CHILTON®

ASIAN
SERVICE MANUAL
2008 EDITION
VOLUME I
Acura
Honda
Isuzu

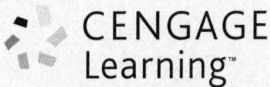
CENGAGE
Learning™

Australia • Brazil • Japan • Korea • Mexico • Singapore • Spain • United Kingdom • United States

CENGAGE
Learning™

CHILTON®
Asian Service Manual
2008 Edition
Volume I
Acura, Honda, Isuzu

Vice President,
Technology & Trades Professional
Business Unit:
  Gregory L. Clayton

Publisher,
Technology & Trades Professional
Business Unit:
  David Koontz

Director of Marketing:
  Beth A. Lutz

Marketing Manager:
  Jennifer Stall

Marketing Assistant:
  Rachael Conover

Production Director:
  Carolyn Miller

Editorial Assistant:
  Jason Yager

Production Manager:
  Andrew Crouth

Publishing Coordinator:
  Paula Baillie

Sr. Content Project Manager:
  Elizabeth C. Hough

Managing Editor:
  Terry L. Blomquist

Editors:
  Dennis Bailey
  Nick D'Andrea
  James R. Marotta
  David G. Olson
  Ryan Price
  Christine Sheeky
  Jon Wallace

Graphical Designer:
  Melinda Possinger

For more information contact:
Cengage Learning
Executive Woods
5 Maxwell Drive, PO Box 8007,
Clifton Park, NY 12065-8007
Visit us at **www.chilton.cengage.com**
Visit our corporate website at **www.cengage.com**
For permission to use material from
the text or product, contact us by
Tel. (800) 730-2214
Fax (800) 730-2215
www.cengage.com/permissions

Cengage Learning products are represented in Canada by Nelson Education, Ltd.

ISBN 10: 1-4283-2215-9
ISBN 13: 978-1-4283-2215-8
ISSN: 1939-621X

NOTICE TO THE READER

Publisher does not warrant or guarantee any of the products described herein or perform any independent analysis in connection with any of the product information contained herein. Publisher does not assume, and expressly disclaims, any obligation to obtain and include information other than that provided to it by the manufacturer.

The reader is expressly warned to consider and adopt all safety precautions that might be indicated by the activities herein and to avoid all potential hazards. By following the instructions contained herein, the reader willingly assumes all risks in connection with such instructions.

The publisher makes no representation or warranties of any kind, including but not limited to, the warranties of fitness for particular purpose or merchantability, nor are any such representations implied with respect to the material set forth herein, and the publisher takes no responsibility with respect to such material. The publisher shall not be liable for any special, consequential, or exemplary damages resulting, in whole or part, from the readers' use of, or reliance upon, this material.

Printed in the United States of America
1 2 3 4 5 xx 13 12 11 10 09 08

Table of Contents

Model Index

USING THIS INFORMATION

Organization

To find where a particular model section or procedure is located, look in the Table of Contents. Main topics are listed with the page number on which they may be found. Following the main topics is an alphabetical listing of all of the procedures within the section and their page numbers.

Manufacturer and Model Coverage

This product covers 2007–2008 Asian models that are produced in sufficient quantities to warrant coverage, and which have technical content available from the vehicle manufacturers before our publication date. Although this information is as complete as possible at the time of publication, some manufacturers may make changes which cannot be included here. While striving for total accuracy, the publisher cannot assume responsibility for any errors, changes, or omissions that may occur in the compilation of this data.

Part Numbers & Special Tools

Part numbers and special tools are recommended by the publisher and vehicle manufacturer to perform specific jobs. Before substituting any part or tool for the one recommended, you must be completely satisfied that neither your personal safety, nor the performance of the vehicle will be endangered.

ACKNOWLEDGEMENT

The publisher would like to express appreciation to the following vehicle manufacturers for their assistance in producing this manual. No further reproduction or distribution of the material in this manual is allowed without the expressed written permission of the vehicle manufacturers and the publisher. Portions of materials contained herein have been reprinted under license from American Honda Corporation, License Agreement 07201AH. Additional appreciation to Isuzu Motors of America, Inc.

PRECAUTIONS

Before servicing any vehicle, please be sure to read all of the following precautions, which deal with personal safety, prevention of component damage, and important points to take into consideration when servicing a motor vehicle:

• Always wear safety glasses or goggles when drilling, cutting, grinding or prying.

• Steel-toed work shoes should be worn when working with heavy parts. Pockets should not be used for carrying tools. A slip or fall can drive a screwdriver into your body.

• Work surfaces, including tools and the floor should be kept clean of grease, oil or other slippery material.

• When working around moving parts, don't wear loose clothing. Long hair should be tied back under a hat or cap, or in a hair net.

• Always use tools only for the purpose for which they were designed. Never pry with a screwdriver.

• Keep a fire extinguisher and first aid kit handy.

• Always properly support the vehicle with approved stands or lift.

• Always have adequate ventilation when working with chemicals or hazardous material.

• Carbon monoxide is colorless, odorless and dangerous. If it is necessary to operate the engine with vehicle in a closed area such as a garage, always use an exhaust collector to vent the exhaust gases outside the closed area.

• When draining coolant, keep in mind that small children and some pets are attracted by ethylene glycol antifreeze, and are quite likely to drink any left in an open container, or in puddles on the ground. This will prove fatal in sufficient quantity. Always drain the coolant into a sealable container.

• To avoid personal injury, do not remove the coolant pressure relief cap while the engine is operating or hot. The cooling system is under pressure; steam and hot liquid can come out forcefully when the cap is loosened slightly. Failure to follow these instructions may result in personal injury. The coolant must be recovered in a suitable, clean container for reuse. If the coolant is contaminated it must be recycled or disposed of correctly.

• When carrying out maintenance on the starting system be aware that heavy gauge leads are connected directly to the battery. Make sure the protective caps are in place when maintenance is completed. Failure to follow these instructions may result in personal injury.

• Do not remove any part of the engine emission control system. Operating the engine without the engine emission control system will reduce fuel economy and engine ventilation. This will weaken engine performance and shorten engine life. It is also a violation of Federal law.

• Due to environmental concerns, when the air conditioning system is drained, the refrigerant must be collected using refrigerant recovery/recycling equipment. Federal law requires that refrigerant be recovered into appropriate recovery equipment and the process be conducted by qualified technicians who have been certified by an approved organization, such as MACS, ASI, etc. Use of a recovery machine dedicated to the appropriate refrigerant is necessary to reduce the possibility of oil and refrigerant incompatibility concerns. Refer to the instructions provided by the equipment manufacturer when removing refrigerant from or charging the air conditioning system.

• Always disconnect the battery ground when working on or around the electrical system.

• Batteries contain sulfuric acid. Avoid contact with skin, eyes, or clothing. Also, shield your eyes when working near batteries to protect against possible splashing of the acid solution. In case of acid contact with skin or eyes, flush immediately with water for a minimum of 15 minutes and get prompt medical attention. If acid is swallowed, call a physician immediately. Failure to follow these instructions may result in personal injury.

• Batteries normally produce explosive gases. Therefore, do not allow flames, sparks or lighted substances to come near the battery. When charging or working near a battery, always shield your face and protect your eyes. Always provide ventilation. Failure to follow these instructions may result in personal injury.

• When lifting a battery, excessive pressure on the end walls could cause acid to spew through the vent caps, resulting in personal injury, damage to the vehicle or battery. Lift with a battery carrier or with your hands on opposite corners. Failure to follow these instructions may result in personal injury.

• Observe all applicable safety precautions when working around fuel. Whenever servicing the fuel system, always work in a well-ventilated area. Do not allow fuel spray or vapors to come in contact with a spark, open flame, or excessive heat (a hot drop light, for example). Keep a dry chemical fire extinguisher near the work area. Always keep fuel in a container specifically designed for fuel storage; also, always properly seal fuel containers to avoid the possibility of fire or explosion. Do not smoke or carry lighted tobacco or open flame of any type when working on or near any fuel-related components.

• Fuel injection systems often remain pressurized, even after the engine has been turned OFF. The fuel system pressure must be relieved before disconnecting any fuel lines. Failure to do so may result in fire and/or personal injury.

• The evaporative emissions system contains fuel vapor and condensed fuel vapor. Although not present in large quantities, it still presents the danger of explosion or fire. Disconnect the battery ground cable from the battery to minimize the possibility of an electrical spark occurring, possibly causing a fire or explosion if fuel vapor or liquid fuel is present in the area. Failure to follow these instructions can result in personal injury.

• The EPA warns that prolonged contact with used engine oil may cause a number of skin disorders, including cancer! You should make every effort to minimize your exposure to used engine oil. Protective gloves should be worn when changing oil. Wash your hands and any other exposed skin areas as soon as possible after exposure to used engine oil. Soap and water, or waterless hand cleaner should be used.

• Some vehicles are equipped with an air bag system, often referred to as a Supple-mental Restraint System (SRS) or Supple-mental Inflatable Restraint (SIR) system. The system must be disabled before performing service on or around system components, steering column, instrument panel components, wiring and sensors. Failure to follow safety and disabling procedures could result in accidental air bag deployment, possible personal injury and unnecessary system repairs.

• Always wear safety goggles when working with, or around, the air bag system. When carrying a non-deployed air bag, be sure the bag and trim cover are pointed away from your body. When placing a non-deployed air bag on a work surface, always face the bag and trim cover upward, away from the surface. This will reduce the motion of the module if it is accidentally deployed.

• Electronic modules are sensitive to electrical charges. The ABS module can be damaged if exposed to these charges.

• Brake pads and shoes may contain asbestos, which has been determined to be a cancer-causing agent. Never clean brake surfaces with compressed air. Avoid inhaling brake dust. Clean all brake surfaces with a commercially available brake cleaning fluid.

• When replacing brake pads, shoes, discs or drums, replace them as complete axle sets.

• When servicing drum brakes, disassemble and assemble one side at a time, leaving the remaining side intact for reference.

• Brake fluid often contains polyglycol ethers and polyglycols. Avoid contact with the eyes and wash your hands thoroughly after handling brake fluid. If you do get brake fluid in your eyes, flush your eyes with clean, running water for 15 minutes. If eye irritation persists, or if you have taken brake fluid internally, immediately seek medical assistance.

• Clean, high quality brake fluid from a sealed container is essential to the safe and proper operation of the brake system. You should always buy the correct type of brake fluid for your vehicle. If the brake fluid becomes contaminated, completely flush the system with new fluid. Never reuse any brake fluid. Any brake fluid that is removed from the system should be discarded. Also, do not allow any brake fluid to come in contact with a painted or plastic surface; it will damage the paint.

• Never operate the engine without the proper amount and type of engine oil; doing so will result in severe engine damage.

• Timing belt maintenance is extremely important! Many models utilize an interference-type, non-freewheeling engine. If the timing belt breaks, the valves in the cylinder head may strike the pistons, causing potentially serious (also time-consuming and expensive) engine damage.

• Disconnecting the negative battery cable on some vehicles may interfere with the functions of the on-board computer system (s) and may require the computer to undergo a relearning process once the negative battery cable is reconnected.

• Steering and suspension fasteners are critical parts because they affect performance of vital components and systems and their failure can result in major service expense. They must be replaced with the same grade or part number or an equivalent part if replacement is necessary. Do not use a replacement part of lesser quality or substitute design. Torque values must be used as specified during reassembly to ensure proper retention of these parts.

ACURA

RL • TL • TSX

1

SPECIFICATIONS AND MAINTENANCE CHARTS

ENGINE AND VEHICLE IDENTIFICATION

Engine							Model Year	
Code	Liters (cc)	Cu. In.	Cyl.	Fuel Sys.	Engine Type	Eng. Mfg.	Code ①	Year
K24A2	2.4 (2354)	144	4	PGM-FI	DOHC	Honda	7	2007
J32A3	3.2 (3210)	196	6	PGM-FI	SOHC	Honda	8	2008
J35A8	3.5 (3471)	222	6	PGM-FI	SOHC	Honda		

PGM-FI: Programmed Fuel Injection

DOHC: Double Overhead Camshaft

SOHC: Single Overhead Camshaft

① 10th digit of the Vehicle Identification Number (VIN)

22140_ACUR_C0001

GENERAL ENGINE SPECIFICATIONS

Year	Model	Engine Displacement Liters	Engine ID	Net Horsepower @ rpm	Net Torque @ rpm (ft. lbs.)	Bore x Stroke (in.)	Com-pression Ratio	Oil Pressure @ rpm
2007	TSX	2.4	K24A2	200@6800	166@4500	3.42X3.89	10.5:1	44@3000
	TL	3.2	J32A3	258@6200	233@5000	3.50x3.39	11.0:1	71@3000
	TL ②	3.5	J35A8	286@6200	256@5000	3.50x3.66	11.0:1	71@3000
	RL	3.5	J35A8	290@6200	256@5000	3.50x3.66	11.0:1	71@3000
2008	TSX	2.4	K24A2	200@6800	166@4500	3.42X3.89	10.5:1	44@3000
	TL	3.2	J32A3	258@6200	233@5000	3.50x3.39	11.0:1	71@3000
	TL ②	3.5	J35A8	286@6200	256@5000	3.50x3.66	11.0:1	71@3000
	RL	3.5	J35A8	290@6200	256@5000	3.50x3.66	11.0:1	71@3000

PGM-FI: Programmed Fuel Injection

① DB7: 4 door

   DC4: 3 door

② Type-S

22140_ACUR_C0002

GASOLINE ENGINE TUNE-UP SPECIFICATIONS

Year	Engine Displacement Liters	Engine ID/VIN	Spark Plug Gap (in.)	Ignition Timing (deg.) MT	Ignition Timing (deg.) AT	Fuel Pump (psi)	Idle Speed (rpm) MT	Idle Speed (rpm) AT	Valve Clearance In.	Valve Clearance Ex.
2007	2.4	K24A2	0.039-0.043	6-10 B	6-10 B	48-55 ①	700-800	750-850	0.008-0.010	0.010-0.011
	3.2	J32A3	0.039-0.043	8-10B	8-10B	57-64 ①	—	700-800	0.008-0.009	0.011-0.013
	3.5	J35A8	0.039-0.043	—	8-12B	55-63 ①	—	630-730	0.008-0.009	0.011-0.013
2008	2.4	K24A2	0.039-0.043	6-10 B	6-10 B	48-55 ①	700-800	750-850	0.008-0.010	0.010-0.011
	3.2	J32A3	0.039-0.043	8-10B	8-10B	57-64 ①	—	700-800	0.008-0.009	0.011-0.013
	3.5	J35A8	0.039-0.043	—	8-12B	55-63 ①	—	630-730	0.008-0.009	0.011-0.013

NOTE: The Vehicle Emission Control Information label reflects specification changes during production and must be used if they differ from this chart.

B: Before Top Dead Center

① At idle, pressure regulator vacuum hose disconnected

22140_ACUR_C0003

CAPACITIES

Year	Model	Engine Displacement Liters	Engine ID/VIN	Engine Oil with Filter (qts.)	Transmission (pts.) 5-Spd	Transmission (pts.) 6-Spd	Transmission (pts.) Auto.	Drive Axle Front (pts.)	Drive Axle Rear (pts.)	Fuel Tank (gal.)	Cooling System (qts.)
2007	TSX	2.4	K24A2	4.4	—	4.2	6.0	—	—	17.1	③
	TL	3.2	J32A3	4.5	—	4.4	6.6	—	—	17.1	5.9
	TL ②	3.5	J35A8	4.5	—	4.4	6.6	—	—	17.1	5.9
	RL	3.5	J35A8	4.5	—	—	5.8	—	1.5	19.4	6.4
2008	TSX	2.4	K24A2	4.4	—	4.2	6.0	—	—	17.1	③
	TL	3.2	J32A3	4.5	—	4.4	6.6	—	—	17.1	5.9
	TL ②	3.5	J35A8	4.5	—	4.4	6.6	—	—	17.1	5.9
	RL	3.5	J35A8	4.5	—	—	5.8	—	1.5	19.4	6.4

NOTE: All capacities are approximate. Add fluid gradually and ensure a proper fluid level is obtained.

NOTE: Capacities given are service, not overhaul capacities

① Automatic transmission: 5.3
   Manual Transmission: 5.4

② Type-S

③ Automatic transmission: 5.6
   Manual Transmission: 5.7

22140_ACUR_C0004

FLUID SPECIFICATIONS

Year	Model	Engine Displ. Liters	Engine Oil	Man. Trans.	Auto. Trans.	Drive Axle Front	Drive Axle Rear	Transfer Case	Power Steering Fluid	Brake Master Cylinder	Cooling System
2007	TSX	2.4	5W-30 Acura	Acura MTF	Acura ATF-Z1	—	—	—	Acura PS Fluid	Acura DOT 3	①
	TL	3.2	5W-20 Acura	Acura MTF	Acura ATF-Z1	—	—	—	Acura PS Fluid	Acura DOT 3	①
	TL①	3.5	5W-20 Acura	Acura MTF	Acura ATF-Z1	—	—	—	Acura PS Fluid	Acura DOT 3	①
	RL	3.5	5W-20 Honda	—	Acura ATF-Z1	—	Acura ATF-Z1	③	Acura PS Fluid	Acura DOT 3	①
2008	TSX	2.4	5W-30 Acura	Acura MTF	Acura ATF-Z1	—	—	—	Acura PS Fluid	Acura DOT 3	①
	TL	3.2	5W-20 Acura	Acura MTF	Acura ATF-Z1	—	—	—	Acura PS Fluid	Acura DOT 3	①
	TL①	3.5	5W-20 Acura	Acura MTF	Acura ATF-Z1	—	—	—	Acura PS Fluid	Acura DOT 3	①
	RL	3.5	5W-20 Honda	—	Acura ATF-Z1	—	Acura ATF-Z1	③	Acura PS Fluid	Acura DOT 3	①

DOT: Department Of Transportation

① Acura Long Life Antifreeze/Coolant-Type2

② Type-S

③ Hypoid gear oil SAE 90 or SAE 80W-90 viscosity, /

22140_ACUR_C0005

VALVE SPECIFICATIONS

Year	Engine Displacement Liters	Engine ID/VIN	Seat Angle (deg.)	Face Angle (deg.)	Spring Test Pressure (lbs. @ in.)	Spring Installed Height (in.)	Stem-to-Guide Clearance (in.) Intake	Stem-to-Guide Clearance (in.) Exhaust	Stem Diameter (in.) Intake	Stem Diameter (in.) Exhaust
2007	2.4	K24A2	45	45	NA	NA	0.0012-0.0022	0.0022-0.0031	0.2156-0.2159	0.2146-0.2150
	3.2	J32A3	45	45	NA	NA	0.0008-0.0018	0.0022-0.0031	0.2159-0.2163	0.2146-0.2150
	3.5	J35A8	45	45	NA	NA	0.0008-0.0018	0.0022-0.0031	0.2159-0.2163	0.2146-0.2150
2008	2.4	K24A2	45	45	NA	NA	0.0012-0.0022	0.0022-0.0031	0.2156-0.2159	0.2146-0.2150
	3.2	J32A3	45	45	NA	NA	0.0008-0.0018	0.0022-0.0031	0.2159-0.2163	0.2146-0.2150
	3.5	J35A8	45	45	NA	NA	0.0008-0.0018	0.0022-0.0031	0.2159-0.2163	0.2146-0.2150

NA: Not Available

22140_ACUR_C0006

CAMSHAFT AND BEARING SPECIFICATIONS

All measurements are given in inches.

Year	Engine Displacement Liters	Engine VIN	Journal Diameter	Brg. Oil Clearance	Shaft End-play	Runout	Journal Bore	Lobe Height Intake	Lobe Height Exhaust
2007	2.4	K24A2	NA	①	0.0020-0.0080	0.0010	NA	⑤	⑥
	3.2	J32A3	NA	0.0020-0.0035	0.0020-0.0080	0.0010	NA	⑦	1.4302
	3.5	J35A8	NA	0.0020-0.0035	0.0020-0.0080	0.0010	NA	⑧	1.4326
2008	2.4	K24A2	NA	①	0.0020-0.0080	0.0010	NA	⑤	⑥
	3.2	J32A3	NA	0.0020-0.0035	0.0020-0.0080	0.0010	NA	⑦	1.4302
	3.5	J35A8	NA	0.0020-0.0035	0.0020-0.0080	0.0010	NA	⑧	1.4326

NA: Information not available

① No. 1 Journal: 0.0010-0.0030 in.
   Other Journals: 0.0020-0.0040 in.
② Primary: 1.3356 in.
   Secondary: 1.1168 in.
③ Primary: 1.2910 in.
   Mid: 1.3990 in.
   Secondary: 1.2865 in.
④ Primary: 1.2902 in.
   Mid: 1.3688 in.
   Secondary: 1.2859 in.

⑤ Primary: 1.3156 in.
   Mid: 1.3792 in.
   Secondary: 1.3156 in.
⑥ Primary: 1.2990 in.
   Mid: 1.3815 in.
   Secondary: 1.2990 in.
⑦ Primary: 1.3796 in.
   Mid: 1.43480 in.
⑧ Primary: 1.3824 in.
   Mid: 1.4328 in.
   Secondary: 1.3824 in.

22140_ACUR_C0009

CRANKSHAFT AND CONNECTING ROD SPECIFICATIONS

All measurements are given in inches.

Year	Engine Displacement Liters	Engine ID/VIN	Crankshaft Main Brg. Journal Dia.	Crankshaft Main Brg. Oil Clearance	Crankshaft Shaft End-play	Thrust on No.	Connecting Rod Journal Diameter	Connecting Rod Oil Clearance	Connecting Rod Side Clearance
2007	2.4	K24A2	②	①	0.0040-0.0140	4	1.8888-1.8898	0.0013-0.0026	0.0060-0.0160
	3.2	J32A3	2.8337-2.8346	0.0008-0.0017	0.0040-0.0140	3	2.1644-2.1654	0.0008-0.0017	0.0060-0.0140
	3.5	J35A8	2.8337-2.8346	0.0008-0.0017	0.0040-0.0140	3	2.1644-2.1654	0.0008-0.0017	0.0060-0.0140
2008	2.4	K24A2	②	①	0.0040-0.0140	4	1.8888-1.8898	0.0013-0.0026	0.0060-0.0160
	3.2	J32A3	2.8337-2.8346	0.0008-0.0017	0.0040-0.0140	3	2.1644-2.1654	0.0008-0.0017	0.0060-0.0140
	3.5	J35A8	2.8337-2.8346	0.0008-0.0017	0.0040-0.0140	3	2.1644-2.1654	0.0008-0.0017	0.0060-0.0140

① Nos. 1, 2, 4 and 5: 0.0007-0.0016
   No. 3: 0.0010-0.0019
② Nos. 1, 2, 4 and 5: 2.1648-2.1657
   No. 3: 2.1644-2.1654

22140_ACUR_C0007

PISTON AND RING SPECIFICATIONS

All measurements are given in inches

Year	Engine Displacement Liters	Engine ID/VIN	Piston Clearance	Ring Gap			Ring Side Clearance		
				Top Compression	Bottom Compression	Oil Control	Top Compression	Bottom Compression	Oil Control
2007	2.4	K24A2	0.0008-0.0016	0.0080-0.0014	0.0200-0.0260	0.0080-0.0280	0.0018-0.0028	0.0016-0.0026	NA
	3.2	J32A3	0.0006-0.0016	0.0080-0.0140	0.0160-0.0220	0.0080-0.0280	0.0022-0.0031	0.0012-0.0022	NA
	3.5	J35A8	0.0006-0.00160	0.0080-0.0140	0.0160-0.0220	0.0080-0.0280	0.0022-0.0031	0.0012-0.0022	NA
2008	2.4	K24A2	0.0008-0.0016	0.0080-0.0014	0.0200-0.0260	0.0080-0.0280	0.0018-0.0028	0.0016-0.0026	NA
	3.2	J32A3	0.0006-0.0016	0.0080-0.0140	0.0160-0.0220	0.0080-0.0280	0.0022-0.0031	0.0012-0.0022	NA
	3.5	J35A8	0.0006-0.00160	0.0080-0.0140	0.0160-0.0220	0.0080-0.0280	0.0022-0.0031	0.0012-0.0022	NA

NA; Not Applicable

22140_ACUR_C0008

TORQUE SPECIFICATIONS

All readings in ft. lbs.

Year	Engine Displacement Liters	Engine ID/VIN	Cylinder Head Bolts	Main Bearing Bolts	Rod Bearing Bolts	Crankshaft Damper Bolts	Flywheel Bolts	Manifold		Spark Plugs	Oil Pan Drain Plug
								Intake	Exhaust		
2007	2.4	K24A2	①	②	④	④	76	16	33	13	33
	3.2	J32A3	⑧	⑨	⑩	⑪	⑫	16	23	13	29
	3.5	J35A8	⑬	⑭	⑩	④	54	16	23	13	29
2008	2.4	K24A2	①	②	④	④	76	16	33	13	33
	3.2	J32A3	⑧	⑨	⑩	⑪	⑫	16	23	13	29
	3.5	J35A8	⑬	⑭	⑩	④	54	16	23	13	29

① Step 1: 29 ft. lbs.
Step 2: Rotate 90 degrees
Step 3: Rotate an additional 90 degrees
Step 4 (new bolts only): additional 90 degree

② Step 1: 22 ft. lbs.
Step 2: 63 ft. lbs.

③ Step 1: 14 ft. lbs.
Step 2: 23 ft. lbs.

④ Manual transmission: 76 ft. lbs.
Automatic transmission: 54 ft. lbs.

⑤ DB8: 4 door (Except Type R)
DC2: 3 door

⑥ Step 1: 14 ft. lbs.
Step 2: 33 ft. lbs.

⑦ NGK type IFRG-11KS & Denso type SK22PRM11S: 18 ft. lbs.
All others: 13 ft. lbs.

⑧ Cap bolts: 29 ft. lbs.
Cap bridge bolts: 48 ft. lbs.

⑨ 14 ft. lbs. plus 116 degrees

⑩ Step 1: Cap bolts 48 ft. lbs.
Step 2: Side bolts 36 ft. lbs.

⑪ Step 1: Outer (9mm) 29 ft. lbs.
Step 2: Inner (11mm) 56 ft. lbs.
Step 3: Side (10mm) 36 ft. lbs.

⑫ Step 1: 29 ft. lbs.
Step 2: 51 ft. lbs.
Step 3: 72.3 ft. lbs.

⑬ Step 1: Cap bolts 56 ft. lbs.
Step 2: Side bolts 36 ft. lbs.

⑭ Step 1: 14 ft. lbs.
Step 2: Rotate 90 degrees

22140_ACUR_C0006

WHEEL ALIGNMENT

Year	Model		Caster Range (+/-Deg.)	Caster Preferred Setting (Deg.)	Camber Range (+/-Deg.)	Camber Preferred Setting (Deg.)	Toe-in (in.)
2007	TL	F	0.75	+3.28	0.50	-0.50	0 +/- 0.08
		R	—	—	0.50	-1.00	0 +/- 0.08
	TL Type S	F	0.75	+3.28	0.50	-0.50	0 +/- 0.08
		R	—	—	0.50	-1.00	0 +/- 0.08
	TSX	F	0.75	+3.22	0.75	0	0 +/- 0.08
		R	—	—	0.50	-1.00	0.08 +/- 0.08
	RL	F	0.50	+2.17	0.50	-0.13	0 +/- 0.08
		R	—	—	0.50	-1.25	0.08 +/- 0.08
2008	TL	F	0.75	+3.28	0.50	-0.50	0 +/- 0.08
		R	—	—	0.50	-1.00	0 +/- 0.08
	TL Type S	F	0.75	+3.28	0.50	-0.50	0 +/- 0.08
		R	—	—	0.50	-1.00	0 +/- 0.08
	TSX	F	0.75	+3.22	0.75	0	0 +/- 0.08
		R	—	—	0.50	-1.00	0.08 +/- 0.08
	RL	F	0.50	+2.17	0.50	-0.13	0 +/- 0.08
		R	—	—	0.50	-1.25	0.08 +/- 0.08

22140_ACUR_C0011

TIRE, WHEEL AND BALL JOINT SPECIFICATIONS

Year	Model	OEM Tires Standard	OEM Tires Optional	Tire Pressures (psi) Front	Tire Pressures (psi) Rear	Wheel Size	Ball Joint Inspection	Lug Nut (ft. lbs.)
2007	TL	P235/45R17	None	NA	NA	NA	NS	80
	TL Type S	P235/45R17	None	NA	NA	NA	NS	80
	TSX	P215/50/R17	None	NA	NA	NA	NS	80
	RL	P245/50/R17	None	NA	NA	NA	NS	94
2008	TL	P235/45R17	None	NA	NA	NA	NS	80
	TL Type S	P235/45R17	None	NA	NA	NA	NS	80
	TSX	P215/50/R17	None	NA	NA	NA	NS	80
	RL	P245/50/R17	None	NA	NA	NA	NS	94

OEM: Original Equipment Manufacturer

PSI: Pounds Per Square Inch

NS: Not Specified by manufacturer

NA: Not Available

22140_ACUR_C0012

BRAKE SPECIFICATIONS
All measurements in inches unless noted

Year	Model		Brake Disc			Brake Drum Diameter			Minimum Lining Thickness		Brake Caliper	
			Original Thickness	Minimum Thickness	Maximum Runout	Original Inside Diameter	Max. Wear Limit	Maximum Machine Diameter	Front	Rear	Bracket Bolts (ft. lbs.)	Mounting Bolts (ft. lbs.)
2007	TL	F	1.100-1.100	1.020	0.004	—	—	—	0.06	—	80	37
		R	0.350-0.358	0.300	0.004	—	—	—	—	0.06	41	17
	TL Type S	F	0.980-0.990	0.910	0.004	—	—	—	0.06	—	80	37
		R	0.350-0.358	0.300	0.004	—	—	—	—	0.06	41	17
	TSX	F	1.100-1.110	1.020	0.002	—	—	—	0.06	—	80	37
		R	0.350-0.360	0.310	0.002	—	—	—	—	0.06	54	17
	RL	F	1.10-1.11	1.020	0.004	—	—	—	0.06	—	—	58
		R	0.625-0.634	0.550	0.006	—	—	—	—	0.06	79	17
2008	TL	F	1.100-1.100	1.020	0.004	—	—	—	0.06	—	80	37
		R	0.350-0.358	0.300	0.004	—	—	—	—	0.06	41	17
	TL Type S	F	0.980-0.990	0.910	0.004	—	—	—	0.06	—	80	37
		R	0.350-0.358	0.300	0.004	—	—	—	—	0.06	41	17
	TSX	F	1.100-1.110	1.020	0.002	—	—	—	0.06	—	80	37
		R	0.350-0.360	0.310	0.002	—	—	—	—	0.06	54	17
	RL	F	1.10-1.11	1.020	0.004	—	—	—	0.06	—	—	58
		R	0.625-0.634	0.550	0.006	—	—	—	—	0.06	79	17

NA: Not Available

F: Front

R: Rear

① A/T: 1.10-1.11 in.
M/T: 0.98-0.99 in.

② A/T: 1.02 in.
M/T: 0.91 in.

22140_ACUR_C0013

SCHEDULED MAINTENANCE INTERVALS
Acura TL, TSX & RL

TO BE SERVICED	TYPE OF SERVICE	SERVICE INTERVALS						
		Symbol A	Symbol B	1	2	3	4	5
Air cleaner element	Replace				✓			
Brake fluid	Inspect		✓					
Brake hoses and lines	Inspect		✓					
Clutch fluid (if equipped)	Inspect		✓					
Drive belt	Inspect				✓			
Driveshaft boots	Inspect		✓					
Dust & pollen filter	Replace				✓			
Engine coolant	Inspect		✓					
Engine Coolant	Replace							✓
Engine oil	Replace	✓						
Engine oil & filter	Replace		✓					
Exhaust system	Inspect		✓					
Front & Rear brakes	Inspect		✓					
Fuel lines	Inspect		✓					
Parking brake adjustment	Inspect		✓					
Power steering fluid	Inspect		✓					
Spark plugs	Replace						✓	
Suspension components	Inspect		✓					
Tie-rod ends, steering gearbox and boots	Inspect		✓					
Timing belt	Replace						✓	
Tires	Rotate			✓				
Transmission fluid	Inspect		✓					
Transmission fluid	Replace					✓		
Valve clearance	Inspect						✓	
Windshield washer fluid	Inspect		✓					

22140_ACUR_C0016

PRECAUTIONS

Before servicing any vehicle, please be sure to read all of the following precautions, which deal with personal safety, prevention of component damage, and important points to take into consideration when servicing a motor vehicle:

• Never open, service or drain the radiator or cooling system when the engine is hot; serious burns can occur from the steam and hot coolant.

• Observe all applicable safety precautions when working around fuel. Whenever servicing the fuel system, always work in a well-ventilated area. Do not allow fuel spray or vapors to come in contact with a spark, open flame, or excessive heat (a hot drop light, for example). Keep a dry chemical fire extinguisher near the work area. Always keep fuel in a container specifically designed for fuel storage; also, always properly seal fuel containers to avoid the possibility of fire or explosion. Refer to the additional fuel system precautions later in this section.

• Fuel injection systems often remain pressurized, even after the engine has been turned **OFF**. The fuel system pressure must be relieved before disconnecting any fuel lines. Failure to do so may result in fire and/or personal injury.

• Brake fluid often contains polyglycol ethers and polyglycols. Avoid contact with the eyes and wash your hands thoroughly after handling brake fluid. If you do get brake fluid in your eyes, flush your eyes with clean, running water for 15 minutes. If eye irritation persists, or if you have taken brake fluid internally, IMMEDIATELY seek medical assistance.

• The EPA warns that prolonged contact with used engine oil may cause a number of skin disorders, including cancer. You should make every effort to minimize your exposure to used engine oil. Protective gloves should be worn when changing oil. Wash your hands and any other exposed skin areas as soon as possible after exposure to used engine oil. Soap and water, or waterless hand cleaner should be used.

• All new vehicles are now equipped with an air bag system, often referred to as a Supplemental Restraint System (SRS) or Supplemental Inflatable Restraint (SIR) system. The system must be disabled before performing service on or around system components, steering column, instrument panel components, wiring and sensors. Failure to follow safety and disabling procedures could result in accidental air bag deployment, possible personal injury and unnecessary system repairs.

• Always wear safety goggles when working with, or around, the air bag system. When carrying a non-deployed air bag, be sure the bag and trim cover are pointed away from your body. When placing a non-deployed air bag on a work surface, always face the bag and trim cover upward, away from the surface. This will reduce the motion of the module if it is accidentally deployed. Refer to the additional air bag system precautions later in this section.

• Clean, high quality brake fluid from a sealed container is essential to the safe and proper operation of the brake system. You should always buy the correct type of brake fluid for your vehicle. If the brake fluid becomes contaminated, completely flush the system with new fluid. Never reuse any brake fluid. Any brake fluid that is removed from the system should be discarded. Also, do not allow any brake fluid to come in contact with a painted surface; it will damage the paint.

• Never operate the engine without the proper amount and type of engine oil; doing so WILL result in severe engine damage.

• Timing belt maintenance is extremely important. Many models utilize an interference-type, non-freewheeling engine. If the timing belt breaks, the valves in the cylinder head may strike the pistons, causing potentially serious (also time-consuming and expensive) engine damage. Refer to the maintenance interval charts for the recommended replacement interval for the timing belt, and to the timing belt section for belt replacement and inspection.

• Disconnecting the negative battery cable on some vehicles may interfere with the functions of the on-board computer system(s) and may require the computer to undergo a relearning process once the negative battery cable is reconnected.

• When servicing drum brakes, only disassemble and assemble one side at a time, leaving the remaining side intact for reference.

• Only an MVAC-trained, EPA-certified automotive technician should service the air conditioning system or its components.

BRAKES

ANTI-LOCK BRAKE SYSTEM (ABS)

GENERAL INFORMATION

When conventional brakes are applied in an emergency stop or on ice, one or more wheels may lock. This may result in loss of steering control and vehicle stability. The purpose of the Anti-lock Brake System (ABS) is to prevent lock up when traction is marginal or under heavy braking conditions. This system offers many benefits allowing the driver increased safety and control during braking. Anti-lock braking operates only at speeds above 3 mph (5 km/h).

Under normal braking conditions, the ABS functions the same as a standard brake system with a diagonally split master cylinder and conventional vacuum assist.

If wheel lock is detected during the brake application, the system will enter anti-lock mode. During anti-lock mode, hydraulic pressure in the four wheel circuits is modulated to prevent any one wheel from locking. Each wheel circuit is designed with a set of electrical valves and hydraulic line to provide modulation, although for vehicle stability, both rear wheel valves receive the same electrical signal. The system can build or reduce pressure at each wheel, depending on signals generated by the Wheel Speed Sensors (WSS) at each wheel and received at the Controller Anti-lock Brake (CAB).

Anti-lock Braking Systems (ABS) are available on all Honda models. When this system engages, some audible noise as well as pulses in the brake pedal may occur. Do not be alarmed; this is normal system operation.

BRAKES   BLEEDING THE BRAKE SYSTEM

BLEEDING PROCEDURE

See Figure 1.

1. Make sure the brake fluid level in the reservoir (A) is at the MAX (upper) level line (B).

2. Slide a piece of clear plastic hose over the first bleed screw, and submerge the other end in a container of new brake fluid.

3. Have someone slowly pump the brake pedal several times, then apply steady pressure.

4. Loosen the left-front brake bleed screw, loosen the brake bleed screw to allow air to escape from the system. Then tighten the bleed screw securely.

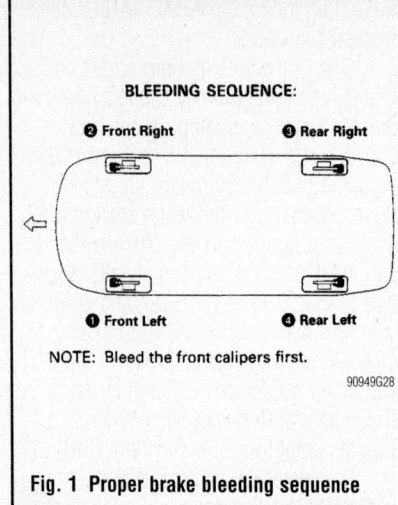

BLEEDING SEQUENCE:

❷ Front Right   ❸ Rear Right

❶ Front Left   ❹ Rear Left

NOTE: Bleed the front calipers first.

90949G28

Fig. 1 Proper brake bleeding sequence

5. Type-S Models only: First bleed the outside piston of the front caliper. Then bleed the inside piston of the front caliper.

➡**The bleed valve uses an 11 mm hex wrench on the Type-S model (4-piston caliper type).**

6. Repeat the procedure for each caliper until no air bubbles are in the fluid. Bleed the calipers in the sequence shown.

7. Refill the master cylinder reservoir to the MAX (upper) level line.

8. Test-drive the vehicle, and make at least two slow ABS-activating stops.

9. Check for pedal feel.

10. If the pedal feel is hard, you are done.

11. If the pedal feel is soft, repeat the brake system bleeding procedure.

BRAKES   FRONT DISC BRAKES

BRAKE CALIPER

REMOVAL & INSTALLATION

TL Models

1. Raise safely support the vehicle.
2. Remove the front wheel.
3. Remove the brake hose bracket mounting bolts.
4. Remove the banjo bolt and disconnect the brake line from the caliper.
5. Remove the caliper mounting bolts and caliper.

To install:

6. Install the caliper and tighten the mounting bolts to 80 ft. lbs. (109 Nm).

7. Install the brake hose. Replace the crush washers and tighten the banjo bolt to 25 ft. lbs. (35 Nm).

8. Install the brake hose bracket.

9. Bleed the brake system. Torque the bleed screws to 84 inch lbs. (9 Nm).

10. Install the front wheel and tighten the lug nuts to 84 inch lbs. (7 Nm).

RL Models

1. Raise and safely support the vehicle.
2. Remove or disconnect the following:
   - Wheels
   - Brake line fitting and disconnect the brake line from the caliper
   - Caliper mounting bolts
   - Caliper
   - Brake pads and shims

To install:

3. Install or connect the following:
   - Brake pads and shims
   - Caliper bolts to 58 ft. lbs. (78 Nm)
4. Install or connect the following:
   - Brake line and fitting. Torque the fitting to 11 ft. lbs. (15 Nm).
5. Bleed the brake system. Torque the bleed screws to 6 ft. lbs. (8 Nm).
   - Front wheels

TSX Models

1. Raise and safely support the vehicle.
2. Remove or disconnect the following:
   - Wheels
   - Banjo bolt and disconnect the brake line from the caliper
   - Caliper mounting bolts
   - Caliper
   - Brake pads and shims
   - Pad spring from the caliper body, if equipped
   - Caliper bracket mounting bolts and bracket

To install:

3. Install or connect the following:
   - Bracket and torque the bolts to 80 ft. lbs. (109 Nm)
   - Pad spring, brake pads, shims, caliper and slide mounting bolts
   - Caliper slide mounting bolt and torque to 36 ft. lbs. (49 Nm)
   - Brake line and the banjo bolt. Replace the crush washers and torque the banjo bolt to 25 ft. lbs. (35 Nm).
4. Bleed the brake system.
   - Front wheels

DISC BRAKE PADS

REMOVAL & INSTALLATION

TL Models

Except Type-S Models

1. Remove some brake fluid from the master cylinder.

2. Raise and safely support the vehicle.

3. Remove the front wheel.

4. Remove the brake hose bracket mounting bolts.

5. Remove the flange bolt while holding the caliper pin with a wrench being careful not to damage the pin boot, and pivot the caliper up out of the way. Check the hose and pin boots for damage and deterioration.

6. Remove the brake pads and the pad shims.

7. Remove the pad retainers.

To install:

8. Install the pad retainers. Wipe excess assembly paste off the retainers. Keep the assembly paste off the brake disc and brake pads.

9. Mount the brake caliper piston compressor on the caliper body.

10. Press in the piston with the brake caliper piston compressor so the caliper will fit over the brake pads. Make sure the piston boot is in position to prevent damaging it when pivoting the caliper down.

➡**Be careful when pressing in the piston; brake fluid might overflow from the master cylinder reservoir.**

❈❈ WARNING

If brake fluid gets on any painted surface, wash it off immediately with water.

11. Remove the brake caliper piston compressor.

12. Apply a thin coat of high temperature brake grease to the pad side of the shims, the back of the brake pads and the other areas indicated by the arrows. Wipe excess assembly paste off the pad shims and brake pads. Contaminated brake discs or brake pads reduce stopping ability. Keep grease and assembly paste off the brake discs and brake pads.

13. Install the brake pads and pad shims correctly. Install the brake pad with the wear indicator on the upper inside. If you are reusing the brake pads, always reinstall the brake pads in their original positions to prevent a momentary loss of braking efficiency.

14. Pivot the caliper down into position. Install the flange bolt, and tighten it to 37 ft. lbs. (50 Nm) while holding the caliper pin with a wrench being careful not to damage the pin boot.

15. Install the brake hose bracket mounting bolts, and tighten them to the specified torque.

16. Clean the mating surfaces of the brake disc and the inside of the wheel, then install the front wheels.

17. Press the brake pedal several times to make sure the brakes work.

➡**Engagement may require a greater pedal stroke immediately after the brake pads have been replaced as a set. Several applications of the brake pedal will restore the normal pedal stroke.**

18. Add brake fluid as needed.

19. After installation, check for leaks at hose and line joints or connections, and retighten if necessary. Test-drive the vehicle, then check for leaks.

Type-S Models

See Figure 2.

1. Remove some brake fluid from the master cylinder.

2. Raise the front of the vehicle, and support it with safety stands in the proper locations.

3. Remove the front wheels. Take care not to scratch the calipers.

4. Depress the pad spring. Remove pad pins from the caliper by pushing from the outside to the inside with a commercially available 5/32" pin punch.

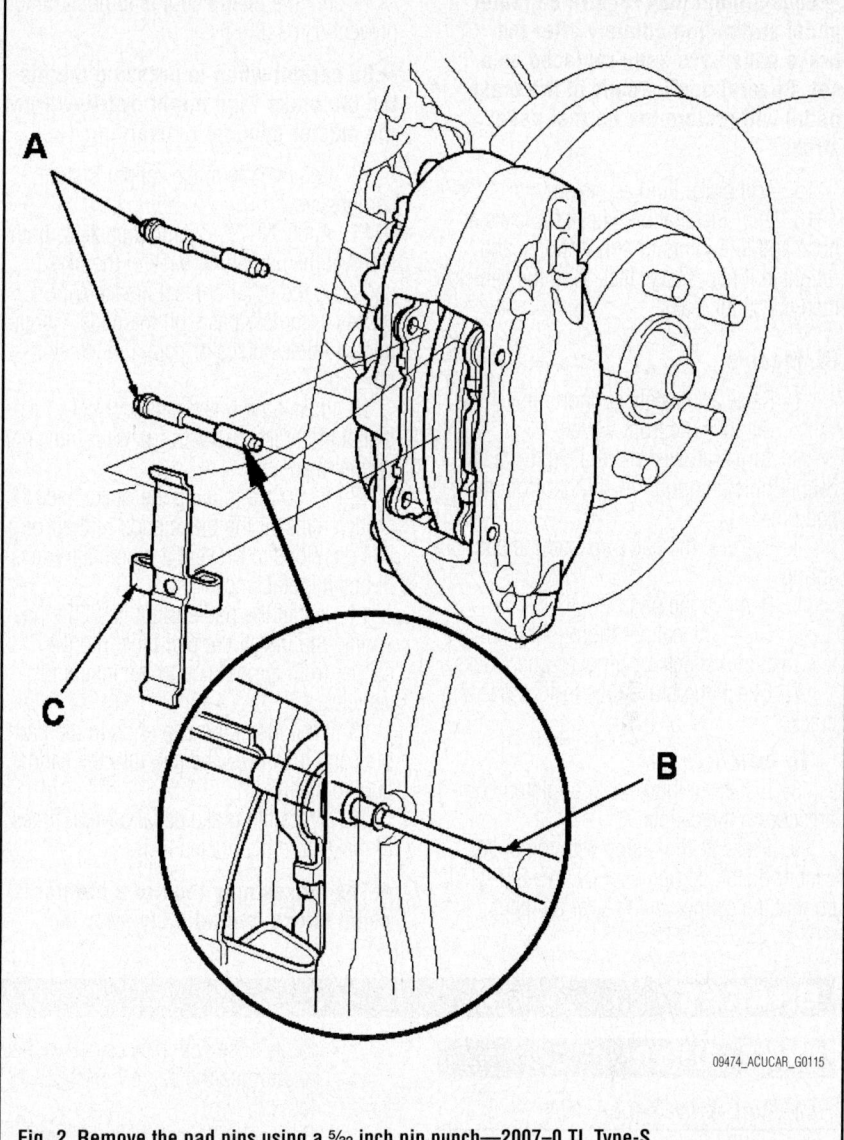

09474_ACUCAR_G0115

Fig. 2 Remove the pad pins using a 5/32 inch pin punch—2007–0 TL Type-S

5. Remove the pad shims and the brake pads.

To install:

6. Mount the brake caliper piston compressor on the caliper.

7. Press in the outside and inside pistons with the brake caliper piston compressor so the caliper will fit over the brake pads.

➡**Be careful when pressing in the piston; brake fluid might overflow from the master cylinder reservoir. If brake fluid gets on any painted surface, wash it off immediately with water.**

8. Remove the brake caliper piston compressor.

9. Apply Molykote CU-7439 PLUS PASTE (P/N 08798-9027) to the pad side of the shims, the back of the brake pads and the other areas indicated by the arrows. Wipe excess grease off the pad shims and brake pads. Contaminated brake discs or brake pads reduce stopping ability. Keep grease off the brake disc and pads.

10. Install the brake pads and pad shims correctly. Install the brake pad with the wear indicator on the upper outside. If you are reusing the brake pads, always reinstall the brake pads in their original positions to prevent a momentary loss of braking efficiency.

11. Install the pad spring.

12. Hold the pad spring, and install the pad pins into the caliper from the inside to the outside of vehicle using the commercially available 5/32" pin punch.

13. Clean the mating surfaces of the brake disc and the inside of the wheel, then install the front wheels.

14. Press the brake pedal several times to make sure the brakes work.

➡Engagement may require a greater pedal stroke immediately after the brake pads have been replaced as a set. Several applications of the brake pedal will restore the normal pedal stroke.

15. Add brake fluid as needed.

16. After installation, check for leaks at hose and line joints or connections, and retighten if necessary. Test-drive the vehicle, then check for leaks.

RL Models

1. Raise and safely support the vehicle.

2. Remove the front wheels.

3. Turn and twist out the clip from the caliper hole and pull the clip out from the pad pins.

4. Remove the pad pins and the pad spring.

5. Remove the pad.

6. Clean the caliper thoroughly; remove any rust, and check for grooves and cracks.

7. Check the brake disc for damage and cracks.

To install:

8. Install the brake caliper piston compressor on the caliper.

9. Press in the piston with the brake caliper piston compressor 07AAE-SEPA101 so that the caliper will fit over the pads.

Make sure the piston boot is in position to prevent damaging it.

➡Be careful when in pressing the piston the brake fluid might overflow from the master cylinder's reservoir.

10. Remove the brake caliper piston compressor.

11. Apply M-77 assembly paste to both sides of the pad shim, back of the brake pads and the other contact areas. Wipe excess assembly paste off the pads. Contaminated brake discs or pads reduce stopping ability.

12. Install the brake pads correctly. Install the brake pad with the wear indicator on the inside.

13. If you are reusing the brake pads, always reinstall the brake pads in their original positions to prevent a momentary loss of braking efficiency.

14. Install the pad spring. Hold the pad spring and install the pad pins into the caliper from the outside to the inside of vehicle.

15. First insert the clip ends to the pad pins and then twist the clip into the caliper hole to stabilize.

16. Press the brake pedal several times to make sure the brakes work.

➡The brakes may require a greater pedal stroke immediately after the

brake pads have been replaced as a set. Several applications of the brake pedal will restore the normal pedal stroke.

17. After installation, check for leaks at the brake hose and line joints or connections, and retighten if necessary.

18. Reinstall the front wheels, then test-drive the vehicle.

19. Check for leaks.

TSX Models

1. Raise and safely support the vehicle.

2. Install or connect the following:
- Pad retainers in position on the caliper bracket
- High temperature brake grease to the back side of the pads and both sides of shims and wipe off the excess
- Pads and shims
- Inner brake pad with the wear indicator facing upward

3. Loosen the bleed screw slightly and push in the caliper piston to allow mounting of the caliper over the rotor. Torque the bleed screw to 84 inch lbs. (9 Nm).

4. Pivot the caliper down over the rotor and install the caliper bolts. Torque the bolts to 36 ft. lbs. (50 Nm)

5. If disconnected, install the brake pad wear indicator connector. Install the wheels.

BRAKES

BRAKE CALIPER

REMOVAL & INSTALLATION

TL Models

1. Remove or disconnect the following:
- Wheels
- Caliper dust shield
- Parking brake cable from the caliper arm, if equipped
- Brake line from the caliper
- Caliper mounting bolts and pull the caliper off the bracket
- Pads, shim, and pad retainer spring
- Caliper bracket mounting bolts
- Bracket from the rotor

To install:

2. Install or connect the following:
- Caliper bracket. Torque the mounting bolts to 28 ft. lbs. (39 Nm).
- Pads, shims, and pad retainer springs
- Caliper. Torque the mounting bolts to 17 ft. lbs. (23 Nm).

- Brake hose with new crush washers and torque the banjo bolt to 25 ft. lbs. (35 Nm)
- Parking brake cable, if equipped

3. Bleed the brake system.
- Caliper dust shield and tighten the bolts to 84 inch lbs. (10 Nm)
- Rear wheels and torque the wheel nuts to 80 ft. lbs. (109 Nm)

RL Models

1. Raise and safely support the vehicle.

2. Remove or disconnect the following:
- Wheels
- Parking brake cable from the caliper arm, if equipped
- Brake line from the caliper
- Caliper mounting bolts and pull the caliper off the bracket
- Pads, shim, and pad retainer spring.
- Caliper bracket mounting bolts
- Bracket from the rotor

REAR DISC BRAKES

To install:

3. Install or connect the following:
- Caliper bracket. Torque the mounting bolts to 79 ft. lbs. (108 Nm).
- Pads, shims, and pad retainer springs
- Caliper. Torque the mounting bolts to 17 ft. lbs. (23 Nm).
- Brake hose.
- Parking brake cable, if equipped

4. Bleed the brake system.
- Rear wheels

TSX Models

1. Remove or disconnect the following:
- Wheels
- Caliper dust shield
- Parking brake cable from the caliper arm, if equipped
- Brake line from the caliper
- Caliper mounting bolts and pull the caliper off the bracket
- Pads, shim, and pad retainer spring.
- Caliper bracket mounting bolts
- Bracket from the rotor

To install:

2. Install or connect the following:
   - Caliper bracket. Torque the mounting bolts to 28 ft. lbs. (39 Nm).
   - Pads, shims, and pad retainer springs
   - Caliper. Torque the mounting bolts to 17 ft. lbs. (23 Nm).
   - Brake hose with new crush washers and torque the banjo bolt to 25 ft. lbs. (35 Nm).
   - Parking brake cable
3. Bleed the brake system.
4. Install the rear wheel.

DISC BRAKE PADS

REMOVAL & INSTALLATION

TL Models

1. Remove some brake fluid from the master cylinder.
2. Raise and safely support the vehicle.
3. Remove the rear wheel.
4. Release the parking brake.
5. Remove the brake hose from the bracket by removing the mounting bolt.
6. Remove the flange bolts while holding the caliper pins with a wrench being careful not to damage the pin boot, and remove the caliper. Check the hose and pin boots for damage and deterioration. Thoroughly clean the outside of the caliper to prevent dust and dirt from entering inside. Support the caliper with a piece of wire so it does not hang from the brake hose.
7. Remove the pad shims, brake pads and pad retainers.

To install:

8. Install the pad retainers.
9. Mount a brake caliper piston compressor on the caliper body (B).
10. Press in the piston with the brake caliper piston compressor so the caliper will fit over the brake pads. Make sure the piston boot is in position to prevent damaging it when installing the caliper.

➡**Be careful when pressing in the piston; brake fluid might overflow from the master cylinder's reservoir. If brake fluid gets on any painted surface, wash it off immediately with water.**

11. Remove the brake caliper piston compressor.
12. Apply a thin coat of M-77 assembly paste (P/N 08798-9010) to the brake pad side of the shims, and the back of the brake pads, and the other areas indicated by the arrows. Wipe excess assembly paste off the pad shims and brake pads. Contaminated brake discs or pads reduce stopping ability. Keep grease and assembly paste off the brake discs and brake pads.

13. Install the brake pads and pad shims on the caliper bracket. Install the inner brake pad with its wear indicator facing downward.
14. If you are reusing the brake pads, always reinstall the brake pads in their original positions to prevent a momentary loss of braking efficiency.
15. Push in the piston so the caliper will fit over the brake pads. Make sure the piston boot is in position to prevent damaging it when installing the caliper.
16. Install the brake caliper.
17. Install the flange bolts, and tighten it to the specified torque while holding the caliper pin with a wrench being careful not to damage the pin boot.
18. Install the brake hose onto the bracket with the mounting bolt.
19. Clean the mating surfaces of the brake disc/drum and the inside of the wheel, then install the rear wheels.
20. Press the brake pedal several times to make sure the brakes work.

➡**Engagement may require a greater pedal stroke immediately after the brake pads have been replaced as a set. Several applications of the brake pedal will restore the normal pedal stroke.**

21. Add brake fluid as needed.
22. After installation, check for leaks at hose and line joints or connections, and retighten if necessary.
23. Test-drive the vehicle, then check for leaks.

RL Models

1. Raise and safely support the vehicle.
2. Remove the rear wheels.
3. Release the parking brake.
4. Remove the caliper bracket mounting bolts while holding the caliper pins) with a wrench being careful not to damage the pin boot, and remove the caliper. Check the hose and pin boots for damage and deterioration. Thoroughly clean the outside of the caliper to prevent dust and dirt from entering inside.
5. Support the caliper with a piece of wire so it does not hang from the brake hose.
6. Remove the pad shims and brake pads.
7. Remove the pad retainers.

To install:

8. Apply M-77 assembly paste to the retainers on their mating surfaces against the caliper bracket.
9. Install the pad retainers. Wipe excess assembly paste off the retainers. Contaminated brake discs and pads reduce stopping ability. Keep assembly paste off the discs and pads.
10. Install the pad retainers.
11. Apply M-77 assembly paste to the pad side of the shims and back of the brake pads and the all other contact. Wipe excess assembly paste off the pad shims and brake pads. Contaminated brake discs or pads reduce stopping ability.
12. Install the brake pads and pad shims on the caliper bracket. Install the inner brake pad with its wear indicator facing on top.
13. If you are reusing the brake pads, always reinstall the brake pads in their original positions to prevent a momentary loss of braking efficiency

Push in the piston (A) so the caliper will fit over the brake pads. Make sure the piston boot is in position to prevent damaging it when installing the caliper.

➡**Be careful when pushing in the caliper, brake fluid might overflow from the master cylinder's reservoir.**

14. Apply M-77 assembly paste to the piston edges on their mating surfaces against the inner pad shim.
15. Install the brake caliper.
16. Install the caliper bolts, and torque them to 17 ft. lbs. (23 Nm) while holding the caliper pins with a wrench being and careful not to damage the pin boot.
17. Press the brake pedal several times to make sure the brakes work, then road-test the vehicle.

➡**The brake may require a greater pedal stroke immediately after the brake pads have been replaced as a set. Several applications of the brake pedal will restore the normal pedal stroke.**

18. After installation, check for leaks at the hose and line joints and connections, and retighten if necessary.

TSX Models

1. Remove some brake fluid from the master cylinder.
2. Raise and safely support the vehicle.
3. Remove the rear wheel.
4. Remove the flange bolts while holding the caliper pin with a wrench being

careful not to damage the pin boot, and remove the caliper. Check the hose and pin boots for damage and deterioration.

5. Remove the brake pads, shims and pad retainers.

To install:

6. Apply a thin coat of M-77 assembly paste to the retainers on their mating surfaces against the caliper.

7. Install the pad retainers.

8. Apply a thin coat of M-77 assembly paste to both sides of the pad shim, the back of brake pads, and the other areas indicated by the arrows. Wipe excess assembly paste off the pad shims and pads. Contaminated brake discs or pads reduce stopping ability. Keep grease and paste off the brake discs and pads.

9. Install the brake pads and pad shims correctly. Install the brake pad with the wear indicator on the bottom inside. If you are reusing the brake pads, always reinstall the brake pads in their original positions to prevent a momentary loss of braking efficiency.

10. Rotate the caliper piston clockwise into the cylinder, then align the cutout in the piston with the tab on the inner pad by turning the piston back. Lubricate the boot with rubber grease to avoid twisting the piston boot. If the piston boot is twisted, back it out so it is positioned properly.

✳✳ WARNING

Be careful when moving the piston back in the caliper; brake fluid might overflow from the master cylinder's reservoir.

11. Install the caliper. Install the flange bolts, and tighten it to the specified torque while holding the caliper pin (E) with a wrench being careful not to damage the pin boot.

12. Clean the mating surfaces of the brake disc and the inside of the wheel, then install the rear wheels.

13. Press the brake pedal several times to make sure the brakes work.

➡**Engagement may require a greater pedal stroke immediately after the brake pads have been replaced as a set. Several applications of the brake pedal will restore the normal pedal stroke.**

14. Add brake fluid as needed.

15. After installation, check for leaks at hose and line joints or connections, and retighten if necessary. Test-drive the vehicle, then recheck for leaks.

BRAKES

PARKING BRAKE CABLES

ADJUSTMENT

TL Models

1. Raise and safely support the vehicle.

2. Release the parking brake lever fully.

3. Open the console box lid, and remove the console mat.

4. Remove the lid to access the adjusting nut.

5. Pull the parking brake lever one notch.

6. Tighten the parking brake adjusting nut until the parking brakes drag slightly when the rear wheels are turned.

7. Release the parking brake lever fully, and check that the parking brakes do not drag when the rear wheels are turned. Readjust if necessary.

8. Make sure the parking brakes are fully applied when the parking brake lever is pulled up all the way.

9. Reinstall the center console lid and console mat.

RL Models

1. Raise and safely support the vehicle.

2. Release the parking brake pedal fully.

3. Press the parking brake pedal one notch.

4. Tighten the adjusting nut until the parking brakes drag slightly when the rear wheels are rotated.

5. Release the parking brake pedal fully, and check that the parking brakes do not drag when the rear wheels are rotated. Readjust if necessary.

6. Make sure the parking brakes are fully applied when the parking brake pedal is pressed all the way.

TSX Models

1. Release the parking brake lever fully.

2. Remove the center console.

3. Loosen the parking brake adjusting nut, start the engine, and press the brake

PARKING BRAKE

pedal several times to set the self-adjusting brake before adjusting the parking brake.

4. Raise and safely support the vehicle.

5. Remove the rear wheels.

6. Make sure the parking brake arm on the rear brake caliper contacts the brake caliper pin.

7. Pull the parking brake lever one click.

8. Clean the mating surfaces of the brake disc and the inside of the wheel, then install the rear wheels.

9. Tighten the adjusting nut until the parking brakes drag slightly when the rear wheels are turned.

10. Release the parking brake lever fully, and check that the parking brakes do not drag when the rear wheels are turned. Readjust if necessary.

11. Make sure the parking brakes are fully applied when the parking brake lever is pulled all the way.

12. Install the center console.

CHASSIS ELECTRICAL   AIR BAG (SUPPLEMENTAL RESTRAINT SYSTEM)

GENERAL INFORMATION

SERVICE PRECAUTIONS

Disconnect and isolate the battery negative cable before beginning any airbag system component diagnosis, testing, removal, or installation procedures. Allow system capacitor to discharge for two minutes before beginning any component service. This will disable the airbag system. Failure to disable the airbag system may result in accidental airbag deployment, personal injury, or death.

Do not place an intact undeployed airbag face down on a solid surface. The airbag will propel into the air if accidentally deployed and may result in personal injury or death.

When carrying or handling an undeployed airbag, the trim side (face) of the airbag should be pointing towards the body to minimize possibility of injury if accidental deployment occurs. Failure to do this may result in personal injury or death.

Replace airbag system components with OEM replacement parts. Substitute parts may appear interchangeable, but internal differences may result in inferior occupant protection. Failure to do so may result in occupant personal injury or death.

Wear safety glasses, rubber gloves, and long sleeved clothing when cleaning powder residue from vehicle after an airbag deployment. Powder residue emitted from a deployed airbag can cause skin irritation. Flush affected area with cool water if irritation is experienced. If nasal or throat irritation is experienced, exit the vehicle for fresh air until the irritation ceases. If irritation continues, see a physician.

Do not use a replacement airbag that is not in the original packaging. This may result in improper deployment, personal injury, or death.

The factory installed fasteners, screws and bolts used to fasten airbag components have a special coating and are specifically designed for the airbag system. Do not use substitute fasteners. Use only original equipment fasteners listed in the parts catalog when fastener replacement is required.

During, and following, any child restraint anchor service, due to impact event or vehicle repair, carefully inspect all mounting hardware, tether straps, and anchors for proper installation, operation, or damage. If a child restraint anchor is found damaged in any way, the anchor must be replaced. Failure to do this may result in personal injury or death.

Deployed and non-deployed airbags may or may not have live pyrotechnic material within the airbag inflator.

Do not dispose of driver/passenger/curtain airbags or seat belt tensioners unless you are sure of complete deployment. Refer to the Hazardous Substance Control System for proper disposal.

Dispose of deployed airbags and tensioners consistent with state, provincial, local, and federal regulations.

After any airbag component testing or service, do not connect the battery negative cable. Personal injury or death may result if the system test is not performed first.

If the vehicle is equipped with the Occupant Classification System (OCS), do not connect the battery negative cable before performing the OCS Verification Test using the scan tool and the appropriate diagnostic information. Personal injury or death may result if the system test is not performed properly.

Never replace both the Occupant Restraint Controller (ORC) and the Occupant Classification Module (OCM) at the same time. If both require replacement, replace one, then perform the Airbag System test before replacing the other.

Both the ORC and the OCM store Occupant Classification System (OCS) calibration data, which they transfer to one another when one of them is replaced. If both are replaced at the same time, an irreversible fault will be set in both modules and the OCS may malfunction and cause personal injury or death.

If equipped with OCS, the Seat Weight Sensor is a sensitive, calibrated unit and must be handled carefully. Do not drop or handle roughly. If dropped or damaged, replace with another sensor. Failure to do so may result in occupant injury or death.

If equipped with OCS, the front passenger seat must be handled carefully as well. When removing the seat, be careful when setting on floor not to drop. If dropped, the sensor may be inoperative, could result in occupant injury, or possibly death.

If equipped with OCS, when the passenger front seat is on the floor, no one should sit in the front passenger seat. This uneven force may damage the sensing ability of the seat weight sensors. If sat on and damaged, the sensor may be inoperative, could result in occupant injury, or possibly death.

DISARMING THE SYSTEM

Disconnect the negative battery cable and wait at least 3 minutes before beginning to work.

ARMING THE SYSTEM

Reconnect the negative battery cable. Turn the ignition switch to **ON**. The SRS indicator should come on for about 6 seconds and then go off.

CLOCKSPRING CENTERING

See Figure 3.

1. Rotate the cable reel clockwise until it stops.
2. Then rotate it counterclockwise (about 2½–3 turns) until the arrow mark on the cable reel label points straight up.

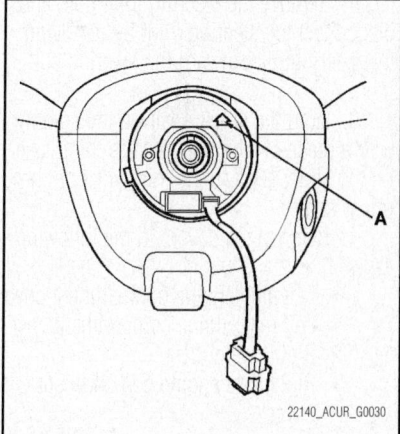

22140_ACUR_G0030

Fig. 3 After rotating the cable reel clockwise, rotate it counterclockwise (about 2½–3 turns) until the arrow mark on the cable reel label points straight up as shown.

DRIVETRAIN

AUTOMATIC TRANSAXLE ASSEMBLY

REMOVAL & INSTALLATION

TL Models

See Figures 4 through 16.

1. Set the wheels in the straight ahead position.

2. Lock the steering wheel.

3. Disconnect the support strut from both sides of the pivot ball (bolted to the hood). Secure the hood in a vertical position. Remove the right side pivot ball and install it into the lower threaded hole, then reattach the support strut.

➡**Do not attempt to close the hood with the support strut in the vertical position; it will damage the support strut and the hood.**

4. Set the wheels in the straight ahead position.

5. Lock the steering wheel.

6. Drain the power steering system fluid from the reservoir.

7. Remove the steering joint cover.

8. Make a reference mark across the steering joint and steering gearbox pinion shaft.

9. Remove the steering joint bolt, and disconnect the steering joint by removing the steering joint toward the steering column.

10. Hold the slider shaft on the column with a piece of wire between the joint yoke on the slider shaft to the joint yoke on the upper shaft.

11. Remove the covers in the following order:
   - Left side engine compartment cover
   - Left rear engine compartment cover
   - Right fender trim
   - Right side engine compartment cover
   - Right rear engine compartment cover
   - Left fender trim
   - Front bulkhead cover
   - Intake manifold cover

12. Remove the harness clamp, two 6 mm bolts and the strut brace.

13. Remove the power steering pump outlet line from the pump, and remove the hose from its clamp.

14. Remove the transmission under cover.

15. Remove the splash shield.

16. Drain the transmission fluid.

17. Reinstall the drain plug with a new sealing washer.

18. Disconnect the battery cables, negative cable first.

19. Remove the battery hold-down bracket, the battery and battery tray.

20. Remove the resonator cover and resonator.

21. Remove the intake air duct and air cleaner housing.

22. Remove the bolts retaining the battery base from under the vehicle, and in the engine compartment, then remove the base.

23. Remove the starter.

24. Remove the transmission upper mount bracket and bracket plate.

25. Remove the transmission sub-harness connector from its bracket and disconnect it.

26. Disconnect the input shaft speed sensor connector and 4th clutch transmission fluid pressure switch connector.

27. Disconnect the vacuum hose from the vacuum line.

28. Disconnect shift solenoid valve connectors and the automatic transmission clutch pressure control solenoid valve connector, then remove the harness clamps from the clamp brackets.

29. Disconnect the transmission clutch pressure control solenoid valve connectors, then remove the harness clamp from the clamp bracket.

30. Disconnect the connectors from the torque converter clutch solenoid valve and shift solenoid valve.

31. Remove the transmission range switch connector from its bracket, and disconnect it.

32. Disconnect the output shaft speed sensor connector, and remove the harness clamps from the brackets.

33. Disconnect the vacuum hose from the vacuum line.

34. Remove the harness clamps from the brackets.

35. Remove the harness cover from the bracket.

36. Remove the bolt retaining the bracket.

37. Remove the transmission ground cable, and disconnect the breather tube.

38. Unfasten the bolts retaining the ATF warmer and the bracket.

39. Remove the ATF warmer from the transmission housing. Cover the fluid passages on the transmission and ATF warmer with tape. Do not disconnect the hoses.

40. Remove the connector bracket from the engine front cylinder head; use the

bracket bolt to attach engine balancer bar front arm.

41. Remove the harness clamp bracket from the engine rear cylinder head; use the bracket bolt to attach engine balancer bar rear arm.

42. Lift and support the engine with engine hanger and engine balancer bar. Attach the front arm to the front cylinder head with a spacer and the connector bracket 10 x 1.25 mm bolt. Attach the rear arm to the rear cylinder head with the harness clamp bracket 8 x 1.25 mm bolt.

43. Remove the front mount stop, and remove the front mount bolt.

44. Insert a 6 mm Allen wrench in the top of the ball joint pin, remove the nuts and separate the stabilizer link from the lower arms.

45. Separate the ball joints from the lower arms.

46. Separate the tie-rod end ball joints from the knuckles.

47. Remove exhaust pipe and its mount.

48. Remove the steering gearbox heat shield.

49. Remove the power steering fluid hose from its line on the front sub-frame.

50. Disconnect the power steering pressure switch connector.

51. Remove the torque converter cover and the drive plate bolts while rotating the crankshaft pulley.

52. Remove the engine-to-torque converter housing mounting bolts.

53. Remove the bolts attaching the shift cable holder and the shift cable cover.

➡**To prevent damage to the control lever joint, remove the bolts retaining the holder before removing the bolts retaining the cover.**

54. Remove the lock bolt retaining the selector control lever, then remove the shift cable and the control lever. Do not bend the shift cable excessively.

55. Install a 6 x 1.0 x 14 mm bolt and nut on the shift cable cover, then reinstall the shift cable cover to the torque converter housing. If you do not perform this step, the bolt head of the cable cover may prevent you from removing the torque converter during transmission removal.

56. Remove the rear mount base bracket bolts.

57. Remove the transmission lower mount nuts.

58. Remove the both mid-mounts.

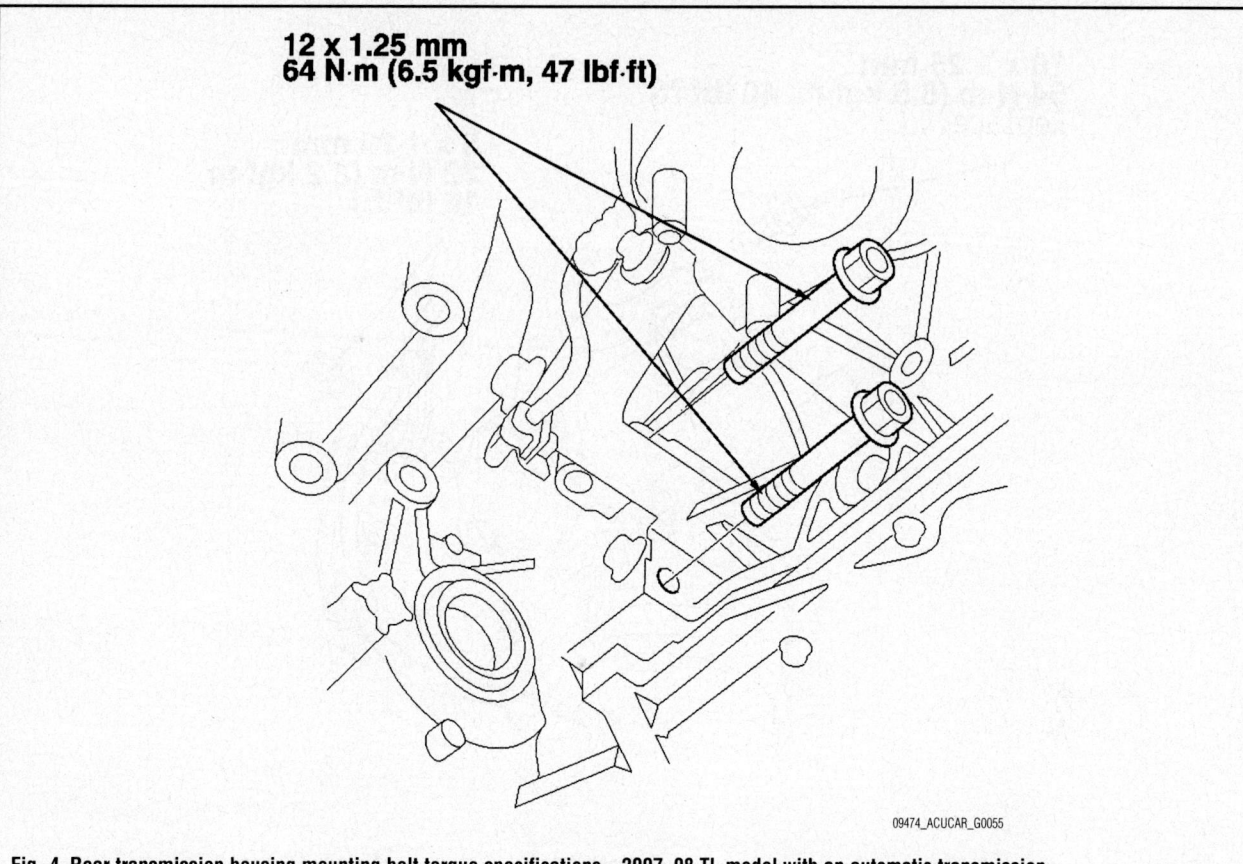

12 x 1.25 mm
64 N·m (6.5 kgf·m, 47 lbf·ft)

09474_ACUCAR_G0055

Fig. 4 Rear transmission housing mounting bolt torque specifications—2007–08 TL model with an automatic transmission

12 x 1.25 mm
64 N·m
(6.5 kgf·m, 47 lbf·ft)

09474_ACUCAR_G0056

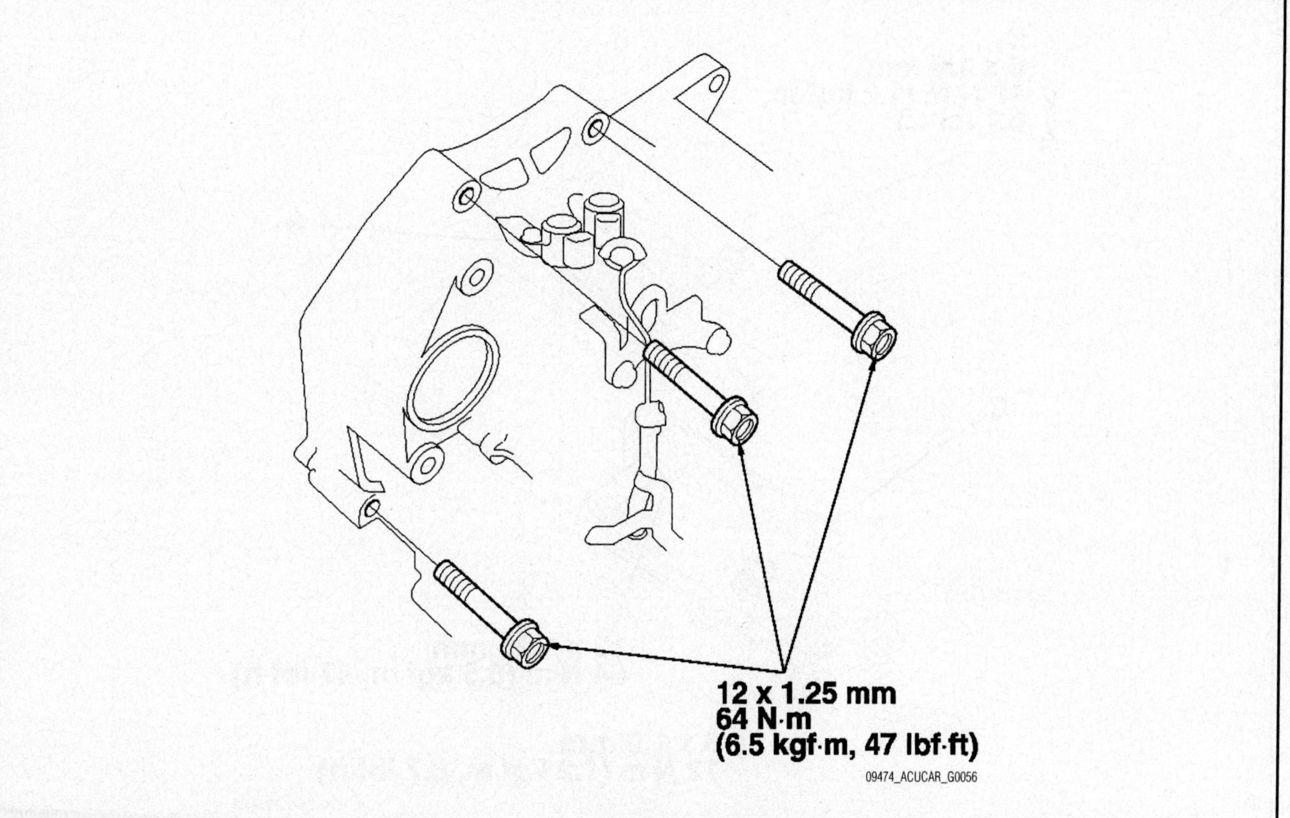

Fig. 5 Upper and front transmission housing mounting bolt torque specifications—2007–08 TL model with an automatic transmission

10 x 1.25 mm
54 N·m (5.5 kgf·m, 40 lbf·ft)
Replace.

8 x 1.25 mm
22 N·m (2.2 kgf·m,
16 lbf·ft)

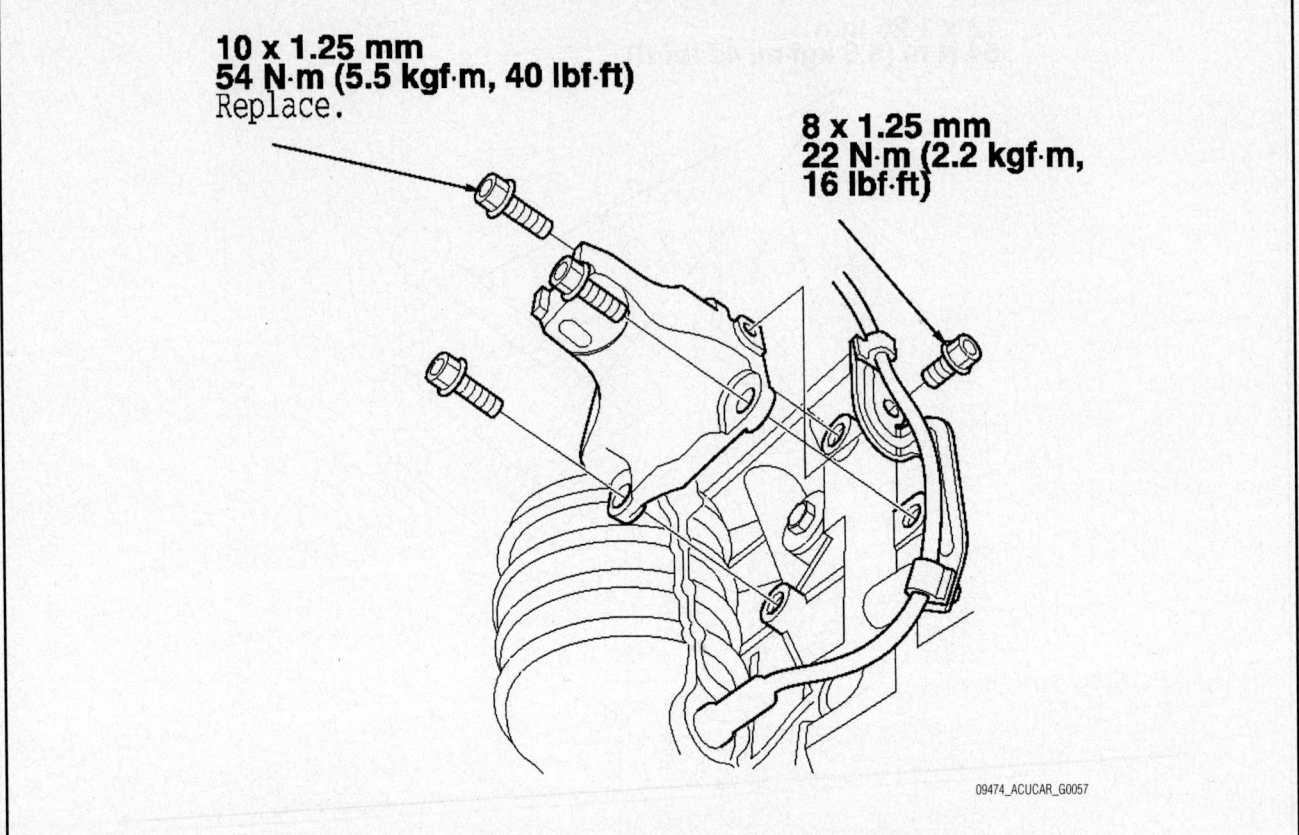

09474_ACUCAR_G0057

Fig. 6 Front mount bracket and harness clamp bolt torque specifications—2007–08 TL model with an automatic transmission

B
6 x 1.0 mm
12 N·m (1.2 kgf·m,
8.7 lbf·ft)

A

C

12 x 1.25 mm
64 N·m (6.5 kgf·m, 47 lbf·ft)

6 x 1.0 mm
12 N·m (1.2 kgf·m, 8.7 lbf·ft)

09474_ACUCAR_G0058

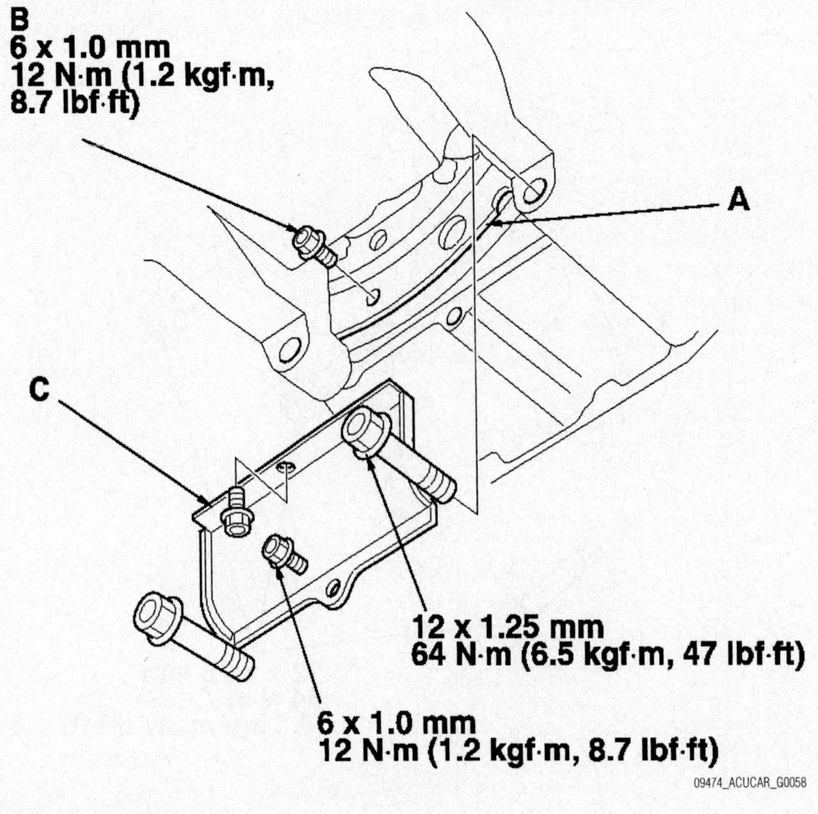

Fig. 7 Engine-to-torque converter housing mounting bolt torque specifications—2007–08 TL model with an automatic transmission

59. Make the appropriate reference lines at both ends of the sub-frame that line up with the edge of the stiffeners.

60. Attach the tool VSB02C000016to the sub-frame by hanging the strap of the tool over the front of the sub-frame, then secure the strap with its stop.

61. Raise the jack and line up the slots in the arms with the bolt holes on the corner of the jack base, then attach them with bolts securely.

62. Remove the four bolts retaining the stiffeners, the four bolts retaining the front sub-frame and lower the front sub-frame.

63. Remove the transmission lower mounts.

64. Remove the driveshafts from the differential and intermediate shaft.

65. Remove the exhaust manifold bracket and heat shield.

Remove the intermediate shaft.

66. Coat all precision finished surfaces with clean engine oil, then tie plastic bags over both ends of driveshaft and intermediate shaft.

67. Place a jack under the transmission.

68. Remove the transmission housing mounting bolts.

69. Remove the bolt retaining the harness clamp bracket, and remove the front mount bracket.

70. Remove the transmission housing mounting bolts.

71. Slide the transmission away from the engine to remove it from the vehicle.

To install:

72. Place the transmission on the jack, and raise the transmission to the engine level.

73. Attach the transmission to the engine, then install the transmission housing mounting bolts.

74. Install the transmission housing mounting bolts and tighten to 47 ft. lbs. (64 Nm).

75. Remove the jack from the transmission.

76. Install the front mount bracket with the new bolts, tighten the bolts to 40 ft. lbs. (54 Nm).

77. Install the harness clamp on the mount bracket and tighten to 16 ft. lbs. (22 Nm).

78. Install the engine-to-torque converter housing mounting bolts and tighten to 47 ft. lbs. (64 Nm).

79. Attach the torque converter to the drive plate with the eight bolts. Rotate the crankshaft pulley as necessary to tighten the bolts to 4.5 ft. lbs. (6 Nm), then to the final torque in a crisscross pattern to 9 ft. lbs. (12 Nm). After tightening the last bolt, check that the crankshaft rotates freely.

80. Install the torque converter cover.

81. Install the new set ring on the intermediate shaft.

82. Install the exhaust manifold bracket tighten the bolts to 20 ft. lbs. (26 Nm).

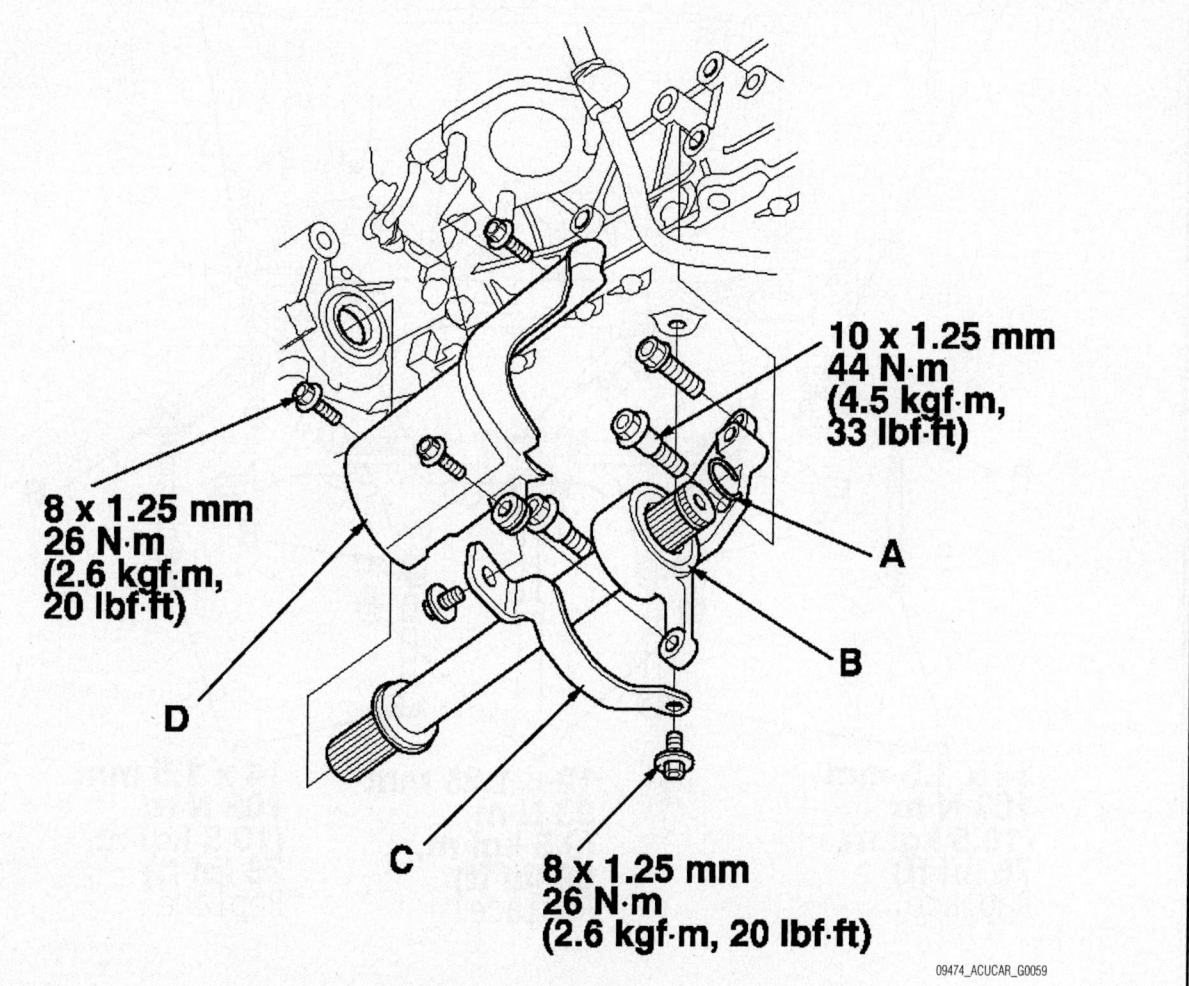

09474_ACUCAR_G0059

Fig. 0 Install the new set ring (A) on the shaft (B) and tighten the exhaust manifold bracket (C) and heat shield (D) bolts—2007–08 TL model with an automatic transmission

Install the heat shield tighten the bolts to 20 ft. lbs. (26 Nm).

83. Install the new set ring on the left driveshaft, then install the left driveshaft in the differential. While installing the driveshaft in the differential, make sure not to allow dirt or other particles to enter the transmission. Install the left driveshaft over the intermediate shaft.

➡ **Clean the areas where the driveshaft and intermediate shaft contact the transmission with solvent and dry with compressed air. Turn the right and left steering knuckle fully out, and slide the driveshaft and intermediate shaft into the differential and intermediate shaft until you feel the set ring engage the side gear.**

84. Install the transmission lower front mount tighten the bolts to 33 ft. lbs. (44 Nm).

85. Install the transmission lower rear mount with the new bolts. Tighten the bolts to 33 ft. lbs. (44 Nm).

86. Support the front sub-frame with the tool and a jack, and lift it up to body.

87. Loosely install the new sub-frame mounting bolts, and new rear stiffener mounting bolts, and the front stiffener mounting bolts.

88. Loosely install both of the new both mid-mount mounting bolts.

89. Align the reference marks with edge of both rear stiffeners, and tighten the rear sub-frame mounting bolts, the front bolts and the stiffener bolts in that order to the torque shown in the accompanying illustration.

90. Tighten the mid-mount mounting bolts to the torque shown in the accompanying illustration.

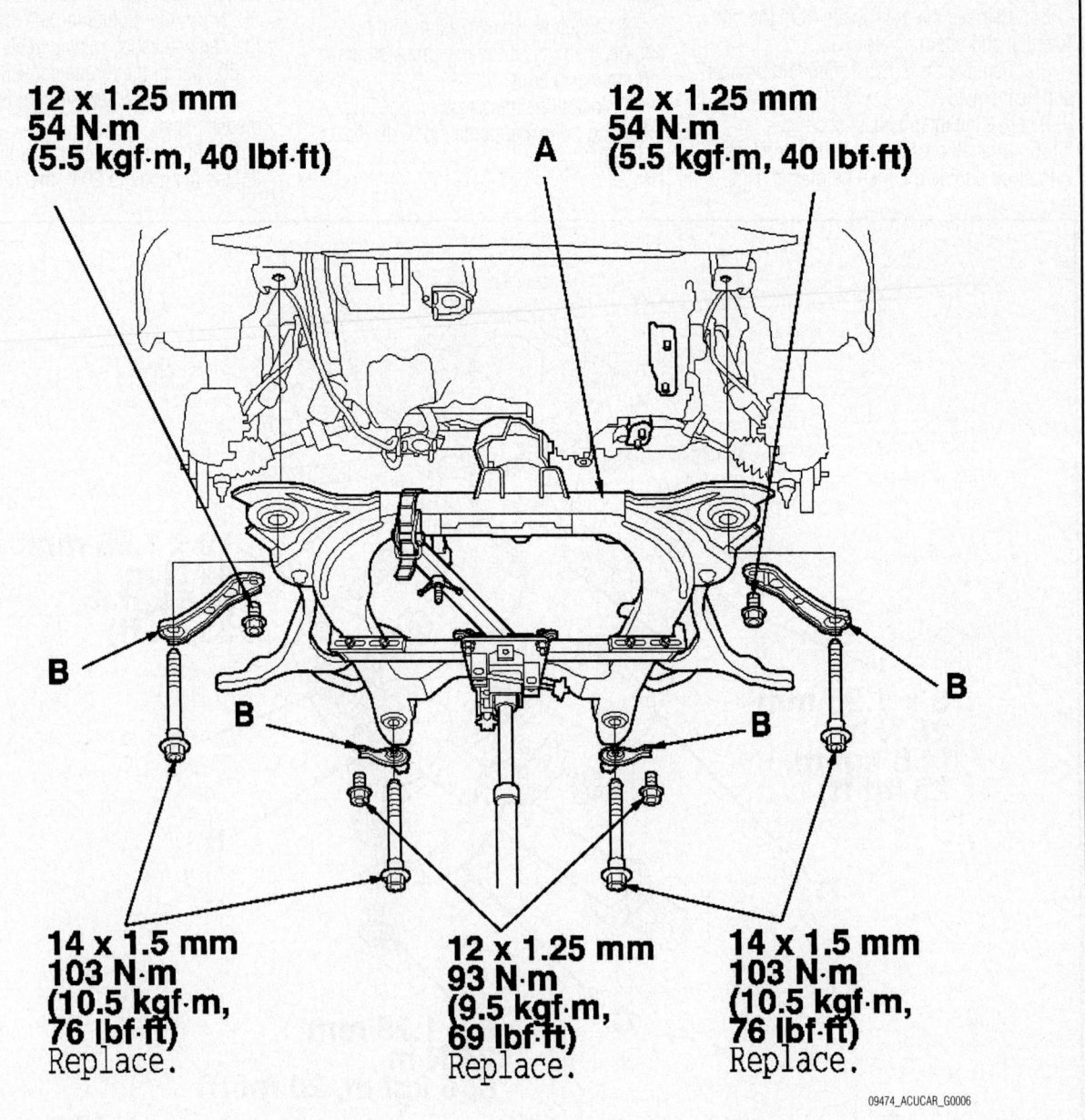

12 x 1.25 mm
54 N·m
(5.5 kgf·m, 40 lbf·ft)

A

12 x 1.25 mm
54 N·m
(5.5 kgf·m, 40 lbf·ft)

B

B

B

B

14 x 1.5 mm
103 N·m
(10.5 kgf·m, 76 lbf·ft)
Replace.

12 x 1.25 mm
93 N·m
(9.5 kgf·m, 69 lbf·ft)
Replace.

14 x 1.5 mm
103 N·m
(10.5 kgf·m, 76 lbf·ft)
Replace.

09474_ACUCAR_G0006

Fig. 9 Sub-frame bolt locations and torque specifications—2007–08 TL model

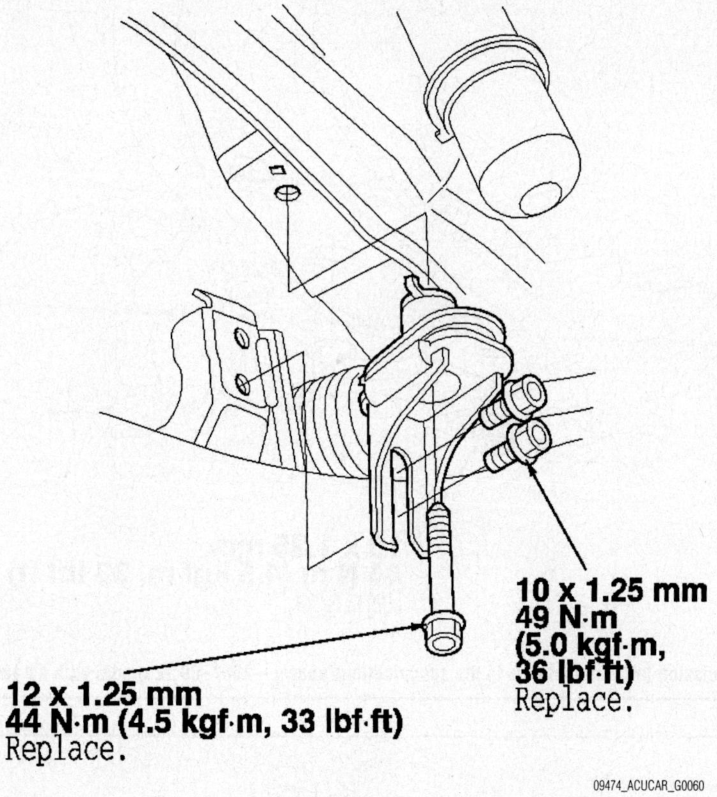

10 x 1.25 mm
49 N·m
(5.0 kgf·m,
36 lbf·ft)
Replace.

12 x 1.25 mm
44 N·m (4.5 kgf·m, 33 lbf·ft)
Replace.

09474_ACUCAR_G0060

Fig. 10 Tighten the mid-mount mounting bolts to the torque shown—2007–08 TL model with an automatic transmission

10 x 1.25 mm
44 N·m (4.5 kgf·m, 33 lbf·ft)
Replace.

09474_ACUCAR_G0061

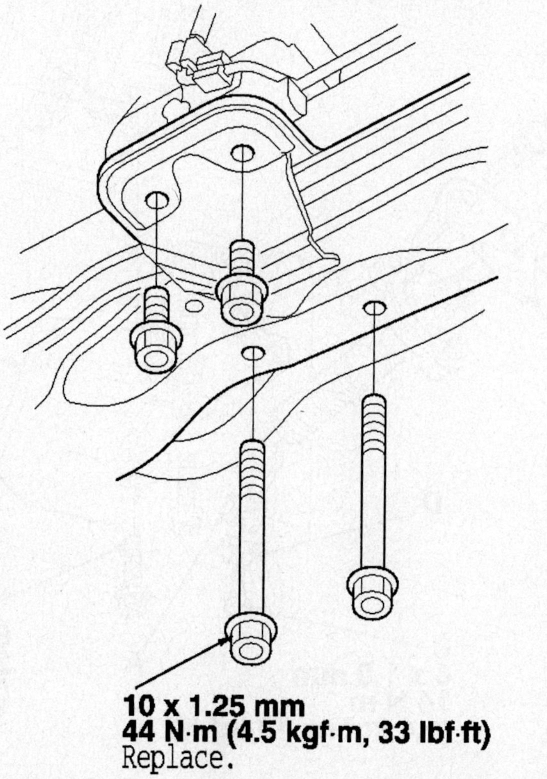

Fig. 11 Tighten the new rear mount bracket bolts to the torque shown 2007–08 TL model with an automatic transmission

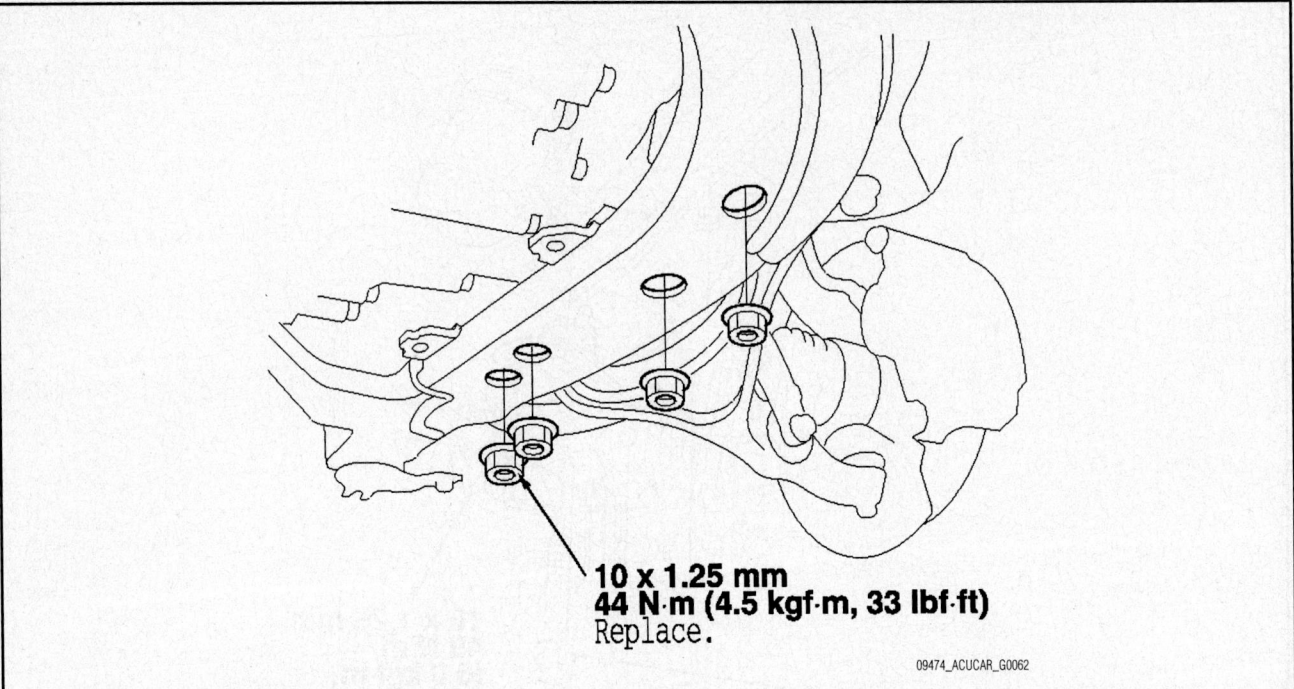

10 x 1.25 mm
44 N·m (4.5 kgf·m, 33 lbf·ft)
Replace.

09474_ACUCAR_G0062

Fig. 12 Tighten the new transmission lower mount nuts to the specifications shown—2007–08 TL model with an automatic transmission

6 x 1.0 mm
9.8 N·m
(1.0 kgf·m, 7.2 lbf·ft)

B

E

F

D

C

A

6 x 1.0 mm
14 N·m
(1.4 kgf·m, 10 lbf·ft)

8 x 1.25 mm
22 N·m
(2.2 kgf·m,
16 lbf·ft)

09474_ACUCAR_G0063

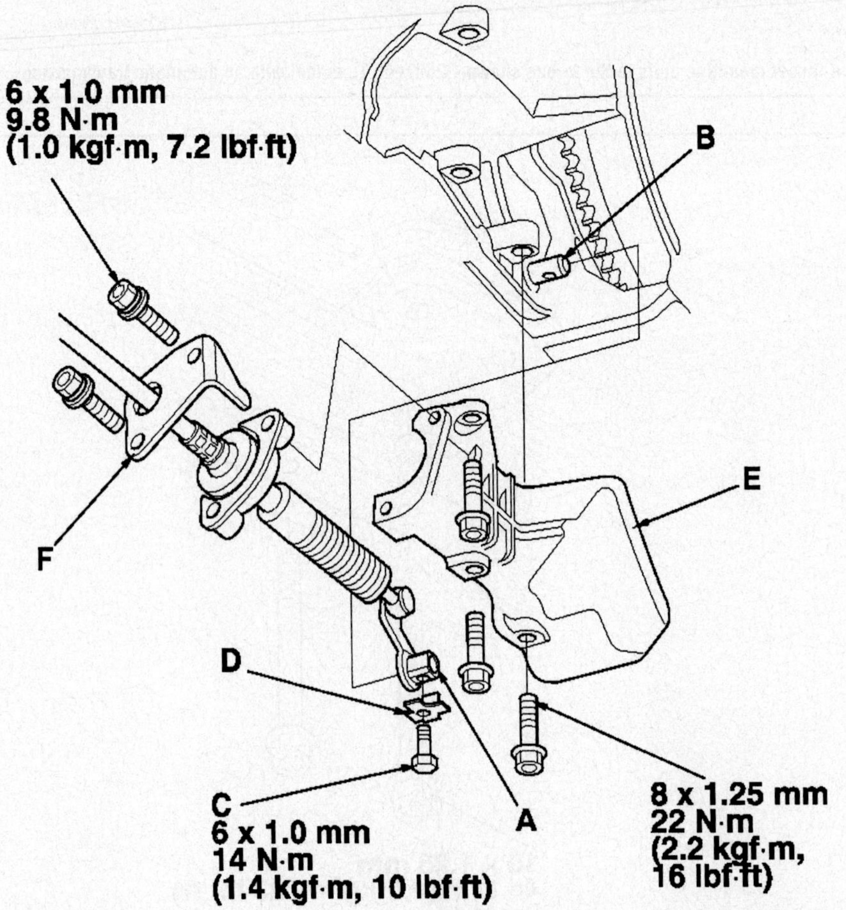

Fig. 13 Install selector control lever on the selector control shaft. Tighten all fasteners to the specifications shown—2007–08 TL model with an automatic transmission

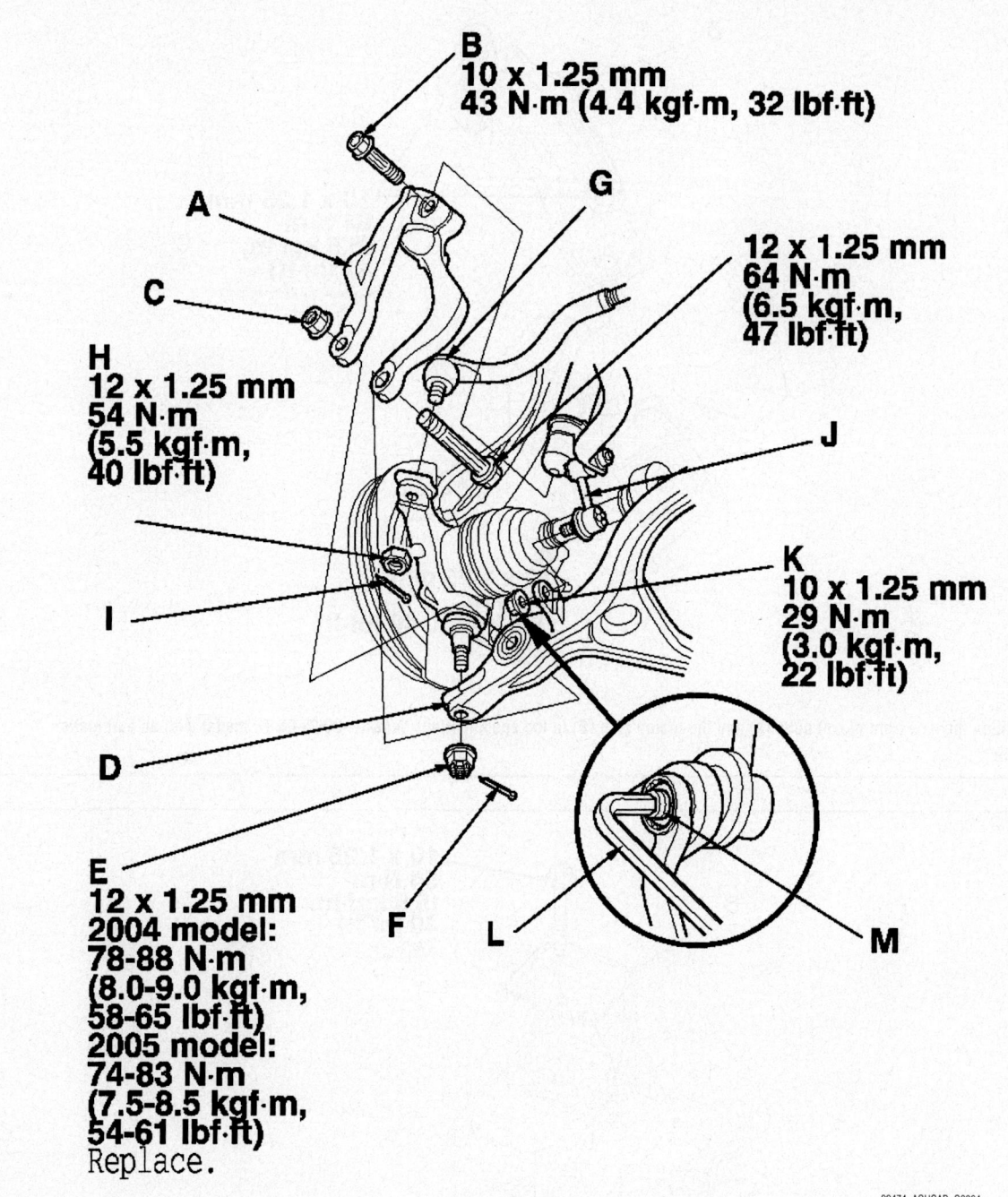

B
10 x 1.25 mm
43 N·m (4.4 kgf·m, 32 lbf·ft)

G

A

C

12 x 1.25 mm
64 N·m
(6.5 kgf·m,
47 lbf·ft)

H
12 x 1.25 mm
54 N·m
(5.5 kgf·m,
40 lbf·ft)

J

K
10 x 1.25 mm
29 N·m
(3.0 kgf·m,
22 lbf·ft)

I

D

E
12 x 1.25 mm
2004 model:
78-88 N·m
(8.0-9.0 kgf·m,
58-65 lbf·ft)
2005 model:
74-83 N·m
(7.5-8.5 kgf·m,
54-61 lbf·ft)
Replace.

F   **L**

M

09474_ACUCAR_G0064

Fig. 14 Exploded view of the damper fork assembly and related components—2007–08 TL model with an automatic transmission

91. Install the new rear mount bracket bolts to the torque shown in the accompanying illustration.

92. Install the new transmission lower mount nuts to the torque shown in the accompanying illustration.

93. Connect the power steering pressure switch connector.

94. Connect the power steering hose to the power steering line.

95. Install selector control lever on the selector control shaft. Do not bend the shift cable excessively.

96. Install the lock bolt with a new lock washer, then bend the lock washer tab against the bolt head.

97. Install the shift cable cover, then secure the shift cable holder on the cover with the bolts.

➡**To prevent damage to the control lever joint, be sure install the shift cable holder after installing the shift**

cable cover to the torque converter housing.

98. Install the remaining components in the reverse order of removal keeping in mind the following component torque specifications:

99. Install the damper forks (A) with damper pinch bolts (B) and the new self-locking nuts (C), then install the ball joints on the lower arms (D) with the ball joint nuts (E) and new cotter pins (F). Install the

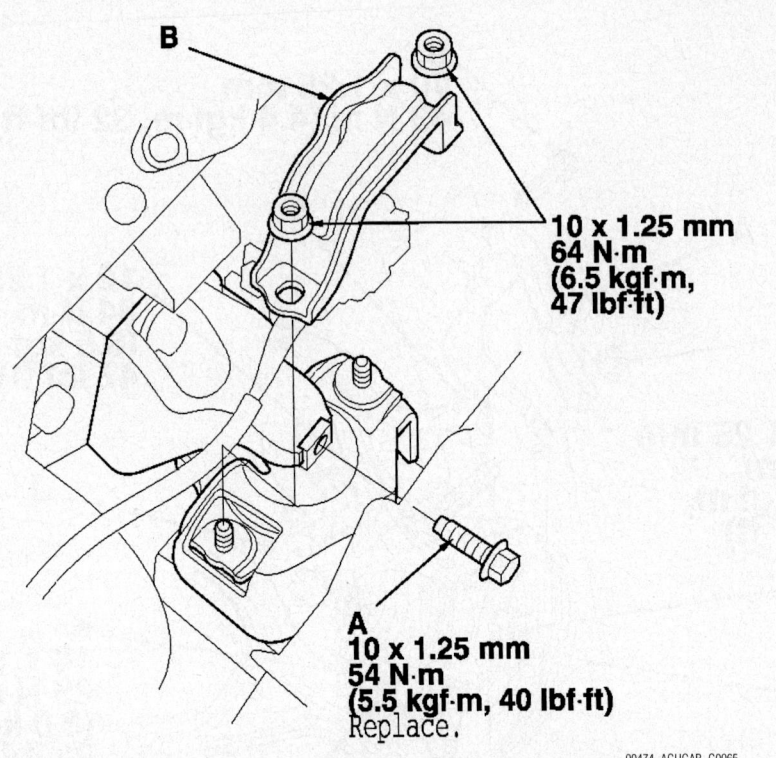

B

10 x 1.25 mm
64 N·m
(6.5 kgf·m,
47 lbf·ft)

A
10 x 1.25 mm
54 N·m
(5.5 kgf·m, 40 lbf·ft)
Replace.

09474_ACUCAR_G0065

Fig. 15 Tighten the new front mount bolts (A) and the mount stop (B) to the specifications shown—2007–08 TL model with an automatic transmission

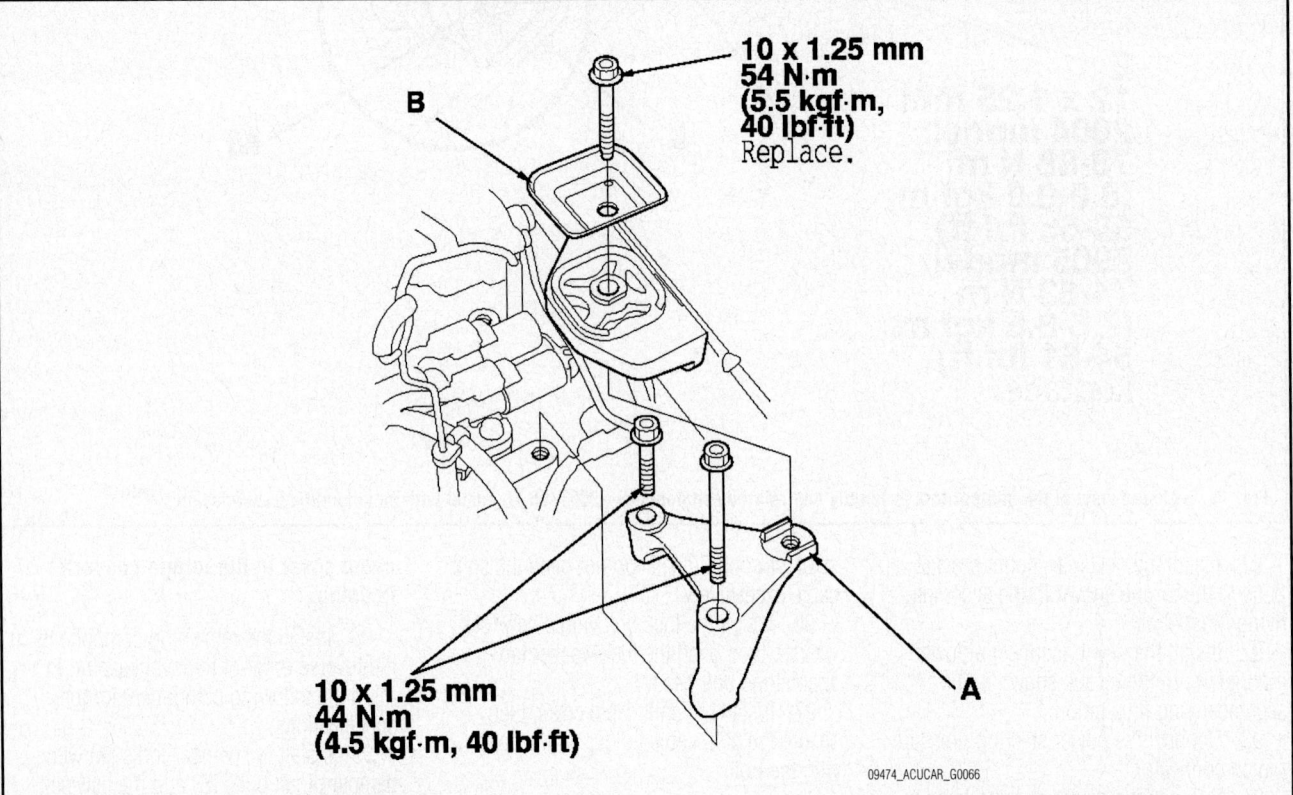

10 x 1.25 mm
54 N·m
(5.5 kgf·m,
40 lbf·ft)
Replace.

B

10 x 1.25 mm
44 N·m
(4.5 kgf·m, 40 lbf·ft)

A

09474_ACUCAR_G0066

Fig. 16 Tighten the front mount stop transmission upper mount bracket (A) and new upper mount bolt and bracket (B) to the torque shown—2007–08 TL model with an automatic transmission

tie-rod end ball joints (G) to each knuckle with the nuts (H) and new cotter pins (I). Install the stabilizer links (J) to the lower arms, and install the nuts (K). Insert a 6 mm Allen wrench (L) in the ball joint (M), and tighten the nuts. Refer to the accompanying illustration for component location and torque value.

100. Install the new front mount bolt to the torque shown in the accompanying illustration.

101. Install the front mount stop to the torque shown in the accompanying illustration.

102. Install the transmission upper mount bracket on the transmission with two bolts to the torque shown in the accompanying illustration.

103. Install the new upper mount bolt and bracket plate to the torque shown in the accompanying illustration.

104. Connect the steering joint to the steering gearbox pinion shaft by aligning the reference mark, and remove the wire from the joint yoke. Tighten the bolt to 20 ft. lbs. (26 Nm).

105. The remainder of the installation is the reverse order of removal.

RL Models

See Figures 17 through 21.

Make sure you have the customer's radio and navigation anti-theft codes, and write down the radio presets.

1. Disconnect the negative battery cable.

2. Remove the battery trim, left upper fender trim, cowl cover upper trim, right upper fender trim, upper grille cover and intake manifold cover.

3. Remove the windshield wiper arms and cowl cover.

4. Disconnect the support strut from the pivot ball on both sides.

5. Raise and secure the hood in a vertical position. Remove the right side pivot ball and install it into the lower threaded hole, then reattach the support strut.

➡**Do not attempt to close the hood with the support strut in the vertical position; it will damage the support strut and the hood.**

6. Set the wheels in the straight ahead position, and lock the steering wheel.

7. Drain the power steering system fluid from the reservoir.

8. Make sure the ignition switch is OFF. Disconnect the negative terminal from the battery, then disconnect the positive terminal.

9. Remove the battery hold-down bracket, and remove the battery cover, battery and battery tray.

10. Remove the air intake duct and air cleaner housing.

11. Remove the battery base.

12. Remove the splash shield.

13. Drain the automatic transmission fluid (ATF).

14. Reinstall the drain plug and a new sealing washer.

15. Remove the strut brace.

16. Disconnect the steering joint.

17. Remove the power steering pump outlet line from the power steering pump and remove the hose from its clamp.

18. Disconnect the power steering pressure switch connector.

19. Remove the starter.

20. Disconnect the solenoid harness connector.

21. Remove the bolt securing the radiator hose clamp.

22. Disconnect the vacuum hose from the vacuum line and remove the vacuum line bolt.

23. Remove the nuts securing the shift cable bracket.

24. Remove the spring clip/washer and the control pin, then separate the shift cable end from the control lever.

25. Disconnect the A/T clutch pressure control solenoid valve A connector, A/T clutch pressure control solenoid valve B connector, and 4th clutch transmission fluid pressure switch connectors, and remove the harness cover mounting bolt.

26. Remove the harness cover mounting bolt, and remove the harness clamps from the clamp brackets.

27. Disconnect the A/T clutch pressure control solenoid valve C connector, and remove the connector bracket and the harness clamp bracket.

28. Disconnect the transmission range switch connector and remove the harness clamp from the clamp bracket.

29. Remove the transmission ground cable.

30. Disconnect the output shaft speed sensor connector, input shaft speed sensor connector, 3rd clutch transmission fluid pressure switch connector and ATF temperature sensor connector.

31. Disconnect the 2nd clutch transmission fluid pressure switch connector and remove the vacuum line bolt.

32. Remove the ATF cooler hoses from the ATF cooler lines. Turn the ends of the cooler hoses up to prevent ATF from flowing out, then plug the cooler hoses and lines.

33. Remove the connector bracket from the engine front cylinder head, use the bracket bolt hole to attach the engine balancer bar front arm.

34. Disconnect the solenoid connector, disconnect the vacuum tube from the joint, then remove the bracket from the engine rear cylinder head; use the bracket bolt hole to attach the engine balancer bar rear arm.

35. Install the engine balancer bar No. VSB02C000019), attach the front arm to the front cylinder head with the spacer and the 10 mm bolt, and attach the rear arm to the rear cylinder head with the 8 mm bolt.

36. Install the engine support hanger No. AAR-T-12566) to the vehicle, and attach the hook to the engine balancer bar slot. Tighten the wing nut by hand, and lift and support the engine.

37. Remove the front mount stop and the front mount bolt.

38. Remove the exhaust pipe.

39. Make a reference mark across the propeller shaft and the transfer companion flange.

40. Separate the propeller shaft from the transfer companion flange.

41. Remove the propeller shaft.

42. Insert a 6 mm Allen wrench in the top of the ball joint pin and remove the nuts, then separate the stabilizer link from the lower arms.

43. Remove the damper pinch bolts, damper fork bolts and self-locking nuts, knuckle holder bolts and nuts. Remove the damper forks.

44. Remove the cotter pins and nuts and separate the steering tie-rod end ball joints from the knuckle.

45. Remove the bolt securing the transfer assembly breather tube bracket and disconnect the breather tube from the breather pipe on the transfer assembly.

46. Remove the transfer assembly from the transmission.

47. Remove the steering gearbox heat shield.

48. Loosen the hose clamp bolt, then disconnect the power steering fluid hose from the line at the right front of the subframe.

49. Remove the ATF cooler hose from the ATF cooler. Turn the end of the cooler hose up to prevent ATF from flowing out, then plug the hose.

50. Remove the torque converter cover and remove the drive plate bolts while rotating the crankshaft pulley.

51. Remove the transmission lower mount bolts.

52. Disconnect the power steering angle sensor connector.

53. Remove the shift cable bracket on the steering gearbox stiffener.

54. Remove the rear mount stop, and remove the rear mount bolt.

55. Attach the front subframe adapter No. VSB02C000016 to the subframe by looping the strap over the front of the sub-frame, then secure the strap.

56. Raise the jack and line up the slots in the arms with the bolt holes on the corner of the jack base, then tighten the bolts.

57. Remove the four bolts securing the stiffeners and the four bolts securing the front subframe, then lower the front sub-frame.

58. Remove the transmission lower mount.

59. Pry out the driveshafts from the dif-ferential and the intermediate shaft.

60. Remove the exhaust manifold bracket and the heat shield.

61. Remove the intermediate shaft.

62. Remove the upper transmission housing mounting bolts.

63. Remove the front mount bracket.

64. Remove the sensor harness from the harness clamp.

65. Remove the transmission housing mounting bolt using a socket 22 mm in length.

66. Remove the rear transmission hous-ing mounting bolts.

67. Lower the transmission by loosening the wing nut of the engine support hanger, and tilt the engine just enough for the trans-mission to clear the side frame.

68. Place a jack under the transmission.

69. Remove the lower transmission housing mounting bolts.

70. Slide the transmission away from the engine to remove it from the vehicle.

To install:

71. Place the transmission on the jack, and raise it to engine level.

72. Attach the transmission to the engine, and install the lower transmission housing mounting bolts. Tighten to 47 ft. lbs. (64 Nm).

73. Install the rear and upper transmis-sion housing mounting bolts. Tighten to 47 ft. lbs. (64 Nm).

74. Install the transmission housing mounting bolt. Tighten to 47 ft. lbs. (64 Nm).

75. Install the sensor harness clamps on the clamp bracket.

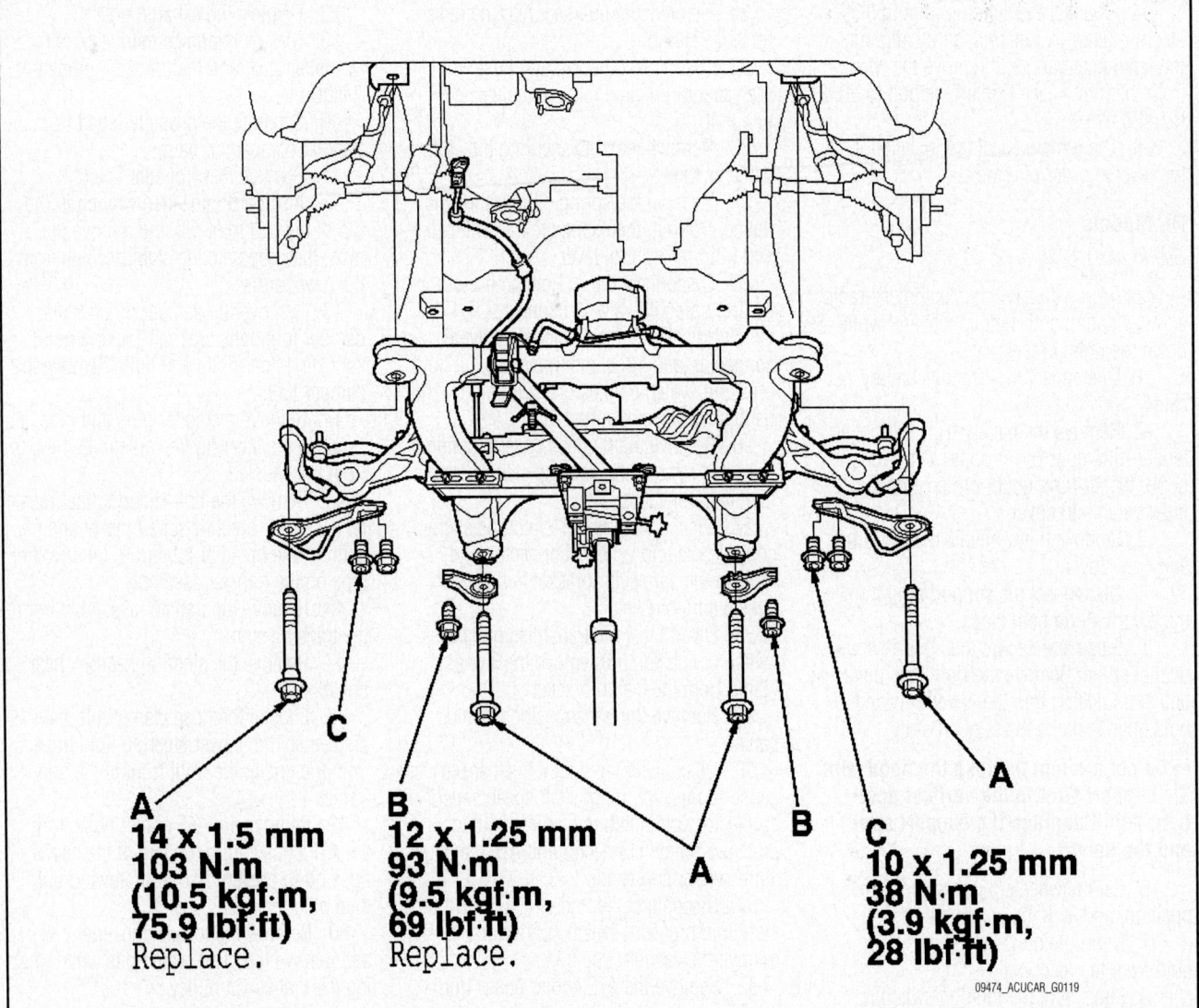

A
14 x 1.5 mm
103 N·m
(10.5 kgf·m,
75.9 lbf·ft)
Replace.

B
12 x 1.25 mm
93 N·m
(9.5 kgf·m,
69 lbf·ft)
Replace.

C
10 x 1.25 mm
38 N·m
(3.9 kgf·m,
28 lbf·ft)

09474_ACUCAR_G0119

Fig. 17 Sub-frame bolt locations and torque specifications—2007–08 RL model

76. Install the front mount bracket with the new mounting bolts. Tighten to 40 ft. lbs. (54 Nm).

77. Install the new set ring on the intermediate shaft and install the intermediate shaft in the differential.

➡ **While installing the intermediate shaft in the differential, be sure not to allow dust or other foreign particles to enter the transmission.**

78. Install the exhaust manifold bracket and heat shield.

79. Install the new set ring on the left driveshaft, then install the left driveshaft in the differential. While installing the driveshaft in the differential, be sure not to allow dust or other foreign particles to enter the transmission. Install the right driveshaft over the intermediate shaft.

➡ **Turn the right and left steering knuckle fully outward, and slide the driveshaft and intermediate shaft into the differential and intermediate shaft until you feel its set ring engage the side gear.**

80. Support the front subframe with the front subframe adapter and a jack, and lift it up to the body.

81. Loosely install the new subframe mounting bolts (A), new rear stiffener mounting bolts (B), front stiffener mounting bolts (C), and front and rear stiffeners. Refer to the illustration for bolt location and its torque specification.

82. Loosely tighten the right rear subframe mounting bolt (A); insert the special tool through the positioning slot (B) on the rear stiffener, through the positioning hole (C) on the subframe, and into the positioning hole (D) on the body, then tighten the subframe mounting bolt. Refer to the illustration for bolt location and its torque specification.

83. Loosely tighten the left rear subframe mounting bolt in the same manner.

84. Loosely tighten the right and left front subframe mounting bolts.

85. Loosen the right rear mounting bolt, then tighten the bolt.

86. Tighten the left rear mounting bolt.

87. Tighten the right and left front mounting bolts.

88. Check that the positioning holes and slot are aligned using the special tool.

89. Tighten the rear and front stiffener mounting bolts to the torque specified in the accompanying illustration.

90. Remove the jack and front subframe adapter. Tighten to 33 ft. lbs. (44 Nm).

91. Attach the torque converter to the drive plate with the eight bolts. Rotate the crankshaft pulley as necessary to tighten the bolt to 4 ft. lbs. (6 Nm), then to the final torque of 9 ft. lbs. (12 Nm), in a crisscross pattern. After tightening the last bolt, check that the crankshaft rotates freely.

92. Install the torque converter cover.

93. Connect the power steering fluid hose to the line at the right front of the front subframe, and secure the hose with its hose clamp.

94. Connect the ATF cooler hose to the ATF cooler, and secure the hose with the clip.

95. Install the steering gearbox heat shield

96. Install the dowel pin in the transmission, and install the transfer assembly on the transmission. Tighten the bolts to 38 ft. lbs. (51 Nm).

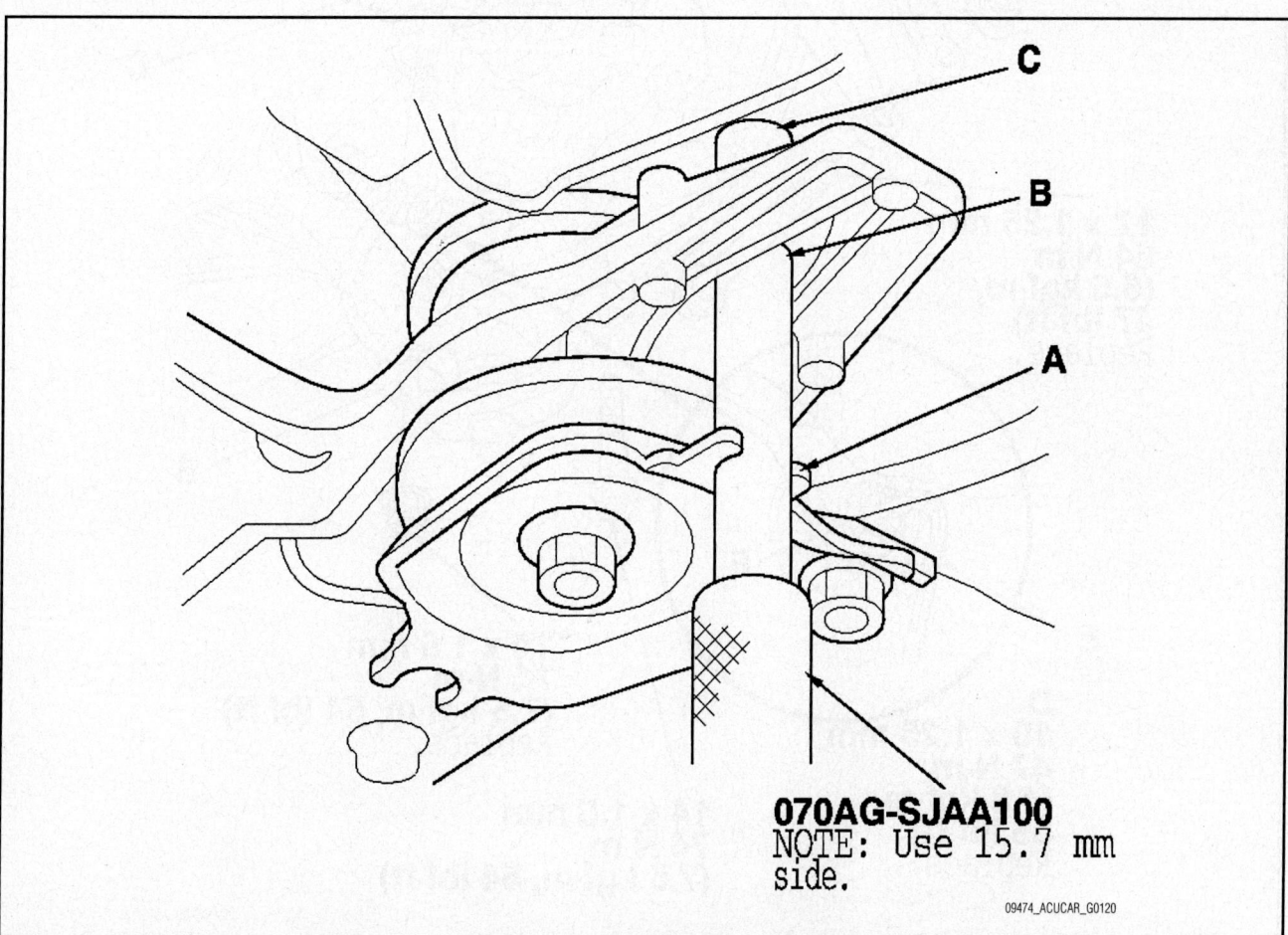

070AG-SJAA100
NOTE: Use 15.7 mm side.

09474_ACUCAR_G0120

Fig. 18 Incert the special tool through the positioning slot (B) on the rear stiffener, through the positioning hole (C) on the subframe—2007–08 RL model

97. Secure the transfer assembly breather tube bracket on the transfer assembly with the bolt, and install the breather tube over the breather pipe. If the breather tube was removed from the clamp, install the tube at the dot on the clamp.

98. Install the propeller shaft.

99. Install the propeller shaft to the transfer companion flange by aligning the previously made reference mark. Tighten the bolts to 54 ft. lbs. (74 Nm).

100. Install exhaust pipe with the new self-locking nuts and the new gaskets.

101. Install the damper forks (A) with the damper pinch bolts, the new damper fork bolts, and the new self-locking nuts. Attach the knuckle holder (B) to the knuckle, and secure the knuckle holder with the bolt and new nut. Refer to the illustration for bolt location and its torque specification.

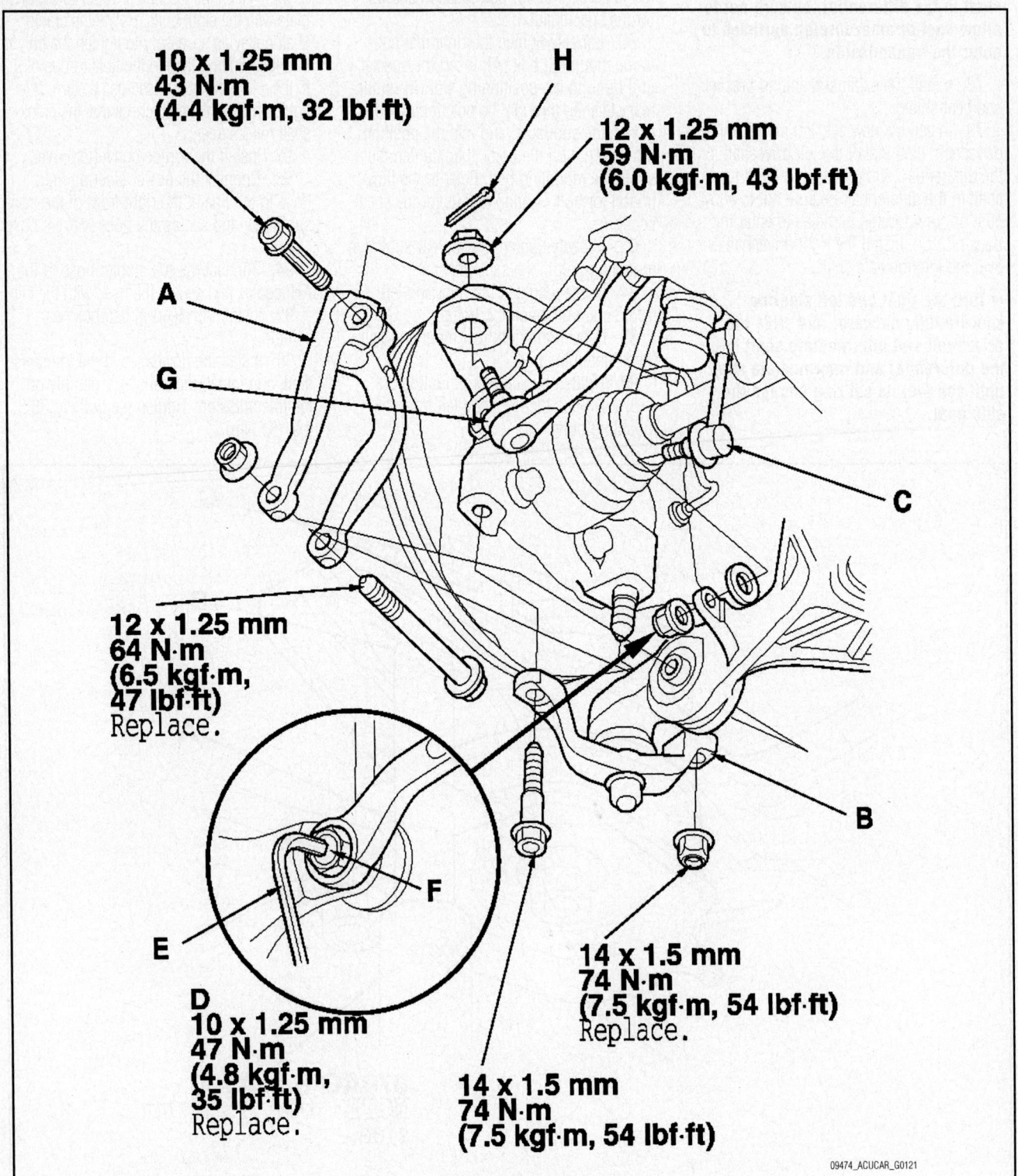

10 x 1.25 mm
43 N·m
(4.4 kgf·m, 32 lbf·ft)

H

12 x 1.25 mm
59 N·m
(6.0 kgf·m, 43 lbf·ft)

A

G

C

12 x 1.25 mm
64 N·m
(6.5 kgf·m,
47 lbf·ft)
Replace.

B

F

E

D
10 x 1.25 mm
47 N·m
(4.8 kgf·m,
35 lbf·ft)
Replace.

14 x 1.5 mm
74 N·m
(7.5 kgf·m, 54 lbf·ft)
Replace.

14 x 1.5 mm
74 N·m
(7.5 kgf·m, 54 lbf·ft)

09474_ACUCAR_G0121

Fig. 19 Exploded view of the damper fork assembly and related components—2007–08 RL model

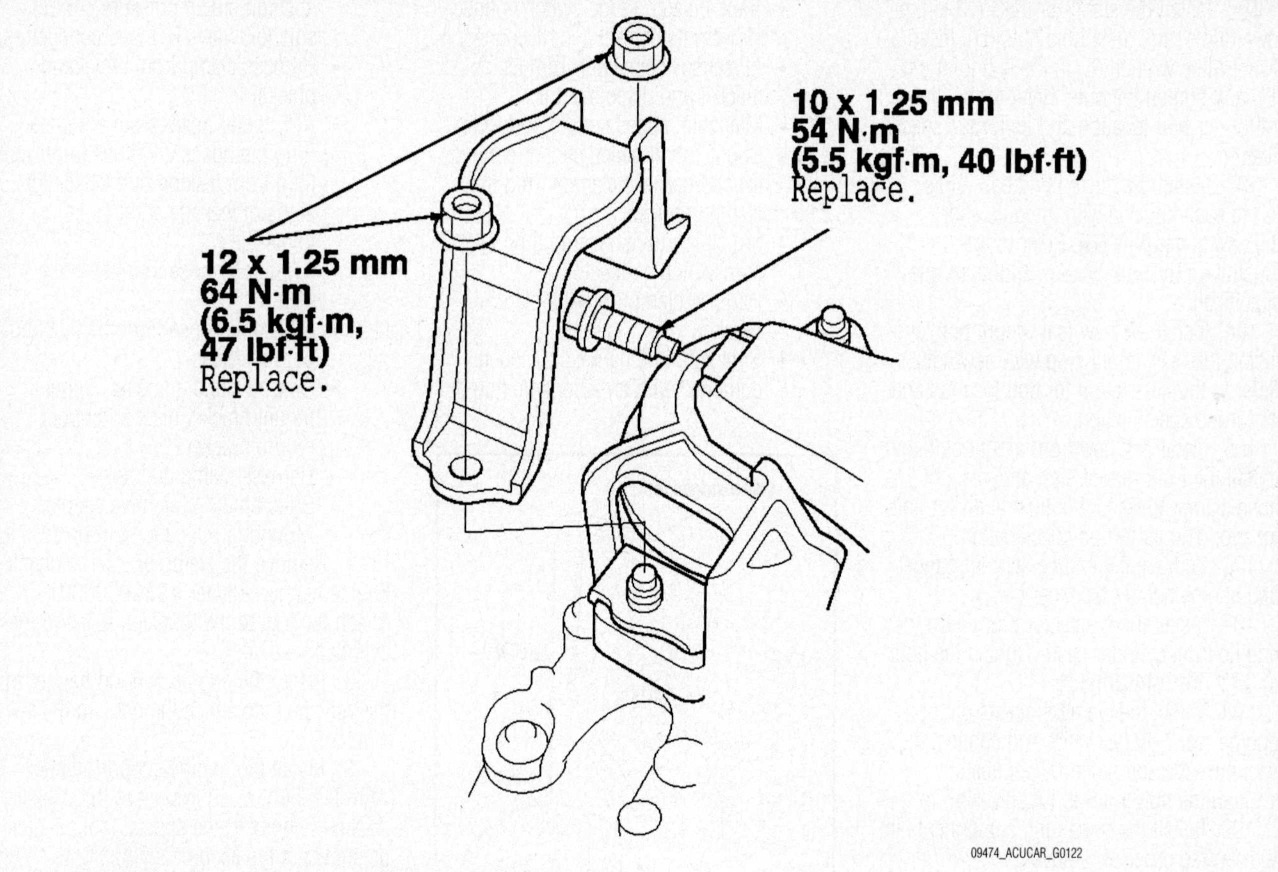

10 x 1.25 mm
54 N·m
(5.5 kgf·m, 40 lbf·ft)
Replace.

12 x 1.25 mm
64 N·m
(6.5 kgf·m,
47 lbf·ft)
Replace.

09474_ACUCAR_G0122

Fig. 20 Exploded view of the rear mount—2007–08 RL model

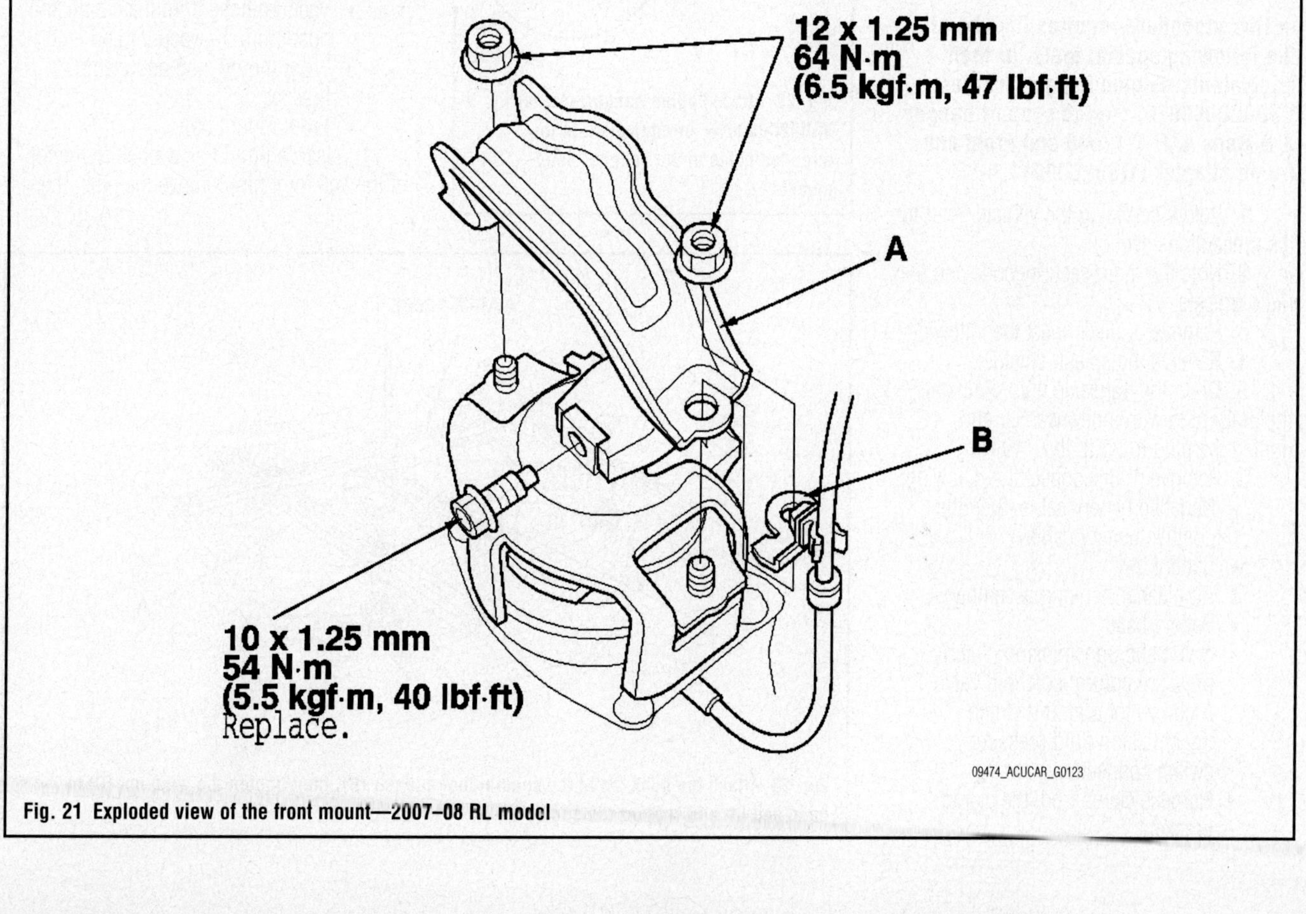

12 x 1.25 mm
64 N·m
(6.5 kgf·m, 47 lbf·ft)

A

B

10 x 1.25 mm
54 N·m
(5.5 kgf·m, 40 lbf·ft)
Replace.

09474_ACUCAR_G0123

Fig. 21 Exploded view of the front mount—2007–08 RL model

102. Install the stabilizer links (C) to the lower arms, and install the nuts (D). Insert a 6 mm Allen wrench (E) in the ball joint pin (F), and tighten the nuts. Refer to the illustration for bolt location and its torque specification.

103. Install the tie-rod end ball joints (G) to each knuckle with the nuts and the new cotter pins (H). Refer to the illustration for bolt location and its torque specification.

104. Install the new rear mount bolt, and install the rear mount stop with new nuts. Refer to the illustration for bolt location and its torque specification.

105. Install the new front mount bolt, and install the front mount stop and vacuum hose clamp. Refer to the illustration for bolt location and its torque specification.

106. Remove the engine support hanger and engine hanger balancer bar.

107. Install the connector bracket on the engine front cylinder head. Tighten the bolt to 33 ft. lbs. (44 Nm).

108. Install the clamp brackets on the engine rear cylinder head, and connect the vacuum tube and solenoid connector. Tighten the bolt to 20 ft. lbs. (26 Nm).

109. Install the remaining components in the reverse order of removal.

TSX Models

See Figures 22 through 26.

➡**This procedure requires the use of the following special tools, or their equivalents: Engine hanger/adapter VSB02C000015, Engine support hanger A & Reds AAR-T-12566 and Front subframe adapter EQS02C00011.**

1. Before servicing the vehicle, refer to the precautions section.

2. Note the radio security code and the radio presets.

3. Remove or disconnect the following:

4. Remove the splash shield.

5. Drain the transaxle fluid. Reinstall the drain plug with a new washer and tighten the plug to 36 ft. lbs. (49 Nm).

6. Remove or disconnect the following:
- Negative battery cable, then the positive battery cable
- Battery tray
- Air cleaner housing assembly
- Battery base
- Automatic transmission clutch pressure control solenoid valve A connector and 2nd clutch transmission fluid pressure switch connector
- Harness clams from the clamp brackets

- Transmission range switch connector from its bracket and disconnect it
- AF sensor connector from its bracket and disconnect it
- Mainshaft speed sensor connector
- Countershaft speed sensor connector and harness clamps from the clamp brackets
- 3rd clutch transmission fluid pressure switch connector
- Harness clamp from the clamp bracket
- Shift solenoid harness connector equipped with an automatic trans-

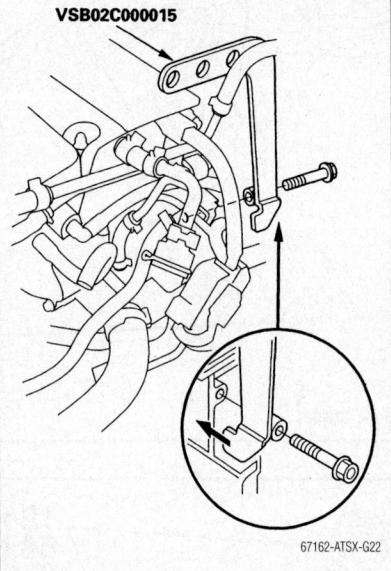

Fig. 22 Attach Engine hanger/adapter VSB02C000015, or equivalent to the threaded holes in the cylinder head—TSX models

mission clutch pressure control solenoid valve B and C connectors
- Harness clamp from the clamp bracket
- ATF cooler hoses from the lines. Plug the hoses and lines to prevent fluid from leaking out. Check for signs of leakage at the hose joints
- ATF cooler hose from the hose clamp
- ATF cooler hose from the line and plug the hose
- Ground cable, transaxle upper mount bracket plate and upper mount bracket
- Harness clamp and hose
- Bolts holding the hose clamps
- Mounting nuts and strut brace

7. Remove the clamp bracket to attach Engine hanger/adapter VSB02C000015. Attach the tool to the threaded holes in the cylinder head.

8. Install the engine support hanger to the vehicle, then attach the hook to the special tool.

9. Install the Engine support hanger AAR-T-12566, or equivalent to the vehicle. Attach the hook to the special tool adapter as shown in the accompanying figure. Tighten the wing nut by hand and lift and support the engine.

10. Remove or disconnect the following:
- Vacuum hose from the clamp and hose from the vacuum line
- Front mount stop and clamp bracket
- Front mount bolt

11. Insert a 6mm hex wrench in the top of the ball joint pin, remove the nuts, then

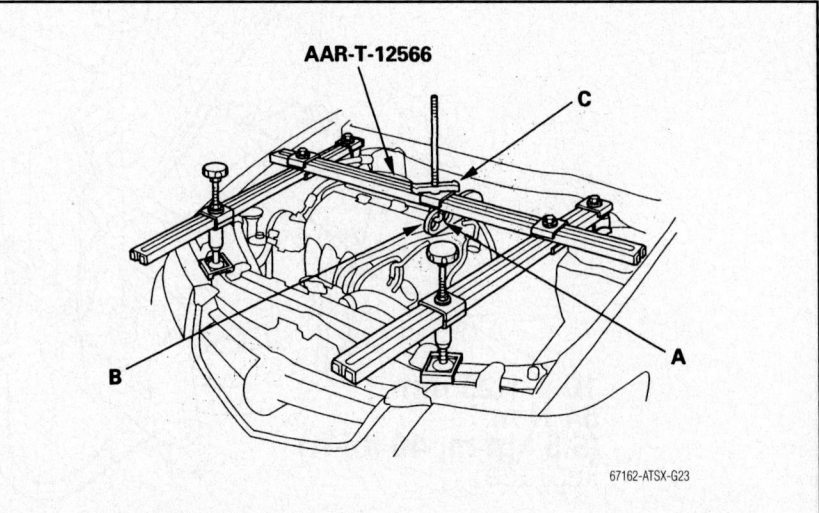

Fig. 23 Attach the hook (A) to the special tool adapter (B), then tighten the wing nut (C) by hand and lift and support the engine

separate the stabilizer link from the lower control arms.

- Cotter pins, castle nuts, damper pinch bolt, self-locking nut, bolt and damper forks, then separate the ball joints from the lower control arms
- Exhaust pipe "A" and its mount
- Steering gearbox heat shield and bolt securing the power steering fluid line bracket
- Power steering line from the clips from the front subframe
- Engine stiffener
- Driveplate bolts, while rotating the crankshaft pulley
- 3 bolts from the shift cable holder, then remove the cover
- Spring clip and control pin and separate the cable from the control lever. Do not bend the cable more than necessary
- Rear mount stop
- Power steering fluid line bracket from the front subframe
- Steering gearbox mounting bolts and stiffener
- Steering gearbox mounting bracket bolts
- Damper and rear mount base bracket bolts
- Transmission lower mount nuts
- Both mid mounts

12. Matchmark the ends of the subframe to the edge of the stiffeners.

13. Attach the special tool to the subframe with hanging the hook of the special tool over the front of the subframe, then tighten the special tool screw.

14. Raise the jack and line up the slots in the arms with the bolt holes on the corner of the jack base, then attach them with the bolts securely.

15. Remove the 4 bolts holding the stiffeners and 4 bolts holding the front subframe. Lower the front subframe while sliding the steering gearbox out to clear the gearbox mounting bracket on the subframe.

16. Securely suspend the gearbox from the body with a piece of wire or rope.

17. Remove or disconnect the following:
- Transaxle lower mounts
- Driveshafts from the differential and intermediate shaft by prying them out
- Intermediate shaft

➡ **Coat all machined surfaces with clean engine oil, then put plastic bags**

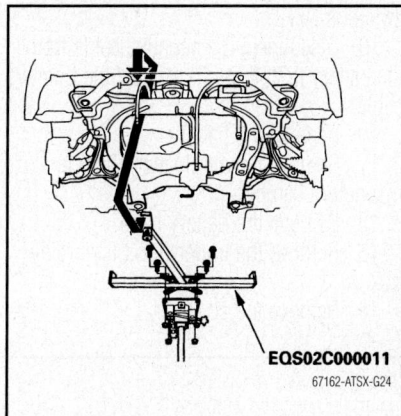

EQS02C000011

67162-ATSX-G24

Fig. 25 Support the subframe with the subframe adapter tool and a suitable jack—TSX models

over the driveshaft and intermediate shaft ends to protect them.

- Rear mount base/bracket
- Rear mount bracket

18. Put a jack under the transmission.
- Transmission housing mounting bolts
- Front mount bracket
- Transmission housing mounting bolts from the front and rear of the transmission

19. Slide the transmission away from the engine to remove it from the vehicle.

20. Remove the torque converter. Check the drive plate and replace if necessary.

21. Inspect the transaxle lower front mount and lower rear mount. If the mount rubber is worn or damage, it must be replaced.

To install:

22. Installation is the reverse of the removal procedure, while using the following torque values:
- Drive plate mounting bolts: 54 ft. lbs. (74 Nm)
- Transmission housing mounting bolts: 47 ft. lbs. (64 Nm)
- New front mount bracket bolts: 47 ft. lbs. (64 Nm)
- New rear mount bracket A bolts: 65 ft. lbs. (88 Nm)
- New rear mount brace bracket bolts: 40 ft. lbs. (54 Nm)

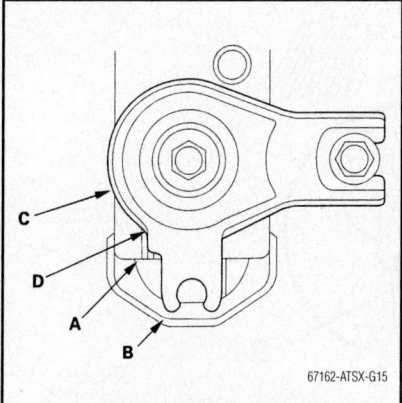

67162-ATSX-G15

Fig. 24 Matchmark (A) the ends of the subframe (B) to the edge (C) of the stiffeners (D)—TSX models

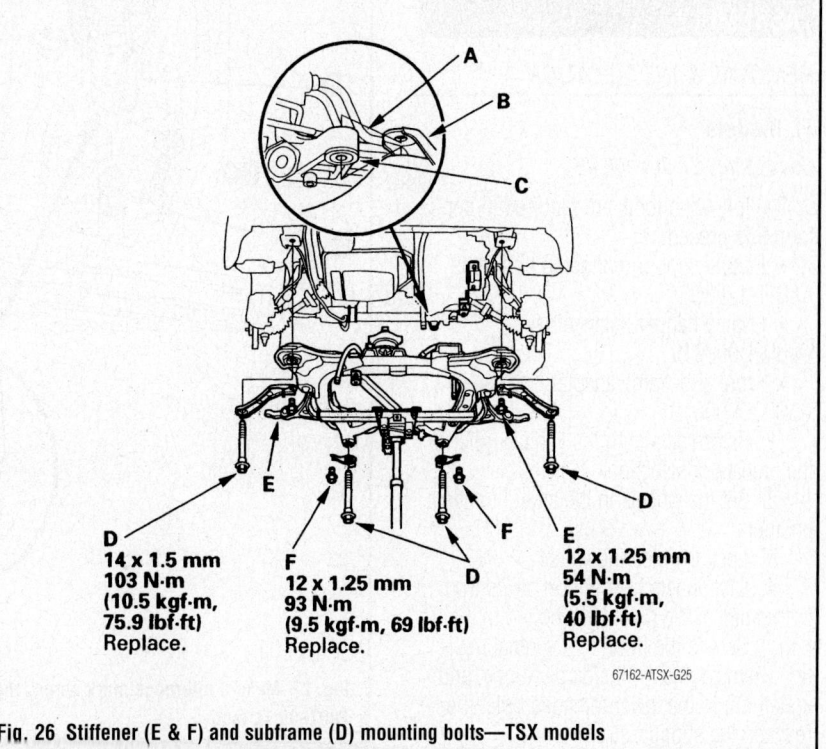

D
14 x 1.5 mm
103 N·m
(10.5 kgf·m, 75.9 lbf·ft)
Replace.

F
12 x 1.25 mm
93 N·m
(9.5 kgf·m, 69 lbf·ft)
Replace.

E
12 x 1.25 mm
54 N·m
(5.5 kgf·m, 40 lbf·ft)
Replace.

67162-ATSX-G25

Fig. 26 Stiffener (E & F) and subframe (D) mounting bolts—TSX models

- Intermediate shaft bolts: 29 ft. lbs. (39 Nm)
- Transmission lower mount bolts: 33 ft. lbs. (45 Nm)
- New subframe mounting bolts: 75.9 ft. lbs. (103 Nm)
- Stiffener mounting bolts (after aligning matchmarks made during removal): Refer to accompanying figure
- Mid mount short bolts: 36 ft. lbs. (49 Nm)
- Mid mount long bolt 33 ft. lbs. (44 Nm)
- Rear mount base bracket short bolts: 16 ft. lbs. (22 Nm)
- Rear mount base bracket long bolts: 36 ft. lbs. (49 Nm)
- Transaxle lower mount nuts: 33 ft. lbs. (44 Nm)
- Steering gearbox mounting bolts: 43 ft. lbs. (60 Nm)
- Steering gearbox mounting bracket bolts: 28 ft. lbs. (38 Nm)
- Rear mount stop: 51 ft. lbs. (69 Nm)
- Torque converter-to-drive plate bolts (in 2 steps): 8.7 ft. lbs. (12 Nm)
- Engine stiffener: 33 ft. lbs. (45 Nm)
- Transaxle upper mount bracket bolts: 40 ft. lbs. (54 Nm)
- Transmission upper mount bracket plate: 40 ft. lbs. (54 Nm)

MANUAL TRANSAXLE ASSEMBLY

REMOVAL & INSTALLATION

TL Models

See Figures 27 through 43.

The following tools are required to perform this procedure:
- Engine support hanger, A and Reds AAR-T-12566
- Engine hanger balance bar VSB02C000019
- Front sub-frame adapter VSB02C000016

1. Before servicing the vehicle, refer to the precautions section.
2. Set the wheels in the straight ahead position.
3. Lock the steering wheel.
4. Disconnect the support struts from both sides of the pivot ball (bolted to the hood). Secure the hood in a vertical position. Remove the right side pivot ball, and install it into the lower threaded hole, then reattach the support strut.

→Do not attempt to close the hood with the support strut in the vertical position; it will damage the support strut and hood.

5. Remove the left rear engine compartment cover and the left side engine compartment cover.
6. Remove the right side engine compartment cover and the right rear engine compartment cover.
7. Remove the intake manifold cover and the bulkhead cover.
8. Make sure you have the anti-theft codes for the radio and the navigation system, then write down the customer's radio channel presets. Make sure the ignition switch is OFF.
9. Disconnect the negative cable from the battery first, then disconnect the positive cable.
10. Remove the battery.
11. Remove the air cleaner housing and resonator chamber.
12. Remove the battery base.
13. Remove the under-hood fuse/relay box.
14. Remove the strut bar.

15. Disconnect the back-up light switch connector and reverse lockout solenoid connector, then remove the harness clips.
16. Disconnect the input shaft speed sensor connector and output shaft speed sensor connector.
17. Remove the harness clips, then disconnect the vacuum hose.
18. Disconnect the starter cable, then remove the starter motor.
19. Remove the harness bracket.
20. Remove the cable bracket, then disconnect the cables from the top of the transmission housing. Carefully remove both cables and the bracket together to avoid bending the cables.
21. Disconnect the vacuum hose.
22. Carefully remove the slave cylinder without bending the clutch line. Do not press the clutch pedal once the slave cylinder has been removed.
23. Remove the engine front mount mounting nuts, engine front mount stop, and the engine front mount mounting bolt.
24. Remove the engine rear mount mounting nuts and engine rear mount stop.

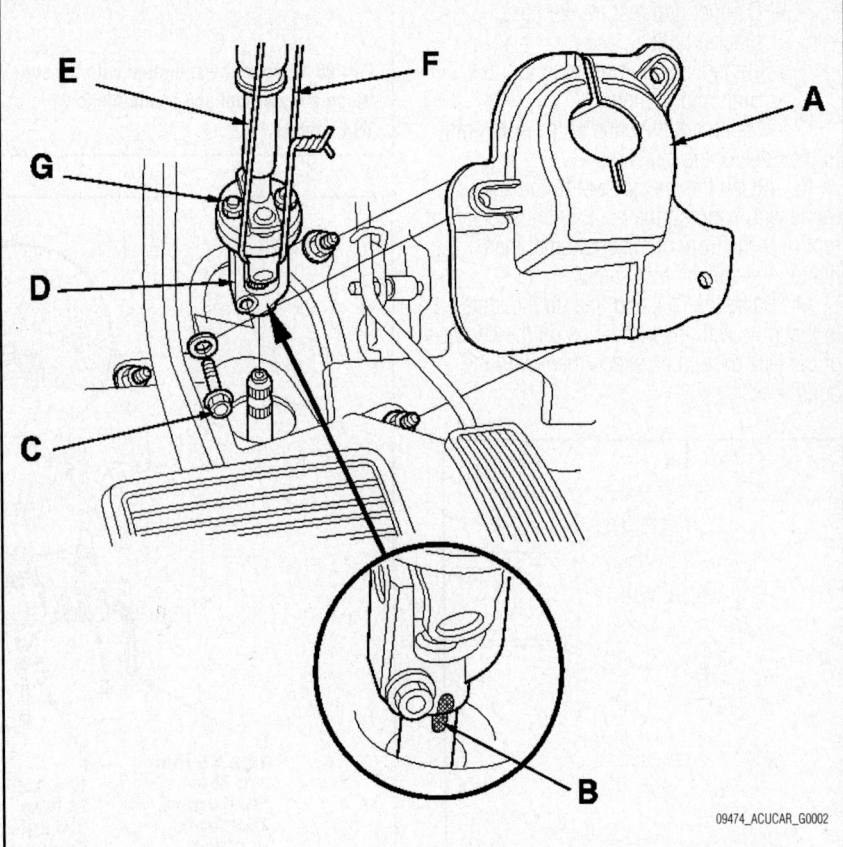

Fig. 27 Make a reference mark across the steering joint and steering gearbox pinion shaft— 2007–08 TL model

09474_ACUCAR_G0002

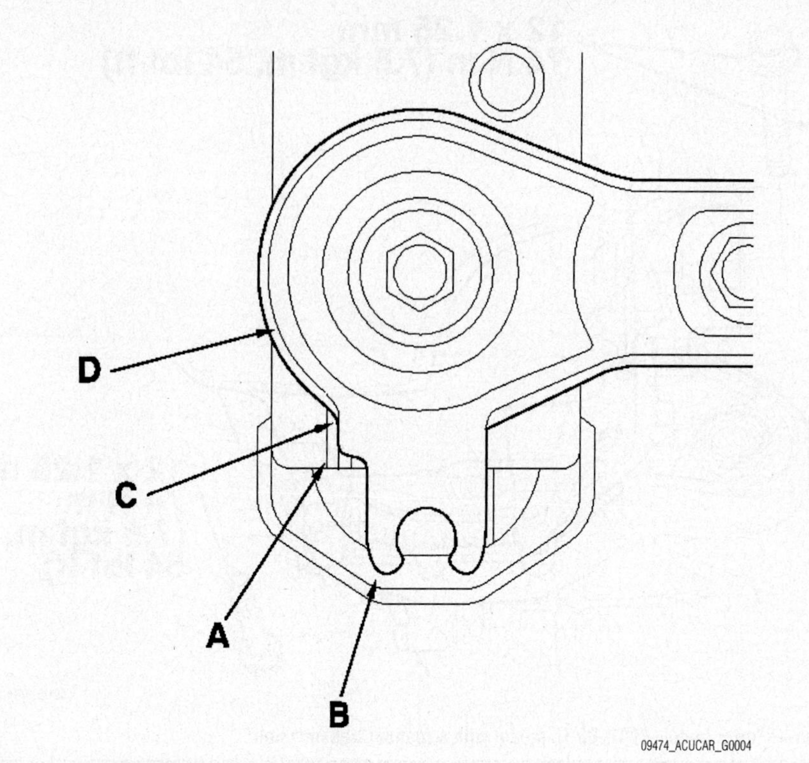

Fig. 28 Make the appropriate reference lines at both ends of the sub-frame that line up—2007–08 TL model

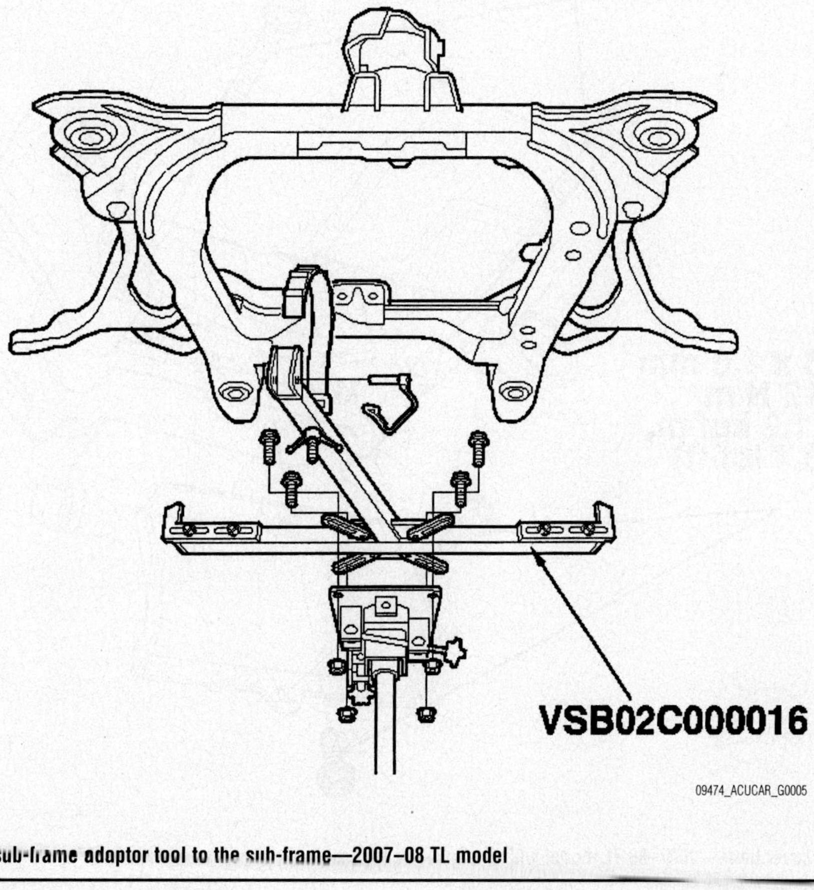

VSB02C000016

Fig. 29 Attach an engine sub-frame adaptor tool to the sub-frame—2007–08 TL model

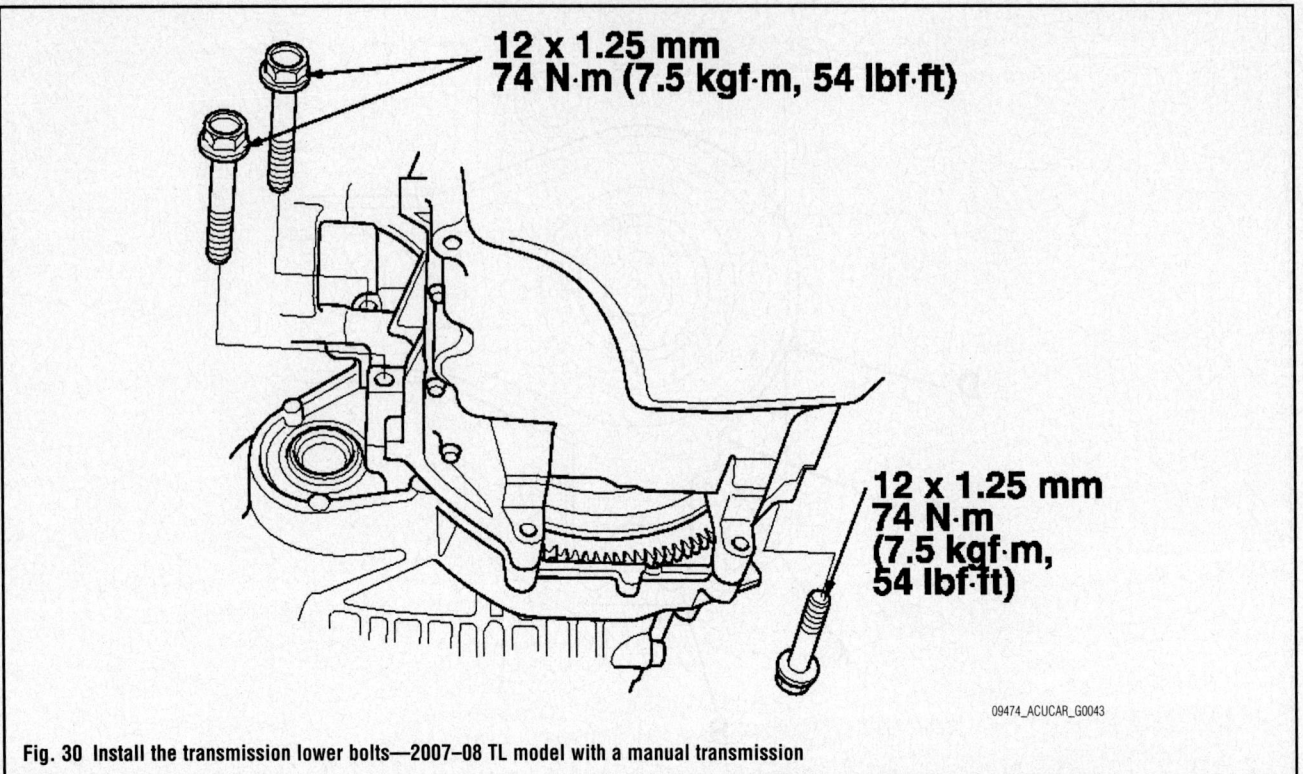

12 x 1.25 mm
74 N·m (7.5 kgf·m, 54 lbf·ft)

12 x 1.25 mm
74 N·m
(7.5 kgf·m,
54 lbf·ft)

09474_ACUCAR_G0043

Fig. 30 Install the transmission lower bolts—2007–08 TL model with a manual transmission

12 x 1.25 mm
74 N·m (7.5 kgf·m, 54 lbf·ft)

6 x 1.0 mm
12 N·m
(1.2 kgf·m,
8.7 lbf·ft)

09474_ACUCAR_G0044

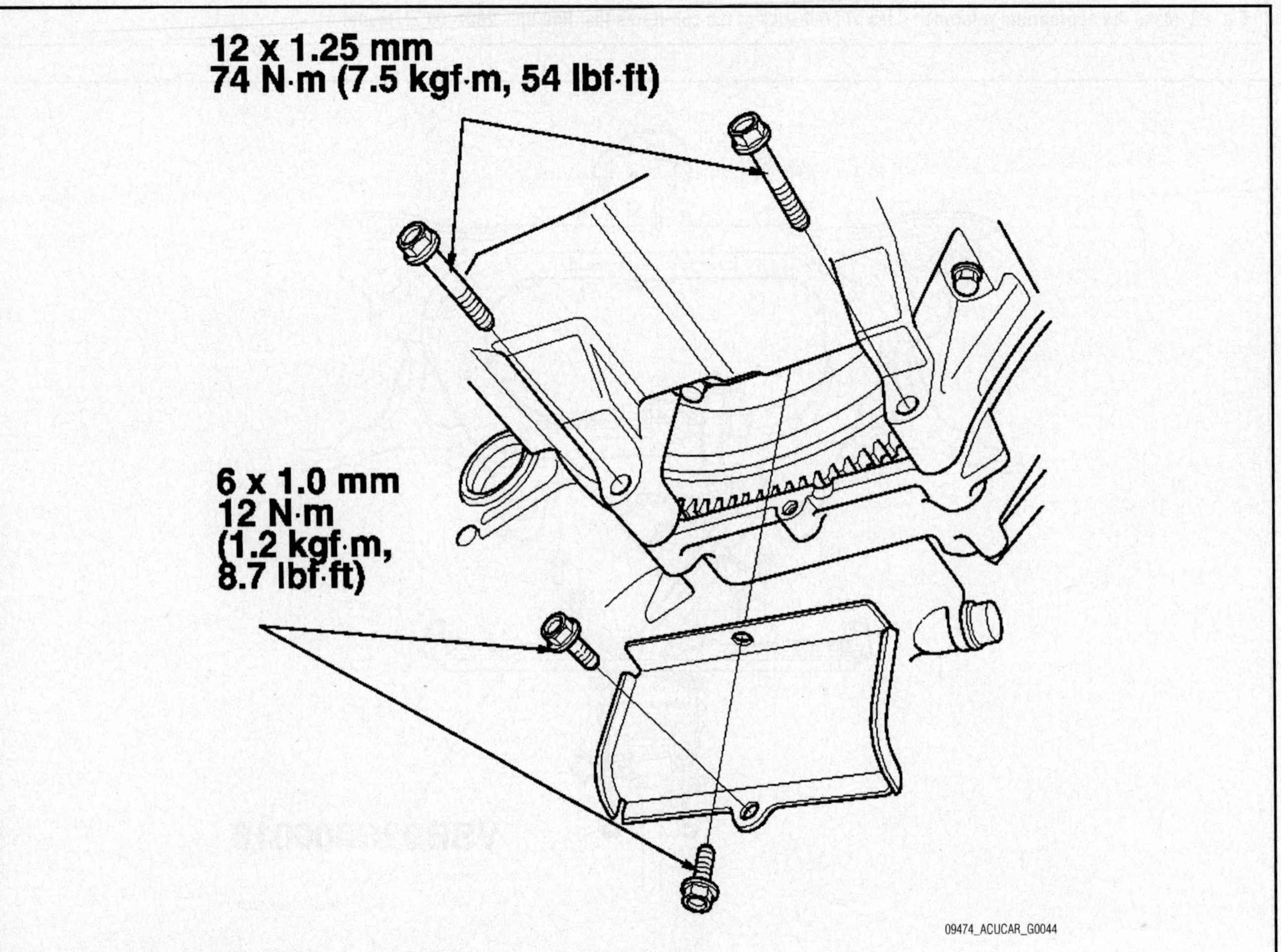

Fig. 31 Install the clutch cover bolts—2007–08 TL model with a manual transmission

25. Remove the transmission upper mounting bolts.

26. Remove and plug the return hose from the power steering fluid reservoir. Wipe off any spilled fluid at once.

27. Remove the power steering pump outlet line from the power steering pump, and remove the hose from its clamp.

28. Lift and support the engine with the engine support hanger and the engine hanger balance bar:

   a. Attach the front arm to the front cylinder head with a spacer and the 10 x 1.25 mm bolt.

   b. Attach the rear arm to the rear cylinder head with the 8 x 1.25 mm bolt.

29. Remove the steering joint cover.

30. Make a reference mark across the steering joint and steering gearbox pinion shaft. Remove the steering joint bolt, and disconnect the steering joint by removing the steering joint toward the steering column. Hold the slider shaft on the column with a piece of wire between the joint yoke on the slider shaft to the joint yoke on the upper shaft.

31. Remove the transmission mount bracket and ground cable.

32. Remove the undercover.

33. Drain the transmission fluid. Install the drain bolt with a new washer.

34. Remove the exhaust pipe.

35. Separate the front stabilizer link.

36. Remove the damper fork.

37. Separate the knuckle from the lower arm.

38. Separate the tie-rod ball joint.

39. Remove the left and right driveshaft inboard joints.

40. Remove the intermediate shaft.

41. Remove the rear engine lower mount mounting bolts.

42. Disconnect the power steering pressure switch connector.

43. Remove the transmission lower mount mounting nuts.

44. Remove the middle sub-frame mounting bolts.

45. Make the appropriate reference lines at both ends of the sub-frame that line up with the edge of the stiffeners.

46. Support the sub-frame with the front sub-frame adapter and a jack.

47. Remove the front suspension sub-frame stays and front suspension sub-frame.

48. Remove the front engine mount upper bracket.

49. Remove the transmission lower front mount and the transmission lower rear mount.

50. Remove the clutch cover.

51. Place the transmission jack under the transmission, and remove the transmission lower mounting bolts.

52. Pull the transmission away from the engine until the transmission mainshaft clears the clutch pressure plate, then lower the transmission on the transmission jack.

To install:

53. Check that the two dowel pins are installed in the clutch housing.

54. Apply super high temp urea grease to the release fork and the release bearing. Install the release fork, the release bearing, and the boot.

55. Place the transmission on the transmission jack, and raise it to the engine level.

56. Install the transmission lower mounting bolts.

57. Install the clutch cover

58. Install the front engine mount upper bracket.

59. Support the subframe with the front subframe adapter and a jack.

60. Install the front suspension subframe (A) and front suspension subframe stays (B).

61. Align the reference marks (A) with edge (B) of both rear stiffer (C), and tighten the rear subframe mounting bolts, then the front bolts, and tighten the stiffener bolts to the specified torque.

62. Install the middle subframe mounting bolts.

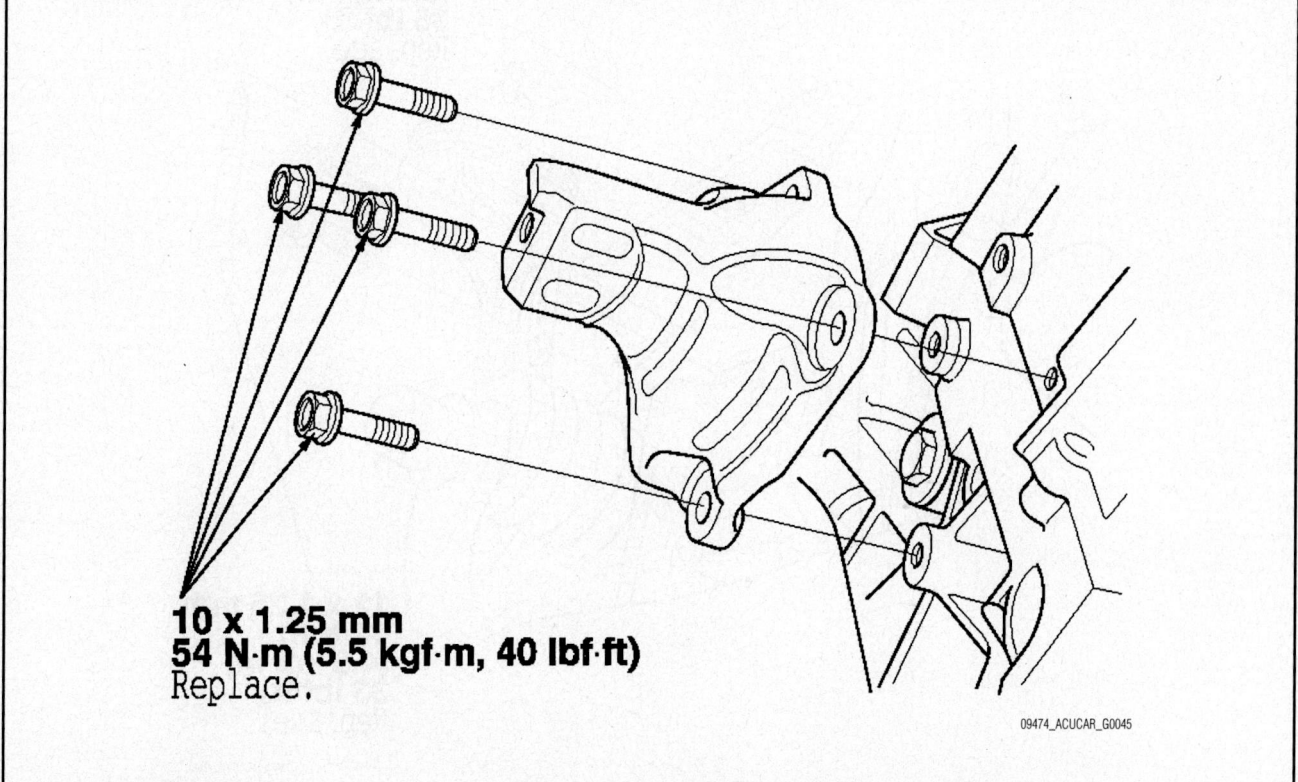

10 x 1.25 mm
54 N·m (5.5 kgf·m, 40 lbf·ft)
Replace.

09474_ACUCAR_G0045

Fig. 32 Install the front engine mount bolts 2007–08 TL model with a manual transmission

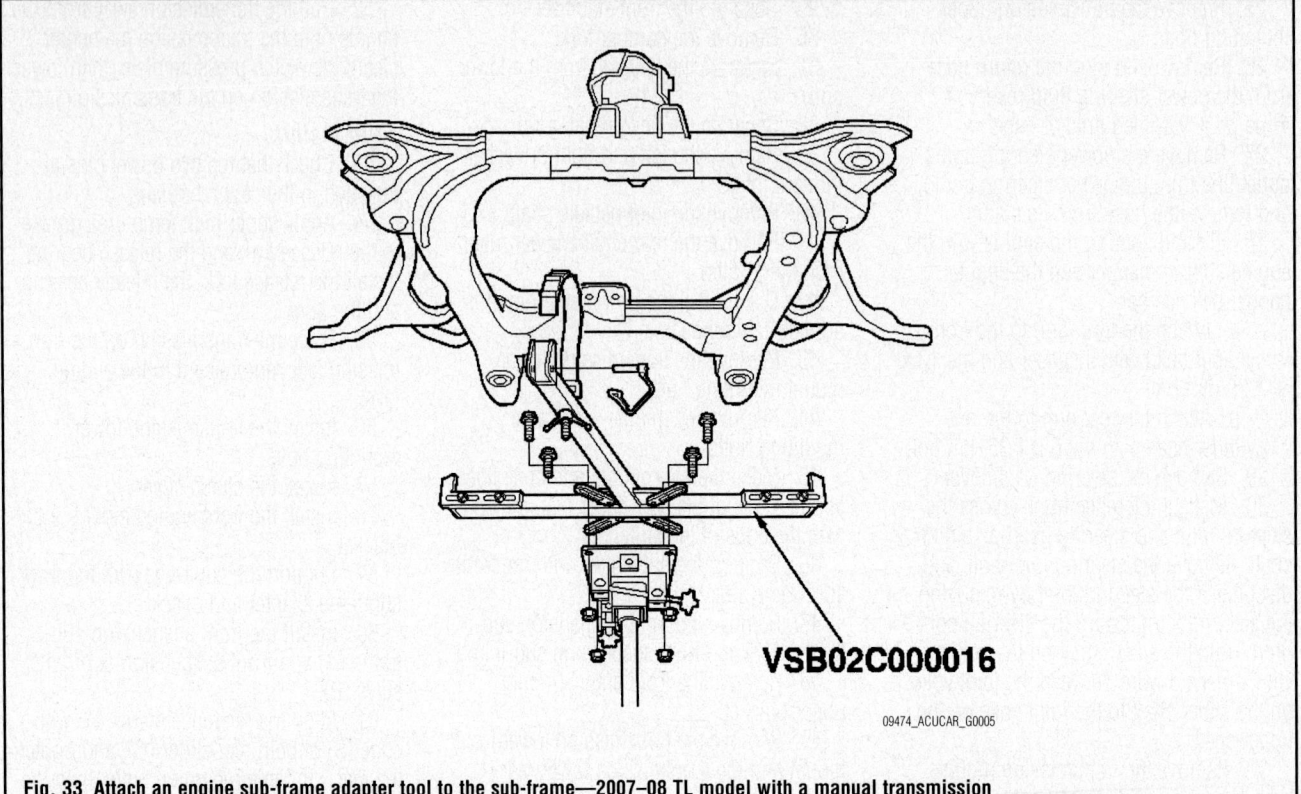

VSB02C000016

09474_ACUCAR_G0005

Fig. 33 Attach an engine sub-frame adapter tool to the sub-frame—2007–08 TL model with a manual transmission

10 x 1.25 mm
49 N·m
(5.0 kgf·m,
36 lbf·ft)
Replace.

12 x 1.25 mm
44 N·m
(4.5 kgf·m,
33 lbf·ft)
Replace.

09474_ACUCAR_G0007

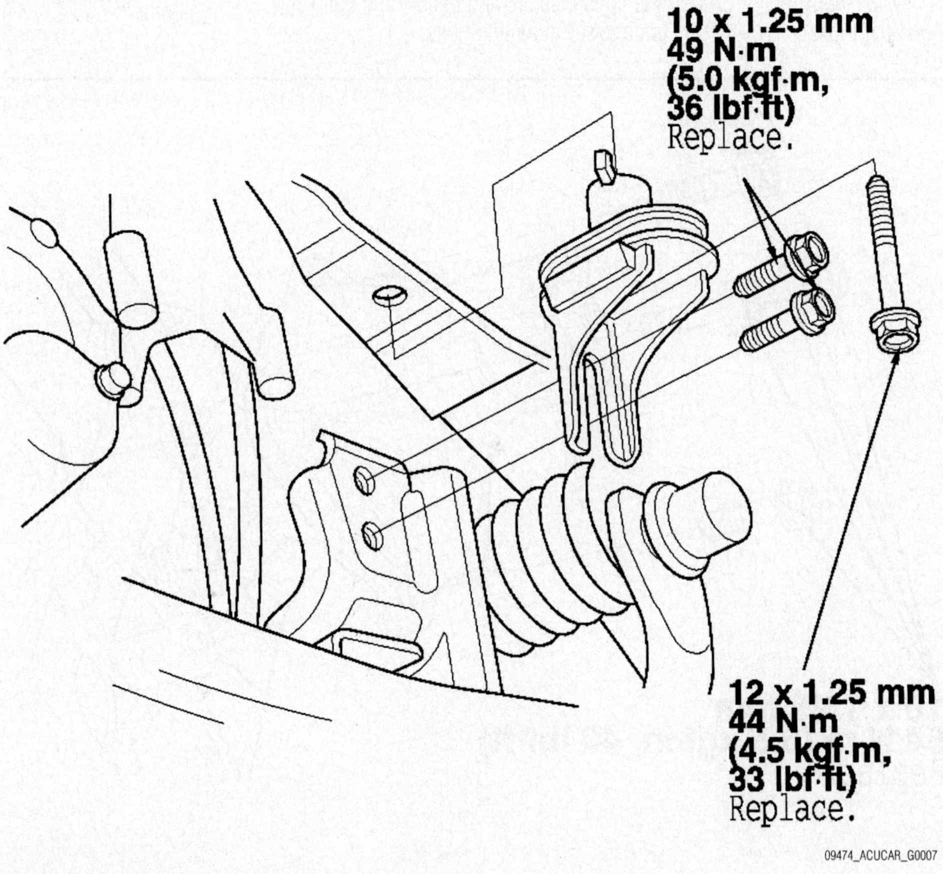

Fig. 34 Sub-frame mid-mounts bolt specifications—2007–08 TL model with a manual transmission

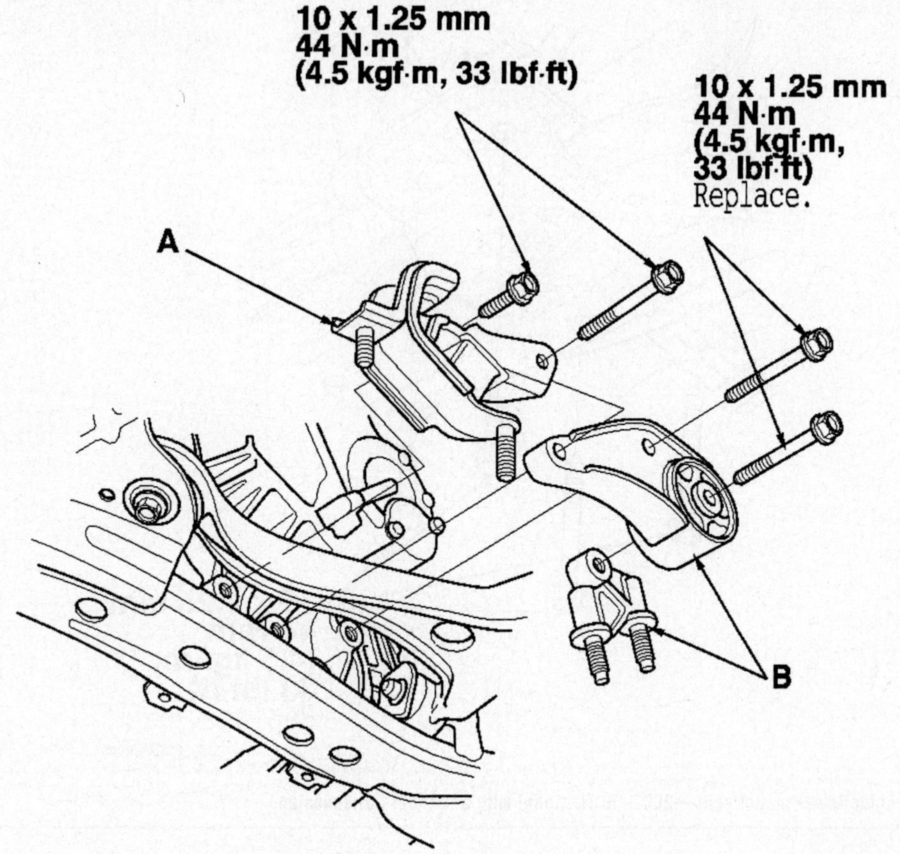

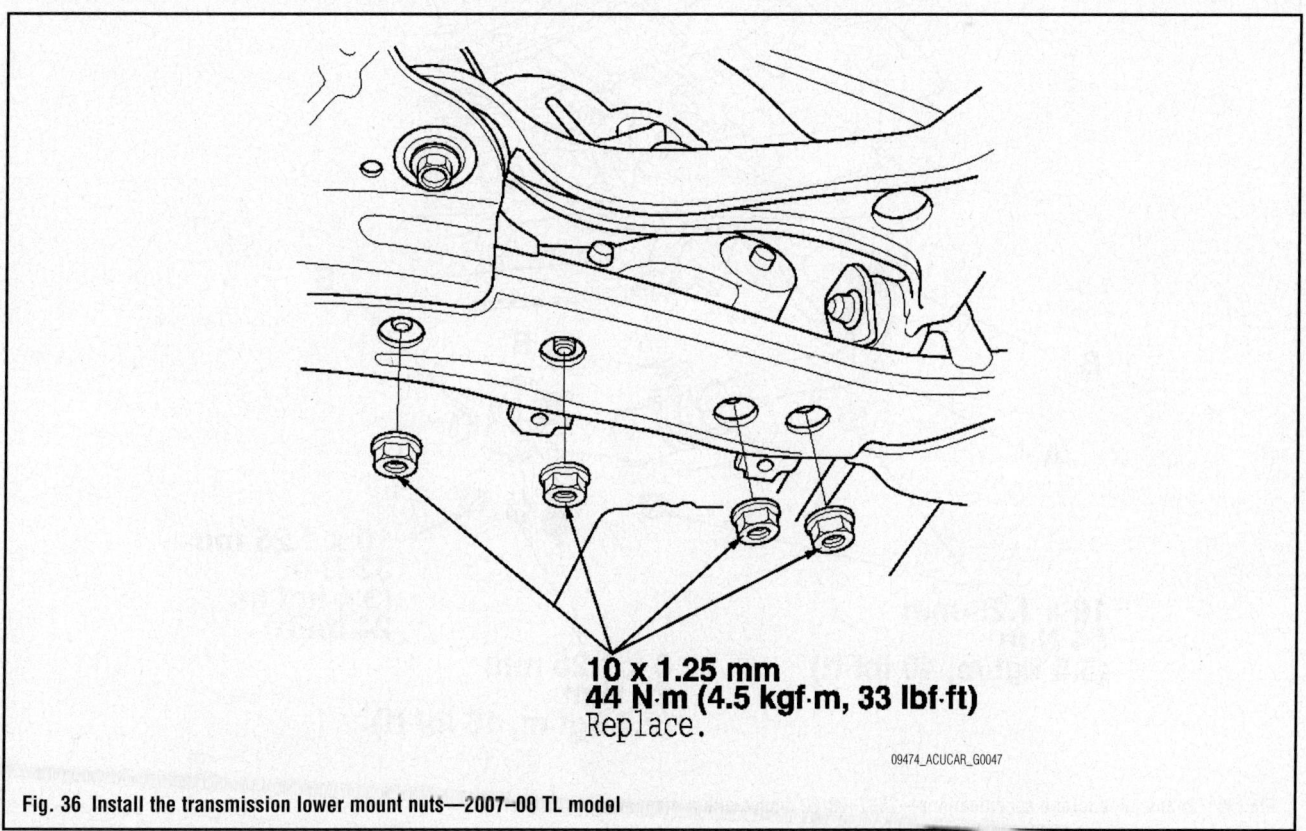

**10 x 1.25 mm
44 N·m
(4.5 kgf·m, 33 lbf·ft)**

**10 x 1.25 mm
44 N·m
(4.5 kgf·m,
33 lbf·ft)
Replace.**

A

B

09474_ACUCAR_G0046

Fig. 35 Install the transmission lower front mount (A) and transmission lower rear mount bolts—2007–08 TL model with a manual transmission

**10 x 1.25 mm
44 N·m (4.5 kgf·m, 33 lbf·ft)
Replace.**

09474_ACUCAR_G0047

Fig. 36 Install the transmission lower mount nuts—2007–00 TL model

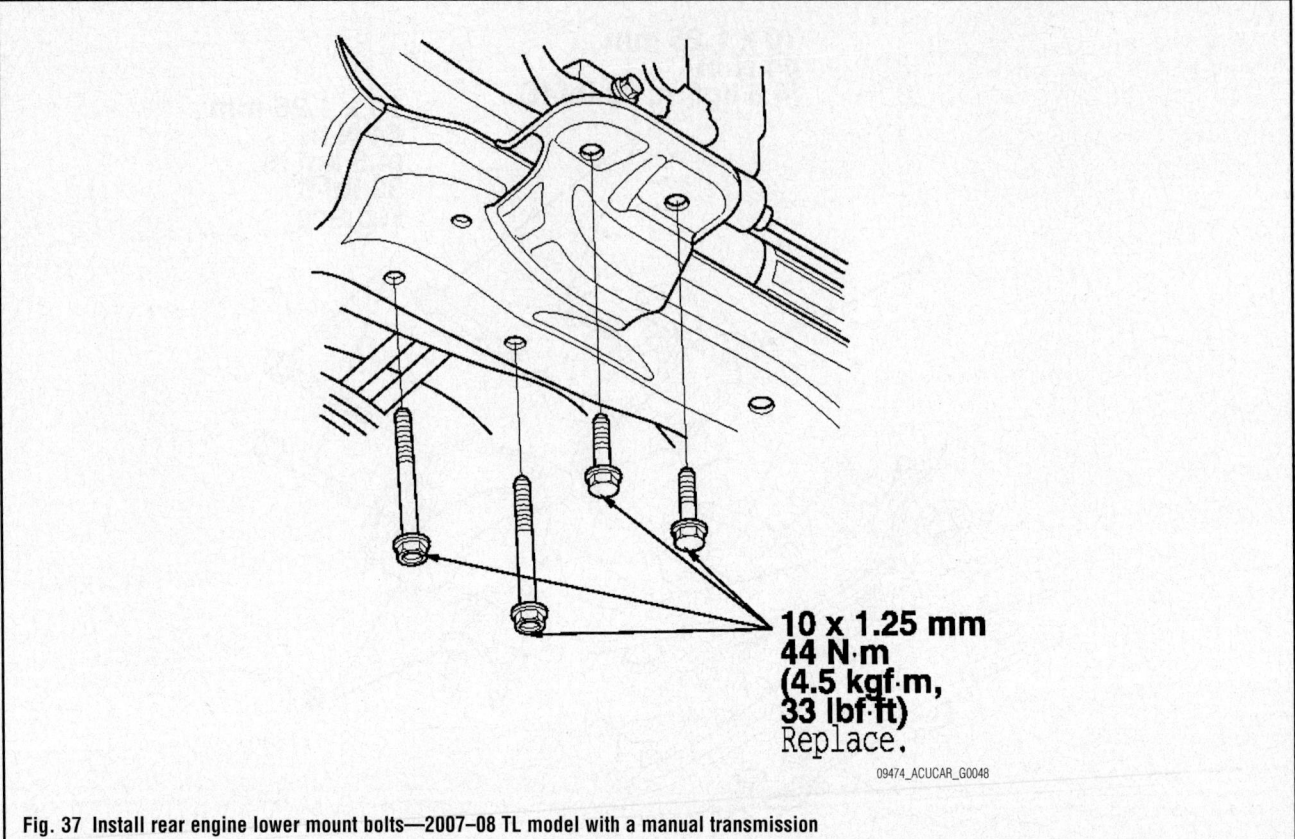

10 x 1.25 mm
44 N·m
(4.5 kgf·m,
33 lbf·ft)
Replace.

09474_ACUCAR_G0048

Fig. 37 Install rear engine lower mount bolts—2007–08 TL model with a manual transmission

B

B

B

B

A

10 x 1.25 mm
33 N·m
(3.4 kgf·m,
24 lbf·ft)

10 x 1.25 mm
54 N·m
(5.5 kgf·m, 40 lbf·ft)

8 x 1.25 mm
22 N·m
(2.2 kgf·m, 16 lbf·ft)

09474_ACUCAR_G0049

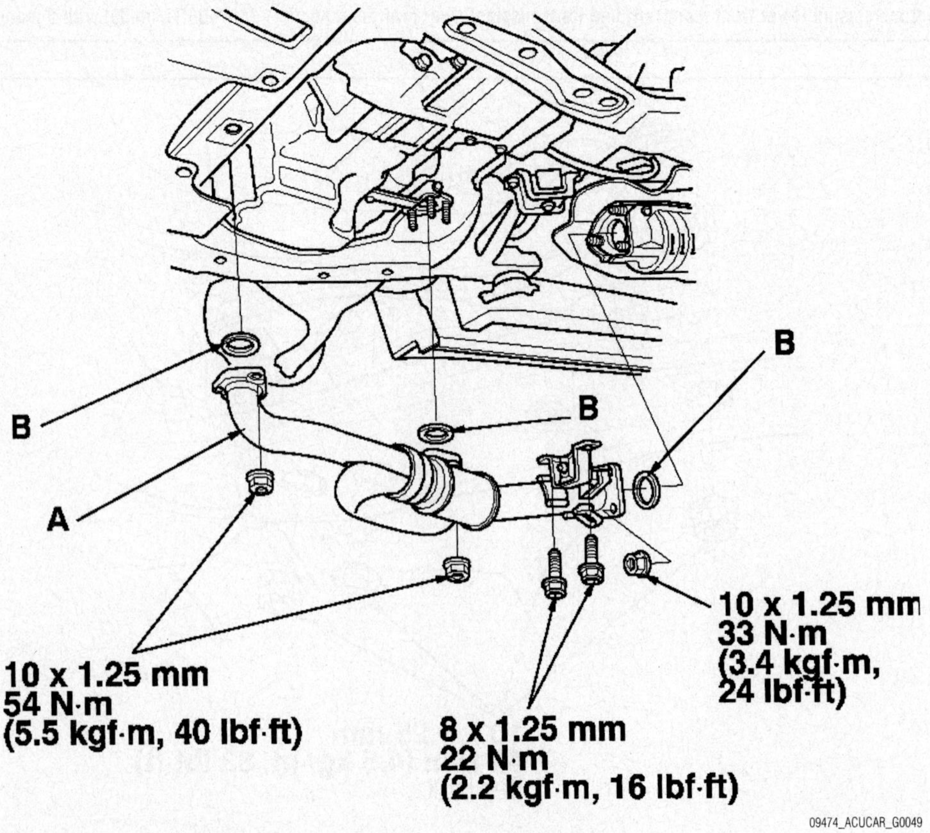

Fig. 38 Exhaust pipe torque specifications—2007–08 TL model with a manual transmission

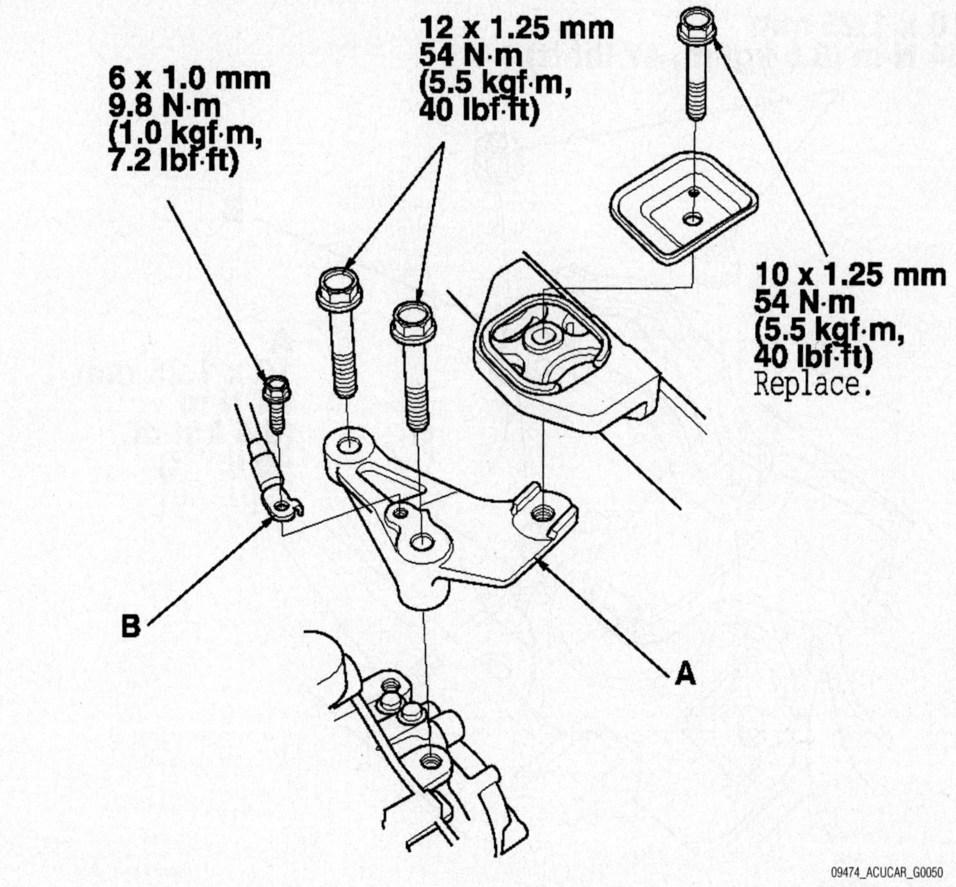

**6 x 1.0 mm
9.8 N·m
(1.0 kgf·m,
7.2 lbf·ft)**

**12 x 1.25 mm
54 N·m
(5.5 kgf·m,
40 lbf·ft)**

**10 x 1.25 mm
54 N·m
(5.5 kgf·m,
40 lbf·ft)
Replace.**

B

A

09474_ACUCAR_G0050

Fig. 39 Transmission mount bracket (A) and ground cable (B) torque specs—2007–08 TL model with a manual transmission

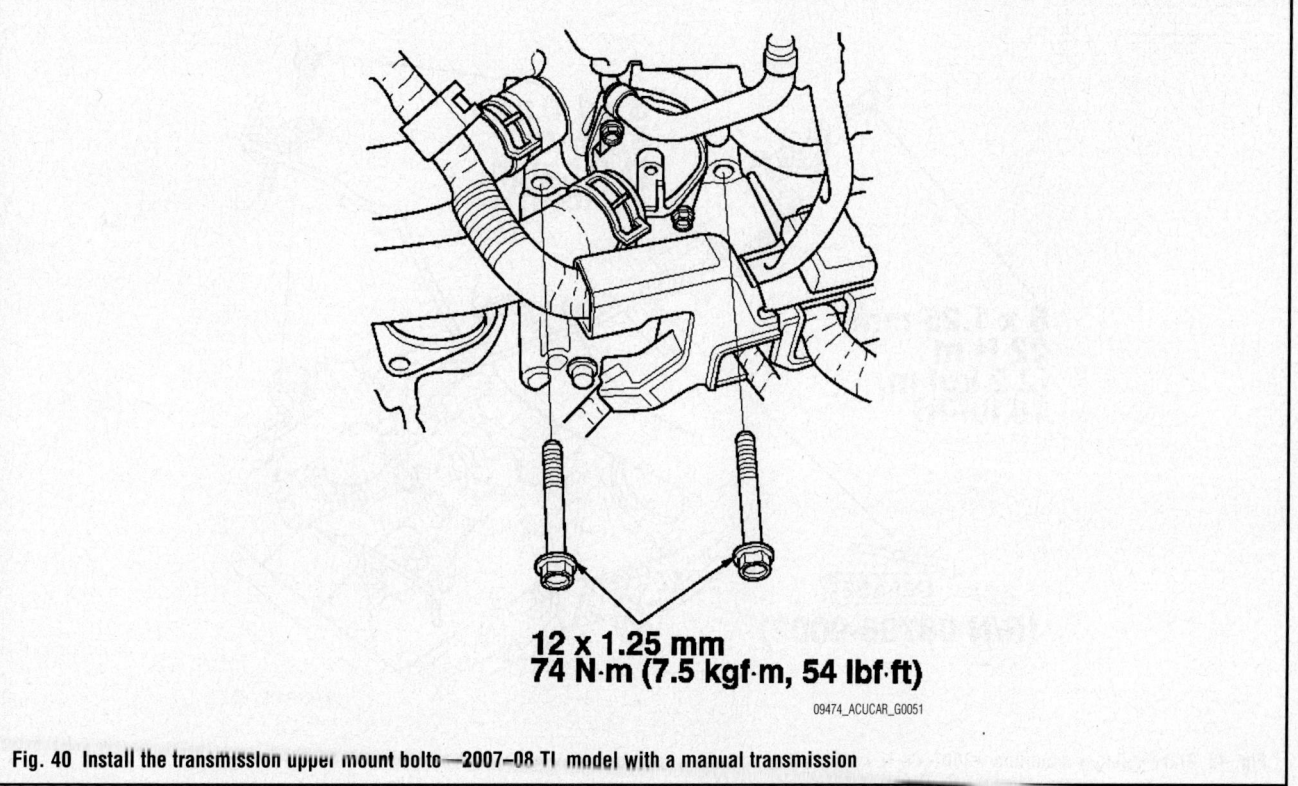

**12 x 1.25 mm
74 N·m (7.5 kgf·m, 54 lbf·ft)**

09474_ACUCAR_G0051

Fig. 40 Install the transmission upper mount bolts—2007–08 TL model with a manual transmission

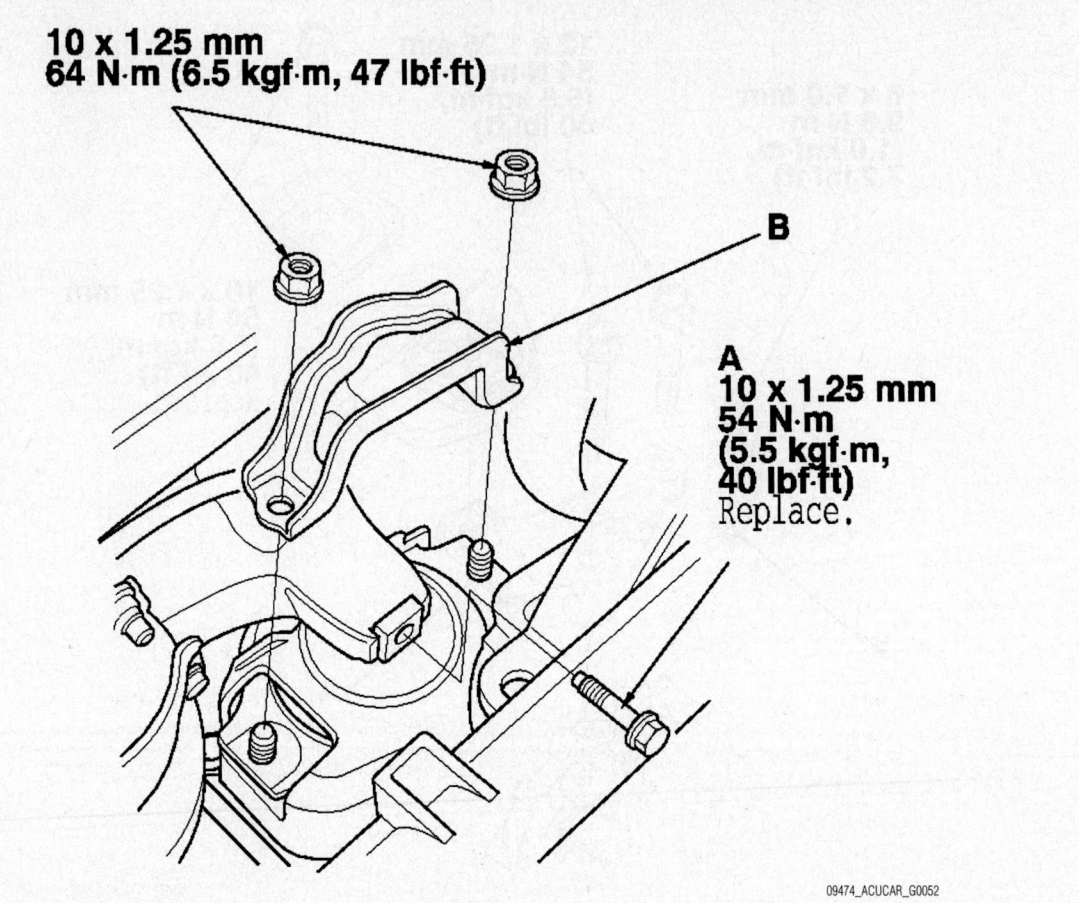

10 x 1.25 mm
64 N·m (6.5 kgf·m, 47 lbf·ft)

B

A
10 x 1.25 mm
54 N·m
(5.5 kgf·m,
40 lbf·ft)
Replace.

09474_ACUCAR_G0052

Fig. 41 Install the front engine mount bolts (A) the front mount stop (B) and the mounting nuts (C)—2007–08 TL model with a manual transmission

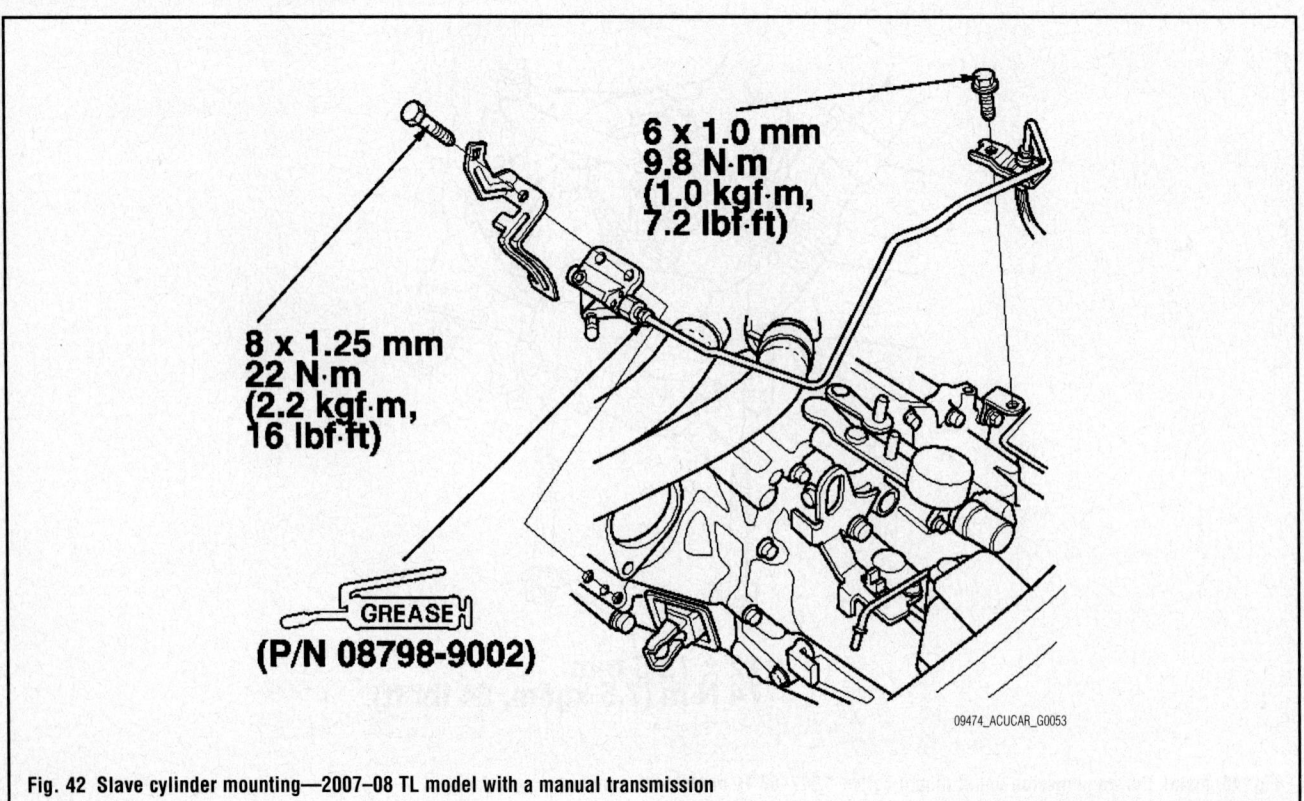

6 x 1.0 mm
9.8 N·m
(1.0 kgf·m,
7.2 lbf·ft)

8 x 1.25 mm
22 N·m
(2.2 kgf·m,
16 lbf·ft)

GREASE
(P/N 08798-9002)

09474_ACUCAR_G0053

Fig. 42 Slave cylinder mounting—2007–08 TL model with a manual transmission

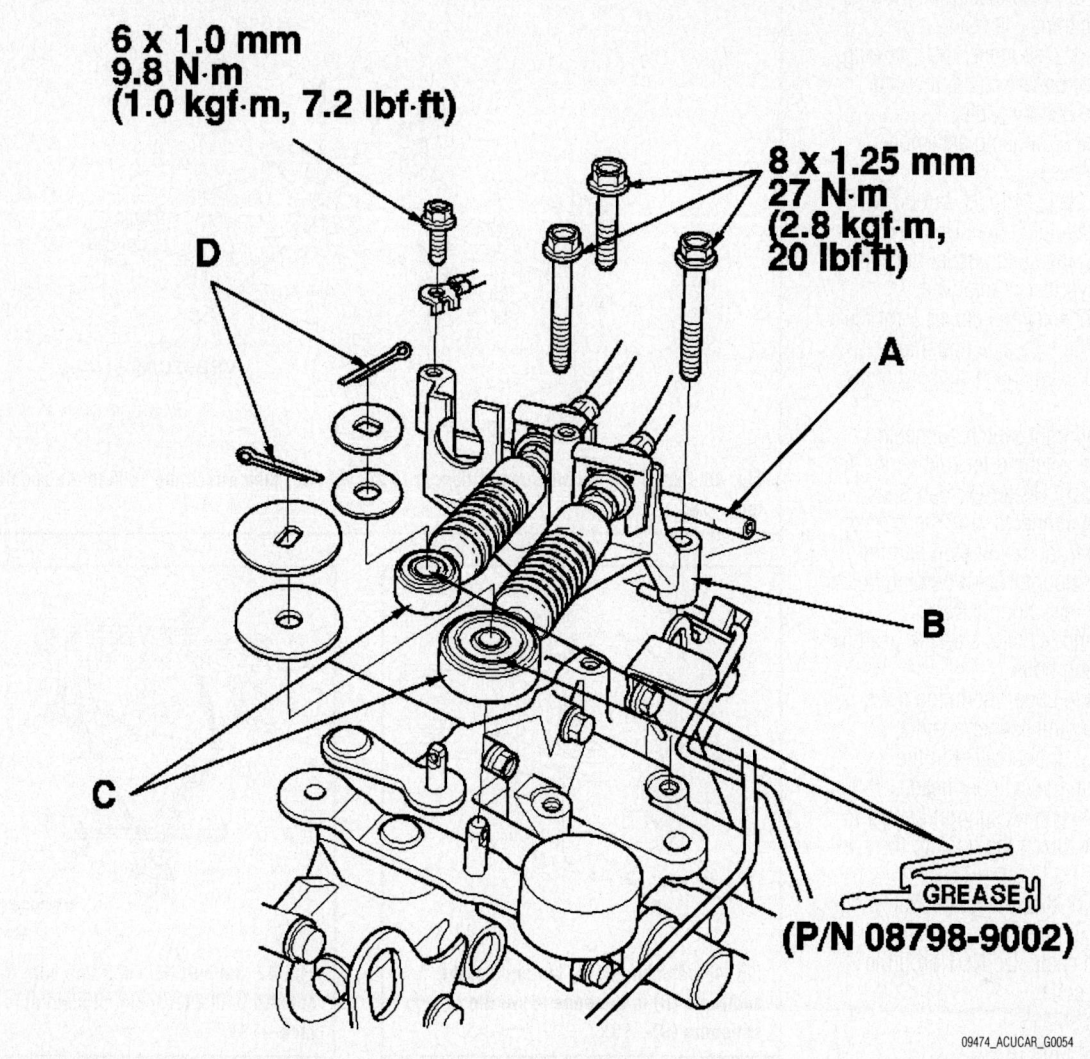

**6 x 1.0 mm
9.8 N·m
(1.0 kgf·m, 7.2 lbf·ft)**

**8 x 1.25 mm
27 N·m
(2.8 kgf·m,
20 lbf·ft)**

D

A

B

C

GREASE

(P/N 08798-9002)

09474_ACUCAR_G0054

Fig. 43 Exploded view cable bracket assembly, cables and the retainer torque values—2007–08 TL model with a manual transmission

63. Install the transmission lower front mount (A) and the transmission lower rear mount (B).

64. Install the transmission lower mount mounting nuts.

65. Install the rear engine lower mount mounting bolts.

66. Connect the power steering pressure switch connector.

67. Connect the return line to the power steering fluid reservoir.

68. Install the tie-rod ball joint.

69. Install the front wheels, and set them in the straight ahead position.

70. Connect the steering joint to the steering gearbox pinion shaft by aligning the reference mark, and remove the wire from the joint yoke.

71. Install the steering joint cover.

72. Install the intermediate shaft.

73. Install the left and right driveshaft inboard joints.

74. Install the knuckle onto the lower arm.

75. Install the damper fork.

76. Connect the front stabilizer.

77. Install the exhaust pipe A and new gaskets (B).

78. Install the splash shield.

79. Install the under cover.

80. Lay the fender cover over the fender, then install the transmission mount bracket (A) and ground cable (B).

81. Remove the engine hanger and tool illustrated from the engine.

82. Install the transmission upper mounting bolts.

83. Install the engine rear mount stop.

84. Install the engine front mount mounting bolt (A) and the engine front mount stop (B).

85. Apply super high temp urea grease to the end of the cylinder rod. Install the slave cylinder. Be careful not to bend the clutch line.

86. Connect the vacuum hose.

87. Install the cable bracket and cables.

88. Apply a light coat of super high temp urea grease to the cable ends, and install new cotter pins.

89. Install the harness bracket.

TSX Models

See Figures 44 through 49.

➡This procedure requires the use of the following special tools, or their equivalents: Engine hanger/adapter VSB02C000015, Engine support hanger A & Reds AAR-T-12566 and Subframe adapter VSB02C000016.

1. Before servicing the vehicle, refer to the precautions in the beginning of this section.

2. Note the radio security code and the radio presets.

3. Remove or disconnect the following:

4. Drain the transaxle fluid.

5. Remove or disconnect the following:
- Negative battery cable, then the positive battery cable
- Air cleaner housing assembly
- Battery base
- Clutch slave cylinder, but do NOT bend the clutch line. Do NOT depress the clutch pedal once the slave cylinder is removed.
- Cable bracket and cables from the top of the transaxle housing
- Countershaft speed sensor connector
- Back-up light switch connector
- Reverse solenoid lockout connector
- Secondary Heater Oxygen Sensor (HO2S) connector and the bracket
- Engine front mount stop and the engine mount front mounting bolt
- Engine rear mount stop
- Right and left side steering gearbox mounting bolts
- Transaxle upper mounting bolts

6. Attach Engine hanger/adapter VSB02C000015, or equivalent to the threaded holes in the cylinder head.

7. Install the engine support hanger to the vehicle, then attach the hook to the special tool.

8. Remove or disconnect the following:
- Transaxle mount bracket and ground cable. Do NOT drop the mount collar.
- Splash shield
- Exhaust pipe "A"
- Front stabilizer link

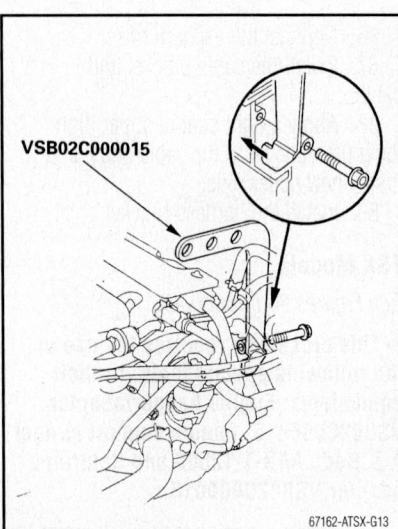

Fig. 44 Attach Engine hanger/adapter VSB02C000015, or equivalent to the threaded holes in the cylinder head—TSX

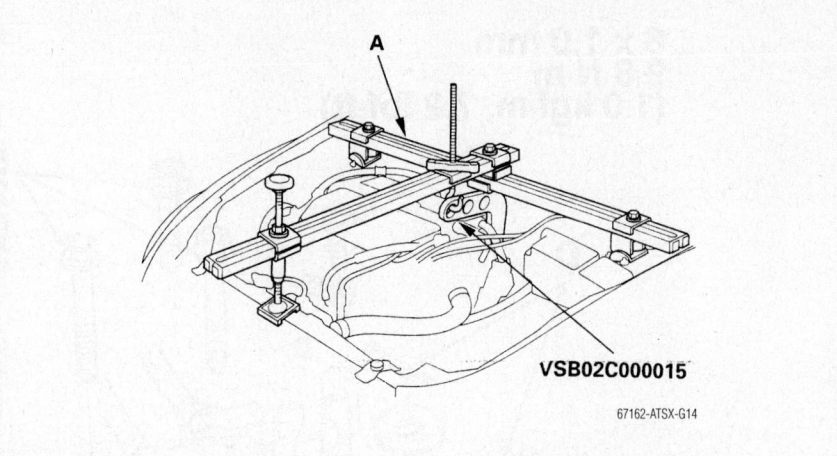

Fig. 45 Install the engine support hanger to the vehicle, then attach the hook to the special tool—TSX

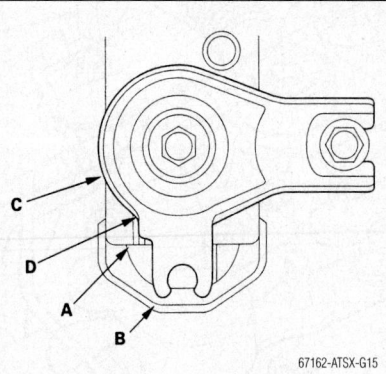

Fig. 46 Matchmark (A) the ends of the subframe (B) to the edge (C) of the stiffeners (D)—TSX

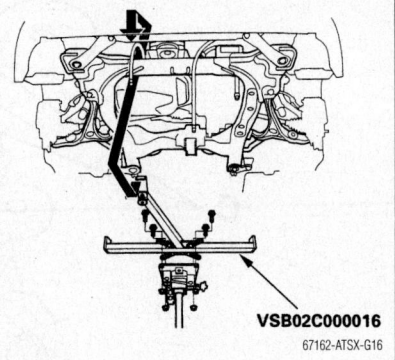

Fig. 47 Support the subframe with the subframe adapter tool and a suitable jack—TSX

- Damper fork
- Knuckle from the lower control arm
- Inboard joint of the driveshaft
- Intermediate shaft
- Heat shield
- Power steering line bracket mounting bolt
- Power steering line from the subframe. There are 2 clamps and 1 mounting bracket bolt.
- Rear engine mount lower mounting bolts
- Transaxle lower mount mounting nuts
- Middle subframe mounting bolts

9. Matchmark the ends of the subframe to the edge of the stiffeners.

10. Support the subframe with the subframe adapter tool and a suitable jack.

11. Suspend the steering gearbox with a suitable support, then remove the front suspension subframe stays and subframe.

12. Remove or disconnect the following:
- Front engine mount upper bracket
- Shift cable bracket, rear engine mount upper mounting bolt and rear engine mount
- Rear engine mount upper bracket
- Clutch cover

13. Place a transmission jack under the transaxle, then remove the lower transaxle mounting bolts.

14. Pull the transaxle away from the engine until the mainshaft clears the clutch pressure plate, then carefully lower the transaxle using the jack.

15. Remove or disconnect the following:
- Transmission lower front mount and lower rear mount
- Release fork boot from the clutch housing
- Release fork from the clutch housing by squeezing the release fork set spring with pliers
- Release bearing

To install:

16. Make sure the 2 dowel pins are installed in the clutch housing.

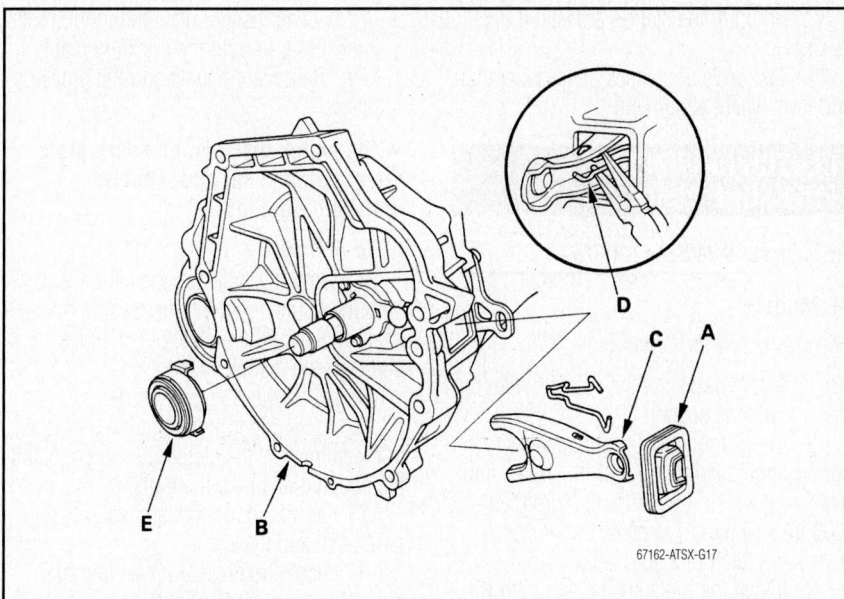

Fig. 48 View of the release fork boot (A), clutch housing (B), release fork (C), set spring (D) and release bearing (E)—TSX

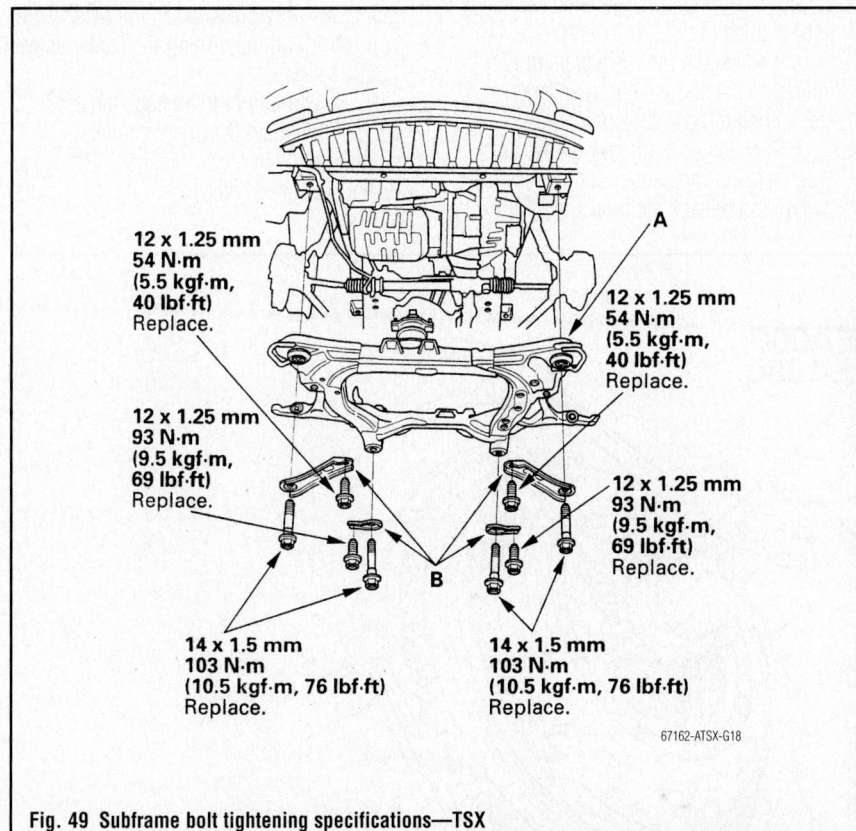

12 x 1.25 mm
54 N·m
(5.5 kgf·m,
40 lbf·ft)
Replace.

12 x 1.25 mm
54 N·m
(5.5 kgf·m,
40 lbf·ft)
Replace.

12 x 1.25 mm
93 N·m
(9.5 kgf·m,
69 lbf·ft)
Replace.

12 x 1.25 mm
93 N·m
(9.5 kgf·m,
69 lbf·ft)
Replace.

14 x 1.5 mm
103 N·m
(10.5 kgf·m, 76 lbf·ft)
Replace.

14 x 1.5 mm
103 N·m
(10.5 kgf·m, 76 lbf·ft)
Replace.

Fig. 49 Subframe bolt tightening specifications—TSX

17. Apply super high temp urea grease (P/N 087989002) to the release fork and release bearing, then install the release fork, release bearing and boot.

18. Make sure the mount bracket collars are installed in the transaxle, then install the transmission lower front and rear mounts. Tighten the front mount bolts to 43 ft. lbs.

(59 Nm) and the lower rear mount bolts to 33 ft. lbs. (45 Nm).

19. Place the transaxle on the transaxle jack and raise it to engine level.

20. Install or connect the following:
- Transmission mounting lower bolts and tighten to 47 ft. lbs. (64 Nm)

- Clutch cover and tighten the retainers to 33 ft. lbs. (45 Nm)
- Rear engine mount upper bracket and tighten NEW bolts to 65 ft. lbs. (90 Nm)
- Rear engine mount
- New rear engine mount upper mounting bolt and tighten to 40 ft. lbs. (54 Nm)
- Shift cable bracket and tighten the bolt to 16 ft. lbs. (22 Nm)
- Front engine mount upper bracket and tighten NEW bolts to 47 ft. lbs. (64 Nm)

21. Support the subframe with the subframe adapter tool and a jack.

22. Install the front suspension subframe and subframe stays.

23. Align the matchmarks with the edge of both rear stiffeners and tighten the rear subframe mounting bolts, then the front bolts, as shown in the accompanying figure.

24. Install or connect the following:
- New middle subframe mounting bolts. Torque the side (short) bolts to 36 ft. lbs. (49 Nm) and the bottom (long) bolt to 33 ft. lbs. (44 Nm).
- Transaxle lower mount mounting nuts and tighten to 33 ft. lbs. (45 Nm)
- New rear engine mount lower mounting bolts and tighten to 33 ft. lbs. (45 Nm)
- Power steering line to the subframe with the 2 clamps and bolt. Torque the bolt to 7.2 ft. lbs. (9.8 Nm).
- Power steering line bracket mounting bolt and tighten to 7.2 ft. lbs. (9.8 Nm)
- Heat shield. Tighten the retainers to 7.2 ft. lbs. (9.8 Nm).
- Intermediate shaft
- Driveshaft inboard joint
- Knuckle onto the lower arm
- Damper fork
- Front stabilizer
- Exhaust pipe "A" with new gaskets. Torque the bolt to 16 ft. lbs. (22 Nm) and the nut to 25 ft. lbs. (33 Nm).
- Splash shield

25. Make sure the mount bracket collars are installed in the transaxle, then install the transaxle mount bracket and ground cable. Torque the mount bracket bolts to 40 ft. lbs. (54 Nm) and the ground cable bolt to 7.2 ft. lbs. (9.8 Nm).

26. Install or connect the following:
- Transmission upper mounting bolts and tighten to 47 ft. lbs. (64 Nm)
- New steering gearbox mounting bolts. Tighten the right side bolts to

28 ft. lbs. (38 Nm) and the left side bolts to 43 ft. lbs. (59 Nm).

- Engine rear mount stop and tighten the nuts to 51 ft. lbs. (69 Nm)
- New engine mount front mounting bolt and tighten to 40 ft. lbs. (54 Nm)
- Engine front mount stop and tighten the nuts to 58 ft. lbs. (78 Nm)
- Bracket and secondary HO2S sensor connector
- Reverse lockout solenoid, back-up light switch and countershaft speed sensor connectors
- Cable bracket and cables. Tighten the bolts to 20 ft. lbs. (27 Nm). Apply a light coat of super high temp grease to the cable ends, then install new cotter pins.

27. Apply super high temp urea grease to the end of the cylinder rod, then install the slave cylinder. Tighten the slave cylinder mounting bolts to 16 ft. lbs. (22 Nm) and the fluid line bolt to 7.2 ft. lbs. (9.8 Nm). Do not bend the clutch line.

- Battery base
- Air cleaner housing
- Battery

28. Fill the transaxle with fluid.

29. Connect the positive and negative battery cables.

30. Program the station presets, if necessary.

31. Test drive and check clutch operation and front wheel alignment.

CLUTCH DRIVEN DISC & PRESSURE PLATE

REMOVAL & INSTALLATION

TL Models

See Figures 50 through 53.

1. Before servicing the vehicle, refer to the precautions section.

2. Check the height of the diaphragm spring fingers using the tool illustrated and a feeler gauge. If the height is more than 0.00 inch (0.8 mm), replace the pressure plate and clutch disc as a set.

3. Install the tools shown and perform the following steps:

   a. Turn the center screw clockwise by hand to apply pressure on the diaphragm spring. Continue turning the center screw until it stops.

   b. Loosen the pressure plate mounting bolts in a star pattern in several steps, then remove the bolts.

   c. Turn the center screw on the pressure plate compressor counterclockwise by hand to release the pressure, then

install two pressure plate mounting bolts, hand-tight, to hold the pressure plate.

   d. Remove the tools and the pressure plate.

➡ The clutch disc and pressure plate are a matched set and must be replaced together.

To install:

4. Temporarily install the clutch disc onto the splines of the transmission mainshaft. Make sure the clutch disc slides freely on the mainshaft.

5. Install the ring gear holder.

6. Apply super high temp urea grease to the splines of the clutch disc, then install the clutch disc using the tools.

7. Align a point mark across the pressure plate and flywheel.

8. Install the pressure plate and the mounting bolts, finger-tight.

9. Install the tools shown and perform the following steps:

   a. Turn the center screw clockwise by hand to apply pressure on the diaphragm spring. Continue turning the center screw until it stops.

10. Be careful not to damage the pressure plate. Tighten the pressure plate mounting bolts in a star pattern to 19 ft. lbs. (25 Nm) in several steps.

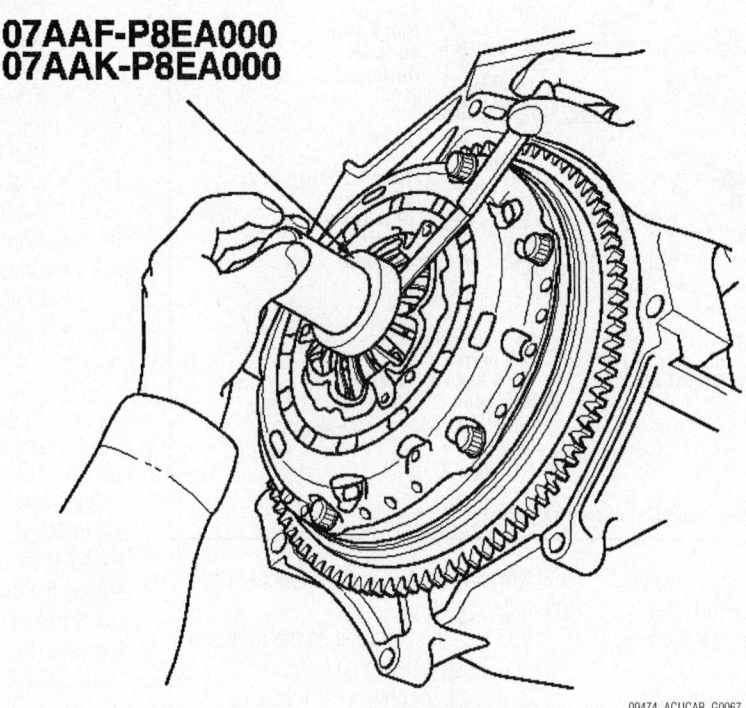

07AAF-P8EA000
07AAK-P8EA000

09474_ACUCAR_G0067

Fig. 50 Checking the height of the diaphragm spring fingers—2007–08 TL model

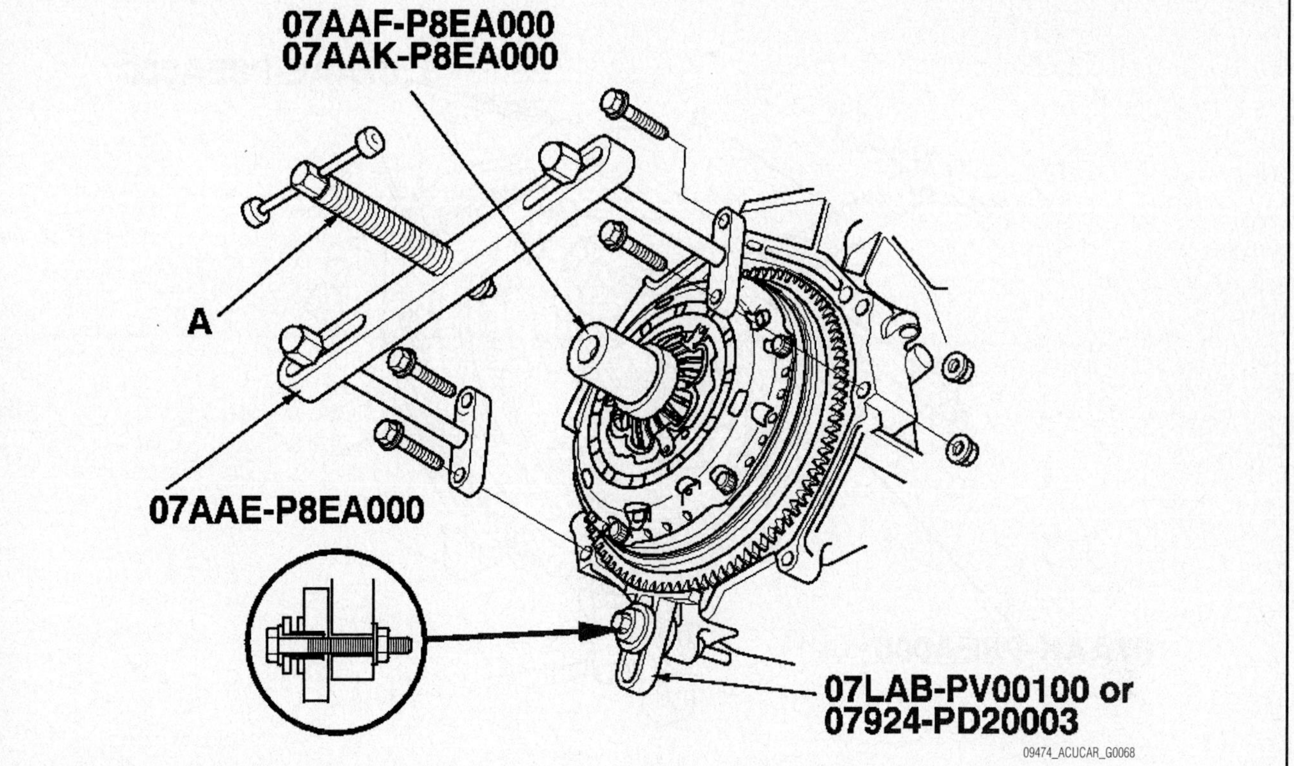

07AAF-P8EA000
07AAK-P8EA000

A

07AAE-P8EA000

07LAB-PV00100 or
07924-PD20003

09474_ACUCAR_G0068

Fig. 51 The tools illustrated are required to remove the clutch assembly—2007–08 TL model

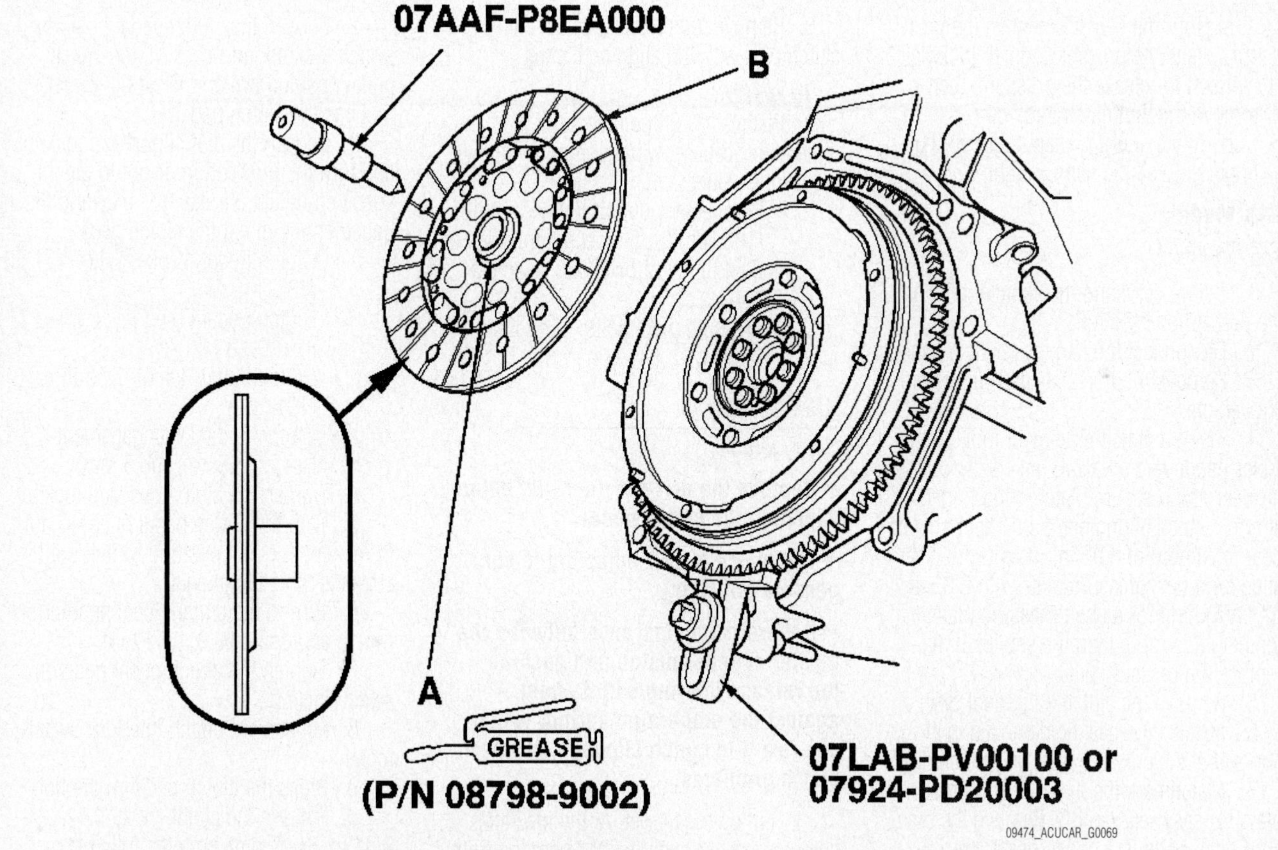

07AAF-P8EA000

B

A

GREASE

(P/N 08798-9002)

07LAB-PV00100 or
07924-PD20003

09474_ACUCAR_G0069

Fig. 52 Remove the tools and the pressure plate—2007–08 TL model

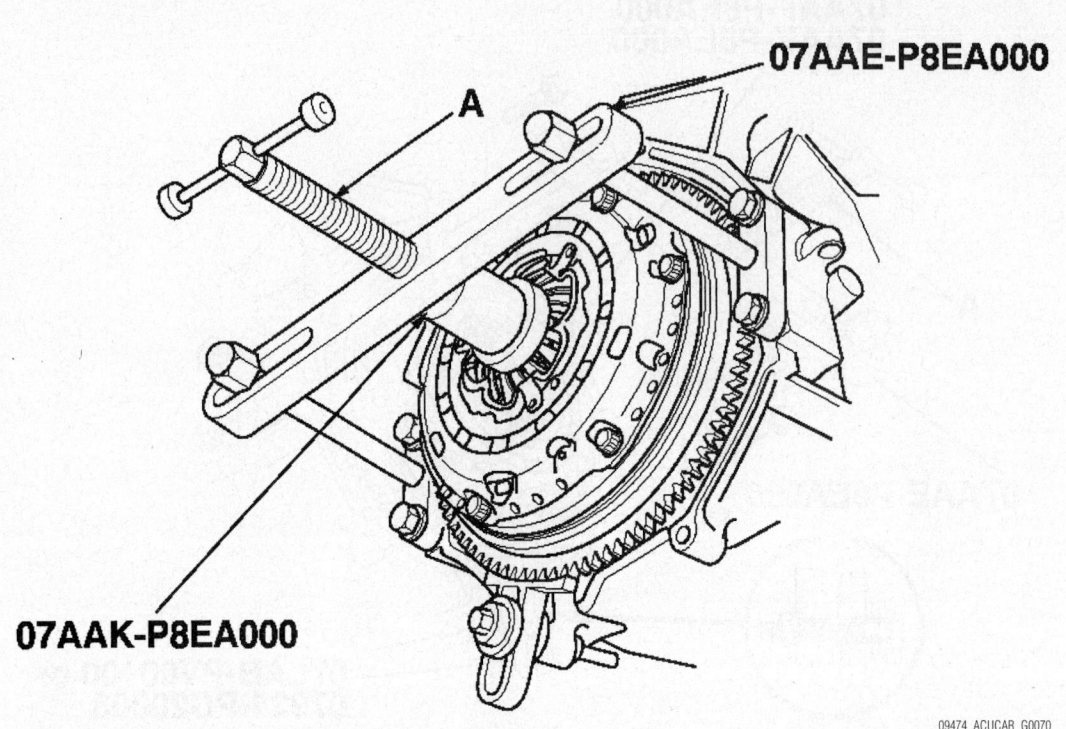

07AAE-P8EA000

A

07AAK-P8EA000

09474_ACUCAR_G0070

Fig. 53 Turn the center screw clockwise by hand to apply pressure on the diaphragm spring—2007–08 TL model

a. Turn the center screw on the pressure plate compressor counterclockwise by hand to release the pressure, then remove the tool illustrated.

b. Make sure the diaphragm spring fingers are all the same height.

TSX Models

See Figure 54.

1. Before servicing the vehicle, refer to the precautions section.

2. Disconnect the negative battery cable.

3. Remove the transaxle assembly from the vehicle.

4. Insert a clutch alignment tool. Use a feeler gauge and measure the clearance between the pressure plate spring fingers and the clutch alignment disc. There should be a maximum of 0.02 in. (0.6mm) of clearance for a new pressure plate with 0.03 in. (0.8mm) limit for a used pressure plate. If the height is more than the service limit, replace the pressure plate.

5. Remove the clutch alignment disc.

6. Install a flywheel holder to aid in the removal of the pressure plate and clutch disc.

7. Matchmark the flywheel and pressure plate for easy reassembly. Remove the pressure plate bolts in a crisscross pattern 2 turns at a time to prevent warping the plate.

8. Remove the pressure plate, then the clutch disc with the alignment shaft.

To install:

9. Installation is the reverse of the removal procedure, while using the following torque values:

- Flywheel mounting bolts: 76 ft. lbs. (103 Nm) for 5–speed transaxles or 90 ft. lbs. (122 Nm) for 6–speed transaxles
- Pressure plate bolts: 19 ft. lbs. (25 Nm)

ADJUSTMENTS

See Figure 55.

➡**Remove the driver's floor mat before adjusting the clutch pedal.**

➡**The clutch is self-adjusting to compensate for wear.**

➡**If there is no clearance between the master cylinder piston and pushrod, the release bearing will be held against the diaphragm spring, which can result in clutch slippage or other clutch problems.**

1. Lift up the carpet. At the insulator cutout, measure pedal height from the right side of the pedal pad.

2. Loosen the clutch pedal position switch locknut, and back off the clutch pedal position switch until it no longer touches the clutch pedal.

3. Loosen the clutch pushrod locknut, and turn the pushrod in or out to get the specified height, stroke, free play and disengagement height at the clutch pedal.

- Clutch Pedal Stroke: 5.12–5.51 in. (130-140 mm)
- Clutch Pedal Free Play: 0.39–0.71 in. (10-18 mm)
- Clutch Pedal Height: 7.52 in. (191 mm)
- Clutch Pedal Disengagement Height: 3.56 in. (90.5 mm)

4. Tighten the clutch pushrod locknut.

5. With the clutch pedal released, turn in the clutch pedal position switch until it contacts the clutch pedal.

6. Turn in the clutch pedal position switch an additional ¾ to 1 turn.

7. Tighten the clutch pedal position switch locknut.

8. Loosen the clutch interlock switch locknut.

9. Press the clutch pedal to the floor.

10. Release the clutch pedal 0.12–0.31 in. (3–8 mm) from the fully pressed position, and hold it there. Adjust the

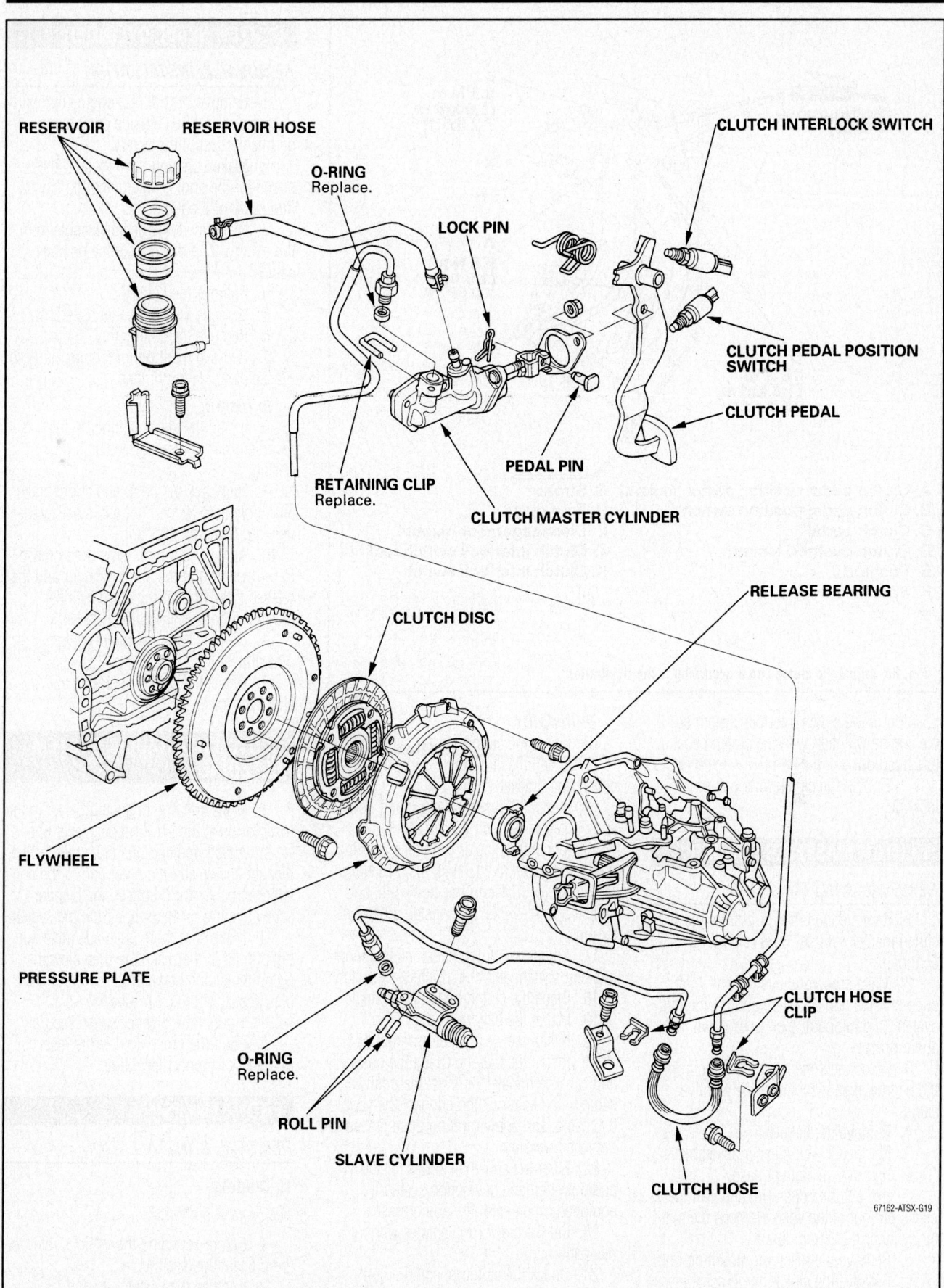

RESERVOIR **RESERVOIR HOSE**

O-RING
Replace.

LOCK PIN

CLUTCH INTERLOCK SWITCH

CLUTCH PEDAL POSITION SWITCH

CLUTCH PEDAL

PEDAL PIN

RETAINING CLIP
Replace.

CLUTCH MASTER CYLINDER

RELEASE BEARING

CLUTCH DISC

FLYWHEEL

PRESSURE PLATE

O-RING
Replace.

ROLL PIN

SLAVE CYLINDER

CLUTCH HOSE CLIP

CLUTCH HOSE

67162-ATSX-G19

Fig. 54 Exploded view of the clutch system components—TSX models

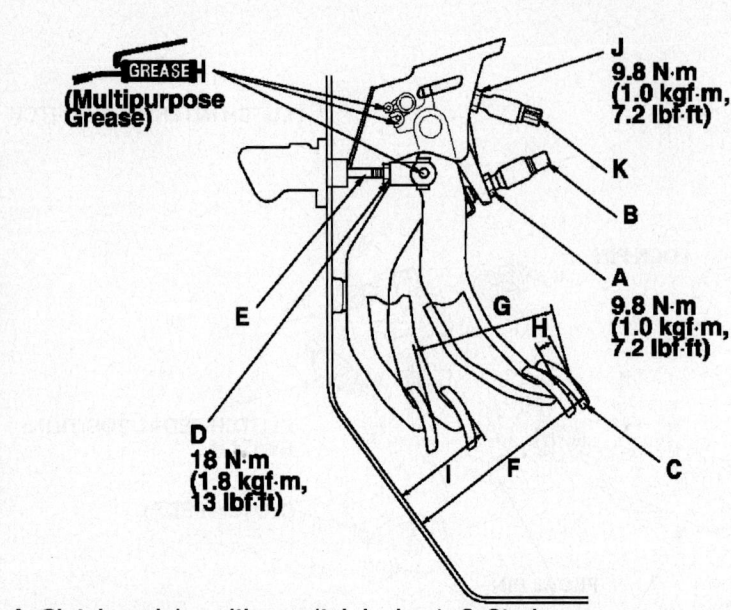

(Multipurpose Grease)

J.
9.8 N·m
(1.0 kgf·m,
7.2 lbf·ft)

A.
9.8 N·m
(1.0 kgf·m,
7.2 lbf·ft)

D.
18 N·m
(1.8 kgf·m,
13 lbf·ft)

A. Clutch pedal position switch locknut
B. Clutch pedal position switch
C. Clutch pedal
D. Clutch pushrod locknut
E. Pushrod
F. Specified height
G. Stroke
H. Free play
I. Disengagement height
J. Clutch interlock switch locknut
K. Clutch interlock switch

22140_ACUR_G0032

Fig. 55 Adjust the clutch pedal according to the illustration

position of the clutch interlock switch so the engine will start with the clutch pedal in this position.

11. Tighten the clutch interlock switch locknut.

CLUTCH MASTER CYLINDER

REMOVAL & INSTALLATION

1. Remove the brake fluid from the clutch master cylinder reservoir with a syringe.
2. Make sure you have the anti-theft codes for the audio system and navigation system (if equipped), then write down the audio presets.
3. Disconnect the negative cable from the battery, then disconnect the positive cable.
4. Remove the battery.
5. Remove the air cleaner assembly.
6. Remove the battery base.
7. Pry out the cotter pin, and pull the clevis pin out of the yoke. Remove the master cylinder mounting nuts.
8. Remove the reservoir mounting bolt.
9. Remove the clutch line bracket.
10. Remove the clutch line clips.
11. Remove the clutch master cylinder.

To install:
12. Install the clutch master cylinder.
13. Install the clutch line clips and clutch line bracket.
14. Install the reservoir mounting bolt and tighten to 88 inch lbs. (10 Nm).
15. Install the master cylinder mounting nuts and tighten to 115 inch lbs. (13 Nm).
16. Apply grease to the clevis pin, and slide it into the yoke, then install a new cotter pin.
17. Adjust the clutch pedal, clutch pedal position switch, and clutch interlock switch.
18. Bleed the clutch hydraulic system.
19. Install the battery base.
20. Install the air cleaner assembly.
21. Install the battery. Clean the posts and cable terminals. Connect the positive cable to the battery, then connect the negative cable, and apply multipurpose grease to prevent corrosion.
22. Enter the anti-theft codes for the audio system and navigation system (if equipped), then enter the audio presets.
23. Set the clock (on vehicles without navigation).
24. Check the clutch operation, and check for leaks.
25. Test-drive the vehicle.

CLUTCH SLAVE CYLINDER

REMOVAL & INSTALLATION

1. Remove the left rear engine compartment cover and the left side engine compartment cover, if necessary.
2. Make sure you have the anti-theft codes for the audio system and the navigation system (if equipped).
3. Disconnect the negative cable from the battery, then disconnect the positive cable.
4. Remove the battery.
5. Remove the air cleaner assembly.
6. Remove the battery tray.
7. Remove the mounting bolts, harness bracket, and slave cylinder.

To install:
8. Install the slave cylinder in the reverse order of removal. Install a new O-ring.
9. Pull back the boot, and apply brake assembly lube to the boot and slave cylinder rod. Reinstall the boot.
10. Apply super high temp urea grease to the pushrod of the slave cylinder and the release fork. Tighten the slave cylinder mounting bolts to 16 ft. lbs. (22 Nm).
11. Do the clutch hydraulic system bleeding.
12. Check the clutch operation and check for leaks.

CLUTCH HYDRAULIC SYSTEM BLEEDING

1. Make sure the brake fluid level in the reservoir is at the MAX (upper) level line.
2. Attach one end of a clear tube to the bleeder screw, and the other end to the container of brake fluid. Loosen the bleeder screw to allow air to escape from the system.
3. Make sure there is an adequate supply of fluid in the reservoir, then slowly pump the clutch pedal until no more bubbles appear at the clear tube.
4. Tighten the bleeder screw securely.
5. Refill the brake fluid in the reservoir to the MAX (upper) level line.

FRONT HALFSHAFT

REMOVAL & INSTALLATION

TL Models

See Figures 56 and 57.

1. Before servicing the vehicle, refer to the precautions section.
2. Remove the wheel nuts and front wheels.

3. Lift up the locking tab on the spindle nut, then remove the nut.

4. Drain the transmission fluid, then reinstall the drain plug using a new washer:

5. Remove exhaust pipe.

6. Hold the stabilizer ball joint pin with a hex wrench, and remove the flange nut. Separate the front stabilizer link from the lower arm.

7. Remove the self-locking nut, 12 mm flange bolt, and 10 mm flange bolt, then remove the damper fork.

8. Remove the cotter pin from the lower arm ball joint, and remove the nut.

➡ **To avoid damaging the ball joint, install the tool illustrated on the threads of the ball joint. Be careful not to damage the ball joint boot when installing the remover. Do not force or hammer on the lower arm, or pry between the lower arm and the knuckle. You could damage the ball joint.**

9. Disconnect the lower ball joint from the lower arm using the tool illustrated.

➡ **The collar on the lower arm is removed with ball joint, the lower arm must be replaced.**

10. Pull the knuckle outward, and remove the outboard joint from the front wheel hub using a plastic hammer.

11. On the left driveshaft, pry the inboard joint from the differential case with a pry bar.

12. On the right driveshaft, drive the inboard joint off of the intermediate shaft with a drift and hammer.

13. Remove the driveshaft as an assembly.

➡ **Do not pull on the driveshaft, because the inboard joint may come apart. Pull the driveshaft straight out to avoid damaging the oil seal.**

To install:

14. Install a new set ring in the set ring groove of the left driveshaft.

15. Apply 0.02–0.04 ounces (0.5–1 gram) of grease to the whole splined surface of the right driveshaft.

16. After applying the grease, remove the grease from the splined grooves at intervals of 2-3 splines and from the set ring groove so that air can bleed from the intermediate shaft.

17. Clean the areas where the driveshaft contacts the differential thoroughly with solvent and dry with compressed air.

18. Insert the inboard end of the driveshaft into the differential or intermediate shaft until the new set ring locks in the groove.

19. Install the outboard joint into the front hub.

20. Clean off any grease from the ball joint tapered section and threads, then install the knuckle onto the lower arm. Torque the new castle nut to the 58 ft. lbs. (78 Nm) on 2004 models or 54 ft. lbs. (74 Nm) on 2005–06 models, then tighten it only far enough to align the slot with the ball joint pin hole. Do not align the nut by loosening it.

➡ **Make sure the ball joint boot is not damaged or cracked.**

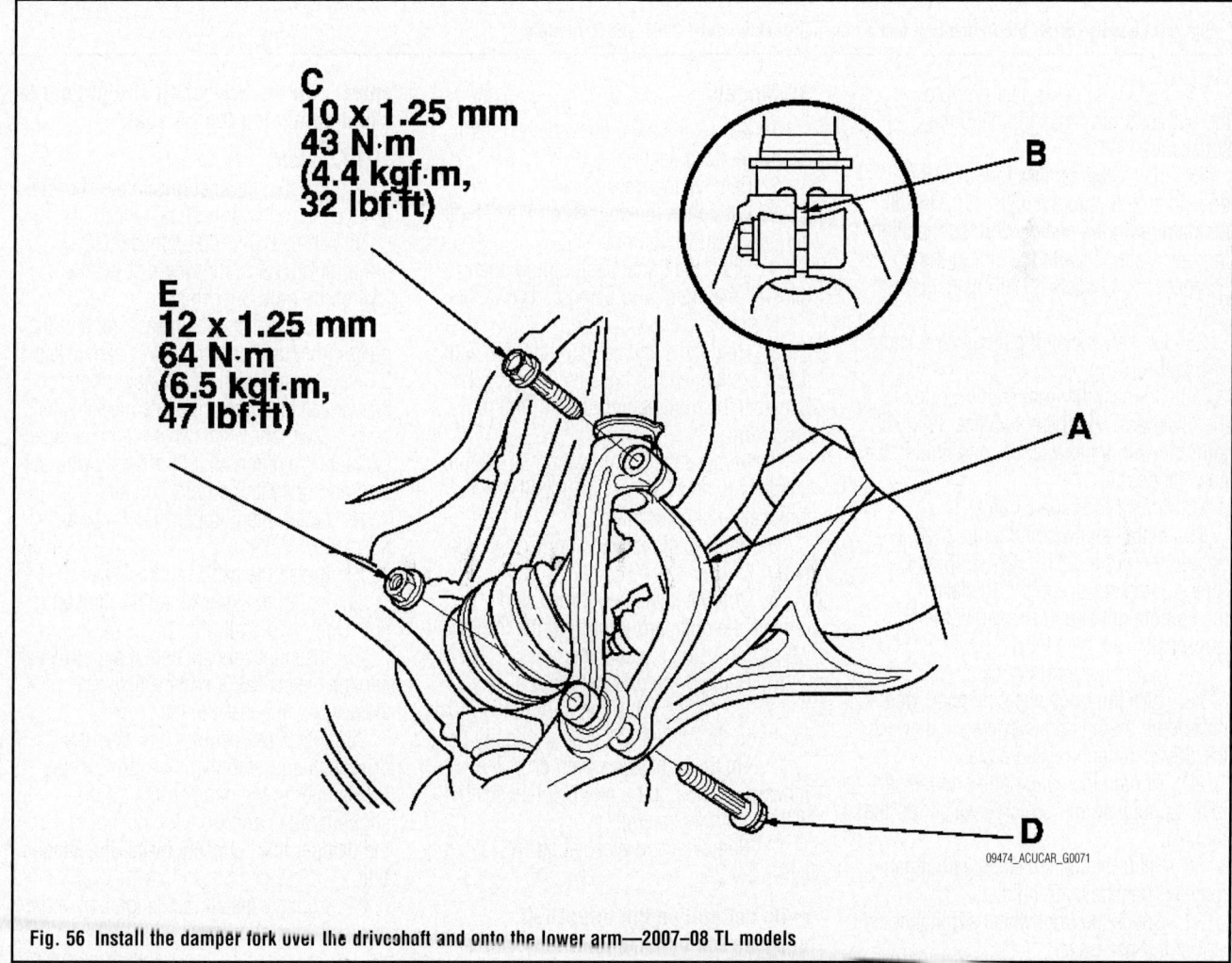

Fig. 56 Install the damper fork over the driveshaft and onto the lower arm—2007–08 TL models

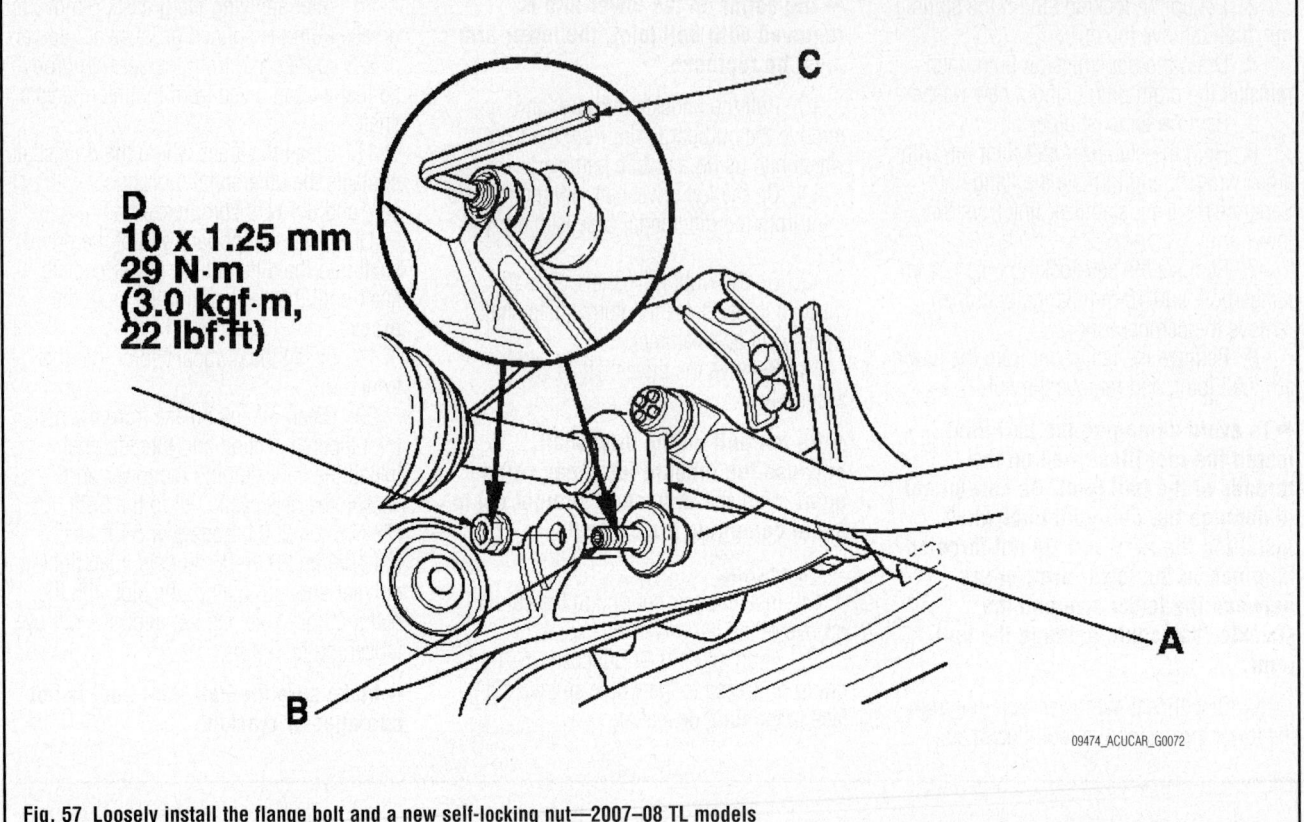

D
10 x 1.25 mm
29 N·m
(3.0 kgf·m,
22 lbf·ft)

C

A

B

09474_ACUCAR_G0072

Fig. 57 Loosely install the flange bolt and a new self-locking nut—2007–08 TL models

21. Install the new cotter pin into the ball joint pin hole, and bend the cotter pin.

22. Install the damper fork over the driveshaft and onto the lower arm. Install the damper in the damper fork so the aligning tab is aligned with the slot in the damper fork. Loosely install the flange bolt.

23. Loosely install the flange bolt and a new self-locking nut.

24. Connect the front stabilizer link to the lower arm. Hold the stabilizer link ball joint pin with a hex wrench, and tighten the new flange nut.

25. Install the exhaust pipe.

26. Install a new spindle nut, then tighten the nut to 181 ft. lbs. (245 Nm). After tightening, use a drift to stake the spindle nut shoulder against the driveshaft.

27. Install the front wheel.

28. Turn the front wheel by hand, and make sure there is no interference between the driveshaft and related parts.

29. Tighten the flange bolt and the self-locking nut with the vehicle's weight on the damper.

30. Refill the transmission with recommended transmission fluid:

31. Check the front wheel alignment, and adjust it if necessary.

RL Models

1. Before servicing the vehicle, refer to the precautions section.

2. Remove the front wheels.

3. Lift up the locking tab on the spindle nut, then remove the nut.

4. Drain the transmission fluid, then reinstall the drain plug using a new washer:

5. Remove exhaust pipe.

6. Hold the stabilizer ball joint pin with a hex wrench, and remove the flange nut. Separate the front stabilizer link from the lower arm.

7. Remove the self-locking nut, 12 mm flange bolt, and 10 mm flange bolt, then remove the damper fork.

8. Remove the knuckle holder bolt and nut.

9. Pull the knuckle outward, and remove the outboard joint from the front wheel hub using a plastic hammer.

10. Remove the exhaust pipe.

11. On the left driveshaft, pry the inboard joint from the differential case with a pry bar.

12. On the right driveshaft, drive the inboard joint off of the intermediate shaft with a drift and hammer.

13. Remove the driveshaft as an assembly.

➡ **Do not pull on the driveshaft, because the inboard joint may come apart. Pull the driveshaft straight out to avoid damaging the oil seal.**

To install:

14. Install a new set ring in the set ring groove of the left driveshaft.

15. Apply 0.02–0.04 ounces (0.5–1 gram) of grease to the whole splined surface of the right driveshaft.

16. After applying the grease, remove the grease from the splined grooves at intervals of 2-3 splines and from the set ring groove so that air can bleed from the intermediate shaft.

17. Clean the areas where the driveshaft contacts the differential thoroughly with solvent and dry with compressed air.

18. Install the outboard joint into the front hub.

19. Install the exhaust pipe.

20. Install the knuckle holder bolt and nut. Tighten to 54 ft. lbs. (74 Nm).

21. Insert the inboard end of the driveshaft into the differential or intermediate shaft until the new set ring locks in the groove.

22. Install the damper fork over the driveshaft and onto the lower arm. Install the damper in the damper fork so the aligning tab is aligned with the slot in the damper fork. Loosely install the flange bolt.

23. Loosely install the flange bolt and a new self-locking nut.

24. Connect the front stabilizer link to the lower arm. Hold the stabilizer link ball joint pin with a hex wrench, and tighten the new flange nut to 40 ft. lbs. (54 Nm).

25. Install a new spindle nut, then tighten the nut to 242 ft. lbs. (349 Nm). After tightening, use a drift to stake the spindle nut shoulder against the driveshaft.

26. Install the front wheel.

27. Turn the front wheel by hand, and make sure there is no interference between the driveshaft and related parts.

28. Tighten the flange bolt and the self-locking nut with the vehicle's weight on the damper.

29. Refill the transmission with recommended transmission fluid:

30. Check the front wheel alignment, and adjust it if necessary

TSX Models

See Figures 58 and 59.

1. Before servicing the vehicle, refer to the precautions section.

2. Drain the differential or transmission lubricant.

3. Remove or disconnect the following:
   - Negative battery cable
   - Wheel(s)
   - Axle nut
   - Strut fork
   - Lower ball joint from the control arm

4. Pull the knuckle outward and remove the halfshaft outboard CV–joint from the knuckle using a plastic mallet.

5. Using a small pry bar carefully pry out the inboard CV–joint approximately ½ in. (13mm) in order to force the spring clip out of the groove in the differential side gears.

➡**Be careful not to damage the oil seal. Do not pull on the inboard CV–joint, it may come apart.**

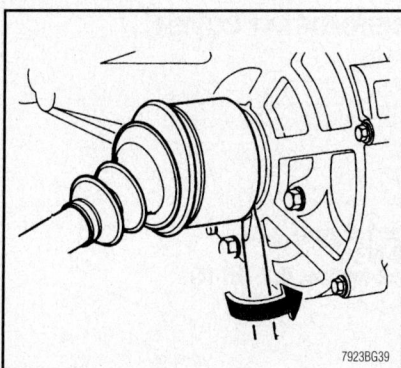

Fig. 58 Carefully pry the inboard joint from the transaxle

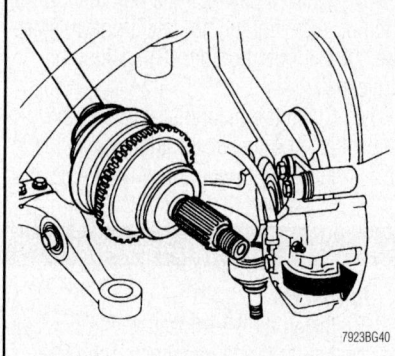

Fig. 59 Pull the hub assembly from the outboard joint

6. Pull the halfshaft out of the differential or the intermediate shaft.

7. Remove the halfshaft from the wheel hub.

To install:

➡**Always use a new set ring whenever the driveshaft is being installed. Be sure the driveshaft locks in the differential side gear groove and that the CV–joint stub–axle bottoms in the differential or the intermediate shaft.**

8. Install or connect the following:
   - Outboard joint to the wheel hub
   - Inboard joint with a new set ring into the differential or intermediate shaft until the set ring locks in the groove
   - Ball joint to the control arm and torque the nut to 36–43 ft. lbs. (49–59 Nm)
   - Strut fork and torque the pinch bolt to 32 ft. lbs. (43 Nm) and the lower bolt to 47 ft. lbs. (64 Nm)
   - Axle nut and torque it to 181 ft. lbs. (245 Nm)
   - Wheel(s)

9. Refill the transmission or differential with the correct amount and type of fluid.

10. Reconnect the battery cable and enter the radio security code.

11. Measure and adjust the wheel alignment.

REAR DIFFERENTIAL

REMOVAL & INSTALLATION

RL Models

1. Drain the differential fluid.

2. Remove the propeller shaft.

3. Disconnect the rear differential harness connectors.

4. Disconnect the breather hose.

5. Using a pair of screwdrivers, pry out both inboard joints from the differential.

6. Place the transmission jack under the rear differential.

7. Remove the differential's rear mounting bolts.

8. Remove the differential's front mounting bolts.

9. As you lower the rear differential (A) with the transmission jack, remove the rear driveshafts (B) from the rear differential.

To install:

10. Raise the rear differential with the transmission jack.

11. If the original differential is being reinstalled, replace the set rings.

12. Apply multipurpose grease to the splines of the rear halfshafts, then install the rear driveshafts to the rear differential.

13. Install the new front mounting bolts and tighten to 22 ft. lbs. (30 Nm).

14. Install the new rear mounting bolts and tighten to 54 ft. lbs. (74 Nm).

15. Connect the breather hose .

16. Connect the rear differential harness connectors.

17. Install the propeller shaft.

18. Refill the differential fluid.

REAR AXLE SHAFT, BEARING & SEAL

REMOVAL & INSTALLATION

1. Raise the vehicle on a lift.

2. Remove the wheel nuts and rear wheels.

3. Lift up the locking tab (A) on the spindle nut (B), then remove the nut.

4. Remove the rear differential, and disconnect the inboard joint from the differential.

5. Remove the rear driveshaft outboard joint from the trailing arm and rear hub using a plastic hammer or a puller if necessary.

6. Remove the rear driveshaft.

To install:

7. Apply 1.5–2.0 g (0.05–0.07 oz) of grease to the whole splined surface (A). After applying grease, remove the grease from the splined grooves at intervals of 2-3 splines and from the set ring groove (B) so that air can bleed from the differential.

8. Install the outboard joint (A) into the rear hub (B).

9. Seat a new set ring in the set ring groove of the differential.

10. Clean the areas where the driveshaft contacts the differential thoroughly with brake cleaner, and dry with compressed air. Do not wash the rubber parts in solvent.

Insert the inboard end (A) of the driveshaft into the differential (B) until the set ring (C) locks in the groove (D).

11. Install the rear differential.

12. Apply a small amount of engine oil to the seating surface of the new spindle nut (A).

13. Install a new spindle nut, then torque the nut. After tightening, use a drift to stake the spindle nut shoulder (B) against the driveshaft.

14. Clean the mating surfaces of the brake disc and the wheel, then install the rear wheels and torque the wheels nuts.

15. Turn the rear wheel by hand, and make sure there is no interference between the driveshaft and surrounding parts.

16. Check the rear wheel alignment, and adjust it if necessary.

ENGINE COOLING

THERMOSTAT

REMOVAL & INSTALLATION

TL Models

See Figure 60.

1. Make sure you have the anti-theft codes for the radio and navigation system, then write down the audio presets. Make sure the ignition switch is OFF.

2. Disconnect the negative cable from the battery first, then disconnect the positive cable.

3. Remove the battery.

4. Drain the engine coolant.

5. Remove the thermostat cover, then remove the thermostat.

To install:

6. Install the thermostat with a new rubber seal and the pin installed on the top side.

7. Install the battery.

8. Refill the radiator with engine coolant, then bleed air from the cooling system.

9. Clean up any spilled engine coolant.

10. Enter the anti-theft codes for the radio and the navigation system, then enter the audio presets.

RL Models

See Figure 61.

1. Make sure you have the anti-theft codes for the radio and navigation system, then write down the audio presets. Make sure the ignition switch is OFF.

2. Disconnect the negative cable from the battery first, then the positive cable, then remove the battery.

3. Drain the engine coolant.

4. Remove the ground cable and thermostat cover, then remove the thermostat.

To install:

5. Install the thermostat with a new rubber seal with the pin on the top side.

6. Install the battery. Clean the battery posts and cable terminals with sandpaper, then assemble them and apply grease to prevent corrosion.

7. Refill the radiator with engine coolant, then bleed air from the cooling system.

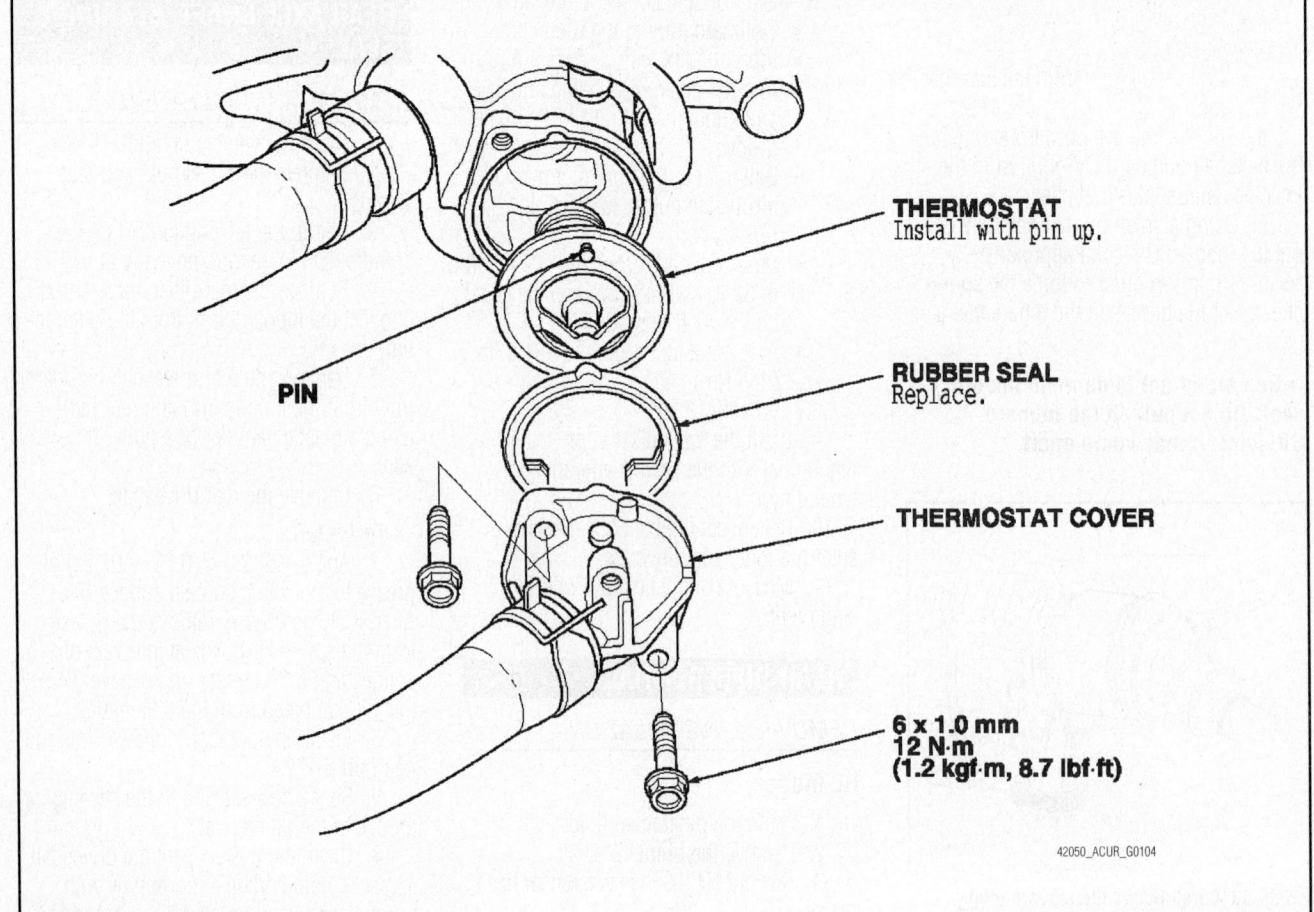

THERMOSTAT
Install with pin up.

PIN

RUBBER SEAL
Replace.

THERMOSTAT COVER

6 x 1.0 mm
12 N·m
(1.2 kgf·m, 8.7 lbf·ft)

42050_ACUR_G0104

Fig. 60 Thermostat mounting—2007–08 TL models

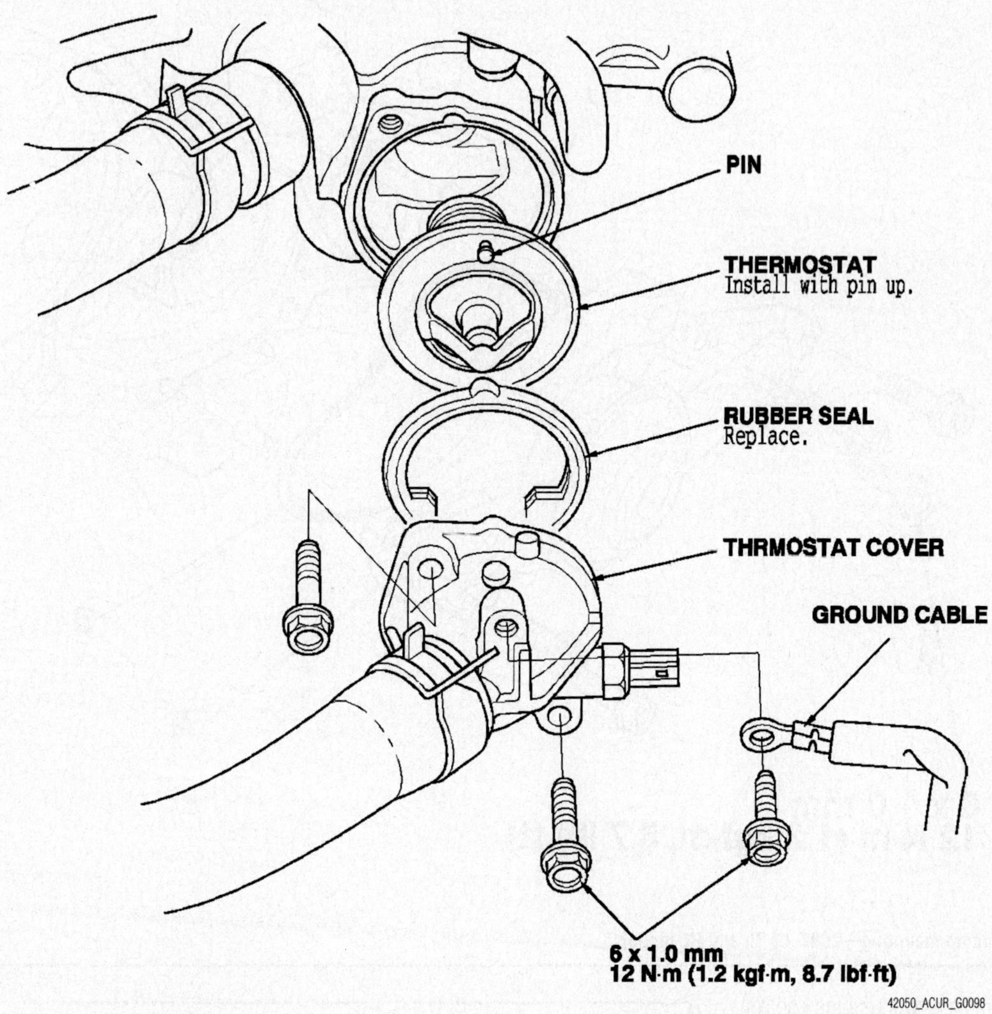

PIN

THERMOSTAT
Install with pin up.

RUBBER SEAL
Replace.

THRMOSTAT COVER

GROUND CABLE

6 x 1.0 mm
12 N·m (1.2 kgf·m, 8.7 lbf·ft)

42050_ACUR_G0098

Fig. 61 Exploded view of the thermostat mounting—2007–08 RL models

8. Clean up any spilled engine coolant.
9. Enter the anti-theft codes for the radio and the navigation system, then enter the audio presets.

TSX Models

See Figure 62.

1. Drain the engine cooling system.
2. Remove the splash shield.
3. Remove the lower radiator hose, then remove the thermostat.

To install:

4. Install the thermostat with a new O-ring.
5. Install the lower radiator hose.
6. Install the splash shield.
7. Refill the radiator with engine coolant, and bleed air from the cooling system with the heater valve open.

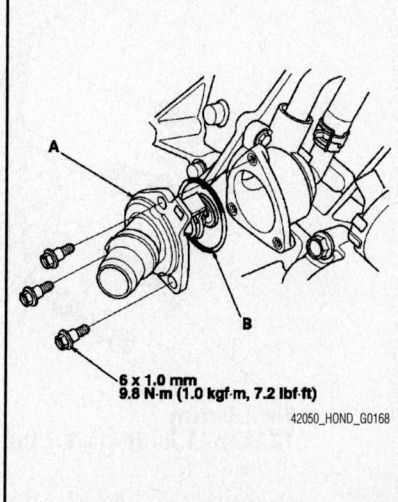

6 x 1.0 mm
9.8 N·m (1.0 kgf·m, 7.2 lbf·ft)

42050_HOND_G0168

Fig. 62 Always use a new O-ring (B) when installing the thermostat (A)

WATER PUMP

REMOVAL & INSTALLATION

TL and RL Models

See Figure 63.

1. Disconnect the negative battery cable.
2. Remove the timing belt.
3. Remove the timing belt adjuster.
4. Remove the water pump by removing the five bolts.
5. Inspect and clean the O-ring groove and the mating surface of the engine block.

To install:

6. Install the water pump with a new O-ring in the reverse order of removal. Tighten the bolts to 8.7 ft. lbs. (12 Nm)
7. Install the timing belt adjuster.
8. Install the timing belt.

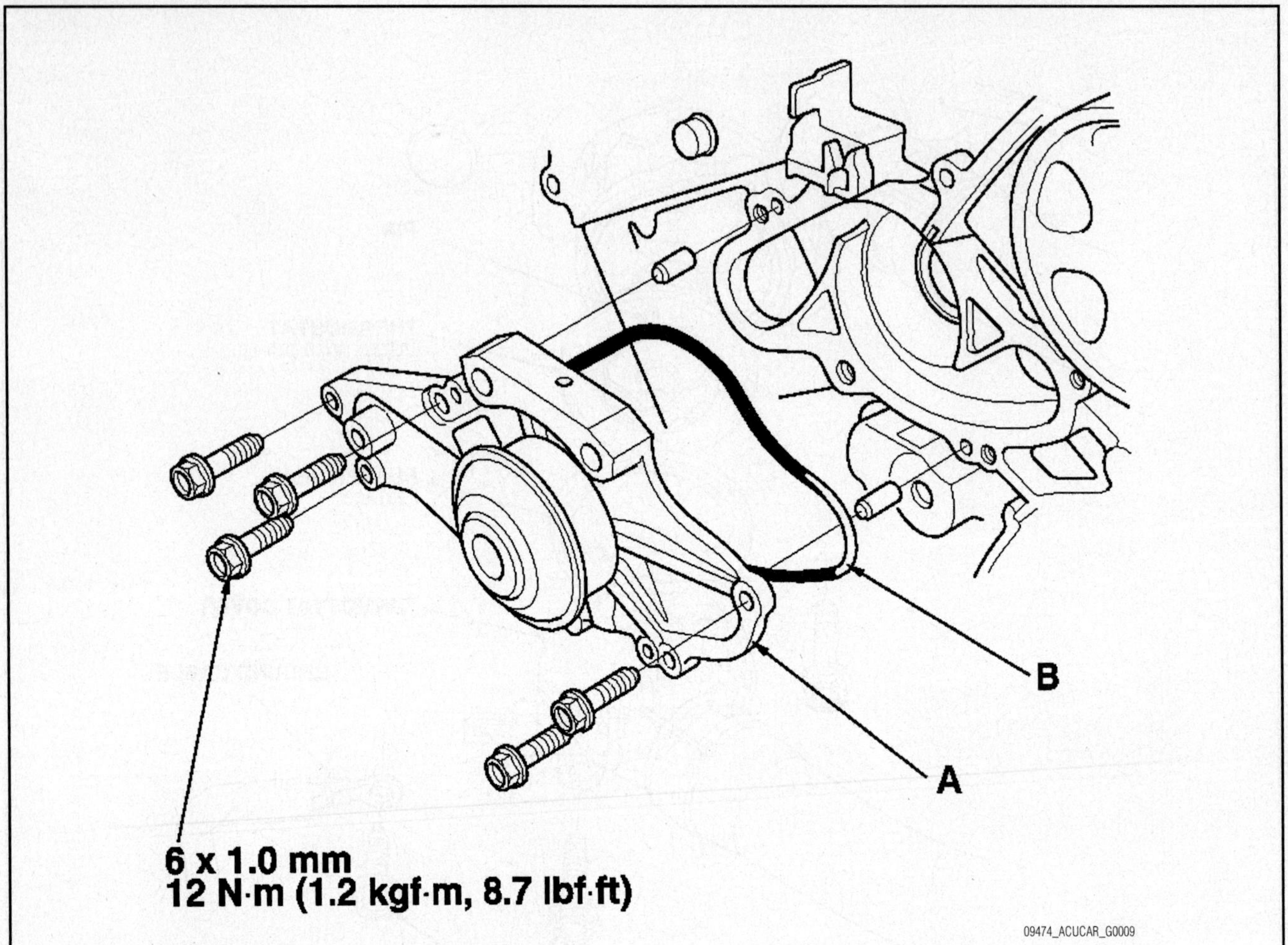

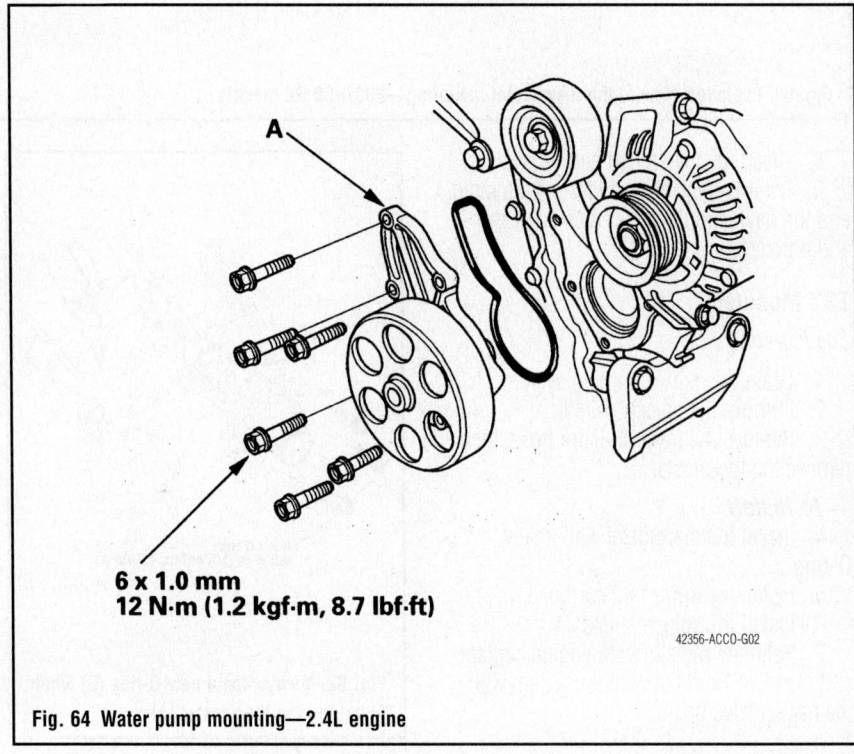

6 x 1.0 mm
12 N·m (1.2 kgf·m, 8.7 lbf·ft)

09474_ACUCAR_G0009

Fig. 63 Water pump mounting—2007–08 TL and RL models

9. Refill the radiator with engine coolant, then bleed the air from the cooling system.

TSX Models

See Figure 64.

1. Disconnect the negative battery cable.
2. Remove the accessory drive belt.
3. Drain the engine cooling system.
4. Remove the drive belt auto-tensioner.
5. Remove the six mounting bolts securing the water pump, then remove the water pump.

To install:

6. Inspect and clean the O-ring groove and mating surface with the water passage.
7. Install the water pump with new O-rings and tighten the mounting bolts 104 inch lbs. (12 Nm).
8. Clean up any spilled engine coolant.
9. Install the drive belt auto-tensioner.
10. Install the accessory drive belt.
11. Refill the cooling system to the correct level.
12. Start the engine and check for leaks.

6 x 1.0 mm
12 N·m (1.2 kgf·m, 8.7 lbf·ft)

42356-ACCO-G02

Fig. 64 Water pump mounting—2.4L engine

ENGINE ELECTRICAL

CHARGING SYSTEM

ALTERNATOR

REMOVAL & INSTALLATION

TL Models

See Figure 65.

1. Before servicing the vehicle, refer to the precautions section.

➡Make sure you have the anti-theft code for the radio, and the navigation system, then write down the radio channel presets. Make sure the ignition switch is OFF. Disconnect the negative cable from the battery first, then disconnect the positive cable.

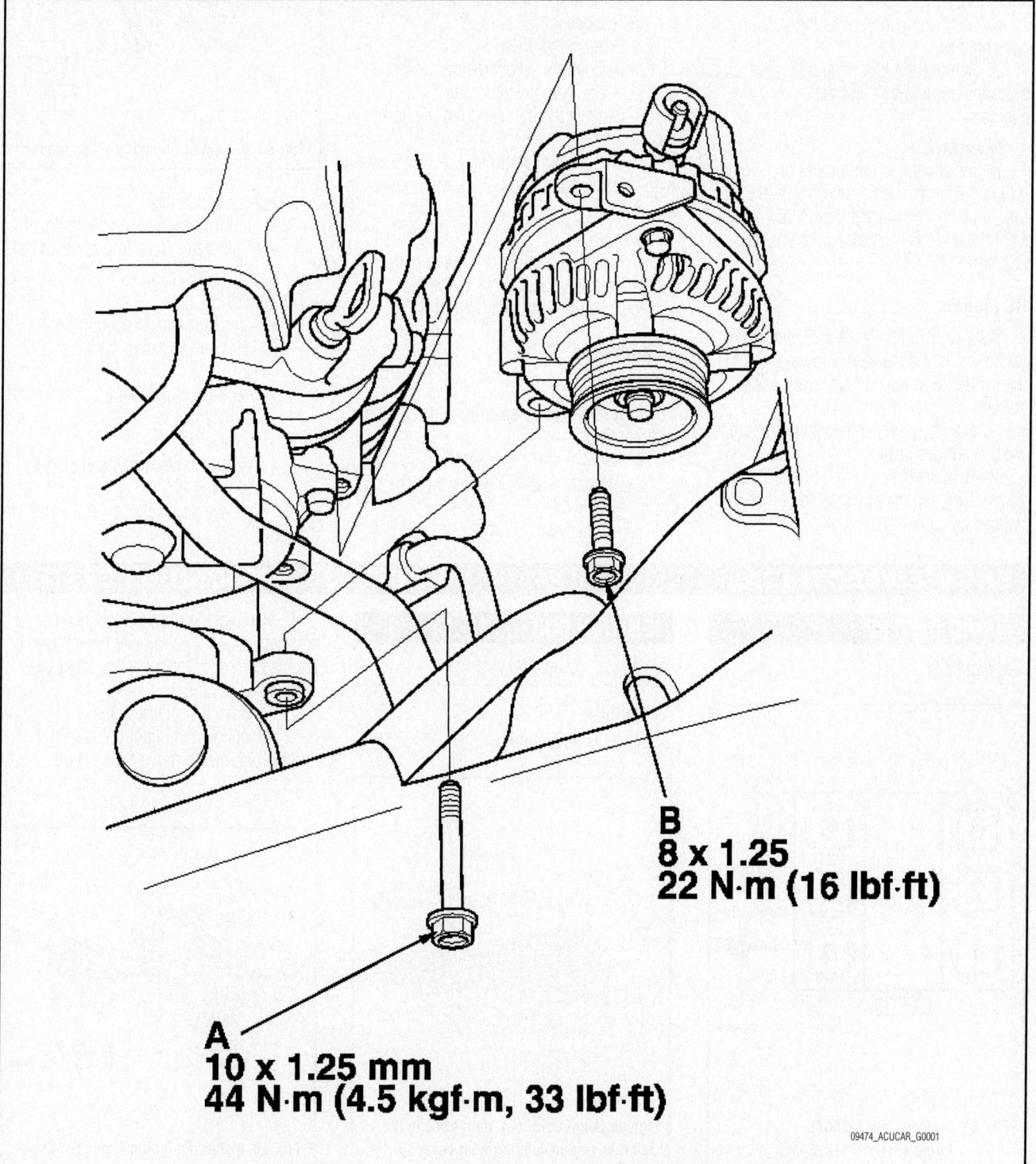

B
8 x 1.25
22 N·m (16 lbf·ft)

A
10 x 1.25 mm
44 N·m (4.5 kgf·m, 33 lbf·ft)

09474_ACUCAR_G0001

Fig. 65 Alternator mounting—2007–08 TL model

2. Disconnect the A/C condenser fan motor connector and the A/C compressor clutch connector, then remove the reserve tank.

3. Remove the two bolts, loosen bolt, and then remove the A/C condenser fan shroud.

4. Remove the drive belt.

5. Disconnect the alternator connector and the BLK wire from the alternator.

6. Remove the bolt retaining the harness bracket.

7. Remove the mounting bolt, alternator bracket mounting bolt and the remove the alternator.

To install:

8. Installation is the reverse of removal. Tighten the mounting bolt to 33 ft. lbs. (44 Nm), the bracket bolt to 16 ft. lbs. (22 Nm) and the harness bolt to 9 ft. lbs. (12 Nm).

RL Models

Make sure you have the anti-theft codes for the radio and the navigation system, then write down the audio presets. Make sure the ignition switch is OFF.

1. Before servicing the vehicle, refer to the precautions section.

2. Remove the right upper fender trim, battery trim, left upper fender trim, then remove the upper grille cover.

3. Disconnect the negative cable from the battery, then disconnect the positive cable.

4. Remove the splash shield.

5. Remove the harness clamps and connector from the A/C condenser fan shroud.

6. Loosen the two bolts securing the A/C condenser fan shroud.

7. Disconnect the fan motor connector, and remove the reserve tank .

8. Remove the two bolts, then remove the A/C condenser fan shroud.

9. Remove the drive belt.

10. Disconnect the alternator connector and BLK wire from the alternator.

11. Remove the bolt securing the harness holder.

12. Remove the mounting bolt and alternator bracket mounting bolt, then remove the alternator.

To install:

13. Installation is the reverse of removal. Tighten the upper bolt to 16 ft. lbs. (22 Nm) and the lower bolt to 33 ft. lbs. (44 Nm).

TSX Models

See Figure 66.

1. Note the radio security code and the radio presets.

2. Remove or disconnect the following:
- Negative battery cable, then the positive
- Drive belt

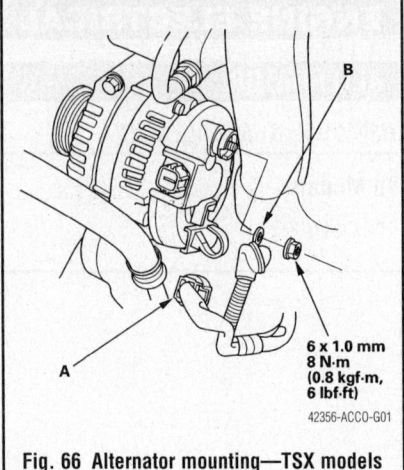

6 x 1.0 mm
8 N·m
(0.8 kgf·m, 6 lbf·ft)
42356-ACCO-G01

Fig. 66 Alternator mounting—TSX models

- Auto-tensioner
- Connectors from the alternator
- 3 alternator mounting bolts and the alternator

To install:

- Alternator and 3 mounting bolts. Torque the bolts to 16 ft. lbs. (22 Nm).
- Electrical connectors
- Auto-tensioner
- Drive belt
- Positive, then negative battery cables

3. Enter the security code and radio presets

ENGINE ELECTRICAL   DISTRIBUTORLESS IGNITION SYSTEM

FIRING ORDERS

See Figure 67.

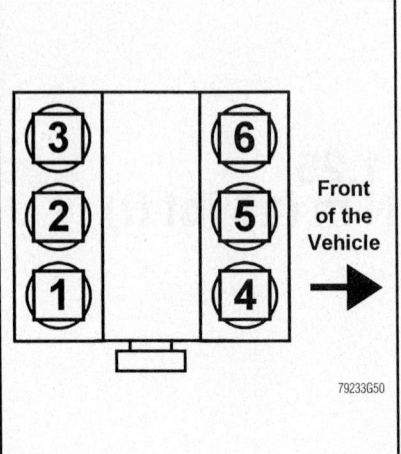

Front of the Vehicle →

79233G50

Fig. 67 3.5L engines (J35A8)
Firing order: 1–4–2–5–3–6
Distributorless ignition system
(one coil per cylinder)

IGNITION COIL

REMOVAL & INSTALLATION

TL Models

See Figures 68 and 69.

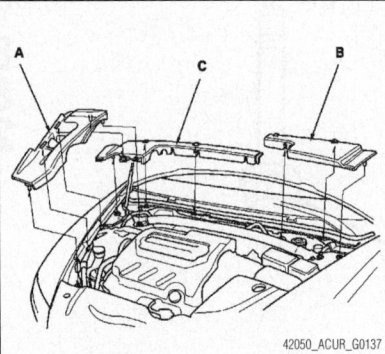

42050_ACUR_G0137

Fig. 68 Remove the right side engine compartment cover (A), then remove the left rear engine compartment cover (B) and right rear engine compartment cover (C)

1. Remove the right side engine compartment cover, then remove the left rear engine compartment cover and right rear engine compartment cover.

2. Remove the engine cover.

3. Disconnect the ignition coil connectors, then remove the ignition coils.

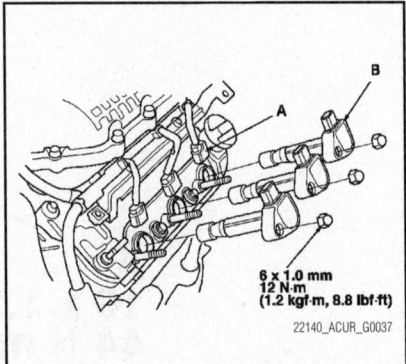

6 x 1.0 mm
12 N·m
(1.2 kgf·m, 8.8 lbf·ft)
22140_ACUR_G0037

Fig. 69 Install the ignition coils (B) and connect the coil connectors (A)—2007–08 TL Models

To install:

4. Install the ignition coils in the reverse order of removal. Tighten the coil mounting bolts to 8.8 ft. lbs. (12 Nm).

5. Install the engine covers.

RL Models

1. Disconnect the negative battery cable.
2. Remove the engine cover.
3. Remove the right upper fender trim.
4. Remove the battery trim cover.
5. Remove the left upper fender trim.
6. Remove the grille cover.
7. Disconnect the ignition coil connectors, the remove the front bank ignition coils.
8. Disconnect the ignition coil connectors, the remove the rear bank ignition coils.

To install:

9. Installation is the reverse order of removal. Tighten the ignition coil nuts to 104 inch lbs. (12 Nm).

TSX Models

See Figure 70.

1. Disconnect the negative battery cable.
2. Remove the ignition coil cover.
3. Disconnect the ignition coil connectors and remove the ignition coils.

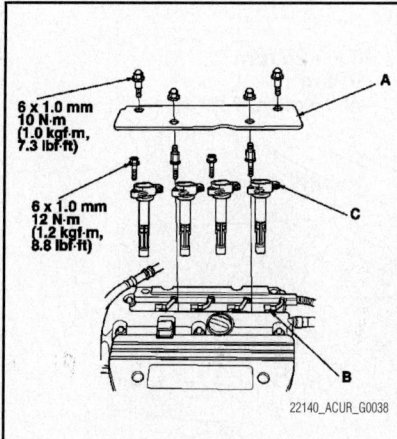

22140_ACUR_G0038

Fig. 70 Remove the ignition coil cover (A) and disconnect the connectors (B) to remove the ignition coils (C)—2007–08 TSX Models

To install:

4. Installation is the reverse order of removal.

IGNITION TIMING

INSPECTION

TL and RL Models

1. Before servicing the vehicle, refer to the precautions section.
2. Connect the Honda Diagnostic System (HDS) to the Data Link Connector (DLC), and check for DTCs.
3. If a DTC is present, diagnose and repair the cause before inspecting the ignition timing.
4. Start the engine. Hold the engine at 3,000 rpm with no load (in Neutral) until the radiator fan comes on, then let it idle.
5. Check the idle speed.
6. Select "SCS" mode using the HDS.
7. Remove the right side engine compartment cover.
8. Connect the timing light to the service loop.
9. Aim the light toward the pointer on the timing belt cover. Check the ignition timing under a no load condition. Headlights, blower fan, rear window defogger, and air conditioner are turned off.
10. The timing should be 8–12degrees Before Top Dead Center (TDC) in park or neutral.
11. If the ignition timing differs from the specification, check the cam timing.
12. Disconnect the HDS and the timing light.

TSX Models

➥**The ignition timing is controlled by the Powertrain Control Module (PCM) and can be checked for diagnostic purposes. If the timing is out of specification, all mechanical and electrical systems should be checked for proper operation before replacing the PCM.**

1. Before servicing the vehicle, refer to the precautions section.

2. To check the ignition timing, start the engine and allow it to fast idle at 3000 rpm with all electrical accessories off and the transmission in **N** or **P**. Allow the engine to warm up and reach normal operating temperature. The engine cooling fan should cycle at least 1 time.

3. Locate the Service Check (SCS) connector under the glove box. Connect the service connector tool part number 07PAZ–0010100 to the SCS terminals.

4. Check the idle speed and adjust if necessary.

5. Connect a timing light to the No. 1 plug wire. While engine idles, point the light toward the pointer on the timing belt cover.

6. Inspect the ignition timing at idle. The specification is 6–10 degrees Before Top Dead Center (BTDC) at idle.

➥**All mechanical and electrical systems should checked for proper operation before replacing the PCM.**

7. If the ignition timing is incorrect, replace the PCM.

8. Remove the service connector.

ADJUSTMENT

The ignition timing is control by the Powertrain Control Module (PCM) and no adjustment is possible.

SPARK PLUGS

REMOVAL & INSTALLATION

All models use a coil over plug ignition system. Each of the spark plugs has its own ignition coil which mounts directly above the spark plug and eliminates the need for a distributor, distributor cap, rotor and spark plug wires. Because the ignition coils are placed above the spark plugs, the coils must be removed before the spark plugs can be accessed.

1. Remove the ignition coils.
2. Remove the spark plugs.

To install:

3. Install the spark plug and tighten to 13 ft. lbs. (18 Nm).

4. Install the ignition coils.

STARTER

REMOVAL & INSTALLATION

TL Models

➡ **Make sure you have the anti-theft code for the radio, and the navigation system, then write down the radio channel presets. Make sure the ignition switch is OFF.**

1. Before servicing the vehicle, refer to the precautions section.
2. Remove the left side engine compartment cover.
3. Disconnect the negative cable and positive cable.
4. Remove the battery hold-down bracket, then remove the battery and battery tray.
5. Remove the harness clamp.
6. Disconnect the starter cable from the B terminal, then disconnect the BLK/WHT wire from the S terminal.
7. Remove the two bolts holding the starter, then remove the starter.

To install:

8. Installation is the reverse of the removal procedure. Tighten the starter bolts to 47 ft. lbs. (64 Nm).
9. Make sure the crimped side of the ring terminal is facing out.
10. Connect the battery positive cable to the battery first, then connect the negative cable.
11. Start the engine to make sure the starter works properly.
12. Enter the anti-theft codes for the radio and the navigation system, then enter the customer's radio channel presets.

RL Models

Make sure you have the anti-theft codes for the radio and the navigation system, then write down the customer's radio presets. Make sure the ignition switch is OFF.

1. Before servicing the vehicle, refer to the precautions section.
2. Remove the battery trim.
3. Disconnect the negative cable from the battery first, then disconnect the positive cable.

4. Remove the intake manifold cover.
5. Remove the vacuum hose and transmission dipstick.
6. Remove the harness clamp.
7. Disconnect the starter cable from the B terminal, then disconnect the BLK/WHT wire from the S terminal.
8. Remove the two bolts holding the starter, then remove the starter.

To install:

9. Installation is the reverse of the removal procedure. Tighten the starter bolts to 33 ft. lbs. (44 Nm).
10. Make sure the crimped side of the ring terminal is facing out.
11. Connect the battery positive cable to the battery first, then connect the negative cable.
12. Start the engine to make sure the starter works properly.
13. Enter the anti-theft codes for the radio and the navigation system, then enter the customer's radio channel presets.

TSX Models

See Figure 71.

➡ **The factory sound system has a coded theft protection system. It is recommended that you know your reset code before you begin.**

1. Before servicing the vehicle, refer to the precautions section.
2. Remove or disconnect the following:
   - Negative then the positive battery cables
   - Intake manifold
   - Starter cable from the B terminal
   - Black/white wire from the S (solenoid) terminal
   - Harness clamp and holder
   - Two bolts that mount the starter to the transaxle assembly
   - Starter

To install:

3. Installation is the reverse of the removal procedure.

➡ **When installing the heavy gauge starter cable, make sure the crimped side of the terminal end is facing out.**

4. Enter the anti-theft code and radio presets.

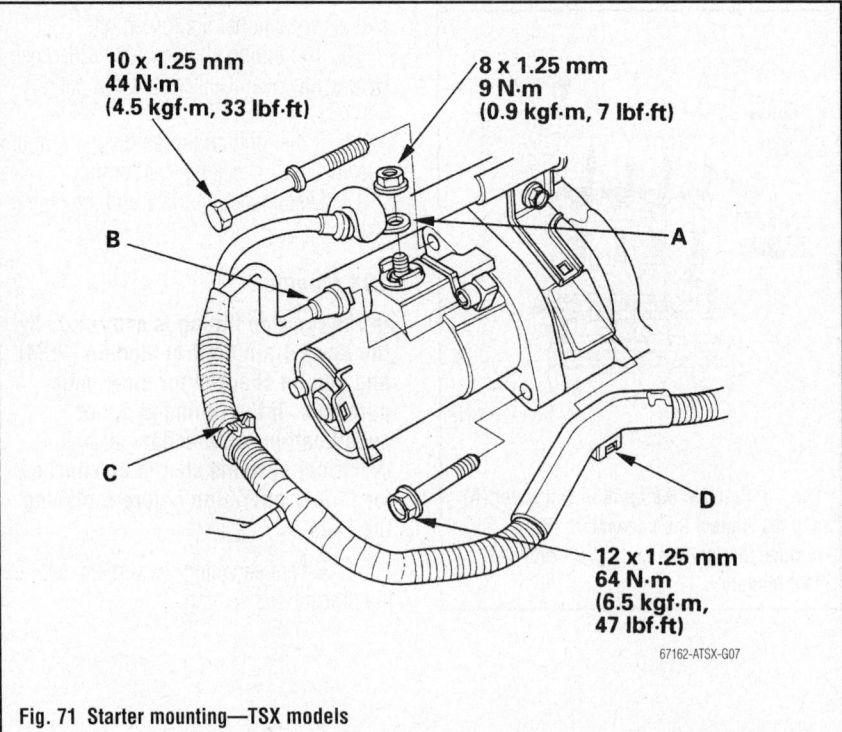

10 x 1.25 mm
44 N·m
(4.5 kgf·m, 33 lbf·ft)

8 x 1.25 mm
9 N·m
(0.9 kgf·m, 7 lbf·ft)

B

A

C

D

12 x 1.25 mm
64 N·m
(6.5 kgf·m, 47 lbf·ft)

67162-ATSX-G07

Fig. 71 Starter mounting—TSX models

ENGINE MECHANICAL

ACCESSORY DRIVE BELTS

ACCESSORY BELT ROUTING

See Figure 72 and 73.

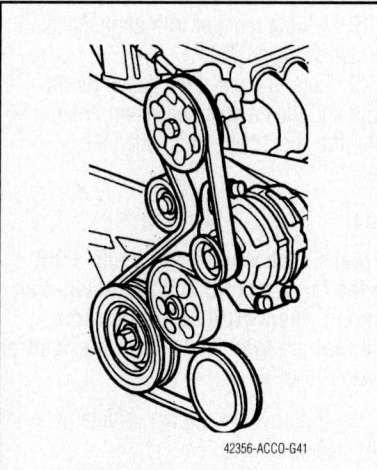

Fig. 72 Accessory drive belt routing—2.4L engines

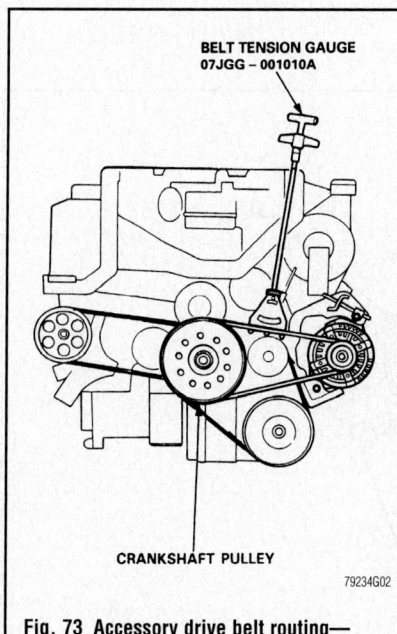

Fig. 73 Accessory drive belt routing—3.5L engines

ADJUSTMENT

Belt tension is maintained by an automatic tensioner. No adjustment is necessary or possible.

REMOVAL & INSTALLATION

TL Models

See Figures 74 and 75.

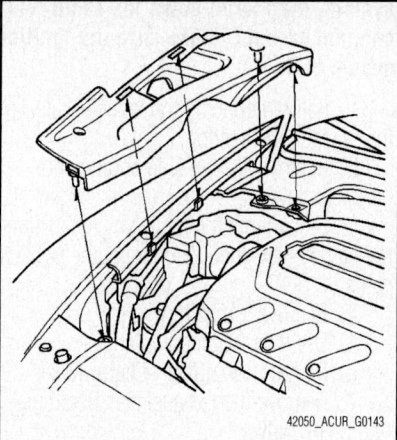

42050_ACUR_G0143

Fig. 74 Remove the right side engine compartment cover

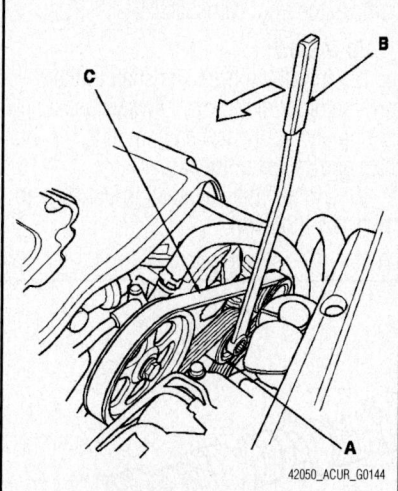

42050_ACUR_G0144

Fig. 75 Move the auto-tensioner (A) with the belt tension release tool (B) to relieve tension from the drive belt (C), then remove the drive belt

➡**This procedure requires Snap-On Belt tension release tool YA9317 or equivalent tool.**

1. Remove the right side engine compartment cover.
2. Move the auto-tensioner with the belt tension release tool to relieve tension from the drive belt, then remove the drive belt.
3. Install the new belt in the reverse order of removal.

RL Models

See Figures 76 and 77.

➡**This procedure requires Snap-On Belt tension release tool YA9317 or equivalent tool.**

42050_ACUR_G0145

Fig. 76 Remove the right upper fender trim

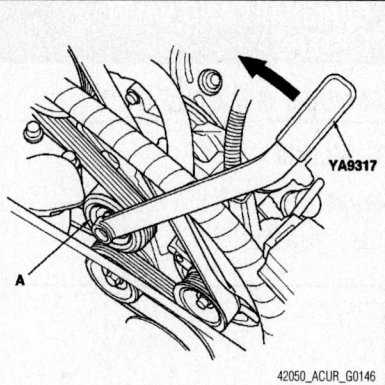

42050_ACUR_G0146

Fig. 77 Move the auto-tensioner (A) with the belt tension release tool in the direction shown to relieve tension from the drive belt, then remove the drive belt

1. Remove the right upper fender trim.
2. Remove the intake manifold cover.
3. Move the auto-tensioner (A) with the belt tension release tool in the direction shown to relieve tension from the drive belt, then remove the drive belt.
4. Install the new belt in the reverse order of removal.

TSX Models

See Figure 78.

➡**This procedure requires Snap-On Belt tension release tool YA9317 or equivalent tool.**

1. Move the auto-tensioner with the belt tension release tool to relieve tension from the drive belt, and remove the drive belt.
2. Install the new belt in the reverse order of removal.

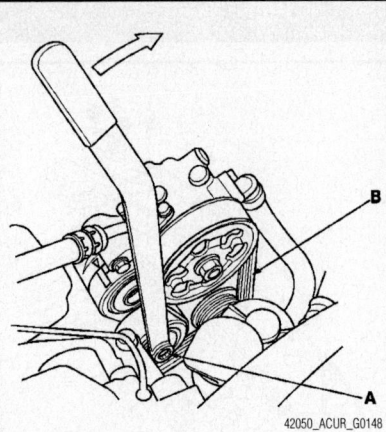

Fig. 78 Move the auto-tensioner (A) with the belt tension release tool to relieve tension from the drive belt (B), and remove the drive belt

CAMSHAFT AND VALVE LIFTERS

REMOVAL & INSTALLATION

TL Models

Front

See Figure 79.

➥**Make sure you have the anti-theft codes for the radio and the navigation system, then write down the radio channel presets. Make sure the ignition switch is OFF.**

1. Before servicing the vehicle, refer to the precautions section.
2. Remove the left side engine compartment cover.
3. Disconnect the negative cable from the battery first, then disconnect the positive cable.
4. Remove the battery.
5. Drain the engine coolant.
6. Remove the upper radiator hose.
7. Remove the Exhaust Gas Recirculation (EGR) valve.
8. Remove the timing belt.
9. Remove the rocker arm assembly.
10. Remove the front camshaft pulley.
11. Remove the thrust cover, then remove the front camshaft.

To install:

12. Install the front camshaft in the reverse order of removal. Always use a new O-ring and apply new engine oil to the journals and camshaft lobes.
13. Tighten the camshaft thrust plate to 16 ft. lbs. (22 Nm).

14. Apply new engine oil to the threads of the camshaft pulley mounting bolt, then install the front camshaft pulley. Tighten the bolt to 67 ft. lbs. (90 Nm).
15. Install the rocker arm assembly.
16. Install the timing belt.
17. Adjust the valve clearance.
18. Install the battery.
19. Fill the radiator with engine coolant and bleed the air out.
20. Enter the anti-theft codes for the radio and the navigation system, then enter the customer's radio channel presets.

Rear

➥**Make sure you have the anti-theft codes for the radio and the navigation system, then write down the radio channel presets. Make sure the ignition switch is OFF.**

1. Before servicing the vehicle, refer to the precautions section.
2. Relieve the fuel system pressure.
3. Disconnect the negative battery cable.
4. Remove the fuse/relay box from under the hood.
5. Drain the cooling system.

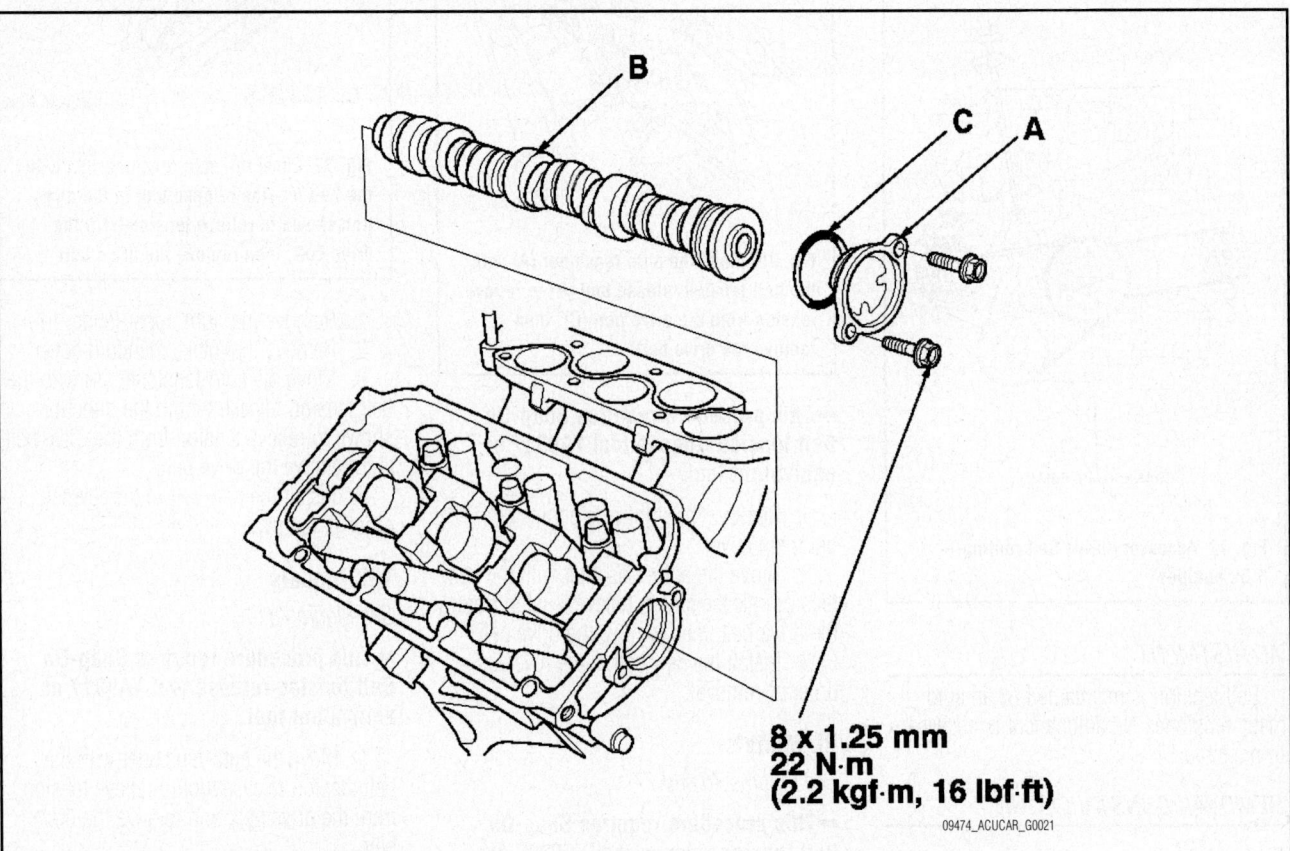

8 x 1.25 mm
22 N·m
(2.2 kgf·m, 16 lbf·ft)

Fig. 79 Typical camshaft mounting—2007–08 TL and 2007–08 RL models

6. Remove the heater hose and two nuts retaining the fuel line.

7. Remove the timing belt.

8. Remove the rocker arm assembly.

9. Remove the rear camshaft pulley.

10. Remove the thrust cover, then remove the rear camshaft.

To install:

11. Install the rear camshaft in the reverse order of removal. Always use a new O-ring and apply new engine oil to the journals and camshaft lobes.

12. Tighten the camshaft thrust plate to 16 ft. lbs. (22 Nm).

13. Apply new engine oil to the threads of the camshaft pulley mounting bolt, then install the rear camshaft pulley. Tighten the bolt to 67 ft. lbs. (90 Nm).

14. Install the rocker arm assembly.

15. Install the timing belt.

16. Adjust the valve clearance.

17. Install the battery.

18. Fill the radiator with engine coolant and bleed the air out.

19. Enter the anti-theft codes for the radio and the navigation system, then enter the customer's radio channel presets.

RL Models

Front

➡**Make sure you have the anti-theft codes for the radio and the navigation system, then write down the radio channel presets. Make sure the ignition switch is OFF.**

1. Before servicing the vehicle, refer to the precautions section.

2. Remove battery trim and the left upper fender trim.

3. Disconnect the negative cable from the battery first, then disconnect the positive cable.

4. Remove the battery.

5. Drain the engine coolant.

6. Remove the radiator hoses.

7. Remove the Exhaust Gas Recirculation (EGR) valve.

8. Remove the timing belt.

9. Remove the rocker arm assembly.

10. Remove the front camshaft pulley.

11. Remove the thrust cover then remove the front camshaft.

To install:

12. Install the front camshaft in the reverse order of removal. Always use a new O-ring and apply new engine oil to the journals and camshaft lobes.

13. Tighten the camshaft thrust plate to 16 ft. lbs. (22 Nm).

14. Apply new engine oil to the threads of the camshaft pulley mounting bolt, then install the front camshaft pulley. Tighten the bolt to 67 ft. lbs. (90 Nm).

15. Install the rocker arm assembly.

16. Install the timing belt.

17. Adjust the valve clearance.

18. Install the battery.

19. Fill the radiator with engine coolant and bleed the air out.

20. Enter the anti-theft codes for the radio and the navigation system, then enter the customer's radio channel presets.

Rear

➡**Make sure you have the anti-theft codes for the radio and the navigation system, then write down the radio channel presets. Make sure the ignition switch is OFF.**

1. Before servicing the vehicle, refer to the precautions section.

2. Relieve the fuel system pressure.

3. Disconnect the negative battery cable.

4. Drain the cooling system.

5. Remove the timing belt.

6. Remove the rocker arm assembly.

7. Remove the rear camshaft pulley.

8. Remove the two nuts securing the purge joint.

9. Remove the thrust cover, then remove the rear camshaft.

To install:

10. Install the rear camshaft in the reverse order of removal. Always use a new O-ring and apply new engine oil to the journals and camshaft lobes.

11. Tighten the camshaft thrust plate to 16 ft. lbs. (22 Nm).

12. Apply new engine oil to the threads of the camshaft pulley mounting bolt, then install the rear camshaft pulley. Tighten the bolt to 67 ft. lbs. (90 Nm).

13. Install the rocker arm assembly.

14. Install the timing belt.

15. Adjust the valve clearance.

16. Install the battery.

17. Fill the radiator with engine coolant and bleed the air out.

18. Enter the anti-theft codes for the radio and the navigation system, then enter the customer's radio channel presets.

TSX Models

See Figures 80 through 84.

1. Disconnect the negative battery cable.

2. Remove or disconnect the following:

- Timing chain
- Loosen the rocker arm adjusting screws
- Camshaft holder bolts, two turns at a time in sequence
- Timing chain guide (B), camshaft and camshafts

3. Insert the bolts (A) into the rocker shaft holder, then remove the rocker arm assembly (B)

- Camshafts by carefully lifting them out of the cylinder head

To install:

4. Clean and dry the No. 5 rocker shaft holding mating surface.

5. Apply a suitable liquid gasket P/N 08718-0009, or equivalent, evenly to the cylinder head mating surface of the No. 5 rocker shaft holder.

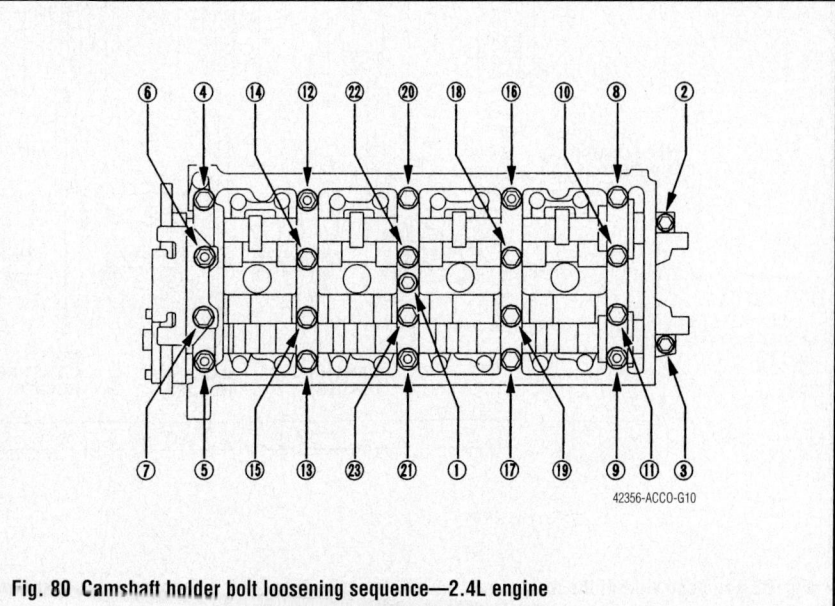

42356-ACCO-G10

Fig. 80 Camshaft holder bolt loosening sequence—2.4L engine

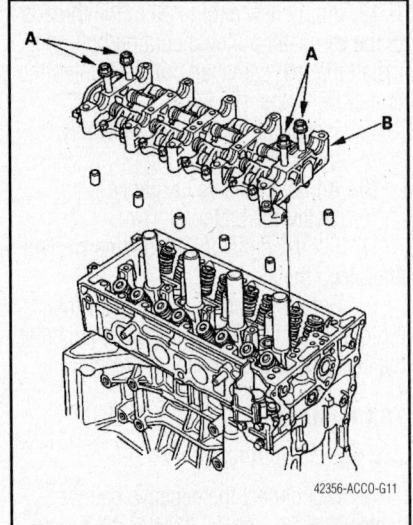

Fig. 81 Insert the bolts (A) into the rocker shaft holder, then remove the rocker arm assembly (B)—2.4L engine

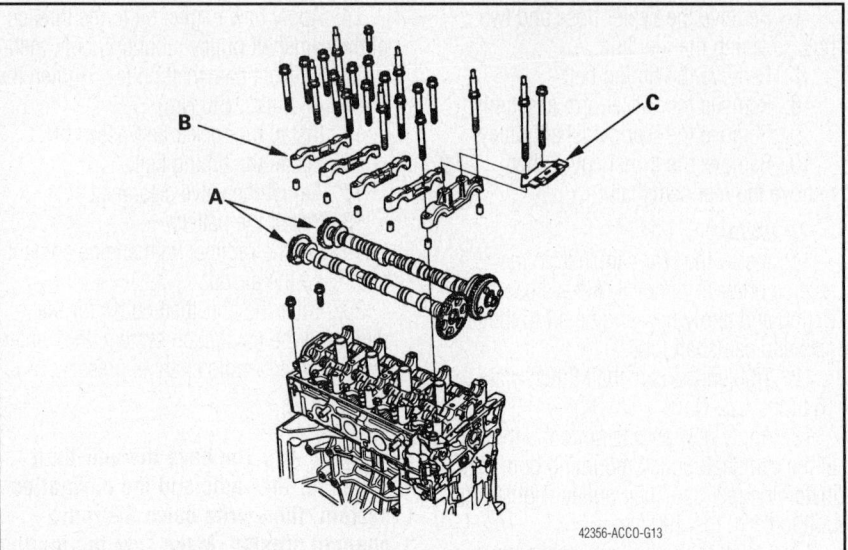

Fig. 83 When installing the camshafts (A) make sure the punch marks on the VTC actuator and exhaust cam sprockets are facing up—2.4L engine

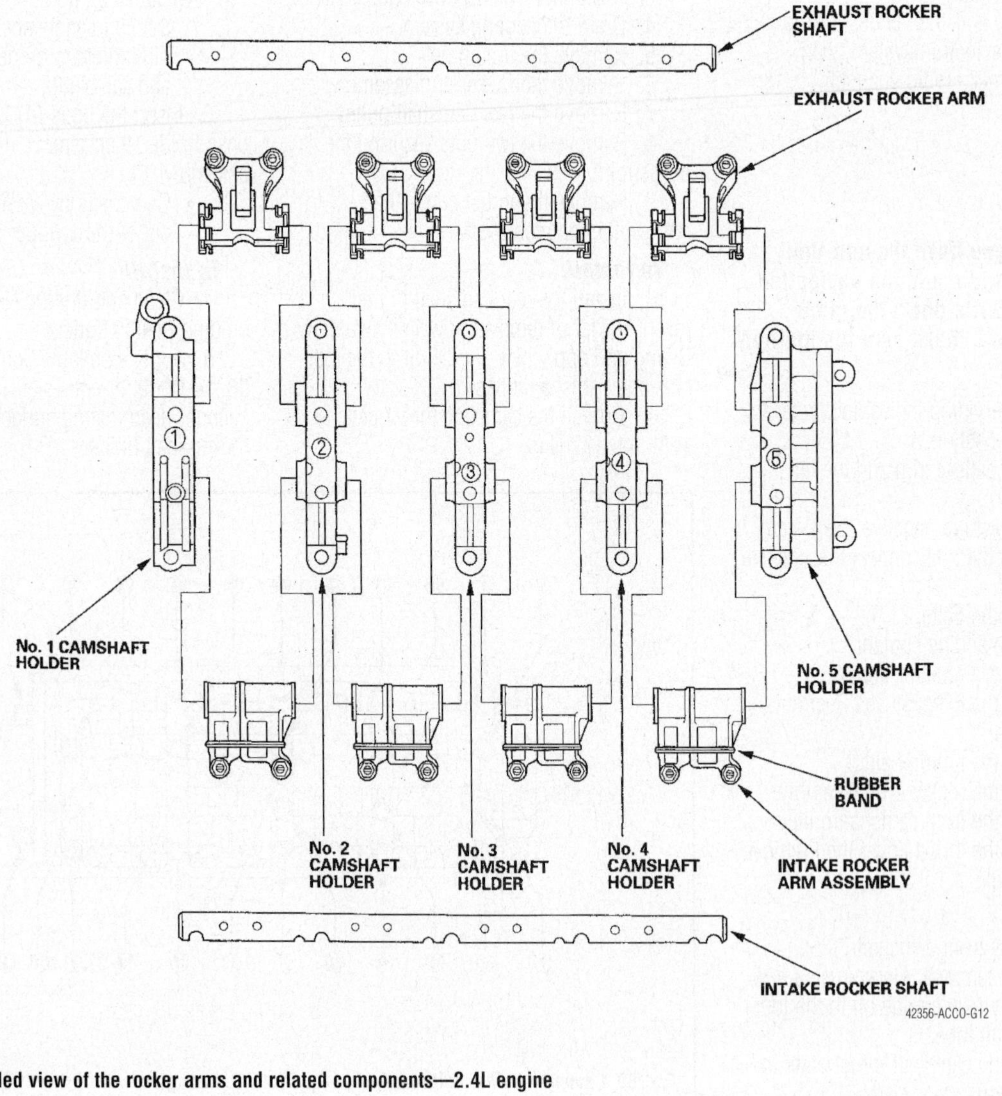

EXHAUST ROCKER SHAFT

EXHAUST ROCKER ARM

No. 1 CAMSHAFT HOLDER

No. 5 CAMSHAFT HOLDER

No. 2 CAMSHAFT HOLDER

No. 3 CAMSHAFT HOLDER

No. 4 CAMSHAFT HOLDER

RUBBER BAND

INTAKE ROCKER ARM ASSEMBLY

INTAKE ROCKER SHAFT

Fig. 82 Exploded view of the rocker arms and related components—2.4L engine

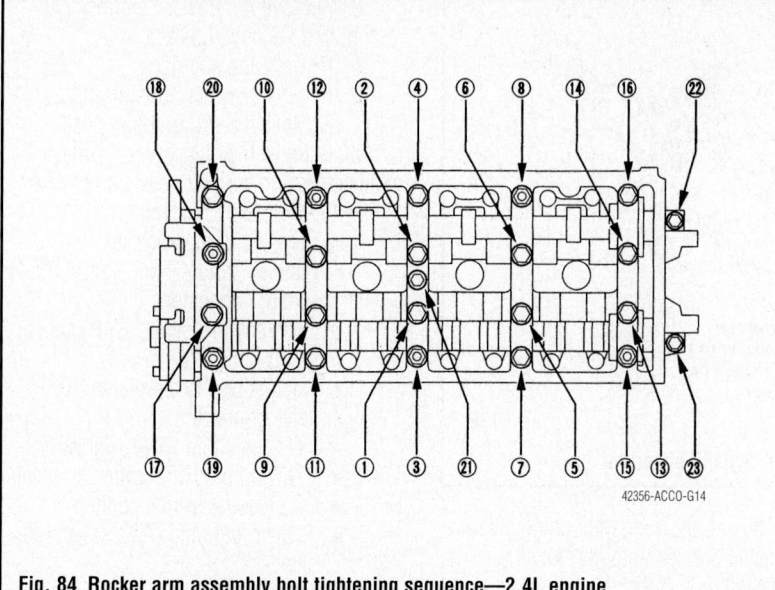

Fig. 84 Rocker arm assembly bolt tightening sequence—2.4L engine

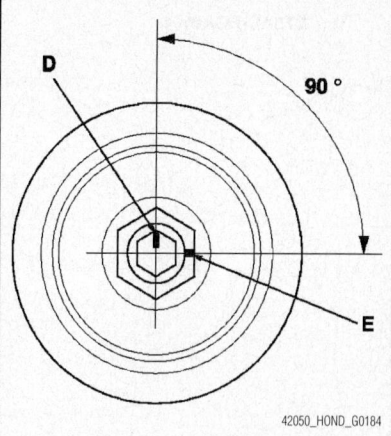

Fig. 87 Mark the bolt head (D) and the crankshaft pulley (E) as shown, then tighten the bolt an additional 90°. The mark on the bolt head should line up with the mark on the crankshaft pulley—TSX Models

➡**The parts must be installed within 5 minutes of applying the liquid gasket.**

6. Reassemble the rocker arm assembly, as necessary.

7. Install or connect the following
- Bolts (A) into the rocker shaft holder, then the rocker arm assembly on the cylinder head. Remove the bolts from the rocker shaft holder.

8. Make sure the punch marks on the variable valve timing control (VTC) actuator and exhaust camshaft sprocket are facing up, then set the camshafts (A) in the holder.

9. Set the camshaft holders (B) and timing chain guide B (C) in place.

10. Tighten the bolts, in sequence, to the following specification:
- a. 8mm bolts: 16 ft. lbs. (22 Nm)
- b. 6mm bolts: 104 inch lbs. (12 Nm)

11. Install the timing chain and adjust the valve lash.

CRANKSHAFT DAMPER

REMOVAL & INSTALLATION

See Figures 85 through 87.

➡**This procedure requires the use of the following special tools or their equivalents: Holder handle Holder handle 07JAB-001020A, 50mm Holder attachment 07NAB-001040A and 19mm Socket 07JAA-001020A.**

1. Raise and safely support the vehicle.
2. Remove the front wheels.
3. Remove the splash shield.
4. Remove the drive belt and other component necessary to access the crankshaft pulley (damper).

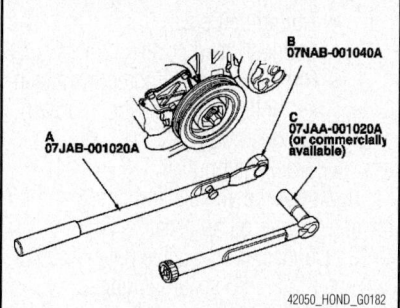

Fig. 85 While holding the pulley with holder handle (A) and holder attachment (B), remove the bolt with a heavy duty 19mm socket (C) and breaker bar

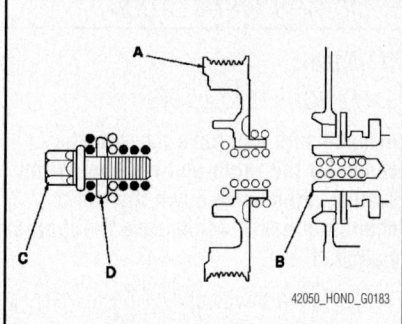

Fig. 86 Clean the crankshaft pulley (A), crankshaft (B), bolt (C), and washer (D). Lubricate with new engine oil as shown—TSX shown, others similar

5. Hold the pulley with holder handle (A) and holder attachment (B).

6. Remove the bolt with a 19 mm socket (C) and breaker bar, then remove the crankshaft.

To install:

7. Clean the crankshaft pulley (A), crankshaft (B), bolt (C), and washer (D). Lubricate as shown.

8. Tighten the crankshaft pulley bolt, as follows: Do not use an impact wrench.
- a. Hold the pulley with the holder handle (A) and pulley holder attachment (B). Torque the bolt, with a torque wrench and heavy duty 19mm socket, as follows
- b. 2007–08 TSX models: to 37 ft. lbs. (50 Nm), then mark the bolt head (D) and the crankshaft pulley (E) as shown, then tighten the bolt an additional 90°. The mark on the bolt head should line up with the mark on the crankshaft pulley.
- c. 2007–08 TL and RL models: 48 ft. lbs. (65 Nm), then mark the bolt head and the crankshaft pulley, then tighten the bolt an additional 60°. The mark on the bolt head should line up with the mark on the crankshaft pulley.

9. Install the drive belt and other components, as necessary.

10. Install the splash shield.
11. Install the front wheels.
12. Carefully lower the vehicle.

CRANKSHAFT FRONT SEAL

REMOVAL & INSTALLATION

TL & RL Models

See Figure 88.

1. Remove the Crankshaft Position (CKP) sensor A and B, timing belt, and timing belt drive pulley.

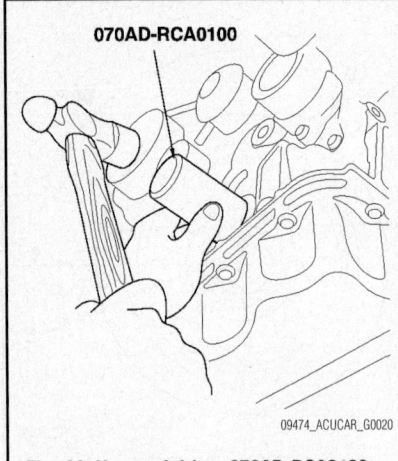

Fig. 88 Use seal driver 070AD-RCA0100 to install the crankshaft oil seal—2007–08 TL and RL models

2. Remove the pulley end crankshaft oil seal.

3. Clean and dry the crankshaft oil seal housing.

4. Apply a light coat of multipurpose grease to the crankshaft and to the lip of the seal.

5. Using the seal driver 070AD-RCA0100, drive in the crankshaft oil seal until the driver bottoms against the oil pump.

6. When the seal is in place, clean any excess grease off the crankshaft, and check that the oil seal lip is not distorted.

7. Install the timing belt drive pulley, CKP sensor A and B, and timing belt.

TSX Models

See Figure 89.

1. Disconnect negative cable at the battery.

2. Raise and safely support the vehicle. Drain the engine oil and properly dispose of it.

3. Be sure the crankshaft is at TDC on No. 1 cylinder by aligning the white mark on the crankshaft pulley with the pointer on the lower timing belt cover.

4. Remove or disconnect the following:
   • Crankshaft pulley
   • Cylinder head cover
   • Timing belt cover
   • Timing belt
   • Crankshaft Speed Fluctuation (CKF) sensor (if equipped)
   • Timing belt drive gear from the crankshaft

5. Using a suitable pry tool, carefully remove the seal.

To install:

6. Apply a light coat of oil to the seal lip.

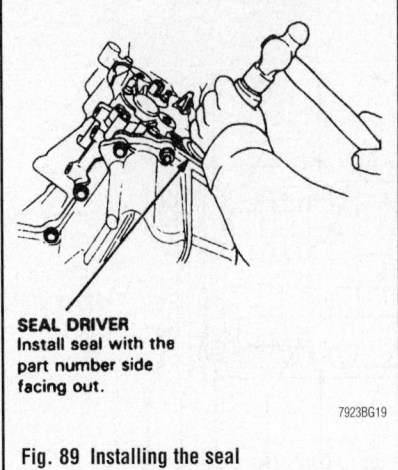

SEAL DRIVER
Install seal with the part number side facing out.

Fig. 89 Installing the seal

7. Position the seal, then using a seal driver, install the seal into the housing.

8. Install or connect the following:
   • Timing belt drive gear
   • Timing belt
   • Timing belt cover
   • Cylinder head cover
   • CKF sensor and tighten the attaching bolts to 96 inch lbs. (11 Nm) (if equipped)
   • Crankshaft pulley

9. Lower the vehicle and check and fill the engine with oil as necessary.

10. Connect the negative battery cable and enter the radio security code.

11. Run the engine and check for leaks.

12. Turn off engine and check the oil level. Top off the oil level if necessary.

CYLINDER HEAD

REMOVAL & INSTALLATION

TL Models

See Figures 90 through 92.

➡Make sure you have the anti-theft codes for the radio and the navigation system, then write down the radio channel presets. Make sure the ignition switch is OFF.

1. Before servicing the vehicle, refer to the precautions section.

2. Relieve the fuel system pressure.

3. Disconnect the battery negative cable from the battery.

4. Drain the engine coolant.

5. Remove the front warm up three way catalytic converter (front WU-TWC) and rear warm up three way catalytic converter (rear WU-TWC).

6. Remove the drive belt.

7. Remove the timing belt.

8. Remove the Power Steering (PS) pump, and PS hose bracket.

9. Remove the alternator.

10. Remove the intake manifold.

11. Remove the six ignition coils.

12. Remove the engine wire harness connectors and wire harness clamps from the cylinder head as follows:
   • Six injector connectors
   • Engine Coolant Temperature (ECT) sensor connector
   • Crankshaft Position (CKP) sensor A and B connector
   • Exhaust Gas Recirculation (EGR) valve connector
   • VTEC solenoid valve connector
   • VTEC oil pressure switch connector
   • Oil pressure switch connector
   • Two air fuel ratio (A/F) sensor connectors
   • Two secondary heated oxygen sensor (secondary HO2S) connectors

13. Remove the upper radiator hose and lower radiator hose.

14. Remove the heater hose and water bypass hose(s).

15. Remove the bolt retaining the harness holder.

16. Remove the bolt retaining the harness bracket.

17. Remove the fuel rails

18. Remove the bolt retaining the harness bracket.

19. Remove the water passage

20. Remove the front and rear camshaft pulleys and front and rear back covers.

21. Remove the cylinder head covers.

22. Remove the cylinder head bolts. To prevent warpage, unscrew the bolts in sequence 1/3 turn at a time; repeat the sequence until all bolts are loosened.

23. Remove the cylinder heads.

To install:

24. Clean the cylinder head and block surface.

25. Clean and install the oil control orifices with new O-rings.

26. Install the dowel pins and new cylinder head gaskets.

27. Put the cylinder head onto the engine block.

28. Clean the timing belt pulleys, timing belt guide plate, and the upper and lower covers.

29. Set the timing belt drive pulley to Top Dead Center (TDC) by aligning the TDC mark on the tooth of the timing belt drive pulley with the pointer on the oil pump.

30. Set the camshaft pulleys to TDC by aligning the TDC marks on the camshaft pulleys with the pointers on the back covers.

3
5
7
1
2
8
6
4
09474_ACUCAR_G0010

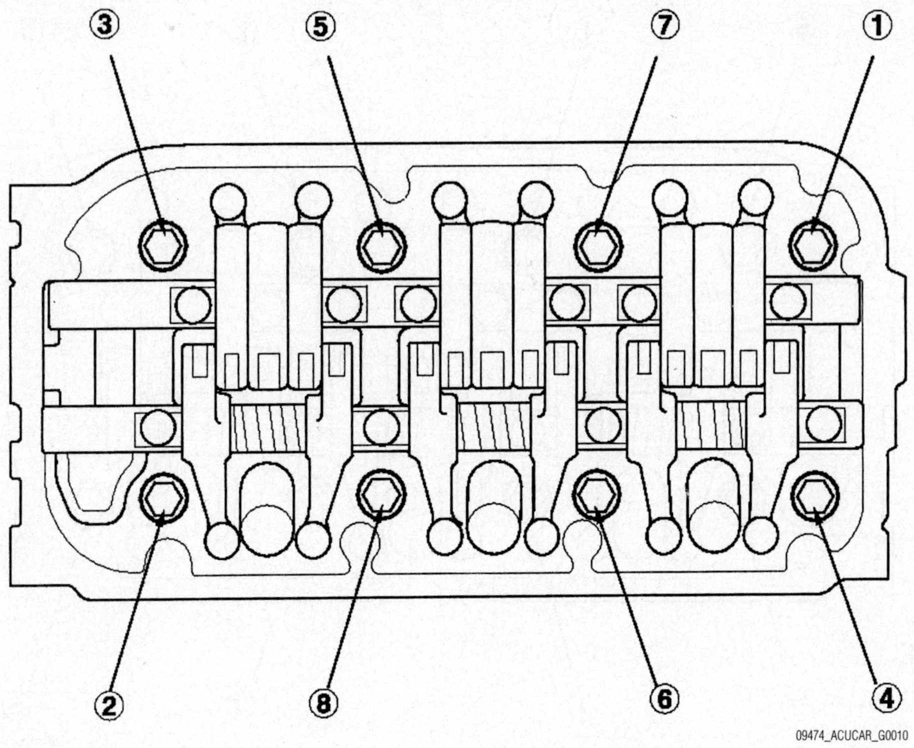

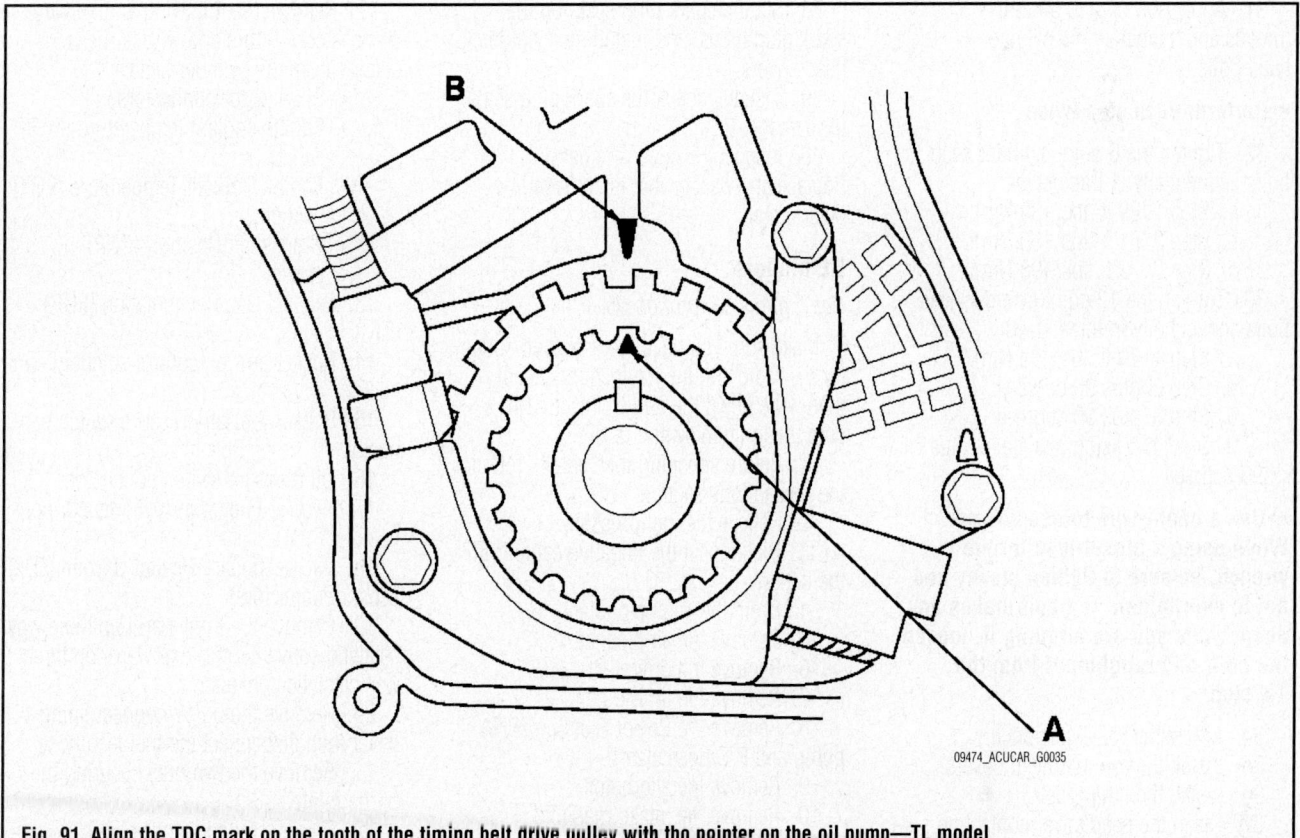

Fig. 90 Remove the cylinder head bolts in the sequence shown using several passes—TL model

B
A
09474_ACUCAR_G0035

Fig. 91 Align the TDC mark on the tooth of the timing belt drive pulley with tho pointer on the oil pump—TL model

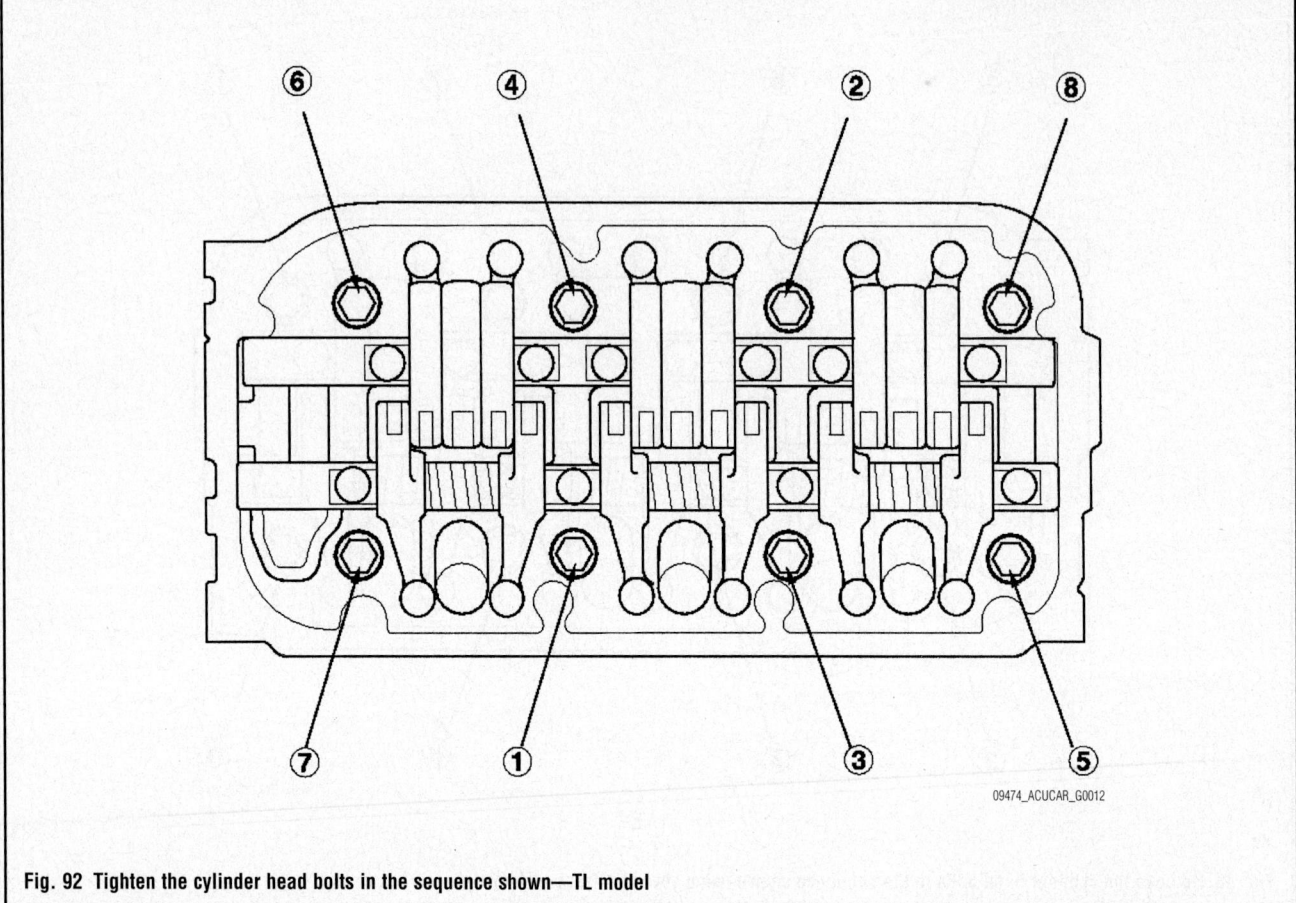

Fig. 92 Tighten the cylinder head bolts in the sequence shown—TL model

09474_ACUCAR_G0012

31. Apply new engine oil to the threads and flanges of the cylinder head bolts.

➡**Perform each step twice.**

32. Tighten the 6 point cylinder head bolts sequentially in three steps:
    a. Step 1: 29 ft. lbs. (39 Nm).
    b. Step 2: 51 ft. lbs. (69 Nm).
    c. Step 3: 72 ft. lbs. (98 Nm).

33. Tighten the 12 point cylinder head bolts sequentially in three steps:
    a. Step 1: 29 ft. lbs. (39 Nm).
    b. Step 2: plus 90 degrees
    c. Step 3: plus 90 degrees
    d. Step 4: if using new bolts: plus 90 degrees

➡**Use a beam-type torque wrench. When using a preset-type torque wrench, be sure to tighten slowly and not to overtighten. If a bolt makes any noise while you are torquing it, loosen the bolt, and retighten it from the 1st step.**

34. Adjust the valve clearance.
35. Install the cylinder head covers.
36. Install the timing belt.
37. Clean the head cover contacting surfaces with a shop towel.

38. Set the spark plug seals on the spark plug tubes, and install the cylinder head covers.
39. Visually check the spark plug seals for damage.
40. Inspect the cover washers. Replace any washer that is damaged or deteriorated.

RL Models

See Figures 93 through 95.

Make sure you have the anti-theft codes for the radio and the navigation system, then write down the audio presets. Make sure the ignition switch is OFF.
1. Before servicing the vehicle, refer to the precautions section.
2. Relieve the fuel pressure.
3. Disconnect the negative cable from the battery.
4. Drain the engine coolant.
5. Remove the air cleaner.
6. Remove the drive belt.
7. Remove the timing belt.
8. Remove the Power Steering (P/S) pump and P/S hose clamp.
9. Remove the alternator.
10. Remove the intake manifold.
11. Remove the six ignition coils.

12. Remove the following engine wire harness connectors and wire harness clamps from the cylinder head:
    • Six injector connectors
    • Engine coolant temperature (ECT) sensor 1
    • Engine Coolant Temperature (ECT) sensor 2
    • Crankshaft Position (CKP) sensor
13. Exhaust Gas Recirculation (EGR) valve.
14. Rocker arm oil control solenoid connector.
15. Rocker arm oil pressure switch connector.
16. Oil pressure switch connector.
17. Two Air Fuel ratio (A/F) sensor connectors.
18. Two secondary Heated Oxygen (O2S) sensor connectors.
19. Remove the front warm up three way catalytic converter and rear warm up three way catalytic converter.
20. Remove the quick-connect fitting cover then disconnect the fuel feed hose.
21. Remove the two nuts securing the purge joint.
22. Remove the radiator hoses.

③
⑤
⑦
①
②
⑧
⑥
④
09474_ACUCAR_G0010

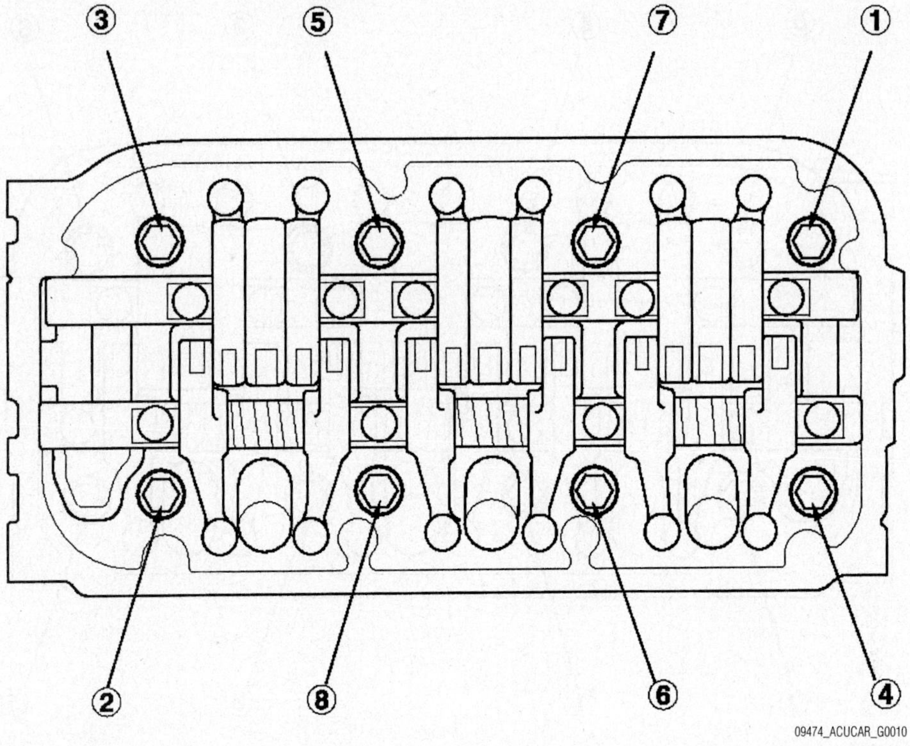

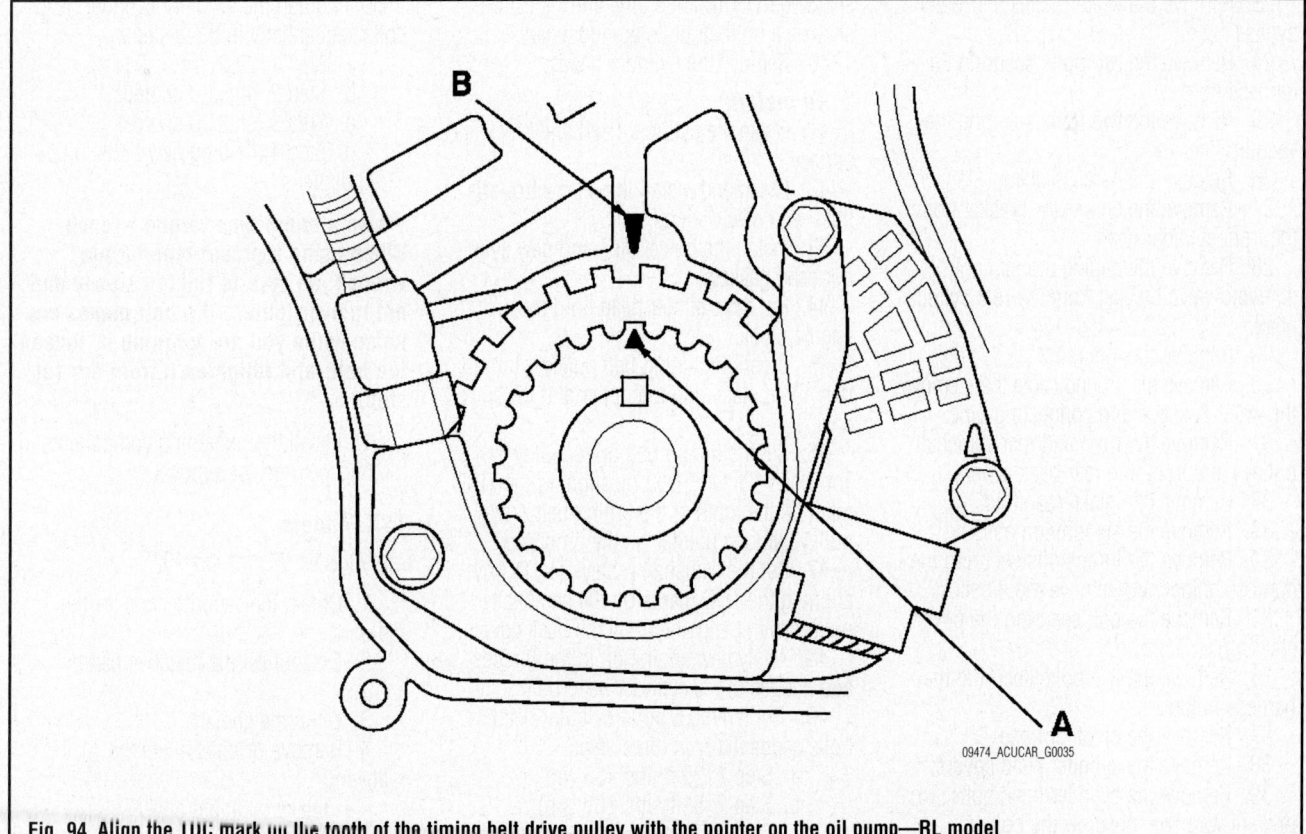

Fig. 93 Remove the cylinder head bolts in the sequence shown using several passes—RL model

B
A
09474_ACUCAR_G0035

Fig. 94 Align the TDC mark on the tooth of the timing belt drive pulley with the pointer on the oil pump—RL model

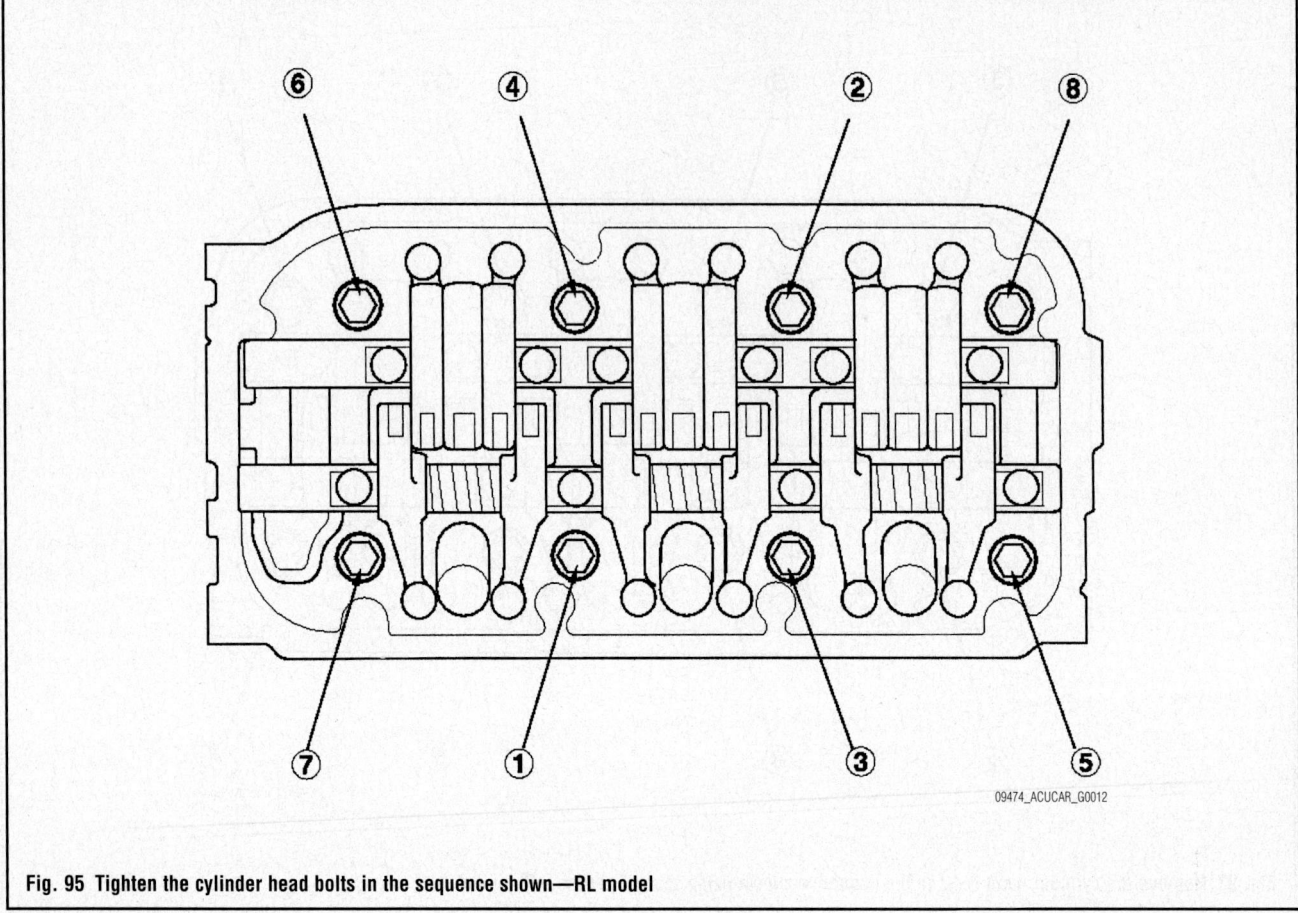

09474_ACUCAR_G0012

Fig. 95 Tighten the cylinder head bolts in the sequence shown—RL model

23. Remove the heater hoses and water bypass hose.

24. Remove the two bolts securing the harness holder.

25. Remove the two bolts securing the vacuum line.

26. Remove the harness clamp.

27. Remove the connector bracket from the front cylinder head.

28. Remove the engine mount control solenoid valve bracket from the rear cylinder head.

29. Remove the fuel rails.

30. Remove the ground cable then remove the water passage and connecting pipe.

31. Remove the front and rear camshaft pulleys and front and rear back covers.

32. Remove the intake manifold.

33. Remove the six ignition coils.

34. Remove the three bolts securing the harness holder, and remove the dipstick.

35. Remove the bolt securing the power steering hose clamp.

36. Remove the two bolts securing the harness holder.

37. Remove the breather hose.

38. Remove the cylinder head covers.

39. Remove the cylinder head bolts. To prevent warpage, unscrew the bolts in sequence 1/3 turn at a time and repeat the sequence until all bolts are loosened.

40. Remove the cylinder heads.

To install:

41. Clean the cylinder head and block surface.

42. Clean and install the oil control orifices with new O-rings.

43. Install the dowel pins and new cylinder head gaskets.

44. Put the cylinder head onto the engine block.

45. Clean the timing belt pulleys, timing belt guide plate, and the upper and lower covers.

46. Set the timing belt drive pulley to Top Dead Center (TDC) by aligning the TDC mark on the tooth of the timing belt drive pulley with the pointer on the oil pump.

47. Set the camshaft pulleys to TDC by aligning the TDC marks on the camshaft pulleys with the pointers on the back covers.

48. Apply new engine oil to the threads and flanges of the cylinder head bolts.

49. Tighten the 6 point cylinder head bolts sequentially in three steps:
   a. Step 1: 29 ft. lbs. (39 Nm).
   b. Step 2: 51 ft. lbs. (69 Nm).
   c. Step 3: 72 ft. lbs. (98 Nm).

50. Tighten the 12 point cylinder head bolts sequentially in three steps:
   a. Step 1: 29 ft. lbs. (39 Nm).
   b. Step 2: plus 90 degrees
   c. Step 3: plus 90 degrees
   d. Step 4: if using new bolts: plus 90 degrees

➡**Use a beam-type torque wrench. When using a preset-type torque wrench, be sure to tighten slowly and not to overtighten. If a bolt makes any noise while you are torquing it, loosen the bolt, and retighten it from the 1st step.**

51. Install the remaining components in the reverse order of removal.

TSX Models

See Figures 96 through 102.

1. Obtain the security code for the radio.

2. Disconnect the negative battery cable.

3. Drain the coolant.

4. Remove or disconnect the following:
   • Intake manifold cover
   • 4 ignition coils

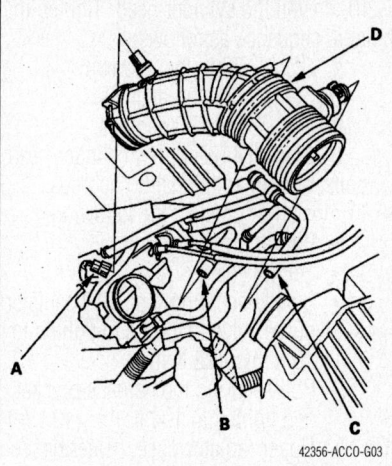

Fig. 96 Remove the vacuum hose (B), breather pipe (C), then remove the intake air duct (D)—2.4L engine

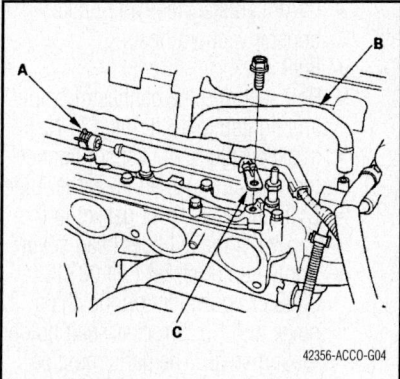

Fig. 97 Disconnect the PCV hose (A), vacuum hose (B) and ground cable (C)—2.4L engine

- 2 bolts securing the vacuum line
- Bolt securing the power steering hose bracket
- Dipstick and breather hose
- Retainers and cylinder head cover
- Fuel line
- Drive belt
- Intake Air Temperature (IAT) sensor connector
- Vacuum hose (B) and breather pipe (C), then the intake air duct (D)
- Bolt securing the connecting pipe
- Evaporative emission (EVAP) canister hose and brake booster vacuum hose
- Intake manifold
- Exhaust manifold
- Positive Crankcase Ventilation (PCV) hose, vacuum hose and ground cable

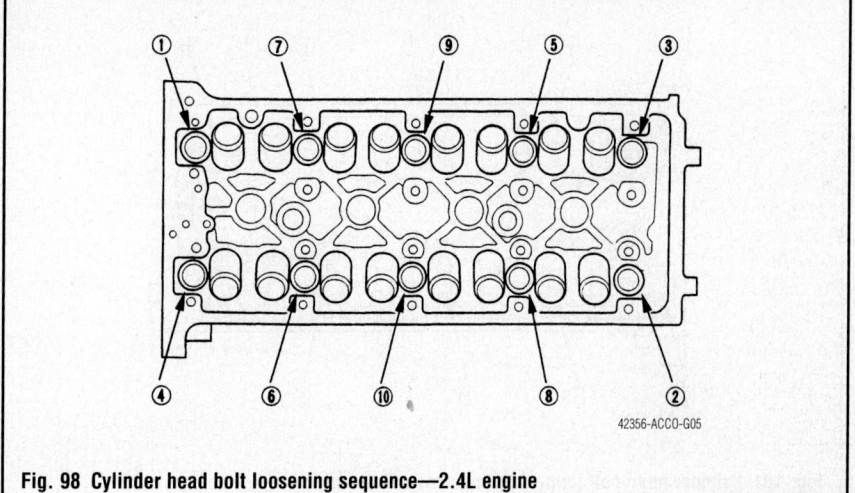

Fig. 98 Cylinder head bolt loosening sequence—2.4L engine

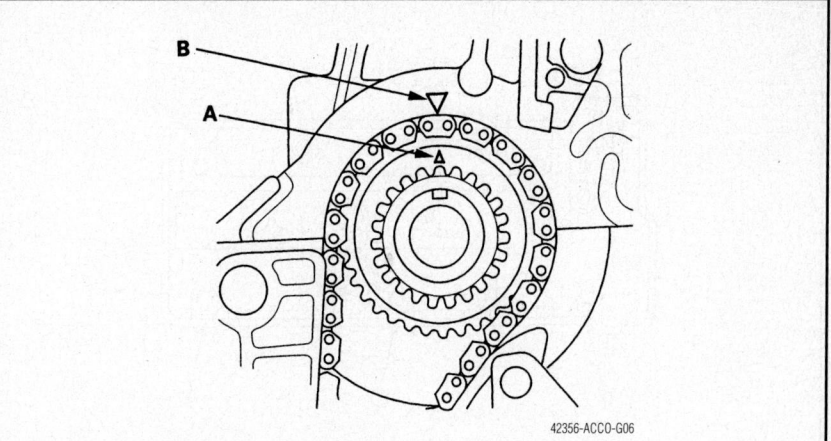

Fig. 99 Set the crankshaft to TDC by aligning the mark (A) on the crankshaft sprocket with the pointer (B) on the cylinder block—2.4L engine

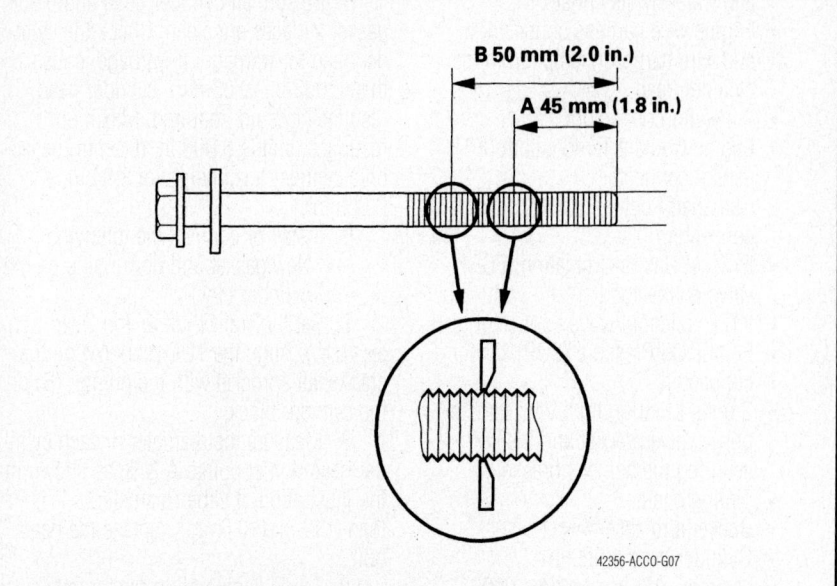

Fig. 100 You must measure the cylinder head bolts to see if they can be reused or need to be replaced—2.4L engine

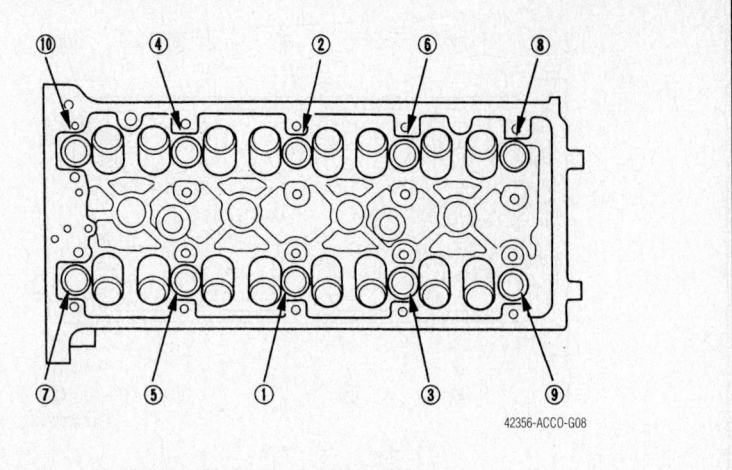

Fig. 101 Cylinder head bolt tightening sequence—2.4L engine

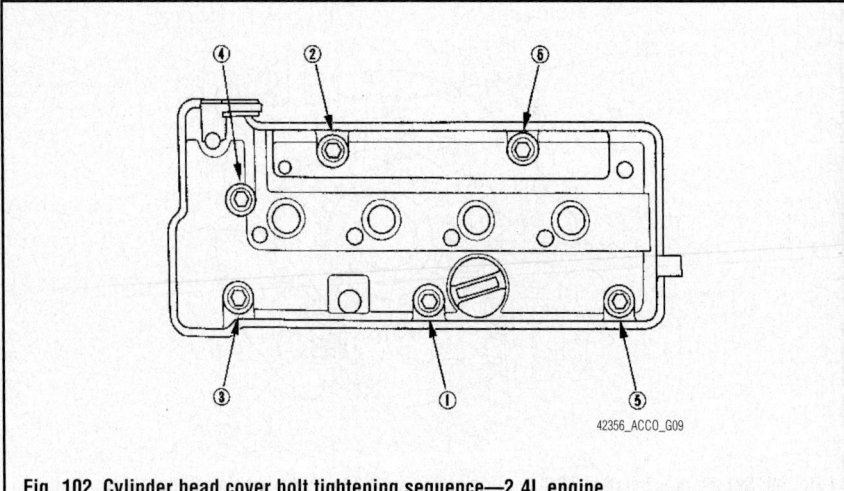

Fig. 102 Cylinder head cover bolt tightening sequence—2.4L engine

- Upper radiator hose, heater hoses and water bypass hose
- Engine wire harness connectors and wire harness clamps from the cylinder head
- 4 injector connectors
- Engine Coolant Temperature (ECT) sensor connector
- Camshaft Position (CMP) sensor connectors
- Exhaust Gas Recirculation (EGR) valve connector
- VTEC solenoid valve connector
- Engine Oil Pressure (EOP) sensor connector
- 2 bolts securing the EVAP canister purge valve bracket and the bolt securing the harness bracket
- Timing chain
- Rocker arm assembly
- Cylinder head bolts, in sequence, ⅓turn at a time until completely loosened
- Cylinder head. Discard the gasket.

To install:

5. Be sure all cylinder head and block gasket surfaces are clean. Check the cylinder head for warpage. If warpage is less than 0.002 in. (0.05mm), cylinder head resurfacing is not required. Maximum resurface limit is 0.008 in. (0.2mm) based on a cylinder head height of 3.94 in. (100mm).

6. Install or connect the following:
- New gasket and dowel pins on the cylinder block

7. Set the crankshaft to Top Dead Center (TDC). Align the TDC mark (A) on the crankshaft sprocket with the pointer (B) on the cylinder block.

8. Measure the diameter of each cylinder head bolt at points A & B, as shown in the illustration. If either diameter is less than 0.42 in. (10.6mm), replace the head bolt

9. Apply engine oil to the threads and under the bolt heads of all of the bolts.

10. Install the cylinder head. Tighten the bolts in sequence as follows:
    a. Step 1: 29 ft. lbs. (39 Nm).
    b. Step 2: Plus 90 degrees.
    c. Step 3: Plus 90 degrees.
    d. Step 4: If using new cylinder head bolts, add an additional 90 degrees.

11. Install or connect the following:
- Rocker arm assembly
- Timing chain
- 2 bolts securing the EVAP canister purge valve bracket and tighten to 16 ft. lbs. (22 Nm)
- Bolt securing the harness bracket and tighten to 104 ft. lbs. (12 Nm)
- Upper radiator hose, heater hoses and water bypass hose
- PCV hose, vacuum hose and ground cable
- Exhaust manifold
- Intake manifold
- EVAP canister hose and brake booster vacuum hose
- Fuel line
- Bolt securing the connecting pipe and tighten to 16 ft. lbs. (22 Nm)
- Intake air duct, IAT sensor connector, vacuum hose and breather pipe
- Cylinder head cover gasket in the groove of the cylinder head cover
- Apply liquid gasket P/N 08718-0009 or equivalent on the chain cover and No. 5 rocker shaft holder mating areas. The parts must be installed within 5 minutes of applying liquid gasket.
- Spark plug seals on the spark plug tubes
- Cylinder head cover on the cylinder head, then slide the cover back and forth gently to seat it
- Cover washers
- Cylinder head cover bolts. Torque, in sequence, in 2 or 3 steps to 104 inch lbs. (12 Nm).
- Dipstick and breather hose
- Bolt securing the power steering hose bracket
- 2 bolts securing the vacuum line
- 4 ignition coils
- Intake manifold cover, and tighten the retainers to 104 inch lbs. (12 Nm)
- All of the remaining hoses, tubes, and connectors are installed correctly.

12. Fill the cooling system.

13. Connect the negative battery cable and enter the radio security code.

14. Start the engine and check carefully for any leaks.

ENGINE ASSEMBLY

REMOVAL & INSTALLATION

TL Models

See Figures 103 through 109.

➡ **Be sure you have the anti-theft codes for the radio and navigation system, then write down the audio channel presets. Make sure the ignition switch is OFF.**

1. Before servicing the vehicle, refer to the precautions section.

2. Disconnect the support struts from both sides of the pivot ball (bolted to the hood).

3. Secure the hood in a vertical position. Remove the right side pivot ball and install it into the lower threaded hole, then reattach the support strut.

➡ **Do not attempt to close the hood with the support strut in the vertical position, as it will damage the support strut and hood.**

4. Remove the left side engine compartment cover and the left rear engine compartment cover.

5. Remove the right side engine compartment cover, then remove the right rear engine compartment cover.

6. Remove the intake manifold cover.

7. Drain the power steering system fluid, then plug the fluid reservoir and return hose.

8. Relieve the fuel system pressure.

9. Disconnect the negative cable from the battery first, then disconnect the positive cable.

10. Remove the battery.

11. Remove the air cleaner housing.

12. Remove the harness clamp.

13. Remove the four bolts, then remove the battery base.

14. Remove the battery cables from the under-hood fuse/relay box, then disconnect the harness connector.

15. Remove the under-hood fuse/relay box from the body.

16. Remove the harness clamp, and remove the two 6 mm bolts, then remove the strut brace.

17. Remove the brake booster vacuum hose and evaporative emission (EVAP) canister hose.

18. Remove the harness clamp, and disconnect the engine wire harness connectors on the left side of the engine compartment.

19. Remove the three bolts retaining the shift cable holder, then remove the shift cable and select cable.

➡ **Do not bend the cables excessively if equipped with a manual transmission.**

20. Remove the clutch slave cylinder and clutch line bracket mounting bolt.

➡ **Do not operate the clutch pedal once the slave cylinder has been removed if equipped with a manual transmission.**

21. Remove the drive belt.

22. Remove the Power Steering (PS) pump outlet line from the PS pump, then plug the outlet line and the PS pump.

23. Remove the PS hose from the clamp.

24. Remove the PS fluid reservoir from the clamp.

25. Remove the steering joint cover.

26. Lock the steering wheel. Make a reference mark across the steering joint and steering gearbox pinion shaft.

27. Remove the steering joint bolt, and disconnect the steering joint from the steering gearbox pinion shaft. To prevent damage to the cable reel, do not turn the steering wheel once the steering joint has been removed.

28. Disconnect the A/C condenser fan motor connector and A/C compressor clutch connector, and remove the reserve tank. Wipe up any spilled engine coolant immediately.

29. Remove the two bolts, loosen the bottom bolt, and then remove the A/C condenser fan shroud.

30. Remove the four bolts retaining the A/C compressor.

31. Remove the radiator cap.

32. Raise the vehicle on the hoist to full height.

33. Remove the front tires/wheels.

34. Remove the engine under cover.

35. Remove the splash shield.

36. Loosen the drain plug in the radiator, and drain the engine coolant.

37. Drain the transmission fluid:

38. Drain the engine oil.

39. Disconnect the stabilizer links.

40. Remove the damper fork.

41. Separate the tie-rod end ball joints from the knuckles.

42. Separate the knuckles from the lower arms.

43. Remove the driveshafts and coat all precision-finished surfaces with clean engine oil. Tie plastic bags over the driveshaft ends.

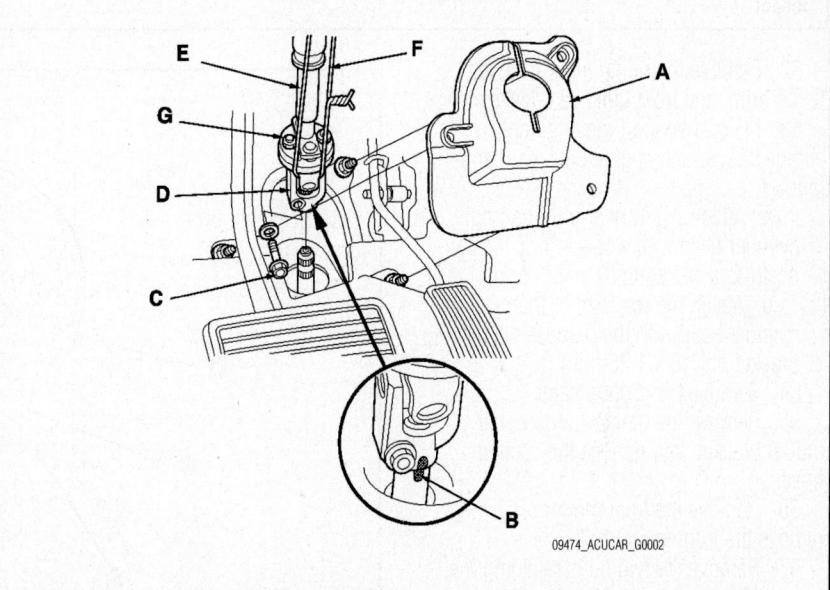

Fig. 103 Make a reference mark across the steering joint and steering gearbox pinion shaft—2007–08 TL model

44. Remove the shift cable. Do not bend the shift cable excessively if equipped with an automatic transmission.

45. Remove exhaust pipe.

46. Remove the PS hose, then plug the line and the hose.

47. Disconnect the power steering pressure switch connector.

48. Remove the nuts retaining the transmission lower front mount and transmission lower rear mount.

49. Remove the upper radiator hose and lower radiator hose.

50. Remove the heater hoses.

51. Remove the two bolts retaining the side engine mount bracket.

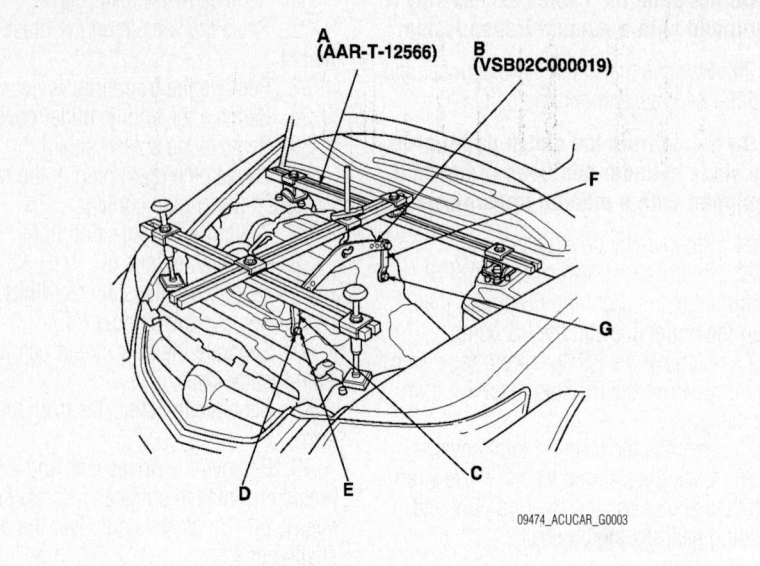

A
(AAR-T-12566)

B
(VSB02C000019)

F

G

C

D

E

09474_ACUCAR_G0003

Fig. 104 Lift and support the engine with engine hanger and engine balancer bar—2007–08 TL model

and wires are disconnected from the engine/transmission assembly.

71. Raise the vehicle all the way.

72. Remove the engine/transmission assembly from under the vehicle.

To install:

73. Position the engine/transmission assembly under the vehicle.

74. Lift and support the engine with engine hanger and engine balancer bar.

75. Attach the front arm to the front cylinder head with a spacer and the connector bracket bolt (10 x 1.25 mm).

76. Attach the rear arm to the rear cylinder head with the harness clamp bracket bolt (8 x 1.25 mm).

77. Lift the engine into position in the vehicle.

78. Reinstall all mounting bolts/support nuts in the sequences given. Failure to follow this may cause excessive noise and vibration, and reduce bushing life.

52. Remove the right fender trim, left fender trim, and front bulkhead cover.

53. Lift and support the engine with engine hanger and engine balancer bar as follows:

a. Attach the front arm to the front cylinder head with a spacer and the connector bracket bolt (10 x 1.25 mm).

b. Attach the rear arm to the rear cylinder head with the harness clamp bracket bolt (8 x 1.25 mm).

54. Remove the ground cable.

55. Remove the transmission upper mount bracket, and remove the vacuum hose.

56. Remove the front mount stop, then remove the front mount bolt.

57. Remove the two bolts retaining the rear engine damper on models with a manual transmission.

58. Remove the rear mount stop, then remove the rear mount bolt.

59. Remove the vacuum hose.

60. Remove the two bolts retaining the shift cable bracket if equipped with a manual transmission.

61. Make sure the hoist brackets are positioned properly.

62. Make the appropriate reference lines at both ends of the sub-frame that line up.

63. Attach an engine sub-frame adapter tool to the sub-frame by hanging the belt over the front of the sub-frame, then secure the belt with its stop.

64. Raise the jack and line up the slots in the engine sub-frame adapter arms with the bolt holes on the jack base, then securely attach them with four bolts.

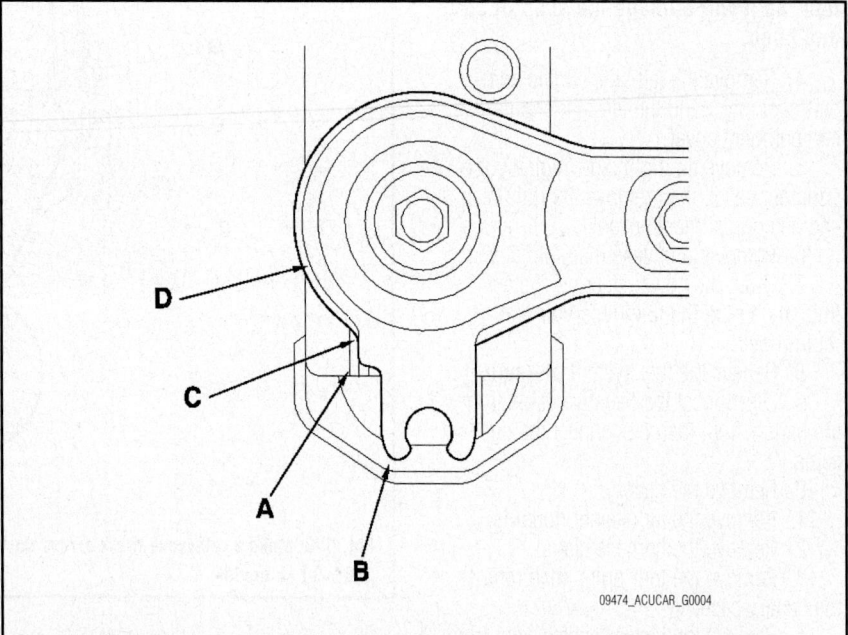

D

C

A

B

09474_ACUCAR_G0004

Fig. 105 Make the appropriate reference lines at both ends of the sub-frame that line up—2007–08 TL model

65. Remove the sub-frame mid-mounts

66. Remove the sub-frame.

67. Check that the engine/transmission is completely free of vacuum hoses, fuel and coolant hoses, and electrical wiring.

68. Lower the vehicle and securely support the engine and transmission assembly.

69. When the engine and transmission are securely supported, and there is no tension on the engine support hanger, remove the engine hanger from the engine.

70. Slowly raise the vehicle about 150 mm (6 in.). Check once again that all hoses

79. Attach the engine sub-frame adapter tool to the sub-frame by hanging the belt over the front of the sub-frame, then secure the belt with its stop. Raise the sub-frame up to the body with a jack.

80. Loosely install the four sub-frame mounting bolts and four 12 x 1.25 mm bolts, with the stiffeners.

81. Loosely install the sub-frame mid mounts, then remove the tool.

82. Align the reference marks with edge of both rear stiffener, and tighten the rear

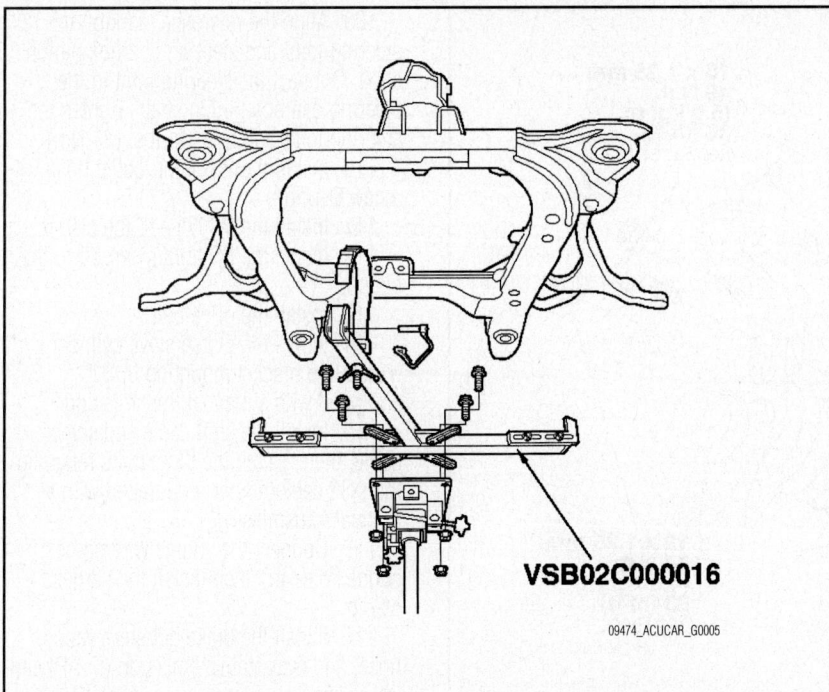

VSB02C000016

09474_ACUCAR_G0005

Fig. 106 Attach an engine sub-frame adapter tool to the sub-frame—2007–08 TL model

Make sure each ring clicks into place in the differential and intermediate shaft.

97. Connect the suspension lower arm ball joints.

98. Connect the tie-rod end ball joints.

99. Install the damper fork.

100. Connect the stabilizer links

101. Install the splash shield

102. Install the engine under cover.

103. Install the front tires/wheels.

➡**On models with a manual transmission, be careful not to damage or chip the paint of the brake calipers when installing the wheels.**

104. Lower the vehicle on the hoist.

105. Install the heater hoses.

106. Install the upper radiator hose and lower radiator hose.

107. Install the A/C compressor.

108. Install the A/C condenser fan shroud and coolant reserve tank, then connect the A/C condenser fan motor connector and A/C compressor clutch connector.

sub-frame mounting bolts, then front bolts, and tighten the stiffener bolts to the specified torque shown in the accompanying illustration.

83. Tighten the bolts retaining the sub-frame mid-mounts to the specified torque shown in the accompanying illustration.

84. Tighten the nuts retaining the transmission lower front mount and transmission lower rear mount to 33 ft. lbs. (44 Nm).

85. Remove the jack and front sub-frame adapter.

86. Install the shift cable, and tighten the two bolts retaining the shift cable bracket if equipped with a manual transmission.

87. Tighten the two bolts retaining the rear engine damper if equipped with a manual transmission.

88. Tighten the front mount bolt, then install the front mount stop.

89. Tighten the two bolts retaining the side engine mount bracket.

90. Install the transmission upper mount bracket, then tighten the bolts in the numbered sequence shown.

91. Install the vacuum hose.

92. Install the ground cable.

93. Connect the power steering pressure switch connector.

94. Install exhaust pipe A using new gaskets and new self locking nuts.

95. Install the shift cable if equipped with an automatic transmission.

96. Install a new set ring on the end of each driveshaft, then install the driveshafts.

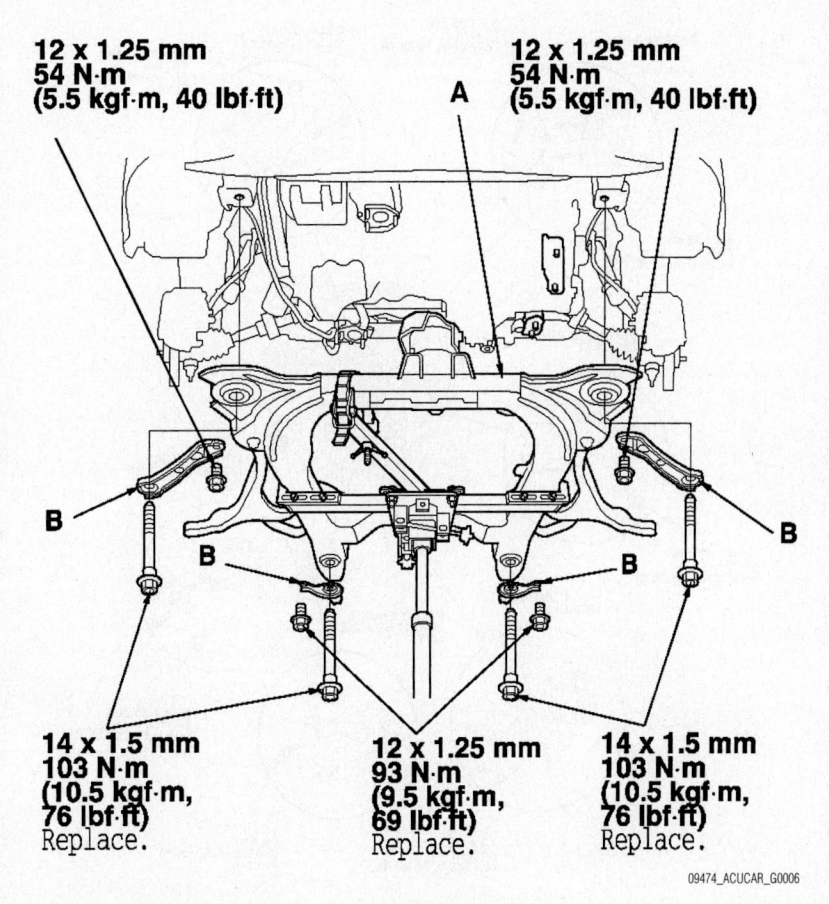

12 x 1.25 mm
54 N·m
(5.5 kgf·m, 40 lbf·ft)

12 x 1.25 mm
54 N·m
(5.5 kgf·m, 40 lbf·ft)

A

B

B

B

B

B

14 x 1.5 mm
103 N·m
(10.5 kgf·m,
76 lbf·ft)
Replace.

12 x 1.25 mm
93 N·m
(9.5 kgf·m,
69 lbf·ft)
Replace.

14 x 1.5 mm
103 N·m
(10.5 kgf·m,
76 lbf·ft)
Replace.

09474_ACUCAR_G0006

Fig. 107 Sub-frame bolt locations and torque specifications—2007–08 TL model

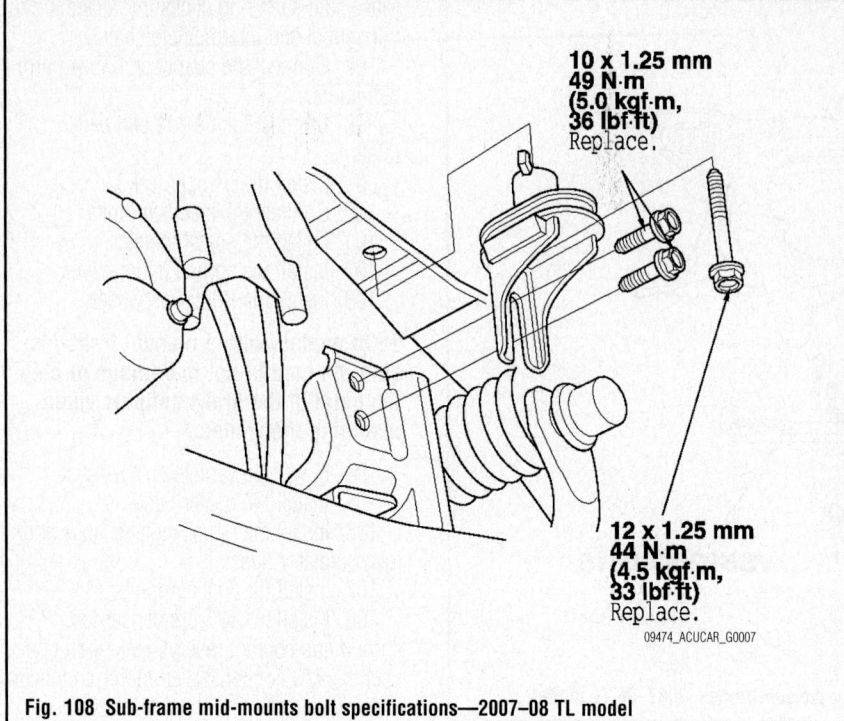

10 x 1.25 mm
49 N·m
(5.0 kgf·m,
36 lbf·ft)
Replace.

12 x 1.25 mm
44 N·m
(4.5 kgf·m,
33 lbf·ft)
Replace.

09474_ACUCAR_G0007

Fig. 108 Sub-frame mid-mounts bolt specifications—2007–08 TL model

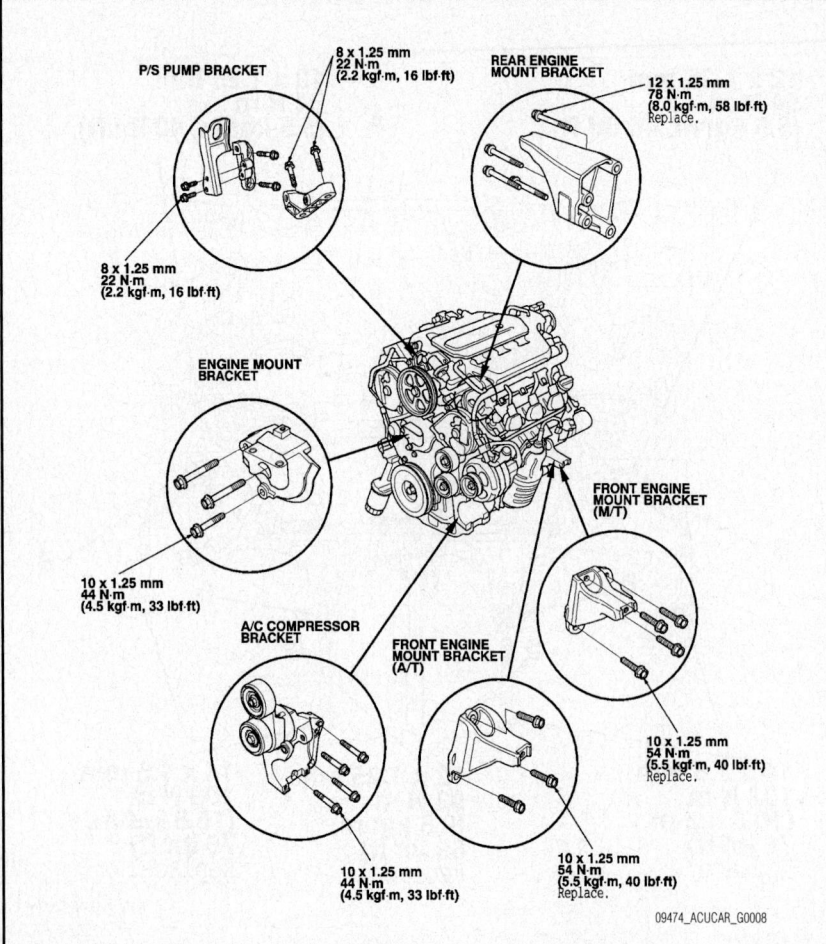

P/S PUMP BRACKET

8 x 1.25 mm
22 N·m
(2.2 kgf·m, 16 lbf·ft)

REAR ENGINE
MOUNT BRACKET

12 x 1.25 mm
78 N·m
(8.0 kgf·m, 58 lbf·ft)
Replace.

8 x 1.25 mm
22 N·m
(2.2 kgf·m, 16 lbf·ft)

ENGINE MOUNT
BRACKET

FRONT ENGINE
MOUNT BRACKET
(M/T)

10 x 1.25 mm
44 N·m
(4.5 kgf·m, 33 lbf·ft)

A/C COMPRESSOR
BRACKET

FRONT ENGINE
MOUNT BRACKET
(A/T)

10 x 1.25 mm
54 N·m
(5.5 kgf·m, 40 lbf·ft)
Replace.

10 x 1.25 mm
44 N·m
(4.5 kgf·m, 33 lbf·ft)
Replace.

10 x 1.25 mm
54 N·m
(5.5 kgf·m, 40 lbf·ft)
Replace.

09474_ACUCAR_G0008

Fig. 109 Exploded view of the engine mounts and their bolt torque specifications—2007–08 TL model

109. Align the reference mark on the steering joint and steering gearbox pinion shaft. Connect the steering joint to the steering gearbox pinion shaft. Tighten the steering joint bolt to 21 ft. lbs. (28 Nm).

110. Install the PS pump outlet hose with a new O-ring.

111. Install the PS hose to the clamp.

112. Install the PS fluid reservoir to the clamp.

113. Install the drive belt.

114. Install the clutch slave cylinder and clutch line bracket mounting bolt if equipped with a manual transmission.

115. Install the shift cable and select cable, then tighten the three bolts retaining the shift cable holder if equipped with a manual transmission.

116. Connect the engine wire harness connectors, and then install the harness clamp.

117. Install the brake booster vacuum hose, and evaporative emission (EVAP) canister hose.

118. Connect the fuel feed hose, then install the quick-connect fitting cover.

119. Install the strut brace, then install the harness clamp, and install the two 6 mm bolts.

120. Install the under-hood fuse/relay box, and connect the harness connector.

121. Install the battery cables.

122. Install the battery base, then install the harness clamp.

123. Install the air cleaner housing.

124. Install the resonator.

125. Install the intake manifold cover.

126. Install the battery.

127. Move the shift lever to each gear, and verify that the automatic transmission gear position indicator follows the transmission range switch if equipped with an automatic transmission.

128. Check that the transmission shifts into gear smoothly if equipped with a manual transmission.

129. Turn the ignition switch ON (II) (do not operate the starter) so the fuel pump runs for about 2 seconds and pressurizes the fuel line. Repeat this operation two or three times, then check for fuel leakage at any point in the fuel line.

130. Refill the engine with engine oil.

131. Refill the transmission with fluid.

132. Refill the radiator with engine coolant, and bleed the air from the cooling system with the heater valve open.

133. Refill the power steering system fluid.

134. Do the Engine Control Module (ECM)/Powertrain Control Module (PCM) reset procedure.

135. Do the Crankshaft Position (CKP) pattern clear/CKP pattern learn procedure.

136. Inspect the idle speed.

137. Inspect the ignition timing.

138. Install the right rear engine compartment cover, then install the right side engine compartment cover.

139. Install the left rear engine compartment cover and the left side engine compartment cover.

140. Enter the anti-theft codes for the radio and the navigation system, then enter the customer's radio channel presets.

RL Models

See Figures 110 through 112.

Make sure you have the anti-theft codes for the radio and navigation system, then write down the audio presets. Make sure the ignition switch is OFF.

1. Before servicing the vehicle, refer to the precautions section.

2. Remove the windshield wiper arms.

3. Remove the intake manifold cover, right upper fender trim, battery trim and left upper fender trim.

4. Remove the upper grille cover and cowl cover.

5. Disconnect the support struts from both sides of the pivot ball. Secure the hood in a vertical position.

6. Remove the right side pivot ball and install it into the lower threaded hole, then reattach the support strut.

➡**Do not attempt to close the hood with the support strut in the vertical position, as it will damage the support strut and hood.**

7. Properly relieve the fuel system pressure.

8. Drain the power steering system fluid, then plug the fluid reservoir and return hose.

9. Disconnect the negative cable from the battery first, then disconnect the positive cable.

10. Remove the battery.

11. Remove the air cleaner.

12. Remove the air intake duct cover.

13. Remove the harness clamps, starter cable and ground cable.

14. Remove the battery base.

15. Remove the battery cable from the under-hood fuse/relay box.

16. Remove the harness clamp, then disconnect the harness connector.

17. Remove the two bolts securing the under-hood fuse/relay box, then remove the under-hood fuse/relay box.

18. Remove the harness clamp and disconnect the engine wire harness connectors on the left side of the engine compartment.

19. Remove the connector and transfer breather hose, then remove the strut brace.

20. Remove the quick-connect fitting cover, then disconnect the fuel feed hose.

21. Remove the brake booster vacuum hose and Evaporative Emission canister hose.

22. Remove the shift cable.

23. Remove the drive belt.

24. Remove the Power Steering (P/S) pump outlet line from the P/S pump, then plug the outlet line and P/S pump.

25. Remove the P/S hose from the clamp.

26. Remove the power steering system fluid reservoir from the clamp.

27. Remove the heat shield.

28. Remove the steering wheel.

29. Remove the steering joint cover.

30. Make a reference mark across the steering joint and steering gearbox pinion shaft. Remove the steering joint bolt and disconnect the steering joint from the steering gearbox pinion shaft.

31. Remove the radiator cap.

32. Raise the vehicle on the hoist to full height.

33. Remove the front wheels.

34. Remove the splash shield.

35. Drain and recycle the engine coolant.

36. Drain the automatic transmission fluid.

37. Drain the engine oil.

38. Disconnect the stabilizer links.

39. Remove the damper fork.

40. Separate the tie-rod end ball joints from the knuckles.

41. Separate the knuckles from the lower arms.

42. Remove the driveshafts.

43. Remove the exhaust pipe assembly.

44. Remove the propeller shaft .

45. Remove the transfer assembly.

46. Remove the ATF cooler hoses from the transmission.

47. Remove the P/S hose, then plug the line and hose.

48. Disconnect the power steering pressure switch connector.

49. Disconnect the connector from the steering gearbox.

50. Remove the bolts securing the transmission lower mount.

51. Lower the vehicle on the hoist.

52. Remove the radiator hoses.

53. Remove the heater hoses.

54. Remove the radiator.

55. Remove the four bolts securing the A/C compressor.

56. Remove the connector bracket from the front cylinder head; use the bracket bolt hole to attach engine balancer bar front arm.

57. Remove the engine mount control solenoid valve bracket from the rear cylinder head; use the bracket bolt hole to attach engine balancer bar rear arm.

58. Lift and support the engine with engine hanger and engine balancer bar. Attach the front arm to the front cylinder head with a spacer and the 10 x 1.25 mm bolt. Attach the rear arm to the rear cylinder head with the 8 x 1.25 mm bolt.

59. Remove the vacuum hose and front mount stop, then remove the front mount bolt.

60. Remove the vacuum hose and rear mount stop, then remove the rear mount bolt.

61. Remove the bolt securing the shift cable bracket.

62. Raise the vehicle on the hoist to full height.

63. Attach the tool to the subframe by hanging the belt over the front of the subframe, then secure the belt with its stop.

64. Raise the jack and line up the slots in the special tool arms with the bolt holes on the jack base, then securely attach them with four bolts.

65. Remove the subframe.

66. Lower the vehicle and attach a chain hoist to the engine hook and transmission hook. Lift up on the engine/transmission assembly until it's securely supported by the chain hoist, and remove the engine support hanger from the engine and vehicle.

67. Remove the two bolts securing the upper bracket.

68. Check that the engine/transmission is completely free of vacuum hoses, fuel and coolant hoses, and electrical wiring.

69. Slowly lower the engine/transmission assembly about 150 mm (6 in.). Check once again that all hoses and electrical wiring are disconnected and free from the engine/transmission, then lower it all the way.

70. Disconnect the chain hoist from the engine/transmission assembly.

71. Raise the vehicle all the way, and remove the engine/transmission assembly from under the vehicle.

To install:

72. Position the engine/transmission assembly under the vehicle. Be sure that they are properly aligned. Carefully lower the vehicle until the engine and transmission are properly positioned in the engine compartment.

73. Make sure the vehicle is not resting on any part of the engine or transmission.

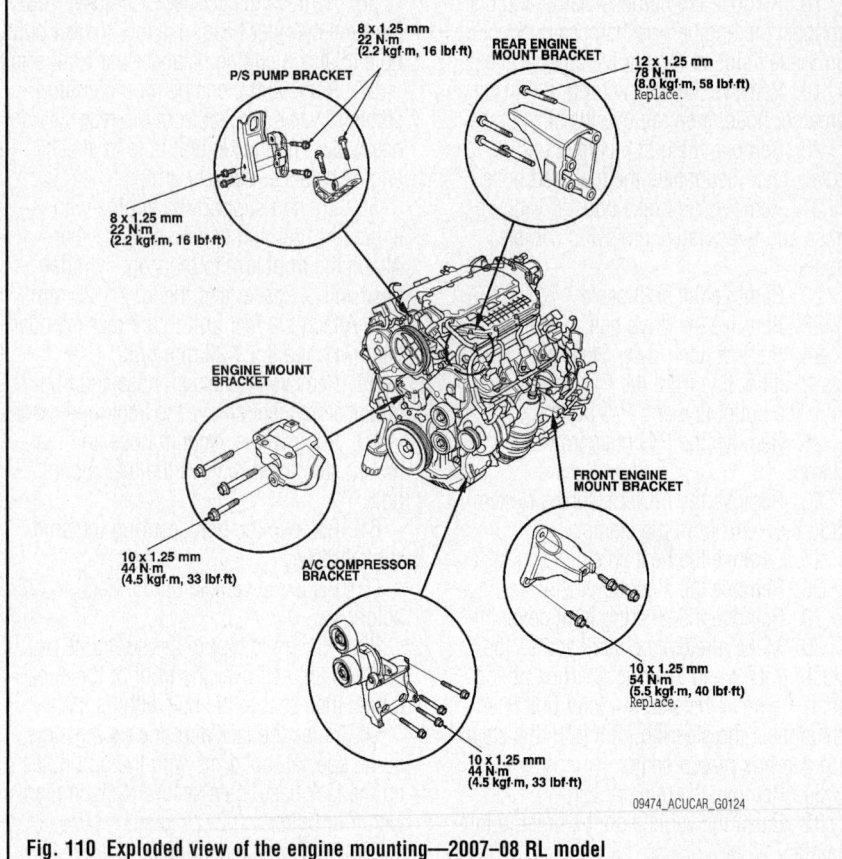

Fig. 110 Exploded view of the engine mounting—2007–08 RL model

P/S PUMP BRACKET
8 x 1.25 mm
22 N·m
(2.2 kgf·m, 16 lbf·ft)

REAR ENGINE MOUNT BRACKET
12 x 1.25 mm
78 N·m
(8.0 kgf·m, 58 lbf·ft)
Replace.

8 x 1.25 mm
22 N·m
(2.2 kgf·m, 16 lbf·ft)

ENGINE MOUNT BRACKET
10 x 1.25 mm
44 N·m
(4.5 kgf·m, 33 lbf·ft)

A/C COMPRESSOR BRACKET
10 x 1.25 mm
44 N·m
(4.5 kgf·m, 33 lbf·ft)

FRONT ENGINE MOUNT BRACKET
10 x 1.25 mm
54 N·m
(5.5 kgf·m, 40 lbf·ft)
Replace.

09474_ACUCAR_G0124

Lift and support the engine with a chain hoist and carefully raise the engine/transmission assembly into place.

➡**Reinstall the mounting bolts/support nuts in the sequence given in the following steps. Failure to follow this sequence may cause excessive noise and vibration, and reduce engine mount life.**

74. Install the two bolts securing the upper bracket. Tighten the bolts 40 ft. lbs. (54 Nm).

75. Install the engine balancer bar No. VSB02C000019; attach the front arm to the front cylinder head with a spacer and 10 x 1.25 mm bolt, attach the rear arm to the rear cylinder head with 8 x 1.25 mm bolt.

76. Install the engine support hanger No. AAR-T-12566 to the vehicle, and attach the hook to the slotted hole in the engine balancer bar. Tighten the wing nut by hand to lift and support the engine/transmission assembly.

77. Remove the chain hoist, then raise the vehicle on the hoist to full height.

78. Attach the front subframe adapter to the subframe by hanging the belt over the front of the subframe, then secure the belt with its stop. Raise the subframe up to the body with a jack.

79. Loosely install the four subframe mounting bolts, four 10 x 1.25 mm bolts and two 12 x 1.25 mm bolts with the stiffeners.

80. Insert the tool through the positioning slot on the right rear stiffener, through the positioning hole on the subframe, and into the positioning hole on the body, then loosely tighten the subframe right rear mounting bolt.

81. Insert the tool through the positioning slot on the left rear stiffener, through the positioning hole on the subframe, and into the positioning hole on the body, then loosely tighten the subframe left rear mounting bolt.

82. Loosely tighten the subframe right/left front mounting bolts.

83. Loosen the subframe right rear mounting bolt, then retighten the subframe right rear mounting bolt to the torque specified in the accompanying illustration.

84. Tighten the subframe left rear mounting bolt to the torque specified in the accompanying illustration.

85. Tighten the subframe right/left front mounting bolts to the torque specified in the accompanying illustration.

86. Check that the positioning slots on the right/left rear stiffener, the positioning holes on the subframe, and the positioning holes on the body are aligned using the special tool.

87. Tighten the stiffener mounting bolts to the torque specified in the accompanying illustration.

88. Remove the jack and front subframe adapter.

89. Tighten the bolts securing the transmission lower mount to 33 ft. lbs. (44 Nm).

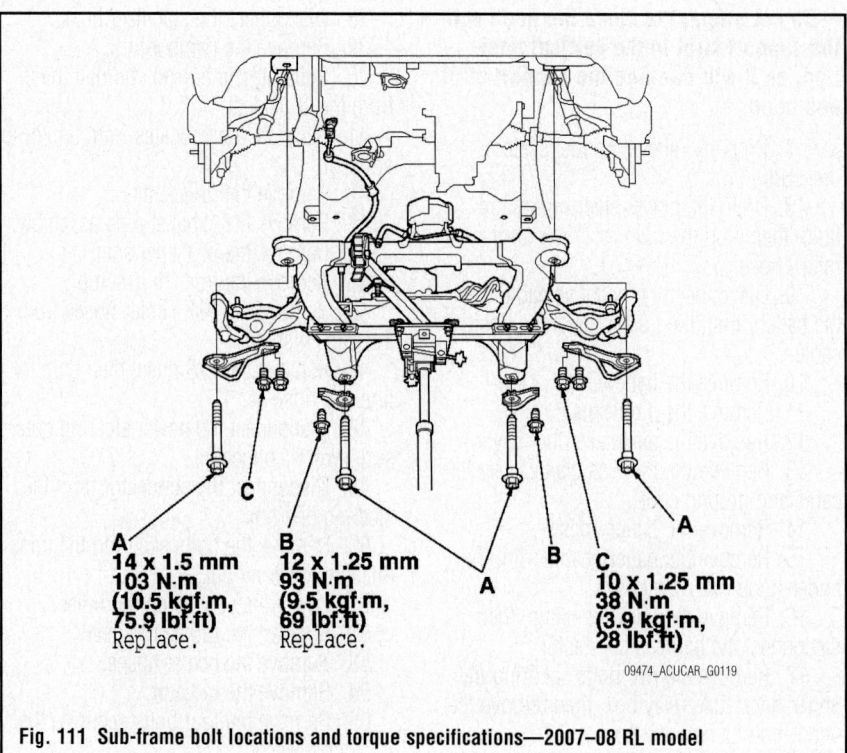

A
14 x 1.5 mm
103 N·m
(10.5 kgf·m, 75.9 lbf·ft)
Replace.

B
12 x 1.25 mm
93 N·m
(9.5 kgf·m, 69 lbf·ft)
Replace.

C
10 x 1.25 mm
38 N·m
(3.9 kgf·m, 28 lbf·ft)

09474_ACUCAR_G0119

Fig. 111 Sub-frame bolt locations and torque specifications—2007–08 RL model

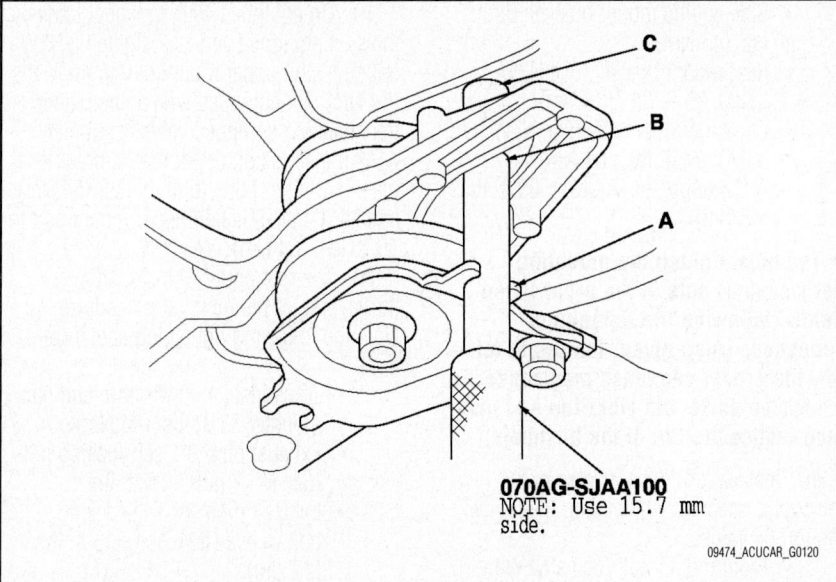

070AG-SJAA100
NOTE: Use 15.7 mm side.

09474_ACUCAR_G0120

Fig. 112 Insert the special tool through the positioning slot (B) on the rear stiffener, through the positioning hole (C) on the subframe—2007–08 RL model

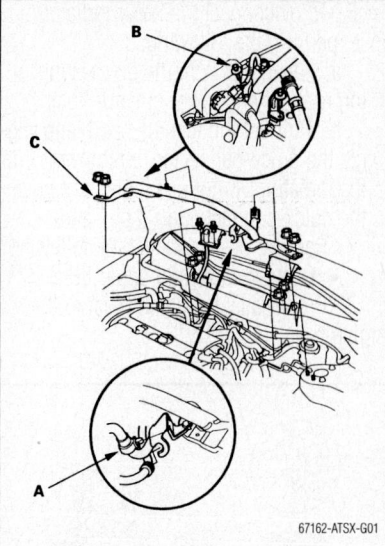

67162-ATSX-G01

Fig. 113 To remove the strut brace (C), you must remove the harness clamp (A) and bolt (B)—2007–08 TSX Models

90. Lower the vehicle on the hoist, then remove the engine hanger and balancer bar.

91. Install the shift cable and tighten the bolt securing the shift cable bracket.

92. Install the vacuum hose.

93. Tighten the rear mount bolt to 40 ft. lbs. (54 Nm), then install the rear mount stop to 47 ft. lbs. (64 Nm).

94. Install the vacuum hose.

95. Tighten the front mount bolt to 40 ft. lbs. (54 Nm), then install the front mount stop to 47 ft. lbs. (64 Nm).

96. Loosen the mounting bolts for the upper half of the side engine mount bracket, then retighten them to 40 ft. lbs. (54 Nm).

97. Install the remaining components in the reverse order of removal. Keeping in mind the following torque specifications:

a. Tighten the A/C compressor bolts to 16 ft. lbs. (22 Nm).

b. Tighten the connector bracket to the front cylinder head bolts to 16 ft. lbs. (22 Nm).

c. Tighten the engine mount control solenoid valve bracket to the rear cylinder head bolts to 16 ft. lbs. (22 Nm).

d. Tighten the steering joint bolt to 21 ft. lbs. (28 Nm).

98. Refill the engine with engine oil.

99. Refill the transmission with fluid.

100. Refill the radiator with engine coolant, and bleed the air from the cooling system with the heater valve open.

101. Refill the power steering system fluid.

102. Do the Engine Control Module (ECM)/Powertrain Control Module (PCM) reset procedure.

103. Do the Crankshaft Position (CKP) pattern clear/CKP pattern learn procedure.

104. Inspect the idle speed.

105. Inspect the ignition timing.

106. Enter the anti-theft codes for the radio and the navigation system, then enter the customer's radio channel presets.

TSX Models

See Figures 113 through 115.

1. Obtain the anti–theft code for the radio.

2. Remove the splash shield in order to drain the vehicle's fluids.

3. Drain the engine oil, coolant, and transmission oil (or fluid) into sealable containers and carefully reinstall the drain plugs using new sealing washers.

4. Properly relieve the fuel system pressure.

5. Remove or disconnect the following:

- Negative and positive battery cables
- Battery
- Air cleaner housing
- Harness clamps and harness bracket
- Battery base retainers and base
- Battery cables from the underhood fuel/relay box and detach the electrical connector
- 2 bolts holding the underhood fuse/relay box
- Harness clamp and bolt and strut brace
- Quick-connect fitting cover and fuel feed hose

- Evaporative emission (EVAP) canister hose and brake booster vacuum hose
- Engine Control Module/Powertrain Control Module (ECM/PCM) connectors and main wire harness
- Accelerator Pedal Position (APP) sensor connector
- Harness clamps and grommet, then pull the engine wire harness through the bulkhead
- Clutch slave cylinder, shift cable and select cable, if equipped with a manual transmission
- Drive belt
- Power steering pump and the hose clamp, but do NOT disconnect the fluid lines
- A/C compressor, but do not disconnect the A/C hoses
- Front wheels and tires
- Stabilizer links
- Damper fork
- Lower control arm ball joints
- Driveshafts. Coat all machined surfaces with clean engine oil, then tie plastic bags over the driveshaft ends.
- Shift control cable, if equipped with an automatic transmission
- Exhaust pipe "A"
- Nuts holding the transmission lower front mount and transmission lower rear mount
- Automatic Transmission Fluid cooler hoses
- Upper radiator hose and heater hoses

6. To disconnect the lower radiator hose, perform the following:

a. Clean dirt from the quick-connector, radiator and lower radiator hose.

b. Pull out the lock by hand, and wiggle the quick-connector to remove it from the radiator. Do not use tools to remove the quick-connector, do it by hand.

7. Remove or disconnect the following:
• Ground cable and upper bracket

8. Proper attach a chain hoist to the engine.

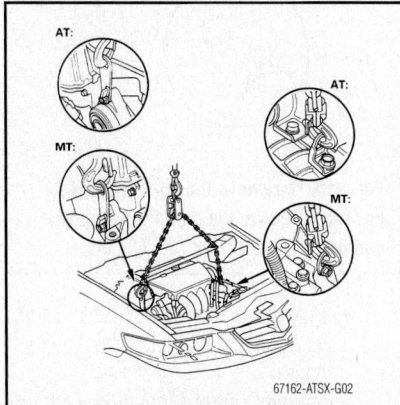

Fig. 114 Make sure the chain hoist is properly attached to the engine—2007–08 TSX Models

9. For vehicles with a equipped with a manual transmission, remove the ground cable, transmission upper mount/bracket assembly and clutch line clamp bracket

10. For vehicles with an equipped with an automatic transmission, remove the ground cable, then remove the bracket plate, transmission upper bracket and transmission upper mount.

11. Remove or disconnect the following:
• Vacuum hose from the vacuum line
• Front mount stop and vacuum hose clamp bracket and the front mount bolt
• Rear mount stop and the rear mount bolt

12. Make sure that the engine/transmission assembly is totally free of all vacuum lines, wiring and fuel and coolant hoses.

13. Slowly raise the engine about 6 in. (150mm), then recheck that all wires/hoses are disconnected from the engine/transmission assembly.

14. Raise the engine/transmission assembly all the way and remove it from the vehicle.

To install:

15. Install the accessory brackets to the engine and tighten to the following torque specifications:

• Side engine mount bracket: 33 ft. lbs. (45 Nm)
• Rear mount bracket (use NEW bolts): 65 ft. lbs. (88 Nm)
• Front mount bracket (use NEW bolts): 47 ft. lbs. (64 Nm)
• A/C compressor bracket: 33 ft. lbs. (45 Nm)

➥You must tighten the mounting bolts/support nuts in the order listed below, following the tightening sequences when given. Failure to follow the proper sequence may cause excessive noise and vibration and may also reduce the life of the bushings.

16. Installation is the reverse of the removal procedure, while using the following torque values:

• Front mount bolt (use a NEW bolt): 40 ft. lbs. (54 Nm)
• Front mount stop nuts: 58 ft. lbs. (78 Nm).
• Rear mount bolt (use a NEW bolt): 40 ft. lbs. (54 Nm)
• Rear mount stop nuts: 51 ft. lbs. (69 Nm)
• 2 top upper bracket bolts: 40 ft. lbs. (54 Nm)
• Side upper bracket bolt: 47 ft. lbs. (64 Nm)
• Ground cable bolt: 104 inch lbs. (12 Nm)

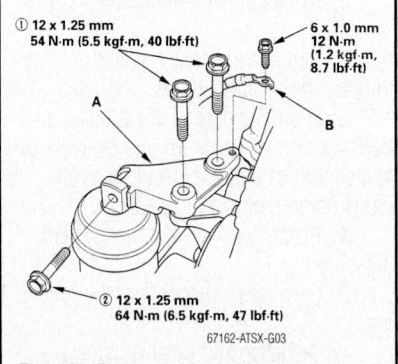

Fig. 115 Proper tightening sequence for the upper bracket bolts—2007–08 TSX Models

17. If equipped with a equipped with a manual transmission, loosen the transmission upper mount bolt, then install the transmission upper mount/bracket assembly, clutch line clamp bracket and ground cable.

18. If equipped with an equipped with an automatic transmission, install the transmission upper mount, transmission upper bracket and ground cable, then loosely install the transmission upper bracket plate.

19. On equipped with a manual transmission equipped vehicles, tighten a NEW transmission upper mount bolt to 40 ft. lbs. (54 Nm). On equipped with an automatic transmission equipped vehicles, tighten NEW transmission upper mount bracket plate mounting bolts to 40 ft. lbs. (54 Nm) and NEW transmission upper mount bolt to 40 ft. lbs. (54 Nm).

20. The remainder of installation is the reverse of the removal procedure. Note the following tightening specifications:

• Transmission lower front and rear mounts: 33 ft. lbs. (44 Nm)
• Exhaust pipe "A" self-locking nuts (use NEW nuts): 25 ft. lbs.
• Exhaust pipe "A" self-locking bolt (use a NEW bolt): 16 ft. lbs. (22 Nm)
• A/C compressor mounting bolts: 16 ft. lbs. (22 Nm)
• Power steering pump mounting bolts: 16 ft. lbs. (22 Nm)
• Power steering hose clamp bolt: 104 inch lbs. (12 Nm)
• Strut brace nuts: 16 ft. lbs. (22 Nm)
• Harness clamp bolt: 104 inch lbs. (12 Nm)
• Underhood fuse/relay box bolts: 104 inch lbs. (12 Nm)
• Battery base bolts: 16 ft. lbs. (22 Nm)
• Harness bracket bolt: 104 inch lbs. (12 Nm)

21. Fill the engine with oil and the transmission with fluid.

22. Fill and bleed (if necessary) the air from the cooling system.

23. Connect the positive, then the negative battery cable and enter the radio security code.

24. Switch the ignition **ON** but do not engage the starter. The fuel pump should run for approximately 2 seconds, building pressure within the lines. Switch the ignition **OFF**, then **ON** 2 or 3 more times to build full system pressure. Check for fuel leaks.

25. Start the engine, allowing it to idle. Check the hoses and lines carefully for any sign of leakage.

26. Check the timing and idle speed.

27. After the engine has warmed up fully and the fan(s) have come on at least once, recheck the engine for fluid leaks. Switch the engine OFF.

28. Adjust the belts and throttle cable as necessary.

EXHAUST MANIFOLD

REMOVAL & INSTALLATION

TL Models

1. Remove or disconnect the following:
   - Negative battery cable
   - Exhaust manifold covers
   - Small heat shields from the cylinder heads (if equipped)
   - Exhaust pipe from the manifold
   - Heated Oxygen (HO₂S) sensors
   - Exhaust manifold nuts in a crisscross pattern starting from the center of the manifold
   - Exhaust manifold

To install:

2. Install or connect the following:
   - Exhaust manifold with a new gasket and new nuts and tighten the nuts in a crisscross pattern starting from the center to 22 ft. lbs. (30 Nm)
   - Small heat shields and tighten the attaching bolts to 16 ft. lbs. (22 Nm) (if equipped)
   - Exhaust pipe to the manifold with a new gasket and tighten the nuts to 40 ft. lbs. (55 Nm)
   - HO₂S sensor and tighten it to 33 ft. lbs. (45 Nm)
   - Manifold covers and tighten the bolts to 16 ft. lbs. (22 Nm)
3. Verify that all vacuum lines and wiring are properly connected.
4. Reconnect the negative battery cable.
5. Start the engine and check for leaks.

RL Models

1. Before servicing the vehicle, refer to the precautions section.
2. Remove or disconnect the following:
   - Negative battery cable
   - Exhaust manifold covers
   - Small heat shields from the cylinder heads (if equipped)
   - Exhaust pipe from the manifold
   - Heated Oxygen (HO₂S) sensors
   - Exhaust manifold nuts in a crisscross pattern starting from the center of the manifold
   - Exhaust manifold

To install:

3. Install or connect the following:
   - Exhaust manifold with a new gasket and new nuts and tighten the nuts in a crisscross pattern starting from the center to 22 ft. lbs. (30 Nm)
   - Small heat shields and tighten the attaching bolts to 16 ft. lbs. (22 Nm) (if equipped)

- Exhaust pipe to the manifold with a new gasket and tighten the nuts to 40 ft. lbs. (55 Nm)
- HO₂S sensor and tighten it to 33 ft. lbs. (45 Nm)
- Manifold covers and tighten the bolts to 16 ft. lbs. (22 Nm)
4. Verify that all vacuum lines and wiring are properly connected.
5. Reconnect the negative battery cable.
6. Start the engine and check for leaks.

TSX Models

See Figure 116.

1. Before servicing the vehicle, refer to the precautions in the beginning of this section.
2. Raise and safely support the vehicle.
3. Remove or disconnect the following:
   - VTEC solenoid valve
   - Driveshaft heat shield
   - Cover and exhaust manifold bracket
   - Exhaust manifold

To install:

4. Clean the mounting surfaces.
5. Install or connect the following:
   - New gasket on the cylinder head
   - Exhaust manifold. Tighten the nuts, in a criss-cross pattern starting with the inner nut, to 33 ft. lbs. (45 Nm).
   - Exhaust manifold bracket and cover
   - Driveshaft heat shield
   - VTEC solenoid valve

FLYWHEEL

REMOVAL & INSTALLATION

See Figures 117 and 118.

The flywheel on cars with manual transaxles serves as the forward clutch engagement surface. It also serves as the ring gear with which the starter pinion engages to crank the engine. The most common reasons to replace the flywheel are:
- Broken teeth on the flywheel ring gear
- Excessive driveline chatter when engaging the clutch
- Excessive wear, scoring or cracking of the clutch surface

On cars with automatic transaxles, the torque converter actually forms part of the flywheel. It is bolted to a thin driveplate which, in turn, is bolted to the crankshaft. The driveplate also serves as the ring gear with which the starter pinion engages in engine cranking. The driveplate occasionally cracks; the teeth on the ring gear may also break, especially if the starter is often engaged while the pinion is still spinning. The torque converter and driveplate must be separated, and the converter and transaxle are be removed together.

1. Remove the transaxle from the vehicle. For more information, refer to the Drive Train Section.
2. On vehicles equipped with a manual transaxle, remove the clutch assembly from the flywheel, as described in the Drive Train Section.

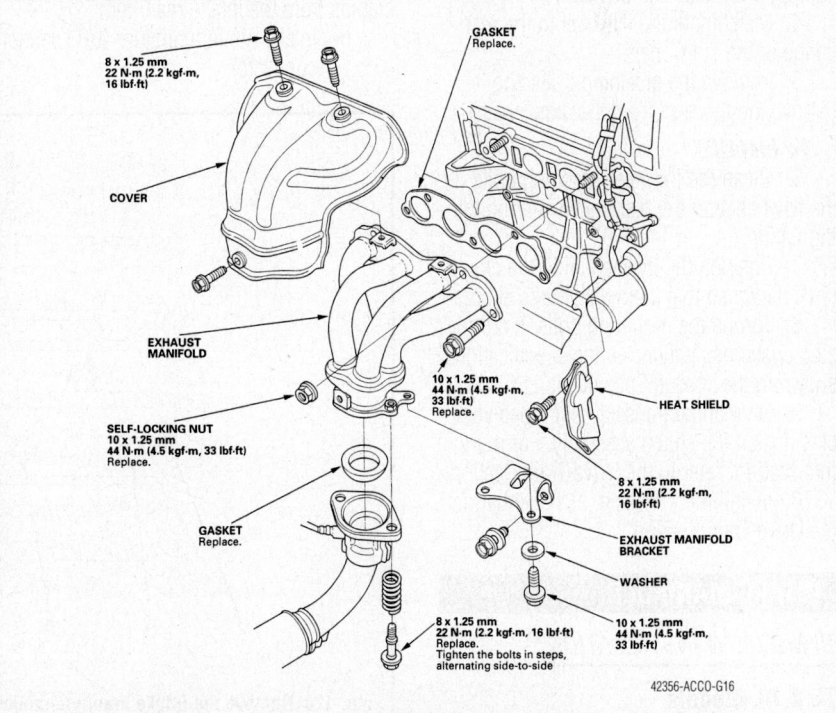

Fig. 116 Exploded view of the exhaust manifold and related components—2.4L engine

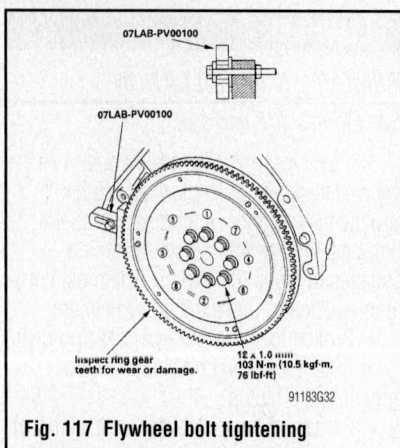

Fig. 117 Flywheel bolt tightening sequence—with manual transaxles

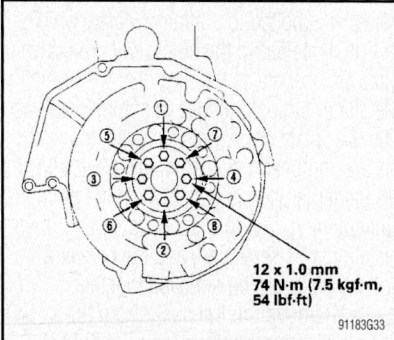

Fig. 118 Driveplate bolt tightening sequence—with automatic transaxles

3. Support the flywheel in a secure manner (the flywheel on manual transaxle-equipped vehicles can be heavy).

4. Matchmark the flywheel to the rear flange of the crankshaft.

5. Remove the attaching bolts and remove the flywheel from the crankshaft.

To install:

6. Clean the flywheel attaching bolts, the flywheel and the rear crankshaft mounting flange.

7. Position the flywheel onto the crankshaft flange so that the matchmarks align.

8. Torque the mounting bolts in a three step crisscross pattern, to the specifications shown in the Torque Specifications Chart.

9. On manual transaxle-equipped vehicles, install the clutch assembly. For more information, refer to the Drive Train Section.

10. Install the transaxle, as described in the Drive Train Section.

INTAKE MANIFOLD

REMOVAL & INSTALLATION

TL & RL Models

See Figures 119 through 122.

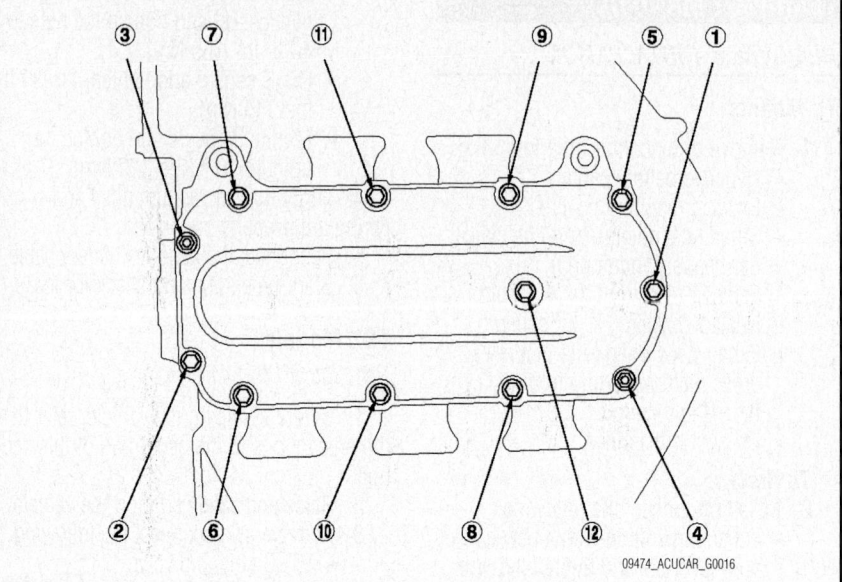

Fig. 119 Remove the upper cover mounting bolts and nuts sequentially in two or three steps—2007–08 TL and RL models

1. Remove the intake manifold cover.

2. Remove the air intake duct.

3. Remove the engine mount control solenoid valve, Positive Crankcase Ventilation (PCV) hose, brake booster vacuum hose, and vacuum hose.

4. Remove the evaporative emission (EVAP) canister purge hose and water bypass hoses, then plug the water bypass hoses.

5. Remove the following engine wire harness connectors and wire harness clamps from the intake manifold:

- Intake Air Temperature (IAT) sensor connector

- Throttle actuator connector
- Manifold Absolute Pressure (MAP) sensor connector
- Evaporative Emission (EVAP) canister purge valve connector
- Intake Manifold Tuning (IMT) valve connector

6. Remove the upper cover mounting bolts and nuts sequentially in two or three steps.

7. Remove the intake manifold mounting bolts and nuts sequentially in two or three steps.

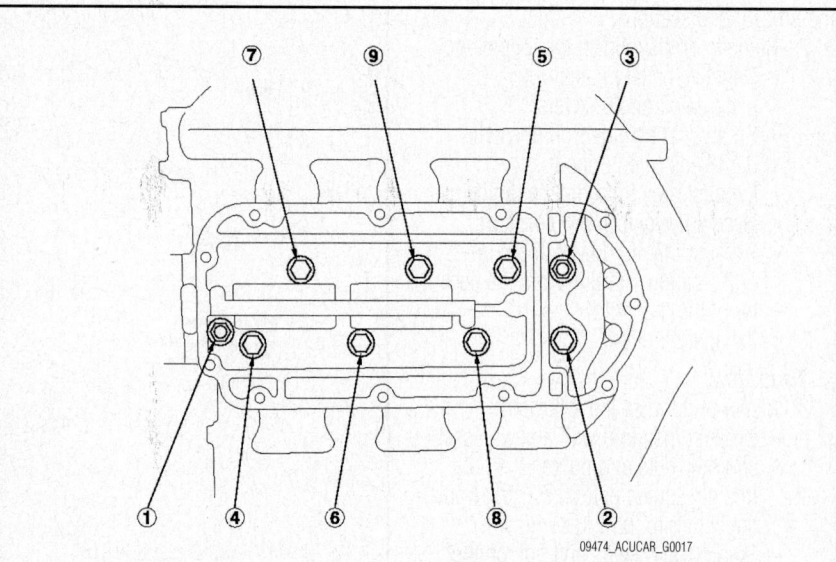

Fig. 120 Remove the intake manifold mounting bolts and nuts sequentially in two or three steps—2007–08 TL and RL models

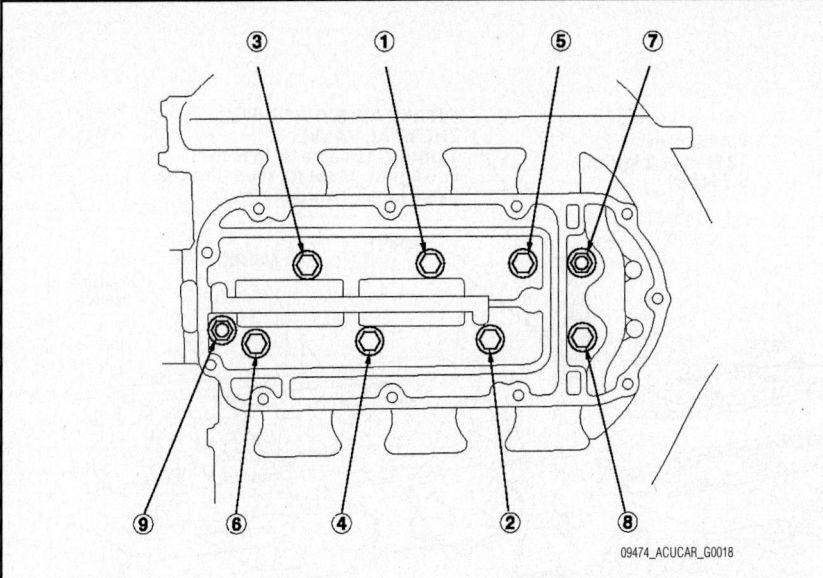

Fig. 121 Tighten the intake manifold mounting bolts and nuts sequentially in two or three steps—2007–08 TL and RL models

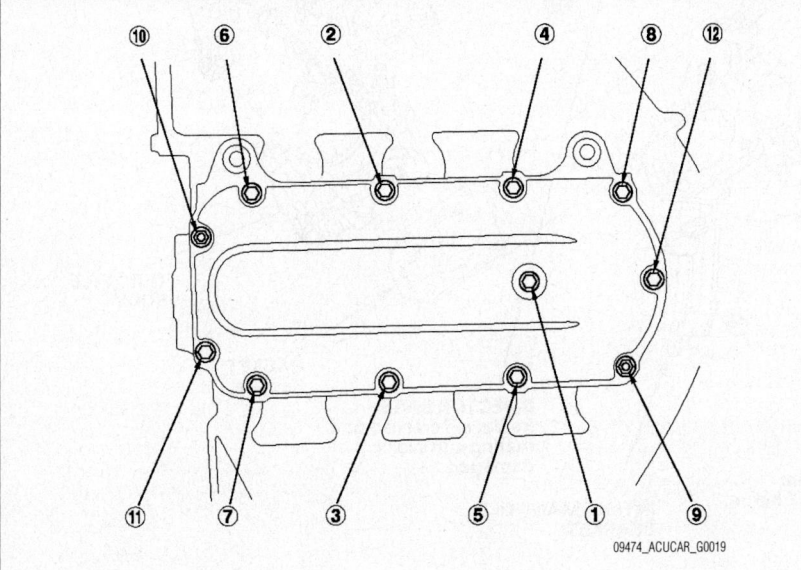

Fig. 122 Tighten the upper cover mounting bolts and nuts sequentially in two or three steps— 2007–08 TL and RL models

TSX Models

See Figure 123.

1. Disconnect the negative battery cable.

2. Drain the engine coolant into a sealable container.

3. Remove or disconnect the following:
- Intake Air Temperature (IAT) sensor electrical connector
- Vacuum hose and breather pipe and the air intake duct
- Intake manifold cover
- Throttle and cruise control cables by loosening the locknuts, then slipping the cable ends out of the accelerator linkage.

➡**Do not bend the cables during removal. Always replace any throttle or cruise control cables that get kinked during removal.**

- Evaporative emission (EVAP) canister hose and brake booster vacuum hose
- Idle Air Control (IAC) valve connectors
- Throttle Position (TP) sensor connector
- Manifold Absolute Pressure (MAP) sensor connector
- Necessary engine wire harness connectors and wire harness clamps from the intake manifold
- Bolt securing the harness holder and remove the harness clamps
- Water bypass hoses, then plug them
- Harness clamp and harness connector from the intake manifold bracket
- Intake manifold bracket
- A/T vacuum hose
- Retainer and intake manifold

To install:

4. Clean the mounting surfaces.

5. Install or connect the following:
- New gasket
- Intake manifold. Tighten the bolts, in a criss-cross pattern beginning with the inner bolt, to 16 ft. lbs. (22 Nm).
- A/T vacuum hose
- Intake manifold bracket
- Harness clamp and connector to the intake manifold bracket
- Water bypass hoses
- Bolt securing the harness holder and tighten to 104 inch lbs. (12 Nm)

To install:

8. Install a new gasket and the intake manifold. Tighten the bolts and nuts sequentially in two or three steps to 16 ft. lbs. (22 Nm).

9. Install a new gasket and the upper cover. Tighten the bolts and nuts sequentially in two or three steps to 9 ft. lbs. (12 Nm).

10. Install the engine wire harness connectors and wire harness clamps to the Intake manifold that were removed earlier.

11. Install the EVAP canister purge hose and water bypass hoses.

12. Install the engine mount control solenoid valve, PCV hose, brake booster vacuum hose, and vacuum hose.

13. Install the intake air duct.

14. After installation, check that all tubes, hoses, and connectors are installed correctly.

15. Install the intake manifold cover.

16. Refill the radiator with engine coolant, then bleed air from the cooling system with the heater valve open.

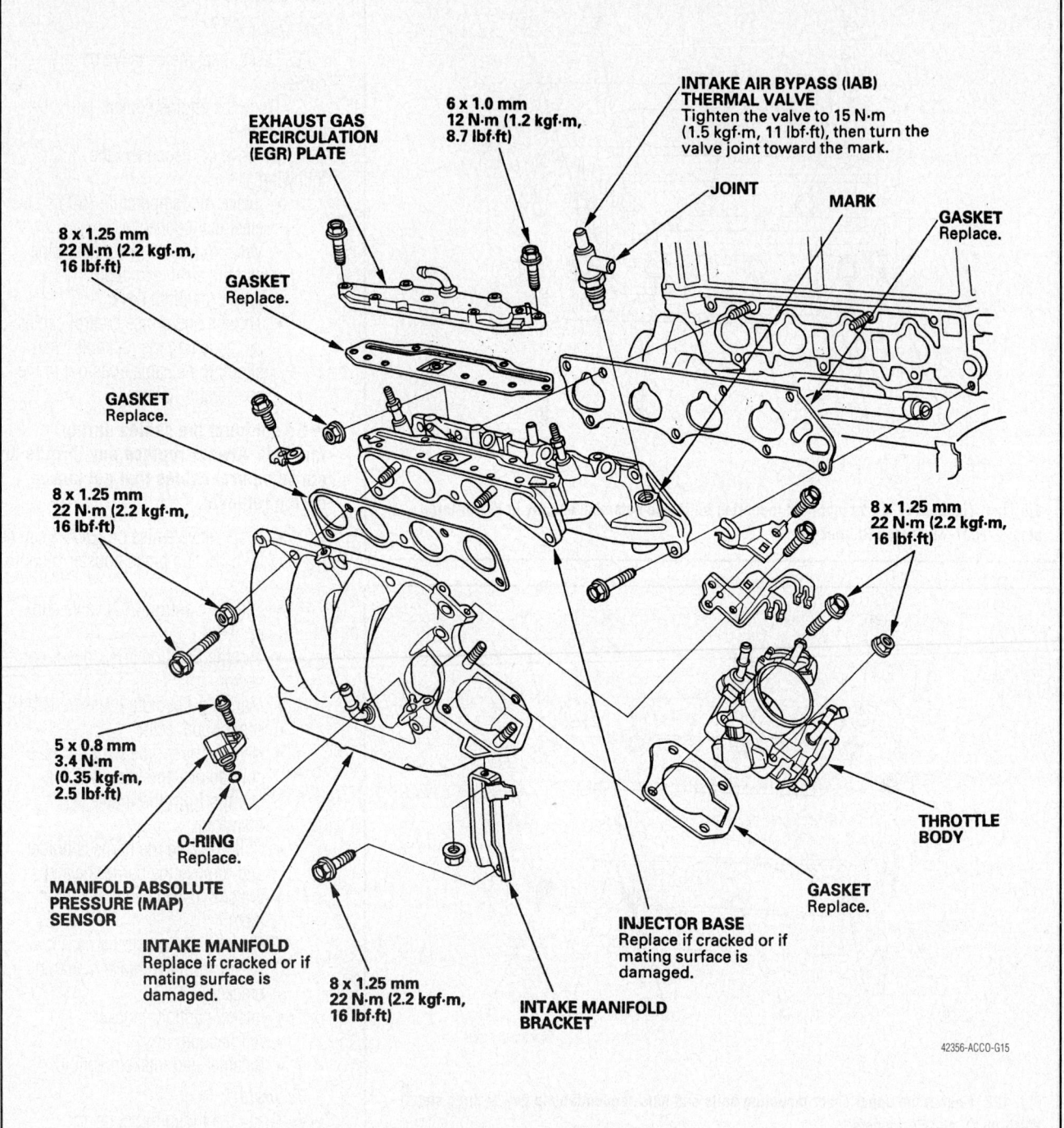

EXHAUST GAS RECIRCULATION (EGR) PLATE

6 x 1.0 mm
12 N·m (1.2 kgf·m, 8.7 lbf·ft)

INTAKE AIR BYPASS (IAB) THERMAL VALVE
Tighten the valve to 15 N·m (1.5 kgf·m, 11 lbf·ft), then turn the valve joint toward the mark.

JOINT

MARK

GASKET
Replace.

8 x 1.25 mm
22 N·m (2.2 kgf·m, 16 lbf·ft)

GASKET
Replace.

GASKET
Replace.

8 x 1.25 mm
22 N·m (2.2 kgf·m, 16 lbf·ft)

8 x 1.25 mm
22 N·m (2.2 kgf·m, 16 lbf·ft)

5 x 0.8 mm
3.4 N·m (0.35 kgf·m, 2.5 lbf·ft)

O-RING
Replace.

MANIFOLD ABSOLUTE PRESSURE (MAP) SENSOR

INTAKE MANIFOLD
Replace if cracked or if mating surface is damaged.

8 x 1.25 mm
22 N·m (2.2 kgf·m, 16 lbf·ft)

INTAKE MANIFOLD BRACKET

INJECTOR BASE
Replace if cracked or if mating surface is damaged.

THROTTLE BODY

GASKET
Replace.

42356-ACCO-G15

Fig. 123 Exploded view of the intake manifold and related components—2007–08 TSX Models

- Harness clamps
- EVAP canister hose and brake booster vacuum hose
- Throttle and cruise control cables
- Intake manifold cover
- Intake air duct
- IAT sensor connector, vacuum hose and breather pipe
6. Refill the cooling system.
7. Connect the negative battery cable, start the engine, and check for leaks.

OIL PAN

REMOVAL & INSTALLATION

TL & RL Models

See Figure 124.

1. Drain the engine oil.
2. Remove the splash shield.
3. Remove the undercover.
4. Remove exhaust pipe A.

5. Remove the rear warm up three way catalytic converter bracket.
6. Remove the torque converter cover and the bolts retaining the transmission.
7. Remove the bolts retaining the oil pan.
8. Using a flat blade screwdriver, separate the oil pan from the block in the places shown.
9. Remove the oil pan.

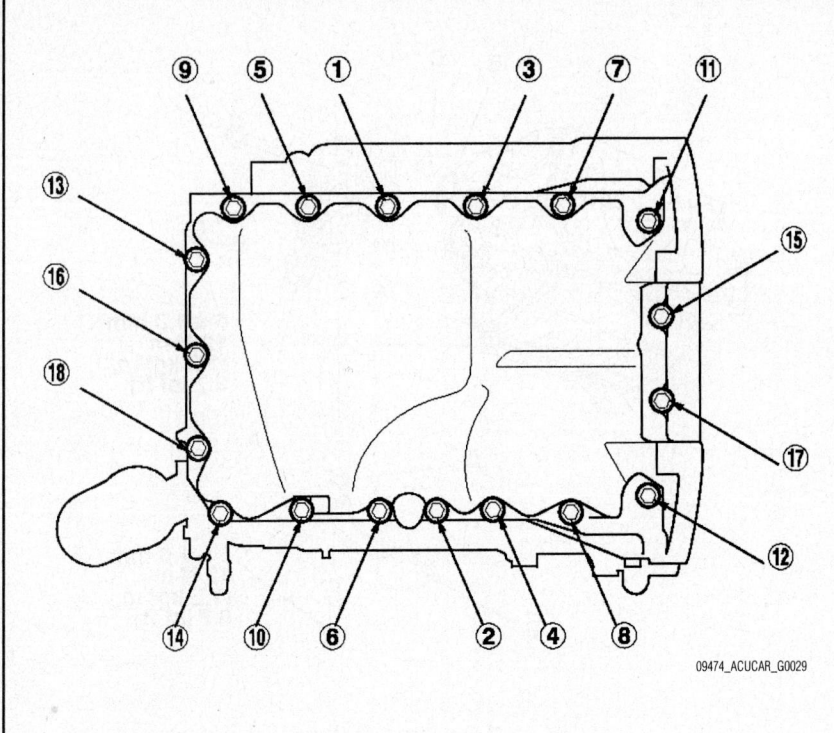

Fig. 124 Oil pan bolt torque sequence—2007–08 TL and RL models

To install:

10. Remove any old liquid gasket from the oil pan mating surfaces, bolts, and bolt holes.

11. Clean and dry the oil pan mating surfaces.

12. Apply liquid gasket, P/N 08717-0004, 08718-0001, 08718-0003, or 08718-0009, evenly to the oil pan mating surface of the engine block.

➡**Do not install the parts if 4 minutes or more have elapsed since applying liquid gasket. Instead, re-apply liquid gasket after removing the old residue.**

13. Install the oil pan on the engine block. Tighten the bolts in two or three steps. In the final step, tighten all bolts, in sequence, to 9 ft. lbs. (12 Nm).

➡**After assembly, wait at least 30 minutes before filling the engine with oil.**

14. Tighten the bolts retaining the transmission to 47 ft. lbs. (64 Nm), then install the torque converter cover.

15. Remove the rear warm up three way catalytic converter bracket. Tighten to 16 ft. lbs. (22 Nm).

16. Install the remaining components in the reverse order of removal, after 30 minutes has elapsed, fill the engine with new oil.

TSX Models

See Figures 125 and 126.

1. Remove or disconnect the following:
- Negative battery cable
- Engine oil
- Front tire and wheel assemblies
- Splash shield
- Stabilizer links
- Right side damper fork
- Right side lower ball joint
- Right side halfshaft. Coat all machined surfaces with clean

engine oil and secure a plastic bag over the end of the halfshaft.

2. From the engine compartment, remove the front mount stop and front mount bolt
- Rear mount stop and rear mount bolt
- Ground cable and upper bracket
- Bolt holding the side engine mount bracket and attach Engine Hanger Plate EQS00BRSX0, or equivalent.

3. Use a jack to lift the engine 1.2–2.4 in. (30–60mm).
- Stiffener
- Oil pan bolts and nuts

4. Hammer a seal cutter between the engine block and oil pan to break the seal.

5. Remove the oil pan.

To install:

6. Clean the oil pan flange and engine block mounting surface.

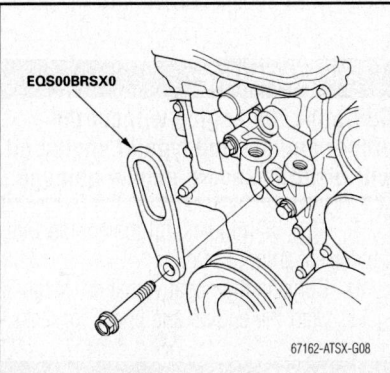

Fig. 125 Proper installation of Engine Hanger Plate EQS00BRSX0—2.4L engine

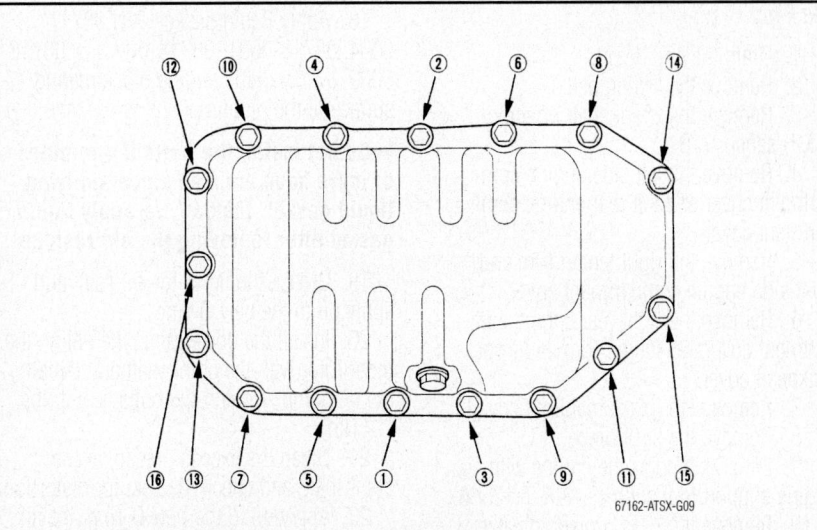

Fig. 126 Oil pan bolt tightening sequence—2.4L engine

7. Install or connect the following:
- Sealant to the oil pan flange. Be sure to apply sealant toward the inside of the bolt holes.
- Oil pan on the engine. Tighten the bolts in sequence, in 2 or 3 steps, to 104 inch lbs. (12 Nm).
- Stiffener. Torque the retaining bolts to 33 ft. lbs. (45 Nm)

8. Remove the engine hanger tool and tighten the side engine mount bracket bolt to 33 ft. lbs. (45 Nm).
- Upper bracket and ground cable
- New set ring on the end of the driveshaft, then install the driveshaft. Make sure each ring "clicks" into plate in the differential
- Right side lower ball joint
- Right side damper fork
- Stabilizer links
- Splash shield
- Tire and wheel assemblies
- Front mount bolt and front mount stop
- Rear mount bolt and rear mount stop

9. After 30 minutes, fill the engine with the correct amount of oil.
10. Connect the negative battery cable.
11. Start the engine and check for leaks.

OIL PUMP

REMOVAL & INSTALLATION

TL Models

See Figure 127.

1. Drain the engine oil.
2. Remove the timing belt.
3. Remove the Crankshaft Position (CKP) sensor A/B.
4. Remove the left side engine compartment cover and left rear engine compartment cover.
5. Remove the right fender trim and right side engine compartment cover.
6. Remove the right rear engine compartment cover, left fender trim and front bulkhead cover.
7. Remove the intake manifold cover.
8. Remove the strut brace.
9. Lift and support the engine with an engine support tool such as AAR-T-12566.
10. Remove the VTEC solenoid valve and oil filter assembly.

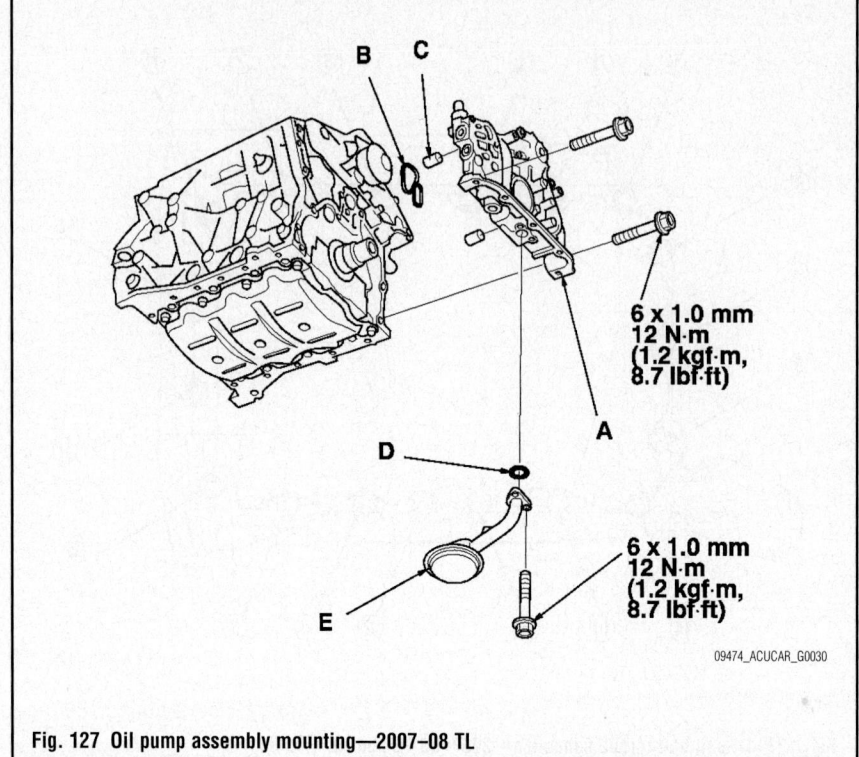

Fig. 127 Oil pump assembly mounting—2007–08 TL

6 x 1.0 mm
12 N·m
(1.2 kgf·m,
8.7 lbf·ft)

6 x 1.0 mm
12 N·m
(1.2 kgf·m,
8.7 lbf·ft)

09474_ACUCAR_G0030

11. Remove the oil pan.
12. Remove the oil screen.
13. Remove the mounting bolts and the oil pump assembly.

To install:
14. Remove the old oil seal from the oil pump.
15. Gently tap in the new oil seal until the seal driver bottoms on the pump.
16. Remove any old liquid gasket from the oil pump mating surfaces, bolts, and bolt holes.
17. Clean and dry the oil pump mating surfaces.
18. Apply liquid gasket, P/N 08717-0004, 08718-0001, 08718-0003, or 08718-0009, evenly to the engine block mating surface of the oil pump.

➡**Do not install the parts if 4 minutes or more have elapsed since applying liquid gasket. Instead, re-apply liquid gasket after removing the old residue.**

19. Grease the lip of the oil seal, and apply oil to the new O-ring.
20. Install the dowel pins, then align the inner rotor with the crankshaft and install the oil pump. Tighten the bolts to 9 ft. lbs. (12 Nm).
21. Clean the excess grease off the crankshaft, and check the seal for distortion.
22. Apply oil to the new O-ring, install the oil screen.
23. Install the oil pan.

24. Install the remaining components in the reverse order of removal.
25. After assembly, wait at least 30 minutes before filling the engine with oil.

RL Models

See Figure 128.

1. Drain the engine oil.
2. Remove the windshield wiper arms.
3. Remove the intake manifold cover, right upper fender trim, battery trim, and left upper fender trim, then remove the upper grille cover and cowl cover.
4. Remove the under-hood fuse/relay box, connector, and transfer breather hose from the strut brace, then remove the strut brace.
5. Remove the connector bracket from the front cylinder head; use the bracket bolt hole to attach engine balancer bar front arm.
6. Remove the engine mount control solenoid valve bracket from the rear cylinder head; use the bracket bolt hole to attach engine balancer bar rear arm.
7. Lift and support the engine with engine hanger and engine balancer bar Attach the front arm to the front cylinder head with a spacer and the 10 x 1.25 mm bolt. Attach the rear arm to the rear cylinder head with the 8 x 1.25 mm bolt.
8. Remove the timing belt.

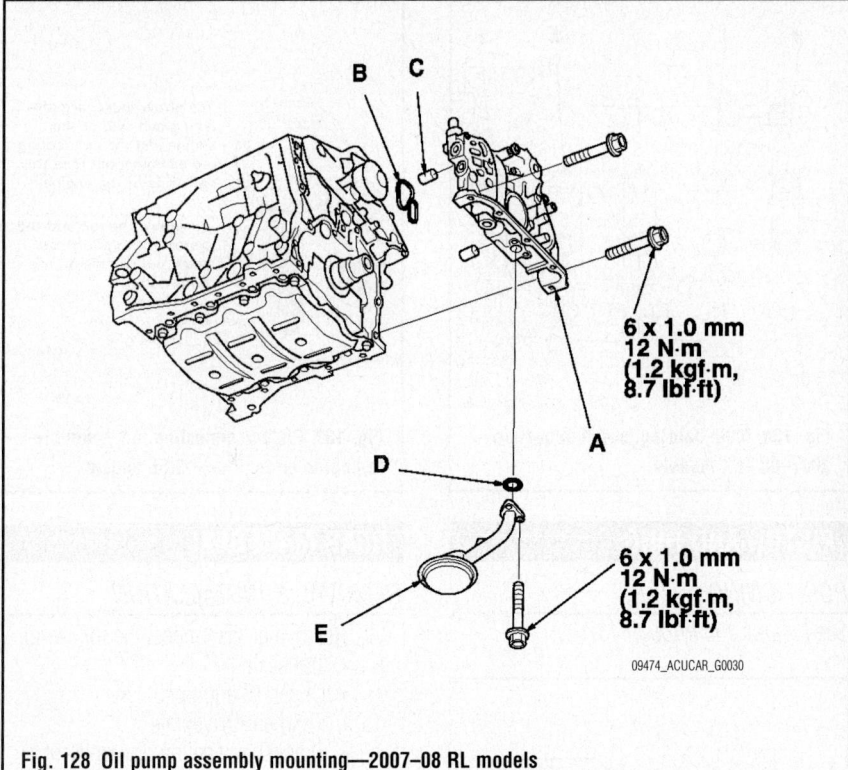

6 x 1.0 mm
12 N·m
(1.2 kgf·m,
8.7 lbf·ft)

6 x 1.0 mm
12 N·m
(1.2 kgf·m,
8.7 lbf·ft)

09474_ACUCAR_G0030

Fig. 128 Oil pump assembly mounting—2007–08 RL models

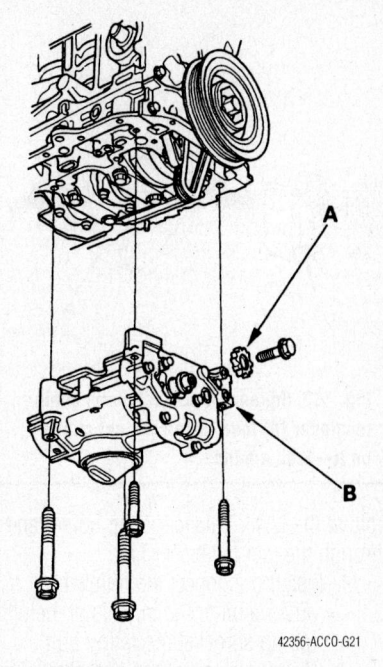

42356-ACCO-G21

Fig. 130 Exploded view of the oil pump sprocket (A) and oil pump (B)—2.4L engine

9. Remove the Crankshaft Position (CKP) sensor.

10. Remove the rocker arm oil control solenoid/oil filter assembly.

11. Remove the oil pan.

12. Remove the oil screen then remove the oil pump.

To install:

13. Remove the old oil seal from the oil pump.

14. Gently tap in the new oil seal until the seal driver bottoms on the pump.

15. Remove any old liquid gasket from the oil pump mating surfaces, bolts, and bolt holes.

16. Clean and dry the oil pump mating surfaces.

17. Apply liquid gasket evenly to the engine block mating surface of the oil pump.

➡**Do not install the parts if 4 minutes or more have elapsed since applying liquid gasket. Instead, re-apply liquid gasket after removing the old residue.**

18. Grease the lip of the oil seal, and apply oil to the new O-ring.

19. Install the dowel pins, then align the inner rotor with the crankshaft and install the oil pump. Tighten the bolts to 9 ft. lbs. (12 Nm).

20. Clean the excess grease off the crankshaft, and check the seal for distortion.

21. Apply oil to the new O-ring, install the oil screen.

22. Install the oil pan.

23. Install the remaining components in the reverse order of removal.

24. After assembly, wait at least 30 minutes before filling the engine with oil.

TSX Models

See Figures 129 through 132.

1. Raise and safely support the vehicle.

2. Drain the engine oil.

3. Turn the crankshaft to position the No. 1 piston at Top Dead Center (TDC) on the compression stroke.

4. Remove or disconnect the following:
- Oil pan
- Oil pump chain tensioner, and discard

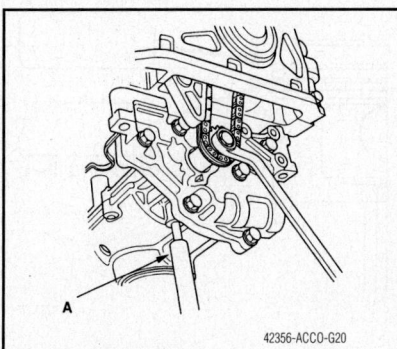

42356-ACCO-G20

Fig. 129 Insert a 6mm pin into the maintenance hole in the lower balancer shaft holder and through the rear balancer shaft

5. Secure the rear balancer shaft by inserting a 6mm pin into the maintenance hole in the lower balancer shaft holder and through the rear balancer shaft
- Oil pump sprocket mounting bolt
- Oil pump sprocket and oil pump

To install:

6. make sure the No. 1 piston is still at TDC.

7. Align the dowel pin on the rear balancer shaft with the mark on the oil pump.

8. Secure the rear balancer shaft by inserting a 6mm pin into the maintenance

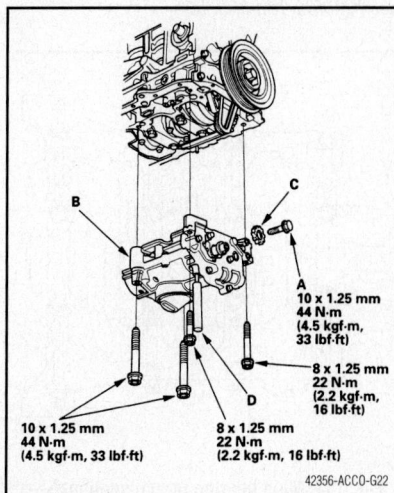

A
10 x 1.25 mm
44 N·m
(4.5 kgf·m,
33 lbf·ft)

8 x 1.25 mm
22 N·m
(2.2 kgf·m,
16 lbf·ft)

10 x 1.25 mm
44 N·m
(4.5 kgf·m, 33 lbf·ft)

8 x 1.25 mm
22 N·m
(2.2 kgf·m, 16 lbf·ft)

42356-ACCO-G22

Fig. 131 Oil pump tightening specifications—2.4L engine

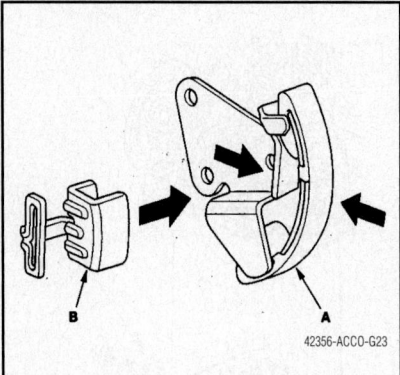

Fig. 132 Squeeze a new oil pump chain tensioner (A) then install the set clip (B) on it—2.4L engine

hole in the lower balancer shaft holder and through the rear balancer shaft.

9. Install or connect the following:
   - Engine oil to the threads of the oil pump sprocket mounting bolt
   - Oil pump and pump sprocket, loosely. Remove the 6mm pin.
   - Oil pump mounting bolts and tighten as shown in the illustration
10. Squeeze a new oil pump chain tensioner then install the set clip on it. The set clip is provided with the oil pump chain tensioner
    - New oil pump chain tensioner and tighten the bolts to 8.7 ft. lbs. (12 Nm). Remove the set clip from the tensioner.
    - Oil pan
11. Fill the crankcase with the proper amount of new engine oil.

MAIN BEARING TORQUE SEQUENCE

See Figures 133 and 134.

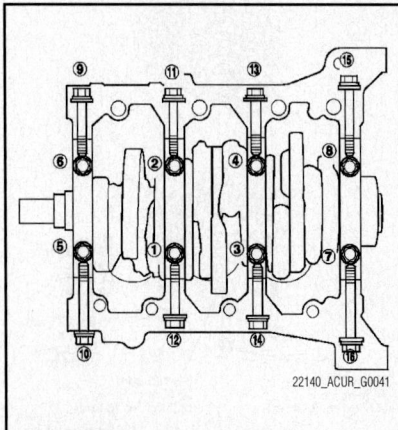

Fig. 133 Main bearing torque sequence— 2007–08 TL and RL Models

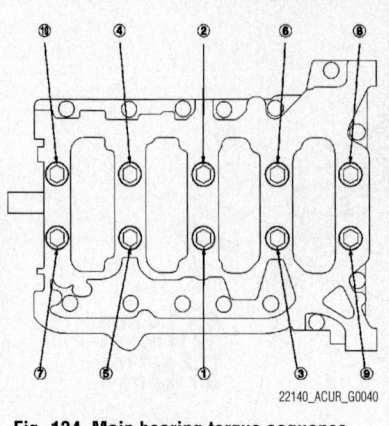

Fig. 134 Main bearing torque sequence— 2007–08 TSX Models

PISTON AND RING

POSITIONING

See Figures 135 through 137.

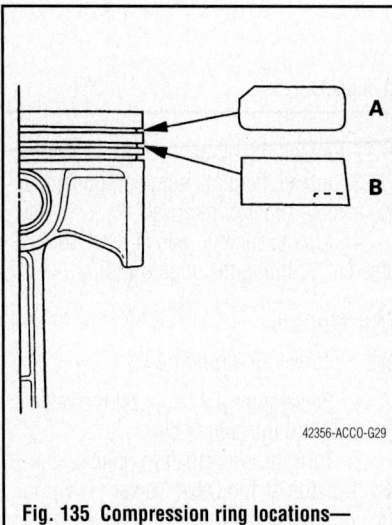

Fig. 135 Compression ring locations— 2.4L (K24A2) engines

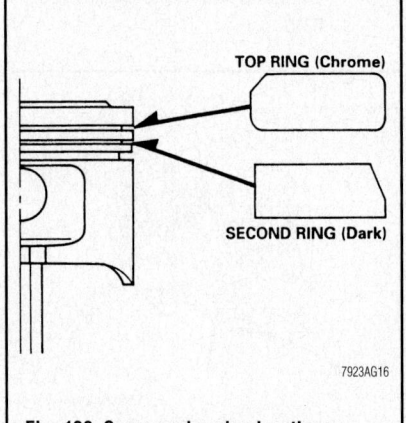

Fig. 136 Compression ring locations— 3.5L engine

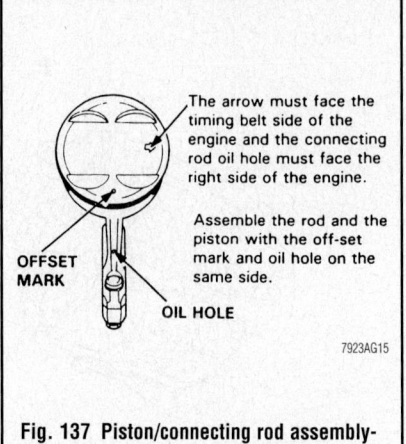

The arrow must face the timing belt side of the engine and the connecting rod oil hole must face the right side of the engine.

Assemble the rod and the piston with the off-set mark and oil hole on the same side.

OFFSET MARK

OIL HOLE

Fig. 137 Piston/connecting rod assembly-to-engine orientation—3.5L engine

REAR MAIN SEAL

REMOVAL & INSTALLATION

1. Remove or disconnect the following:
   - Transaxle
   - Clutch (if equipped)
   - Flywheel/Driveplate
   - Crankshaft seal by prying it out of the retainer

To install:

2. Install or connect the following:
   - Clean engine oil to the lip of the new seal
   - New seal into the retainer using an appropriate seal driver
   - Flywheel/Driveplate
   - Clutch (if equipped)
   - Transaxle

ROCKER ARMS/SHAFTS

REMOVAL & INSTALLATION

TL & RL Models

See Figures 138 through 140.

1. Remove the cylinder head cover.
2. Loosen the adjusting screws (A).
3. Remove the bolts and the rocker arm assembly as follows:

   a. Step 1: Unscrew the rocker shaft mounting bolts two turns at a time, in a crisscross pattern, to prevent damaging the valves or rocker arm assembly.

   b. Step 2: When removing the rocker arm assembly, do not remove the rocker shaft mounting bolts. The bolts will keep the springs and the rocker arms on the shafts

To install:

4. Set the rocker arm assembly in place, and loosely install the bolts. Make sure that

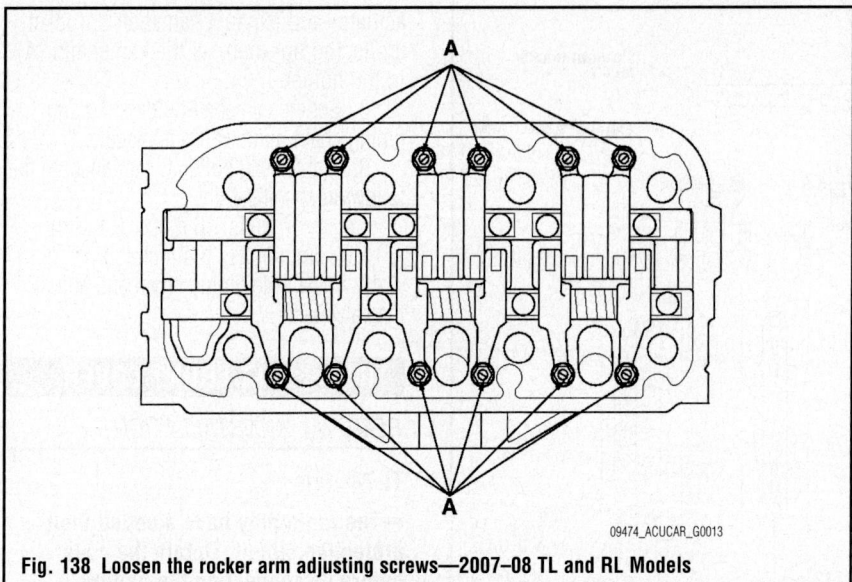

Fig. 138 Loosen the rocker arm adjusting screws—2007–08 TL and RL Models

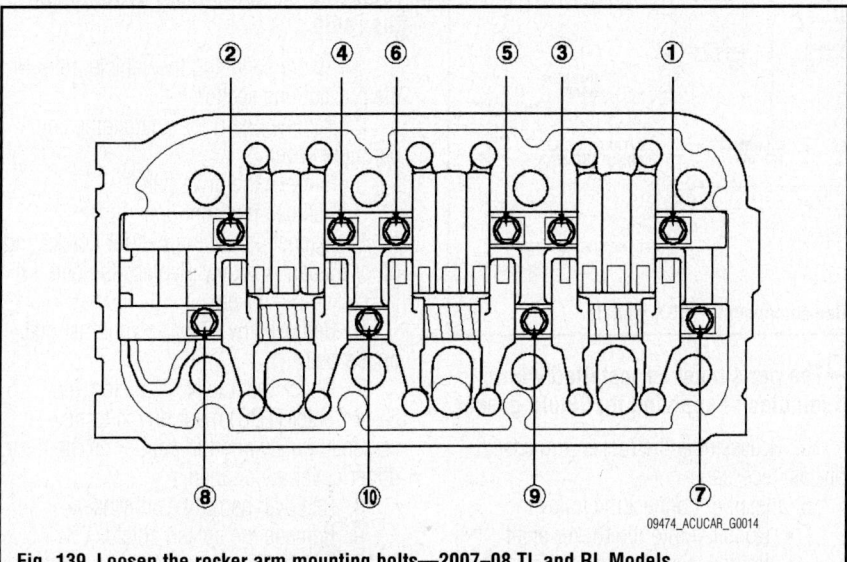

Fig. 139 Loosen the rocker arm mounting bolts—2007–08 TL and RL Models

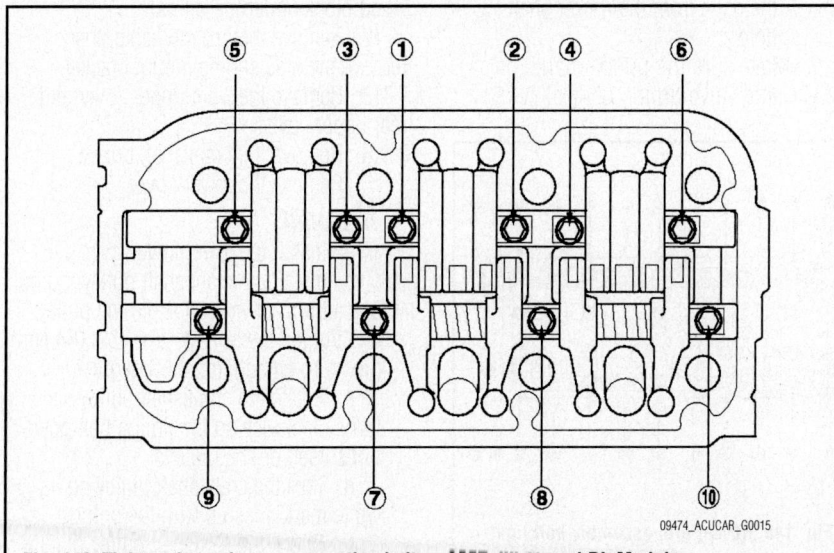

Fig. 140 Tighten the rocker arm mounting bolts—2007–08 TL and RL Models

the rocker arms are properly positioned on the valve stems.

5. Tighten each bolt two turns at a time in the sequence shown to ensure that the rockers do not bind on the valves. Tighten to a final torque of 17 ft. lbs. (24 Nm).

➡ **Apply new engine oil to the threads and flange of the exhaust rocker shaft mounting bolts.**

6. Install the cylinder head cover.

TSX Models

See Figures 141 through 145.

1. Remove or disconnect the following:
   • Timing chain
   • Loosen the rocker arm adjusting screws
   • Camshaft holder bolts, two turns at a time in sequence
   • Timing chain guide (B), camshaft and camshafts

2. Insert the bolts (A) into the rocker shaft holder, then remove the rocker arm assembly (B)

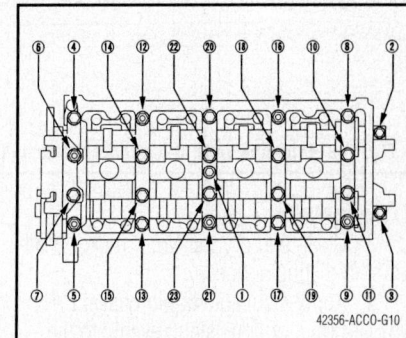

Fig. 141 Camshaft holder bolt loosening sequence—TSX models

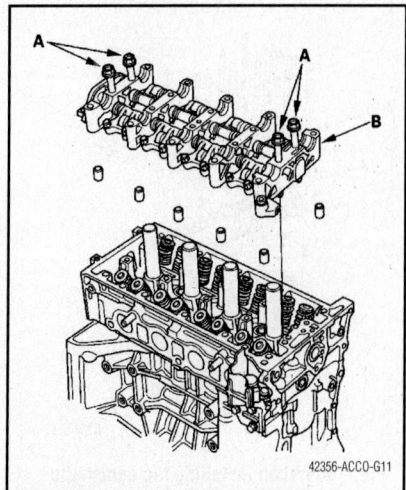

Fig. 142 Insert the bolts (A) into the rocker shaft holder, then remove the rocker arm assembly (B)—TSX models

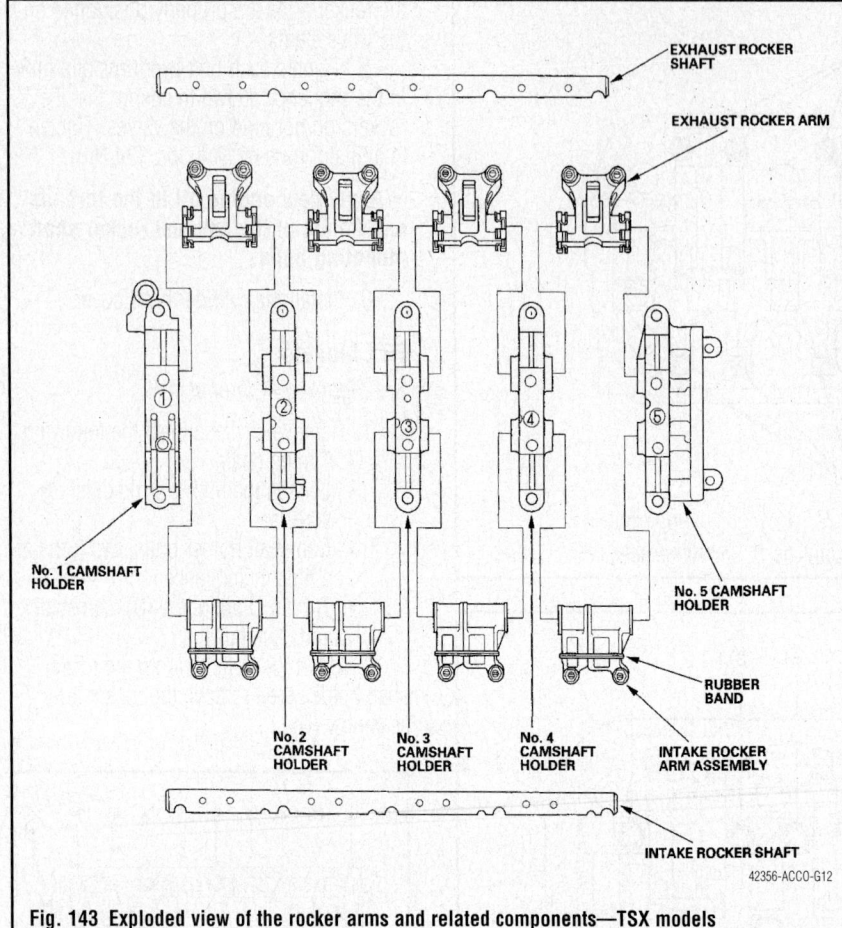

Fig. 143 Exploded view of the rocker arms and related components—TSX models

To install:

3. Clean and dry the No. 5 rocker shaft holding mating surface.

4. Apply a suitable liquid gasket P/N 08718-0009, or equivalent, evenly to the cylinder head mating surface of the No. 5 rocker shaft holder.

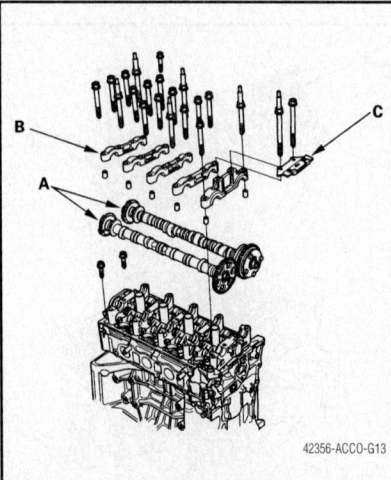

Fig. 144 When installing the camshafts (A) make sure the punch marks on the VTC actuator and exhaust cam sprockets are facing up—TSX models

➡**The parts must be installed within 5 minutes of applying the liquid gasket.**

5. Reassemble the rocker arm assembly, as necessary.

6. Install or connect the following
- Bolts (A) into the rocker shaft holder, then the rocker arm assembly on the cylinder head. Remove the bolts from the rocker shaft holder.

7. Make sure the punch marks on the variable valve timing control (VTC)

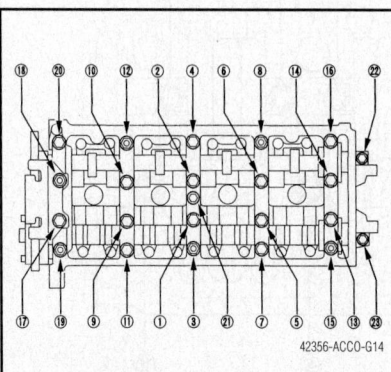

Fig. 145 Rocker arm assembly bolt tightening sequence—TSX models

actuator and exhaust camshaft sprocket are facing up, then set the camshafts (A) in the holder.

8. Set the camshaft holders (B) and timing chain guide B (C) in place.

9. Tighten the bolts, in sequence, to the following specification:
   a. 8mm bolts: 16 ft. lbs. (22 Nm)
   b. 6mm bolts: 104 inch lbs. (12 Nm)
10. Install the timing chain and adjust the valve lash.

TIMING BELT FRONT COVER

REMOVAL & INSTALLATION

TL Models

➡**The radio may have a coded theft protection circuit. Obtain the code before disconnecting the battery, removing the radio fuse, or removing the radio.**

1. Before servicing the vehicle, refer to the precautions section.

2. Remove the right side engine compartment cover.

3. Remove the drive belt.

4. Remove the front upper cover.

5. Inspect the timing belt for cracks and oil or coolant soaking. Replace the belt if it is cracked or soaked.

6. Remove any oil or solvent that gets on the belt.

7. Check that the No. 1 piston Top Dead Center (TDC) mark on the front camshaft pulley and the pointer on the front upper cover are aligned.

8. Remove the right front wheel.

9. Remove the splash shield.

10. Remove the drive belt.

11. Support the engine with a jack and wood block under the oil pan.

12. Remove the ground cable, then remove the side engine mount bracket.

13. Remove the front upper cover and rear upper cover.

14. Remove the crankshaft pulley.

15. Remove the lower cover.

To install:

a. Install the lower cover.

b. Install the crankshaft pulley, do not use an impact wrench. Hold the pulley and tighten the bolt to 47 ft. lbs. (64 Nm) plus turn an additional 60 degrees.

c. Rotate the crankshaft pulley six turns clockwise so the timing belt positions itself on the pulleys.

d. Turn the crankshaft pulley so its white mark lines up with the pointer.

e. Check the camshaft pulley marks, they should be as illustrated.

f. Install the front upper cover and rear upper cover.

g. Install the side engine mount bracket, then tighten the bolts in the numbered sequence shown.

16. Install the remaining components in the reverse order of removal.

RL Models

The radio may have a coded theft protection circuit. Obtain the code before disconnecting the battery, removing the radio fuse, or removing the radio.

1. Before servicing the vehicle, refer to the precautions section.

2. Remove the right upper fender trim.

3. Check that the No. 1 piston Top Dead Center (TDC) mark on the front camshaft pulley and the pointer on the front upper cover are aligned.

4. Remove the right front wheel.

5. Remove the splash shield.

6. Remove the drive belt.

7. Remove the drive belt auto-tensioner.

8. Support the engine with a jack and wood block under the oil pan.

9. Remove the ground cable, then remove the side engine mount bracket.

10. Remove the front upper cover and rear upper cover.

11. Remove the crankshaft pulley.

12. Remove the lower cover.

To install:

13. Install the lower cover.

14. Install the crankshaft pulley, do not use an impact wrench. Hold the pulley and tighten the bolt to 47 ft. lbs. (64 Nm) plus turn an additional 60 degrees.

15. Rotate the crankshaft pulley six turns clockwise so the timing belt positions itself on the pulleys.

16. Turn the crankshaft pulley so its white mark lines up with the pointer.

17. Check the camshaft pulley marks, they should be as illustrated.

18. Install the front upper cover and rear upper cover.

19. Install the side engine mount bracket, then tighten the bolts in the numbered sequence shown.

20. Install the remaining components in the reverse order of removal.

TIMING BELT AND SPROCKETS

REMOVAL & INSTALLATION

TL Models

See Figures 146 through 158.

➡The radio may have a coded theft protection circuit. Obtain the code before

disconnecting the battery, removing the radio fuse, or removing the radio.

1. Before servicing the vehicle, refer to the precautions section.

2. Remove the right side engine compartment cover.

3. Remove the drive belt.

4. Remove the front upper cover.

5. Inspect the timing belt for cracks and oil or coolant soaking. Replace the belt if it is cracked or soaked.

6. Remove any oil or solvent that gets on the belt.

7. Check that the No. 1 piston Top Dead Center (TDC) mark on the front camshaft pulley and the pointer on the front upper cover are aligned.

8. Remove the right front wheel.

9. Remove the splash shield.

10. Remove the drive belt.

11. Support the engine with a jack and wood block under the oil pan.

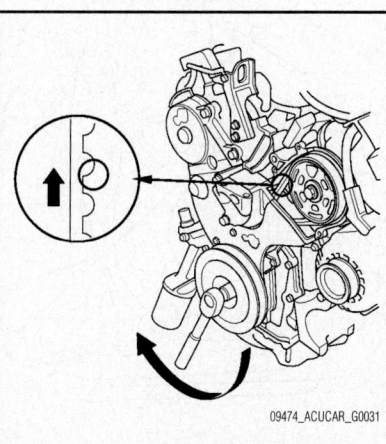

Fig. 146 Inspect the timing belt for cracks and oil or coolant soaking—2007–08 TL model

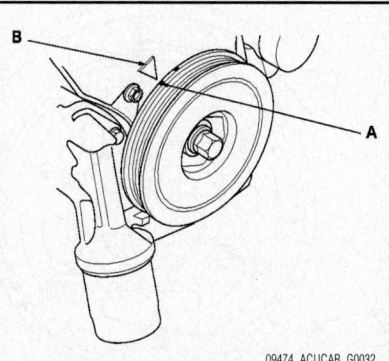

Fig. 147 Check that the No. 1 piston Top Dead Center (TDC) mark on the front camshaft pulley and the pointer on the front upper cover are aligned—2007–08 TL model

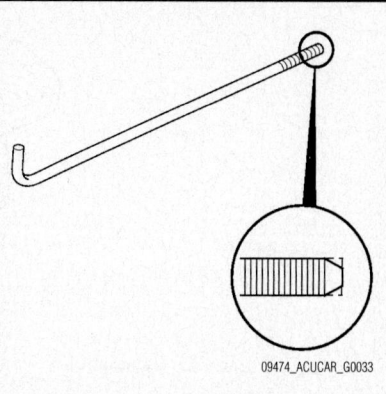

Fig. 148 Grind the battery tray clamp bolt as shown—2007–08 TL model

12. Remove the ground cable, then remove the side engine mount bracket.

13. Remove the front upper cover and rear upper cover.

14. Remove the crankshaft pulley.

15. Remove the lower cover.

16. Remove one of the battery clamp bolts from the battery tray, and grind the end of it as shown.

17. Screw the battery clamp bolt in as shown to hold the timing belt adjuster in its current position. Only tighten it by hand, do not use a wrench.

18. Remove the engine mount bracket.

19. Remove the idler pulley bolt and idler pulley, then remove the timing belt.

To install:

20. If reusing the old belt, perform the following steps.

a. Clean the timing belt pulleys, timing belt guide plate, and the upper and lower covers.

b. Set the timing belt drive pulley to TDC by aligning the TDC mark on the tooth of the timing belt drive pulley with the pointer on the oil pump.

c. Set the camshaft pulleys to TDC by aligning the TDC marks on the camshaft pulleys with the pointers on the back covers.

d. Apply liquid thread lock (P/N 08713-0001) to the idler pulley bolt, then loosely install the idler pulley bolt as the pulley can move and does not come off.

e. Install the timing belt in a counterclockwise sequence as follows:

- Drive pulley
- Idler pulley
- Front camshaft pulley
- Water pump pulley
- Rear camshaft pulley
- Adjusting pulley

f. Tighten the idler pulley bolt 33 ft. lbs. (44 Nm).

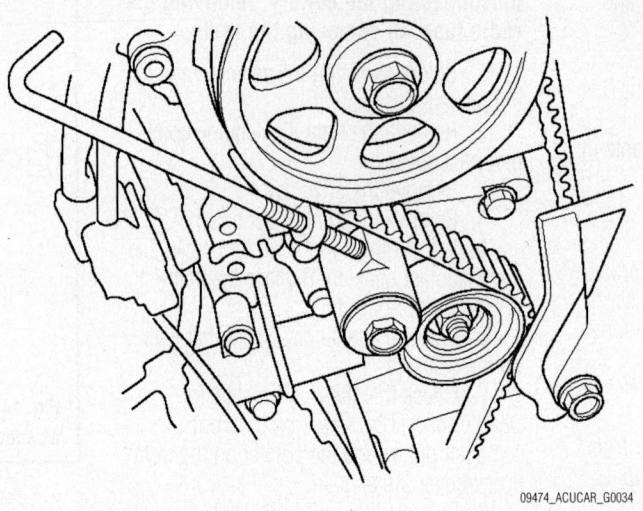

09474_ACUCAR_G0034

Fig. 149 Screw the battery clamp bolt in as shown to hold the timing belt adjuster in its current position—2007–08 TL model

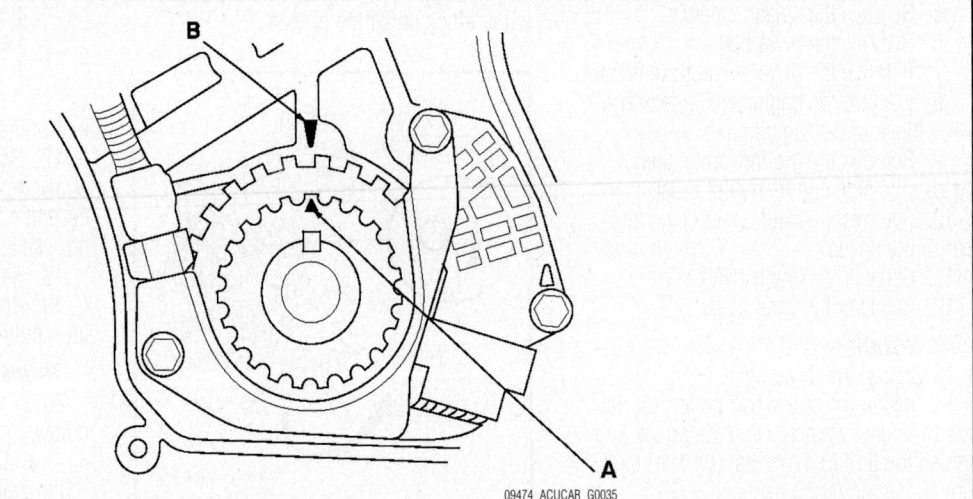

09474_ACUCAR_G0035

Fig. 150 Set the timing belt drive pulley to TDC by aligning the TDC mark on the tooth of the timing belt drive pulley with the pointer on the oil pump—2007–08 TL model

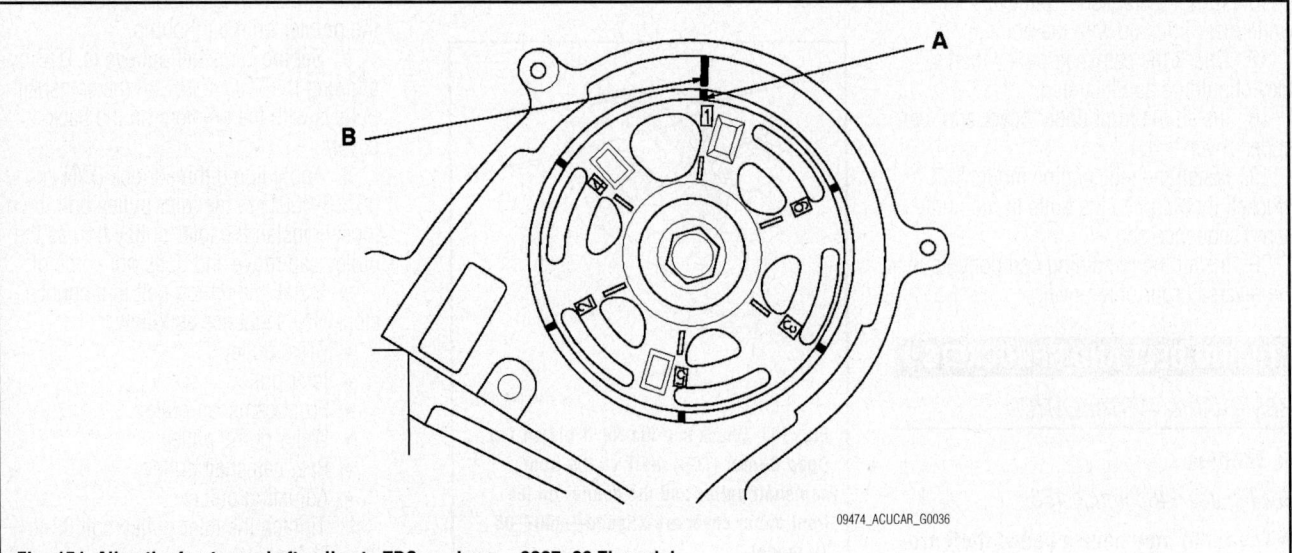

09474_ACUCAR_G0036

Fig. 151 Align the front camshaft pulley to TDC as shown—2007–08 TL model

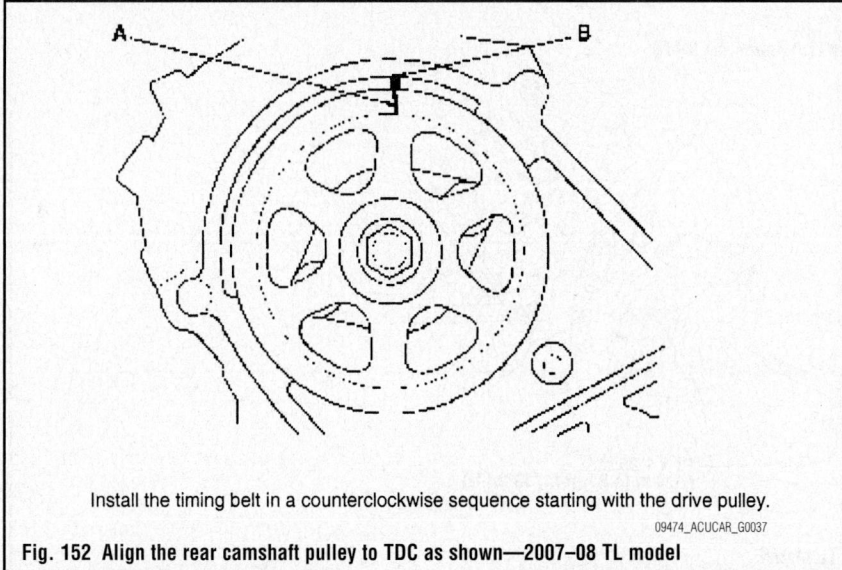

Install the timing belt in a counterclockwise sequence starting with the drive pulley.

09474_ACUCAR_G0037

Fig. 152 Align the rear camshaft pulley to TDC as shown—2007–08 TL model

g. Remove the battery clamp bolt from the back cover.

h. Install the engine mount bracket.

i. Install the timing belt guide plate as shown.

j. Install the lower cover.

k. Install the crankshaft pulley, do not use an impact wrench. Hold the pulley and tighten the bolt to 47 ft. lbs. (64 Nm) plus turn an additional 60 degrees.

l. Rotate the crankshaft pulley six turns clockwise so the timing belt positions itself on the pulleys.

m. Turn the crankshaft pulley so its white mark lines up with the pointer.

n. Check the camshaft pulley marks, they should be as illustrated.

o. Install the front upper cover and rear upper cover.

p. Install the side engine mount bracket, then tighten the bolts in the numbered sequence shown.

21. If installing a new belt, perform the following steps.

a. Clean the timing belt pulleys, timing belt guide plate, and the upper and lower covers.

b. Set the timing belt drive pulley to TDC by aligning the TDC mark on the tooth of the timing belt drive pulley with the pointer on the oil pump.

c. Set the camshaft pulleys to TDC by aligning the TDC marks on the camshaft pulleys with the pointers on the back covers.

d. Remove the auto-tensioner.

e. Align the holes on the rod and body on the tensioner.

f. Use a press to slowly compress the tensioner and insert a 0.08 inch (2 mm) pin through the body and rod. Make sure the pressure to compress the tensioner does not exceed 2,200 lbs. ft. (9,800 N).

g. Install the tensioner with the pin still in place.

h. Screw the battery clamp bolt in as shown to hold the timing belt adjuster in

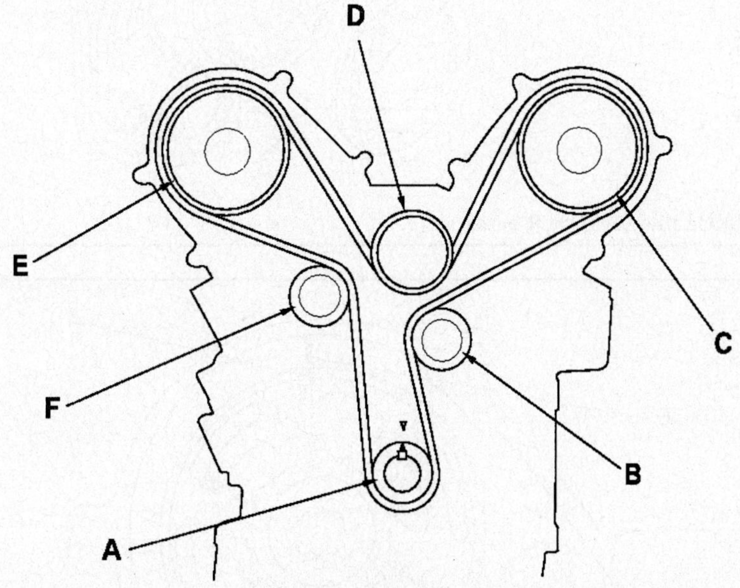

A. Drive pulley
B. Idler pulley
C. Front camshaft pulley
D. Water pump pulley
E. Rear camshaft pulley
F. Adjusting pulley

09474_ACUCAR_G0038

Fig. 153 Timing belt routing—2007–08 TL model

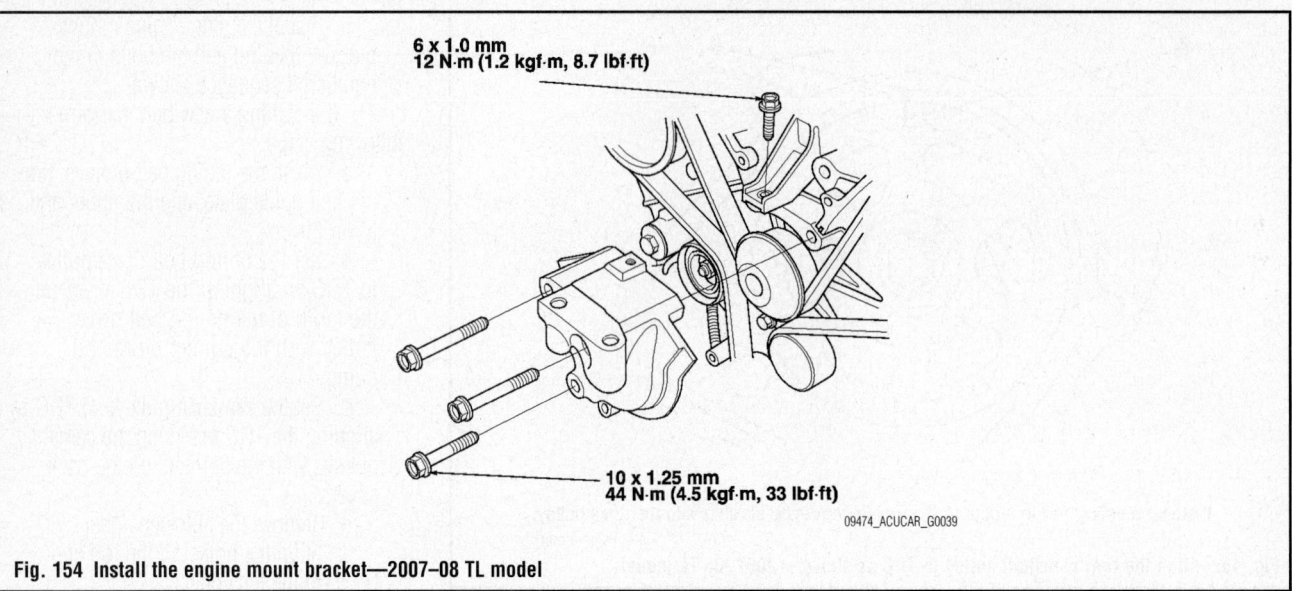

Fig. 154 Install the engine mount bracket—2007–08 TL model

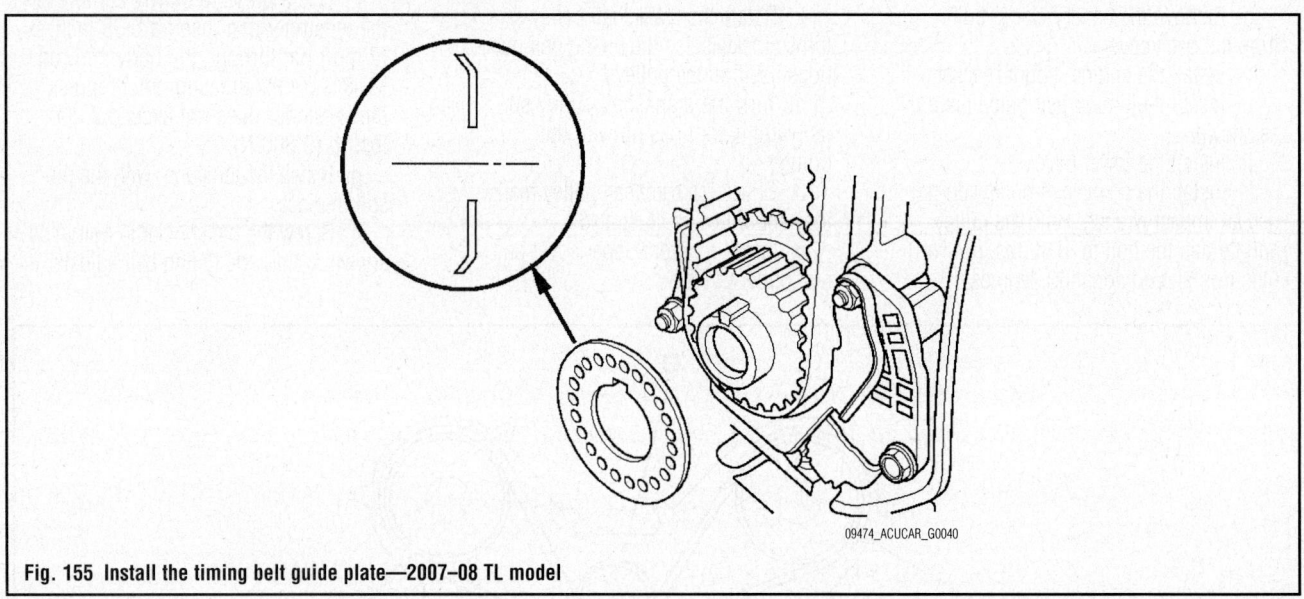

Fig. 155 Install the timing belt guide plate—2007–08 TL model

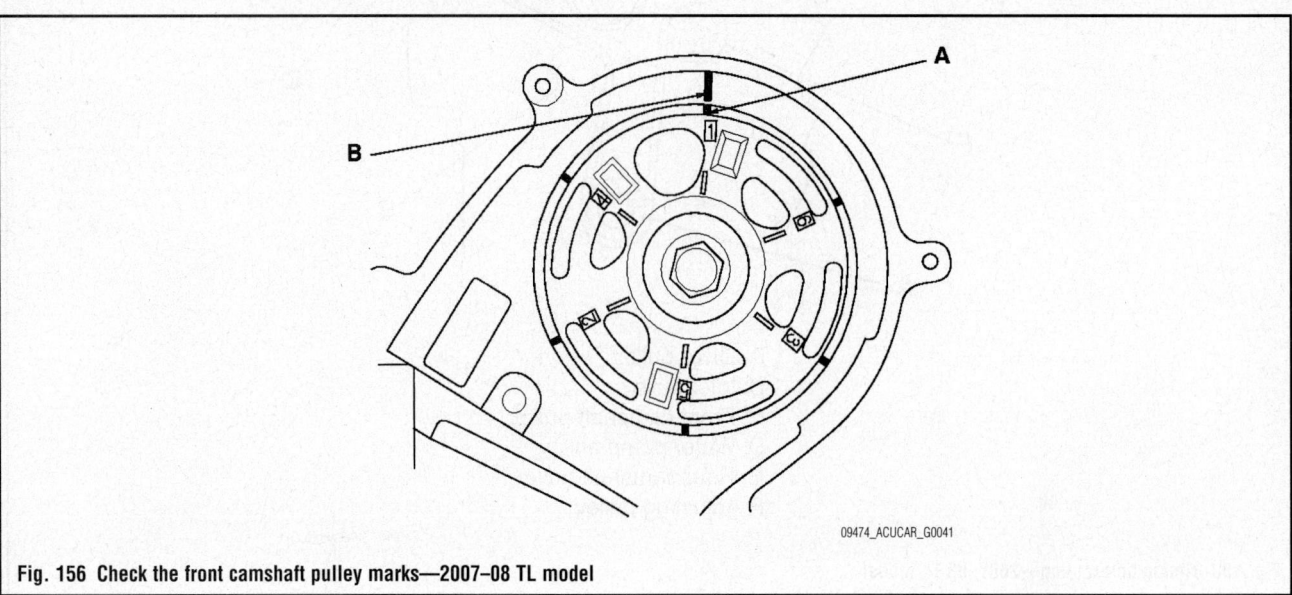

Fig. 156 Check the front camshaft pulley marks—2007–08 TL model

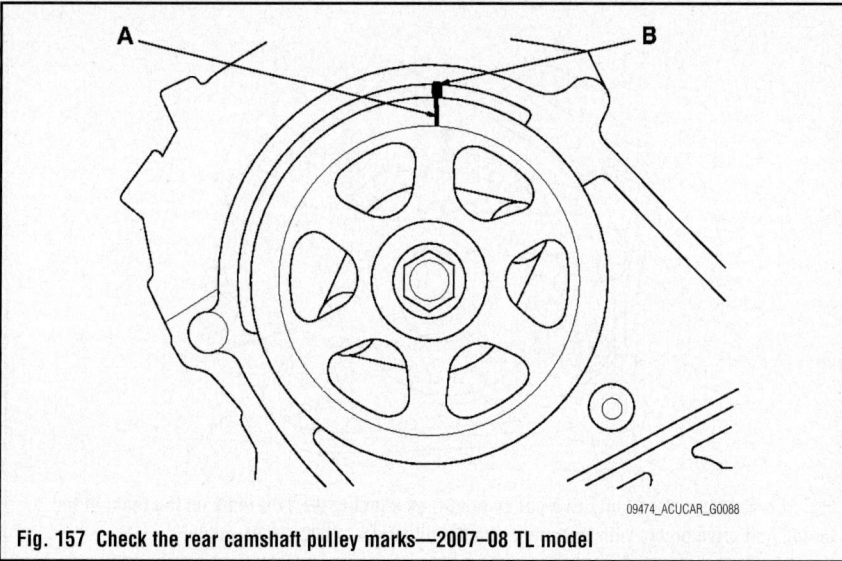

Fig. 157 Check the rear camshaft pulley marks—2007–08 TL model

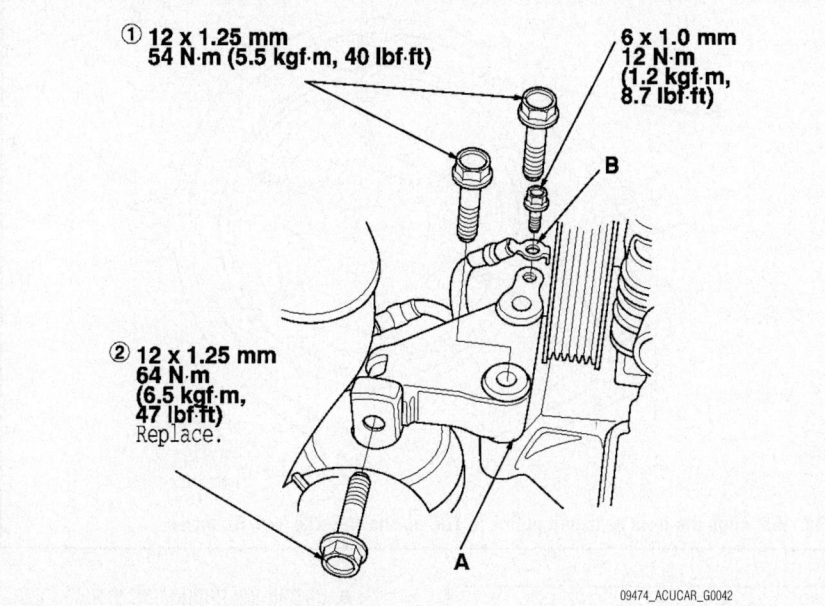

Fig. 158 Install the side engine mount bracket, then tighten the bolts in the numbered sequence shown—2007–08 TL model

its current position. Only tighten it by hand, do not use a wrench.

i. Loosely install the idler pulley with a new bolt so the pulley is free to move but does not come off.

j. Install the timing belt in a counter-clockwise sequence as follows:
- Drive pulley
- Idler pulley
- Front camshaft pulley
- Water pump pulley
- Rear camshaft pulley
- Adjusting pulley

k. Tighten the idler pulley bolt 33 ft. lbs. (44 Nm).

l. Remove the pin from the ten-sioner.

m. Remove the battery clamp bolt from the back cover.

n. Install the engine mount bracket.

o. Install the timing belt guide plate as shown.

p. Install the lower cover.

q. Install the crankshaft pulley, do not use an impact wrench. Hold the pulley and tighten the bolt to 47 ft. lbs. (64 Nm) plus turn an additional 60 degrees.

r. Rotate the crankshaft pulley six turns clockwise so the timing belt positions itself on the pulleys.

s. Turn the crankshaft pulley so its white mark lines up with the pointer.

t. Check the camshaft pulley marks, they should be as illustrated.

u. Install the front upper cover and rear upper cover.

v. Install the side engine mount bracket, then tighten the bolts in the numbered sequence shown.

22. Install the remaining components in the reverse order of removal.

RL Models

See Figures 159 through 170.

The radio may have a coded theft protection circuit. Obtain the code before disconnecting the battery, removing the radio fuse, or removing the radio.

1. Before servicing the vehicle, refer to the precautions section.

2. Remove the right upper fender trim.

3. Check that the No. 1 piston Top Dead Center (TDC) mark on the front camshaft pulley and the pointer on the front upper cover are aligned.

4. Remove the right front wheel.

5. Remove the splash shield.

6. Remove the drive belt.

7. Remove the drive belt auto-tensioner.

8. Support the engine with a jack and wood block under the oil pan.

9. Remove the ground cable, then remove the side engine mount bracket.

10. Remove the front upper cover and rear upper cover.

11. Remove the crankshaft pulley.

12. Remove the lower cover.

13. Remove one of the battery clamp bolts from the battery tray, and grind the end of it as shown.

14. Screw the battery clamp bolt in as shown to hold the timing belt adjuster in its current position. Only tighten it by hand, do not use a wrench.

15. Remove the engine mount bracket.

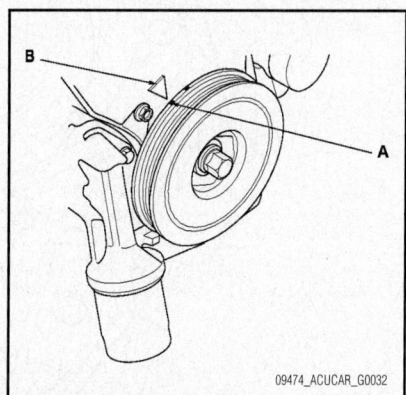

Fig. 159 Check that the No. 1 piston Top Dead Center (TDC) mark on the front camshaft pulley and the pointer on the front upper cover are aligned—2007–08 RL model

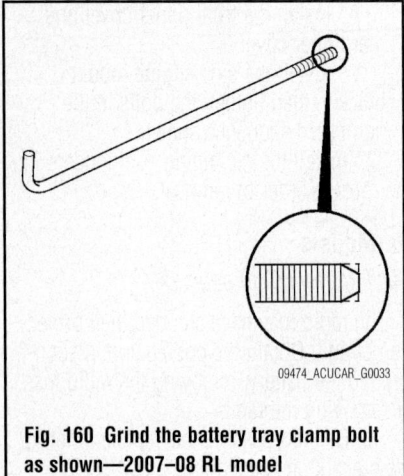

Fig. 160 Grind the battery tray clamp bolt as shown—2007–08 RL model

16. Remove the idler pulley bolt and idler pulley, then remove the timing belt.

To install:

17. If reusing the old belt, perform the following steps.

a. Clean the timing belt pulleys, timing belt guide plate, and the upper and lower covers.

b. Set the timing belt drive pulley to TDC by aligning the TDC mark on the tooth of the timing belt drive pulley with the pointer on the oil pump.

c. Set the camshaft pulleys to TDC by aligning the TDC marks on the camshaft pulleys with the pointers on the back covers.

d. Apply liquid thread lock (P/N 08713-0001) to the idler pulley bolt, then loosely install the idler pulley bolt as the pulley can move and does not come off.

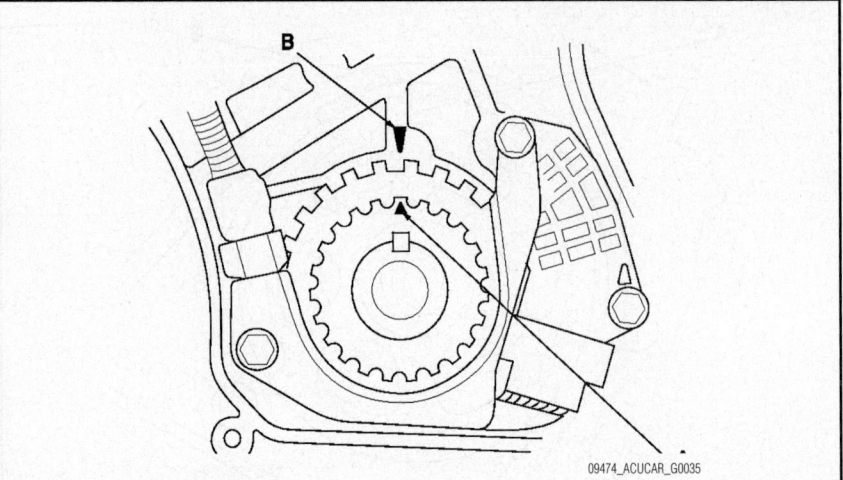

Fig. 162 Set the timing belt drive pulley to TDC by aligning the TDC mark on the tooth of the timing belt drive pulley with the pointer on the oil pump—2007–08 RL model

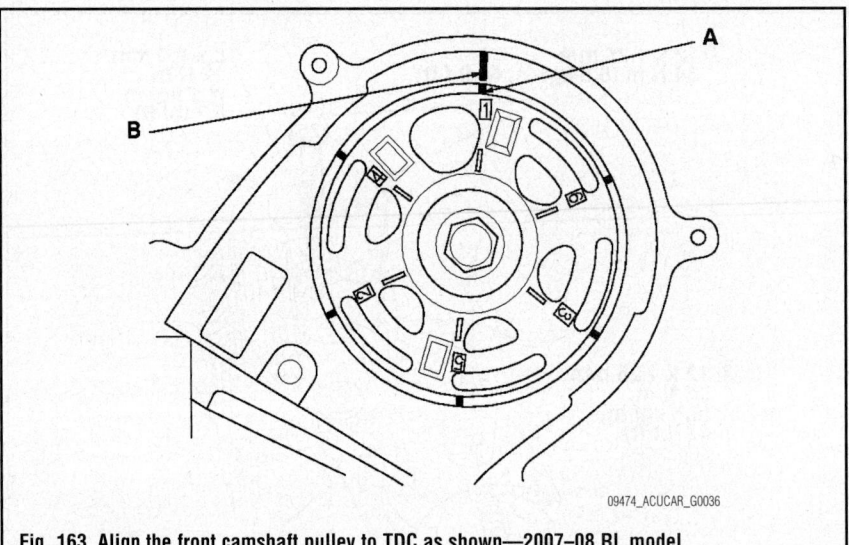

Fig. 163 Align the front camshaft pulley to TDC as shown—2007–08 RL model

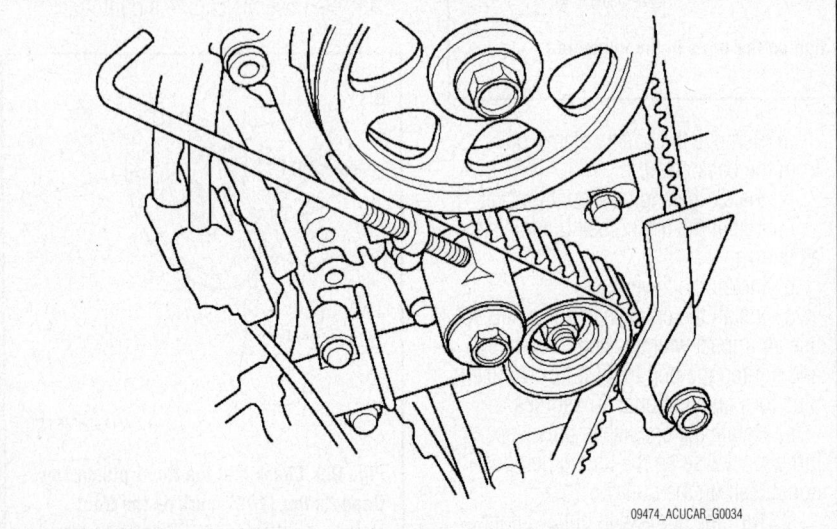

Fig. 161 Screw the battery clamp bolt in as shown to hold the timing belt adjuster in its current position—2007–08 RL model

e. Install the timing belt in a counter-clockwise sequence as follows:
- Drive pulley
- Idler pulley
- Front camshaft pulley
- Water pump pulley
- Rear camshaft pulley
- Adjusting pulley

f. Tighten the idler pulley bolt 33 ft. lbs. (44 Nm).

g. Remove the battery clamp bolt from the back cover.

h. Install the engine mount bracket and tighten the upper bolt to 9 ft. lbs. (12 Nm) and the lower bolts to 33 ft. lbs. (44 Nm).

i. Install the timing belt guide plate as shown.

j. Install the crankshaft pulley, do not use an impact wrench. Hold the pulley and tighten the bolt to 47 ft. lbs. (64 Nm) plus turn an additional 60 degrees.

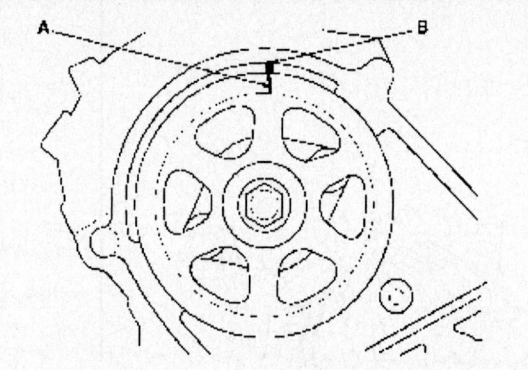

Install the timing belt in a counterclockwise sequence starting with the drive pulley.

09474_ACUCAR_G0037

Fig. 164 Align the rear camshaft pulley to TDC as shown—2007–08 RL model

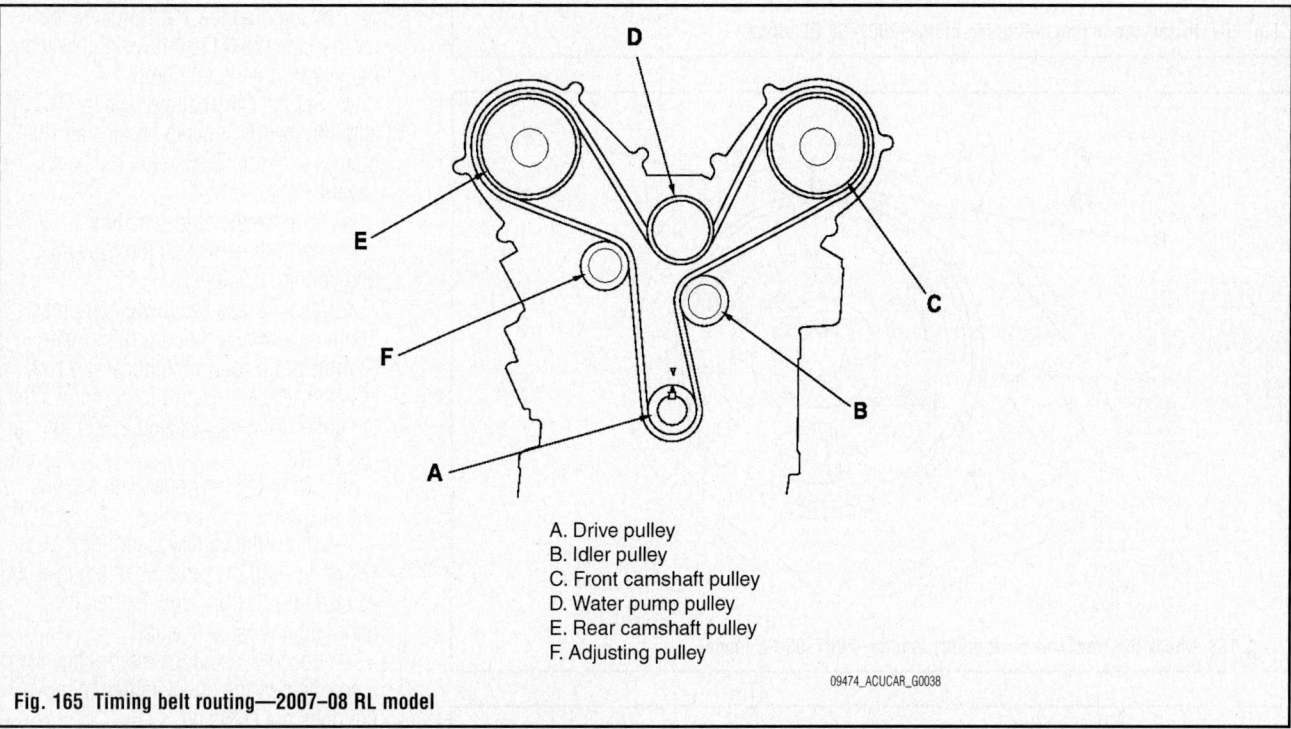

A. Drive pulley
B. Idler pulley
C. Front camshaft pulley
D. Water pump pulley
E. Rear camshaft pulley
F. Adjusting pulley

09474_ACUCAR_G0038

Fig. 165 Timing belt routing—2007–08 RL model

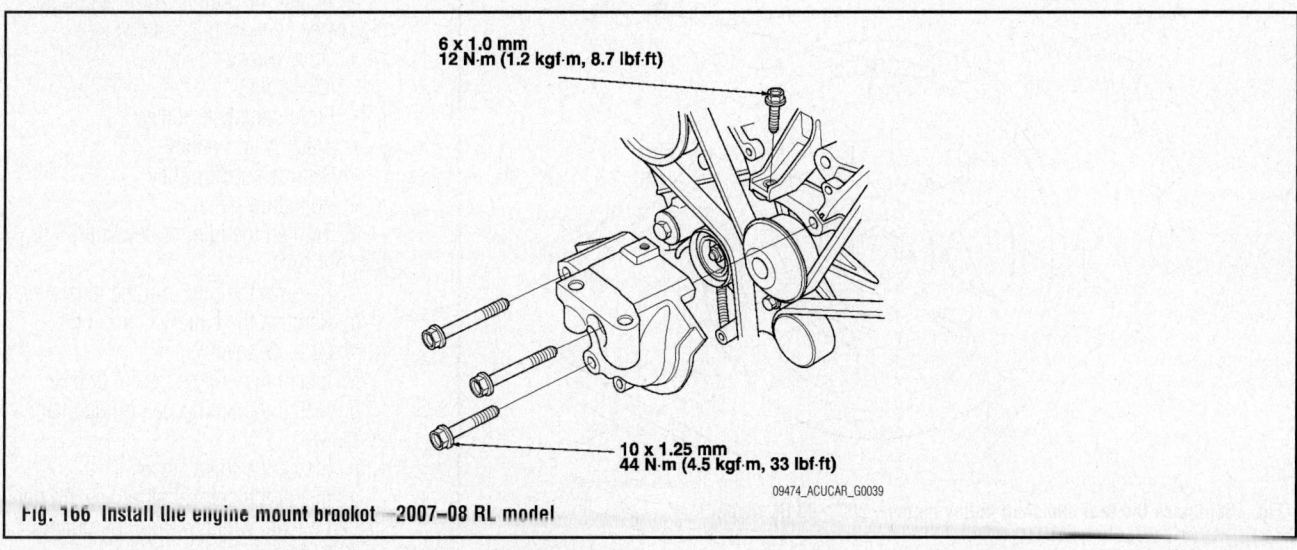

6 x 1.0 mm
12 N·m (1.2 kgf·m, 8.7 lbf·ft)

10 x 1.25 mm
44 N·m (4.5 kgf·m, 33 lbf·ft)

09474_ACUCAR_G0039

Fig. 166 Install the engine mount bracket—2007–08 RL model

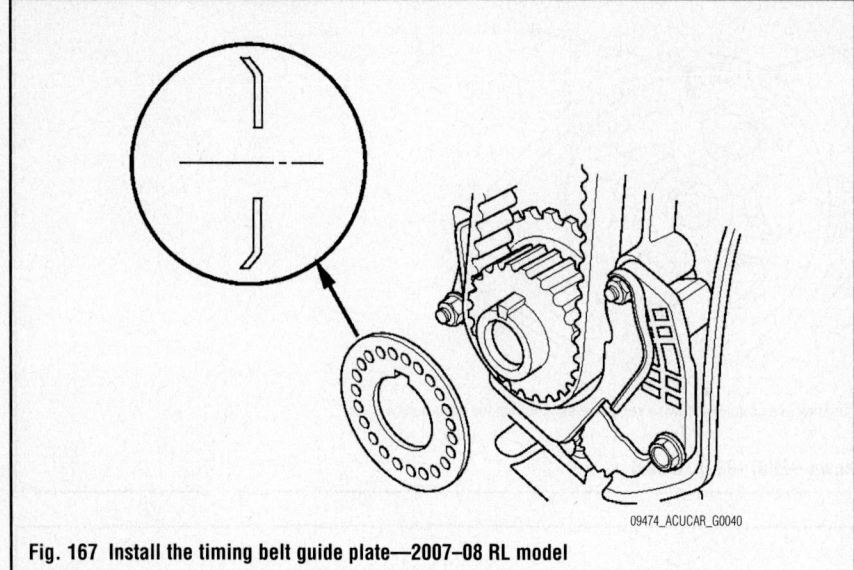

Fig. 167 Install the timing belt guide plate—2007–08 RL model

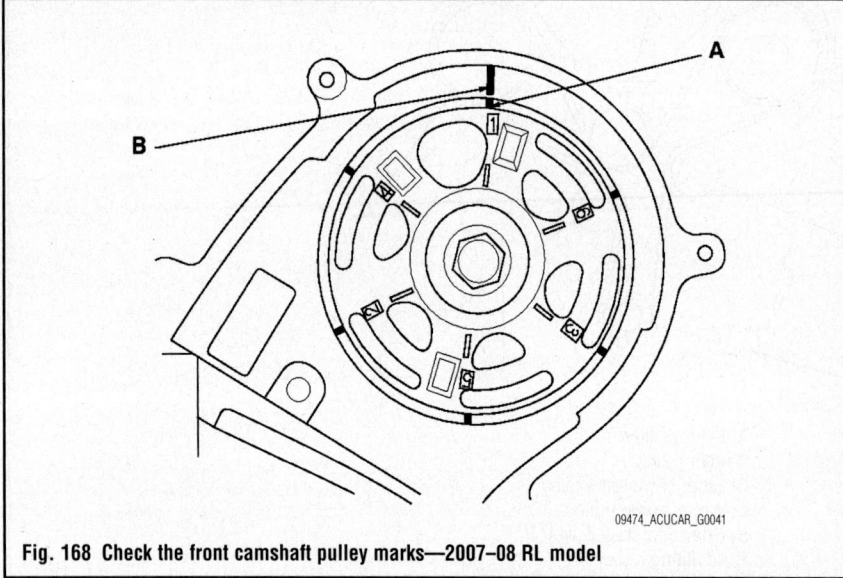

Fig. 168 Check the front camshaft pulley marks—2007–08 RL model

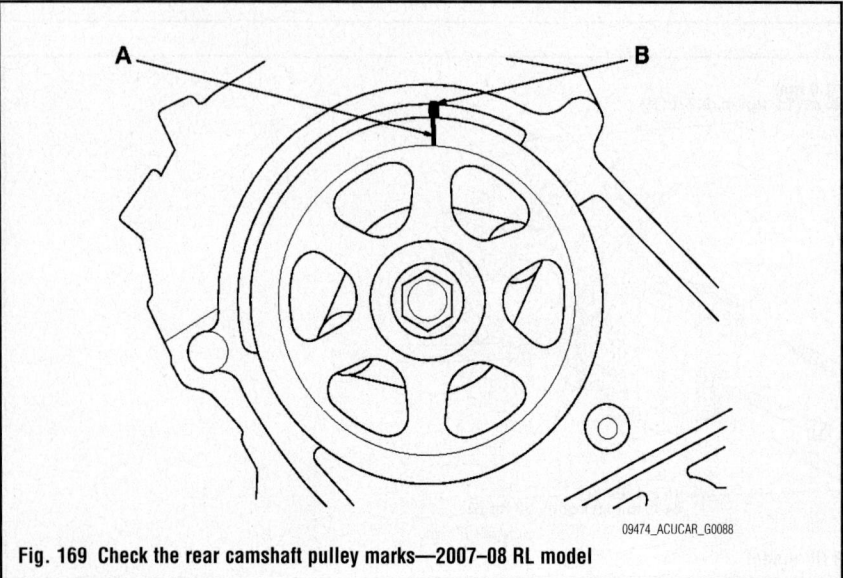

Fig. 169 Check the rear camshaft pulley marks—2007–08 RL model

k. Install the front upper cover and rear upper cover.

l. Rotate the crankshaft pulley six turns clockwise so the timing belt positions itself on the pulleys.

m. Turn the crankshaft pulley so its white mark lines up with the pointer.

n. Check the camshaft pulley marks, they should be as illustrated.

o. Install the side engine mount bracket, then tighten the bolts in the numbered sequence shown.

18. If installing a new belt, perform the following steps.

a. Clean the timing belt pulleys, timing belt guide plate, and the upper and lower covers.

b. Set the timing belt drive pulley to TDC by aligning the TDC mark on the tooth of the timing belt drive pulley with the pointer on the oil pump.

c. Set the camshaft pulleys to TDC by aligning the TDC marks on the camshaft pulleys with the pointers on the back covers.

d. Remove the auto-tensioner.

e. Align the holes on the rod and body on the tensioner.

f. Use a press to slowly compress the tensioner and insert a 0.08 inch (2 mm) pin through the body and rod. Make sure the pressure to compress the tensioner does not exceed 2,200 lbs. ft. (9,800 N).

g. Install the tensioner with the pin still in place.

h. Screw the battery clamp bolt in as shown to hold the timing belt adjuster in its current position. Only tighten it by hand, do not use a wrench.

i. Loosely install the idler pulley with a new bolt so the pulley is free to move but does not come off.

j. Install the timing belt in a counter-clockwise sequence as follows:

• Drive pulley
• Idler pulley
• Front camshaft pulley
• Water pump pulley
• Rear camshaft pulley
• Adjusting pulley

k. Tighten the idler pulley bolt 33 ft. lbs. (44 Nm).

l. Remove the pin from the tensioner.

m. Remove the battery clamp bolt from the back cover.

n. Install the engine mount bracket.

o. Install the timing belt guide plate as shown.

p. Install the lower cover.

q. Install the crankshaft pulley, do not use an impact wrench. Hold the pulley

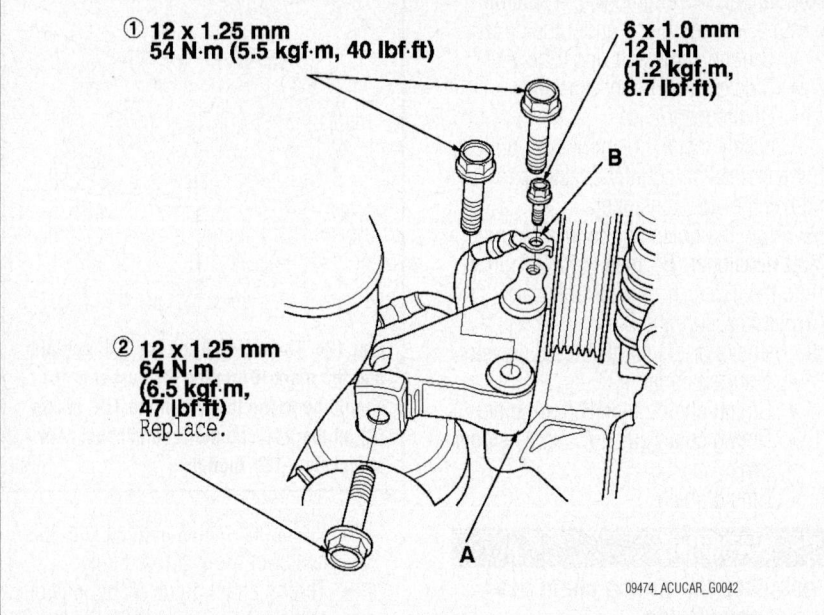

Fig. 170 Install the side engine mount bracket, then tighten the bolts in the numbered sequence shown—2007–08 RL model

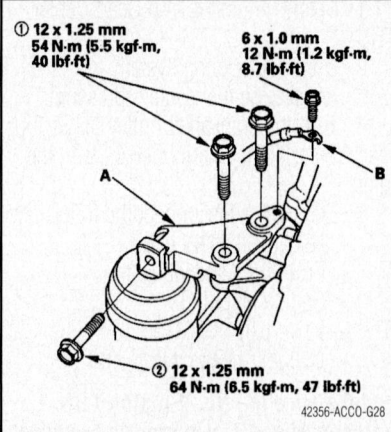

Fig. 171 Tighten the upper bracket upper bolt/nuts in the proper order to the correct specification—TSX models

and tighten the bolt to 47 ft. lbs. (64 Nm) plus turn an additional 60 degrees.

r. Rotate the crankshaft pulley six turns clockwise so the timing belt positions itself on the pulleys.

s. Turn the crankshaft pulley so its white mark lines up with the pointer.

t. Check the camshaft pulley marks, they should be as illustrated.

u. Install the front upper cover and rear upper cover.

v. Install the side engine mount bracket, then tighten the bolts in the numbered sequence shown.

19. Install the remaining components in the reverse order of removal.

TIMING CHAIN COVER AND SEAL

REMOVAL & INSTALLATION

TSX Models

See Figure 171.

1. Remove or disconnect the following:
   - Negative battery cable
   - Front tires and wheels
   - Splash shield
   - Drive belt
   - Cylinder head cover

➡ **Make sure the No. 1 piston TDC marks on the VTC actuator and exhaust camshaft are aligned.**

   - Crankshaft pulley
   - Crankshaft Position (CKP) sensor connector

   - Variable Valve Timing Control (VTC) oil control solenoid valve connector
   - VTC oil control solenoid valve

2. Support the engine with a suitable jack with a wooden block under the oil pan.
   - Ground cable and upper bracket
   - Side engine mount bracket
   - Chain cover/case

To install:

3. Inspect the chain cover seal for damage and replace if necessary. Clean and dry the chain cover mating surfaces.

4. Install or connect the following:
   - Liquid gasket, P/N 08718-0009 evenly to the cylinder block mating surface of the timing chain cover and the inner threads of the holes
   - Liquid gasket to the cylinder block upper surface contact areas on the chain cover and the oil pan mating surface of the chain cover in the inner threads of the holes

➡ **Make sure to install the components within 5 minutes of applying the sealer.**

   - New O-ring the timing chain cover. Set the edge of the cover to the edge of the oil pan, then install the cover on the engine block. Tighten the retainers to 104 inch lbs. (12 Nm).

➡ **When installing the chain case, do not slide the bottom surface on the oil pan mounting surface.**

   - Side engine mounting bracket and tighten the retainers to 33 ft. lbs. (44 Nm)
   - Upper bracket, then tighten the bolts/nuts as shown in the illustration
   - Ground cable
   - VTC oil control solenoid valve
   - CKP sensor and VTC oil control solenoid valve connectors
   - Crankshaft pulley
   - Cylinder head cover
   - Drive belt
   - Splash shield
   - Wheels and tires

5. Fill the engine cooling system and connect the negative battery cable.

TIMING CHAIN AND SPROCKETS

REMOVAL & INSTALLATION

TSX Models

See Figures 172 through 177.

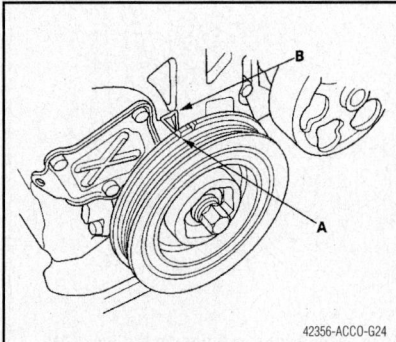

Fig. 172 Turn the crankshaft pulley so the TDC mark (A) is aligned with the pointer (B)—TSX models

1. Set the engine to Top Dead Center (TDC).
2. Drain the cooling system.
3. Relieve the fuel system pressure.
4. Turn the crankshaft pulley so its Top Dead Center (TDC) mark lines up with the pointer.
5. Remove or disconnect the following:
- Negative battery cable
- Front tires and wheels
- Splash shield
- Drive belt
- Cylinder head cover

➡Make sure the No. 1 piston TDC marks on the VTC actuator and exhaust camshaft are aligned.

- Crankshaft pulley
- Crankshaft Position (CKP) sensor connector
- Variable Valve Timing Control (VTC) oil control solenoid valve connector
- VTC oil control solenoid valve

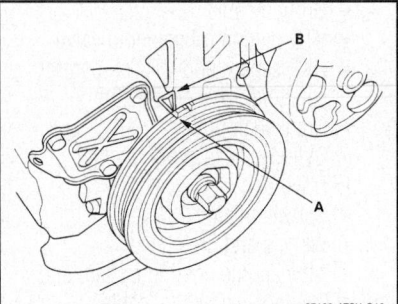

Fig. 173 Turn the crankshaft pulley so its Top Dead Center (TDC) mark (A) line up with the pointer (B)—TSX models

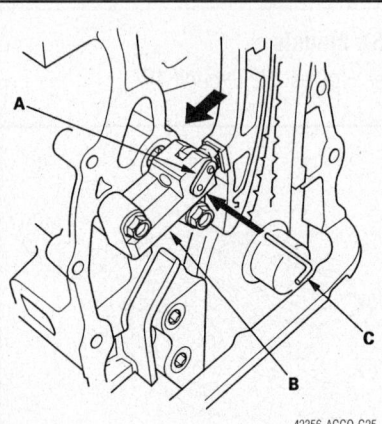

Fig. 174 Align the holes on the lock (A) and the auto-tensioner (B), then place a 1.5mm pin into the holes. Turn the crank-shaft clockwise to secure the pin—TSX models

6. Support the engine with a suitable jack with a wooden block under the oil pan.
- Ground cable and upper bracket
- Side engine mount bracket
- Chain cover/case
7. Loosely install the crankshaft pulley. Turn the crankshaft counterclockwise to compress the auto-tensioner.
8. Align the holes on the lock (A) and the auto-tensioner (B), then place a 1.5mm pin into the holes. Turn the crankshaft clockwise to secure the pin.
9. Remove or disconnect the following:
- Auto-tensioner
- Timing chain guide B (top guide)
- Timing chain guide A and tensioner arm
- Timing chain

✳✳ WARNING

Do not place the timing chain near any magnetic fields.

To install:

10. Set the crankshaft to TDC. Align the TDC mark (A) on the crankshaft sprocket with the pointer (B) on the cylinder block.
11. Set the camshafts to TDC. The punch mark (A) on the VTC actuator and the punch mark (B) on the exhaust camshaft (C) should be at the top. Align the TDC marks (C) on the VTC actuator and exhaust camshaft sprockets.
12. Install or connect the following:
- Timing chain the crankshaft sprocket with the colored link of the chain aligned with the mark on the crank sprocket
- Timing chain on the VTC actuator and exhaust camshaft sprocket with

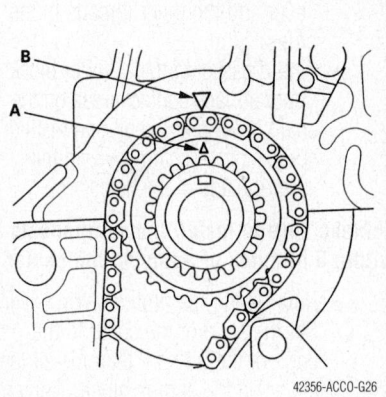

Fig. 175 Set the crankshaft to TDC. Align the TDC mark (A) on the crankshaft sprocket with the pointer (B) on the cylinder block—TSX models

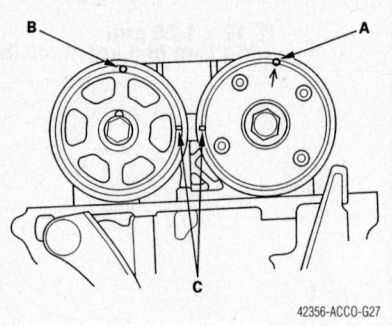

Fig. 176 The mark (A) on the VTC actuator and the mark (B) on the exhaust cam (C) should be at the top. Align the TDC marks (C) on the VTC actuator and exhaust cam sprockets—TSX models

the punch marks aligned with the center of the 2 colored links
- Timing chain guide A and tensioner arm. Tighten the guide bolts to 104 inch lbs. (12 Nm) and the tensioner arm retainer to 16 ft. lbs. (22 Nm).
- Auto-tensioner and tighten the bolts to 104 inch lbs. (12 Nm)
- Timing chain guide B and tighten the retainers to 16 ft. lbs. (22 Nm)

13. Remove the pin from the auto-tensioner.
14. Inspect the chain cover seal for damage and replace if necessary. Clean and dry the chain cover mating surfaces.
15. Install or connect the following:
- Liquid gasket, P/N 08718-0009 evenly to the cylinder block mating surface of the timing chain cover and the inner threads of the holes
- Liquid gasket to the cylinder block upper surface contact areas on the chain cover and the oil pan mating surface of the chain cover in the inner threads of the holes

➡Make sure to install the components within 5 minutes of applying the sealer.

- New O-ring the timing chain cover. Set the edge of the cover to the edge of the oil pan, then install the cover on the engine block. Tighten the retainers to 104 inch lbs. (12 Nm).

➡When installing the chain case, do not slide the bottom surface on the oil pan mounting surface.

- Side engine mounting bracket and tighten the retainers to 33 ft. lbs. (44 Nm)
- Upper bracket, then tighten the bolts/nuts as shown in the illustration

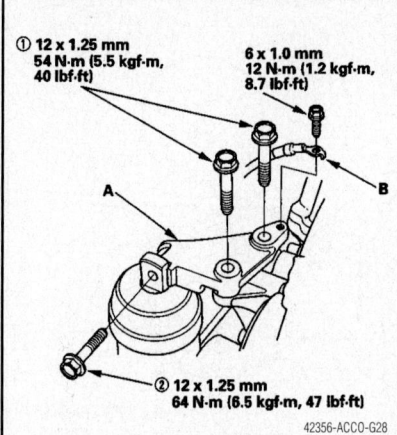

Fig. 177 Tighten the upper bracket upper bolt/nuts in the proper order to the correct specification—TSX models

- Ground cable
- VTC oil control solenoid valve
- CKP sensor and VTC oil control solenoid valve connectors
- Crankshaft pulley
- Cylinder head cover
- Drive belt
- Splash shield
- Wheels and tires

16. Fill the engine cooling system and connect the negative battery cable.

VALVE COVERS

REMOVAL & INSTALLATION

TSX Models

See Figures 178 and 179.

1. Disconnect the negative battery cable.
2. Drain the engine oil.
3. Remove the engine cover.
4. Remove the four ignition coils, as outlined in the Engine Electrical Section.
5. Remove the two bolts securing the vacuum line.
6. Remove the bolt securing the power steering hose bracket.
7. Remove the dipstick and breather hose.
8. Remove the cylinder head cover.

To install:

9. Thoroughly clean the head cover gasket and the groove.
10. Install the head cover gasket in the groove of the cylinder head cover.
11. Check that the mating surfaces are clean and dry.
12. Apply liquid gasket, P/N 08717-0004, 08710-0001, 08718 0002, 08718-0003, or 08718-0009, on the chain case

and the No. 5 rocker shaft holder mating areas.

➡**Do not install components if too much time has passed after applying the liquid gasket (for P/N 08718-0002, no more than 4 minutes, for all others, no more than 5 minutes). Instead, remove the old residue and reapply the liquid gasket.**

13. Set the spark plug seals on the spark plug tubes. Place the cylinder head cover on the cylinder head, then slide the cover slightly back and forth to seat the head cover gasket.
14. Inspect the cover washers. Replace any washer that is damaged or deteriorated.
15. Tighten the cylinder head cover bolts in two or three steps. In the final step tighten all bolts, in sequence, to 8.7 ft. lbs. (12 Nm).

✳✳ WARNING

Wait at least 30 minutes before filling the engine with oil. Do not run the engine for at least 3 hours after installing the head cover.

16. Install the dipstick and breather hose.
17. Tighten the bolt securing the power steering hose bracket.
18. Tighten the two bolts securing the vacuum line.
19. Install the four ignition coils.
20. Check that all tubes, hoses, and connectors are installed correctly.
21. Install the engine cover.

VALVE LASH

ADJUSTMENT

TL & RL Models

See Figures 180 through 186.

➡**Adjust the valves only when the cylinder head temperature is less than 100°F (38°C).**

1. Before servicing the vehicle, refer to the precautions section.
2. Remove the right side engine compartment cover.
3. Remove the cylinder head covers.

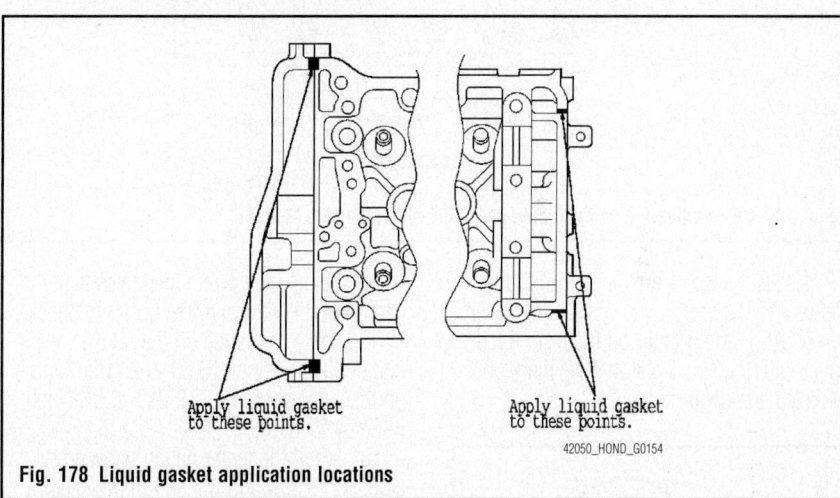

Fig. 178 Liquid gasket application locations

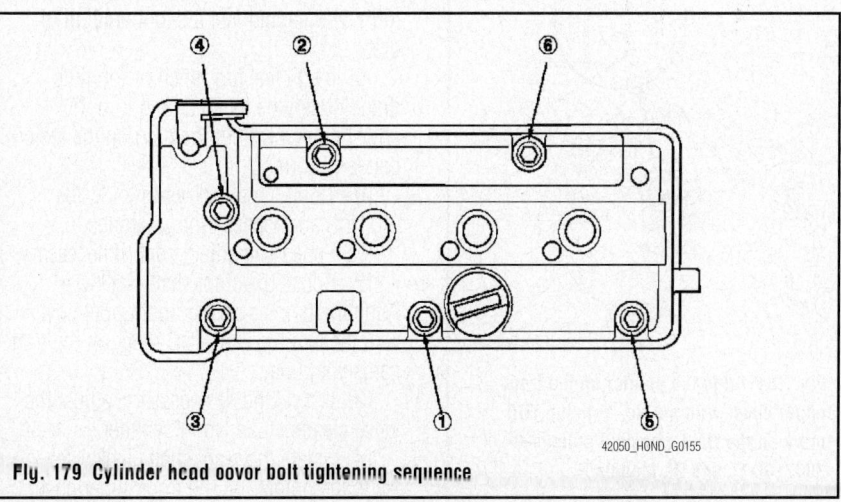

Fig. 179 Cylinder head cover bolt tightening sequence

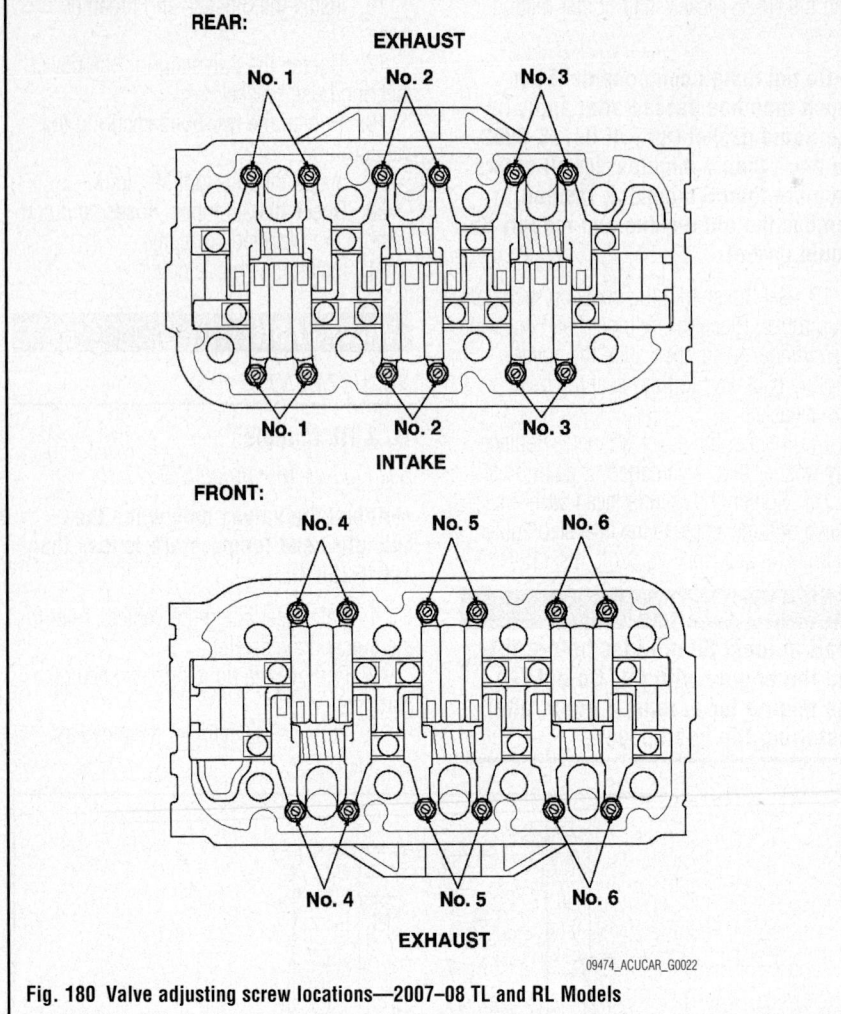

REAR:

EXHAUST

No. 1   No. 2   No. 3

No. 1   No. 2   No. 3

INTAKE

FRONT:

No. 4   No. 5   No. 6

No. 4   No. 5   No. 6

EXHAUST

09474_ACUCAR_G0022

Fig. 180 Valve adjusting screw locations—2007–08 TL and RL Models

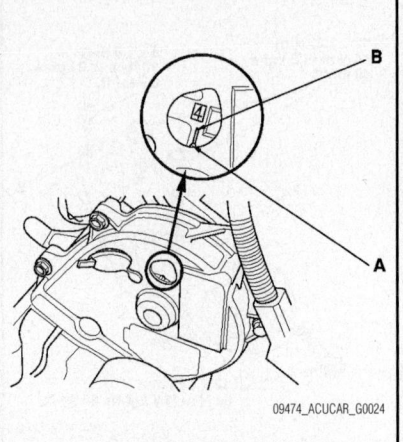

09474_ACUCAR_G0024

Fig. 182 Align the pointer on the front upper cover with the No. 4 piston TDC mark on the front camshaft pulley— 2007–08 TL and RL Models

4. Set the No. 1 piston at Top Dead Center (TDC).

5. Align the pointer on the front upper cover with the No. 1 piston TDC mark on the front camshaft pulley.

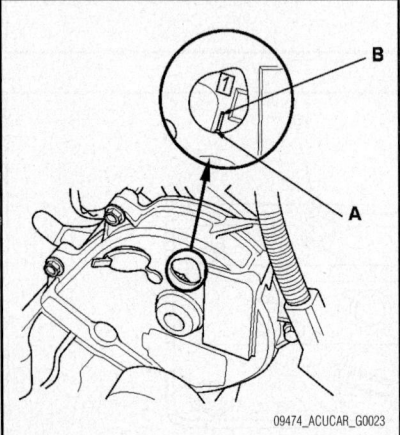

09474_ACUCAR_G0023

Fig. 181 Align the pointer on the front upper cover with the No. 1 piston TDC mark on the front camshaft pulley— 2007–08 TL and RL Models

6. Select the correct thickness feeler gauge for the valves you're going to check.

7. Valve clearance on the intake valves is 0.008–0.009 inch (0.20–0.24 mm) and on the exhaust side it is 0.011–0.013 inch (0.28–0.32 mm).

8. Insert the feeler gauge between the adjusting screw and the end of the valve stem on No. 1 cylinder and slide it back and forth; you should feel a slight amount of drag.

9. If you feel too much or too little drag, loosen the locknut, and turn the adjusting screw until the drag on the feeler gauge is correct.

10. Tighten the locknut to 14 ft. lbs. (20 Nm) and recheck the clearance.

11. Repeat the adjustment, if necessary.

12. Rotate the crankshaft clockwise. Align the pointer on the front upper cover with the No. 4 piston TDC mark on the front camshaft pulley.

13. Check and, if necessary, adjust the valve clearance on No. 4 cylinder.

14. Rotate the crankshaft clockwise. Align the pointer on the front upper cover

with the No. 2 piston TDC mark on the front camshaft pulley.

15. Check and, if necessary, adjust the valve clearance on No. 2 cylinder

16. Rotate the crankshaft clockwise. Align the pointer on the front upper cover with the No. 5 piston TDC mark on the front camshaft pulley.

17. Check and, if necessary, adjust the valve clearance on No. 5 cylinder.

18. Rotate the crankshaft clockwise. Align the pointer on the front upper cover with the No. 3 piston TDC mark on the front camshaft pulley.

Check and, if necessary, adjust the valve clearance on No. 3 cylinder

19. Rotate the crankshaft clockwise. Align the pointer on the front upper cover with the No. 6 piston TDC mark on the front camshaft pulley.

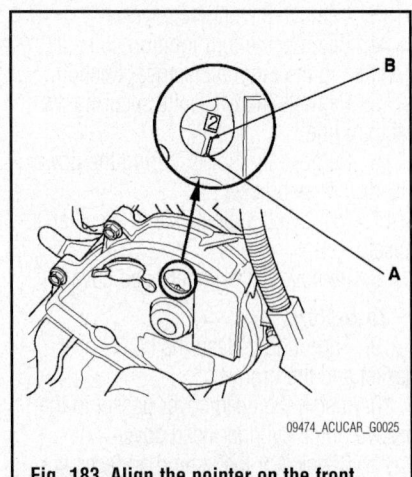

09474_ACUCAR_G0025

Fig. 183 Align the pointer on the front upper cover with the No. 2 piston TDC mark on the front camshaft pulley— 2007–08 TL and RL Models

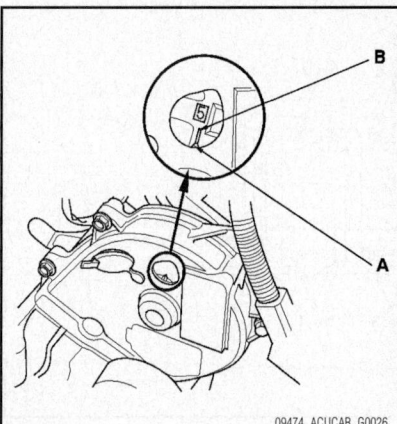

Fig. 184 Align the pointer on the front upper cover with the No. 5 piston TDC mark on the front camshaft pulley—2007–08 TL and RL Models

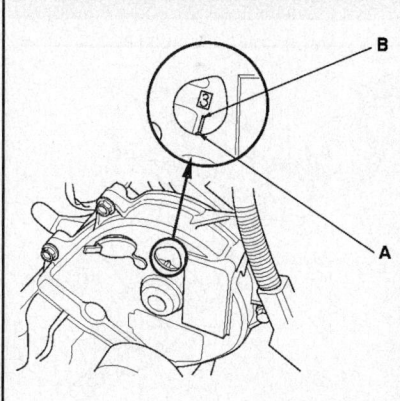

Fig. 185 Align the pointer on the front upper cover with the No. 3 piston TDC mark on the front camshaft pulley—2007–08 TL and RL Models

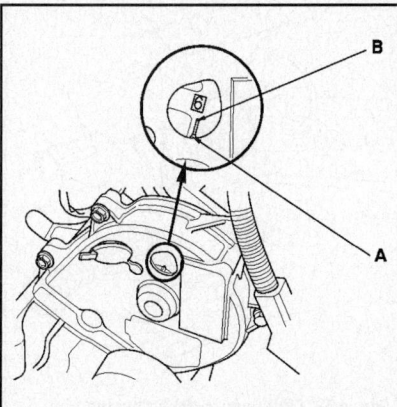

Fig. 186 Align the pointer on the front upper cover with the No. 6 piston TDC mark on the front camshaft pulley—2007–08 TL and RL Models

20. Check and, if necessary, adjust the valve clearance on No. 6 cylinder.

21. Install the cylinder head covers.

22. Install the right side engine compartment cover.

TSX Models

See Figures 187 and 188.

➡**The valve clearance should be adjusted when the engine is cold, the cylinder head temperature should be less than 100°F (38°C).**

➡**The radio may contain a coded theft protection circuit. Always obtain the code number before disconnecting the battery.**

1. Before servicing the vehicle, refer to the precautions in the beginning of this section.

2. Remove or disconnect the following:
   - Negative battery cable

➡**Label the wires before disconnecting them.**

   - Spark plug wires from the spark plugs
   - Positive Crankcase Ventilation (PCV) hose
   - Cylinder head cover. Replace the rubber seals if damaged or deteriorated.

3. Turn the engine to align the timing marks and set cylinder No.1 to TDC. The white mark on the crankshaft pulley should align with the pointer on the timing belt cover. The words **UP** embossed on the camshaft pulley should be aligned in the upward position. The marks on the edge of the pulley should be aligned with the cylinder head or the back cover upper edge.

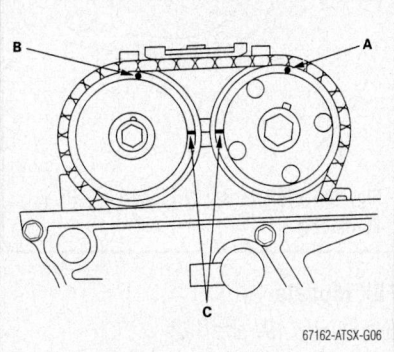

Fig. 187 The punch mark (A) on the VTC actuator and the punch mark (B) on the exhaust cam sprocket should be at the top. Align the TDC marks (C) on the VTC actuator and exhaust cam sprockets—2.4L engine

4. Adjust the valves on cylinder No. 1 by performing the following:
   a. Insert a feeler gauge in between the camshaft lobe and the rocker arm.

➡**The intake valve clearance specification is 0.008–0.010 in. (0.21–0.25mm). The exhaust valve clearance specification is 0.010–0.011 in. (0.25–0.29mm).**

   b. Loosen the locknut and turn the adjusting screw until the feeler gauge slides back and forth with a slight amount of drag.
   c. Tighten the locknut to 14 ft. lbs. (20 Nm) and recheck the valve clearance. Repeat the valve adjustment if necessary.

5. Rotate the crankshaft 180° counterclockwise (the camshaft pulleys will turn 90°) The **UP** arrow marks should be pointing to the exhaust side of the cylinder head.

6. Adjust the valves on cylinder No. 3 by performing the following:
   a. Insert a feeler gauge in between the camshaft lobe and the rocker arm.
   b. Loosen the locknut and turn the adjusting screw until the feeler gauge slides back and forth with a slight amount of drag.
   c. Tighten the locknut to 14 ft. lbs. (20 Nm) and recheck the valve clearance. Repeat the valve adjustment if necessary.

7. Rotate the crankshaft 180° counterclockwise (the camshaft pulleys will turn 90°) to bring No. 4 piston to TDC. The **UP** arrow marks should be pointing down, toward the crankshaft.

8. Adjust the valves on cylinder No. 4 by performing the following:
   a. Insert a feeler gauge in between the camshaft lobe and the rocker arm.
   b. Loosen the locknut and turn the adjusting screw until the feeler gauge slides back and forth with a slight amount of drag.
   c. Tighten the locknut to 14 ft. lbs. (20 Nm) and recheck the valve clearance. Repeat the valve adjustment if necessary.

9. Rotate the crankshaft 180° counterclockwise (the camshaft pulleys will turn 90°) to bring piston No. 2 to TDC. The **UP** arrow marks should be pointing to the intake side of the cylinder head.

10. Adjust the valves on cylinder No. 2 by performing the following:
   a. Insert a feeler gauge in between the camshaft lobe and the rocker arm.
   b. Loosen the locknut and turn the adjusting screw until the feeler gauge slides back and forth with a slight amount of drag.

c. Tighten the locknut to 14 ft. lbs. (20 Nm) and recheck the valve clearance. Repeat the valve adjustment if necessary.

11. Install the cylinder head cover gasket cover to the groove of the cylinder head cover. Before installing the gasket, thoroughly clean the seal and the groove. Seat the recesses for the camshaft first, then work it into the groove around the outside edges. Be sure the gasket is seated securely in the corners of the recesses.

12. Apply liquid gasket to the 4 corners of the recesses of the cylinder head cover gasket. Do not install the parts if 5 minutes or more have elapsed since applying liquid gasket. After assembly, wait at least 20 minutes before filling the engine with oil.

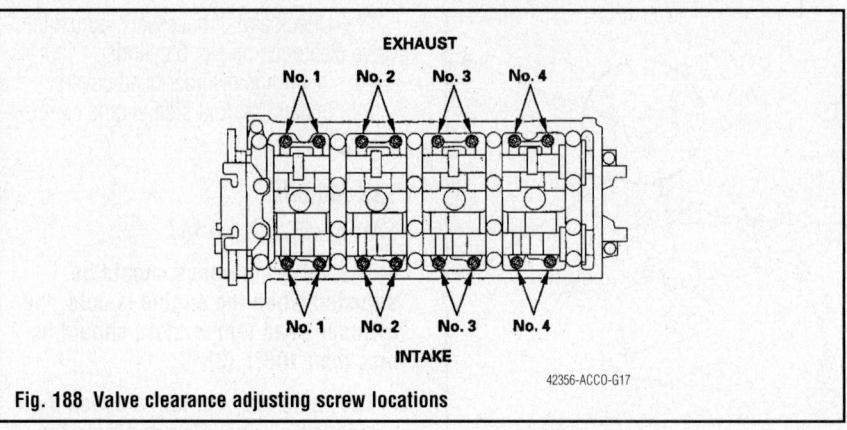

Fig. 188 Valve clearance adjusting screw locations

13. Install or connect the following:
- Cylinder head (valve) cover. Tighten the bolts attaching to 84 inch lbs. (10 Nm).

- Spark plug wires to the correct spark plugs.
- Positive, then the negative battery cable and enter the radio security code.

ENGINE PERFORMANCE & EMISSION CONTROL

ACCELERATOR PEDAL POSITION (APP) SENSOR

LOCATION

See Figure 189.

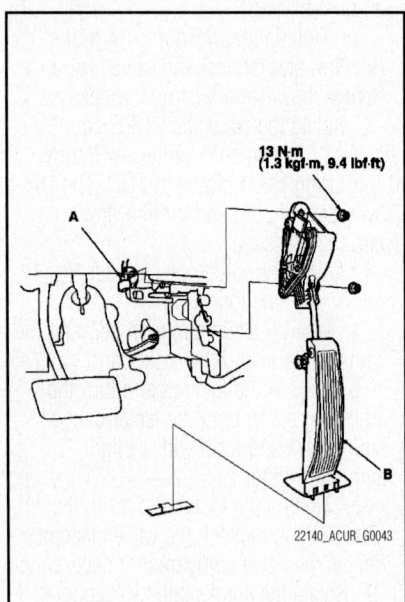

Fig. 189 Showing APP sensor (A) above the accelerator pedal module (B)

REMOVAL & INSTALLATION

1. Disconnect the APP sensor connector (A).
2. Release the lock tab on the base of the accelerator pedal, then remove the accelerator pedal module (B).
3. Install the parts in the reverse order of removal.

CAMSHAFT POSITION (CMP) SENSOR

LOCATION

TL & RL Models

See Figure 190.

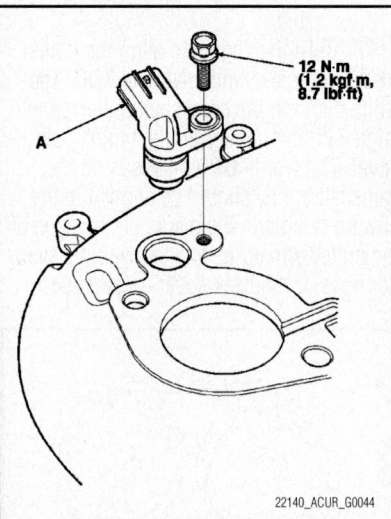

Fig. 190 CMP sensor (A) location—RL & TL models

TSX Models

See Figures 191 and 192.

REMOVAL & INSTALLATION

TL & RL Models

1. Set the No 1 piston at top dead center.
2. Remove the upper covers from the engine.

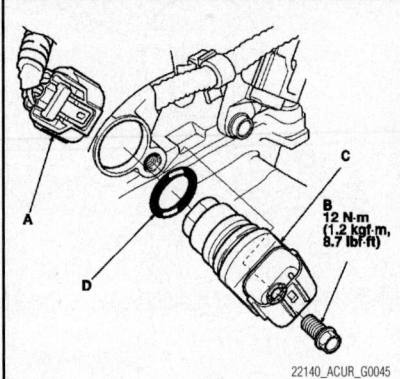

Fig. 191 CMP sensor (C) on the intake camshaft side of the cylinder head—Sensor A

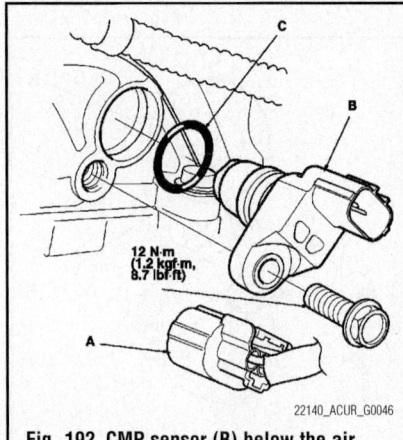

Fig. 192 CMP sensor (B) below the air cleaner assembly—Sensor B

3. To hold the timing belt adjuster in its current position, thread in the battery clamp bolt hand-tight.

4. Loosen the idler pulley bolt about five or six turns, then remove the timing belt from the front camshaft pulley.

5. Remove the front camshaft pulley.

6. Disconnect the CMP sensor connector (B), then remove the back cover (C).

7. Remove the CMP sensor (A) from the back cover.

8. Installation is the reverse order of removal.

TSX Models

Sensor A

1. Remove the air cleaner.

2. Disconnect the CMP sensor A 3P connector (A).

3. Remove the bolt (B).

4. Remove CMP sensor A (C) from the intake camshaft side of the cylinder head.

5. Install the parts in the reverse order of removal with a new O-ring (D).

Sensor B

1. Remove the air cleaner.

2. Remove the EVAP canister purge valve.

3. Disconnect the CMP sensor B 3P connector.

4. Remove CMP sensor B.

5. Install the parts in the reverse order of removal with a new O-ring.

CRANKSHAFT POSITION (CKP) SENSOR

LOCATION

TL & RL Models

See Figure 193.

TSX Models

The CKP sensor is located on the front, passenger side of the engine block.

REMOVAL & INSTALLATION

TL & RL Models

1. Remove the crankshaft pulley.

2. Remove the drive belt auto-tensioner.

3. Remove the upper cover and lower front covers from the engine.

4. Remove the CKP sensor from the oil pump.

5. Install the parts in the reverse order of removal.

TSX Models

1. Disconnect the CKP sensor 3P connector.

2. Remove the CKP sensor.

3. Install the parts in the reverse order of removal with a new O-ring.

EGR VALVE POSITION (EVP) SENSOR

LOCATION

See Figure 194.

REMOVAL & INSTALLATION

1. Remove the engine cover.

2. Disconnect the EGR valve wiring harness.

3. Remove the EGR valve.

4. Install the parts in the reverse order of removal with a new gasket.

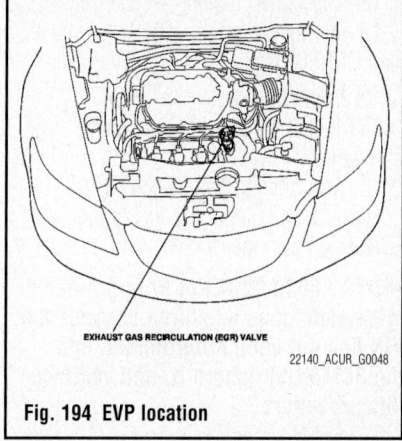

EXHAUST GAS RECIRCULATION (EGR) VALVE

22140_ACUR_G0048

Fig. 194 EVP location

ELECTRONIC CONTROL MODULE (ECM)

LOCATION

TL & TSX Models

The ECM is located under the dashboard behind the center console on all vehicles equipped with a manual transaxle.

REMOVAL & INSTALLATION

TL & TSX Models

1. Connect the HDS to the data link connector (DLC) (A) located under the driver's side of the dashboard.

2. Turn the ignition switch ON (II).

3. Make sure the HDS communicates with the ECM/PCM and other vehicle systems.

4. Select the PGM-FI system with the HDS.

5. Select the INSPECTION MENU with the HDS.

6. Select the ETCS TEST, then select the TP POSITION CHECK, and follow the screen prompts.

➡**If the TP POSITION CHECK indicates FAILED, continue with this procedure.**

7. Select the REPLACE ECM/PCM MENU, then READ DATA and follow the screen prompts.

➡**Doing this step copies (READS) the engine oil life data from the original ECM/PCM so you can later download (WRITES) it into the new ECM/PCM.**

➡**If READ DATA indicates FAILED, continue with this procedure.**

8. Turn the ignition switch OFF.

9. Jump the SCS line with the HDS.

10. Remove the center lower covers (A).

11. Remove the duct (B).

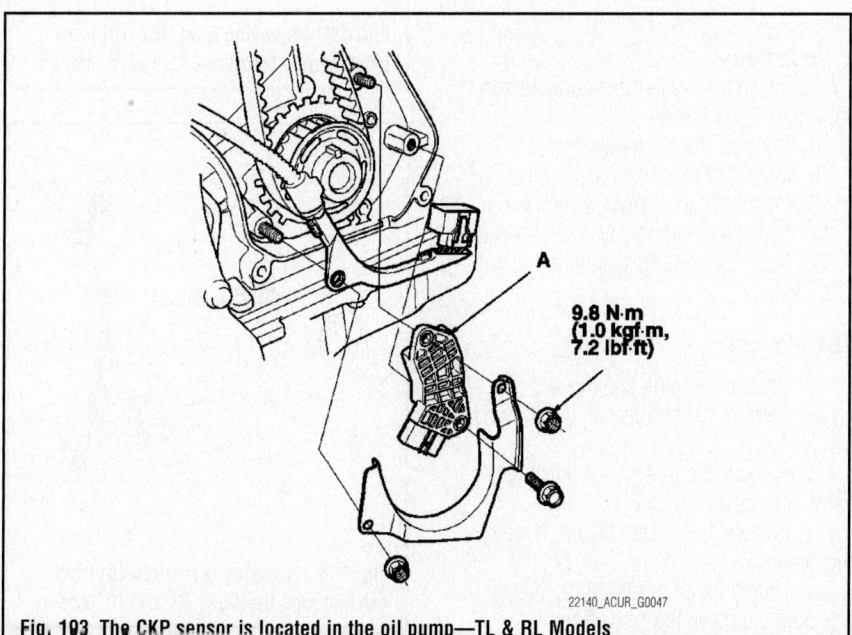

9.8 N·m (1.0 kgf·m, 7.2 lbf·ft)

A

22140_ACUR_G0047

Fig. 193 The CKP sensor is located in the oil pump—TL & RL Models

12. Disconnect ECM/PCM connectors (C).

13. Remove the bolts (D), then remove the ECM/PCM (E).

To install:

14. Install the parts in the reverse order of removal.

15. Turn the ignition switch ON (II).

16. Manually input the VIN to the ECM/PCM with the HDS.

➡ **DTC P0630 "VIN Not Programmed or Mismatch" may be stored because the VIN has not been programmed into the ECM/PCM; ignore it, and continue this procedure.**

17. If the READ DATA (engine oil life) failed in step 7, go to step 20. Otherwise, go to step 18.

18. Select the PGM-FI system with the HDS.

19. Select the REPLACE ECM/PCM MENU, then WRITE DATA and follow the screen prompts.

➡ **If the WRITE DATA indicates FAILED, continue with this procedure.**

20. Select IMMOBI system with the HDS.

21. Enter the immobilizer code with the ECM/PCM replacement procedure in the HDS; it allows you to start the engine.

22. If the TP POSITION CHECK failed in step 6 clean the throttle body, then go to step 23.

23. If the READ DATA failed in step 7 or the WRITE DATA failed in step 19, replace the engine oil and engine oil filter, then go to step 24.

24. Select PGM-FI system and reset the ECM/PCM with the HDS.

25. Update the ECM/PCM if it does not have the latest software.

26. Do the ECM/PCM idle learn procedure.

27. Do the CKP pattern learn procedure.

ENGINE COOLANT TEMPERATURE (ECT) SENSOR

LOCATION

TL Models

ECT sensor 1 is located below the air cleaner assembly. ECT sensor 2 is accessed from the bottom of the vehicle, in the engine block.

RL Models

ECT sensor 1 is located below the air cleaner assembly. ECT sensor 2 is located below the throttle body.

TSX Models

ECT sensor 1 is located below the underhood fuse box. ECT sensor 2 is accessed from the bottom of the vehicle, in the engine block.

REMOVAL & INSTALLATION

TL Models

1. Drain the engine cooling system.

2. Remove the air cleaner, if removing sensor 1.

3. Remove the splash shield, if removing sensor 2.

4. Disconnect the coolant temperature sensor wiring harness.

5. Remove the coolant temperature sensor.

To install:

6. Install the coolant temperature sensor using a new O-ring.

7. Connect the wiring harness.

8. Install the air cleaner, if removed.

9. Install the splash shield, if removed.

10. Refill the cooling system to the correct level.

11. Start the engine and check for leaks.

RL Models

1. Drain the cooling system.

2. Remove the engine cover, if removing sensor 1.

3. Remove the air cleaner, if removing sensor 1.

4. Remove the throttle body, if removing sensor 2.

5. Disconnect the coolant temperature sensor wiring harness.

6. Remove the coolant temperature sensor.

To install:

7. Install the coolant temperature sensor using a new O-ring.

8. Connect the wiring harness.

9. Install the throttle body, if removed.

10. Install the air cleaner, if removed.

11. Install the engine cover, if removed.

12. Refill the cooling system to the correct level.

TSX Models

1. Drain the engine cooling system.

2. Remove the air cleaner, if removing sensor 1.

3. Remove the EVAP canister purge valve, if removing sensor 1.

4. Remove the splash shield, if removing sensor 2.

5. Unbolt the under-hood fuse/relay box bolt and move the box aside.

6. Disconnect the coolant temperature sensor wiring harness.

7. Remove the coolant temperature sensor.

To install:

8. Install the coolant temperature sensor using a new O-ring.

9. Connect the wiring harness.

10. Install the EVAP canister purge valve, if removed.

11. Install the air cleaner, if removed.

12. Install the splash shield, if removed.

13. Refill the cooling system to the correct level.

HEATED OXYGEN (HO2S) SENSOR

LOCATION

TL & RL Models

See Figures 195 and 196.

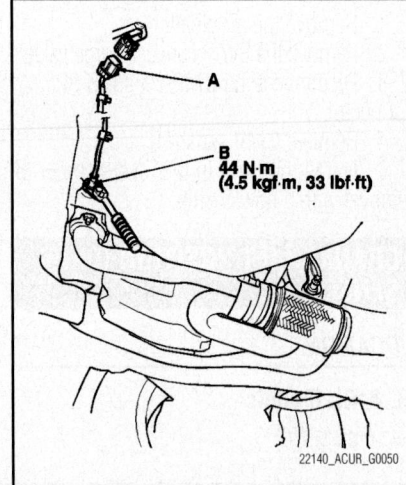

22140_ACUR_G0050

Fig. 195 Removing front HO2S(B) from exhaust pipe location—TL and RL Models

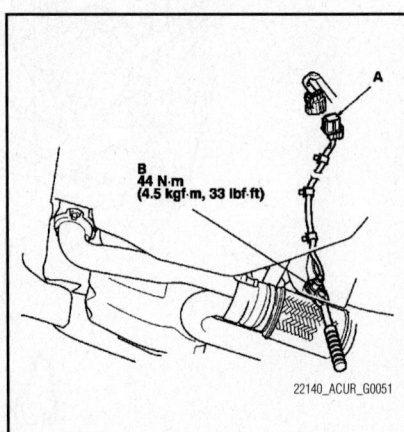

22140_ACUR_G0051

Fig. 196 Removing rear HO2S (B) from exhaust pipe location—TL and RL Models

TSX Models

See Figure 197.

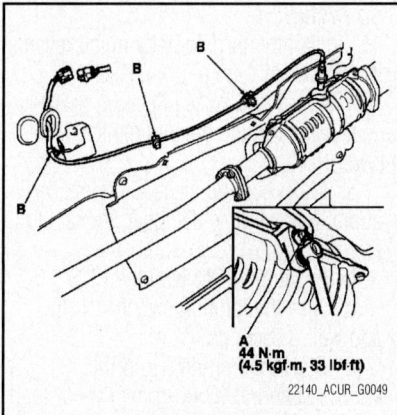

Fig. 197 Removing HO2S (A) from exhaust pipe location—TSX Models

REMOVAL & INSTALLATION

TL & RL Models

Front Bank

Disconnect the front secondary HO2S 4P connector, then remove the front secondary HO2S with an O2 sensor socket wrench.

Install the parts in the reverse order of removal.

Rear Bank

1. Disconnect the rear secondary HO2S 4P connector (A), then remove the rear secondary HO2S (B) with an O2 sensor socket wrench.
2. Install the parts in the reverse order of removal.

TSX Models

1. Pull back the carpet from the floor rail under the front edge of the front passenger seat to expose the secondary HO2S 4P connector that is attached to the floor rail.
2. Disconnect the secondary HO2S 4P connector under the passenger's seat.
3. Remove the secondary HO2S and the clips.
4. Install the parts in the reverse order of removal.

INTAKE AIR TEMPERATURE (IAT) SENSOR

LOCATION

The IAT sensor is located on the air intake assembly.

REMOVAL & INSTALLATION

TL & RL Models

1. Remove the engine cover.
2. Disconnect the IAT sensor 2P connector .
3. Remove the IAT sensor.
4. Install the parts in the reverse order of removal with a new O-ring.

TSX Models

1. Disconnect the IAT sensor 2P connector..
2. Remove the clamp (B) and the IAT sensor.
3. Install the parts in the reverse order of removal.

KNOCK SENSOR (KS)

LOCATION

TL & RL Models

See Figure 198.

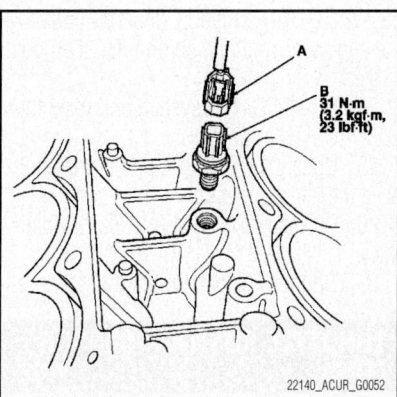

Fig. 198 Indicating the knock sensor location—TL and RL Models

TSX Models

The KS is located below the intake manifold in the engine block.

REMOVAL & INSTALLATION

TL & RL Models

1. Remove the intake manifold.
2. Remove the fuel rails and the intake runner base.
3. Disconnect the knock sensor connector, then remove the knock sensor.
4. Install the parts in the reverse order of removal.

TSX Models

1. Remove the intake manifold.
2. Disconnect the knock sensor 1P connector.

3. Remove the knock sensor.
4. Install the parts in the reverse order of removal.

MALFUNCTION INDICATOR LIGHT (MIL)

RESET PROCEDURES

Clearing MIL's requires the use of the Honda Diagnostic System (HDS) Clear command. MIL's can also be cleared by disconnecting the battery.

MANIFOLD ABSOLUTE PRESSURE (MAP) SENSOR

LOCATION

TL & RL Models

See Figure 199.

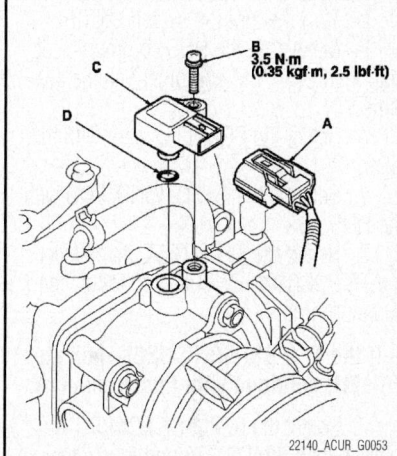

Fig. 199 Removing the MAF sensor from the air cleaner housing—TL and RL Models

TSX Models

See Figure 200.

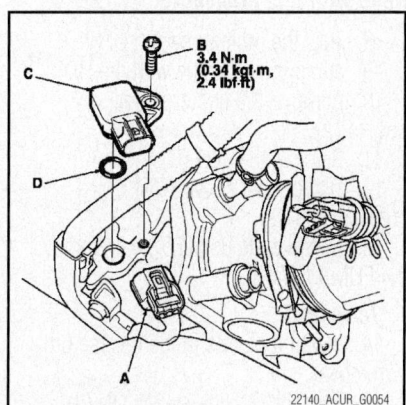

Fig. 200 Removing the MAF sensor from the air cleaner housing—TSX Models

REMOVAL & INSTALLATION

1. Remove the engine cover.
2. Disconnect the MAP sensor 3P connector.
3. Remove the screw.
4. Remove the MAP sensor.
5. Install the parts in the reverse order of removal with a new O-ring.

POWERTRAIN CONTROL MODULE (PCM)

LOCATION

The PCM is located under the dashboard behind the center console on all vehicles equipped with an automatic transaxle.

REMOVAL & INSTALLATION

1. Connect the HDS to the data link connector (DLC) (A) located under the driver's side of the dashboard.
2. Turn the ignition switch ON (II).
3. Make sure the HDS communicates with the ECM/PCM and other vehicle systems.
4. Select the PGM-FI system with the HDS.
5. Select the INSPECTION MENU with the HDS.
6. Select the ETCS TEST, then select the TP POSITION CHECK, and follow the screen prompts.

➡**If the TP POSITION CHECK indicates FAILED, continue with this procedure.**

7. Select the REPLACE ECM/PCM MENU, then READ DATA and follow the screen prompts.

➡**Doing this step copies (READS) the engine oil life data from the original ECM/PCM so you can later download (WRITES) it into the new ECM/PCM.**

➡**If READ DATA indicates FAILED, continue with this procedure.**

8. Turn the ignition switch OFF.
9. Jump the SCS line with the HDS.
10. Remove the center lower covers (A).
11. Remove the duct (B).
12. Disconnect ECM/PCM connectors (C).
13. Remove the bolts (D), then remove the ECM/PCM (E).

To install:

14. Install the parts in the reverse order of removal.
15. Turn the ignition switch ON (II).
16. Manually input the VIN to the ECM/PCM with the HDS.

➡**DTC P0630 "VIN Not Programmed or Mismatch" may be stored because the VIN has not been programmed into the ECM/PCM; ignore it, and continue this procedure.**

17. If the READ DATA (engine oil life) failed in step 7, go to step 20. Otherwise, go to step 18.
18. Select the PGM-FI system with the HDS.
19. Select the REPLACE ECM/PCM MENU, then WRITE DATA and follow the screen prompts.

➡**If the WRITE DATA indicates FAILED, continue with this procedure.**

20. Select IMMOBI system with the HDS.
21. Enter the immobilizer code with the ECM/PCM replacement procedure in the HDS; it allows you to start the engine.
22. If the TP POSITION CHECK failed in step 6 clean the throttle body, then go to step 23.
23. If the READ DATA failed in step 7 or the WRITE DATA failed in step 19, replace the engine oil and engine oil filter, then go to step 24.
24. Select PGM-FI system and reset the ECM/PCM with the HDS.
25. Update the ECM/PCM if it does not have the latest software.
26. Do the ECM/PCM idle learn procedure.
27. Do the CKP pattern learn procedure.

THROTTLE POSITION SENSOR (TPS)

LOCATION

The Throttle Position Sensors are located in the throttle body assembly.

REMOVAL & INSTALLATION

TL & RL Models

➡**If you are replacing the throttle body, begin at step 1. If you are removing the throttle body temporarily, begin at step 4.**

1. Connect the HDS while the engine is stopped.
2. Select the INSPECTION MENU with the HDS.
3. Do the TP POSITION CHECK in the ETCS TEST.
4. Disconnect the MAP sensor connector.
5. Remove the air intake duct.
6. Disconnect the throttle body connector.

7. Disconnect and plug the water bypass hoses.
8. Remove the throttle body.

To install:

9. Install the throttle body in the reverse order of removal with a new gasket.
10. Perform the PCM idle learn procedure after the throttle body is replaced, as follows:

   a. Make sure all electrical items (A/C, audio, rear window defogger, lights, etc.) are off.
   b. Reset the PCM with the HDS.
   c. Turn the ignition switch ON (II), and wait 2 seconds.
   d. Start the engine. Hold the engine speed at 3,000 rpm without load (in Park or neutral) until the radiator fan comes on, or until the engine coolant temperature reaches 194°F (90°C).
   e. Let the engine idle about 5 minutes with the throttle fully closed.

➡**If the radiator fan comes on, do not include its running time in the 5 minutes.**

11. Refill the radiator with engine coolant.

TSX Models

See Figure 201.

➡**If you are replacing the throttle body, begin at step 1. If you are removing the throttle body temporarily, begin at step 4.**

1. Connect the HDS while the engine is stopped.
2. Select the INSPECTION MENU with the HDS.
3. Do the TP POSITION CHECK in the ETCS TEST.
4. Turn the ignition switch OFF.
5. Disconnect the IAT sensor 2P connector, and remove the intake air duct.
6. Disconnect the throttle body 6P connector.
7. Disconnect the water bypass hoses and plug them.
8. Disconnect the vacuum hose.
9. Remove the harness clip and the throttle body.

To install:

10. Install the parts in the reverse order of removal with a new gasket, then do this:
11. Do the ECM/PCM idle learn procedure.
12. Refill the radiator with engine coolant.

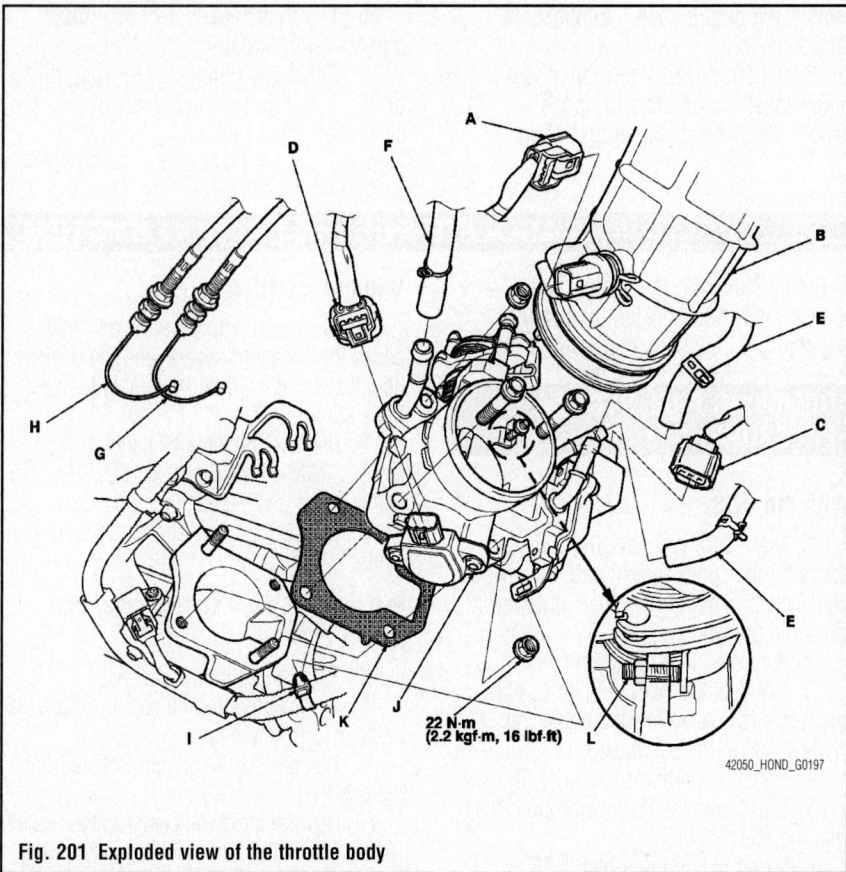

Fig. 201 Exploded view of the throttle body

VEHICLE SPEED SENSOR (VSS)

LOCATION

TL & RL Models

See Figures 202 and 203.

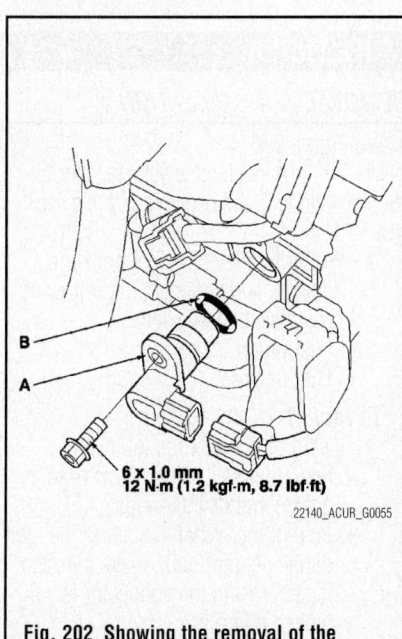

Fig. 202 Showing the removal of the speed sensor from the automatic transaxle— TL and RL Models

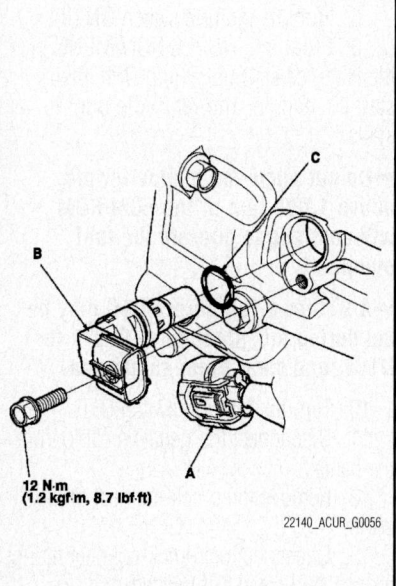

Fig. 203 Showing the removal of the speed sensor from the manual transaxle— TL Models only

TSX Models

See Figures 204 and 205.

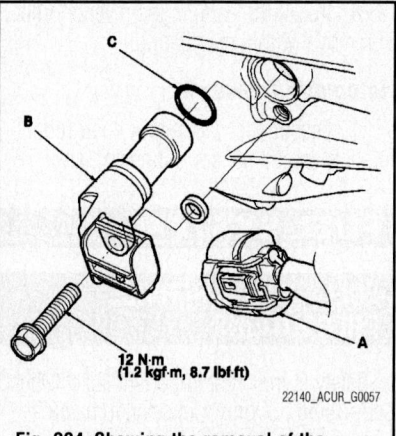

Fig. 204 Showing the removal of the speed sensor from the manual transaxle— TSX Models

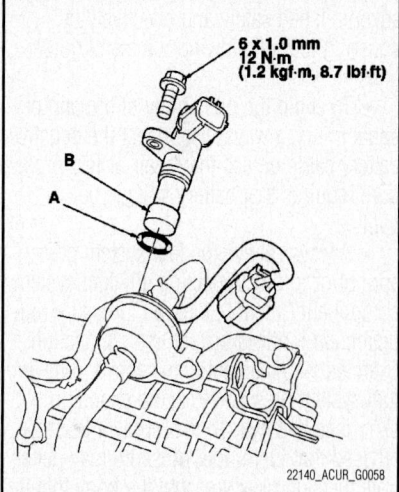

Fig. 205 Showing the removal of the speed sensor from the automatic transaxle—TSX Models

REMOVAL & INSTALLATION

TL & RL Models

1. Disconnect the output shaft (countershaft) speed sensor connector, and remove the output shaft (countershaft) speed sensor.

2. Install a new O-ring on a new output shaft (countershaft) speed sensor, then install the output shaft (countershaft) speed sensor in the transmission housing.

3. Check the connector for rust, dirt, or oil, then connect the connector securely.

TSX Models

Manual Transaxle

1. Remove the air cleaner.

2. Disconnect the output shaft (countershaft) speed sensor 3P connector.

3. Remove the output shaft (countershaft) speed sensor.

4. Install the parts in the reverse order of removal with a new O-ring.

Automatic Transaxle

1. Disconnect the output shaft (countershaft) speed sensor connector, and remove the output shaft (countershaft) speed sensor.

2. Install a new O-ring on a new output shaft (countershaft) speed sensor, then install the output shaft (countershaft) speed sensor in the transmission housing.

3. Check the connector for rust, dirt, or oil, then connect the connector securely.

FUEL SYSTEM   GASOLINE FUEL INJECTION SYSTEM

FUEL SYSTEM SERVICE PRECAUTIONS

Safety is the most important factor when performing not only fuel system maintenance but any type of maintenance. Failure to conduct maintenance and repairs in a safe manner may result in serious personal injury or death. Maintenance and testing of the vehicle's fuel system components can be accomplished safely and effectively by adhering to the following rules and guidelines.

• To avoid the possibility of fire and personal injury, always disconnect the negative battery cable unless the repair or test procedure requires that battery voltage be applied.

• Always relieve the fuel system pressure prior to disconnecting any fuel system component (injector, fuel rail, pressure regulator, etc.), fitting or fuel line connection. Exercise extreme caution whenever relieving fuel system pressure to avoid exposing skin, face and eyes to fuel spray. Please be advised that fuel under pressure may penetrate the skin or any part of the body that it contacts.

• Always place a shop towel or cloth around the fitting or connection prior to loosening to absorb any excess fuel due to spillage. Ensure that all fuel spillage (should it occur) is quickly removed from engine surfaces. Ensure that all fuel soaked cloths or towels are deposited into a suitable waste container.

• Always keep a dry chemical (Class B) fire extinguisher near the work area.

• Do not allow fuel spray or fuel vapors to come into contact with a spark or open flame.

• Always use a back-up wrench when loosening and tightening fuel line connection fittings. This will prevent unnecessary stress and torsion to fuel line piping.

• Always replace worn fuel fitting O-rings with new. Do not substitute fuel hose or equivalent where fuel pipe is installed.

Before servicing the vehicle, make sure to also refer to the precautions in the beginning of this section as well.

RELIEVING FUEL SYSTEM PRESSURE

With the HDS

1. Make sure you have the anti-theft codes for the audio system and the navigation system (if equipped), then write down the audio presets.

2. Turn the ignition switch OFF.

3. Connect the HDS to the data link connector (DLC) (A) located under the driver's side of the dashboard.

4. Turn the ignition switch ON (II).

5. Make sure the HDS communicates with the ECM/PCM.

6. Turn the ignition switch OFF.

7. Remove the fuel fill cap to relieve the pressure in the fuel tank.

8. Turn the ignition switch ON (II).

9. From the INSPECTION MENU of the HDS, select Fuel Pump OFF, then start the engine, and let it idle until it stalls.

➡**Do not allow the engine to idle above 1,000 rpm or the ECM/PCM will continue to operate the fuel pump.**

➡**A DTC or a Temporary DTC may be set during this procedure. Check for DTCs, and clear them as needed.**

10. Turn the ignition switch OFF.

11. Disconnect the negative cable from the battery.

12. Remove the quick-connect fitting cover.

13. Check the fuel quick-connect fitting for dirt, and clean it if needed.

14. Place a rag or shop towel over the quick-connect fitting.

15. Disconnect the quick-connect fitting: Hold the connector with one hand, and squeeze the retainer tabs with the other hand to release them from the locking tabs. Pull the connector off.

Without the HDS

1. Make sure you have the anti-theft codes for the audio system and the navigation system (if equipped), then write down the audio presets.

2. Remove the left kick panel, then remove PGM-FI main relay 2 (FUEL PUMP) (A) from the under-dash fuse/relay box.

3. Start the engine, and let it idle until it stalls.

➡**If any DTCs are stored, clear and ignore them.**

4. Turn the ignition switch OFF.

5. Remove the fuel fill cap to relieve the pressure in the fuel tank.

6. Disconnect the negative cable from the battery.

7. Remove the quick-connect fitting cover.

8. Check the fuel quick-connect fitting for dirt, and clean it if needed.

9. Place a rag or shop towel over the quick-connect fitting.

10. Disconnect the quick-connect fitting: Hold the connector with one hand, and squeeze the retainer tabs with the other hand to release them from the locking tabs. Pull the connector off.

FUEL FILTER

REMOVAL & INSTALLATION

See Figure 206.

1. Before servicing the vehicle, refer to the precautions in the beginning of this section.

2. Properly relieve the fuel pressure.

3. Remove or disconnect the following:
   • Negative battery cable
   • Fuel pump
   • Fuel filter set

To install:

4. Install or connect the following:
   • Fuel filter set, using a new base gasket and new O-rings
   • Fuel pump. When installing the fuel gauge sending unit, make sure the is secure and the connector is locked into place
   • Negative battery cable

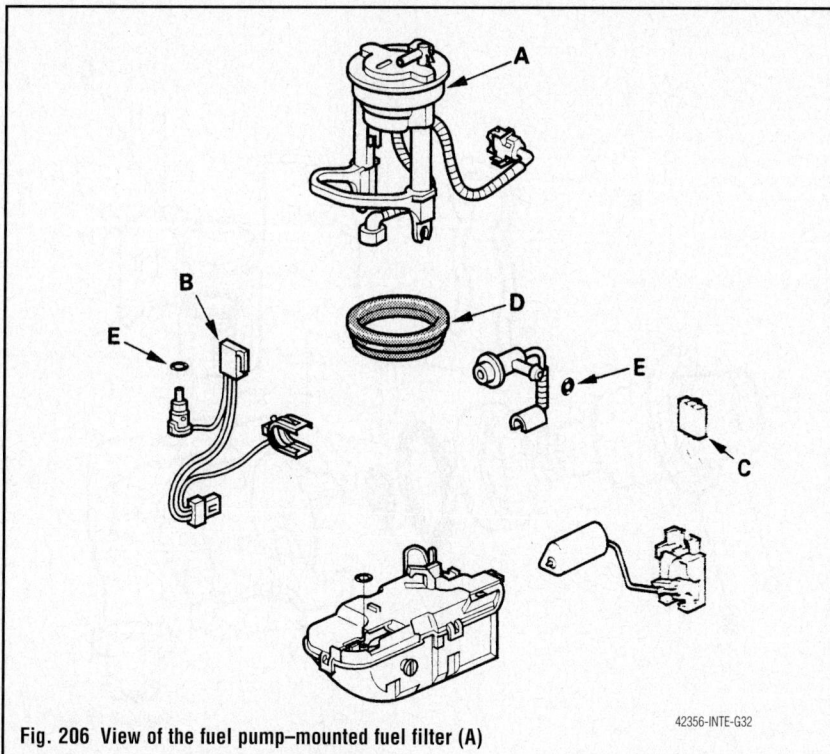

Fig. 206 View of the fuel pump–mounted fuel filter (A)

42356-INTE-G32

FUEL PUMP

REMOVAL & INSTALLATION

TL Models

1. Properly relieve the fuel pressure.
2. Remove the fuel fill cap.
3. Remove the trunk floor.
4. Remove the access panel from the floor.
5. Disconnect fuel pump 5P connector.
6. Disconnect the quick-connect fitting from the fuel tank unit.
7. Using Fuel sender wrench 07AAA-S0XA100, loosen the locknut.
8. Remove the locknut and the fuel tank unit.
9. Remove the fuel filter, the fuel gauge sending unit, the case, the wire harness, and the fuel pressure regulator.

To install:

10. Installation is the reverse of removal, use a new base gasket, new O-rings, and a new locknut

➥**After installation, make sure the base gasket is not pinched.**

RL Models

1. Disconnect the negative battery cable.
2. Properly relieve the fuel system pressure.
3. Remove the rear seat to gain access to the fuel pump access panel.

4. Remove the maintenance access cover.
5. Disconnect the electrical connector from the fuel pump.
6. If equipped with quick–connect fittings, hold the fuel line connector with one hand and press down the retainer tabs with the other hand, and then pull the connector off. Check the contact area of the pipe for dirt or damage, clean or replace the pipe or pump as required. Remove the old retainer from the pipe and discard. Cover the connector and pipe with plastic bags to prevent damage and keep foreign material out.
7. Remove the fuel pump mounting nuts, then remove the fuel pump from the fuel tank.

To install:

8. Installation is the reverse of the removal procedure. Tighten the fuel pump mounting nuts to 53 inch lbs. (6 Nm).

TSX Models

See Figure 207.

1. Properly relieve the fuel pressure.
2. Remove or disconnect the following:
   • Negative battery cable, if not already done
   • Fuel fill cap
   • Trunk floor
   • Access panel from the floor
   • Fuel pump connector
   • Quick-connect fitting from the fuel tank unit
   • Fuel pump/sender assembly locknut using tool 07XAA-001010A
   • Locknut and fuel pump/sender assembly
   • Fuel filter, fuel gauge sending unit, case, wire harness and fuel pressure regulator

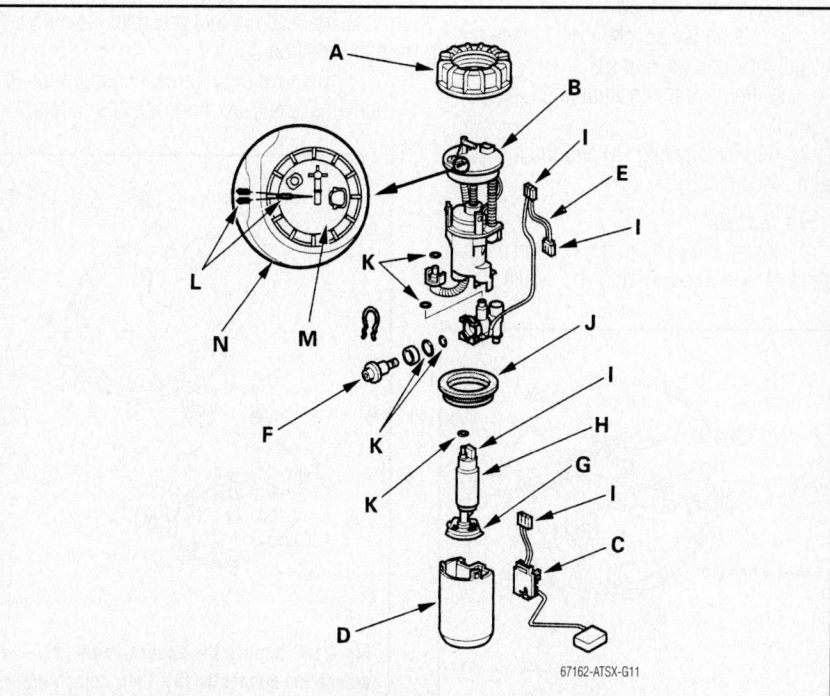

67162-ATSX-G11

Fig. 207 Exploded view of the fuel filter (B), fuel gauge sending unit (C), case (D), wire harness (E), fuel pressure regulator (F) and related components—TSX models

To install:

3. Installation is the reverse of the removal procedure, noting the following points:

a. When connecting the wire harness, make sure the connection is secure and the connectors are firmly locked into place. Do not bend or twist the fuel gauge sending unit more than necessary.

b. Install the pump assembly components in the reverse order of removal, with a new base gasket and new O-rings. Tighten the fuel pump/sender locknut to 69 ft. lbs. (93 Nm) with special tool 07XAA-001010A.

c. When installing the fuel tank unit, align the marks on the unit and the fuel tank.

FUEL PRESSURE REGULATOR

REMOVAL & INSTALLATION

TL Models

See Figures 208 and 209.

1. Relieve the fuel system pressure, as outlined in this section.

2. Remove the fuel tank unit, as follows:

a. Remove the fuel fill cap.

b. Remove the trunk floor.

c. Remove the access panel (A) from the floor.

d. Disconnect the fuel pump 5P connector (B).

e. Disconnect the quick-connect fitting (C) from the fuel tank unit.

f. Using the special tool, loosen the fuel tank unit locknut (A).

g. Remove the locknut and fuel tank unit.

3. Remove the clip (A) and the fuel pressure regulator (B).

To install:

4. Install the parts in the reverse order of removal with a new O-ring (C). Make sure

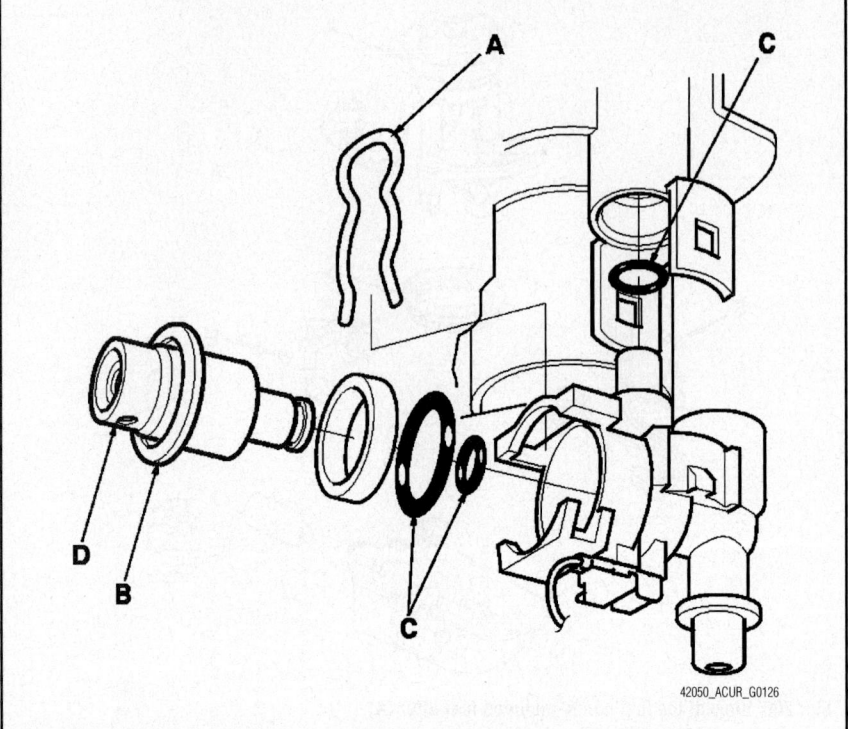

Fig. 209 Exploded view of the fuel pressure regulator (B) and related components

the regulator is installed with the drain hole (D) facing down.

→Coat the O-ring with clean engine oil.

RL Models

See Figures 210 through 213.

1. If the fuel tank is full, drain the fuel.

2. Relieve the fuel pressure, as outlined in this section.

3. Remove the rear seat cushion for RL models or the trunk floor for TL models.

4. Remove the access panel from the left side of the floor.

5. Disconnect the fuel pump electrical connector (B).

6. Disconnect the quick-connect fitting from the fuel tank unit.

7. Remove the fuel tank unit, and disconnect the transfer tube.

8. Remove the strainer case, the fuel gauge sending unit, the wire harness and the fuel pressure regulator.

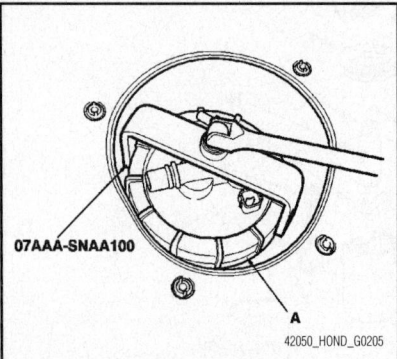

Fig. 208 Use a fuel pump module locknut wrench to loosen the locknut (A)

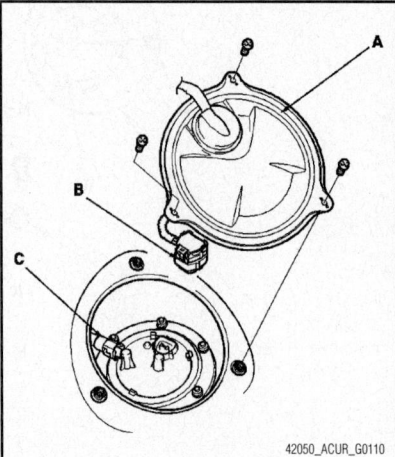

Fig. 210 Remove the access panel (A), detach the connector (B), then detach the quick-connect fitting (C) from the fuel tank unit—2005–06 RL models

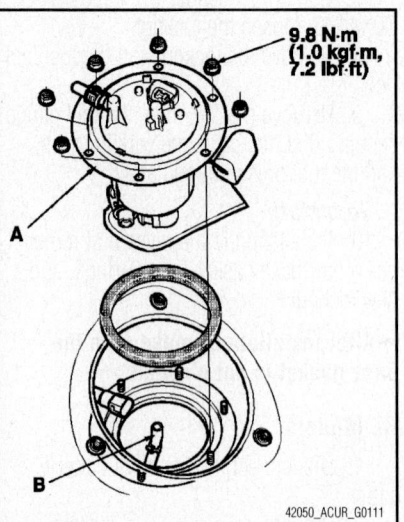

Fig. 211 Remove the fuel tank unit (A), and disconnect the transfer tube (B)

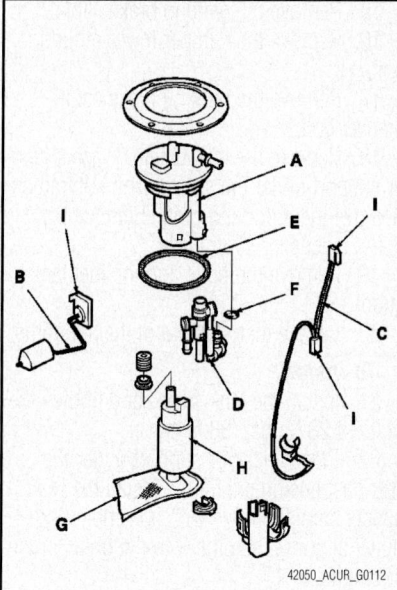

Fig. 212 Remove the strainer case (A), the fuel gauge sending unit (B), the wire harness (C) and the fuel pressure regulator (D)

To install:

9. Install the regulator in the reverse order of removal with a new O-ring (B). Coat the O-ring with clean engine oil before installing it.

10. Install the fuel tank unit in the reverse of the removal procedure.

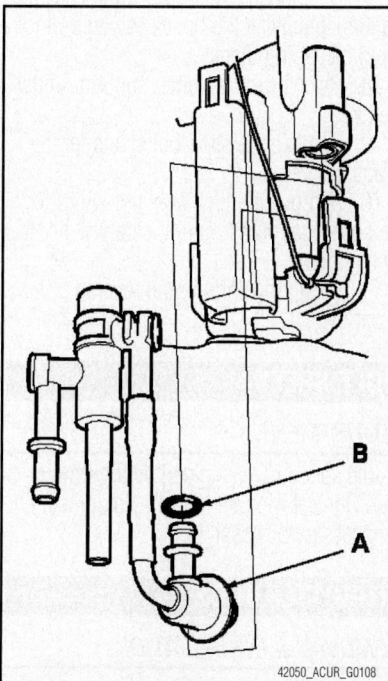

Fig. 213 Install the fuel pressure regulator (A) with a new O-ring (B)—2005–06 RL models

TSX Models

See Figures 208 and 214.

1. Relieve the fuel system pressure, as outlined in this section.

2. Remove the fuel tank unit, as follows:
   a. Remove the fuel fill cap.
   b. Remove the spare tire cover.
   c. Remove the access panel from the floor.
   d. Disconnect the fuel pump 5P connector.
   e. Disconnect the quick-connect fitting from the fuel tank unit.
   f. Using the special tool, loosen the fuel tank unit locknut.
   g. Remove the locknut and fuel pump module.

3. Remove the clip.

4. Remove the fuel pressure regulator.

To install:

5. Install the parts in the reverse order of removal with new O-rings.

6. Make sure the regulator is installed with the drain hole facing down.

➡ Coat the O-ring with clean engine oil.

FUEL RAIL & INJECTORS

REMOVAL & INSTALLATION

TL & RL Models

1. Before servicing the vehicle, refer to the precautions section.

2. Properly relieve the fuel pressure.

3. Remove the intake manifold.

4. Disconnect the connectors from the injectors.

5. Disconnect the quick-connect fittings.

6. Remove the fuel rail mounting bolts from the fuel rail.

7. Remove the injector clip from the injector.

8. Remove the injector from the fuel rail.

To install:

9. Coat the new O-rings with clean engine oil, and insert the injectors into the fuel rail.

10. Install the injector clip.

11. Coat the new injector O-rings with clean engine oil.

12. Install the injectors in the injector base.

13. Install the fuel rail mounting nuts.

14. Connect the connectors on the injectors.

15. Connect the quick-connect fittings.

16. Turn the ignition switch ON, but do not operate the starter. After the fuel pump runs for about 2 seconds, the fuel rail will be pressurized. Repeat this two or three times, then check for fuel leakage.

17. Install the intake manifold.

TSX Models

See Figure 215.

1. Disconnect the negative battery cable.

2. Relieve the fuel system pressure.

3. Remove or disconnect the following:
   - Intake manifold cover, if equipped
   - Fuel injector electrical connectors
   - Ground cable bolt, if necessary
   - Fuel line quick-connect fittings
   - Fuel feed line from the fuel rail
   - Vacuum hose and fuel return line from the fuel pressure regulator
   - Fuel rail mounting nuts and fuel rail
   - Fuel injector retaining clip

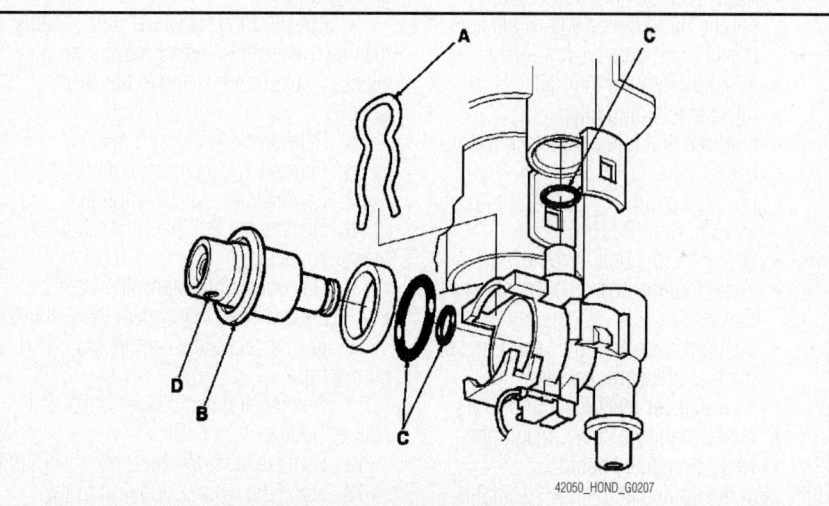

Fig. 214 Exploded view of the fuel pressure regulator (B) and related components—2007–08 TSX Models

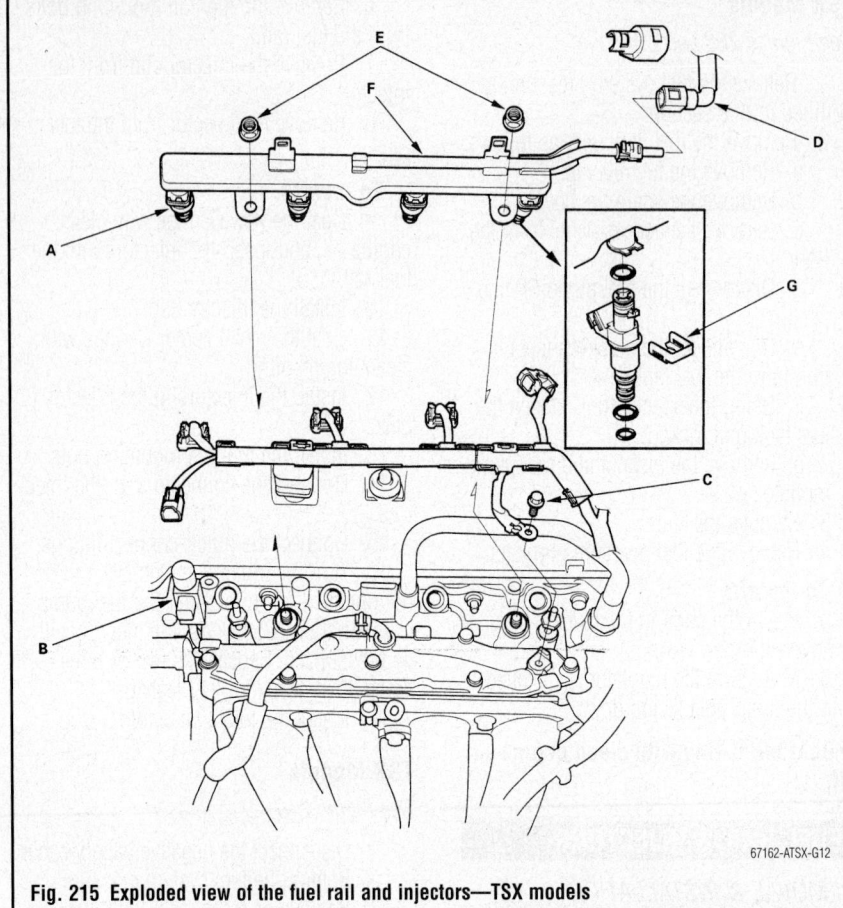

Fig. 215 Exploded view of the fuel rail and injectors—TSX models

67162-ATSX-G12

4. Grasp the fuel injectors body and pull up while gently rocking the fuel injector from side to side.

5. Once removed, inspect the fuel injector cap and body for signs of deterioration. Replace as required.

6. Remove the O-rings and discard.

To install:

7. Install or connect the following:
- Apply a small amount of clean engine oil to the new O–rings and install them onto each injector
- Injectors into the fuel rail
- Injector retaining clips
- Fuel rail and injectors into injector base
- Fuel rail mounting nuts and tighten to 16 ft. lbs. (22 Nm)
- Ground cable bolt, if removed
- Fuel feed line and quick-connect fittings
- Vacuum hose and fuel return line to the fuel pressure regulator
- Fuel injector electrical connectors
- Intake manifold cover, if equipped
- Negative battery cable

8. Run the engine at idle for 2 minutes, then turn the engine **OFF** and check for fuel leaks and proper operation.

FUEL TANK

REMOVAL & INSTALLATION

TL & TSX Models

1. Properly relieve the fuel system pressure.

2. Drain the fuel tank.

3. Install the fuel tank unit, if not already done.

4. Loosen the rear wheel nuts slightly, then raise the vehicle, and make sure it is securely supported. Remove the rear wheels.

5. Release the parking brake.

6. Remove the fuel tank covers.

7. Remove the exhaust muffler.

8. Disconnect the fuel fill neck tube and breather hose.

9. Disconnect the vapor line from the EVAP canister. Then disconnect the fuel line.

10. Disconnect the wheel sensor 2P connector.

11. Remove the two caliper bolts and caliper body.

12. Remove the brake hose mounting bolt.

13. Hook the caliper body on to the damper spring.

14. Remove the brake pads.

15. Remove the parking brake cable.

16. Remove the damper lower mounting bolt.

17. Remove the parking brake cable bracket bolt.

18. Remove the heat shield. Place a jack or support under the suspension subframe. Remove the mounting bolts. Remove the rear suspension subframe.

19. Remove the bolts and the fuel tank straps.

20. Lift the fuel tank out of the subframe.

To install:

21. Install the tank straps and tighten the bolts to 28 ft. lbs. (38 Nm).

22. Place a jack or support under the rear suspension subframe. Install the rear suspension subframe with new mounting bolts, aligning the pins with the holes in the subframe.

23. Install the parts in the reverse order of removal.

24. After installation, perform a wheel alignment.

RL Models

1. Drain the fuel tank.

2. Reinstall the fuel tank unit.

3. Remove the propeller shaft.

4. Remove the rear differential.

5. Jack up the vehicle, and support it with jackstands.

6. Remove the fuel tank covers and parking brake cable bolts.

7. Disconnect the hoses. Slide back the clamps, then twist the hoses as you pull to avoid damaging them.

8. Place a jack, or other support, under the tank.

9. Remove the strap bolts, and let the straps fall free.

10. Remove the fuel tank. If it sticks to the undercoat on its mount, carefully pry it off the mount.

11. Install the parts in the reverse order of removal.

IDLE SPEED

ADJUSTMENT

Idle speed is maintained by the Powertrain Control Module (PCM). No adjustment is necessary or possible.

THROTTLE BODY

REMOVAL & INSTALLATION

TL & RL Models

➡**If you are replacing the throttle body, begin at step 1. If you are removing the**

throttle body temporarily, begin at step 4.

1. Connect the HDS while the engine is stopped.
2. Select the INSPECTION MENU with the HDS.
3. Do the TP POSITION CHECK in the ETCS TEST.
4. Disconnect the MAP sensor connector.
5. Remove the air intake duct.
6. Disconnect the throttle body connector.
7. Disconnect and plug the water bypass hoses.
8. Remove the throttle body.

To install:
9. Install the throttle body in the reverse order of removal with a new gasket.
10. Perform the PCM idle learn procedure after the throttle body is replaced, as follows:
   a. Make sure all electrical items (A/C, audio, rear window defogger, lights, etc.) are off.

b. Reset the PCM with the HDS.
c. Turn the ignition switch ON (II), and wait 2 seconds.
d. Start the engine. Hold the engine speed at 3,000 rpm without load (in Park or neutral) until the radiator fan comes on, or until the engine coolant temperature reaches 194°F (90°C).
e. Let the engine idle about 5 minutes with the throttle fully closed.

➡**If the radiator fan comes on, do not include its running time in the 5 minutes.**

11. Refill the radiator with engine coolant.

TSX Models
See Figure 201.

➡**If you are replacing the throttle body, begin at step 1. If you are removing the throttle body temporarily, begin at step 4.**

1. Connect the HDS while the engine is stopped.
2. Select the INSPECTION MENU with the HDS.
3. Do the TP POSITION CHECK in the ETCS TEST.
4. Turn the ignition switch OFF.
5. Disconnect the IAT sensor 2P connector, and remove the intake air duct.
6. Disconnect the throttle body 6P connector.
7. Disconnect the water bypass hoses and plug them.
8. Disconnect the vacuum hose.
9. Remove the harness clip and the throttle body.

To install:
10. Install the parts in the reverse order of removal with a new gasket, then do this:
11. Do the ECM/PCM idle learn procedure.
12. Refill the radiator with engine coolant.

HEATING & AIR CONDITIONING SYSTEM

BLOWER MOTOR

REMOVAL & INSTALLATION
See Figure 216.

The blower motor is located in the front passenger's right side foot well area.
1. If additional access is needed, remove the front passenger's right side lower kick panel.
2. If necessary, remove the blower motor cover.
3. Detach the electrical connectors from the blower motor.

4. Remove the fasteners on the blower motor mounting flange.
5. Remove the blower motor downward from the blower unit.
6. Installation is the reverse of the removal procedure.

HEATER CORE

REMOVAL & INSTALLATION

TL Models
See Figures 217 through 224.

1. Disconnect the battery negative cable, and wait at least 3 minutes before beginning work.

➡**Make sure you have the anti-theft codes for the radio and the navigation system, then write down the radio channel presets.**

2. Disconnect the suction and receiver lines from the evaporator core.
3. From under the hood, open the cable clamp (A), then disconnect the heater valve

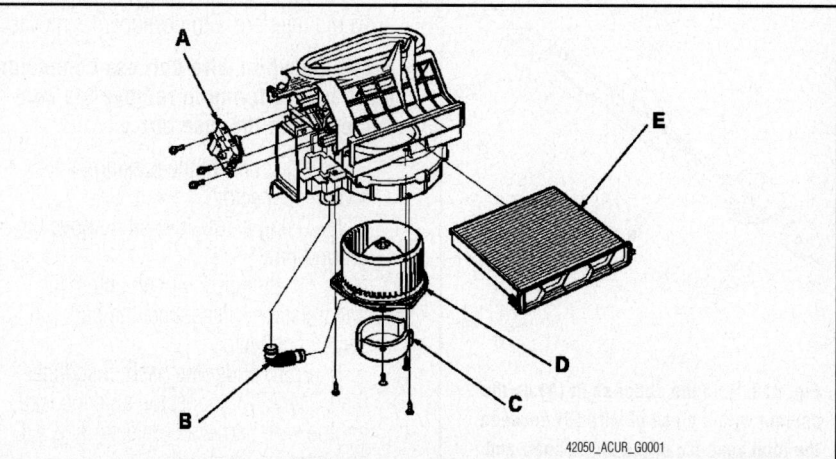

Fig. 216 The recirculation control motor (A), blower motor cooling hose (B), blower motor cover (C), the blower motor (D), and the dust and pollen filter (E) can be replaced without removing the blower unit—TL shown, other models similar

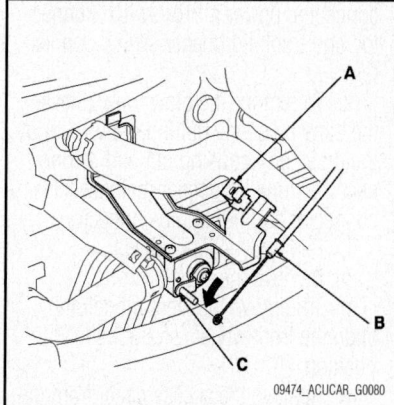

09474_ACUCAR_G0080

Fig. 217 Open the cable clamp (A), then disconnect the heater valve cable (B) from the heater valve arm (C). Turn the heater valve arm to the fully opened position as shown—2007–08 TL model

cable (B) from the heater valve arm (C). Turn the heater valve arm to the fully opened position as shown.

4. Drain and recycle the engine coolant.

5. Disconnect the heater hoses from the heater unit.

6. Remove the mounting nut from the heater unit. Take care not to damage or bend the fuel lines and the brake lines, etc.

7. Remove the dashboard as follows:

a. Remove the driver' and passenger seats.

b. If equipped with a manual transmission, lower the shift lever boot to release the hooks from the boot, then remove the shift knob.

c. Detach the clips and release the hooks by carefully inserting a trim tool between the HVAC panel and the center console trim and prying gently on both sides.

d. Open the console box lid then remove the console mat and screws.

e. Slide the rear section of the center console rearward to release the hooks.

f. Lift up the rear of the console.

g. Release the harness retainer clip from the air duct, and disconnect the accessory power socket connector and the light bulb socket connector, and then remove the rear section of the center console.

h. Remove the driver's side lower cover, by adjusting the steering column upward.

i. Remove the clip, and disconnect the in-car temperature sensor connector and air hose. Gently pull out the bottom of the dashboard lower cover to detach the clips and release the pin and hooks.

j. Disconnect the VSA OFF switch connector, power mirror switch connector, and trunk lid opener switch connector.

k. To remove the glove box, release the clips by gently pulling the bottom of the glove box housing out and disconnect the trunk lid opener main switch connector and light bulb connector.

l. Remove the kick panels.

m. Remove the A-pillar trim

n. To remove the steering column, align the front wheels straight ahead position

o. Remove the access panel from the steering wheel, then disconnect the driver's airbag 4P connector from the cable reel.

p. Remove the two Torx bolts using a Torx T30 bit.

q. Disconnect the horn switch connector, then remove the driver's airbag.

r. Disconnect the cruise control set/resume switch, audio remote switch and HFL/HFL-voice control switch connectors (if equipped).

s. Loosen the steering wheel bolt.

t. Install a commercially available steering wheel puller on the steering and remove the wheel.

✳✳ CAUTION

Do not tap on the steering wheel or the steering column shaft when removing the steering wheel. If you thread the puller bolts into the wheel hub more than five threads, the bolts will hit the cable reel and damage it. To prevent this, install a pair of jam nuts five threads up on each puller bolt.

u. Remove the column covers.

v. Remove the steering joint cover.

w. Release the tilt/telescopic lever, and adjust the steering column to full tilt up position, and to the full telescopic in position. Tighten the tilt/telescopic lever.

x. Hold the slider shaft (A) on the column with a piece of wire (B) between the joint yoke (C) of the slider shaft and joint yoke (D) of the upper shaft to prevent the slider shaft from pulling out.

y. Release the tilt/telescopic lever, and adjust the steering column to the full telescopic out position, then tighten the tilt/telescopic lever.

z. Disconnect the wire harness connectors from the combination switch assembly.

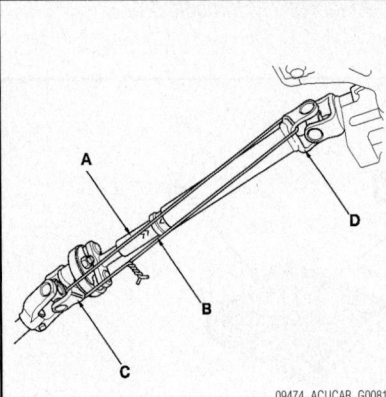

Fig. 218 Hold the slider shaft (A) on the column with a piece of wire (B) between the joint yoke (C) of the slider shaft and joint yoke (D) of the upper shaft to prevent the slider shaft from pulling out—2007–08 TL model

aa. Remove the combination switch assembly from the steering column shaft by removing the screws.

bb. Disconnect the connectors from the ignition switch, and release the wire harness clips from the steering column.

cc. Remove the steering joint bolt, then disconnect the steering joint from the pinion shaft.

dd. Remove the steering column by removing the attaching nuts and bolts. If the lower slide shaft is removed, slip it into the upper shaft by aligning the marks.

ee. If equipped with an automatic transmission, shift the transmission into the R position and remove the nut securing the shift cable end.

ff. Press the holder lock release (A), and pull out the socket holder (B) to remove the shift cable (C) from the shift lever bracket base (D). Do not remove the shift cable by pulling on the shift cable guide (E).

gg. Disconnect shift lock solenoid connector and transmission gear selection switch/park pin switch connector.

hh. Remove the harness clamps.

ii. Remove the shift lever assembly.

jj. Detach the clips, then remove the rear vent duct

kk. Remove the clips, then pull back the carpet.

ll. Remove the rear joint vent ducts on both sides.

mm. Disconnect the connector and remove the bolts, then remove the bracket.

nn. Remove the heater ducts.

oo. Disconnect the SRS connectors and remove the Torx bolts, then pull out the SRS unit.

pp. From under the dash, disconnect any wiring harness connectors that would interfere with dashboard removal.

➡**Lift the white wire harness connector locks before trying to remove the connectors from the fuse box.**

qq. Disconnect the parking brake switch connector.

rr. Using a T30 Torx bit, remove the ground bolt.

ss. If equipped with an automatic transmission, disconnect the park pin switch connector.

tt. From under the dash, disconnect the ECM/PCM connector, antenna lead, engine wire harness connector and A/C sub harness connector.

uu. If equipped with an automatic transmission, disconnect the shift lock solenoid connector.

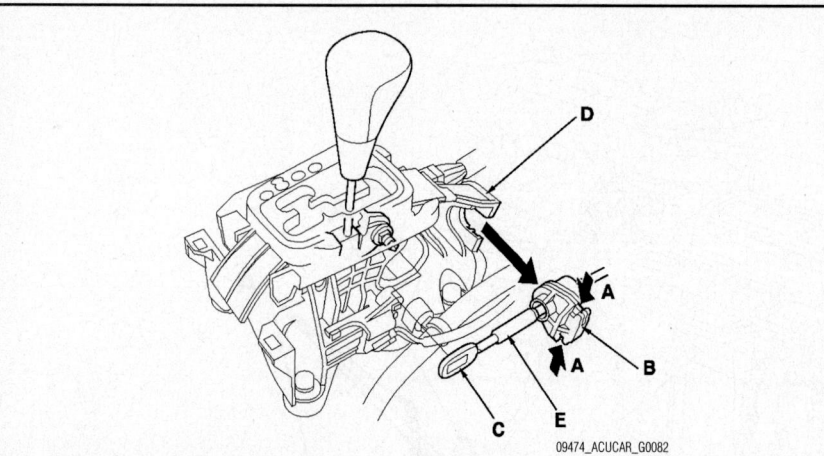

Fig. 219 Press the holder lock release (A), and pull out the socket holder (B) to remove the shift cable (C) from the shift lever bracket base (D). Do not remove the shift cable by pulling on the shift cable guide (E)—2007–08 TL model

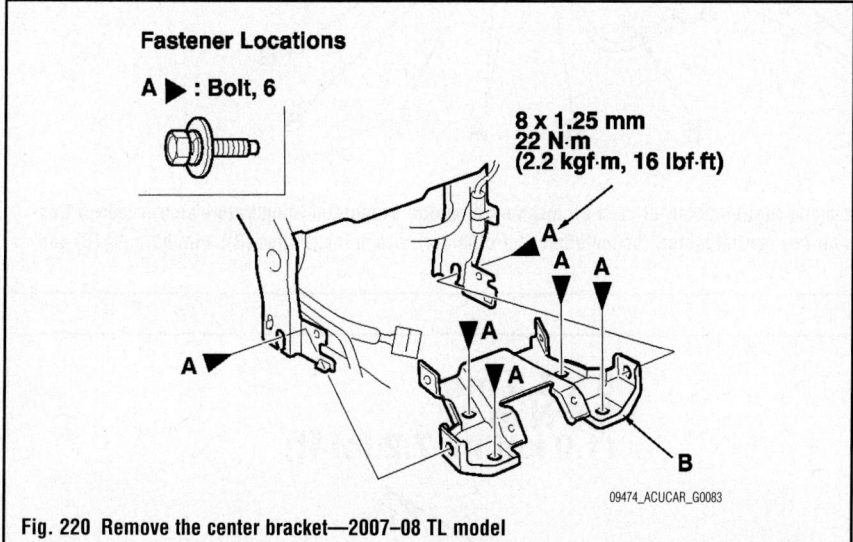

Fig. 220 Remove the center bracket—2007–08 TL model

vv. From under the dash, disconnect the passenger's door wire harness connectors, the three floor wire harness connectors (if the vehicle has a navigation system) or the one floor wire harness connector (if the vehicle does not have a navigation system), stereo amplifier connectors, and ETC unit connector.

ww. Detach all of the harness and connector clips.

xx. Remove bolts, then remove the brake pedal support member.

yy. Unbolt the parking brake handle assembly and the center console bracket. Do not disconnect the parking brake cable; move the assembly to the side of the shifter.

zz. Remove the bolts, then remove the center bracket.

aaa. From outside the driver's door, remove the caps, then remove the bolts

and clips, then lift up on the dashboard to release it from the guide pins.

➡Before removing the dashboard, make sure all the harnesses have been disconnected.

bbb. Carefully remove the dashboard through the front door opening.

8. Disconnect the connectors (A) from the driver's mode control motor, the driver's air mix control motor, the evaporator temperature sensor, the power transistor, the passenger's mode control motor, passenger's air mix control motor, and the recirculation control motor, then remove the wire harness clips (B), the wire harness (C), and the carpet clip (D). Refer to the illustration for component locations.

9. Remove the heater ducts. then remove the mounting nuts and the blower-heater unit.

10. Remove the self-tapping screws, the evaporator temperature sensor (A) and the joint duct (B). Remove the self-tapping screws, then remove the passenger's heater outlet (C), and the heater core cover (D). Remove the self-tapping screws and the passenger's air mix control motor (E). Remove the self-tapping screws, the heater pipe brackets (F), the grommets (G) and carefully pull out the heater core (H) so you don't bend the inlet and outlet pipes. Refer to the illustration for component locations.

To install:

11. Installation is the reverse of removal. refer to the illustrations accompanying the removal procedure for component locations and torque specifications.

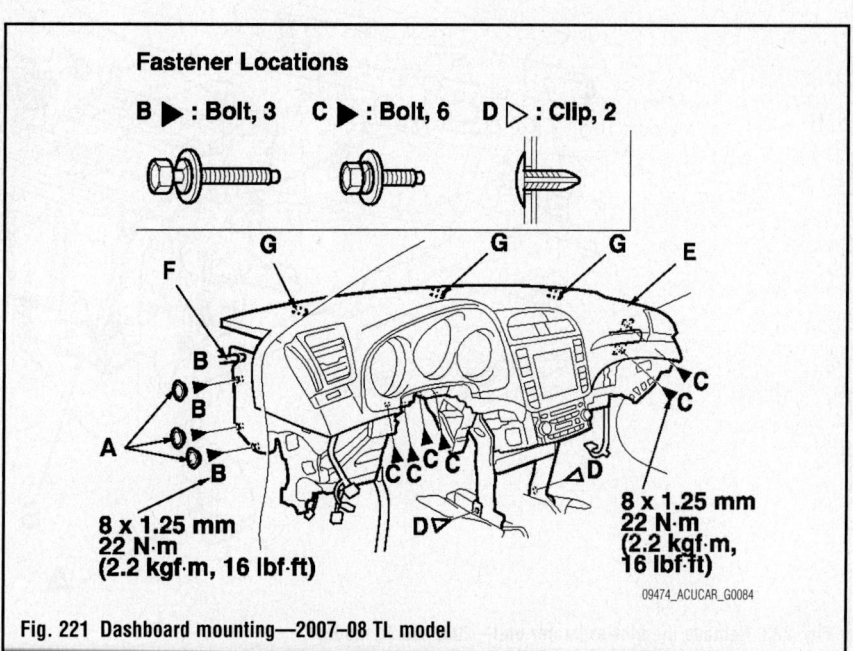

Fig. 221 Dashboard mounting—2007–08 TL model

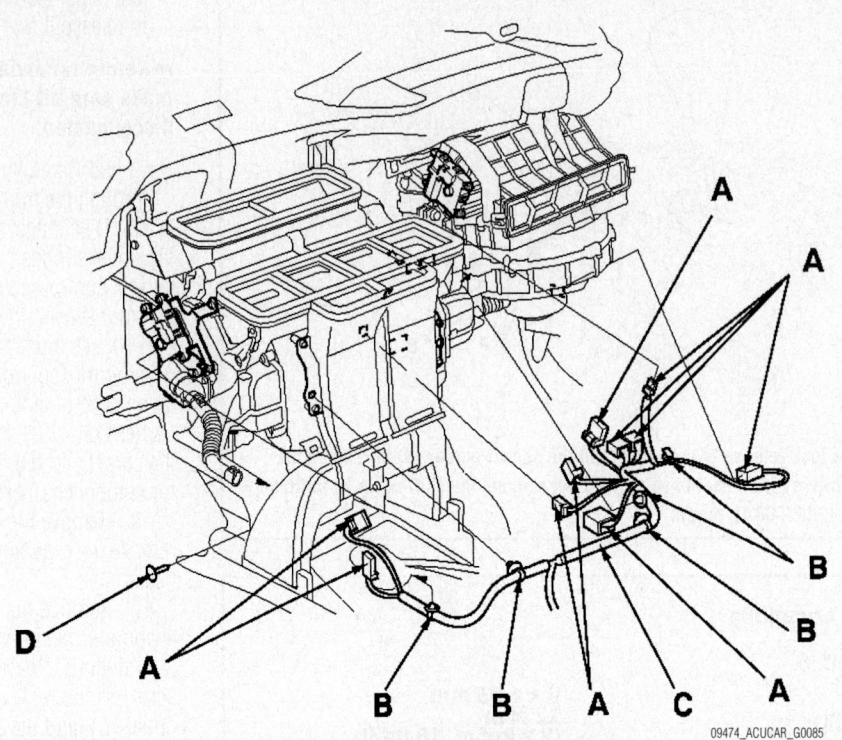

A
A
B
B
A
C
A
B
B
A
D
09474_ACUCAR_G0085

Fig. 222 Unplug the connectors (A) from the driver's mode control motor, driver's air mix control motor, evaporator temperature sensor, power transistor, passenger's mode control motor, passenger's air mix control motor, recirculation control motor, wire harness clips (B), wire harness (C) and carpet clip (D)—2007–08 TL model

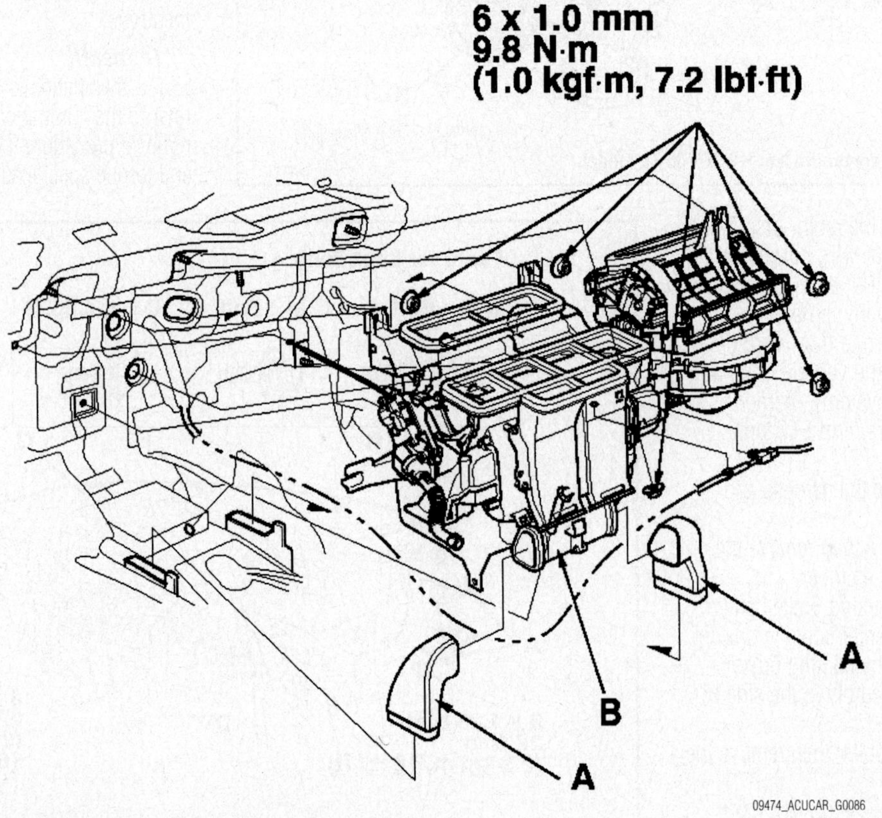

6 x 1.0 mm
9.8 N·m
(1.0 kgf·m, 7.2 lbf·ft)
A
B
A
09474_ACUCAR_G0086

Fig. 223 Remove the blower/heater unit—2007–08 TL model

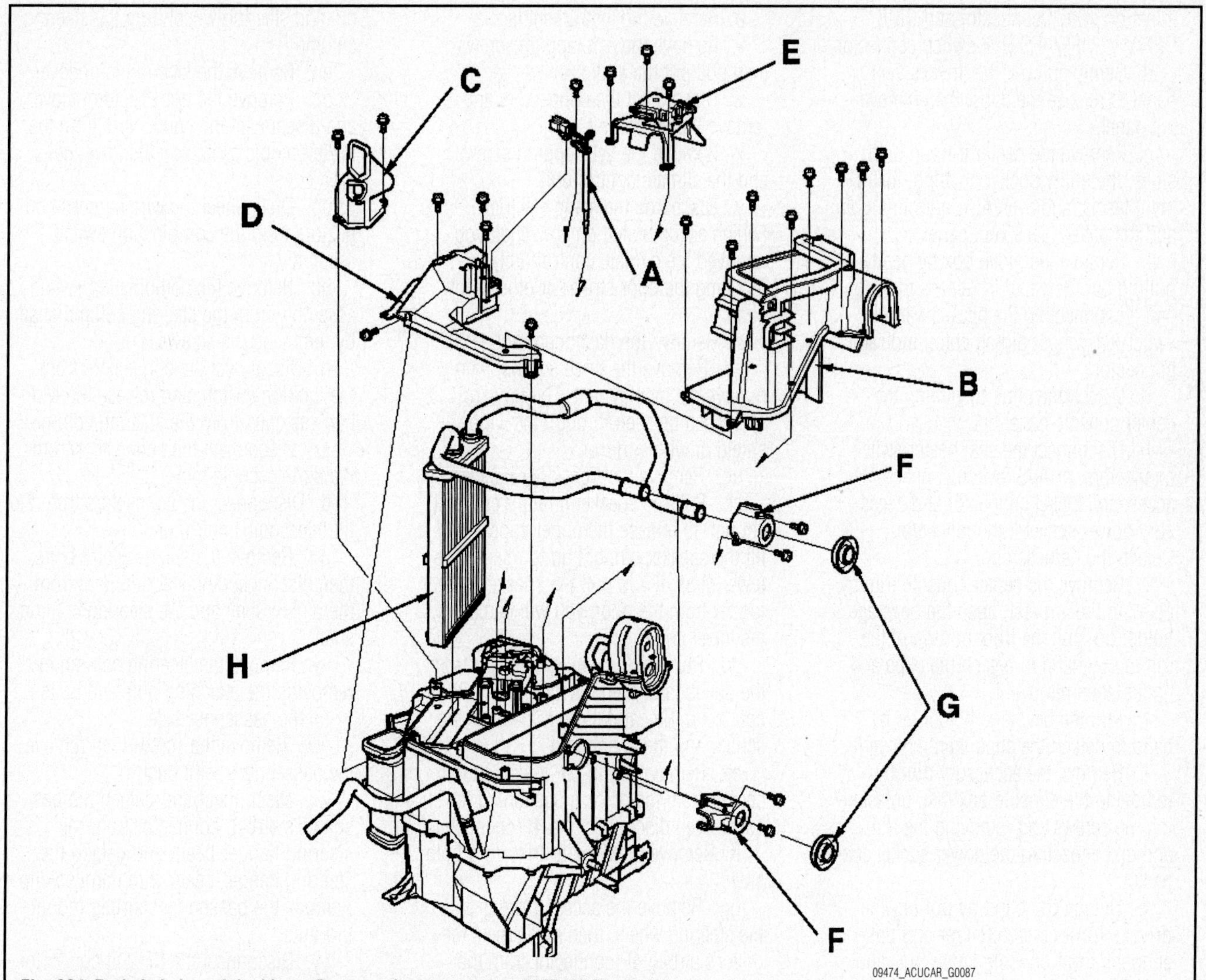

Fig. 224 Exploded view of the blower/heater unit components—2007–08 TL model

09474_ACUCAR_G0087

RL Models

See Figure 225.

1. Disconnect the battery negative cable, and wait at least 3 minutes before beginning work.

➡**Make sure you have the anti-theft codes for the radio and the navigation system, then write down the radio channel presets.**

2. Disconnect the suction and receiver lines from the evaporator core.

3. From under the hood, open the cable clamp (A), then disconnect the heater valve cable (B) from the heater valve arm (C). Turn the heater valve arm to the fully opened position as shown.

4. Drain and recycle the engine coolant.

5. Disconnect the heater hoses from the heater unit.

6. Remove the mounting nut from the heater unit. Take care not to damage or bend the fuel lines and the brake lines, etc.

7. Remove the dashboard as follows:

a. Gently pull out the driver's switch panel to release the hooks and detach the clips. Disconnect the power mirror switch connector then remove the panel.

b. Tilt the steering column down, and telescope it out.

c. Remove the instrument fascia by pulling out the bottom to release the clips gently and disconnecting the illu-

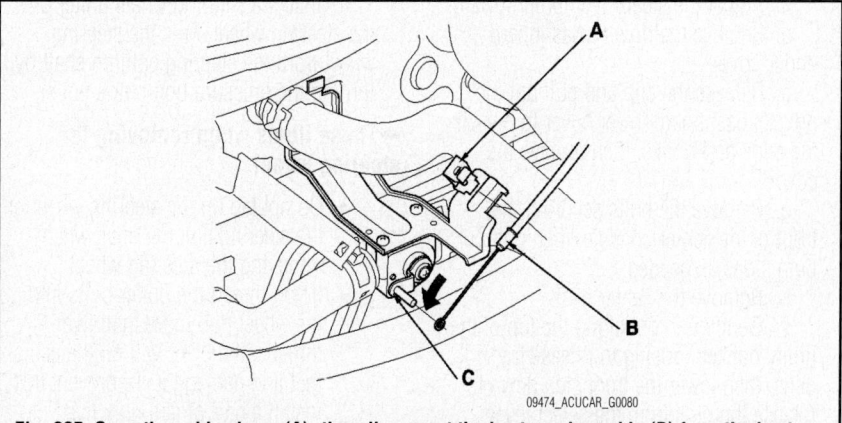

Fig. 225 Open the cable clamp (A), then disconnect the heater valve cable (B) from the heater valve arm (C). Turn the heater valve arm to the fully opened position as shown—2007–08 RL model

09474_ACUCAR_G0080

mination switch connector and CMS OFF/VSA OFF/AFS OFF switch connector

d. Gently pull out the driver's vent panel to release the clips, then remove the panel.

e. Remove the center trim on both sides, the utility pocket housing, instrument fascia, audio-HVAC-control module and the passenger's vent panel.

f. Remove the glove box by gently pulling out the panel to release the clips and disconnecting the hazard warning switch/passenger airbag cutoff indicator connector.

g. Detach the clips by pulling the center console panel up.

h. Disconnect the seat heater switch connectors, or AVS switch connectors, accessory socket connector and accessory power socket light connector. Detach the harness clip.

i. Remove the center console trim by opening the armrest, open the beverage holder lid, pull the front portion of the trim up by hand to detach the clips and hooks, then remove it.

j. Pull the beverage holder up by hand to detach the clips then remove it

k. Remove the accessory trim by removing the console box mat, unfastening the screws and removing the trim and disconnecting the power socket connector.

l. Detach the clips by pulling the driver's front console cover and passenger's front console cover out, and pull them backward to release their pins from the heater unit, then remove them.

m. Remove the center console by first disconnecting the sub harness connector, and detach it from the vent duct. Remove the bolts and pull the console backwards to remove it

n. Adjust the steering column upward.

o. Remove the driver's dashboard under cover.

p. Release the clip and pull out the driver's dashboard lower cover to detach the clips and hooks, then remove the cover.

q. Remove the bolts securing the front of the center console, then pull out both sides as needed.

r. Remove the center trim.

s. Gently pull out along the top of the utility pocket housing to release the clips, then lower the front side down to release the pins from the select lever bracket.

t. Tilt the steering column down, and telescope it out.

u. Remove the instrument fascia.

v. Remove the self tapping screws, then pull out the audio unit.

w. Disconnect the connectors and remove the audio unit.

x. Remove the self-tapping screws and the climate control unit.

y. Discharge the static electricity (which accumulated on you when you removed the climate control unit) by touching the door striker or other body parts

z. Remove the dashboard upper visor.

aa. Remove the three screws, then remove the display unit. The forward screw can be seen through the windshield or with a mirror.

bb. Remove the front door panel.

cc. Pull the speaker straight out, just enough to release the upper clips. Then lift the speaker straight up to release the lower clips. If you pull the speaker out too far from the door, you will damage the lower clips.

dd. Remove the sunlight sensor from the dashboard, then disconnect the 2P connector. Be careful not to damage the sensor and the dashboard.

ee. Remove the in-car temperature sensor from the driver's dashboard lower cover, then disconnect the 4P connector.

ff. Remove the A-pillar trim from both sides.

gg. Remove the access panel from the steering wheel, then disconnect the driver's airbag 4P connector from the cable reel.

hh. Remove the two Torx bolts using a Torx T30 bit.

ii. Disconnect the horn switch connector, then remove the driver's airbag.

jj. Disconnect the cable reel connector.

kk. Loosen the steering wheel nut.

ll. Install a steering wheel puller on the steering wheel. Free the steering wheel from the steering column shaft by turning the pressure bolt of the puller.

➡**These items when removing the steering wheel:**

- Do not tap on the steering wheel or the steering column shaft when removing the steering wheel.
- If you thread the puller bolts into the wheel hub more than five threads, the bolts will hit the cable reel and damage it. To prevent this, install a pair of jam nuts five threads up on each puller bolt.

mm. Remove the steering wheel puller, then remove the steering wheel

nut and steering wheel from the steering column

nn. Remove the steering joint cover.

oo. Remove the upper column cover and disconnect the connectors from the tilt/telescopic switch on the lower column cover.

pp. Disconnect the wire harness connectors from the combination switch assembly.

qq. Remove the combination switch assembly from the steering column shaft by removing the screws.

rr. Disconnect the connectors from the ignition switch, and release the wire harness clips from the steering column.

ss. Disconnect the connectors from the tilt/telescopic motors.

tt. Disconnect the connectors from the tilt/telescopic control unit.

uu. Remove the steering joint bolts, then disconnect the steering joint from the pinion shaft and the steering column shaft.

vv. Remove the steering column by removing the attaching nuts and bolts on the passenger side.

ww. Remove the screw, then remove the passenger's joint duct.

xx. Disconnect and detach the passenger's airbag connector from the steering hanger beam and detach the steering hanger beam wire harness clip. Remove the passenger's airbag mounting nuts.

yy. Disconnect the RF unit connector and detach the harness clip from the unit bracket. Tie a string to the sunlight sensor harness connector then pull the connector out through the hole in the dashboard. Remove the bolts, then remove the RF unit.

zz. From the front of the dashboard, remove the screws and bolts securing the dashboard/steering hanger beam.

aaa. Lift up on the dashboard to release it from the guide pins then carefully remove the dashboard through the front door opening.

➡**Lay the dashboard on its front or back. Do not rest it on the lower console opening or you may damage it.**

8. Disconnect the connectors from the power tilt/telescopic steering control unit, the adaptive front lighting control unit, the power steering control unit, and the daytime running lights control unit.

9. Remove the relay and the power tilt/telescopic steering control unit, then remove the self-tapping screws and the bracket.

10. Disconnect the connectors from the driver's cool vent control motor, the driver's air mix control motor, the mode control motor, and the evaporator temperature sensor, then remove the wire harness clips.

11. Disconnect the connectors from the blower motor, the power transistor, the control motor relay, the throttle actuator control module sub harness, and dashboard wire harnesses.

12. Remove the connector clip and the wire harness clips.

13. Disconnect the connectors from the recirculation control motor, the passenger's cool vent control motor, the passenger's air mix control motor, and the rear vent control motor, and then remove the wire harness clip and the wire harness.

14. Remove the clips, the mounting nuts, and the blower-heater unit.

15. Remove the self-tapping screws and the passenger's heater duct and the expansion valve cover, then remove the bolts and the expansion valve.

16. Remove the self-tapping screws the joint duct and the heater core cover.

17. Remove the self-tapping screws the heater pipe brackets and the grommets and carefully pull out the heater core.

To install:

18. Installation is the reverse of removal. refer to the illustrations accompanying the removal procedure for component locations and torque specifications.

TSX Models

See Figures 226 and 227.

➡Be sure to acquire the anti–theft code for the radio; then, write down the frequencies for the preset buttons.

1. Before servicing the vehicle, refer to the precautions in the beginning of this section.

2. Disconnect the negative battery cable.

⁂ CAUTION

After disconnecting the negative battery cable, wait for at least 3 minutes for the SRS module to deplete its energy.

3. Properly drain the cooling system. Properly discharge the air conditioning system using an approved refrigerant recover/recycling system.

4. Remove or disconnect the following:
   - Suction and receiver lines from the evaporator core

5. From under the hood, open the cable clamp and disconnect the heater valve cable from the heater valve arm. Turn the heater valve arm to the fully opened position.

6. Position a drain pan under the hoses. Slide the hose clamps back Remove the bolt and water valve bracket, then disconnect the inlet and outlet heater hoses from the heater unit.

⁂ WARNING

Do not let any coolant that drains from the hoses to contact any painted surfaces or electrical parts. Immediately wipe up and coolant that spills.

7. Remove or disconnect the following:
   - Mounting nut from the heater unit. Do not damage or bend the fuel lines or brake lines
   - Dashboard
   - Footrest and footrest bracket

8. Remove or disconnect the connectors, wire harness clips, connector clips and wire harness, after disconnecting the following:
   - Driver's side mix control motor
   - Evaporator temperature sensor
   - Power transistor
   - Mode control
   - Passenger's air mix control motor

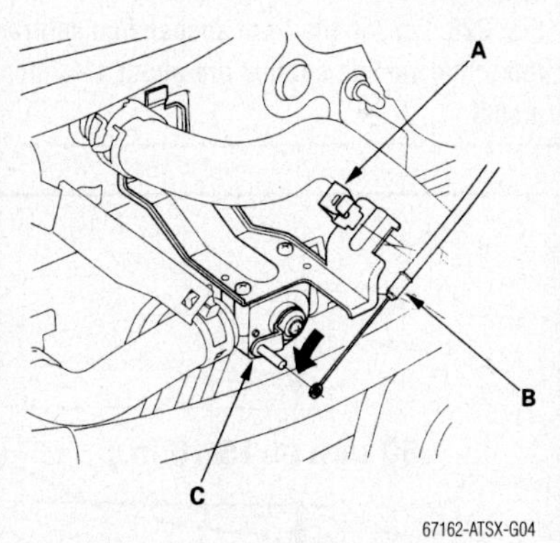

Fig. 226 Open the heater valve cable clamp (A), then disconnect the cable (B) from the heater valve arm. Turn the heater valve arm (C) to the fully opened position—TSX

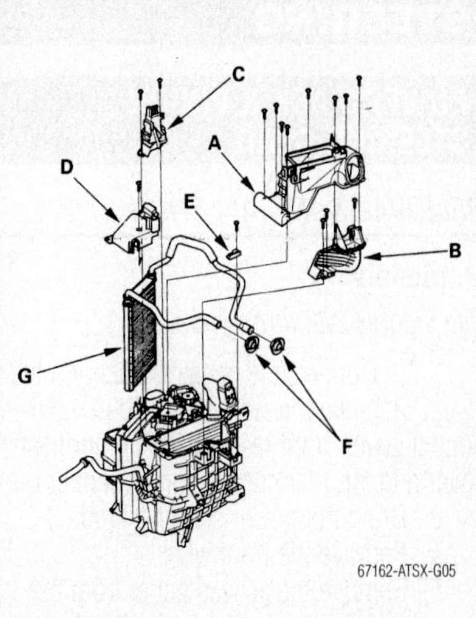

Fig. 227 Exploded view of the passenger heater outlet (C), heater core cover (D), heater pipe brackets (E), grommets (F) and heater core (G)—TSX

9. Remove or disconnect the following:
   - Heater ducts
   - Mounting nuts and blower/heater unit
   - Self-tapping screws and the joint ducts "A" and "B"
   - Self-tapping screws and the passenger's heater outlet and heater core cover
   - Self-tapping screws, heater pipe brackets, grommets and heater pipe brackets
   - Heater core, carefully, being sure not to bend the inlet and outlet pipes.

To install:

10. Installation is the reverse of the removal procedure, noting the following points:
    - Do not interchange the inlet and outlet heater hoses
    - Adjust the heater valve cable
    - Perform the PCM idle learn procedure
    - Perform the power window control unit reset procedure

11. Refill the cooling system.

12. Connect the negative battery cable.

13. Evacuate, charge and leak test the air conditioning system.

14. Operate the engine to normal operating temperatures; then, check the climate control operation and check for leaks.

STEERING

POWER RACK & PINION STEERING GEAR

REMOVAL & INSTALLATION

TL Models

See Figures 228 through 232.

1. Disconnect the negative battery cable.
2. Remove the right side engine component cover, right rear engine component cover and left rear engine component cover.
3. Drain the power steering fluid.
4. Remove the front wheels.
5. Remove the access panel from the steering wheel, then disconnect the driver's airbag 4P connector from the cable reel.
6. Remove the two Torx bolts using a Torx T30 bit.
7. Disconnect the horn switch connector, then remove the driver's airbag.
8. Disconnect the cruise control set/resume switch, audio remote switch and HFL/HFL-voice control switch connectors (if equipped).
9. Loosen the steering wheel bolt.
10. Install a commercially available steering wheel puller on the steering and remove the wheel.

✳✳ CAUTION

Do not tap on the steering wheel or the steering column shaft when removing the steering wheel.

11. If you thread the puller bolts into the wheel hub more than five threads, the bolts will hit the cable reel and damage it. To prevent this, install a pair of jam nuts five threads up on each puller bolt.
12. Remove the steering joint cover.
13. Remove the steering joint bolt, and disconnect the steering joint by moving the steering joint toward the column. Hold the slider shaft on the column with a piece of wire between the joint yoke on the slider shaft to the joint yoke on the upper shaft.
14. Remove the center guide and discard it.
15. Remove the steering joint cover. Be careful not to damage the mating surface on the joint cover and pinion shaft grommet. Replace the cover seal if necessary.
16. Remove and discard the cotter pin from the tie-rod ball joint nut, and loosen the nut.
17. Separate the tie-rod ball joint from the knuckle.
18. Disconnect the power steering pressure switch connector.

19. Remove the splash shield.
20. Remove the engine under cover.
21. Attach the tool illustrated to the front suspension subframe by hanging the hook of the tool over the front of the subframe, then tighten the screw.
22. Raise the jack and line up the slots in the arms with the bolt holes on the corner of the jack base, then attach them with bolts securely.
23. Remove the front suspension subframe right middle mounting bolts.
24. Remove the front suspension subframe left middle mounting.
25. Loosen the front suspension subframe front bracket mounting bolts on the right and left of the vehicle so they are

about 1³⁄₁₆ inch (30 mm) from the mounting surface.
26. Remove the front suspension subframe rear bracket on the right and left of the vehicle from the front suspension subframe.
27. Lower the jack supporting the front suspension subframe with the tool illustrated slowly until the front suspension subframe has dropped about 1¹⁵⁄₁₆ inch (50 mm).
28. Remove the P/S heat baffle plate.
29. Remove the feed line holder mounting bolt on the front suspension subframe.
30. Remove the feed line holder mounting bolt and return hose from the gearbox mounting bracket.

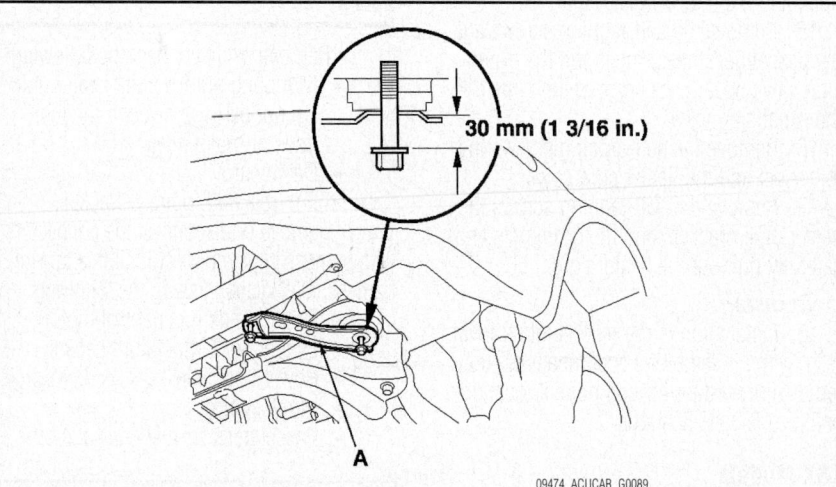

Fig. 228 Loosen the front suspension subframe front bracket mounting bolts on the right and left of the vehicle so they are about 1³⁄₁₆ inch (30 mm) from the mounting surface—2007–08 TL model

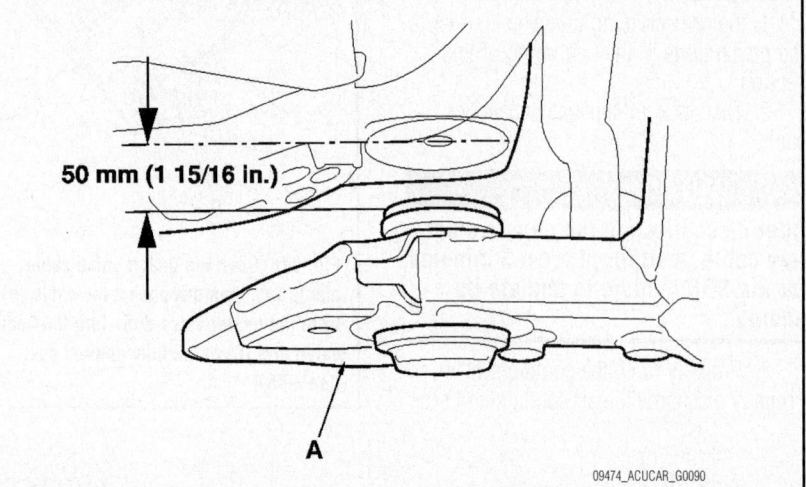

Fig. 229 Lower the jack supporting the front suspension subframe with the tool illustrated slowly until the front suspension subframe has dropped about 1¹⁵⁄₁₆ inch (50 mm)—2007–08 TL model

31. Place several shop towels under the line connections, and cover the gearbox mounting part to protect it from the power steering fluid. Loosen the flare nut and disconnect the feed line.

32. Loosen the flare nut and disconnect the return line on 2004 models or the pipe on 2005–06 models.

33. After disconnecting the lines or pipe, plug or seal them with a piece of tape or equivalent to prevent foreign materials from entering.

➡**Do not loosen the cylinder line A and B between the valve body unit and the cylinder.**

34. Remove the steering gearbox mounting bolts and washers on the left gearbox mount.

35. Remove the two flange bolts from the right side of the gearbox, then remove the gearbox mounting bracket and cushion.

36. Move the steering gearbox toward the front, and remove the pinion shaft grommet from the top of the valve body unit.

37. Apply vinyl tape to the splines on the pinion shaft.

38. Apply vinyl tape or equivalent material to brake lines to protect it from the pinion shaft.

39. Move the steering gearbox to the driver's side, and rotate it so the pinion shaft points toward the front of the vehicle.

40. Carefully move the steering gearbox as an assembly toward the left side of the vehicle until the pinion shaft clears the wheel well opening. Be careful not to damage the brake lines with the pinion shaft.

To install:

41. Before installing the steering gearbox, make sure that no power steering fluid is on the mating surface of the gearbox and front suspension subframe. To prevent the gearbox mounting bolts from loosening after the installation, remove any power steering fluid from the mount cushions and bolt holes.

42. Apply a mild soap and water solution to both sides of the mount cushion mating surfaces.

43. Pass the cylinder of the steering gearbox through the wheel well opening on the driver's side.

44. Carefully move the steering gearbox toward the passenger's side until the pinion shaft clears the wheel well opening on the body.

45. Rotate the steering gearbox so the pinion shaft points upward.

46. Continue moving the gearbox toward the passenger's side until the steering gearbox is in position.

47. Make sure the power steering return line and feed line are routed above the gearbox

48. Remove the vinyl tape from the pinion shaft, and install the pinion shaft grommet. Align the slot in the pinion shaft grommet with the lug portion on the valve housing.

49. Position the cutout on the mounting cushion as shown, and install it on the cylinder of the gearbox securely.

50. Install the new gearbox mounting bracket over the mounting cushion, and loosely install the two flange bolts.

51. Install the new gearbox mounting bolts and washers on the left side of the

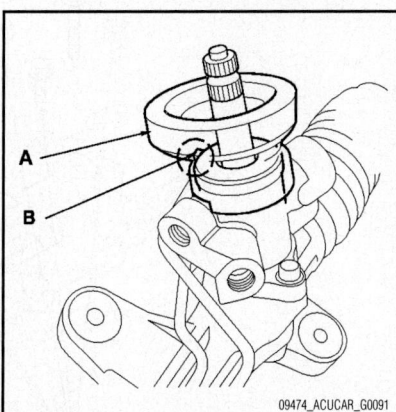

Fig. 230 Install the pinion shaft grommet. Align the slot in the pinion shaft grommet with the lug portion on the valve housing—2007–08 TL model

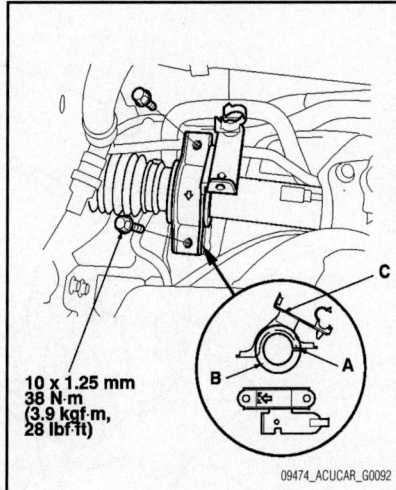

Fig. 231 Install the new gearbox mounting bolts and washers on the left side of the gearbox—2007–08 TL model

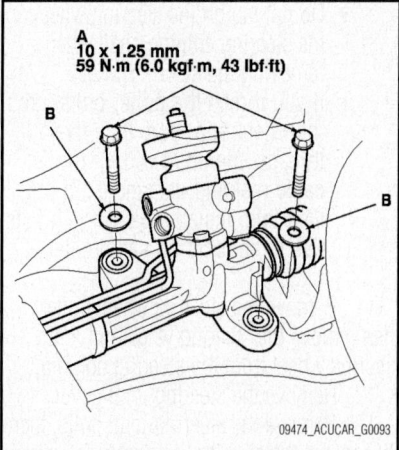

Fig. 232 Tighten the flange bolts on the right side of the gearbox—2007–08 TL model

gearbox, then tighten them to the 28 ft. lbs. (38 Nm).

52. Tighten the flange bolts on the right side of the gearbox to 43 ft. lbs. (59 Nm) alternately in two or more steps.

53. Loosely connect the return line/pipe and feed line by hand.

54. Install the feed line holder and return hose on the gearbox mounting bracket.

55. Install the feed line holder on the front suspension subframe. Make sure that there is no interference between the feed and return lines and any other parts.

56. Install the remaining components in the reverse order of removal.

RL Models

See Figures 233 through 236.

1. Remove the right and left front fender trim.

2. Drain the power steering fluid.

3. Remove the relay box and release the wire harness clips.

4. Remove the front strut brace.

5. Remove the access panel from the steering wheel, then disconnect the driver's airbag 4P connector from the cable reel.

6. Remove the two Torx bolts using a Torx T30 bit.

7. Disconnect the horn switch connector, then remove the driver's airbag.

8. Disconnect the cable reel connector.

9. Loosen the steering wheel nut.

10. Install a steering wheel puller on the steering wheel. Free the steering wheel from the steering column shaft by turning the pressure bolt of the puller.

➡**These items when removing the steering wheel:**

- Do not tap on the steering wheel or the steering column shaft when removing the steering wheel.
- If you thread the puller bolts into the wheel hub more than five threads, the bolts will hit the cable reel and damage it. To prevent this, install a pair of jam nuts five threads up on each puller bolt.

11.   Remove the steering wheel puller, then remove the steering wheel nut and steering wheel from the steering column.

12.   Remove the steering joint cover.

13.   Remove the steering joint bolts, then disconnect the steering joint from the pinion shaft and the steering column shaft.

14.   Remove the center guide.

15.   Remove steering joint cover. Be careful not to damage the mating surface on joint cover and the pinion shaft grommet. Replace the cover seal if necessary.

16.   Remove the pinion shaft grommet from the top of the valve body unit.

17.   Apply vinyl tape to the splines on the pinion shaft.

18.   Remove and discard the cotter pin from the tie-rod ball joint nut and loosen the nut.

19.   Separate the tie-rod ball joint and knuckle using the ball joint remover.

20.   Remove the nuts from the rear engine mount.

21.   Remove the rear engine mount and the engine mount pressure hose holder mounting bolt on the gearbox stiffener.

22.   Disconnect the engine mount pressure hose from the rear engine mount.

23.   Remove the feed line holder mounting bolt and A/T wire line holder mounting bolt on the gearbox stiffener.

24.   Disconnect the return hose from the clamp.

25.   Remove the feed line holder mounting bolt and return line holder mounting bolt on the gearbox.

26.   Disconnect the stepping valve motor connector.

27.   Place shop towels under the line connections, and cover the gearbox mounting part to protect it from the power steering fluid. Loosen the adjustable hose clamp and disconnect the return hose.

28.   Loosen the 14 mm flare nut and disconnect the feed line.

29.   Remove the return line joint from the steering gearbox.

30.   Remove the P/S heat baffle plate.

31.   Remove the gearbox stiffener.

32.   Remove the dynamic damper between the gearbox stiffener and front suspension subframe.

33.   Remove the steering gearbox mounting bolts and washers on the right, then left gearbox mount.

34.   Remove the 4-way brake line joint bolts.

35.   Move the steering gearbox to the passenger's side, and rotate it so the pinion shaft points toward the front of the vehicle.

36.   Carefully move the steering gearbox as an assembly toward the left side of the vehicle until the pinion shaft clears the wheel well opening. Be careful not to damage the brake lines with the pinion shaft.

37.   Lift the passenger's side, and remove the steering gearbox through the wheel well opening on the driver's side.

38.   After removing the steering gearbox, make sure that no power steering fluid gets on the gearbox mount cushions, gearbox housing, surface of the front suspension subframe, and stiffener. Wipe off any spilled fluid at once.

To install:

39.   Apply a mild soap and water solution to both sides of the mount cushion mating surfaces (A).

40.   Lift and pass the passenger's side of the steering gearbox through the wheel well opening on the driver's side.

41.   Carefully move the steering gearbox toward the passenger's side until the pinion shaft clears the wheel well opening on the body.

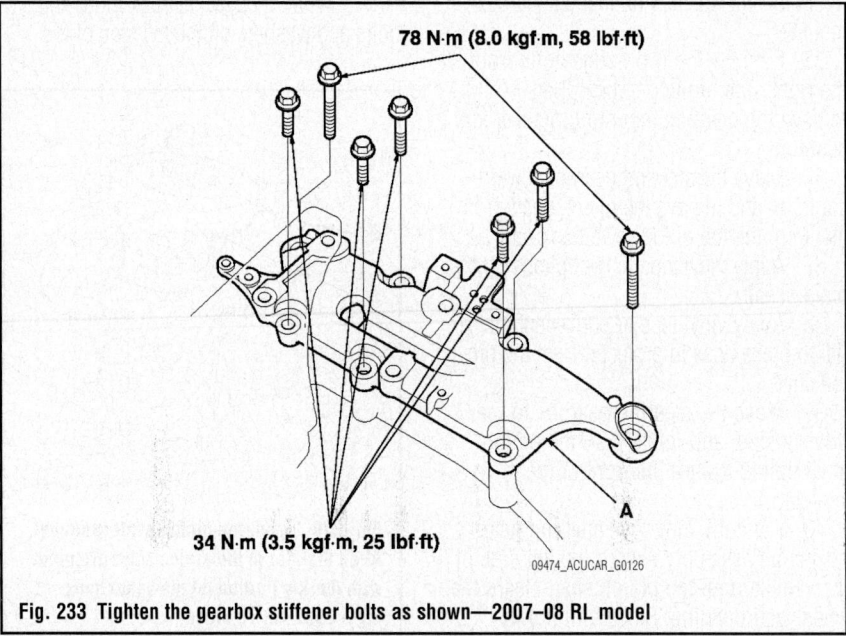

Fig. 233 Tighten the gearbox stiffener bolts as shown—2007–08 RL model

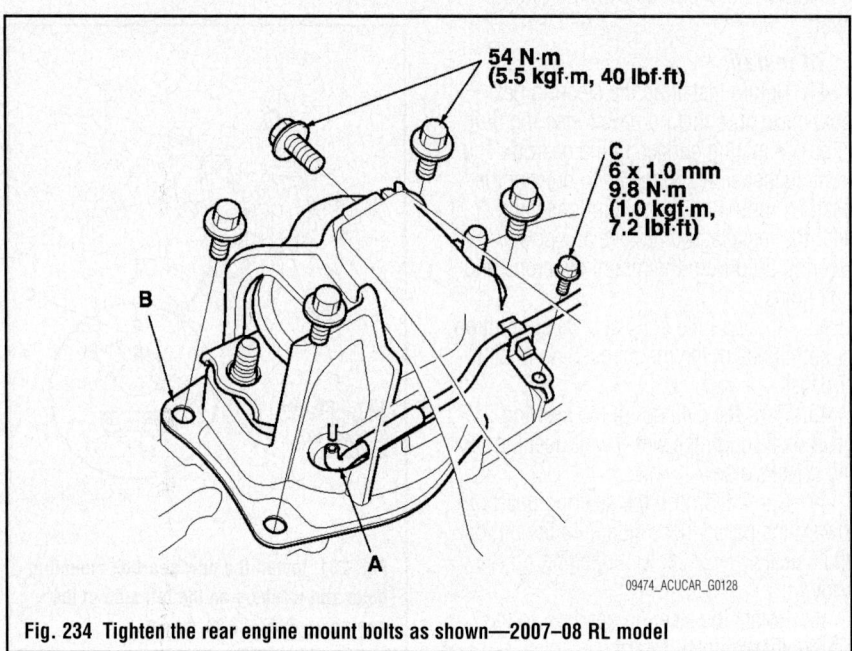

Fig. 234 Tighten the rear engine mount bolts as shown—2007–08 RL model

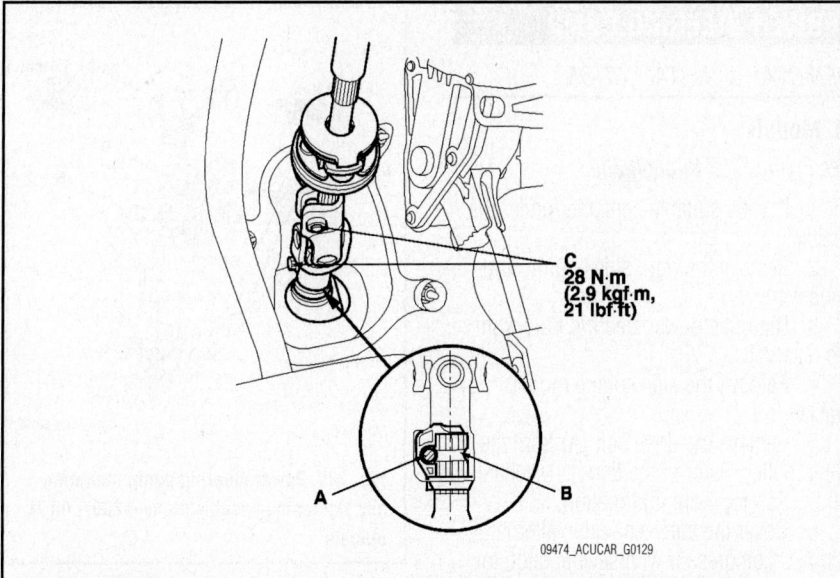

Fig. 235 Align the bolt hole on the steering joint with the groove around the pinion shaft—2007–08 RL model

42. Rotate the steering gearbox so the pinion shaft points upward, then move the steering gearbox to the driver's side.

43. Install the 4-way brake line joint.

44. Install the steering gearbox mounting bolts and washers on the mounts, then tighten them to 58 ft. lbs. (78 Nm).

45. Install the gearbox stiffener, then tighten the mounting bolts to the torque shown in the illustration.

46. Install a dynamic damper in the reverse order of removal.

47. Install the remaining components keeping in mind the following:

a. Tighten the rear engine mount bolt to the specifications shown in the illustration.

b. Install the nuts on the rear engine mount, then tighten them to 47 ft. lbs. (64 Nm).

c. Install the tie-rod end ball joint nut, and tighten it 43 ft. lbs. (59 Nm). Install the new cotter pin.

Install the pinion shaft grommet. Align the slot in the pinion shaft grommet with the lug portion on the valve housing.

d. Align the bolt hole on the steering joint with the groove around the pinion shaft, and loosely install the joint bolts. Be sure that the joint bolt is securely in the groove in the pinion shaft. Pull on the steering joint to make sure that the steering joint is fully seated. Tighten the steering joint bolt to 21 ft. lbs. (28 Nm).

e. Before installing the steering wheel, make sure the front wheels are aligned straight ahead, then center the cable reel (A). Do this by first rotating the

cable reel clockwise until it stops. Then rotate it counterclockwise about three full turns. The arrow mark (B) on the cable reel label point should point straight up. Refer to the illustration for locations.

f. Position the two tabs of the turn signal canceling sleeve. Route the airbag connector through the steering wheel and install the steering wheel on to the steering column shaft, making sure the steering wheel hub engages the pins of the cable reel and tabs of the turn signal canceling sleeve. Do not tap on the steering wheel or steering column shaft when installing the steering wheel.

g. Install the steering wheel nut and tighten it to 36 ft. lbs. (49 Nm). Connect the cable reel connector.

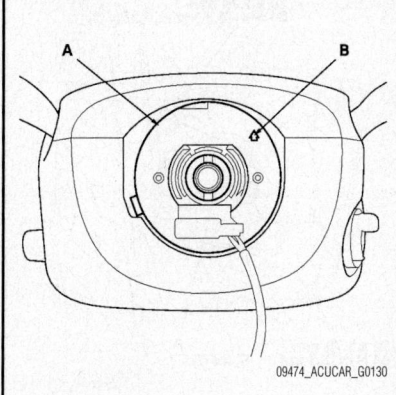

Fig. 236 Center the cable reel (A), the arrow mark (B) on the cable reel label point should point straight up—2007–08 RL model

h. Make sure the wire harness is routed and fastened properly.

i. Fill the system with power steering fluid, and bleed air from the system.

j. Start the engine, allow it to idle, and turn the steering wheel from lock-to-lock several times to warm up to the fluid. Check the gearbox for leaks.

TSX Models

1. Note the radio security code and the radio presets.

2. Disconnect the negative battery cable and wait at least 3 minutes.

3. Make sure the front wheels are in the straight ahead position.

4. Drain the power steering fluid.

5. Remove or disconnect the following:

- Wheels
- Steering wheel access panel
- Supplemental Restraint System (SRS) electrical connector
- 2 Torx®screws
- Horn electrical connector
- Air bag
- Cruise control electrical connector
- Audio remote switch and navigation guide switch, if equipped
- Steering wheel
- Steering joint cover A
- Steering joint lower bolt and pull the joint toward the column. Hold the slider shaft on the column with a piece of wire between the yoke joint on the slider shaft to the joint yoke on the upper shaft
- Center guide and discard
- Steering joint cover B. Do not damage the mating surfaces of the cover and pinion shaft grommet. Replace the cover seal if necessary
- Tie rods from the steering knuckles
- Power steering heat baffle
- Feed line holder mounting from the front suspension subframe
- Feed line holder mounting bolt and return hose from the gearbox mounting bracket
- Shift linkage (if equipped with a manual transmission)

6. Place some rags under the fluid line connections and cover the gearbox mounting part to protect it from power steering fluid. Loosen the flare nut and disconnect the fluid and return lines.

❊❊ CAUTION

After disconnecting the fluid and return lines, plug or cap the hose and pipe to prevent foreign materials from entering the valve body unit.

➡ Do not loosen the cylinder pipes between the valve body unit and the cylinder.

7. Remove or disconnect the following:
- Front suspension subframe left mid mount
- Steering gear mounting bolts from the left gearbox mount, then remove the steering stiffener plate
- 2 flange bolts from the right side of the gearbox, then remove the gearbox mounting bracket, and cushion

8. Move the steering gear toward the front, and remove the pinion shaft grommet from the top of the valve body unit.

9. Put vinyl tape on the brake lines to protect them from the pinion shaft.

10. Move the steering gear to the driver's side and rotate it so the pinion shaft points toward the front of the vehicle.

11. Carefully move the steering gearbox toward the driver's side of the vehicle until the pinion shaft clears the wheel well opening. Do not damage the brake lines with the pinion shaft.

12. Remove the steering gear from the driver's side wheel well opening.

➡ Make sure no power steering fluid gets on the gearbox mount cushions, gearbox housing, surfaces of the subframe or stiffener. Wipe up any spilled fluid immediately.

To install:
13. Installation is the reverse of the removal procedure, while using the following torque values:
- Left steering gear mounting bracket flange bolts: 28 ft. lbs. (38 Nm)
- Right steering gear mounting bolts: 43 ft. lbs. (58 Nm)

➡ After installing the steering gear, check the air hose connections for interference with adjacent parts.

- Return line flare nut: 21 ft. lbs. (28 Nm)
- Feed line flare nut: 31 ft. lbs. (42 Nm)
- Front suspension subframe mid mount bottom (long) bolt: 33 ft. lbs. (44 Nm)
- Front suspension subframe mid mount side (short) bolts: 36 ft. lbs. (49 Nm)
- Steering joint to pinion shaft bolt: 21 ft. lbs. (29 Nm)
- Tie rod end nuts: 32 ft. lbs. (43 Nm)
- Steering wheel nut: 29 ft. lbs. (39 Nm)
- Air bag bolts: 86 inch lbs. (9.8 Nm)

POWER STEERING PUMP

REMOVAL & INSTALLATION

TL Models
See Figures 237 through 239.

1. Place a suitable container under the vehicle.

2. Remove the right side engine component cover.

3. Drain the power steering fluid from the reservoir.

4. Remove the side engine mount bracket.

5. Remove the drive belt (A) from the pump pulley. Refer to the Engine Mechanical Section for more information.

6. Cover the auto-tensioner, alternator, and A/C compressor with several shop towels to protect them from spilled power steering fluid. Disconnect the pump inlet hose (B) and pump outlet hose (C) from the pump (D), and plug them. Take care not to spill the fluid on the body or parts. Wipe off

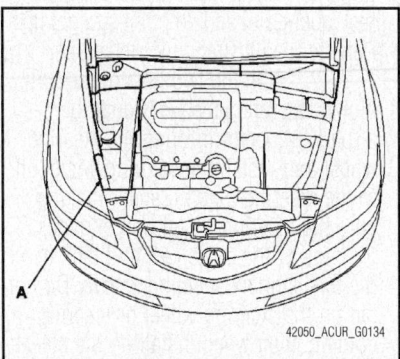

Fig. 237 Remove the right side engine component cover (A)

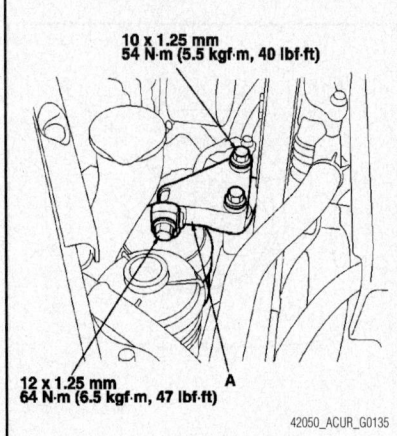

Fig. 238 Side engine mount bracket (A) and tightening specifications—2007–08 TL models

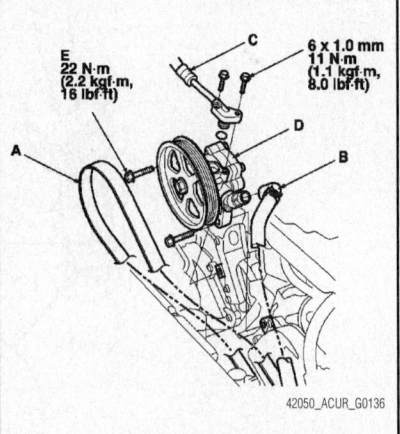

Fig. 239 Power steering pump mounting and tightening specifications—2007–08 TL models

any spilled fluid at once. Do not turn the steering wheel with the pump removed.

7. Remove the pump mounting bolts (E).

To install:
8. Cover the opening of the pump with a piece of tape to prevent foreign material from entering the pump.

9. Connect the pump inlet hose and pump outlet hose onto the new pump.

10. Loosely install the pump in the pump bracket with the mounting bolts, then tighten the pump fittings securely.

11. Tighten the pump mounting bolts to the specified torque.

12. Install the drive belt (A). Make sure that the belt is properly positioned on the pulleys (B).

✳✳ WARNING

Do not get power steering fluid or grease on the auto-tensioner, alternator, A/C compressor, and drive belt or pulley faces. Clean off any fluid or grease before installation.

13. Install the side engine mount bracket. Tighten the bolts to the specified torque.

14. Fill the reservoir to the upper level line and bleed the system.

15. Install the right side engine component cover.

RL Models
See Figures 240 and 241.

1. Place a suitable container under the vehicle.

2. Remove the right front fender trim.

3. Drain the power steering fluid from the reservoir.

4. Remove the drive belt (A) from the pump pulley.

5. Cover the auto-tensioner, alternator, and A/C compressor with several shop towels to protect them from spilled power steering fluid. Disconnect the pump inlet hose (B) and pump outlet hose (C) from the pump (D), and plug them. Take care not to spill the fluid on the body or parts. Wipe off any spilled fluid at once. Do not turn the steering wheel with the pump removed.

6. Remove the pump mounting bolts (E).

7. Cover the opening of the pump with a piece of tape to prevent foreign material from entering the pump.

To install:

8. Connect the pump inlet hose and pump outlet hose onto the new pump.

9. Loosely install the pump in the pump bracket with the mounting bolts, then tighten the pump fittings securely.

10. Tighten the pump mounting bolts to

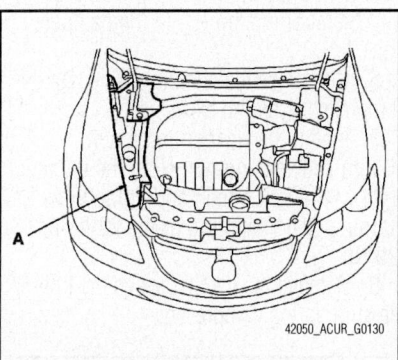

Fig. 240 Remove the right front fender trim (A)

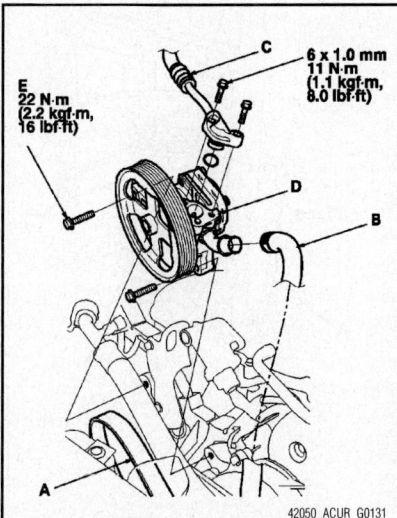

Fig. 241 Exploded view of the power steering pump (C) and related components—2007–08 RL models

the specified torque shown in the accompanying illustration.

11. Install the drive belt (A). Make sure that the belt is properly positioned on the pulleys (B).

❋❋ WARNING

Do not get power steering fluid or grease on the auto-tensioner, alternator, A/C compressor, and drive belt or pulley faces. Clean off any fluid or grease before installation.

12. Fill the reservoir to the upper level line and bleed the system.

13. Install the right side engine component cover.

TSX Models

See Figure 242.

1. Place a suitable container under the vehicle.

2. Drain the power steering fluid from the reservoir.

3. Remove the drive belt (A) from the pump pulley.

4. Remove the pump mounting bolts (B).

5. Cover the auto-tensioner, alternator, and A/C compressor with several shop towels to protect them from spilled power steering fluid. Disconnect the pump inlet hose (C) and pump outlet hose (D) from the pump (E), and plug them. Take care not to spill the fluid on the body or parts. Wipe off any spilled fluid at once. Do not turn the steering wheel with the pump removed.

6. Cover the opening of the pump with a piece of tape to prevent foreign material from entering the pump.

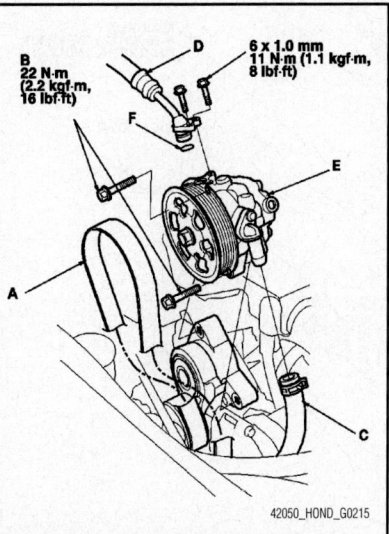

Fig. 242 Power steering pump mounting and tightening specifications—2007–08 TSX Models

To install:

7. Connect the pump inlet hose and the pump outlet hose onto the new pump with the new O-ring (F).

8. Loosely install the pump in the pump bracket with the mounting bolts, then tighten the pump fittings securely.

9. Install the drive belt (A). Make sure that the belt is properly positioned on the pulleys.

❋❋ WARNING

Do not get power steering fluid or grease on the auto-tensioner, alternator, A/C compressor, and drive belt or pulley faces. Clean off any fluid or grease before installation.

10. Tighten the pump mounting bolts to the specified torque, as shown in the accompanying illustration.

11. Fill the reservoir to the upper level line and bleed the system, as outlined in this section.

BLEEDING

Check the reservoir at regular intervals, and add the recommended fluid as necessary. Always use Honda Power Steering Fluid. Using any other type of power steering fluid or automatic transmission fluid can cause increased wear and poor steering in cold weather.

➡**If the fluid is contaminated, the screen in the reservoir may be partially blocked. Replace the reservoir if necessary.**

1. Remove any panels necessary to access the power steering fluid reservoir.

2. Remove the reservoir from its holder. Raise the reservoir, then disconnect the return hose to drain the reservoir. Take care not to spill the fluid on the body and parts. Wipe off any spilled fluid at once.

➡**Inspect the reservoir screen for any debris. If the reservoir screen is clogged, replace the reservoir.**

3. Connect a hose of suitable diameter to the disconnected return hose, and put the hose end in a suitable container.

4. Start the engine, let it run at idle, and turn the steering wheel from lock-to-lock several times. When fluid stops running out of the hose, shut off the engine. Discard the fluid.

5. Reinstall the return hose on the reservoir.

6. Fill the reservoir to the upper level line.

7. Start the engine and run it at fast idle, then turn the steering from lock-to-lock several times to bleed air from the system.

8. Recheck the fluid level and add some if necessary. Do not fill the reservoir beyond the upper level line. If the fluid is contaminated, dark, or discolored, repeat the procedure as necessary.

STEERING LINKAGE

REMOVAL & INSTALLATION

Tie Rod Ends

1. Raise and safely support the vehicle. Remove the front wheels.

2. Remove the cotter pin and the nut from the tie rod end. Use a press type ball joint remover tool, separate the tie rod from the steering knuckle. Be careful to not damage the threads on the joint.

3. Disconnect the air tube at the dust seal joint. Remove the tie rod dust seal bellows clamps and move the rubber bellows back on the tie rod rack joints.

4. Straighten the tie rod lockwasher tabs at the tie rod-to-rack joint and remove the tie rod by turning it with a wrench. On some models, the lock washer is staked.

To install:

5. Reverse the removal procedure. Always use a new tie rod lockwasher and cotter pin during reassembly.

6. Torque the tie rod end-to-power steering gear to 40 ft. lbs. (55 Nm) and the tie rod end-to-steering knuckle nut to 32 ft. lbs. (44 Nm). Install a new cotter pin.

7. Fit the locating lugs into the slots on the rack and bend the outer edge of the washer over the flat part of the rod, after the tie rod nut has been properly tightened.

SUSPENSION

See Figure 243.

COIL SPRING

REMOVAL & INSTALLATION

TL & RL Models

See Figures 244 and 245.

1. Compress the damper spring with a strut spring compressor then remove the self-locking nut while holding the damper shaft with a hex wrench. Do not compress the spring more than necessary to remove the nut.

2. Release the pressure from the strut spring compressor, then disassemble the damper.

To install:

3. Reassemble the damper except for the washer and self-locking nut.

Install the damper assembly on a strut

FRONT SUSPENSION

spring compressor and compress the spring lightly.

4. Align the bottom of the spring and the stepped part of the lower spring seat.

5. Position the damper mounting base (A) so the stud bolt (B) is aligned with the aligning tab (C) in the damper unit.

6. Compress the damper spring. Do not compress the spring excessively.

7. Install the washer and a new 10 mm self-locking nut. Hold the damper shaft with a hex wench and tighten the 10 mm self-locking nut to the 22 ft. lbs. (29 Nm).

8. Remove the damper assembly from the strut spring compressor.

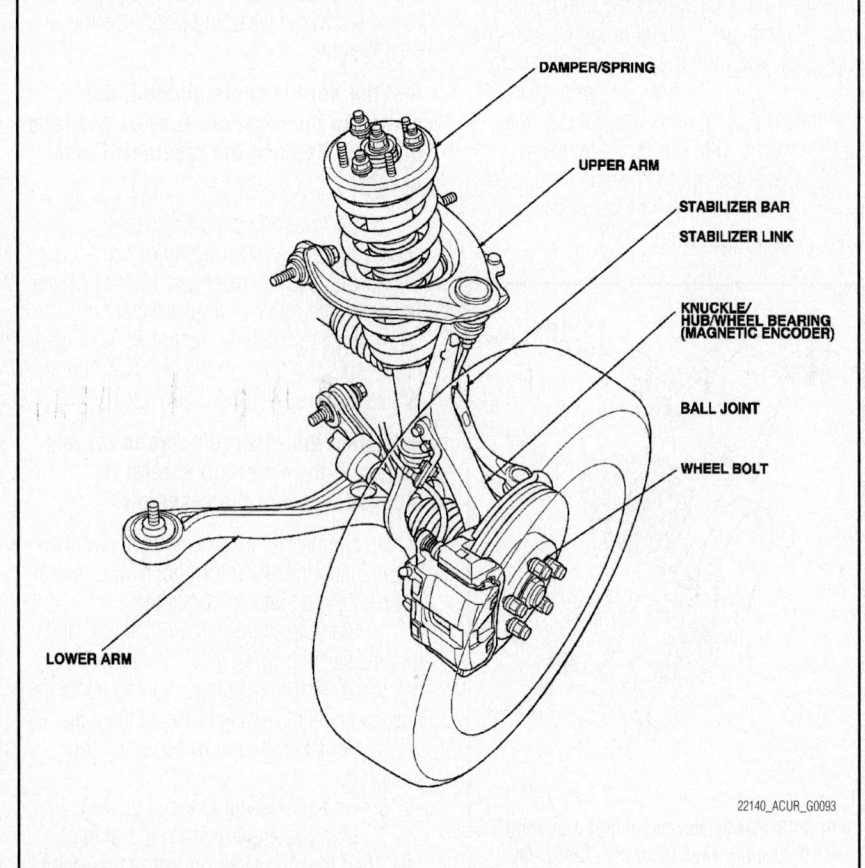

DAMPER/SPRING

UPPER ARM

STABILIZER BAR

STABILIZER LINK

KNUCKLE/
HUB/WHEEL BEARING
(MAGNETIC ENCODER)

BALL JOINT

WHEEL BOLT

LOWER ARM

22140_ACUR_G0093

Fig. 243 Front suspension component locations—All Models

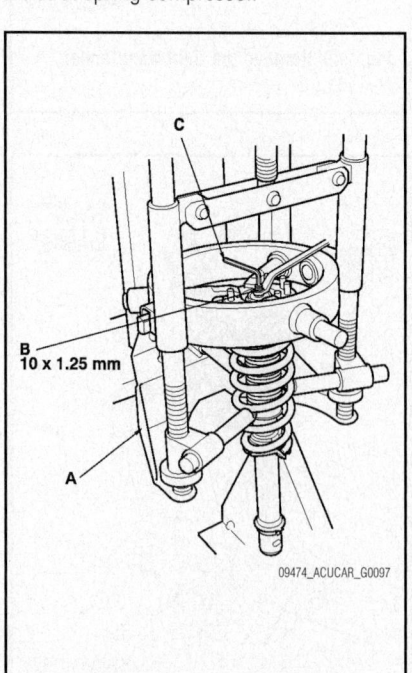

C

B
10 x 1.25 mm

A

09474_ACUCAR_G0097

Fig. 244 Compress the spring assembly and remove the locking nut—2007–08 TL and RL Models

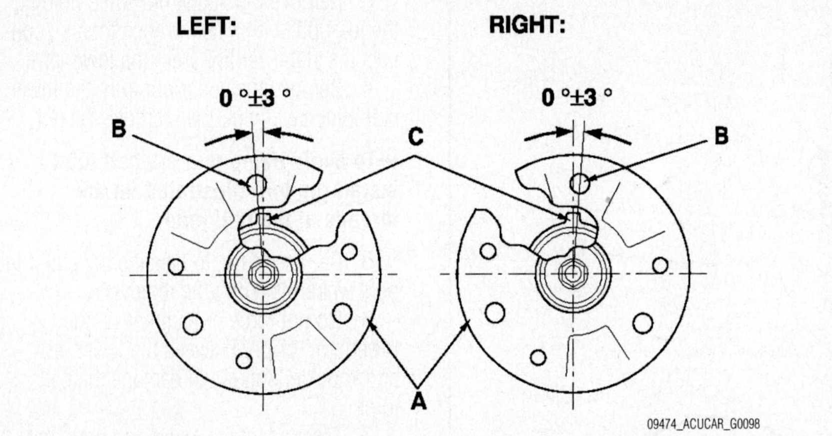

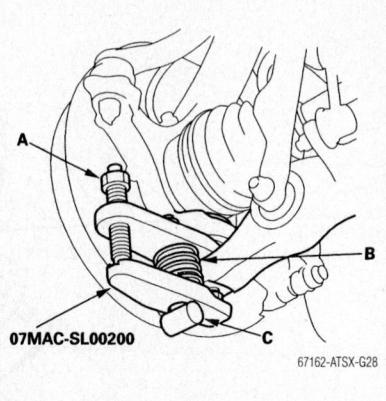

Fig. 245 Position the damper mounting base (A) so the stud bolt (B) is aligned with the aligning tab (C) in the damper unit—2007–08 TL and RL Models

Fig. 247 Pressure bolt (A), adjusting bolt (B) and proper position of the head of the adjusting bolt (C)—TSX

TSX Models

1. Before servicing the vehicle, refer to the precautions section.

2. Raise and support the vehicle and remove the front wheels.

3. Remove the strut (damper).

4. Place the strut assembly in a coil spring compressor.

5. Compress the coil spring and remove the locking nut from the top of the strut.

6. Release the pressure from the spring compressor.

7. Remove the coil spring and related pieces from the strut.

To install:

➡**Use new self–locking nuts and bolts when assembling the strut.**

8. Install the strut, coil spring and related components on the spring compressor.

9. Compress the spring.

10. Install the mounting washer, and loosely install a new self–locking nut.

11. Hold the strut piston rod with a hex wrench and tighten the self–locking nut to 22 ft. lbs. (30 Nm).

12. Install the strut in the vehicle.

13. Check and adjust the vehicle's front wheel alignment.

LOWER BALL JOINT

REMOVAL & INSTALLATION

See Figures 246 and 247.

1. Raise and safely support the vehicle
2. Remove or disconnect the following:
   - Wheels
   - Axle nut
   - Brake hose from the knuckle

- Brake caliper mounting from the knuckle
- Wheel sensor wire bracket and the sensor from the knuckle
- Tie rod end from the knuckle
- Lower ball joint from the control arm
- Upper ball joint from the knuckle
- Knuckle and hub by sliding the assembly off the halfshaft. Tap the end of the halfshaft with a plastic mallet to release it from the knuckle.
- Hub and rotor assembly from the knuckle
- Splash guard from the knuckle

To install:

3. Install or connect the following:
- Splash guard and torque the bolts to 84 inch lbs. (9.5 Nm)
- Hub assembly and tighten the self–locking bolts to 33 ft. lbs. (45 Nm)

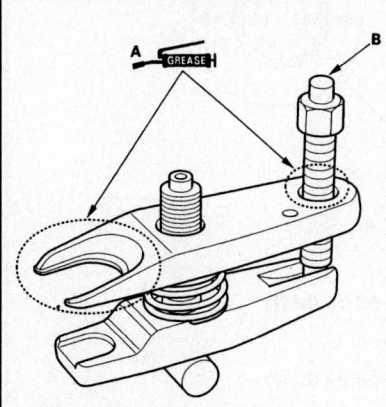

Fig. 246 Apply grease to the A & B areas on the ball joint separator tool—TSX

➡**Be sure that all the hub bolts are properly tightened to avoid warpage of the brake disc.**

- Knuckle and hub assembly onto the halfshaft
- Tie rod and torque the nut to 36–43 ft. lbs. (49–59 Nm)
- Upper ball joint and torque the nut to 29–35 ft. lbs. (39–47 Nm)
- Lower ball joint and torque the nut to 36–43 ft. lbs. (49–59 Nm)
- Wheel sensor and torque the bolts to 16 ft. lbs. (22 Nm)
- Wheel sensor wire and torque the bolts to 84 inch lbs. (9.5 Nm)
- Brake caliper and torque the bolts to 80 ft. lbs. (108 Nm)
- Brake hoses and torque the bolts to 84 inch lbs. (9.5 Nm)
- New axle nut and torque it to 181 ft. lbs. (245 Nm)
- Wheel

4. Measure and adjust the wheel alignment.

LOWER CONTROL ARM

REMOVAL & INSTALLATION

TL Models

See Figures 248 through 252.

1. Remove the front wheels.

2. Remove the damper fork from the damper and lower arm.

➡**During installation, insert the damper fork into the damper lower end so the aligning tab (D) is aligned with the slot (E) in the damper fork. Replace the damper fork mounting nut (F) with a new one.**

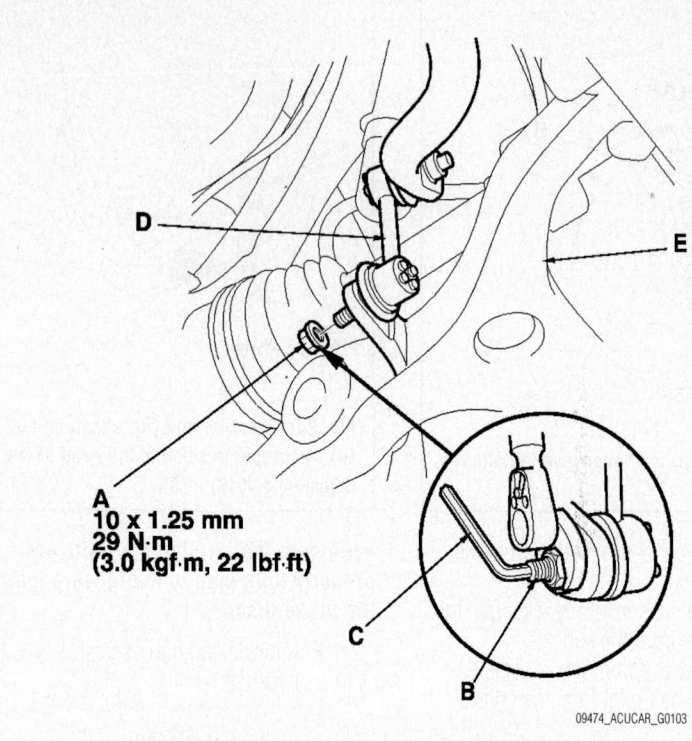

A
10 x 1.25 mm
29 N·m
(3.0 kgf·m, 22 lbf·ft)

09474_ACUCAR_G0103

Fig. 248 Remove the flange nut while holding the joint pin with a hex wrench—2007–08 TL model

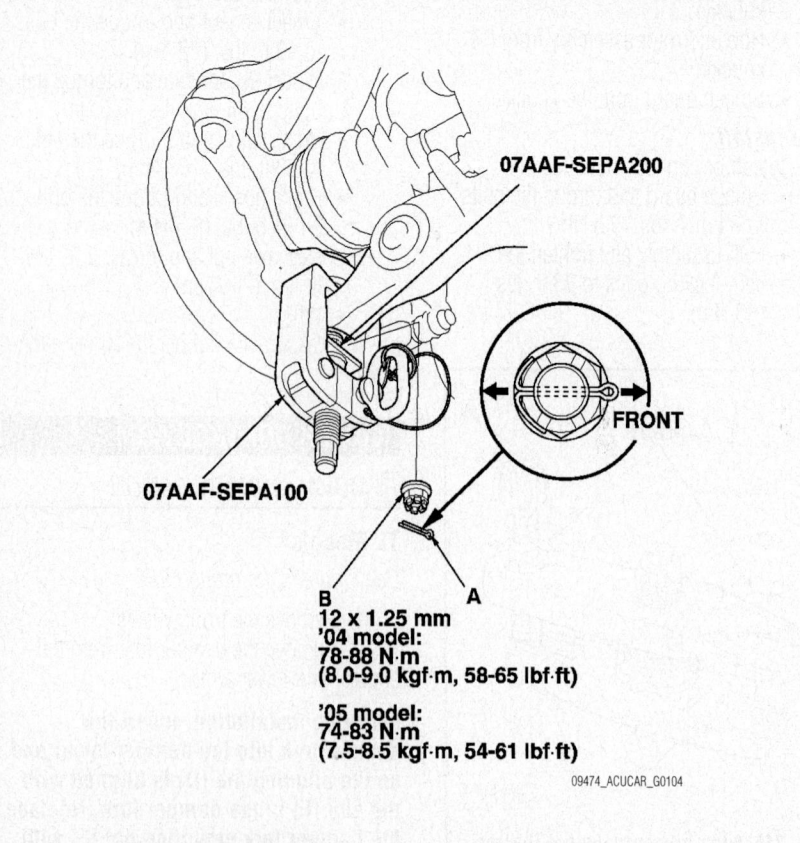

07AAF-SEPA200

07AAF-SEPA100

FRONT

B   A
12 x 1.25 mm
'04 model:
78-88 N·m
(8.0-9.0 kgf·m, 58-65 lbf·ft)

'05 model:
74-83 N·m
(7.5-8.5 kgf·m, 54-61 lbf·ft)

09474_ACUCAR_G0104

Fig. 249 Removing/installing the lower ball joint—2007–08 TL model

3. Remove the flange nut while holding the joint pin with a hex wrench then disconnect the stabilizer link from the lower arm.

4. Remove the cotter pin from the lower ball joint castle nut, then remove the nut.

➡ **To avoid damaging the ball joint, install the tool illustrated on the threads of the ball joint.**

5. Be careful not to damage the ball joint boot when installing the remover.

6. Do not force or hammer on the lower arm, or pry between the lower arm and knuckle. You could damage the ball joint.

7. Disconnect the lower arm ball joint from the knuckle using the tools shown.

8. Remove the flange bolts, nuts and washers, then remove the lower arm.

To install:

9. Install the lower arm in the reverse order of removal, and note the following:

  a. Check the collar sleeve (A) on the lower arm (B). Replace the lower arm if the collar/sleeve is loose or damaged.

  b. Be careful not to damage the ball joint boot when installing the knuckle.

  c. Tighten all mounting hardware to the torque specified in the accompanying illustrations.

  d. Before connecting the lower ball joint to the lower arm, degrease the threaded section and tapered portion of the ball joint pin, the lower arm connecting hole, the threaded section and mating surface of the castle nut.

  e. First install all the components and lightly tighten the bolts and nuts, then raise the suspension to load it with the vehicle's weight before fully tightening to the specified torque values. Do not place the jack against the ball joint pin of the knuckle.

  f. Torque the castle nut to the lower torque specification, then tighten it only far enough to align the slot with the ball joint pin hole. Do not align the castle nut by loosening it.

  g. Install a new cotter pin on the castle nut after torquing.

  h. Check the wheel alignment, and adjust if necessary.

RL Models

See Figures 253 through 256.

1. Remove the front wheels.

2. Remove the holder from the knuckle.

3. Remove the cotter pin from the lower ball joint castle nut and then remove the nut.

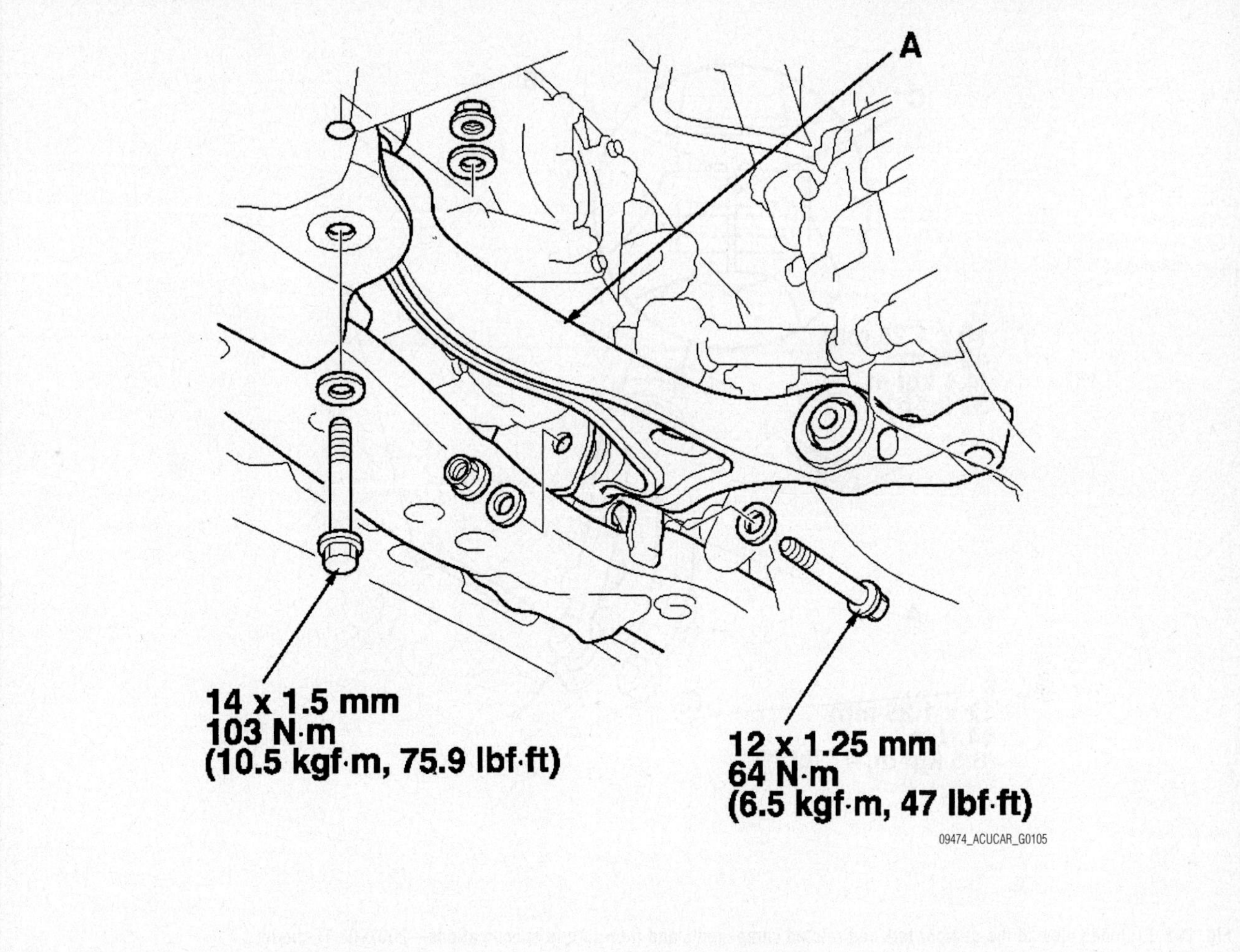

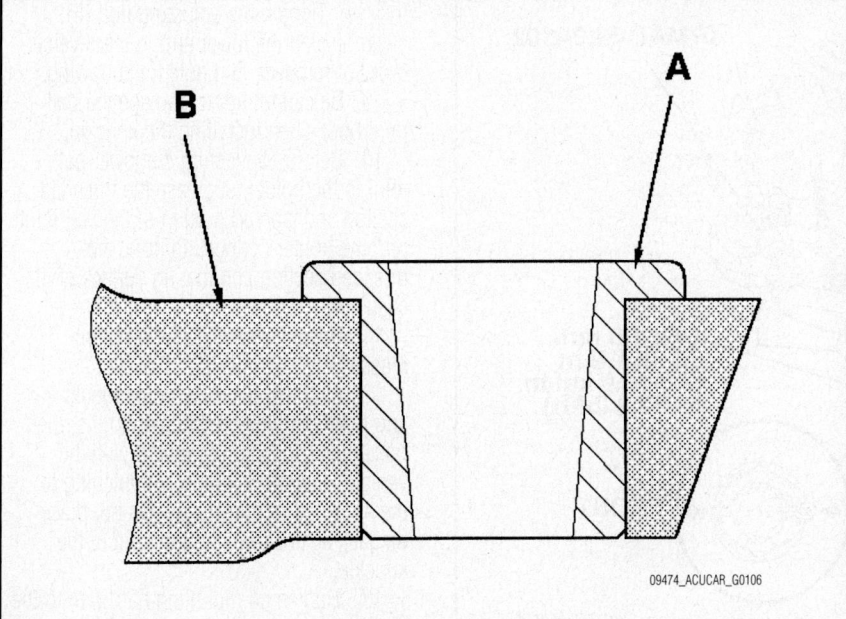

Fig. 250 Removing/installing the flange bolts, nuts, washers and lower arm—2007–08 TL model

14 x 1.5 mm
103 N·m
(10.5 kgf·m, 75.9 lbf·ft)

12 x 1.25 mm
64 N·m
(6.5 kgf·m, 47 lbf·ft)

09474_ACUCAR_G0105

Fig. 251 Check the collar sleeve (A) on the lower arm (B). Replace the lower arm if the collar/sleeve is loose or damaged—2007–08 TL model

09474_ACUCAR_G0106

→To avoid damaging the ball joint, install a hex nut on the threads of the ball joint. Be careful not to damage the ball joint boot when installing the remover. Do not force or hammer on the lower arm, or pry between the lower arm and knuckle. You could damage the ball joint. Insert the new cotter pin into the ball joint pin hole from the rear to the front of vehicle, and bend its end as shown.

4. Disconnect the lower arm ball joint from the holder using the ball joint remover.

5. Remove the self-locking nut and washer, then remove the damper fork mounting nut and bolt.

6. Remove the flange bolts and self-locking nut, then remove the lower arm.

To install:

7. In case of loosening lower arm bracket, follow the steps to retighten its bolts, refer to the illustration for bolt locations:

   a. Lightly tighten the three bolts.

   b. Align the center of the cam (B) to the edge (C)

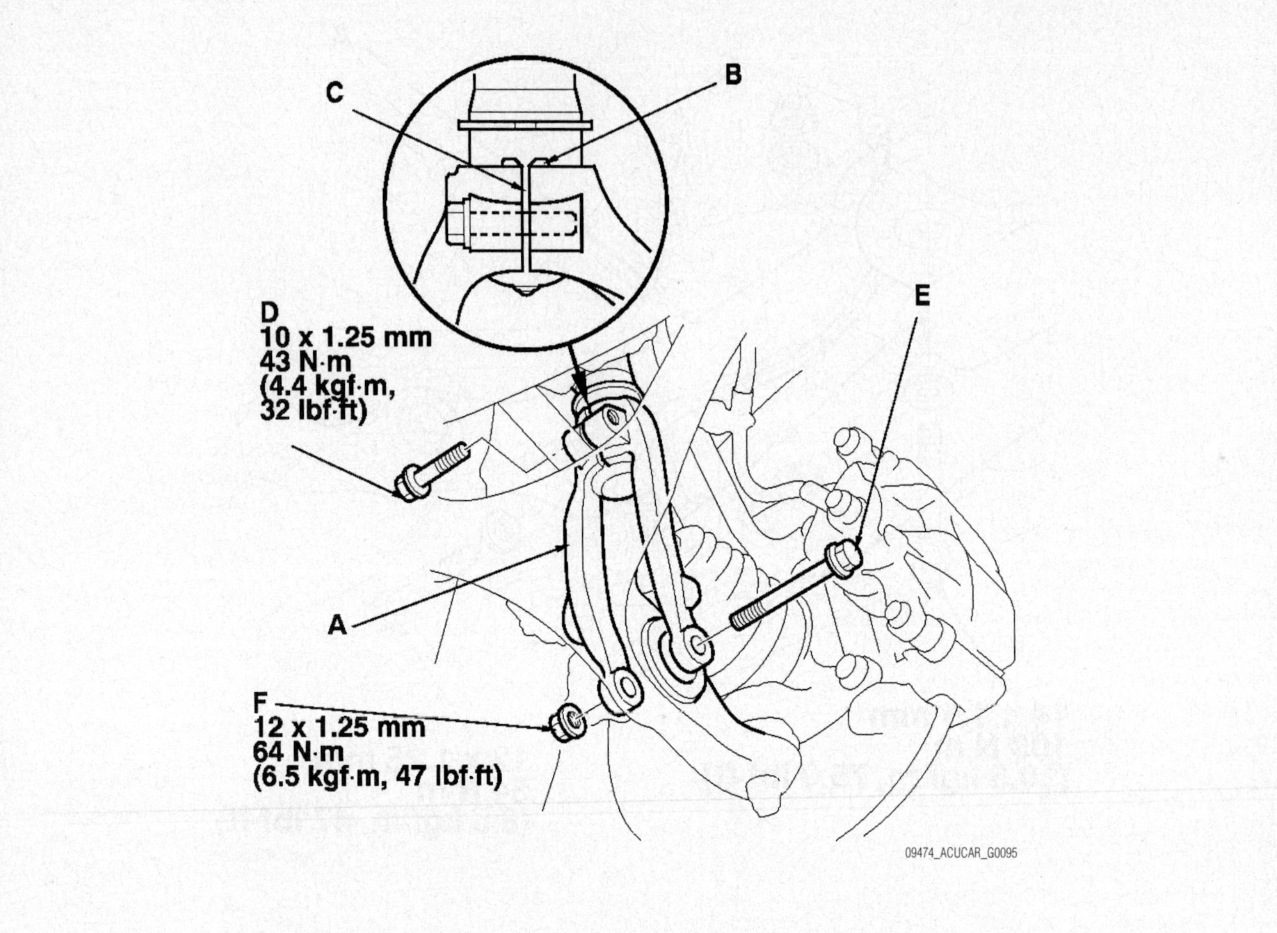

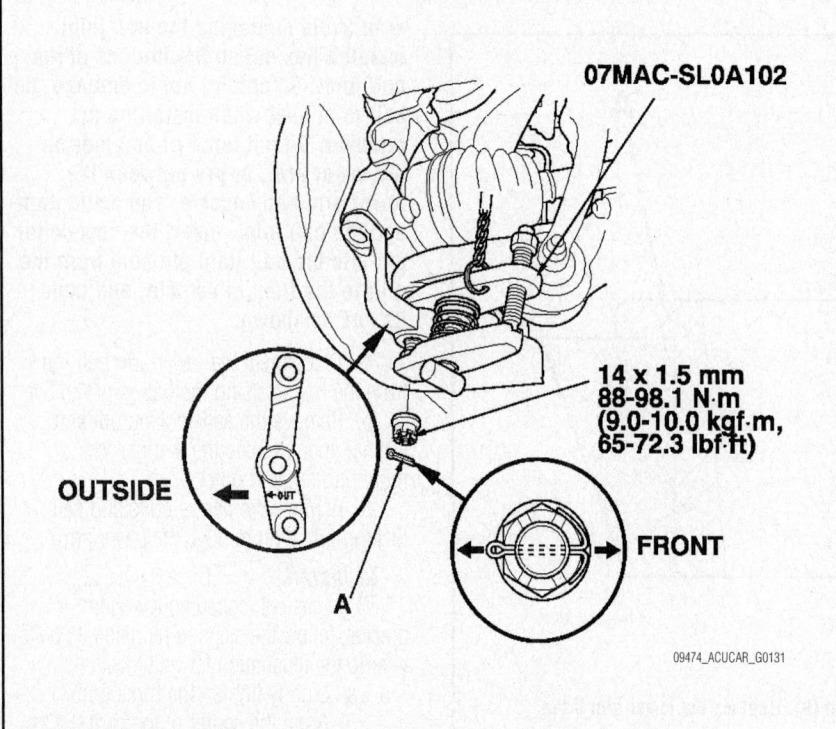

D
10 x 1.25 mm
43 N·m
(4.4 kgf·m,
32 lbf·ft)

F
12 x 1.25 mm
64 N·m
(6.5 kgf·m, 47 lbf·ft)

09474_ACUCAR_G0095

Fig. 252 Exploded view of the damper fork and related components and their torque specifications—2007–08 TL model

07MAC-SL0A102

14 x 1.5 mm
88-98.1 N·m
(9.0-10.0 kgf·m,
65-72.3 lbf·ft)

OUTSIDE

FRONT

09474_ACUCAR_G0131

Fig. 253 Ball joint removal/installation, front lower arm—2007–08 RL model

c. Tighten the adjusting bolt (D).
d. Tighten the adjusting bolt (E).
e. Tighten the adjusting bolt (F).

8. Install the lower arm in the reverse order of removal, and note the following:

9. Be careful not to damage the ball joint boot when installing the knuckle.

10. Before connecting the lower ball joint to the holder, degrease the threaded section and tapered portion of the ball joint pin, the holder connecting hole, the threaded section and mating surface of the castle nut.

11. Use a new self-locking nut on reassembly.

12. First install all the components and lightly tighten the bolts and nuts, then raise the suspension to load it with the vehicle's weight before fully tightening to the specified torque values. Do not place the jack against the ball joint pin of the knuckle.

13. Tighten all mounting hardware to the specified torque values.

14. Torque the castle nut to the lower torque specification, then tighten it only far enough to align the slot with the ball joint

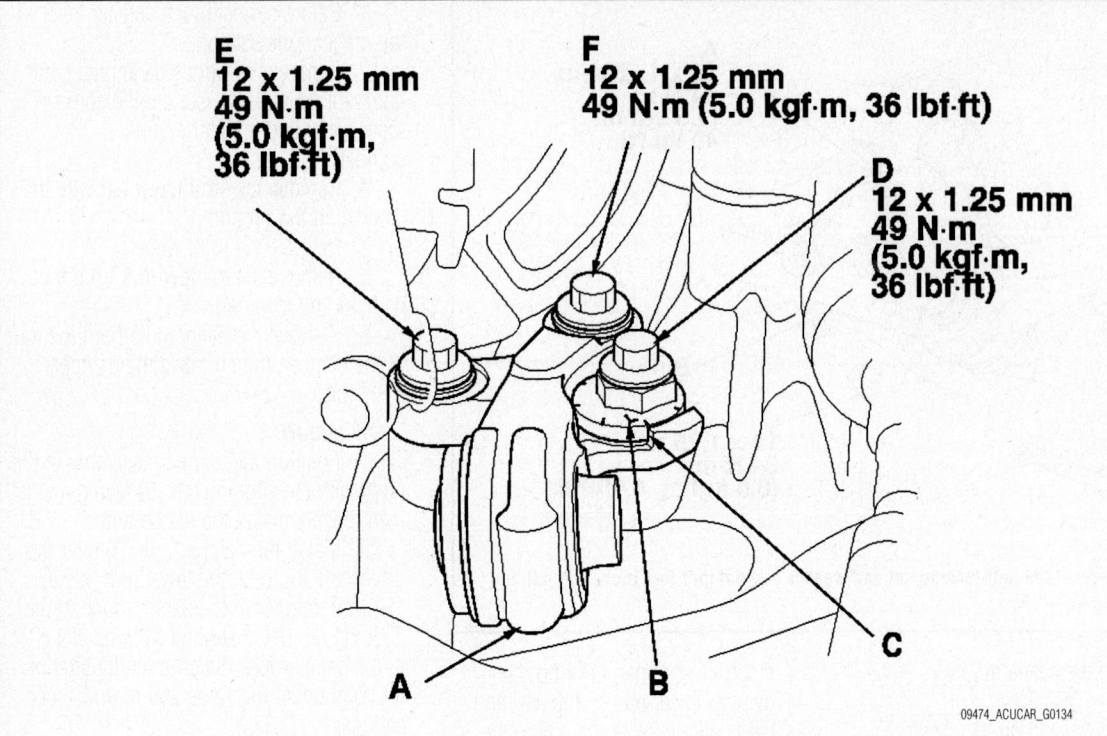

E
12 x 1.25 mm
49 N·m
(5.0 kgf·m,
36 lbf·ft)

F
12 x 1.25 mm
49 N·m (5.0 kgf·m, 36 lbf·ft)

D
12 x 1.25 mm
49 N·m
(5.0 kgf·m,
36 lbf·ft)

A

B

C

09474_ACUCAR_G0134

Fig. 254 In case of loosening the front lower arm bracket, follow the steps to retighten its bolts—2007–08 RL model

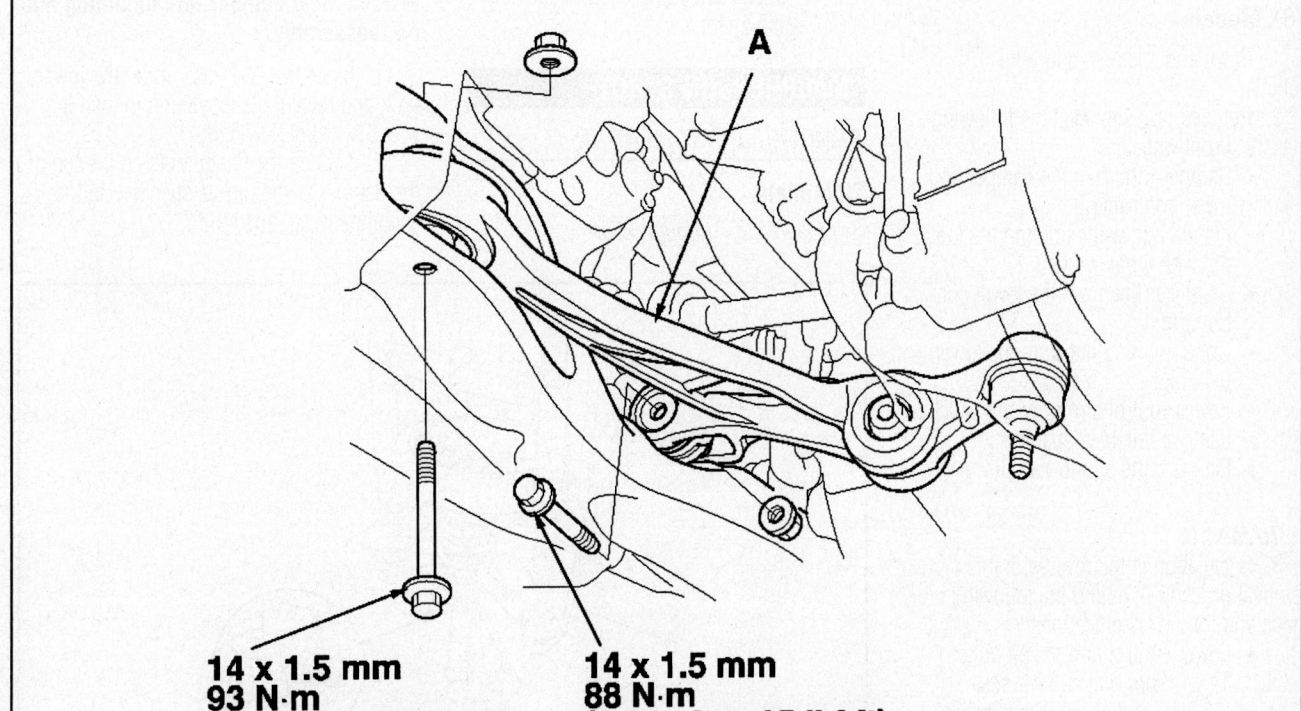

A

14 x 1.5 mm
93 N·m
(9.5 kgf·m, 69 lbf·ft)

14 x 1.5 mm
88 N·m
(9.0 kgf·m, 65 lbf·ft)

09474_ACUCAR_G0133

Fig. 255 Exploded view of the front lower arm and the flange bolts—2007–08 RL model

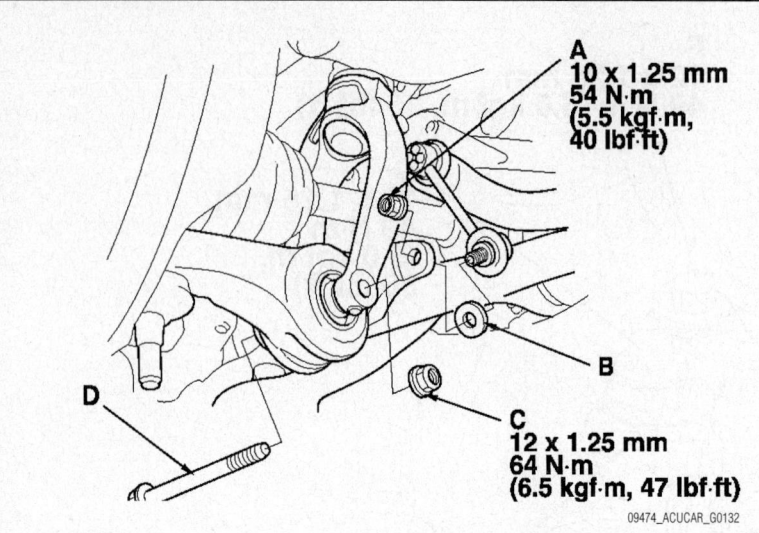

Fig. 256 Exploded view of the self-locking nut and washer and damper fork mounting nut and bolt—2007–08 RL model

pin hole. Do not align the castle nut by loosening it.

15. Install a new cotter pin on the castle nut after torquing.

16. Before installing the wheel, clean the mating surfaces of the brake disc and the inside of the wheel.

17. Check the wheel alignment, and adjust if necessary.

TSX Models

1. Raise and safely support the vehicle.

2. Remove or disconnect the following:
   - Front wheels
   - Damper fork from the damper and lower control arm
   - Flange nut, while holding the joint pin with a hex wrench
   - Stabilizer link from the lower control arm
   - Cotter pin and nut from the lower ball joint
   - Lower control arm from the knuckle using a suitable separator tool
   - Flange bolts and lower control arm

To install:

3. Installation is the reverse of the removal procedure, noting the following steps and torque specifications:
   - Lower control arm flange bolts: 14 x 1.5mm bolt to 61 ft. lbs. (83 Nm) and 12 x 1.25mm bolt to 47 ft. lbs. (64 Nm)
   - Lower ball joint castle nut: 58–65 ft. lbs. (78–88 Nm)
   - Stabilizer link-to-control arm flange nut: 22 ft. lbs. (29 Nm)

- Insert the damper fork into the damper lower end so the aligning lab is aligned with the slot in the damper fork. Use a new damper fork mounting nut
- Lower damper fork mounting bolt: 47 ft. lbs. (64 Nm)
- Upper damper fork mounting bolt: 32 ft. lbs. (44 Nm)
- Check and adjust the front wheel alignment

MACPHERSON STRUT

REMOVAL & INSTALLATION

TL Models

See Figures 257 through 259.

1. Before servicing the vehicle, refer to the precautions section.

2. Remove the right side engine compartment cover, right rear engine compartment cover and left rear engine compartment cover.

3. Remove the strut tower bar with the vehicle on the ground.

4. Remove the front wheel.

5. Remove the damper fork from the damper and lower arm.

6. Remove the flange nuts from the top of the damper, then remove the damper assembly.

To install:

7. Position the damper assembly in the body with the aligning tab (B) facing inside, then loosely install the flange nuts.

8. Install the damper fork (A) over the driveshaft and onto the lower arm. Install the front damper in the damper fork so the aligning tab (B) is aligned with the slot (C) in the damper fork. Refer to the illustration for component locations and torque specifications.

9. Loosely install the damper pinch bolt into the damper fork.

10. Install the flange bolt to the damper fork and lower arm, then lightly tighten the new damper fork mounting nut.

➡**Use a new damper fork mounting nut on reassembly.**

11. Place the floor jack under the lower arm, and raise the suspension to load it with the vehicle's weight.

12. Tighten the flange nuts on the top of the damper to the torque specified in the accompanying illustration.

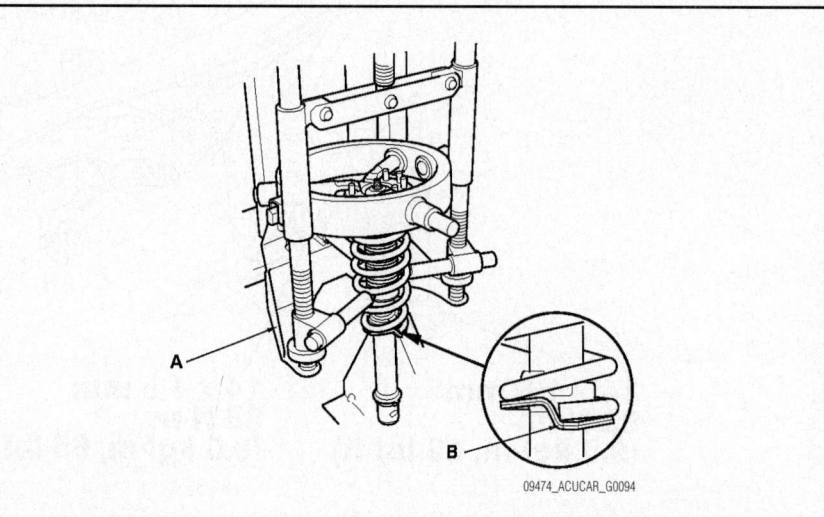

Fig. 257 Position the damper assembly in the body with the aligning tab (B) facing inside—2007–08 TL Models

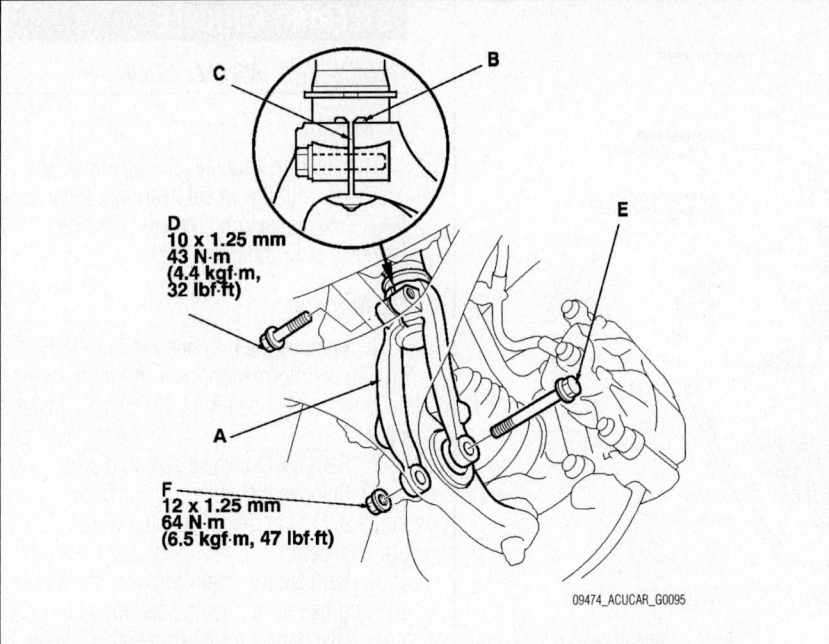

Fig. 258 Exploded view of the damper fork and related components and their torque specifications—2007–08 TL Models

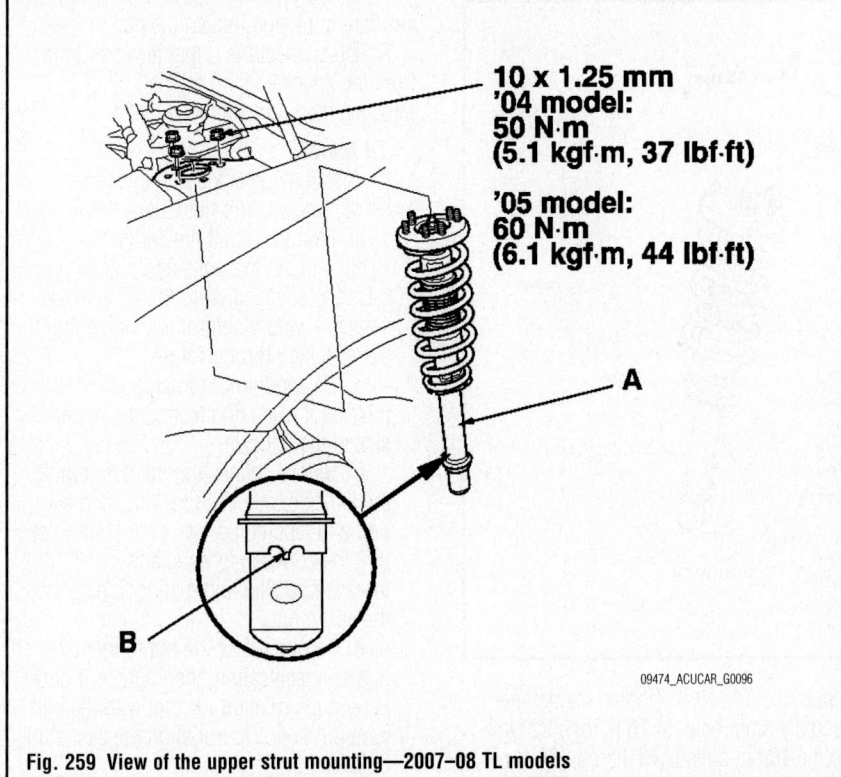

Fig. 259 View of the upper strut mounting—2007–08 TL models

13. Tighten the damper pinch bolts to the torque specified in the accompanying illustration.

14. Tighten the flange nut on the damper fork to the torque specified in the accompanying illustration.

15. Install the strut tower bar with the vehicle on the ground.

16. Install the engine compartment covers.

17. Install the front wheel.

RL Models

1. Before servicing the vehicle, refer to the precautions section.

2. Remove covers from the engine compartment cover.

3. Remove the strut tower bar with the vehicle on the ground.

4. Remove the front wheel.

5. Remove the damper fork from the damper and lower arm.

6. Remove the flange nuts from the top of the damper, then remove the damper assembly.

To install:

7. Position the damper assembly in the body with the aligning tab (B) facing inside, then loosely install the flange nuts.

8. Loosely install the damper pinch bolt into the damper fork.

9. Install the flange bolt to the damper fork and lower arm, then lightly tighten the new damper fork mounting nut.

➡Use a new damper fork mounting nut on reassembly.

10. Place the floor jack under the lower arm, and raise the suspension to load it with the vehicle's weight.

11. Tighten the flange nuts on the top of the damper to 25 ft. lbs. (34 Nm) on the 8 mm nuts and 42 ft. lbs. (55 Nm) on the 10 mm nuts.

12. Tighten the damper pinch bolts to the torque specified in the accompanying illustration.

13. Tighten the flange nut on the damper fork to the torque specified in the accompanying illustration.

14. Install the strut tower bar with the vehicle on the ground.

15. Install the engine compartment covers.

16. Install the front wheel.

TSX Models

See Figures 260 and 261.

1. Before servicing the vehicle, refer to the precautions in the beginning of this section.

2. Raise and safely support the vehicle.

3. Remove or disconnect the following:
   • Front wheels
   • Damper fork bolts, then the damper fork from the damper and lower arm
   • 2 8mm flange nuts and 3 10mm flange nuts
   • Strut (damper assembly) from the vehicle

To install:

➡Use new self-locking bolts when installing the struts and assembling the damper forks.

4. Install or connect the following:
   • Strut into the vehicle with the aligning tab facing inside, if equipped. Hand-tighten the mounting nuts.

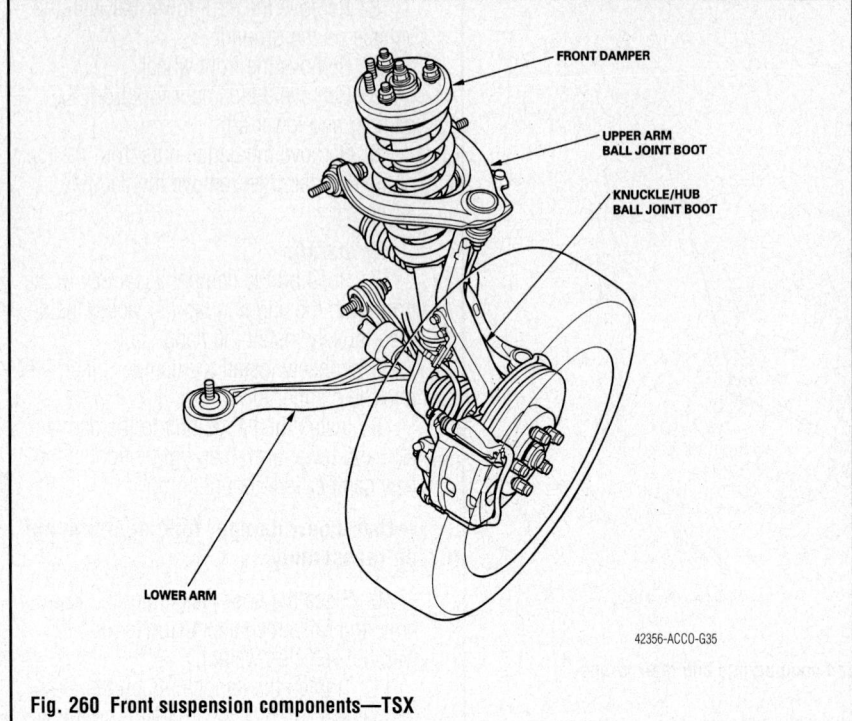

Fig. 260 Front suspension components—TSX

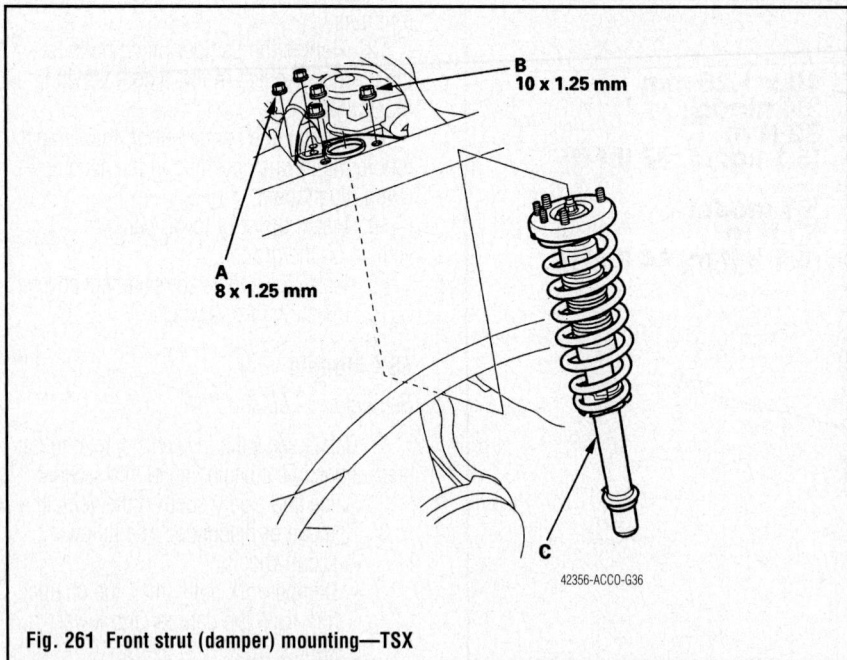

Fig. 261 Front strut (damper) mounting—TSX

- Strut into the damper fork. The alignment mark on the strut tube fits into the slot on the damper fork.
- Pinch bolt and damper fork bolt/nut. Only hand-tighten these bolts.
- Front wheels and lower the vehicle.

5. With all 4 of the vehicle's wheels on the ground, tighten the damper fork nut to 47 ft. lbs. (65 Nm) while holding the damper fork bolt. Tighten the damper fork pinch bolt to 32 ft. lbs. (44 Nm). Tighten the damper assembly 8mm bolts to 16 ft. lbs. (22 Nm) and the 10mm bolts to 37 ft. lbs. (50 Nm).

6. Tighten the wheel nuts to 80 ft. lbs. (110 Nm).

7. Check and adjust the vehicle's front end alignment.

OVERHAUL

For disassembly procedures, refer to the following section, "Front Suspension, Coil Spring, Removal & Installation."

STEERING KNUCKLE

REMOVAL & INSTALLATION

TL Models

The steering knuckle is removed as an assembly with the wheel bearings. Refer to the following section, "Wheel Bearings, Removal & Installation."

RL Models

1. Remove the hub bearing unit. For additional information, refer to the following section, "Wheel bearings, Removal & Installation."

2. Remove the wheel splash guard.

3. Remove the wheel sensor from the knuckle. Do not disconnect the wheel sensor connector.

4. Remove the cotter pin from the tie-rod end ball joint, then loosen the nut.

5. Disconnect the tie-rod end ball joint from the knuckle using the ball joint remover.

6. Remove the holder from the knuckle.

7. Remove the lock pin from the upper arm ball joint, and loosen the nut.

8. Disconnect the upper arm ball joint from the knuckle using the ball joint remover, then remove the knuckle.

To install:

9. Install the knuckle in the reverse order of removal, and note these items:

  a. First install all the components and lightly tighten the bolts and nuts, then raise the suspension to load it with the vehicle's weight before fully tightening to the specified torque values.

  b. Be careful not to damage the ball joint boot when connecting the upper arm to the knuckle.

  c. Before connecting the ball joint to the knuckle, degrease the threaded section and tapered portion of the ball joint pin, the connecting hole, and the threaded section and mating surface of the castle nut.

  d. Torque the castle nut to the lower torque specification, then tighten it only far enough to align the slot with the ball joint pin hole. Do not align the castle nut by loosening it.

  e. Before installing the wheel, clean the mating surfaces on the brake disk and the inside of the wheel.

  f. Check the wheel alignment, and adjust it if necessary.

TSX Models

The steering knuckle is removed as an assembly with the wheel bearings. Refer to the following section, "Wheel Bearings, Removal & Installation."

STABILIZER BAR

REMOVAL & INSTALLATION

See Figure 262.

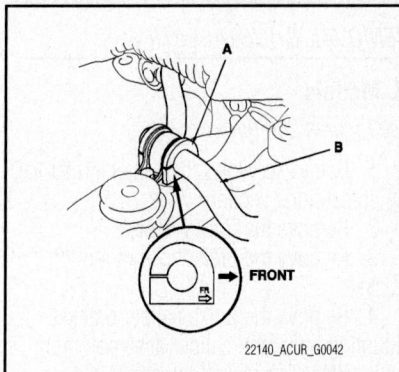

22140_ACUR_G0042

Fig. 262 Orientation of the bushing when installing the front stabilizer bar

1. Raise the front of the vehicle, and support it with safety stands in the proper locations.
2. Remove the front wheels, taking care not to scratch the calipers on the Type-S model.
3. Disconnect the stabilizer links from the stabilizer bar on the right and left sides.
4. Remove the front suspension subframe (A) from the body.
5. Remove the flange bolts (B) and the bushing holders (C), then remove the bushings (D) and the stabilizer bar (E).

To install:

6. Install the stabilizer bar in the reverse order of removal, and note these items:
   a. Note the right and left direction of the stabilizer bar.
   b. Do not set the bushings (A) on the bent or curved part of the stabilizer bar (B).
   c. Note the direction of installation for the bushings.
   d. Refer to stabilizer link removal/installation to connect the stabilizer bar to the links.
   e. Before installing the wheel, clean the mating surfaces of the brake disc and inside of the wheel.
   f. Check the wheel alignment, and adjust it if necessary.

UPPER BALL JOINT

REMOVAL & INSTALLATION

The upper ball joint cannot be removed from the control arm. If the ball joint is

damaged, the upper arm assembly must be replaced.

UPPER CONTROL ARM

REMOVAL & INSTALLATION

TL Models

See Figures 263 and 264.

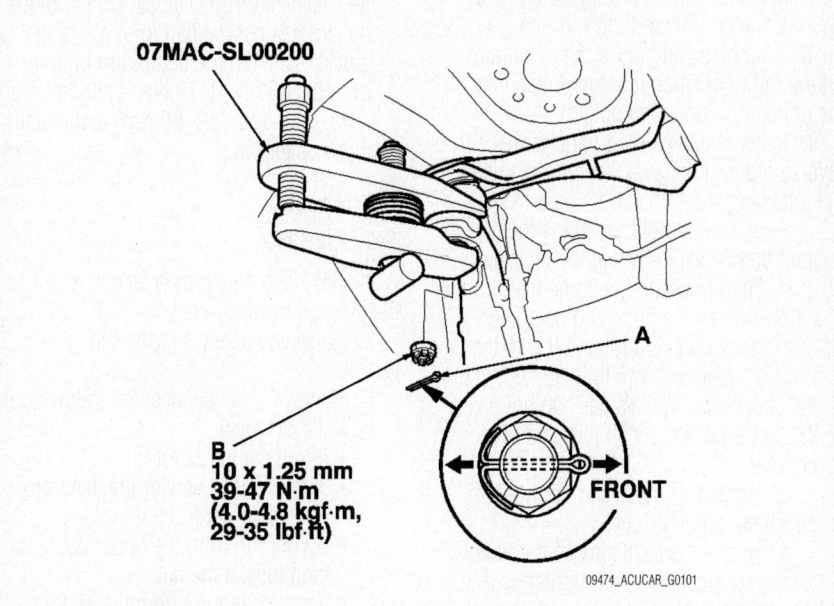

07MAC-SL00200

B
10 x 1.25 mm
39-47 N·m
(4.0-4.8 kgf·m,
29-35 lbf·ft)

FRONT

09474_ACUCAR_G0101

Fig. 263 Use the following tool to separate the ball joint and tighten the nut as shown—2007–08 TL model

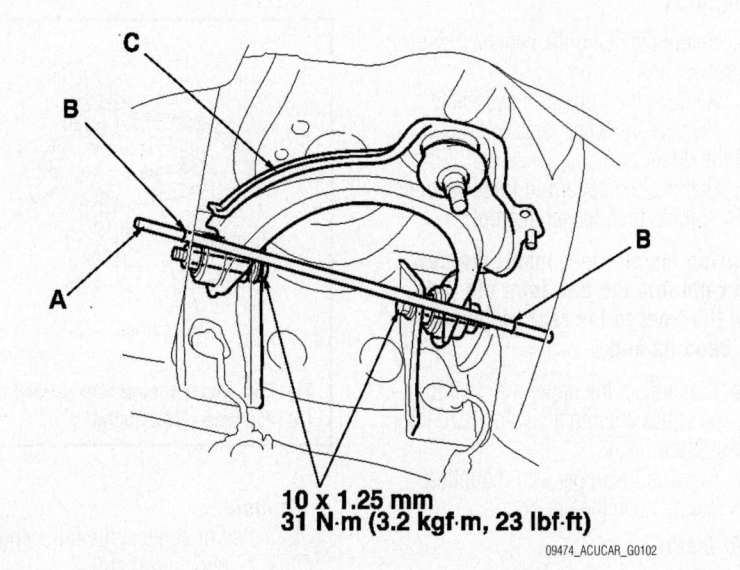

10 x 1.25 mm
31 N·m (3.2 kgf·m, 23 lbf·ft)

09474_ACUCAR_G0102

Fig. 264 Insert a 6mm diameter 300mm long rod into the positioning holes and place the upper arm on the rod to position it before tightening the upper arm mounting bolts—2007–08 TL model

1. Before servicing the vehicle, refer to the precautions section.
2. Remove the front strut assembly.
3. Remove the wheel sensor bracket from the upper arm.
4. Remove the cotter pin from the upper arm ball joint, then loosen the nut.

➡During installation, insert the new cotter pin into the ball joint pin hole from the front to the rear of vehicle, and bend its end as shown.

5. Disconnect the upper arm ball joint from the knuckle using the tool illustrated.

6. Remove the upper arm mounting bolts then remove the upper arm.

To install:

7. Install the upper arm by inserting a 6mm diameter 300mm long rod into the positioning holes and place the upper arm on the rod to position it before tightening the upper arm mounting bolts.

8. Install the remaining parts in the reverse order of removal, keeping in mind the following:

a. Be careful not to damage the ball joint boot when installing the knuckle.

b. First install all the components and lightly tighten the bolts and nuts, then raise the suspension to load it with the vehicle's weight before fully tightening to the specified torque values. Do not place the jack against the ball joint pin of the knuckle.

c. Tighten all mounting hardware to specified torque values.

d. Torque the castle nut to the lower torque specification, then tighten it only far enough to align the slot with the ball joint pin hole. Do not align the castle nut by loosening it.

e. Install a new cotter pin on the castle nut after tightening.

RL Models

1. Before servicing the vehicle, refer to the precautions section.

2. Remove the front strut assembly.

3. Remove the wheel sensor bracket from the upper arm.

4. Remove the cotter pin from the upper arm ball joint, then loosen the nut.

➡During installation, insert the new cotter pin into the ball joint pin hole from the front to the rear of vehicle, and bend its end.

5. Disconnect the upper arm ball joint from the knuckle using a suitable ball joint separator tool.

6. Remove the upper arm mounting bolts then remove the upper arm.

To install:

7. Install the upper arm by inserting a 6mm diameter 300mm long rod into the positioning holes and place the upper arm on the rod to position it before tightening the upper arm mounting bolts.

8. Install the remaining parts in the reverse order of removal, keeping in mind the following:

a. Be careful not to damage the ball joint boot when installing the knuckle.

b. First install all the components and lightly tighten the bolts and nuts, then raise the suspension to load it with the vehicle's weight before fully tightening to the specified torque values. Do not place the jack against the ball joint pin of the knuckle. Tighten the upper arm bolts to 23 ft. lbs. (31 Nm). Tighten ball joint nut to 43–51 ft. lbs. (59–69 Nm) and install a new cotter pin.

TSX Models

See Figure 265.

1. Before servicing the vehicle, refer to the precautions section.

2. Raise and safely support the vehicle.

3. Remove or disconnect the following:
- Front wheel
- Front damper/strut
- Wheel speed sensor bracket from the upper control arm
- Cotter pin from the upper ball joint and loosen the nut
- Upper ball joint from the knuckle, using a suitable separator tool
- Upper control arm bolts or nuts, as applicable
- Upper control arm from the vehicle

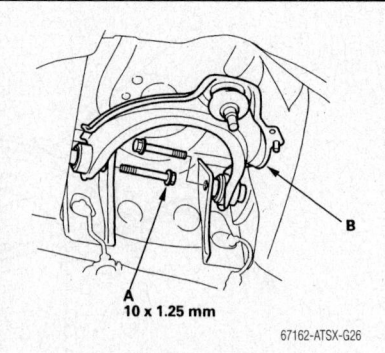

10 x 1.25 mm

67162-ATSX-G26

Fig. 265 Upper control arm (B) and bolt (A) mounting—TSX model

To install:

4. Install or connect the following:
- Upper control arm
- Upper control arm bolts: 23 ft. lbs. (31 Nm)
- Upper control arm-to-chassis nuts and torque them to 47 ft. lbs. (64 Nm)

- Ball joint to the steering knuckle and torque the nut to 29–35 ft. lbs. (39–47 Nm)
- Front wheel

5. Check the front wheel alignment and adjust if necessary.

WHEEL BEARINGS

REMOVAL & INSTALLATION

TL Models

See Figures 266 through 271.

1. Before servicing the vehicle, refer to the precautions section.

2. Remove the front wheels.

3. Remove the brake hose mounting bracket.

4. Remove the brake caliper bracket mounting bolts and caliper assembly from the knuckle. To prevent damage to the caliper assembly or brake hose, use a short piece of wire to hang the caliper assembly from the undercarriage. Do not twist the brake hose with force.

5. Remove the washers on models with a manual transmission.

6. Remove the wheel sensor from the knuckle. Do not disconnect the wheel sensor connector.

7. Remove the spindle nut.

8. Remove the brake rotor.

9. Remove the cotter pin from the tie-rod end ball joint, then loosen the nut.

➡During installation, install the new cotter pin after tightening the nut, and bend its end as shown.

10. Disconnect the tie-rod end ball joint from the knuckle.

11. Remove the cotter pin from the lower arm ball joint, and remove the nut.

➡To avoid damaging the ball joint, install the tool shown in the control arm procedures on the threads of the ball joint.

➡Be careful not to damage the ball joint boot when installing the remover. Do not force or hammer on the lower arm, or pry between the lower arm and the knuckle. You could damage the ball joint.

12. Disconnect the lower ball joint from the lower arm using the tool illustrated.

➡If the collar/sleeve on the lower arm is removed with ball joint, the lower arm must be replaced.

13. Remove the cotter pin from the upper arm ball joint, then loosen the nut.

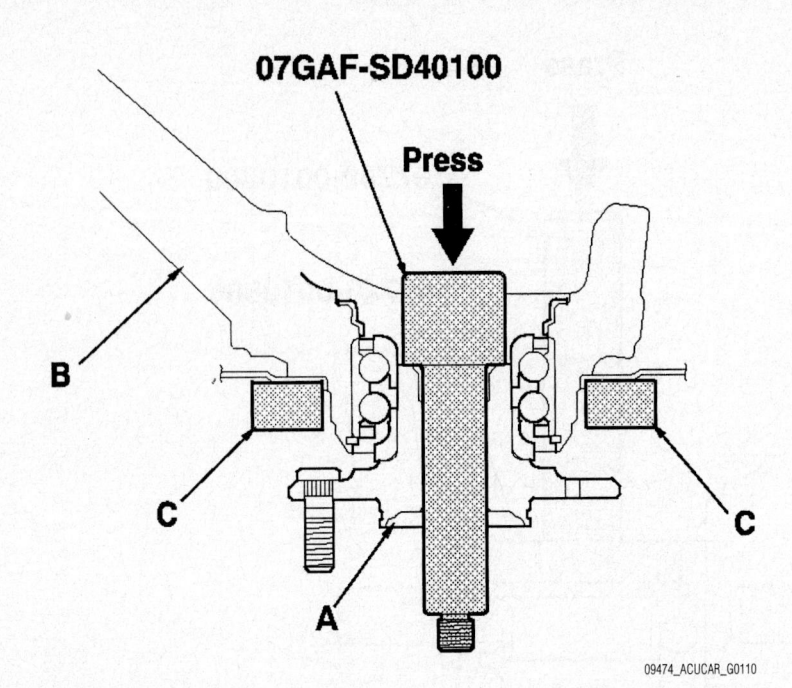

Fig. 266 Separate the hub from the knuckle using the tool illustrated and a hydraulic press. Hold the knuckle with the attachment of the hydraulic press or equivalent tool—Front wheel bearing on 2007–08 TL models

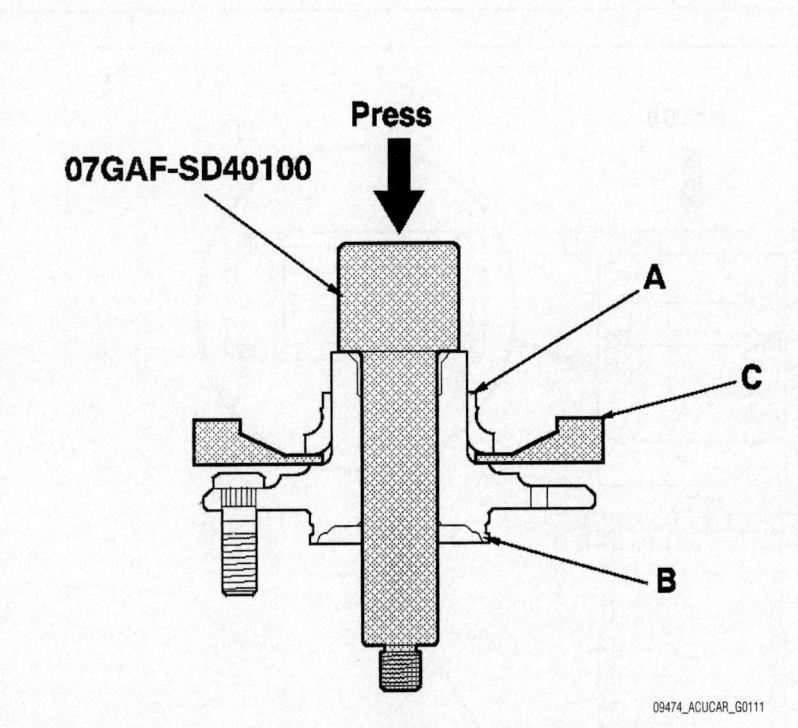

Fig. 267 Press the wheel bearing inner race off of the hub using the tool illustrated, a commercially available bearing separator, and a press—Front wheel bearing on 2007–08 TL models

➥During installation, insert the new cotter pin into the ball joint pin from the front to the rear of the vehicle, and bend its end as shown.

14. Disconnect the upper arm ball joint from the knuckle.

15. Remove the driveshaft outboard joint from the, then remove the knuckle.

➥Do not pull the driveshaft end outward. The inner driveshaft joint may come apart.

16. Separate the hub from the knuckle using the tool illustrated and a hydraulic press. Hold the knuckle with the attachment of the hydraulic press or equivalent tool. Be careful not to deform the splash guard.

17. Hold onto the hub to keep it from falling when pressed clear

18. Press the wheel bearing inner race off of the hub using the tool illustrated, a commercially available bearing separator, and a press.

19. Remove the snap ring and the splash guard from the knuckle.

20. Check the front knuckle ring for damage or deformation, and replace it if necessary.

21. Press the wheel bearing out of the knuckle using the tool illustrated and a press.

To install:

22. Wash the knuckle and hub thoroughly in high flash point solvent before reassembly.

23. Press a new wheel bearing into the knuckle using the old bearing, a steel plate, the tool illustrated, and a press.

➥Install the wheel bearing with the wheel sensor magnetic encoder (brown color), toward the inside of the knuckle.

24. Remove any oil, grease, dust, metal debris, and other foreign material from the encoder surface.

25. Keep all magnetic tools away from the encoder surface.

➥Be careful not to damage the encoder surface when you insert the wheel bearing.

26. Install the new front knuckle ring on the inside of the knuckle by aligning the cutout portion on the ring with the wheel sensor hole in the knuckle. Be careful not to damage or deform the ring when installing it.

27. Install the snap ring securely in the knuckle.

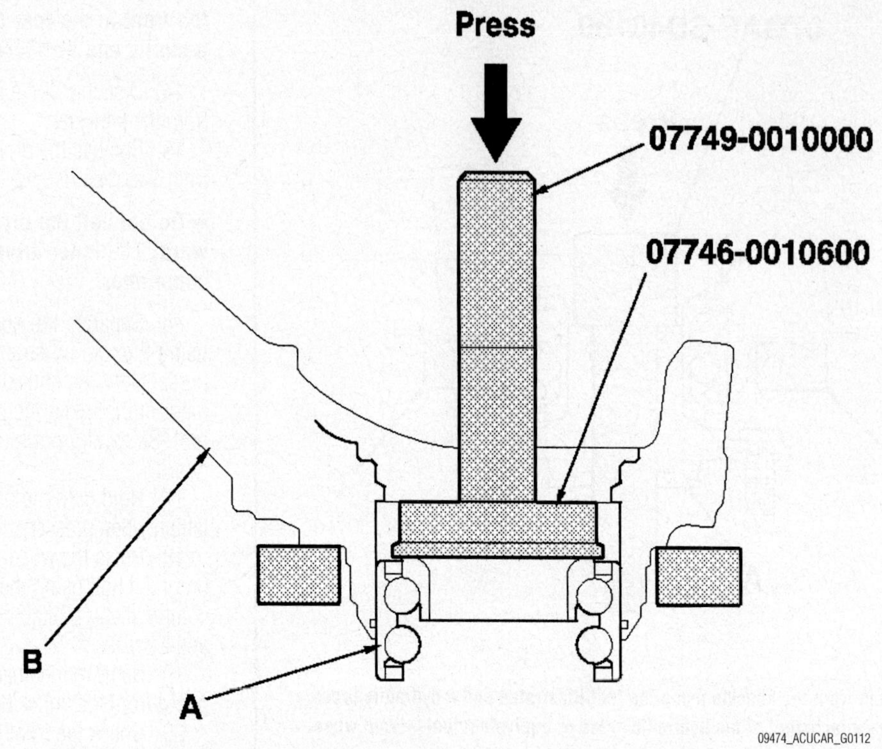

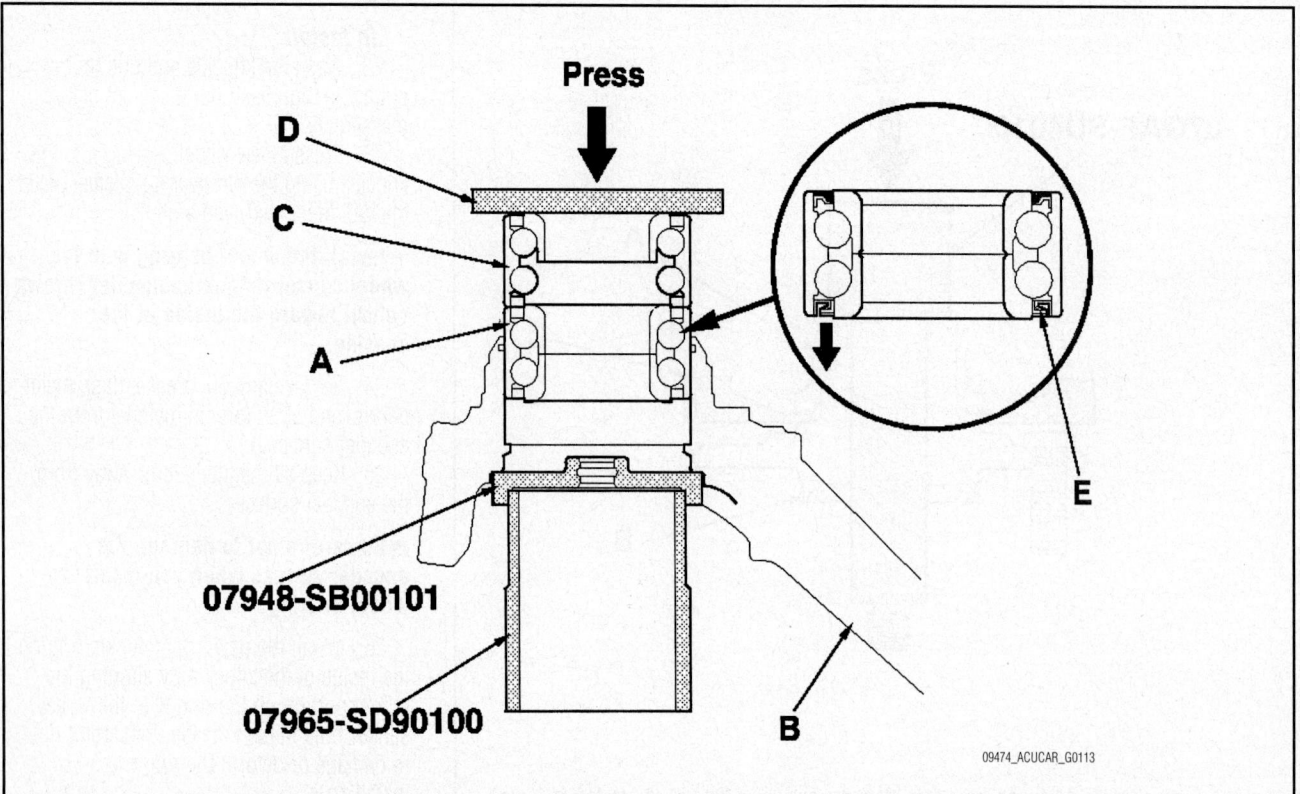

Press

07749-0010000

07746-0010600

B

A

09474_ACUCAR_G0112

Fig. 268 Press the wheel bearing out of the knuckle using the tool illustrated and a press—Front wheel bearing on 2007–08 TL models

Press

D

C

A

07948-SB00101

07965-SD90100

B

E

09474_ACUCAR_G0113

Fig. 269 Press a new wheel bearing into the knuckle using the old bearing, a steel plate, the tool illustrated, and a press—Front wheel bearing on 2007–08 TL models

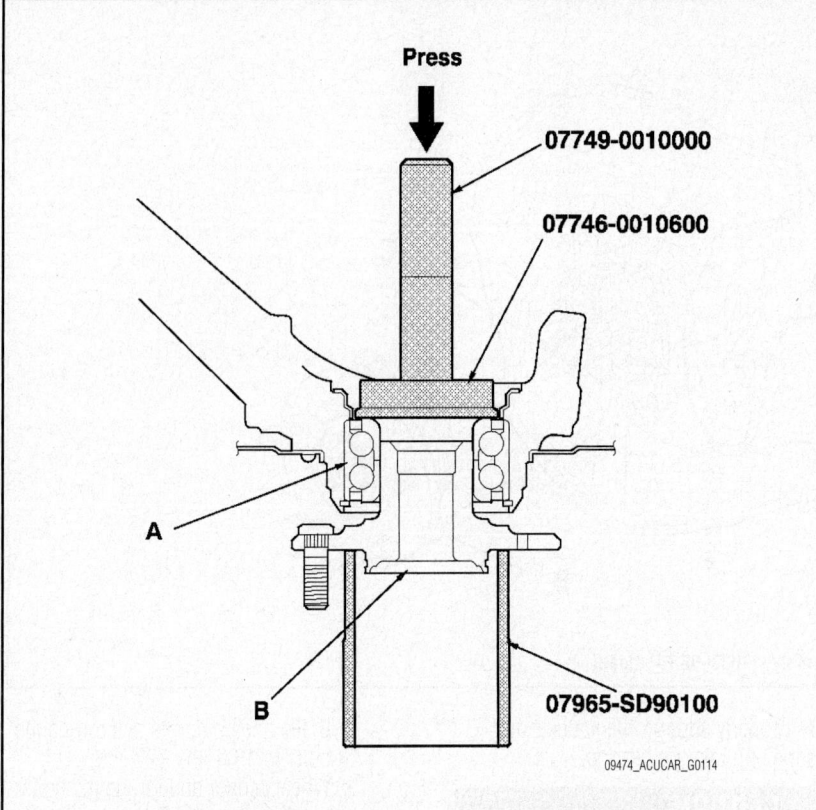

Fig. 270 Press the wheel bearing onto the hub using the tool illustrated and a press—Front wheel bearing on 2007–08 TL models

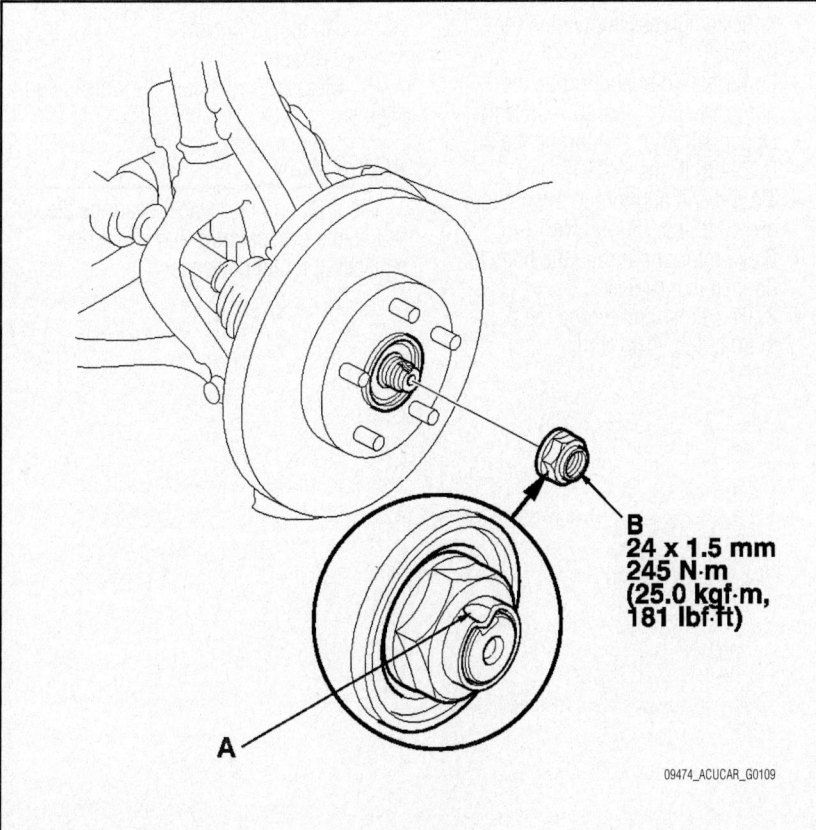

Fig. 271 Tighten the spindle nut—2007–08 TL model

28. Install the splash guard and tighten the screws.

29. Press the wheel bearing onto the hub using the tool illustrated and a press.

30. Install the remaining components in the reverse order of removal.

RL Models

See Figure 272.

➡ **The front wheel bearings on these vehicles are part of the wheel hub and are not serviceable. If the bearing is bad the wheel hub must be replaced.**

1. Before servicing the vehicle, refer to the precautions section.

2. Remove the front wheels.

3. Remove the brake hose mounting bracket.

4. Remove the brake caliper bracket mounting bolts and caliper assembly from the knuckle. To prevent damage to the caliper assembly or brake hose, use a short piece of wire to hang the caliper assembly from the undercarriage. Do not twist the brake hose with force.

5. Remove the wheel sensor and o-ring from the knuckle. Do not disconnect the wheel sensor connector. use a new O-ring during installation.

6. Remove the spindle nut.

7. Remove the brake rotor.

8. Remove the cotter pin from the tie-rod end ball joint, then loosen the nut.

9. Remove the hub bearing unit mounting bolts.

10. Remove the hub bearing unit by tapping the driveshaft end with plastic hammer while drawing the hub bearing unit outward.

To install:

11. Installation is the reverse of removal. Tighten the mounting bolts to 72 ft. lbs. (93 Nm).

TSX Models

1. Before servicing the vehicle, refer to the precautions section.

2. Raise and safely support the vehicle.

3. Remove or disconnect the following:
- Wheel
- Axle nut
- Wheel sensor and wire brackets from the knuckle
- Brake caliper from the knuckle
- Brake rotor from the knuckle
- Tie rod from the knuckle
- Lower control arm from the knuckle
- Upper arm from the knuckle
- Knuckle/hub assembly from the halfshaft

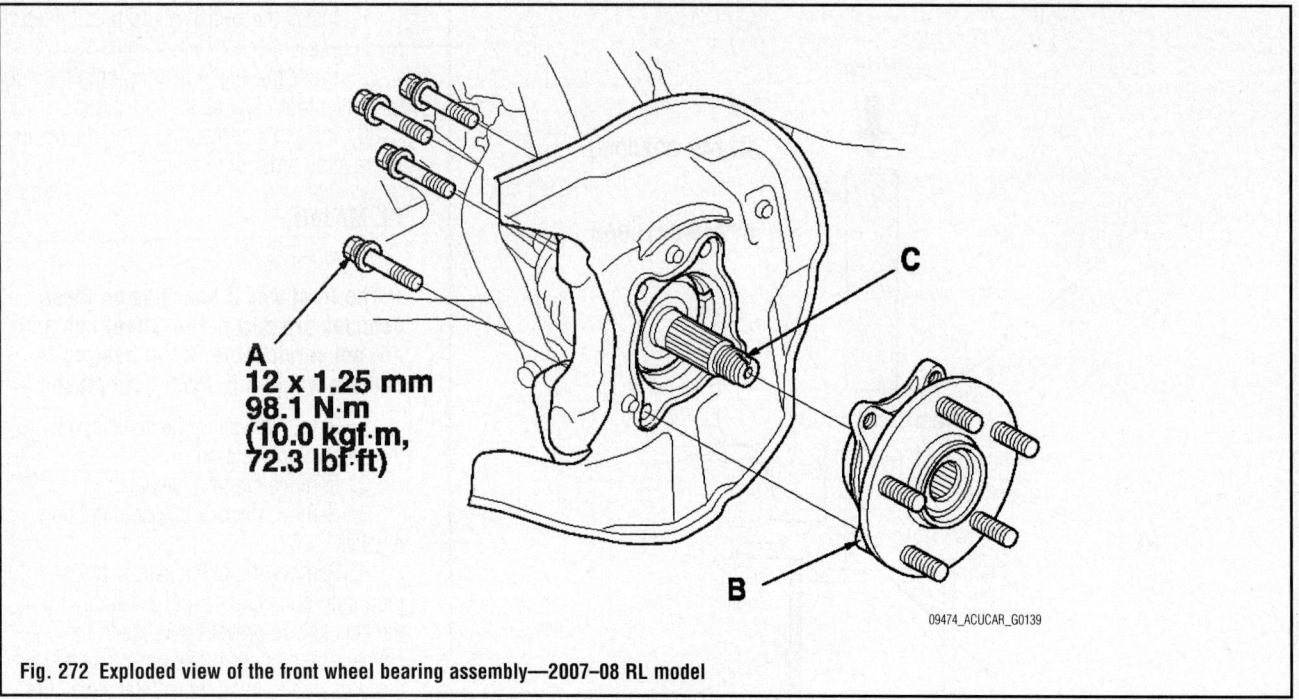

A
12 x 1.25 mm
98.1 N·m
(10.0 kgf·m,
72.3 lbf·ft)

09474_ACUCAR_G0139

Fig. 272 Exploded view of the front wheel bearing assembly—2007–08 RL model

4. Clamp the knuckle in a vise and secure a slide hammer to the wheel studs to separate the hub from the knuckle.

5. Remove the splash guard.

6. Remove the snapring from the knuckle.

7. Support the knuckle and press the bearing out towards the wheel side.

8. If the inner bearing race stayed on the hub, use a puller to remove it.

To install:

9. Press a new inner race on the hub.

10. Press a new bearing into the knuckle.

11. Install the outer snapring.

12. Install the splash guard and torque the bolts to 4 ft. lbs. (4.9 Nm).

13. Properly support the knuckle and press the hub into the bearing.

> ❊❊ **CAUTION**
>
> **Do not press on the wheel studs or they will press out of the hub.**

14. Install or connect the following:
- Knuckle/hub assembly onto the halfshaft
- Lower ball joint and torque the nut to 51–58 ft. lbs. (69–78 Nm)
- Upper ball joint and torque the nut to 29–35 ft. lbs. (40–48 Nm)
- Tie rod end and torque the nut to 36–43 ft. lbs. (50–60 Nm)
- Brake rotor and torque the bolts to 84 inch lbs. (9.5 Nm)
- Brake caliper and torque the bolts to 80 ft. lbs. (108 Nm)
- Brake hose brackets and torque the bolts to 16 ft. lbs. (22 Nm)
- Wheel sensor and torque the bolts to 84 inch lbs. (9.5 Nm)
- Wheel sensor wire brackets and torque the bolts to 84 inch lbs. (9.5 Nm)
- New axle nut and torque it to 181 ft. lbs. (245 Nm)
- Wheel

15. Measure and adjust the wheel alignment

ADJUSTMENT

The front and rear wheel bearings are not adjustable or repairable and should be replaced if found defective.

REAR SUSPENSION

See Figures 273 and 274.

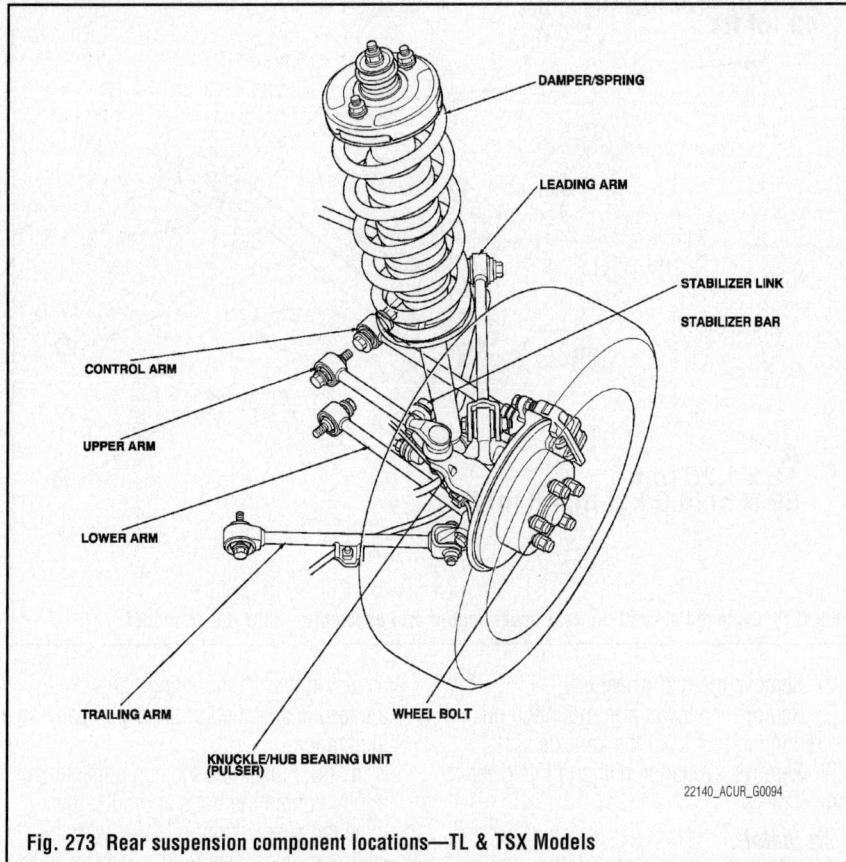

Fig. 273 Rear suspension component locations—TL & TSX Models

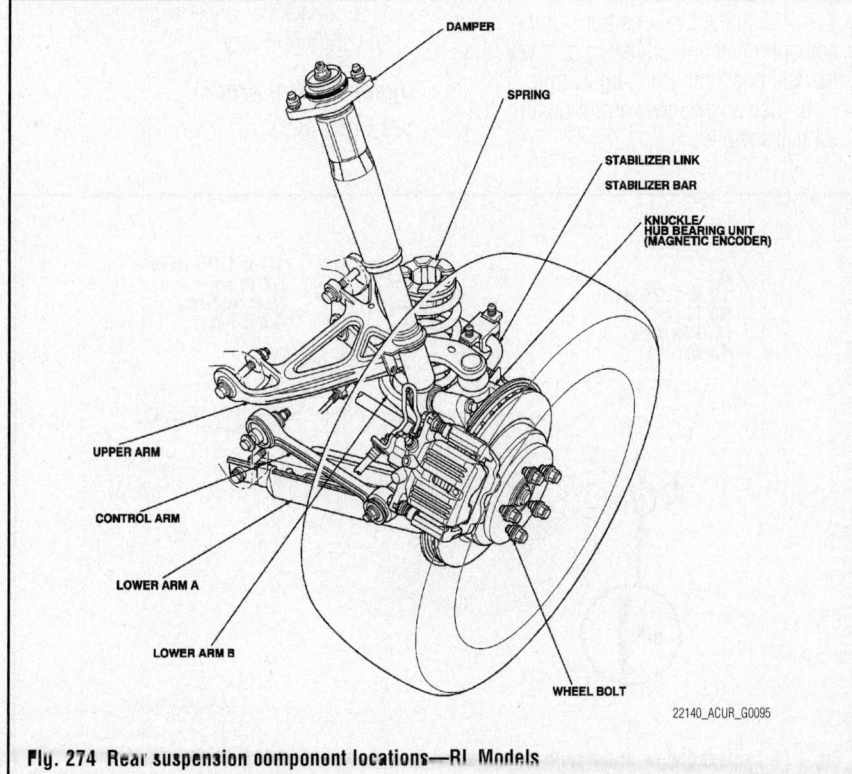

Fig. 274 Rear suspension component locations—RL Models

COIL SPRING

REMOVAL & INSTALLATION

TL & TSX Models

See Figure 275.

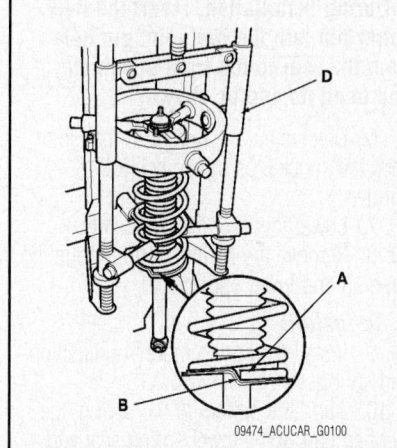

09474_ACUCAR_G0100

Fig. 275 Align the bottom of the spring (A) and the stepped part of the lower spring seat (B), and align the damper mounting base (C)—2007–08 TL and TSX models

1. Before servicing the vehicle, refer to the precautions section.
2. Compress the damper spring with a strut spring, then remove the self-locking nut while holding the damper shaft with a hex wrench. Do not compress the spring more than necessary to remove the nut.
3. Release the pressure from the strut spring compressor, then disassemble the damper.

To install:

4. Install all parts except the self-locking nut and washer onto the.
5. Align the bottom of the spring (A) and the stepped part of the lower spring seat (B), and align the damper mounting base (C).
6. Install the damper assembly on a strut spring compressor.
7. Compress the damper spring with the spring compressor.
8. Install the washer and a new self-locking nut. Hold the damper shaft with a hex wench and tighten the 10 mm self-locking nut to the 22 ft. lbs. (29 Nm).
9. Remove the damper assembly from the strut spring compressor.

RL Models

1. Before servicing the vehicle, refer to the precautions section.

2. Remove the rear wheel.

3. Remove the self-locking nut and the washer while holding the joint pin with hex wrench and disconnect the stabilizer link from lower arm.

4. Place a floor jack at the connecting point of lower arm and the stabilizer link.

5. Remove the cotter pin from the lower arm ball joint, and loosen the nut.

➡ **During installation, insert the new cotter pin into the ball joint pin hole from the rear to the front of vehicle, and bend its end as shown.**

6. Disconnect the lower arm ball joint from the knuckle using the ball joint remover.

7. Lower the floor jack gradually.

8. Remove the spring, spring mounting cushion and lower spring seat.

To install:

9. Install the spring mounting cushion and spring.

10. Align the bottom of the spring, the stepped part of the lower spring seat and lower arm.

11. Place the floor jack at the connecting point of lower arm and the stabilizer link.

12. Raise the jack slowly until you can align the bolt hole of lower arm and the knuckle ball joint pin, then loosely install the castle nut.

13. Install the stabilizer link on the lower arm with the washer and the new self-locking nut, and lightly tighten them.

14. Raise the rear suspension with a floor jack to load it with the vehicle's weight.

15. Tighten the castle nut and self-locking nut to the 54–61 ft. lbs. (74–83 Nm).

➡ **Torque the castle nut to the lower torque specification, then tightens if only far enough to align the slot with the ball joint pin hole. Do not align the castle nut by loosening it.**

16. Insert a new cotter pin into the ball joint pin from the rear to the front of the vehicle, and bend its end.

17. Install the rear wheel.

18. Check the rear wheel alignment, and adjust it if necessary.

CONTROL ARMS/LINKS

REMOVAL & INSTALLATION

TL Models

Lower Control Arm

See Figure 276.

1. Before servicing the vehicle, refer to the precautions section.

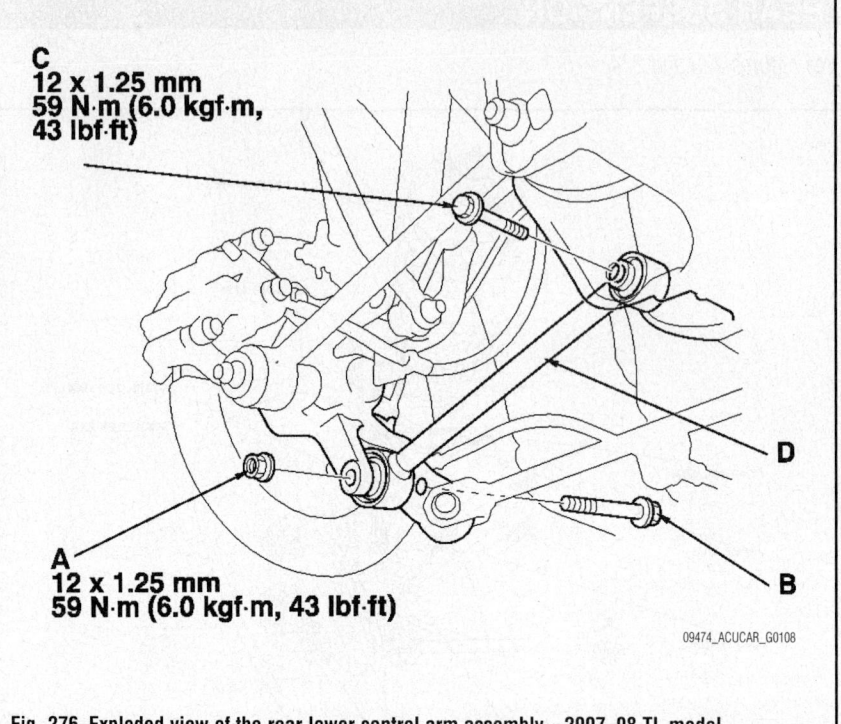

**C
12 x 1.25 mm
59 N·m (6.0 kgf·m,
43 lbf·ft)**

**A
12 x 1.25 mm
59 N·m (6.0 kgf·m, 43 lbf·ft)**

D

B

09474_ACUCAR_G0108

Fig. 276 Exploded view of the rear lower control arm assembly—2007–08 TL model

2. Remove the rear wheel.

3. Remove the lower arm mounting nut and mounting bolt from the knuckle side.

4. Remove the flange bolt and the lower arm.

To install:

5. Installation is the reverse of removal, keep in mind the following:

a. Align the cam positions of the adjusting bolt and adjusting cam with the marked positions when tightening.

b. Use a new lower arm mounting nut on reassembly.

c. Tighten all mounting hardware to the torque specified in the accompanying illustration.

d. First install all the components and lightly tighten the bolts and nuts, then raise the suspension to load it with the vehicle's weight before fully tightening to the final torque specs.

e. Check the wheel alignment, and adjust if necessary.

Upper Control Arm

See Figure 277.

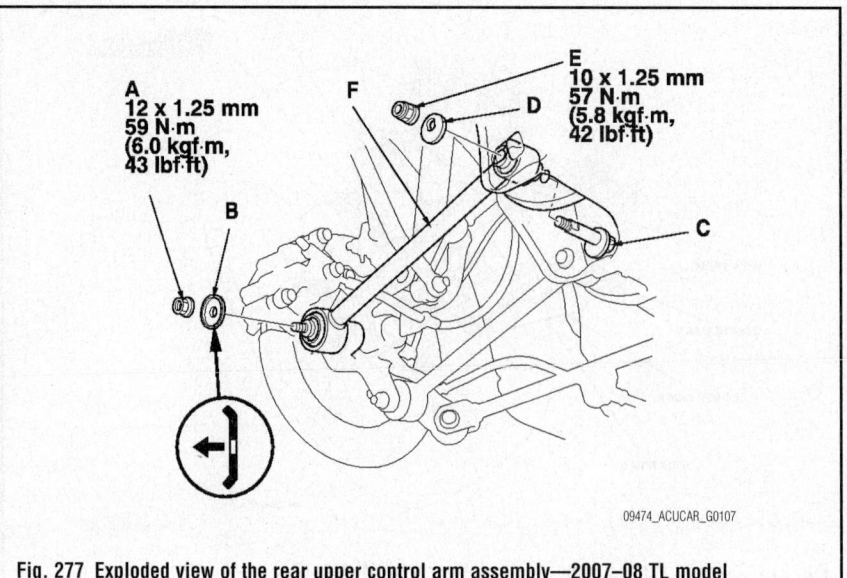

**A
12 x 1.25 mm
59 N·m
(6.0 kgf·m,
43 lbf·ft)**

F

**E
10 x 1.25 mm
57 N·m
(5.8 kgf·m,
42 lbf·ft)**

D

B

C

09474_ACUCAR_G0107

Fig. 277 Exploded view of the rear upper control arm assembly—2007–08 TL model

1. Before servicing the vehicle, refer to the precautions section.

2. Remove the rear wheel.

3. Remove the control arm mounting nut) and washer from the knuckle side.

4. Mark the cam positions of the adjusting bolt and adjusting cam, then remove the self-locking nut, adjusting cam, and adjusting bolt. Discard the self-locking nut and control arm mounting nut.

5. Remove the control arm.

To install:

6. Installation is the reverse of removal, keep in mind the following:

a. Align the cam positions of the adjusting bolt and adjusting cam with the marked positions when tightening.

b. Use a new self-locking nut and control arm mounting nut on reassembly.

c. Tighten all mounting hardware to the torque specified in the accompanying illustration.

d. First install all the components and lightly tighten the bolts and nuts, then raise the suspension to load it with the vehicle's weight before fully tightening to the final torque specs.

e. Check the wheel alignment, and adjust if necessary.

RL Models

Lower Control Arm A

See Figure 278.

1. Before servicing the vehicle, refer to the precautions section.

2. Remove the rear wheel.

3. Remove the lower arm A mounting nut, washers and mounting bolt from the knuckle side.

4. Remove the self-locking nut, washers and the flange bolt, then remove the lower arm A.

To install:

5. Install the lower arm in the reverse order of removal, please note the following:

a. Use a new lower arm A mounting nut and self-locking nut on reassembly.

b. First, install the components and lightly tighten the bolts and nuts, then raise the suspension to load it with the vehicle's weight before fully tightening to the specified torque.

c. Tighten all mounting hardware to the specified torque.

d. Check the wheel alignment, and adjust it if necessary.

Lower Control Arm B

See Figure 279

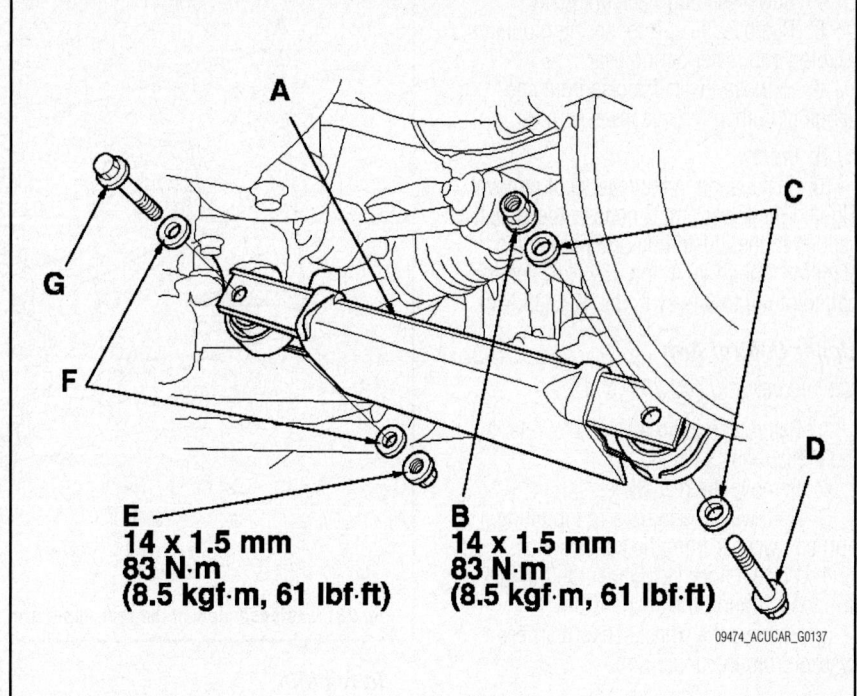

E
14 x 1.5 mm
83 N·m
(8.5 kgf·m, 61 lbf·ft)

B
14 x 1.5 mm
83 N·m
(8.5 kgf·m, 61 lbf·ft)

09474_ACUCAR_G0137

Fig. 278 Exploded view of the rear lower control arm A assembly—2007–08 RL model

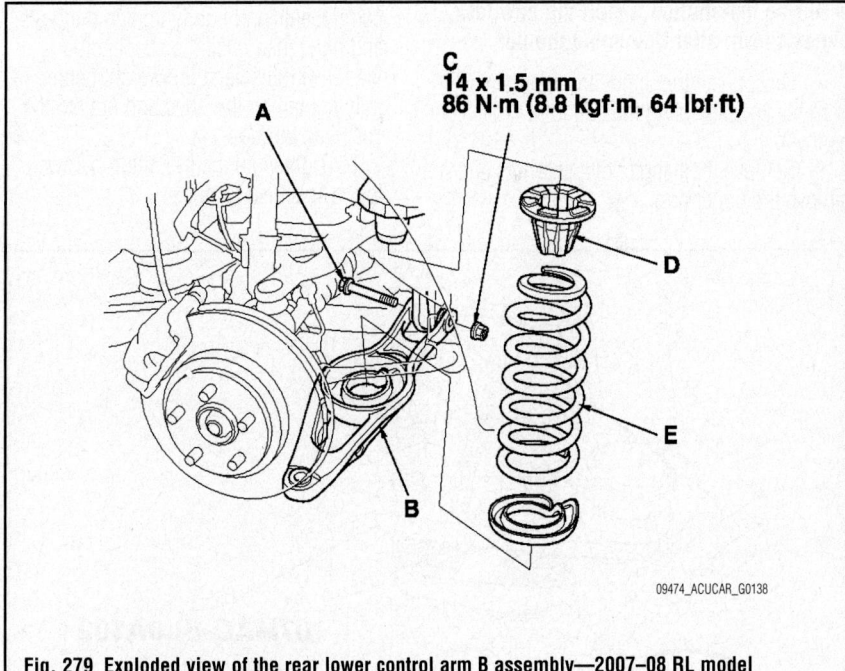

C
14 x 1.5 mm
86 N·m (8.8 kgf·m, 64 lbf·ft)

09474_ACUCAR_G0138

Fig. 279 Exploded view of the rear lower control arm B assembly—2007–08 RL model

1. Before servicing the vehicle, refer to the precautions section.

2. Remove the rear wheel.

3. Remove the self-locking nut and the washer while holding the joint pin with a hex wrench and disconnect the stabilizer link from lower arm B.

4. Place a floor jack at the connecting point of lower arm B and the stabilizer link.

5. Remove the cotter pin from the lower arm B ball joint, and loosen the nut.

➡**During installation, insert the new cotter pin into the ball joint pin hole from the rear to the front of vehicle, and bend its end.**

6. Disconnect the lower arm ball joint from the knuckle using a separator tool.

7. Lower the floor jack gradually.

8. Remove the spring, spring mounting cushion and lower spring seat.

9. Remove the self-locking nut and flange bolt, then remove lower arm B.

To install:

10. Installation is the reverse of removal. Tighten the lower arm B bolts to the specifications in the illustration and tighten the stabilizer link to 40 ft. lbs. (54 Nm) and the ball joint nut to 54–61 ft. lbs. (74–83 Nm).

Upper Control Arm

See Figures 280 and 281.

1. Before servicing the vehicle, refer to the precautions section.

2. Remove the rear wheel.

3. Remove the control arm mounting nut) and washer from the knuckle side.

4. Place a floor jack under the trailing arm, and support the suspension.

5. Remove the wheel sensor harness bracket from the upper arm.

➡ **Use a new bracket on reassembly.**

6. Remove the lock pin from the upper arm ball joint, and loosen the nut.

➡ **During installation, insert the new lock pin as shown after tightening the nut.**

7. Disconnect the upper arm ball joint from the knuckle using the ball joint remover.

8. Remove the flange bolts and nuts and remove the upper arm.

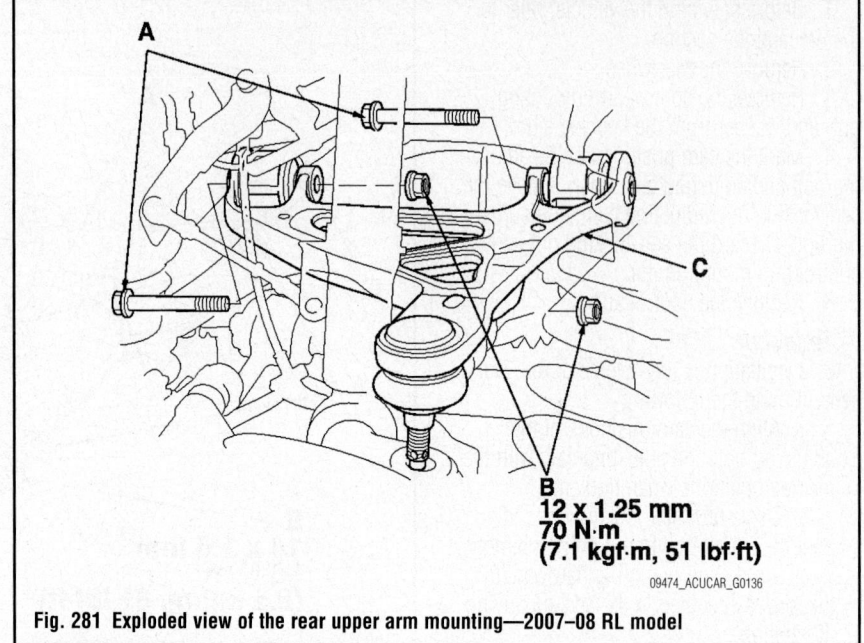

Fig. 281 Exploded view of the rear upper arm mounting—2007–08 RL model

To install:

9. Install the upper arm in the reverse order of removal, and please note the following:

a. First install all the suspension components and lightly tighten the bolts and nuts, then raise the suspension to load it with the vehicle's weight before fully tightening the bolts and nuts to the specified torque.

b. Tighten all the mounting hardware to the specified torque.

c. Be careful not to damage the ball joint boot when installing the knuckle.

d. Before connecting the ball joint to the knuckle, degrease the threaded section and tapered portion of the ball joint pin, the connecting hole, and the threaded section and mating surface of the castle nut.

e. Torque the castle nut to the lower torque specification, then tighten it only far enough to align the slot with the ball joint pin hole. Do not align the castle nut by loosening it.

f. Check the wheel alignment, and adjust it if necessary.

TSX Models

Lower Control Arm A

See Figure 282.

1. Before servicing the vehicle, refer to the precautions section.

2. Raise and safely support the vehicle.

3. Remove or disconnect the following:

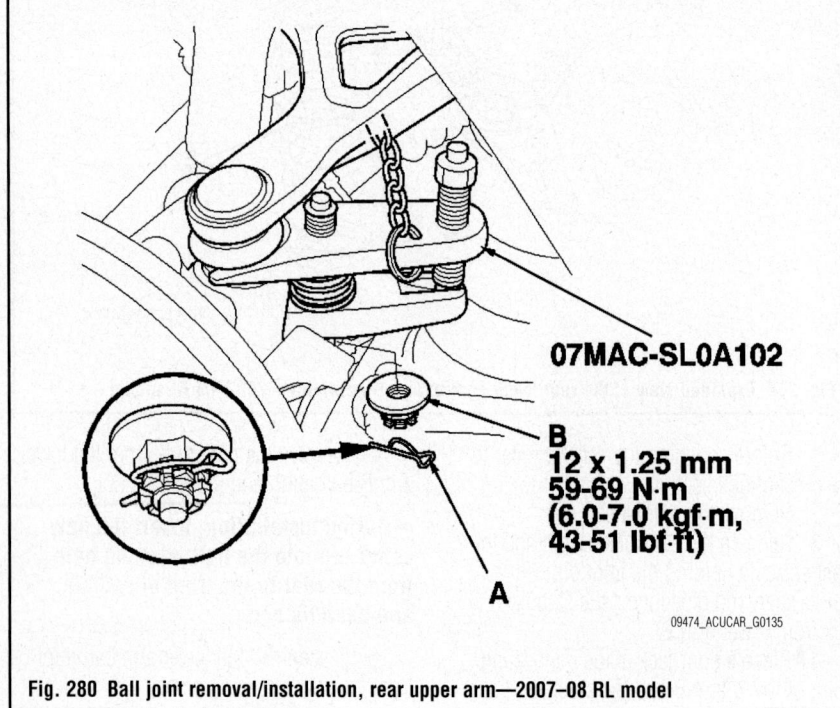

Fig. 280 Ball joint removal/installation, rear upper arm—2007–08 RL model

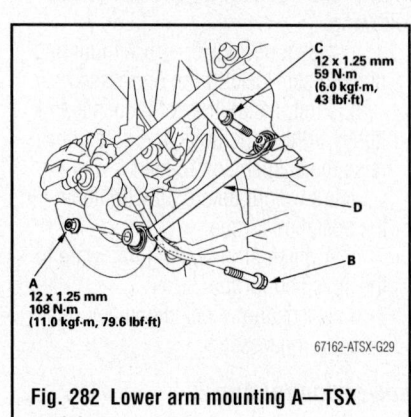

Fig. 282 Lower arm mounting A—TSX models

- Rear wheels
- Lower arm mounting nut and mounting bolt from the knuckle side
- Flange bolt
- Lower arm

4. Installation is the reverse of the removal procedure. Tighten the flange bolt to 43 ft. lbs. (59 Nm) and the mounting nut to 79.6 ft. lbs. (108 Nm).

Lower Control Arm B

See Figure 283.

1. Before servicing the vehicle, refer to the precautions section.
2. Raise and safely support the vehicle.
3. Remove or disconnect the following:
   - Rear wheels
   - Control arm mounting nut and washer from the knuckle side

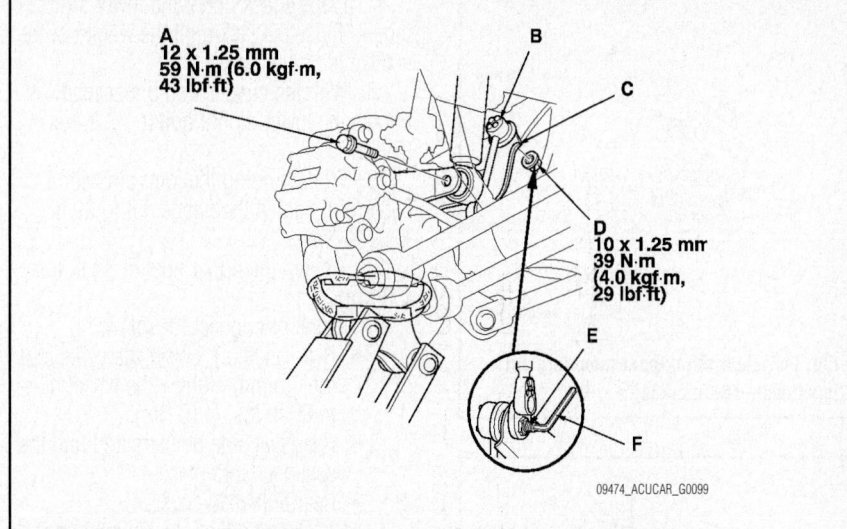

Fig. 284 Install the flange bolt (A) on the bottom of the damper. Connect the stabilizer link (B) on the bracket (C), then loosely install the flange nut (D)—2007–08 TL models

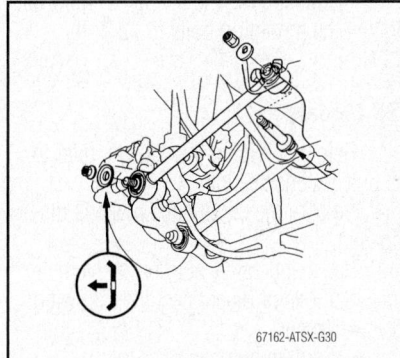

67162-ATSX-G30

Fig. 283 Control arm mounting B—TSX

4. Matchmark the cam positions of the adjusting bolt and adjusting cam, then remove the self-locking nut and adjusting cam and adjusting bolt. Discard the self-locking nut and control arm mounting nut.
   - Lower control arm
5. Installation is the reverse of the removal procedure. Tighten the self-locking nut to 40 ft. lbs. (54 Nm) and the mounting nut to 51 ft. lbs. (69 Nm).

MACPHERSON STRUTS

REMOVAL & INSTALLATION

TL Models

See Figure 284.

1. Before servicing the vehicle, refer to the precautions section.
2. Remove the rear wheel.
3. Remove the rear shelf.
4. Remove the two flange nuts.
5. Remove the flange nut while holding the joint pin with a hex wrench, then disconnect the stabilizer link from the stabilizer link bracket.

6. Remove the flange bolt from the knuckle.
7. Lower the rear suspension, then remove the damper from the vehicle. Damper springs are different, left and right. Mark the springs L and R before you continue.

To install:

8. Lower the rear suspension, and position the damper (A) in the body.
9. Loosely install the flange nuts onto the top of the damper.
10. Loosely install the flange bolt (A) on the bottom of the damper. Connect the stabilizer link (B) on the bracket (C), then loosely install the flange nut (D).
11. Raise the rear suspension with a floor jack to load the vehicle weight, and tighten the flange bolt to the torque specified in the accompanying illustration.
12. Tighten the flange nut while holding the joint pin with a hex wrench.
13. Tighten the two flange nuts on top of the damper to 37 ft. lbs. (50 Nm).
14. Install the rear shelf.
15. Install the rear wheel.
16. Check the rear wheel alignment, and adjust if necessary.

RL Models

1. Before servicing the vehicle, refer to the precautions section.
2. Remove the rear wheel.
3. Place a floor jack at the connecting point of lower arm and the stabilizer link to support them.
4. Disconnect the headlight leveling sensor linkage from the damper.
5. Remove the rear shelf.
6. Remove the two flange nuts.

7. Remove the damper lower mounting bolt from the knuckle.
8. Lower the rear suspension, then remove the damper from the vehicle.

To install:

9. Place a floor jack at the connecting point of lower arm and the stabilizer link.
10. Compress the damper by hand, and move it into position.
11. Loosely tighten the damper lower mounting bolt.
12. Loosely install the flange nuts.
13. Raise the rear suspension with a floor jack to load it with the vehicle's weight.
14. Tighten the flange nuts and the damper lower mounting bolt to 28 ft. lbs. (38 Nm).
15. Tighten the damper lower mounting bolt to 43 ft. lbs. (59 Nm).
16. Connect the headlight leveling sensor linkage to the damper.
17. Install the rear wheel.
18. Check the rear wheel alignment, and adjust it if necessary.

TSX Models

See Figures 285 and 286.

1. Before servicing the vehicle, refer to the precautions section.
2. Fold the rear seat forward.
3. Remove or disconnect the following:
   - Rear shelf cover
   - Seat side bolster cushions. The side bolster cushions are secured by 1 screw at the bottom, and 2 clips at the top.
   - Strut mount cap, if equipped, and upper strut mounting nuts
4. Raise and safely support the vehicle.

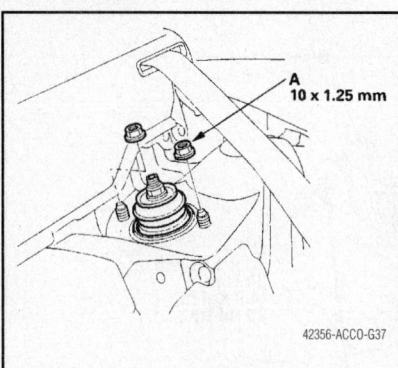

Fig. 285 Rear strut upper mounting nut locations—TSX models

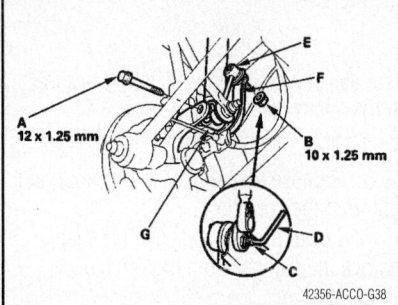

Fig. 286 Remove the lower flange (A) bolt from the knuckle, then remove the flange nut (B) while holding the joint pin (C) with a hex wrench (D) and disconnect the stabilizer link (E) from the bracket (F)—TSX models

- Rear wheels, then support the knuckle with a floor jack
- Strut lower flange bolt from the knuckle
- Strut flange nut while holding the joint pin with a hex wrench
- Stabilizer link from the bracket
- Strut, while lowering rear suspension

➡ **The left and right struts are different, so be sure to mark them L & R if you are removing both struts before continuing.**

To install:

➡ **Use new self-locking nuts when installing the strut.**

5. Lower the rear suspension.
6. Install or connect the following:
- Strut into the upper mount. Only hand-tighten the upper mounting nuts.
- Strut into position on the knuckle, then loosely install the flange bolt on the bottom of the strut.
- Stabilizer link on the bracket and loosely install the flange nut.

7. Place a jack under the lower strut mount. Raise the jack until the weight of the vehicle is on the jack.
8. With the suspension under load, tighten the lower mount bolt to 43 ft. lbs. (59 Nm).
9. While holding the joint pin with a cotter pin, tighten the flange nut to 29 ft. lbs. (39 Nm).
10. Tighten the upper nuts to 37 ft. lbs. (50 Nm).
11. Install or connect the following:
- Rear wheel(s). Lower the vehicle to the ground. Tighten the wheel nuts to 80 ft. lbs. (110 Nm).
- Rear seat side bolsters and fold the seat back into place.
- Rear bulkhead cover
12. Check and adjust the vehicle's rear wheel alignment.

OVERHAUL

For disassembly procedures, refer to the following section, "Rear Suspension, Coil Spring, Removal & Installation."

WHEEL BEARINGS

REMOVAL & INSTALLATION

TL Models

➡ **The rear wheel bearings on these vehicles are part of the wheel hub and are not serviceable. If the bearing is bad the wheel hub must be replaced.**

1. Before servicing the vehicle, refer to the precautions section.
2. Remove the rear wheels.
3. Remove the brake hose mounting bracket.
4. Remove the brake caliper bracket mounting bolts and caliper assembly from the knuckle. To prevent damage to the caliper assembly or brake hose, use a short piece of wire to hang the caliper assembly from the undercarriage. Do not twist the brake hose with force.
5. Remove the hub cap (A).
6. Remove the spindle nut.
7. Remove the brake disc.
8. Remove the hub bearing unit.
9. Installation is the reverse of removal. Tighten the hub bearing unit nuts to 28 ft. lbs. (38 Nm).

RL Models

➡ **The rear wheel bearings on these vehicles are part of the wheel hub and are not serviceable. If the bearing is bad the wheel hub must be replaced.**

1. Before servicing the vehicle, refer to the precautions section.
2. Remove the rear wheels.
3. Remove the brake caliper bracket mounting bolts and caliper assembly from the knuckle. To prevent damage to the caliper assembly or brake hose, use a short piece of wire to hang the caliper assembly from the undercarriage. Do not twist the brake hose with force.
4. Remove the 2 washers.
5. Remove the wheel sensor and o-ring from the knuckle. Do not disconnect the wheel sensor connector. use a new O-ring during installation.
6. Remove the spindle nut.
7. Remove the brake rotor.
8. Remove the hub bearing unit mounting bolts.
9. Remove the hub bearing unit.
10. Installation is the reverse of removal. Tighten the mounting bolts to 72 ft. lbs. (93 Nm).

TSX Models

1. Before servicing the vehicle, refer to the precautions section.
2. Be sure the emergency brake is disengaged.
3. Raise and safely support the vehicle.
4. Remove or disconnect the following:
- Wheel
- Spindle hub cap
- Spindle nut
- Caliper shield, if equipped
- Brake hose mounting bolts from the knuckle
- Brake caliper
- Brake rotor
- Hub assembly from the knuckle

To install:
5. Install or connect the following:
- Hub/bearing assembly
- Brake disc and tighten the bolts to 84 inch lbs. (9.5 Nm)
- Brake caliper and torque the bolts to 41 ft. lbs. (56 Nm)
- Brake hose clamps and torque the bolts to 16 ft. lbs. (22 Nm)
- Brake caliper shield and torque the mounting bolts to 84 inch lbs. (9.5 Nm), if equipped
- New spindle nut and torque it to 134 ft. lbs. (181 Nm)
- Spindle hub cap
- Rear wheels

ADJUSTMENT

The front and rear wheel bearings are not adjustable or repairable and should be replaced if found defective.

ACURA

MDX

SPECIFICATIONS AND MAINTENANCE CHARTS

ENGINE AND VEHICLE IDENTIFICATION

Engine							Model Year	
Code	Liters (cc)	Cu. In.	Cyl.	Fuel Sys.	Engine Type	Eng. Mfg.	Code ①	Year
J37A1	3.7 (3664)	224	6	PGM-FI	SOHC	Honda	7	2007
							8	2008

PGM-FI: Programmed Fuel Injection

SOHC: Single Overhead Camshaft

① 10th digit of the Vehicle Identification Number (VIN)

22140_AMDX_C0001

GENERAL ENGINE SPECIFICATIONS

Year	Model	Engine Displacement Liters	Engine ID	Net Horsepower @ rpm	Net Torque @ rpm (ft. lbs.)	Bore x Stroke (in.)	Com- pression Ratio	Oil Pressure @ rpm
2007	MDX	3.7	J37A1	300@6000	275@5000	3.54x3.78	11.0:1	71@3000
2008	MDX	3.7	J37A1	300@6000	275@5000	3.54x3.78	11.0:1	71@3000

22140_AMDX_C0002

GASOLINE ENGINE TUNE-UP SPECIFICATIONS

Year	Engine Displacement Liters	Engine ID/VIN	Spark Plug Gap (in.)	Ignition Timing (deg.) MT	Ignition Timing (deg.) AT	Fuel Pump (psi)	Idle Speed (rpm) MT	Idle Speed (rpm) AT	Valve Clearance In.	Valve Clearance Ex.
2007	3.7	J37A1	0.039-0.043	—	8-12B	57-64 ①	—	660-760	0.008-0.009	0.011-0.013
2008	3.7	J37A1	0.039-0.043	—	8-12B	57-64 ①	—	660-760	0.008-0.009	0.011-0.013

NOTE: The Vehicle Emission Control Information label reflects specification changes during production and must be used if they differ from this chart.

B: Before Top Dead Center

HYD: Hydraulic

① At idle, pressure regulator vacuum hose disconnected

22140_AMDX_C0003

CAPACITIES

| Year | Model | Engine Displacement Liters | Engine ID/VIN | Engine Oil with Filter (qts.) | Transmission (pts.) | | | Drive Axle | | Fuel Tank (gal.) | Cooling System (qts.) |
					5-Spd	6-Spd	Auto.	Front (pts.)	Rear (pts.)		
2007	MDX	3.5	J35A5	4.5	—	—	6.0	—	5.34	21.0	7.7
2008	MDX	3.5	J35A5	4.5	—	—	6.0	—	5.34	21.0	7.7

NOTE: All capacities are approximate. Add fluid gradually and ensure a proper fluid level is obtained.

NOTE: Capacities given are service, not overhaul capacities

22140_AMDX_C0004

FLUID SPECIFICATIONS

| Year | Model | Engine Displ. Liters | Engine Oil | Man. Trans. | Auto. Trans. | Drive Axle | | Transfer Case | Power Steering Fluid | Brake Master Cylinder | Cooling System |
						Front	Rear				
2007	MDX	3.7	5W-20 Acura	Acura MTF	Acura ATF-Z1	—	—	①	Acura PS Fluid	Acura DOT 3	②
2008	MDX	3.7	5W-20 Acura	Acura MTF	Acura ATF-Z1	—	—	①	Acura PS Fluid	Acura DOT 3	②

① Hypoid gear oil SAE 90 or SAE 80W-90 viscosity, API classified GL4 or GL5 only

② Acura Long Life Antifreeze/Coolant-Type2

22140_AMDX_C0005

VALVE SPECIFICATIONS

| Year | Engine Displacement Liters | Engine ID/VIN | Seat Angle (deg.) | Face Angle (deg.) | Spring Test Pressure (lbs. @ in.) | Spring Installed Height (in.) | Stem-to-Guide Clearance (in.) | | Stem Diameter (in.) | |
							Intake	Exhaust	Intake	Exhaust
2007	3.7	J37A1	45	45	NA	NA	0.0008-0.0018	0.0022-0.0031	0.2159-0.2163	0.2146-0.2150
2008	3.7	J37A1	45	45	NA	NA	0.0008-0.0018	0.0022-0.0031	0.2159-0.2163	0.2146-0.2150

NA: Not Available

22140_AMDX_C0006

CAMSHAFT AND BEARING SPECIFICATIONS
All measurements are given in inches.

Year	Engine Displacement Liters	Engine VIN	Journal Diameter	Brg. Oil Clearance	Shaft End-play	Runout	Journal Bore	Lobe Height Intake	Lobe Height Exhaust
2007	3.7	J37A1	NA	0.0020-0.0035	0.0020-0.0080	0.0010	NA	①	1.4326
2008	3.7	J37A1	NA	0.0020-0.0035	0.0020-0.0080	0.0010	NA	①	1.4326

NA: Information not available
① Primary: 1.3824 in.
  Mid: 1.4328 in.
  Secondary: 1.3824 in.

22140_AMDX_C0009

CRANKSHAFT AND CONNECTING ROD SPECIFICATIONS
All measurements are given in inches.

Year	Engine Displacement Liters	Engine ID/VIN	Crankshaft Main Brg. Journal Dia.	Crankshaft Main Brg. Oil Clearance	Crankshaft Shaft End-play	Crankshaft Thrust on No.	Connecting Rod Journal Diameter	Connecting Rod Oil Clearance	Connecting Rod Side Clearance
2007	3.7	J37A1	2.8337-2.8346	0.0007-0.0018	0.0040-0.0140	3	2.2431-2.2441	0.0008-0.0017	0.0060-0.0140
2008	3.7	J37A1	2.8337-2.8346	0.0007-0.0018	0.0040-0.0140	3	2.2431-2.2441	0.0008-0.0017	0.0060-0.0140

22140_AMDX_C0007

PISTON AND RING SPECIFICATIONS
All measurements are given in inches

Year	Engine Displacement Liters	Engine ID/VIN	Piston Clearance	Ring Gap Top Compression	Ring Gap Bottom Compression	Ring Gap Oil Control	Ring Side Clearance Top Compression	Ring Side Clearance Bottom Compression	Ring Side Clearance Oil Control
2007	3.7	J37A1	0.0002-0.0013	0.0120-0.0160	0.0160-0.0220	0.0080-0.0280	0.0022-0.0033	0.0012-0.0024	NA
2008	3.7	J37A1	0.0002-0.0013	0.0120-0.0160	0.0160-0.0220	0.0080-0.0280	0.0022-0.0033	0.0012-0.0024	NA

NA; Not Applicable

22140_AMDX_C0008

TORQUE SPECIFICATIONS
All readings in ft. lbs.

Year	Engine Displacement Liters	Engine ID/VIN	Cylinder Head Bolts	Main Bearing Bolts	Rod Bearing Bolts	Crankshaft Damper Bolts	Flywheel Bolts	Manifold Intake	Manifold Exhaust	Spark Plugs	Oil Pan Drain Plug
2007	3.7	J37A1	①	②	③	④	54	16	40	13	29
2008	3.7	J37A1	①	②	③	④	54	16	40	13	29

① Step 1: 22 ft. lbs.
  Step 2: Rotate 90 degrees
  Step 3: Rotate an additional 90 degrees
  Step 4 (new bolts only): additional 90 degrees

② Step 1: Cap bolts 56 ft. lbs.
  Step 2: Side bolts 36 ft. lbs.

③ Step 1: 14 ft. lbs.
  Step 2: Rotate 90 degrees

④ Step 1: 47 ft. lbs.
  Step 2: Rotate 60 degrees

22140_AMDX_C0010

WHEEL ALIGNMENT

Year	Model		Caster Range (+/-Deg.)	Caster Preferred Setting (Deg.)	Camber Range (+/-Deg.)	Camber Preferred Setting (Deg.)	Toe-in (in.)
2007	MDX	F	0.58	+4.20	1.00	-0.50	0 +/- 0.08
		R	—	—	0.75	-0.50	0.08 +/- 0.08
2008	MDX	F	0.58	+4.20	1.00	-0.50	0 +/- 0.08
		R	—	—	0.75	-0.50	0.08 +/- 0.08

22140_AMDX_C0011

TIRE, WHEEL AND BALL JOINT SPECIFICATIONS

Year	Model	OEM Tires Standard	OEM Tires Optional	Tire Pressures (psi) Front	Tire Pressures (psi) Rear	Wheel Size	Ball Joint Inspection	Lug Nut (ft. lbs.)
2007	MDX	P255/55R18	None	NA	NA	NA	NS	80
2008	MDX	P255/55R18	None	NA	NA	NA	NS	80

OEM: Original Equipment Manufacturer

PSI: Pounds Per Square Inch

NS: Not Specified by manufacturer

NA: Not available

22140_AMDX_G0012

BRAKE SPECIFICATIONS
All measurements in inches unless noted

Year	Model		Brake Disc			Brake Drum Diameter			Minimum Lining Thickness		Brake Caliper	
			Original Thickness	Minimum Thickness	Maximum Runout	Original Inside Diameter	Max. Wear Limit	Maximum Machine Diameter	Front	Rear	Bracket Bolts (ft. lbs.)	Mounting Bolts (ft. lbs.)
2007	MDX	F	1.100-1.111	1.020	0.004	—	—	—	0.06	—	101	27
		R	0.430-0.440	0.350	0.004	—	—	—	—	0.06	41	27
2008	MDX	F	1.100-1.111	1.020	0.004	—	—	—	0.06	—	101	27
		R	0.430-0.440	0.350	0.004	—	—	—	—	0.06	41	27

NA: Not Available

F: Front

R: Rear

22140_AMDX_C0013

SCHEDULED MAINTENANCE INTERVALS
Acura MDX

TO BE SERVICED	TYPE OF SERVICE	SERVICE INTERVALS						
		Symbol A	Symbol B	1	2	3	4	5
Air cleaner element	Replace				✓			
Brake fluid	Inspect		✓					
Brake hoses and lines	Inspect		✓					
Clutch fluid (if equipped)	Inspect		✓					
Drive belt	Inspect				✓			
Driveshaft boots	Inspect		✓					
Dust & pollen filter	Replace				✓			
Engine coolant	Inspect		✓					
Engine coolant	Replace							✓
Engine oil	Replace	✓						
Engine oil & filter	Replace		✓					
Exhaust system	Inspect		✓					
Front & Rear brakes	Inspect		✓					
Fuel lines	Inspect		✓					
Parking brake adjustment	Inspect		✓					
Power steering fluid	Inspect		✓					
Spark plugs	Replace						✓	
Suspension components	Inspect		✓					
Tie-rod ends, steering gearbox and boots	Inspect		✓					
Timing belt	Replace						✓	
Tires	Rotate			✓				
Transmission fluid	Inspect		✓					
Transmission fluid	Replace					✓		
Valve clearance	Inspect						✓	
Windshield washer fluid	Inspect		✓					

Symbol A and B represent the maintenance symbol that will appear on the dash maintenance reminder.
Numbers 1through 5 indicate when the service is performed. See the owners manual for specifics.

22140_AMDX_C0014

PRECAUTIONS

Before servicing any vehicle, please be sure to read all of the following precautions, which deal with personal safety, prevention of component damage, and important points to take into consideration when servicing a motor vehicle:

• Never open, service or drain the radiator or cooling system when the engine is hot; serious burns can occur from the steam and hot coolant.

• Observe all applicable safety precautions when working around fuel. Whenever servicing the fuel system, always work in a well-ventilated area. Do not allow fuel spray or vapors to come in contact with a spark, open flame, or excessive heat (a hot drop light, for example). Keep a dry chemical fire extinguisher near the work area. Always keep fuel in a container specifically designed for fuel storage; also, always properly seal fuel containers to avoid the possibility of fire or explosion. Refer to the additional fuel system precautions later in this section.

• Fuel injection systems often remain pressurized, even after the engine has been turned **OFF**. The fuel system pressure must be relieved before disconnecting any fuel lines. Failure to do so may result in fire and/or personal injury.

• Brake fluid often contains polyglycol ethers and polyglycols. Avoid contact with the eyes and wash your hands thoroughly after handling brake fluid. If you do get brake fluid in your eyes, flush your eyes with clean, running water for 15 minutes. If eye irritation persists, or if you have taken

brake fluid internally, IMMEDIATELY seek medical assistance.

• The EPA warns that prolonged contact with used engine oil may cause a number of skin disorders, including cancer. You should make every effort to minimize your exposure to used engine oil. Protective gloves should be worn when changing oil. Wash your hands and any other exposed skin areas as soon as possible after exposure to used engine oil. Soap and water, or waterless hand cleaner should be used.

• All new vehicles are now equipped with an air bag system, often referred to as a Supplemental Restraint System (SRS) or Supplemental Inflatable Restraint (SIR) system. The system must be disabled before performing service on or around system components, steering column, instrument panel components, wiring and sensors. Failure to follow safety and disabling procedures could result in accidental air bag deployment, possible personal injury and unnecessary system repairs.

• Always wear safety goggles when working with, or around, the air bag system. When carrying a non-deployed air bag, be sure the bag and trim cover are pointed away from your body. When placing a non-deployed air bag on a work surface, always face the bag and trim cover upward, away from the surface. This will reduce the motion of the module if it is accidentally deployed. Refer to the additional air bag system precautions later in this section.

• Clean, high quality brake fluid from a sealed container is essential to the safe and

proper operation of the brake system. You should always buy the correct type of brake fluid for your vehicle. If the brake fluid becomes contaminated, completely flush the system with new fluid. Never reuse any brake fluid. Any brake fluid that is removed from the system should be discarded. Also, do not allow any brake fluid to come in contact with a painted surface; it will damage the paint.

• Never operate the engine without the proper amount and type of engine oil; doing so WILL result in severe engine damage.

• Timing belt maintenance is extremely important. Many models utilize an interference-type, non-freewheeling engine. If the timing belt breaks, the valves in the cylinder head may strike the pistons, causing potentially serious (also time-consuming and expensive) engine damage. Refer to the maintenance interval charts for the recommended replacement interval for the timing belt, and to the timing belt section for belt replacement and inspection.

• Disconnecting the negative battery cable on some vehicles may interfere with the functions of the on-board computer system(s) and may require the computer to undergo a relearning process once the negative battery cable is reconnected.

• When servicing drum brakes, only disassemble and assemble one side at a time, leaving the remaining side intact for reference.

• Only an MVAC-trained, EPA-certified automotive technician should service the air conditioning system or its components.

BRAKES

ANTI-LOCK BRAKE SYSTEM (ABS)

GENERAL INFORMATION

PRECAUTIONS

• Certain components within the ABS system are not intended to be serviced or repaired individually.

• Do not use rubber hoses or other parts not specifically specified for and ABS system. When using repair kits, replace all parts included in the kit. Partial or incorrect repair may lead to functional problems and require the replacement of components.

• Lubricate rubber parts with clean, fresh brake fluid to ease assembly. Do not

use shop air to clean parts; damage to rubber components may result.

• Use only DOT 3 brake fluid from an unopened container.

• If any hydraulic component or line is removed or replaced, it may be necessary to bleed the entire system.

• A clean repair area is essential. Always clean the reservoir and cap thoroughly before removing the cap. The slightest amount of dirt in the fluid may plug an orifice and impair the system function. Perform repairs after components have been thoroughly cleaned; use only denatured alcohol

to clean components. Do not allow ABS components to come into contact with any substance containing mineral oil; this includes used shop rags.

• The Anti-Lock control unit is a microprocessor similar to other computer units in the vehicle. Ensure that the ignition switch is **OFF** before removing or installing controller harnesses. Avoid static electricity discharge at or near the controller.

• If any arc welding is to be done on the vehicle, the control unit should be unplugged before welding operations begin.

BRAKES

BLEEDING PROCEDURE

BLEEDING PROCEDURE

See Figure 1.

1. Make sure the brake fluid level in the reservoir is at the MAX (upper) level line.

2. Slide a piece of clear plastic hose over the first bleed screw, and submerge the other end in a container of new brake fluid.

3. Have someone slowly pump the brake pedal several times, then apply steady pressure.

4. Loosen the left-front brake bleed screw, loosen the brake bleed screw to allow air to escape from the system. Then tighten the bleed screw securely.

5. Repeat the procedure for each caliper until no air bubbles are in the fluid. Bleed the calipers in the sequence shown.

6. Refill the master cylinder reservoir to the MAX (upper) level line.

7. Test-drive the vehicle, and make at least two slow ABS-activating stops.

8. Check for pedal feel.

9. If the pedal feel is hard, you are done.

10. If the pedal feel is soft, repeat the brake system bleeding procedure.

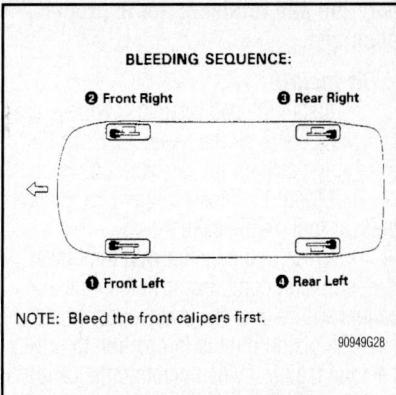

BLEEDING SEQUENCE:

❷ Front Right   ❸ Rear Right

❶ Front Left   ❹ Rear Left

NOTE: Bleed the front calipers first.

90949G28

Fig. 1 Proper brake bleeding sequence

BRAKES

✳✳ CAUTION

Dust and dirt accumulating on brake parts during normal use may contain asbestos fibers from production or aftermarket brake linings. Breathing excessive concentrations of asbestos fibers can cause serious bodily harm. Exercise care when servicing brake parts. Do not sand or grind brake lining unless equipment used is designed to contain the dust residue. Do not clean brake parts with compressed air or by dry brushing. Cleaning should be done by dampening the brake components with a fine mist of water, then wiping the brake components clean with a dampened cloth. Dispose of cloth and all residue containing asbestos fibers in an impermeable container with the appropriate label. Follow practices prescribed by the Occupational Safety and Health Administration (OSHA) and the Environmental Protection Agency (EPA) for the handling, processing, and disposing of dust or debris that may contain asbestos fibers.

BRAKE CALIPER

REMOVAL & INSTALLATION

See Figure 2.

1. Raise and safely support the vehicle.
2. Remove the front wheel.
3. Remove the brake hose bracket mounting bolt and banjo bolt to disconnect the brake hose from the caliper.

Plug the brake hose to prevent excessive fluid loss.

4. Remove the brake caliper bracket mounting bolts, then remove the caliper assembly from the knuckle.

To install:

5. Install the caliper assembly and tighten the mounting bolts to 101 ft. lbs. (137 Nm).

6. Connect the brake hose using new sealing washers and tighten the banjo bolt to 25 ft. lbs. (34 Nm).

7. The remainder of the installation is the reverse order of removal.

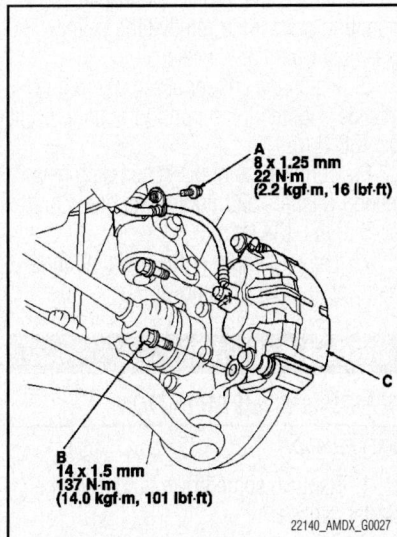

A
8 x 1.25 mm
22 N·m
(2.2 kgf·m, 16 lbf·ft)

B
14 x 1.5 mm
137 N·m
(14.0 kgf·m, 101 lbf·ft)

22140_AMDX_G0027

Fig. 2 Remove the caliper mounting bolts (B) to remove the caliper assembly (C) from the vehicle—MDX

DISC BRAKE PADS

REMOVAL & INSTALLATION

See Figure 3.

1. Remove some brake fluid from the master cylinder.

2. Raise and safely support the vehicle.

3. Remove the front wheel.

4. Remove the flange bolt, and pivot the caliper up out of the way.

5. Remove the pad shims and brake pads.

6. Remove the pad retainers.

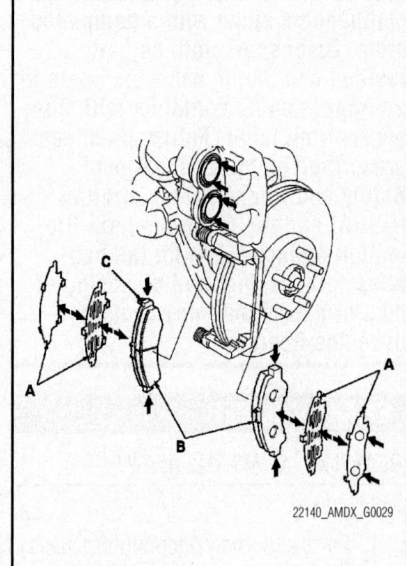

22140_AMDX_G0029

Fig. 3 Apply assembly to the pad side of the shims (A) and back of the brake pads (B). Install the pad with the wear indicator (C) on the upper side—MDX

➡The upper and lower pad retainers are different. During installation, make sure the pad retainers are in proper positions.

To install:

7. Install the pad retainers. Wipe excess assembly paste off the retainers. Keep the assembly paste off the discs and pads.

8. Mount the brake caliper piston compressor tool on the caliper body.

9. Press in the piston with the brake caliper piston compressor tool so the caliper will fit over the brake pads. Make sure the piston boot is in position to prevent damaging it when pivoting the caliper down.

➡Be careful when pressing in the piston; brake fluid might overflow from the master cylinder's reservoir.

10. Remove the brake caliper piston compressor tool.

11. Apply a thin coat of M-77 assembly paste (P/N 08798-9010) to the pad side of the shims, the back of the brake pads and the other areas indicated by the arrows. Wipe excess assembly paste off the pad shims and brake pads. Contaminated brake discs or brake pads reduce stopping ability. Keep grease and assembly paste off the brake discs and brake pads.

12. Install the brake pads and pad shims correctly. Install the brake pad with the wear indicator on the upper inside. If you are reusing the brake pads, always reinstall the brake pads in their original positions to prevent a momentary loss of braking efficiency.

13. Pivot the caliper down into position. Install the flange bolt and tighten it to 53 ft. lbs. (72 Nm).

14. Clean the mating surfaces of the brake disc and the inside of the wheel, then install the front wheels.

15. Press the brake pedal several times to make sure the brakes work.

16. Add brake fluid as needed.

17. After installation, check for leaks at hose and line joints or connections, and retighten if necessary.

18. Test-drive the vehicle, then recheck for leaks.

BRAKES

❉❉ CAUTION

Dust and dirt accumulating on brake parts during normal use may contain asbestos fibers from production or aftermarket brake linings. Breathing excessive concentrations of asbestos fibers can cause serious bodily harm. Exercise care when servicing brake parts. Do not sand or grind brake lining unless equipment used is designed to contain the dust residue. Do not clean brake parts with compressed air or by dry brushing. Cleaning should be done by dampening the brake components with a fine mist of water, then wiping the brake components clean with a dampened cloth. Dispose of cloth and all residue containing asbestos fibers in an impermeable container with the appropriate label. Follow practices prescribed by the Occupational Safety and Health Administration (OSHA) and the Environmental Protection Agency (EPA) for the handling, processing, and disposing of dust or debris that may contain asbestos fibers.

BRAKE CALIPER

REMOVAL & INSTALLATION

See Figure 4.

1. Raise and safely support the vehicle.
2. Remove the rear wheel.
3. Remove the brake hose bracket mounting bolt and banjo bolt to disconnect the brake hose from the caliper. Plug the brake hose to prevent excessive fluid loss.

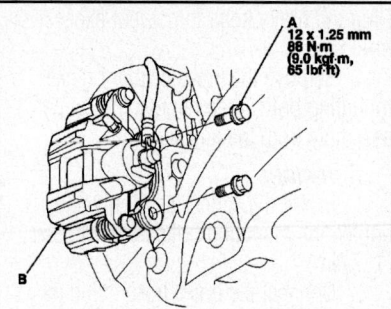

A
12 x 1.25 mm
88 N·m
(9.0 kgf·m,
65 lbf·ft)

22140_AMDX_G0030

Fig. 4 Remove the brake caliper mounting bolts (A) to remove the caliper assembly—MDX

4. Remove the brake caliper bracket mounting bolts, then remove the caliper assembly from the knuckle.

5. Install the caliper assembly to the knuckle. Tighten the mounting bolts to 65 ft. lbs. (88 Nm).

6. Connect the brake hose using new sealing washers and tighten the banjo bolt to 25 ft. lbs. (34 Nm).

7. The remainder of the installation is the reverse order of removal.

DISC BRAKE PADS

REMOVAL & INSTALLATION

See Figure 5.

1. Remove some brake fluid from the master cylinder.
2. Raise and safely support the vehicle.
3. Remove the rear wheel.
4. Remove the flange bolt, and pivot the caliper up out of the way.
5. Remove the pad shims and brake pads.
6. Remove the pad retainers.

REAR DISC BRAKES

To install:

7. Install the pad retainers. Wipe excess assembly paste off the retainers. Keep the assembly paste off the discs and pads.

8. Mount the brake caliper piston compressor tool on the caliper body.

9. Press in the piston with the brake caliper piston compressor tool so the caliper will fit over the brake pads. Make sure the piston boot is in position to prevent damaging it when pivoting the caliper down.

❉❉ CAUTION

Be careful when pressing in the piston; brake fluid might overflow from the master cylinder's reservoir. If brake fluid gets on any painted surface, wash it off immediately with water.

10. Remove the brake caliper piston compressor tool.

11. Apply a thin coat of M-77 assembly paste to the pad side of the shims, the back of the brake pads, and the other areas

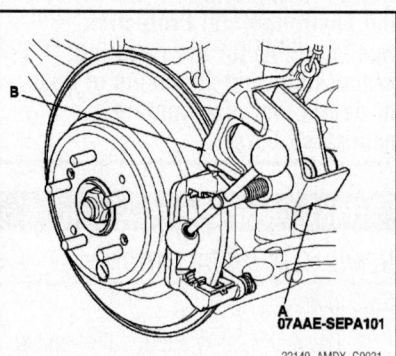

07AAE-SEPA101

22140_AMDX_G0031

Fig. 5 Use the brake caliper piston compressor tool (A) on the caliper body (B)—MDX

indicated by the arrows. Wipe excess assembly paste off the pad shims and brake pads. Contaminated brake disc/drums or brake pads reduce stopping ability. Keep grease and assembly paste off the brake disc/drums and brake pads.

12. Install the brake pads and pad shims correctly. Install the brake pad with the wear indicator on the bottom inside. If you are reusing the brake pads, always reinstall the brake pads in their original positions to prevent a momentary loss of braking efficiency.

13. Pivot the caliper down into position. Install the flange bolt and tighten it to 27 ft. lbs. (37 Nm).

14. Clean the mating surfaces of the brake disc/drum and the inside of the wheel, then install the rear wheels.

15. Press the brake pedal several times to make sure the brakes work.

16. Add brake fluid as needed.

17. After installation, check for leaks at hose and line joints or connections, and retighten if necessary.

18. Test-drive the vehicle, then check for leaks.

BRAKES   REAR DRUM BRAKES

✳✳ CAUTION

Dust and dirt accumulating on brake parts during normal use may contain asbestos fibers from production or aftermarket brake linings. Breathing excessive concentrations of asbestos fibers can cause serious bodily harm. Exercise care when servicing brake parts. Do not sand or grind brake lin-ing unless equipment used is designed to contain the dust residue. Do not clean brake parts with compressed air or by dry brushing. Cleaning should be done by dampening the brake components with a fine mist of water, then wiping the brake components clean with a dampened cloth. Dispose of cloth and all residue containing asbestos fibers in an impermeable container with the appropriate label. Follow practices prescribed by the Occupational Safety and Health Administration (OSHA) and the Environmental Protection Agency (EPA) for the handling, processing, and disposing of dust or debris that may contain asbestos fibers.

BRAKES   PARKING BRAKE

PARKING BRAKE SHOES

REMOVAL & INSTALLATION

See Figures 6 through 14.

1. Raise the rear of the vehicle, and make sure it is securely supported. Remove the rear wheels.

2. Release the parking brake, and remove the rear brake caliper and brake disc/drum.

3. Disconnect and remove the upper return springs.

4. Remove the tension pins by pushing and turning the retainers.

5. Remove the connecting rod.

6. Lower the parking brake shoe assembly.

7. Remove the forward brake shoe by removing the lower return spring and adjuster assembly.

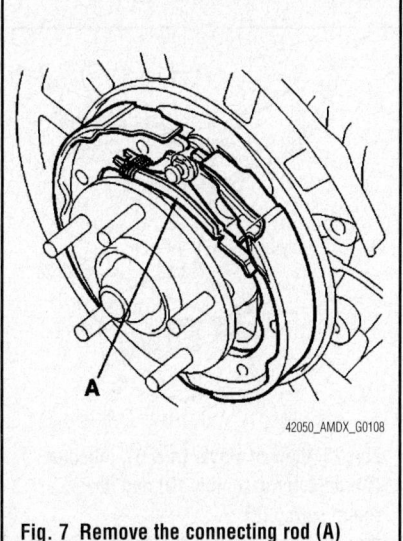

Fig. 7 Remove the connecting rod (A)

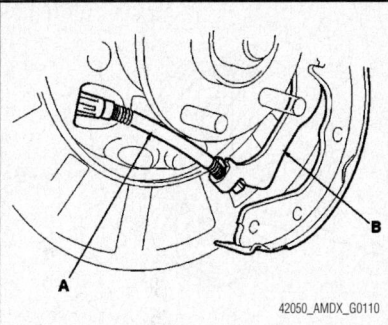

Fig. 9 Remove the rearward brake shoe by disconnecting the parking brake cable (A) from the parking brake lever (B)

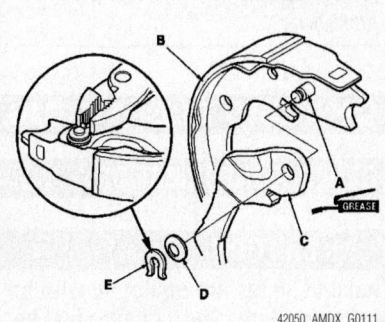

Fig. 10 Apply Molykote® 44 MA grease to the sliding surface of the pivot pin (A) of the rearward brake shoe (B). Install the parking brake lever (C) and wave washer (D) on the pivot pin, and secure with a new U-clip (E)

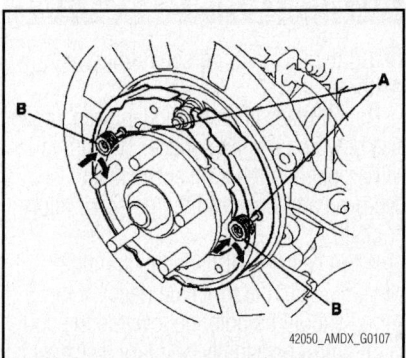

Fig. 6 Remove the tension pins (A) by pushing and turning the retainers (B)

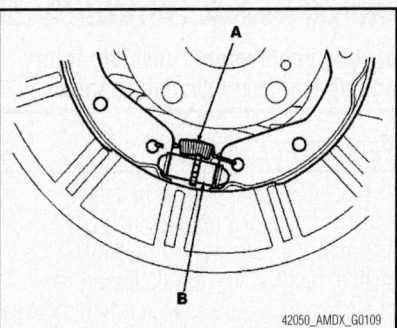

Fig. 8 Remove the forward brake shoe by removing the lower return spring (A) and adjuster assembly (B)

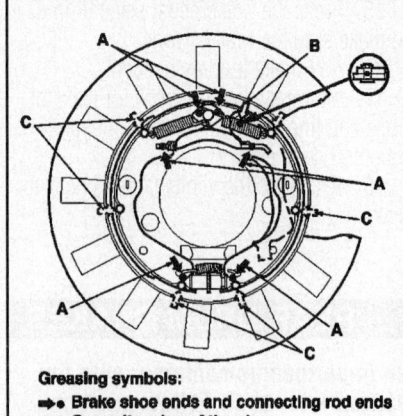

Greasing symbols:
➡● Brake shoe ends and connecting rod ends
⟷○ Opposite edge of the shoe
⟹● Sliding surface

42050_AMDX_G0112

Fig. 11 Apply Molykote 44 MA grease to the shoe ends and connecting rod ends (A), sliding surfaces (B), and opposite edges of the parking brake shoe (C) as shown

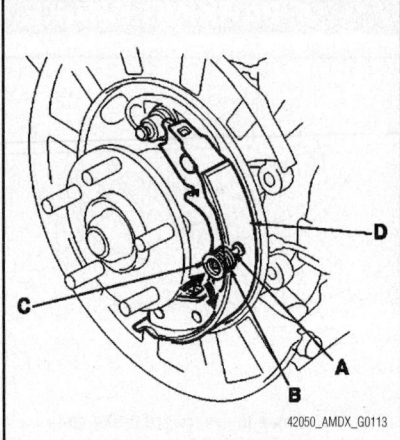

42050_AMDX_G0113

Fig. 12 Install the tension pin (A), retainer spring (B), and retainer (C) of the rearward brake shoe (D). Make sure the tension pin does not contact the parking brake lever

8. Remove the rearward brake shoe by disconnecting the parking brake cable from the parking brake lever.

9. Remove the U-clip, wave washer, and parking brake lever from the brake shoe.

To install:

10. Apply Molykote® 44 MA grease to the sliding surface of the pivot pin of the rearward brake shoe.

11. Install the parking brake lever and wave washer on the pivot pin, and secure with a new U-clip, noting the following:

    a. Install the wave washer with its convex side facing out.

    b. Pinch the U-clip securely to prevent the parking brake lever from coming out from the brake shoe.

12. Connect the parking brake cable to the parking brake lever.

13. Apply Molykote® 44 MA grease to the shoe ends and connecting rod ends, sliding surfaces, and opposite edges of the parking brake shoe as shown. Wipe off

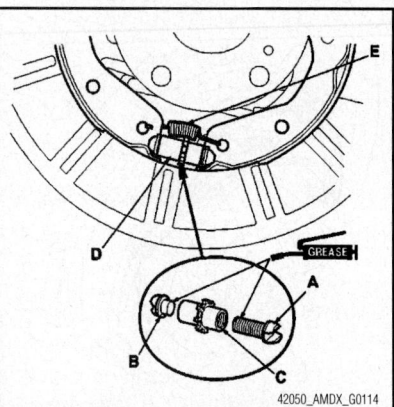

42050_AMDX_G0114

Fig. 13 View of clevis (A & B), adjuster (C), adjuster assembly (D) and lower return spring (E)

any excess. Keep grease off the brake linings.

14. Install the tension pin, retainer spring, and retainer of the rearward brake shoe. Make sure the tension pin does not contact the parking brake lever.

15. Clean the threaded portions of clevis A, and coat the threads of clevis A with grease. Clean the sliding surface of clevis B, and coat the sliding surface of clevis B with grease. Install clevis A and B on the adjuster, and shorten clevis A by turning the adjuster.

16. Reinstall the brake shoe adjuster assembly, and hook the lower return spring on the parking brake shoes.

17. Install the rod spring to the connecting rod first. Then install the connecting rod on the parking brake shoes.

18. Install the tension pin, retainer spring, and retainer of the forward brake shoe.

19. Install the upper return springs.

20. Install the rear brake disc/drum and rear brake caliper.

21. Adjust the parking brake.

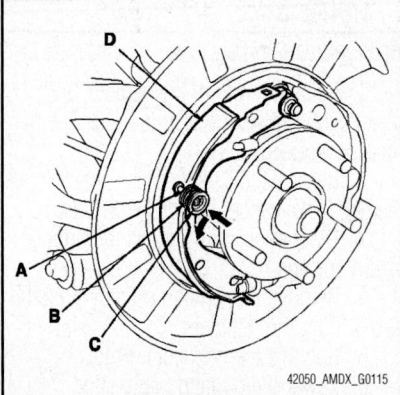

42050_AMDX_G0115

Fig. 14 Install the tension pin (A), retainer spring (B), and retainer (C) of the forward brake shoe (D)

CHASSIS ELECTRICAL

AIR BAG (SUPPLEMENTAL RESTRAINT SYSTEM)

GENERAL INFORMATION

✳✳ CAUTION

These vehicles are equipped with an air bag system. The system must be disarmed before performing service on, or around, system components, the steering column, instrument panel components, wiring and sensors. Failure to follow the safety precautions and the disarming procedure could result in accidental

air bag deployment, possible injury and unnecessary system repairs.

SERVICE PRECAUTIONS

Disconnect and isolate the battery negative cable before beginning any airbag system component diagnosis, testing, removal, or installation procedures. Allow system capacitor to discharge for two minutes before beginning any component service. This will disable the airbag system. Failure to disable the airbag system may result in accidental

airbag deployment, personal injury, or death.

Do not place an intact undeployed airbag face down on a solid surface. The airbag will propel into the air if accidentally deployed and may result in personal injury or death.

When carrying or handling an undeployed airbag, the trim side (face) of the airbag should be pointing towards the body to minimize possibility of injury if accidental deployment occurs. Failure to do this may result in personal injury or death.

Replace airbag system components with OEM replacement parts. Substitute parts may appear interchangeable, but internal differences may result in inferior occupant protection. Failure to do so may result in occupant personal injury or death.

Wear safety glasses, rubber gloves, and long sleeved clothing when cleaning powder residue from vehicle after an airbag deployment. Powder residue emitted from a deployed airbag can cause skin irritation. Flush affected area with cool water if irritation is experienced. If nasal or throat irritation is experienced, exit the vehicle for fresh air until the irritation ceases. If irritation continues, see a physician.

Do not use a replacement airbag that is not in the original packaging. This may result in improper deployment, personal injury, or death.

The factory installed fasteners, screws and bolts used to fasten airbag components have a special coating and are specifically designed for the airbag system. Do not use substitute fasteners. Use only original equipment fasteners listed in the parts catalog when fastener replacement is required.

During, and following, any child restraint anchor service, due to impact event or vehicle repair, carefully inspect all mounting hardware, tether straps, and anchors for proper installation, operation, or damage. If a child restraint anchor is found damaged in any way, the anchor must be replaced. Failure to do this may result in personal injury or death.

Deployed and non-deployed airbags may or may not have live pyrotechnic material within the airbag inflator.

Do not dispose of driver/passenger/curtain airbags or seat belt tensioners unless you are sure of complete deployment. Refer to the Hazardous Substance Control System for proper disposal.

Dispose of deployed airbags and tensioners consistent with state, provincial, local, and federal regulations.

After any airbag component testing or service, do not connect the battery negative cable. Personal injury or death may result if the system test is not performed first.

If the vehicle is equipped with the Occupant Classification System (OCS), do not connect the battery negative cable before performing the OCS Verification Test using the scan tool and the appropriate diagnostic information. Personal injury or death may result if the system test is not performed properly.

Never replace both the Occupant Restraint Controller (ORC) and the Occupant Classification Module (OCM) at the same time. If both require replacement, replace one, then perform the Airbag System test before replacing the other.

Both the ORC and the OCM store Occupant Classification System (OCS) calibration data, which they transfer to one another when one of them is replaced. If both are replaced at the same time, an irreversible fault will be set in both modules and the OCS may malfunction and cause personal injury or death.

If equipped with OCS, the Seat Weight Sensor is a sensitive, calibrated unit and must be handled carefully. Do not drop or handle roughly. If dropped or damaged, replace with another sensor. Failure to do so may result in occupant injury or death.

If equipped with OCS, the front passenger seat must be handled carefully as well. When removing the seat, be careful when setting on floor not to drop. If dropped, the sensor may be inoperative, could result in occupant injury, or possibly death.

If equipped with OCS, when the passenger front seat is on the floor, no one should sit in the front passenger seat. This uneven force may damage the sensing ability of the seat weight sensors. If sat on and damaged, the sensor may be inoperative, could result in occupant injury, or possibly death.

DISARMING THE SYSTEM

Disconnect and isolate the negative battery cable. Wait 3 minutes for the system capacitor to discharge before performing any service.

ARMING THE SYSTEM

Connect the negative battery cable.

CLOCKSPRING CENTERING

See Figure 15.

1. Rotate the cable reel clockwise until it stops. Then rotate it counterclockwise (about three turns) until the arrow mark on the cable reel label points straight up.

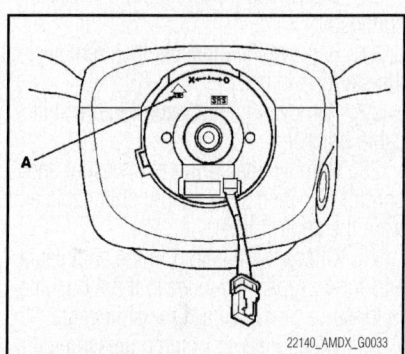

22140_AMDX_G0033

Fig. 15 Rotate the cable reel counterclockwise (about three turns) until the arrow mark (A) on the cable reel label points straight up—MDX

DRIVETRAIN

AUTOMATIC TRANSAXLE ASSEMBLY

REMOVAL & INSTALLATION

See Figures 16 and 17.

1. Secure the hood in the wide open position.
2. Set the wheels in the straight ahead position and lock the steering wheel.
3. Make sure you have the anti-theft code for the radio and the navigation system, then write down the frequencies for the radio's preset buttons.
4. Disconnect the negative cable from the battery first, then disconnect the positive cable. Wait at least 3 minutes before proceeding.
5. Drain the power steering system.
6. Remove the battery.
7. Remove the intake manifold cover, intake air duct, and air cleaner housing.
8. Remove the under-hood sub-fuse box from its bracket, and remove the bracket from the battery base.
9. Remove the battery base and battery base bracket.
10. Remove the front bulkhead cover.
11. Raise and safely support the vehicle.
12. Remove the front wheels.
13. Remove the transmission undercover and splash shield.

14. Drain the transaxle and reinstall the drain plug.
15. Disconnect the steering joint.
16. Remove the power steering pump outlet line from the power steering pump, and remove the hose clamp bolt.
17. Disconnect the transmission breather hose from the breather pipe at the transmission housing.
18. Disconnect the vacuum hose from the vacuum line and remove the vacuum line bolt.
19. Disconnect the starter cables from the starter, and remove the harness clamp from the clamp bracket.

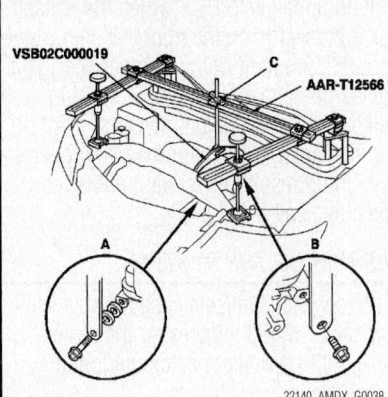

VSB02C000019

C

AAR-T12566

A

B

22140_AMDX_G0038

Fig. 16 Attach the front arm (A) to the front cylinder head and attach the rear arm (B) to the rear cylinder head. Install the engine support hanger (AAR-T12566) to the vehicle, and attach the hook to the slotted hole in the engine balance bar. Tighten the wing nut (C) by hand, to lift and support the engine—MDX

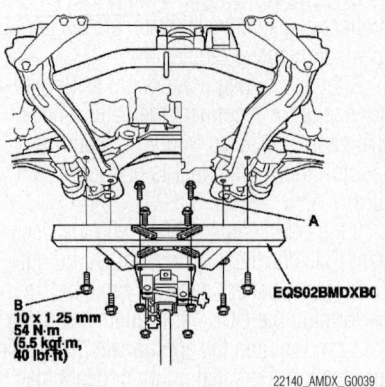

A

B
10 x 1.25 mm
54 N·m
(5.5 kgf·m,
40 lbf·ft)

EQS02BMDXB0

22140_AMDX_G0039

Fig. 17 Line up the slots in the arms with the bolt holes on the corner of the jack base, then attach the adapter (EQS02BMDXB0) to the jack base with the bolts (A) that came with the jack. Raise the jack to the vehicle height, then attach the adapter to the front subframe using the subframe stiffener mounting bolts (B) and bolt holes—MDX

20. Disconnect the solenoid harness connector.

21. Remove the dipstick, then remove the starter and gasket.

22. Remove the nuts securing the shift cable bracket.

23. Remove the spring clip/washer and control pin, then separate the shift cable end from the control lever.

24. Check the bushing in the shift cable end for a proper fit and wear. If the bushing is loose or worn, replace the shift cable.

25. Disconnect A/T clutch pressure control solenoid valve A connector, A/T clutch pressure control solenoid valve B connector, and 4th clutch transmission fluid pressure switch connector, and remove the harness cover mounting bolt.

26. Disconnect A/T clutch pressure control solenoid valve C connector.

27. Remove the bolts securing the harness cover.

28. Disconnect the transmission range switch connector, and remove the harness clamp from the clamp bracket.

29. Remove the ATF cooler hoses from the ATF cooler lines. Turn the ends of the cooler hoses up to prevent ATF from flowing out, then plug the cooler hoses and lines.

30. Disconnect the input shaft (mainshaft) speed sensor connector, output shaft (countershaft) speed sensor connector, ATF temperature sensor connector, and 3rd clutch transmission fluid pressure switch connector.

31. Disconnect the 2nd clutch transmission fluid pressure switch connector, and remove the vacuum line bolt.

32. Remove the bolt securing the engine mount control solenoid valve mounting bracket.

33. Remove the connector bracket from the engine front cylinder head; use the bracket bolt hole to attach engine hanger balancer bar front arm.

34. Remove the transfer breather bracket from the engine rear cylinder head; use the bracket bolt hole to attach engine hanger balancer bar rear arm.

35. Remove the service caps for the front damper flange nuts from the cowl cover. Position the engine hanger adapters (VSB02C000031) over the damper flange nuts.

36. Install the engine balance bar (VSB02C000019); attach the front arm to the front cylinder head with a spacer and the 10 mm bolt, and attach the rear arm to the rear cylinder head with the 8 mm bolt.

37. Install the engine support hanger (AAR-T12566) to the vehicle, and attach the hook to the slotted hole in the engine balance bar. Tighten the wing nut by hand, to lift and support the engine.

38. Remove the front mount stop and front mount bolt.

39. Remove the front subframe stiffener.

40. Remove the exhaust pipe and its mount.

41. Disconnect the headlight adjuster leveling sensor connector on both lower arms, and remove the harness clamps from the front subframe for models equipped with the headlight adjuster leveling system.

42. Remove the lock pins and castle nuts, and separate the lower arms from the knuckles.

43. Insert a 6 mm Allen wrench in the top of the ball joint pin, and remove the nuts, then separate the stabilizer link.

44. Remove the cotter pins and nuts, and separate the tie-rod end ball joints from the knuckles.

45. Remove the bolt securing the transfer breather hose bracket, and disconnect the breather hose from the breather pipe on the transfer assembly.

46. Make a reference mark across the propeller shaft and the transfer companion flange, and separate the propeller shaft from the transfer companion flange.

47. Remove the transfer assembly from the transaxle.

48. Remove the torque converter cover, and remove the drive plate bolts (8) while rotating the crankshaft pulley.

49. Disconnect the power steering pressure switch connector.

50. Loosen the hose clamp bolt, then disconnect the power steering fluid hose from the fluid line at the left front of the subframe.

51. Remove the transmission lower mounting bolt.

52. Remove the four rear mounting bracket bolts.

53. Remove the bolt securing the power steering fluid line clamp bracket on the rear mount bracket, and turn the bracket away from the rear mount bracket.

54. Loosen the four bolts holding the adjustable arms of the front subframe adapter (EQS02BMDXSB0) to its center plate.

55. Line up the slots in the arms with the bolt holes on the corner of the jack base, then attach the adapter (EQS02BMDXB0) to the jack base with the bolts that came with the jack. Tighten the bolts securely.

56. Raise the jack to the vehicle height, then attach the adapter to the front subframe using the subframe stiffener mounting bolts and bolt holes.

57. Remove the six bolts securing the front stiffeners and rear stiffeners, and four bolts securing the front subframe, and lower the front subframe.

58. Remove the transmission lower mount.

59. Remove the transmission ground terminal.

60. Remove the driveshafts from the differential and the intermediate shaft. Coat all precision machined surfaces with clean engine oil, then put plastic bags over driveshaft ends.

61. Remove the exhaust manifold bracket and heat shield.

62. Remove the intermediate shaft. Coat all precision machined surfaces with clean engine oil, then put plastic bags over intermediate shaft ends.

63. Remove the upper transaxle housing mounting bolts.

64. Remove the lower transaxle housing mounting bolts.

65. Remove the harness clamp bracket from the front mount bracket, and remove the mount bracket.

66. Remove the transaxle housing mounting bolt using a socket 22 mm in length.

67. Remove the rear transaxle housing mounting bolts.

68. Place a jack under the transaxle.

69. Lower the transaxle by loosening the wing nut of the engine support hanger, and tilt the engine just enough for the transaxle to clear its end from the side frame.

70. Slide the transaxle away from the engine to remove it from the vehicle.

To install:

71. Place the transmission on the jack, and raise it to engine level.

72. Attach the transaxle to the engine, and install the lower transmission housing mounting bolts and tighten to 47 ft. lbs. (64 Nm).

73. Install the rear transaxle mounting bolts and tighten to 47 ft. lbs. (64 Nm).

74. Install the upper transaxle housing mounting bolts and tighten to 47 ft. lbs. (64 Nm).

75. Install the transmission housing mounting bolt and tighten to 47 ft. lbs. (64 Nm).

76. Install the front mount bracket with the new mounting bolts and tighten to 9 ft. lbs. (12 Nm).

77. The remainder of the installation is the reverse order of installation.

78. Refill the transaxle with fluid to the correct level.

79. Refill the power steering system to the correct level.

80. Start the engine and check for leaks.

TRANSFER CASE ASSEMBLY

REMOVAL & INSTALLATION

See Figures 18 and 19.

1. Raise and safely support the vehicle.
2. Shift the transmission into neutral.
3. Remove the transaxle undercover.
4. Drain the transaxle fluid and replace the drain plug.
5. Remove the front subframe stiffener.

6. Remove the exhaust pipe.

7. Remove the bolt securing the transfer breather hose bracket, and disconnect the breather hose from the breather pipe on the transfer assembly.

8. Make a reference mark across the driveshaft and the transfer companion flange.

9. Separate the driveshaft from the transfer companion flange.

10. Remove the transfer case assembly from the transaxle.

To install:

11. Install the dowel pin in the transaxle, and install the transfer case assembly on the transaxle. Tighten the mounting bolts to 38 ft. lbs. (51 Nm).

12. Align the matchmarks and install the driveshaft to the transfer case companion flange. Tighten the mounting bolts to 53 ft. lbs. (72 Nm).

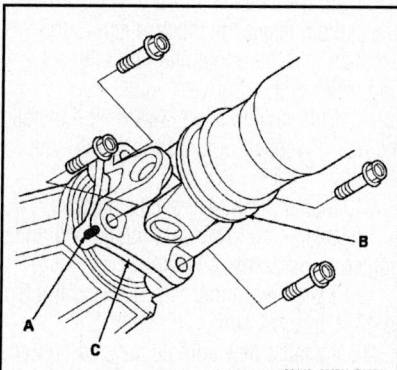

22140_AMDX_G0050

Fig. 18 Make a reference mark (A) across the driveshaft (B) and the transfer companion flange (C) before separating—MDX

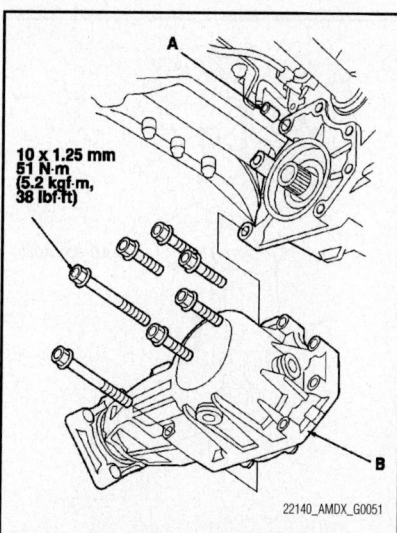

10 x 1.25 mm
51 N·m
(5.2 kgf·m,
38 lbf·ft)

22140_AMDX_G0051

Fig. 19 Install the dowel pin (A) in the transaxle before installing the transfer case (B) to the transaxle—MDX

13. Secure the transfer breather hose bracket on the transfer assembly with the bolt, and install the breather hose over the breather pipe with the dot facing out.

14. Install the exhaust pipe, using new nuts and gaskets.

15. Install the front subframe stiffener with new mounting bolts. Tighten the mounting bolts to 40 ft. lbs. (54 Nm).

16. Refill the transfer case with hypoid gear oil to the correct level.

17. Refill the transaxle with fluid to the correct level.

18. Install the transaxle undercover and tighten the mounting bolts to 87 inch lbs. (9.8 Nm).

FRONT HALFSHAFT

REMOVAL & INSTALLATION

See Figures 20 and 21.

1. Raise and safely support the vehicle.
2. Remove the front wheel.
3. Pry up the locking tab on the spindle nut and remove the nut.
4. Remove the splash shield.
5. Drain the transaxle fluid.
6. Remove the suspension stroke sensor, if equipped.
7. Separate the lower ball joint from the wheel knuckle. For additional information, refer to the following section, "Lower Ball Joint, Removal & Installation."
8. Loosen the halfshaft outboard joint from the front hub using a plastic hammer.
9. Pull the wheel knuckle outward and separate the outboard joint from the front hub.

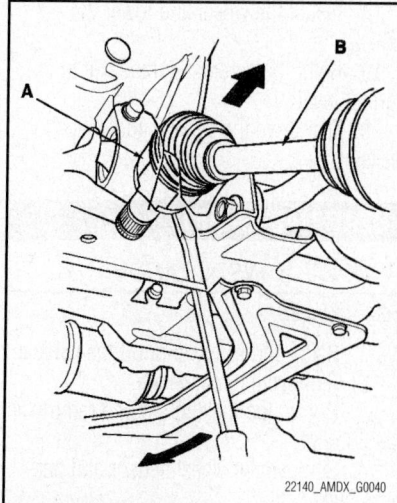

22140_AMDX_G0040

Fig. 20 To remove the left halfshaft (B), pry the inboard joint (A) from the differential with a prybar—MDX

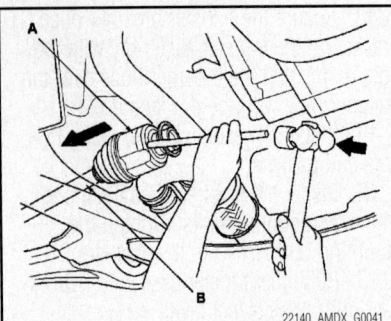

Fig. 21 If removing the right halfshaft (B), drive the inboard joint (A) off of the intermediate shaft using a drift and a hammer—MDX

10. If removing the left halfshaft, pry the inboard joint from the differential with a prybar. Remove the halfshaft as an assembly.

11. If removing the right halfshaft, drive the inboard joint off of the intermediate shaft using a drift and a hammer. Remove the halfshaft as an assembly.

To install:

➡**Use new circlips and cotter pins during installation.**

12. Insert the inboard end of the halfshaft into the differential (or intermediate shaft) until the set ring locks in the groove.

13. Install the outboard joint into the front hub.

14. Install the lower ball joint and tighten the nut to 76–83 ft. lbs. (103–113 Nm).

15. Install the suspension stroke sensor, if equipped.

16. Install the splash shield.

17. Install the new hub nut and tighten to 242 ft. lbs. (328 Nm) and stake the nut.

18. Install the wheel and lower the vehicle.

19. Refill the automatic transaxle to the correct level.

20. Test drive the vehicle and check for leaks.

REAR HALFSHAFT

REMOVAL & INSTALLATION

See Figure 22.

1. Raise and safely support the vehicle.
2. Remove the rear wheel.
3. Pry up the locking tab and remove the hub nut.
4. Remove the wheel sensor and harness clip.
5. Remove the lock pin from the upper arm ball joint castle nut, and remove the nut.

6. Separate the ball joint from the upper arm with the ball joint remover.

7. Remove the lower control arm.

8. Place a transmission jack under lower arm and remove the flange bolt.

9. Loosen the rear driveshaft outboard joint from the rear hub using a plastic hammer.

10. Pull the knuckle outward, then separate the rear driveshaft outboard joint.

11. Using the driveshaft remover and the hammer, pry out the inboard joint from the rear differential.

12. Remove the rear halfshaft, then remove the set ring.

To install:

13. Apply grease to the whole splined surface of the halfshaft.

14. Install a new set ring in the set ring groove of the differential.

15. Clean the areas where the driveshaft contacts the differential thoroughly with solvent or brake cleaner, and dry with compressed air. Insert the inboard end of the halfshaft into the differential until the set ring locks in the groove.

16. Pull the knuckle outward, and install the rear driveshaft outboard joint into the rear hub.

17. Install the wheel sensor bracket.

18. Install the lower arm flange bolt and tighten to 54 ft. lbs. (74 Nm).

19. Install the upper arm bolt and tighten to 47 ft. lbs. (64 Nm).

20. Install a new spindle nut and tighten to 181 ft. lbs. (245 Nm). Use a drift to stake the spindle nut shoulder once the nut is tightened to specification.

21. Install the wheel

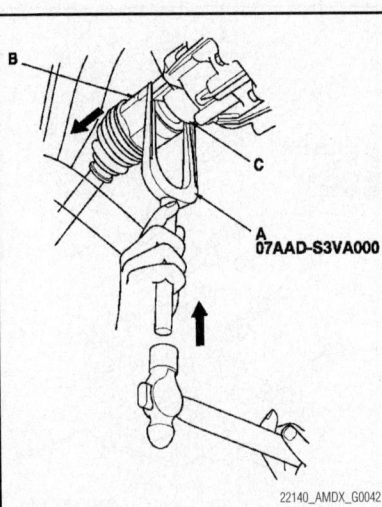

Fig. 22 Using the driveshaft remover (A) and the hammer, pry out the inboard joint (B) from the rear differential (C)—MDX

22. Turn the wheel to make sure there is no binding between the driveshaft and wheel.

23. Refill the differential until the fluid level is at the bottom of the fill hole. A complete oil change would require 2.79 quarts of VTM–4 differential fluid. Install the plug and tighten to 35 ft. lbs. (47 Nm).

24. Check and adjust the wheel alignment.

REAR PINION SEAL

REMOVAL & INSTALLATION

See Figures 23 through 29.

1. Drain the rear differential fluid.

2. Remove the exhaust system.

3. Remove the right rear halfshaft. For additional information, refer to the following section, "Halfshafts, Removal & Installation, Rear."

4. Matchmark the driveshaft to the rear differential companion flange. Separate the driveshaft from the rear differential.

5. Place the transmission jack under the rear differential.

6. Using the driveshaft remover and a

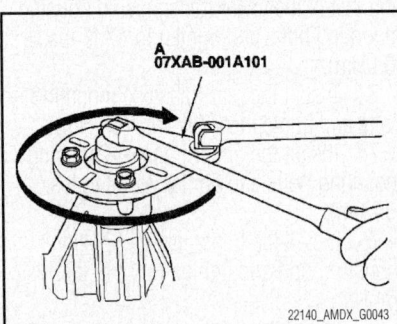

Fig. 23 Install the companion flange holder (A) on the companion flange to loosen the locknut—MDX

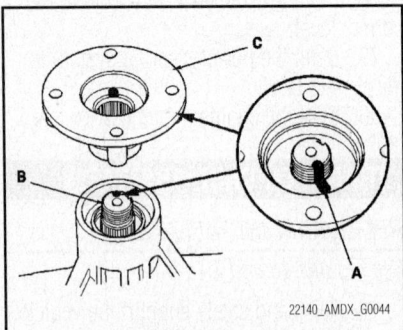

Fig. 24 Make a reference mark (A) across the input shaft (B) and companion flange (C), before removing the companion flange.

hammer, and disconnect the left rear inboard joint from the rear differential.

7. Remove the rear differential mounting bolts.

8. Lower the rear differential a little on the transmission jack, then remove the left rear driveshaft inboard joint from the rear differential.

9. Disconnect the right solenoid wiring harness and the rear differential fluid temperature sensor wiring harness, then remove the harness clip (C).

10. Disconnect the breather tube from the breather pipe.

11. Slowly lower the rear differential a little on the transmission jack.

12. Disconnect the remaining wiring harnesses and remove the differential.

13. Remove the rear differential front mounting brackets from the differential.

14. Remove the rear differential fluid temperature sensor cover, the rear differential fluid temperature sensor, the O-ring, and the rear differential harness bracket with bolts.

15. Remove the left and right side case from the center differential.

16. Remove the bearing set plate and shim, by removing the mounting bolts in a crisscross pattern.

17. Remove the mounting bolts in a crisscross pattern and remove the differential housing assembly.

18. Remove the ring gear assembly, the bearing outer races, the 75 mm shim, and the 8 x 14 mm dowel pins.

19. Raise the locknut tab from the groove of the input shaft, making sure that the tab completely clears the groove to prevent damaging the input shaft.

20. Install the companion flange holder on the companion flange. Loosen the 27 mm locknut (left-hand threads).

21. Loosen the locknut clockwise so that its tab comes out from the groove in the input shaft.

22. Tighten the locknut until its tab aligns with the groove.

23. Remove any dirt from inside of the groove in the input shaft, then loosen the locknut.

24. Remove the 27 mm locknut, the 27 mm spring washer, the 28 mm back-up ring, and the 28 mm O-ring..

25. Make a reference mark across the input shaft and companion flange, then remove the companion flange.

26. Remove the drive pinion, pinion spacer, and the thrust washer by tapping on the drive pinion with a plastic hammer.

27. Remove the pinion oil seal with a commercially available seal remover.

To install:

28. Apply ATF to the tapered roller bearing, then install the new front case oil seal with the dust seal driver and the 39 x 47.3 mm fork seal driver weight.

29. Apply ATF to the tapered roller bearing, then install the drive pinion, pinion spacer, and the thrust washer into the differential carrier.

30. Align the match mark and install the companion flange on the input shaft.

31. Apply ATF to the new 28 mm O-ring, then install the new 28 mm O-ring, the 28 mm backup ring, the 27 mm spring washer, and the new 27 mm locknut.

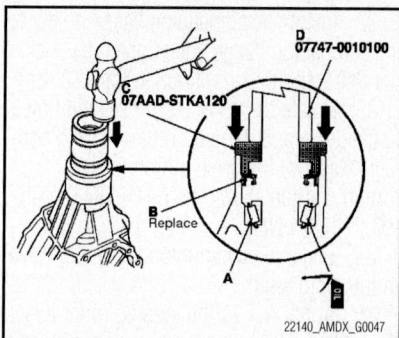

Fig. 27 Apply ATF to the tapered roller bearing (A), then install the new front case oil seal (B) with the dust seal driver (C) and the 39 x 47.3 mm fork seal driver weight (D).

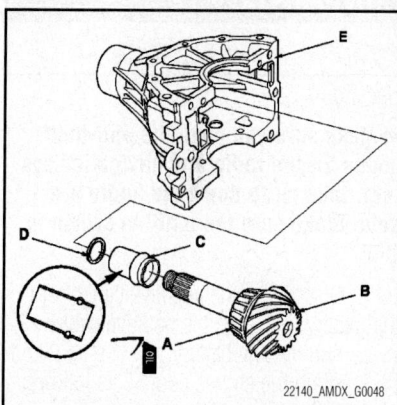

Fig. 28 Apply ATF to the tapered roller bearing (A), then install the drive pinion (B), pinion spacer (C), and the thrust washer (D) into the differential carrier (E).

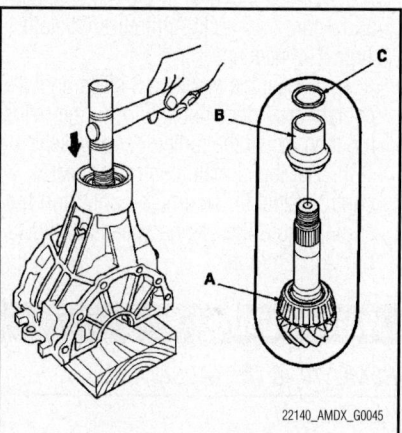

Fig. 25 Remove the drive pinion (A), pinion spacer (B), and the thrust washer (C) by tapping on the drive pinion with a plastic hammer.

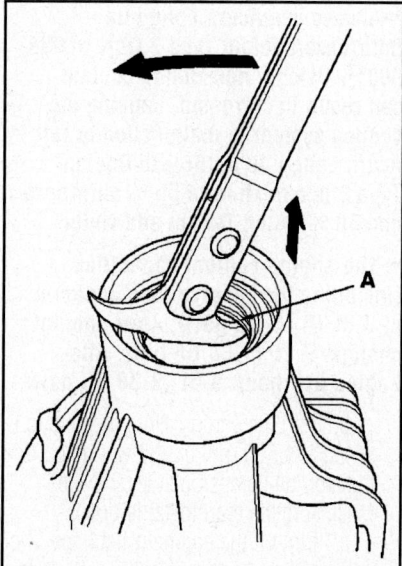

Fig. 26 Remove the front case oil seal (A) with a commercially available seal remover.

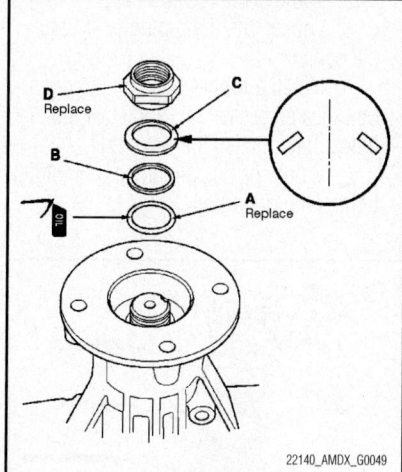

Fig. 29 Apply ATF to the new 28 mm O-ring (A), then install the new 28 mm O-ring, the 28 mm backup ring (B), the 27 mm spring washer (C), and the new 27 mm locknut (D).

32. Install the companion flange holder to the companion flange, then tighten the locknut (left-hand threads) to 53 ft. lbs. (72 Nm).

33. Rotate the drive pinion several times to assure proper tapered roller bearing contact. Measure the drive pinion turning torque. Torque should be 69.0–134.0 ft. lbs. (93.2–181.4 Nm).

34. Stake the locknut tab into the groove in the input shaft.

35. Apply ATF to the tapered roller bearings, then install the ring gear assembly, the bearing outer races, the 75 mm shim, and the 8 x 14 mm dowel pins.

36. Apply sealant the mating surface of the differential housing install the differential housing. Tighten the mounting bolts in a crisscross pattern to 32 ft. lbs. (44 Nm).

37. Install the 83 mm shim and bearing set plate. Install the bolts in a crisscross pattern to 20 ft. lbs. (27 Nm).

38. Apply sealant the mating surfaces of the left and right cases. Install the left and right cases and tighten the mounting bolts in a crisscross pattern to 20 ft. lbs. (27 Nm).

39. Install the rear differential fluid temperature sensor, a new O-ring, the rear differential harness bracket, and the rear differential fluid temperature sensor cover with bolts.

40. Install the rear differential mounting brackets and tighten the mounting bolts to 63 ft. lbs. (85 Nm).

41. The remainder of the installation is the reverse order of removal.

42. Refill the differential with fluid to the correct level.

43. Test drive the vehicle and check for leaks.

ENGINE COOLING

THERMOSTAT

REMOVAL & INSTALLATION

See Figure 30.

➡**Make sure you have the anti-theft codes for the radio and navigation system, then write down the audio presets. Make sure the ignition switch is OFF.**

1. Disconnect the negative cable from the battery first, then the positive cable.

2. Remove the battery.

3. Drain the engine coolant as follows:

 a. Turn the ignition switch ON (II). Set the displayed temperature on the climate control system to 90∞F (32∞C), then turn the ignition switch OFF. Make sure the engine and radiator are cool to the touch.

 b. Remove the radiator cap.

 c. Loosen the drain plug, and drain the coolant.

 d. Install a rubber hose on the drain bolt located at the rear of the cylinder block, then loosen the drain bolt.

 e. When the coolant stops draining, tighten the drain bolt.

 f. Tighten the radiator drain plug securely.

 g. Remove, drain, and reinstall the coolant reservoir.

4. Remove the thermostat cover, then remove the thermostat.

To install:

5. Install the thermostat with a new rubber seal.

6. Install the battery. Clean the battery posts and cable terminals with sandpaper, then assemble them and apply grease to prevent corrosion.

7. Fill the radiator with engine coolant and bleed the air from the system, as follows:

 a. Pour Acura Long Life Antifreeze/Coolant Type 2 (P/N OL999-9001) into the radiator up to the base of the filler neck.

➡**Always use Acura Long Life Antifreeze/Coolant Type 2 (P/N OL999-9001). Using a non-Honda coolant can result in corrosion, causing the cooling system to malfunction or fail. Acura Long Life Antifreeze/Coolant Type 2 is a mixture of 50 % antifreeze and 50 % water. Do not add water.**

➡**The engine coolant capacities (including the reserve tank capacity of 0.6L (0.16 US gal)): After coolant change: 7.1L (1.88 US gal). After engine overhaul: 9.0L (2.38 US gal).**

 b. Start the engine. Hold the engine speed at 1,500 rpm until it warms up (the radiator fan comes on at least twice). Make sure the thermostat is open.

 c. Turn off the engine. Check the level in the radiator, and add Acura Long Life Antifreeze/Coolant Type 2, if needed.

 d. Set the climate control or heater control panel to maximum cool. Start the engine. Hold the engine speed at 1,500 rpm for 5 minutes, then turn off the engine.

 e. Check the level in the radiator, and add Acura Long Life Antifreeze/Coolant Type 2, if needed.

 f. Set the climate control or heater control panel to maximum heat. Start the engine. Hold the engine speed at 1,500 rpm for 5 minutes, then turn off the engine.

 g. Check the level in the radiator, and add Acura Long Life Antifreeze/Coolant Type 2, if needed.

 h. Set the climate control or heater control panel to maximum cool. Start the engine. Hold the engine speed at 1,500 rpm for 3 minutes, then turn off the engine.

 i. Check the level in the radiator, and add Acura Long Life Antifreeze/Coolant Type 2, if needed.

 Set the climate control or heater control panel to maximum heat. Start the engine. Hold the engine speed at 1,500 rpm for 3 minutes, then turn off the engine.

 j. Check the level in the radiator, and add Acura Long Life Antifreeze/Coolant Type 2, if needed.

 k. Repeat the previous 3 steps until the coolant level does not change in the radiator, then install the radiator cap loosely.

 l. Set the climate control or heater control panel to maximum cool. Start the engine. Hold the engine speed at 2,500 rpm for 1 minute.

WATER PUMP

REMOVAL & INSTALLATION

See Figure 31.

1. Disconnect the negative battery cable.

2. Drain the cooling system.

3. Remove the timing belt. For additional information, refer to the following

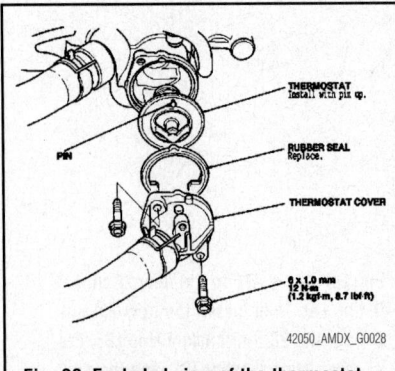

THERMOSTAT
Install with pin up.

PIN

RUBBER SEAL
Replace.

THERMOSTAT COVER

6 x 1.0 mm
12 N·m
(1.2 kgf·m, 8.7 lbf·ft)

42050_AMDX_G0028

Fig. 30 Exploded view of the thermostat and related components—MDX

section, "Timing Belt, Removal & Installation."

4. Remove the timing belt adjuster.

5. Remove the water pump mounting bolts.

6. Remove the water pump.

To install:

7. Install the water pump with a new O-ring. Tighten the mounting bolts to 105 inch lbs. (12 Nm).

8. The remainder of the installation is the reverse order of removal.

9. Refill the cooling system to the correct level.

10. Start the engine and check for leaks.

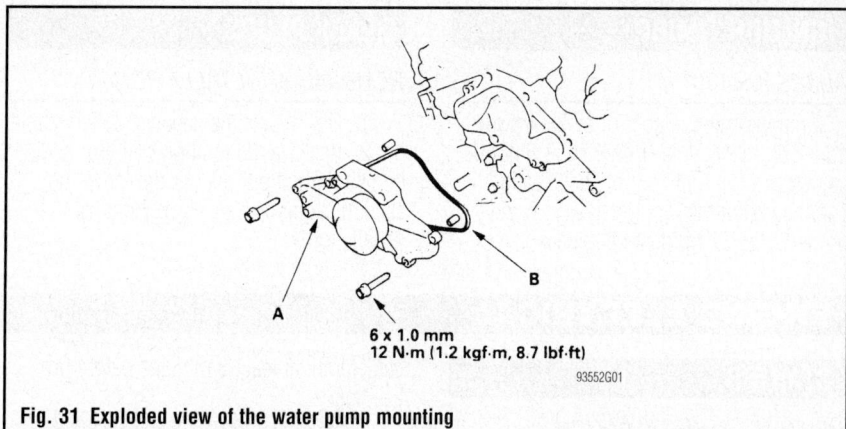

Fig. 31 Exploded view of the water pump mounting

ENGINE ELECTRICAL CHARGING SYSTEM

ALTERNATOR

REMOVAL & INSTALLATION

See Figure 32.

1. Make sure you have the anti-theft code for the radio and the navigation system, then write down the frequencies for the radio's preset buttons.

2. Disconnect the negative cable from the battery first, then disconnect the positive cable. Wait at least 3 minutes before proceeding.

3. Remove the engine appearance cover.

4. Remove the accessory drive belt.

5. Remove the A/C suction line from the brackets.

6. Remove the coolant reservoir, then remove power steering fluid reservoir from the bracket.

7. Remove the harness bracket, then disconnect the A/C compressor clutch wiring harness.

8. Disconnect the alternator wiring harness and the BLK wire from the alternator.

9. Remove the mounting bolt, the bracket mounting bolt, then remove the alternator.

To install:

10. Install the alternator. Tighten the mounting bolt to 33 ft. lbs. (45 Nm) and bracket mounting bolt to 16 ft. lbs. (22 Nm).

11. Connect the alternator wiring harness and BLK wire to the alternator.

12. Install the harness bracket, then connect the A/C compressor clutch wiring harness

13. Install the power steering fluid reservoir to the bracket, then install the coolant reservoir.

14. Install the A/C suction line to the brackets.

15. Install the accessory drive belt.

16. Install the engine appearance cover.

17. Connect the negative battery cable.

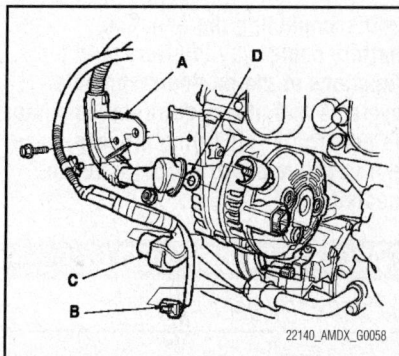

Fig. 32 Remove the following connectors to remove the alternator—MDX

ENGINE ELECTRICAL IGNITION SYSTEM

FIRING ORDER

The firing order is 1–4–2–5–3–6.

IGNITION COIL PACK

REMOVAL & INSTALLATION

See Figure 33.

1. Disconnect the negative battery cable.

2. Remove the engine appearance cover.

3. Disconnect the ignition coil wiring harness.

4. Remove the ignition coil.

To install:

5. Installation is the reverse order of removal. Tighten the ignition coil mounting nut to 108 inch lbs. (12 Nm).

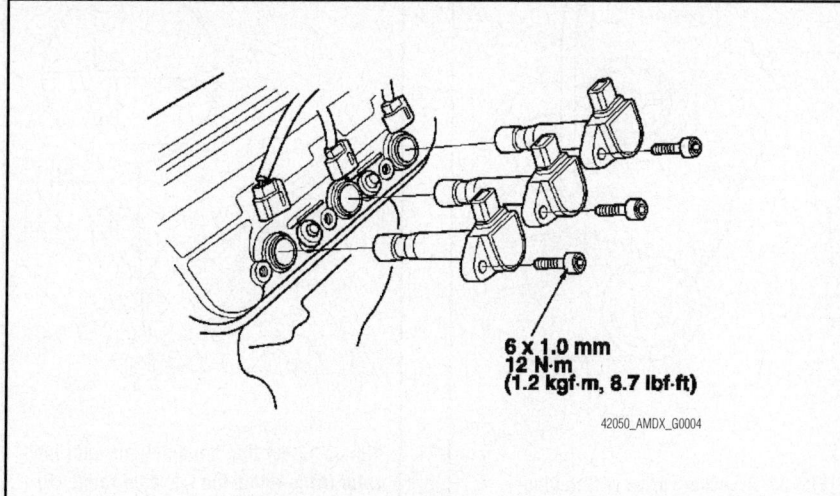

Fig. 33 Exploded view of the ignition coils —MDX

IGNITION TIMING

ADJUSTMENT

Ignition timing is control by the Powertrain Control Module (PCM) and cannot be adjusted. If timing is out of specification and the rest of the ignition system work properly, the PCM must be replaced.

ENGINE ELECTRICAL

STARTER

REMOVAL & INSTALLATION

1. Make sure you have the anti-theft codes for the radio, and the navigation system, then write down the audio presets. Make sure the ignition switch is OFF.
2. Disconnect the negative cable from the battery first, then disconnect the positive cable. Wait at least 3 minutes before proceeding.
3. Remove the air intake assembly.
4. Remove the battery and battery tray.
5. Remove the starter wiring harness clamp.
6. Disconnect the positive starter cable from the B terminal. Disconnect the wiring harness from the S terminal.

SPARK PLUGS

REMOVAL & INSTALLATION

1. Disconnect the negative battery cable.
2. Remove the ignition coil. For additional information, refer to the following section, "Ignition Coil Pack, Removal & Installation."
3. Remove the spark plug.
4. Inspect the spark plug.

To install:

5. Install the spark plug and tighten to 13 ft. lbs. (18 Nm).
6. Install the ignition coil.
7. Connect the negative battery cable.

STARTING SYSTEM

7. Remove the starter mounting bolts and remove the starter.

To install:

8. Install the starter using a new gasket. Tighten the mounting bolts to 33 ft. lbs. (44 Nm).
9. The remainder of the installation is the reverse order of removal.
10. Start the engine to engine proper operation.

ENGINE MECHANICAL

➡**Disconnecting the negative battery cable may interfere with the functions of the on board computer systems and may require the computer to undergo a relearning process, once the negative battery cable is reconnected.**

ACCESSORY DRIVE BELTS

ACCESSORY BELT ROUTING

See Figure 34.

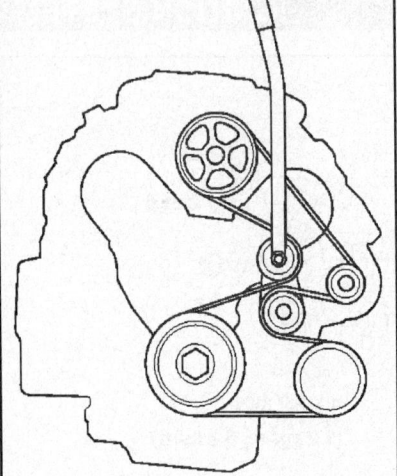

Fig. 34 Accessory drive belt routing—MDX

INSPECTION

See Figure 35.

1. Inspect the belt for cracks and damage. If the belt is cracked or damaged, replace it.
2. Check that the auto-tensioner indicator is within the standard range. If it is out of the standard range, replace the drive belt.

ADJUSTMENT

Belt tension is automatically maintained by a belt tensioner. No adjustments are necessary.

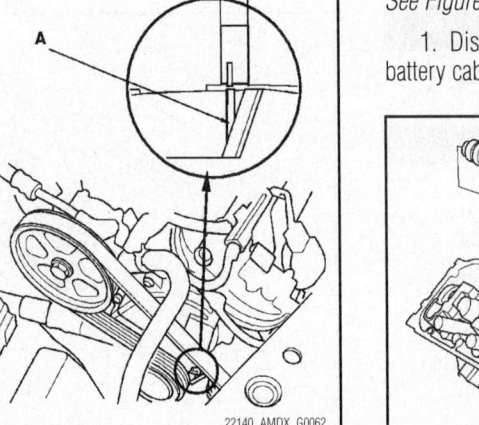

Fig. 35 Check that the auto-tensioner indicator (A) is within the standard range (B) as shown

REMOVAL & INSTALLATION

➡**For this procedure, you will need a Belt tension release tool (Snap-on YA9317) or equivalent.**

1. Move the auto-tensioner with the belt tension release tool to relieve tension from the drive belt, then remove the drive belt.
2. Install the new belt in the reverse order of removal.

CAMSHAFT AND VALVE LIFTERS

REMOVAL & INSTALLATION

Front

See Figure 36.

1. Disconnect the negative and positive battery cable.

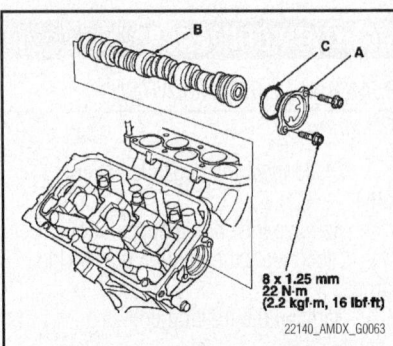

Fig. 36 Remove the thrust cover (A) to remove the camshaft (B). Replace the O-ring (C) when reinstalling—MDX

2. Remove the battery and battery box.

3. Drain the engine cooling system.

4. Disconnect the upper radiator hose.

5. Remove the exhaust gas recirculation valve.

6. Remove the timing belt. For additional information, refer to the following section, "Timing Belt, Removal & Installation."

7. Remove the intake manifold. For additional information, refer to the following section, "Intake Manifold, Removal & Installation."

8. Remove the cylinder head cover. For additional information, refer to the following section, "Cylinder Head, Removal & Installation."

9. Loosen the locknuts and adjusting screws.

10. Remove the bolts and the rocker arm assembly.

✳✳ CAUTION

Loosen the rocker shaft mounting bolts two turns at a time, to prevent damaging the valves or rocker arm assembly.

➡**When removing the rocker arm assembly, do not remove the rocker shaft mounting bolts. The bolts will keep the springs and the rocker arms on the shafts.**

11. Remove the front camshaft pulley.

12. Remove the thrust cover, the remove the camshaft.

To install:

13. Apply clean engine oil to the journals and cam lobes of the camshaft. Install the camshaft using a new O-ring. Tighten the thrust plate to 16 ft. lbs. (22 Nm).

14. Apply clean engine oil to the threads of the camshaft pulley mounting bolt, then install the front camshaft pulley.

15. Install the rocker arm assembly. Tighten the mounting bolts two turns at a time in the sequence shown to 17 ft. lbs. (24 Nm).

16. Install the timing belt.

17. Adjust the valve clearance. For additional information, refer to the following section, "Valve Lash, Adjustment."

18. The remainder of the installation is the reverse order of removal.

19. Refill the cooling system to the correct level.

Rear

1. Disconnect the negative battery cable.

2. Drain the cooling system.

3. Remove the intake manifold. For additional information, refer to the following

section, "Intake Manifold, Removal & Installation."

4. Remove the purge joint.

5. Remove the brake lines from the master cylinder.

6. Remove the timing belt. For additional information, refer to the following section, "Timing Belt, Removal & Installation."

7. Remove the cylinder head cover. For additional information, refer to the following section, "Cylinder Head, Removal & Installation."

8. Loosen the locknuts and adjusting screws.

9. Remove the bolts and the rocker arm assembly.

✳✳ CAUTION

Loosen the rocker shaft mounting bolts two turns at a time, to prevent damaging the valves or rocker arm assembly.

➡**When removing the rocker arm assembly, do not remove the rocker shaft mounting bolts. The bolts will keep the springs and the rocker arms on the shafts.**

10. Remove the rear camshaft pulley.

11. Remove the thrust cover, then remove the camshaft.

To install:

12. Apply clean engine oil to the journals and cam lobes of the camshaft. Install the camshaft using a new O-ring. Tighten the thrust plate to 16 ft. lbs. (22 Nm).

13. Apply clean engine oil to the threads of the camshaft pulley mounting bolt, then install the rear camshaft pulley.

14. Install the rocker arm assembly. Tighten the mounting bolts two turns at a time in the sequence shown to 17 ft. lbs. (24 Nm).

15. Install the timing belt.

16. Adjust the valve clearance. For additional information, refer to the following section, "Valve Lash, Adjustment."

17. Install the brake lines to the master cylinder and bleed the brake system.

18. The remainder of the installation is the reverse order of removal.

19. Refill the cooling system to the correct level.

CRANKSHAFT DAMPER

REMOVAL & INSTALLATION

See Figures 37 through 40.

➡**This procedure requires the use of the following special tools or their equivalents:**

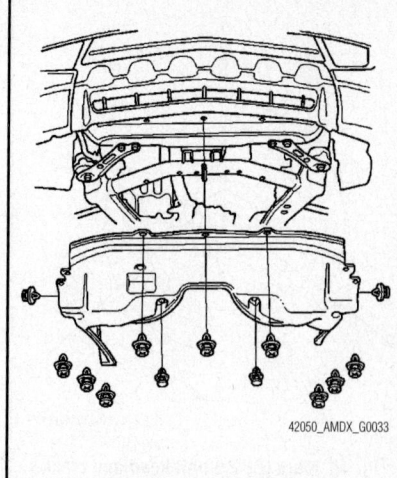

Fig. 37 Remove the splash shield

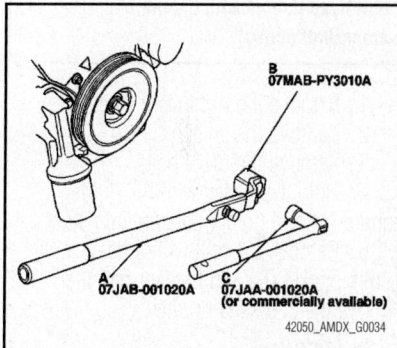

Fig. 38 Hold the pulley with the holder handle (A) and holder attachment (B). Remove the bolt with a heavy duty 19mm socket (C) and breaker bar, then remove the crankshaft pulley

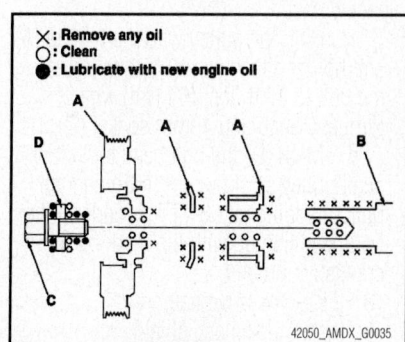

Fig. 39 Remove any oil or clean the pulleys (A), crankshaft (B), bolt (C), and washer (D). Lubricate new engine oil as shown—

- Holder handle 07JAB-001020A
- Holder attachment, 50mm, offset 07MAB-PY3010A
- Socket, 19mm 07JAA-001020A or a commercially available 19mm socket

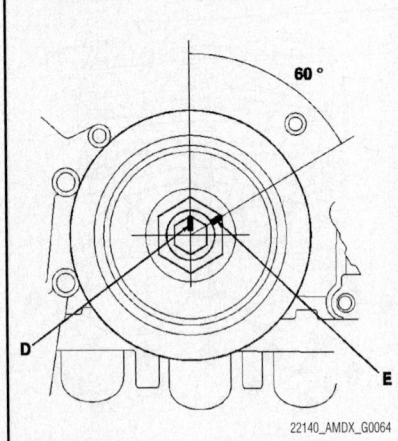

Fig. 40 Mark (D) the bolt head and crankshaft pulley as shown, then tighten the bolt an additional 60°. (The mark on the bolt head line up with the mark on the crankshaft pulley).

1. Remove the right front wheel.
2. Remove the splash shield.
3. Remove the drive belt.
4. Hold the pulley with the holder handle (A) and holder attachment (B).
5. Remove the bolt with a heavy duty 19mm socket (C) and breaker bar, then remove the crankshaft pulley.

To install:

6. Remove any oil or clean the pulleys (A), crankshaft (B), bolt (C), and washer (D). Lubricate new engine oil as shown in the accompanying illustration.
7. Install the crankshaft pulley, and tighten the bolt as follows. Do not use an impact wrench.
   a. Hold the pulley with the handle (A) and holder attachment (B), then tighten the bolt to 47 ft. lbs. (64 Nm) with a torque wrench and 19mm socket (C).
   b. Mark (D) the bolt head and crankshaft pulley as shown, then tighten the bolt an additional 60°. (The mark on the bolt head line up with the mark on the crankshaft pulley).
8. Install the drive belt.
9. Install the splash shield.
10. Install the right front wheel.

CRANKSHAFT FRONT SEAL

REMOVAL & INSTALLATION

See Figure 41.

1. Remove the crankshaft position (CKP) sensor, the timing belt, and the timing belt drive pulley. For additional information, refer to the following section, "Timing Belt, Removal & Installation."

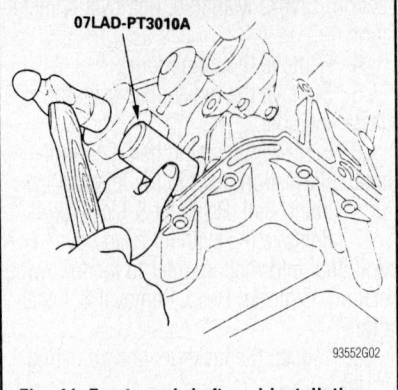

Fig. 41 Front crankshaft seal installation

2. Remove the crankshaft front seal.

To install:

3. Clean and dry the crankshaft oil seal housing.
4. Apply a light coat of multipurpose grease to the crankshaft and to the lip of the seal.
5. Using the seal driver, drive in the crankshaft oil seal until the driver bottoms against the oil pump.
6. When the seal is in place, clean any excess grease off the crankshaft, and check that the oil seal lip is not distorted.
7. Install the timing belt drive pulley, the timing belt, and the CKP sensor.

CYLINDER HEAD

REMOVAL & INSTALLATION

See Figures 42 through 45.

1. Properly relieve the fuel system pressure.
2. Make sure you have the anti-theft code for the radio and the navigation system, then write down the frequencies for the radio's preset buttons.
3. Disconnect the negative cable from the battery first, then disconnect the positive cable. Wait at least 3 minutes before proceeding.

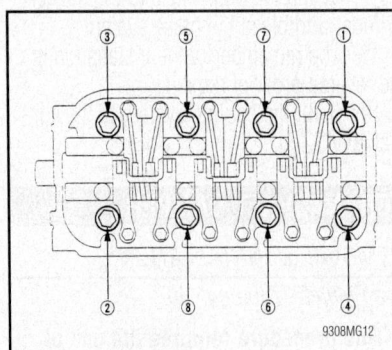

Fig. 42 Cylinder head bolt loosening sequence—MDX

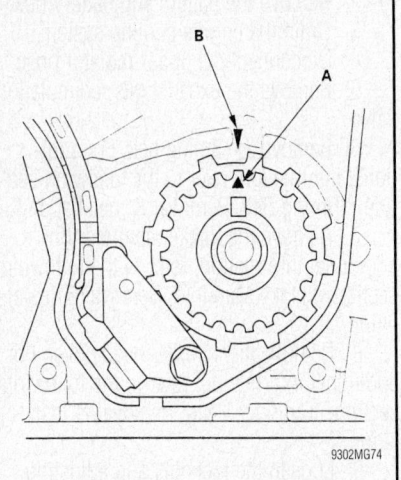

Fig. 43 Crankshaft timing belt sprocket TDC marks. Align sprocket mark (A) with pointer (B)—MDX

4. Drain the engine cooling system.
5. Remove the accessory drive belt.
6. Remove the power steering pump and power steering hose clamp.
7. Remove the alternator.

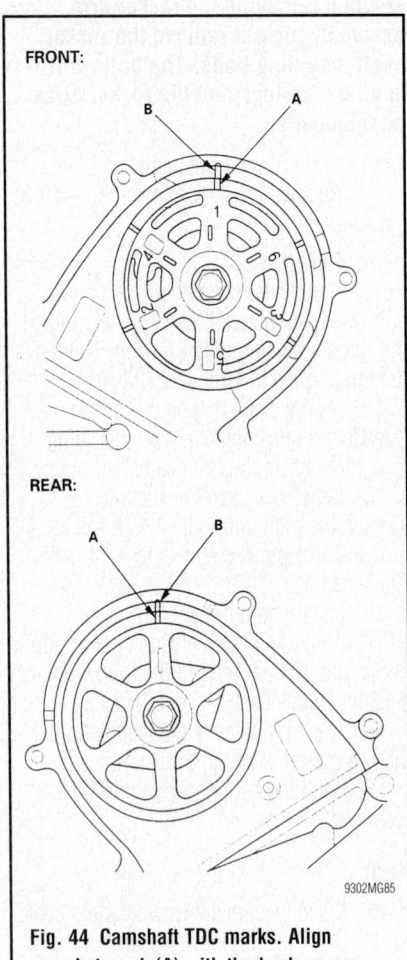

Fig. 44 Camshaft TDC marks. Align sprocket mark (A) with the back cover pointer (B)—MDX

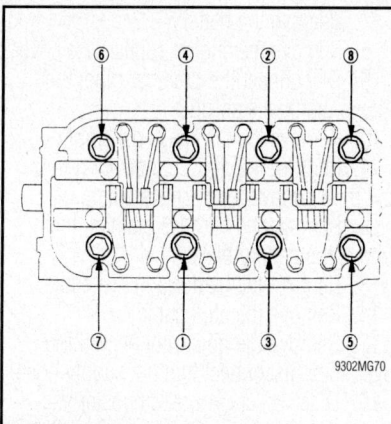

Fig. 45 Cylinder head bolt tightening sequence—MDX

8. Remove the timing belt.

9. Remove the intake manifold.

10. Remove the ignition coils.

11. Disconnect the upper and lower radiator hoses from the engine assembly.

12. Remove the quick-connect cover, then disconnect the fuel supply hose.

13. Remove the purge joint.

14. Remove the following engine wiring harnesses:

- Six injector wiring harnesses
- Engine coolant temperature (ECT) sensor 1 wiring harness
- Crankshaft position (CKP) sensor wiring harness
- Exhaust gas recirculation (EGR) valve wiring harness
- Rocker arm oil control solenoid wiring harness
- Rocker arm oil pressure switch wiring harness
- Oil pressure switch wiring harness
- Two air fuel ratio (A/F) sensor wiring harnesses
- Two secondary heated oxygen sensor (secondary HO2S) wiring harnesses

15. Remove the front warm up three way catalytic converter (front WU-TWC) and rear warm up three way catalytic converter (rear WU-TWC).

16. Remove the connector bracket from the front cylinder head.

17. Remove the bracket from the rear cylinder head.

18. Remove the fuel rails.

19. Remove the water passage.

20. Remove the front and rear camshaft pulleys and back covers.

21. Remove the cylinder head cover as follows:

a. Remove the dipstick.

h. Remove the two bolts securing the harness holder, and disconnect the front

air fuel ratio (A/F) sensor connector, front secondary heated oxygen sensor (secondary HO2S) connector, exhaust gas recirculation (EGR) valve connector and engine coolant temperature (ECT) sensor 1 connector.

c. Disconnect the three injector connectors from the injectors on the rear side cylinder head.

d. Remove the power steering hose bracket mounting bolt, the harness holder mounting bolts, and the engine ground cable bolt.

e. Remove the engine ground cable and disconnect the breather hose.

f. Remove the cylinder head covers.

22. Loosen the cylinder head bolts in sequence and ⅓ turns until all bolts are loose.

23. Remove the cylinder head.

To install:

24. Clean the cylinder head and engine block surface.

25. Clean and install the oil control orifices with new O-rings.

26. Install the dowel pins and new cylinder head gaskets.

27. Clean the timing belt pulleys, timing belt guide plate, and the upper and lower covers.

28. Align the crankshaft and camshaft sprocket TDC marks as shown.

29. Put the cylinder head onto the engine block.

30. Apply clean engine oil to the cylinder head bolt threads and flanges.

31. Tighten the cylinder head bolts in sequence as follows:

a. Step 1: 22 ft. lbs. (29 Nm)

b. Step 2: Plus 90 degrees

c. Step 3: Plus an additional 90 degrees

32. Install the timing belt.

33. Adjust the valve clearance.

34. Install the cylinder head covers and tighten to 108 inch lbs. (12 Nm).

35. Install the water passage.

36. Install the front warm up three way catalytic converter (front WU-TWC) and rear warm up three way catalytic converter (rear WU-TWC).

37. Install the fuel rails.

38. Install the connector bracket to the front cylinder head.

39. Install the bracket to the rear cylinder head.

40. Install the purge joint and tighten the mounting nuts to 109 inch lbs. (12 Nm).

41. Connect the fuel supply hose and replace the quick-connect fitting cover.

42. Reconnect the engine wiring harnesses that were previously removed.

43. Connect the upper and lower radiator hoses.

44. Install the intake manifold.

45. Install the alternator.

46. Install the power steering pump and hose bracket.

47. Install the accessory drive belt.

48. Reconnect the battery cables.

49. After installation, check that all tubes, hoses and connectors are installed correctly.

50. Refill the engine cooling system to the correct level.

51. Start the engine and check for leaks.

ENGINE ASSEMBLY

REMOVAL & INSTALLATION

See Figures 46 through 48.

➡**The engine and transaxle are removed as an assembly.**

1. Remove the support struts from the engine hood. Move the engine hood to a vertical position, then reinstall the support strut in the lower bolt hole.

2. Properly relieve the fuel system pressure.

3. Drain the power steering system fluid, then plug the fluid reservoir and the return hose.

4. Remove the bulkhead cover.

5. Make sure you have the anti-theft code for the radio and the navigation system, then write down the frequencies for the radio's preset buttons.

6. Disconnect the negative cable from the battery first, then disconnect the positive cable. Wait at least 3 minutes before proceeding.

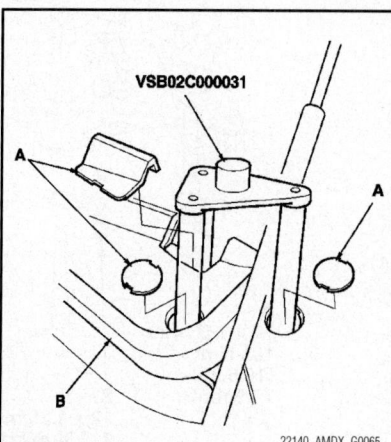

Fig. 46 Remove the service caps (A) for the front damper flange nuts from the cowl cover (B)—MDX

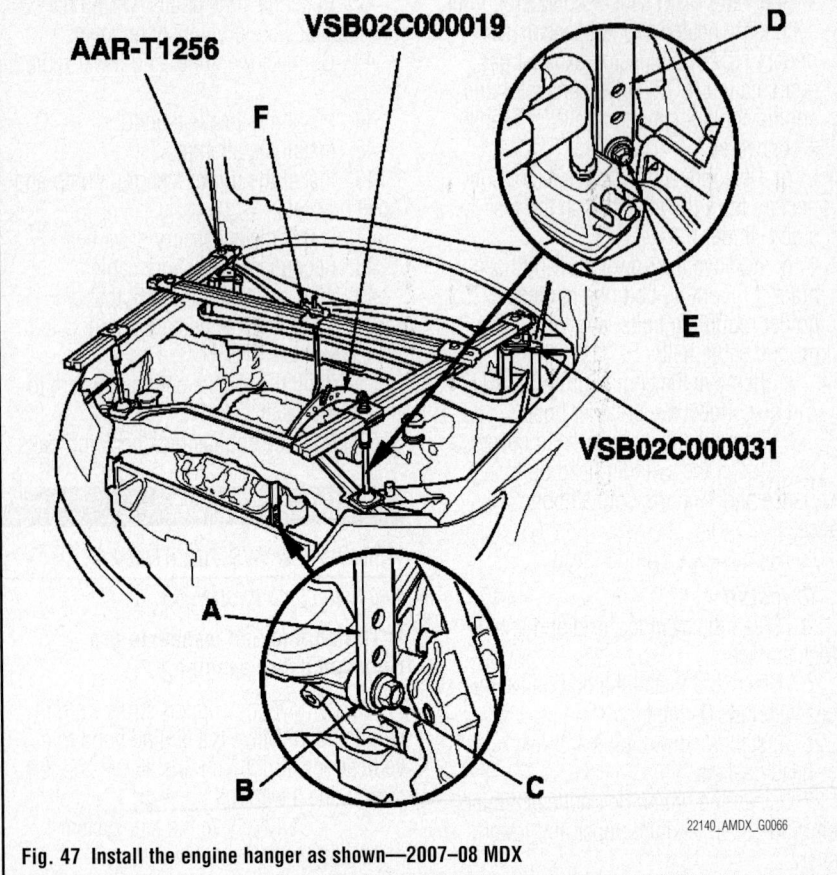

AAR-T1256   VSB02C000019   D

F

E

VSB02C000031

A

B   C

22140_AMDX_G0066

Fig. 47 Install the engine hanger as shown—2007–08 MDX

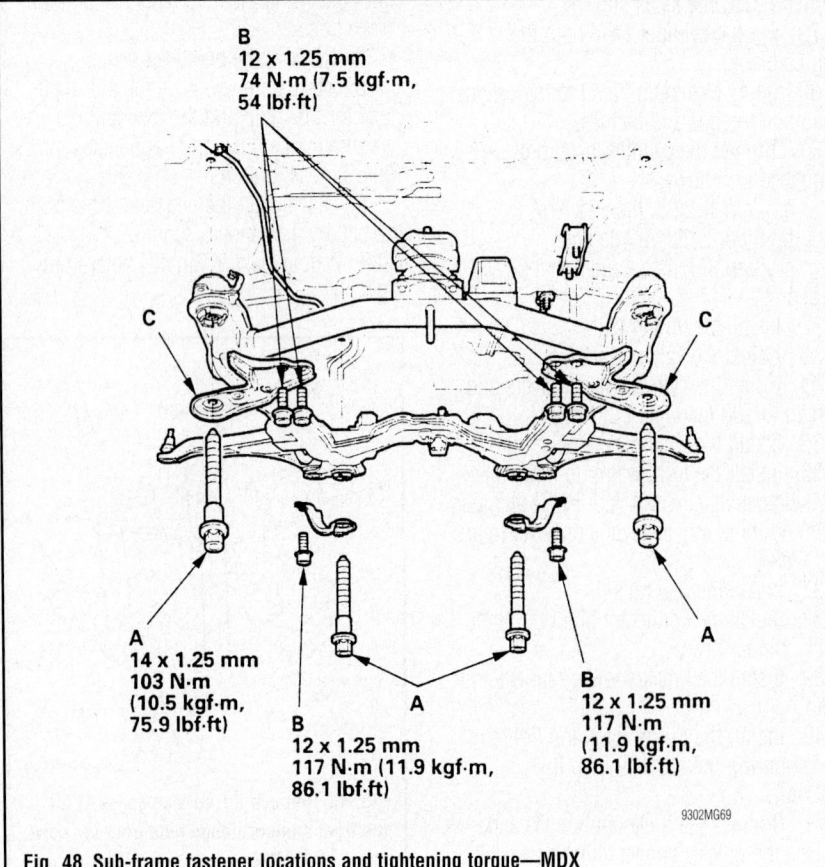

B
12 x 1.25 mm
74 N·m (7.5 kgf·m,
54 lbf·ft)

C   C

A
14 x 1.25 mm
103 N·m
(10.5 kgf·m,
75.9 lbf·ft)

A   A

B   B
12 x 1.25 mm   12 x 1.25 mm
117 N·m (11.9 kgf·m,   117 N·m
86.1 lbf·ft)   (11.9 kgf·m,
                                              86.1 lbf·ft)

9302MG69

Fig. 48 Sub-frame fastener locations and tightening torque—MDX

7. Remove the battery.

8. Remove the engine appearance cover.

9. Disconnect the breather pipe, then remove the intake air duct.

10. Remove the air intake assembly.

11. Remove the main under-hood fuse box, then remove the harness clamps.

12. Remove the harness clamps, then remove the battery base.

13. Remove the starter positive cable.

14. Remove the shift cable.

15. Remove the quick-connect fitting cover, then disconnect the fuel supply hose.

16. Disconnect the brake booster vacuum hose, and evaporative emission (EVAP) canister hose.

17. Disconnect the EVAP canister purge valve connector.

18. Remove the steering wheel.

19. Remove the steering joint cover.

20. Make a reference mark across the steering joint and steering gearbox pinion shaft. Remove the steering joint bolt, and loosen the steering joint bolt, then disconnect the steering joint from the steering gearbox pinion shaft.

21. Remove the alternator cable from the main under-hood fuse box.

22. Remove the coolant reservoir from the bracket.

23. Remove the support bracket, then remove the Powertrain Control Module (PCM) cover.

24. Remove the harness clamps, then disconnect the two PCM connectors and engine wire harness connector.

25. Remove the accessory drive belt.

26. Remove the power steering (P/S) pump inlet hose and P/S pump outlet hose from the P/S pump, then plug the P/S lines and P/S pump.

27. Remove the P/S hose clamp from the cylinder head cover.

28. Remove the radiator cap.

29. Raise and safely support the vehicle.

30. Remove the front wheels.

31. Remove the splash shield.

32. Loosen the drain plug in the radiator, and drain the engine coolant.

33. Drain the automatic transmission fluid (ATF).

34. Drain the engine oil.

35. Remove the front subframe stiffener.

36. Remove exhaust pipe A.

37. Separate the stabilizer links from the dampers.

38. Separate the tie-rod end ball joints from the knuckles.

39. Separate the knuckles from the lower arms.

40. Remove the halfshafts. Coat all precision-finished surfaces with new

engine oil. Tie plastic bags over the half-shaft ends.

41. Remove the driveshaft from the transfer shaft flange.

42. Remove the transfer case assembly.

43. Disconnect the P/S hose, then plug the line and hose.

44. Disconnect the power steering pressure switch connector.

45. Remove the transmission lower mount bolts and ground cable.

46. Remove the rear mount bolts, and power steering line bracket.

47. Disconnect the A/C compressor clutch connector, then remove the A/C compressor without disconnecting the A/C hoses.

48. Disconnect the heater hoses.

49. Disconnect the upper and lower radiator hoses from the engine.

50. Disconnect the ATF cooler hoses from the transmission, then plug the ATF cooler hoses and lines.

51. Remove the radiator.

52. Remove the connector bracket from the front cylinder head; use the bracket bolt hole to attach the engine balance bar front arm.

53. Remove the bracket from the rear cylinder head; use the bracket bolt hole to attach the engine balance bar rear arm.

54. Remove the service caps for the front damper flange nuts from the cowl cover. Position the engine hanger adapters (VSB02C000031) with the "FRONT" mark facing forward over the damper flange nuts.

55. Install the engine hanger balance bar (VSB02C000019); attach the front arm (A) to the front cylinder head with a spacer (B) and 10 x 1.25 mm bolt (C), attach the rear arm (D) to the rear cylinder head with an 8 x 1.25 mm bolt (E).

56. Install the engine support hanger (AAR-T1256) to the vehicle, and attach the hook to the slotted hole in the engine hanger balance bar. Tighten the wing nut (F) by hand to lift and support the engine/transmission.

57. Disconnect the front engine mount control solenoid valve tube, then remove the front engine mount bolts.

58. Disconnect the front suspension stroke sensor connectors and remove the harness clamps, if equipped.

59. Loosen the four bolts holding the adjustable arms of the front subframe adapter (EQS02BMDXSBO) to its center plate.

60. Line up the slots in the arms with the bolt holes on the corner of the jack base, then attach the front subframe adapter to the jack base with the bolts that came with the jack. Tighten all bolts securely.

61. Raise the jack to vehicle height, then attach the front subframe adapter to the front subframe using the subframe stiffener mounting bolts and bolt holes.

62. Remove the six 12 x 1.25 mm bolts (A) securing the subframe stiffeners, the four subframe mounting bolts, and the stiffeners, then lower the front subframe

63. Lower the vehicle and attach a chain hoist to the engine hook and transmission hook. Lift up on the engine/transmission until it's securely supported by the chain hoist, then remove the engine support hanger and the engine balance bar from the engine and vehicle.

64. Remove the mounting bolts from the upper half of the side engine mount bracket.

65. Check that the engine/transmission is completely free of vacuum hoses, fuel and coolant hoses, and electrical wiring.

66. Slowly lower the engine/transmission about 150 mm (6 in.). Check once again that all hoses and electrical wiring are disconnected and free from the engine/transmission, then lower it all the way.

67. Disconnect the chain hoist from the engine/transmission.

68. Raise the vehicle all the way on the lift, then remove the engine/transmission from under the vehicle.

To install:

69. Position the engine/transmission under the vehicle. Be sure that they are properly aligned. Carefully lower the vehicle until the engine and transmission are properly positioned in the engine compartment. Make sure the vehicle is not resting on any part of the engine or transmission. Lift and support the engine with a chain hoist, and carefully raise the engine/transmission into place.

70. Install new mounting bolts into the upper half of the side engine mount bracket. Tighten the bolts to 33 ft. lbs. (44 Nm).

71. Remove the chain hoist, then raise the vehicle on the lift.

72. Install the front sub-frame and tighten the bolts as shown.

73. The remainder of the installation is the reverse order of removal. Note the following torque values:

- A/C compressor bolts to 16 ft. lbs. (22 Nm)
- Transaxle lower front mount nuts to 28 ft. lbs. (38 Nm)
- Transaxle lower rear mount bolts to 28 ft. lbs. (38 Nm)
- Front mount bracket support nut to 40 ft. lbs. (54 Nm)
- Side engine mount bracket bolts to

33 ft. lbs. (44 Nm) and the through bolt to 40 ft. lbs. (54 Nm)

74. Refill the engine with oil to the correct level.

75. Refill the transaxle with ATF to the correct level.

76. Refill the cooling system to the correct level.

77. Refill the power steering system to the correct level.

78. Start the engine and check for leaks.

79. Check the wheel alignment and adjust if necessary.

EXHAUST MANIFOLD

REMOVAL & INSTALLATION

1. Remove the radiator and fan assembly.

2. Disconnect the front air fuel ratio (A/F) sensor connector and front secondary heated oxygen sensor (secondary HO2S) connector.

3. Carefully remove the exhaust manifold

To install:

4. Carefully install the exhaust manifold, and tighten the nuts in a crisscross pattern in two or three steps to 108 inch lbs. (12 Nm).

5. The remainder of the installation is the reverse order of removal.

INTAKE MANIFOLD

REMOVAL & INSTALLATION

See Figures 49 through 52.

1. Disconnect the negative battery cable.

2. Remove the engine appearance cover.

3. Remove the air intake assembly.

4. Disconnect the positive crankcase ventilation (PCV) hose, the brake booster vacuum hose, vacuum hose, and

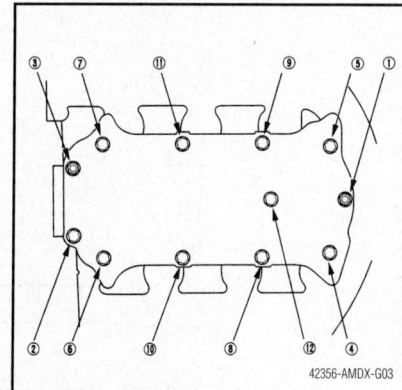

42356-AMDX-G03

Fig. 49 Upper cover loosening sequence—MDX

the intake manifold tuning (IMT) actuator connector.

5. Disconnect the evaporative emission (EVAP) canister hose, EVAP canister purge valve connector, throttle actuator connector, and manifold absolute pressure (MAP) sensor connector.

6. Disconnect and plug the water bypass hoses.

7. Remove the upper cover mounting bolts and nuts sequentially in three steps, then remove the upper cover.

8. Remove the intake manifold mounting bolts and nuts sequentially in three steps, then remove the intake manifold.

To install:

9. Install the new intake manifold gasket.

10. Install the intake manifold and tighten the bolts in sequence to 16 ft. lbs. (22 Nm).

11. Install the upper cover and new gasket. Tighten the bolts and nuts in sequence to 9 ft. lbs. (12 Nm).

12. The remainder of the installation is the reverse order of removal.

13. Start the engine and check for proper operation.

OIL PAN

REMOVAL & INSTALLATION

See Figures 53 and 54.

1. Drain the engine oil.
2. Disconnect the negative battery cable.
3. Remove the front splash shield.
4. Remove the exhaust front pipe.
5. Remove the exhaust manifold.
6. Remove the torque converter cover and the four bolts securing the transaxle.
7. Remove the oil pan mounting bolts.
8. Use a flat bladed screwdriver to separate the oil pan from the block.

To install:

9. Remove all of the old liquid gasket material.

10. Apply liquid gasket to the mating surfaces of the oil pan as shown.

11. Install the oil pan on the engine block. Tighten the mounting bolts in sequence to 9 ft. lbs. (12 Nm).

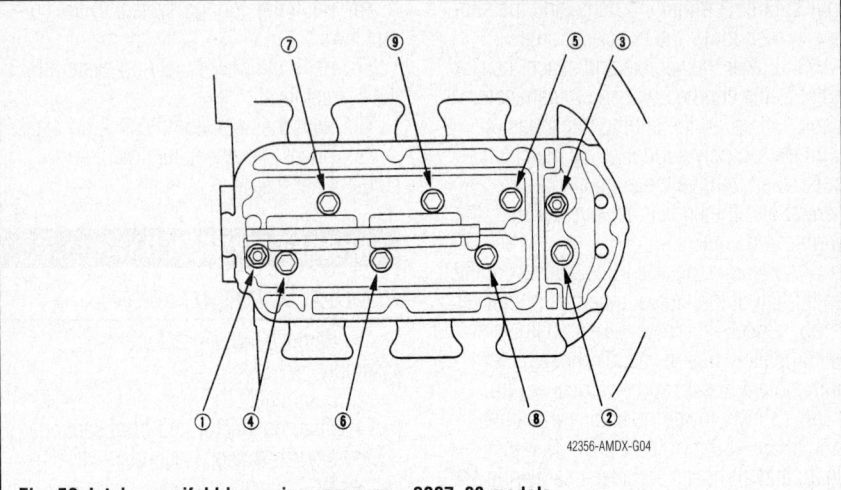

Fig. 50 Intake manifold loosening sequence—2007–08 models

42356-AMDX-G04

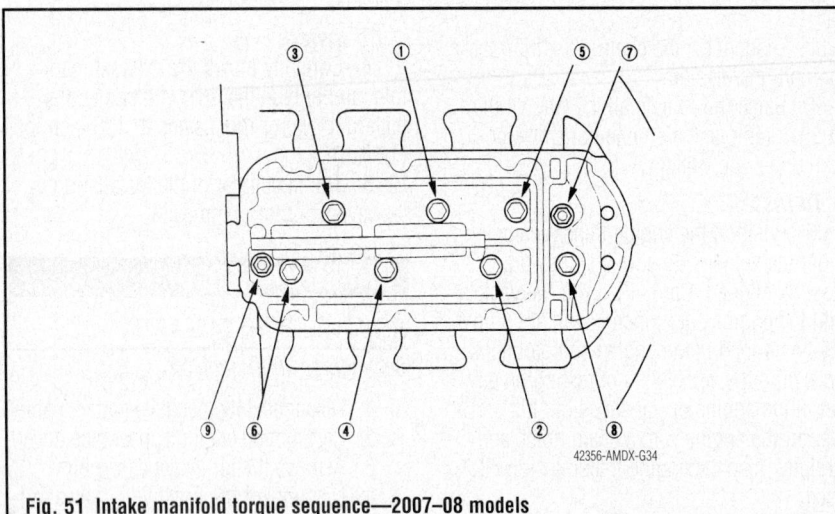

Fig. 51 Intake manifold torque sequence—2007–08 models

42356-AMDX-G34

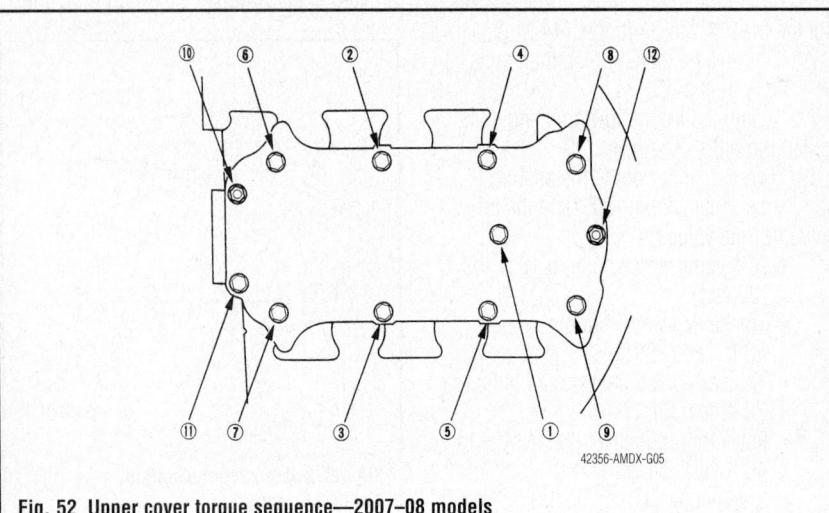

Fig. 52 Upper cover torque sequence—2007–08 models

42356-AMDX-G05

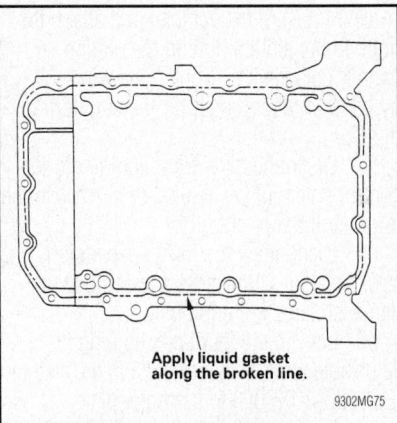

Apply liquid gasket along the broken line.

9302MG75

Fig. 53 Apply liquid gasket to the inner threads of the bolt holes and the engine block along the area indicated by the broken line—MDX

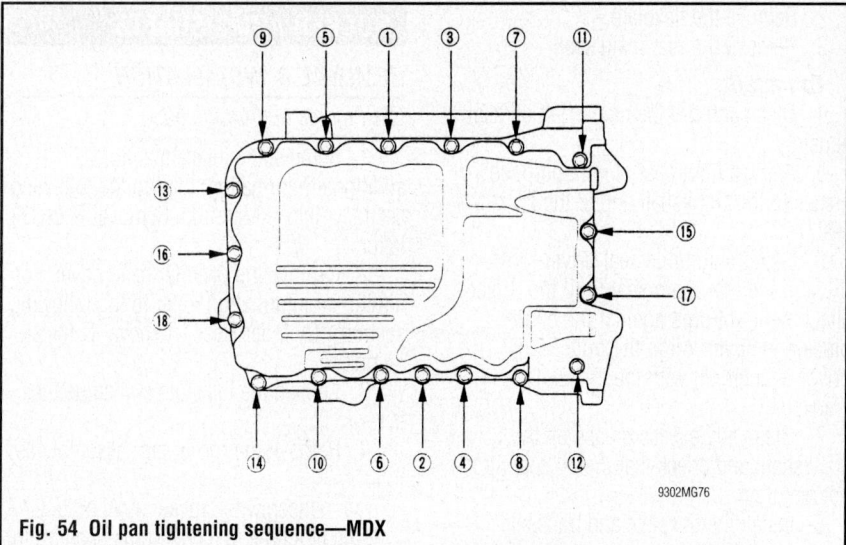

Fig. 54 Oil pan tightening sequence—MDX

12. Tighten the four bolts securing the transmission, then install the torque converter cover.

13. Install the exhaust manifold and exhaust front pipe.

14. Install the splash shield.

15. Connect the negative battery cable.

➡**Wait at least 30 minutes before adding oil to the engine.**

16. Refill the engine with oil to the correct level.

OIL PUMP

REMOVAL & INSTALLATION

See Figures 55 and 56.

1. Remove the timing belt. For additional information, refer to the following section, "Timing Belt, Removal & Installation."

2. Remove the crankshaft position sensor.

3. Remove the rocker arm control solenoid/oil filter assembly.

4. Remove the oil pan.

5. Remove the oil screen before removing the oil pump.

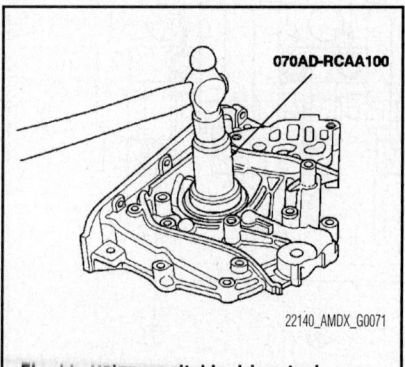

Fig. 55 Using a suitable driver tool, gently tap in a new oil—MDX

To install:

6. Remove the old oil seal from the oil pump.

7. Gently tap in the new oil seal until the oil seal driver bottoms on the pump.

8. Remove all of the old liquid gasket from the oil pump mating surfaces, bolts, and bolt holes.

9. Clean and dry the oil pump mating surfaces.

10. Apply liquid gasket evenly to the engine block mating surface of the oil pump. Install the component within 5 minutes of applying the liquid gasket.

11. Grease the lip of the oil seal, and apply oil to the new O-ring.

12. Install the dowel pins, then align the inner rotor with the crankshaft, and install the oil pump.

13. Clean the excess grease off the crankshaft, and check the seal for distortion.

14. Install the oil screen with new O-ring.

15. The remainder of the installation is the reverse order of removal.

16. Refill the engine with oil to the correct level.

MAIN BEARING TORQUE SEQUENCE

See Figure 57.

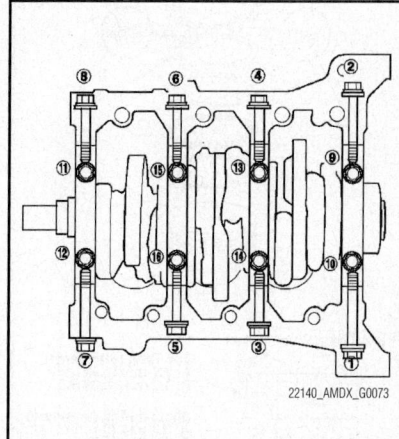

Fig. 57 Main bearing torque sequence— MDX

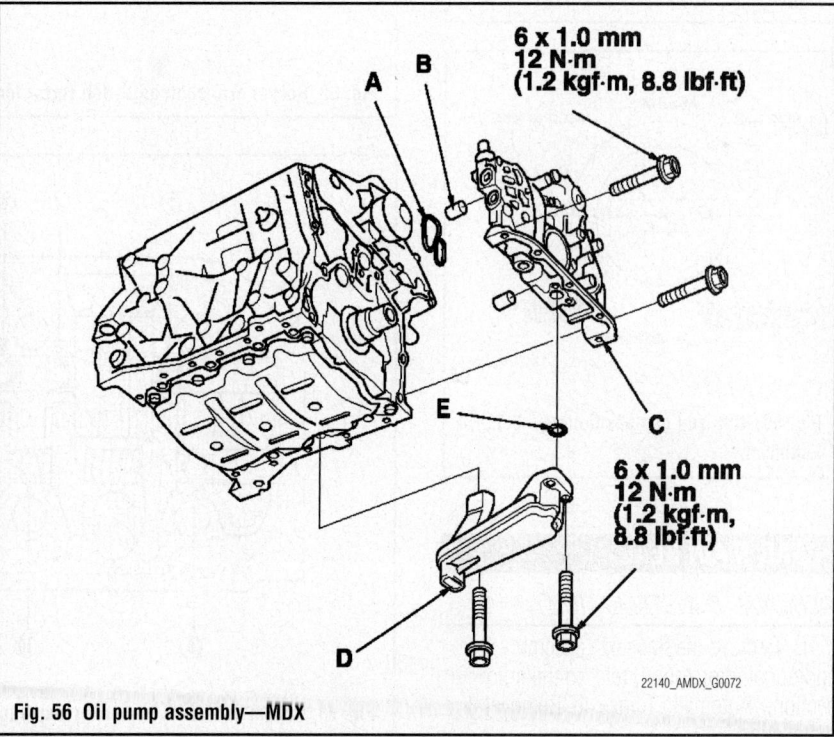

Fig. 56 Oil pump assembly—MDX

PISTON AND RING

POSITIONING

See Figures 58 and 59.

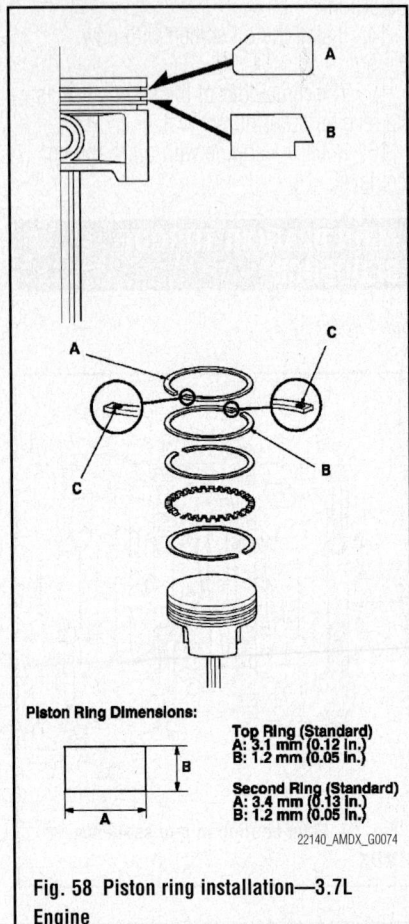

Piston Ring Dimensions:

Top Ring (Standard)
A: 3.1 mm (0.12 in.)
B: 1.2 mm (0.05 in.)

Second Ring (Standard)
A: 3.4 mm (0.13 in.)
B: 1.2 mm (0.05 in.)

22140_AMDX_G0074

Fig. 58 Piston ring installation—3.7L Engine

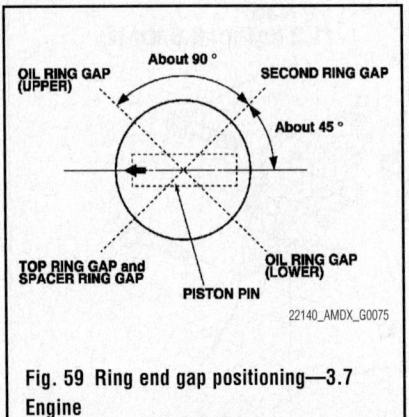

22140_AMDX_G0075

Fig. 59 Ring end gap positioning—3.7 Engine

REAR MAIN SEAL

REMOVAL & INSTALLATION

1. Remove the transaxle assembly. For additional information, refer to the following section, "Automatic Transaxle, Removal & Installation."

2. Remove the flexplate.
3. Remove the rear main seal.

To install:

4. Clean and dry the crankshaft oil seal housing.

5. Apply a light coat of multipurpose grease to the crankshaft and to the lip of the seal.

6. Using a suitable seal driver tool, drive in the rear main seal until the driver attachment bottoms against the engine block end cover. Align the hole in the driver attachment with the pin on the crankshaft.

7. Clean any excess grease off the crankshaft, and check that the oil seal lip is not distorted.

8. Install the flexplate and transaxle assembly.

ROCKER ARMS/SHAFTS

REMOVAL & INSTALLATION

See Figures 60 through 62.

1. Remove the intake manifold. For additional information, refer to the following section, "Intake Manifold, Removal & Installation."

2. Remove the cylinder head cover. For additional information, refer to the following section, "Cylinder Head, Removal & Installation."

3. Loosen the locknuts and adjusting screws.

4. Remove the rocker arm assembly as follows:

   a. Unscrew the rocker shaft bolts 2 turns at a time in a criss-cross pattern to

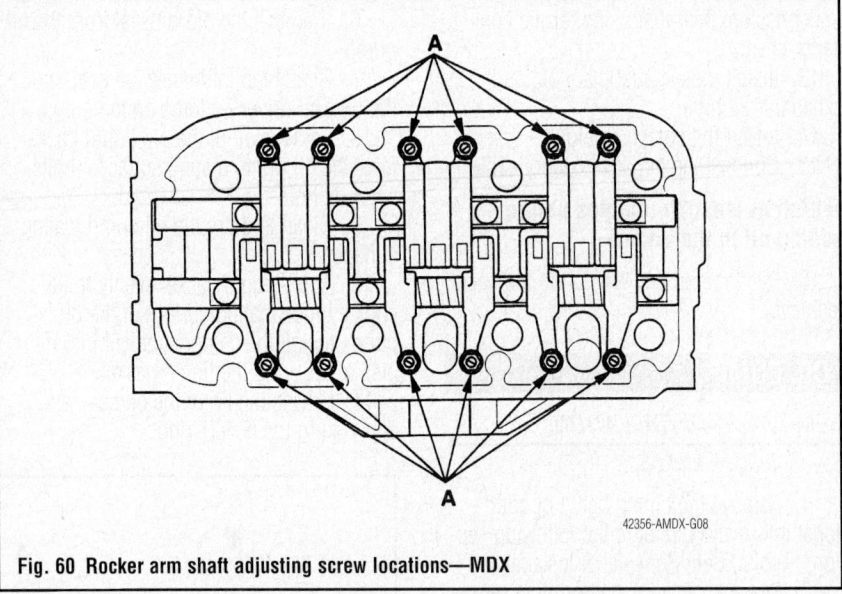

42356-AMDX-G08

Fig. 60 Rocker arm shaft adjusting screw locations—MDX

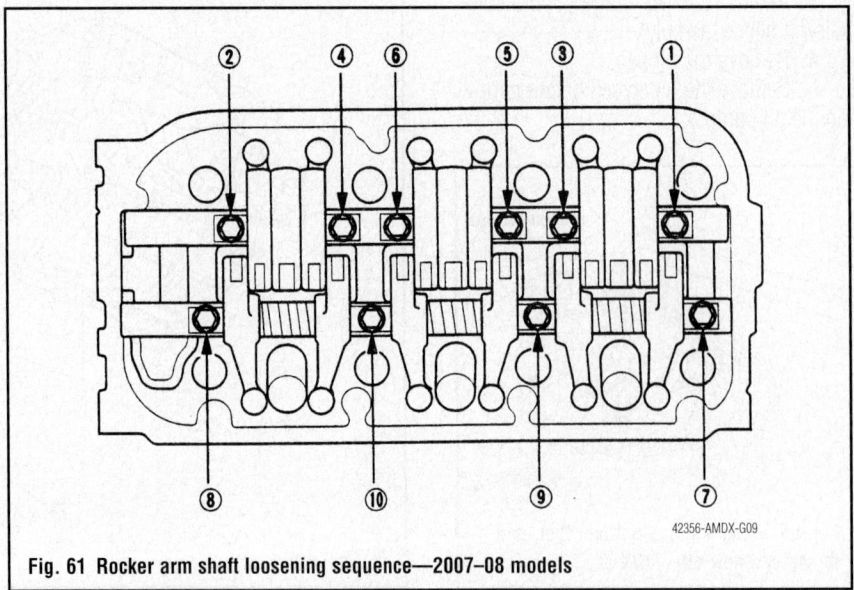

42356-AMDX-G09

Fig. 61 Rocker arm shaft loosening sequence—2007–08 models

avoid damaging the vales or rocker assembly.

b. Do not remove the rocker shaft bolts. These bolts keep the springs and rocker arms on the shafts.

5. Remove the rocker arms and shafts from the vehicle as an assembly.

➡**Keep all valve train components in order for assembly.**

6. Remove the rocker arms and springs from the rocker arm shafts.

To install:

7. Assemble the rocker arms and springs to the rocker arm shafts in their original positions.

8. Install the rocker arm assemblies. Tighten the bolts in sequence and in multiple passes to 17 ft. lbs. (24 Nm).

9. Adjust the valve clearance.

10. Install the cylinder head cover.

11. Install the intake manifold.

TIMING BELT FRONT COVER

REMOVAL & INSTALLATION

See Figures 63 and 64.

1. Disconnect the negative battery cable.

2. Remove the splash shield.

3. Remove the accessory drive belt.

4. Remove the accessory drive belt auto-tensioner.

5. Support the engine assembly with a suitable jack.

6. Remove the ground cable, then remove the upper half of the side engine mount bracket.

7. Remove the front upper and rear upper cover.

8. Remove the crankshaft pulley.

9. Remove the lower cover.

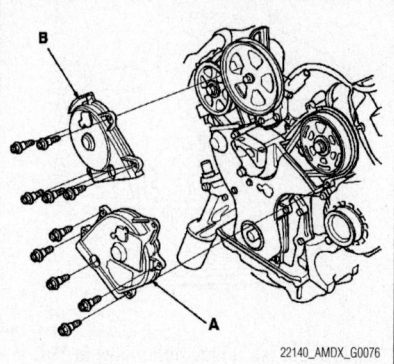

Fig. 63 Removing the front upper (A) and rear upper (B) cover—MDX

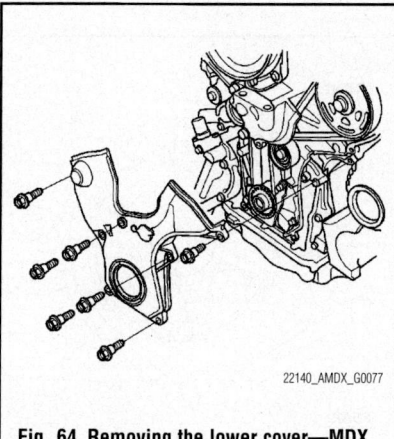

Fig. 64 Removing the lower cover—MDX

To install:

10. Install the lower cover and tighten the mounting bolts to 9 ft. lbs. (12 Nm).

11. Install the front upper and rear upper cover and tighten the mounting bolts to 9 ft. lbs. (12 Nm).

12. Install the crankshaft pulley.

13. Install the accessory drive belt auto-tensioner.

14. Install the accessory drive belt.

15. Install the splash shield.

16. Connect the negative battery cable.

17. Start the engine and check for leaks.

TIMING BELT AND SPROCKETS

REMOVAL & INSTALLATION

See Figures 65 through 73.

1. Turn the crankshaft so the white mark aligns with the pointer.

2. Check that the No. 1 piston top dead center (TDC) mark on the front camshaft pulley and the pointer on the front upper cover are aligned.

➡**If the marks are not aligned, rotate the crankshaft 360 degrees, and recheck the camshaft pulley mark.**

3. Raise and safely support the vehicle, then remove the right front wheel.

4. Remove the timing belt front covers.

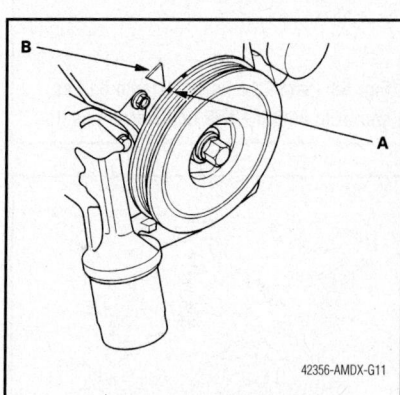

Fig. 65 Turn the crankshaft so the white mark (A) aligns with the pointer (B)

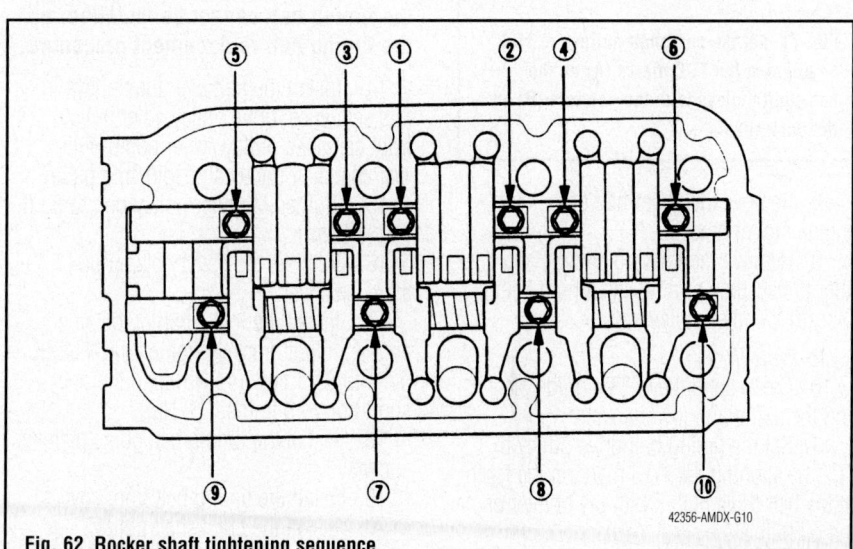

Fig. 62 Rocker shaft tightening sequence

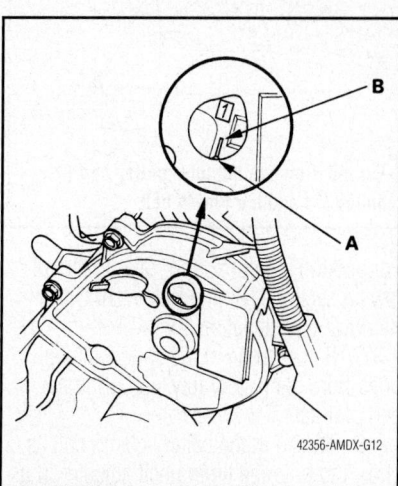

Fig. 66 Make sure the number 1 piston is at top dead center (A) on the front camshaft pulley and pointer (B)

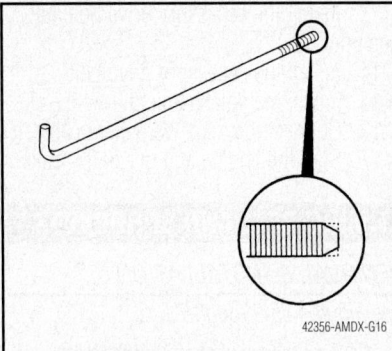

Fig. 67 Remove a battery clamp bolt and grind the end as shown

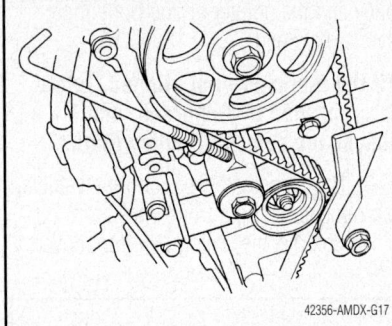

Fig. 68 Install the battery clamp bolt as shown to hold the belt adjuster in position

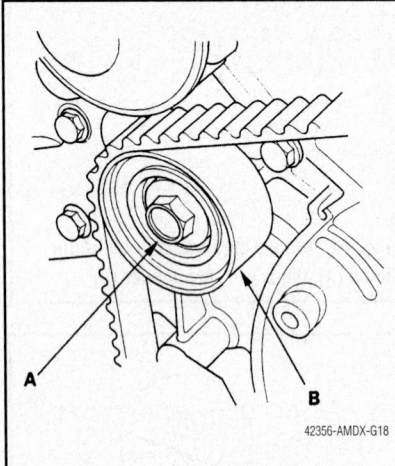

Fig. 69 Remove the idler pulley bolt (A), pulley (B) and the timing belt

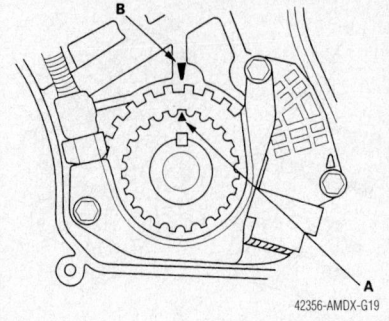

Fig. 70 Set the timing belt pulley to TDC by aligning the TDC mark (A) on the tooth of the belt pulley with the pointer (B) on the oil pump

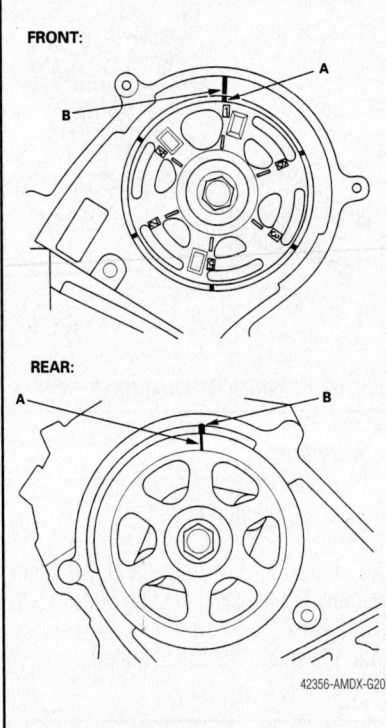

Fig. 71 Set the camshaft pulleys to TDC by aligning the TDC marks (A) on the camshaft pulleys with the pointers (B) on the back covers

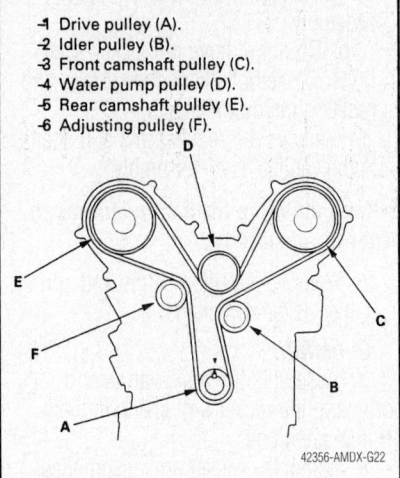

-1 Drive pulley (A).
-2 Idler pulley (B).
-3 Front camshaft pulley (C).
-4 Water pump pulley (D).
-5 Rear camshaft pulley (E).
-6 Adjusting pulley (F).

Fig. 72 Route the belt as shown in the sequence listed

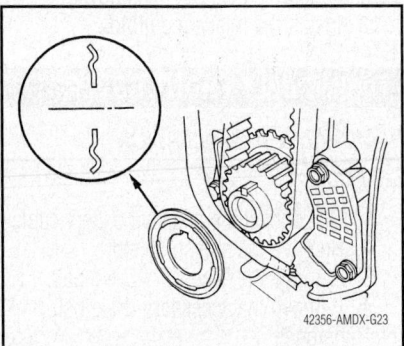

Fig. 73 Install the timing belt guide plate as shown

For additional information, refer to the following section, "Timing Belt Front Cover, Removal & Installation."

5. Remove one of the battery clamp bolts from the battery tray, and grind the end of it flat.

6. Thread in the battery clamp bolt as shown to hold the timing belt adjuster in its current position. Tighten it by hand; do not use a wrench.

7. Remove the timing belt guide plate.

8. Remove the lower half of the side engine mount bracket.

9. Remove the idler pulley bolt and idler pulley, then remove the timing belt. Discard the idler pulley bolt.

To install:

10. Clean the pulleys, belt guide plate and the upper and lower covers.

11. Set the timing belt drive pulley to TDC by aligning the TDC mark on the tooth of the belt drive pulley with the pointer on the oil pump.

12. Set the camshaft pulleys to TDC by

aligning the TDC marks on the camshaft pulleys with the pointers on the back covers.

13. Loosely install the idler pulley with a new idler pulley bolt so the pulley can move but does not come off.

➡ **If the auto-tensioner has extended and the timing belt cannot be installed, do the timing belt replacement procedure.**

14. Install the belt over the pulleys in this sequence; drive pulley, idler pulley, front camshaft pulley, water pump pulley, rear camshaft pulley and adjusting pulley.

15. Tighten the idler pulley bolt to 33 ft. lbs. (44 Nm).

16. Remove the battery clamp bolt from the back cover.

17. Install the lower half of the side engine mount bracket. Tighten the 3 long bolts to 33 ft. lbs. (44 Nm) and the one short bolt to 9 ft. lbs. (12 Nm).

18. Install the timing belt guide plate as shown.

19. Install the timing belt front covers.

20. Install the crankshaft pulley.

21. Rotate the crankshaft pulley about

six turns clockwise so the timing belt positions itself on the pulleys.

22. Turn the crankshaft pulley so the white mark lines up with the pointer.

23. Check the camshaft pulley marks are aligned. If the marks are aligned, proceed to the next step. If the marks are not aligned, remove the timing belt and reinstall using the steps outlined before this step.

24. Install the upper half of the side engine mount bracket, and tighten the new mounting bolts to 33 ft. lbs. (44 Nm), then tighten the mass damper mounting bolt to 40 ft. lbs. (54 Nm).

25. Install the accessory drive belt and auto-tensioner.

26. Install the splash shield.

27. Install the right front wheel.

28. Start the engine and check for leaks.

TIMING BELT REAR COVER

REMOVAL & INSTALLATION

1. Remove the timing belt. For additional information, refer to the following section, "Timing Belt, Removal & Installation."

2. Remove the camshaft pulley.

3. Remove the rear cover.

To install:

4. Install the rear cover.

5. Install the camshaft pulley.

6. Install the timing belt.

7. Start the engine and check for leaks.

VALVE LASH

ADJUSTMENT

See Figure 74.

1. Remove the cylinder head cover. For additional information, refer to the following section, "Valve Covers, Removal & Installation."

2. Set the No. 1 piston at top dead center (TDC).

3. Measure the valve clearance. If adjustment is necessary, loosen the locknut and turn the adjusting screw as necessary to achieve the correct valve clearance.

4. The correct valve clearance is:
   • Intake valves: 0.008–0.009 inches (0.20–0.24mm)
   • Exhaust valves: 0.011–0.013 inches (0.28–0.32mm)

5. After adjustment, tighten the locknuts to 14 ft. lbs. (20 Nm).

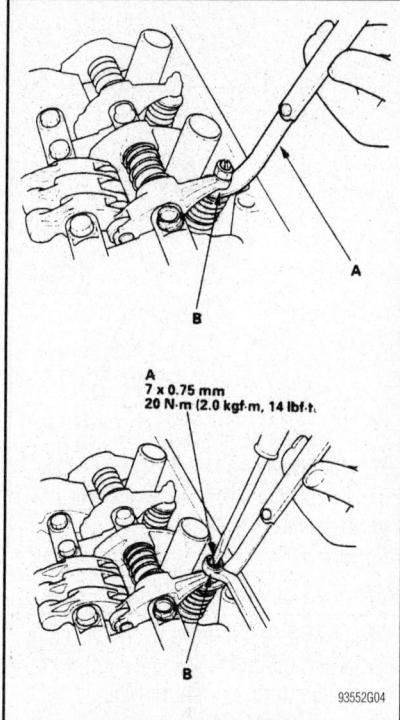

A
7 x 0.75 mm
20 N·m (2.0 kgf·m, 14 lbf·t.

93552G04

Fig. 74 Inspect the valve clearance, adjust to specification and tighten the retainer to specification

ENGINE PERFORMANCE & EMISSION CONTROL

COMPONENT LOCATIONS

See Figure 75.

ACCELERATOR PEDAL POSITION (APP) SENSOR

LOCATION

The accelerator pedal position sensor is located on the accelerator pedal module. The accelerator pedal position sensor is an integrated part of the accelerator pedal module. If the sensor is faulty, the entire module must be replaced.

REMOVAL & INSTALLATION

See Figure 76.

1. Disconnect the accelerator pedal position (APP) sensor 6P connector.

2. Remove the accelerator pedal module.

3. Installation is the reverse order of removal.

BAROMETRIC PRESSURE (BARO) SENSOR

LOCATION

The barometric pressure sensor is located inside the Powertrain Control Module (PCM).

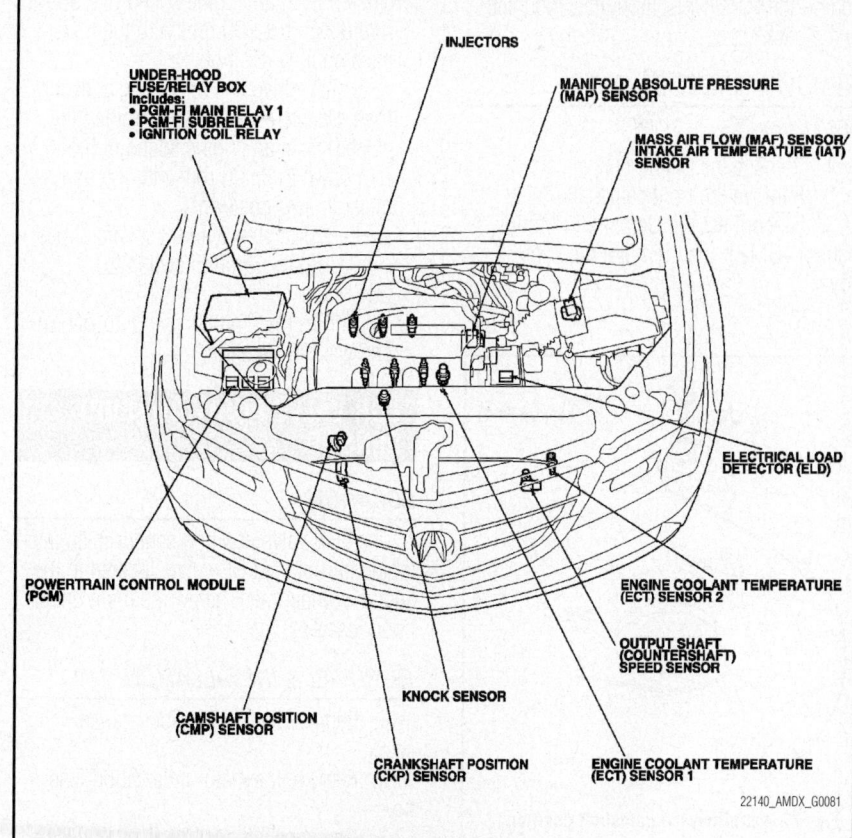

INJECTORS

UNDER-HOOD FUSE/RELAY BOX
Includes:
• PGM-FI MAIN RELAY 1
• PGM-FI SUBRELAY
• IGNITION COIL RELAY

MANIFOLD ABSOLUTE PRESSURE (MAP) SENSOR

MASS AIR FLOW (MAF) SENSOR/ INTAKE AIR TEMPERATURE (IAT) SENSOR

ELECTRICAL LOAD DETECTOR (ELD)

POWERTRAIN CONTROL MODULE (PCM)

ENGINE COOLANT TEMPERATURE (ECT) SENSOR 2

OUTPUT SHAFT (COUNTERSHAFT) SPEED SENSOR

CAMSHAFT POSITION (CMP) SENSOR

KNOCK SENSOR

CRANKSHAFT POSITION (CKP) SENSOR

ENGINE COOLANT TEMPERATURE (ECT) SENSOR 1

22140_AMDX_G0081

Fig. 75 Fuel Injection system component locations—MDX

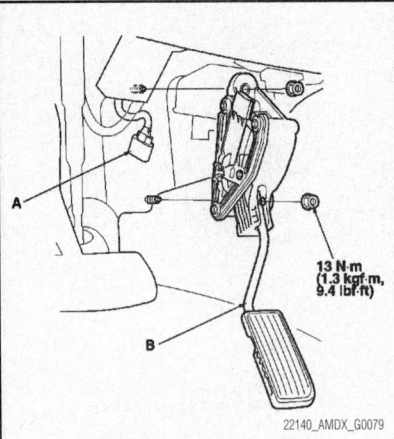

13 N·m
(1.3 kgf·m,
9.4 lbf·ft)

Fig. 76 Disconnect the APP sensor wiring harness (A) to remove the accelerator pedal module (B)—MDX

22140_AMDX_G0079

REMOVAL & INSTALLATION

The BARO sensor is an integrated component of the PCM. Refer to the following section, "Powertrain Control Module, Removal & Installation."

CAMSHAFT POSITION (CMP) SENSOR

LOCATION

The camshaft position sensor is located behind the front camshaft pulley, mounted in the back cover.

REMOVAL & INSTALLATION

See Figure 77.

1. Remove the timing belt.
2. Remove the front camshaft pulley.
3. Disconnect the Camshaft Position Sensor (CMP), then remove the back cover.

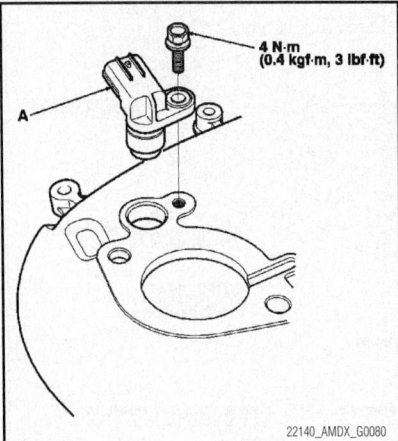

4 N·m
(0.4 kgf·m, 3 lbf·ft)

A

22140_AMDX_G0080

Fig. 77 Installing the camshaft position sensor into the back cover—MDX

4. Remove the CMP from the back cover.
5. Installation is the reverse order of removal.

CRANKSHAFT POSITION (CKP) SENSOR

LOCATION

The crankshaft position sensor is mounted on the oil pump.

REMOVAL & INSTALLATION

1. Move the auto-tensioner to remove tension from the drive belt, then remove the accessory drive belt.
2. Remove the crankshaft pulley.
3. Remove the upper and lower front covers from the engine.
4. Remove the Crankshaft Position CKP sensor from the oil pump.
5. Installation is the reverse order of removal.
6. After installation, perform the CKP learn procedure as follows:
   a. Start the engine. Hold the engine speed at 3,000 rpm without load (in Park or neutral) until the radiator fan comes on.
   b. Test-drive the vehicle on a level road: Decelerate (with the throttle fully closed) from an engine speed of 2,500 rpm down to 1,000 rpm with the transmission in D position.
   c. Test-drive the vehicle on a level road: Decelerate (with the throttle fully closed) from an engine speed of 5,000 rpm down to 3,000 rpm with the transmission in D position.
   d. Repeat step 2 and 3 several times.
   e. Turn the ignition switch to LOCK (0).
   f. Turn the ignition switch to ON (II), and wait 30 seconds.

EGR VALVE POSITION (EVP) SENSOR

LOCATION

The EGR valve position sensor is an integrated part of the EGR valve, located in the engine compartment below the main under-hood fuse box.

REMOVAL & INSTALLATION

1. Remove the engine appearance cover.
2. Remove the main under-hood fuse box.
3. Disconnect the EGR valve connector.
4. Remove the EGR valve.

5. Installation is the reverse order of removal, using a new gasket.

ENGINE COOLANT TEMPERATURE (ECT) SENSOR

LOCATION

Sensor 1 is located in the engine compartment below the main underhood fuse box. Sensor 2 is located at the bottom of the radiator.

REMOVAL & INSTALLATION

Sensor 1

1. Drain the engine cooling system.
2. Remove the engine appearance cover.
3. Remove the main under-hood fuse/relay box.
4. Disconnect the ECT sensor 1 connector.
5. Remove ECT sensor 1.
6. Install the parts in the reverse order of removal with a new O-ring, then refill the radiator with engine coolant.

Sensor 2

1. Drain the engine cooling system.
2. Remove the splash shield.
3. Disconnect the ECT sensor 2 connector.
4. Remove ECT sensor 2.
5. Install the parts in the reverse order of removal with a new O-ring, then refill the radiator with engine coolant.

HEATED OXYGEN (HO2S) SENSOR

LOCATION

There are front and rear bank HO2S sensors, one located in each of catalytic converters.

REMOVAL & INSTALLATION

➡The following procedures require the use of an O2 sensor socket wrench.

Front Bank

1. Disconnect the front secondary HO2S wiring harness, then remove the front secondary HO2S.
2. Install the parts in the reverse order of removal.

Rear Bank

1. Disconnect the rear secondary HO2S wiring harness, then remove the rear secondary HO2S.
2. Install the parts in the reverse order of removal.

INTAKE AIR TEMPERATURE (IAT) SENSOR

LOCATION

The IAT sensor is integrated with the Mass Air Flow (MAF) sensor. Refer the following section for additional information, "Mass Air Flow (MAF) sensor."

KNOCK SENSOR (KS)

LOCATION

The knock sensor is location in the top of the engine block, below the fuel rail.

REMOVAL & INSTALLATION

1. Remove the intake manifold.
2. Remove the injector rails and the injector base.
3. Disconnect the knock sensor wiring harness, then remove the knock sensor.
4. Installation is the reverse order of removal. Tighten the knock sensor to 23 ft. lbs. (31 Nm).

MALFUNCTION INDICATOR LIGHT (MIL)

RESET PROCEDURES

➡ If you are using a generic scan tool to clear commands, be aware that there is only one setting for clearing the PCM, and it clears all commands at the same time (crank (CKP) pattern learn, idle learn, readiness codes, freeze data, on-board snapshot, and DTCs). After you clear all commands, you then need to do these procedures, in this order: PCM idle learn procedure; crank (CKP) pattern learn procedure; Test-drive to set readiness codes to complete.

1. Clear the DTC with the Honda Diagnostic System (HDS) while the engine is stopped.
2. Turn the ignition switch to LOCK (0).
3. Turn the ignition switch to ON (II), and wait 30 seconds.
4. Turn the ignition switch to LOCK (0), and disconnect the HDS from the DLC.

MASS AIR FLOW (MAF) SENSOR

LOCATION

The MAF/IAT sensor is located in the intake tube of the air intake assembly.

REMOVAL & INSTALLATION

1. Disconnect the MAF sensor/IAT senor connector.
2. Remove the screw.
3. Remove the MAF sensor/IAT senor.
4. Install the parts in the reverse order of removal with a new gasket.

MANIFOLD ABSOLUTE PRESSURE (MAP) SENSOR

LOCATION

The MAP sensor is mounted on top of the throttle body assembly.

REMOVAL & INSTALLATION

1. Remove the engine appearance cover.
2. Disconnect the MAP sensor connector.
3. Remove the screw.
4. Remove the MAP sensor .
5. Install the parts in the reverse order of removal with a new O-ring.

POWERTRAIN CONTROL MODULE (PCM)

LOCATION

The PCM is located in the right side engine compartment, in front of the under-hood fuse/relay box.

REMOVAL & INSTALLATION

➡ This procedure requires the use of a Honda diagnostic system (HDS) tablet tester, Honda interface module (HIM) and an iN workstation with HDS and CM update software and HDS pocket tester

If you are replacing the PCM after substituting a known-good PCM, reinstall the original PCM, then do this procedure.

➡ During the procedure, if any READ DATA, WRITE DATA, or other data checks fail, note the failure, then continue.

1. Connect the HDS to the data link connector (DLC) (A) located under the driver's side of the dashboard.
2. Turn the ignition switch to ON (II).
3. Select the PGM-FI system with the HDS.
4. Select the INSPECTION MENU with the HDS.
5. Select the ETCS TEST, then select the TP POSITION CHECK, and follow the screen prompts.

➡ If the TP POSITION CHECK indicates FAILED, continue with this procedure.

6. Select the REPLACE PCM MENU, then select READ DATA and follow the screen prompts.

➡ Doing this step copies (READS) the engine oil life data from the original PCM so you can later download (WRITES) it into the new PCM.

7. If READ DATA indicates FAILED, continue with this procedure.
8. Select the A/T system with the HDS.
9. Select the REPLACE TCM/PCM MENU, then select READ DATA and follow the screen prompts.

➡ Doing this step copies (READS) the ATF life data from the original PCM so you can later download (WRITES) it into the new PCM.

10. If READ DATA indicates FAILED, continue with this procedure.
11. Turn the ignition switch to LOCK (0).
12. Jump the SCS line with the HDS.
13. Remove the bracket (D), then free the A/C discharge line (E) from the clip (F) and remove the A/C suction line mounting bracket bolt (G).
14. Remove the cover (H), then disconnect PCM connectors A, B, and C.

➡ PCM connectors A, B, and C have symbols (A=?, B=? , C=?) embossed on them for identification.

15. Remove the bolts (I), then remove the PCM (J).

To install:

16. Installation is the reverse order of removal.
17. Turn the ignition switch to ON (II).
18. Manually input the VIN to the PCM with the HDS.

➡ DTC P0630 "VIN Not Programmed or Mismatch" may be stored because the VIN has not been programmed into the PCM; ignore it, and continue this procedure.

19. Select the PGM-FI system with the HDS.
20. Select the REPLACE PCM MENU, then select WRITE DATA and follow the screen prompts.

➡ If the WRITE DATA indicates FAILED, continue with this procedure.

21. If the READ DATA (ATF life) failed, skip the next two steps.
22. Select the A/T SYSTEM with the HDS.
23. Select the REPLACE TCM/PCM MENU, then select WRITE DATA and follow the screen prompts.

→**If the WRITE DATA indicates FAILED, continue with this procedure.**

24. Select IMMOBI system with the HDS.

25. Enter the immobilizer code with the PCM replacement procedure in the HDS; it allows you to start the engine.

26. If the TP POSITION CHECK failed, clean the throttle body.

27. If the READ DATA or the WRITE DATA failed for engine oil life, replace the engine oil and engine oil filter.

28. If the READ DATA or the WRITE DATA failed for the A/T system, replace the ATF.

29. Select PGM-FI system and reset the PCM with the HDS.

30. Reconnect all connectors, then update the PCM if it does not have the latest software.

31. Do the PCM idle learn procedure.

32. Do the CKP pattern learn procedure.

THROTTLE POSITION SENSOR (TPS)

LOCATION

The throttle position sensor is located on the throttle body, and is part of the throttle actuator control module.

REMOVAL & INSTALLATION

The TPS is integrated into the throttle body assembly. For additional information, refer to the following section, "Throttle Body, Removal & Installation."

VEHICLE SPEED SENSOR (VSS)

LOCATION

The vehicle speed sensor is located on the underside of the transaxle assembly.

REMOVAL & INSTALLATION

1. Raise the vehicle up on a lift, or apply the parking brake, block both rear wheels, and raise the front of the vehicle. Make sure it is securely supported.

2. Remove the transmission undercover and splash shield.

3. Remove the damper from the front subframe.

4. Disconnect the output shaft (countershaft) speed sensor connector, and remove the output shaft (countershaft) speed sensor and sensor washer.

To install:

5. Install the new O-ring on the new output shaft (countershaft) speed sensor, then install the output shaft (countershaft) speed sensor and sensor washer.

6. Check the connector for rust, dirt, or oil, then connect it securely.

7. Install the damper on the front sub-frame.

8. Install the splash shield and transmission undercover.

FUEL   GASOLINE FUEL INJECTION SYSTEM

FUEL SYSTEM SERVICE PRECAUTIONS

Safety is the most important factor when performing not only fuel system maintenance but any type of maintenance. Failure to conduct maintenance and repairs in a safe manner may result in serious personal injury or death. Maintenance and testing of the vehicle's fuel system components can be accomplished safely and effectively by adhering to the following rules and guidelines.

• To avoid the possibility of fire and personal injury, always disconnect the negative battery cable unless the repair or test procedure requires that battery voltage be applied.

• Always relieve the fuel system pressure prior to disconnecting any fuel system component (injector, fuel rail, pressure regulator, etc.), fitting or fuel line connection. Exercise extreme caution whenever relieving fuel system pressure to avoid exposing skin, face and eyes to fuel spray. Please be advised that fuel under pressure may penetrate the skin or any part of the body that it contacts.

• Always place a shop towel or cloth around the fitting or connection prior to loosening to absorb any excess fuel due to spillage. Ensure that all fuel spillage (should it occur) is quickly removed from engine surfaces. Ensure that all fuel soaked cloths or towels are deposited into a suitable waste container.

• Always keep a dry chemical (Class B) fire extinguisher near the work area.

• Do not allow fuel spray or fuel vapors to come into contact with a spark or open flame.

• Always use a back-up wrench when loosening and tightening fuel line connection fittings. This will prevent unnecessary stress and torsion to fuel line piping.

• Always replace worn fuel fitting O-rings with new Do not substitute fuel hose or equivalent where fuel pipe is installed.

Before servicing the vehicle, make sure to also refer to the precautions in the beginning of this section as well.

RELIEVING FUEL SYSTEM PRESSURE

1. Remove the driver's side dashboard lower cover and disconnect the PGM-FI main relay.

2. Remove the fuel filler cap.

3. Start the engine and let it stall.

→**A temporary code may be set during this procedure and the codes must be cleared after repairs are completed.**

4. Turn the ignition **OFF**.

5. Disconnect the negative battery cable.

6. Remove the quick connect fitting cover.

7. Place a shop towel over the quick connect fitting and disconnect the fitting.

8. After the pressure is release, reconnect the fitting and install the cover.

9. After repairs are complete install the relay and connect the negative battery cable.

FUEL FILTER

REMOVAL & INSTALLATION

1. Properly relieve the fuel system pressure.

2. Remove the fuel pump module. For additional information, refer to the following section, "Fuel Pump, Removal & Installation."

3. Disassemble the fuel pump module and remove the fuel filter.

4. Installation is the reverse order of removal. Use new O-rings when installing.

FUEL RAIL & INJECTORS

REMOVAL & INSTALLATION

See Figure 78.

1. Properly relieve the fuel system pressure.

2. Remove the intake manifold. For additional information, refer to the following section, "Intake Manifold, Removal & Installation."

3. Disconnect the connectors from the injectors.

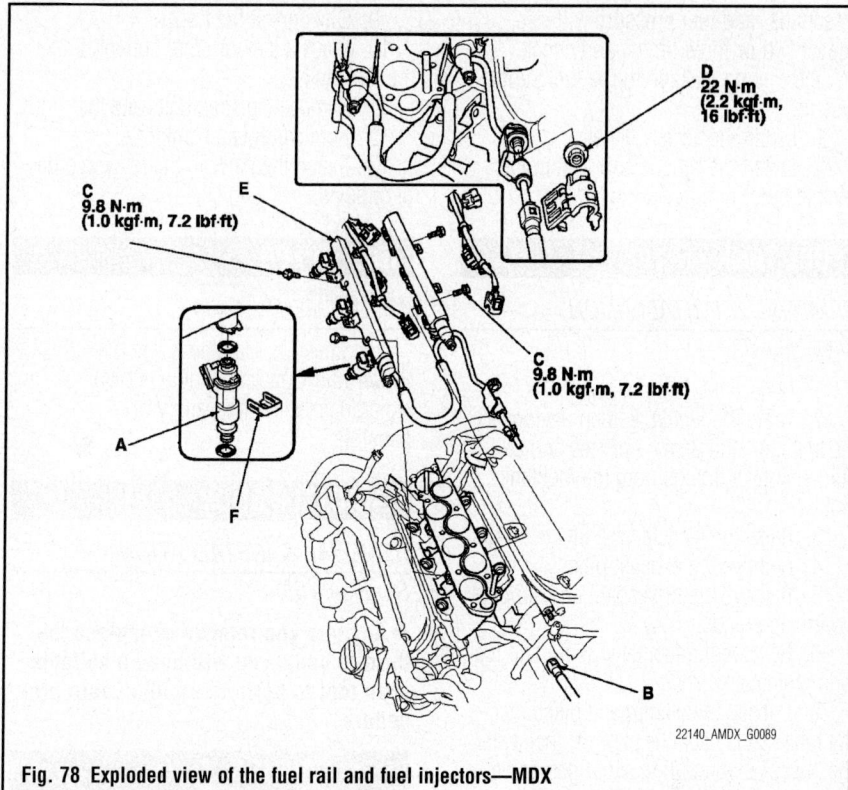

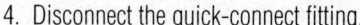

Fig. 78 Exploded view of the fuel rail and fuel injectors—MDX

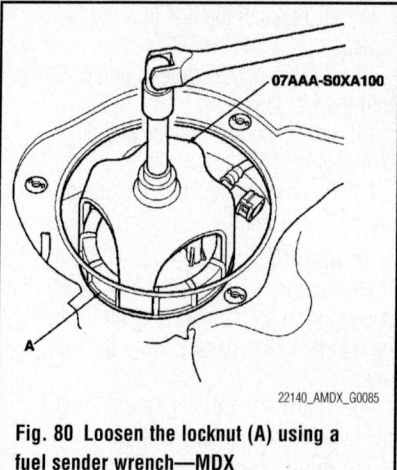

Fig. 80 Loosen the locknut (A) using a fuel sender wrench—MDX

4. Disconnect the quick-connect fitting.

5. Remove the fuel rail mounting bolts and nut from the fuel rails.

6. Remove the injector clips from the fuel rail.

7. Remove the injectors from the rails.

To install:

8. Coat the new O-ring with clean engine oil, and insert the injectors into the fuel rails.

9. Install the injector clips.

10. Coat the new injector O-ring with clean engine oil.

11. Install the injectors in the injector base.

12. Install the fuel rail mounting bolts and nut.

13. Install the injector connectors.

14. Connect the quick-connect fitting.

15. Turn the ignition switch to ON (II), but do not operate the starter. After the fuel pump runs for about 2 seconds, the fuel pressure in the fuel line rises. Repeat this two or three times, then check for fuel leaks.

16. Install the intake manifold.

FUEL PUMP

REMOVAL & INSTALLATION

See Figures 79 through 81.

1. Properly relieve the fuel system pressure.

2. Raise the second row seat-back.

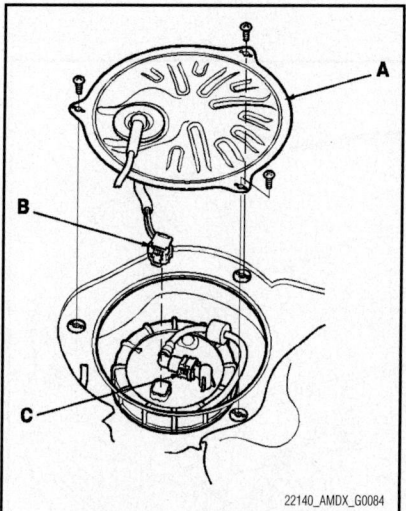

Fig. 79 Remove the access panel (A), wiring harness (B) and quick-connect fitting (C)—MDX

3. Remove the front foot inner cover and front foot outer cover from the front of the seat tracks.

4. Fold the second row seat-back forward.

5. Remove the rear ISO-FIX bracket cover cap (left second row seat), rear foot inner cover, and rear foot outer cover from the back of the seat tracks.

6. From the front of the seat tracks, remove the bolts securing the second row seat.

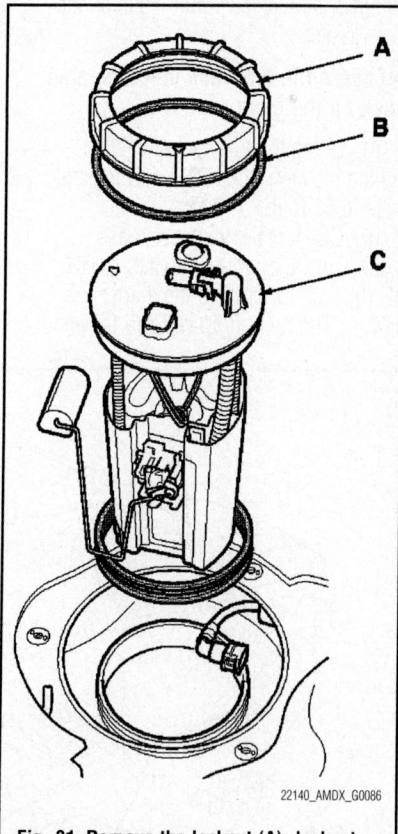

Fig. 81 Remove the locknut (A), locknut plate (B), to remove the fuel tank unit (C)—MDX

7. From the back of the seat tracks, remove the bolts securing the second row seat.

8. If equipped with a seat heater, lift up the front seat, then detach the harness clip, and disconnect the seat-back heater connector.

9. With the help of an assistant, carefully remove the second row seat through the rear door opening.

10. Remove the access panel from the floor.

11. Disconnect the fuel tank unit wiring harness.

12. Disconnect the quick-connect fitting from the fuel tank unit.

13. Using the fuel sender wrench 07AAA-S0XA100, loosen the locknut.

14. Remove the locknut, locknut plate, and fuel tank unit.

To install:

15. Temporarily attach a new base gasket to the fuel tank unit, then insert the fuel tank unit partially into the fuel tank.

16. Transfer the base gasket from the fuel tank unit to the fuel tank.

17. Align the marks on the fuel tank and the fuel tank unit, then insert the fuel tank unit into the fuel tank until it sits on the base gasket.

➡**Ensure the fuel tank unit is seated flush on the base gasket.**

18. Using the fuel sender wrench, tighten the new lock with a new locknut plate to 51 ft. lbs. (70 Nm).

19. Connect the wiring harness.

20. Turn the ignition switch to ON (II) (but do not operate the starter motor). The fuel pump will run for about 2 seconds, and fuel pressure will rise. Repeat two or three times, and check that there is no leakage in the fuel supply system.

21. Install the access panel.

22. Install the second row seat in the reverse order of removal.

FUEL TANK

REMOVAL & INSTALLATION

See Figure 82.

1. Relieve the fuel pressure.

2. Drain the fuel tank, then disconnect the fuel tank unit wiring harness and the quick-connect fittings from the fuel tank unit.

3. Raise the vehicle on a lift.

4. Remove the exhaust pipe.

5. Remove the driveshaft, and support it with jackstands.

6. Remove the fuel tank guard and fuel tank protector.

7. Loosen the clamp, and disconnect the tube. Slide back the clamps, then twist the hoses as you pull to avoid damaging them.

8. Open the clamps.

9. Disconnect the hoses.

10. Place a jack or other support under the fuel tank.

11. Remove the strap bolts and the straps.

12. Remove the fuel tank.

13. Install the parts in the reverse order of removal

IDLE SPEED

ADJUSTMENT

Idle speed is maintained by the Powertrain Control Module (PCM). No adjustment is necessary or possible.

THROTTLE BODY

REMOVAL & INSTALLATION

See Figure 83.

➡**Anytime you remove or replace the throttle body, you will need a suitable scan tool to perform an Idle Learn procedure.**

✷✷ CAUTION

Do not insert your fingers into the installed throttle body when you turn the ignition switch ON (II) or while the ignition switch is ON (II). If you do, you will seriously injure your fingers if the throttle valve is activated.

➡**If you are replacing the throttle body, begin at step 1. If you are removing the throttle body temporarily, begin at step 4.**

1. Connect the HDS while the engine is stopped.

2. Select the INSPECTION MENU with the HDS.

3. Do the TP POSITION CHECK in the ETCS TEST.

4. Disconnect the MAP sensor connector.

5. Remove the intake air duct.

6. Disconnect the throttle body connector.

7. Disconnect the water bypass hoses, and plug the water bypass hoses.

8. Remove the throttle body.

9. Clean the throttle body and intake manifold surface.

To install:

10. Installation is the reverse of the removal procedure. Make sure to use a new gasket (E). Tighten the throttle body mounting bolts to 16 ft. lbs. (22 Nm).

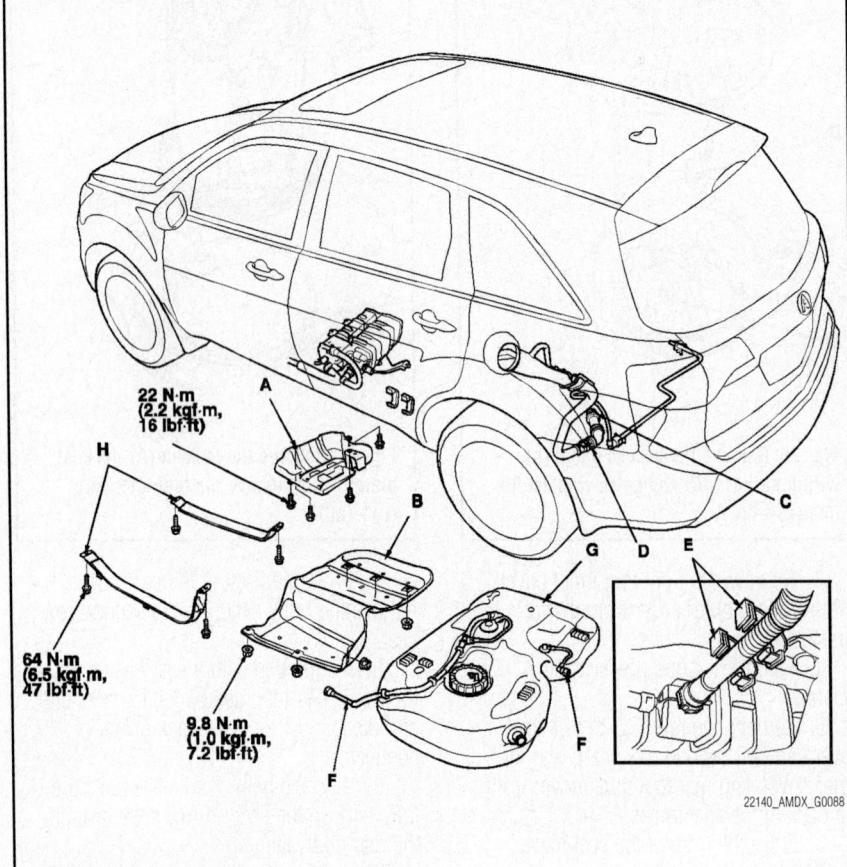

22 N·m (2.2 kgf·m, 16 lbf·ft)

64 N·m (6.5 kgf·m, 47 lbf·ft)

9.8 N·m (1.0 kgf·m, 7.2 lbf·ft)

22140_AMDX_G0088

Fig. 82 Removing the fuel tank assembly—MDX

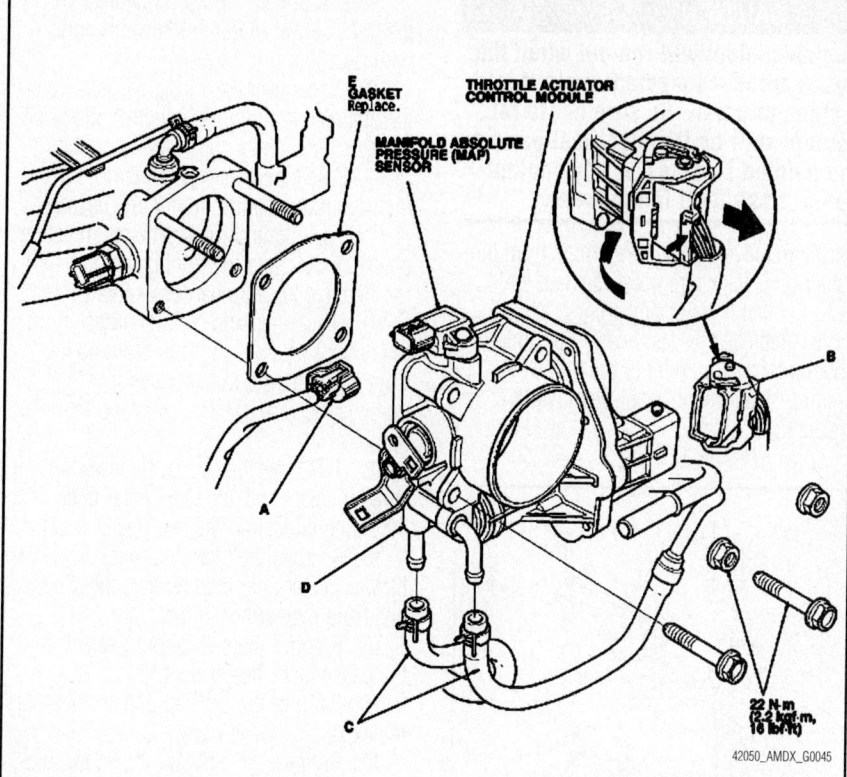

Fig. 83 View of the MAP sensor connector (A), throttle body connector (B), water bypass hoses (C), throttle body (D) and gasket (E)

➡The idle learn procedure must be done so the PCM can learn the engine idle characteristics. Perform the idle learn procedure whenever you do any of these actions: Replace PCM, Reset PCM, Update PCM or Replace or clean the throttle body.

11. Perform the PCM idle learn procedure, as follows:

➡Erasing DTCs with the HDS does not require you to do the idle learn procedure.

    a. Make sure all electrical items (A/C, audio, rear window defogger, lights, etc.) are off.
    b. Reset the PCM with the HDS.
    c. Turn the ignition switch ON (II), and wait 2 seconds.
    d. Start the engine. Hold the engine speed at 3,000 rpm without load (in Park or neutral) until the radiator fan comes on, or until the engine coolant temperature reaches 194 °F (90 °C).
    e. Let the engine idle for about 5 minutes with the throttle fully closed.

➡If the radiator fan comes on, do not include its running time in the 5 minutes.

12. Refill the engine with coolant.

HEATING & AIR CONDITIONING SYSTEM

BLOWER MOTOR

REMOVAL & INSTALLATION

See Figures 84 and 85.

1. Remove the passenger's dashboard undercover.
2. Open the glove box.
3. Remove the cap, then disconnect the glove box damper from the pivot on the glove box.
4. Close the glove box.
5. Remove the mounting bolts.
6. While holding the glove box, release the glove box stop on each side from the dashboard by pushing them in, then remove the glove box.
7. Remove the harness clips, the bolts, and the glove box frame.
8. Cut the plastic cross brace in the glove box opening with diagonal cutters in the area shown, and discard it.
9. Remove the wire harness clips, the self-tapping screws, and the passenger's heater duct.
10. Disconnect the connector from the front blower motor. Remove the wire harness clips.
11. Disconnect the connectors from the recirculation control motor and adaptive front lighting control unit, then remove the harness clip. Remove the self-tapping screws, the mounting nuts, and the blower unit.
12. Install the unit in the reverse order of removal. Make sure that there is no air leakage.

Fig. 84 Cut the plastic cross brace (A) in the glove box opening with diagonal cutters in the area shown, and discard it.

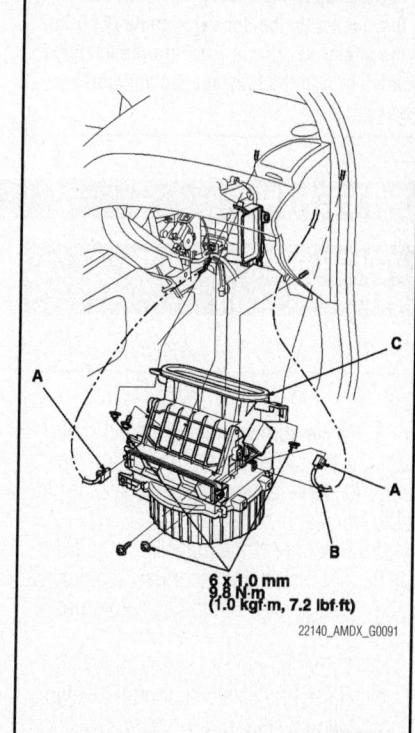

Fig. 85 Removing the blower unit—MDX

HEATER CORE

REMOVAL & INSTALLATION

See Figures 86 and 87.

1. Disconnect the negative battery cable.
2. Recover the refrigerant with a recovery/recycling/charging station, as outlined in this section.
3. Disconnect the front receiver line and front suction line from the front evaporator core.
4. From under the hood, open the cable clamp, then disconnect the heater valve cable from the heater valve arm. Turn the heater valve arm to the fully opened position.
5. Drain the engine cooling system.
6. Remove the clamp and slide the hose clamps back. Remove the nut and the water valve, then disconnect the inlet heater hose and the outlet heater hose from the heater unit.

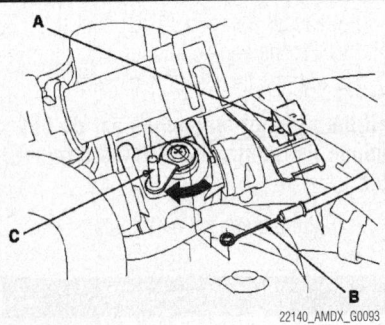

22140_AMDX_G0093

Fig. 86 Open the cable clamp (A), then disconnect the heater valve cable (B) from the heater valve arm (C). Turn the heater valve arm to the fully opened position as shown.

✳✳ CAUTION

Engine coolant will run out when the hoses are disconnected; drain it into a clean drip pan. Be sure not to let coolant spill on the electrical parts or the painted surfaces. If any coolant spills, rinse it off immediately.

7. Remove the mounting nuts from the heater unit. Take care not to damage or bend the fuel lines or brake lines, etc.
8. Remove the dashboard. For additional information, refer to the following section, "Instrument Panel, Removal & Installation."

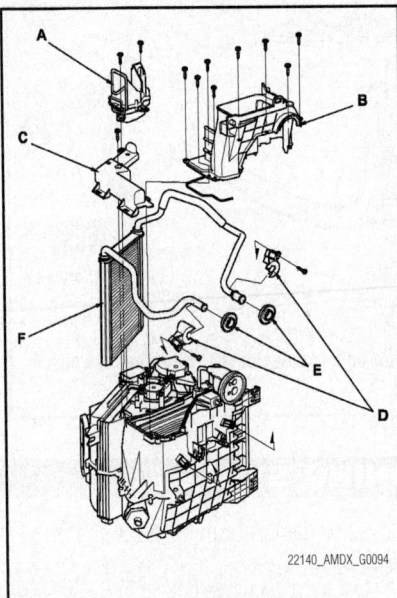

22140_AMDX_G0094

Fig. 87 Removing the heater core from the blower-heater unit—MDX

9. Disconnect the wiring harness from the front blower motor. Remove the wire harness clips.
10. Disconnect the wiring harness from the adaptive front lighting control unit, then remove the harness clip.
11. Disconnect the wiring harness from the front mode control motor, the passenger's air mix control motor, the recirculation control motor, and the front power transistor. Remove the wire harness clips.
12. Disconnect the wiring harness from the driver's air mix control motor and the front evaporator temperature sensor. Remove the wire harness clips and the wire harness.
13. Turn over the carpet. Remove the wire harness clips, the rear heater duct mounting clips, and the rear heater duct.
14. Remove the mounting nuts. Slide the blower-heater unit, then remove the drain hose and blower-heater unit.
15. Remove the self-tapping screws and the passenger's heater duct.
16. Remove the self-tapping screws and the expansion valve cover.
17. Remove the self-tapping screw and the front heater core cover.
18. Remove the self-tapping screws, the heater pipe brackets, the grommets, and carefully pull out the front heater core.

To install:

19. Installation is the reverse order of removal. Tighten the heater mounting nuts to 9 ft. lbs. (12 Nm).
20. Refill the cooling system to the correct level.
21. Adjust the heater valve cable if necessary.

STEERING

POWER RACK & PINION STEERING GEAR

REMOVAL & INSTALLATION

See Figures 88 through 93.

1. Make sure you have the anti-theft code for the radio and the navigation system, then write down the frequencies for the radio's preset buttons.
2. Disconnect the negative cable from the battery first, then disconnect the positive cable. Wait at least 3 minutes before proceeding.
3. Drain the power steering fluid.
4. Raise and safely support the vehicle.
5. Remove the front wheels.
6. Remove the driver's dashboard undercover.

7. Remove the steering wheel. For additional information, refer to the following section, "Steering Wheel, Removal & Installation."
8. Remove steering joint cover.
9. Remove the steering joint bolts, disconnect the steering joint by moving the steering joint toward the column
10. Remove the center guide (if equipped) from the top of the pinion shaft, and discard it.

➡ **The center guide is for factory assembly use only.**

11. Apply vinyl tape to the splines on the pinion shaft.
12. Remove the air cleaner housing.
13. Remove the service caps for the front damper flange nuts from the cowl cover.

Position the engine hanger adapters (VSB02C000031) with the "FRONT" mark facing forward over the damper flange nuts.

14. Install the engine hanger balancer bar (VSB02C000019); attach the front arm to the front cylinder head with a spacer and 10 x 1.25 mm bolt, attach the rear arm to the rear cylinder head with 8 x 1.25 mm bolt.
15. Install the engine support hanger (AAR-T-12566) to the vehicle, and attach the hook to the slotted hole in the engine hanger balancer bar. Tighten the wing nut by hand to lift and support the engine/transmission.
16. Loosen the adjustable hose clamp, and disconnect the return hose.
17. Loosen the 16 mm flare nut, and disconnect the inlet line.

18. Remove the cotter pins from the tie-rod ball joints and remove the nuts.

19. Separate the tie-rod ball joints and knuckles using the ball joint remover.

20. Disconnect the lower arm ball joints from the knuckles.

21. Disconnect the stabilizer links from the lower arms.

22. Remove the front undercover and the front splash shield.

23. Remove the front subframe stiffener plate.

24. Remove exhaust pipe A.

25. Remove the driveshaft and driveshaft protectors.

26. Remove the inlet line clamp bolts.

27. Remove the rear mount stop from the rear engine mount.

28. Remove the rear engine mount from the base bracket, the remove the base bracket from the front subframe.

29. Remove the ground cable bolt from the transmission.

30. Make reference marks on the body across the marks on the edge of the front subframe.

31. Loosen the four bolts holding the adjustable arms of the front subframe adapter to its center plate.

32. Line up the slots in the arms with the bolt holes on the corner of the jack base, then attach the front subframe adapter to the jack base with the bolts that came with the jack. Tighten all bolts securely.

33. Raise the jack to vehicle height, then attach the front subframe adapter to the front subframe using the subframe stiffener mounting bolts and bolt holes.

34. Support the front subframe securely by raising the transmission jack.

35. Remove the mounting bolts from the front subframe front brackets.

36. Loosen the special bolts on the front subframe front brackets so they are about 40 mm (1 9/16 in.) from the mounting surface. Do not loosen the special bolts more than necessary.

37. Remove the mounting bolts from the front subframe rear brackets.

38. Loosen the special bolts on the front subframe rear brackets so they are about 40 mm (1 9/16 in.) from the mounting surface. Do not loosen the special bolts more than necessary.

39. Lower the transmission jack slowly until the front subframe has dropped about 40 mm (1 9/16 in.).

40. Remove the mounting bolts from the left side of the steering gearbox.

41. Remove the gearbox stiffener bracket from the left side of the front subframe.

42. Remove the mounting bolts from the right side of the steering gearbox,

then remove the mounting bracket and cushion.

43. Carefully move the steering gearbox toward the driver's side until the pinion shaft clears the wheel well opening on the frame.

44. Remove the steering gearbox through the wheel well opening on the driver's side.

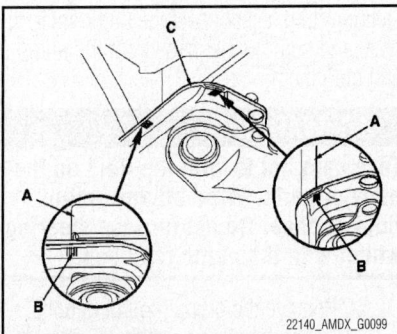

Fig. 88 Make reference marks (A) on the body across the marks (B) on the edge of the front subframe (C).

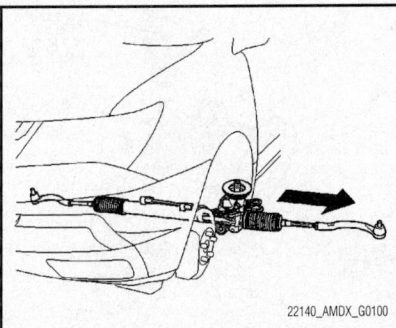

Fig. 89 Remove the steering gearbox through the driver's side wheel well opening—MDX

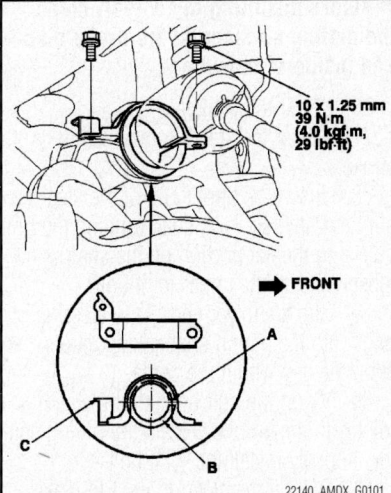

Fig. 90 Position the cutout (A) on the mounting cushion (B) as shown, and install it on the right side of the steering gearbox. Install the gearbox mounting bracket (C) over the mounting cushion, and loosely install the mounting bolts.

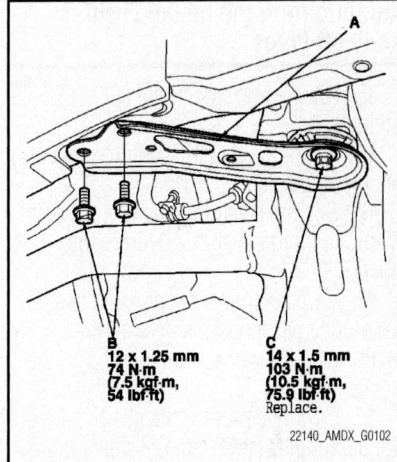

Fig. 91 Front subframe front bracket mounting bolt torques—MDX

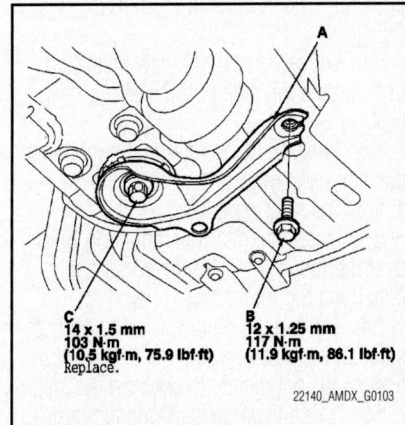

Fig. 92 Rear subframe front bracket mounting bolt torques—MDX

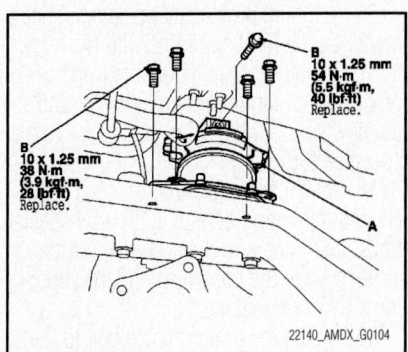

Fig. 93 Rear engine mount bolt torques—MDX

To install:

✳✳ CAUTION

Before installing the steering gearbox, make sure that no power steering fluid is on the mating surface of the steering gearbox and front subframe. To prevent the gearbox mounting bolts from loosening after the installation, remove any power steering fluid from the mount cushions and bolt holes.

45. Apply vinyl tape to the splines on the pinion shaft.
46. Install the pinion shaft grommet on the top of the valve housing.
47. Slide the steering gearbox between the front subframe and body from the driver's side.
48. Carefully move the steering gearbox toward the passenger's side until the pinion shaft clears the wheel well opening on the body.
49. Continue moving the gearbox toward the passenger's side until the steering gearbox is in position.
50. Install the gearbox stiffener bracket on the left side of the front subframe, and tighten the bolts and nut to 28 ft. lbs. (38 Nm).
51. Loosely install the new mounting bolts on the left side of the steering gearbox.
52. Position the cutout on the mounting cushion as shown, and install it on the right side of the steering gearbox.
53. Install the gearbox mounting bracket over the mounting cushion, and loosely install the mounting bolts.
54. Tighten the mounting bolts on both sides of the steering gearbox to the specified torque alternately in two or more steps.
55. Carefully raise the front subframe with the subframe adapter and the transmission jack or the powertrain lift until the subframe is it position.
56. Align all previously made reference marks on the front subframe with the body.
57. Install the front subframe front brackets (A) with mounting bolts (B) and the new special bolts (C), and tighten to the specified torque.
58. Install the front subframe rear brackets with mounting bolts and the new special bolts, and tighten to the specified torque.
59. Lower the transmission jack supporting the front subframe.
60. Install the ground cable bolt to the transmission and tighten to 33 ft. lbs. (44 Nm).
61. Install the base bracket with new

short mounting bolts and tighten them all to 31 ft. lbs. (42 Nm).
62. Install the rear engine mount with new mounting bolts, and tighten to the specified torque.
63. Install the rear mount stop with new mounting nuts, and tighten to 54 ft. lbs. (74 Nm).
64. Install the inlet line clamp bolts (A), and tighten to 7 ft. lbs. (9.8 Nm).
65. Install the exhaust pipe.
66. Install the driveshaft and driveshaft protectors.
67. Install the front subframe stiffener plate with new mounting bolts (B), and tighten to 40 ft. lbs. (54 Nm).
68. Install the front undercover and the front splash shield.
69. Connect the stabilizer links to the lower arms.
70. Connect the lower arm ball joints to the knuckles.
71. Wipe off any grease contamination from the ball joint tapered section and threads. Reconnect the tie-rod ball joints to the knuckles. Install the 12 mm nuts and tighten it to 40 ft. lbs. (54 Nm). Install new cotter pins.
72. Remove the engine support hanger, the hanger balance bar, and the hanger adapter set.
73. Connect the return hose securely, and tighten the adjustable hose clamp.
74. Connect the inlet line and tighten the 16 mm flare nut to 31 ft. lbs. (42 Nm).
75. Install the front wheel, then set the wheels in the straight ahead position.

➡**Before installing the wheel, clean the mating surfaces of the brake disc and inside of the wheel.**

76. Lower the vehicle.
77. Center the steering rack within its stroke.
78. Insert the upper end of the steering joint onto the steering shaft (line up the bolt hole with the flat portion on the shaft), and loosely install the upper joint bolt.
79. Slip the lower end of the steering joint onto the pinion shaft taking care to align the gap within the angle.
80. Align the bolt hole on the steering joint with the groove around the pinion shaft then loosely install the joint bolt.
81. Pull on the steering joint to make sure that the steering joint is fully seated, then tighten the lower joint bolt to the specified torque.
   - 2007 Models: 21 ft. lbs. (28 Nm)
   - 2008 Models: 16 ft. lbs. (22 Nm)
82. Tighten the upper joint bolt to the specified torque.

   - 2007 Models: 21 ft. lbs. (28 Nm)
   - 2008 Models: 16 ft. lbs. (22 Nm)
83. Install the steering joint cover.
84. Install the steering wheel.
85. Connect the battery cables.
86. Turn the ignition switch to ON (II); the SRS indicator should come on for about 6 seconds and then go off.
87. Make sure the horn and turn signal switches work properly.
88. Make sure the steering wheel switches work properly.
89. Refill the power steering system to the correct level.
90. Start the engine, turn the steering wheel from lock-to-lock several times and check for leaks.

POWER STEERING PUMP

REMOVAL & INSTALLATION

See Figure 94.

1. Drain the power steering fluid from the reservoir.
2. Remove the engine appearance cover.
3. Remove the accessory drive belt from the pump pulley.
4. Cover the auto-tensioner, alternator, and A/C compressor with several shop towels to protect them from spilled power steering fluid. Disconnect the pump inlet hose and pump outlet hose from the pump, and plug them.

✳✳ WARNING

Take care not to spill the fluid on the body or parts. Wipe off any spilled fluid at once. Do not turn the steering wheel with the pump removed.

5. Remove the pump mounting bolts.
6. Cover the opening of the pump with a piece of tape to prevent foreign material from entering the pump.

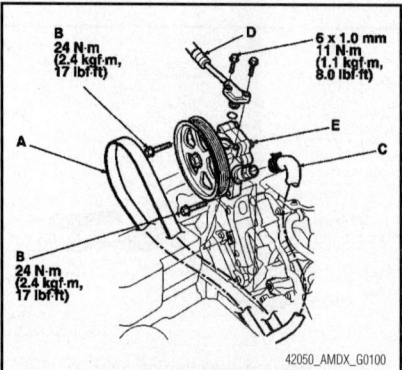

42050_AMDX_G0100

Fig. 94 Exploded view of the power steering pump and tightening specifications

To install:

7. Connect the pump inlet hose and pump outlet hose onto the new pump with the new O-ring.

8. Loosely install the pump in the pump bracket with the mounting bolts, then tighten the pump fittings securely.

9. Tighten the pump mounting bolts to 16 ft. lbs. (22 Nm).

10. Install the accessory drive belt.

11. Refill the power steering system to the correct level.

BLEEDING

1. Fill the reservoir to the upper level line.

2. Start the engine and run it at idle, then turn the steering from lock-to-lock several times to bleed air from the system.

3. Recheck the fluid level and add some if necessary. Do not fill the reservoir beyond the upper level line.

4. If the fluid is contaminated, dark, or discolored, repeat the procedure as necessary.

STEERING LINKAGE

REMOVAL & INSTALLATION

1. Disconnect the tie-rod ball joint from the knuckle.

2. Remove the tie-rod end from the rack end.

3. Remove the tie-rod ball joint boot from the tie-rod end, and wipe the old grease off the ball pin.

4. Pack the lower area of the ball pin with fresh multipurpose grease.

5. Pack the interior of the new tie-rod ball joint boot and lip with fresh multipurpose grease.

➡Keep grease off the boot mounting area and the tapered section of the ball pin.

✳✳ CAUTION

Do not allow dust, dirt, or other foreign materials to enter the boot.

6. Install the new tie-rod ball joint boot using the front hub disassembly tool. The boot must not have a gap at the boot installation sections. After installing the boot, check the ball pin tapered section for grease contamination, and wipe it if necessary.

7. Install the tie-rod end to the rack end.

8. Connect the tie-rod ball joint to the knuckle.

9. Check the wheel alignment, and adjust it if necessary.

SUSPENSION

See Figure 95.

COIL SPRING

REMOVAL & INSTALLATION

The coil spring is removed during the MacPherson strut overhaul process. For additional information, refer to the following section, "MacPherson Strut, Overhaul."

FRONT SUSPENSION

LOWER BALL JOINT

REMOVAL & INSTALLATION
See Figure 96.

1. Remove the lower control arm. For additional information, refer to the following section, "Lower Control Arm, Removal & Installation."

2. Install a hex nut onto the threads of the ball joint. Make sure the nut is flush with the ball joint pin end to prevent damage to the threaded end of the ball joint pin.

3. Apply grease to the ball joint remover on the areas shown. This will ease installation of the tool and prevent damage to the pressure bolt threads.

4. Loosen the pressure bolt, and install the ball joint remover. Insert the jaws carefully, making sure not to damage the ball joint boot. Adjust the jaw spacing by turning the adjusting bolt.

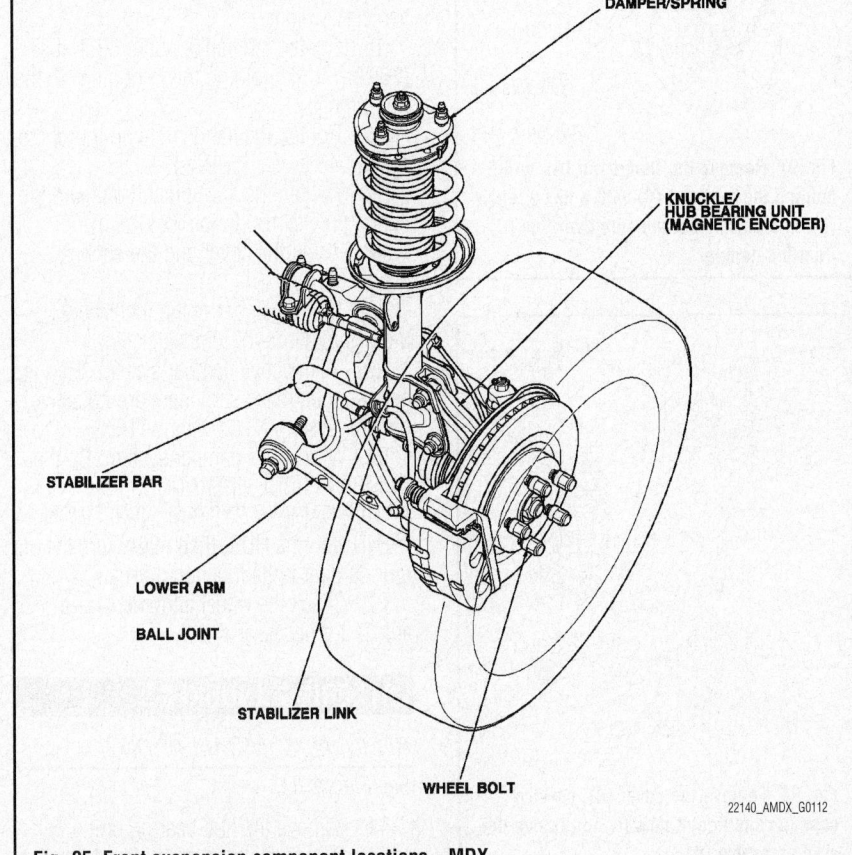

DAMPER/SPRING

KNUCKLE/
HUB BEARING UNIT
(MAGNETIC ENCODER)

STABILIZER BAR

LOWER ARM

BALL JOINT

STABILIZER LINK

WHEEL BOLT

22140_AMDX_G0112

Fig. 95 Front suspension component locations—MDX

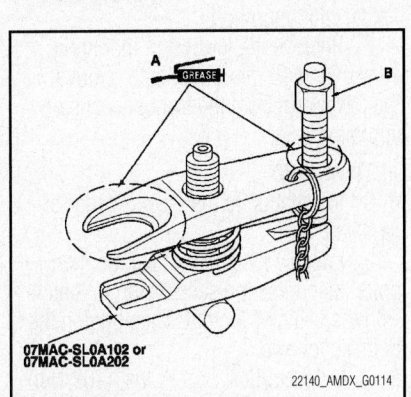

A GREASE
B
07MAC-SL0A102 or
07MAC-SL0A202

22140_AMDX_G0114

Fig. 96 Apply grease to the ball joint remover on the areas shown (A)

✴✴ CAUTION

Fasten the safety chain securely to a suspension arm or the subframe. Do not fasten it to a brake line or wire harness.

5. After adjusting the adjusting bolt, make sure the head of the adjusting bolt is in the position shown to allow the jaw to pivot.

6. With a wrench, tighten the pressure bolt until the ball joint pin pops loose from the ball joint connecting hole. If necessary, apply penetrating type lubricant to loosen the ball joint pin.

✴✴ CAUTION

Do not use pneumatic or electric tools on the pressure bolt.

7. Remove the ball joint remover, then remove the nut from the end of the ball joint pin, and pull the ball joint out of the ball joint connecting hole. Inspect the ball joint boot, and replace it if damaged.

LOWER CONTROL ARM

REMOVAL & INSTALLATION

1. Raise and safely support the vehicle.

2. Remove the front wheel.

3. Remove the suspension stroke sensor from the lower control arm, if equipped with the active damper system.

4. Remove the cotter pin from the lower arm ball joint, then remove the castle nut.

5. Disconnect the lower arm ball joint from the knuckle using the ball joint thread protector and the ball joint remover.

6. Remove the stabilizer bar bushing holder mounting bolt.

7. Remove the lower arm mounting 14 mm and 16 mm bolts, then remove the lower arm from the front suspension subframe.

To install:

8. Installation is the reverse order of removal.

9. Loosely install all of the mounting bolts, then raise the suspension to load it with the vehicle's weight. Then tighten the bolts as follows:

- 16mm bolt: 119 ft. lbs. (162 Nm)
- 14mm bolts: 69 ft. lbs. (93 Nm)
- Lower ball joint nut: 76–83 ft. lbs. (103–113 Nm)

MACPHERSON STRUT

REMOVAL & INSTALLATION

See Figures 97 through 99.

1. Raise and safely support the vehicle.

2. Remove the front wheel.

3. If equipped with the active damper system: Disconnect the damper coil connector, and remove the flange bolt, and the harness clip.

4. Remove the wheel speed sensor harness and the brake hose from the damper. Do not disconnect the wheel speed sensor connector.

5. Remove the flange nut, while holding the joint pin with a hex wrench, and disconnect the stabilizer link from the damper.

6. Remove the damper pinch bolts and flange nuts from the strut assembly.

7. Remove the top mount cover and the service caps .

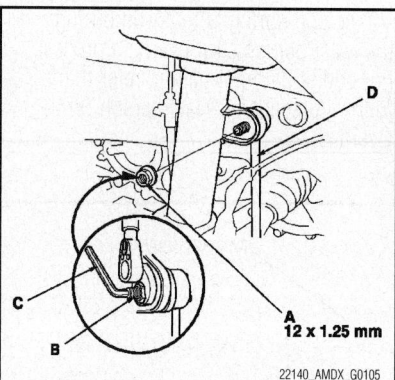

Fig. 97 Remove the flange nut (A), while holding the joint pin (B) with a hex wrench (C), and disconnect the stabilizer link (D) from the damper

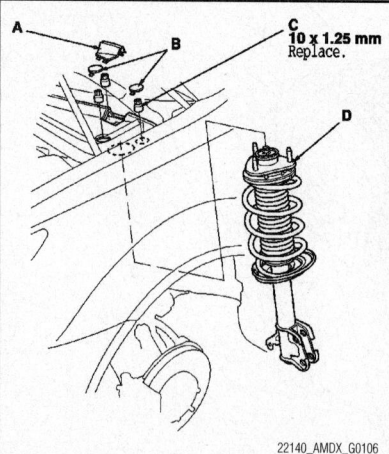

Fig. 98 Remove the cover (A), service caps (B) and flange nuts (C) to remove the strut assembly (D)

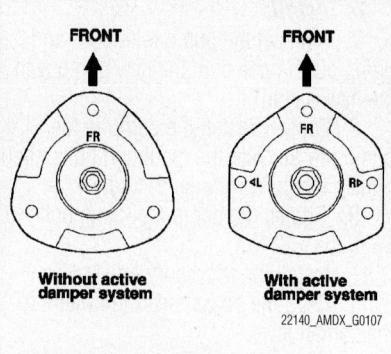

Fig. 99 Note the direction of the strut mounting base during installation

8. Remove the flange nuts from the top of the strut.

9. Remove the strut assembly.

To install:

10. Install the strut assembly on to the frame. Note the direction of the strut mounting base as shown.

11. Loosely install the new flange nuts to the top of the strut.

12. Loosely install new strut pinch bolts and new flange nuts to the damper strut.

13. Connect the stabilizer link (A) to the damper (B), and loosely install a new flange nut (C).

14. Raise the front suspension with a floor jack to load the suspension with the vehicle's weight.

15. Tighten the flange nut to 58 ft. lbs. (78 Nm) while holding the joint pin with the hex wrench.

16. Tighten the flange nuts on top of the strut to 43 ft. lbs. (59 Nm).

17. Tighten the strut pinch bolts and the flange nuts to 156 ft. lbs. (211 Nm).

18. Install the cover and the service caps.

19. The remainder of the installation is the reverse order of removal.

20. With active damper system: Start the engine, then make sure there are no active damper system DTCs with the HDS.

21. With active damper system: Do the DAMPER FORCE OPERATION in the ACTIVE DAMPER SYSTEM INSPECTION MENU with the HDS, then make sure the all four damper units function normally.

22. Check the wheel alignment, and adjust it if necessary.

STEERING KNUCKLE

REMOVAL & INSTALLATION

See Figure 100.

1. Remove the hub bearing unit. For additional information, refer to the following

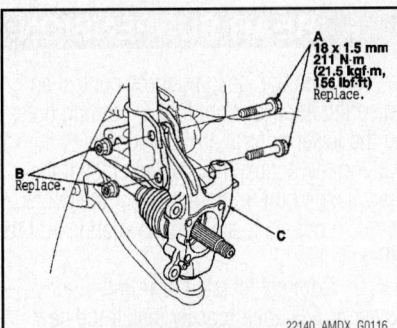

Fig. 100 Remove the damper pinch bolts (A) and flange nuts (B) from the damper, then remove the knuckle (C)

section, "Wheel Bearings, Removal & Installation."

2. Remove the wheel speed sensor from the knuckle. Do not disconnect the wheel speed sensor connector.

3. Remove the cotter pin from the tie-rod end ball joint, then remove the nut.

➡**During installation, install the new cotter pin after tightening the nut, and bend its end as shown.**

4. Disconnect the tie-rod end ball joint from the knuckle using the ball joint remover.

5. Remove the lock pin from the lower arm ball joint, then remove the castle nut.

6. Disconnect the lower arm ball joint from the knuckle using the ball joint thread protector, and the ball joint remover.

7. Remove the damper pinch bolts and flange nuts from the damper, then remove the knuckle.

8. Installation is the reverse order of removal.

9. Loosely install all of the mounting bolts, then raise the suspension to load it with the vehicle's weight. Then tighten the damper pinch bolts to 156 ft. lbs. (211 Nm).

STABILIZER BAR

REMOVAL & INSTALLATION
See Figures 101 and 102.

1. Raise and safely support the vehicle.

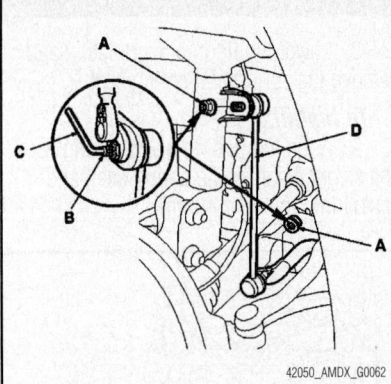

Fig. 101 Remove the flange nuts (A) while holding the respective joint pin (B) with a hex wrench (C), and remove the stabilizer link (D)

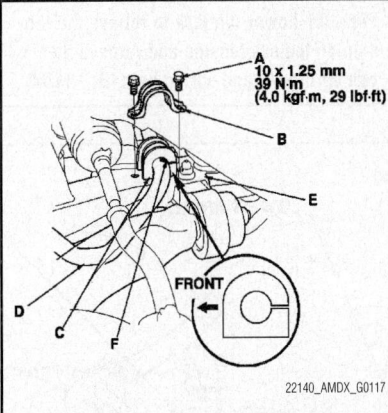

Fig. 102 Note the direction of the stabilizer bar bushings when reinstalling— MDX

2. Remove the front wheels.

3. Remove the flange nuts (A) while holding the respective joint pin (B) with a hex wrench (C), and remove the stabilizer link (D).

4. Remove the flange bolts and the bushing holders, then remove the bushings and the stabilizer bar from the front suspension subframe.

5. Installation is the reverse order of removal. Tighten the bushing mounting bolts to 29 ft. lbs. (39 Nm).

WHEEL HUB AND BEARING

REMOVAL & INSTALLATION
See Figure 103.

1. Raise and safely support the vehicle.

2. Remove the front wheel.

3. Remove the brake hose mounting bolt from the damper.

4. Remove the brake caliper bracket mounting bolts, then remove the caliper assembly from the knuckle.

➡**To prevent damage to the caliper assembly or brake hose, use a short piece of wire to hang the caliper assembly from the undercarriage. Do not twist the brake hose excessively.**

5. Raise the stake, then remove the spindle nut.

6. Remove the brake rotor.

7. Remove the hub bearing unit and the splash guard from the knuckle.

8. Installation is the reverse order of removal. Note the following torque values:
- Hub bearing unit: 72 ft. lbs. (98 Nm)
- New spindle nut: 242 ft. lbs. (328 Nm)

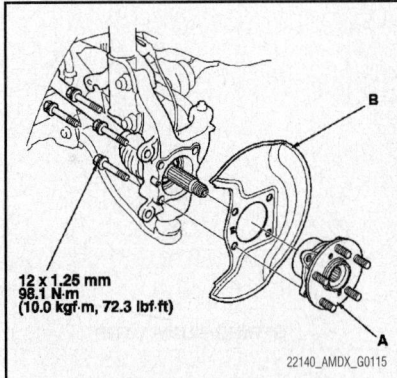

Fig. 103 Remove the mounting bolts to remove the hub bearing unit (A) and splash guard (B)

ADJUSTMENT

1. No adjustment is possible. If the hub bearing unit end play is outside of the service limits, the hub bearing unit must be replaced.

See Figure 104.

COIL SPRING

REMOVAL & INSTALLATION

See Figures 105 and 106.

1. Raise and safely support the vehicle.
2. Remove the rear wheel.
3. Remove the muffler from the muffler hanger.
4. Remove the flange nut while holding the joint pin with a hex wrench, and disconnect the stabilizer link from the lower control arm B. Discard the flange nut.
5. Position a floor jack under the lower arm B and raise the floor jack until the suspension begins to compress.
6. Remove the flange bolt from the bottom of the shock.
7. Remove the flange bolt from the knuckle.

8. Lower the floor jack gradually and remove the spring and spring seat.

To *install:*

9. Align the bottom of the spring with the stepped part of the spring seat, and install into the control arm.

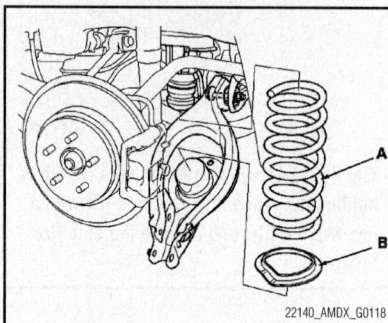

Fig. 105 Lower the jack to relieve the tension on the suspension and remove the coil spring (A) and spring seat (B)—MDX

22140_AMDX_G0118

10. With the jack under the control arm, raise the floor jack until the mounting hole in the lower control arm B aligns with the hole in the shock, then loosely install the new flange bolt to the bottom of the shock.
11. Loosely install the new flange bolt to the knuckle.
12. Connect the stabilizer link to the lower arm B, then loosely install the new flange nut.
13. Raise the rear suspension with the floor jack to load the suspension with the vehicle's weight.
14. Tighten the flange nut 36 ft. lbs. (49 Nm), while holding the joint pin with a hex wrench.
15. Tighten the shock flange mounting bolt to 47 ft. lbs. (64 Nm). Tighten the knuckle flange mounting bolt to 83 ft. lbs. (113 Nm).
16. Install the muffler to the muffler hanger.
17. Install the rear wheel.

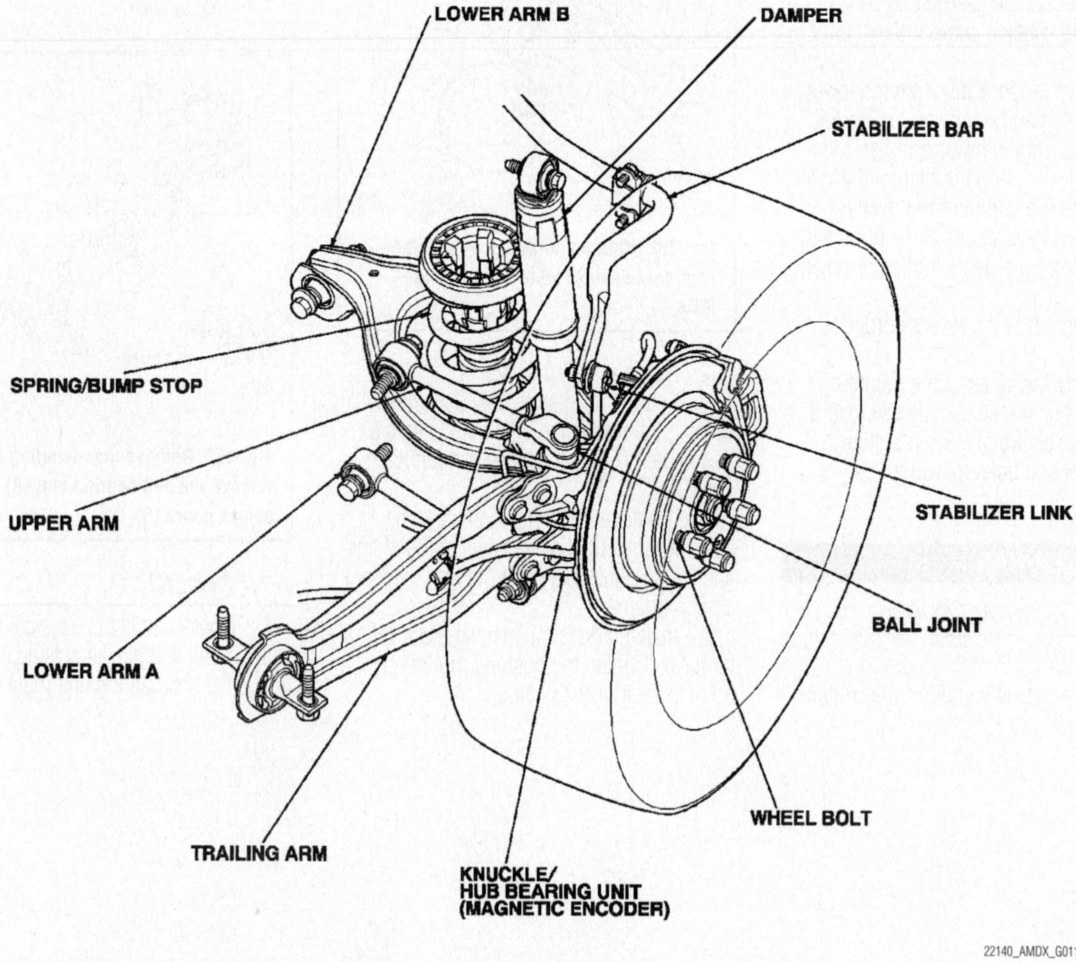

LOWER ARM B   DAMPER

STABILIZER BAR

SPRING/BUMP STOP

UPPER ARM

STABILIZER LINK

BALL JOINT

LOWER ARM A

WHEEL BOLT

TRAILING ARM

KNUCKLE/
HUB BEARING UNIT
(MAGNETIC ENCODER)

22140_AMDX_G0113

Fig. 104 Rear suspension component locations—MDX

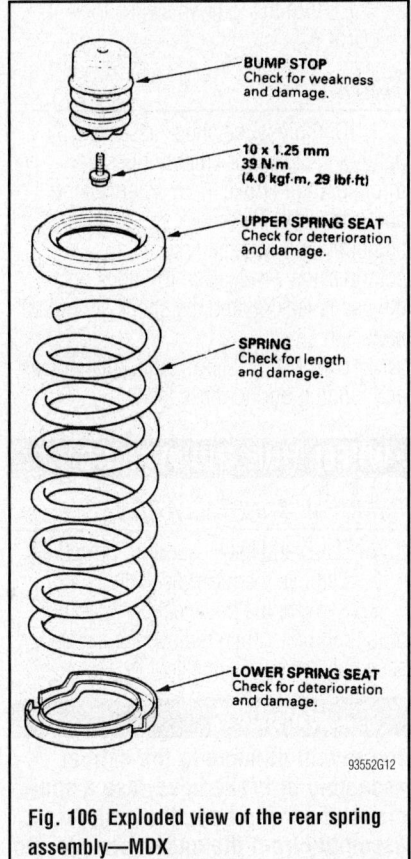

Fig. 106 Exploded view of the rear spring assembly—MDX

CONTROL ARMS/LINKS

REMOVAL & INSTALLATION

Lower Control Arm A

See Figure 107.

1. Raise and safely support the vehicle.

2. Remove the rear wheel.

3. Using a floor jack under lower control arm B, raise the jack until the suspension begins to compress.

4. Remove the flange bolts from the trailing arm.

5. Remove the nut and washer, and flange bolt, then remove the lower control arm.

6. Installation is the reverse order of removal. Position the paint mark on the control arm toward the outside of the vehicle.

Lower Control Arm B

See Figure 108.

1. Raise and safely support the vehicle.

2. Remove the coil spring. For additional information, refer to the following section, Coil Spring, Removal & Installation."

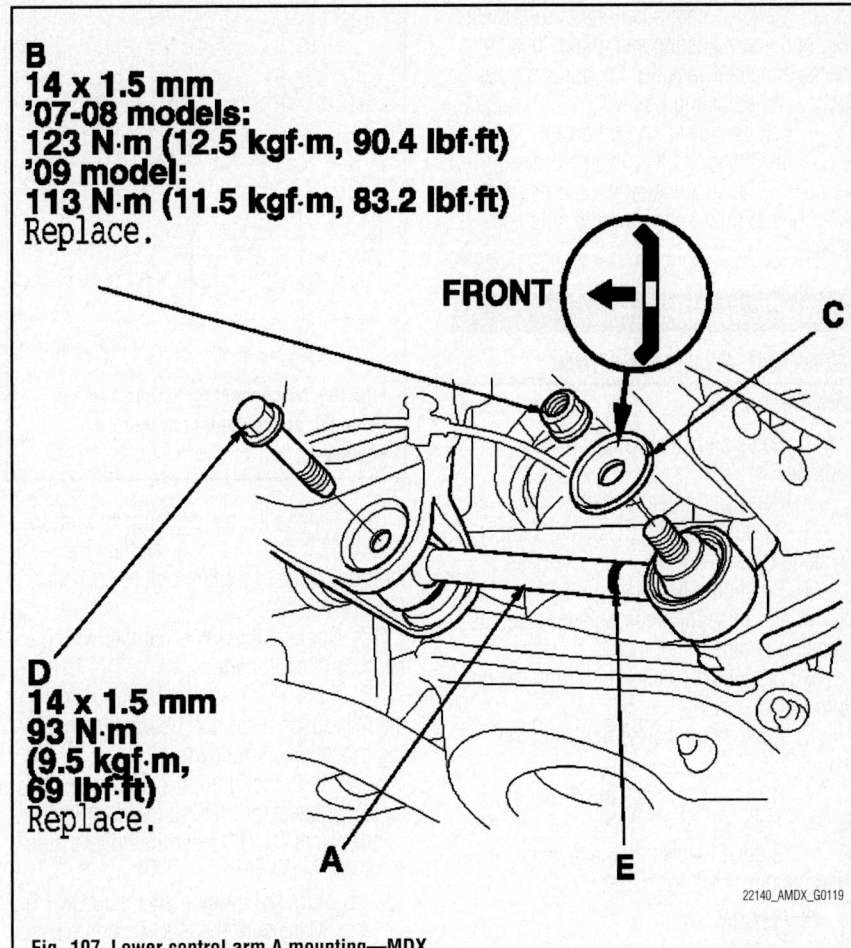

Fig. 107 Lower control arm A mounting—MDX

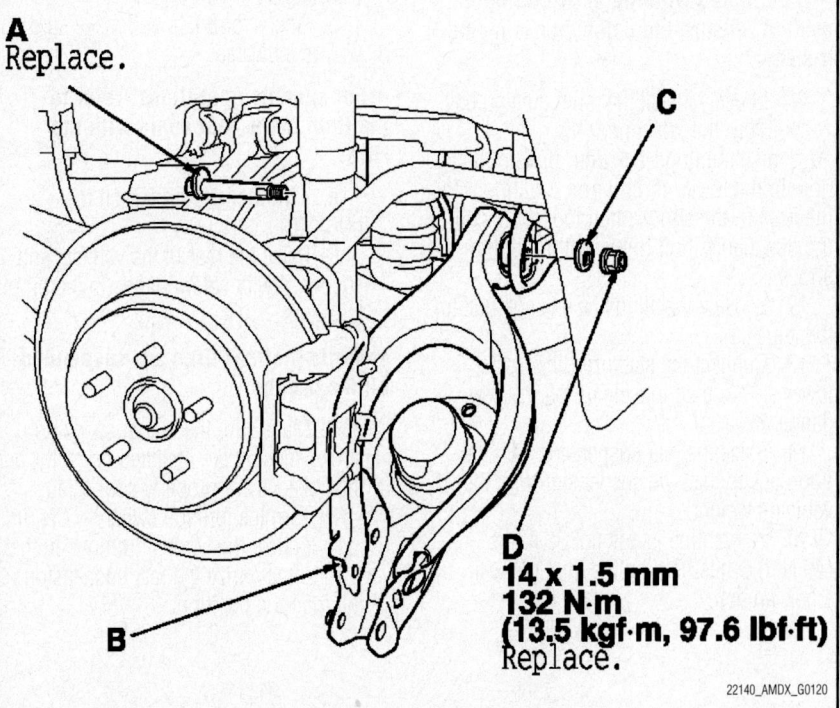

Fig. 108 Matchmark the adjusting bolt (A) and adjusting cam plate (C) before removing the nuts (D)—MDX

3. Mark the cam positions of the adjusting bolt and adjusting cam plate, then remove the self-locking nut, adjusting cam plate, and adjusting bolt.

4. Remove lower control arm B.

5. Installation is the reverse order of removal. The suspension should be under load before fully tightening the nuts and bolts.

SHOCK ABSORBER

REMOVAL & INSTALLATION

See Figure 109.

1. Raise and safely support the vehicle.

2. Remove the rear wheel.

3. If equipped with the active damper system, disconnect the wiring harness.

4. Position a floor jack under the lower arm B and raise the floor jack until the suspension begins to compress.

5. Remove the flange bolt from the bottom of the shock.

6. Remove the flange bolt from the knuckle.

7. Remove the shock top mounting bolt.

8. Gradually lower the jack, then remove the shock from the vehicle.

To install:

9. Install the shock between the body and the lower control arm.

➡**If equipped with the active damper system, ensure the connector is facing the rear.**

10. Loosely install the top mounting bolt.

11. With the jack under the control arm, raise the floor jack until the mounting hole in the lower control arm B aligns with the hole in the shock, then loosely install the new flange bolt to the bottom of the shock.

12. Loosely install the new flange bolt to the knuckle.

13. Connect the stabilizer link to the lower arm B, then loosely install the new flange nut.

14. Raise the rear suspension with the floor jack to load the suspension with the vehicle's weight.

15. Tighten the flange nut 36 ft. lbs. (49 Nm), while holding the joint pin with a hex wrench.

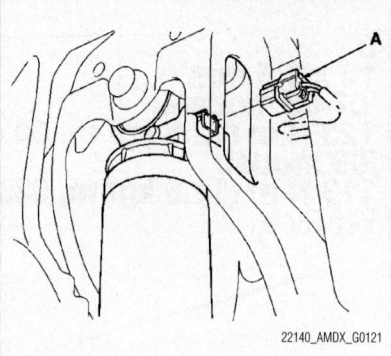

22140_AMDX_G0121

Fig. 109 Disconnect the wiring harness (A) of the active damper system, if equipped.

16. Tighten the shock flange mounting bolt to 47 ft. lbs. (64 Nm). Tighten the knuckle flange mounting bolt to 83 ft. lbs. (113 Nm).

17. Connect the active damper wiring harness, if equipped.

18. If the shock was replaced, perform the rear suspension full rebound memorizing procedure as follows:

a. With the ignition switch in LOCK (0), connect the HDS to the data link connector (DLC) under the driver's side of the dashboard.

b. Turn the ignition switch to ON (II).

c. Make sure the HDS communicates with the vehicle and the active damper control unit. If it doesn't, troubleshoot the DLC circuit.

d. Check for DTCs and make sure no DTC is indicated.

➡**For specific operations, refer to the Help menu that came with the HDS.**

e. Turn the ignition switch to LOCK (0).

f. Raise the rear of the vehicle, and support it with safety stands in the proper locations.

➡**Verify the rear tires are suspended off the ground.**

g. Make sure the rear suspension stroke sensors are installed correctly, and the linkages are properly connected.

h. Turn the ignition switch to ON (II).

i. Follow the screen prompts on the HDS to memorize the rear suspension full rebound position.

j. Turn the ignition switch to LOCK (0).

TESTING

1. Compress the shock assembly by hand, and check for smooth operation through a full stroke, both compression and extension. The shock should extend smoothly and constantly when compression is released. If it does not, the gas is leaking and the shock should be replaced.

2. Check for oil leaks, abnormal noises, and binding during these tests.

WHEEL HUB AND BEARING

REMOVAL & INSTALLATION

1. Raise and safely support the vehicle.

2. Remove the rear wheel.

3. Remove the brake caliper bracket mounting bolts, then remove the caliper assembly from the knuckle.

✳✳ WARNING

To prevent damage to the caliper assembly or brake hose, use a short piece of wire to hang the caliper assembly from the undercarriage. Do not twist the brake hose excessively.

4. Remove the two mounting washers.

5. Raise the stake, then remove the spindle nut.

6. Release the parking brake, and remove the brake disc/drum.

7. Remove the mounting bolts, then remove the hub bearing unit.

8. Installation is the reverse order of removal. Note the following during installation:

a. Tighten hub mounting bolts to 72 ft. lbs. (98 Nm).

b. Using a new spindle nut, apply a small amount of clean engine oil to the seating surface of the nut. Tighten the nut to 181 ft. lbs. (245 Nm), then use a drift to stake the spindle nut shoulder against the driveshaft.

ADJUSTMENT

1. No adjustment is possible. If the hub bearing unit end play is outside of the service limits, the hub bearing unit must be replaced.

ACURA

RDX

HEATING & AIR CONDITIONING SYSTEM......................3-67

PRECAUTIONS.................3-9

SPECIFICATIONS AND MAINTENANCE CHARTS 3-3

STEERING 3-75

SUSPENSION.................3-78

FRONT SUSPENSION..........3-78

REAR SUSPENSION 3-86

SPECIFICATIONS AND MAINTENANCE CHARTS

ENGINE AND VEHICLE IDENTIFICATION CHART

		Engine Code					Model Year	
Code	Liters (cc)	Cu. In.	Cyl.	Fuel Sys.	Engine Type	Eng. Mfg.	Code ①	Year
K23A1	2.3 (2350)	140	4	SMFI	DOHC	Honda	7	2007
							8	2008

DOHC: Dual Overhead Cam

SMFI: Sequential Multi-port Fuel Injection

① 10th position of VIN

22140_ARDX_C0001

GENERAL ENGINE SPECIFICATIONS

Year	Model	Engine Displacement Liters (VIN)	Net Horsepower @ rpm	Net Torque @ rpm (ft. lbs.)	Bore x Stroke (in.)	Com-pression Ratio	Oil Pressure @ rpm
2007	RDX	2.3 (K23A1)	NA	NA	3.39x3.90	8.8:1	44@3000
2008	RDX	2.3 (K23A1)	NA	NA	3.39x3.90	8.8:1	44@3000

SMFI: Sequential Multi-port Fuel Injection

NA: Not Available

22140_ARDX_C0002

ENGINE TUNE-UP SPECIFICATIONS

Year	Engine Displacement Liters (VIN)	Spark Plug Gap (in.)	Ignition Timing (deg.) MT	AT	Fuel Pump (psi)	Idle Speed (rpm) MT	AT	Valve Clearance (in.) In.	Ex.
2007	2.3 (K23A1)	0.028-0.031	—	12-14B	47-54	—	670-770	0.008-0.010	0.011-0.013
2008	2.3 (K23A1)	0.028-0.031	—	12-14B	47-54	—	670-770	0.008-0.010	0.011-0.013

NOTE: The Vehicle Emission Control Information label often reflects changes made during production and must be used if they differ from this chart.

NOTE: The fuel pressure readings are given with the vacuum hose disconnected

B: Before top dead center

22140_ARDX_C0003

CAPACITIES

Year	Model	Engine Displacement Liters (VIN)	Engine Oil with Filter (qts.)	Transmission (pts.) 5-Spd	Transmission (pts.) Auto.	Transfer Case (pts.)	Drive Axle Front (pts.)	Drive Axle Rear (pts.)	Fuel Tank (gal.)	Cooling System (qts.)
2007	RDX	2.3 (K23A1)	4.7	—	①	②	—	③	18.0	④
2008	RDX	2.3 (K23A1)	4.7	—	①	②	—	③	18.0	④

NOTE: All capacities are approximate. Add fluid gradually and check to be sure a proper fluid level is obtained.

① Fluid change: 3.5 quarts
 Overhaul: 8.2 quarts

② Fluid change: .45 quart
 Overhaul: .48 quart

③ Fluid change: 2.67 quarts
 Overhaul: 2.93 quarts

④ Fluid change: 1.85 gallons
 Overhaul (engine): 2.22 gallons

22140_ARDX_C0004

FLUID SPECIFICATIONS

Year	Model	Engine Displacement Liters (VIN)	Engine Oil	Auto. Trans.	Drive Axle	Power Steering Fluid	Brake Master Cylinder
2007	RDX	2.3 (K23A1)	①	②	③	④	DOT 3
2008	RDX	2.3 (K23A1)	①	②	③	④	DOT 3

DOT: Department Of Transportation

Note: If specification disagrees with specification in owners manual, use specification in owners manaual

① Mobil 1 (P/N 5w-30-MB1-000) or equivalent that meets Acure HTO-06 standard

② Acura ATF-Z1 fluid

③ Transfer case: API classified GL4 or GL5 only. SAE 90 or SAE 80W-90 viscosity.
 Rear differential: Acura ATF-Z1 fluid

④ Acura power steering fluid

22140_ARDX_C0014

VALVE SPECIFICATIONS

Year	Engine Displacement Liters (VIN)	Seat Angle (deg.)	Face Angle (deg.)	Spring Test Pressure (lbs. @ in.)	Spring Installed Height (in.)	Stem-to-Guide Clearance (in.) Intake	Stem-to-Guide Clearance (in.) Exhaust	Stem Diameter (in.) Intake	Stem Diameter (in.) Exhaust
2007	2.3 (K23A1)	NA	NA	NA	①	0.0012-0.0022	0.0022-0.0031	0.2156-0.2159	0.2146-0.2150
2008	2.3 (K23A1)	NA	NA	NA	①	0.0012-0.0022	0.0022-0.0031	0.2156-0.2159	0.2146-0.2150

NA: Not Available

① Valve spring free length:
 Intake: 1.874 in.
 Exhaust: 1.954 in.

22140_ARDX_C0005

CAMSHAFT SPECIFICATIONS
All measurements in inches unless noted

Year	Model	Engine Displacement Liters (VIN)	Journal Dia.	Brg. Oil Clearance	Shaft End-play	Circle Runout	Lobe Height Intake	Lobe Height Exhaust
2007	RDX	2.3 (K23A1)	NA	①	0.0020-0.0080	NA	②	1.3422
2008	RDX	2.3 (K23A1)	NA	①	0.0020-0.0080	NA	②	1.3422

NA: Not Available

① No. 1 Journal: 0.001-0.003 inch
All others: 0.002-0.004 inch

② Intake primary: 1.3356 inch. Intake secondary: 1.1668 inch.
All others: 0.002-0.004 inch

22140_ARDX_C0006

CRANKSHAFT AND CONNECTING ROD SPECIFICATIONS
All measurements are given in inches

Year	Engine Displacement Liters (VIN)	Crankshaft Main Brg. Journal Dia.	Crankshaft Main Brg. Oil Clearance	Crankshaft Shaft End-play	Crankshaft Thrust on No.	Connecting Rod Journal Diameter	Connecting Rod Oil Clearance	Connecting Rod Side Clearance
2007	2.3 (K23A1)	①	②	0.0040-0.0140	NA	1.8888 1.8898	0.0013-0.0025	0.0060-0.0140
2008	2.3 (K23A1)	①	②	0.0040-0.0140	NA	1.8888 1.8898	0.0013-0.0025	0.0060-0.0140

NA: Not Available

① Except No. 3: 2.1648-2.1657 inches
No. 3: 2.1644-2.1654 inches

② Except No. 3: 0.0007-0.0016 inches
No. 3: 0.0010-0.0019 inches

22140_ARDX_C0007

PISTON AND RING SPECIFICATIONS
All measurements are given in inches

Year	Engine Displacement Liters (VIN)	Piston Clearance	Ring Gap Top Compression	Ring Gap Bottom Compression	Ring Gap Oil Control	Ring Side Clearance Top Compression	Ring Side Clearance Bottom Compression	Ring Side Clearance Oil Control
2007	2.3 (K23A1)	0.0008-0.0016	0.0604-0.0616	0.0484-0.0488	0.0791-0.0795	①	②	③
2008	2.3 (K23A1)	0.0008-0.0016	0.0604-0.0616	0.0484-0.0488	0.0791-0.0795	①	②	③

① Ring to groove: 0.0018-0.0035 inch
Ring end gap: 0.008-0.012 inch

② Ring to groove: 0.0016-0.0026 inch
Ring end gap: 0.016-0.020 inch

③ Ring to groove: 0.0008-0.0020 inch
Ring end gap: 0.0004-0.012 inch

22140_ARDX_C0008

TORQUE SPECIFICATIONS
All readings in ft. lbs.

Year	Engine Displacement Liters (VIN)	Cylinder Head Bolts	Main Bearing Bolts	Rod Bearing Bolts	Crankshaft Damper Bolts	Flywheel Bolts	Manifold		Spark Plugs	Oil Pan Drain Plug
							Intake	Exhaust		
2007	2.3 (K23A1)	①	②	③	④	NA	⑤	⑥	13	29
2008	2.3 (K23A1)	①	②	③	④	NA	⑤	⑥	13	29

NA: Not Available

① Step 1: 29 ft. lbs.

Step 2: After tightening all bolts, in two steps (90 degrees per step). If using a new bolt tighten an extra 90 degrees

② Step 1: 22 ft. lbs.

Step 2: additional 67 degrees

③ Step 1: 22 ft. lbs.

Step 2: 90 degrees

④ Used bolt: 36 ft. lbs.

New bolt: 130 ft. lbs., loosen and retighten to 36 ft. lbs.

⑤ Tighten bolts and nuts in a crisscross pattern beginning with the inner bolt, in three stages to 16 ft. lbs.

⑥ Tighten bolts and nuts in a crisscross pattern beginning with the inner bolt, in three stages to 33 ft. lbs.

22140_ARDX_C0009

WHEEL ALIGNMENT

Year	Model		Caster		Camber		Toe-in (in.)
			Range (+/-Deg.)	Preferred Setting (Deg.)	Range (+/-Deg.)	Preferred Setting (Deg.)	
2007	RDX	F	1.00	1.00	1.00	30	0+/-1/16
		R	—	—	0	30	0+/-1/16
2008	RDX	F	1.00	1.00	1.00	30	0+/-1/16
		R	—	—	0	30	0+/-1/16

22140_ARDX_C0010

TIRE, WHEEL AND BALL JOINT SPECIFICATIONS

Year	Model	OEM Tires		Tire Pressures (psi)		Wheel Size	Ball Joint Inspection	Lug Nut (ft. lbs.)
		Standard	Optional	Front	Rear			
2007	RDX	P235/65R18	None	32	32	R18	NA	80
2008	RDX	P235/65R18	None	32	32	R18	NA	80

OEM: Original Equipment Manufacturer

PSI: Pounds Per Square Inch

NA: Not Available

22140_ARDX_C0011

BRAKE SPECIFICATIONS

All measurements in inches unless noted

Year	Model		Brake Disc			Brake Drum Diameter			Minimum Lining Thickness		Brake Caliper	
			Original Thickness	Minimum Thickness	Maximum Runout	Original Inside Diameter	Max. Wear Limit	Maximum Machine Diameter	Front	Rear	Bracket Bolts (ft. lbs.)	Mounting Bolts (ft. lbs.)
2007	RDX	F	①	1.020	0.0016	—	—	—	0.060	—	101	37
		R	②	0.300	NA	—	—	—	—	0.060	80	17
2008	RDX	F	①	1.020	0.0016	—	—	—	0.060	—	101	37
		R	②	0.300	NA	—	—	—	—	0.060	80	17

NA: Not Available

F: Front

R: Rear

① 1.09-1.11 inch

② 0.35-0.36 inch

22140_ARDX_C0012

SCHEDULED MAINTENANCE INTERVALS
ACURA—RDX

TO BE SERVICED	TYPE OF SERVICE	VEHICLE MILEAGE INTERVAL (x1000)															
		7.5	15	22.5	30	37.5	45	52.5	60	67.5	75	82.5	90	97.5	105	112.5	120
Accessory drive belts	I & A				✓				✓				✓				✓
Air cleaner element	R				✓				✓				✓				✓
Brake fluid	R	Every 3 years															
Brake hoses & lines (incl. ABS)	I		✓		✓		✓		✓		✓		✓		✓		✓
Cooling system hoses & connections	I		✓		✓		✓		✓		✓		✓		✓		✓
Engine coolant ①	R						✓						✓				
Engine oil	R	✓	✓	✓	✓	✓	✓	✓	✓	✓	✓	✓	✓	✓	✓	✓	✓
Engine oil and coolant levels	I	Inspect at each fuel stop															
Engine oil filter	R		✓		✓		✓		✓		✓		✓		✓		✓
Exhaust system	I		✓		✓		✓		✓		✓		✓		✓		✓
Fluid levels and condition	I		✓		✓		✓		✓		✓		✓		✓		✓
Front and rear brakes	I		✓		✓		✓		✓		✓		✓		✓		✓
Fuel lines & connection	I		✓		✓		✓		✓		✓		✓		✓		✓
Halfshaft boots	I		✓		✓		✓		✓		✓		✓		✓		✓
Idle speed	I & A														✓		
Parking brake system	I & A		✓		✓		✓		✓		✓		✓		✓		✓
Rear differential fluid	R	✓			✓		✓		✓				✓				✓
Rotate and inspect tires	I	✓	✓	✓	✓	✓	✓	✓	✓	✓	✓	✓	✓	✓	✓	✓	✓
Spark plugs	R														✓		
Supplemental Restrain System	I	Inspect the SRS 10 years after production															
Suspension components	I		✓		✓		✓		✓		✓		✓		✓		✓
Tie rod ends, steering gear box & boots	I		✓		✓		✓		✓		✓		✓		✓		✓
Timing belt	R														✓		
Transmission fluid	R						✓				✓				✓		
Valve clearance	I	Adjust if valves are noisy															
Water pump	S/I														✓		

R: Replace   I: Inspect   A: Adjust

① Every 12,000 miles or 10 years, then every 60,000 miles or 5 years

FREQUENT OPERATION MAINTENANCE (SEVERE SERVICE)

If a vehicle is operated under any of the following conditions it is considered severe service:

- Towing a trailer or using a camper or car-top carrier.
- Repeated short trips of less than 5 miles in temperatures below freezing, or trips of less than 10 miles in any temperature.
- Extensive idling or low-speed driving for long distances as in heavy commercial use, such as delivery, taxi or police cars.
- Operating on rough, muddy or salt-covered roads.
- Operating on unpaved or dusty roads.
- Driving in extremely hot (over 90°) conditions.

Air cleaner element: replace every 15,000 miles

Engine oil and filter: replace every 3750 miles or 6 months, whichever occurs first.

Timing belt: replace every 60,000 miles if the vehicle is regularly driven in temperatures above 110°F or below -20°F, or if frequently towing a trailer.

Transmission fluid: replace every 30,000 miles.

Rear differential fluid: replace every 60,000 miles.

Front and rear brakes: inspect every 7500 miles or 6 months, whichever occurs first.

Locks and hinges: lubricate every 15,000 miles.

Tie rods, steering gear box, boots: inspect every 7500 miles or 6 months, whichever occurs first.

Suspension components: inspect every 7500 miles or 6 months, whichever occurs first.

Halfshaft boots: inspect every 7500 miles or 6 months, whichever occurs first.

PRECAUTIONS

Before servicing any vehicle, please be sure to read all of the following precautions, which deal with personal safety, prevention of component damage, and important points to take into consideration when servicing a motor vehicle:

• Never open, service or drain the radiator or cooling system when the engine is hot; serious burns can occur from the steam and hot coolant.

• Observe all applicable safety precautions when working around fuel. Whenever servicing the fuel system, always work in a well-ventilated area. Do not allow fuel spray or vapors to come in contact with a spark, open flame, or excessive heat (a hot drop light, for example). Keep a dry chemical fire extinguisher near the work area. Always keep fuel in a container specifically designed for fuel storage; also, always properly seal fuel containers to avoid the possibility of fire or explosion. Refer to the additional fuel system precautions later in this section.

• Fuel injection systems often remain pressurized, even after the engine has been turned **OFF**. The fuel system pressure must be relieved before disconnecting any fuel lines. Failure to do so may result in fire and/or personal injury.

• Brake fluid often contains polyglycol ethers and polyglycols. Avoid contact with the eyes and wash your hands thoroughly after handling brake fluid. If you do get brake fluid in your eyes, flush your eyes with clean, running water for 15 minutes. If eye irritation persists, or if you have taken

brake fluid internally, IMMEDIATELY seek medical assistance.

• The EPA warns that prolonged contact with used engine oil may cause a number of skin disorders, including cancer. You should make every effort to minimize your exposure to used engine oil. Protective gloves should be worn when changing oil. Wash your hands and any other exposed skin areas as soon as possible after exposure to used engine oil. Soap and water, or waterless hand cleaner should be used.

• All new vehicles are now equipped with an air bag system, often referred to as a Supplemental Restraint System (SRS) or Supplemental Inflatable Restraint (SIR) system. The system must be disabled before performing service on or around system components, steering column, instrument panel components, wiring and sensors. Failure to follow safety and disabling procedures could result in accidental air bag deployment, possible personal injury and unnecessary system repairs.

• Always wear safety goggles when working with, or around, the air bag system. When carrying a non-deployed air bag, be sure the bag and trim cover are pointed away from your body. When placing a non-deployed air bag on a work surface, always face the bag and trim cover upward, away from the surface. This will reduce the motion of the module if it is accidentally deployed. Refer to the additional air bag system precautions later in this section.

• Clean, high quality brake fluid from a sealed container is essential to the safe and

proper operation of the brake system. You should always buy the correct type of brake fluid for your vehicle. If the brake fluid becomes contaminated, completely flush the system with new fluid. Never reuse any brake fluid. Any brake fluid that is removed from the system should be discarded. Also, do not allow any brake fluid to come in contact with a painted surface; it will damage the paint.

• Never operate the engine without the proper amount and type of engine oil; doing so WILL result in severe engine damage.

• Timing belt maintenance is extremely important. Many models utilize an interference-type, non-freewheeling engine. If the timing belt breaks, the valves in the cylinder head may strike the pistons, causing potentially serious (also time-consuming and expensive) engine damage. Refer to the maintenance interval charts for the recommended replacement interval for the timing belt, and to the timing belt section for belt replacement and inspection.

• Disconnecting the negative battery cable on some vehicles may interfere with the functions of the on-board computer system(s) and may require the computer to undergo a relearning process once the negative battery cable is reconnected.

• When servicing drum brakes, only disassemble and assemble one side at a time, leaving the remaining side intact for reference.

• Only an MVAC-trained, EPA-certified automotive technician should service the air conditioning system or its components.

BRAKES

ANTI-LOCK BRAKE SYSTEM (ABS)

GENERAL INFORMATION

PRECAUTIONS

• Certain components within the ABS system are not intended to be serviced or repaired individually.

• Do not use rubber hoses or other parts not specifically specified for and ABS system. When using repair kits, replace all parts included in the kit. Partial or incorrect repair may lead to functional problems and require the replacement of components.

• Lubricate rubber parts with clean, fresh brake fluid to ease assembly. Do not

use shop air to clean parts; damage to rubber components may result.

• Use only DOT 3 brake fluid from an unopened container.

• If any hydraulic component or line is removed or replaced, it may be necessary to bleed the entire system.

• A clean repair area is essential. Always clean the reservoir and cap thoroughly before removing the cap. The slightest amount of dirt in the fluid may plug an orifice and impair the system function. Perform repairs after components have been thoroughly cleaned; use only denatured alcohol

to clean components. Do not allow ABS components to come into contact with any substance containing mineral oil; this includes used shop rags.

• The Anti-Lock control unit is a microprocessor similar to other computer units in the vehicle. Ensure that the ignition switch is **OFF** before removing or installing controller harnesses. Avoid static electricity discharge at or near the controller.

• If any arc welding is to be done on the vehicle, the control unit should be unplugged before welding operations begin.

BRAKES

BLEEDING THE BRAKE SYSTEM

BLEEDING PROCEDURE

BLEEDING PROCEDURE

See Figure 1.

➡Do not reuse the drained fluid. Use only clean Honda DOT 3 Brake Fluid from an unopened container. Using a non-Honda brake fluid can cause corrosion and shorten the life of the system.

❊ WARNING

Make sure no dirt or other foreign matter is allowed to contaminate the brake fluid.

❊ WARNING

Do not spill brake fluid on the vehicle, it may damage the paint; if brake fluid does contact the

BLEEDING SEQUENCE:

② Front Right   ③ Rear Right

① Front Left   ④ Rear Left

42050_PILO_G0102

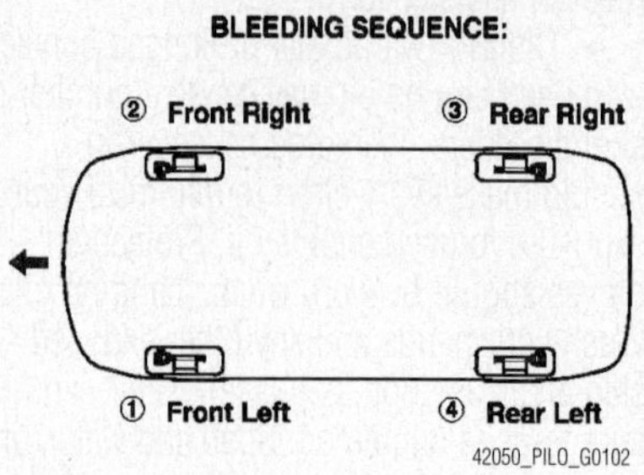

Fig. 1 Brake bleeding sequence

paint, wash it off immediately with water.

1. The reservoir on the master cylinder must be at the MAX (upper) level mark at

the start of the bleeding procedure and checked after bleeding each brake caliper. Add fluid as required.

2. Make sure the brake fluid level in the reservoir is at the MAX (upper) level line.

3. Slide a piece of clear plastic hose over the first bleed screw, and submerge the other end in a container of new brake fluid.

4. Have someone slowly pump the brake pedal several times, and then apply steady pressure.

5. Starting at the left-front, loosen the brake bleed screw to allow air to escape from the system. Then tighten the bleed screw securely.

6. Repeat the procedure for each wheel in the sequence shown following until air bubbles no longer appear in the fluid.

7. Refill the master cylinder reservoir to the MAX (upper) level line.

BRAKES

FRONT DISC BRAKES

❊ CAUTION

Dust and dirt accumulating on brake parts during normal use may contain asbestos fibers from production or aftermarket brake linings. Breathing excessive concentrations of asbestos fibers can cause serious bodily harm. Exercise care when servicing brake parts. Do not sand or grind brake lining unless equipment used is designed to contain the dust residue. Do not clean brake parts with compressed air or by dry brushing. Cleaning should be done by dampening the brake components with a fine mist of water, then wiping the brake components clean with a dampened cloth. Dispose of cloth and all residue containing asbestos fibers in an impermeable container with the appropriate label. Follow practices prescribed by the Occupational Safety and Health Administration (OSHA) and the Environmental Protection Agency (EPA) for the handling, processing, and disposing of dust or debris that may contain asbestos fibers.

BRAKE CALIPER

REMOVAL & INSTALLATION

1. Before servicing the vehicle, refer to the Precautions Section.

2. As required, remove a small amount of brake fluid from the reservoir using a suction pump.

3. Raise and safely support the vehicle.

4. Remove the tire and wheel assembly.

5. Remove the brake hose mounting bolt. Disconnect and plug the brake line hose at the caliper.

6. Remove the brake caliper mounting bolts.

7. Remove the caliper assembly from its mounting.

➡Do not allow the caliper to hang by the brake line hose, as damage to the hose may result.

8. As required, remove the disc brake pads and shims from the caliper.

To install:

9. Clean the caliper thoroughly; remove any dirt or dust. Check the brake rotor for grooves or cracks and machine or replace, as necessary.

10. Installation is the reverse of the removal procedure.

11. Be sure to fill the brake system using the proper grade and type brake fluid.

12. Bleed the brake system.

13. Check for leaks and correct as required.

DISC BRAKE PADS

REMOVAL & INSTALLATION

See Figure 2.

1. Before servicing the vehicle, refer to the Precautions Section.

2. As required, remove a small amount of brake fluid from the reservoir using a suction pump.

3. Raise and safely support the vehicle.

4. Remove the tire and wheel assembly.

5. Remove the flange bolt while holding the caliper pin with a wrench.

➡Be careful not to damage the pin boot, and pivot the caliper up and out of the way.

6. Remove the pad shims and pads. Remove the pad retainers.

To install:

7. Install the pad retainers.

➡Apply molybdenum brake grease to both surfaces of the shims and the back of the disc brake pads, prior to installation.

8. Install the pads and shims. See illustration for proper positioning.

9. Use a suitable tool to push caliper piston into its bore and enable the caliper to fit over the pads. Lubricate the piston boot with silicon grease. Avoid twisting the boot.

10. Continue the installation in the reverse order of the removal procedure.

11. Add brake fluid to the master cylinder reservoir. Depress the brake pedal several times to seat the pads. Bleed the brakes if necessary.

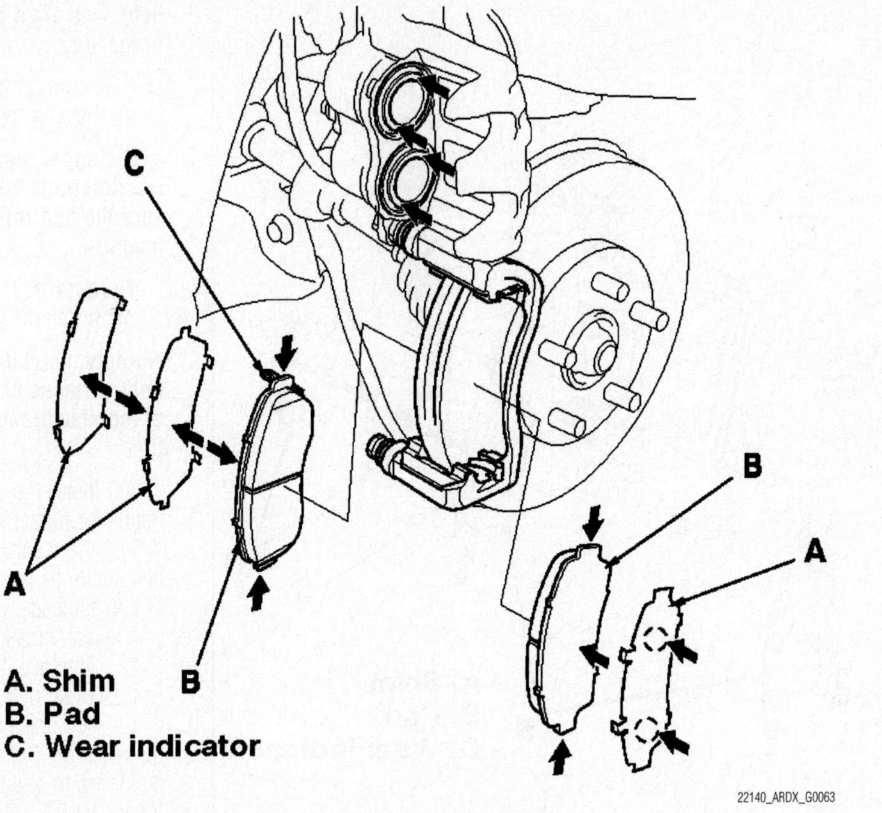

A. Shim
B. Pad
C. Wear indicator

22140_ARDX_G0063

Fig. 2 Front brake pad positioning

BRAKES   **REAR DISC BRAKES**

✳✳ CAUTION

Dust and dirt accumulating on brake parts during normal use may contain asbestos fibers from production or aftermarket brake linings. Breathing excessive concentrations of asbestos fibers can cause serious bodily harm. Exercise care when servicing brake parts. Do not sand or grind brake lining unless equipment used is designed to contain the dust residue. Do not clean brake parts with compressed air or by dry brushing. Cleaning should be done by dampening the brake components with a fine mist of water, then wiping the brake components clean with a dampened cloth. Dispose of cloth and all residue containing asbestos fibers in an impermeable container with the appropriate label. Follow practices prescribed by the Occupational Safety and Health Administration (OSHA) and the Environmental Protection Agency (EPA) for the handling, processing, and disposing of dust or

debris that may contain asbestos fibers.

DISC BRAKE PADS

REMOVAL & INSTALLATION

See Figures 3 and 4.

1. Before servicing the vehicle, refer to the Precautions Section.

2. As required, remove a small amount of brake fluid from the reservoir using a suction pump.
3. Raise and safely support the vehicle.
4. Remove the tire and wheel assembly.
5. Remove the flange nuts and remove the brake hose mounting bracket.
6. Remove the flange bolt while holding the caliper pin with a wrench.

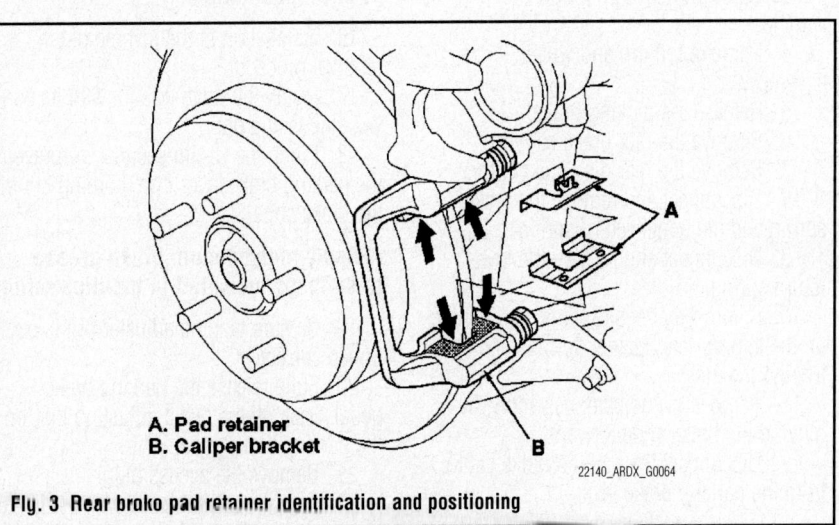

A. Pad retainer
B. Caliper bracket

22140_ARDX_G0064

Fig. 3 Rear brake pad retainer identification and positioning

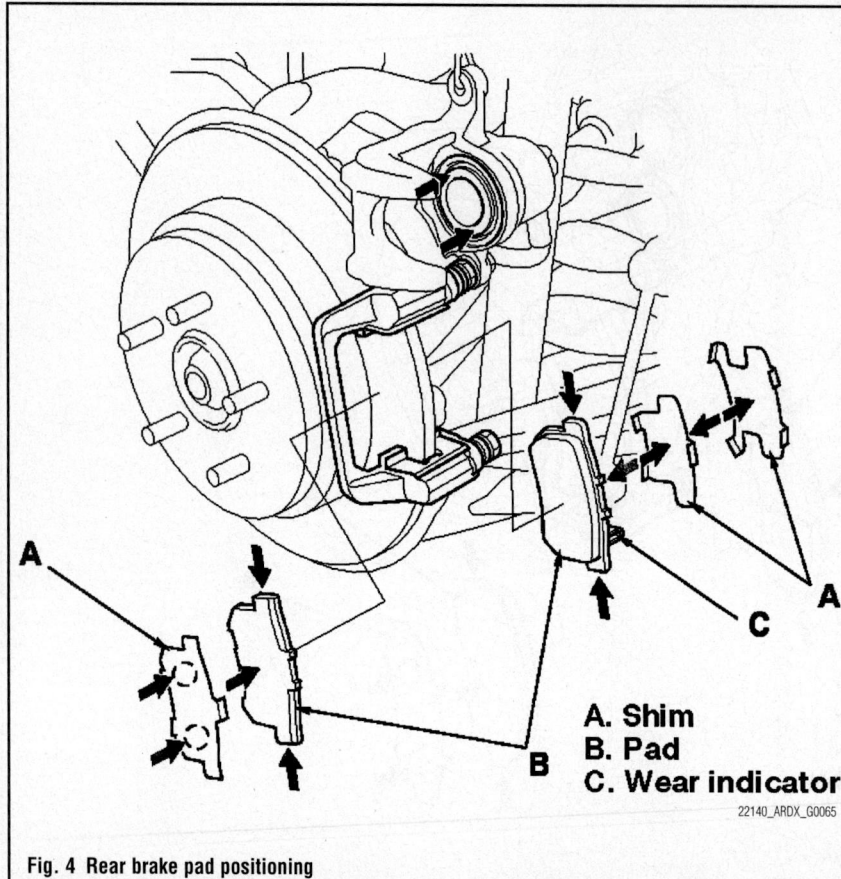

A. Shim
B. Pad
C. Wear indicator

22140_ARDX_G0065

Fig. 4 Rear brake pad positioning

➡ Be careful not to damage the pin boot, and pivot the caliper up and out of the way.

7. Remove the pad shims and pads.
8. Remove the pad retainers.

➡ The upper and lower pad retainers are different. During installation, make sure the pad retainers are in the proper position.

To install:

9. Install the pad retainers.

➡ Apply molybdenum brake grease to both surfaces of the shims and the back of the disc brake pads, prior to installation.

10. Install the pads and shims. See illustration for proper positioning.
11. Use a suitable tool to push caliper piston into its bore and enable the caliper to fit over the pads. Lubricate the piston boot with silicon grease. Avoid twisting the boot.
12. Continue the installation in the reverse order of the removal procedure.
13. Add brake fluid to the master cylinder reservoir. Depress the brake pedal several times to seat the pads. Bleed the brakes if necessary.

BRAKES
PARKING BRAKE

PARKING BRAKE SHOES

REMOVAL & INSTALLATION

See Figures 5 and 6.

1. Before servicing the vehicle, refer to the Precautions Section.
2. As required, remove a small amount of brake fluid from the reservoir using a suction pump.
3. Raise and safely support the vehicle.
4. Remove the tire and wheel assembly.
5. Release the parking brake.
6. Remove the rear brake caliper.
7. Remove the rear disc brake rotor.
8. Disconnect and remove the brake spring and the upper return spring.
9. Disconnect and remove the lower return spring.
10. Remove the tension pins, by pushing on the appropriate retainer springs and turning the pin.
11. Remove the adjuster assembly, by moving the brake shoe forward.
12. Disconnect the parking brake cable from the parking brake lever.

13. Disconnect the rod spring and remove the strut.
14. Remove the parking brake shoes.
15. Remove the U-clip, wave washer, parking brake lever and the pivot pin from the shoe.

To install:

➡ Apply molybdenum brake grease to the sliding surface of the pivot pin, prior to installation.

16. Installation is the reverse of the removal procedure.
17. Install the wave washer with its convex side facing out.
18. Pinch the U-clip securely to prevent the parking brake lever from coming out of the brake shoe.

➡ Apply molybdenum brake grease to surfaces indicated in the illustration.

19. Be sure that the adjuster pin is positioned correctly.
20. Fully release the parking brake pedal. Back off the pedal adjusting nut, on the parking brake pedal.
21. Remove the access plug.
22. Turn the ratchet teeth on the adjuster

nut until the shoes lock against the parking brake drum.
23. Back off the adjuster eight clicks. Install the access plug.

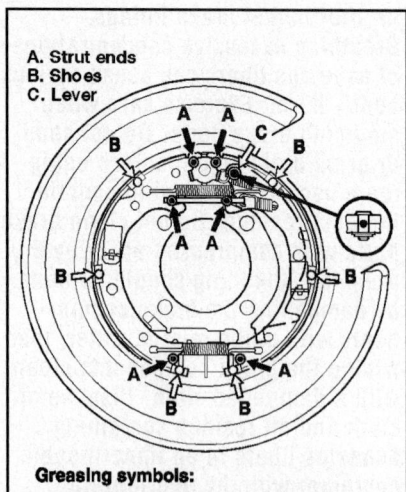

A. Strut ends
B. Shoes
C. Lever

Greasing symbols:
➡ ● Brake shoe ends and strut ends
⇨ ○ Sliding surfaces of the shoe
⇨ ● Pivot of parking brake lever

22140_ARDX_G0066

Fig. 5 Rear parking brake grease application points

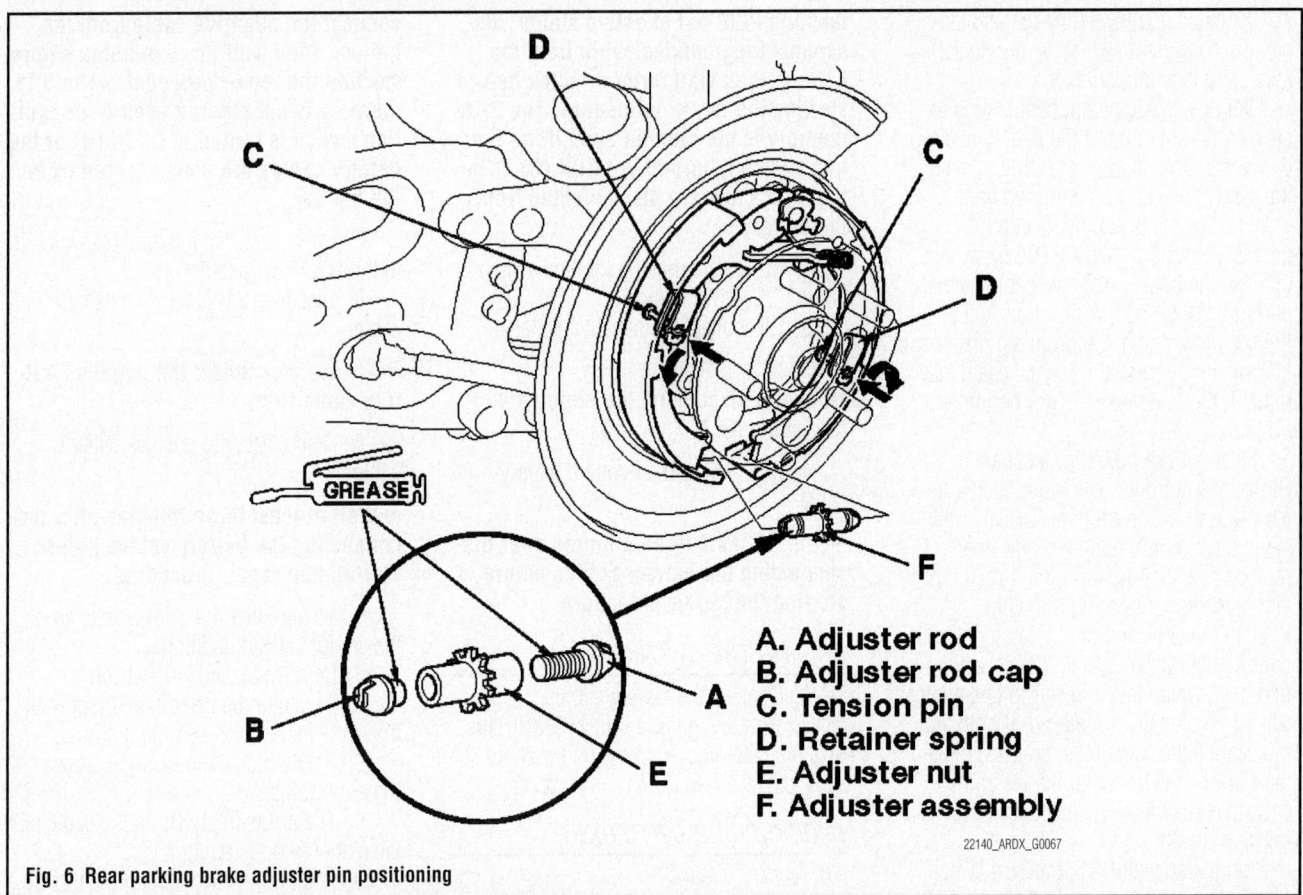

A. Adjuster rod
B. Adjuster rod cap
C. Tension pin
D. Retainer spring
E. Adjuster nut
F. Adjuster assembly

22140_ARDX_G0067

Fig. 6 Rear parking brake adjuster pin positioning

CHASSIS ELECTRICAL | AIR BAG (SUPPLEMENTAL RESTRAINT SYSTEM)

GENERAL INFORMATION

✳✳ CAUTION

These vehicles are equipped with an air bag system. The system must be disarmed before performing service on, or around, system components, the steering column, instrument panel components, wiring and sensors. Failure to follow the safety precautions and the disarming procedure could result in accidental air bag deployment, possible injury and unnecessary system repairs.

SERVICE PRECAUTIONS

Disconnect and isolate the battery negative cable before beginning any airbag system component diagnosis, testing, removal, or installation procedures. Allow system capacitor to discharge for two minutes before beginning any component service. This will disable the airbag system. Failure to disable the airbag system may result in accidental airbag deployment, personal injury, or death.

Do not place an intact undeployed airbag face down on a solid surface. The airbag will propel into the air if accidentally deployed and may result in personal injury or death.

When carrying or handling an undeployed airbag, the trim side (face) of the airbag should be pointing towards the body to minimize possibility of injury if accidental deployment occurs. Failure to do this may result in personal injury or death.

Replace airbag system components with OEM replacement parts. Substitute parts may appear interchangeable, but internal differences may result in inferior occupant protection. Failure to do so may result in occupant personal injury or death.

Wear safety glasses, rubber gloves, and long sleeved clothing when cleaning powder residue from vehicle after an airbag deployment. Powder residue emitted from a deployed airbag can cause skin irritation. Flush affected area with cool water if irritation is experienced. If nasal or throat irritation is experienced, exit the vehicle for fresh air until the irritation ceases. If irritation continues, see a physician.

Do not use a replacement airbag that is not in the original packaging. This may result in improper deployment, personal injury, or death.

The factory installed fasteners, screws and bolts used to fasten airbag components have a special coating and are specifically designed for the airbag system. Do not use substitute fasteners. Use only original equipment fasteners listed in the parts catalog when fastener replacement is required.

During, and following, any child restraint anchor service, due to impact event or vehicle repair, carefully inspect all mounting hardware, tether straps, and anchors for proper installation, operation, or damage. If a child restraint anchor is found damaged in any way, the anchor must be replaced. Failure to do this may result in personal injury or death.

Deployed and non-deployed airbags may or may not have live pyrotechnic material within the airbag inflator.

Do not dispose of driver/passenger/curtain airbags or seat belt tensioners unless you are sure of complete deployment. Refer to the Hazardous Substance Control System for proper disposal.

Dispose of deployed airbags and tensioners consistent with state, provincial, local, and federal regulations.

After any airbag component testing or service, do not connect the battery negative cable. Personal injury or death may result if the system test is not performed first.

If the vehicle is equipped with the Occupant Classification System (OCS), do not connect the battery negative cable before performing the OCS Verification Test using the scan tool and the appropriate diagnostic information. Personal injury or death may result if the system test is not performed properly.

Never replace both the Occupant Restraint Controller (ORC) and the Occupant Classification Module (OCM) at the same time. If both require replacement, replace one, then perform the Airbag System test before replacing the other.

Both the ORC and the OCM store Occupant Classification System (OCS) calibration data, which they transfer to one another when one of them is replaced. If both are replaced at the same time, an irreversible fault will be set in both modules and the OCS may malfunction and cause personal injury or death.

If equipped with OCS, the Seat Weight Sensor is a sensitive, calibrated unit and must be handled carefully. Do not drop or handle roughly. If dropped or damaged, replace with another sensor. Failure to do so may result in occupant injury or death.

If equipped with OCS, the front passenger seat must be handled carefully as well. When removing the seat, be careful when setting on floor not to drop. If dropped, the sensor may be inoperative, could result in occupant injury, or possibly death.

If equipped with OCS, when the passenger front seat is on the floor, no one should sit in the front passenger seat. This uneven force may damage the sensing ability of the seat weight sensors. If sat on and damaged, the sensor may be inoperative, could result in occupant injury, or possibly death.

DISARMING THE SYSTEM

1. Before servicing the vehicle, refer to the Precautions Section.

➡ **Before disconnecting the battery cables make sure you have the anti theft codes for the audio/navigation system.**

➡ **Except when doing electrical inspections, always turn the ignition switch to LOCK (0), ground the SCS line with the Honda Diagnostic Service (HDS) tool to** take the PCM out of active status, disconnect the negative cable from the battery, then wait three minutes before starting the repair procedure. The SRS memory is not cleared even if the ignition switch is turned to LOCK (0) or the battery cables are disconnected from the battery.

2. Make sure that the ignition switch is in the LOCK (0) position.
3. Disconnect the negative battery cable.

➡ **Always disconnect the negative battery cable first.**

4. Disconnect the positive battery cable.

➡ **Wait at least three minutes after disconnecting the battery cables before starting the repair procedure.**

ARMING THE SYSTEM

1. Connect the battery cables.
2. Turn the ignition switch ON (II), the SRS indicator should come on for about six seconds, and then go off.

CLOCKSPRING CENTERING

See Figures 7 and 8.

1. Before servicing the vehicle, refer to the Precautions Section.

➡ **Before disconnecting the battery cables make sure you have the anti theft codes for the audio/navigation system.**

➡ **Except when doing electrical inspections, always turn the ignition switch to LOCK (0), ground the SCS line with the Honda Diagnostic Service (HDS) tool to take the PCM out of active status, dis-** connect the negative cable from the battery, then wait three minutes before starting the repair procedure. The SRS memory is not cleared even if the ignition switch is turned to LOCK (0) or the battery cables are disconnected from the battery.

2. Make sure that the ignition switch is in the LOCK (0) position.
3. Disconnect the negative battery cable.

➡ **Always disconnect the negative battery cable first.**

4. Disconnect the positive battery cable.

➡ **Wait at least three minutes after disconnecting the battery cables before starting the repair procedure.**

5. Be sure that the front wheels are in the straight ahead position.
6. Remove the steering wheel.
7. To center the cable reel, first rotate the cable reel clockwise until it stops.
8. Rotate it counterclockwise about three full turns.
9. The arrow (mark B) on the cable reel label should point straight up.
10. Position the two tabs of the turn signal canceling sleeve, as shown in the illustration.
11. Install the steering wheel onto the steering column shaft, making sure that the steering wheel hub engages the pins of the cable reel and the tabs of the turn signal canceling sleeve.

➡ **Do not tap on the steering wheel of the steering column shaft when installing the steering wheel.**

12. Install the steering wheel. Tighten the retaining bolt to 29 ft. lbs.

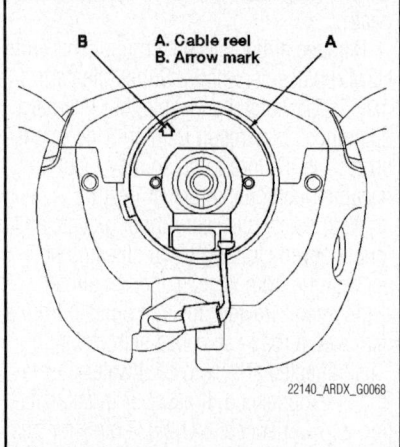

Fig. 7 Cable reel arrow location and positioning

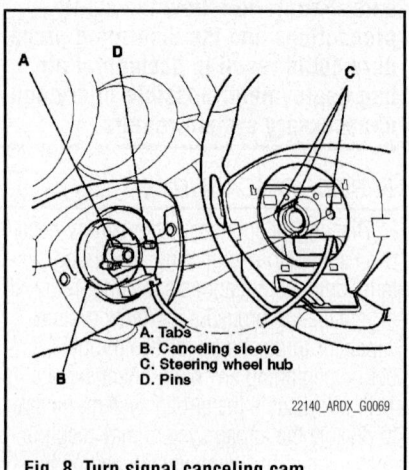

Fig. 8 Turn signal canceling cam positioning

DRIVETRAIN

AUTOMATIC TRANSMISSION ASSEMBLY

REMOVAL & INSTALLATION

See Figures 9 through 14.

1. Before servicing the vehicle, refer to the Precautions Section.

➡**Before disconnecting the battery cables make sure you have the anti theft codes for the audio/navigation system.**

➡**Except when doing electrical inspections, always turn the ignition switch to LOCK (O), ground the SCS line with the Honda Diagnostic Service (HDS) tool to take the PCM out of active status, dis-**

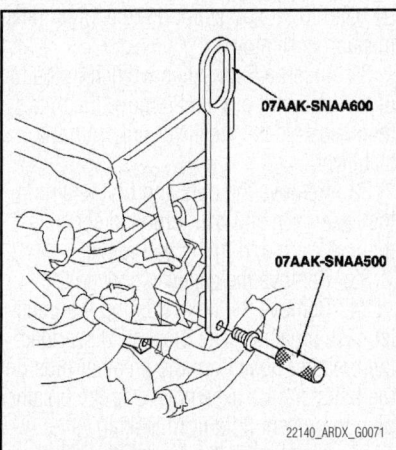

22140_ARDX_G0071

Fig. 9 Engine support tool installation— part one

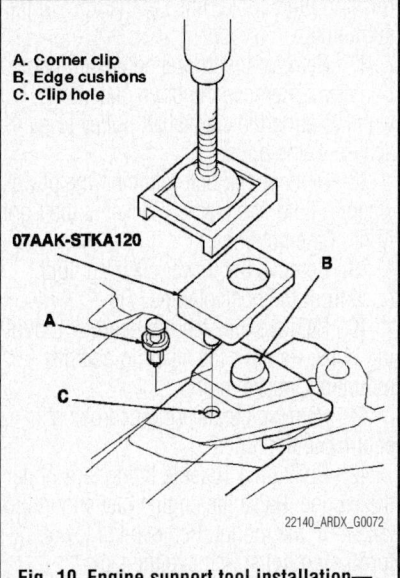

A. Corner clip
B. Edge cushions
C. Clip hole

07AAK-STKA120

22140_ARDX_G0072

Fig. 10 Engine support tool installation— part two

connect the negative cable from the battery, then wait three minutes before starting the repair procedure. The SRS memory is not cleared even if the ignition switch is turned to LOCK (O) or the battery cables are disconnected from the battery.

2. Make sure that the ignition switch is in the LOCK (O) position.
3. Disconnect the negative battery cable.

➡**Always disconnect the negative battery cable first.**

4. Disconnect the positive battery cable.

➡**Wait at least three minutes after disconnecting the battery cables before starting the repair procedure.**

5. Position the hood in the wide open position.
6. Remove the battery from the vehicle.
7. Remove the charge air cooler cover.
8. Remove the air cleaner. Remove the battery base.
9. Raise and support the vehicle safely.
10. Remove the front tire and wheel assemblies.
11. Remove the splash shield.
12. Drain and properly dispose of the transmission fluid. Reinstall the drain plug, using a new washer.
13. On 2007 vehicles, disconnect the transmission range switch connector.
14. Remove the transmission ground cable. Remove the harness clamp from the bracket.

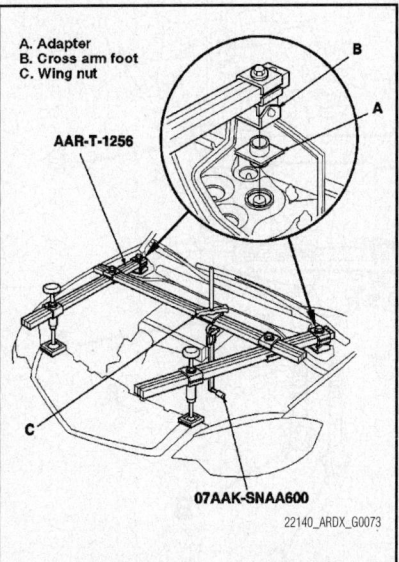

A. Adapter
B. Cross arm foot
C. Wing nut

AAR-T-1256

07AAK-SNAA600

22140_ARDX_G0073

Fig. 11 Engine support tool installation— part three

15. Remove the nuts securing the shift cable bracket.
16. On 2007 vehicles remove the spring clip/washer and the control pin. Then separate the shift cable end from the control lever.
17. On 2008 vehicles remove the spring clip and the control pin. Then separate the shift cable end from the control lever.
18. Check the bushing in the shift cable end for proper fit and wear, replace as required.
19. On 2008 vehicles, disconnect the transmission range switch connector.
20. Disconnect the input shaft speed sensor connector, the output shaft speed sensor connector, the third clutch transmission fluid pressure switch connector and the ATF temperature sensor connector.
21. Disconnect the A/T clutch pressure control solenoid valve connector, the solenoid valve connectors, the fourth clutch transmission fluid pressure switch connector and the shift solenoid harness connector.
22. Remove the bolt retaining the harness cover.
23. Disconnect the transmission fluid hoses from the transmission lines. Plug the lines to prevent fluid loss.
24. Remove the breather hose from the hose clip. Remove the cooler bypass hoses from the clamps. Remove the bolt retaining the clamp bracket.
25. Remove the bolts retaining the harness clamp/junction box bracket from the transmission housing.

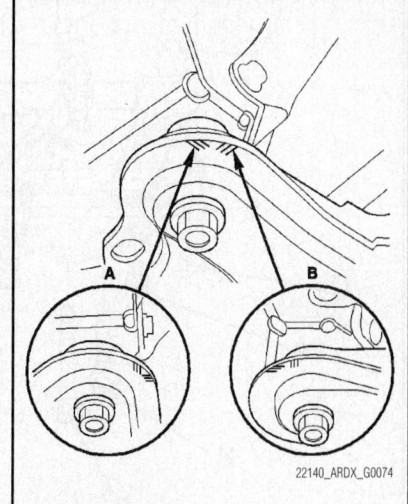

A   B

22140_ARDX_G0074

Fig. 12 Subframe reference mark locations

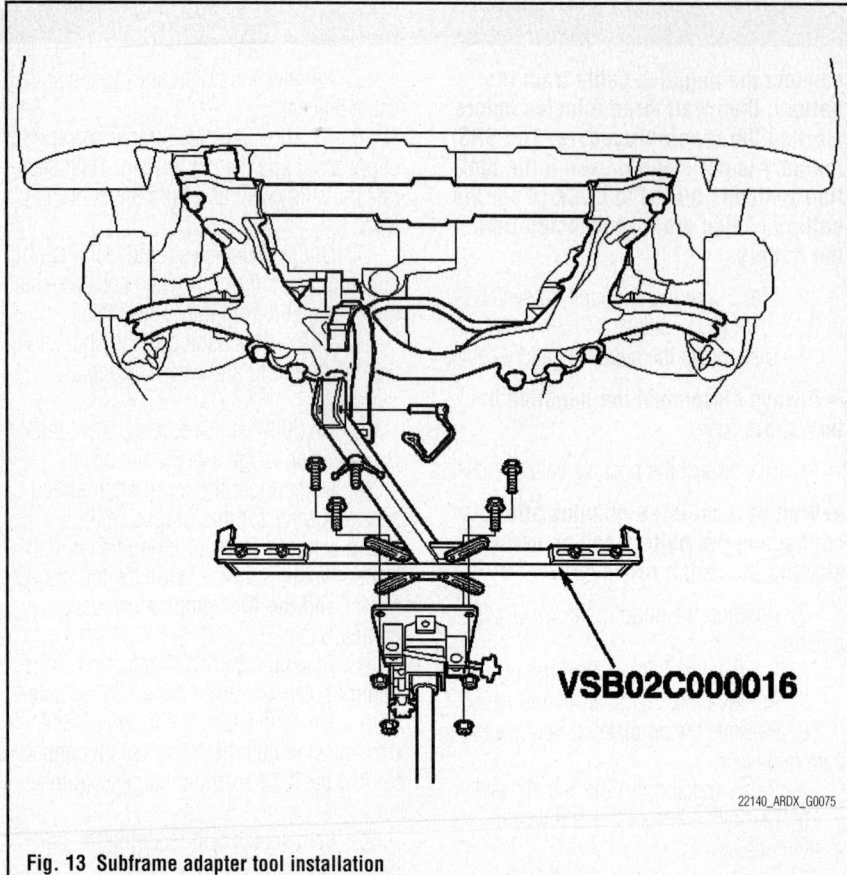

VSB02C000016

22140_ARDX_G0075

Fig. 13 Subframe adapter tool installation

A. Bracket mounting bolt
B. Bolt (B)
C. Bolt (C)
D. Subframe

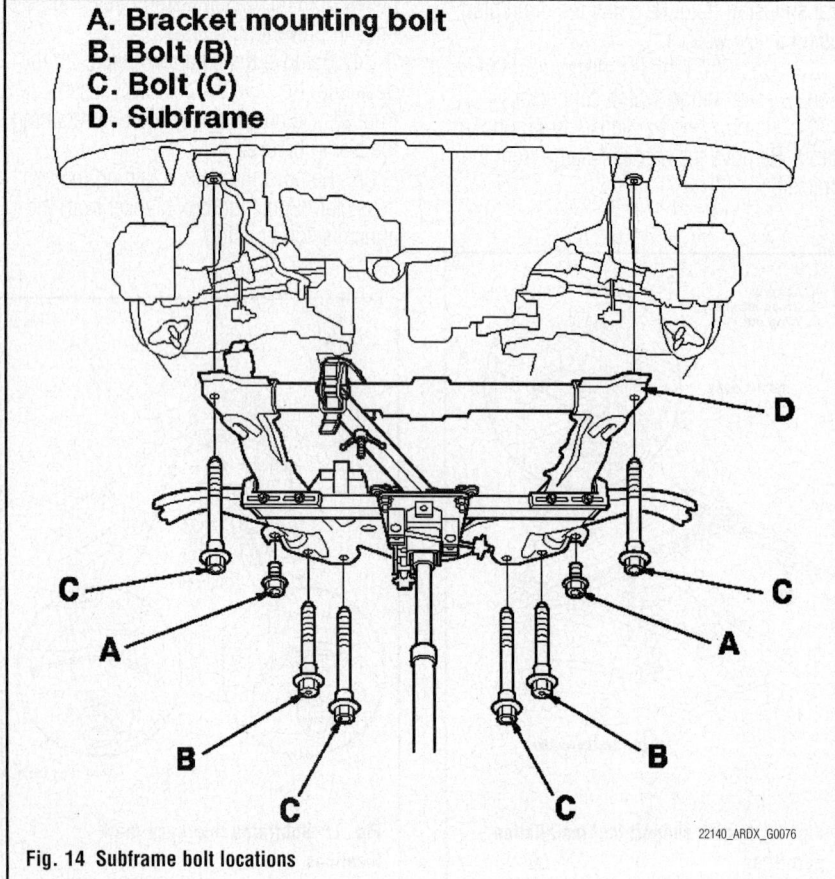

22140_ARDX_G0076

Fig. 14 Subframe bolt locations

26. Remove the hose from the clamp on the intake air duct. Remove the air intake duct.

27. Attach the engine support eyelet (07AAK-SNAA600) to the cylinder head, using the support bolt (07AAK-SNAA500). Hand tighten the bolt.

28. Remove the corner clip, from the front grille on both sides. Place the engine support mount pads (07AAK-STKA120) over the hood edge cushions and align the pins with the clip hole.

29. Remove both lids from the front damper flange nuts from the cowl cover. Use the adapter between the cross arm foot and the body when positioning the engine support hanger.

30. Carefully position the engine support hanger (AAR-T-1256) on the vehicle and attach the hook to the slotted hole in the engine support eyelet. Tighten the wing nut by hand to lift and support the engine/transmission assembly.

31. Insert a 6mm Allen wrench in the top of the ball joint pin, and remove the nuts, then separate the stabilizer link from the stabilizer.

32. Remove the nuts and bolt retaining the lower arm and ball joint and separate the lower arms from the ball joints.

33. Remove the catalytic converter.

34. Remove the bolt retaining the transfer case breather hose bracket. Disconnect the breather hose from the breather pipe on the transfer case assembly. Cap the breather pipe to prevent fluid from leaking.

35. Matchmark the driveshaft. Remove the retaining bolts and remove the driveshaft from the companion flange.

36. Remove the bolts retaining the transmission fluid cooler lines at the front of the subframe.

37. Remove the torque converter inspection cover. Remove the drive plate bolts, while rotating the crankshaft pulley to gain access to the bolts.

38. Remove the bolt retaining the power steering fluid line clamp bracket at the right front of the subframe.

39. Remove the power steering fluid hose from the hose clamps.

40. Remove the steering gearbox mounting bolts. Remove the steering gearbox mounting bracket bolts.

41. Remove the steering gearbox stiffener mounting bolts.

42. Position a suitable lifting jack under the engine. Raise the engine just enough to take it off the mount. Remove the lower torque rod bolts. Remove the jack.

43. Remove the driveshaft protector.

44. Place reference marks on the body

across the marks (A) and (B) on the edge of the subframe as indicated in the illustration.

45. Attach the front subframe adapter (VSB02C000016) to the subframe, by looping the strap over the front of the subframe. Secure the strap.

46. Raise the jack and line up the slots in the arms with the bolt holes on the corner of the jack base. Tighten the bolts.

47. Remove the lower arm bracket mounting bolts (A). Remove the six bolts (B) and (C) securing the front subframe, and lower the subframe.

➡**Hang the steering box from the body with a strap.**

48. Disconnect the second clutch transmission fluid pressure switch connector.

49. Remove the halfshafts from the differential and intermediate shaft. Place a plastic bag around the components to prevent debris from sticking to the splines.

50. Remove the intermediate shaft heat shield. Remove the intermediate shaft.

➡**Coat all machined surfaces with clean engine oil. Put plastic bags over the components to prevent debris from sticking to the shaft ends.**

51. Remove the upper transmission housing retaining bolts.

52. Position a suitable jack under the transmission. Raise the transmission just enough to take it off the mount. Remove the transmission mount bracket bolt and nuts. Remove the jack.

53. Remove the front transmission housing retaining bolts.

54. Remove the rear transmission housing retaining bolts.

55. Lower the transmission by loosening the wing nut on the engine support hanger. Tilt the engine just enough for the transmission to clear the side frame.

56. Position a jack under the transmission.

57. Remove the lower transmission housing retaining bolts.

58. Slide the transmission away from the engine to remove it from the vehicle.

59. Install the driveshaft protector, as required.

60. Remove the torque converter and dowel pins, as required.

➡**Inspect the drive plate, replace as required.**

61. Remove the clamp bracket from the transmission.

62. Remove the transfer case from the transmission, as required.

To install:

63. Installation is the reverse of the removal procedure.

64. Tighten the transfer case to transmission retaining bolts to 38 ft. lbs.

65. Tighten the lower transmission housing retaining bolts to 47 ft. lbs.

66. Tighten the rear transmission housing retaining bolts to 47 ft. lbs.

67. Tighten the front transmission housing retaining bolts to 47 ft. lbs.

68. Tighten the upper transmission housing retaining bolt to 47 ft. lbs.

69. When securing the transmission mount bracket on the transmission housing be sure to use a new mounting bolt. Tighten the bolt and nuts to 61 ft. lbs.

70. Be sure to use new set rings when installing the halfshafts.

71. Loosely install the new subframe bolts. When installing the subframe tighten (B) bolts to 43 ft. lbs. and (A) bolts to 76 ft. lbs. Be sure to use new bolts. Align all reference marks on the front subframe with the body, and then tighten the mounting bolts on the subframe to specification.

72. Tighten the lower torque rod retaining bolts to 47 ft. lbs. Be sure to use new bolts.

73. Tighten the steering gearbox bracket retaining bolts to 54 ft. lbs.

74. Tighten the eight torque converter to drive plate retaining bolts to 9 ft. lbs, using a crisscross tightening sequence.

75. Tighten the driveshaft to transfer case companion flange retaining bolts to 53 ft. lbs. Be sure to use the reference mark made during the removal procedure for proper alignment.

76. Refill the transmission with the proper grade and type transmission fluid.

77. Check and adjust the front end alignment.

78. Perform the power window reset procedure.

79. Road test the vehicle, correct problems as required.

TRANSFER CASE ASSEMBLY

REMOVAL & INSTALLATION

1. Before servicing the vehicle, refer to the Precautions Section.

➡**Before disconnecting the battery cables make sure you have the anti theft codes for the audio/navigation system.**

➡**Except when doing electrical inspections, always turn the ignition switch to**

LOCK (O), ground the SCS line with the Honda Diagnostic Service (HDS) tool to take the PCM out of active status, disconnect the negative cable from the battery, then wait three minutes before starting the repair procedure. The SRS memory is not cleared even if the ignition switch is turned to LOCK (O) or the battery cables are disconnected from the battery.

2. Make sure that the ignition switch is in the LOCK (O) position.

3. Disconnect the negative battery cable.

➡**Always disconnect the negative battery cable first.**

4. Disconnect the positive battery cable.

➡**Wait at least three minutes after disconnecting the battery cables before starting the repair procedure.**

5. Raise and support the vehicle safely.

6. Shift the transmission into Neutral.

7. Remove the drain plug and drain the transmission fluid. Dispose of used fluid properly. Install the drain plug using a new sealing washer.

8. Remove the catalytic converter.

9. Remove the bolt retaining the transfer case breather hose bracket. Disconnect the breather hose from the breather pipe on the transfer case assembly. Cap the breather pipe to prevent fluid from leaking.

10. Matchmark the driveshaft. Remove the retaining bolts and remove the driveshaft from the companion flange.

11. Remove the transfer case retaining bolts. Remove the transfer case from the vehicle.

To install:

12. Installation is the reverse of the removal procedure.

13. Tighten the transfer case to transmission retaining bolts to 38 ft. lbs.

14. Tighten the driveshaft to transfer case companion flange retaining bolts to 53 ft. lbs. Be sure to use the reference mark made during the removal procedure for proper alignment.

15. Refill the transmission with the proper grade and type transmission fluid.

16. Refill the transfer case with the proper grade and type fluid, as required.

17. Road test the vehicle, correct problems as required.

FRONT HALFSHAFT

REMOVAL & INSTALLATION

See Figures 15 through 20.

1. Before servicing the vehicle, refer to the Precautions Section.

➡**Before disconnecting the battery cables make sure you have the anti theft codes for the audio/navigation system.**

➡**Except when doing electrical inspections, always turn the ignition switch to LOCK (O), ground the SCS line with the Honda Diagnostic Service (HDS) tool to take the PCM out of active status, disconnect the negative cable from the battery, then wait three minutes before starting the repair procedure. The SRS memory is not cleared even if the ignition switch is turned to LOCK (O) or the battery cables are disconnected from the battery.**

2. Make sure that the ignition switch is in the LOCK (O) position.
3. Disconnect the negative battery cable.

➡**Always disconnect the negative battery cable first.**

4. Disconnect the positive battery cable.

➡**Wait at least three minutes after disconnecting the battery cables before starting the repair procedure.**

5. Raise and support the vehicle safely.
6. Remove the drain plug and drain the

transmission fluid. Dispose of used fluid properly. Install the drain plug using a new sealing washer.

7. Pry up the locking tab on the spindle nut and remove the nut.
8. Remove the nuts and bolts and separate the lower arm, using a prybar.
9. Loosen the halfshaft outboard joint from the front hub, using a plastic hammer.

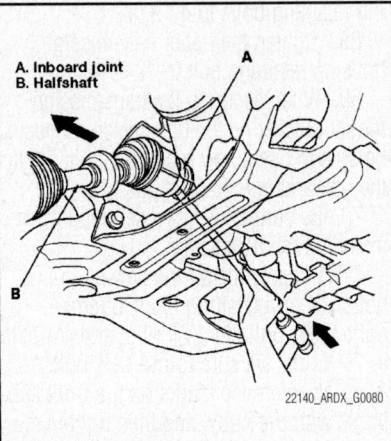

Fig. 16 Right halfshaft removal locating points

A. Inboard joint
B. Halfshaft

22140_ARDX_G0080

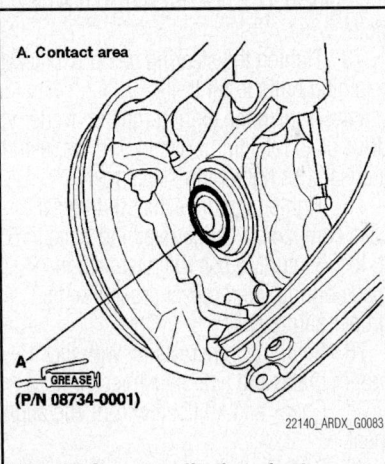

A. Contact area

GREASE
(P/N 08734-0001)

22140_ARDX_G0083

Fig. 17 Grease application—front halfshaft

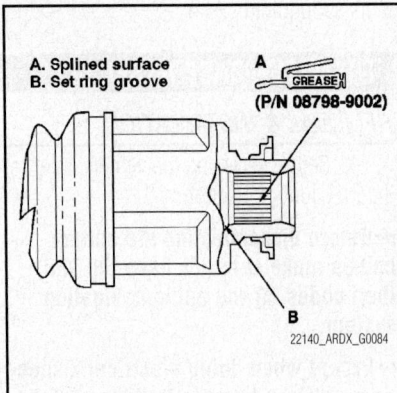

A. Splined surface
B. Set ring groove

GREASE
(P/N 08798-9002)

22140_ARDX_G0084

Fig. 18 Right halfshaft grease application

10. Pull the knuckle outward and separate the outboard joint from the front hub.
11. If removing the left halfshaft, install the prybar through the reference hole (A) of the front subframe. Pry the inboard joint (B) from the differential using a prybar. See the illustration for locating points.
12. If removing the right halfshaft, drive the inboard joint off of the intermediate shaft using a drift and hammer. Remove the halfshaft as an assembly.

➡**Do not pull the halfshaft or the inboard joint may come apart.**

13. Remove the set ring from the inboard ring.
14. Remove the set ring from the intermediate shaft.

To install:

15. Installation is the reverse of the removal procedure.
16. Be sure that the mating surfaces of the joint and splined section are clean and free of debris.
17. Apply about 0.18 ounce of moly 60 paste (PN 08734-0001 or equivalent) to the contact area of the outboard joint and front wheel bearing.
18. Be sure to use new set rings on both shafts.
19. Apply 0.02–0.04 ounce of super high grease to the whole splined surface of the right halfshaft.

➡**After applying the grease, remove the grease from the splined grooves at intervals of 2 to 3 splines and from the set ring groove so that air can bleed from the intermediate shaft.**

20. Clean the areas where the halfshaft contacts the differential thoroughly with solvent and dry with compressed air.

➡**Do not wash rubber parts with solvent.**

21. Insert the inboard end of the halfshaft into the differential or intermediate shaft until the set ring locks in the groove.

➡**Insert the halfshaft horizontally to prevent damage to the oil seal.**

22. Continue the installation in the reverse order of the removal procedure.
23. Tighten the lower arm to knuckle retaining bolt and nuts to 39 ft. lbs. Be sure to use new bolt and nuts. Use the sequence in the illustration.
24. Use a new spindle nut and tighten to 242 ft. lbs.
25. Refill the transmission with the correct grade and type transmission fluid.

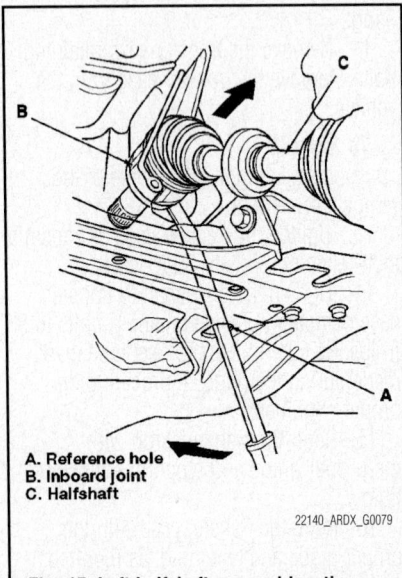

A. Reference hole
B. Inboard joint
C. Halfshaft

22140_ARDX_G0079

Fig. 15 Left halfshaft removal locating points

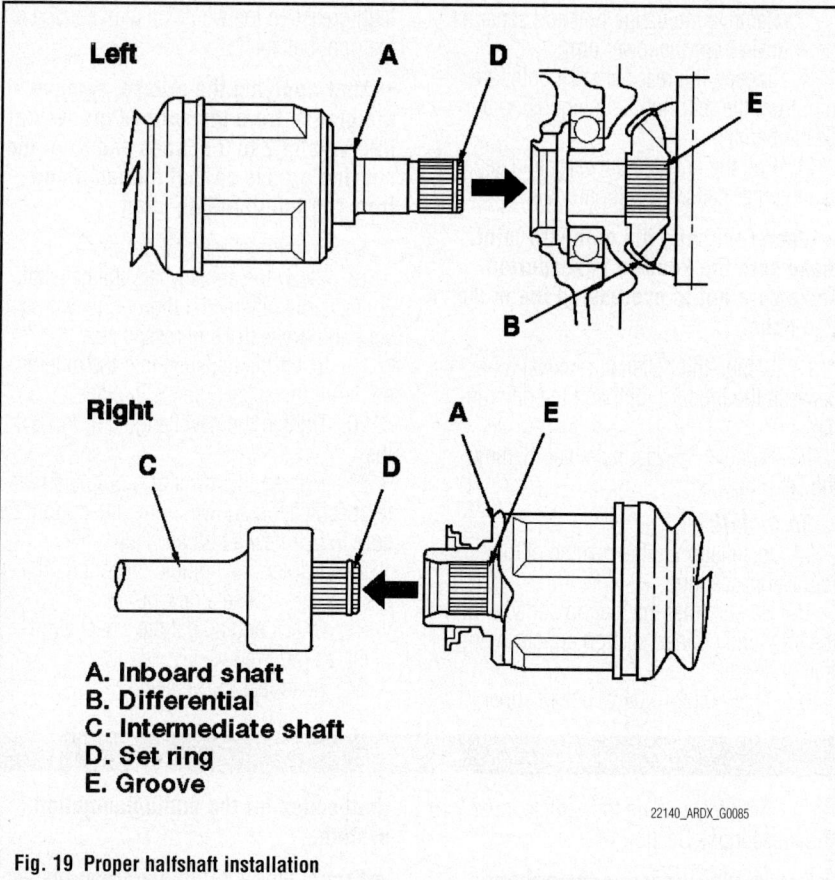

A. Inboard shaft
B. Differential
C. Intermediate shaft
D. Set ring
E. Groove

22140_ARDX_G0085

Fig. 19 Proper halfshaft installation

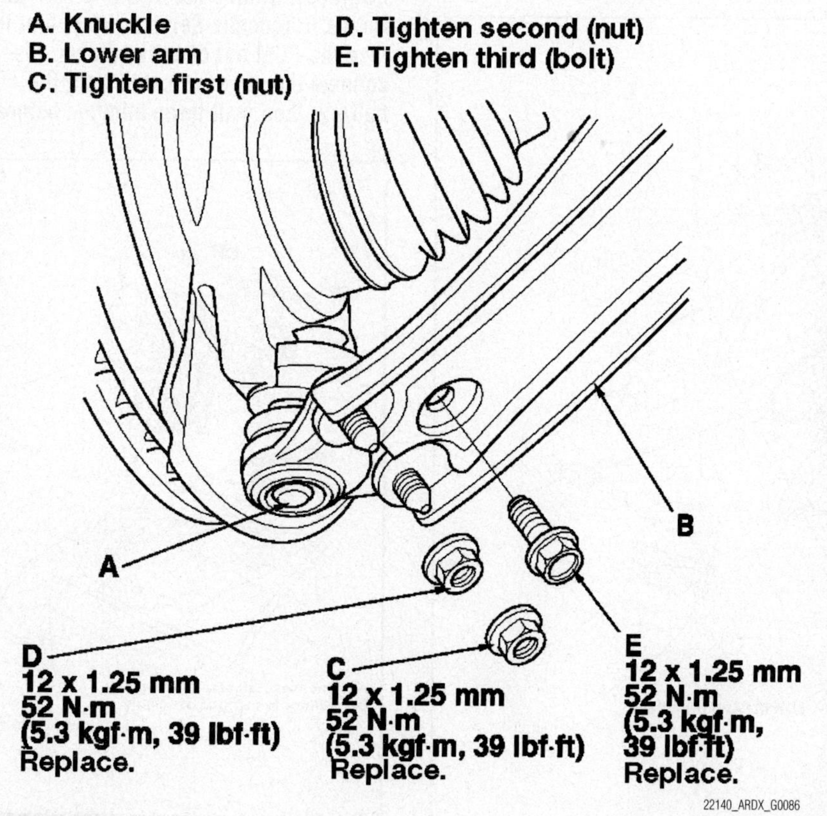

A. Knuckle
B. Lower arm
C. Tighten first (nut)
D. Tighten second (nut)
E. Tighten third (bolt)

D
12 x 1.25 mm
52 N·m
(5.3 kgf·m, 39 lbf·ft)
Replace.

C
12 x 1.25 mm
52 N·m
(5.3 kgf·ft, 39 lbf·ft)
Replace.

E
12 x 1.25 mm
52 N·m
(5.3 kgf·m,
39 lbf·ft)
Replace.

22140_ARDX_G0086

Fig. 20 Lower arm to knuckle tightening sequence

26. Check and adjust the wheel alignment, as required.
27. Road test the vehicle.

REAR HALFSHAFT

REMOVAL & INSTALLATION

See Figures 21 and 22.

1. Before servicing the vehicle, refer to the Precautions Section.

➡ **Before disconnecting the battery cables make sure you have the anti theft codes for the audio/navigation system.**

➡ **Except when doing electrical inspections, always turn the ignition switch to LOCK (O), ground the SCS line with the Honda Diagnostic Service (HDS) tool to take the PCM out of active status, disconnect the negative cable from the battery, then wait three minutes before starting the repair procedure. The SRS**

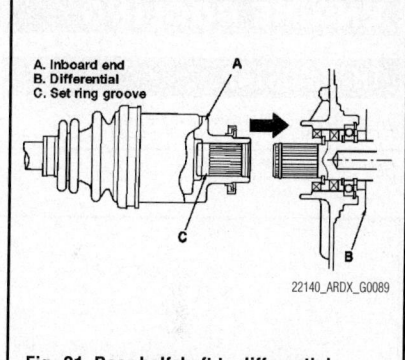

A. Inboard end
B. Differential
C. Set ring groove

22140_ARDX_G0089

Fig. 21 Rear halfshaft to differential alignment

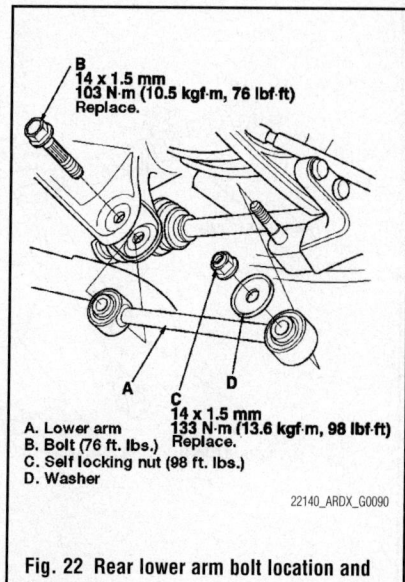

B
14 x 1.5 mm
103 N·m (10.5 kgf·m, 76 lbf·ft)
Replace.

14 x 1.5 mm
133 N·m (13.6 kgf·m, 98 lbf·ft)
Replace.

A. Lower arm
B. Bolt (76 ft. lbs.)
C. Self locking nut (98 ft. lbs.)
D. Washer

22140_ARDX_G0090

Fig. 22 Rear lower arm bolt location and torque data

memory is not cleared even if the ignition switch is turned to LOCK (O) or the battery cables are disconnected from the battery.

2. Make sure that the ignition switch is in the LOCK (O) position.

3. Disconnect the negative battery cable.

➡**Always disconnect the negative battery cable first.**

4. Disconnect the positive battery cable.

➡**Wait at least three minutes after disconnecting the battery cables before starting the repair procedure.**

5. Raise and support the vehicle safely.

6. Remove the rear tire and wheel assemblies.

7. Remove the locking tab on the spindle nut. Remove the nut.

8. Remove the lower arm.

9. Remove the flange bolt and separate the knuckle from the lower arm.

10. Loosen the rear halfshaft outboard joint from the rear hub housing, using a plastic hammer.

11. Pull the knuckle outward, and separate the rear halfshaft outboard joint.

➡**When removing the outboard joint, make sure the knuckle is supported. Make sure not to overextend the brake line hose.**

12. Wedge the halfshaft removal tool between the inboard joint and the differential.

13. Remove the rear halfshaft. Remove the set ring.

To install:

14. Installation is the reverse of the removal procedure.

15. Be sure that the mating surfaces of the joint and splined section are clean and free of debris.

16. Apply 0.02–0.04 ounce of super

high grease to the whole splined surface of the right halfshaft.

➡**After applying the grease, remove the grease from the splined grooves at intervals of 2 to 3 splines and from the set ring groove so that air can bleed from the intermediate shaft.**

17. Be sure to use new set rings.

18. Clean the areas where the halfshaft contacts the differential thoroughly with solvent and dry with compressed air.

19. Insert the halfshaft into the differential, until the set ring locks in place.

20. Tighten the new flange bolt to 76 ft. lbs.

21. Tighten the lower arm bolt and nuts to specification shown in the illustration. Be sure to use new a bolt and nut.

22. Tighten the spindle locknut to 181 ft. lbs. Be sure to use a new nut.

23. Check and adjust the wheel alignment, as required.

24. Road test the vehicle.

ENGINE COOLING

THERMOSTAT

REMOVAL & INSTALLATION

See Figures 23 and 24.

1. Before servicing the vehicle, refer to the Precautions Section.

➡**Before disconnecting the battery cables make sure you have the anti**

theft codes for the audio/navigation system.

➡**Except when doing electrical inspections, always turn the ignition switch to LOCK (O), ground the SCS line with the Honda Diagnostic Service (HDS) tool to take the PCM out of active status, disconnect the negative cable from the battery, then wait three minutes before**

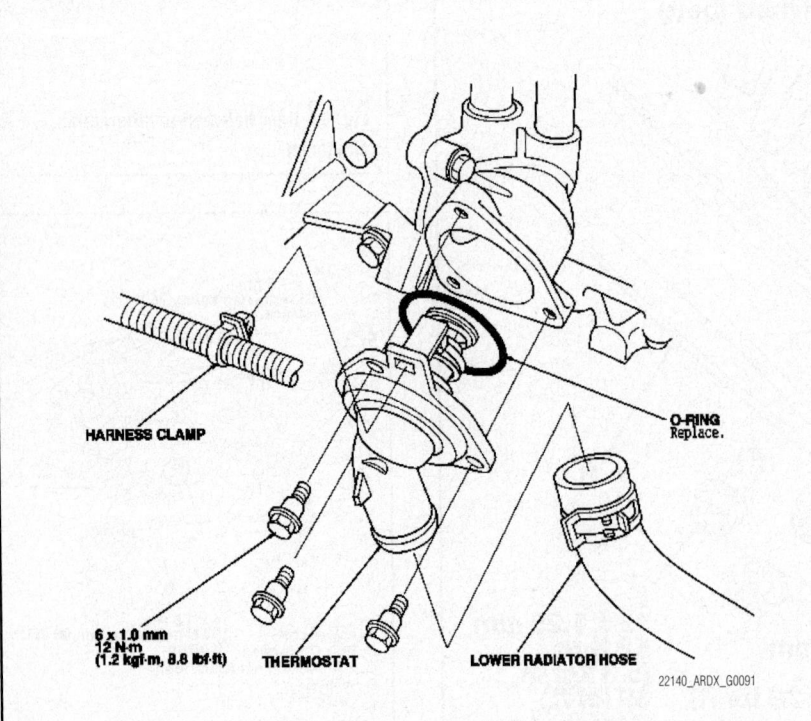

HARNESS CLAMP

O-RING
Replace.

6 x 1.0 mm
12 N·m
(1.2 kgf·m, 8.8 lbf·ft)   THERMOSTAT   LOWER RADIATOR HOSE

22140_ARDX_G0091

Fig. 23 Thermostat and related components

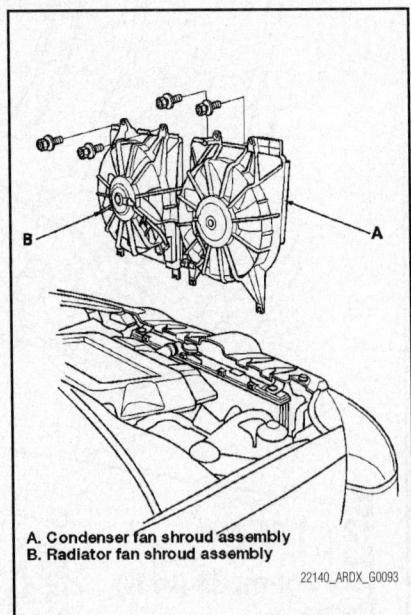

A. Condenser fan shroud assembly
B. Radiator fan shroud assembly

22140_ARDX_G0093

Fig. 24 Condenser fan shroud and related components

starting the repair procedure. The SRS memory is not cleared even if the ignition switch is turned to LOCK (O) or the battery cables are disconnected from the battery.

2. Make sure that the ignition switch is in the LOCK (O) position.
3. Disconnect the negative battery cable.

➡**Always disconnect the negative battery cable first.**

4. Disconnect the positive battery cable.

➡**Wait at least three minutes after disconnecting the battery cables before starting the repair procedure.**

5. Drain the cooling system. Be sure to properly dispose of used coolant.
6. Remove the splash shield.
7. Disconnect the fan motor connectors and remove the harness clamp. Remove the coolant reservoir from the holder.
8. Remove the clips and the support rod clamp bracket.
9. Remove the condenser fan shroud assembly, then remove the radiator fan shroud assembly from the condenser fan shroud side.
10. Remove the lower hose.
11. Remove the thermostat retaining bolts. Remove the thermostat from its mounting.

To install:

12. Installation is the reverse of the removal procedure.
13. Be sure to use a new O-ring.
14. Tighten the retaining bolts to 8.8 ft. lbs.
15. Fill the cooling system with the proper grade and type coolant.
16. Bleed the air from the cooling system with the heater valve open.
17. Loosely install the radiator cap.
18. Start the engine and allow it to reach operating temperature.

➡**Until the radiator fan comes on twice.**

19. Turn the engine off. Check the coolant level. Correct as required.

➡**Removing the radiator cap while the engine is hot can cause the coolant to spray out. Always let the engine and radiator cool before removing the radiator cap.**

20. Install the radiator cap. Run the engine and check for leaks.
21. Connect the Honda Diagnostic System (HDS) tool to the DLC.
22. Turn the ignition switch ON (II).

23. Select BODY ELECTRICAL on the HDS tool.
24. Select ADJUSTMENT in the GAUGE MENU.
25. Select RESET in the MAINTENANCE MINDER.
26. Select MAINTENANCE SUB ITEM 5 RESET.

WATER PUMP

REMOVAL & INSTALLATION

See Figure 25.

1. Before servicing the vehicle, refer to the Precautions Section.

➡**Before disconnecting the battery cables make sure you have the anti theft codes for the audio/navigation system.**

➡**Except when doing electrical inspections, always turn the ignition switch to LOCK (O), ground the SCS line with the Honda Diagnostic Service (HDS) tool to take the PCM out of active status, disconnect the negative cable from the battery, then wait three minutes before starting the repair procedure. The SRS memory is not cleared even if the ignition switch is turned to LOCK (O) or the battery cables are disconnected from the battery.**

2. Make sure that the ignition switch is in the LOCK (O) position.
3. Disconnect the negative battery cable.

➡**Always disconnect the negative battery cable first.**

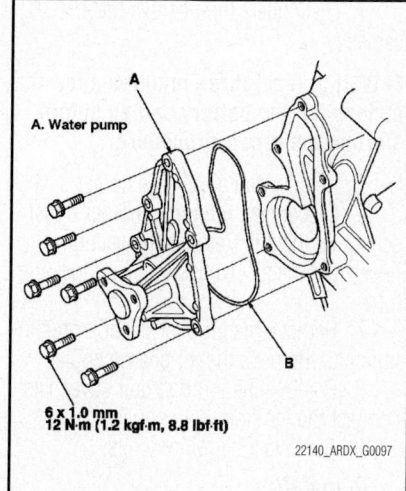

A. Water pump

6 x 1.0 mm
12 N·m (1.2 kgf·m, 8.8 lbf·ft)

22140_ARDX_G0097

Fig. 25 Water pump and related components

4. Disconnect the positive battery cable.

➡**Wait at least three minutes after disconnecting the battery cables before starting the repair procedure.**

5. Drain the cooling system. Be sure to properly dispose of used coolant.
6. Remove the splash shield.
7. Move the drive belt auto tensioner using a belt tension release tool, to relieve the tension on the drive belt. Remove the drive belt.
8. Hold the water pump pulley using an air conditioning clutch holder tool.
9. Remove the water pump pulley mounting bolts. Remove the water pump pulley.
10. Remove the water pump retaining bolts. Remove the water pump from its mounting.

To install:

11. Installation is the reverse of the removal procedure.
12. Be sure to use a new O-ring.
13. Tighten the retaining bolts to 8.8 ft. lbs.
14. Fill the cooling system with the proper grade and type coolant.
15. Bleed the air from the cooling system with the heater valve open.
16. Loosely install the radiator cap.
17. Start the engine and allow it to reach operating temperature.

➡**Until the radiator fan comes on twice.**

18. Turn the engine off. Check the coolant level. Correct as required.

➡**Removing the radiator cap while the engine is hot can cause the coolant to spray out. Always let the engine and radiator cool before removing the radiator cap.**

19. Install the radiator cap. Run the engine and check for leaks.
20. Connect the Honda Diagnostic System (HDS) tool to the DLC.
21. Turn the ignition switch ON (II).
22. Select BODY ELECTRICAL on the HDS tool.
23. Select ADJUSTMENT in the GAUGE MENU.
24. Select RESET in the MAINTENANCE MINDER.
25. Select MAINTENANCE SUB ITEM 5 RESET.

ENGINE ELECTRICAL

CHARGING SYSTEM

ALTERNATOR

REMOVAL & INSTALLATION

See Figure 26.

1. Before servicing the vehicle, refer to the Precautions Section.

➡ Before disconnecting the battery cables make sure you have the anti theft codes for the audio/navigation system.

➡ Except when doing electrical inspections, always turn the ignition switch to LOCK (0), ground the SCS line with the Honda Diagnostic Service (HDS) tool to take the PCM out of active status, disconnect the negative cable from the battery, then wait three minutes before starting the repair procedure. The SRS memory is not cleared even if the ignition switch is turned to LOCK (0) or the battery cables are disconnected from the battery.

2. Make sure that the ignition switch is in the LOCK (0) position.
3. Disconnect the negative battery cable.

➡ Always disconnect the negative battery cable first.

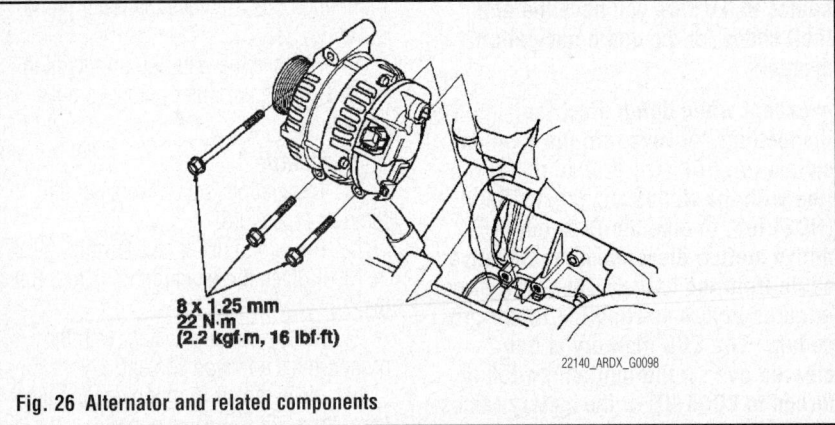

8 x 1.25 mm
22 N·m
(2.2 kgf·m, 16 lbf·ft)

22140_ARDX_G0098

Fig. 26 Alternator and related components

4. Disconnect the positive battery cable.

➡ Wait at least three minutes after disconnecting the battery cables before starting the repair procedure.

5. Move the drive belt auto tensioner using a belt tension release tool, to relieve the tension on the drive belt. Remove the drive belt.
6. Remove the clip and remove the coolant reservoir from its holder.
7. Remove the clips and the support rod clamp bracket.
8. Disconnect the condenser fan motor connector. Remove the condenser fan shroud assembly.
9. Disconnect the alternator electrical connectors.
10. Remove the harness clamp from the alternator.
11. Remove the alternator retaining bolts. Remove the component from its mounting.

To install:

12. Installation is the reverse of the removal procedure.
13. Tighten the retaining bolts to 16 ft. lbs.

ENGINE ELECTRICAL

IGNITION SYSTEM

IGNITION COIL

REMOVAL & INSTALLATION

See Figures 27 through 29.

1. Before servicing the vehicle, refer to the Precautions Section.

➡ Before disconnecting the battery cables make sure you have the anti theft codes for the audio/navigation system.

➡ Except when doing electrical inspections, always turn the ignition switch to LOCK (0), ground the SCS line with the Honda Diagnostic Service (HDS) tool to take the PCM out of active status, disconnect the negative cable from the battery, then wait three minutes before starting the repair procedure. The SRS memory is not cleared even if the ignition switch is turned to LOCK (0) or the battery cables are disconnected from the battery.

2. Make sure that the ignition switch is in the LOCK (0) position.

3. Disconnect the negative battery cable.

➡ Always disconnect the negative battery cable first.

4. Disconnect the positive battery cable.

➡ Wait at least three minutes after disconnecting the battery cables before starting the repair procedure.

5. Remove the air charge cooler cover.
6. Disconnect the turbocharger boost sensor connector. Remove the vacuum hoses and the air bypass outlet connecting tube.
7. Remove the air charge cooler retaining bolts. Remove the air charge cooler.
8. Remove the ignition coil cover. Disconnect the ignition coil connectors.
9. Remove the ignition coils.

To install:

10. Tighten the coil retaining bolt to 8.8 ft. lbs.
11. Install the air charge cooler and connect the intake air ducts.

12. Tighten the hose bands until edge (B) of the band aligns with the mark (C) on the band.

➡ If the hose band edge exceeds the mark, replace the hose band.

13. Continue the installation in the reverse order of the removal procedure.

IGNITION TIMING

ADJUSTMENT

The ignition timing is not adjustable. It is controlled by the PCM.

SPARK PLUGS

REMOVAL & INSTALLATION

1. Before servicing the vehicle, refer to the Precautions Section.

➡ Before disconnecting the battery cables make sure you have the anti theft codes for the audio/navigation system.

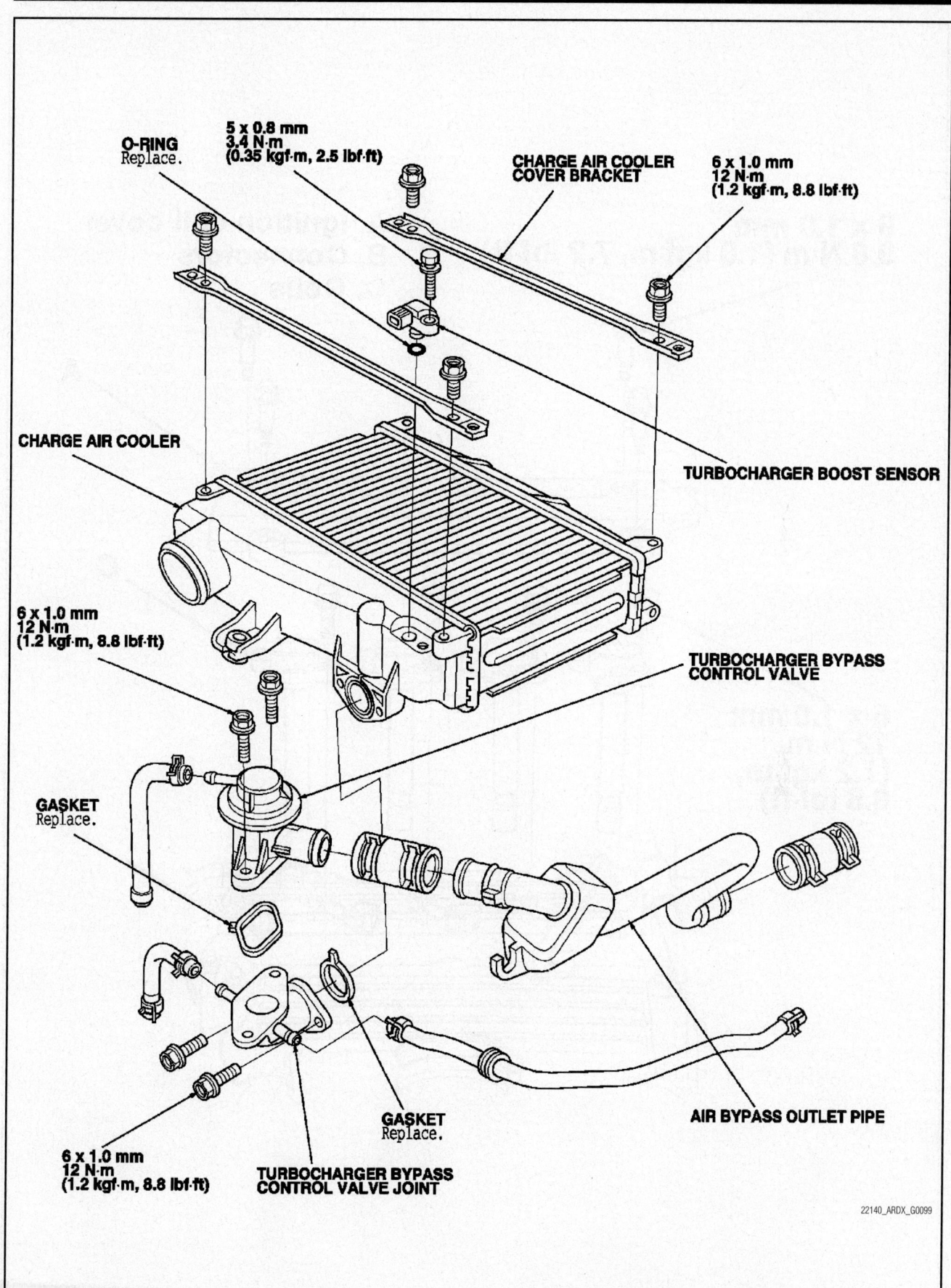

O-RING
Replace.

5 x 0.8 mm
3.4 N·m
(0.35 kgf·m, 2.5 lbf·ft)

CHARGE AIR COOLER
COVER BRACKET

6 x 1.0 mm
12 N·m
(1.2 kgf·m, 8.8 lbf·ft)

CHARGE AIR COOLER

TURBOCHARGER BOOST SENSOR

6 x 1.0 mm
12 N·m
(1.2 kgf·m, 8.8 lbf·ft)

TURBOCHARGER BYPASS
CONTROL VALVE

GASKET
Replace.

GASKET
Replace.

AIR BYPASS OUTLET PIPE

6 x 1.0 mm
12 N·m
(1.2 kgf·m, 8.8 lbf·ft)

TURBOCHARGER BYPASS
CONTROL VALVE JOINT

22140_ARDX_G0099

Fig. 27 Air charge cooler and related components

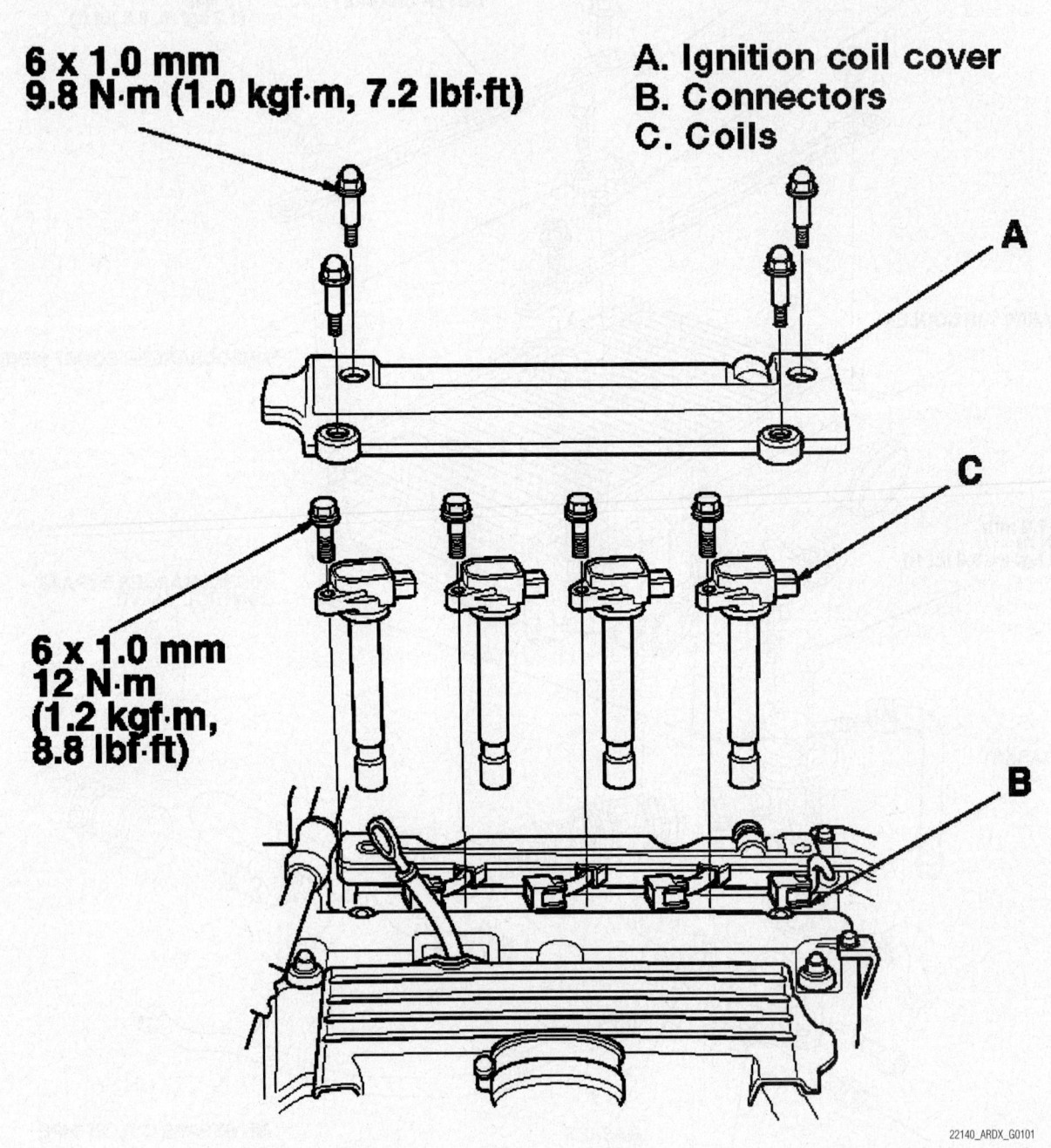

6 x 1.0 mm
9.8 N·m (1.0 kgf·m, 7.2 lbf·ft)

A. Ignition coil cover
B. Connectors
C. Coils

A

C

6 x 1.0 mm
12 N·m
(1.2 kgf·m,
8.8 lbf·ft)

B

22140_ARDX_G0101

Fig. 28 Ignition coils and related components

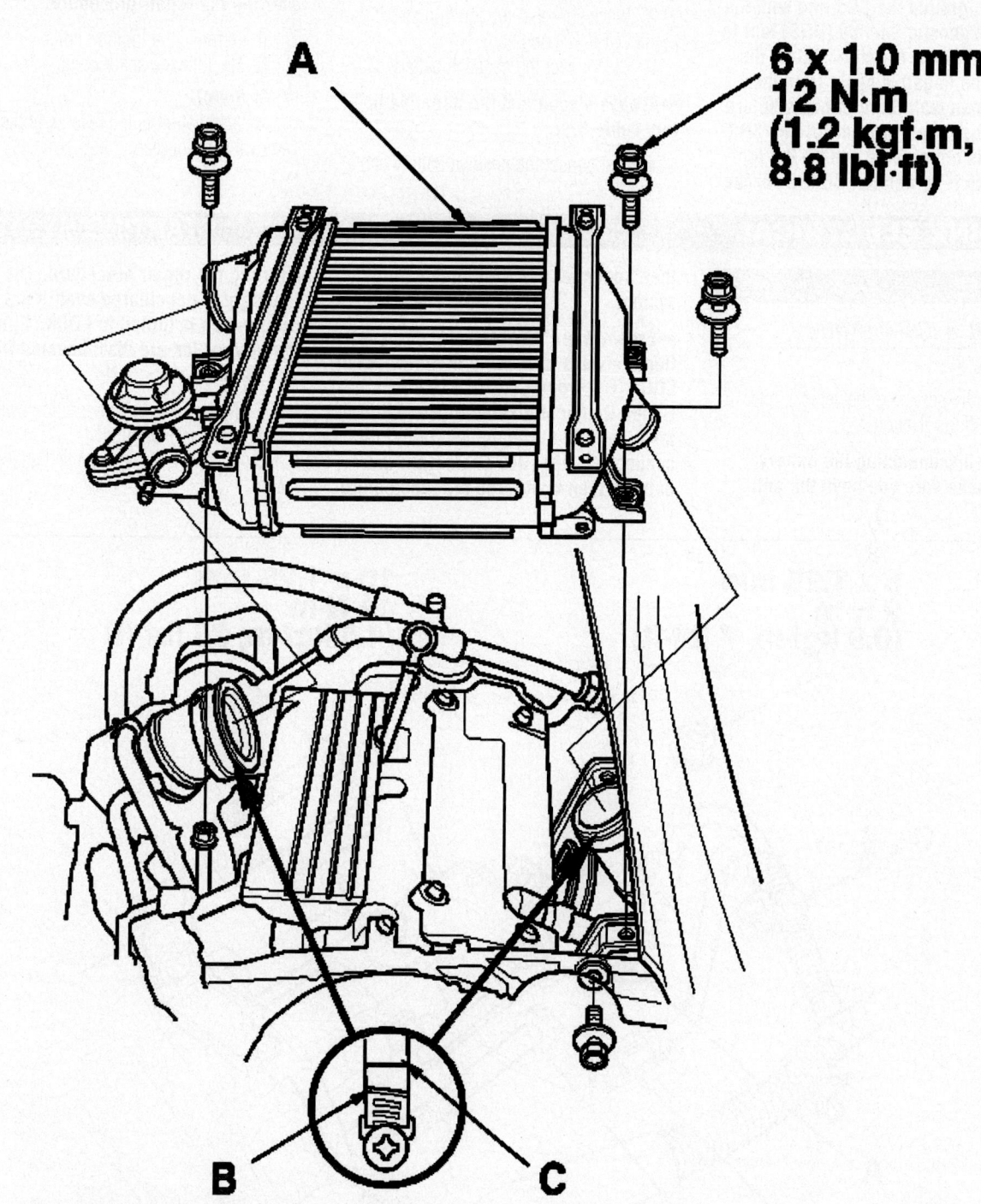

**6 x 1.0 mm
12 N·m
(1.2 kgf·m,
8.8 lbf·ft)**

**A. Air charge cooler
B. Edge (B)
C. Mark (C)**

22140_ARDX_G0100

Fig. 29 Air charge cooler band tightening

➡Except when doing electrical inspections, always turn the ignition switch to LOCK (0), ground the SCS line with the Honda Diagnostic Service (HDS) tool to take the PCM out of active status, disconnect the negative cable from the battery, then wait three minutes before starting the repair procedure. The SRS memory is not cleared even if the ignition switch is turned to LOCK (0) or the

battery cables are disconnected from the battery.

2. Make sure that the ignition switch is in the LOCK (0) position.
3. Disconnect the negative battery cable.

➡Always disconnect the negative battery cable first.

4. Disconnect the positive battery cable.

➡Wait at least three minutes after disconnecting the battery cables before starting the repair procedure.

5. Remove the ignition coils.
6. Remove the spark plugs.

To install:
7. Installation is the reverse of the removal procedure.

ENGINE ELECTRICAL

STARTER

REMOVAL & INSTALLATION
See Figure 30.

1. Before servicing the vehicle, refer to the Precautions Section.

➡Before disconnecting the battery cables make sure you have the anti

theft codes for the audio/navigation system.

➡Except when doing electrical inspections, always turn the ignition switch to LOCK (0), ground the SCS line with the Honda Diagnostic Service (HDS) tool to take the PCM out of active status, disconnect the negative cable from the battery, then wait three minutes before

STARTING SYSTEM

starting the repair procedure. The SRS memory is not cleared even if the ignition switch is turned to LOCK (0) or the battery cables are disconnected from the battery.

2. Make sure that the ignition switch is in the LOCK (0) position.
3. Disconnect the negative battery cable.

8 x 1.25 mm
9 N·m
(0.9 kgf·m, 7 lbf·ft)

10 x 1.25 mm
44 N·m
(4.5 kgf·m, 33 lbf·ft)

12 x 1.25 mm
64 N·m
(6.5 kgf·m, 47 lbf·ft)

A. Starter cable
B. Connector
C. Harness clamps

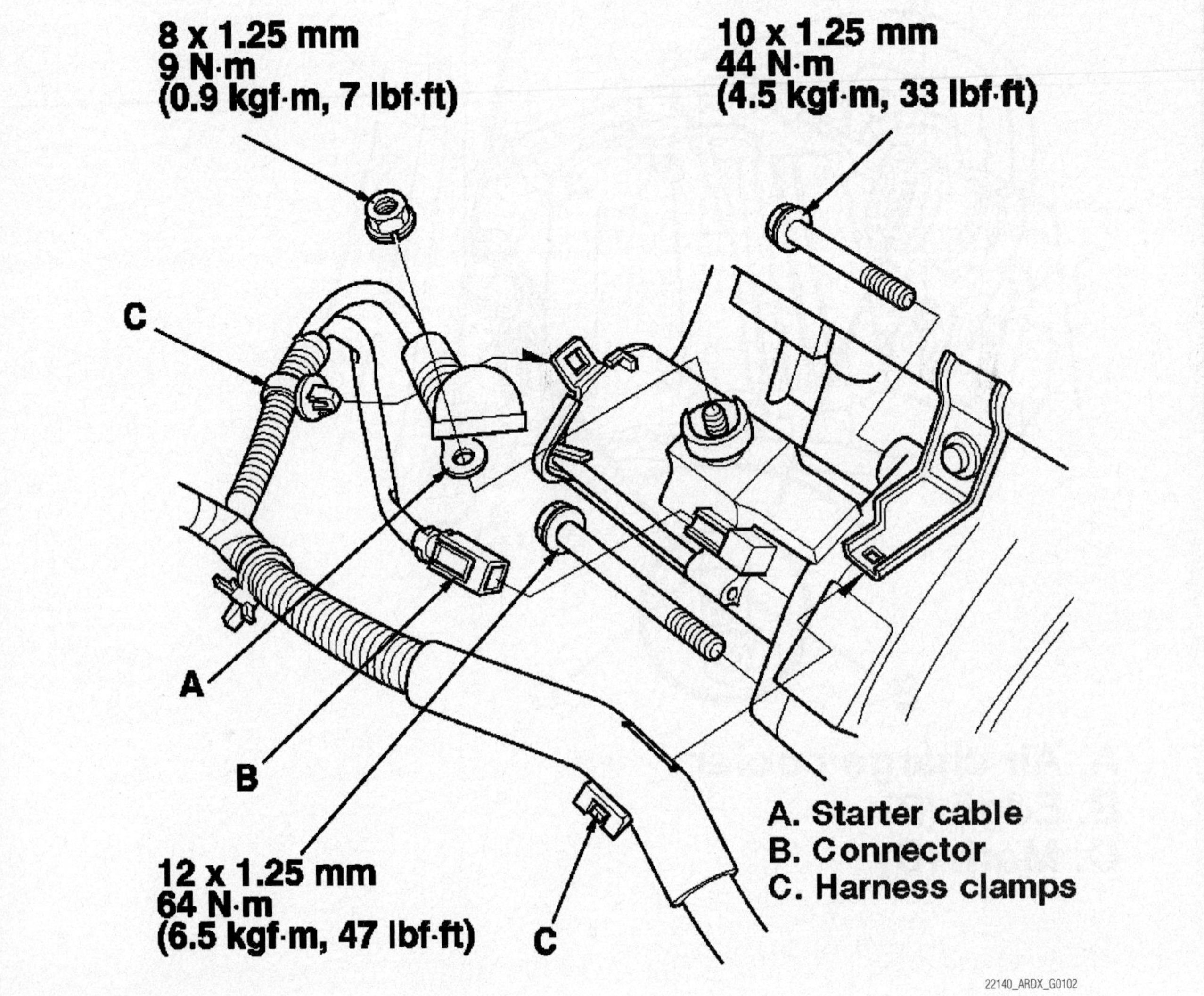

Fig. 30 Starter and related components

➥**Always disconnect the negative battery cable first.**

4. Disconnect the positive battery cable.

➥**Wait at least three minutes after disconnecting the battery cables before starting the repair procedure.**

5. Remove the clip and remove the coolant reservoir from its holder.

6. Remove the clips and the support rod clamp bracket.

7. Disconnect the condenser fan motor connector. Remove the condenser fan shroud assembly.

8. Remove the harness clamp and connector. Remove the intake manifold bracket.

9. Disconnect the electrical connectors

from the starter. Disconnect the harness clamps.

10. Remove the starter retaining bolts. Remove the starter from the vehicle.

To install:

11. Installation is the reverse of the removal procedure.

12. Tighten the starter retaining bolts as shown in the illustration.

ENGINE MECHANICAL

➥**Disconnecting the negative battery cable may interfere with the functions of the on board computer systems and may require the computer to undergo a relearning process, once the negative battery cable is reconnected.**

ACCESSORY DRIVE BELTS

ACCESSORY BELT ROUTING

See Figure 31.

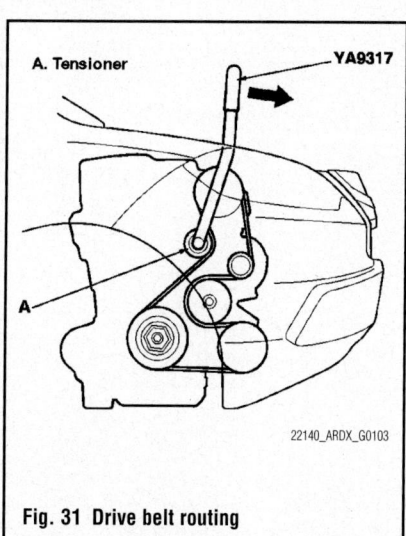

A. Tensioner — YA9317

22140_ARDX_G0103

Fig. 31 Drive belt routing

INSPECTION

See Figure 32.

1. Before servicing the vehicle, refer to the Precautions Section.
2. Inspect the belt for cracks and/or damage.
3. If damage exists, replace the belt.
4. Check that the auto tensioner indicator is within range. If not replace the belt.

ADJUSTMENT

The drive belt is adjusted automatically. If out of specification, it must be replaced.

REMOVAL & INSTALLATION

1. Before servicing the vehicle, refer to the Precautions Section.

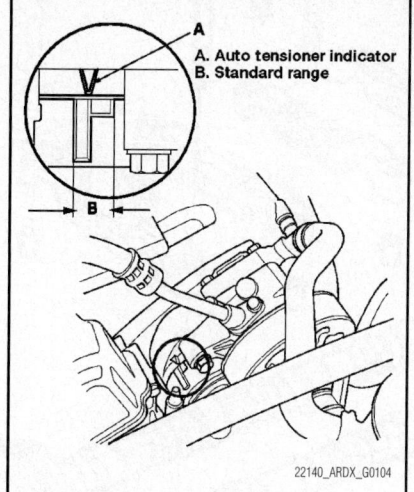

A. Auto tensioner indicator
B. Standard range

22140_ARDX_G0104

Fig. 32 Drive belt replacement specification

➥**Before disconnecting the battery cables make sure you have the anti theft codes for the audio/navigation system.**

➥**Except when doing electrical inspections, always turn the ignition switch to LOCK (O), ground the SCS line with the Honda Diagnostic Service (HDS) tool to take the PCM out of active status, disconnect the negative cable from the battery, then wait three minutes before starting the repair procedure. The SRS memory is not cleared even if the ignition switch is turned to LOCK (O) or the battery cables are disconnected from the battery.**

2. Make sure that the ignition switch is in the LOCK (O) position.

3. Disconnect the negative battery cable.

➥**Always disconnect the negative battery cable first.**

4. Disconnect the positive battery cable.

➥**Wait at least three minutes after disconnecting the battery cables before starting the repair procedure.**

5. Move the drive belt auto tensioner using a belt tension release tool, to relieve the tension on the drive belt. Remove the drive belt.

To install:

6. Installation is the reverse of the removal procedure.

CAMSHAFT AND VALVE LIFTERS

REMOVAL & INSTALLATION

See Figures 33 through 39.

1. Before servicing the vehicle, refer to the Precautions Section.

➥**Before disconnecting the battery cables make sure you have the anti theft codes for the audio/navigation system.**

➥**Except when doing electrical inspections, always turn the ignition switch to LOCK (O), ground the SCS line with the Honda Diagnostic Service (HDS) tool to take the PCM out of active status, disconnect the negative cable from the battery, then wait three minutes before starting the repair procedure. The SRS memory is not cleared even if the ignition switch is turned to LOCK (O) or the battery cables are disconnected from the battery.**

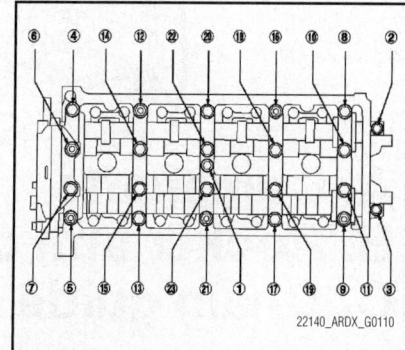

22140_ARDX_G0110

Fig. 33 Rocker arm cap loosening sequence

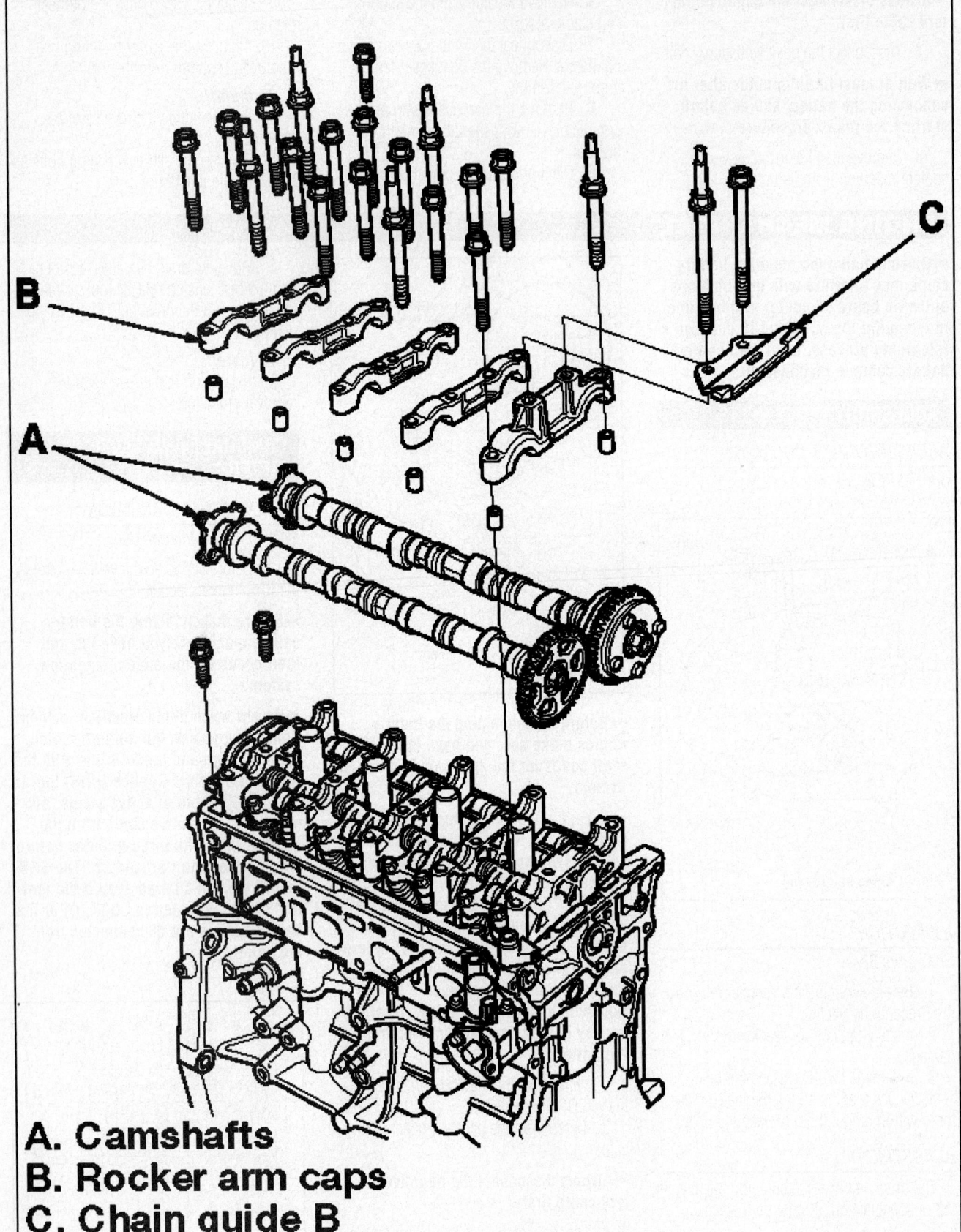

A. Camshafts
B. Rocker arm caps
C. Chain guide B

22140_ARDX_G0111

Fig. 34 Camshafts and related components

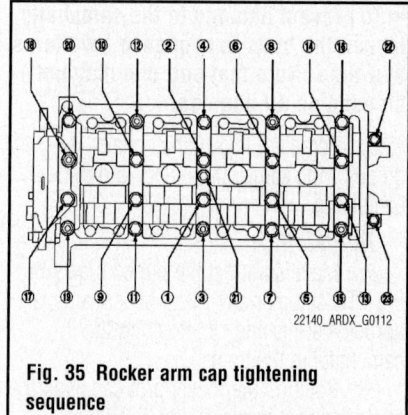

Fig. 35 Rocker arm cap tightening sequence

2. Make sure that the ignition switch is in the LOCK (O) position.

3. Disconnect the negative battery cable.

➡**Always disconnect the negative battery cable first.**

4. Disconnect the positive battery cable.

➡**Wait at least three minutes after disconnecting the battery cables before starting the repair procedure.**

5. Remove the air charge cooler cover.

6. Disconnect the turbocharger boost sensor connector. Remove the vacuum hoses and the air bypass outlet connecting tube.

7. Remove the air charge cooler retaining bolts. Remove the air charge cooler.

8. Remove the fuel injector cover.

9. Remove the ignition coil cover. Disconnect the ignition coil connectors.

10. Remove the ignition coils.

11. Remove the power steering hose bracket, the breather hose and the dipstick.

12. Remove the bracket mounting bolts, from the valve cover.

13. Disconnect the rocker arm oil control solenoid connector, the rocker are oil pressure switch connector and the variable valve timing control (VTC) oil control solenoid valve connector. Remove the harness clamp.

14. Disconnect the air fuel ration (AF) sensor connector, the oil pressure switch connector and the crankshaft position (CKP) sensor. Remove the harness clamps.

15. Remove the valve cover retaining bolts.

16. Remove the valve cover from the engine. Discard the gasket.

17. Remove the timing chain.

18. Loosen the rocker arm adjusting screws.

19. Remove the camshaft retaining bolts.

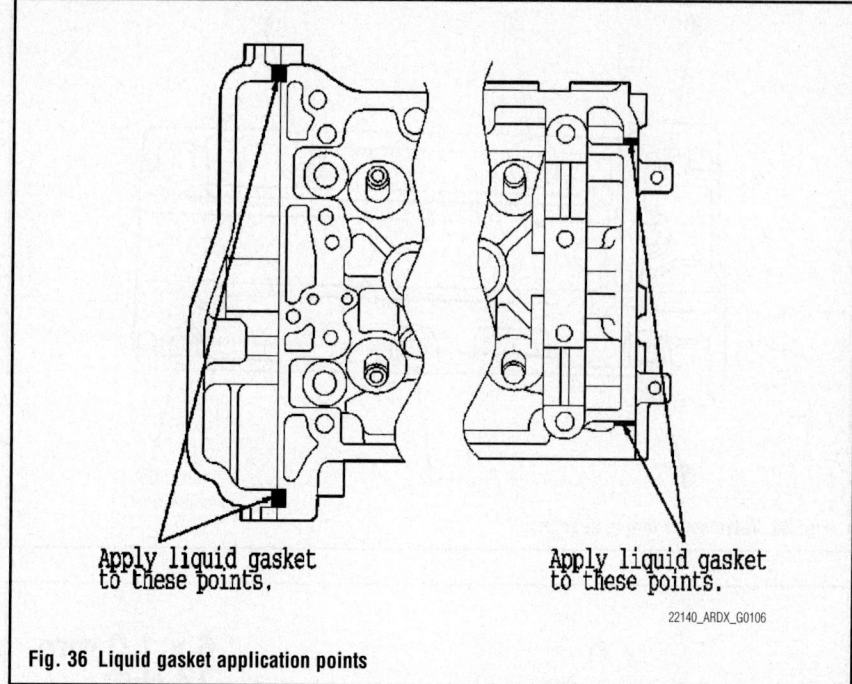

Apply liquid gasket to these points.   Apply liquid gasket to these points.

Fig. 36 Liquid gasket application points

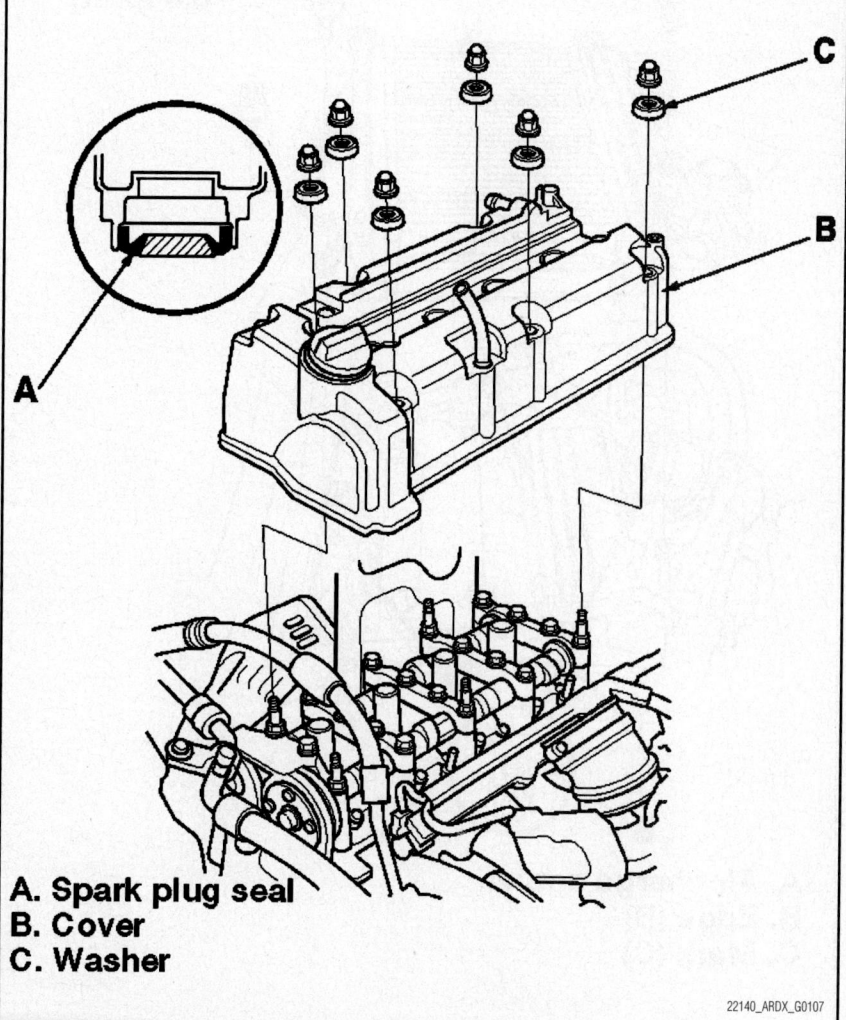

A. Spark plug seal
B. Cover
C. Washer

Fig. 37 Spark plug seal placement

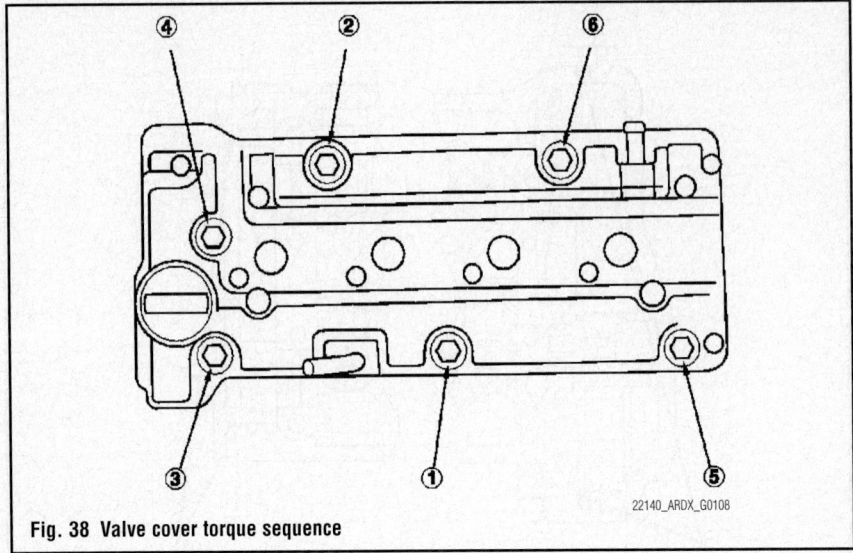

Fig. 38 Valve cover torque sequence

22140_ARDX_G0108

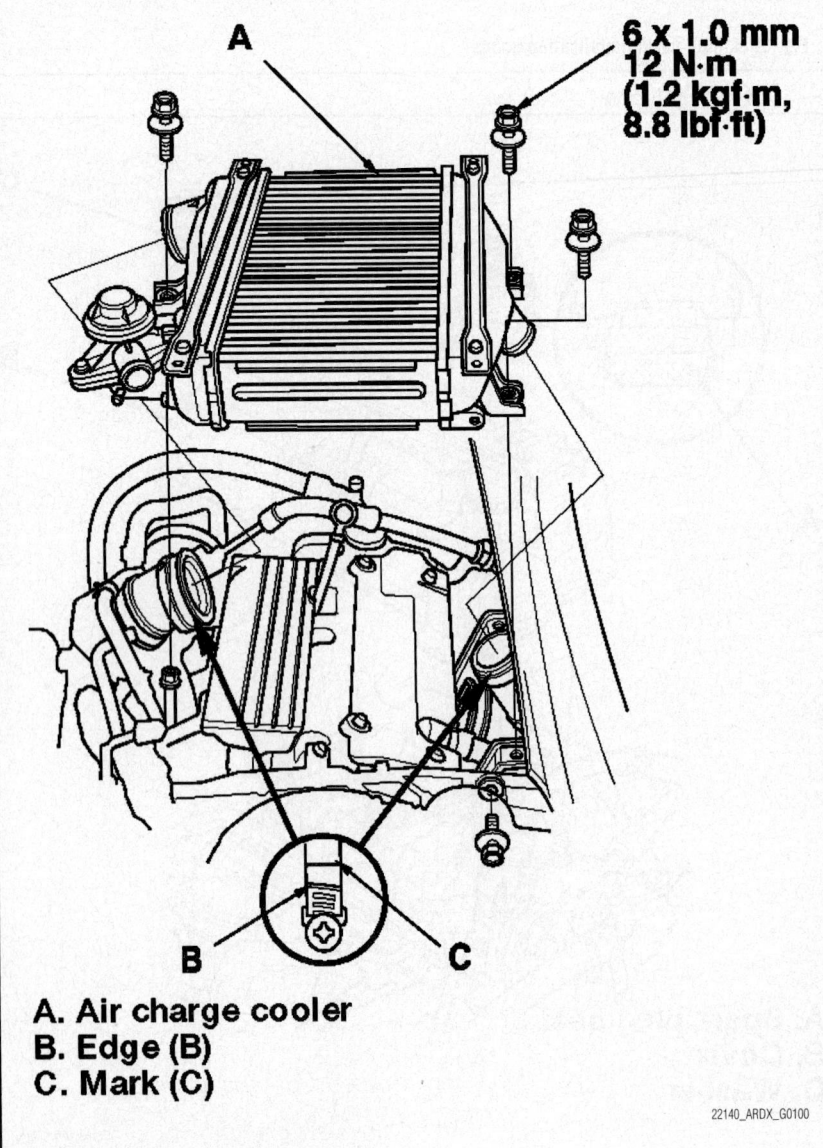

6 x 1.0 mm
12 N·m
(1.2 kgf·m,
8.8 lbf·ft)

A
B   C

A. Air charge cooler
B. Edge (B)
C. Mark (C)

22140_ARDX_G0100

Fig. 39 Air charge cooler band tightening

➡ To prevent damage to the camshafts, loosen the bolts in sequence, two turns at a time. Note that bolt one may not be used on all engines.

20. Remove the camshaft chain guide B, the camshaft holders and the camshafts.

To install:

21. Make sure that the punch marks on the VTC actuator and the exhaust camshaft sprocket are facing UP. Position the camshafts in the holder.

22. Position the rocker arm caps and the chain guide B in place.

23. Tighten the camshaft retaining bolts to specification and in the sequence shown in the illustration. Tighten 8x1.25mm bolts to 16 ft. lbs. Tighten 6x1.0mm bolts to 8.8 ft lbs (21, 22 and 23).

➡ If the engine does not use bolt number 21, skip it and continue the torque sequence.

24. Continue the installation in the reverse order of the removal procedure.

25. Before installing the valve cover, clean the mating surfaces of the cylinder head and valve cover.

26. Install a new gasket in the groove of the valve cover.

27. Be sure that the mating surfaces are clean and dry.

28. Apply liquid gasket part number 08717-0004 or equivalent on the chain case and the number five rocker shaft holder mating surface areas.

➡ Install the component within five minutes of applying the liquid gasket. Some liquid gasket materials require installation within four minutes. Be sure to read and follow the manufacturer's instruction. If too much time has passed before installing the component remove the liquid gasket and residue, then reapply new liquid gasket.

29. Position the spark plug seals on the spark plug tubes.

30. Position the valve cover on the cylinder head. Slide the cover back and forth to seat the gasket.

31. Inspect the valve cover washers, replace as required.

32. Tighten the retaining bolts in three steps to 8.8 ft. lbs., using the sequence shown in the illustration.

➡ Wait at least 30 minutes to allow the gasket to cure before filling the engine with oil.

➡**Do not run the engine for at least three hours after installing the valve cover.**

33. Continue the installation in the reverse order of the removal procedure.

34. Install the air charge cooler and connect the intake air ducts.

35. Tighten the hose bands until edge (B) of the band aligns with the mark (C) on the band.

➡**If the hose band edge exceeds the mark, replace the hose band.**

36. Start the engine and check for leaks, correct as required.

37. Adjust the valves.

38. Turn the ignition switch ON (II), the SRS indicator should come on for about six seconds, and then go off.

CRANKSHAFT FRONT SEAL

REMOVAL & INSTALLATION

See Figure 40.

1. Before servicing the vehicle, refer to the Precautions Section.

➡**Before disconnecting the battery cables make sure you have the anti**

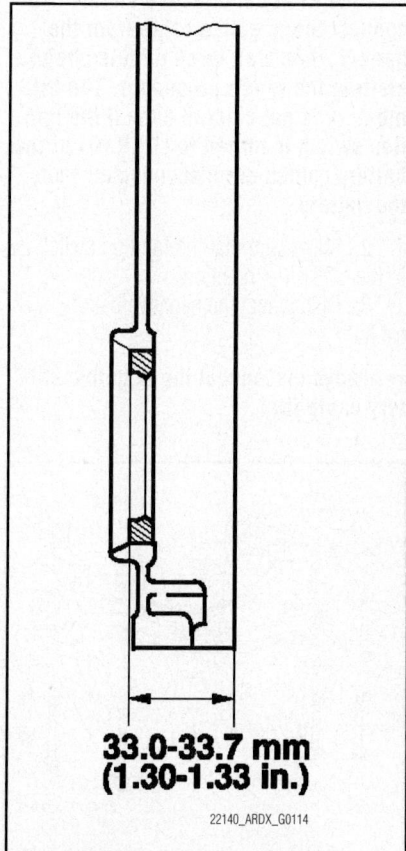

**33.0-33.7 mm
(1.30-1.33 in.)**

22140_ARDX_G0114

Fig. 40 Timing cover oil seal height specification

theft codes for the audio/navigation system.

➡**Except when doing electrical inspections, always turn the ignition switch to LOCK (O), ground the SCS line with the Honda Diagnostic Service (HDS) tool to take the PCM out of active status, disconnect the negative cable from the battery, then wait three minutes before starting the repair procedure. The SRS memory is not cleared even if the ignition switch is turned to LOCK (O) or the battery cables are disconnected from the battery.**

2. Make sure that the ignition switch is in the LOCK (O) position.

3. Disconnect the negative battery cable.

➡**Always disconnect the negative battery cable first.**

4. Disconnect the positive battery cable.

➡**Wait at least three minutes after disconnecting the battery cables before starting the repair procedure.**

5. Remove the timing chain cover.

6. Carefully pry the old seal out of its mounting, using a suitable tool.

To install:

7. Position a new seal on the cover. Coat it with clean engine oil.

8. Use a seal driver tool to drive the new seal squarely into the chain case to the specified installed height.

9. Measure the distance between the chain case surface and the oil seal.

10. The oil seal installed height should be 1.30–1.33 inch.

11. Continue the installation in the reverse order of the removal procedure.

12. Start the engine and check for leaks, correct as required.

13. Turn the ignition switch ON (II), the SRS indicator should come on for about six seconds, and then go off.

CYLINDER HEAD

REMOVAL & INSTALLATION

See Figures 41 through 46.

1. Before servicing the vehicle, refer to the Precautions Section.

➡**Before disconnecting the battery cables make sure you have the anti theft codes for the audio/navigation system.**

➡**Except when doing electrical inspections, always turn the ignition**

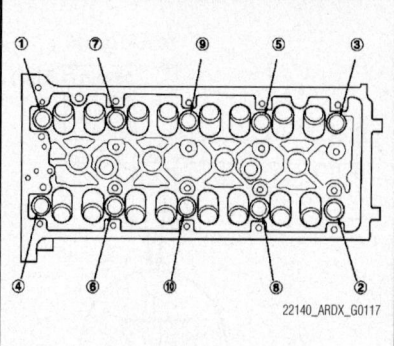

22140_ARDX_G0117

Fig. 41 Cylinder head bolt removal sequence

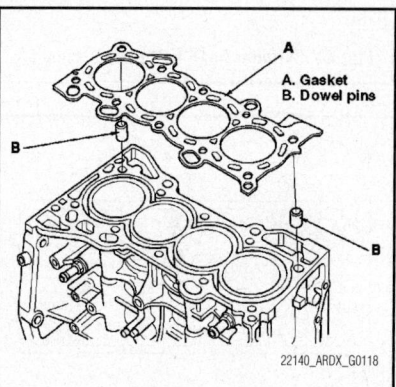

A. Gasket
B. Dowel pins

22140_ARDX_G0118

Fig. 42 Cylinder head gasket positioning

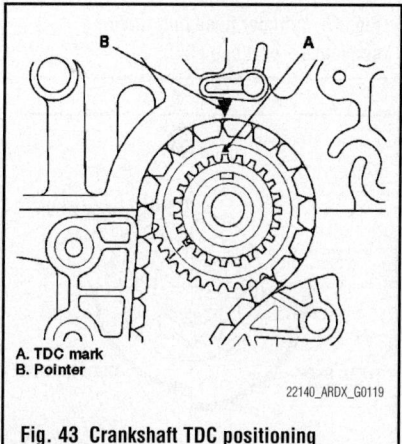

A. TDC mark
B. Pointer

22140_ARDX_G0119

Fig. 43 Crankshaft TDC positioning

switch to LOCK (O), ground the SCS line with the Honda Diagnostic Service (HDS) tool to take the PCM out of active status, disconnect the negative cable from the battery, then wait three minutes before starting the repair procedure. The SRS memory is not cleared even if the ignition switch is turned to LOCK (O) or the battery cables are disconnected from the battery.

2. Make sure that the ignition switch is in the LOCK (O) position.

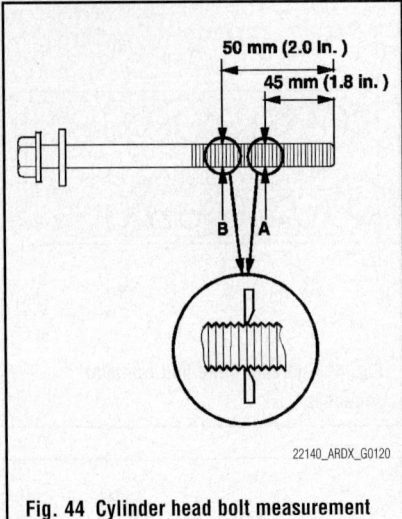

22140_ARDX_G0120

Fig. 44 Cylinder head bolt measurement

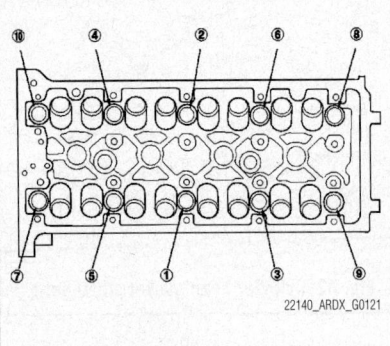

22140_ARDX_G0121

Fig. 45 Cylinder head bolt torquing sequence—part one

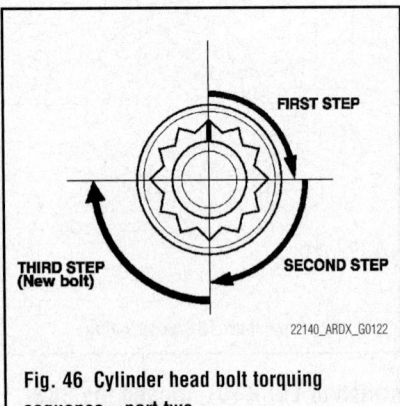

22140_ARDX_G0122

Fig. 46 Cylinder head bolt torquing sequence—part two

3. Disconnect the negative battery cable.

➡**Always disconnect the negative battery cable first.**

4. Disconnect the positive battery cable.

➡**Wait at least three minutes after disconnecting the battery cables before starting the repair procedure.**

5. Relieve the fuel system pressure.
6. Drain the engine coolant.
7. Remove the air cleaner housing assembly.
8. Remove the drive belt.
9. Remove the intake manifold.
10. Remove the exhaust manifold.
11. Remove the engine wire harness connectors and harness clamps from the cylinder head (CMP sensor A, ECT sensor, CMP sensor B and EVAP).
12. Remove the quick connect fitting cover. Disconnect the fuel feed hose.
13. Disconnect the EVAP canister hose and the brake booster hoses.
14. Remove the two bolts retaining the EVAP canister purge control valve mounting bracket. Remove the harness clamps.
15. Disconnect the upper radiator hose, the heater hose and the water bypass hose.
16. Remove the timing chain.
17. Remove the camshafts.
18. Remove the cylinder head bolts.

➡**To prevent damage to the head loosen the bolts, in the sequence shown in the illustration, a third turn at a time. Repeat the sequence until all bolts are loosened.**

19. Remove the cylinder head from the engine.

To install:

20. Clean the cylinder head and block mating surfaces.
21. Install the new cylinder head gasket and dowel pins on the engine block.
22. Position the crankshaft to TDC. Align the TDC mark on the crankshaft sprocket with the pointer on the engine block.
23. Install the cylinder head.
24. Measure the diameter of each cylinder head bolt at points "A" and "B", as shown in the illustration. If either diameter is less than 0.42 inch, replace the bolt.

➡**Apply engine oil to the threads and under the bolt heads, prior to installation.**

25. Tighten the cylinder head bolts to 29 ft. lbs, and in the sequence shown in the illustration.

➡**Use a beam type torque wrench. If using a preset torque wench be sure to tighten slowly and do not over tighten. If the bolt makes any noise while tightening it, loosen it and retighten it.**

26. After tightening, tighten all bolts in two steps (90 degrees per step) in the same sequence used for tightening the cylinder head bolts in the step above.

➡**If you are using new bolts tighten an extra 90 degrees.**

27. If the bolt you tightened is beyond the specified angle, re-measure the bolt. Do not loosen it back to the specified angle.
28. Continue the installation in the reverse order of the removal procedure.
29. Start the engine and check for leaks, correct as required.
30. Inspect the idle speed. Inspect the ignition timing.
31. Turn the ignition switch ON (II), the SRS indicator should come on for about six seconds, and then go off.

ENGINE ASSEMBLY

REMOVAL & INSTALLATION
See Figures 47 through 56.

1. Before servicing the vehicle, refer to the Precautions Section.

➡**Before disconnecting the battery cables make sure you have the anti theft codes for the audio/navigation system.**

➡**Except when doing electrical inspections, always turn the ignition switch to LOCK (0), ground the SCS line with the Honda Diagnostic Service (HDS) tool to take the PCM out of active status, disconnect the negative cable from the battery, then wait three minutes before starting the repair procedure. The SRS memory is not cleared even if the ignition switch is turned to LOCK (0) or the battery cables are disconnected from the battery.**

2. Make sure that the ignition switch is in the LOCK (0) position.
3. Disconnect the negative battery cable.

➡**Always disconnect the negative battery cable first.**

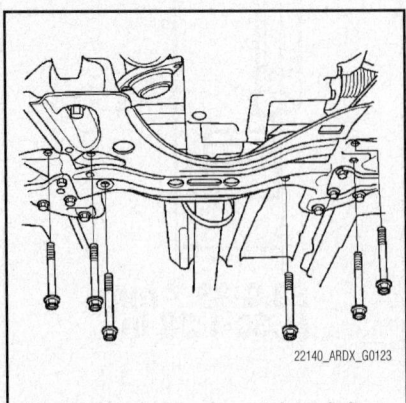

22140_ARDX_G0123

Fig. 47 Steering gearbox retaining bolts

4. Disconnect the positive battery cable.

➤**Wait at least three minutes after disconnecting the battery cables before starting the repair procedure.**

5. Relieve the fuel system pressure.
6. Drain the engine coolant.
7. Remove the air cleaner housing assembly.
8. Remove the relay block and the harness clamps. Loosen the lower bolts. Remove the upper bolts. Remove the battery base.
9. Remove the charge air cooler cover. Remove the charge air cooler cover front bracket. Remove the air bypass outlet pipe.
10. Remove the battery cables from the under hood fuse/relay box.
11. Disconnect the harness connector.
12. Disconnect the PCM connectors and the engine wire harness connector. Remove the harness clamps.
13. Remove the under hood fuse/relay box from the bracket.
14. Remove the quick connect fitting cover. Disconnect the fuel feed hose.
15. Remove the heater hoses from the clamps, and then remove the bolt retaining the heater hose clamp bracket.
16. Disconnect the EVAP canister hose and the brake booster hoses.
17. Remove the shift cable. Remove the drive belt.
18. Remove the power steering pump retaining bolts. Remove the power steering pump from its mounting. Do not disconnect the hoses. Position the component to the side.
19. Remove the radiator cap.
20. Raise and support the vehicle safely. Remove the front tire and wheel assemblies.

21. Remove the splash shield.
22. Drain the engine oil. Drain the coolant. Drain the transmission fluid.
23. Remove the catalytic converter.
24. Separate the stabilizer links from the stabilizer bar.
25. Separate the knuckles from the lower arms.
26. Remove the halfshafts.
27. Remove the driveshaft from the transfer shaft flange.
28. Remove the bolt securing the power steering fluid line and bracket. Unclamp the lines on the front subframe.
29. Remove the bolts retaining the steering gearbox mounting brackets.
30. Lower the vehicle on the lift.

31. Remove the radiator.
32. Disconnect the air conditioning compressor electrical connectors. Remove the retaining bolts. Remove the compressor from it mounting. Do not disconnect the refrigerant lines.
33. Disconnect the heater hoses.
34. Disconnect and plug the transmission fluid lines. Remove the ground cable.
35. Attach the engine support eyelet tool 07AAK-SNAA600 or equivalent.
36. Remove the corner clip, from the front grille on both sides. Place the engine support mount pads (07AAK-STKA120) over the hood edge cushions and align the pins with the clip hole.

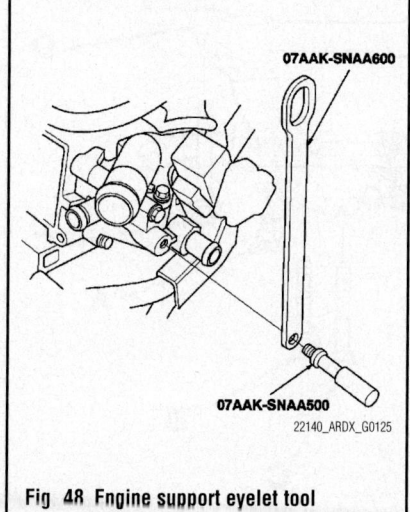

Fig 48 Engine support eyelet tool installation

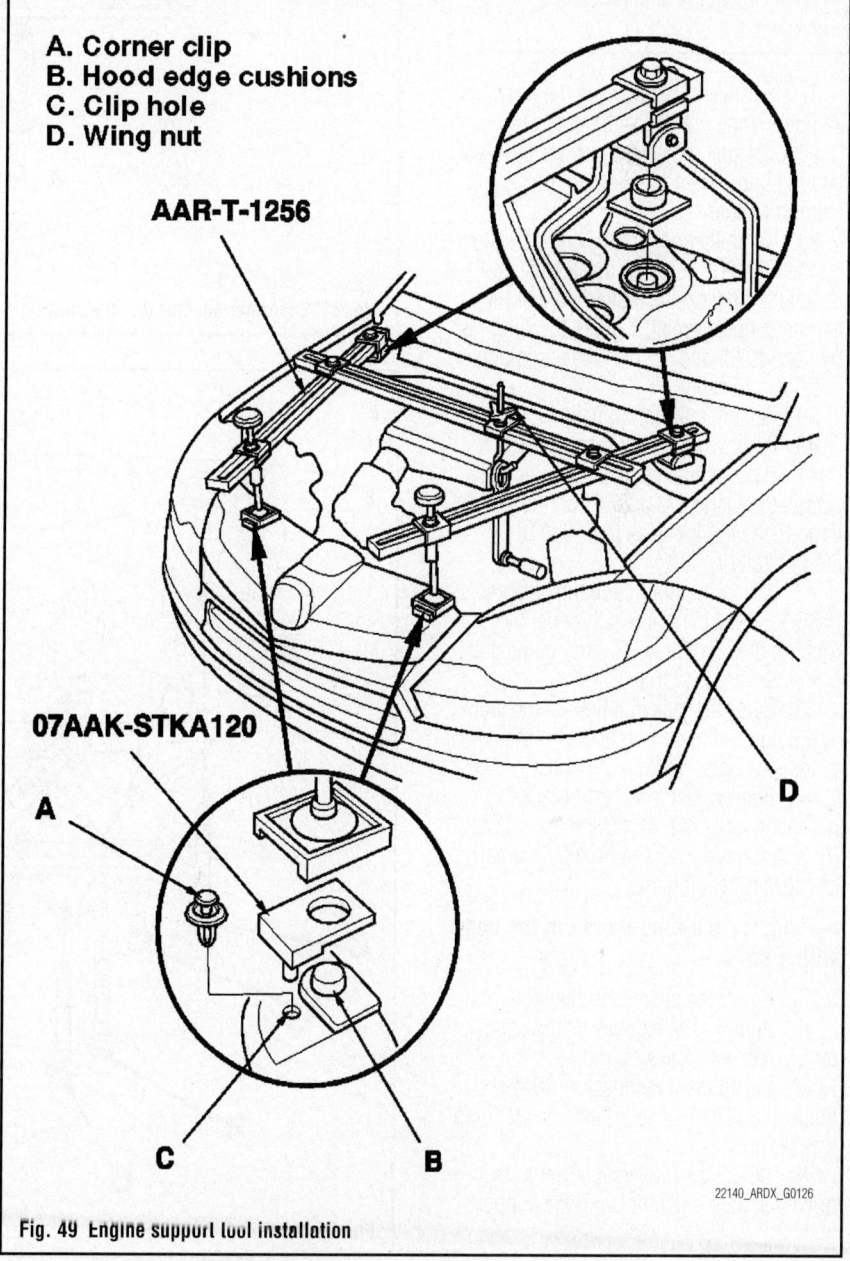

A. Corner clip
B. Hood edge cushions
C. Clip hole
D. Wing nut

AAR-T-1256

07AAK-STKA120

22140_ARDX_G0126

Fig. 49 Engine support tool installation

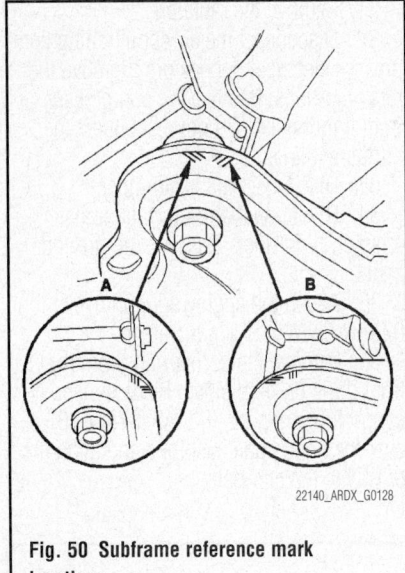

Fig. 50 Subframe reference mark locations

37. Remove both lids from the front damper flange nuts from the cowl cover. Use the adapter between the cross arm foot and the body when positioning the engine support hanger.

38. Carefully position the engine support hanger (AAR-T-1256) on the vehicle and attach the hook to the slotted hole in the engine support eyelet. Tighten the wing nut by hand to lift and support the engine/transmission assembly.

39. Raise the lift to full height.

40. Remove the lower torque rod.

41. Place reference marks on the body across the marks (A) and (B) on the edge of the subframe as indicated in the illustration.

42. Attach the front subframe adapter (VSB02C000016) to the subframe, by looping the strap over the front of the subframe. Secure the strap.

43. Raise the jack and line up the slots in the arms with the bolt holes on the corner of the jack base. Tighten the bolts.

44. Remove the lower arm bracket mounting bolts (A). Remove the six bolts (B) and (C) securing the front subframe, and lower the subframe.

➡**Hang the steering box from the body with a strap.**

45. Lower the vehicle on the lift.

46. Remove the harness clamps and remove the transmission mount.

47. Install the transmission hanger bracket (21232-RCT-A00) and washer onto the transmission.

48. Attach the universal eyelet to the drive belt auto tensioner with the support bolt.

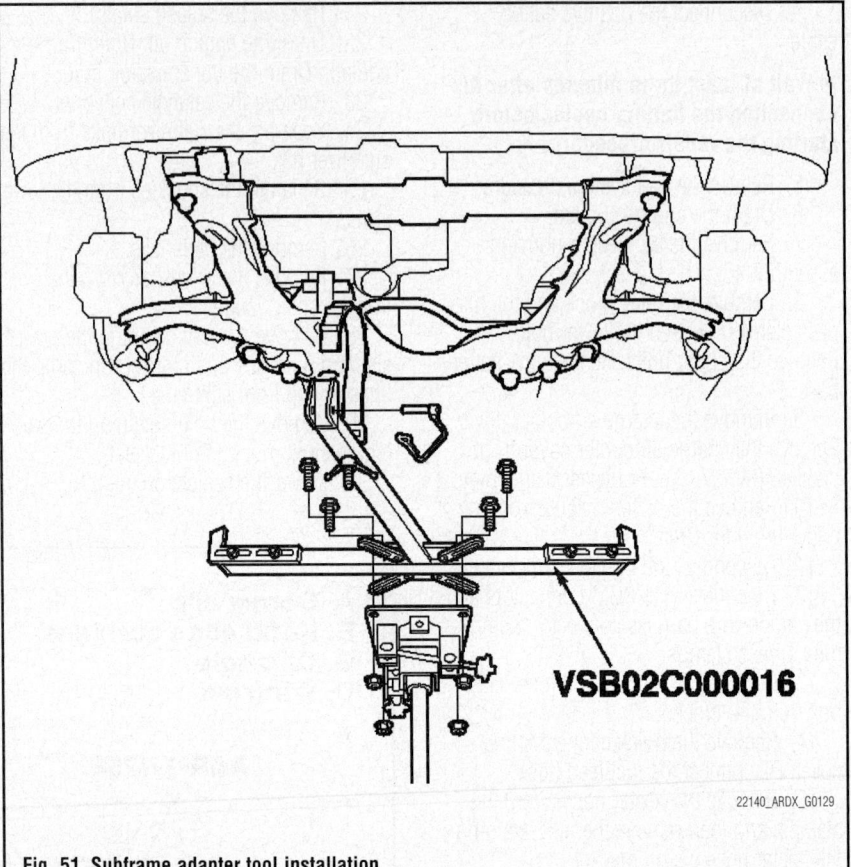

VSB02C000016

Fig. 51 Subframe adapter tool installation

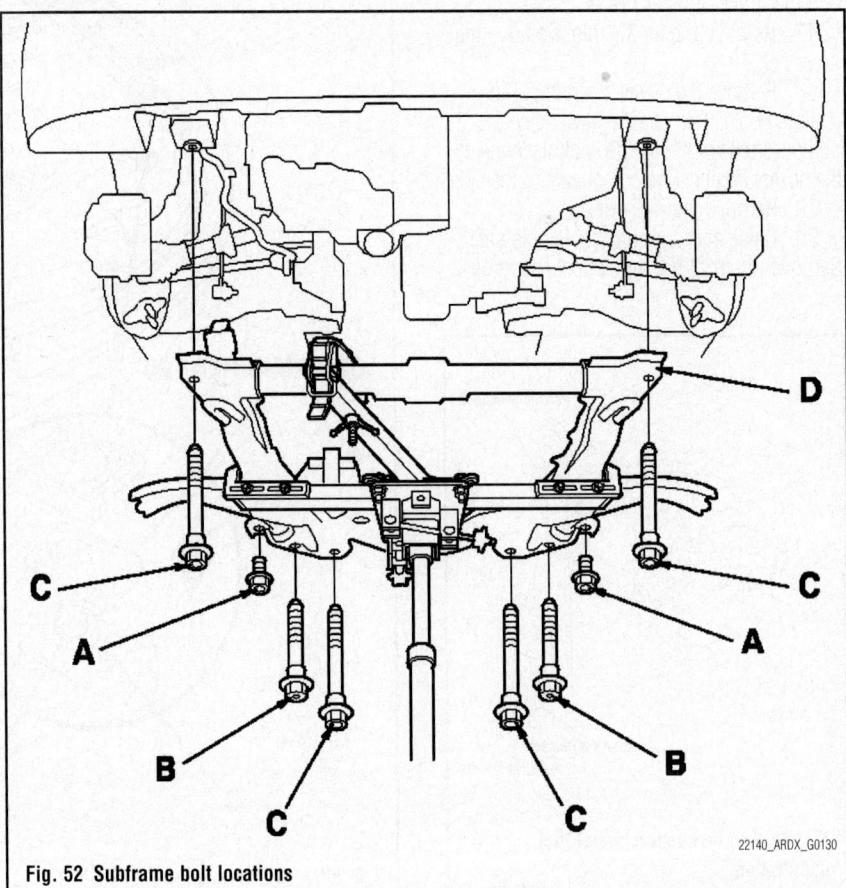

Fig. 52 Subframe bolt locations

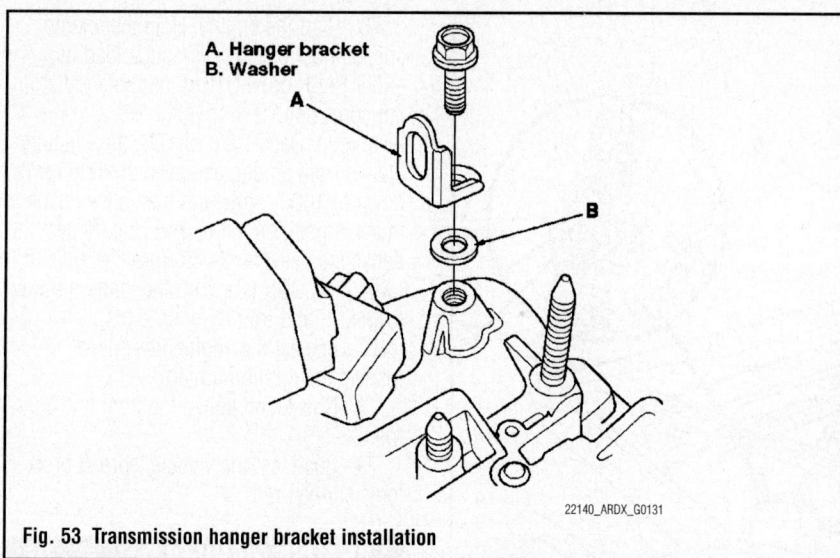

Fig. 53 Transmission hanger bracket installation

A. Hanger bracket
B. Washer

22140_ARDX_G0131

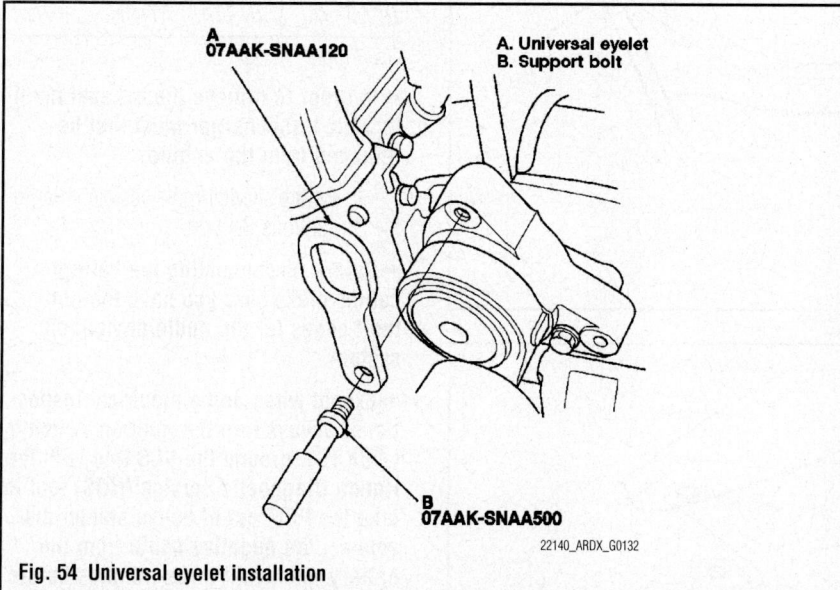

Fig. 54 Universal eyelet installation

A
07AAK-SNAA120

A. Universal eyelet
B. Support bolt

B
07AAK-SNAA500

22140_ARDX_G0132

49. Attach a chain hoist to the universal eyelet and the transmission hook. Lift up on the engine/transmission assembly until it is securely supported by the chain hoist.

Remove the engine hanger.

50. Remove the side engine mount bracket mounting bolt.

51. Check that the engine/transmission assembly is free from all wires, lines hoses etc.

52. Slowly lower the engine/transmission assembly about six inches. Check again that the engine/transmission assembly is free from all wires, lines hoses etc.

53. Disconnect the chain hoist from the assembly.

54. Raise the vehicle all the way on the lift. Remove the engine/transmission assembly from under the vehicle.

To install:

55. Install the accessory brackets.

56. Raise the vehicle on the lift. Position the engine/transmission assembly under the vehicle. Lower the vehicle and attach the universal eyelet and transmission hook to the chain hoist. Lift the assembly into position.

➡**Reinstall the mounting bolts and nuts in the correct sequence, reverse the removal procedure. Failure to follow the proper sequence may cause excessive noise and vibration and reduce engine mount life. Always use new mounting bolts and nuts.**

57. Tighten the new side engine mounting bracket bolt and nut to 47 ft. lbs.

58. Tighten the new transmission mounting bolts to 61 ft. lbs. and the new stiffener bolts to 16 ft. lbs.

59. Loosely install the new subframe bolts. When installing the subframe tighten (B) bolts to 40 ft. lbs., (A) bolts to 76 ft. lbs. and (C) bolts to 43 ft. lbs. Be sure to use new bolts. Align all reference marks on the front subframe with the body, and then tighten the mounting bolts on the subframe to specification.

60. Install the lower torque rod, then tighten the new lower torque rod mounting bolts in the sequence shown in the illustration, to the specified torque.

61. Lower the vehicle on the lift.

62. Remove the special tools.

63. Continue the installation in the reverse order of the removal procedure.

64. Tighten the steering gearbox mounting bracket bolts to 54 ft. lbs.

65. Tighten the driveshaft to transfer case companion flange retaining bolts to 53 ft. lbs. Be sure to use the reference mark made during the removal procedure for proper alignment.

66. Refill the transmission with the proper grade and type transmission fluid.

67. Refill the engine with the proper grade and type engine oil.

68. Refill the cooling system with the proper grade and fluid.

➡**The PCM relearn procedure must be performed whenever you replace the PCM, Reset the PCM, update the PCM, remove the engine or replace/clean the throttle body. Erasing DTC's with the HDS tool does not require you to perform the relearn procedure.**

69. Reset the PCM with the Honda Diagnostic System (HDS) tool while the engine is stopped. Turn the ignition switch to LOCK (0). Turn the ignition switch to ON (II), and wait thirty seconds. Turn the ignition switch to LOCK (0) and disconnect the tool from the DLC.

70. Make sure that all electrical items are turned off.

71. Reset the PCM with the HDS tool.

72. Turn the ignition switch ON (II) and wait two seconds.

73. Start the engine. Hold the engine speed at 3000 rpm's, without a load in either PARK or NEUTRAL until the radiator fan comes on or until the coolant temperature reaches 194 degrees F.

74. Let the engine idle for about five minutes with the throttle fully closed.

➡**If the radiator fan comes on, do not include its running time in the five minutes.**

75. Perform the CKP pattern clear/relearn procedure, as follows.

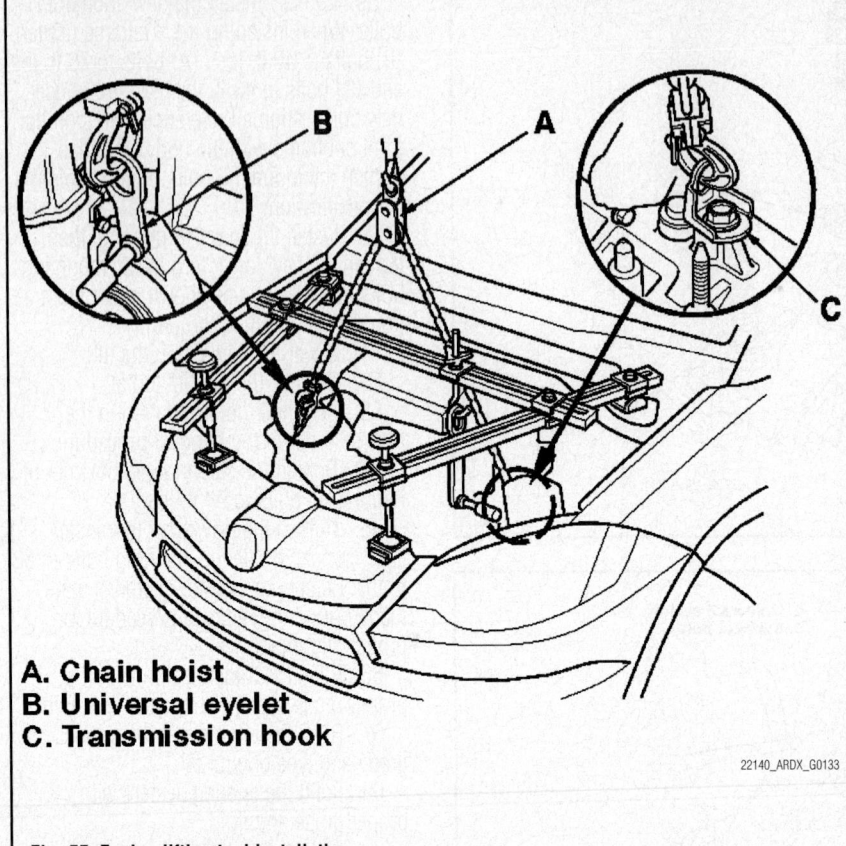

A. Chain hoist
B. Universal eyelet
C. Transmission hook

22140_ARDX_G0133

Fig. 55 Engine lifting tool installation

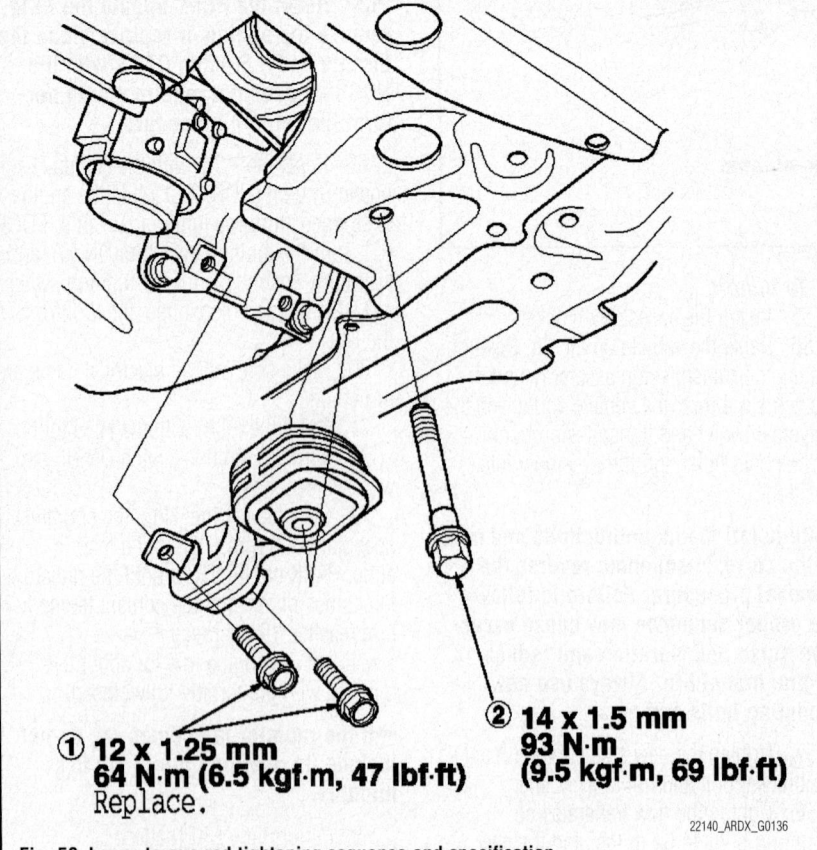

① 12 x 1.25 mm
64 N·m (6.5 kgf·m, 47 lbf·ft)
Replace.

② 14 x 1.5 mm
93 N·m
(9.5 kgf·m, 69 lbf·ft)

22140_ARDX_G0136

Fig. 56 Lower torque rod tightening sequence and specification

76. Start the engine. Hold the engine speed at 3000 rpm's, without a load in either PARK or NEUTRAL until the radiator fan comes on. Test drive the vehicle on a level road. Decelerate with the throttle fully closed from an engine speed of 2500 rpm's down to 1000 rpm's with the transmission in the S position 1st or 2nd gear. Repeat the above step several times. Turn the ignition switch to LOCK (0). Turn the ignition switch to ON (II) and wait thirty seconds.

77. Inspect the engine idle speed. Inspect the ignition timing.

78. Check and adjust the front end alignment.

79. Road test the vehicle, correct problems as required.

EXHAUST MANIFOLD

REMOVAL & INSTALLATION

See Figure 57.

➡ In order to remove the exhaust manifold the turbocharger must first be removed from the engine.

1. Before servicing the vehicle, refer to the Precautions Section.

➡ Before disconnecting the battery cables make sure you have the anti theft codes for the audio/navigation system.

➡ Except when doing electrical inspections, always turn the ignition switch to LOCK (0), ground the SCS line with the Honda Diagnostic Service (HDS) tool to take the PCM out of active status, disconnect the negative cable from the battery, then wait three minutes before starting the repair procedure. The SRS memory is not cleared even if the ignition switch is turned to LOCK (0) or the battery cables are disconnected from the battery.

2. Make sure that the ignition switch is in the LOCK (0) position.

3. Disconnect the negative battery cable.

➡ Always disconnect the negative battery cable first.

4. Disconnect the positive battery cable.

➡ Wait at least three minutes after disconnecting the battery cables before starting the repair procedure.

5. Remove the turbocharger.
6. Remove the water bypass hoses.
7. Remove the exhaust manifold retaining bolts.

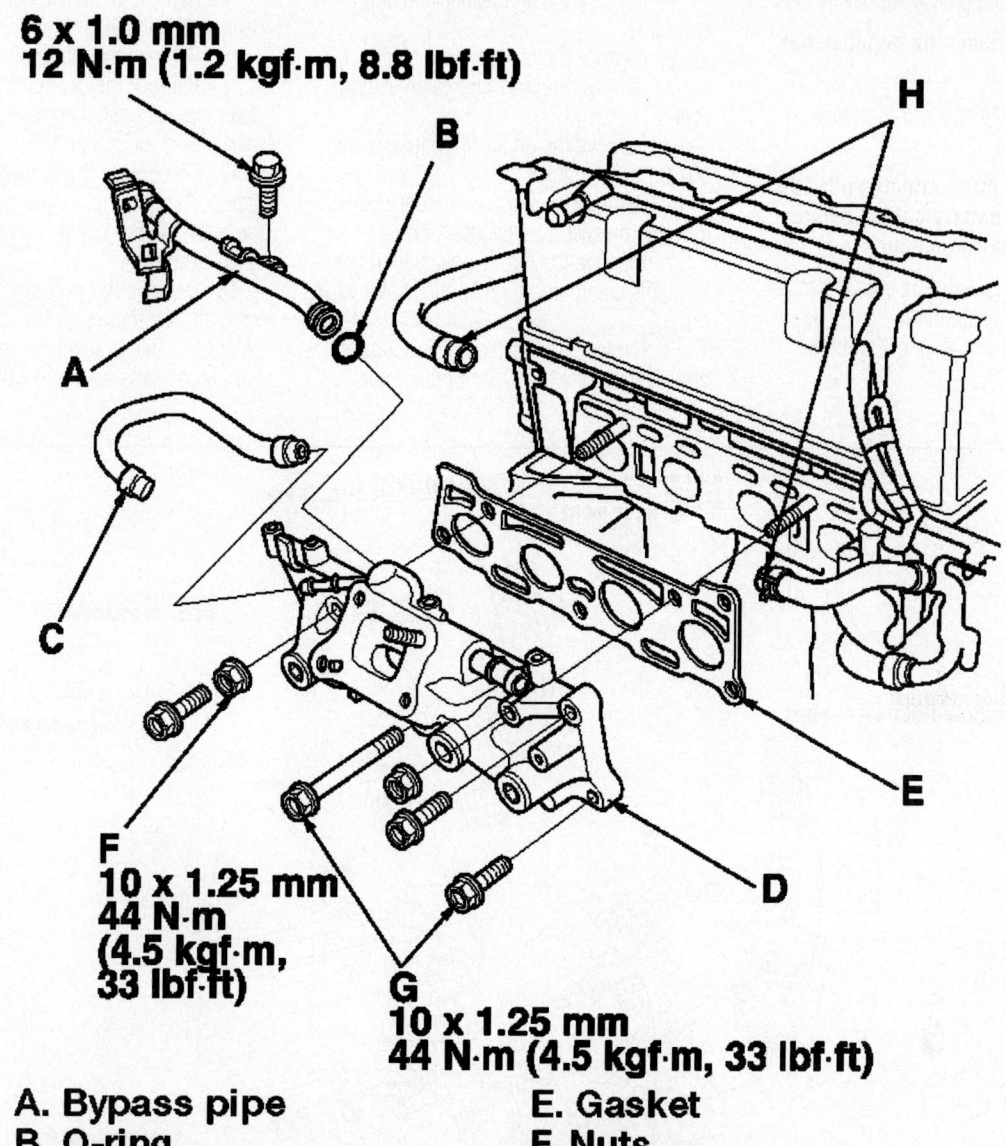

6 x 1.0 mm
12 N·m (1.2 kgf·m, 8.8 lbf·ft)

B

H

A

C

E

F
10 x 1.25 mm
44 N·m
(4.5 kgf·m,
33 lbf·ft)

G
10 x 1.25 mm
44 N·m (4.5 kgf·m, 33 lbf·ft)

D

A. Bypass pipe
B. O-ring
C. Water bypass hose
D. Exhaust manifold
E. Gasket
F. Nuts
G. Bolts
H. Water bypass hoses

22140_ARDX_G0138

Fig. 57 Exhaust manifold and related components

8. Remove the exhaust manifold from the engine.

To install:

9. Clean the mating surfaces of the engine and the exhaust manifold of dirt and debris.

10. Be sure to use a new gasket.

11. Tighten the retaining bolts and nuts to 33 ft. lbs.

12. Continue the installation in the reverse order of the removal procedure.

13. Turn the ignition switch ON (II), the SRS indicator should come on for about six seconds, and then go off.

14. Road test the vehicle, correct problems as required.

TURBOCHARGER

REMOVAL & INSTALLATION

See Figures 58 and 59.

1. Before servicing the vehicle, refer to the Precautions Section.

➡Before disconnecting the battery cables make sure you have the anti theft codes for the audio/navigation system.

➡Except when doing electrical inspections, always turn the ignition switch to LOCK (O), ground the SCS line with the Honda Diagnostic Service (HDS) tool to take the PCM out of active status, disconnect the negative cable from the battery, then wait three minutes before starting the repair procedure. The SRS memory is not cleared even if the ignition switch is turned to LOCK (O) or the battery cables are disconnected from the battery.

2. Make sure that the ignition switch is in the LOCK (O) position.

3. Disconnect the negative battery cable.

➡**Always disconnect the negative battery cable first.**

4. Disconnect the positive battery cable.

➡**Wait at least three minutes after disconnecting the battery cables before starting the repair procedure.**

5. Drain the coolant. Be sure to properly dispose of used coolant.

6. Remove the cowl panel and the under cowl panel.

7. Remove the air cleaner housing assembly.

8. Remove the charge air cooler.

9. Remove the warm up catalytic converter.

10. Remove the under hood fuse/relay box from the bracket.

11. Remove the master cylinder reservoir from the bracket. Remove the bracket.

12. Remove the heater hoses, then open the cable clamp and remove the heater valve cable.

13. Remove the heater valve mounting bolt. Remove the air bypass outlet pipe.

14. Remove the EVAP canister hose. Remove the intake air duct.

15. Remove the intake air ducts.

16. Remove the breather hose and the water bypass hose then remove the water bypass line mounting bolt.

17. Disconnect the turbocharger wastegate control solenoid valve connector and the turbocharger boost control solenoid valve connector.

18. Remove the vacuum hose, the vacuum line mounting bolt and the solenoid valve bracket mounting bolt, then loosen the solenoid valve bracket

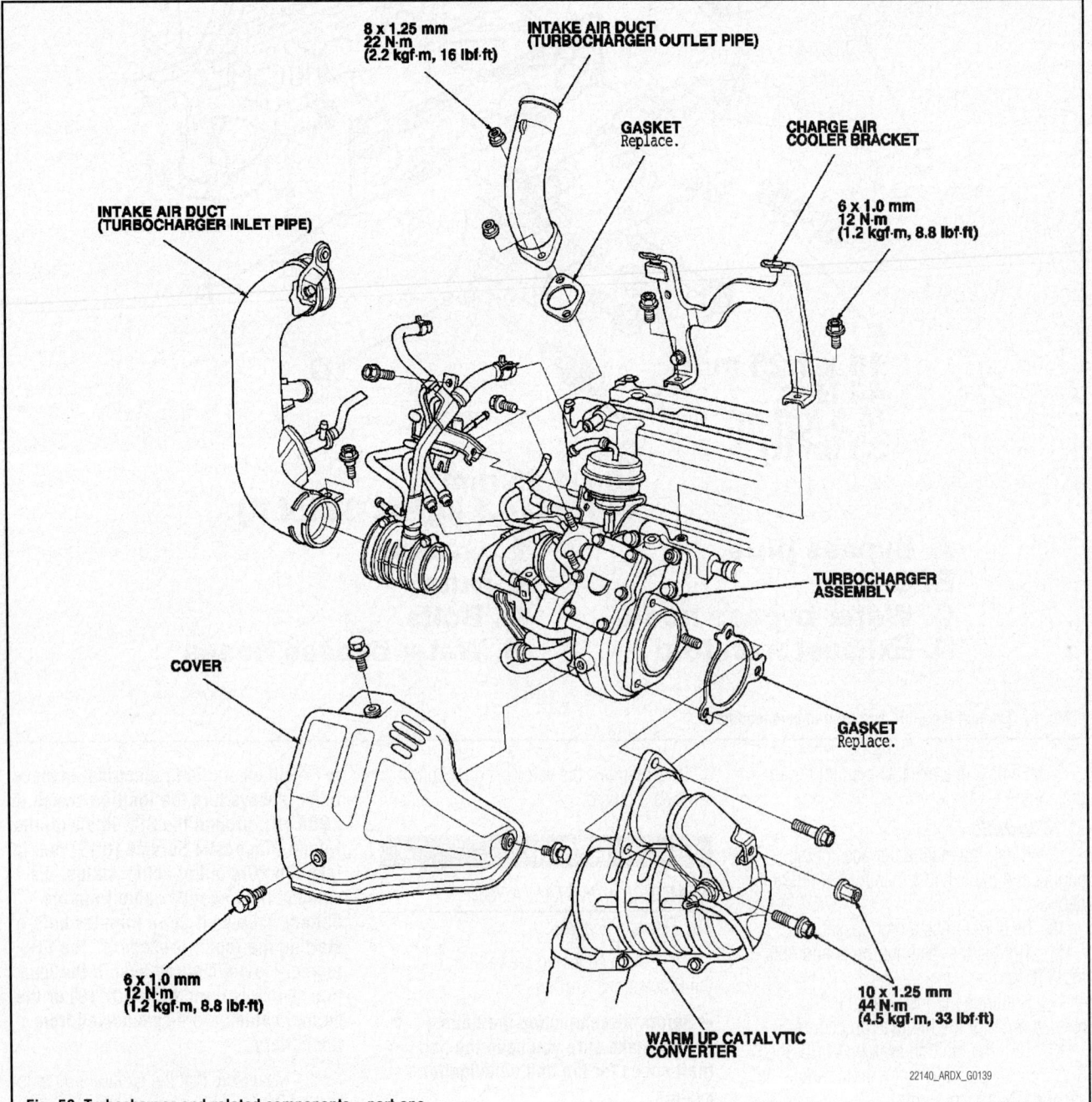

Fig. 58 Turbocharger and related components—part one

22140_ARDX_G0139

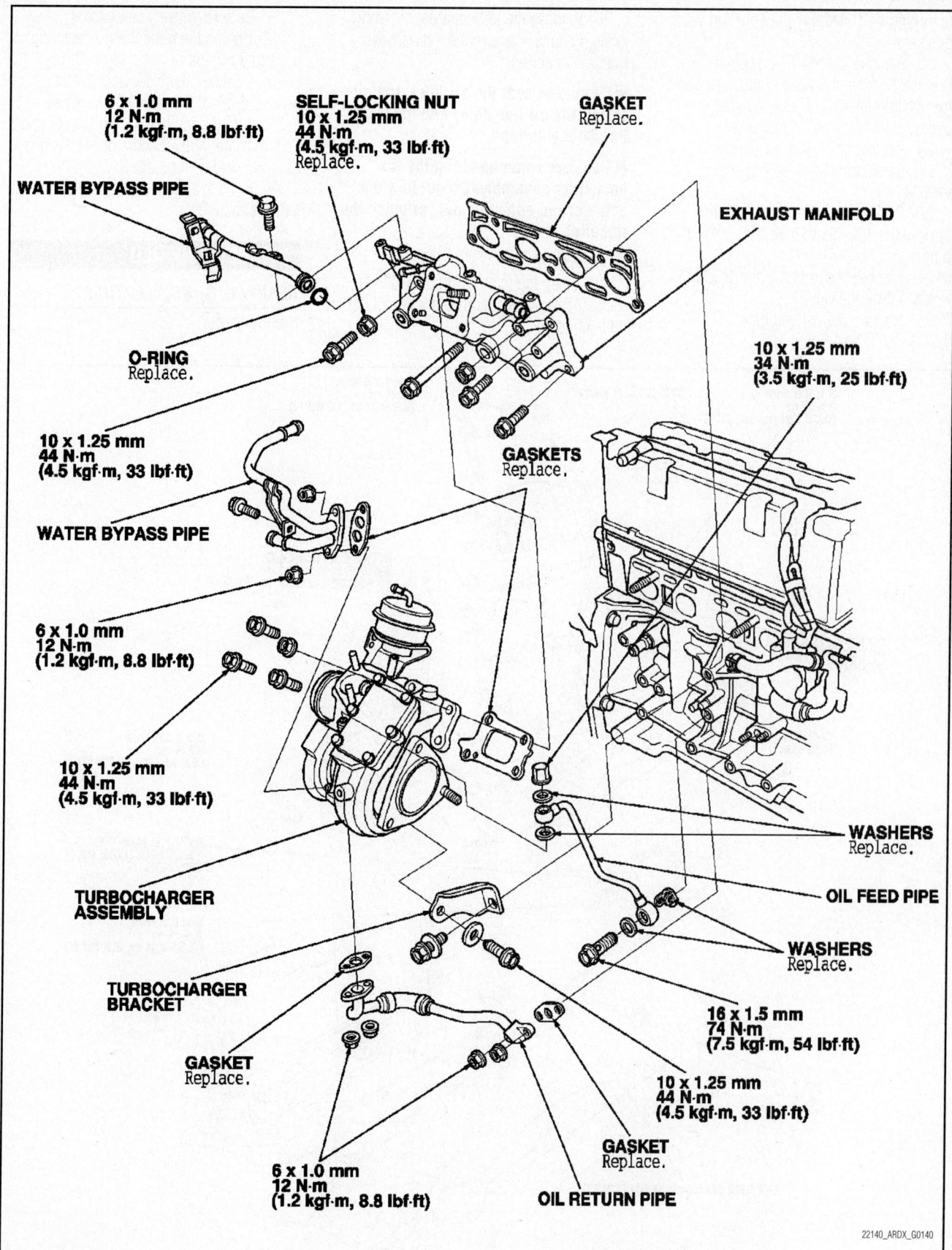

6 x 1.0 mm
12 N·m
(1.2 kgf·m, 8.8 lbf·ft)

WATER BYPASS PIPE

SELF-LOCKING NUT
10 x 1.25 mm
44 N·m
(4.5 kgf·m, 33 lbf·ft)
Replace.

GASKET
Replace.

EXHAUST MANIFOLD

O-RING
Replace.

10 x 1.25 mm
34 N·m
(3.5 kgf·m, 25 lbf·ft)

10 x 1.25 mm
44 N·m
(4.5 kgf·m, 33 lbf·ft)

WATER BYPASS PIPE

GASKETS
Replace.

6 x 1.0 mm
12 N·m
(1.2 kgf·m, 8.8 lbf·ft)

10 x 1.25 mm
44 N·m
(4.5 kgf·m, 33 lbf·ft)

WASHERS
Replace.

OIL FEED PIPE

TURBOCHARGER
ASSEMBLY

WASHERS
Replace.

TURBOCHARGER
BRACKET

16 x 1.5 mm
74 N·m
(7.5 kgf·m, 54 lbf·ft)

GASKET
Replace.

10 x 1.25 mm
44 N·m
(4.5 kgf·m, 33 lbf·ft)

GASKET
Replace.

6 x 1.0 mm
12 N·m
(1.2 kgf·m, 8.8 lbf·ft)

OIL RETURN PIPE

22140_ARDX_G0140

Fig. 59 Turbocharger and related components—part two

mounting bolt. Remove the solenoid valve bracket.

19. Remove the charge air cooler bracket. Remove the water bypass hose and the vacuum hoses. Remove the water bypass pipe from the turbocharger.

20. Remove the intake air duct.

21. Raise and safely support the vehicle.

22. Remove the oil return pipe, then remove the 16x1.5mm bolt at the oil feed pipe.

23. Remove the turbocharger bracket.

24. Lower the vehicle.

25. Remove the oil feed pipe.

26. Remove the turbocharger assembly retaining bolts. Remove the turbocharger from its mounting.

➡ **Be sure to seal the air inlet, the air outlet, the oil line hole, and the water line hole with tape.**

➡ **Use care when handling the turbocharger assembly. Do not turn the actuator rod adjusting nut, or touch the impeller.**

To install:

27. Clean the mating surfaces of the components to be sure they are free of dirt and debris.

28. Be sure to use a new gasket.

29. Tighten the turbocharger retaining bolts to 33 ft. lbs.

30. Continue the installation in the reverse order of the removal procedure.

31. Turn the ignition switch ON (II), the SRS indicator should come on for about six seconds, and then go off.

32. Road test the vehicle, correct problems as required.

INTAKE MANIFOLD

REMOVAL & INSTALLATION

See Figure 60.

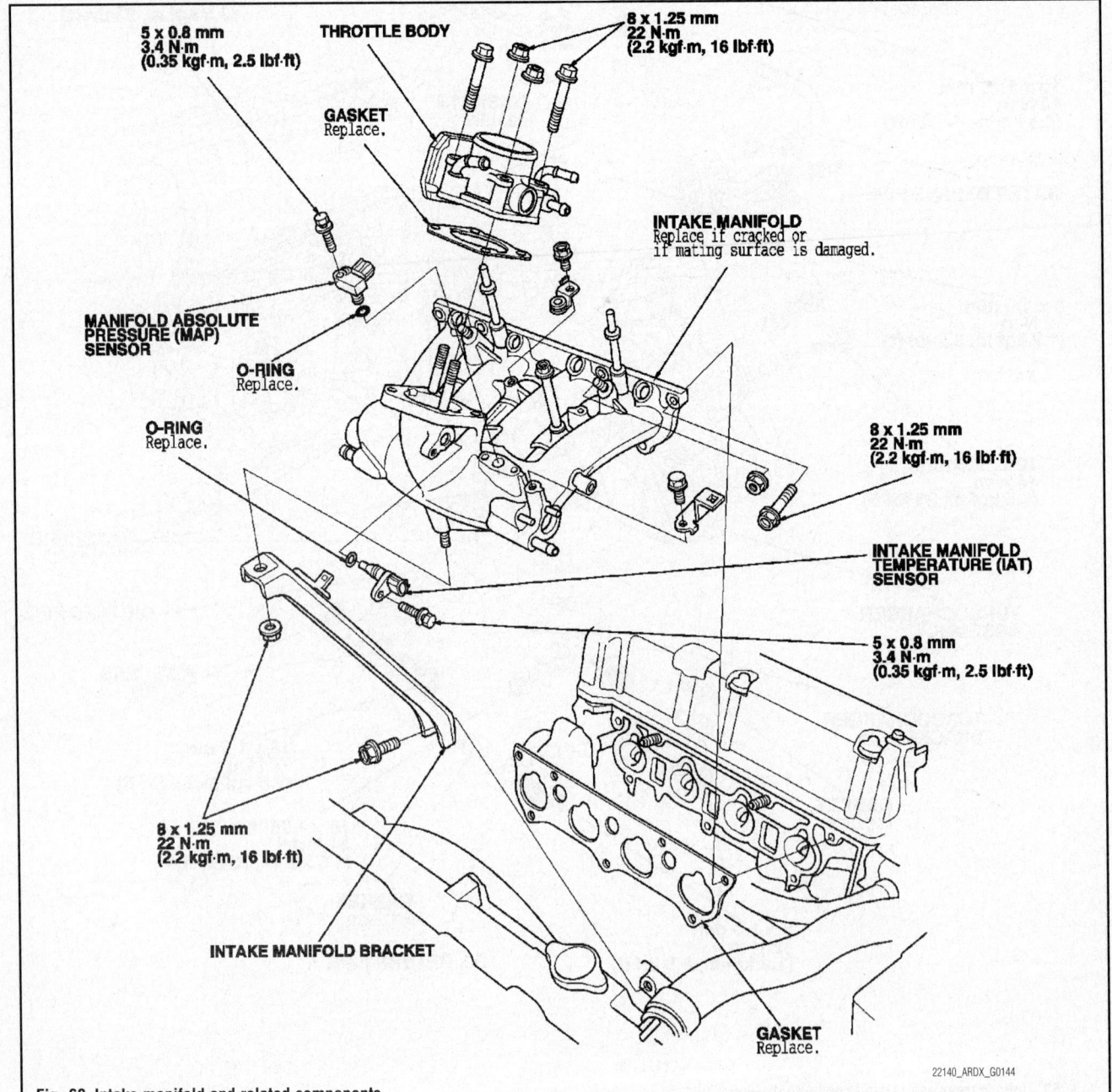

Fig. 60 Intake manifold and related components

22140_ARDX_G0144

1. Before servicing the vehicle, refer to the Precautions Section.

➡**Before disconnecting the battery cables make sure you have the anti theft codes for the audio/navigation system.**

➡**Except when doing electrical inspections, always turn the ignition switch to LOCK (O), ground the SCS line with the Honda Diagnostic Service (HDS) tool to take the PCM out of active status, disconnect the negative cable from the battery, then wait three minutes before starting the repair procedure. The SRS memory is not cleared even if the ignition switch is turned to LOCK (O) or the battery cables are disconnected from the battery.**

2. Make sure that the ignition switch is in the LOCK (O) position.
3. Disconnect the negative battery cable.

➡**Always disconnect the negative battery cable first.**

4. Disconnect the positive battery cable.

➡**Wait at least three minutes after disconnecting the battery cables before starting the repair procedure.**

5. Relieve the fuel system pressure.
6. Remove the air charge cooler cover.
7. Disconnect the turbocharger boost sensor connector. Remove the vacuum hoses and the air bypass outlet connecting tube.
8. Remove the air charge cooler retaining bolts. Remove the air charge cooler.
9. Remove the fuel injector cover. Remove the injector electrical connectors.
10. Remove the throttle actuator connector. Remove the MAP sensor connector. Remove the IAT sensor connector. Remove the turbocharger bypass control solenoid valve connector.
11. Remove the quick connect fitting cover. Disconnect the fuel feed hose.
12. Remove the ground cable, the vacuum hoses, the brake booster vacuum hoses and the turbocharger bypass control solenoid valve bracket mounting bolt.
13. Remove the EVAP canister hose and the water bypass line bracket mounting bolt.
14. Remove the water bypass hoses, plug the lines.
15. Remove the PCV hose and the harness holder.
16. Remove the harness clamp and the connector, and then remove the intake manifold bracket.

17. Remove the intake manifold retaining bolts. Remove the intake manifold from the engine.

To install:
18. Clean the mating surfaces of the components to be sure they are free of dirt and debris.
19. Be sure to use a new gasket.
20. Tighten the retaining bolts in a criss-cross pattern, beginning with the inner bolt to 16 ft. lbs.
21. Continue the installation in the reverse order of the removal procedure.
22. Turn the ignition switch ON (II), the SRS indicator should come on for about six seconds, and then go off.
23. Road test the vehicle, correct problems as required.

OIL PAN

REMOVAL & INSTALLATION
See Figures 61 through 66.

1. Before servicing the vehicle, refer to the Precautions Section.

➡**Before disconnecting the battery cables make sure you have the anti theft codes for the audio/navigation system.**

➡**Except when doing electrical inspections, always turn the ignition switch to LOCK (O), ground the SCS line with the Honda Diagnostic Service (HDS) tool to take the PCM out of active status, disconnect the negative cable from the battery, then wait three minutes before starting the repair procedure. The SRS memory is not cleared even if the ignition switch is turned to LOCK (O) or the battery cables are disconnected from the battery.**

2. Make sure that the ignition switch is in the LOCK (O) position.
3. Disconnect the negative battery cable.

➡**Always disconnect the negative battery cable first.**

4. Disconnect the positive battery cable.

➡**Wait at least three minutes after disconnecting the battery cables before starting the repair procedure.**

5. Raise and support the vehicle safely. Remove the front tire and wheel assemblies.
6. Drain the engine oil.
7. Remove the splash shield.
8. Remove the bolt securing the power

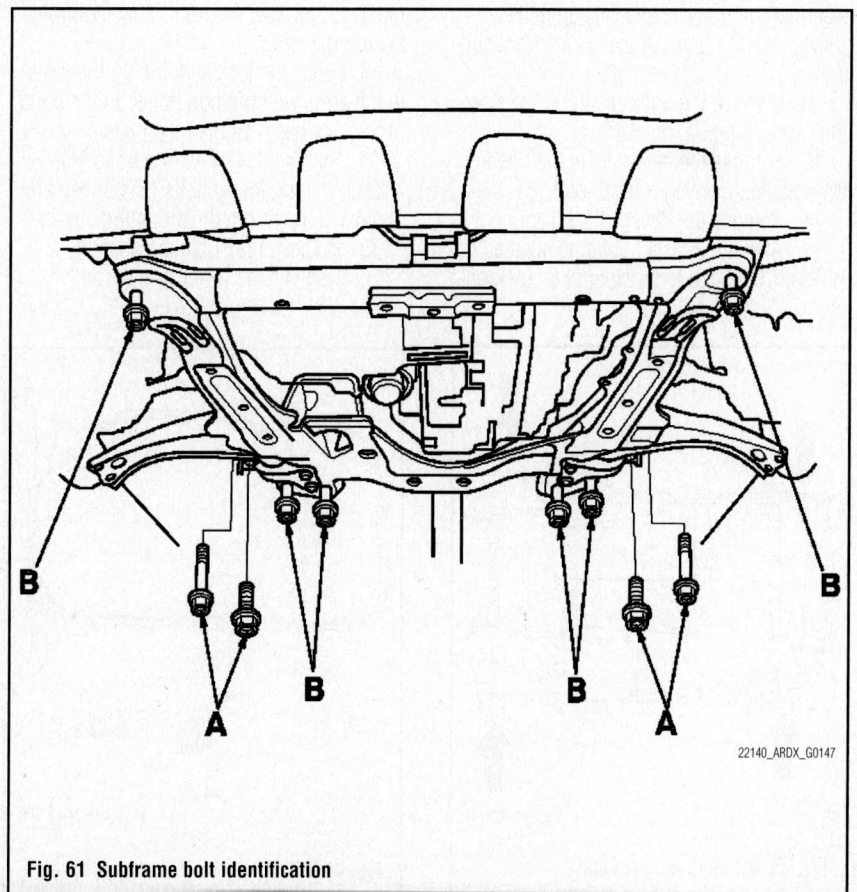

Fig. 61 Subframe bolt identification

22140_ARDX_G0147

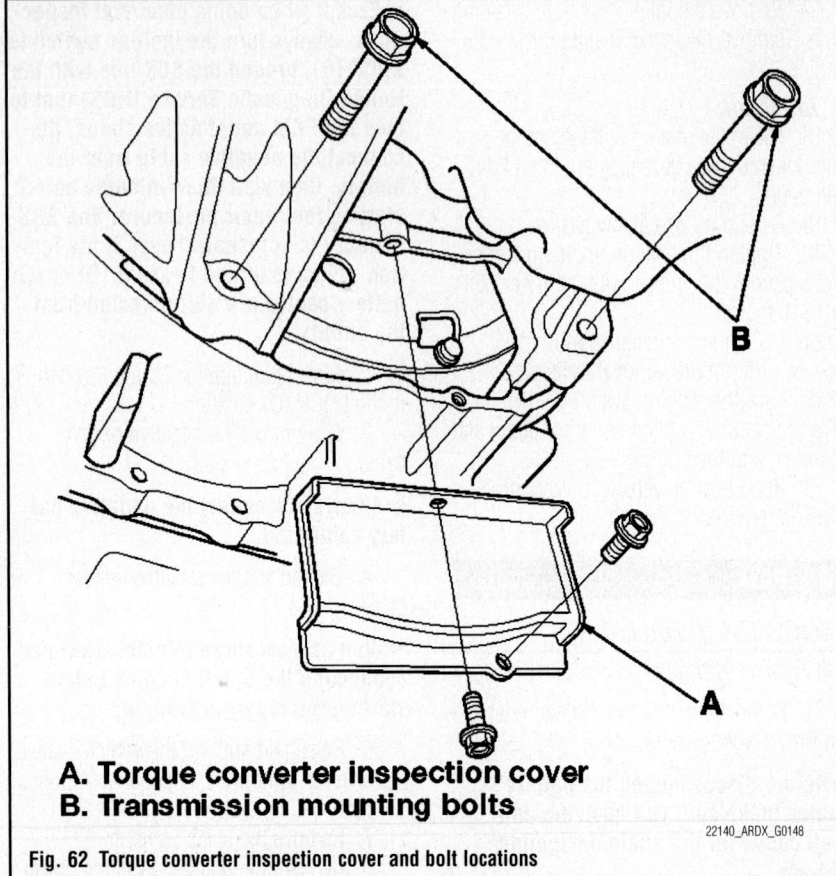

A. Torque converter inspection cover
B. Transmission mounting bolts

22140_ARDX_G0148

Fig. 62 Torque converter inspection cover and bolt locations

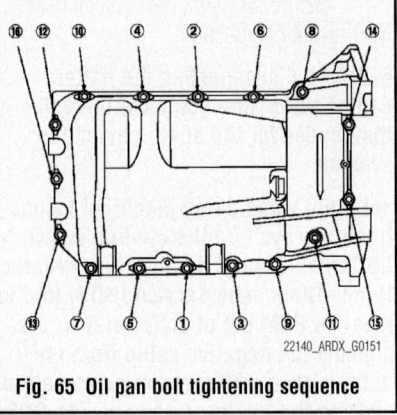

22140_ARDX_G0151

Fig. 65 Oil pan bolt tightening sequence

steering line bracket, and unclamp the power steering line clamps on the front subframe.

9. Remove the bolts retaining the steering gearbox mounting brackets.

10. Remove the two bolts at the three way catalytic converter front joint.

11. Remove the lower torque rod.

12. Attach the front subframe adapter (VSB02C000016) to the subframe, by looping the strap over the front of the subframe. Secure the strap.

13. Raise the jack and line up the slots in the arms with the bolt holes on the corner of the jack base. Tighten the bolts.

14. Remove the mounting bolts (A). Loosen the mounting bolts (B) so they are about 1/2 inch from the mounting surface.

15. Remove the subframe adapter.

16. Remove the lower torque rod

bracket. Remove the torque converter cover inspection plate. Remove the transmission mounting bolts.

17. Remove the oil pan retaining bolts.

18. Using a suitable pry tool, separate the oil pan from the block, as shown in the illustration.

To install:

19. Clean the mating surfaces of the components to be sure they are free of dirt and debris.

20. Apply liquid gasket part number 08717-0004 or equivalent to the engine block mating surface of the oil pan and to the inside edge of the threaded bolt holes.

➡**Install the component within five minutes of applying the liquid gasket. Some liquid gasket materials require installation within four minutes. Be sure to read and follow the manufacturer's instruction. If too much time has passed before installing the component remove the liquid gasket and residue, then reapply new liquid gasket.**

21. Apply liquid gasket about 0.12 inch diameter bead along the broken line (A).

22. Install the oil pan.

23. Tighten the bolts in three steps. In the final step, tighten all bolts in the sequence shown in the illustration, to 8.8 ft. lbs. Wipe off the excess liquid gasket on each side of the crankshaft pulley and drive plate.

➡**Wait at least 30 minutes to allow the gasket to cure before filling the engine with oil.**

➡**Do not run the engine for at least three hours after installing the oil pan.**

24. Continue the installation in the reverse order of the removal procedure.

25. When installing the subframe be sure to tighten the bolts as shown in the illustration.

26. Fill the engine with the proper grade and type engine oil.

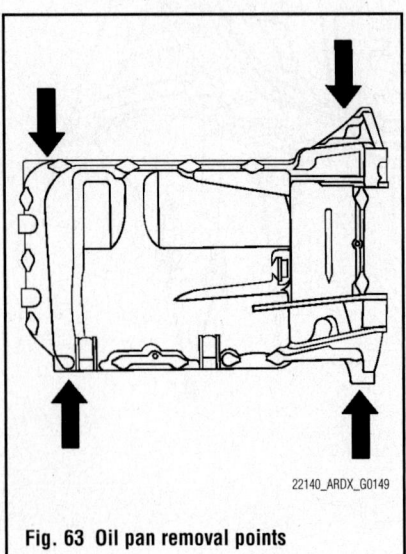

22140_ARDX_G0149

Fig. 63 Oil pan removal points

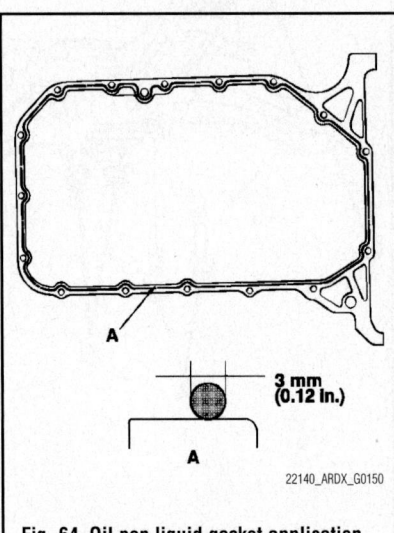

3 mm
(0.12 In.)

A

22140_ARDX_G0150

Fig. 64 Oil pan liquid gasket application

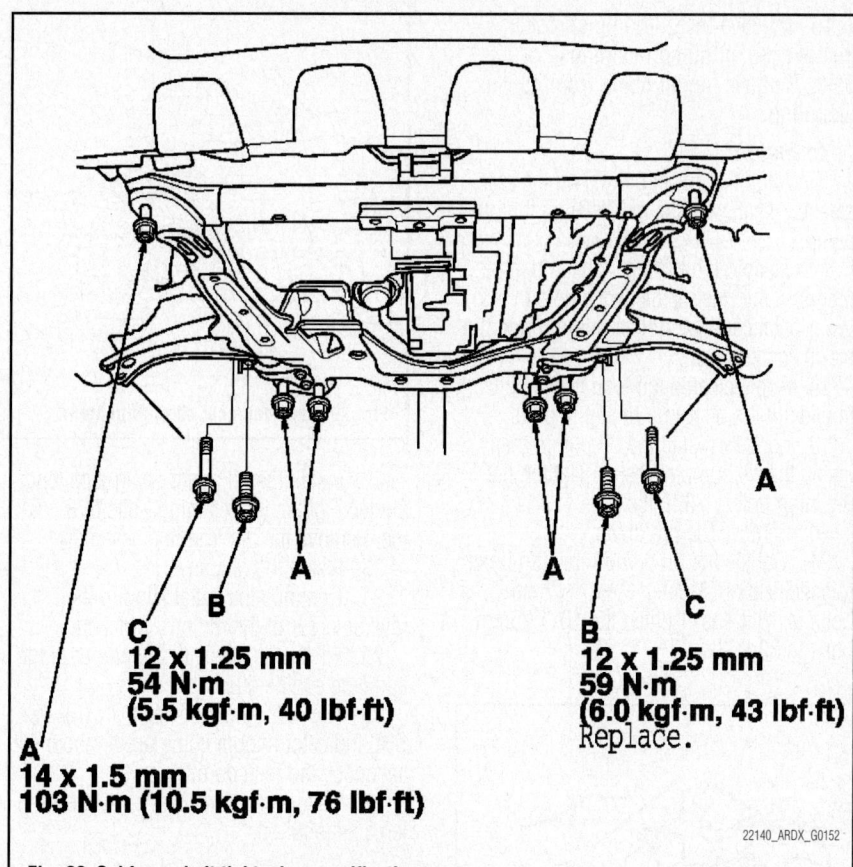

C
12 x 1.25 mm
54 N·m
(5.5 kgf·m, 40 lbf·ft)

B
12 x 1.25 mm
59 N·m
(6.0 kgf·m, 43 lbf·ft)
Replace.

A
14 x 1.5 mm
103 N·m (10.5 kgf·m, 76 lbf·ft)

22140_ARDX_G0152

Fig. 66 Subframe bolt tightening specifications

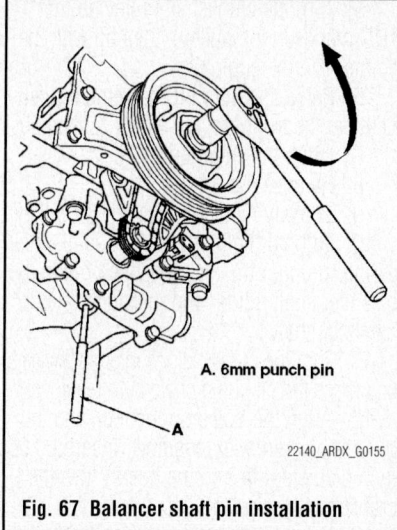

A. 6mm punch pin

22140_ARDX_G0155

Fig. 67 Balancer shaft pin installation

2. Make sure that the ignition switch is in the LOCK (0) position.

3. Disconnect the negative battery cable.

➡**Always disconnect the negative battery cable first.**

4. Disconnect the positive battery cable.

➡**Wait at least three minutes after disconnecting the battery cables before starting the repair procedure.**

27. Turn the ignition switch ON (II), the SRS indicator should come on for about six seconds, and then go off.

28. Road test the vehicle, correct problems as required.

OIL PUMP

REMOVAL & INSTALLATION

See Figures 67 through 70.

1. Before servicing the vehicle, refer to the Precautions Section.

➡**Before disconnecting the battery cables make sure you have the anti theft codes for the audio/navigation system.**

➡**Except when doing electrical inspections, always turn the ignition switch to LOCK (0), ground the SCS line with the Honda Diagnostic Service (HDS) tool to take the PCM out of active status, disconnect the negative cable from the battery, then wait three minutes before starting the repair procedure. The SRS memory is not cleared even if the ignition switch is turned to LOCK (0) or the battery cables are disconnected from the battery.**

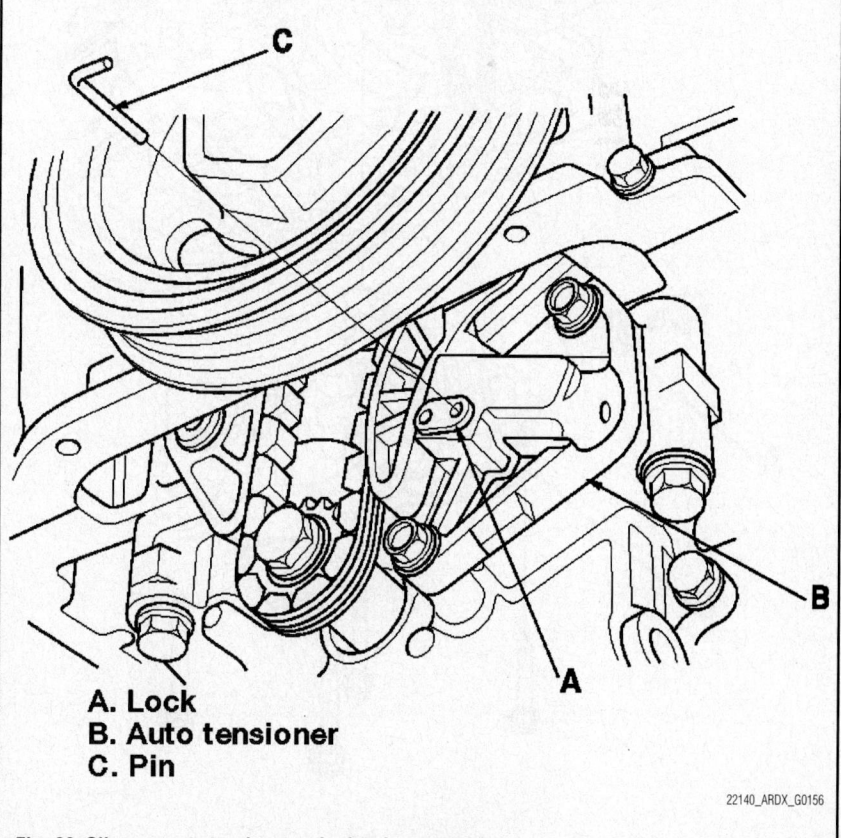

A. Lock
B. Auto tensioner
C. Pin

22140_ARDX_G0156

Fig. 68 Oil pump auto tensioner and related components

5. Turn the crankshaft pulley so that TDC mark (on the pulley) lines up with the pointer (on the engine).

6. Raise and support the vehicle safely. Remove the front tire and wheel assemblies.

7. Drain the engine oil.

8. Remove the oil pan.

9. Loosely install the crankshaft pulley.

10. To hold the rear balancer shaft, insert a 6mm pin into the maintenance hole in the balancer shaft holder and through the rear balancer shaft.

11. Turn the crankshaft counterclockwise to compress the oil pump chain auto tensioner.

12. Align the holes on the lock and the oil pump chain auto tensioner. Insert a 0.06 inch diameter pin into the holes. Turn the crankshaft clockwise to secure the pin.

13. Remove the oil pump chain auto tensioner.

14. Loosen the oil pump sprocket mounting bolt.

15. Remove the oil pump sprocket. Remove the oil pump retaining bolts. Remove the oil pump from its mounting.

To install:

16. Align the dowel pin (A) on the rear balancer shaft with the mark (B) on the oil pump.

17. To hold the balancer shaft, insert a 6mm pin into the maintenance hole in the lower balancer shaft holder and through the rear balancer shaft.

18. Apply clean engine oil to the bolt threads of the oil pump retaining bolts.

19. Loosely install the oil pump, then install the oil pump sprocket. Tighten the retaining bolt to 33 ft. lbs.

20. Remove the pin driver.

21. Tighten the oil pump retaining bolts to specification. Tighten the 8x1.25mm bolts to 16 ft. lbs. Tighten the 10x1.25mm bolts to 33 ft. lbs.

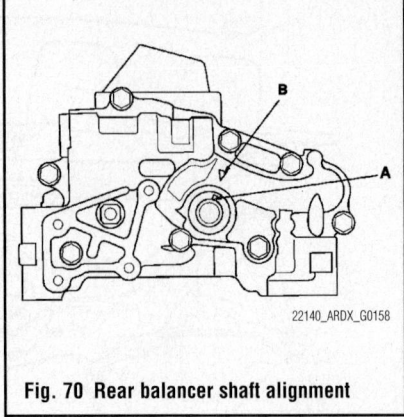

Fig. 70 Rear balancer shaft alignment

22. Install the oil pump chain auto tensioner. Tighten the retaining bolts to 8.8 ft. lbs. Remove the pin from the assembly.

23. Install the oil pan.

24. Continue the installation in the reverse order of the removal procedure.

25. Fill the engine with the proper grade and type engine oil.

26. Turn the ignition switch ON (II), the SRS indicator should come on for about six seconds, and then go off.

27. Road test the vehicle, correct problems as required.

PISTON AND RING

POSITIONING

See Figure 71.

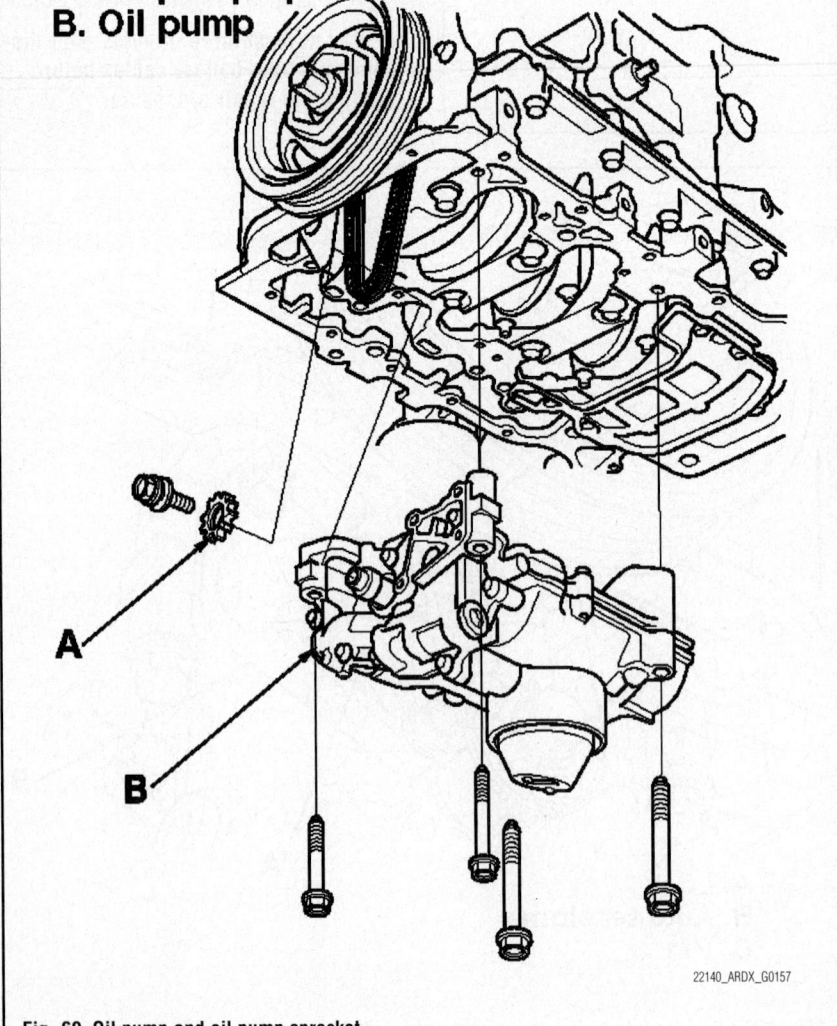

A. Oil pump sprocket
B. Oil pump

Fig. 69 Oil pump and oil pump sprocket

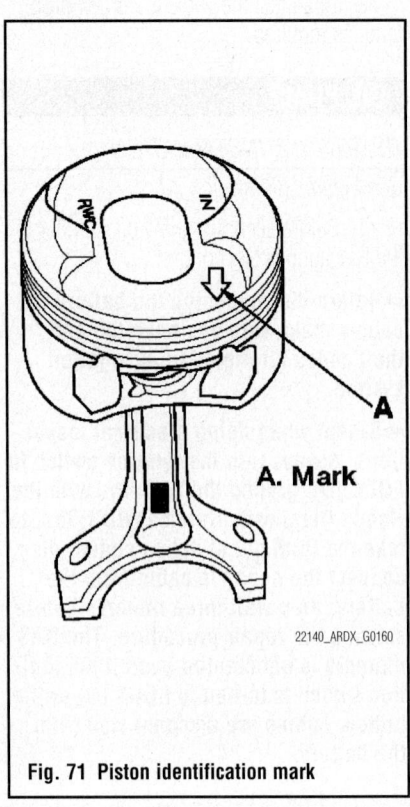

A. Mark

Fig. 71 Piston identification mark

REAR MAIN SEAL

REMOVAL & INSTALLATION

See Figures 72 and 73.

1. Before servicing the vehicle, refer to the Precautions Section.

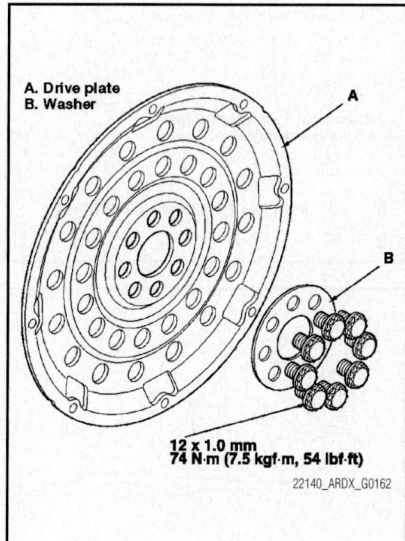

A. Drive plate
B. Washer

12 x 1.0 mm
74 N·m (7.5 kgf·m, 54 lbf·ft)

22140_ARDX_G0162

Fig. 72 Drive plate and related components

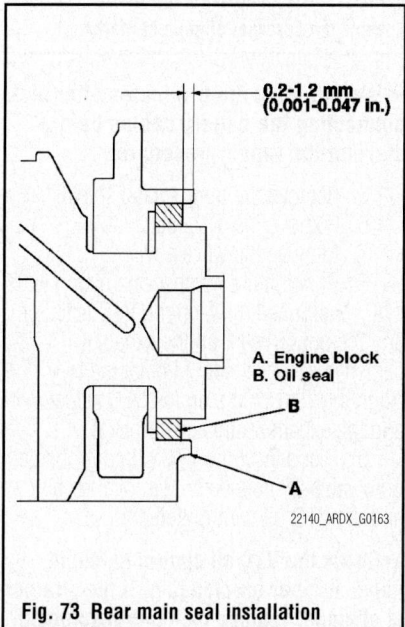

0.2-1.2 mm
(0.001-0.047 in.)

A. Engine block
B. Oil seal

22140_ARDX_G0163

Fig. 73 Rear main seal installation specification

➡**Before disconnecting the battery cables make sure you have the anti theft codes for the audio/navigation system.**

➡**Except when doing electrical inspections, always turn the ignition switch to** LOCK (O), ground the SCS line with the Honda Diagnostic Service (HDS) tool to take the PCM out of active status, disconnect the negative cable from the battery, then wait three minutes before starting the repair procedure. **The SRS memory is not cleared even if the ignition switch is turned to LOCK (O) or the battery cables are disconnected from the battery.**

2. Make sure that the ignition switch is in the LOCK (O) position.
3. Disconnect the negative battery cable.

➡**Always disconnect the negative battery cable first.**

4. Disconnect the positive battery cable.

➡**Wait at least three minutes after disconnecting the battery cables before starting the repair procedure.**

5. Remove the transmission.
6. Remove the drive plate.
7. Remove the seal from its mounting, using the proper seal removal tool.

To install:

8. Clean and dry the crankshaft oil seal housing.
9. Use the driver tool and the attachment to drive a new seal squarely into the block, to the specified installed height.
10. Measure the distance between the engine block and the oil seal. Specified height should be 0.001–0.047 inch.
11. Install the drive plate. Tighten the bolts to specification and in a crisscross pattern. Tightening specification is 54 ft. lbs.
12. Continue the installation in the reverse order of the removal procedure.
13. Turn the ignition switch ON (II), the SRS indicator should come on for about six seconds, and then go off.
14. Road test the vehicle, correct problems as required.

TIMING CHAIN, SPROCKETS, FRONT COVER AND SEAL

REMOVAL & INSTALLATION

See Figures 74 through 89.

1. Before servicing the vehicle, refer to the Precautions Section.

➡**Before disconnecting the battery cables make sure you have the anti theft codes for the audio/navigation system.**

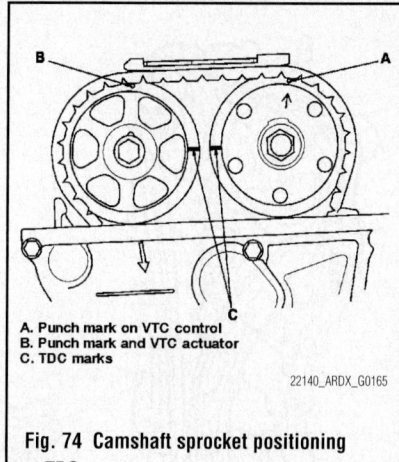

A. Punch mark on VTC control
B. Punch mark and VTC actuator
C. TDC marks

22140_ARDX_G0165

Fig. 74 Camshaft sprocket positioning at TDC

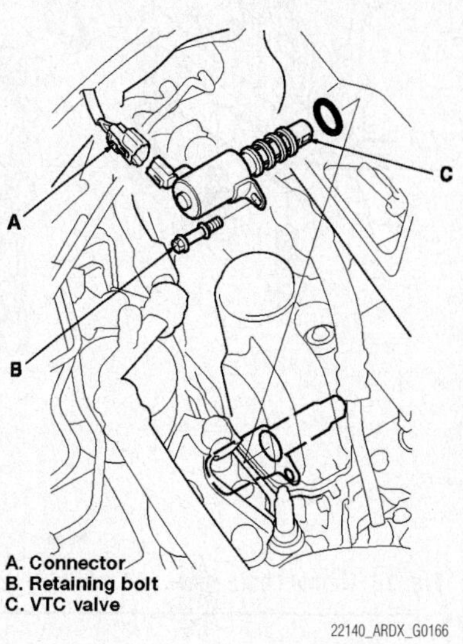

A. Connector
B. Retaining bolt
C. VTC valve

22140_ARDX_G0166

Fig. 75 VTC solenoid valve location

➡**Except when doing electrical inspections, always turn the ignition switch to LOCK (O), ground the SCS line with the Honda Diagnostic Service (HDS) tool to take the PCM out of active status, disconnect the negative cable from the battery, then wait three minutes before starting the repair procedure. The SRS memory is not cleared even if the ignition switch is turned to LOCK (O) or the battery cables are disconnected from the battery.**

2. Make sure that the ignition switch is in the LOCK (O) position.
3. Disconnect the negative battery cable.

➡**Always disconnect the negative battery cable first.**

4. Disconnect the positive battery cable.

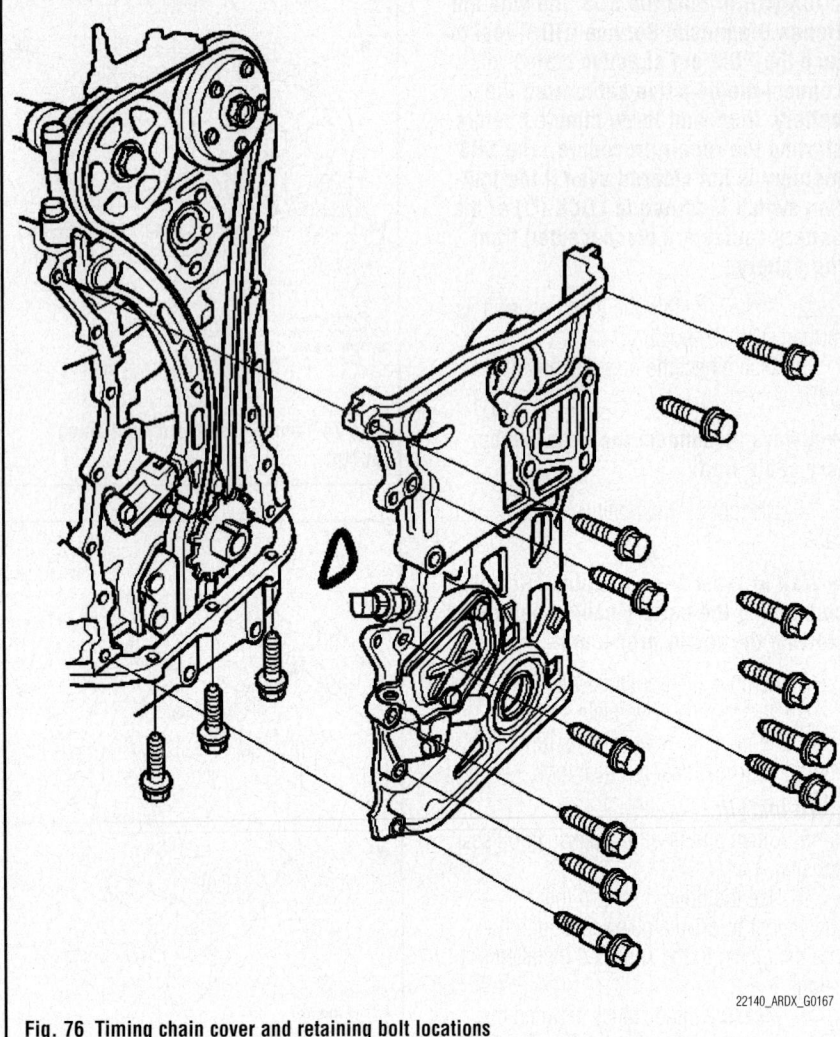

Fig. 76 Timing chain cover and retaining bolt locations

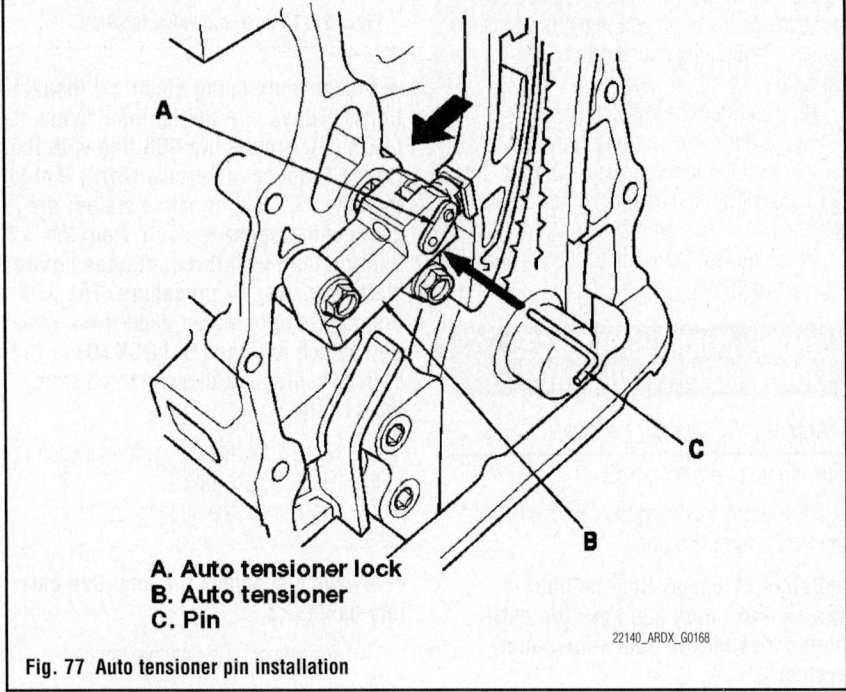

A. Auto tensioner lock
B. Auto tensioner
C. Pin

Fig. 77 Auto tensioner pin installation

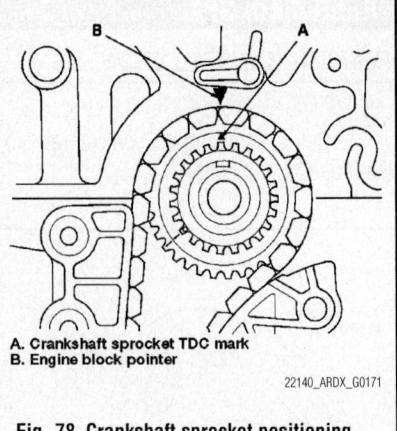

A. Crankshaft sprocket TDC mark
B. Engine block pointer

Fig. 78 Crankshaft sprocket positioning at TDC

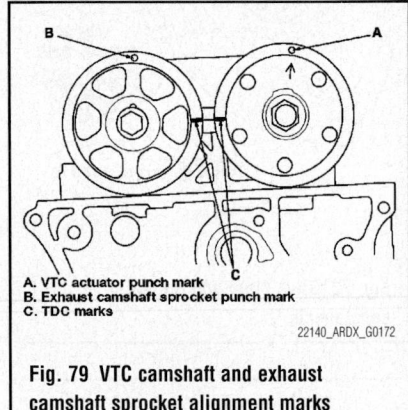

A. VTC actuator punch mark
B. Exhaust camshaft sprocket punch mark
C. TDC marks

Fig. 79 VTC camshaft and exhaust camshaft sprocket alignment marks

➡**Wait at least three minutes after disconnecting the battery cables before starting the repair procedure.**

5. Remove the front splash shield.
6. Remove the drive belt.
7. Remove the valve cover.
8. Position the number one piston to TDC. The punch mark on the VTC actuator and the punch mark on the exhaust camshaft sprocket should be at the top. Align the TDC marks on the VTC actuator and the exhaust camshaft sprocket.
9. Disconnect the VTC oil control solenoid valve 2P connector. Remove the bolt and the VTC oil control solenoid valve.

➡**Check the VTC oil control solenoid valve strainer for clogging. If the strainer is clogged, replace the control solenoid.**

10. Remove the crankshaft pulley retaining bolt. Remove the crankshaft pulley.
11. Properly support the engine with a suitable jack and block of wood under the oil pan.
12. Remove the torque rod stiffener and upper torque rod.
13. Remove the ground cable. Remove the side engine mount bracket.

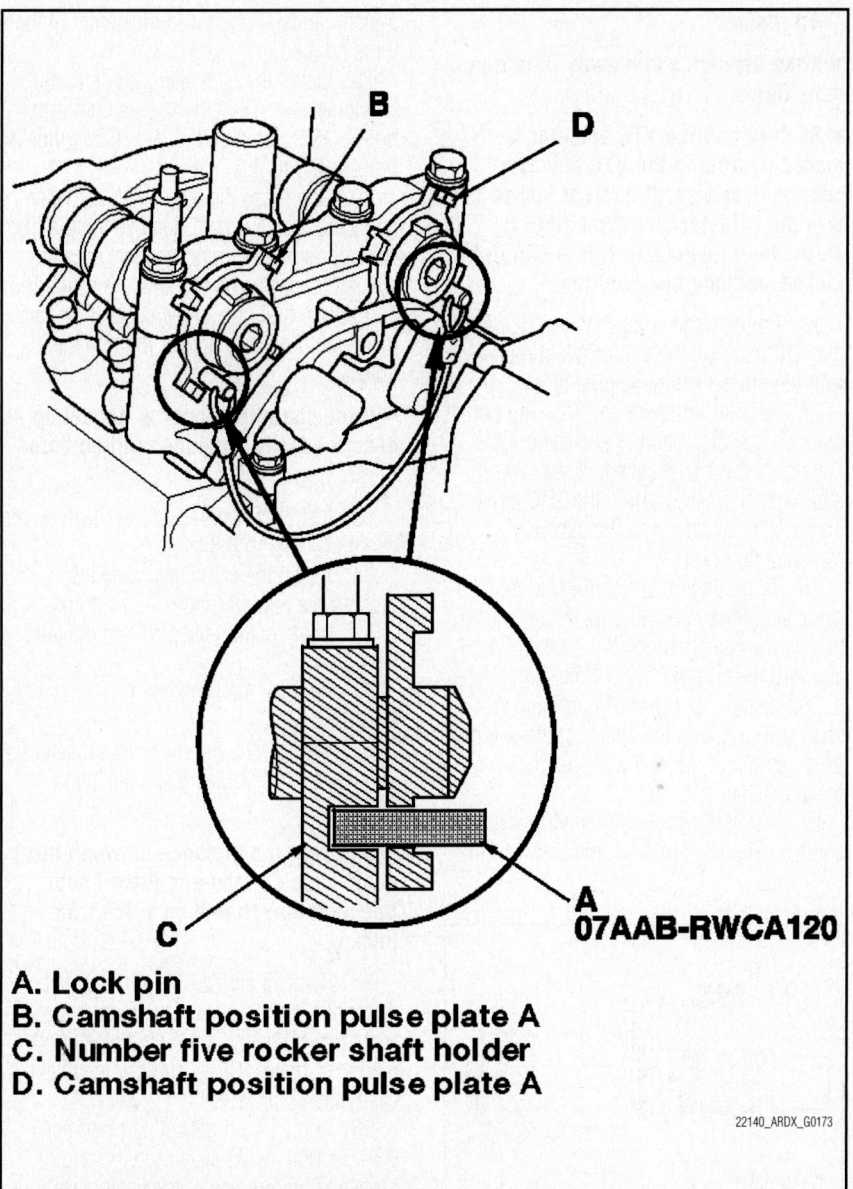

A. Lock pin
B. Camshaft position pulse plate A
C. Number five rocker shaft holder
D. Camshaft position pulse plate A

07AAB-RWCA120

22140_ARDX_G0173

Fig. 80 Holding the intake camshaft in position

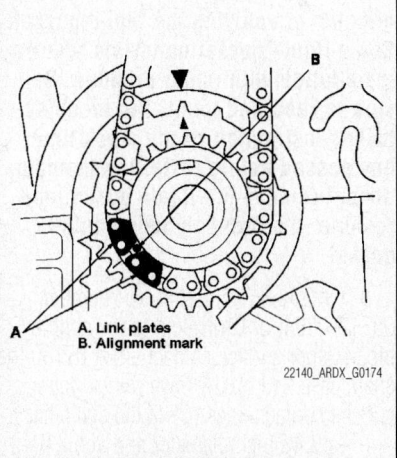

A. Link plates
B. Alignment mark

22140_ARDX_G0174

Fig. 81 Timing chain link plate location and identification at the crankshaft

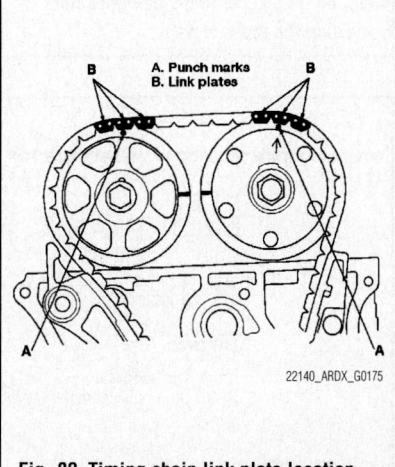

A. Punch marks
B. Link plates

22140_ARDX_G0175

Fig. 82 Timing chain link plate location and identification at the camshaft

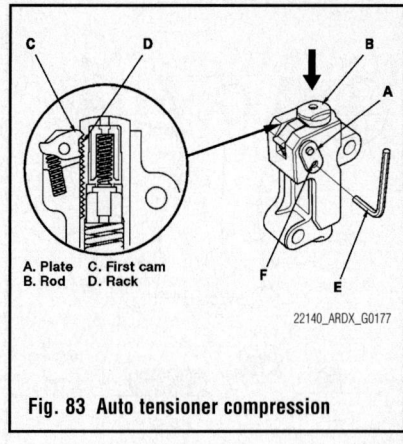

A. Plate   C. First cam
B. Rod   D. Rack

22140_ARDX_G0177

Fig. 83 Auto tensioner compression

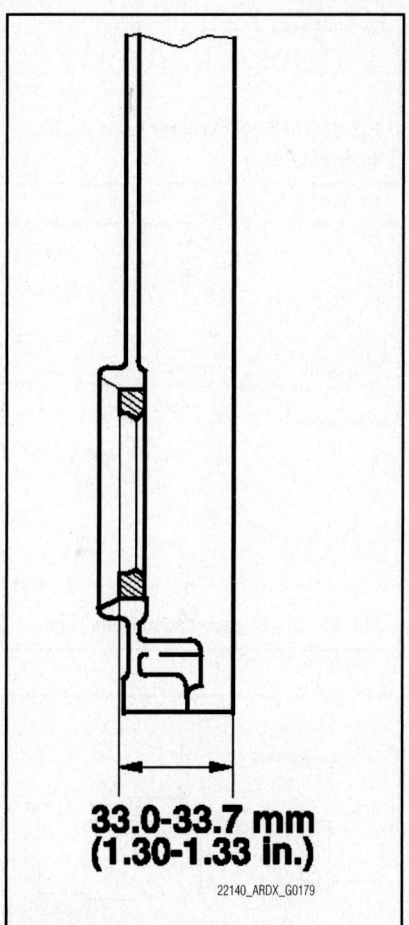

**33.0-33.7 mm
(1.30-1.33 in.)**

22140_ARDX_G0179

Fig. 84 Timing chain cover seal positioning

14. Remove the water bypass mounting bolt. Remove the engine mount bracket.

15. Remove the timing chain case cover retaining bolts. Remove the cover from its mounting on the engine.

16. Loosely install the crankshaft pulley.

17. Turn the crankshaft counterclockwise to compress the auto tensioner.

18. Align the holes on the lock and the auto tensioner. Insert a 0.06 inch diameter pin into the holes. Turn the crankshaft clockwise to secure the pin.

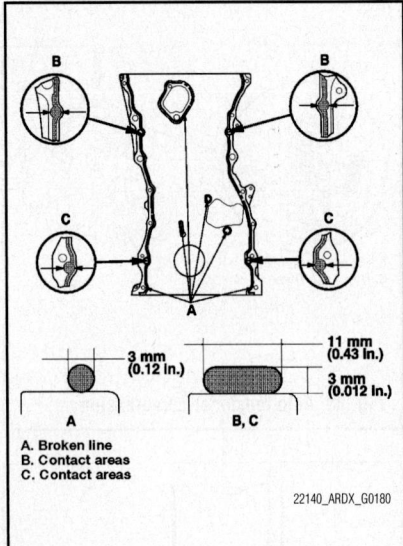

Fig. 85 Timing chain cover liquid gasket application points

A. Broken line
B. Contact areas
C. Contact areas

22140_ARDX_G0180

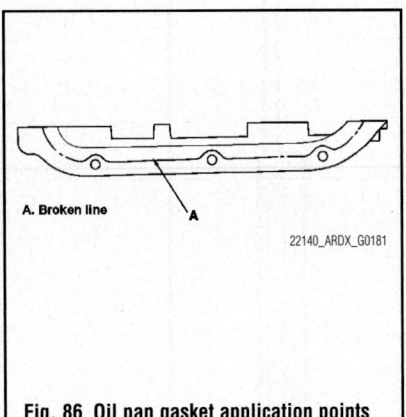

A. Broken line

22140_ARDX_G0181

Fig. 86 Oil pan gasket application points

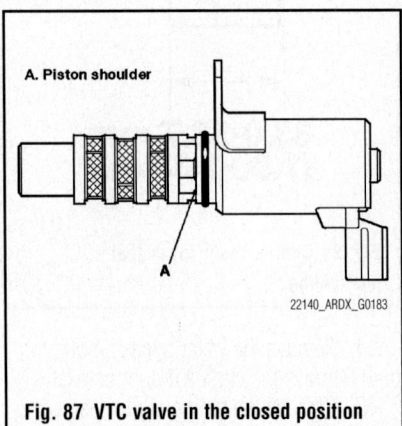

A. Piston shoulder

22140_ARDX_G0183

Fig. 87 VTC valve in the closed position

19. Remove the auto tensioner assembly retaining bolts. Remove the auto tensioner from its mounting.
20. Remove the cam chain guide B.
21. Remove the cam chain guide A and the tensioner arm.
22. Remove the cam chain.

To install:

➡ **Keep the cam chain away from magnetic fields.**

➡ **Be sure that the VTC actuator is locked by turning the VTC actuator counterclockwise. If it is not locked, turn the actuator clockwise until it stops, then recheck it. If it is still not locked, replace the actuator.**

23. Position the crankshaft at TDC. Align the TDC mark on the crankshaft sprocket with the pointer on the engine block.
24. Set the camshafts to TDC. The punch mark on the VTC actuator and the punch mark on the exhaust camshaft sprocket should be at the top. Align the TDC marks on the VTC actuator and the exhaust camshaft sprocket.
25. To hold the intake camshaft, insert a camshaft lock pin (A) into the maintenance hole in the camshaft position pulse plate A and thru the number five rocker shaft holder.
26. Install the cam chain on the crankshaft sprocket with the center of the two colored link plates aligned with the mark on the crankshaft sprocket.
27. Install the chain on the VTC actuator and the exhaust camshaft sprocket with the

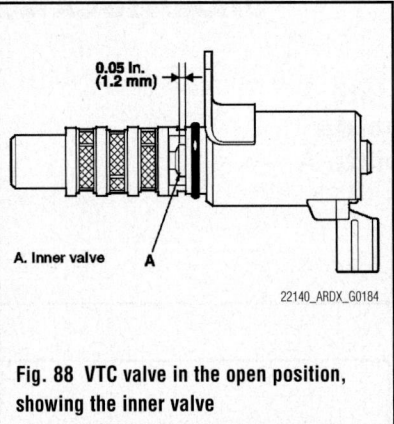

0.05 in. (1.2 mm)

A. Inner valve

22140_ARDX_G0184

Fig. 88 VTC valve in the open position, showing the inner valve

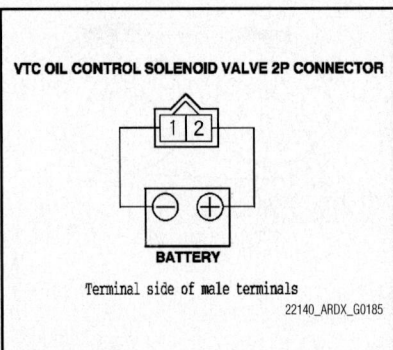

VTC OIL CONTROL SOLENOID VALVE 2P CONNECTOR

1 2

BATTERY

Terminal side of male terminals

22140_ARDX_G0185

Fig. 89 VTC terminal positioning and connection points

punch marks aligned with the center of the three colored link plates.

28. Install the cam chain guide A and tensioner arm. Tighten the tensioner arm bolt to 16 ft. lbs and the cam chain guide A bolts to 8.8 ft. lbs.
29. Compress the auto tensioner when replacing the camshaft. Remove the pin from the auto tensioner. Turn the plate counterclockwise, to release the lock. Press the rod and set the first cam to the edge of the rack. Insert a 0.05 inch diameter pin into the holes.

➡ **If the chain tensioner is not set up as described, the tensioner will be damaged.**

30. Install the auto tensioner. Tighten the retaining bolts to 8.8 ft. lbs.
31. Install the cam chain guide B. Tighten the retaining bolts to 16 ft. lbs.
32. Remove the lock pin from the auto tensioner.
33. Remove the camshaft lock pin set (tool).
34. Replace the timing cover oil seal. Be sure to use the proper seal installation tools.

➡ **Measure the distance between the chain case surface and the oil seal. Specification should be 1.30–1.33 inch.**

35. Remove all the old liquid gasket material from the chain case mating surfaces, the bolts and the bolts holes of all mating surfaces. Clean and dry the chain case mating surfaces.
36. Apply liquid gasket part number 08717-0004, or equivalent to the engine block mating surface of the chain case. Apply a 0.12 inch diameter bead along the broken line (A), shown in the illustration.

➡ **Install the component within five minutes of applying the liquid gasket. Some liquid gasket materials require installation within four minutes. Be sure to read and follow the manufacturer's instruction. If too much time has passed before installing the component remove the liquid gasket and residue, then reapply new liquid gasket.**

37. Apply liquid gasket part number 08717-0004, or equivalent to the engine block upper surface contact areas (B) on the chain case and on the lower block upper surface contact areas (C) of the chain case. Apply a 0.43 inch diameter and about 0.12 inch thickness of gasket to areas (B) and (C), shown in the illustration.

38. Apply liquid gasket part number 08717-0004, or equivalent to the oil pan mating surface of the chain case. Apply a 0.12 inch diameter bead along the broken line (A), shown in the illustration.

➡**Install the component within five minutes of applying the liquid gasket. Some liquid gasket materials require installation within four minutes. Be sure to read and follow the manufacturer's instruction. If too much time has passed before installing the component remove the liquid gasket and residue, then reapply new liquid gasket.**

39. Install a new O-ring on the chain case. Set the edge of the chain case to the edge of the oil pan. Install the chain case on the engine block. Wipe any excess liquid gasket on the oil pan and chain case mating surfaces. When installing the chain case, do not slide the bottom surface onto the oil pan mounting surface.

➡**Wait at least 30 minutes to allow the gasket to cure before filling the engine with oil.**

➡**Do not run the engine for at least three hours after installing the valve cover.**

40. Tighten the cover retaining bolts to specification. Tightening specification is 8.8 ft. lbs.

41. Continue the installation in the reverse order of the removal procedure.

42. When installing the VTC valve note the amount of valve opening by observing the position of the shoulder piston, thru the valve retard drain valve. If you see the shoulder of the piston, the valve is open and must be replaced.

43. Connect the battery positive terminal of the VTC terminal to the solenoid valve 2P connector terminal number two. Connect the battery negative terminal to the VTC solenoid valve 2P connector terminal number one. Refer to the accompanying illustration.

44. Appearance of the inner valve in the port should be at least 0.05 inch. If the inner valve does not open, replace it and repeat the procedure.

45. Perform the CKP pattern clear/ relearn procedure as follows. Start the engine. Hold the engine speed at 3000 rpm's, without a load in either PARK or NEUTRAL until the radiator fan comes on. Test drive the vehicle on a level road. Decelerate with the throttle fully closed from an engine speed of 2500 rpm's down to 1000 rpm's with the transmission in the

S position 1st or 2nd gear. Repeat the above step several times. Turn the ignition switch to LOCK (0). Turn the ignition switch to ON (II) and wait thirty seconds.

VALVE LASH

ADJUSTMENT

See Figures 90 through 100.

1. Before servicing the vehicle, refer to the Precautions Section.

➡**Before disconnecting the battery cables make sure you have the anti theft codes for the audio/navigation system.**

➡**Except when doing electrical inspections, always turn the ignition switch to LOCK (0), ground the SCS line with the Honda Diagnostic Service (HDS) tool to**

take the PCM out of active status, disconnect the negative cable from the battery, then wait three minutes before starting the repair procedure. The SRS memory is not cleared even if the ignition switch is turned to LOCK (0) or the battery cables are disconnected from the battery.

2. Make sure that the ignition switch is in the LOCK (0) position.
3. Disconnect the negative battery cable.

➡**Always disconnect the negative battery cable first.**

4. Disconnect the positive battery cable.

➡**Wait at least three minutes after disconnecting the battery cables before starting the repair procedure.**

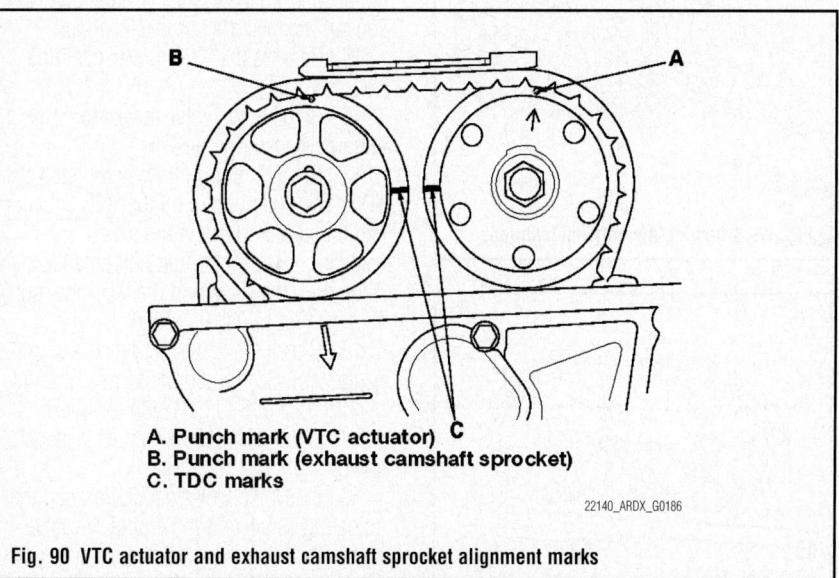

A. Punch mark (VTC actuator)
B. Punch mark (exhaust camshaft sprocket)
C. TDC marks

22140_ARDX_G0186

Fig. 90 VTC actuator and exhaust camshaft sprocket alignment marks

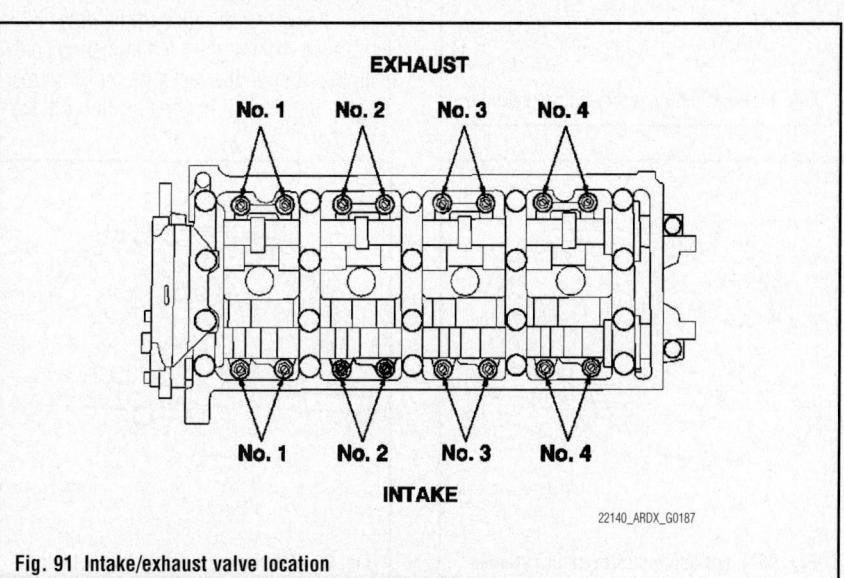

22140_ARDX_G0187

Fig. 91 Intake/exhaust valve location

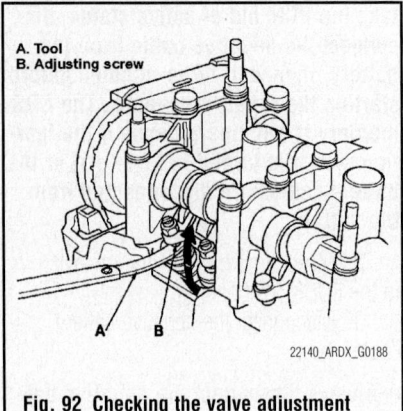

Fig. 92 Checking the valve adjustment

A. Tool
B. Adjusting screw

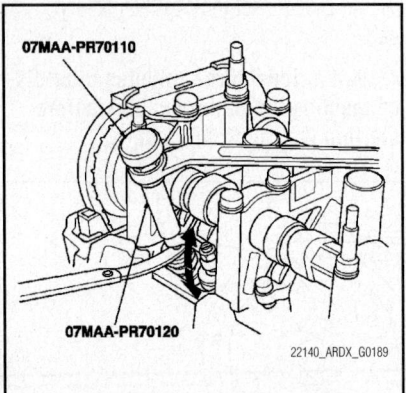

07MAA-PR70110

07MAA-PR70120

Fig. 93 Adjusting the valve adjustment

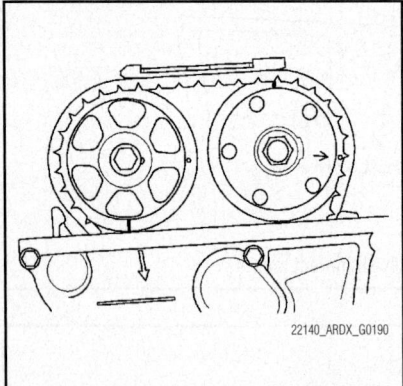

Fig. 94 Positioning number three cylinder

5. Remove the air charge cooler cover.

6. Disconnect the turbocharger boost sensor connector. Remove the vacuum hoses and the air bypass outlet connecting tube.

7. Remove the air charge cooler retaining bolts. Remove the air charge cooler.

8. Remove the fuel injector cover.

9. Remove the ignition coil cover. Disconnect the ignition coil connectors.

10. Remove the ignition coils.

11. Remove the power steering hose bracket, the breather hose and the dipstick.

12. Remove the bracket mounting bolts, from the valve cover.

13. Disconnect the rocker arm oil control solenoid connector, the rocker are oil pressure switch connector and the variable valve timing control (VTC) oil control solenoid valve connector. Remove the harness clamp.

14. Disconnect the air fuel ration (AF) sensor connector, the oil pressure switch connector and the crankshaft position (CKP) sensor. Remove the harness clamps.

15. Remove the valve cover retaining bolts.

16. Remove the valve cover from the engine. Discard the gasket.

17. Position the number one piston at TDC. The punch mark on the VTC actuator and the punch mark on the exhaust camshaft sprocket should be at the top. Align the TDC marks on the VTC actuator and exhaust camshaft sprocket.

18. Check the valve clearance, using a gauge tool. See illustration. Specification should be 0.008–0.010 inch for intake valves and 0.011–0.013 inch for exhaust valves.

19. Insert the feeler gauge between the adjusting screw and the end of the valve stem. Slide the gauge back and forth, you should feel a slight drag.

20. If you feel too much/little drag, loosen the locknut, turn the adjusting screw until the drag on the feeler gauge is correct. Tighten the locknut to specification (intake

14 ft. lbs. exhaust 10 ft. lbs.). Be sure to apply clean engine oil to the nut threads.

21. Repeat the adjustment as required.

22. Rotate the crankshaft 180 degrees clockwise. Check and adjust the valve clearance on number three cylinder.

23. Rotate the crankshaft 180 degrees clockwise. Check and adjust the valve clearance on number four cylinder.

24. Rotate the crankshaft 180 degrees clockwise. Check and adjust the valve clearance on number two cylinder.

25. Clean the mating surfaces of the cylinder head and valve cover.

26. Install a new gasket in the groove of the valve cover.

27. Be sure that the mating surfaces are clean and dry.

28. Apply liquid gasket part number 08717-0004 or equivalent on the chain case and the number five rocker shaft holder mating surface areas.

➡**Install the component within five minutes of applying the liquid gasket. Some liquid gasket materials require installation within four minutes. Be sure to read and follow the manufacturer's instruction. If too much time has passed before installing the component remove the liquid gasket and residue, then reapply new liquid gasket.**

29. Position the spark plug seals on the spark plug tubes.

30. Position the valve cover on the cylinder head. Slide the cover back and forth to seat the gasket.

31. Inspect the valve cover washers, replace as required.

32. Tighten the retaining bolts in three steps to 8.8 ft. lbs., using the sequence shown in the illustration.

➡**Wait at least 30 minutes to allow the gasket to cure before filling the engine with oil.**

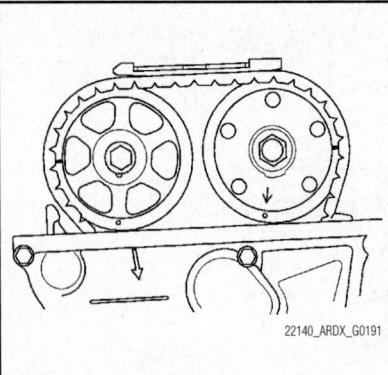

Fig. 95 Positioning number four cylinder

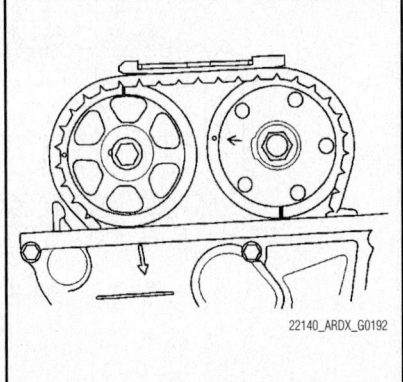

Fig. 96 Positioning number two cylinder

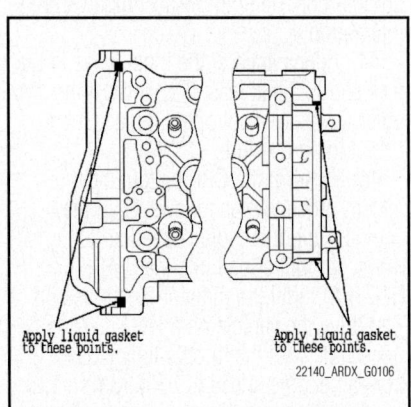

Apply liquid gasket to these points.

Apply liquid gasket to these points.

Fig. 97 Liquid gasket application points

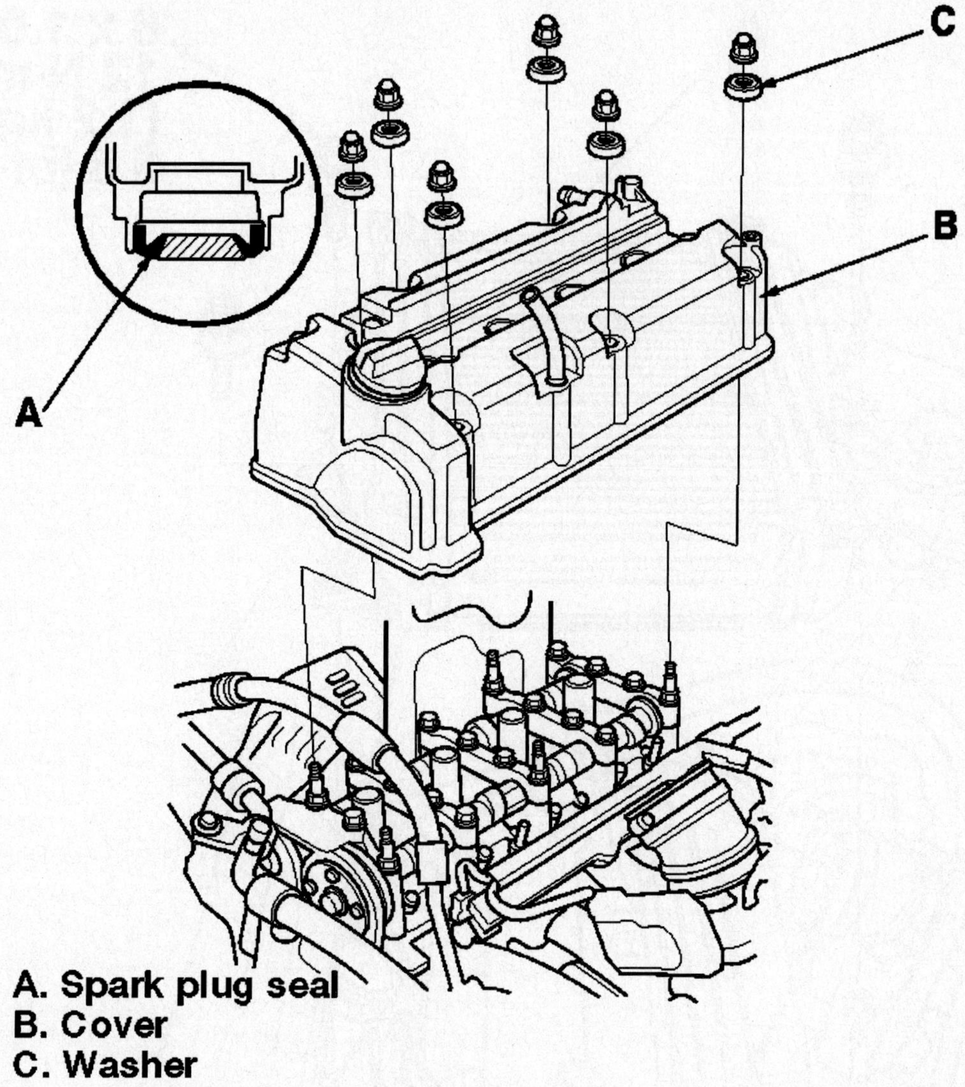

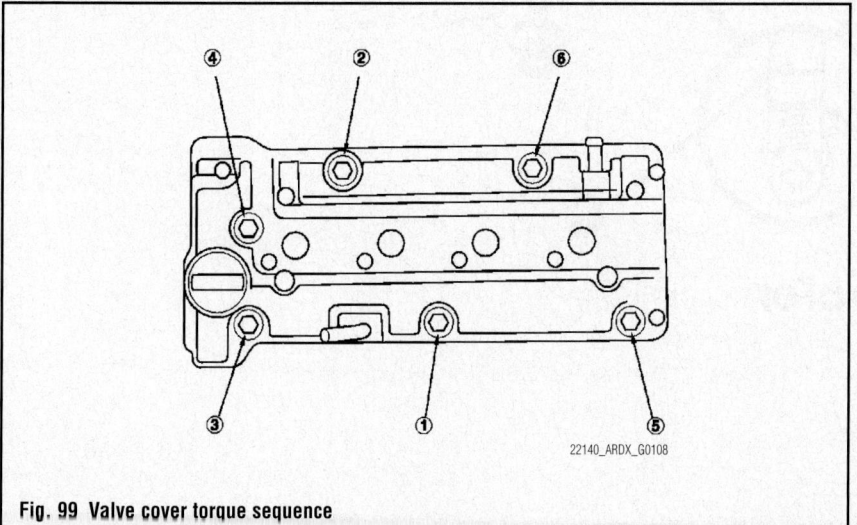

A. Spark plug seal
B. Cover
C. Washer

22140_ARDX_G0107

Fig. 98 Spark plug seal placement

22140_ARDX_G0108

Fig. 99 Valve cover torque sequence

➡**Do not run the engine for at least three hours after installing the valve cover.**

33. Continue the installation in the reverse order of the removal procedure.

34. Install the air charge cooler and connect the intake air ducts.

35. Tighten the hose bands until edge (B) of the band aligns with the mark (C) on the band.

➡**If the hose band edge exceeds the mark, replace the hose band.**

36. Start the engine and check for leaks, correct as required.

37. Turn the ignition switch ON (II), the SRS indicator should come on for about six seconds, and then go off.

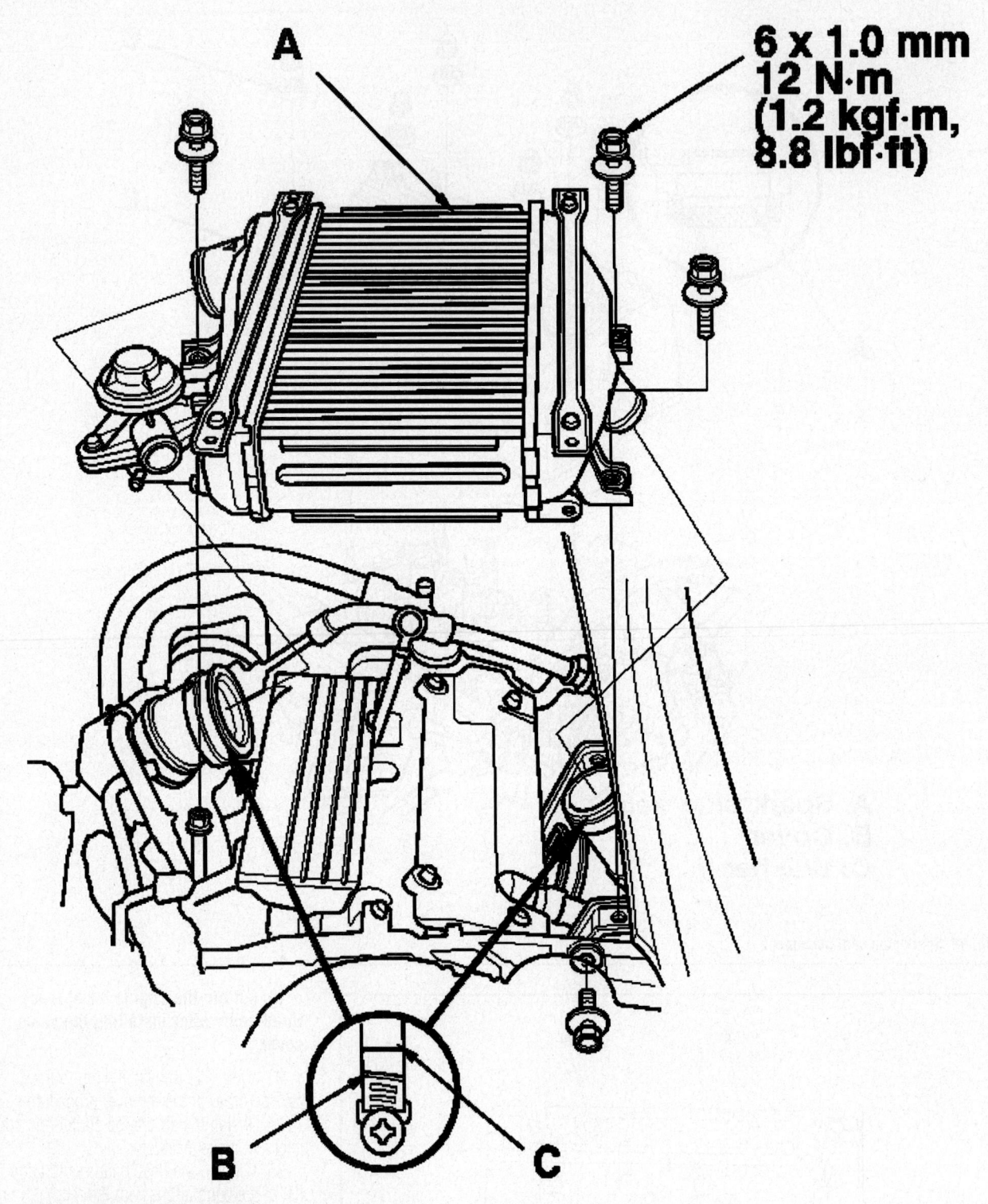

6 x 1.0 mm
12 N·m
(1.2 kgf·m,
8.8 lbf·ft)

A. Air charge cooler
B. Edge (B)
C. Mark (C)

22140_ARDX_G0100

Fig. 100 Air charge cooler band tightening

ENGINE PERFORMANCE & EMISSION CONTROL

COMPONENT LOCATIONS

See Figures 101 through 106.

AIR FUEL (A/F) SENSOR

REMOVAL & INSTALLATION

See Figure 107.

1. Before servicing the vehicle, refer to the Precautions Section.

➡️Before disconnecting the battery cables make sure you have the anti theft codes for the audio/navigation system.

➡️Except when doing electrical inspections, always turn the ignition switch to

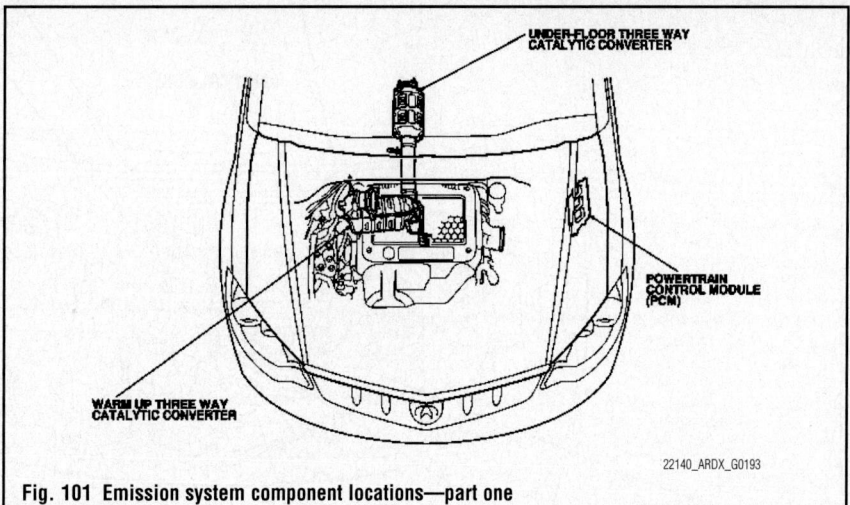

Fig. 101 Emission system component locations—part one

22140_ARDX_G0193

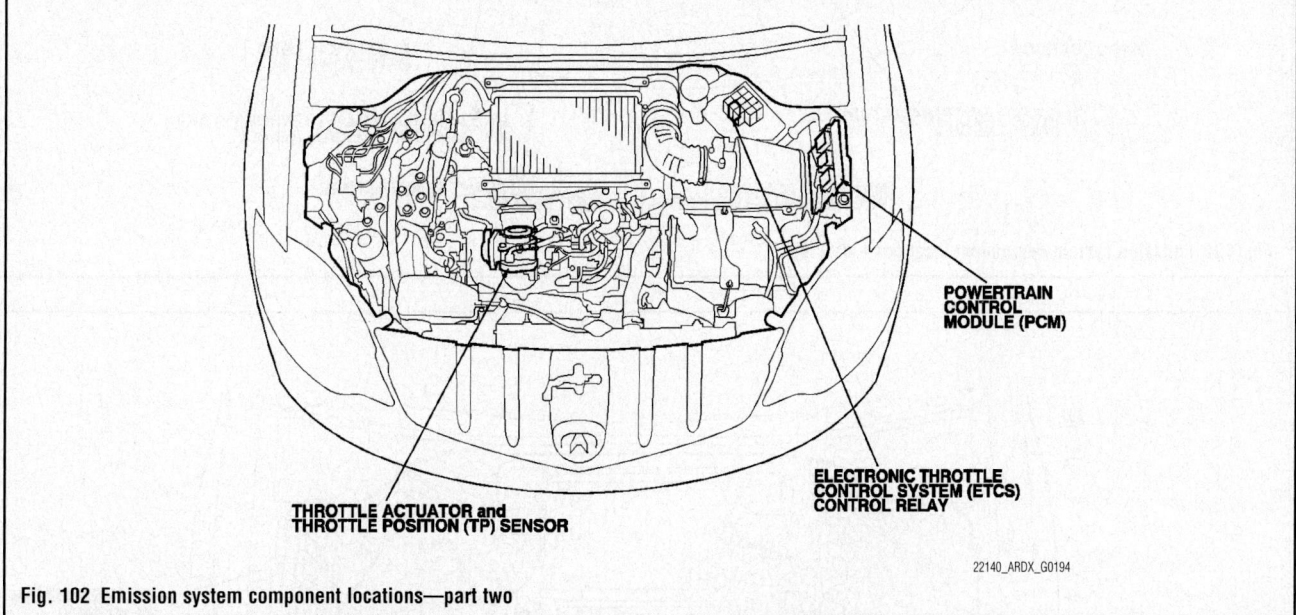

Fig. 102 Emission system component locations—part two

22140_ARDX_G0194

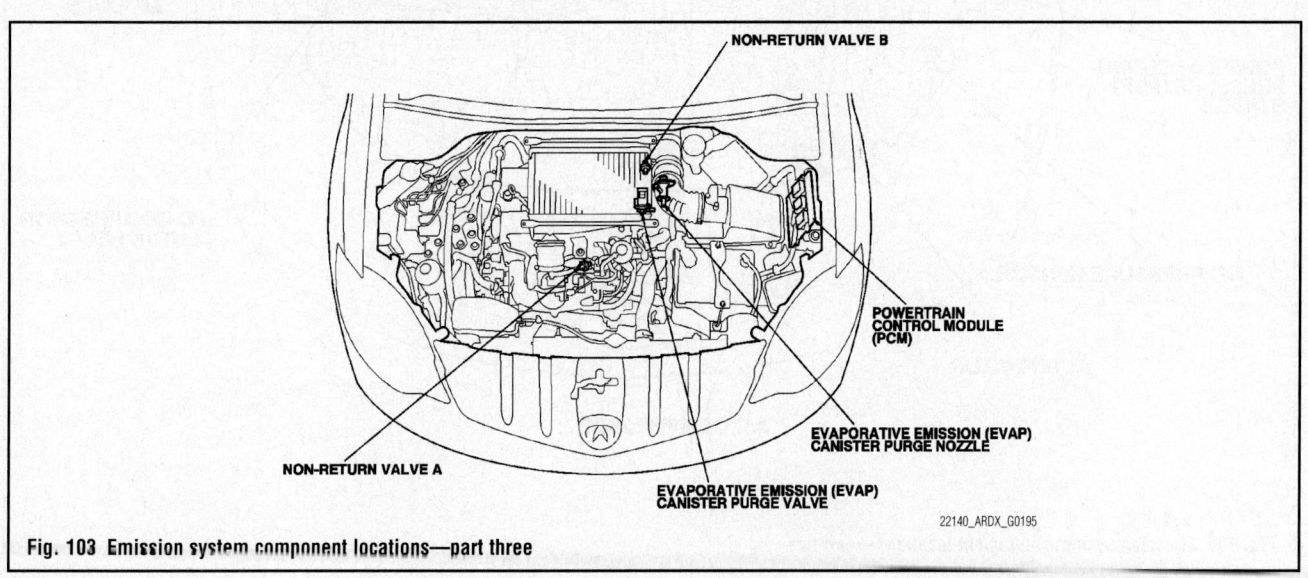

Fig. 103 Emission system component locations—part three

22140_ARDX_G0195

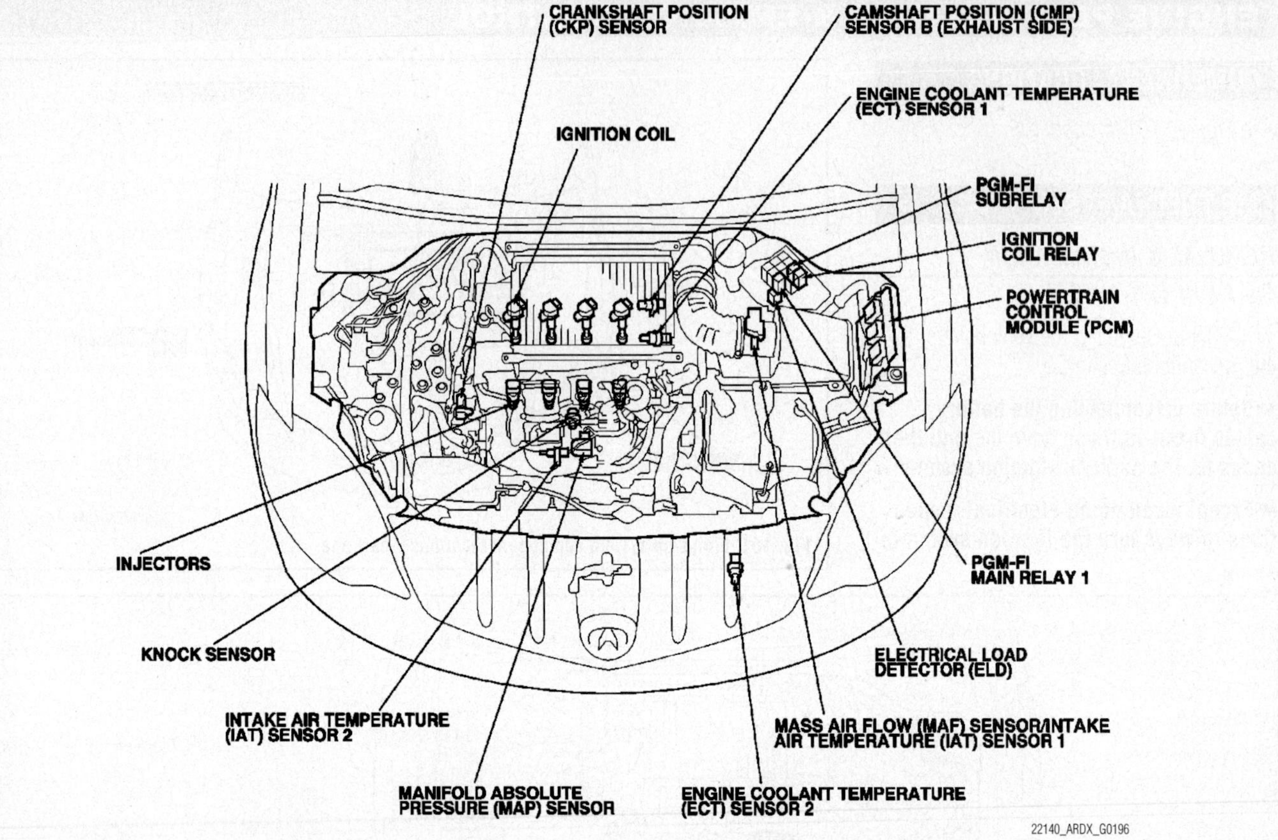

CRANKSHAFT POSITION (CKP) SENSOR

CAMSHAFT POSITION (CMP) SENSOR B (EXHAUST SIDE)

ENGINE COOLANT TEMPERATURE (ECT) SENSOR 1

IGNITION COIL

PGM-FI SUBRELAY

IGNITION COIL RELAY

POWERTRAIN CONTROL MODULE (PCM)

INJECTORS

PGM-FI MAIN RELAY 1

KNOCK SENSOR

ELECTRICAL LOAD DETECTOR (ELD)

INTAKE AIR TEMPERATURE (IAT) SENSOR 2

MASS AIR FLOW (MAF) SENSOR/INTAKE AIR TEMPERATURE (IAT) SENSOR 1

MANIFOLD ABSOLUTE PRESSURE (MAP) SENSOR

ENGINE COOLANT TEMPERATURE (ECT) SENSOR 2

22140_ARDX_G0196

Fig. 104 Emission system component locations—part four

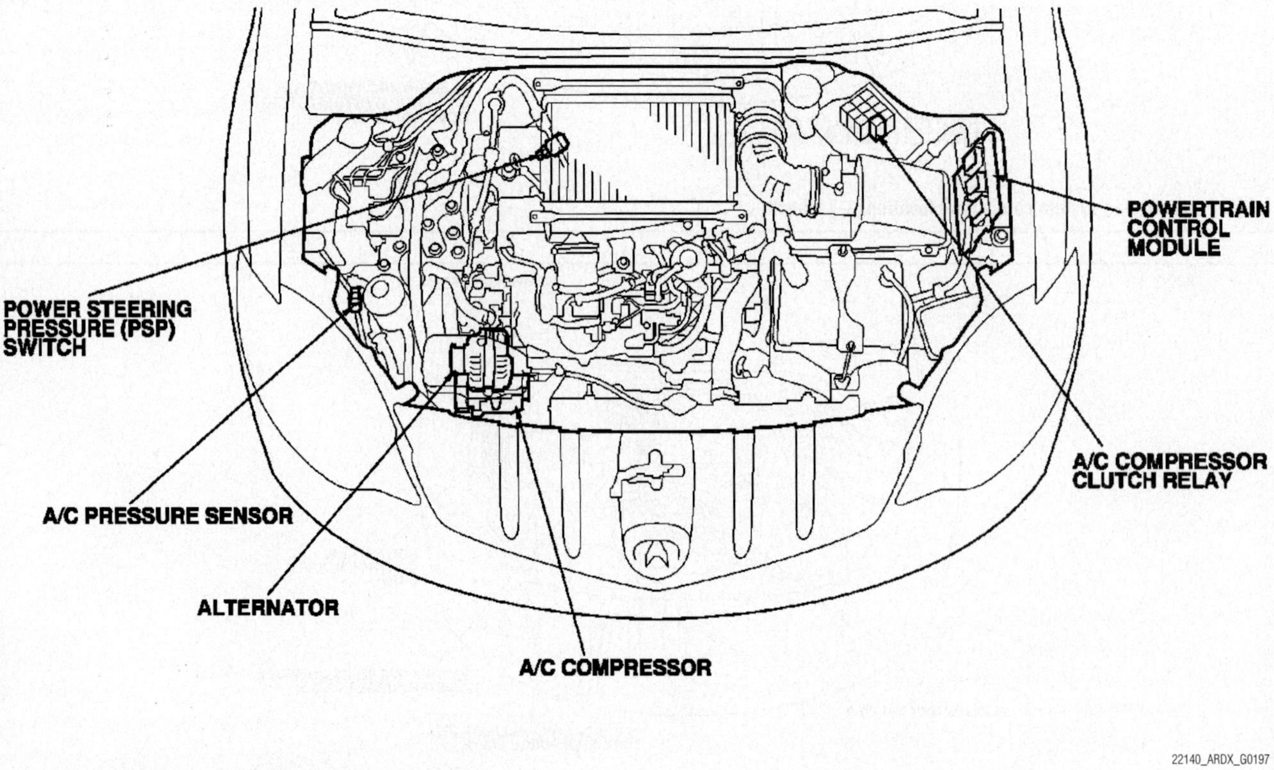

POWERTRAIN CONTROL MODULE

POWER STEERING PRESSURE (PSP) SWITCH

A/C COMPRESSOR CLUTCH RELAY

A/C PRESSURE SENSOR

ALTERNATOR

A/C COMPRESSOR

22140_ARDX_G0197

Fig. 105 Emission system component locations—part five

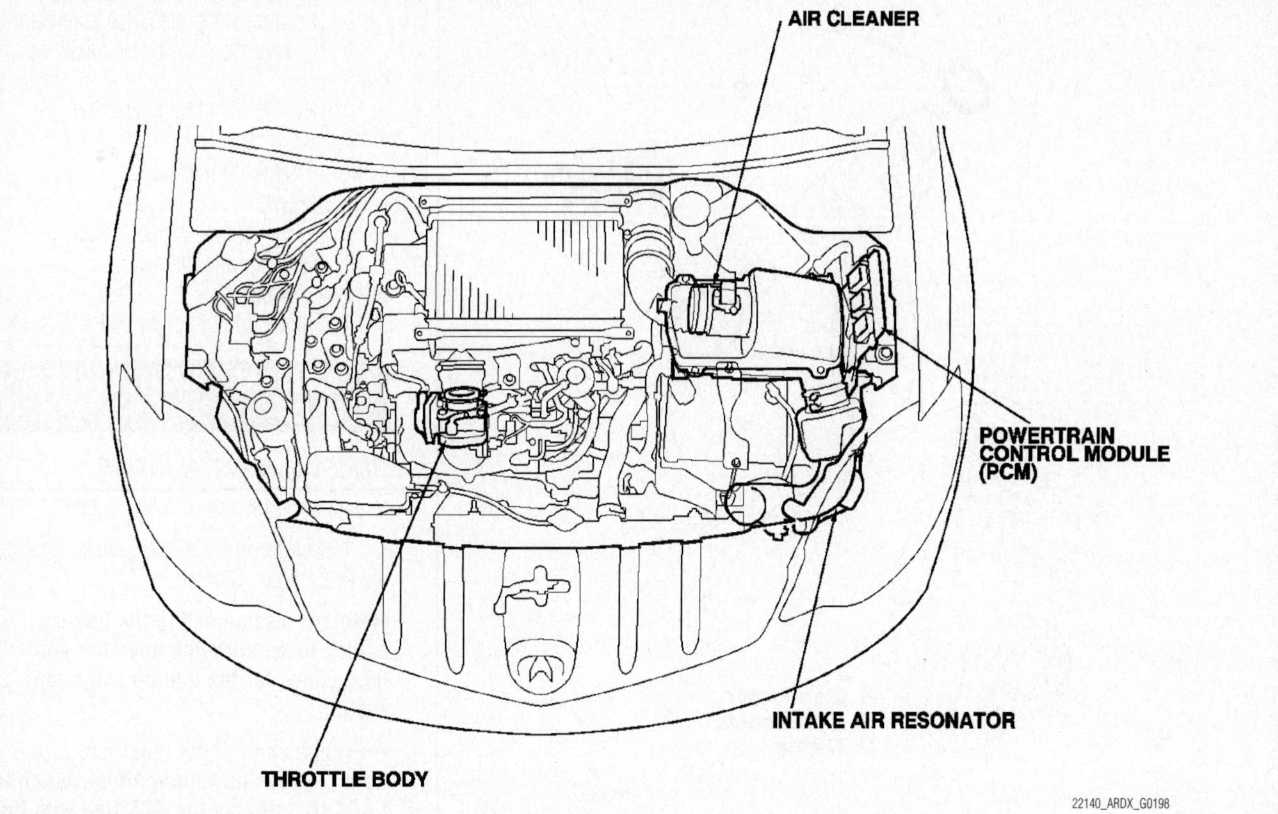

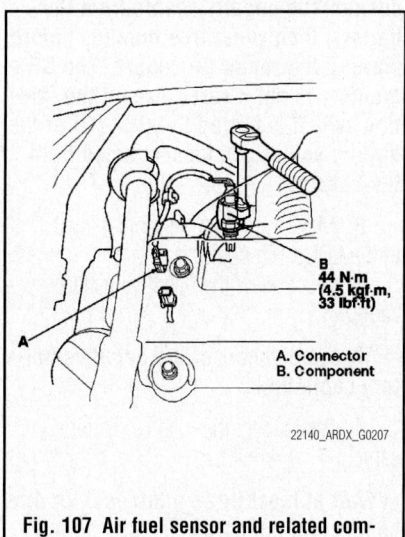

Fig. 106 Emission system component locations—part six

Fig. 107 Air fuel sensor and related components

A. Connector
B. Component

B 44 N·m (4.5 kgf·m, 33 lbf·ft)

22140_ARDX_G0207

LOCK (O), ground the SCS line with the Honda Diagnostic Service (HDS) tool to take the PCM out of active status, disconnect the negative cable from the battery, then wait three minutes before starting the repair procedure. The SRS memory is not cleared even if the ignition switch is turned to LOCK (O) or the battery cables are disconnected from tho battery.

2. Make sure that the ignition switch is in the LOCK (O) position.

3. Disconnect the negative battery cable.

➡Always disconnect the negative battery cable first.

4. Disconnect the positive battery cable.

➡Wait at least three minutes after disconnecting the battery cables before starting the repair procedure.

5. Disconnect the electrical connector.

6. Remove the component from its mounting, using a sensor removal tool.

To install:

7. Installation is the reverse of the removal procedure.

8. Tighten the sensor to 33 ft. lbs.

CAMSHAFT POSITION (CMP) SENSOR

REMOVAL & INSTALLATION
See Figures 108 and 109.

1. Before servicing the vehicle, refer to the Precautions Section.

➡Before disconnecting the battery cables make sure you have the anti

theft codes for the audio/navigation system.

➡Except when doing electrical inspections, always turn the ignition switch to LOCK (O), ground the SCS line with the Honda Diagnostic Service (HDS) tool to take the PCM out of active status, disconnect the negative cable from the battery, then wait three minutes before starting the repair procedure. The SRS memory is not cleared even if the ignition switch is turned to LOCK (O) or the battery cables are disconnected from the battery.

2. Make sure that the ignition switch is in the LOCK (O) position.

3. Disconnect the negative battery cable.

➡Always disconnect the negative battery cable first.

4. Disconnect the positive battery cable.

➡Wait at least three minutes after disconnecting the battery cables before starting the repair procedure.

5. Remove the air cleaner, if removing sensor A.

6. Remove the air charge cooler, if removing sensor B.

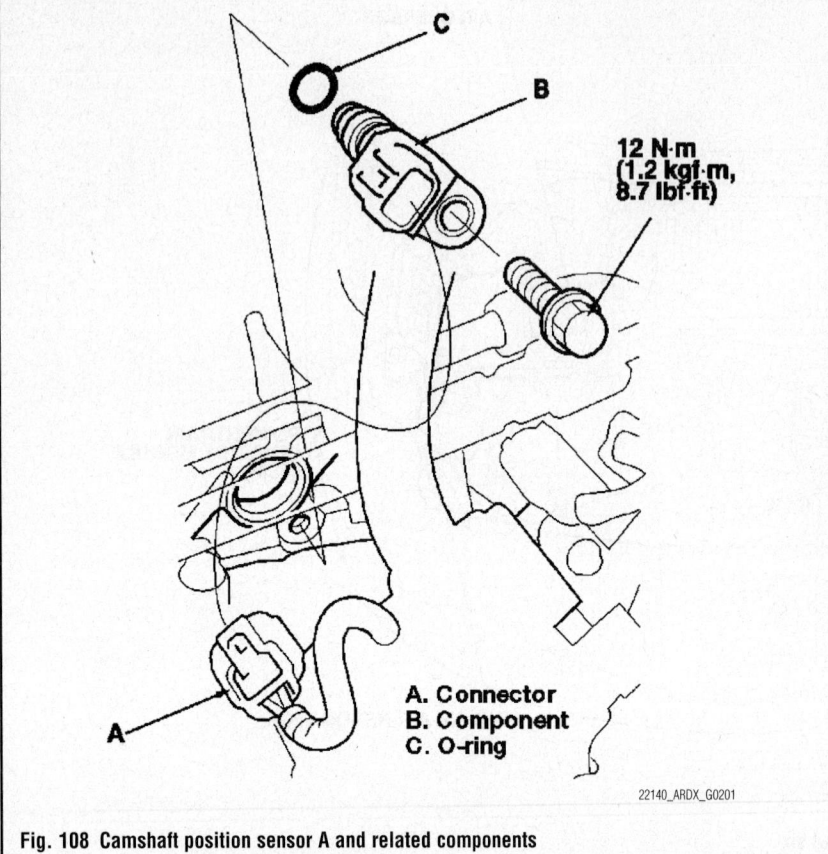

A. Connector
B. Component
C. O-ring

12 N·m
(1.2 kgf·m,
8.7 lbf·ft)

22140_ARDX_G0201

Fig. 108 Camshaft position sensor A and related components

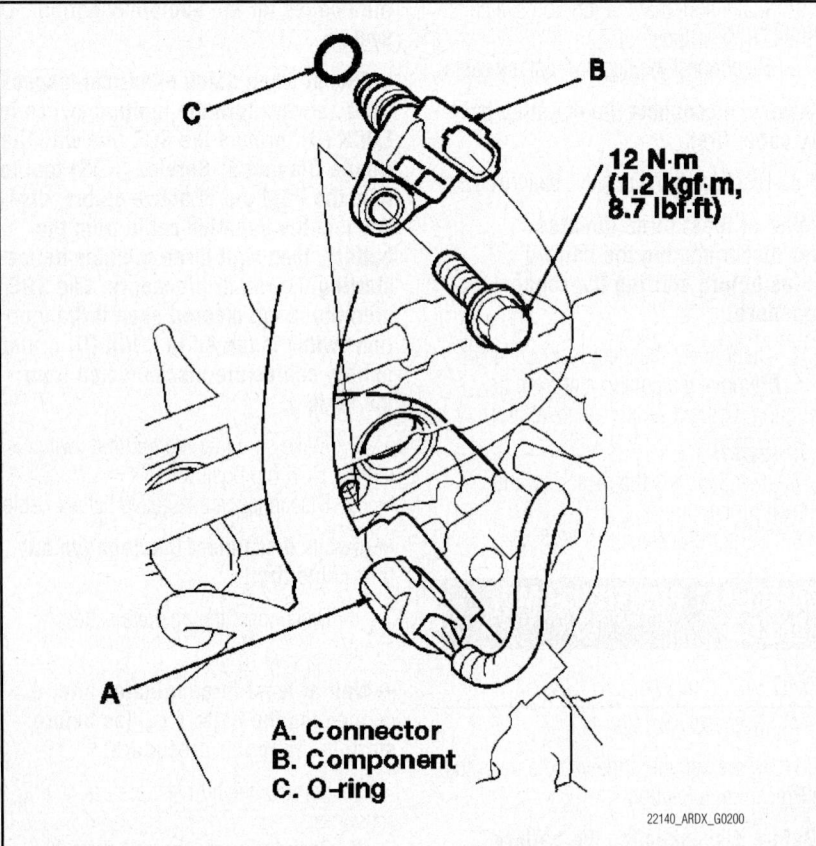

A. Connector
B. Component
C. O-ring

12 N·m
(1.2 kgf·m,
8.7 lbf·ft)

22140_ARDX_G0200

Fig. 109 Camshaft position sensor B and related components

7. Remove the air bypass outlet pipe.
8. Disconnect the electrical connector.
9. Remove the component retaining bolt.
10. Remove the component from its mounting.
11. Discard the O-ring.

To install:
12. Be sure to use a new O-ring.
13. Installation is the reverse of the removal procedure.
14. Tighten the retaining bolt to 9 ft. lbs.

CRANKSHAFT POSITION (CKP) SENSOR

REMOVAL & INSTALLATION

See Figure 110.

1. Before servicing the vehicle, refer to the Precautions Section.

➡**Before disconnecting the battery cables make sure you have the anti theft codes for the audio/navigation system.**

➡**Except when doing electrical inspections, always turn the ignition switch to LOCK (0), ground the SCS line with the Honda Diagnostic Service (HDS) tool to take the PCM out of active status, disconnect the negative cable from the battery, then wait three minutes before starting the repair procedure. The SRS memory is not cleared even if the ignition switch is turned to LOCK (0) or the battery cables are disconnected from the battery.**

2. Make sure that the ignition switch is in the LOCK (0) position.
3. Disconnect the negative battery cable.

➡**Always disconnect the negative battery cable first.**

4. Disconnect the positive battery cable.

➡**Wait at least three minutes after disconnecting the battery cables before starting the repair procedure.**

5. Disconnect the electrical connector.
6. Remove the component retaining bolt.
7. Remove the component from its mounting.
8. Discard the O-ring.

To install:
9. Be sure to use a new O-ring.
10. Installation is the reverse of the removal procedure.

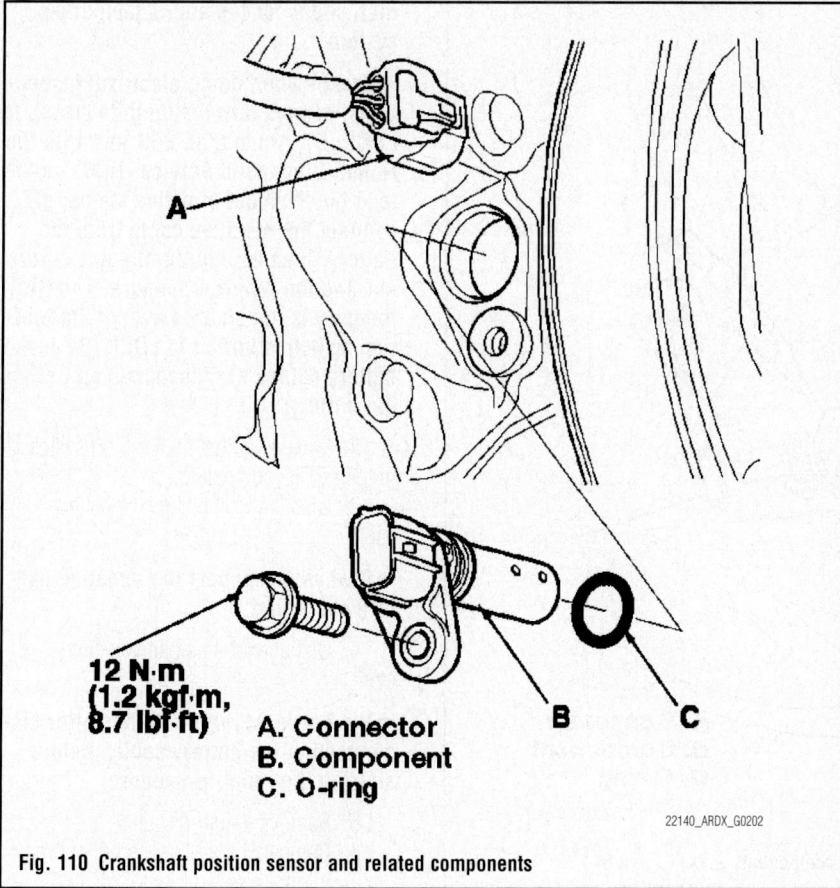

12 N·m
(1.2 kgf·m,
8.7 lbf·ft)

A. Connector
B. Component
C. O-ring

22140_ARDX_G0202

Fig. 110 Crankshaft position sensor and related components

Honda Diagnostic Service (HDS) tool to take the PCM out of active status, disconnect the negative cable from the battery, then wait three minutes before starting the repair procedure. The SRS memory is not cleared even if the ignition switch is turned to LOCK (O) or the battery cables are disconnected from the battery.

2. Make sure that the ignition switch is in the LOCK (O) position.
3. Disconnect the negative battery cable.

➡**Always disconnect the negative battery cable first.**

4. Disconnect the positive battery cable.

➡**Wait at least three minutes after disconnecting the battery cables before starting the repair procedure.**

5. Drain the coolant. Be sure to properly dispose of used coolant.
6. Remove the splash shield, if removing ECT 2.
7. Remove the air cleaner, if removing ECT 1.
8. Remove the air charge cooler, if removing ECT 1.

11. Tighten the retaining bolt to 9 ft. lbs.
12. Perform the CKP pattern clear/relearn procedure as follows. Start the engine. Hold the engine speed at 3000 rpm's, without a load in either PARK or NEUTRAL until the radiator fan comes on. Test drive the vehicle on a level road. Decelerate with the throttle fully closed from an engine speed of 2500 rpm's down to 1000 rpm's with the transmission in the S position 1st or 2nd gear. Repeat the above step several times. Turn the ignition switch to LOCK (O). Turn the ignition switch to ON (II) and wait thirty seconds.

ENGINE COOLANT TEMPERATURE (ECT) SENSOR

REMOVAL & INSTALLATION
See Figures 111 and 112.

1. Before servicing the vehicle, refer to the Precautions Section.

➡**Before disconnecting the battery cables make sure you have the anti theft codes for the audio/navigation system.**

➡**Except when doing electrical inspections, always turn the ignition switch to LOCK (O), ground the SCS line with the**

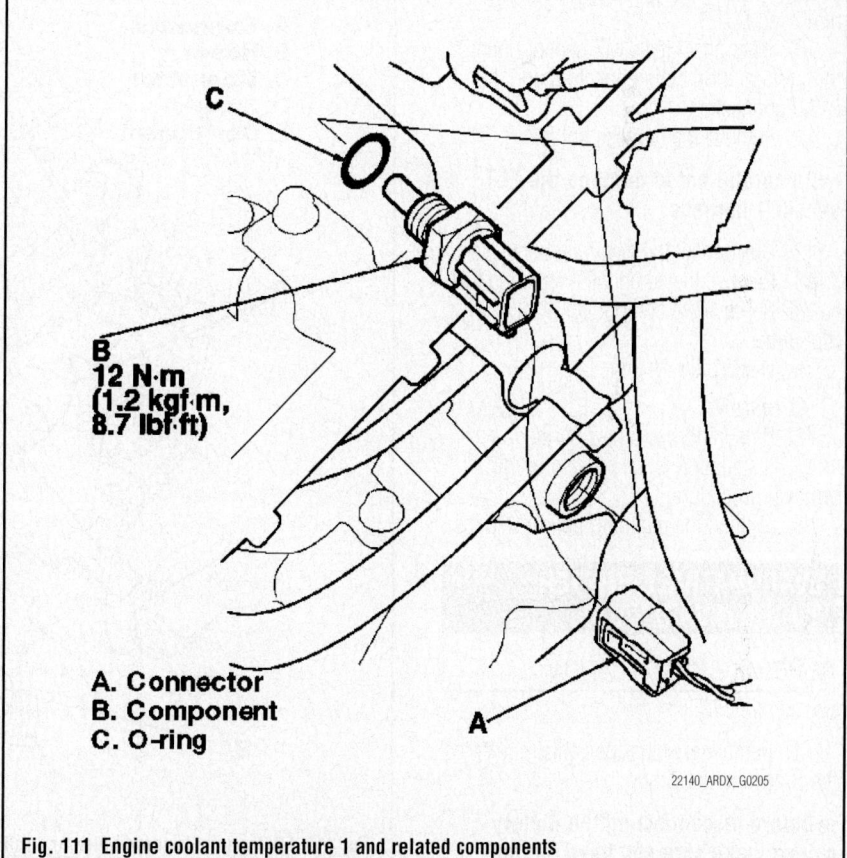

C

B
12 N·m
(1.2 kgf·m,
8.7 lbf·ft)

A

A. Connector
B. Component
C. O-ring

22140_ARDX_G0205

Fig. 111 Engine coolant temperature 1 and related components

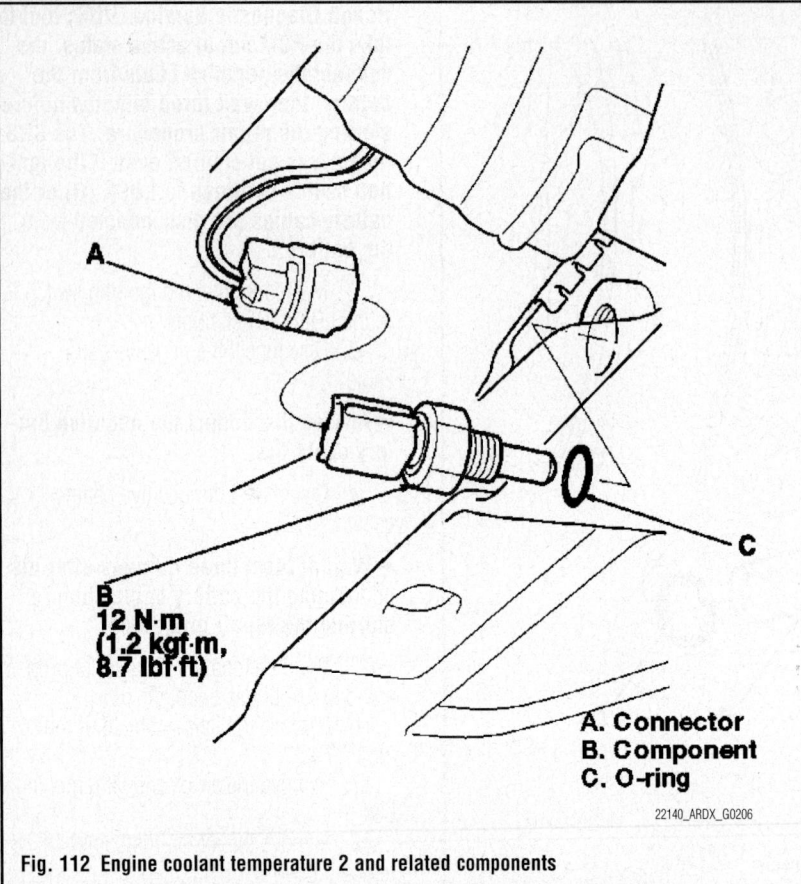

A. Connector
B. Component
C. O-ring

22140_ARDX_G0206

Fig. 112 Engine coolant temperature 2 and related components

theft codes for the audio/navigation system.

➡Except when doing electrical inspections, always turn the ignition switch to LOCK (O), ground the SCS line with the Honda Diagnostic Service (HDS) tool to take the PCM out of active status, disconnect the negative cable from the battery, then wait three minutes before starting the repair procedure. The SRS memory is not cleared even if the ignition switch is turned to LOCK (O) or the battery cables are disconnected from the battery.

2. Make sure that the ignition switch is in the LOCK (O) position.
3. Disconnect the negative battery cable.

➡Always disconnect the negative battery cable first.

4. Disconnect the positive battery cable.

➡Wait at least three minutes after disconnecting the battery cables before starting the repair procedure.

5. Remove the air cleaner.
6. Remove the air bypass outlet pipe.

9. Remove the ignition coil cover, if removing ECT 1.
10. Disconnect the CMP sensor connectors, the ignition coil connectors and the EVAP connector.
11. Remove the harness holder.

➡Be careful not to damage the ECT sensor 1 harness.

12. Disconnect the electrical connector.
13. Remove the component retaining bolt.
14. Remove the component from its mounting.
15. Discard the O-ring.

To install:
16. Be sure to use a new O-ring.
17. Installation is the reverse of the removal procedure.
18. Tighten the retaining bolt to 9 ft. lbs.

EVAPORATIVE CANISTER PURGE (EVAP) VALVE

REMOVAL & INSTALLATION

See Figure 113.

1. Before servicing the vehicle, refer to the Precautions Section.

➡Before disconnecting the battery cables make sure you have the anti

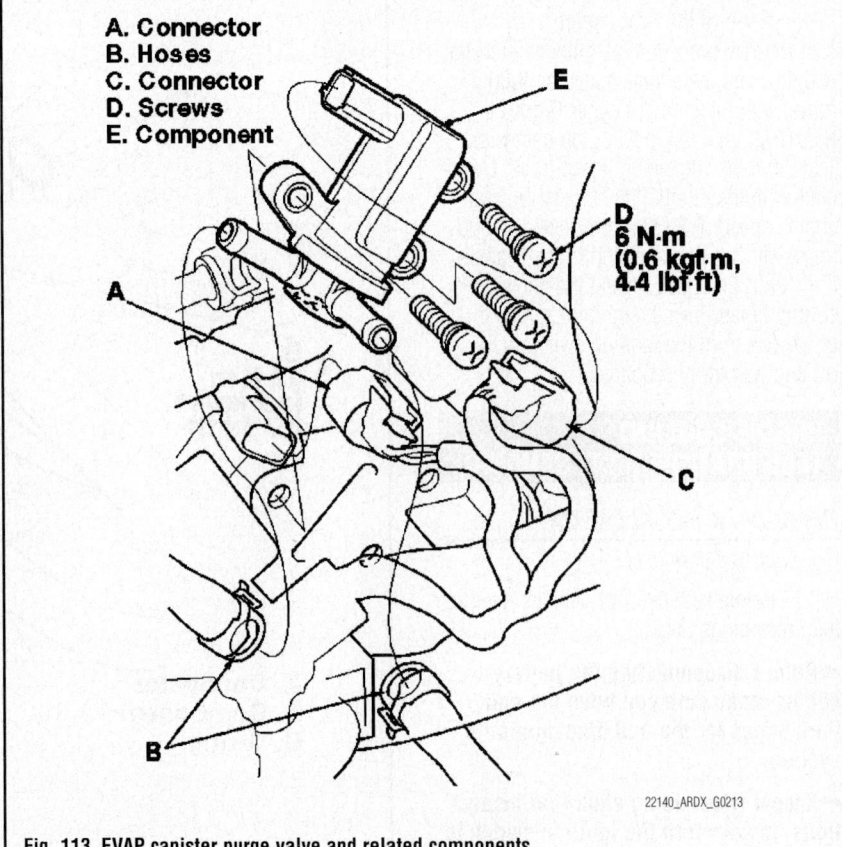

A. Connector
B. Hoses
C. Connector
D. Screws
E. Component

D
6 N·m
(0.6 kgf·m,
4.4 lbf·ft)

22140_ARDX_G0213

Fig. 113 EVAP canister purge valve and related components

7. Disconnect the CMP sensor B connector.

8. Disconnect the hoses at the EVAP canister.

9. Disconnect the electrical connector.

10. Remove the component retaining screws.

11. Remove the component from its mounting.

To install:

12. Installation is the reverse of the removal procedure.

13. Tighten the retaining screws to 4.4 ft. lbs.

HEATED OXYGEN (HO2S) SENSOR

REMOVAL & INSTALLATION

See Figure 114.

1. Before servicing the vehicle, refer to the Precautions Section.

➡**Before disconnecting the battery cables make sure you have the anti theft codes for the audio/navigation system.**

➡**Except when doing electrical inspections, always turn the ignition switch to LOCK (O), ground the SCS line with the Honda Diagnostic Service (HDS) tool to take the PCM out of active status, disconnect the negative cable from the battery, then wait three minutes before starting the repair procedure. The SRS memory is not cleared even if the ignition switch is turned to LOCK (O) or the battery cables are disconnected from the battery.**

2. Make sure that the ignition switch is in the LOCK (O) position.

3. Disconnect the negative battery cable.

➡**Always disconnect the negative battery cable first.**

4. Disconnect the positive battery cable.

➡**Wait at least three minutes after disconnecting the battery cables before starting the repair procedure.**

5. Remove the WU-TWC stay.

6. Disconnect the electrical connector.

7. Remove the component from its mounting, using a sensor removal tool.

To install:

8. Installation is the reverse of the removal procedure.

9. Tighten the sensor to 33 ft. lbs.

INTAKE AIR TEMPERATURE (IAT) SENSOR

REMOVAL & INSTALLATION

See Figure 115.

1. Before servicing the vehicle, refer to the Precautions Section.

➡**Before disconnecting the battery cables make sure you have the anti theft codes for the audio/navigation system.**

➡**Except when doing electrical inspections, always turn the ignition switch to LOCK (O), ground the SCS line with the Honda Diagnostic Service (HDS) tool to take the PCM out of active status, disconnect the negative cable from the battery, then wait three minutes before starting the repair procedure. The SRS memory is not cleared even if the ignition switch is turned to LOCK (O) or the battery cables are disconnected from the battery.**

2. Make sure that the ignition switch is in the LOCK (O) position.

3. Disconnect the negative battery cable.

➡**Always disconnect the negative battery cable first.**

4. Disconnect the positive battery cable.

➡**Wait at least three minutes after disconnecting the battery cables before starting the repair procedure.**

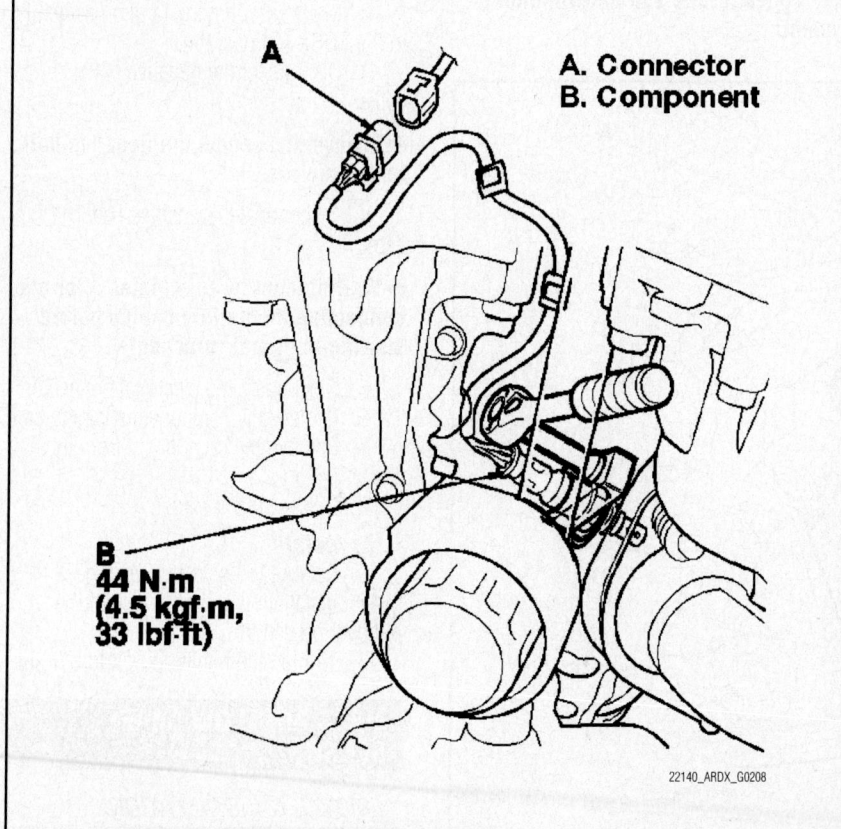

A. Connector
B. Component

B
44 N·m
(4.5 kgf·m,
33 lbf·ft)

22140_ARDX_G0208

Fig. 114 Heated oxygen sensor and related components

A. Connector
B. Component
C. O-ring

C
B

3.4 N·m (0.35 kgf·m, 2.5 lbf·ft)

22140_ARDX_G0209

Fig. 115 IAT sensor and related components

5. Remove the air charge cooler cover.

6. Disconnect the electrical connector.

7. Remove the component retaining bolt.

8. Remove the component from its mounting.

9. Discard the O-ring.

To install:

10. Be sure to use a new O-ring.

11. Installation is the reverse of the removal procedure.

12. Tighten the retaining bolt to 3 ft. lbs.

KNOCK SENSOR (KS)

REMOVAL & INSTALLATION

See Figure 116.

1. Before servicing the vehicle, refer to the Precautions Section.

➡ **Before disconnecting the battery cables make sure you have the anti theft codes for the audio/navigation system.**

➡ **Except when doing electrical inspections, always turn the ignition switch to LOCK (O), ground the SCS line with the Honda Diagnostic Service (HDS) tool to take the PCM out of active status, disconnect the negative cable from the battery, then wait three minutes before starting the repair procedure. The SRS memory is not cleared even if the ignition switch is turned to LOCK (O) or the** battery cables are disconnected from the battery.

2. Make sure that the ignition switch is in the LOCK (O) position.

3. Disconnect the negative battery cable.

➡ **Always disconnect the negative battery cable first.**

4. Disconnect the positive battery cable.

➡ **Wait at least three minutes after disconnecting the battery cables before starting the repair procedure.**

5. Disconnect the electrical connector.

6. Remove the component from its mounting.

To install:

7. Installation is the reverse of the removal procedure.

8. Tighten the retaining bolt to 23 ft. lbs.

MANIFOLD ABSOLUTE PRESSURE (MAP) SENSOR

REMOVAL & INSTALLATION

See Figure 117.

1. Before servicing the vehicle, refer to the Precautions Section.

➡ **Before disconnecting the battery cables make sure you have the anti theft codes for the audio/navigation system.**

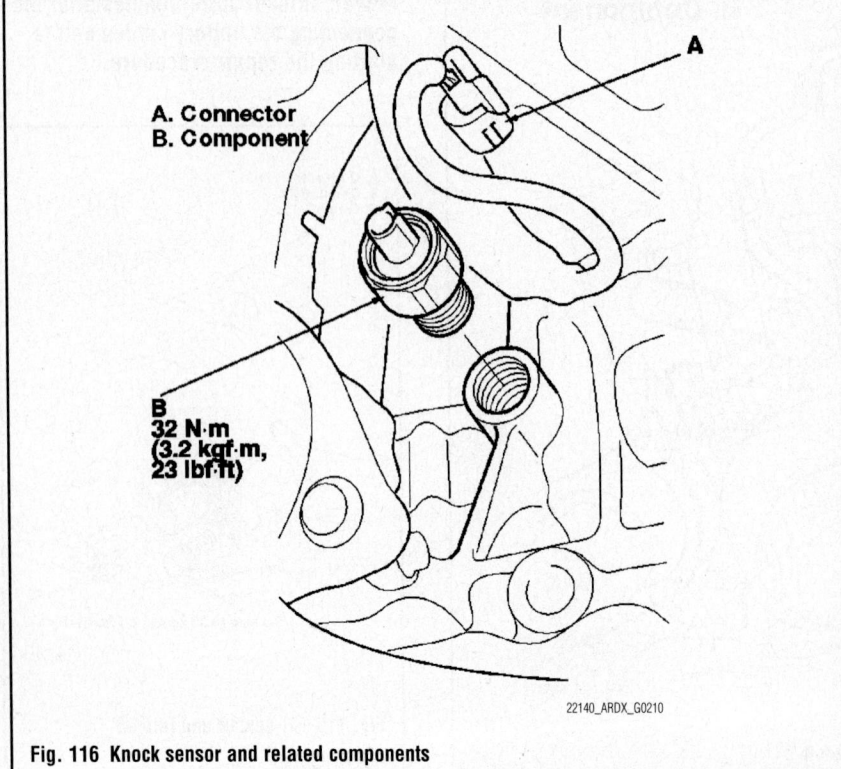

A. Connector
B. Component

B
32 N·m
(3.2 kgf·m,
23 lbf·ft)

22140_ARDX_G0210

Fig. 116 Knock sensor and related components

3.4 N·m
(0.35 kgf·m,
2.5 lbf·ft)

A. Connector
B. Component
C. O-ring

22140_ARDX_G0211

Fig. 117 MAP sensor and related components

➡ **Except when doing electrical inspections, always turn the ignition switch to LOCK (O), ground the SCS line with the Honda Diagnostic Service (HDS) tool to take the PCM out of active status, disconnect the negative cable from the battery, then wait three minutes before starting the repair procedure. The SRS memory is not cleared even if the ignition switch is turned to LOCK (O) or the battery cables are disconnected from the battery.**

2. Make sure that the ignition switch is in the LOCK (O) position.

3. Disconnect the negative battery cable.

➡ **Always disconnect the negative battery cable first.**

4. Disconnect the positive battery cable.

➡ **Wait at least three minutes after disconnecting the battery cables before starting the repair procedure.**

5. Disconnect the electrical connector.

6. Remove the component retaining bolt.

7. Remove the component from its mounting.

8. Discard the O-ring.

To install:

9. Be sure to use a new O-ring.

10. Installation is the reverse of the removal procedure.

11. Tighten the retaining bolt to 3 ft. lbs.

MASS AIR FLOW (MAF) SENSOR

REMOVAL & INSTALLATION

See Figure 118.

1. Before servicing the vehicle, refer to the Precautions Section.

➡Before disconnecting the battery cables make sure you have the anti theft codes for the audio/navigation system.

➡Except when doing electrical inspections, always turn the ignition switch to LOCK (O), ground the SCS line with the Honda Diagnostic Service (HDS) tool to take the PCM out of active status, disconnect the negative cable from the battery, then wait three minutes before starting the repair procedure. The SRS memory is not cleared even if the ignition switch is turned to LOCK (O) or the battery cables are disconnected from the battery.

2. Make sure that the ignition switch is in the LOCK (O) position.

3. Disconnect the negative battery cable.

➡Always disconnect the negative battery cable first.

4. Disconnect the positive battery cable.

➡Wait at least three minutes after disconnecting the battery cables before starting the repair procedure.

5. Disconnect the electrical connector.

6. Remove the component retaining screws.

7. Remove the component from its mounting.

8. Discard the O-ring.

To install:

9. Be sure to use a new O-ring.

10. Installation is the reverse of the removal procedure.

11. Tighten the retaining screws to 1 ft. lbs.

POWERTRAIN CONTROL MODULE (PCM)

REMOVAL & INSTALLATION

See Figure 119.

1. Before servicing the vehicle, refer to the Precautions Section.

➡Before disconnecting the battery cables make sure you have the anti theft codes for the audio/navigation system.

➡Except when doing electrical inspections, always turn the ignition switch to LOCK (O), ground the SCS line with the Honda Diagnostic Service (HDS) tool to take the PCM out of active status, disconnect the negative cable from the battery, then wait three minutes before starting the repair procedure. The SRS memory is not cleared even if the ignition switch is turned to LOCK (O) or the battery cables are disconnected from the battery.

2. Make sure that the ignition switch is in the LOCK (O) position.

3. Disconnect the negative battery cable.

➡Always disconnect the negative battery cable first.

4. Disconnect the positive battery cable.

➡Wait at least three minutes after disconnecting the battery cables before starting the repair procedure.

➡Refer to the HDS tool instructions for the particular model you are repairing. Use those procedures to position the PCM for replacement. Diagnostics for engine oil replacement, ATF fluid replacement, and throttle body cleaning will be addressed on both US and Canadian vehicles.

5. Disconnect the electrical connectors.

➡These connectors have symbols embossed in them for identification, see illustration.

6. Remove the component retaining bolts.

7. Remove the component from its mounting.

8. Remove the cover and the bracket from the PCM.

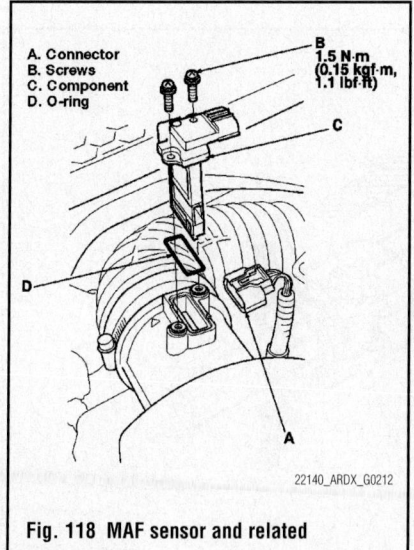

A. Connector
B. Screws
C. Component
D. O-ring
B. 1.5 N·m (0.15 kgf·m, 1.1 lbf·ft)

22140_ARDX_G0212

Fig. 118 MAF sensor and related components

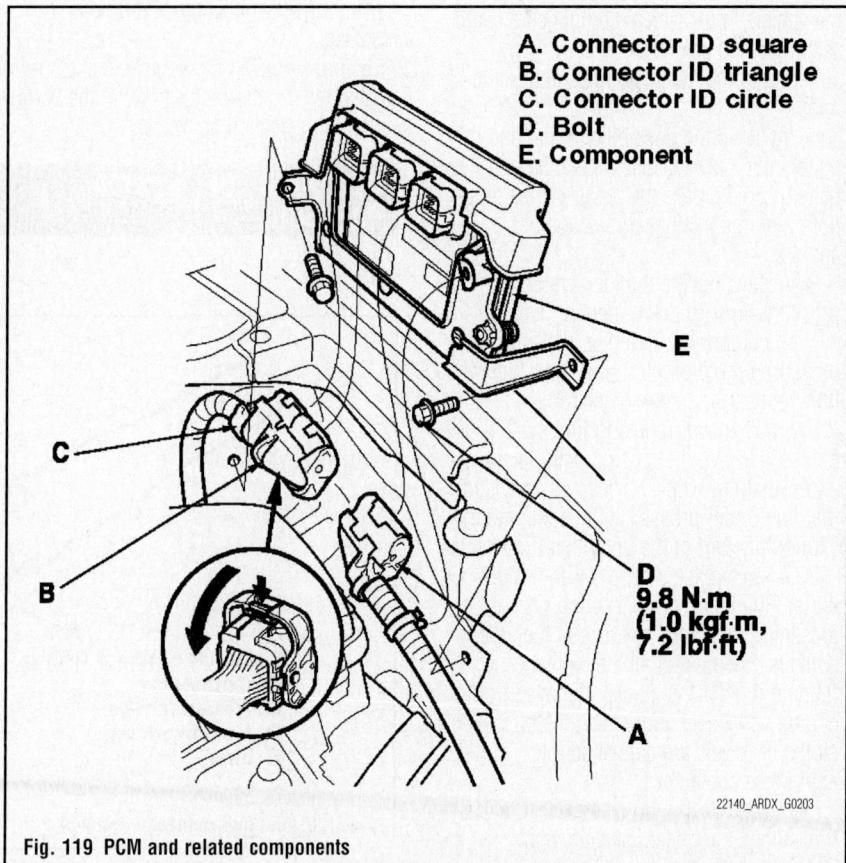

A. Connector ID square
B. Connector ID triangle
C. Connector ID circle
D. Bolt
E. Component

D. 9.8 N·m (1.0 kgf·m, 7.2 lbf·ft)

22140_ARDX_G0203

Fig. 119 PCM and related components

To install:

9. Installation is the reverse of the removal procedure.

10. Tighten the retaining bolt to 7 ft. lbs.

➡Refer to the HDS tool instructions for the particular model you are repairing. Use those procedures to check the PCM after replacement. Diagnostics for engine oil replacement, ATF fluid replacement, and throttle body cleaning will be addressed on both US and Canadian vehicles.

➡Update the PCM if it does not have the latest software.

➡The PCM relearn procedure must be performed whenever you replace the PCM, Reset the PCM, update the PCM, remove the engine or replace/clean the throttle body. Erasing DTC's with the HDS tool does not require you to perform the relearn procedure.

11. Reset the PCM with the Honda Diagnostic System (HDS) tool while the engine is stopped. Turn the ignition switch to LOCK (O). Turn the ignition switch to ON (II), and wait thirty seconds. Turn the ignition switch to LOCK (O) and disconnect the tool from the DLC.

12. Make sure that all electrical items are turned off.

13. Reset the PCM with the HDS tool.

14. Turn the ignition switch ON (II) and wait two seconds.

15. Start the engine. Hold the engine speed at 3000 rpm's, without a load in either PARK or NEUTRAL until the radiator fan comes on or until the coolant temperature reaches 194 degrees F.

16. Let the engine idle for about five minutes with the throttle fully closed

➡If the radiator fan comes on, do not include its running time in the five minutes.

17. Perform the CKP pattern clear/relearn procedure as follows. Start the engine. Hold the engine speed at 3000 rpm's, without a load in either PARK or NEUTRAL until the radiator fan comes on. Test drive the vehicle on a level road. Decelerate with the throttle fully closed from an engine speed of 2500 rpm's down to 1000 rpm's with the transmission in the S position 1st or 2nd gear. Repeat the above step several times. Turn the ignition switch to LOCK (O). Turn the ignition switch to ON (II) and wait thirty seconds.

FUEL

GASOLINE FUEL INJECTION SYSTEM

FUEL SYSTEM SERVICE PRECAUTIONS

Safety is the most important factor when performing not only fuel system maintenance but any type of maintenance. Failure to conduct maintenance and repairs in a safe manner may result in serious personal injury or death. Maintenance and testing of the vehicle's fuel system components can be accomplished safely and effectively by adhering to the following rules and guidelines.

• To avoid the possibility of fire and personal injury, always disconnect the negative battery cable unless the repair or test procedure requires that battery voltage be applied.

• Always relieve the fuel system pressure prior to disconnecting any fuel system component (injector, fuel rail, pressure regulator, etc.), fitting or fuel line connection. Exercise extreme caution whenever relieving fuel system pressure to avoid exposing skin, face and eyes to fuel spray. Please be advised that fuel under pressure may penetrate the skin or any part of the body that it contacts.

• Always place a shop towel or cloth around the fitting or connection prior to loosening to absorb any excess fuel due to spillage. Ensure that all fuel spillage (should it occur) is quickly removed from engine surfaces. Ensure that all fuel soaked cloths or towels are deposited into a suitable waste container.

• Always keep a dry chemical (Class B) fire extinguisher near the work area.

• Do not allow fuel spray or fuel vapors to come into contact with a spark or open flame.

• Always use a back-up wrench when loosening and tightening fuel line connection fittings. This will prevent unnecessary stress and torsion to fuel line piping.

• Always replace worn fuel fitting O-rings with new Do not substitute fuel hose or equivalent where fuel pipe is installed.

Before servicing the vehicle, make sure to also refer to the precautions in the beginning of this section as well.

RELIEVING FUEL SYSTEM PRESSURE

See Figure 120.

➡Before relieving the fuel system pressure be sure the engine is cold. Before disconnecting fuel lines or hoses, relieve pressure from the system by disabling the fuel pump, running the engine until it stalls, then disconnecting the fuel line/quick connect fitting in the engine compartment.

1. Before servicing the vehicle, refer to the Precautions Section.

2. Remove the driver's dashboard undercover.

3. Remove the PGM-FI main relay 2 (fuel pump) from the under dash fuse panel.

4. Start the engine and let it run until it stalls.

➡If any DTC's are stored, clear and ignore them.

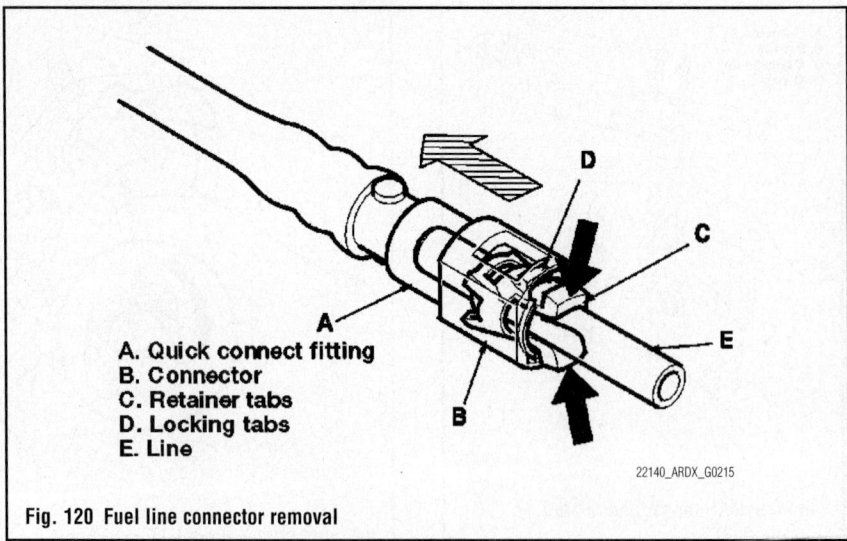

A. Quick connect fitting
B. Connector
C. Retainer tabs
D. Locking tabs
E. Line

22140_ARDX_G0215

Fig. 120 Fuel line connector removal

5. Turn the ignition switch to the LOCK (0) position.

6. Remove the fuel fill cap to relieve the pressure in the tank.

7. Disconnect the negative battery terminal.

8. Disconnect the positive battery terminal.

9. Remove the quick connect fitting cover.

➡**Check the fitting for dirt and debris, clean as required.**

10. Place a shop towel or rag over the fitting.

11. To disconnect the fitting hold the connector with one hand and squeeze the retainer tabs with the other hand to release them from the locking tabs. Pull the connector off.

➡**Do not use tools. Be careful not to damage any parts. Do not remove the retainer from the line, once it is removed it must be replaced. If the connector does not move, keep the retainer tabs pressed down and alternately pull and push the connector until it comes off easily.**

FUEL FILTER

REMOVAL & INSTALLATION

1. Before servicing the vehicle, refer to the Precautions Section.

➡**Before disconnecting the battery cables make sure you have the anti theft codes for the audio/navigation system.**

➡**Except when doing electrical inspections, always turn the ignition switch to LOCK (0), ground the SCS line with the Honda Diagnostic Service (HDS) tool to take the PCM out of active status, disconnect the negative cable from the battery, then wait three minutes before starting the repair procedure. The SRS memory is not cleared even if the ignition switch is turned to LOCK (0) or the battery cables are disconnected from the battery.**

2. Make sure that the ignition switch is in the LOCK (0) position.

3. Disconnect the negative battery cable.

➡**Always disconnect the negative battery cable first.**

4. Disconnect the positive battery cable.

➡**Wait at least three minutes after disconnecting the battery cables before starting the repair procedure.**

5. Properly relieve the fuel system pressure.

6. Remove the fuel tank.

7. Remove the fuel strainer set.

To install:

8. Installation is the reverse of the removal procedure.

9. Be sure to use new O-rings. Coat the O-rings with clean engine oil.

FUEL INJECTORS

REMOVAL & INSTALLATION

See Figure 121.

1. Before servicing the vehicle, refer to the Precautions Section.

➡**Before disconnecting the battery cables make sure you have the anti theft codes for the audio/navigation system.**

➡**Except when doing electrical inspections, always turn the ignition switch to LOCK (0), ground the SCS line with the Honda Diagnostic Service (HDS) tool to take the PCM out of active status, disconnect the negative cable from the battery, then wait three minutes before starting the repair procedure. The SRS**

memory is not cleared even if the ignition switch is turned to LOCK (0) or the battery cables are disconnected from the battery.

2. Make sure that the ignition switch is in the LOCK (0) position.

3. Disconnect the negative battery cable.

➡**Always disconnect the negative battery cable first.**

4. Disconnect the positive battery cable.

➡**Wait at least three minutes after disconnecting the battery cables before starting the repair procedure.**

5. Properly relieve the fuel system pressure.

6. Remove the air charge cooler cover.

7. Disconnect the turbocharger boost sensor connector. Remove the vacuum hoses and the air bypass outlet connecting tube.

8. Remove the air charge cooler retaining bolts. Remove the air charge cooler.

9. Disconnect the throttle body connector. Remove the injector cover.

10. Disconnect the connectors from the injectors. Remove the harness holder from the fuel rail.

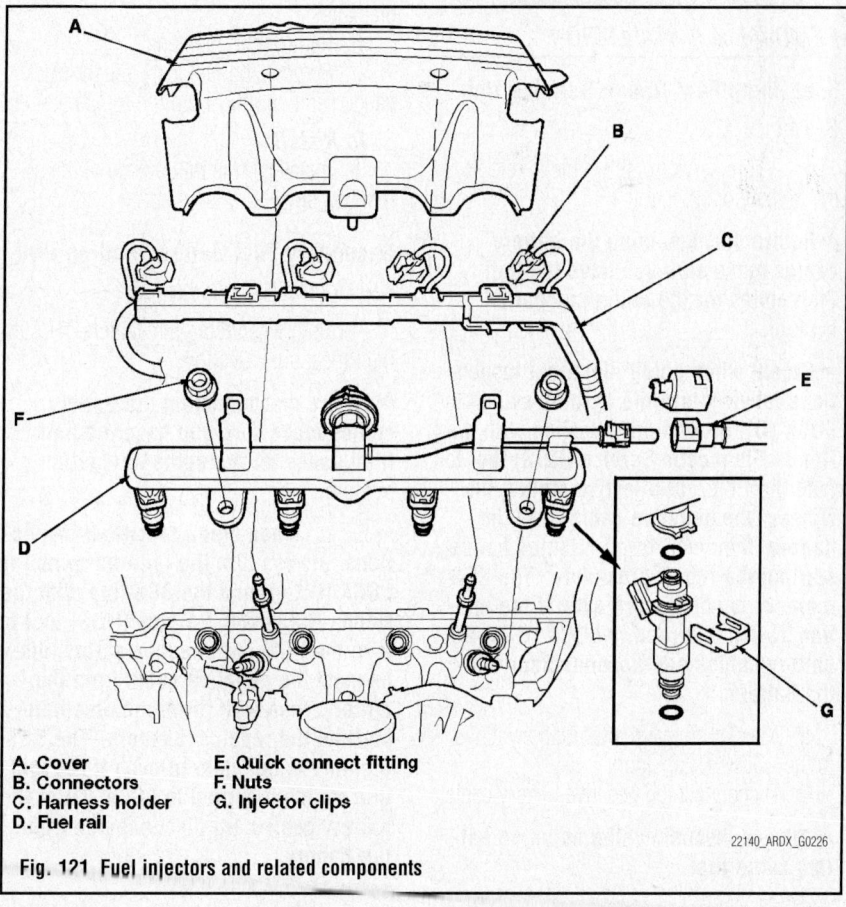

A. Cover
B. Connectors
C. Harness holder
D. Fuel rail
E. Quick connect fitting
F. Nuts
G. Injector clips

Fig. 121 Fuel injectors and related components

22140_ARDX_G0226

11. Disconnect the quick connectors.

12. Remove the fuel retaining nuts. Remove the fuel rail from its mounting.

13. Remove the injector clips from the injectors. Remove the injectors from the fuel rail.

To install:

14. Coat the new O-rings (black) with clean engine oil, and then insert the injectors into the fuel rail. Install the injector clips.

15. Coat the new O-rings (green) with clean engine oil, and then install the fuel rail and the injectors in the injector base.

16. Install the fuel rail retaining nuts. Tighten them to 16 ft. lbs.

17. Connect the quick connect fitting.

18. Turn the ignition switch ON (II). The fuel pump will run for about two seconds and fuel pressure will rise.

➡**Do not operate the starter motor.**

19. Repeat the above Step two or three times and check that there is no leakage in the fuel supply system.

20. Continue the installation in the reverse order of the removal procedure.

21. Clear any stored DTC codes, as required.

FUEL PUMP

REMOVAL & INSTALLATION

Fuel Pump/Fuel Gauge Sending Unit

See Figure 122.

1. Before servicing the vehicle, refer to the Precautions Section.

➡**Before disconnecting the battery cables make sure you have the anti theft codes for the audio/navigation system.**

➡**Except when doing electrical inspections, always turn the ignition switch to LOCK (0), ground the SCS line with the Honda Diagnostic Service (HDS) tool to take the PCM out of active status, disconnect the negative cable from the battery, then wait three minutes before starting the repair procedure. The SRS memory is not cleared even if the ignition switch is turned to LOCK (0) or the battery cables are disconnected from the battery.**

2. Make sure that the ignition switch is in the LOCK (0) position.

3. Disconnect the negative battery cable.

➡**Always disconnect the negative battery cable first.**

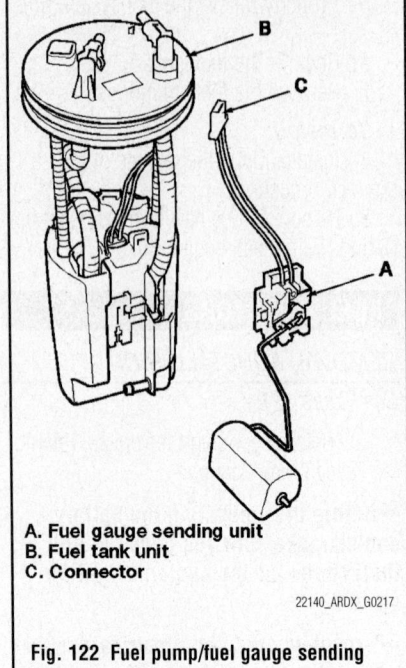

A. Fuel gauge sending unit
B. Fuel tank unit
C. Connector

22140_ARDX_G0217

Fig. 122 Fuel pump/fuel gauge sending unit related components

4. Disconnect the positive battery cable.

➡**Wait at least three minutes after disconnecting the battery cables before starting the repair procedure.**

5. Properly relieve the fuel system pressure.

6. Remove the fuel tank.

7. Remove the fuel level sensor (fuel gauge sending unit) from the fuel tank unit.

To install:

8. Installation is the reverse of removal procedure.

Secondary Fuel Gauge Sending Unit

See Figures 123 through 125.

1. Before servicing the vehicle, refer to the Precautions Section.

➡**Before disconnecting the battery cables make sure you have the anti theft codes for the audio/navigation system.**

➡**Except when doing electrical inspections, always turn the ignition switch to LOCK (0), ground the SCS line with the Honda Diagnostic Service (HDS) tool to take the PCM out of active status, disconnect the negative cable from the battery, then wait three minutes before starting the repair procedure. The SRS memory is not cleared even if the ignition switch is turned to LOCK (0) or the battery cables are disconnected from the battery.**

2. Make sure that the ignition switch is in the LOCK (0) position.

3. Disconnect the negative battery cable.

➡**Always disconnect the negative battery cable first.**

4. Disconnect the positive battery cable.

➡**Wait at least three minutes after disconnecting the battery cables before starting the repair procedure.**

5. Properly relieve the fuel system pressure.

6. Remove the fuel cap.

7. Fold the rear seat down and remove the right rear door seal trim.

8. Pull back the carpet and expose the access panel.

9. Remove the access panel from the right side of the floor.

10. Disconnect the secondary fuel gauge sending unit connector.

11. Using tool 07AAA-SOXA100 or equivalent, loosen the locknut.

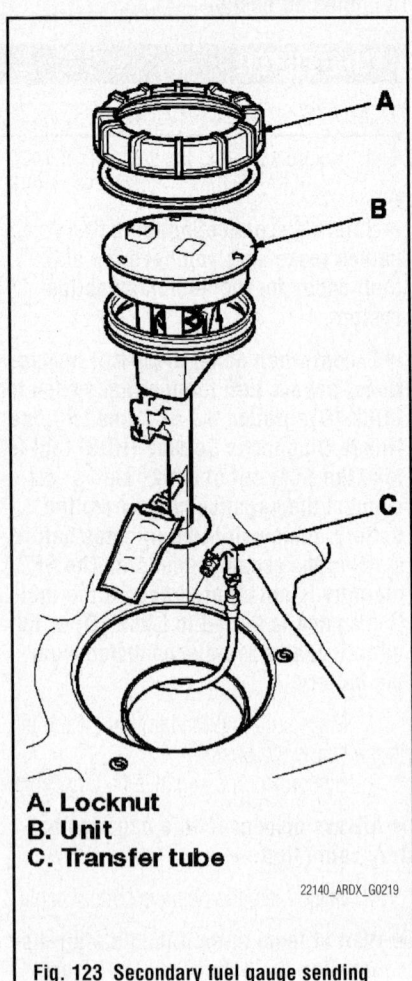

A. Locknut
B. Unit
C. Transfer tube

22140_ARDX_G0219

Fig. 123 Secondary fuel gauge sending unit

12. Remove the locknut. Lift up the secondary fuel gauge sending unit from the fuel tank. Disconnect the transfer tube.

To install:

13. Temporarily attach a new base gasket to the assembly.

14. Position the mark on the transfer tube up.

15. Insert the assembly partially into the fuel tank.

➡ **Be careful not to damage the new base gasket. Be careful not to bend the sending unit. Do not coat the base gasket with oil.**

16. Transfer the base gasket from the sending unit to the fuel tank.

17. Align the marks (B) on the fuel tank and the fuel gauge sending unit. Insert the sending unit and allow it to contact the base gasket. See the illustration for proper alignment.

➡ **To avoid a fuel leak, check the base gasket visually or with your hand to be sure it is not pinched. Correct as required.**

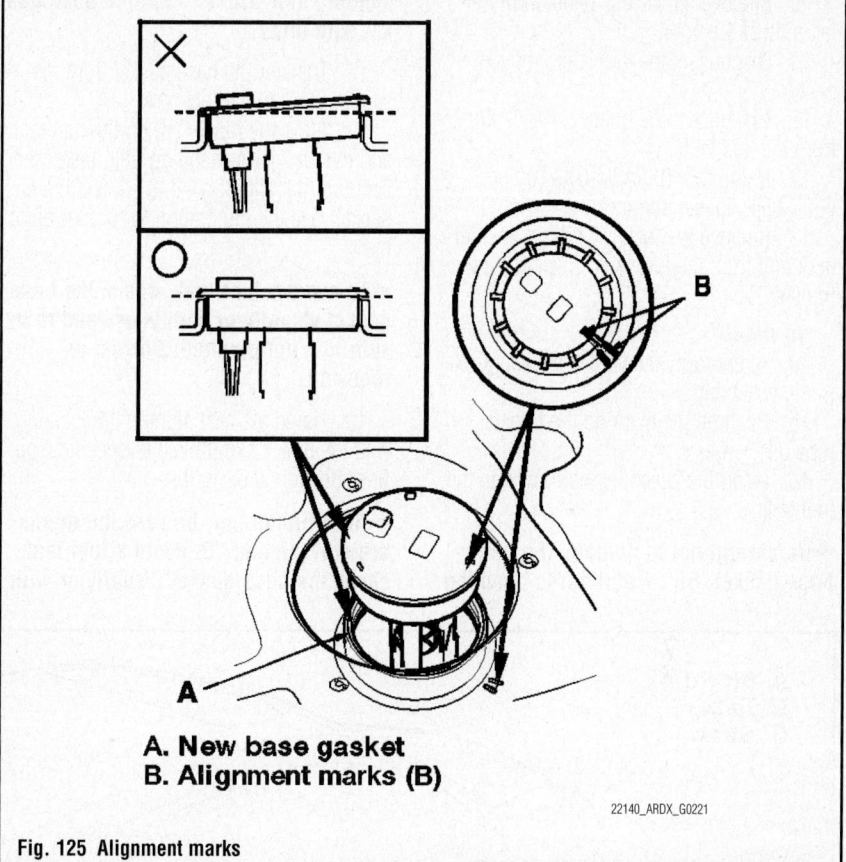

A. New base gasket
B. Alignment marks (B)

22140_ARDX_G0221

Fig. 125 Alignment marks

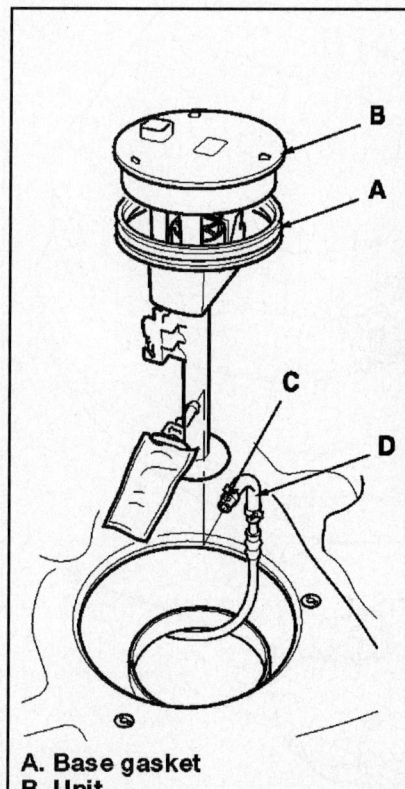

A. Base gasket
B. Unit
C. Transfer tube mark location
D. Transfer tube

22140_ARDX_G0220

Fig. 124 Transfer tube alignment

18. Using the tool, tighten the locknut with the new locknut plate to specification. Specification is 52 ft. lbs.

➡ **After tightening, be sure the marks are still aligned. To avoid a fuel leak, check the base gasket visually or with your hand to be sure it is not pinched. Correct as required.**

19. Connect the electrical connector.

20. Reconnect the negative battery cable and turn the ignition switch ON (II). The fuel pump will run for about two seconds and fuel pressure will rise.

➡ **Do not operate the starter motor.**

21. Repeat the above Step two or three times and check that there is no leakage in the fuel supply system.

22. Continue the installation in the reverse order of the removal procedure.

Fuel Tank Unit

1. Before servicing the vehicle, refer to the Precautions Section.

➡ **Before disconnecting the battery cables make sure you have the anti theft codes for the audio/navigation system.**

➡ **Except when doing electrical inspections, always turn the ignition switch to** LOCK (O), ground the SCS line with the **Honda Diagnostic Service (HDS) tool to take the PCM out of active status, disconnect the negative cable from the battery, then wait three minutes before starting the repair procedure. The SRS memory is not cleared even if the ignition switch is turned to LOCK (O) or the battery cables are disconnected from the battery.**

2. Make sure that the ignition switch is in the LOCK (O) position.

3. Disconnect the negative battery cable.

➡ **Always disconnect the negative battery cable first.**

4. Disconnect the positive battery cable.

➡ **Wait at least three minutes after disconnecting the battery cables before starting the repair procedure.**

5. Properly relieve the fuel system pressure.

6. Remove the fuel cap.

7. Fold the rear seat down and remove the left rear door seal trim.

8. Pull back the carpet and expose the access panel.

9. Remove the access panel from the left side of the floor.

10. Disconnect the fuel tank unit connector.

11. Disconnect the quick connect fittings from the fuel tank.

12. Using tool 07AAA-SOXA100 or equivalent, loosen the locknut.

13. Remove the locknut. Lift up the fuel tank unit from the fuel tank. Disconnect the transfer tube.

To install:

14. Temporarily attach a new base gasket to the assembly.

15. Position the mark on the transfer tube up.

16. Insert the assembly partially into the fuel tank.

➡ **Be careful not to damage the new base gasket. Be careful not to bend the**

sending unit. Do not coat the base gasket with oil.

17. Transfer the base gasket from the sending unit to the fuel tank.

18. Align the marks (B) on the fuel tank and the fuel gauge sending unit. Insert the sending unit and allow it to contact the base gasket. See the illustration for proper alignment.

➡ **To avoid a fuel leak, check the base gasket visually or with your hand to be sure it is not pinched. Correct as required.**

19. Using the tool, tighten the locknut with the new locknut plate to specification. Specification is 52 ft. lbs.

➡ **After tightening, be sure the marks are still aligned. To avoid a fuel leak, check the base gasket visually or with**

your hand to be sure it is not pinched. Correct as required.

20. Connect the electrical connector.

21. Reconnect the negative battery cable and turn the ignition switch ON (II). The fuel pump will run for about two seconds and fuel pressure will rise.

➡ **Do not operate the starter motor.**

22. Repeat the above Step two or three times and check that there is no leakage in the fuel supply system.

23. Continue the installation in the reverse order of the removal procedure.

FUEL TANK

REMOVAL & INSTALLATION

See Figure 126.

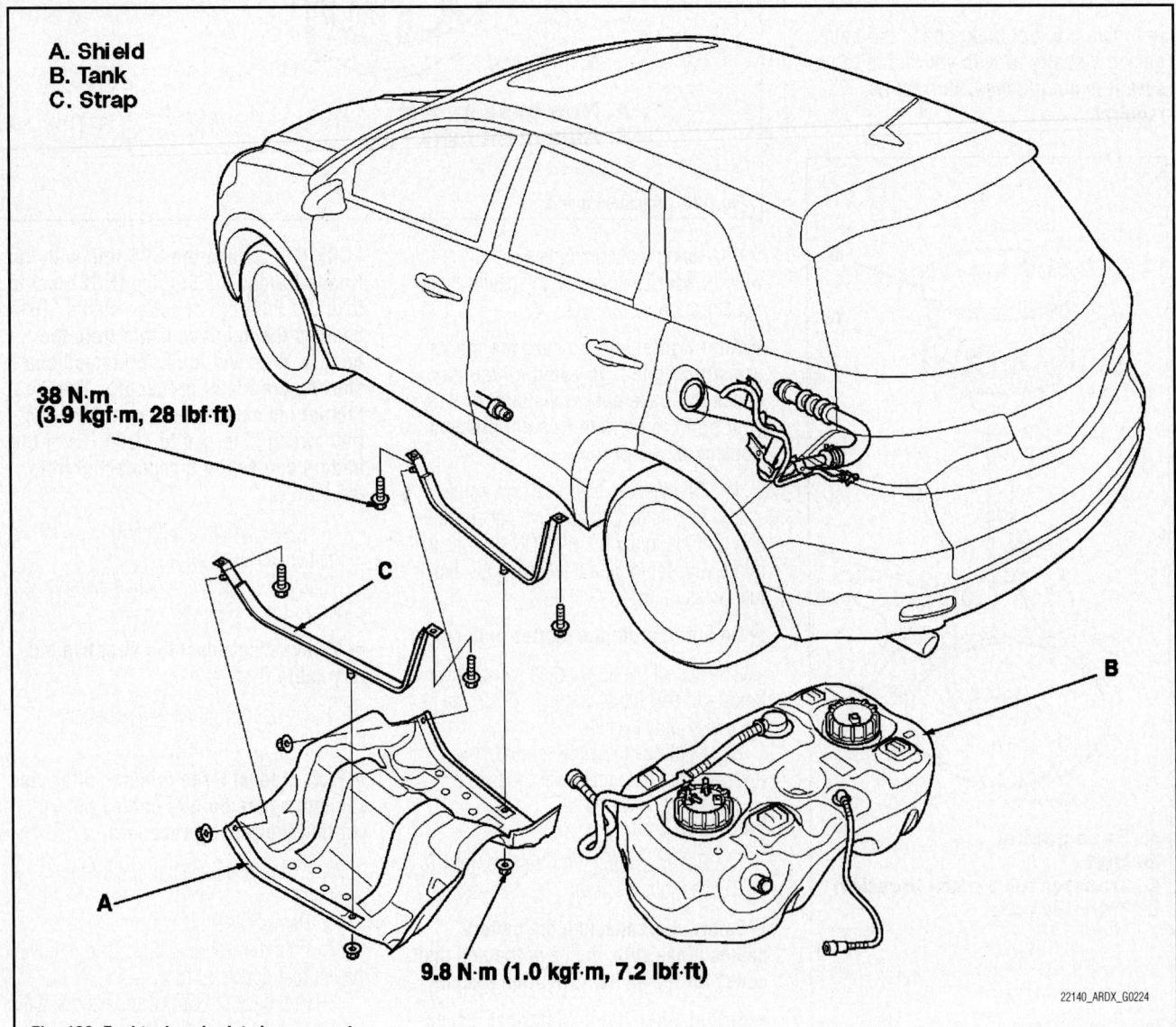

A. Shield
B. Tank
C. Strap

38 N·m
(3.9 kgf·m, 28 lbf·ft)

9.8 N·m (1.0 kgf·m, 7.2 lbf·ft)

22140_ARDX_G0224

Fig. 126 Fuel tank and related components

1. Before servicing the vehicle, refer to the Precautions Section.

➡**Before disconnecting the battery cables make sure you have the anti theft codes for the audio/navigation system.**

➡**Except when doing electrical inspections, always turn the ignition switch to LOCK (O), ground the SCS line with the Honda Diagnostic Service (HDS) tool to take the PCM out of active status, disconnect the negative cable from the battery, then wait three minutes before starting the repair procedure. The SRS memory is not cleared even if the ignition switch is turned to LOCK (O) or the battery cables are disconnected from the battery.**

2. Make sure that the ignition switch is in the LOCK (O) position.

3. Disconnect the negative battery cable.

➡**Always disconnect the negative battery cable first.**

4. Disconnect the positive battery cable.

➡**Wait at least three minutes after disconnecting the battery cables before starting the repair procedure.**

5. Properly relieve the fuel system pressure.

6. Remove the fuel tank unit. Using a hand pump, hose and container drain the fuel from the tank.

➡**Be sure to properly dispose of the drained fuel.**

7. Reinstall the fuel tank unit without connecting the quick connect fittings and the electrical connector.

8. Disconnect the secondary fuel gauge electrical connector.

9. Remove the fuel fill pipe cover. Disconnect the quick connect fitting.

10. Raise and support the vehicle safely.

11. Disconnect the fuel fill tube from the fuel tank. Slide back the clamp, then twist the tube as you pull to avoid damage. Disconnect the fuel vent tube.

12. Remove the exhaust pipe.

13. Remove the driveshaft.

14. Remove the fuel tank cover.

15. Properly support the fuel tank assembly by positioning a jack or support under the fuel tank.

16. Remove the fuel tank retaining straps.

17. Carefully lower the tank. Check to be sure nothing is attached. Remove the tank from the vehicle.

To install:

18. Installation is the reverse of the removal procedure.

19. Tighten the retaining strap bolts to 28 ft. lbs.

IDLE SPEED

ADJUSTMENT

The idle speed is controlled by the PCM and is not adjustable.

HEATING & AIR CONDITIONING SYSTEM

BLOWER MOTOR

REMOVAL & INSTALLATION
See Figure 127.

1. Before servicing the vehicle, refer to the Precautions Section.

➡**Before disconnecting the battery cables make sure you have the anti theft codes for the audio/navigation system.**

➡**Except when doing electrical inspections, always turn the ignition switch to LOCK (O), ground the SCS line with the Honda Diagnostic Service (HDS) tool to take the PCM out of active status, disconnect the negative cable from the battery, then wait three minutes before starting the repair procedure. The SRS memory is not cleared even if the ignition switch is turned to LOCK (O) or the battery cables are disconnected from the battery.**

2. Make sure that the ignition switch is in the LOCK (O) position.

3. Disconnect the negative battery cable.

➡**Always disconnect the negative battery cable first.**

4. Disconnect the positive battery cable.

➡**Wait at least three minutes after disconnecting the battery cables before starting the repair procedure.**

5. Remove the necessary trim panels.

6. Remove the necessary components to gain access to the blower motor assembly.

7. Disconnect the blower motor electrical connectors.

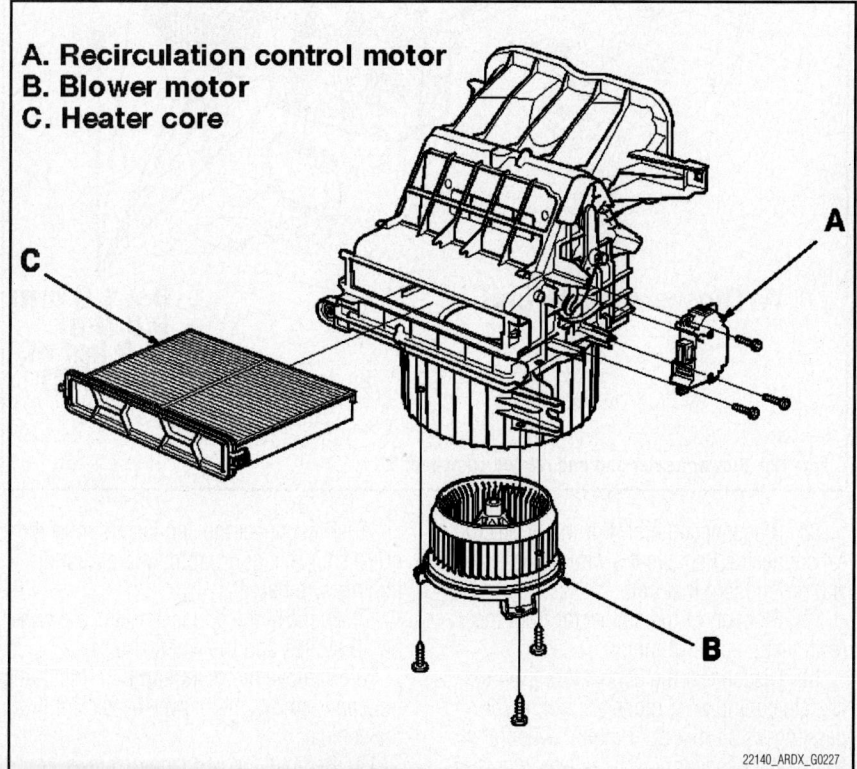

A. Recirculation control motor
B. Blower motor
C. Heater core

22140_ARDX_G0227

Fig. 127 Blower motor and related components

8. Remove the blower motor retaining screws.

9. Remove the blower motor from the blower unit.

To install:

10. Installation is the reverse of the removal procedure.

HEATER CORE

REMOVAL & INSTALLATION

See Figures 128 and 129.

1. Before servicing the vehicle, refer to the Precautions Section.

➡**Before disconnecting the battery cables make sure you have the anti theft codes for the audio/navigation system.**

➡**Except when doing electrical inspections, always turn the ignition switch to LOCK (0), ground the SCS line with the Honda Diagnostic Service (HDS) tool to take the PCM out of active status, disconnect the negative cable from the battery, then wait three minutes before starting the repair procedure. The SRS memory is not cleared even if the ignition switch is turned to LOCK (0) or the battery cables are disconnected from the battery.**

2. Make sure that the ignition switch is in the LOCK (0) position.

3. Disconnect the negative battery cable.

➡**Always disconnect the negative battery cable first.**

4. Disconnect the positive battery cable.

➡**Wait at least three minutes after disconnecting the battery cables before starting the repair procedure.**

5. Properly discharge the air conditioning system. Properly recycle used refrigerant.

6. Drain the coolant. Properly dispose of used coolant.

7. Remove the air cleaner housing assembly.

8. From under the hood open the cable clamp then disconnect the heater valve cable from the heater valve arm. Turn the heater valve arm to the fully open position as shown in the illustration.

9. Disconnect and plug the heater hoses at the heater core.

10. Remove the mounting nut from the heater unit.

11. Remove the lower instrument panel.

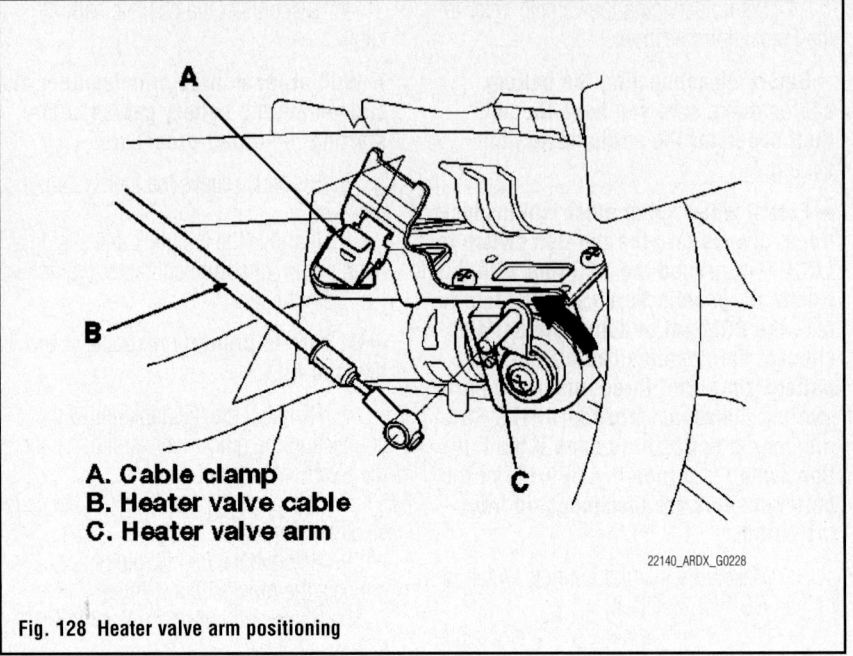

A. Cable clamp
B. Heater valve cable
C. Heater valve arm

22140_ARDX_G0228

Fig. 128 Heater valve arm positioning

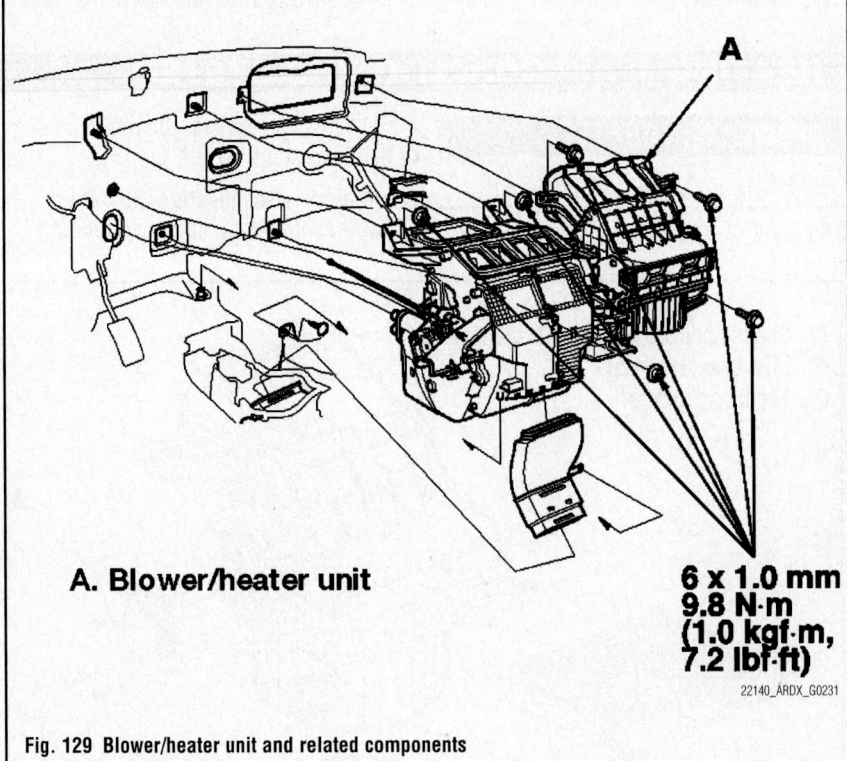

A. Blower/heater unit

6 x 1.0 mm
9.8 N·m
(1.0 kgf·m,
7.2 lbf·ft)

22140_ARDX_G0231

Fig. 129 Blower/heater unit and related components

12. Disconnect the blower motor electrical connector. Remove the wire harness clip and ground terminal bolt.

13. Disconnect the connector from the recirculation control motor.

14. Disconnect the connectors from the climate control unit, mode control motor, passenger's air mix door motor, evaporator temperature sensor and the power transistor.

15. Remove the wire harness.

16. Disconnect the connectors from the driver's air mix door motor and air conditioning wire harness.

17. Remove the connector clip, the wire harness clips and the wire harness.

18. Remove the mounting bolt, mounting nuts and remove the blower/heater unit from the vehicle.

19. Remove the self taping screws. Remove the heater core cover.

20. Remove the grommet and carefully pull the heater core out of its mounting.

To install:

21. Installation is the reverse of the removal procedure.

22. Do not interchange the heater inlet and outlet hoses.

23. Be sure to use new O-rings, as required.

24. Properly recharge the air conditioning system.

25. Fill the engine with the proper grade and type engine coolant.

26. Start the engine and check for leaks. Correct as required.

27. Be sure that the air conditioning system is functioning properly.

REMOVAL & INSTALLATION

Lower Instrument Panel

See Figures 130 through 138.

➡️**According to the manufacturer the steering column must be removed from**

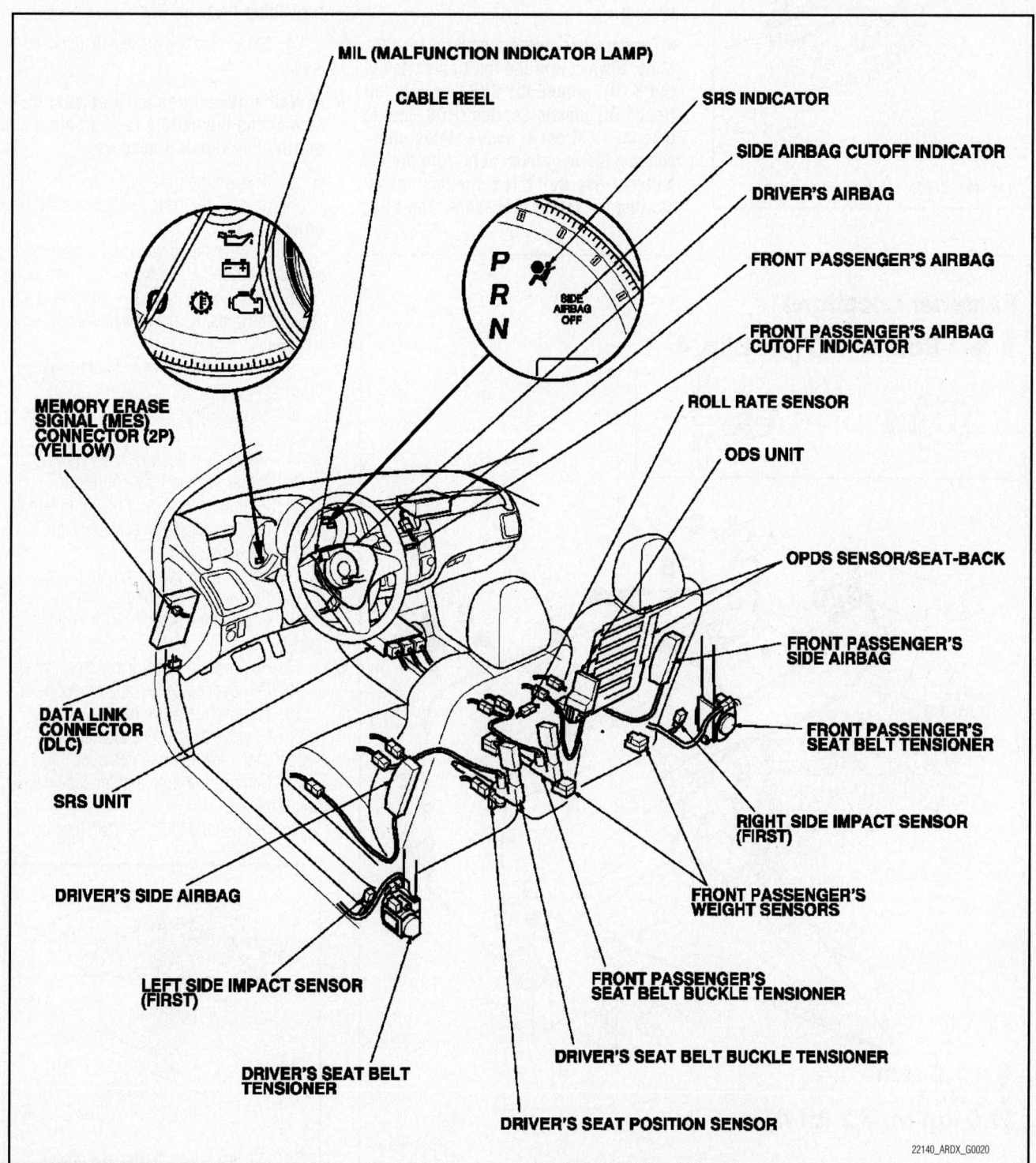

MIL (MALFUNCTION INDICATOR LAMP)

CABLE REEL

SRS INDICATOR

SIDE AIRBAG CUTOFF INDICATOR

DRIVER'S AIRBAG

FRONT PASSENGER'S AIRBAG

FRONT PASSENGER'S AIRBAG CUTOFF INDICATOR

ROLL RATE SENSOR

ODS UNIT

MEMORY ERASE SIGNAL (MES) CONNECTOR (2P) (YELLOW)

OPDS SENSOR/SEAT-BACK

FRONT PASSENGER'S SIDE AIRBAG

FRONT PASSENGER'S SEAT BELT TENSIONER

DATA LINK CONNECTOR (DLC)

SRS UNIT

RIGHT SIDE IMPACT SENSOR (FIRST)

DRIVER'S SIDE AIRBAG

FRONT PASSENGER'S WEIGHT SENSORS

LEFT SIDE IMPACT SENSOR (FIRST)

FRONT PASSENGER'S SEAT BELT BUCKLE TENSIONER

DRIVER'S SEAT BELT BUCKLE TENSIONER

DRIVER'S SEAT BELT TENSIONER

DRIVER'S SEAT POSITION SENSOR

22140_ARDX_G0020

Fig. 130 Front seats and related components

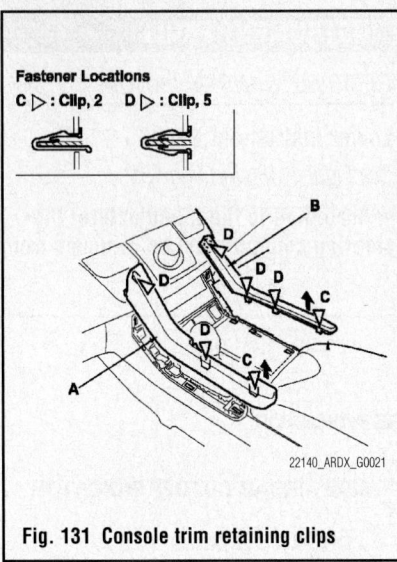

Fastener Locations

C ▷ : Clip, 2   D ▷ : Clip, 5

Fig. 131 Console trim retaining clips

the vehicle. However, at this time the manufacturer does not provide Removal and Installation procedures for the steering column.

1. Before servicing the vehicle, refer to the Precautions Section.

➡ Before disconnecting the battery cables make sure you have the anti theft codes for the audio/navigation system.

➡ Except when doing electrical inspections, always turn the ignition switch to LOCK (0), ground the SCS line with the Honda Diagnostic Service (HDS) tool to take the PCM out of active status, disconnect the negative cable from the battery, then wait three minutes before starting the repair procedure. The SRS

memory is not cleared even if the ignition switch is turned to LOCK (0) or the battery cables are disconnected from the battery.

2. Make sure that the ignition switch is in the LOCK (0) position.

3. Disconnect the negative battery cable.

➡ Always disconnect the negative battery cable first.

4. Disconnect the positive battery cable.

➡ Wait at least three minutes after disconnecting the battery cables before starting the repair procedure.

5. Remove the front seats.

6. Remove the gear selector indicator cover.

7. Remove the driver's and passenger's side trim covers, at the top of the console.

8. Open the console arm rest and detach the clips. Remove the center console trim panel.

9. Remove the driver's side console cover.

10. Remove the passenger's side console cover.

11. Disconnect the accessory power socket connector and light connector.

12. Remove the console retaining bolts.

13. Lift the rear of the assembly and detach the clips.

14. Remove the center console.

15. Remove the steering column.

16. Remove the power mirror switch panel.

17. Remove the upper instrument panel.

18. Remove the center instrument panel.

19. Remove the glove box.

20. Remove the driver's side and passenger side dashboard undercover.

21. Remove the right and left side dashboard panels.

22. Release the tilt/telescopic lock lever,

Fastener Locations

B ▶ : Bolt, 6   C ▷ : Clip, 4

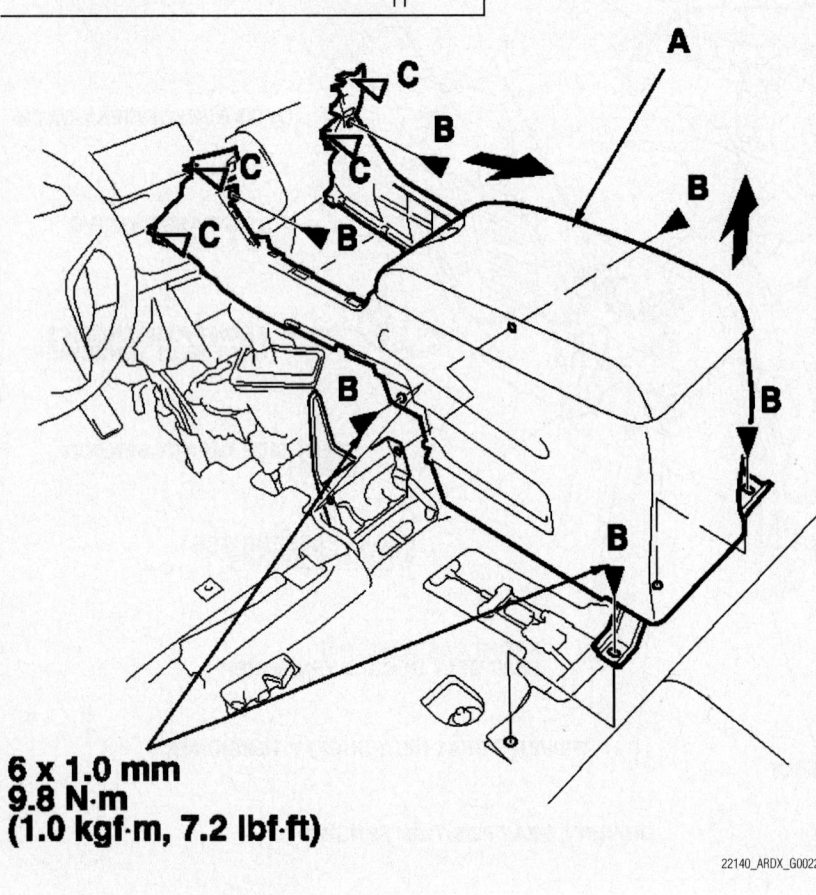

6 x 1.0 mm
9.8 N·m
(1.0 kgf·m, 7.2 lbf·ft)

Fig. 132 Console retaining bolt and clip locations

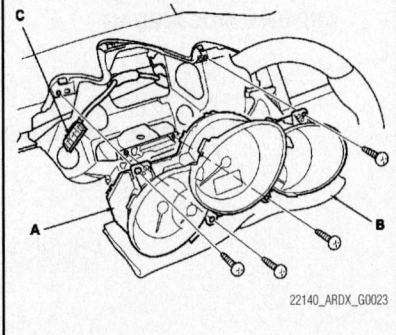

Fig. 133 Instrument cluster and related components

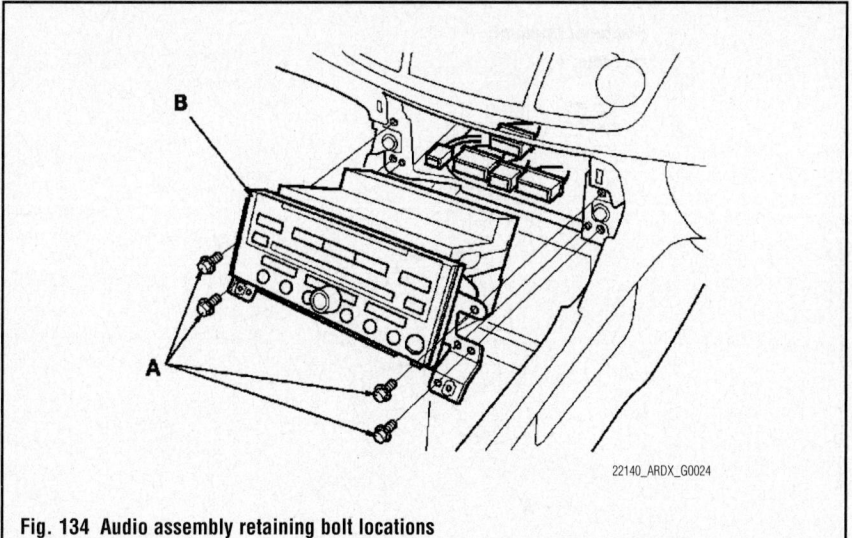

22140_ARDX_G0024

Fig. 134 Audio assembly retaining bolt locations

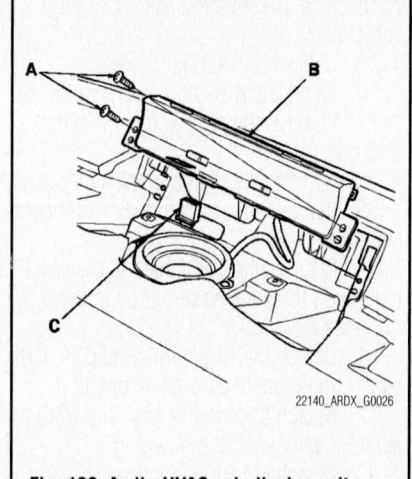

22140_ARDX_G0026

Fig. 136 Audio-HVAC sub-display unit retaining screw locations

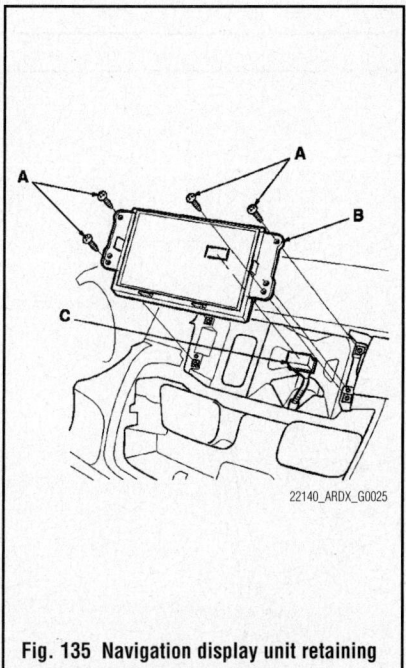

22140_ARDX_G0025

Fig. 135 Navigation display unit retaining bolt locations (audio-HVAC display unit similar)

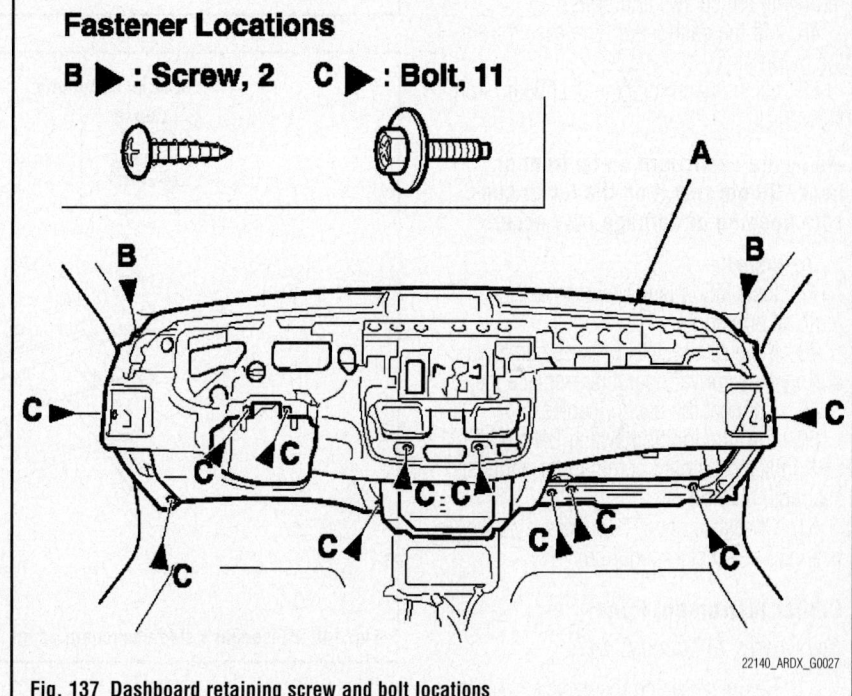

Fastener Locations

B ▶ : Screw, 2   C ▶ : Bolt, 11

22140_ARDX_G0027

Fig. 137 Dashboard retaining screw and bolt locations

and adjust the steering wheel so that it is fully tilted downward and fully pulled out.

23. Remove the instrument cluster retaining screws.

24. Gently pull out the cluster and disconnect the electrical connectors.

25. Remove the assembly from the vehicle.

26. Remove the audio pocket retaining screws. Pull the assembly forward and disconnect the electrical connectors. Remove the audio pocket from the vehicle.

27. Remove the audio assembly retaining bolts. Pull the assembly forward and disconnect the electrical connectors.

28. Remove the component from the vehicle.

29. Remove the navigation display or audio-HVAC display unit retaining bolts. Pull the assembly forward and disconnect the electrical connectors.

30. Remove the component from the vehicle.

31. Remove the front center speaker retaining screws. Disconnect the electrical connector.

32. Remove the front center speaker.

33. To remove the audio-HVAC sub-display unit, first remove the center upper dashboard panel.

34. Remove the audio-HVAC sub-display unit retaining screws. Pull the assembly forward and disconnect the

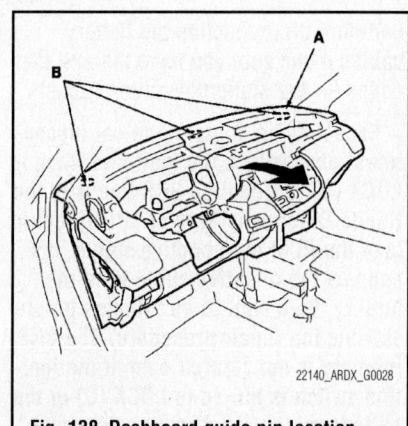

22140_ARDX_G0028

Fig. 138 Dashboard guide pin location points

connectors. Remove the assembly from the vehicle.

35. Remove the sunlight sensor.

36. Remove the in-car temperature sensor.

37. Remove the right and left A-pillar trim covers.

38. Remove the right and left kick panels.

39. Remove the right and left dashboard side lids.

40. Remove the GPS antenna bracket, then push the antenna out through the hole in the dashboard.

41. Detach the wire harness clip from the center upper dashboard panel opening.

42. Detach the wire harness clip from the center dashboard panel opening.

43. Detach the wire harness clip from the glove box opening. Detach the antenna connector.

44. Remove the dashboard by first removing the screws and bolts.

45. Lift the dashboard to release it from the guide pins.

46. Carefully remove the dashboard from the vehicle.

➡Lay the dashboard on its front or back. Do not rest it on the lower console opening or damage may occur.

To install:

47. Installation is the reverse of the removal procedure.

48. If the retaining clips were damaged during the removal operation, replace them.

49. Connect the battery cables.

50. Turn the ignition switch ON (II), the SRS indicator should come on for about six seconds, then go off.

51. Check for any DTC's that may have been set. Correct as required.

Center Instrument Panel

See Figures 139 through 142

1. Before servicing the vehicle, refer to the Precautions Section.

➡Before disconnecting the battery cables make sure you have the anti theft codes for the audio/navigation system.

➡Except when doing electrical inspections, always turn the ignition switch to LOCK (O), ground the SCS line with the Honda Diagnostic Service (HDS) tool to take the PCM out of active status, disconnect the negative cable from the battery, then wait three minutes before starting the repair procedure. The SRS memory is not cleared even if the ignition switch is turned to LOCK (O) or the battery cables are disconnected from the battery.

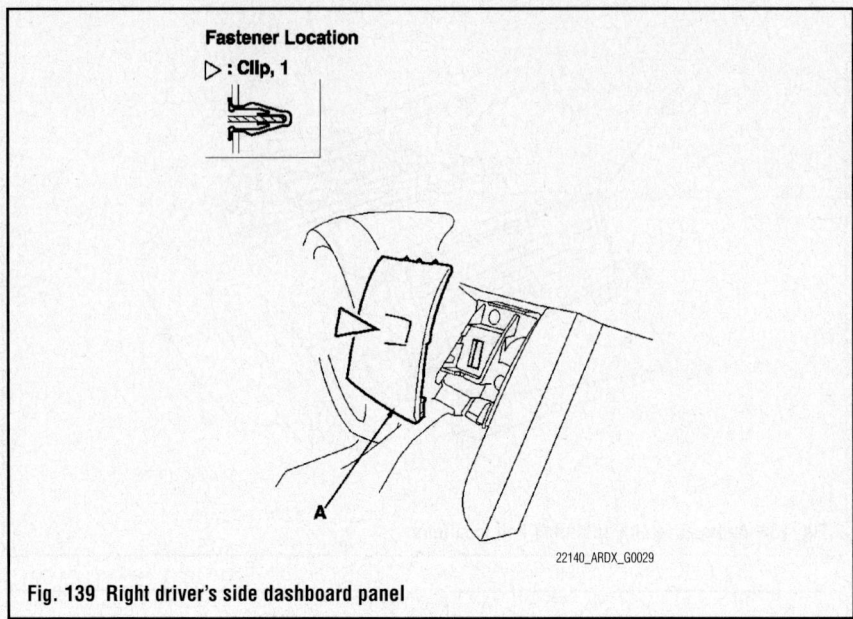

Fastener Location

▷ : Clip, 1

22140_ARDX_G0029

Fig. 139 Right driver's side dashboard panel

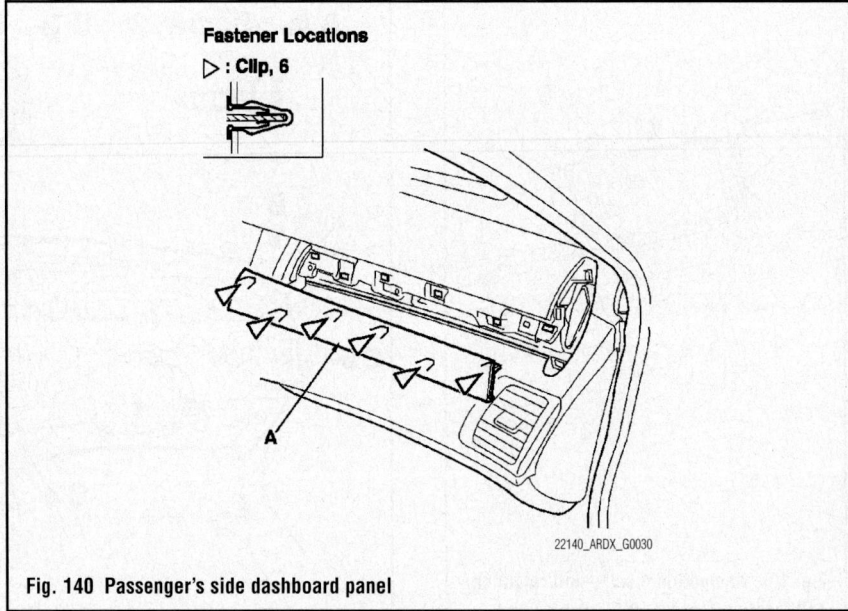

Fastener Locations

▷ : Clip, 6

22140_ARDX_G0030

Fig. 140 Passenger's side dashboard panel

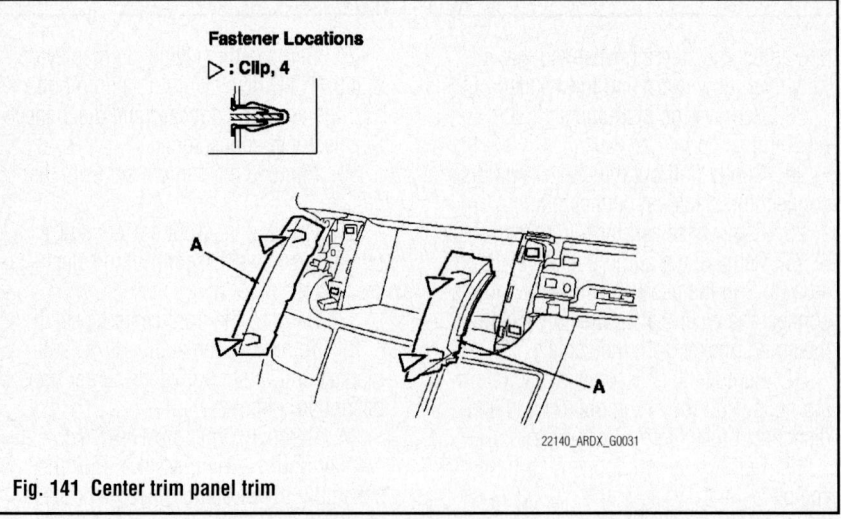

Fastener Locations

▷ : Clip, 4

22140_ARDX_G0031

Fig. 141 Center trim panel trim

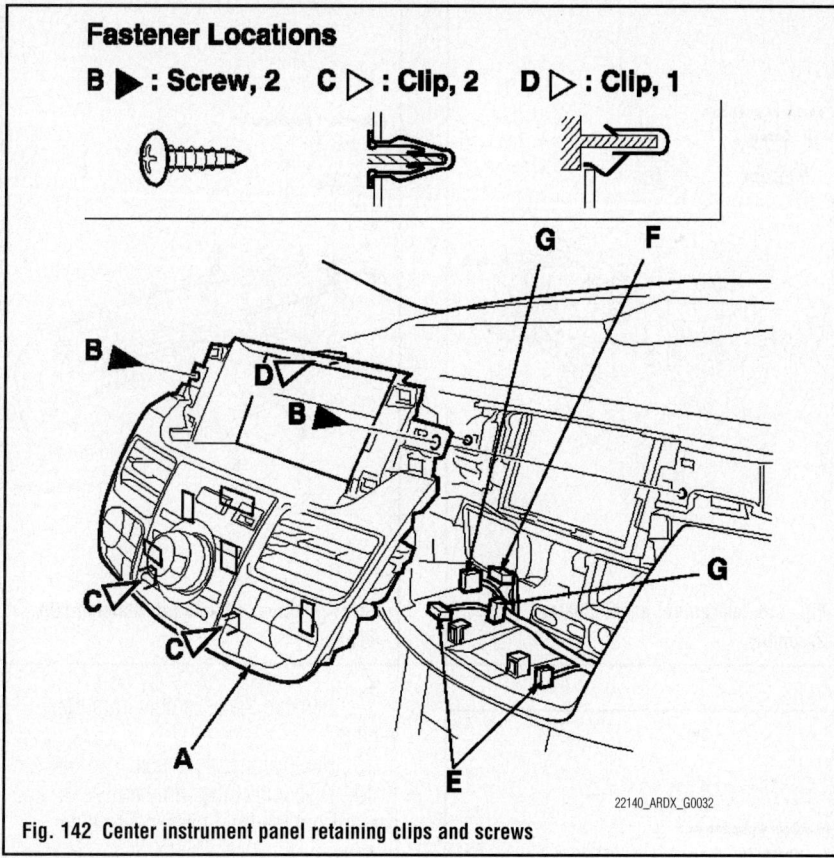

Fastener Locations

B ▶ : Screw, 2   C ▷ : Clip, 2   D ▷ : Clip, 1

22140_ARDX_G0032

Fig. 142 Center instrument panel retaining clips and screws

2. Make sure that the ignition switch is in the LOCK (0) position.

3. Disconnect the negative battery cable.

➡**Always disconnect the negative battery cable first.**

4. Disconnect the positive battery cable.

➡**Wait at least three minutes after disconnecting the battery cables before starting the repair procedure.**

5. Remove the right driver's side dashboard panel.

6. Remove the passenger's side dashboard trim panel.

7. Remove the center panel trim.

8. Remove the center panel retaining screws.

9. Gently pull the panel to release the clips.

10. Disconnect the electrical connectors.

11. Remove the center panel from the vehicle.

To install:

12. Installation is the reverse of the removal procedure.

13. If the retaining clips were damaged during the removal operation, replace them.

14. Connect the battery cables.

15. Turn the ignition switch ON (II), the

SRS indicator should come on for about six seconds, then go off.

Upper Instrument Panel
See Figures 143 through 148.

➡**Before disconnecting the battery cables make sure you have the anti theft codes for the audio/navigation system.**

➡**Except when doing electrical inspections, always turn the ignition switch to**

LOCK (0), ground the SCS line with the Honda Diagnostic Service (HDS) tool to take the PCM out of active status, disconnect the negative cable from the battery, then wait three minutes before starting the repair procedure. The SRS memory is not cleared even if the ignition switch is turned to LOCK (0) or the battery cables are disconnected from the battery.

1. Make sure that the ignition switch is in the LOCK (0) position.

2. Disconnect the negative battery cable.

➡**Always disconnect the negative battery cable first.**

3. Disconnect the positive battery cable.

➡**Wait at least three minutes after disconnecting the battery cables before starting the repair procedure.**

4. Remove the center upper dashboard panel.

5. Remove the right side driver's dashboard panel.

6. Remove the dashboard side trim, both sides.

7. Remove the passenger's side dashboard trim panel.

8. Remove the center instrument panel.

9. Remove the instrument cluster assembly.

➡**Release the tilt/telescopic lock lever, and adjust the steering wheel so that it is fully tilted downward and fully pulled out.**

10. Remove the instrument cluster retaining screws.

11. Gently pull out the cluster and disconnect the electrical connectors.

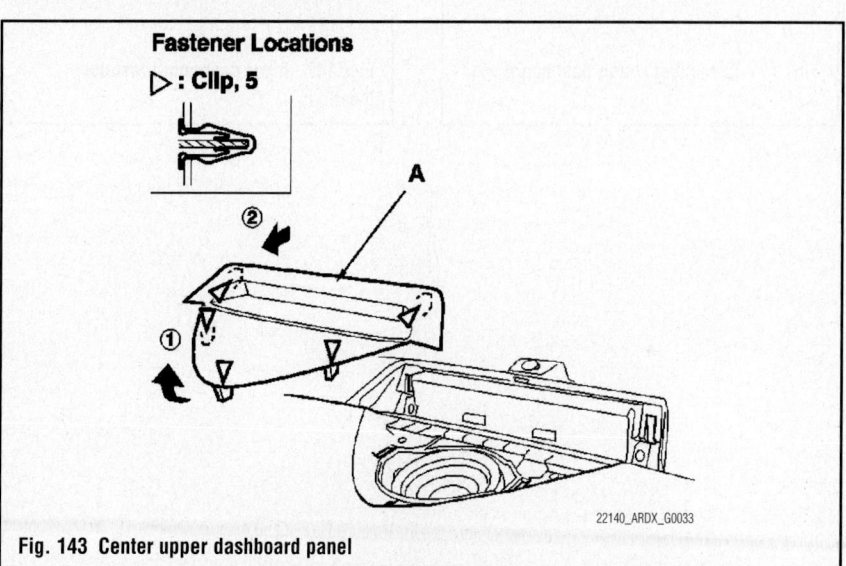

Fastener Locations

▷ : Clip, 5

22140_ARDX_G0033

Fig. 143 Center upper dashboard panel

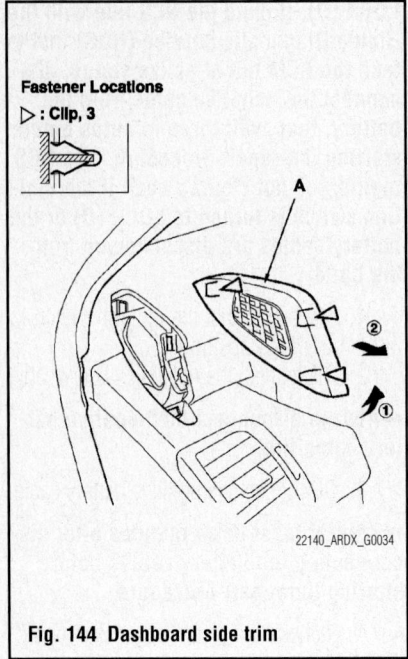

Fig. 144 Dashboard side trim

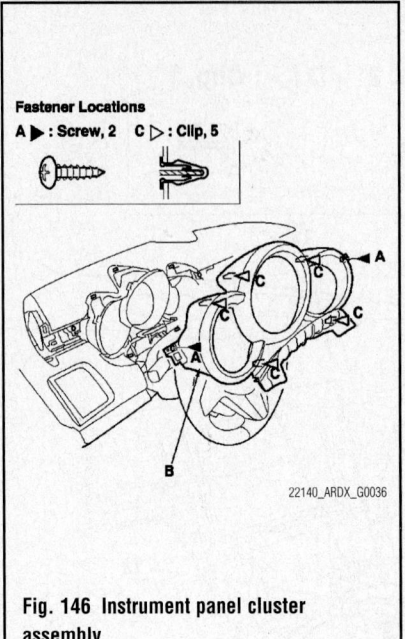

**Fig. 146 Instrument panel cluster
assembly**

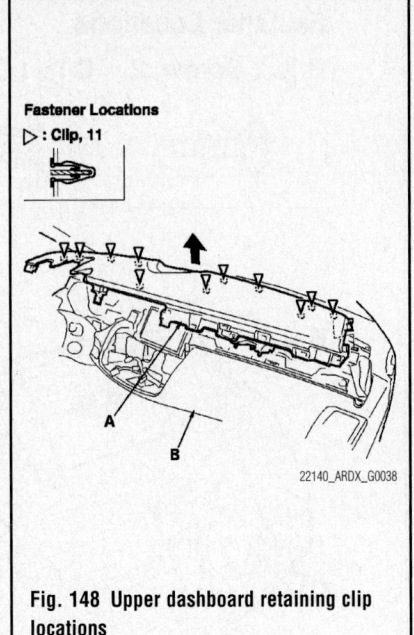

**Fig. 148 Upper dashboard retaining clip
locations**

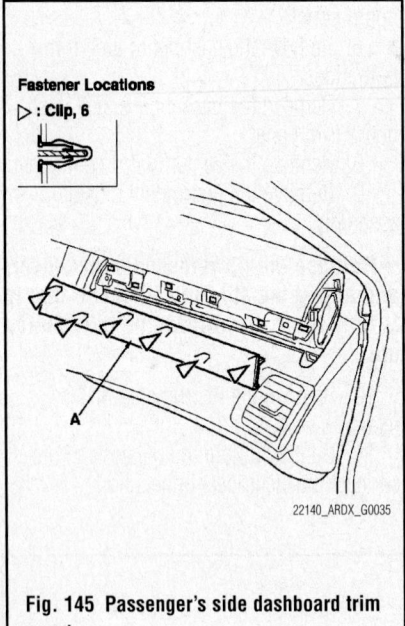

**Fig. 145 Passenger's side dashboard trim
panel**

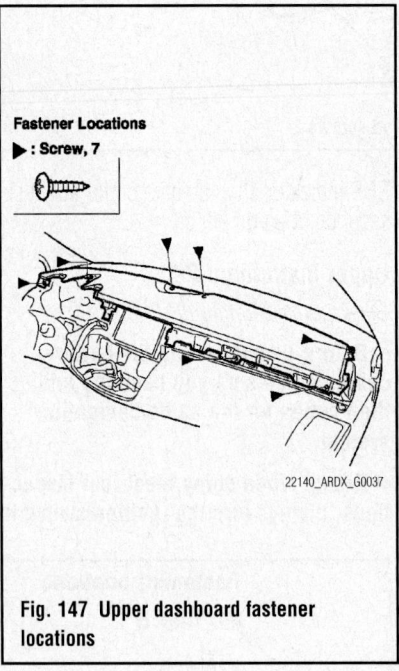

**Fig. 147 Upper dashboard fastener
locations**

12. Remove the assembly from the vehicle.

13. Remove the glove box.

14. Disconnect the wire harness 4P connector from the front passenger's side airbag.

15. Remove the three retaining nuts.

16. Remove the upper dashboard panel retaining clips.

17. Remove the component from the vehicle.

18. If required, remove the airbag from the dashboard assembly.

To install:

19. Installation is the reverse of the removal procedure.

20. If the retaining clips were damaged during the removal operation, replace them.

21. Connect the battery cables.

22. Turn the ignition switch ON (II), the SRS indicator should come on for about six seconds, then go off.

STEERING

POWER STEERING GEAR

REMOVAL & INSTALLATION

See Figures 149 through 155.

1. Before servicing the vehicle, refer to the Precautions Section.

➡**Before disconnecting the battery cables make sure you have the anti theft codes for the audio/navigation system.**

➡**Except when doing electrical inspections, always turn the ignition switch to LOCK (0), ground the SCS line with the Honda Diagnostic Service (HDS) tool to take the PCM out of active status, disconnect the negative cable from the battery, then wait three minutes before starting the repair procedure. The SRS memory is not cleared even if the ignition switch is turned to LOCK (0) or the battery cables are disconnected from the battery.**

2. Make sure that the ignition switch is in the LOCK (0) position.

3. Disconnect the negative battery cable.

➡**Always disconnect the negative battery cable first.**

4. Disconnect the positive battery cable.

➡**Wait at least three minutes after disconnecting the battery cables before starting the repair procedure.**

5. Drain the power steering reservoir. Be sure to properly dispose of used power steering fluid.

6. Raise and support the vehicle safely. Remove the front tire and wheel assemblies.

7. Remove the access panel from the steering wheel. Disconnect the driver's side airbag connector from the cable reel.

8. Remove the two Torx® head bolts. Discard the bolts.

9. Disconnect the horn switch connector. Remove the driver's airbag module.

10. Disconnect the cable reel sub harness connector.

11. Loosen the steering wheel retaining bolt.

12. Using a steering wheel puller, free the steering wheel from the steering shaft.

13. Remove the steering wheel retaining nut.

14. Remove the steering wheel from its mounting.

15. Remove the driver's dashboard undercover.

16. Remove the steering joint cover.

17. Remove the steering joint bolt. Disconnect the steering joint by moving the joint toward the column. Hold the lower

slide shaft on the column using a piece of mechanics wire, between the joint yoke on the lower slide shaft to the joint yoke on the upper shaft. See illustration for proper positioning.

18. Remove the center guide, if equipped and discard it

➡**The center guide is for factory assembly only**

19. Remove the tie rod ball joint cotter pin and remove the nut.

20. Separate the tie rod ball joint and knuckle using tool 07MAC-SL0A202, or equivalent.

21. Remove the air cleaner housing.

22. Remove the under hood fuse relay box.

23. Loosen the adjustable hose clamp and disconnect the return hose. Remove the inlet clamp bolt. Loosen the flare nut and disconnect the inlet line.

24. Remove the flange bolt from the driver's side of the steering gearbox.

25. Disconnect the return line holder and remove the return line clamp bolt.

26. Remove the pump outlet hose clamp from the steering gearbox.

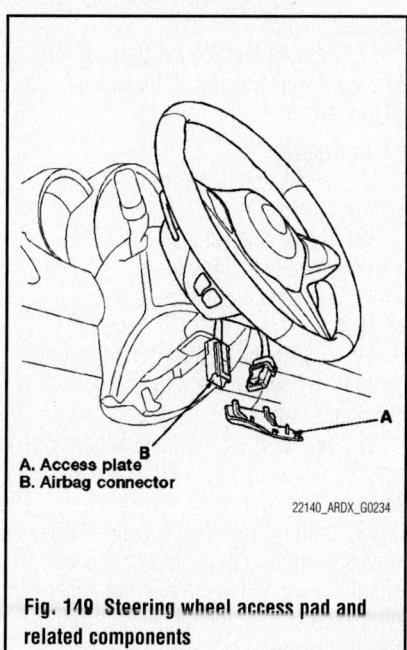

A. Access plate
B. Airbag connector

22140_ARDX_G0234

Fig. 149 Steering wheel access pad and related components

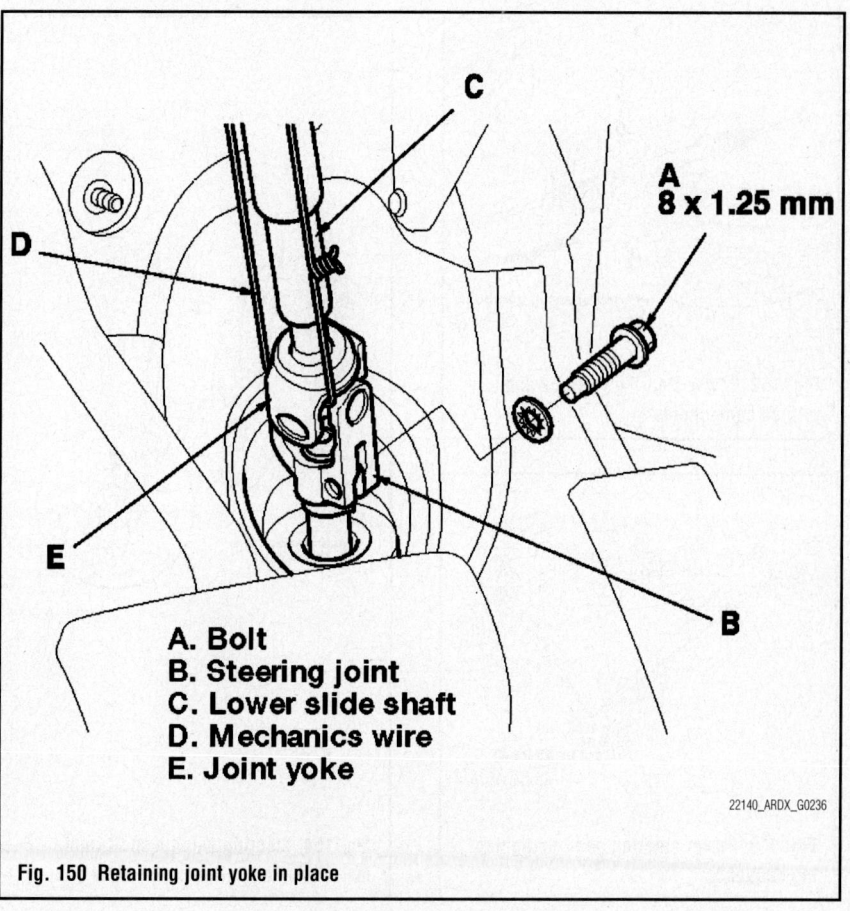

A. Bolt
B. Steering joint
C. Lower slide shaft
D. Mechanics wire
E. Joint yoke

22140_ARDX_G0236

Fig. 150 Retaining joint yoke in place

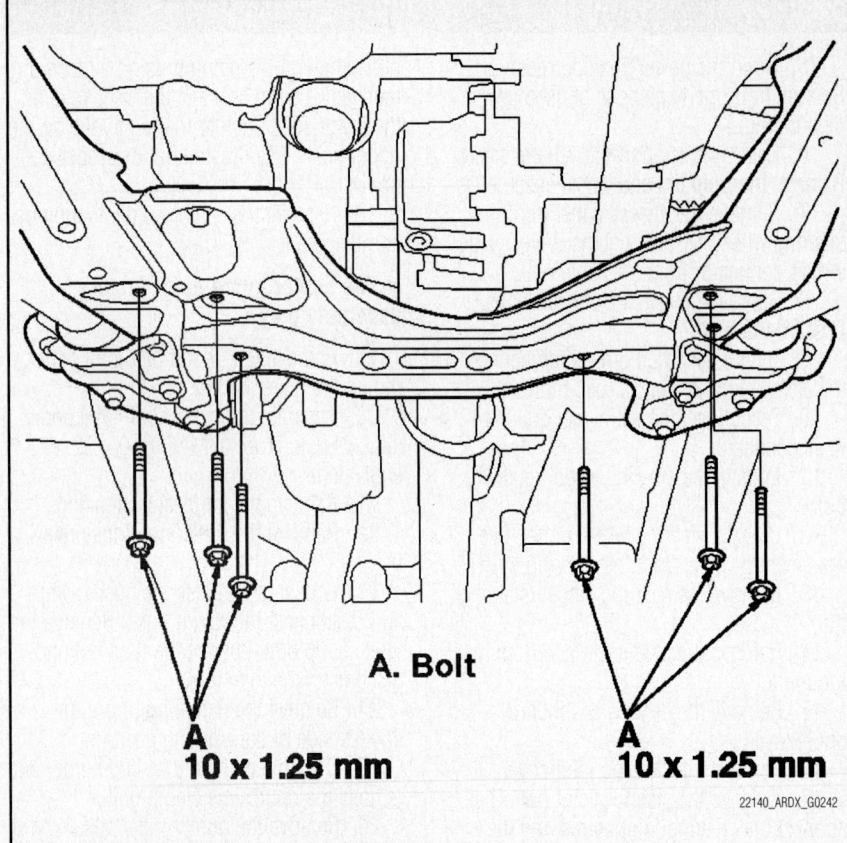

A. Bolt

A
10 x 1.25 mm

A
10 x 1.25 mm

22140_ARDX_G0242

Fig. 151 Steering gearbox mounting bolt locations

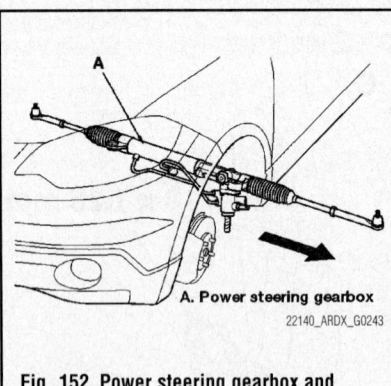

A. Power steering gearbox

22140_ARDX_G0243

Fig. 152 Power steering gearbox and related components

A

10 x 1.25 mm
59 N·m
(6.0 kgf·m, 43 lbf·ft)

22140_ARDX_G0244

Fig. 153 Power steering gearbox tightening sequence

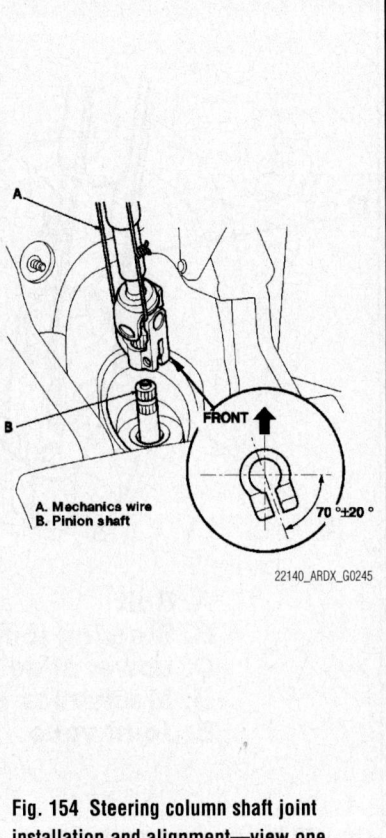

A. Mechanica wire
B. Pinion shaft

FRONT

70 °±20 °

22140_ARDX_G0245

Fig. 154 Steering column shaft joint installation and alignment—view one

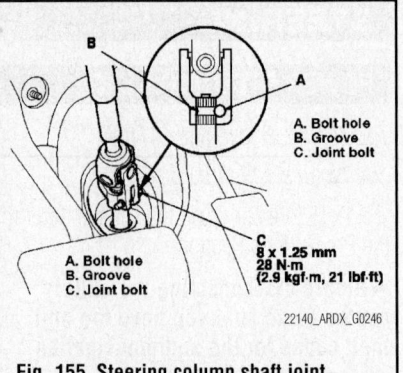

A. Bolt hole
B. Groove
C. Joint bolt

A. Bolt hole
B. Groove
C. Joint bolt

C
8 x 1.25 mm
28 N·m
(2.9 kgf·m, 21 lbf·ft)

22140_ARDX_G0246

Fig. 155 Steering column shaft joint installation and alignment—view two

27. Disconnect the return hose clips.

28. Remove the power steering heat shield.

29. Remove the flange bolt and washer from the driver's side of the steering gearbox.

30. On the driver's side, remove the flange bolts and remove the stabilizer bar bushing holder.

31. On the passenger's side, remove the flange bolts and remove the stabilizer bar bushing holder.

32. Remove the steering gearbox mounting bolts from both sides of the subframe.

33. Remove the steering gearbox brackets.

34. Move the steering gearbox toward the front, and remove the pinion shaft grommet from the top of the valve housing.

35. Apply tape to the splines of the pinion shaft.

36. Carefully move the steering gearbox and tie rods as an assembly toward the front side until the pinion shaft clears the wheel well opening on the frame.

37. Remove the steering gearbox thru the wheel well opening on the driver's side of the vehicle.

To install:

38. Installation is the reverse of the removal procedure.

39. Tighten the subframe gearbox mounting bolts to 54 ft. lbs.

40. Tighten the stabilizer bushing bolts to 16 ft. lbs.

41. Tighten the gearbox retaining bolts to 43 ft. lbs. and in the sequence shown in the illustration

42. Tighten the tie rod end to steering knuckle retaining bolt to 40 ft. lbs. Be sure to use a new cotter pin.

43. With the rack in the straight ahead driving position, cut the mechanics wire and slip the lower end of the steering joint onto the pinion shaft in the range as shown in the illustration.

44. Align the bolt hole on the steering joint with the groove around the pinion shaft, then loosely install the joint bolt.

45. Be sure that the joint bolt is securely in the groove in the pinion shaft.

46. Pull on the steering joint to make sure that the joint is fully seated. Tighten the bolt to 21 ft. lbs.

47. When installing the airbag module be sure to tighten the new Torx® bolts to 7 ft. lbs.

48. Fill the power steering system with the proper grade and type power steering fluid.

49. Bleed the system.

50. Start the engine and check for leaks, correct as required.

51. Check and adjust the front end alignment, as required.

POWER STEERING PUMP

REMOVAL & INSTALLATION

See Figure 156.

1. Before servicing the vehicle, refer to the Precautions Section.

➡**Before disconnecting the battery cables make sure you have the anti theft codes for the audio/navigation system.**

➡**Except when doing electrical inspections, always turn the ignition switch to LOCK (O), ground the SCS line with the Honda Diagnostic Service (HDS) tool to take the PCM out of active status, disconnect the negative cable from the battery, then wait three minutes before starting the repair procedure. The SRS memory is not cleared even if the ignition switch is turned to LOCK (O) or the battery cables are disconnected from the battery.**

2. Make sure that the ignition switch is in the LOCK (O) position.

3. Disconnect the negative battery cable.

➡**Always disconnect the negative battery cable first.**

4. Disconnect the positive battery cable.

➡**Wait at least three minutes after disconnecting the battery cables before starting the repair procedure.**

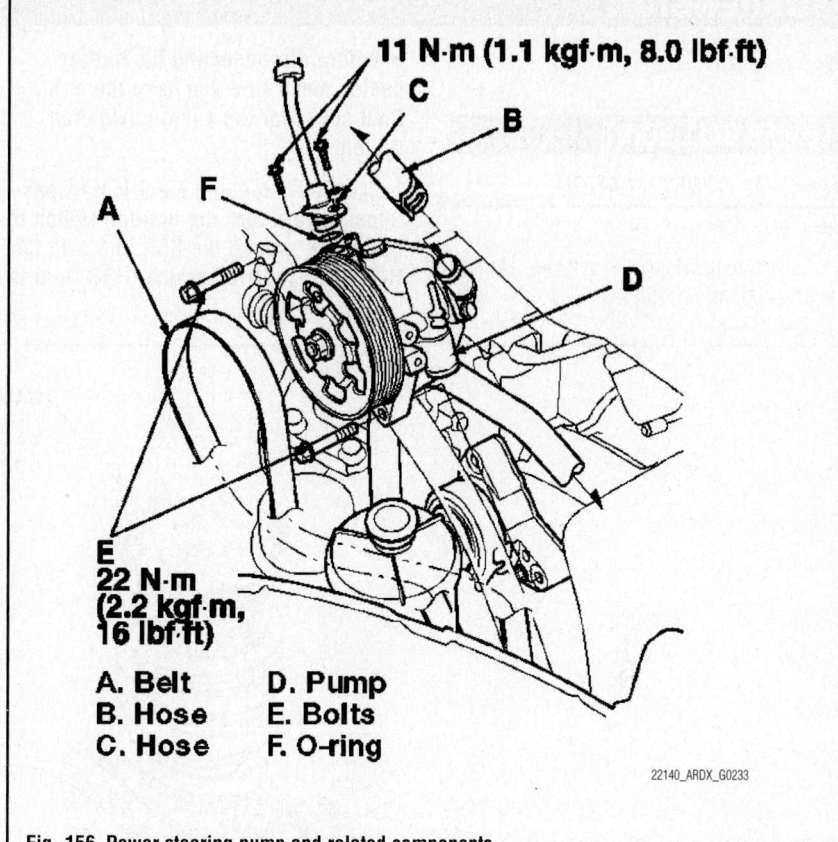

11 N·m (1.1 kgf·m, 8.0 lbf·ft)

E
**22 N·m
(2.2 kgf·m,
16 lbf·ft)**

A. Belt   D. Pump
B. Hose   E. Bolts
C. Hose   F. O-ring

22140_ARDX_G0233

Fig. 156 Power steering pump and related components

5. Drain the power steering reservoir. Be sure to properly dispose of used power steering fluid.

6. Remove the drive belt from the pump pulley.

7. Disconnect and plug the power steering pump fluid lines.

➡**Be sure to cover the auto tensioner, alternator and compressor with a shop towel in order to prevent fluid from spilling on these components.**

8. Remove the power steering pump retaining bolts. Remove the power steering pump from its mounting.

➡**Cover the opening of the pump with a piece of tape to prevent dirt and debris from entering the pump.**

To install:

9. Connect the power steering pump hoes using new O-rings.

10. Loosely install the pump in the pump bracket with the mounting bolts. Tighten to specification shown in graphic.

11. Install the drive belt. Make sure that the belt is properly positioned on the pulleys.

12. Fill the reservoir with the proper grade and type power steering fluid.

BLEEDING

1. Before servicing the vehicle, refer to the Precautions Section.

2. Fill the power steering reservoir with the proper grade and type power steering fluid.

3. Start the engine and run it at fast idle.

4. Turn the steering wheel from lock-to-lock several times to bleed the air from the system.

5. Recheck the fluid level, correct as required.

6. Do not fill the reservoir past the upper level line.

See Figure 157.

LOWER BALL JOINT

REMOVAL & INSTALLATION
See Figure 158.

1. Before servicing the vehicle, refer to the Precautions Section.

→Before disconnecting the battery cables make sure you have the anti theft codes for the audio/navigation system.

→Except when doing electrical inspections, always turn the ignition switch to LOCK (0), ground the SCS line with the Honda Diagnostic Service (HDS) tool to take the PCM out of active status, disconnect the negative cable from the battery, then wait three minutes before starting the repair procedure. The SRS memory is not cleared even if the ignition switch is turned to LOCK (0) or the battery cables are disconnected from the battery.

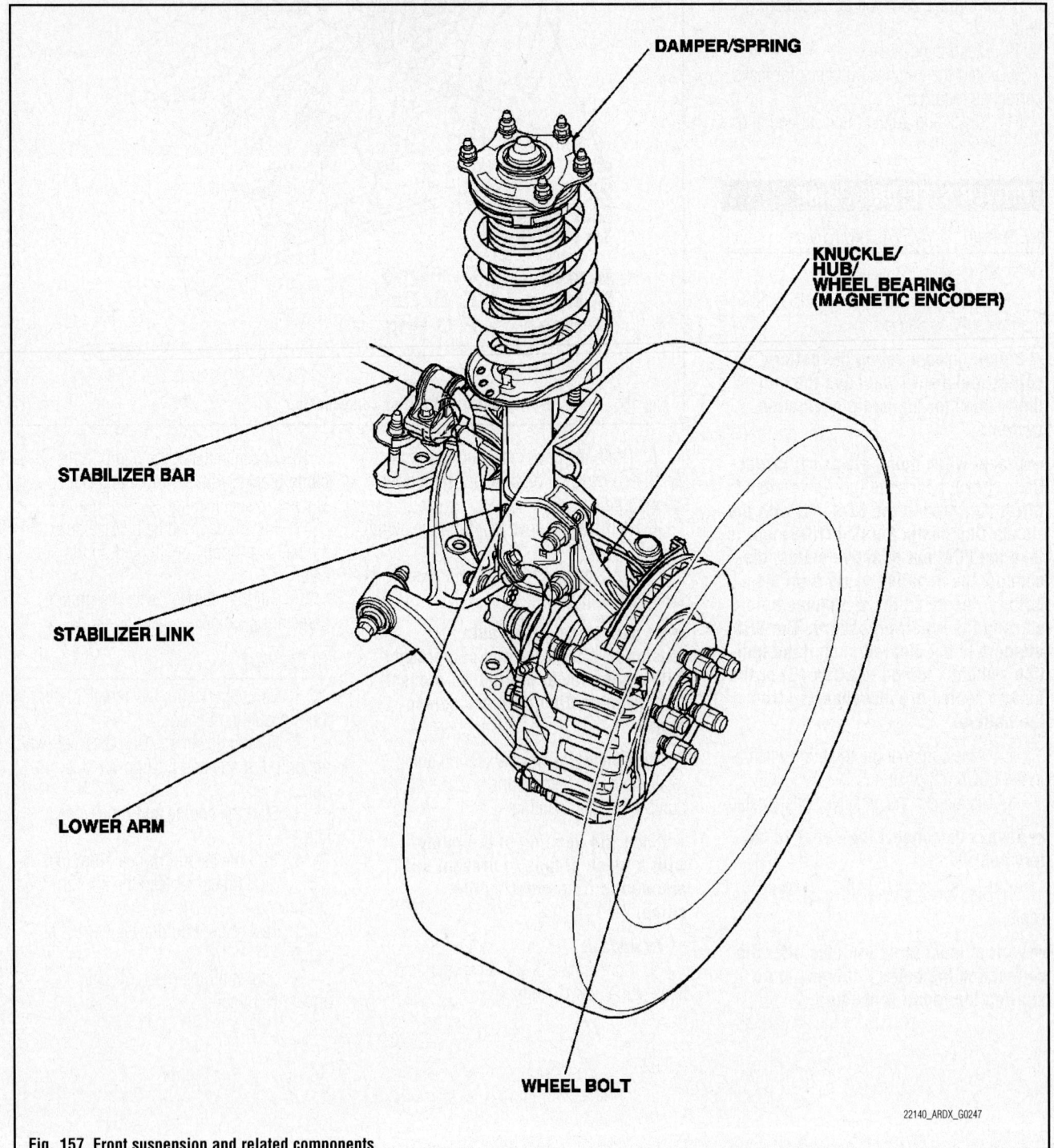

Fig. 157 Front suspension and related components

22140_ARDX_G0247

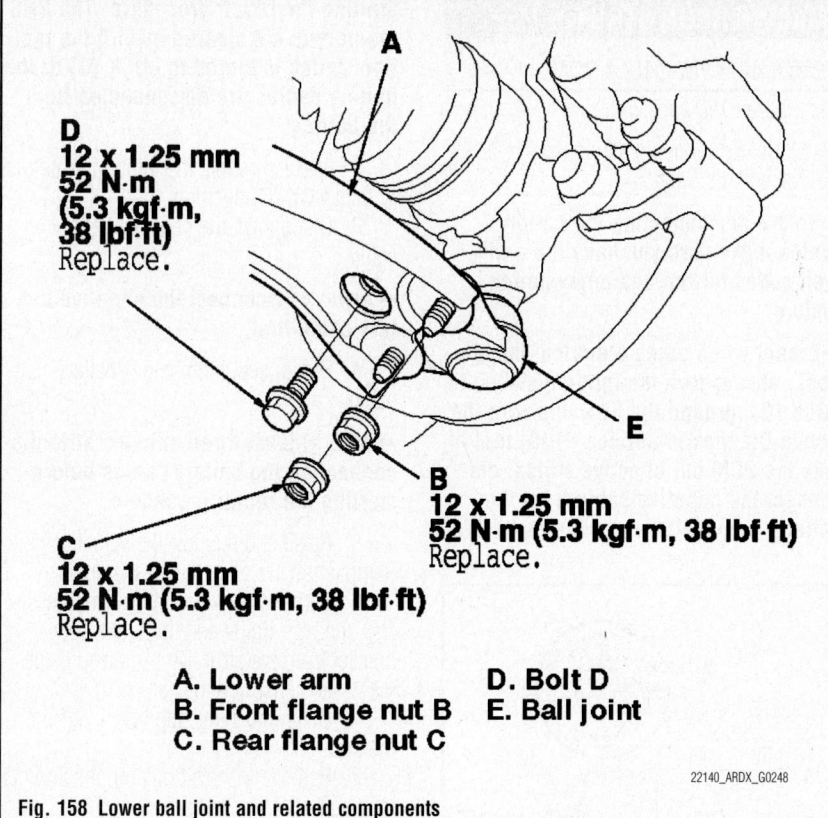

D
**12 x 1.25 mm
52 N·m
(5.3 kgf·m,
38 lbf·ft)
Replace.**

B
**12 x 1.25 mm
52 N·m (5.3 kgf·m, 38 lbf·ft)
Replace.**

C
**12 x 1.25 mm
52 N·m (5.3 kgf·m, 38 lbf·ft)
Replace.**

A. Lower arm	**D. Bolt D**
B. Front flange nut B	**E. Ball joint**
C. Rear flange nut C	

22140_ARDX_G0248

Fig. 158 Lower ball joint and related components

2. Make sure that the ignition switch is in the LOCK (O) position.

3. Disconnect the negative battery cable.

➡**Always disconnect the negative battery cable first.**

4. Disconnect the positive battery cable.

➡**Wait at least three minutes after disconnecting the battery cables before starting the repair procedure.**

5. Raise and support the vehicle safely. Remove the tire and wheel assemblies.

6. Remove the flange bolt and flange nuts from the lower arm.

7. Disconnect the lower ball joint from the lower arm.

8. Remove the lock pin from the lower ball joint. Remove the castle nut.

9. Install the ball joint thread protector. Using the ball joint removal tool, remove the lower ball joint.

To install:

10. Installation is the reverse of the removal procedure.

11. Loosely install the new flange bolt and flange nuts. Tighten to specification and in the following order, front nut (B), rear nut (C) and bolt (D). See the illustration.

12. Tighten the castle nut to 43–54 ft lbs. then tighten it only enough to install the

cotter pin. Do not align the castle nut by loosening it.

➡**First install all the components, and lightly tighten the bolts and nuts, then tighten the lower ball joint to the lower arm to specification. Raise the suspension to load it with the vehicles weight before fully tightening the lower ball joint to the knuckle to specification.**

13. Check and adjust the wheel alignment, as required.

LOWER CONTROL ARM

REMOVAL & INSTALLATION
See Figure 159.

1. Before servicing the vehicle, refer to the Precautions Section.

➡**Before disconnecting the battery cables make sure you have the anti theft codes for the audio/navigation system.**

➡**Except when doing electrical inspections, always turn the ignition switch to LOCK (O), ground the SCS line with the Honda Diagnostic Service (HDS) tool to**

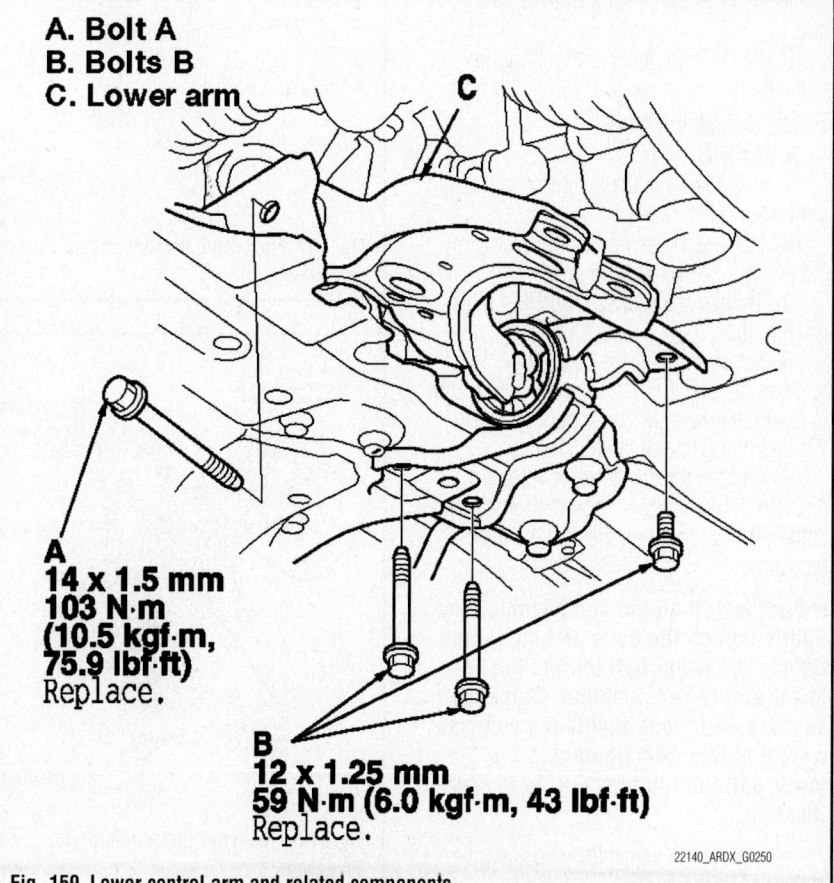

A. Bolt A
B. Bolts B
C. Lower arm

A
**14 x 1.5 mm
103 N·m
(10.5 kgf·m,
75.9 lbf·ft)
Replace.**

B
**12 x 1.25 mm
59 N·m (6.0 kgf·m, 43 lbf·ft)
Replace.**

22140_ARDX_G0250

Fig. 159 Lower control arm and related components

take the PCM out of active status, disconnect the negative cable from the battery, then wait three minutes before starting the repair procedure. The SRS memory is not cleared even if the ignition switch is turned to LOCK (O) or the battery cables are disconnected from the battery.

2. Make sure that the ignition switch is in the LOCK (O) position.
3. Disconnect the negative battery cable.

➡️**Always disconnect the negative battery cable first.**

4. Disconnect the positive battery cable.

➡️**Wait at least three minutes after disconnecting the battery cables before starting the repair procedure.**

5. Raise and support the vehicle safely. Remove the tire and wheel assemblies.
6. Remove the flange bolts and the bushing holder bolts from the front stabilizer bar.
7. Remove the flange bolt and flange nuts from the lower arm.
8. Disconnect the lower ball joint from the lower arm.
9. Remove the lower arm mounting bolt.
10. Remove the lower arm mounting bolts. Remove the lower arm from the front suspension subframe.

To install:

11. Installation is the reverse of the removal procedure.
12. Be sure to use new bolts when installing the lower control arm.
13. Tighten the stabilizer bushing bolts to 16 ft. lbs.
14. Loosely install the new flange bolt and flange nuts. Tighten to specification and in the following order, front nut (B), rear nut (C) and bolt (D). See the illustration.
15. Tighten the castle nut to 43–54 ft. lbs. then tighten it only enough to install the cotter pin. Do not align the castle nut by loosening it.

➡️**First install all the components, and lightly tighten the bolts and nuts, then tighten the lower ball joint to the lower arm to specification. Raise the suspension to load it with the vehicles weight before fully tightening the lower ball joint to the knuckle to specification.**

16. Check and adjust the wheel alignment, as required.

MACPHERSON STRUT

REMOVAL & INSTALLATION

See Figures 160 and 161.

1. Before servicing the vehicle, refer to the Precautions Section.

➡️**Before disconnecting the battery cables make sure you have the anti theft codes for the audio/navigation system.**

➡️**Except when doing electrical inspections, always turn the ignition switch to LOCK (O), ground the SCS line with the Honda Diagnostic Service (HDS) tool to take the PCM out of active status, disconnect the negative cable from the battery, then wait three minutes before**

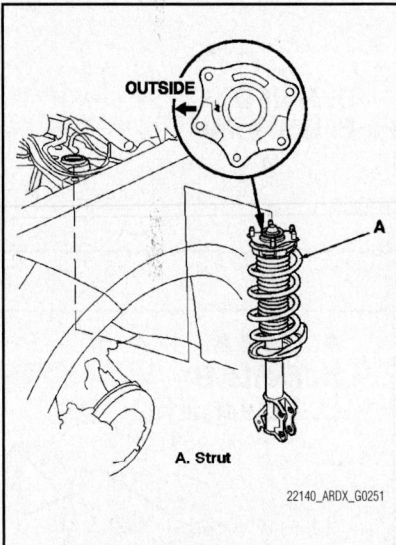

Fig. 160 Front strut installation and positioning

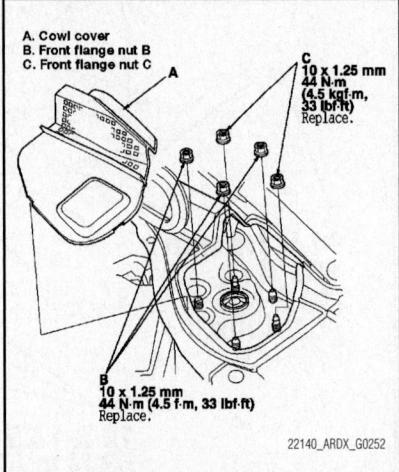

Fig. 161 Front strut flange nut torque sequence

starting the repair procedure. The SRS memory is not cleared even if the ignition switch is turned to LOCK (O) or the battery cables are disconnected from the battery.

2. Make sure that the ignition switch is in the LOCK (O) position.
3. Disconnect the negative battery cable.

➡️**Always disconnect the negative battery cable first.**

4. Disconnect the positive battery cable.

➡️**Wait at least three minutes after disconnecting the battery cables before starting the repair procedure.**

5. Raise and support the vehicle safely. Remove the tire and wheel assemblies.
6. Remove the wheel sensor harness clip, the wire guide and the brake hose bracket from the strut. Do not disconnect the wheel sensor connector.
7. Disconnect the stabilizer link from the strut.
8. Remove the strut pinch bolts and self locking nuts from the strut.
9. Remove the lid from the cowl cover.
10. Remove the flange nuts from the top of the strut.

➡️**Strut springs are different, left and right. Mark the springs L and R before removing them.**

11. Remove the strut from the vehicle.

To install:

12. Installation is the reverse of the removal procedure.
13. Be sure to use new bolts and nuts, as required.
14. Loosely install the new upper strut retaining flange nuts.
15. Loosely install the new front strut retaining bolts and new self locking nuts.
16. Connect the stabilizer link to the strut. Tighten the new nut to 58 ft. lbs.
17. Install the wheel speed sensor harness clip, wire guide and brake hose bracket to the strut.
18. Raise the front suspension with a floor jack to load the suspension with the vehicles weight.
19. Tighten the strut bolts and self locking nuts to 122 ft. lbs.
20. Tighten the upper strut nuts to specification and in the order shown in the illustration. Specification is 33 ft. lbs. Install the cowl cover.
21. Check and adjust the wheel alignment, as required.

STABILIZER BAR

REMOVAL & INSTALLATION

See Figure 162.

1. Before servicing the vehicle, refer to the Precautions Section.

➡ **Before disconnecting the battery cables make sure you have the anti theft codes for the audio/navigation system.**

➡ **Except when doing electrical inspections, always turn the ignition switch to LOCK (O), ground the SCS line with the Honda Diagnostic Service (HDS) tool to take the PCM out of active status, disconnect the negative cable from the battery, then wait three minutes before starting the repair procedure. The SRS memory is not cleared even if the ignition switch is turned to LOCK (O) or the battery cables are disconnected from the battery.**

2. Make sure that the ignition switch is in the LOCK (O) position.
3. Disconnect the negative battery cable.

➡ **Always disconnect the negative battery cable first.**

4. Disconnect the positive battery cable.

➡ **Wait at least three minutes after disconnecting the battery cables before starting the repair procedure.**

5. Raise and support the vehicle safely. Remove the tire and wheel assemblies.
6. Disconnect both stabilizer links from the stabilizer bar.
7. Remove the flange bolts and the bushing holders. Remove the bushings.

8. Disconnect both tie rod ball joints from the steering knuckles.
9. Remove the stabilizer bar thru the wheel well opening on the passenger's side of the vehicle.

To install:

10. Installation is the reverse of the removal procedure.
11. Be sure to use new bolts and nuts, as required.
12. Note the right and left direction of the stabilizer bar.
13. Align the paint marks on the stabilizer bar with the sides of the bushings.
14. Note the fore/aft direction of the bushing.
15. Check and adjust the wheel alignment, as required.

STABILIZER LINK

REMOVAL & INSTALLATION

See Figure 163.

1. Before servicing the vehicle, refer to the Precautions Section.

➡ **Before disconnecting the battery cables make sure you have the anti theft codes for the audio/navigation system.**

➡ **Except when doing electrical inspections, always turn the ignition switch to LOCK (O), ground the SCS line with the Honda Diagnostic Service (HDS) tool to take the PCM out of active status, disconnect the negative cable from the battery, then wait three minutes before starting the repair procedure. The SRS memory is not cleared even if the ignition switch is turned to LOCK (O) or the battery cables are disconnected from the battery.**

2. Make sure that the ignition switch is in the LOCK (O) position.
3. Disconnect the negative battery cable.

➡ **Always disconnect the negative battery cable first.**

4. Disconnect the positive battery cable.

➡ **Wait at least three minutes after disconnecting the battery cables before starting the repair procedure.**

5. Raise and support the vehicle safely. Remove the tire and wheel assemblies.
6. Remove the flange nuts while holding the respective joint pin, with the hex wrench. Remove the stabilizer link.

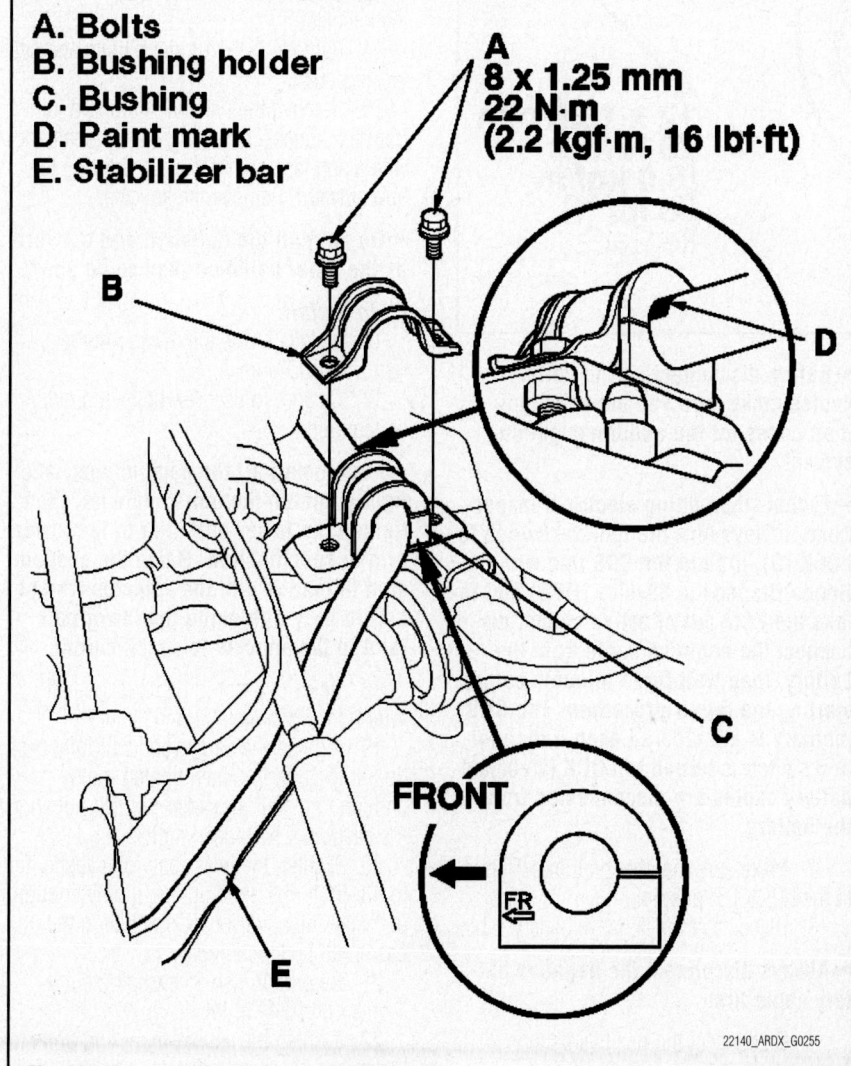

A. Bolts
B. Bushing holder
C. Bushing
D. Paint mark
E. Stabilizer bar

A
8 x 1.25 mm
22 N·m
(2.2 kgf·m, 16 lbf·ft)

FRONT

FR

22140_ARDX_G0255

Fig. 162 Front stabilizer and related components

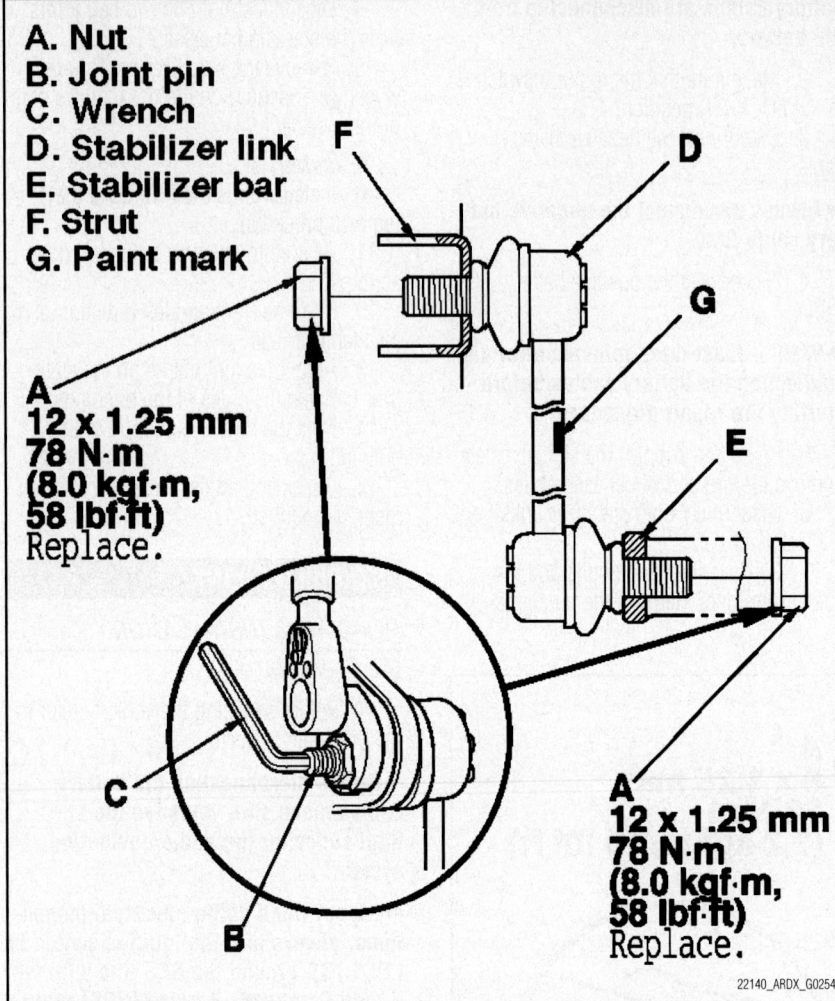

A. Nut
B. Joint pin
C. Wrench
D. Stabilizer link
E. Stabilizer bar
F. Strut
G. Paint mark

A
12 x 1.25 mm
78 N·m
(8.0 kgf·m,
58 lbf·ft)
Replace.

A
12 x 1.25 mm
78 N·m
(8.0 kgf·m,
58 lbf·ft)
Replace.

22140_ARDX_G0254

Fig. 163 Front stabilizer link

To install:

7. Installation is the reverse of the removal procedure.

8. Be sure to use new bolts and nuts, as required.

9. Install the stabilizer link on the stabilizer bar and the strut with the joint pins set at the center of their range movement.

➡ **The stabilizer link has a paint mark. Align the paint mark on the stabilizer link facing inward.**

10. Install the new flange nuts and lightly tighten them. Tighten them to specification while holding the respective joint pin with the hex wrench. Specification is 58 ft. lbs.

11. Check and adjust the wheel alignment, as required.

STEERING KNUCKLE

REMOVAL & INSTALLATION

See Figure 164.

1. Before servicing the vehicle, refer to the Precautions Section.

➡ **Before disconnecting the battery cables make sure you have the anti theft codes for the audio/navigation system.**

➡ **Except when doing electrical inspections, always turn the ignition switch to LOCK (0), ground the SCS line with the Honda Diagnostic Service (HDS) tool to take the PCM out of active status, disconnect the negative cable from the battery, then wait three minutes before starting the repair procedure. The SRS memory is not cleared even if the ignition switch is turned to LOCK (0) or the battery cables are disconnected from the battery.**

2. Make sure that the ignition switch is in the LOCK (0) position.

3. Disconnect the negative battery cable.

➡ **Always disconnect the negative battery cable first.**

4. Disconnect the positive battery cable.

➡ **Wait at least three minutes after disconnecting the battery cables before starting the repair procedure.**

5. Raise and support the vehicle safely. Remove the tire and wheel assemblies.

6. Remove the brake hose mounting bolt. Remove the caliper and position it to the side. Do not allow the caliper to hang by the brake hose.

7. Remove the wheel speed sensor from the knuckle. Do not disconnect the wheel speed sensor connector.

8. Remove the spindle nut. Remove the rotor.

9. Check the hub for damage and cracks.

10. Remove the cotter pin from the tie rod end ball joint. Remove the nut.

11. Disconnect the tie rod ball joint from the steering knuckle using a ball joint removal tool.

12. Remove the lock pin from the lower ball joint. Remove the castle nut.

13. Disconnect the lower ball joint from the knuckle.

14. Remove the strut pinch bolts and self locking nuts.

15. Remove the halfshaft outboard joint from the knuckle by taping the halfshaft end with a soft face hammer while drawing the hub outward. Remove the knuckle.

➡ **Do not pull the halfshaft end outward as the inner halfshaft may come apart.**

To install:

16. Installation is the reverse of the removal procedure.

17. Be sure to use new bolts and nuts, as required.

➡ **First install all the components, and lightly tighten the bolts and nuts, then tighten the lower ball joint to the lower arm to specification. Raise the suspension to load it with the vehicles weight before fully tightening the lower ball joint to the knuckle to specification.**

18. Be careful not to damage the ball joint boot when installing the knuckle.

19. Tighten the tie rod end ball joint castle nut to 40 ft. lbs. then tighten it only enough to install the cotter pin. Do not align the castle nut by loosening it.

20. Tighten the lower ball joint castle nut to 43–51 ft. lbs. then tighten it only enough to install the cotter pin. Do not align the castle nut by loosening it.

21. Be sure to use a new spindle nut. Tighten it to 242 ft. lbs.

22. Tighten the caliper mounting bolts to 101 ft. lbs.

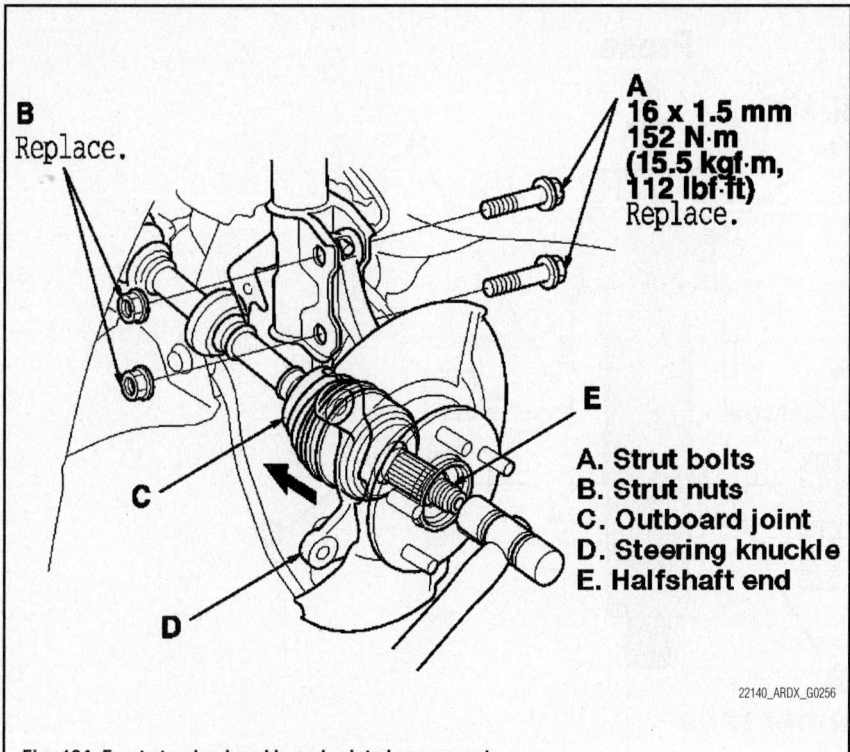

Fig. 164 Front steering knuckle and related components

A
**16 x 1.5 mm
152 N·m
(15.5 kgf·m,
112 lbf·ft)
Replace.**

B
Replace.

A. Strut bolts
B. Strut nuts
C. Outboard joint
D. Steering knuckle
E. Halfshaft end

22140_ARDX_G0256

23. Check and adjust the wheel alignment, as required.

WHEEL BEARINGS

REMOVAL & INSTALLATION

See Figures 165 through 169.

1. Before servicing the vehicle, refer to the Precautions Section.

➡ Before disconnecting the battery cables make sure you have the anti theft codes for the audio/navigation system.

➡ Except when doing electrical inspections, always turn the ignition switch to LOCK (O), ground the SCS line with the Honda Diagnostic Service (HDS) tool to take the PCM out of active status, disconnect the negative cable from the battery, then wait three minutes before starting the repair procedure. The SRS memory is not cleared even if the ignition switch is turned to LOCK (O) or the battery cables are disconnected from the battery.

2. Make sure that the ignition switch is in the LOCK (O) position.

3. Disconnect the negative battery cable.

➡ Always disconnect the negative battery cable first.

4. Disconnect the positive battery cable.

➡ Wait at least three minutes after disconnecting the battery cables before starting the repair procedure.

5. Raise and support the vehicle safely. Remove the tire and wheel assemblies.

6. Remove the steering knuckle.

7. Separate the hub from the knuckle using a hub disassembly tool and a hydraulic press.

➡ Be careful not to deform the splash shield. Hold the hub to prevent it from falling after it is pressed free from the tool.

8. Press the inner wheel bearing race off of the hub.

9. Remove the splash guard and the snapring from the knuckle.

10. Press the wheel bearing out of the knuckle, using the driver attachment and the press.

To install:

11. Press a new wheel bearing into the knuckle using the old bearing, a steel plate,

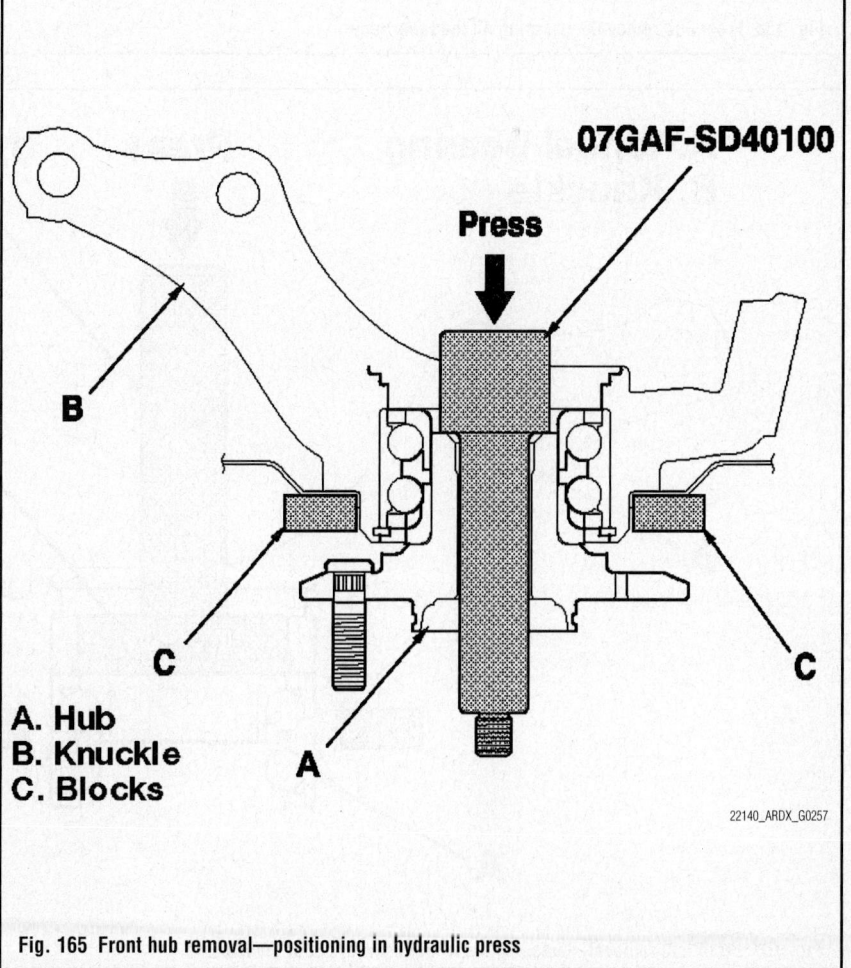

07GAF-SD40100

Press

A. Hub
B. Knuckle
C. Blocks

22140_ARDX_G0257

Fig. 165 Front hub removal—positioning in hydraulic press

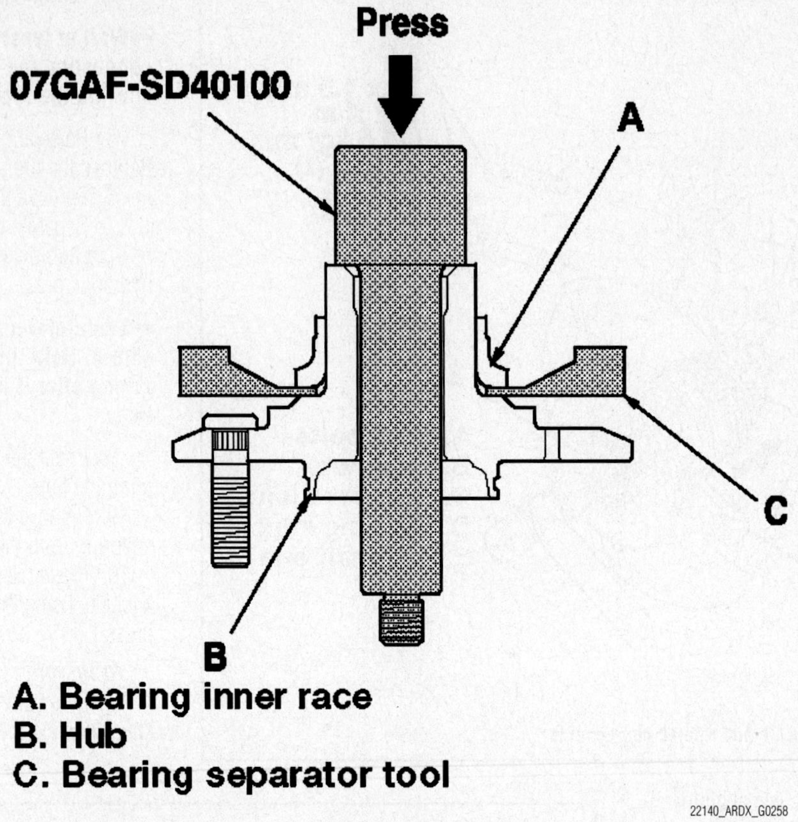

Press

07GAF-SD40100

A

B

C

A. Bearing inner race
B. Hub
C. Bearing separator tool

22140_ARDX_G0258

Fig. 166 Front hub removal—pressing off the inner race

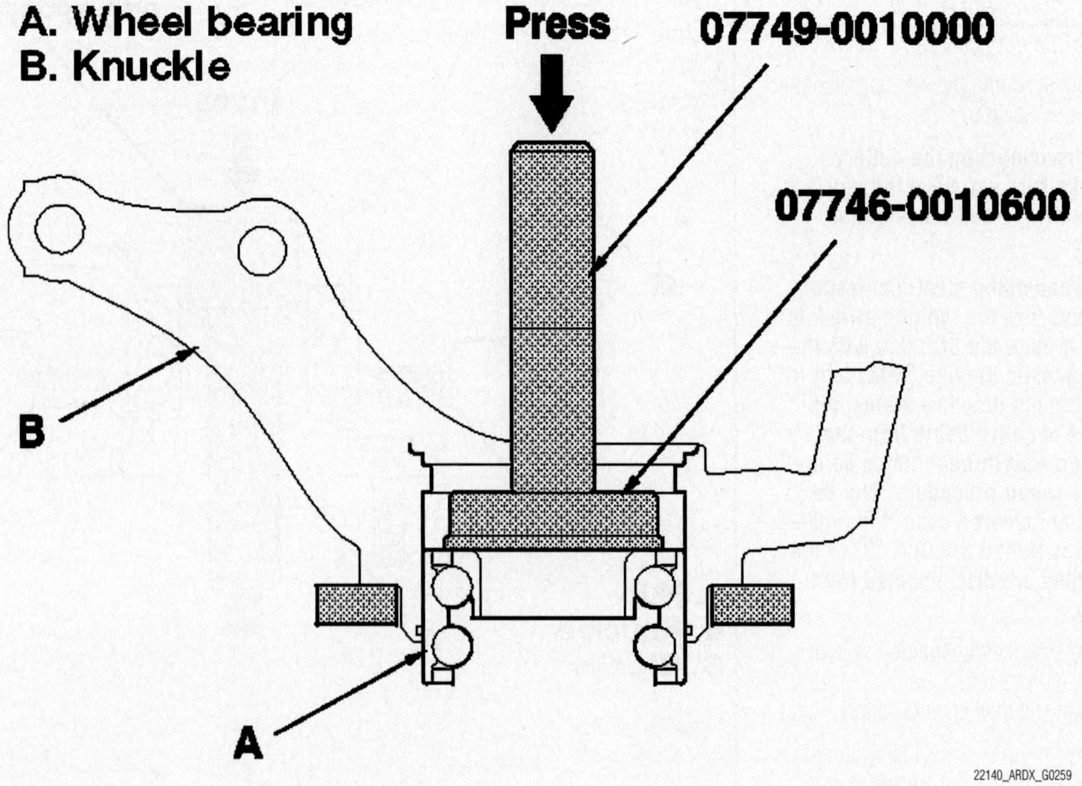

A. Wheel bearing
B. Knuckle

Press

07749-0010000

07746-0010600

B

A

22140_ARDX_G0259

Fig. 167 Front hub removal—pressing the wheel bearing out of the knuckle

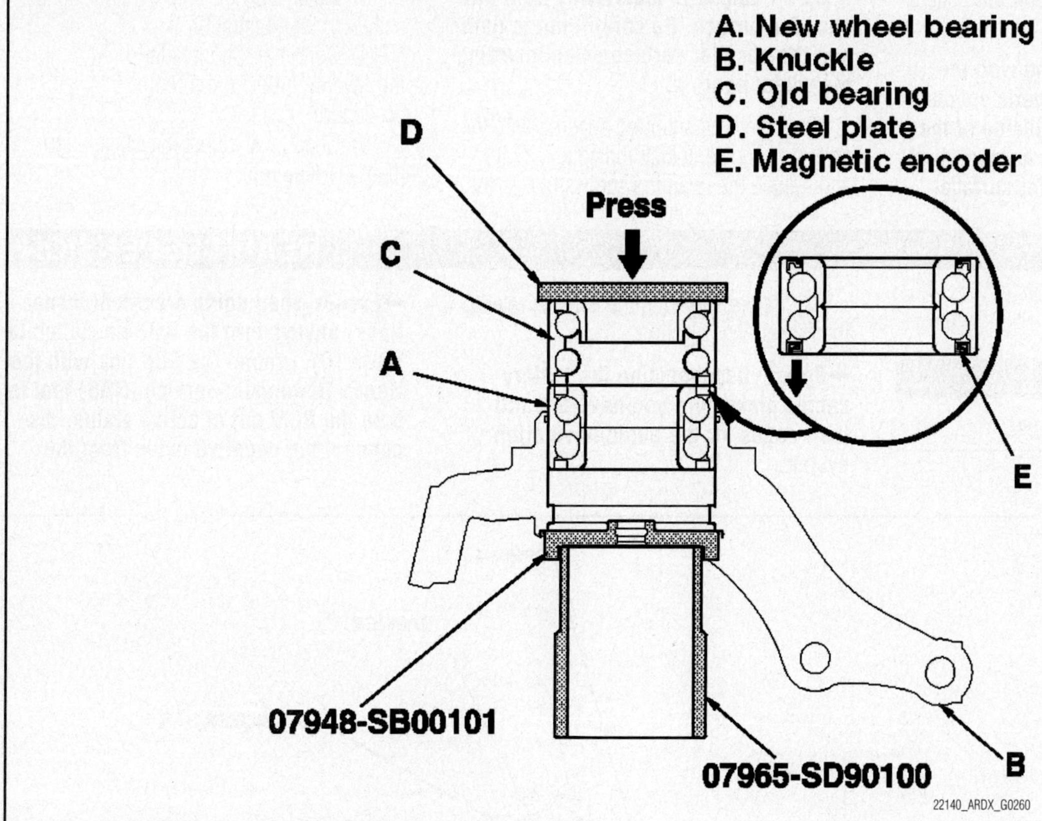

A. New wheel bearing
B. Knuckle
C. Old bearing
D. Steel plate
E. Magnetic encoder

Press

D
C
A

E

07948-SB00101

07965-SD90100

B

22140_ARDX_G0260

Fig. 168 Front hub bearing installation and encoder positioning

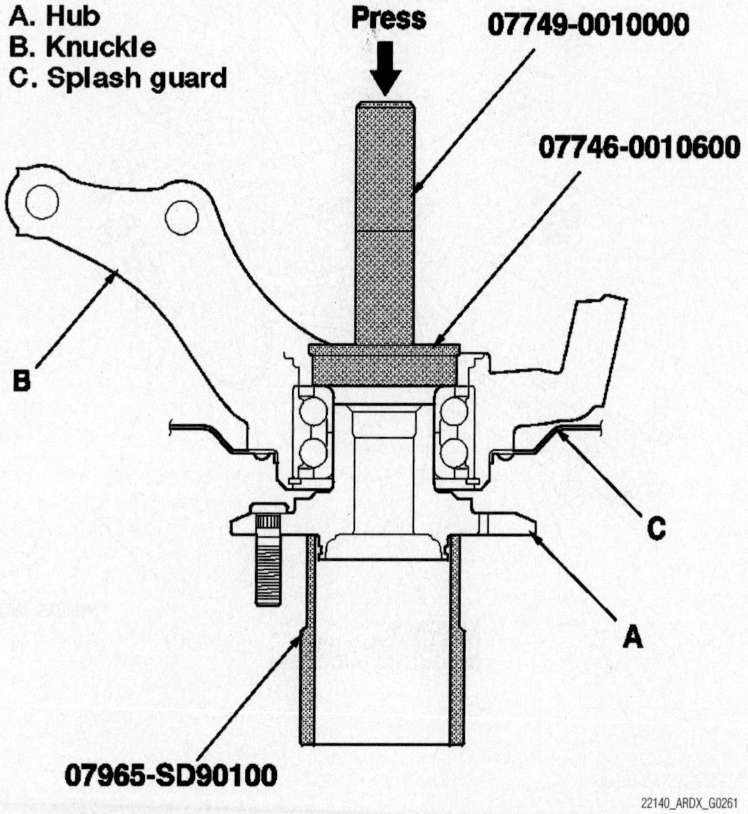

A. Hub
B. Knuckle
C. Splash guard

Press   07749-0010000

07746-0010600

B

C

A

07965-SD90100

22140_ARDX_G0261

Fig. 169 Front hub to knuckle installation

the tool attachment, the support base and a press.

➡Install the wheel bearing with the wheel speed sensor magnetic encoder (brown color) toward the inside of the knuckle. Remove any oil, grease, dust, debris etc from the encoder surface.

Keep all magnetic tools away from the encoder surface. Be careful not to damage the encoder surface when inserting the wheel bearing.

12. Install the snapring securely into the steering knuckle. Install the splash guard and tighten the retaining screws to 4 ft. lbs.

13. Install the hub onto the knuckle, as shown in the illustration.

14. Continue the installation in the reverse order of the removal procedure.

15. Check and adjust the wheel alignment, as required.

SUSPENSION

See Figure 170.

COIL SPRING

REMOVAL & INSTALLATION
See Figures 171 and 172.

1. Before servicing the vehicle, refer to the Precautions Section.

➡Before disconnecting the battery cables make sure you have the anti theft codes for the audio/navigation system.

REAR SUSPENSION

➡Except when doing electrical inspections, always turn the ignition switch to LOCK (O), ground the SCS line with the Honda Diagnostic Service (HDS) tool to take the PCM out of active status, disconnect the negative cable from the

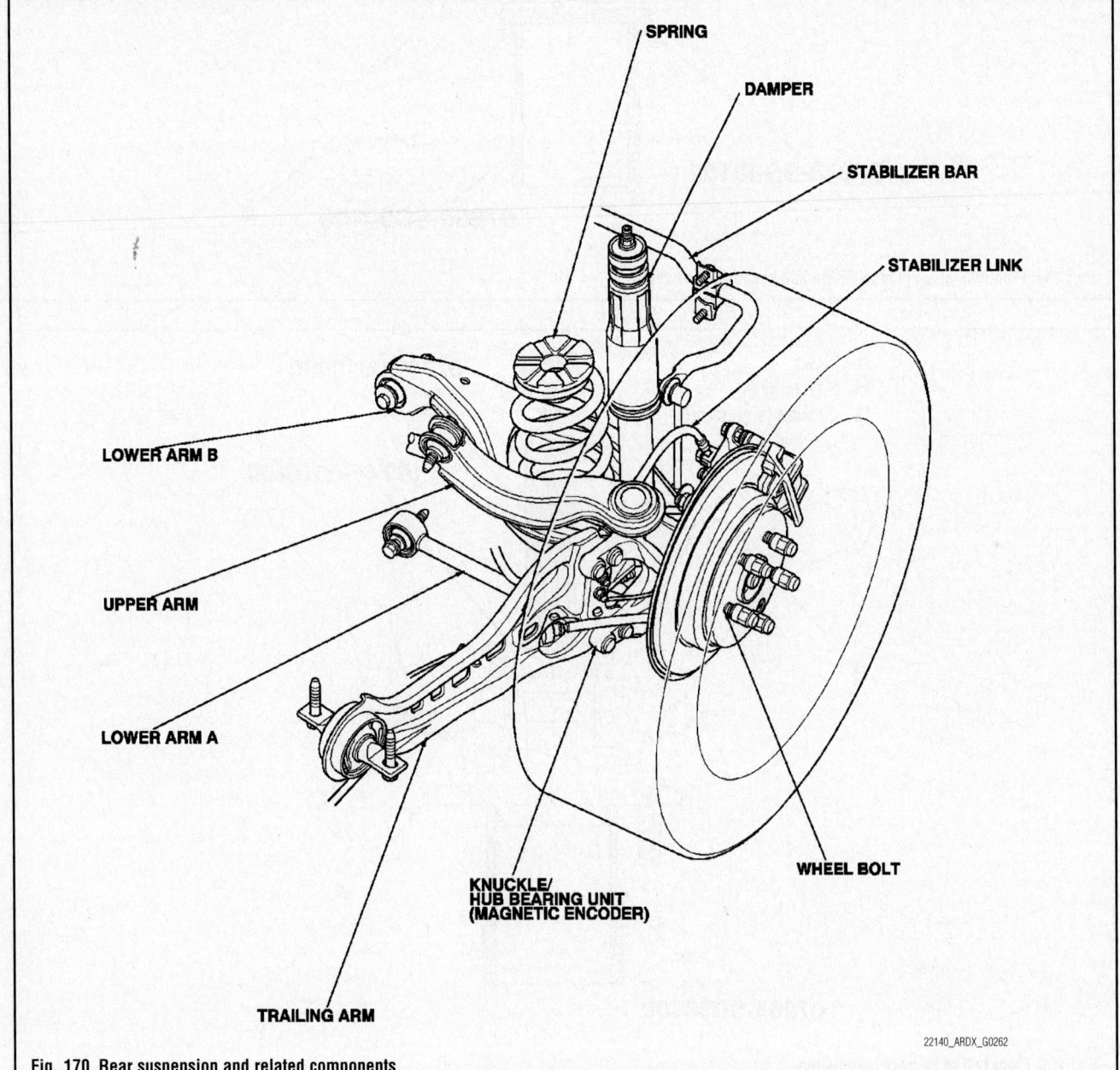

Fig. 170 Rear suspension and related components

22140_ARDX_G0262

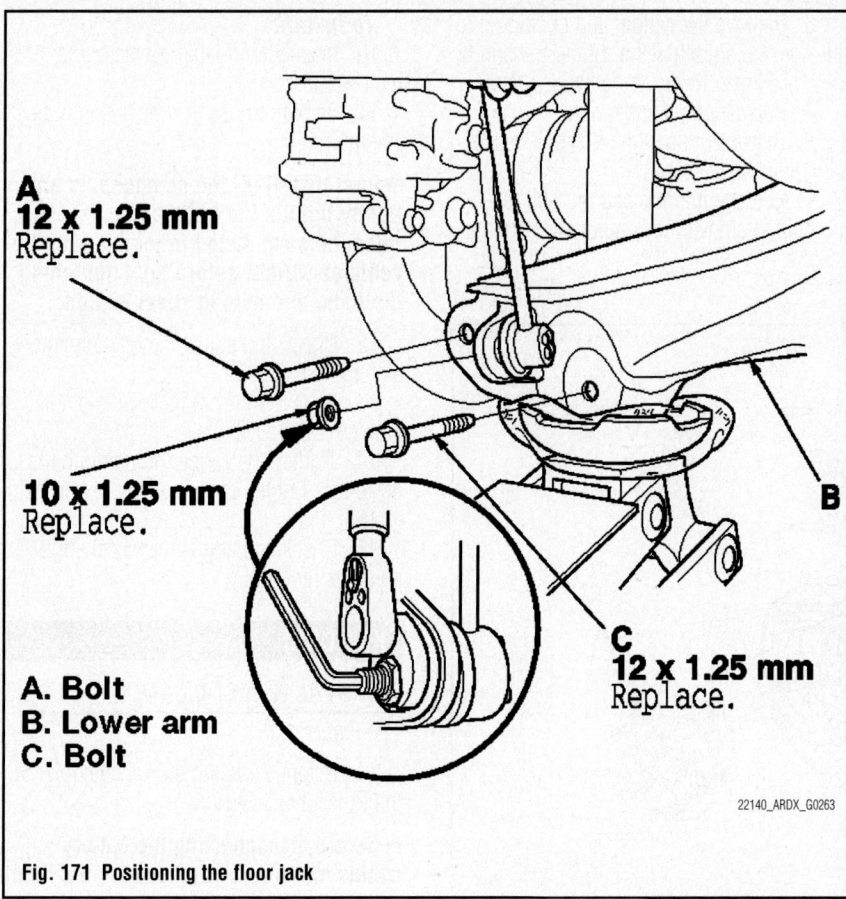

A
12 x 1.25 mm
Replace.

10 x 1.25 mm
Replace.

C
12 x 1.25 mm
Replace.

A. Bolt
B. Lower arm
C. Bolt

22140_ARDX_G0263

Fig. 171 Positioning the floor jack

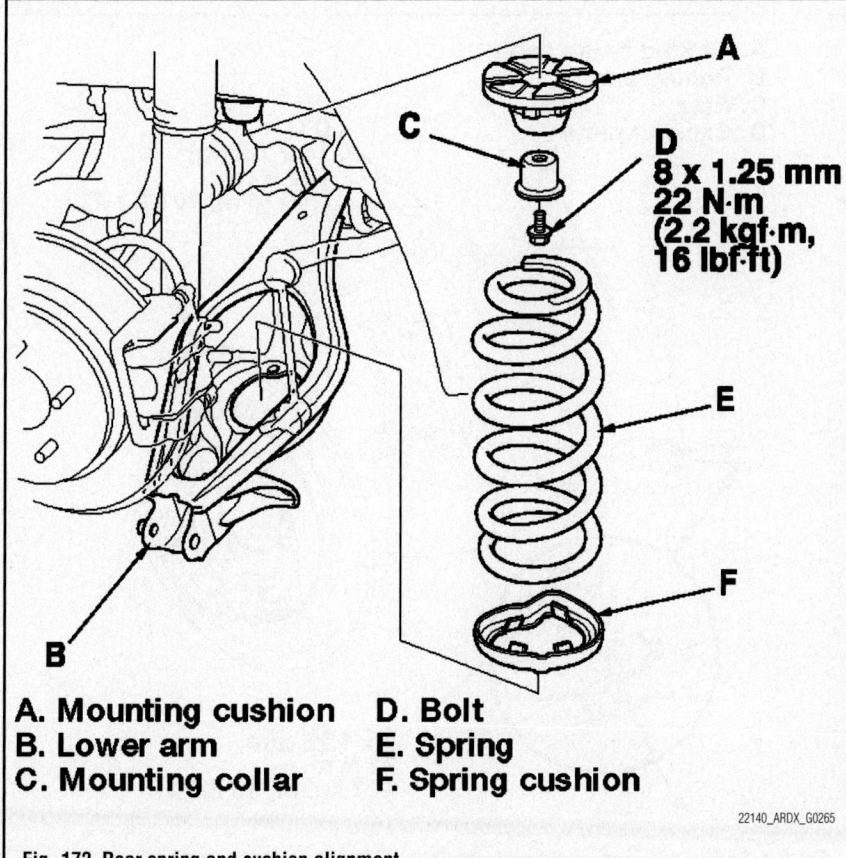

A

C

D
8 x 1.25 mm
22 N·m
(2.2 kgf·m,
16 lbf·ft)

E

F

B

A. Mounting cushion   D. Bolt
B. Lower arm   E. Spring
C. Mounting collar   F. Spring cushion

22140_ARDX_G0265

Fig. 172 Rear spring and cushion alignment

battery, then wait three minutes before starting the repair procedure. The SRS memory is not cleared even if the ignition switch is turned to LOCK (0) or the battery cables are disconnected from the battery.

2. Make sure that the ignition switch is in the LOCK (0) position.

3. Disconnect the negative battery cable.

➡**Always disconnect the negative battery cable first.**

4. Disconnect the positive battery cable.

➡**Wait at least three minutes after disconnecting the battery cables before starting the repair procedure.**

5. Raise and support the vehicle safely. Remove the tire and wheel assemblies.

6. Position the floor jack at the connecting point of the lower control arm and the lower strut mount.

7. Disconnect the stabilizer link from the lower arm.

8. Remove the flange bolt that connects the lower arm and the knuckle.

9. Remove the flange bolt that connects the lower arm and the bottom of the strut.

10. Lower the floor jack, slowly.

11. Remove the spring and the lower spring cushion.

12. Remove the flange bolt that connects to the body. Remove the spring mounting collar and the spring mounting cushion, as required.

To install:

13. Install the spring and the lower spring cushion. Align the bottom of the spring and the lower spring cushion, as shown in the illustration.

14. Position the floor jack at the connecting point of the lower control arm.

15. Slowly raise the jack until you align the bolt hole with the holes in the lower arm and the knuckle.

16. Loosely install a new flange bolt.

17. Compress the strut by hand until you can align the bolt hole with the holes in the lower arm and the strut.

18. Loosely install a new flange bolt.

19. Connect the stabilizer link to the lower arm.

20. Raise the rear suspension with the floor jack until the vehicle just lifts off the safety stands.

21. Tighten the flange bolts and the self locking nut to specification. Bolt specification is 76 ft. lbs. Self locking nut specification is 29 ft. lbs.

22. Continue the installation in the reverse order of the removal procedure.

23. Check and adjust the rear alignment, as required.

HUB AND BEARING

REMOVAL & INSTALLATION

See Figure 173.

1. Before servicing the vehicle, refer to the Precautions Section.

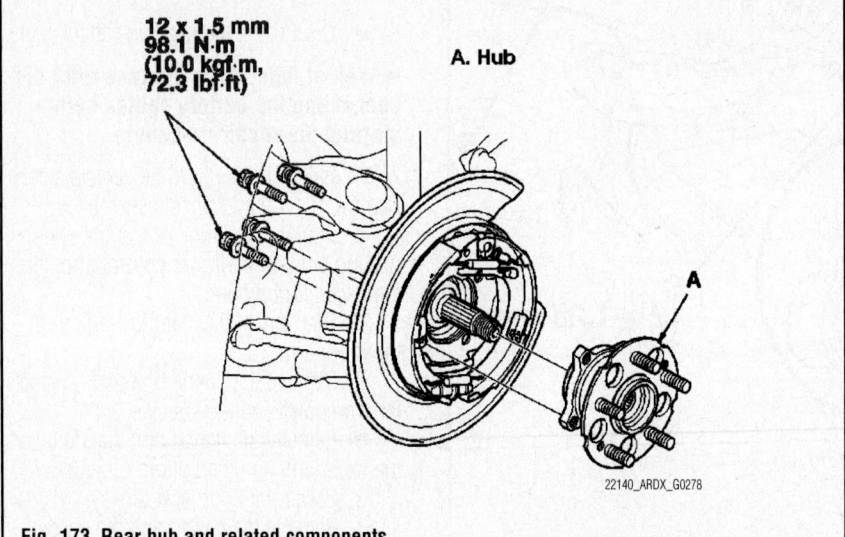

Fig. 173 Rear hub and related components

➤Before disconnecting the battery cables make sure you have the anti theft codes for the audio/navigation system.

➤Except when doing electrical inspections, always turn the ignition switch to LOCK (O), ground the SCS line with the Honda Diagnostic Service (HDS) tool to take the PCM out of active status, disconnect the negative cable from the battery, then wait three minutes before starting the repair procedure. The SRS memory is not cleared even if the ignition switch is turned to LOCK (O) or the battery cables are disconnected from the battery.

2. Make sure that the ignition switch is in the LOCK (O) position.

3. Disconnect the negative battery cable.

➤Always disconnect the negative battery cable first.

4. Disconnect the positive battery cable.

➤Wait at least three minutes after disconnecting the battery cables before starting the repair procedure.

5. Raise and support the vehicle safely. Remove the tire and wheel assemblies.

6. Remove the caliper and position it to the side. Do not allow the caliper to hang by the brake hose. Remove the two washers.

7. Remove the spindle nut.

8. Release the parking brake, if necessary. Remove the rotor.

9. Remove the hub bearing unit retaining bolts. Remove the component from the vehicle.

To install:

10. Installation is the reverse of the removal procedure.

11. Be sure to use new bolts and nuts, as required.

➤First install all the components and lightly tighten the bolts and nuts, then raise the suspension to load it with the vehicles weight before fully tightening the bolts and nuts to specification.

12. Check and adjust the rear alignment, as required.

13. Be sure to use a new spindle nut. Tighten it to 181 ft. lbs.

14. Tighten the caliper mounting bolts to 80 ft. lbs. Tighten the caliper mounting nuts to 16 ft. lbs.

15. Check and adjust the wheel alignment, as required.

KNUCKLE

REMOVAL & INSTALLATION

See Figure 174.

1. Before servicing the vehicle, refer to the Precautions Section.

➤Before disconnecting the battery cables make sure you have the anti theft codes for the audio/navigation system.

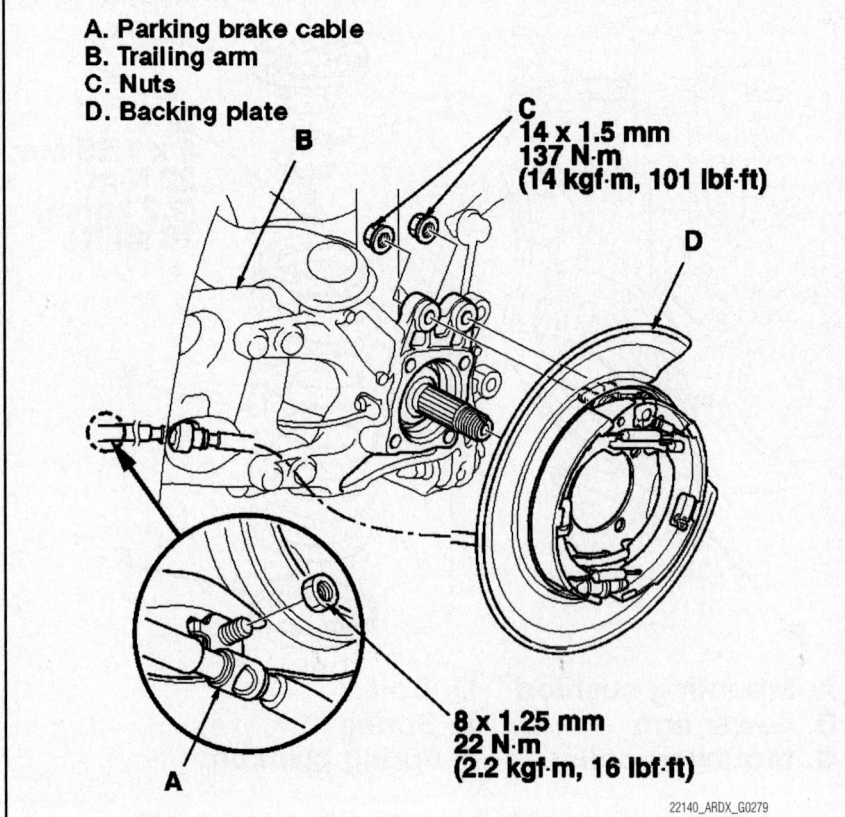

Fig. 174 Rear knuckle and related components

➡Except when doing electrical inspections, always turn the ignition switch to LOCK (O), ground the SCS line with the Honda Diagnostic Service (HDS) tool to take the PCM out of active status, disconnect the negative cable from the battery, then wait three minutes before starting the repair procedure. The SRS memory is not cleared even if the ignition switch is turned to LOCK (O) or the battery cables are disconnected from the battery.

2. Make sure that the ignition switch is in the LOCK (O) position.

3. Disconnect the negative battery cable.

➡Always disconnect the negative battery cable first.

4. Disconnect the positive battery cable.

➡Wait at least three minutes after disconnecting the battery cables before starting the repair procedure.

5. Raise and support the vehicle safely. Remove the tire and wheel assemblies.

6. Remove the hub bearing unit.

7. Remove the parking brake cable from the trailing arm.

8. Remove the flange nuts and remove the backing plate.

➡Hang the backing plate, using mechanics wire, to prevent damage to the parking brake cable and the backing plate.

9. Remove the wheel speed sensor from the knuckle. Do not disconnect the sensor at the connector.

➡Before installation apply multi grease to the inside hole on the knuckle.

10. Remove the lock pin from the upper ball joint. Loosen the castle nut.

11. Using the proper removal tool disconnect the upper arm ball joint from the knuckle.

12. Remove the lower control arm link.

13. Remove the locking nut and separate the knuckle from the trailing arm.

14. Position a floor jack under the lower arm. Remove the flange bolt. Remove the knuckle from the vehicle.

To install:

15. Installation is the reverse of the removal procedure.

16. Be sure to use new bolts and nuts, as required.

17. Tighten the new knuckle flange bolt to 76 ft. lbs. Tighten the trailing arm to knuckle retaining nuts to 87 ft. lbs.

➡First install all the components and lightly tighten the bolts and nuts, then raise the suspension to load it with the vehicles weight before fully tightening the bolts and nuts to specification.

18. Check and adjust the rear alignment, as required.

LOWER CONTROL ARM

REMOVAL & INSTALLATION

See Figure 175.

1. Before servicing the vehicle, refer to the Precautions Section.

➡Before disconnecting the battery cables make sure you have the anti theft codes for the audio/navigation system.

➡Except when doing electrical inspections, always turn the ignition switch to LOCK (O), ground the SCS line with the Honda Diagnostic Service (HDS) tool to take the PCM out of active status, disconnect the negative cable from the battery, then wait three minutes before starting the repair procedure. The SRS memory is not cleared even if the ignition switch is turned to LOCK (O) or the battery cables are disconnected from the battery.

2. Make sure that the ignition switch is in the LOCK (O) position.

3. Disconnect the negative battery cable.

➡Always disconnect the negative battery cable first.

4. Disconnect the positive battery cable.

➡Wait at least three minutes after disconnecting the battery cables before starting the repair procedure.

5. Raise and support the vehicle safely. Remove the tire and wheel assemblies.

6. Position the floor jack at the connecting point of the lower control arm and the lower strut mount.

7. Disconnect the stabilizer link from the lower arm.

8. Remove the flange bolt that connects the lower arm and the knuckle.

9. Remove the flange bolt that connects the lower arm and the bottom of the strut.

10. Lower the floor jack, slowly.

11. Remove the spring and the lower spring cushion.

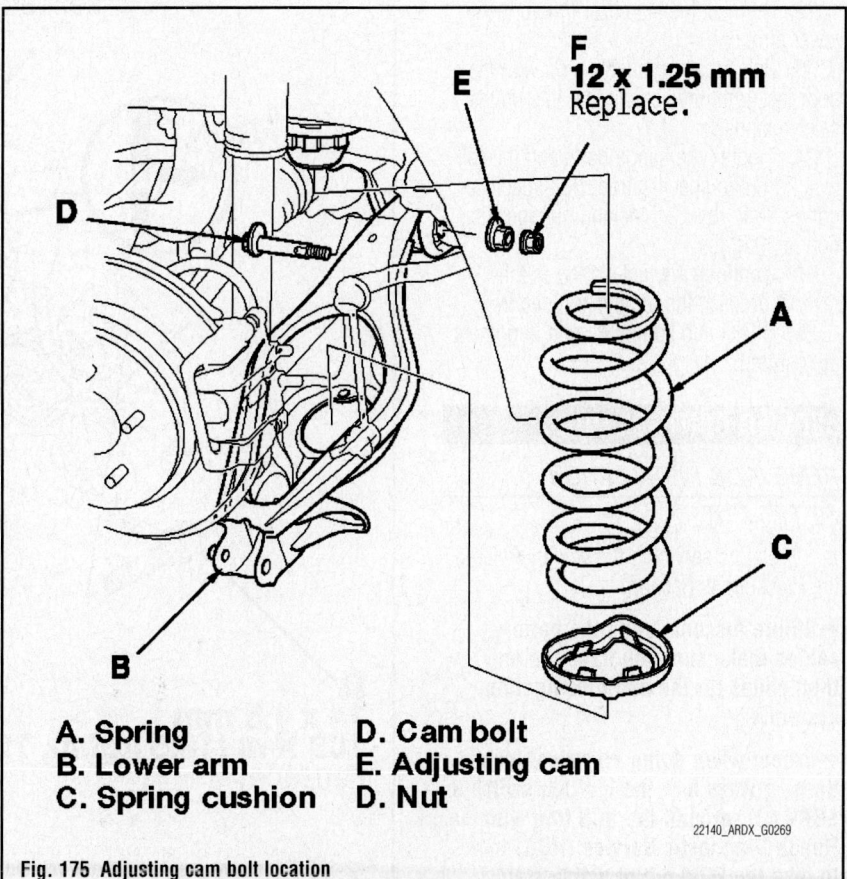

A. Spring
B. Lower arm
C. Spring cushion
D. Cam bolt
E. Adjusting cam
D. Nut

Fig. 175 Adjusting cam bolt location

22140_ARDX_G0269

12. Remove the flange bolt that connects to the body. Remove the spring mounting collar and the spring mounting cushion, as required.

13. Mark the cam positions of the adjusting bolt and the adjusting cam. Remove the nut, adjusting cam and adjusting bolt.

14. Set aside the nut and control arm mounting nut. Remove the lower arm.

To install:

15. Position the lower arm. Loosely install a new adjusting bolt, adjusting cam and self locking nut.

➡**At final tightening, torque the nut to 43 ft. lbs.**

16. Install the spring and the lower spring cushion. Align the bottom of the spring and the lower spring cushion, as shown in the illustration.

17. Position the floor jack at the connecting point of the lower control arm.

18. Slowly raise the jack until you align the bolt hole with the holes in the lower arm and the knuckle.

19. Loosely install a new flange bolt.

20. Compress the strut by hand until you can align the bolt hole with the holes in the lower arm and the strut.

21. Loosely install a new flange bolt.

22. Connect the stabilizer link to the lower arm.

23. Raise the rear suspension with the floor jack until the vehicle just lifts off the safety stands.

24. Tighten the flange bolts and the self locking nut to specification. Bolt specification is 76 ft. lbs. Self locking nut specification is 29 ft. lbs.

25. Continue the installation in the reverse order of the removal procedure.

26. Check and adjust the rear alignment, as required.

LOWER CONTROL ROD

REMOVAL & INSTALLATION

See Figure 176.

1. Before servicing the vehicle, refer to the Precautions Section.

➡**Before disconnecting the battery cables make sure you have the anti theft codes for the audio/navigation system.**

➡**Except when doing electrical inspections, always turn the ignition switch to LOCK (O), ground the SCS line with the Honda Diagnostic Service (HDS) tool to take the PCM out of active status,** disconnect the negative cable from the battery, then wait three minutes before starting the repair procedure. The SRS memory is not cleared even if the ignition switch is turned to LOCK (O) or the battery cables are disconnected from the battery.

2. Make sure that the ignition switch is in the LOCK (O) position.

3. Disconnect the negative battery cable.

➡**Always disconnect the negative battery cable first.**

4. Disconnect the positive battery cable.

➡**Wait at least three minutes after disconnecting the battery cables before starting the repair procedure.**

5. Raise and support the vehicle safely. Remove the tire and wheel assemblies.

6. Position the floor jack at the connecting point of the lower control arm and the knuckle.

7. Remove the self locking nut, the washer and the flange bolt.

8. Remove the lower control rod from its mounting.

To install:

9. Installation is the reverse of the removal procedure.

10. Tighten the new flange bolt to 76 ft. lbs. Tighten the new nut to 98 ft. lbs.

➡**First install all the components and lightly tighten the bolts and nuts, then raise the suspension to load it with the vehicles weight before fully tightening the bolts and nuts to specification.**

11. Check and adjust the rear alignment, as required.

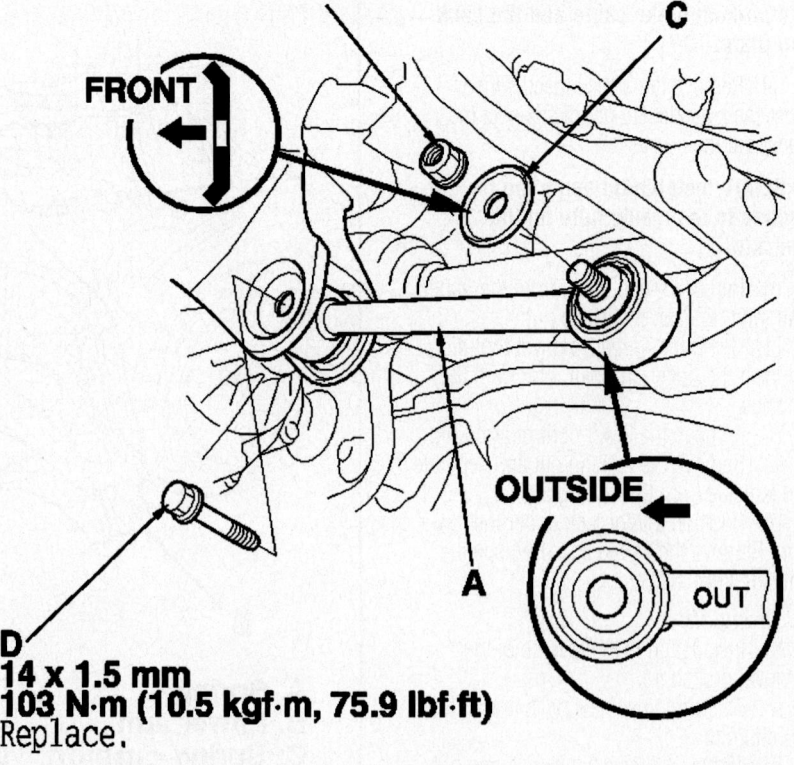

A. Lower control rod
B. Nut
C. Washer
D. Bolt

B
12 x 1.25 mm
132 N·m
(13.5 kgf·m, 97.6 lbf·ft)
Replace.

D
14 x 1.5 mm
103 N·m (10.5 kgf·m, 75.9 lbf·ft)
Replace.

FRONT

OUTSIDE

OUT

22140_ARDX_G0267

Fig. 176 Rear lower control rod and related components

STABILIZER BAR

REMOVAL & INSTALLATION

See Figure 177.

1. Before servicing the vehicle, refer to the Precautions Section.

➡Before disconnecting the battery cables make sure you have the anti theft codes for the audio/navigation system.

➡Except when doing electrical inspections, always turn the ignition switch to LOCK (0), ground the SCS line with the Honda Diagnostic Service (HDS) tool to take the PCM out of active status, disconnect the negative cable from the battery, then wait three minutes before starting the repair procedure. The SRS memory is not cleared even if the ignition switch is turned to LOCK (0) or the battery cables are disconnected from the battery.

2. Make sure that the ignition switch is in the LOCK (0) position.

3. Disconnect the negative battery cable.

➡Always disconnect the negative battery cable first.

4. Disconnect the positive battery cable.

➡Wait at least three minutes after disconnecting the battery cables before starting the repair procedure.

5. Raise and support the vehicle safely. Remove the tire and wheel assemblies.

6. Disconnect both stabilizer links from the stabilizer bar.

7. Remove the flange bolts and the bushing holders.

8. Remove the stabilizer bar from its mounting.

To install:

9. Installation is the reverse of the removal procedure.

10. Tighten the stabilizer bar bushing bolts to 29 ft. lbs.

➡Note the right and left direction of the stabilizer bar. The bar has a paint mark on the center facing downward.

11. Check and adjust the rear alignment, as required.

STABILIZER LINK

REMOVAL & INSTALLATION

See Figure 178.

1. Before servicing the vehicle, refer to the Precautions Section.

➡Before disconnecting the battery cables make sure you have the anti theft codes for the audio/navigation system.

➡Except when doing electrical inspections, always turn the ignition switch to LOCK (0), ground the SCS line with the Honda Diagnostic Service (HDS) tool to take the PCM out of active status, disconnect the negative cable from the battery, then wait three minutes before starting the repair procedure. The SRS memory is not cleared even if the ignition switch is turned to LOCK (0) or the battery cables are disconnected from the battery.

2. Make sure that the ignition switch is in the LOCK (0) position.

3. Disconnect the negative battery cable.

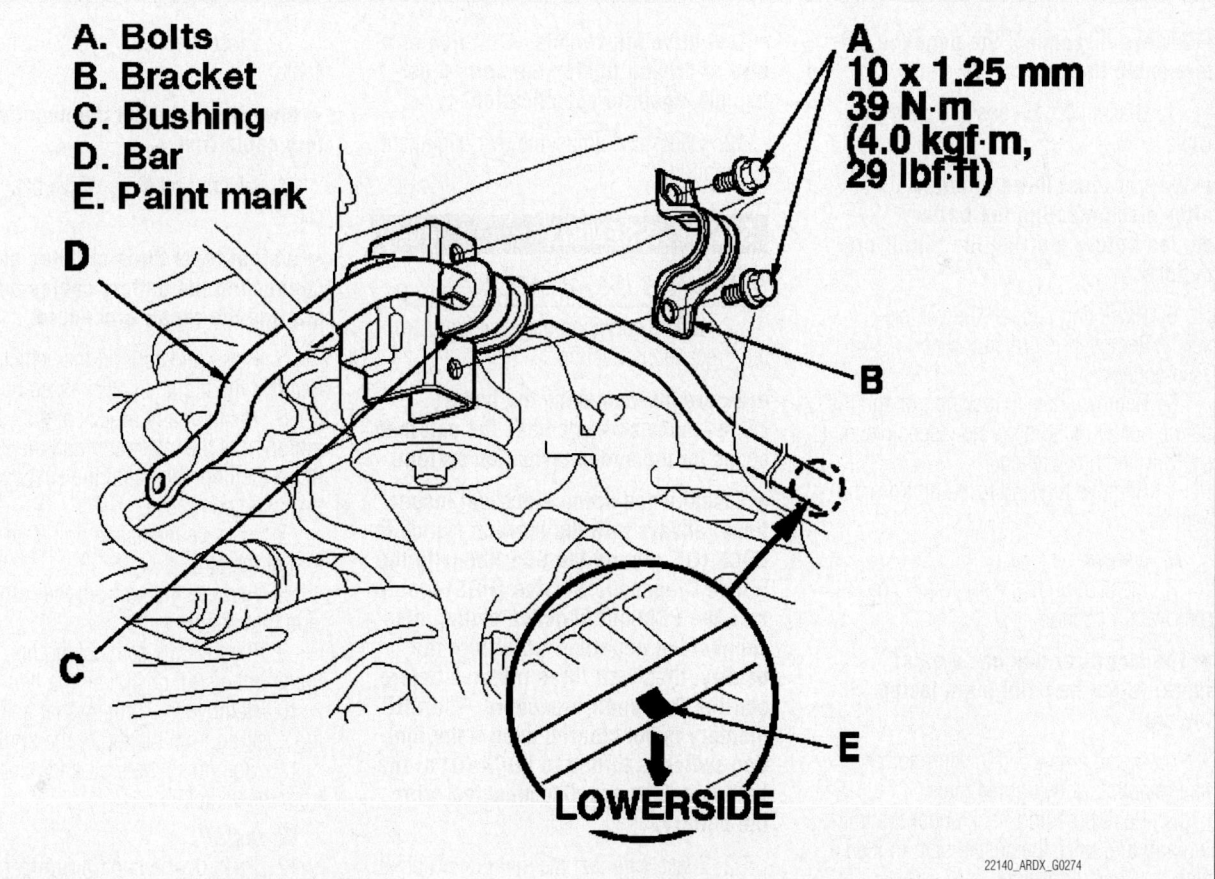

A. Bolts
B. Bracket
C. Bushing
D. Bar
E. Paint mark

A
10 x 1.25 mm
39 N·m
(4.0 kgf·m,
29 lbf·ft)

LOWERSIDE

22140_ARDX_G0274

Fig. 177 Rear stabilizer bar and related components

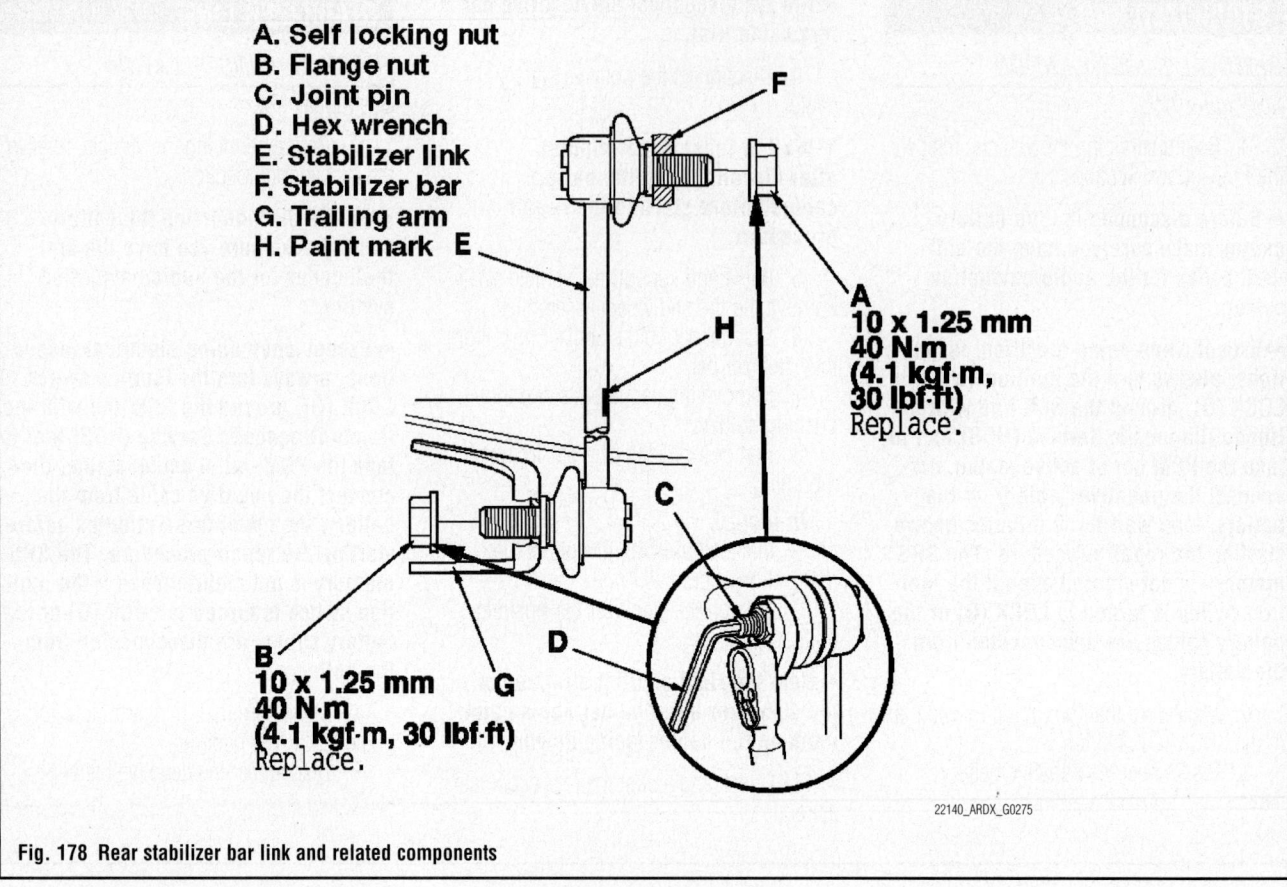

A. Self locking nut
B. Flange nut
C. Joint pin
D. Hex wrench
E. Stabilizer link
F. Stabilizer bar
G. Trailing arm
H. Paint mark

A
10 x 1.25 mm
40 N·m
(4.1 kgf·m,
30 lbf·ft)
Replace.

B
10 x 1.25 mm
40 N·m
(4.1 kgf·m, 30 lbf·ft)
Replace.

22140_ARDX_G0275

Fig. 178 Rear stabilizer bar link and related components

➡**Always disconnect the negative battery cable first.**

4. Disconnect the positive battery cable.

➡**Wait at least three minutes after disconnecting the battery cables before starting the repair procedure.**

5. Raise and support the vehicle safely. Remove the tire and wheel assemblies.

6. Remove the self locking nut and the flange nut while holding the respective joint pin with the hex wrench.

7. Remove the stabilizer link from the vehicle.

To install:

8. Installation is the reverse of the removal procedure.

➡**The stabilizer link has a paint mark. Align the paint mark facing inward.**

9. Install a new self locking nut and flange bolt. Lightly tighten them.

10. Position a floor jack under the trailing arm and raise the suspension to load it with the vehicle's weight.

11. Tighten the nuts to 30 ft. lbs.

➡**Test drive the vehicle. After five minutes of driving tighten the self adjusting nut, again to specification.**

12. Check and adjust the rear alignment, as required.

STRUT ASSEMBLY

REMOVAL & INSTALLATION

1. Before servicing the vehicle, refer to the Precautions Section.

➡**Before disconnecting the battery cables make sure you have the anti theft codes for the audio/navigation system.**

➡**Except when doing electrical inspections, always turn the ignition switch to LOCK (O), ground the SCS line with the Honda Diagnostic Service (HDS) tool to take the PCM out of active status, disconnect the negative cable from the battery, then wait three minutes before starting the repair procedure. The SRS memory is not cleared even if the ignition switch is turned to LOCK (O) or the battery cables are disconnected from the battery.**

2. Make sure that the ignition switch is in the LOCK (O) position.

3. Disconnect the negative battery cable.

➡**Always disconnect the negative battery cable first.**

4. Disconnect the positive battery cable.

➡**Wait at least three minutes after disconnecting the battery cables before starting the repair procedure.**

5. Raise and support the vehicle safely. Remove the tire and wheel assemblies.

6. Position a floor jack at the connecting point of the lower arm and the knuckle. Raise the floor jack until the suspension begins to compress.

7. Remove the flange bolt from the bottom of the strut.

8. Remove the lid from the cargo area side trim panel.

9. Remove the self locking nut, while holding the strut shaft, using a hex wrench.

10. Remove the strut washer and mounting cushion from the top of the strut.

11. Compress the strut and remove it from the vehicle.

To install:

12. Position the floor jack under the lower arm, to support the suspension.

13. Slowly raise the floor jack until you align the bolt hole with the holes in the lower arm and the strut. Loosely install the new flange bolt.

➡**First install the component and lightly tighten the new flange bolt, then raise the suspension to load it with the vehicles weight before fully tightening the bolt to specification.**

14. Tighten the new flange bolt to 76 ft. lbs.

15. Continue the installation in the reverse order of the removal procedure.

16. Tighten the new self locking nut to 22 ft. lbs.

17. Check and adjust the rear alignment, as required.

TRAILING ARM

REMOVAL & INSTALLATION

See Figure 179.

1. Before servicing the vehicle, refer to the Precautions Section.

➡**Before disconnecting the battery cables make sure you have the anti** theft codes for the audio/navigation system.

➡**Except when doing electrical inspections, always turn the ignition switch to LOCK (O), ground the SCS line with the Honda Diagnostic Service (HDS) tool to take the PCM out of active status, disconnect the negative cable from the battery, then wait three minutes before starting the repair procedure. The SRS memory is not cleared even if the ignition switch is turned to LOCK (O) or the battery cables are disconnected from the battery.**

2. Make sure that the ignition switch is in the LOCK (O) position.

3. Disconnect the negative battery cable.

➡**Always disconnect the negative battery cable first.**

4. Disconnect the positive battery cable.

➡**Wait at least three minutes after disconnecting the battery cables before starting the repair procedure.**

5. Raise and support the vehicle safely. Remove the tire and wheel assemblies.

6. Remove the parking brake shoe. Remove the parking brake cable from the backing plate.

7. Remove the parking brake cable from the trailing arm.

8. Remove the brake hose mounting bracket from the trailing arm.

9. Position a floor jack at the connecting point of the lower arm and knuckle.

10. Remove the self locking nuts and flange bolts from the trailing arm.

11. Remove the trailing arm from its mounting.

To install:

12. Installation is the reverse of the removal procedure.

➡**First install all the component and lightly tighten the new flange bolts and nuts, then raise the suspension to load it with the vehicles weight before fully tightening the bolts and nuts to specification.**

13. Tighten the new flange bolt to 76 ft. lbs. Tighten the new nuts to 87 ft. lbs.

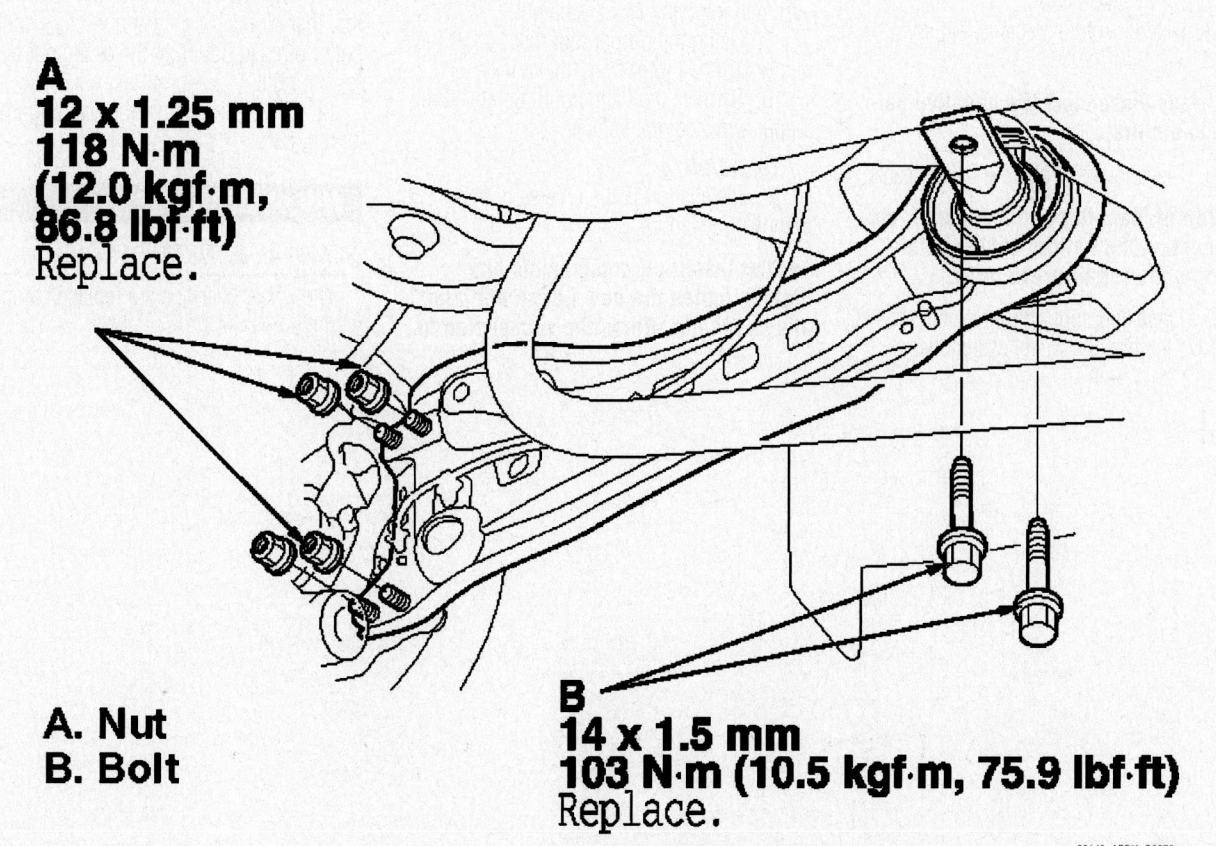

A
12 x 1.25 mm
118 N·m
(12.0 kgf·m,
86.8 lbf·ft)
Replace.

A. Nut
B. Bolt

B
14 x 1.5 mm
103 N·m (10.5 kgf·m, 75.9 lbf·ft)
Replace.

22140_ARDX_G0276

Fig. 179 Rear trailing arm and related components

14. Check the brake hose for interference and twisting, correct as required.

15. Check and adjust the rear alignment, as required.

UPPER CONTROL ARM

REMOVAL & INSTALLATION

See Figure 180.

1. Before servicing the vehicle, refer to the Precautions Section.

➡ **Before disconnecting the battery cables make sure you have the anti theft codes for the audio/navigation system.**

➡ **Except when doing electrical inspections, always turn the ignition switch to LOCK (O), ground the SCS line with the Honda Diagnostic Service (HDS) tool to take the PCM out of active status, disconnect the negative cable from the battery, then wait three minutes before starting the repair procedure. The SRS memory is not cleared even if the ignition switch is turned to LOCK (O) or the battery cables are disconnected from the battery.**

2. Make sure that the ignition switch is in the LOCK (O) position.

3. Disconnect the negative battery cable.

➡ **Always disconnect the negative battery cable first.**

4. Disconnect the positive battery cable.

➡ **Wait at least three minutes after disconnecting the battery cables before starting the repair procedure.**

5. Raise and support the vehicle safely. Remove the tire and wheel assemblies.

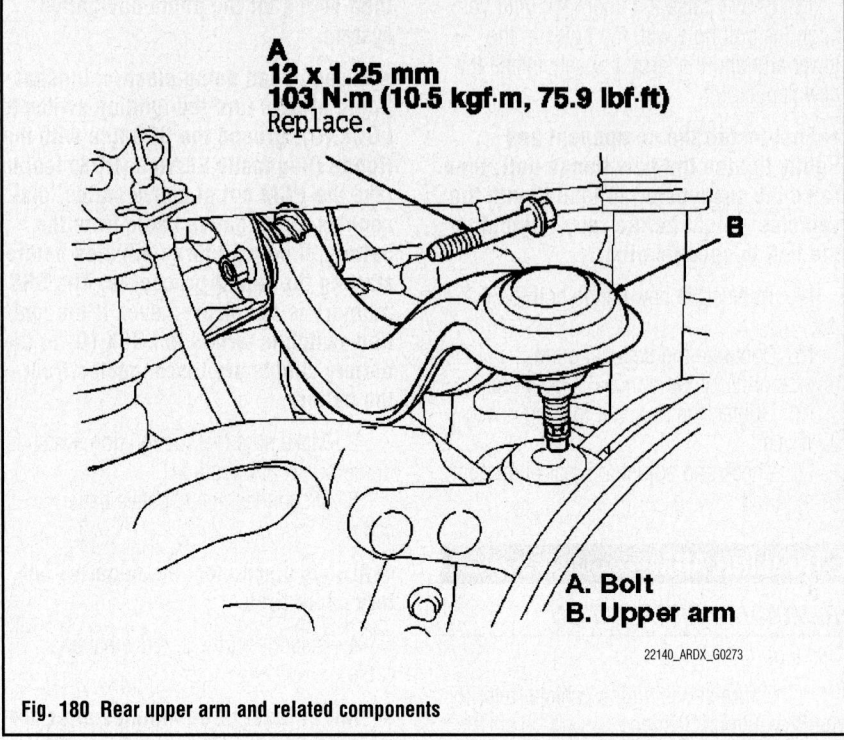

A
12 x 1.25 mm
103 N·m (10.5 kgf·m, 75.9 lbf·ft)
Replace.

B

A. Bolt
B. Upper arm

22140_ARDX_G0273

Fig. 180 Rear upper arm and related components

6. Position a floor jack at the connecting point of the lower arm and knuckle.

7. Remove the lock pin from the upper arm ball joint and loosen the nut.

8. Using the proper tool disconnect the upper arm ball joint from the knuckle.

9. Remove the flange bolt. Remove the upper arm from the vehicle.

To install:

10. Installation is the reverse of the removal procedure.

➡ **First install all components and lightly tighten the new flange bolt and the castle nut. Raise the suspension to load it with the vehicle's weight before fully tightening.**

11. Tighten the castle nut to 43–47 ft. lbs. then tighten it only enough to install the cotter pin. Do not align the castle nut by loosening it.

12. Check and adjust the rear alignment, as required.

WHEEL BEARINGS

REMOVAL & INSTALLATION

The wheel bearings are replaced along with the hub, as an assembly.

HONDA

Accord • Civic • Civic Hybrid • S2000

SPECIFICATIONS AND MAINTENANCE CHARTS

ENGINE AND VEHICLE IDENTIFICATION

Engine						Model Year	
Code	Liters	Cu. In. (cc)	Cyl.	Fuel Sys.	Eng. Mfg.	Code	Year
F22C1	2.2	132.0 (2157)	4	PGM-FI	Honda	7	2007
J30A5	3.0	183.0 (2997)	6	PGM-FI	Honda	8	2008
J35Z2	3.5	212.0 (3471)	6	PGM-FI	Honda		
J35Z3	3.5	212.0 (3471)	6	PGM-FI	Honda		
K20Z3	2.0	121.9 (1997)	4	PGM-FI	Honda		
K24A8	2.4	144.0 (2354)	4	PGM-FI	Honda		
K24Z2	2.4	144.0 (2354	4	PGM-FI	Honda		
K24Z3	2.4	144.0 (2354	4	PGM-FI	Honda		
LDA2	1.3	82.0 (1339)	4	PGM-FI	Honda		
R18A1	1.8	110.0 (1798)	4	PGM-FI	Honda		

PGM-FI: Programmed Fuel Injection

22140_HOND_C0001

GENERAL ENGINE SPECIFICATIONS

Year	Model	Engine Displacement Liters	Engine ID/VIN	Net Horsepower @ rpm	Net Torque @ rpm (ft. lbs.)	Bore X Stroke (in.)	Compression Ratio	Oil Pressure @ rpm
2007	Civic	1.8	R18A1	140@6300	128@4800	3.19x3.44	10.5:1	50@3000
	Civic	2.0	K20Z3	197@7800	139@6100	3.39x3.39	11.0:1	44@3000
	Civic Hybrid	1.3	LDA2	110@6000	123@2500	2.87x3.15	10.8:1	50@3000
	Accord	2.4	K24A8	166@5800	160@4000	3.43x3.90	9.7:1	44@3000
	Accord	3.0	J30A5	244@6250	211@5000	3.39x3.39	10.0:1	71@3000
	S2000	2.2	F22C1	237@7800	162@6800	3.43X3.57	11.1:1	85@3000
2008	Civic	1.8	R18A1	140@6300	128@4800	3.19x3.44	10.5:1	50@3000
	Civic	2.0	K20Z3	197@7800	139@6100	3.39x3.39	11.0:1	44@3000
	Civic Hybrid	1.3	LDA2	110@6000	123@2500	2.87x3.15	10.8:1	50@3000
	Accord	2.4	K24Z2	177@6500	161@4300	3.43x3.90	10.5:1	44@3000
	Accord	2.4	K24Z3	190@7000	162@4400	3.43x3.90	10.5:1	44@3000
	Accord	3.5	J35Z2	268@6200	248@5000	3.50x3.66	10.5:1	71@3000
	Accord	3.5	J35Z3	268@6200	248@5000	3.50x3.66	10.0:1	71@3000
	S2000	2.2	F22C1	237@7800	162@6800	3.43X3.57	11.1:1	85@3000

22140_HOND_C0002

ENGINE TUNE-UP SPECIFICATIONS

Year	Engine Displacement Liters	Engine ID/VIN	Spark Plugs Gap (in.)	Ignition Timing (deg.) MT	AT	Fuel Pump (psi)	Idle Speed (rpm) MT	AT	Valve Clearance In.	Ex.
2007	1.8	R18A1	0.039-0.043	6-10B	6-10B	48-55	620-720	620-720	0.007-0.009	0.009-0.011
	2.0	K20Z3	0.039-0.043	6-10B	—	48-55	700-800	—	0.008-0.010	0.010-0.011
	1.3	LDA2	0.039-0.043	—	8-12B	38-46	—	770-870	0.006-0.007	0.009-0.011
	2.2	F22C1	0.039-0.043	3-7B	—	55-63	850-950	—	0.008-0.010	0.010-0.011
	2.4	K24A8	0.039-0.043	6-10B	6-10B	47-54	670-770	750-850	0.008-0.010	0.011-0.013
	3.0	J30A5	0.039-0.043	8-12B	8-12B	55-63	700-800	740-840	0.008-0.009	0.011-0.013
2008	1.8	R18A1	0.039-0.043	6-10B	6-10B	48-55	620-720	620-720	0.007-0.009	0.009-0.011
	2.0	K20Z3	0.039-0.043	6-10B	—	48-55	700-800	—	0.008-0.010	0.010-0.011
	1.3	LDA2	0.039-0.043	—	8-12B	38-46	—	770-870	0.006-0.007	0.009-0.011
	2.2	F22C1	0.039-0.043	3-7B	—	55-63	850-950	—	0.008-0.010	0.010-0.011
	2.4	K24Z2/3	0.039-0.043	6-10B	6-10B	48-55	730-830	730-830	0.008-0.010	0.010-0.011
	3.5	J35Z2	0.039-0.043	3-7B	8-12B	57-64	600-700	600-700	0.008-0.009	0.011-0.013
	3.5	J35Z3	0.039-0.043	3-7B	8-12B	57-64	700-800	700-800	0.008-0.009	0.011-0.013

NOTE: The Vehicle Emission Control Information label often reflects specification changes made during production.

The label figures must be used if they differ from those in this chart

B: Before Top Dead Center

22140_HOND_C0003

CAPACITIES

Year	Model	Engine Displacement Liters	Engine ID	Engine Oil with Filter	Transmission (pts.)		Drive Axle		Fuel Tank (gal.)	Cooling System (qts.)
					5-Spd	Auto.	Front (pts.)	Rear (pts.)		
2007	Accord	2.4	K24A8	4.4	4.0	5.2	—	—	18.5	①
	Accord	3.0	J30A5	4.5	4.4	7.0	—	—	18.5	7.0
	Civic	1.8	R18A1	3.9	3.0	5.0	—	—	13.2	②
	Civic	2.0	K20Z3	4.6	3.2	—	—	—	13.2	7.2
	Civic Hybrid	1.3	LDA2	3.4	3.2	6.0	—	—	12.3	5.0
	S2000	2.2	F22C1	5.1	3.1	—	—	—	13.2	6.9
2008	Accord	2.4	K24Z2	4.4	4.0	5.2	—	—	18.5	③
	Accord	2.4	K24Z3	4.4	4.0	5.2	—	—	18.5	③
	Accord	3.5	J35Z2	4.5	—	7.0	—	—	18.5	7.0
	Accord	3.5	J35Z3	4.5	4.4	—	—	—	18.5	7.0
	Civic	1.8	R18A1	3.9	3.0	5.0	—	—	13.2	②
	Civic	2.0	K20Z3	4.6	3.2	—	—	—	13.2	7.2
	Civic Hybrid	1.3	LDA2	3.4	3.2	6.0	—	—	12.3	5.0
	S2000	2.2	F22C1	5.1	3.1	—	—	—	13.2	6.9

NOTE: All capacities are approximate. Add fluid gradually and ensure a proper fluid level is obtained.

NOTE: Capacities given are service, not overhaul capacities

① Automatic Transaxle: 6.2
Manual Transaxle: 6.7

② Manual Transaxle 4 door: 5.4
Automatic transaxle 4 door: 5.4
Manual transaxle 2 door: 5.4
Automatic transaxle 4 door: 5.8

③ Automatic Transaxle: 6.2
Manual Transaxle: 6.4

22140_HOND_C0004

FLUID SPECIFICATIONS

Year	Model	Engine Displ. Liters	Engine Oil	Man. Trans.	Auto. Trans.	Drive Axle Front	Drive Axle Rear	Transfer Case	Power Steering Fluid	Brake Master Cylinder	Cooling System
2007	Accord	2.4	5W-20 Honda	Honda MTF	Honda ATF-Z1	—	—	—	Honda PS Fluid	Honda DOT 3	①
	Accord	3.0	5W-20 Honda	Honda MTF	Honda ATF-Z1	—	—	—	Honda PS Fluid	Honda DOT 3	①
	Civic	1.8	5W-20 Honda	Honda MTF	Honda ATF-Z1	—	—	—	Honda PS Fluid	Honda DOT 3	①
	Civic	2.0	5W-20 Honda	Honda MTF	—	—	—	—	Honda PS Fluid	Honda DOT 3	①
	Civic Hybrid	1.3	0W-20 Honda	—	Honda CVTF	—	—	—	NA	Honda DOT 3	①
	S2000	2.2	10W-30 Honda	Honda MTF	—	—	②	—	NA	Honda DOT 3	①
2008	Accord	2.4	5W-20 Honda	Honda MTF	Honda ATF-Z1	—	—	—	Honda PS Fluid	Honda DOT 3	①
	Accord	3.5	5W-20 Honda	Honda MTF	Honda ATF-Z1	—	—	—	Honda PS Fluid	Honda DOT 3	①
	Civic	1.8	5W-20 Honda	Honda MTF	Honda ATF-Z1	—	—	—	Honda PS Fluid	Honda DOT 3	①
	Civic	2.0	5W-20 Honda	Honda MTF	—	—	—	—	Honda PS Fluid	Honda DOT 3	①
	Civic Hybrid	1.3	0W-20 Honda	—	Honda CVTF	—	—	—	NA	Honda DOT 3	①
	S2000	2.2	10W-30 Honda	Honda MTF	—	—	—	—	NA	Honda DOT 3	①

DOT: Department Of Transpotation

① Honda Long Life Antifreeze/Coolant-Type2

② Hypoid gear oil SAE 90, API classified GL4 or GL5 only

22140_HOND_C0005

VALVE SPECIFICATIONS

Year	Engine Displacement Liters	Engine ID/VIN	Seat Angle (deg.)	Face Angle (deg.)	Spring Test Pressure (lbs. @ in.)	Spring Installed Height (in.)	Stem-to-Guide Clearance (in.)		Stem Diameter (in.)	
							Intake	Exhaust	Intake	Exhaust
2007	2.4	K24A8	45	45	NA	NA	0.0012-0.0022	0.0022-0.0031	0.2156-0.2159	0.2146-0.2150
	3.0	J30A5	45	45	NA	NA	0.0008-0.0018	0.0022-0.0031	0.2159-0.2163	0.2146-0.2150
	1.8	R18A1	45	45	NA	NA	0.0008-0.0020	0.0020-0.0031	0.2157-0.2161	0.2146-0.2150
	2.0	K20Z3	45	45	NA	NA	0.0012-0.0022	0.0022-0.0031	0.2156-0.2159	0.2146-0.2150
	1.3	LDA2	45	45	NA	NA	0.0008-0.0020	0.0020-0.0031	0.2157-0.2161	0.2146-0.2150
	2.2	F22C1	45	45	NA	NA	0.0010-0.0020	0.0020-0.0030	0.2157-0.2161	0.2146-0.2150
2008	2.4	K24Z2	45	45	NA	NA	0.0012-0.0022	0.0022-0.0031	0.2156-0.2159	0.2146-0.2150
	2.4	K24Z3	45	45	NA	NA	0.0012-0.0022	0.0022-0.0031	0.2156-0.2159	0.2146-0.2150
	3.5	J35Z2	45	45	NA	NA	0.0008-0.0018	0.0022-0.0031	0.2159-0.2163	0.2146-0.2150
	3.5	J35Z3	45	45	NA	NA	0.0008-0.0018	0.0022-0.0031	0.2159-0.2163	0.2146-0.2150
	1.8	R18A1	45	45	NA	NA	0.0008-0.0020	0.0020-0.0031	0.2157-0.2161	0.2146-0.2150
	2.0	K20Z3	45	45	NA	NA	0.0012-0.0022	0.0022-0.0031	0.2156-0.2159	0.2146-0.2150
	1.3	LDA2	45	45	NA	NA	0.0008-0.0020	0.0020-0.0031	0.2157-0.2161	0.2146-0.2150
	2.2	F22C1	45	45	NA	NA	0.0010-0.0020	0.0020-0.0030	0.2157-0.2161	0.2146-0.2150

NA: Information not available

22140_HOND_C0007

CAMSHAFT AND BEARING SPECIFICATIONS
All measurements are given in inches.

Year	Engine Displacement Liters	Engine VIN	Journal Diameter	Brg. Oil Clearance	Shaft End-play	Runout	Journal Bore	Lobe Height Intake	Exhaust
2007	2.4	K24A8	NA	①	0.0020-0.0080	0.0010	NA	②	1.3422
	3.0	J30A5	NA	0.0020-0.0035	0.0020-0.0080	0.0010	NA	③	1.4302
	1.8	R18A1	NA	0.0018-0.0033	0.0020-0.0100	0.0010	NA	④	1.4100
	2.0	K20Z3	NA	①	0.0020-0.0080	0.0010	NA	⑤	⑥
	1.3	LDA2	NA	0.0020-0.0035	0.0020-0.0060	0.0010	NA	⑦	⑧
	2.2	F22C1	NA	0.0020-0.0040	0.0020-0.0060	0.0010	NA	⑨	⑩
2008	2.4	K24Z2	NA	⑪	0.0020-0.0080	0.0012	NA	⑫	⑬
	2.4	K24Z3	NA	⑪	0.0020-0.0080	0.0012	NA	⑫	⑬
	3.5	J35Z2	NA	0.0020-0.0035	0.0020-0.0080	0.0012	NA	⑭	⑮
	3.5	J35Z3	NA	0.0020-0.0035	0.0020-0.0080	0.0012	NA	⑯	1.4472
	1.8	R18A1	NA	0.0018-0.0033	0.0020-0.0100	0.0010	NA	④	1.4100
	2.0	K20Z3	NA	①	0.0020-0.0080	0.0010	NA	⑤	⑥
	1.3	LDA2	NA	0.0020-0.0035	0.0020-0.0060	0.0010	NA	⑦	⑧
	2.2	F22C1	NA	0.0020-0.0040	0.0020-0.0060	0.0010	NA	⑨	⑩

NA: Information not available

① No. 1 journal: 0.001-0.003 in.
No. 2-5 journals: 0.002-0.004 in

② Primary: 1.3356 in.
Secondary: 1.1668 in.

③ Primary: 1.3796 in.
Mid: 1.4348 in.
Secondary: 1.3891 in.

④ Primary: 1.4076 in.
Secondary A: 1.3920 in.
Secondary B: 1.4184 in.

⑤ Primary: 1.2910 in.
Mid: 1.3990 in.
Secondary: 1.2865 in.

⑥ Primary: 1.2902 in.
Mid: 1.3688 in.
Secondary: 1.2859 in.

⑦ 1st: 1.1692 in.
2nd: 1.4003 in.
3rd: 1.4196 in.

⑧ 1st: 1.1771 in.
2nd: 1.4054 in.

⑨ Primary: 1.3370 in.
Mid: 1.4340 in.
Secondary: 1.3370 in.

⑩ Primary: 1.2902 in.
Mid: 1.3688 in.
Secondary: 1.2859 in.

⑪ No. 1 journal: 0.0012-0.0030 in.
No. 2-5 journals: 0.002-0.004 in

⑫ Primary: 1.3285 in.
Mid: 1.3959 in.
Secondary: 1.3285 in.

⑬ Except PZEV: 1.3500 in.
PZEV: 1.3477 in.

⑭ Nos. 1-4 cylinders: 1.3965 in.
Nos 5-6 cylinders: 1.3964 in.

⑮ Nos. 1-4 cylinders: 1.4481 in.
Nos 5-6 cylinders: 1.4472 in.

⑯ Primary: 1.4024 in.
Secondary: 1.3504 in.

22140_HOND_C0006

CRANKSHAFT AND CONNECTING ROD SPECIFICATIONS

All measurements are given in inches.

| Year | Engine Displacement Liters | Engine ID/VIN | Crankshaft | | | | Connecting Rod | | |
			Main Brg. Journal Dia.	Main Brg. Oil Clearance	Shaft End-play	Thrust on No.	Journal Diameter	Oil Clearance	Side Clearance
2007	2.4	K24A8	①	②	0.0040-0.0140	4	1.8888-1.8898	0.0008-0.0020	0.0060-0.0140
	3.0	J30A5	2.8337-2.8346	0.0008-0.0017	0.0040-0.0140	3	2.0857-2.0866	0.0008-0.0017	0.0060-0.0140
	1.8	R18A1	2.1644-2.1654	0.0007-0.0013	0.0040-0.0140	4	1.7707-1.7716	0.0009-0.0017	0.0060-0.0140
	2.0	K20Z3	①	②	0.0040-0.0140	4	1.7707-1.7717	0.0013-0.0026	0.0060-0.0120
	1.3	LDA2	1.9676-1.9685	0.0007-0.0014	0.0040-0.0140	4	1.5739-1.5748	0.0008-0.0015	0.0060-0.0120
	2.2	F22C1	2.1644-2.1654	0.0007-0.0016	0.0040-0.0140	4	1.8888-1.8898	0.0012-0.0021	0.0060-0.0120
2008	2.4	K24Z2	③	②	0.0040-0.0140	4	1.8888-1.8898	0.0013-0.0026	0.0060-0.0140
	2.4	K24Z3	③	②	0.0040-0.0140	4	1.8888-1.8898	0.0013-0.0026	0.0060-0.0140
	3.5	J35Z2	2.8337-2.8346	0.0007-0.0018	0.0040-0.0140	3	2.1644-2.1654	0.0008-0.0017	0.0060-0.0140
	3.5	J35Z3	2.8337-2.8346	0.0007-0.0018	0.0040-0.0140	3	2.1644-2.1654	0.0008-0.0017	0.0060-0.0140
	1.8	R18A1	2.1644-2.1654	0.0007-0.0013	0.0040-0.0140	4	1.7707-1.7716	0.0009-0.0017	0.0060-0.0140
	2.0	K20Z3	①	②	0.0040-0.0140	4	1.7707-1.7717	0.0013-0.0026	0.0060-0.0120
	1.3	LDA2	1.9676-1.9685	0.0007-0.0014	0.0040-0.0140	4	1.5739-1.5748	0.0008-0.0015	0.0060-0.0120
	2.2	F22C1	2.1644-2.1654	0.0007-0.0016	0.0040-0.0140	4	1.8888-1.8898	0.0012-0.0021	0.0060-0.0120

① Journals 1, 2 4 and 5: 2.1648-2.1657
Journal 3: 2.1644-2.1654

② Journals 1, 2 4 and 5: 0.0007-0.0016
Journal 3: 0.0010-0.0019

③ Journals 1, 2 4 and 5: 2.1647-2.1
Journal 3: 2.1644-2.1654

22140_HOND_C0008

PISTON AND RING SPECIFICATIONS

All measurements are given in inches.

Year	Engine Displacement Liters	Engine ID/VIN	Piston Clearance	Ring Gap			Ring Side Clearance		
				Top Compression	Bottom Compression	Oil Control	Top Compression	Bottom Compression	Oil Control
2007	2.4	K24A8	0.0008-0.0016	0.0080-0.0140	0.0160-0.022	0.0080-0.0028	0.0018-0.0028	0.0020-0.0030	NA
	3.0	J30A5	0.0006-0.0016	0.0080-0.0140	0.0016-0.0220	0.0080-0.0280	0.0022-0.0031	0.0012-0.0022	NA
	1.8	R18A1	0.0004-0.0014	0.0080-0.0140	0.0160-0.0220	①	0.0018-0.0028	0.0014-0.0024	NA
	2.0	K20Z3	0.0008-0.0016	0.0080-0.0140	0.0200-0.0260	0.0080-0.0280	0.0018-0.0028	0.0016-0.0026	NA
	1.3	LDA2	0.0004-0.0016	0.0060-0.0120	0.0140-0.0200	0.0080-0.0280	0.0025-0.0035	0.0012-0.0022	NA
	2.2	F22C1	0.0002-0.0011	0.0100-0.0140	0.0240-0.0300	0.0080-0.0280	0.0018-0.0035	0.0016-0.0028	NA
2008	2.4	K24Z2	0.0008-0.0016	0.0080-0.0140	0.0200-0.0260	0.0080-0.0280	0.0024-0.0033	0.0016-0.0026	NA
	2.4	K24Z3	0.0008-0.0016	0.0080-0.0140	0.0200-0.0260	0.0080-0.0280	0.0024-0.0033	0.0016-0.0026	NA
	3.5	J35Z2	0.0006-0.0016	0.0080-0.0140	0.0160-0.0220	0.0080-0.0280	0.0022-0.0031	0.0012-0.0022	NA
	3.5	J35Z3	0.0006-0.0016	0.0080-0.0140	0.0160-0.0020	0.0080-0.0280	0.0022-0.0031	0.0012-0.0022	NA
	1.8	R18A1	0.0004-0.0014	0.0080-0.0140	0.0160-0.0020	①	0.0018-0.0028	0.0014-0.0024	NA
	2.0	K20Z3	0.0008-0.0016	0.0080-0.0140	0.0200-0.0260	0.0080-0.0280	0.0018-0.0028	0.0016-0.0026	NA
	1.3	LDA2	0.0004-0.0016	0.0060-0.0120	0.0140-0.0200	0.0080-0.0280	0.0025-0.0035	0.0012-0.0022	NA
	2.2	F22C1	0.0002-0.0011	0.0100-0.0140	0.0240-0.0300	0.0080-0.0280	0.0018-0.0035	0.0016-0.0028	NA

NA: Information not available

① Riken: 0.008-0.020 in.
  Except RIKEN: 0.008-0.028 in.

22140_HOND_C0009

TORQUE SPECIFICATIONS
All readings in ft. lbs.

Year	Engine Displacement Liters	Engine ID/VIN	Cylinder Head Bolts	Main Bearing Bolts	Rod Bearing Bolts	Crankshaft Damper Bolts	Flywheel Bolts	Manifold Intake	Manifold Exhaust	Spark Plugs	Oil Pan Drain Plug
2007	2.4	K24A8	①	②	③	④	⑤	16	33	13	33
	3.0	J30A5	⑥	⑦	③	⑧	⑤	16	23	13	29
	1.8	R18A1	①	⑨	③	⑩	⑤	17	23	13	30
	2.0	K20Z3	①	②	⑪	⑩	⑤	16	23	13	30
	1.3	LDA2	⑫	⑬	⑭	⑩	33	17	16	13	29
	2.2	F22C1	⑮	⑯	⑰	192	94	16	23	13	29
2008	2.4	K24Z2	①	⑱	⑲	⑳	㉑	16	25	13	29
	2.4	K24Z3	①	⑱	⑲	⑳	㉑	16	25	13	29
	3.5	J35Z2	⑮	⑦	③	⑧	⑤	16	23	13	29
	3.5	J35Z3	⑮	⑦	③	⑧	⑤	16	23	13	29
	1.8	R18A1	①	⑨	③	⑩	⑤	17	23	13	30
	2.0	K20Z3	①	②	⑪	⑩	⑤	16	23	13	30
	1.3	LDA2	⑫	⑬	⑭	⑩	33	17	16	13	29
	2.2	F22C1	⑮	⑯	⑰	192	94	16	23	13	29

① Step 1: 29 ft. lbs.
  Step 2: Rotate 90 degrees
  Step 3: Rotate 90 degrees
  Step 4: If new bolt rotate
  additional 90 degrees

② Step 1: 22 ft. lbs.
  Step 2: Rotate 56 degrees

③ Step 1: 14 ft. lbs.
  Step 2: Rotate 90 degrees

④ Step 1: 36 ft. lbs.
  Step 2: Rotate 90 degrees

⑤ Automatic transaxle: 54 ft. lbs.
  Manual transaxle: 76 ft. lbs.

⑥ 6-point bolts:
  Step 1: 29 ft. lbs.
  Step 2: 51 ft. lbs.
  Step 3: 72.3 ft. lbs.
  12-point bolts:
  Step 1: 29 ft. lbs.
  Step 2: Rotate 90 degrees
  Step 3: Rotate 90 degrees
  Step 4: If new bolt rotate
  additional 90 degrees

⑦ Step 1: Cap bolts, 56 ft. lbs.
  Step 2: Side bolts, 36 ft. lbs.

⑧ Step 1: 47 ft. lbs.
  Step 2: Rotate 60 degrees

⑨ Step 1: 18 ft. lbs.
  Step 2: Rotate 57 degrees

⑩ Old bolt:
  Step 1: 27 ft. lbs
  Step 2: Plus 90 degrees
  New bolt:
  Step 1: 130 ft. lbs
  Step 2: Loosen fully
  Step 3: 27 ft. lbs
  Step 4: Plus 90 degrees

⑪ Step 1: 22 ft. lbs.
  Step 2: Rotate 90 degrees

⑫ Step 1: 22 ft. lbs.
  Step 2: Rotate 130 degrees

⑬ Step 1: 18 ft. lbs.
  Step 2: Rotate 40 degrees

⑭ Step 1: 7.2 ft. lbs.
  Step 2: Rotate 90 degrees

⑮ Step 1: 22 ft. lbs.
  Step 2: Rotate 90 degrees
  Step 3: Rotate 90 degrees
  Step 4: If new bolt rotate
  additional 90 degrees

⑯ Step 1: 22 ft. lbs.
  Step 2: Rotate 60 degrees
  Step 3: 8mm bolts to 16 ft. lbs.

⑰ Step 1: 18 ft. lbs.
  Step 2: Rotate 90 degrees

⑱ Step 1: 22 ft. lbs.
  Step 2: Rotate 48 degrees

⑲ Step 1: 30 ft. lbs.
  Step 2: Rotate 120 degrees

⑳ Old bolt:
  Step 1: 36 ft. lbs
  Step 2: Plus 90 degrees
  New bolt:
  Step 1: 130 ft. lbs
  Step 2: Loosen fully
  Step 3: 36 ft. lbs
  Step 4: Plus 90 degrees

㉑ Automatic transaxle: 54 ft. lbs.
  Manual transaxle: 90 ft. lbs.

22140_HOND_C0010

WHEEL ALIGNMENT

Year	Model		Caster Range (+/-Deg.)	Caster Preferred Setting (Deg.)	Camber Range (+/-Deg.)	Camber Preferred Setting (Deg.)	Toe-in (in.)
2007	Accord	F	0.75	+3.25	0.75	0.00	0.00 +/-0.06
		R	—	—	0.50	-1 00'	0.06 +/-0.06
	Civic	F	1.00	+7.00	0.50	0.00	0.00+/-0.08
		R	—	—	0 - +1.83	-0.75	0.00+/-0.08
	Civic	F	1.00	+7.10	0.50	0.05	0.00+/-0.08
	Hybrid	R	—	—	0.183	-0.90	0.00+/-0.08
	S2000	F	0.75	+6.00	0.50	-0.50	0.00+/-0.08
		R	—	—	0.50	-1.50	0.14+/-0.08
2008	Accord	F	-1.08 - +0.42	+3.8	0.75	0.00	0.00+/-0.08
		R	—	—	0.75	-1 00'	0.00+/-0.08
	Civic	F	1.00	+7.00	0.50	0.00	0.00+/-0.08
		R	—	—	0 - +1.83	-0.75	0.00+/-0.08
	Civic	F	1.00	+7.10	0.50	0.05	0.00+/-0.08
	Hybrid	R	—	—	0.183	-0.90	0.00+/-0.08
	S2000	F	0.75	+6.00	0.50	-0.50	0.00+/-0.08
		R	—	—	0.50	-1.50	0.14+/-0.08

22140_HOND_C0011

TIRE, WHEEL AND BALL JOINT SPECIFICATIONS

Year	Model	OEM Tires Standard	OEM Tires Optional	Tire Pressures (psi) Front	Tire Pressures (psi) Rear	Wheel Size	Ball Joint Inspection	Lug Nut (ft. lbs.)
2007	Accord	P195/65R15	P2065R15	32	30	NA	NA	80
	Civic	①	None	②	②	NA	NA	80
	Civic Hybrid	195/65R15	None	32	32	NA	NA	80
	S2000	③	None	32	32	NA	NA	80
2008	Accord	P215/60R16	P225/50R17	30	30	NA	NA	80
	Civic	①	None	②	②	NA	NA	80
	Civic Hybrid	195/65R15	None	32	32	NA	NA	80
	S2000	③	None	32	32	NA	NA	80

OEM: Original Equipment Manufacturer

PSI: Pounds Per Square Inch

NA: Information not available

① DX, DX-VP, DX-G: P195/65R15
LX, LX-S, EX, EX-L: P205/55R16
Si: P215/45R17

② 4-door:
DX, DX-VP, DX-G: 30 psi
LX, LX-S, EX, EX-L: 32 psi
Si with ABS: Front 32, Rear 29 psi
Si with VSA: Front 32, Rear 33 psi
2-door:
Except Si Model: 32 psi
Si with ABS: Front 32, Rear 29 psi
Si with VSA: 32 psi

③ Front: 215/45R17
Rear, Except CR: 245/40R17
Rear, CR Model: 255/40R17

22140_HOND_C0012

BRAKE SPECIFICATIONS

All measurements in inches unless noted

Year	Model		Brake Disc			Brake Drum Diameter			Minimum Lining Thickness		Brake Caliper	
			Original Thickness	Minimum Thickness	Maximum Runout	Original Inside Diameter	Max. Wear Limit	Maximum Machine Diameter	Front	Rear	Bracket Bolts (ft. lbs.)	Mounting Bolts (ft. lbs.)
2007	Accord	F	①	②	0.004	—	—	—	0.060	—	80	③
		R	④	0.310	0.004	8.66	8.70	8.70	—	0.060	41	17
	Civic	F	⑤	⑥	0.004	—	—	—	0.060	—	80	25
		R	0.360	0.310	0.004	7.87	7.91	7.91	—	0.080	41	17
	Civic Hybrid	F	⑤	⑥	0.002	—	—	—	0.060	—	80	25
		R	—	—	—	8.66	8.70	8.7	—	②	—	—
	S2000	F	0.990	0.910	0.002	—	—	—	0.060	—	84	24
		R	0.476	0.390	0.002	—	—	—	—	0.060	41	17
2008	Accord	F	①	②	0.002	—	—	—	0.060	—	80	③
		R	④	0.310	0.002	—	—	—	—	⑦	41	17
	Civic	F	⑤	⑥	0.004	—	—	—	0.060	—	80	25
		R	0.360	0.310	0.004	7.87	7.91	7.91	—	0.080	41	17
	Civic Hybrid	F	⑤	⑥	0.002	—	—	—	0.060	—	80	25
		R	—	—	—	8.66	8.70	8.7	—	②	—	—
	S2000	F	0.990	0.910	0.002	—	—	—	0.060	—	84	24
		R	0.476	0.390	0.002	—	—	—	—	0.060	41	17

NA: Information not available

F: Front

R: Rear

① Except V6 w/manual transaxle: 0.910
   V6 w/automatic transaxle: 1.110

② Except V6 w/manual transaxle: 0.830
   V6 w/automatic transaxle: 1.020

③ V6: 37 ft. lbs.
   4 cyl. 26 ft. lbs.

④ Except V6 w/manual transaxle: 0.360
   V6 w/automatic transaxle: 0.040

⑤ 4 cyl (R18A1) engine: 0.83
   4 cyl (K20Z3) engine: 0.99

⑥ 4 cyl (R18A1) engine: 0.75
   4 cyl (K20Z3) engine: 0.91

⑦ 4 cyl: 0.06 in.
   6 cyl: 0.04 in.

22140_HOND_C0013

MAINTENANCE MINDER SCHEDULE
Honda—Civic, Accord & S2000

All Honda's displays engine oil life and maintenance service items in the information display to indicate when to perform maintenance service. If the engine oil life is 15% or less, based on the onboard computer's caluculations, you will see SERVICE DUE SOON in the information display every time the ignition key is turned to ON. The maintenance minder indicator will also come on and the maintenance code(s) for other scheduled maintenance items needing service will be displayed below the message.

Symbol	Item	Service
A	Engine oil ①	Change
B	Engine oil and filter	Change
	Fluid levels	Inspect
	Brakes	Inspect
	Parking brake adjustment	Check
	Steering gear and linkage	Inspect
	Suspension components	Inspect
	Driveshaft boots	Inspect
	Brake hoses and lines	Inspect
	Exhaust system	Inspect
	Fuel lines and connections	Inspect
1	Tires	Rotate
2	Engine air filter ②	Replace
	Dust and pollen filter ③	Replace
	Accessory drive belt	Inspect
3	Transmission fluid ④	Replace
	Transfer case fluid ④	Replace
4	Spark plugs	Replace
	Timing belt ⑤	Replace
	Water pump	Inspect
	Valve clearance ⑥	Inspect
5	Engine coolant	Replace
6	VTM-4 rear differential fluid	Replace

① If the message SERVICE DUE NOW does not appear more than 12 months after the display is reset, change every year.

② If driven in dusty conditions, replace every 15,000 miles.

③ If driven in urban areas that have a high concentration of soot from industry and diesel, replace every 15,000 miles

④ If regularly driven in mountainous areas at very low speed or trailer towing, change the fluid every 30,000 miles.

⑤ If driven regularly in temperatures over 110 deg.F or below -20 deg.F, or towing a trailer, replace every 60,000 miles.

⑥ Adjust if necessary.

Additionally, replace the brake fluid every 3 years, and inspect the idle speed every 160,000 miles.
To reset the Engine Oil Life Display:

1. Turn the ignition switch to ON.

2. Press the SELECT button repeatedly until the engine oil life display or the service message is displayed.

3. Press the RESET button for about 10 seconds. You will see a MAINT RESET message.

4. Select the appropriate answer, MAINT RESET >N (NO) or MAINT RESET > y (YES) by pressing the SELECT button repeatedly.

>N or >Y is displayed on the outside temperature >N or >Y is displayed on the outside temperature display.

5. Select the MAINT RESET > Y (YES), and press and hold the RESET button again to reset the engine oil life to 100%.

PRECAUTIONS

Before servicing any vehicle, please be sure to read all of the following precautions, which deal with personal safety, prevention of component damage, and important points to take into consideration when servicing a motor vehicle:

• Never open, service or drain the radiator or cooling system when the engine is hot; serious burns can occur from the steam and hot coolant.

• Observe all applicable safety precautions when working around fuel. Whenever servicing the fuel system, always work in a well-ventilated area. Do not allow fuel spray or vapors to come in contact with a spark, open flame, or excessive heat (a hot drop light, for example). Keep a dry chemical fire extinguisher near the work area. Always keep fuel in a container specifically designed for fuel storage; also, always properly seal fuel containers to avoid the possibility of fire or explosion. Refer to the additional fuel system precautions later in this section.

• Fuel injection systems often remain pressurized, even after the engine has been turned **OFF**. The fuel system pressure must be relieved before disconnecting any fuel lines. Failure to do so may result in fire and/or personal injury.

• Brake fluid often contains polyglycol ethers and polyglycols. Avoid contact with the eyes and wash your hands thoroughly after handling brake fluid. If you do get brake fluid in your eyes, flush your eyes with clean, running water for 15 minutes. If eye irritation persists, or if you have taken brake fluid internally, IMMEDIATELY seek medical assistance.

• The EPA warns that prolonged contact with used engine oil may cause a number of skin disorders, including cancer. You should make every effort to minimize your exposure to used engine oil. Protective gloves should be worn when changing oil. Wash your hands and any other exposed skin areas as soon as possible after exposure to used engine oil. Soap and water, or waterless hand cleaner should be used.

• All new vehicles are now equipped with an air bag system, often referred to as a Supplemental Restraint System (SRS) or Supplemental Inflatable Restraint (SIR) system. The system must be disabled before performing service on or around system components, steering column, instrument panel components, wiring and sensors. Failure to follow safety and disabling procedures could result in accidental air bag deployment, possible personal injury and unnecessary system repairs.

• Always wear safety goggles when working with, or around, the air bag system. When carrying a non-deployed air bag, be sure the bag and trim cover are pointed away from your body. When placing a non-deployed air bag on a work surface, always face the bag and trim cover upward, away from the surface. This will reduce the motion of the module if it is accidentally deployed. Refer to the additional air bag system precautions later in this section.

• Clean, high quality brake fluid from a sealed container is essential to the safe and proper operation of the brake system. You should always buy the correct type of brake fluid for your vehicle. If the brake fluid becomes contaminated, completely flush the system with new fluid. Never reuse any brake fluid. Any brake fluid that is removed from the system should be discarded. Also, do not allow any brake fluid to come in contact with a painted surface; it will damage the paint.

• Never operate the engine without the proper amount and type of engine oil; doing so WILL result in severe engine damage.

• Timing belt maintenance is extremely important. Many models utilize an interference-type, non-freewheeling engine. If the timing belt breaks, the valves in the cylinder head may strike the pistons, causing potentially serious (also time-consuming and expensive) engine damage. Refer to the maintenance interval charts for the recommended replacement interval for the timing belt, and to the timing belt section for belt replacement and inspection.

• Disconnecting the negative battery cable on some vehicles may interfere with the functions of the on-board computer system(s) and may require the computer to undergo a relearning process once the negative battery cable is reconnected.

• When servicing drum brakes, only disassemble and assemble one side at a time, leaving the remaining side intact for reference.

• Only an MVAC-trained, EPA-certified automotive technician should service the air conditioning system or its components.

BRAKES   ANTI-LOCK BRAKE SYSTEM (ABS)

GENERAL INFORMATION

PRECAUTIONS

• Certain components within the ABS system are not intended to be serviced or repaired individually.

• Do not use rubber hoses or other parts not specifically specified for and ABS system. When using repair kits, replace all parts included in the kit. Partial or incorrect repair may lead to functional problems and require the replacement of components.

• Lubricate rubber parts with clean, fresh brake fluid to ease assembly. Do not use shop air to clean parts; damage to rubber components may result.

• Use only DOT 3 brake fluid from an unopened container.

• If any hydraulic component or line is removed or replaced, it may be necessary to bleed the entire system.

• A clean repair area is essential. Always clean the reservoir and cap thoroughly before removing the cap. The slightest amount of dirt in the fluid may plug an orifice and impair the system function. Perform repairs after components have been thoroughly cleaned; use only denatured alcohol to clean components. Do not allow ABS components to come into contact with any substance containing mineral oil; this includes used shop rags.

• The Anti-Lock control unit is a microprocessor similar to other computer units in the vehicle. Ensure that the ignition switch is **OFF** before removing or installing controller harnesses. Avoid static electricity discharge at or near the controller.

• If any arc welding is to be done on the vehicle, the control unit should be unplugged before welding operations begin.

BRAKES **BLEEDING THE BRAKE SYSTEM**

BLEEDING PROCEDURE

BLEEDING PROCEDURE

See Figure 1.

➡Do not reuse the drained fluid. Use only clean Honda DOT 3 Brake Fluid from an unopened container.

Using a non-Honda brake fluid can cause corrosion and shorten the life of the system.

➡Do not mix different brands of brake fluid; they may not be compatible.

➡Make sure no dirt or other foreign matter is allowed to contaminate the brake fluid.

❋ WARNING

Do not spill brake fluid on the vehicle, it may damage the paint; if brake fluid does contact the paint,

wash it off immediately with water.

➡The reservoir on the master cylinder must be at the MAX (upper) level mark at the start of the bleeding procedure and checked after bleeding each brake caliper. Add fluid as required.

1. Make sure the brake fluid level in the reservoir is at the MAX (upper) level line.
2. Attach a length of clear drain tube to the bleed screw.
3. Have someone slowly pump the brake pedal several times, and then apply steady pressure.
4. Starting at the left-front, loosen the brake bleed screw to allow air to escape from the system. Then tighten the bleed screw securely.
5. Repeat the procedure for each wheel in the sequence shown following until air bubbles no longer appear in the fluid.
6. Refill the master cylinder reservoir to the MAX (upper) level line.

BLEEDING THE ABS SYSTEM

The bleeding procedure for the ABS System is the same as the Conventional Bleeding Procedure.

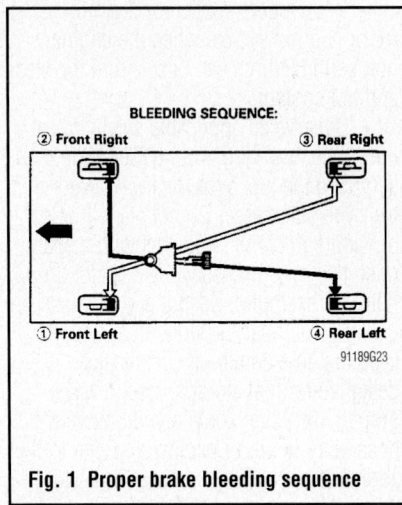

BLEEDING SEQUENCE:
② Front Right ③ Rear Right
① Front Left ④ Rear Left
91189G23

Fig. 1 Proper brake bleeding sequence

BRAKES **FRONT DISC BRAKES**

❋❋ CAUTION

Dust and dirt accumulating on brake parts during normal use may contain asbestos fibers from production or aftermarket brake linings. Breathing excessive concentrations of asbestos fibers can cause serious bodily harm. Exercise care when servicing brake parts. Do not sand or grind brake lining unless equipment used is designed to contain the dust residue. Do not clean brake parts with compressed air or by dry brushing. Cleaning should be done by dampening the brake components with a fine mist of water, then wiping the brake components clean with a dampened cloth. Dispose of cloth and all residue containing asbestos fibers in an impermeable container with the appropriate label. Follow practices prescribed by the Occupational Safety and Health Administration (OSHA) and the Environmental Protection Agency (EPA) for the handling, processing, and disposing of dust or debris that may contain asbestos fibers.

BRAKE CALIPER

REMOVAL & INSTALLATION

Accord

See Figure 2.

1. Before servicing the vehicle, refer to the precautions in the beginning of this section.
2. Remove and discard brake fluid from the master cylinder, as necessary.
3. Raise and support the vehicle safely. Remove the front tires.
4. Disconnect and plug the brake line hose.
5. Remove the brake hose bracket mounting bolt.
6. Remove the caliper flange bolts.
7. Remove the brake pads and shims.
8. Remove the pad retainers and check the caliper pins for free movement.

To install:

9. Apply a thin coat of M-77 assembly paste part number 08798-9010 to the brake pad sides of the pad shims and the back of the brake pads, wipe excess paste off the shim.
10. Continue the installation in the reverse order of the removal procedure.

11. Check and refill the master cylinder, as necessary.
12. Road test the vehicle.

Civic

See Figure 3.

1. Remove and discard brake fluid from the master cylinder, as necessary.
2. Raise and support the vehicle safely. Remove the front tires.
3. Disconnect and plug the brake line hose.
4. If necessary, remove the brake hose bracket mounting bolt.
5. Remove the caliper flange bolts.
6. Remove the brake pads and shims.
7. Remove the pad retainers and check the caliper pins for free movement.

To install:

8. Installation is the reverse of the removal procedure.
9. Apply a thin coat of M-77 assembly paste part number 08798-9010 to the brake pad sides of the pad shims and the back of the brake pads, wipe excess paste off the shim.
10. Check and refill the master cylinder, as necessary.
11. Road test the vehicle.

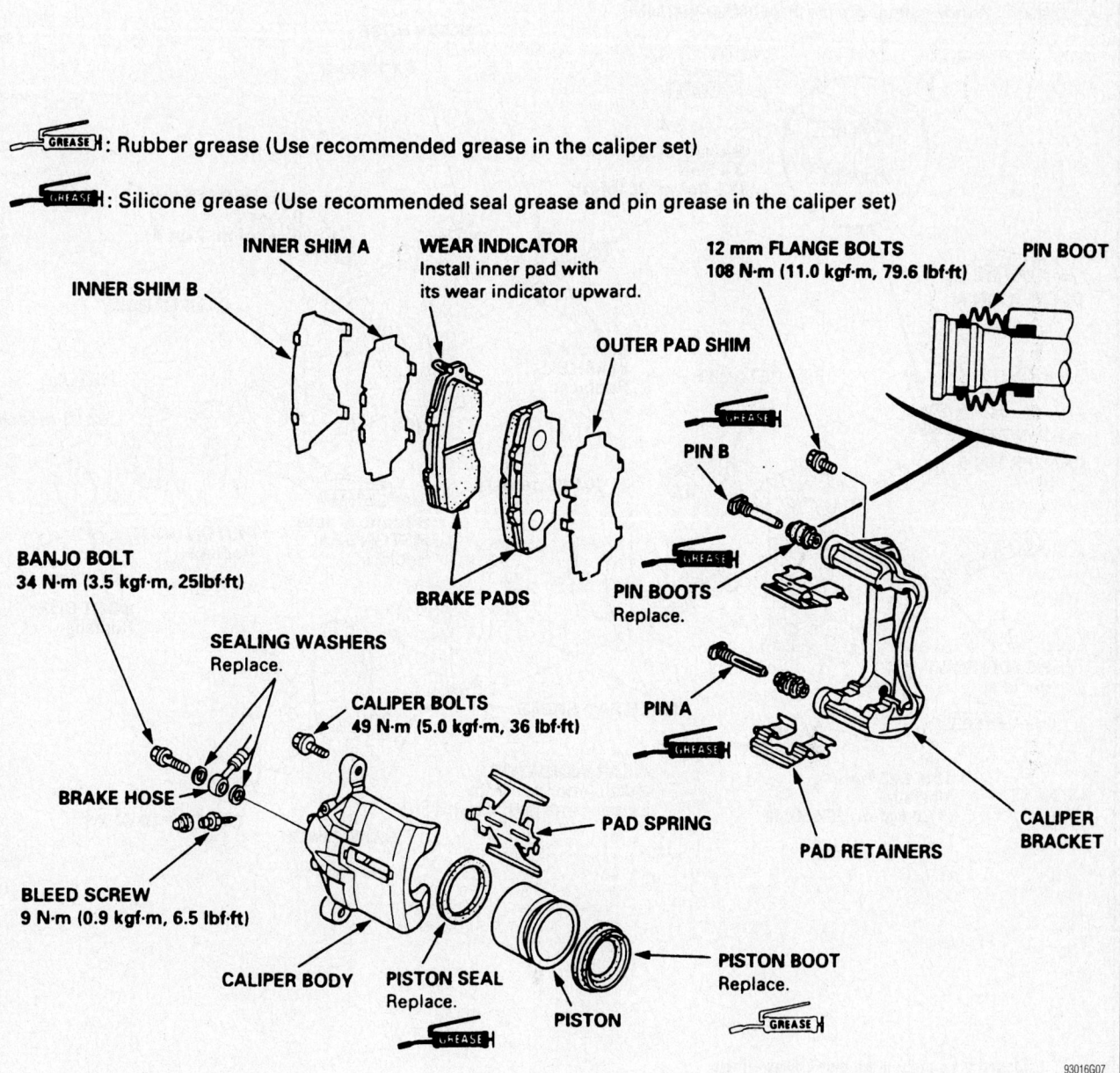

GREASE : Rubber grease (Use recommended grease in the caliper set)

GREASE : Silicone grease (Use recommended seal grease and pin grease in the caliper set)

INNER SHIM B

INNER SHIM A

WEAR INDICATOR
Install inner pad with
its wear indicator upward.

OUTER PAD SHIM

12 mm FLANGE BOLTS
108 N·m (11.0 kgf·m, 79.6 lbf·ft)

PIN BOOT

PIN B

BRAKE PADS

PIN BOOTS
Replace.

BANJO BOLT
34 N·m (3.5 kgf·m, 25 lbf·ft)

SEALING WASHERS
Replace.

CALIPER BOLTS
49 N·m (5.0 kgf·m, 36 lbf·ft)

PIN A

BRAKE HOSE

PAD SPRING

PAD RETAINERS

CALIPER
BRACKET

BLEED SCREW
9 N·m (0.9 kgf·m, 6.5 lbf·ft)

CALIPER BODY

PISTON SEAL
Replace.

PISTON

PISTON BOOT
Replace.

GREASE

GREASE

93016G07

Fig. 2 Exploded view of the front disc brakes—Accord V6 shown

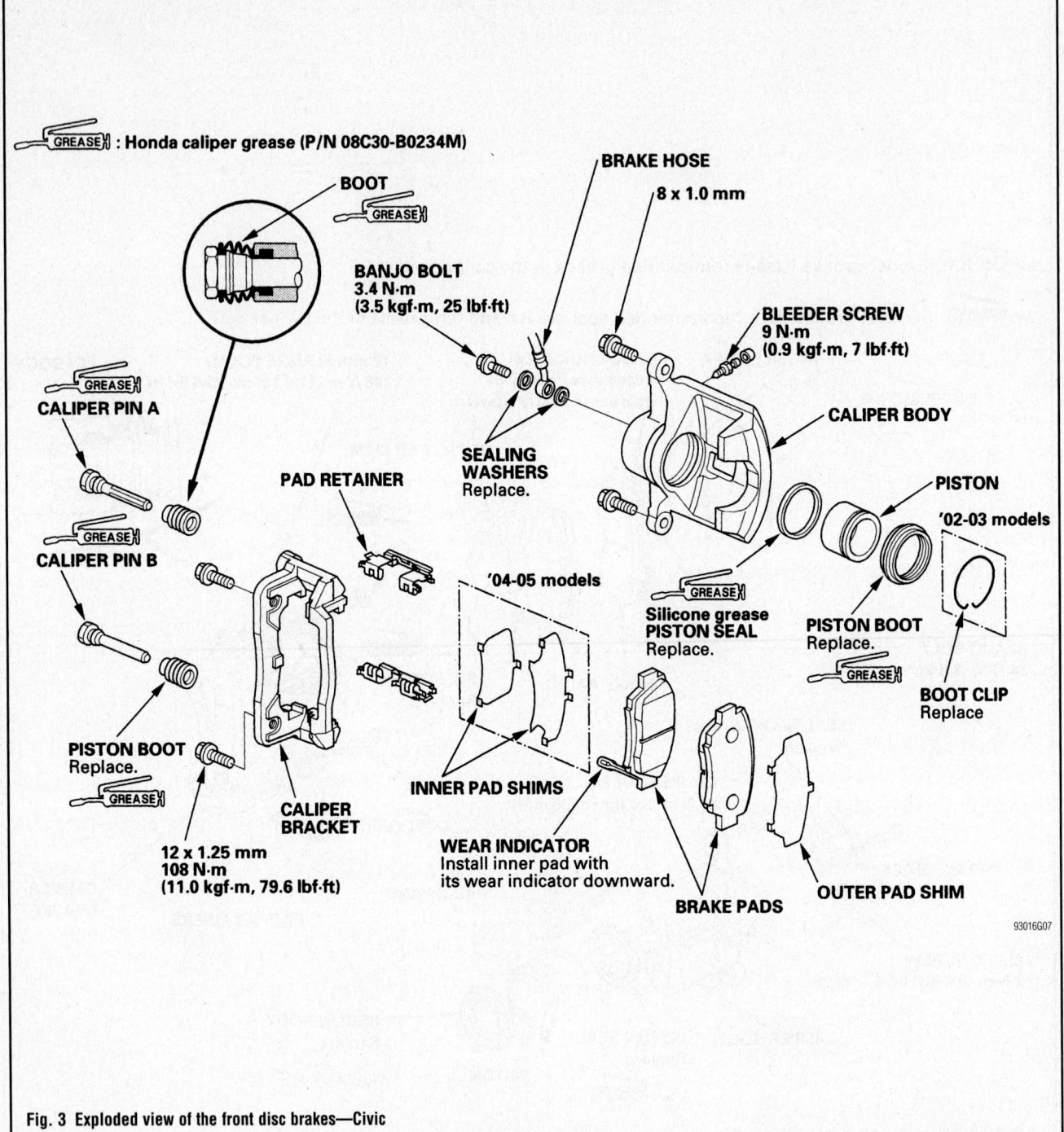

Fig. 3 Exploded view of the front disc brakes—Civic

S2000

See Figure 4.

1. Remove and discard brake fluid from the master cylinder, as necessary.

2. Raise and support the vehicle safely. Remove the front tires.

3. Disconnect and plug the brake line hose.

4. Remove the brake hose bracket mounting bolt.

5. Remove the caliper flange bolts.

6. Remove the brake pads and shims.

7. Remove the pad retainers and check the caliper pins for free movement.

To install:

8. Apply a thin coat of M-77 assembly paste part number 08798-9010 to the brake pad sides of the pad shims and the back of the brake pads, wipe excess paste off the shim.

9. Continue the installation in the reverse order of the removal procedure.

10. Check and refill the master cylinder, as necessary.

11. Road test the vehicle.

DISC BRAKE PADS

REMOVAL & INSTALLATION

Accord

See Figures 5 and 6.

1. Remove some brake fluid from the master cylinder.

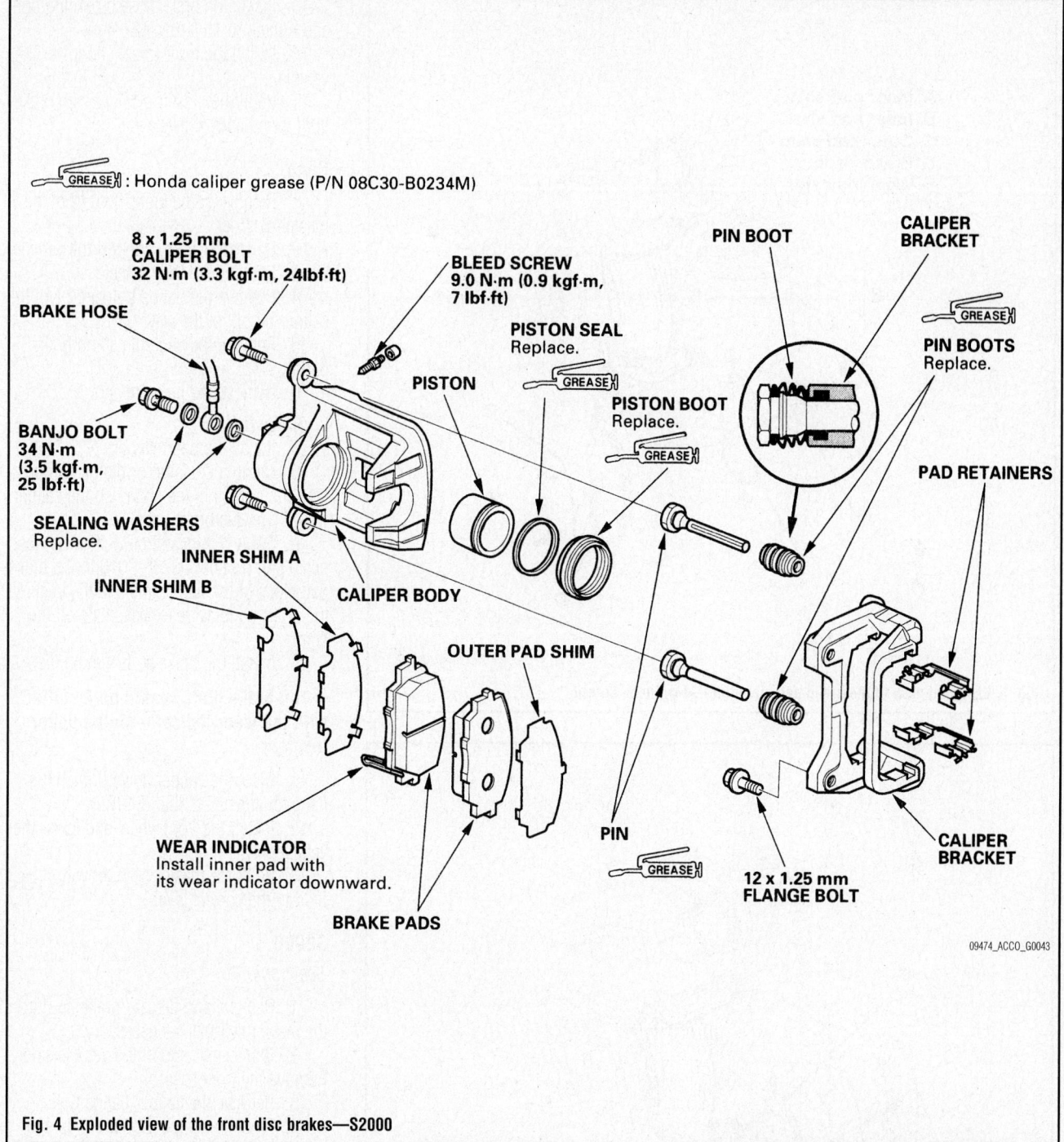

GREASE : Honda caliper grease (P/N 08C30-B0234M)

8 x 1.25 mm
CALIPER BOLT
32 N·m (3.3 kgf·m, 24lbf·ft)

BLEED SCREW
9.0 N·m (0.9 kgf·m, 7 lbf·ft)

BRAKE HOSE

PISTON SEAL
Replace.

PISTON

GREASE

PISTON BOOT
Replace.

GREASE

BANJO BOLT
34 N·m
(3.5 kgf·m,
25 lbf·ft)

SEALING WASHERS
Replace.

INNER SHIM A

INNER SHIM B

CALIPER BODY

PIN BOOT

CALIPER BRACKET

GREASE

PIN BOOTS
Replace.

PAD RETAINERS

OUTER PAD SHIM

WEAR INDICATOR
Install inner pad with
its wear indicator downward.

BRAKE PADS

PIN

GREASE

12 x 1.25 mm
FLANGE BOLT

CALIPER BRACKET

09474_ACCO_G0043

Fig. 4 Exploded view of the front disc brakes—S2000

2. Raise and safely support the vehicle.
3. Remove the front wheel.
4. Remove the brake hose mounting bolt.
5. Remove the flange bolt, while holding the caliper pin, and pivot the caliper up out of the way.
6. Remove the brake pads and pad shims.
7. Remove the pad retainers.

To install:

8. Apply a thin coat of M-77 assembly paste part number 08798-9010 to the brake pad sides of the pad shims and the back of the brake pads, wipe excess paste off the shim.
9. Install the pad retainers.
10. Install the brake pads and shims.

➡**The brake pads should be installed with the wear indicator on top (4-cylinder), or on the inside (6-cylinder)**

11. Push in the piston, so the caliper will fit over the brake pads.

※※ **WARNING**

Check the brake fluid level. The brake fluid may overflow if the reservoir is too full.

12. Pivot the caliper down into position. Install the flange bolt and tighten it to:

- 4 cylinder models: 26 ft. lbs. (35 Nm)
- 6-cylinder models: 37 ft. lbs. (50 Nm)

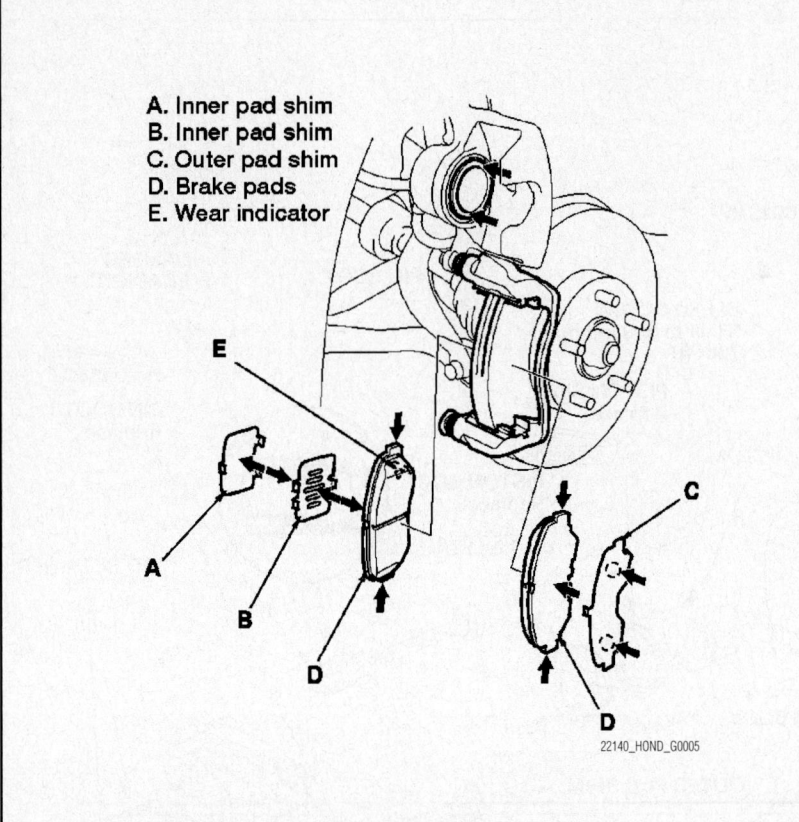

A. Inner pad shim
B. Inner pad shim
C. Outer pad shim
D. Brake pads
E. Wear indicator

22140_HOND_G0005

Fig. 5 Exploded view of the brake pad mounting—4-cylinder Accord

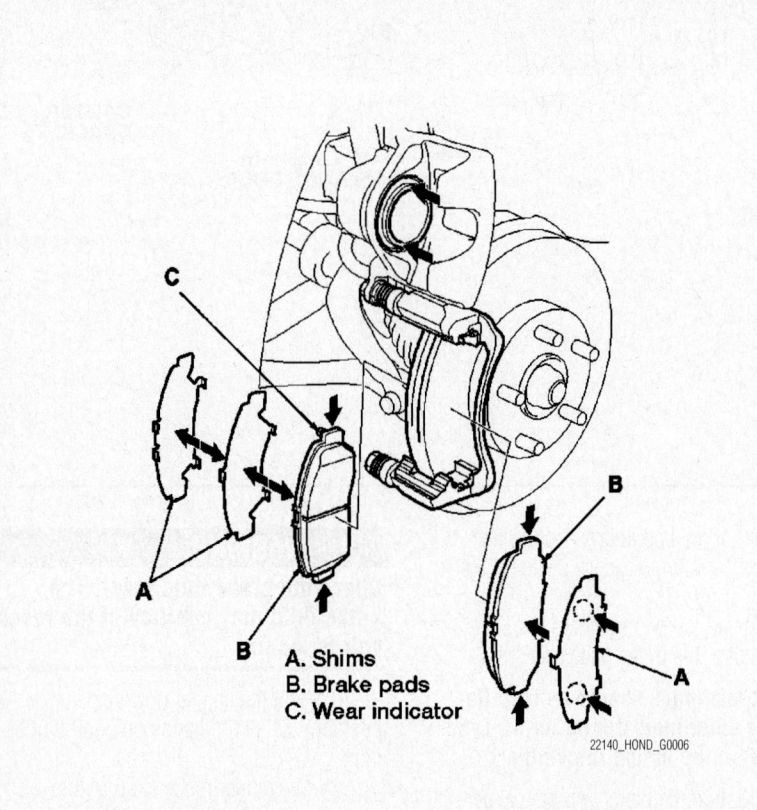

A. Shims
B. Brake pads
C. Wear indicator

22140_HOND_G0006

Fig. 6 Exploded view of the brake pad mounting—6-cylinder Accord

13. Install the brake hose mounting bolt and tighten to 16 ft. lbs. (22 Nm).

14. Install the front wheel and lower the vehicle.

15. Refill the master cylinder with brake fluid to the correct level.

Civic

1. Remove some brake fluid from the master cylinder.

2. Raise and safely support the vehicle.

3. Remove the front wheel.

4. Remove the flange bolt and pivot the caliper up out of the way.

5. Remove the pad shims and brake pads.

6. Remove the pad retainers.

To install:

7. Install the pad retainers.

8. Using a C-clamp or piston compression tool, press in the piston so the caliper will fit over the brake pads.

9. Apply a thin coat of M-77 assembly paste part number 08798-9010 to the brake pad sides of the pad shims and the back of the brake pads, wipe excess paste off the shim.

10. Install the brake pads and shims.

➡**The brake pads should be installed with the wear indicator on the upper inside.**

11. Pivot the caliper down tighten the flange bolt to 25 ft. lbs. (34 Nm).

12. Install the front wheel and lower the vehicle.

13. Refill the master cylinder with brake fluid to the correct level.

S2000

See Figure 7.

1. Remove and discard brake fluid from the master cylinder, as necessary.

2. Raise and support the vehicle safely. Remove the front tires.

3. Remove the caliper flange bolt.

4. Pivot the caliper assembly up and out of the way. Remove the brake pads and shims.

5. Remove the pad retainers and check the caliper pins for free movement.

To install:

6. Apply a thin coat of M-77 assembly paste part number 08798-9010 to the brake pad sides of the pad shims and the back of the brake pads, wipe excess paste off the shim.

7. Continue the installation in the reverse order of the removal procedure.

8. Check and refill the master cylinder, as necessary.

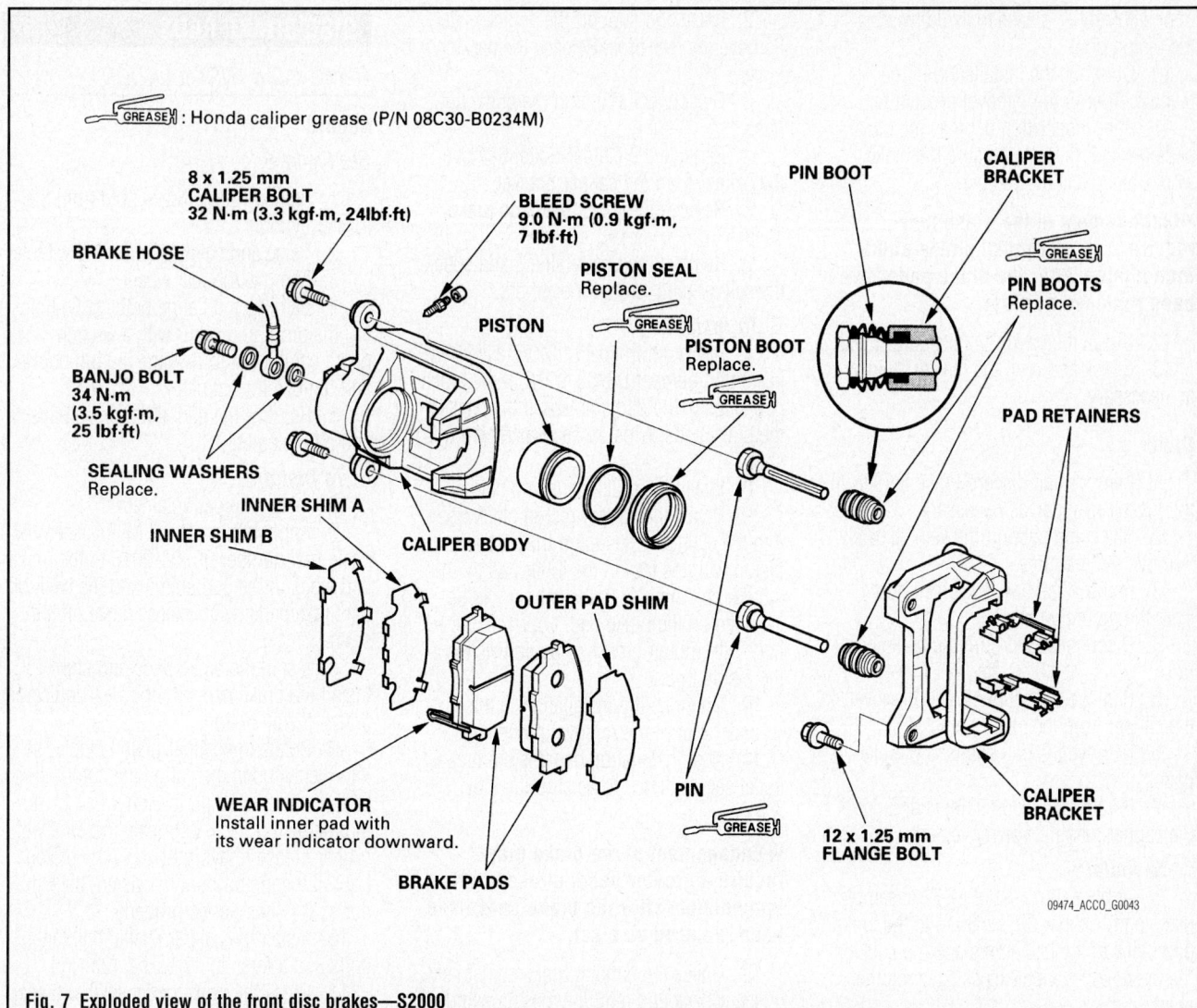

GREASE : Honda caliper grease (P/N 08C30-B0234M)

8 x 1.25 mm
CALIPER BOLT
32 N·m (3.3 kgf·m, 24lbf·ft)

BLEED SCREW
9.0 N·m (0.9 kgf·m,
7 lbf·ft)

PISTON SEAL
Replace.

BRAKE HOSE

PISTON

GREASE

PISTON BOOT
Replace.

GREASE

BANJO BOLT
34 N·m
(3.5 kgf·m,
25 lbf·ft)

SEALING WASHERS
Replace.

INNER SHIM A

INNER SHIM B

CALIPER BODY

OUTER PAD SHIM

PIN BOOT

CALIPER
BRACKET

GREASE

PIN BOOTS
Replace.

PAD RETAINERS

CALIPER
BRACKET

PIN

GREASE

12 x 1.25 mm
FLANGE BOLT

WEAR INDICATOR
Install inner pad with
its wear indicator downward.

BRAKE PADS

09474_ACCO_G0043

Fig. 7 Exploded view of the front disc brakes—S2000

BRAKES

REAR DISC BRAKES

✳✳ CAUTION

Dust and dirt accumulating on brake parts during normal use may contain asbestos fibers from production or aftermarket brake linings. Breathing excessive concentrations of asbestos fibers can cause serious bodily harm. Exercise care when servicing brake parts. Do not sand or grind brake lining unless equipment used is designed to contain the dust residue. Do not clean brake parts with compressed air or by dry brushing. Cleaning should be done by dampening the brake components with a fine mist of water, then wiping the brake components clean with a dampened cloth. Dispose of cloth and all residue containing asbestos fibers in

an impermeable container with the appropriate label. Follow practices prescribed by the Occupational Safety and Health Administration (OSHA) and the Environmental Protection Agency (EPA) for the handling, processing, and disposing of dust or debris that may contain asbestos fibers.

BRAKE CALIPER

REMOVAL & INSTALLATION

Accord

1. Remove and discard brake fluid from the master cylinder, as necessary.
2. Raise and support the vehicle safely. Remove the rear tires. Release the parking brake.

3. Disconnect and plug the brake line hose.
4. Remove the caliper bolts. Remove the caliper from the caliper bracket.
5. Remove the pad shims and brake pads.
6. Remove the pad retainers and check the caliper pins for free movement.

To install:

7. Apply a thin coat of M-77 assembly paste part number 08798-9010 to the brake pad sides of the pad shims and the back of the brake pads, wipe excess paste off the shim.
8. Install the brake pads and shims.
9. Rotate the caliper piston clockwise into the cylinder, then align the cutout in the piston with the tab on the inner pad by turning the piston back so the caliper can be installed on the brake pad. Lubricate the

boot with rubber grease to avoid twisting the piston boot.

10. Continue the installation in the reverse order of the removal procedure.

11. After installation depress the brake pedal several times to be sure the brake works. Road test the vehicle.

➡ **Engagement of the brake may require a greater pedal stroke effort immediately after the brake pads have been replaced as a set.**

12. Check the parking brake adjustment.

13. Check and refill the master cylinder, as necessary.

Civic

1. Remove and discard brake fluid from the master cylinder, as necessary.

2. Raise and support the vehicle safely. Remove the rear tires.

3. Remove the bolt and brake hose from the mounting bracket.

4. Disconnect and plug the brake line hose.

5. Remove the caliper bolts. Remove the caliper from the caliper bracket.

6. Remove the pad shims and brake pads.

7. Remove the pad retainers and check the caliper pins for free movement.

To install:

8. Apply a thin coat of M-77 assembly paste part number 08798-9010 to the brake pad sides of the pad shims and the back of the brake pads, wipe excess paste off the shim.

9. Install the brake pads and shims.

10. Rotate the caliper piston clockwise into the cylinder, then align the cutout in the piston with the tab by turning the piston back so the caliper can be installed on the brake pad. Lubricate the boot with rubber grease to avoid twisting the piston boot.

11. Continue the installation in the reverse order of the removal procedure.

12. After installation depress the brake pedal several times to be sure the brake works. Road test the vehicle.

➡ **Engagement of the brake may require a greater pedal stroke effort immediately after the brake pads have been replaced as a set.**

13. Check and refill the master cylinder, as necessary.

14. Check the parking brake adjustment.

S2000

1. Remove and discard brake fluid from the master cylinder, as necessary.

2. Raise and support the vehicle safely. Remove the rear tires. Release the parking brake.

3. Disconnect and plug the brake line hose.

4. Remove the caliper bolts. Remove the caliper from the caliper bracket.

5. Remove the pad shims and brake pads.

6. Remove the pad retainers and check the caliper pins for free movement.

To install:

7. Apply a thin coat of M-77 assembly paste part number 08798-9010 to the brake pad sides of the pad shims and the back of the brake pads, wipe excess paste off the shim.

8. Install the brake pads and shims.

9. Rotate the caliper piston clockwise into the cylinder, then align the cutout in the piston with the tab on the inner pad by turning the piston back so the caliper can be installed on the brake pad. Lubricate the boot with rubber grease to avoid twisting the piston boot.

10. Continue the installation in the reverse order of the removal procedure.

11. After installation depress the brake pedal several times to be sure the brake works.

➡ **Engagement of the brake may require a greater pedal stroke effort immediately after the brake pads have been replaced as a set.**

12. Check the parking brake adjustment.

13. Check and refill the master cylinder, as necessary.

DISC BRAKE PADS

REMOVAL & INSTALLATION

Accord

See Figure 8.

1. Remove some brake fluid from the master cylinder.

2. Raise and safely support the vehicle.

3. Remove the rear wheel.

4. Remove the flange bolts while holding the pin A and pin B with a wrench, being careful not to damage the pin boots, and remove the caliper.

5. Remove the pad shim, brake pads and pad retainers.

To install:

6. Install the pad retainers.

7. Apply a thin coat of M-77 assembly paste part number 08798-9010 to the brake pad sides of the pad shims and the back of the brake pads, wipe excess paste off the shim.

8. Install the brake pads and shim. Install the brake pad with the wear indicator on the inside bottom.

9. Rotate the caliper piston clockwise into the cylinder, then align the cutout in the piston with the tab on the inner pad by turning the piston back. Lubricate the boot with rubber grease to avoid twisting the piston boot. If the piston boot is twisted, back it out so it is positioned properly.

10. Install the caliper. Install the flange bolts and tighten to 17 ft. lbs. (23 Nm).

11. Press the brake pedal several times to make sure the brakes work.

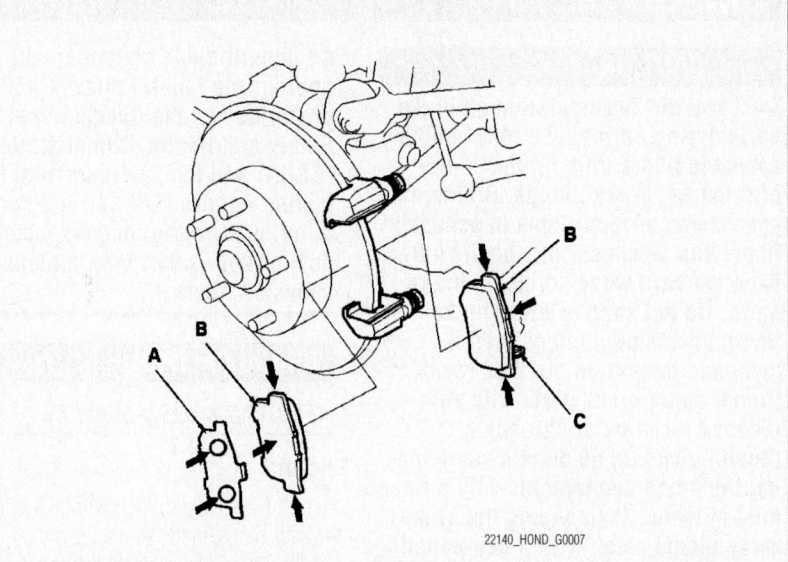

Fig. 8 Install the shim (A) and brake pads (B) correctly, with the wear indicator (C) on the bottom—2007–08 Accord

22140_HOND_G0007

➡Engagement of the brake may require a greater pedal stroke immediately after the brake pads have been replaced as a set. Several applications of the brake pedal will restore the normal pedal stroke.

12. The remainder of the installation is the reverse order of removal.

13. Add brake fluid as needed.

Civic

1. Remove some brake fluid from the master cylinder.

2. Raise and safely support the vehicle.

3. Remove the rear wheel.

4. Remove the brake hose mounting bolt.

5. Remove the flange bolts while holding respective caliper pin with a wrench. Be careful not to damage the pin boot, and remove the caliper.

6. Remove the pad shims, brake pads and pad retainers.

To install:

7. Install the pad retainers.

8. Apply a thin coat of M-77 assembly paste part number 08798-9010 to the brake pad sides of the pad shims and the back of the brake pads, wipe excess paste off the shim.

9. Install the brake pads and shim. Install the brake pad with the wear indicator on the inside bottom.

10. Rotate the caliper piston clockwise into the cylinder, then align the cutout in the piston with the tab on the inner pad by turning the piston back. Lubricate the boot with rubber grease to avoid twisting the piston boot. If the piston boot is twisted, back it out so it is positioned properly.

11. Install the caliper. Install the flange bolts and tighten to 17 ft. lbs. (23 Nm).

12. Press the brake pedal several times to make sure the brakes work.

➡Engagement of the brake may require a greater pedal stroke immediately after the brake pads have been replaced as a set. Several applications of the brake pedal will restore the normal pedal stroke.

13. The remainder of the installation is the reverse order of removal. Tighten the brake hose mounting bolt to 16 ft. lbs. (22 Nm).

14. Add brake fluid as needed.

S2000

See Figure 9.

1. Remove some brake fluid from the master cylinder.

2. Raise and safely support the vehicle.

3. Remove the rear wheel.

4. Remove the brake hose mounting bolt.

5. Remove the flange bolts while holding respective caliper pin with a wrench. Be careful not to damage the pin boot, and remove the caliper.

6. Remove the pad shims, brake pads and pad retainers.

To install:

7. Install the pad retainers.

8. Apply a thin coat of M-77 assembly paste part number 08798-9010 to the brake pad sides of the pad shims and the back of the brake pads, wipe excess paste off the shim.

9. Install the brake pads and shims. Install the brake pad with the wear indicator on the upper inside.

10. Rotate the caliper piston clockwise into the cylinder, then align the cutout in the piston with the tab on the inner pad by turning the piston back. Lubricate the boot with rubber grease to avoid twisting the piston boot. If the piston boot is twisted, back it out so it is positioned properly.

11. Install the caliper. Install the flange bolts and tighten to 16 ft. lbs. (22 Nm).

12. Press the brake pedal several times to make sure the brakes work.

➡Engagement of the brake may require a greater pedal stroke immediately after the brake pads have been replaced as a set. Several applications of the brake pedal will restore the normal pedal stroke.

13. The remainder of the installation is the reverse order of removal.

14. Add brake fluid as needed.

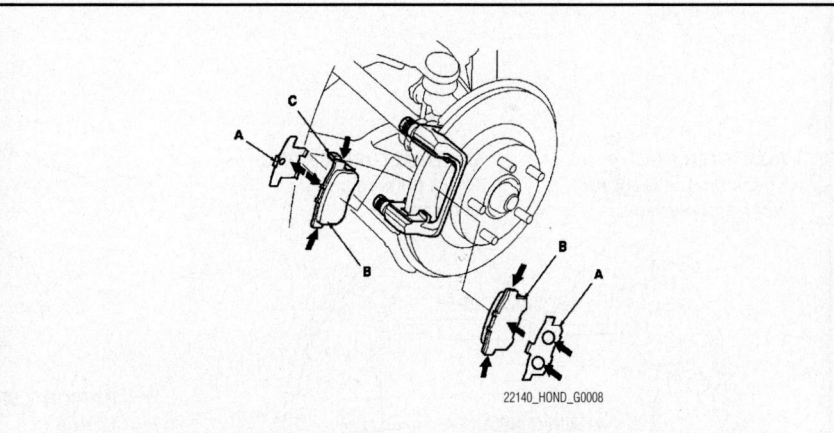

22140_HOND_G0008

Fig. 9 Install the shim (A) and brake pads (B) correctly, with the wear indicator (C) on the upper inside—2007–08 S2000

BRAKES   REAR DRUM BRAKES

❄❄ CAUTION

Dust and dirt accumulating on brake parts during normal use may contain asbestos fibers from production or aftermarket brake linings. Breathing excessive concentrations of asbestos fibers can cause serious bodily harm. Exercise care when servicing brake parts. Do not sand or grind brake lin-

ing unless equipment used is designed to contain the dust residue. Do not clean brake parts with compressed air or by dry brushing. Cleaning should be done by dampening the brake components with a fine mist of water, then wiping the brake components clean with a dampened cloth. Dispose of cloth and all residue containing asbestos fibers in an

impermeable container with the appropriate label. Follow practices prescribed by the Occupational Safety and Health Administration (OSHA) and the Environmental Protection Agency (EPA) for the handling, processing, and disposing of dust or debris that may contain asbestos fibers.

BRAKE DRUM

REMOVAL & INSTALLATION

1. Raise and safely support the vehicle.
2. Remove the rear tire.
3. Release the parking brake.
4. Remove the brake drum.

➡**If necessary, turn the adjuster bolt with a flat-blade screwdriver until the shoes become loose.**

➡**If the brake drum is stuck to the hub bearing unit, thread two 8 x 1.25 mm bolts into the brake drum to push it away from the hub bearing unit. Turn each bolt 90 degrees at a time to prevent the brake drum from binding.**

To install:

5. Installation is the reverse of the removal procedure.
6. Make certain the brake shoes are adjusted to allow the drum clearance during installation.

BRAKE SHOES

REMOVAL & INSTALLATION

See Figure 10.

1. Raise and safely support the vehicle.
2. Remove the brake drum. For additional information, refer to the following section, "Brake Drum, Removal & Installation."
3. Remove the tension pins by pushing the respective retainer spring and turning the pin.
4. Remove the lower return spring, and remove the brake shoe assembly over the hub.
5. Remove the forward brake shoe by removing the upper return spring, and disassemble the brake shoe assembly.
6. Remove the rearward brake shoe by disconnecting the parking brake cable from the parking brake lever.
7. Remove the U-clip, wave washer, and pivot pin, and separate the parking brake lever from the brake shoe.

To install:

8. Apply Molykote 44MA to the sliding surface of the pivot pin for the rearward brake shoe.
9. Install the parking brake lever and the wave washer on the pivot pin, and secure with a new U-clip.

➡**Pinch the U-clip securely to prevent the parking brake lever from coming out of the brake shoe.**

10. Connect the parking brake cable to the parking brake lever.
11. Apply a thin coat of Molykote 44MA to the connecting rod ends, and the sliding surfaces. Wipe off any excess. Keep grease off the brake linings.
12. Apply a thin coat of Molykote 44MA to the shoe ends and to the edge of the shoe surfaces that make contact with the backing plate. Wipe off any excess. Keep grease off the brake linings.
13. Install connecting rods A and B on the adjuster bolt.

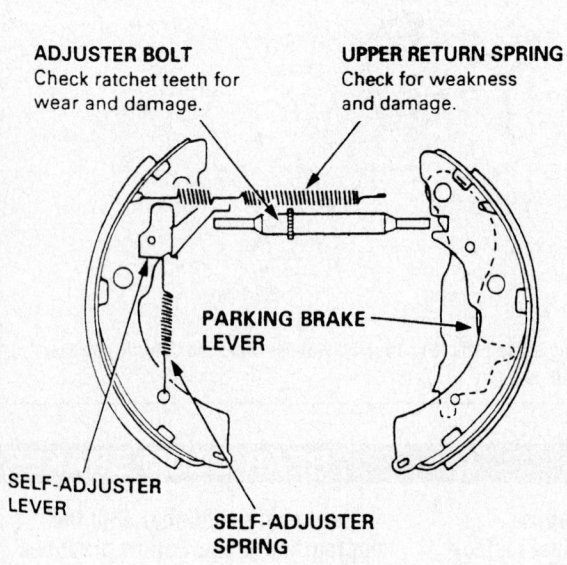

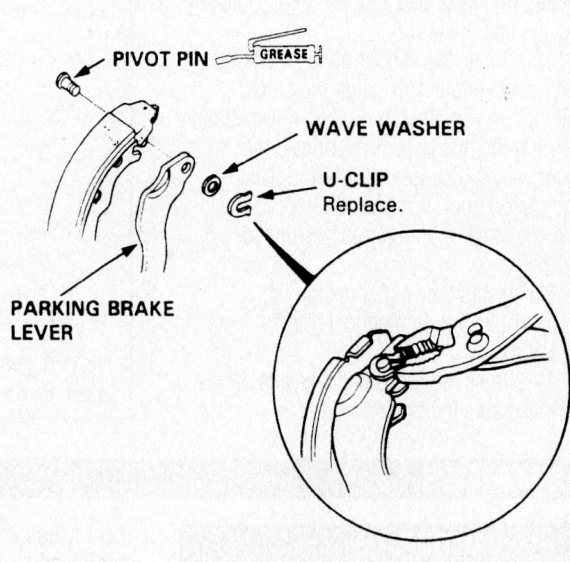

Fig. 10 Exploded view of the rear drum brakes—2007–08 Civic, Accord similar

➡**Clean the threaded portions of connecting rod A and the sliding surface of connecting rod B, then coat them with Molykote 44MA.**

➡**Shorten connecting rod A by fully turning the adjuster bolt.**

14. Assemble the brake shoes, the upper return spring, and the connecting rods with the adjuster bolt against the backing plate, then install the self-adjuster lever and the self-adjuster spring on the forward brake shoe.

15. Install the tension pins and the retainer springs by pushing in respective spring and turning each pin.

16. Install the lower return spring.

➡**Make sure the brake shoes positioning on the brake shoe bosses of the backing plate, and fitting the top of the brake shoes onto the wheel cylinder pistons.**

17. Install the brake drum.

➡**Before installing the brake drum, clean the mating surfaces between the rear hub and the inside of the brake drum.**

18. Clean the mating surfaces between the brake drum and the inside of the wheel, then install the rear wheels.

19. Press the brake pedal several times to make sure the brakes work and to set the self-adjusting brake.

➡**Engagement of the brakes may require a greater pedal stroke immediately after the brake shoes have been replaced as a set. Several applications of the brake pedal will restore the normal pedal stroke.**

20. Adjust the parking brake.

ADJUSTMENT

1. Remove the center console panel.
2. Release the parking brake lever fully.

3. Loosen the parking brake adjusting nut.

4. Press the brake pedal several times to set the self-adjusting brake before adjusting the parking brake.

5. Pull the parking brake lever 1 click. Raise the rear of the vehicle, and support it with safety stands in the proper locations.

6. Tighten the parking brake adjusting nut until the parking brakes drag slightly when the rear wheels are turned.

7. Release the parking brake lever fully, and check that the parking brakes do not drag when the rear wheels are turned. Readjust if necessary.

8. Make sure the parking brakes are fully applied when the parking brake lever is pulled all the way (8 to 10 clicks).

9. Install the center console panel.

BRAKES

PARKING BRAKE CABLES

ADJUSTMENT

Accord

1. Raise and safely support the vehicle.
2. Remove the center console.
3. Release the parking brake lever fully.
4. Loosen the parking brake adjusting nut and remove the rear wheels.

5. Make sure the parking brake arm on the rear brake caliper contacts the brake caliper pin.

➡**The parking brake arm will only contact the brake caliper pin when the parking brake adjusting nut is loosened.**

6. Pull the parking brake lever up one click.

PARKING BRAKE

7. Tighten the adjusting nut until the parking brakes drag slightly when the rear wheels are turned.

8. Release the parking brake lever fully, and check that the parking brakes do not drag when the rear wheels are turned. Readjust if necessary.

9. Pull the parking brake all the way up, and make sure the parking brakes are fully applied.

10. Reinstall the center console.

CHASSIS ELECTRICAL

GENERAL INFORMATION

✳✳ CAUTION

These vehicles are equipped with an air bag system. The system must be disarmed before performing service on, or around, system components, the steering column, instrument panel components, wiring and sensors. Failure to follow the safety precautions and the disarming procedure could result in accidental air bag deployment, possible injury and unnecessary system repairs.

SERVICE PRECAUTIONS

Disconnect and isolate the battery negative cable before beginning any airbag system component diagnosis, testing, removal,

AIR BAG (SUPPLEMENTAL RESTRAINT SYSTEM)

or installation procedures. Allow system capacitor to discharge for two minutes before beginning any component service. This will disable the airbag system. Failure to disable the airbag system may result in accidental airbag deployment, personal injury, or death.

Do not place an intact undeployed airbag face down on a solid surface. The airbag will propel into the air if accidentally deployed and may result in personal injury or death.

When carrying or handling an undeployed airbag, the trim side (face) of the airbag should be pointing towards the body to minimize possibility of injury if accidental deployment occurs. Failure to do this may result in personal injury or death.

Replace airbag system components with OEM replacement parts. Substitute parts may appear interchangeable, but internal

differences may result in inferior occupant protection. Failure to do so may result in occupant personal injury or death.

Wear safety glasses, rubber gloves, and long sleeved clothing when cleaning powder residue from vehicle after an airbag deployment. Powder residue emitted from a deployed airbag can cause skin irritation. Flush affected area with cool water if irritation is experienced. If nasal or throat irritation is experienced, exit the vehicle for fresh air until the irritation ceases. If irritation continues, see a physician.

Do not use a replacement airbag that is not in the original packaging. This may result in improper deployment, personal injury, or death.

The factory installed fasteners, screws and bolts used to fasten airbag components have a special coating and are specifically designed for the airbag system. Do not use substitute fasteners. Use only original equipment fasteners listed in the parts catalog when fastener replacement is required.

During, and following, any child restraint anchor service, due to impact event or vehicle repair, carefully inspect all mounting hardware, tether straps, and anchors for proper installation, operation, or damage. If a child restraint anchor is found damaged in any way, the anchor must be replaced. Failure to do this may result in personal injury or death.

Deployed and non-deployed airbags may or may not have live pyrotechnic material within the airbag inflator.

Do not dispose of driver/passenger/curtain airbags or seat belt tensioners unless you are sure of complete deployment. Refer to the Hazardous Substance Control System for proper disposal.

Dispose of deployed airbags and tensioners consistent with state, provincial, local, and federal regulations.

After any airbag component testing or service, do not connect the battery negative cable. Personal injury or death may result if the system test is not performed first.

If the vehicle is equipped with the Occupant Classification System (OCS), do not connect the battery negative cable before performing the OCS Verification Test using the scan tool and the appropriate diagnostic information. Personal injury or death may result if the system test is not performed properly.

Never replace both the Occupant Restraint Controller (ORC) and the Occupant Classification Module (OCM) at the same time. If both require replacement, replace one, then perform the Airbag System test before replacing the other.

Both the ORC and the OCM store Occupant Classification System (OCS) calibration data, which they transfer to one another when one of them is replaced. If both are replaced at the same time, an irreversible fault will be set in both modules and the OCS may malfunction and cause personal injury or death.

If equipped with OCS, the Seat Weight Sensor is a sensitive, calibrated unit and must be handled carefully. Do not drop or handle roughly. If dropped or damaged, replace with another sensor. Failure to do so may result in occupant injury or death.

If equipped with OCS, the front passenger seat must be handled carefully as well. When removing the seat, be careful when setting on floor not to drop. If dropped, the sensor may be inoperative, could result in occupant injury, or possibly death.

If equipped with OCS, when the passenger front seat is on the floor, no one should sit in the front passenger seat. This uneven force may damage the sensing ability of the seat weight sensors. If sat on and damaged, the sensor may be inoperative, could result in occupant injury, or possibly death.

DISARMING THE SYSTEM

Disconnect and isolate the negative battery cable. Wait 3 minutes for the system capacitor to discharge before performing any service.

ARMING THE SYSTEM

Connect the negative battery cable.

CLOCKSPRING CENTERING

See Figure 11.

1. Rotate the cable reel clockwise until it stops.
2. Then rotate it counterclockwise (about 2½–3 turns) until the arrow mark on the cable reel label points straight up.

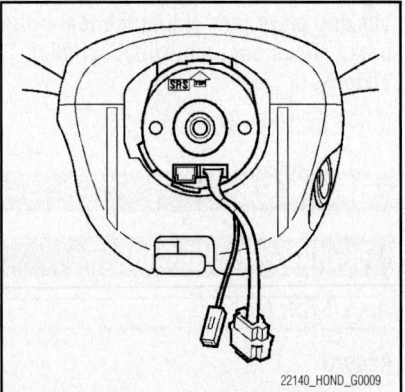

22140_HOND_G0009

Fig. 11 Ensure the arrow mark is facing straight up to center the clockspring—Civic shown, others similar

DRIVETRAIN

AUTOMATIC TRANSAXLE ASSEMBLY

REMOVAL & INSTALLATION

Accord

2007 2.4L Engine

See Figures 12 through 17.

1. Make sure you have the anti-theft code for the radio and the navigation system, then write down the frequencies for the radio's preset buttons.
2. Set the steering wheel in the straight-ahead position, and lock in position.
3. Remove the steering joint cover.
4. Make a reference mark across the steering joint and steering gearbox pinion shaft. Remove the steering joint bolt, and disconnect the steering joint by removing the steering joint toward the steering column. Hold the slider shaft on the column

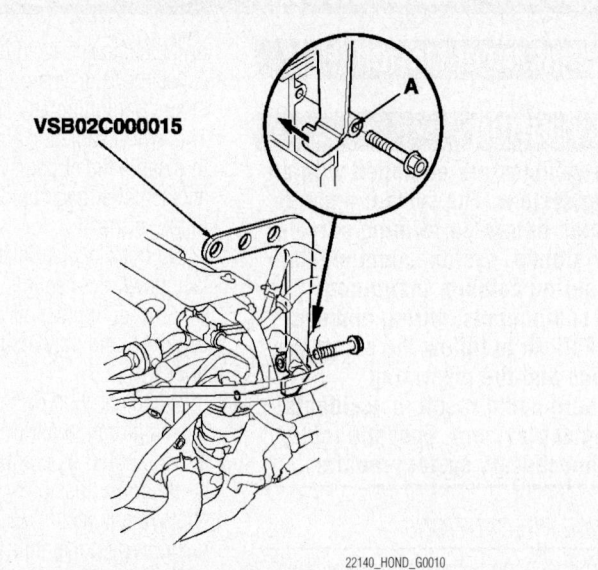

VSB02C000015

22140_HOND_G0010

Fig. 12 Attach the special tool adapter (VSB02C000015) to the threaded hole (A) in the cylinder head—2007 2.4L Accord

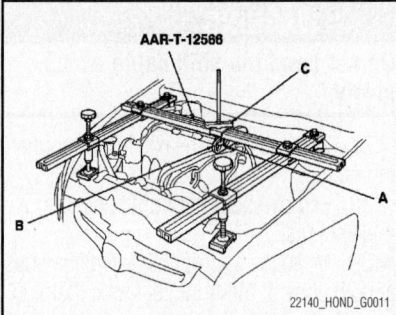

Fig. 13 Install the engine support hanger and attach the hook (A) to the special tool adapter (B). Tighten the wing nut (C) by hand—2007 2.4L Accord

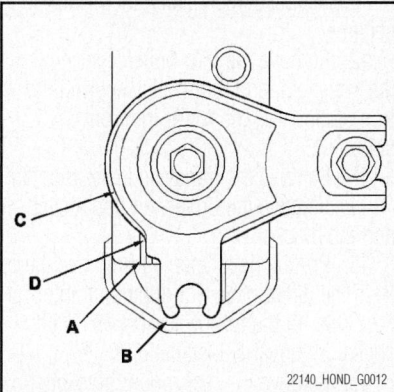

Fig. 14 Matchmark (A) both ends of the front subframe (B) that line up with the edge (C) of the stiffeners (D)—2007 2.4L Accord

with a piece of wire between the joint yoke on the slider shaft to the joint yoke on the upper shaft.

5. Remove and plug the return hose from the power steering fluid reservoir.

6. Remove the power steering pump outlet line from the power steering pump.

7. Remove the bolt securing the power steering hose clamp.

8. Raise the vehicle, and make sure it is supported securely.

9. Remove the splash shield.

10. Remove the drain plug, and drain the automatic transmission fluid (ATF).

11. Reinstall the drain plug with a new sealing washer.

12. Disconnect the battery negative terminal, then disconnect the battery positive terminal.

13. Remove the battery hold-down bracket, then remove the battery cover, the battery, and the battery tray.

14. Remove the intake air duct and the air cleaner assembly.

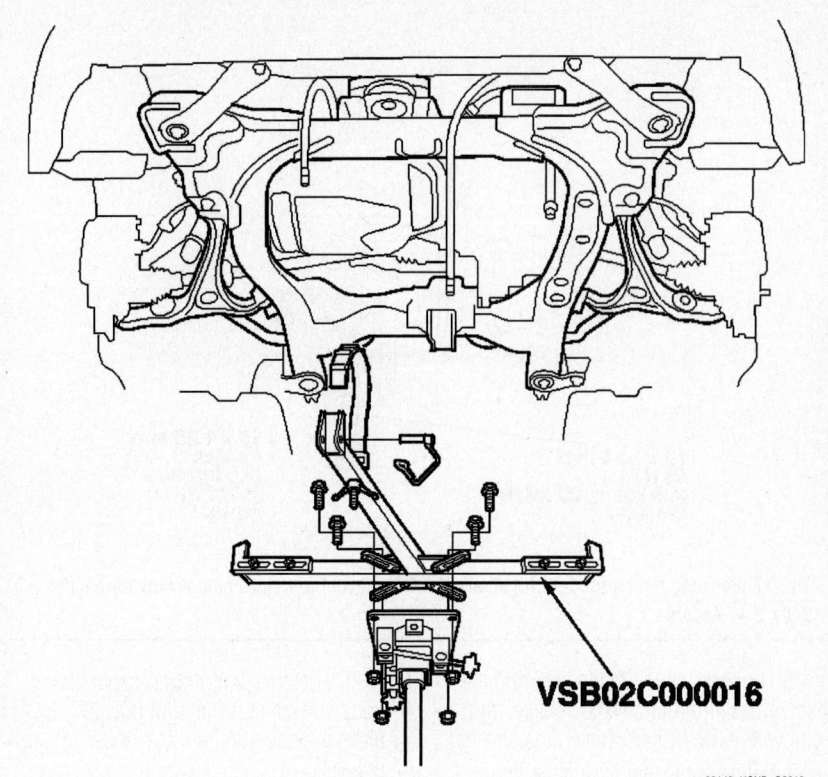

Fig. 15 Attach the special tool to the front subframe with hanging the belt of the special tool over the front of the subframe—2007 2.4L Accord

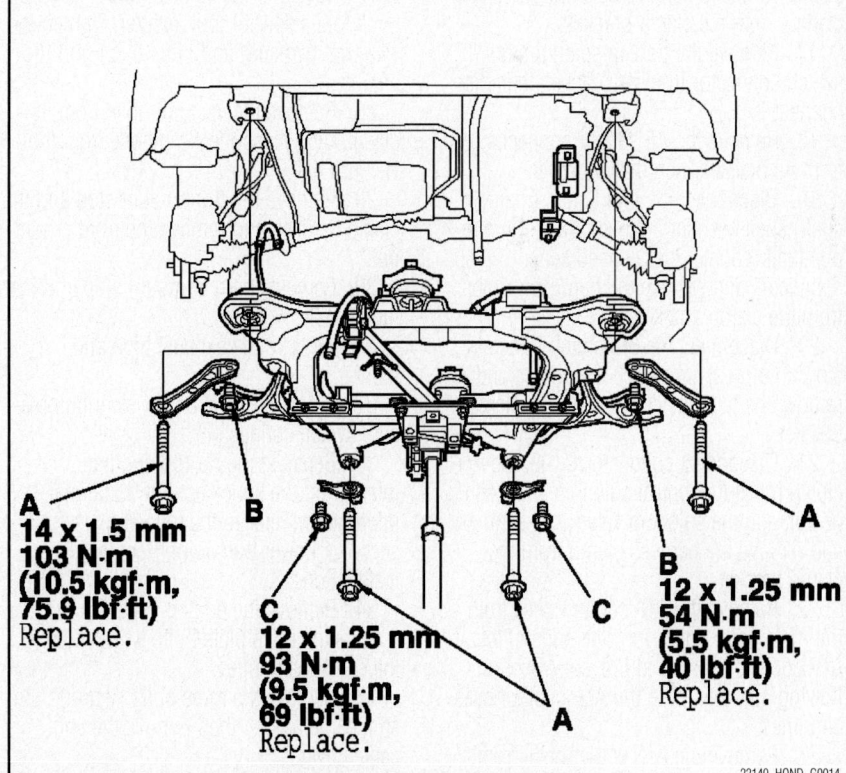

Fig. 16 Showing the torque specifications of the front subframe mounting bolts (A) and stiffener mounting bolts (B, C)—2007 2.4L Accord

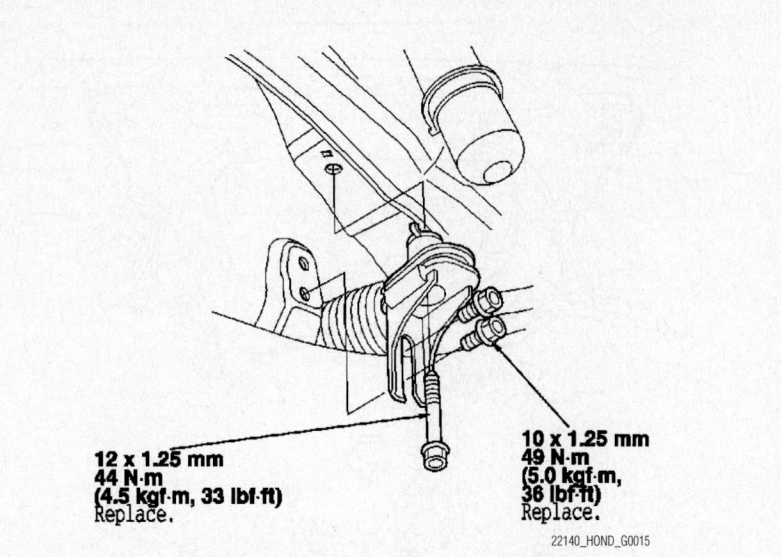

12 x 1.25 mm
44 N·m
(4.5 kgf·m, 33 lbf·ft)
Replace.

10 x 1.25 mm
49 N·m
(5.0 kgf·m,
36 lbf·ft)
Replace.

22140_HOND_G0015

Fig. 17 Showing the torque specifications of the front subframe mid-mount mounting bolts—2007 2.4L Accord

15. Loosen the two bolts securing the battery base from under the vehicle, and remove the two bolts securing the battery base in the engine compartment, then remove the battery base.

16. Disconnect the A/T clutch pressure control solenoid valve connector and the 2nd clutch transmission fluid pressure switch connector, and remove the harness clamps from the clamp brackets.

17. Remove the transmission range switch connector from its bracket, then disconnect it.

18. Remove the AF sensor connector from its bracket, then disconnect it.

19. Disconnect the input shaft (mainshaft) speed sensor connector and the output shaft (countershaft) speed sensor connector, and remove the harness clamps from the clamp brackets.

20. Disconnect the 3rd clutch transmission fluid pressure switch connector, and remove the harness clamp from the clamp bracket

21. Disconnect connector C108, the A/T clutch pressure control solenoid valve connector, and the solenoid valve connector, and remove the harness clamp from the clamp bracket.

22. Remove the ATF cooler hoses from the ATF cooler lines. Turn the ends of the ATF cooler hoses up to prevent ATF from flowing out, then plug the ATF cooler hoses and lines.

23. Remove the ATF cooler hose from the hose clamp.

24. Disconnect the ATF cooler hose from the ATF cooler line, then plug the hose end.

25. Remove the ground cable, the transmission upper mount bracket plate, and the transmission upper mount bracket.

26. Attach the special tool adapter VSB02C000015 to the threaded hole in the cylinder head.

27. Install the engine support hanger AAR-T-12566 to the vehicle, and attach the hook to the special tool adapter. Tighten the wing nut by hand, and lift and support the engine.

28. Remove the vacuum hose from its clamp, then disconnect the hose from the vacuum line.

29. Remove the front mount stop and the clamp bracket, and remove the front mount bolt.

30. Remove the rear mount stop and rear mount bolt.

31. Remove the exhaust pipe and mount.

32. Disconnect the power steering pressure switch connector.

33. Separate the tie-rod end ball joints from the knuckles. For additional information, refer to the following section, "Lower Ball Joint, Removal & Installation."

34. Remove the engine stiffener, and remove the drive plate bolts while rotating the crankshaft pulley.

35. Remove the three bolts securing the shift cable holder, then remove the shift cable cover.

36. Remove the spring clip and the control pin, then separate the shift cable from the selector control lever.

⁂ **WARNING**

Do not bend the shift cable excessively.

37. Remove the transaxle lower mounting nuts.

38. Remove both the front subframe mid-mounts.

39. Make the appropriate reference lines at both ends of the front subframe that line up with the edge of the stiffeners.

40. Attach Special Tool VSB02C000016 to the front subframe with hanging the belt of the special tool over the front of the subframe, then secure the belt with its stop.

41. Raise the jack and line up the slots in the arms with the bolt holes on the corner of the jack base, then attach them with bolts securely.

42. Remove the four bolts securing the stiffeners, and four bolts securing the front subframe, and lower the front subframe.

43. Remove the transaxle lower mounts.

44. Remove the driveshaft boot cover and the bracket.

45. Pry the halfshafts, and remove them from the differential. For additional information, refer to the following section, "Halfshafts, Removal & Installation."

46. Remove the rear mount bracket.

47. Place a suitable jack under the transaxle assembly.

48. Remove the upper transaxle housing mounting bolts.

49. Remove the front mount bracket, the remove the front housing mounting bolts.

50. Remove the rear transaxle housing mounting bolts.

51. Slide the transaxle assembly away from the engine to remove it from the vehicle.

To install:

52. Inspect the drive plate and replace if damaged.

53. Inspect the rubber of the mounts and replace if worn or damaged.

54. Place the transaxle assembly on the jack, and raise it to engine level.

55. Attach the transaxle to the engine assembly, then tighten the rear housing mounting bolts to 47 ft. lbs. (64 Nm).

56. Install the front housing mounting bolts and tighten to 47 ft. lbs. (64 Nm).

57. Install the upper housing mounting bolts and tighten to 47 ft. lbs. (64 Nm).

58. Remove the jack from the transaxle assembly.

59. Install the front mount bracket and, using new bolts, tighten to 47 ft. lbs. (64 Nm).

60. Install the rear mount bracket and tighten the bolts to 65 ft. lbs. (88 Nm).

61. Install the left and right halfshafts. For additional information, refer to the following section, "Halfshafts, Removal & Installation."

62. Install the transaxle lower mounts and tighten the bolts to 33 ft. lbs. (44 Nm).

63. Support the front subframe using a suitable jack on Special Tool VSB02C000016, and lift the assembly up to the body.

64. Loosely install the new front subframe mounting bolts, stiffener mounting bolts, and new subframe mid-mount mounting bolts.

65. Align the reference marks with edge of both rear stiffeners, and tighten the rear subframe mounting bolts, then front bolts, then the stiffener bolts, and then the front subframe mid-mount mounting to the specified torque.

66. Install the transaxle lower mount nuts and tighten to 33 ft. lbs. (44 Nm).

67. The remainder of the installation is the reverse order of removal.

68. Refill the power steering system with fluid to the correct level.

69. Refill the transaxle with fluid to the correct level.

70. Remove the steering wheel and center the clockspring. For additional information, refer to the following section, "AIR BAG (SUPPLEMENTAL RESTRAINT SYSTEM), Clockspring Centering."

71. Check and adjust the wheel alignment as necessary.

72. Start the engine and check for leaks.

2008 2.4L Engine

See Figures 18 through 22.

1. Secure the hood in the wide open position with the support rod.

2. Do the battery removal procedure.

3. Remove the front grille cover.

4. Remove the strut brace.

5. Remove the air cleaner assembly and the intake air duct.

6. Remove the nut securing the underhood fuse/relay box, and swing it out of the way.

7. Loosen the two bolts located behind the battery base, and remove the two bolts securing the battery base, then remove the battery base.

8. Raise the vehicle on a lift, and make sure it is securely supported.

9. Remove the front wheels.

10. Remove the splash shield.

11. Remove the drain plug and drain the automatic transmission fluid (ATF). Reinstall

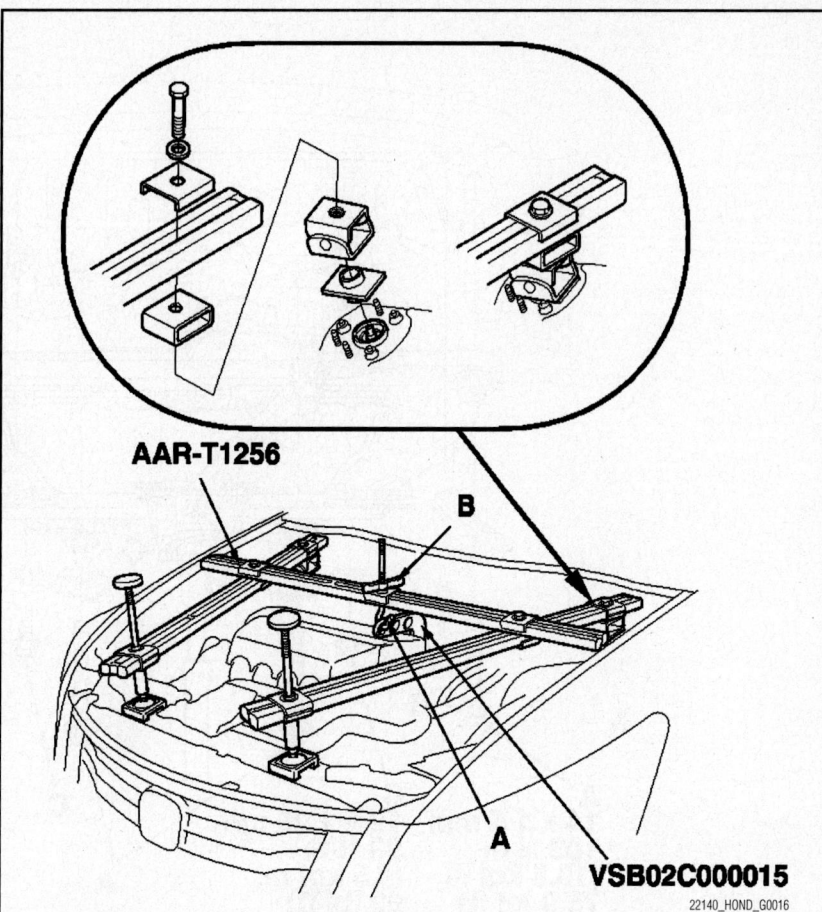

Fig. 18 Install the engine support hanger (AAR-T1256) to the vehicle, and attach the hook (A) to the slotted hole in the engine hanger adapter (VSB02C000015), and turn the wing nut (B) to lift—2008 2.4L Accord

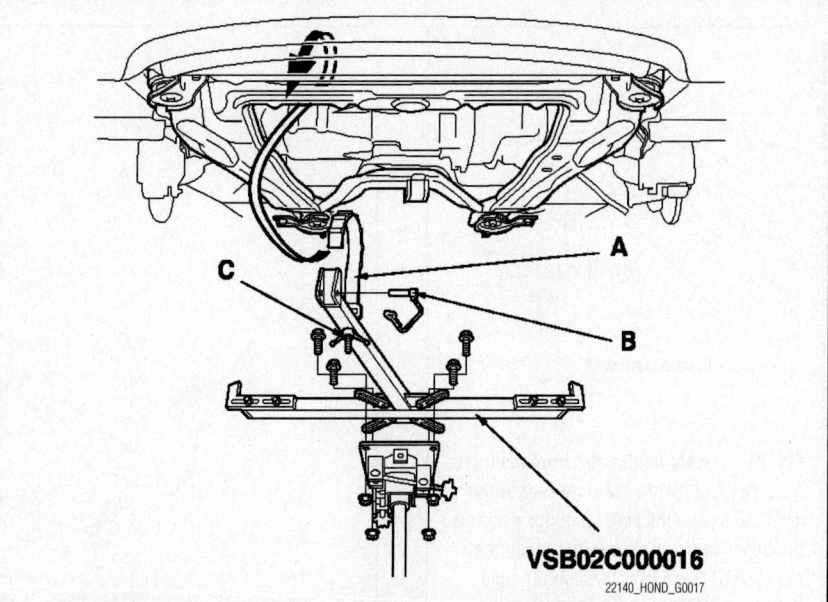

Fig. 19 Attach the front subframe adapter (VSB02C000016) to the front subframe by looping the strap (A) over the front of the front subframe, then secure the strap with the stop (B), then tighten the wing nut (C)—2007 2.4L Accord

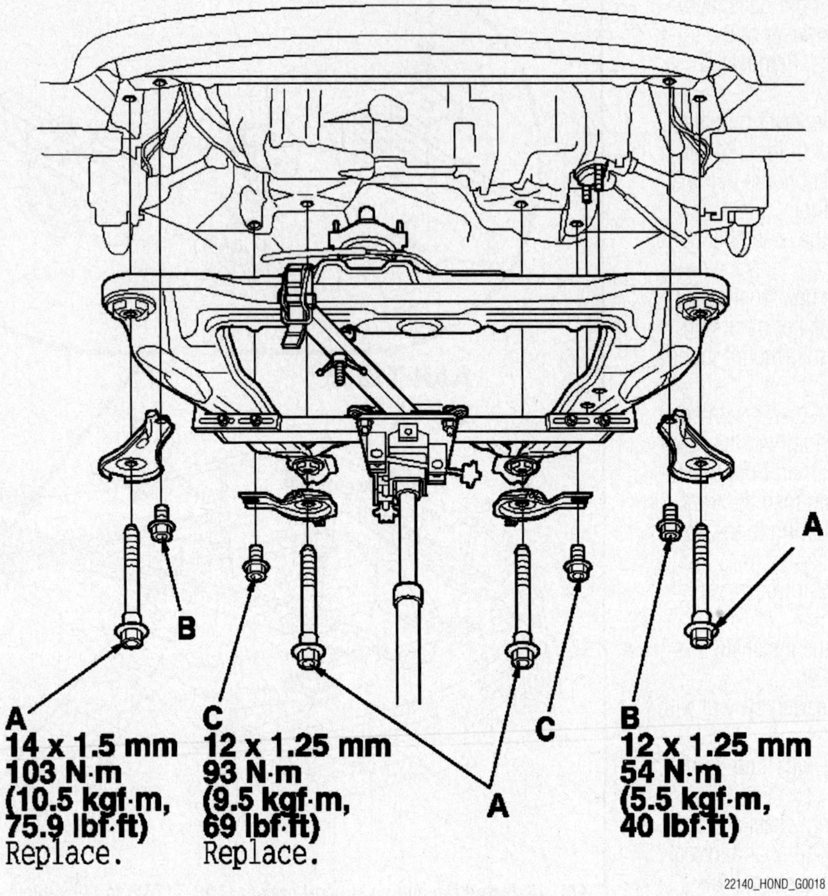

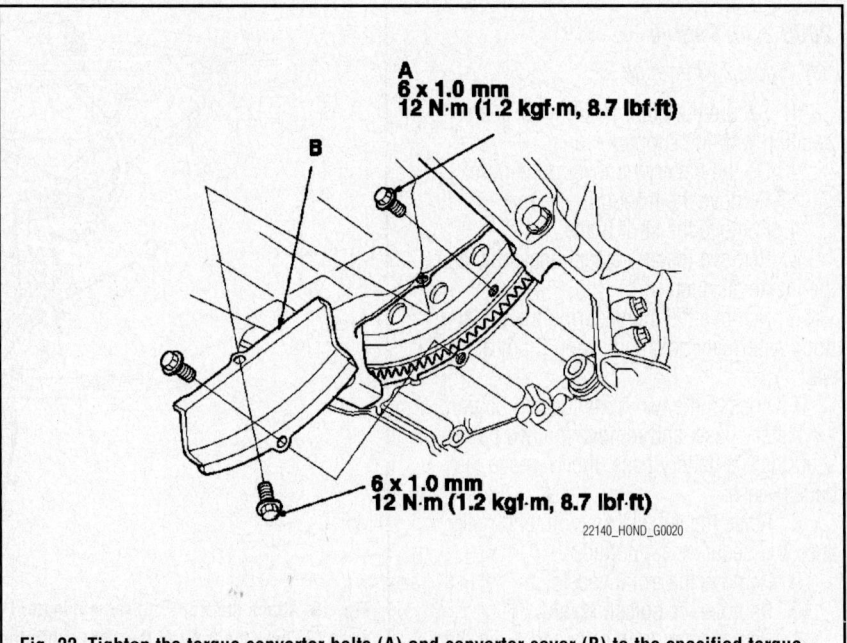

A
14 x 1.5 mm
103 N·m
(10.5 kgf·m,
75.9 lbf·ft)
Replace.

C
12 x 1.25 mm
93 N·m
(9.5 kgf·m,
69 lbf·ft)
Replace.

B
12 x 1.25 mm
54 N·m
(5.5 kgf·m,
40 lbf·ft)

22140_HOND_G0018

Fig. 20 Location of the front subframe mounting bolts (A), the stiffener mounting bolts (B), and new stiffener mounting bolts (C)—2008 2.4L Accord

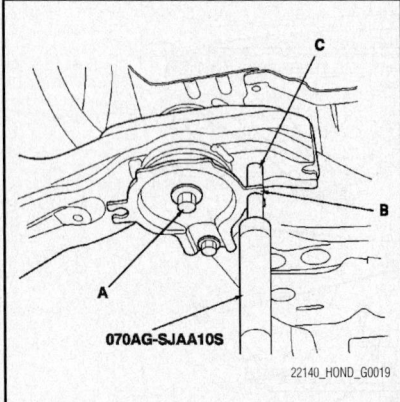

070AG-SJAA10S

22140_HOND_G0019

Fig. 21 Loosely tighten the front subframe mounting bolt (A) in the right rear stiffener until the front subframe insulator contacts the body; insert the subframe alignment pin (070AG-SJAA10S) through the positioning slot (B) on the right rear stiffener, through the positioning hole (C) on the front subframe, and into the positioning hole on the body.

A
6 x 1.0 mm
12 N·m (1.2 kgf·m, 8.7 lbf·ft)

B

6 x 1.0 mm
12 N·m (1.2 kgf·m, 8.7 lbf·ft)

22140_HOND_G0020

Fig. 22 Tighten the torque converter bolts (A) and converter cover (B) to the specified torque—2008 2.4L Accord

the drain plug with a new sealing washer. Tighten to 36 ft. lbs. (49 Nm).

12. Disconnect the A/T clutch pressure control solenoid valve connector and the 2nd clutch transmission fluid pressure switch connector, and remove the harness clamps from the clamp brackets.

13. Remove the transmission range switch sub-harness connector from its bracket and disconnect it.

14. Remove the A/F sensor connector from its bracket and disconnect it.

15. Disconnect the input shaft (mainshaft) speed sensor connector and the output shaft (countershaft) speed sensor connector.

16. Remove the harness cover mounting bolt, and remove the engine wire harness cover from the ATF filter bracket.

17. Disconnect the shift solenoid wire harness connector, the A/T clutch pressure control solenoid valve connector, the A/T clutch pressure control solenoid valve connector, and remove the harness clamp from the clamp bracket.

18. Remove the ATF cooler hoses from the ATF cooler lines. Turn the ends of the ATF cooler hoses up to prevent ATF from flowing out, and then plug the ATF cooler hoses and the lines.

➡**Check for any signs of leakage at the hose joints.**

19. Disconnect the 3rd clutch transmission fluid pressure switch connector, and remove the harness clamp from the clamp bracket.

20. Remove the vacuum hose.

21. Disconnect the ATF cooler hose from the ATF line. Turn the end of the ATF cooler hose up to prevent ATF from flowing out, and then plug the hose and line.

22. Remove the ATF cooler hose from the hose clamp.

23. Remove the upper transaxle mount bracket bolts.

✳✳ WARNING

Do not remove the TORX® bolt from the upper transmission mount. If the TORX® bolt is removed, the upper transmission mount must be replaced as an assembly.

24. Attach the engine hanger adapter VSB02C000015 to the threaded hole located on the rear side of the cylinder head.

25. Install the engine support hanger AAR-T1256 to the vehicle, and attach the hook to the slotted hole in the engine hanger adapter VSB02C000015. Tighten the wing nut by hand to lift and support the engine.

26. Remove the vacuum hose from its clamp.

27. Remove the front mount stop and the clamp bracket, and remove the front mount bolt

28. Remove the heat shield mounting bolts, then remove the heat shield.

29. Remove the steering gearbox mounting bracket bolts.

30. Remove the power steering hose clamp bolts.

31. Remove the rear engine mount bolts, and then remove the rear engine mount.

32. Remove the mounting bolts, and then remove the rear engine mount upper bracket.

33. Remove the steering gearbox stiffeners.

34. Remove the remaining power steering fluid line clamp bolt.

35. Remove the power steering fluid line from its clamp

36. Remove the exhaust pipe.

37. Separate the tie-rod end ball joints from the knuckles. For additional information, refer to the following sections, "Lower Ball Joint, Removal & Installation" and "Upper Ball Joint, Removal & Installation."

38. Remove the torque converter cover. Remove the eight drive plate bolts while rotating the crankshaft pulley.

39. Separate the shift cable from the selector control lever.

40. Remove the transaxle lower mounting nuts.

41. Remove both front subframe mid-mount bolts.

42. Hang the steering gearbox to the body with a strap.

43. Attach the front subframe adapter VSB02C000016 to the front subframe by looping the strap over the front of the front subframe, then secure the strap with the stop, then tighten the wing nut.

44. Raise a suitable jack and line up the slots in the arms with the bolt holes on the corner of the jack base, and then tighten the bolts.

45. Remove the four bolts securing the stiffeners, and remove the four bolts securing the front subframe, and then lower the front subframe.

46. Place a suitable jack under the transaxle assembly.

47. Remove the transaxle lower mount.

48. Remove the left side halfshaft from the differential and the right side halfshaft from the intermediate shaft. Coat all precision machined surfaces with clean engine oil, and then put plastic bags over the driveshaft ends. Hang the driveshafts to the body with a strap.

49. Remove the intermediate shaft. Coat all precision machined surfaces with clean engine oil, then put plastic bags over the intermediate shaft ends.

50. Remove the rear engine mount bracket.

51. Remove the front engine mount bracket.

52. Remove the jack.

53. Remove the upper transaxle housing mounting bolts.

54. Lower the transaxle by loosening the wing nut on the engine support hanger, and tilt the engine just enough for the transaxle to clear the side frame. Check that the transaxle is completely free of the ATF and coolant hoses, vacuum hoses, and the electrical wiring.

55. Place the jack under the transaxle.

56. Remove the front transmission housing mounting bolts.

57. Remove the crankshaft position (CKP) sensor cover.

58. Remove the rear transaxle housing mounting bolts.

✳✳ WARNING

Be careful not to damage the CKP sensor and the sensor harness.

59. Slide the transaxle assembly away from the engine to remove it from the vehicle.

To install:

60. Inspect the drive plate and replace if damaged.

61. Place the transaxle assembly on a suitable jack, and raise the transaxle to the engine level, then fit the transaxle to the engine assembly.

62. Install the rear transaxle housing mounting bolts and tighten to 47 ft. lbs. (64 Nm).

63. Install the CKP sensor cover and tighten to 108 inch lbs. (12 Nm).

64. Install the front transaxle housing mounting bolts and tighten to 47 ft. lbs. (64 Nm).

65. Remove the jack.

66. Install the upper transaxle housing mounting bolts and tighten to 47 ft. lbs. (64 Nm).

67. Place the jack under the transmission.

68. Install the front engine mount bracket with new bolts, and tighten to 47 ft. lbs. (64 Nm).

69. Install the rear engine mount bracket with new bolts, and tighten to 65 ft. lbs. (88 Nm).

70. Install the intermediate shaft.

71. Connect the left and right halfshafts. For additional information, refer to the following section, "Halfshaft, Removal & Installation."

72. Install the transaxle lower mounts with new bolts, and tighten to 40 ft. lbs. (54 Nm).

73. Set the front subframe adapter (VSB02C000016) to the front subframe by looping the strap over the front of the front subframe, then secure the strap with the stop, then tighten the wing nut.

74. Raise the front subframe up to the body, then loosely install new front subframe mounting bolts, the stiffener mounting bolts, and new stiffener mounting bolts.

75. Loosely tighten the front subframe mounting bolt in the right rear stiffener until the front subframe insulator contacts the body; insert the subframe alignment pin 070AG-SJAA10S through the positioning slot on the right rear stiffener, through the positioning hole on the front subframe, and into the positioning hole on the body.

76. Loosely tighten the front subframe mounting bolt in the left rear stiffener in the same manner.

77. Reinsert the subframe alignment pin through the positioning slot on the right rear stiffener, through the positioning hole on the front subframe, and into the positioning hole on the body, then tighten the front subframe mounting bolt to the specified torque.

78. Tighten the subframe mounting bolt in the left rear stiffener in the same manner.

79. Tighten the front subframe mounting bolts in the right front stiffener and the left front stiffener to the specified torque.

80. Check that the positioning holes and slots are aligned using the subframe alignment pin.

81. Tighten the rear and front stiffener mounting bolts to the specified torque.

82. Remove the jack and the front subframe adapter.

83. Replace the front subframe midmount mounting bolts and tighten to 36 ft. lbs. (49 Nm).

84. Install the transaxle mounting nuts and tighten to 33 ft. lbs. (44 Nm).

85. Connect the shift cable to the selector.

86. Attach the torque converter to the drive plate with the eight bolts. Rotate the crankshaft pulley as necessary to tighten the bolts to ½ of the specified torque, then to the final torque, in a crisscross pattern. After tightening the last bolt, check that the crankshaft rotates freely.

87. Install the torque converter cover.

88. The remainder of the installation is the reverse order of removal.

89. Refill the transaxle with fluid to the correct level.

90. Check and adjust the wheel alignment as necessary.

91. Start the engine and check for leaks.

Civic, Except Hybrid

1. The radio may contain a coded theft protection circuit. Always obtain the code number before disconnecting the battery.

2. Remove the cowl cover and the under cowl panel. Remove the front grille cover.

3. Disconnect the negative battery cable. Disconnect the positive battery cable. Remove the battery.

4. Remove the air cleaner housing and air intake duct. Remove the battery base and resonator.

Raise and support the vehicle safely. Remove the splash shield. Drain the transaxle fluid.

5. Lower the vehicle. Position the hood in the vertical position.

6. Remove the mounting bolts securing the harness cover and remove the harness clamp.

7. Disconnect the transaxle pressure control switch solenoid valve, solenoid valve B connector and solenoid valve C connector.

8. Remove the harness clamp from its bracket. Remove the air cleaner housing mounting bracket. Disconnect the transaxle range switch connector.

9. Disconnect the output shaft speed sensor connector and remove the harness clamps from the clamp brackets.

10. Disconnect the input shaft speed sensor connector and the second clutch transaxle fluid pressure switch connector.

11. Disconnect the ATF warmer hoses from the transaxle fluid lines. Plug the hoses and lines to prevent fluid leakage.

12. Remove the bolts securing the ATF warmer. Do not disconnect the ATF hoses and water by pass hose from the ATF warmer.

13. Disconnect the shift solenoid harness connector and the third clutch transaxle fluid pressure switch connector. Remove the harness clamps from the clamp brackets.

14. Remove the harness clamp from its bracket. Remove the radiator hose from the clamp. Remove the air cleaner hose mounting bracket.

15. Install the support eyelet part number 07AAK-SNAA400, or equivalent behind

the breather pipe and down to the threaded hole on the cylinder head. Attach another support eyelet to the cylinder head with the support bolt part number 07AAK-SNAA500, or equivalent. Hand tighten the bolt.

16. Install tool VSB02C000025, or equivalent onto the engine hanger. Carefully position the engine hanger on the vehicle and support the engine and transaxle assembly.

17. Remove the nuts and bolt securing the lower arm and ball joint. Separate the lower arms from the ball joints. Remove both body mount brackets. Remove the steering gearbox mounting bracket bolts.

18. Remove the rear steering gearbox mounting bolt, stiffener mounting bolt and stiffener. Remove the bolt securing the power steering fluid line clamp bracket.

19. Remove the steering gearbox mounting bolt, stiffener mounting bolt and stiffener. Remove the bolt securing the power steering fluid line bracket on the right of the front subframe. Remove the power steering line from the clamp. Remove the lower torque rod bolts.

20. Make reference marks on the front subframe that line up with the edge of the body. Attach the front subframe adapter tool, VSB02000016 or equivalent, to the subframe. Secure the strap with the stop and tighten the wing nut. Raise the jack and line up the slots in the arms with the bolt holes on the corner of the jack base. Tighten the bolts.

21. Remove the four bolts securing the front subframe. Lower the subframe. Hang the steering gearbox to the side, with rope or wire.

22. Remove the driveshafts from the differential. Remove the shift cable cover. Remove the three bolts securing the shift cable holder. Pry up on the lock tab of the lock washer. Remove the lock bolt and lock washer. Separate the shift cable from the control shaft.

23. Remove the shift cable holder bracket from the transaxle. Remove the torque converter cover. Remove the drive plate bolts. Remove the upper transaxle housing mounting bolts.

24. Remove the transaxle mount bracket bolts. Remove the front transaxle housing mounting bolts. Remove the rear transaxle housing mounting bolts.

25. Lower the transaxle assembly by loosening the wing nut of the engine support hanger. Lift the engine just enough for the transaxle to clear the side frame.

26. Position a jack under the transaxle. Remove the lower transaxle housing mounting bolts.

27. Slide the transaxle away from the engine and remove it from the vehicle.

28. Remove the torque converter and dowel pins.

To install:

29. Install the torque converter on the mainshaft, using a new O-ring.

30. Install the dowel pins in the torque converter housing. Place the transaxle on the transaxle jack. Raise the unit to engine level.

31. Install the lower transaxle housing mounting bolt part way in the bolt hole on the engine. Attach the transaxle to the engine.

32. Install the lower transaxle housing mounting bolt, tighten the bolts.

33. Continue the installation in the reverse order of the removal procedure.

34. Check and adjust the front wheel alignment.

35. Properly fill and check the automatic transaxle fluid level.

36. Reprogram the power window control unit as follows.

37. Turn the ignition switch ON. Lower the window, all the way down. Open the driver's side door.

38. Turn the ignition switch OFF. Push and hold the driver's window DOWN switch. Turn the ignition switch ON. Release the driver's window DOWN switch. This must be done within five seconds.

39. Repeat the above step three more times. Wait one second.

40. Confirm that the 'AUTO UP' and 'AUTO DOWN' do not work. If they work repeat the procedure, paying close attention to the five second time limit.

41. Move the driver's window all the way down by holding the driver's window DOWN switch to the AUTO DOWN position.

42. Pull up and hold the driver's window UP switch to the AUTO UP position until the window reaches the fully closed position. Hold the switch for one second.

43. Confirm that the power window master switch has been reset by using the driver's window AUTO UP and DOWN function.

44. If the AUTO UP and DOWN feature is still not working, repeat the complete procedure, paying close attention to the five second time limit.

45. Set the clock.

46. Road test the vehicle.

Civic Hybrid

See Figures 23 through 27.

1. Disconnect the negative battery cable.

2. Remove the cowl cover and under-cowl panel.

3. Move and fix the hood in the vertical position.

4. Remove the front grille cover.

5. Disconnect the positive battery cable, and remove the mounting nuts and remove the battery.

6. Remove the reservoir, the reservoir bracket, the under-hood fuse/relay box bracket, the PCM, and the PCM bracket.

7. Remove the battery base and the resonator.

8. Raise the vehicle on a lift, and make sure it is securely supported.

9. Remove the engine undercover and splash shield.

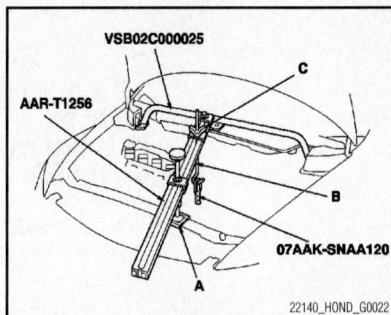

Fig. 24 Install the front leg assembly (A), the hook (B), and the wing nut (C) from an A and Reds engine support hanger (AAR-T1256) onto the engine hanger (VSB02C000025)—2007–08 Civic Hybrid

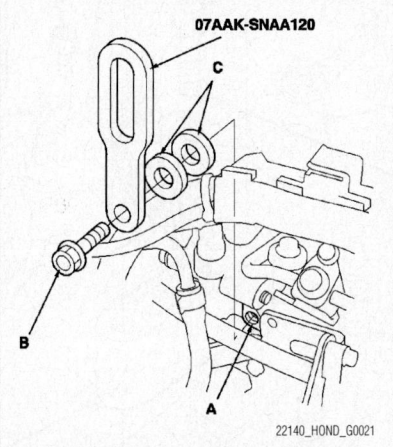

Fig. 23 Installing the universal eyelet (07AAK-SNAA120) to the bolt hole (A) with 8 x 1.25 mm bolt (B) and spacers (C)—2007–08 Civic Hybrid

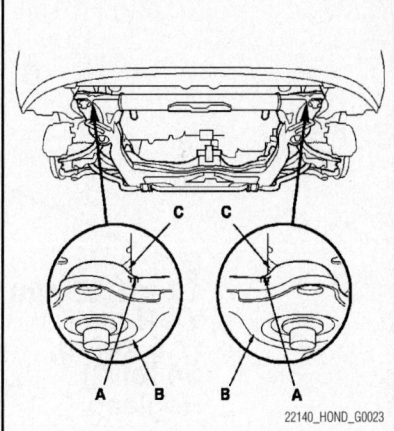

Fig. 25 Matchmark (A) the front subframe (B) that line up with the edge of the body—2007–08 Civic Hybrid

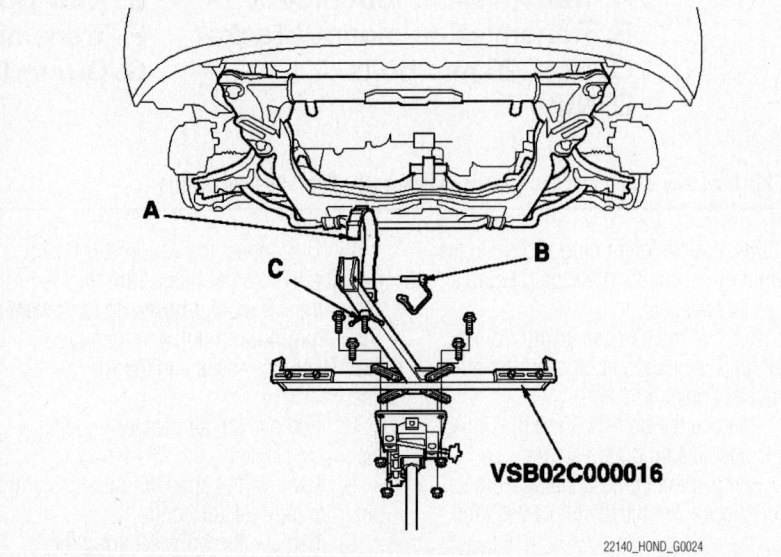

Fig. 26 Attach the front subframe adapter (VSB02C000016) to the front subframe by looping the strap (A) over the front of the front subframe, then secure the strap with the stop (B), then tighten the wing nut (C)—2007–08 Civic Hybrid

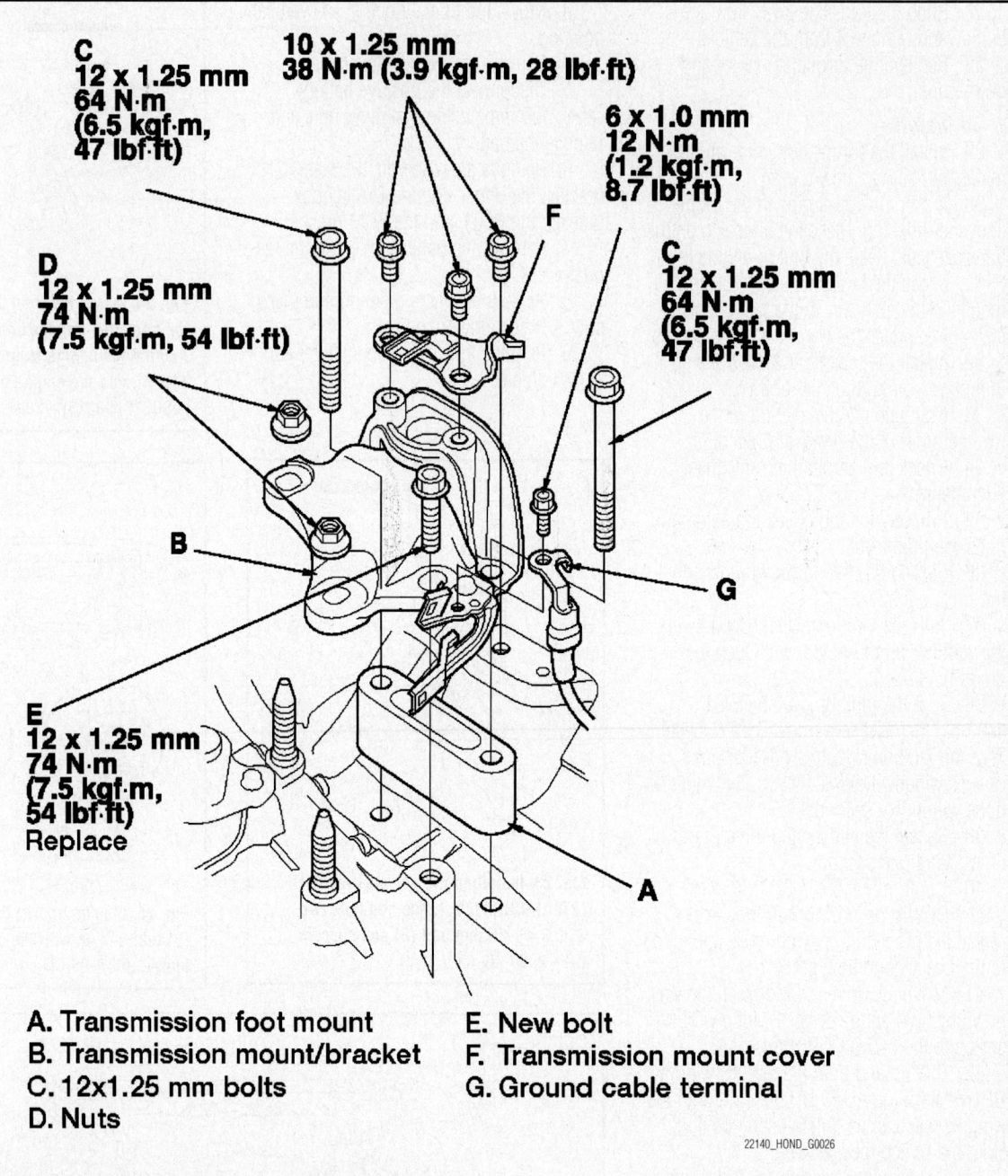

C
12 x 1.25 mm
64 N·m
(6.5 kgf·m,
47 lbf·ft)

10 x 1.25 mm
38 N·m (3.9 kgf·m, 28 lbf·ft)

6 x 1.0 mm
12 N·m
(1.2 kgf·m,
8.7 lbf·ft)

D
12 x 1.25 mm
74 N·m
(7.5 kgf·m, 54 lbf·ft)

F

C
12 x 1.25 mm
64 N·m
(6.5 kgf·m,
47 lbf·ft)

B

G

E
12 x 1.25 mm
74 N·m
(7.5 kgf·m,
54 lbf·ft)
Replace

A

A. Transmission foot mount
B. Transmission mount/bracket
C. 12x1.25 mm bolts
D. Nuts
E. New bolt
F. Transmission mount cover
G. Ground cable terminal

22140_HOND_G0026

Fig. 27 Transaxle mount foot and various components—2007–08 Civic Hybrid

10. Remove the drain plug and drain the CVT fluid. Reinstall the drain plug using a new sealing washer.

11. Remove the harness clamp from its bracket, and remove the bolt securing the IMA harness cover bracket.

12. Remove the bolt securing the upper radiator hose clamp on the starter.

13. Remove the harness clamp from its bracket, disconnect the starter cables, and remove the starter.

14. Disconnect the solenoid harness connector and the CVT input shaft (drive pulley) speed sensor connector.

15. Disconnect the ATF cooler hoses from the ATF cooler lines. Turn the end of the ATF cooler hoses up to prevent CVTF from flowing out, and then plug the ATF cooler hoses and the ATF cooler lines.

16. Remove the air cleaner housing mounting bracket.

17. Remove the snap pin and the control pin from the selector control lever.

18. Remove the bolts securing the shift cable bracket, and then separate the shift cable from the selector control lever. Do not bend the shift cable excessively.

19. Disconnect the transmission range switch connector, and remove the harness clamp from its bracket.

20. Disconnect the CVT output shaft (driven pulley) speed sensor connector, and the CVT speed sensor connector.

21. Remove the harness clamps from the clamp bracket.

22. Remove the drive plate bolts (six) at the opening of the starter while rotating the engine crankshaft pulley.

23. Disconnect the EVAP canister purge valve connector, and remove the bolts securing the connector bracket.

24. Remove the bolts securing the harness cover, and remove the harness cover from its bracket.

25. Install the universal eyelet 07AAK-SNAA120 to the bolt hole with 8 x 1.25 mm bolt and spacers.

26. Install the front leg assembly, the hook, and the wing nut from an A and Reds engine support hanger AAR-T1256 onto the engine hanger VSB02C000025. Carefully position the engine hanger on the vehicle, and attach the hook to the slotted hole in the universal eyelet 07AAK-SNAA120. Tighten the wing nut by hand to lift and support the engine.

※※ CAUTION

Be careful when working around the windshield.

27. Remove the nuts and the bolt securing both lower arm and ball joint, and separate the lower arms from the ball joints.

28. Remove both mid stiffeners.

29. Remove the exhaust pipe mount rubber.

30. Remove the bolt securing the steering gearbox heat shield.

31. Remove the steering gearbox bracket mounting bolts.

32. Remove the steering gearbox mounting bolt, the stiffener mounting bolt, and the stiffener.

33. Remove the lower torque rod bolts.

34. Remove the bolt securing the steering gearbox heat shield.

35. Remove the bolts securing the ATF filter bracket and the IMA harness cover.

36. Make reference marks on the front subframe that line up with the edge of the body.

37. Attach the front subframe adapter VSB02C000016 to the front subframe by looping the strap over the front of the front subframe, then secure the strap with the stop, then tighten the wing nut.

38. Raise the jack and line up the slots in the arms with the bolt holes on the corner of the jack base, and then tighten the bolts.

39. Remove the four bolts securing the front subframe, and lower the front subframe.

40. Secure the steering gearbox to the body with a strap.

41. Remove the halfshafts from the differential and intermediate shaft. For additional information, refer to the following section, "Halfshafts, Removal & Installation."

42. Remove the heat shield and the intermediate shaft.

43. Coat all precision machined surfaces with clean engine oil, and then put plastic bags over both ends of the intermediate shaft.

44. Remove the ground cable terminal.

45. Remove the bolts and the nuts, and remove the transmission mount cover, the transmission mount/bracket, and the transmission mount foot.

46. Remove the upper and front transmission housing mounting bolts.

47. Place a jack under the transmission.

48. Remove the lower torque rod bracket.

49. Remove the rear and lower transmission housing mounting bolts.

50. Lower the transmission by loosening the wing nut of the engine support hanger, and tilt the engine just enough for the transmission to clear the side frame.

51. Slide the transmission away from the motor to remove it from the vehicle.

To install:

52. Place the transmission on a jack, and raise the transmission to the engine level, then fit the transmission to the motor housing/engine.

53. Install the rear and lower transmission housing mounting bolts and tighten to 47 ft. lbs. (64 Nm).

54. Install the upper and front transmission housing mounting bolts and tighten to 47 ft. lbs. (64 Nm).

55. Install the transmission mount foot and the transmission mount/bracket on the body with 12 x 1.25 mm bolts loosely.

56. Secure the transmission mount bracket on the transmission with the nuts and new bolt, and tighten the 12 x 1.25 mm bolts to the specified torque.

57. Install the transmission mount cover and the ground cable terminal.

58. Install the lower torque rod bracket with new bolts and tighten to 54 ft. lbs. (74 Nm).

59. Install a new set ring on the intermediate shaft.

60. Install the intermediate shaft and the heat shield.

61. Install the halfshafts. For additional information, refer to the following section, "Halfshafts, Removal & Installation."

62. Set the front subframe adapter VSB02C000016 to the front subframe by looping the strap over the front of the front subframe, then secure the strap with the stop, then tighten the wing nut.

63. Line up the slots in the arms with the bolt holes on the corner of the jack base, and tighten the bolts, then lift the front subframe up to the body.

64. Loosely install new front subframe mounting bolts.

65. Align the reference marks on the front subframe with the edge of the body, and tighten the mounting bolts to 76 ft. lbs. (103 Nm).

66. Secure the lower torque rod with the bolts.

67. Secure the steering gearbox heat shield with the bolt.

68. Remove the engine support hanger.

69. Position the steering gearbox on the gearbox mounting bracket of the front subframe.

70. Install the gearbox stiffener and the stiffener mounting bolts, and tighten the bolts loosely.

71. Install the steering gearbox mounting bracket bolts.

72. Tighten the steering gearbox mounting bolts and the stiffener mounting bolts to the specified torque.

73. Install both mid stiffeners with new bolts and tighten to 47 ft. lbs. (64 Nm).

74. Install both of the lower arms to the ball joints, and loosely install new mounting nuts and bolts.

75. Loosely tighten the nuts and the bolts in the following order; the nut on the front, the nut on the rear, then the bolt to 43 ft. lbs. (59 Nm).

76. Secure the steering gearbox heat shield with the bolt.

77. Install the exhaust pipe mount rubber.

78. Secure the ATF filter bracket and the IMA harness cover with the bolts.

79. Remove the universal eyelet and spacers.

80. Install the harness cover on its bracket, and secure it with the bolts.

81. Secure the connector bracket with the bolts, and connect the EVAP canister purge valve connector.

82. Attach the flywheel to the drive plate with six bolts. Rotate the crankshaft pulley to tighten the bolts to ½ of 9 ft. lbs. (12 Nm), then to the final torque, in a crisscross pattern. After tightening the last bolt, check that the crankshaft rotates freely.

83. The remainder of the installation is the reverse order of removal.

84. Refill the transmission with CVT fluid.

85. Check and adjust the wheel alignment as necessary.

86. Start the engine and check for leaks.

MANUAL TRANSAXLE ASSEMBLY

REMOVAL & INSTALLATION

Accord

2007 2.4L Engine

See Figures 28 through 33.

1. Make sure you have the anti-theft code for the radio and the navigation system, and then write down the frequencies for the radio's preset buttons.

2. Set the steering wheel in the straight-ahead position, and lock in position.

3. Disconnect the battery cables, then remove the battery.

4. Remove the air intake assembly.

5. Remove the battery tray.

6. Carefully remove the slave cylinder to avoid bending the clutch line.

✳✳ WARNING

Do not press the clutch pedal after the slave cylinder has been removed.

7. Remove the change cable bracket, then disconnect the change cables from the top of the transmission housing.

➡**Carefully remove both cables and the bracket together to avoid bending the cables.**

8. Disconnect the output shaft (countershaft) speed sensor connector and back-up light switch connector.

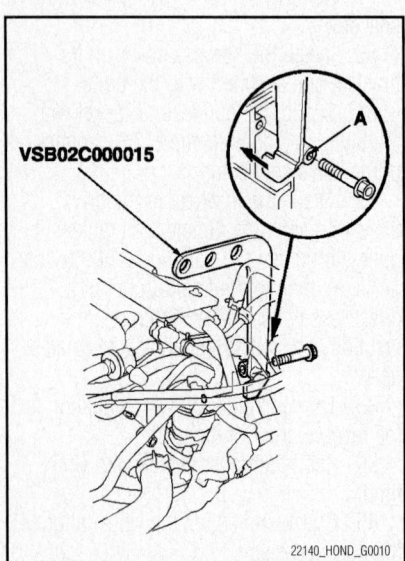

Fig. 28 Attach the engine hanger/adapter to the threaded hole in the cylinder head

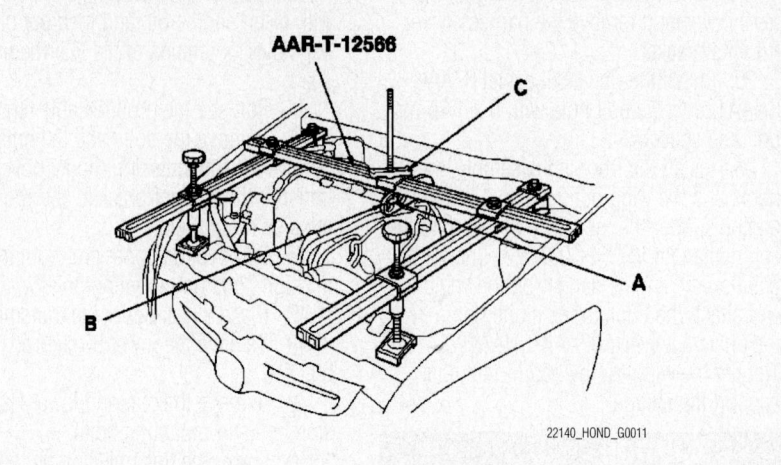

Fig. 29 Install the engine support hanger and attach the hook (A) to the special tool adapter (B). Tighten the wing nut (C) by hand—2007 2.4L Accord

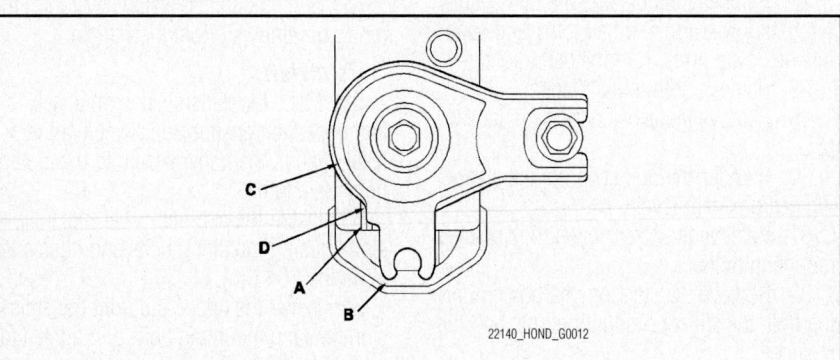

Fig. 30 Matchmark (A) both ends of the front subframe (B) that line up with the edge (C) of the stiffeners (D)—2007 2.4L Accord

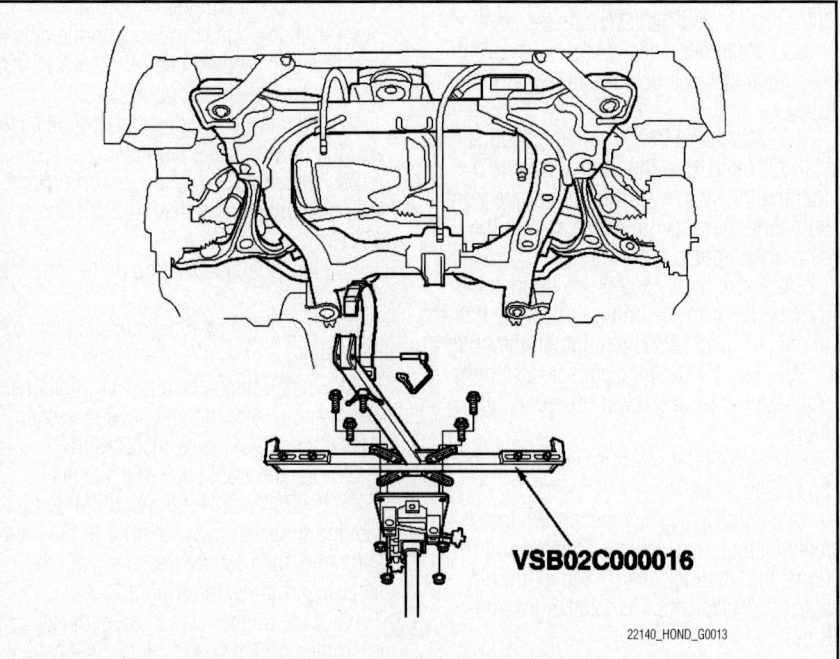

Fig. 31 Attach the special tool to the front subframe with hanging the belt of the special tool over the front of the subframe—2007 2.4L Accord

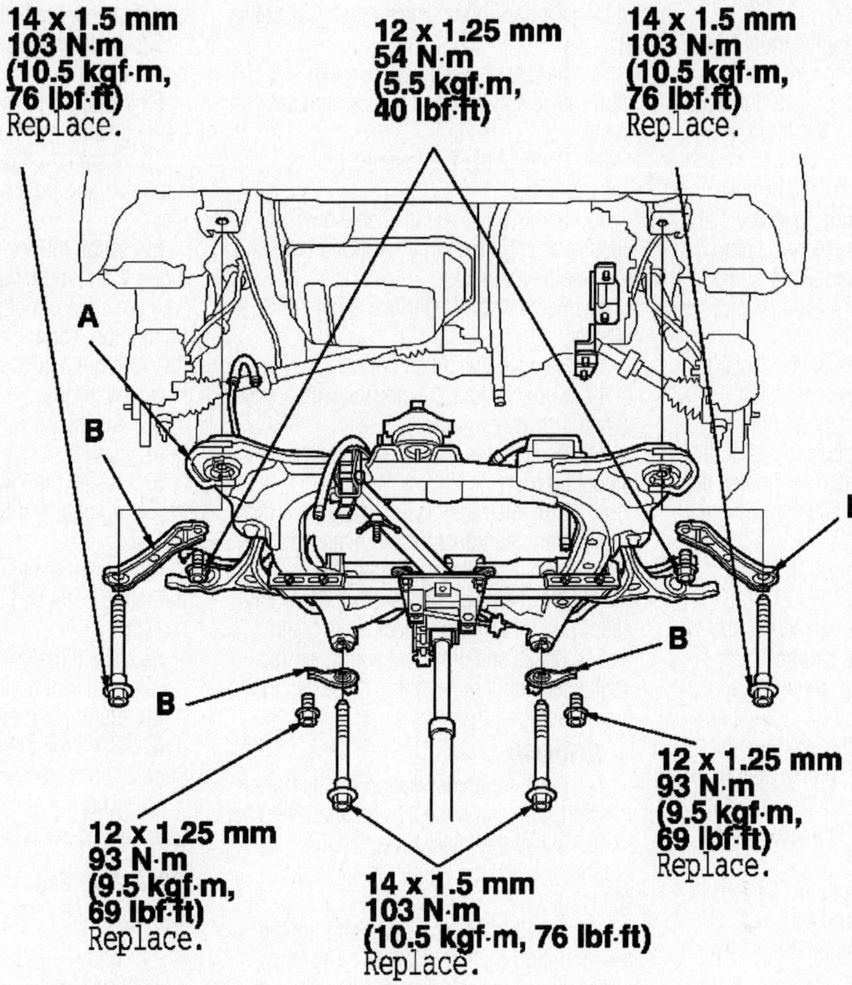

14 x 1.5 mm
103 N·m
(10.5 kgf·m,
76 lbf·ft)
Replace.

12 x 1.25 mm
54 N·m
(5.5 kgf·m,
40 lbf·ft)

14 x 1.5 mm
103 N·m
(10.5 kgf·m,
76 lbf·ft)
Replace.

A

B

B

B

B

B

12 x 1.25 mm
93 N·m
(9.5 kgf·m,
69 lbf·ft)
Replace.

14 x 1.5 mm
103 N·m
(10.5 kgf·m, 76 lbf·ft)
Replace.

12 x 1.25 mm
93 N·m
(9.5 kgf·m,
69 lbf·ft)
Replace.

22140_HOND_G0027

Fig. 32 Showing the torque specifications of the front subframe mounting bolts (A) and subframe stays (B)—2007 2.4L Accord

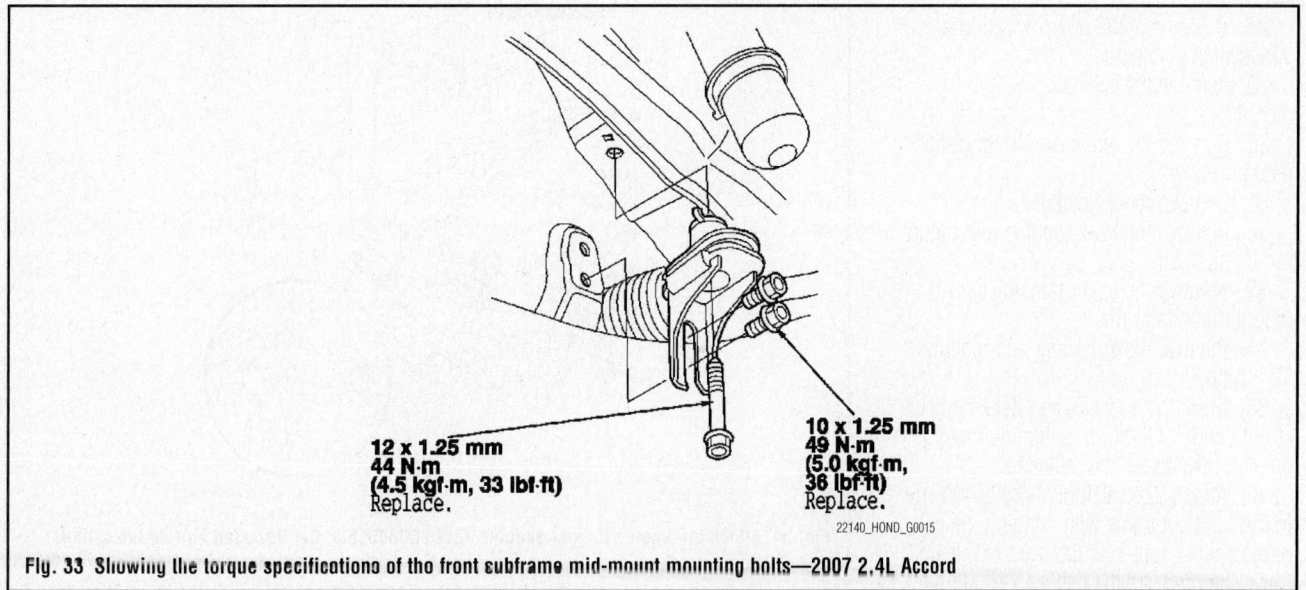

12 x 1.25 mm
44 N·m
(4.5 kgf·m, 33 lbf·ft)
Replace.

10 x 1.25 mm
49 N·m
(5.0 kgf·m,
36 lbf·ft)
Replace.

22140_HOND_G0015

Fig. 33 Showing the torque specifications of the front subframe mid-mount mounting bolts—2007 2.4L Accord

9. Disconnect the secondary heated oxygen sensor (secondary HO2S) connector, then remove the bracket.

10. Remove the front engine mount stop, and then remove the front mount bolt.

11. Remove the rear engine mount stop.

12. Remove the upper transmission mounting bolts.

13. Remove and plug the return hose from the power steering fluid reservoir.

14. Remove the power steering pump outlet line from the power steering pump.

15. Remove the power steering pump outlet line bracket.

16. Attach the engine hanger/adapter to the threaded hole in the cylinder head.

17. Install the engine support hanger AAR-T-12566 to the vehicle, and attach the hook to the special tool adapter. Tighten the wing nut by hand, and lift and support the engine.

18. Remove the transmission mount bracket and ground cable.

19. Make a reference mark across the steering joint and steering gearbox pinion shaft. Remove the steering joint bolt, and disconnect the steering joint by removing the steering joint toward the steering column. Hold the slider shaft on the column with a piece of wire between the joint yoke on the slider shaft to the joint yoke on the upper shaft.

20. Raise and safely support the vehicle.

21. Remove the front wheels.

22. Drain the transmission fluid. Reinstall the drain plug.

23. Remove the splash shield.

24. Separate the tie-rod ball joint.

25. Separate the front stabilizer link.

26. Remove the damper fork.

27. Separate the knuckle from the lower arm.

28. Disconnect the power steering pressure switch connector.

29. Remove the exhaust pipe and the mount.

30. Remove the left and right driveshaft inboard joints.

31. Remove the intermediate shaft.

32. Remove the lower engine rear mount mounting bolts.

33. Remove the lower transmission mount mounting nuts.

34. Remove the subframe mid mounts from both sides.

35. Make the appropriate reference lines at both ends of the front subframe that line up with the edge of the stiffeners.

36. Attach Special Tool VSB02C000016 to the front subframe with hanging the belt of the special tool over the front of the subframe, then secure the belt with its stop.

37. Raise the jack and line up the slots in the arms with the bolt holes on the corner of the jack base, then attach them with bolts securely.

38. Remove the front suspension subframe stays and the front suspension subframe.

39. Remove the front engine mount upper bracket.

40. Remove the change cable bracket, upper rear engine mount mounting bolt, and rear engine mount.

41. Remove the upper rear engine mount bracket.

42. Remove the clutch cover.

43. Support the transmission with a suitable transmission jack.

44. Remove the lower transmission mounting bolts.

45. Pull the transmission away from the engine until the transmission mainshaft clears the clutch pressure plate.

46. Slowly lower the transaxle assembly about 6 in. (151 mm). Check once again that all hoses and electrical wiring are disconnected and free from the transaxle, and then lower it all the way.

To install:

47. Place the transmission on the transmission jack, and raise it to engine level.

48. Install the lower transmission mounting bolts and tighten to 47 ft. lbs. (64 Nm).

49. Install the clutch cover and tighten the mounting bolts to 33 ft. lbs. (44 Nm).

50. Install the upper rear engine mount bracket and tighten the bolts to 65 ft. lbs. (88 Nm).

51. Install the rear engine mount, upper rear engine mount mounting bolt, and change cable bracket.

52. Install the front engine mount upper bracket.

53. Support the subframe with the subframe adapter and a suitable jack.

54. Install the front suspension subframe and front suspension subframe stays.

55. Align the reference marks that you noted when you removed the transmission with the edge of both rear stiffeners, then tighten the rear subframe mounting bolts, the front bolts, and the stiffener bolts to the specified torque.

56. Install the subframe mid mounts on both sides

57. Install the lower transmission mount mounting nuts and tighten to 33 ft. lbs. (44 Nm).

58. Install the lower engine rear mount mounting bolts and tighten to 36 ft. lbs. (49 Nm).

59. The remainder of the installation the reverse order of removal.

60. Refill the transaxle to correct level.

61. Refill the power steering system with fluid to the correct level.

62. Check and adjust the wheel alignment as necessary.

2008 2.4L Engine

See Figures 34 through 42.

1. Make sure you have the anti-theft code for the radio and the navigation system, and then write down the frequencies for the radio's preset buttons.

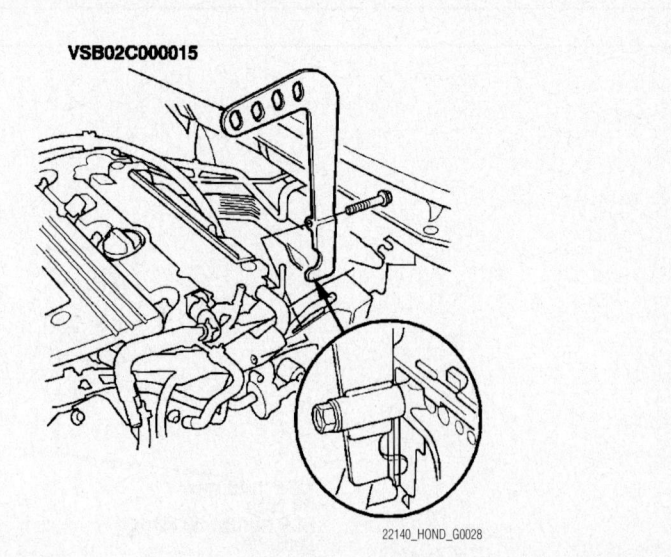

VSB02C000015

22140_HOND_G0028

Fig. 34 Attach the engine hanger adapter VSB02C000015 to the threaded hole in the cylinder head—2008 2.4L Accord

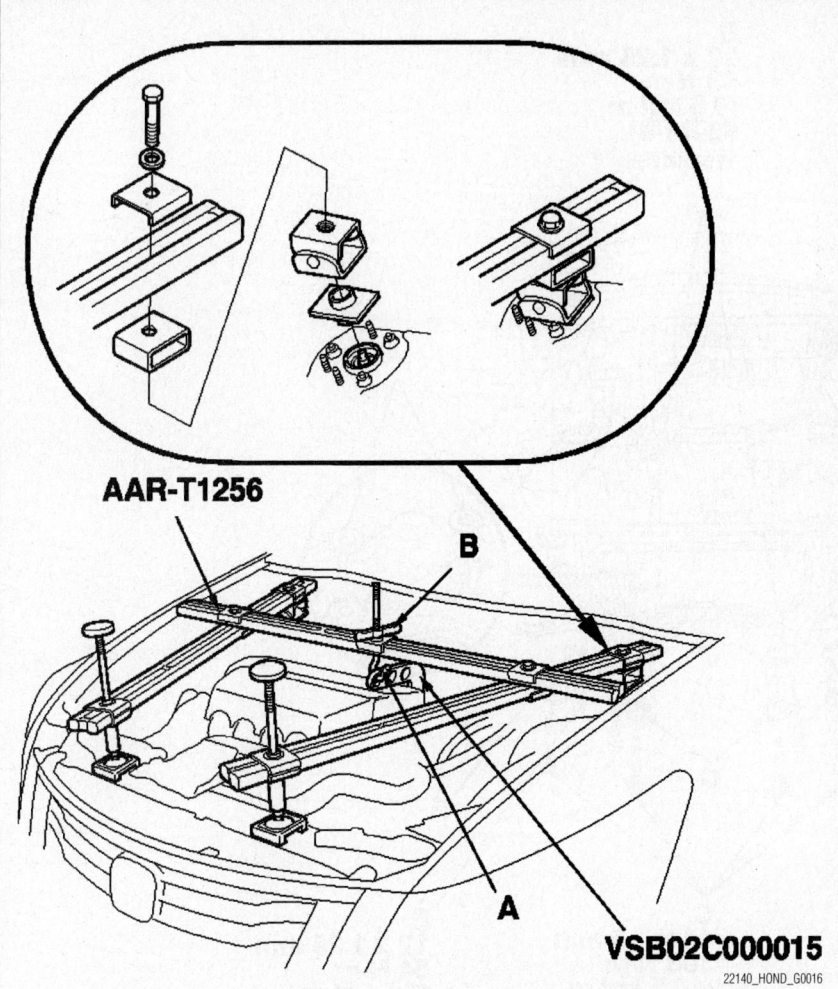

AAR-T1256

B

A

VSB02C000015

22140_HOND_G0016

Fig. 35 Install the engine support hanger (AAR-T1256) to the vehicle, and attach the hook (A) to the slotted hole in the engine hanger adapter (VSB02C000015), and turn the wing nut (B) to lift—2008 2.4L Accord

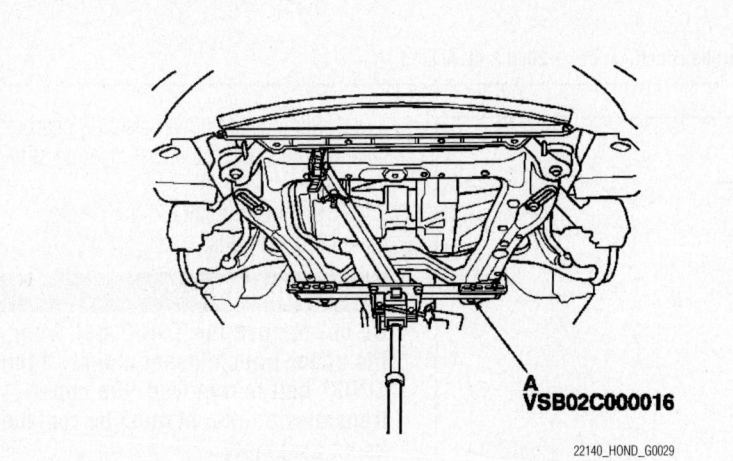

A
VSB02C000016

22140_HOND_G0029

Fig. 36 Attach the front subframe adapter (A) to the front suspension subframe and hang the belt of the subframe adapter over the front of the subframe, then secure the belt with its stop—2008 2.4L Accord

2. Secure the hood in the wide open position with the support strut.

3. Remove the front grille cover.

4. Disconnect the battery cables.

5. Remove the battery and battery tray.

6. Remove the air intake assembly.

7. Remove the battery base bolts, loosen the two bolts, remove the cable clamp and the harness bracket bolt, and then remove the battery base.

8. Remove the slave cylinder mounting bolts and the bracket mounting nut, then carefully move the slave cylinder out of the way to avoid bending the clutch line.

> ✳✳ **WARNING**
>
> **Do not press the clutch pedal after the slave cylinder has been removed.**

9. Remove the mounting nut and the clamp, and then move the under-hood fuse/relay box out of the way.

10. Remove the lock pins, the shift cable bracket bolt, and the shift cable bracket, then disconnect the shift cables from the change lever assembly. Carefully remove both cables and the bracket together to avoid bending the cables.

11. Disconnect the output shaft (counter-shaft) speed sensor connector, the back-up light switch connector, and the harness clamp.

12. Remove the bolts, the harness clamp, and the bracket.

13. Remove the hose from the clamp and the nuts, and then remove the strut brace.

14. Remove the evaporative emission (EVAP) canister hose (A) and the brake booster vacuum hose.

15. Attach the engine hanger adapter VSB02C000015 to the threaded hole in the cylinder head.

16. Install the engine support hanger AAR-T1256 to the vehicle, and attach the hook to the slotted hole in the engine hanger adapter VSB02C000015. Tighten the wing nut by hand to lift and support the engine.

17. Remove the front engine mount stop nuts, the front engine mount stop, and the vacuum hose bracket, then remove the front engine mount bolt, and disconnect the vacuum hose.

18. Remove the two rear engine mount bracket bolts and three rear engine mount bolts.

19. Remove the power steering line holder mounting bolts.

20. Remove the two heat shield bolts, the heat shield, and the two power steering gearbox mounting bracket bolts, then remove the power steering gearbox mounting bracket.

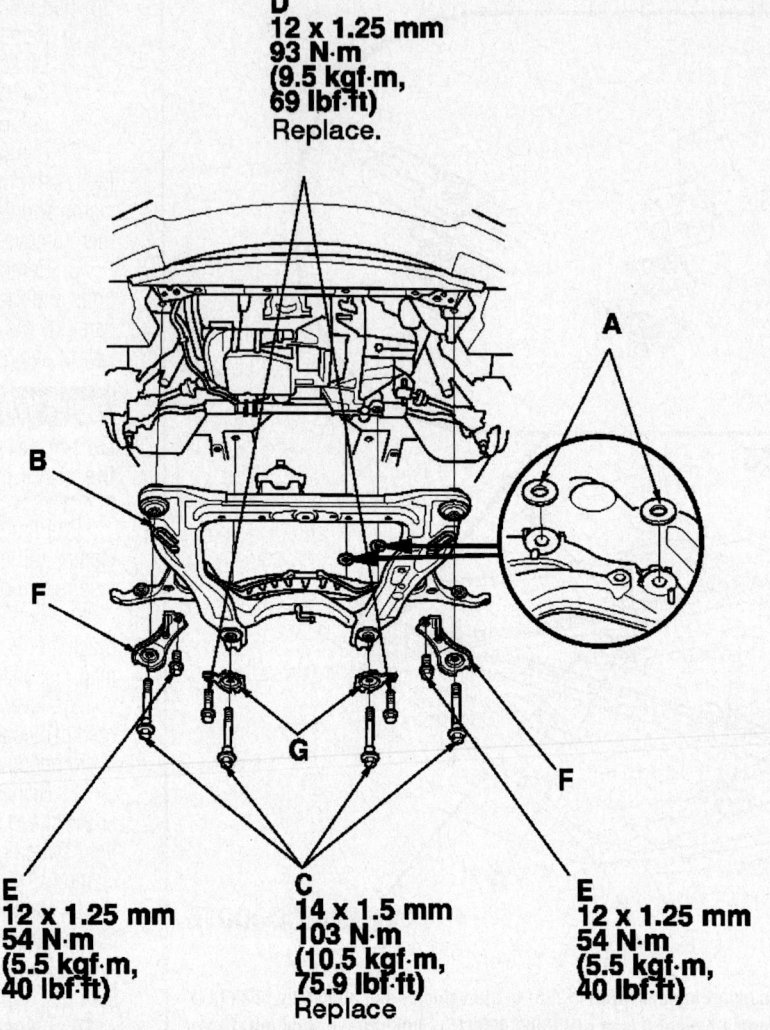

D
12 x 1.25 mm
93 N·m
(9.5 kgf·m,
69 lbf·ft)
Replace.

A

B

F

G

F

E
12 x 1.25 mm
54 N·m
(5.5 kgf·m,
40 lbf·ft)

C
14 x 1.5 mm
103 N·m
(10.5 kgf·m,
75.9 lbf·ft)
Replace

E
12 x 1.25 mm
54 N·m
(5.5 kgf·m,
40 lbf·ft)

A. Steering gearbox washers   E. Front stiffener mounting bolts
B. Front suspension subframe   F. Front stiffeners
C. Subframe mounting bolts   G. Rear stiffeners
D. Rear stiffener mounting bolts

22140_HOND_G0030

Fig. 37 Install the subframe and stiffener bolts to the correct torque specifications—2008 2.4L Accord

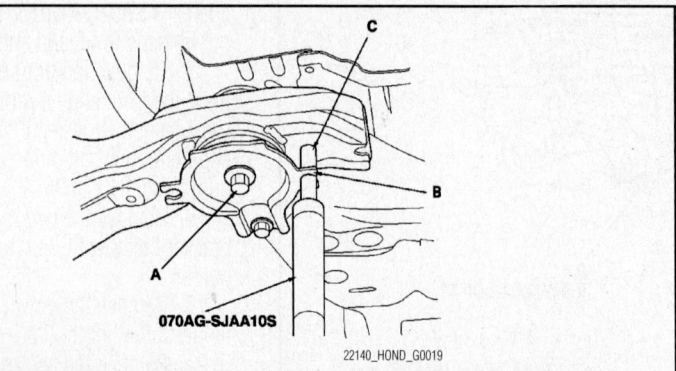

C

B

A

070AG-SJAA10S

22140_HOND_G0019

Fig. 38 Loosely tighten the front subframe mounting bolt (A) in the right rear stiffener until the front subframe insulator contacts the body; insert the subframe alignment pin (070AG-SJAA10S) through the positioning slot (B) on the right rear stiffener, through the positioning hole (C) on the front subframe, and into the positioning hole on the body

21. Remove the power steering gearbox stiffener bolts and the power steering stiffener plates.

22. Remove the upper transmission mount bracket bolts.

✳✳ WARNING

Do not remove the TORX® bolt from the upper transmission mount. If the TORX® bolt is removed, the upper transmission mount must be replaced as an assembly.

23. Remove the three upper transmission mount bracket bolts, the upper transmission mount bracket, the ground cable mount bolt, and the ground cable.

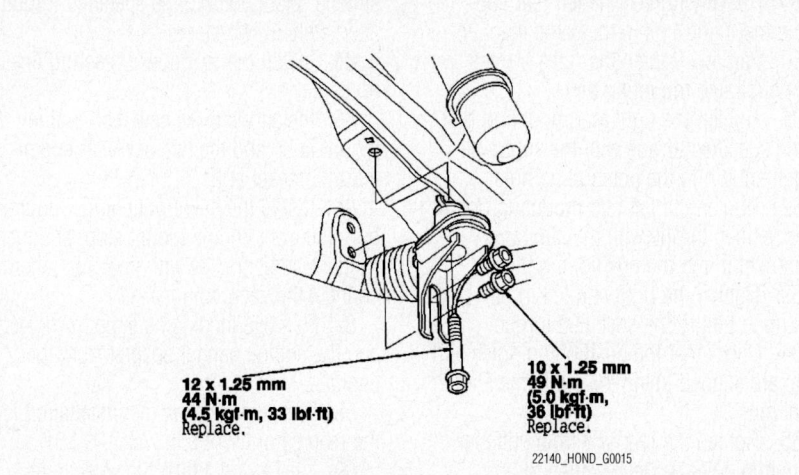

Fig. 39 Showing the torque specifications of the front subframe mid-mount mounting bolts—2007 2.4L Accord

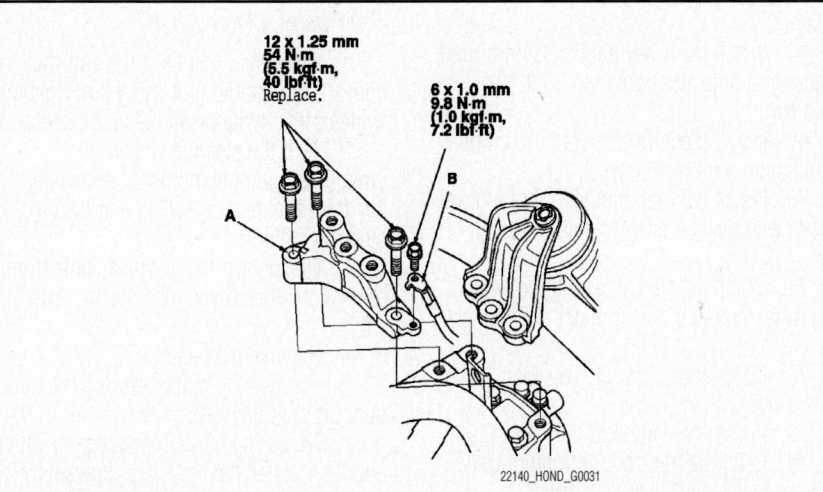

Fig. 40 Install the upper transmission mount bracket (A) with new bolts, and connect the ground cable (B) by installing its mounting bolt—2008 2.4L Accord

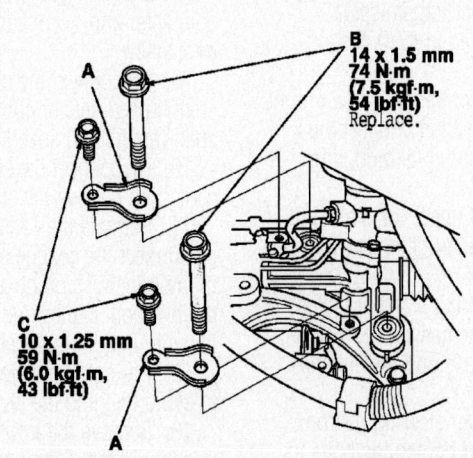

Fig. 41 When installing the steering stiffener plates (A), make sure the lower washers are correctly positioned before installing the power steering gear box mounting bolts (B) and stiffener plate bolts (C)—2008 2.4L Accord

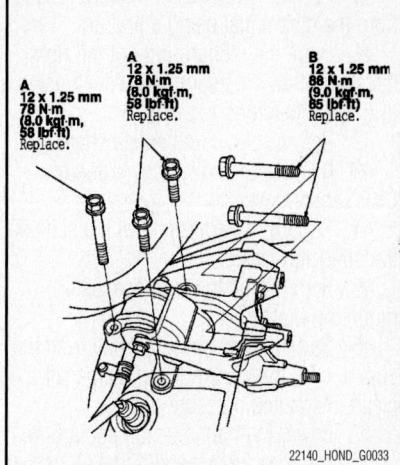

Fig. 42 Install the three new rear engine mount bolts (A) and the two new rear engine mount bracket bolts (B)—2008 2.4L Accord

24. Remove the upper transmission mount bolts.

25. Raise and safely support the vehicle.

26. Remove the front wheels.

27. Remove the front splash shield.

28. Drain the transmission fluid. Reinstall the drain plug using a new sealing washer.

29. Remove the damper fork.

30. Separate the knuckle ball joint from the lower arm.

31. Remove the exhaust pipe and the gaskets.

32. Remove the rear engine mount bracket bolts.

33. Remove the subframe mid mount bolts and the subframe mid mount from both sides.

34. Remove the lower transmission mount mounting nuts.

35. Remove the power steering line holder bolt and the power steering line from the clamp.

36. Attach front subframe adapter VSB02C000016 to the front suspension subframe and hang the belt of the subframe adapter over the front of the subframe, then secure the belt with its stop.

37. Raise the jack and line up the slots in the front subframe adapter arms with the bolt holes on the jack base, and then securely attach them with four bolts.

38. Remove the four front stiffener bolts, the front stiffeners, the four rear stiffeners bolts, and the rear stiffeners, then remove the front suspension subframe.

39. Remove the lower transmission mount bolts and the lower transmission mount.

40. Pry the left halfshaft inboard joint from the differential using a prybar.

41. Drive the inboard joint of the right halfshaft off of the intermediate shaft using a drift and a hammer.

42. Remove the intermediate shaft.

43. Remove the clutch cover and the CKP sensor cover.

44. Support the transaxle with a suitable transmission jack.

45. Remove the lower transmission mounting bolts.

46. Pull the transmission away from the engine until the transmission mainshaft clears the clutch pressure plate.

47. Lower the transmission about 6 in. (151 mm). Check once again that all hoses and harnesses are disconnected and free from the transmission, and then lower it completely.

To install:

48. With the transaxle assembly on the transmission jack, and raise it to engine level.

49. Align the transmission mainshaft and the clutch pressure plate, then move the transmission inward until there is no gap between the transmission housing and the engine block.

50. Install the lower transmission mounting bolts and tighten to 47 ft. lbs. (64 Nm).

51. Install the clutch cover and the CKP sensor cover. Tighten the bolts to 108 inch lbs. (12 Nm).

52. Install the intermediate shaft.

53. Install both halfshafts. For additional information, refer to the following section, "Halfshafts, Removal & Installation."

54. Install the lower transmission mount with new bolts and tighten to 40 ft. lbs. (54 Nm). Remove the transmission jack.

55. Support the front suspension subframe with front subframe adapter VSB02C000016 and a jack.

56. Position the steering gearbox washers on the front suspension subframe, and lift the subframe up to the body.

57. Loosely install the new subframe mounting bolts, the new rear stiffener mounting bolts, the front stiffener mounting bolts, the front stiffeners, and the rear stiffeners.

58. Partially tighten the right rear subframe mounting bolt; insert the subframe alignment pin through the positioning slot on the rear stiffener, through the positioning hole on the subframe, and into the positioning hole on the body, then tighten the subframe mounting bolt.

59. Partially tighten the left rear subframe mounting bolt in the same manner.

60. Partially tighten the right and left front subframe mounting bolts.

61. Tighten the right rear mounting bolt to the specified torque with the subframe alignment pin in the positioning hole.

62. Tighten the left rear mounting bolt to the specified torque with the subframe alignment pin in the positioning hole.

63. Tighten the right and left front mounting bolt to the specified torque.

64. Check that the positioning holes and slots are aligned using the subframe alignment pin.

65. Tighten the rear and front stiffener mounting bolts to the specified torque.

66. Remove the jack and front subframe adapter.

67. Install the power steering line holder bolt and the power steering line to the clamp.

68. Install the lower transmission mount mounting nuts and tighten to 33 ft. lbs. (44 Nm).

69. Install the subframe mid mount on both sides with new bolts.

70. Install the rear engine mount bracket bolts with new bolts and tighten to 65 ft. lbs. (88 Nm).

71. Install exhaust pipe with the new gaskets, the bolts, the springs, and the new nuts.

72. Connect the knuckle ball joint onto the lower arm.

73. Install the damper fork.

74. Refill the transaxle with fluid to the correct level.

75. Install the front splash shield.

76. Install the front wheels, and set them in the straight-ahead position.

77. Lower the vehicle.

78. Install the upper transmission mount bolts and tighten to 47 ft. lbs. (64 Nm).

79. Install the upper transmission mount bracket with new bolts, and connect the ground cable by installing its mounting bolt.

80. Install the new upper transmission mount bracket bolts and tighten to 43 ft. lbs. (59 Nm).

81. Install the steering stiffener plates, and loosely tighten the new power steering gear box mounting bolts and the stiffener plate bolts.

82. Install the power steering gearbox mounting bracket, and tighten the bolts to the specified torque, then install the heat shield.

83. Tighten the driver's side of the power steering gear box mounting bolts and the stiffener plate bolts to the specified torque alternately in two steps.

84. Install the rear power steering line holders.

85. Install the three new rear engine mount bolts and the two new rear engine mount bracket bolts.

86. Install the new front engine mount bolt, the front engine mount stop, and the vacuum hose bracket with new nuts, then connect the vacuum hose.

87. Remove the engine support hanger and the engine hanger adapter from the engine.

88. The remainder of the installation is the reverse order of removal.

89. Check and adjust the wheel alignment as necessary.

90. Start the engine and check for leaks.

2007 3.0L Engine

See Figures 43 through 46.

1. Make sure you have the anti-theft codes for the audio unit and the navigation system, then write down the audio presets.

2. Turn the steering wheel to the straight-ahead position, then remove the key from the ignition switch and lock the steering column.

3. Disconnect the negative cable from the battery, then disconnect the positive cable.

4. Remove the battery.

5. Remove the bolts securing the hood support strut brackets on both sides of the hood. Secure the hood in a vertical position, and then install the right strut bracket by turning it over with its bolt in the lower position.

6. Remove the battery base.

7. Remove the air intake assembly.

8. Disconnect the back-up light switch connector and reverse lockout solenoid connector.

9. Disconnect the output shaft (countershaft) speed sensor connector and input shaft (mainshaft) speed sensor connector.

10. Disconnect the starter cable and wire, then remove the starter motor.

11. Remove the cable bracket, and then disconnect the change cables from the top of the transmission housing. Carefully remove both cables and the bracket together to avoid bending the cables.

12. Carefully remove the slave cylinder to avoid bending the clutch line.

13. Remove the front engine mount stop, and then remove the front mount bolt.

14. Remove the rear engine mount stop.

15. Remove the upper transmission mounting bolts.

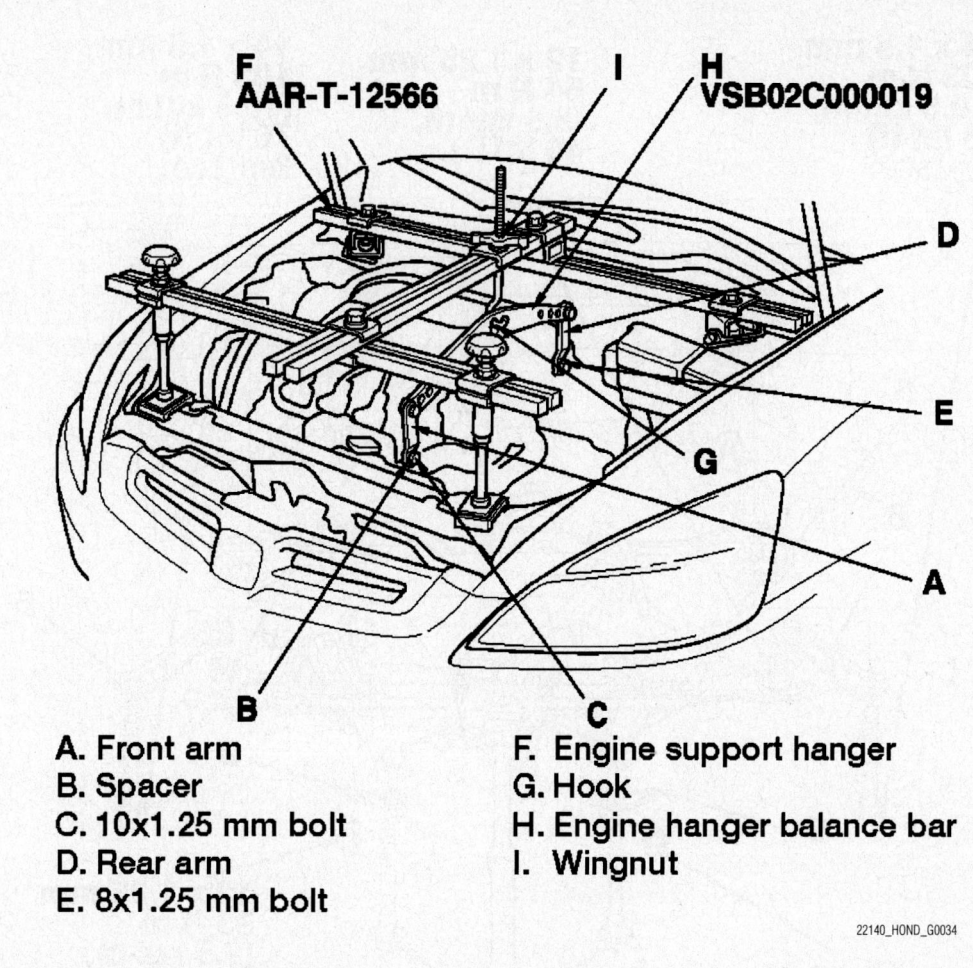

F. AAR-T-12566
I.
H. VSB02C000019
D
E
G
A
B
C

A. Front arm
B. Spacer
C. 10x1.25 mm bolt
D. Rear arm
E. 8x1.25 mm bolt

F. Engine support hanger
G. Hook
H. Engine hanger balance bar
I. Wingnut

22140_HOND_G0034

Fig. 43 Installation of the engine hanger to support the engine assembly—2007 3.0L Accord

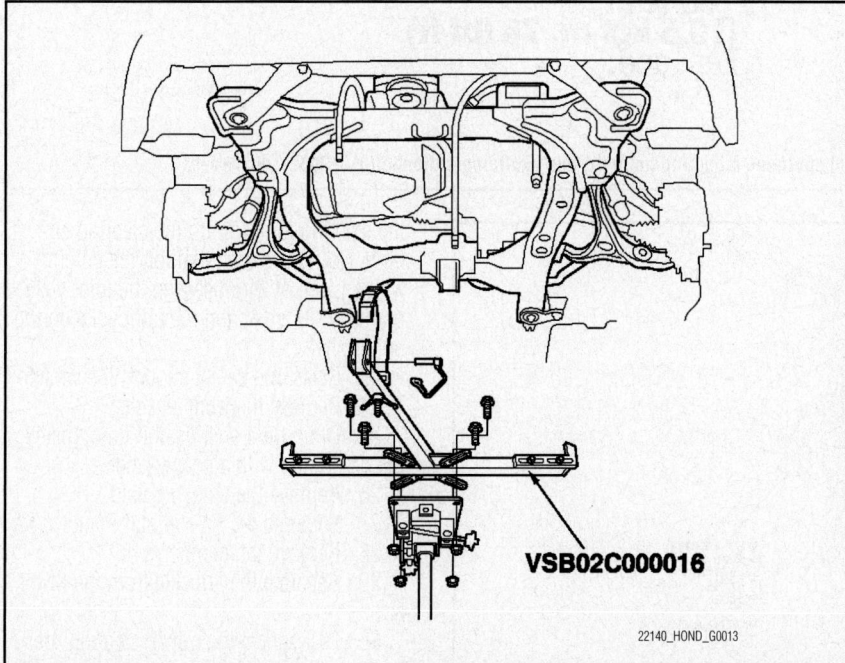

VSB02C000016

22140_HOND_G0013

Fig. 44 Attach the special tool to the front subframe with hanging the belt of the special tool over the front of the subframe—2007 2.4L Accord

16. Remove and plug the return hose from the power steering fluid reservoir.

17. Remove the power steering pump outlet line from the power steering pump, and remove the hose from its clamp.

18. Remove the front bulkhead cover:

19. Install the engine hanger as follows:

   a. Attach the front arm to the front cylinder head with a spacer and the 10 x 1.25 mm bolt.

   b. Attach the rear arm to the rear cylinder head with the 8 x 1.25 mm bolt.

   c. Install the engine support hanger to the vehicle, and attach the hook to the engine hanger balance bar. Tighten the wing nut by hand, and lift and support the engine.

20. Remove the transmission mount bracket and ground cable.

21. Remove the steering joint cover.

22. Make a reference mark across the steering joint and steering gearbox pinion shaft. Remove the steering joint bolt, and disconnect the steering joint by removing

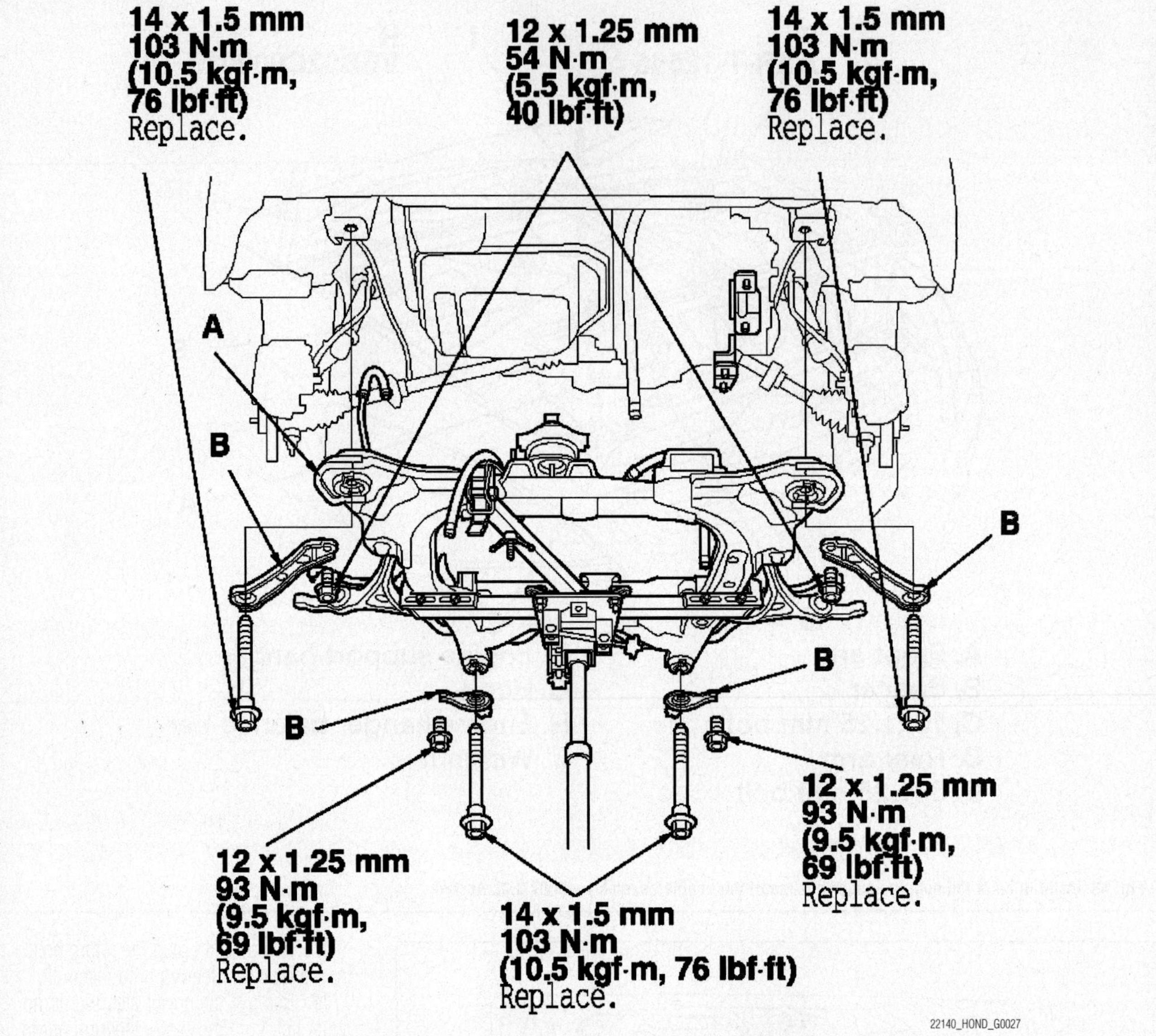

14 x 1.5 mm
103 N·m
(10.5 kgf·m,
76 lbf·ft)
Replace.

12 x 1.25 mm
54 N·m
(5.5 kgf·m,
40 lbf·ft)

14 x 1.5 mm
103 N·m
(10.5 kgf·m,
76 lbf·ft)
Replace.

A

B

B

B

B

B

12 x 1.25 mm
93 N·m
(9.5 kgf·m,
69 lbf·ft)
Replace.

12 x 1.25 mm
93 N·m
(9.5 kgf·m,
69 lbf·ft)
Replace.

14 x 1.5 mm
103 N·m
(10.5 kgf·m, 76 lbf·ft)
Replace.

22140_HOND_G0027

Fig. 45 Showing the torque specifications of the front subframe mounting bolts (A) and subframe stiffeners(B)—2007 3.0L Accord

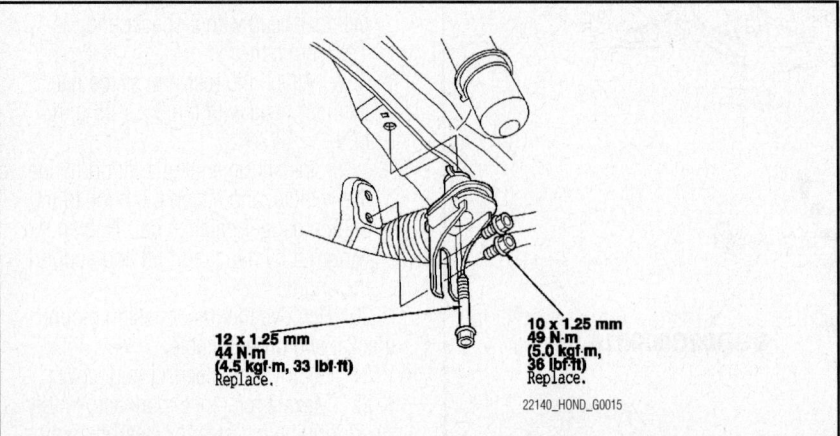

12 x 1.25 mm
44 N·m
(4.5 kgf·m, 33 lbf·ft)
Replace.

10 x 1.25 mm
49 N·m
(5.0 kgf·m,
36 lbf·ft)
Replace.

22140_HOND_G0015

Fig. 46 Showing the torque specifications of the front subframe mid-mount mounting bolts—2007 3.0L Accord

the steering joint toward the steering column. Hold the slider shaft on the column with a piece of wire between the joint yoke on the slider shaft and the joint yoke on the upper shaft.

23. Raise and safely support the vehicle.

24. Remove the front wheels.

25. Drain the transmission fluid. Install the drain plug with a new washer.

26. Remove the splash shield.

27. Separate the tie-rod ball joint.

28. Remove the damper fork.

29. Separate the knuckle from the lower arm.

30. Disconnect the power steering pressure switch connector.

31. Remove the exhaust pipe.

32. Remove the left and right halfshaft inboard joints.

33. Remove the intermediate shaft.

34. Remove the lower engine rear mount mounting bolts.

35. Remove the lower transmission mount mounting nuts.

36. Remove the subframe mid mounts from both sides.

37. Attach the subframe adapter to the subframe by wrapping the band over the subframe and attaching the end of the band to the subframe adapter with the pin.

38. Raise the jack, and line up the slots in the arms with the bolt holes on the corner of the jack base, then attach them securely.

39. Remove the four bolts securing the stiffeners, and four bolts securing the front subframe, and lower the front subframe.

40. Remove the front engine mount upper bracket.

41. Remove the clutch cover.

42. Support the transaxle assembly with a transmission jack.

43. Remove the lower transmission mounting bolts.

44. Pull the transmission away from the engine until the transmission mainshaft clears the clutch pressure plate.

45. Slowly lower the transmission about 6 in. (151 mm). Check once again that all hoses and electrical wiring are disconnected and free from the transmission, and then lower it all the way.

To install:

46. With the transaxle assembly on the transmission jack, raise it to engine level.

47. Install the lower transmission mounting bolts and tighten to 54 ft. lbs. (74 Nm).

48. Install the clutch cover and tighten the mounting bolts to 54 ft. lbs. (74 Nm).

49. Install the front engine mount upper bracket and tighten to 40 ft. lbs. (54 Nm).

50. Support the subframe with the subframe adapter and a jack.

51. Install the front suspension subframe and front suspension subframe stiffeners.

52. Align the reference marks that you noted when you removed the transaxle with the edge of both rear stiffeners, then tighten the rear subframe mounting bolts, the front

bolts, and the stiffener bolts to the specified torque.

53. Install the subframe mid mounts on both sides.

54. Install the lower transmission mount mounting nuts and tighten to 33 ft. lbs. (44 Nm).

55. Install the lower engine rear mount mounting bolts and tighten to 36 ft. lbs. (39 Nm).

56. The remainder of the installation is the reverse order of removal.

57. Refill the power steering reservoir to the correct level.

58. Refill the transaxle fluid to the correct level.

59. Check and adjust the wheel alignment as necessary.

60. Start the engine and check for leaks.

2008 3.5L Engine

See Figures 47 through 52.

1. Remove the hood support struts from both sides of the engine hood. Secure the hood in a vertical position, and then reinstall the support strut.

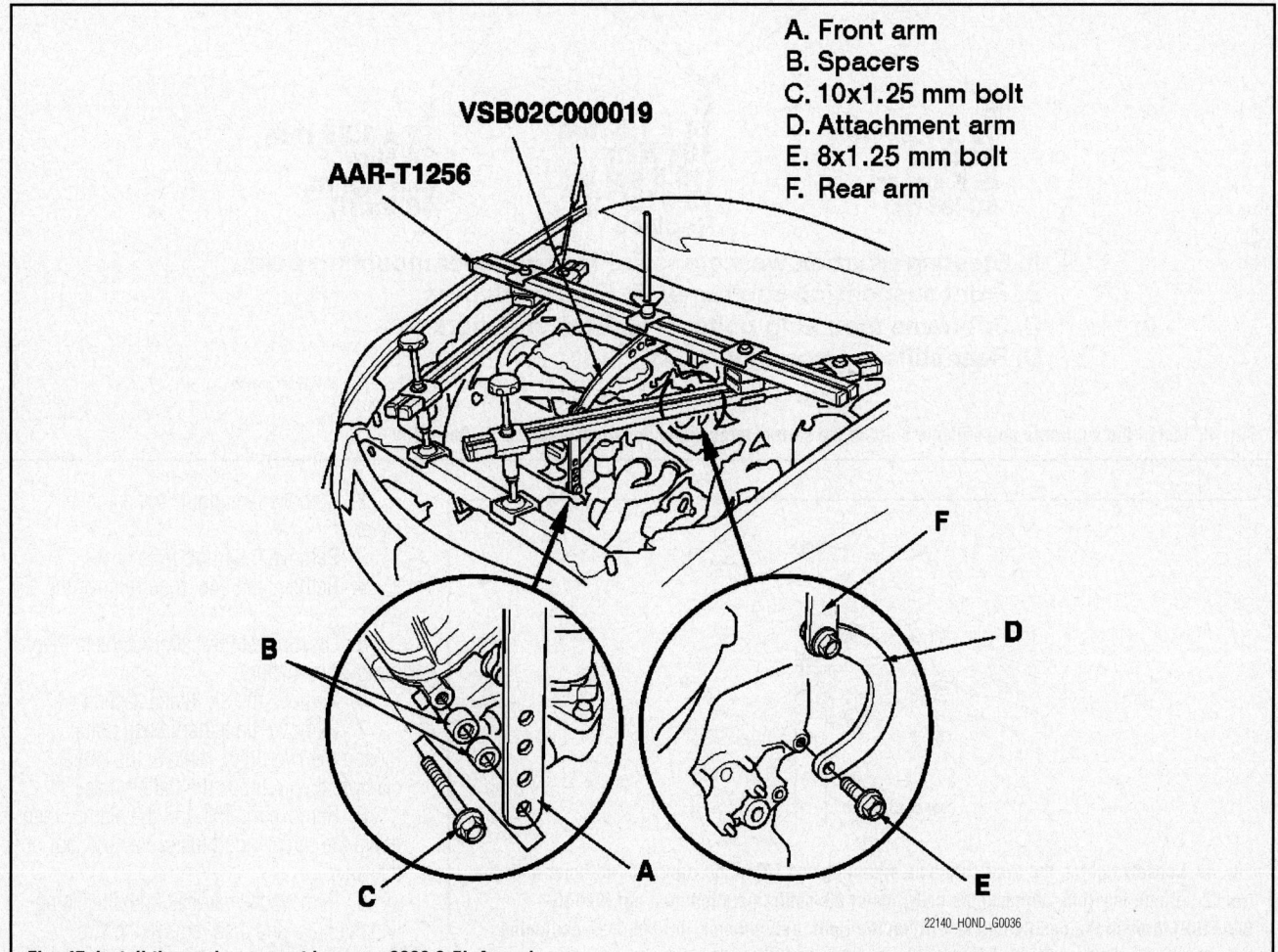

A. Front arm
B. Spacers
C. 10x1.25 mm bolt
D. Attachment arm
E. 8x1.25 mm bolt
F. Rear arm

VSB02C000019

AAR-T1256

22140_HOND_G0036

Fig. 47 Install the engine support hanger—2008 3.5L Accord

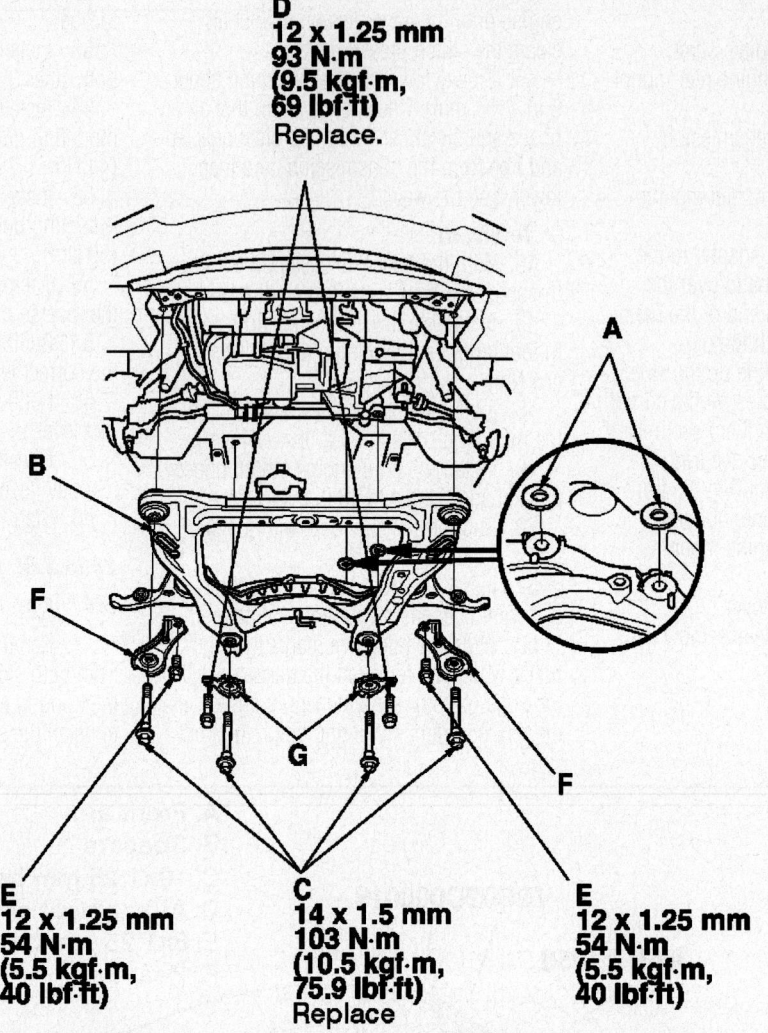

D
12 x 1.25 mm
93 N·m
(9.5 kgf·m,
69 lbf·ft)
Replace.

A

B

F

G

F

E
12 x 1.25 mm
54 N·m
(5.5 kgf·m,
40 lbf·ft)

C
14 x 1.5 mm
103 N·m
(10.5 kgf·m,
75.9 lbf·ft)
Replace

E
12 x 1.25 mm
54 N·m
(5.5 kgf·m,
40 lbf·ft)

A. Steering gearbox washers   E. Front stiffener mounting bolts
B. Front suspension subframe   F. Front stiffeners
C. Subframe mounting bolts   G. Rear stiffeners
D. Rear stiffener mounting bolts

22140_HOND_G0030

Fig. 48 Install the subframe and stiffener bolts to the correct torque specifications—2008 3.5L Accord

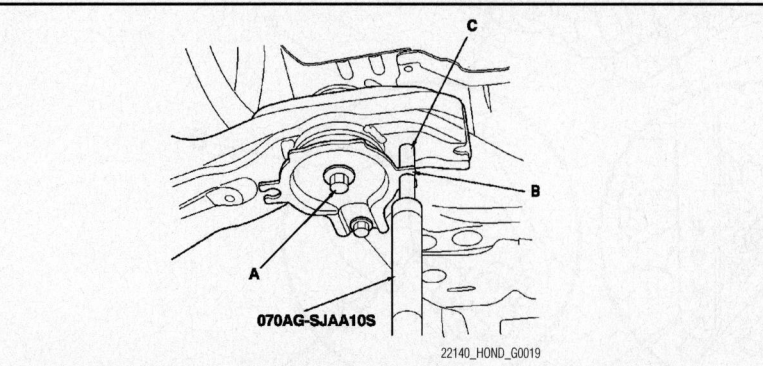

C

A

B

070AG-SJAA10S

22140_HOND_G0019

Fig. 49 Loosely tighten the front subframe mounting bolt (A) in the right rear stiffener until the front subframe insulator contacts the body; insert the subframe alignment pin (070AG-SJAA10S) through the positioning slot (B) on the right rear stiffener, through the positioning hole (C) on the front subframe, and into the positioning hole on the body.

2. Remove the engine appearance cover.

3. Remove the front grille cover.

4. Remove the clip, then remove the air duct.

5. Disconnect the battery cables, then remove the battery.

6. Remove the air intake assembly.

7. Remove the battery base bolts, loosen the two bolts, remove the cable clamps, then remove the battery base.

8. Remove the nut and the clamp, then move the under-hood fuse/relay box out of the way.

9. Remove the harness from the clamp and the nuts, and then remove the strut brace.

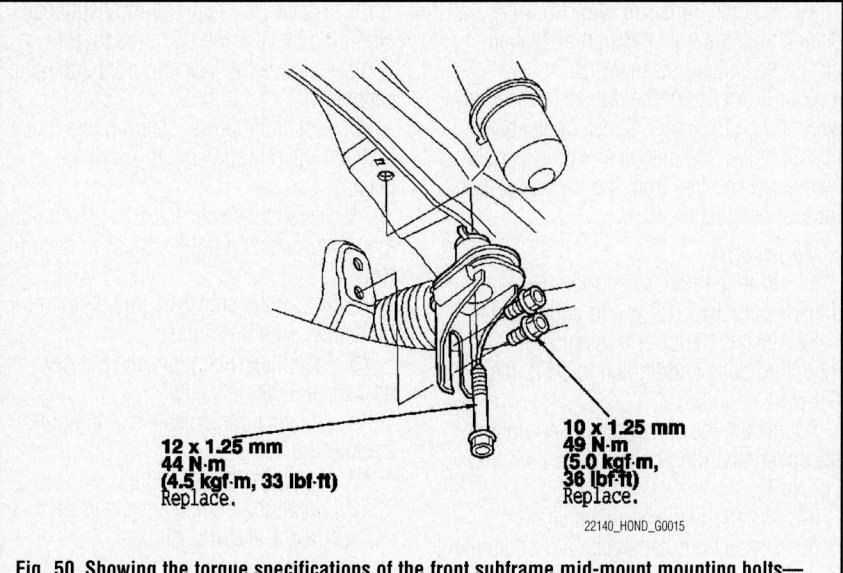

12 x 1.25 mm
44 N·m
(4.5 kgf·m, 33 lbf·ft)
Replace.

10 x 1.25 mm
49 N·m
(5.0 kgf·m,
36 lbf·ft)
Replace.

22140_HOND_G0015

Fig. 50 Showing the torque specifications of the front subframe mid-mount mounting bolts—2008 3.5L Accord

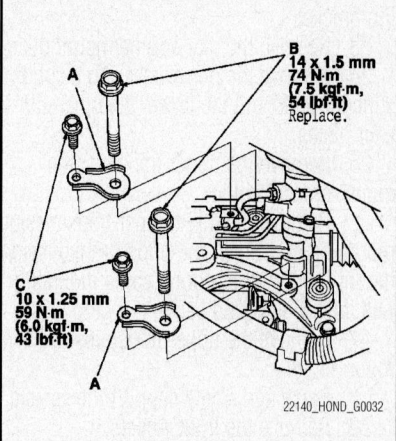

B
14 x 1.5 mm
74 N·m
(7.5 kgf·m,
54 lbf·ft)
Replace.

A

C
10 x 1.25 mm
59 N·m
(6.0 kgf·m,
43 lbf·ft)

A

22140_HOND_G0032

Fig. 52 When installing the steering stiffener plates (A), make sure the lower washers are correctly positioned before installing the power steering gear box mounting bolts (B) and stiffener plate bolts (C)—2008 3.5L Accord

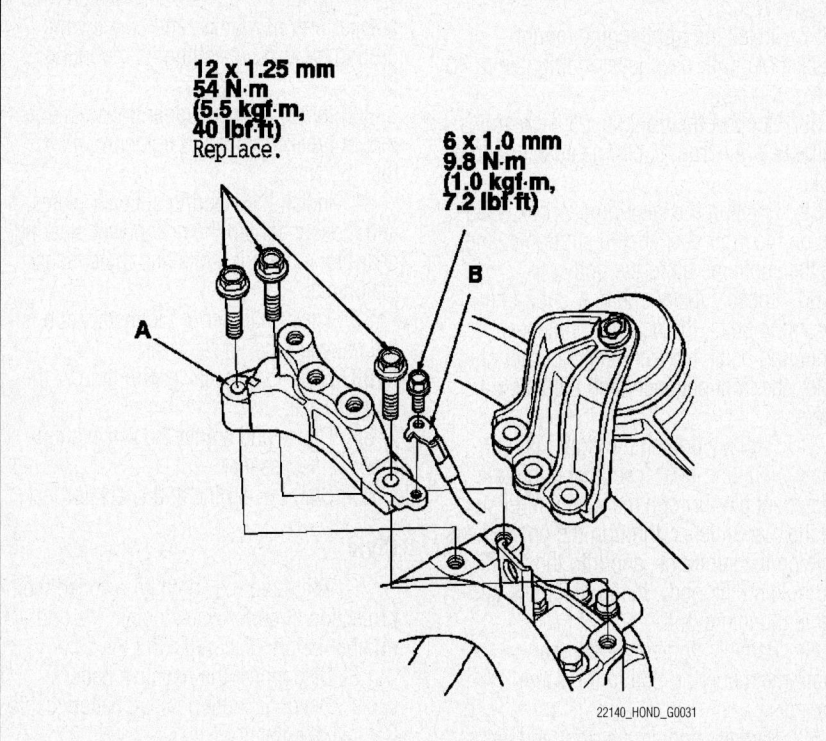

12 x 1.25 mm
54 N·m
(5.5 kgf·m,
40 lbf·ft)
Replace.

6 x 1.0 mm
9.8 N·m
(1.0 kgf·m,
7.2 lbf·ft)

B

A

22140_HOND_G0031

Fig. 51 Install the upper transmission mount bracket (A) with new bolts, and connect the ground cable (B) by installing its mounting bolt—2008 3.5L Accord

10. Disconnect the back-up light switch connector, the input shaft (mainshaft) speed sensor connector and the output shaft (countershaft) speed sensor connector, then remove the harness clips.

11. Disconnect the reverse lockout solenoid connector, remove the harness mounting bolt and the harness clamps, then remove the harness.

12. Remove the starter cable nut and the harness clamp, disconnect the starter cable, and disconnect the connector.

13. Remove the starter and the vacuum hose bracket.

14. Remove the harness bracket mounting bolt and the harness bracket.

15. Remove the lock pins, the shift cable bracket bolts, and the shift cable bracket,

then disconnect the shift cables from the select lever and the change lever. Carefully remove both cables and the bracket together to avoid bending the cables.

16. Remove the slave cylinder mounting bolts and the bracket mounting nut, then carefully move the slave cylinder out of the way to avoid bending the clutch line.

17. Remove the connector bracket from the front cylinder head; use the bracket bolt hole to attach the engine hanger balance bar front arm.

18. Remove the harness clamp bracket from the rear cylinder head; use the bracket bolt hole to attach the 2008 V6 attachment arm.

19. Lift and support the engine with engine support hanger (AAR-T1256) and the engine hanger balance bar (VSB02C000019). Attach the front arm to the front cylinder head with several spacers and the bolt (10 x 1.25 mm). Attach the 2008 V6 attachment arm (SIL02C000033) to the rear cylinder head and the bolt (8 x 1.25mm), then attach the rear arm to the 2008-2009 V6 attachment arm.

20. Remove the front engine mount stop nuts, the front engine mount stop, the vacuum hose bracket, and the harness bracket, then remove the front engine mount bolt, and disconnect the vacuum hose.

21. Remove the power steering line holder mounting bolt, and the power steering hose clamp.

22. Remove the two heat shield bolts, the heat shield and the power steering gearbox mounting bracket bolts, and then

remove the power steering gearbox mounting bracket.

23. Remove the rear engine mount bolts.

24. Remove the power steering gearbox stiffener bolts and the power steering stiffener plates.

25. Remove the upper transmission mount bracket bolts.

26. Remove the three upper transmission mount bracket bolts, the upper transmission mount bracket, the ground cable mounting bolt, and the ground cable.

27. Remove the upper transmission mount bolts.

28. Raise and safely support the vehicle.

29. Remove the front wheels.

30. Remove the front splash shield.

31. Drain the transaxle fluid, then reinstall the drain plug.

32. Remove the damper fork.

33. Separate the knuckle ball joint from the lower arm.

34. Remove the exhaust pipe and gaskets.

35. Remove the subframe mid mount bolts and the subframe mid mount from both sides.

36. Remove the lower transmission mount mounting nuts.

37. Remove the power steering line holder bolt and the power steering line from the clamp.

38. Attach the front subframe adapter to the front suspension subframe and hang the belt of the subframe adapter over the front of the subframe, then secure the belt with its stop.

39. Raise the jack and line up the slots in the front subframe adapter arms with the bolt holes on the jack base, then securely attach them with four bolts.

40. Remove the four front stiffeners bolts, the front stiffeners, the four rear stiffeners bolts, the rear stiffeners, then remove the front suspension subframe.

41. Remove the front engine mount bracket bolts and the front engine mount bracket.

42. Remove the lower transmission mount bolts and lower transmission mount.

43. Pry the left halfshaft inboard joint from the differential using a prybar.

44. Drive the inboard joint of the right driveshaft off of the intermediate shaft using a drift and a hammer.

45. Remove the intermediate shaft.

46. Remove the clutch cover and the harness cover.

47. Support the transaxle assembly with a suitable transmission jack.

48. Remove the lower transmission mounting bolts.

49. Pull the transaxle away from the engine until the transmission mainshaft clears the clutch pressure plate.

50. Slowly lower the transaxle assembly about 6 in. (151 mm). Check once again that all hoses and electrical wiring are disconnected and free from the transaxle, then lower it completely.

To install:

51. With the transaxle assembly on the transmission jack, raise it to engine level.

52. Install the lower transmission mounting bolts and tighten to 54 ft. lbs. (74 Nm).

53. Install the clutch cover and tighten the cover mounting bolts to 108 inch lbs. (12 Nm).

54. Install the intermediate shaft.

55. Install both halfshafts. For additional information, refer to the following section, "Halfshafts, Removal & Installation."

56. Install the lower transmission mount with new bolts and tighten them to 40 ft. lbs. (54 Nm).

57. Install the front engine mount bracket (A), with new bolts and tighten to 40 ft. lbs. (54 Nm).

58. Support the front suspension subframe with the front subframe adapter and a jack.

59. Position the steering gearbox washers on the front suspension subframe, and lift the subframe up to the body.

60. Loosely install the new subframe mounting bolts, the new rear stiffener mounting bolts, the front stiffener mounting bolts, the front stiffeners and the rear stiffeners.

61. Partially tighten the right rear subframe mounting bolt; insert the subframe alignment pin through the positioning slot on the rear stiffener, through the positioning hole on the subframe, and into the positioning hole on the body, then tighten the subframe mounting bolt.

62. Partially tighten the left rear subframe mounting bolt in the same manner.

63. Partially tighten the right and left front subframe mounting bolts.

64. Tighten the right rear mounting bolt to the specified torque with the subframe alignment pin in the positioning hole.

65. Tighten the left rear mounting bolt to the specified torque with the subframe alignment pin in the positioning hole.

66. Tighten the right and left front mounting bolt to the specified torque.

67. Check that the positioning holes and slots are aligned using the subframe alignment pin.

68. Tighten the rear and front stiffener mounting bolts to the specified torque.

69. Remove the jack and front subframe adapter.

70. Install the power steering line holder bolt and the power steering line to the clamp.

71. Install the lower transmission mount mounting nuts and tighten to 33 ft. lbs. (44 Nm).

72. Install the subframe mid mount on both sides with new bolts.

73. Install exhaust pipe with the new gaskets and the new nuts.

74. Connect the knuckle ball joint onto the lower arm.

75. Install the damper fork.

76. Install the front wheels, and set them in the straight-ahead position.

77. Lower the vehicle.

78. Install the upper transmission mounting bolts and tighten to 54 ft. lbs. (74 Nm).

79. Install the upper transmission mount bracket with new bolts, and connect the ground cable by installing its mounting bolt.

80. Install the new upper transmission mount bracket bolts and tighten to 43 ft. lbs. (59 Nm).

81. Install the steering stiffener plates, and loosely tighten the new power steering gear box mounting bolts and the stiffener plate bolts.

82. The remainder of the installation is the reverse order of removal.

83. Refill the transaxle with fluid to the correct level.

84. Check and adjust the wheel alignment as necessary.

85. Start the engine and check for leaks.

Civic

1. The radio may contain a coded theft protection circuit. Always obtain the code number before disconnecting the battery.

2. Disconnect the negative battery cable. Disconnect the positive battery cable. Remove the battery.

3. Remove the cowl cover and under cowl panel.

4. Remove the air cleaner assembly. Remove the harness clips and the intake air duct. Remove the battery base with the reservoir tank.

5. Remove the clutch bracket and carefully remove the slave cylinder. Do not depress the clutch pedal once the slave cylinder has been removed.

6. Disconnect the backup switch electrical connector, the vehicle speed sensor,

the reverse solenoid lockout connector and the harness clips.

7. Remove the cable bracket, then disconnect the cables from the top of the transaxle housing. Carefully remove the cables and the bracket.

8. Remove the harness clips from the clutch cable bracket and the harness bracket. Remove the engine wire harness cover. Slide the harness forward off of the air cleaner housing mounting bracket.

9. Attach tool VSP02C000015, or equivalent to the holes in the cylinder head. Install the front leg of the tool on to the engine hanger in the front of the vehicle. This operation will support the engine and transaxle assembly.

10. Remove the two upper transaxle upper mounting bolts. Remove the under hood fuse/relay box and position it to the side.

11. Remove the ECM stay and position it to the side. Remove the clutch line clamp. Disconnect the ground cable. Remove the transaxle mount bracket bolts and nuts. Remove the transaxle mount bracket.

12. Raise and safely support the vehicle. Drain the transaxle fluid.

13. Remove the splash shield. Separate the lower arm. Remove the stiffener plate and mounting bracket from the steering gearbox. Disconnect the exhaust mounting rubber.

14. Remove the stiffener plate and harness clip. Remove the front engine mount bracket mounting bolt and nut. Remove the lower radiator hose from the front mount bracket.

15. Remove the front engine mount bracket from the transaxle and engine. Remove the rear engine mount bracket mounting bolt.

16. Remove the middle subframe mounting bolts. Install tool VSB02C000016, or equivalent, to the subframe. Raise the jack and line up the slots in the arms with the bolt holes on the corner of the jack base. Attach them securely.

17. Remove the front suspension subframe mounting bolts and front suspension subframe. Suspend the steering gearbox to the side.

18. Pry out the driveshafts inboard joint. Remove the intermediate shafts.

19. Remove the clutch cover. Position a transaxle jack under the transaxle assembly. Remove the transaxle mounting bolts. Carefully remove the transaxle from the vehicle.

To install:

20. Apply high temperature grease to the mainshaft splines, release fork contact points, and throw-out bearing. The manu-

facturer recommends part No. 08798-9002, Honda Super High temp Urea Grease.

21. Place the transaxle on a transaxle jack and raise it to the level of the engine.

22. Align the transaxle and engine. Be sure the transaxle case dowel pins are securely seated. Install the transaxle retaining bolts.

23. Continue the installation in the reverse order of the removal procedure.

24. Reprogram the power window control unit as follows.

25. Turn the ignition switch ON. Lower the window, all the way down. Open the driver's side door.

26. Turn the ignition switch OFF. Push and hold the driver's window DOWN switch. Turn the ignition switch ON. Release the driver's window DOWN switch. This must be done within five seconds.

27. Repeat the above step three more times. Wait one second.

28. Confirm that the AUTO UP and AUTO DOWN do not work. If they work repeat the procedure, paying close attention to the five second time limit.

29. Move the driver's window all the way down by holding the driver's window DOWN switch to the AUTO DOWN position.

30. Pull up and hold the driver's window UP switch to the AUTO UP position until the window reaches the fully closed position. Hold the switch for one second.

31. Confirm that the power window master switch has been reset by using the driver's window AUTO UP and DOWN function.

32. If the AUTO UP and DOWN feature is still not working, repeat the complete procedure, paying close attention to the five second time limit.

MANUAL TRANSMISSION ASSEMBLY

REMOVAL & INSTALLATION

S2000

1. Before servicing the vehicle, refer to the precautions in the beginning of this section.

2. Remove or disconnect the following:
   - Battery cables and the battery
   - Shift lever knob
   - Center console
   - Shift lever boot
   - Shift lever
   - Air cleaner housing
   - Steering shaft from the steering gear box
   - Alternator

- A/C compressor
- Exhaust manifold heat shields
- Upper starter mounting bolt
- Upper intake manifold bracket mounting bolt
- Suction valve hose
- Camshaft Position (CMP) sensor connectors
- Splash shield
- Steering gear box electrical connector
- Torque sensor connector
- Intake manifold bracket
- Heated Oxygen (HO$_2$S) sensor connectors
- Catalytic converter
- Exhaust manifold
- Driveshaft
- Shifter boot holder bolts
- Clutch slave cylinder
- Clutch release fork
- Lower transmission flange bolts

3. Support the front subframe with a floor jack and remove the two center mounting bolts.

4. Loosen the four outer mounting bolts 3 inches (75 mm).

5. Lower the front subframe until it is supported by the loosened bolts.

6. Support the transmission with the floor jack.

7. Remove or disconnect the following:
   - Rear transmission mount
   - Speed sensor connector and wiring harness
   - Upper transmission flange bolts
   - Transaxle

To install:

➡**Use new subframe bolts for assembly.**

8. Installation is the reverse of the removal procedure, while using the following torque values:
   - Transaxle flange bolts: 47 ft. lbs. (64 Nm)
   - Rear transmission mount bolts: 28 ft. lbs. (38 Nm)
   - Subframe mounting bolts: 14mm bolts to 85 ft. lbs. (116 Nm) and 12mm bolts to 43 ft. lbs. (59 Nm)
   - Clutch slave cylinder bolts: 16 ft. lbs. (22 Nm)
   - A/C compressor bolts: 33 ft. lbs. (44 Nm)
   - Steering shaft pinch bolt: 16 ft. lbs. (22 Nm)
   - Shift lever bolts: 86 inch lbs. (10 Nm)

CLUTCH

REMOVAL & INSTALLATION

See Figures 53 and 54.

1. Remove the transaxle assembly from the vehicle. For additional information, refer to the following section, "Manual Transaxle Assembly, Removal & Installation."

2. Install the ring gear holder, clutch alignment shaft and remover handle.

3. To prevent warping, unscrew the pressure plate mounting bolts in a criss-cross pattern in several steps, then remove the pressure plate.

4. Remove the clutch disc, clutch alignment shaft, and remover handle.

To install:

5. Temporarily install the clutch disc onto the splines of the transmission mainshaft. Make sure the clutch disc slides freely on the mainshaft.

6. Install the ring gear holder.

7. Apply a light coat of super high temp urea grease (P/N 08798-9002) to the crankshaft pilot bushing.

8. Apply super high temp urea grease (P/N 08798-9002) to the splines of the clutch disc, then install the clutch disc using the clutch alignment shaft and remover handle.

9. Install the pressure plate and the mounting bolts finger-tight.

10. Torque the mounting bolts to 19 ft.

lbs. (25 Nm) in a crisscross pattern. Tighten the bolts in several steps to prevent warping the diaphragm spring.

11. Remove the ring gear holder, clutch alignment shaft, and remover handle.

BLEEDING

1. Make sure the DOT 3 brake fluid level in the clutch reservoir is at the MAX (upper) level line Attach one end of a clear tube to the bleeder screw, and put the other end to the container. Loosen the bleeder screw to allow air to escape from the system.

2. Make sure there is an adequate supply of fluid in the reservoir, then slowly push the clutch pedal all the way down. Before releasing the pedal, have an assistant temporarily tighten the bleeder screw. Loosen the bleeder screw, and push the clutch pedal down again. Repeat this step until no more bubbles appear at the clear tube.

❄❄ WARNING

Make sure the fluid level on the reservoir does not go below MIN.

3. Tighten the bleeder screw securely.

4. Refill the brake fluid in the reservoir to the MAX (upper) level line.

FRONT DRIVESHAFT

REMOVAL & INSTALLATION

S2000

See Figures 55 and 56.

1. Raise and safely support the vehicle.

2. Remove the driveshaft protector.

3. Matchmark the driveshaft to the transmission companion flange.

4. Separate the driveshaft from the transmission.

5. Matchmark the driveshaft to the rear differential companion flange.

6. Separate the driveshaft form the rear differential, then remove the driveshaft.

To install:

7. Install the driveshaft onto the rear differential by aligning the matchmarks. Tighten the bolts to 36 ft. lbs. (49 Nm).

➡**If the driveshaft is replaced, align the white marks on the new driveshaft with the white mark on the differential.**

8. Install the driveshaft onto the transmission by aligning the matchmarks. Tighten the bolts to 36 ft. lbs. (49 Nm).

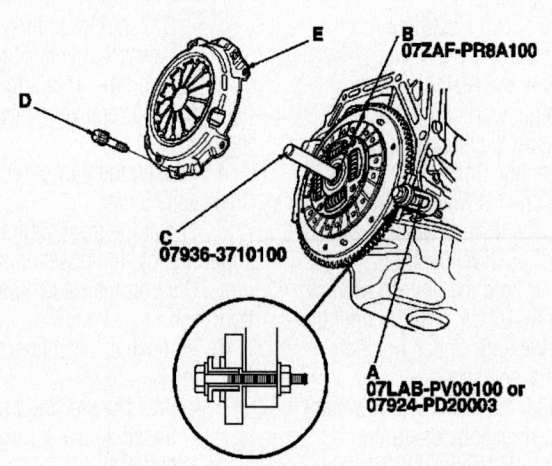

22140_HOND_G0037

Fig. 53 Install the ring gear holder (A), clutch alignment shaft (B), and remover handle (C)— Accord

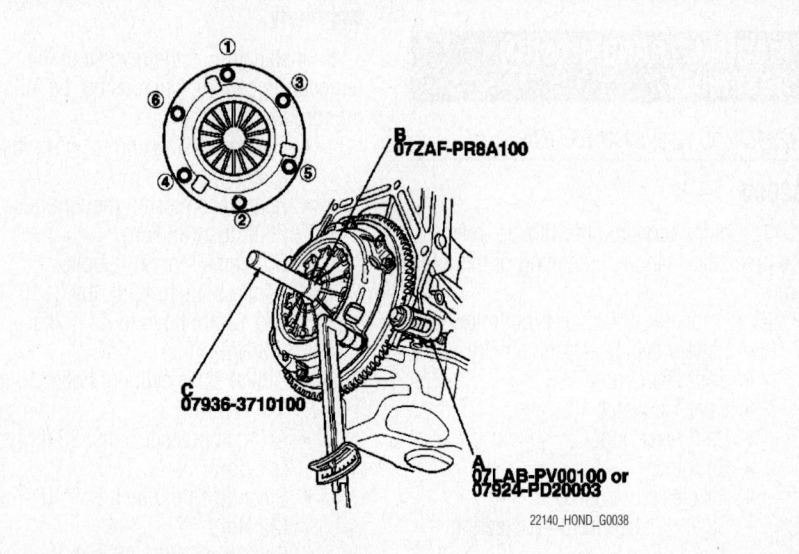

22140_HOND_G0038

Fig. 54 Torque sequence of the pressure plate—Accord

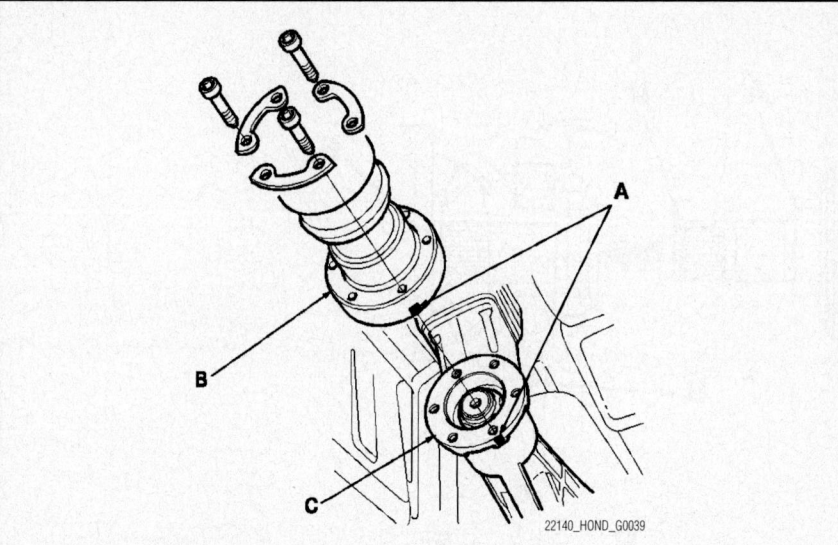

Fig. 55 Make reference marks (A) across the propeller shaft (B) and the transmission companion flange (C)—2007–08 S2000

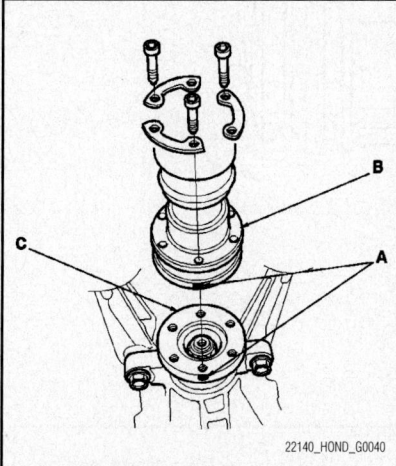

Fig. 56 Make reference marks (A) across the propeller shaft (B) and the rear differential companion flange (C)—2007–08 S2000

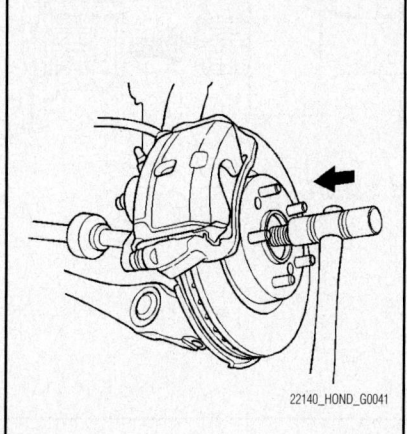

Fig. 57 Pull the knuckle outward, and remove the outboard joint from the front wheel hub using a plastic hammer—2007 2.4L Accord

➡️**If the driveshaft is replaced, align the white marks on the new driveshaft with the white mark on the transmission.**

9. Install the driveshaft protector, and tighten the bolts to 16 ft. lbs. (22 Nm).

FRONT HALFSHAFT

REMOVAL & INSTALLATION

Accord

See Figures 57 and 58.

1. Raise and safely support the vehicle.
2. Remove the front wheels.

3. Lift up the locking tab and remove the spindle nut.
4. Drain the transaxle fluid and reinstall the drain plug.
5. Separate the front stabilizer link from the lower control arm. For additional information, refer to the following section, "Stabilizer Bar, Removal & Installation."
6. Remove the self-locking nut, 12 mm flange bolt, and 10 mm flange bolt, then remove the damper fork.
7. Remove the cotter pin from the lower arm ball joint castle nut, and remove the nut, then separate the ball joint from the lower arm using the ball joint thread protector and remover.

8. Pull the knuckle outward, and remove the outboard joint from the front wheel hub using a plastic hammer.
9. Remove the heat shield, if equipped with automatic transmission.
10. If removing the left halfshaft:
    a. Pry the inboard joint from the transmission housing with a prybar.
    b. Remove the halfshaft as an assembly.
11. If removing the right halfshaft:
    a. If equipped with manual transmission: Drive the inboard joint off of the intermediate shaft using a drift and hammer.
    b. If equipped with automatic transmission: Pry the inboard joint from the transmission housing with a prybar.
    c. Remove the halfshaft as an assembly.

To install:

12. Apply about 0.18 oz moly 60 paste (P/N 08734-0001) to the contact area of the outboard joint and the front wheel bearing.
13. Install a new set ring onto the set ring groove of the halfshaft.
14. If equipped with manual transmission: Apply 0.02–0.04 oz of grease to the whole splined surface of the right halfshaft. After applying grease, remove the grease from the splined grooves at intervals of 2–3 splines and from the set ring groove so that air can bleed from the intermediate shaft.
15. Clean the areas where the halfshaft contacts the differential thoroughly with solvent or brake cleaner, and dry with compressed air.
16. Insert the inboard end of the halfshaft into the differential or intermediate shaft until the set ring locks in the groove.
17. Install the outboard joint into the front hub.
18. If equipped with automatic transmission, install the heat shield and tighten the mounting bolts to 16 ft. lbs. (22 Nm).
19. Clean off any grease contamination from the ball joint tapered section and threads, then install the knuckle onto the lower arm.

✳✳ WARNING

Be careful not to damage the ball joint boot.

20. Wipe off the grease before tightening the nut at the ball joint. Tighten the new castle nut to 65–72 ft. lbs. (88–98 Nm), then tighten it only far enough to align the slot with the ball joint pin hole.
21. Install a new cotter pin into the ball joint pin hole.

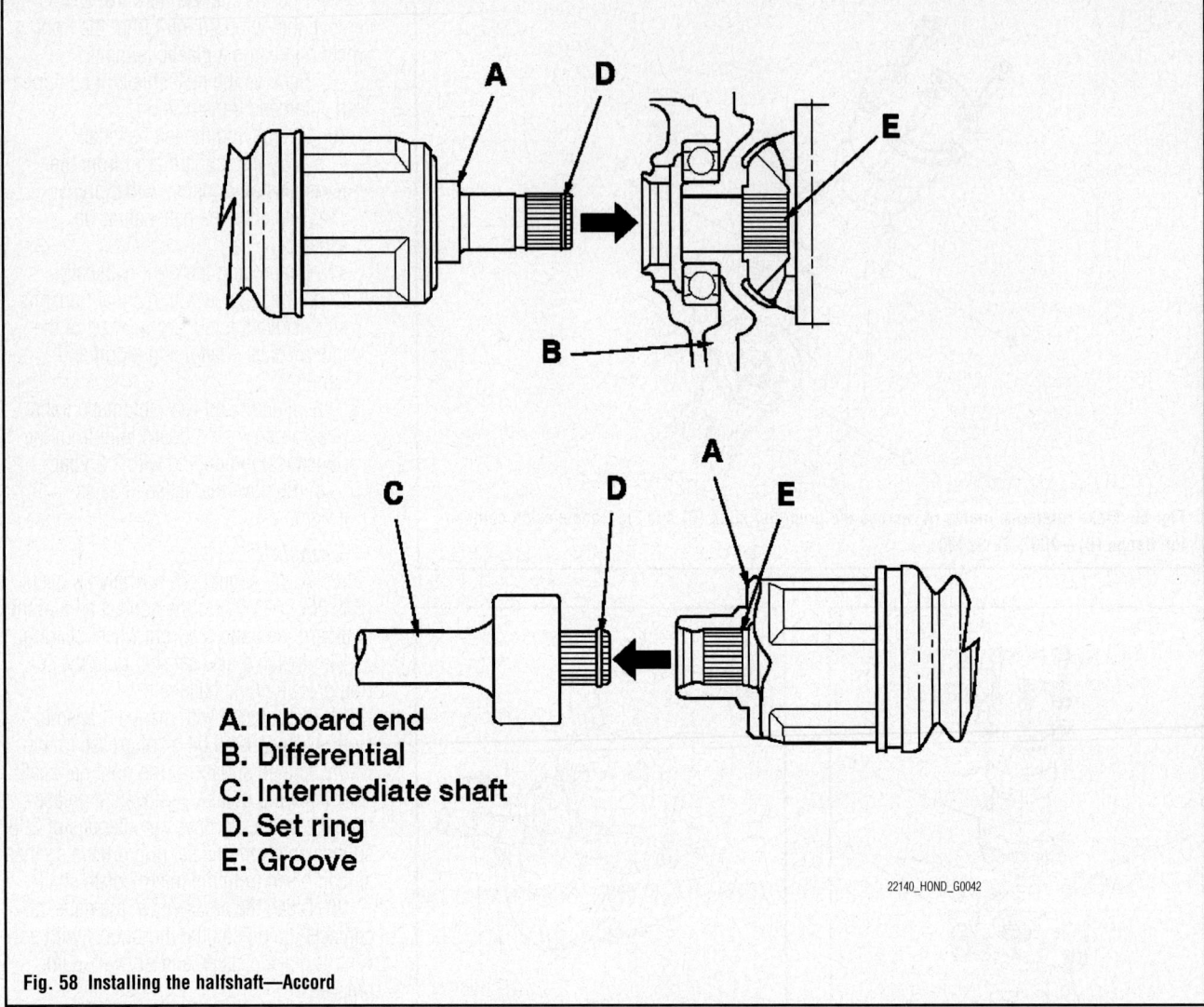

A. Inboard end
B. Differential
C. Intermediate shaft
D. Set ring
E. Groove

22140_HOND_G0042

Fig. 58 Installing the halfshaft—Accord

22. Install the damper fork over the half-shaft and onto the lower arm. Install the damper in the damper fork so the aligning tab is aligned with the slot in the damper fork. Loosely install the flange bolt.

23. Loosely install the flange bolt and a new self-locking nut.

24. Connect the front stabilizer link to the lower arm.

25. Apply a small amount of engine oil to the seating surface of the new spindle nut.

26. Install a new spindle nut, then tighten the nut as follows:
- 2007 Automatic transmission: 134 ft. lbs. (181 Nm)
- 2007 Manual transmission: 181 ft. lbs. (245 Nm)
- 2008 Models: 242 ft. lbs. (328 Nm)

27. After tightening, use a drift to stake the spindle nut shoulder against the half-shaft.

28. Install the front wheels.

29. Turn the front wheel by hand, and make sure there is no interference between the halfshaft and surrounding parts.

30. Tighten the flange bolt and the self-locking nut with the vehicle's weight on the damper.

31. Refill the transmission with the recommended transmission fluid:

32. Check and adjust the front wheel alignment as necessary.

Civic

1. Raise and safely support the vehicle.

2. Remove the front wheels.

3. Pry up the locking tab on the spindle nut, and then remove the nut.

4. Drain the transaxle fluid and reinstall the drain plug.

5. Remove the nuts and the bolt, and then separate the lower arm using a prybar.

6. Separate the halfshaft outboard joint from the front hub using a plastic hammer.

7. Pull the knuckle outward, and remove the halfshaft outboard joint from the front hub.

8. If removing the left halfshaft:
   a. Pry the inboard joint from the differential using a prybar.
   b. Remove the halfshaft as an assembly.

9. If removing the right halfshaft:
   a. Drive the inboard joint off of the intermediate shaft using a drift and a hammer, if equipped with manual transmission.
   b. Pry the inboard joint from the differential using a prybar, if equipped with automatic transmission.
   c. Remove the halfshaft as an assembly.

To install:

10. Apply about 0.18 oz moly 60 paste (P/N 08734-0001) to the contact area of the outboard joint and the front wheel bearing.

11. Install a new set ring onto the set ring groove of the halfshaft and intermediate shaft.

12. If equipped with manual transmission: Apply 0.02–0.04 oz of grease to the whole splined surface of the right halfshaft. After applying grease, remove the grease from the splined grooves at intervals of 2–3 splines and from the set ring groove so that air can bleed from the intermediate shaft.

13. Clean the areas where the halfshaft contacts the differential thoroughly with solvent or brake cleaner, and dry with compressed air.

14. Insert the inboard end of the halfshaft into the differential or intermediate shaft until the set ring locks in the groove.

15. Install the outboard joint into the front hub.

16. Install the knuckle onto the lower arm using a new flange bolt and new self-locking nuts. After lightly tightening all three fasteners, tighten them in the following order to 43 ft. lbs. (59 Nm).
- Nut on the front
- Nut on the rear
- Bolt

17. Apply a small amount of engine oil to the seating surface of new spindle nut.

18. Install the spindle nut, then tighten the nut as follows:
- All models except Si: 134 ft. lbs. (181 Nm)
- Si models: 180 ft. lbs. (245 Nm)

19. After tightening, use a drift to stake the spindle nut shoulder against the halfshaft.

20. Install the front wheels.

21. Turn the front wheels by hand, and make sure there is no interference between the halfshaft and surrounding parts.

22. Refill the transaxle with fluid to the correct level.

23. Check and adjust the wheel alignment as necessary.

CV-JOINTS OVERHAUL

Accord & Civic

Inboard Side

See Figures 59 through 62.

1. Remove the set ring.
2. Remove the boot bands as follows:
- If the boot band is a welded type, cut the boot band.
- If the boot band is a double loop type, lift up the band end, and push it into the clip.
- If the boot band is a low profile type, pinch the boot band using commercially available boot band pliers.

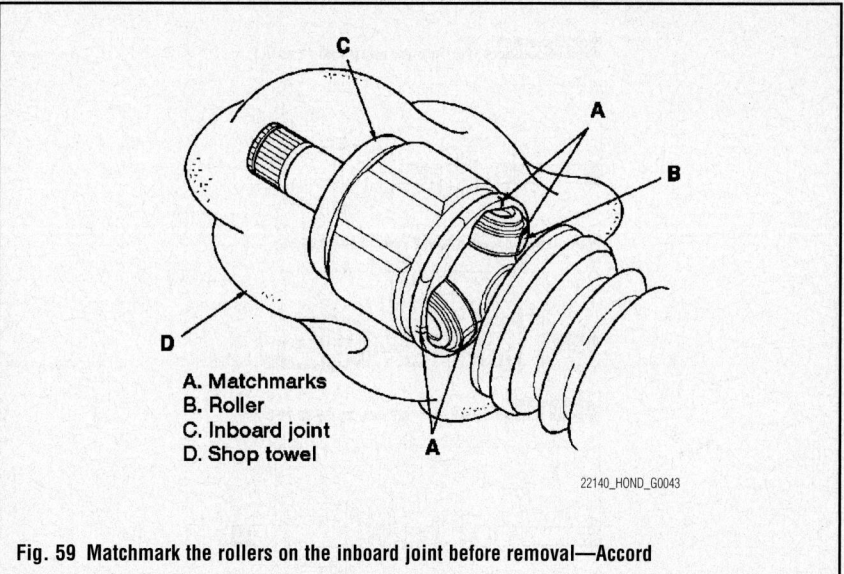

A. Matchmarks
B. Roller
C. Inboard joint
D. Shop towel

22140_HOND_G0043

Fig. 59 Matchmark the rollers on the inboard joint before removal—Accord

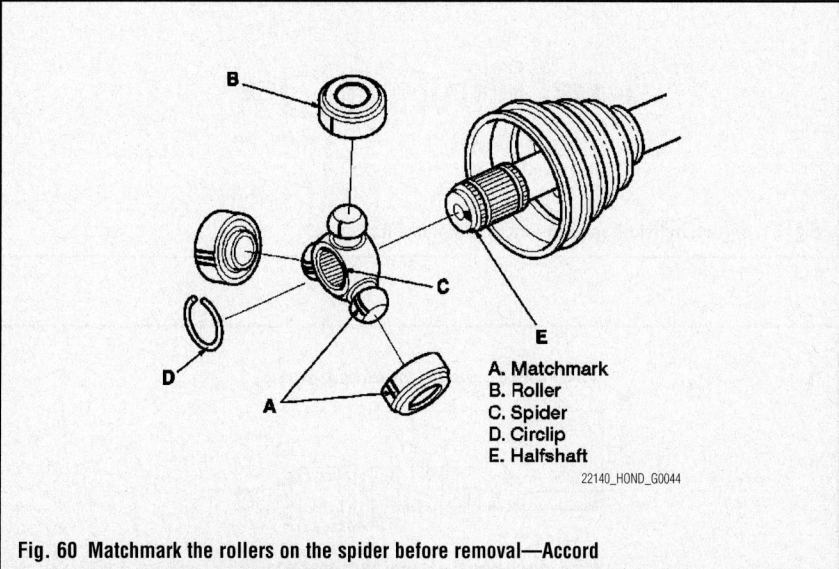

A. Matchmark
B. Roller
C. Spider
D. Circlip
E. Halfshaft

22140_HOND_G0044

Fig. 60 Matchmark the rollers on the spider before removal—Accord

✳✳ WARNING

Be careful not to damage the boot and dynamic damper.

3. Make a mark on each roller and inboard joint to identify the locations of rollers and grooves in the inboard joint. Then remove the inboard joint on the shop towel. Be careful not to drop the rollers when separating them from the inboard joint.

4. Matchmark the rollers and spider to identify the locations of the rollers on the spider, the remove the rollers.

5. Remove the circlip.

6. Matchmark the spider to the halfshaft to identify the position of the spider on the shaft.

7. Remove the spider.

8. Wrap the splines on the halfshaft with vinyl tape to prevent damage to the boot.

9. Remove the inboard boot. Be careful not to damage the boot.

10. Remove the vinyl tape.

To install:

11. Wrap the splines with vinyl tape to prevent damage to the inboard boot.

12. Install the inboard boot onto the halfshaft and then remove the vinyl tape. Be careful not to damage the inboard boot.

13. Install the spider onto the halfshaft by aligning the marks you made on the spider and the end of the halfshaft.

14. Install a new circlip into the halfshaft groove. Always rotate the circlip in its groove to make sure it is fully seated.

15. Fit the rollers onto the spider with the high shoulders facing outward, aligning the matchmarks.

16. Pack the inboard joint with the joint grease included in the new inboard boot set.

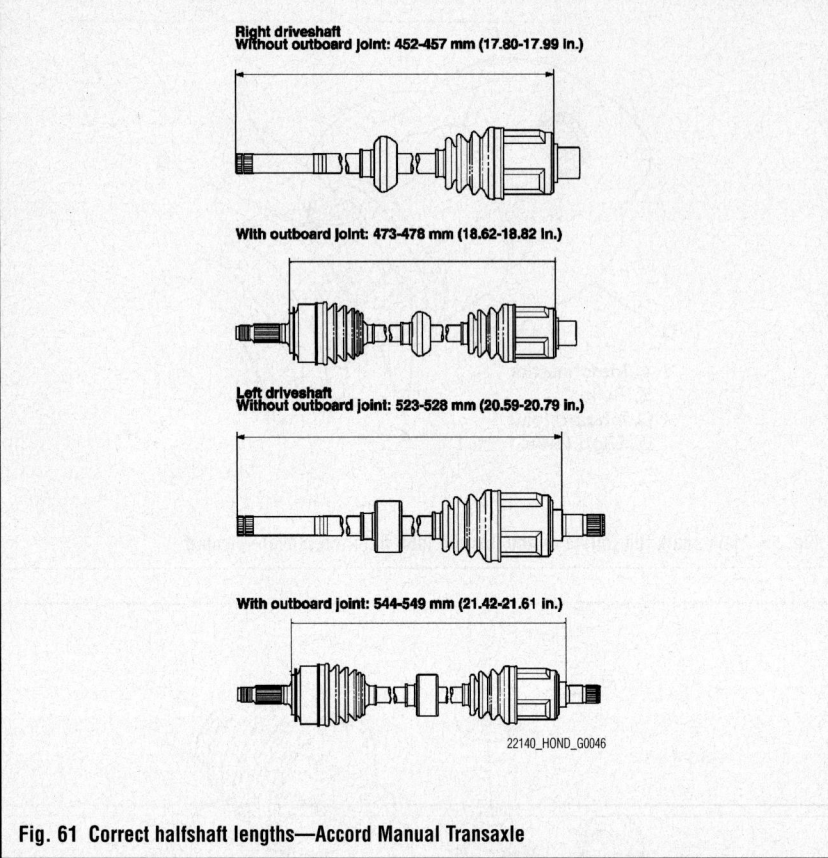

Right driveshaft
Without outboard joint: 452-457 mm (17.80-17.99 in.)

With outboard joint: 473-478 mm (18.62-18.82 in.)

Left driveshaft
Without outboard joint: 523-528 mm (20.59-20.79 in.)

With outboard joint: 544-549 mm (21.42-21.61 in.)

22140_HOND_G0046

Fig. 61 Correct halfshaft lengths—Accord Manual Transaxle

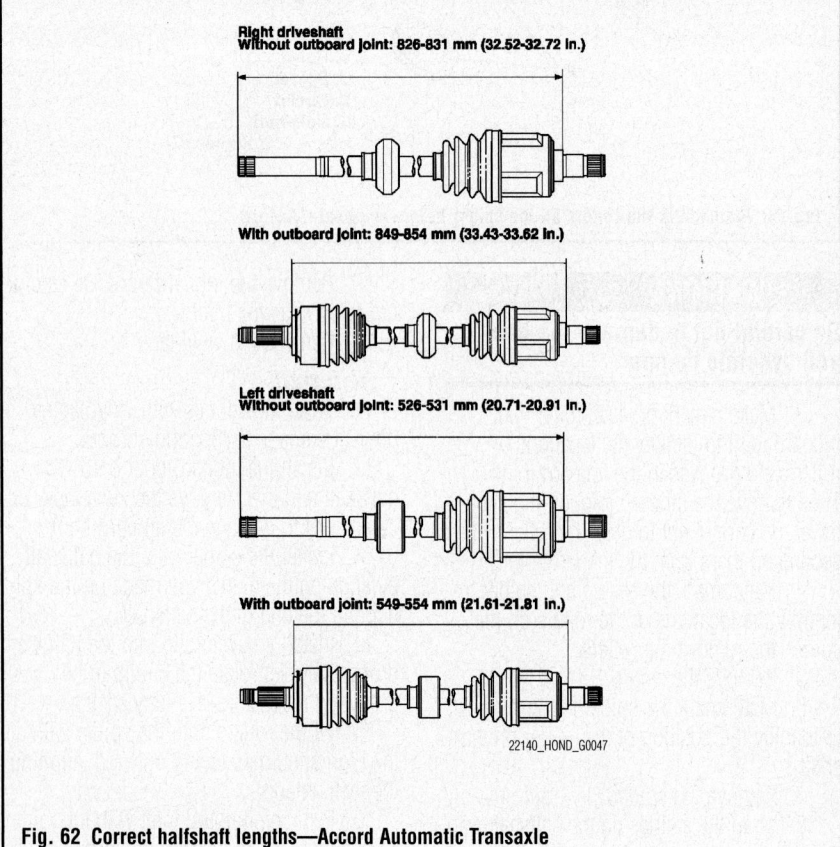

Right driveshaft
Without outboard joint: 826-831 mm (32.52-32.72 in.)

With outboard joint: 849-854 mm (33.43-33.62 in.)

Left driveshaft
Without outboard joint: 526-531 mm (20.71-20.91 in.)

With outboard joint: 549-554 mm (21.61-21.81 in.)

22140_HOND_G0047

Fig. 62 Correct halfshaft lengths—Accord Automatic Transaxle

17. Fit the inboard joint onto the half-shaft aligning the matchmarks.

18. Fit the boot ends onto the halfshaft and the inboard joint.

19. Adjust the length of the halfshafts to the figure as shown, then adjust the boots to halfway between full compression and full extension.

20. Install the boot bands.

Outboard Side

See Figures 63 and 64.

1. Remove the boot bands by lifting up the three tabs with a screwdriver. Be careful not to damage the boot and dynamic damper.

2. Slide the outboard boot partially to the inboard joint side. Be careful not to damage the boot.

3. Wipe off the grease to expose the halfshaft and the outboard joint inner race.

4. Matchmark the halfshaft at the same level as the outboard joint rim.

5. Securely clamp the halfshaft in a bench vise with a shop towel.

6. Remove the outboard joint using the threaded adapter and a commercially available 5/8"-18 UNF slide hammer.

7. Remove the halfshaft from the bench vise.

8. Remove the stop ring from the half-shaft.

9. Wrap the splines on the halfshaft with vinyl tape to prevent damaging the boot.

10. Remove the outboard boot. Be careful not to damage the boot.

11. Remove the vinyl tape.

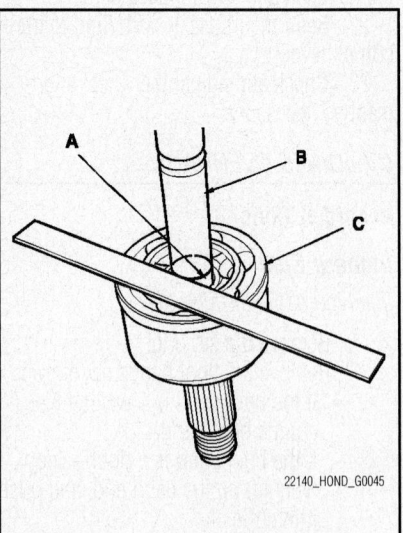

22140_HOND_G0045

Fig. 63 Make a mark (A) on the halfshaft (B) at the same level as the outboard joint rim (C)—Accord

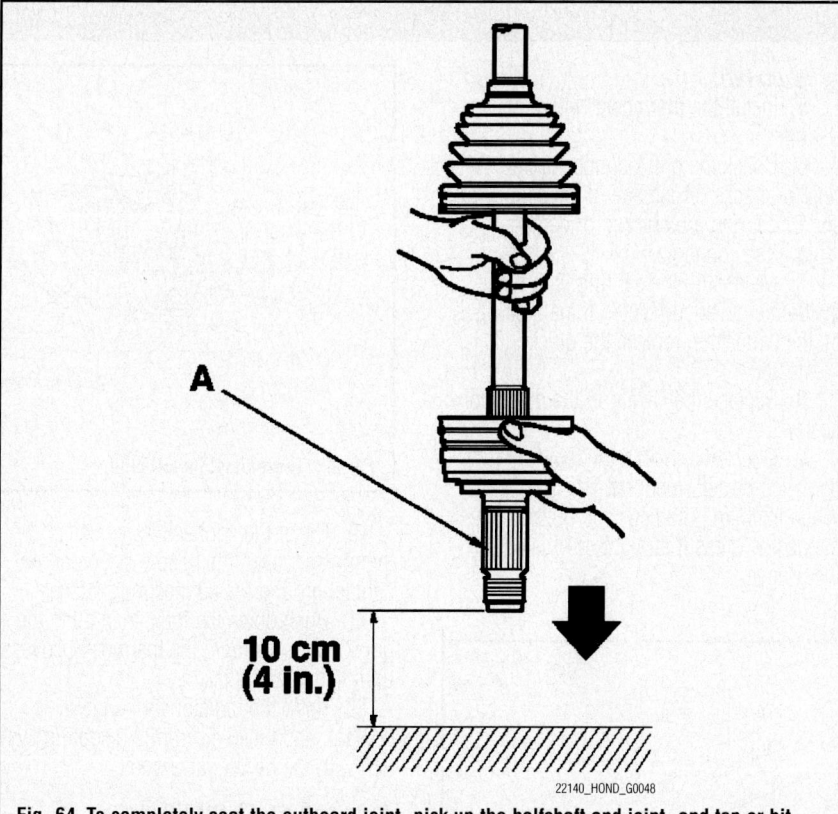

Fig. 64 To completely seat the outboard joint, pick up the halfshaft and joint, and tap or hit them from a height of about 4 in. (10 cm) onto a hard surface—Accord

22140_HOND_G0048

To install:

12. Wrap the splines with vinyl tape to prevent damaging the outboard boot.

13. Install the new ear clamp bands and outboard boot, then remove the vinyl tape. Be careful not to damage the outboard boot.

14. Install the new stop ring in the halfshaft groove.

15. Pack about half of the grease included in the new outboard boot set into the halfshaft hole in the outboard joint.

16. Insert the halfshaft into the outboard joint until the stop ring is close to the joint.

17. To completely seat the outboard joint, pick up the halfshaft and joint, and tap or hit them from a height of about 4 in. (10 cm) onto a hard surface.

18. Check the alignment of the paint mark you made with the outboard joint rim.

19. Pack the outboard joint with the remaining joint grease included in the new outboard boot set.

20. Fit the boot ends onto the halfshaft and outboard joint.

21. Close the ear portion of the band with commercially available boot band pliers.

22. Check the clearance between the closed ear portion of the band. If the clearance is not within the 0.12 in. (3 mm), close the ear portion of the band tighter.

23. Repeat the process for the band on the other side of the boot.

REAR HALFSHAFT

REMOVAL & INSTALLATION

S2000

1. Raise and safely support the vehicle.
2. Remove the rear tires.
3. Lift up the locking tab on the spindle nut. Remove the nut.
4. Remove the cotter pin from the lower arm ball joint castle nut. Remove the nut. Separate the ball joint from the lower arm.
5. Remove the wheel sensor harness from the upper arm.
6. Make reference marks across the inboard joint and the rear differential.
7. Remove the six inboard joint mounting bolts and nuts. Remove the inboard joint from the rear differential.
8. Pull the knuckle outward and remove the outboard joint from the wheel hub using a plastic hammer.
9. Remove the halfshaft.

To install:

10. Installation is the reverse of the removal procedure.

11. After the front tire is installed, turn the front wheel by hand and make sure there is no interference between the halfshaft and surrounding parts.

12. Check the rear wheel alignment. Adjust if necessary.

REAR PINION SEAL

REMOVAL & INSTALLATION

S2000

1. Raise and safely support the vehicle.
2. Remove the rear differential.
3. Mount the rear differential in a bench vise.
4. Remove the output shafts from the differential using pry bars.
5. Remove the ten mounting bolts in a crisscross pattern in several steps, then remove the differential case.
6. Make marks on the bearing cap, the adjustment screw, and the differential carrier.
7. Remove the lock plates and the bearing caps.
8. Remove the adjustment screws, the bearing outer races, and the Torsen LSD assembly.
9. Install the holder handle and the companion flange holder on the companion flange, then remove the locknut and the pinion washer.
10. Remove the oil seal from the differential carrier.

To install:

11. Using an oil seal driver tool, install the oil seal into differential carrier.

12. Apply molybdenum grease to the surface end of the companion flange, then install the companion flange, the drive pinion washer, and a new locknut.

13. Using a holder and companion flange holder, tighten the locknut to 14 ft. lbs. (20 Nm).

14. Rotate the drive pinion several times to ensure proper tapered roller bearing contact. Measure the drive pinion turning torque before tightening the locknut to the specified torque.

15. Tighten the locknut to 94 ft. lbs. (127 Nm)., then remove the holder handle and the companion flange holder.

16. Rotate the drive pinion several times to assure proper tapered roller bearing contact. Measure the drive pinion turning torque. If the drive pinion turning torque exceeds the standard, replace the pinion spacer.

17. The remainder of the installation is the reverse order of removal.

ENGINE COOLING

THERMOSTAT

REMOVAL & INSTALLATION

Accord

2.4L Engine

See Figures 65 through 67.

1. Drain the engine coolant.
2. Clean any dirt off the quick connector, thermostat cover, and lower radiator hose.
3. Pull out lock by hand, then wiggle the quick connector loose, and remove it from the thermostat cover. Do not use any tools to remove the quick connector.
4. Remove the thermostat.

To install:

5. Install the thermostat with a new O-ring.
6. Check the quick connector and set ring for cracks or damage. If the connector and/or set ring are cracked or damaged, replace the connector.
7. Make sure the set ring is in place inside the quick connector. If the set ring is off the connector, replace the quick connector.
8. Replace the O-ring in the quick connector.
9. Check the lock. If the lock is damaged or deformed, replace it. When installing the new lock on the connector, press it straight down along the groove.

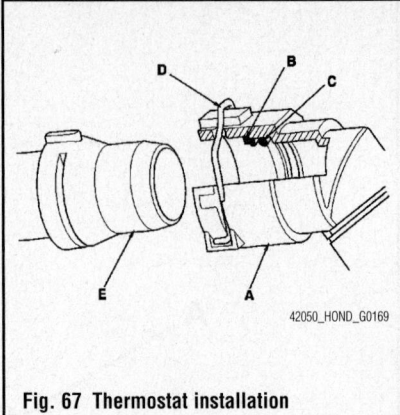

Fig. 67 Thermostat installation

10. Clean the connecting surface of the thermostat cover, then apply clean engine coolant around the connecting surface.
11. Push down the lock, then push the quick connector onto the thermostat cover until you hear it click.
12. Refill the radiator with engine coolant, and bleed air from the cooling system with the heater valve open.

3.0L & 3.5L ENGINES

See Figure 68.

1. Make sure you have the anti-theft codes for the audio unit (and navigation code if equipped), then write down the audio presets. Make sure the ignition switch is OFF.
2. Disconnect the negative cable from the battery first, then the positive cable, then remove the battery.
3. Drain the engine coolant.
4. Remove the ground cable and thermostat cover, then remove the thermostat.

To install:

5. Install the thermostat with new rubber seal.
6. Install the battery. Clean the battery posts and cable terminals with sandpaper, then assemble them and apply grease to prevent corrosion.
7. Refill the radiator with engine coolant, then bleed air from the cooling system.

Civic

1.8L ENGINE

See Figure 69.

1. Drain the engine coolant.
2. Remove the harness clamp bracket and thermostat cover, then remove the thermostat.
3. Install the thermostat with a new rubber seal.

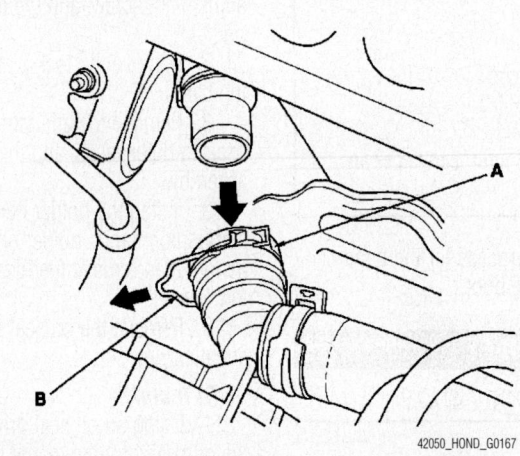

Fig. 65 Clean the dirt off of the quick connector (A), then pull out lock (B) by hand, then wiggle the quick connector loose, and remove it from the thermostat cover

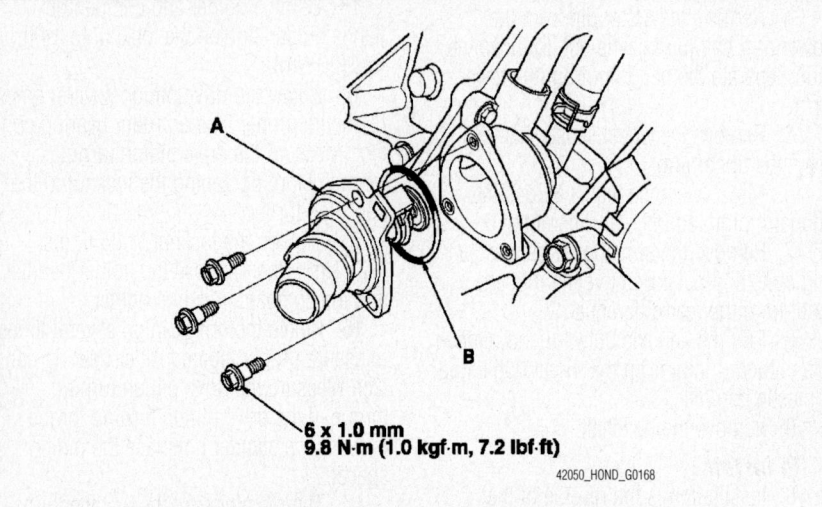

6 x 1.0 mm
9.8 N·m (1.0 kgf·m, 7.2 lbf·ft)

Fig. 66 Always use a new O-ring (B) when installing the thermostat (A)

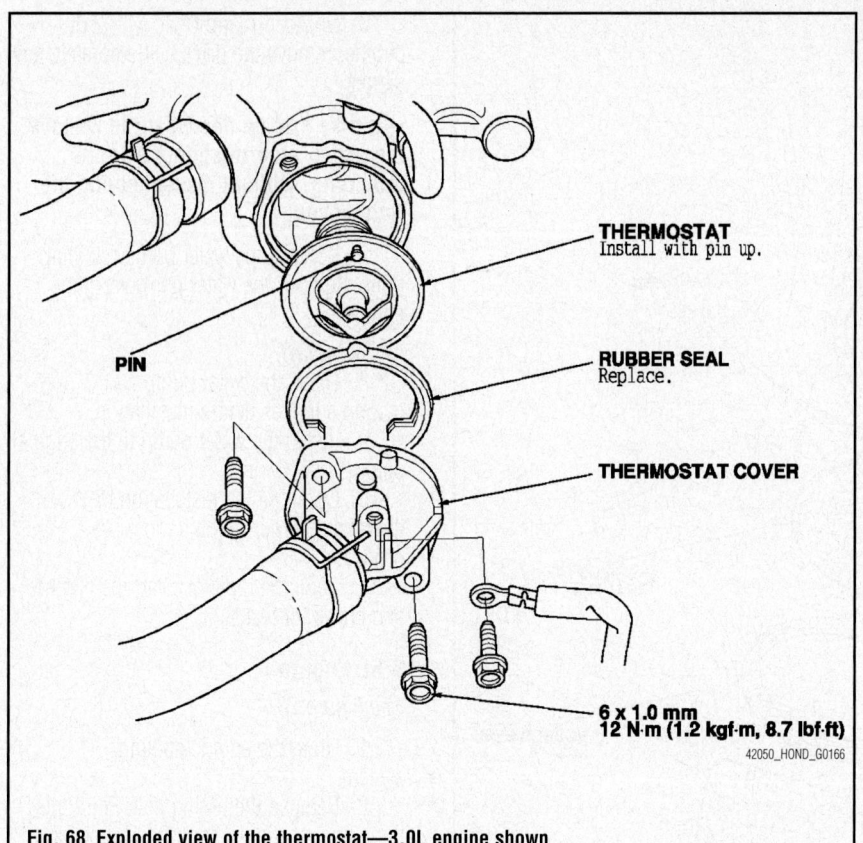

THERMOSTAT
Install with pin up.

PIN

RUBBER SEAL
Replace.

THERMOSTAT COVER

6 x 1.0 mm
12 N·m (1.2 kgf·m, 8.7 lbf·ft)

42050_HOND_G0166

Fig. 68 Exploded view of the thermostat—3.0L engine shown

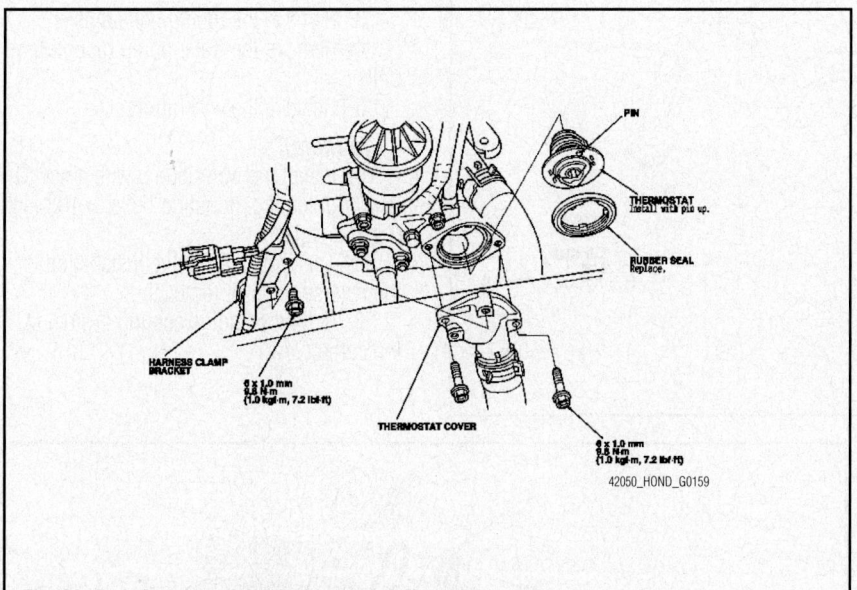

HARNESS CLAMP
BRACKET

6 x 1.0 mm
9.8 Nm
(1.0 kgf·m, 7.2 lbf·ft)

THERMOSTAT COVER

PIN

THERMOSTAT
Install with pin up.

RUBBER SEAL
Replace.

6 x 1.0 mm
9.8 Nm
(1.0 kgf·m, 7.2 lbf·ft)

42050_HOND_G0159

Fig. 69 Exploded view of the thermostat—1.8L engine

4. Refill the radiator with engine coolant, then bleed air from the cooling system.

5. Clean up any spilled engine coolant.

2.0L Engine

See Figure 70.

1. Drain the engine coolant.
2. Remove the splash shield.

3. Remove the lower hose, then remove the thermostat.

Civic Hybrid

1. Drain the engine cooling system.
2. Remove the air intake assembly.
3. Disconnect the upper radiator hose, lower radiator hose, heater hoses and water bypass hose.

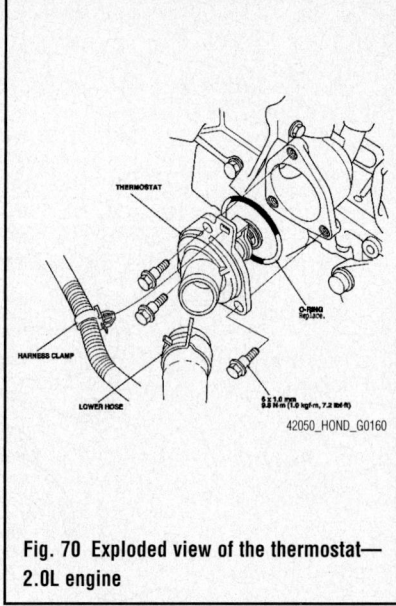

THERMOSTAT

O-RING
Replace.

HARNESS CLAMP

LOWER HOSE

6 x 1.0 mm
9.8 Nm (1.0 kgf·m, 7.2 lbf·ft)

42050_HOND_G0160

Fig. 70 Exploded view of the thermostat—2.0L engine

4. Remove the water passage, then remove the thermostat.

To install:

5. Remove all of the old liquid gasket from the water passage, thermostat housing mating surfaces, the bolts, and the bolt holes.

6. Apply a 1.5 mm wide bead of the liquid gasket on the water passage, P/N 08717-0004, 08718-0001, 08718-0003, or 08718-0009, along the broken lines. Install the component within 5 minutes of applying the liquid gasket.

7. Install the rubber seal on the thermostat, then install the thermostat with the pin up.

8. Install the water passage and tighten the mounting bolts to 108 inch lbs. (12 Nm).

9. The remainder of the installation is the reverse order of removal.

10. Refill the engine cooling system to the correct level.

S2000

See Figure 71.

1. Drain the engine coolant.
2. Remove the lower radiator hose and the thermostat cover, then remove the thermostat.

To install:

3. Install the thermostat with a new rubber seal.

4. Install the thermostat cover and the lower radiator hose.

5. Refill the radiator with engine coolant, then bleed air from the cooling system.

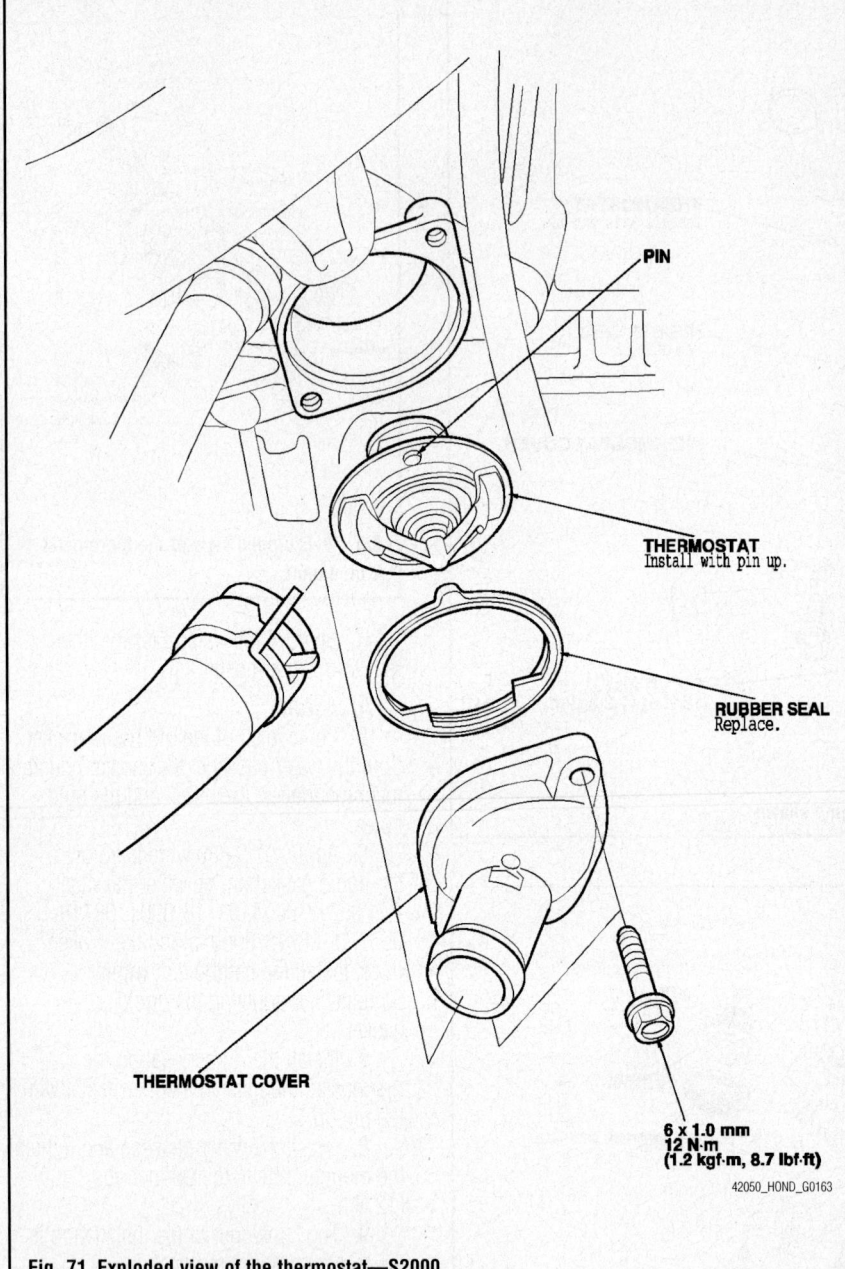

PIN

THERMOSTAT
Install with pin up.

RUBBER SEAL
Replace.

THERMOSTAT COVER

6 x 1.0 mm
12 N·m
(1.2 kgf·m, 8.7 lbf·ft)

42050_HOND_G0163

Fig. 71 Exploded view of the thermostat—S2000

6. On Accord and Civic with 2.0L engine, remove the drive belt automatic tensioner.

→It may first be necessary to remove the power steering pump from its mounting, without disconnecting the fluid hoses.

7. Remove the water pump retaining bolts. Remove the water pump from the engine.

To install:

8. Clean the water pump and O-ring mating surfaces before installation.

9. Install the water pump to the engine using a new O-ring.

10. Continue the installation in the reverse order of the removal procedure.

11. Refill the radiator. Start the engine and check for leaks.

3.5L Engine

See Figure 73.

1. Drain the engine cooling system.

2. Remove the timing belt. For additional information, refer to the following section, "Timing Belt, Removal & Installation."

3. Remove the timing belt adjuster.

4. Remove the water pump mounting bolts.

5. Remove the water pump.

To install:

6. Install the water pump with a new O-ring. Tighten the mounting bolts to 108 inch lbs. (12 Nm).

7. The remainder of the installation is the reverse order of removal.

8. Refill the engine cooling system to the correct level.

WATER PUMP

REMOVAL & INSTALLATION

1.8L, 2.0L, 2.2L and 2.4L Engines

See Figure 72.

1. Before servicing the vehicle, refer to the precautions in the beginning of this section.

2. Drain the cooling system.

3. Remove the drive belt.

4. On 2007 Civic with 2.0L engine, remove the crankshaft pulley.

5. On S2000, remove the water pump pulley.

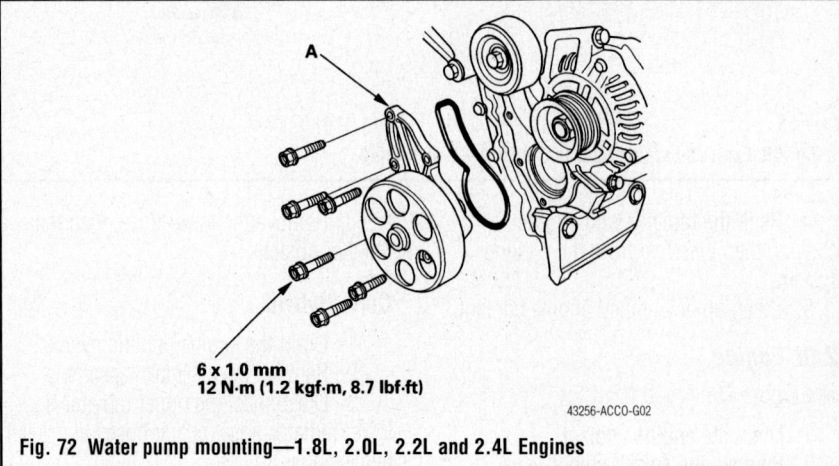

A

6 x 1.0 mm
12 N·m (1.2 kgf·m, 8.7 lbf·ft)

43256-ACCO-G02

Fig. 72 Water pump mounting—1.8L, 2.0L, 2.2L and 2.4L Engines

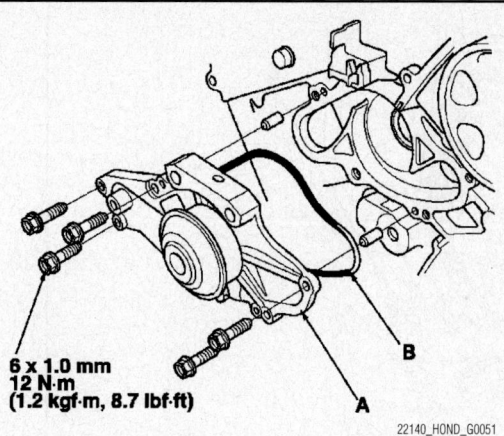

Fig. 73 Install the water pump (A) with a new O-ring (B)—3.5L Engine

Civic Hybrid

See Figure 74.

1. Drain the engine cooling system.
2. Remove the engine undercover and splash shield.
3. Remove the accessory drive belt.
4. Remove the water pump pulley mounting bolts.
5. Remove the auto tensioner.
6. Remove the five bolts securing the water pump, then remove the water pump with water pump pulley.

To install:

7. Install the water pump with a new O-ring and tighten the mounting bolts to 108 inch lbs. (12 Nm).
8. The remainder of the installation is the reverse order of removal.
9. Refill the engine cooling system to the correct level.
10. Start the engine and check for leaks.

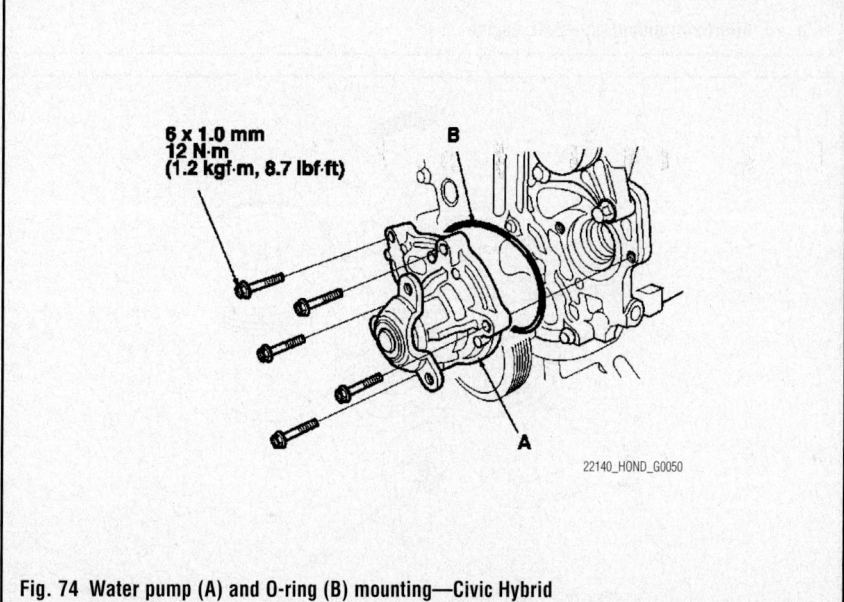

Fig. 74 Water pump (A) and O-ring (B) mounting—Civic Hybrid

ENGINE ELECTRICAL

ALTERNATOR

REMOVAL & INSTALLATION

Accord

2.4L Engine

See Figure 75.

1. Before servicing the vehicle, refer to the precautions in the beginning of this section.
2. Note the radio security code and the radio presets.
3. Remove or disconnect the following:
   • Negative battery cable, then the positive

• Drive belt
• Auto-tensioner, 2007 models only
• Connectors from the alternator
• Alternator mounting bolts and the alternator

To install:
• Alternator and mounting bolts. Torque the bolts to 16 ft. lbs. (22 Nm).
• Electrical connectors
• Auto-tensioner
• Drive belt
• Positive, then negative battery cables

4. Enter the security code and radio presets.

CHARGING SYSTEM

3.0L Engine

See Figures 76 through 78.

1. Before servicing the vehicle, refer to the precautions in the beginning of this section.
2. Note the radio security code and the radio presets.
3. Remove or disconnect the following:
   • Negative battery cable, then the positive
   • Alternator belt tension by pulling back on the adjuster and then remove the belt
   • Condenser fan motor connector from the shroud
   • Condenser fan assembly
   • Four prong connector from the rear of the alternator

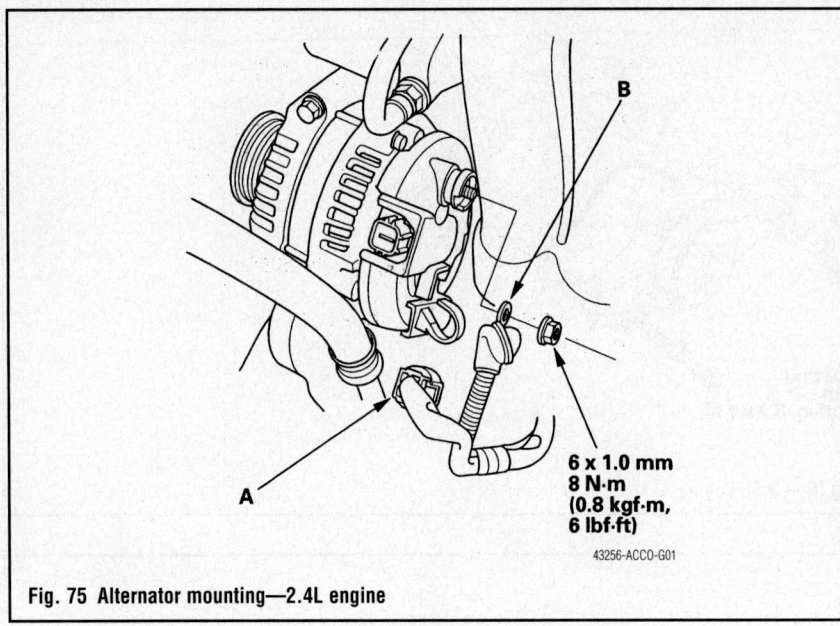

Fig. 75 Alternator mounting—2.4L engine

6 x 1.0 mm
8 N·m
(0.8 kgf·m,
6 lbf·ft)

43256-ACCO-G01

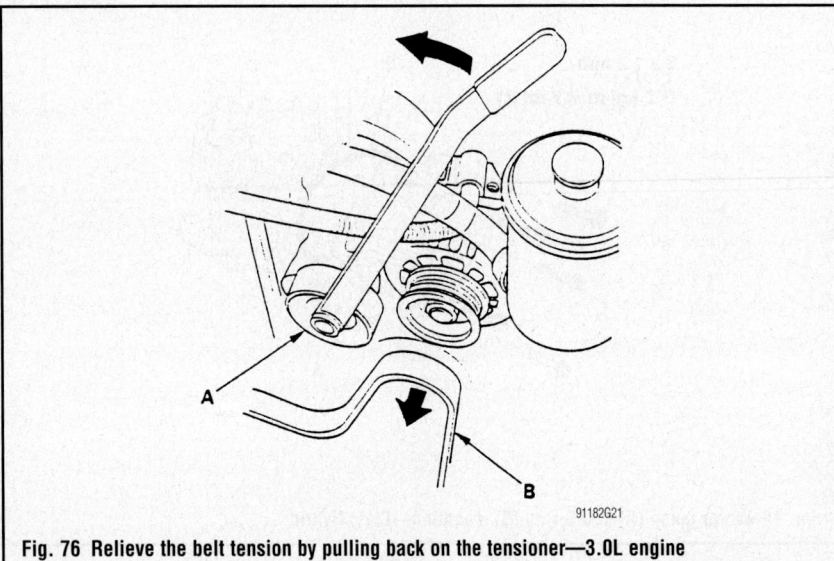

Fig. 76 Relieve the belt tension by pulling back on the tensioner—3.0L engine

91182G21

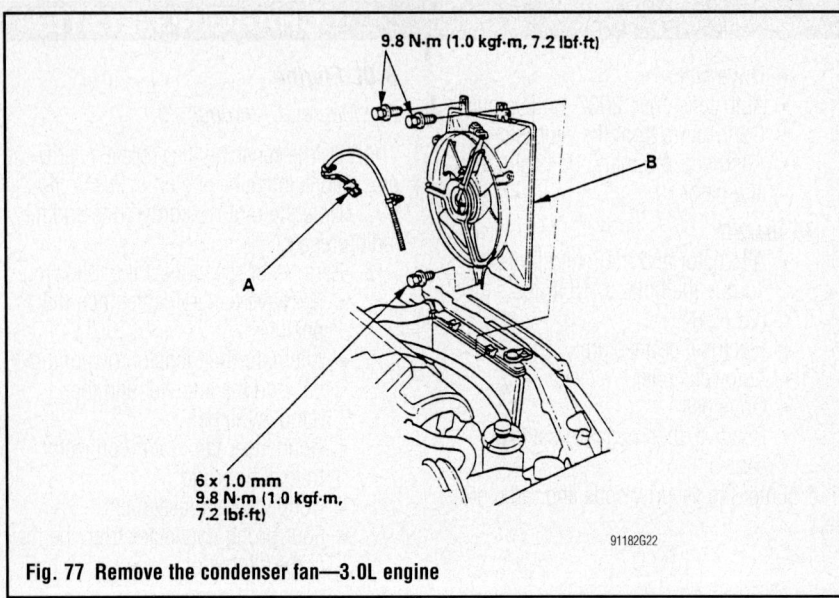

9.8 N·m (1.0 kgf·m, 7.2 lbf·ft)

6 x 1.0 mm
9.8 N·m (1.0 kgf·m,
7.2 lbf·ft)

91182G22

Fig. 77 Remove the condenser fan—3.0L engine

- Alternator mounting bolts
- Wiring harness clamp
- Alternator assembly

To install:

4. Alternator installation is the reverse of the removal procedure.

5. Connect the positive battery cable, then the negative battery cable. Enter the radio security code and station presets.

3.5L Engine

See Figure 79.

1. Disconnect the battery cables.

2. Remove the engine splash shield.

3. Disconnect the A/C condenser fan motor connector and remove the harness clamp.

4. Loosen the A/C condenser fan shroud mounting bolts

5. Remove the A/C condenser fan shroud assembly and the coolant reservoir.

6. Remove the accessory drive belt.

7. Disconnect the alternator connector and the positive alternator cable from the alternator.

8. Remove the harness clamp from the alternator and disconnect the A/C compressor clutch connector from the A/C compressor.

9. Remove the bolt securing the harness holder.

10. Remove the mounting bolt and the alternator bracket mounting bolt, then remove the alternator.

To install:

11. Install the alternator. Tighten the mounting bolt to 33 ft. lbs. (44 Nm). Tighten the bracket mounting bolt to 16 ft. lbs. (22 Nm).

12. The remainder of the installation is the reverse order of removal.

Civic

1.8L Engine

1. Before servicing the vehicle, refer to the precautions in the beginning of this section.

2. Remove or disconnect the following:
- Negative battery cable, then the positive
- Accessory drive belt
- 4P connector and battery terminal wire
- Alternator bolts
- Alternator

To install:

3. Installation is the reverse of removal.

4. Adjust the alternator belt tension.

5. Connect the negative battery cable, then the positive cable.

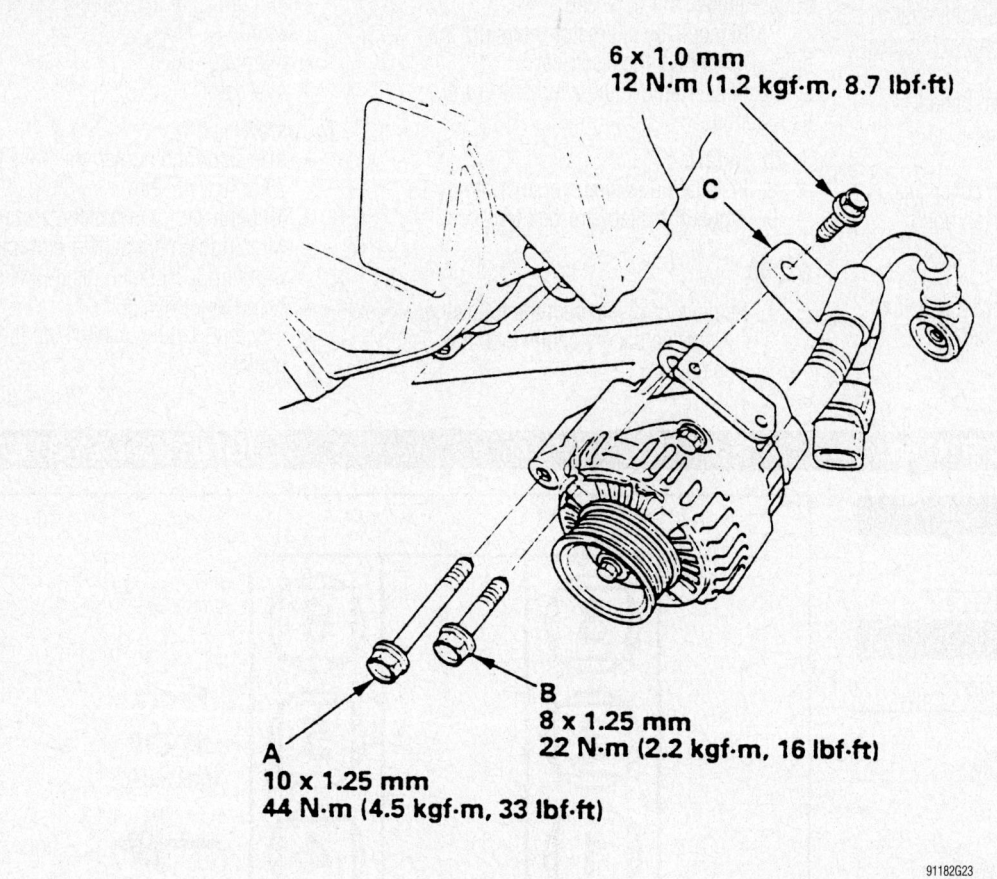

6 x 1.0 mm
12 N·m (1.2 kgf·m, 8.7 lbf·ft)

C

B
8 x 1.25 mm
22 N·m (2.2 kgf·m, 16 lbf·ft)

A
10 x 1.25 mm
44 N·m (4.5 kgf·m, 33 lbf·ft)

91182G23

Fig. 78 Torque the alternator bolts to the specs shown—3.0L engine

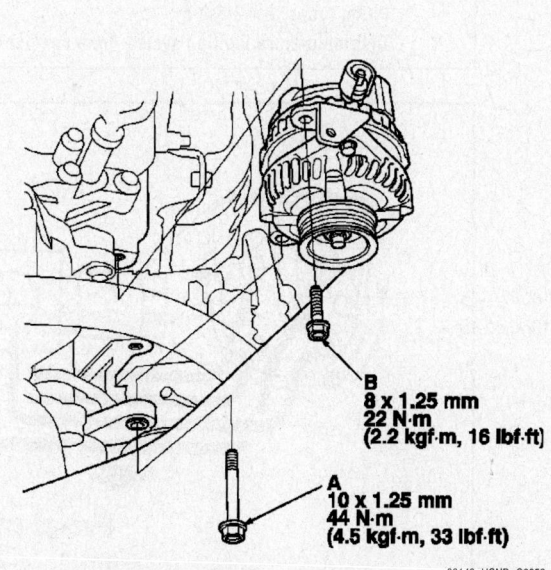

B
8 x 1.25 mm
22 N·m
(2.2 kgf·m, 16 lbf·ft)

A
10 x 1.25 mm
44 N·m
(4.5 kgf·m, 33 lbf·ft)

22140_HOND_G0052

Fig. 79 Remove the mounting bolt (A) and the alternator bracket mounting bolt (B) to remove
the alternator—3.5L Engines

2.0L Engine

1. Before servicing the vehicle, refer to the precautions in the beginning of this section.

2. Disconnect the negative battery cable, then the positive cable.

3. Remove the drive belt.

4. Remove the front grille cover. Disconnect the fan motor connector, hood switch connector. Remove the harness clamps.

5. Remove the reservoir hose, radiator cap mounting bolts and upper radiator brackets.

6. Remove the condenser mounting bolts. Remove the bulkhead.

7. Remove the alternator retaining bolts. Disconnect the electrical connections. Remove the alternator from its mounting.

To install:

8. Installation is the reverse of removal.

9. Adjust the alternator belt tension.

S2000

1. Remove or disconnect the following:
   - Negative battery cable, then the positive
   - Accessory drive belt
   - 4P connector and battery terminal wire
   - Alternator bolts
   - Alternator

To install:
   - Alternator and tighten the bolts to 33 ft. lbs. (44 Nm)
   - 4P connector and battery terminal wire. Tighten the battery terminal wire nut to 108 inch lbs. (12 Nm).
   - Accessory drive belt
   - Negative battery cable, then the positive

ENGINE ELECTRICAL   IGNITION SYSTEM

FIRING ORDER

See Figures 80 and 81.

IGNITION COIL

REMOVAL & INSTALLATION

Accord

2.4L Engine

See Figure 82.

1. Disconnect the negative battery cable.

2. Remove the ignition coil cover, disconnect the ignition coil connectors, then remove the ignition coils.

3. Install the ignition coils in the reverse order of removal.

3.0L & 3.5L Engines

See Figures 83 and 84.

1. Disconnect the negative battery cable.

2. Remove the engine cover.

3. Disconnect the ignition coil connectors, then remove the front bank ignition coils.

4. Disconnect the ignition coil connectors, then remove the rear bank ignition coils.

5. Installation of the ignition coils is the reverse of the removal procedure.

Civic

1.8L Engine

See Figure 85.

1. Disconnect the negative battery cable.

2. Disconnect the ignition coil connectors, then remove the ignition coils.

3. Installation of the ignition coils is the reverse of the removal procedure.

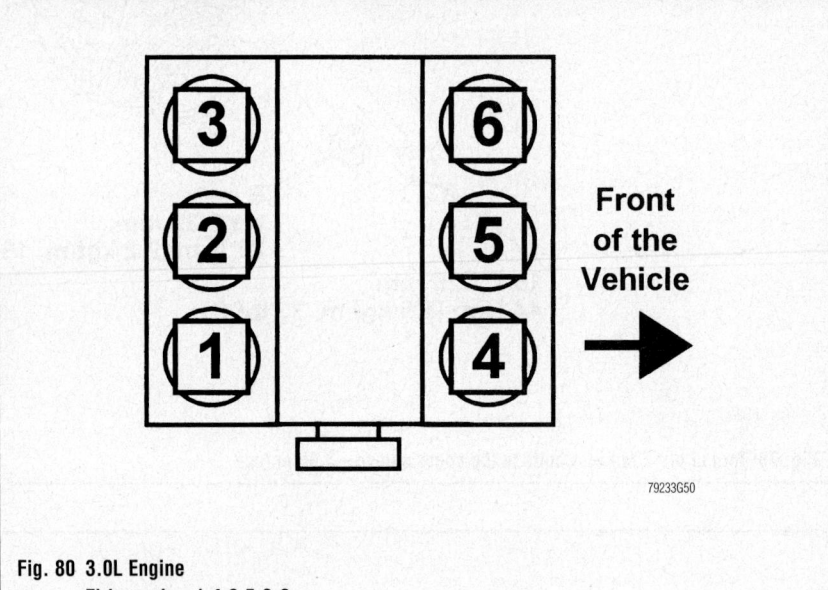

Fig. 80 3.0L Engine
Firing order: 1-4-2-5-3-6
Distributorsless ignition system (one coil per cylinder)

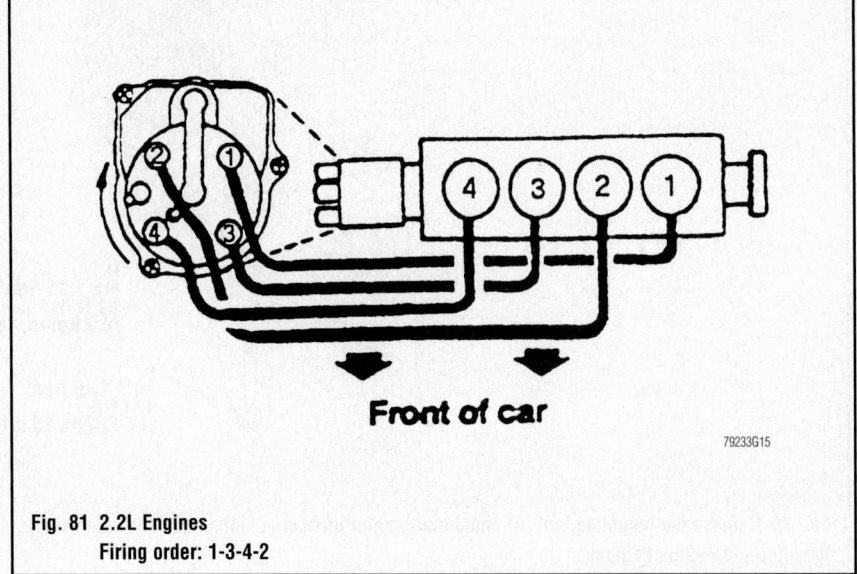

Fig. 81 2.2L Engines
Firing order: 1-3-4-2

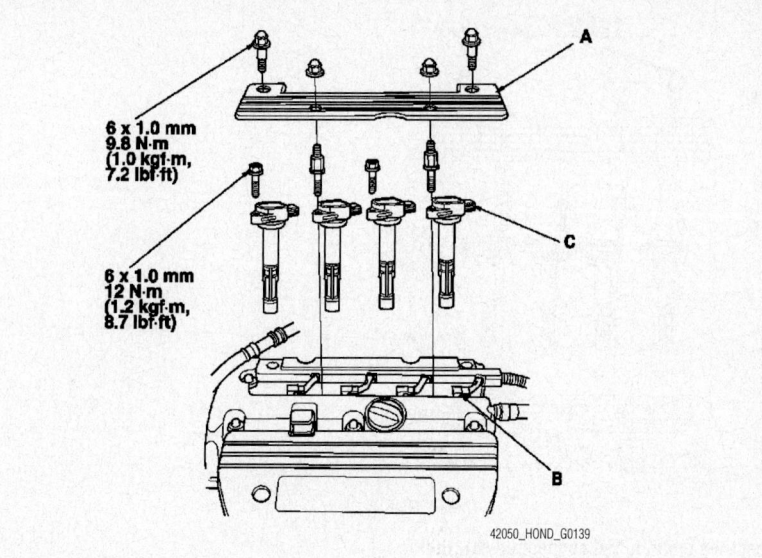

Fig. 82 Remove the ignition coil cover (A), disconnect the ignition coil connectors (B), then remove the ignition coils (C)

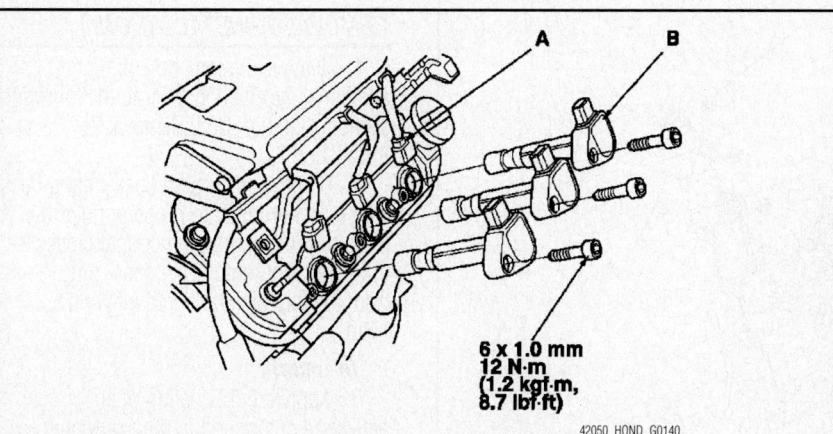

Fig. 83 Disconnect the ignition coil connectors (A), then remove the front bank ignition coils (B)

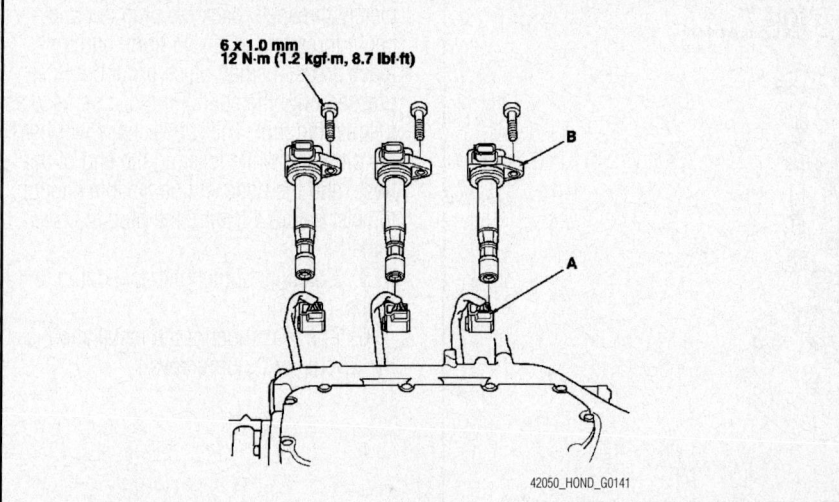

Fig. 84 Disconnect the ignition coil connectors (A), then remove the rear bank ignition coils (B)

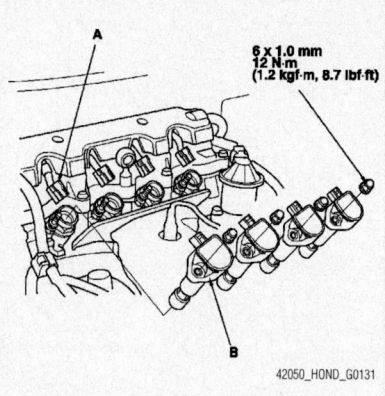

Fig. 85 Disconnect the ignition coil connectors (A), then remove the ignition coils (B)

2.0L Engine

See Figure 86.

1. Disconnect the negative battery cable.
2. Remove the under-cowl panel.
3. Remove the ignition coil cover, disconnect the ignition coil connectors, then remove the ignition coils.
4. Installation of the ignition coils is the reverse of the removal procedure.

Civic Hybrid

See Figure 87.

1. Remove the engine appearance cover.
2. Disconnect the ignition coil connectors, then remove the intake side ignition coils and the exhaust side ignition coils.
3. Install the ignition coils in the reverse order of removal.

S2000

See Figure 88.

1. Disconnect the negative battery cable.
2. Remove the ignition coil cover (A).
3. Disconnect the ignition coil connectors, then remove the ignition coils (B).
4. Installation of the ignition coils is the reverse of the removal procedure.

IGNITION TIMING

ADJUSTMENT

Ignition timing is control by the Power Control Module (PCM). No adjustment is necessary.

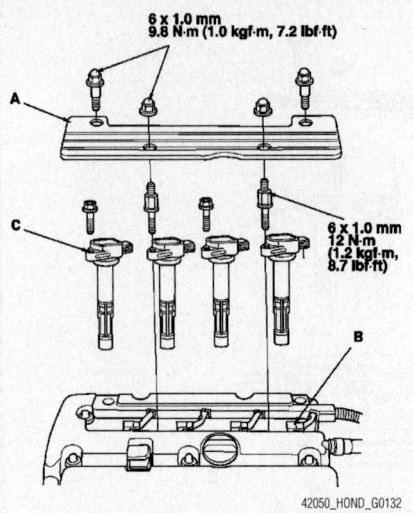

6 x 1.0 mm
9.8 N·m (1.0 kgf·m, 7.2 lbf·ft)

6 x 1.0 mm
12 N·m
(1.2 kgf·m,
8.7 lbf·ft)

42050_HOND_G0132

Fig. 86 Remove the ignition coil cover (A), disconnect the ignition coil connectors (B), then remove the ignition coils (C)

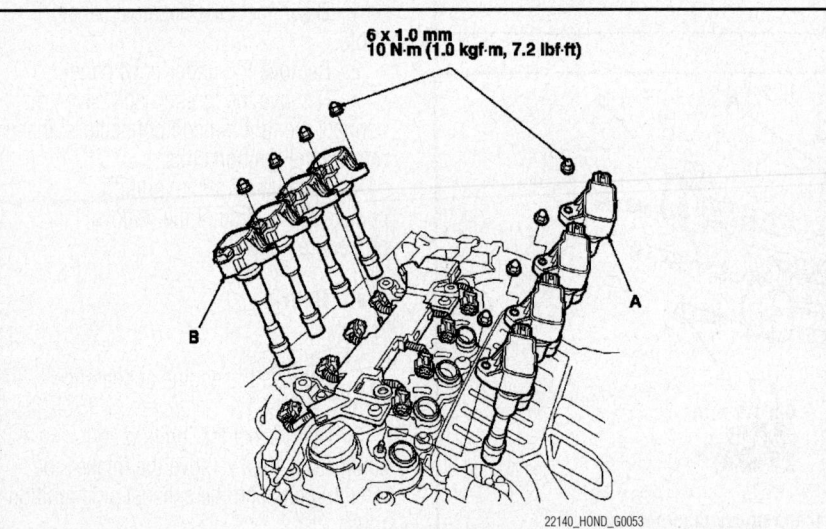

6 x 1.0 mm
10 N·m (1.0 kgf·m, 7.2 lbf·ft)

22140_HOND_G0053

Fig. 87 Exploded view of the intake side (A) and exhaust side (B) ignition coils—1.3L Engine

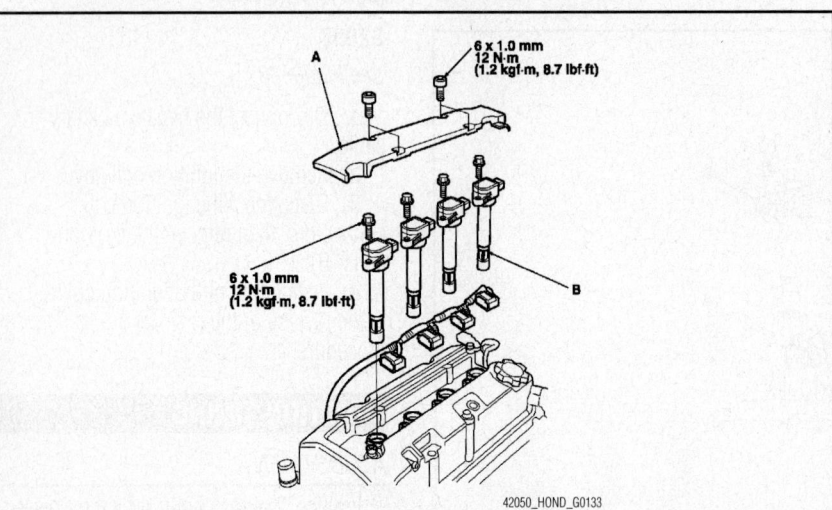

6 x 1.0 mm
12 N·m
(1.2 kgf·m, 8.7 lbf·ft)

6 x 1.0 mm
12 N·m
(1.2 kgf·m, 8.7 lbf·ft)

42050_HOND_G0133

Fig. 88 Remove the ignition coil cover (A), detach the connectors, then remove the ignition coils (B)

SPARK PLUGS

REMOVAL & INSTALLATION

1. Remove the ignition coil. For additional information, refer to the following section, "Ignition Coil, Removal & Installation."

2. Using a spark plug socket equipped with a rubber insert to properly hold the plug, turn the spark plug counterclockwise to loosen and remove the spark plug from the threaded hole in the cylinder head.

To install:

3. Apply a light coating of an anti-seize compound to the spark plug threads.

4. Carefully thread the plug into the threaded spark plug hole by hand. If resistance is felt before the plug is almost completely threaded, back the plug out and begin threading again. In tight, hard to reach areas, a small piece of rubber hose pressed onto the spark plug can be used as a threading tool. The rubber hose will hold the plug and while twisting the end of the hose, and the hose will be flexible enough to twist before allowing the plug to cross thread.

5. Carefully tighten the spark plug to 13 ft. lbs. (18 Nm).

6. The remainder of the installation is the reverse order of removal.

STARTER

REMOVAL & INSTALLATION

Accord

2.4L Engine

See Figure 89.

1. Note the radio security code and the radio presets.
2. Disconnect the negative battery cable. Disconnect the positive battery cable.
3. Drain the cooling system.
4. Remove the intake manifold.
5. Disconnect the starter cable from the B terminal on the solenoid. Disconnect the BLK/WHT wire from the S terminal.
6. Remove the harness clamp. Remove the harness holder.
7. Remove the starter retaining bolts.
8. Remove the starter from the vehicle.
9. Installation is the reverse order of removal. Refer to the illustration for the torque specifications.

3.0L Engine

See Figure 90.

1. Note the radio security code and the radio presets.
2. Disconnect the negative battery cable. Disconnect the positive battery cable.
3. Remove the battery. Remove the harness clamp.
4. Disconnect the starter cable from the B terminal on the solenoid. Disconnect the BLK/WHT wire from the S terminal.
5. Remove the starter retaining bolts.
6. Remove the starter from the vehicle.
7. Installation is the reverse order of removal.

3.5L Engine

1. Disconnect the negative battery cable. Disconnect the positive battery cable.
2. Remove the battery. Remove the harness clamp.
3. Remove the intake air ducts and splash separator.
4. Remove the starter wiring harness clamp.
5. Disconnect the starter wiring harness.
6. If equipped with automatic transmission, remove the upper radiator hose bracket and dipstick.
7. Remove the starter mounting bolts.
8. Remove the starter.

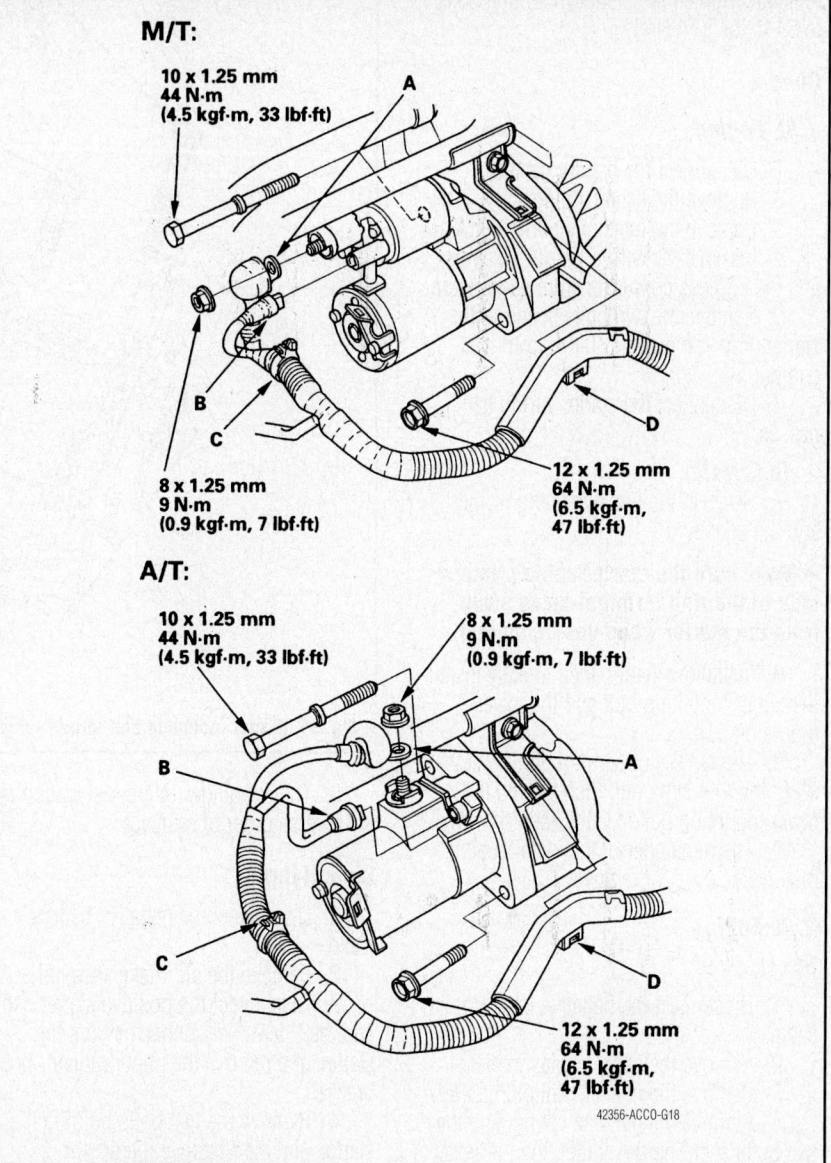

M/T:

10 x 1.25 mm
44 N·m
(4.5 kgf·m, 33 lbf·ft)

8 x 1.25 mm
9 N·m
(0.9 kgf·m, 7 lbf·ft)

12 x 1.25 mm
64 N·m
(6.5 kgf·m, 47 lbf·ft)

A/T:

10 x 1.25 mm
44 N·m
(4.5 kgf·m, 33 lbf·ft)

8 x 1.25 mm
9 N·m
(0.9 kgf·m, 7 lbf·ft)

12 x 1.25 mm
64 N·m
(6.5 kgf·m, 47 lbf·ft)

42356-ACCO-G18

Fig. 89 Starter mounting—2.4L engine

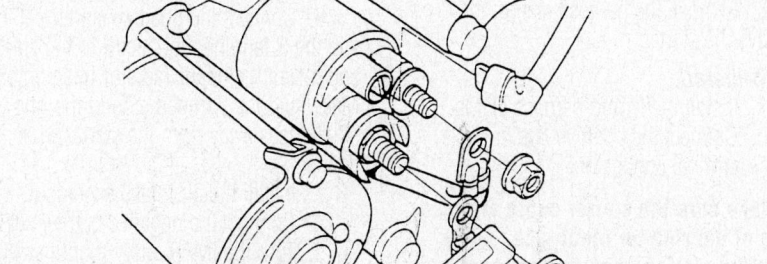

91182G27

Fig. 90 Location of starter wiring—3.0L engine

To install:

9. Installation is the reverse order of removal. Tighten the starter mounting bolts to 54 ft. lbs. (74 Nm).

Civic

1.8L Engine

1. Disconnect the battery cables.
2. Remove the exhaust pipe.
3. Remove the intake manifold bracket.
4. Remove the wiring harness clamps and the harness connector from each clamp.
5. Remove the two bolts securing the starter, then remove the starter from the engine.
6. Disconnect the starter wiring harnesses.

To install:

7. Connect the positive starter cable and the connector.

➡**Make sure the starter cable crimped side of the ring terminal faces away from the starter when you connect it.**

8. Install the starter, then loosely install the upper mounting bolt and the lower mounting bolt.
9. Tighten the upper mounting bolt to 33 ft. lbs. (44 Nm), and then tighten the lower mounting bolt to 33 ft. lbs. (44 Nm).
10. The remainder of the installation is the reverse order of removal.

2.0L Engine

See Figure 91.

1. Disconnect the negative battery cable.
2. Remove the engine splash shield.
3. Remove the intake manifold bracket.
4. Remove the harness clamp, and the two bolts securing the starter, then remove the starter from the engine.
5. Disconnect the positive starter cable and the S terminal connector.
6. Remove the harness clamp, then remove the starter.

To install:

7. Install the wiring harness clamp.
8. Connect the positive starter cable and S terminal connector.

➡**Make sure the starter cable crimped side of the ring terminal faces away from the starter when you connect it.**

9. Install the starter and tighten the mounting bolts.

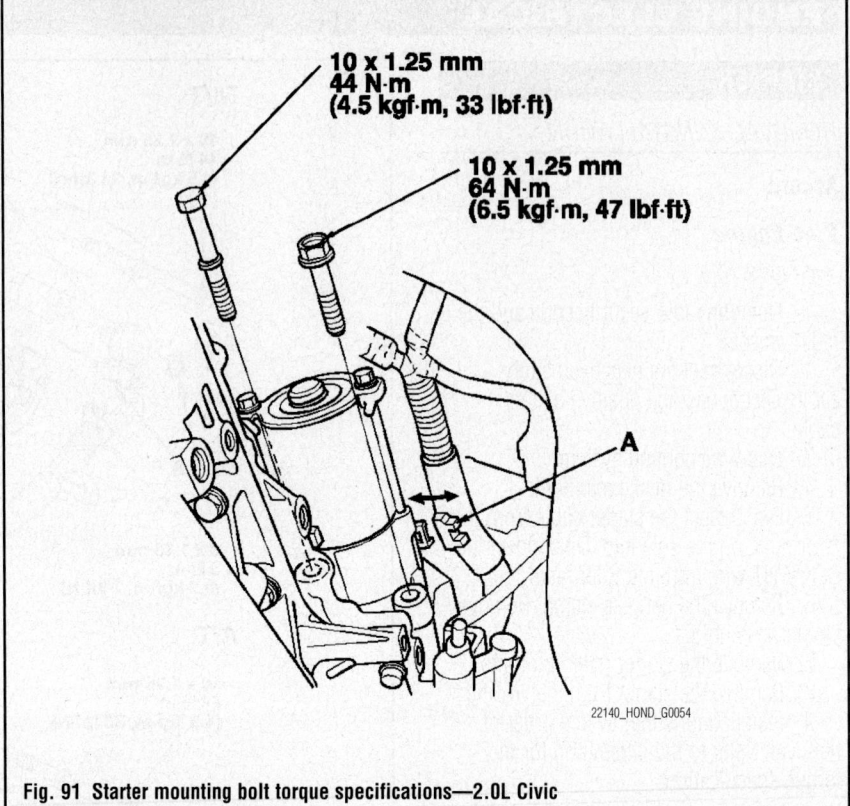

10 x 1.25 mm
44 N·m
(4.5 kgf·m, 33 lbf·ft)

10 x 1.25 mm
64 N·m
(6.5 kgf·m, 47 lbf·ft)

A

22140_HOND_G0054

Fig. 91 Starter mounting bolt torque specifications—2.0L Civic

10. The remainder of the installation is the reverse order of removal.

Civic Hybrid

1. Disconnect the negative battery cable.
2. Remove the air intake assembly.
3. Disconnect the positive starter cable and the S terminal connector from the starter, and remove the upper radiator hose bracket.
4. Remove the two bolts holding the starter, and then remove the starter.

To install:

5. Install the starter, then tighten the mounting bolts to 33 ft. lbs. (44 Nm).
6. Connect the positive starter cable and the S terminal connector to the starter, and install the upper radiator hose bracket. Make sure the crimped side of the ring terminal faces away from the starter when you connect it.
7. Install the air intake assembly.
8. Connect the negative battery cable.
9. Turn the IMA battery module switch OFF.
10. Start the engine to make sure the starter works properly.

11. Turn the IMA battery module switch ON.
12. If the IMA battery level gauge (BAT) displays no segments, start the engine, and hold it between 3,500 RPM and 4,000 RPM without load (in N or P) until the BAT displays at least three segments.

S2000

1. Disconnect the negative battery cable. Disconnect the positive battery cable.
2. Disconnect the intake air temperature sensor connector and breather pipe. Remove the manifold absolute sensor harness from the holder. Remove the air cleaner housing cover and the air cleaner assembly. Remove the IAT sensor harness clamps and the intake air housing.
3. Remove the drive belt. Remove the alternator.
4. Disconnect the starter cable from the B terminal on the solenoid. Disconnect the BLK/WHT wire from the S terminal.
5. Remove the starter retaining bolts. Remove the starter from the vehicle.
6. Installation is the reverse of removal.

ENGINE MECHANICAL

➡Disconnecting the negative battery cable may interfere with the functions of the on board computer systems and may require the computer to undergo a relearning process, once the negative battery cable is reconnected.

ACCESSORY DRIVE BELTS

ACCESSORY BELT ROUTING

See Figures 92 through 96.

ADJUSTMENT

Belt tension is maintained by an automatic tensioner. No adjustment is necessary or possible.

REMOVAL & INSTALLATION

➡Refer to the Accessory Belt Routing Illustrations located earlier in this section for routing diagrams.

1. Place a long-handled, boxed-end wrench or a belt tension release tool on the drive belt auto-tensioner from above the engine. Slowly turn the wrench in the direction shown to release the tension, then remove the drive belt.

✳✳ WARNING

This is a hydraulic type auto-tensioner; you must turn the wrench slowly.

2. Install the new belt in the reverse order of removal.

CAMSHAFT AND VALVE LIFTERS

REMOVAL & INSTALLATION

Accord

2.4L Engine

See Figures 97 through 100.

1. Disconnect the negative battery cable.
2. Remove or disconnect the following:
   • Timing chain
   • Loosen the rocker arm adjusting screws
   • Camshaft holder bolts, two turns at a time in sequence
   • Timing chain guide (B), camshaft and camshafts

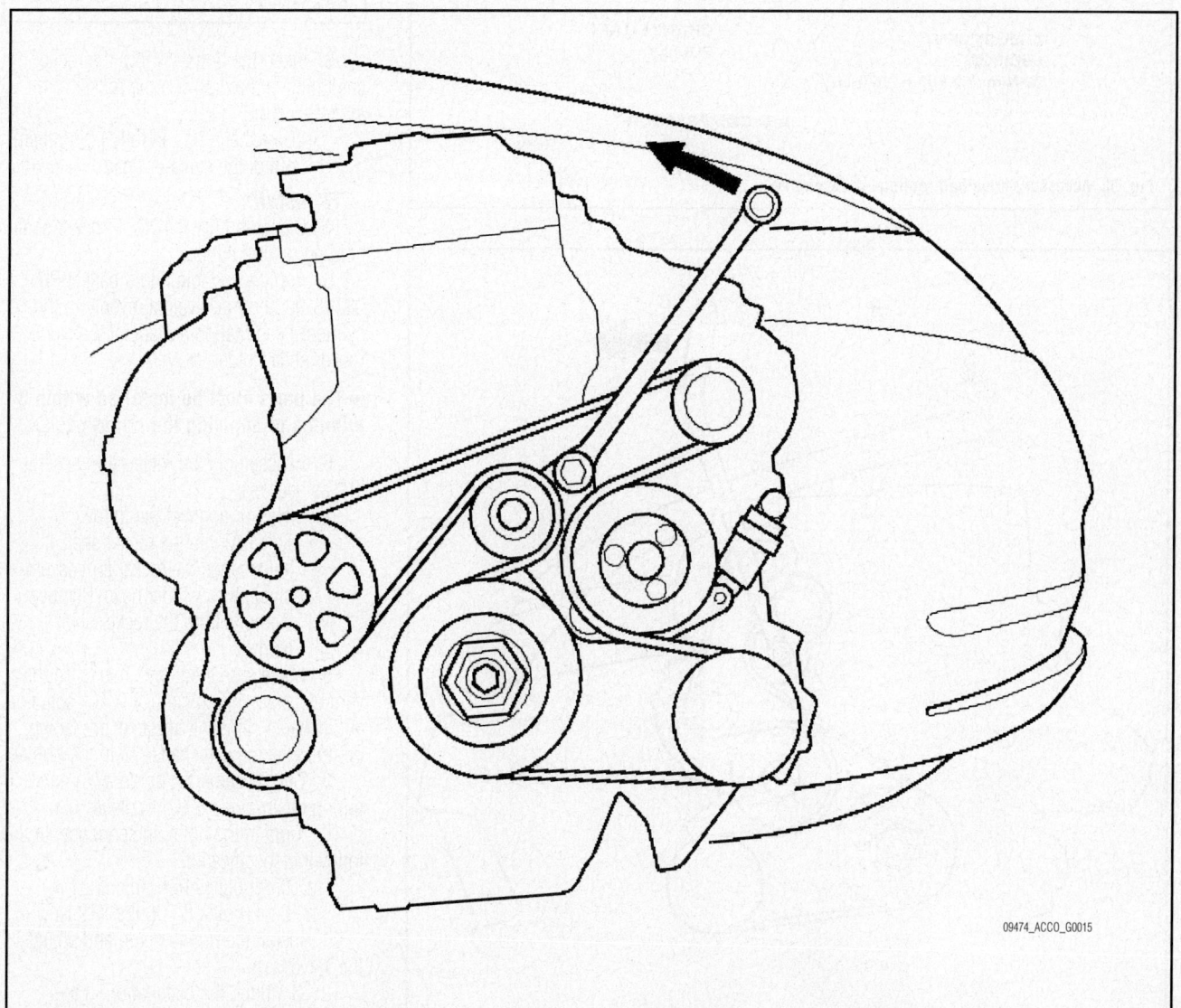

09474_ACCO_G0015

Fig. 92 Accessory drive belt routing   1.8L ongine

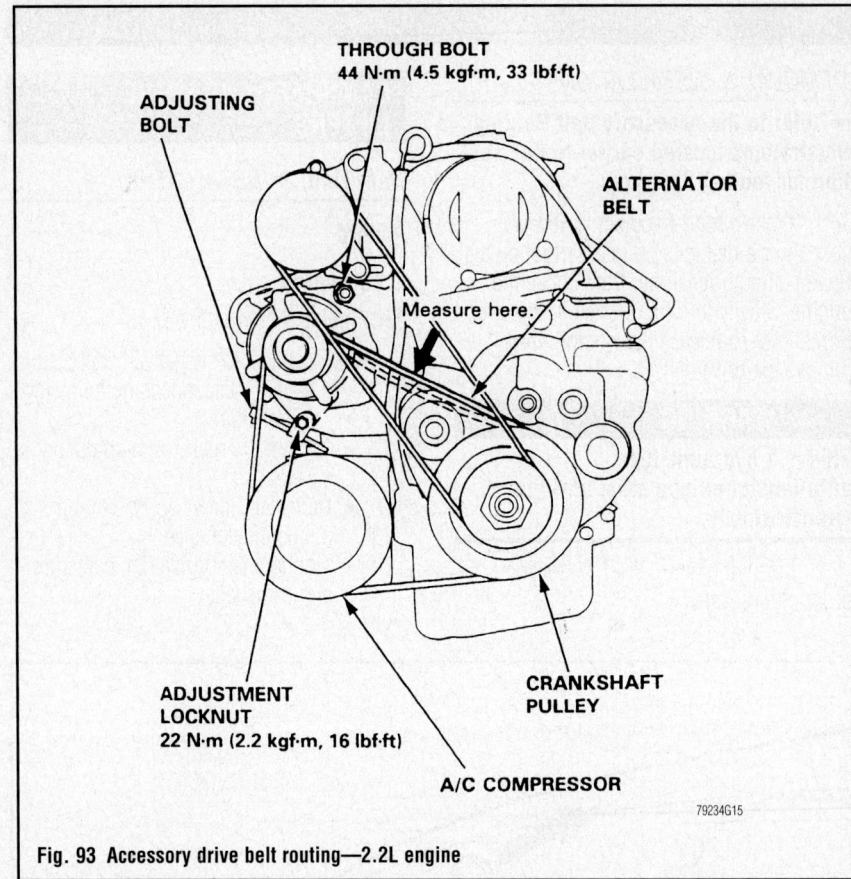

ADJUSTING BOLT

THROUGH BOLT
44 N·m (4.5 kgf·m, 33 lbf·ft)

ALTERNATOR BELT

Measure here.

ADJUSTMENT LOCKNUT
22 N·m (2.2 kgf·m, 16 lbf·ft)

CRANKSHAFT PULLEY

A/C COMPRESSOR

79234G15

Fig. 93 Accessory drive belt routing—2.2L engine

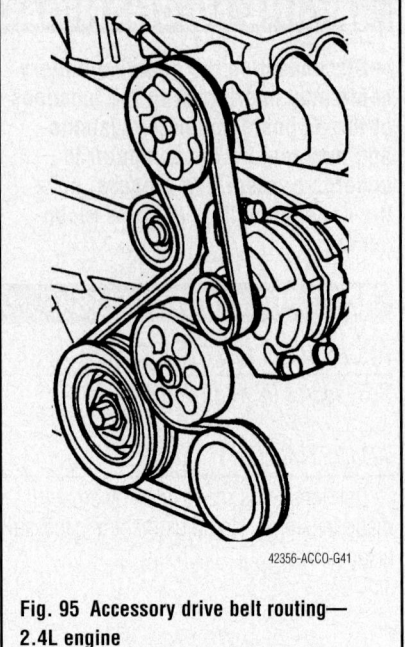

42356-ACCO-G41

Fig. 95 Accessory drive belt routing—2.4L engine

3. Insert the bolts (A) into the rocker shaft holder, then remove the rocker arm assembly (B)
  • Camshafts by carefully lifting them out of the cylinder head

To install:

4. Clean and dry the No. 5 rocker shaft holding mating surface.

5. Apply a suitable liquid gasket P/N 08718-0009, or equivalent, evenly to the cylinder head mating surface of the No. 5 rocker shaft holder.

➡**The parts must be installed within 5 minutes of applying the liquid gasket.**

6. Reassemble the rocker arm assembly, as necessary.

7. Install or connect the following
  • Bolts (A) into the rocker shaft holder, then the rocker arm assembly on the cylinder head. Remove the bolts from the rocker shaft holder.

8. Make sure the punch marks on the variable valve timing control (VTC) actuator and exhaust camshaft sprocket are facing up, then set the camshafts (A) in the holder.

9. Set the camshaft holders (B) and timing chain guide B (C) in place.

10. Tighten the bolts, in sequence, to the following specification:
  a. 8mm bolts: 16 ft. lbs. (22 Nm)
  b. 6mm bolts: 8.7 ft. lbs. (12 Nm)

11. Install the timing chain and adjust the valve lash.

12. Continue the installation in the reverse order of the removal procedure.

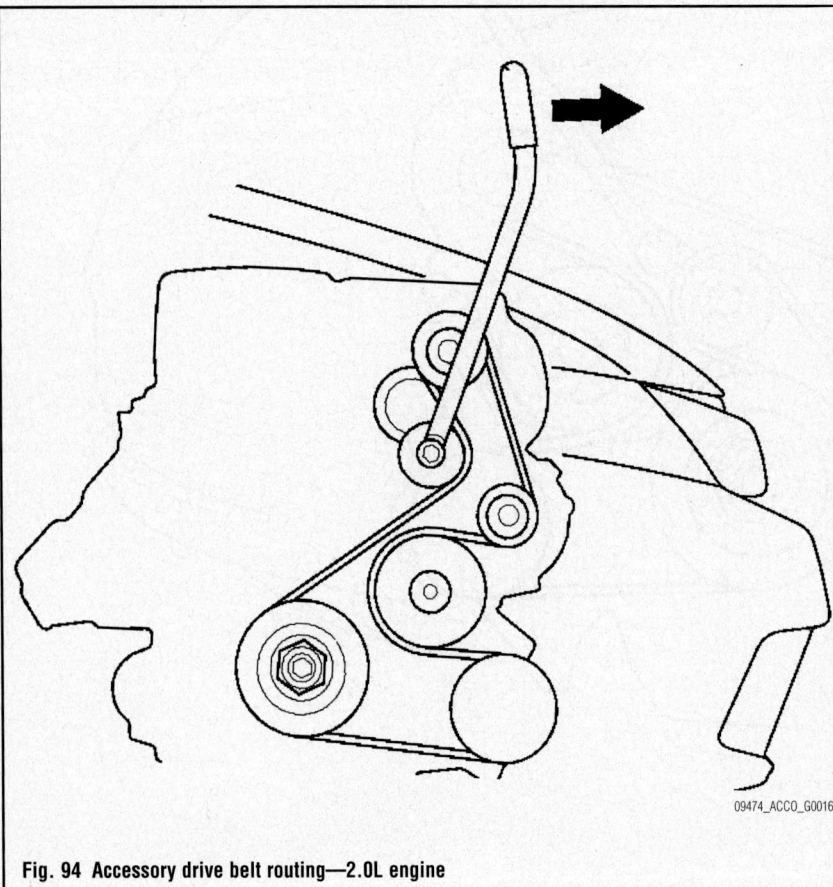

09474_ACCO_G0016

Fig. 94 Accessory drive belt routing—2.0L engine

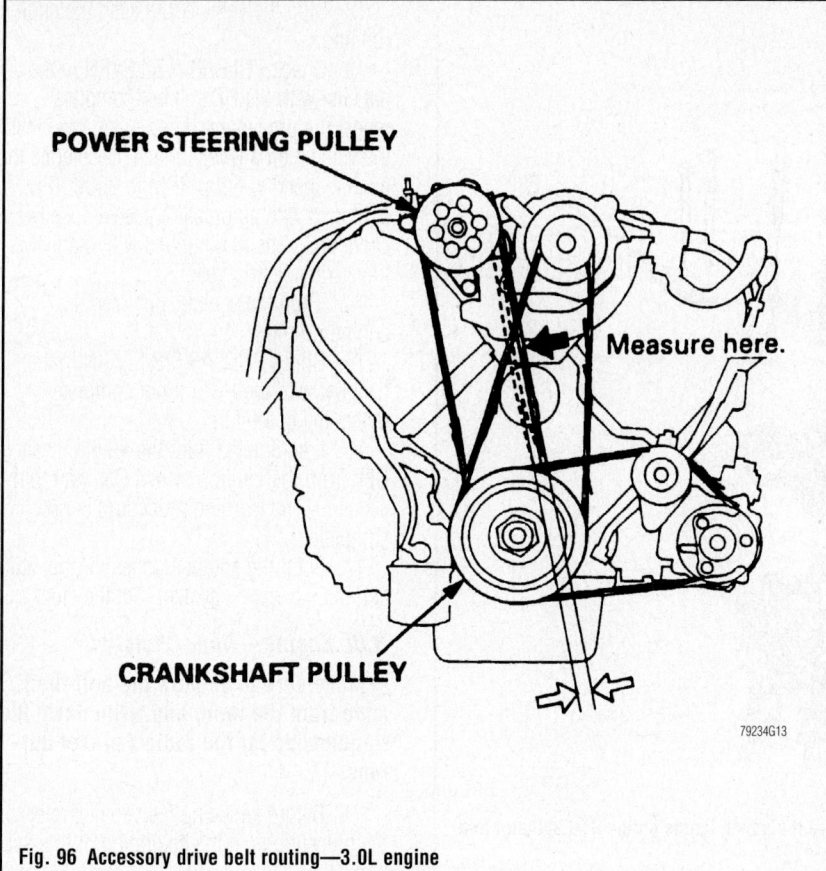

Fig. 96 Accessory drive belt routing—3.0L engine

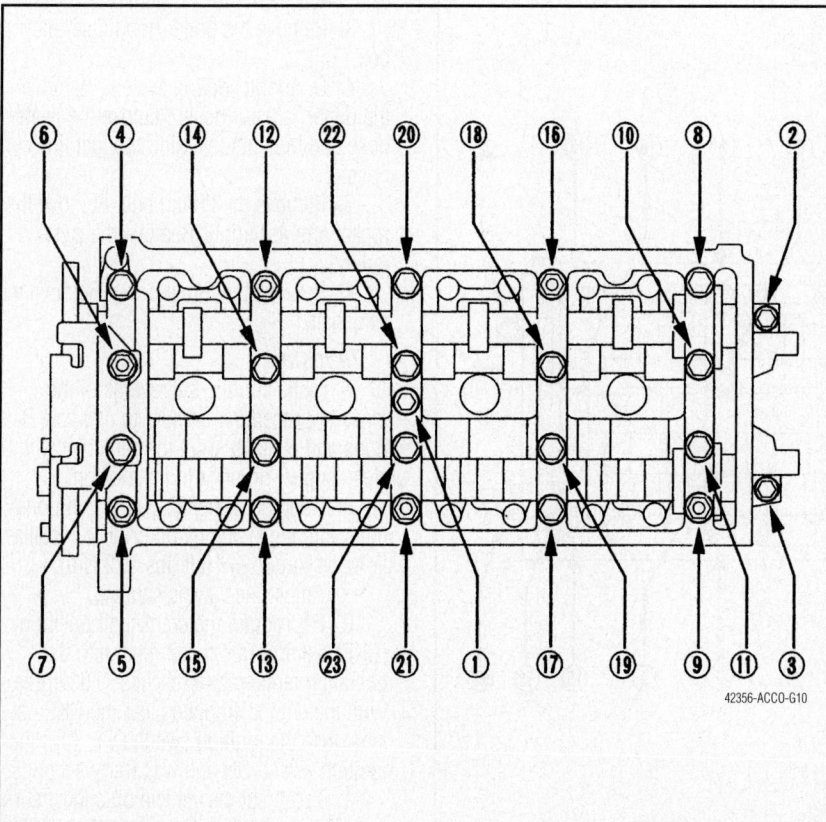

Fig. 97 Camshaft holder bolt loosening sequence—2.4l engine

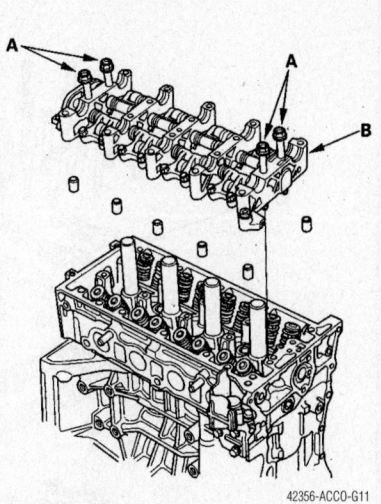

Fig. 98 Insert the bolts (A) into the rocker shaft holder, then remove the rocker arm assembly (B)—2.4L engine

3.0L Engine—Front Camshaft

➡ Make sure to acquire the anti-theft code from the radio and write down the frequencies for the radio's preset buttons.

1. Before servicing the vehicle, refer to the precautions in the beginning of this section.

2. Disconnect the negative battery cable. Remove the battery from the vehicle.

3. Drain the cooling system. Remove the upper radiator hose.

4. Remove the EGR valve. Remove the timing belt.

5. Remove the rocker arm assembly. Remove the camshaft pulley.

6. Remove the thrust cover. Remove the camshaft.

To install:

7. Installation is the reverse of the removal procedure. Be sure to use new O-rings and gaskets. Coat the camshaft with clean engine oil prior to installation.

8. Coat the camshaft pulley mounting bolts with clean engine oil prior to installation and torque to 16 ft. lbs. (22 Nm).

9. Adjust the valve clearance.

10. Reprogram the crankshaft position (CKP) pattern. Run the engine until the operating temperature reaches 176 degree. With the engine stopped clear the CKP pattern. Turn the ignition switch OFF. Turn the ignition switch ON and wait thirty seconds.

11. Road test the vehicle on a level surface. Decelerate the engine speed of 2500 RPM to 1000 RPM. If equipped with automatic transaxle use two Drive positions. If equipped with manual transaxle use first gear.

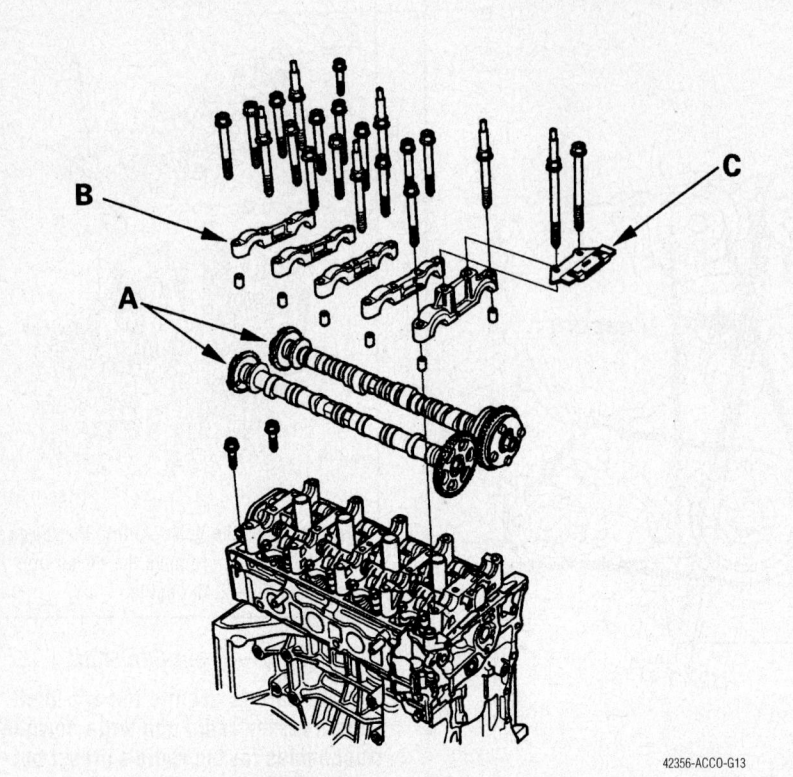

Fig. 99 When installing the camshafts (A) make sure the punch marks on the VTC actuator and exhaust cam sprockets are facing up—2.4L engine

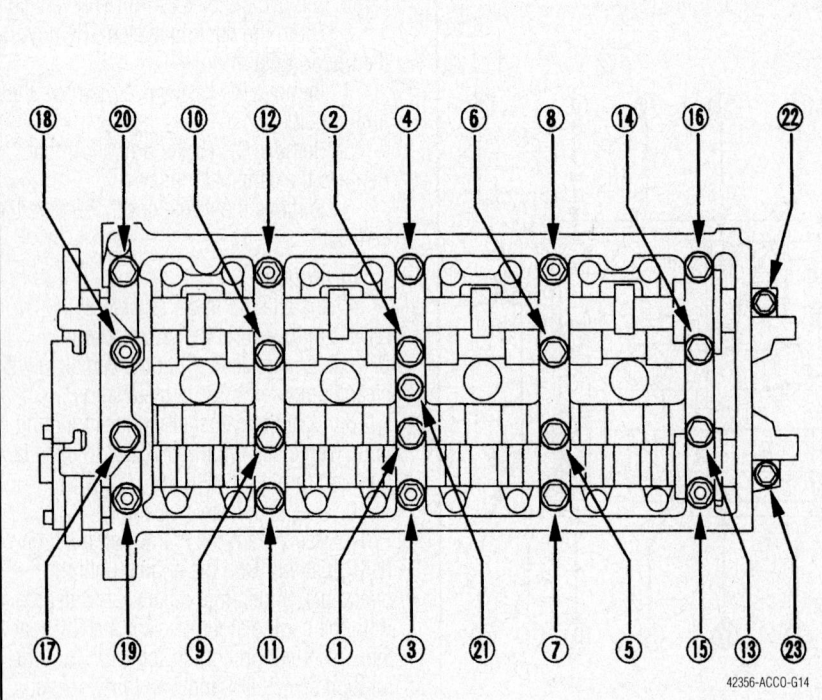

Fig. 100 Rocker arm assembly bolt tightening sequence—2.4L engine

12. Stop the vehicle, but keep the engine running.

13. Check PULSAR F/B LEARN in the data list with the HDS. If not complete repeat the procedure. If complete, road test the vehicle on a level surface. Decelerate the engine speed of 5000 RPM to 3000 RPM. If equipped with automatic transaxle use two Drive positions. If equipped with manual transaxle use first gear.

14. Stop the vehicle, but keep the engine running.

15. Check PULSAR F/B LEARN in the data list with the HDS. If not complete repeat the procedure.

16. If completed, turn the ignition switch OFF. Turn the ignition switch ON, wait thirty seconds. The learning procedure is now complete.

17. Enter the antitheft codes for the radio and the navigation system. Set the clock.

3.0L Engine—Rear Camshaft

➡ **Make sure to acquire the anti-theft code from the radio and write down the frequencies for the radio's preset buttons.**

1. Before servicing the vehicle, refer to the precautions in the beginning of this section.

2. Disconnect the negative battery cable. Relieve the fuel system pressure.

3. Remove the under hood fuse/relay box.

4. Drain the cooling system. Remove the upper radiator hose. Remove the heater hose and the nuts retaining the fuel line hose.

5. Remove the timing belt. Remove the rocker arm assembly. Remove the camshaft pulley.

6. Remove the thrust cover. Remove the camshaft.

To install:

7. Installation is the reverse of the removal procedure. Be sure to use new O-rings and gaskets. Coat the camshaft with clean engine oil prior to installation.

8. Coat the camshaft pulley mounting bolts with clean engine oil prior to installation and torque to 16 ft. lbs. (22 Nm).

9. Adjust the valve clearance.

10. Reprogram the crankshaft position (CKP) pattern. Run the engine until the operating temperature reaches 176 degree. With the engine stopped clear the CKP pattern. Turn the ignition switch OFF. Turn the ignition switch ON and wait thirty seconds.

11. Road test the vehicle on a level surface. Decelerate the engine speed of 2500 RPM to 1000 RPM. If equipped with

automatic transaxle use two Drive positions. If equipped with manual transaxle use first gear.

12. Stop the vehicle, but keep the engine running.

13. Check PULSAR F/B LEARN in the data list with the HDS. If not complete repeat the procedure. If complete, road test the vehicle on a level surface. Decelerate the engine speed of 5000 RPM to 3000 RPM. If equipped with automatic transaxle use two Drive positions. If equipped with manual transaxle use first gear.

14. Stop the vehicle, but keep the engine running.

15. Check PULSAR F/B LEARN in the data list with the HDS. If not complete repeat the procedure.

16. If completed, turn the ignition switch OFF. Turn the ignition switch ON, wait thirty seconds. The learning procedure is now complete.

17. Enter the antitheft codes for the radio and the navigation system. Set the clock.

3.5L Engine—Front Camshaft

See Figure 101.

1. Disconnect the battery cables.

2. Remove the battery clamp and battery.

3. Drain the engine cooling system.

4. Disconnect the radiator hoses.

5. Remove the Exhaust Gas Recirculation (EGR) valve.

6. Remove the EGR valve stud bolts.

7. Remove the timing belt. For additional information, refer to the following section, "Timing Belt, Removal & Installation."

8. Remove the rocker arm assembly.

9. Remove the front camshaft pulley.

10. Remove the thrust cover, then remove the front camshaft.

To install:

11. Install the front camshaft in the reverse order of removal. Always use a new O-ring. Apply new engine oil to the journals and the cam lobes.

12. Tighten the rocker arm assembly bolts in the sequence shown to 16 ft. lbs. (22 Nm).

13. Install the timing belt. For additional information, refer to the following section, "Timing Belt, Removal & Installation."

14. Install the EGR valve stud bolt, and then install the EGR valve.

15. Connect the radiator hoses.

16. Install the battery and connect the battery cables.

17. Refill the engine cooling system to the correct level.

3.5L Engine—Rear Camshaft

See Figure 102.

1. Properly relieve the fuel system pressure.

2. Disconnect the negative, then positive battery cable.

3. Remove the under-hood fuse/relay box.

4. Drain the engine cooling system.

5. Disconnect the fuel supply hose.

6. Disconnect the heater hoses and remove the mounting bracket.

7. Remove the timing belt. For additional information, refer to the following section, "Timing Belt, Removal & Installation."

8. Loosen the rocker arm assembly locknuts and adjusting screws.

9. Remove the rocker shaft bridge mounting bolts, the rocker shaft holder mounting bolts, and the rocker arm assembly.

10. Remove the rear camshaft pulley.

11. Remove the thrust cover, and then remove the rear camshaft.

To install:

12. Install the rear camshaft in the reverse order of removal. Always use a new O-ring. Apply new engine oil to the journals and cam lobes.

13. Apply new engine oil to the threads of the camshaft pulley mounting bolt, then install the rear camshaft pulley.

14. Install the rocker arm assembly, then tighten the mounting bolts.

15. Install the timing belt. For additional information, refer to the following section, "Timing Belt, Removal & Installation."

16. Adjust the valve clearance. For additional information, refer to the following section, "Valve Lash, Removal & Installation."

17. Connect the heater hoses and install the bracket.

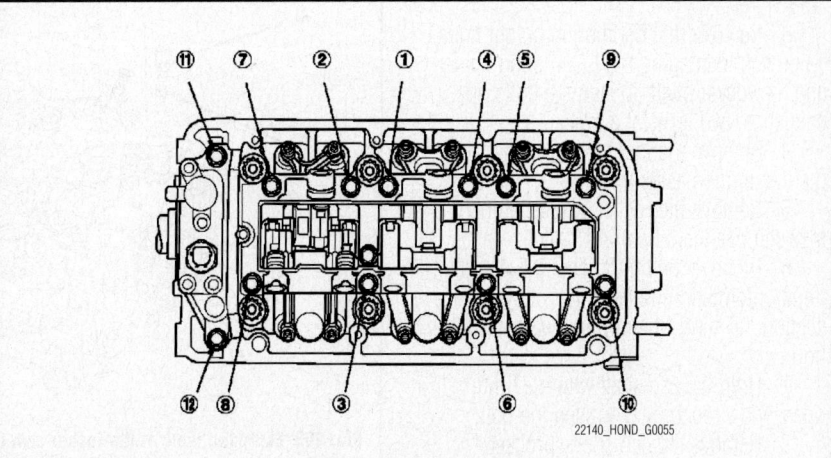

22140_HOND_G0055

Fig. 101 Front bank rocker arm assembly torque sequence—3.5L Engine

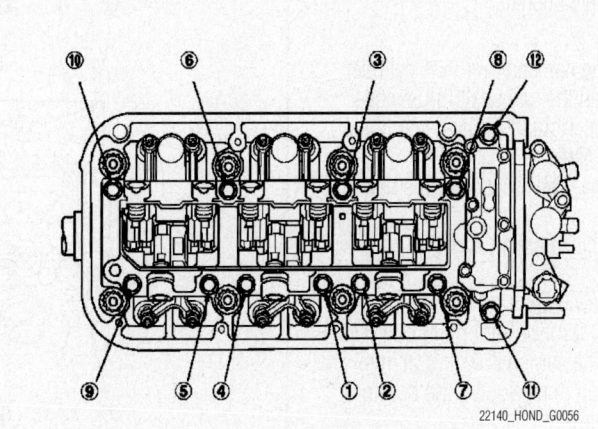

22140_HOND_G0056

Fig. 102 Rear bank rocker arm assembly torque sequence—3.5L Engine

18. Connect the fuel feed hose, then install the quick-connect fitting cover.

19. Install the under-hood fuse/relay box.

20. Connect the battery cables.

21. Inspect for fuel leaks. Turn the ignition switch to ON (II) (do not operate the starter) so the fuel pump runs for about 2 seconds and pressurizes the fuel line. Repeat this operation three times, then check for fuel leakage at any point in the fuel line.

22. Refill the engine cooling system to the correct level.

Civic

1.8L Engine

See Figures 103 and 104.

1. Disconnect the negative battery cable.

2. Remove the cylinder head. For additional information, refer to the following section, "Cylinder Head, Removal & Installation."

3. Remove the lost motion holder bolts. To prevent damaging the lost motion holder and the rocker shaft, loosen the bolts, in sequence, two turns at a time.

4. Remove the lost motion holder and the lost motion assemblies.

5. Remove the rocker arm assembly, then remove the oil control orifice.

6. Remove the timing chain. For additional information, refer to the following section, "Timing Chain, Removal & Installation."

7. Hold the camshaft with a 27 mm open-end wrench, then loosen the bolt.

8. Remove the camshaft sprocket.

9. Remove the camshaft position (CMP) sensor.

10. Remove the camshaft thrust cover, then pull out the camshaft.

To install:

11. Install the camshaft into the cylinder head, then install the camshaft thrust cover with new O-ring. Tighten thrust cover bolts to 7 ft. lbs. (10 Nm).

12. Install the CMP sensor with a new O-ring.

13. Mount the camshaft sprocket onto the camshaft.

14. Apply new engine oil to the threads of the camshaft sprocket mounting bolt, and install it. Hold the camshaft with a 27 mm open-end wrench, then tighten the bolt to 41 ft. lbs. (56 Nm).

15. Install the oil control orifice with a new O-ring, then install the rocker arm assembly.

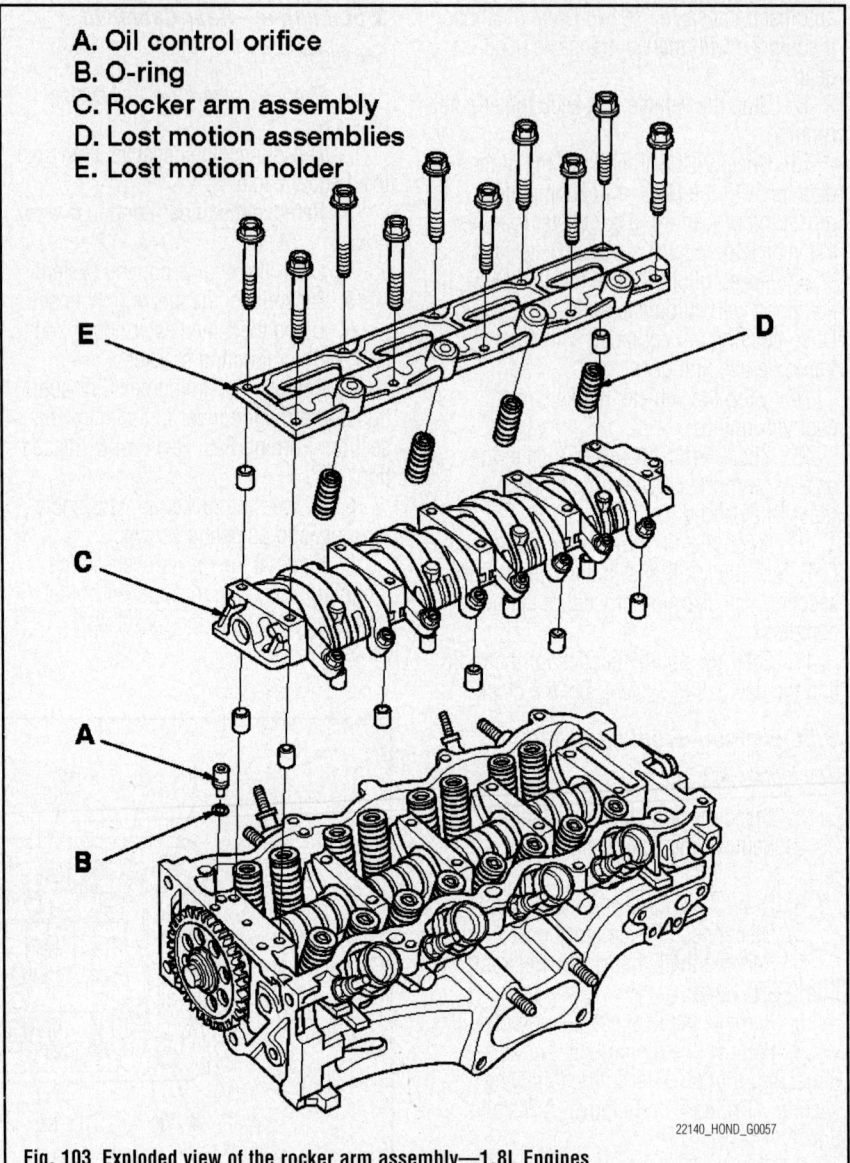

A. Oil control orifice
B. O-ring
C. Rocker arm assembly
D. Lost motion assemblies
E. Lost motion holder

22140_HOND_G0057

Fig. 103 Exploded view of the rocker arm assembly—1.8L Engines

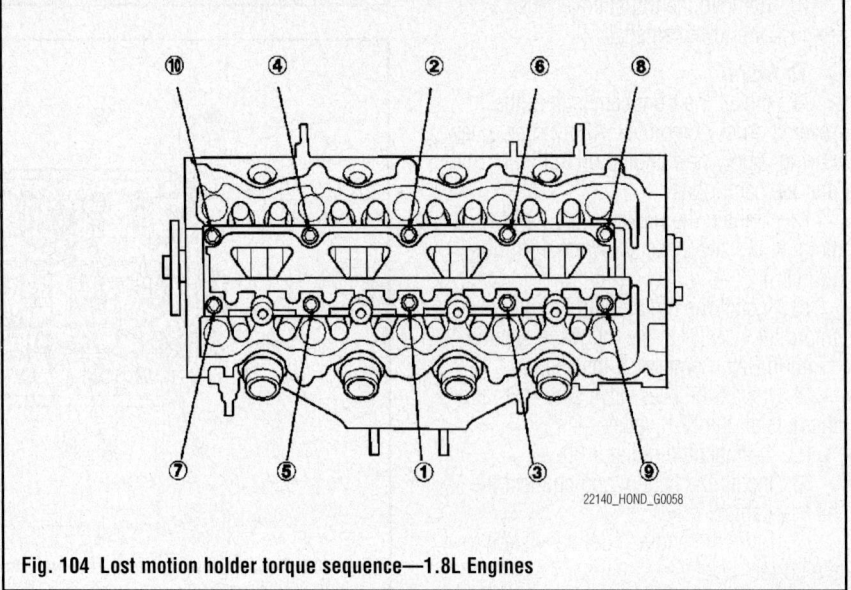

22140_HOND_G0058

Fig. 104 Lost motion holder torque sequence—1.8L Engines

16. Install the lost motion assembles and the lost motion holder in sequence to 11 ft. lbs. (15 Nm).

17. Install the cylinder head. For additional information, refer to the following section, "Cylinder Head, Removal & Installation."

18. Connect the negative battery cable.

2.0L Engine

See Figures 105 and 106.

1. Disconnect the negative battery cable.

2. Remove the timing chain. For additional information, refer to the following section, "Timing Chain, Removal & Installation."

3. Loosen the rocker arm adjusting screws.

4. Remove the camshaft holder bolts. To prevent damaging the camshafts, loosen the bolts, in sequence, two turns at a time.

5. Remove the timing chain guide, the camshaft holders, and the camshafts.

To install:

6. Install the camshafts, camshaft holders and timing chain guide.

7. Tighten the camshaft holders in sequence as follows:
- 8x1.25 mm bolts: 16 ft. lbs. (22 Nm)
- 6x1.0 mm bolts: 9 ft. lbs. (12 Nm)

8. The remainder of the installation is the reverse order of removal.

Civic Hybrid

See Figures 107 and 108.

1. Disconnect the negative battery cable.

2. Remove the engine appearance cover.

3. Remove the ignition coils.

4. Remove the ignition coil wiring harness holder, then disconnect the breather hose.

5. Remove the cylinder head cover.

6. Make a reference mark across the camshaft sprocket and the timing chain.

7. Apply new engine oil to the sliding surface of the cam chain tensioner through the oil return hole in the cylinder head.

8. Remove the cylinder head plug.

9. Hold the crankshaft pulley and set the socket wrench on the camshaft sprocket bolt.

10. Remove the maintenance bolt, and turn the camshaft clockwise to compress the cam chain tensioner, then install the 6 x 1.0 mm bolt in the bolt hole in the engine block through the maintenance hole and the cam chain tensioner

11. Hold the camshaft with a 27 mm open-end wrench, then loosen the camshaft sprocket bolt.

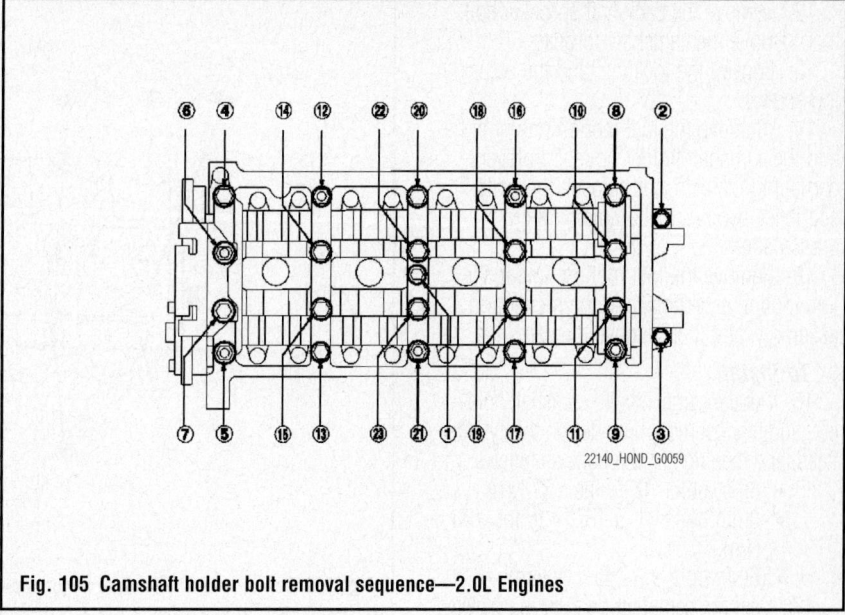

Fig. 105 Camshaft holder bolt removal sequence—2.0L Engines

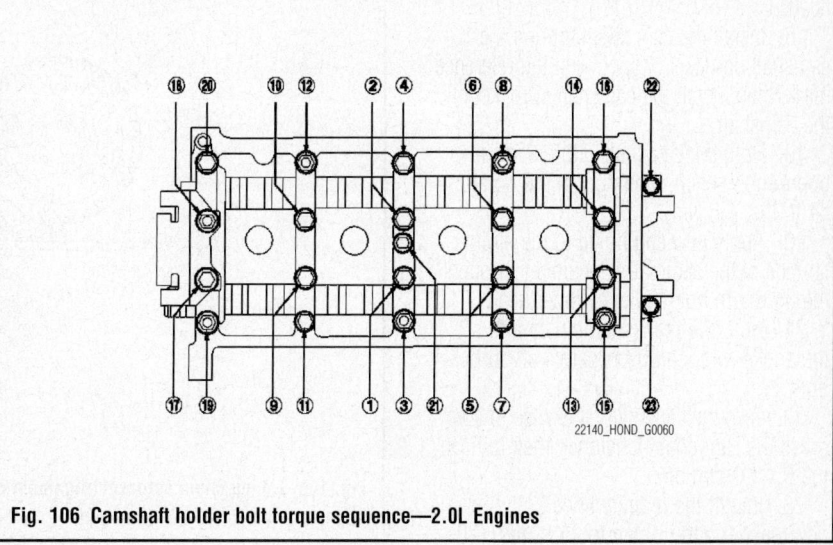

Fig. 106 Camshaft holder bolt torque sequence—2.0L Engines

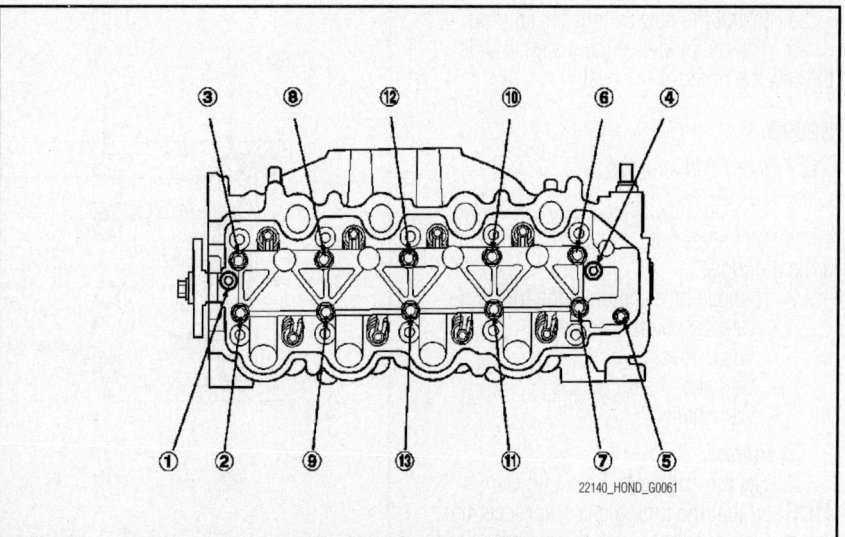

Fig. 107 Lost motion holder bolt removal sequence—1.3L Engine

12. Remove the camshaft sprocket bolt, then remove the camshaft sprocket.

13. Loosen the locknuts and the adjusting screws.

14. Remove the lost motion holder bolts and the camshaft holder bolts. To prevent damaging the camshaft, loosen the bolts in sequence two turns at a time, in a criss-cross pattern.

15. Remove the lost motion holder, the lost motion assembly, and the rocker arm assembly, and then remove the camshaft.

To install:

16. Put the camshaft, the camshaft holders, and the lost motion holder on the cylinder head, and then tighten the bolts as follows:
   • 8mm bolts: 16 ft. lbs. (22 Nm)
   • 8mm bolts 11 & 13: 14 ft. lbs. (20 Nm)
   • 6mm bolt: 9 ft. lbs. (12 Nm)

17. Seat the camshaft by pushing it away from the camshaft pulley end of the cylinder head.

18. Install the cam chain around the camshaft sprocket aligned with the reference mark, then install the camshaft sprocket on the camshaft.

19. Hold the camshaft with a 27 mm open-end wrench, then tighten the bolt to 41 ft. lbs. (56 Nm).

20. Apply new engine oil to the sliding surface of the cam chain tensioner through the oil return hole in the cylinder head.

21. Hold the crankshaft pulley and set the socket wrench on the camshaft sprocket bolt.

22. Turn the camshaft clockwise to compress the cam chain tensioner, then remove the 6 x 1.0 mm bolt.

23. Install the maintenance bolt and new washer and tighten to 15 ft. lbs. (20 Nm).

24. Install the new cylinder head plug.

25. The remainder of the installation is the reverse order of removal.

S2000

See Figures 109 and 110.

1. Loosen the valve adjustment screws so that all valves are closed and all rocker arms are loose.

2. Remove or disconnect the following:
   • Negative battery cable
   • Valve cover
   • Camshaft bearing caps
   • Camshafts

To install:

3. Set the engine to Top Dead Center (TDC) so that the timing chain sprocket timing marks are aligned with the cylinder head surface as shown.

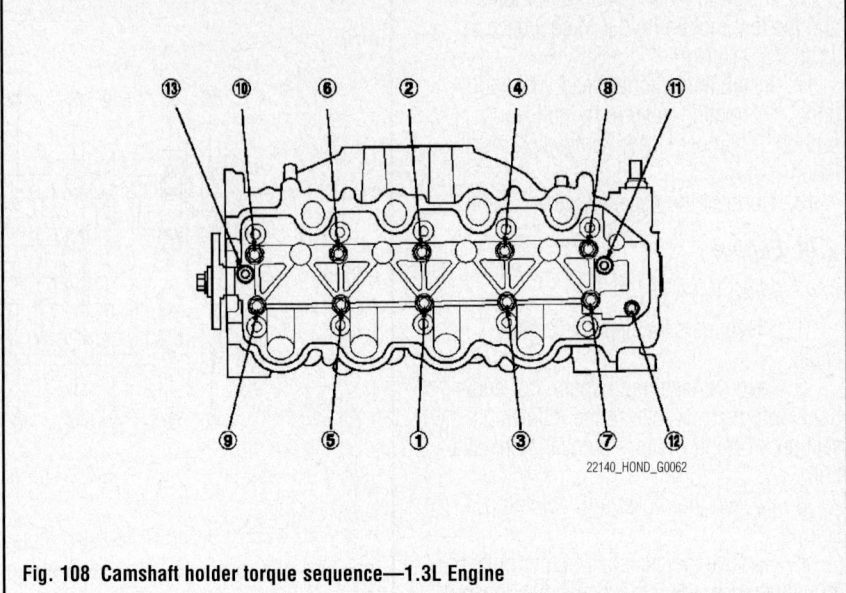

Fig. 108 Camshaft holder torque sequence—1.3L Engine

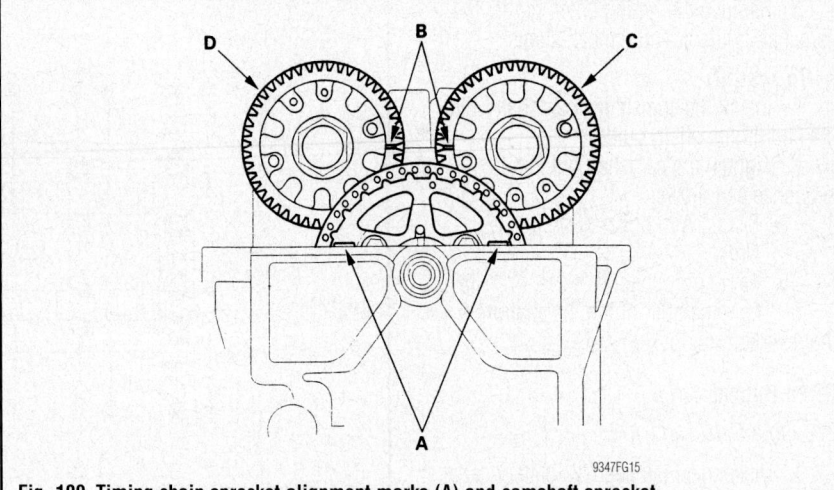

Fig. 109 Timing chain sprocket alignment marks (A) and camshaft sprocket alignment marks (B)—2.2L engine

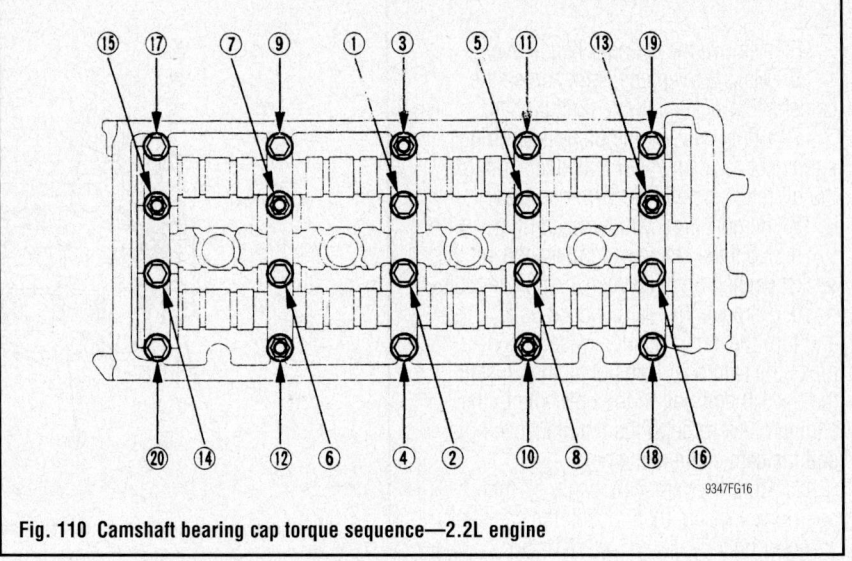

Fig. 110 Camshaft bearing cap torque sequence—2.2L engine

4. Install or connect the following:
- Camshafts with the sprocket timing marks aligned as shown
- Camshaft bearing caps and tighten the bolts in sequence to 16 ft. lbs. (22 Nm). Adjust the valve clearance.
- Valve cover
- Negative battery cable

CRANKSHAFT FRONT SEAL

REMOVAL & INSTALLATION

Accord

1. Disconnect the negative battery cable.
2. Remove the crankshaft position (CKP) sensor.
3. Remove the timing belt and timing belt drive pulley. For additional information, refer to the following section, "Timing Belt, Removal & Installation."
4. Remove the crankshaft pulley and crankshaft oil seal.

To install:

5. Clean and dry the crankshaft oil seal housing.
6. Apply a light coat of multipurpose grease to the crankshaft and to the lip of the seal.
7. Using the seal driver, drive in the crankshaft oil seal until the driver bottoms against the oil pump. When the seal is in place, clean any excess grease off the crankshaft, and check that the oil seal lip is not distorted.
8. The remainder of the installation is the reverse order of removal.

Civic

1. Raise and safely support the vehicle.
2. Remove the right front wheel.
3. Remove the accessory drive belt.
4. Hold the pulley with the holder handle and crankshaft pulley holder.
5. Remove the bolt with a socket, 19 mm and a breaker bar, then remove the crankshaft pulley.
6. Remove the pulley end crankshaft oil seal.

To install:

7. Clean and dry the crankshaft oil seal housing.
8. Apply a light coat of new engine oil around the crankshaft oil seal.
9. Apply a light coat of new engine oil to the crankshaft and to the lip of the crankshaft oil seal.
10. Using the oil seal driver, drive in the crankshaft oil seal until the driver bottoms against the oil pump. When the seal is in place, clean any excess grease off the

crankshaft, and check that the oil seal lip is not distorted.
11. Clean the crankshaft pulley, crankshaft, bolt, and washer.
12. Install the crankshaft pulley onto the crankshaft by aligning the flat sides of the pulley with the flat sides of the inner oil pump gear.
13. Hold the crankshaft pulley with the holder handle and the holder attachment. Tighten the bolt to 51 ft. lbs. (69 Nm) with a torque wrench and 19 mm socket. Do not use an impact wrench. If the pulley bolt or crankshaft are new, tighten the bolt to 130 ft. lbs. (177 Nm) then remove the bolt and tighten it to 51 ft. lbs. (69 Nm). Tighten the bolt an additional 90°.
14. Install the accessory drive belt.
15. Install the right front wheel.

CYLINDER HEAD

REMOVAL & INSTALLATION

Accord

2.4L Engine

See Figures 111 through 114.

➡**Be sure the cylinder head is cool to the touch before beginning the removal procedure. The coolant temperature must be below 100°F (38°C).**

1. Relieve the fuel system pressure. Disconnect the negative battery cable. Drain the engine coolant. Remove the drive belt.
2. Disconnect the intake air temperature sensor. Remove the vacuum hose and breather pipe. Remove the intake air duct.

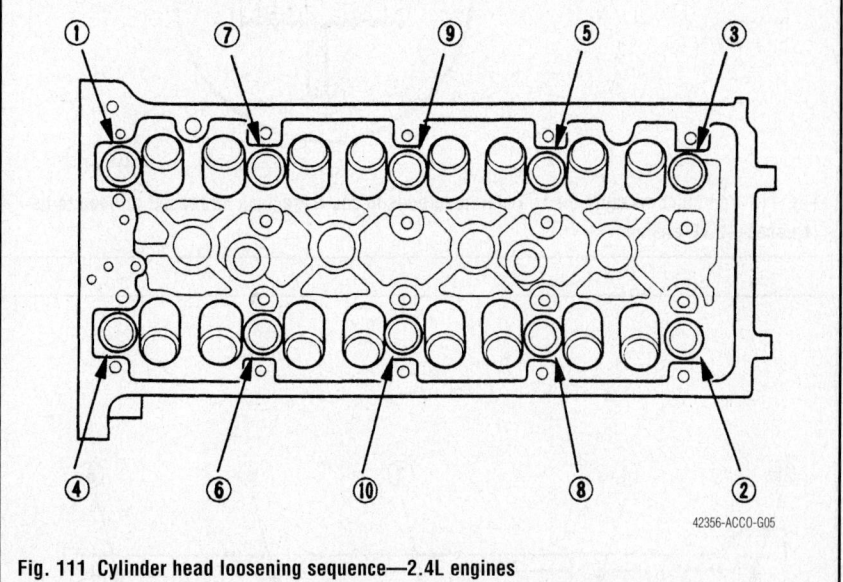

42356-ACCO-G05
Fig. 111 Cylinder head loosening sequence—2.4L engines

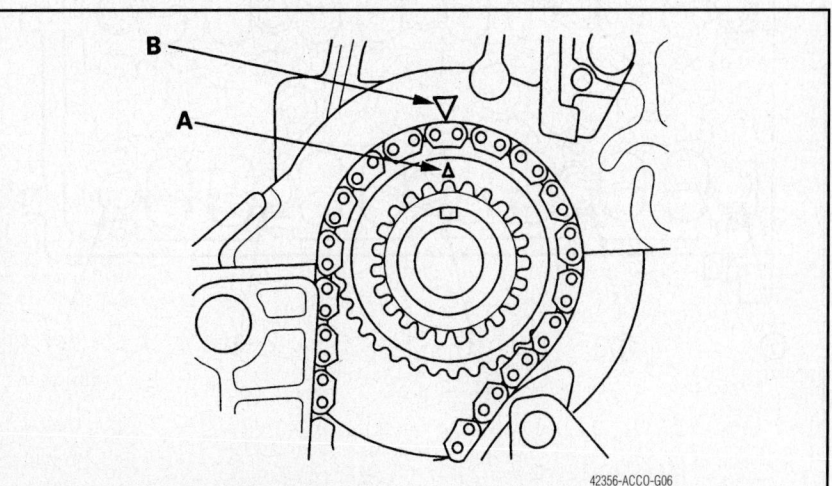

42356-ACCO-G06
Fig. 112 Set the crankshaft to TDC by aligning the mark (A) on the crankshaft sprocket with the pointer (D) on the cylinder block—2.4L engine

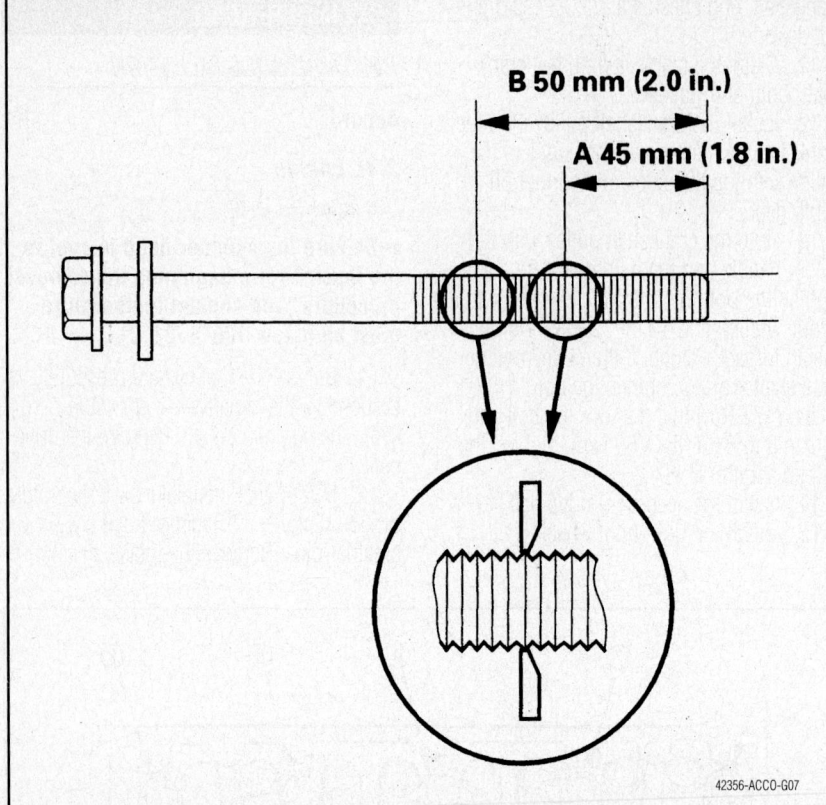

Fig. 113 You must measure the cylinder head bolts to see if they can be reused or need to be replaced—2.4L engine

B 50 mm (2.0 in.)

A 45 mm (1.8 in.)

42356-ACCO-G07

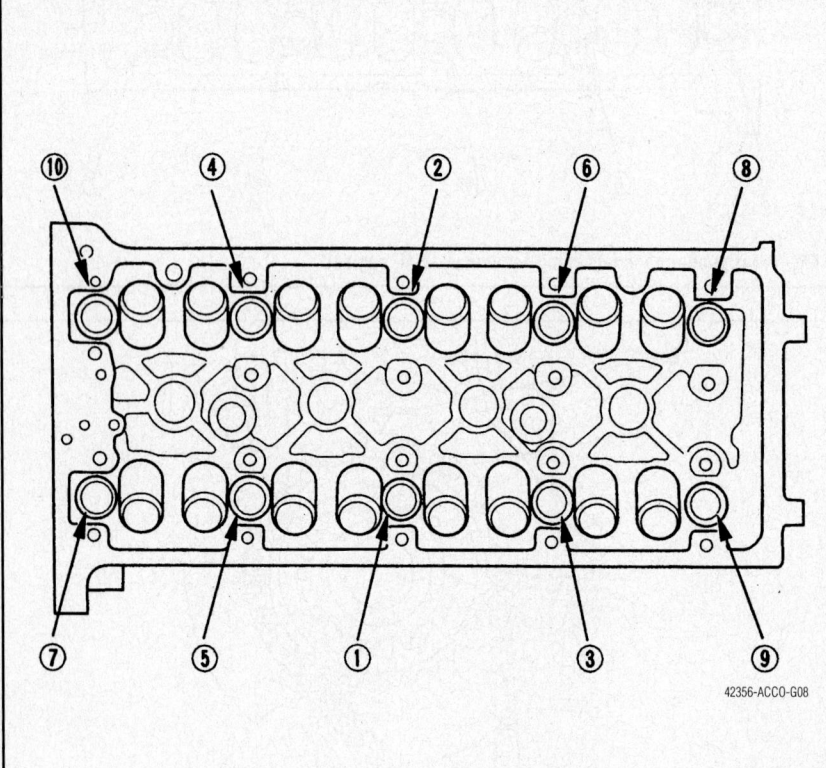

Fig. 114 Cylinder head bolt tightening sequence—2.4L engine

42356-ACCO-G08

Remove the quick connect fitting cover. Disconnect the fuel feed hose.

3. Remove the intake manifold. Remove the exhaust manifold.

4. Remove the bolt securing the connecting pipe. Remove the water bypass hose. Remove the evaporative emission canister hose and power brake vacuum hose.

5. Remove the PCV valve hose. Remove the ground cable. Remove the upper radiator hose and the heater hoses.

6. Remove the engine wiring harness connectors and the wire harness clamps from the cylinder head.

7. Disconnect the fuel injector connectors, engine coolant temperature sensor connector, camshaft position sensor connectors and the EGR valve connector.

8. Disconnect the rocker arm oil control solenoid connector and engine oil pressure sensor connector.

9. Remove the two bolts securing the EVAP canister purge valve bracket and remove the bolt securing the harness bracket.

10. Remove the cam chain. Remove the rocker arm assembly.

11. Remove the cylinder head retaining bolts. To prevent warpage, loosen the cylinder head bolts in a 3-step crisscross pattern in the reverse order of the tightening sequence.

12. Remove the cylinder head from the engine.

To install:

➡**Use new O-ring, seals, and gaskets when installing the cylinder head and its components.**

13. Be sure the cylinder head and the engine block surfaces are clean, level, and straight.

14. Be sure the cylinder head dowel pins and control orifice are aligned. Clean the oil control orifice and reinstall it with a new O-ring.

15. Set the crankshaft to TDC. Align the TDC mark on the crankshaft sprocket with the pointer on the engine block.

16. Position the cylinder head on the engine.

17. Measure the diameter of each cylinder head bolt at points A & B, as shown in the illustration. If either diameter is less than 0.42 in. (10.6mm), replace the head bolt

18. Coat the threads of the cylinder head retaining bolts with clean engine oil. Tighten the cylinder head bolts sequentially to the proper torque.

19. Continue the installation in the reverse order of the removal procedure.

20. Reprogram the ECM/PCM with the HDS, turn the ignition switch the OFF position. Turn the ignition switch to the ON position. Wait thirty seconds. Turn the ignition switch OFF and disconnect the HDS from the DLC.

21. Reprogram the ECM engine idle characteristics. Be sure all electrical items are OFF.

22. Start the engine. Hold the idle speed at 3000 RPM's in neutral until the radiator fan comes on or the temperature reached 176 degrees.

23. Let the engine idle for about five minutes with the throttle fully closed.

24. If the radiator fan comes on during the five minutes, do not count this toward the five minute programming time.

25. Enter the antitheft codes for the radio and the navigation system. Set the clock.

3.0L Engine

See Figures 115 and 116.

➥Be sure the cylinder head is cool to the touch before beginning the removal procedure. The coolant temperature must be below 100°F (38°C).

1. Relieve the fuel system pressure. Disconnect the negative battery cable. Drain the engine coolant. Remove the drive belt. Remove the air cleaner.

2. Remove the timing belt. Remove the power steering pump and hose clamp. Remove the alternator.

3. Remove the intake manifold. Remove the six ignition coils.

4. Remove the engine wire harness connectors and wire harness from the cylinder head.

5. Disconnect the six fuel injector connectors, engine coolant temperature sensor connector, crankshaft position sensor connector, and the EGR valve sensor connector.

6. Disconnect the rocker arm oil control solenoid connector, rocker arm oil pressure switch connector, and the two air fuel ratio sensor connectors. Disconnect the secondary heated oxygen sensor connectors.

7. Remove the upper and lower radiator hoses and the heater hoses. Remove the water bypass hose.

8. Remove the ground cable. Remove the bolt securing the harness bracket. Remove the fuel rails. Remove the bolt securing the harness bracket.

9. Remove the front and rear three way catalytic converter. Remove the water passage.

10. Remove the front and rear camshaft pulleys and front and rear back covers.

11. Remove the cylinder head covers.

12. Remove the cylinder head retaining bolts. To prevent warpage, loosen the cylinder head bolts in a 3-step crisscross pattern in the reverse order of the tightening sequence.

13. Remove the cylinder head from the engine.

To install:

➥Use new O-ring, seals, and gaskets when installing the cylinder head and its components.

14. Be sure the cylinder head and the engine block surfaces are clean, level, and straight.

15. Be sure the cylinder head dowel pins and control orifice are aligned. Clean the oil control orifice and reinstall it with a new O-ring.

16. Set the timing belt drive pulley to TDC by aligning the TDC mark on the tooth of the timing belt drive pulley with the pointer on the oil pump.

17. Set the camshaft pulleys to TDC by aligning the TDC marks on the camshaft pulleys with the pointers on the back covers.

18. Position the cylinder head on the engine.

19. If the head bolts are Dodecagon type, measure the diameter of each cylinder head bolt between point A and B, see illustration. If either diameter is less than 0.42 inch, replace the bolt.

20. Coat the threads of the cylinder head retaining bolts with clean engine oil. Tighten the cylinder head bolts sequentially to the proper torque.

21. Continue the installation in the reverse order of the removal procedure.

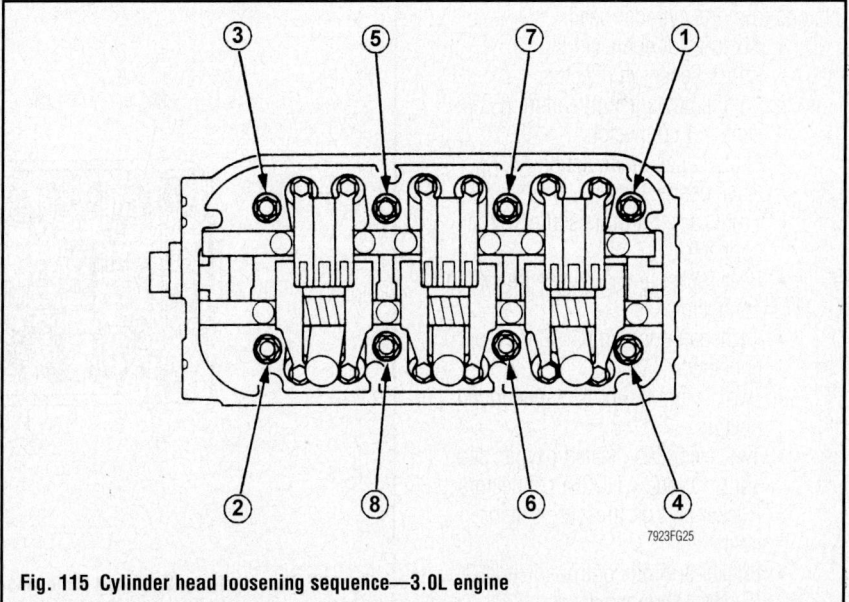

Fig. 115 Cylinder head loosening sequence—3.0L engine

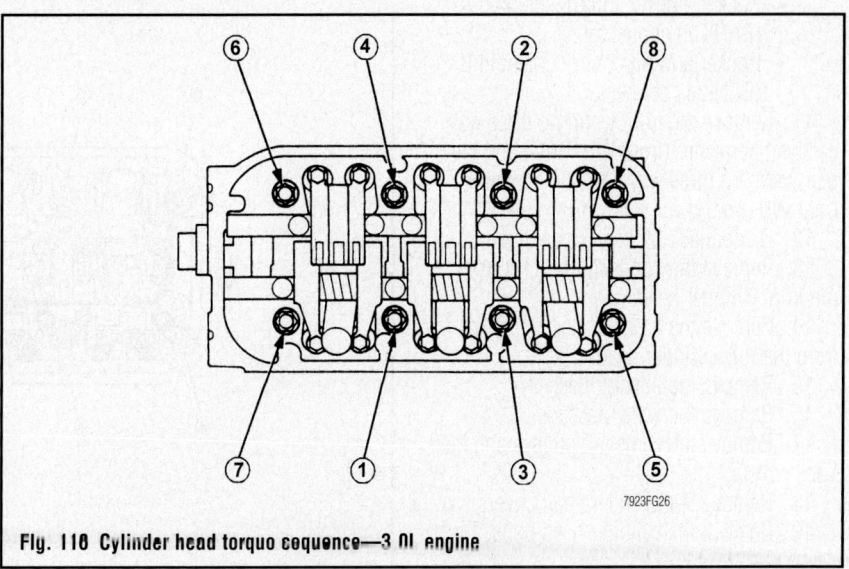

Fig. 116 Cylinder head torque sequence—3.0L engine

3.5L Engine

See Figures 117 and 118.

➡ **Be sure the cylinder head is cool to the touch before beginning the removal procedure. The coolant temperature must be below 100°F (38°C).**

1. Properly relieve the fuel system pressure.
2. Disconnect the battery cables.
3. Drain the engine cooling system.
4. Remove the accessory drive belt.
5. Remove the power steering (P/S) pump and the bolt securing the P/S hose bracket.
6. Remove the alternator.
7. Remove the intake manifold.
8. Remove the six ignition coils.
9. Remove the timing belt.
10. Disconnect the following engine wire harness connectors and wire harness clamps from the cylinder head:
  - Six injector connectors
  - Knock sensor connector
  - Engine coolant temperature (ECT) sensor 1 connector
  - Exhaust gas recirculation (EGR) valve connector
  - Front rocker arm pressure switch connector
  - Rear rocker arm oil pressure switch connector
  - Camshaft position (CMP) sensor connector
  - Two air fuel ratio (A/F) sensor connectors
  - Two secondary heated oxygen sensor (secondary HO2S) connectors
  - Rocker arm oil pressure sensor connector
  - Rocker arm oil control solenoid A (BANK 1) connector
  - Rocker arm oil control solenoid A (BANK 2) connector
  - Rocker arm oil control solenoid B (BANK 1) connector
11. Remove the front warm up three way catalytic converter (front WU-TWC) and the rear warm up three way catalytic converter (rear WU-TWC).
12. Disconnect the fuel supply hose.
13. Remove the connector bracket from the front cylinder head.
14. Remove the harness clamp bracket from the rear cylinder head.
15. Remove the injector bases.
16. Remove the water passage.
17. Remove the camshaft pulleys and the back covers.
18. Remove the cylinder head covers.
19. Remove the cylinder head bolts. To prevent warpage, loosen the bolts in sequence 1/3 turn at a time; repeat the sequence until all bolts are loosened.
20. Remove the cylinder heads.

To install:

21. Clean the cylinder head and the engine block surface.
22. Clean and install the oil control orifices with new O-rings.
23. Install the dowel pins and the new cylinder head gaskets.
24. Clean the timing belt pulleys, the timing belt guide plate, and the upper and lower covers.
25. Set the timing belt drive pulley to top dead center (TDC) by aligning the TDC mark on the tooth of the timing belt drive pulley with the pointer on the oil pump.
26. Set the camshaft pulleys to TDC by aligning the TDC marks on the camshaft pulleys with the pointers on the back covers.
27. Install the cylinder heads on the engine block.
28. Measure the diameter of each cylinder head bolt at point A and point B.
29. If either diameter is less than 10.6 mm (0.42 in.), replace the cylinder head bolt.
30. Apply new engine oil to the threads and under the bolt heads of all cylinder head bolts.
31. Tighten the cylinder head bolts in sequence to 22 ft. lbs. (29 Nm) using a beam-type torque wrench. When using a preset click-type torque wrench, be sure to tighten slowly and do not overtighten. If a bolt makes any noise while you are torquing it, loosen the bolt and retighten it from the first step.

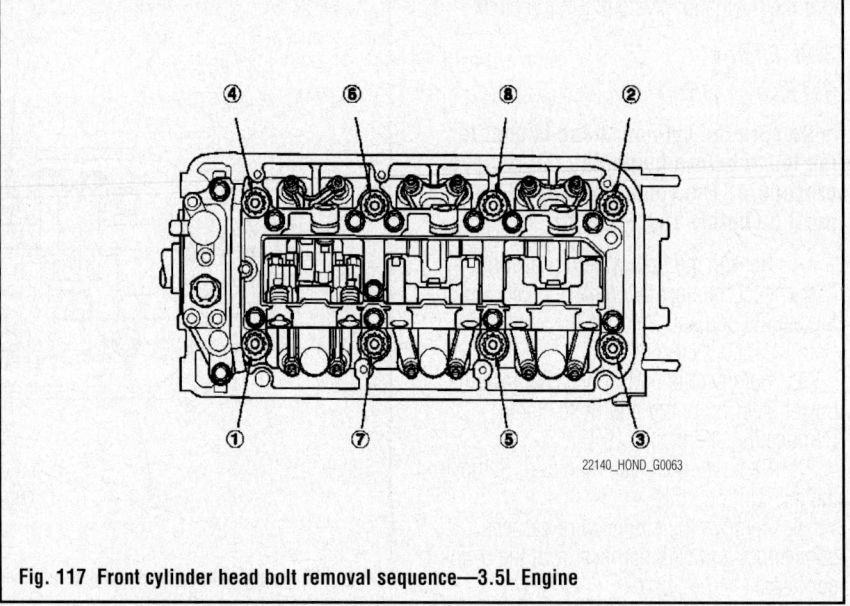

Fig. 117 Front cylinder head bolt removal sequence—3.5L Engine

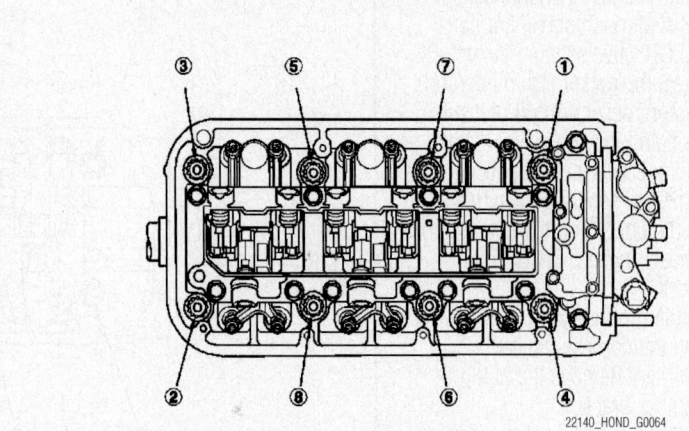

Fig. 118 Rear cylinder head bolt removal sequence—3.5L Engine

32. Tighten the cylinder head bolts in sequence an additional 90 degrees.

33. If using new bolts, tighten the bolts in sequence an additional 90 degrees.

34. Install the timing belt.

35. Adjust the valve clearance.

36. Install the cylinder head covers.

37. Install the water passage.

38. Install the injector bases.

39. The remainder of the installation is the reverse order of removal.

40. Refill the engine cooling system to the correct level.

41. Start the engine and check for leaks.

Civic

1.8L Engine

See Figures 119 and 120.

1. Before servicing the vehicle, refer to the precautions in the beginning of this section.

2. Be sure the cylinder head is cool to the touch before beginning the removal procedure. The coolant temperature must be below 100°F (38°C).

3. Relieve the fuel system pressure. Disconnect the negative battery cable. Drain the engine coolant.

4. Remove the drive belt. Remove the intake manifold.

5. Remove the harness clamps. Remove the PCV valve hose from the clamp. Remove the air cleaner housing bracket and the harness holder from the cylinder head.

6. If equipped with manual transaxle, remove the upper radiator hose and heater hoses.

7. If equipped with automatic transaxle, remove the upper radiator hose, heater hoses and water bypass hose.

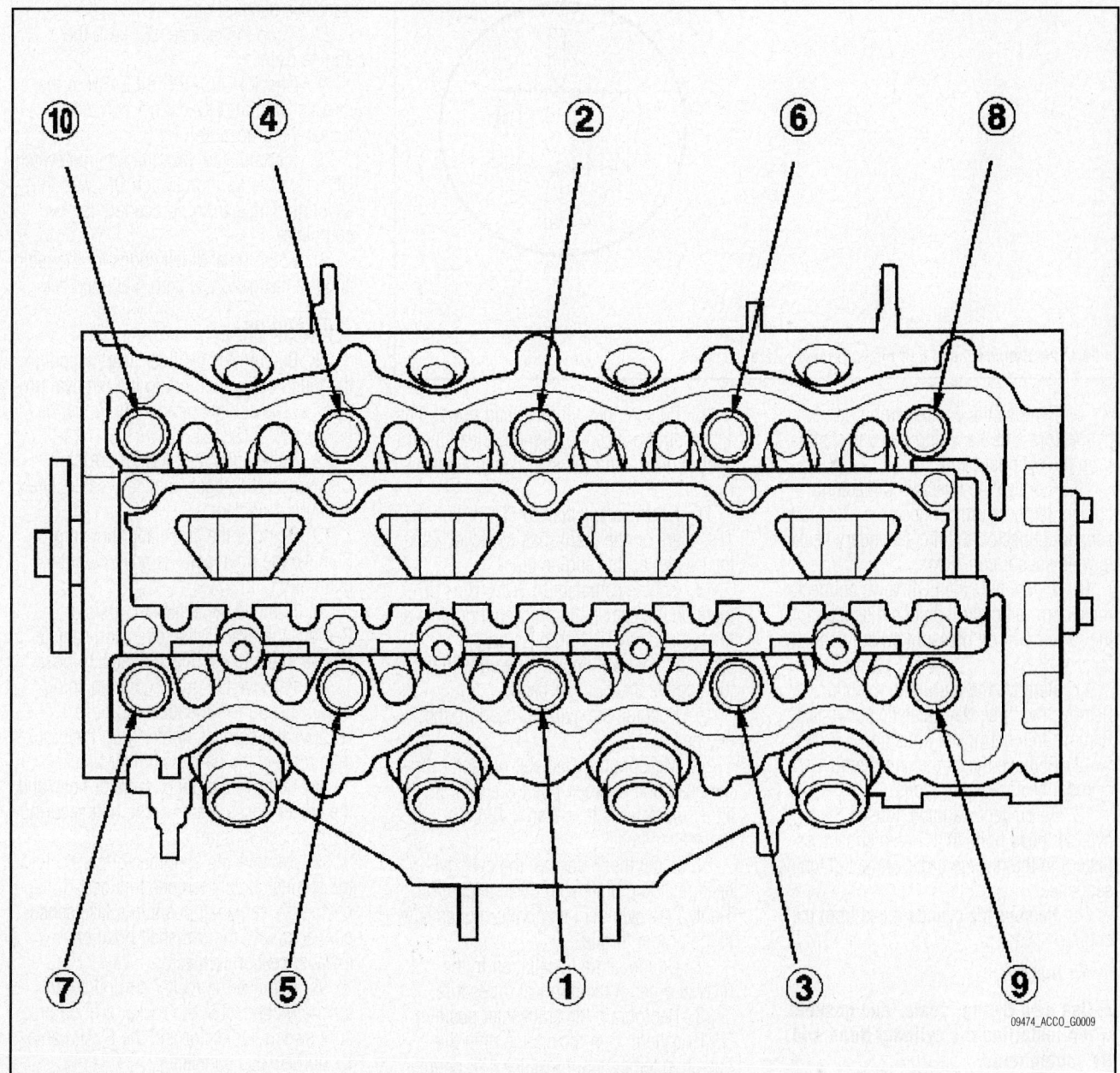

Fig. 119 Cylinder head torque sequence—1.8L engine

09474_ACCO_G0009

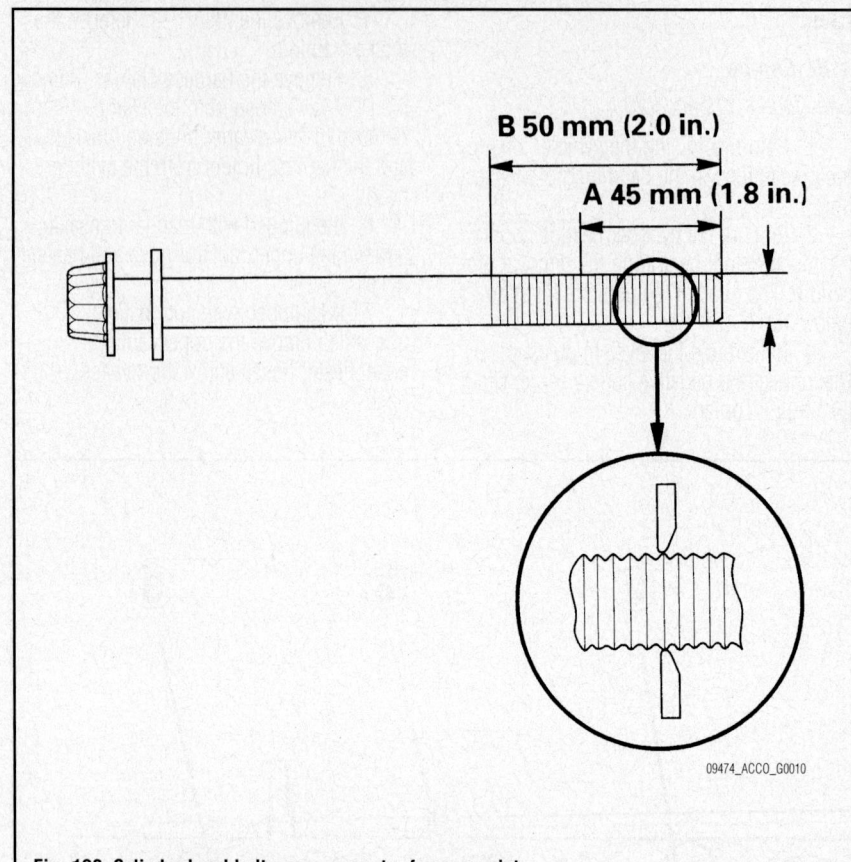

B 50 mm (2.0 in.)

A 45 mm (1.8 in.)

09474_ACCO_G0010

Fig. 120 Cylinder head bolt measurement reference points

8. Remove the engine wiring harness connectors and the wiring harness clamps from the cylinder head.

9. Remove the injector connectors, coolant temperature connector, air fuel ratio sensor connector and the secondary heated oxygen sensor connector.

10. Disconnect the EGR valve connector, rocker arm oil control solenoid sensor and the rocker arm oil pressure switch connector.

11. Remove the four ignition coils. Remove the three way catalytic converter. Remove the thermostat housing.

12. Remove the cam chain. Remove the cylinder head retaining bolts.

13. To prevent warpage, loosen the cylinder head bolts in a 3-step crisscross pattern in the reverse order of the tightening sequence.

14. Remove the cylinder head from the engine.

To install:

➡**Use new O-ring, seals, and gaskets when installing the cylinder head and its components.**

15. Be sure the cylinder head and the engine block surfaces are clean, level, and straight.

16. Be sure the cylinder head dowel pins and control orifice are aligned. Clean the oil control orifice and reinstall it with a new O-ring.

17. Set the crankshaft to TDC. Align the TDC mark on the crankshaft sprocket with the pointer on the engine block.

18. Set the camshaft to TDC. The "UP" mark on the camshaft sprocket should be at the top and the TDC grooves on the camshaft sprocket should line up with the top edge of the cylinder head.

19. Position the cylinder head on the engine.

20. Measure the diameter of each cylinder head bolt between point A and B. If either diameter is less than 0.42 inch, replace the bolt.

21. Coat the threads of the cylinder head retaining bolts with clean engine oil. Tighten the cylinder head bolts sequentially to the proper torque.

22. Continue the installation in the reverse order of the removal procedure.

23. Reprogram the crankshaft position (CKP) pattern. Run the engine until the operating temperature reaches 176 degree. With the engine stopped clear the CKP pattern. Turn the ignition switch OFF. Turn the ignition switch ON and wait thirty seconds.

24. Road test the vehicle on a level surface. Decelerate the engine speed of 2500 RPM to 1000 RPM. If equipped with automatic transaxle use two Drive positions. If equipped with manual transaxle use first gear.

25. Stop the vehicle, but keep the engine running.

26. Check PULSAR F/B LEARN in the data list with the HDS. If not complete repeat the procedure. If complete, road test the vehicle on a level surface. Decelerate the engine speed of 5000 RPM to 3000 RPM. If equipped with automatic transaxle use two Drive positions. If equipped with manual transaxle use first gear.

27. Stop the vehicle, but keep the engine running.

28. Check PULSAR F/B LEARN in the data list with the HDS. If not complete repeat the procedure.

29. If completed, turn the ignition switch OFF. Turn the ignition switch ON, wait thirty seconds. The learning procedure is now complete.

30. Enter the antitheft codes for the radio and the navigation system. Set the clock.

2.0L Engine

1. Be sure the cylinder head is cool to the touch before beginning the removal procedure. The coolant temperature must be below 100°F (38°C).

2. Relieve the fuel system pressure. Disconnect the negative battery cable. Drain the engine coolant.

3. Remove the air cleaner housing. Remove the drive belt. Remove the intake manifold.

4. Remove the exhaust manifold. Remove the evaporative emission canister hose and the brake booster vacuum hose.

5. Remove the quick-connect fitting cover. Disconnect the fuel feed hose. Remove the harness holder from the bracket then remove the harness holder bracket.

6. Remove the upper radiator hose and the heater hoses. Remove the bolt securing the connecting pipe.

7. Remove the fuel injector connectors, the engine coolant temperature sensor connector, the camshaft position intake sensor connector and the camshaft position exhaust sensor connector.

8. Remove the rocker arm oil control solenoid connector, the rocker arm oil pressure switch connector and the EVAP canister purge valve connector.

9. Remove the cam chain. Remove the rocker arm cover. Remove the rocker arm assembly. Remove the cylinder head retaining bolts.

10. To prevent warpage, loosen the cylinder head bolts in a 3-step crisscross pattern in the reverse order of the tightening sequence.

11. Remove the cylinder head from the engine.

To install:

➡️**Use new O-ring, seals, and gaskets when installing the cylinder head and its components.**

12. Be sure the cylinder head and the engine block surfaces are clean, level, and straight.

13. Be sure the cylinder head dowel pins and control orifice are aligned. Clean the oil control orifice and reinstall it with a new O-ring.

14. Set the crankshaft to TDC. Align the TDC mark on the crankshaft sprocket with the pointer on the engine block.

15. Position the cylinder head on the engine.

16. Measure the diameter of each cylinder head bolt between point A and B. If either diameter is less than 0.42 inch, replace the bolt.

17. Coat the threads of the cylinder head retaining bolts with clean engine oil. Tighten the cylinder head bolts sequentially to the proper torque.

18. Continue the installation in the reverse order of the removal procedure.

Civic Hybrid

See Figures 121 and 122.

➡️**Connect the Honda Diagnostic System (HDS) to the data link connector (DLC), and monitor the engine coolant temperature (ECT) sensor 1. To avoid damaging the cylinder head, wait until the ECT 1 temperature drops below 100°F (38°C) before loosening the cylinder head bolts.**

1. Properly relieve the fuel system pressure.

2. Remove the engine appearance cover.

3. Drain the engine cooling system.

4. Remove the air intake assembly.

5. Remove the intake manifold. For additional information, refer to the following section, "Intake Manifold, Removal & Installation."

6. Disconnect the knock sensor connector, then remove the harness clamp and the connecting pipe.

7. Disconnect the camshaft position (CMP) sensor and the ECT sensor 1 connector, then remove the harness clamp.

8. Remove the eight ignition coils.

9. Remove the harness holder and disconnect the breather hose.

10. Disconnect the evaporative emission (EVAP) canister purge valve connector, then remove the harness clamp.

11. Disconnect the purge joint from the bracket, then remove the fuel pipe nut and remove the EVAP canister purge valve bracket bolts.

12. Disconnect the upper radiator hose, the lower radiator hose, the water bypass hose, and the heater hoses.

13. Remove the accessory drive belt.

14. Turn the crankshaft pulley so its top dead center (TDC) mark lines up with the pointer.

15. Remove the water pump.

16. Remove the cylinder head cover.

17. Remove the warm-up three way catalytic converter (WU-TWC).

18. Remove the crankshaft pulley.

19. Support the engine with a suitable jack and a wood block under the oil pan.

20. Remove the ground cable, and then remove the side engine mount bracket.

21. Disconnect the crankshaft position (CKP) sensor and the harness clamp, then remove the dipstick tube.

22. Remove the timing chain case.

23. Make a reference mark across the camshaft sprocket and the timing chain.

24. Loosely install the crankshaft pulley.

25. Apply new engine oil to the sliding surface of the cam chain tensioner through the oil return hole in the cylinder head.

26. Hold the crankshaft pulley and set the socket wrench on the camshaft sprocket bolt.

27. Turn the camshaft clockwise to compress the cam chain tensioner, then install the

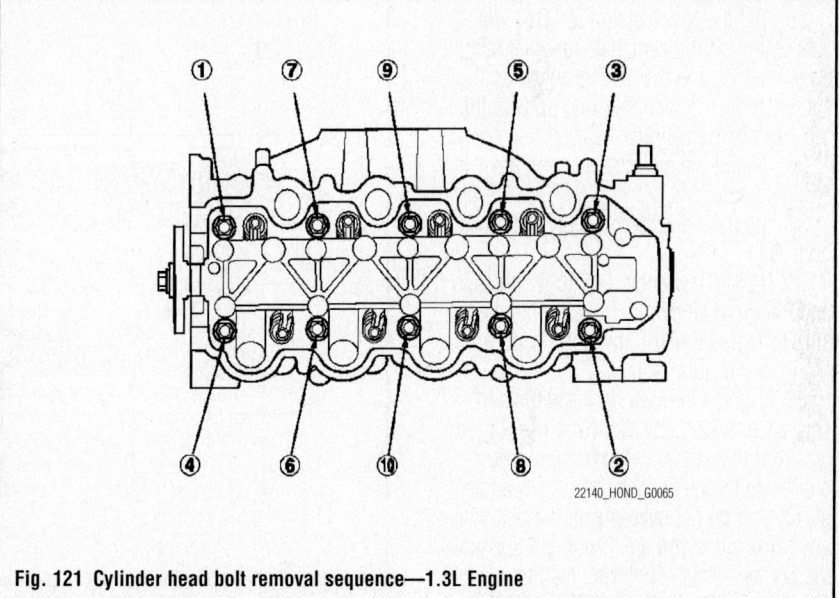

Fig. 121 Cylinder head bolt removal sequence—1.3L Engine

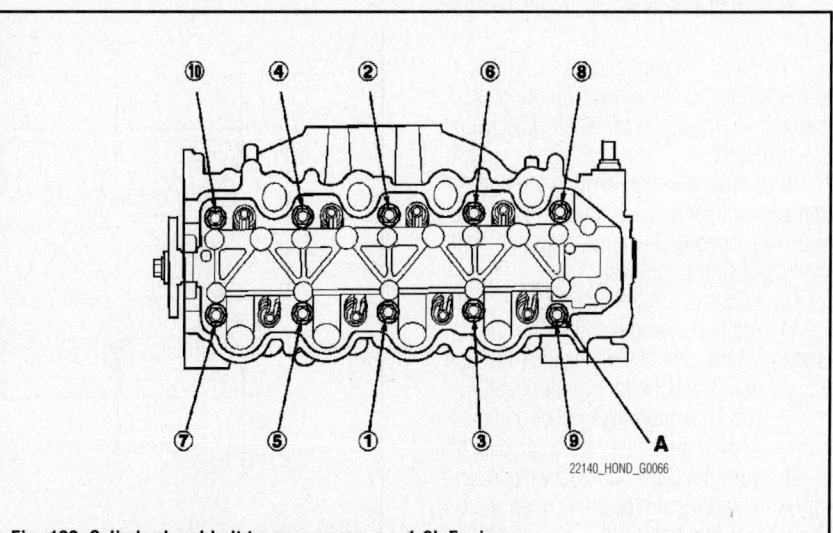

Fig. 122 Cylinder head bolt torque sequence—1.3L Engine

6 x 1.0 mm bolt in the bolt hole in the engine block through the timing chain tensioner.

28. Hold the camshaft with a 27 mm open-end wrench, then remove the camshaft sprocket.

29. Remove the top bolt that secures the timing chain guide.

30. Remove the cylinder head bolts. To prevent warpage, loosen the bolts in sequence ⅓ turn at a time; repeat the sequence until all bolts are loosened.

31. Remove the cylinder head.

To install:

32. Clean the cylinder head and the engine block surface.

33. Install the new cylinder head gasket and the dowel pins on the engine block. Always use a new cylinder head gasket.

34. Set the crankshaft to top dead center (TDC). Align the TDC mark on the crankshaft sprocket with the pointer on the oil pump.

35. Set the No. 1 piston at TDC. The "UP" mark on the camshaft sprocket should be at the top, and the TDC grooves on the camshaft sprocket should line up with the top edge of the cylinder head.

36. Install the cylinder head on the engine block.

37. Replace any stretched cylinder head bolts.

38. Apply new engine oil to the threads and flange of all cylinder head bolts. Be sure to install the 165 mm long head bolt (A) in the location shown.

39. Tighten the cylinder head bolts in sequence to 22 ft. lbs. (29 Nm), use a beam-type torque wrench. When using a preset-click-type torque wrench, be sure to tighten slowly and do not overtighten. If a bolt makes any noise while you are torquing it, loosen the bolt and retighten it from the first step. Then tighten the bolts an additional 130°.

40. Install the timing chain guide mounting top bolt.

41. Install the cam chain around the camshaft sprocket aligned with the reference mark, then install the camshaft sprocket on the camshaft.

42. Apply new engine oil to the bolt threads and flange. Hold the camshaft with a 27 mm open-end wrench, then tighten the bolt to 41 ft. lbs. (56 Nm).

43. Loosely install the crankshaft pulley.

44. Apply new engine oil to the sliding surface of the cam chain tensioner through the oil return hole in the cylinder head.

45. Hold the crankshaft pulley and set the socket wrench on the camshaft sprocket bolt.

46. Turn the camshaft clockwise to compress the cam chain tensioner, then remove the 6 x 1.0 mm bolt.

47. Remove all of the old liquid gasket from the chain case mating surfaces, the bolts, and the bolt holes.

48. Clean and dry the chain case mating surfaces.

49. Apply liquid gasket, P/N 08717-0004, 08718-0001, 08718-0003, or 08718-0009, evenly to the cylinder head and engine block mating surface of the timing chain case. Install the component within 5 minutes of applying the liquid gasket.

50. Apply liquid gasket, P/N 08718-0001, or 08718-0009 evenly to the oil pan mating surface of the timing chain case. Install the component within 5 minutes of applying the liquid gasket.

51. Set the edge of the chain case to the edge of the oil pan, then install the chain case on the engine block. Wipe off the excess liquid gasket on the oil pan and chain case mating area.

52. The remainder of the installation is the reverse order of removal.

53. Inspect for fuel leaks.

54. Refill the engine cooling system to the correct level.

55. Start the engine and check for leaks.

S2000

See Figures 123 and 124.

➡Be sure the cylinder head is cool to the touch before beginning the removal procedure. The coolant temperature must be below 100°F (38°C).

1. Relieve the fuel system pressure. Disconnect the negative battery cable. Drain the engine coolant. Drain the engine oil.

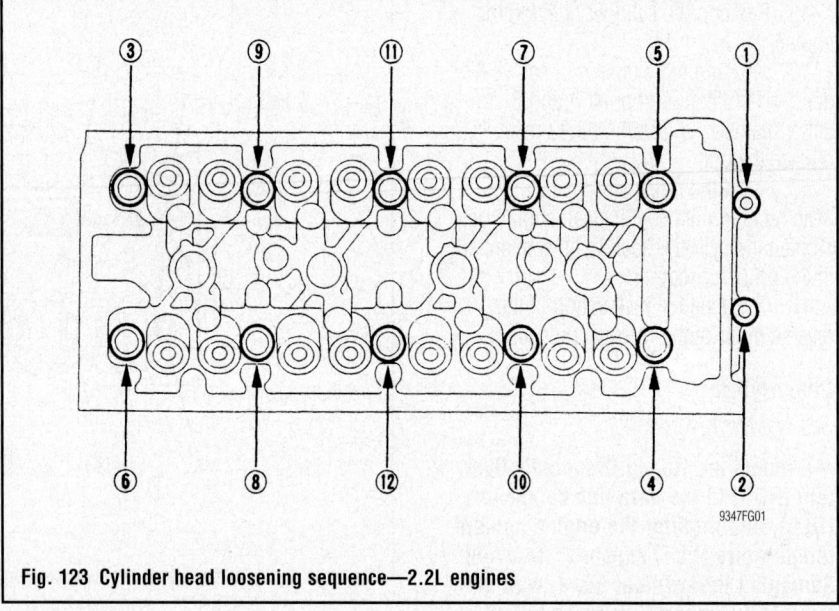

Fig. 123 Cylinder head loosening sequence—2.2L engines

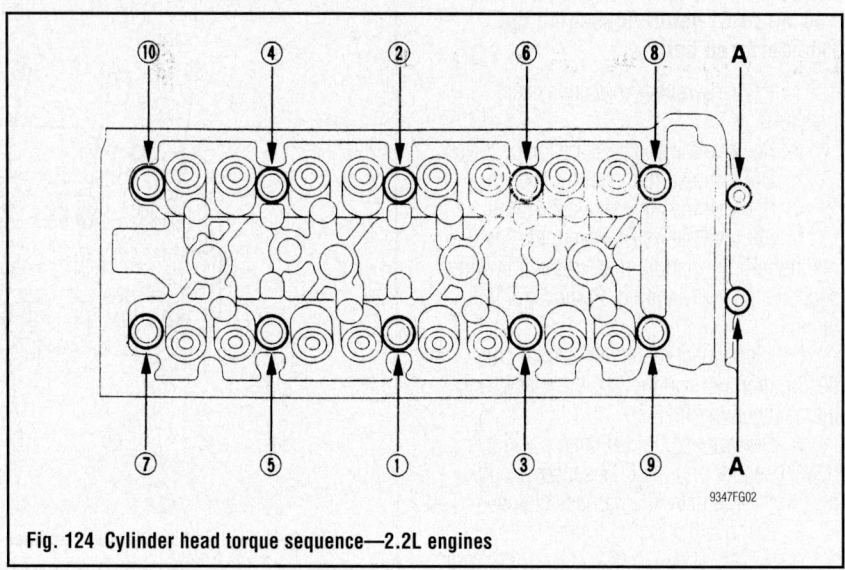

Fig. 124 Cylinder head torque sequence—2.2L engines

2. Remove the air cleaner assembly. Remove the intake air cleaner housing. Remove the drive belt.

3. Remove the intake manifold cover, power brake booster hose and the quick-connect fitting cover, then disconnect the fuel feed hose.

4. Disconnect the evaporative emission canister hose. Remove the intake manifold bracket retaining bolt. Remove the water outlet cover.

5. Remove the engine wire harness connectors and the wire harness clamps from the cylinder head and intake manifold. Remove the four fuel injector connectors

6. Disconnect the engine coolant temperature sensor connector, throttle body connector, and the air/fuel sensor connector.

7. Disconnect the manifold absolute pressure connector, rocker arm oil control solenoid connector, rocker arm oil pressure switch connector and the crankshaft position sensor connector.

8. Remove the water bypass hose, EVAP hose and bracket and the intake manifold bracket.

9. Remove the four bolts retaining the exhaust manifold cover. Remove the heat shield retaining bolts. Remove the heat shield.

10. Remove the exhaust manifold cover. Remove the exhaust manifold retaining bolts and remove the exhaust manifold.

11. Remove the intake manifold bracket clips. Remove the intake manifold retaining bolts. Remove the intake manifold.

12. Remove the cylinder head cover retaining bolts. Remove the cylinder head cover.

13. Position the No. 1 piston at TDC. The TDC marks on the cam chain sprocket should align with the cylinder head surface.

14. Remove the end cover and nozzle from the cam chain auto tensioner. Thread a nut onto a 5 × 0.8 mm bolt at is at least 40 mm long. Thread the bolt into the maintenance hole in the cam chain auto tensioner.

15. Turn the bolt clockwise to compress the cam chain auto tensioner and lock it in place with the nut. Remove the cam chain auto tensioner.

16. Loosen the rocker arm adjusting screws. Remove the camshaft holders and camshafts.

17. Insert the bolts into the rocker shaft holder and remove the rocker arm assembly.

18. Remove the idler gear/cam chain sprocket assembly, idler gear collar and washer.

19. Remove the cylinder head retaining bolts. To prevent warpage, loosen the cylinder

head bolts in a 3-step crisscross pattern in the reverse order of the tightening sequence.

20. Remove the cylinder head from the engine.

To install:

➡**Use new O-ring, seals, and gaskets when installing the cylinder head and its components.**

21. Be sure the cylinder head and the engine block surfaces are clean, level, and straight.

22. Be sure the cylinder head dowel pins and control orifice are aligned. Clean the oil control orifice and reinstall it with a new O-ring.

23. Apply liquid gasket, part number 08717-0004, 08718-0001, 08718-0003 or 08718-0009 to the cylinder head mating surface of the block and chain case within 5mm of the edge of the cylinder head gasket.

➡**Do not install the parts if more than five minutes have elapsed since applying the liquid gasket. Instead, reapply the liquid gasket after removing the old residue.**

24. Position the cylinder head on the engine.

25. Coat the threads of the cylinder head retaining bolts with clean engine oil. Tighten the cylinder head bolts sequentially to the proper torque.

26. Continue the installation in the reverse order of the removal procedure.

27. Reprogram the crankshaft position (CKP) pattern. Run the engine until the operating temperature reaches 176 degree. With the engine stopped clear the CKP pattern. Turn the ignition switch OFF. Turn the ignition switch ON and wait thirty seconds.

28. Road test the vehicle on a level surface. Decelerate the engine speed of 2500 RPM to 1000 RPM. If equipped with automatic transaxle use two Drive positions. If equipped with manual transaxle use first gear.

29. Stop the vehicle, but keep the engine running.

30. Check PULSAR F/B LEARN in the data list with the HDS. If not complete repeat the procedure. If complete, road test the vehicle on a level surface. Decelerate the engine speed of 5000 RPM to 3000 RPM. If equipped with automatic transaxle use two Drive positions. If equipped with manual transaxle use first gear.

31. Stop the vehicle, but keep the engine running.

32. Check PULSAR F/B LEARN in the data list with the HDS. If not complete repeat the procedure.

33. If completed, turn the ignition switch OFF. Turn the ignition switch ON, wait thirty seconds. The learning procedure is now complete.

34. Enter the antitheft codes for the radio and the navigation system. Set the clock.

ENGINE ASSEMBLY

REMOVAL & INSTALLATION

Accord

2.4L Engine

1. Before servicing the vehicle, refer to the precautions in the beginning of this section.

2. Note the radio security code and the radio presets.

3. Properly relieve the fuel system pressure.

4. Disconnect the negative battery cable, than the positive cable. Remove the hood support from the driver's side of the vehicle. Lock the hood in the full up position.

5. Remove the battery.

6. Disconnect the air intake sensor. Remove the vacuum hose, and breather pipe then remove the intake air duct.

7. Remove the air cleaner housing. Remove the harness clamp.

8. Remove the harness terminal wires. Remove the bolts retaining the battery base and remove it. Remove the battery cables from the under hood fuse/relay box. Disconnect the harness connector. Remove the two retaining bolts holding the under hood fuse/relay box in place. Remove the box.

9. Remove the harness clamp and the strut brace.

10. Remove the quick connect fitting cover. Disconnect the fuel feed hose. Remove the evaporative emission canister hose and brake booster vacuum hose.

11. Disconnect the ECM/PCM connectors and the main wire harness connectors. Disconnect the accelerator pedal position (APP) sensor connector. Remove the harness clamps and grommet. Pull the wire harness thru the bulkhead.

12. If equipped with manual transaxle, remove the clutch slave cylinder and clutch line bracket mounting bolts. Remove the shift and select cables.

13. Disconnect the air fuel sensor connector. Remove the drive belt.

14. Remove the power steering pump. Remove the air conditioning compressor

without disconnecting the air conditioning hoses. Position the compressor to the side.

15. Remove the radiator cap. Raise and support the vehicle safely. Remove the front tires. Drain the radiator. Drain the transaxle. Drain the engine oil.

16. Disconnect the stabilizer links. Remove the damper fork. Separate the knuckles from the lower arm ball joints. Remove the driveshafts. Remove the exhaust pipe.

17. On automatic transaxle remove the bolts securing the shift cable holder. Remove the shift cable cover. Remove the spring clip and control pin then separate the shift cable from the control lever.

18. Remove the transaxle lower front and lower rear mounts. Lower the vehicle. On automatic transaxle models, remove the fluid cooler lines.

19. Remove the upper radiator hose and heater hoses. Remove the lower radiator hoses. Remove the battery ground cable and upper bracket.

20. Properly attach the engine lifting device.

21. If equipped with manual transaxle, remove the ground cable. Remove the transaxle upper mount/bracket assembly and the clutch line clamp bracket. Remove the front stop then the front mount bolt.

22. If equipped with automatic transaxle, remove the ground cable. Remove the transaxle upper mount/bracket assembly. Remove the vacuum hose. Remove the front mount stop and vacuum hose clamp bracket. Remove the front transaxle mount bolt.

23. Remove the rear mount stop retaining nuts. Remove the rear mount stop. Remove the rear mount bolt.

24. Be sure that the engine/transaxle assembly is free from all hoses, vacuum lines and electrical wires and connectors.

25. Remove the engine/transaxle assembly from under the vehicle.

To install:

26. Installation is the reverse of the removal procedure. Be sure to check and adjust all required fluid levels.

27. Reprogram the ECM/PCM with the HDS, turn the ignition switch the OFF position. Turn the ignition switch to the ON position. Wait thirty seconds. Turn the ignition switch OFF and disconnect the HDS from the DLC.

28. Reprogram the ECM engine idle characteristics. Be sure all electrical items are OFF.

29. Start the engine. Hold the idle speed at 3000 RPM's in neutral until the radiator

fan comes on or the temperature reached 176 degrees.

30. Let the engine idle for about five minutes with the throttle fully closed.

31. If the radiator fan comes on during the five minutes, do not count this toward the five minute programming time.

32. Enter the antitheft codes for the radio and the navigation system. Set the clock.

3.0L Engine

1. Note the radio security code and the radio presets.

2. Properly relieve the fuel system pressure.

3. Disconnect the negative battery cable, than the positive cable. Remove the battery.

4. Remove the windshield wiper arms. Remove the cowl cover. Remove the bulkhead cover.

5. Remove the support struts from the engine hood. Move the hood to a vertical position. Install the right side hood support.

6. Drain the power steering fluid. Plug the reservoir and return hose. Remove the air cleaner housing.

7. Remove the harness terminal wires. Remove the bolts retaining the battery base and remove it. Remove the battery cables from the under hood fuse/relay box. Disconnect the harness connector. Remove the two retaining bolts holding the under hood fuse/relay box in place. Remove the box. Remove the ground cable.

8. Remove the brake booster vacuum hose, evaporative emission canister hose and vacuum hose. Remove the harness clamp and disconnect the engine wiring harness from the left side of the engine compartment.

9. Remove the drive belt. Remove the power steering pump outlet line from the power steering pump. Remove the power steering hose from the clamp. Remove the power steering system fluid reservoir from the clamp.

10. Remove the steering gear protective cover. Lock the steering wheel in position. Make a reference mark across the steering joint and steering gear box pinion shaft. Remove the steering joint bolt. Disconnect the steering joint from the steering gearbox pinion shaft.

➡**To prevent damage to the cable reel, do not turn the steering wheel once the steering joint has been removed.**

11. Disconnect the fan motor connector. Disconnect the compressor clutch connector. Remove the coolant tank.

12. Remove the condenser fan shroud retaining bolts. Remove the fan shroud. Remove the four bolts retaining the air conditioning compressor.

13. If equipped with manual transaxle, remove the ground cable from the shift cable holder. Remove the bolts retaining the shift cable holder. Remove the shift cable and the select cable. Remove the clutch slave cylinder mounting bolt. Remove the slave cylinder.

➡**Do not operate the clutch pedal once the slave cylinder has been removed.**

14. Remove the radiator cap. Raise and safely support the vehicle. Remove the front tires. Remove the splash pan.

15. Drain the radiator. Drain the engine oil. Drain the transaxle.

16. Disconnect the stabilizer links. Remove the damper fork. Separate the tie rod end ball joints from the knuckles. Separate the knuckles from the lower arms.

17. Remove the driveshafts. Remove the exhaust pipe.

18. On automatic transaxle, Remove the shift cable cover retaining bolt. Remove the cover. Remove the lock bolt retaining then control lever, remove the lever assembly.

➡**To prevent damage to the control lever joint, be sure to remove the bolts retaining the shift cable holder before removing the bolts retaining the shift cable cover.**

19. Remove and plug the power steering line hose. Disconnect the power steering pressure switch connector.

20. Remove the nuts retaining the transaxle lower front mount and lower rear mount. Lower the vehicle to the ground.

21. Remove the radiator hoses. Remove the heater hoses. Remove the connector clamp from the front and rear cylinder head.

➡**Removal of these bolts is necessary if using engine support tool AAR-T-12566 and VSB02C000919.**

22. Lift and support the engine, using the engine support tool.

23. If equipped with manual transaxle, remove the ground cable. If equipped with automatic transaxle, remove the transaxle upper mount bracket and vacuum hose.

24. Remove the front mount stop. Remove the front mount bolt. Remove the rear mount stop. Remove the rear mount bolt.

25. If equipped with manual transaxle, remove the two bolts retaining the shift cable bracket. Raise and safely support the vehicle.

26. Matchmark the subframe with the edge of the subframe stiffener. Attach special tool VSB02C000016 by hanging the belt over the front of the subframe. Secure the belt to the stop.

27. Raise the jack, line up the slots in the tool arms with the bolt holes on the corner of the jack base. Attach the four bolts.

28. Remove the subframe mid mount. Remove the subframe. Lower the vehicle to the ground.

29. Attach a lifting device. Raise the engine/transaxle assembly up slightly. Remove the engine support tool.

30. Remove the two bolts retaining the side engine mount bracket.

31. Be sure that the engine/transaxle assembly is free from all hoses, vacuum lines and electrical wires and connectors.

32. Remove the engine/transaxle assembly from under the vehicle.

To install:

33. Installation is the reverse of the removal procedure. Be sure to check and adjust all required fluid levels.

34. Reprogram the ECM/PCM with the HDS, turn the ignition switch the OFF position. Turn the ignition switch to the ON position. Wait thirty seconds. Turn the ignition switch OFF and disconnect the HDS from the DLC.

35. Reprogram the crankshaft position (CKP) pattern. Run the engine until the operating temperature reaches 176 degree. With the engine stopped clear the CKP pattern. Turn the ignition switch OFF. Turn the ignition switch ON and wait thirty seconds.

36. Road test the vehicle on a level surface. Decelerate the engine speed of 2500 RPM to 1000 RPM. If equipped with automatic transaxle use two Drive positions. If equipped with manual transaxle use first gear.

37. Stop the vehicle, but keep the engine running.

38. Check PULSAR F/B LEARN in the data list with the HDS. If not complete repeat the procedure. If complete, road test the vehicle on a level surface. Decelerate the engine speed of 5000 RPM to 3000 RPM. If equipped with automatic transaxle use two Drive positions. If equipped with manual transaxle use first gear.

39. Stop the vehicle, but keep the engine running.

40. Check PULSAR F/B LEARN in the data list with the HDS. If not complete repeat the procedure.

41. If completed, turn the ignition switch OFF. Turn the ignition switch ON, wait thirty seconds. The learning procedure is now complete.

42. Enter the antitheft codes for the radio and the navigation system. Set the clock.

3.5L Engine

See Figures 125 through 127.

1. Remove the hood support rod, then use it as shown to prop the hood in the wide-open position.

2. Remove the strut brace.

3. Drain the power steering fluid.

4. Properly relieve the fuel pressure.

5. Remove the engine appearance cover.

6. Disconnect the battery cables, then remove the battery.

7. Remove the splash separator.

8. Remove the front grille cover.

9. Remove the air intake assembly.

10. Remove the harness clamp and the bolt, then remove the battery base.

11. Disconnect the battery cables from the under-hood fuse/relay box.

12. Remove the engine control module (ECM)/ powertrain control module (PCM) cover, then remove the three bolts securing the ECM/PCM.

13. Disconnect the ECM/PCM connectors, the engine wire harness connector, and the harness clamps, and remove the bracket.

14. If equipped with manual transmission, disconnect the engine wire harness connector.

15. Disconnect the evaporative emission (EVAP) canister hose and the brake booster vacuum hose.

16. Remove the quick-connect fitting cover, then disconnect the fuel feed hose.

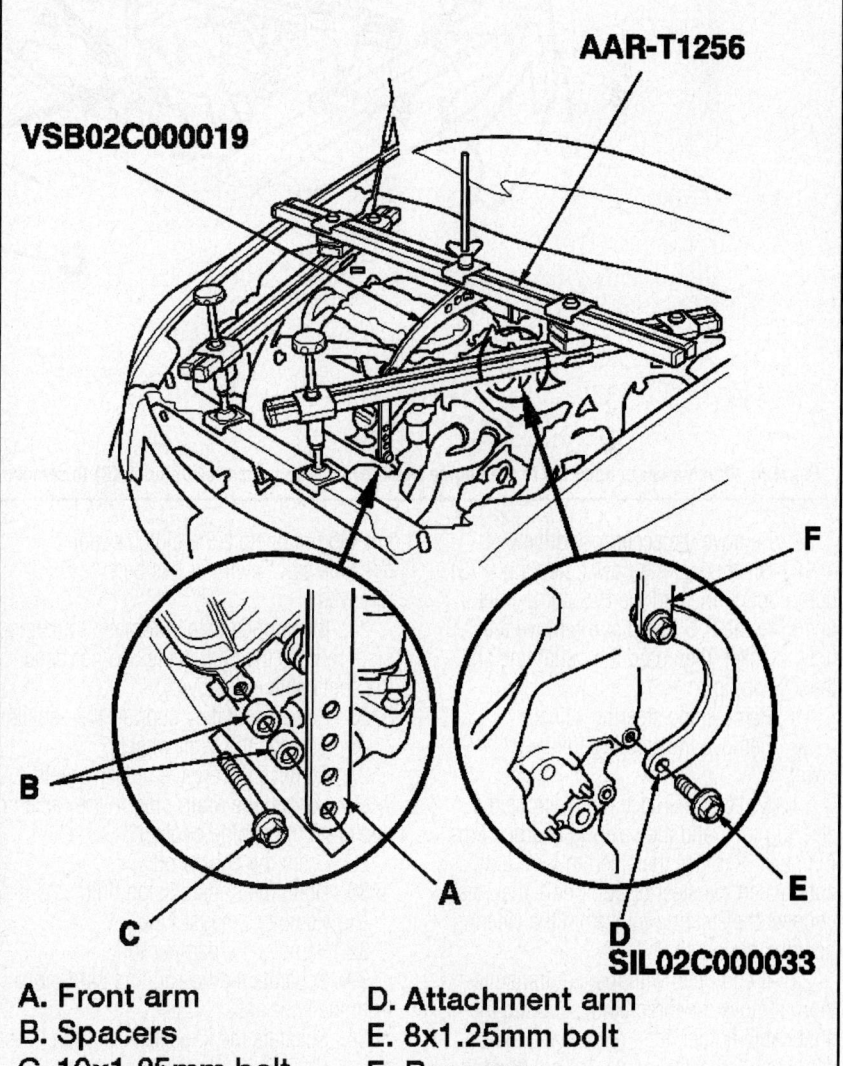

A. Front arm
B. Spacers
C. 10x1.25mm bolt
D. Attachment arm
E. 8x1.25mm bolt
F. Rear arm

22140_HOND_G0067

Fig. 125 Install the engine support hanger—3.5L Engine

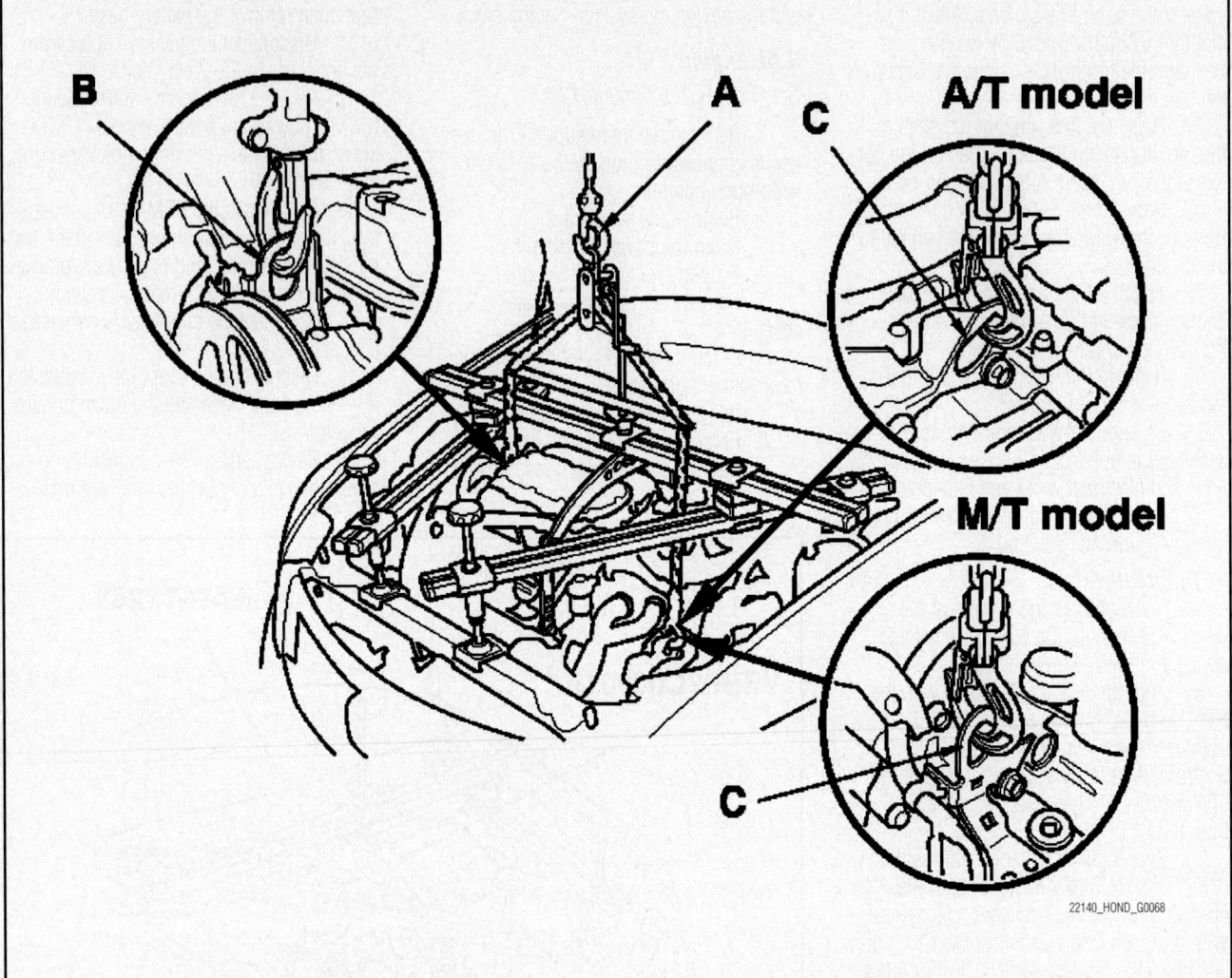

Fig. 126 Attach a chain hoist (A) to the engine hook (B) and the transmission hook (C) to remove the engine assembly—3.5L Engine

17. Remove the accessory drive belt.

18. Disconnect the power steering (P/S) pump outlet line and the P/S pump inlet line from the P/S pump and remove the P/S hose bracket, then plug the outlet line and the P/S pump.

19. Remove the steering wheel.

20. Remove the steering joint cover.

21. Make a reference mark across the steering joint and the steering gearbox pinion shaft. Remove the steering joint bolt, and loosen the steering joint bolt, then disconnect the steering joint from the steering gearbox pinion shaft.

22. If equipped with manual transmission, remove the three bolts securing the shift cable holder, then remove the shift cable and the select cable. Do not bend the cables excessively.

23. If equipped with manual transmission, remove the clutch slave cylinder and the clutch line bracket mounting

nut. Do not operate the clutch pedal once the slave cylinder has been removed.

24. If equipped with automatic transmission, remove the shift cable. Do not bend the shift cable excessively.

25. Raise and safely support the vehicle.

26. Remove the front wheels.

27. Remove the engine splash shield.

28. Loosen the drain plug in the radiator, and drain the engine coolant.

29. Drain the engine oil.

30. Drain the transmission fluid.

31. Remove exhaust pipe.

32. Remove the damper fork.

33. Separate the tie-rod end ball joints from the knuckles.

34. Separate the knuckles from the lower arms.

35. Remove the halfshafts. Coat all precision-finished surfaces with new engine oil. Tie plastic bags over the driveshaft ends.

36. Disconnect the P/S hose, then plug the line and the hose.

37. Disconnect the power steering pressure switch connector.

38. Lower the vehicle.

39. Remove the radiator.

40. Disconnect the heater hoses.

41. Disconnect the A/C compressor clutch connector, then remove the A/C compressor without disconnecting the A/C hoses. Do not bend the A/C hoses excessively.

42. If equipped with automatic transmission, remove the ATF cooler hoses from the transmission, then plug the ATF cooler hoses and the lines.

43. Remove the harness holder and the connectors from the front cylinder head.

44. Remove the connector bracket from the front cylinder head; use the bracket bolt hole to attach the engine hanger balance bar front arm.

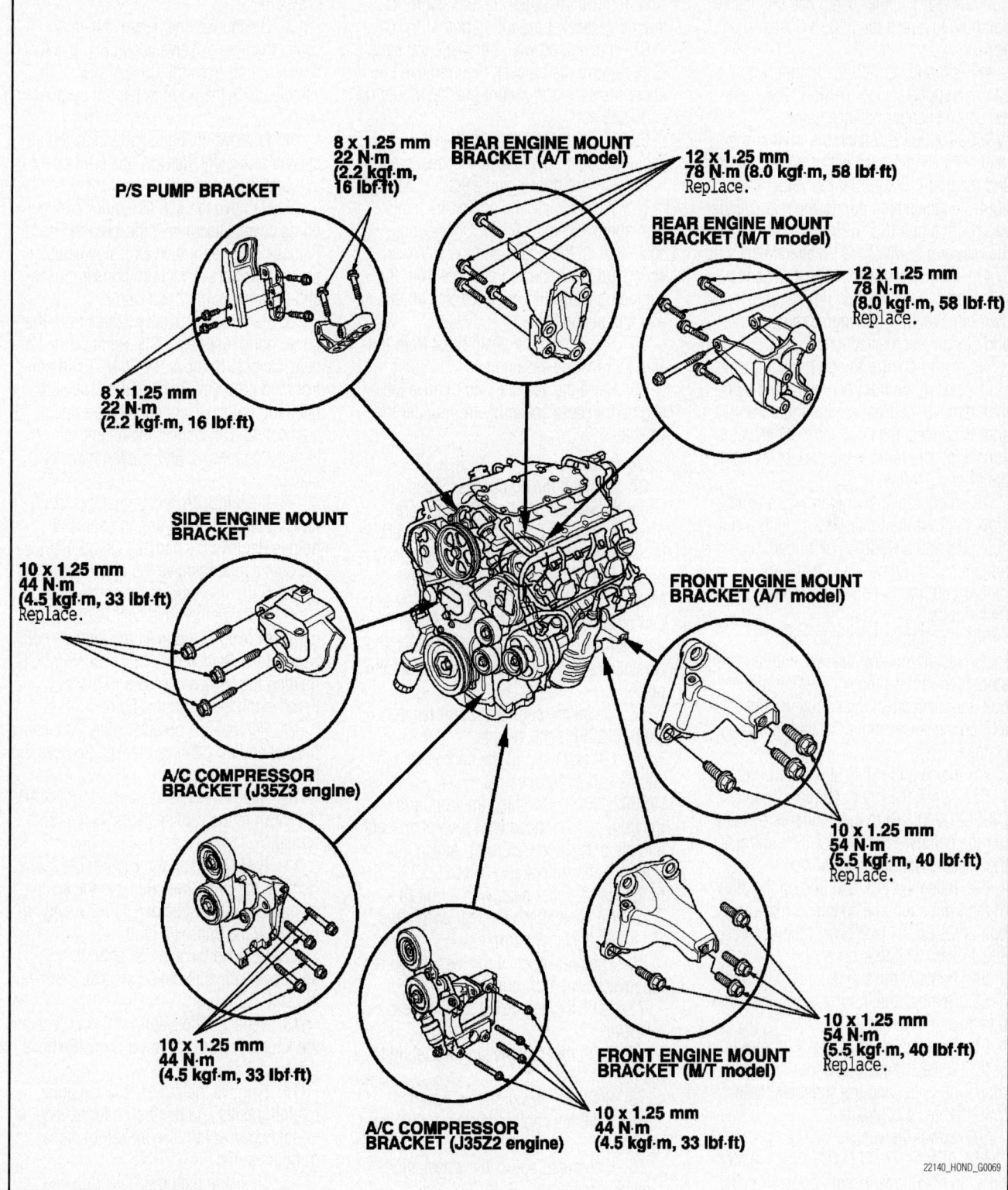

P/S PUMP BRACKET

8 x 1.25 mm
22 N·m
(2.2 kgf·m,
16 lbf·ft)

8 x 1.25 mm
22 N·m
(2.2 kgf·m, 16 lbf·ft)

REAR ENGINE MOUNT BRACKET (A/T model)

12 x 1.25 mm
78 N·m (8.0 kgf·m, 58 lbf·ft)
Replace.

REAR ENGINE MOUNT BRACKET (M/T model)

12 x 1.25 mm
78 N·m
(8.0 kgf·m, 58 lbf·ft)
Replace.

SIDE ENGINE MOUNT BRACKET

10 x 1.25 mm
44 N·m
(4.5 kgf·m, 33 lbf·ft)
Replace.

FRONT ENGINE MOUNT BRACKET (A/T model)

10 x 1.25 mm
54 N·m
(5.5 kgf·m, 40 lbf·ft)
Replace.

A/C COMPRESSOR BRACKET (J35Z3 engine)

10 x 1.25 mm
44 N·m
(4.5 kgf·m, 33 lbf·ft)

A/C COMPRESSOR BRACKET (J35Z2 engine)

10 x 1.25 mm
44 N·m
(4.5 kgf·m, 33 lbf·ft)

FRONT ENGINE MOUNT BRACKET (M/T model)

10 x 1.25 mm
54 N·m
(5.5 kgf·m, 40 lbf·ft)
Replace.

22140_HOND_G0069

Fig. 127 Engine assembly mounting bracket torque specifications—3.5L Engine

45. Remove the harness clamp bracket from the rear cylinder head; use the bracket bolt hole to attach the 2008 V6 attachment arm.

46. Lift and support the engine with the engine support hanger (AAR-T1256) and the engine hanger balance bar (VSB02C000019). Attach the front arm to the front cylinder head with several spacers and the bolt (10 x 1.25 mm). Attach the 2008 V6 attachment arm to the rear cylinder head with a bolt (8 x 1.25 mm), then attach the rear arm to the 2008 V6 attachment arm.

47. If equipped with manual transmission, remove the front engine mount stop, then remove the front engine mount bolt and disconnect the vacuum hose.

48. If equipped with automatic transmission, remove the front engine mount stop, then remove the front engine mount bolt and disconnect the engine mount actuator connector and remove the connector from the front subframe.

49. If equipped with manual transmission, support the transmission with a suitable jack and a wood block under the transmission. Remove the rear engine mount bolt, then remove the mount-to-subframe bolts.

50. If equipped with automatic transmission, remove the rear engine mount stop, then remove the rear engine mount bolt and disconnect the engine mount actuator connector. Remove the wiring harness clamps.

51. Raise and safely support the vehicle.

52. Attach the front subframe adapter (VSB02C000016) to the subframe by hanging the belt over the front of the subframe, then secure the belt with its stop.

53. Raise the jack and line up the slots in the front subframe adapter arms with the bolt holes on the jack base, then securely attach them with four bolts.

54. Remove the subframe middle mount.

55. Remove the nuts securing the lower transmission mount.

56. Remove the four 12 x 1.25 mm bolts securing the stiffeners, the four subframe mounting bolts, and the stiffeners, then lower the front subframe.

57. Lower the vehicle.

58. Attach a chain hoist to the engine hook and the transmission hook, then lift the engine/transmission until it is securely supported by the chain hoist, and remove the engine support hanger and the engine hanger balance bar.

59. Remove the mounting bolts from the upper half of the side engine mount bracket.

60. Remove the transmission ground cable.

61. If equipped with manual transmission, remove the upper transmission mounting bracket mounting bolts.

62. If equipped with automatic transmission, remove the bracket, then remove the upper transmission mount bracket mounting bolts and nut.

63. Check that the engine/transmission is completely free of vacuum hoses, fuel hoses, coolant hoses, and electrical wiring.

64. Slowly lower the engine/transmission about 150 mm (6 in.). Check once again that all the hoses and the electrical wiring are disconnected and free from the engine/transmission, then lower it all the way and support it.

65. Disconnect the chain hoist from the engine/transaxle assembly.

66. Raise the vehicle, and remove the engine/transaxle assembly from under the vehicle.

To install:

67. Position the engine/transmission under the vehicle. Be sure that they are properly aligned. Carefully lower the vehicle until the engine/transmission are properly positioned in the engine compartment. Make sure the vehicle is not resting on any part of the engine/transmission. Support the engine/transmission with a chain hoist and carefully raise the engine/transmission into place.

68. Install the engine support hanger (AAR-T1256) to the vehicle.

69. Lift and support the engine with the engine hanger balance bar (VSB02C000019). Attach the front arm to the front cylinder head with several spacers and the bolt (10 x 1.25 mm). Attach the 2008 V6 attachment arm (SIL02C000033) to the rear cylinder head with a bolt (8 x 1.25 mm), then attach the rear arm to the 2008 V6 attachment arm.

70. The remainder of the installation is the reverse order of removal.

71. Refill the engine with oil to the correct level.

72. Refill the transmission with fluid to the correct level.

73. Refill the power steering system.

74. Refill the cooling system to the correct level.

75. Check and adjust the wheel alignment as necessary.

76. Start the engine and check for leaks.

Civic

1.8L Engine

1. Note the radio security code and the radio presets.

2. Properly relieve the fuel system pressure.

3. Disconnect the negative battery cable, than the positive cable. Remove the hood support from the driver's side of the vehicle. Lock the hood in the full up position.

4. Remove the battery. Remove the air cleaner assembly. Remove the cowl cover and undercover panel.

5. Remove the fuel line quick connect fitting cover. Disconnect the fuel feed hose. Remove the evaporative emission canister hose, power brake vacuum booster hoses and power steering hose clamp.

6. Remove the battery cables from the under hood fuse/relay box. Remove the harness clamps. Remove the ECM /PCM control module cover. Remove the assembly retaining bolts. Disconnect the electrical connectors. Disconnect the engine wiring harness connector. Remove the harness clamps.

7. If equipped with manual transaxle, remove the bolts retaining the harness holder. Remove the harness clamp. Remove the shift cable. Remove the clutch slave cylinder and line bracket mounting bolt.

8. Remove the drive belt. Remove the harness cover. Disconnect the ignition coil connectors. Remove the four retaining bolts holding the harness holders in place. Remove the front harness holder.

9. Remove the breather pipe, air cleaner bracket and air harness bracket. Remove the harness holder bracket.

10. Attach the engine support tool (AAR-T-12566) to the vehicle. Remove the radiator cap.

11. Raise and support the vehicle safely. Remove the front tires. Remove the splash shield. Drain the radiator. Drain the engine oil. Drain the transaxle fluid.

12. Remove the exhaust pipe. If equipped with automatic transaxle, remove the shift cable.

13. Separate the stabilizer links. Separate the knuckles from the lower arms. Remove the driveshafts.

14. Remove the power steering pump retaining bolts and position it to the side. It is not necessary to disconnect the hoses from the pump.

15. Unclamp the power steering line clamp from the subframe. Remove the power steering line mounting bolts.

16. Remove the gearbox mounting bolt. Remove the steering gearbox stiffener

17. Lower the vehicle to the ground. Remove the radiator.

18. Disconnect the air conditioning compressor clutch connector and remove the

harness clamp. Remove the compressor assembly, without disconnecting the refrigerant lines.

19. Remove the heater hoses. Raise and support the vehicle safely. Remove the lower torque rod.

20. Matchmark the subframe with the edge of the subframe stiffener. Attach special tool VSB02C000016 by hanging the belt over the front of the subframe. Secure the belt to the stop.

21. Raise the jack, line up the slots in the tool arms with the bolt holes on the corner of the jack base. Attach the four bolts.

22. Remove the subframe mid mount. Remove the subframe. Lower the vehicle to the ground. Remove the ground cable.

23. Attach a lifting device. Raise the engine/transaxle assembly up slightly. Remove the engine support tool.

24. Remove the side engine mount bracket. Remove the transaxle mounting bolt and nuts. Remove the ground cable.

25. Be sure that the engine/transaxle assembly is free from all hoses, vacuum lines and electrical wires and connectors.

26. Remove the engine/transaxle assembly from under the vehicle.

To install:

27. Installation is the reverse of the removal procedure. Be sure to check and adjust all required fluid levels.

28. Reprogram the ECM engine idle characteristics. Be sure all electrical items are OFF.

29. Start the engine. Hold the idle speed at 3000 RPM's in neutral until the radiator fan comes on or the temperature reached 194 degrees.

30. Let the engine idle for about five minutes with the throttle fully closed.

31. If the radiator fan comes on during the five minutes, do not count this toward the five minute programming time.

32. Reprogram the crankshaft position (CKP) pattern. Run the engine until the operating temperature reaches 176 degree. With the engine stopped clear the CKP pattern. Turn the ignition switch OFF. Turn the ignition switch ON and wait thirty seconds.

33. Road test the vehicle on a level surface. Decelerate the engine speed of 2500 RPM to 1000 RPM. If equipped with automatic transaxle use two Drive positions. If equipped with manual transaxle use first gear.

34. Stop the vehicle, but keep the engine running.

35. Check PULSAR F/B LEARN in the data list with the HDS. If not complete repeat the procedure. If complete, road test

the vehicle on a level surface. Decelerate the engine speed of 5000 RPM to 3000 RPM. If equipped with automatic transaxle use two Drive positions. If equipped with manual transaxle use first gear.

36. Stop the vehicle, but keep the engine running.

37. Check PULSAR F/B LEARN in the data list with the HDS. If not complete repeat the procedure.

38. If completed, turn the ignition switch OFF. Turn the ignition switch ON, wait thirty seconds. The learning procedure is now complete.

39. Enter the antitheft codes for the radio and the navigation system. Set the clock.

2.0L Engine

1. Note the radio security code and the radio presets.

2. Properly relieve the fuel system pressure.

3. Disconnect the negative battery cable, than the positive cable. Remove the hood support from the driver's side of the vehicle. Lock the hood in the full up position.

4. Remove the battery. Remove the air cleaner assembly. Remove the cowl cover and undercover panel.

5. Disconnect the harness clamp. Remove the resonator unit.

6. Remove the battery cables from the under hood fuse/relay box. Remove the harness clamps. Remove the ECM /PCM control module cover. Remove the assembly retaining bolts. Disconnect the electrical connectors. Disconnect the engine wiring harness connector. Remove the harness clamps.

7. Remove the intake manifold cover. Remove the fuel line quick connect fitting cover. Disconnect the fuel feed hose. Remove the evaporative emission canister hose, power brake vacuum booster hoses and power steering hose clamp.

8. If equipped with manual transaxle, remove the bolts retaining the harness holder. Remove the harness clamp. Remove the shift cable. Remove the clutch slave cylinder and line bracket mounting bolt.

9. Remove the air cleaner housing bracket. Remove the drive belt. Remove the idler pulley base. Remove the radiator cap.

10. Raise and safely support the vehicle. Remove the front tires. Remove the splash shield. Drain the radiator. Drain the engine oil. Drain the transaxle fluid.

11. Disconnect the air fuel ratio (A/F) sensor connector. Remove the grommet and disconnect the oxygen sensor connector. Remove the three way catalytic converter.

12. Separate the stabilizer links. Separate the knuckles from the lower arms. Remove the driveshafts.

13. Remove the steering gearbox bracket. Remove the steering gearbox mounting bolt, stiffener mounting bolt and stiffener. Remove the harness clamp from the subframe.

14. Disconnect the air conditioning compressor clutch connector and remove the harness clamp. Remove the compressor assembly, without disconnecting the refrigerant lines. Lower the vehicle to the ground.

15. Remove the radiator. Remove the heater hoses. Attach the engine support tool (AAR-T-12566) to the vehicle.

16. Raise and support the vehicle safely. Remove the lower torque rod. Remove the lower radiator hose from the clamp. Remove the front mount retaining bolt.

17. Matchmark the subframe with the edge of the subframe stiffener. Attach special tool VSB02C000016 by hanging the belt over the front of the subframe. Secure the belt to the stop.

18. Raise the jack, line up the slots in the tool arms with the bolt holes on the corner of the jack base. Attach the four bolts.

19. Remove the subframe. Lower the vehicle to the ground. Remove the side engine mount bracket retaining bolt and nut. Remove the transaxle mount bracket and retaining bolt and nuts.

20. Attach a lifting device. Raise the engine/transaxle assembly up slightly. Remove the engine support tool.

21. Remove the side engine mount bracket. Remove the transaxle mounting bolt and nuts. Remove the ground cable.

22. Be sure that the engine/transaxle assembly is free from all hoses, vacuum lines and electrical wires and connectors.

23. Remove the engine/transaxle assembly from under the vehicle.

To install:

24. Installation is the reverse of the removal procedure. Be sure to check and adjust all required fluid levels.

25. Reprogram the ECM engine idle characteristics. Be sure all electrical items are OFF.

26. Start the engine. Hold the idle speed at 3000 RPM's in neutral until the radiator fan comes on or the temperature reached 194 degrees.

27. Let the engine idle for about five minutes with the throttle fully closed.

28. If the radiator fan comes on during the five minutes, do not count this toward the five minute programming time.

29. Reprogram the crankshaft position (CKP) pattern. Run the engine until the operating temperature reaches 176 degree. With the engine stopped clear the CKP pattern. Turn the ignition switch OFF. Turn the ignition switch ON and wait thirty seconds.

30. Road test the vehicle on a level surface. Decelerate the engine speed of 2500 RPM to 1000 RPM. If equipped with automatic transaxle use two Drive positions. If equipped with manual transaxle use first gear.

31. Stop the vehicle, but keep the engine running.

32. Check PULSAR F/B LEARN in the data list with the HDS. If not complete repeat the procedure. If complete, road test the vehicle on a level surface. Decelerate the engine speed of 5000 RPM to 3000 RPM. If equipped with automatic transaxle use two Drive positions. If equipped with manual transaxle use first gear.

33. Stop the vehicle, but keep the engine running.

34. Check PULSAR F/B LEARN in the data list with the HDS. If not complete repeat the procedure.

35. If completed, turn the ignition switch OFF. Turn the ignition switch ON, wait thirty seconds. The learning procedure is now complete.

36. Enter the antitheft codes for the radio and the navigation system. Set the clock.

Civic Hybrid

See Figure 128.

1. Secure the hood in the wide open position (support rod in the lower hole).

2. Do the battery removal procedure.

3. Remove the cowl cover and the under-cowl panel.

4. Turn the IMA battery module switch OFF.

5. Properly relieve the fuel system pressure.

6. Remove the engine cover.

7. Remove the resonator.

8. Remove the intake air duct.

9. Remove the air cleaner assembly.

10. Disconnect the positive starter cable, then remove the harness clamp and the ground cable.

11. Remove the powertrain control module (PCM) cover, then remove the bolts.

12. Disconnect the PCM connectors.

13. Disconnect the engine wire harness connectors, then remove the harness clamp on the left side of the engine compartment.

14. Disconnect the vacuum hose.

15. Disconnect the evaporative emission (EVAP) canister purge valve connector, then remove the harness clamp.

16. Disconnect the purge joint from the bracket, then remove the fuel pipe nut and the EVAP canister purge valve bracket bolts.

17. Remove the accessory drive belt.

18. Wait until the engine is cool, then carefully remove the radiator cap.

19. Raise and safely support the vehicle.

20. Remove the front wheels.

21. Remove the front undercover and the splash shield.

22. Loosen the drain plug in the radiator, and drain the engine coolant.

23. Drain the CVT fluid (CVTF).

24. Drain the engine oil.

25. Lower the vehicle.

26. Remove the radiator.

27. Remove the A/C compressor, then disconnect the A/C compressor clutch connector without disconnecting the A/C hoses. Do not bend the A/C hoses excessively.

28. Remove the steering wheel.

29. Remove the steering joint cover.

30. Make reference marks across the steering joint and the steering gearbox pinion shaft. Remove the steering joint bolt, and then disconnect the steering joint from the steering gearbox pinion shaft.

31. Remove the IMA motor power cable clamps, then remove the bolt.

32. Disconnect the water bypass hose and the heater hoses.

33. Disconnect the IMA motor power cable connector.

✳✳ WARNING

If the motor power cable terminals are wet, dry them with a clean towel. Do not use compressed air.

34. Remove the shift cable Do not bend the shift cable excessively.

35. Disconnect the solenoid harness connector and the CVT input shaft (drive pulley) speed sensor connector, then remove the clamp.

36. Remove the IMA motor power cable mounting bolts

37. Raise and safely support the vehicle.

38. Remove the under-floor three way catalytic converter (TWC).

39. Disconnect the front stabilizer link.

40. Disconnect the suspension lower arm ball joints.

41. Remove the halfshafts. Coat all precision-finished surfaces with new engine oil. Tie plastic bags over the driveshaft ends.

42. Disconnect the steering gearbox harness connectors.

43. Lower the vehicle.

44. Install the universal lifting eyelet (07AAK-SNAA120) to the threaded hole on the cylinder head with a 10 x 1.25 mm bolt.

45. Install the front leg assembly, the hook, and the wing nut to an A and Reds engine support hanger (AAR-T1256) onto the Civic engine hanger (VSB02C000025). Carefully position the engine hanger on the

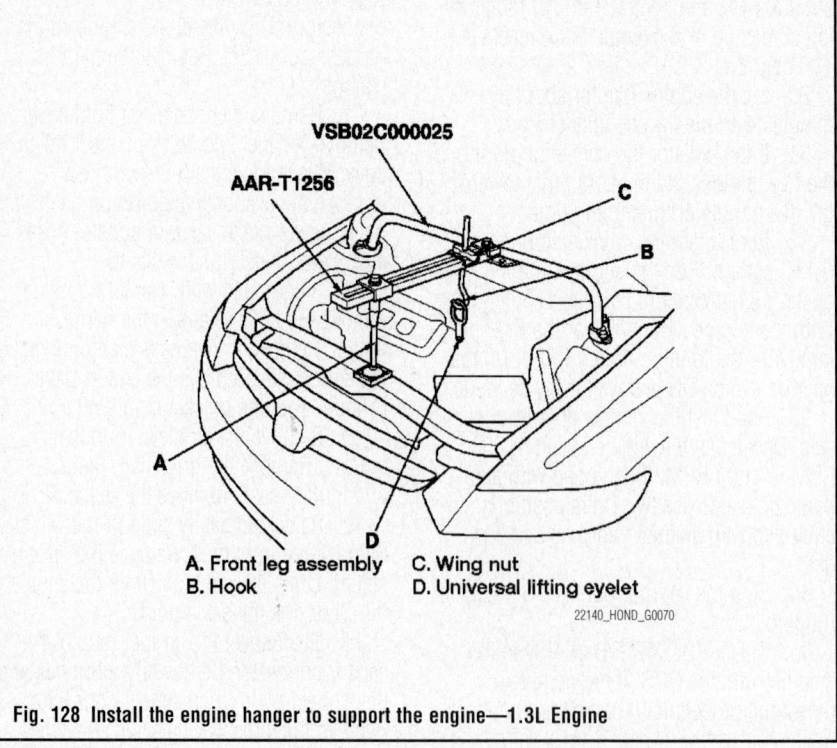

VSB02C000025
AAR-T1256
C
B
A
D

A. Front leg assembly   C. Wing nut
B. Hook   D. Universal lifting eyelet

22140_HOND_G0070

Fig. 128 Install the engine hanger to support the engine—1.3L Engine

vehicle, and attach the hook to the universal lifting eyelet. Tighten the wing nut by hand to lift and support the engine/IMA motor/transmission.

46. Make sure the hoist brackets are positioned properly. Raise and safely support the vehicle.

47. Remove the lower torque rod mounting bolts.

48. Make the appropriate reference lines at the both sides of the front subframe that line up with the edge on the body.

49. Loosen the mid stiffener mounting bolts on both sides.

50. Attach the front subframe adapter (VSB02C000016) to the front subframe by hanging the belt over the front of the front subframe, then secure the belt with its stop, then tighten the wing nut.

51. Raise the jack and line up the slots in the front subframe arms with the bolt holes on the jack base, then carefully attach them with four bolts.

52. Remove the four bolts securing the front subframe, and lower the front subframe.

53. Lower the vehicle.

54. Install the universal lifting eyelet to the threaded hole on the cylinder head with a 8 x 1.25 mm bolt.

55. Attach a chain hoist to the universal lifting eyelet, and the transmission hook. Lift the engine/IMA motor/transmission until its securely supported by the chain hoist, and remove the engine support hanger and the Civic engine hanger.

56. Remove the transmission mount bracket support bolt and nuts.

57. Remove the ground cable, then remove the side engine mount bracket.

58. Check that the engine/IMA motor/transmission is completely free of vacuum hoses, fuel hoses, coolant hoses, and electrical wiring.

59. Slowly lower the engine/IMA motor/transmission about 150 mm (6 in.). Check once again that all hoses and electrical wiring are disconnected and free from the engine/IMA motor/transmission, then lower it all the way and support it.

60. Disconnect the chain hoist from the engine/IMA motor/transmission.

61. Raise the vehicle, and remove the engine/IMA motor/transmission from under the vehicle.

To install:

62. Raise the vehicle on the lift, and position the engine/IMA motor/transmission under the vehicle. Be sure that they are properly aligned. Carefully lower the vehicle until the engine/IMA motor/transmission is

properly positioned in the engine compartment. Make sure the vehicle is not resting on any part of the engine/IMA motor/transmission.

63. Attach a chain hoist to the universal lifting eyelet (07AAK-SNAA120) and the transmission hook. Carefully raise the engine/IMA motor/transmission with the chain hoist into place.

64. Install the universal lifting eyelet to the threaded hole on the cylinder head with a 10 x 1.25 mm bolt.

65. Install the front leg assembly, the hook, and the wing nut to an A and Reds engine support hanger (AAR-T1256) onto the Civic engine hanger (VSB02C000025). Carefully position the engine hanger on the vehicle, and attach the hook to the universal lifting eyelet. Tighten the wing nut by hand to lift and support the engine/IMA motor/transmission.

66. The remainder of the installation is the reverse order of removal.

67. Turn the IMA battery module switch ON.

68. Inspect for fuel leaks. Turn the ignition switch to ON (II) (do not operate the starter) so the fuel pump runs for about 2 seconds and pressurizes the fuel lines. Repeat this operation three times, then check fuel leakage at any point in the fuel line.

69. Refill the engine with oil to the correct level.

70. Refill the transmission with CVT fluid (CVTF).

71. Move the shift lever to each gear, and verify that the A/T gear position indicator follows the transmission range switch.

72. Refill the radiator with engine coolant, and bleed the air from the cooling system with the heater valve open.

73. Check for fluid leaks.

74. Do the PCM reset procedure.

75. Do the PCM idle learn procedure.

76. Do the crankshaft position (CKP) pattern clear/CKP pattern learn procedure.

77. Inspect the idle speed.

78. Inspect the ignition timing.

79. Check the wheel alignment.

80. Do the start clutch pressure calibration procedures.

81. If the IMA battery level gauge (BAT) displays no segment, start the engine, and hold it between 3,500 RPM and 4,000 RPM without load (in N or P) until the BAT displays at least three segments.

S2000

1. Before servicing the vehicle, refer to the precautions in the beginning of this section.

2. Note the radio security code and the radio presets.

3. Properly relieve the fuel system pressure.

4. Disconnect the negative battery cable, than the positive cable. Remove the hood support from the driver's side of the vehicle. Lock the hood in the full up position.

5. Remove the battery.

6. Raise and support the vehicle safely. Drain the engine oil.

7. Remove the transmission. Lower the vehicle.

8. Disconnect the dashboard wiring harness connector.

9. Remove the electrical power steering (EPS) retaining bolts. Remove the control unit from its mounting.

10. Remove the battery cable from the main under hood fuse/relay box. Remove the harness clamps.

11. Remove the battery cable from the auxiliary under hood fuse box. Remove the ground cable and harness clamps.

12. Remove the harness retaining grommet from its mounting and pull out the ECM connectors.

13. Remove the radiator cap. Raise and safely support the vehicle. Drain the engine coolant.

14. Remove the engine stop bracket cushion and stop bracket. Lower the vehicle.

15. Disconnect and remove the heater hoses. Remove the lower radiator hose. Remove the upper radiator hose.

16. Remove the intake manifold cover. Remove the quick connect fitting cover, then disconnect the fuel feed hose and brake booster vacuum hose.

17. Disconnect the evaporative emission (EVAP) canister hose.

18. Properly attach the engine lifting device.

19. Remove the support nut from the left side engine mount bracket. Remove the support nut and the four mounting bolts. Remove the right side engine mount bracket.

20. Raise the engine. Check to insure that no wires, hoses etc are attached. Remove the engine from the vehicle.

To install:

21. Installation is the reverse of the removal procedure, while using the following torque values:

- Right motor mount bracket bolts: 28 ft. lbs. (38 Nm)
- Motor mount nuts: 40 ft. lbs. (54 Nm)
- Front motor mount bolts: 16 ft. lbs. (22 Nm)

22. Reprogram the ECM engine idle characteristics. Be sure all electrical items are OFF.

23. Start the engine. Hold the idle speed at 3000 RPM's in neutral until the radiator fan comes on or the temperature reached 176 degrees.

24. Let the engine idle for about five minutes with the throttle fully closed.

25. If the radiator fan comes on during the five minutes, do not count this toward the five minute programming time.

26. Set the clock.

EXHAUST MANIFOLD

REMOVAL & INSTALLATION

Accord

2.4L Engine

See Figure 129.

1. Raise and safely support the vehicle.
2. Remove or disconnect the following:
   - VTEC solenoid valve
   - Driveshaft heat shield

- Cover and exhaust manifold bracket
- Exhaust manifold

To install:

3. Clean the mounting surfaces. Be sure to use a new gasket.
4. Install or connect the following:
   - Exhaust manifold. Tighten the nuts, in a crisscross pattern starting with the inner nut, to specification.
   - Exhaust manifold bracket and cover
   - Driveshaft heat shield
   - VTEC solenoid valve

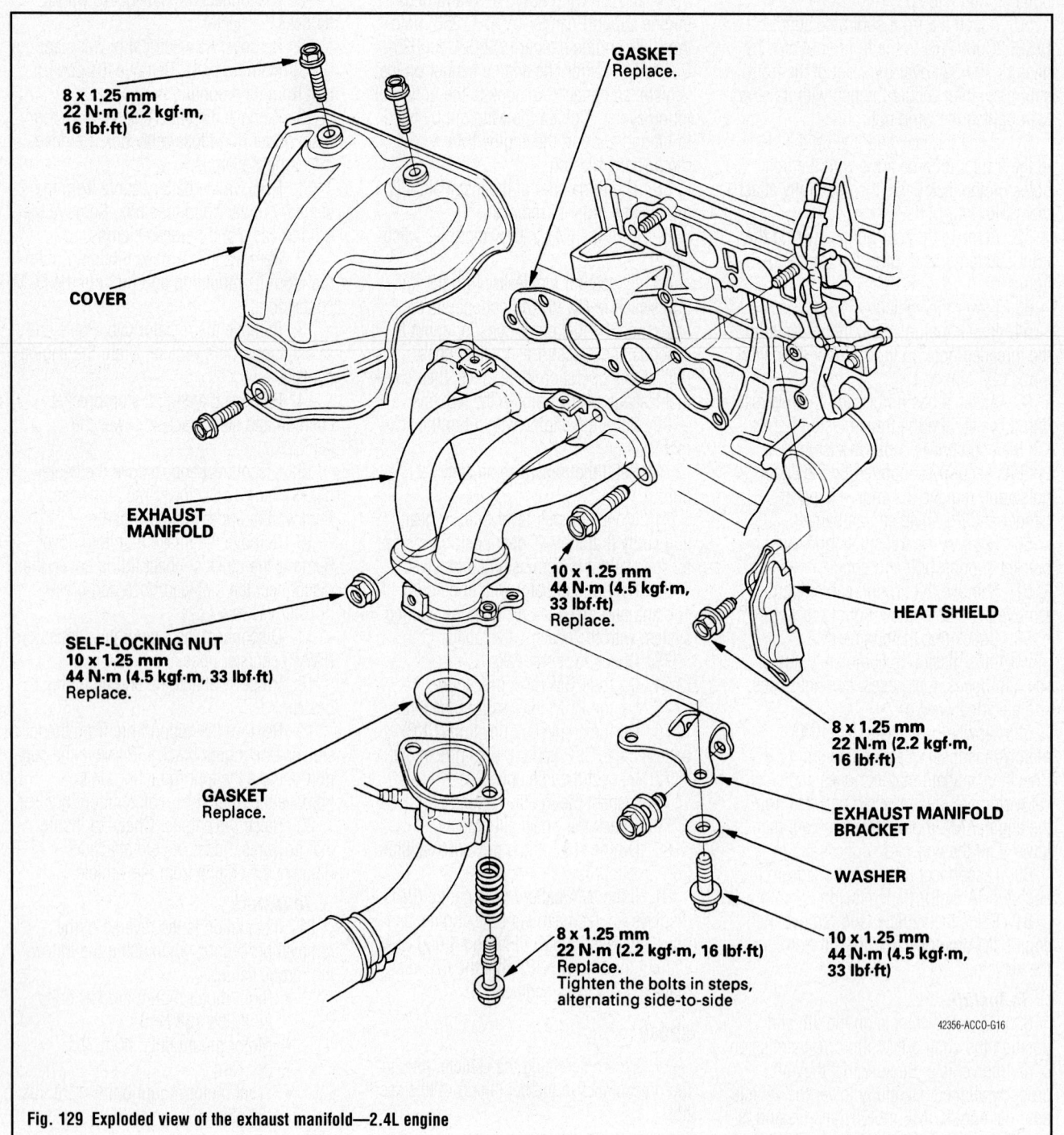

GASKET
Replace.

8 x 1.25 mm
22 N·m (2.2 kgf·m, 16 lbf·ft)

COVER

EXHAUST MANIFOLD

SELF-LOCKING NUT
10 x 1.25 mm
44 N·m (4.5 kgf·m, 33 lbf·ft)
Replace.

GASKET
Replace.

10 x 1.25 mm
44 N·m (4.5 kgf·m, 33 lbf·ft)
Replace.

HEAT SHIELD

8 x 1.25 mm
22 N·m (2.2 kgf·m, 16 lbf·ft)

EXHAUST MANIFOLD BRACKET

WASHER

8 x 1.25 mm
22 N·m (2.2 kgf·m, 16 lbf·ft)
Replace.
Tighten the bolts in steps, alternating side-to-side

10 x 1.25 mm
44 N·m (4.5 kgf·m, 33 lbf·ft)

42356-ACCO-G16

Fig. 129 Exploded view of the exhaust manifold—2.4L engine

3.0L Engine

See Figure 130.

1. Raise and safely support the vehicle.
2. Remove or disconnect the following:
   - Engine undercover
   - Exhaust pipe from the manifold to be removed
3. Lower the vehicle.
4. Remove or disconnect the following:
   - Exhaust manifold heat shield
   - Mounting nuts and the exhaust manifold.

To install:

5. Use new gaskets when installing the exhaust manifold.
6. Clean all gasket mating surfaces.
7. Install or connect the following:
   - Exhaust manifold. Tighten the nuts to specification.
   - Heat shield. Tighten the bolts to 16 ft. lbs. (22 Nm).

8. Raise the vehicle and connect the exhaust pipe to the manifold using a new gasket. Tighten the nuts to 40 ft. lbs. (54 Nm).

3.5L Engine—Front

1. Remove the engine appearance cover.
2. Remove the No. 2 the ignition coil and the ignition coil heat insulator.
3. Remove the front A/F sensor (Sensor 1) and the front secondary HO2S (Sensor 2).
4. Remove the exhaust pipe A mounting nuts (front WU-TWC side).
5. Remove the EGR pipe.
6. Remove the A/C condenser fan assembly and the radiator upper bracket/cushion.
7. Carefully remove the front exhaust manifold.

To install:

8. Carefully install the front exhaust manifold with a new gasket and new self-locking nuts. Tighten the nuts in a criss-cross pattern in two or three steps to 23 ft. lbs. (31 Nm).
9. Install the parts in the reverse order of removal.

3.5L Engine—Rear

1. Remove the strut brace.
2. Remove the P/S feed hose clamp.
3. Remove the cowl cover.
4. Remove the No. 4 ignition coil and the ignition coil heat insulator.
5. Remove the rear A/F sensor (Sensor 1) and the rear secondary HO2S (Sensor 2).
6. Remove the exhaust pipe A mounting nuts (rear WU-TWC side).
7. Carefully remove the rear exhaust manifold.

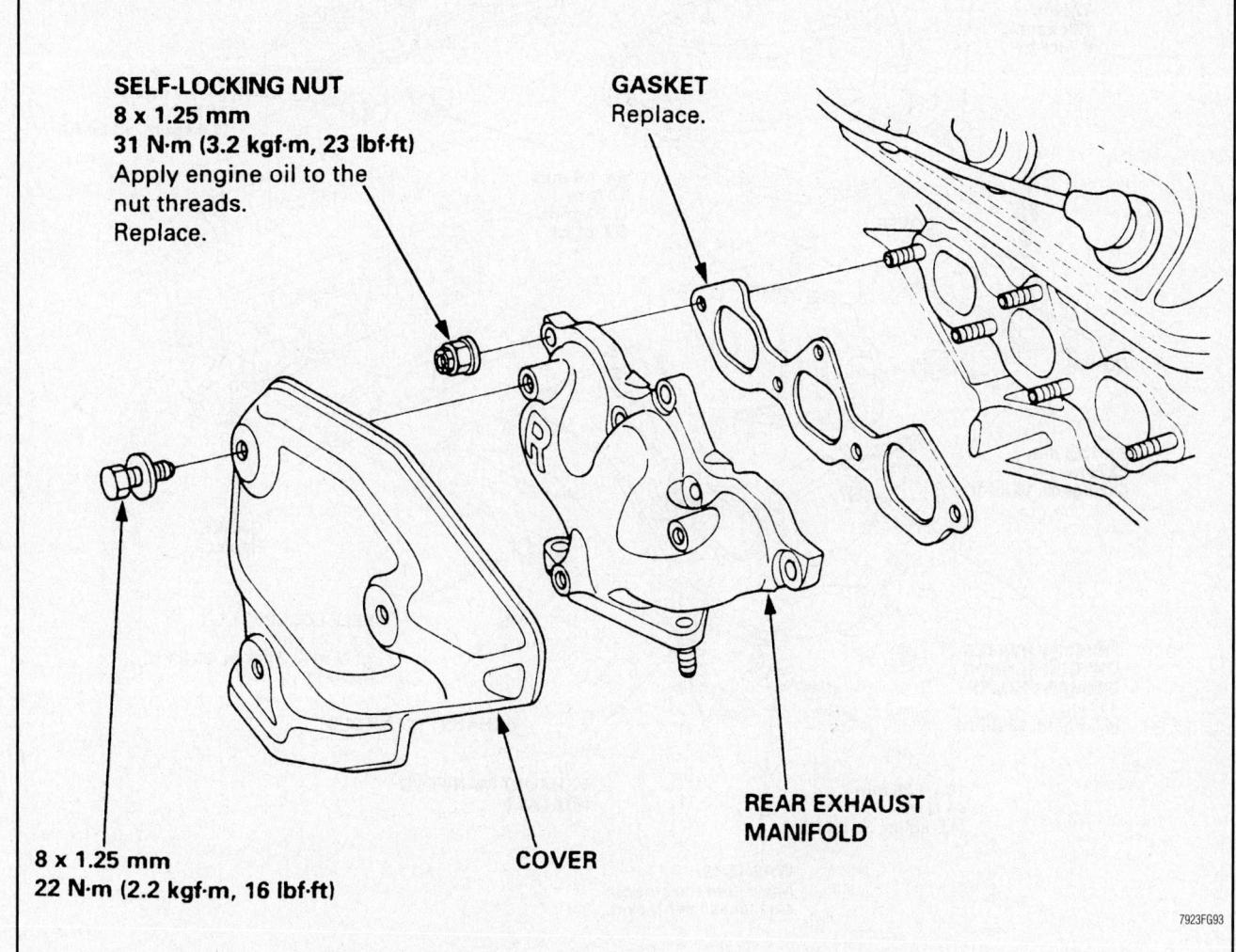

SELF-LOCKING NUT
8 x 1.25 mm
31 N·m (3.2 kgf·m, 23 lbf·ft)
Apply engine oil to the
nut threads.
Replace.

GASKET
Replace.

8 x 1.25 mm
22 N·m (2.2 kgf·m, 16 lbf·ft)

COVER

REAR EXHAUST MANIFOLD

7923FG93

Fig. 130 Exploded view of the exhaust manifold—3.0L engine

To install:

8. Carefully install the rear exhaust manifold with a new gasket and new self-locking nuts. Tighten the nuts in a crisscross pattern in two or three steps to 23 ft. lbs. (31 Nm).

Install the parts in the reverse order of removal.

Civic

2.0L Engine

See Figure 131.

1. Before servicing the vehicle, refer to the precautions in the beginning of this section.

2. Drain the cooling system and relieve the fuel system pressure.

3. Remove or disconnect the following:
- Negative battery cable
- Catalytic converter
- Exhaust manifold heat shields
- Heated Oxygen (HO2S) sensor connector

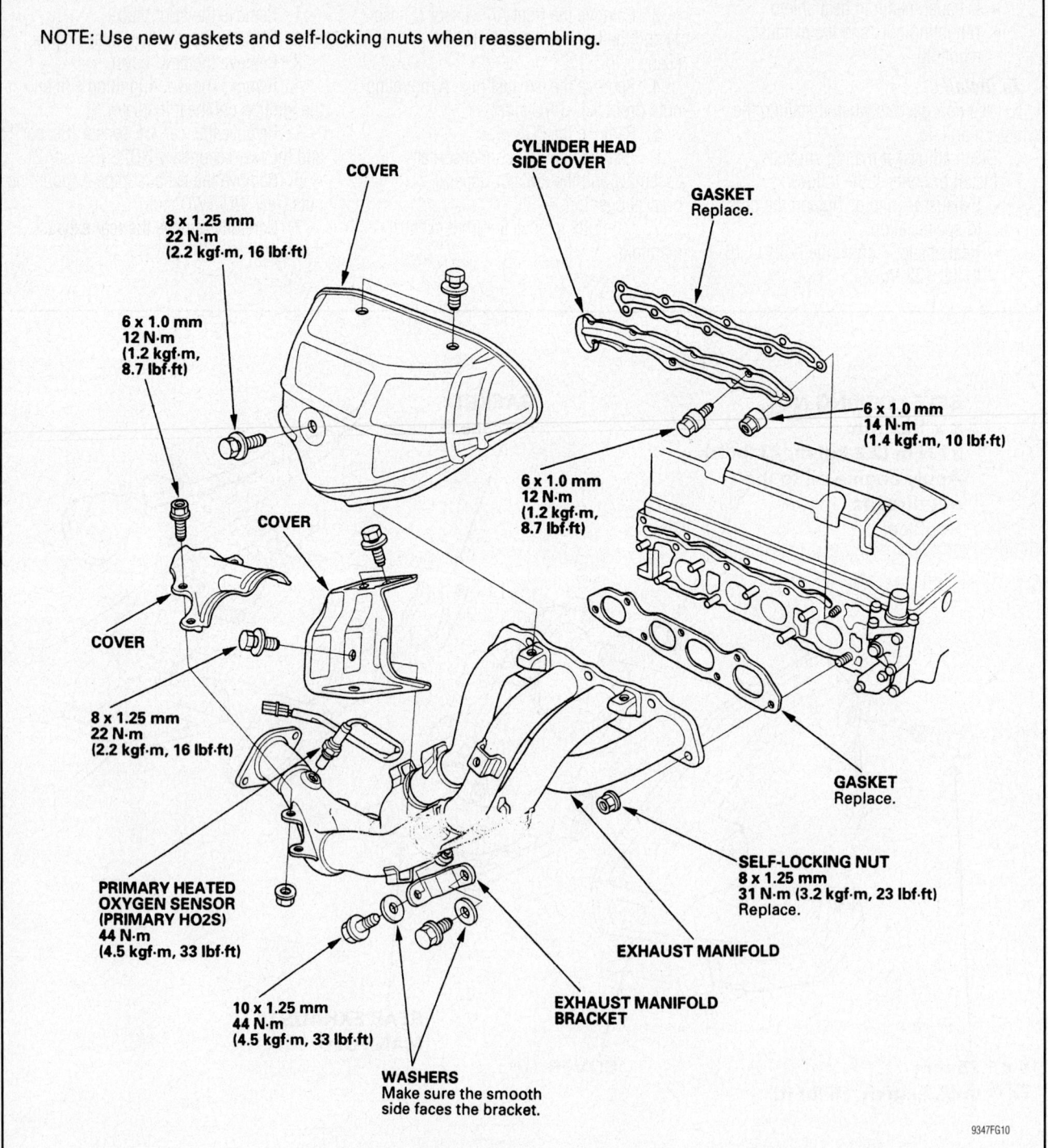

NOTE: Use new gaskets and self-locking nuts when reassembling.

COVER

CYLINDER HEAD SIDE COVER

GASKET
Replace.

8 x 1.25 mm
22 N·m
(2.2 kgf·m, 16 lbf·ft)

6 x 1.0 mm
12 N·m
(1.2 kgf·m, 8.7 lbf·ft)

6 x 1.0 mm
14 N·m
(1.4 kgf·m, 10 lbf·ft)

6 x 1.0 mm
12 N·m
(1.2 kgf·m, 8.7 lbf·ft)

COVER

COVER

GASKET
Replace.

8 x 1.25 mm
22 N·m
(2.2 kgf·m, 16 lbf·ft)

SELF-LOCKING NUT
8 x 1.25 mm
31 N·m (3.2 kgf·m, 23 lbf·ft)
Replace.

EXHAUST MANIFOLD

PRIMARY HEATED OXYGEN SENSOR
(PRIMARY HO2S)
44 N·m
(4.5 kgf·m, 33 lbf·ft)

EXHAUST MANIFOLD BRACKET

10 x 1.25 mm
44 N·m
(4.5 kgf·m, 33 lbf·ft)

WASHERS
Make sure the smooth side faces the bracket.

9347FG10

Fig. 131 Exploded view of the exhaust manifold—2.0L and 2.2L engines

- Exhaust manifold bracket
- Exhaust manifold

4. Installation is the reverse of the removal procedure.

S2000

1. Remove or disconnect the following:
   - Negative battery cable
2. Safely raise and support the vehicle.
3. Remove or disconnect the following:
   - Oxygen Sensor (O2S) connector, if it is located in the manifold.
   - Exhaust manifold upper cover
   - Heat insulator from the manifold, if equipped with air conditioning.

- Nuts attaching the exhaust manifold to the front exhaust pipe.
- Pipe from the manifold and discard the gasket. Support the pipe with wire; do not allow it to hang by itself.
- Exhaust manifold bracket(s) bolts and bracket(s).
- Exhaust manifold attaching nuts, using a crisscross pattern (starting from the center).
- Manifold and discard the gasket. Clean the manifold and cylinder head mating surfaces.
- Lower manifold cover from the manifold, if equipped.

4. Installation is the reverse order of removal. Tighten the nuts to 23 ft. lbs. (31 Nm).

INTAKE MANIFOLD

REMOVAL & INSTALLATION

Accord

2.4L Engine

See Figure 132.

→**Make sure to acquire the anti-theft code from the radio and write down the frequencies for the radio's preset buttons.**

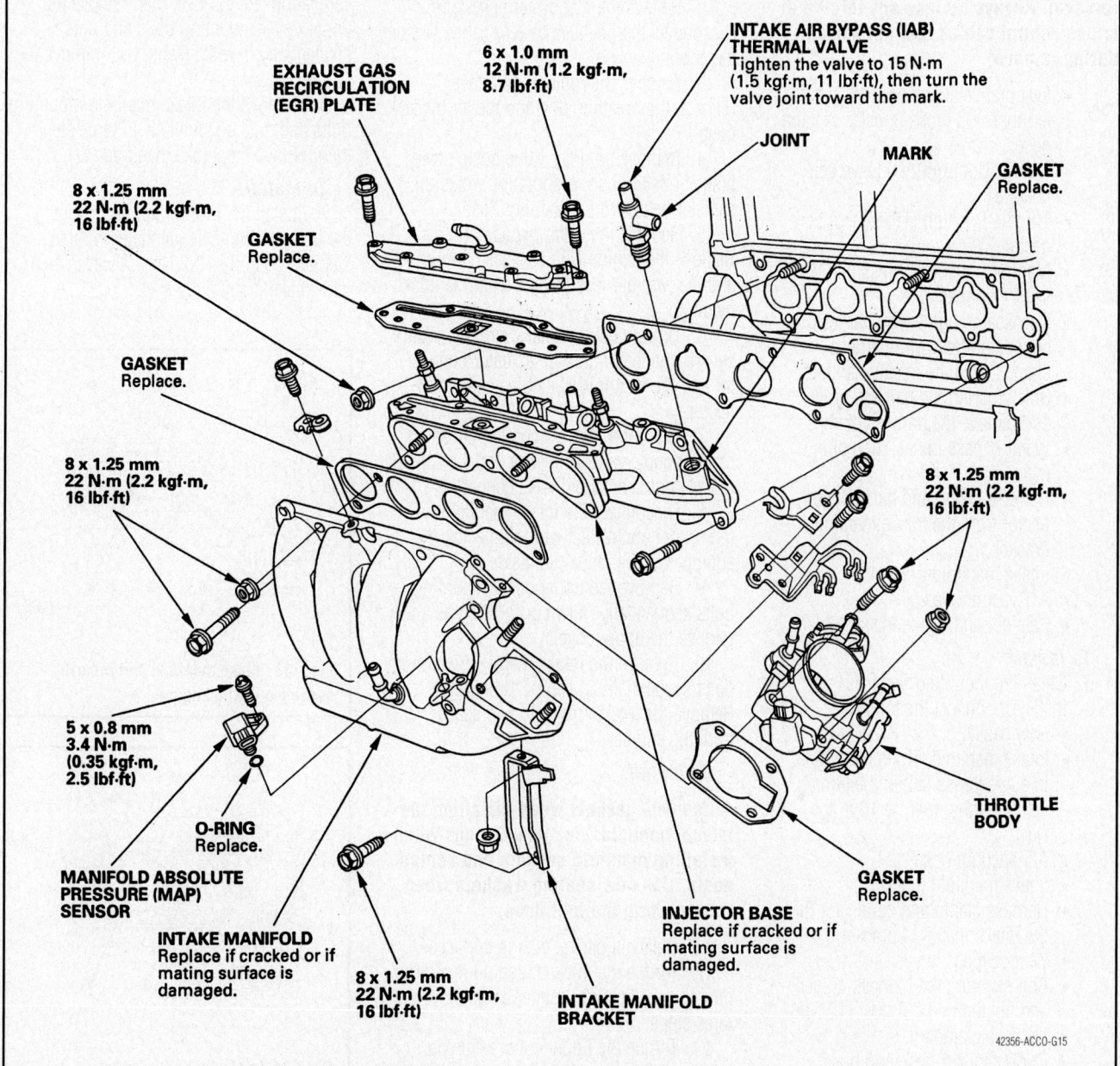

EXHAUST GAS
RECIRCULATION
(EGR) PLATE

6 x 1.0 mm
12 N·m (1.2 kgf·m,
8.7 lbf·ft)

INTAKE AIR BYPASS (IAB)
THERMAL VALVE
Tighten the valve to 15 N·m
(1.5 kgf·m, 11 lbf·ft), then turn the
valve joint toward the mark.

JOINT

MARK

GASKET
Replace.

8 x 1.25 mm
22 N·m (2.2 kgf·m,
16 lbf·ft)

GASKET
Replace.

GASKET
Replace.

8 x 1.25 mm
22 N·m (2.2 kgf·m,
16 lbf·ft)

8 x 1.25 mm
22 N·m (2.2 kgf·m,
16 lbf·ft)

5 x 0.8 mm
3.4 N·m
(0.35 kgf·m,
2.5 lbf·ft)

O-RING
Replace.

MANIFOLD ABSOLUTE
PRESSURE (MAP)
SENSOR

INTAKE MANIFOLD
Replace if cracked or if
mating surface is
damaged.

8 x 1.25 mm
22 N·m (2.2 kgf·m,
16 lbf·ft)

INTAKE MANIFOLD
BRACKET

INJECTOR BASE
Replace if cracked or if
mating surface is
damaged.

THROTTLE
BODY

GASKET
Replace.

42356-ACCO-G15

Fig. 132 Exploded view of the intake manifold and related components—2.4L engine

1. Before servicing the vehicle, refer to the precautions in the beginning of this section.

2. Disconnect the negative battery cable.

3. Drain the engine coolant into a sealable container.

4. Remove or disconnect the following:
- Intake Air Temperature (IAT) sensor electrical connector
- Vacuum hose and breather pipe and the air intake duct
- Intake manifold cover
- Throttle and cruise control cables by loosening the locknuts, then slipping the cable ends out of the accelerator linkage.

➡**Do not bend the cables during removal. Always replace any throttle or cruise control cables that get kinked during removal.**

- Evaporative emission (EVAP) canister hose and brake booster vacuum hose
- Idle Air Control (IAC) valve connectors
- Throttle Position (TP) sensor connector
- Manifold Absolute Pressure (MAP) sensor connector
- Necessary engine wire harness connectors and wire harness clamps from the intake manifold
- Bolt securing the harness holder and remove the harness clamps
- Water bypass hoses, then plug them
- Harness clamp and harness connector from the intake manifold bracket
- Intake manifold bracket
- A/T vacuum hose
- Retainer and intake manifold

To install:

5. Clean the mounting surfaces.

6. Install or connect the following:
- New gasket
- Intake manifold. Tighten the bolts, in a crisscross pattern beginning with the inner bolt, to 16 ft. lbs. (22 Nm).
- A/T vacuum hose
- Intake manifold bracket
- Harness clamp and connector to the intake manifold bracket
- Water bypass hoses
- Bolt securing the harness holder and tighten to 8.7 ft. lbs. (12 Nm)
- Harness clamps
- EVAP canister hose and brake booster vacuum hose

- Throttle and cruise control cables
- Intake manifold cover
- Intake air duct
- IAT sensor connector, vacuum hose and breather pipe

7. Refill the cooling system.

8. Connect the negative battery cable, start the engine, and check for leaks.

3.0L Engine

➡**Make sure to acquire the anti-theft code from the radio and write down the frequencies for the radio's preset buttons.**

1. Before servicing the vehicle, refer to the precautions in the beginning of this section.

2. Relieve the fuel system pressure. Disconnect the negative battery cable. Drain the cooling system.

3. Remove the ignition coil cover. Remove the breather pipe and the air intake duct.

4. Remove the PCV valve hose, power brake booster hose, evaporative emission canister hose and the vacuum hose.

5. Remove the water bypass hoses. Remove the engine wire harness connectors and the wire harness clamps from the intake manifold

6. Disconnect the intake air temperature sensor connector, throttle actuator connector and the manifold absolute sensor connector.

7. Disconnect the evaporative emission canister purge valve connector and the engine mount control solenoid valve connector.

8. Disconnect the intake manifold tuning runner control actuator connector, if equipped with manual transaxle.

9. Remove the upper cover mounting bolts sequentially in two or three steps and remove the upper cover.

10. Remove the intake manifold retaining bolts sequentially in two or three steps. Remove the intake manifold and spacer from the engine.

To install:

➡**Use new gaskets when installing the intake manifold. Use new O-rings when installing manifold sensors and components. Use new sealing washers when reconnecting the fuel lines.**

11. Clean all gasket mating surfaces.

12. Torque the intake manifold retaining nuts to specification, sequentially in two or three steps.

13. Torque the upper cover retaining bolts to 8.7 ft. lbs. (12 Nm), sequentially in two or three steps.

14. Continue the installation in the reverse order of the removal procedure.

3.5L Engine

See Figures 133 and 134.

1. Remove the strut brace.

2. Remove the engine appearance cover.

3. Disconnect the manifold absolute pressure (MAP) sensor connector and the breather pipe, and remove the intake air duct.

4. Disconnect the throttle actuator connector, the evaporative emission (EVAP) canister purge valve connector, the water bypass hoses, the EVAP hose, and the brake booster vacuum hose.

5. Disconnect the positive crankcase ventilation (PCV) hose, then remove the upper cover mounting bolts and nuts sequentially in three steps, then remove the upper cover.

6. Remove the intake manifold mounting bolts and nuts sequentially in three steps, then remove the intake manifold.

To install:

7. Install the intake manifold. Tighten the bolts and nuts sequentially in three steps to 16 ft. lbs. (22 Nm). Always use a new intake manifold gasket.

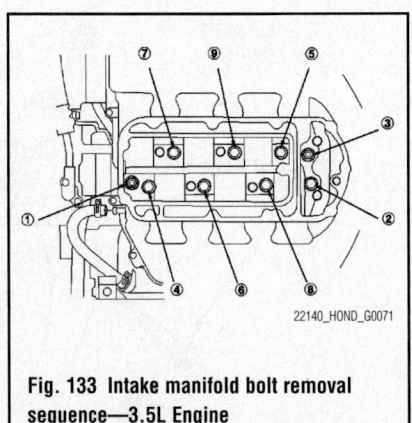

22140_HOND_G0071

Fig. 133 Intake manifold bolt removal sequence—3.5L Engine

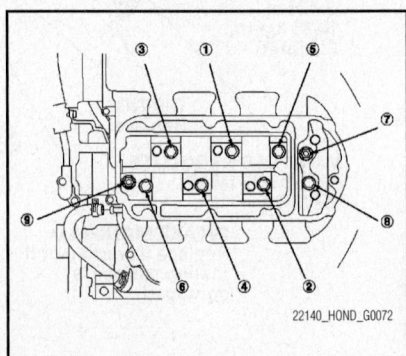

22140_HOND_G0072

Fig. 134 Intake manifold torque sequence—3.5L Engine

8. Install the upper cover and tighten the bolts to 9 ft. lbs. (12 Nm).

9. The remainder of the installation is the reverse order of removal.

Civic

1.8L Engine

➡ **Make sure to acquire the anti-theft code from the radio and write down the frequencies for the radio's preset buttons.**

1. Before servicing the vehicle, refer to the precautions in the beginning of this section.

2. Disconnect the negative battery cable. Drain the cooling system.

3. Remove the cowl cover and the under cowl panel. Remove the air cleaner housing assembly. Remove the air intake duct.

4. Remove the injector cover. Remove the evaporative emission canister hose, power brake booster hose and the power steering pump hose clamp.

5. Remove the quick connect fitting cover. Disconnect the fuel feed hose.

6. Remove the engine wire harness connectors and the wire harness clamps from the intake manifold.

7. Disconnect the throttle actuator connector, the manifold absolute pressure connector, the evaporative emission control canister purge connector and the intake manifold tuning valve actuator connector.

8. Remove the water bypass hoses. Remove the throttle body. Remove the heater hose clamp bracket.

9. Raise and safely support the vehicle. Remove the intake manifold bracket. Lower the vehicle.

10. Remove the intake manifold retaining bolts. Remove the intake manifold from the engine.

To install:

➡ **Use new gaskets when installing the intake manifold. Use new O-rings when installing manifold sensors and components. Use new sealing washers when reconnecting the fuel lines.**

11. Clean all gasket mating surfaces.

12. Torque the intake manifold retaining nuts to 17 ft. lbs. (24 Nm).

13. Continue the installation in the reverse order of the removal procedure.

Civic Hybrid

See Figure 135.

1. Disconnect the negative battery cable.

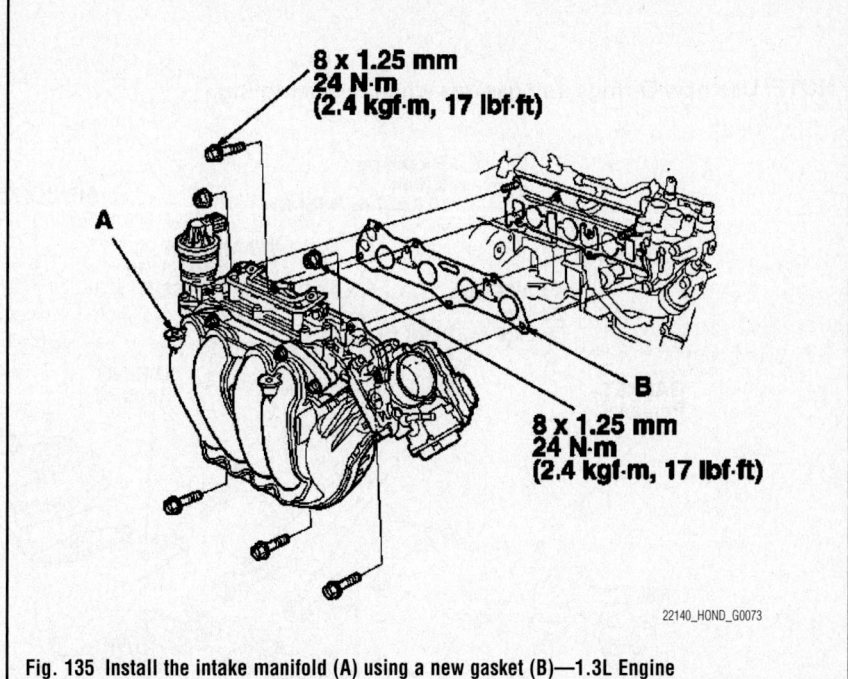

8 x 1.25 mm
24 N·m
(2.4 kgf·m, 17 lbf·ft)

A

B

8 x 1.25 mm
24 N·m
(2.4 kgf·m, 17 lbf·ft)

22140_HOND_G0073

Fig. 135 Install the intake manifold (A) using a new gasket (B)—1.3L Engine

2. Drain the engine cooling system.

3. Remove the engine appearance cover.

4. Remove the resonator.

5. Remove the intake air duct.

6. Remove the air cleaner assembly.

7. Remove the front grille cover.

8. Raise the vehicle on the lift.

9. Remove the front wheels.

10. Remove the front undercover and the splash shield.

11. Remove the intake manifold bracket bolts and the A/C compressor harness clamp.

12. Lower the vehicle on the lift.

13. Disconnect the water bypass hose and the vacuum hose, and the throttle actuator connector.

14. Disconnect the water bypass hose.

15. Disconnect the EGR valve connector, the intake manifold sub-harness connector, and the clamp, then remove the ground cables.

16. Disconnect the fuel injector connectors, rocker arm oil control valve connectors, and the rocker arm oil pressure sensor connector, then remove the engine wire harness from the brackets.

17. Disconnect the positive crankcase ventilation (PCV) hose.

18. Remove the intake manifold assembly.

To install:

19. Install the intake manifold assembly and tighten the bolts/nuts in a crisscross pattern in three steps, beginning with the inner bolt. Use a new gasket.

20. The remainder of the installation is the reverse order of removal.

21. Refill the engine cooling system to the correct level.

S2000

See Figure 136.

➡ **Make sure to acquire the anti-theft code from the radio and write down the frequencies for the radio's preset buttons.**

1. Before servicing the vehicle, refer to the precautions in the beginning of this section.

2. Drain the cooling system.

3. Relieve the fuel system pressure.

4. Remove or disconnect the following:
- Negative battery cable
- Cooling hoses from the intake manifold
- Vacuum hoses and electrical connectors from the manifold and throttle body
- Throttle cable from the throttle body
- Fuel rail and fuel injectors
- Intake manifold support brackets
- Intake manifold

To install:

➡ **Use new gaskets when installing the intake manifold. Use new O-rings when installing manifold sensors and components. Use new sealing washers when reconnecting the fuel lines.**

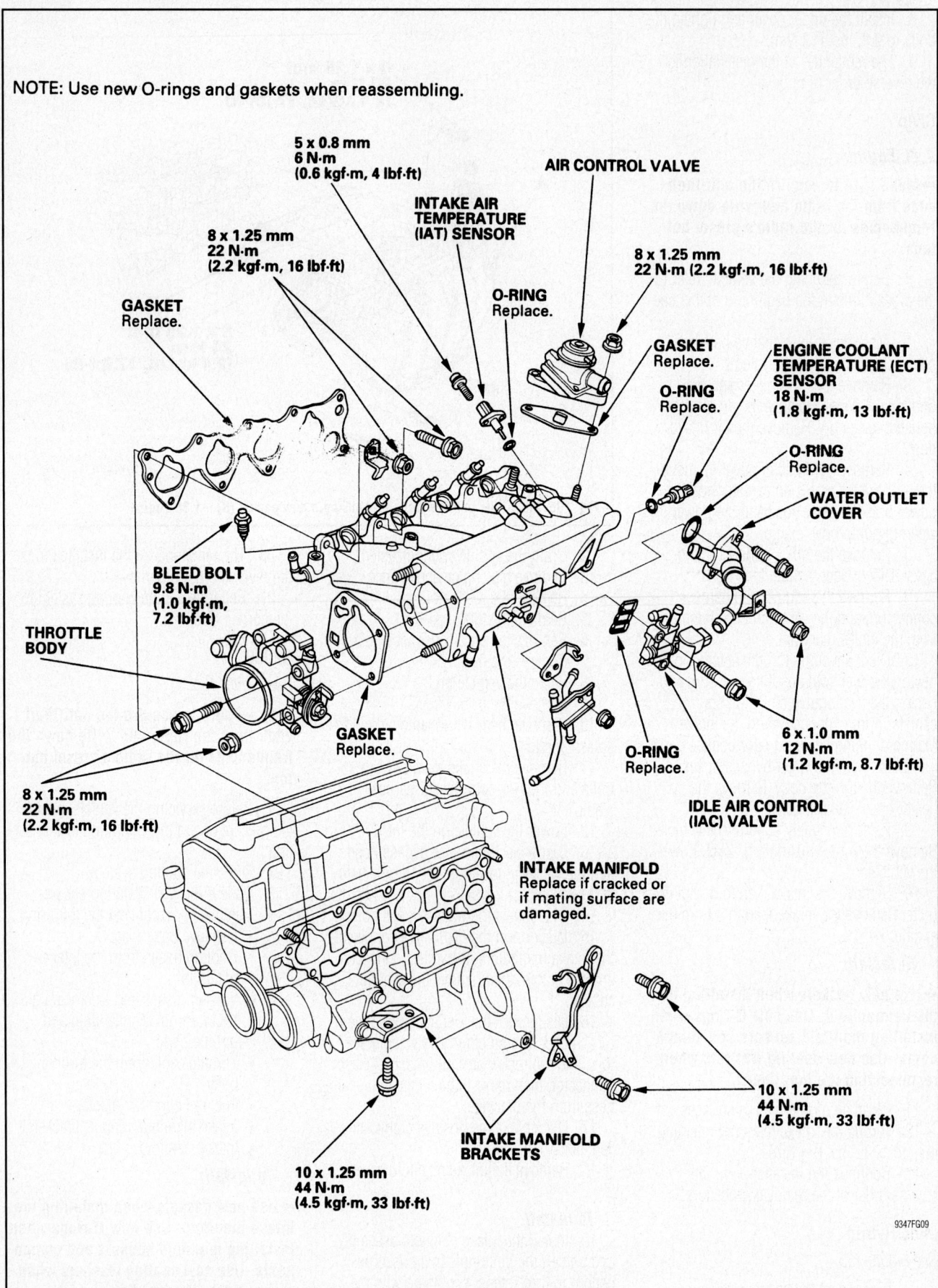

NOTE: Use new O-rings and gaskets when reassembling.

5 x 0.8 mm
6 N·m
(0.6 kgf·m, 4 lbf·ft)

AIR CONTROL VALVE

INTAKE AIR
TEMPERATURE
(IAT) SENSOR

8 x 1.25 mm
22 N·m (2.2 kgf·m, 16 lbf·ft)

8 x 1.25 mm
22 N·m
(2.2 kgf·m, 16 lbf·ft)

O-RING
Replace.

GASKET
Replace.

O-RING
Replace.

ENGINE COOLANT
TEMPERATURE (ECT)
SENSOR
18 N·m
(1.8 kgf·m, 13 lbf·ft)

GASKET
Replace.

O-RING
Replace.

WATER OUTLET
COVER

BLEED BOLT
9.8 N·m
(1.0 kgf·m,
7.2 lbf·ft)

THROTTLE
BODY

GASKET
Replace.

8 x 1.25 mm
22 N·m
(2.2 kgf·m, 16 lbf·ft)

O-RING
Replace.

6 x 1.0 mm
12 N·m
(1.2 kgf·m, 8.7 lbf·ft)

IDLE AIR CONTROL
(IAC) VALVE

INTAKE MANIFOLD
Replace if cracked or
if mating surface are
damaged.

10 x 1.25 mm
44 N·m
(4.5 kgf·m, 33 lbf·ft)

INTAKE MANIFOLD
BRACKETS

10 x 1.25 mm
44 N·m
(4.5 kgf·m, 33 lbf·ft)

9347FG09

Fig. 136 Exploded view of the intake manifold and related components—S2000 with 2.0L engine

5. Clean all gasket mating surfaces.

6. Torque the intake manifold retaining nuts to specification.

7. Continue the installation in the reverse order of the removal procedure.

OIL PAN

REMOVAL & INSTALLATION

Accord

2.4L Engine

See Figures 137 and 138.

1. Before servicing the vehicle, refer to the precautions in the beginning of this section.

2. Note the radio security code and the radio presets. Disconnect the negative battery cable.

3. Disconnect the battery positive cable. Remove the battery. Remove the air cleaner housing. Remove the battery base.

4. If equipped with manual transaxle remove the clutch slave cylinder and clutch line bracket mounting bolt.

5. Remove the ground cable. Remove the transaxle upper mount/bracket assembly.

6. Remove the front mount stop. Remove the front mount bolt.

7. Remove the rear mount stop. Remove the rear mount bolt.

8. Raise and support the vehicle safely. Drain the engine oil. Remove the front tires.

9. Disconnect the stabilizer links. Remove the left side damper fork. Disconnect the left side suspension lower arm ball joint.

10. Remove the left side driveshaft. Remove the nuts securing the transaxle lower front mount and transaxle lower rear mount.

11. Use a transaxle jack and support the transaxle assembly. Remove the stiffener.

12. Remove the oil pan retaining bolts. Drive an oil pan seal cutter tool between the oil pan and engine block. Cut the oil pan seal by striking the side of the tool along the oil pan

13. Remove the oil pan from the vehicle.

To install:

14. Remove any old gasket from the mating surfaces. Be sure these surfaces are clean and dry.

15. Apply liquid gasket, part number 08717-004, 08718-0001, 08718-0003 or 08718-0009 evenly to the engine block mating surface of the oil pan.

➡**Do not install the parts if more than four minutes have elapsed since applying the liquid gasket. Instead, reapply after removing the previous coating material.**

16. Position the oil pan in place. Tighten the oil pan retaining bolts in two or three steps to specification and in the proper sequence.

17. Continue the installation in the reverse order of the removal procedure.

18. After assembly, wait at least thirty minutes before filling the engine with clean engine oil.

19. Do not run the engine for at least three hours after installing the oil pan.

3.0L Engine

See Figure 139.

1. Note the radio security code and the radio presets. Disconnect the negative battery cable.

2. Raise and support the vehicle safely. Drain the engine oil.

3. Remove the exhaust pipe. Remove the torque converter cover. Remove the two lower transaxle retaining bolts.

4. Remove the oil pan retaining bolts. Using a flat bladed tool, carefully separate the oil pan from the engine block. Remove the oil pan from the vehicle.

To install:

5. Remove any old gasket from the mating surfaces. Be sure these surfaces are clean and dry.

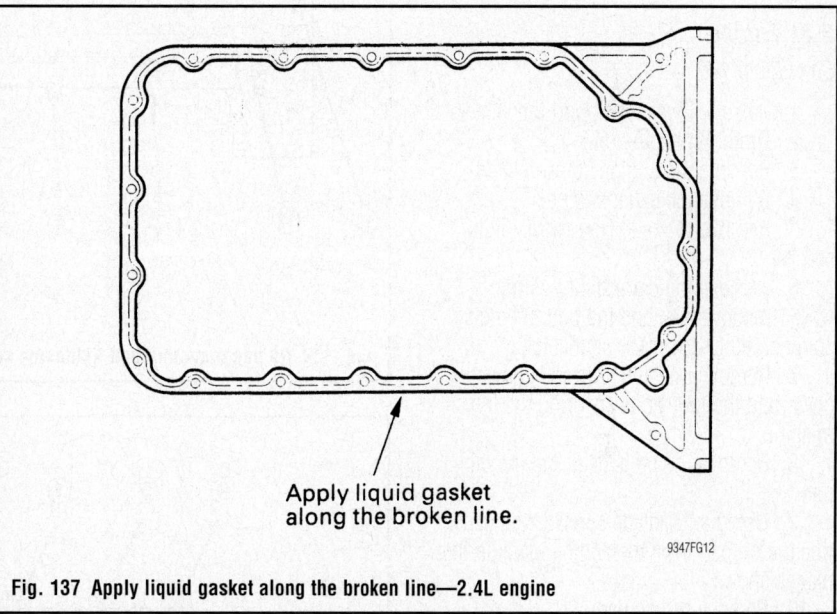

Apply liquid gasket along the broken line.

9347FG12

Fig. 137 Apply liquid gasket along the broken line—2.4L engine

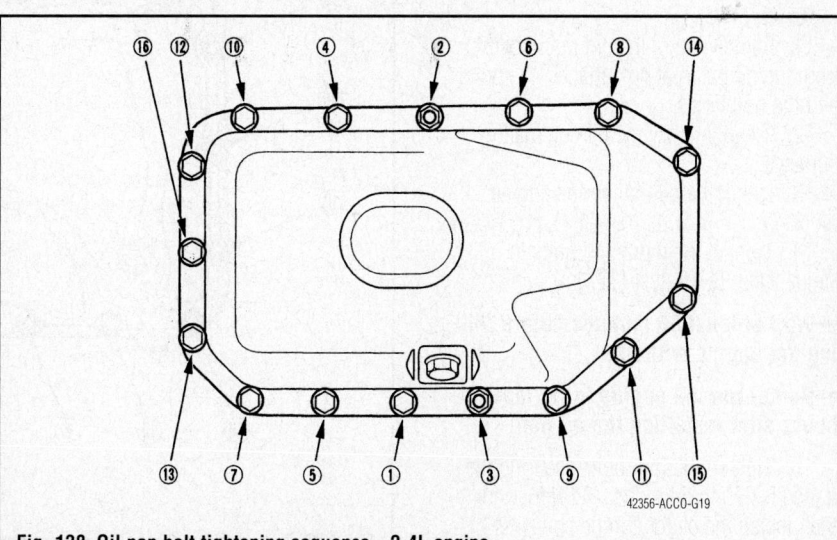

42356-ACCO-G19

Fig. 138 Oil pan bolt tightening sequence—2.4L engine

6. Apply liquid gasket, part number 08717-004, 08718-0001, 08718-0003 or 08718-0009 evenly to the engine block mating surface of the oil pan.

➡**Do not install the parts if more than four minutes have elapsed since applying the liquid gasket. Instead, reapply after removing the previous coating material.**

7. Position the oil pan in place. Tighten the oil pan retaining bolts in two or three steps to specification and in the proper sequence.

8. Continue the installation in the reverse order of the removal procedure.

9. After assembly, wait at least thirty minutes before filling the engine with clean engine oil.

10. Do not run the engine for at least three hours after installing the oil pan.

3.5L Engine

See Figure 140.

1. Raise and safely support the vehicle.
2. Drain the engine oil.
3. Remove the engine splash shield.
4. Remove the exhaust pipe.
5. Remove the rear exhaust manifold bracket.
6. Remove the crankshaft position (CKP) sensor cover and the bolt, then disconnect the CKP sensor connector.
7. Remove the clutch/torque converter cover and the four bolts securing the transmission.
8. Remove the bolts securing the oil pan.
9. Using a flat blade screwdriver, separate the oil pan from the engine block in the places shown.
10. Remove the oil pan.

To install:

11. Remove all of the old liquid gasket from the oil pan mating surfaces, the bolts, and the bolt holes.
12. Clean and dry the oil pan mating surfaces.
13. Install the oil pan on the engine block.
14. Tighten the mounting bolts in sequence to 8.7 ft. lbs. (12 Nm).

➡**Wait at least 30 minutes before filling the engine with oil.**

➡**Do not run the engine for at least 3 hours after installing the oil pan.**

15. Tighten the four bolts securing the transmission to 54 ft. lbs. (74 Nm), and then install the clutch/torque converter cover.

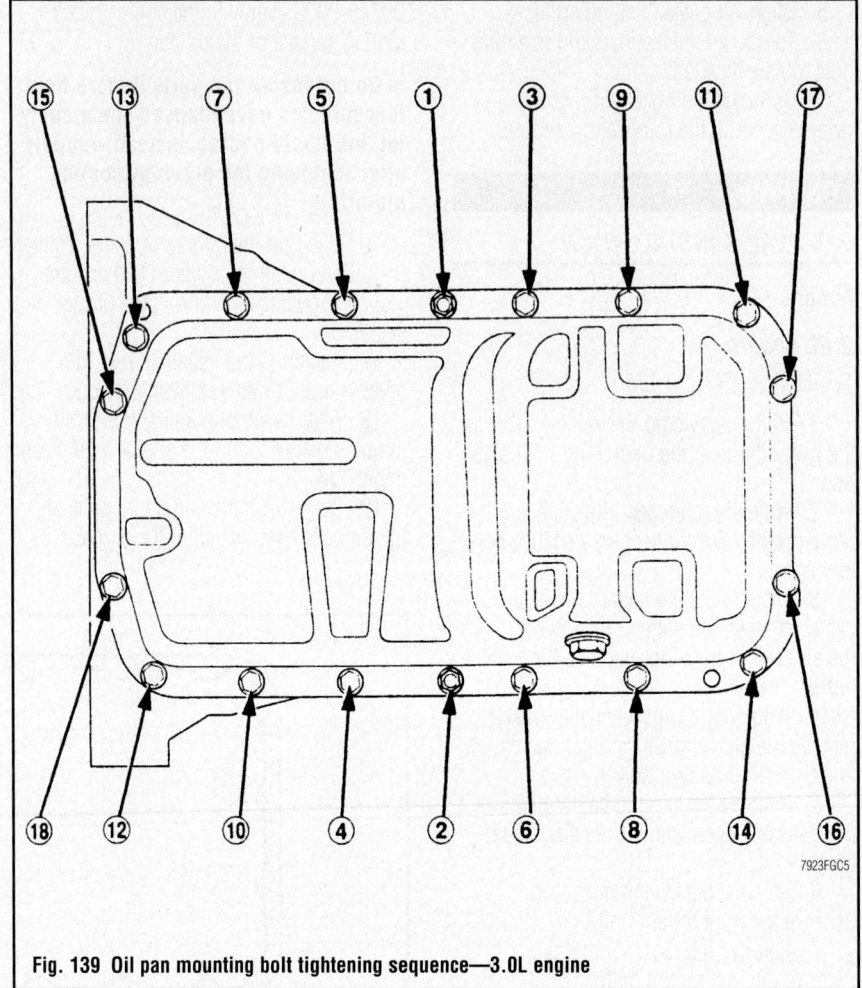

Fig. 139 Oil pan mounting bolt tightening sequence—3.0L engine

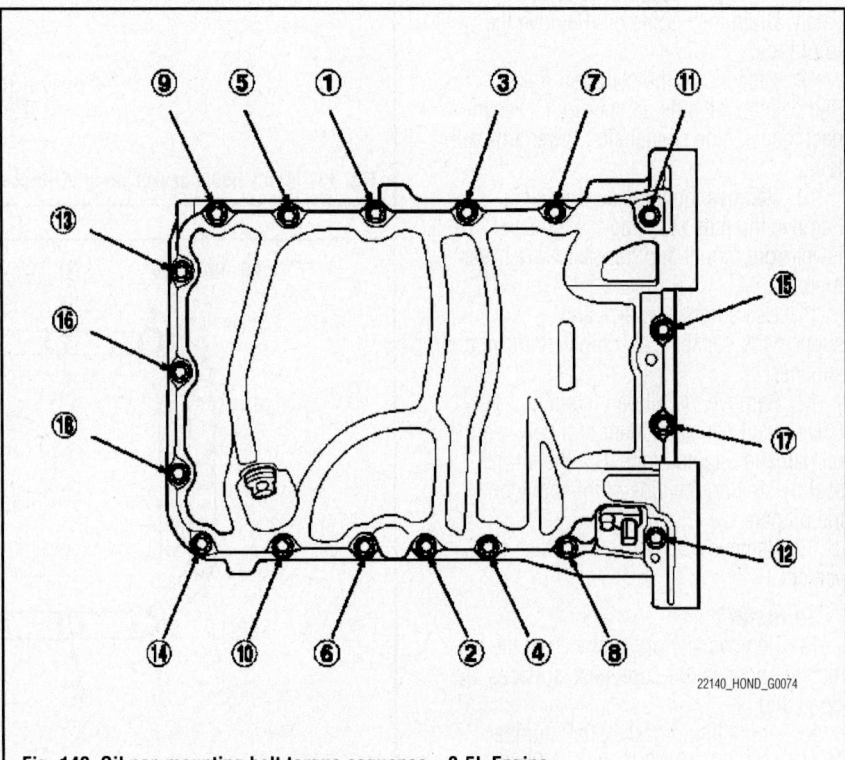

Fig. 140 Oil pan mounting bolt torque sequence—3.5L Engine

16. Connect the crankshaft position (CKP) sensor connector, then install the CKP sensor cover and the bolt.

17. The remainder of the installation is the reverse order of removal.

18. Refill the engine with oil to the correct level.

Civic

1.8L Engine

See Figure 141.

1. Note the radio security code and the radio presets. Disconnect the negative battery cable.

2. Remove the drive belt. Remove the air conditioning condenser fan shroud.

3. Disconnect the compressor electrical connectors. Remove the compressor retaining bolts, and position it to the side without discharging the system.

4. Raise and support the vehicle safely. Remove the splash shield. Drain the engine oil.

5. Remove the exhaust pipe. Properly support the oil pan. Remove the lower torque rod. Remove the oil pan support tool.

6. Remove the lower torque rod bracket. Remove the air conditioning compressor bracket.

7. If equipped with automatic transaxle, remove the shift cable cover. Remove the torque converter cover.

8. Remove the clutch cover if equipped with manual transaxle.

9. Remove the oil pan retaining bolts. Using a flat bladed tool, carefully separate the oil pan from the engine block. Remove the oil pan from the vehicle.

To install:

10. Remove any old gasket from the mating surfaces. Be sure these surfaces are clean and dry.

11. Apply liquid gasket, part number 08717-004, 08718-0001, 08718-0003 or 08718-0009 evenly to the engine block mating surface of the oil pan.

➡ **Do not install the parts if more than five minutes have elapsed since applying the liquid gasket. Instead, reapply after removing the previous coating material.**

12. Install the dowel pins, using new O-rings.

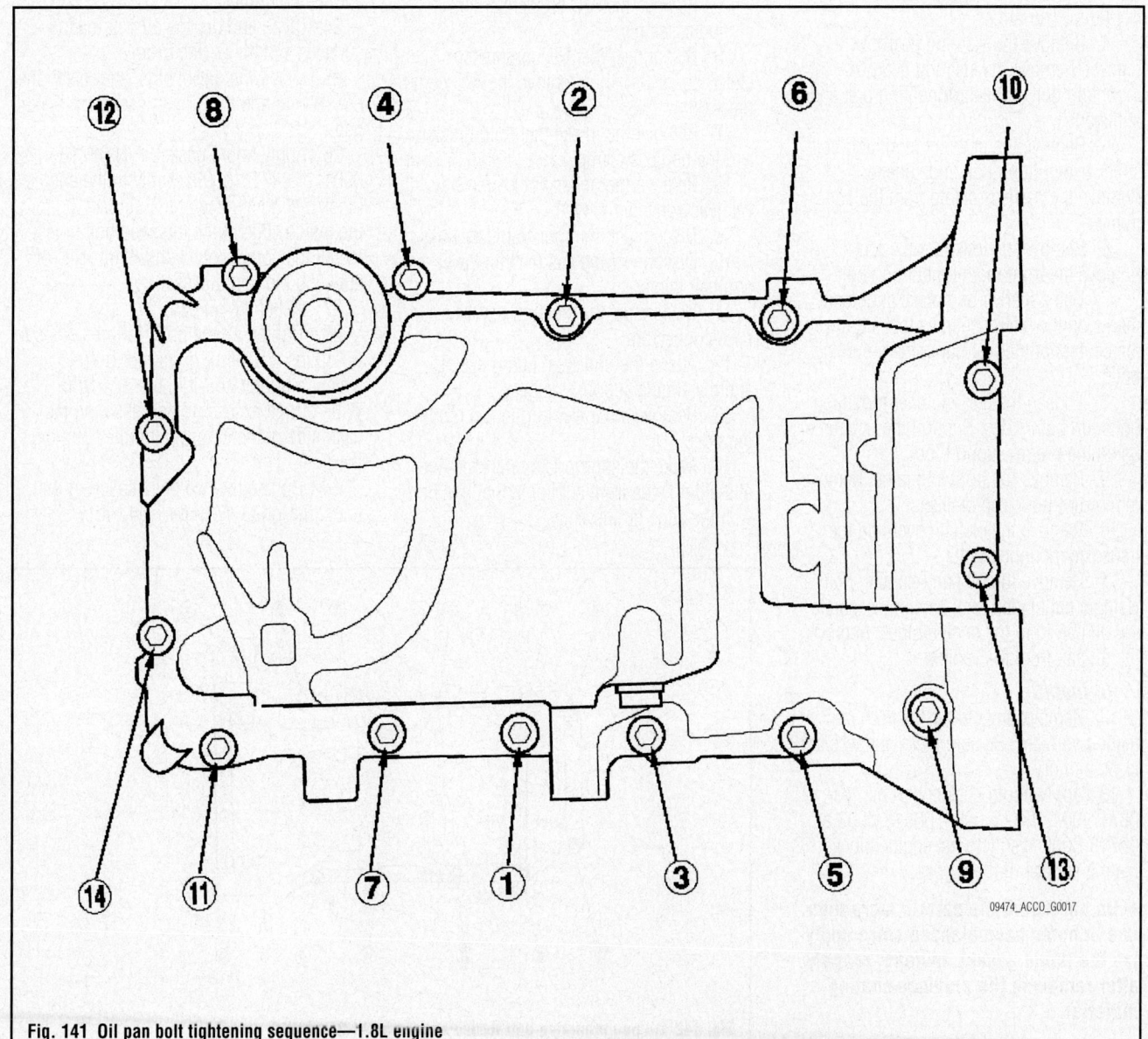

Fig. 141 Oil pan bolt tightening sequence—1.8L engine

09474_ACCO_G0017

13. Position the oil pan in place. Tighten the oil pan retaining bolts in two or three steps to specification and in the proper sequence.

14. Continue the installation in the reverse order of the removal procedure.

15. After assembly, wait at least thirty minutes before filling the engine with clean engine oil.

16. Do not run the engine for at least three hours after installing the oil pan.

2.0L Engine

1. Note the radio security code and the radio presets. Disconnect the negative battery cable.

2. Raise and support the vehicle safely. Drain the engine oil. Remove the front tires.

3. Remove the splash shield. Separate the stabilizer links. Separate the knuckles from the lower arms.

4. Remove the steering gearbox bracket. Remove the steering gearbox mounting bolt, stiffener mounting bolt and stiffener.

5. Remove the gearbox mounting bolt, stiffener mounting bolt and stiffener. Remove the harness clamp from the subframe

6. Remove the lower torque rod. Remove the front mount mounting bolt.

7. Use a marker and make alignment marks on the reference lines that align with the centers of the rear subframe mounting bolts.

8. Loosen the mid-stiffener mounting bolts, on both sides. Support the subframe using the proper support tool.

9. Remove the front subframe. Remove the lower torque rod bracket.

10. Remove the clutch cover and the transaxle mounting bolts.

11. Remove the oil pan retaining bolts. Using a flat bladed tool, carefully separate the oil pan from the engine block. Remove the oil pan from the vehicle.

To install:

12. Remove any old gasket from the mating surfaces. Be sure these surfaces are clean and dry.

13. Apply liquid gasket, part number 08717-004, 08718-0001, 08718-0003 or 08718-0009 evenly to the engine block mating surface of the oil pan.

➡**Do not install the parts if more than five minutes have elapsed since applying the liquid gasket. Instead, reapply after removing the previous coating material.**

14. Position the oil pan in place. Tighten the oil pan retaining bolts in two or three steps to specification and in the proper sequence.

15. Continue the installation in the reverse order of the removal procedure.

16. After assembly, wait at least thirty minutes before filling the engine with clean engine oil.

17. Do not run the engine for at least three hours after installing the oil pan.

Civic Hybrid

See Figure 142.

1. Drain the engine oil.

2. Remove the steering wheel, then remove the steering joint bolt.

3. Remove the front wheels.

4. Remove the front undercover and the splash shield.

5. Remove the A/C condenser fan shroud assembly.

6. Disconnect the A/C compressor clutch connector, then remove the A/C compressor.

7. Remove the intake manifold bracket and the harness clamp.

8. Remove the under-floor three way catalytic converter (TWC).

9. Disconnect the front stabilizer links.

10. Disconnect the suspension lower arm ball joints.

11. Disconnect the steering gearbox harness connectors.

12. Attach the universal lifting eyelet, and the engine support hanger.

13. Remove the lower torque rod mounting bolts.

14. Make the appropriate reference line at the both side front subframe that line up with the edge on the body.

15. Loosen the mid stiffener mounting bolts.

16. Using the front subframe adapter, then remove the front subframe.

17. Remove the harness clamps and the mounting bolt, then remove the dipstick tube.

18. Remove the lower torque rod bracket mounting bolts.

19. Remove the two bolts securing the transmission.

20. Remove the bolts securing the oil pan.

21. Insert a flat blade screwdriver where shown, and separate the oil pan from the engine block.

22. Remove the oil pan.

To install:

23. Remove all of the old liquid gasket from the oil pan mating surfaces, the bolts, and the bolt holes.

24. Clean and dry the oil pan mating surfaces and the O-ring groove.

25. Install the dowel pins, and install the new O-ring and the oil pan gasket on the oil pan.

26. Apply liquid gasket, P/N 08718-0001, or 08718-0009, evenly to the engine block mating surface of the oil pan and to the inside edge of the threaded bolt holes. Install the component within 5 minutes of applying the liquid gasket.

27. Install the oil pan.

28. Tighten all the bolts in three steps to 8.7 ft. lbs. (12 Nm), except for bolt #1 which is tighten to 17 ft. lbs. (24 Nm). Wipe off the excess liquid gasket on the each side of the crankshaft pulley and the flywheel.

29. Tighten the two bolts securing the transmission to 47 ft. lbs. (64 Nm).

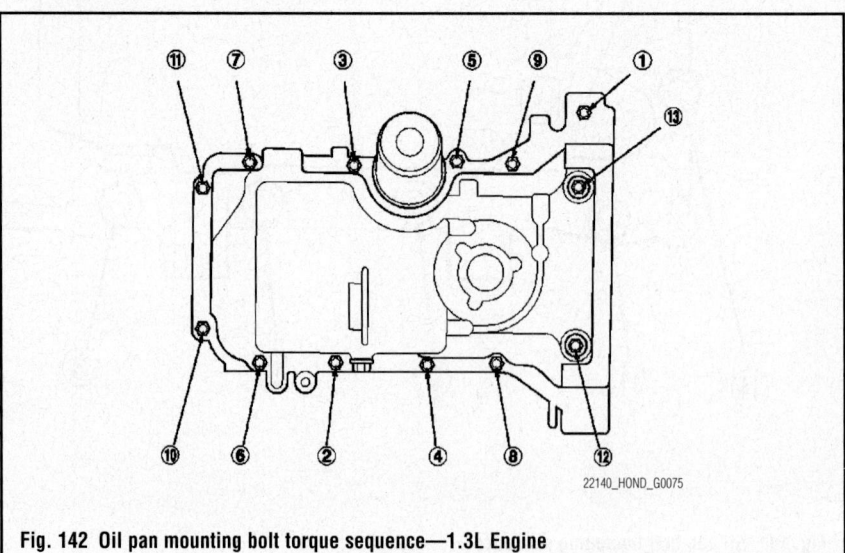

22140_HOND_G0075

Fig. 142 Oil pan mounting bolt torque sequence—1.3L Engine

30. Install the lower torque rod bracket mounting bolts and tighten to 54 ft. lbs. (74 Nm).

31. The remainder of the installation is the reverse order of removal.

32. Wait at least 30 minutes before filling the engine oil.

33. Check and adjust the wheel alignment as necessary.

S2000

See Figure 143.

1. Note the radio security code and the radio presets. Disconnect the negative battery cable.

2. Drain the engine oil. Raise and support the vehicle safely.

3. Remove the oil pan retaining bolts. Drive an oil pan seal cutter tool between the oil pan and engine block. Cut the oil pan seal by striking the side of the tool along the oil pan

4. Remove the oil pan from the vehicle.

To install:

5. Remove any old gasket from the mating surfaces. Be sure these surfaces are clean and dry.

6. Apply liquid gasket, part number 08717-004, 08718-0001, 08718-0003 or 08718-0009 evenly to the engine block mating surface of the oil pan.

➡**Do not install the parts if more than four minutes have elapsed since applying the liquid gasket. Instead, reapply after removing the previous coating material.**

7. Position the oil pan in place. Tighten the oil pan retaining bolts in two or three steps to specification and in the proper sequence.

8. Continue the installation in the reverse order of the removal procedure.

9. After assembly, wait at least thirty minutes before filling the engine with clean engine oil.

10. Do not run the engine for at least three hours after installing the oil pan.

OIL PUMP

REMOVAL & INSTALLATION

Accord

2.4L Engine

1. Note the radio security code and the radio presets. Disconnect the negative battery cable.

2. Position the number one piston at TDC. Remove the oil pan.

3. Remove and discard the oil pump chain tensioner.

4. To hold the rear balancer shaft, insert a 6mm pin driver into the maintenance hole in the lower balancer shaft holder and through the rear balancer shaft.

5. Loosen the oil pump sprocket mounting bolt.

6. Remove the oil pump sprocket. Remove the oil pump.

To install:

7. Remove any old gasket from the mating surfaces. Be sure these surfaces are clean and dry.

8. Apply clean engine oil to the threads of the oil pump sprocket mounting bolt.

9. loosely install the oil pump, and then install the oil pump sprocket. Remove the pin driver.

10. Tighten the pump retaining bolts to specification.

11. Squeeze the new oil pump chain tensioner and then install the set clip.

12. Install the oil pump chain tensioner. Remove the set clip from the pump chain tensioner.

13. Continue the installation in the reverse order of the removal procedure.

14 After assembly, wait at least thirty minutes before filling the engine with clean engine oil.

➡**Do not run the engine for at least three hours after installing the oil pan.**

15. Refill the engine with oil to the correct level.

3.0L Engine

See Figures 144 and 145.

1. Note the radio security code and the radio presets.

2. Disconnect the negative battery cable.

3. Raise and support the vehicle safely.

4. Drain the engine oil.

5. Remove the bulkhead cover. Lower the vehicle.

6. Remove the windshield wiper arms.

7. Remove the cowl cover.

8. Remove the connector bracket from the front of the cylinder head. Remove the harness clamp bracket from the rear of the cylinder head. Attach engine support tool, VSB02C0000019. Lift and support the engine.

9. Remove the timing belt. Remove the crankshaft position sensor (CKP). Remove the rocker arm oil control solenoid and oil filter assembly.

10. Raise and support the vehicle safely. Remove the oil pan.

11. Remove the oil pump screen. Remove the oil pump retaining bolts. Remove the oil pump from the engine.

To install:

12. Remove any old gasket from the mating surfaces. Be sure these surfaces are clean and dry.

13. Apply liquid gasket, part number 08717-004, 08718-0001, 08718-0003 or 08718-0009 evenly to the block mating surface of the oil pump.

➡**Do not install the parts if more than four minutes have elapsed since applying the liquid gasket. Instead, reapply after removing the previous coating material.**

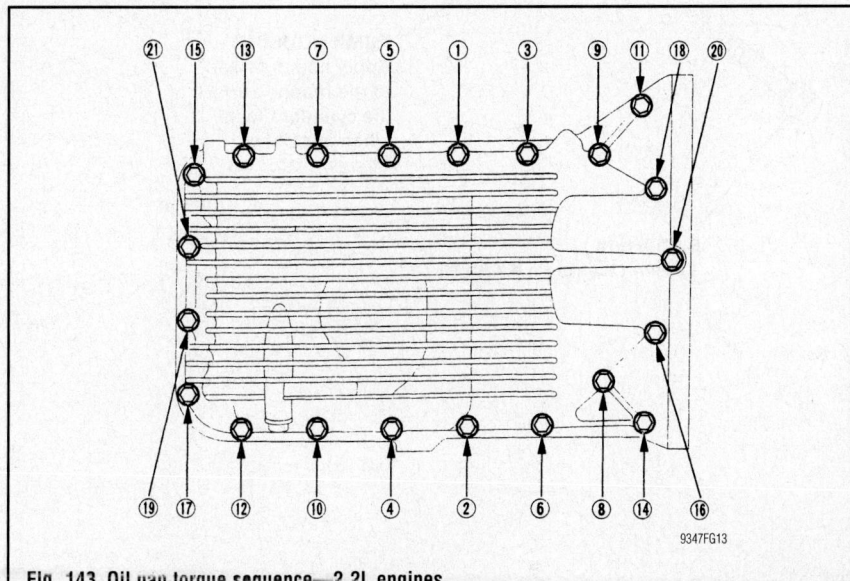

Fig. 143 Oil pan torque sequence—2.2L engines

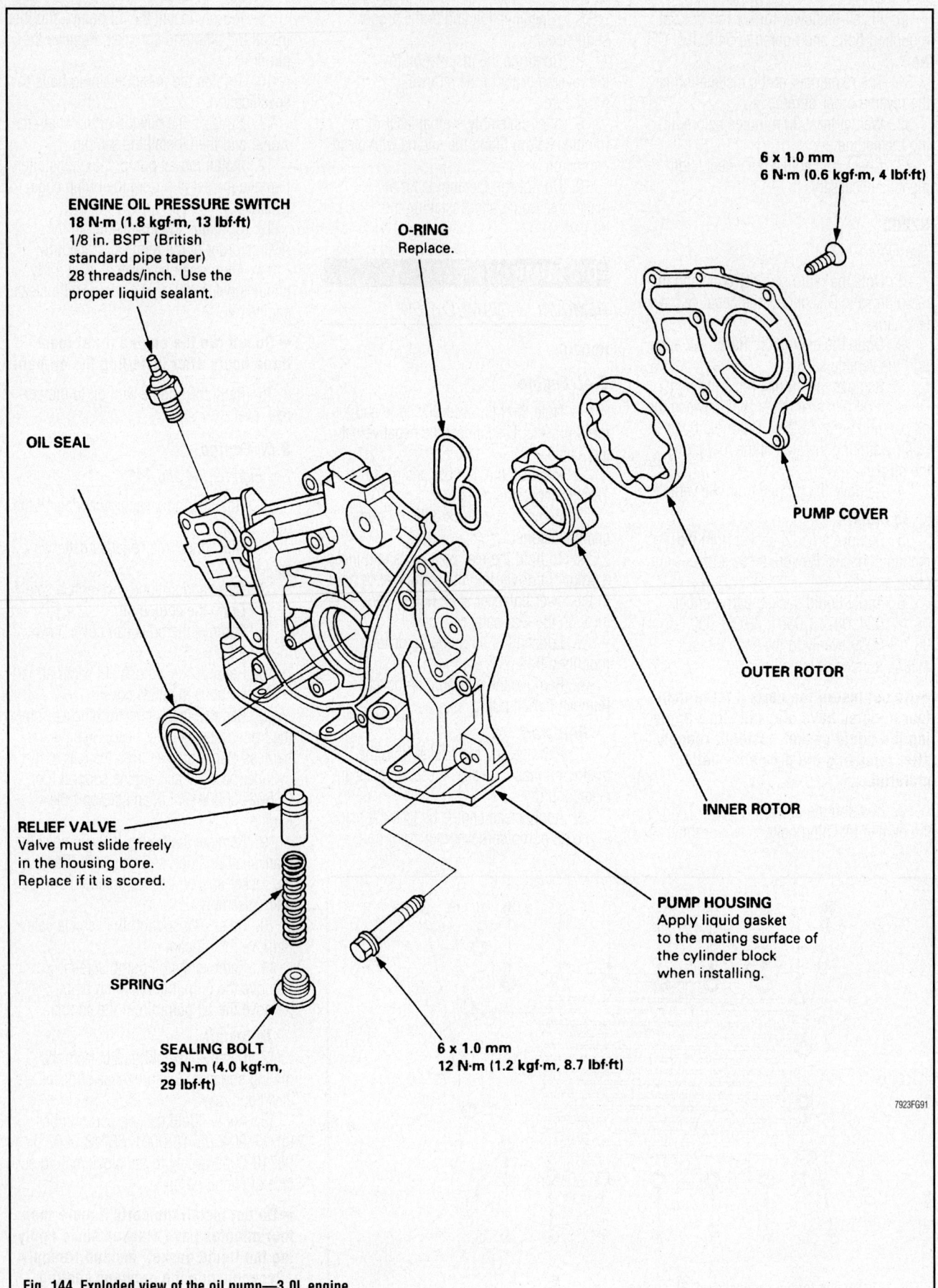

ENGINE OIL PRESSURE SWITCH
18 N·m (1.8 kgf·m, 13 lbf·ft)
1/8 in. BSPT (British
standard pipe taper)
28 threads/inch. Use the
proper liquid sealant.

O-RING
Replace.

6 x 1.0 mm
6 N·m (0.6 kgf·m, 4 lbf·ft)

PUMP COVER

OIL SEAL

OUTER ROTOR

INNER ROTOR

RELIEF VALVE
Valve must slide freely
in the housing bore.
Replace if it is scored.

PUMP HOUSING
Apply liquid gasket
to the mating surface of
the cylinder block
when installing.

SPRING

SEALING BOLT
39 N·m (4.0 kgf·m,
29 lbf·ft)

6 x 1.0 mm
12 N·m (1.2 kgf·m, 8.7 lbf·ft)

7923FG91

Fig. 144 Exploded view of the oil pump—3.0L engine

14. Grease the lip of the oil seal and apply clean oil to the new O-rings.

15. Install the dowel pins then align the inner rotor with the crankshaft. Install the oil pump.

16. Continue the installation in the reverse order of the removal procedure.

➡**After assembly, wait at least thirty minutes before filling the engine with clean engine oil.**

➡**Do not run the engine for at least three hours after installing the oil pan.**

17. Refill the engine with oil to the correct level.

3.5L Engine

See Figure 146.

1. Disconnect the negative battery cable.
2. Drain the engine oil.
3. Remove the timing belt. For additional information, refer to the following section, "Timing Belt, Removal & Installation."
4. Remove the timing belt drive pulley from the crankshaft.
5. Attach a chain hoist to the engine hook on the power steering pump bracket.
6. Remove the oil filter base/oil filter assembly.
7. Remove the oil pan. For additional information, refer to the following section, "Oil Pan, Removal &Installation."
8. Remove the oil screen.
9. Remove the mounting bolts, then remove the oil pump assembly.

To install:

10. Remove the old oil seal from the oil pump.
11. Apply a light coat of new engine oil around the crankshaft oil seal.
12. Gently tap in the new oil seal until the oil seal driver bottoms on the pump.
13. Remove all of the old liquid gasket from the oil pump mating surfaces, the bolts, and the bolt holes.
14. Clean and dry the oil pump mating surfaces.
15. Apply liquid gasket, P/N 08717-0004, 08718-0001, 08718-0003, or 08718-0009, evenly to the engine block mating surface of the oil pump and to the inside edge of the threaded bolt holes. Install the component within 5 minutes of applying the liquid gasket.
16. Apply a light coat of new engine oil to the lip of the crankshaft oil seal, and apply new engine oil to the new O-ring.

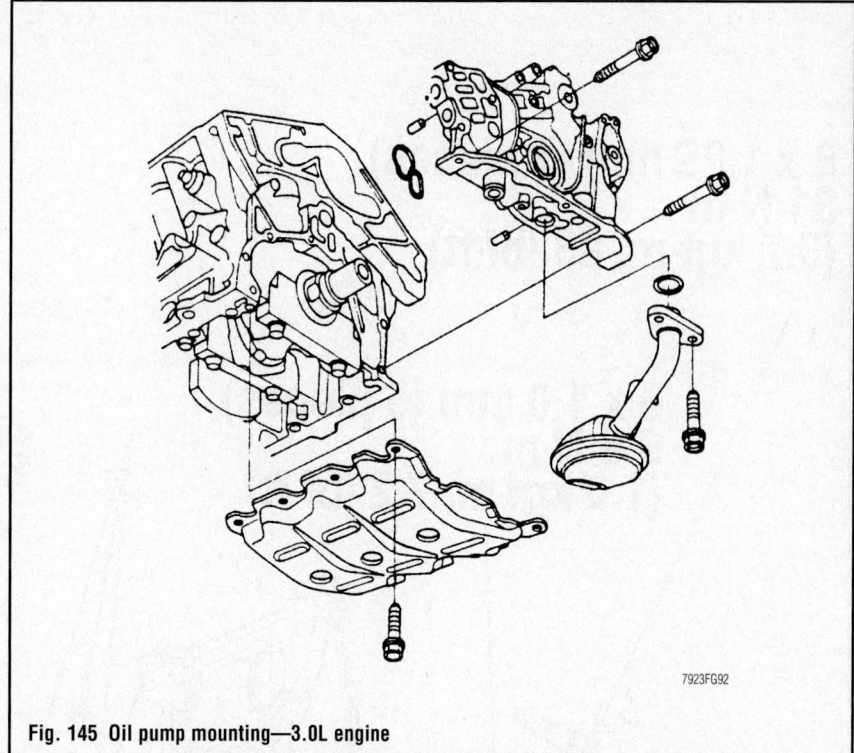

Fig. 145 Oil pump mounting—3.0L engine

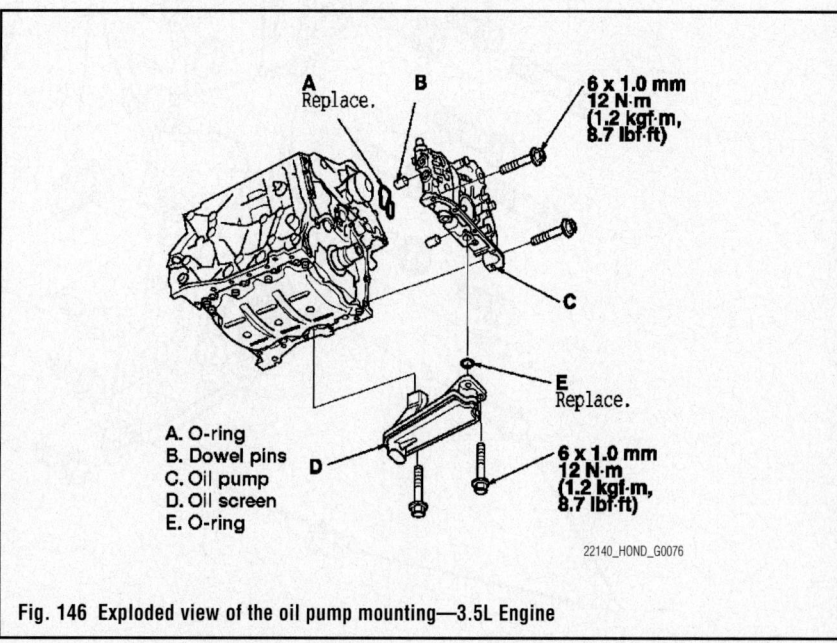

A. O-ring
B. Dowel pins
C. Oil pump
D. Oil screen
E. O-ring

Fig. 146 Exploded view of the oil pump mounting—3.5L Engine

17. Install the dowel pins, then align the inner rotor with the crankshaft, and install the oil pump.
18. Clean any excess grease off the crankshaft, and check the seal for distortion.
19. Install the oil screen with a new O-ring.
20. Install the oil pan.
21. Install the oil filter base/oil filter assembly.
22. Install the timing belt.

➡**After assembly, wait at least 30 minutes before filling the engine with oil.**

23. Refill the engine with oil to the correct level.
24. Connect the negative battery cable.

Civic

1.8L Engine

See Figures 147 and 148.

1. Note the radio security code and the radio presets.

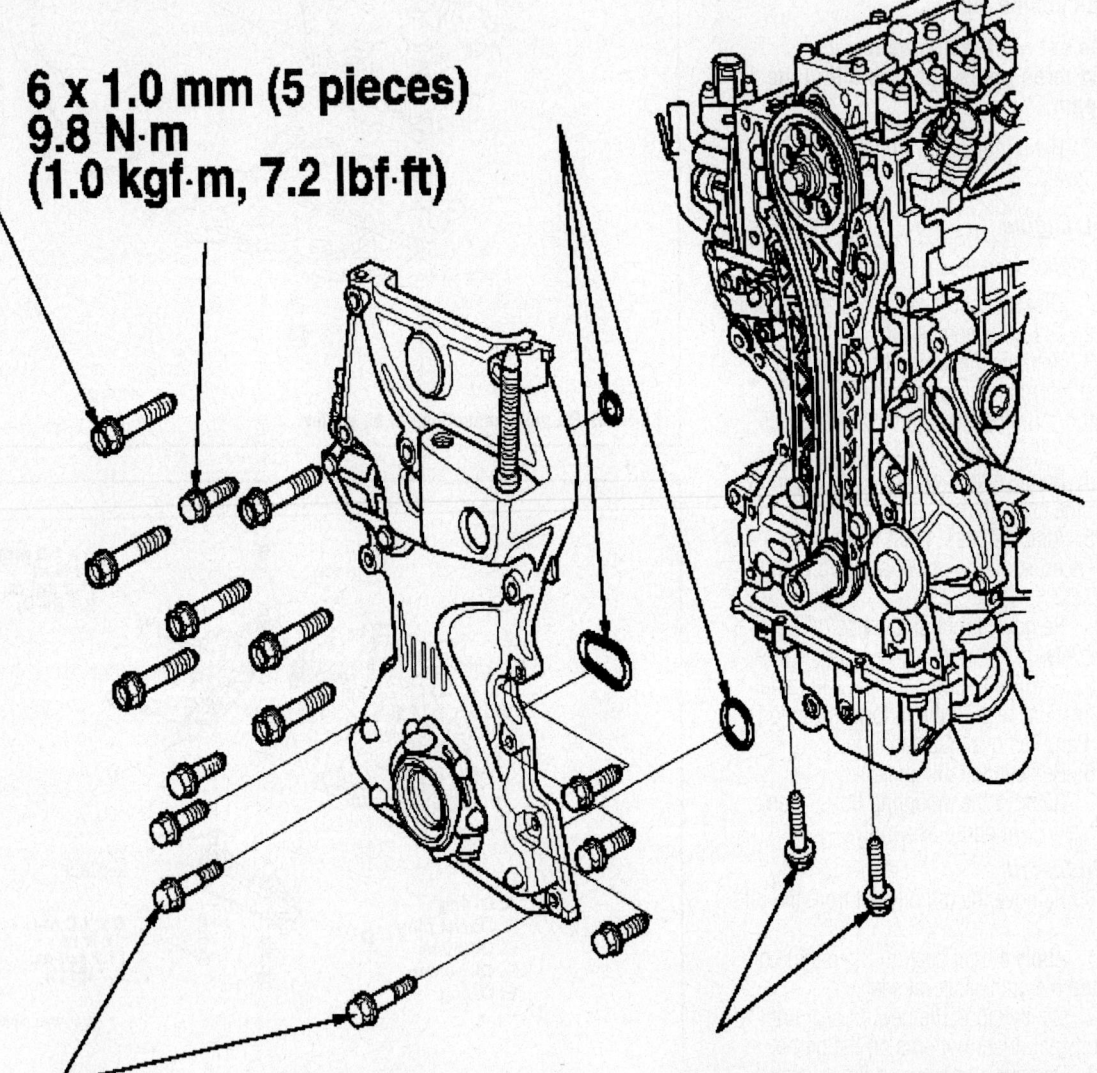

8 x 1.25 mm (7 pieces)
31 N·m
(3.2 kgf·m, 23 lbf·ft)

6 x 1.0 mm (5 pieces)
9.8 N·m
(1.0 kgf·m, 7.2 lbf·ft)

6 x 1.0 mm (2 pieces)
12 N·m
(1.2 kgf·m, 8.7 lbf·ft)

6 x 1.0 mm (2 pieces)
18 N·m
(1.8 kgf·m, 13 lbf·ft)

09474_ACCO_G0020

Fig. 147 Oil pump mounting—1.8L engine

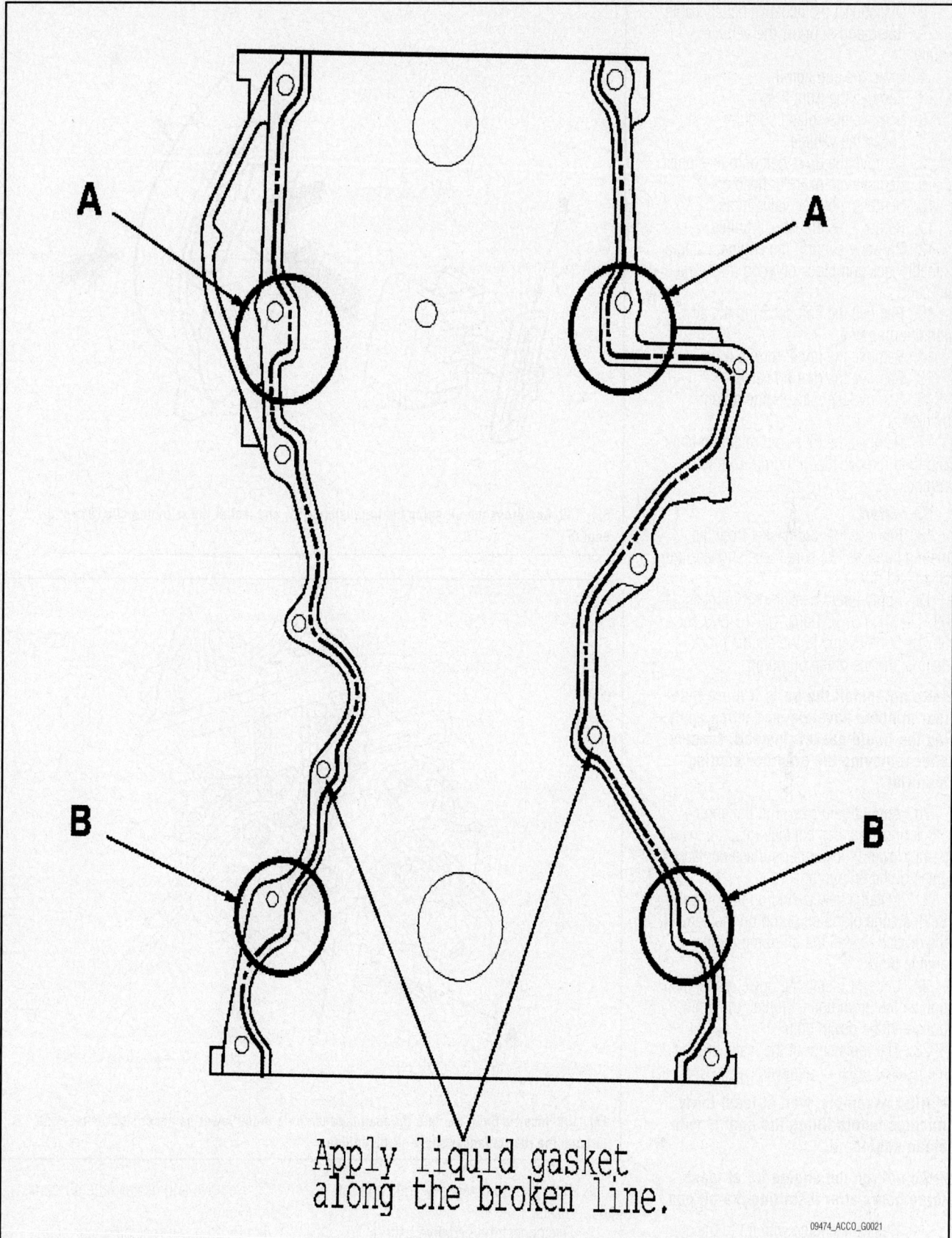

A A

B B

Apply liquid gasket
along the broken line.

09474_ACCO_G0021

Fig. 148 Apply liquid gasket to the engine block upper surface contact areas (A) on the oil pump, lower block upper surface contact areas (B) of the oil pump—1.8L engine

2. Disconnect the negative battery cable.

3. Raise and support the vehicle safely.

4. Drain the engine oil.

5. Remove the front tires.

6. Remove the splash shield.

7. Lower the vehicle.

8. Remove the drive belt auto tensioner.

9. Remove the cylinder head cover.

10. Remove the PCV valve hose.

11. Remove the crankshaft pulley.

12. Properly support the engine, using a suitable jack and block of wood under the oil pan.

13. Remove the bolt securing the air conditioning line.

14. Remove the upper torque rod.

15. Remove the ground cable.

16. Remove the side engine mount bracket.

17. Remove the oil pump retaining bolts, and then remove the oil pump from the engine.

To install:

18. Remove any old gasket from the mating surfaces. Be sure these surfaces are clean and dry.

19. Apply liquid gasket, part number 08717-004, 08718-0001, 08718-0003 or 08718-0009 evenly to the engine block mating surface of the oil pump.

➡**Do not install the parts if more than four minutes have elapsed since applying the liquid gasket. Instead, reapply after removing the previous coating material.**

20. Apply liquid gasket to the engine block upper surface contact areas on the oil pump, lower block upper surface contact areas of the oil pump.

21. Install a new O-ring on the oil pump. Set the edge of the oil pump on the edge of the oil pan. Install the oil pump on the engine block.

22. Loosely install the dowel bolts, then tighten the 8mm bolts. Tighten the 6mm bolts and the dowel bolts.

23. The remainder of the installation is the reverse order of removal.

➡**After assembly, wait at least thirty minutes before filling the engine with clean engine oil.**

➡**Do not run the engine for at least three hours after installing the oil pan.**

24. Refill the engine with oil to the correct level.

2.0L Engine

See Figures 149 through 152.

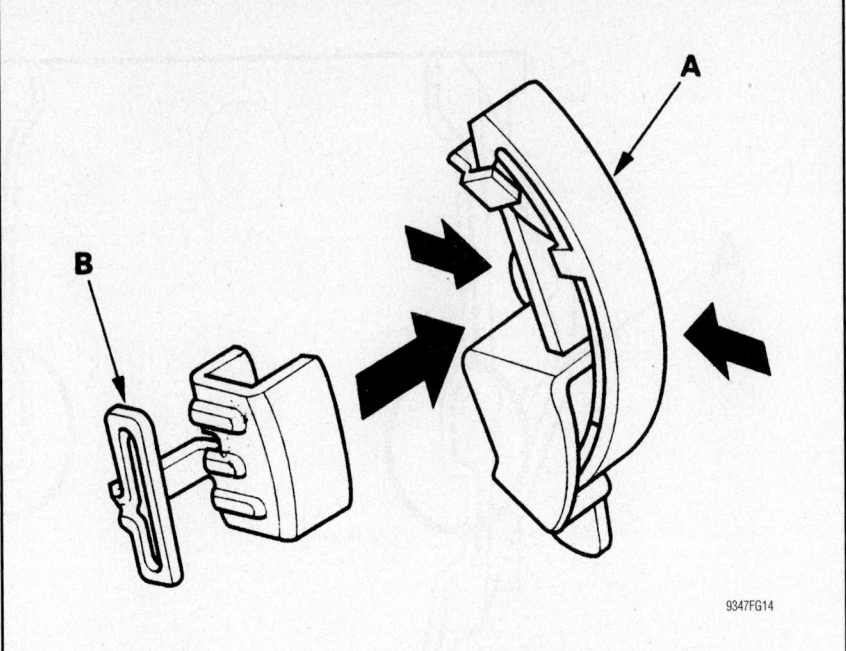

Fig. 149 Compress the oil pump chain tensioner (A) and install the retaining clip (B)—2.0L engine

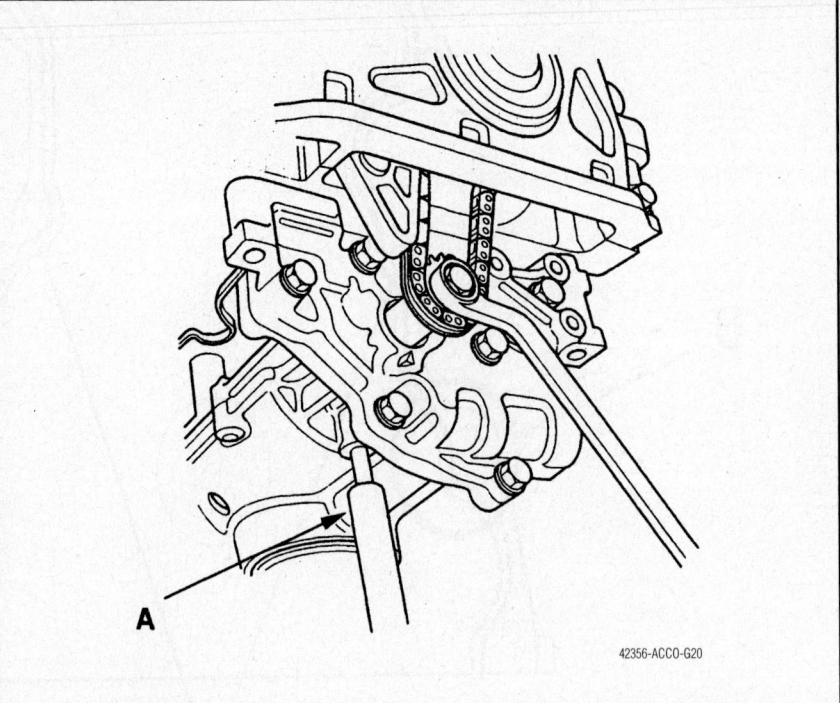

Fig. 150 Insert a 6mm pin into the maintenance hole in the lower balancer shaft holder and through the rear balancer shaft—2.0L engine

1. Note the radio security code and the radio presets.

2. Disconnect the negative battery cable.

3. Position the number one piston at TDC.

4. Remove the oil pan.

5. Remove and discard the oil pump chain tensioner.

6. To hold the rear balancer shaft, insert a 6mm pin driver into the maintenance hole in the lower balancer shaft holder and through the rear balancer shaft.

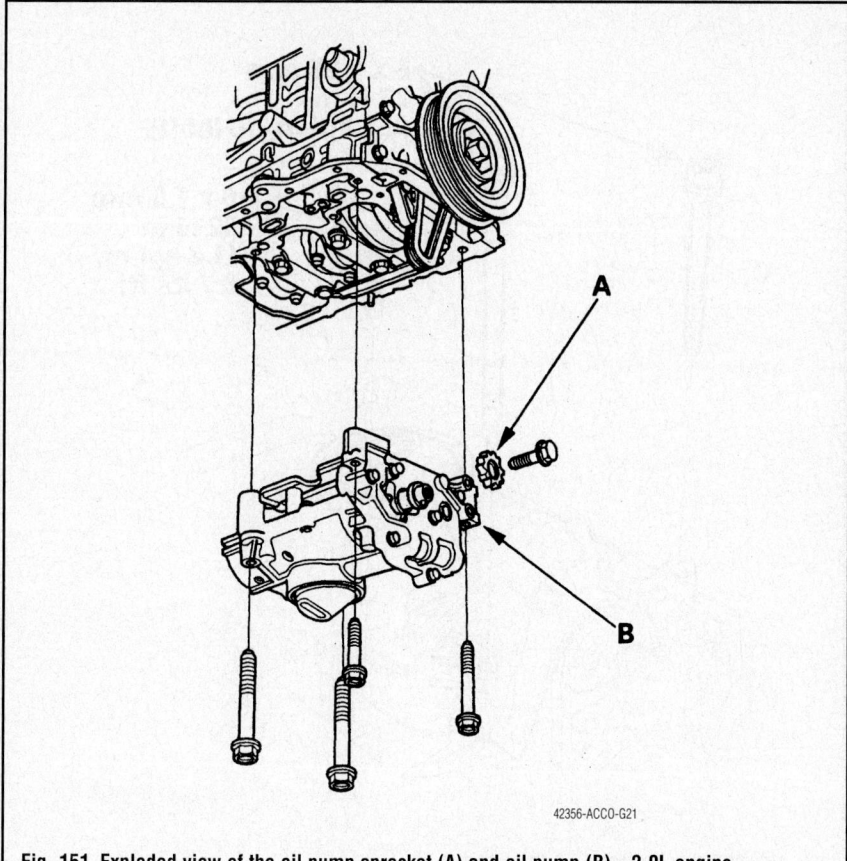

Fig. 151 Exploded view of the oil pump sprocket (A) and oil pump (B)—2.0L engine

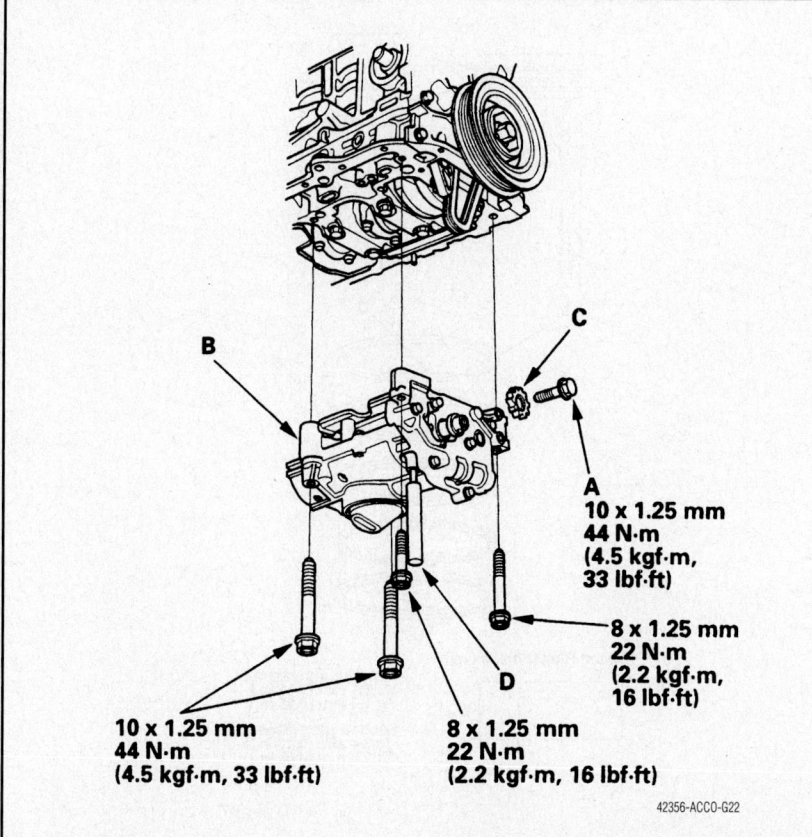

**A
10 x 1.25 mm
44 N·m
(4.5 kgf·m,
33 lbf·ft)**

**8 x 1.25 mm
22 N·m
(2.2 kgf·m,
16 lbf·ft)**

**10 x 1.25 mm
44 N·m
(4.5 kgf·m, 33 lbf·ft)**

**8 x 1.25 mm
22 N·m
(2.2 kgf·m, 16 lbf·ft)**

42356-ACCO-G22

Fig. 152 Oil pump tightening specifications—2.0L engine

7. Loosen the oil pump sprocket mounting bolt.

8. Remove the oil pump sprocket. Remove the oil pump.

To install:

9. Remove any old gasket from the mating surfaces. Be sure these surfaces are clean and dry.

10. Apply clean engine oil to the threads of the oil pump sprocket mounting bolt.

11. Loosely install the oil pump, and then install the oil pump sprocket. Remove the pin driver.

12. Tighten the pump retaining bolts to specification.

13. Squeeze the new oil pump chain tensioner and then install the set clip.

14. Install the oil pump chain tensioner. Remove the set clip from the pump chain tensioner.

15. Continue the installation in the reverse order of the removal procedure.

➡**After assembly, wait at least thirty minutes before filling the engine with clean engine oil.**

➡**Do not run the engine for at least three hours after installing the oil pan.**

16. Refill the engine with oil to the correct level.

Civic Hybrid

1. Disconnect the negative battery cable.

2. Drain the engine oil.

3. Remove the timing chain. For additional information, refer to the following section, "Timing Chain, Removal & Installation."

4. Remove the oil screen.

5. Remove the oil pump.

To install:

6. Install the dowel pins and a new O-ring on the oil pump, then align the inner rotor with the crankshaft, and install the oil pump. Tighten the mounting bolts to 86 inch lbs. (10 Nm)

7. Install the oil screen with a new gasket.

8. Install the timing chain.

9. Refill the engine with oil to the correct level.

10. Connect the negative battery cable.

S2000

See Figure 153.

1. Note the radio security code and the radio presets. Disconnect the negative battery cable.

2. Remove the timing chain. For additional information, refer to the following section, "Timing Chain, Removal & Installation."

3. Remove the oil pan.

4. Remove the oil pump chain tensioner. Remove the baffle plate.

5. Remove the oil pump retaining bolts. Remove the oil pump, oil pump chain and crankshaft sprocket.

To install:

6. Remove any old gasket from the mating surfaces. Be sure these surfaces are clean and dry.

7. Squeeze the new oil pump chain tensioner and then install the set clip.

8. Install the crankshaft sprocket, oil pump chain and oil pump. Install the baffle plate.

9. Set the crankshaft sprocket so that the number one piston is at TDC. Align the key on the sprocket and crankshaft with the pointer on the engine block.

10. Move the cam chain so that the colored piece aligns with the punched mark on the crankshaft.

11. Install the oil pump chain guide and the oil pump chain tensioner with the seat clip.

12. Remove the set clip from the oil pump chain tensioner.

13. Continue the installation in the reverse order of the removal procedure.

➡**After assembly, wait at least thirty minutes before filling the engine with clean engine oil.**

➡**Do not run the engine for at least three hours after installing the oil pan.**

14. Refill the engine with oil to the correct level.

15. Connect the negative battery cable.

PISTON AND RING

POSITIONING

See Figures 154 and 155.

REAR MAIN SEAL

REMOVAL & INSTALLATION

1. Remove the transaxle from the vehicle.

2. Remove the IMA motor, if equipped.

3. Remove the driveplate from the crankshaft.

4. Carefully pry the crankshaft seal out of the retainer.

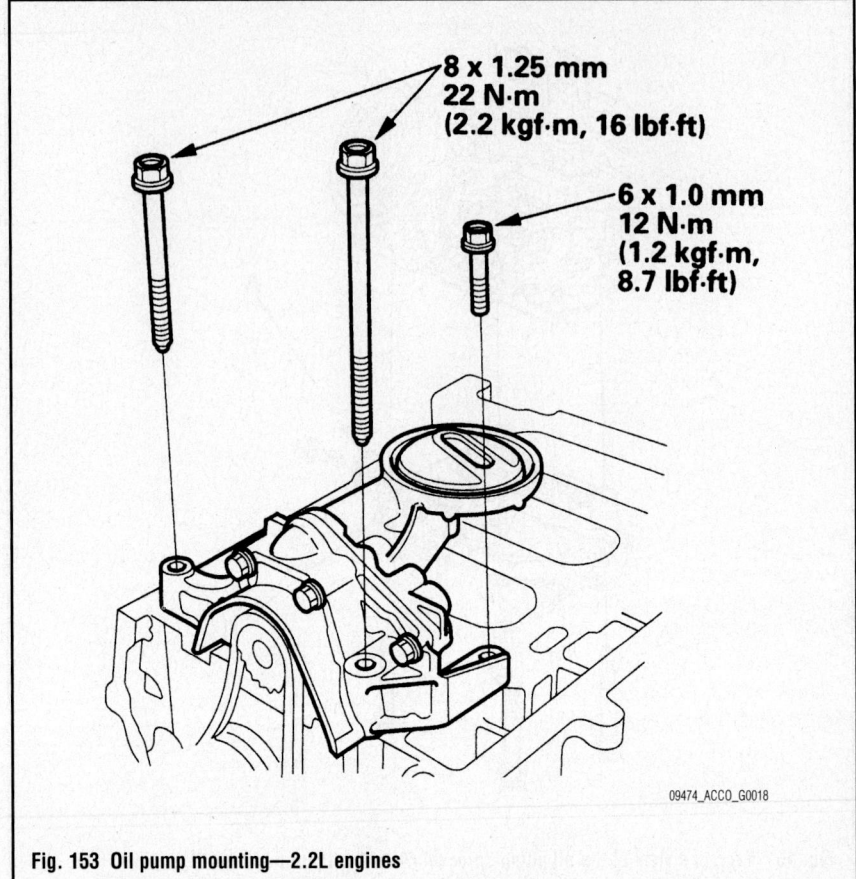

8 x 1.25 mm
22 N·m
(2.2 kgf·m, 16 lbf·ft)

6 x 1.0 mm
12 N·m
(1.2 kgf·m, 8.7 lbf·ft)

09474_ACCO_G0018

Fig. 153 Oil pump mounting—2.2L engines

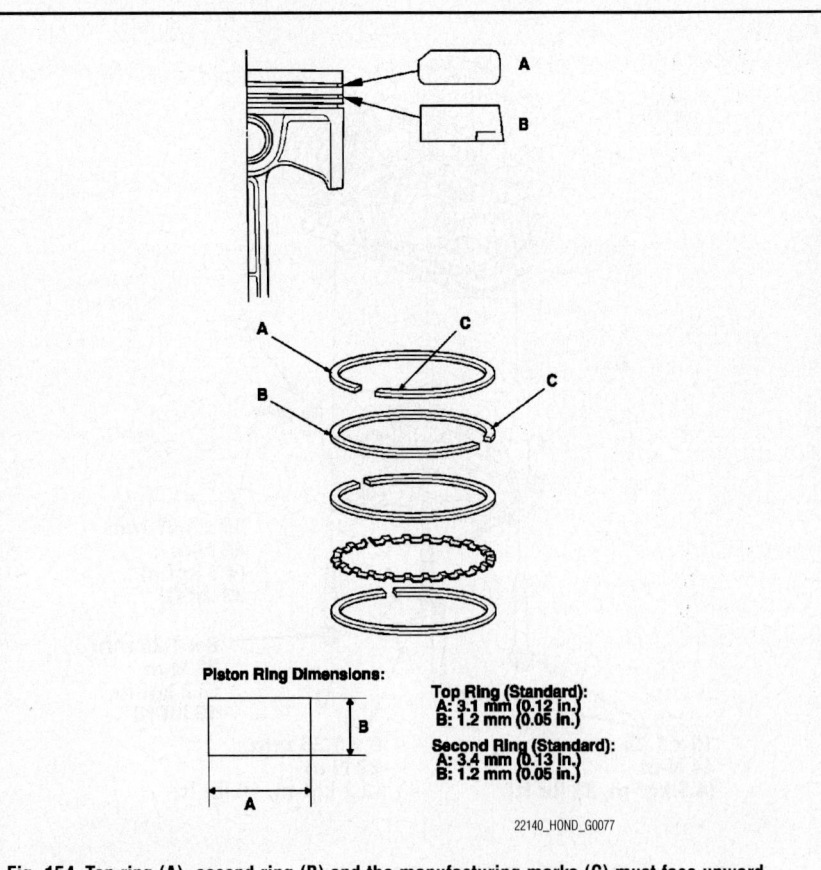

Piston Ring Dimensions:

Top Ring (Standard):
A: 3.1 mm (0.12 in.)
B: 1.2 mm (0.05 in.)

Second Ring (Standard):
A: 3.4 mm (0.13 in.)
B: 1.2 mm (0.05 in.)

22140_HOND_G0077

Fig. 154 Top ring (A), second ring (B) and the manufacturing marks (C) must face upward

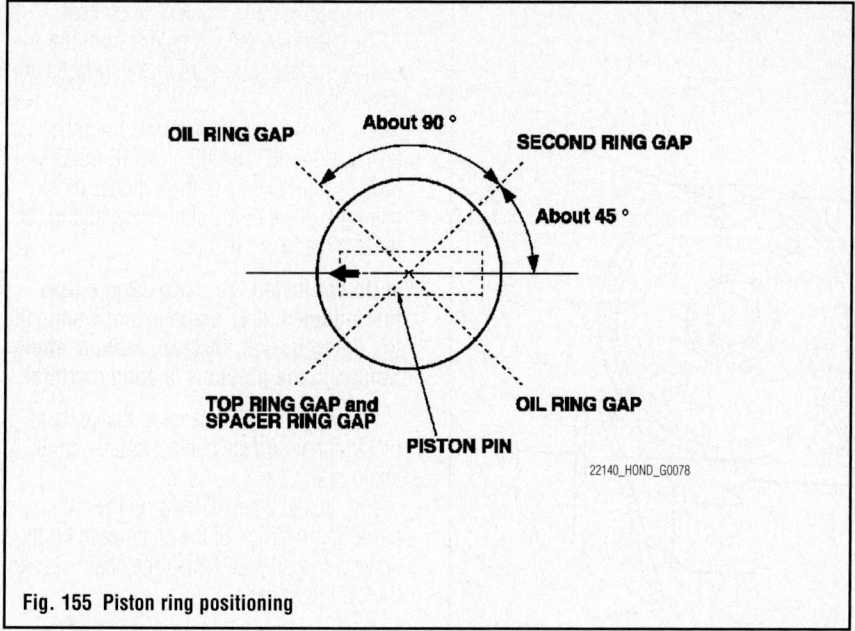

Fig. 155 Piston ring positioning

To install:

5. Apply clean engine oil to the lip of the new seal.

6. Install the seal onto the crankshaft and into the retainer using the appropriate seal driver.

7. Install the IMA motor, if equipped.

8. Install the driveplate and the transmission.

TIMING CHAIN, SPROCKETS, FRONT COVER AND SEAL

REMOVAL & INSTALLATION

Accord

See Figures 156 through 158.

➡**Keep the cam chain away from magnetic fields.**

1. Before servicing the vehicle, refer to the precautions in the beginning of this section.

2. Note the radio security code and the radio presets. Disconnect the negative battery cable.

3. Rotate the crankshaft to set the engine at Top Dead Center (TDC) on the compression stroke for the No. 1 piston. The TDC mark on the crankshaft pulley should line up with the pointer.

4. Raise and support the vehicle safely. Remove the front tires. Remove the splash shield. Lower the vehicle.

5. Remove the drive belt. Remove the cylinder head cover. Remove the crankshaft pulley.

6. Check that the number one piston TDC marks on the variable timing control

(VTC) actuator and exhaust camshaft sprocket are aligned.

7. Disconnect the crankshaft position (CKP) sensor connector and the variable valve timing control (VTC) oil control solenoid valve connector.

8. Remove the VTC oil control solenoid valve.

9. Properly support the engine using a block of wood under the oil pan. Remove the ground cable. Remove the upper bracket. Remove the side engine mount bracket.

10. Remove the chain case cover.

11. Loosely install the crankshaft pulley. Turn the crankshaft counterclockwise to compress the auto tensioner. Align the holes on the lock and the auto tensioner. Insert a 0.05 inch diameter pin into the holes. Turn the crankshaft clockwise to secure the pin. Remove the auto tensioner.

12. Remove the cam chain guide. Remove the Remove the tensioner arm.

13. Remove the cam chain from the engine.

To install:

➡**Check that the VTC actuator is locked by turning the actuator counterclockwise. If not locked, turn the actuator clockwise until it stops. Recheck it. If it is still not locked, replace it.**

14. Set the crankshaft to TDC. Align the TDC mark on the crankshaft sprocket with the pointer on the engine block.

15. Set the camshafts to TDC. The punch mark on the VTC actuator and the punch mark on the exhaust camshaft sprocket should be at the top. Align the TDC marks on the VTC actuator and the exhaust camshaft sprocket.

16. Install the cam chain on the crankshaft sprocket with the colored piece aligned with the punch mark on the crankshaft sprocket.

17. Install the cam chain on the VTC actuator and exhaust camshaft sprocket with the punch marks aligned with the two colored pieces.

18. Install the cam chain guide and tensioner arm. Install the auto tensioner. Install the cam chain guide. Remove the pin from the auto tensioner.

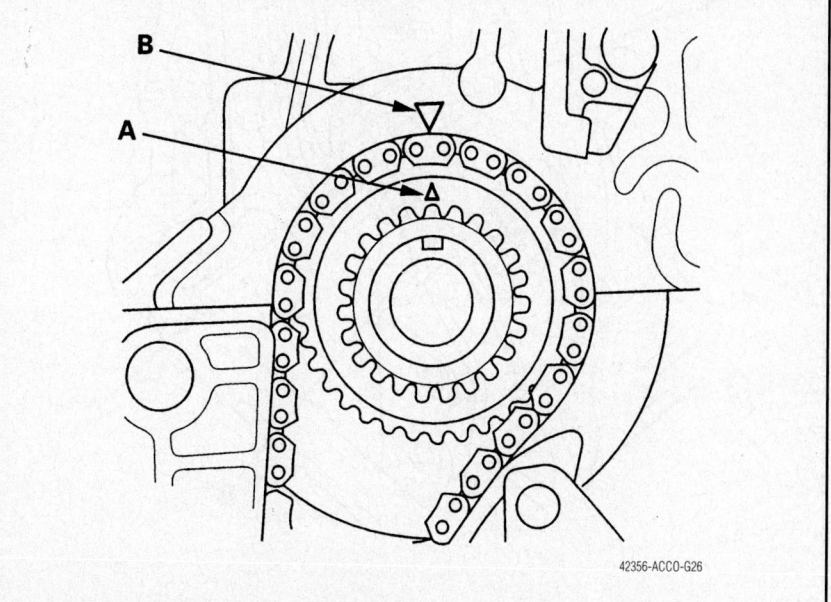

Fig. 156 Set the crankshaft to TDC. Align the TDC mark (A) on the crankshaft sprocket with the pointer (B) on the cylinder block—2.4L engine

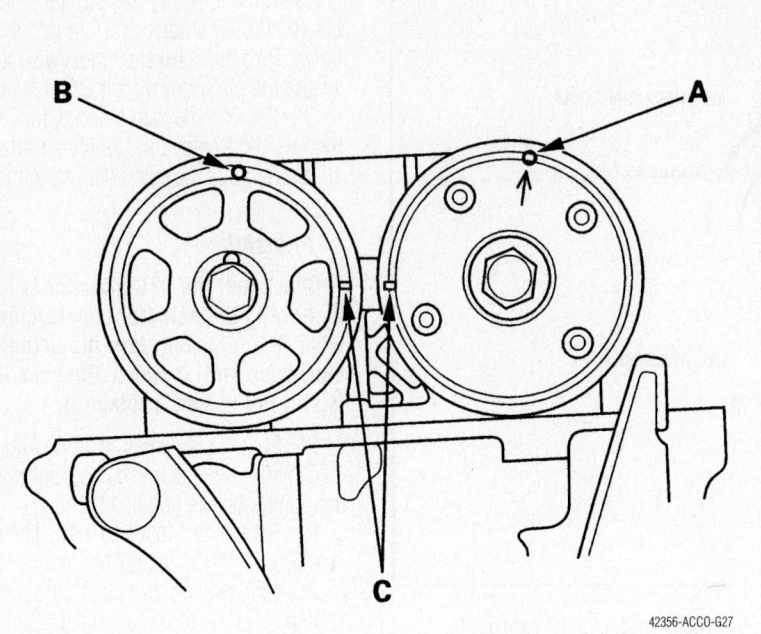

Fig. 157 The mark (A) on the VTC actuator and the mark (B) on the exhaust cam (C) should be at the top. Align the TDC marks (C) on the VTC actuator and exhaust cam sprockets—2.4L engine

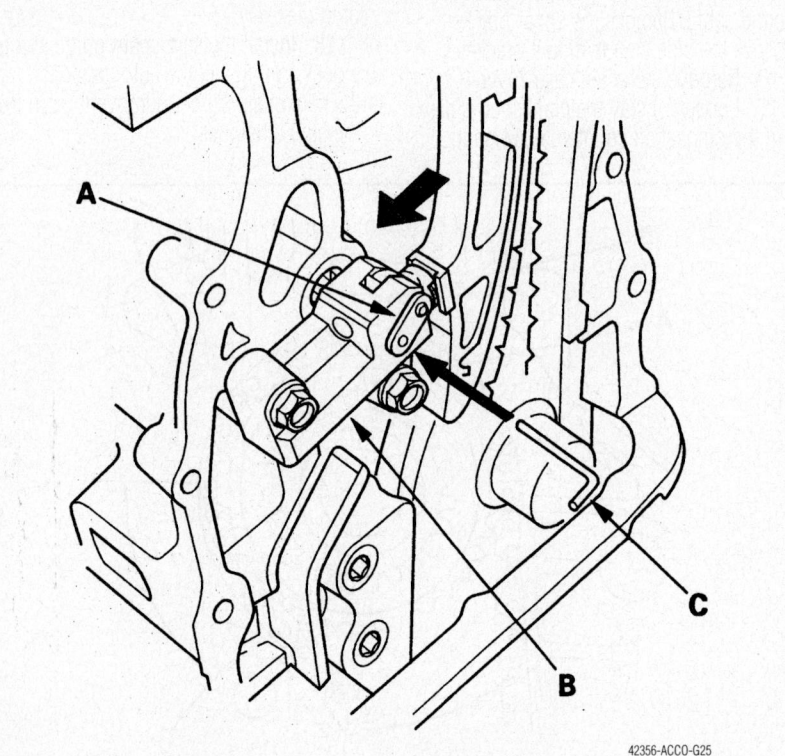

Fig. 158 Align the holes on the lock (A) and the auto-tensioner (B), then place a 1.5mm pin into the holes. Turn the crankshaft clockwise to secure the pin—2.4L engine

19. Replace the chain case oil seal.

20. Remove any old gasket from the mating surfaces. Be sure these surfaces are clean and dry.

21. Apply liquid gasket, part number 08717-004, 08718-0001, 08718-0003 or 08718-0009 evenly to the cylinder block mating surface of the chain case and to the inner threads of the holes.

➡**Do not install the parts if more than four minutes have elapsed since applying the liquid gasket. Instead, reapply after removing the previous coating material.**

22. Apply liquid gasket to the cylinder block upper surface contact areas on the chain case.

23. Install a new O-ring on the chain case. Set the edge of the chain case on the edge of the oil pan. Install the chain case to the cylinder block.

➡**When installing the chain case, do not slide the bottom surface on the oil pan mounting surface.**

24. Continue the installation in the reverse order of the removal procedure.

25. After assembly, wait at least thirty minutes before filling the engine with clean engine oil.

26. Do not run the engine for at least three hours after installing the oil pan.

27. Reprogram the crankshaft position (CKP) pattern. Run the engine until the operating temperature reaches 176 degree. With the engine stopped clear the CKP pattern. Turn the ignition switch OFF. Turn the ignition switch ON and wait thirty seconds.

28. Road test the vehicle on a level surface. Decelerate the engine speed of 2500 RPM to 1000 RPM. If equipped with automatic transaxle use two Drive positions. If equipped with manual transaxle use first gear.

29. Stop the vehicle, but keep the engine running.

30. Check PULSAR F/B LEARN in the data list with the HDS. If not complete repeat the procedure. If complete, road test the vehicle on a level surface. Decelerate the engine speed of 5000 RPM to 3000 RPM. If equipped with automatic transaxle use two Drive positions. If equipped with manual transaxle use first gear.

31. Stop the vehicle, but keep the engine running.

32. Check PULSAR F/B LEARN in the data list with the HDS. If not complete repeat the procedure.

33. If completed, turn the ignition switch OFF. Turn the ignition switch ON, wait thirty seconds. The learning procedure is now complete.

Civic

1.8L Engine

See Figures 159 through 162.

1. Before servicing the vehicle, refer to the precautions in the beginning of this section.

2. Note the radio security code and the radio presets. Disconnect the negative battery cable.

3. Rotate the crankshaft to set the engine at Top Dead Center (TDC) on the compression stroke for the No. 1 piston. The UP mark on the camshaft sprocket should be at the top, and the TDC grooves on the camshaft sprocket should line up with the top edge of the cylinder head.

4. Raise and support the vehicle safely. Remove the front tires. Remove the splash shield. Lower the vehicle.

5. Remove the drive belt auto tensioner. Remove the cylinder head cover. Remove the PCV valve hose. Remove the crankshaft pulley.

6. Properly support the engine using a block of wood under the oil pan. Remove the bolt securing the air conditioning line and remove the upper torque rod. Remove the ground cable. Remove the engine mount bracket.

7. Remove the oil pump.

8. Loosely install the crankshaft pulley. Turn the crankshaft counterclockwise to compress the auto tensioner. Align the holes on the lock and the auto tensioner. Insert a 0.04 inch diameter pin into the holes. Turn the crankshaft clockwise to secure the pin.

9. Remove the auto tensioner. Remove the crankshaft pulley.

10. Remove the cam chain guide and cam chain tensioner arm. Remove the cam chain from the engine.

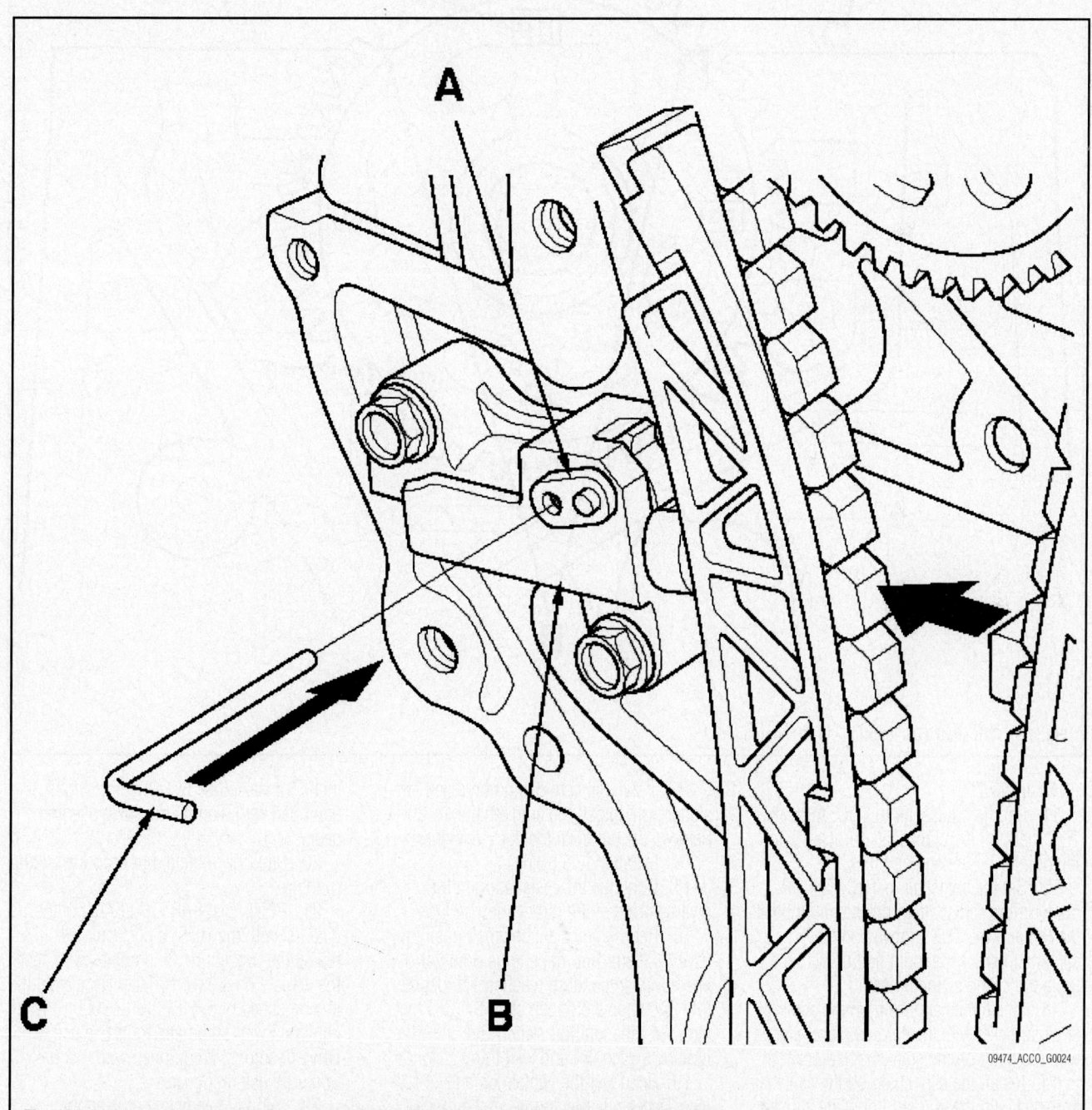

09474_ACCO_G0024

Fig. 159 Locking auto tensioner: (A) alignment hole, (B) tensioner and (C) lock pin—1.8L engine

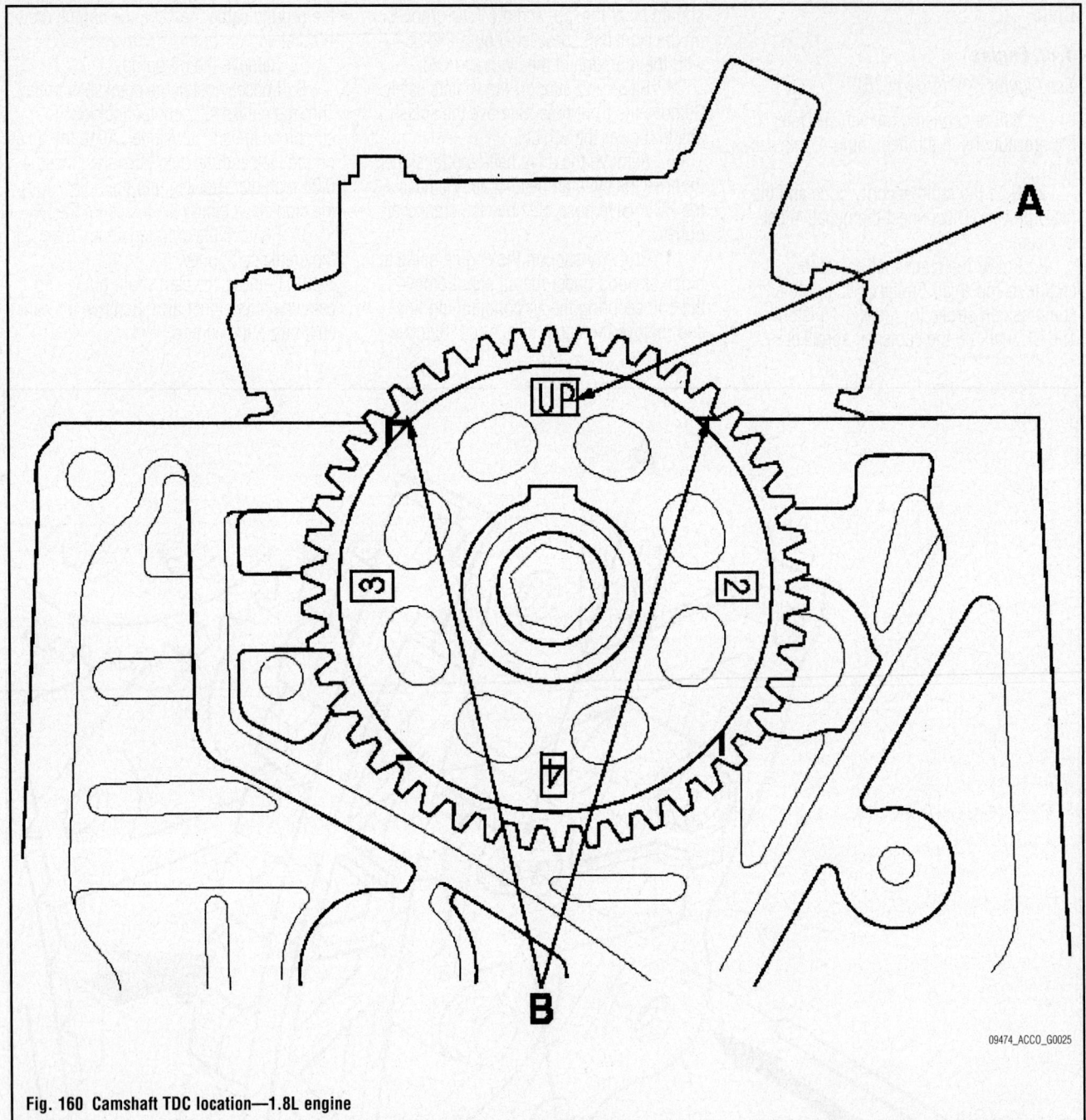

Fig. 160 Camshaft TDC location—1.8L engine

To install:

11. Set the crankshaft to TDC. Align the TDC mark on the crankshaft sprocket with the pointer on the engine block.

12. Set the camshaft to TDC. The UP mark on the camshaft sprocket should be at the top and the TDC grooves on the camshaft sprocket should line up with the top edge of the cylinder head.

13. Install the cam chain on the crankshaft sprocket with the colored piece aligned with the mark on the crankshaft sprocket.

14. Install the cam chain on the camshaft sprocket with the colored link plate aligned with the mark on the crankshaft sprocket.

15. Install the cam chain guide and tensioner arm. Install the auto tensioner. Remove the pin from the auto tensioner.

16. Install the oil pump.

17. Continue the installation in the reverse order of the removal procedure.

18. Reprogram the crankshaft position (CKP) pattern. Run the engine until the operating temperature reaches 176 degree. With the engine stopped clear the CKP pattern. Turn the ignition switch OFF. Turn the ignition switch ON and wait thirty seconds.

19. Road test the vehicle on a level surface. Decelerate the engine speed of 2500 RPM to 1000 RPM. If equipped with auto-matic transaxle use two Drive positions. If equipped with manual transaxle use first gear.

20. Stop the vehicle, but keep the engine running.

21. Check PULSAR F/B LEARN in the data list with the HDS. If not complete repeat the procedure. If complete, road test the vehicle on a level surface. Decelerate the engine speed of 5000 RPM to 3000 RPM. If equipped with automatic transaxle use two Drive positions. If equipped with manual transaxle use first gear.

22. Stop the vehicle, but keep the engine running.

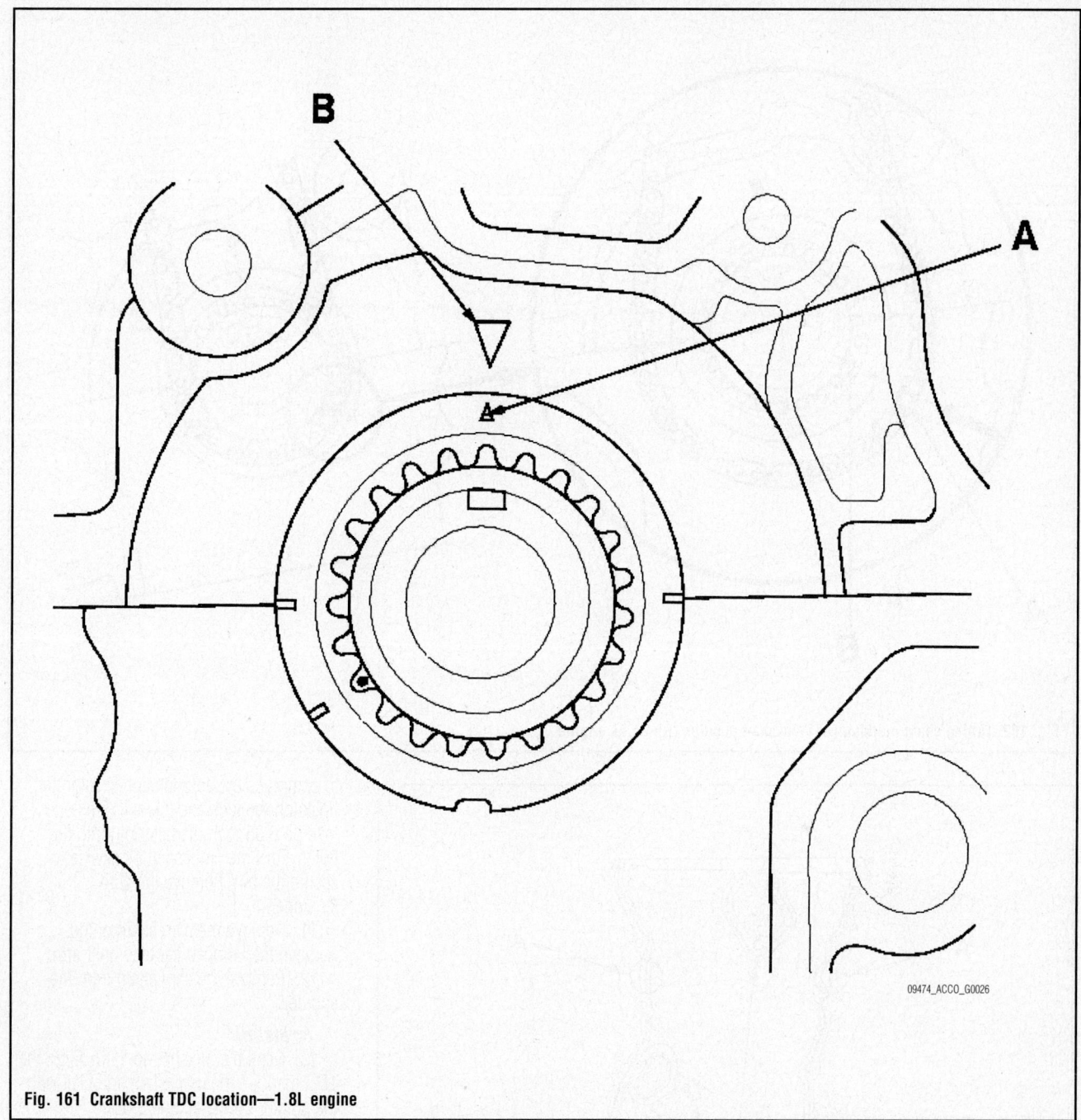

Fig. 161 Crankshaft TDC location—1.8L engine

23. Check PULSAR F/B LEARN in the data list with the HDS. If not complete repeat the procedure.

24. If completed, turn the ignition switch OFF. Turn the ignition switch ON, wait thirty seconds. The learning procedure is now complete.

2.0L Engine

See Figure 163.

➡**Keep the cam chain away from magnetic fields.**

1. Before servicing the vehicle, refer to the precautions in the beginning of this section.

2. Note the radio security code and the radio presets. Disconnect the negative battery cable.

3. Rotate the crankshaft to set the engine at Top Dead Center (TDC) on the compression stroke for the No. 1 piston. The TDC mark on the crankshaft pulley should line up with the pointer.

4. Raise and support the vehicle safely. Remove the front tires. Remove the splash shield. Lower the vehicle.

5. Remove the drive belt. Remove the cylinder head cover. Remove the crankshaft pulley.

6. Disconnect the crankshaft position (CKP) sensor connector and the variable valve timing control (VTC) oil control solenoid valve connector.

7. Remove the VTC oil control solenoid valve.

8. Properly support the engine using a block of wood under the oil pan. Remove the ground cable. Remove the upper bracket. Remove the side engine mount bracket.

9. Remove the chain case cover.

10. Loosely install the crankshaft pulley. Turn the crankshaft counterclockwise

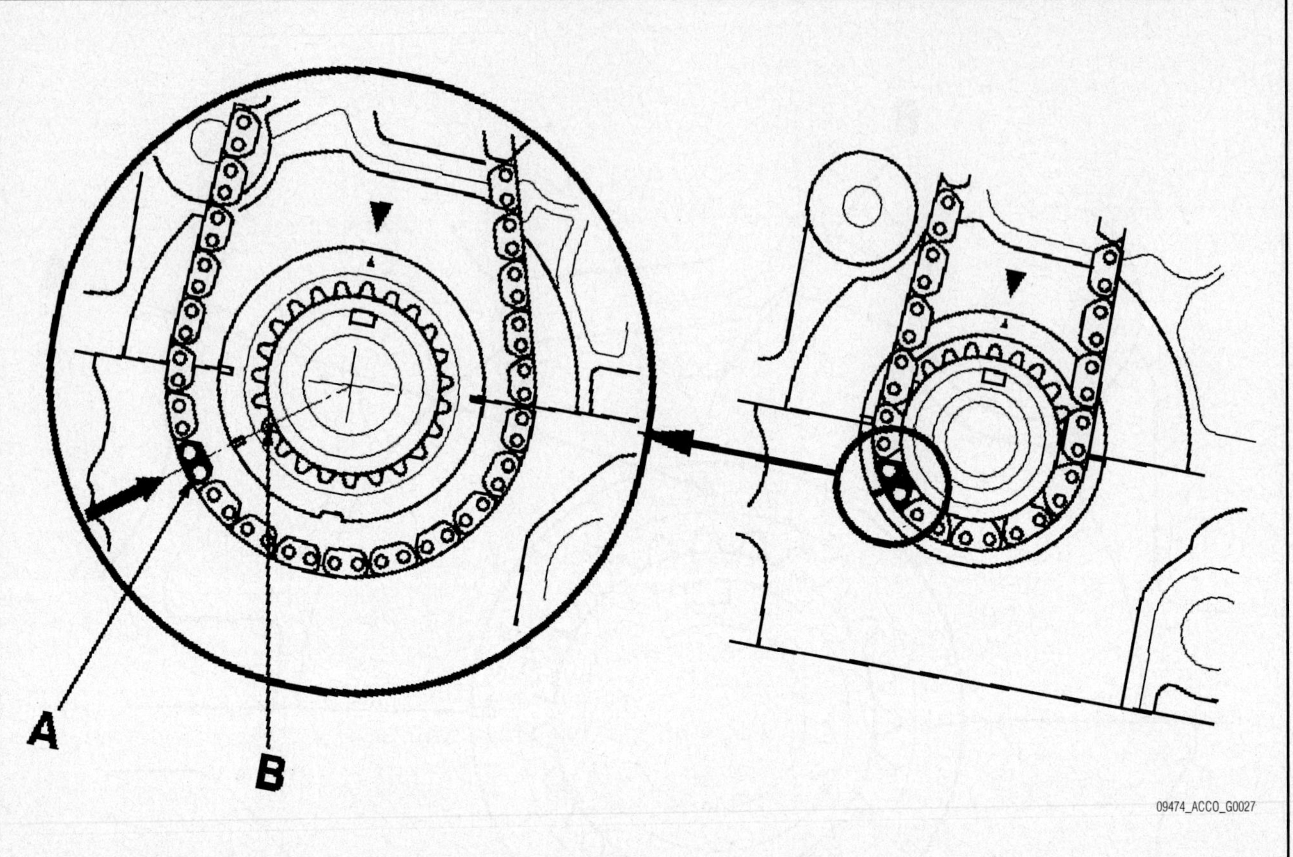

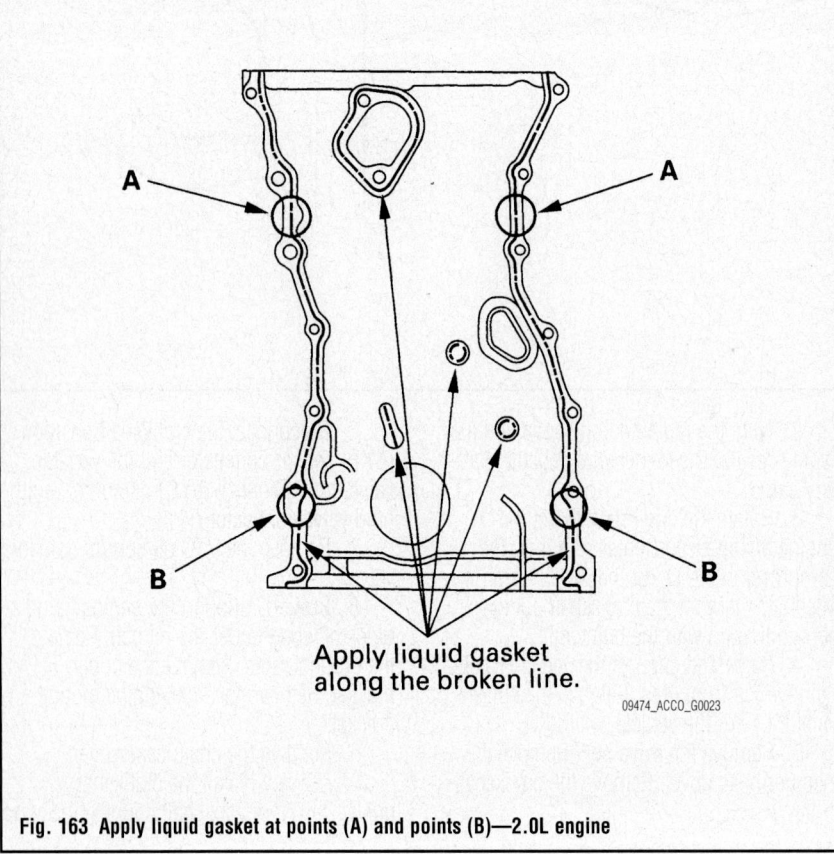

Fig. 162 Timing chain marking (A) crankshaft position (B)—1.8L engine

Fig. 163 Apply liquid gasket at points (A) and points (B)—2.0L engine

Apply liquid gasket
along the broken line.

09474_ACCO_G0023

to compress the auto tensioner. Align the holes on the lock and the auto tensioner. Insert a 0.05 inch diameter pin into the holes. Turn the crankshaft clockwise to secure the pin. Remove the auto tensioner.

11. Remove the cam chain guide. Remove the Remove the tensioner arm.

12. Remove the cam chain from the engine.

To install:

13. Set the crankshaft to TDC. Align the TDC mark on the crankshaft sprocket with the pointer on the engine block.

14. Set the camshafts to TDC. The punch mark on the VTC actuator and the punch mark on the exhaust camshaft sprocket should be at the top. Align the TDC marks on the VTC actuator and the exhaust camshaft sprocket.

15. Install the cam chain on the crankshaft sprocket with the colored piece aligned with the punch mark on the crankshaft sprocket.

16. Install the cam chain on the VTC actuator and exhaust camshaft sprocket with the punch marks aligned with the two colored pieces.

17. Install the cam chain guide and tensioner arm. Install the auto tensioner. Install

the cam chain guide. Remove the pin from the auto tensioner.

18. Replace the chain case oil seal.

19. Remove any old gasket from the mating surfaces. Be sure these surfaces are clean and dry.

20. Apply liquid gasket, part number 08717-004, 08718-0001, 08718-0003 or 08718-0009 evenly to the cylinder block mating surface of the chain case and to the inner threads of the holes.

➡**Do not install the parts if more than four minutes have elapsed since applying the liquid gasket. Instead, reapply after removing the previous coating material.**

21. Apply liquid gasket to the cylinder block upper surface contact areas on the chain case.

22. Install a new O-ring on the chain case. Set the edge of the chain case on the edge of the oil pan. Install the chain case to the cylinder block.

➡**When installing the chain case, do not slide the bottom surface on the oil pan mounting surface.**

23. Continue the installation in the reverse order of the removal procedure.

24. After assembly, wait at least thirty minutes before filling the engine with clean engine oil.

25. Do not run the engine for at least three hours after installing the oil pan.

26. Reprogram the crankshaft position (CKP) pattern. Run the engine until the operating temperature reaches 176 degree. With the engine stopped clear the CKP pattern. Turn the ignition switch OFF. Turn the ignition switch ON and wait thirty seconds.

27. Road test the vehicle on a level surface. Decelerate the engine speed of 2500 RPM to 1000 RPM. If equipped with automatic transaxle use two Drive positions. If equipped with manual transaxle use first gear.

28. Stop the vehicle, but keep the engine running.

29. Check PULSAR F/B LEARN in the data list with the HDS. If not complete repeat the procedure. If complete, road test the vehicle on a level surface. Decelerate the engine speed of 5000 RPM to 3000 RPM. If equipped with automatic transaxle use two Drive positions. If equipped with manual transaxle use first gear.

30. Stop the vehicle, but keep the engine running.

31. Check PULSAR F/B LEARN in the data list with the HDS. If not complete repeat the procedure.

32. If completed, turn the ignition switch OFF. Turn the ignition switch ON, wait thirty seconds. The learning procedure is now complete.

Civic Hybrid

See Figures 164 through 166.

1. Disconnect the negative battery cable.

2. Raise and safely support the vehicle.

3. Remove the front wheels.

4. Remove the front undercover and the splash shield.

5. Remove the accessory drive belt.

6. Turn the crankshaft pulley so its top dead center (TDC) mark lines up with the pointer.

7. Remove the water pump pulley.

8. Remove the cylinder head cover.

9. Remove the crankshaft pulley.

10. Remove the oil pan.

11. Support the engine with a suitable jack and a wood block under the engine block.

12. Remove the ground cable, and then remove the side engine mount bracket.

13. Disconnect the crankshaft position (CKP) sensor connector, then remove the dipstick tube mounting bolt and the harness clamps.

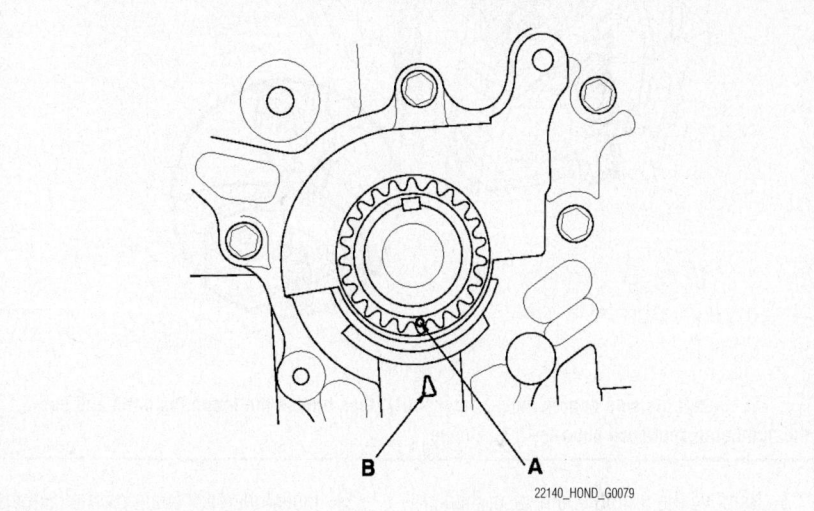

22140_HOND_G0079

Fig. 164 Set the crankshaft to top dead center (TDC). Align the TDC mark (A) on the crankshaft sprocket with the pointer (B) on the oil pump—1.3L Engine

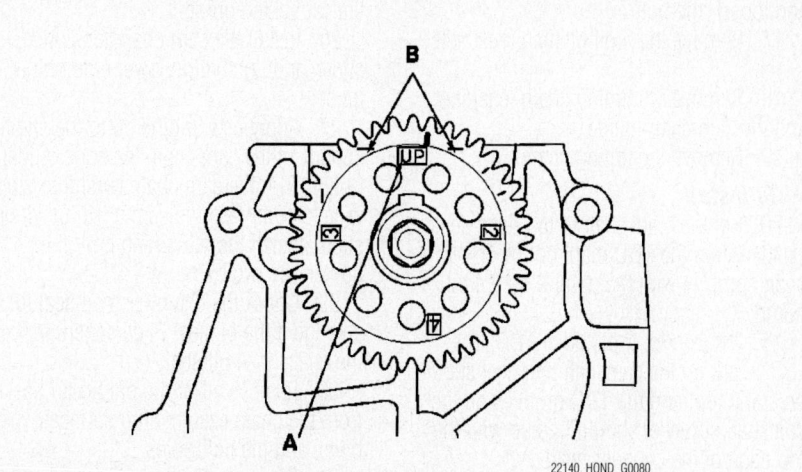

22140_HOND_G0080

Fig. 165 Set the No. 1 piston at TDC. The "UP" mark (A) on the camshaft sprocket should be at the top, and the TDC grooves (B) on the camshaft sprocket should line up with the top edge of the cylinder head—1.3L Engine

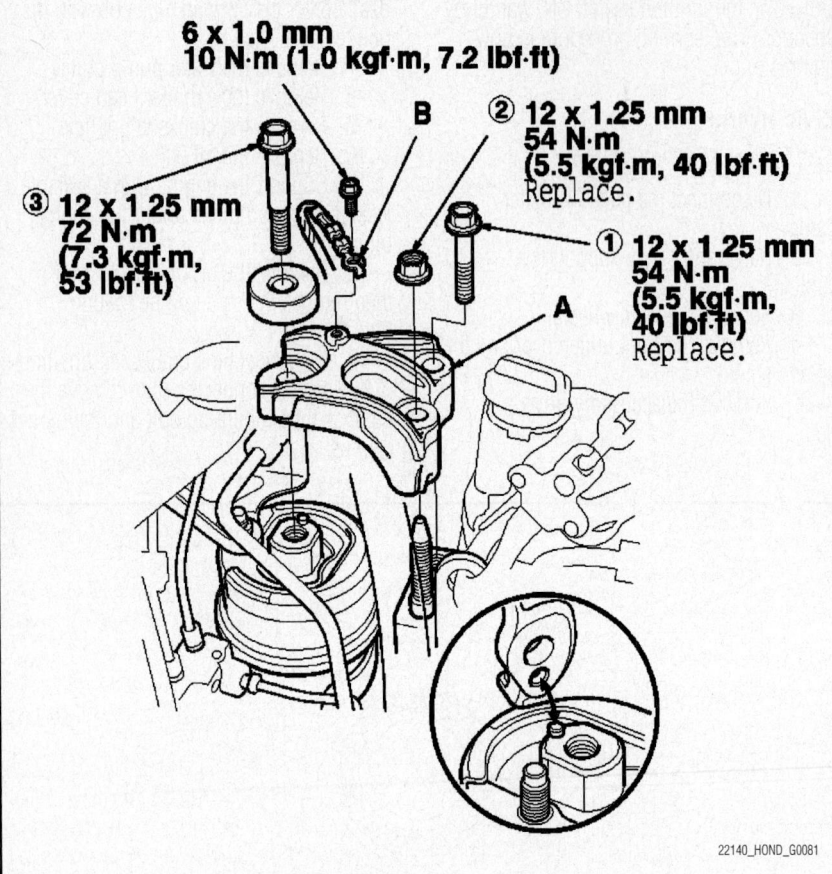

6 x 1.0 mm
10 N·m (1.0 kgf·m, 7.2 lbf·ft)

B

② **12 x 1.25 mm**
54 N·m
(5.5 kgf·m, 40 lbf·ft)
Replace.

③ **12 x 1.25 mm**
72 N·m
(7.3 kgf·m,
53 lbf·ft)

① **12 x 1.25 mm**
54 N·m
(5.5 kgf·m,
40 lbf·ft)
Replace.

A

22140_HOND_G0081

Fig. 166 Install the side engine mount bracket (A), then tighten the mounting bolts and nut in the numbered sequence shown—1.3L Engine

14. Remove the timing chain case, then remove the CKP pulse plate.

15. Apply new engine oil to the sliding surface of the cam chain tensioner slider.

16. Hold the cam chain tensioner slider with a screwdriver, then remove the bolt, and loosen the bolt.

17. Remove the timing chain tensioner slider.

18. Remove the timing chain tensioner and the cam chain guide.

19. Remove the timing chain.

To install:

20. Set the crankshaft to top dead center (TDC). Align the TDC mark on the crankshaft sprocket with the pointer on the oil pump.

21. Set the No. 1 piston at TDC. The "UP" mark on the camshaft sprocket should be at the top, and the TDC grooves on the camshaft sprocket should line up with the top edge of the cylinder head.

22. Install the cam chain on the crankshaft sprocket with the colored piece aligned with the TDC mark on the crankshaft sprocket.

23. Install the cam chain on the camshaft sprocket with the pointer aligned with the center of the two colored pieces.

24. Apply new engine oil to the threads of the cam chain tensioner mounting bolt.

25. Install the cam chain tensioner and the cam chain guide.

26. Install the cam chain tensioner slider, and tighten the lower side bolt loosely.

27. Apply new engine oil to the sliding surface of the cam chain tensioner slider.

28. Turn the cam chain tensioner clockwise to compress the cam chain tensioner slider. Install the remaining bolt, then tighten the two bolts.

29. Check the chain case oil seal for damage If the oil seal is damaged, replace the chain case oil seal..

30. Remove all of the old liquid gasket from the chain case mating surfaces, the bolts, and the bolt holes.

31. Clean and dry the chain case mating surfaces.

32. Apply liquid gasket, P/N 08717-0004, 08718-0001, 08718-0003, or 08718-0009, evenly to the cylinder head and the engine block mating surface of the chain case. Install the component within 5 minutes of applying the liquid gasket.

33. Install the crankshaft position (CKP) pulse plate and the timing chain case. Tighten the mounting bolts to 23 ft. lbs. (31 Nm).

34. Install the harness clamps and the dipstick tube mounting bolt, then connect the CKP sensor connector.

35. Install the side engine mount bracket, then tighten the mounting bolts and nut in the numbered sequence shown.

36. Install the ground cable.

37. Remove the jack and the wood block.

38. Install the oil pan.

39. Install the crankshaft pulley.

40. Install the cylinder head cover.

41. Install the water pump pulley.

42. Install the accessory drive belt.

43. Install the splash shield and the front undercover.

44. Install the front wheels.

45. Connect the negative battery cable.

S2000

See Figures 167 through 169.

➡**Keep the cam chain away from magnetic fields.**

1. Before servicing the vehicle, refer to the precautions in the beginning of this section.

2. Note the radio security code and the radio presets. Relieve the fuel system pressure.

3. Disconnect the negative battery cable. Disconnect the positive battery cable.

4. Drain the engine coolant. Drain the engine oil.

5. Loosen the water pump pulley bolts. Remove the drive belt.

6. Remove the cylinder head.

7. Remove the water bypass hose. Remove the water bypass tube retaining bolts and tube. Remove the water pump pulley. Remove the auto tensioner.

8. Remove the alternator. Remove the idler pulley. Remove the idler pulley base.

9. Remove the oil pan.

10. Remove the crankshaft pulley. Remove the chain case retaining bolts. Remove the chain case.

11. Remove the CKP pulse plate. Remove the oil pump chain guide. Remove the cam chain.

To install:

12. Set the crankshaft to TDC. Align the key on the sprocket and the crankshaft with the pointer on the engine block.

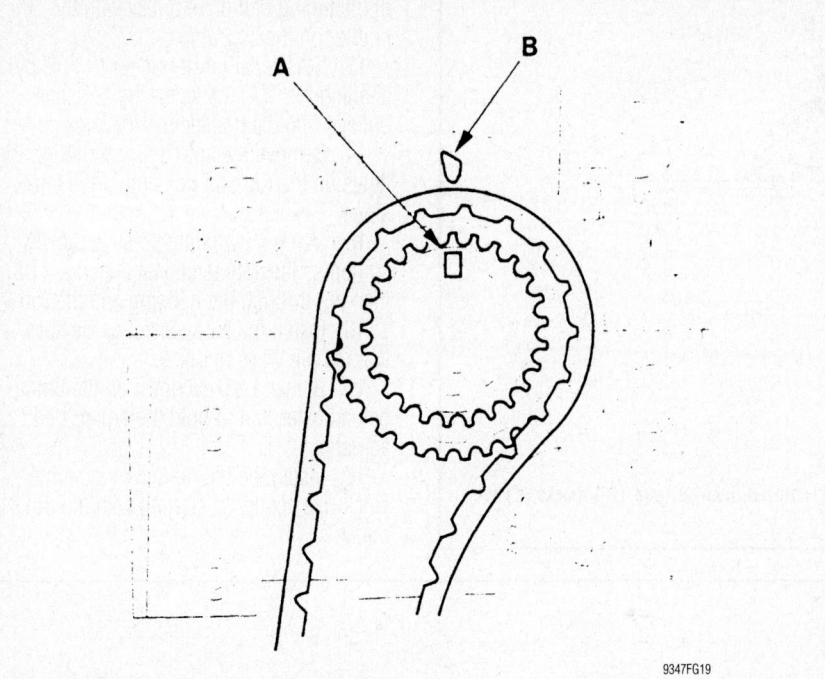

9347FG19

Fig. 167 Align the sprocket key (A) with the cylinder block pointer (B) to set the engine to TDC—S2000

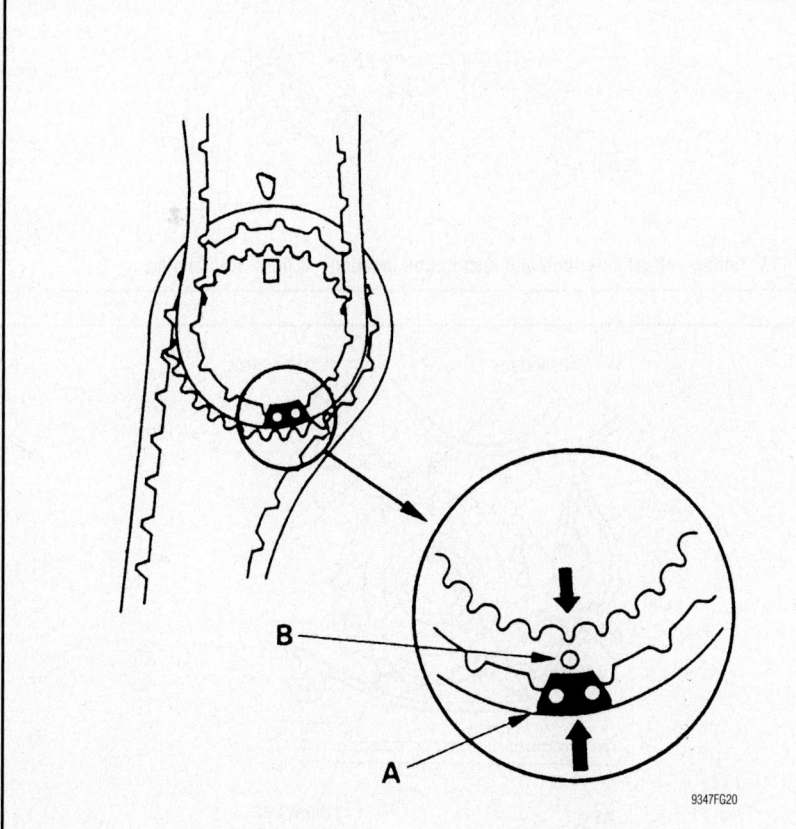

9347FG20

Fig. 168 Install the timing chain with the colored link (A) aligned with the crankshaft sprocket punch mark (B)—S2000

13. Install the cam chain with the colored piece aligned with the punch mark on the crankshaft sprocket.

14. Install the oil pump chain guide. Install the CKP pulse plate.

15. Replace the chain case oil seal.

16. Remove any old gasket from the mating surfaces. Be sure these surfaces are clean and dry.

17. Apply liquid gasket, part number 08717-004, 08718-0001, 08718-0003 or 08718-0009 evenly to the cylinder block mating surface of the chain case.

➡**Do not install the parts if more than four minutes have elapsed since applying the liquid gasket. Instead, reapply after removing the previous coating material.**

18. Install the dowel pins and the chain case using a new O-ring.

19. Continue the installation in the reverse order of the removal procedure.

20. After assembly, wait at least thirty minutes before filling the engine with clean engine oil.

21. Do not run the engine for at least three hours after installing the oil pan.

22. Reprogram the crankshaft position (CKP) pattern. Run the engine until the operating temperature reaches 176 degree. With the engine stopped clear the CKP pattern. Turn the ignition switch OFF. Turn the ignition switch ON and wait thirty seconds.

23. Road test the vehicle on a level surface. Decelerate the engine speed of 2500 RPM to 1000 RPM. If equipped with automatic transaxle use two Drive positions. If equipped with manual transaxle use first gear.

24. Stop the vehicle, but keep the engine running.

25. Check PULSAR F/B LEARN in the data list with the HDS. If not complete repeat the procedure. If complete, road test the vehicle on a level surface. Decelerate the engine speed of 5000 RPM to 3000 RPM. If equipped with automatic transaxle use two Drive positions. If equipped with manual transaxle use first gear.

26. Stop the vehicle, but keep the engine running.

27. Check PULSAR F/B LEARN in the data list with the HDS. If not complete repeat the procedure.

28. If completed, turn the ignition switch OFF. Turn the ignition switch ON, wait thirty seconds. The learning procedure is now complete.

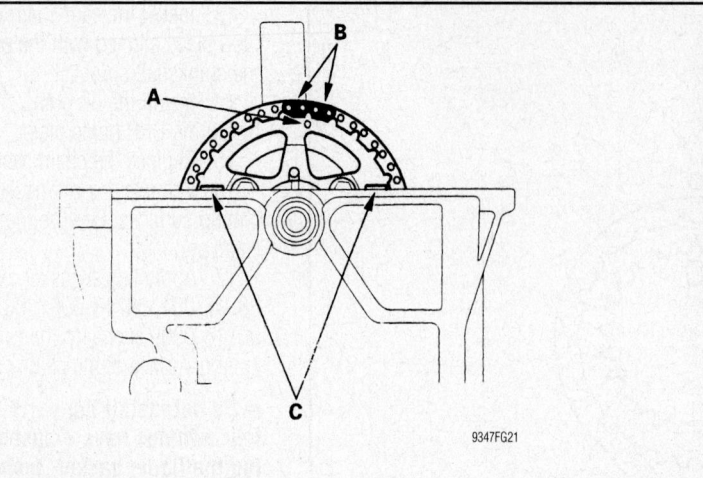

Fig. 169 Timing chain idler sprocket punch mark (A), colored links (B) and TDC marks (C) in proper alignment—S2000

of the timing belt drive pulley with the pointer on the oil pump.

13. Set the camshaft pulleys to TDC by aligning the TDC marks on the camshaft pulleys with the pointers on the back covers.

14. Remove the auto tensioner. Align the holes on the rod and housing of the tensioner.

15. Using a hydraulic press to slowly compress the auto tensioner, insert a 0.08 inch pin through the housing and the rod.

16. Install the auto tensioner. Be sure that the pin stays in place.

17. By hand, screw down on the timing belt adjuster bolt to hold the timing belt adjuster.

18. Install the timing belt in a counter-clockwise sequence starting with the drive pulley.

TIMING BELT, SPROCKETS, FRONT COVER AND SEAL

REMOVAL & INSTALLATION

3.0L and 3.5L Engine

See Figures 170 through 173.

1. Before servicing the vehicle, refer to the precautions in the beginning of this section.

2. Note the radio security code and the radio presets. Disconnect the negative battery cable.

3. Turn the crankshaft so that its white mark lines up with the pointer. Check to insure that number one piston is at TDC. Be sure that the mark on the front camshaft pulley and the pointer on the front upper cover are aligned.

4. Raise and support the vehicle safely. Remove the front tires. Remove the splash shield.

5. Remove the drive belt. Remove the drive belt auto tensioner.

6. Properly support the engine using a block of wood under the oil pan. Remove the ground cable. Remove the side engine mount bracket.

7. Remove the front upper cover. Remove the rear upper cover.

8. Remove the crankshaft pulley. Remove the lower cover.

9. Using the proper size bolt, screw it into the timing belt adjuster. Tighten by hand, do not use a wrench.

10. Remove the engine mount bracket.

11. Remove the idler pulley bolt and idler pulley. Remove the timing belt.

To install:

12. Set the timing belt drive pulley to TDC by aligning the TDC mark on the tooth

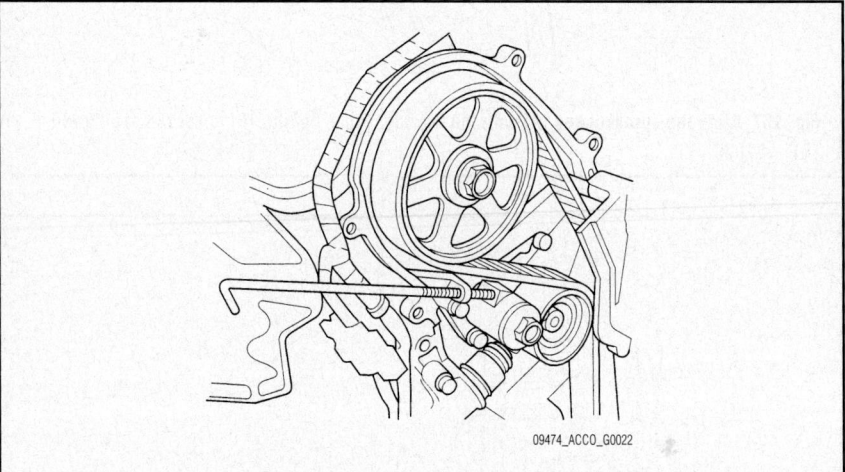

Fig. 170 Timing belt adjuster bolt and installation location—3.0L & 3.5L engine

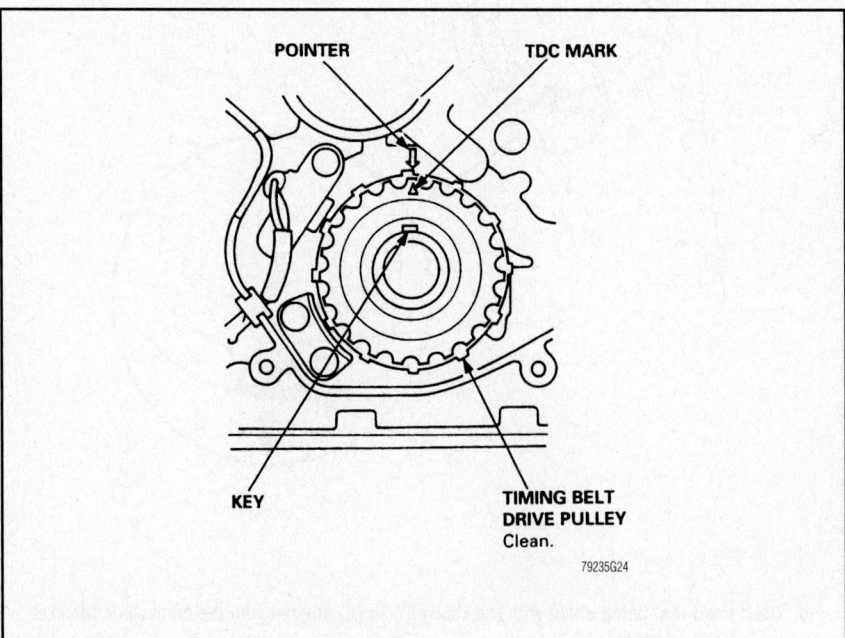

POINTER TDC MARK

KEY TIMING BELT DRIVE PULLEY
Clean.

Fig. 171 Crankshaft timing belt sprocket alignment mark locations—3.0L & 3.5L engine

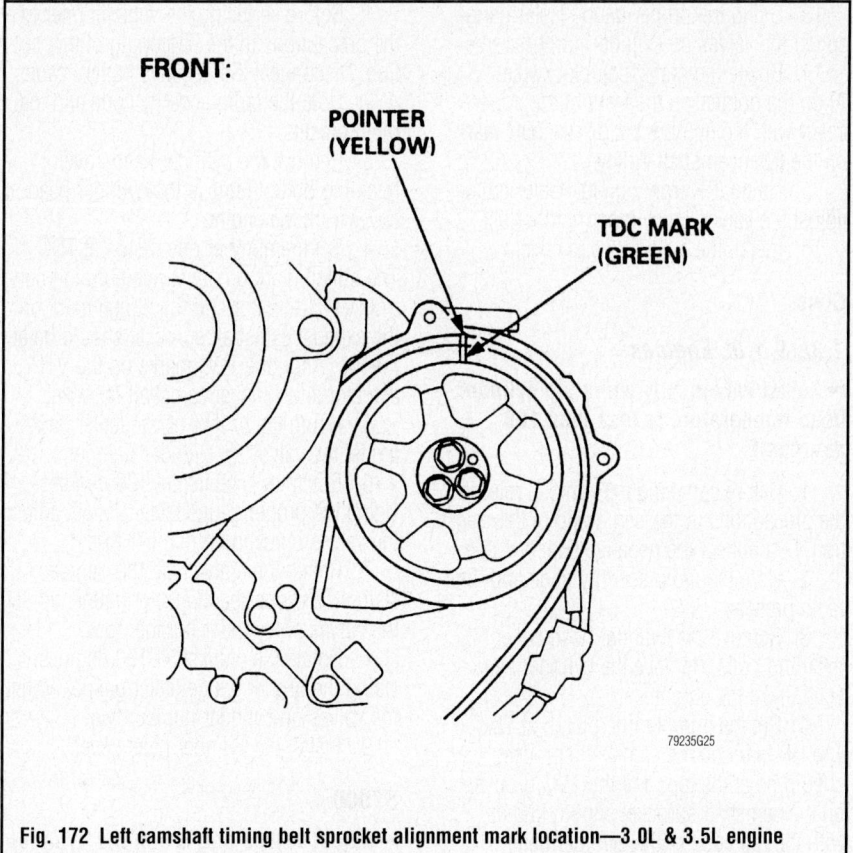

Fig. 172 Left camshaft timing belt sprocket alignment mark location—3.0L & 3.5L engine

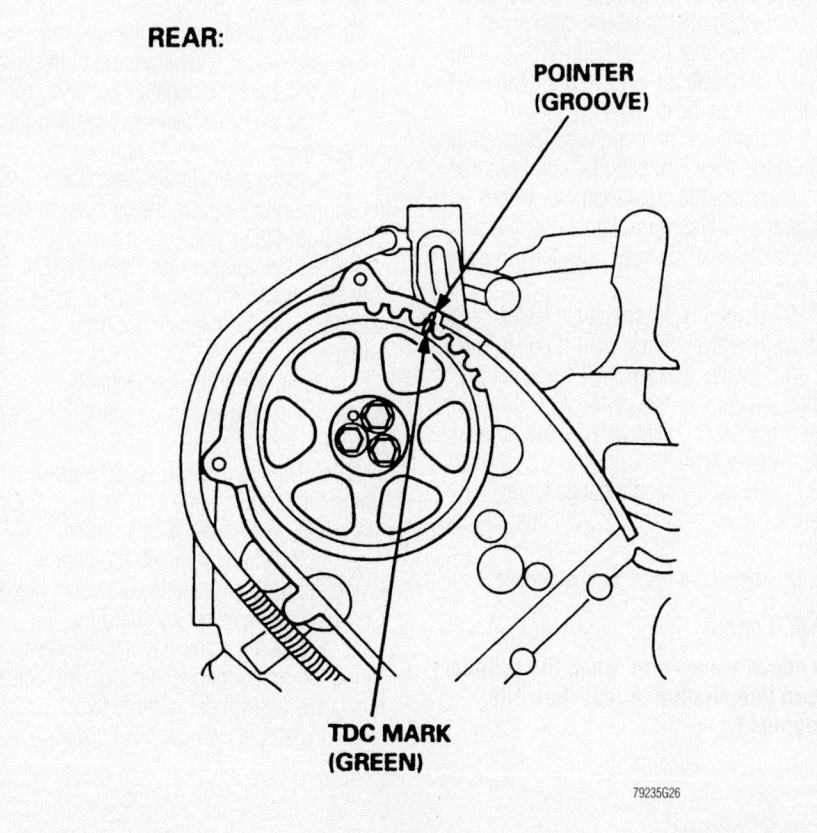

Fig. 173 Rear camshaft timing belt sprocket alignment mark location—3.0L & 3.5L engine

19. Remove the pin from the auto tensioner. Remove the bolt from the timing belt adjuster.

20. Continue the installation in the reverse order of the removal procedure.

21. Reprogram the crankshaft position (CKP) pattern. Run the engine until the operating temperature reaches 176 degree. With the engine stopped clear the CKP pattern. Turn the ignition switch OFF. Turn the ignition switch ON and wait thirty seconds.

22. Road test the vehicle on a level surface. Decelerate the engine speed of 2500 RPM to 1000 RPM. If equipped with automatic transaxle use two Drive positions. If equipped with manual transaxle use first gear.

23. Stop the vehicle, but keep the engine running.

24. Check PULSAR F/B LEARN in the data list with the HDS. If not complete repeat the procedure. If complete, road test the vehicle on a level surface. Decelerate the engine speed of 5000 RPM to 3000 RPM. If equipped with automatic transaxle use two Drive positions. If equipped with manual transaxle use first gear.

25. Stop the vehicle, but keep the engine running.

26. Check PULSAR F/B LEARN in the data list with the HDS. If not complete repeat the procedure.

27. If completed, turn the ignition switch OFF. Turn the ignition switch ON, wait thirty seconds. The learning procedure is now complete.

VALVE LASH

ADJUSTMENT

Accord

2.4L Engine

➡**Adjust valves only when the cylinder head temperature is less than 100 degrees F.**

1. Before servicing the vehicle, refer to the precautions in the beginning of this section. Disconnect the negative battery cable.

2. Note the radio security code and the radio presets.

3. Remove the cylinder head cover retaining bolts. Remove the cylinder head cover from the engine.

4. Set the number one piston at TDC. The punch mark on the variable timing control (VTC) actuator and the punch mark on the exhaust camshaft sprocket should be at the top. Align the TDC marks on the VTC actuator and exhaust camshaft sprocket.

5. Using the proper gauge feeler gauge, adjust the valves on cylinder number one.

6. Rotate the crankshaft 180 degrees. Using the proper gauge feeler gauge, adjust the valves on cylinder number three.

7. Rotate the crankshaft 180 degrees. Using the proper gauge feeler gauge, adjust the valves on cylinder number four.

8. Rotate the crankshaft 180 degrees. Using the proper gauge feeler gauge, adjust the valves on cylinder number two.

9. Install the cylinder head cover.

3.0L & 3.5L Engines

➡**Adjust valves only when the cylinder head temperature is less than 100 degrees F.**

1. Before servicing the vehicle, refer to the precautions in the beginning of this section. Disconnect the negative battery cable.

2. Note the radio security code and the radio presets.

3. Remove the cylinder head cover retaining bolts. Remove the cylinder head cover from the engine.

4. Set the number one piston at TDC. Align the pointer on the front of the upper cover with the number one piston TDC mark on the front camshaft pulley.

5. Using the proper gauge feeler gauge, adjust the valves on cylinder number one.

6. Rotate the crankshaft clockwise. Align the pointer on the front of the upper cover with the number four piston TDC mark on the front camshaft pulley.

7. Using the proper gauge feeler gauge, adjust the valves on cylinder number four.

8. Rotate the crankshaft clockwise. Align the pointer on the front of the upper cover with the number two piston TDC mark on the front camshaft pulley.

9. Using the proper gauge feeler gauge, adjust the valves on cylinder number two.

10. Rotate the crankshaft clockwise. Align the pointer on the front of the upper cover with the number five piston TDC mark on the front camshaft pulley.

11. Using the proper gauge feeler gauge, adjust the valves on cylinder number five.

12. Rotate the crankshaft clockwise. Align the pointer on the front of the upper cover with the number three piston TDC mark on the front camshaft pulley.

13. Using the proper gauge feeler gauge, adjust the valves on cylinder number three.

14. Rotate the crankshaft clockwise. Align the pointer on the front of the upper cover with the number six piston TDC mark on the front camshaft pulley.

15. Using the proper gauge feeler gauge, adjust the valves on cylinder number six.

16. Install the cylinder head cover.

Civic

1.3L & 1.8L Engines

➡**Adjust valves only when the cylinder head temperature is less than 100 degrees F.**

1. Before servicing the vehicle, refer to the precautions in the beginning of this section. Disconnect the negative battery cable.

2. Note the radio security code and the radio presets.

3. Remove the cylinder head cover retaining bolts. Remove the cylinder head cover from the engine.

4. Set the number one piston at TDC. The UP mark on the camshaft sprocket should be at the top, and the TDC grooves on the camshaft sprocket should line up with the top edge of the cylinder head.

5. Using the proper gauge feeler gauge, adjust the valves on cylinder number one.

6. Rotate the crankshaft clockwise. Align the number three piston TDC groove on the camshaft sprocket with the top edge of the cylinder head.

7. Using the proper gauge feeler gauge, adjust the valves on cylinder number three.

8. Rotate the crankshaft clockwise. Align the number four piston TDC groove on the camshaft sprocket with the top edge of the cylinder head.

9. Using the proper gauge feeler gauge, adjust the valves on cylinder number four.

10. Rotate the crankshaft clockwise. Align the number two piston TDC groove on the camshaft sprocket with the top edge of the cylinder head.

11. Using the proper gauge feeler gauge, adjust the valves on cylinder number two.

12. Install the cylinder head cover.

2.0L Engine

➡**Adjust valves only when the cylinder head temperature is less than 100 degrees F.**

1. Before servicing the vehicle, refer to the precautions in the beginning of this section. Disconnect the negative battery cable.

2. Note the radio security code and the radio presets.

3. Remove the cylinder head cover retaining bolts. Remove the cylinder head cover from the engine.

4. Set the number one piston at TDC. The punch mark on the variable timing control (VTC) actuator and the punch mark on the exhaust camshaft sprocket should be at the top. Align the TDC marks on the VTC actuator and exhaust camshaft sprocket.

5. Using the proper gauge feeler gauge, adjust the valves on cylinder number one.

6. Rotate the crankshaft 180 degrees. Using the proper gauge feeler gauge, adjust the valves on cylinder number three.

7. Rotate the crankshaft 180 degrees. Using the proper gauge feeler gauge, adjust the valves on cylinder number four.

8. Rotate the crankshaft 180 degrees. Using the proper gauge feeler gauge, adjust the valves on cylinder number two.

9. Install the cylinder head cover.

S2000

➡**Adjust valves only when the cylinder head temperature is less than 100 degrees F.**

1. Before servicing the vehicle, refer to the precautions in the beginning of this section. Disconnect the negative battery cable.

2. Note the radio security code and the radio presets.

3. Remove the cylinder head cover retaining bolts. Remove the cylinder head cover from the engine.

4. Set the number one piston at TDC. The TDC marks on the cam chain sprocket should align with the cylinder head surface.

5. Using the proper gauge feeler gauge, adjust the valves on cylinder number one.

6. Rotate the crankshaft 180 degrees. Using the proper gauge feeler gauge, adjust the valves on cylinder number three.

7. Rotate the crankshaft 180 degrees. Using the proper gauge feeler gauge, adjust the valves on cylinder number four.

8. Rotate the crankshaft 180 degrees. Using the proper gauge feeler gauge, adjust the valves on cylinder number two.

9. Install the cylinder head cover.

ENGINE PERFORMANCE & EMISSION CONTROL

CAMSHAFT POSITION (CMP) SENSOR

LOCATION

The CMP sensor is located in the cylinder head.

REMOVAL & INSTALLATION

Accord

2.4L Engine—Sensor A

1. Remove the air cleaner.
2. Remove the EGR valve.
3. Disconnect the CMP sensor A connector.
4. Remove the bolt.
5. Remove the CMP sensor A from the intake camshaft side of the cylinder head.
6. Install the parts in the reverse order of removal with a new O-ring.

2.4L Engine—Sensor B

1. Remove the air cleaner.
2. Remove the EVAP canister purge valve.
3. Disconnect the Camshaft Position (CMP) sensor B connector.
4. Remove the CMP sensor B from the exhaust camshaft side of the cylinder head.
5. Install the parts in the reverse order of removal with a new O-ring.

3.0L & 3.5L Engines

1. Set the No 1 piston at top dead center.
2. Remove the upper covers from the engine.
3. To hold the timing belt adjuster in its current position, thread in the battery clamp bolt hand-tight.
4. Loosen the idler pulley bolt about five or six turns, then remove the timing belt from the front camshaft pulley.
5. Remove the front camshaft pulley.
6. Disconnect the Camshaft Position (CMP) sensor connector, then remove the back cover
7. Remove the CMP sensor from the back cover.
8. Installation is the reverse order of removal.

Civic

1.8L Engine

1. Remove the cowl cover and the under-cowl panel.
2. Disconnect the Camshaft Position (CMP) sensor 3P connector.

3. Remove the CMP sensor.
4. Install the parts in the reverse order of removal with a new O-ring.

2.0L Engine—Sensor A

1. Remove the air intake assembly.
2. Disconnect the Camshaft Position (CMP) sensor A 3P connector.
3. Remove the CMP sensor A from the intake camshaft side of the cylinder head.
4. Install the parts in the reverse order of removal with a new O-ring.

2.0L Engine—Sensor B

1. Remove the air intake assembly.
2. Disconnect the Camshaft Position (CMP) sensor B 3P connector.
3. Remove the CMP sensor B.
4. Install the parts in the reverse order of removal with a new O-ring.

Civic Hybrid

1. Remove the air intake assembly.
2. Disconnect the Camshaft Position (CMP) sensor 3P connector.
3. Remove the CMP sensor.
4. Install the parts in the reverse order of removal with a new O-ring.

S2000

1. Disconnect the Camshaft Position (CMP) sensor 3P connector.
2. Remove the CMP sensor from the exhaust camshaft side of the cylinder head.
3. Install the parts in the reverse order of removal with a new O-ring

CRANKSHAFT POSITION (CKP) SENSOR

LOCATION

The CKP sensor is located in the engine block.

REMOVAL & INSTALLATION

Accord

2.4L Engine

1. Disconnect the Crankshaft Position (CKP) sensor connector.
2. Remove the CKP sensor.
3. Install the parts in the reverse order of removal with a new O-ring.

3.0L Engine

1. Move the auto-tensioner to remove tension from the accessory drive belt, then remove the belt.

2. Remove the crankshaft pulley.
3. Remove the upper and lower timing belt covers from the engine.
4. Remove the Crankshaft Position (CKP) sensor A/B from the oil pump.
5. Install the parts in the reverse order of removal.

3.5L Engine

1. Raise and safely support the vehicle.

✻✻ WARNING

Make the vehicle the horizontal because engine oil comes out when the sensor removed.

2. Remove the Crankshaft Position (CKP) sensor cover.
3. Disconnect the CKP sensor connector.
4. Remove the CKP sensor.
5. Install the parts in the reverse order of removal with a new O-ring.

Civic

1.8L Engine

1. Remove the engine splash shield.
2. Disconnect the Crankshaft Position (CKP) sensor 3P connector.
3. Remove the CKP sensor.
4. Install the parts in the reverse order of removal with a new O-ring.

2.0L Engine

1. Disconnect the Crankshaft Position (CKP) sensor 3P connector.
2. Remove the CKP sensor.
3. Install the parts in the reverse order of removal with a new O-ring

Civic Hybrid

1. Disconnect the Crankshaft Position (CKP) 3P connector.
2. Remove the CKP sensor.
3. Install the parts in the reverse order of removal with a new O-ring.

S2000

1. Remove the Crankshaft Position (CKP) sensor.
2. Disconnect the CKP sensor 3P connector.
3. Install the parts in the reverse order of removal with a new O-ring

ENGINE COOLANT TEMPERATURE (ECT) SENSOR

LOCATION

Except S2000

There are two ECT sensor, one located in the upper engine block and one underneath the engine in the bottom part of the engine block.

S2000

The ECT sensor is located in the radiator assembly.

REMOVAL & INSTALLATION

Accord

2.4L Engine—Sensor 1

See Figure 174.

1. Remove the air intake assembly.
2. Remove the EVAP canister purge valve.
3. Unbolt the under-hood fuse/relay box, and move the box aside.
4. Disconnect the ECT sensor/ECT sensor 1 connector.
5. Remove the ECT sensor/ECT sensor 1.
6. Install the parts in the reverse order of removal with a new O-ring.
7. Refill the radiator with engine coolant.

2.4L Engine—Sensor 2

See Figure 175.

1. Drain the engine coolant.
2. Remove the splash shield.
3. Disconnect the ECT sensor 2 connector (A), then remove ECT sensor 2.
4. Install the ECT sensor 2 with a new O-ring.
5. Install the splash shield.
6. Refill the radiator with engine coolant.

3.0L Engine—Sensor 1

See Figure 176.

1. Remove the air intake assembly.
2. Disconnect the ECT sensor/ECT sensor 1 connector.
3. Remove the ECT sensor/ECT sensor 1.
4. Install the parts in the reverse order of removal with a new O-ring.
5. Refill the radiator with engine coolant.

3.0L Engine—Sensor 2

See Figure 177.

1. Remove the throttle body.
2. Disconnect the ECT sensor 2 connector.

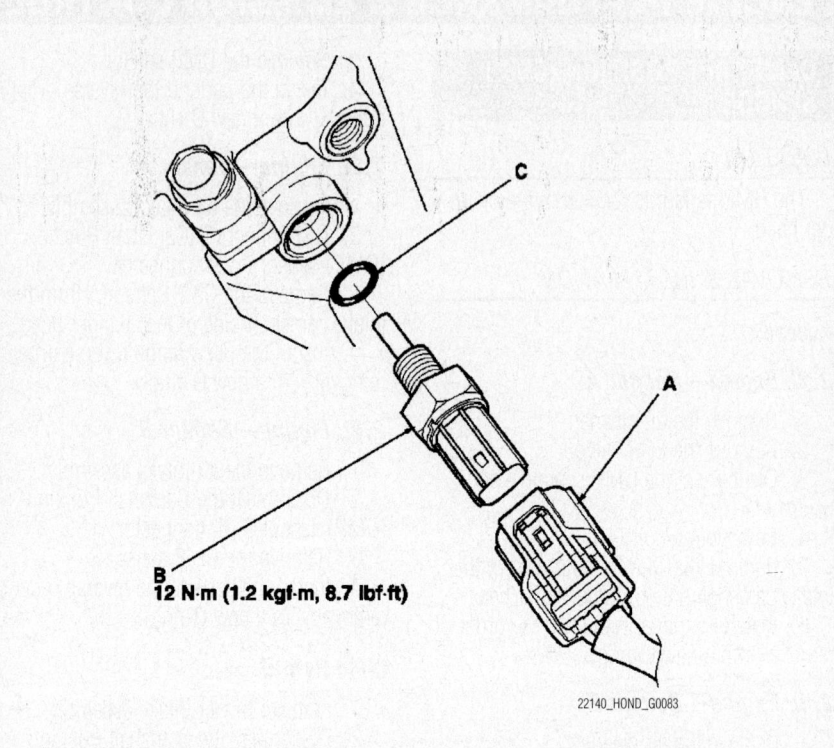

Fig. 174 Connect the electrical connector (A) and use a new O-ring (C) to install ECT sensor 1 (B)—2.4L Engine

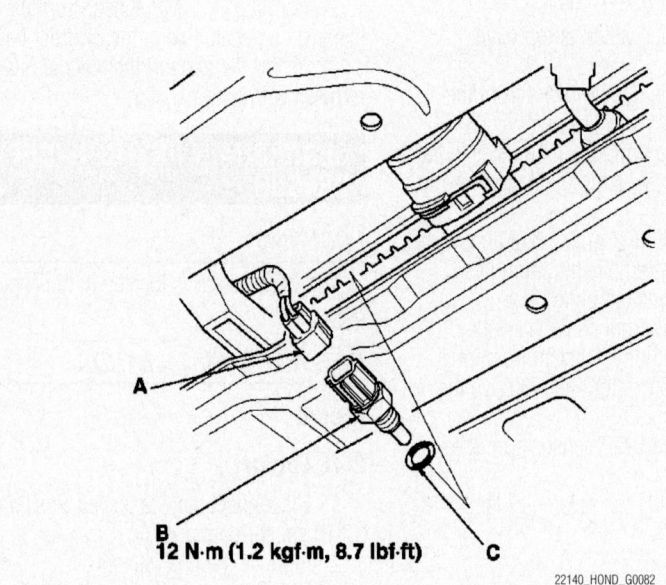

Fig. 175 Connect the electrical connector (A) and use a new O-ring (C) to install ECT sensor 2 (B)—2.4L Engine

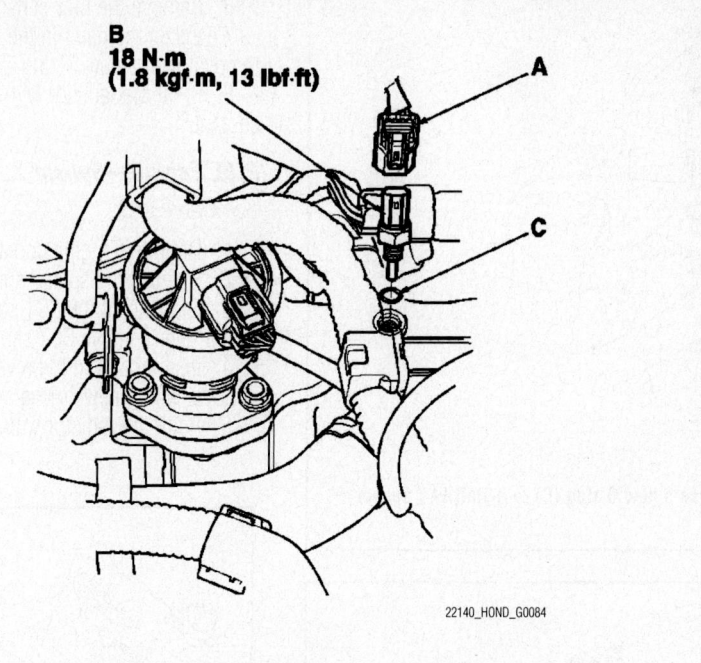

Fig. 176 Connect the electrical connector (A) and use a new O-ring (C) to install ECT sensor 1 (B)—3.0L Engine

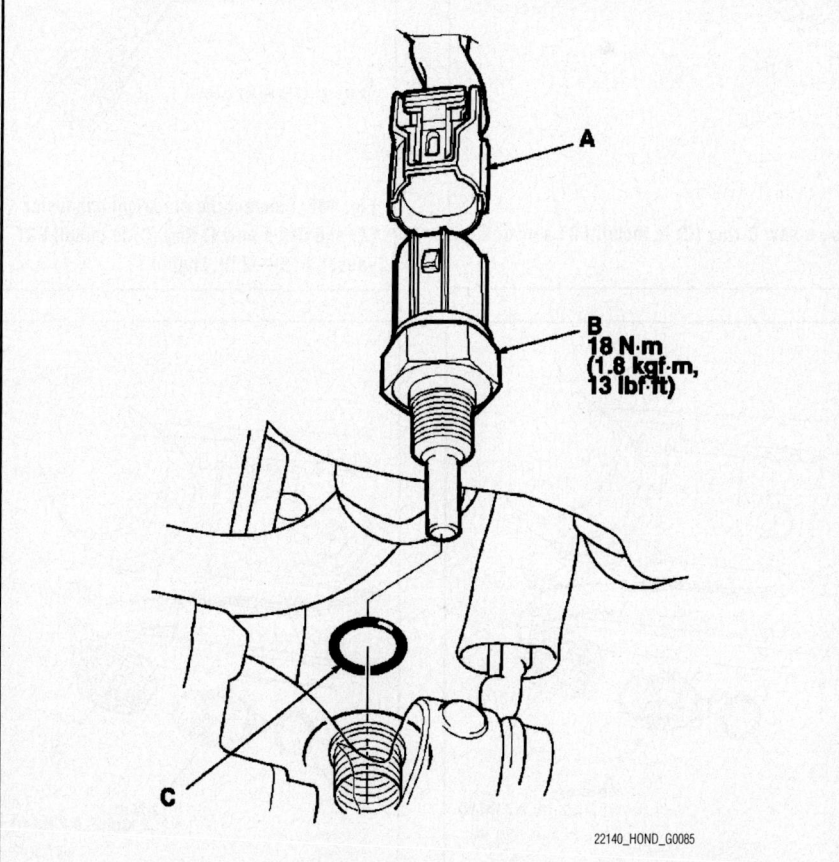

Fig. 177 Connect the electrical connector (A) and use a new O-ring (C) to install ECT sensor 2 (B)—3.0L Engine

3. Remove the ECT sensor 2.
4. Install the parts in the reverse order of removal with a new O-ring.
5. Refill the radiator with engine coolant.

3.5L Engine—Sensor 1

See Figure 178.

1. Drain the engine coolant.
2. Remove the engine appearance cover.
3. Disconnect the ECT sensor 1 2P connector.
4. Remove the ECT sensor 1.
5. Install the parts in the reverse order of removal with a new O-ring.
6. Refill the radiator with engine coolant.

3.5L Engine—Sensor 2

See Figure 179.

1. Remove the engine splash shield.
2. Drain the engine coolant.
3. Disconnect the ECT sensor 2 2P connector.
4. Remove the ECT sensor 2.
5. Install the parts in the reverse order of removal with a new O-ring.
6. Refill the radiator with engine coolant.

Civic

1.8L Engine—Sensor 1

See Figure 180.

1. Drain the engine coolant.
2. Disconnect the ECT sensor 1 2P connector.
3. Remove the ECT sensor 1.
4. Install the parts in the reverse order of removal with a new O-ring
5. Refill the radiator with engine coolant.

1.8L Engine—Sensor 2

See Figure 181.

1. Drain the engine coolant.
2. Remove the engine splash shield.
3. Disconnect the ECT sensor 2 2P connector.
4. Remove the ECT sensor 2.
5. Install the parts in the reverse order of removal with a new O-ring.
6. Refill the radiator with engine coolant.

2.0L Engine—Sensor 1

See Figure 182.

1. Drain the engine coolant.
2. Remove the air intake assembly.
3. Disconnect the CMP sensor A 3P connector, and the CMP sensor B 3P connector.
4. Remove the harness cover.
5. Disconnect the ECT sensor 1 2P connector.

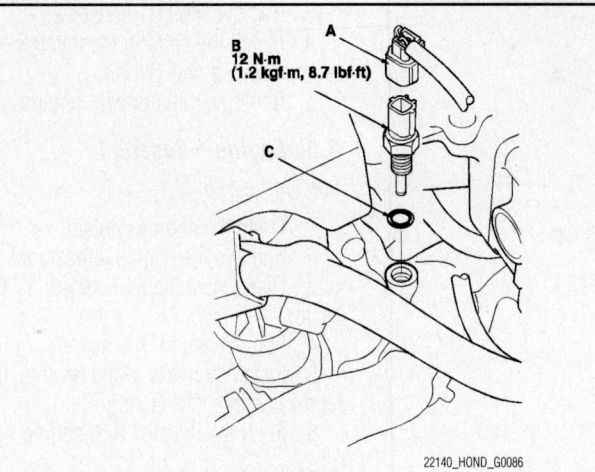

Fig. 178 Connect the electrical connector (A) and use a new O-ring (C) to install ECT sensor 1 (B)—3.5L Engine

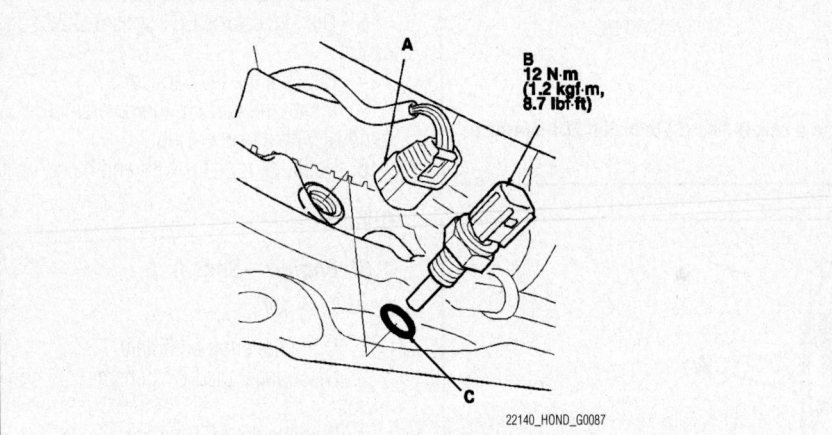

Fig. 179 Connect the electrical connector (A) and use a new O-ring (C) to install ECT sensor 2 (B)—3.5L Engine

6. Remove the ECT sensor 1.

7. Install the parts in the reverse order of removal with a new O-ring.

8. Refill the radiator with engine coolant.

2.0L Engine—Sensor 2

See Figure 183.

1. Drain the engine coolant.

2. Remove the splash shield.

3. Disconnect the ECT sensor 2 2P connector, then remove ECT sensor 2.

4. Install parts in the reverse order of removal with a new O-ring.

5. Refill the radiator with engine coolant.

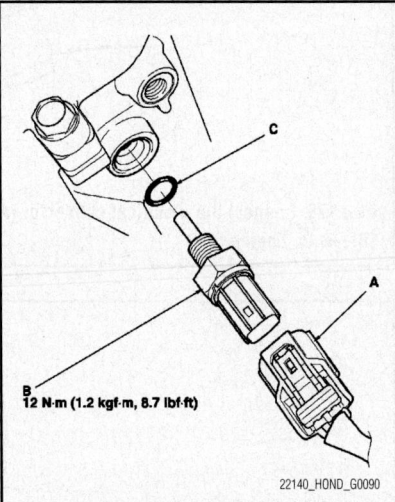

Fig. 182 Connect the electrical connector (A) and use a new O-ring (C) to install ECT sensor 1 (B)—2.0L Engine

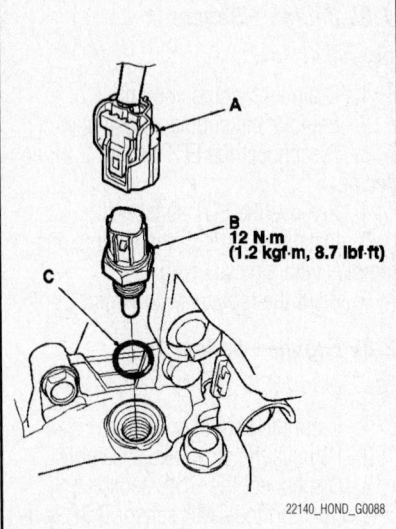

Fig. 180 Connect the electrical connector (A) and use a new O-ring (C) to install ECT sensor 1 (B)—1.8L Engine

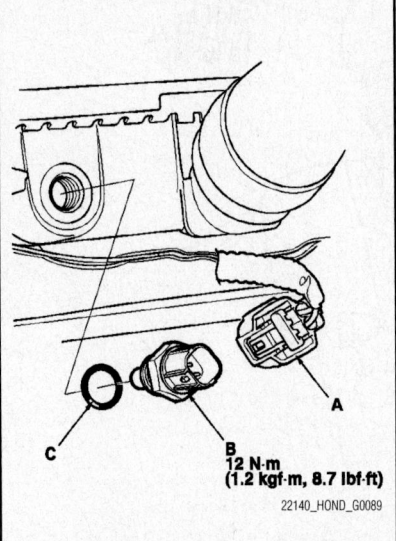

Fig. 181 Connect the electrical connector (A) and use a new O-ring (C) to install ECT sensor 2 (B)—1.8L Engine

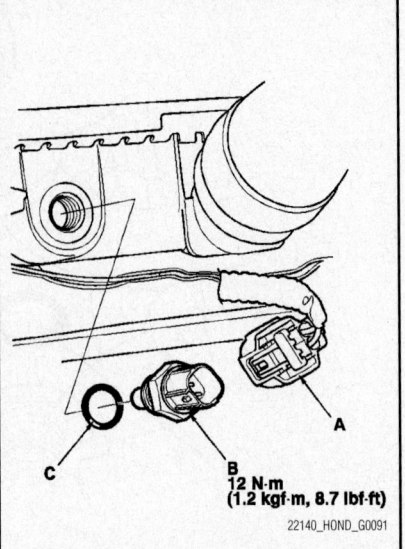

Fig. 183 Connect the electrical connector (A) and use a new O-ring (C) to install ECT sensor 2 (B)—2.0L Engine

Civic Hybrid

Sensor 1

See Figure 184.

1. Drain the engine coolant.
2. Remove the air intake assembly.
3. Disconnect the ECT sensor 1 2P connector.
4. Remove the ECT sensor 1.
5. Install the parts in the reverse order of removal with a new O-ring.
6. Refill the radiator with engine coolant.

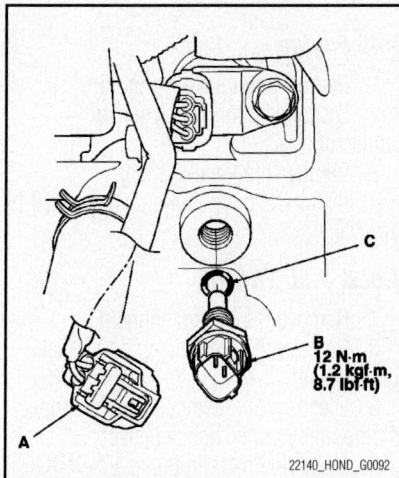

Fig. 184 Connect the electrical connector (A) and use a new O-ring (C) to install ECT sensor 1 (B)—1.3L Engine

Sensor 2

See Figure 185.

1. Drain the engine coolant.
2. Remove the splash shield.

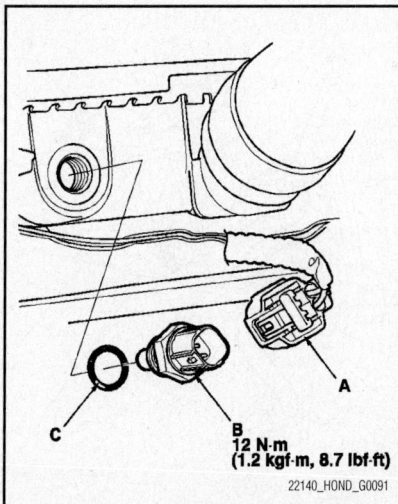

Fig. 185 Connect the electrical connector (A) and use a new O-ring (C) to install ECT sensor 2 (B)—1.3L Engine

3. Disconnect the ECT sensor 2 2P connector.
4. Remove the ECT sensor 2.
5. Install the parts in the reverse order of removal with a new O-ring.
6. Refill the radiator with engine coolant.
7. Install the splash shield.

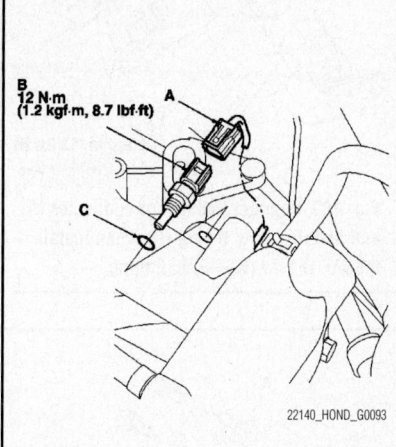

Fig. 186 Connect the electrical connector (A) and use a new O-ring (C) to install ECT sensor (B)—2.2L Engine

S2000

See Figure 186.

1. Drain the engine coolant.
2. Disconnect the ECT sensor 1 2P connector.
3. Remove the ECT sensor 1.
4. Install the parts in the reverse order of removal with a new O-ring.
5. Refill the radiator with engine coolant.

HEATED OXYGEN (HO2S) SENSOR

LOCATION

The HO2S sensors are located in the exhaust pipe assembly.

REMOVAL & INSTALLATION

Accord

2.4L Engine

1. Pull back the carpet from the floor rail under the front edge of the front passenger seat to expose the secondary HO2S 4P connector.
2. Disconnect the secondary HO2S 4P connector
3. Remove the secondary HO2S, and the covers.

4. Install the parts in the reverse order of removal.

3.0L & 3.5L Engines—Front Bank

1. Disconnect the front secondary HO2S (Bank 2) 4P connector, then remove the front secondary HO2S.
2. Install the parts in the reverse order of removal.

3.0L & 3.5L Engines—Rear Bank

1. Disconnect the rear secondary HO2S (Bank 1) 4P connector, then remove the rear secondary HO2S.
2. Install the parts in the reverse order of removal.

Civic

1. Disconnect the secondary HO2S 4P connector, then remove the secondary HO2S.
2. Install the parts in the reverse order of removal.

Civic Hybrid

1. Disconnect the secondary HO2S 4P connector, then remove the secondary HO2S.
2. Install the parts in the reverse order of removal.

S2000

See Figure 187.

1. Disconnect the secondary HO2S (Sensor 2) 4P connector.
2. Remove the secondary HO2S (Sensor 2).
3. Install the parts in the reverse order of removal.

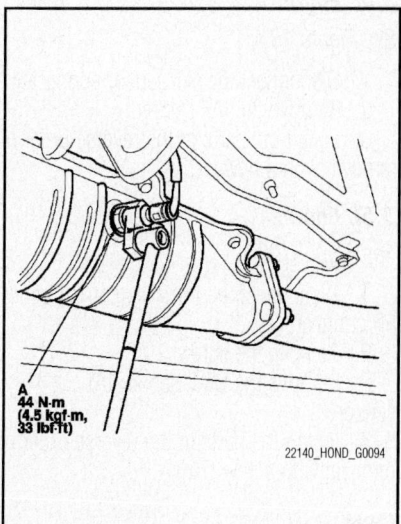

Fig. 187 Installing the HO2S sensor— S2000

INTAKE AIR TEMPERATURE (IAT) SENSOR

LOCATION

The IAT sensor is located in the throttle body assembly.

REMOVAL & INSTALLATION

Accord

2.4L Engine

See Figure 188.

1. Disconnect the MAF sensor/IAT sensor connector.
2. Remove the bolts.
3. Remove the MAF sensor/IAT sensor.
4. Install the parts in the reverse order of removal.

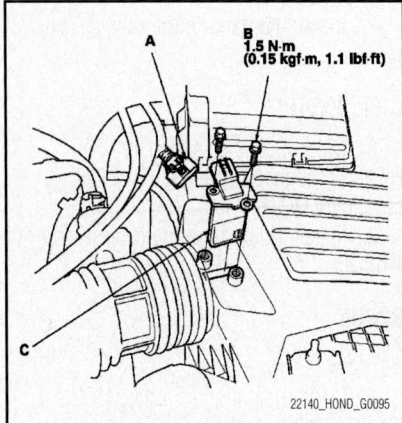

Fig. 188 Disconnect the sensor connector (A) and remove the bolts (B) to remove the IAT/MAF sensor—2.4L Engine

3.0L Engine

See Figure 189.

1. Disconnect the IAT sensor connector.
2. Remove the IAT sensor.
3. Install the parts in the reverse order of removal with a new O-ring

3.5L Engine

See Figure 190.

1. Disconnect the MAF sensor/IAT senor 5P connector.
2. Remove the screw.
3. Remove the MAF sensor/IAT sensor.
4. Install the parts in the reverse order of removal with a new O-ring.

Civic

1. Disconnect the MAF sensor/IAT sensor 5P connector.

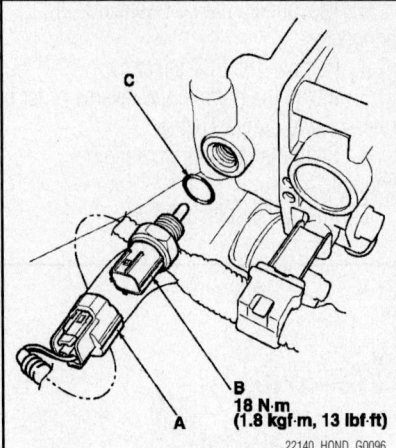

Fig. 189 Connect the sensor connector (A) and install a new O-ring (C) when install the IAT sensor (B)—3.0L Engine

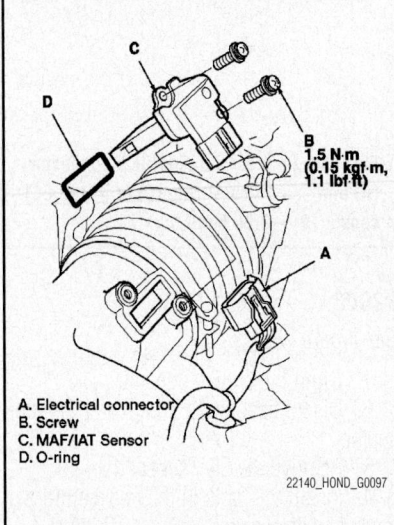

A. Electrical connector
B. Screw
C. MAF/IAT Sensor
D. O-ring

Fig. 190 IAT/MAF sensor mounting—3.5L Engine

2. Remove the bolts.
3. Remove the MAF sensor/IAT sensor.
4. Install the parts in the reverse order of removal with a new O-ring.

Civic Hybrid

1. Disconnect the MAF sensor/IAT sensor 5P connector.
2. Remove the bolts.
3. Remove the MAF sensor/IAT sensor.
4. Install the parts in the reverse order of removal with a new O-ring.

S2000

See Figure 191.

1. Disconnect the IAT sensor 2P connector.

2. Remove the clip (B) and the IAT sensor.
3. Install the parts in the reverse order of removal.

KNOCK SENSOR (KS)

LOCATION

The knock sensor is located in the top of the engine block.

REMOVAL & INSTALLATION

Accord

2.4L Engine

1. Remove the intake manifold.
2. Disconnect the knock sensor connector.
3. Remove the knock sensor.
4. Install the parts in the reverse order of removal.

3.0L & 3.5L ENGINE

1. Remove the intake manifold.
2. Remove the fuel rails and the intake runner base.
3. Disconnect the knock sensor connector, then remove the knock sensor.
4. Install the parts in the reverse order of removal.

Civic, Civic Hybrid & S2000

1. Remove the intake manifold.
2. Disconnect the knock sensor 1P connector.
3. Remove the knock sensor.
4. Install the parts in the reverse order of removal.

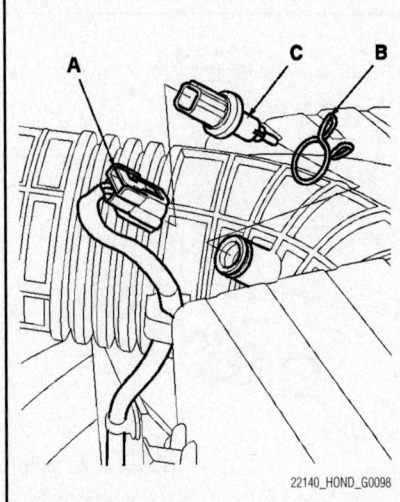

Fig. 191 Remove the electrical connector (A) and clip (B) to remove the IAT sensor (C)—S2000

MANIFOLD ABSOLUTE PRESSURE (MAP) SENSOR

LOCATION

The MAP sensor is located on the air intake assembly.

REMOVAL & INSTALLATION

Accord & S2000

1. Disconnect the MAP sensor connector.
2. Remove the screw.
3. Remove the MAP sensor.
4. Install the parts in the reverse order of removal with a new O-ring.

Civic

1.8L Engine

1. Remove the cowl cover and the under-cowl panel.
2. Disconnect the MAP sensor 3P connector.
3. Remove the MAP sensor.
4. Install the parts in the reverse order of removal with a new O-ring.

2.0L Engine

1. Disconnect the MAP sensor 3P connector.
2. Remove the MAP sensor.
3. Install the parts in the reverse order of removal with a new O-ring.

Civic Hybrid

1. Disconnect the MAP sensor 3P connector.
2. Remove the MAP sensor.
3. Install the parts in the reverse order of removal with a new O-ring

THROTTLE POSITION SENSOR (TPS)

LOCATION

The TPS is located under the dashboard as part of the throttle actuator module.

REMOVAL & INSTALLATION

Accord

1. Remove the right kick panel.
2. Disconnect the throttle actuator control module 16P connector.
3. Remove the bolts and the throttle actuator control module.
4. Install the parts in the reverse order of removal.

S2000

1. Remove the passenger's dashboard lower cover.
2. Push the tab, and disconnect the throttle actuator control module 16P connector.
3. Remove the nuts and the throttle actuator control module.
4. Install the parts in the reverse order of removal.

VEHICLE SPEED SENSOR (VSS)

LOCATION

The vehicle speed sensor is located on the rear of the transaxle assembly.

REMOVAL & INSTALLATION

Accord

2.4L Engine

1. Disconnect the output shaft (countershaft) speed sensor connector, and remove the output shaft (countershaft) speed sensor.
2. Install a new O-ring on the new output shaft (countershaft) speed sensor, then install the output shaft (countershaft) speed sensor in the transmission housing.
3. Check the connector for rust, dirt, or oil, and clean if necessary, then connect the connector securely.

3.0L Engine

1. Disconnect the negative cable from the battery, then disconnect the positive cable from the battery.
2. Unbolt the under-hood fuse/relay box.
3. Reposition the under-hood fuse/relay box to gain access to the output shaft (countershaft) speed sensor.
4. Remove the battery hold-down bracket, then remove the battery cover, battery, and battery tray.
5. Loosen the two bolts securing the battery base from under the vehicle, and remove the two bolts securing the battery base in the engine compartment, then remove the battery base.
6. Disconnect output shaft (countershaft) speed sensor connector.
7. Remove the bolt securing the output shaft (countershaft) speed sensor from transmission housing.
8. Install the new O-ring on the output shaft (countershaft) speed sensor, then install the output shaft (countershaft) speed sensor in the transmission housing. Do not allow dust or foreign particles to enter the transmission.

9. Check for rust, dirt, or oil, then connect the connector securely.
10. Install the battery base and battery, then connect the battery terminals.

3.5L Engine

1. Disconnect the battery cables and remove the battery.
2. Remove the 6 x 1.0 mm bolt securing the resonator bracket under the battery base bracket, and remove the battery base and the battery base bracket in the engine compartment.
3. Disconnect the input shaft (mainshaft) speed sensor connector, and remove the input shaft (mainshaft) speed sensor.
4. Install a new O-ring on a new input shaft (mainshaft) speed sensor, then install the input shaft (mainshaft) speed sensor.
5. Check the connector for rust, dirt, or oil, then connect the connector securely.
6. Install the battery base bracket and the battery base, and secure the resonator bracket under the battery base bracket with the bolt.
7. Install the battery and connect the battery cables.

Civic

1. Remove the air intake assembly.
2. Disconnect the output shaft (countershaft) speed sensor connector, and remove the output shaft (countershaft) speed sensor.
3. Install a new O-ring on a new output shaft (countershaft) speed sensor, then install the output shaft (countershaft) speed sensor in the transmission housing.
4. Check the connector for rust, dirt, or oil, then connect the connector securely.
5. Install the air intake assembly.

Civic Hybrid

1. Remove the air intake assembly.
2. Disconnect the CVT speed sensor connector, then remove the CVT speed sensor.
3. Install a new O-ring on a new CVT speed sensor, then install the CVT speed sensor in the transmission housing.
4. Check the connector for rust, dirt, or oil, then connect the connector securely.
5. Install the air cleaner assembly and the intake air duct.

S2000

1. Disconnect the output shaft (countershaft) speed sensor 3P connector.
2. Remove the output shaft (countershaft) speed sensor.
3. Install the parts in the reverse order of removal with a new O-ring.

FUEL SYSTEM SERVICE PRECAUTIONS

Safety is the most important factor when performing not only fuel system maintenance but any type of maintenance. Failure to conduct maintenance and repairs in a safe manner may result in serious personal injury or death. Maintenance and testing of the vehicle's fuel system components can be accomplished safely and effectively by adhering to the following rules and guidelines.

• To avoid the possibility of fire and personal injury, always disconnect the negative battery cable unless the repair or test procedure requires that battery voltage be applied.

• Always relieve the fuel system pressure prior to disconnecting any fuel system component (injector, fuel rail, pressure regulator, etc.), fitting or fuel line connection. Exercise extreme caution whenever relieving fuel system pressure to avoid exposing skin, face and eyes to fuel spray. Please be advised that fuel under pressure may penetrate the skin or any part of the body that it contacts.

• Always place a shop towel or cloth around the fitting or connection prior to loosening to absorb any excess fuel due to spillage. Ensure that all fuel spillage (should it occur) is quickly removed from engine surfaces. Ensure that all fuel soaked cloths or towels are deposited into a suitable waste container.

• Always keep a dry chemical (Class B) fire extinguisher near the work area.

• Do not allow fuel spray or fuel vapors to come into contact with a spark or open flame.

• Always use a back-up wrench when loosening and tightening fuel line connection fittings. This will prevent unnecessary stress and torsion to fuel line piping.

• Always replace worn fuel fitting O-rings with new Do not substitute fuel hose or equivalent where fuel pipe is installed.

Before servicing the vehicle, make sure to also refer to the precautions in the beginning of this section as well.

RELIEVING FUEL SYSTEM PRESSURE

Accord
See Figure 192.

✳✳ CAUTION

The fuel injection system remains under pressure after the engine has been turned OFF. Properly relieve fuel pressure before disconnecting any fuel lines. Failure to do so may result in fire or personal injury.

✳✳ CAUTION

Do not allow fuel spray or fuel vapors to come in contact with a spark or open flame. Keep a dry chemical fire extinguisher nearby. Never store fuel in an open container due to risk of fire or explosion.

➡ **The radio may contain a coded theft protection circuit. Always obtain the code number before disconnecting the battery.**

1. Before servicing the vehicle, refer to the precautions in the beginning of this section.
2. Turn the ignition switch OFF.
3. Remove the left side kick panel. Remove the PGM-FI main relay.
4. Start the engine and let it idle until it stalls.

➡ **If any codes are stored, ignore them.**

5. Turn the ignition OFF.
6. If necessary, remove the engine cover.
7. Disconnect the negative battery cable.
8. Remove the fuel filler cap to relieve the fuel pressure in the tank.
9. On vehicles equipped with a quick connect cover, remove it. Clean any dirt from the quick-connect fitting. Place a rag or shop towel over quick-connect fitting.
10. Detach the quick-connect fitting by holding the connector with one hand, then squeeze the retainer tabs with the other hand to release them from the locking pawls. Pull the connector off.
11. If the connector does not move, keep the retainer tabs pressed down and alternately pull and push the connector until it comes off easily.
12. Do not move the retainer from the line; once removed, the retainer must be replaced with a new one.

Civic

✳✳ CAUTION

The fuel injection system remains under pressure after the engine has been turned OFF. Properly relieve fuel pressure before disconnecting any fuel lines. Failure to do so may result in fire or personal injury.

✳✳ CAUTION

Do not allow fuel spray or fuel vapors to come in contact with a spark or open flame. Keep a dry chemical fire extinguisher nearby. Never store fuel in an open container due to risk of fire or explosion.

➡ **The radio may contain a coded theft protection circuit. Always obtain the code number before disconnecting the battery.**

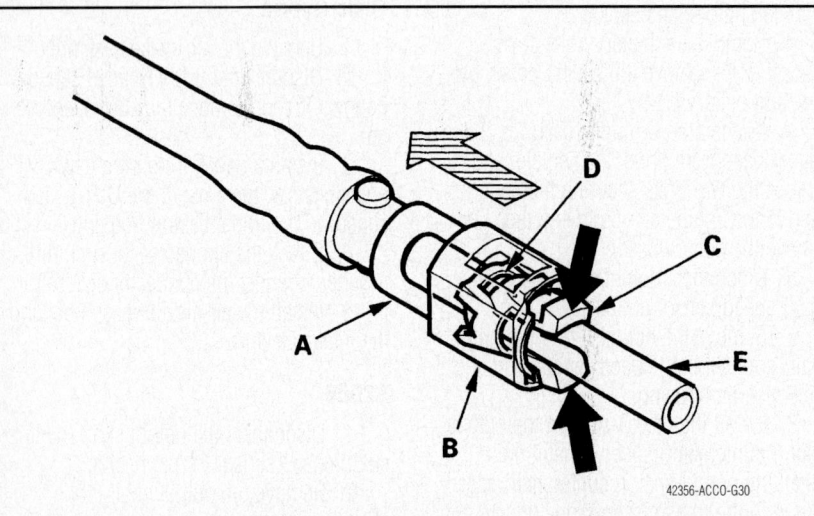

42356-ACCO-G30

Fig. 192 Hold the quick-connect (A) connector (B) with one hand, then squeeze the retainer tabs (C) with the other hand to release them from the locking pawls (D)

1. Remove the under dash fuse/relay box.
2. Remove the PGM-FI main relay. Reinstall the under dash fuse/relay box.
3. Start the engine and let it idle until it stalls.

➡**If any codes are stored, ignore them.**

4. Turn the ignition OFF.
5. Disconnect the negative battery cable.
6. Remove the fuel filler cap to relieve the fuel pressure in the tank.
7. On vehicles equipped with a quick connect cover, remove it. Clean any dirt from the quick-connect fitting. Place a rag or shop towel over quick-connect fitting.
8. Detach the quick-connect fitting by holding the connector with one hand, then squeeze the retainer tabs with the other hand to release them from the locking pawls. Pull the connector off.
9. If the connector does not move, keep the retainer tabs pressed down and alternately pull and push the connector until it comes off easily.
10. Do not move the retainer from the line; once removed, the retainer must be replaced with a new one.

Civic Hybrid & S2000

1. Remove PGM-FI main relay 2 from the auxiliary under-hood fuse/relay box.
2. Start the engine, and let it idle until it stalls.

➡**If any DTCs are stored, clear and ignore them.**

3. Turn the ignition switch to LOCK (0).
4. Remove the fuel fill cap.
5. Do the battery terminal reconnection procedure.
6. Remove the air intake assembly.
7. Remove the cover and quick-connect fitting cover.
8. Check the fuel quick-connect fitting for dirt, and clean it if needed.
9. Place a rag or shop towel over the quick-connect fitting.
10. Disconnect the quick-connect fitting: Hold the connector with one hand, and squeeze the retainer tabs with the other hand to release them from the locking tabs. Pull the connector off.

FUEL FILTER

REMOVAL & INSTALLATION

The fuel filter is an integrated part of the fuel pump assembly. For additional information, refer to the following section, "Fuel Pump, Removal & Installation."

FUEL INJECTORS

REMOVAL & INSTALLATION

Accord

2.4L Engine

1. Properly relieve the fuel system pressure.
2. Remove the engine appearance cover.
3. Disconnect the connectors from the injectors. On A/T models, disconnect the engine mount control solenoid valve.
4. Remove the ground cable bolt (G101).
5. Disconnect the quick-connect fittings.
6. Remove the fuel rail mounting nuts from the fuel rail.
7. Remove the injector clip from the injector.
8. Remove the injector from the fuel rail.

To install:

9. Coat the new O-rings with clean engine oil, and insert the injectors into the fuel rail.
10. Install the injector clip.
11. Coat the injector O-rings with clean engine oil.
12. Install the injectors into the injector base.
13. Install the fuel rail mounting nuts and the ground cable bolt (G101).
14. Connect the connectors on the injectors. On A/T models, connect the engine mount control solenoid valve.
15. Connect the quick-connect fittings.
16. Turn the ignition switch ON (II), but do not operate the starter. After the fuel pump runs for about 2 seconds, the fuel rail will be pressurized. Repeat this two or three times, then check for fuel leakage.

3.0L & 3.5L Engines

1. Properly relieve the fuel system pressure.
2. Remove the intake manifold.
3. Disconnect the connectors from the injectors.
4. Disconnect the quick-connect fittings.
5. Remove the fuel rail mounting bolts from the fuel rail.
6. Remove the injector clip from the injector.
7. Remove the injector from the fuel rail.

To install:

8. Coat the new O-rings with clean engine oil, and insert the injectors into the fuel rail.
9. Install the injector clip.
10. Coat the injector O-rings with clean engine oil.
11. Install the injectors in the injector base.
12. Install the fuel rail mounting nuts.
13. Connect the connectors on the injectors.
14. Connect the quick-connect fittings.
15. Turn the ignition switch ON (II), but do not operate the starter. After the fuel pump runs for about 2 seconds, the fuel rail will be pressurized. Repeat this two or three times, then check for fuel leakage.

Civic

1.8L Engine

1. Properly relieve the fuel system pressure.
2. Remove the cowl cover and the under-cowl panel.
3. Remove the fuel line cover.
4. Disconnect the injector connectors from the injectors and the rocker arm oil control valve connector.
5. Disconnect the quick-connect fitting.
6. Remove the fuel rail mounting nuts from the fuel rail.
7. Remove the injector clips from the injectors.
8. Remove the injectors from the fuel rail.

To install:

9. Coat the new O-rings with clean engine oil, and insert the injectors into the fuel rail.
10. Install the injector clips.
11. Coat the new injector O-rings with clean engine oil.
12. Install the fuel rail and the injectors into the cylinder head.
13. Install the fuel rail mounting nuts.
14. Connect the connectors on the injectors.
15. Connect the quick-connect fittings.
16. Turn the ignition switch to ON (II), but do not operate the starter. After the fuel pump runs for about 2 seconds, the fuel rail is pressurized. Repeat this two or three times, then check for fuel leakage.
17. Install the fuel line cover.
18. Install the cowl cover and the under-cowl panel.

2.0L Engine

1. Properly relieve the fuel system pressure.

2. Remove the engine appearance cover.

3. Disconnect the connectors from the injectors

4. Remove the ground cable bolt (G101).

5. Disconnect the quick-connect fittings.

6. Remove the fuel rail mounting nuts from the fuel rail.

7. Remove the injector clip from the injector.

8. Remove the injector from the fuel rail.

To install:

9. Coat the new O-rings with clean engine oil, and insert the injectors into the fuel rail.

10. Install the injector clip.

11. Coat the injector O-rings with clean engine oil.

12. Install the injectors into the injector base.

13. Install the fuel rail mounting nuts and the ground cable bolt (G101).

14. Connect the connectors on the injectors

15. Connect the quick-connect fittings.

16. Turn the ignition switch ON (II), but do not operate the starter. After the fuel pump runs for about 2 seconds, the fuel rail will be pressurized. Repeat this two or three times, then check for fuel leakage.

Civic Hybrid

1. Properly relieve the fuel system pressure.

2. Remove the air intake assembly.

3. Remove the mounting nut and bolt.

4. Disconnect the connectors from the injectors, the intake side ignition coils, the EGR valve, the rocker arm oil control valve, the rocker arm oil pressure sensor, the MAP sensor, and the rocker arm oil pressure switch.

5. Remove the ground terminals.

6. Disconnect the hoses from the fuel rail.

7. Remove the fuel rail mounting nuts from the fuel rail, then remove the injectors and fuel rail together.

8. Remove the injector clips from the injectors.

9. Remove the injectors from the fuel rail.

To install:

10. Coat the new O-ring with clean engine oil, and insert the injectors into the fuel rail.

11. Install the injector clips.

12. Coat the injector O-rings with clean engine oil.

13. Install the injectors into the injector base.

14. Install the fuel rail mounting nuts and the bolt.

15. Install the nut with a new O-ring.

16. Reconnect the fuel hoses.

17. Connect the ground terminal.

18. Connect the connectors on the injectors, the intake side ignition coils, the EGR valve, the rocker arm oil control valve, the rocker arm oil pressure sensor, the MAP sensor, and the rocker arm oil pressure switch.

S2000

1. Properly relieve the fuel system pressure.

2. Remove the fuel rail cover.

3. Disconnect the injector connectors from the injectors, then remove the harness holder from the fuel rail.

4. Disconnect the quick-connect fitting from the fuel rail.

5. Remove the retainer nuts and the bolts from the fuel rail.

6. Disconnect the fuel rail from the intake manifold.

7. Remove the injector clips from the fuel rail.

8. Remove the injectors from the fuel rail.

9. Remove the O-rings from the injectors.

To install:

10. Coat new O-rings with clean engine oil, and put them on the injectors.

11. Insert the injectors into the fuel rail.

12. Install the injector clips.

13. Install the injectors in the intake manifold.

14. Install and tighten the retainer nuts and bolts.

15. Connect the quick-connect fitting.

16. Connect the injector connectors and the harness holder.

17. Install the injector cover.

18. Turn the ignition switch to ON (II), but do not operate the starter. After the fuel pump runs for about 2 seconds, the fuel pressure in the fuel line rises. Repeat this 2 or 3 times, then check for fuel leakage.

FUEL PUMP

REMOVAL & INSTALLATION

Accord

2.4L Engine

1. Properly relieve the fuel system pressure.

2. Remove the fuel fill cap.

3. Remove the spare tire lid.

4. Remove the access panel from the floor.

5. Disconnect the fuel pump 5P connector.

6. Disconnect the quick-connect fitting from the fuel pump assembly.

7. Remove the fuel pump assembly.

8. Installation is the reverse order of removal. Tighten the mounting nuts to 36 inch lbs. (4 Nm).

3.0L Engine

See Figure 193.

1. Properly relieve the fuel system pressure.

2. Remove the fuel fill cap.

3. Remove the spare tire lid.

4. Remove the access panel from the floor.

5. Disconnect the fuel pump 5P connector.

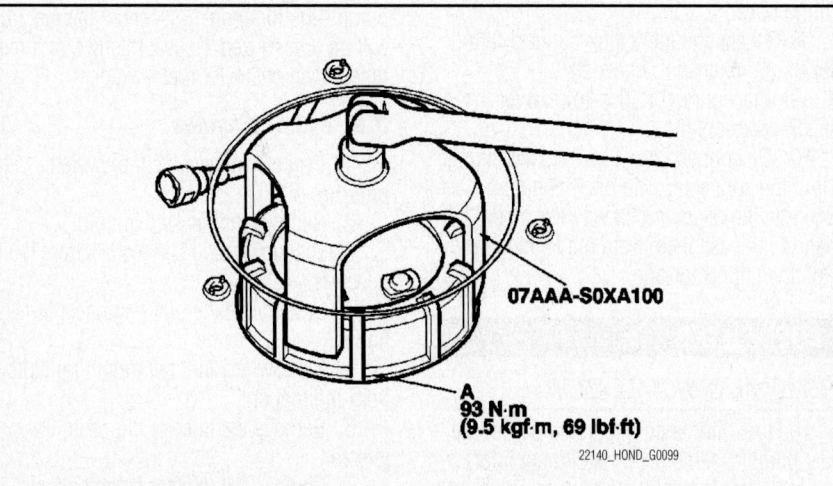

07AAA-S0XA100

A
93 N·m
(9.5 kgf·m, 69 lbf·ft)

22140_HOND_G0099

Fig. 193 Using a suitable removal tool, remove the fuel pump assembly locknut—Accord 3.0L Models

6. Disconnect the quick-connect fitting from the fuel pump assembly.

7. Using suitable fuel sender wrench, remove the pump assembly.

8. Installation is the reverse order of removal. Note the following:
- Use a new base gasket and locknut
- Line up the matchmarks on the fuel tank and top of fuel pump assembly
- Tighten the locknut to 69 ft. lbs. (93 Nm)

3.5L Engine

See Figure 194.

1. Properly relieve the fuel pressure.
2. Remove the fuel fill cap.
3. Remove the rear seat cushion.
4. Remove the access panel from the floor.
5. Disconnect the fuel pump 4P connector.
6. Disconnect the quick-connect fitting from the fuel pump assembly.
7. Using suitable fuel sender wrench, remove the pump assembly.
8. Remove the locknut and fuel pump assembly.

To install:

9. Temporarily attach a new base gasket to the fuel pump assembly, then insert the assembly partially into the fuel tank.

10. Transfer the base gasket from the fuel pump assembly to the fuel tank.

11. Align the marks on the fuel tank and fuel tank unit, then insert the fuel tank unit into the fuel tank until the fuel tank unit rests on top of the base gasket.

✳✳ WARNING

Ensure the base gasket is sitting flush to avoid any fuel leaks.

12. Using a suitable fuel sender wrench, tighten a new locknut to 51 ft. lbs. (70 Nm).

13. The remainder of the installation is the reverse order of removal.

Civic & Civic Hybrid

See Figure 195.

✳✳ CAUTION

The fuel injection system remains under pressure, even after the engine has been turned OFF. The fuel system pressure must be relieved before disconnecting any fuel lines. Failure to follow this procedure may result in fire, explosion, or personal injury.

1. Before servicing the vehicle, refer to the precautions in the beginning of this section.

2. Disconnect the negative battery cable.

3. Relieve the fuel pressure.

4. Remove or disconnect the following:
- Rear seat cushions
- Remove the fuel filler cap
- Rear floor upper crossmember

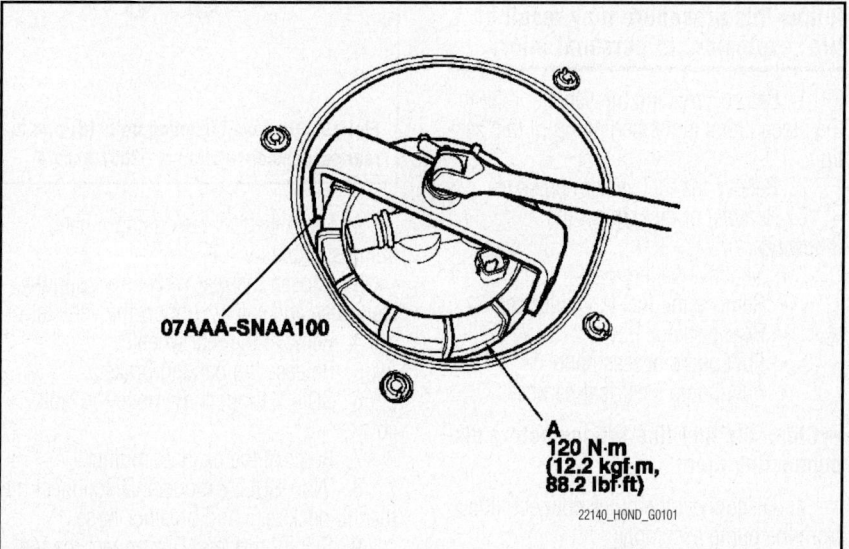

07AAA-SNAA100

A
120 N·m
(12.2 kgf·m,
88.2 lbf·ft)

22140_HOND_G0101

Fig. 195 Using a suitable fuel sender wrench, tighten a new locknut to 88 ft. lbs. (120 Nm)—Civic models

- Fuel pump access panel
- Fuel pump electrical harness

➡**Clean the fuel line fittings before disconnecting them.**

5. Disconnect the quick connect fitting from the pump assembly.

6. Using Special Tool 07AAA-S0XA100, or equivalent, loosen the fuel tank unit locknut. Remove the locknut and the fuel tank unit. Allow the fuel in the pump drain into the tank before removing the pump from the vehicle.

To install:

7. Be sure to use a new base gasket and new O-rings.

8. When installing the fuel tank unit, align the marks on the fuel tank and the fuel tank unit.

9. Make sure the electrical connections are secure and the connector firmly locked in place.

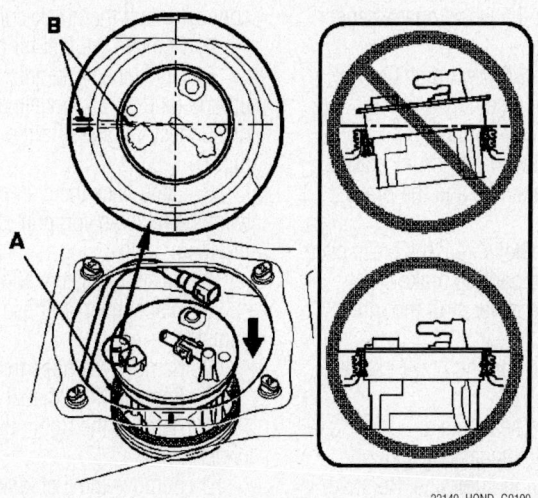

22140_HOND_G0100

Fig. 194 After transfer the gasket (A) to the fuel tank, align the marks (B) on the tank and fuel pump assembly—Accord 3.5L Models

10. Make sure the fuel line connections are secure and the connector firmly locked in place.

11. Using Special Tool 07AAA-S0XA100, or equivalent, tighten the fuel tank unit locknut to 88 ft. lbs. (120 Nm).

12. The remainder of the installation is the reverse order of removal.

S2000

> **⁂ CAUTION**
>
> **The fuel injection system remains under pressure, even after the engine has been turned OFF. The fuel system pressure must be relieved before disconnecting any fuel lines. Failure to follow this procedure may result in fire, explosion, or personal injury.**

1. Before servicing the vehicle, refer to the precautions in the beginning of this section.

2. Relieve the fuel system pressure.

3. Remove or disconnect the following:

- Negative battery cable
- Remove the fuel tank filler cap
- Rear package tray
- Fuel pump access panel
- Fuel pump electrical harness

➡Clean the fuel line fittings before disconnecting them.

4. Disconnect the quick connect fitting from the pump assembly.

5. Remove the fuel tank retaining bolts. Remove the fuel tank unit from the vehicle.

To install:

6. Be sure to use a new base gasket.

7. When installing the fuel tank unit, align the marks on the fuel tank and the fuel tank unit.

8. Make sure the electrical connections are secure and the connector firmly locked in place.

9. Make sure the fuel line connections are secure and the connector firmly locked in place.

10. The remainder of the installation is the reverse order of removal.

FUEL TANK

REMOVAL & INSTALLATION

Accord

2007 Models

See Figure 196.

1. Properly relieve the fuel system pressure.

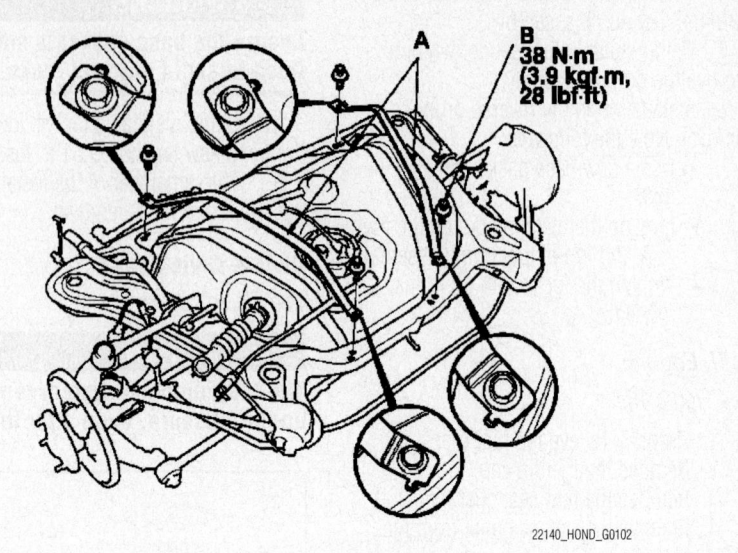

Fig. 196 Remove mounting bolts (B) and fuel tank straps (A) to remove the fuel tank from the rear suspension subframe—2007 Accord

2. Drain the fuel tank using a hand pump suitable for fuel.

3. Loosen the rear wheel nuts slightly, then raise and safely support the vehicle.

4. Remove the rear wheels.

5. Release the parking brake.

6. SULEV models: Remove the tank covers.

7. Remove the exhaust muffler.

8. Non-SULEV models: Disconnect the fuel fill neck tube and breather hose.

9. SULEV models: Disconnect the fuel fill neck tube, fuel tank vapor recirculation tube, and fuel tank vapor signal tube.

10. Disconnect the vapor line from the EVAP canister. Then disconnect the fuel line.

11. Disconnect the wheel speed sensor electrical connector.

12. If equipped with rear disc brakes, remove the caliper assembly and hang securely from the coil spring.

13. If equipped with rear drum brakes, disconnect the brake line and the brake hose clip.

14. Remove the bolts from the brake plate.

15. Remove the parking brake cable.

16. Remove the lower strut mounting bolt.

17. Remove the parking brake cable bracket.

18. Remove the heat shield.

19. Place a suitable jack or support under the suspension subframe. Remove the mounting bolts. Remove the rear suspension subframe.

20. Remove the fuel tank mounting bolts and the fuel tank straps.

21. Lift the fuel tank out of the subframe.

22. Installation is the reverse order of assembly. Tighten the fuel tank mounting strap bolts to 28 ft. lbs. (38 Nm).

23. If equipped with rear drum brakes, bleed the brake system.

24. Check and adjust the rear wheel alignment as necessary.

2008 Models

1. Properly relieve the fuel system pressure.

2. Remove the fuel pump assembly and drain the fuel tank using a hand pump suitable for fuel.

3. Reinstall the fuel pump assembly without connecting the fuel tank unit 4P connector and the quick-connect fitting.

4. Remove the fuel fill pipe cover.

5. Disconnect the quick-connect fittings (and on all models except PZEV, disconnect the fuel fill tube from the fuel fill pipe).

6. Slide back the clamps, then twist the hose as you pull to avoid damaging them.

7. Raise and safely support the vehicle.

8. Disconnect the hose from the EVAP canister.

9. Remove the hose from the clamp.

10. Remove the exhaust pipe.

11. Remove the right side middle floor undercover.

12. Remove the fuel tank protector.

13. Remove the right parking brake cable mounting bolts.

14. Place a suitable jack or other support under the fuel tank.

15. Remove the strap bolts and the straps.

16. Remove the fuel tank.

17. Installation is the reverse order of removal. Tighten the fuel tank mounting strap bolts to 28 ft. lbs. (38 Nm).

Civic & Civic Hybrid

See Figure 197.

1. Properly relieve the fuel system pressure.

2. Drain the fuel tank, then disconnect the fuel tank unit 4P connector and the quick-connect fitting.

3. Raise and safely support the vehicle.

4. Remove the cover, and the EVAP canister guard pipe.

5. Disconnect the fuel fill tube. Slide back the clamps, then twist the hoses as you pull to avoid damaging them.

6. Disconnect the quick-connect fitting and the fuel tank vapor recirculation tube.

7. Place a suitable jack or other support under the fuel tank and secure the tank with a winch strap.

8. Remove the strap bolts and the strap.

9. Remove the fuel tank.

10. Installation is the reverse order of removal. Tighten the fuel tank mounting strap bolts to 28 ft. lbs. (38 Nm).

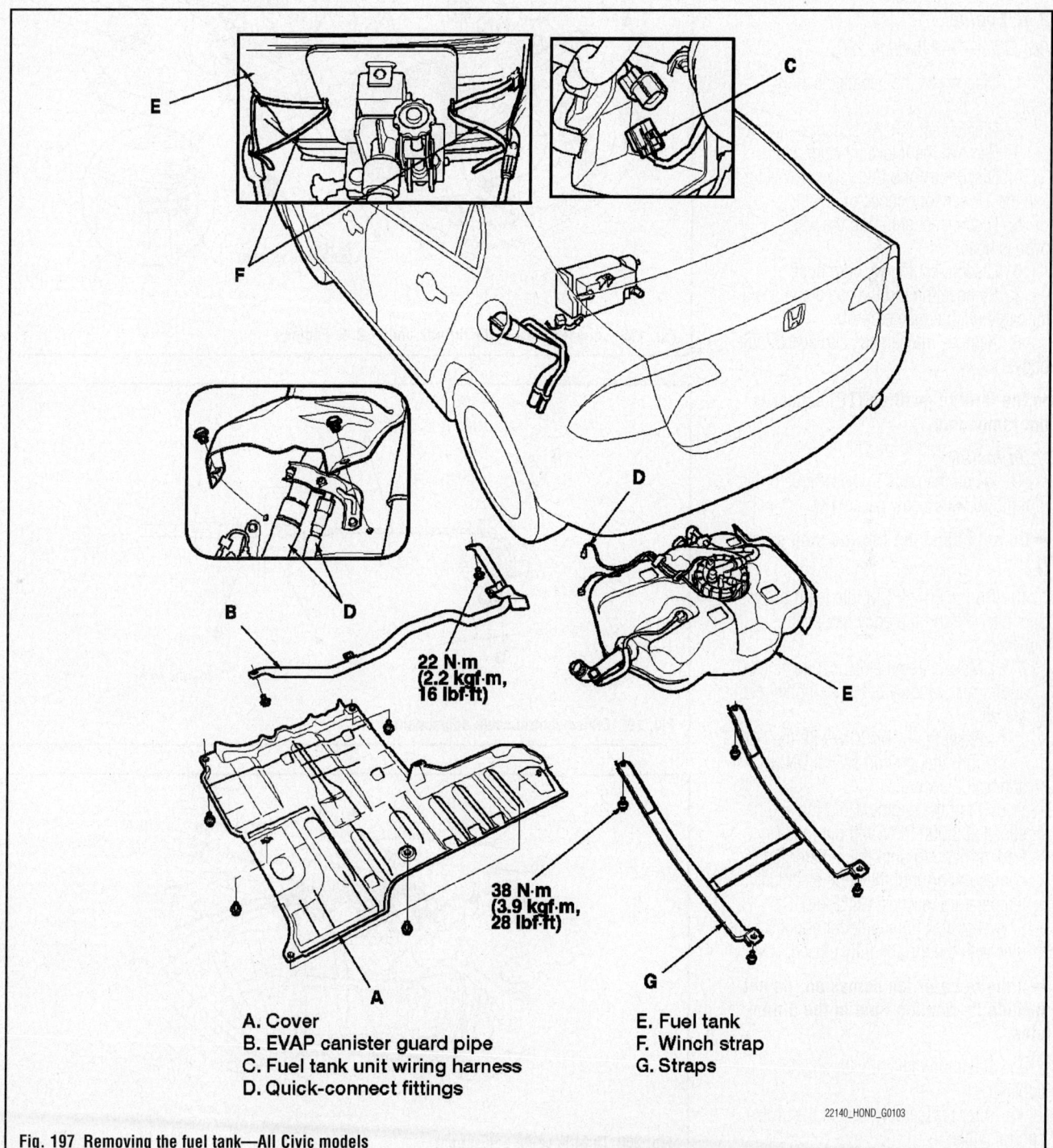

A. Cover
B. EVAP canister guard pipe
C. Fuel tank unit wiring harness
D. Quick-connect fittings
E. Fuel tank
F. Winch strap
G. Straps

22140_HOND_G0103

Fig. 197 Removing the fuel tank—All Civic models

IDLE SPEED

ADJUSTMENT

Idle speed is maintained by the Power-train Control Module (PCM). No adjustment is necessary or possible.

THROTTLE BODY

REMOVAL & INSTALLATION

Accord

2.4L Engine

See Figures 198 through 200.

1. Disconnect the negative battery cable.
2. Disconnect the IAT sensor connector.
3. Remove the intake air duct.
4. Disconnect the IAC valve connector, and the TP sensor connector.
5. Disconnect and plug the water bypass hoses.
6. Disconnect the vacuum hose.
7. Remove the throttle cable and actuator cable (with cruise control).
8. Remove the harness clip and throttle body.

➡The throttle position (TP) sensor is not removable.

To install:

9. Install the parts in the reverse order of removal with a new gasket (K).

➡Do not adjust the throttle stop screw (L).

10. Do the ECM/PCM idle learn procedure after the throttle body has been replaced:

a. Make sure all electrical items (A/C, audio, rear window defogger, lights, etc.) are off.

b. Reset the ECM/PCM with the HDS.

c. Turn the ignition switch ON (II), and wait 2 seconds.

d. Start the engine. Hold the engine speed at 3,000 RPM without load (in Park or neutral) until the radiator fan comes on, or until the engine coolant temperature reaches 194°F (90°C).

e. Let the engine idle for about 5 minutes with the throttle fully closed.

➡If the radiator fan comes on, do not include its running time in the 5 minutes.

11. Refill the radiator with engine coolant.
12. After reassembly, adjust the cruise control cable:

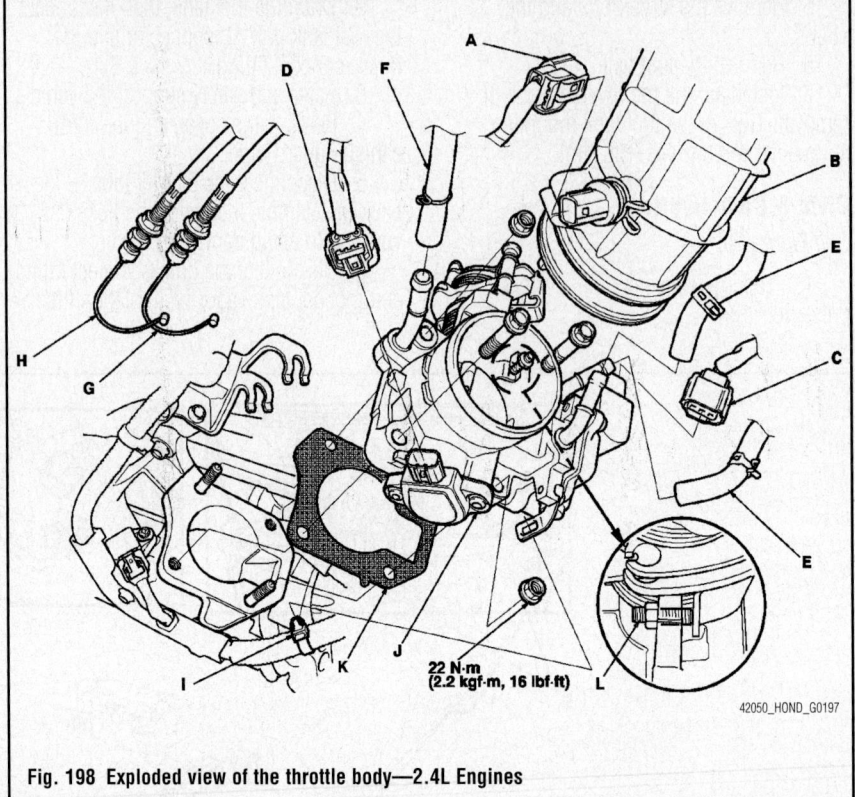

Fig. 198 Exploded view of the throttle body—2.4L Engines

22 N·m
(2.2 kgf·m, 16 lbf·ft)

42050_HOND_G0197

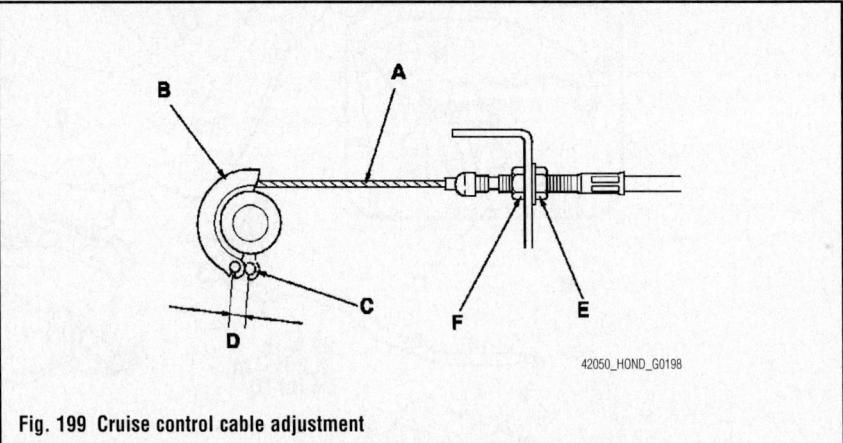

Fig. 199 Cruise control cable adjustment

42050_HOND_G0198

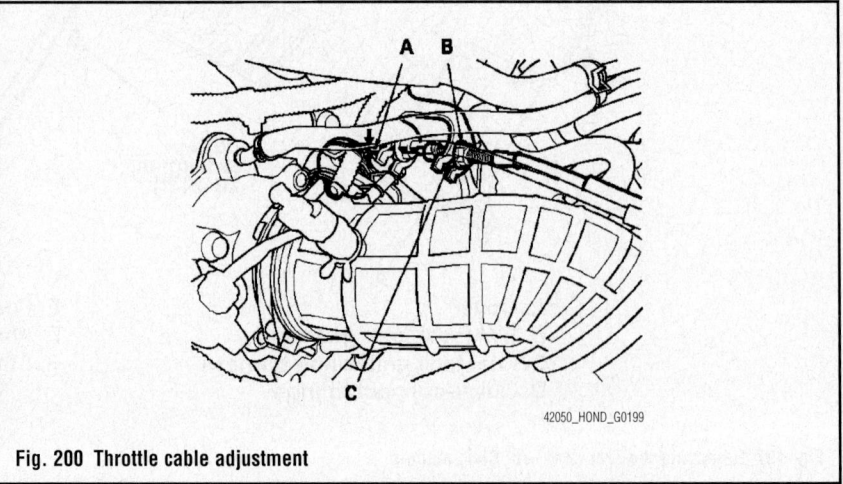

Fig. 200 Throttle cable adjustment

42050_HOND_G0199

a. Check that the actuator cable (A) moves smoothly with no binding or sticking.

b. Measure the amount of movement of the output linkage (B) until the engine speed starts to increase.

c. At first, the output linkage should be located at the fully closed position (C). The free play (D) should be 3.75±0.5 mm (0.15±0.02 in.).

d. If the free play is not within specification, loosen the locknut (E), and turn the adjusting nut (F) until the free play is as specified, then retighten the locknut.

e. Adjust the throttle cable as follows:

f. Check cable free play at the throttle linkage. Cable free play (A) should be 3/8–1/2 in. (10–12 mm).

g. If the free play is not within spec (3/8–1/2 in., 10–12 mm), loosen the locknut (B), turn the adjusting nut (C) until the free play is as specified, then retighten the locknut.

h. With the cable properly adjusted, check the throttle valve to be sure it opens fully when you push the accelerator pedal to the floor. Also check the throttle valve to be sure it returns to the idle position whenever you release the accelerator pedal.

3.0L & 3.5L Engines

See Figure 201.

➡**If you are replacing or cleaning the throttle body, begin from step 1. If you are removing the throttle body temporarily, begin from step 4.**

1. Connect the HDS while the engine is stopped.

2. Select the INSPECTION MENU with the HDS.

3. Do the TP POSITION CHECK in the ETCS TEST.

4. Disconnect the MAP sensor connector (A).

5. Remove the intake air duct (B).

6. Disconnect the throttle body connector (C).

7. Disconnect and plug the water bypass hoses (D).

8. Remove the harness clip (E) and the throttle body (F).

9. Install the parts in the reverse order of removal with a new gasket (G).

10. Do the ECM/PCM idle learn procedure after the throttle body has been replaced, as follows:

a. Make sure all electrical items (A/C, audio, rear window defogger, lights, etc.) are off.

b. Reset the ECM/PCM with the HDS.

c. Turn the ignition switch ON (II), and wait for 2 seconds.

d. Start the engine. Hold the engine speed at 3,000 RPM without load (in Park or neutral) until the radiator fan comes on, or until the engine coolant temperature reaches 194°F (90°C).

e. Let the engine idle for about 5 minutes with the throttle fully closed.

➡**If the radiator fan comes on, do not include its running time in the 5 minutes.**

11. Refill the radiator with engine coolant.

Civic

1.8L Engine

➡**If you are replacing or cleaning the throttle body, start at step 1. If you are removing the throttle body, start at step 4.**

1. Connect the HDS to the DLC while the engine is stopped.

2. Select the INSPECTION MENU on the HDS.

3. Do the TP POSITION CHECK in the ETCS TEST.

4. Remove the cowl cover and under-cowl panel.

5. Remove the air cleaner.

6. Disconnect the throttle body connector and the EVAP purge control solenoid valve connector.

7. Disconnect and plug the water bypass hoses.

8. Remove the throttle body.

9. Install the parts in the reverse order of removal with a new gasket.

10. Do the ECM/PCM idle learn procedure after replacing throttle body, as follows:

a. Make sure all electrical items (A/C, audio, lights, etc.) are off.

b. Reset the ECM/PCM with the HDS.

c. Turn the ignition switch ON (II), and wait 2 seconds.

d. Start the engine. Hold the engine speed at 3,000 RPM without load (in Park or neutral) until the radiator fan comes on, or until the engine coolant temperature reaches 194°F (90°C).

e. Let the engine idle for about 5 minutes with the throttle fully closed.

➡**If the radiator fan comes on, do not include its running time in the 5 minutes.**

f. Verify on the HDS data list that the idle learn procedure is complete.

11. Refill the radiator with engine coolant.

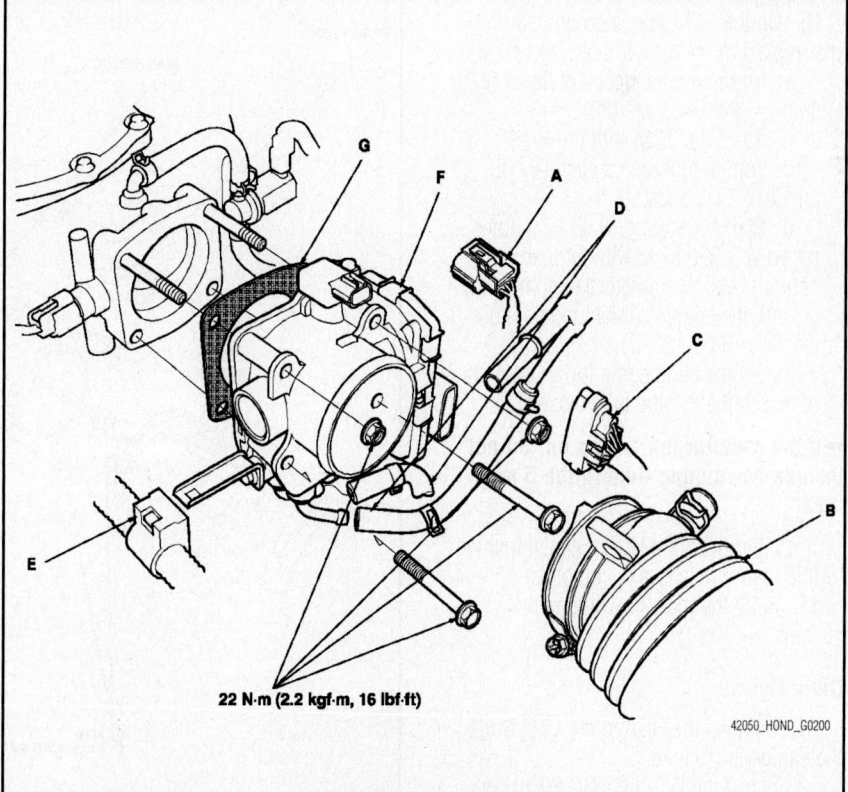

22 N·m (2.2 kgf·m, 16 lbf·ft)

42050_HOND_G0200

Fig. 201 Exploded view of the throttle body—3.0L engine shown

2.0L Engine

See Figure 202.

✳✳ WARNING

Do not insert your fingers into the installed throttle body when you turn the ignition switch ON (II) or while the ignition switch is ON (II). If you do, there will be serious injury to your fingers if the throttle valve is activated.

➡️**If you are replacing or cleaning the throttle body, start at step 1. If you are removing the throttle body, start at step 4.**

1. Connect the HDS to the DLC while the engine is stopped.
2. Select the INSPECTION MENU on the HDS.
3. Do the TP POSITION CHECK in the ETCS TEST.
4. Remove the air intake duct.
5. Disconnect the throttle body connector.
6. Disconnect the water bypass hoses, and plug the water bypass hoses.
7. Disconnect the vacuum hose and clamp.
8. Remove the throttle body.
9. Install the parts in the reverse order of removal with a new gasket, then check these items:
10. Do the ECM idle learn procedure after replacing the throttle body, as follows:
   a. Make sure all electrical items (A/C, audio, lights, etc.) are off.
   b. Reset the ECM with the HDS.
   c. Turn the ignition switch ON (II), and wait 2 seconds.
   d. Start the engine. Hold the engine speed at 3,000 RPM without load (in neutral) until the radiator fan comes on, or until the engine coolant temperature reaches 194°F (90°C).
   e. Let the engine idle for about 5 minutes with the throttle fully closed.

➡️**If the radiator fan comes on, do not include its running time in the 5 minutes.**

   f. Verify on the HDS data list that the idle learn procedure is complete.
11. Refill the radiator with engine coolant.

Civic Hybrid

1. Connect the HDS to the DLC while the engine is stopped.
2. Select the INSPECTION MENU on the HDS.

3. Do the TP POSITION CHECK in the ETCS TEST.
4. Turn the ignition switch to LOCK (0).
5. Remove the cowl cover and under-cowl panel.
6. Remove the air cleaner, and the intake air duct.
7. Disconnect the throttle body connector.
8. Disconnect and plug the water bypass hoses and vacuum hose.

9. Remove the throttle body.
10. Install the parts in the reverse order of removal with a new gasket.

S2000

See Figures 203 through 207.

1. Refer to the accompanying illustration for Throttle Body replacement.

➡️**Do not adjust the throttle stop screw. The TP sensor is not removable.**

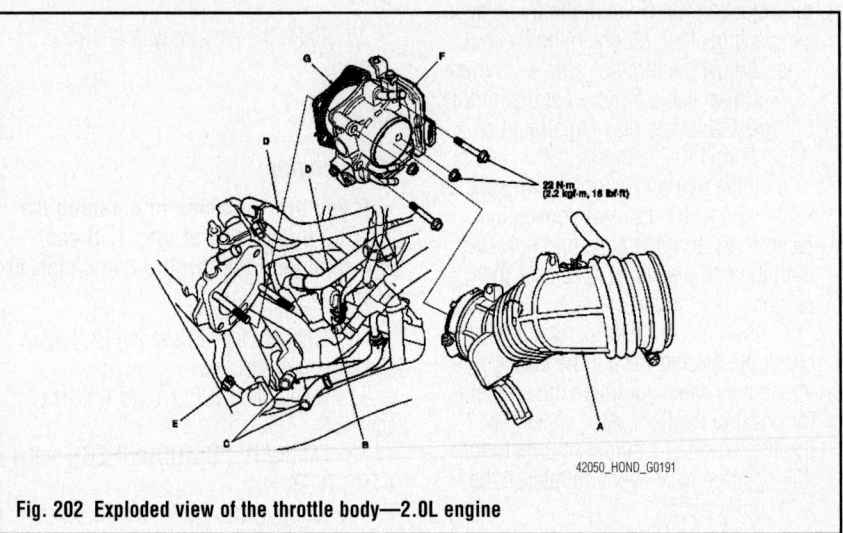

Fig. 202 Exploded view of the throttle body—2.0L engine

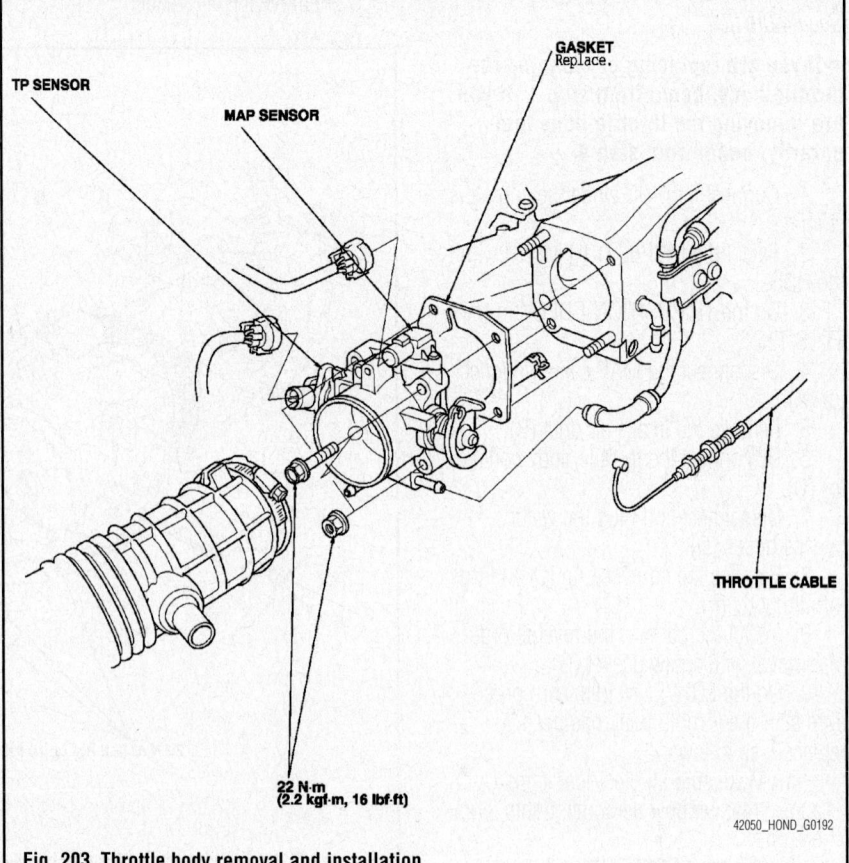

Fig. 203 Throttle body removal and installation

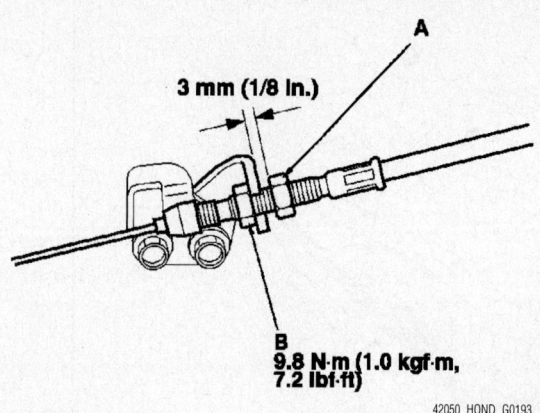

Fig. 204 Turn the adjusting nut (A) until it is 3 mm (1/8 in.) away from the throttle cable bracket. Tighten the locknut (B). The throttle cable deflection should now be 4–6 mm (0.16–0.24 in.)

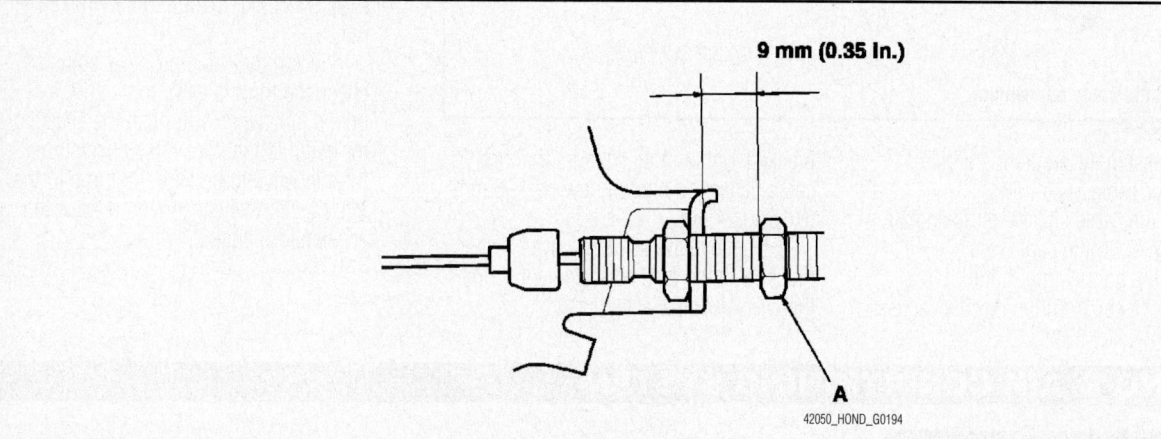

Fig. 205 Turn the adjusting nut (A) until it is 9 mm (0.35 in.) away from the actuator cable bracket when the throttle linkage starts open

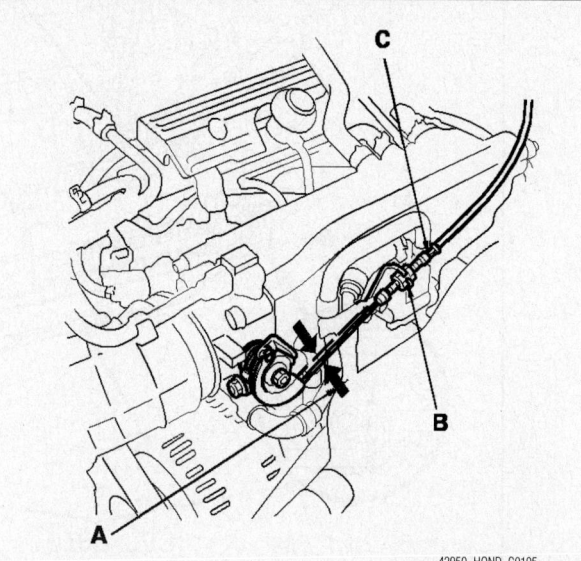

Fig. 206 Pull the cable so the adjusting nut (A) touches the bracket, and tighten the locknut (B). Make sure the throttle linkage starts open when the actuator cable is pulled 9 mm (0.35 in.) from the starting point by rotating the actuator linkage

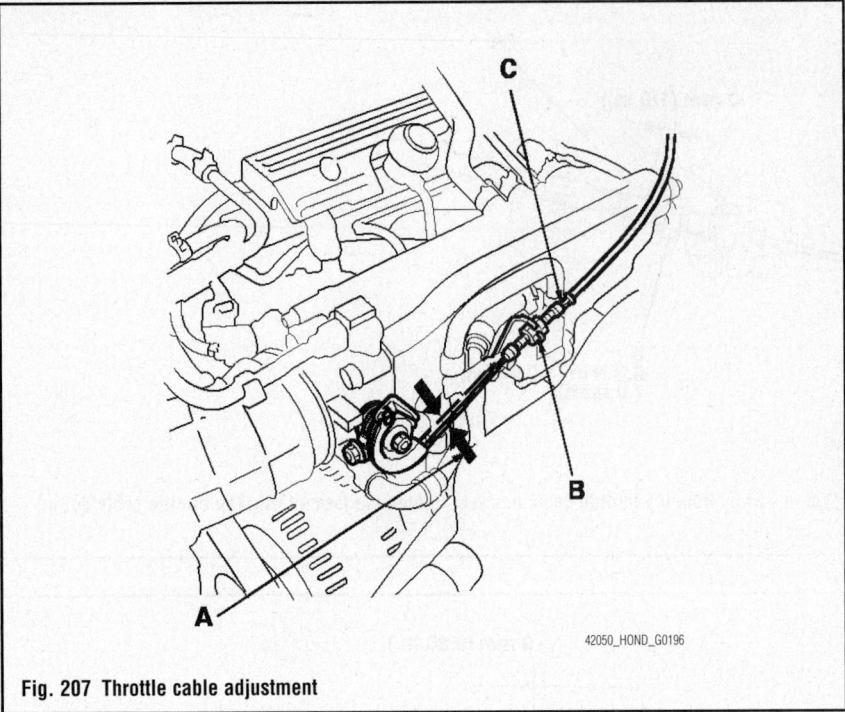

Fig. 207 Throttle cable adjustment

2. After reassembly, adjust the cruise control actuator cable, as follows:

a. Hold the cable sheath, removing all slack from the throttle cable.

b. Turn the adjusting nut until it is 3 mm (1/8 in.) away from the throttle cable

bracket. Tighten the locknut. The throttle cable deflection should now be 4–6 mm (0.16–0.24 in.).

c. Remove the actuator cover, then disconnect the actuator cable end from the cruise control actuator.

d. Turn the adjusting nut until it is 9 mm (0.35 in.) away from the actuator cable bracket when the throttle linkage starts open.

e. Pull the cable so the adjusting nut touches the bracket, and tighten the locknut.

f. Make sure the throttle linkage starts open when the actuator cable is pulled 9 mm (0.35 in.) from the starting point by rotating the actuator linkage.

3. Adjust the throttle cable as follows:

a. Check cable free play at the throttle linkage. Cable free play should be 4–6 mm (3/16–1/4 in.).

b. If free play is not within spec (4–6 mm, 3/16–1/4 in.) loosen the locknut, turn the adjusting nut until the free play is as specified, then retighten the locknut.

c. With the cable properly adjusted, check the throttle valve to be sure it opens fully when you push the accelerator pedal to the floor. Also check the throttle valve to be sure it returns to the idle position whenever you release the accelerator pedal.

HEATING & AIR CONDITIONING SYSTEM

BLOWER MOTOR

REMOVAL & INSTALLATION

Accord

2007 Models

See Figure 208.

1. Remove the glove box, and the passenger's kick panel.

2. Remove the connector clips, the wire harness clip, and bolt.

3. Cut the plastic cross brace in the glove box opening with diagonal cutters in the area shown. Retain these parts to be reinstalled later.

4. Disconnect the connectors from the blower motor, and the recirculation control motor, then remove the wire harness clip.

5. Remove the mounting nuts, the mounting bolts and the blower unit

6. Installation is the reverse order of removal.

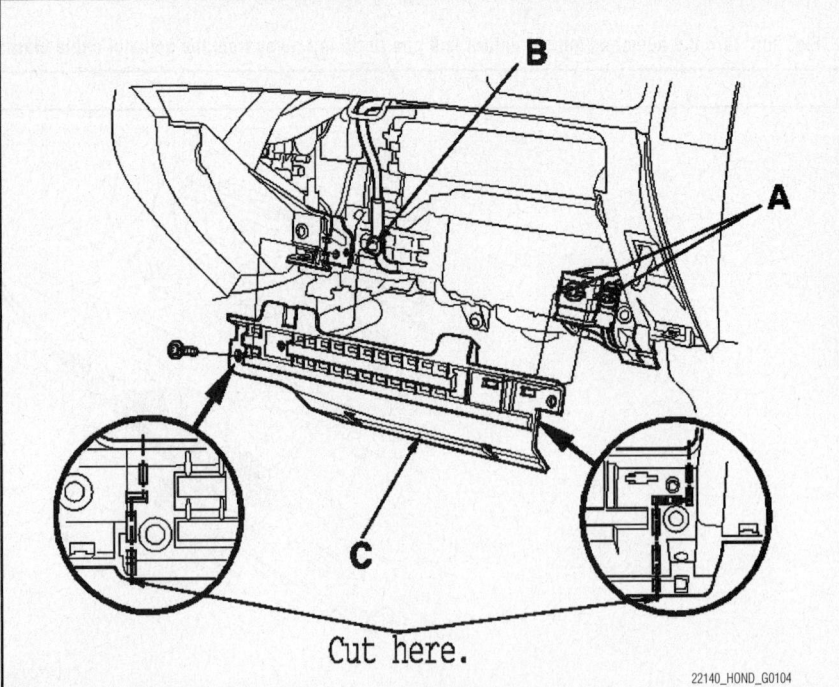

Cut here.

22140_HOND_G0104

Fig. 208 Cut the plastic cross brace where indicated to remove the blower motor—2007 Accord Models

2008 Models

See Figures 209 and 210.

1. Remove the glove box.
2. Remove the passenger's undercover.
3. Remove the right kick panel.
4. Remove the dust and pollen filter assembly from the blower unit.
5. Remove the bolts and the wire harness clip. Then cut the plastic cross brace in the glove box opening with diagonal cutters in the area shown. Retain plastic cross brace to be reinstalled later.
6. Disconnect these connectors: Passenger's under-dash fuse/relay box connector D (28P), stereo amplifier (with premium audio system), AM/FM antenna lead, and right side wire harness 20P connector (C410). Remove the wire harness clips.
7. Remove the two screws, then remove the cover.
8. Disconnect the connector from the blower motor and the wire harness clip. Remove the self-tapping screw and the mounting nut.
9. Remove the self-tapping screws, and the passenger's heater duct.
10. Disconnect the connector from the recirculation control motor. Remove the mounting nuts.
11. Pull the blower unit out while rotating it clockwise as shown, so that the glove box bracket passes through the dust and pollen filter area.
12. Installation is the reverse order of removal. Make sure that there is no air leakage.

Civic & Civic Hybrid

See Figure 211.

1. Remove the glove box.
2. Cut the plastic cross brace in the glove box opening with diagonal cutters in the area shown, and discard it.
3. Remove the bolts and the glove box frame.
4. Remove the wire harness clip, the self-tapping screws, and the passenger's heater duct.
5. Disconnect the connector from the blower motor. Remove the wire harness clip.
6. Disconnect the connector from the recirculation control motor. Remove the self-tapping screws, the bolt, the mounting nuts, and the blower unit.

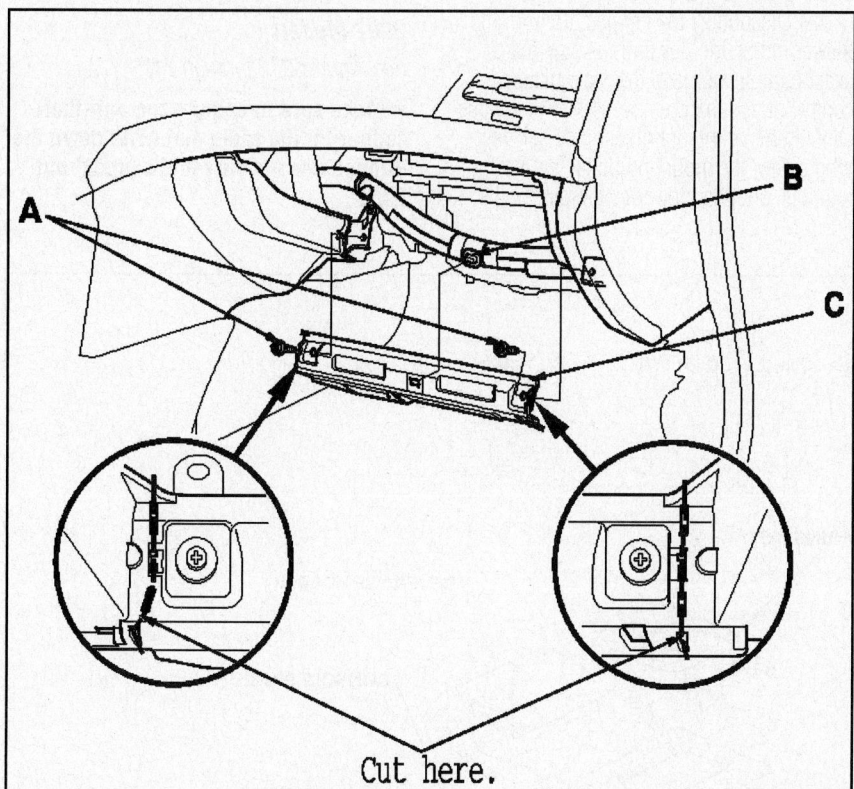

Cut here.

22140_HOND_G0105

Fig. 209 Cut the plastic cross brace where indicated to remove the blower motor—2008 Accord Models

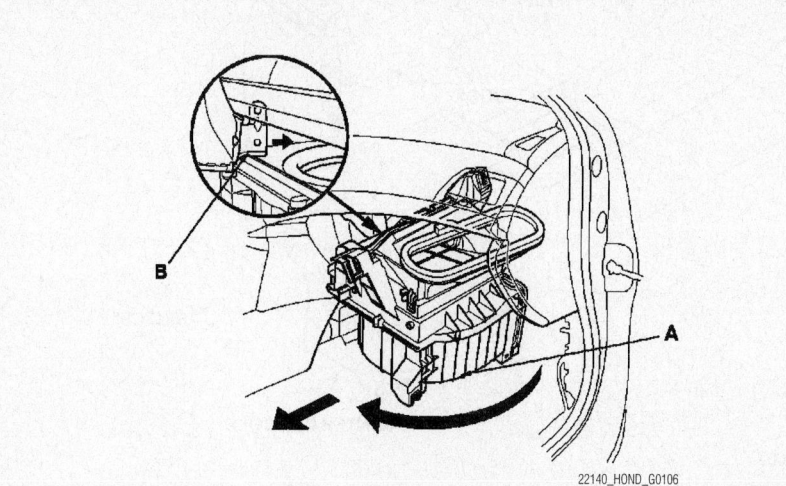

22140_HOND_G0106

Fig. 210 Pull the blower unit (A) out while rotating it clockwise as shown, so that the glove box bracket (B) passes through the dust and pollen filter area—2008 Accord Models

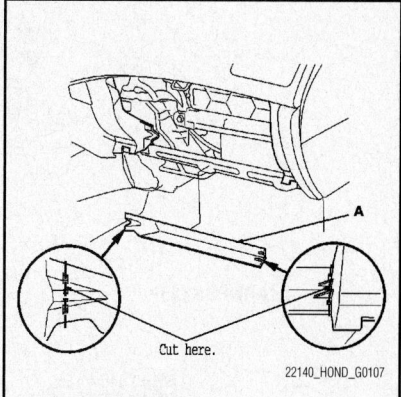

Cut here.

22140_HOND_G0107

Fig. 211 Cut the plastic cross brace (A) in the glove box opening with diagonal cutters in the area shown, and discard it—All Civic Models

7. Installation is the reverse order of removal. Make sure that there is no air leakage.

S2000

1. Discharge and recover the A/C refrigerant.
2. Disconnect the battery cables and remove the battery.
3. Pull out the grommets of the A/C lines, then carefully separate the upper grommet from the lower grommet by releasing the lock tabs.
4. Remove the bolt, then disconnect the suction line and the receiver line from the blower/evaporator unit. Plug or cap the lines immediately after disconnecting

them to avoid moisture and dust contamination.

5. Remove the passenger's dashboard lower cover and the right kick panel.
6. Disconnect the dashboard wire harness connector from the passenger's door wire harness connector, then remove it. Remove the wire harness connectors and the convertible top control unit from the steering hanger beam.
7. Disconnect the connectors from the blower motor, the power transistor, the evaporator sensor, and the recirculation control motor, then remove the wire harness clips. Remove the drain hose, the self-tapping screw, the mounting bolts, the mounting nuts, and the blower/evaporator unit.

8. Installation is the reverse order of removal.
9. Replace any O-ring for the A/C lines.
10. Recharge the A/C system refrigerant.

HEATER CORE

REMOVAL & INSTALLATION

Accord

2007 Models

See Figures 212 through 215.

➡**Make sure to acquire the anti-theft code from the radio and write down the frequencies for the radio's preset buttons.**

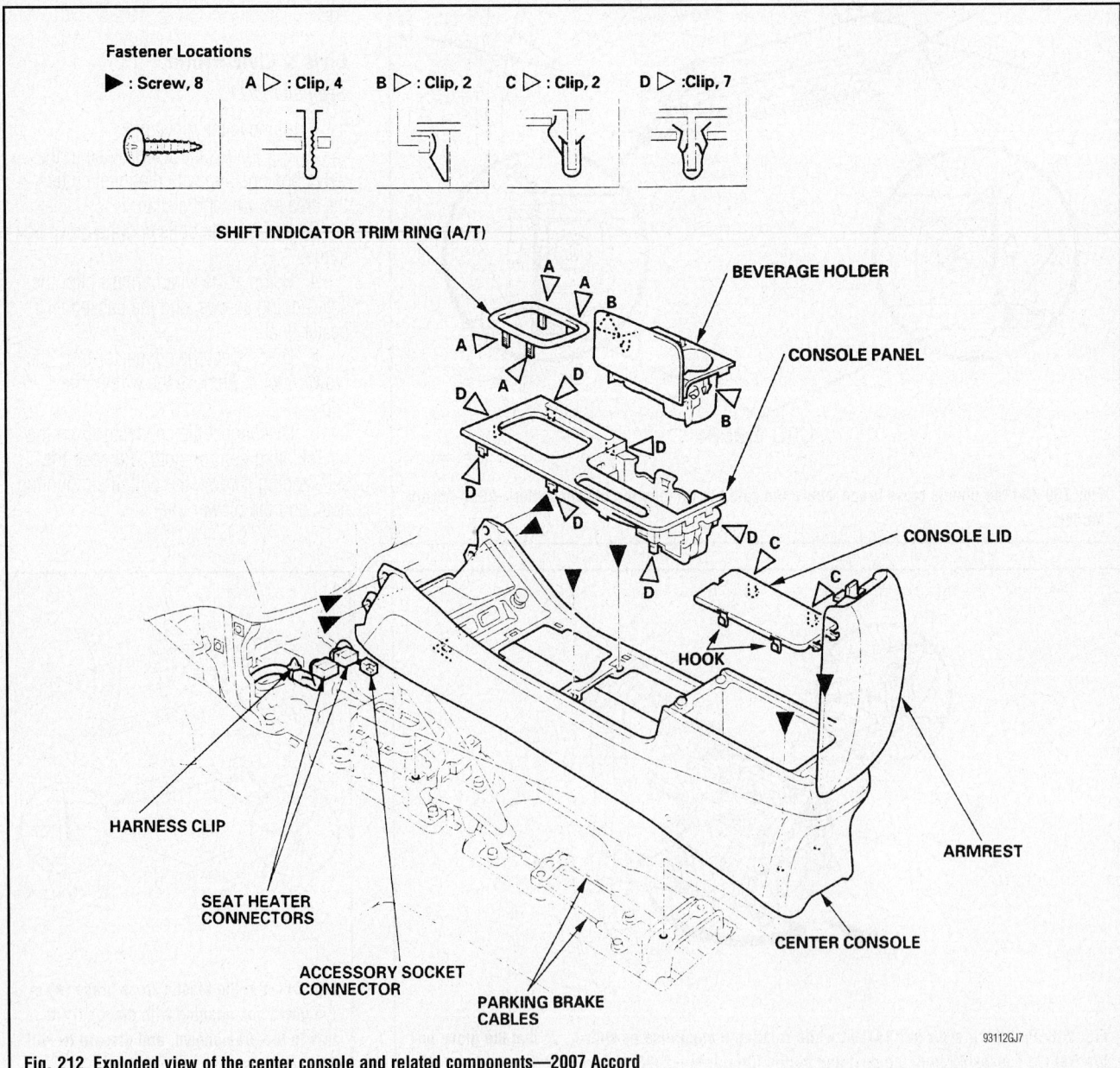

Fastener Locations

▶ : Screw, 8   A ▷ : Clip, 4   B ▷ : Clip, 2   C ▷ : Clip, 2   D ▷ : Clip, 7

SHIFT INDICATOR TRIM RING (A/T)

BEVERAGE HOLDER

CONSOLE PANEL

CONSOLE LID

HOOK

HARNESS CLIP

ARMREST

SEAT HEATER CONNECTORS

ACCESSORY SOCKET CONNECTOR

PARKING BRAKE CABLES

CENTER CONSOLE

93112GJ7

Fig. 212 Exploded view of the center console and related components—2007 Accord

09474_ACCO_G0001

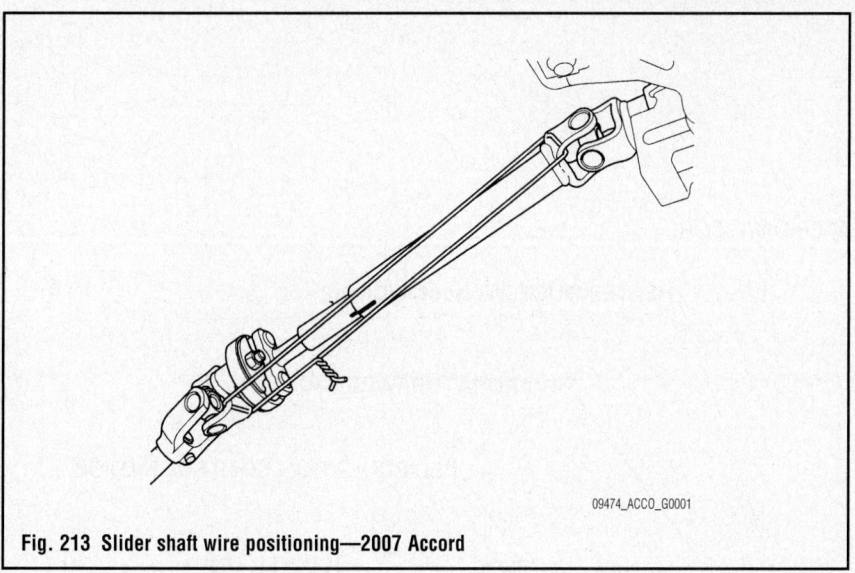

Fig. 213 Slider shaft wire positioning—2007 Accord

1. Disconnect the negative battery cable.

⁂ CAUTION

After disconnecting the negative battery cable, wait for at least 3 minutes for the air bag module to deplete its energy before working the on the instrument panel or steering wheel.

2. Discharge the air conditioning system. Disconnect the suction and the receiver lines from the evaporator core.
3. Drain the engine coolant.
4. From under the hood, open the cable clamp. Disconnect the heater control cable from the valve. Turn the valve to the fully opened position. Disconnect the heater hoses from heater core assembly. Remove the mounting nut from the heater unit.

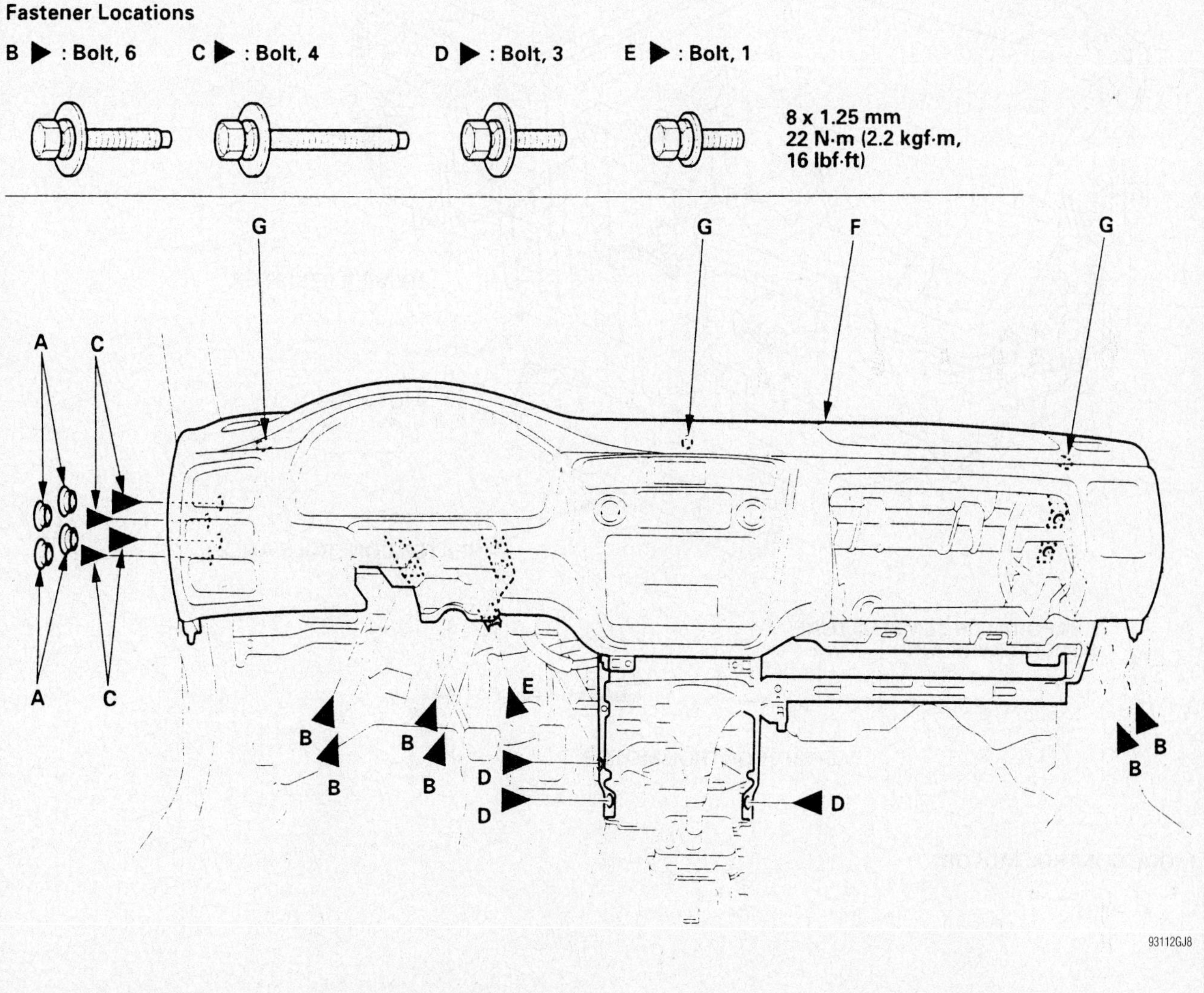

Fastener Locations
B ▶ : Bolt, 6
C ▶ : Bolt, 4
D ▶ : Bolt, 3
E ▶ : Bolt, 1
8 x 1.25 mm
22 N·m (2.2 kgf·m,
16 lbf·ft)
G
G
F
G
A
C
A
C
B
B
B
B
E
D
D
D
B
B
93112GJ8

Fig. 214 View of the instrument panel bolt locations—2007 Accord

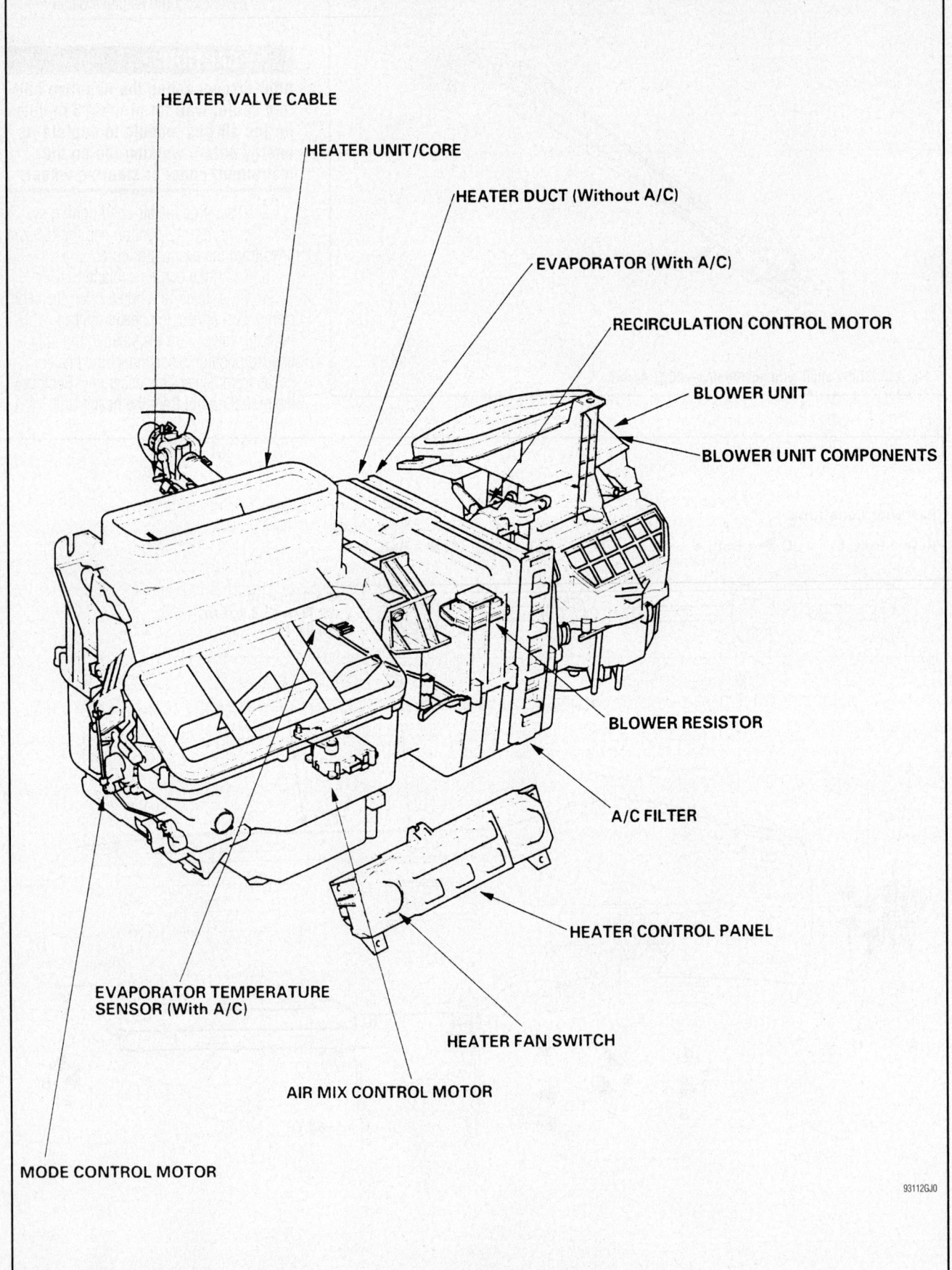

HEATER VALVE CABLE

HEATER UNIT/CORE

HEATER DUCT (Without A/C)

EVAPORATOR (With A/C)

RECIRCULATION CONTROL MOTOR

BLOWER UNIT

BLOWER UNIT COMPONENTS

BLOWER RESISTOR

A/C FILTER

HEATER CONTROL PANEL

HEATER FAN SWITCH

AIR MIX CONTROL MOTOR

EVAPORATOR TEMPERATURE
SENSOR (With A/C)

MODE CONTROL MOTOR

93112GJ0

Fig. 215 View of the heater housing, evaporator housing and related components—2007 Accord

5. Remove the center console as follows.

- If equipped with manual transaxle, remove the shifter knob
- Detach the center console retaining clips and pull the assembly upward
- If equipped with heated seats, detach the heat switch connectors
- Pry up on the right side of the console pocket, detach the clips and remove the pocket
- Remove the screws, detach the clips and disconnect the front power accessory connector
- Open the console lid and remove the console mat
- If equipped, disconnect the rear accessory power socket electrical connector
- Remove the console retaining screws and remove the console from the vehicle

6. Remove the driver's side lower dashboard cover.

7. Remove the passenger's side lower dashboard cover.

8. Remove the center pocket assembly.

9. Remove the glovebox.

10. Remove the driver's side under cover panel.

11. Remove the drivers and passengers kick panels.

12. Remove the A-pillar trim panels.

13. Remove the steering column as follows.

- Remove the driver's air bag assembly
- Remove the steering wheel
- Remove the steering column covers
- Remove the steering joint cover
- Release the tilt/telescopic lever and adjust the column in the full up position
- Tighten the tilt/telescopic lever
- Hold the slider shaft on the column using a piece of wire
- Release the tilt/telescopic lever and adjust the column to the full up position
- Tighten the tilt/telescopic lever
- Disconnect the wire harness from the combination switch and remove the switch
- Disconnect the connectors from the ignition switch and release the wire harness clips from the column
- Remove the steering joint bolt and disconnect the steering joint from the pinion shaft
- Remove the steering column retaining nuts and remove the column from the vehicle

14. Remove the driver's side center lower cover. Remove the passenger's side center lower cover and rear duct vent. Remove the SRS control unit.

15. On the driver's side disconnect all electrical connectors, air hoses and harness that interfere with the removal of the dashboard.

16. In the middle section of the dashboard disconnect the parking brake switch connector, radio antenna connector and lead. Disconnect all electrical connectors, air hoses and harness that interfere with the removal of the dashboard. Disconnect the ECM/PCM connector.

17. On the passenger's side disconnect all electrical connectors, air hoses and harness that interfere with the removal of the dashboard.

18. Remove the brake pedal support member.

19. Open the driver's side door remove the bolts and clips after remove their protective caps. Lift upward on the dashboard/steering hanger beam to release if from the guide pins. Be sure all electrical connectors are unplugged.

20. Carefully remove the dashboard assembly from the vehicle.

21. Disconnect the connectors from the air mix control motor, evaporator temperature sensor, power transistor and recirculation control motor.

22. Remove the wiring harness clips, connector clip and wire harness. Remove the heater ducts.

23. Remove the mounting nuts and remove the blower-heater unit from the vehicle.

24. Remove the heater core cover. Remove the heater core.

To install:

25. Installation is the reverse of the removal procedure.

26. Evacuate, charge and leak test the air conditioning system refrigerant.

27. Operate the engine to normal operating temperatures; then, check the climate control operation and check for leaks.

28. Enter the antitheft codes for the radio and the navigation system. Set the clock.

2008 Models

See Figures 216 through 218.

➡**Make sure to acquire the anti-theft code from the radio and write down the frequencies for the radio's preset buttons.**

1. Disconnect the negative battery cable.

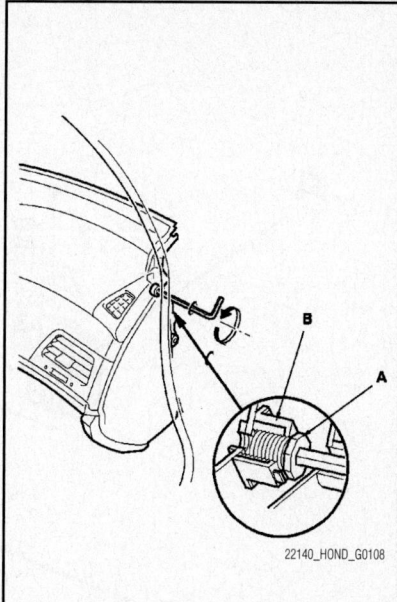

22140_HOND_G0108

Fig. 216 Slightly screw the collar bolt (A) into the space adjuster (B) with 8 mm hexagonal wench—2008 Accord Models

✳✳ **CAUTION**

After disconnecting the negative battery cable, wait for at least 3 minutes for the air bag module to deplete its energy before working the on the instrument panel or steering wheel.

2. Discharge the air conditioning system. Disconnect the suction and the receiver lines from the evaporator core.

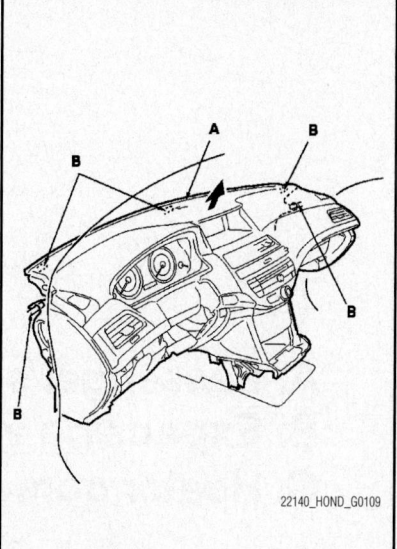

22140_HOND_G0109

Fig. 217 Lift up on the dashboard (A) to release it from the guide pins (B)—2008 Accord Models

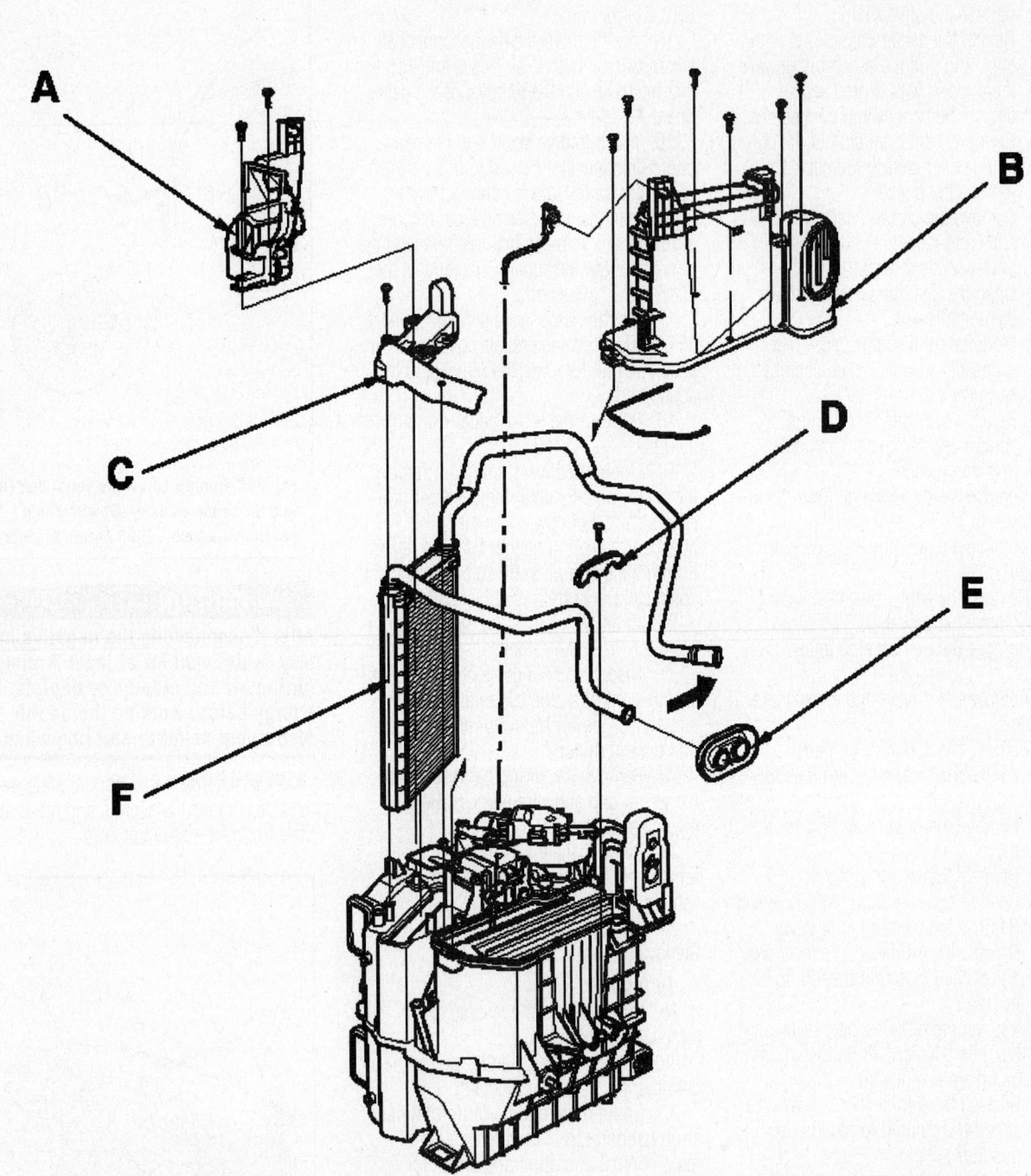

A. Passenger's heater duct
B. Expansion valve cover
C. Heater core cover
D. Heater pipe bracket
E. Grommet
F. Heater core

22140_HOND_G0110

Fig. 218 Exploded view of the heater unit—2008 Accord Models

3. Drain the engine coolant.

4. From under the hood, open the cable clamp. Disconnect the heater control cable from the valve. Turn the valve to the fully opened position. Disconnect the heater hoses from heater core assembly. Remove the mounting nut from the heater unit.

5. Remove the driver's dashboard lower cover as follows:

a. Adjust the steering column upward.

b. Pull out the bottom of the cover to detach the clips.

c. Pull out along the edge of the cover to detach the clips.

d. Release the bottom hooks.

e. Disconnect the VSA OFF switch connector

f. If necessary, remove the screws, and detach the hooks, then remove the pocket trim from the driver's dashboard lower cover.

6. Remove the center console as follows:

a. Remove the center console panel.

b. Remove the center console rear trim.

c. If equipped with heated seats, detach the harness clips fastening the front seat heater switch harnesses from the center console.

d. Disconnect the console accessory power socket connector and the auxiliary jack assembly connector, and detach the harness clip.

e. Remove the screws and the bolts securing the center console.

f. Detach the clips by pulling the front bottom edges of the center console out from both sides, and pull up the console to release the hooks, then remove the console.

7. Remove the dashboard center lower cover, both sides, as follows:

a. Release both front portions of the center console from the dashboard.

b. Remove the driver's dashboard center lower cover.

c. Remove the passenger's dashboard center lower cover.

8. Remove the glove box as follows:

a. Open the glove box.

b. Disconnect the glove box damper from the pivot on the glove box.

c. Close the glove box.

d. Remove the bolts.

e. While holding the glove box, release the glove box stop on each side from the dashboard by pushing them in, then remove the glove box.

f. Open the front door, and remove the passenger's dashboard side lid.

9. Remove the kick panels as follows:

a. Remove the front side cap from the door sill trim.

b. Driver's side: Remove the trunk lid/fuel fill door opener lock cylinder, and loosen the opener mounting bolt.

c. Driver's side: Remove the screw securing the door sill trim and trunk lid opener/fuel fill door opener.

d. Detach the hooks and the tabs from the kick panel and the rear side trim panel, and pull the door sill trim up by hand to detach the clips, then remove it.

e. Remove the driver's and passenger's kick panels.

10. Remove the A-pillar trim as follows:

a. Pull the door opening seal away from the A-pillar as needed.

b. Hit the upper clip in the A-pillar trim with a rubber mallet. The clip is located under the point where the triangle mark on the edge of the trim indicates. Hitting the clip breaks the projections on the pin and pushes it into the grommet on the body.

c. Pull the top of the A-pillar trim back by hand to remove the upper clip from the body.

d. Pull the A-pillar trim by hand to detach the clips. Pull the trim up from the dashboard, then remove it.

11. Remove the steering column as follows:

a. Adjust the steering column to the full tilt down position, and to the full telescopic out position.

b. Remove the driver's airbag, and the steering wheel.

c. Remove the steering column covers.

d. Remove the steering joint cover.

e. Loosen the upper steering joint bolt, and remove the lower steering joint bolt. Disconnect the steering joint by moving the steering joint toward the column. Do not disconnect the steering joint from the column shaft.

f. Disconnect the wire harness connectors from the combination switch assembly/cable reel.

g. Remove the combination switch assembly/cable reel from the steering column shaft by removing the three screws.

h. Disconnect the connectors from the ignition switch, and release the wire harness clips from the steering column.

i. Remove the steering column by removing the attaching nuts and bolts.

j. Remove the shift lever housing (M/T Models) or shift liver (A/T Models).

12. On the driver's side: From under the dash, disconnect the left engine compartment wire harness connectors, the driver's door wire harness connector, the roof wire harness connectors, and the left side wire harness connector, and disconnect the left engine compartment wire harness connectors and the left side wire harness connector from the driver's under-dash fuse/relay box.

13. If equipped with climate control: From under the dash, disconnect the air hose, then remove it.

14. From the shift lever portion: Remove the clip, and remove the left rear heater joint duct. Disconnect the floor wire harness connector, the SRS unit connector, the parking brake switch connector, and the yaw rate-lateral acceleration sensor connector, and detach the wire harness clips. Using a TORX T30 bit, remove the ground bolt.

15. From the middle portion: From under the dash, disconnect the A/C wire harness connector.

16. From the passenger's side: From under the dash, disconnect the passenger's door wire harness connector, the right side wire harness connectors, the antenna lead connector, and the right engine compartment wire harness connectors. Disconnect the dashboard wire harness connectors and the audio wire harness connector from the passenger's under-dash fuse/relay box. Detach the harness clip. With premium sound system: Disconnect the stereo amplifier connectors.

17. Detach all of the harness and the connector clips.

18. Detach the harness clip and the clip fastening the relay from the brake pedal support member. Remove the bolts then remove the member, and disconnect the TPMS control unit connector.

19. Remove bolts, then remove the center joint bracket.

20. Remove the special bolts from outside the passenger's door.

21. Slightly screw the collar bolt into the space adjuster with 8 mm hexagonal wench.

22. If the collar bolts are not screwed fully into the fixed space adjuster when removing the special bolts, screw collar bolt into the fixed space adjuster.

23. From outside the driver's door, remove the caps, then remove the bolts from outside the driver's door.

24. Lift up on the dashboard to release it from the guide pins. Carefully remove the dashboard through the front door opening. Take care not to scratch the body with the collar nuts on the passenger's side.

25. Disconnect the connector. Remove the clips, ducts, and the drain hose. Then remove the mounting bolt, mounting nuts, and the blower-heater unit.

26. Remove the two screws, then remove the cover.

27. Disconnect the connector from the blower motor. Remove the wire harness clip.

28. Disconnect these connectors: The mode control motor, the power transistor, the evaporator temperature sensor, the passenger's air mix control motor (with climate control), and the recirculation control motor. Remove the wire harness clips.

29. Disconnect the connector from the air mix control motor. Remove the wire harness clips, the connector clip, and the wire harness.

30. Remove the self-tapping screws and the passenger's heater duct. Remove the self-tapping screws and the expansion valve cover. Remove the self-tapping screw and the heater core cover. Remove the self-tapping screws, the heater pipe bracket, and

the grommet, and carefully pull out the heater core.

31. Installation is the reverse order of removal.

32. Refill the engine cooling system to the correct level.

33. Recharge the A/C refrigerant.

Civic

See Figures 219 and 220.

➡**Make sure to acquire the anti-theft code from the radio and write down the frequencies for the radio's preset buttons.**

1. Disconnect the negative battery cable.

✳✳ CAUTION

After disconnecting the negative battery cable, wait for at least 3 minutes for the air bag module to deplete its energy before working the on the instrument panel or steering wheel.

2. Discharge the air conditioning system.

3. Remove the air cleaner assembly. Disconnect the heater hoses. Remove the mounting nut from the heater unit.

4. Remove the subdisplay visor. Remove the navigation system, if equipped. On vehicles without navigation system remove the radio.

5. Remove the dashboard retaining screws. Detach the retaining clips along the lower edge of the instrument panel.

6. Detach the retaining clips along the upper edge of the instrument panel. Gently pull forward to release the hooks from the holder of the gauge module.

7. Remove the instrument panel.

8. Disconnect the electrical connector from the blower motor. Remove the wire harness clip. Disconnect the connector from the recirculation control motor.

9. Disconnect the connectors from the mode control motor, the evaporator temperature sensor and the power transistor. Remove the wire clip harness.

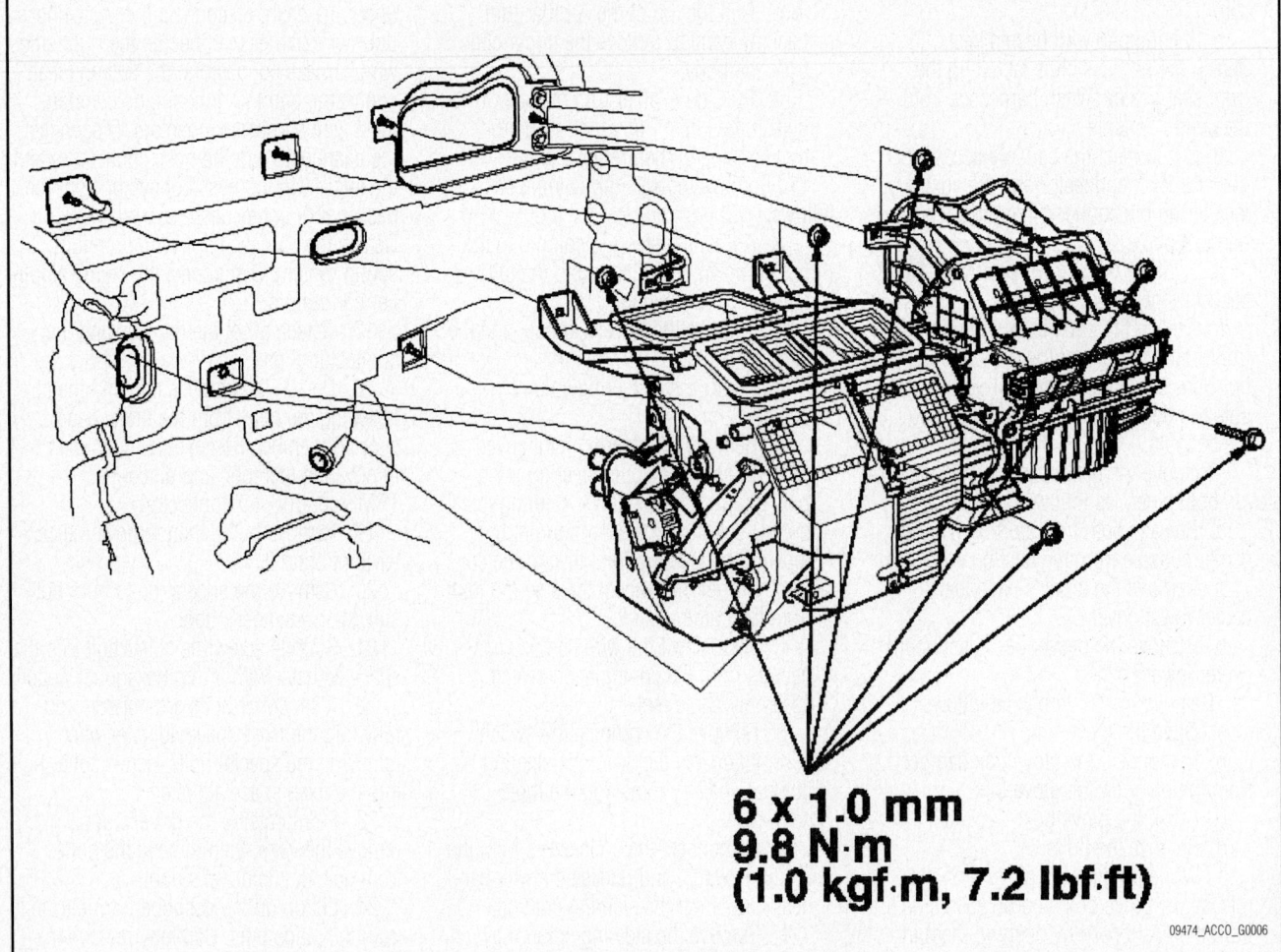

6 x 1.0 mm
9.8 N·m
(1.0 kgf·m, 7.2 lbf·ft)

09474_ACCO_G0006

Fig. 219 View of the heater housing, evaporator housing and related components—Civic

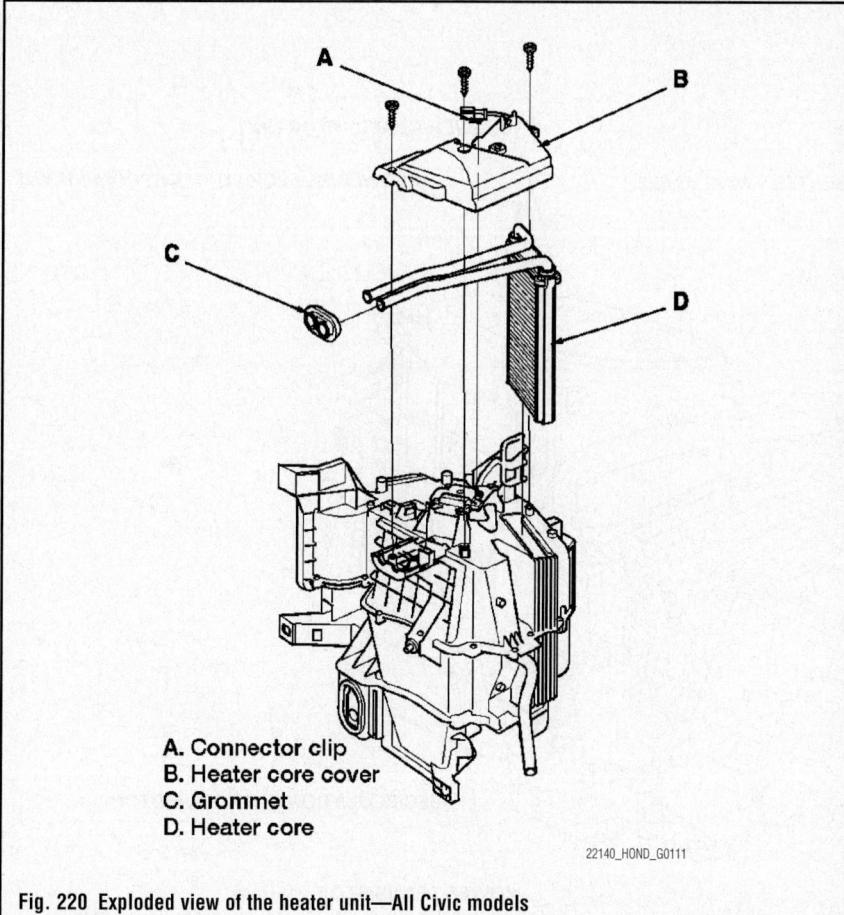

A. Connector clip
B. Heater core cover
C. Grommet
D. Heater core

22140_HOND_G0111

Fig. 220 Exploded view of the heater unit—All Civic models

10. Disconnect the connectors from the air mix control motor and the air conditioning wire harness. Remove the connector clip and the wire harness clips.

11. Remove the mounting bolt, mounting nuts and the heater/blower unit from the vehicle.

12. Remove the self taping screws, remove the grommet and carefully pry out the heater core.

To install:

13. Installation is the reverse of the removal procedure.

14. Evacuate, charge and leak test the air conditioning system refrigerant, as required.

15. Operate the engine to normal operating temperatures; then, check the climate control operation and check for leaks.

16. Enter the antitheft codes for the radio and the navigation system.

S2000

See Figure 221.

➡**Make sure to acquire the anti-theft code from the radio and write down the frequencies for the radio's preset buttons.**

1. Disconnect the negative battery cable.

✳✳ CAUTION

After disconnecting the negative battery cable, wait for at least 3 minutes for the air bag module to deplete its energy before working the on the instrument panel or steering wheel.

2. Drain the engine coolant. Disconnect the heat shield from the exhaust manifold. Remove the battery.

3. Discharge the air conditioning system. Disconnect the suction and the receiver lines from the evaporator core.

4. From under the hood, open the cable clamp. Disconnect the heater control cable from the valve. Turn the valve arm to the fully open position.

5. Remove the mounting bolt from the heater valve. Disconnect the heater hoses from the heater unit. Remove the mounting nut from the heater unit. Disconnect the air conditioning lines from the evaporator housing.

6. Remove the radio panel retaining clips and remove the panel. Remove the radio retaining screws and pull the unit

forward. Disconnect the antenna lead and electrical connectors and remove the radio.

7. Remove the driver's side air bag. Remove the steering wheel and cable reel. Remove the steering column covers.

8. Remove the combination switch assembly from the column shaft by disconnecting the connectors and removing the screws. Disconnect the ignition switch connectors from the under dash fuse/relay box.

9. Remove the steering joint bolt, at the base of the column. Remove the steering column retaining bolts and remove the column.

10. Remove the driver's side lower dash cover. Remove the passenger's side lower dash cover. Remove the drivers and passengers side front console cover.

11. Remove the passenger's side air bag assembly.

12. Remove the kick panels and A-pillar trim from both the drivers and passenger's side.

13. Disconnect all electrical connectors, air hoses and harness that interfere with the removal of the dashboard. Remove the ground bolts.

14. Open the driver's side door remove the bolts and clips after remove their protective caps. Lift upward on the dashboard/steering hanger beam to release if from the guide pins. Be sure all electrical connectors are unplugged.

15. Remove the blower/evaporator housing from the vehicle.

16. Remove the mounting bolts, the center brackets and the radio mounting brackets.

17. Remove the SRS unit. Remove the defroster outlet and wire harness clips.

18. Disconnect the connectors from the mode control motor and the air mix control motor. Remove the wire harness clip.

19. Remove the mounting nuts, the mounting bolts and remove the heater unit from the vehicle.

20. Remove the heater core cover. Remove the heater core.

To install:

21. Installation is the reverse of the removal procedure.

22. Evacuate, charge and leak test the air conditioning system refrigerant.

23. Operate the engine to normal operating temperatures; then, check the climate control operation and check for leaks.

24. Enter the antitheft codes for the radio and the navigation system. Set the clock.

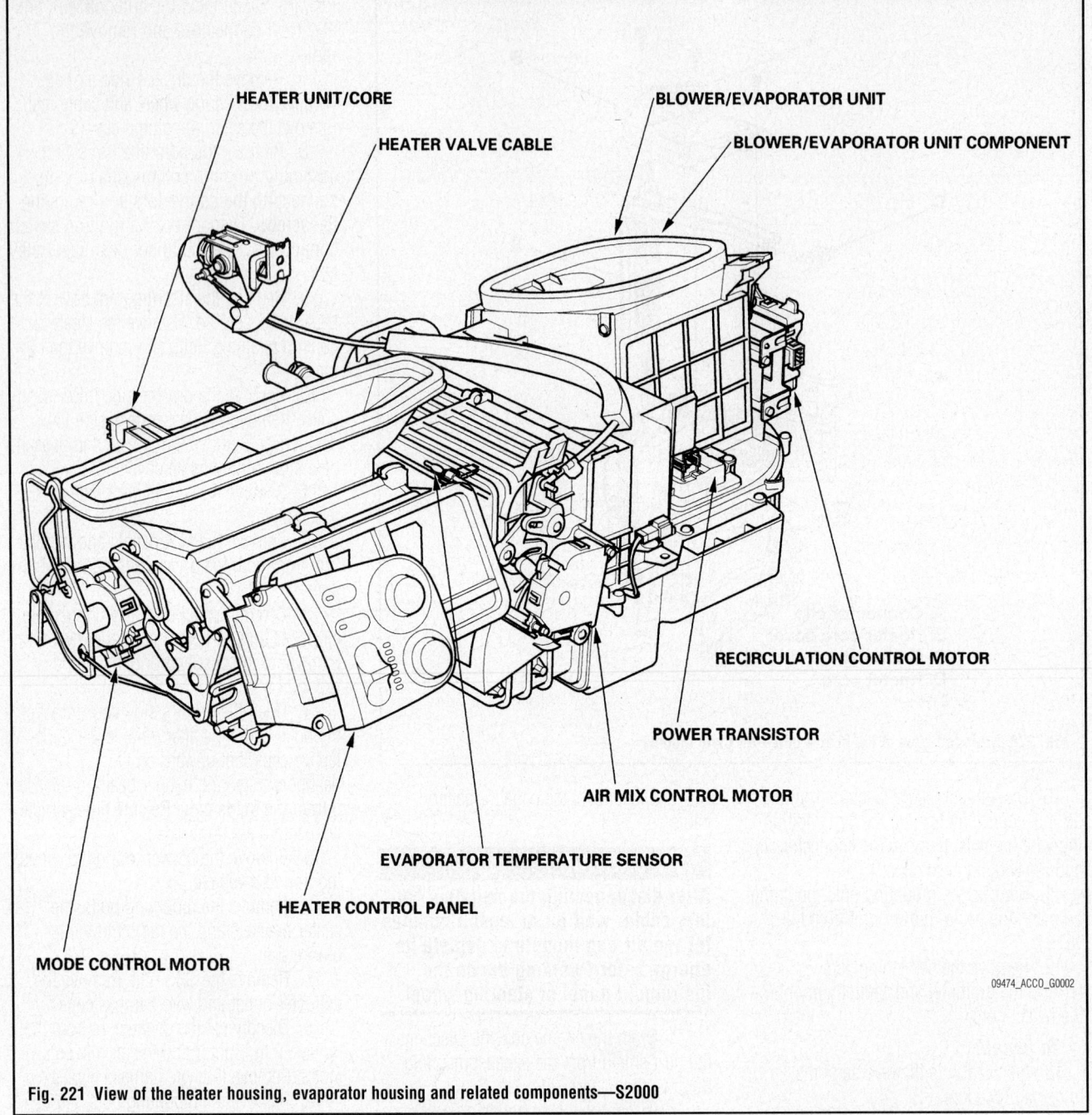

HEATER UNIT/CORE

HEATER VALVE CABLE

BLOWER/EVAPORATOR UNIT

BLOWER/EVAPORATOR UNIT COMPONENT

RECIRCULATION CONTROL MOTOR

POWER TRANSISTOR

AIR MIX CONTROL MOTOR

EVAPORATOR TEMPERATURE SENSOR

HEATER CONTROL PANEL

MODE CONTROL MOTOR

09474_ACCO_G0002

Fig. 221 View of the heater housing, evaporator housing and related components—S2000

25. Reprogram the ECM engine idle characteristics. Be sure all electrical items are OFF.

26. Reset the ECM with the HDS. Turn the ignition switch to the ON position. Wait two seconds.

27. Start the engine. Hold the idle speed at 3000 RPM's in neutral until the radiator fan comes on or the temperature reached 194 degrees.

28. Let the engine idle for about five minutes with the throttle fully closed.

29. If the radiator fan comes on during the five minutes, do not count this toward the five minute programming time.

STEERING

POWER STEERING GEAR

REMOVAL & INSTALLATION

Accord

✳✳ CAUTION

The air bag must be disabled before removing the steering wheel to center the cable reel. Failure to disarm the air bag system may cause accidental air bag deployment, resulting in unnecessary air bag system repairs and the risk of personal injury.

➡**The radio may contain a coded theft protection circuit. Always obtain the code number before disconnecting the battery.**

1. Before servicing the vehicle, refer to the precautions in the beginning of this section.
2. Disconnect the negative and positive battery cables. Wait 3 minutes before working around the air bags.
3. Raise and support the vehicle safely. Remove the front tires. Drain the power steering fluid.
4. Remove the driver's side airbag. Remove the steering wheel.

➡**Be sure to remove the steering wheel before disconnecting the steering joint as damage to the cable reel can occur.**

5. Remove the top half of the steering joint cover.
6. Remove the steering joint bolts. Disconnect the steering joint by moving the joint toward the column. Hold the lower slide shaft on the column with a piece of wire between the joint yoke on the lower slide shaft to the joint yoke on the upper shaft.
7. Remove the lower half of the steering joint cover.
8. Remove the cotter pin from the tie rod ball joint and loosen the nut. Separate the tie rod ball joint and the knuckle, using the proper tool.
9. Remove the power steering heat baffle plate. Remove the feed line holder mounting bolt on the front suspension subframe.
10. Remove the feed line holder mounting bolt and return hose from the gearbox mounting bracket.
11. Place shop towels under the fluid lines and disconnect them. Plug the lines to prevent dirt from entering the system.

➡**Do not loosen the cylinder line between the valve body unit and the cylinder.**

12. Attach tool EQS02C000016 to the front subframe.
13. Remove the front subframe right middle mount bolt. Remove the front subframe left middle mount bolt.
14. Remove the steering gearbox mounting bolts on the left gearbox mount. Remove the steering stiffener plate.
15. Remove the front suspension subframe rear bracket and mounting bolt.
16. Loosen the two front subframe rear mounting bolts on the right and left. Slowly lower the special tool supporting the subframe until the subframe has dropped 1 1/8 in.
17. Remove the two flange bolts from the right side of the gearbox. Remove the gearbox mounting bracket and cushion.
18. Move the steering gearbox toward the front and remove the pinion shaft grommet from the top of the valve body unit.
19. Apply vinyl tape to the splines on the pinion shaft. Tape the brake lines to protect them from the pinion shaft.
20. Move the steering gearbox to the driver's side, and rotate it so that the pinion shaft points toward the front of the vehicle.
21. Carefully move the steering gearbox as an assembly toward the driver's side of the vehicle until the pinion shaft clears the wheel well opening.
22. Remove the steering gearbox through the wheel well opening on the driver's side of the vehicle.

To install:

23. Position the assembly to its mounting on the vehicle.
24. Install the mounting bolts.
25. Continue the installation in the reverse order of the removal procedure.
26. Be sure to center the cable reel by first rotating it clockwise until it stops. Then rotate it counterclockwise (approximately two and a half turns) until the arrow mark on the label points straight up. Install the steering wheel.
27. Fill the system with the proper grade and type power steering fluid.
28. Start the engine, allow it to idle, and turn the steering wheel from lock-to-lock several times to warm up the fluid. Check and adjust the fluid level.
29. Check the gearbox for fluid leaks, correct as necessary.
30. Check front end alignment.
31. Check the steering wheel spoke angle. Adjust by turning the right and left tie rods equally, if necessary.

Civic & Civic Hybrid

See Figure 222.

✳✳ CAUTION

The air bag must be disabled before removing the steering wheel to center the cable reel. Failure to disarm the air bag system may cause accidental air bag deployment, resulting in unnecessary air bag system repairs and the risk of personal injury.

➡**The radio may contain a coded theft protection circuit. Always obtain the code number before disconnecting the battery.**

1. Before servicing the vehicle, refer to the precautions in the beginning of this section.
2. Disconnect the negative and positive battery cables. Wait 3 minutes before working around the air bags.
3. Raise and support the vehicle safely. Remove the front tires. Drain the power steering fluid.
4. Remove the driver's side airbag. Remove the steering wheel.

➡**Be sure to remove the steering wheel before disconnecting the steering joint as damage to the cable reel can occur.**

5. Remove the driver's dashboard lower cover. Remove the driver's dashboard under cover.
6. Remove the steering joint bolts. Disconnect the steering joint by moving the joint toward the column. Hold the lower slide shaft on the column with a piece of wire between the joint yoke on the lower slide shaft to the joint yoke on the upper shaft.
7. Remove the center guide, if equipped and discard it. The center guide is for factory assembly only.
8. Remove the harness holder and the air cleaner housing bracket. Remove the breather pipe and the harness holder bracket.
9. Attach special tool 07AAK-SNAA120, or equivalent to the cylinder head. Install the engine support hanger, AAR-12566) or equivalent. Attach the hook to the tool and tighten the wing nut. Lift and support the engine.

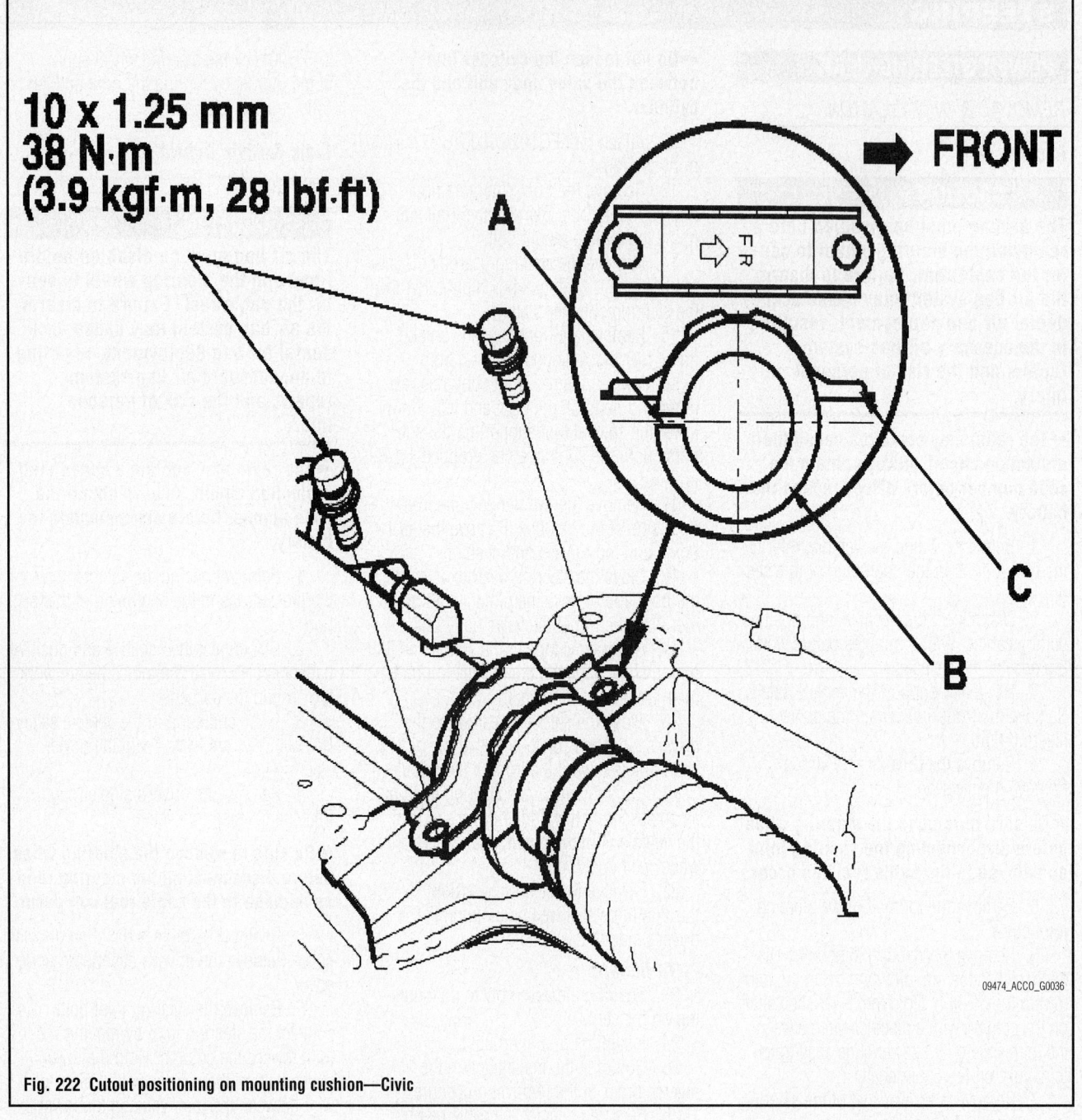

10 x 1.25 mm
38 N·m
(3.9 kgf·m, 28 lbf·ft)

A

➡ FRONT

C

B

09474_ACCO_G0036

Fig. 222 Cutout positioning on mounting cushion—Civic

10. Remove the cotter pin from the tie rod ball joint and loosen the nut. Separate the tie rod ball joint and the knuckle, using the proper tool.

11. Remove the pump outlet hose clamp from the intake manifold. Remove the inlet line clamp bolt. Open the return line holder and remove the return line clamp bolt.

12. Loosen the adjustable hose clamp and disconnect the return hose. Loosen the flare nut and disconnect the inlet line.

13. Remove the splash shield. Remove the lower ball joint mounting bolt and flange nuts from the lower arm. Disconnect the lower arm from the lower ball housing.

14. Attach tool VSB02C000016 to the front subframe and a transaxle jack or powertrain lift tool. Make sure that the subframe is securely supported.

15. Remove the front subframe middle mount bolt from the left side. Remove the front subframe middle mount bolt from the right side. Remove the two 12mm flange bolts from the lower torque rod and bracket.

16. Remove the front subframe front mounting bolts from the right and left sides of the vehicle. Discard them.

17. Remove the front subframe rear mounting bolts from the right and left sides of the vehicle. Discard them.

18. Lower the front subframe and steering gearbox as an assembly. Remove the pinion shaft grommet from the top of the valve body unit.

19. Remove the two 10mm bolts from the right side of the steering gearbox. Remove the mounting bracket and mounting cushion.

20. Remove the four 10mm flange bolts from the left side of the steering gearbox. Remove the stiffener plates.

21. Remove the steering gearbox from the front subframe

To install:

22. Position the assembly to its mounting on the subframe.

23. Loosely install the stiffener plates and gearbox mounting bolts on the left side of the steering gearbox.

24. Position the cutout on the mounting cushion and install it on the right side of the steering gearbox.

25. Install the gearbox mounting bracket over the mounting cushion and loosely install the two 10mm bolts.

26. Tighten the 10mm bolts on both sides of the steering gearbox to 40 ft. lbs. right side and 28 ft. lbs. left side, alternately in two or more steps.

27. Install the pinion shaft grommet. Align the slot in the pinion shaft grommet with the lug portion on the valve housing. The grommet must not have a gap at the mating surface of the grommet and valve housing.

28. Carefully raise the front subframe in position.

29. Continue the installation in the reverse order of the removal procedure.

30. Be sure to center the cable reel by first rotating it clockwise until it stops. Then rotate it counterclockwise (approximately two and a half turns) until the arrow mark on the label points straight up. Install the steering wheel.

31. Fill the system with the proper grade and type power steering fluid.

32. Start the engine, allow it to idle, and turn the steering wheel from lock-to-lock several times to warm up the fluid. Check and adjust the fluid level.

33. Check the gearbox for fluid leaks, correct as necessary.

34. Check front end alignment.

35. Check the steering wheel spoke angle. Adjust by turning the right and left tie rods equally, if necessary.

S2000

See Figure 223.

> ※※ **CAUTION**
>
> **The air bag must be disabled before removing the steering wheel to center the cable reel. Failure to disarm the air bag system may cause accidental air bag deployment, resulting in unnecessary air bag system repairs and the risk of personal injury.**

➡**The radio may contain a coded theft protection circuit. Always obtain the code number before disconnecting the battery.**

1. Remove or disconnect the following:
   • Negative battery cable
   • Front wheels
   • Driver's air bag
   • Steering wheel
   • Steering coupler
   • Outer tie rod ends
   • Splash shield
   • Stabilizer bar brackets
   • Steering gear wiring connectors
   • Steering gear mounting bolts

2. Move the steering gear forward and to the right to remove the steering gear.

To install:

3. Installation is the reverse of the removal procedure, while using the following torque values:
   • Steering gear mounting bolts: 33 ft. lbs. (44 Nm)
   • Steering gear ground cable bolt: 88 inch lbs. (10 Nm)

• Stabilizer bar bracket bolts: 61 ft. lbs. (83 Nm)
• Splash shield bolts: 88 inch lbs. (10 Nm)
• Outer tie rod end nuts: 40 ft. lbs. (54 Nm)
• Steering coupler pinch bolts: 16 ft. lbs. (22 Nm)

POWER STEERING PUMP

REMOVAL & INSTALLATION

Accord

2.4L Engine

See Figure 224.

1. Place a suitable container under the vehicle.

2. Drain the power steering fluid from the reservoir.

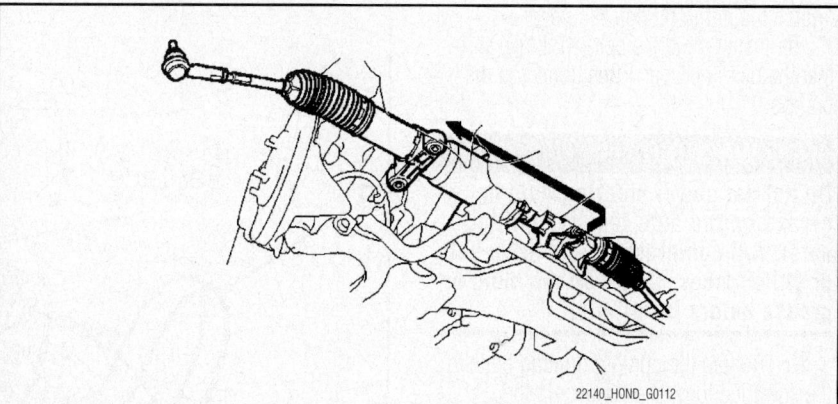

Fig. 223 Move the steering gear forward and to the right to remove the steering gear—S2000 Models

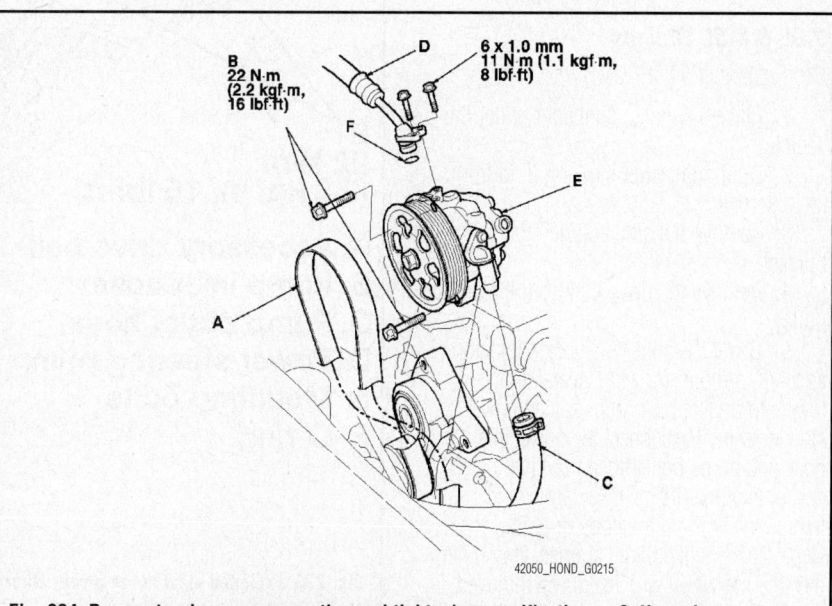

Fig. 224 Power steering pump mounting and tightening specifications—2.4L engine

3. Remove the drive belt (A) from the pump pulley.

4. Remove the pump mounting bolts (B).

5. Cover the auto-tensioner, alternator, and A/C compressor with several shop towels to protect them from spilled power steering fluid. Disconnect the pump inlet hose (C) and pump outlet hose (D) from the pump (E), and plug them. Take care not to spill the fluid on the body or parts. Wipe off any spilled fluid at once. Do not turn the steering wheel with the pump removed.

6. Cover the opening of the pump with a piece of tape to prevent foreign material from entering the pump.

To install:

7. Connect the pump inlet hose and the pump outlet hose onto the new pump with the new O-ring (F).

8. Loosely install the pump in the pump bracket with the mounting bolts, then tighten the pump fittings securely.

9. Install the drive belt (A). Make sure that the belt is properly positioned on the pulleys.

❋❋ WARNING

Do not get power steering fluid or grease on the auto-tensioner, alternator, A/C compressor, and drive belt or pulley faces. Clean off any fluid or grease before installation.

10. Tighten the pump mounting bolts to the specified torque, as shown in the accompanying illustration.

11. Fill the reservoir to the upper level line and bleed the system, as outlined in this section.

3.0L & 3.5L Engines

See Figures 225 and 226.

1. Place a suitable container under the vehicle.

2. Drain the power steering fluid from the reservoir.

3. Remove the side engine mount bracket.

4. Remove the drive belt from the pump pulley.

5. Cover the auto-tensioner, alternator, and A/C compressor with several shop towels to protect them from spilled power steering fluid. Disconnect the pump inlet hose and pump outlet hose from the pump, and plug them. Take care not to spill the fluid on the body or parts. Wipe off any spilled fluid at once. Do not turn the steering wheel with the pump removed.

6. Remove the pump mounting bolts.

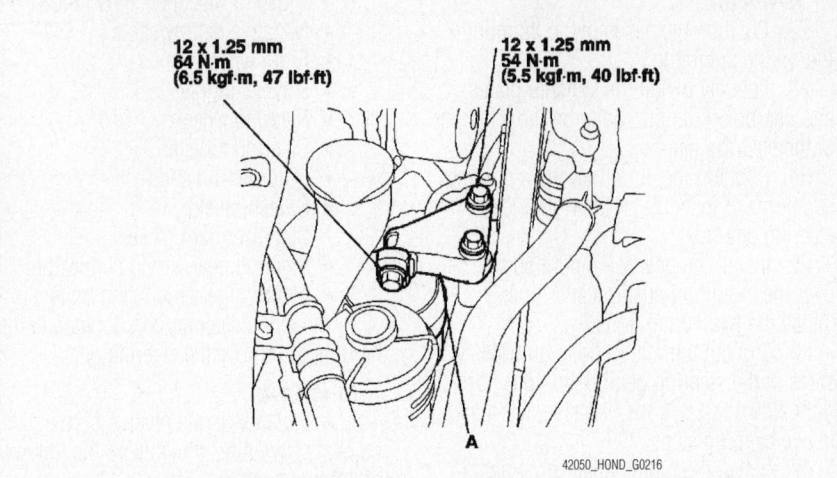

12 x 1.25 mm
64 N·m
(6.5 kgf·m, 47 lbf·ft)

12 x 1.25 mm
54 N·m
(5.5 kgf·m, 40 lbf·ft)

A

42050_HOND_G0216

Fig. 225 Side engine mount bracket (A) mounting and tightening specifications—3.0L engine shown, 3.5L engine similar

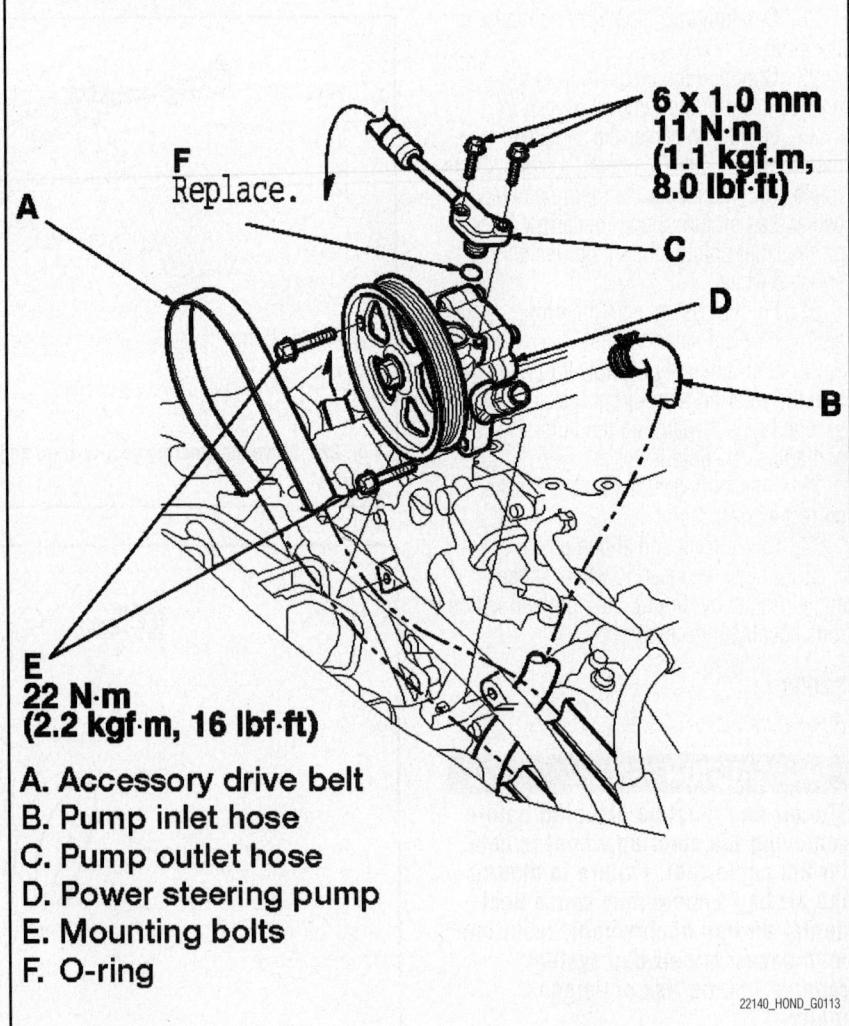

F
Replace.

A

6 x 1.0 mm
11 N·m
(1.1 kgf·m, 8.0 lbf·ft)

C

D

B

E
22 N·m
(2.2 kgf·m, 16 lbf·ft)

A. Accessory drive belt
B. Pump inlet hose
C. Pump outlet hose
D. Power steering pump
E. Mounting bolts
F. O-ring

22140_HOND_G0113

Fig. 226 Exploded view of the power steering pump and tightening specifications—3.5L engine shown, 3.0L engine similar

7. Cover the opening of the pump with a piece of tape to prevent foreign material from entering the pump.

To install:

8. Connect the pump inlet hose and pump outlet hose onto the new pump with the new O-ring. Loosely install the pump in the pump bracket with the mounting bolts, then tighten the pump fittings securely.

9. Tighten the pump mounting bolts to the specified torque.

10. Install the drive belt. Make sure that the belt is properly positioned on the pulleys.

✳✳ WARNING

Do not get power steering fluid or grease on the auto-tensioner, alternator, A/C compressor, and drive belt or pulley faces. Clean off any fluid or grease before installation.

11. Install the side engine mount bracket. Tighten the bolts to the specified torque.

12. Fill the reservoir to the upper level line and bleed the system as outlined in this section.

Civic

See Figures 227 and 228.

1. Place a suitable container under the vehicle.

2. Drain the power steering fluid from the reservoir.

3. Remove the cowl cover and under-cowl panel.

4. Remove the air cleaner.

5. Remove the front splash shield.

6. If equipped with a manual transaxle, remove the shift cable bracket.

7. If equipped with an automatic transaxle, disconnect the shift cable from the control lever.

8. Remove the upper torque rod from the body.

9. Remove the drive belt from the pump pulley.

10. Cover the parts around the power steering pump with several shop towels to protect them from spilled power steering fluid.

11. Disconnect the pump inlet hose and pump outlet hose from the pump, and plug them.

✳✳ WARNING

Take care not to spill the fluid on the body or any parts. Wipe off any spilled fluid at once. Do not turn the steering wheel with the pump removed.

12. Remove the pump outlet hose O-ring, and discard it.

13. Remove the pump mounting bolts.

14. Cover the opening of the pump with a piece of tape to prevent foreign material from entering the pump.

To install:

15. Move the power steering pump toward the driver's side, then raise it.

16. Connect the pump inlet hose and pump outlet hose onto the new pump with new O-ring.

17. Loosely install the pump in the pump bracket with the mounting bolts, then tighten the pump fittings securely.

18. Tighten the pump mounting bolts to the specified torque shown in the accompanying illustration.

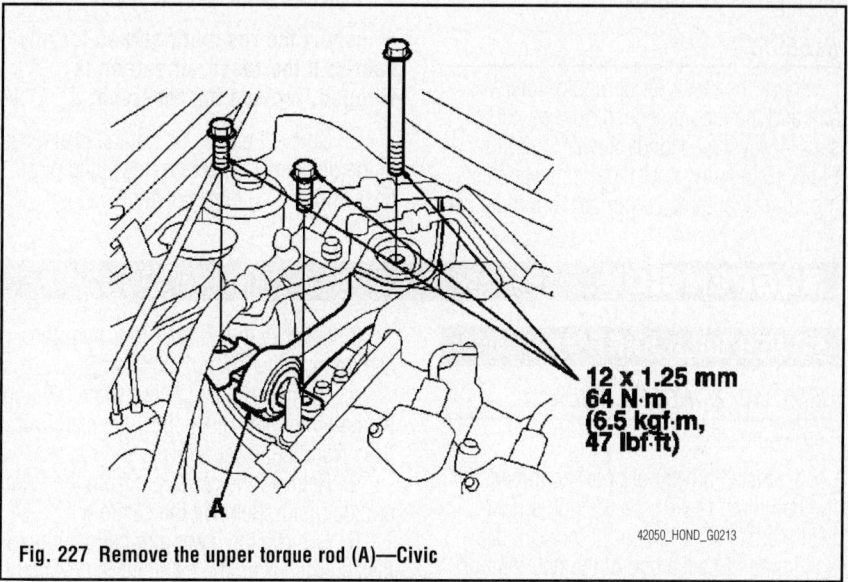

Fig. 227 Remove the upper torque rod (A)—Civic

42050_HOND_G0213

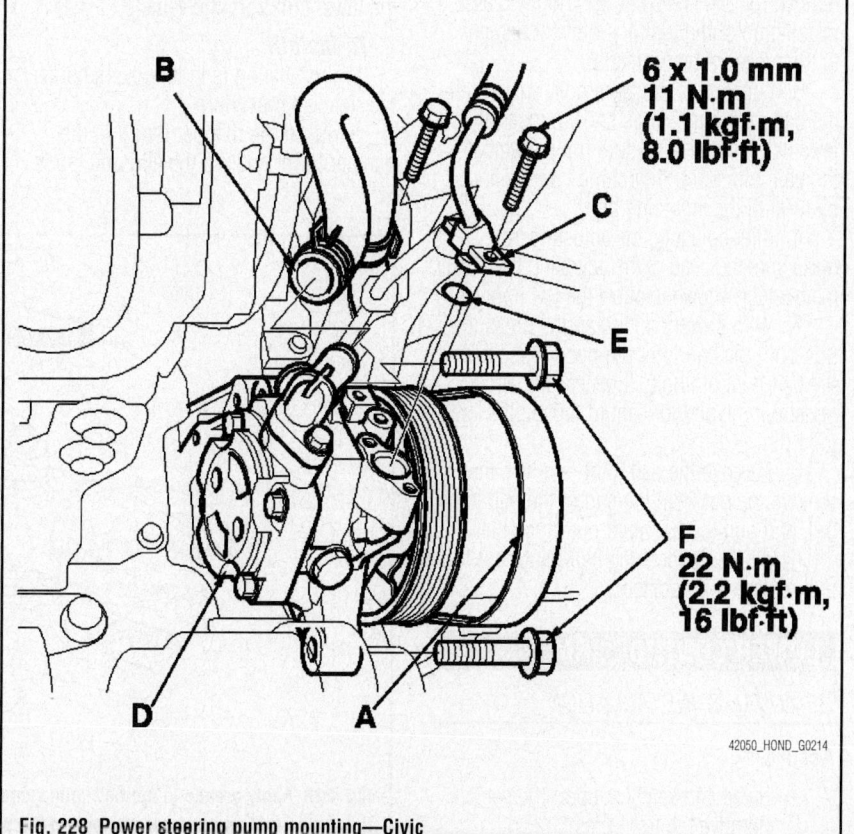

Fig. 228 Power steering pump mounting—Civic

42050_HOND_G0214

19. Install the drive belt. Make sure that the belt is properly positioned on the pulleys.

❈ WARNING

Do not get power steering fluid or grease on any parts around the power steering pump, drive belt, or pulley faces. Clean off any fluid or grease before installation.

20. Fill the reservoir to the upper level line and bleed the system.

BLEEDING

Check the reservoir at regular intervals, and add the recommended fluid as necessary. Always use Honda Power Steering Fluid. Using any other type of power steering fluid or automatic transmission fluid can cause increased wear and poor steering in cold weather.

➡**If the fluid is contaminated, the screen in the reservoir may be partially blocked. Replace the reservoir if necessary.**

1. Remove the reservoir from its holder. Raise the reservoir, then disconnect the return hose to drain the reservoir. Take care not to spill the fluid on the body and parts. Wipe off any spilled fluid at once.

➡**Inspect the reservoir screen for any debris. If the reservoir screen is clogged, replace the reservoir.**

2. Connect a hose of suitable diameter to the disconnected return hose, and put the hose end in a suitable container.

3. Start the engine, let it run at idle, and turn the steering wheel from lock-to-lock several times. When fluid stops running out of the hose, shut off the engine. Discard the fluid.

4. Reinstall the return hose on the reservoir.

5. Fill the reservoir to the upper level line.

6. Start the engine and run it at fast idle, then turn the steering from lock-to-lock several times to bleed air from the system.

7. Recheck the fluid level and add some if necessary. Do not fill the reservoir beyond the upper level line.

8. If the fluid is contaminated, dark, or discolored, repeat the procedure as necessary.

SUSPENSION

LOWER BALL JOINT

REMOVAL & INSTALLATION

See Figure 229.

1. Install a hex nut onto the threads of the ball joint. Make sure the nut is flush with the ball joint pin end to prevent damage to the threaded end of the ball joint pin.

2. Apply grease to the ball joint remover on the areas shown. This will ease installation of the tool and prevent damage to the pressure bolt threads.

3. Loosen the pressure bolt, and install the ball joint remover as shown. Insert the jaws carefully, making sure not to damage the ball joint boot. Adjust the jaw spacing by turning the adjusting bolt.

4. After adjusting the adjusting bolt, make sure the head of the adjusting bolt is in the position shown to allow the jaw to pivot.

5. With a wrench, tighten the pressure bolt until the ball joint pin pops loose from the ball joint pin hole. If necessary, apply penetrating type lubricant to loosen the ball joint pin.

6. Remove the ball joint remover, then remove the nut from the end of the ball joint pin, and pull the ball joint out of the ball joint pin hole. Inspect the ball joint boot, and replace it if damaged.

LOWER CONTROL ARM

REMOVAL & INSTALLATION

Accord

1. Raise and safely support the vehicle.
2. Remove the front tires.

3. Remove the damper fork from the damper and the lower control arm.

4. Remove the flange nut while holding the ball joint pin. Disconnect the stabilizer links from the lower arm.

5. Remove the cotter pin from the lower arm ball joint. Remove the castle nut.

6. Remove the lower arm ball joint from the knuckle, using the ball joint removal tool.

7. Remove the flange bolts and remove the lower arm from the vehicle.

To install:

8. Installation is in the reverse order of the removal procedure.

9. Insert the damper fork into the damper lower end, so the aligning tab is

FRONT SUSPENSION

aligned with the slot in the damper fork. Replace the damper fork retaining nut, with a new one.

10. Be sure to use new flange bolts, castle nut and lock pin.

11. Check and adjust the wheel alignment, as necessary.

Civic

1. Raise and support the vehicle safely. Remove the front tires.

2. Remove the flange nut while holding the ball joint pin. Disconnect the stabilizer links from the lower arm.

3. Turn the stabilizer bar backward to gain access to the front side of the lower arm mounting bolt.

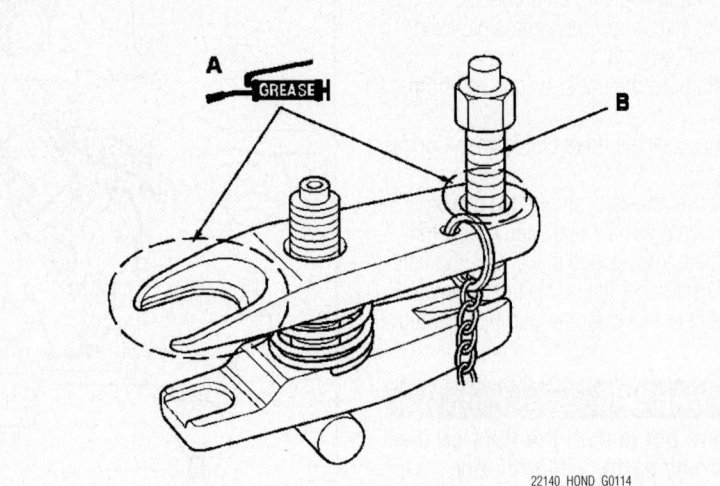

22140_HOND_G0114

Fig. 229 Apply grease to the ball joint remover on the areas shown (A). This will ease installation of the tool and prevent damage to the pressure bolt (B) threads.

4. Remove the flange bolt and nuts from the lower arm.

5. Disconnect the lower ball joint from the lower arm.

6. Remove the lower arm mounting bolts. Remove the lower arm from the front suspension subframe.

To install:

7. Installation is in the reverse order of the removal procedure.

8. Be sure to use new flange bolts, castle nut and lock pin. Tighten the flange nut to 25 ft. lbs. (34 Nm).

9. Check and adjust the wheel alignment, as necessary.

S2000

See Figure 230.

1. Raise and support the vehicle safely. Remove the front tires.

2. Remove the flange nut while holding the ball joint pin. Disconnect the stabilizer links from the lower arm.

3. Remove the flange bolt and nuts from the lower arm.

4. Remove the cotter pin from the lower arm ball joint. Remove the castle nut.

5. Remove the lower arm ball joint from the knuckle, using the ball joint removal tool.

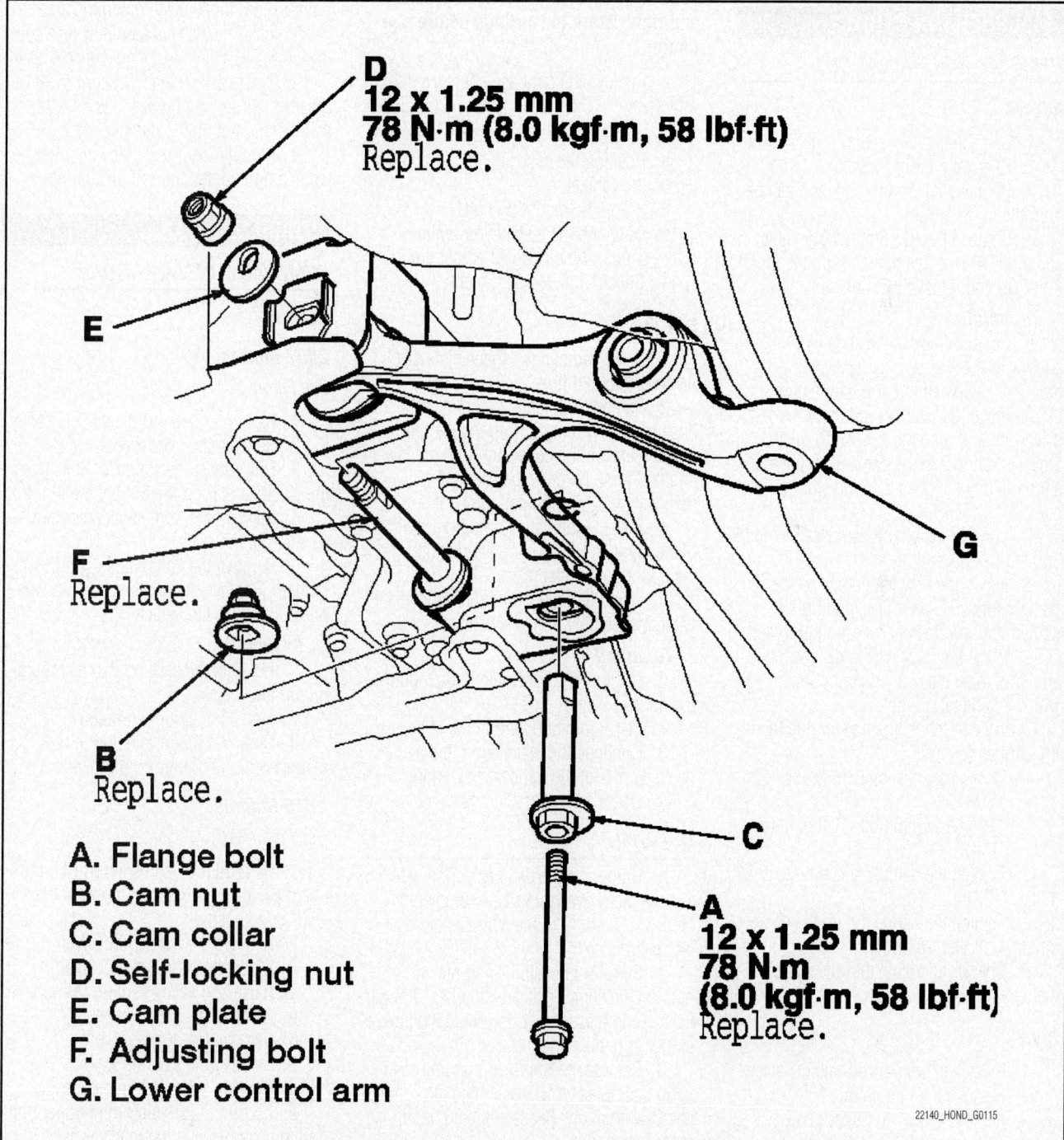

D
12 x 1.25 mm
78 N·m (8.0 kgf·m, 58 lbf·ft)
Replace.

E

F
Replace.

B
Replace.

G

C

A
12 x 1.25 mm
78 N·m
(8.0 kgf·m, 58 lbf·ft)
Replace.

A. Flange bolt
B. Cam nut
C. Cam collar
D. Self-locking nut
E. Cam plate
F. Adjusting bolt
G. Lower control arm

22140_HOND_G0115

Fig. 230 Exploded view of the lower control arm mounting—S2000

6. Remove the self locking nut and self locking cam nut. Remove the cam plate adjusting bolt, cam collar, flange bolt and the lower arm.

To install:

7. Installation is in the reverse order of the removal procedure.

8. Be sure to use new flange bolts, castle nut and lock pin.

9. Check and adjust the wheel alignment, as necessary.

MACPHERSON STRUT

REMOVAL & INSTALLATION

Accord

1. Raise and safely support the vehicle.
2. Remove the front wheel.
3. Remove the damper fork and damper from the lower arm.
4. Remove the two 8 mm flange nuts and three 10 mm flange nuts from the top of the strut, and remove the strut assembly.

To install:

5. Position the strut assembly in the body with the aligning tab facing inside, then loosely install the new flange nuts.

6. Install the damper fork over the driveshaft and onto the lower arm. Install the front damper in the damper fork so the aligning tab is aligned with the slot in the damper fork.

7. Loosely install the damper pinch bolt into the damper fork.

8. Install the new flange bolt to the damper fork and lower arm, and lightly tighten the new damper fork mounting nut.

9. Place the floor jack under the lower arm, and raise the suspension to load it with the vehicle's weight.

10. Tighten the flange nuts on the top of the strut as follows:

- 8 × .1.25 mm bolt: 16 ft. lbs. (22 Nm)
- 10 × 1.25mm bolt: 37 ft. lbs. (50 Nm)

11. Tighten the damper pinch bolts to 32 ft. lbs. (43 Nm).

12. Tighten the flange nut on the damper fork to 47 ft. lbs. (64 Nm).

13. Check and adjust the wheel alignment as necessary.

Civic

1. Raise and support the vehicle safely.
2. Remove the front tires.
3. Remove the wheel sensor harness bracket and brake hose bracket from the damper. Do not disconnect the wheel sensor connector.

4. Remove the damper pinch bolts from the bottom of the damper. Do not allow the knuckle to rotate too far outward as this may cause the inner CV joint bearing to unseat.

5. Turn the ignition switch to the ON position. Turn the windshield wipers on; turn the ignition switch off leaving the wipers near the A-pillars.

6. Remove the service cap and lid. Remove the three flange nuts at the top of the damper.

➡ Damper springs are different, left and right. Mark the springs before continuing.

7. Remove the strut assembly from the vehicle.

To install:

8. Position the assembly to its mounting on the vehicle.

9. Install the mounting bolts.

10. Continue the installation in the reverse order of the removal procedure.

11. Check front end alignment.

S2000

1. Raise and support the vehicle safely. Remove the front tires.

2. Remove the flange bolt and brake hose mounting bracket from the damper.

3. Remove the cotter pin from the lower arm ball joint. Remove the castle nut. Remove the lower arm ball joint.

4. Remove the flange nuts from the top of the damper. Remove the flange bolt from the bottom of the damper.

5. Lower the lower arm and remove the damper assembly from the vehicle.

To install:

6. Position the assembly to its mounting on the vehicle.

7. Install the mounting bolts.

8. Continue the installation in the reverse order of the removal procedure.

9. Check front end alignment.

OVERHAUL

1. Compress the damper spring with a commercially available strut spring compressor according to the manufacturer's instructions.

2. Remove the self locking nut while holding the damper shaft with a hex wrench. Do not compress the spring more than necessary to remove the nut.

3. Release the pressure from the strut spring compressor. Disassemble the damper assembly. Reassemble all parts, except the spring.

4. Compress the damper assembly by hand and check for smooth operation thru

full stroke, both compression and extension.

5. The damper should extend smoothly and constantly when compression is released. If not, gas is leaking and the damper should be replaced.

6. Install all parts except the self locking nut onto the damper unit.

7. Align the bottom of the spring and the stepped part of the lower spring seat. The hole in the upper spring seat and the arrow on the damper mounting base must point toward the knuckle mounting area.

8. Install the damper assembly on a commercially available strut spring compressor.

9. Compress the damper spring with the strut spring compressor. Install a new self locking nut on the damper shaft.

10. Hold the damper shaft with a hex wrench and tighten the self locking nut.

STABILIZER BAR

REMOVAL & INSTALLATION

Accord

2007 Models

See Figure 231.

1. Raise and support the vehicle safely.
2. Remove the front tires.
3. Disconnect the stabilizer links from the stabilizer bar on the right and left sides.
4. Remove the front suspension subframe from the body.
5. Remove the flange bolts and bushing holders. Remove the bushings and the stabilizer bar from the vehicle.

To install:

6. Position the assembly to its mounting on the vehicle.

7. Install the mounting bolts.

8. Continue the installation in the reverse order of the removal procedure.

2008 Models

See Figure 232.

1. Remove the hood support rod, then use it as shown to prop the hood in the wide-open position.

2. Remove the front grille cover.
3. Disconnect the battery cables.
4. Raise and safely support the vehicle.
5. Remove the front wheels.
6. Remove the driver's airbag.
7. Remove the steering wheel.
8. Remove the steering joint cover.
9. Loosen the steering joint upper bolt, and remove the steering joint lower bolt, then disconnect the steering joint from the pinion shaft.

After removing the subframe mounting bolts, front suspension subframe middle mounting rubber mounting bolts, front suspension subframe rear bracket mounting bolts, and front suspension subframe rear damper front mounting bolt, be sure to replace them with new ones.

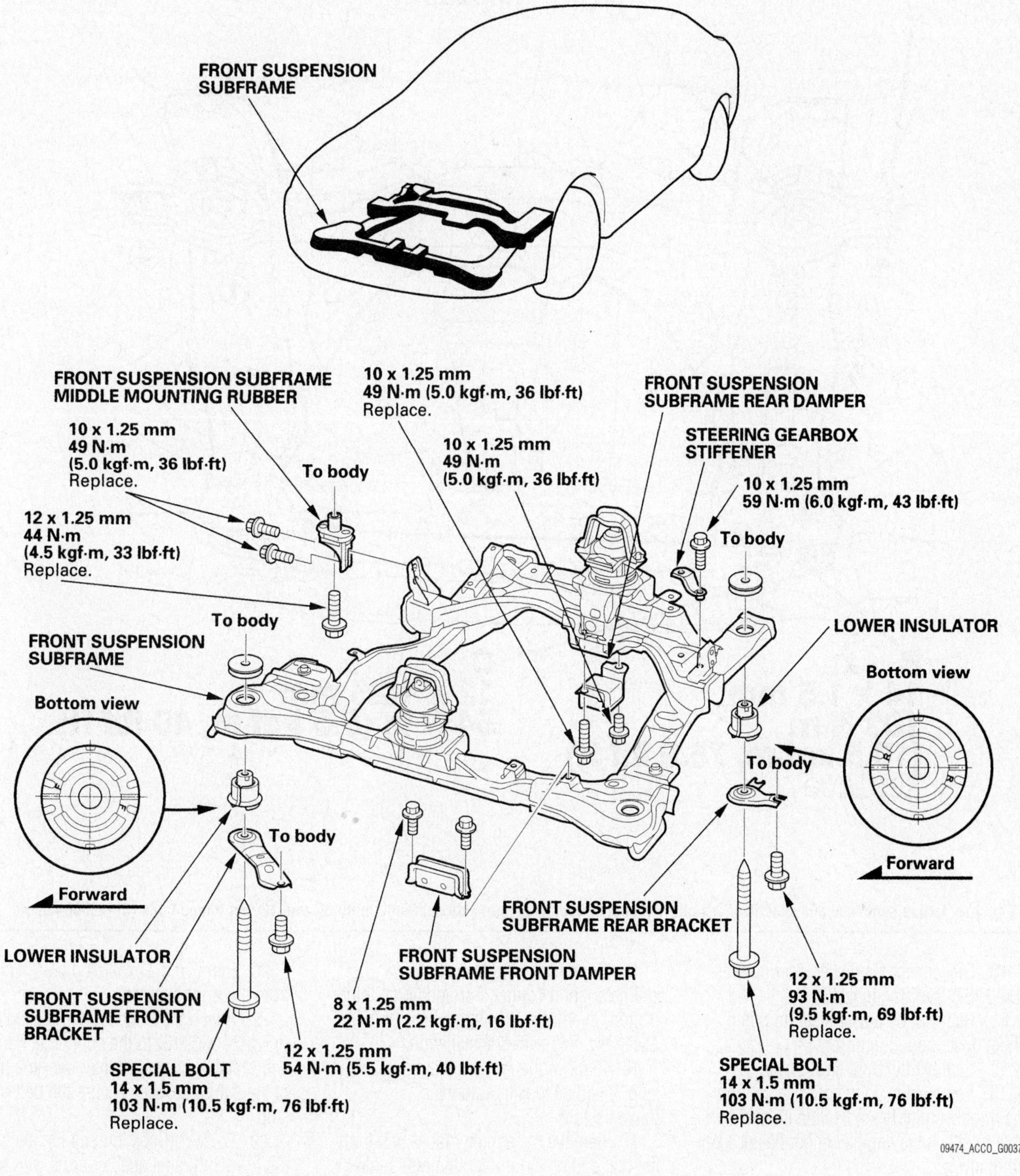

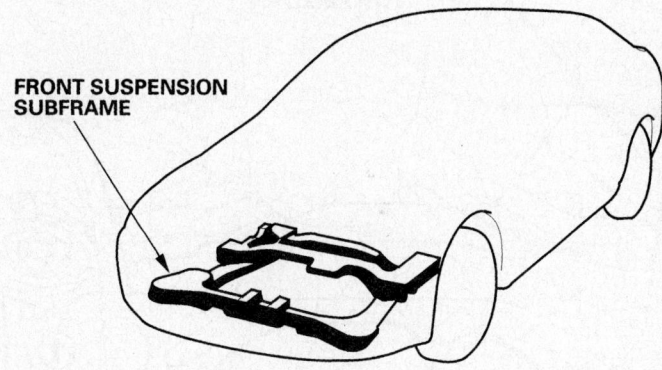

FRONT SUSPENSION SUBFRAME

FRONT SUSPENSION SUBFRAME MIDDLE MOUNTING RUBBER

10 x 1.25 mm
49 N·m
(5.0 kgf·m, 36 lbf·ft)
Replace.

12 x 1.25 mm
44 N·m
(4.5 kgf·m, 33 lbf·ft)
Replace.

To body

10 x 1.25 mm
49 N·m (5.0 kgf·m, 36 lbf·ft)
Replace.

10 x 1.25 mm
49 N·m
(5.0 kgf·m, 36 lbf·ft)

FRONT SUSPENSION SUBFRAME REAR DAMPER

STEERING GEARBOX STIFFENER

10 x 1.25 mm
59 N·m (6.0 kgf·m, 43 lbf·ft)

To body

LOWER INSULATOR

Bottom view

To body

FRONT SUSPENSION SUBFRAME

Bottom view

To body

Forward

Forward

LOWER INSULATOR

To body

FRONT SUSPENSION SUBFRAME FRONT BRACKET

SPECIAL BOLT
14 x 1.5 mm
103 N·m (10.5 kgf·m, 76 lbf·ft)
Replace.

8 x 1.25 mm
22 N·m (2.2 kgf·m, 16 lbf·ft)

12 x 1.25 mm
54 N·m (5.5 kgf·m, 40 lbf·ft)

FRONT SUSPENSION SUBFRAME FRONT DAMPER

FRONT SUSPENSION SUBFRAME REAR BRACKET

12 x 1.25 mm
93 N·m
(9.5 kgf·m, 69 lbf·ft)
Replace.

SPECIAL BOLT
14 x 1.5 mm
103 N·m (10.5 kgf·m, 76 lbf·ft)
Replace.

09474_ACCO_G0037

Fig. 231 Front subframe and related components—Accord

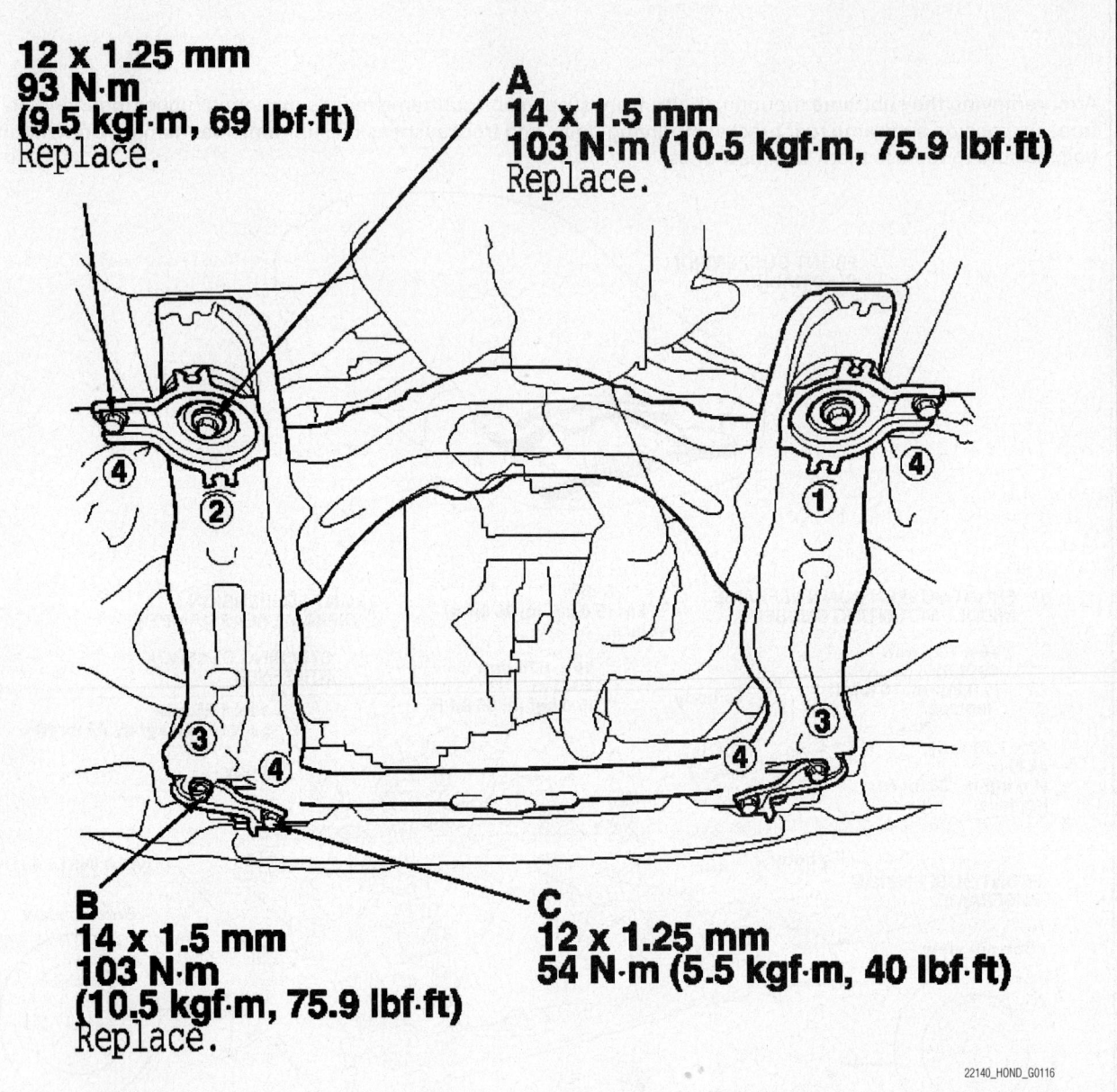

12 x 1.25 mm
93 N·m
(9.5 kgf·m, 69 lbf·ft)
Replace.

A
14 x 1.5 mm
103 N·m (10.5 kgf·m, 75.9 lbf·ft)
Replace.

B
14 x 1.5 mm
103 N·m
(10.5 kgf·m, 75.9 lbf·ft)
Replace.

C
12 x 1.25 mm
54 N·m (5.5 kgf·m, 40 lbf·ft)

22140_HOND_G0116

Fig. 232 Torque sequence and specifications of the rear side bolts (A), front side subframe bolts (B) and 12 mm flange bolts (C)—2008 Accord

10. Disconnect the power steering pressure (PSP) switch connector.

11. Remove the power steering pump outlet hose mounting bolt.

12. Remove the front strut brace.

13. Remove the connector bracket from the front cylinder head; use the bracket bolt hole to attach the engine hanger balance bar front arm.

14. Remove the harness clamp bracket from the rear cylinder head; use the bracket bolt hole to attach the 2008 attachment arm Special Tool.

15. Support the engine with a suitable engine support hanger. For additional information, refer the following section, "Engine Assembly, Removal & Installation."

16. Remove the rear engine mount stop, if equipped with automatic transmission.

17. Remove the engine mount bolt from the rear engine mount and the rear engine mount bracket.

18. Remove the front splash shield.

19. Remove the nuts securing the lower transmission mount.

20. Remove the exhaust pipe A hanger from the front subframe.

21. Attach the front subframe adapter (VSB02C000016) to the subframe, hang the belt of the subframe adapter over the front of the subframe, then secure the belt with its stop.

22. Raise the jack, line up the slots in the front subframe adapter arms with the bolt holes on the jack base, then securely attach them with four bolts.

23. Remove the front subframe mounting bolts on both sides of the middle mount.

24. Disconnect both sides of the stabilizer link from the stabilizer bar.

25. Remove the flange bolts on both sides of the front subframe front bracket.

26. Loosen the front side of the front subframe mounting bolts until there is 20 mm (¹³⁄₁₆ in.) distance between the bolt seat and the mounting surface. Do not loosen the mounting bolts more than necessary.

27. Remove the flange bolts on both sides of the front subframe rear bracket.

28. Loose the rear side of the front subframe mounting bolts until there is 30 mm (1³⁄₁₆ in.) distance between the bolt seat and the mounting surface. Do not loosen the mounting bolts more than necessary.

29. Lower the transmission jack with the front subframe adapter slowly until the front subframe has dropped about 30 mm (1³⁄₁₆ in.).

30. Remove the flange bolts and the bushing holders, then remove the bushings.

31. Move the stabilizer bar toward the passenger's side, and remove the stabilizer bar.

To install:

32. Install the stabilizer bar.

33. Align the front subframe using the subframe alignment pin. Vertically install the subframe alignment pin, and align the right-rear corner of the front subframe and vehicle frame holes, then loosely tighten the new subframe mounting bolt until the front subframe contacts the body frame.

34. Loosely tighten the left-rear subframe mounting bolt using the same procedure as the right-rear with the subframe alignment pin.

35. Loosely install the new 12 mm flange bolts to the subframe rear bracket.

36. Torque the subframe mounting bolts to the specified torque values starting with the right-rear bolt. Use the subframe alignment pin when tightening the rear side bolts.

37. Check all of the front subframe mounting bolts, and retighten if necessary.

38. The remainder of the installation is the reverse order of removal.

39. Check and adjust the wheel alignment as necessary.

Civic

1. Raise and support the vehicle safely. Remove the front tires.

2. Disconnect both stabilizer links from the stabilizer bar.

3. Remove the flange bolts and the bushing holders. Remove the bushings and the stabilizer bar from the front suspension subframe.

To install:

4. Position the assembly to its mounting on the vehicle.

5. Install the mounting bolts.

6. Continue the installation in the reverse order of the removal procedure.

S2000

1. Raise and support the vehicle safely.

2. Remove the front tires.

3. Remove the splash shield.

4. Remove the self locking nuts while holding the joint pin. Disconnect the stabilizer links from the stabilizer bar on the right and left sides.

5. Remove the flange bolts and bushing holders. Remove the bushings and the stabilizer bar from the vehicle.

To install:

6. Position the assembly to its mounting on the vehicle.

7. Install the mounting bolts.

8. Continue the installation in the reverse order of the removal procedure.

STEERING KNUCKLE

REMOVAL & INSTALLATION

The steering knuckle is removed with the wheel bearings as an assembly. Refer to the following section for additional information, "Wheel Bearings, Removal & Installation."

UPPER CONTROL ARM

REMOVAL & INSTALLATION

Accord

See Figure 233.

1. Remove the front strut assembly.

2. Remove the wheel sensor bracket from the upper arm.

3. Remove the cotter pin from the upper arm ball joint, and loosen the nut.

4. Disconnect the upper arm ball joint from the knuckle using the ball joint remover.

5. Remove the upper arm mounting bolts, and remove the upper arm.

To install:

6. 2007 Models: Install the upper arm by inserting a rod of appropriate size (O.D. 6 mm/L: 300 mm) into the positioning holes, and place the upper arm on the rod to position it before tightening the upper arm mounting bolts.

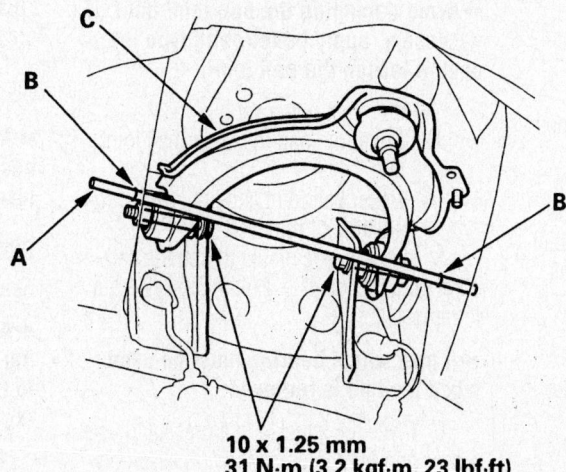

10 x 1.25 mm
31 N·m (3.2 kgf·m, 23 lbf·ft)

42356-ACCO-G39

Fig. 233 Insert a rod (A) into the positioning holes (B) and place the upper control arm (C) on the rod to position it before tightening the bolts— 2007 Accord

7. The remainder of the installation is the reverse order of removal.

S2000

1. Raise and safely support the vehicle.
2. Remove the front wheel.
3. Remove the flange bolts and wheel speed sensor harness from the upper arm.
4. Remove the lock pin from the upper arm ball joint, then remove the castle nut.
5. Disconnect the upper arm ball joint from the knuckle using the ball joint thread protector and the ball joint remover.
6. Remove the flange bolts, and remove the upper arm.
7. Installation is the reverse order of assembly. Tighten the mounting bolts to 76 ft. lbs. (103 Nm).
8. Check and adjust the wheel alignment as necessary.

WHEEL BEARINGS

REMOVAL & INSTALLATION

Accord

1. Raise and safely support the vehicle.
2. Remove the front wheel.
3. Remove the brake hose mounting bracket.
4. Remove the brake caliper bracket mounting bolts, and remove the caliper assembly from the knuckle. To prevent damage to the caliper assembly or brake hose, use a short piece of wire to hang the caliper assembly from the undercarriage. Do not twist the brake hose with force.
5. Remove the wheel sensor from the knuckle.
6. Raise the stake, and remove the spindle nut.
7. Remove the brake rotor.
8. Disconnect the tie-rod end ball joint from the knuckle using the ball joint remover.
9. Remove the cotter pin from the upper arm ball joint, and loosen the nut.
10. Disconnect the upper arm ball joint from the knuckle using the ball joint remover.
11. Remove the driveshaft outboard joint from the knuckle by tapping the driveshaft end with a plastic hammer while drawing the hub outward, then remove the knuckle/bearing assembly.
12. Separate the hub from the knuckle using the hub disassembly tool and a hydraulic press. Hold the knuckle with the attachment of the hydraulic press or equivalent tool. Be careful not to deform the splash guard. Hold onto the hub to keep it from falling when pressed clear.

To install:

13. Install the hub onto the knuckle using the attachment, the driver support base, and a hydraulic press. Be careful not to distort the splash guard.
14. The remainder of the installation is the reverse order of removal.

Civic

See Figure 234.

➡A hydraulic press and several bearing drivers and attachments are needed to remove and install the hub and bearing.

1. Before servicing the vehicle, refer to the precautions in the beginning of this section.
2. Pry the spindle nut stake away from the spindle, then loosen the nut.
3. Raise and safely support the vehicle.
4. Remove or disconnect the following:
   - Front wheel and the spindle nut
   - Wheel sensor wire bracket from the knuckle, but don't disconnect it.
   - Caliper mounting bolts and the caliper. Support the caliper out of the way with a length of wire. Do not let the caliper hang from the brake hose.
   - 6mm brake disc retaining screws. Screw 2, 12mm bolts into the disc to push it away from the hub.
   - Tie rod castle nut
   - Tie rod ball joint using a suitable ball joint remover.
   - Cotter pin and loosen the lower arm ball joint nut half the length of the joint threads.
   - Ball joint and lower arm using a suitable puller with the pawls applied to the lower arm.

➡Avoid damaging the ball joint boot. If necessary, apply penetrating type lubricant to loosen the ball joint.

   - Ball joint nut cover
   - Cotter pin and the upper ball joint nut.
   - Upper ball joint and knuckle using a ball joint remover.
5. Use a plastic mallet to free the half-shaft from the knuckle. Pull the knuckle out to remove it.

➡A new wheel bearing must be used when the hub is removed.

6. Place the knuckle in a press and use a base and pilot to press the hub assembly out of the wheel bearing.
7. Remove the knuckle ring seal and circlip. Remove the splash guard from the knuckle.

8. Press the wheel bearing out of the knuckle using a driving attachment.

To install:

9. Clean the knuckle and hub assembly and inspect them for damage.
10. Install or connect the following:
    - New wheel bearing into the hub using a driving tool.
    - Circlip in the outer groove of the knuckle.
    - Splash guard
    - Hub assembly into the steering knuckle using a base and a driving and guide tool.
    - Knuckle ring seal
    - Knuckle onto the spindle
    - Knuckle onto the upper and lower ball joints and tighten the castle nuts.
    - Tie rod ball joint onto the steering knuckle.
11. Tighten the upper ball joint nut and tie rod nut to 29–35 ft. lbs. (40–48 Nm) and the lower ball joint castle nut to 36–43 ft. lbs. (50–60 Nm).
12. Install or connect the following:
    - Anti-lock Brake System (ABS) wheel sensor wire brackets onto the knuckle. Tighten the mounting bolts to 84 inch lbs. (10 Nm).
    - Brake disc; use 2 lug nuts to evenly draw the disc onto the hub.
    - Retainer screws: 84 inch lbs. (10 Nm)
    - Spindle washer and nut. Don't tighten the nut until the vehicle is on the ground.
    - Brake caliper and tighten the bolts to 80 ft. lbs. (110 Nm).
    - Front wheels and lower the vehicle.
13. Tighten the spindle nut to 134 ft. lbs. (185 Nm), stake the nut, and install the grease cap.
14. Check and adjust the vehicle's front wheel alignment.

➡Avoid damaging the ball joint boot. If necessary, apply penetrating-type lubricant to loosen the ball joint.

S2000

See Figure 235.

➡A hydraulic press and several bearing drivers and attachments are needed to remove and install the hub and bearing.

1. Before servicing the vehicle, refer to the precautions in the beginning of this section.
2. Remove or disconnect the following:
   - Front wheel
   - Brake hose bracket mounting bolts

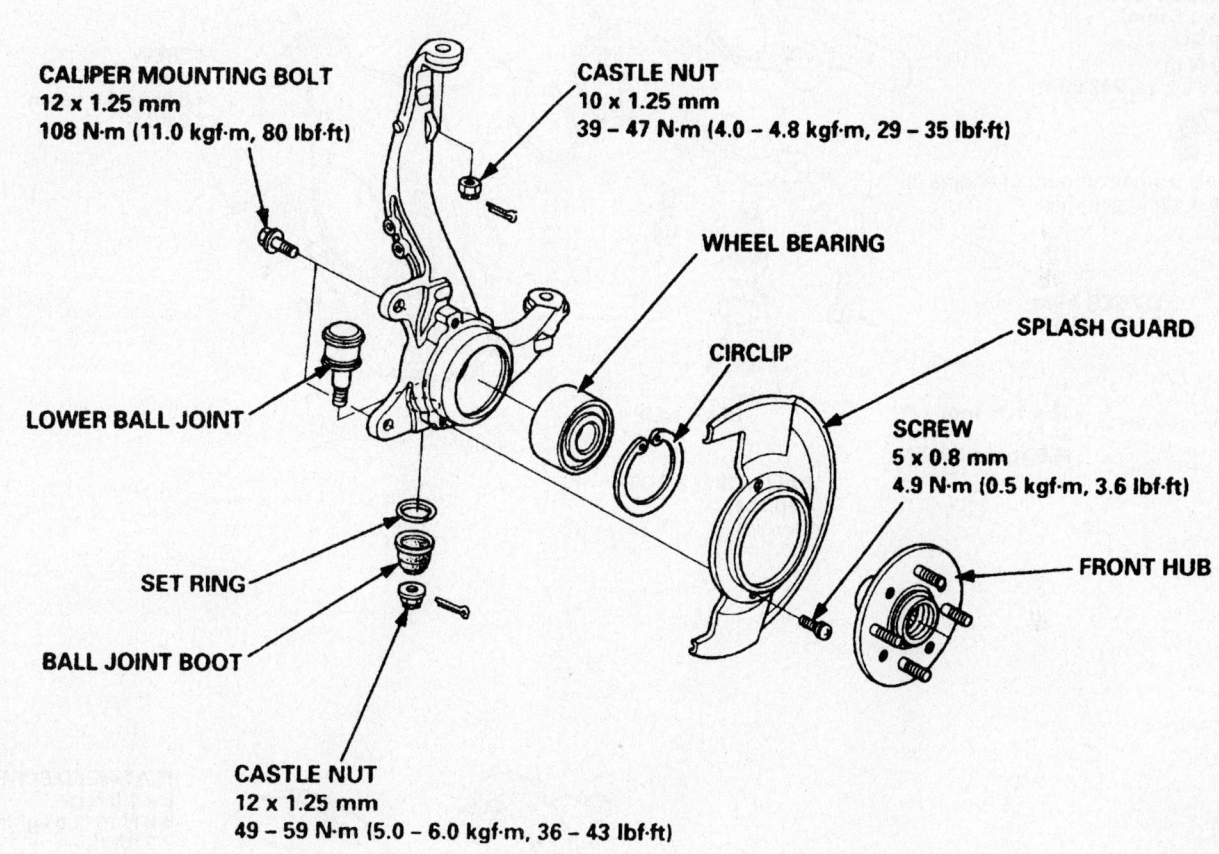

CALIPER MOUNTING BOLT
12 x 1.25 mm
108 N·m (11.0 kgf·m, 80 lbf·ft)

CASTLE NUT
10 x 1.25 mm
39 – 47 N·m (4.0 – 4.8 kgf·m, 29 – 35 lbf·ft)

WHEEL BEARING

CIRCLIP

SPLASH GUARD

LOWER BALL JOINT

SCREW
5 x 0.8 mm
4.9 N·m (0.5 kgf·m, 3.6 lbf·ft)

SET RING

FRONT HUB

BALL JOINT BOOT

CASTLE NUT
12 x 1.25 mm
49 – 59 N·m (5.0 – 6.0 kgf·m, 36 – 43 lbf·ft)

7923FGB4

Fig. 234 Knuckle components—Civic

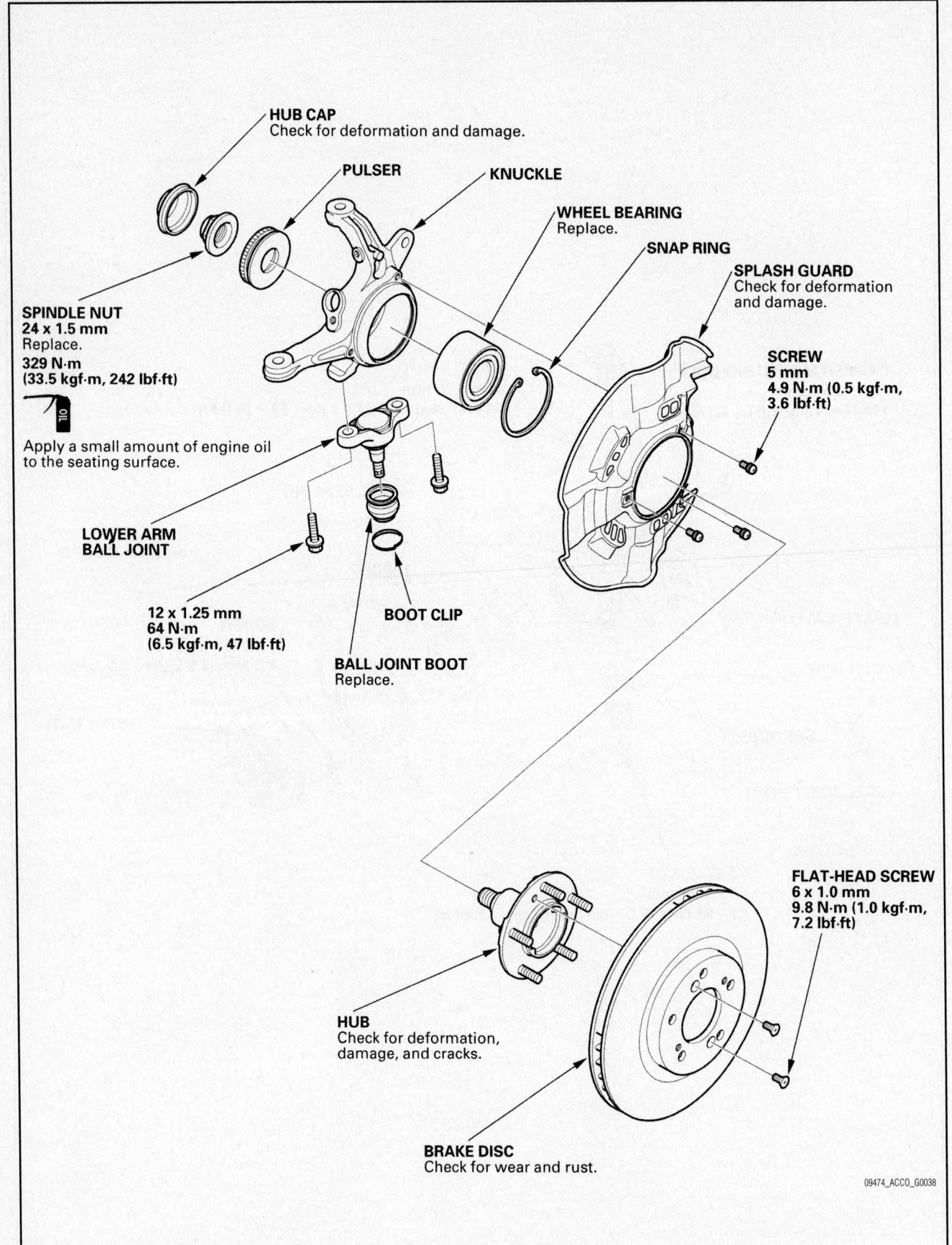

HUB CAP
Check for deformation and damage.

PULSER

KNUCKLE

WHEEL BEARING
Replace.

SNAP RING

SPLASH GUARD
Check for deformation
and damage.

SPINDLE NUT
24 x 1.5 mm
Replace.
329 N·m
(33.5 kgf·m, 242 lbf·ft)

SCREW
5 mm
4.9 N·m (0.5 kgf·m,
3.6 lbf·ft)

Apply a small amount of engine oil
to the seating surface.

**LOWER ARM
BALL JOINT**

12 x 1.25 mm
64 N·m
(6.5 kgf·m, 47 lbf·ft)

BOOT CLIP

BALL JOINT BOOT
Replace.

FLAT-HEAD SCREW
6 x 1.0 mm
9.8 N·m (1.0 kgf·m,
7.2 lbf·ft)

HUB
Check for deformation,
damage, and cracks.

BRAKE DISC
Check for wear and rust.

09474_ACCO_G0038

Fig. 235 Knuckle components—S2000

- Brake caliper and caliper support
- Wheel speed sensor
- Brake rotor
- Outer tie rod end
- Upper and lower ball joints
- Steering knuckle from the vehicle
- Dust cover
- Spindle nut
- Wheel speed pulse ring

3. Mount the steering knuckle in a press and press the hub out of the wheel bearing.

4. Remove the splash guard and the wheel bearing snapring.

5. Press the wheel bearing out of the steering knuckle.

To install:

6. Installation is the reverse of the removal procedure, while using the following torque values:
- Splash guard screws: 48 inch lbs. (5 Nm)
- Spindle nut: 242 ft. lbs. (329 Nm)

- Upper ball joint nut: 36–43 ft. lbs. (49–59 Nm)
- Lower ball joint nut: 43–51 ft. lbs. (56–69 Nm)
- Outer tie rod end nut: 40 ft. lbs. (54 Nm)
- Brake caliper support bolts: 83 ft. lbs. (113 Nm)

ADJUSTMENT

The wheel bearings are not adjustable. If they are not within specification, the wheel bearings must be replaced.

SUSPENSION   REAR SUSPENSION

LOWER CONTROL ARM

REMOVAL & INSTALLATION

Accord

2007 Models

1. Raise and support the vehicle safely. Remove the rear tires.
2. Remove the lower arm mounting nut and mounting bolt from the knuckle side.
3. Remove the flange bolt. Remove the lower arm from the vehicle.

To install:

4. Position the lower arm on the vehicle. Be sure to use new retaining bolts and nuts.
5. Install all the suspension components and lightly tighten the bolts and nuts. Position a jack under the trailing arm and raise the suspension to load it with the vehicles weight, before fully tightening the bolts and nuts.
6. Continue the installation in the reverse order of the removal procedure.
7. Check and adjust the wheel alignment, as required.

2008 Models—Lower Arm A

See Figure 236.

1. Raise and safely support the vehicle.
2. Remove the rear wheel.
3. Remove the parking brake cable mounting bolt.
4. Remove the lower arm A mounting bolts, then remove lower arm A.

To install:

5. Install lower arm A using new bolts to 43 ft. lbs. (59 Nm).
6. The remainder of the installation is the reverse order of removal.

2008 Models—Lower Arm B

See Figure 237.

1. Raise and safely support the vehicle.
2. Remove the rear wheel.

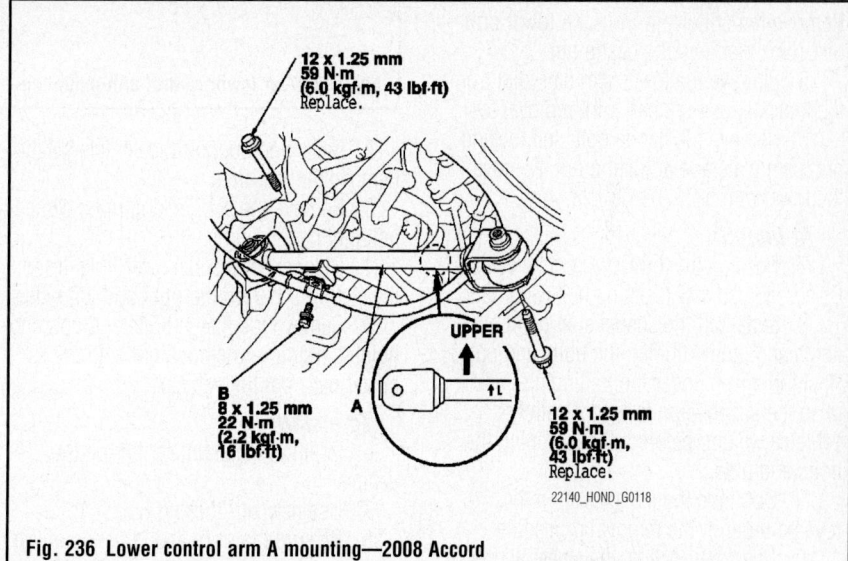

Fig. 236 Lower control arm A mounting—2008 Accord

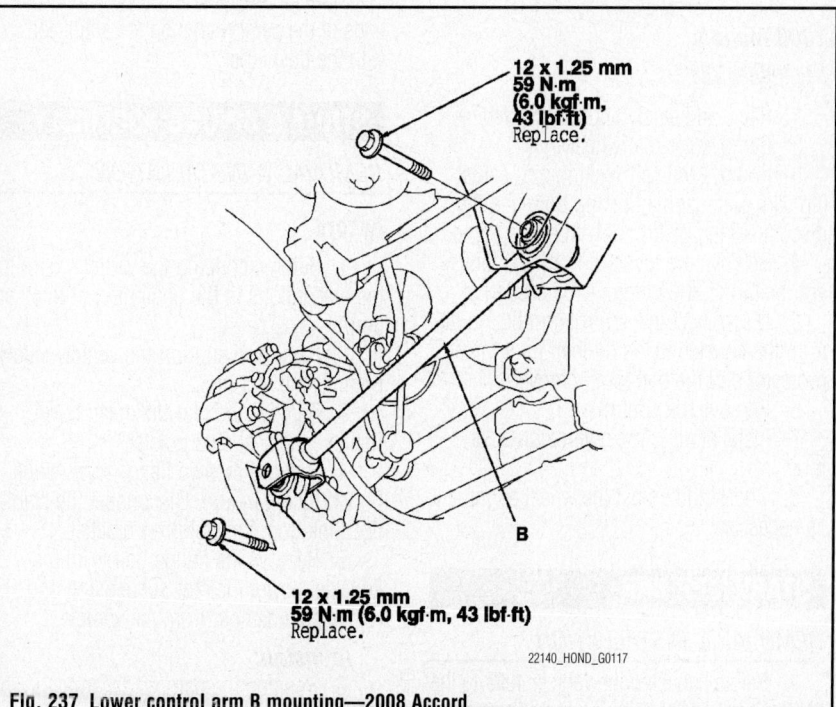

Fig. 237 Lower control arm B mounting—2008 Accord

3. Remove the lower arm B mounting bolts, then remove lower arm B.

4. Install lower arm B using new bolts to 43 ft. lbs. (59 Nm).

5. The remainder of the installation is the reverse order of removal.

S2000

2007 Models

1. Raise and support the vehicle safely.

2. Remove the rear tires.

3. Remove the flange nut while holding the joint pin. Disconnect the stabilizer link from the lower arm.

4. Remove the flange bolt and disconnect the damper from the lower arm. Remove the cotter pin from the lower arm ball joint. Remove the castle nut.

5. Remove the lower arm ball joint from the knuckle, using a ball joint removal tool.

6. Remove the flange bolt, self locking nut, cam plate and adjusting nut. Remove the lower arm from the vehicle.

To install:

7. Position the lower arm on the vehicle. Be sure to use new retaining bolts and nuts.

8. Install all the suspension components and lightly tighten the bolts and nuts. Position a jack under the trailing arm and raise the suspension to load it with the vehicles weight, before fully tightening the bolts and nuts.

9. Continue the installation in the reverse order of the removal procedure.

10. Check and adjust the wheel alignment, as required.

2008 Models

See Figure 238.

1. Raise and safely support the vehicle.

2. Remove the rear wheel.

3. Remove the self-locking nut, adjusting cam plate, and adjusting bolt, and then disconnect the control arm from the frame.

4. Remove the lock pin from the control arm ball joint, then remove the castle nut.

5. Disconnect the control arm ball joint from the knuckle using the ball joint thread protector and the ball joint remover.

6. Remove the control arm.

7. Installation is the reverse order of removal.

8. Check and adjust the wheel alignment as necessary.

STABILIZER BAR

REMOVAL & INSTALLATION

1. Before servicing the vehicle, refer to the precautions in the beginning of this section.

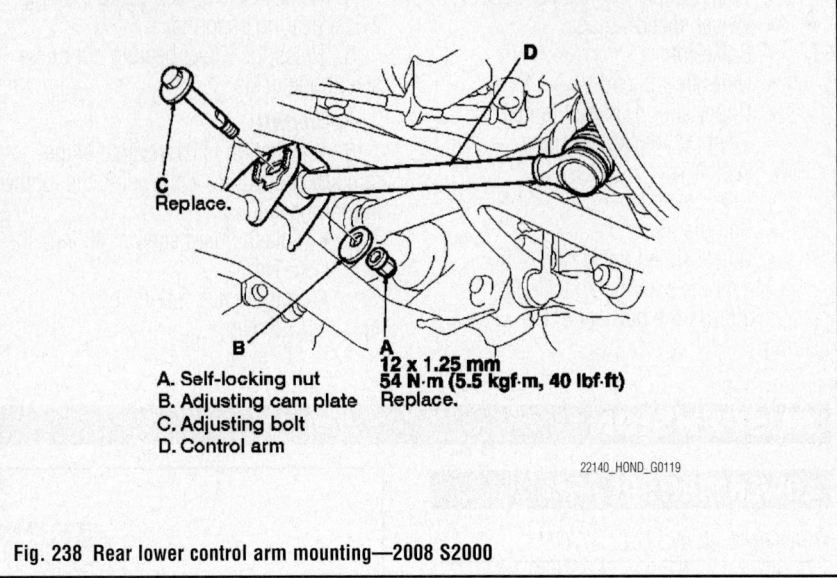

A. Self-locking nut
B. Adjusting cam plate
C. Adjusting bolt
D. Control arm

12 x 1.25 mm
54 N·m (5.5 kgf·m, 40 lbf·ft)
Replace.

22140_HOND_G0119

Fig. 238 Rear lower control arm mounting—2008 S2000

2. Raise and support the vehicle safely. Remove the rear tires.

3. Remove the self locking nuts, while holding the joint pins.

4. Disconnect the stabilizer links from the stabilizer bar on the right and left sides.

5. Remove the flange bolts and bushing holders. Remove the bushings and the stabilizer bar from the vehicle.

To install:

6. Position the stabilizer bar on the vehicle.

7. Use new self locking nuts.

8. Be sure the right and left ends of the stabilizer bar are installed on their respective sides of the vehicle. Properly align the ends of the paint marks on the stabilizer bar with the bushings

STRUT & SPRING ASSEMBLY

REMOVAL & INSTALLATION

Accord

1. Before servicing the vehicle, refer to the precautions in the beginning of this section.

2. Raise and support the vehicle safely. Remove the rear tires.

3. Remove the rear bulkhead cover. Remove the seat side bolster.

4. Remove the two flange nuts, while holding the joint pin. Disconnect the stabilizer link from the stabilizer bracket.

5. Remove the flange bolt from the knuckle. Lower the rear suspension and remove the damper from the vehicle.

To install:

6. Position the damper assembly in the vehicle.

➡**Damper springs are different for the left and right side. Mark the springs prior to removal.**

7. Loosely install the flange nuts onto the top of the damper.

8. Loosely install the flange bolt on the bottom of the damper.

9. Connect the stabilizer link on the bracket and loosely install the flange nut.

10. Raise the floor jack until the suspension begins to compress. Tighten the bolts.

11. Continue the installation in the reverse order of the removal procedure.

Civic

1. Raise and support the vehicle safely. Remove the rear tires.

2. Position a floor jack at the connecting point of the trailing arm and the knuckle. Raise the floor jack until the suspension begins to compress.

3. Remove the flange bolt and discard it.

4. Remove the trunk side trim panel.

5. Remove the self locking nut while holding the damper shaft. Compress the damper unit, by hand, and remove it from the vehicle.

To install:

6. Position the damper assembly in the vehicle.

7. Position a floor jack under the trailing to support the suspension. Install a new damper mounting bolt.

8. Loosely tighten the damper mounting bolt. Raise the floor jack until the suspension begins to compress. Tighten the damper mounting bolt.

9. Continue the installation in the reverse order of the removal procedure.

S2000

See Figure 239.

1. Before servicing the vehicle, refer to the precautions in the beginning of this section.

2. Remove the spare tire from the trunk.

3. Raise and support the vehicle safely. Remove the rear tires.

4. Remove the flange nuts from the top of the damper.

5. Remove the flange bolt from the bottom of the damper.

6. Lower the lower arm and remove the damper assembly.

To install:

7. Position the damper assembly in the vehicle.

8. Loosely install new flange nuts onto the damper studs.

9. Position the bottom of the damper assembly on the lower arm. Install a new flange bolt.

10. Raise the floor jack until the suspension begins to compress. Tighten the damper mounting bolt.

11. Continue the installation in the reverse order of the removal procedure.

UPPER CONTROL ARM

REMOVAL & INSTALLATION

Accord

1. Before servicing the vehicle, refer to the precautions in the beginning of this section.

2. Raise and support the vehicle safely. Remove the rear tires.

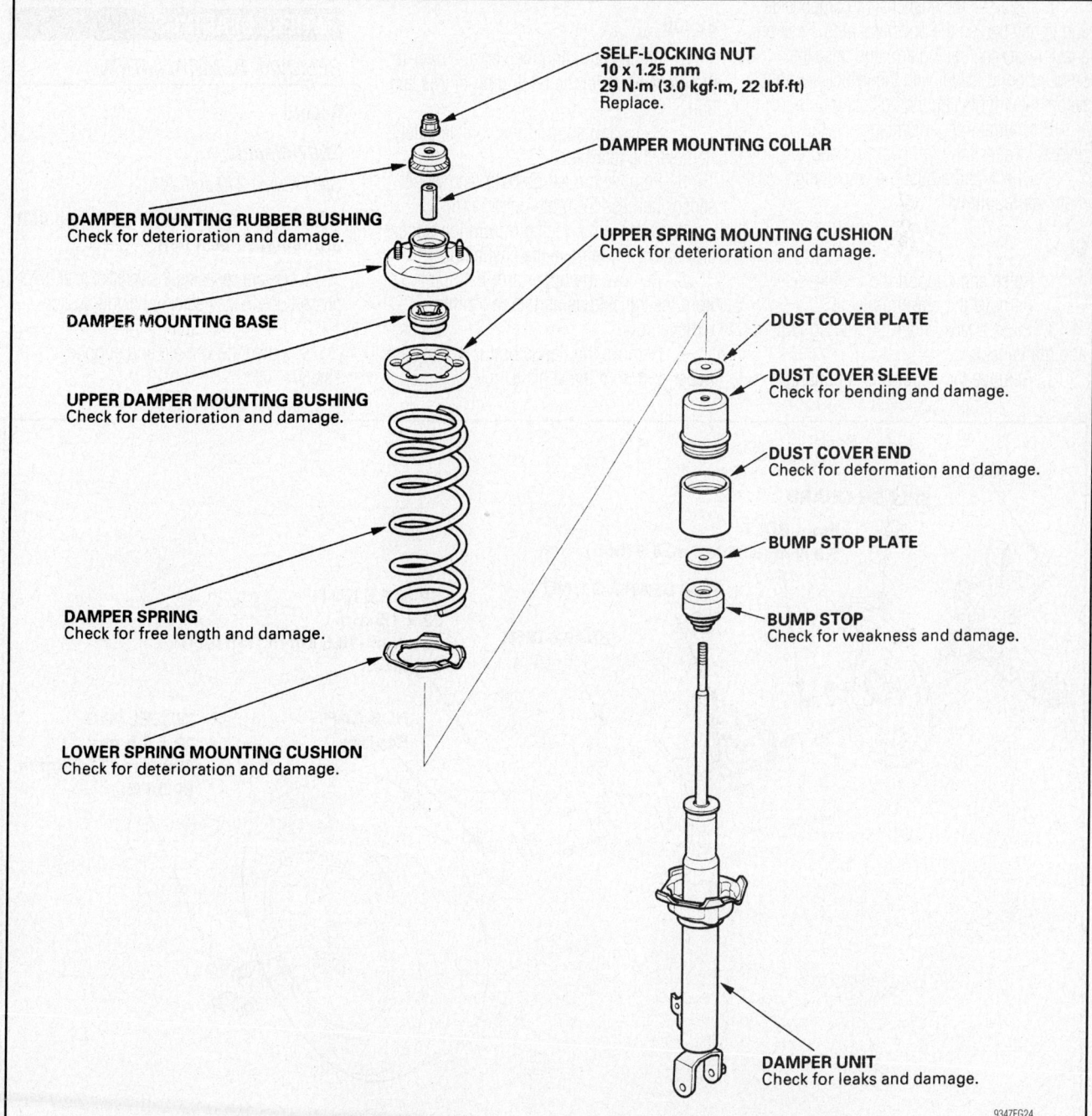

SELF-LOCKING NUT
10 x 1.25 mm
29 N·m (3.0 kgf·m, 22 lbf·ft)
Replace.

DAMPER MOUNTING COLLAR

DAMPER MOUNTING RUBBER BUSHING
Check for deterioration and damage.

UPPER SPRING MOUNTING CUSHION
Check for deterioration and damage.

DAMPER MOUNTING BASE

DUST COVER PLATE

DUST COVER SLEEVE
Check for bending and damage.

UPPER DAMPER MOUNTING BUSHING
Check for deterioration and damage.

DUST COVER END
Check for deformation and damage.

BUMP STOP PLATE

DAMPER SPRING
Check for free length and damage.

BUMP STOP
Check for weakness and damage.

LOWER SPRING MOUNTING CUSHION
Check for deterioration and damage.

DAMPER UNIT
Check for leaks and damage.

9347FG24

Fig. 239 Exploded view of the strut and spring assembly—S2000

3. Remove the lock pin from the upper arm ball joint, and loosen the castle nut.

4. Disconnect the upper arm ball joint from the knuckle using the ball joint removal tool.

5. Remove the brake hose mounting bracket. Do not disconnect the brake line.

6. Remove the wheel sensor harness mounting bracket.

7. Remove the flange bolt. Remove the upper arm from the vehicle.

To install:

8. Position the upper arm on the vehicle. Be sure to use new retaining bolts and nuts.

9. Install all the suspension components and lightly tighten the bolts and nuts. Position a jack under the trailing arm and raise the suspension to load it with the vehicles weight, before fully tightening the bolts and nuts.

10. Continue the installation in the reverse order of the removal procedure.

11. Check and adjust the wheel alignment, as required.

Civic

1. Raise and support the vehicle safely.
2. Remove the rear tires.
3. Place a jack under the trailing arm and the knuckle.
4. Remove the upper arm mounting

bolts and the flange bolt. Remove the upper arm from the vehicle.

To install:

5. Position the upper arm on the vehicle. Be sure to use new retaining bolts and nuts.

6. Install all the suspension components and lightly tighten the bolts and nuts. Position a jack under the trailing arm and raise the suspension to load it with the vehicles weight, before fully tightening the bolts and nuts.

7. Continue the installation in the reverse order of the removal procedure.

8. Check and adjust the wheel alignment, as required.

S2000

1. Before servicing the vehicle, refer to the precautions in the beginning of this section.

2. Raise and support the vehicle safely. Remove the rear tires.

3. Remove the flange bolts and wheel sensor harness from the upper arm.

4. Remove the lock pin from the upper arm ball joint, remove the castle nut.

5. Remove the upper arm ball joint from the knuckle using the ball joint removal tool.

6. Remove the flange bolts. Remove the upper arm from the vehicle.

To install:

7. Position the upper arm on the vehicle. Be sure to use new retaining bolts and nuts.

8. Install all the suspension components and lightly tighten the bolts and nuts. Position a jack under the trailing arm and raise the suspension to load it with the vehicles weight, before fully tightening the bolts and nuts.

9. Continue the installation in the reverse order of the removal procedure.

10. Check and adjust the wheel alignment, as required.

WHEEL BEARINGS

REMOVAL & INSTALLATION

Accord

2007 Models

See Figures 240 and 241.

➡ **The rear wheel bearing and hub unit are replaced as a unit.**

1. Before servicing the vehicle, refer to the precautions in the beginning of this section.

2. Loosen the spindle nut.

3. Raise the vehicle and support it safely.

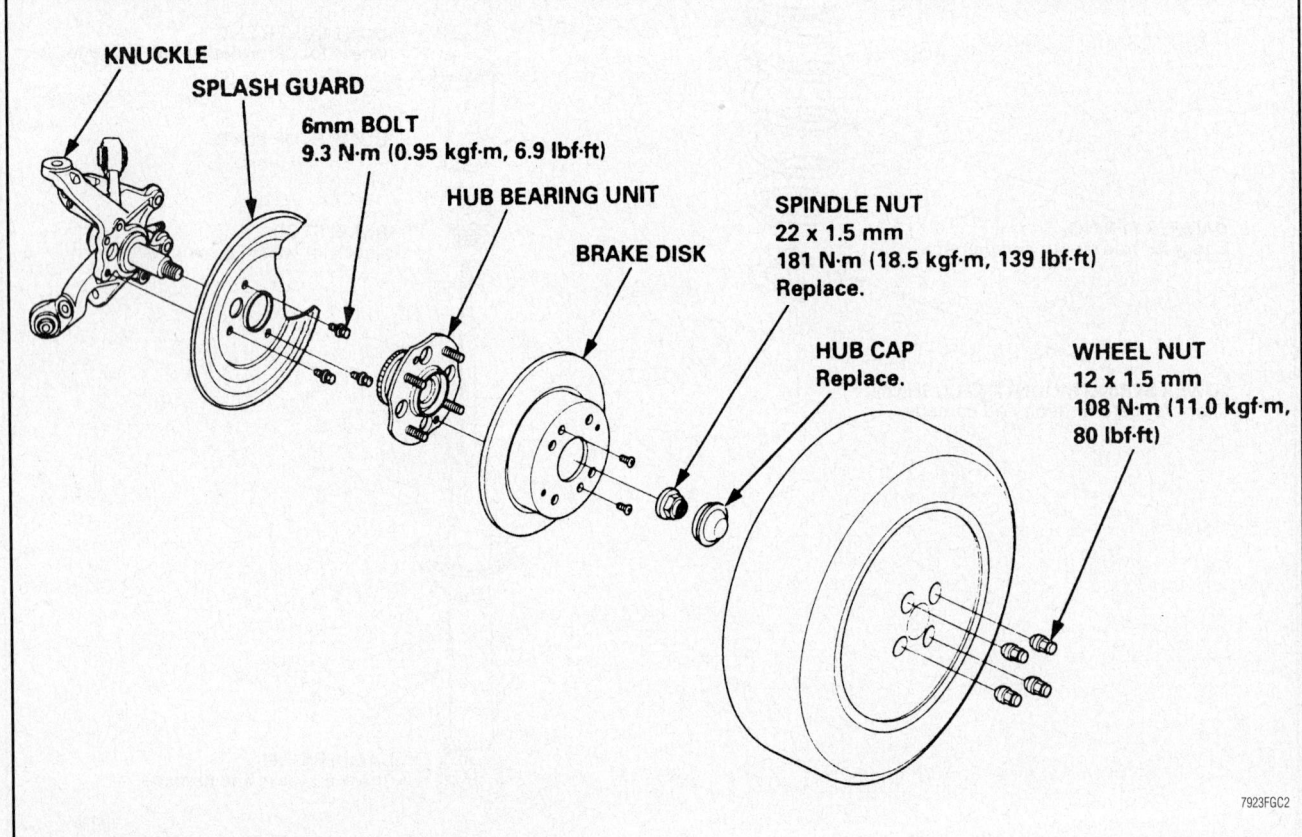

KNUCKLE

SPLASH GUARD

6mm BOLT
9.3 N·m (0.95 kgf·m, 6.9 lbf·ft)

HUB BEARING UNIT

BRAKE DISK

SPINDLE NUT
22 x 1.5 mm
181 N·m (18.5 kgf·m, 139 lbf·ft)
Replace.

HUB CAP
Replace.

WHEEL NUT
12 x 1.5 mm
108 N·m (11.0 kgf·m, 80 lbf·ft)

7923FGC2

Fig. 240 Knuckle and related components—Accord with disc brakes

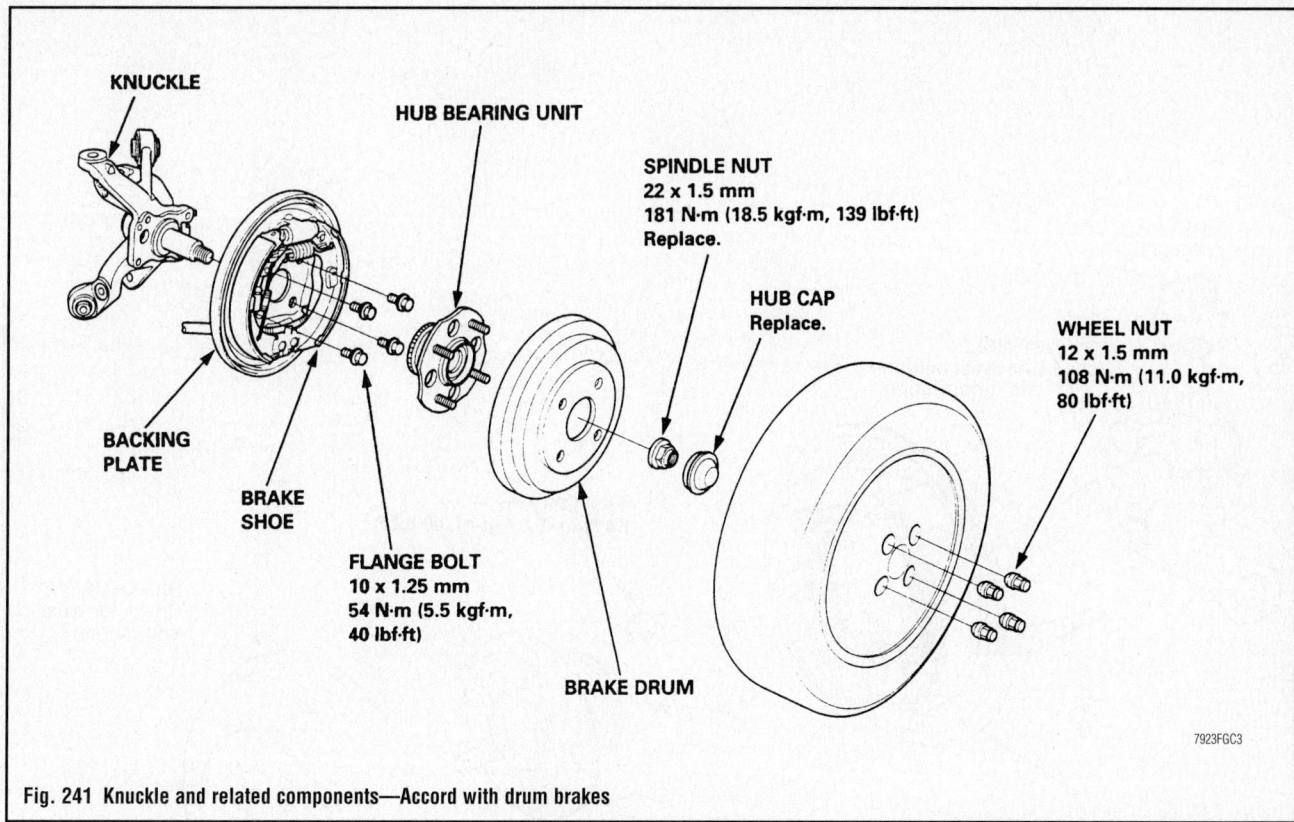

Fig. 241 Knuckle and related components—Accord with drum brakes

4. Remove or disconnect the following:
- Rear wheels
- Brake disc retaining screws
- Brake hose brackets from the knuckle
- Caliper bracket mounting bolts and hang the caliper out of the way with a piece of wire.
- Brake disc. If the disc is frozen on the hub, screw 2, 8 x 1.25mm bolts evenly into the disc to push it away from the hub.
- Spindle nut and pull the hub unit off of the spindle.

➡Clean the backing plate and the mating surfaces of the brake disc and hub with brake cleaner. Clean the spindle, washer, and hub with solvent.

To install:

5. Inspect the hub unit for signs of damage or wear. If the bearings are worn, the entire unit must be replaced.

6. Install or connect the following:
- Hub unit and spindle washer onto the spindle.
- Spindle nut but do not tighten it.
- Brake disc and tighten the retaining screws to 84 inch lbs. (10 Nm).
- Brake caliper and tighten the mounting bolts to 28 ft. lbs. (39 Nm).

- Brake hose brackets onto the knuckle and tighten the bolts to 16 ft. lbs. (22 Nm).
- Rear wheels and lower the vehicle.

7. With the vehicle on the ground, tighten the new spindle nut to 134 ft. lbs. (185 Nm), then stake the nut with a punch.

8. Tighten the wheel nuts to 80 ft. lbs. (110 Nm).

9. Test the operation of the brakes.

2008 Models

See Figure 242.

➡The rear wheel bearing and hub unit are replaced as a unit.

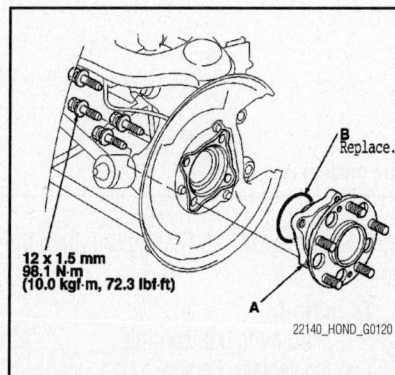

Fig. 242 Hub bearing unit (A) mounting. Replace the O ring—2008 Accord

1. Raise and safely support the vehicle.
2. Remove the rear wheel.
3. Fully release the parking brake lever.
4. Loosen the parking brake cable adjusting nut.
5. Remove the flange bolt from the arm. Then disconnect the parking brake cable from the lever.
6. Remove the brake hose mounting bolt.
7. Remove the brake caliper bracket mounting bolts, then remove the caliper assembly from the knuckle. To prevent damage to the caliper assembly or the brake hose, use a short piece of wire to hang the caliper assembly from the undercarriage.

✳✳ WARNING

Do not twist the brake hose excessively.

8. Remove the brake rotor.
9. Remove the hub bearing unit and O-Ring.
10. Installation is the reverse order of removal.

Civic

With Drum Brakes

See Figure 243.

1. Raise and support the vehicle safely.
2. Remove the rear tires.

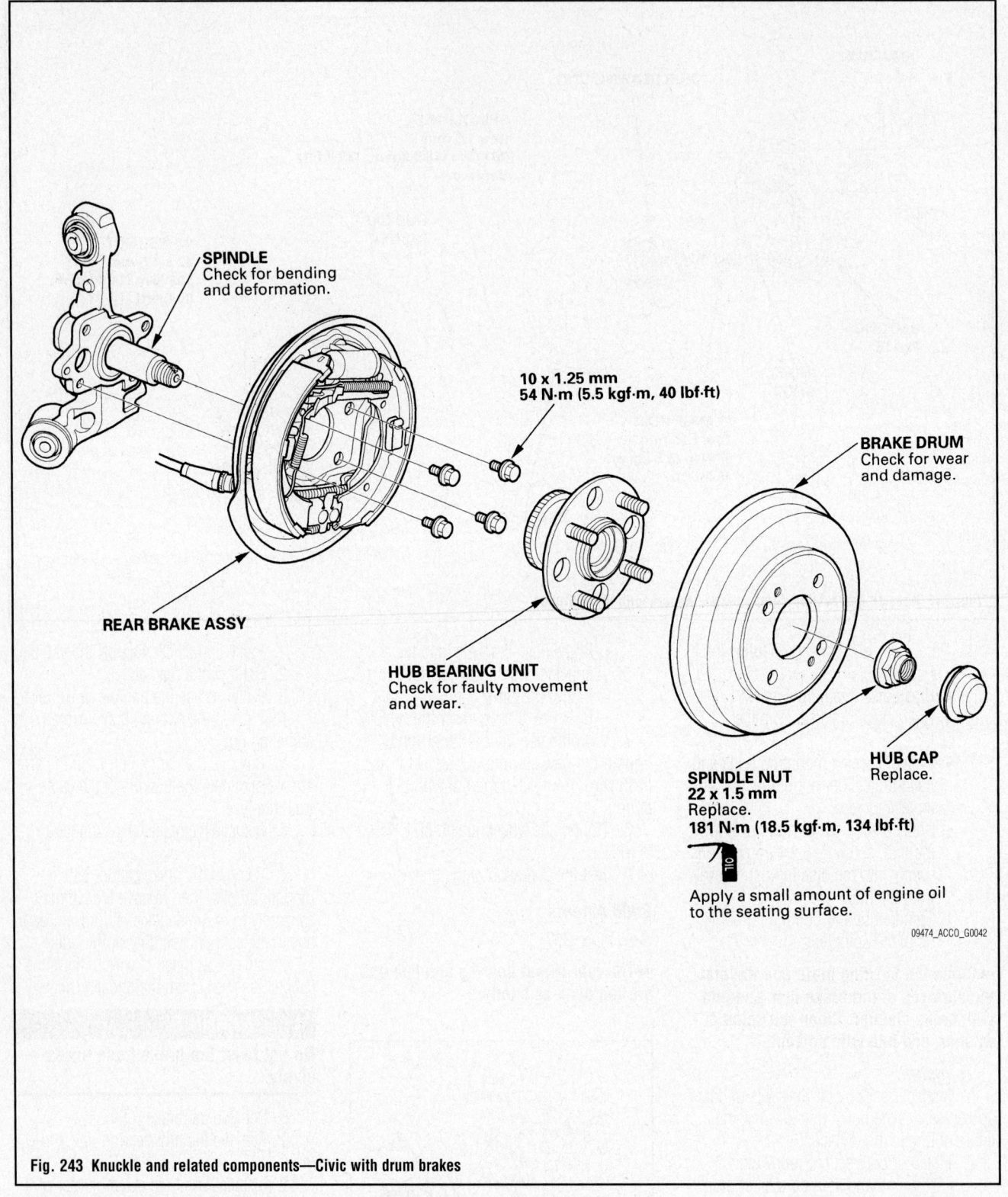

SPINDLE
Check for bending
and deformation.

10 x 1.25 mm
54 N·m (5.5 kgf·m, 40 lbf·ft)

BRAKE DRUM
Check for wear
and damage.

REAR BRAKE ASSY

HUB BEARING UNIT
Check for faulty movement
and wear.

HUB CAP
Replace.

SPINDLE NUT
22 x 1.5 mm
Replace.
181 N·m (18.5 kgf·m, 134 lbf·ft)

Apply a small amount of engine oil
to the seating surface.

09474_ACCO_G0042

Fig. 243 Knuckle and related components—Civic with drum brakes

3. Remove the hub cap, raise the stake and remove the spindle nut.

4. Screw two 8 × 1.25mm bolts into the brake drum to push it away from the hub. Turn each bolt two turns at a time to prevent cocking the drum excessively. Remove the brake drum.

5. Remove the hub bearing unit from the spindle.

To install:

6. Position the hub bearing unit on the vehicle. Be sure to use new bolts and nuts, as necessary. Use a new hub cap.

7. Continue the installation in the reverse order of the removal procedure.

With Disc Brakes

See Figure 244.

1. Raise and support the vehicle safely.
2. Remove the rear tires.

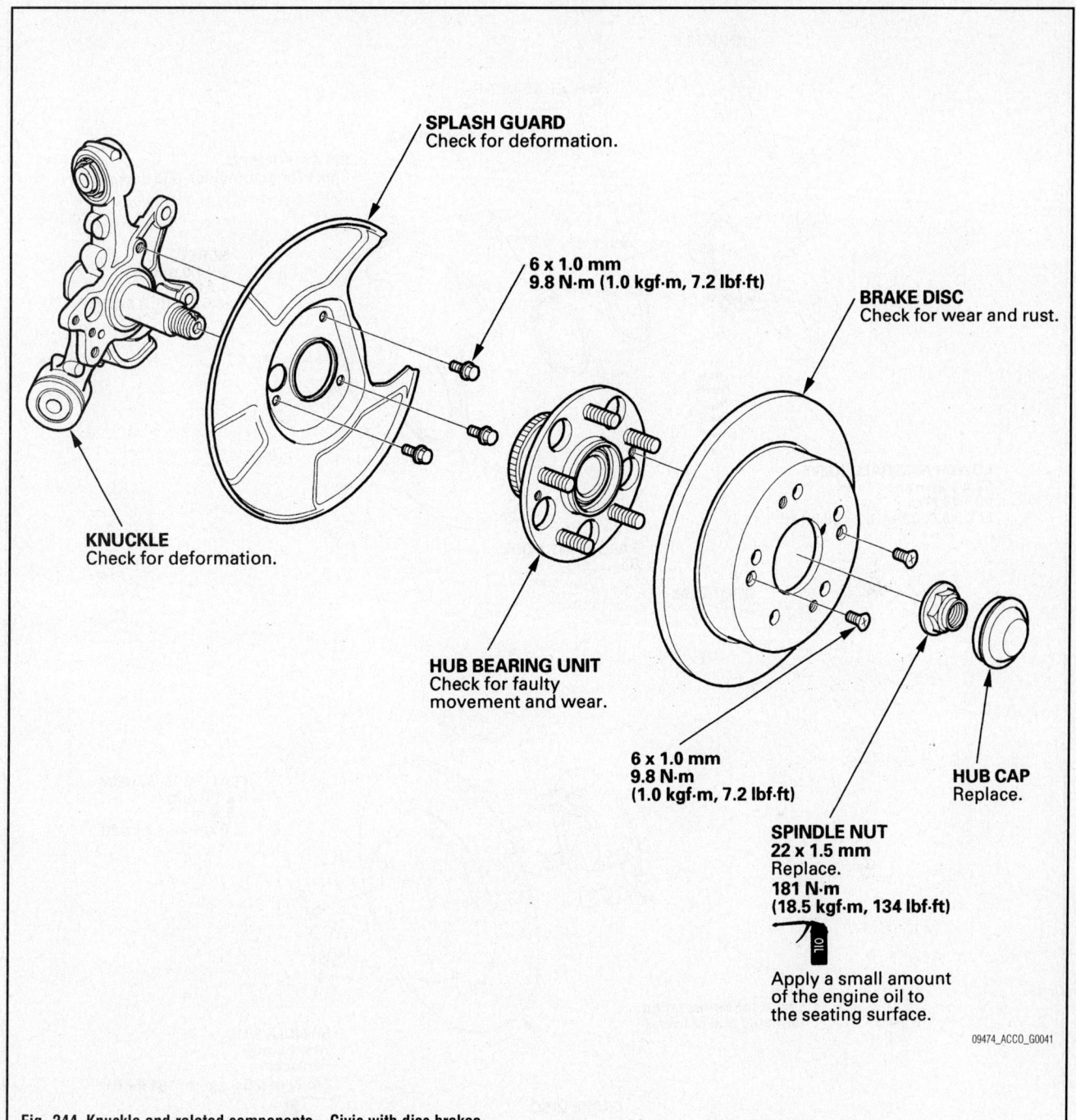

SPLASH GUARD
Check for deformation.

6 x 1.0 mm
9.8 N·m (1.0 kgf·m, 7.2 lbf·ft)

BRAKE DISC
Check for wear and rust.

KNUCKLE
Check for deformation.

HUB BEARING UNIT
Check for faulty
movement and wear.

6 x 1.0 mm
9.8 N·m
(1.0 kgf·m, 7.2 lbf·ft)

HUB CAP
Replace.

SPINDLE NUT
22 x 1.5 mm
Replace.
181 N·m
(18.5 kgf·m, 134 lbf·ft)

Apply a small amount
of the engine oil to
the seating surface.

09474_ACCO_G0041

Fig. 244 Knuckle and related components—Civic with disc brakes

3. Remove the brake hose bracket mounting bolt from the knuckle.

4. Remove the brake caliper mounting bolts. Remove the caliper assembly from the knuckle. Remove the two washers.

5. Remove the 6mm brake disc retaining screws. Remove the rotor from the hub bearing nut.

➡ If the brake disc has clung to the hub bearing unit, screw two 8x1.25MM bolts into the rotor to push It away from the hub bearing unit. Turn each bolt 90

degrees at a time to prevent cocking the rotor.

6. Remove the hub bearing unit and the O-ring.

To install:

7. Position the hub bearing unit on the vehicle. Be sure to use new bolts and nuts, as necessary. Use a new hub cap.

8. Continue the installation in the reverse order of the removal procedure.

9. Check the wheel alignment, adjust as required.

S2000

See Figure 245.

1. Remove or disconnect the following:
   - Rear wheel
   - Brake caliper support bracket
   - Wheel speed sensor
   - Spindle nut
   - Brake rotor
   - Control arm
   - Upper and lower ball joints
   - Spindle from the vehicle

2. Mount the steering knuckle in a press and press the hub out of the wheel bearing.

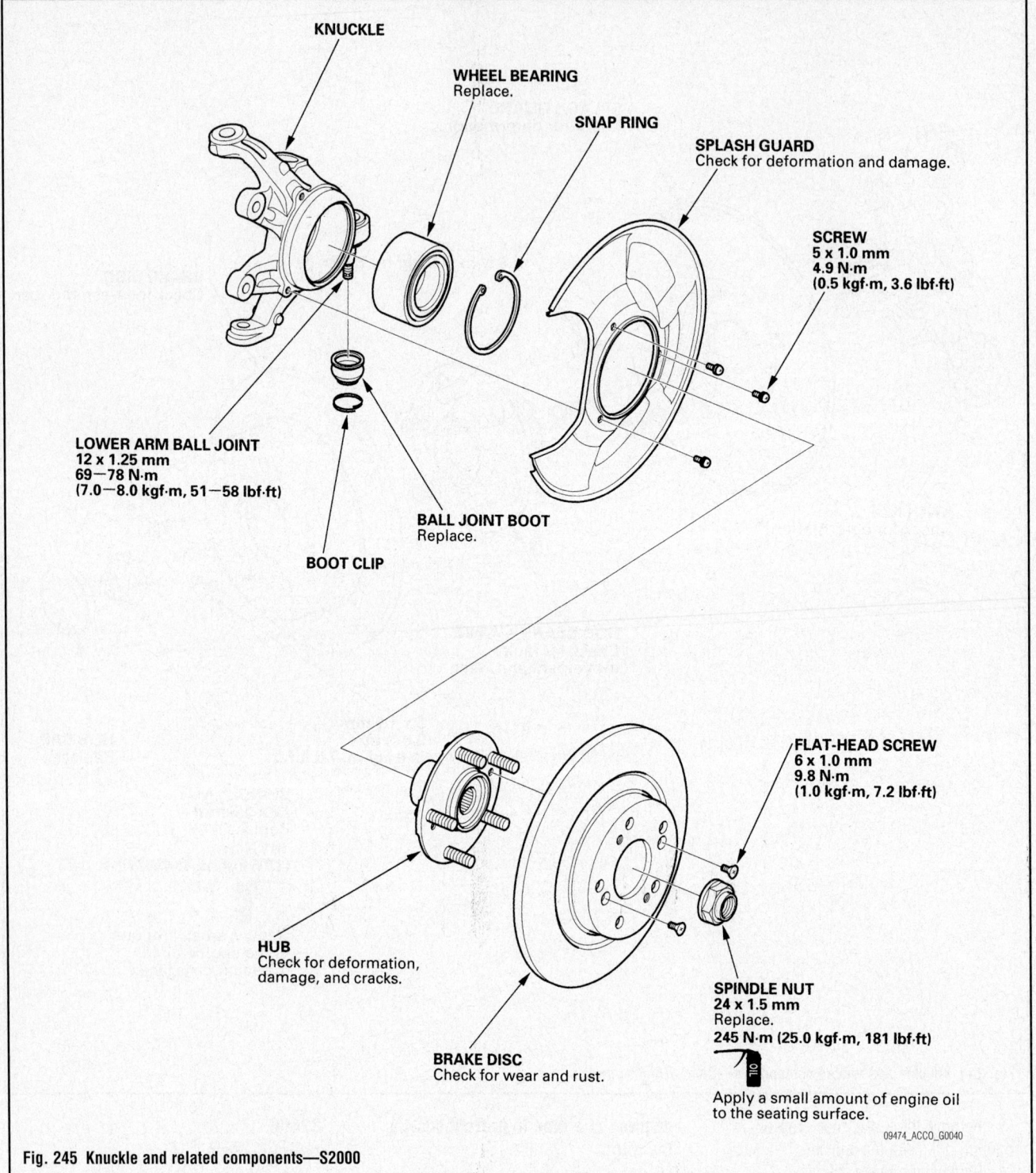

KNUCKLE

WHEEL BEARING
Replace.

SNAP RING

SPLASH GUARD
Check for deformation and damage.

SCREW
5 x 1.0 mm
4.9 N·m
(0.5 kgf·m, 3.6 lbf·ft)

LOWER ARM BALL JOINT
12 x 1.25 mm
69—78 N·m
(7.0—8.0 kgf·m, 51—58 lbf·ft)

BALL JOINT BOOT
Replace.

BOOT CLIP

HUB
Check for deformation,
damage, and cracks.

FLAT-HEAD SCREW
6 x 1.0 mm
9.8 N·m
(1.0 kgf·m, 7.2 lbf·ft)

SPINDLE NUT
24 x 1.5 mm
Replace.
245 N·m (25.0 kgf·m, 181 lbf·ft)

Apply a small amount of engine oil
to the seating surface.

BRAKE DISC
Check for wear and rust.

09474_ACCO_G0040

Fig. 245 Knuckle and related components—S2000

3. Remove the splash guard and the wheel bearing snapring.

4. Press the wheel bearing out of the steering knuckle.

To install:

5. Installation is the reverse of the removal procedure, while using the following torque values:

- Splash guard screws: 48 inch lbs. (5 Nm)
- Spindle nut: 181 ft. lbs. (245 Nm)
- Upper ball joint nut: 36–43 ft. lbs. (49–59 Nm)
- Lower ball joint nut: 51–58 ft. lbs. (68–78 Nm)
- Control arm ball joint nut: 36–43 ft. lbs. (49–59 Nm)
- Brake caliper support bolts: 41 ft. lbs. (55 Nm)

ADJUSTMENT

The wheel bearings are not adjustable. If they are not within specification, the wheel bearings must be replaced.

HONDA

CR-V

5

SPECIFICATIONS AND MAINTENANCE CHARTS

ENGINE AND VEHICLE IDENTIFICATION

	Engine						Model Year	
Code ①	Liters (cc)	Cu. In.	Cyl.	Fuel Sys.	Engine Type	Eng. Mfg.	Code ②	Year
K24Z1	2.4 (2,354)	144	4	SMFI	DOHC	Honda	7	2007
							8	2008

SMFI: Sequential Multi-port Fuel Injection

DOHC: Double Overhead Camshaft

① Stamped into the front of the engine block and can be seen through the window next to the "H" logo on the front grill

② 10th digit of the Vehicle Identification Number (VIN)

22140_HCRV_C0001

GENERAL ENGINE SPECIFICATIONS
All measurements are given in inches.

Year	Model	Engine Displacement Liters	Engine Series ID/VIN	Net Horsepower @ rpm	Net Torque @ rpm (ft. lbs.)	Bore x Stroke (in.)	Compression Ratio	Oil Pressure @ rpm
2007	CR-V	2.4	K24Z1	166@5800	161@4200	3.43 x 3.90	9.7:1	44@3000
2008	CR-V	2.4	K24Z1	166@5800	161@4200	3.43 x 3.90	9.7:1	44@3000

22140_HCRV_C0002

GASOLINE ENGINE TUNE-UP SPECIFICATIONS

Year	Engine Displacement Liters	Engine Series ID/VIN	Spark Plug Gap (in.)	Ignition Timing (deg.) MT	Ignition Timing (deg.) AT	Fuel Pump (psi)	Idle Speed (rpm) MT	Idle Speed (rpm) AT	Valve Clearance (in.) Intake	Valve Clearance (in.) Exhaust
2007	2.4	K24Z1	0.039-0.043	—	6-10B	47-54	—	600-700	0.008-0.010	0.011-0.013
2008	2.4	K24Z1	0.039-0.043	—	6-10B	47-54	—	600-700	0.008-0.010	0.011-0.013

NOTE: The Vehicle Emission Control Information label reflects specification changes made during production.

Follow the figures on the label if they differ from those in this chart.

B: Before top dead center

22140_HCRV_C0003

CAPACITIES

Year	Model	Engine Displacement Liters	Engine Series ID/VIN	Engine Oil with Filter (qts.)	Transmission (pts.) Manual	Transmission (pts.) Auto. ①	Transfer Case (pts.)	Drive Axle Front (pts.)	Drive Axle Rear (pts.) ①	Fuel Tank (gal.)	Cooling System (qts.)
2007	CR-V	2.4	K24Z1	4.4	—	②	③	③	2.6	15.3	5.3
2008	CR-V	2.4	K24Z1	4.4	—	②	③	③	2.6	15.3	5.3

NOTE: All capacities are approximate. Add fluid gradually and check to be sure a proper fluid level is obtained.

① Drain and refill

② 2WD: 5.4 pts.
   4WD: 5.2 pts.

③ Included in transaxle refill figure

22140_HCRV_C0004

FLUID SPECIFICATIONS

Year	Model	Engine Displacement Liters	Engine Series ID/VIN	Engine Oil	Manual Trans.	Auto. Trans.	Drive Axle	Transfer Case	Power Steering Fluid	Brake Master Cylinder	Cooling System
2007	CR-V	2.4	K24Z1	①	—	②	③	②	④	⑤	⑥
2008	CR-V	2.4	K24Z1	①	—	②	③	②	④	⑤	⑥

DOT: Department Of Transportation

① Honda Motor Oil: American Honda P/N 08798-9023 (5W-20), Honda Canada P/N CA66806 (5W-20)

② Honda Automatic Transmission Fluid (ATF-Z1): American Honda P/N 08200-9001, Honda Canada P/N CA66689

③ Rear differential (4WD) Honda Dual Pump Fluid II: P/N 08200-9007

④ Honda Power Steering Fluid: P/N 08206-9002

⑤ Honda DOT 3 Brake Fluid: P/N 08798-9008

⑥ Honda Long Life Antifreeze/Coolant Type 2: P/N OL 999-9001; Honda Coolant Concentrate: P/N OL 999-9020

22140_HCRV_C0005

VALVE SPECIFICATIONS

Year	Engine Displacement Liters	Engine Series ID/VIN	Seat Angle (deg.)	Face Angle (deg.)	Spring Test Pressure (lbs. @ in.)	Spring Installed Height (in.)	Stem-to-Guide Clearance (in.) Intake	Stem-to-Guide Clearance (in.) Exhaust	Stem Diameter (in.) Intake	Stem Diameter (in.) Exhaust
2007	2.4	K24Z1	NA	NA	NA	①	0.0012-0.0022	0.0022-0.0031	0.2156-0.2159	0.2146-0.2150
2008	2.4	K24Z1	NA	NA	NA	①	0.0012-0.0022	0.0022-0.0031	0.2156-0.2159	0.2146-0.2150

NA: Information not available

① Valve spring free length:
   Intake: 1.873 in.
   Exhaust: 1.954 in.

22140_HCRV_C0006

CAMSHAFT AND BEARING SPECIFICATIONS CHART

All measurements are given in inches.

Year	Engine Displ. Liters	Engine Series ID/VIN	Journal Dia.	Brg. Oil Clearance	Shaft End-play	Runout	Journal Bore	Lobe Height Intake	Lobe Height Exhaust
2007	2.4	K24Z1	NA	①	0.002-0.008	0.001 max.	NA	②	1.3422
2008	2.4	K24Z1	NA	①	0.002-0.008	0.001 max.	NA	②	1.3422

NA: Information not available

① Journal No. 1: 0.002-0.003 in.
Journals No. 2, 3, 4, 5: 0.002-0.004 in.

② Intake, primary: 1.3489 in.
Intake, secondary: 1.1668 in.

22140_HCRV_C0007

CRANKSHAFT AND CONNECTING ROD SPECIFICATIONS

All measurements are given in inches.

Year	Engine Displacement Liters	Engine Series ID/VIN	Crankshaft Main Brg. Journal Dia.	Crankshaft Main Brg. Oil Clearance	Crankshaft Shaft End-play	Crankshaft Thrust on No.	Connecting Rod Journal Diameter	Connecting Rod Oil Clearance	Connecting Rod Side Clearance
2007	2.4	K24Z1	①	②	0.0040-0.0140	3	1.8888-1.8898	0.0008-0.0020	0.0060-0.0120
2008	2.4	K24Z1	①	②	0.0040-0.0140	3	1.8888-1.8898	0.0008-0.0020	0.0060-0.0120

① Except No. 3: 2.1648-2.1657
No. 3: 2.1644-2.1654

② Except No. 3: 0.0007-0.0016
No. 3: 0.0010-0.0019

22140_HCRV_C0008

PISTON AND RING SPECIFICATIONS

All measurements are given in inches.

Year	Engine Displ. Liters	Engine Series ID/VIN	Piston Clearance	Ring Gap Top Compression	Ring Gap Bottom Compression	Ring Gap Oil Control	Ring Side Clearance Top Compression	Ring Side Clearance Bottom Compression	Ring Side Clearance Oil Control
2007	2.4	K24Z1	0.0008-0.0016	0.0080-0.0140	0.0160-0.0220	0.0100-0.0260	0.0014-0.0024	0.0012-0.0022	NA
2008	2.4	K24Z1	0.0008-0.0016	0.0080-0.0140	0.0160-0.0220	0.0100-0.0260	0.0014-0.0024	0.0012-0.0022	NA

22140_HCRV_C0009

TORQUE SPECIFICATIONS
All readings in ft. lbs.

Year	Engine Displacement Liters	Engine Series ID/VIN	Cylinder Head Bolts	Main Bearing Bolts	Rod Bearing Bolts	Crankshaft Damper Bolts	Flywheel Bolts	Manifold Intake	Manifold Exhaust	Spark Plugs	Oil Pan Drain Plug
2007	2.4	K24Z1	①	②	③	④	⑤	⑥	⑦	13	29
2008	2.4	K24Z1	①	②	③	④	⑤	⑥	⑦	13	29

NOTE: Dip cylinder head bolts, main bearing bolts, and crankshaft damper bolt in clean engine oil prior to tightening.

① Step 1: 29 ft. lbs.
   Step 2: plus 90 degrees
   Step 3: plus 90 degrees
   Step 4: NEW BOLT ONLY plus 90 degrees

② Tighten bearing cap bolts in sequence:
   Step 1: 22 ft. lbs.
   Step 2: plus 56 degrees
   Tighten the 8mm bolts to 16 ft. lbs., in sequence

③ Step 1: Tighten to 14 ft. lbs.
   Step 2: plus 90 degrees

④ Step 1: Tighten to 36 ft. lbs.
   Step 2: plus 90 degrees

⑤ Tighten in sequence to 54 ft. lbs. in at least 2 steps

⑥ Tighten in sequence to 16 ft. lbs. in 3 steps

⑦ Tighten in sequence to 33 ft. lbs. in 3 steps

22140_HCRV_C0010

WHEEL ALIGNMENT

Year	Model		Caster Range (+/-Deg.)	Caster Preferred Setting (Deg.)	Camber Range (+/-Deg.)	Camber Preferred Setting (Deg.)	Toe-in (Deg.)
2007	CR-V	Front	1.00	+3.02	0.30	0	0+/-0.08
		Rear	—	—	0.45	-1.00	0.08+/-0.08
2008	CR-V	Front	1.00	+3.02	0.30	0	0+/-0.08
		Rear	—	—	0.45	-1.00	0.08+/-0.08

NOTE: Measurements are given for unladen vehicle: fuel, engine coolant, and fluid levels are full. Spare tire, jack, hand tools, and mats are in designated positio

22140_HCRV_C0011

TIRE, WHEEL AND BALL JOINT SPECIFICATIONS

| Year | Model | OEM Tires | | Tire Pressures (psi) | | Wheel Size | Ball Joint Inspection | Lug Nut Torque (ft. lbs.) |
		Standard	Optional	Front	Rear			
2007	CR-V	P225/65R17	—	30	30	NA	NA	80
2008	CR-V	P225/65R17	—	30	30	NA	NA	80

OEM: Original Equipment Manufacturer

PSI: Pounds Per Square Inch

NA: Information not available

22140_HCRV_C0012

BRAKE SPECIFICATIONS

All measurements in inches unless noted

| Year | Model | | Brake Disc | | | Brake Drum Diameter | | | Minimum Lining Thickness | Brake Caliper | |
			Original Thickness	Minimum Thickness	Maximum Runout	Original Inside Diameter	Max. Wear Limit	Maximum Machine Diameter		Bracket Bolts (ft. lbs.)	Mounting Bolts (ft. lbs.)
2007	CR-V	F	1.10-1.11	1.020	0.0016	—	—	—	0.060	80	25
		R	0.35-0.36	0.300	0.0016	—	—	—	0.060	41	16
2008	CR-V	F	1.10-1.11	1.020	0.0016	—	—	—	0.060	80	25
		R	0.35-0.36	0.300	0.0016	—	—	—	0.060	41	16

F: Front

R: Rear

22140_HCRV_C0013

SCHEDULED MAINTENANCE INTERVALS
HONDA—CR-V

TO BE SERVICED	TYPE OF SERVICE	VEHICLE MILEAGE INTERVAL (x1000) ①											
		10	20	30	40	50	60	70	80	90	100	110	120
Accessory drive belts	S/I			✓			✓			✓			✓
Air cleaner element (engine)	R			✓			✓			✓			✓
Air conditioner system	S/I	Inspect system operation annually											
Air conditioner filter	R			✓			✓			✓			✓
Automatic transmission fluid	R												✓
Brake fluid level	S/I	Once a month											
Brake fluid	R	Every 3 years											
Brake hoses/lines (incl. ABS)	S/I		✓		✓		✓		✓		✓		✓
Cooling system hoses/connections	S/I		✓		✓		✓		✓		✓		✓
Engine coolant	R	Every 120,000 miles or 10 years, then every 60,000 miles or 5 years											
Engine oil and filter	R	✓	✓	✓	✓	✓	✓	✓	✓	✓	✓	✓	✓
Engine oil and coolant levels	I	Inspect at each fuel stop											
Exhaust system	S/I		✓		✓		✓		✓		✓		✓
Fluid levels and condition	S/I		✓		✓		✓		✓		✓		✓
Front and rear brakes	S/I		✓		✓		✓		✓		✓		✓
Fuel lines and connections	S/I		✓		✓		✓		✓		✓		✓
Halfshaft boots	S/I		✓		✓		✓		✓		✓		✓
Idle speed	S/I	Every 160,000 miles											
Lights	S/I	Once a month											
Parking brake system	S/I		✓		✓		✓		✓		✓		✓
Power steering fluid level	S/I		✓		✓		✓		✓		✓		✓
Rear differential fluid level	S/I		✓		✓		✓		✓		✓		✓
Rear differential fluid	R	Every 90,000 miles or 5 years, whichever comes first											
Rotate and inspect tires	S/I	✓	✓	✓	✓	✓	✓	✓	✓	✓	✓	✓	✓
Spark plugs	R											✓	
Suspension components	S/I		✓		✓		✓		✓		✓		✓
Tie rod ends, steering gear box, and boots	S/I		✓		✓		✓		✓		✓		✓
Tire inflation and condition	S/I	Once a month											
Transmission fluid level	S/I		✓		✓		✓		✓		✓		✓
Valve clearance (when noisy)	S/I	Inspect at 110,000 miles or when making noise											

R: Replace   S/I: Service or Inspect

FREQUENT OPERATION MAINTENANCE (SEVERE SERVICE)

If a vehicle is operated under any of the following conditions it is considered severe service:

- Extremely dusty areas.

- 50% or more of the vehicle operation is in 90°F (32°C) or higher temperatures, or constant operation in temperatures below 32°F (0°C).

- Prolonged idling (vehicle operation in stop and go traffic).

- Frequent short running periods (engine does not warm to normal operating temperatures).

- Police, taxi, delivery usage, or trailer towing usage.

Air cleaner element (engine) replace every 15,000 miles

Automatic transmission fluid replace every 60,000 miles

Engine oil and filter replace every 5,000 miles

Front and rear brakes inspect every 10,000 miles

Halfshaft boots inspect every 10,000 miles

Rear differential fluid replace every 60,000 miles

Suspension components inspect every 10,000 miles

Tires rotate every 10,000 miles

① Some of the maintenance intervals are determined by the on-board computer system and are subject to individual driving habits.

Follow the owner's manual to appropriately respond to the "SERVICE" messages which appear in the display area.

PRECAUTIONS

Before servicing any vehicle, please be sure to read all of the following precautions, which deal with personal safety, prevention of component damage, and important points to take into consideration when servicing a motor vehicle:

• Never open, service or drain the radiator or cooling system when the engine is hot; serious burns can occur from the steam and hot coolant.

• Observe all applicable safety precautions when working around fuel. Whenever servicing the fuel system, always work in a well-ventilated area. Do not allow fuel spray or vapors to come in contact with a spark, open flame, or excessive heat (a hot drop light, for example). Keep a dry chemical fire extinguisher near the work area. Always keep fuel in a container specifically designed for fuel storage; also, always properly seal fuel containers to avoid the possibility of fire or explosion. Refer to the additional fuel system precautions later in this section.

• Fuel injection systems often remain pressurized, even after the engine has been turned **OFF**. The fuel system pressure must be relieved before disconnecting any fuel lines. Failure to do so may result in fire and/or personal injury.

• Brake fluid often contains polyglycol ethers and polyglycols. Avoid contact with the eyes and wash your hands thoroughly after handling brake fluid. If you do get brake fluid in your eyes, flush your eyes with clean, running water for 15 minutes. If eye irritation persists, or if you have taken brake fluid internally, IMMEDIATELY seek medical assistance.

• The EPA warns that prolonged contact with used engine oil may cause a number of skin disorders, including cancer. You should make every effort to minimize your exposure to used engine oil. Protective gloves should be worn when changing oil. Wash your hands and any other exposed skin areas as soon as possible after exposure to used engine oil. Soap and water, or waterless hand cleaner should be used.

• All new vehicles are now equipped with an air bag system, often referred to as a Supplemental Restraint System (SRS) or Supplemental Inflatable Restraint (SIR) system. The system must be disabled before performing service on or around system components, steering column, instrument panel components, wiring and sensors. Failure to follow safety and disabling procedures could result in accidental air bag deployment, possible personal injury and unnecessary system repairs.

• Always wear safety goggles when working with, or around, the air bag system. When carrying a non-deployed air bag, be sure the bag and trim cover are pointed away from your body. When placing a non-deployed air bag on a work surface, always face the bag and trim cover upward, away from the surface. This will reduce the motion of the module if it is accidentally deployed. Refer to the additional air bag system precautions later in this section.

• Clean, high quality brake fluid from a sealed container is essential to the safe and proper operation of the brake system. You should always buy the correct type of brake fluid for your vehicle. If the brake fluid becomes contaminated, completely flush the system with new fluid. Never reuse any brake fluid. Any brake fluid that is removed from the system should be discarded. Also, do not allow any brake fluid to come in contact with a painted surface; it will damage the paint.

• Never operate the engine without the proper amount and type of engine oil; doing so WILL result in severe engine damage.

• Timing belt maintenance is extremely important. Many models utilize an interference-type, non-freewheeling engine. If the timing belt breaks, the valves in the cylinder head may strike the pistons, causing potentially serious (also time-consuming and expensive) engine damage. Refer to the maintenance interval charts for the recommended replacement interval for the timing belt, and to the timing belt section for belt replacement and inspection.

• Disconnecting the negative battery cable on some vehicles may interfere with the functions of the on-board computer system(s) and may require the computer to undergo a relearning process once the negative battery cable is reconnected.

• When servicing drum brakes, only disassemble and assemble one side at a time, leaving the remaining side intact for reference.

• Only an MVAC-trained, EPA-certified automotive technician should service the air conditioning system or its components.

BRAKES

ANTI-LOCK BRAKE SYSTEM (ABS)

GENERAL INFORMATION

PRECAUTIONS

• Certain components within the ABS system are not intended to be serviced or repaired individually.

• Do not use rubber hoses or other parts not specifically specified for and ABS system. When using repair kits, replace all parts included in the kit. Partial or incorrect repair may lead to functional problems and require the replacement of components.

• Lubricate rubber parts with clean, fresh brake fluid to ease assembly. Do not use shop air to clean parts; damage to rubber components may result.

• Use only DOT 3 brake fluid from an unopened container.

• If any hydraulic component or line is removed or replaced, it may be necessary to bleed the entire system.

• A clean repair area is essential. Always clean the reservoir and cap thoroughly before removing the cap. The slightest amount of dirt in the fluid may plug an orifice and impair the system function. Perform repairs after components have been thoroughly cleaned; use only denatured alcohol to clean components. Do not allow ABS components to come into contact with any substance containing mineral oil; this includes used shop rags.

• The Anti-Lock control unit is a microprocessor similar to other computer units in the vehicle. Ensure that the ignition switch is **OFF** before removing or installing controller harnesses. Avoid static electricity discharge at or near the controller.

• If any arc welding is to be done on the vehicle, the control unit should be unplugged before welding operations begin.

BLEEDING PROCEDURE

➡The brake fluid should be replaced at least every 3 years or 45,000 miles (72,000 km).

⚙ WARNING

Avoid spilling brake fluid on the vehicle's paint. It will damage the finish. If a spill does occur, wash it immediately with water.

The purpose of bleeding the brakes is to expel air trapped in the hydraulic system. The system should be checked for fluid condition, brake hose condition, and air infiltration whenever the pedal feels spongy. The system must be bled whenever it has been opened, repaired, or a hydraulic component is replaced. If you are not using a pressure bleeder, you will need an assistant for this job.

⚙ WARNING

Never reuse brake fluid which has been bled from the brake system. Brake fluid absorbs moisture which can lower its boiling point, therefore re-using old, used, or contaminated fluid can decrease the effectiveness of the braking system.

BLEEDING PROCEDURE

See Figures 1 and 2.

➡If the ABS modulator has been opened or replaced, it may be necessary to bleed each line at the modulator unit.

When bleeding the brakes, air may be trapped in the brake lines or valves far upstream, as much as 10 feet from the bleeder screw. Therefore, it is very important to have a fast flow of a large volume of brake fluid when bleeding the brakes to make sure all of the air is expelled from the system.

➡Proper manual bleeding of the hydraulic brake system will require the use of an assistant unless a suitable self-bleeding tool is available. If using a self-bleeding tool, refer to the manufacturer's directions for tool use and follow the proper bleeding sequence listed in this section.

✳✳ WARNING

Avoid spilling brake fluid on the vehicle's paint. It will damage the

finish. If a spill does occur, wash it with water immediately.

✳✳ CAUTION

Brake fluid contains polyglycol ethers and polyglycols. Avoid contact with the eyes and wash your hands thoroughly after handling brake fluid. If you do get brake fluid in your eyes, flush your eyes with clean, running water for 15 minutes. If eye irritation persists, or if you have taken brake fluid internally, IMMEDIATELY seek medical assistance.

1. To bleed the brakes, if the ABS modulator has not been opened or replaced, proceed as follows.

➡If the master cylinder reservoir runs dry during the bleeding process, restart from the first fitting.

2. Remove the old brake fluid and clean the brake master cylinder reservoir with a clean lint-free cloth.

3. Refill the brake master cylinder reservoir.

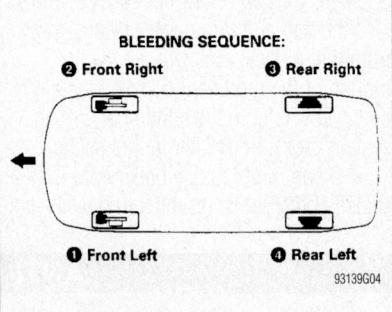

BLEEDING SEQUENCE:
❷ Front Right   ❸ Rear Right

❶ Front Left   ❹ Rear Left

93139G04

Fig. 1 Always follow the proper bleeding sequence when bleeding the brake hydraulic system

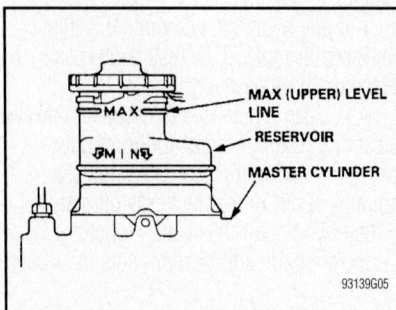

MAX (UPPER) LEVEL LINE
RESERVOIR
MASTER CYLINDER

93139G05

Fig. 2 Check the fluid level often, and top off the master cylinder reservoir as necessary as the system is being bled

4. Bleed the brake system at each fitting. Do not proceed to the next fitting until all air bubbles are removed from the previous fitting. Bleed the brakes, making sure to follow this sequence:
- Left Front
- Right Front
- Right Rear
- Left Rear

5. Attach a clear plastic hose to the bleeder screw, then place the hose into a clean jar that has enough fresh brake fluid to submerge the end of the hose.

6. Have an assistant pump the brake pedal 3–4 times, and hold pressure on it, then open the bleeder screw at least ¼ turn. When the bleeder screw opens, the brake pedal will drop. Have the assistant hold it there until the bleed valve is closed.

7. Close the bleeder screw and have the assistant slowly release the brake pedal AFTER the bleeder screw is closed, then check the master cylinder fluid level and top off as necessary.

8. Repeat the bleeding procedure until there are no air bubbles, or a minimum of 4 or 5 times at each bleeder screw, then check the pedal for travel and feel. If the pedal travel is excessive, or feels spongy, it is possible that not enough fluid has passed through the system to expel all of the trapped air.

➡Constantly check and top off the master cylinder. Do not allow the master cylinder to run dry, otherwise air will re-enter the brake system.

9. Once completed, test drive the vehicle to be sure the brakes are operating correctly and that the pedal feel is firm.

MASTER CYLINDER BLEEDING

1. Before servicing the vehicle, refer to the Precautions Section.

2. If removed from the vehicle, clamp the master cylinder in a vise with soft-jaw caps.

3. Attach the special tools for bleeding the master cylinder in the following fashion:

   a. Thread the bleeder tube adapters into the primary and secondary outlet ports of the master cylinder and tighten the adapters.

   b. Thread a bleeder tube into each adapter and tighten the tube nuts.

   c. Flex each bleeder tube and place the open ends into the neck of the master cylinder reservoir. Position the open ends of the tubes into the reservoir so their outlets are below the surface of the brake fluid in the reservoir when filled.

➡Make sure the ends of the bleeder tubes stay below the surface of the brake fluid in the reservoir at all times during the bleeding procedure.

4. Fill the brake fluid reservoir with fresh brake fluid (DOT 3).

5. Using an appropriately sized wooden dowel as a pushrod, slowly press the pistons inward discharging brake fluid through the bleeder tubes, then release the pressure, allowing the pistons to return to the released position. Repeat this several times until all air bubbles are expelled from the master cylinder bore and bleeder tubes.

6. Remove the bleeder tubes and adapters from the master cylinder and plug the master cylinder outlet ports.

7. Install the fill cap on the reservoir.

8. Remove the master cylinder from the vise.

9. Install the master cylinder on the vehicle.

BRAKE LINE BLEEDING

Refer to Bleeding the Brake System, Bleeding Procedure.

BRAKES

✳✳ CAUTION

Dust and dirt accumulating on brake parts during normal use may contain asbestos fibers from production or aftermarket brake linings. Breathing excessive concentrations of asbestos fibers can cause serious bodily harm. Exercise care when servicing brake parts. Do not sand or grind brake lining unless equipment used is designed to contain the dust residue. Do not clean brake parts with compressed air or by dry brushing. Cleaning should be done by dampening the brake components with a fine mist of water, then wiping the brake components clean with a dampened cloth. Dispose of cloth and all residue containing asbestos fibers in an impermeable container with the appropriate label. Follow practices prescribed by the Occupational Safety and Health Administration (OSHA) and the Environmental Protection Agency (EPA) for the handling, processing, and disposing of dust or debris that may contain asbestos fibers.

BRAKE CALIPER

REMOVAL & INSTALLATION
See Figure 3.

✳✳ CAUTION

Frequent inhalation of brake pad dust, regardless of material composition, could be hazardous to your health. Avoid breathing dust particles. Never use an air hose or brush to clean brake assemblies. Use an OSHA-approved vacuum cleaner.

1. Before servicing the vehicle, refer to the Precautions Section.

2. Remove the wheel nuts and front wheel.

3. Remove the brake hose mounting bolt.

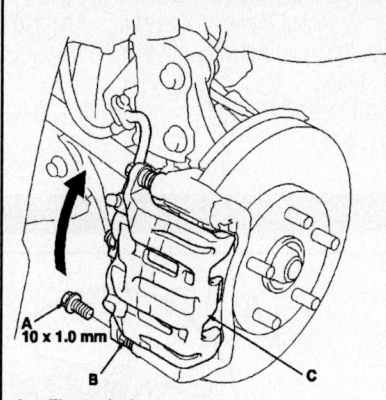

A. Flange bolt
B. Caliper pin
C. Caliper assembly

22140_HCRV_G0132

Fig. 3 Remove the flange bolt while holding the caliper pin with a wrench. Remove the caliper assembly from the knuckle

4. Remove the flange bolt (A) while holding the caliper pin (B) with a wrench. Be careful not to damage the pin boot.

5. Remove the caliper assembly (C) from the knuckle.

➡Check the hose and the pin boots for damage and deterioration.

6. Detach the brake caliper from the brake hose, if the caliper is being replaced.

7. If the caliper is not being replaced, it can be moved out of the way and suspended with wire so the hose is not under strain.

✳✳ WARNING

To prevent damage to the caliper assembly or brake hose, use a short piece of wire to hang the caliper assembly from the undercarriage. Do not twist the brake hose excessively.

To install:

8. Position the caliper in place.

9. If removed, reattach the brake line with the banjo bolt to the caliper and tighten to 26 ft. lbs. (35 Nm).

FRONT DISC BRAKES

10. Install and tighten the caliper mounting bolts to 37 ft. lbs. (50 Nm).

11. Install the brake hose clip retaining bolt.

12. If the brake hose was removed from the caliper, bleed the brake system. Refer to Bleeding the Brake System.

13. Install the wheels.

14. Before attempting to move the vehicle, pump the brake pedal to seat the pads against the rotors. Make sure the vehicle has a firm brake pedal. Check the level of the brake fluid and add fluid if necessary.

DISC BRAKE PADS

REMOVAL & INSTALLATION
See Figures 4 and 5.

1. Before servicing the vehicle, refer to the Precautions Section.

2. Remove a small amount of brake fluid from the reservoir using a suction pump.

3. Remove the front wheels.

4. Remove the lower caliper retaining bolt and pivot the caliper upward, off of the pads.

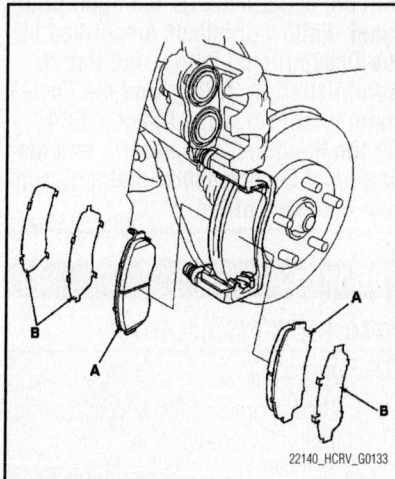

22140_HCRV_G0133

Fig. 4 Remove the pad shims (B) and pad retainers. Remove the disc brake pads (A) from the caliper

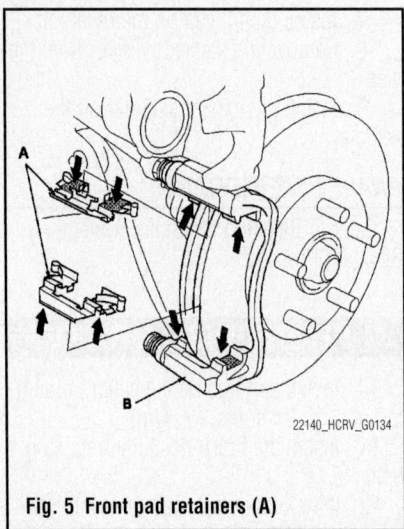

Fig. 5 Front pad retainers (A)

5. Remove the pad shims (B) and pad retainers.

6. Remove the disc brake pads (A) from the caliper.

To install:

7. Clean the caliper thoroughly; remove any rust from the lip of the rotor. Check the brake rotor for grooves or cracks. If any heavy scoring is present, the rotor must be replaced.

8. Install the pad retainers. Apply molybdenum brake grease to both surfaces of the shims and the back of the disc brake pads.

9. Install the pads and shims. The pad with the wear indicator goes in the inboard position.

10. Install the pad springs while holding the pads.

→Push in the caliper piston so the caliper will fit over the pads. This is most easily accomplished with a pad spreader or large C-clamp.

11. Install the caliper down into position and tighten the mounting bolt to 37 ft. lbs. (50 Nm).

12. Install the wheels.

13. Add brake fluid to the master cylinder reservoir and install the cap.

14. Depress the brake pedal several times and make sure that the movement feels normal. The first brake pedal application may result in a very long pedal action due to the pistons being retracted. Always make several brake applications before starting the vehicle. Bleed the system if necessary.

BRAKES

REAR DISC BRAKES

✳✳ CAUTION

Dust and dirt accumulating on brake parts during normal use may contain asbestos fibers from production or aftermarket brake linings. Breathing excessive concentrations of asbestos fibers can cause serious bodily harm. Exercise care when servicing brake parts. Do not sand or grind brake lining unless equipment used is designed to contain the dust residue. Do not clean brake parts with compressed air or by dry brushing. Cleaning should be done by dampening the brake components with a fine mist of water, then wiping the brake components clean with a dampened cloth. Dispose of cloth and all residue containing asbestos fibers in an impermeable container with the appropriate label. Follow practices prescribed by the Occupational Safety and Health Administration (OSHA) and the Environmental Protection Agency (EPA) for the handling, processing, and disposing of dust or debris that may contain asbestos fibers.

BRAKE CALIPER

REMOVAL & INSTALLATION

See Figure 6.

1. Before servicing the vehicle, refer to the Precautions Section.

2. Remove some fluid from the reservoir with a suction pump.

3. Raise and safely support the vehicle.

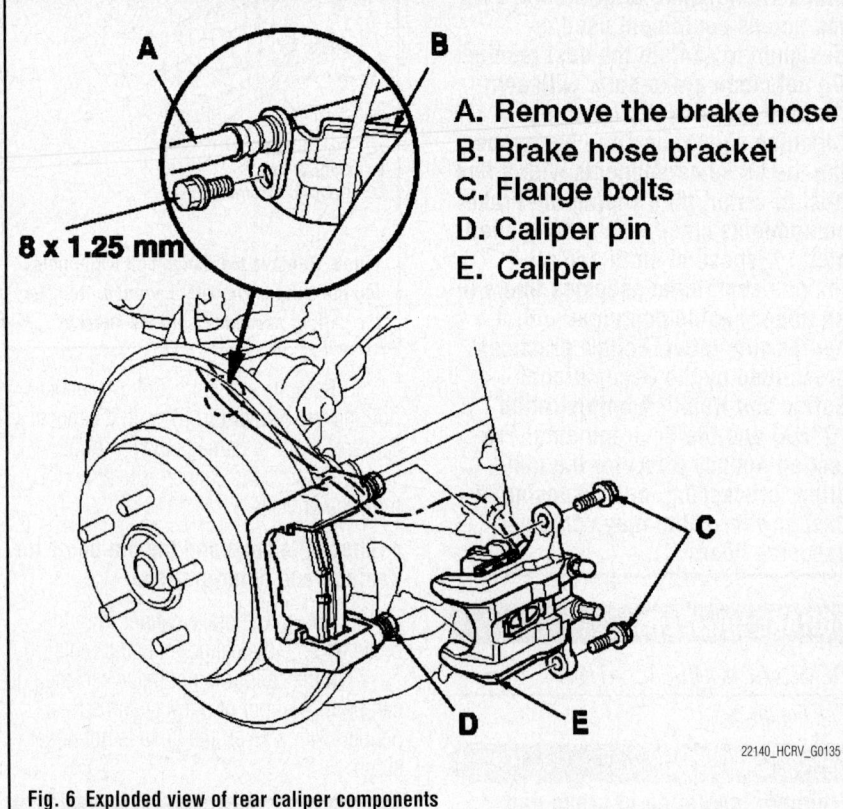

A. Remove the brake hose
B. Brake hose bracket
C. Flange bolts
D. Caliper pin
E. Caliper

8 x 1.25 mm

Fig. 6 Exploded view of rear caliper components

4. Remove the rear wheels.

5. Remove the brake hose (A) from the brake hose bracket (B).

6. Remove the banjo bolt and disconnect the brake hose from the caliper. Plug the hose to prevent fluid loss and contamination.

7. Remove the flange bolts (C) while holding the caliper pin (D) with a wrench. Be careful not to damage the pin boot.

8. Remove the caliper (E).

9. Check the hose and pin boots for damage and deterioration.

To install:

10. Install the caliper over the pads and onto its mounting bracket. Tighten the caliper bolts to 17 ft. lbs. (23 Nm).

11. Install the brake hose with new sealing washers. Tighten the banjo bolt to 25 ft. lbs. (34 Nm).

12. Fill the reservoir with fluid and bleed

the brake system. Adjust the parking brake if necessary.

13. Install the rear wheels.

DISC BRAKE PADS

REMOVAL & INSTALLATION

1. Before servicing the vehicle, refer to the Precautions Section.

2. Remove a small amount of brake fluid from the reservoir using a suction pump.

3. Remove the lower caliper pin bolt and pivot the caliper upward.

4. Remove the pads, shims, and pad retainers.

To install:

5. Clean the caliper thoroughly; remove any dirt or dust. Check the brake rotor for grooves or cracks and machine or replace, as necessary.

6. Install the pad retainers. Apply molybdenum brake grease to both surfaces of the shims and the back of the disc brake pads.

7. Install the pads and shims. The wear retainer on the inboard pad faces down.

8. Use a suitable tool to push caliper piston into its bore and enable the caliper to fit over the pads. Lubricate the piston boot with silicon grease. Avoid twisting the boot.

9. Rotate the caliper down and tighten the mounting bolts to 17 ft. lbs. (23 Nm)

10. Install the rear wheels.

11. Add brake fluid to the master cylinder reservoir. Depress the brake pedal several times to seat the pads. Bleed the brakes if necessary.

BRAKES

PARKING BRAKE SHOES

REMOVAL & INSTALLATION

See Figures 7 through 13.

1. Before servicing the vehicle, refer to the Precautions Section.

2. Raise and safely support the vehicle.

3. Remove the rear wheels.

4. Release the parking brake and remove the rear brake caliper and brake disc.

5. Disconnect and remove the upper brake spring.

6. Disconnect and remove the lower brake spring.

7. Remove the tension pins (A) by pushing the retainer springs (B) and turning the pins.

8. Remove the adjuster assembly (C) by moving the forward brake shoe (D).

9. Remove the rearward brake shoe by disconnecting the parking brake cable from the parking brake lever.

10. Disconnect the rod spring (A), and remove the strut (B).

11. Remove the parking brake shoes (C).

12. Remove the U-clip (A), wave washer (B), parking brake lever (C), and pivot pin (D) from the brake shoe.

To install:

13. Apply Molykote 44 MA grease to the sliding surface of the pivot pin (A) and insert the pin into the rearward brake shoe (B).

14. Install the parking brake lever (C) and wave washer (D) on the pivot pin and secure with a new U-clip (E).

15. Install the wave washer with its convex side facing out.

16. Pinch the U-clip securely to prevent the pivot pin from coming out of the brake shoe.

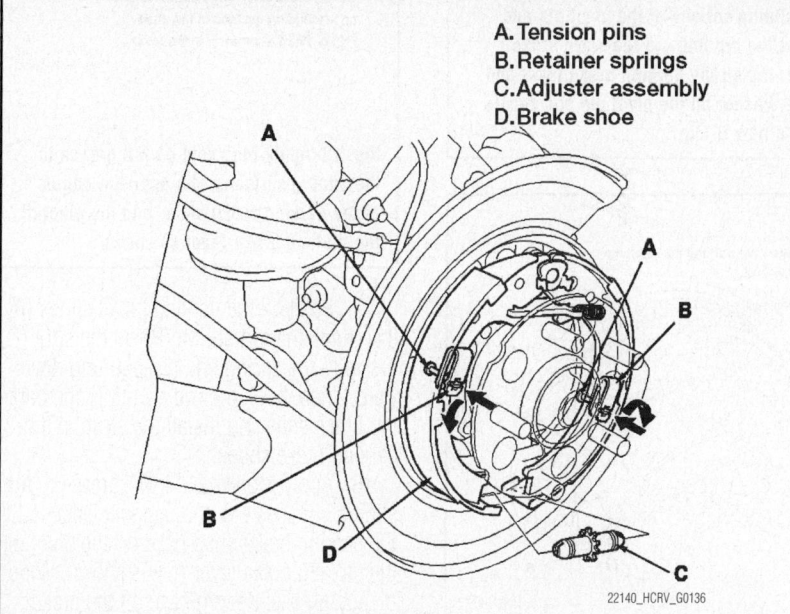

A. Tension pins
B. Retainer springs
C. Adjuster assembly
D. Brake shoe

22140_HCRV_G0136

Fig. 7 Remove the tension pins by pushing the retainer springs and turning the pins. Remove the adjuster assembly by moving the forward brake shoe.

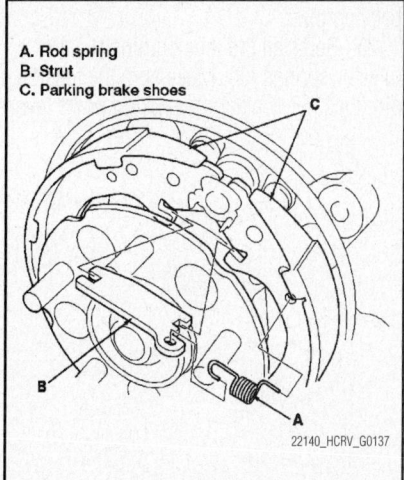

A. Rod spring
B. Strut
C. Parking brake shoes

22140_HCRV_G0137

Fig. 8 Disconnect the rod spring and remove the strut. Remove the parking brake shoes.

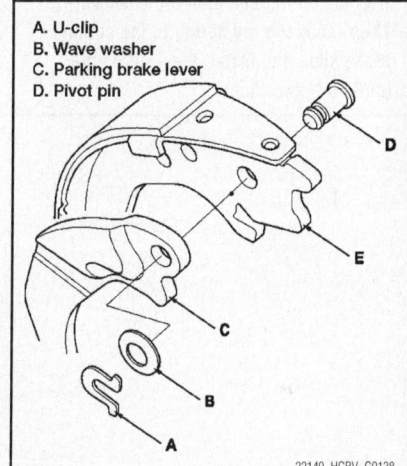

A. U-clip
B. Wave washer
C. Parking brake lever
D. Pivot pin

22140_HCRV_G0138

Fig. 9 Remove the U-clip, wave washer, parking brake lever, and pivot pin from the brake shoe.

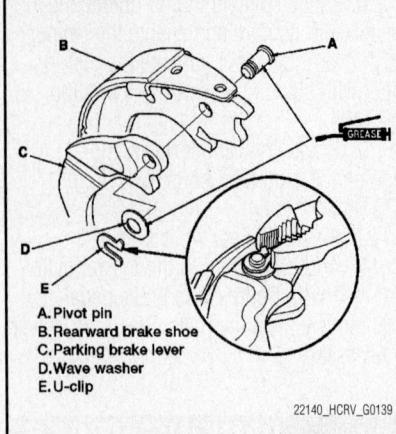

A. Pivot pin
B. Rearward brake shoe
C. Parking brake lever
D. Wave washer
E. U-clip

22140_HCRV_G0139

Fig. 10 Apply Molykote 44 MA grease to the sliding surface of the pivot pin and insert the pin into the rearward brake shoe. Install the parking brake lever and wave washer on the pivot pin and secure with a new U-clip.

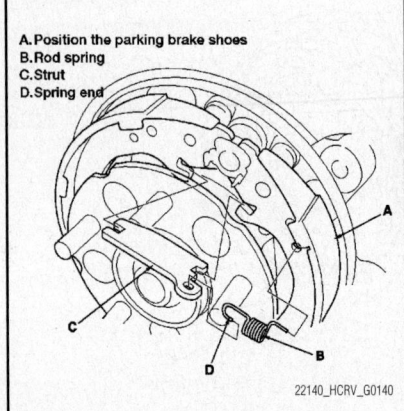

A. Position the parking brake shoes
B. Rod spring
C. Strut
D. Spring end

22140_HCRV_G0140

Fig. 11 Position the parking brake shoes, then hook the rod spring on the strut first with the spring end pointing downward. Then, hook the rod spring to the parking brake shoe and install the strut on the parking brake shoes.

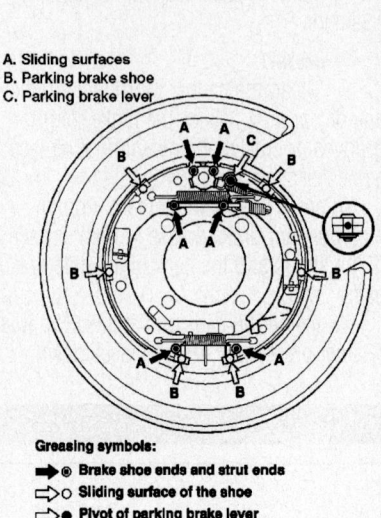

A. Sliding surfaces
B. Parking brake shoe
C. Parking brake lever

Greasing symbols:
➡️⊛ Brake shoe ends and strut ends
⇨○ Sliding surface of the shoe
⇨● Pivot of parking brake lever

22140_HCRV_G0141

Fig. 12 Apply Molykote 44 MA grease to the sliding surfaces, the opposite edges of the parking brake shoe, and the pivot of the parking brake lever as shown

17. Position the parking brake shoes (A), then hook the rod spring (B) on the strut (C) first with the spring end (D) pointing downward. Then, hook the rod spring to the parking brake shoe and install the strut on the parking brake shoes.

18. Apply Molykote 44 MA grease to the sliding surfaces (A), the opposite edges of the parking brake shoe (B), and the pivot of the parking brake lever (C) as shown. Wipe off any excess. Keep grease off the brake linings.

19. Connect the parking brake cable to the parking brake lever. Apply silicone grease to the cable contact surface on the backing plate.

20. Reinstall the tension pins (C) and retainer springs (D). Make sure the tension pin does not contact the parking brake lever.

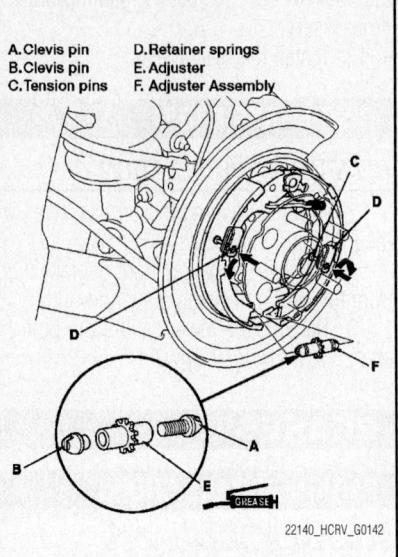

A. Clevis pin   D. Retainer springs
B. Clevis pin   E. Adjuster
C. Tension pins   F. Adjuster Assembly

22140_HCRV_G0142

Fig. 13 Reinstall the tension pins and retainer springs. Make sure the tension pin does not contact the parking brake lever. Clean the threaded portions of the clevis A, and coat the threads of the clevis with grease. Clean the sliding surface of the clevis B, and coat the sliding surface of the clevis B with grease. Install the clevis A and B on the adjuster and shorten the clevis A by turning the adjuster.

21. Clean the threaded portions of the clevis A, and coat the threads of the clevis with grease. Clean the sliding surface of the clevis B, and coat the sliding surface of the clevis B with grease. Install the clevis A and B on the adjuster (E) and shorten the clevis A by turning the adjuster.

22. Position the brake shoe adjuster assembly (F) on the parking brake shoes.

23. Reinstall the lower brake spring.

24. Reinstall the upper brake spring.

25. Install the rear brake disc/drum and rear brake caliper.

26. Adjust the parking brake.

CHASSIS ELECTRICAL — AIR BAG (SUPPLEMENTAL RESTRAINT SYSTEM)

GENERAL INFORMATION

✳✳ CAUTION

These vehicles are equipped with an air bag system. The system must be disarmed before performing service on, or around, system components, the steering column, instrument panel components, wiring and sensors. Failure to follow the safety precautions and the disarming procedure could result in accidental air bag deployment, possible injury and unnecessary system repairs.

SERVICE PRECAUTIONS

Disconnect and isolate the battery negative cable before beginning any airbag system component diagnosis, testing, removal, or installation procedures. Allow system capacitor to discharge for two minutes before beginning any component service. This will disable the airbag system. Failure to disable the airbag system may result in accidental airbag deployment, personal injury, or death.

Do not place an intact undeployed airbag face down on a solid surface. The airbag will propel into the air if accidentally deployed and may result in personal injury or death.

When carrying or handling an undeployed airbag, the trim side (face) of the airbag should be pointing towards the body to minimize possibility of injury if accidental deployment occurs. Failure to do this may result in personal injury or death.

Replace airbag system components with OEM replacement parts. Substitute parts may appear interchangeable, but internal differences may result in inferior occupant protection. Failure to do so may result in occupant personal injury or death.

Wear safety glasses, rubber gloves, and long sleeved clothing when cleaning powder residue from vehicle after an airbag deployment. Powder residue emitted from a deployed airbag can cause skin irritation. Flush affected area with cool water if irritation is experienced. If nasal or throat irritation is experienced, exit the vehicle for fresh air until the irritation ceases. If irritation continues, see a physician.

Do not use a replacement airbag that is not in the original packaging. This may result in improper deployment, personal injury, or death.

The factory installed fasteners, screws and bolts used to fasten airbag components have a special coating and are specifically designed for the airbag system. Do not use substitute fasteners. Use only original equipment fasteners listed in the parts catalog when fastener replacement is required.

During, and following, any child restraint anchor service, due to impact event or vehicle repair, carefully inspect all mounting hardware, tether straps, and anchors for proper installation, operation, or damage. If a child restraint anchor is found damaged in any way, the anchor must be replaced. Failure to do this may result in personal injury or death.

Deployed and non-deployed airbags may or may not have live pyrotechnic material within the airbag inflator.

Do not dispose of driver/passenger/curtain airbags or seat belt tensioners unless you are sure of complete deployment. Refer to the Hazardous Substance Control System for proper disposal.

Dispose of deployed airbags and tensioners consistent with state, provincial, local, and federal regulations.

After any airbag component testing or service, do not connect the battery negative cable. Personal injury or death may result if the system test is not performed first.

If the vehicle is equipped with the Occupant Classification System (OCS), do not connect the battery negative cable before performing the OCS Verification Test using the scan tool and the appropriate diagnostic information. Personal injury or death may result if the system test is not performed properly.

Never replace both the Occupant Restraint Controller (ORC) and the Occupant Classification Module (OCM) at the same time. If both require replacement, replace one, then perform the Airbag System test before replacing the other.

Both the ORC and the OCM store Occupant Classification System (OCS) calibration data, which they transfer to one another when one of them is replaced. If both are replaced at the same time, an irreversible fault will be set in both modules and the OCS may malfunction and cause personal injury or death.

If equipped with OCS, the Seat Weight Sensor is a sensitive, calibrated unit and must be handled carefully. Do not drop or handle roughly. If dropped or damaged, replace with another sensor. Failure to do so may result in occupant injury or death.

If equipped with OCS, the front passenger seat must be handled carefully as well. When removing the seat, be careful when setting on floor not to drop. If dropped, the sensor may be inoperative, could result in occupant injury, or possibly death.

If equipped with OCS, when the passenger front seat is on the floor, no one should sit in the front passenger seat. This uneven force may damage the sensing ability of the seat weight sensors. If sat on and damaged, the sensor may be inoperative, could result in occupant injury, or possibly death.

DISARMING THE SYSTEM

1. Before servicing the vehicle, refer to the Precautions Section.
2. Turn the ignition switch to **OFF**.
3. Disconnect the negative battery cable and isolate it from accidental reconnection. Insulate the cable end with high-quality electrical tape or a similar non-conductive wrapping.
4. Wait at least 3 minutes for the system capacitor to discharge before performing any service. The air bag system is designed to retain enough voltage to deploy the air bag for a short period of time after the battery has been disconnected.

ARMING THE SYSTEM

1. Before servicing the vehicle, refer to the Precautions Section.
2. Reconnect the negative battery cable.
3. To confirm proper system operation, turn the ignition switch to the **ON** position. The SRS indicator light should light for at least 7 seconds and then go off.

CLOCKSPRING CENTERING

See Figures 14 through 18.

✳✳ CAUTION

Before servicing, or working around, the SRS system, turn the ignition switch OFF, disconnect both battery cables and wait at least 3 minutes. When servicing, or working around, the SRS system, do not work directly in front of the air bag module.

1. Before servicing the vehicle, refer to the Precautions Section.
2. Position the front wheels in the straight ahead position.
3. Remove the driver's airbag.
4. Disconnect the connector from the clockspring (also called a cable reel), then remove the steering wheel bolt.

A. Dashboard wire harness 4P connector
B. Clockspring 4P connector
C. Dashboard wire harness 13P or 5P connector
D. Clockspring

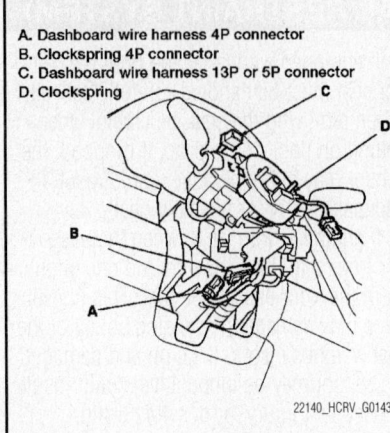

22140_HCRV_G0143

Fig. 14 Disconnect the dashboard wire harness 4P connector from the clockspring 4P connector, then disconnect the dashboard wire harness 13P or 5P connector from the clockspring

A. Lock tab
B. 90 degree hook shaped tool
C. Lower lock tab

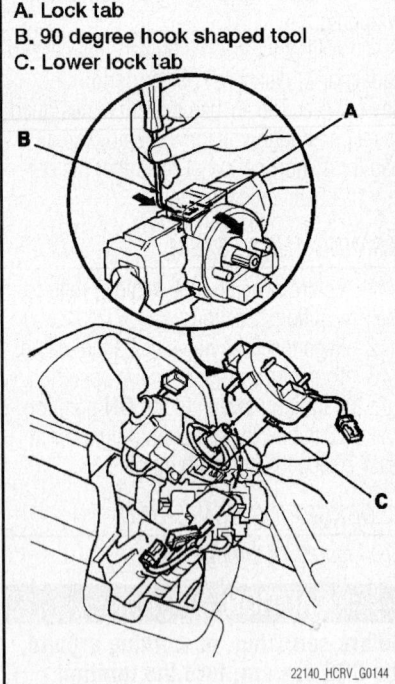

22140_HCRV_G0144

Fig. 15 Release the lock tab under the clockspring connector with a 90° hook shaped tool. Slide the tool below the clockspring connector just above the lock tab. Release the lower lock tab and slide the clockspring off the column.

5. Confirm that the front wheels point straight ahead, then remove the steering wheel with a steering wheel puller .Do not tap on the steering wheel or steering column shaft when removing the steering wheel.

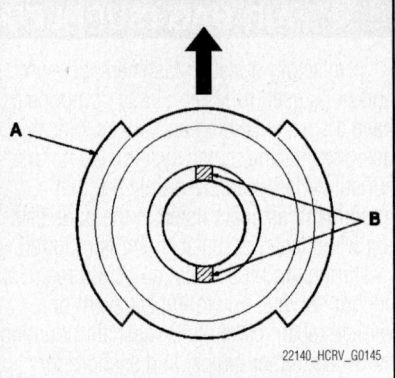

22140_HCRV_G0145

Fig. 16 Set the turn signal canceling sleeve (A) so that the projections (B) are aligned vertically

6. Remove the driver's dashboard undercover.
7. Remove the column cover screws, then remove the column covers.
8. Disconnect the dashboard wire harness 4P connector (A) from the clockspring 4P connector (B), then disconnect the dashboard wire harness 13P or 5P connector (C) from the clockspring (D).
9. Release the lock tab (A) under the clockspring connector with a 90° hook shaped tool (B). Slide the tool below the clockspring connector just above the lock tab. Release the lower lock tab (C) and slide the clockspring off the column.

To install:
10. Before installing the steering wheel, make sure the front wheels are aligned straight ahead.
11. If not already done, perform the battery terminal disconnection procedure. Some systems store data in memory that is lost when the battery is disconnected. Do the following steps before disconnecting the battery:
    a. Make sure to record the anti-theft code(s) for the audio and/or the navigation system (if equipped).
    b. Write down the audio presets (AM and FM) and the XM audio presets (if equipped), because the audio unit does not retain the presets after the battery is disconnected.
    c. Make sure the ignition switch is in LOCK position.
    d. Disconnect and isolate the negative cable from the battery.

➡Always disconnect the negative cable from the battery first.

    e. Disconnect the positive cable from the battery.
12. Wait for 3 minutes for the SRS system to discharge before continuing the procedure.

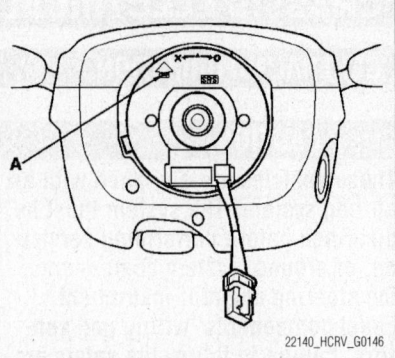

22140_HCRV_G0146

Fig. 17 Center the clockspring by rotating the clockspring clockwise until it stops. Then, rotate it counterclockwise (about 3 turns) until the arrow mark (A) on the clockspring label points straight up.

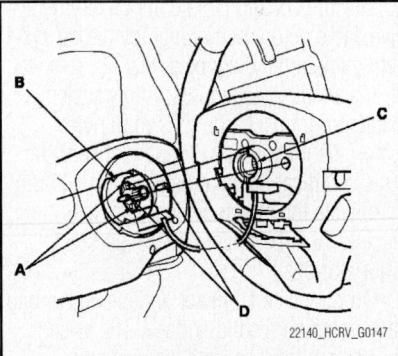

22140_HCRV_G0147

Fig. 18 Position the 2 tabs (A) of the turn signal canceling sleeve (B) as shown, and install the steering wheel on to the steering column shaft, making sure the steering wheel hub (C) engages the pins (D) of the clockspring and tabs of the turn signal canceling sleeve.

13. Set the turn signal canceling sleeve (A) so that the projections (B) are aligned vertically.
14. Carefully install the clockspring (D) on the steering column shaft. Then connect 13P or 5P connector (C) to the clockspring, and connect the 4P connector (B) to the dashboard wire harness 4P connector (A).
15. Install the steering column covers.
16. If necessary, center the clockspring. Do this by first rotating the clockspring clockwise until it stops. Then, rotate it counterclockwise (about 3 turns) until the arrow mark (A) on the clockspring label points straight up.

➡New replacement clocksprings come centered.

17. Position the 2 tabs (A) of the turn signal canceling sleeve (B) as shown, and install the steering wheel on to the

steering column shaft, making sure the steering wheel hub (C) engages the pins (D) of the clockspring and tabs of the turn signal canceling sleeve. Do not tap on the steering wheel or steering column shaft when installing the steering wheel.

18. Install a new steering wheel bolt and tighten to 29 ft. lbs. (39 Nm).

19. Reconnect the connectors.

20. Install the driver's airbag.
21. Connect the battery terminals.
22. Clear any DTC's.
23. After installing the clockspring, confirm proper system operation:
   • Turn the ignition switch to ON (II); the SRS indicator should come on for about 6 seconds and then go off

• After the SRS indicator has turned off, turn the steering wheel fully left and right to confirm the SRS indicator does not come on
• Make sure the horn works
• Make sure the cruise control buttons work
• Make sure the switches near the steering wheel all work properly

DRIVETRAIN

AUTOMATIC TRANSAXLE ASSEMBLY

REMOVAL & INSTALLATION
See Figures 19 through 28.

1. Before servicing the vehicle, refer to the Precautions Section.
   Special tools required:
   • Engine hanger adapter VSB02C000015
   • CR-V engine hanger adapter VSB02C000032
   • Front subframe adapter VSB02C000016
   • Engine support hanger AAR-T-1256

✳ WARNING
Use fender covers to avoid damaging painted surfaces. The special tool engine hanger must be used with the side engine mount installed.

2. Remove the front bulkhead cover.
3. Disconnect and remove the battery.
4. Remove the air cleaner housing.
5. Remove the air intake duct.
6. Remove the battery base.
7. Raise the vehicle and make sure it is securely supported.
8. Remove the front wheels.
9. Remove the splash shield.
10. Remove the drain plug and drain the Automatic Transmission Fluid (ATF).
11. Reinstall the drain plug with a new sealing washer.
12. Fix the hood in the vertical position.
13. Remove the harness clamp from the clamp bracket and remove the air cleaner housing bracket.
14. Disconnect the A/T clutch pressure control solenoid valve A connector, the 2nd clutch transaxle fluid pressure switch connector, and remove the harness clamps from the clamp brackets.
15. Disconnect the transaxle range switch connector and remove the connector from its bracket.
16. Disconnect the output shaft (countershaft) speed sensor connector and the input

shaft (mainshaft) speed sensor connector and remove the harness clamp from the clamp bracket.

17. Disconnect the 3rd clutch transaxle fluid pressure switch connector and remove the harness clamp from its bracket.

18. Disconnect the A/T clutch pressure control solenoid valve B connector, the A/T clutch pressure control solenoid valve C connector, and the shift solenoid harness connector, then remove the harness clamp from the clamp bracket.

19. Disconnect the ATF cooler hoses from the ATF lines. Turn the end of the ATF cooler hoses up to prevent ATF from flowing out, then plug the hoses and lines.

20. Attach the engine hanger adapter (VSB02C000015) to the threaded hole in the cylinder head.

21. Remove both lids (A) for the front strut flange nuts from the cowl cover. Position the engine hanger adapters (VSB02C000032) over the strut flange nuts.

22. Carefully position the engine support hanger (AAR-T-1256) on the vehicle, and attach the hook (A) to the engine hanger adapter (VSB02C000015). Tighten the wing nut (B) by hand, and lift and support the engine.

23. Insert a 6 mm Allen wrench (A) in the top of the ball joint pin (B), and remove the nut (C), then separate the stabilizer link (D) from the stabilizer (E). Repeat this for the other stabilizer link.

24. Remove the nuts (F) and the bolt (G) securing the lower arm (H) and the ball joint (I), and separate the lower arms from the ball joints.

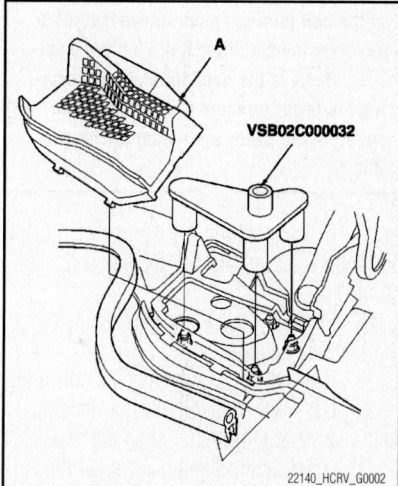
Fig. 20 Remove both lids (A) for the front strut flange nuts from the cowl cover and position the engine hanger adapters (VSB02C000032) over the strut flange nuts

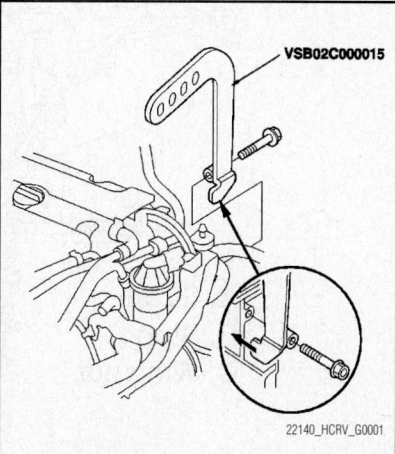
Fig. 19 Attach the engine hanger adapter (VSB02C000015) to the threaded hole in the cylinder head

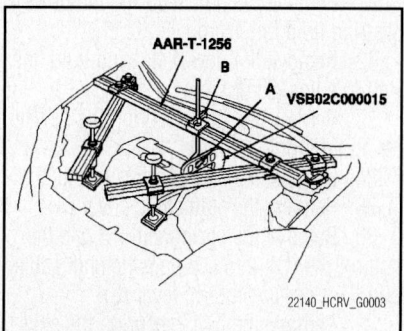
Fig. 21 Position the engine support hanger (AAR-T-1256) on the vehicle, and attach the hook (A) to the engine hanger adapter (VSB02C000015). Tighten the wing nut (B) by hand.

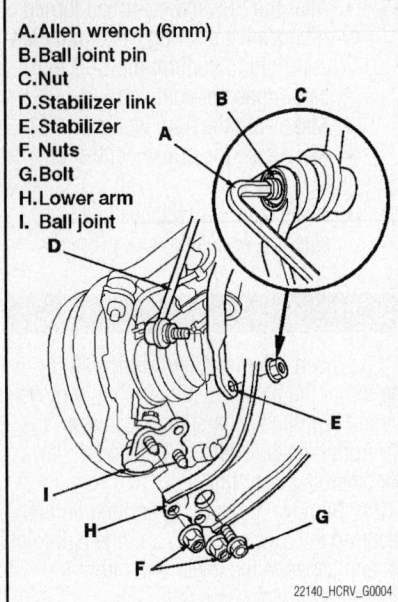

A. Allen wrench (6mm)
B. Ball joint pin
C. Nut
D. Stabilizer link
E. Stabilizer
F. Nuts
G. Bolt
H. Lower arm
I. Ball joint

22140_HCRV_G0004

Fig. 22 Insert an Allen wrench in the top of the ball joint pin and remove the nut to separate the stabilizer link from the stabilizer. Remove the nuts and the bolt securing the lower arm and the ball joint and separate the lower arms from the ball joints.

25. Disconnect the A/F sensor connector and the secondary heated oxygen sensor connector.

26. Remove the sensor harnesses from the harness clamps.

27. Remove the 3-way catalytic converter.

28. For 4WD models, make a reference mark across the propeller shaft and the transfer companion flange. Separate the propeller shaft from the transfer companion flange.

29. Remove the steering gearbox bracket mounting bolts.

30. Remove the steering gearbox bracket mounting bolts.

31. Remove the bolt securing the power steering fluid line clamp.

32. Remove the power steering fluid line from the line clamps.

33. Remove the bolt securing the ATF filter on the front subframe.

34. Remove the lower torque rod bolts.

35. Remove the shift cable cover (A).

36. Remove the spring clip (B) and the control pin (C), and separate the shift cable end (D) from the control lever (E).

37. Remove the bolts securing the shift cable brackets (F) (G). Do not bend the shift cable excessively.

38. Check bushing (H) in the shift cable end for proper fit and wear. If the bushing is loose or worn, replace the shift cable.

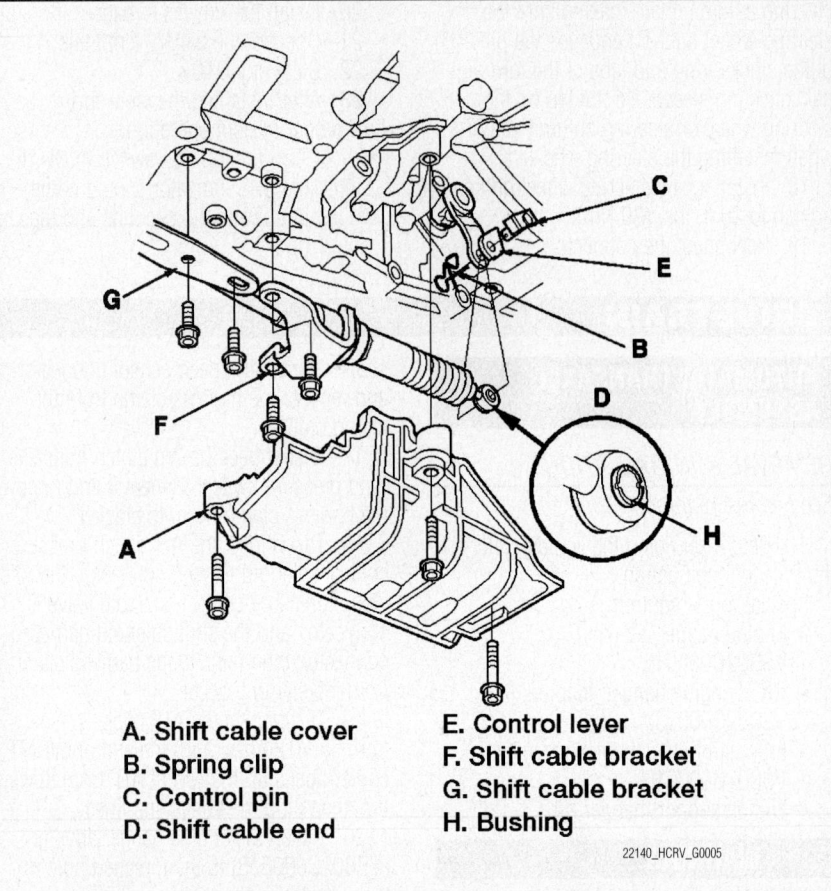

A. Shift cable cover
B. Spring clip
C. Control pin
D. Shift cable end
E. Control lever
F. Shift cable bracket
G. Shift cable bracket
H. Bushing

22140_HCRV_G0005

Fig. 23 Remove the shift cable cover, spring clip, and control pin. Separate the shift cable end from the control lever and remove the bolts securing the shift cable brackets.

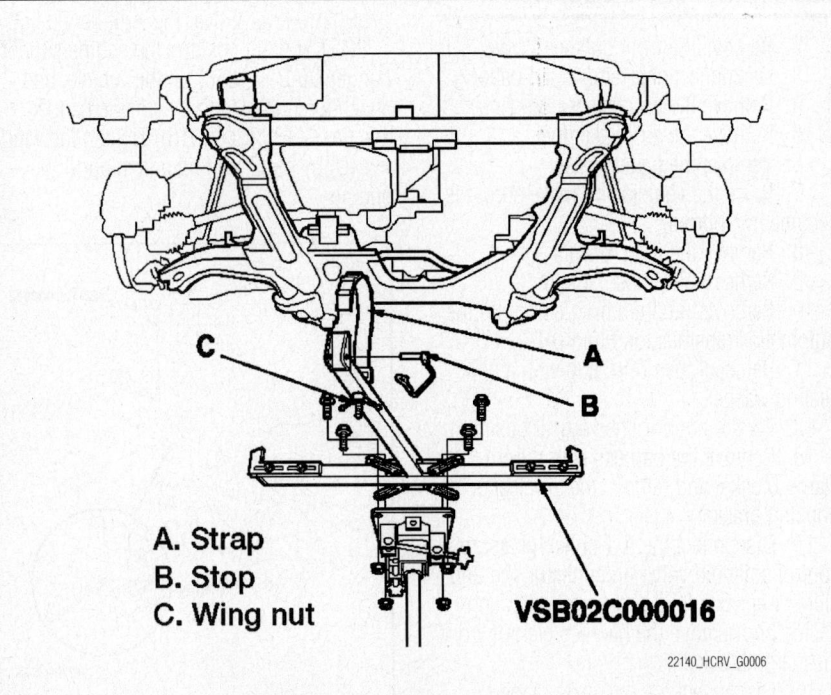

A. Strap
B. Stop
C. Wing nut

VSB02C000016

22140_HCRV_G0006

Fig. 24 Attach the front subframe adapter (VSB02C000016) to the subframe by looping the strap over the front of the subframe, then secure the strap with the stop and tighten the wing nut

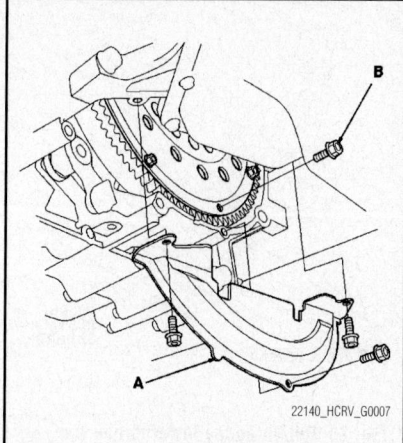

Fig. 25 Remove the torque converter cover (A) and remove the 8 drive plate bolts (B)

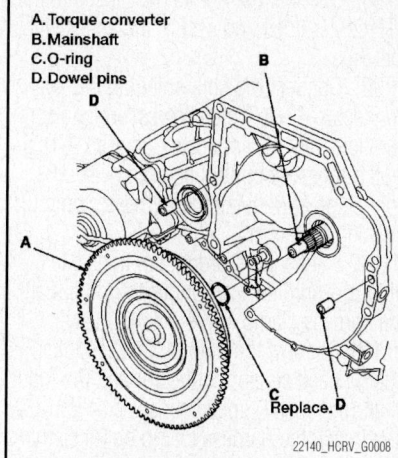

A. Torque converter
B. Mainshaft
C. O-ring
D. Dowel pins

Fig. 26 Install the torque converter on the mainshaft with a new O-ring

39. Remove the shift cable holder bracket from the transaxle.

40. For 4WD models, remove the propeller shaft protector.

41. Remove both mid-bracket bolts.

42. Make reference marks on the body across the marks on the edge of the front subframe.

43. Attach the front subframe adapter (VSB02C000016) to the subframe by looping the strap (A) over the front of the subframe, then secure the strap with the stop (B), then tighten the wing nut (C).

44. Raise the jack and line up the slots in the arms with the bolt holes on the corner of the jack base, then tighten the bolts.

45. Remove the 6 bolts securing the front subframe and lower the subframe.

46. Secure the steering gearbox from the body with a strap.

47. Remove the torque converter cover (A), and remove the drive plate bolts (B) (8) while rotating the crankshaft pulley.

48. Remove the driveshafts from the differential and the intermediate shaft. Clean then coat all precision machined surfaces with clean engine oil, and put plastic bags over the driveshaft ends.

49. Remove the intermediate shaft. Clean then coat all precision machined surfaces with clean engine oil, and put plastic bags over the intermediate shaft ends.

50. Remove the upper transaxle housing mounting bolts.

51. Remove the transaxle mount bracket bolts.

52. Remove the front transaxle housing mounting bolts.

53. Lower the transaxle by loosening the wing nut on the engine support hanger, and tilt the engine just enough for the transaxle to clear the side frame.

54. Place a jack under the transaxle.

55. Remove the rear transaxle housing mounting bolts.

56. Slide the transaxle away from the engine to remove it from the vehicle.

57. Remove the torque converter and the dowel pins.

58. Inspect the drive plate, and replace it if it is damaged.

59. For 4WD models, install the propeller shaft protector.

60. If installing an overhauled or remanufactured transaxle, clean the ATF cooler.

To install:

61. For 4WD models, remove the propeller shaft protector.

62. Install the torque converter (A) on the mainshaft (B) with a new O-ring (C).

63. Install the 14 x 20mm dowel pins (D) in the torque converter housing.

64. Place the transaxle on the jack and raise the transaxle to the engine level, then fit the transaxle to the engine.

65. Install the rear transaxle housing mounting bolts and tighten to 47 ft. lbs. (64 Nm).

66. Install the front transaxle housing mounting bolts and tighten to 47 ft. lbs. (64 Nm).

67. Install the upper transaxle housing mounting bolts and tighten to 47 ft. lbs. (64 Nm).

68. Secure the transaxle mount bracket on the transaxle housing with new mounting bolts and tighten to 54 ft. lbs. (74 Nm).

69. Install a new set ring on the intermediate shaft.

70. Clean the areas where the intermediate shaft contacts the transaxle (differential) with solvent, and dry with compressed air. Apply ATF to the intermediate shaft splines, then install the intermediate shaft.

✳✳ WARNING

Be sure not to allow dust or other foreign particles to enter the transaxle.

71. Install a new set ring on the left driveshaft.

72. Clean the areas where the left driveshaft contacts the transaxle (differential) with solvent, and dry with compressed air. Then install the left driveshaft, be sure not to allow dust or other foreign particles to enter the transaxle. Turn the steering knuckle fully outward, and slide the driveshaft into the differential until you feel the set ring fully engage the side gear.

73. Apply the recommended grease to the right driveshaft inboard-joint splines.

74. Slide the right driveshaft over the intermediate shaft splines until the driveshaft fully engages the intermediate shaft set ring.

75. Attach the torque converter to the drive plate with the 8 bolts. Rotate the crankshaft pulley, as necessary, to tighten the bolts to ½ of the specified torque, then to the final torque, in a crisscross pattern. After tightening the last bolt, check that the crankshaft rotates freely. Tightening torque: 108 inch lbs. (12 Nm).

76. Install the torque converter cover and tighten to 108 inch lbs. (12 Nm).

77. Set the front subframe adapter (VSB02C000016) to the subframe by looping the strap (A) over the front of the subframe, then secure the strap with the stop (B) and tighten the wing nut (C).

78. Loosely install new subframe mounting bolts.

79. Align all reference marks on the front subframe with the marks you made on the body, then tighten the mounting bolts on the subframe to 76 ft. lbs. (103 Nm).

80. Install new mid-bracket bolts and tighten to 47 ft. lbs. (64 Nm).

81. Secure the lower torque rod with new bolts and tighten to 65 ft. lbs. (88 Nm).

82. For 4WD models, install the propeller shaft protector and tighten the bolts to 16 ft. lbs. (22 Nm).

83. For 4WD models, install the propeller shaft to the transfer companion flange by aligning the reference marks made earlier and tighten the bolts to 24 ft. lbs. (32 Nm).

84. Loosely install the steering gearbox bracket mounting bolts, then tighten the bracket mounting bolts to 48 ft. lbs. (66 Nm).

85. Install both lower arms to both ball joints and loosely install new mounting nuts and bolts. Tighten the nuts and the bolts to

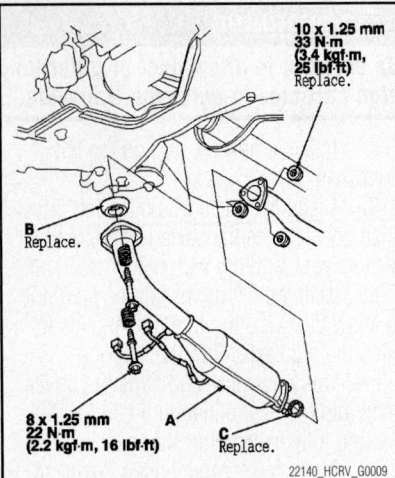

Fig. 27 Install the 3-way catalytic converter (A) with the bolts, new self-locking nuts, and new gaskets (B) (C)

43 ft. lbs. (59 Nm) in the following order: the nut on the front, the nut on the rear, then the bolt.

86. Install the stabilizer links to the stabilizer and install the nuts. Insert a 6mm Allen wrench in the ball joint pin, and tighten the nuts to 58 ft. lbs. (78 Nm).

87. Install the 3-way catalytic converter (A) with the bolts, new self-locking nuts, and new gaskets (B) (C). Tighten as illustrated.

88. Connect the A/F sensor connector and the secondary heated oxygen sensor connector.

89. Install the sensor harnesses in the harness clamps.

90. Install the shift cable holder bracket on the transaxle.

91. Apply molybdenum grease to the hole in the bushing in the shift cable end. Attach the shift cable end to the control lever, then insert the control pin into the control lever hole through the shift cable end, and secure the control pin with the spring clip. Do not bend the shift cable excessively.

92. Secure the shift cable brackets with the bolts and tighten to 108 inch lbs. (12 Nm).

93. Install the shift cable cover and tighten the bolts to 108 inch lbs. (12 Nm).

94. Secure the ATF filter with the bolt on the front subframe and tighten the bolt to 108 inch lbs. (12 Nm).

95. Secure the power steering fluid line clamp with the bolt and the power steering fluid line with the clamps.

96. Remove the engine support hanger and the engine hanger adapters.

97. Remove the engine hanger adapter (VSB02C000015) from the cylinder head.

98. Connect the ATF cooler hoses to the ATF cooler lines and secure the hoses with the clips.

99. Connect the shift solenoid harness connector, the A/T clutch pressure control solenoid valve C connector, and the A/T clutch pressure control solenoid valve B connector, and install the harness clamp on the clamp bracket.

100. Connect the 3rd clutch transaxle fluid pressure switch connector and install the harness clamp on its clamp bracket.

101. Connect the output shaft (countershaft) speed sensor connector and the input shaft (mainshaft) speed sensor connector, and install the harness clamp on its clamp bracket.

102. Install the transaxle range switch connector on the connector bracket and connect the connector.

103. Connect the A/T clutch pressure control solenoid valve A connector and the 2nd clutch transaxle fluid pressure switch connector and install the harness clamps on the clamp brackets.

104. Install the air cleaner housing mounting bracket, and install the harness clamp on the harness clamp bracket.

105. Refill the transaxle with the proper amount and type of ATF.

106. Install the battery base.

107. Install the air intake duct and the air cleaner.

108. Install the front bulkhead cover.

109. Install both lids on the cowl cover.

110. Install and connect the battery.

111. Set the parking brake. Start the engine and shift the transaxle through all positions 3 times.

112. Check the shift lever operation, the A/T gear position indicator operation, and the shift cable adjustment.

113. Install the splash shield.

114. Install the wheels, check and adjust the front wheel alignment.

115. Check the ATF level.

116. Road-test the vehicle.

117. After the road test, raise and safely support the vehicle.

118. Loosen the upper torque rod bolt.

119. Loosen the transaxle mount base bracket bolt and the nuts.

120. Loosen the lower torque rod bolts.

121. Retighten the lower torque rod bolts (B) to 65 ft. lbs. (88 Nm).

122. Retighten the bolt (A) to 69 ft. lbs. (93 Nm).

123. Tighten the transaxle mount base bracket bolt and the nuts to 54 ft. lbs. (74 Nm).

124. Tighten the upper torque rod bolt to 40 ft. lbs. (54 Nm).

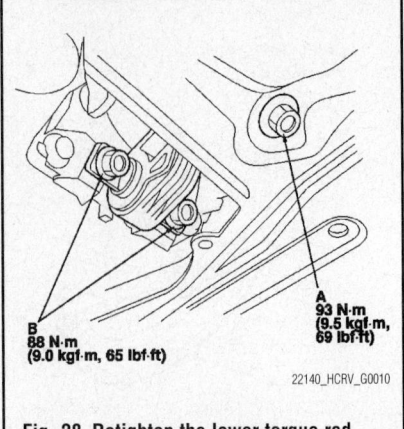

Fig. 28 Retighten the lower torque rod bolts to the specified torque

TRANSFER CASE ASSEMBLY

REMOVAL & INSTALLATION

4WD Models

See Figures 29 through 31.

1. Before servicing the vehicle, refer to the Precautions Section.

2. Shift the transaxle to **N**.

3. Raise the vehicle on a lift, and make sure it is supported securely.

4. Remove the drain plug and drain the Automatic Transmission Fluid (ATF).

5. Reinstall the drain plug with a new sealing washer and tighten to 36 ft. lbs. (49 Nm).

6. Disconnect the A/F sensor connector and the secondary heated oxygen sensor connector.

7. Remove the sensor harnesses from the harness clamps.

8. Remove the 3-way catalytic converter.

9. Make a reference mark across the propeller shaft and the transfer companion flange.

10. Separate the propeller shaft from the transfer companion flange.

11. Remove the 4 transfer assembly mounting bolts (A), and pull out the bolt (B) to the limit of travel.

12. Remove the transfer assembly (C) and the dowel pin (D) from the transaxle.

To install:

13. Clean the areas where the transfer assembly contacts the transaxle with solvent and dry with compressed air. Then apply transmission fluid to the seal contact area.

14. Install a new O-ring (A) on the transfer assembly (B).

15. Install the dowel pin (C) in the transfer housing.

16. Install the one bolt (D) part-way in the rear lower of the transfer housing, and install the transfer assembly on the transaxle.

17. Install the remaining bolts through the transfer case and tighten to 33 ft. lbs. (44 Nm).

18. Install the propeller shaft to the transfer companion flange by aligning the reference mark and tighten the bolts to 24 ft. lbs. (32 Nm).

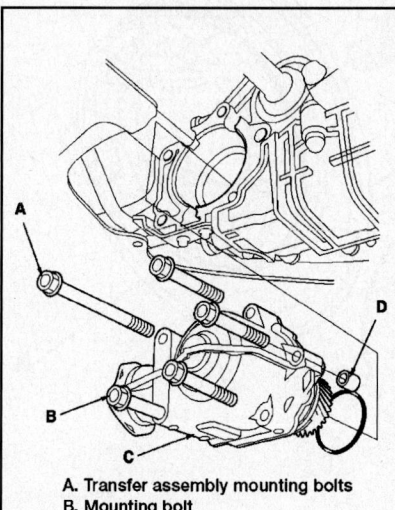

A. Transfer assembly mounting bolts
B. Mounting bolt
C. Transfer assembly
D. Dowel pin

22140_HCRV_G0011

Fig. 29 Remove the transfer assembly and the dowel pin from the transaxle

A. O-ring
B. Transfer assembly
C. Dowel pin
D. Bolt

10 x 1.25 mm
44 N·m
(4.5 kgf·m,
33 lbf·ft)

A Replace

22140_HCRV_G0012

Fig. 30 Install a new O-ring on the transfer assembly and the dowel pin in the transfer housing. Install the dowel pin in the transfer housing and install the one bolt (D) part-way in the rear lower of the transfer housing

19. Install the 3-way catalytic converter with the bolts, new self-locking nuts, and new gaskets. Tighten the bolts and nuts as illustrated.

20. Connect the A/F sensor connector and the secondary heated oxygen sensor connector.

21. Install the sensor harnesses in the harness clamps.

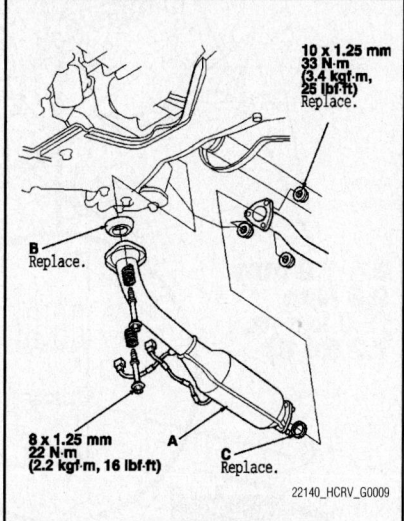

10 x 1.25 mm
33 N·m
(3.4 kgf·m,
25 lbf·ft)
Replace.

B
Replace.

8 x 1.25 mm
22 N·m
(2.2 kgf·m, 16 lbf·ft)

A

C
Replace.

22140_HCRV_G0009

Fig. 31 Install the 3-way catalytic converter (A) with the bolts, new self-locking nuts, and new gaskets (B) (C)

22. Refill the transaxle with the proper type and amount of ATF.

FRONT HALFSHAFT

REMOVAL & INSTALLATION

See Figure 32.

1. Before servicing the vehicle, refer to the Precautions Section.
2. Drain the transaxle.
3. Remove or disconnect the following:
   - Negative battery cable
   - Front wheels
   - Stabilizer bar
   - Lower ball joint
   - Spindle nut
4. On the left side, pry the inboard joint from the case with a prybar.
5. On the right side, drive the inboard shaft off the intermediate shaft with a drift and hammer.

To install:

6. Installation is the reverse of the removal procedure.
7. Observe the following torques:
   - Ball stud nuts: 40 ft. lbs. (54 Nm)
   - Stabilizer link nuts: 29 ft. lbs. (39 Nm)
   - Spindle nut: 181 ft. lbs. (245 Nm)

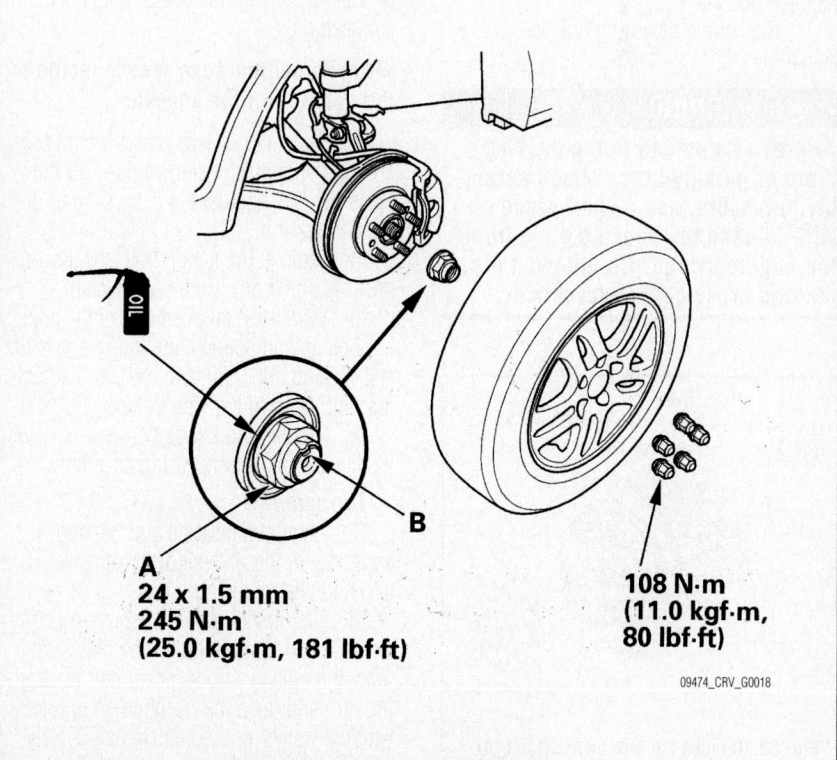

A
24 x 1.5 mm
245 N·m
(25.0 kgf·m, 181 lbf·ft)

B

108 N·m
(11.0 kgf·m,
80 lbf·ft)

09474_CRV_G0018

Fig. 32 View of the front halfshaft locknut (A) and shoulder (B)

REAR AXLE SHAFT, BEARING & SEAL

REMOVAL & INSTALLATION

See Figures 33 and 34.

1. Before servicing the vehicle, refer to the Precautions Section.
2. Raise and safely support the vehicle.
3. Remove the wheel nuts and the rear wheel.
4. Remove the brake hose bracket mounting bolt from the knuckle.
5. Remove the brake caliper bracket mounting bolts and remove the caliper assembly from the knuckle.

⚠ WARNING

To prevent damage to the caliper assembly or brake hose, use a short piece of wire to hang the caliper assembly from the undercarriage. Do not twist the brake hose excessively.

6. Remove the 2 washers from the caliper assembly.
7. On 4WD models, raise the stake and remove the spindle nut.
8. Release the parking brake and remove the brake disc.
9. Remove the hub bearing unit (A) from the spindle.
10. Check the hub bearing unit for damage and cracks.
11. Remove the parking brake assembly (A).

⚠ WARNING

To prevent damage to the backing plate or parking brake shoes assembly and cable, use a short piece of wire to hang the backing plate from the undercarriage. Do not twist the parking brake cable excessively.

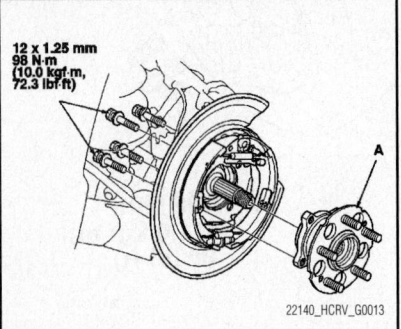

Fig. 33 Remove the hub bearing unit (A) from the spindle

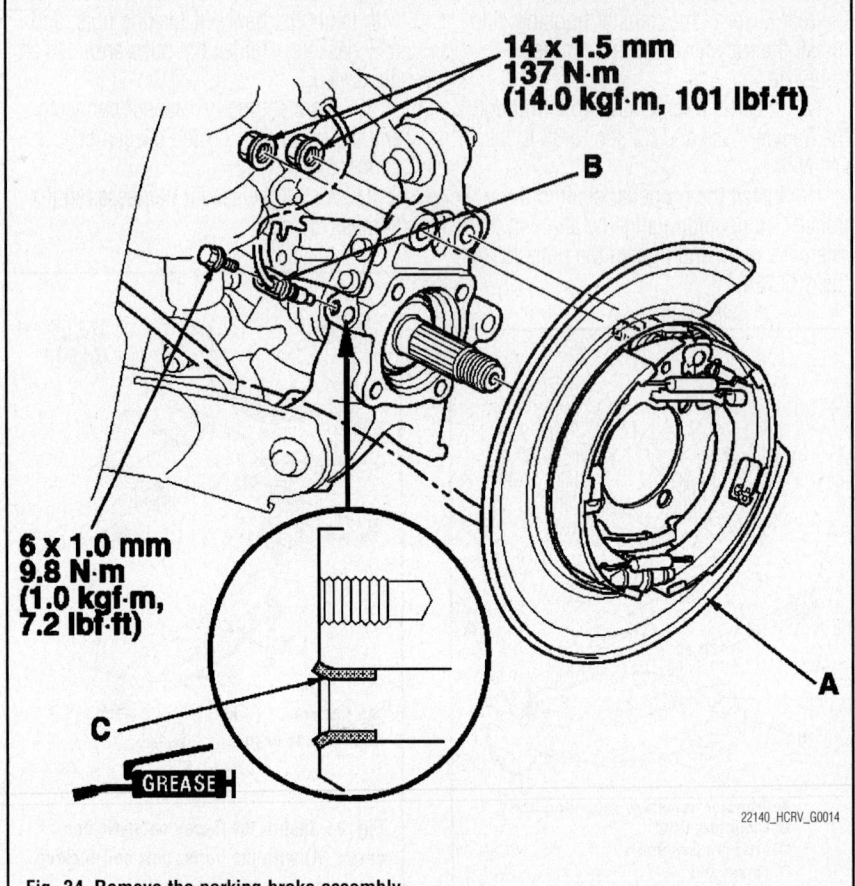

Fig. 34 Remove the parking brake assembly

12. Remove the wheel speed sensor (B). Do not disconnect the wheel speed sensor connector.

➡**Apply multipurpose grease inside of the hole (C) on the knuckle.**

13. Place a floor jack under the trailing arm and support the suspension. Do not place the jack against the plate section of the trailing arm.
14. Remove the flange bolt and disconnect the upper arm from the knuckle.
15. Mark the cam positions of the adjusting bolt and adjusting cam, then remove the self-locking nut, adjusting cam, and adjusting bolt. Discard the self-locking nut.
16. Remove the flange bolt and remove the knuckle and hub bearing assembly.

To install:

17. Install the knuckle and bearing assembly in the reverse order of removal, and note the following:
18. First install all the suspension components and lightly tighten the bolts and nuts, then place a floor jack under the trailing arm and raise the suspension to load it with the weight of the vehicle before fully tightening the bolts and nuts to the specified torque values.

19. Align the cam positions of the adjusting bolt and adjusting cam with the marked positions when tightening.
20. Use a new self-locking nut during reassembly.
21. Use a new spindle nut during reassembly.
22. Before installing the spindle nut, apply a small amount of engine oil to the seating surface of the nut. After tightening, use a drift to stake the spindle nut shoulder against the driveshaft. Tighten the nut to 181 ft. lbs. (245 Nm).
23. Before installing the brake disc, clean the mating surface of the hub bearing unit and the inside of the brake disc.
24. Before installing the wheel, clean the mating surface of the brake disc and the inside of the wheel.
25. Check the wheel alignment and adjust if necessary.

REAR HALFSHAFT

REMOVAL & INSTALLATION

4WD Models

1. Before servicing the vehicle, refer to the Precautions Section.

2. Raise and safely support the vehicle.

3. Remove the rear wheels.

4. Pry up the locking tab on the spindle nut, then remove the nut.

5. Drain the differential fluid.

6. Remove the rear halfshaft inboard joint from the rear differential assembly.

7. Pull up and outward on the knuckle and separate the outboard joint from the rear wheel hub using a plastic hammer.

8. Remove the rear halfshaft.

✴ WARNING

Be careful not to damage the wheel speed sensor. Pull on the outer joint. Do not pull on the driveshaft because the joint may come apart.

9. Remove the set ring from inboard joint.

To install:

10. Make sure the mating surfaces of the joint and the splined section are free from dirt or dust.

11. Install a new set ring onto the set ring groove of the rear driveshaft inboard joint.

12. Install the outboard joint into the rear hub.

✴ WARNING

Be careful not to damage the wheel speed sensor.

13. Clean the areas where the driveshaft contacts the differential thoroughly with solvent, or brake cleaner, and dry with compressed air. Do not wash the rubber parts with solvent.

14. Insert the inboard end of the driveshaft into the differential until the set ring locks in the groove.

➡ **Insert the driveshaft horizontally to prevent damaging the oil seal.**

15. Install the rear differential.

16. Apply a small amount of engine oil to the seating surface of a new spindle nut.

17. Install the new spindle nut, then tighten the nut to 181 ft. lbs. (245 Nm). After tightening, use a drift to stake the spindle nut shoulder against the driveshaft.

18. Clean the mating surfaces of the brake disc and the wheel, then install the rear wheels.

19. Turn the rear wheel by hand and make sure there is no interference between the driveshaft and surrounding parts.

20. Refill the differential fluid with the proper type and amount of fluid.

21. Check for any grease or fluid leakage.

REAR PINION SEAL

REMOVAL & INSTALLATION

See Figure 35.

1. Before servicing the vehicle, refer to the Precautions Section.

2. Remove or disconnect the following:
   - Driveshaft
   - Companion flange
   - Pinion seal

To install:

➡ **Use a new locknut and O-ring for assembly.**

3. Install the pinion seal. Drive the seal square into the bore.

4. Install the companion flange. Tighten the locknut to 87 ft. lbs. (118 Nm).

5. Install the driveshaft. Tighten the flange bolts to 24 ft. lbs. (32 Nm).

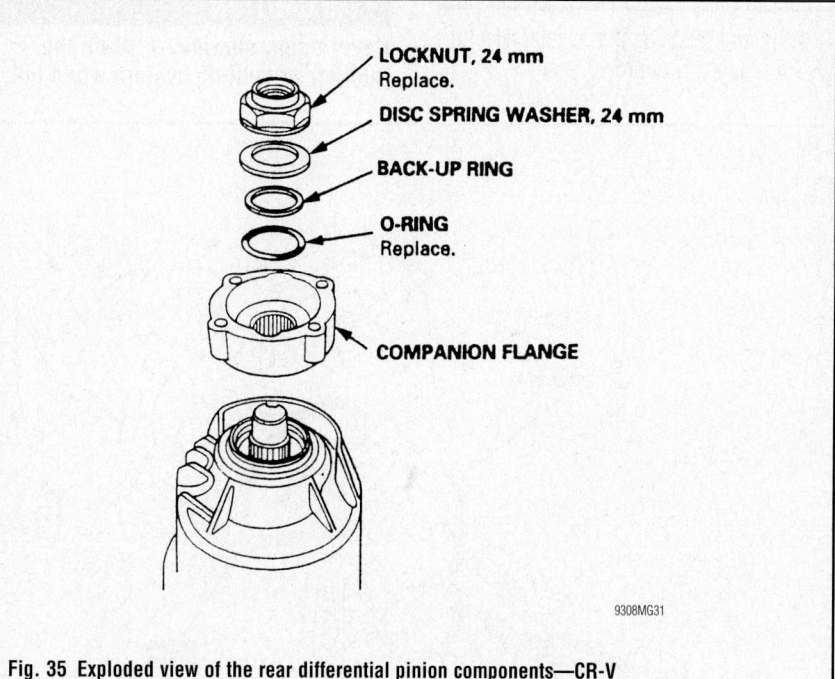

9308MG31

Fig. 35 Exploded view of the rear differential pinion components—CR-V

ENGINE COOLING

THERMOSTAT

REMOVAL & INSTALLATION

See Figure 36.

❋❋ CAUTION

Never open, service, or drain the radiator or cooling system when hot; serious burns can occur from the steam and hot coolant. Also, when draining engine coolant, keep in mind that cats and dogs are attracted to ethylene glycol antifreeze and could drink any that is left in an uncovered container or in puddles on the ground. This will prove fatal in sufficient quantities. Always drain coolant into a sealable container. Coolant should be reused unless it is contaminated or is several years old.

1. Before servicing the vehicle, refer to the Precautions Section.

2. Drain the engine coolant.
3. Remove the splash shield.
4. Remove the lower coolant hose.
5. Remove the thermostat.

To install:

6. Install the thermostat with a new O-ring. Tighten the bolts to 84 inch lbs. (10 Nm).
7. Install the lower coolant hose.
8. Install the splash shield.
9. Refill the radiator with engine coolant and bleed the air from the cooling system while running the engine with the heater valve open.

WATER PUMP

REMOVAL & INSTALLATION

See Figure 37.

❋❋ CAUTION

Never open, service, or drain the radiator or cooling system when hot;

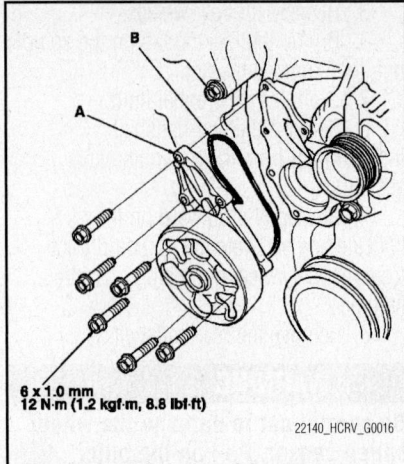

6 x 1.0 mm
12 N·m (1.2 kgf·m, 8.8 lbf·ft)

22140_HCRV_G0016

Fig. 37 Exploded view of the water pump

serious burns can occur from the steam and hot coolant. Also, when draining engine coolant, keep in mind that cats and dogs are attracted to ethylene glycol antifreeze and could drink any that is left in an uncovered container or in puddles on the ground. This will prove fatal in sufficient quantities. Always drain coolant into a sealable container. Coolant should be reused unless it is contaminated or is several years old.

1. Before servicing the vehicle, refer to the Precautions Section.
2. Remove the drive belt.
3. Drain the engine coolant.
4. Remove the drive belt auto-tensioner pulley.
5. Remove the 6 bolts securing the water pump, then remove the water pump (A).

To install:

6. Inspect and clean the O-ring groove and mating surface of the water passage.
7. Install the water pump with new O-ring (B) and tighten the 6 bolts to 106 inch lbs. (12 Nm).
8. Clean up any spilled engine coolant.
9. Install the drive belt auto-tensioner pulley.
10. Refill the radiator with engine coolant and bleed the air from the cooling system while running the engine with the heater valve open.

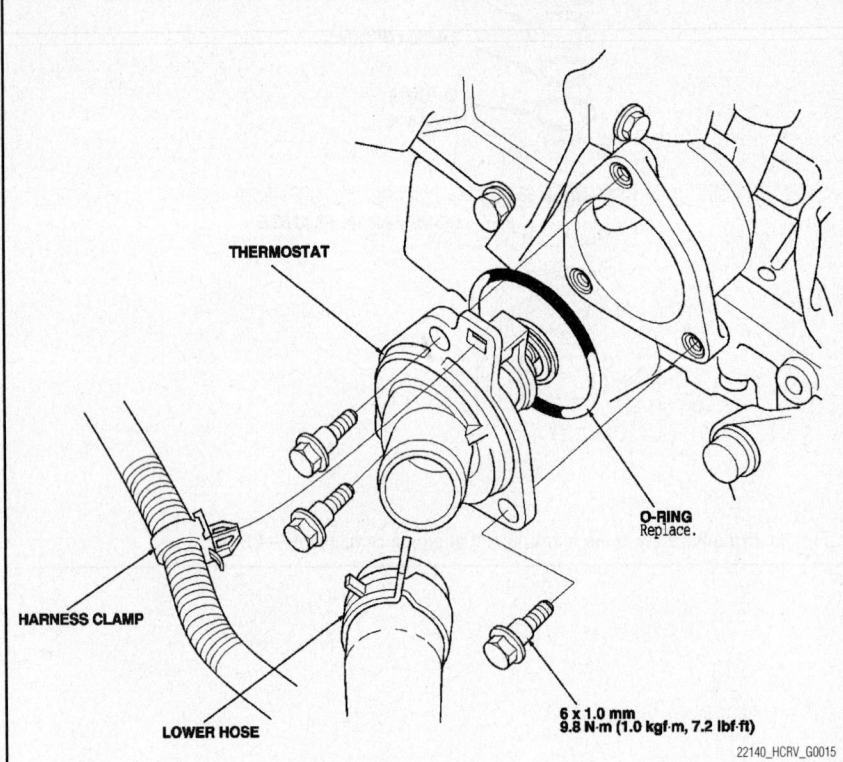

THERMOSTAT

O-RING
Replace.

HARNESS CLAMP

LOWER HOSE

6 x 1.0 mm
9.8 N·m (1.0 kgf·m, 7.2 lbf·ft)

22140_HCRV_G0015

Fig. 36 Exploded view of the thermostat and related components

ENGINE ELECTRICAL

ALTERNATOR

REMOVAL & INSTALLATION

See Figure 38.

1. Before servicing the vehicle, refer to the Precautions Section.
2. Disconnect the negative battery terminal.
3. Remove the accessory drive belt. Refer to Accessory Drive Belts, removal & installation.
4. Remove the drive belt auto-tensioner.
5. Remove the 3 bolts securing the alternator.
6. Disconnect the alternator electrical

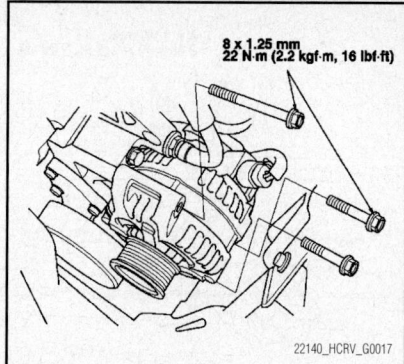

8 x 1.25 mm
22 N·m (2.2 kgf·m, 16 lbf·ft)

22140_HCRV_G0017

Fig. 38 Install the 3 bolts securing the alternator

CHARGING SYSTEM

connectors and the harness clamp from the alternator.
7. Remove the alternator.

To install:

8. Place the alternator into position.
9. Connector the alternator electrical connectors and harness clamp to the alternator.
10. Install the 3 bolts securing the alternator and tighten to 16 ft. lbs. (22 Nm).
11. Install the drive belt auto-tensioner.
12. Install the drive belt. Refer to Accessory Drive Belts, removal & installation.
13. Connect the negative battery terminal.

ENGINE ELECTRICAL

FIRING ORDER

2.4L Engine, Firing order: 1–3–4–2

IGNITION COIL

REMOVAL & INSTALLATION

See Figure 39.

1. Before servicing the vehicle, refer to the Precautions Section.
2. Disconnect the negative battery cable.
3. Remove the ignition coil cover (A).
4. Disconnect the ignition coil connectors (B).
5. Remove the ignition coil mounting bolts.
6. Remove the ignition coils (C).

To install:

7. Place the ignition coils (C) into position.
8. Tighten the ignition coil mounting bolts to 106 inch lbs. (12 Nm).
9. Connect the ignition coil connectors (B).
10. Install the ignition coil cover (A) and tighten the bolts and nuts to 86 inch lbs. (10 Nm).
11. Connect the negative battery cable.

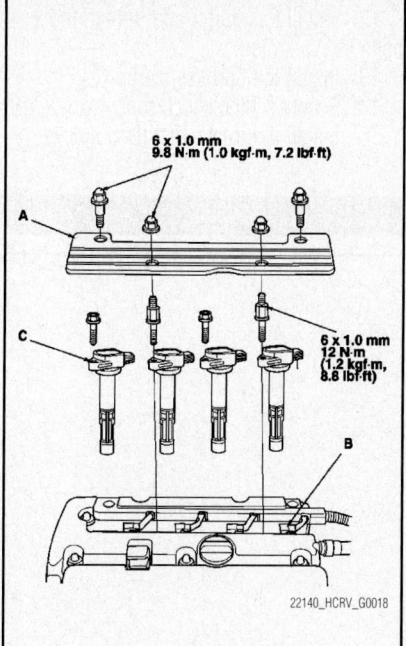

6 x 1.0 mm
9.8 N·m (1.0 kgf·m, 7.2 lbf·ft)

6 x 1.0 mm
12 N·m (1.2 kgf·m, 8.8 lbf·ft)

22140_HCRV_G0018

Fig. 39 Exploded view of ignition cover and coils

IGNITION SYSTEM

IGNITION TIMING

ADJUSTMENT

The ignition timing is controlled by the Powertrain Control Module (PCM). No adjustment is necessary or possible.

SPARK PLUGS

REMOVAL & INSTALLATION

1. Before servicing the vehicle, refer to the Precautions Section.
2. Disconnect the negative battery cable.
3. Remove the ignition coils. Refer to Ignition Coil, removal & installation.
4. Remove the spark plug using a spark plug socket and wrench.

To install:

5. Be sure the spark plug gap is to specification (0.039–0.043 inch).
6. Carefully install the spark plug and torque to specification: 13 ft. lbs. (18 Nm).
7. Install the ignition coils. Refer to Ignition Coil, removal & installation.
8. Connect the negative battery cable.

STARTER

REMOVAL & INSTALLATION

See Figures 40 and 41.

1. Before servicing the vehicle, refer to the Precautions Section.
2. Disconnect the negative battery cable.
3. Remove the splash shield.
4. Remove the intake manifold bracket.
5. Disconnect the knock sensor connector.
6. Remove the harness clamp (A) and remove the 2 bolts securing the starter.
7. Remove the starter from the engine.
8. Disconnect the starter cable (A) from the B terminal, then disconnect the BLK/WHT wire (B) from the S terminal.
9. Remove the harness clamp (C) and remove the starter.

To install:

10. Install the starter cable (A) and BLK/WHT (B) wire. Make sure the starter cable crimped side of the ring terminal faces away from the starter when you connect it.

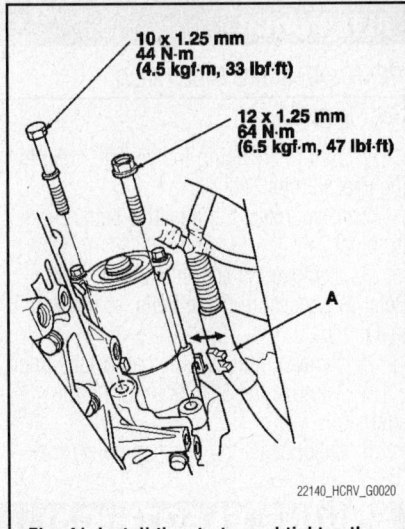

Fig. 40 Install the starter cable connections as illustrated

11. Install the harness clamp (C).
12. Install the starter and tighten the 2 bolts as illustrated.
13. Install the harness clamp (A).
14. Reconnect the knock sensor connector.
15. Install the intake manifold bracket.

Fig. 41 Install the starter and tighten the 2 bolts as illustrated

16. Install the splash shield.
17. Connect the negative battery cable.
18. Start the engine to make sure the starter works properly.

ENGINE MECHANICAL

➡ Disconnecting the negative battery cable may interfere with the functions of the on board computer systems and may require the computer to undergo a relearning process, once the negative battery cable is reconnected.

ACCESSORY DRIVE BELTS

ACCESSORY BELT ROUTING

See Figure 42.

INSPECTION

See Figure 43.

Inspect the drive belt for signs of glazing or cracking. A glazed belt will be perfectly smooth from slippage, while a good belt will have a slight texture of fabric visible. Cracks will usually start at the inner edge of the belt and run outward. All worn or damaged drive belts should be replaced immediately.

Check that the auto-tensioner indicator (A) is within the standard range (B) as shown. If it is out of the standard range, the belt must be replaced.

ADJUSTMENT

The 2.4L engine uses 1 belt to drive the alternator, air conditioner compressor, and

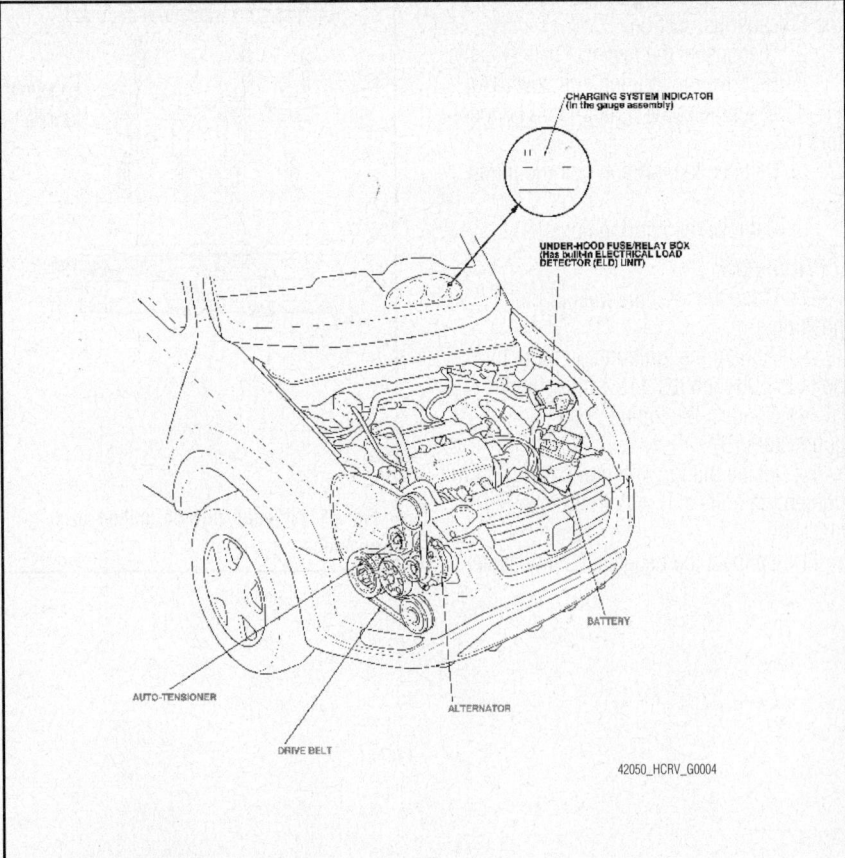

Fig. 42 Accessory drive belt routing

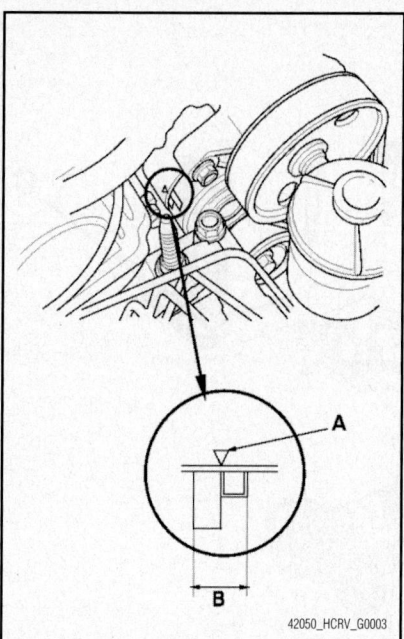

Fig. 43 Check that the auto-tensioner indicator (A) is within the standard range (B) as shown. If it is out of the standard range, the belt must be replaced

the power steering pump. The belt tension is automatically maintained by a tensioner. No adjustment is necessary or possible.

REMOVAL & INSTALLATION

See Figure 44.

The 2.4L engine uses 1 belt to drive the alternator, air conditioner compressor, and the power steering pump.

1. Before servicing the vehicle, refer to the Precautions Section.

2. Remove the splash shield.

3. Use a suitable belt tension release tool to move the auto-tensioner (A) to relieve tension from the drive belt (B) and remove the drive belt.

4. Install the new belt in the reverse order of the removal procedure.

CAMSHAFT AND VALVE LIFTERS

REMOVAL & INSTALLATION

See Figures 45 through 49.

1. Before servicing the vehicle, refer to the Precautions Section.

2. Remove the timing chain and sprockets. Refer to Timing Chain, Sprockets, Front Cover and Seal, removal & installation.

3. Loosen the rocker arm adjusting screws.

4. Remove the camshaft holder bolts. To prevent damaging the camshafts, loosen the bolts, in sequence, 2 turns at a time.

➡**Mark the camshafts, camshaft brackets, and bolts so they are placed in the same position and direction at installation.**

5. Remove the camshaft chain guide, the camshaft holders, and the camshafts.

To install:

6. Reassemble the rocker arm assembly.

7. Clean and dry the No. 5 rocker shaft holder mating surface.

8. Apply liquid gasket (P/N 08717-0004, 08718-0001, 08718-0003, or 08718-0009) to the cylinder head mating surface of the No. 5 rocker shaft holder and to the inside edge of the bolt holes. Install the component within 5 minutes of applying the liquid gasket.

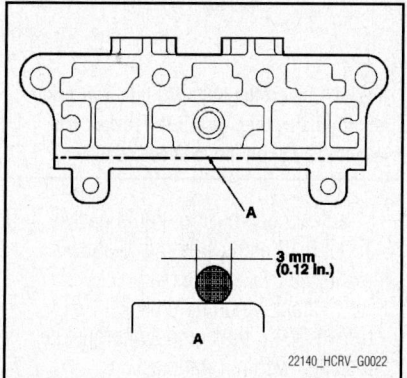

Fig. 46 Apply a bead of liquid gasket about 0.12 inch (3 mm) in diameter along the broken line (A)

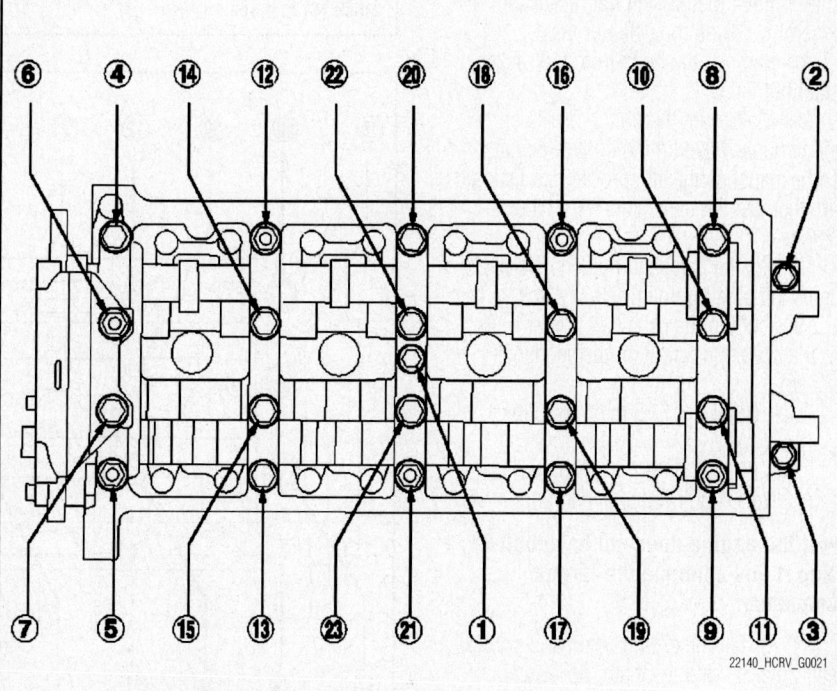

Fig. 44 Use a suitable belt tension release tool to move the auto-tensioner (A) in order to remove the belt (B)

Fig. 45 Remove the camshaft holder bolts. To prevent damage to the camshafts, loosen the bolts 2 turns at a time in the sequence illustrated

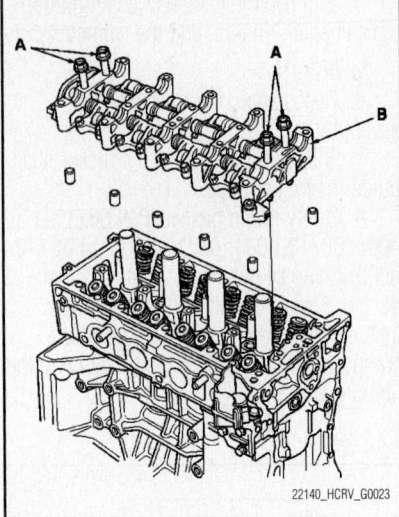

Fig. 47 Insert the bolts (A) into the rocker shaft holder, then install the rocker arm assembly (B) on the cylinder head

a. Apply a bead of liquid gasket about 0.12 inch (3 mm) in diameter along the broken line (A).

b. If applying liquid gasket P/N 08718-0012, the component must be installed within 4 minutes.

c. If too much time has passed after applying the liquid gasket, remove the old liquid gasket and residue, then reapply new liquid gasket.

9. Insert the bolts (A) into the rocker shaft holder, then install the rocker arm assembly (B) on the cylinder head.

10. Remove the bolts from the rocker shaft holder.

11. Make sure the punch marks on the variable valve timing control actuator and exhaust camshaft sprocket are facing up, then set the camshafts (A) in the holder.

12. Set the camshaft holders (B) and camshaft chain guide B (C) in place.

13. Tighten the camshaft holder bolts to the specified torque in the sequence shown.

a. Bolts 8 x 1.25 mm: 16 ft. lbs. (22 Nm).

b. Bolts 6 x 1.0 mm: 106 inch lbs. (12 Nm)—bolts 21, 22, and 23.

➡**If the engine does not have bolt 21, skip it and continue the torque sequence.**

14. Install the timing chain and sprockets. Refer to Timing Chain, Sprockets, Front Cover and Seal, removal & installation.

15. Adjust the valve clearance as needed.

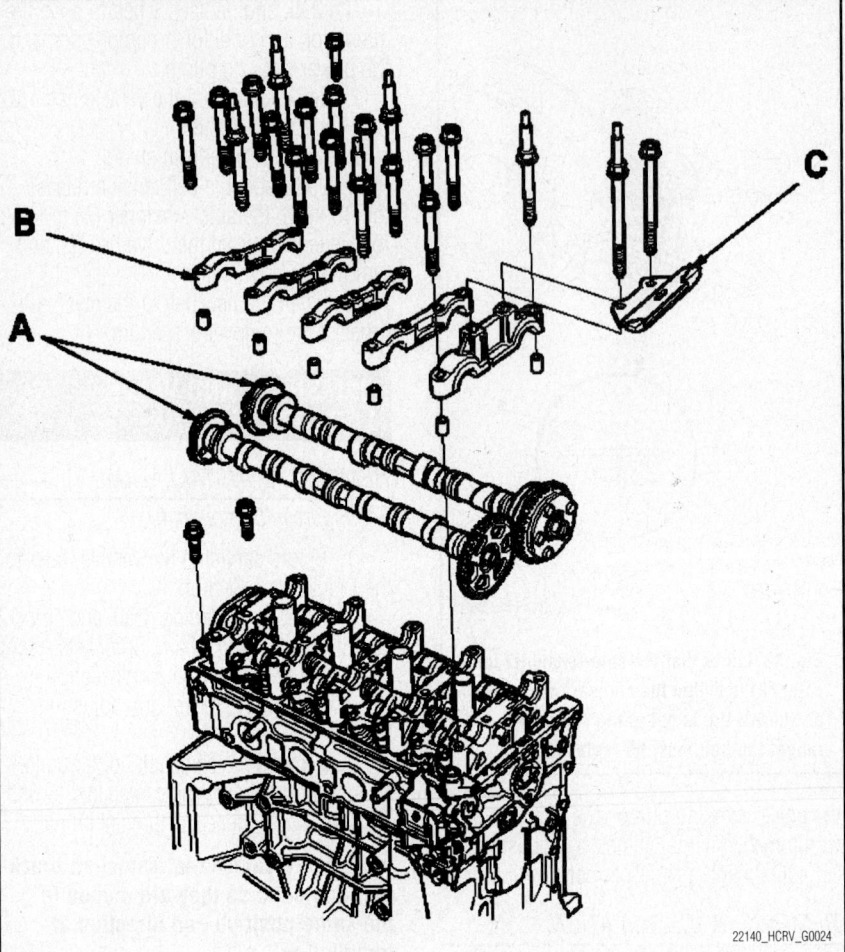

Fig. 48 Set the camshafts (A) in the holder. Set the camshaft holders (B) and camshaft chain guide (C) in place

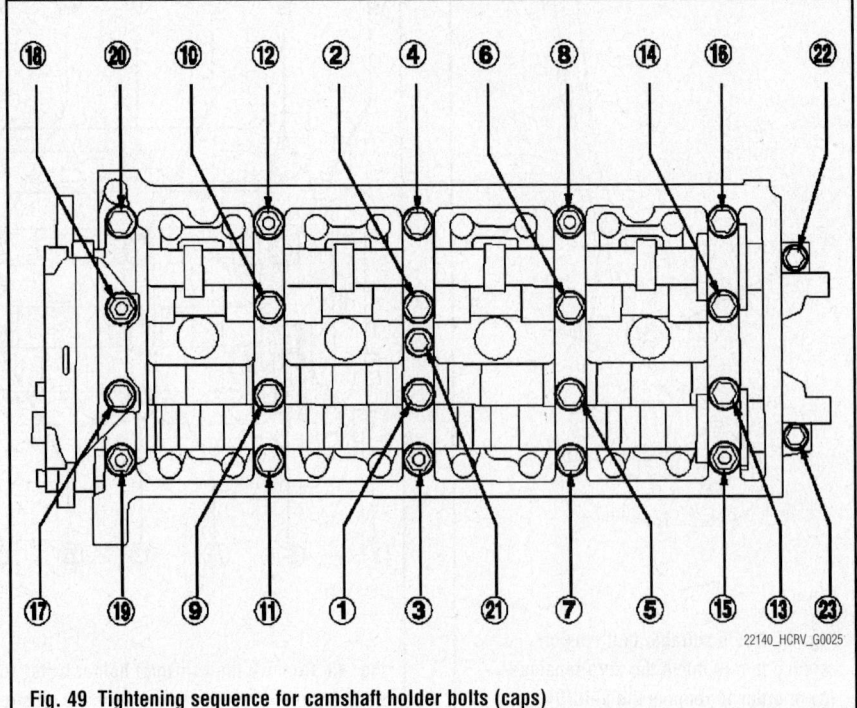

Fig. 49 Tightening sequence for camshaft holder bolts (caps)

CRANKSHAFT FRONT SEAL

REMOVAL & INSTALLATION

The 2.4L engine utilizes a camshaft timing chain case seal. Refer to Timing Chain, Sprockets, Front Cover and Seal, removal & installation.

CYLINDER HEAD

REMOVAL & INSTALLATION

See Figures 50 through 53.

1. Before servicing the vehicle, refer to the Precautions Section.

✱✱ WARNING

Use fender covers to avoid damaging painted surfaces. To avoid damage, unplug the wiring connectors carefully while holding the connector portion. To avoid damaging the cylinder head, wait until the engine coolant temperature drops below 100°F (38°C) before loosening the cylinder head bolts.

➡**Mark all wiring and hoses to avoid misconnection.**

2. Relieve fuel pressure.
3. Drain the engine coolant.
4. Remove the air cleaner housing.
5. Remove the accessory drive belt.
6. Remove the intake manifold.
7. Remove the exhaust manifold.

8. Disconnect the Evaporative Emission (EVAP) canister hose and the brake booster vacuum hose.
9. Remove the quick-connect fitting cover, then disconnect the fuel feed hose.
10. Disconnect the positive crankcase ventilation (PCV) hose and remove the ground cable.
11. Remove the harness holder from the bracket, then remove the harness holder bracket.
12. Disconnect the upper radiator hose and the heater hoses.
13. Remove the bolt securing the connecting pipe.
14. Disconnect the water bypass hose.
15. Remove the following engine wire harness connectors and wire harness clamps from the cylinder head:
   - Four fuel injector connectors
   - Engine Coolant Temperature (ECT) sensor 1 connector
   - Camshaft Position (CMP) sensor A (Intake) connector
   - Camshaft Position (CMP) sensor B (Exhaust) connector
   - Rocker arm oil control solenoid connector
   - Rocker arm oil pressure switch connector
   - EVAP canister purge valve connector
   - Exhaust Gas Recirculation (EGR) valve connector
16. Remove the camshaft chain.
17. Remove the rocker arm assembly.

18. Remove the cylinder head bolts. To prevent warpage, loosen the bolts in sequence ⅓ turn at a time; repeat the sequence until all bolts are loosened.
19. Remove the cylinder head.

To install:

20. Install a new coolant separator in the engine block whenever the engine block is replaced.
21. Clean the cylinder head and block surface.
22. Install a new cylinder head gasket and dowel pins on the engine block.

➡**Always use a new cylinder head gasket.**

23. Set the crankshaft to Top Dead Center (TDC). Align the TDC mark (A) on the crankshaft sprocket with the pointer (B) on the engine block.
24. Install the cylinder head on the block.

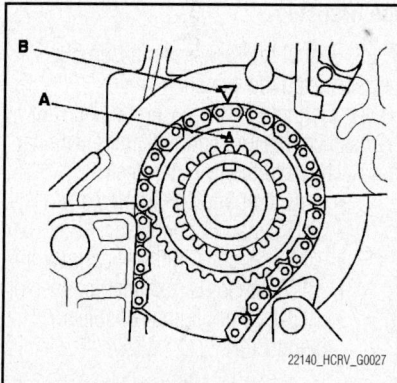

Fig. 51 Align the TDC mark (A) on the crankshaft sprocket with the pointer (B) on the engine block

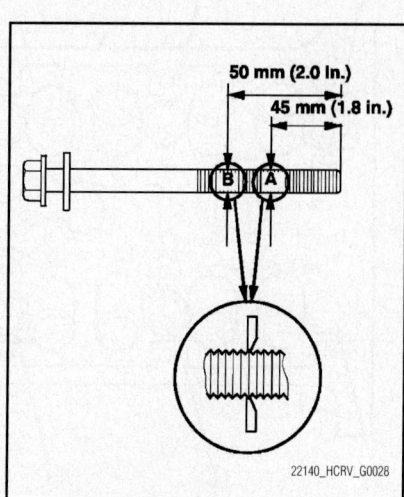

Fig. 52 Measure the diameter of each cylinder head bolt at point A and point B. Replace with new bolts if required

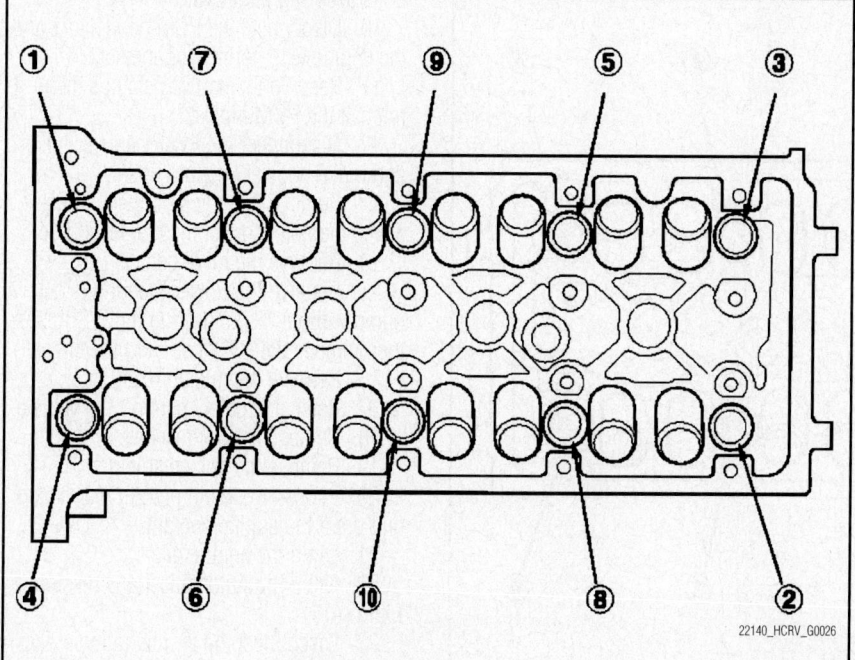

Fig. 50 Loosen the cylinder head bolts in sequence

25. If reusing the cylinder head bolts:
   a. Measure the diameter of each cylinder head bolt at point A and point B.
   b. If either diameter is less than 0.42 inch (10.6 mm), replace the cylinder head bolt.

26. Apply engine oil to the threads and under the bolt heads of all cylinder head bolts.

27. Tighten the cylinder head bolts in sequence:
   a. Step 1: Tighten to 29 ft. lbs. (39 Nm).
   b. Step 2: Angle-tighten 90°.
   c. Step 3: Angle-tighten another 90 °.
   d. Step 4: (Only if using new bolts) Angle-tighten another 90 °.

➡Use a beam-type torque wrench. When using a preset click-type torque wrench, be sure to tighten slowly and do not over-tighten. If a bolt makes any noise while you are tightening it, loosen the bolt and retighten it from the first step.

28. Install the rocker arm assembly.
29. Install the camshaft chain.
30. Connect the following engine wire harness connectors and install the wire harness clamps to the cylinder head:
   • Four fuel injector connectors
   • ECT sensor 1 connector
   • CMP sensor A (Intake) connector
   • CMP sensor B (Exhaust) connector
   • Rocker arm oil control solenoid connector

   • Rocker arm oil pressure switch connector
   • EVAP canister purge valve connector
   • EGR valve connector

31. Install the bolt securing the connecting pipe.
32. Install the water bypass hose.
33. Install the upper radiator hose and the heater hoses.
34. Install the harness holder bracket, then install the harness holder.
35. Install the PCV hose and the ground cable.
36. Connect the fuel feed hose, then install the quick-connect fitting cover.
37. Install the EVAP canister hose and the brake booster vacuum hose.
38. Install the exhaust manifold.
39. Install the intake manifold.
40. Install the accessory drive belt.
41. Install the air cleaner housing.
42. After installation, check that all tubes, hoses, and connectors are installed correctly.
43. Inspect for fuel leaks. Turn the ignition switch **ON** (do not operate the starter) so the fuel pump runs for about 2 seconds and pressurizes the fuel line. Repeat this operation 3 times, then check for fuel leakage at all points in the fuel line.
44. Refill the radiator with engine coolant and bleed the air from the cooling system with the heater valve open and the engine running.
45. Inspect the idle speed.
46. Inspect the ignition timing.

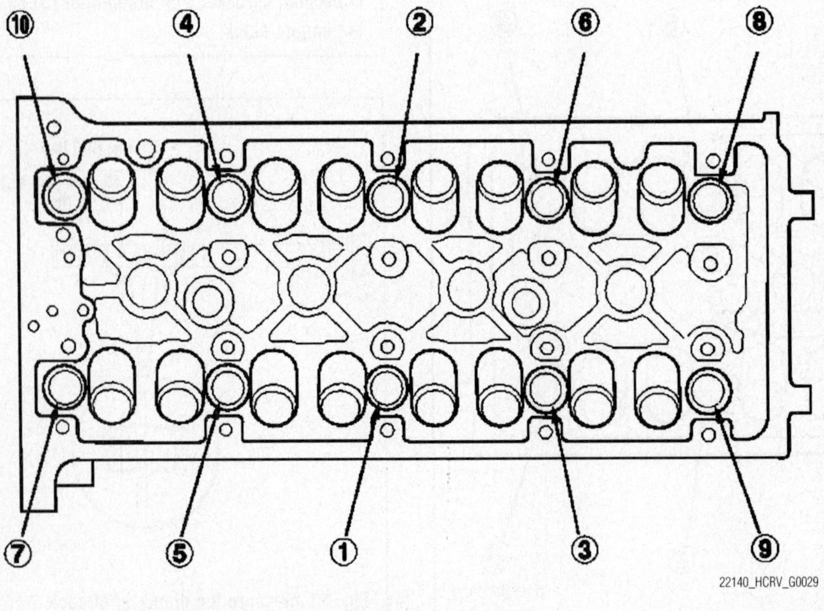

ENGINE ASSEMBLY

REMOVAL & INSTALLATION

See Figures 54 through 68.

Special Tools Required:
• Universal lifting eyelet 07AAK-SNAA120
• Engine hanger adapter VSB02C000015
• Front subframe adapter VSB02C000016
• CR-V engine hanger adapter VSB02C000032
• Engine support hanger, A and Reds AAR-T1256

➡Use fender covers to avoid damaging painted surfaces. To avoid damaging the wiring and terminals, unplug the wiring connectors carefully while holding the connector portion. Mark all wiring and hoses to avoid misconnection.

1. Before servicing the vehicle, refer to the Precautions Section.
2. Remove the hood support rod, then use it to prop the hood in the wide-open position.
3. Relieve the fuel pressure.
4. Disconnect and remove the battery.
5. Remove the air cleaner housing assembly.
6. Remove the harness clamp and the ground cable beneath the battery holder.
7. Remove the battery cables from the under-hood fuse/relay box.
8. Disconnect the harness connector and remove the harness clamp.
9. Remove the Powertrain Control Module (PCM) cover, then remove the 3 bolts securing the PCM.
10. Disconnect the PCM connectors and the engine wire harness connector.
11. Remove the harness clamps then remove the PCM bracket.
12. Disconnect the EVAP canister hose and brake booster vacuum hose.
13. Remove the quick-connect fitting cover, then disconnect the fuel feed hose.
14. Remove the drive belt.
15. Remove the Power Steering (P/S) pump without disconnecting the P/S hoses, then remove the P/S hose from the clamp.
16. Remove the radiator cap.
17. Raise and safely support the vehicle.
18. Remove the front wheels.
19. Remove the splash shield.
20. Loosen the drain plug in the radiator and drain the engine coolant.
21. Drain the engine oil.
22. Drain the Automatic Transmission Fluid (ATF).
23. Disconnect the air fuel ratio sensor connector and the secondary heated oxygen sensor connector.

22140_HCRV_G0029

Fig. 53 Cylinder head bolt tightening sequence

24. Remove the 3-way catalytic converter.

25. Remove the shift cable.

26. Separate the stabilizer links from the stabilizer bar.

27. Separate the knuckles from the lower arms.

28. Remove the driveshafts. Coat all precision-finished surfaces with clean engine oil. Tie plastic bags over the driveshaft ends.

29. Remove the propeller shaft.

30. Remove the bolt securing the P/S fluid line bracket, and unclamp the P/S fluid line clamps on the front subframe.

31. Remove the bolts securing the left steering gearbox mounting bracket.

32. Remove the bolts securing the right steering gearbox mounting brackets.

33. Disconnect the A/C compressor clutch connector, then remove the A/C compressor without disconnecting the A/C hoses. Do not bend the A/C hoses excessively.

34. Lower the vehicle.

35. Remove the radiator.

36. Disconnect the ATF cooler hose, then plug the line and the hose.

37. Disconnect the heater hoses.

38. Install the front bulkhead.

39. Attach the engine hanger adapter (VSB02C000015) to the threaded hole in the cylinder head.

40. Remove both lids (A) for the front strut flange nuts from the cowl cover. Position the engine hanger adapters (VSB02C000032) over the strut flange nuts.

41. Carefully position the engine support hanger (AAR-T-1256) on the vehicle, and attach the hook (A) to the engine hanger adapter (VSB02C000015). Tighten the wing nut (B) by hand, and lift and support the engine.

✸✸ WARNING

Be careful when working around the windshield as breakage is possible.

42. Raise and safely support the vehicle.

43. Remove the lower torque rod.

44. Make the appropriate reference lines at both ends of the subframe that line up with the body.

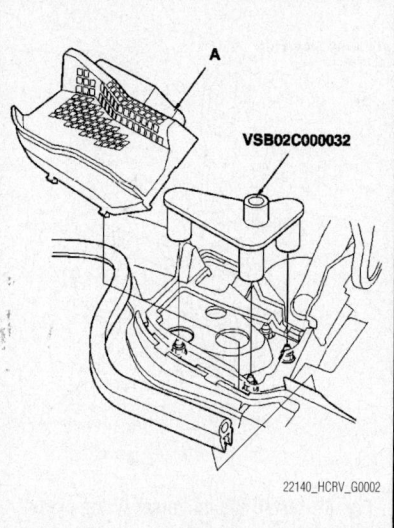

Fig. 55 Remove both lids (A) for the front strut flange nuts from the cowl cover and position the engine hanger adapters (VSB02C000032) over the strut flange nuts

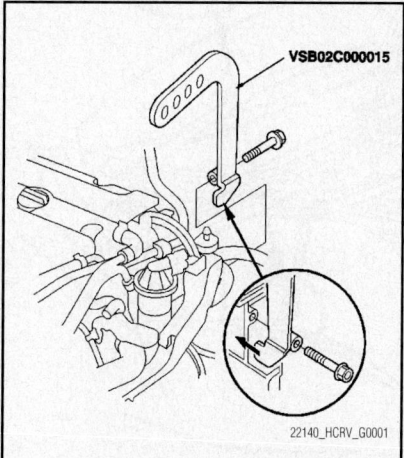

Fig. 54 Attach the engine hanger adapter (VSB02C000015) to the threaded hole in the cylinder head

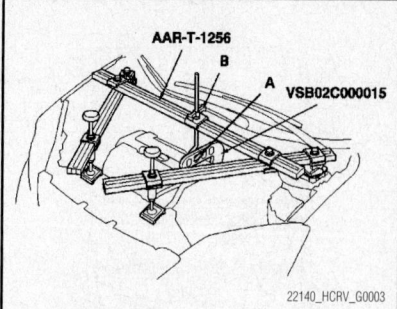

Fig. 56 Position the engine support hanger (AAR-T-1256) on the vehicle, and attach the hook (A) to the engine hanger adapter (VSB02C000015). Tighten the wing nut (B) by hand

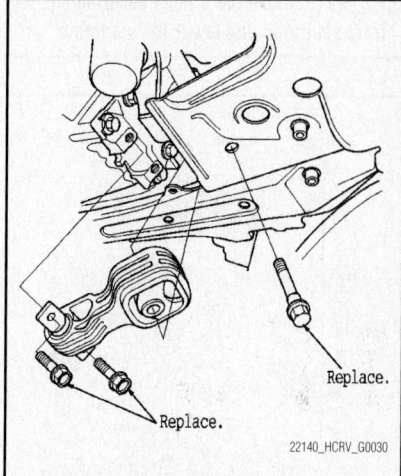

Fig. 57 Remove the lower torque rod

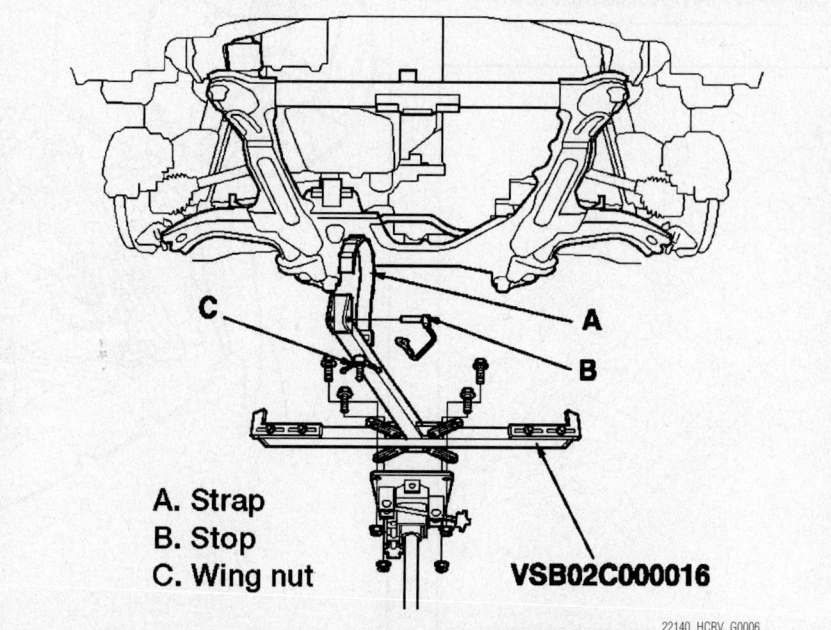

A. Strap
B. Stop
C. Wing nut

Fig. 58 Attach the front subframe adapter (VSB02C000016) to the subframe by looping the strap over the front of the subframe, then secure the strap with the stop and tighten the wing nut

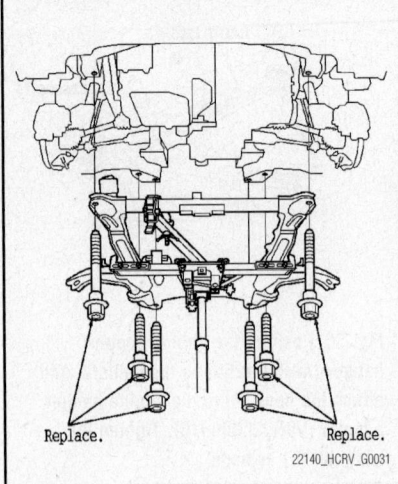

Fig. 59 Remove the 6 bolts securing the front subframe and lower the subframe

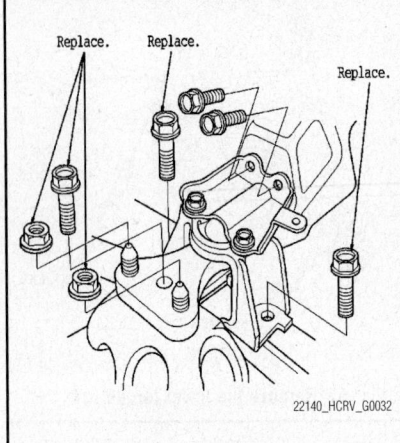

Fig. 60 Remove the transaxle mount

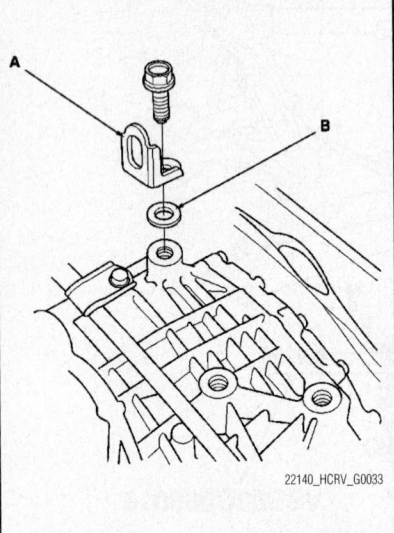

Fig. 61 Install the transmission hanger bracket (P/N 21232-RCT-A00) (A) and the washer (B) on the transaxle

45. Remove the subframe mounting bolts on both sides.

46. Attach the front subframe adapter (VSB02C000016) to the subframe by looping the strap (A) over the front of the subframe, then secure the strap with the stop (B), then tighten the wing nut (C).

47. Raise the jack and line up the slots in the arms with the bolt holes on the corner of the jack base, then tighten the bolts.

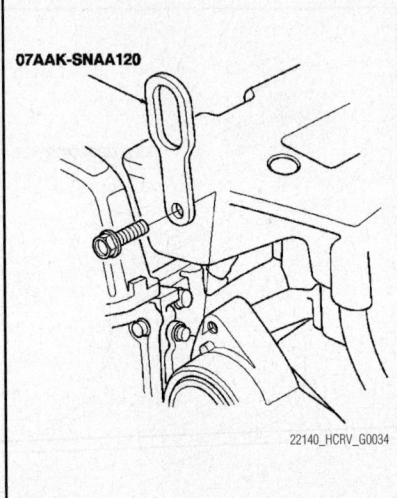

Fig. 62 Install the universal lifting eyelet to the drive belt auto-tensioner

48. Remove the 6 bolts securing the front subframe and lower the subframe.

49. Remove the subframe.

50. Lower the vehicle.

51. Remove the transaxle mount.

52. Remove the ground cable, then remove the transaxle mount bracket.

53. Install the transmission hanger bracket (P/N 21232-RCT-A00) (A) and the washer (B) on the transaxle.

54. Install the universal lifting eyelet to the drive belt auto-tensioner.

55. Attach a chain hoist (A) to the universal lifting eyelet (B), and the transmission hook (C). Lift up on the engine/transmission assembly until it's securely supported by the chain hoist, then remove the engine support hanger.

56. Remove the side engine mount bracket mounting bolt and nut.

57. Check that the engine/transaxle is completely free of vacuum hoses, fuel hoses, coolant hoses, and electrical wiring.

58. Slowly lower the engine/transaxle assembly about 6 inches (150 mm). Check once again that all the hoses and the electrical wiring are disconnected and free from the engine/transaxle, then lower it all the way.

59. Disconnect the chain hoist from the engine/transaxle assembly.

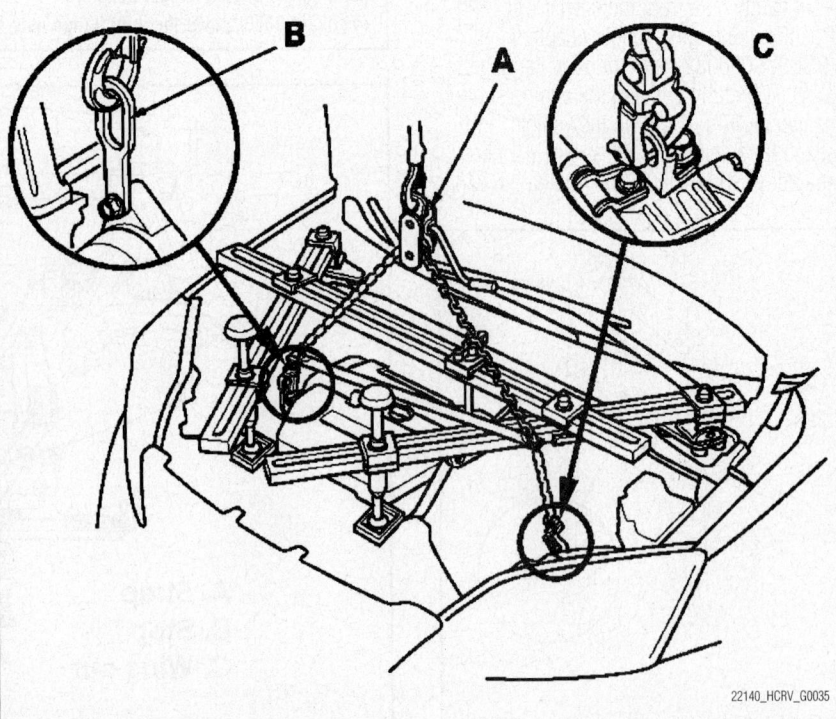

Fig. 63 Attach a chain hoist (A) to the universal lifting eyelet (B) and the transmission hook (C), and lift up on the engine/transaxle assembly

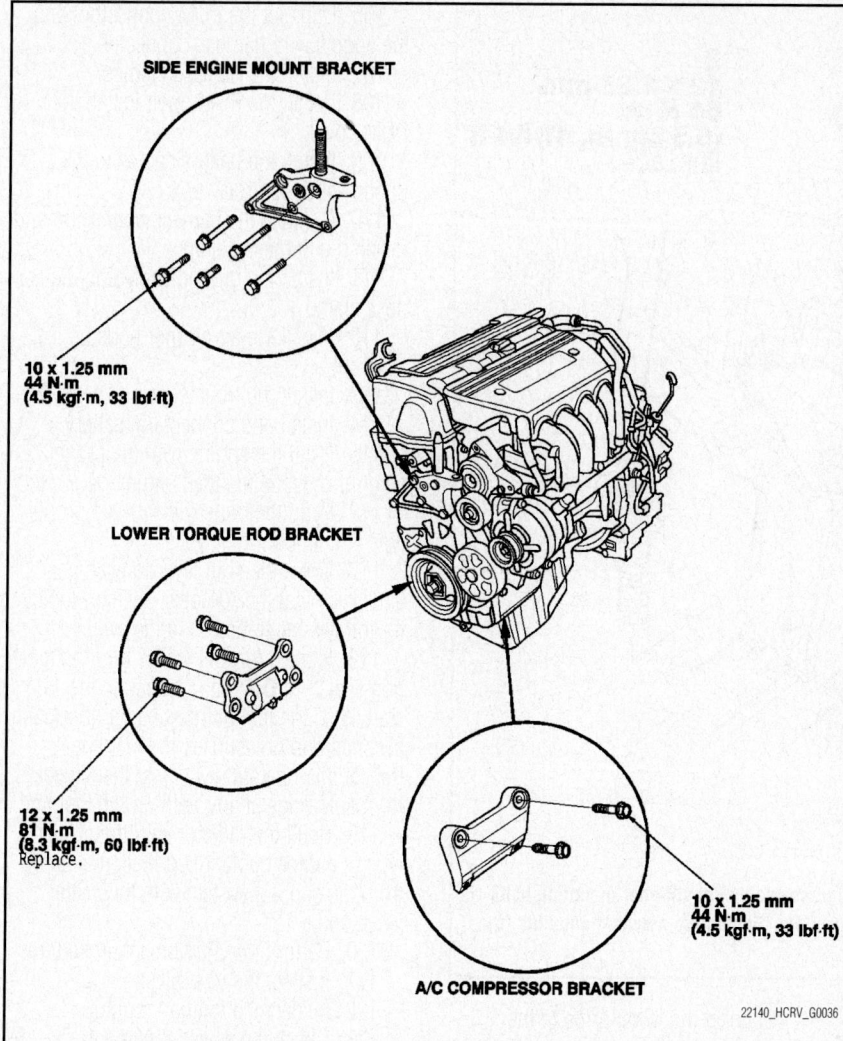

SIDE ENGINE MOUNT BRACKET

10 x 1.25 mm
44 N·m
(4.5 kgf·m, 33 lbf·ft)

LOWER TORQUE ROD BRACKET

12 x 1.25 mm
81 N·m
(8.3 kgf·m, 60 lbf·ft)
Replace.

10 x 1.25 mm
44 N·m
(4.5 kgf·m, 33 lbf·ft)

A/C COMPRESSOR BRACKET

22140_HCRV_G0036

Fig. 64 Install the accessory brackets and tighten bolts to the specified torque, as illustrated

60. Raise the vehicle, then remove the engine/transaxle assembly from under the vehicle.

To install:

61. Install the accessory brackets and tighten bolts to the specified torque, as illustrated.

62. Raise the vehicle and position the engine/transaxle assembly under the vehicle. Lower the vehicle and attach the universal lifting eyelet (A) and transmission hook (B) to the chain hoist, then lift the engine into position.

➡**Reinstall the mounting bolts and support nuts in the sequence given in the following steps. Failure to follow this sequence may cause excessive noise and vibration and reduce engine mount life.**

63. Attach the special tool adapter (VSB02C000015) to the threaded hole in the cylinder head.

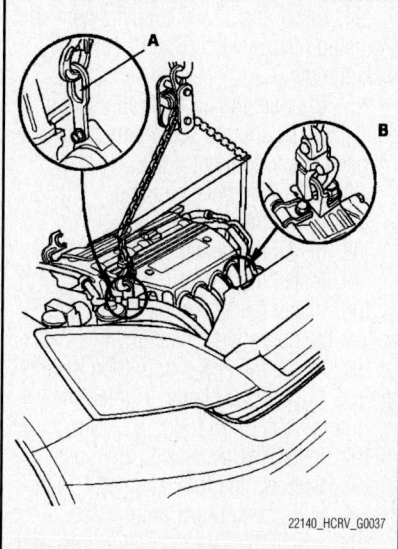

22140_HCRV_G0037

Fig. 65 Attach the universal lifting eyelet (A) and transmission hook (B) to the chain hoist

64. Remove both lids (A) from the cowl cover. Position the engine hanger adapters (VSB02C000032) over the strut flange nuts.

65. Install the A and Reds engine support hanger (AAR-T1256), then attach the hook to the slotted hole in the hanger adapter. Tighten the wing nut by hand to lift and support the engine/transaxle assembly.

66. Loosen the upper torque rod mounting bolt.

67. Tighten the new side engine mount bracket mounting bolt and nut to 54 ft. lbs. (74 Nm).

68. Remove the chain hoist.

69. Remove the universal lifting eyelet.

70. Remove the transmission hanger bracket (P/N 21232-RCT-A00) (A) and the washer (B).

71. Install the transaxle mount bracket and tighten to 54 ft. lbs. (74 Nm).

72. Install the ground cable and tighten to 86 inch lbs. (10 Nm).

73. Install the transaxle mount, then tighten the transaxle mount stiffener mounting bolts (A) to 16 ft. lbs. (22 Nm) and the new transaxle mount mounting bolts (B) to 47 ft. lbs. (64 Nm).

74. Tighten the new bolt and nuts (C) to 54 ft. lbs. (74 Nm).

75. Raise the vehicle on the lift.

76. Using the subframe adapter (VSB02C000016) and a jack, raise the subframe up to the body.

77. Loosely install the 6 new 14 x 1.5 mm bolts.

78. Align all reference marks on the front subframe with the body, then tighten the bolts on the front subframe to 76 ft. lbs. (103 Nm).

79. Tighten the new subframe mounting bolts to 47 ft. lbs. (64 Nm) on both sides.

80. Install the lower torque rod, then tighten the new lower torque rod mounting bolts in the numbered sequence shown.

81. Lower the vehicle.

82. Remove the engine support hanger from the vehicle. Remove the adapter from the cylinder head.

83. Tighten the upper torque rod mounting bolt to 40 ft. lbs. (54 Nm).

84. Raise the vehicle.

85. Install the A/C compressor, then connect the A/C compressor clutch connector. Tighten the mounting bolts to 16 ft. lbs. (22 Nm).

86. Install the bolts securing the left steering gearbox mounting bracket. Tighten the bolts to 43 ft. lbs. (59 Nm).

87. Install the bolts securing the right steering gearbox mounting bracket. Tighten the bolts to 43 ft. lbs. (59 Nm).

88. Install the power steering fluid line

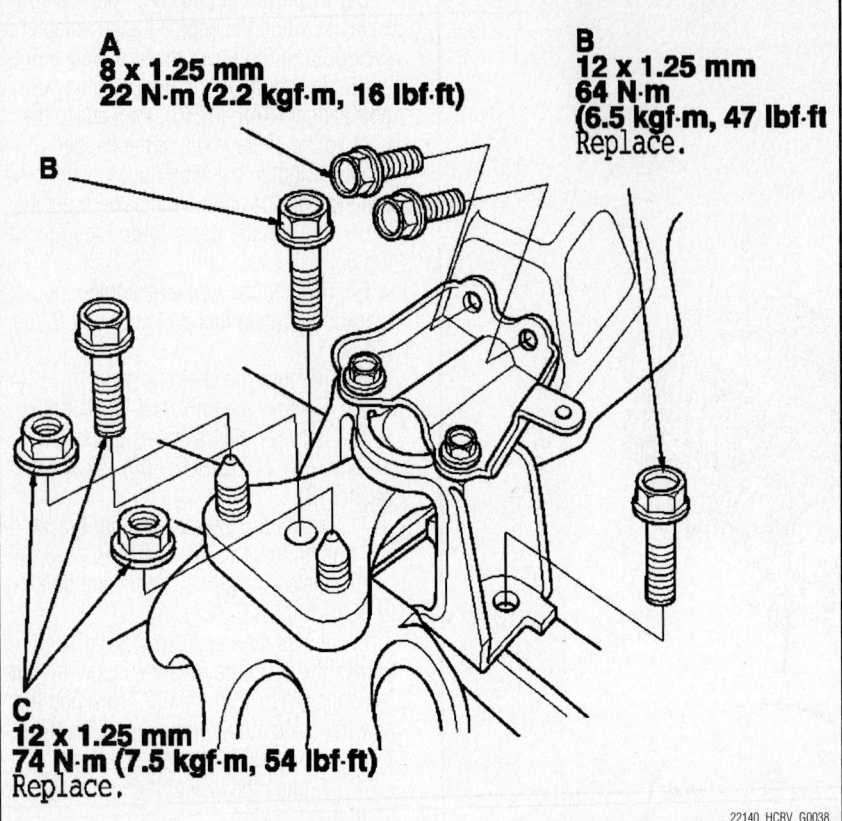

A
8 x 1.25 mm
22 N·m (2.2 kgf·m, 16 lbf·ft)

B

B
12 x 1.25 mm
64 N·m
(6.5 kgf·m, 47 lbf·ft
Replace.

C
12 x 1.25 mm
74 N·m (7.5 kgf·m, 54 lbf·ft)
Replace.

22140_HCRV_G0038

Fig. 66 Install the transaxle mount, then tighten the transaxle mount stiffener mounting bolts (A) and the new transaxle mount mounting bolts (B) and the Tighten the new bolt and nuts (C) as illustrated.

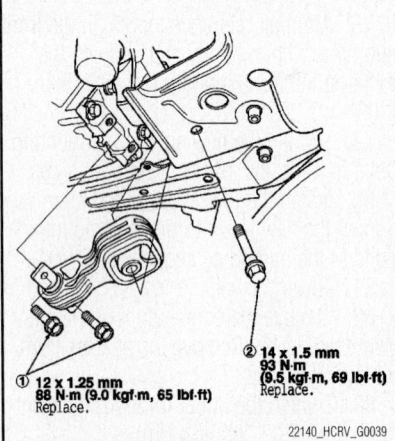

② 14 x 1.5 mm
93 N·m
(9.5 kgf·m, 69 lbf·ft)
Replace.

① 12 x 1.25 mm
88 N·m (9.0 kgf·m, 65 lbf·ft)
Replace.

22140_HCRV_G0039

Fig. 67 Install the lower torque rod, then tighten the new lower torque rod mounting bolts in the numbered sequence shown

bracket and secure the hose with the hose clamps.

89. Install the propeller shaft.

90. Install a new set ring on the end of each driveshaft, then install the driveshafts. Make sure each ring clicks into place in the differential and intermediate shaft.

91. Connect the lower arms to the knuckles.

92. Connect the stabilizer links to the stabilizer bar.

93. Install the shift cable.

94. Install the 3-way catalytic converter (A) using new gaskets (B) and new self-locking nuts (C).

95. Connect the air fuel ratio sensor connector (D) and the secondary heated oxygen sensor connector (E).

96. Install the splash shield.

97. Lower the vehicle.

98. Install the heater hoses.

99. Install the radiator.

100. Install the ATF cooler hose and secure the hoses with the clip.

101. Install the P/S pump, then install the P/S hose to the clamp. Tighten the mounting bolts to 16 ft. lbs. (22 Nm).

102. Install the accessory drive belt.

103. Connect the fuel feed hose, then install the quick-connect fitting cover.

104. Install the EVAP canister hose and the brake booster vacuum hose.

105. Install the PCM bracket, then install the harness clamp. Tighten the bracket bolts to 86 inch lbs. (10 Nm).

106. Connect the PCM connectors and the engine wire harness connector.

107. Install the harness clamps.

108. Install the PCM, then install the PCM cover.

109. Install the battery cables to the under-hood fuse/relay box.

110. Connect the harness connector, and install the harness clamp.

111. Install the ground cable and the harness clamp.

112. Install the air cleaner housing assembly.

113. Install the front wheels.

114. Install and connect the battery.

115. Refill the engine with the proper amount of recommended engine oil.

116. Refill the transaxle with the proper type and amount of ATF.

117. Move the shift lever to each gear and verify that the A/T gear position indicator follows the transaxle range switch.

118. Inspect for fuel leaks: Turn the ignition switch to **ON** (do not operate the starter) so the fuel pump runs for about 2 seconds and pressurizes the fuel line. Repeat this operation 3 times, then check for fuel leakage at any point in the fuel line.

119. Refill the radiator with the proper type of engine coolant and bleed the air from the cooling system with the heater valve open.

120. Perform the PCM reset procedure.

121. Inspect the idle speed.

122. Inspect the ignition timing.

123. Check the wheel alignment.

PCM Reset Procedure

1. Reset the PCM with the Honda Diagnostic System (HDS) while the engine is stopped.

2. Turn the ignition switch to LOCK.

3. Turn the ignition switch to ON and wait 30 seconds.

4. Turn the ignition switch to LOCK and disconnect the HDS from the Data Link Connector (DLC).

5. Perform the PCM idle learn procedure.

PCM Idle Learn Procedure—With HDS

1. Connect the Honda Diagnostic System (HDS) to the Data Link Connector (DLC) located under the driver's side of the dashboard.

2. Turn the ignition switch to ON.

3. Make sure the HDS communicates with the PCM and other vehicle systems. If it does not, go to the DLC circuit troubleshooting.

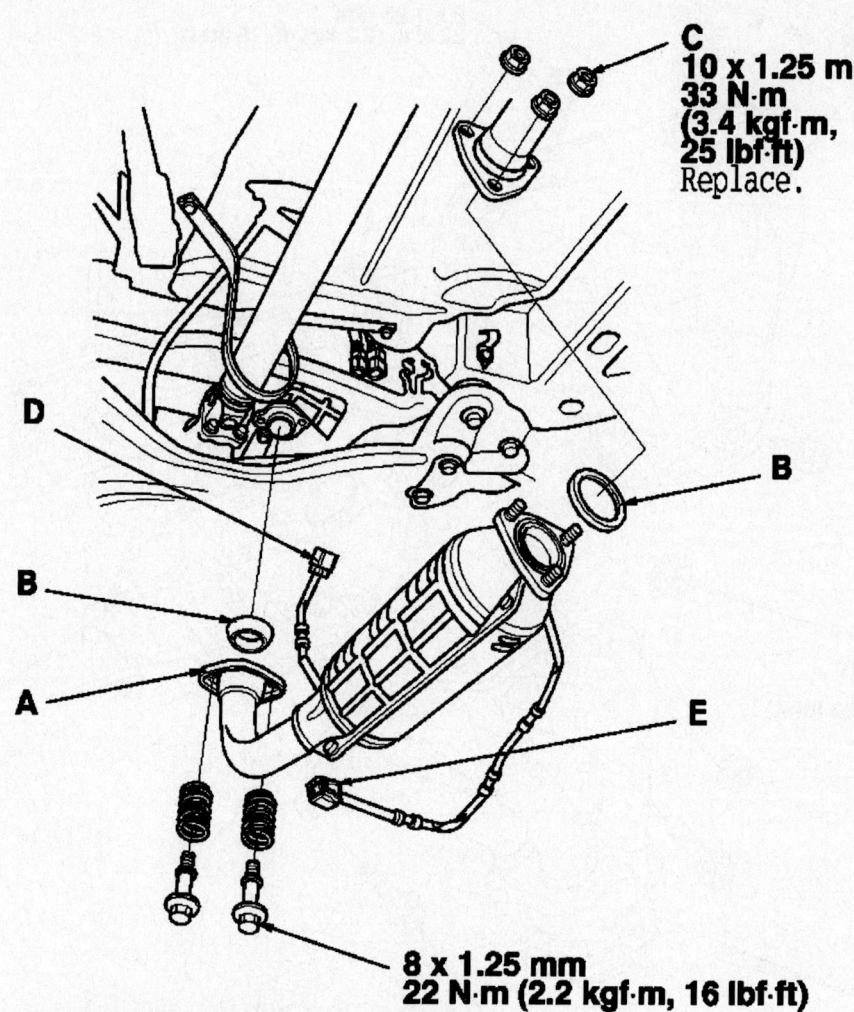

C
10 x 1.25 mm
33 N·m
(3.4 kgf·m,
25 lbf·ft)
Replace.

8 x 1.25 mm
22 N·m (2.2 kgf·m, 16 lbf·ft)

A. 3-way catalytic converter
B. New gaskets
C. New self-locking nuts
D. Air fuel ratio sensor connector
E. Secondary heated oxygen sensor connector

22140_HCRV_G0040

Fig. 68 Install the 3-way catalytic converter using new gaskets and new self-locking nuts

4.Select CRANK PATTERN in the ADJUSTMENT MENU with the HDS.

5. Select CRANK PATTERN LEARNING with the HDS and follow the screen prompts.

PCM Idle Learn Procedure—Without HDS

1. Start the engine.

2. Hold the engine speed at 3,000 RPM without load (in P or N) until the radiator fan comes on.

3. Test-drive the vehicle on a level road: decelerate (with the throttle fully closed) from an engine speed of 2,500 RPM down to 1,000 RPM with the transaxle in 2 position.

4. Repeat step 3 several times.

5. Turn the ignition switch to LOCK.

6. Turn the ignition switch to ON and wait 30 seconds.

EXHAUST MANIFOLD

REMOVAL & INSTALLATION

See Figure 69.

1. Before servicing the vehicle, refer to the Precautions Section.

2. Disconnect the air fuel ratio sensor connector and secondary heated oxygen sensor connector.

3. Remove the 3-way catalytic converter.

4. Remove the under-cowl cover, then remove the strut brace.

5. Remove the rocker arm oil control solenoid.

6. Remove the cover and exhaust manifold bracket.

7. Remove the exhaust manifold.

To install:

8. Install the exhaust manifold with a new gasket.

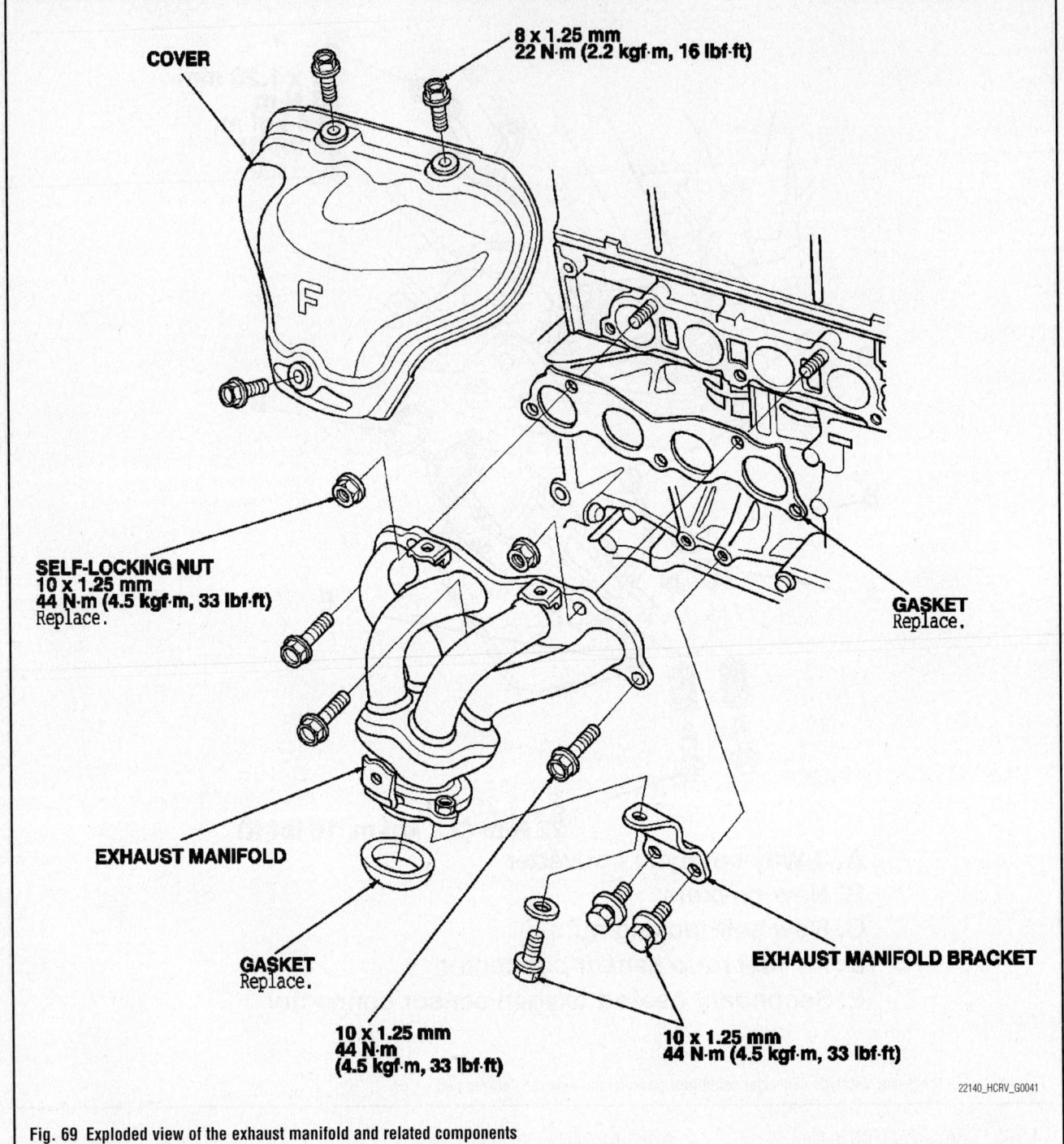

COVER

8 x 1.25 mm
22 N·m (2.2 kgf·m, 16 lbf·ft)

F

SELF-LOCKING NUT
10 x 1.25 mm
44 N·m (4.5 kgf·m, 33 lbf·ft)
Replace.

GASKET
Replace.

EXHAUST MANIFOLD

GASKET
Replace.

10 x 1.25 mm
44 N·m
(4.5 kgf·m, 33 lbf·ft)

10 x 1.25 mm
44 N·m (4.5 kgf·m, 33 lbf·ft)

EXHAUST MANIFOLD BRACKET

22140_HCRV_G0041

Fig. 69 Exploded view of the exhaust manifold and related components

9. Tighten the bolts and nuts in a criss-cross pattern in 3 steps, beginning with the inner nut. Torque specification: 33 ft. lbs. (44 Nm).

10. Install the other parts in the reverse order of removal and torque according to the illustration.

INTAKE MANIFOLD

REMOVAL & INSTALLATION

See Figures 70 through 77.

1. Before servicing the vehicle, refer to the Precautions Section.

2. Remove the hood support rod, then use it to prop the hood in the wide-open position.

3. Remove the bulkhead cover:
   a. Put on gloves to protect your hands. Take care not to scratch the front bumper and the body.
   b. Remove the clips (A, B) by carefully pulling the front grille cover (C) up, then remove the cover by releasing

the front edge of the cover from the grille (D).
   c. To remove the clips, pry the inner clip up at the edge near the line (E) on its head.

4. Disconnect the fan motor connectors and the hood switch connector, then remove the harness clips.

5. Remove the radiator upper brackets (A), then remove the front bulkhead (B).

6. Remove the intake manifold cover.

7. Disconnect the vacuum hose (A) and

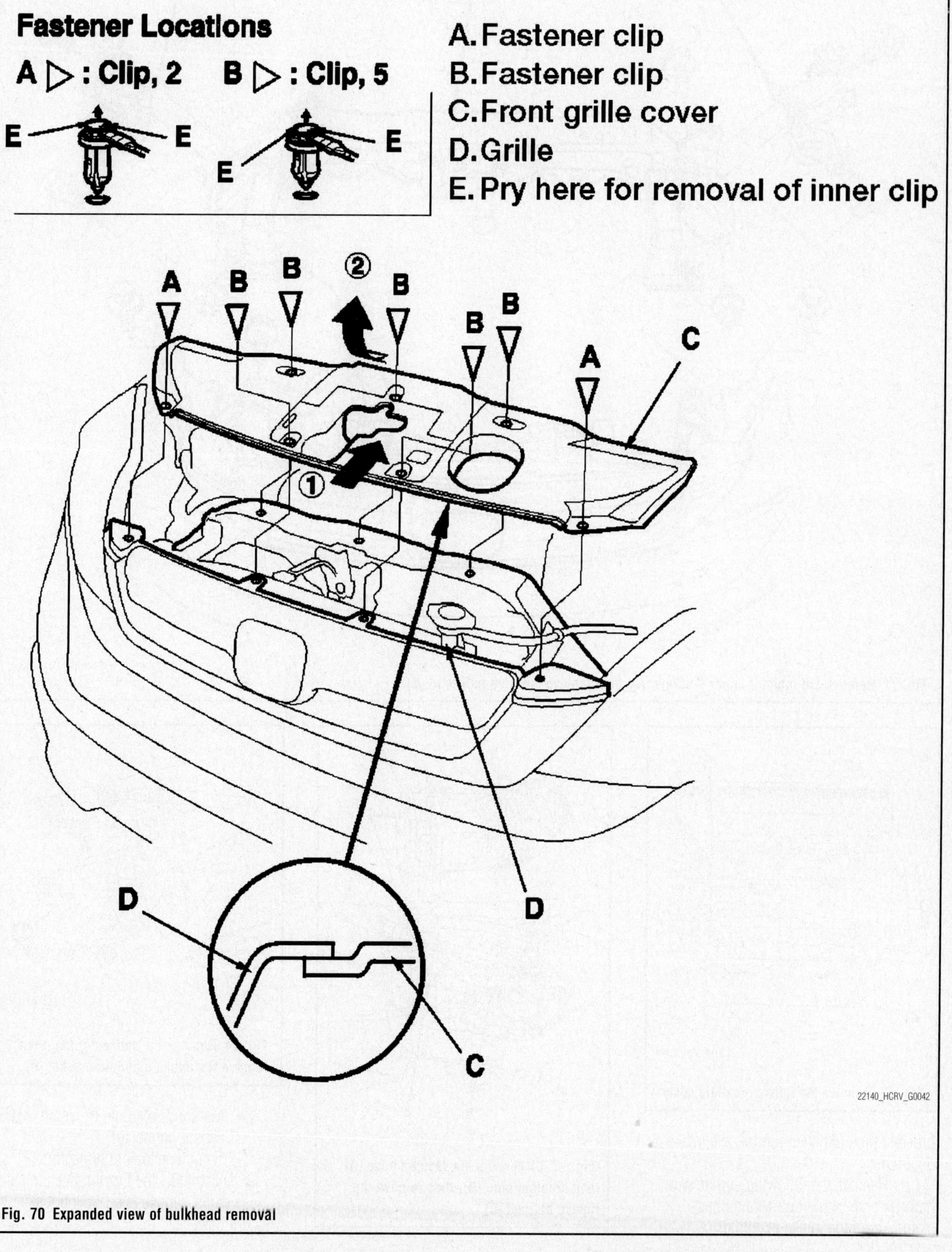

Fastener Locations

A ▷ : Clip, 2   B ▷ : Clip, 5

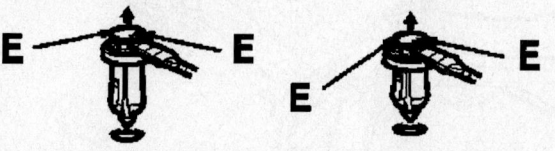

A. Fastener clip
B. Fastener clip
C. Front grille cover
D. Grille
E. Pry here for removal of inner clip

Fig. 70 Expanded view of bulkhead removal

22140_HCRV_G0042

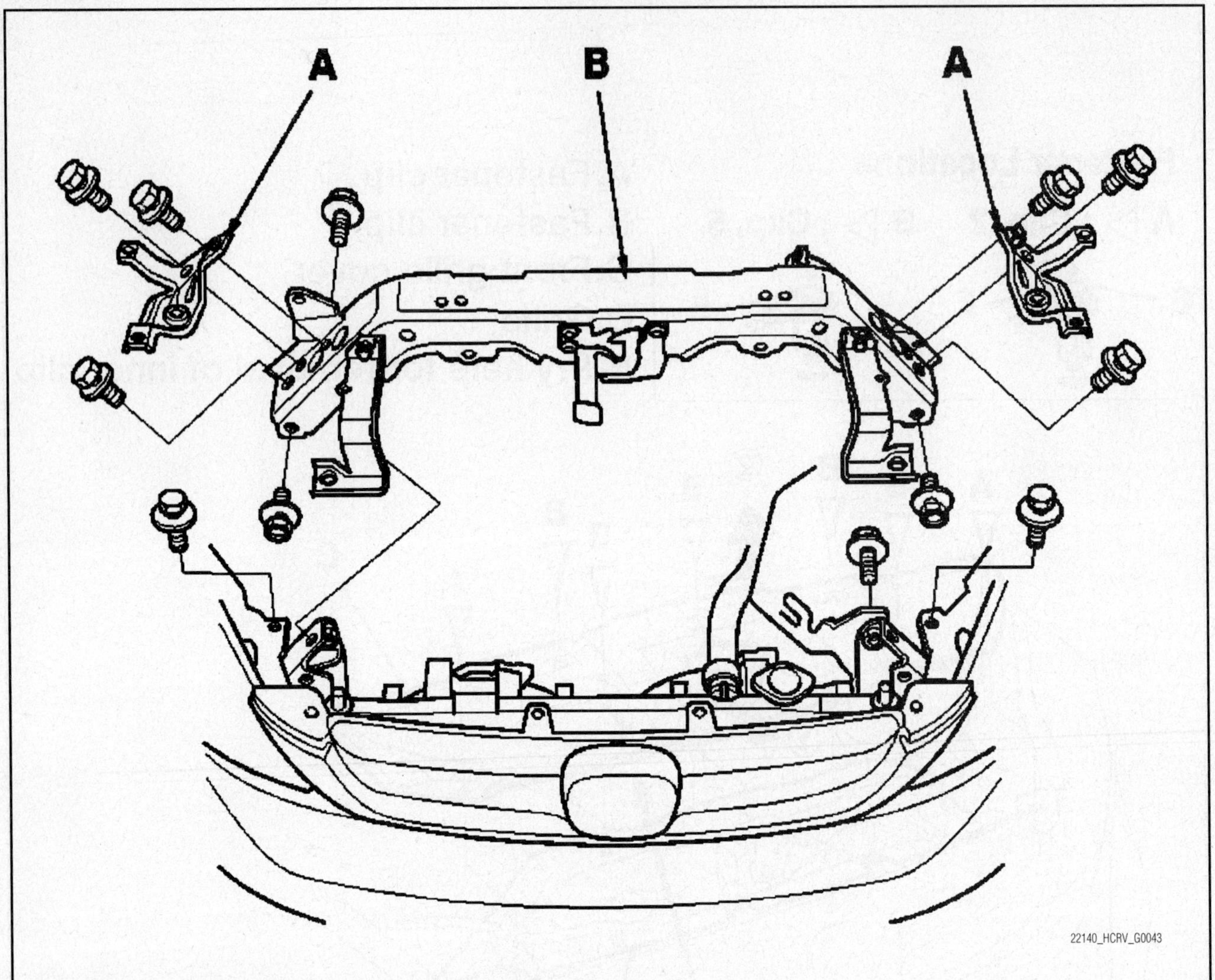

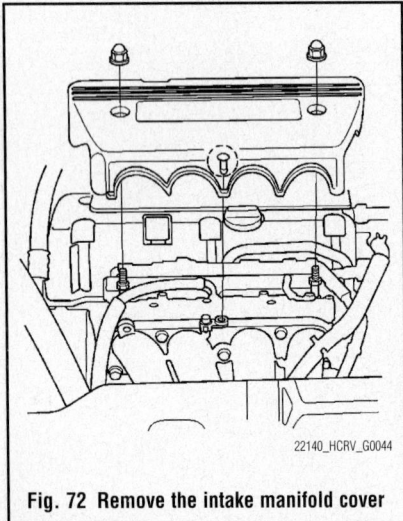

Fig. 71 Remove the radiator upper brackets (A), then remove the front bulkhead (B)

22140_HCRV_G0043

Fig. 72 Remove the intake manifold cover

22140_HCRV_G0044

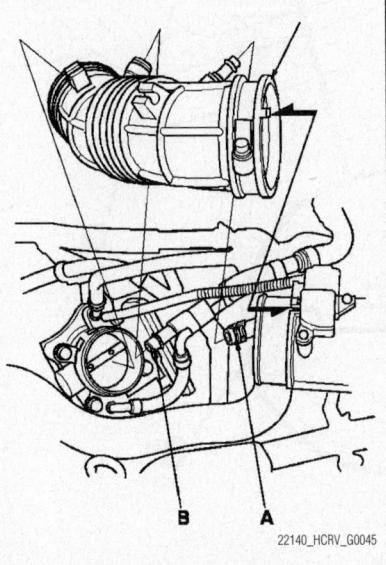

Fig. 73 Disconnect the vacuum hose (A) and breather pipe (B), then remove the intake air duct (C)

22140_HCRV_G0045

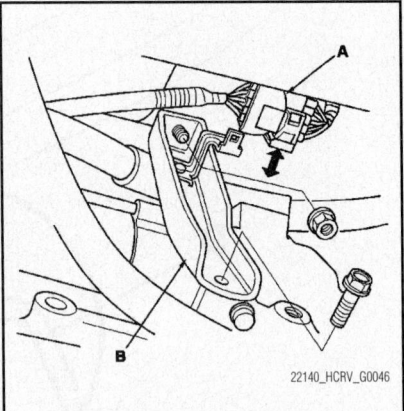

Fig. 74 Remove the connector (A), then remove the intake manifold bracket (B)

22140_HCRV_G0046

breather pipe (B), then remove the intake air duct (C).

8. Remove the following engine wire harness connectors and wire harness clamps from the intake manifold:

- Manifold Absolute Pressure (MAP) sensor connector
- Throttle actuator connector

9. Disconnect the Evaporative Emission (EVAP) canister hose and brake booster vacuum hose.

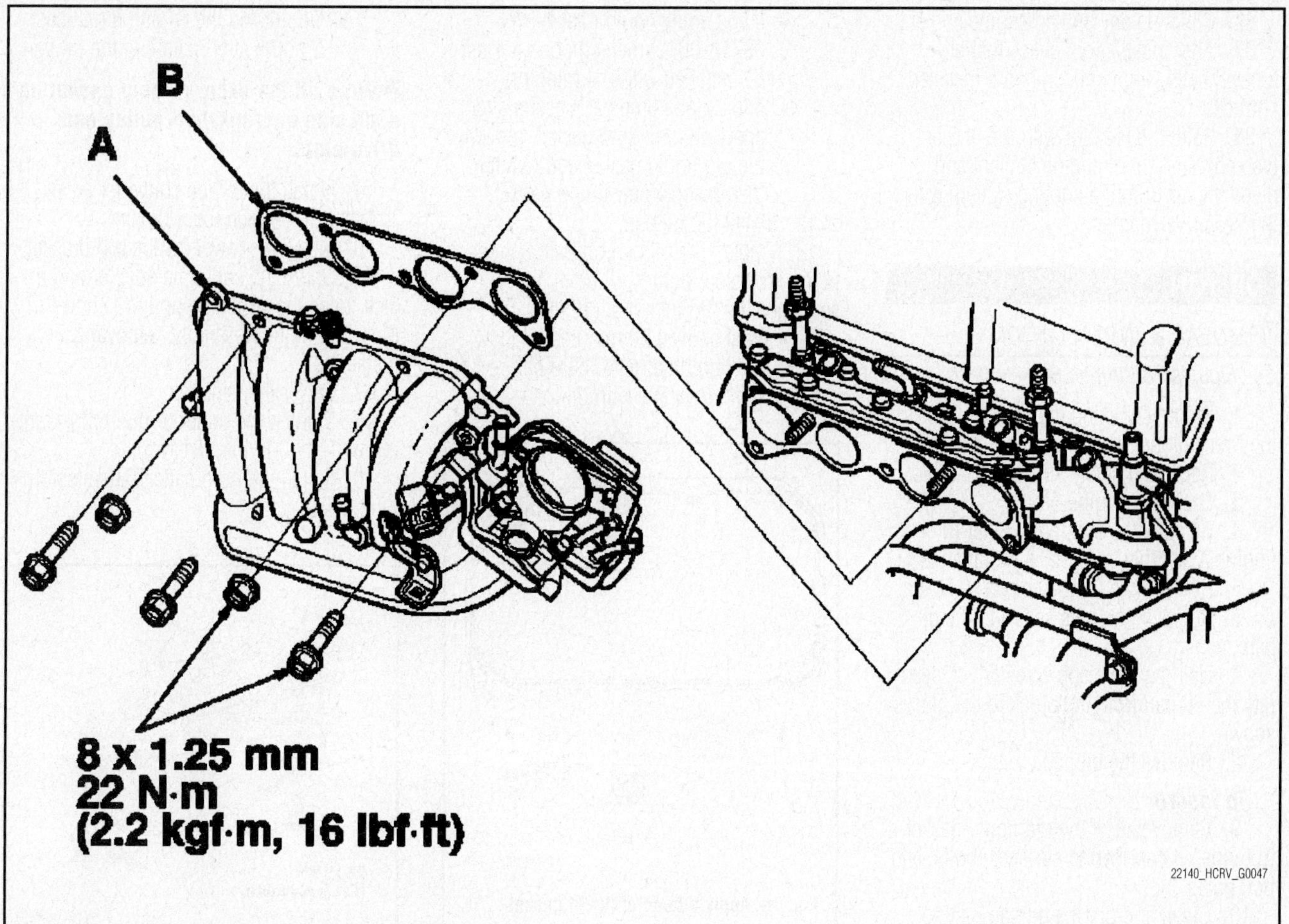

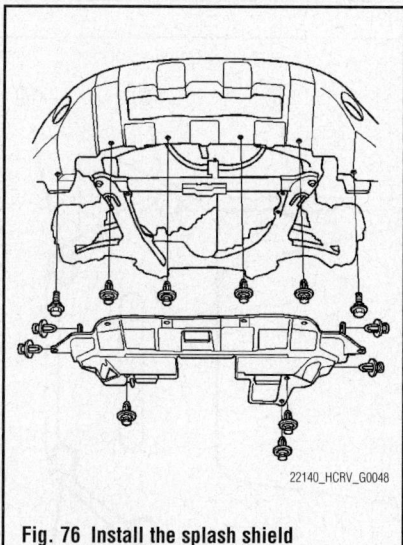

8 x 1.25 mm
22 N·m
(2.2 kgf·m, 16 lbf·ft)

22140_HCRV_G0047

Fig. 75 Install the intake manifold (A) with a new gasket (B)

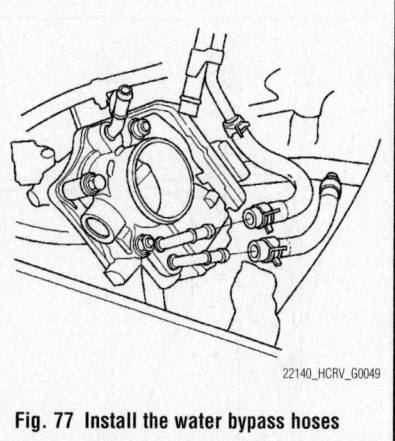

22140_HCRV_G0048

Fig. 76 Install the splash shield

22140_HCRV_G0049

Fig. 77 Install the water bypass hoses

10. Remove the harness bracket mounting bolts.

11. Disconnect the water bypass hoses, then plug the water bypass hoses.

12. Raise and safely support the vehicle.

13. Remove the splash shield.

14. Remove the connector (A), then remove the intake manifold bracket (B).

15. Lower the vehicle.

16. Remove the intake manifold.

To install:

17. Install the intake manifold (A) with a new gasket (B), and tighten the bolts and nuts in a crisscross pattern in 3 steps, beginning with the inner bolt. Tighten the bolts and nuts to 16 ft. lbs. (22 Nm).

18. Raise and safely support the vehicle.

19. Install the intake manifold bracket (B), then install the connector (A). Tighten

the bracket bolt and nut to 16 ft. lbs. (22 Nm).

20. Install the splash shield.

21. Lower the vehicle.

22. Install the water bypass hoses.

23. Install the EVAP canister hose and the brake booster vacuum hose.

24. Install the harness bracket mounting bolts.

25. Connect the engine wire harness connectors and install the wire harness clamps to the intake manifold including the:
- MAP sensor connector
- Throttle actuator connector

26. Install the intake air duct, then install the vacuum hose and the breather pipe.

27. Install the intake manifold cover and tighten the nuts to 106 inch lbs. (12 Nm).

28. Install the front bulkhead, then install the radiator upper brackets. Tighten the bolts to 86 inch lbs. (10 Nm).

29. Apply body paint to the bulkhead mounting bolts.

30. Connect the fan motor connectors and the hood switch connector, then install the harness clips.

31. Install the bulkhead cover.

32. Clean up any spilled engine coolant.

33. After installation, check that all tubes, hoses, and connectors are installed correctly.

34. Refill the radiator with the proper type and amount of engine coolant and bleed the air from the cooling system with the heater valve open.

OIL PAN

REMOVAL & INSTALLATION

See Figures 78 through 81.

1. Before servicing the vehicle, refer to the Precautions Section.

2. Raise and safely support the vehicle.

3. Drain the engine oil.

4. Remove the subframe. Refer to Engine Assembly, removal & installation.

5. Remove the torque converter cover.

6. Remove the bolts securing the oil pan.

7. Using a flat blade screwdriver, separate the oil pan from the block in the places shown.

8. Remove the oil pan.

To install:

9. Remove all of the old liquid gasket from the oil pan mating surfaces, bolts, and bolt holes.

10. Clean and dry the oil pan mating surfaces.

11. Apply liquid gasket (P/N 08717-0004, 08718-0001, 08718-0003, or 08718-0009) to the engine block mating surface of the oil pan and to the inside edge of the bolt holes. Install the component within 5 minutes of applying the liquid gasket.

- Apply a bead of liquid gasket about 0.12 inch (3mm) in diameter along the broken line (A), as illustrated.

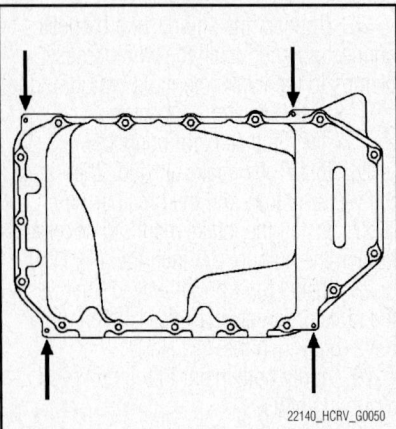

Fig. 78 Using a flat blade screwdriver, separate the oil pan from the block in the places shown

- If you apply liquid gasket P/N 08718-0012, the component must be installed within 4 minutes.
- If too much time has passed after applying the liquid gasket, remove the old liquid gasket and residue, then reapply new liquid gasket.

12. Install the oil pan.

13. Tighten the oil pan bolts in 3 steps. In the final step, tighten all bolts, in sequence, to 106 inch lbs. (12 Nm).

- Wait at least 30 minutes to allow the liquid gasket to cure before filling the engine with oil

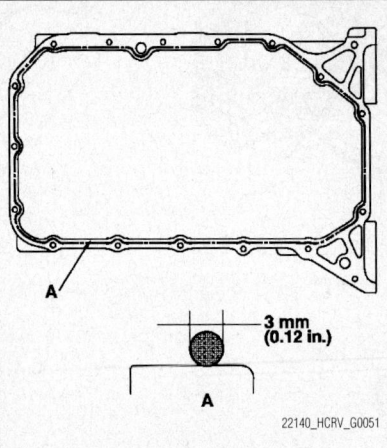

Fig. 79 Apply a bead of liquid gasket about 0.12 inch (3mm) in diameter along the broken line (A), as illustrated

- Do not run the engine for at least 3 hours after installing the oil pan

➡**Wipe off the excess liquid gasket on each side of crankshaft pulley and drive plate.**

14. Install the torque converter cover. Tighten to 106 inch lbs. (12 Nm).

15. Install the lower torque rod bracket.

16. Using the subframe adapter and a jack, raise the subframe up to the body. Refer to Engine Assembly, removal & installation.

17. Lower the vehicle.

18. Tighten the upper torque rod mounting bolt to 60 ft. lbs. (81 Nm).

19. Install the Automatic Transmission Fluid (ATF) filter.

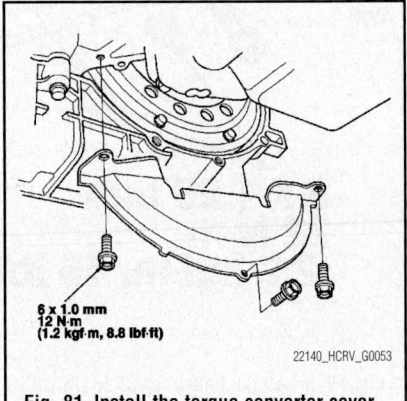

Fig. 81 Install the torque converter cover

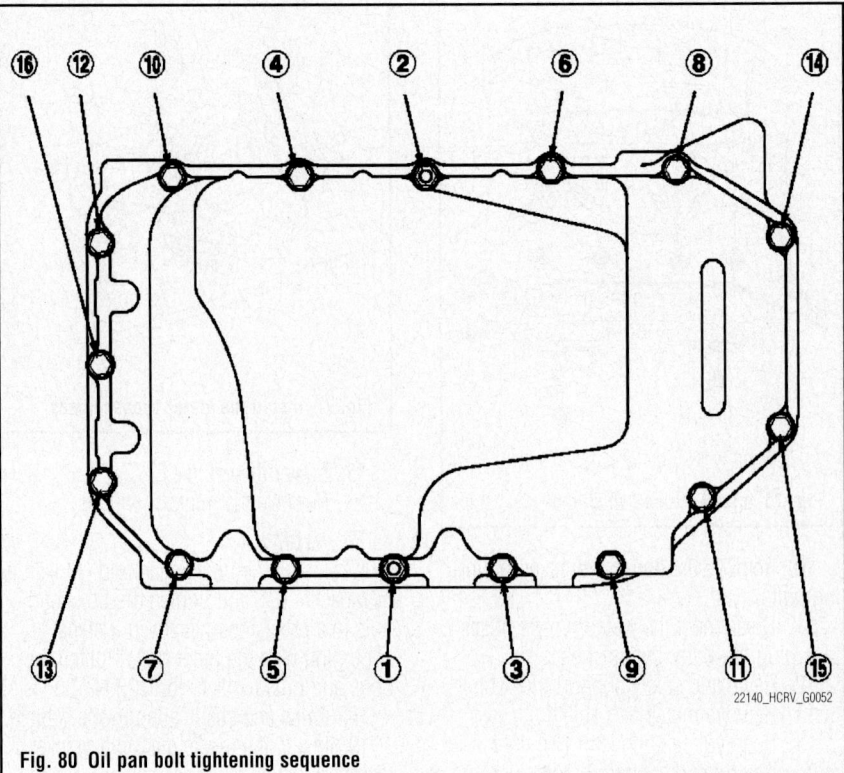

Fig. 80 Oil pan bolt tightening sequence

20. Install the bolts securing the left steering gearbox mounting bracket.

21. Install the bolts securing the right steering gearbox mounting bracket.

22. Install the Power Steering (P/S) fluid line bracket and secure the hose with the hose clamps.

23. Install the propeller shaft.

24. Install a new set ring on the end of each driveshaft, then install the driveshafts. Make sure each ring clicks into place in the differential and intermediate shaft.

25. Connect the lower arms to the knuckles.

26. Connect the stabilizer links to the stabilizer bar.

27. Install the shift cable.

28. Install the 3-way catalytic converter. Use new gaskets and new self-locking nuts. Connect the air fuel ratio sensor connector and secondary heated oxygen sensor connector.

29. Install the splash shield.

30. Install the front wheels.

31. Check the wheel alignment.

OIL PUMP

REMOVAL & INSTALLATION

See Figures 82 through 86.

1. Before servicing the vehicle, refer to the Precautions Section.

2. Turn the crankshaft pulley so the Top Dead Center (TDC) mark (A) lines up with the pointer (B).

3. Remove the oil pan. Refer to Oil Pan, removal & installation.

4. Remove and discard the oil pump chain tensioner.

5. To hold the rear balancer shaft, insert a 6 mm pin driver (A) into the maintenance hole in the lower balancer shaft holder and through the rear balancer shaft.

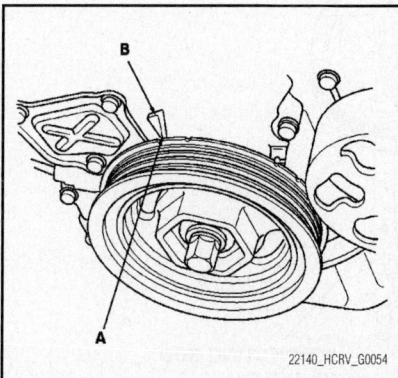

22140_HCRV_G0054

Fig. 82 Turn the crankshaft pulley so the Top Dead Center (TDC) mark (A) lines up with the pointer (B)

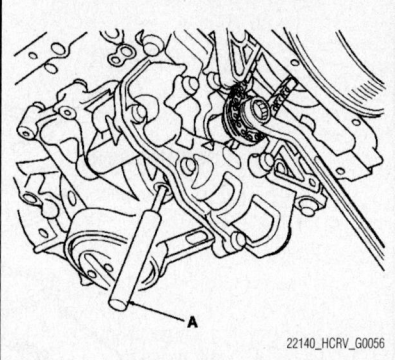

22140_HCRV_G0056

Fig. 83 Hold the rear balancer shaft with a 6 mm pin driver (A) inserted into the maintenance hole in the lower balancer shaft holder and through the rear balancer shaft

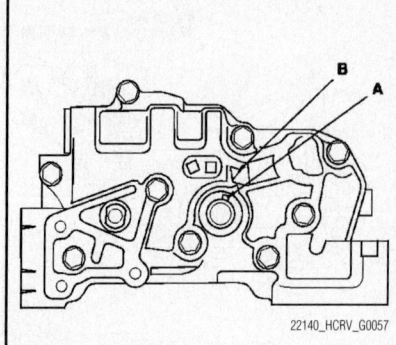

22140_HCRV_G0057

Fig. 84 Align the dowel pin (A) on the rear balancer shaft with the mark (B) on the oil pump

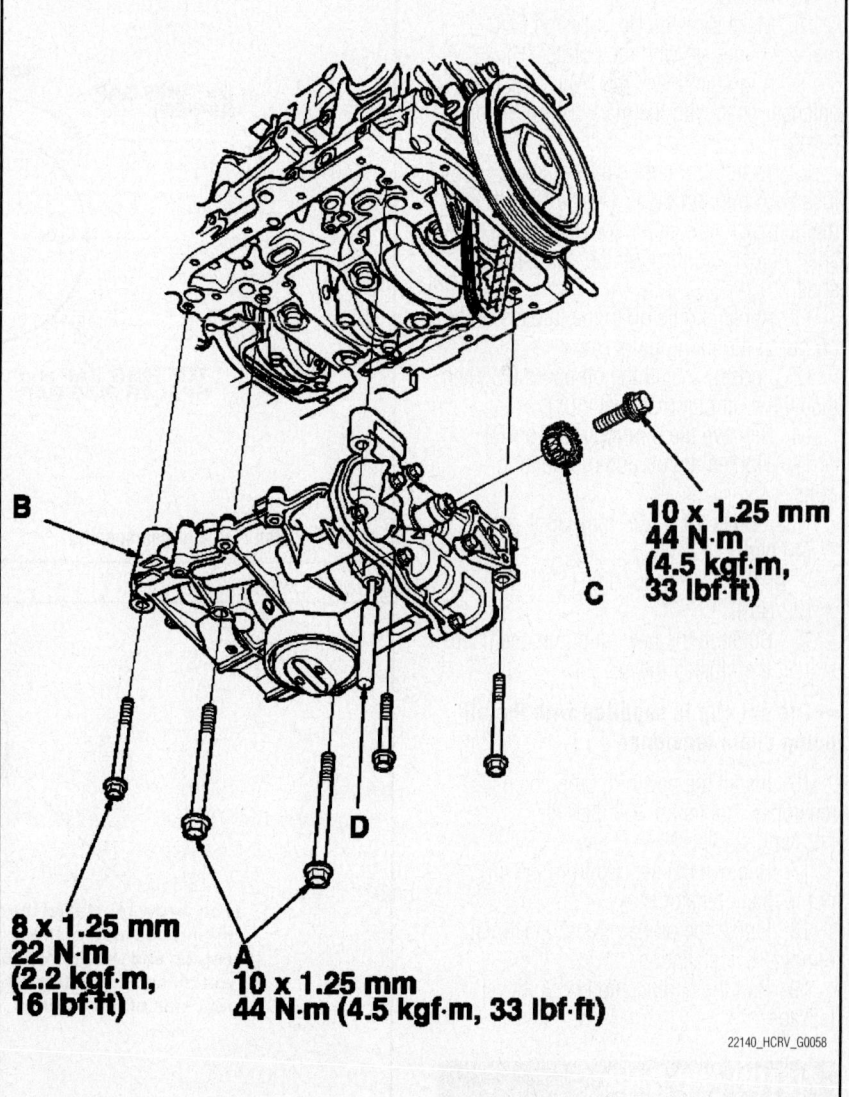

**10 x 1.25 mm
44 N·m
(4.5 kgf·m,
33 lbf·ft)**

**8 x 1.25 mm
22 N·m
(2.2 kgf·m,
16 lbf·ft)**

**10 x 1.25 mm
44 N·m (4.5 kgf·m, 33 lbf·ft)**

22140_HCRV_G0058

Fig. 85 Exploded view of the oil pump and related components

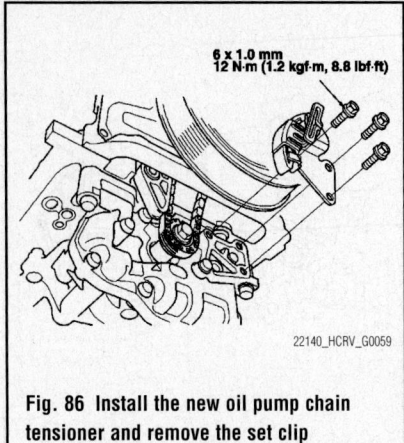

Fig. 86 Install the new oil pump chain tensioner and remove the set clip

6. Loosen the oil pump sprocket mounting bolt.

7. Remove the oil pump sprocket, then remove the oil pump.

To install:

8. Make sure the No. 1 piston TDC mark (A) lines up with the pointer (B).

9. Align the dowel pin (A) on the rear balancer shaft with the mark (B) on the oil pump.

10. To hold the rear balancer shaft, insert a 6 mm pin driver (A) into the maintenance hole in the lower balancer shaft holder and through the rear balancer shaft.

11. Apply engine oil to the threads of the oil pump mounting bolts (A).

12. Loosely install the oil pump (B), then install the oil pump sprocket (C).

13. Remove the 6 mm pin driver (D).

14. Tighten the oil pump mounting bolts:

   a. Bolts 10 x 1.25 mm: 33 ft. lbs. (44 Nm).

   b. Bolts 8 x 1.25 mm: 16 ft. lbs. (22 Nm).

15. Squeeze the new oil pump chain tensioner, then install the set clip.

➡**The set clip is supplied with the oil pump chain tensioner.**

16. Install the new oil pump chain tensioner. Tighten to 106 inch lbs. (12 Nm).

17. Remove the set clip from the oil pump chain tensioner.

18. Install the oil pan. Refer to Oil Pan, removal & installation.

19. Run the engine and check for oil leakage.

PISTON AND RING

POSITIONING

See Figures 87 through 89.

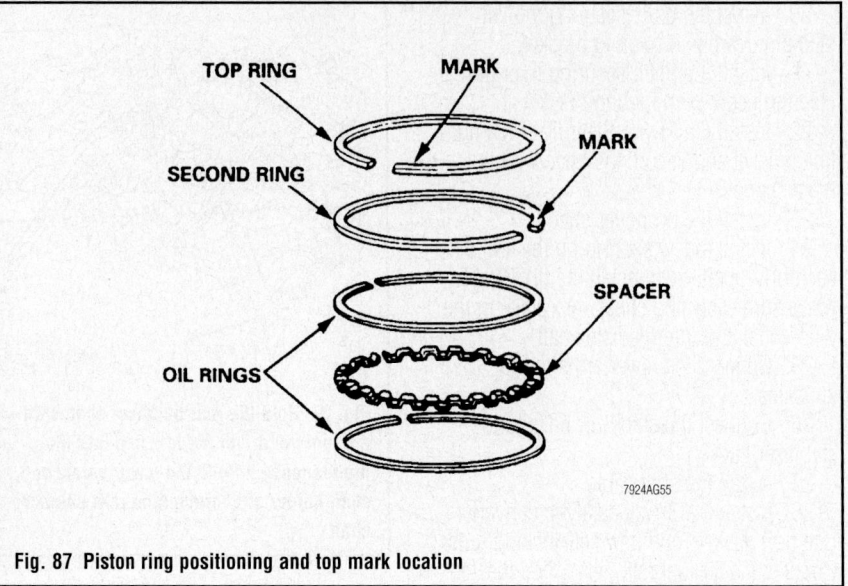

Fig. 87 Piston ring positioning and top mark location

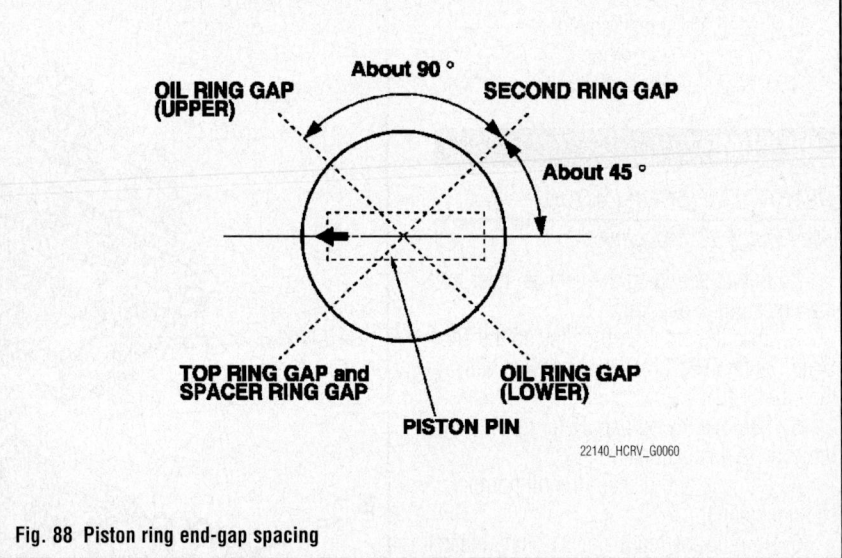

Fig. 88 Piston ring end-gap spacing

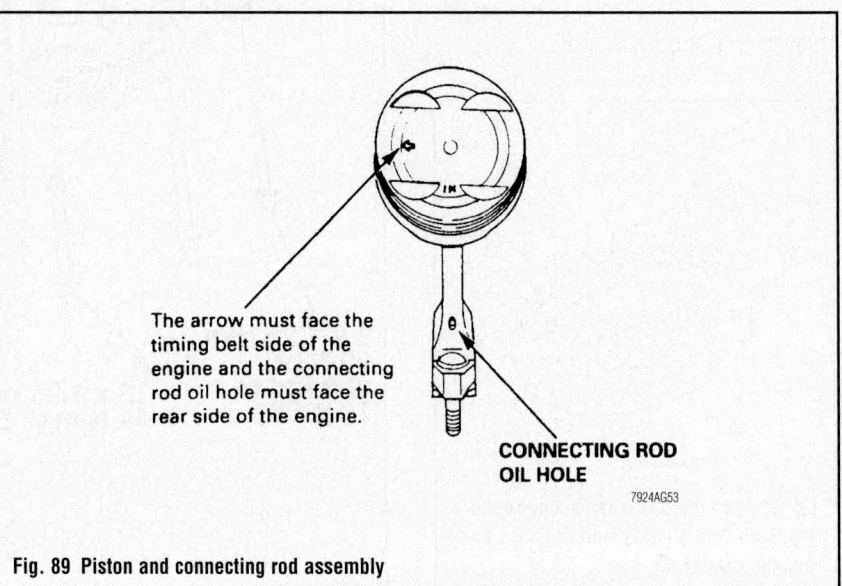

The arrow must face the timing belt side of the engine and the connecting rod oil hole must face the rear side of the engine.

Fig. 89 Piston and connecting rod assembly

REAR MAIN SEAL

REMOVAL & INSTALLATION

Special tools required:
• Driver handle, 15 x 135L 07749-0010000
• Oil seal driver attachment 96 mm 07ZAD-PNAA100

1. Before servicing the vehicle, refer to the Precautions Section.
2. Remove the transaxle. Refer to Automatic Transaxle Assembly, removal & installation.
3. Remove the drive plate.
4. Use a suitable tool to remove the old crankshaft oil seal.

To install:

5. Clean and dry the crankshaft oil seal housing.
6. Use the driver and attachment to drive a new oil seal squarely into the block to the specified installed height of 0.001–0.047 inch (0.2–1.2 mm) from flush.
7. Install the drive plate.
8. Install the transaxle. Refer to Automatic Transaxle Assembly, removal & installation.

TIMING CHAIN, SPROCKETS, FRONT COVER AND SEAL

REMOVAL & INSTALLATION

See Figures 90 through 102.

1. Before servicing the vehicle, refer to the Precautions Section.

➡**Keep the camshaft chain away from magnetic fields.**

2. Remove the front wheels.
3. Remove the splash shield.

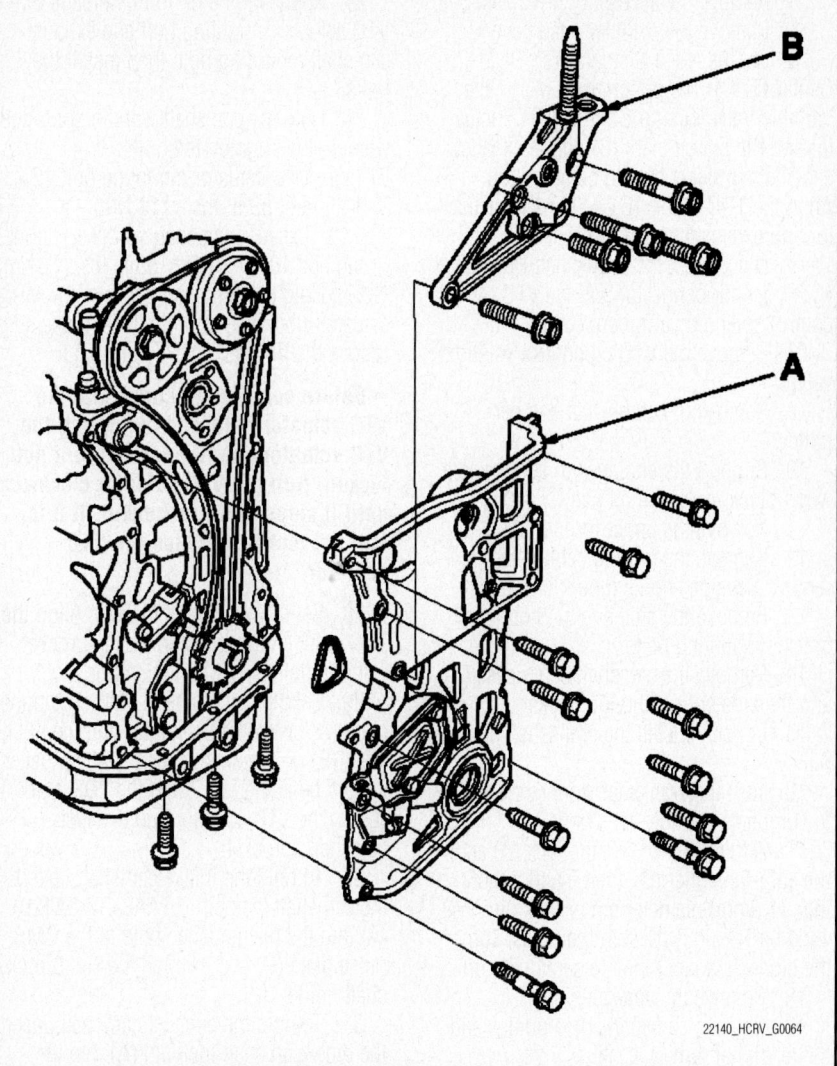

22140_HCRV_G0064

Fig. 91 Remove the camshaft chain case (A) and the side engine mount bracket (B)

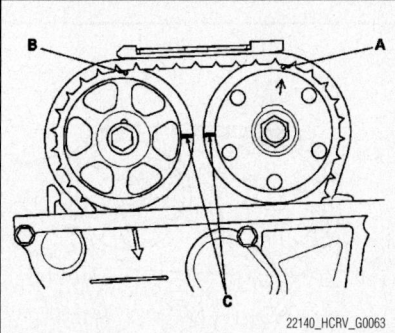

22140_HCRV_G0063

Fig. 90 Set the No. 1 piston at TDC. The punch mark (A) on the VTC actuator and the punch mark (B) on the exhaust camshaft sprocket should be at the top. Align the TDC marks (C) on the VTC actuator and exhaust camshaft sprocket.

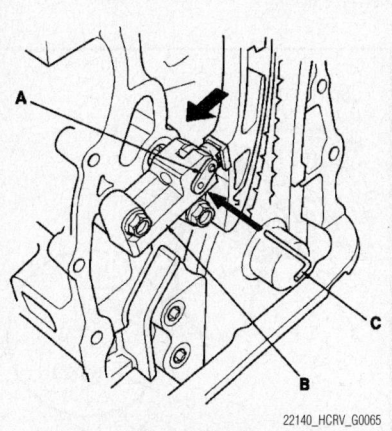

22140_HCRV_G0065

Fig. 92 Align the holes on the lock (A) and the auto-tensioner (B), then insert a 0.05 inch (1.2mm) diameter pin or lock pin (P/N 14511-PNA-003) (C) into the holes. Turn the crankshaft clockwise to secure the pin.

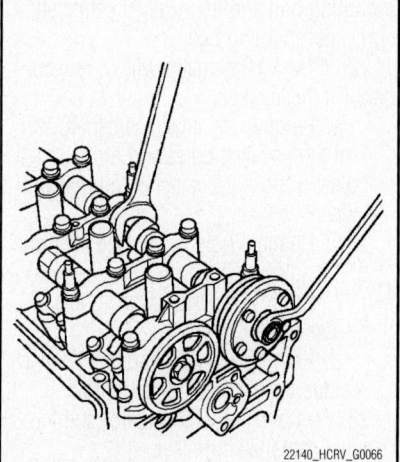

22140_HCRV_G0066

Fig. 93 Removing the camshaft sprockets holding the camshaft with an open-end wrench

4. Remove the accessory drive belt.

5. Remove the cylinder head cover.

6. Set the No. 1 piston at Top Dead Center (TDC). The punch mark (A) on the Variable Valve Timing Control (VTC) actuator and the punch mark (B) on the exhaust camshaft sprocket should be at the top. Align the TDC marks (C) on the VTC actuator and exhaust camshaft sprocket.

7. Disconnect the Crankshaft Position (CKP) sensor connector and the VTC oil control solenoid valve connector.

8. Remove the VTC oil control solenoid valve.

9. Remove the crankshaft damper pulley.

10. Support the engine with a jack and a wood block under the oil pan.

11. Remove the upper torque rod.

12. Remove the ground cable, then remove the side engine mount bracket.

13. Remove the side engine mount bracket mounting bolts.

14. Remove the camshaft chain case (A) and the side engine mount bracket (B).

15. Loosely install the crankshaft damper pulley.

16. Turn the crankshaft counterclockwise to compress the auto-tensioner.

17. Align the holes on the lock (A) and the auto-tensioner (B), then insert a 0.05 inch (1.2mm) diameter pin or lock pin (P/N 14511-PNA-003) (C) into the holes. Turn the crankshaft clockwise to secure the pin.

18. Remove the auto-tensioner.

19. Remove camshaft chain guides and the tensioner arm.

20. Remove the camshaft chain.

21. Hold the camshaft with an open-end wrench, then loosen the VTC actuator mounting bolt and the exhaust camshaft sprocket mounting bolt.

22. If the VTC actuator will be reused, perform these steps:

    a. Remove the intake camshaft, and seal the advance holes and the retard holes in the No 1 camshaft journal with tape.

    b. Punch a hole in the tape over one of the advance holes.

    c. Apply air to the advance hole to release the lock.

    d. Remove the tape and any adhesive residue from the camshaft journal.

23. Remove the VTC actuator and the exhaust camshaft sprocket.

To install:

24. Install the VTC actuator and the exhaust camshaft sprocket.

➡**Install the VTC actuator to unlock position.**

25. Apply engine oil to the threads of the VTC actuator mounting bolt and exhaust camshaft mounting bolt, then install the bolts.

26. Hold the camshaft with an open-end wrench, then tighten the bolts:

    a. VTC actuator mounting bolt 12 x 1.25 mm: 83 ft. lbs. (113 Nm).

    b. Exhaust camshaft sprocket mounting bolt 10 x 1.25 mm: 53 ft. lbs. (72 Nm).

27. Hold the camshaft and turn the VTC actuator clockwise until it clicks. Make sure to lock the VTC actuator by turning it.

➡**Before continuing, check that the VTC actuator is locked by turning the VTC actuator counterclockwise. If not locked, turn the VTC actuator clockwise until it stops, then recheck it. If it is still not locked, replace the VTC actuator.**

28. Set the crankshaft to TDC. Align the TDC mark (A) on the crankshaft sprocket with the pointer (B) on the engine block.

29. Set the camshafts to TDC. The punch mark (A) on the VTC actuator and the punch mark (B) on the exhaust camshaft sprocket should be at the top. Align the TDC marks (C) on the VTC actuator and exhaust camshaft sprocket.

30. To hold the intake camshaft, insert the camshaft lock pin (07AAB-RWCA120) (A) into the maintenance hole in the CMP pulse plate (B) and through the No. 5 rocker shaft holder (C).

31. To hold the exhaust camshaft, insert the other camshaft lock pin (A) into the maintenance hole in the CMP pulse plate (D) and through No. 5 rocker shaft holder (C).

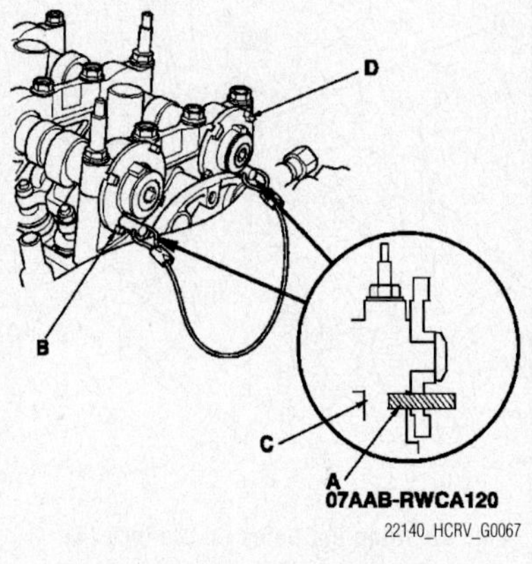

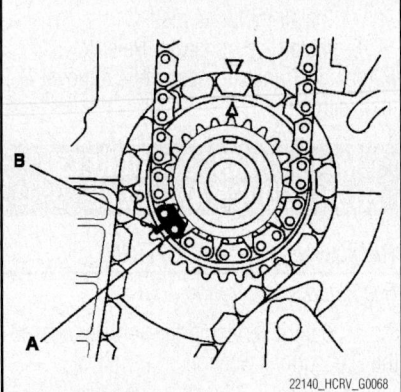

Fig. 94 Insert a camshaft lock pin (07AAB-RWCA120) (A) into the maintenance hole in the CMP pulse plate (B) and through the No. 5 rocker shaft holder (C)

32. Install the camshaft chain on the crankshaft sprocket with the colored link plate (A) aligned with the mark (B) on the crankshaft sprocket.

33. Install the camshaft chain on the VTC actuator and the exhaust camshaft sprocket with the punch marks (A) aligned with the center of the 2 colored link plates (B).

34. Install the camshaft chain guide (A) and the tensioner arm (B). Tighten the mounting bolts:

    a. Bolts 6 x 1.0 mm: 106 inch lbs. (12 Nm).

    b. Bolt 8 x 1.25 mm: 16 ft. lbs. (22 Nm).

35. Compress the auto-tensioner when replacing the camshaft chain. Remove the pin (P/N 14511-PNA-003) (A) from the auto-tensioner that was installed during removal. Turn the plate (B) counterclockwise, to release the lock, then press the rod

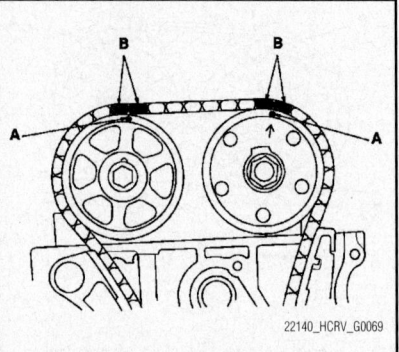

Fig. 95 Install the camshaft chain on the crankshaft sprocket with the colored link plate (A) aligned with the mark (B) on the crankshaft sprocket

Fig. 96 Install the camshaft chain on the VTC actuator and the exhaust camshaft sprocket with the punch marks (A) aligned with the center of the 2 colored link plates (B)

(C), and set the first cam (D) to the edge of the rack (E). Insert the 0.05 inch (1.2mm) diameter pin or lock pin into the holes (F).

✳✳ WARNING

If the chain tensioner is not set up as described, the tensioner will become damaged.

36. Install the auto-tensioner and tighten the bolts to 106 inch lbs. (12 Nm).

37. Install the camshaft chain guide and tighten the mounting bolts to 16 ft. lbs. (22 Nm).

38. Remove the pin or lock pin (P/N 14511-PNA-003) from the auto-tensioner.

39. Remove the camshaft lock pin set.

40. Check the chain case oil seal for damage. If the oil seal is damaged, replace the chain case oil seal.

41. Remove old liquid gasket from the chain case mating surfaces, bolts, and bolt holes.

42. Clean and dry the chain case mating surfaces.

43. Apply liquid gasket (P/N 08717-0004, 08718-0001, 08718-0003, or 08718-0009) to the engine block mating surface of the chain case and to the inside edge of the bolt holes. Install the component within 5 minutes of applying the liquid gasket.

- Apply a bead of liquid gasket about 0.12 inch (3 mm) in diameter along the broken line (A).

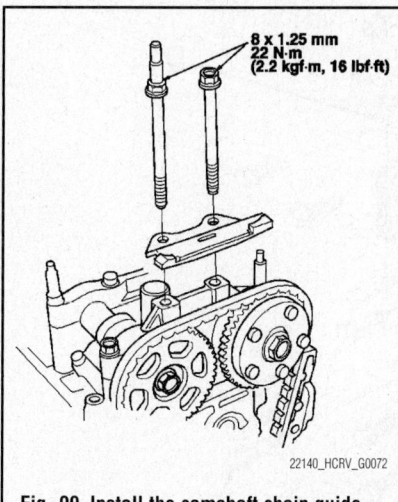

Fig. 99 Install the camshaft chain guide

- If you apply liquid gasket P/N 08718-0012, the component must be installed within 4 minutes.
- If too much time has passed after applying the liquid gasket, remove the old liquid gasket and residue, then reapply new liquid gasket.

44. Apply liquid gasket to the engine block upper surface contact areas (B) and the lower block upper surface contact areas (C) on the chain case.

➡**Apply about 0.43 inch (11 mm) diameter and about 0.12 inch (3 mm) thickness of liquid gasket to the areas (B) and (C).**

45. Apply liquid gasket (P/N 08717-0004, 08718-0001, 08718-0003, or 08718-0009) to the oil pan mating surface of the oil pump. Install the component within 5 minutes of applying the liquid gasket.

46. Install a new O-ring (A), the side engine mount bracket (B), and the mounting bolts (C) on the chain case. Set the edge of the chain case (D) to the edge of the oil pan (E), then install the chain case on the

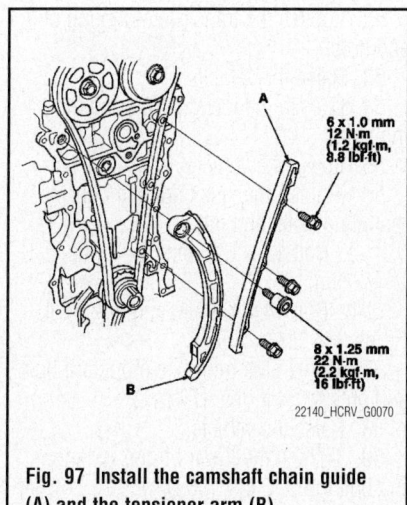

Fig. 97 Install the camshaft chain guide (A) and the tensioner arm (B)

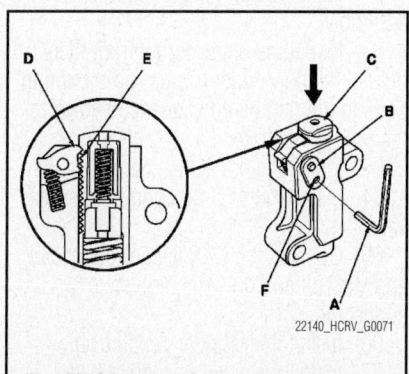

Fig. 98 Compress the auto-tensioner as illustrated using a lock pin

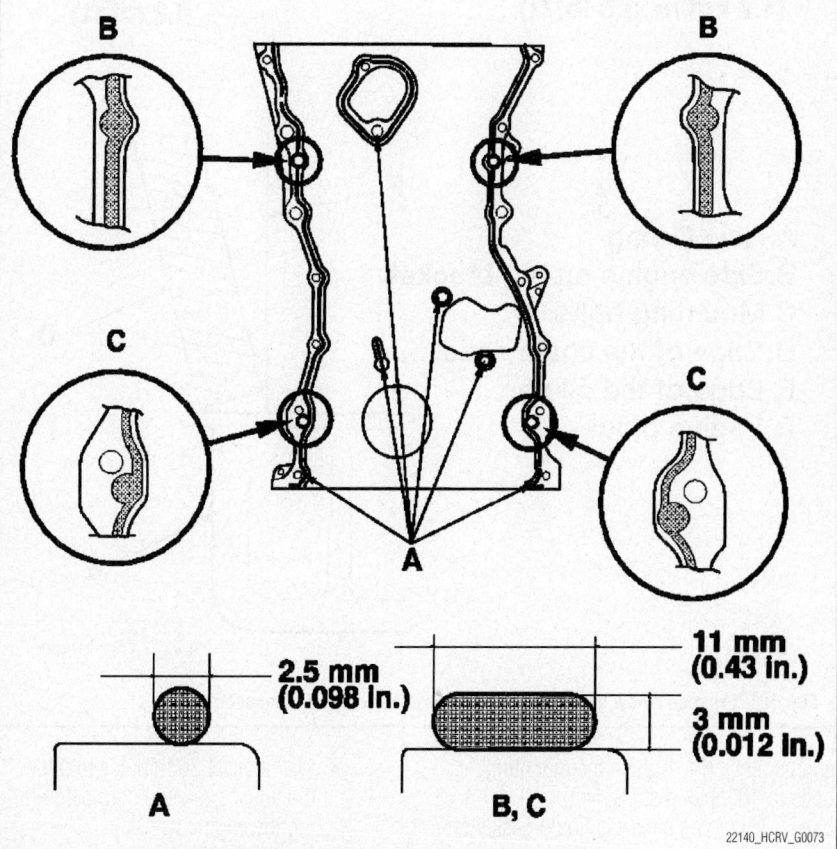

Fig. 100 Apply liquid gasket to the engine block mating surface of the chain case and to the inside edge of the bolt holes as illustrated

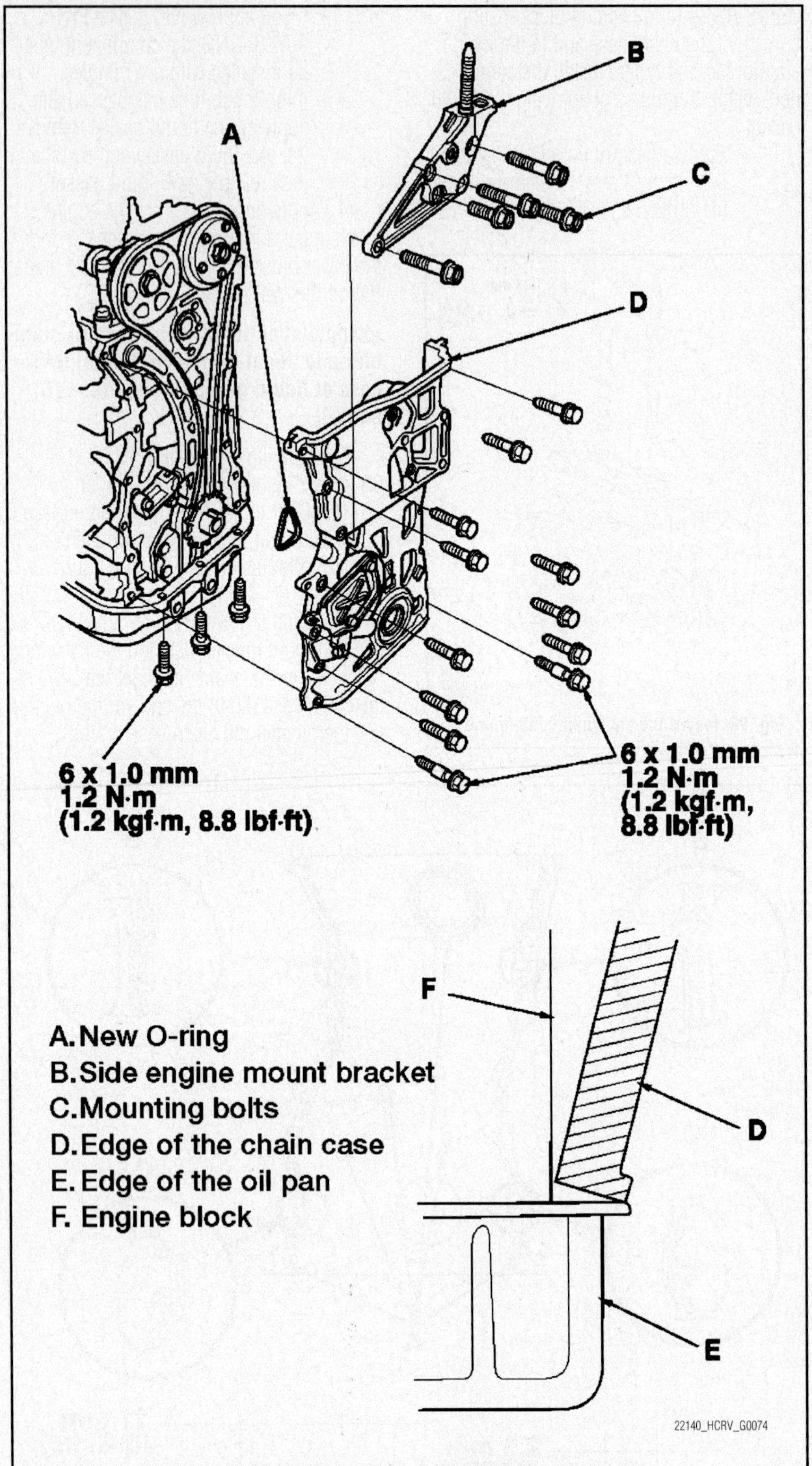

B

A

C

D

**6 x 1.0 mm
1.2 N·m
(1.2 kgf·m, 8.8 lbf·ft)**

**6 x 1.0 mm
1.2 N·m
(1.2 kgf·m,
8.8 lbf·ft)**

F

D

E

A. New O-ring
B. Side engine mount bracket
C. Mounting bolts
D. Edge of the chain case
E. Edge of the oil pan
F. Engine block

22140_HCRV_G0074

Fig. 101 Expanded view of the timing chain case and related components

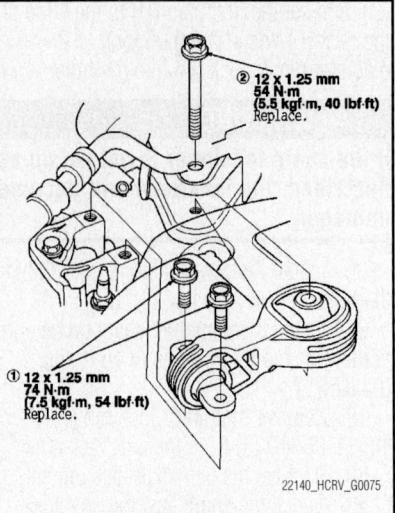

② 12 x 1.25 mm
54 N·m
(5.5 kgf·m, 40 lbf·ft)
Replace.

① 12 x 1.25 mm
74 N·m
(7.5 kgf·m, 54 lbf·ft)
Replace.

22140_HCRV_G0075

Fig. 102 Install the upper torque rod, then tighten the mounting bolts in the numbered sequence shown

49. Install the side engine mount bracket, then loosely tighten the new bolt and nut.

50. Install the ground cable.

51. Remove the air cleaner housing assembly.

52. Loosen the transaxle mounting bolt and nuts.

53. Raise the vehicle.

54. Loosen the lower torque rod mounting bolt.

55. Lower the vehicle.

56. Tighten the side engine mount mounting bolts and nut.
   a. Bolt 12 x 1.25 mm: 52 ft. lbs. (72 Nm).
   b. Bolt 14 x 1.5 mm and nut: 54 ft. lbs. (74 Nm).

57. Tighten the transaxle mounting bolt and nuts to 54 ft. lbs. (74 Nm).

58. Raise the vehicle.

59. Tighten the lower torque rod mounting bolt to 69 ft. lbs. (93 Nm).

60. Lower the vehicle.

61. Install the air cleaner housing assembly.

62. Install the upper torque rod, then tighten the new upper torque rod mounting bolts in the numbered sequence shown to the torque illustrated.

63. Install the crankshaft damper pulley.

64. Install the VTC oil control solenoid valve.

65. Connect the CKP sensor connector and VTC oil control solenoid valve connector.

66. Install the cylinder head cover.

67. Install the accessory drive belt.

68. Perform the CKP pattern clear/CKP learn procedure.

engine block (F). Tighten the mounting bolts to 106 inch lbs. (12 Nm).

47. Wipe off the excess liquid gasket on the oil pan and chain case mating area.

 • When installing the chain case, do not slide the bottom surface onto the oil pan mounting surface.

 • Wait at least 30 minutes to allow the liquid gasket to cure before filling the engine with oil.

 • Do not run the engine for at least 3 hours after installing the chain case.

48. Tighten the side engine mount bracket mounting bolts to 33 ft. lbs. (44 Nm).

CKP Pattern Clear/CKP Pattern Learn

With Honda Diagnostic System (HDS)

1. Connect the Honda Diagnostic System (HDS) to the Data Link Connector (DLC) located under the driver's side of the dashboard.

2. Turn the ignition switch to ON.

3. Make sure the HDS communicates with the PCM and other vehicle systems. If it does not, go to the DLC circuit troubleshooting.

4. Select CRANK PATTERN in the ADJUSTMENT MENU with the HDS.

5. Select CRANK PATTERN LEARNING with the HDS and follow the screen prompts.

Without Honda Diagnostic System (HDS)

1. Start the engine.

2. Hold the engine speed at 3,000 RPM without load (in P or N) until the radiator fan comes on.

3. Test-drive the vehicle on a level road: decelerate (with the throttle fully closed) from an engine speed of 2,500 RPM down to 1,000 RPM with the transaxle in 2 position.

4. Repeat step 3 several times.

5. Turn the ignition switch to LOCK.

6. Turn the ignition switch to ON and wait 30 seconds.

VALVE LASH

ADJUSTMENT

See Figures 103 and 104.

Special tools required:
- Adjuster 07MAA-PR70110
- Locknut wrench 07MAA-PR70120

Adjust the valves only when the cylinder head temperature is less than 100°F (38°C).

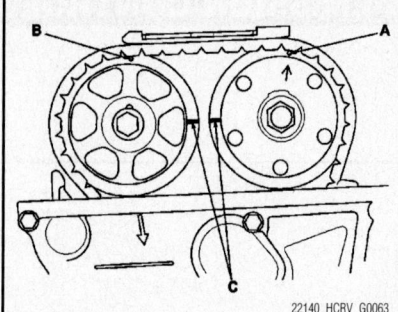

Fig. 103 Set the No. 1 piston at TDC. The punch mark (A) on the VTC actuator and the punch mark (B) on the exhaust camshaft sprocket should be at the top. Align the TDC marks (C) on the VTC actuator and exhaust camshaft sprocket.

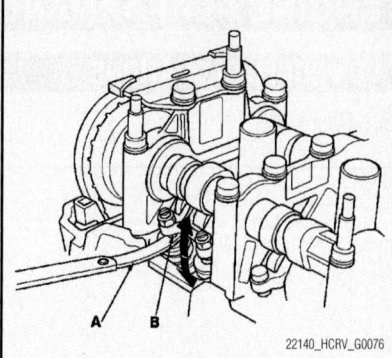

Fig. 104 Insert the feeler gauge (A) between the adjusting screw (B) and the end of the valve stem, and slide it back and forth; there should be a slight amount of drag

1. Before servicing the vehicle, refer to the Precautions Section.

2. Remove the cylinder head cover.

3. Set the No. 1 piston at Top Dead Center (TDC). The punch mark (A) on the Variable Valve Timing Control (VTC) actuator and the punch mark (B) on the exhaust camshaft sprocket should be at the top. Align the TDC marks (C) on the VTC actuator and the exhaust camshaft sprocket.

4. Select the correct feeler gauge for the valves to be checked.

   a. Intake valve clearance: 0.008–0.010 inch (0.21–0.25 mm).

   b. Exhaust valve clearance: 0.011–0.013 inch (0.28–0.32 mm).

5. Insert the feeler gauge (A) between the adjusting screw (B) and the end of the valve stem, and slide it back and forth; there should be a slight amount of drag.

6. If there is too much or too little drag, loosen the locknut with the locknut wrench and adjuster, and turn the adjusting screw until the drag on the feeler gauge is correct.

7. Tighten the locknut to the specified torque and recheck the clearance. Repeat the adjustment if necessary. Apply engine oil to the nut threads.

   a. Intake locknut torque: 14 ft. lbs. (20 Nm).

   b. Exhaust locknut torque: 10 ft. lbs. (14 Nm).

8. Rotate the crankshaft 180° clockwise (the camshaft pulley turns 90°).

9. Check and, if necessary, adjust the valve clearance on the No. 3 cylinder.

10. Rotate the crankshaft 180° clockwise (the camshaft pulley turns 90°).

11. Check and, if necessary, adjust the valve clearance on the No. 4 cylinder.

12. Rotate the crankshaft 180° clockwise (the camshaft pulley turns 90°).

13. Check and, if necessary, adjust the valve clearance on the No. 2 cylinder.

14. Install the cylinder head cover.

ENGINE PERFORMANCE & EMISSION CONTROL

COMPONENT LOCATIONS

See Figure 105.

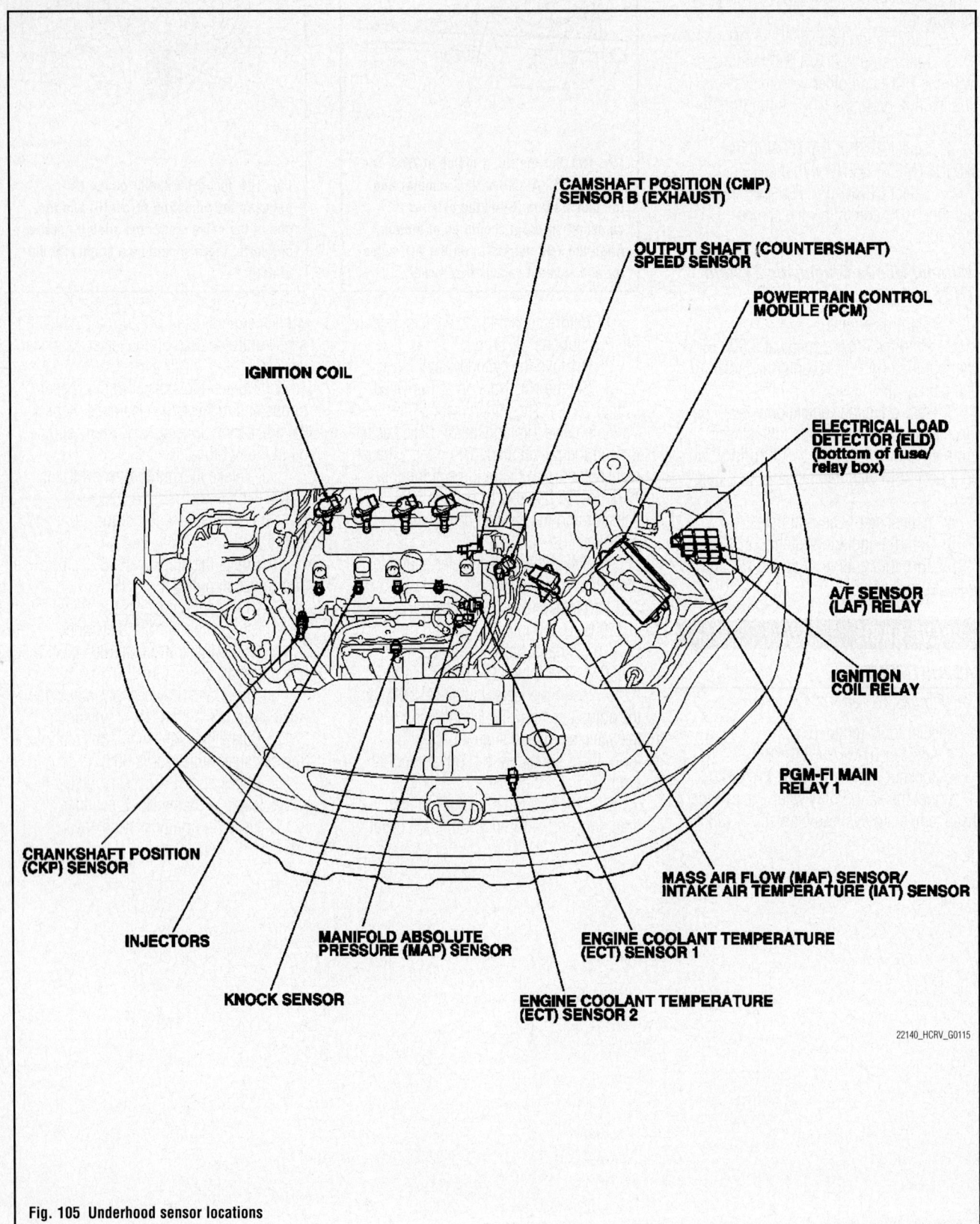

CAMSHAFT POSITION (CMP)
SENSOR B (EXHAUST)

OUTPUT SHAFT (COUNTERSHAFT)
SPEED SENSOR

POWERTRAIN CONTROL
MODULE (PCM)

IGNITION COIL

ELECTRICAL LOAD
DETECTOR (ELD)
(bottom of fuse/
relay box)

A/F SENSOR
(LAF) RELAY

IGNITION
COIL RELAY

PGM-FI MAIN
RELAY 1

CRANKSHAFT POSITION
(CKP) SENSOR

MASS AIR FLOW (MAF) SENSOR/
INTAKE AIR TEMPERATURE (IAT) SENSOR

INJECTORS

MANIFOLD ABSOLUTE
PRESSURE (MAP) SENSOR

ENGINE COOLANT TEMPERATURE
(ECT) SENSOR 1

KNOCK SENSOR

ENGINE COOLANT TEMPERATURE
(ECT) SENSOR 2

22140_HCRV_G0115

Fig. 105 Underhood sensor locations

CAMSHAFT POSITION (CMP) SENSOR

LOCATION

See Figures 106 and 107.

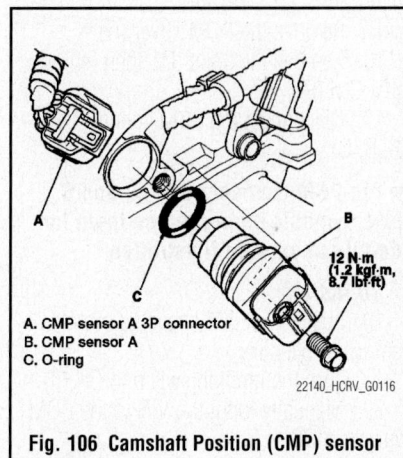

A. CMP sensor A 3P connector
B. CMP sensor A
C. O-ring

22140_HCRV_G0116

Fig. 106 Camshaft Position (CMP) sensor A location

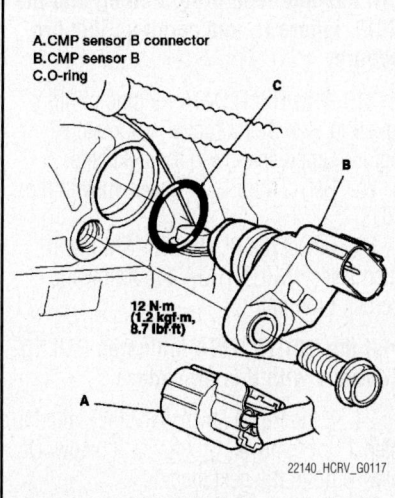

A. CMP sensor B connector
B. CMP sensor B
C. O-ring

22140_HCRV_G0117

Fig. 107 Camshaft Position (CMP) sensor B location

REMOVAL & INSTALLATION

See Figures 106 through 108.

1. Before servicing the vehicle, refer to the Precautions Section.
2. Remove the air cleaner.
3. Relieve the fuel pressure.
4. Remove the fuel line (A) and disconnect the hose (B).
5. Disconnect the CMP sensor A connector (C).
6. Remove the EGR valve 6P connector (D).
7. Remove the EGR valve (E).
8. Disconnect the CMP sensor A 3P connector (A).

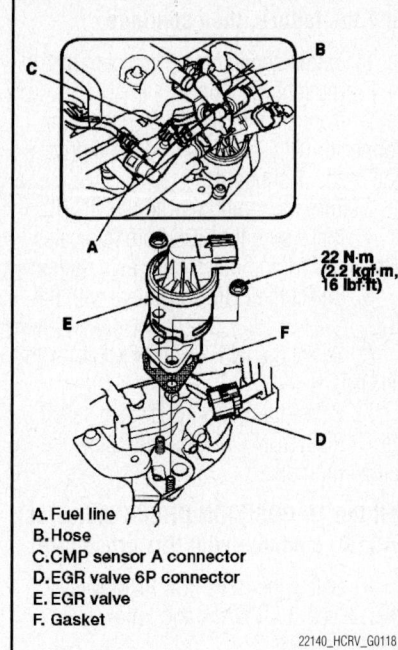

A. Fuel line
B. Hose
C. CMP sensor A connector
D. EGR valve 6P connector
E. EGR valve
F. Gasket

22140_HCRV_G0118

Fig. 108 Exhaust Gas Recirculation (EGR) valve location

9. Remove CMP sensor A (B) from the intake camshaft side of the cylinder head.
10. Disconnect the CMP sensor B connector (A).
11. Remove CMP sensor B (B).

To install:

12. Install the CMP sensor B with a new O-ring (C) and tighten the mounting bolt to 104 inch lbs. (12 Nm).
13. Connect the CMP sensor B connector (A).
14. Install CMP sensor A (B) to the intake camshaft side of the cylinder head with a new O-ring (C) and tighten the mounting bolt to 104 inch lbs. (12 Nm).
15. Connect the CMP sensor A 3P connector (A).
16. Install the EGR valve (E) with a new gasket (F) and tighten the mounting nuts to 16 ft. lbs. (22 Nm).
17. Connect the EGR valve 6P connector (D).
18. Connect the fuel line (A) and the hose (B).
19. Install the air cleaner.

CRANKSHAFT POSITION (CKP) SENSOR

LOCATION

See Figure 109.

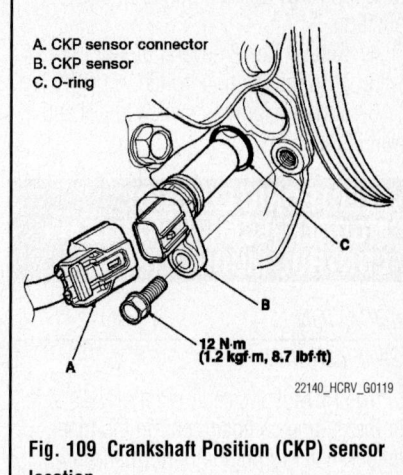

A. CKP sensor connector
B. CKP sensor
C. O-ring

22140_HCRV_G0119

Fig. 109 Crankshaft Position (CKP) sensor location

REMOVAL & INSTALLATION

See Figure 109.

1. Before servicing the vehicle, refer to the Precautions Section.
2. Disconnect the CKP sensor connector (A).
3. Remove the CKP sensor (B).

To install:

4. Install the CKP sensor (B) with a new O-ring (C).
5. Tighten the CKP mounting bolt to 104 inch lbs. (12 Nm).
6. Perform the CKP pattern clear/CKP pattern learn procedure.

CKP Pattern Clear/CKP Pattern Learn

With Honda Diagnostic System (HDS)

1. Connect the Honda Diagnostic System (HDS) to the Data Link Connector (DLC) located under the driver's side of the dashboard.
2. Turn the ignition switch to ON.
3. Make sure the HDS communicates with the PCM and other vehicle systems. If it does not, go to the DLC circuit troubleshooting.
4. Select CRANK PATTERN in the ADJUSTMENT MENU with the HDS.
5. Select CRANK PATTERN LEARNING with the HDS and follow the screen prompts.

Without Honda Diagnostic System (HDS)

1. Start the engine.
2. Hold the engine speed at 3,000 RPM without load (in P or N) until the radiator fan comes on.
3. Test-drive the vehicle on a level road: decelerate (with the throttle fully closed) from an engine speed of 2,500 RPM down

to 1,000 RPM with the transaxle in 2 position.

4. Repeat step 3 several times.
5. Turn the ignition switch to LOCK.
6. Turn the ignition switch to ON and wait 30 seconds.

ELECTRONIC CONTROL MODULE (ECM)/POWERTRAIN CONTROL MODULE (PCM)

LOCATION

See Figure 110.

The PCM is located on the driver's side in the engine compartment on the inner fender. Refer to Engine Performance & Emission Control, Component Locations.

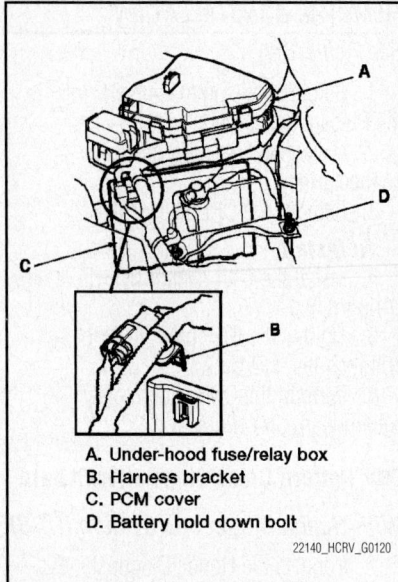

A. Under-hood fuse/relay box
B. Harness bracket
C. PCM cover
D. Battery hold down bolt

22140_HCRV_G0120

Fig. 110 Powertrain Control Module (PCM) location

REMOVAL & INSTALLATION

See Figures 110 and 111.

Special tools required (one of the following):
• Honda Diagnostic System (HDS) tablet tester
• Honda Interface Module (HIM) and an iN workstation with the latest software version
• HDS pocket tester
• GNA600 and an iN workstation with the latest software version

➡**Make sure the HDS is loaded with the latest software version. If you are replacing the PCM after substituting a known-good PCM, reinstall the original PCM, then do this procedure. During the procedure, if any READ DATA,**

WRITE DATA, or other data checks fail, note the failure, then continue.

1. Before servicing the vehicle, refer to the Precautions Section.
2. Connect the HDS to the Data Link Connector (DLC) located under the driver's side of the dashboard.
3. Turn the ignition switch to ON (II).
4. Make sure the HDS communicates with the PCM and other vehicles systems.
5. Select the PGM-FI system with the HDS.
6. Select the INSPECTION MENU with the HDS.
7. Select the ETCS TEST, then select the TP POSITION CHECK, and follow the screen prompts.

➡**If the TP POSITION CHECK indicates FAILED, continue with this procedure.**

8. Select the REPLACE PCM MENU, then select READ DATA and follow the screen prompts.

➡**Doing this step copies (READS) the engine oil life data from the original PCM so you can later download (WRITES) it into the new PCM. If READ DATA indicates FAILED, continue with this procedure.**

9. Select the A/T system with the HDS.
10. Select the REPLACE TCM/PCM MENU, then select READ DATA and follow the screen prompts.

➡**Doing this step copies (READS) the ATF life data from the original PCM so you can later download (WRITES) it into the new PCM. If READ DATA indicates FAILED, continue with this procedure.**

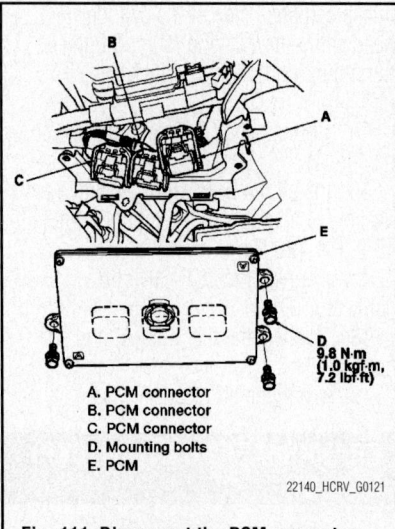

A. PCM connector
B. PCM connector
C. PCM connector
D. Mounting bolts
E. PCM

D 9.8 N·m
(1.0 kgf·m,
7.2 lbf·ft)

22140_HCRV_G0121

Fig. 111 Disconnect the PCM connectors and remove the PCM

11. Turn the ignition switch to LOCK (0).
12. Remove the under-hood fuse/relay box (A).
13. Remove the harness bracket (B).
14. Loosen the battery hold down bolt (D), and reposition the battery away from the PCM.
15. Remove the PCM cover (C).
16. Remove the bolts (D), then remove the PCM (E).
17. Disconnect the PCM connectors (A, B, C).

➡**The PCM connectors A, B, and C have symbols embossed on them for identification. See illustration.**

To install:

18. Installation is the reverse of the removal procedure.
19. Turn the ignition switch to ON (II).
20. Manually input the VIN to the PCM with the HDS.

➡**DTC P0630 VIN Not Programmed or Mismatch may be stored because the VIN has not been programmed into the PCM; ignore it, and continue this procedure.**

21. If the READ DATA (engine oil life) failed in step 9 of removal, go to step 7 below. Otherwise, go to the next step.
22. Select the PGM-FI system with the HDS.
23. Select the REPLACE PCM MENU, then select WRITE DATA and follow the screen prompts.

➡**If the WRITE DATA indicates FAILED, continue with this procedure.**

24. If the READ DATA (ATF life) failed in step 11 of removal, go to step 9 below. Otherwise go to the next step.
25. Select the A/T SYSTEM with the HDS.
26. Select the REPLACE TCM/PCM MENU, then select WRITE DATA and follow the screen prompts.

➡**If the WRITE DATA indicates FAILED, continue with this procedure.**

27. Select IMMOBI system with the HDS.
28. Enter the immobilizer code with the PCM replacement procedure in the HDS; this allows you to start the engine.
29. If the TP POSITION CHECK failed in step 7 of removal, clean the throttle body, then go to the next step.
30. If the READ DATA failed in step 8 of removal or the WRITE DATA failed in step 6 in installation, replace the engine oil and engine oil filter, then go to the next step.
31. If the READ DATA failed in step 11 of removal or the WRITE DATA failed in step 9

of installation, replace the ATF, then go to the next step.

32. Select PGM-FI system and reset the PCM with the HDS.

33. Update the PCM if it does not have the latest software.

34. Perform the PCM idle learn procedure.

35. Perform the CKP pattern learn procedure.

PCM Idle Learn Procedure—With HDS

1. Connect the Honda Diagnostic System (HDS) to the Data Link Connector (DLC) located under the driver's side of the dashboard.

2. Turn the ignition switch to ON.

3. Make sure the HDS communicates with the PCM and other vehicle systems. If it does not, go to the DLC circuit troubleshooting.

4. Select CRANK PATTERN in the ADJUSTMENT MENU with the HDS.

5. Select CRANK PATTERN LEARNING with the HDS and follow the screen prompts.

PCM Idle Learn Procedure—Without HDS

1. Start the engine.

2. Hold the engine speed at 3,000 RPM without load (in P or N) until the radiator fan comes on.

3. Test-drive the vehicle on a level road: decelerate (with the throttle fully closed) from an engine speed of 2,500 RPM down to 1,000 RPM with the transaxle in 2 position.

4. Repeat step 3 several times.

5. Turn the ignition switch to LOCK.

6. Turn the ignition switch to ON and wait 30 seconds.

CKP Pattern Clear/CKP Pattern Learn

With Honda Diagnostic System (HDS)

1. Connect the Honda Diagnostic System (HDS) to the Data Link Connector (DLC) located under the driver's side of the dashboard.

2. Turn the ignition switch to ON.

3. Make sure the HDS communicates with the PCM and other vehicle systems. If it does not, go to the DLC circuit troubleshooting.

4. Select CRANK PATTERN in the ADJUSTMENT MENU with the HDS.

5. Select CRANK PATTERN LEARNING with the HDS and follow the screen prompts.

Without Honda Diagnostic System (HDS)

1. Start the engine.

2. Hold the engine speed at 3,000 RPM without load (in P or N) until the radiator fan comes on.

3. Test-drive the vehicle on a level road: decelerate (with the throttle fully closed) from an engine speed of 2,500 RPM down to 1,000 RPM with the transaxle in 2 position.

4. Repeat step 3 several times.

5. Turn the ignition switch to LOCK.

6. Turn the ignition switch to ON and wait 30 seconds.

ENGINE COOLANT TEMPERATURE (ECT) SENSOR

LOCATION

See Figures 112 and 113.

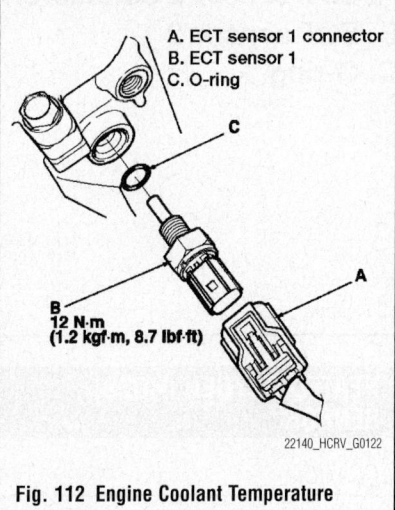

Fig. 112 Engine Coolant Temperature (ECT) sensor 1 location

ECT sensor 1 is located on the engine block. ECT sensor 2 is located on the radiator. Refer to Engine Performance & Emission Control, Component Locations.

REMOVAL & INSTALLATION

ECT Sensor 1

See Figure 112.

1. Before servicing the vehicle, refer to the Precautions Section.

2. Drain the engine coolant.

3. Remove the air cleaner.

4. Disconnect the ECT sensor 1 connector (A).

5. Remove the ECT sensor 1 (B).

To install:

6. Install the ECT sensor 1 (B) with a new O-ring (C) and tighten the sensor to 104 inch lbs. (12 Nm).

7. Refill the radiator with the proper type and amount of engine coolant.

ECT Sensor 2

See Figure 114.

1. Before servicing the vehicle, refer to the Precautions Section.

2. Remove the splash shield.

3. Drain the engine coolant.

4. Disconnect the ECT sensor 2 connector (A), then remove ECT sensor 2 (B).

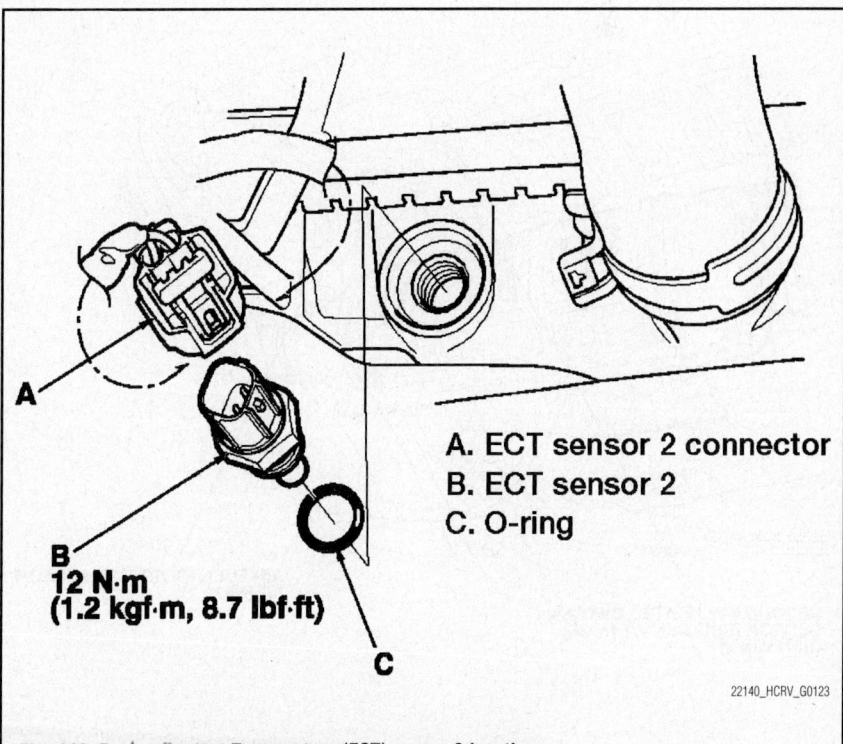

Fig. 113 Engine Coolant Temperature (ECT) sensor 2 location

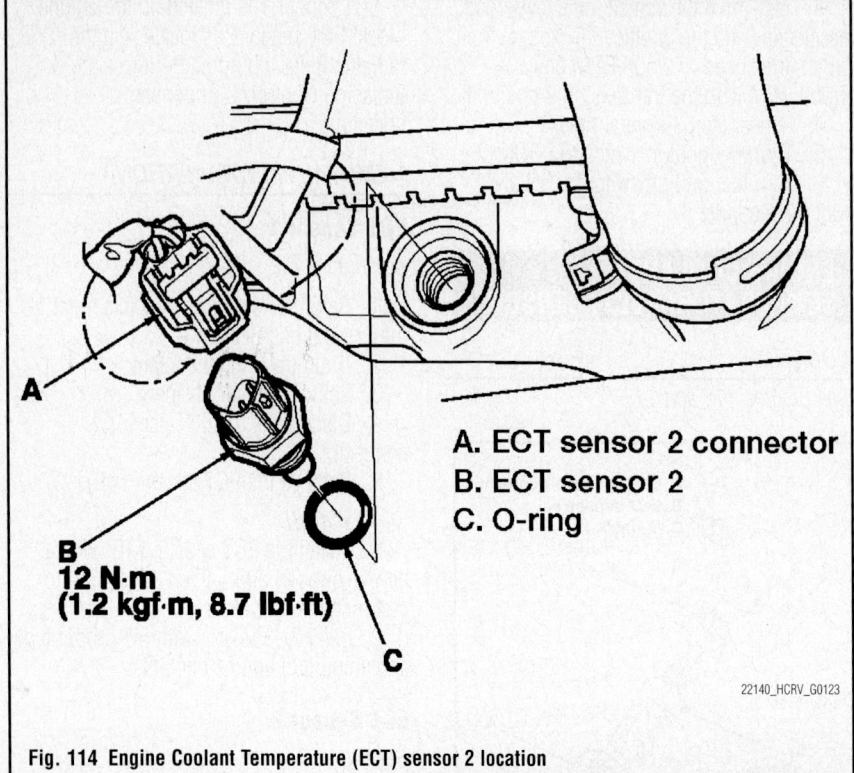

A. ECT sensor 2 connector
B. ECT sensor 2
C. O-ring

12 N·m
(1.2 kgf·m, 8.7 lbf·ft)

22140_HCRV_G0123

Fig. 114 Engine Coolant Temperature (ECT) sensor 2 location

To install:

5. Install ECT sensor 2 with a new O-ring (C) and tighten to 104 inch lbs. (12 Nm).
6. Install the splash shield.
7. Refill the radiator with the proper type and amount of engine coolant.

HEATED OXYGEN (HO2S) SENSOR

LOCATION

See Figure 115.

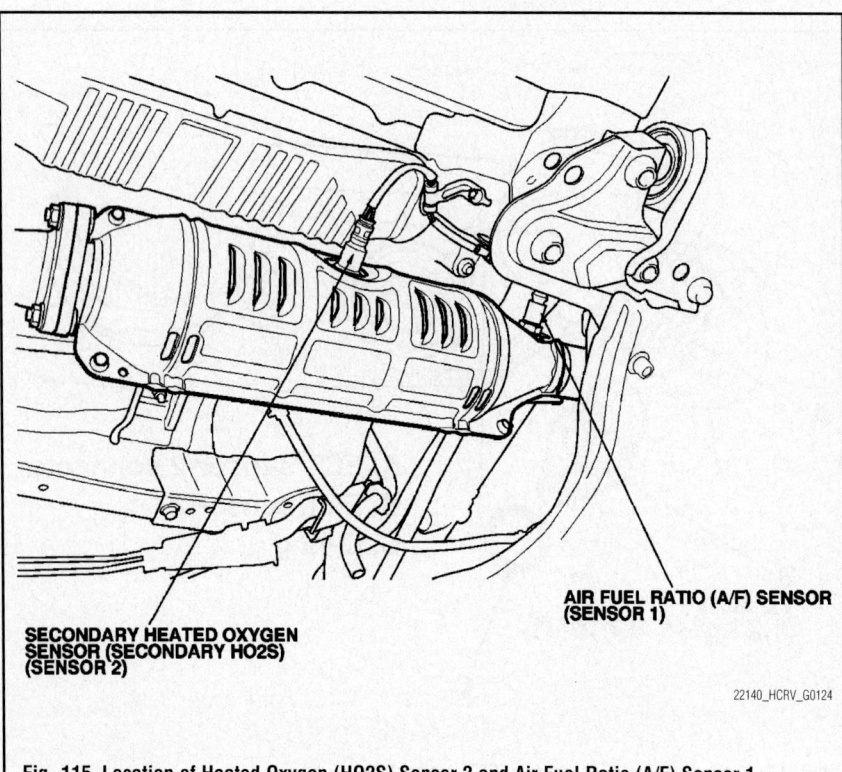

SECONDARY HEATED OXYGEN SENSOR (SECONDARY HO2S) (SENSOR 2)

AIR FUEL RATIO (A/F) SENSOR (SENSOR 1)

22140_HCRV_G0124

Fig. 115 Location of Heated Oxygen (HO2S) Sensor 2 and Air Fuel Ratio (A/F) Sensor 1

Air Fuel Ratio (A/F) Sensor 1

See Figure 116.

Special tool required:
• O2 sensor wrench, Snap-On® S3176, or equivalent

> ✳✳ **CAUTION**
>
> **The temperature of the exhaust system is extremely high after the engine has been run. To prevent personal injury, allow the exhaust system to cool before removing the sensor from the exhaust system.**

1. Before servicing the vehicle, refer to the Precautions Section.
2. Disconnect the A/F sensor 4P connector (A).
3. Remove the A/F sensor (B).

To install:
4. Install the A/F sensor (B) and tighten to 33 ft. lbs. (44 Nm).
5. Connect the A/F sensor 4P connector (A).

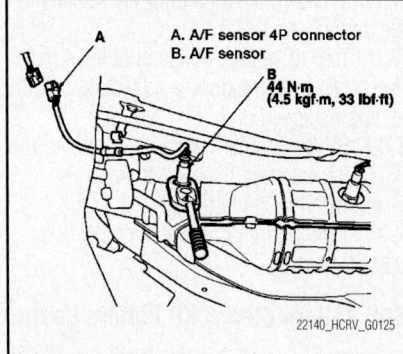

A. A/F sensor 4P connector
B. A/F sensor

44 N·m
(4.5 kgf·m, 33 lbf·ft)

22140_HCRV_G0125

Fig. 116 Removing the Air Fuel Ratio (A/F) Sensor 1

Heated Oxygen (HO2S) Sensor 2

See Figure 117.

Special tool required:
• O2 sensor wrench, Snap-On® S3176, or equivalent

> ✳✳ **CAUTION**
>
> **The temperature of the exhaust system is extremely high after the engine has been run. To prevent personal injury, allow the exhaust system to cool before removing the sensor from the exhaust system.**

1. Before servicing the vehicle, refer to the Precautions Section.

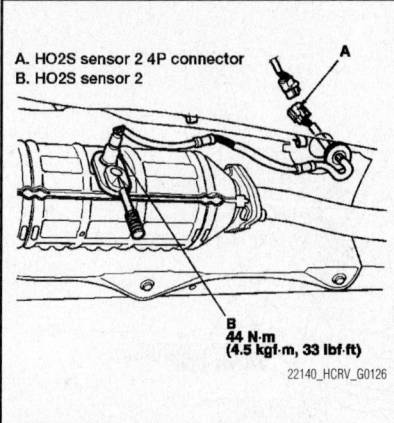

A. HO2S sensor 2 4P connector
B. HO2S sensor 2

B
44 N·m
(4.5 kgf·m, 33 lbf·ft)

22140_HCRV_G0126

**Fig. 117 Removing the Heated Oxygen
(HO2S) Sensor 2**

2. Disconnect the secondary HO2S 4P
connector (A).
3. Remove the secondary HO2S (B).

To install:
4. Install the secondary HO2S (B) and
tighten to 33 ft. lbs. (44 Nm).
5. Connect the secondary HO2S 4P
connector (A).

INTAKE AIR TEMPERATURE (IAT) SENSOR

LOCATION

See Figure 118.

The Intake Air Temperature (IAT) sensor
is mounted in the intake air hose of the air
cleaner assembly and is integrated with the
Mass Air Flow (MAF) sensor.

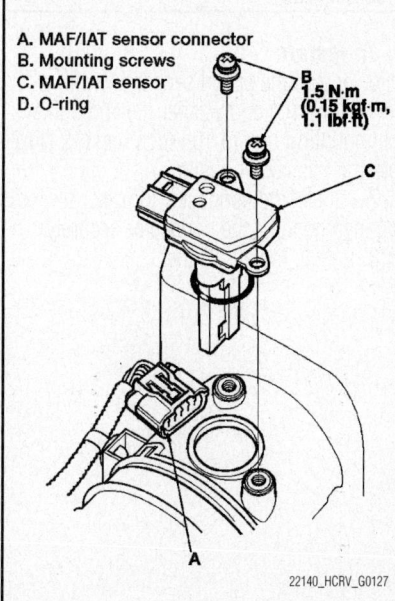

A. MAF/IAT sensor connector
B. Mounting screws
C. MAF/IAT sensor
D. O-ring

B
1.5 N·m
(0.15 kgf·m,
1.1 lbf·ft)

C

A

22140_HCRV_G0127

**Fig. 118 Location of Mass Air Flow (MAF)
and Intake Air Temperature (IAT) sensor**

REMOVAL & INSTALLATION

See Figure 118.

1. Before servicing the vehicle, refer to
the Precautions Section.
2. Disconnect the MAF sensor/IAT sensor connector (A).
3. Remove the screws (B).
4. Remove the MAF sensor/IAT
sensor (C).

To install:
5. Install the MAF sensor/IAT sensor
with a new O-ring (D).
6. Tighten the screws (B) to 13 inch lbs.
(1.5 Nm).

KNOCK SENSOR (KS)

LOCATION

See Figure 119.

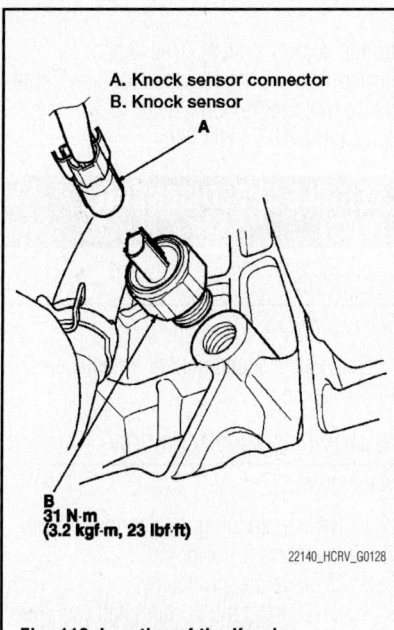

A. Knock sensor connector
B. Knock sensor

B
31 N·m
(3.2 kgf·m, 23 lbf·ft)

22140_HCRV_G0128

**Fig. 119 Location of the Knock
Sensor (KS)**

REMOVAL & INSTALLATION

See Figure 119.

1. Before servicing the vehicle, refer to
the Precautions Section.
2. Raise and safely support the
vehicle.
3. Disconnect the knock sensor
connector (A).
4. Remove the knock sensor (B).

To install:
5. Install the knock sensor (B) and
tighten to 23 ft. lbs. (31 Nm).
6. Connect the knock sensor
connector (A).

MANIFOLD ABSOLUTE PRESSURE (MAP) SENSOR

LOCATION

See Figure 120.

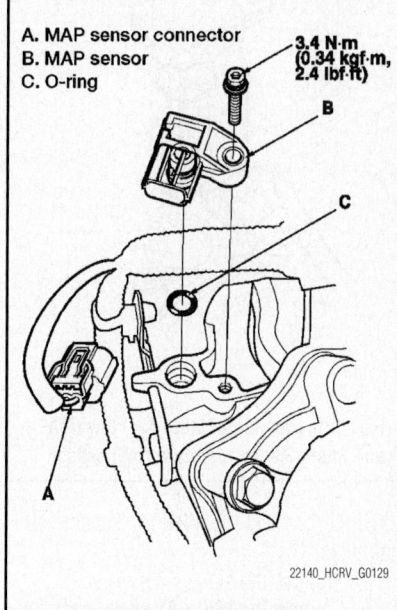

A. MAP sensor connector
B. MAP sensor
C. O-ring

3.4 N·m
(0.34 kgf·m,
2.4 lbf·ft)

B

C

A

22140_HCRV_G0129

**Fig. 120 Location of the Manifold
Absolute Pressure (MAP) Sensor**

REMOVAL & INSTALLATION

See Figure 120.

1. Before servicing the vehicle, refer to
the Precautions Section.
2. Disconnect the MAP sensor
connector (A).
3. Remove the MAP sensor (B).

To install:
4. Install the MAP sensor (B) with a
new O-ring (C) and tighten to 29 inch lbs.
(3 Nm).
5. Connect the MAP sensor
connector (A).

MASS AIR FLOW (MAF) SENSOR

LOCATION

See Figure 121.

The Mass Air Flow (MAF) sensor is
mounted in the intake air hose of the air
cleaner assembly and is integrated with the
Intake Air Temperature (IAT) sensor.

REMOVAL & INSTALLATION

See Figure 121.

1. Before servicing the vehicle, refer to
the Precautions Section.

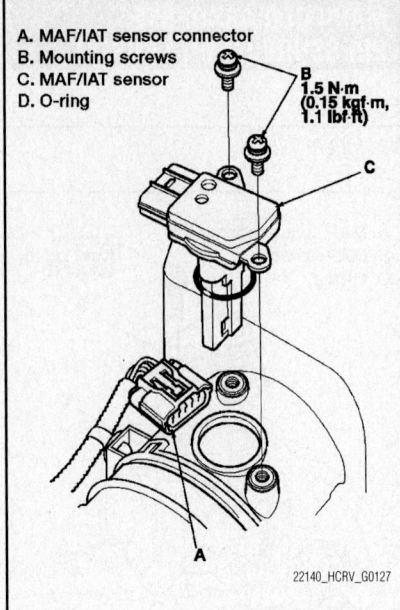

A. MAF/IAT sensor connector
B. Mounting screws
C. MAF/IAT sensor
D. O-ring

B
1.5 N·m
(0.15 kgf·m,
1.1 lbf·ft)

C

A

22140_HCRV_G0127

**Fig. 121 Location of Mass Air Flow (MAF)
and Intake Air Temperature (IAT) sensor**

2. Disconnect the MAF/IAT sensor
connector (A).
3. Remove the screws (B).
4. Remove the MAF/IAT sensor (C).

To install:
5. Install the MAF/IAT sensor with a new
O-ring (D).
6. Tighten the screws (B) to 13 inch lbs.
(1.5 Nm).

THROTTLE POSITION SENSOR (TPS)

LOCATION
See Figure 122.

The Throttle Position Sensor (TPS) is
mounted on the throttle body and is
incorporated into the throttle body
assembly.

REMOVAL & INSTALLATION

The TPS is a component of the electronic

throttle control system. This sensor is
located with the throttle actuator and is integral to the throttle body. Refer to Throttle
Body, removal & installation.

VEHICLE SPEED SENSOR (VSS)

LOCATION
See Figure 123.

The vehicle speed sensor is installed on
the transaxle.

REMOVAL & INSTALLATION
See Figure 123.

1. Before servicing the vehicle, refer to
the Precautions Section.
2. Remove the air cleaner.
3. Disconnect the output shaft (countershaft) speed sensor connector.
4. Remove the mounting bolt.
5. Remove the output shaft (countershaft) speed sensor (A) and the sensor
washer (B).

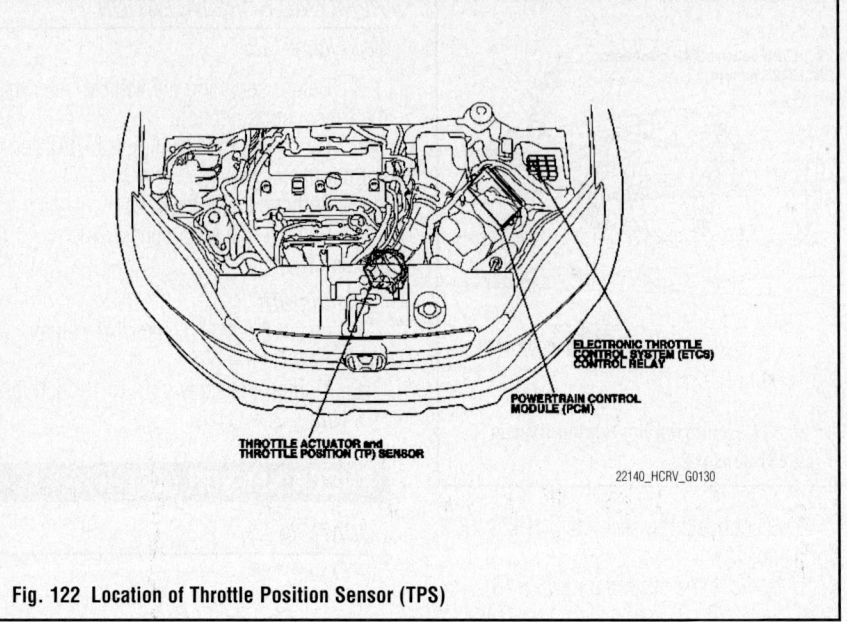

ELECTRONIC THROTTLE
CONTROL SYSTEM (ETCS)
CONTROL RELAY

POWERTRAIN CONTROL
MODULE (PCM)

THROTTLE ACTUATOR and
THROTTLE POSITION (TP) SENSOR

22140_HCRV_G0130

Fig. 122 Location of Throttle Position Sensor (TPS)

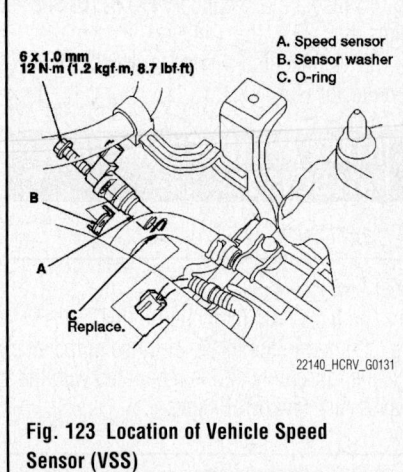

6 x 1.0 mm
12 N·m (1.2 kgf·m, 8.7 lbf·ft)

A. Speed sensor
B. Sensor washer
C. O-ring

B

A

C
Replace.

22140_HCRV_G0131

**Fig. 123 Location of Vehicle Speed
Sensor (VSS)**

To install:
6. Install the speed sensor (A) with a
new O-ring (C) and washer (B) and tighten
the mounting bolt to 104 inch lbs. (12 Nm)
into the transaxle housing.
7. Check the connector for rust, dirt, or
oil, then connect the connector securely.
8. Install the air cleaner.

FUEL SYSTEM SERVICE PRECAUTIONS

Safety is the most important factor when performing not only fuel system maintenance but any type of maintenance. Failure to conduct maintenance and repairs in a safe manner may result in serious personal injury or death. Maintenance and testing of the vehicle's fuel system components can be accomplished safely and effectively by adhering to the following rules and guidelines.

• To avoid the possibility of fire and personal injury, always disconnect the negative battery cable unless the repair or test procedure requires that battery voltage be applied.

• Always relieve the fuel system pressure prior to disconnecting any fuel system component (injector, fuel rail, pressure regulator, etc.), fitting or fuel line connection. Exercise extreme caution whenever relieving fuel system pressure to avoid exposing skin, face and eyes to fuel spray. Please be advised that fuel under pressure may penetrate the skin or any part of the body that it contacts.

• Always place a shop towel or cloth around the fitting or connection prior to loosening to absorb any excess fuel due to spillage. Ensure that all fuel spillage (should it occur) is quickly removed from engine surfaces. Ensure that all fuel soaked cloths or towels are deposited into a suitable waste container.

• Always keep a dry chemical (Class B) fire extinguisher near the work area.

• Do not allow fuel spray or fuel vapors to come into contact with a spark or open flame.

• Always use a back-up wrench when loosening and tightening fuel line connection fittings. This will prevent unnecessary stress and torsion to fuel line piping.

• Always replace worn fuel fitting O-rings with new Do not substitute fuel hose or equivalent where fuel pipe is installed.

Before servicing the vehicle, make sure to also refer to the precautions in the beginning of this section as well.

RELIEVING FUEL SYSTEM PRESSURE

Before disconnecting fuel lines or hoses, relieve pressure from the system by disabling the fuel pump and then disconnecting the fuel tube/quick connect fitting in the engine compartment.

WITH HONDA DIAGNOSTICS SYSTEM (HDS)

See Figure 124.

1. Before servicing the vehicle, refer to the Precautions Section.
2. Remove the fuel fill cap, to relieve the pressure in the fuel tank.
3. Turn the ignition switch to ON.
4. From the INSPECTION MENU of the Honda Diagnostics System (HDS), select Fuel Pump OFF, then start the engine, and let it idle until it stalls.

➡ **Do not allow the engine to idle above 1,000 RPM or the PCM will continue to operate the fuel pump. A DTC or a Temporary DTC may be set during this procedure. Check for DTC's, and clear them as needed.**

5. Turn the ignition switch to LOCK.
6. Perform the battery terminal disconnection procedure. Some systems store data in memory that is lost when the battery is disconnected. Do the following steps before disconnecting the battery:

   a. Make sure to record the anti-theft code(s) for the audio and/or the navigation system (if equipped).

   b. If replacing the audio unit, write down the audio presets (AM and FM), and the XM audio presets (if equipped), because the audio unit does not retain the presets after the battery is disconnected.

   c. Make sure the ignition switch is in LOCK position.

   d. Disconnect and isolate the negative cable from the battery.

➡ **Always disconnect the negative cable from the battery first.**

   e. Disconnect the positive cable from the battery.
7. Remove the quick-connect fitting cover.
8. Check the fuel quick-connect fitting for dirt, and clean it if needed.
9. Place a rag or shop towel over the quick-connect fitting.
10. Disconnect the quick-connect fitting (A): hold the connector (B) with one hand and squeeze the retainer tabs (C) with the other hand to release them from the locking tabs (D). Pull the connector off.

   • Be careful not to damage the line (E) or other parts

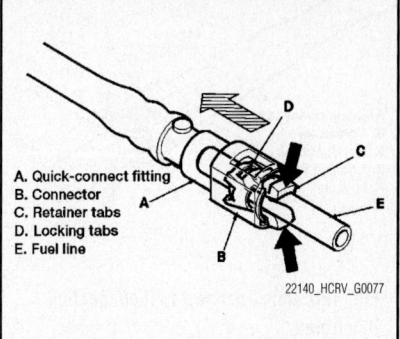

A. Quick-connect fitting
B. Connector
C. Retainer tabs
D. Locking tabs
E. Fuel line

22140_HCRV_G0077

Fig. 124 Quick-connect fuel connection illustrated

   • Do not use tools
   • If the connector does not move, keep the retainer tabs pressed down, and alternately pull and push the connector until it comes off easily
   • Do not remove the retainer from the line; once removed, the retainer must be replaced with a new one
11. After disconnecting the quick-connect fitting, check it for dirt or damage.
12. Reconnect the battery as needed.

WITHOUT HONDA DIAGNOSTICS SYSTEM (HDS)

See Figures 125 and 126.

1. Before servicing the vehicle, refer to the Precautions Section.
2. Remove the under-dash fuse/relay box, then remove the PGM-FI main relay 2 (FUEL PUMP) (A) from the under-dash fuse/relay box.
3. Reinstall the under-dash fuse/relay box.

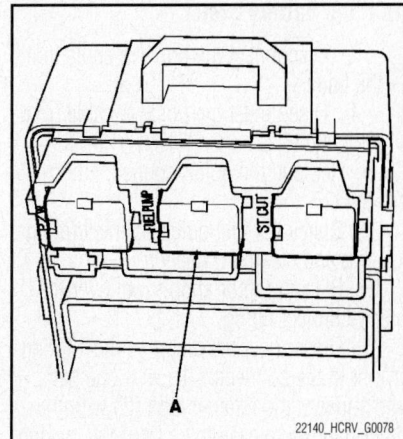

22140_HCRV_G0078

Fig. 125 Remove the PGM-FI main relay 2 (FUEL PUMP) (A) from the under-dash fuse/relay box

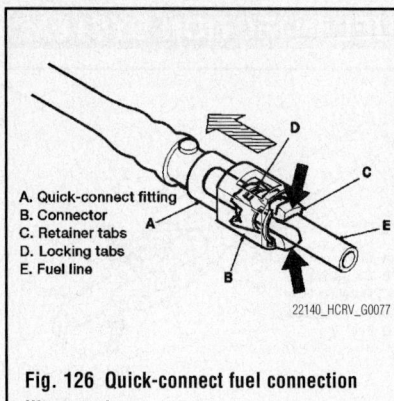

A. Quick-connect fitting
B. Connector
C. Retainer tabs
D. Locking tabs
E. Fuel line

22140_HCRV_G0077

Fig. 126 Quick-connect fuel connection illustrated

4. Start the engine, and let it idle until it stalls.

➡ **If any DTC's are stored, clear and ignore them.**

5. Turn the ignition switch to LOCK.
6. Remove the fuel fill cap.
7. Perform the battery terminal disconnection procedure. Some systems store data in memory that is lost when the battery is disconnected. Do the following steps before disconnecting the battery:

    a. Make sure to record the anti-theft code(s) for the audio and/or the navigation system (if equipped).

    b. If replacing the audio unit, write down the audio presets (AM and FM), and the XM audio presets (if equipped), because the audio unit does not retain the presets after the battery is disconnected.

    c. Make sure the ignition switch is in LOCK position.

    d. Disconnect and isolate the negative cable from the battery.

➡ **Always disconnect the negative cable from the battery first.**

    e. Disconnect the positive cable from the battery.

    f. Disconnect the positive cable from the battery.

8. Remove the quick-connect fitting cover.
9. Check the fuel quick-connect fitting for dirt, and clean it if needed.
10. Place a rag or shop towel over the quick-connect fitting.
11. Disconnect the quick-connect fitting (A): hold the connector (B) with one hand and squeeze the retainer tabs (C) with the other hand to release them from the locking tabs (D). Pull the connector off.

- Be careful not to damage the line (E) or other parts
- Do not use tools

- If the connector does not move, keep the retainer tabs pressed down, and alternately pull and push the connector until it comes off easily
- Do not remove the retainer from the line; once removed, the retainer must be replaced with a new one

12. After disconnecting the quick-connect fitting, check it for dirt or damage.
13. Reconnect the battery as needed.

FUEL FILTER

REMOVAL & INSTALLATION
See Figure 127.

The fuel filter should be replaced whenever the fuel pressure drops below the specified value (47–54 psi), after making sure that the fuel pump and the fuel pressure regulator are functioning properly.

1. Before servicing the vehicle, refer to the Precautions Section.
2. Relieve the fuel pressure. Refer to Relieving Fuel System Pressure.

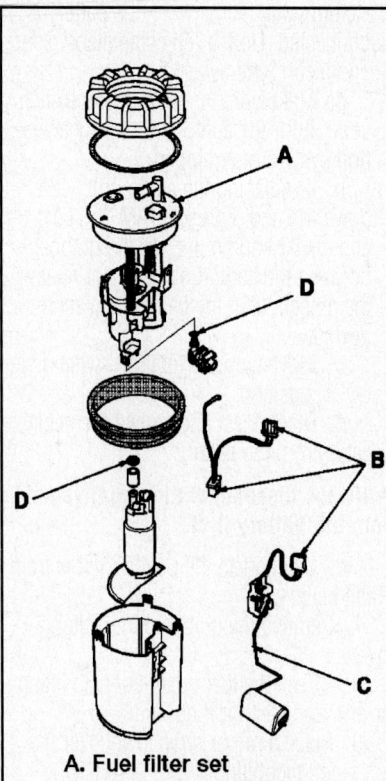

A. **Fuel filter set**
B. **Electrical connectors**
C. **Fuel gauge sending unit**
D. **O-rings**

22140_HCRV_G0079

Fig. 127 Exploded view of the fuel pump tank unit

3. Remove the fuel pump unit. Refer to Fuel Pump, removal & installation.
4. Remove the fuel filter set (A).

To install:
5. Check these items before installing the fuel pump tank unit:

    a. When connecting the wire harness, make sure the connection is secure and the connectors (B) are firmly locked into place.

    b. When installing the fuel gauge sending unit (C), make sure the connection is secure and the connector is firmly locked into place. Be careful not to bend or twist it excessively.

6. Install the parts in the reverse order of removal with new O-rings (D).

➡ **Coat the O-rings with clean engine oil.**

7. Install the fuel pump tank unit.

FUEL INJECTORS

REMOVAL & INSTALLATION
See Figure 128.

1. Before servicing the vehicle, refer to the Precautions Section.
2. Relieve the fuel pressure. Refer to Relieving Fuel System Pressure.
3. Remove the engine cover.
4. Disconnect the connectors from the injectors.
5. Remove the ground cable bolt.
6. Disconnect the quick-connect fitting.
7. Remove the fuel rail mounting nuts from the fuel rail.
8. Remove the injector clips from the injectors.
9. Remove the injectors from the fuel rail.

To install:
10. Coat the new O-rings (A) with clean engine oil and insert the injectors (B) into the fuel rail (C).
11. Install the injector clips (D).
12. Coat the injector O-rings (E) with clean engine oil.
13. Install the fuel rail and the injectors in the injector base (F).
14. Install the fuel rail mounting nuts (G) and tighten to 16 ft. lbs. (22 Nm).
15. Install the ground cable and bolt.
16. Connect the injector connectors.
17. Connect the quick-connect fitting.
18. Turn the ignition switch to **ON**, but do not operate the starter. After the fuel pump runs for about 2 seconds, the fuel rail will be pressurized. Repeat this 2–3 times, then check for fuel leakage.
19. Reinstall the engine cover.

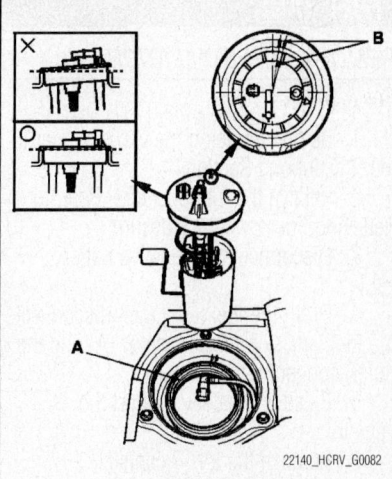

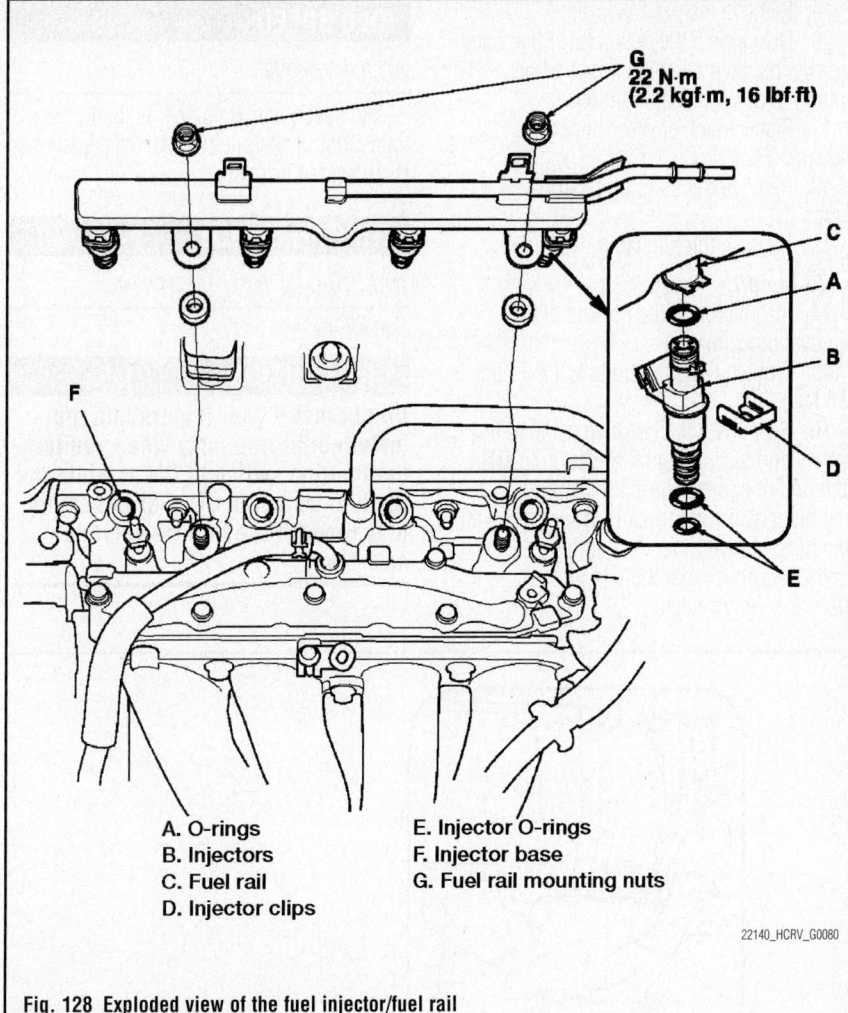

A. O-rings
B. Injectors
C. Fuel rail
D. Injector clips

E. Injector O-rings
F. Injector base
G. Fuel rail mounting nuts

22140_HCRV_G0080

Fig. 128 Exploded view of the fuel injector/fuel rail

FUEL PUMP

REMOVAL & INSTALLATION

See Figures 129 and 130.

Special tools required:
- Fuel sender wrench 07AAA-S0XA100

1. Before servicing the vehicle, refer to the Precautions Section.
2. Relieve the fuel pressure. Refer to Relieving Fuel System Pressure.
3. Disconnect the negative battery cable.
4. Remove the fuel fill cap.
5. Fold the left side rear seat forward and pull back the carpet to expose the access panel.
6. Remove the access panel (A) from the floor.
7. Disconnect the fuel tank unit 4P connector (B).
8. Disconnect the quick-connect fitting (C) from the fuel tank unit.
9. Using the special tool, loosen the locknut.
10. Remove the locknut, the base gasket, and the fuel pump tank unit.

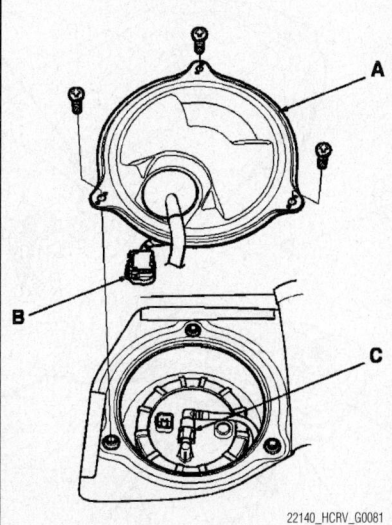

22140_HCRV_G0081

Fig. 129 Remove the access panel (A) from the floor, disconnect the 4P connector (B) and the quick-connect fitting (C) from the fuel pump tank unit

Fig. 130 Install the base gasket (A). Align the marks (B) on the fuel tank and the fuel tank unit, then insert the fuel tank unit until it sits on the base gasket.

To install:

11. Temporarily attach a new base gasket to the fuel pump tank unit, then insert the fuel tank unit partially into the fuel tank.
- Be careful not to damage the new base gasket
- Be careful not to bend the fuel gauge sending unit
- Do not coat the base gasket with oil
12. Transfer the base gasket (A) from the fuel tank unit to the fuel tank.
13. Align the marks (B) on the fuel tank and the fuel tank unit, then insert the fuel tank unit until it sits on the base gasket.

❈❈ CAUTION

To prevent a fuel leak, check the base gasket, visually or by hand, to make sure it is not pinched.

14. Using the special tool, tighten the new locknut to 52 ft. lbs. (70 Nm).
- After tightening, make sure the marks are still aligned
- After installation, check the base gasket visually or by hand to be sure it is not pinched
15. Connect the fuel tank unit 4P connector.
16. Reconnect the negative cable to the battery and turn the ignition switch to **ON** (but do not operate the starter motor). The fuel pump will run for about 2 seconds, and fuel pressure will rise. Repeat this 2–3 times, and check that there is no leakage in the fuel supply system.
17. Install the parts in the reverse order of removal.

FUEL TANK

REMOVAL & INSTALLATION

See Figure 131.

1. Before servicing the vehicle, refer to the Precautions Section.
2. Relieve the fuel pressure. Refer to Relieving Fuel System Pressure.
3. Disconnect the negative battery cable.
4. Drain the fuel tank, then disconnect the fuel tank unit 4P connector (A) and the quick-connect fittings (B).
5. Raise and safely support the vehicle.
6. Remove the EVAP canister.
7. Remove the cover (C) and the fuel tank guard (D).
8. On 4WD models, remove the propeller shaft.

9. Remove the exhaust pipe.
10. Disconnect the hoses (E). Slide back the clamps, then twist the hoses while pulling to avoid damaging them.
11. Place a jack or other support under the tank (F).
12. Remove the strap bolts and the straps (G).
13. Remove the fuel tank.

To install:

14. Installation is the reverse of the removal procedure.
15. Tighten the strap bolts to 28 ft. lbs. (38 Nm).
16. Reconnect the negative cable to the battery and turn the ignition switch to **ON** (but do not operate the starter motor). The fuel pump will run for about 2 seconds, and fuel pressure will rise. Repeat this 2–3 times, and check that there is no leakage in the fuel supply system.

IDLE SPEED

ADJUSTMENT

Idle speed is maintained by the Powertrain Control Module (PCM). No adjustment is necessary or possible.

THROTTLE BODY

REMOVAL & INSTALLATION

See Figure 132.

✳✳ CAUTION

Do not insert your fingers into the installed throttle body when you turn the ignition switch to ON or while the ignition switch is ON. Serious injury may result if the throttle valve is activated.

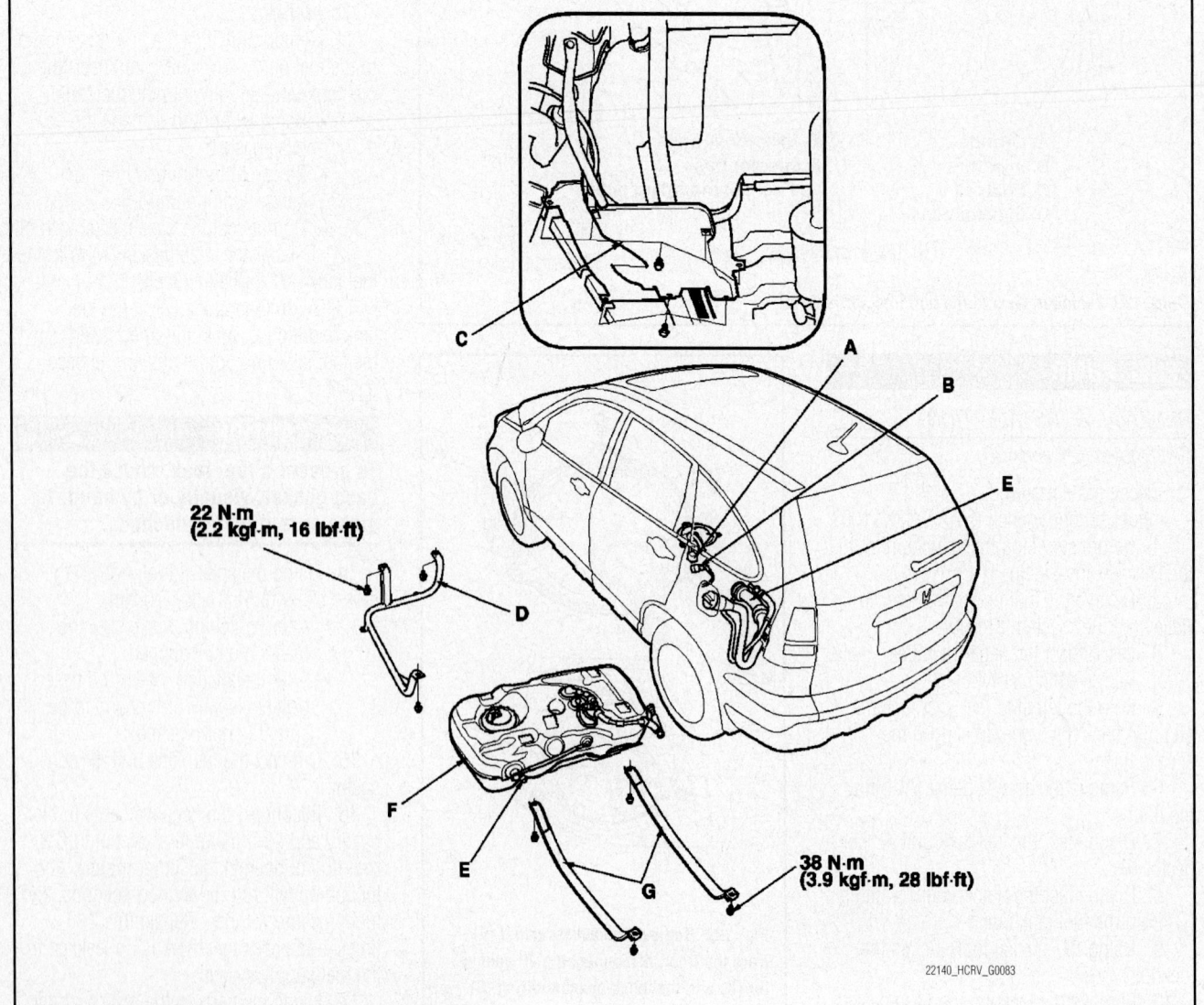

22 N·m
(2.2 kgf·m, 16 lbf·ft)

38 N·m
(3.9 kgf·m, 28 lbf·ft)

22140_HCRV_G0083

Fig. 131 Expanded view of fuel tank and related components

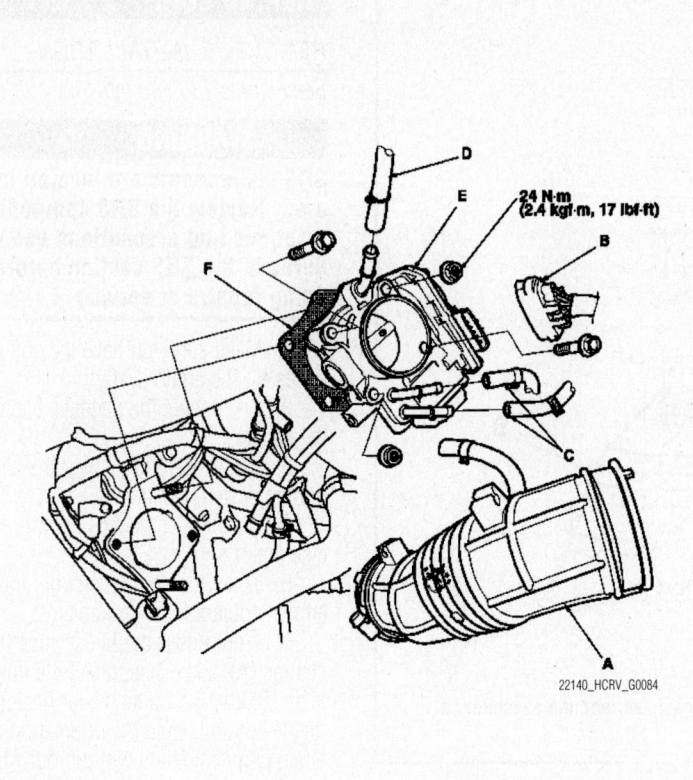

Fig. 132 View of throttle body components

➡**If replacing or cleaning the throttle body, start at step 2. If removing the throttle body, start at step 5.**

1. Before servicing the vehicle, refer to the Precautions Section.
2. Connect the Honda Diagnostics System (HDS) to the Data Link Connector (DLC) while the engine is stopped.
3. Select the INSPECTION MENU on the HDS.
4. Perform the TP POSITION CHECK in the ETCS TEST.
5. Turn the ignition switch to LOCK.
6. Remove the air intake duct (A).
7. Disconnect the throttle body connector (B).
8. Disconnect the water bypass hoses (C) and plug them.
9. Disconnect the vacuum hose (D).
10. Remove the throttle body (E).

To install:
11. Installation is the reverse of the removal procedure.
12. Install a new throttle body gasket (F) and tighten the retaining bolts and nuts to 17 ft. lbs. (24 Nm).
13. Refill the radiator with the proper type and amount of engine coolant.
14. Perform the PCM idle learn procedure. Refer to Engine Assembly, removal & installation.

HEATING & AIR CONDITIONING SYSTEM

BLOWER MOTOR

REMOVAL & INSTALLATION

See Figures 133 through 135.

1. Before servicing the vehicle, refer to the Precautions Section.
2. Remove the glove box.
3. Remove the glove box frame.
4. Cut the plastic cross brace (A) in the glove box opening with diagonal cutters in the area shown. Retain the plastic cross brace as it will be reinstalled.
5. Remove the wire harness clips (A), the self-tapping screws, and the passenger's heater duct (B).
6. Disconnect the connector (A) from the blower motor. Remove the wire harness clip (B).
7. Disconnect the connector from the recirculation control motor. Remove the self-tapping screws, the bolt, the mounting nuts, and the blower unit.

To install:
8. Install the unit in the reverse order of removal.
9. Make sure that there are no air leaks.

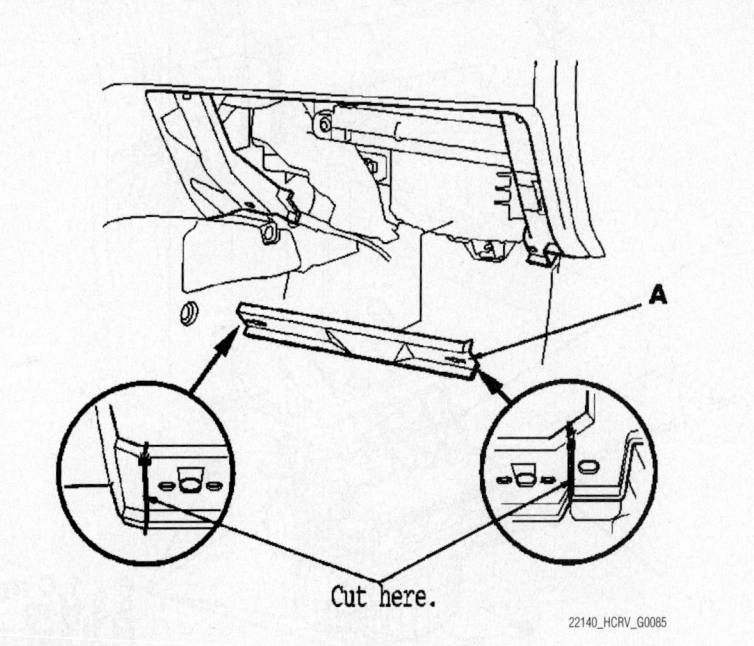

Fig. 133 Cut the plastic cross brace (A) in the glove box opening with diagonal cutters in the area shown

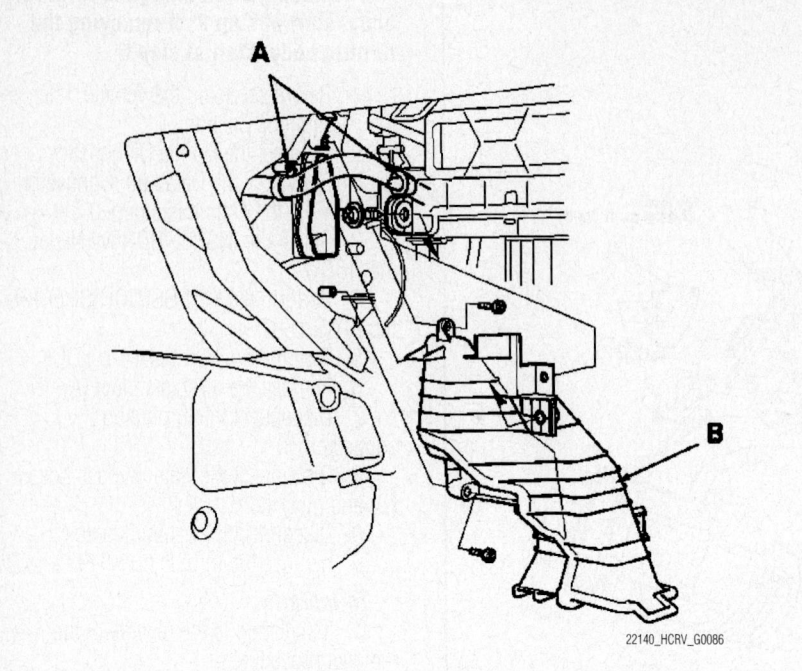

Fig. 134 Remove the wire harness clips (A), the self-tapping screws, and the passenger's heater duct (B)

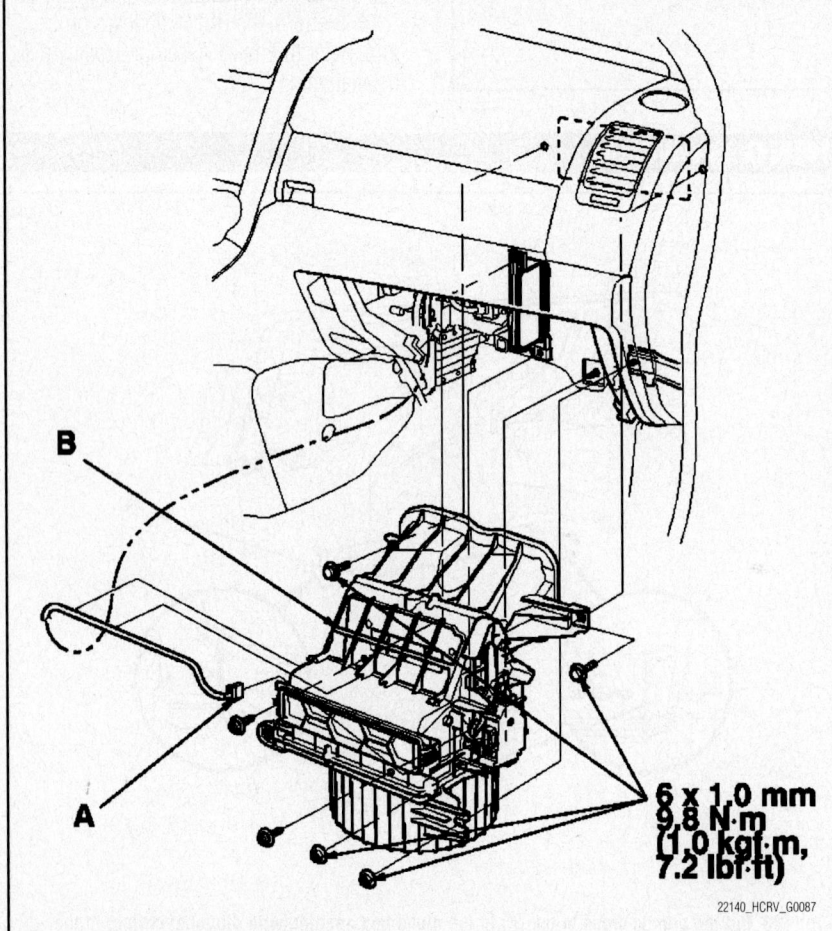

Fig. 135 Blower unit installation

HEATER CORE

REMOVAL & INSTALLATION

See Figures 136 through 139.

※ WARNING

SRS components are located in this area. Review the SRS component locations and precautions and procedures in the SRS section before doing repairs or service.

1. Make sure you have the anti-theft codes for the audio system.
2. Disconnect the negative cable from the battery.
3. Discharge and recover the air conditioning system refrigerant.
4. Disconnect the A/C line from the evaporator core.
5. When the engine is cool, drain the engine coolant from the radiator.
6. From under the hood, slide the hose clamps (A) back. Disconnect the inlet heater hose (B) and the outlet heater hose (C) from the heater unit. Note the orientation of the hose. Engine coolant will run out when the hoses are disconnected; drain it into a clean drip pan. Be sure not to let coolant spill on the electrical parts or the painted surfaces. If any coolant spills, rinse it off immediately.
7. Remove the mounting nut from the heater unit. Take care not to damage or bend the fuel lines or the brake lines.

➡**Have an assistant help when removing and installing the dashboard. Use the appropriate tool from the KTC trim tool set (SOJATP2014) to avoid damage when prying components. Take care not to scratch the dashboard and related parts.**

8. Remove or disconnect the following:
   • Center upper dashboard panel
   • Driver's vent panel

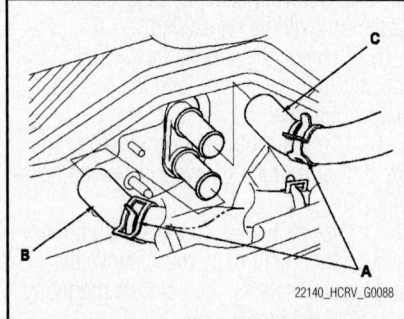

Fig. 136 From under the hood, slide the hose clamps (A) back. Disconnect the inlet heater hose (B) and the outlet heater hose (C) from the heater unit

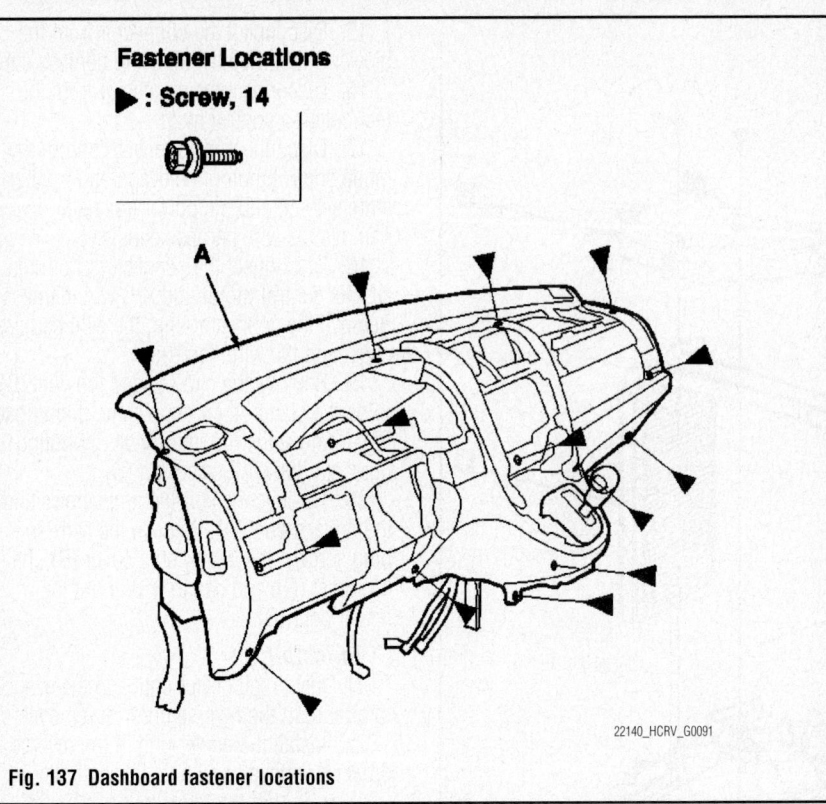

Fastener Locations

▶ : Screw, 14

A

Fig. 137 Dashboard fastener locations

22140_HCRV_G0091

- Dashboard center vent
- Passenger's vent panel
- Passenger's airbag
- Glove box
- Instrument panel
- Gauge control module
- Navigation display unit, with navigation system (if equipped)
- Audio unit, without navigation system (if equipped)

9. On the driver's side: remove the screws, then remove the GPS antenna, then pull the antenna out through the hole in the dashboard.

10. In the center, detach the wire harness clips from the shift lever panel opening.

11. From the back of the dashboard, disconnect the tweeter connectors, then detach the wire harness clips.

12. On the passenger's side, carefully pry the glove box light from the inside of the dashboard, and detach the wire harness clips.

13. From the back of the dashboard, remove the screws and the center joint duct.

14. From the front of the dashboard, remove the screws.

A. Remove the clip
B. Duct
C. Drain hose
D. Blower-heater unit

6 x 1.0 mm
9.8 N·m
(1.0 kgf·m, 7.2 lbf·ft)

6 x 1.0 mm
9.8 N·m
(1.0 kgf·m, 7.2 lbf·ft)

C   D   B   A

22140_HCRV_G0089

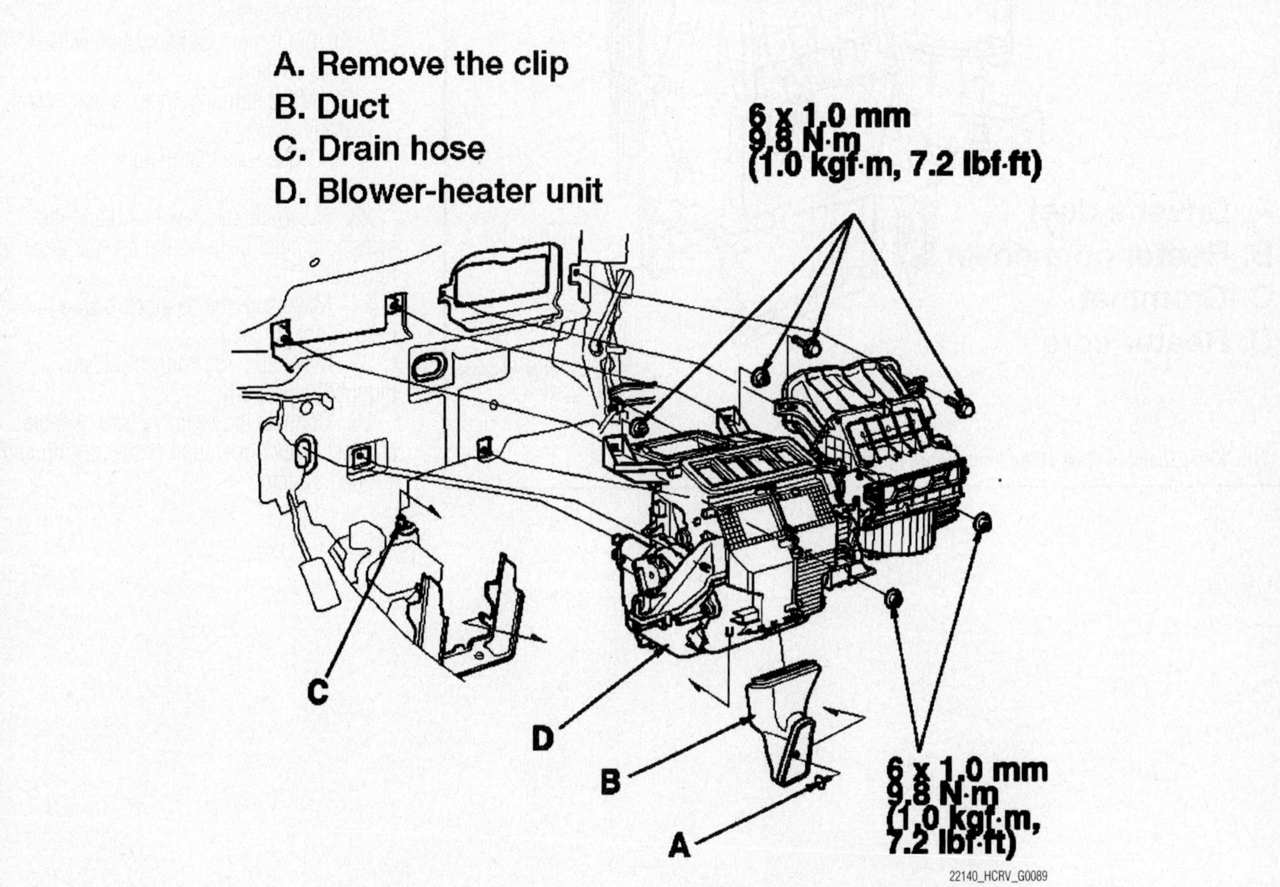

Fig. 138 Remove blower-heater unit

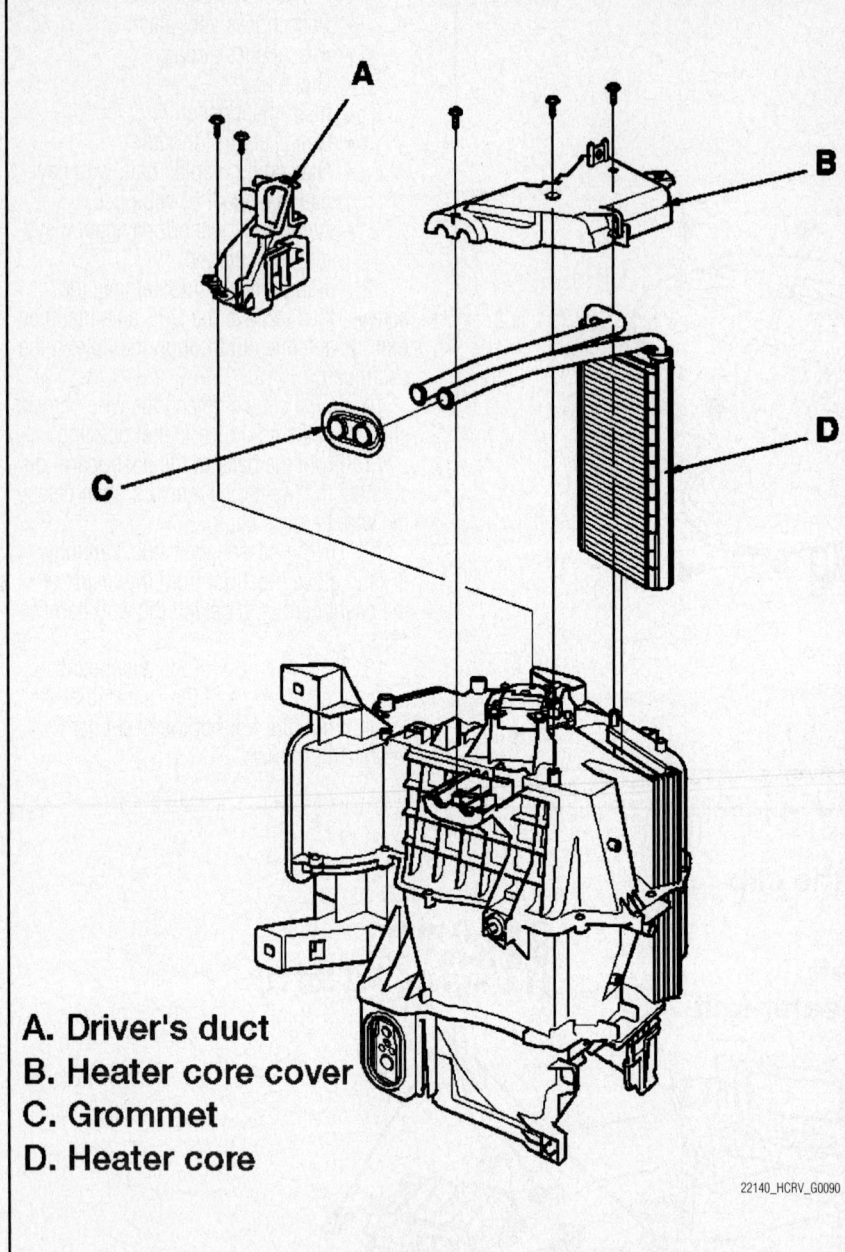

A. Driver's duct
B. Heater core cover
C. Grommet
D. Heater core

22140_HCRV_G0090

Fig. 139 Expanded view of the heater core mounting

15. Disconnect the connector from the blower motor. Remove the wire harness clip.

16. Disconnect the connector from the recirculation control motor.

17. Disconnect the connectors from the mode control motor, the evaporator temperature sensor, and the power transistor. Remove the wire harness clips.

18. Disconnect the connectors from the air mix control motor and A/C wire harness. Remove the connector clip, the wire harness clips, and the wire harness.

19. Remove the clip (A) and the duct (B). Slide the clamp, then remove the drain hose (C). Remove the mounting bolt, mounting nuts, and blower-heater unit (D).

20. Remove the self-tapping screws and the driver's duct (A). Remove the self-tapping screws, the heater core cover (B), the grommet (C), and carefully pull out the heater core (D).

To install:

21. Install the heater core and the evaporator core in the reverse order of removal.

22. Install the heater unit in the reverse order of removal, and note these items:

   a. Do not interchange the inlet and outlet heater hoses. Install the hose clamps securely.

   b. Refill the cooling system with engine coolant.

   c. Make sure that there is no coolant leakage.

   d. Make sure that there is no air leakage.

23. Assemble the dashboard and the steering hanger beam in the reverse order of removal.

24. Make sure the dashboard wire harness is not pinched.

25. Make sure the connectors are plugged in properly.

26. Connect the negative battery cable and test proper operation of the heating and cooling system.

STEERING

POWER STEERING GEAR

REMOVAL & INSTALLATION

See Figure 140.

❊❊ WARNING

Do not permit the steering wheel to turn whenever the steering gear is disconnected from the steering column. Damage to the air bag wiring can result.

1. Before servicing the vehicle, refer to the Precautions Section.
2. Center the steering wheel and lock it in position.
3. Remove or disconnect the following:
   - Negative battery cable
   - Air bag and steering wheel
   - Front wheels
   - Driver's side dashboard lower cover and undercover
   - Air cleaner housing
   - Steering joint bolts
   - Tie rod ends
   - Steering hoses
   - Left side flange bolts
   - Mounting brackets
4. Lower the unit so the pinion shaft points outward. Remove the pinion shaft grommet. The steering gear is removed through the driver's side.

To install:

5. Installation is the reverse of removal.
6. Tighten the mounting bracket and side flange bolts to 46 ft. lbs. (62 Nm).
7. Tighten the supply line flare nut to 27 ft. lbs. (37 Nm).
8. Tighten the tie rod ball stud nuts to 32 ft. lbs. (43 Nm).
9. Tighten the steering joint bolts to 21 ft. lbs. (28 Nm).

POWER STEERING PUMP

REMOVAL & INSTALLATION

See Figure 141.

1. Before servicing the vehicle, refer to the Precautions Section.
2. Place a suitable container under the vehicle.
3. Drain the power steering fluid from the reservoir.
4. Remove the drive belt (A) from the pump pulley.
5. Cover the auto-tensioner, alternator, and A/C compressor with several shop towels to protect them from spilled power steering fluid.
6. Disconnect the pump inlet hose (B) and pump outlet hose (C) from the pump (D), and plug them.

➡**Take care not to spill the fluid on the body or parts. Wipe off any spilled fluid at once. Do not turn the steering wheel with the pump removed.**

7. Remove the pump mounting bolts (E).
8. Cover the opening of the pump with a piece of tape to prevent foreign material from entering the pump.
9. Remove the power steering pump.

To install:

10. Install the pump in the pump bracket with the mounting bolts. Tighten the outlet hose bolts to 96 inch lbs. (11 Nm) and the power steering pump mounting bolts to 16 ft. lbs. (22 Nm).

➡ **Connect the pump inlet hose and pump outlet hose onto new pump with a new O-ring (F).**

11. Connect the pump inlet hose and pump outlet hose.
12. Install the drive belt (A).
13. Note these items during drive belt installation:
    - Make sure that the drive belt is properly positioned on the pulleys.
    - Do not get power steering fluid or grease on the auto-tensioner, alternator, A/C compressor, drive belt, or pulley faces. Clean off any fluid or grease before installation.
14. Fill the power steering pump reservoir with the proper type and amount of fluid.
15. Bleed the power steering system.

BLEEDING

1. Before servicing the vehicle, refer to the Precautions Section.
2. Stop the engine.

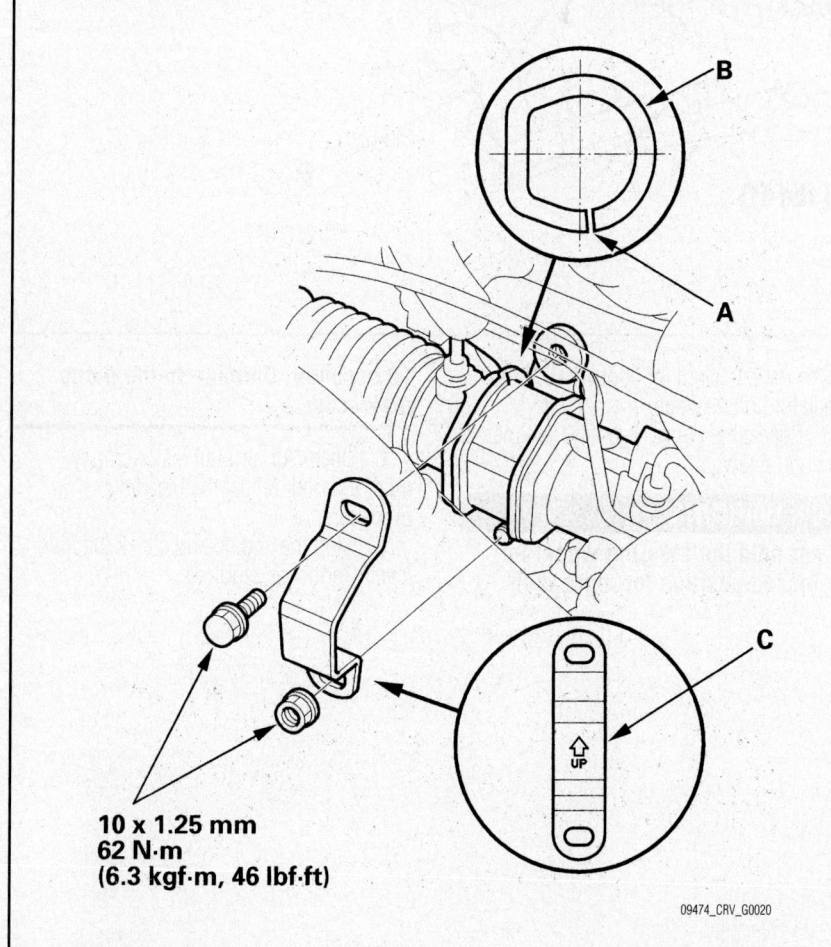

10 x 1.25 mm
62 N·m
(6.3 kgf·m, 46 lbf·ft)

09474_CRV_G0020

Fig. 140 Power steering gear installation

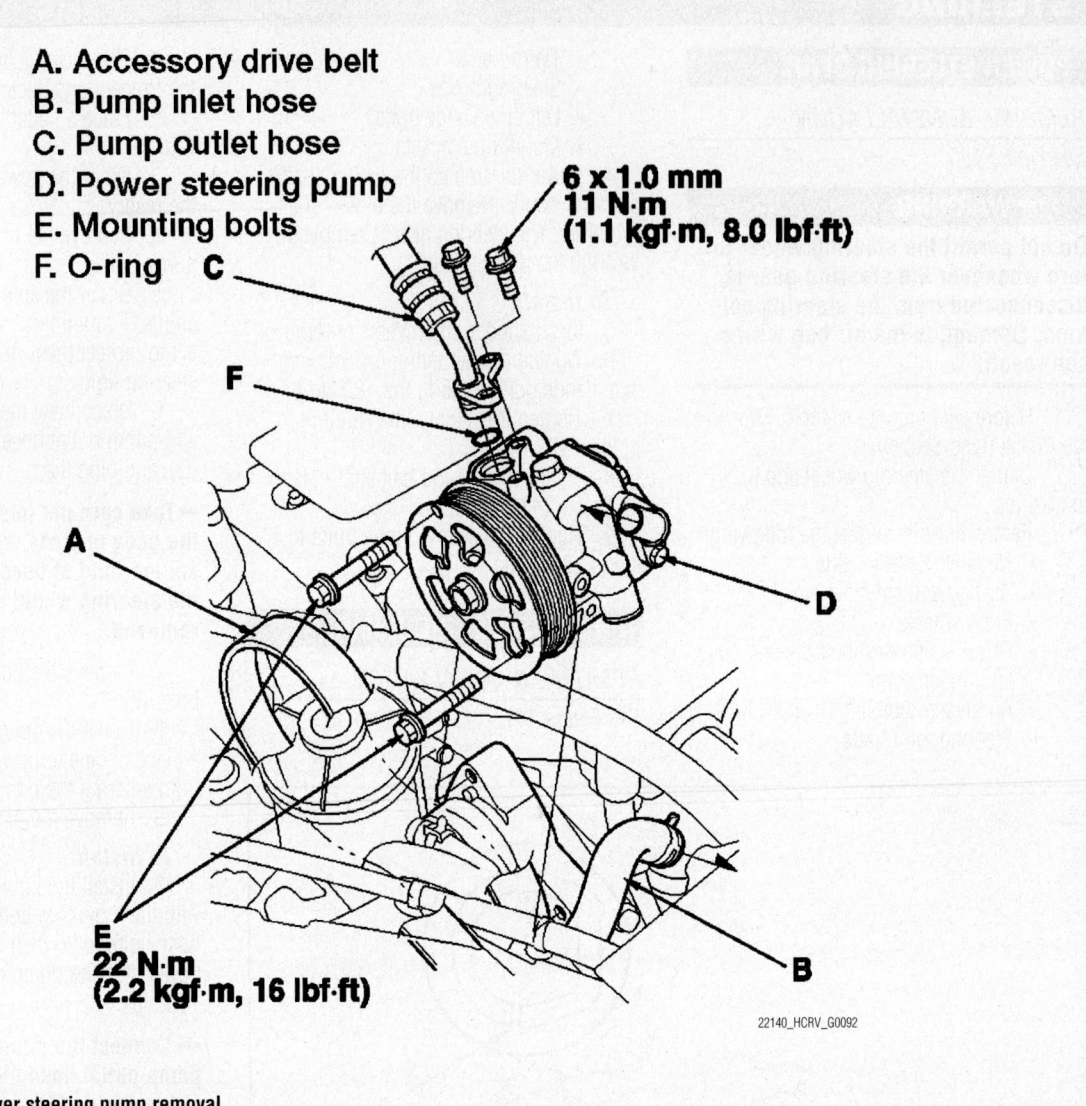

A. Accessory drive belt
B. Pump inlet hose
C. Pump outlet hose
D. Power steering pump
E. Mounting bolts
F. O-ring

6 x 1.0 mm
11 N·m
(1.1 kgf·m, 8.0 lbf·ft)

22 N·m
(2.2 kgf·m, 16 lbf·ft)

22140_HCRV_G0092

Fig. 141 View of power steering pump removal

3. Turn the steering wheel fully to the right and left several times.

➡**Do not allow the fluid level in the reservoir tank to go below the MIN level line. Check and add fluid as needed.**

4. Run the engine at idle speed. Turn the steering wheel fully to the right and then fully to the left. Hold for about 3 seconds. Check for fluid leakage.

5. Repeat the above step several times at 3 second intervals.

❊❊ **WARNING**

Do not hold the steering wheel in the locked position for more than 10 seconds. Damage to the pump may occur.

6. Check for air bubbles or cloudy fluid. If found, repeat the bleeding procedure.

7. Stop the engine and check the fluid level. Correct as required.

See Figure 142.

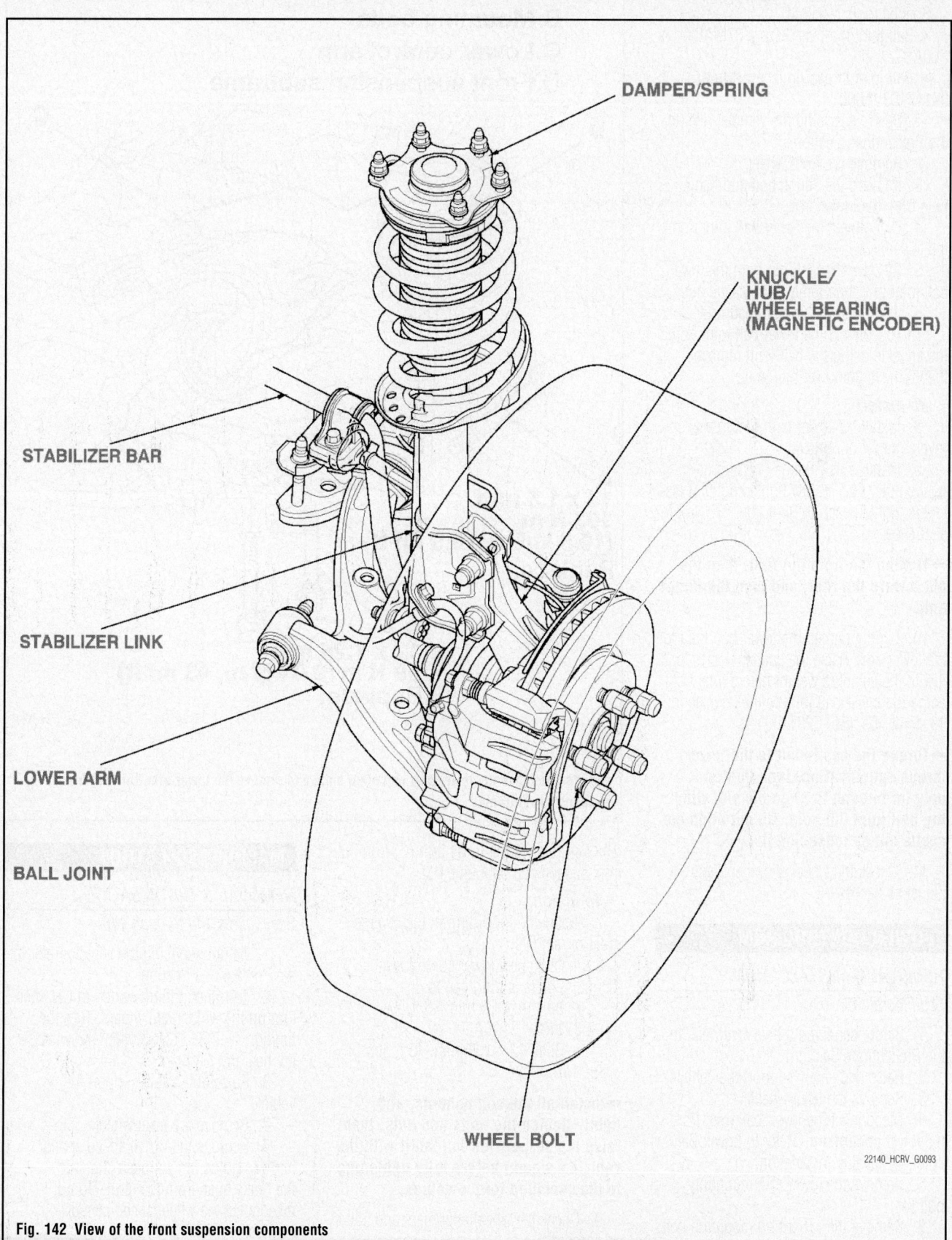

Fig. 142 View of the front suspension components

LOWER BALL JOINT

REMOVAL & INSTALLATION

Special tools required:
• Ball joint remover, 32mm 07MAC-SL0A102
• Ball joint thread protector, 14mm 071AF-S3VA000

1. Before servicing the vehicle, refer to the Precautions Section.
2. Remove the front wheel.
3. Remove the flange bolt and flange nuts from the lower arm.
4. Disconnect the lower ball joint from the lower arm.
5. Remove the lock pin from the lower ball joint pin, then remove the castle nut.
6. Install the ball joint thread protector.
7. Disconnect the lower ball joint from the knuckle using the ball joint remover, then remove the lower ball joint.

To install:

8. Install the lower ball joint in the reverse order of removal.
9. Install a new flange bolt and new flange nuts. After lightly tightening all 3 fasteners, tighten each to 38 ft. lbs. (52 Nm) in sequence.

➡**Tighten the front nut first, then the nut toward the rear, and then the flange bolt.**

10. Lightly tighten the lower ball joint to the lower arm. Raise the suspension to load it with the vehicle's weight before fully tightening the lower ball joint to the knuckle to 43–51 ft. lbs. (59–69 Nm).

➡**Torque the castle nut to the lower torque specification, then tighten it only far enough to align the slot with the ball joint pin hole. Do not align the castle nut by loosening it.**

11. Check the wheel alignment, and adjust as necessary.

LOWER CONTROL ARM

REMOVAL & INSTALLATION

See Figure 143.

1. Before servicing the vehicle, refer to the Precautions Section.
2. Raise and safely support the vehicle.
3. Remove the front wheel.
4. Disconnect the lower ball joint from the lower control arm. Refer to Lower Ball Joint, removal & installation.
5. Remove the lower arm mounting bolt (A).
6. Remove the lower arm mounting bolts

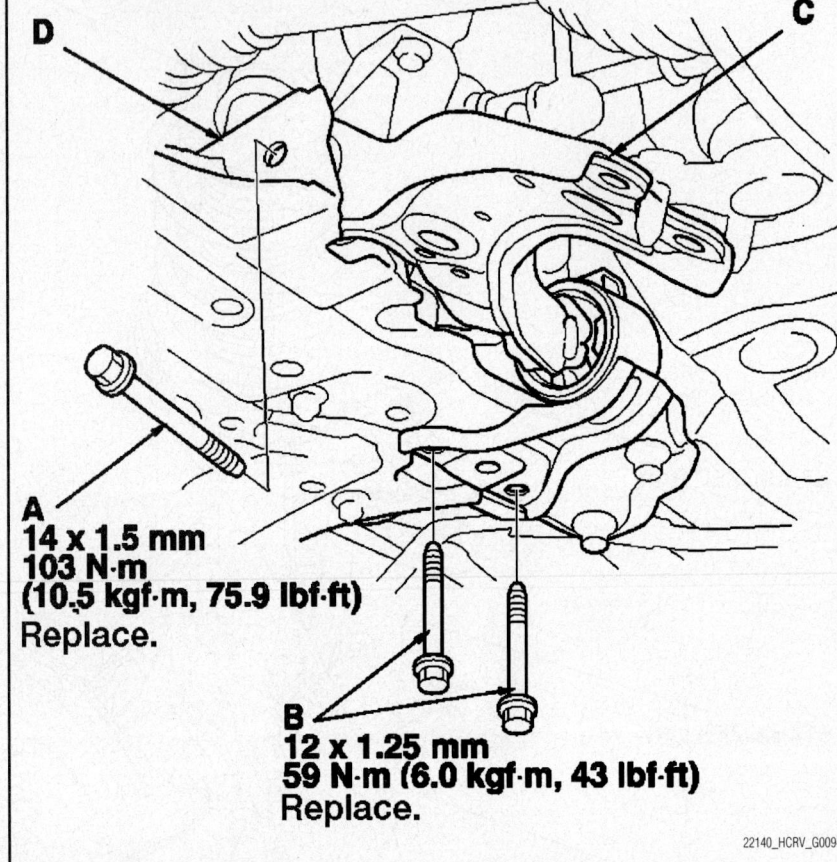

A. Mounting bolt
B. Mounting bolts
C. Lower control arm
D. Front suspension subframe

A
14 x 1.5 mm
103 N·m
(10.5 kgf·m, 75.9 lbf·ft)
Replace.

B
12 x 1.25 mm
59 N·m (6.0 kgf·m, 43 lbf·ft)
Replace.

22140_HCRV_G0094

Fig. 143 Remove the lower arm mounting bolts and remove the lower arm from the front suspension subframe

(B), then remove the lower arm (C) from the front suspension subframe (D).

To install:

7. Install the lower arm in the reverse order of removal.
8. Install a new lower control arm mounting bolts and tighten to:
  a. Bolt 14 x 1.5 mm: 76 ft. lbs. (103 Nm).
  b. Bolts 12 x 1.25 mm: 43 ft. lbs. (59 Nm).

➡**Install all the components, and lightly tighten the bolts and nuts, then raise the suspension to load it with the vehicle's weight before fully tightening to the specified torque values.**

9. Check the wheel alignment and adjust as necessary.

MACPHERSON STRUT

REMOVAL & INSTALLATION

See Figures 144 through 146.

1. Before servicing the vehicle, refer to the Precautions Section.
2. Turn the ignition switch to ON, then turn on the windshield wipers. Turn the ignition switch to LOCK when the wipers are near the A-pillars.
3. Raise and safely support the vehicle.
4. Remove the front wheel.
5. Remove the wheel speed sensor harness guide, the harness clip, and the brake hose from the strut. Do not disconnect the wheel speed sensor connector.

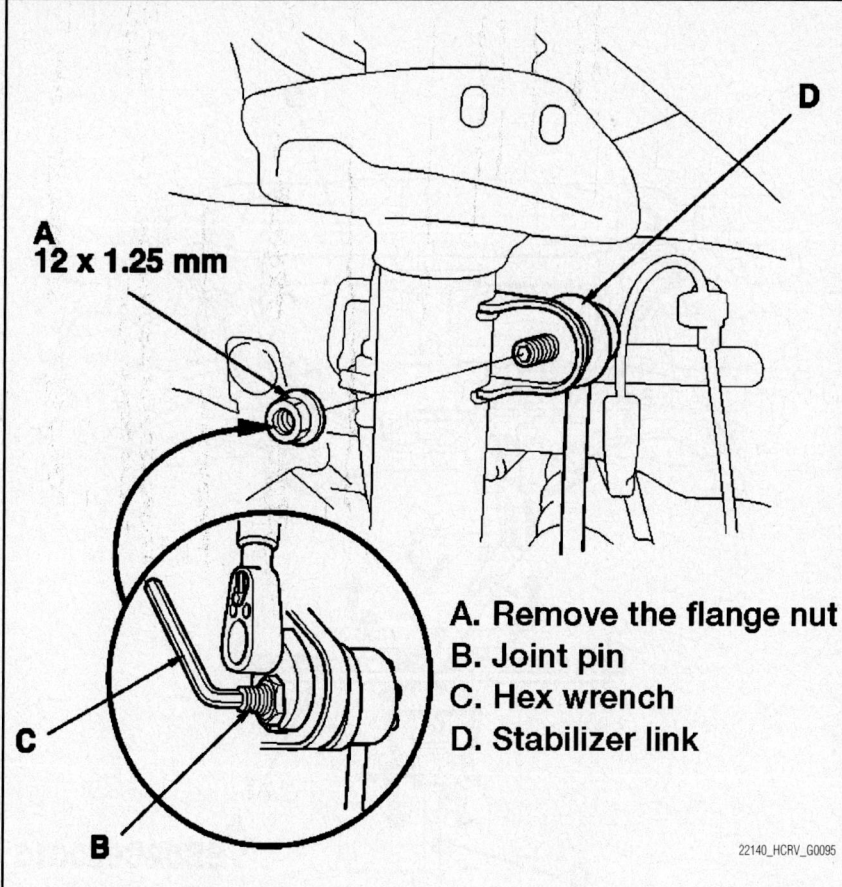

A
12 x 1.25 mm

A. Remove the flange nut
B. Joint pin
C. Hex wrench
D. Stabilizer link

22140_HCRV_G0095

Fig. 144 Remove the flange nut, while holding the joint pin with a hex wrench, and disconnect the stabilizer link from the strut

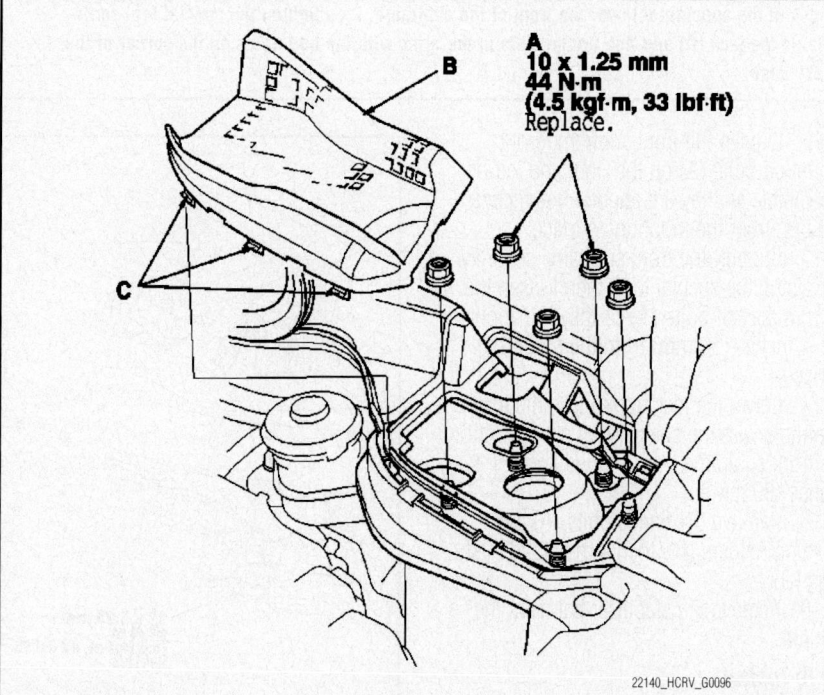

B
A
10 x 1.25 mm
44 N·m
(4.5 kgf·m, 33 lbf·ft)
Replace.

C

22140_HCRV_G0096

Fig. 145 Remove the lid (B) by releasing the hooks (C) and remove the flange nuts (A) from the top of the strut

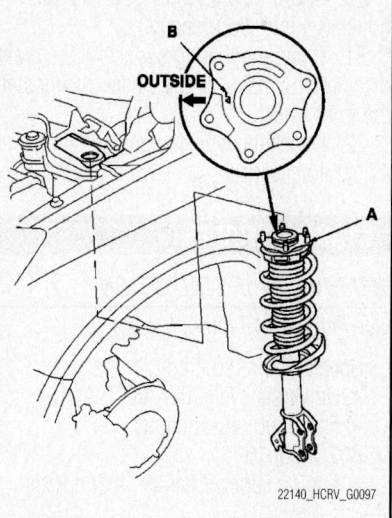

22140_HCRV_G0097

Fig. 146 View of strut assembly (A) and directional stamp for mounting (B)

6. Make an alignment marking on the camber adjusting bolt and strut for approximate installation alignment later.

7. Remove the flange nut (A), while holding the joint pin (B) with a hex wrench (C), and disconnect the stabilizer link (D) from the strut.

8. Remove the strut mounting bolts and self-locking nuts from the strut.

9. Remove the lid (B) by releasing the hooks (C).

10. Remove the flange nuts (A) from the top of the strut.

11. Remove the strut assembly (A).

➡The strut springs are different. Mark the springs L and R before removal.

To install:

12. Install the strut assembly (A) on to the frame. Note the direction of the strut mounting base so that the stamp (B) on it is toward the outside of the vehicle.

13. Loosely install new flange nuts to the top of the strut.

14. Loosely install new strut mounting bolts and new self-locking nuts to the strut.

15. Connect the stabilizer link to the strut, and loosely install the flange nut, while holding the joint pin with a hex wrench.

16. Install the wheel speed sensor harness guide, the harness clip, and the brake hose to the strut.

17. Raise the front suspension with a floor jack to load the suspension with the vehicle's weight.

18. Tighten the strut mounting bolts and the self-locking nuts to 116 ft. lbs. (157 Nm).

19. Tighten the flange nuts on top of the strut to 33 ft. lbs. (44 Nm).

20. Install the lid (B) by pushing the hooks (C) into place securely.

21. Clean the mating surface of the brake disc and the inside of the wheel, then install the front wheel.

22. Check the wheel alignment and adjust if necessary.

STABILIZER BAR

REMOVAL & INSTALLATION

See Figures 147 through 152.

Special tools required:
- Universal eyelet 07AAK-SNAA120
- Front subframe adapter VSB02C000016
- Engine support hanger, A and Reds AAR-T-12566

1. Before servicing the vehicle, refer to the Precautions Section.

2. Matchmark the stabilizer bar for proper reinstallation.

3. Raise and safely support the vehicle.

4. Remove the front wheels.

5. Disconnect both stabilizer control links from the stabilizer bar.

6. Remove the cowl cover.

7. Attach the universal eyelet to the cylinder head.

8. Install the engine support hanger (AAR-T-12566) to the vehicle and attach the hook to the universal eyelet.

9. Attach the front subframe adapter (A) to the front subframe (B) by hanging the hook of the special tool over the front of the subframe, then tighten the special tool screw.

10. Raise the jack (C) and line up the slots in the arms with the bolt holes on the corner of the jack base, then securely attach them with bolts.

11. Remove both mid-bracket bolts.

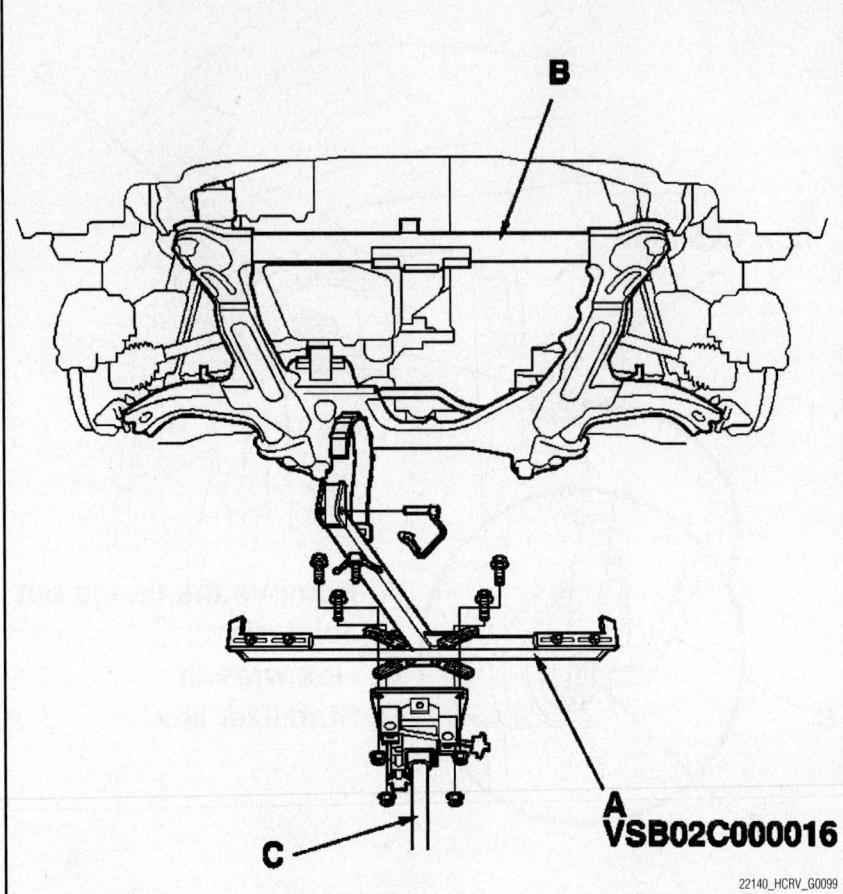

Fig. 148 Attach the front subframe adapter (A) to the front subframe (B) by hanging the hook of the special tool over the front of the subframe, then tighten the special tool screw. Raise the jack (C) and line up the slots in the arms with the bolt holes on the corner of the jack base.

12. Loosen the front subframe front mounting bolts (A) on the right and left of the vehicle so they are about 1 ³⁄₁₆ inches (30mm) from the mounting surface.

13. Support the front subframe securely by raising the special tool, then loosen the 14 mm special bolts (A) so they are about 1 ³⁄₁₆ inches (30 mm) from the mounting surface.

14. Lower the jack supporting the front subframe with the special tool slowly until the front subframe has dropped about 1 ³⁄₁₆ inches (30 mm).

15. Remove the flange bolts (A) and bushing holders (B), then remove the bushings (C).

16. Remove the stabilizer bar from the vehicle.

To install:

17. Install the stabilizer bar into position aligning the paint marks (D) on the stabilizer bar with the sides of the bushings.

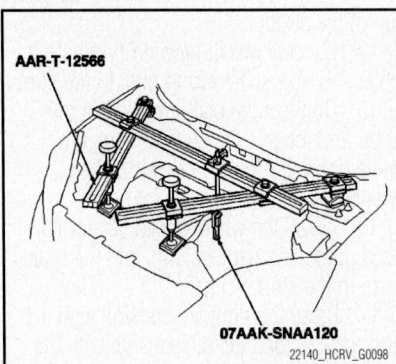

Fig. 147 Install the engine support hanger (AAR-T-12566) to the vehicle and attach the hook to the universal eyelet

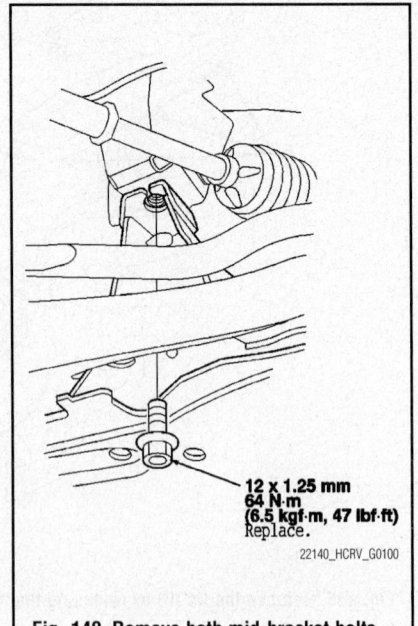

Fig. 149 Remove both mid-bracket bolts

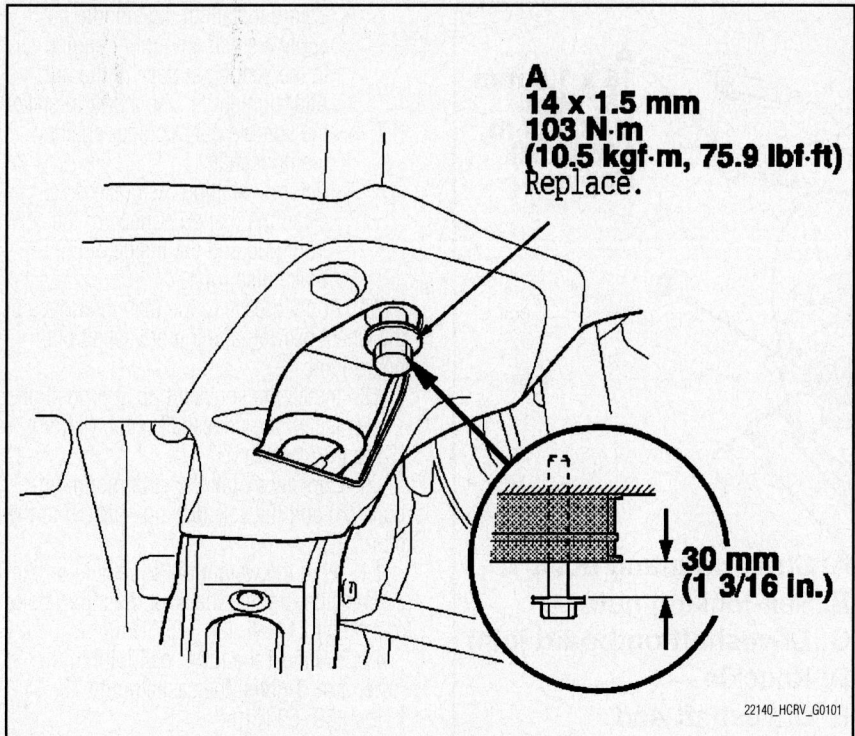

Fig. 150 Loosen the front subframe front mounting bolts (A) on the right and left of the vehicle so they are about 1 ³⁄₁₆ inches (30 mm) from the mounting surface

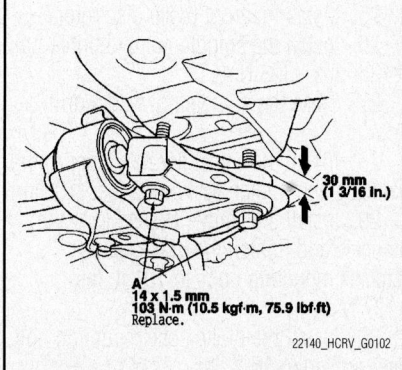

Fig. 151 Support the front subframe securely by raising the special tool, then loosen the 14 mm special bolts (A) so they are about 1 ³⁄₁₆ inches (30 mm) from the mounting surface

➡Ensure the right and left direction of the stabilizer bar is correct before installation. Make sure the fore and aft direction of the bushings is correct.

18. Install the bushings (C), bushing holders (B), and the flange bolts (A). Tighten the bolts to 16 ft. lbs. (22 Nm).

19. Raise the jack supporting the front subframe with the special tool slowly until the front subframe makes contact with the body frame.

20. Installation continues in the reverse of the removal procedure.

21. Tighten the bolts to the specified torque:

   a. Bolts 14 x 1.5mm: 76 ft. lbs. (103 Nm).

   b. Bolts 12 x 1.25mm: 47 ft. lbs. (64 Nm).

22. Check the front wheel alignment and make adjustments as necessary.

STEERING KNUCKLE

REMOVAL & INSTALLATION

See Figure 153.

Special tools required:
- Ball joint remover, 32 mm 07MAC-SL0A102
- Ball joint remover, 28 mm 07MAC-SL0A202
- Ball joint thread protector, 14 mm 071AF-S3VA000

1. Before servicing the vehicle, refer to the Precautions Section.

2. Raise and safely support the vehicle.

3. Remove the front wheels.

4. Remove the brake hose mounting bolt.

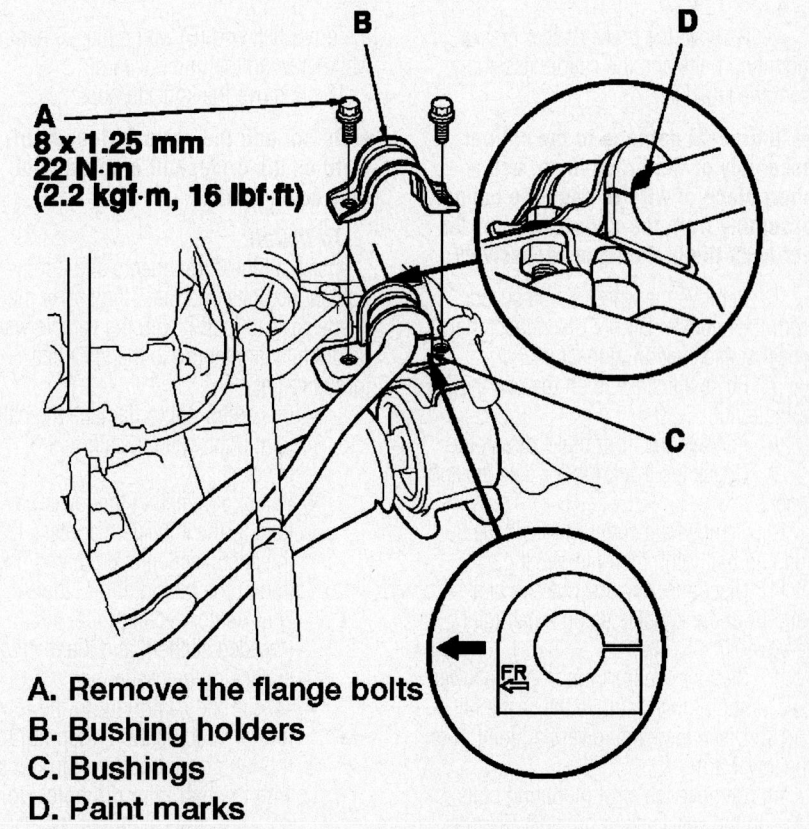

A. Remove the flange bolts
B. Bushing holders
C. Bushings
D. Paint marks

Fig. 152 Remove the flange bolts and bushing holders, then remove the bushings. (Align the paint marks during installation)

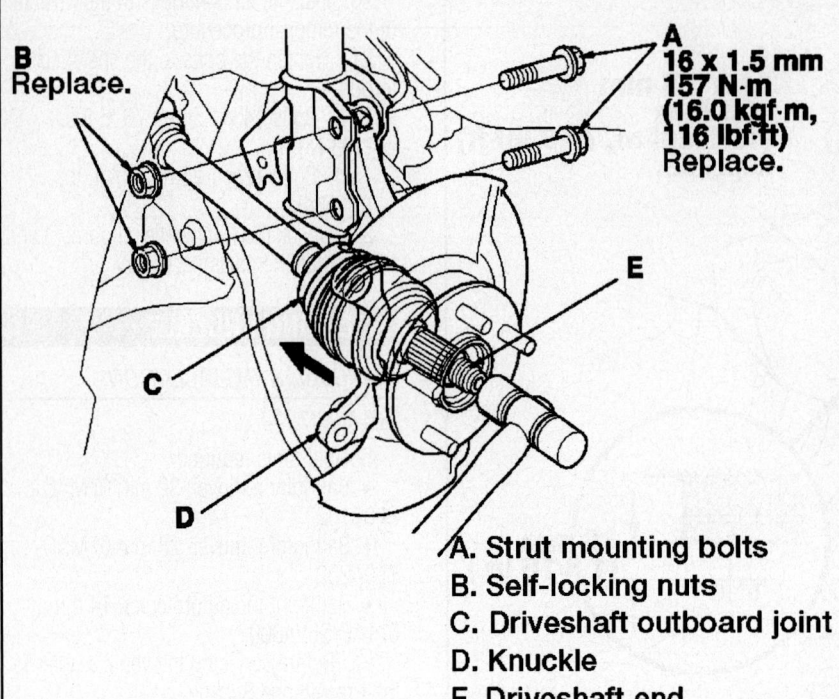

B
Replace.

A
16 x 1.5 mm
157 N·m
(16.0 kgf·m,
116 lbf·ft)
Replace.

A. **Strut mounting bolts**
B. **Self-locking nuts**
C. **Driveshaft outboard joint**
D. **Knuckle**
E. **Driveshaft end**

22140_HCRV_G0104

Fig. 153 Remove the strut mounting bolts and the self-locking nuts from the strut. Remove the driveshaft outboard joint from the knuckle by tapping the driveshaft end with a plastic hammer while drawing the hub outward

5. Remove the brake caliper bracket mounting bolts and the caliper assembly from the knuckle.

→**To prevent damage to the caliper assembly or the brake hose, use a short piece of wire to hang the caliper assembly from the undercarriage. Do not twist the brake hose excessively.**

6. Remove the wheel speed sensor from the knuckle. Do not disconnect the wheel speed sensor connector.

7. Raise the stake, then remove the spindle nut.

8. Remove the front brake disc rotor.

9. Check the front hub for damage and cracks.

10. Remove the cotter pin from the tie-rod end ball joint, then remove the nut.

11. Disconnect the tie-rod end ball joint from the knuckle using a ball joint remover.

12. Remove the lock pin from the lower ball joint pin, then remove the castle nut.

13. Disconnect the lower ball joint from the lower arm.

14. Remove the strut mounting bolts (A) and the self-locking nuts (B) from the strut.

15. Remove the driveshaft outboard joint (C) from the knuckle (D) by tapping

the driveshaft end (E) with a plastic hammer while drawing the hub outward.

16. Remove the knuckle/hub.

→**Do not pull the driveshaft end outward as the driveshaft inboard joint may come apart.**

To install:

Install all the components and lightly tighten the bolts and nuts, then raise the suspension to load it with the vehicle weight before fully tightening to the specified torque values.

- Be careful not to damage the ball joint boot when installing the knuckle
- Before connecting the lower ball joint to the knuckle, degrease the threaded section and tapered portion of the ball joint pin, the knuckle connecting hole, the threaded section, and the mating surface of the castle nut
- Torque the castle nut to the lower torque specification, then tighten it only far enough to align the slot with the ball joint pin hole. Do not align the castle nut by loosening it
- Use a new spindle nut during reassembly

- Before installing the spindle nut, apply a small amount of engine oil to the seating surface of the nut. After tightening, use a drift to stake the spindle nut shoulder against the driveshaft.
- Before installing the brake disc, clean the mating surface of the front hub and the inside of the brake disc rotor.

17. Apply grease to the mating surface of the wheel bearing and the driveshaft outboard joint.

18. Install the knuckle/hub into position.

19. Install the driveshaft outboard joint (C) to the knuckle (D).

20. Loosely install the strut mounting bolts (A) and the self-locking nuts (B) to the strut.

21. With the weight of the vehicle on the suspension, tighten the strut mounting bolts and nuts to 116 ft. lbs. (157 Nm).

22. Connect the lower ball joint to the lower arm. Tighten the castle nut to 43–51 ft. lbs. (59–69 Nm).

23. Install the lock pin into the lower ball joint.

24. Connect the tie-rod end ball joint to the knuckle and tighten the nut to 40 ft. lbs. (54 Nm). Install the cotter pin.

25. Install the front brake disc rotor.

26. Install the spindle nut and tighten to 242 ft. lbs. (328 Nm).

27. Stake the spindle nut with a drift tool.

28. Install the wheel speed sensor to the knuckle and tighten to 86 inch lbs. (10 Nm).

29. Install the caliper assembly to the knuckle and tighten the brake caliper bracket mounting bolts to 101 ft. lbs. (137 Nm).

30. Install the brake hose mounting bolt and tighten to 16 ft. lbs. (22 Nm).

31. Install the wheel and tighten to 80 ft. lbs. (108 Nm).

32. Check the wheel alignment and adjust as necessary.

STRUT & SPRING ASSEMBLY

REMOVAL & INSTALLATION

Refer to MacPherson Strut, removal & installation.

WHEEL BEARINGS

REMOVAL & INSTALLATION

See Figures 154 through 157.

Special tools required:
- Hub disassembly/assembly tool 07GAF-SD40100

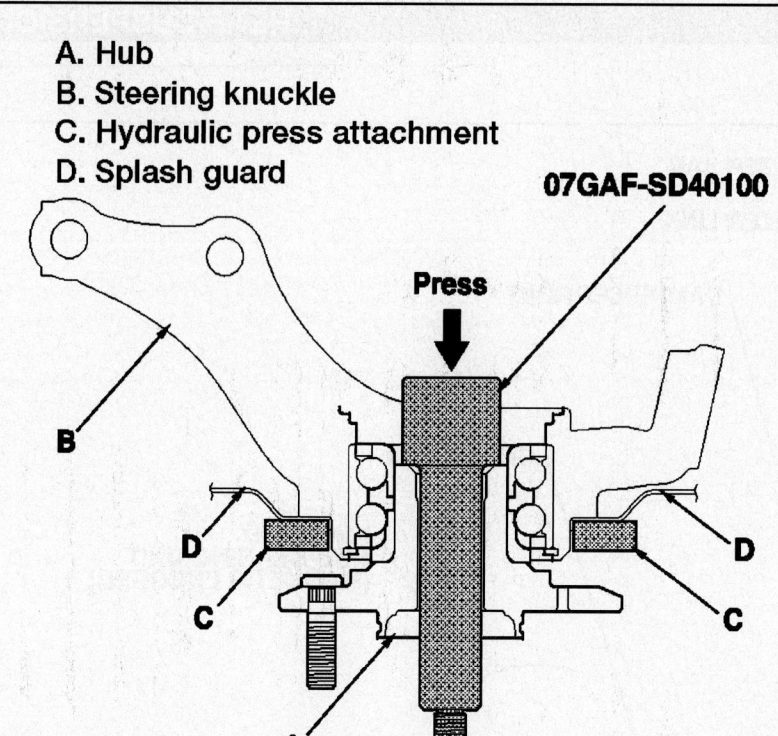

A. Hub
B. Steering knuckle
C. Hydraulic press attachment
D. Splash guard

07GAF-SD40100

Press

22140_HCRV_G0105

Fig. 154 Separate the hub from the knuckle using the hub disassembly/assembly tool and a hydraulic press. Hold the knuckle with the attachment of the hydraulic press or equivalent tool. Be careful not to deform the splash guard.

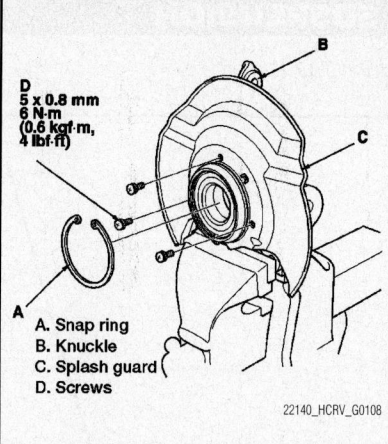

D 5 x 0.8 mm 6 N·m (0.6 kgf·m, 4 lbf·ft)

A. Snap ring
B. Knuckle
C. Splash guard
D. Screws

22140_HCRV_G0108

Fig. 157 Install the snap ring securely in the knuckle. Install the splash guard and tighten the screws

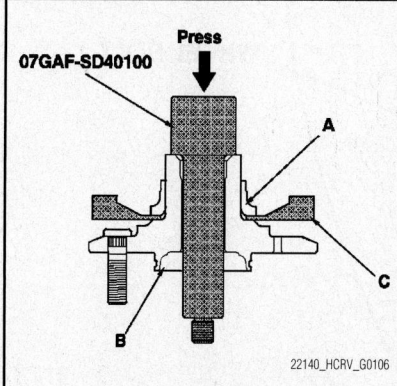

07GAF-SD40100

Press

22140_HCRV_G0106

Fig. 155 Press the wheel bearing inner race (A) off of the hub (B) using the hub disassembly/assembly tool, a commercially available bearing separator (C), and a press

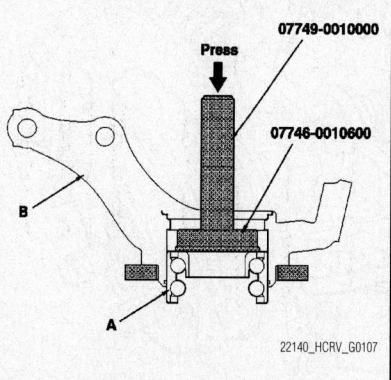

Press

07749-0010000

07746-0010600

22140_HCRV_G0107

Fig. 156 Press the wheel bearing (A) out of the knuckle (B) using the attachment, the driver, and a press

4. Press the wheel bearing inner race (A) off of the hub (B) using the hub disassembly/assembly tool, a commercially available bearing separator (C), and a press.

5. Remove the splash guard and the snap ring from the knuckle.

6. Press the wheel bearing (A) out of the knuckle (B) using the attachment, the driver, and a press.

To install:

7. Wash the knuckle and hub thoroughly in high flash-point solvent before reassembly.

8. Press a new wheel bearing into the knuckle using the old bearing, a steel plate, the attachment, the support base, and a press.

- Install the wheel bearing with the wheel speed sensor magnetic encoder (brown color) toward the inside of the knuckle
- Remove any oil, grease, dust, metal debris, and other foreign material from the encoder surface
- Keep all magnetic tools away from the encoder surface
- Be careful not to damage the encoder surface when inserting the wheel bearing

9. Install the snap ring (A) securely in the knuckle (B).

10. Install the splash guard (C) and tighten the screws (D) to 48 inch lbs. (6 Nm).

11. Install the hub onto the knuckle using the attachment, the driver, the support base, and a hydraulic press. Be careful not to distort the splash guard.

12. Install the knuckle/hub assembly. Refer to Steering Knuckle, removal & installation.

- Attachment, 72 x 75 mm 07746-0010600
- Driver 07749-0010000
- Attachment, 96mm 07948-SB00101
- Support base 07965-SD90100

1. Before servicing the vehicle, refer to the Precautions Section.

2. Remove the knuckle/hub assembly.

Refer to Steering Knuckle, removal & installation.

3. Separate the hub (A) from the knuckle (B) using the hub disassembly/assembly tool and a hydraulic press. Hold the knuckle with the attachment (C) of the hydraulic press or equivalent tool. Be careful not to deform the splash guard (D). Hold onto the hub to keep it from falling when pressed clear.

See Figure 158.

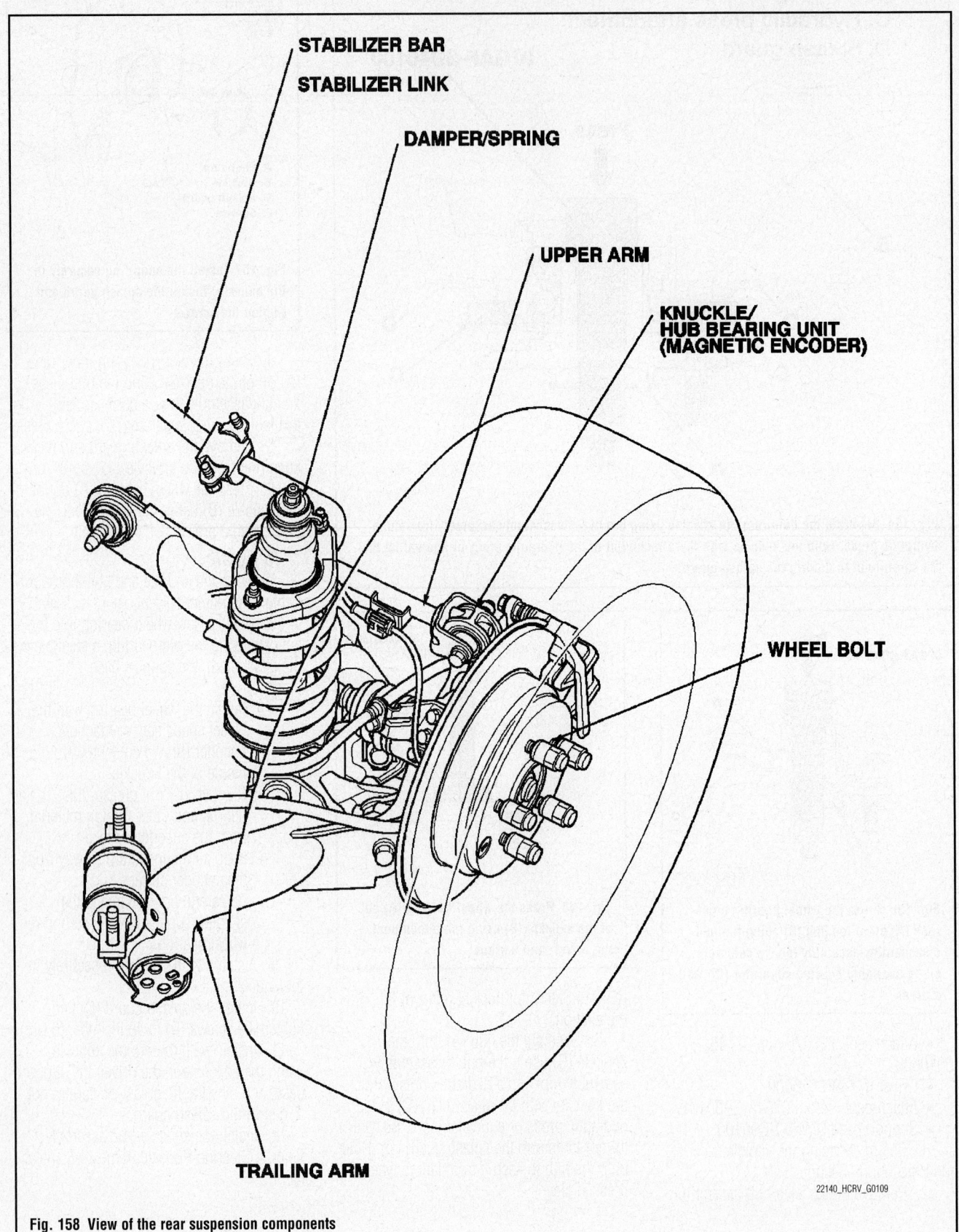

STABILIZER BAR

STABILIZER LINK

DAMPER/SPRING

UPPER ARM

KNUCKLE/
HUB BEARING UNIT
(MAGNETIC ENCODER)

WHEEL BOLT

TRAILING ARM

22140_HCRV_G0109

Fig. 158 View of the rear suspension components

STRUT & SPRING ASSEMBLY

REMOVAL & INSTALLATION

See Figures 159 through 161.

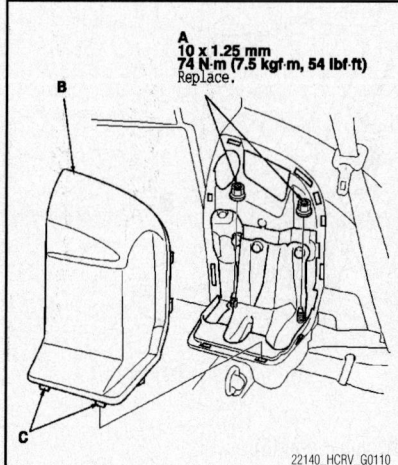

Fig. 159 Remove the lid (B) on the cargo area side trim panel by releasing the hooks (C) and remove the flange nuts (A) from the top of the strut

1. Before servicing the vehicle, refer to the Precautions Section.
2. Raise and safely support the vehicle.
3. Remove the rear wheel.
4. Remove the lid (B) on the cargo area side trim panel by releasing the hooks (C).
5. Remove the flange nuts (A) from the top of the strut.
6. Remove the flange bolts, then remove the brake hose bracket.
7. Position a floor jack under the trailing arm (A). Raise the floor jack until the suspension begins to compress.
8. Disconnect the stabilizer link (D) from the trailing arm.
9. Remove the flange bolt (B) from the bottom of the strut.
10. Remove the flange bolt, then remove the parking brake cable.
11. Remove the trailing arm mounting bolts.
12. Lower the rear suspension, then remove the strut assembly from the vehicle.

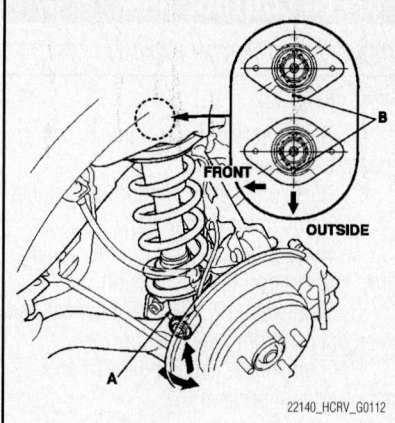

Fig. 161 Position the strut assembly (A) between the body and the trailing arm. Note the direction of the strut mounting base and the hook or the paint mark (B).

To install:

13. Position the strut assembly (A) between the body and the trailing arm. Note the direction of the strut mounting base so that the hook or the paint mark (B) on it is toward the outside of the vehicle.
14. Loosely install new flange nuts on the top of the strut.
15. Loosely install new trailing arm mounting bolts.
16. Install the parking brake cable, then install the flange bolt.
17. Position a floor jack under the trailing arm. Raise the floor jack until the hole in the trailing arm aligns with the hole in the strut.
18. Loosely install a new flange bolt on the bottom of the strut.
19. Connect the stabilizer link to the trailing arm.
20. Raise the rear suspension with a floor jack to load the suspension with the vehicle's weight.
21. Tighten new flange nuts, new flange bolts, and other fasteners to the specified torque values.
   a. Flange nuts and bolts: 54 ft. lbs. (74 Nm).
   b. Parking brake cable bolt: 16 ft. lbs. (22 Nm).
   c. New flange bolt at bottom of strut: 69 ft. lbs. (93 Nm).
   d. Brake hose bracket bolts: 16 ft. lbs. (22 Nm).
22. Install the strut cover lid.
23. Clean the mating surface of the brake disc and the inside of the wheel, then install the rear wheel.
24. Check the wheel alignment and adjust it if necessary.

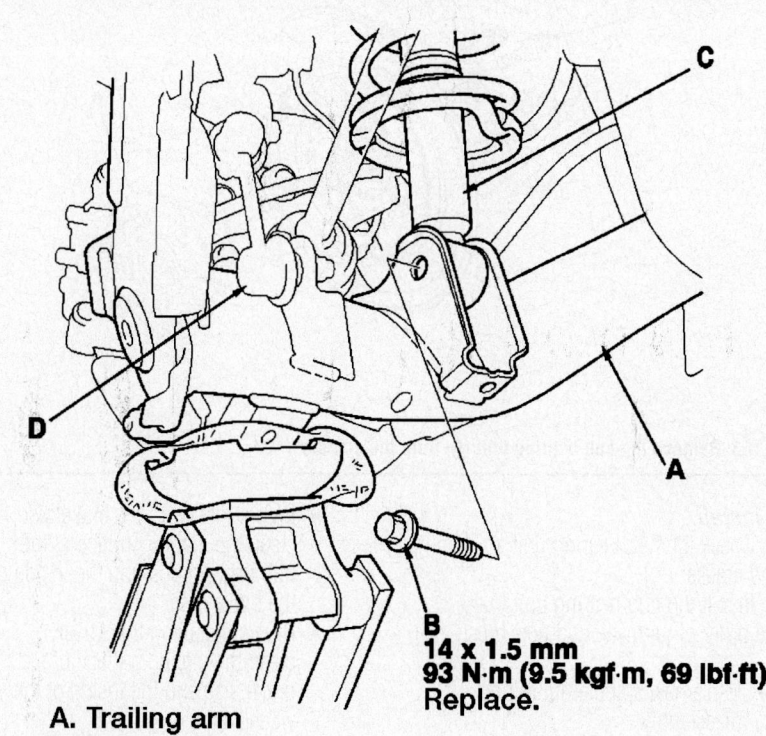

A. Trailing arm
B. Flange bolt
C. Strut
D. Stabilizer link

Fig. 160 Position a floor jack under the trailing arm and raise it until the suspension begins to compress. Disconnect the stabilizer link from the trailing arm. Remove the flange bolt from the bottom of the strut.

UPPER CONTROL ARM

REMOVAL & INSTALLATION

See Figure 162.

1. Before servicing the vehicle, refer to the Precautions Section.
2. Raise and safely support the vehicle.
3. Remove the rear wheel.
4. Place a floor jack under the trailing arm, and support the suspension.
5. Remove the wheel speed sensor harness bracket from the upper arm.
6. Remove the flange bolts (A) and remove the upper arm (B).

To install:

7. Install the upper arm (B) and loosely install the flange bolts (A).
8. Install the wheel speed sensor harness bracket to the upper arm.
9. Place a floor jack under the trailing arm and raise the suspension to load it with the vehicle weight before tightening the flange bolts.
10. Tighten the new flange bolts to 69 ft. lbs. (93 Nm).
11. Install the rear wheel.
12. Check the wheel alignment and adjust it if necessary.

WHEEL BEARINGS

REMOVAL & INSTALLATION

See Figure 163.

1. Before servicing the vehicle, refer to the Precautions Section.
2. Raise and safely support the vehicle.
3. Remove the rear wheel.
4. Remove the disc rotor.
5. Remove the brake hose bracket mounting bolt from the knuckle.
6. Remove the brake caliper bracket mounting bolts, and remove the caliper assembly from the knuckle.

➡**To prevent damage to the caliper assembly or brake hose, use a short piece of wire to hang the caliper assembly from the undercarriage. Do not twist the brake hose excessively.**

7. Remove the 2 washers from the mounting bolt holes.
8. On 4WD models, raise the stake, then remove the spindle nut.
9. Release the parking brake and remove the brake disc.
10. Remove the hub bearing unit (A) from the spindle.

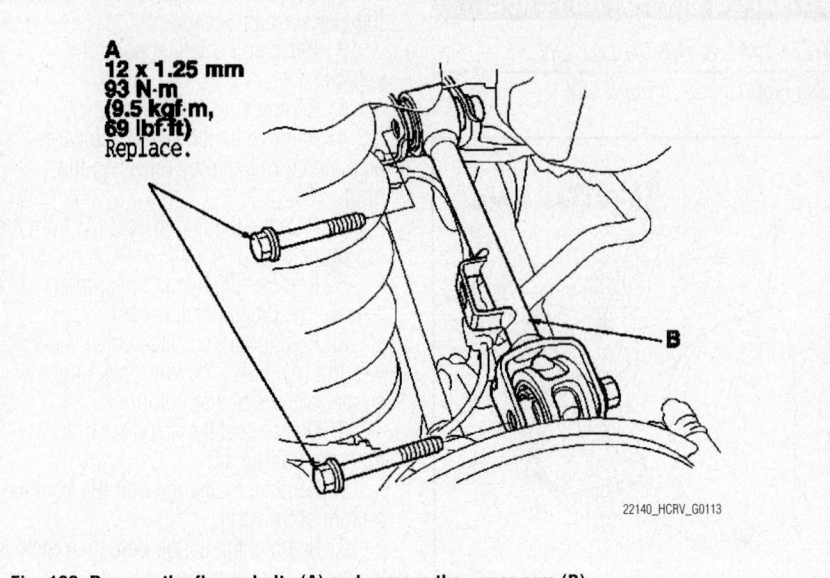

Fig. 162 Remove the flange bolts (A) and remove the upper arm (B)

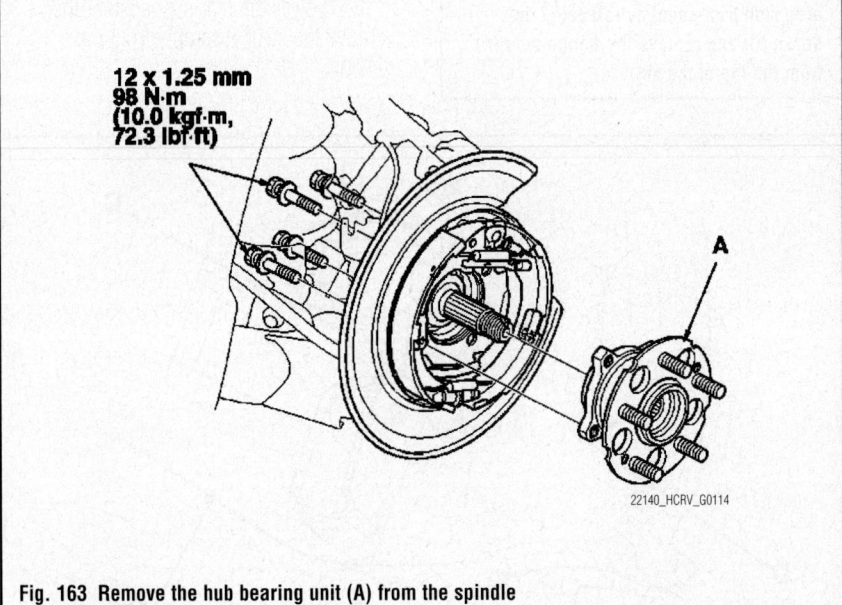

Fig. 163 Remove the hub bearing unit (A) from the spindle

To install:

11. Check the hub bearing unit for damage and cracks.
12. Install the hub bearing unit in the reverse order of removal, and note these items:
   - Use a new spindle nut during reassembly
   - On 4WD models: before installing the spindle nut, apply a small amount of engine oil to the seating surface of the nut. After tightening, use a drift to stake the spindle nut shoulder against the driveshaft
   - Before installing the brake disc, clean the mating surface of the hub bearing unit and the inside of the brake disc
   - Before installing the wheel, clean the mating surface of the brake disc and the inside of the wheel
   - After installation, press the brake pedal several times to make sure the brakes function properly

13. Check the wheel alignment and adjust it if necessary.

HONDA

Element

SPECIFICATIONS AND MAINTENANCE CHARTS

ENGINE AND VEHICLE IDENTIFICATION CHART

		Engine Code						Model Year	
Code	Liters (cc)	Cu. In.	Cyl.	Fuel Sys.	Engine Type	Eng. Mfg.		Code ①	Year
K24A4	2.4 (2354)	144	4	SMFI	DOHC	Honda		7	2007
								8	2008

DOHC: Double Overhead Cam

SMFI: Sequential Multi-port Fuel Injection

① 10th position of VIN

22140_ELEM_C0001

GENERAL ENGINE SPECIFICATIONS

Year	Model	Engine Displacement Liters (VIN)	Net Horsepower @ rpm	Net Torque @ rpm (ft. lbs.)	Bore x Stroke (in.)	Com-pression Ratio	Oil Pressure @ rpm
2007	Element	2.4 (K24A4)	166@5800	161@4500	3.42x3.90	9.7:1	44@3000
2008	Element	2.4 (K24A4)	166@5800	161@4500	3.42x3.90	9.7:1	44@3000

SMFI: Sequential Multi-port Fuel Injection

22140_ELEM_C0002

ENGINE TUNE-UP SPECIFICATIONS

Year	Engine Displacement Liters (VIN)	Spark Plug Gap (in.)	Ignition Timing (deg.) MT	Ignition Timing (deg.) AT	Fuel Pump (psi)	Idle Speed (rpm) MT	Idle Speed (rpm) AT	Valve Clearance (in.) In.	Valve Clearance (in.) Ex.
2007	2.4 (K24A4)	0.039-0.043	6-10B	6-10B	48-55	650-750	650-750	0.008-0.010	0.011-0.013
2008	2.4 (K24A4)	0.039-0.043	6-10B	6-10B	48-55	650-750	650-750	0.008-0.010	0.011-0.013

NOTE: The Vehicle Emission Control Information label often reflects changes made during production and must be used if they differ from this chart.

NOTE: The fuel pressure readings are given with the vacuum hose connected to the regulator and the engine running

B: Before top dead center

HYD: Hydraulic

22140_ELEM_C0003

CAPACITIES

Year	Model	Engine Displacement Liters (VIN)	Engine Oil with Filter (qts.)	Transmission (pts.)		Transfer Case (pts.)	Drive Axle		Fuel Tank (gal.)	Cooling System (qts.)
				5-Spd	Auto.		Front (pts.)	Rear (pts.)		
2007	Element	2.4 (K24A4)	4.4	4.0	①	②	②	2.2	15.9	③
2008	Element	2.4 (K24A4)	4.4	4.0	①	②	②	2.2	15.9	③

NOTE: All capacities are approximate. Add fluid gradually and check to be sure a proper fluid level is obtained.

① 2WD: 6.6 pts. for fluid change, 15.2 for overhaul
　4WD: 6.3 pts. for fluid change, 13.8 for overhaul

② Included in transaxle refill figure

③ Manual trans: 7.6 qts
　Auto trans: 7.5 qts.

22140_ELEM_C0004

VALVE SPECIFICATIONS

Year	Engine Displacement Liters (VIN)	Seat Angle (deg.)	Face Angle (deg.)	Spring Test Pressure (lbs. @ in.)	Spring Installed Height (in.)	Stem-to-Guide Clearance (in.)		Stem Diameter (in.)	
						Intake	Exhaust	Intake	Exhaust
2007	2.4 (K24A4)	NA	NA	NA	①	0.0012-0.0022	0.0022-0.0031	0.2156-0.2159	0.2146-0.2150
2008	2.4 (K24A4)	NA	NA	NA	①	0.0012-0.0022	0.0022-0.0031	0.2156-0.2159	0.2146-0.2150

NA: Not Available

① Valve spring free length:
　Intake: 1.668 in.
　Exhaust: 1.745 in.

22140_ELEM_C0005

CRANKSHAFT AND CONNECTING ROD SPECIFICATIONS

All measurements are given in inches

Year	Engine Displacement Liters (VIN)	Crankshaft				Connecting Rod		
		Main Brg. Journal Dia.	Main Brg. Oil Clearance	Shaft End-play	Thrust on No.	Journal Diameter	Oil Clearance	Side Clearance
2007	2.4 (K24A4)	①	②	0.0040-0.0140	3	1.8888-1.8898	0.0008-0.0019	0.016
2008	2.4 (K24A4)	①	②	0.0040-0.0140	3	1.8888-1.8898	0.0008-0.0019	0.016

① Except No. 3: 2.1648-2.1657
　No. 3: 2.1644-2.1654

② Except No. 3: 0.0007-0.0016
　No. 3: 0.0010-0.0019

22140_ELEM_C0006

PISTON AND RING SPECIFICATIONS

All measurements are given in inches

Year	Engine Displacement Liters (VIN)	Piston Clearance	Ring Gap			Ring Side Clearance		
			Top Compression	Bottom Compression	Oil Control	Top Compression	Bottom Compression	Oil Control
2007	2.4 (K24A4)	0.0008-0.0016	0.0080-0.0140	0.0160-0.0220	0.0080-0.0280	0.0018-0.0028	0.0020-0.0030	NA
2008	2.4 (K24A4)	0.0008-0.0016	0.0080-0.0140	0.0160-0.0220	0.0080-0.0280	0.0018-0.0028	0.0020-0.0030	NA

NA: Not Applicable

22140_ELEM_C0007

TORQUE SPECIFICATIONS

All readings in ft. lbs.

Year	Engine Displacement Liters (VIN)	Cylinder Head Bolts	Main Bearing Bolts	Rod Bearing Bolts	Crankshaft Damper Bolts	Flywheel Bolts	Manifold		Spark Plugs	Oil Pan Drain Plug
							Intake	Exhaust		
2007	2.4 (K24A4)	①	②	③	④	76	16	33	13	33
2008	2.4 (K24A4)	①	②	③	④	76	16	33	13	33

NOTE: Dip main bearing bolts and crankshaft damper bolt in clean engine oil prior to tightening.

① Step 1: 29 ft. lbs.

 Step 2: +90 degrees

 Step 3: +90 degrees

 Step 4: NEW BOLT ONLY +90 degrees

② 22 ft. lbs. +56 degrees

③ 14 ft. lbs. +90 degrees

④ 36 ft. lbs. +90 degrees

22140_ELEM_C0008

WHEEL ALIGNMENT

Year	Model		Caster		Camber		Toe-in (in.)
			Range (+/-Deg.)	Preferred Setting (Deg.)	Range (+/-Deg.)	Preferred Setting (Deg.)	
2007	Element	F	1.00	+1.75	0.75	0	0+/-0.08
		R	—	—	0.75	-1.00	0.08+/-0.08
2008	Element	F	1.00	+1.75	0.75	0	0+/-0.08
		R	—	—	0.75	-1.00	0.08+/-0.08

22140_ELEM_C0009

TIRE, WHEEL AND BALL JOINT SPECIFICATIONS

Year	Model	OEM Tires		Tire Pressures (psi)		Wheel Size	Ball Joint Inspection	Lug Nut (ft. lbs.)
		Standard	Optional	Front	Rear			
2007	Element	P215/70R16	None	26	26	6JJ	NS	80
2008	Element	P215/70R16	None	26	26	6JJ	NS	80

OEM: Original Equipment Manufacturer

PSI: Pounds Per Square Inch

NS: Not specified by manufacturer

22140_ELEM_C0010

BRAKE SPECIFICATIONS
All measurements in inches unless noted

Year	Model		Brake Disc			Brake Drum Diameter			Minimum Lining Thickness		Brake Caliper	
			Original Thickness	Minimum Thickness	Maximum Runout	Original Inside Diameter	Max. Wear Limit	Maximum Machine Diameter	Front	Rear	Bracket Bolts (ft. lbs.)	Mounting Bolts (ft. lbs.)
2007	Element	F	0.910	0.830	0.004	—	—	—	0.060	—	80	25
		R	0.350	0.300	0.004	—	—	—	0.060	—	41	16
2008	Element	F	0.910	0.830	0.004	—	—	—	0.060	—	80	25
		R	0.350	0.300	0.004	—	—	—	0.060	—	41	16

F: Front

R: Rear

22140_ELEM_C0011

SCHEDULED MAINTENANCE INTERVALS

HONDA—ELEMENT

TO BE SERVICED	TYPE OF SERVICE	VEHICLE MILEAGE INTERVAL (x1000)											
		10	20	30	40	50	60	70	80	90	100	110	120
Accessory drive belts	I & A			✓			✓			✓			✓
Air cleaner element	R			✓			✓			✓			✓
Air conditioning filter	R			✓			✓			✓			✓
Brake fluid	R											✓	
Brake hoses & lines (including ABS)	I		✓		✓		✓		✓		✓		
Cooling system hoses & connections	I		✓		✓		✓		✓		✓		
Engine coolant	R												✓
Engine oil	R	✓	✓	✓	✓	✓	✓	✓	✓	✓	✓	✓	✓
Engine oil and coolant levels	I	Inspect at each fuel stop											
Engine oil filter	R		✓		✓		✓		✓		✓		
Exhaust system	I		✓		✓		✓		✓		✓		
Fluid levels and condition	I		✓		✓		✓		✓		✓		
Front and rear brakes	I		✓		✓		✓		✓		✓		
Fuel lines & connection	I		✓		✓		✓		✓		✓		
Halfshaft boots	I		✓		✓		✓		✓		✓		
Idle speed	I & A											✓	
Parking brake system	I & A		✓		✓		✓		✓		✓		
Rear differential fluid	R									✓			
Rotate and inspect tires	I	✓	✓	✓	✓	✓	✓	✓	✓	✓	✓	✓	✓
Spark plugs	R											✓	
Suspension components	I		✓		✓		✓		✓		✓		
Tie rod ends, steering gear box & boots	I		✓		✓		✓		✓		✓		
Transmission fluid	R												✓
Valve clearance	I											✓	

R: Replace   I: Inspect   A: Adjust

FREQUENT OPERATION MAINTENANCE (SEVERE SERVICE)

If a vehicle is operated under any of the following conditions it is considered severe service:

- Towing a trailer or using a camper or car-top carrier.
- Repeated short trips of less than 5 miles in temperatures below freezing, or trips of less than 10 miles in any temperature.
- Extensive idling or low-speed driving for long distances as in heavy commercial use, such as delivery, taxi or police cars.
- Operating on rough, muddy or salt-covered roads.
- Operating on unpaved or dusty roads.
- Driving in extremely hot (over 90°) conditions.

Air cleaner element: replace every 15,000 miles

Engine oil and filter: replace every 3750 miles or 6 months, whichever occurs first.

Timing belt: replace every 60,000 miles if the vehicle is regularly driven in temperatures above 110°F or below -20°F.

Transmission fluid: replace every 30,000 miles.

Rear differential fluid: replace every 60,000 miles.

Front and rear brakes: inspect every 7500 miles or 6 months, whichever occurs first.

Locks and hinges: lubricate every 15,000 miles.

Tie rods, steering gear box, boots: inspect every 7500 miles or 6 months, whichever occurs first.

Suspension components: inspect every 7500 miles or 6 months, whichever occurs first.

Halfshaft boots: inspect every 7500 miles or 6 months, whichever occurs first.

22140_ELEM_C0012

PRECAUTIONS

Before servicing any vehicle, please be sure to read all of the following precautions, which deal with personal safety, prevention of component damage, and important points to take into consideration when servicing a motor vehicle:

• Never open, service or drain the radiator or cooling system when the engine is hot; serious burns can occur from the steam and hot coolant.

• Observe all applicable safety precautions when working around fuel. Whenever servicing the fuel system, always work in a well-ventilated area. Do not allow fuel spray or vapors to come in contact with a spark, open flame, or excessive heat (a hot drop light, for example). Keep a dry chemical fire extinguisher near the work area. Always keep fuel in a container specifically designed for fuel storage; also, always properly seal fuel containers to avoid the possibility of fire or explosion. Refer to the additional fuel system precautions later in this section.

• Fuel injection systems often remain pressurized, even after the engine has been turned **OFF**. The fuel system pressure must be relieved before disconnecting any fuel lines. Failure to do so may result in fire and/or personal injury.

• Brake fluid often contains polyglycol ethers and polyglycols. Avoid contact with the eyes and wash your hands thoroughly after handling brake fluid. If you do get brake fluid in your eyes, flush your eyes with clean, running water for 15 minutes. If eye irritation persists, or if you have taken brake fluid internally, IMMEDIATELY seek medical assistance.

• The EPA warns that prolonged contact with used engine oil may cause a number of skin disorders, including cancer. You should make every effort to minimize your exposure to used engine oil. Protective gloves should be worn when changing oil. Wash your hands and any other exposed skin areas as soon as possible after exposure to used engine oil. Soap and water, or waterless hand cleaner should be used.

• All new vehicles are now equipped with an air bag system, often referred to as a Supplemental Restraint System (SRS) or Supplemental Inflatable Restraint (SIR) system. The system must be disabled before performing service on or around system components, steering column, instrument panel components, wiring and sensors. Failure to follow safety and disabling procedures could result in accidental air bag deployment, possible personal injury and unnecessary system repairs.

• Always wear safety goggles when working with, or around, the air bag system. When carrying a non-deployed air bag, be sure the bag and trim cover are pointed away from your body. When placing a non-deployed air bag on a work surface, always face the bag and trim cover upward, away from the surface. This will reduce the motion of the module if it is accidentally deployed. Refer to the additional air bag system precautions later in this section.

• Clean, high quality brake fluid from a sealed container is essential to the safe and proper operation of the brake system. You should always buy the correct type of brake fluid for your vehicle. If the brake fluid becomes contaminated, completely flush the system with new fluid. Never reuse any brake fluid. Any brake fluid that is removed from the system should be discarded. Also, do not allow any brake fluid to come in contact with a painted surface; it will damage the paint.

• Never operate the engine without the proper amount and type of engine oil; doing so WILL result in severe engine damage.

• Timing belt maintenance is extremely important. Many models utilize an interference-type, non-freewheeling engine. If the timing belt breaks, the valves in the cylinder head may strike the pistons, causing potentially serious (also time-consuming and expensive) engine damage. Refer to the maintenance interval charts for the recommended replacement interval for the timing belt, and to the timing belt section for belt replacement and inspection.

• Disconnecting the negative battery cable on some vehicles may interfere with the functions of the on-board computer system(s) and may require the computer to undergo a relearning process once the negative battery cable is reconnected.

• When servicing drum brakes, only disassemble and assemble one side at a time, leaving the remaining side intact for reference.

• Only an MVAC-trained, EPA-certified automotive technician should service the air conditioning system or its components.

BRAKES

GENERAL INFORMATION

PRECAUTIONS

• Certain components within the ABS system are not intended to be serviced or repaired individually.

• Do not use rubber hoses or other parts not specifically specified for and ABS system. When using repair kits, replace all parts included in the kit. Partial or incorrect repair may lead to functional problems and require the replacement of components.

• Lubricate rubber parts with clean, fresh brake fluid to ease assembly. Do not use shop air to clean parts; damage to rubber components may result.

• Use only DOT 3 brake fluid from an unopened container.

• If any hydraulic component or line is removed or replaced, it may be necessary to bleed the entire system.

• A clean repair area is essential. Always clean the reservoir and cap thoroughly before removing the cap. The slightest amount of dirt in the fluid may plug an orifice and impair the system function. Perform repairs after components have been thoroughly cleaned; use only denatured alcohol

ANTI-LOCK BRAKE SYSTEM (ABS)

to clean components. Do not allow ABS components to come into contact with any substance containing mineral oil; this includes used shop rags.

• The Anti-Lock control unit is a microprocessor similar to other computer units in the vehicle. Ensure that the ignition switch is **OFF** before removing or installing controller harnesses. Avoid static electricity discharge at or near the controller.

• If any arc welding is to be done on the vehicle, the control unit should be unplugged before welding operations begin.

BRAKES

BLEEDING THE BRAKE SYSTEM

BLEEDING PROCEDURE

See Figure 1.

When bleeding the brake system, observe the following:

• Do not reuse the drained fluid. Use only clean Honda DOT 3 Brake Fluid from an unopened container. Using a non-Honda brake fluid can cause corrosion and shorten the life of the system. Do not mix different brands of brake fluid; they may not be compatible.

• Make sure no dirt or other foreign matter is allowed to contaminate the brake fluid.

• Do not spill brake fluid on the vehicle, it may damage the paint; if brake fluid does contact the paint, wash it off immediately with water.

1. The reservoir on the master cylinder

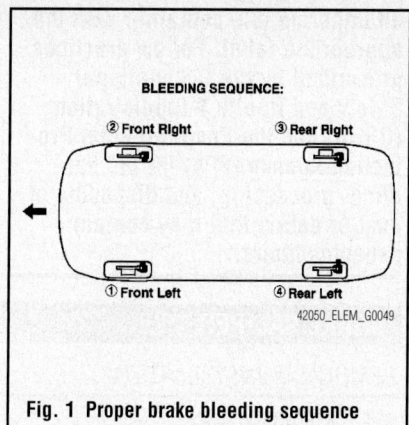

BLEEDING SEQUENCE:

② Front Right   ③ Rear Right

① Front Left   ④ Rear Left

42050_ELEM_G0049

Fig. 1 Proper brake bleeding sequence

must be at the MAX (upper) level mark at the start of the bleeding procedure and checked after bleeding each brake caliper. Add fluid as required.

2. Make sure the brake fluid level in the reservoir is at the MAX (upper) level line.

3. Slide a piece of clear plastic hose over the bleed screw, and submerge the other end in a container of new brake fluid.

4. Have someone slowly pump the brake pedal several times, then apply steady pressure.

5. Loosen the left-front brake bleed screw to allow air to escape from the system. Then tighten the bleed screw securely.

6. Repeat the procedure for each caliper until no air bubbles are in the fluid. Bleed the calipers in the sequence shown.

7. Refill the master cylinder reservoir to the MAX (upper) level line.

BRAKES

FRONT DISC BRAKES

✳✳ CAUTION

Dust and dirt accumulating on brake parts during normal use may contain asbestos fibers from production or aftermarket brake linings. Breathing excessive concentrations of asbestos fibers can cause serious bodily harm. Exercise care when servicing brake parts. Do not sand or grind brake lining unless equipment used is designed to contain the dust residue. Do not clean brake parts with compressed air or by dry brushing. Cleaning should be done by dampening the brake components with a fine mist of water, then wiping the brake components clean with a dampened cloth. Dispose of cloth and all residue containing asbestos fibers in an impermeable container with the
appropriate label. Follow practices prescribed by the Occupational Safety and Health Administration (OSHA) and the Environmental Protection Agency (EPA) for the handling, processing, and disposing of dust or debris that may contain asbestos fibers.

BRAKE CALIPER

REMOVAL & INSTALLATION

1. Remove the wheel
2. Remove the brake hose banjo bolt and washers. Discard the washers.
3. Remove the upper and lower bolts.
4. Lift off the caliper.
5. Remove the pad springs.
6. Remove the pads and shims.
7. Remove the pad retainers.

8. Installation is the reverse of removal. Coat both sides of the shims and the backs of the pads with brake grease. Torque the bolts to 25 ft. lbs. (34 Nm). Install new washers and torque the banjo bolt to 25 ft. lbs. (34 Nm).

DISC BRAKE PADS

REMOVAL & INSTALLATION

1. Remove the lower bolt.
2. Pivot the caliper up and hold the pads.
3. Remove the pad springs.
4. Remove the pads and shims.
5. Remove the pad retainers.
6. Installation is the reverse of removal. Coat both sides of the shims and the backs of the pads with brake grease. Torque the lower bolt to 25 ft. lbs. (34 Nm).

BRAKES

⋇ CAUTION

Dust and dirt accumulating on brake parts during normal use may contain asbestos fibers from production or aftermarket brake linings. Breathing excessive concentrations of asbestos fibers can cause serious bodily harm. Exercise care when servicing brake parts. Do not sand or grind brake lining unless equipment used is designed to contain the dust residue. Do not clean brake parts with compressed air or by dry brushing. Cleaning should be done by dampening the brake components with a fine mist of water, then wiping the brake components clean with a dampened cloth. Dispose of cloth and all residue containing asbestos fibers in an impermeable container with the appropriate label. Follow practices prescribed by the Occupational Safety and Health Administration (OSHA) and the Environmental Protection Agency (EPA) for the handling, processing, and disposing of dust or debris that may contain asbestos fibers.

BRAKE CALIPER

REMOVAL & INSTALLATION

1. Remove the wheel
2. Remove the brake hose banjo bolt and washers. Discard the washers.
3. Remove the caliper pin bolts.
4. Lift off the caliper and suspend it safely.

REAR DISC BRAKES

5. Remove the pads and shims.
6. Remove the pad retainers.
7. Installation is the reverse of removal. Coat both sides of the shims and the backs of the pads with brake grease. Torque the bolts to 16 ft. lbs. (22 Nm). Install new washers and torque the banjo bolt to 25 ft. lbs. (34 Nm).

DISC BRAKE PADS

REMOVAL & INSTALLATION

1. Remove the caliper pin bolts.
2. Lift off the caliper and suspend it safely.
3. Remove the pads and shims.
4. Remove the pad retainers.
5. Installation is the reverse of removal. Coat both sides of the shims and the backs of the pads with brake grease. Torque the bolts to 16 ft. lbs. (22 Nm).

BRAKES

PARKING BRAKE

PARKING BRAKE CABLES

ADJUSTMENT

1. Before servicing the vehicle, refer to the precautions section.
2. Pull the parking brake lever (A) with 44 lbs. (196 N) of force to fully apply the parking brake. The parking brake lever should be locked within 4–7 clicks.
3. Adjust the parking brake if the lever clicks are not within the specification.

Minor Adjustment

See Figure 2.

1. Raise the rear of the vehicle, and support it with safety stands in the proper locations.
2. Release the parking brake lever fully.
3. Remove the center console, by pulling the front edge of the center console

to release the clips, and remove the console by releasing the hooks.

4. Pull the parking brake lever 1 click.
5. Tighten the adjusting nut (A) until the parking brakes drag slightly when the rear wheels are turned.
6. Release the parking brake lever fully, and check that the parking brakes do not drag when the rear wheels are turned. Readjust if necessary.
7. Make sure the parking brakes are fully applied when the parking brake lever is pulled all the way.
8. Install the console in the reverse order of removal. During installation, check for damaged or stress-whitened clips, and replace them with new ones. Push the clip portions into place securely.

Major Adjustment

See Figures 3 and 4.

➡This adjustment should be done when replacing parking brake shoes and after lining surface break-in.

1. Raise the rear of the vehicle, and support it with safety stands in the proper locations.
2. Release the parking brake lever fully.
3. Remove the center console, by pulling the front edge of the center console to release the clips, and remove the console by releasing the hooks.
4. Back off the adjusting nut (A) in the equalizer.

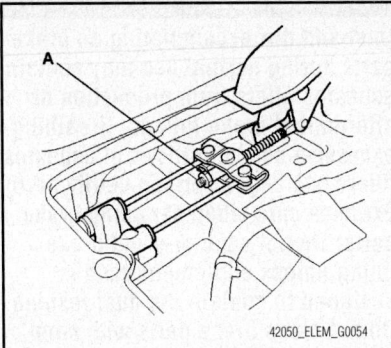

Fig. 3 Back off the adjusting nut (A) in the equalizer

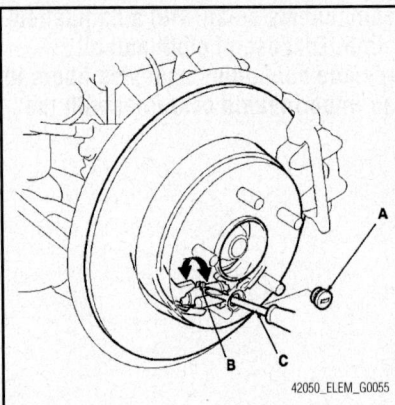

Fig. 4 Remove the access plug (A). Turn the adjuster (B) with a flat-tip screwdriver (C) until the shoes lock against the drum. Then back off 8 clicks, and install the access plug.

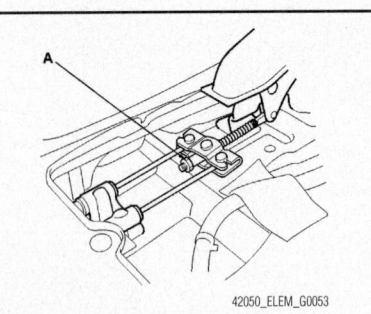

Fig. 2 Tighten the adjusting nut (A) until the parking brakes drag slightly when the rear wheels are turned

5. Remove the rear wheels.

6. Remove the access plug (A).

7. Turn the adjuster (B) with a flat-tip screwdriver (C) until the shoes lock against the drum. Then back off 8 clicks, and install the access plug.

8. Perform the minor adjustment procedure.

9. Install the rear wheels.

10. Install the console in the reverse order of removal. During installation, check for damaged or stress-whitened clips, and replace them with new ones. Push the clip portions into place securely.

PARKING BRAKE SHOES

ADJUSTMENT

1. Before servicing the vehicle, refer to the precautions section.

2. Pull the parking brake lever (A) with 44 lbs. (196 N) of force to fully apply the parking brake. The parking brake lever should be locked within 4–7 clicks.

3. Adjust the parking brake if the lever clicks are not within the specification.

Minor Adjustment

See Figure 5.

1. Raise the rear of the vehicle, and support it with safety stands in the proper locations.

2. Release the parking brake lever fully.

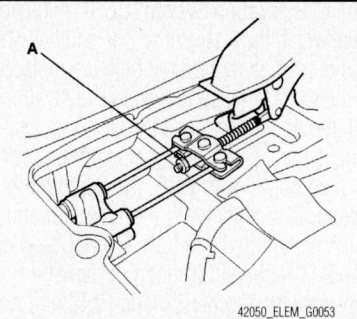

Fig. 5 Tighten the adjusting nut (A) until the parking brakes drag slightly when the rear wheels are turned

3. Remove the center console, by pulling the front edge of the center console to release the clips, and remove the console by releasing the hooks.

4. Pull the parking brake lever 1 click.

5. Tighten the adjusting nut (A) until the parking brakes drag slightly when the rear wheels are turned.

6. Release the parking brake lever fully, and check that the parking brakes do not drag when the rear wheels are turned. Readjust if necessary.

7. Make sure the parking brakes are fully applied when the parking brake lever is pulled all the way.

8. Install the console in the reverse order of removal. During installation, check for damaged or stress-whitened clips, and

replace them with new ones. Push the clip portions into place securely.

Major Adjustment

See Figures 3 and 4.

➡This adjustment should be done when replacing parking brake shoes and after lining surface break-in.

1. Raise the rear of the vehicle, and support it with safety stands in the proper locations.

2. Release the parking brake lever fully.

3. Remove the center console, by pulling the front edge of the center console to release the clips, and remove the console by releasing the hooks.

4. Back off the adjusting nut (A) in the equalizer.

5. Remove the rear wheels.

6. Remove the access plug (A).

7. Turn the adjuster (B) with a flat-tip screwdriver (C) until the shoes lock against the drum. Then back off 8 clicks, and install the access plug.

8. Perform the minor adjustment procedure.

9. Install the rear wheels.

10. Install the console in the reverse order of removal. During installation, check for damaged or stress-whitened clips, and replace them with new ones. Push the clip portions into place securely.

CHASSIS ELECTRICAL   AIR BAG (SUPPLEMENTAL RESTRAINT SYSTEM)

GENERAL INFORMATION

✷✷ CAUTION

These vehicles are equipped with an air bag system. The system must be disarmed before performing service on, or around, system components, the steering column, instrument panel components, wiring and sensors. Failure to follow the safety precautions and the disarming procedure could result in accidental air bag deployment, possible injury and unnecessary system repairs.

SERVICE PRECAUTIONS

Disconnect and isolate the battery negative cable before beginning any airbag system component diagnosis, testing, removal, or installation procedures. Allow system capacitor to discharge for two minutes before beginning any component service. This will

disable the airbag system. Failure to disable the airbag system may result in accidental airbag deployment, personal injury, or death.

Do not place an intact undeployed airbag face down on a solid surface. The airbag will propel into the air if accidentally deployed and may result in personal injury or death.

When carrying or handling an undeployed airbag, the trim side (face) of the airbag should be pointing towards the body to minimize possibility of injury if accidental deployment occurs. Failure to do this may result in personal injury or death.

Replace airbag system components with OEM replacement parts. Substitute parts may appear interchangeable, but internal differences may result in inferior occupant protection. Failure to do so may result in occupant personal injury or death.

Wear safety glasses, rubber gloves, and long sleeved clothing when cleaning powder residue from vehicle after an airbag deployment. Powder residue emitted from a deployed airbag can cause skin irritation.

Flush affected area with cool water if irritation is experienced. If nasal or throat irritation is experienced, exit the vehicle for fresh air until the irritation ceases. If irritation continues, see a physician.

Do not use a replacement airbag that is not in the original packaging. This may result in improper deployment, personal injury, or death.

The factory installed fasteners, screws and bolts used to fasten airbag components have a special coating and are specifically designed for the airbag system. Do not use substitute fasteners. Use only original equipment fasteners listed in the parts catalog when fastener replacement is required.

During, and following, any child restraint anchor service, due to impact event or vehicle repair, carefully inspect all mounting hardware, tether straps, and anchors for proper installation, operation, or damage. If a child restraint anchor is found damaged in any way, the anchor must be replaced. Failure to do this may result in personal injury or death.

Deployed and non-deployed airbags may or may not have live pyrotechnic material within the airbag inflator.

Do not dispose of driver/passenger/curtain airbags or seat belt tensioners unless you are sure of complete deployment. Refer to the Hazardous Substance Control System for proper disposal.

Dispose of deployed airbags and tensioners consistent with state, provincial, local, and federal regulations.

After any airbag component testing or service, do not connect the battery negative cable. Personal injury or death may result if the system test is not performed first.

If the vehicle is equipped with the Occupant Classification System (OCS), do not connect the battery negative cable before performing the OCS Verification Test using the scan tool and the appropriate diagnostic information. Personal injury or death may result if the system test is not performed properly.

Never replace both the Occupant Restraint Controller (ORC) and the Occupant Classification Module (OCM) at the same time. If both require replacement, replace one, then perform the Airbag System test before replacing the other.

Both the ORC and the OCM store Occupant Classification System (OCS) calibration data, which they transfer to one another when one of them is replaced. If both are replaced at the same time, an irreversible fault will be set in both modules and the OCS may malfunction and cause personal injury or death.

If equipped with OCS, the Seat Weight Sensor is a sensitive, calibrated unit and must be handled carefully. Do not drop or handle roughly. If dropped or damaged, replace with another sensor. Failure to do so may result in occupant injury or death.

If equipped with OCS, the front passenger seat must be handled carefully as well. When removing the seat, be careful when setting on floor not to drop. If dropped, the sensor may be inoperative, could result in occupant injury, or possibly death.

If equipped with OCS, when the passenger front seat is on the floor, no one should sit in the front passenger seat. This uneven force may damage the sensing ability of the seat weight sensors. If sat on and damaged, the sensor may be inoperative, could result in occupant injury, or possibly death.

DISARMING THE SYSTEM

1. Disconnect and isolate the negative battery cable. Wait 3 minutes for the system capacitor to discharge before performing any service.

2. To disarm the driver's airbag, remove the access panel from the steering wheel, then disconnect the driver's airbag 4P connector from the cable reel.

3. To disarm the front passenger's airbag, remove the glove box, then disconnect the passenger's airbag 4P connector from dashboard wire harness B.

4. To disarm the side airbag, disconnect the side airbag 2P connector from the floor wire harness.

5. To disarm the seat belt tensioner, disconnect the seat belt tensioner 2P connector from the rear door wire harness.

6. To disarm the seat belt buckle tensioner, disconnect the seat belt buckle tensioner 4P connector.

7. To disarm the SRS unit, disconnect the SRS unit connector A, B or C, as applicable.

ARMING THE SYSTEM

1. To rearm, connect the electrical connector(s) as necessary, then connect the negative battery cable.

DRIVETRAIN

AUTOMATIC TRANSAXLE ASSEMBLY

REMOVAL & INSTALLATION

See Figures 6 and 7.

1. Before servicing the vehicle, refer to the precautions section.
2. Drain the transaxle.
3. Remove or disconnect the following:
- Splash shield
- Battery cables, negative cable first
- Air intake assembly
- Battery
- Battery tray
- Harness clamp from the batter base and remove the base
- Transaxle ground cable
- Clutch pressure switch connectors
- 2nd clutch pressure control solenoid valve connector
- Clutch pressure control valve connector
- Harness clamps from the brackets
- Countershaft speed sensor connector
- Mainshaft speed sensor connector
- Transmission Range Switch (TRS)

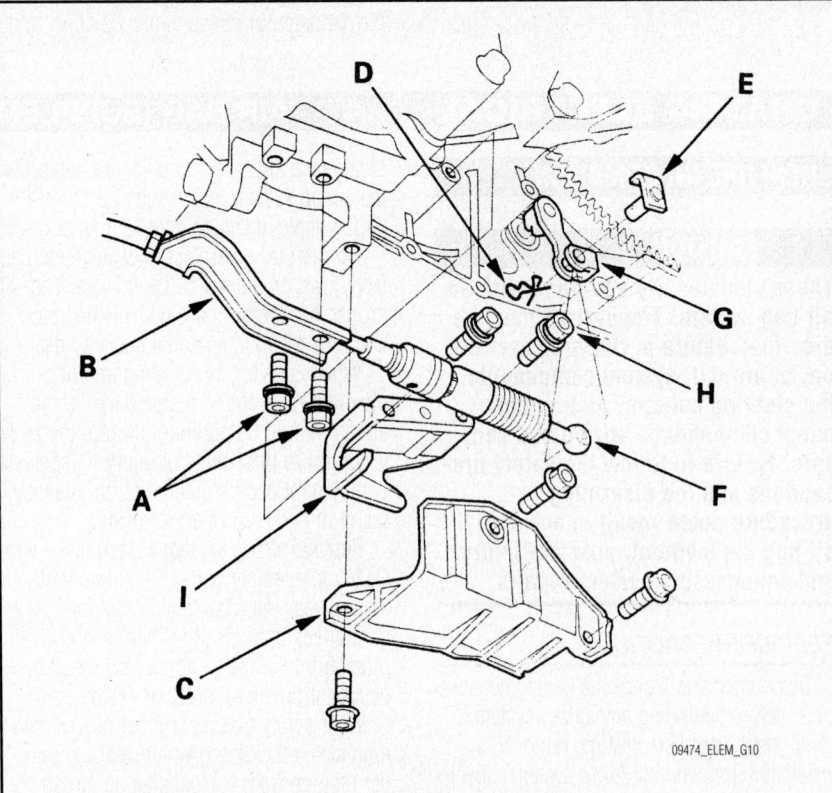

Fig. 6 Exploded view of the shift cable components on models equipped with four wheel drive

09474_ELEM_G10

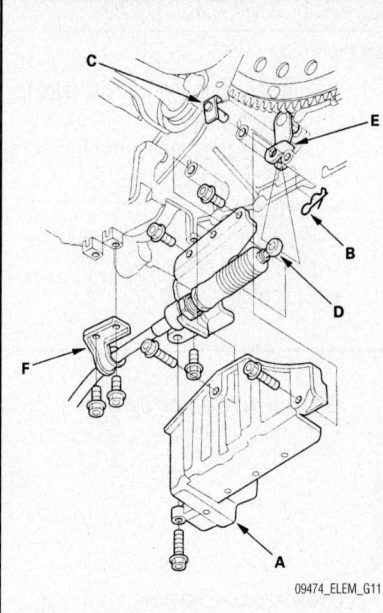

Fig. 7 Exploded view of the shift cable components on models equipped with two wheel drive

connector from the bracket and disconnect it

- 3rd clutch pressure control solenoid valve connector
- Shift control solenoid valve connectors
- Harness clamps from the brackets
- Transaxle oil cooler lines
- Harness clamp from the clamp bracket and harness cover from the bracket
- Water pipe mounting bolt and brackets

4. Attach engine hanger VSB02C000015 to the treaded holes in the cylinder head.

5. Install engine support hanger AAR-T-12566 to the engine and hanger VSB02C000015.

6. Install a 5mm hex wrench in the top ball joint pin and remove the nut. Separate the stabilizer link from the lower arm.

- Lower arms from the knuckles
- Torque converter cover and converter bolts

7. On four wheel drive models, refer to the illustration for location and perform the following:

a. Remove the bolts (A) attaching the shift cable bracket (B) and remove the cable cover (C).

b. Remove the spring clip (D) and control pin (E) and separate the cable (F)

from the selector lever (G) being careful not to bend the cable too much.

c. Remove the bolts (H) attaching the shift cable bracket (I) and remove the bracket from the cable.

8. On two wheel drive models, refer to the illustration for location and perform the following:

a. Remove the shift cable cover (A).

b. Remove the spring clip (B) and control pin (C) the separate the cable (D) from the selector lever (E).

c. Remove the bolts attaching the shift cable bracket (F) and remove the bracket from the cable.

9. Remove or disconnect the following:

- Transmission cooler hose from the cooler line and plug the hose
- Front mount bolt and nut
- Rear mount bolts

10. Matchmark the sub-frame mounting bolt centers.

There is a special tool necessary for sub-frame removal. The Honda tool number is EQS02C000011.

11. Attach the tool as explained in the tool instructions, attach a floor jack with adapter, remove the 4 sub-frame bolts and lower the sub-frame.

- Rear driveshaft, if equipped

12. Separate the inner CV-joints from the transaxle and intermediate shaft and support the axle halfshafts out of the work area with safety wire.

- Intermediate shaft

13. Support the transmission with a suitable jack.

- Transmission mount bolts and nuts
- Transmission housing bolts
- Transaxle

To install:

14. Installation is the reverse of removal. Observe the following torques:

- Air cleaner housing bracket bolt: 16 ft. lbs. (22 Nm)
- Front mount bolts: 47 ft. lbs. (64 Nm)
- Rear mount bracket bolts: 40 ft. lb. (54 Nm)
- Transmission-to-engine bolts: 47 ft. lbs. (64 Nm)
- Upper transmission mount bolt: 40 ft. lbs. (54 Nm)
- Upper transmission mount nuts: 40 ft. lbs. (54 Nm)
- Rear driveshaft bolts: 24 ft. lbs. (32 Nm)
- Subframe bolts: 76 ft. lbs. (103 Nm)

MANUAL TRANSAXLE ASSEMBLY

REMOVAL & INSTALLATION

1. Before servicing the vehicle, refer to the precautions section.

2. Secure the hood in a vertical position.

3. Remove or disconnect the following:

- Air intake assembly
- Battery cables, battery and tray
- Transaxle ground cable
- Vehicle Speed Sensor (VSS) connector
- Back-up light switch connector
- Cable bracket and the cables from the top of the transmission housing. Be careful when removing the cables to avoid bending them.
- Clutch slave cylinder without bending the clutch line. Do not press the clutch pedal after the slave cylinder has been removed.
- Two upper transmission bolts

4. Attach engine hanger VSB02C000015 to the treaded holes in the cylinder head.

5. Install engine support hanger AAR-T-12566 to the engine and hanger VSB02C000015.

- Transmission mount bracket and bolt
- Air cleaner bracket

6. Drain the transmission fluid

- Splash shield
- Driveshafts
- Intermediate shaft
- Front engine mount bracket bolt
- 3 bolts attaching the rear transmission mount

7. Matchmark the sub-frame mounting bolt centers.

There is a special tool necessary for subframe removal. The Honda tool number is EQS02C000011.

8. Attach the tool as explained in the tool instructions, attach a floor jack with adapter, remove the 4 sub-frame bolts and lower the sub-frame.

- Clutch cover
- Front engine mount

9. Support the transmission with a suitable jack and remove the 4 lower mounting bolts.

10. Pull the transmission away from the engine until the mainshaft clears the pressure plate and lower the transmission.

To install:

11. Installation is the reverse of removal. Observe the following torques:

- Transaxle rear mount and bracket. Tighten the bracket bolts to 40 ft. lbs. (54 Nm) and the through bolt to 47 ft. lbs. (64 Nm)
- Transaxle. Tighten the flange bolts to 47 ft. lbs. (64 Nm)
- Front mount and bracket. Tighten the bolts to 47 ft. lbs. (64 Nm).
- Clutch housing cover. Tighten the bolts to 29 ft. lbs. (39 Nm)
- Subframe: 76 ft. lbs. (98 Nm)

12. Fill the transmission with the correct type and amount of fluid.
13. Road test the vehicle.

CLUTCH DRIVEN DISC AND PRESSURE PLATE

ADJUSTMENTS

The Element is equipped with a hydraulic clutch system. No adjustment is necessary.

REMOVAL & INSTALLATION

See Figure 8.

1. Before servicing the vehicle, refer to the precautions section.
2. Remove or disconnect the following:
   - Negative battery cable
   - Transaxle
   - Pressure plate. Loosen the bolts evenly in a crossing pattern.
   - Clutch disc

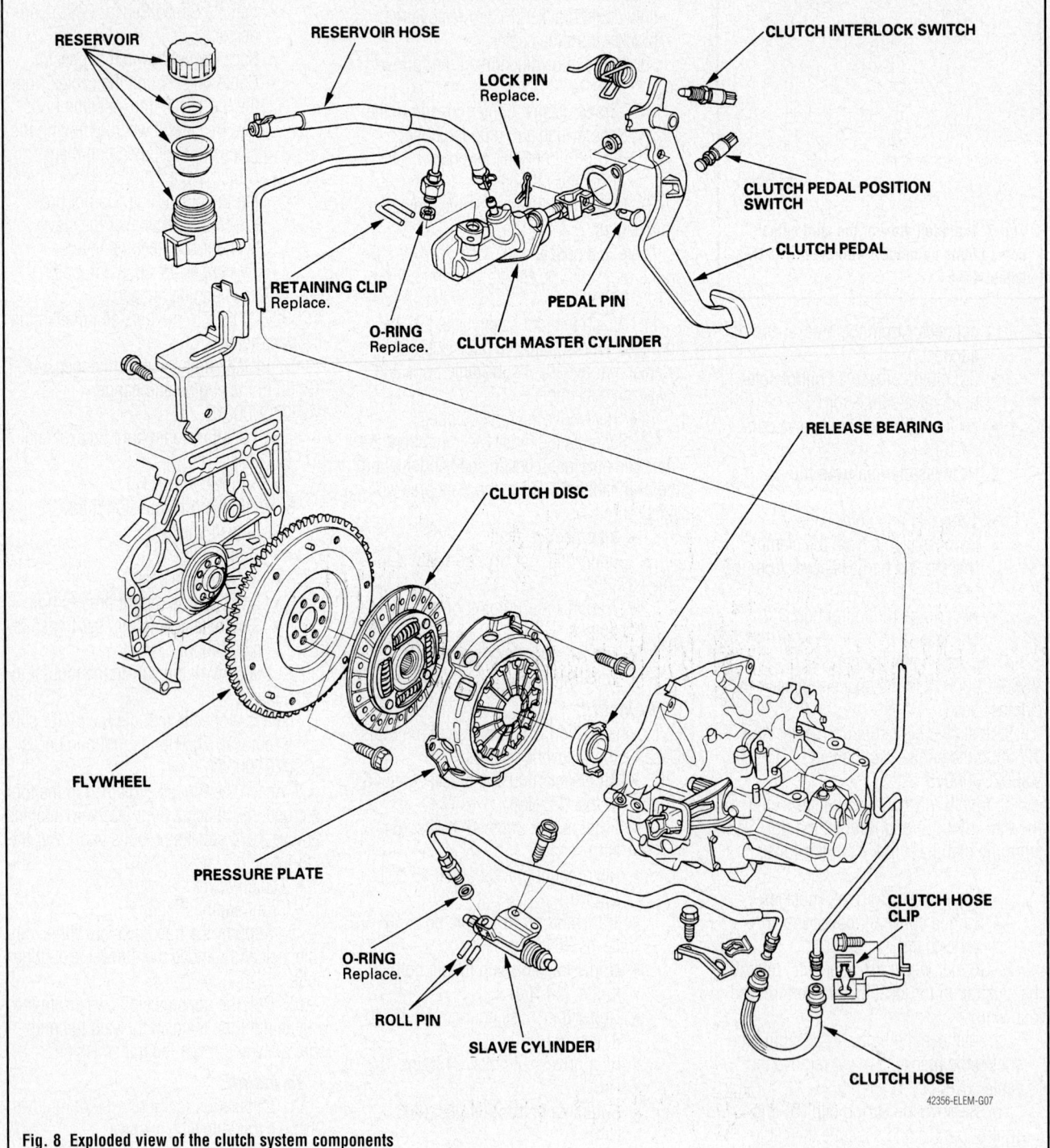

Fig. 8 Exploded view of the clutch system components

42356-ELEM-G07

To install:

3. Install the clutch disc and pressure plate. Tighten the pressure plate bolts in a crisscross pattern, in several steps to 19 ft. lbs. (26 Nm).

4. Install or connect the following:
- Transaxle
- Negative battery cable

CLUTCH MASTER CYLINDER

REMOVAL & INSTALLATION

See Figures 9 through 13.

✳✳ WARNING

Always use fender covers to avoid damaging painted surfaces. Do not spill brake fluid on the vehicle; it may damage the paint; if brake fluid does contact the paint, wash it off immediately with water.

1. Remove the brake fluid from the clutch master cylinder reservoir with a syringe.

2. Make sure you have the anti-theft code for the radio, then write down the audio presets. Disconnect the negative cable from the battery first, then disconnect the positive cable. Remove the battery.

3. Remove the air cleaner housing.

4. Remove the battery base.

5. Pry out the lock pin, and pull the pedal pin out of the yoke. Remove the master cylinder mounting nuts.

6. Remove the reservoir mounting bolt.

7. Remove the clutch line bracket.

8. Remove the clutch master cylinder from the vehicle.

9. Disconnect the reservoir hose, then remove the retaining clip and clutch line from the clutch master cylinder. Plug the end of the reservoir hose and clutch line

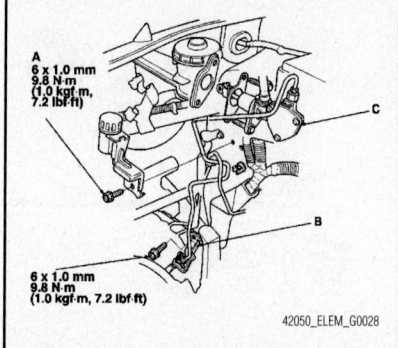

Fig. 10 Remove the reservoir mounting bolt (A), the clutch line bracket (B), then remove the clutch master cylinder (C)

with a shop towel to prevent brake fluid from coming out.

10. Remove the O-ring and clutch master cylinder seal from the clutch master cylinder.

To install:

11. Install the clutch master cylinder in the reverse order of removal, and note these items:

    a. Apply brake assembly lube to the clutch line, and install a new O-ring.

    b. Tighten the master cylinder mounting nuts to 9 ft. lbs. (13 Nm).

12. Install the battery base.

13. Install the air cleaner housing.

14. Install the battery. Clean the battery posts and cable terminals with sandpaper. Connect the positive cable to the battery first, then connect the negative cable, and apply grease to prevent corrosion.

15. Enter the anti-theft code for the radio, then enter the audio presets, and set the clock.

16. Perform the power window control unit reset procedure, as follows.

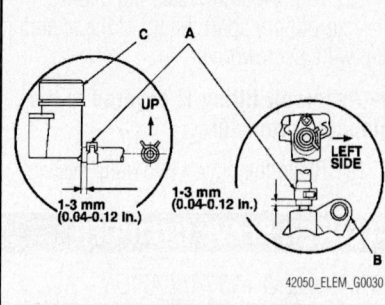

Fig. 12 The hose clamps (A) must be properly positioned on the master cylinder (B) and reservoir (C)

➡Resetting the power window control unit is required after performing the following procedures: Loss of battery power, Loss of power from the No. 23 (20 A) fuse in the under-dash fuse/relay box, Open circuit caused by disconnecting the 14P connector from the power window master switch

    a. Make sure the driver's window does not work in AUTO with the ignition switch ON (II).

    b. Start the engine.

    c. Lower the driver's window all the way down by pushing the driver's power window switch to the second detent (AUTO DOWN); when the window reaches the bottom, hold the switch in the AUTO DOWN position for 2 seconds.

    d. Raise the driver's window all the way up without stopping by pulling the driver's power window switch to the UP position; when the window reaches the top, hold the switch in the UP position for 2 seconds.

    e. If the window does not work in AUTO, repeat steps b through e.

17. Make sure the hose clamps are positioned on the master cylinder and reservoir as shown.

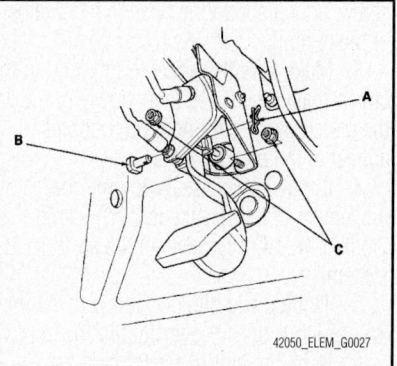

Fig. 9 Pry out the lock pin (A), and pull the pedal pin (B) out of the yoke. Remove the master cylinder mounting nuts (C)

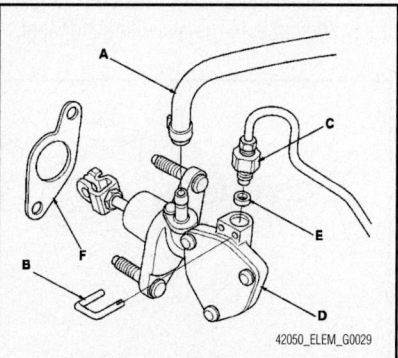

Fig. 11 View of the reservoir hose (A), retaining clips (B), clutch line (C), clutch master cylinder (D), O-ring (E) and clutch master cylinder seal (F)

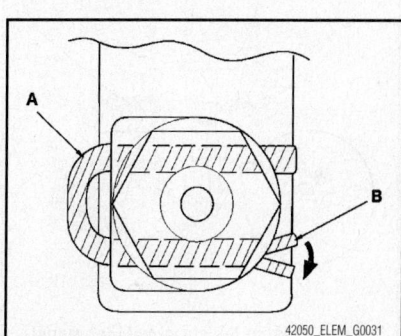

Fig. 13 Pry the tip of the retaining clips (B) apart with a screwdriver to prevent the clip (A) from coming off

18. To prevent the retaining clip from coming off, pry apart the tip of the retaining clip with a screwdriver.

→ **Reservoir filling is covered in the bleeding procedure.**

19. Bleed the clutch hydraulic system.

CLUTCH SLAVE CYLINDER

REMOVAL & INSTALLATION

See Figures 14 through 16.

❊❊ WARNING

Always use fender covers to avoid damaging painted surfaces. Do not spill brake fluid on the vehicle; it may damage the paint; if brake fluid does contact the paint, wash it off immediately with water.

1. Remove the brake fluid from the clutch master cylinder reservoir with a syringe.

2. Make sure you have the anti-theft code for the radio, then write down the audio presets. Disconnect the negative cable from the battery first, then disconnect the positive cable. Remove the battery.

3. Remove the air cleaner housing.

4. Remove the battery base.

5. Remove the clutch line bracket.

6. Remove the mounting bolts and the slave cylinder.

7. Remove the roll pins. Disconnect the clutch line, and remove the O-ring. Plug the end of the clutch line with a shop towel to prevent brake fluid from coming out.

To install:

8. Install the slave cylinder in the reverse order of removal. Install a new O-ring (A).

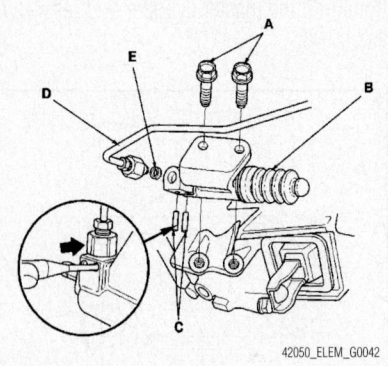

Fig. 14 View of the slave cylinder mounting bolts (A), slave cylinder (B), roll pins (C), clutch line (D) and O-ring (E)

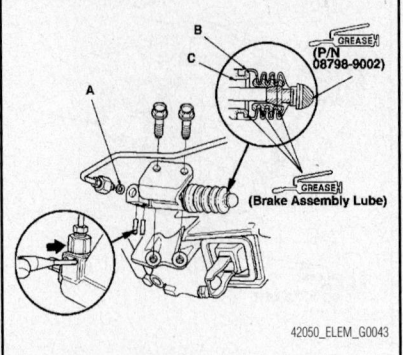

Fig. 15 Slave cylinder installation. Apply the proper type of grease where shown

9. Pull back the boot (B), and apply brake assembly lube to the boot and slave cylinder rod (C). Reinstall the boot.

10. Apply super high temp urea grease (P/N 08798-9002) to the slave cylinder push rod. Tighten the slave cylinder mounting bolts to 16 ft. lbs. (22 Nm).

11. Attach a hose (A) to the bleeder screw, and suspend the hose in a container of brake fluid.

12. Make sure there is enough fluid in the clutch master cylinder, then slowly pump the clutch pedal until no more bubbles appear at the bleeder hose.

13. It may be necessary to limit the movement of the release fork (B) with a block of wood to remove all the air from the system.

14. Tighten the bleeder screw to 70 inch lbs. (8 Nm); do not overtighten the screw.

15. Refill the clutch master cylinder with fluid when done. Use only Honda DOT 3 Brake Fluid from an unopened container.

16. Make sure the fluid level in the reservoir is at the MAX (upper) level line.

17. Install the air cleaner housing.

18. Install the battery base.

19. Install the battery. Clean the battery posts and cable terminals with sandpaper.

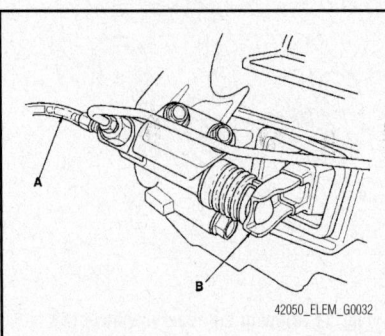

Fig. 16 View of the hose (A) attached to the bleeder screw and the release fork (B)

Connect the positive cable to the battery first, then connect the negative cable, and apply grease to prevent corrosion.

20. Enter the anti-theft code for the radio, then enter the audio presets, and set the clock.

21. Perform the power window control unit reset procedure, as follows.

→**Resetting the power window control unit is required after performing the following procedures: Loss of battery power, Loss of power from the No. 23 (20 A) fuse in the under-dash fuse/relay box, Open circuit caused by disconnecting the 14P connector from the power window master switch**

　a. Make sure the driver's window does not work in AUTO with the ignition switch ON (II).

　b. Start the engine.

　c. Lower the driver's window all the way down by pushing the driver's power window switch to the second detent (AUTO DOWN); when the window reaches the bottom, hold the switch in the AUTO DOWN position for 2 seconds.

　d. Raise the driver's window all the way up without stopping by pulling the driver's power window switch to the UP position; when the window reaches the top, hold the switch in the UP position for 2 seconds.

　e. If the window does not work in AUTO, repeat steps b through e.

HYDRAULIC CLUTCH SYSTEM

BLEEDING

See Figure 16.

1. Before servicing the vehicle, refer to the precautions section.

2. Attach a hose (A) to the bleeder screw, and suspend the hose in a container of brake fluid.

3. Make sure there is enough fluid in the clutch master cylinder, then slowly pump the clutch pedal until no more bubbles appear at the bleeder hose.

4. It may be necessary to limit the movement of the release fork (B) with a block of wood to remove all the air from the system.

5. Tighten the bleeder screw to 70 inch lbs. (8 Nm); do not overtighten the screw.

6. Refill the clutch master cylinder with fluid when done. Use only Honda DOT 3 Brake Fluid from an unopened container.

7. Make sure the fluid level in the reservoir is at the MAX (upper) level line.

TRANSFER ASSEMBLY

REMOVAL & INSTALLATION

See Figure 17.

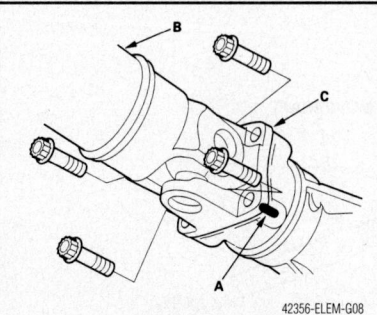

42356-ELEM-G08

Fig. 17 Matchmark (A) the installed position of the propeller shaft (B) and transfer companion flange (C)

1. Before servicing the vehicle, refer to the precautions section.
2. Drain the transaxle fluid. Install the drain plug with a new gasket and tighten to 36 ft. lbs. (49 Nm).
3. Disconnect the negative battery cable.
4. Matchmark the installed position of the propeller shaft and transfer companion flange.
5. Remove or disconnect the following:
   - Propeller shaft from the transfer assembly
   - Mounting bolts and transfer assembly

To install:

6. Clean the transfer assembly mating surfaces, then apply clean transmission fluid to the mating surfaces.
7. Install or connect the following:
   - New O-ring seal on the transfer assembly
   - 4 bolts in the transfer housing, then the transfer assembly with the dowel pin. Tighten the 10mm bolts to 33 ft. lbs. (44 Nm).
   - Propeller shaft to the transfer companion flange, aligning the mark made during removal. Tighten the 8mm bolts to 24 ft. lbs. (33 Nm).
   - Negative battery cable
8. Fill the transaxle to the correct level and check for leaks.

FRONT HALFSHAFT

REMOVAL & INSTALLATION

1. Before servicing the vehicle, refer to the precautions section.
2. Drain the transaxle.

3. Remove or disconnect the following:
   - Negative battery cable
   - Front wheels
   - Spindle nut
   - Stabilizer bar
   - Lower ball joint from the control arm
4. On the left side, pry the inboard joint from the case with a prybar.
5. On the right side, drive the inboard shaft off the intermediate shaft with a drift and hammer.
6. Installation is the reverse of removal. Observe the following torques:
   - Ball stud nuts: 40 ft. lbs. (54 Nm)
   - Stabilizer link nuts: 29 ft. lbs. (39 Nm)
   - Spindle nut: 181 ft. lbs. (245 Nm)

CV-JOINT OVERHAUL

Outboard Joint

1. Before servicing the vehicle, refer to the precautions section.
2. Remove or disconnect the following:
   - Axle halfshaft from the vehicle and place it in a vise
   - Outboard joint boot clamps and push the boot back
   - Outboard joint by driving it off the axle shaft with a brass drift and hammer
   - Outboard joint boot

To install:

➡**Use new circlips and boot clamps for assembly.**

3. Install the outboard joint boot and clamps to the axle shaft.
4. Fill the outboard joint with grease. Install the outboard joint to the axle shaft. Tap the stub shaft with a brass hammer to seat the circlip.
5. Fill the outboard joint boot with grease and install the boot clamps.
6. Install the axle halfshaft to the vehicle.

Inboard Joint

1. Before servicing the vehicle, refer to the precautions section.
2. Remove or disconnect the following:
   - Axle halfshaft from the vehicle.
   - Inboard joint boot clamps and push the boot back
   - Inboard joint housing from the axle
   - Rollers from the spider
   - Snapring and the spider from the axle shaft
   - Inboard joint boot

To install:

➡**Use new circlips and boot clamps for assembly.**

3. Install or connect the following:
   - Inboard joint boot and clamps to the axle shaft
   - Spider with a new snapring
   - Rollers to the spider
4. Fill the joint housing with grease and install it.
5. Fill the inboard joint boot with grease and install the boot clamps.
6. Install the axle halfshaft to the vehicle.

REAR HALFSHAFT

REMOVAL & INSTALLATION

1. Before servicing the vehicle, refer to the precautions section.
2. Drain the differential.
3. Remove or disconnect the following:
   - Negative battery cable
   - Rear wheels
   - Spindle nut
4. Pry the inboard joint from the differential.
5. Remove the outer CV-joint stub shaft from the hub by tapping the stub shaft with a plastic hammer.

To install:

➡**Use new circlips and self-locking nuts for assembly.**

6. Install the outer CV-joint stub shaft into the hub.
7. Install the inner CV-joint to the differential until the circlip locks in the retaining groove.
8. Install or connect the following:
   - Spindle nut. Tighten the nut to 134 ft. lbs. (181 Nm).
   - Rear wheels
   - Negative battery cable
9. Fill the differential to the correct level and check for leaks.

CV-JOINT OVERHAUL

See Figure 18.

1. Before servicing the vehicle, refer to the precautions section.
2. Remove or disconnect the following:
   - Axle halfshaft from the vehicle
   - Joint boot clamps and push the boot back
   - Joint housing from the axle
   - Rollers from the spider
   - Snapring and the spider from the axle shaft
   - Joint boot

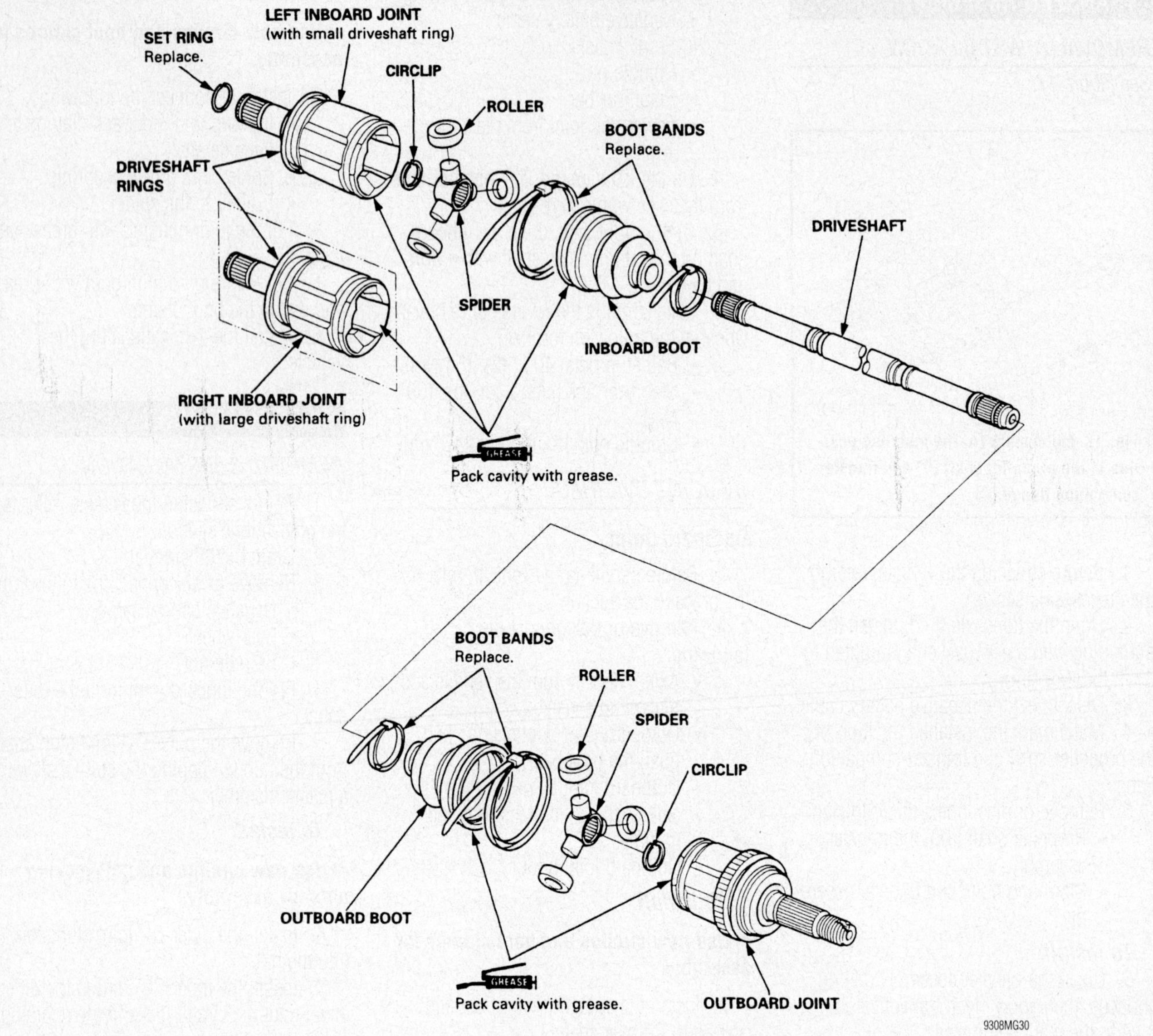

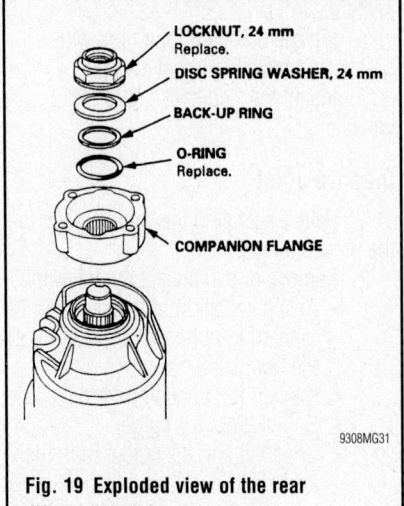

SET RING Replace.

LEFT INBOARD JOINT (with small driveshaft ring)

CIRCLIP

ROLLER

BOOT BANDS Replace.

DRIVESHAFT RINGS

DRIVESHAFT

SPIDER

INBOARD BOOT

RIGHT INBOARD JOINT (with large driveshaft ring)

Pack cavity with grease.

BOOT BANDS Replace.

ROLLER

SPIDER

CIRCLIP

OUTBOARD BOOT

Pack cavity with grease.

OUTBOARD JOINT

9308MG30

Fig. 18 Exploded view of the rear axle

To install:

➡ **Use new circlips and boot clamps for assembly.**

3. Install or connect the following:
   - Joint boot and clamps to the axle shaft
   - Spider with a new snapring
   - Rollers to the spider
4. Fill the joint housing with grease and install it.
5. Fill the joint boot with grease and install the boot clamps.
6. Install the axle halfshaft to the vehicle.

REAR PINION SEAL

REMOVAL & INSTALLATION

See Figure 19.

LOCKNUT, 24 mm Replace.

DISC SPRING WASHER, 24 mm

BACK-UP RING

O-RING Replace.

COMPANION FLANGE

9308MG31

Fig. 19 Exploded view of the rear differential pinion components

1. Before servicing the vehicle, refer to the precautions section.
2. Remove or disconnect the following:
   - Driveshaft
   - Companion flange
   - Pinion seal

To install:

➡ **Use a new locknut and O-ring for assembly.**

3. Install or connect the following:
   - Pinion seal. Drive the seal square into the bore.
   - Companion flange. Tighten the locknut to 108 ft. lbs. (147 Nm).
   - Driveshaft. Tighten the flange bolts to 24 ft. lbs. (32 Nm).

ENGINE COOLING

THERMOSTAT

REMOVAL & INSTALLATION

See Figures 20 and 21.

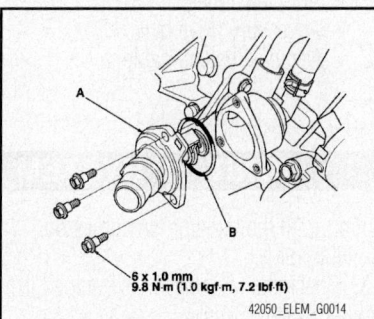

Fig. 20 View of the thermostat (A) and O-ring (B). Always use a new O-ring during thermostat installation

1. Before servicing the vehicle, refer to the precautions section.
2. Drain the engine coolant.
3. Disconnect the negative battery cable.
4. Clean any dirt from quick connector, thermostat cover, and lower radiator hose.

5. Pull the lock out by hand, then wiggle the quick connector loose, and remove it from the thermostat cover. Do not use any tools to remove the quick connector.
6. Remove the thermostat.

To install:

7. Install the thermostat with a new O-ring.
8. Check the quick connector (A) and set ring (B) for cracks or damage. If the connector and/or set ring are cracked or damaged, replace the connector.

➡**Make sure the set ring is in place inside the quick connector. If the set ring is off the connector, replace the quick connector.**

9. Replace the O-ring (C) in the quick connector.
10. Check the lock (D). If the lock is damaged or deformed, replace it. When installing the new lock to the connector, push it straight down along the groove.
11. Clean the connecting surface of the thermostat cover (E), then apply clean engine coolant around the connecting surface.

12. Push the lock down, then push the quick connector onto the thermostat cover until you hear an audible click.
13. Refill the radiator with engine coolant, and bleed air from the cooling system with the heater valve open.

WATER PUMP

REMOVAL & INSTALLATION

See Figure 22.

1. Before servicing the vehicle, refer to the precautions section.
2. Drain the cooling system.
3. Remove or disconnect the following:
   - Negative battery cable
   - Accessory drive belt
   - Crankshaft pulley
   - Water pump (6 bolts)

To install:

4. Clean the water pump mating surfaces.
5. Install or connect the following:
   - Water pump with a new O-ring. Torque the bolts to 8.7 ft. lbs. (12 Nm).
   - Crankshaft pulley
   - Accessory drive belt
   - Negative battery cable
6. Refill the engine cooling system.

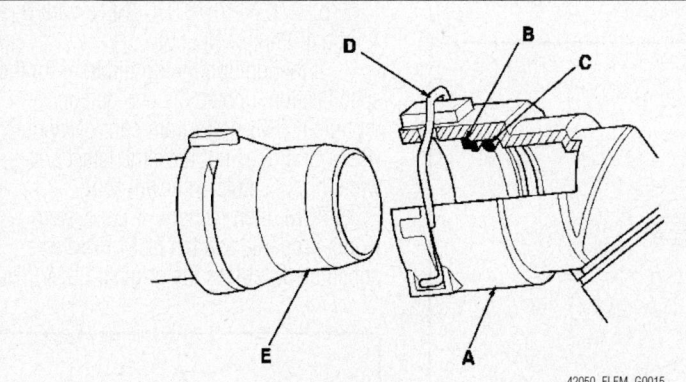

Fig. 21 View of the quick connector (A), set ring (B), O-ring (C), lock (D) and thermostat cover (E)

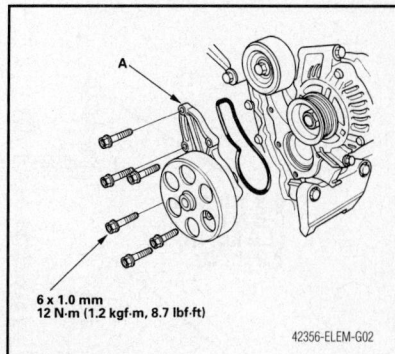

Fig. 22 Exploded view of the water pump mounting

ENGINE ELECTRICAL

ALTERNATOR

REMOVAL & INSTALLATION

1. Before servicing the vehicle, refer to the precautions section.
2. Remove or disconnect the following:
   - Negative, then the positive battery cables
   - Accessory drive belt
   - Auto-tensioner
   - Alternator wiring harness connectors and harness clamp
   - Positive Crankcase Ventilation (PCV) valve
   - 3 bolts holding the alternator
   - Alternator

To install:

3. Install or connect the following:

ENGINE ELECTRICAL

IGNITION COIL

REMOVAL & INSTALLATION
See Figure 23.

1. Before servicing the vehicle, refer to the precautions section.
2. Disconnect the negative battery cable.
3. Remove the ignition coil cover, disconnect the ignition coil connectors, then remove the ignition coils.

To install:

4. Install the ignition coils and tighten the retainers to 8.7 ft. lbs. (12 Nm).
5. Attach the ignition coil electrical connectors.

6. Install the ignition coil cover and tighten the retainers to 7.2 ft. lbs. (9.8 Nm).
7. Connect the negative battery cable.

IGNITION TIMING

INSPECTION****
See Figure 24.

1. Connect the Honda Diagnostic System (HDS) to the data link connector (DLC), and check for DTCs. If a DTC is present, diagnose and repair the cause before inspecting the ignition timing.
2. Start the engine. Hold the engine speed at 3,000 rpm without load (in Park or

CHARGING SYSTEM

- Alternator. Tighten the bolts to 16 ft. lbs. (22 Nm).
- PCV valve
- Alternator wiring harness connectors and harness clamp
- Auto tensioner
- Accessory drive belt
- Negative battery cable

IGNITION SYSTEM

Neutral) until the radiator fan comes on, then let it idle.

3. Check the idle speed, as outlined in the Fuel System Section.
4. Jump the SCS line with the HDS.
5. Free the service loop from the wire harness, then connect the timing light to the service loop.
6. Aim the light toward the pointer (A) on the cam chain case. Check the ignition timing under a no load condition (headlights, blower fan, rear window defogger, and air conditioner are turned off). The ignition timing should be:
   a. M/T: 6–10° BTDC (RED mark B) at idle in Neutral
   b. A/T: 6–10° BTDC (RED mark B) at idle in Park or Neutral
7. If the ignition timing differs from the specification, update the engine control module (ECM)/powertrain control module (PCM) if it does not have the latest software, or substitute a known-good ECM/PCM, then recheck. If the system works properly, and the ECM/PCM was substituted, replace the original ECM/PCM.

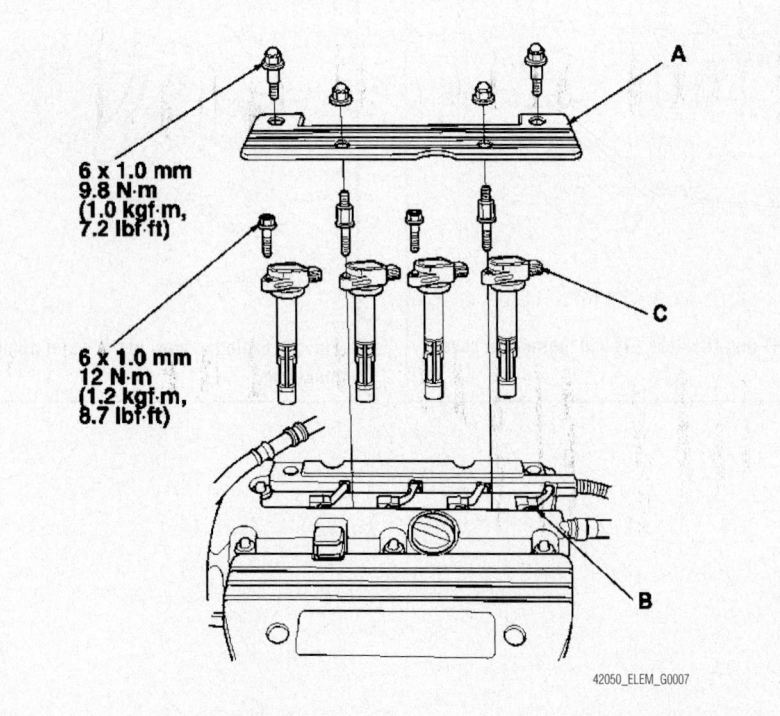

6 x 1.0 mm
9.8 N·m
(1.0 kgf·m,
7.2 lbf·ft)

6 x 1.0 mm
12 N·m
(1.2 kgf·m,
8.7 lbf·ft)

42050_ELEM_G0007

Fig. 23 Exploded view of the ignition coil cover (A), coil connectors (B) and ignition coils (C)

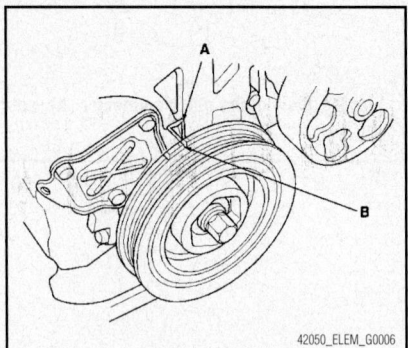

42050_ELEM_G0006

Fig. 24 Aim the light toward the pointer (A) on the cam chain case. Check the ignition timing under a no load condition (headlights, blower fan, rear window defogger, and air conditioner are turned off)

8. Disconnect the HDS and the timing light.

9. Secure the service loop to the wire harness with wire ties.

ADJUSTMENT

The ignition timing is controlled by the Powertrain Control Module (PCM). No adjustment is necessary or possible.

SPARK PLUGS

REMOVAL & INSTALLATION

See Figure 25.

1. Disconnect the negative battery cable.
2. Remove the ignition coil, as outlined later in this section.
3. Remove the spark plug.
4. Inspect the spark plug.

To install:

5. Install the spark plug and tighten to 13 ft. lbs. (18 Nm).
6. Install the ignition coil.
7. Connect the negative battery cable.

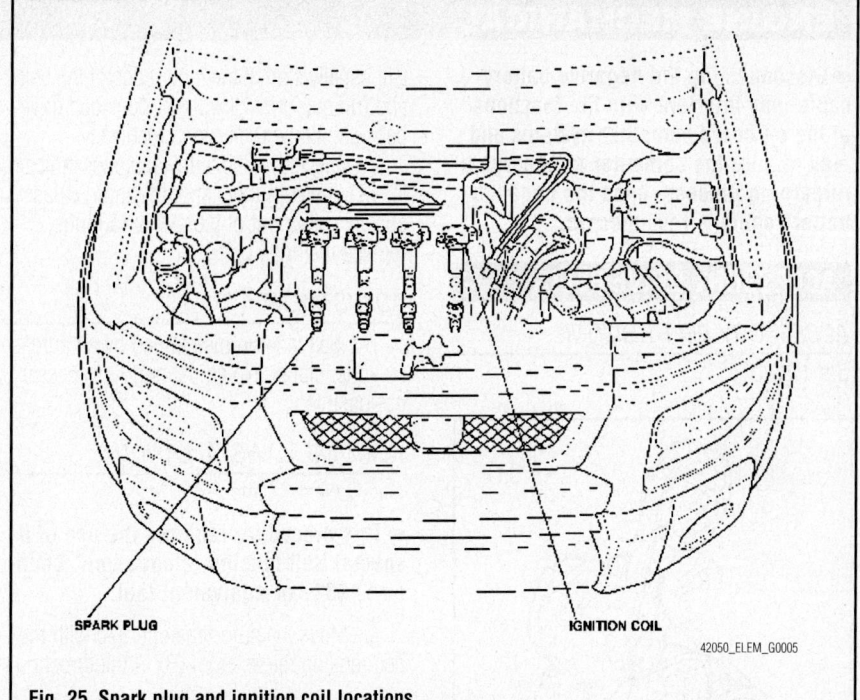

SPARK PLUG   IGNITION COIL

42050_ELEM_G0005

Fig. 25 Spark plug and ignition coil locations

ENGINE ELECTRICAL   STARTING SYSTEM

STARTER

REMOVAL & INSTALLATION

See Figure 26.

➡**The factory sound system has a coded theft protection system. It is recommended that you know your reset code before you begin.**

1. Before servicing the vehicle, refer to the precautions section.
2. Remove or disconnect the following:
   - Negative then the positive battery cables
   - Intake manifold
   - Starter cable from the B terminal
   - Black/white wire from the S (solenoid) terminal
   - Harness clamp and holder
   - Two bolts that mount the starter to the transaxle assembly
   - Starter

To install:

3. Install in the reverse order of removal. Refer to the illustration for torque specifications.

➡**When installing the heavy gauge starter cable, make sure the crimped side of the terminal end is facing out.**

4. Enter the anti-theft code and radio presets.

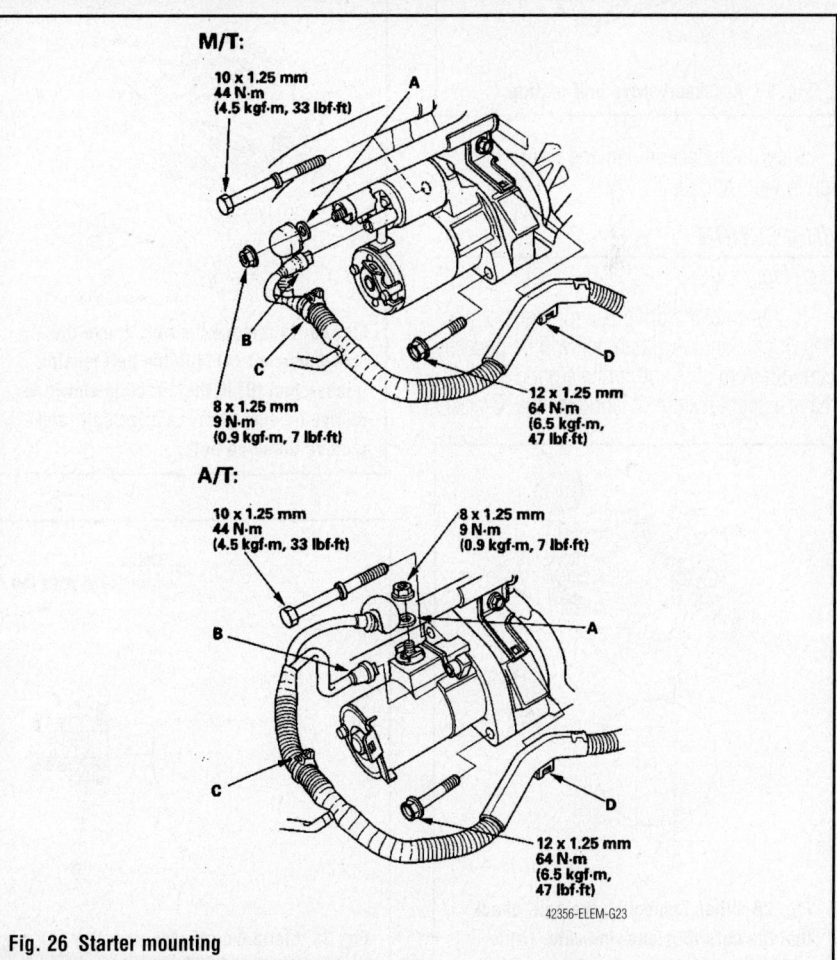

M/T:

10 x 1.25 mm
44 N·m
(4.5 kgf·m, 33 lbf·ft)

8 x 1.25 mm
9 N·m
(0.9 kgf·m, 7 lbf·ft)

12 x 1.25 mm
64 N·m
(6.5 kgf·m, 47 lbf·ft)

A/T:

10 x 1.25 mm
44 N·m
(4.5 kgf·m, 33 lbf·ft)

8 x 1.25 mm
9 N·m
(0.9 kgf·m, 7 lbf·ft)

12 x 1.25 mm
64 N·m
(6.5 kgf·m, 47 lbf·ft)

42356-ELEM-G23

Fig. 26 Starter mounting

ENGINE MECHANICAL

➡Disconnecting the negative battery cable may interfere with the functions of the on board computer systems and may require the computer to undergo a relearning process, once the negative battery cable is reconnected.

ACCESSORY DRIVE BELTS

ACCESSORY BELT ROUTING

See Figure 27.

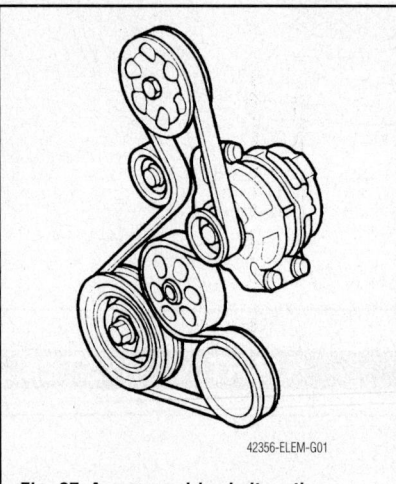

42356-ELEM-G01

Fig. 27 Accessory drive belt routing

Refer to the accompanying figure for drive belt routing.

INSPECTION

See Figure 28.

1. Inspect the drive belt for signs of glazing or cracking. A glazed belt will be perfectly smooth from slippage, while a good belt will have a slight texture of fabric visible. Cracks

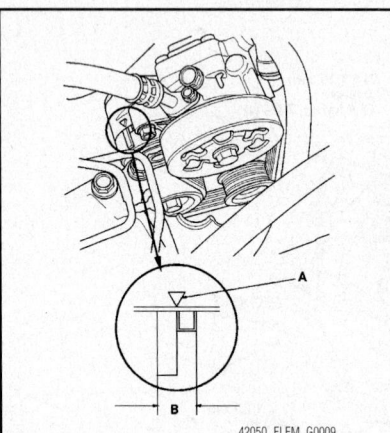

42050_ELEM_G0009

Fig. 28 When inspecting the belt, check that the auto-tensioner indicator (A) is within the standard range (B)

will usually start at the inner edge of the belt and run outward. All worn or damaged drive belts should be replaced immediately.

2. Check that the auto-tensioner indicator (A) is within the standard range (B) as shown. If it is out of the standard range, replace the drive belt.

ADJUSTMENT

The belt tension maintained by an automatic tensioner. No adjustment is necessary or possible.

REMOVAL & INSTALLATION

See Figures 27 and 29.

➡This procedure requires the use of a special Belt tension release tool, Snap-on YA9317 or equivalent tool.

1. Move the auto-tensioner (A) with the belt tension release tool (B) in the direction

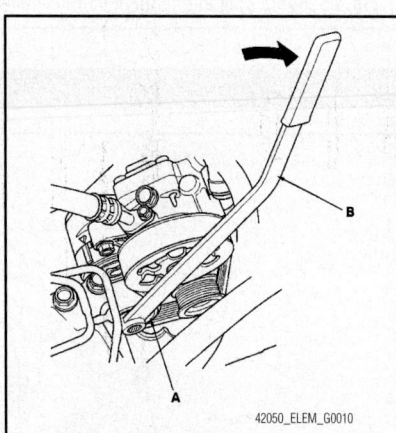

42050_ELEM_G0010

Fig. 29 To remove the belt, move the auto-tensioner (A) with the belt tension release tool (B) in the direction shown to relieve tension from the drive belt, and remove the drive belt

shown to relieve tension from the drive belt, and remove the drive belt.

2. Install the new belt in the reverse order of removal.

CAMSHAFT AND VALVE LIFTERS

REMOVAL & INSTALLATION

See the Rocker Arm Shaft Removal & Installation procedure.

CRANKSHAFT DAMPER

REMOVAL & INSTALLATION

See Figures 30 through 32.

➡This procedure requires the use of the following special tools, or their equivalents: Holder handle 07JAB-001020B, 50mm Holder attachment 07NAB-001040A, and 19mm Socket 07JAA-001020A.

1. Before servicing the vehicle, refer to the precautions section.

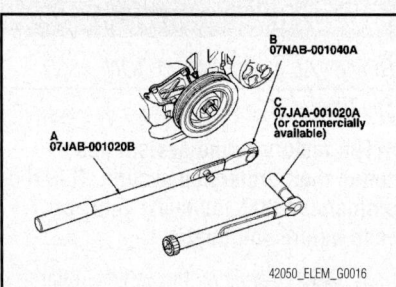

42050_ELEM_G0016

Fig. 30 To remove the crankshaft pulley, hold the pulley with holder handle (A) and holder attachment (B), then remove the bolt with a 19mm socket (C) and breaker bar

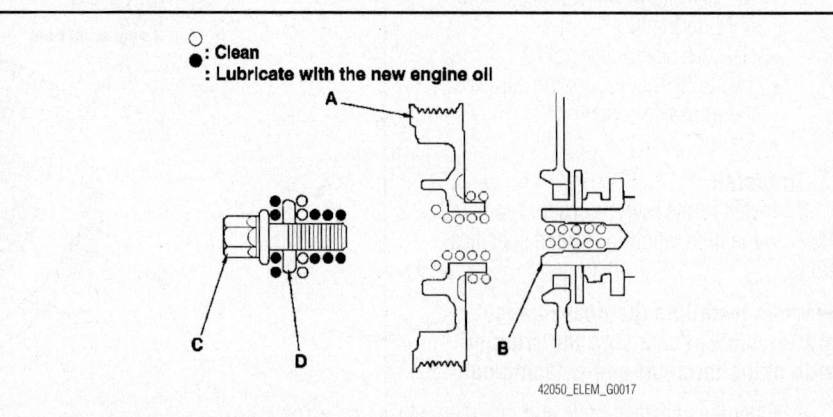

○ : Clean
● : Lubricate with the new engine oil

42050_ELEM_G0017

Fig. 31 Clean the crankshaft pulley (A), crankshaft (B), bolt (C), and washer (D). Lubricate with the new engine oil as shown

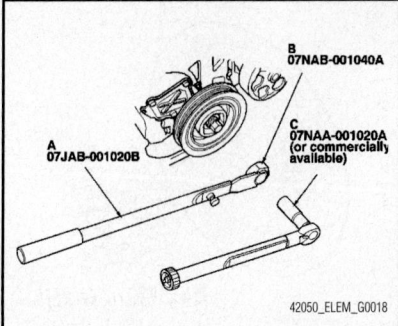

Fig. 32 Install the crankshaft pulley, and hold the pulley with holder handle (A) and holder attachment (B). Tighten the bolt to specifications with a torque wrench and 19mm socket

2. Remove the right front wheel.
3. Remove the splash shield.
4. Remove the drive belt.
5. Hold the pulley with holder handle (A) and holder attachment (B).
6. Remove the bolt with a 19mm socket (C) and breaker bar, then remove the crankshaft pulley.

To install:

7. Clean the crankshaft pulley (A), crankshaft (B), bolt (C), and washer (D). Lubricate with the new engine oil as shown.
8. Install the crankshaft pulley, and hold the pulley with holder handle (A) and holder attachment (B).
9. Tighten the bolt to 36 ft. lbs. (49 Nm) with a torque wrench and 19mm socket (C). Never use an impact wrench to tighten the crankshaft pulley bolt.
10. Tighten the pulley bolt an additional 90°.
11. Install the drive belt.
12. Install the splash shield.
13. Install the right front wheel.

CRANKSHAFT FRONT SEAL

REMOVAL & INSTALLATION

For the 2.4L engine, see the Timing Chain Removal & Installation procedure.

CYLINDER HEAD

REMOVAL & INSTALLATION

See Figures 33 through 36.

1. Before servicing the vehicle, refer to the precautions section.
2. Drain the cooling system.
3. Relieve the fuel system pressure.
4. Remove or disconnect the following:
   • Negative battery cable
   • Accessory drive belt

• Intake Air Temperature (IAT) sensor connector
• Vacuum hoses and breather pipe and air intake duct
• Fuel feed hose
• Bolt securing the connecting pipe support bracket to the engine block
• Evaporative emission (EVAP) canister hose and brake booster vacuum hose
• Intake manifold
• Exhaust manifold
• Cam chain
• Positive Crankcase Ventilation (PCV) hose and ground cable
• Upper radiator hose, heater hoses and water bypass hose

5. Remove the following engine wire harness connectors and wire harness clamps from the cylinder head:
   • Four injector connector
   • Engine Coolant Temperature (ECT) sensor connector
   • Camshaft Position (CMP) sensor A & B (intake & exhaust) connectors
   • VTEC solenoid valve connector
   • Engine Oil Pressure (EOP) sensor connector

6. Remove or disconnect the following:
   • 3 bolts holding the EVAP canister purge valve bracket and remove the two bolts (B) securing the harness bracket
   • Timing (cam) chain
   • Rocker arm assembly

7. Loosen the cylinder head bolts in sequence and ⅓ turns until all bolts are loose.

8. Remove the cylinder head.

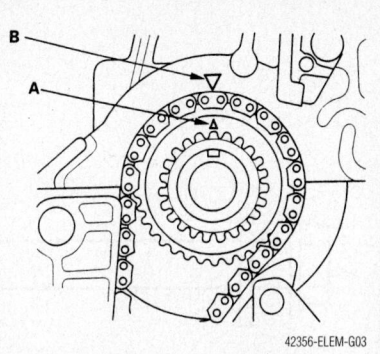

Fig. 34 Set the crankshaft to TDC by aligning the mark (A) on the crankshaft sprocket with the pointer (B) on the cylinder block

To install:

9. Be sure all cylinder head and block gasket surfaces are clean. Check the cylinder head for warpage. If warpage is less than 0.002 in. (0.05mm), cylinder head resurfacing is not required. Maximum resurface limit is 0.008 in. (0.2mm) based on a cylinder head height of 3.94 in. (100mm).

10. Install or connect the following:
    • New gasket and dowel pins on the cylinder block

11. Set the crankshaft to Top Dead Center (TDC). Align the TDC mark (A) on the crankshaft sprocket with the pointer (B) on the cylinder block.

12. Measure the diameter of each cylinder head bolt at points A & B, as shown in the illustration. If either diameter is less than 0.42 in. (10.6mm), replace the head bolt

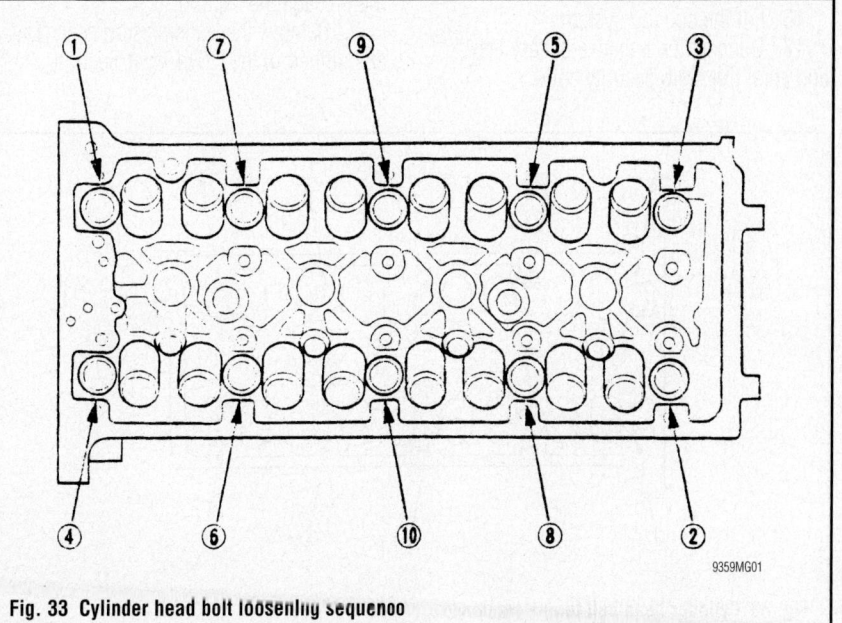

Fig. 33 Cylinder head bolt loosening sequence

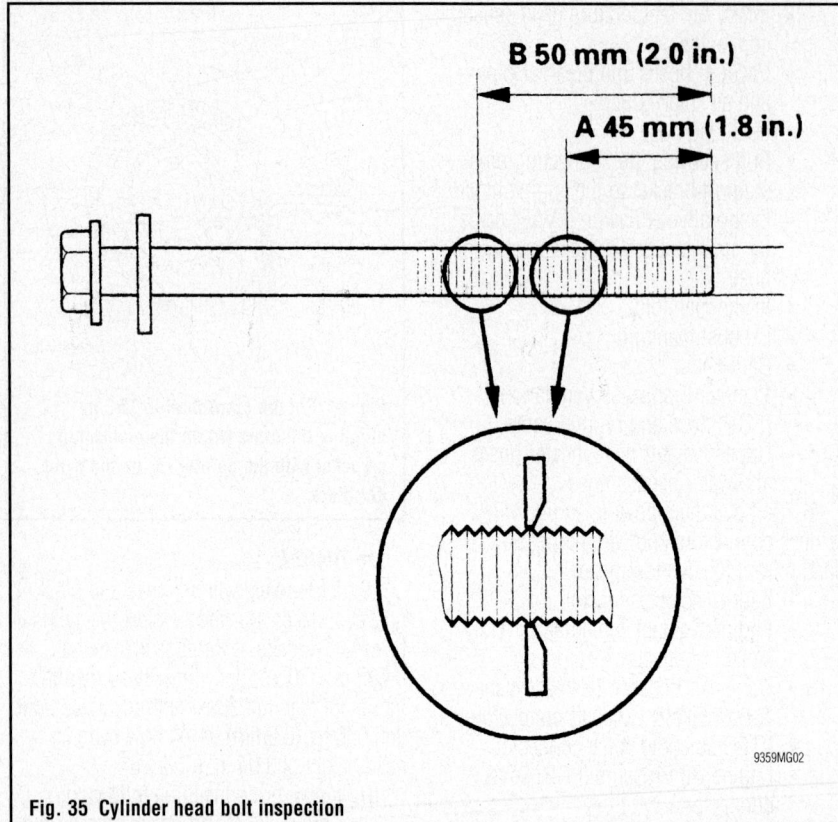

B 50 mm (2.0 in.)

A 45 mm (1.8 in.)

Fig. 35 Cylinder head bolt inspection

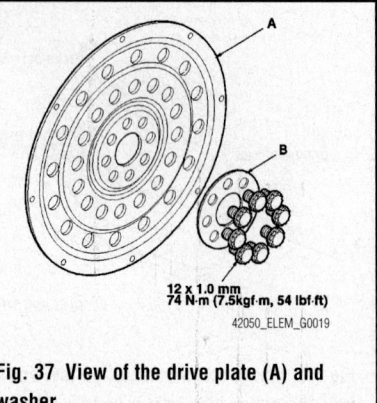

Fig. 37 View of the drive plate (A) and washer

3. Remove the drive plate (A) and washer (B) from the engine crankshaft.

To install:

4. Install the drive plate and washer on the engine crankshaft, and tighten the eight bolts in a crisscross pattern in two or more steps to a final torque of 54 ft. lbs. (74 Nm).

5. Install the transmission assembly, as outlined in the Drive Train Section.

ENGINE ASSEMBLY

REMOVAL & INSTALLATION

➡**The engine and transaxle are removed from the vehicle as a unit.**

1. Before servicing the vehicle, refer to the precautions section.
2. Drain the cooling system.
3. Drain the transaxle fluid.
4. Drain the engine oil.
5. Relieve fuel system pressure.
6. Remove or disconnect the following:

- Negative, then the positive battery cables
- IAT sensor connector
- Intake Air Temperature (IAT) sensor connector
- Breather and vacuum hoses
- Intake duct
- Battery
- Air cleaner housing
- Battery tray
- Cables from the power distribution center
- Ground cable
- Throttle and cruise cables
- Fuel lines
- EVAP canister
- Powertrain Control Module (PCM) connectors and grommet. Pull the PCM harness through the firewall.
- Clutch slave cylinder and line mounting bracket if equipped with a manual transmission

13. Apply engine oil to the threads and under the bolt heads of all of the bolts.

14. Install the cylinder head. Tighten the bolts in sequence as follows:

    a. Step 1: 29 ft. lbs. (39 Nm).
    b. Step 2: Plus 90 degrees.
    c. Step 3: Plus 90 degrees.
    d. Step 4: If using new cylinder head bolts, add an additional 90 degrees.

15. The remainder of installation is the reverse of removal.

16. Fill the cooling system.

17. Connect the negative battery cable and enter the radio security code.

18. Start the engine and check carefully for any leaks.

DRIVEPLATE

REMOVAL & INSTALLATION

With Automatic Transmission

See Figure 37.

1. Before servicing the vehicle, refer to the precautions section.

2. Remove the transmission assembly, as outlined in the Drive Train Section.

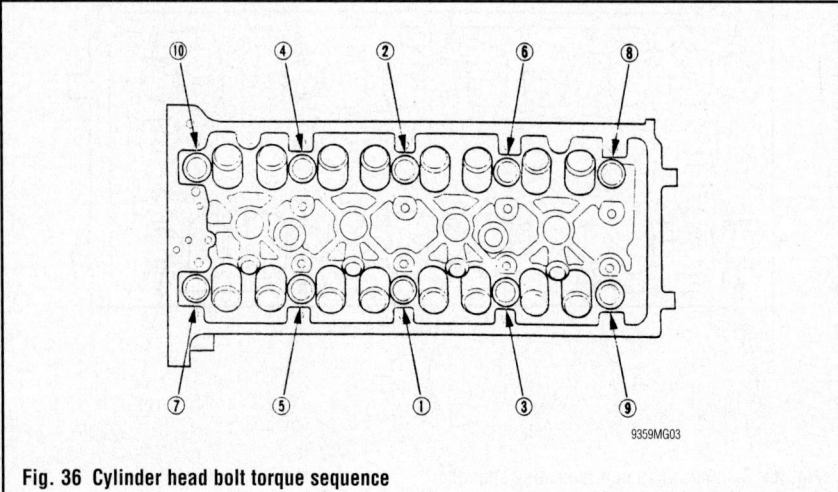

Fig. 36 Cylinder head bolt torque sequence

- Shift cable if equipped with a manual transmission
- Drive belt
- Power steering pump, leaving the hoses connected
- Power steering hose from the bracket on the valve cover
- Wheels
- Splash shield
- Air/Fuel ratio (A/F) sensor connector
- Secondary Heated Oxygen (O2S) sensor connector
- Catalytic converter
- Stabilizer links
- Lower ball joints
- Halfshafts
- Shift cable on models with an automatic transmission
- Propeller shaft on 4 wheel drive models
- Transmission cooler lines and ATF filter mounting if equipped with an automatic transmission
- Radiator and heater hoses
- Bolt attaching the power steering line if equipped with an automatic transmission

7. Support the transmission with a jack and block of wood if equipped with an automatic transmission.

- Transmission mount

8. Attach a hoist to the engine lifting eyes and support the powertrain weight.

9. Remove the jack and block of wood if equipped with an automatic transmission.

- Transmission mount bracket
- Upper engine mount bracket retainers
- Rear mount bracket bolts.
- Front mount bolt

10. Matchmark the sub-frame mounting bolt centers.

There is a special tool necessary for subframe removal. The Honda tool number is EQS02C000011. Attach the tool as explained in the tool instructions, attach a floor jack with adapter, remove the 4 subframe bolts and lower the sub-frame.

11. Remove or disconnect the following:

12. A/C compressor without disconnecting the hoses

13. Check that all hoses and wires are disconnected.

14. Lower the engine about 6 inches and recheck all clearances.

15. Lower the engine all the way.

16. Remove the chain hoist.

To install:

17. Installation is the reverse of removal. Observe the following torques:

- Front engine mount bracket bolts: 33 ft. lbs. (44Nm)
- A/C compressor bracket: 33 ft. lbs. (44Nm)
- Stiffener 10mm bolts: 33 ft. lbs. (44Nm); 6mm bolts 9 ft. lbs. (12 Nm)
- A/C compressor bolts: 16 ft. lbs. (22 Nm)
- Subframe bolts: 76 ft. lbs. (98 Nm)
- Upper bracket bolt and nut: 45 ft. lbs. (61 Nm)
- Transmission mount bracket support bolts/nuts: 40 ft. lbs. (54 Nm)
- PS pump bolts: 16 ft. lbs. (22 Nm)

➡**Use new self-locking nuts and color-coded self-locking bolts when installing the engine mounts and suspension components.**

➡**Do not tighten the engine or transaxle mount fasteners until instructed to do so.**

18. Lower the powertrain into position.

19. Install or connect the following:

- Transaxle mount and bracket. Tighten the frame mounting bolts to 47 ft. lbs. (64 Nm).
- Upper bracket. Tighten the nuts in sequence to 54 ft. lbs. (74 Nm).
- Rear mount bracket through bolt
- Right front mount and bracket
- Left front mount and bracket

20. Tighten the remaining mount fasteners as follows:

a. Transaxle mount fasteners to 47 ft. lbs. (64 Nm) and the through bolt to 54 ft. lbs. (74 Nm).

b. Rear mount bracket through bolt to 43 ft. lbs. (59 Nm).

c. Right front mount 12mm bolts to 47 ft. lbs. (64 Nm) and the 10mm bolts to 33 ft. lbs. (44 Nm).

d. Left front mount 12mm stud bolt to 61 ft. lbs. (83 Nm), 10mm bolts to 33 ft. lbs. (44 Nm), and 12mm nut to 43 ft. lbs. (59 Nm).

e. Right front mount 12mm nut to 43 ft. lbs. (59 Nm).

21. Install or connect the following:

- Propeller shaft, if equipped
- A/C compressor
- Radiator
- A/C hose clamp

22. If equipped with a manual transaxle, install or connect the following:

- Shift cables
- Transaxle ground cable
- Clutch hose bracket
- Clutch slave cylinder

23. If equipped with an automatic transaxle, install or connect the following:

- Transaxle fluid cooler lines
- Transaxle ground cable and hose clamp
- Shift cable
- Shift cable cover

24. For all vehicles, install or connect the following:

- Axle halfshafts
- Lower ball joints
- Right damper fork
- Exhaust front pipe
- HO2S connector
- Heater hoses
- Radiator hoses
- Splash shield
- PSP switch
- Accelerator cable
- Brake booster vacuum line
- Fuel lines
- A/C compressor drive belt
- Power steering pump and belt
- Cruise control actuator
- Left engine wire harness connectors
- PCM connectors and grommet
- Air intake assembly
- Battery and tray
- Fuse/Relay box battery cables
- Negative battery cable

25. Fill the engine crankcase to the correct level.

26. Fill the transaxle to the correct level.

27. Fill the cooling system.

28. Start the engine and check for leaks.

29. Check the wheel alignment and adjust as necessary.

EXHAUST MANIFOLD

REMOVAL & INSTALLATION

See Figure 38.

1. Before servicing the vehicle, refer to the precautions section.

2. Raise and safely support the vehicle.

3. Remove or disconnect the following:

- VTEC solenoid valve
- Intermediate shaft heat cover
- Cover and exhaust manifold bracket
- Exhaust manifold

To install:

4. Clean the mounting surfaces.

5. Install or connect the following:

- New gasket on the cylinder head
- Exhaust manifold. Tighten the nuts, in a criss-cross pattern starting with the inner nut, to 33 ft. lbs. (45 Nm).
- Exhaust manifold bracket and cover
- Intermediate shaft heat cover
- VTEC solenoid valve

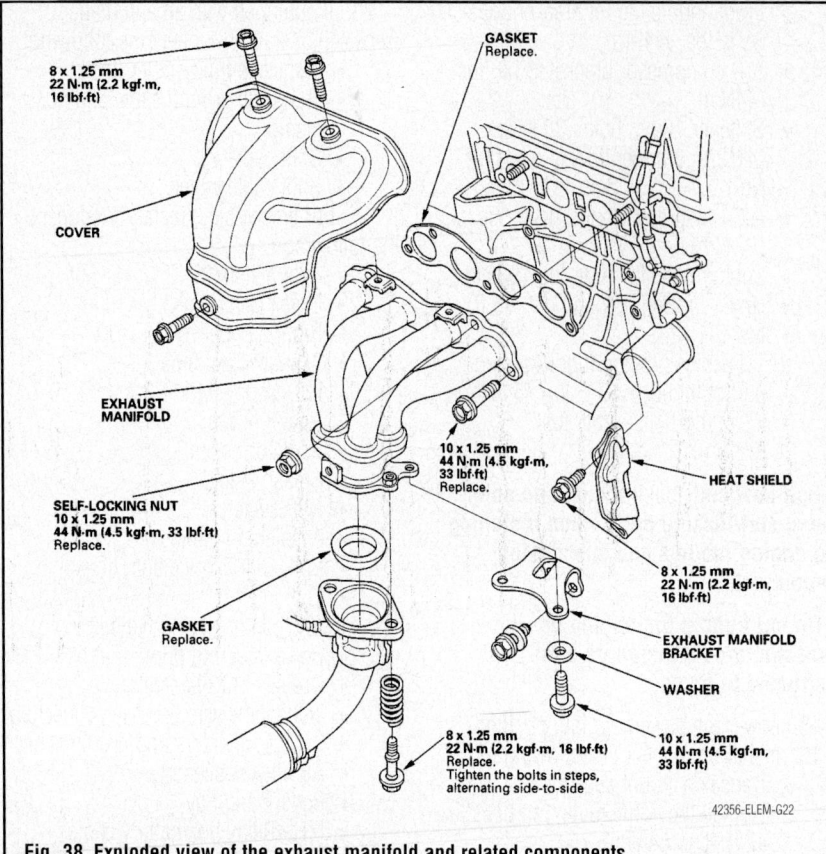

Fig. 38 Exploded view of the exhaust manifold and related components

FLYWHEEL

REMOVAL & INSTALLATION

With Manual Transmission

See Figure 39.

1. Before servicing the vehicle, refer to the precautions section.

2. Remove the pressure plate and clutch disc, as outlined in the Drive Train Section.

3. Install the special tool on the flywheel, as shown in the accompanying illustration.

4. Remove the flywheel mounting bolts in a crisscross pattern in several steps, then remove the flywheel.

To install:

5. Install the flywheel on the crankshaft, and install the mounting bolts, finger-tight.

Install the special tool, then torque the flywheel mounting bolts in a crisscross pattern in several steps to 76 ft. lbs. (103 Nm).

6. Install the clutch disc and pressure plate, as outlined in the Drive Train Section.

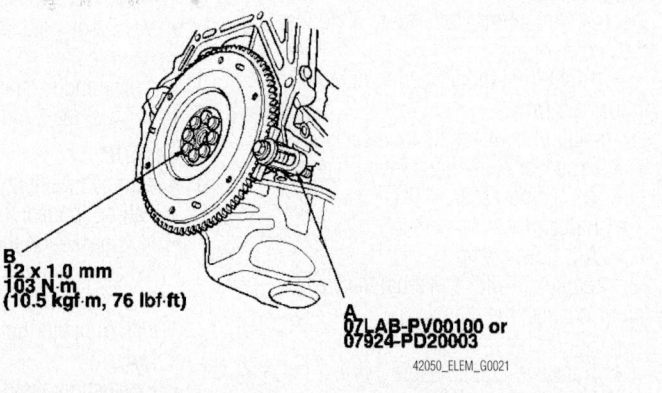

Fig. 39 After installing the special tool (A), tighten the flywheel mounting bolts (B) to specifications, in several steps, using a criss-cross pattern

INTAKE MANIFOLD

REMOVAL & INSTALLATION

See Figure 41.

1. Before servicing the vehicle, refer to the precautions section.

2. Disconnect the negative battery cable.

3. Drain the engine coolant into a sealable container.

4. Remove or disconnect the following:
   - Intake Air Temperature (IAT) sensor electrical connector
   - Vacuum hose and breather pipe and the air intake duct
   - Intake manifold cover
   - Throttle and cruise control cables by loosening the locknuts, then slipping the cable ends out of the accelerator linkage.

➡**Do not bend the cables during removal. Always replace any throttle or cruise control cables that get kinked during removal.**

   - Evaporative emission (EVAP) canister hose and brake booster vacuum hose
   - Idle Air Control (IAC) valve connectors
   - Throttle Position (TP) sensor connector
   - Manifold Absolute Pressure (MAP) sensor connector
   - Necessary engine wire harness connectors and wire harness clamps from the intake manifold
   - Bolt securing the harness holder and remove the harness clamps
   - Water bypass hoses, then plug them
   - Harness clamp and harness connector from the intake manifold bracket
   - Intake manifold bracket
   - A/T vacuum hose
   - Retainer and intake manifold

To install:

5. Clean the mounting surfaces.

6. Install or connect the following:
   - New gasket
   - Intake manifold. Tighten the bolts, in a criss-cross pattern beginning with the inner bolt, to 16 ft. lbs. (22 Nm).
   - A/T vacuum hose
   - Intake manifold bracket
   - Harness clamp and connector to the intake manifold bracket
   - Water bypass hoses
   - Bolt securing the harness holder and tighten to 8.7 ft. lbs. (12 Nm)

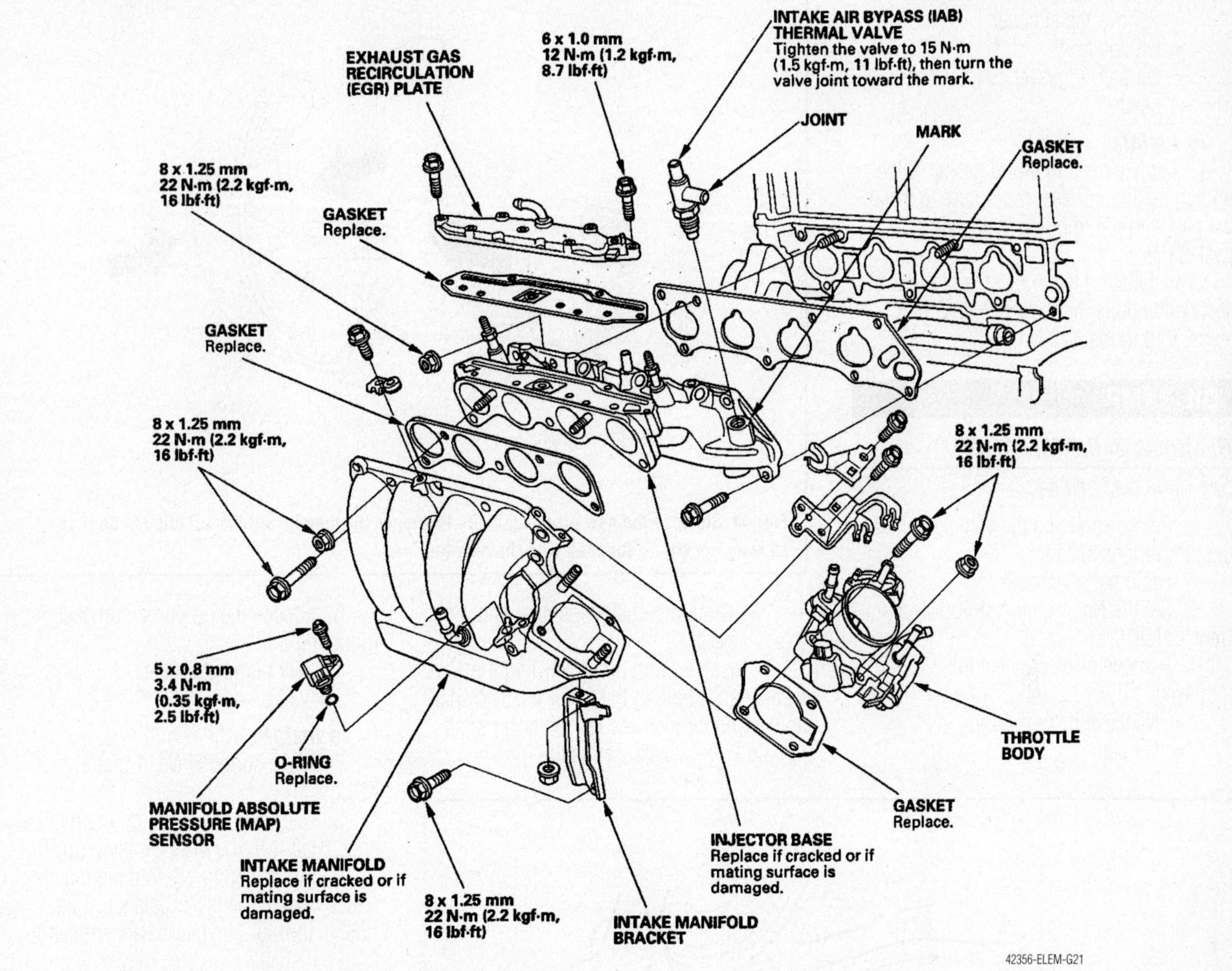

EXHAUST GAS RECIRCULATION (EGR) PLATE

6 x 1.0 mm
12 N·m (1.2 kgf·m, 8.7 lbf·ft)

INTAKE AIR BYPASS (IAB) THERMAL VALVE
Tighten the valve to 15 N·m (1.5 kgf·m, 11 lbf·ft), then turn the valve joint toward the mark.

JOINT

MARK

GASKET Replace.

8 x 1.25 mm
22 N·m (2.2 kgf·m, 16 lbf·ft)

GASKET Replace.

GASKET Replace.

8 x 1.25 mm
22 N·m (2.2 kgf·m, 16 lbf·ft)

8 x 1.25 mm
22 N·m (2.2 kgf·m, 16 lbf·ft)

5 x 0.8 mm
3.4 N·m (0.35 kgf·m, 2.5 lbf·ft)

O-RING Replace.

MANIFOLD ABSOLUTE PRESSURE (MAP) SENSOR

INTAKE MANIFOLD
Replace if cracked or if mating surface is damaged.

8 x 1.25 mm
22 N·m (2.2 kgf·m, 16 lbf·ft)

INTAKE MANIFOLD BRACKET

INJECTOR BASE
Replace if cracked or if mating surface is damaged.

THROTTLE BODY

GASKET Replace.

42356-ELEM-G21

Fig. 41 Exploded view of the intake manifold and related components

- Harness clamps
- EVAP canister hose and brake booster vacuum hose
- Throttle and cruise control cables
- Intake manifold cover
- Intake air duct
- IAT sensor connector, vacuum hose and breather pipe

7. Refill the cooling system.

8. Connect the negative battery cable, start the engine, and check for leaks.

OIL PAN

REMOVAL & INSTALLATION

See Figure 42.

1. Before servicing the vehicle, refer to the precautions section.

2. Drain the engine oil.

3. Remove or disconnect the following:

- Subframe. See Engine Removal and Installation.

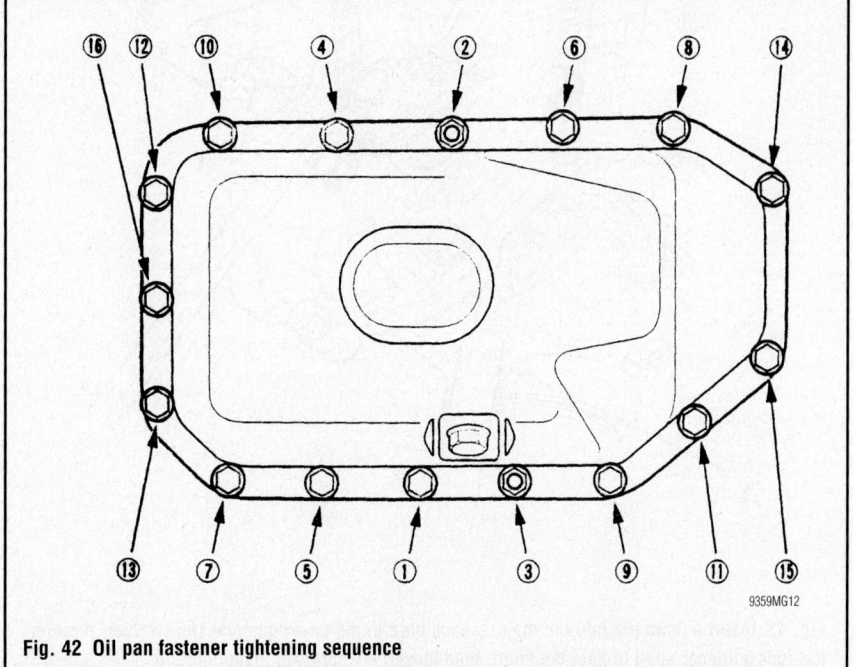

9359MG12

Fig. 42 Oil pan fastener tightening sequence

- With a manual transmission, remove the stiffener
- Oil pan bolts
- Oil pan. A gasket cutter will be needed.

To install:

4. Apply a bead of liquid gasket to the oil pan mating surface. make sure to install the pan within 4 minutes of applying the gasket maker.

5. Installation is the reverse of removal. Torque the bolts, in sequence, in 2 or 3 steps, to 9 ft. lbs. (12 Nm).

OIL PUMP

REMOVAL & INSTALLATION

See Figures 43 and 44.

1. Before servicing the vehicle, refer to the precautions section.
2. Drain the engine oil.
3. Set the No. 1 piston to Top Dead Center (TDC).
4. Remove or disconnect the following:
   - Negative battery cable
   - Oil pan

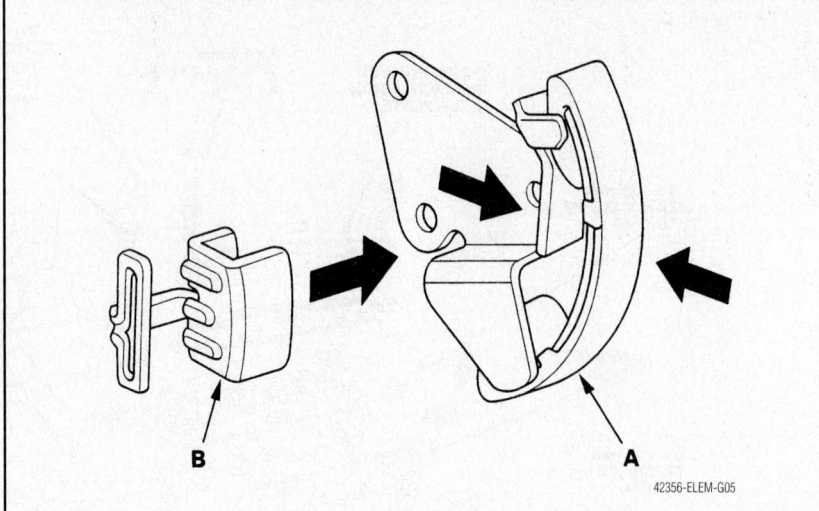

Fig. 44 Squeeze the new oil pump chain tensioner (A) then install the set clip (A) on it as shown. The clip is supplied with the new tensioner

- Oil pump chain tensioner and discard

5. Insert a 6mm pin driver into the maintenance hole in the lower balance shaft holder and through the rear balancer shaft to hold the rear balancer shaft.

6. Loosen the oil pump sprocket mounting bolt.
   - Oil pump sprocket
   - Oil pump

To install:

7. Make sure that No. 1 piston is at TDC.
8. Align the dowel pin on the rear balance shaft with the mark on the pump.
9. Insert a 6mm pin into the maintenance hole in the lower balance shaft holder, through the rear balancer shaft to hold the shaft.
10. Install or connect the following:
   - Engine oil to the threads of the oil pump sprocket mounting bolt
   - Oil pump and sprocket loosely
11. Remove the balance shaft holding pin.
12. Torque the 10mm mounting bolts to 33 ft. lbs. (44 Nm); the 8mm bolts to 16 ft. lbs. (22 Nm).
13. Torque the pulley bolt to 33 ft. lbs. (44 Nm).
14. Squeeze the new oil pump chain tensioner then install the set clip on it as shown in the illustration.
15. Install or connect the following:
   - New oil pump chain tensioner and torque the bolts to 9 ft. lbs. (12 Nm). Remove the set clip from the tensioner.
   - Oil pan
16. Fill the engine with oil.

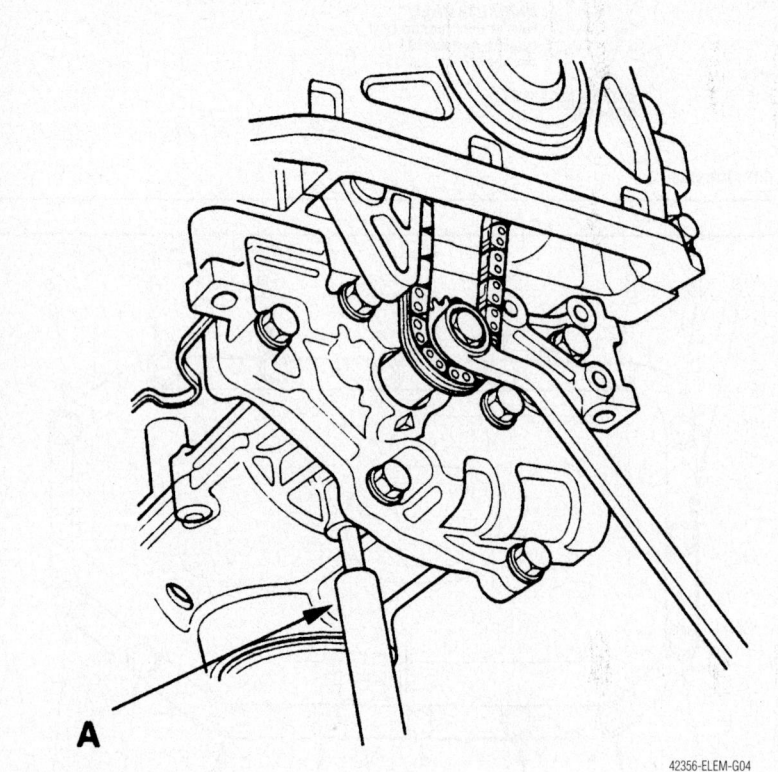

Fig. 43 Insert a 6mm pin into the maintenance hole in the lower balance shaft holder, through the rear balancer shaft to hold the shaft, then loosen the sprocket mounting bolt

PISTON AND RING

POSITIONING

See Figures 45 through 47.

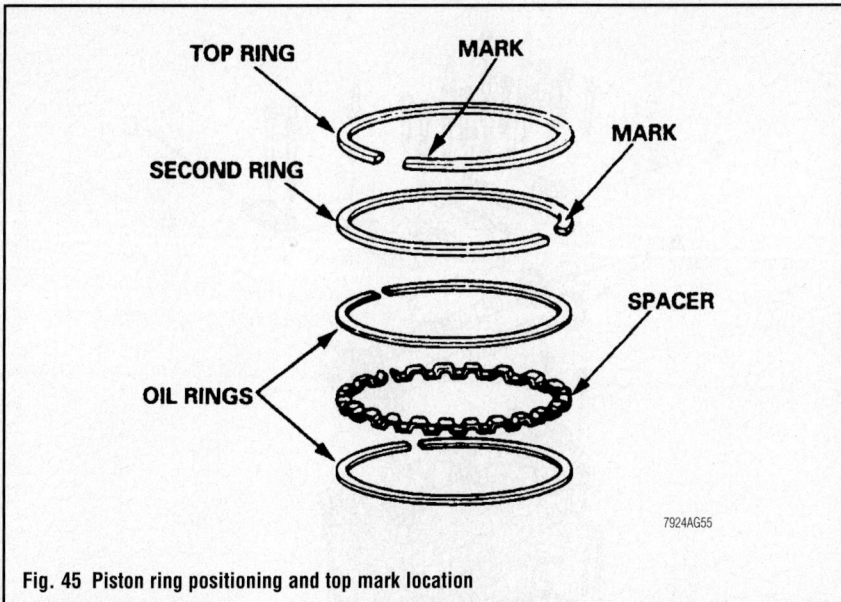

Fig. 45 Piston ring positioning and top mark location

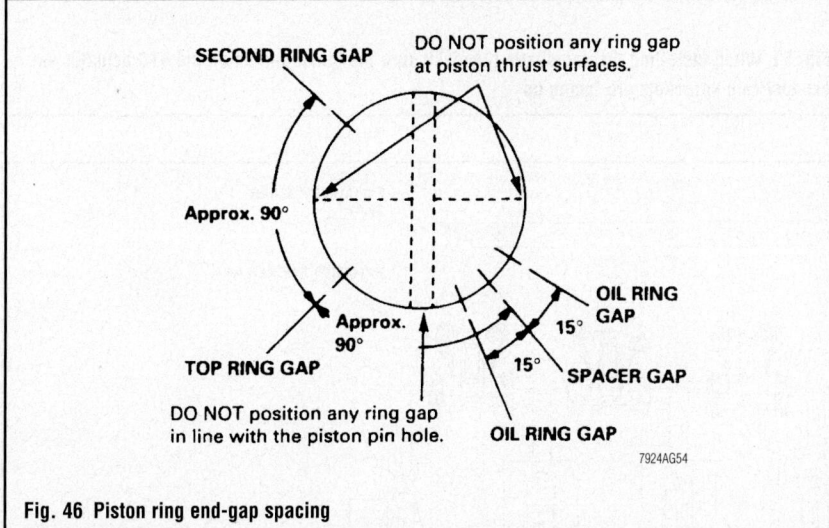

Fig. 46 Piston ring end-gap spacing

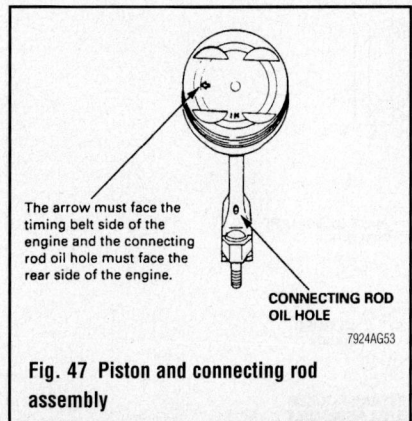

The arrow must face the timing belt side of the engine and the connecting rod oil hole must face the rear side of the engine.

CONNECTING ROD OIL HOLE

Fig. 47 Piston and connecting rod assembly

REAR MAIN SEAL

REMOVAL & INSTALLATION

1. Before servicing the vehicle, refer to the precautions section.

2. Remove or disconnect the following:
   - Transaxle
   - Clutch pressure plate and disc, if equipped
   - Flywheel
   - Oil seal

To install:

3. Install or connect the following:
   - Oil seal. Drive the seal square into the seal case.
   - Flywheel. Tighten the bolts in a crossing pattern to 76 ft. lbs. (103 Nm).
   - Clutch pressure plate and disc, if equipped
   - Transaxle
4. Check the fluid levels.
5. Start the engine and check for leaks.

ROCKER ARMS/SHAFTS

REMOVAL & INSTALLATION

See Figures 48 through 52.

1. Before servicing the vehicle, refer to the precautions section.
2. Remove or disconnect the following:
   - Timing (cam) chain
   - Loosen the rocker arm adjusting screws
   - Camshaft holder bolts, two turns at a time in sequence
   - Timing chain guide (B), camshaft holders and camshafts
3. Insert the bolts (A) into the rocker shaft holder, then remove the rocker arm assembly (B)

To install:

4. Clean and dry the No. 5 rocker shaft holding mating surface.

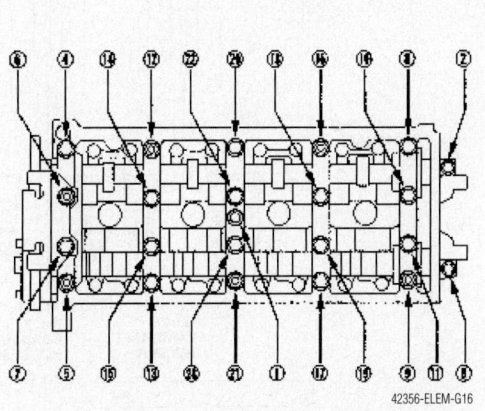

Fig. 48 Camshaft holder bolt loosening sequence. Note that bolt 1 in the illustration is not on all engines.

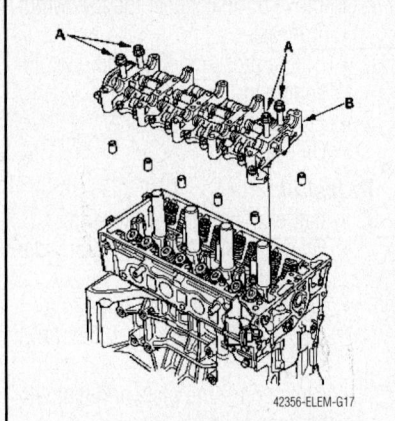

Fig. 49 Insert the bolts (A) into the rocker shaft holder, then remove the rocker arm assembly (B)

5. Apply a suitable liquid gasket P/N 08718-0009, or equivalent, evenly to the cylinder head mating surface of the No. 5 rocker shaft holder.

➡The parts must be installed within 5 minutes of applying the liquid gasket.

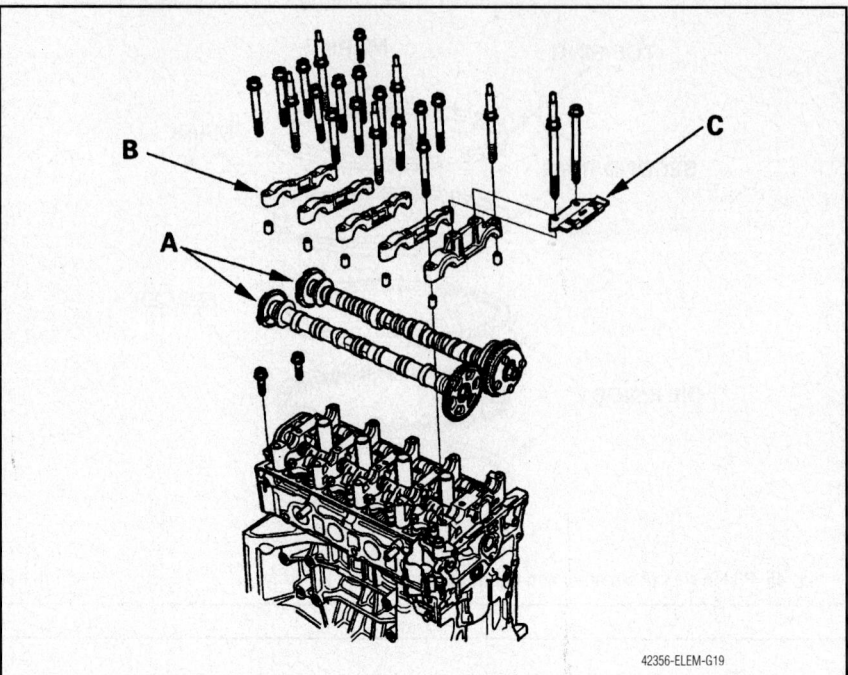

Fig. 51 When installing the camshafts (A) make sure the punch marks on the VTC actuator and exhaust cam sprockets are facing up

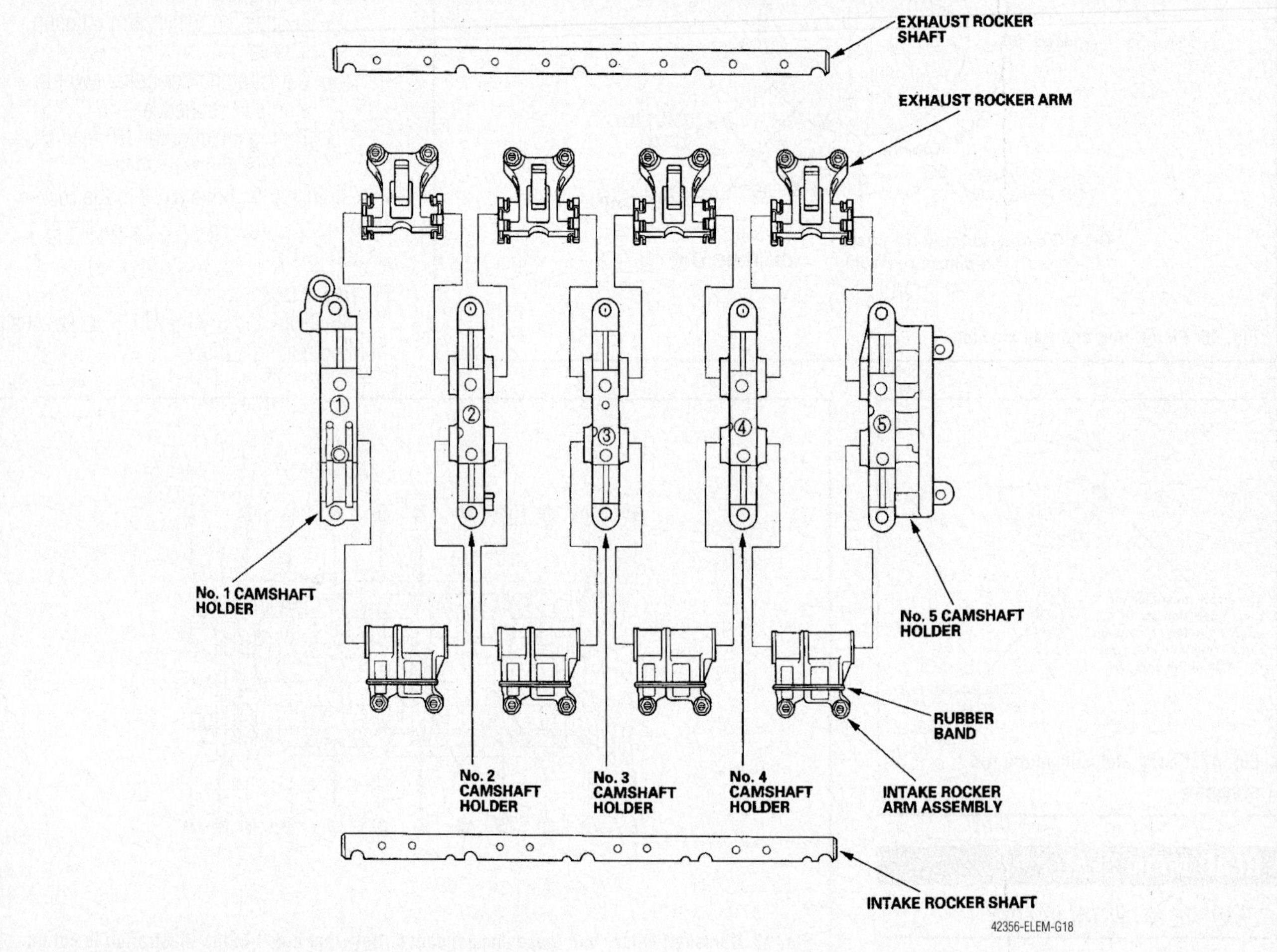

EXHAUST ROCKER SHAFT

EXHAUST ROCKER ARM

No. 1 CAMSHAFT HOLDER

No. 5 CAMSHAFT HOLDER

No. 2 CAMSHAFT HOLDER

No. 3 CAMSHAFT HOLDER

No. 4 CAMSHAFT HOLDER

RUBBER BAND

INTAKE ROCKER ARM ASSEMBLY

INTAKE ROCKER SHAFT

Fig. 50 Exploded view of the rocker arms and related components

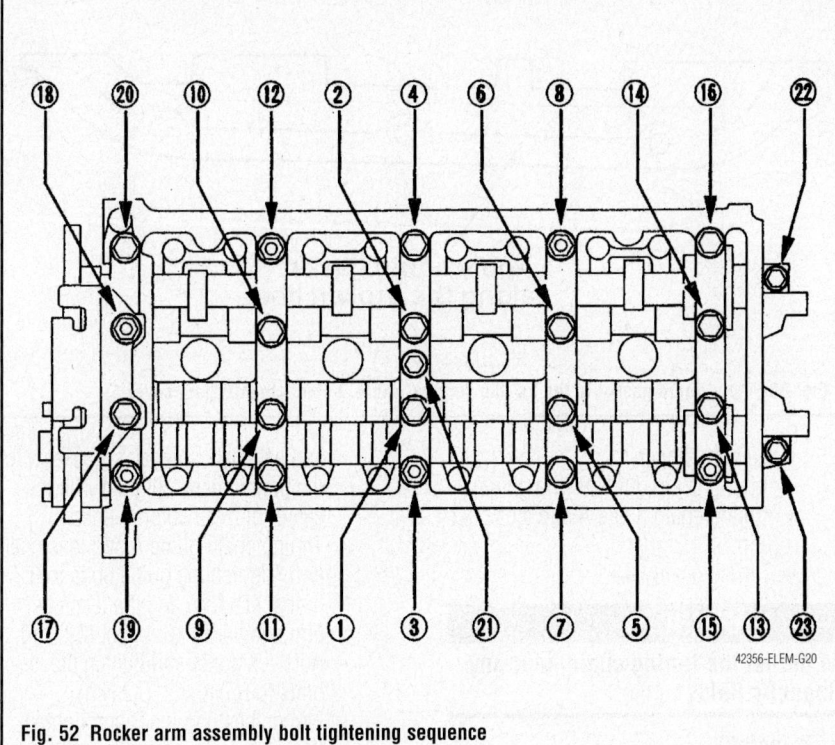

Fig. 52 Rocker arm assembly bolt tightening sequence

6. Reassemble the rocker arm assembly, as necessary.

7. Install or connect the following
- Bolts (A) into the rocker shaft holder, then the rocker arm assembly on the cylinder head. Remove the bolts from the rocker shaft holder.

8. Make sure the punch marks on the variable valve timing control (VTC) actuator and exhaust camshaft sprocket are facing up, then set the camshafts (A) in the holder

9. Set the camshaft holders (B) and timing chain guide B (C) in place.

10. Tighten the bolts, in sequence, to the following specification:
   a. 8mm bolts: 16 ft. lbs. (22 Nm).
   b. 6mm bolts: 8.7 ft. lbs. (12 Nm).
   The 6mm bolts are 21, 22 and 23.

11. Install the timing chain and adjust the valve lash.

TIMING CHAIN & FRONT SEAL

REMOVAL & INSTALLATION

See Figures 53 through 61.

1. Before servicing the vehicle, refer to the precautions section.

2. Set the engine to Top Dead Center (TDC).

3. Drain the cooling system.

4. Relieve the fuel system pressure.

5. Remove or disconnect the following:

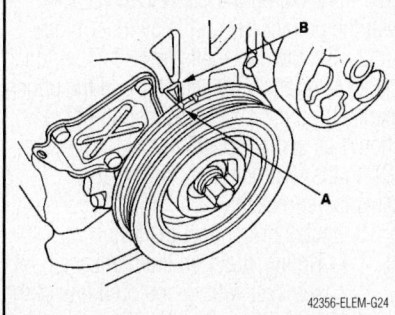

Fig. 53 Turn the crankshaft pulley so the TDC mark (A) is aligned with the pointer (B)

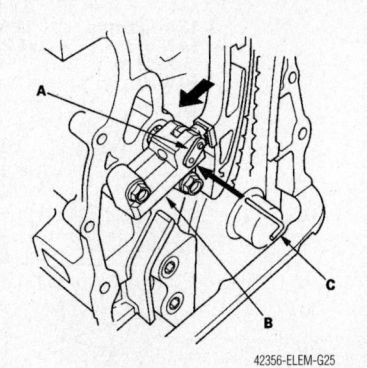

Fig. 54 Align the holes on the lock (A) and the auto-tensioner (B), then place a 1.5mm pin into the holes. Turn the crankshaft clockwise to secure the pin

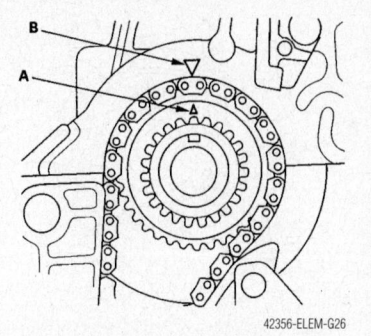

Fig. 55 Set the crankshaft to TDC. Align the TDC mark (A) on the crankshaft sprocket with the pointer (B) on the cylinder block.

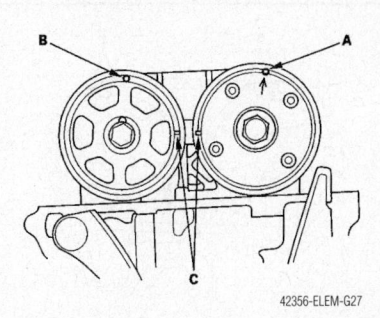

Fig. 56 The mark (A) on the VTC actuator and the mark (B) on the exhaust cam (C) should be at the top. Align the TDC marks (C) on the VTC actuator and exhaust cam sprockets.

- Negative battery cable
- Front tires and wheels
- Splash shield
- Drive belt
- Cylinder head cover.

6. Check that the No. 1 piston TDC marks on the Variable Valve Timing Control (VTC) actuator and exhaust camshaft sprocket are aligned.
- Crankshaft pulley
- Crankshaft Position (CKP) sensor connector
- VTC oil control solenoid valve connector
- VTC oil control solenoid valve

7. Support the engine with a suitable jack with a wooden block under the oil pan.
- Ground cable and upper engine mount bracket
- Side engine mount bracket
- Chain (case) cover

8. Loosely install the crankshaft pulley. Turn the crankshaft counterclockwise to compress the auto-tensioner.

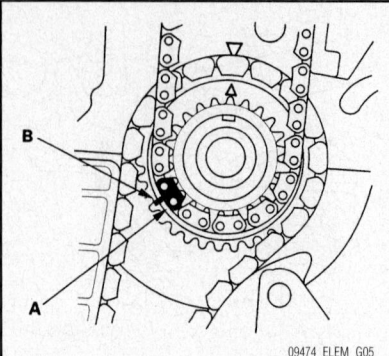

Fig. 57 Install the timing chain on the crankshaft sprocket with the colored link of the chain aligned with the mark on the crank sprocket

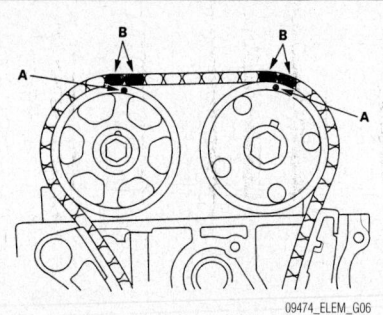

Fig. 58 Install the timing chain on the VTC actuator and exhaust camshaft sprocket with the punch marks aligned with the center of the 2 colored links

9. Align the holes on the lock (A) and the auto-tensioner (B), then place a 1.5mm pin into the holes. Turn the crankshaft clockwise to secure the pin.

10. Remove or disconnect the following:

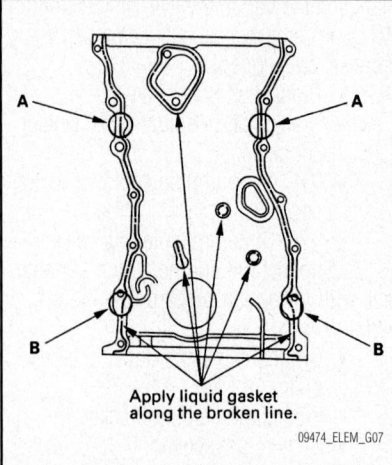

Fig. 59 Apply liquid gasket to the chain cover locations illustrated

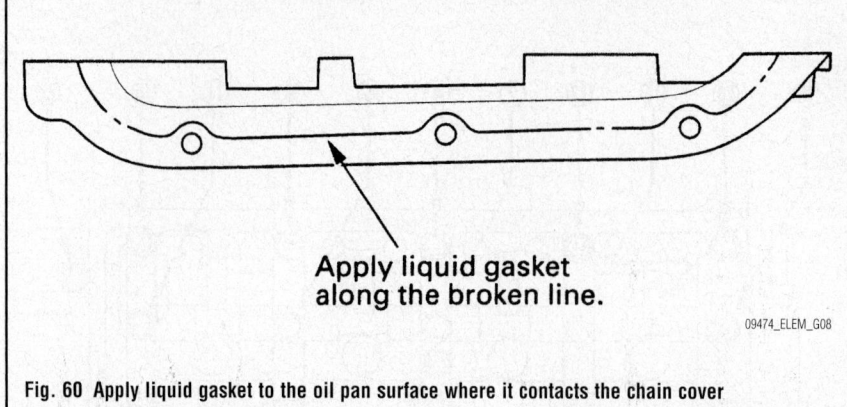

Apply liquid gasket along the broken line.

Fig. 60 Apply liquid gasket to the oil pan surface where it contacts the chain cover

- Auto-tensioner
- Timing chain guide B (top guide)
- Timing chain guide A and tensioner arm
- Timing chain

✳✳ WARNING

Do not let the timing chain near any magnetic fields.

To install:

11. Set the crankshaft to TDC. Align the TDC mark (A) on the crankshaft sprocket with the pointer (B) on the cylinder block.

12. Set the camshafts to TDC. The punch mark (A) on the VTC actuator and the punch mark (B) on the exhaust camshaft (C) should be at the top. Align the TDC marks (C) on the VTC actuator and exhaust camshaft sprockets.

13. Install or connect the following:
- Timing chain on the crankshaft sprocket with the colored link of the chain aligned with the mark on the crank sprocket
- Timing chain on the VTC actuator

and exhaust camshaft sprocket with the punch marks aligned with the center of the 2 colored links
- Timing chain guide A and tensioner arm. Tighten the guide bolts to 8.7 ft. lbs. (12 Nm) and the tensioner arm retainer to 16 ft. lbs. (22 Nm).
- Auto-tensioner and tighten the bolts to 8.7 ft. lbs. (12 Nm)
- Timing chain guide B and tighten the retainers to 16 ft. lbs. (22 Nm)

14. Remove the pin from the auto-tensioner.

15. Inspect the chain cover seal for damage and replace if necessary. Clean and dry the chain cover mating surfaces.

16. Install or connect the following:
- Liquid gasket, P/N 08718-0009 evenly to the cylinder block mating surface of the timing chain cover and the inner threads of the holes
- Liquid gasket to the cylinder block upper surface contact areas on the chain cover and the oil pan mating surface of the chain cover in the inner threads of the holes

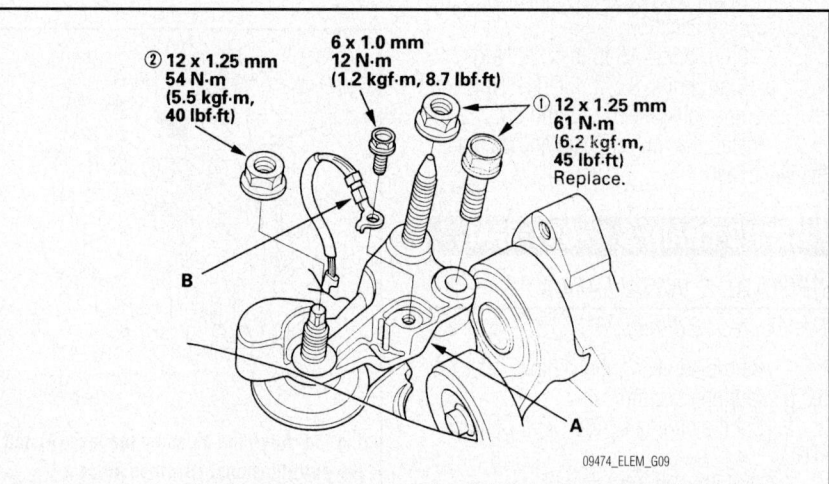

Fig. 61 Tighten the upper bracket upper bolt/nuts in the proper order to the correct specification

- Liquid gasket to the oil pan surface where it contacts the chain cover

➡**Make sure to install the components within 4 minutes of applying the sealer.**

- New O-ring the timing chain cover. Set the edge of the cover to the edge of the oil pan, then install the cover on the engine block. Tighten the retainers to 8.7 ft. lbs. (12 Nm).

➡**When installing the chain case, do not slide the bottom surface on the oil pan mounting surface.**

- Side engine mounting bracket and tighten the retainers to 33 ft. lbs. (44 Nm)
- Upper mount, then tighten the bolts/nuts as shown in the illustration
- Ground cable
- VTC oil control solenoid valve
- CKP sensor and VTC oil control solenoid valve connectors
- Crankshaft pulley. Tighten the bolt to 36 ft. lbs. (49 Nm), then tighten an additional 90 degrees.
- Cylinder head cover
- Drive belt
- Splash shield

17. Fill the engine cooling system and connect the negative battery cable.

VALVE (ROCKER ARM) COVERS

REMOVAL & INSTALLATION

1. Before servicing the vehicle, refer to the precautions section.
2. Disconnect the negative battery cable.
3. Remove the intake manifold cover.
4. Remove the four ignition coils.

5. Remove the two bolts securing the vacuum line.
6. Remove the bolt securing the power steering hose bracket.
7. Remove the dipstick and breather hose.
8. Remove the retainers and the cylinder head cover.
9. Installation is the reverse of the removal procedure.

VALVE LASH

ADJUSTMENT

See Figure 62.

Adjust the valves only when the cylinder head temperature is less than 100°F (38°C).

1. Before servicing the vehicle, refer to the precautions section.
2. Remove or disconnect the following:
   - Negative battery cable
   - Cylinder head cover
3. Set the timing marks as shown in the illustration with NO. 1 at TDC. Check all clearances. Intake should be 0.008–0.010 in.; exhaust should be 0.011–0.013 in. Intake locknut torque is 14 ft. lbs. (19 Nm); exhaust is 10 ft. lbs. (14 Nm).
4. Rotate the crankshaft 180 degrees clockwise and recheck No. 3.
5. Rotate the crankshaft 180 degrees clockwise and recheck No. 4.
6. Rotate the crankshaft 180 degrees clockwise and recheck No. 2.

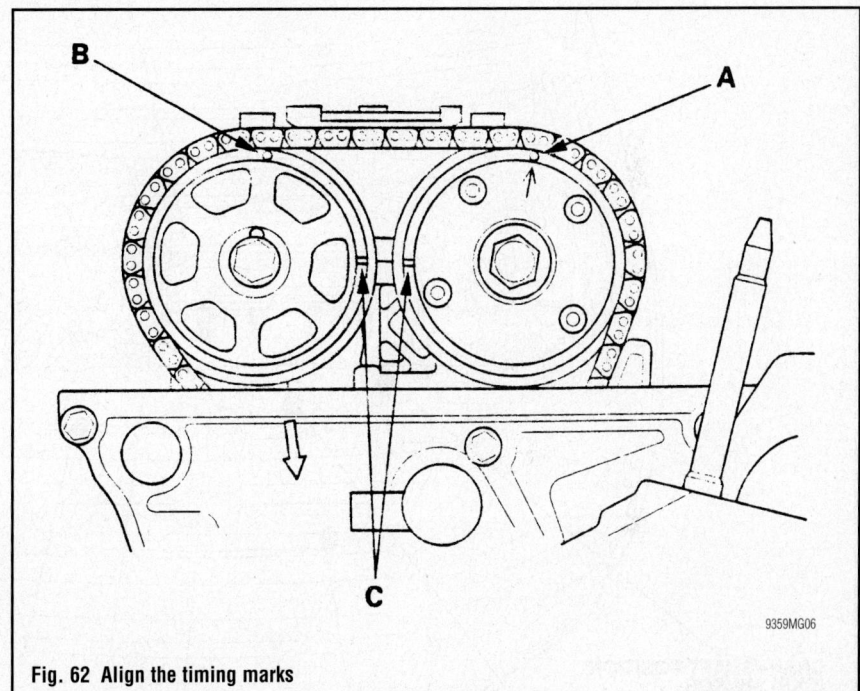

9359MG06

Fig. 62 Align the timing marks

ENGINE PERFORMANCE & EMISSION CONTROL

COMPONENT LOCATIONS

See Figure 63.

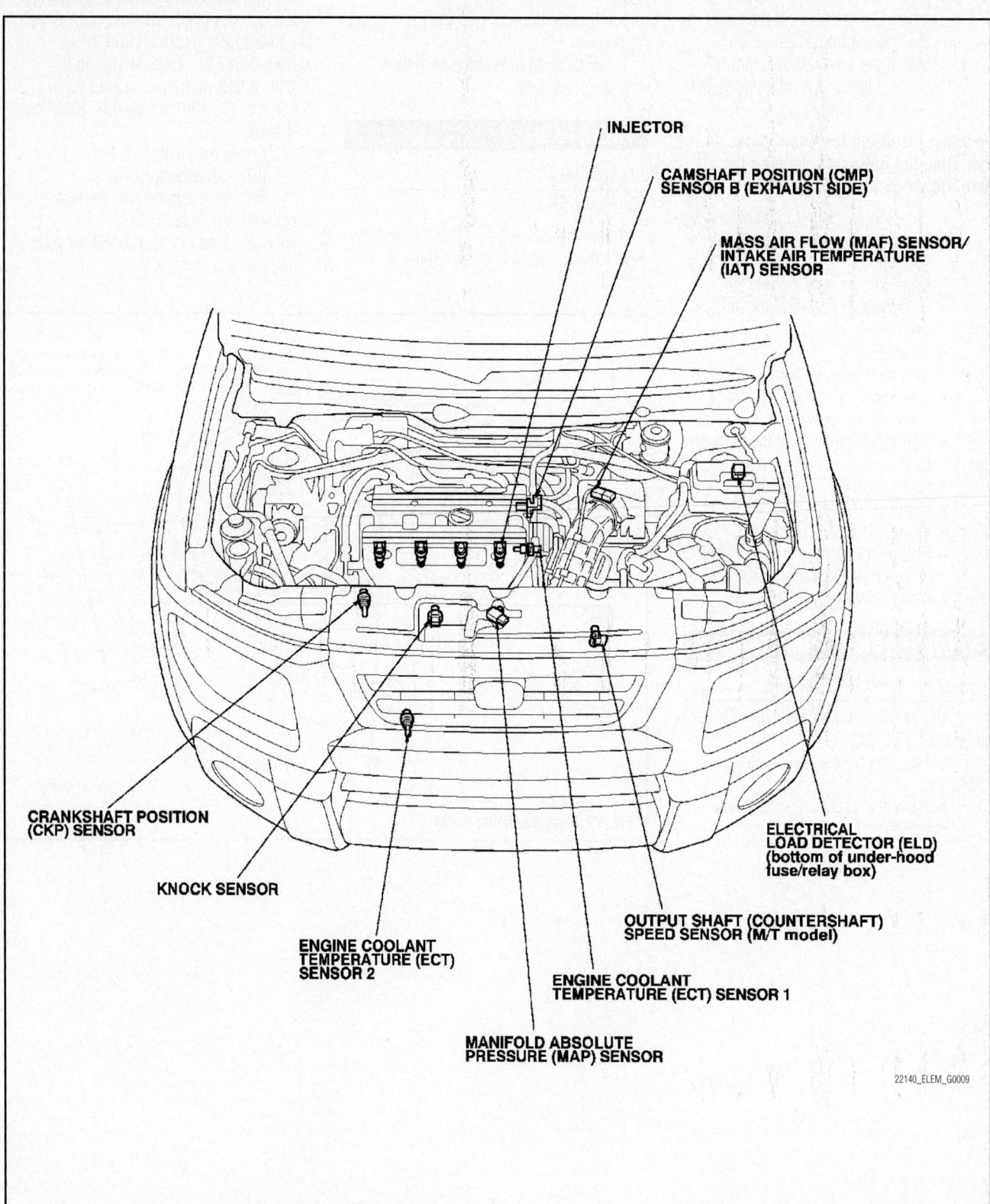

INJECTOR

CAMSHAFT POSITION (CMP) SENSOR B (EXHAUST SIDE)

MASS AIR FLOW (MAF) SENSOR/ INTAKE AIR TEMPERATURE (IAT) SENSOR

CRANKSHAFT POSITION (CKP) SENSOR

KNOCK SENSOR

ENGINE COOLANT TEMPERATURE (ECT) SENSOR 2

MANIFOLD ABSOLUTE PRESSURE (MAP) SENSOR

ENGINE COOLANT TEMPERATURE (ECT) SENSOR 1

OUTPUT SHAFT (COUNTERSHAFT) SPEED SENSOR (M/T model)

ELECTRICAL LOAD DETECTOR (ELD) (bottom of under-hood fuse/relay box)

22140_ELEM_G0009

Fig. 63 Engine performance and emission control component locations

AIR FUEL (A/F) RATIO SENSOR

LOCATION

See Figure 64.

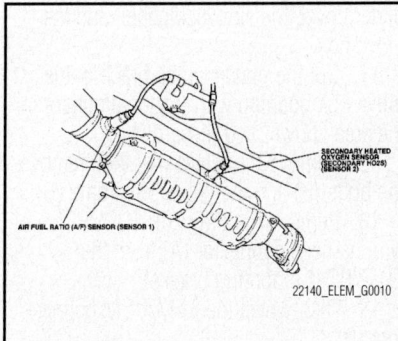

Fig. 64 A/F ratio sensor and Secondary HO2S locations

Refer to the accompanying illustration for sensor location.

REMOVAL & INSTALLATION

See Figure 65.

➡This procedure requires the use of O2 sensor socket wrench, Snap-on YA8875, SP Tools 93750, or equivalent, commercially available O2 sensor removal tool.

1. Disconnect the A/F sensor 4P connector (A), then remove the A/F sensor (B), using the proper tool.
2. Install the parts in the reverse order of removal. Tighten the sensor to 33 ft. lbs. (44 Nm).

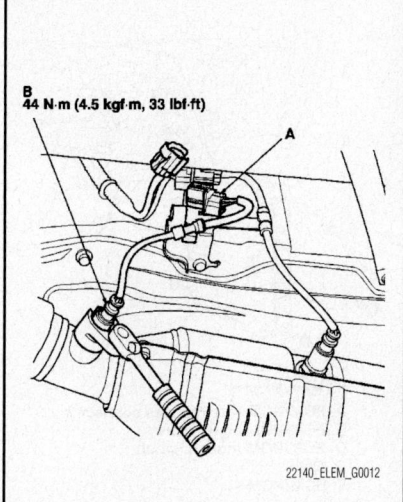

Fig. 65 View of the A/F sensor connector (A) and sensor (B)

CAMSHAFT POSITION (CMP) SENSOR

LOCATION

See Figure 63.

Refer to the illustration under Component Locations for sensor location.

REMOVAL & INSTALLATION

See Figure 66.

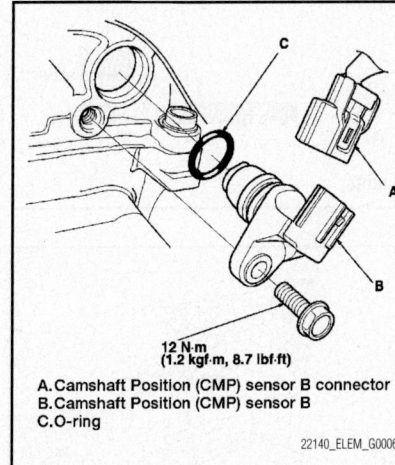

Fig. 66 Exploded view of the CMP sensor B connector (A) and sensor (B)

1. Remove the air cleaner.
2. Remove the EVAP canister purge valve.
3. Disconnect the CMP sensor B connector (A).
4. Remove CMP sensor B (B).
5. Install the parts in the reverse order of removal with a new O-ring (C) coated with clean engine oil. Tighten the sensor bolt to 8.7 ft. lbs. (12 Nm).

CRANKSHAFT POSITION (CKP) SENSOR

LOCATION

See Figure 63.

Refer to the illustration under Component Locations for sensor location.

REMOVAL & INSTALLATION

See Figures 67 and 68.

1. Disconnect the CKP sensor connector (A).
2. Remove the CKP sensor (B).

To install:

3. Install the parts in the reverse order of removal with a new O-ring (C) coated with clean engine oil. Tighten the sensor bolt to 8.7 ft. lbs. (12 Nm).

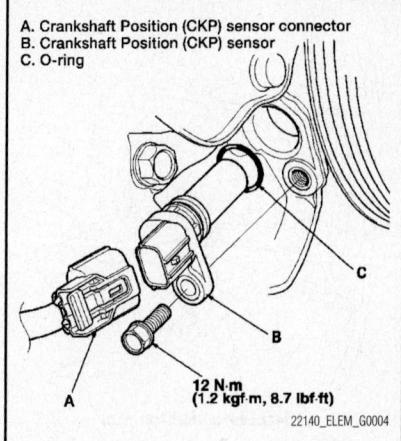

Fig. 67 View of the CKP sensor connector (A) and sensor (B)

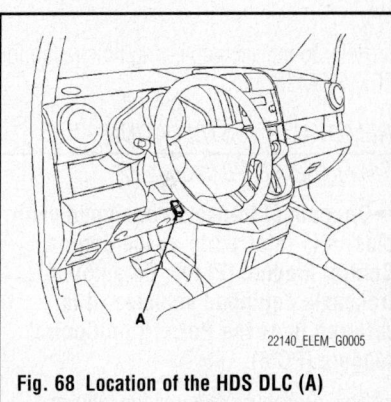

Fig. 68 Location of the HDS DLC (A)

4. Perform the CKP pattern clear/pattern learn procedure:
   a. Connect the HDS to the data link connector (DLC) (A) located under the driver's side of the dashboard.
   b. Turn the ignition switch ON (II).
   c. Make sure the HDS communicates with the ECM/PCM and other vehicle system.
   d. Select CRANK PATTERN in the ADJUSTMENT MENU with the HDS.
   e. Select CRANK PATTERN LEARNING with the HDS, and follow the screen prompts.

ELECTRONIC CONTROL MODULE (ECM) POWERTRAIN CONTROL MODULE

LOCATION

See Figure 69.

➡On manual transaxle equipped vehicles, it is referred to as the Engine Control Module (ECM). On automatic transaxle equipped vehicles, it is referred to as the Powertrain Control Module (PCM).

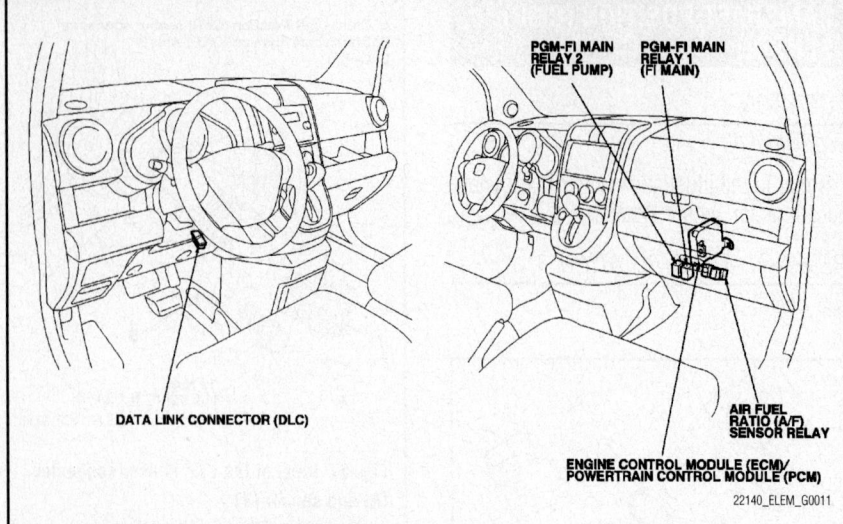

Fig. 69 Location of the ECM/PCM and related components

Refer to the accompanying illustration for ECM/PCM location.

REMOVAL & INSTALLATION

See Figures 68, 70 through 72.

➡**On manual transaxle equipped vehicles, it is referred to as the Engine Control Module (ECM). On automatic transaxle equipped vehicles, it is referred to as the Powertrain Control Module (PCM).**

This procedure requires the following special tools (or their equivalents):
- Honda diagnostics system (HDS) tablet tester
- Honda interface module (HIM) and an iN workstation with HDS and CM update software
- HDS pocket tester
- GNA 600 and an iN workstation with HDS and CM update software

➡**Make sure the HDS is loaded with the latest software version.**

- If you are replacing the ECM/PCM after substituting a known-good ECM/PCM, reinstall the original ECM/PCM, then do this procedure.
- During the procedure, if any READ DATA, WRITE DATA, or other data checks fail, note the failure, then continue.
1. Connect the HDS to the data link connector (DLC) (A) located under the driver's side of the dashboard.
2. Turn the ignition switch ON (II).
3. Make sure the HDS communicates with the ECM/PCM and other vehicle system. If you are returning from DLC circuit troubleshooting, skip steps 4 through 7, 21 through 23, and 26 through

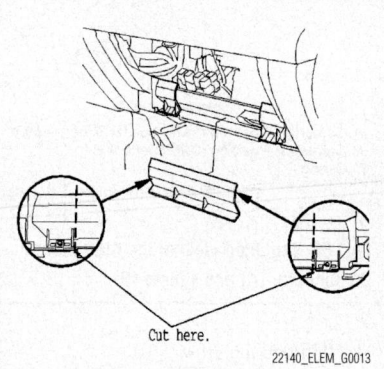

Cut here.

Fig. 70 Cut the plastic cross brace in the glove box opening with diagonal cutters in the area shown, and discard it

27, and do this after replacing the ECM/PCM:
   a. Replace the engine oil and the engine oil filter.
   b. Clean the throttle body.
4. Select the PGM-FI system with the HDS.
5. Select the INSPECTION MENU with the HDS.
6. Select the ETCS TEST, then select the TP POSITION CHECK, and follow the screen prompts.

➡**If the TP POSITION CHECK indicates FAILED, continue with this procedure.**

7. Select the REPLACE ECM/PCM MENU, then select READ DATA and follow the screen prompts.

➡**Doing this step copies (READS) the engine oil life data from the original ECM/PCM so you can later download (WRITE) it into the new ECM/PCM. If**

READ DATA indicates FAILED, continue with this procedure.

8. Turn the ignition switch OFF.
9. Jump the SCS line with the HDS.
10. Remove the passenger's dashboard under cover, the side kick panel, and the glove box.
11. Cut the plastic cross brace in the glove box opening with diagonal cutters in the area shown, and discard it.
12. Remove the relays (A), then remove the bolts (B) and the glove box frame (C).
13. Remove the gray 20P ECM/PCM wire harness connector (A) from the ECM/PCM mounting bracket.
14. Disconnect the ECM/PCM connectors (B).
15. Remove the ECM/PCM mounting bolt (C) and the bracket.
16. Remove the nuts (D), then remove the ECM/PCM (E).

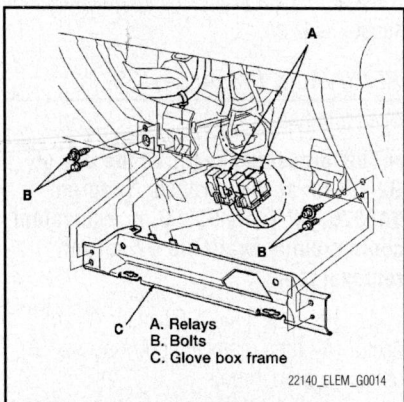

A. Relays
B. Bolts
C. Glove box frame

Fig. 71 Remove the relays (A), then remove the bolts (B) and the glove box frame (C)

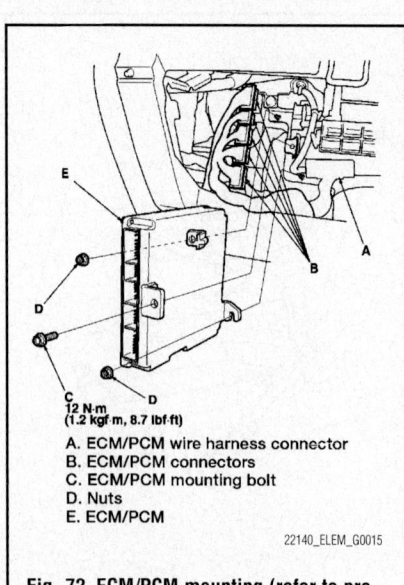

12 N·m
(1.2 kgf·m, 8.7 lbf·ft)
A. ECM/PCM wire harness connector
B. ECM/PCM connectors
C. ECM/PCM mounting bolt
D. Nuts
E. ECM/PCM

Fig. 72 ECM/PCM mounting (refer to procedure for component identification)

To install:

17. Install the parts in the reverse order of removal.
18. Open the SCS line with the HDS.
19. Turn the ignition switch ON (II).
20. Manually input the VIN to the ECM/PCM with the HDS.

➡**DTC P0630 "VIN Not Programmed or Mismatch" may be stored because the VIN has not been programmed into the ECM/PCM; ignore it, and continue this procedure. If the READ DATA (engine oil life) failed in step 7, go to step 24. Otherwise, go to step 22.**

21. Select the PGM-FI system with the HDS.
22. Select the REPLACE ECM/PCM MENU, then select WRITE DATA and follow the screen prompts.

➡ **If the WRITE DATA indicates FAILED, continue with this procedure.**

23. Select IMMOBI system with the HDS.
24. Enter the immobilizer code with the ECM/PCM replacement procedure in the HDS; it allows you to start the engine.
25. If the TP POSITION CHECK failed in step 6 clean the throttle body, then go to step 27.
26. If the READ DATA failed in step 7 or the WRITE DATA failed in step 23, replace the engine oil and engine oil filter, then go to step 28.
27. Select PGM-FI system, and reset the ECM/PCM with the HDS.
28. Update the ECM/PCM if it does not have the latest software.
29. Perform the ECM/PCM idle learn procedure, as follows:
   a. Make sure all electrical items (A/C, audio, lights, etc.) are off.
   b. Reset the ECM/PCM with the HDS.
   c. Turn the ignition switch ON (II), and wait 2 seconds.
   d. Start the engine. Hold the engine speed at 3,000 rpm without load (in Park or neutral) until the radiator fan comes on, or until the engine coolant temperature reaches 194 °F (90 °C).
   e. Let the engine idle for about 5 minutes with the throttle fully closed.

➡**If the radiator fan comes on, do not include its running time in the 5 minutes**

30. Perform the CKP pattern learn procedure, as follows:
   a. Connect the HDS to the data link connector (DLC) (A) located under the driver's side of the dashboard.
   b. Turn the ignition switch ON (II).

   c. Make sure the HDS communicates with the ECM/PCM and other vehicle system.
   d. Select CRANK PATTERN in the ADJUSTMENT MENU with the HDS.
   e. Select CRANK PATTERN LEARNING with the HDS, and follow the screen prompts.

ENGINE COOLANT TEMPERATURE (ECT) SENSOR

LOCATION
See Figure 63.

Refer to the illustration under Component Locations for sensor locations.

REMOVAL & INSTALLATION

Sensor 1
See Figure 73.

1. Remove the air cleaner.
2. Remove the EVAP canister purge valve.
3. Unbolt the under-hood fuse/relay box bolt, and move the assembly aside.
4. Drain the engine coolant.
5. Disconnect the ECT sensor 1 connector (A).
6. Remove the ECT sensor 1 (B).
7. Install the parts in the reverse order of removal with a new O-ring (C) coated with clean engine oil, then refill the radiator with engine coolant.

To install:

8. Install the parts in the reverse order of removal with a new O-ring (C) coated with clean engine oil. Tighten to 8.7 ft. lbs. (12 Nm).
9. Refill the radiator with engine coolant.

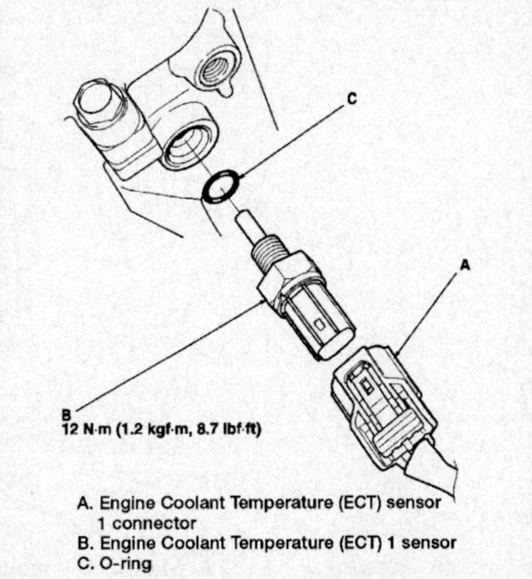

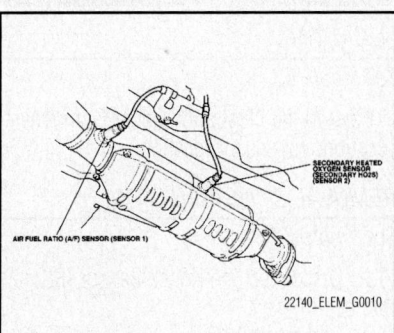

Fig. 73 View of the ECT sensor 1 connector (A) and sensor (B). Use a new O-ring (C) coated with clean engine oil during installation

Sensor 2
See Figure 74.

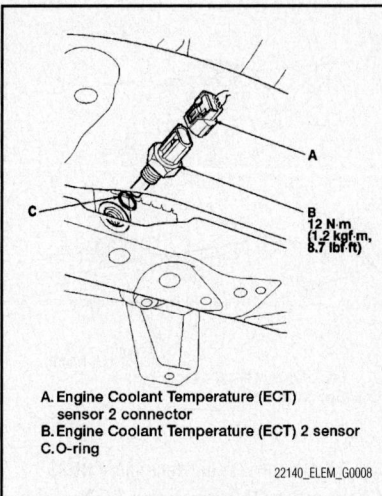

A. Engine Coolant Temperature (ECT) sensor 2 connector
B. Engine Coolant Temperature (ECT) 2 sensor
C. O-ring

22140_ELEM_G0008

Fig. 74 View of the ECT sensor 2 connector (A) and sensor (B). Use a new O-ring (C) coated with clean engine oil during installation

1. Drain the engine coolant.
2. Remove the splash shield.
3. Disconnect the ECT sensor 2 connector (A).
4. Remove ECT sensor 2 (B).

To install:

5. Install the parts in the reverse order of removal with a new O-ring (C) coated with clean engine oil. Tighten to 8.7 ft. lbs. (12 Nm).
6. Refill the radiator with engine coolant.

HEATED OXYGEN (HO2S) SENSOR

LOCATION
See Figure 75.

Refer to the accompanying illustration for sensor location.

22140_ELEM_G0007

Fig. 75 A/F ratio sensor and Secondary HO2S locations

REMOVAL & INSTALLATION

See Figure 76.

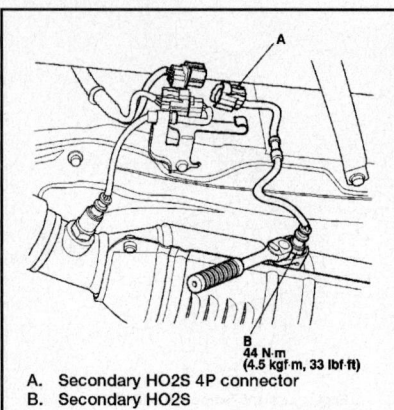

A. Secondary HO2S 4P connector
B. Secondary HO2S

22140_ELEM_G0016

Fig. 76 Disconnect the secondary HO2S 4P connector (A), then remove the secondary HO2S (B)

➡**This procedure requires the use of O2 sensor socket wrench, Snap-on YA8875, SP Tools 93750, or equivalent, commercially available O2 sensor removal tool.**

1. Disconnect the secondary HO2S 4P connector (A), then remove the secondary HO2S (B).
2. Install the parts in the reverse order of removal. Tighten the sensor to 33 ft. lbs. (44 Nm).

INTAKE AIR TEMPERATURE (IAT) SENSOR

LOCATION

Refer to Mass Air Flow (MAF) sensor.

REMOVAL & INSTALLATION

Refer to Mass Air Flow (MAF) sensor.

KNOCK SENSOR (KS)

LOCATION

See Figure 63.

Refer to the illustration under Component Locations for sensor location.

REMOVAL & INSTALLATION

See Figure 77.

1. Disconnect the Knock Sensor (KS) connector (A).
2. Remove the KS (B).
3. Install the parts in the reverse order of removal. Tighten the KS to 23 ft. lbs. (32 Nm).

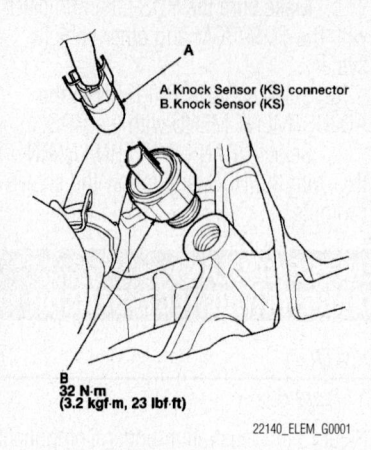

A. Knock Sensor (KS) connector
B. Knock Sensor (KS)

B
32 N·m
(3.2 kgf·m, 23 lbf·ft)

22140_ELEM_G0001

Fig. 77 View of the Knock Sensor (KS) connector (A) and KS (B)

MANIFOLD ABSOLUTE PRESSURE (MAP) SENSOR

LOCATION

See Figure 63.

Refer to the illustration under Component Locations for sensor location.

REMOVAL & INSTALLATION

See Figure 78.

1. Disconnect the MAP sensor connector (A).
2. Remove the MAP sensor (B).
3. Install the parts in the reverse order of removal with a new O-ring (C) coated with clean engine oil. Tighten the sensor retainer to 2.4 ft. lbs. (3.4 Nm).

MASS AIR FLOW (MAF) SENSOR

LOCATION

See Figure 63.

Refer to the illustration under Component Locations for sensor location.

REMOVAL & INSTALLATION

See Figure 79.

➡**The Intake Air Temperature Sensor is integrated with the Mass Air Flow (MAF) sensor.**

1. Disconnect the MAF sensor/IAT sensor connector (A).
2. Remove the bolts (B).

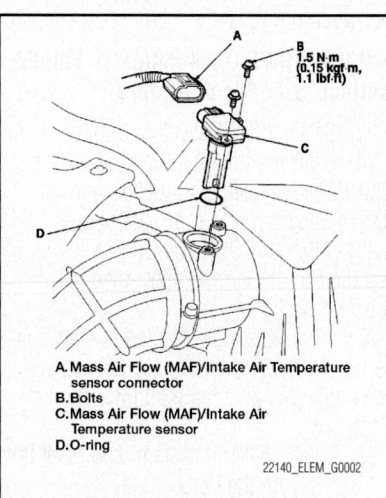

A. Mass Air Flow (MAF)/Intake Air Temperature sensor connector
B. Bolts
C. Mass Air Flow (MAF)/Intake Air Temperature sensor
D. O-ring

22140_ELEM_G0002

Fig. 79 Exploded view of the MAF sensor/IAT sensor

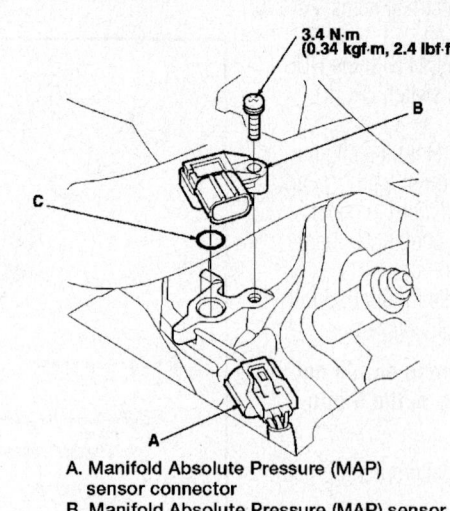

3.4 N·m
(0.34 kgf·m, 2.4 lbf·ft)

A. Manifold Absolute Pressure (MAP) sensor connector
B. Manifold Absolute Pressure (MAP) sensor
C. O-ring

22140_ELEM_G0003

Fig. 78 View of the MAP sensor connector (A) and sensor (B)

3. Remove the MAF sensor/IAT sensor (C).

4. Install the parts in the reverse order of removal with a new O-ring (D) coated with clean engine oil.

OUTPUT SHAFT SPEED (OSS) SENSOR

LOCATION

With Manual Transaxle

See Figure 63.

Refer to the illustration under Component Locations for sensor location.

REMOVAL & INSTALLATION

With Manual Transaxle

See Figure 80.

1. Remove the air cleaner.
2. Disconnect the output shaft (countershaft) speed sensor connector (A).
3. Remove the output shaft (countershaft) speed sensor (B).
4. Install the parts in the reverse order of removal with a new O-ring (C) coated with clean engine oil. Tighten the sensor retainer to 8.7 ft. lbs. (12 Nm).

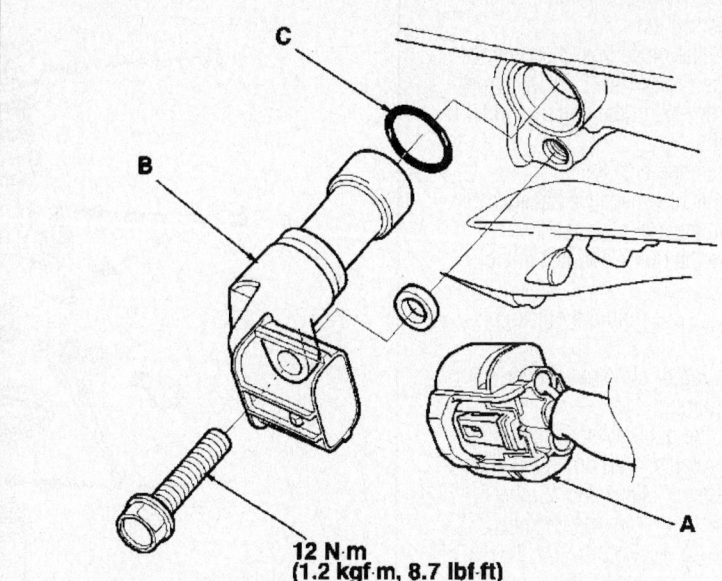

**12 N·m
(1.2 kgf·m, 8.7 lbf·ft)**

**A. Output Shaft Speed (OSS) sensor connector
B. Output Shaft Speed (OSS) sensor
C. O-ring**

22140_ELEM_G0017

Fig. 80 View of the Output Shaft Speed (OSS) connector (A), sensor (B) and O-ring (C)

FUEL

GASOLINE FUEL INJECTION SYSTEM

FUEL SYSTEM SERVICE PRECAUTIONS

Safety is the most important factor when performing not only fuel system maintenance but any type of maintenance. Failure to conduct maintenance and repairs in a safe manner may result in serious personal injury or death. Maintenance and testing of the vehicle's fuel system components can be accomplished safely and effectively by adhering to the following rules and guidelines.

• To avoid the possibility of fire and personal injury, always disconnect the negative battery cable unless the repair or test procedure requires that battery voltage be applied.

• Always relieve the fuel system pressure prior to disconnecting any fuel system component (injector, fuel rail, pressure regulator, etc.), fitting or fuel line connection. Exercise extreme caution whenever relieving fuel system pressure to avoid exposing skin, face and eyes to fuel spray. Please be advised that fuel under pressure may penetrate the skin or any part of the body that it contacts.

• Always place a shop towel or cloth around the fitting or connection prior to loosening to absorb any excess fuel due to spillage. Ensure that all fuel spillage (should it occur) is quickly removed from engine surfaces. Ensure that all fuel soaked cloths or towels are deposited into a suitable waste container.

• Always keep a dry chemical (Class B) fire extinguisher near the work area.

• Do not allow fuel spray or fuel vapors to come into contact with a spark or open flame.

• Always use a back-up wrench when loosening and tightening fuel line connection fittings. This will prevent unnecessary stress and torsion to fuel line piping.

• Always replace worn fuel fitting O-rings with new Do not substitute fuel hose or equivalent where fuel pipe is installed.

Before servicing the vehicle, make sure to also refer to the precautions in the beginning of this section as well.

RELIEVING FUEL SYSTEM PRESSURE

See Figure 81.

✳✳ CAUTION

The fuel injection system remains under pressure after the engine has been turned OFF. Properly relieve fuel pressure before disconnecting any fuel lines. Failure to do so may result in fire or personal injury.

➡**The radio may contain a coded theft protection circuit. Always obtain the code number before disconnecting the battery.**

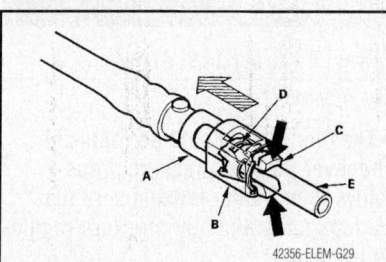

42356-ELEM-G29

Fig. 81 Hold the quick-connect (A) connector (B) with one hand, then squeeze the retainer tabs (C) with the other hand to release them from the locking pawls (D)

1. Before servicing the vehicle, refer to the precautions section.

2. Remove the glove box, then remove the PGM-FI main relay (FUEL PUMP) from the fuse/relay box. Start the engine and let it run until it stalls.

3. Turn the engine OFF.

4. Disconnect the negative battery cable.

5. Remove the fuel filler cap.

6. Remove the quick-connect fitting cover.

7. Clean any dirt from the quick-connect fitting.

8. Place a rag or shop towel over quick-connect fitting.

9. Detach the quick-connect fitting by holding the connector with one hand, then squeeze the retainer tabs with the other hand to release them from the locking pawls. Pull the connector off.

✷✷ CAUTION

Do not allow fuel spray or fuel vapors to come in contact with a spark or open flame. Keep a dry chemical fire extinguisher nearby. Never store fuel in an open container due to risk of fire or explosion.

➡**A fuel pressure gauge may be attached at the quick-connect location.**

10. Connect the quick-connect fitting, making sure the locking pawls are properly engaged,

11. Clean up any fuel spilled on the engine and intake manifold.

12. Install the fuel pump relay to the under dash fuel/relay box and install the glove box.

13. Install the fuel filler cap.

14. Reconnect the negative battery cable.

15. Turn the ignition **ON**, but don't start the engine. Repeat this 2 or 3 times to pressurize the fuel system. Check for fuel leaks.

16. Enter the radio security code.

FUEL FILTER

REMOVAL & INSTALLATION

See Figure 82.

➡**The fuel filter should be replaced whenever the fuel pressure drops below 48 psi, after making sure that the fuel pump and fuel pressure regulator are okay.**

1. Before servicing the vehicle, refer to the precautions section.

2. Relieve the fuel system pressure.

3. Remove or disconnect the following:
  • Negative battery cable

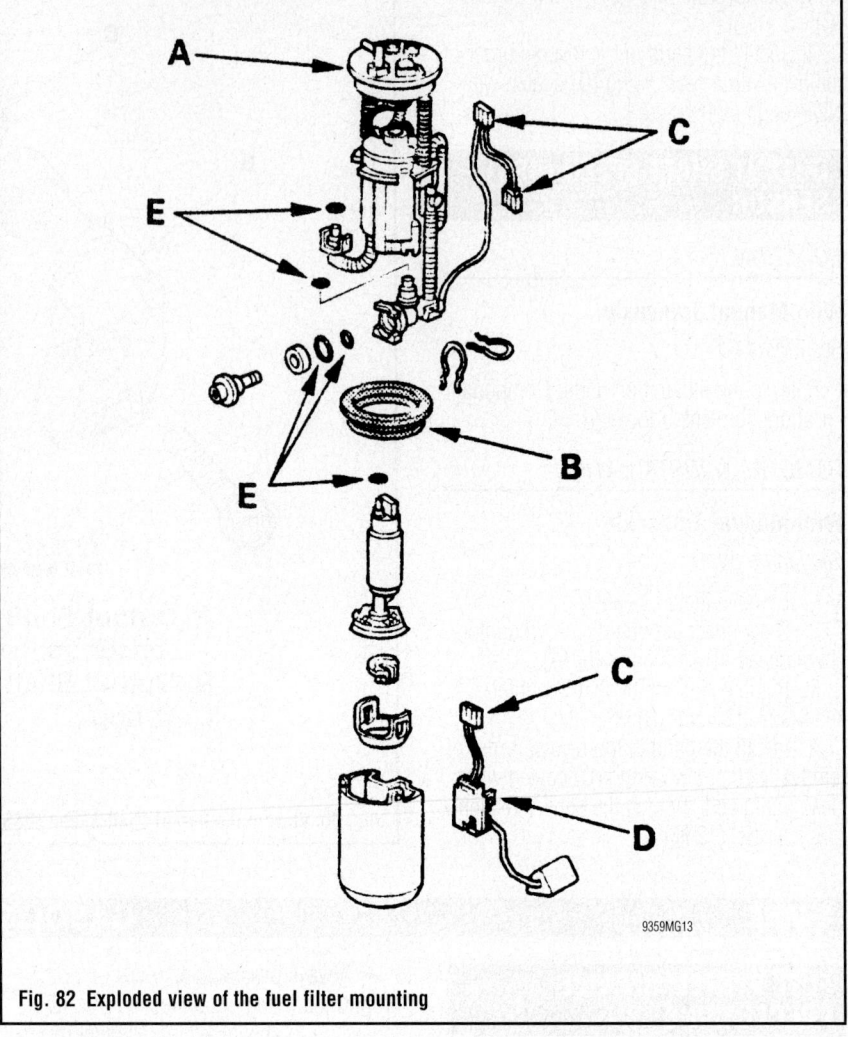

Fig. 82 Exploded view of the fuel filter mounting

9359MG13

  • Fuel pump
  • Fuel filter carrier (A)
  • Fuel filter

To install:

4. Install or connect the following:
  • Fuel filter
  • Fuel lines
  • New gasket (B)
  • New o-rings (E)
  • Connectors (C)
  • Sending unit (D)

5. Start the engine and check for leaks.

FUEL PUMP

REMOVAL & INSTALLATION

1. Before servicing the vehicle, refer to the precautions section.

2. Relieve the fuel system pressure.

3. Remove or disconnect the following:
  • Negative battery cable
  • Fuel filler cap
  • Center console, then both track floor covers and sill trims.

4. Fold back the floor covering until you can get to the access panel
  • Access panel from the floor
  • Fuel pump connector
  • Fuel supply and return line quick-connect fittings
  • Fuel pump locknut, using special tool No. 07XAA-001010A
  • Fuel pump/sending unit assembly

5. Installation is the reverse of removal.

FUEL RAIL & INJECTORS

REMOVAL & INSTALLATION

See Figure 83.

1. Before servicing the vehicle, refer to the precautions section.

2. Relieve the fuel system pressure.

3. Remove or disconnect the following:
  • Negative battery cable
  • Engine cover
  • Injector connectors, ground cable and harness holder

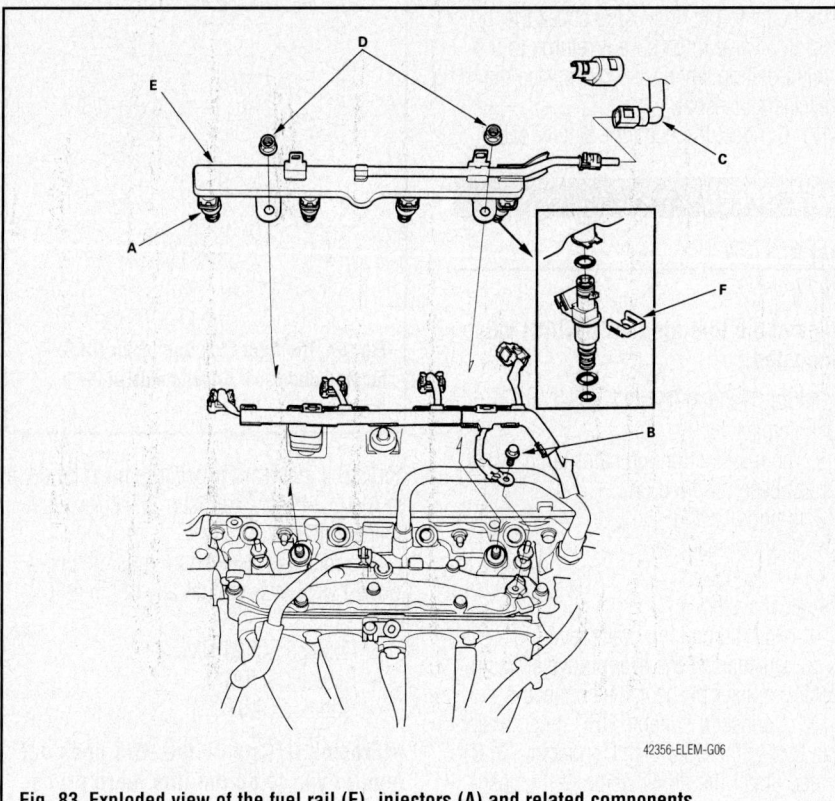

Fig. 83 Exploded view of the fuel rail (E), injectors (A) and related components

- Fuel line quick-connect fittings
- Fuel rail mounting nuts
- Injector clip(s) from the injector(s)
- Fuel injectors from the fuel rail

To install:

4. Install or connect the following:
- Injectors to the fuel rail with new O-rings coated with clean engine oil.
- Injector clips
- Injectors in the injector base
- Fuel rail and injector assembly. Tighten the nuts to 16 ft. lbs. (22 Nm).
- Ground cable bolt
- Injector connectors
- Fuel lines
- Negative battery cable
5. Start the engine and check for leaks.

THROTTLE BODY

REMOVAL & INSTALLATION

See Figures 84 through 86.

1. Before servicing the vehicle, refer to the precautions section.
2. Relieve the fuel system pressure.
3. Disconnect the negative battery cable.
4. Remove the engine cover.
5. Detach the electrical connectors and cables from the throttle body.

6. Remove the retainers, then remove the throttle body.
7. Remove and discard the throttle body gasket.

To install:

8. Position a new throttle body gasket, then install the throttle body. Tighten the retainers to 16 ft. lbs. (22 Nm).
9. Attach the cables and electrical connectors to the throttle body.
10. Adjust the actuator cable, as follows:
    a. Make sure the actuator cable moves smoothly with no binding or sticking.
    b. Measure the amount of movement of the output linkage until the engine speed starts to increase. At first, the output linkage should be located at the fully closed position. The free play should be 0.13–0.17 in. (3.25–4.25mm).
    c. If the free play is not within specs, loosen the locknut, and turn the adjusting nut until the free play is as specified, then retighten the locknut.
11. Adjust the throttle cable, as follows:
    a. Check cable free play at the throttle linkage. Cable free play should be ⅜–½ in. (10–12mm). If the free play is not

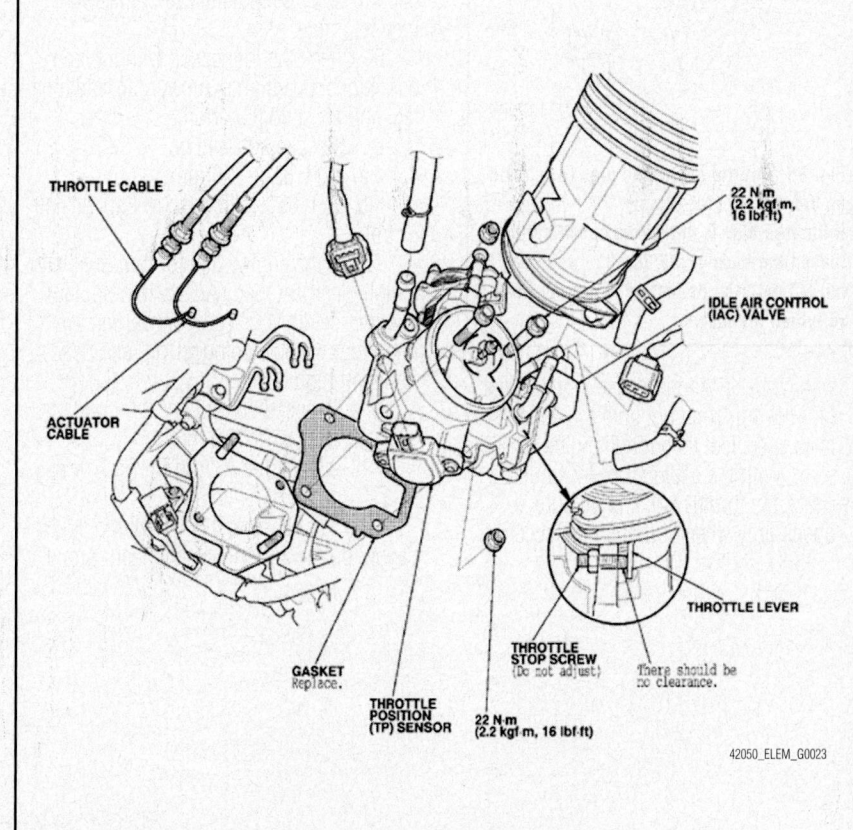

Fig. 84 Exploded view of the throttle body and related components

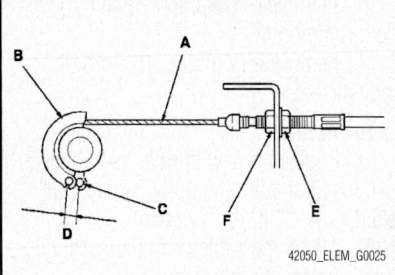

Fig. 85 View of the actuator cable (A). Measure the movement of the output linkage (B) which should first be located at the fully closed position (C). If the freeplay (D) is not within specs, loosen the locknut (E) and turn the adjusting nut (F) until the free play is correct

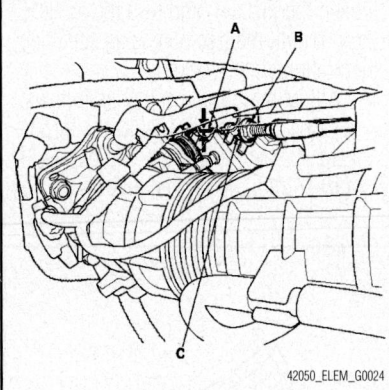

Fig. 86 Throttle cable free play (A) should be 3/8–1/2 in. (10–12mm).
If the free play is not within specifications, loosen the locknut (B), turn the adjusting nut (C) until the deflection is correct, then retighten the locknut

within specifications, loosen the locknut, turn the adjusting nut until the deflection is as specified, then retighten the locknut.

b. With the cable properly adjusted, check the throttle valve to be sure it opens fully when you push the accelera-

tor pedal to the floor. Also check the throttle valve to be sure it returns to the idle position whenever you release the accelerator pedal.

12. Connect the negative battery cable.

IDLE SPEED

INSPECTION

See Figure 87.

➡**Leave the idle air control (IAC) valve connected.**

Before checking the idle speed, check these items:
- The malfunction indicator lamp (MIL) has not been reported on.
- Ignition timing
- Spark plugs
- Air cleaner
- PCV system

1. Pull the parking brake lever up.
2. Disconnect the evaporative emission (EVAP) canister purge valve connector.
3. Connect a suitable OBD II compliant scan tool to the Data Link Connector (DLC) located under the driver's side of the dashboard.
4. Start the engine. Hold the engine speed at 3,000 rpm without load (in Park or neutral) until the radiator fan comes on, then let it idle.
5. Check the idle speed without load conditions: headlights, blower fan, radiator fan, and air conditioner off.
6. Idle speed should be:
   a. M/T: 650–750 rpm
   b. A/T: 650–750 rpm (in Park or neutral)
7. Let the engine idle for 1 minute with a high electrical load (A/C switch on, temperature set to MAX cool, blower fan on high, rear window defogger on, and headlights on high beam).
8. Idle speed should be:
   a. M/T: 670–770 rpm
   b. A/T: 670–770 rpm (in Park or neutral)
9. If the idle speed is not within specification, do the ECM/PCM idle learn proce-

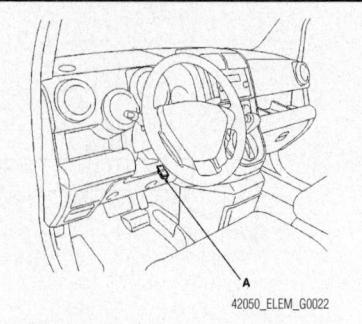

Fig. 87 The Data Link Connector (DLC) is located under the driver's side of the dashboard

dure. The idle learn procedure must be done so the ECM/PCM can learn the engine idle characteristics.

The idle learn procedure must be performed whenever you do any of the following:
- Replace ECM/PCM.
- Reset ECM/PCM.
- Update ECM/PCM.

➡**Erasing DTCs with the HDS does not require you to do the idle learn procedure.**

- Clean or replace the throttle body.
   a. Make sure all electrical items (A/C, audio, rear window defogger, lights, etc.) are off.
   b. Reset the ECM/PCM with the HDS.
   c. Turn the ignition switch ON (II), and wait 2 seconds.
   d. Start the engine. Hold the engine speed at 3,000 rpm without load (in Park or neutral) until the radiator fan comes on, or until the engine coolant temperature reaches 194 °F (90 °C).
   e. Let the engine idle for about 5 minutes with the throttle fully closed.

➡**If the radiator fan comes on, do not include its running time in the 5 minutes.**

10. Reconnect the EVAP canister purge valve connector.

HEATING & AIR CONDITIONING SYSTEM

BLOWER MOTOR

REMOVAL & INSTALLATION

See Figures 88 through 92.

1. Remove the passenger's dashboard under cover, as follows:

   a. Gently pull down the rear edge to release the clips.

   b. Pull the cover away to release the pins (B) from the holders (C).

2. Use a suitable trim panel removal tool to remove the passenger's kick panel.

✳✳ WARNING

Be careful not to scratch or damage the dash when removing the glove box.

3. Remove the glove box, as follows:

   a. While holding the glove box, remove the glove box stop on each side.

   b. Remove the bolts, then remove the glove box.

4. Cut the plastic cross brace in the glove box opening with diagonal cutters in the area shown. Remove and discard the plastic cross brace.

5. Remove the relays (A) and the wire harness clip (B), then remove the bolts and the glove box frame (C).

6. Remove the ECM/PCM.

7. Disconnect the connectors from the blower motor, the power transistor, and the recirculation control motor, then remove the wire harness clips.

8. If blower motor replacement is nec-

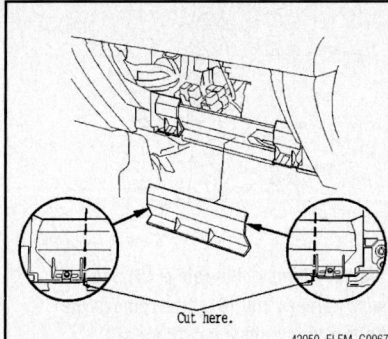

Cut here.

42050_ELEM_G0067

Fig. 89 Cut the plastic cross brace in the glove box opening with diagonal cutters in the area shown. Remove and discard the plastic cross brace

---

Fastener Locations

A ▷ : Clip, 9 (White) B ▷ : Clip, 2 (Orange) C ▷ : Clip, 3 (White) D ▷ : Clip, 3 (Black)

TWEETER CONNECTOR

A-PILLAR TRIM

PASSENGER'S KICK PANEL

[A] Portions

DOOR OPENING TRIM

HOOK

[A]

[A]

HOOK

HOOK

C

DOOR OPENING TRIM

DRIVER'S KICK PANEL

[A]

DOOR SILL TRIM

HOOKS

HOOK

FRONT SEAT TRACK FLOOR COVER

HOOKS

42050_ELEM_G0071

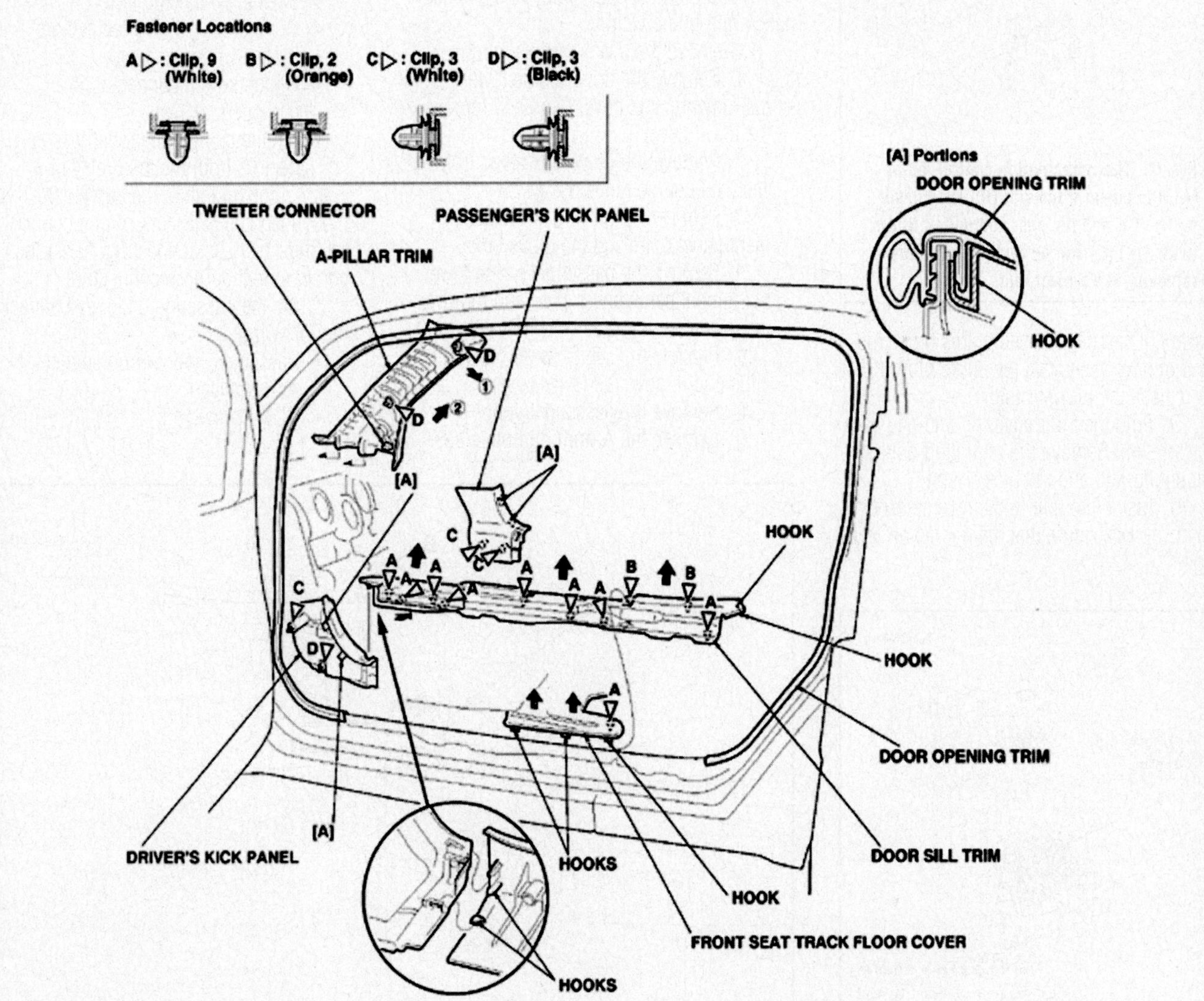

Fig. 88 Passenger side kick panel and related interior trim panels

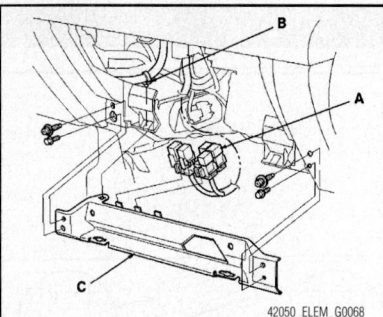

Fig. 90 Remove the relays (A) and the wire harness clip (B), then remove the bolts and the glove box frame (C)

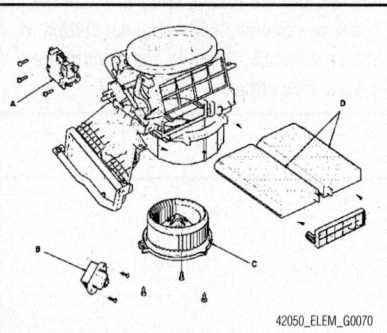

Fig. 91 The recirculation control motor (A), the power transistor (B), the blower motor (C), and the dust and pollen filters (with A/C) (D) can be replaced without removing the blower unit

essary, it can be removed at this time. You do not have to remove the entire blower unit to replace the blower motor.

9. Fold the floor covering and pad back toward you. Remove the mounting bolts, the mounting nut, and the blower unit.

10. Install the unit in the reverse order of removal. Make sure that there is no air leakage.

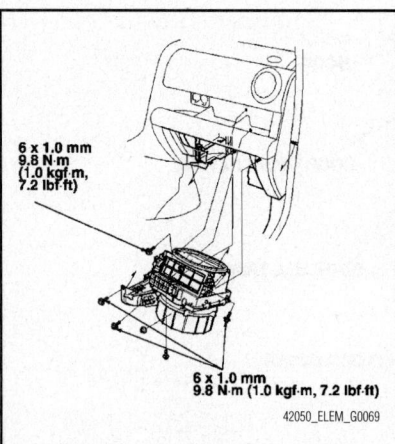

6 x 1.0 mm
9.8 N·m
(1.0 kgf·m,
7.2 lbf·ft)

6 x 1.0 mm
9.8 N·m (1.0 kgf·m, 7.2 lbf·ft)

42050_ELEM_G0069

Fig. 92 Blower unit mounting and fastener tightening specifications.

HEATER CORE

REMOVAL & INSTALLATION

See Figures 93 through 96.

1. Before servicing the vehicle, refer to the precautions section.

2. Disconnect the negative battery cable.

3. Drain the cooling system into a clean container for reuse.

4. Disconnect the A/C lines from the evaporator core if equipped with A/C.

5. Open the cable clamp (A) and disconnect the heater valve cable (B) from the heater valve arm (C). Turn the heater valve arm to the full open position as illustrated.

6. Remove the heater hoses from the heater core

7. Remove the mounting bolt and heater valve. Remove the nut from the heater unit being careful of any lines, hoses and wiring in the vicinity.

8. Remove the dashboard as follows:

a. Remove the driver's side lower instrument panel cover clips and remove the lower cover.

b. Remove the glove box stops from each side of the glove box.

c. Remove the glove box-to-instrument panel bolts and the glove box.

d. Remove the passenger's side lower instrument panel cover clips and remove the lower cover.

e. Remove the center lower cover clips and remove the lower cover.

f. Remove the passenger vent.

g. Remove the A–trim on both sides.

h. Remove the passenger side kick panel.

i. Remove the steering wheel.

j. Remove the steering column covers.

k. Disconnect the wiring from the combination switch and remove the assembly by removing the screw on top of the switch.

l. Disconnect the ignition switch connectors and release the wire harness clips from the column.

m. Disconnect the steering joint bolt and disconnect it from the column shaft.

n. Remove the steering column retainers and the column.

o. Control cable if equipped with an automatic transmission or shift cable if equipped with a manual transaxle.

p. Woofer, if equipped.

q. On the driver's side disconnect the following:

- Tweeter connector
- Drivers door wiring connectors
- Brake switch connector
- Clutch switch connector, if equipped
- Engine compartment harness connectors from the fuse/relay box

r. In the middle of the dashboard, disconnect the SRS control unit connector, floor harness connector and engine compartment harness connectors.

s. On the passenger side, disconnect the following:

- Passenger door wiring connectors
- Antenna lead
- PCM connectors

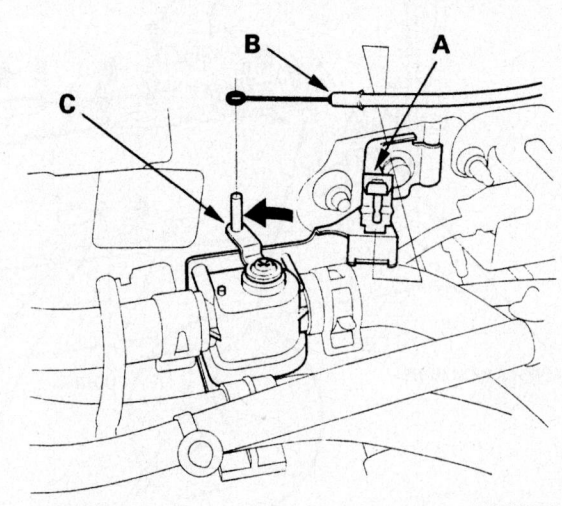

09474_ELEM_G01

Fig. 93 Open the cable clamp (A) and disconnect the heater valve cable (B) from the heater valve arm (C). Turn the heater valve arm to the full open position

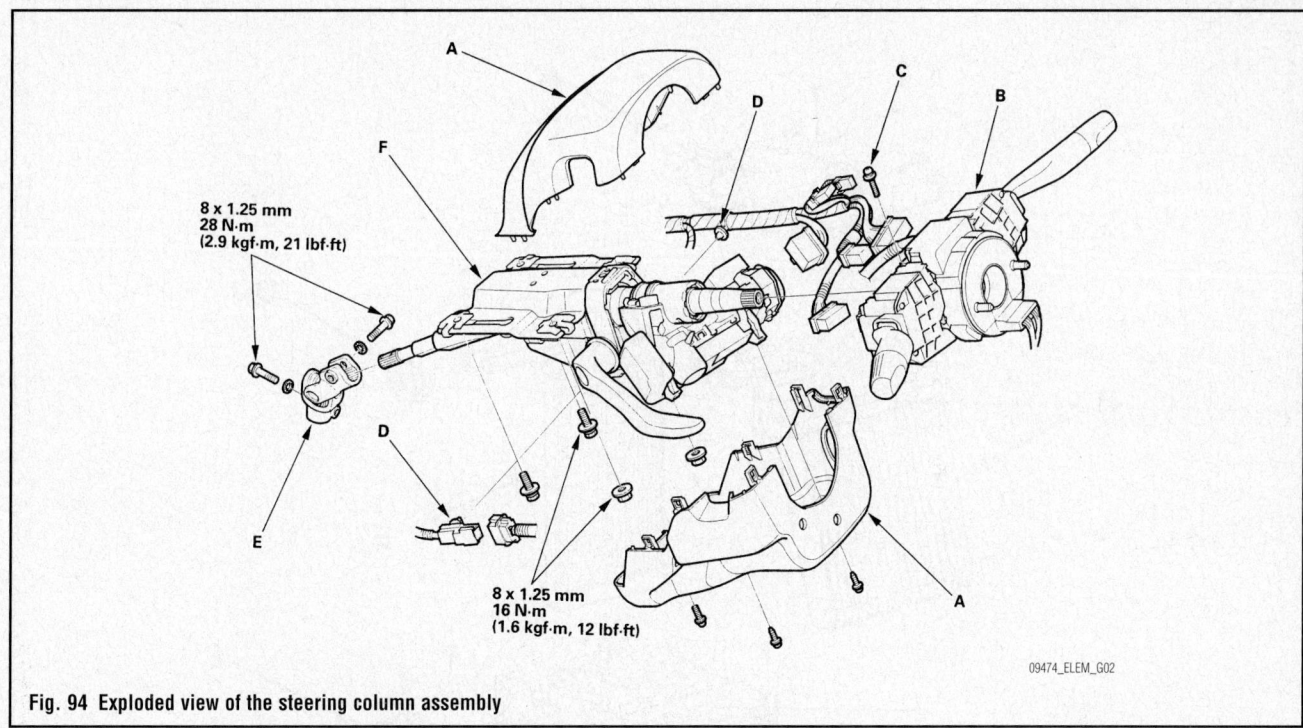

8 x 1.25 mm
28 N·m
(2.9 kgf·m, 21 lbf·ft)

8 x 1.25 mm
16 N·m
(1.6 kgf·m, 12 lbf·ft)

09474_ELEM_G02

Fig. 94 Exploded view of the steering column assembly

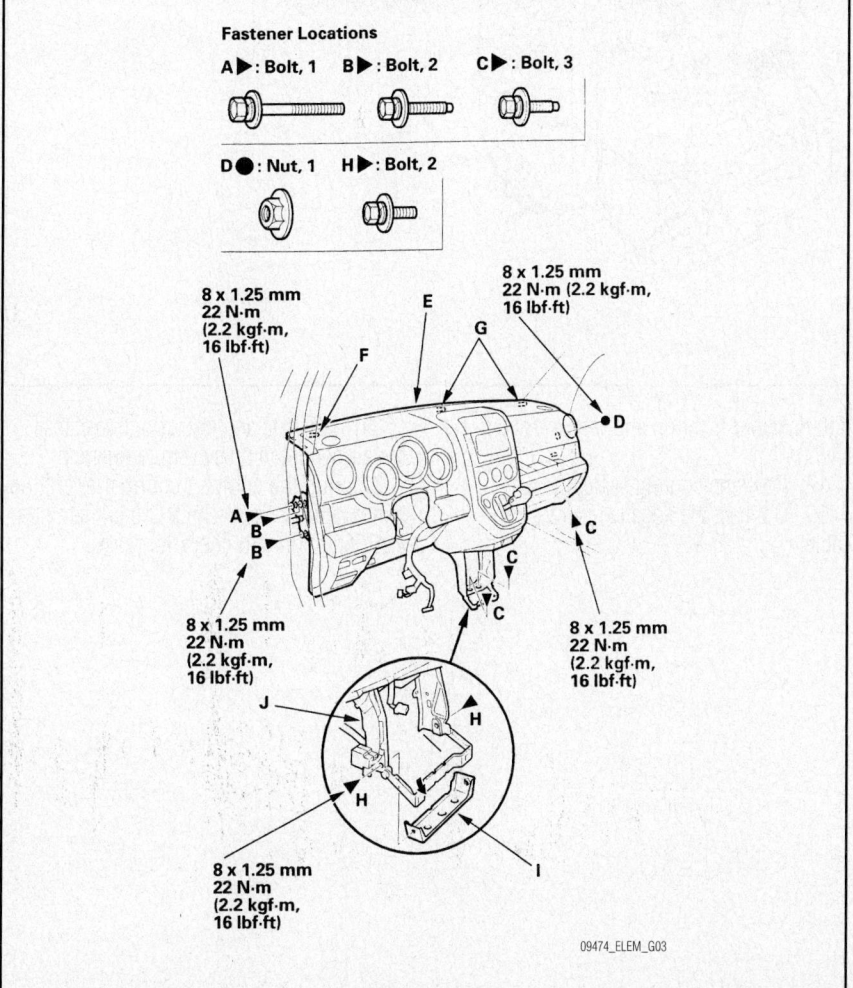

Fastener Locations

A ▶ : Bolt, 1   B ▶ : Bolt, 2   C ▶ : Bolt, 3

D ● : Nut, 1   H ▶ : Bolt, 2

8 x 1.25 mm
22 N·m
(2.2 kgf·m,
16 lbf·ft)

8 x 1.25 mm
22 N·m (2.2 kgf·m,
16 lbf·ft)

8 x 1.25 mm
22 N·m
(2.2 kgf·m,
16 lbf·ft)

8 x 1.25 mm
22 N·m
(2.2 kgf·m,
16 lbf·ft)

8 x 1.25 mm
22 N·m
(2.2 kgf·m,
16 lbf·ft)

09474_ELEM_G03

Fig. 95 Exploded view of the dashboard, retainers and the retainer torque specifications

- Engine wire harness connectors
- Heater sub-harness connectors
- Passenger airbag connectors.
- Amplifier connectors
- Wire harness protector from the amplifier, if equipped

t. Remove any remaining harness and connector clips.

u. Dashboard bolts. Refer to the exploded view for bolt location and torque values.

v. Remove the dashboard.

9. Remove the PCM.

10. Disconnect the following connectors:
- Dashboard wiring harness
- Air mixture control motor
- Evaporator temperature sensor
- Power transistor
- Mode control motor
- Blower motor

11. Disconnect the following clips:
- Wire harness clips
- Connector clips

12. Remove the wire harness, heater duct and clip.

13. Remove the drain hose and the mounting nuts and the heater unit.

14. Remove the screws and the expansion valve cover (A).

15. If equipped with A/C, remove the evaporator core (B).

16. Remove the screws and the flange cover (C).

17. Remove the grommet (D) and the heater core being careful not to damage any lines.

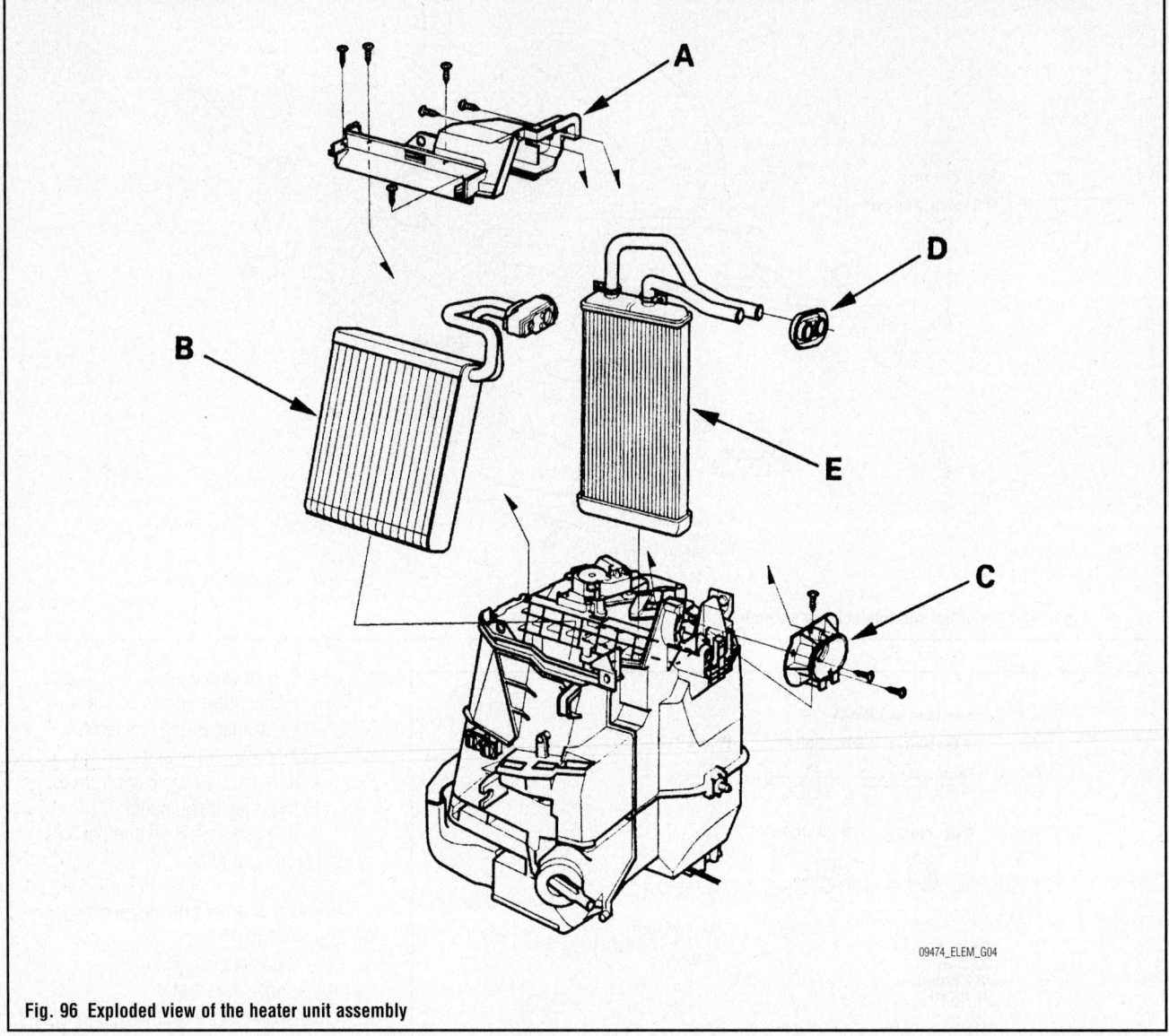

Fig. 96 Exploded view of the heater unit assembly

09474_ELEM_G04

To install:

18. Installation is the reverse of removal. Refer to the exploded views of the heater unit assembly, dashboard and steering column assembly for component location, fastener location and torque specifications.

19. Refill the cooling system.

20. Connect the negative battery cable.

21. Evacuate and charge and leak test the air conditioning system refrigerant.

22. Run the engine to normal operating temperatures; then, check the climate control operation and check for leaks.

STEERING

POWER RACK & PINION STEERING GEAR

REMOVAL & INSTALLATION

See Figure 97.

✳✳ WARNING

Do not permit the steering wheel to turn whenever the steering gear is disconnected from the steering col-umn. Damage to the air bag wiring can result.

1. Before servicing the vehicle, refer to the precautions section.
2. Center the steering wheel and lock it in position.
3. Remove or disconnect the following:
   - Negative battery cable and wait at least 3 minutes before continuing
   - Front wheels
4. Remove the air bag and steering wheel as follows:
   a. Align the front wheels in the straight ahead position
   b. Remove the access panel from the steering wheel and disconnect the drivers airbag 4P connector.
   c. Remove the two Torx® bolts using a T30 bit.
   d. Remove the airbag.

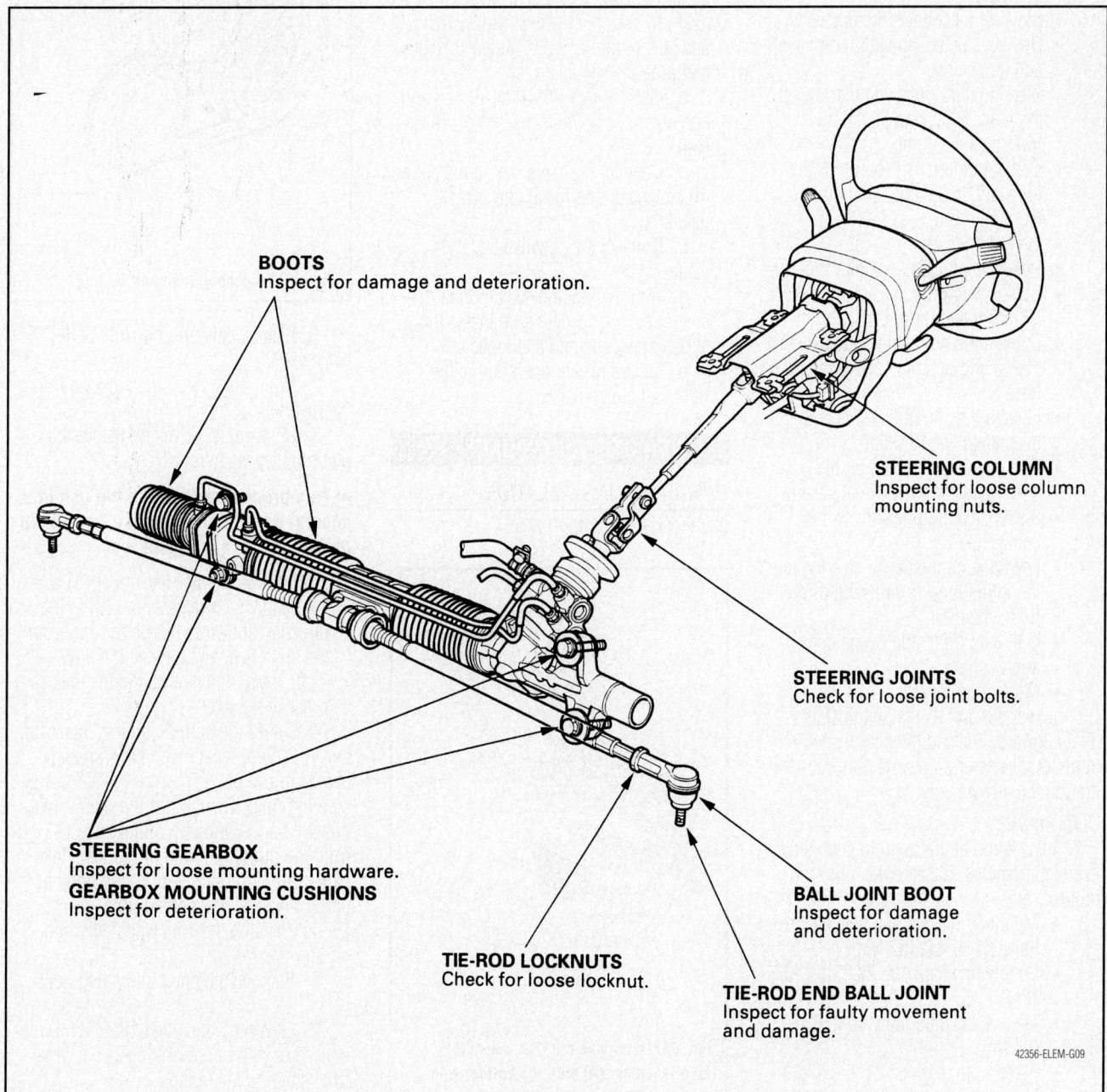

BOOTS
Inspect for damage and deterioration.

STEERING COLUMN
Inspect for loose column mounting nuts.

STEERING JOINTS
Check for loose joint bolts.

STEERING GEARBOX
Inspect for loose mounting hardware.
GEARBOX MOUNTING CUSHIONS
Inspect for deterioration.

TIE-ROD LOCKNUTS
Check for loose locknut.

BALL JOINT BOOT
Inspect for damage and deterioration.

TIE-ROD END BALL JOINT
Inspect for faulty movement and damage.

42356-ELEM-G09

Fig. 97 Power steering gear and related components

e. Disconnect the cruise control connector and horn switch connector.

f. Loosen the steering wheel bolt and using a suitable puller, free the steering wheel.

➡**Do not tap on the steering wheel or column shaft during removal. If you thread the puller bolts more than 5 threads into the wheel hub you will hit the cable reel and damage it. To prevent damage, insert a pair of jam nuts 5 threads up on each puller bolt.**

g. Remove the puller, steering wheel bolt and wheel.

5. Remove or disconnect the following:
- Driver's side dashboard lower cover and undercover
- Steering joint bolts and disconnect the joint by moving the joint towards the column
- Center pin from the top of the pinion shaft, if equipped and discard the pin
- Tie rod ends
- Power steering heat baffle plate
- Engine wiring harness clamp and clip from their brackets
- Loosen the adjustable hose clamp and disconnect the return hose
- Loosen the 14mm flare nut and disconnect the feed line
- Open the hose holders on the return hose and remove the clamp
- Power steering pressure switch connector
- Feed line on the power steering line mounting bracket and set it aside
- Body stiffener
- Left, then right side flange bolts and washers
- Mounting brackets

6. Lower the unit so the pinion shaft points upward. Remove the pinion shaft grommet. The steering gear is removed through the driver's side.

To install:

7. Installation of the steering gear is the reverse of removal. Observe the following torques:
- Mounting bracket and side flange bolts: 46 ft. lbs. (62 Nm)
- Supply line flare nut: 27 ft. lbs. (37 Nm)
- Tie rod ball stud nuts: 40 ft. lbs. (54 Nm)
- Steering joint bolts: 21 ft. lbs. (28 Nm)

8. Install the steering wheel and air bag as follows:

a. First make sure the front wheels are aligned straight ahead. Center the cable reel by rotating the cable reel clockwise until it stops, then rotate it counterclockwise about 2½ turns. The arrow mark on the cable reel should point straight up.

b. Position the tabs on the turn signal canceling sleeve, install the steering wheel and make sure the wheel hub engages the pins of the cable reel and tabs of the canceling sleeve. Do not tap on the wheel or column.

c. Install the steering wheel bolt and tighten to 29 ft. lbs. (39 Nm). Connect the horn switch, cruise control switch and ensure the wiring is routed correctly and properly secured.

d. Install the driver's side air bag and tighten the Torx® bolts to 7 ft. lbs. (9 Nm).

e. Connect the cable reel to the airbag 4P connector and install the access panel.

f. Connect the negative battery cable.

g. Turn the ignition switch on and ensure the airbag light illuminates for about 6 seconds and then goes out.

h. Ensure proper operation of the horn and cruise control.

POWER STEERING PUMP

REMOVAL & INSTALLATION
See Figures 98 through 100.

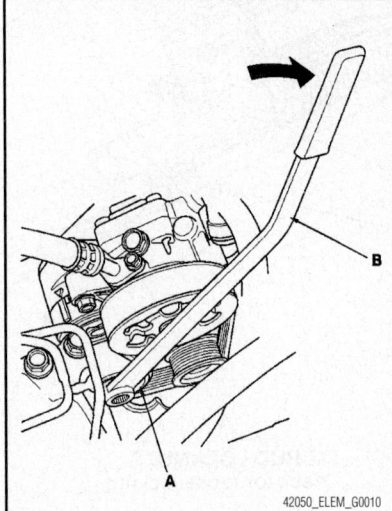

Fig. 98 To remove the belt, move the auto-tensioner (A) with the belt tension release tool (B) in the direction shown to relieve tension from the drive belt, and remove the drive belt

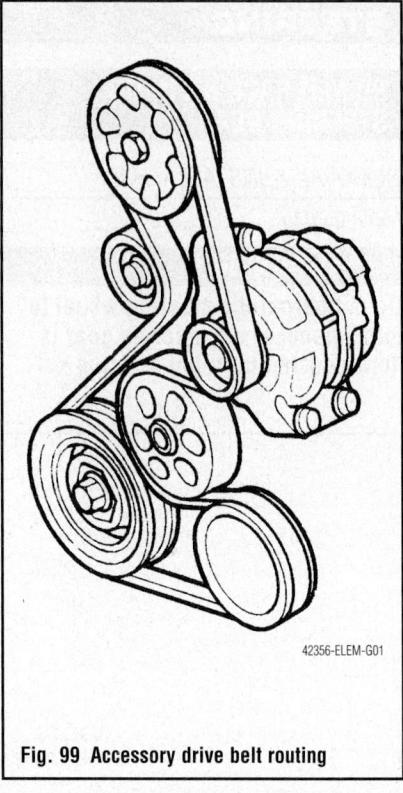

42356-ELEM-G01

Fig. 99 Accessory drive belt routing

1. Place a suitable container under the vehicle.

2. Drain the power steering fluid from the reservoir.

3. Remove the drive belt (A) from the pump pulley, as follows:

➡**This procedure requires the use of a special Belt tension release tool, Snap-on YA9317 or equivalent tool.**

a. Move the auto-tensioner (A) with the belt tension release tool (B) in the direction shown to relieve tension from the drive belt, and remove the drive belt.

b. Install the new belt in the reverse order of removal.

4. Cover the auto-tensioner, alternator, and A/C compressor with several shop towels to protect them from spilled power steering fluid. Disconnect the pump inlet hose (B) and the pump outlet hose (C) from the pump (D), and plug them. Take care not to spill the fluid on the body or parts. Wipe off any spilled fluid at once. Do not turn the steering wheel with the pump removed.

5. Remove the pump mounting bolts (E).

6. Cover the opening of the pump with a piece of tape to prevent foreign material from entering the pump.

To install:

7. Connect the pump inlet hose and the pump outlet hose with a new O-ring.

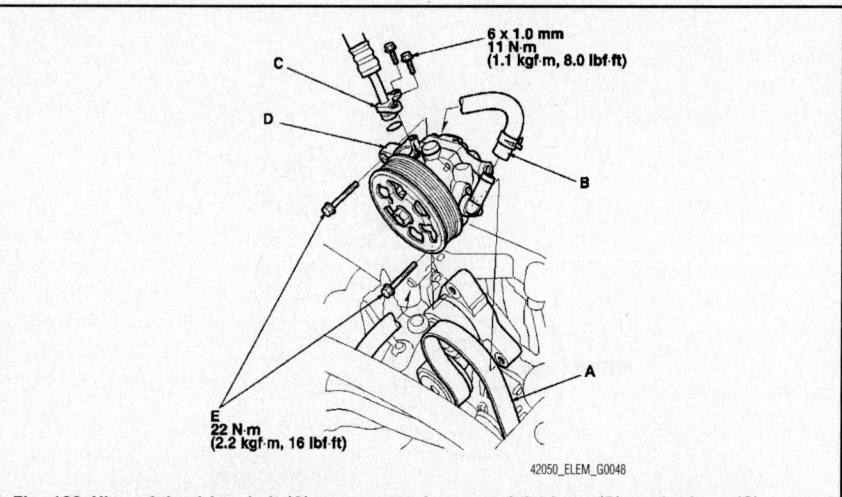

Fig. 100 View of the driver belt (A), power steering pump inlet hose (B), outlet hose (C), power steering pump (D) and mounting bolts (E)

8. Loosely install the pump in the pump bracket with the mounting bolts, then tighten the pump fittings securely.

9. Install the drive belt. Make sure that the belt is properly positioned on the pulleys.
10. Tighten the pump mounting bolts to 16 ft. lbs. (22 Nm).
11. Fill the power steering fluid reservoir to the upper level line.

SUSPENSION   FRONT SUSPENSION

COIL SPRING

REMOVAL & INSTALLATION

See Figure 101.

1. Before servicing the vehicle, refer to the precautions section.
2. Remove the strut from the vehicle and install in a strut spring compressor. Compress the spring until the end of the spring comes away from the spring seat.
3. Remove the upper strut mount, spring seat and related components.
4. Remove the coil spring from the strut spring compressor.

To install:

➡**Use a new self-locking nut.**

5. Compress the spring and position the strut so that the end of the spring aligns with the notch in the spring seat.
6. Install the upper strut mounting components and tighten the nut to 33 ft. lbs. (44 Nm).
7. Install the strut to the vehicle.
8. Check the wheel alignment and adjust as necessary.

LOWER BALL JOINT

REMOVAL & INSTALLATION

The ball joint is not replaceable.

LOWER CONTROL ARM

REMOVAL & INSTALLATION

1. Before servicing the vehicle, refer to the precautions section.

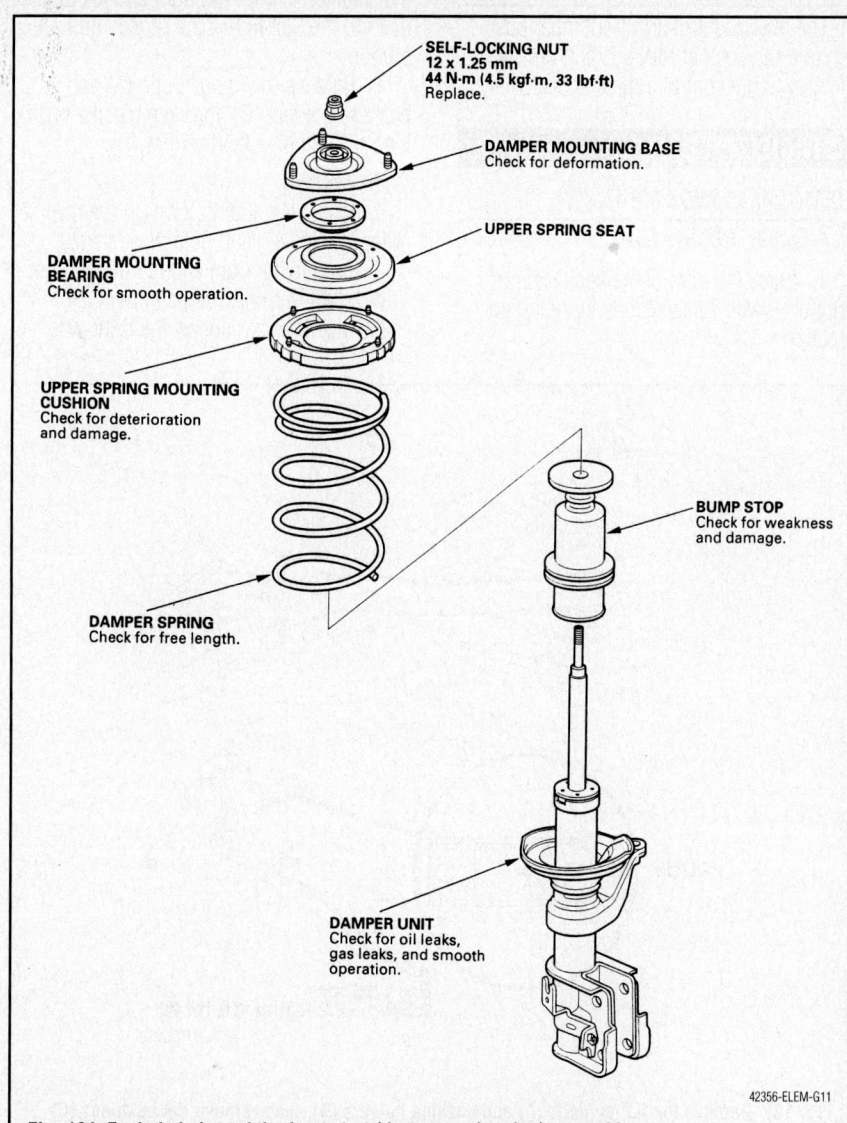

Fig. 101 Exploded view of the front strut (damper and spring) assembly

2. Remove or disconnect the following:
- Front wheel
- Stabilizer link
- Lower arm from the knuckle
- Lower arm

To install:

3. Install all suspension components and fasteners and hand tighten them.. Place a jack under the suspension, raise the suspension with the jack and load the jack with the vehicle weight.

4. Installation is the reverse of removal. Observe the following torques:
- Lower arm bolts: 61 ft. lbs. (83 Nm)
- Ball stud nut: 51 ft. lbs. (69 Nm). Install the cotter pin into the ball joint from the inside to the outside of the vehicle.
- Stabilizer link: 29 ft. lbs. (39 Nm)

CONTROL ARM BUSHING REPLACEMENT

The lower control arm front inner bushing and the damper fork bushing are serviced with the control arm as an assembly.

STABILIZER BAR

REMOVAL & INSTALLATION

See Figures 102 and 103.

1. Raise the front of the vehicle, and support it with safety stands in the proper locations.

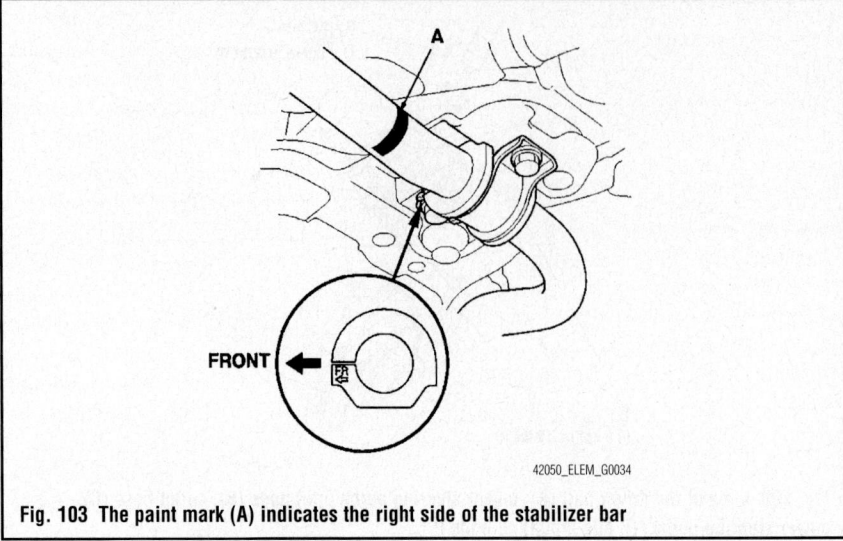

Fig. 103 The paint mark (A) indicates the right side of the stabilizer bar

2. Remove the front wheels.

3. Disconnect the stabilizer links from the stabilizer bar on the right and left sides. Refer to the Stabilizer Link procedure in this section.

4. Remove the flange bolts (A) and bushing holders (B), then remove the bushings (C) and the stabilizer bar (D).

To install:

5. Install the stabilizer bar in the reverse order of removal, and note these items:
   a. Note the right and left direction of the stabilizer bar. The paint mark (A) on the stabilizer bar shows the right side.

   b. Do not set the bushings on the bent or curved part of the stabilizer bar.

   c. Note the fore/aft direction of the bushing holders.

   d. Refer to stabilizer link removal/installation to connect the stabilizer bar to the links.

STABILIZER LINK

REMOVAL & INSTALLATION

See Figures 104 through 106.

1. Raise the front of the vehicle, and support it with safety stands in the proper locations.

2. Remove the front wheel.

3. Remove the self-locking nut (A) and

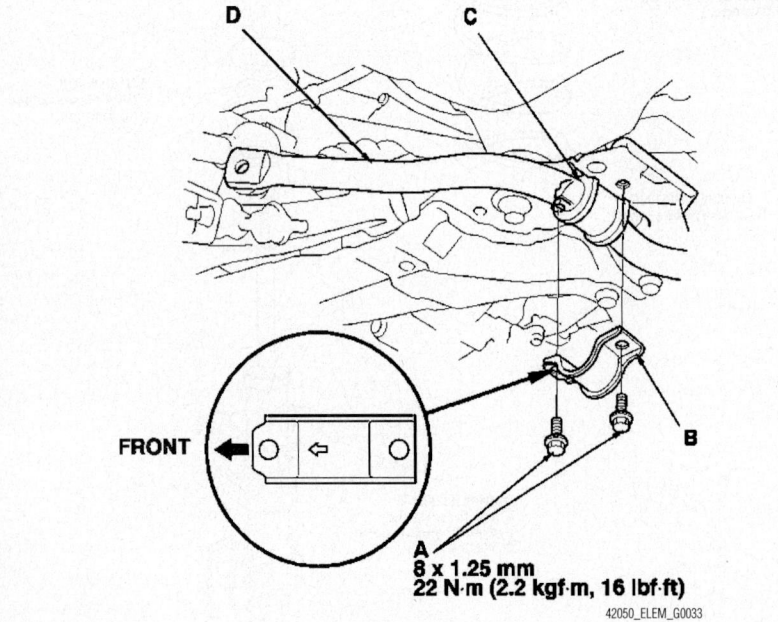

Fig. 102 Remove the flange bolts (A) and bushing holders (B), then remove the bushings (C) and the stabilizer bar (D)

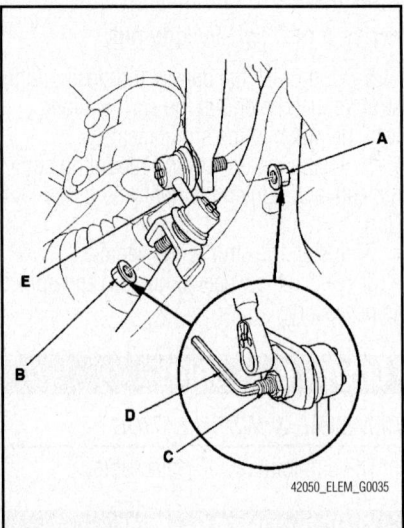

Fig. 104 Remove the self-locking nut (A) and flange nut (B) while holding the respective joint pin (C) with a hex wrench (D), and remove the stabilizer link (E)

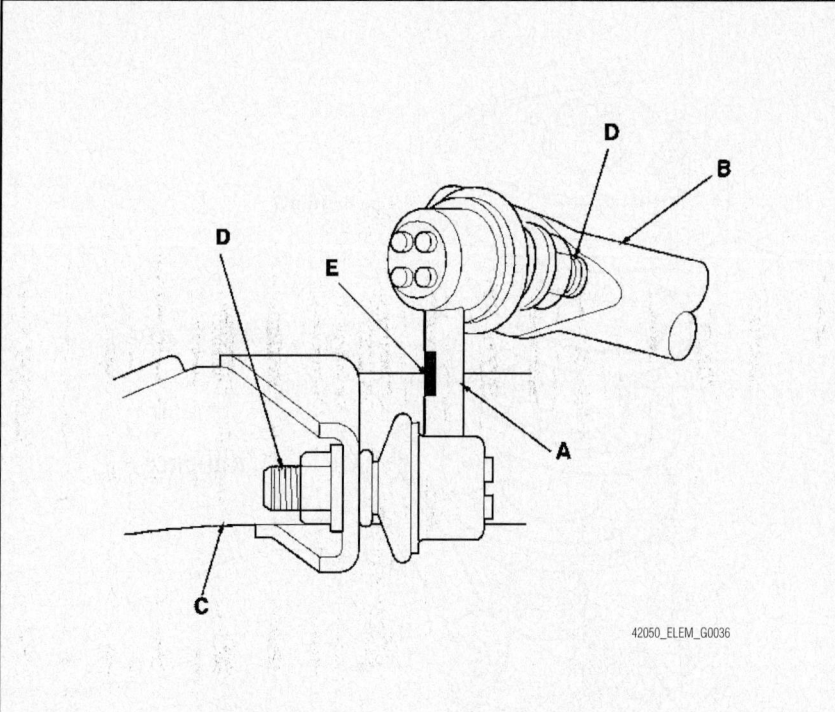

42050_ELEM_G0036

Fig. 105 View of the stabilizer link (A), stabilizer bar (B), lower arm (C), joint pins (D) and left stabilizer yellow paint mark (E)

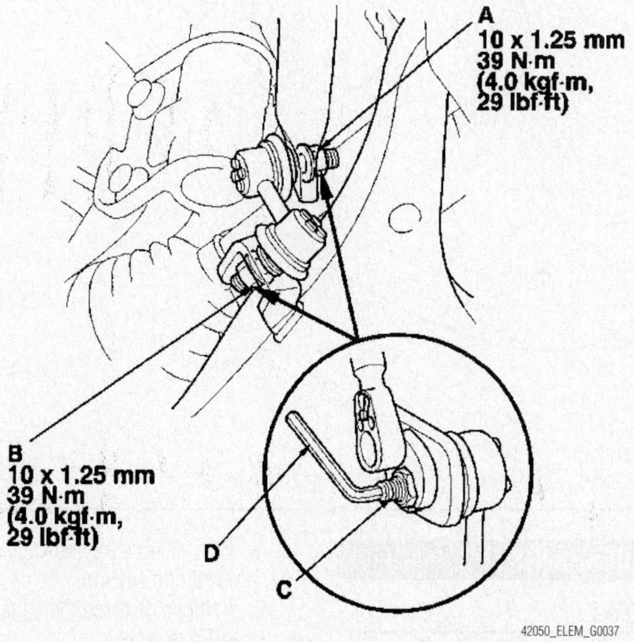

A
10 x 1.25 mm
39 N·m
(4.0 kgf·m,
29 lbf·ft)

B
10 x 1.25 mm
39 N·m
(4.0 kgf·m,
29 lbf·ft)

D

C

42050_ELEM_G0037

Fig. 106 View of the self-locking nut (A) and flange nut (B), joint pin (C) and hex wrench (D)

flange nut (B) while holding the respective joint pin (C) with a hex wrench (D), and remove the stabilizer link (E).

To install:

4. Install the stabilizer link (A) on the stabilizer bar (B) and lower arm (C) with the joint pins (D) set at the center of their range of movement.

➡**The left stabilizer has a yellow paint mark (E), while the right stabilizer link has a white paint mark.**

5. Install a new self-locking nut and flange nut, and lightly tighten them.

➡**Use a new self-locking nut during installation.**

✳✳ WARNING

Do not place the jack against the ball joint pin.

6. Tighten the self-locking nut (A) and flange nut (B) to 29 ft. lbs. (39 Nm) while holding the respective joint pin (C) with a hex wrench (D).

7. Reinstall the front wheel and test-drive the vehicle.

8. After 5 minutes of driving, retighten the self-locking nut to 29 ft. lbs. (39 Nm).

STEERING KNUCKLE

REMOVAL & INSTALLATION

Refer to the Wheel Bearing Removal & Installation procedure.

STRUT/DAMPER

REMOVAL & INSTALLATION

See Figure 107.

1. Before servicing the vehicle, refer to the precautions section.

2. Remove or disconnect the following:
   - Front wheel
   - Tie rod end
   - Brake hose retainer
   - ABS sensor harness bracket and brake hose bracket. Do not disconnect the wheel sensor connector.
   - Pinch bolts from the damper, while holding the nuts
   - Flange nuts from the top of the damper
   - Strut (damper), after lowering the lower control arm

To install:

➡**Use new self-locking fasteners for assembly.**

3. Install or connect the following:

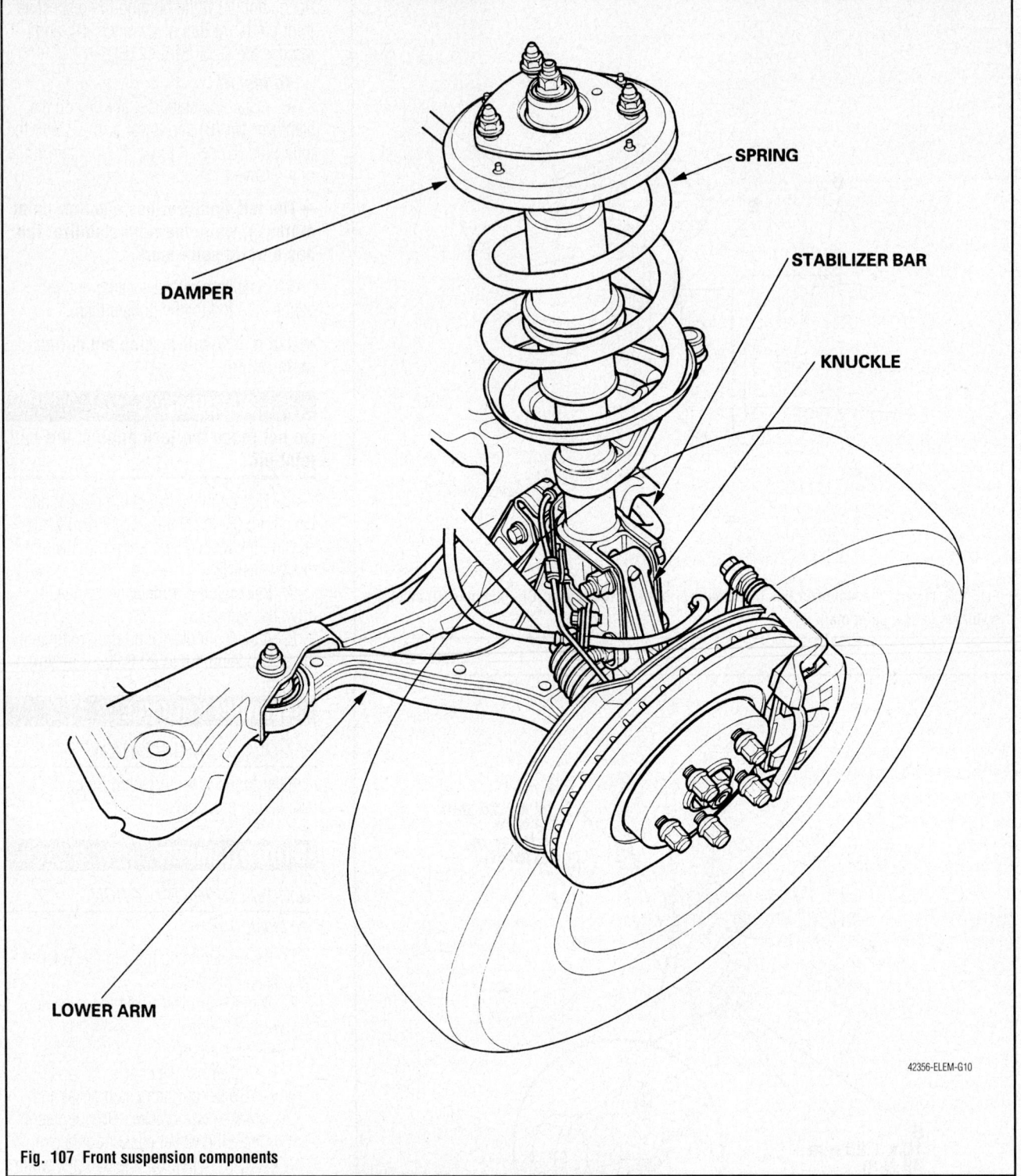

DAMPER

SPRING

STABILIZER BAR

KNUCKLE

LOWER ARM

42356-ELEM-G10

Fig. 107 Front suspension components

- Strut (damper). Tighten the upper mounting nuts to 33 ft. lbs. (44 Nm).
- Tighten the pinch bolts to 116 ft. lbs. (157 Nm)
- ABS sensor
- Tie rod end
- Brake hose retainer
- Front wheel

WHEEL BEARINGS

ADJUSTMENT

The wheel bearings are sealed units and are not adjustable.

REMOVAL & INSTALLATION

See Figure 108.

1. Before servicing the vehicle, refer to the precautions section.
2. Remove or disconnect the following:
   - Front wheel
   - Brake hose bracket
   - Brake caliper and rotor. Forcing screws are needed to remove the rotor.
   - Spindle nut

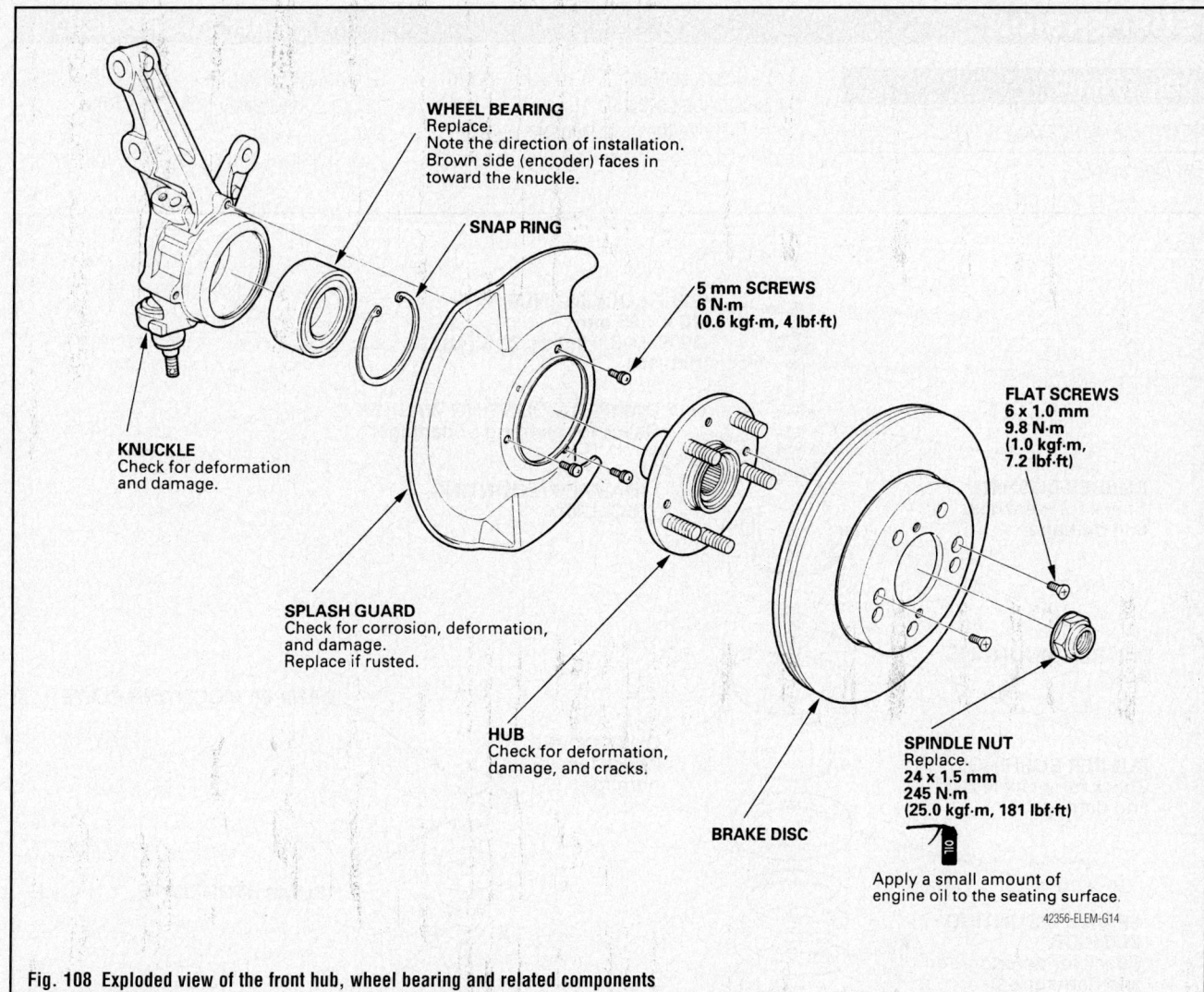

WHEEL BEARING
Replace.
Note the direction of installation.
Brown side (encoder) faces in
toward the knuckle.

SNAP RING

5 mm SCREWS
6 N·m
(0.6 kgf·m, 4 lbf·ft)

FLAT SCREWS
6 x 1.0 mm
9.8 N·m
(1.0 kgf·m,
7.2 lbf·ft)

KNUCKLE
Check for deformation
and damage.

SPLASH GUARD
Check for corrosion, deformation,
and damage.
Replace if rusted.

HUB
Check for deformation,
damage, and cracks.

SPINDLE NUT
Replace.
24 x 1.5 mm
245 N·m
(25.0 kgf·m, 181 lbf·ft)

BRAKE DISC

Apply a small amount of
engine oil to the seating surface.

42356-ELEM-G14

Fig. 108 Exploded view of the front hub, wheel bearing and related components

- ABS sensor
- Stabilizer link
- Lower arm from the knuckle
- Strut-to-knuckle bolts
- Steering hub/knuckle assembly
3. Press the hub from the knuckle. The

bearings and races can now be pressed out
and replaced.

➡**With ABS, install the bearing with
the magnetic encoder (brown color)
toward the inside of the knuckle.**

4. Observe the following torques:
- Strut bolts: 116 ft. lbs. (157 Nm)
- Ball stud nuts: 51 ft. lbs. (69 Nm)
- Stabilizer bar link: 29 ft. lbs. (39 Nm)
- Spindle nut: 181 ft. lbs. (245 Nm)

COIL SPRING

REMOVAL & INSTALLATION

See Figure 109.

1. Before servicing the vehicle, refer to the precautions section.
2. Remove the strut from the vehicle and install in a strut spring compressor.

Compress the spring until the end of the spring comes away from the spring seat.

3. Remove or disconnect the following:

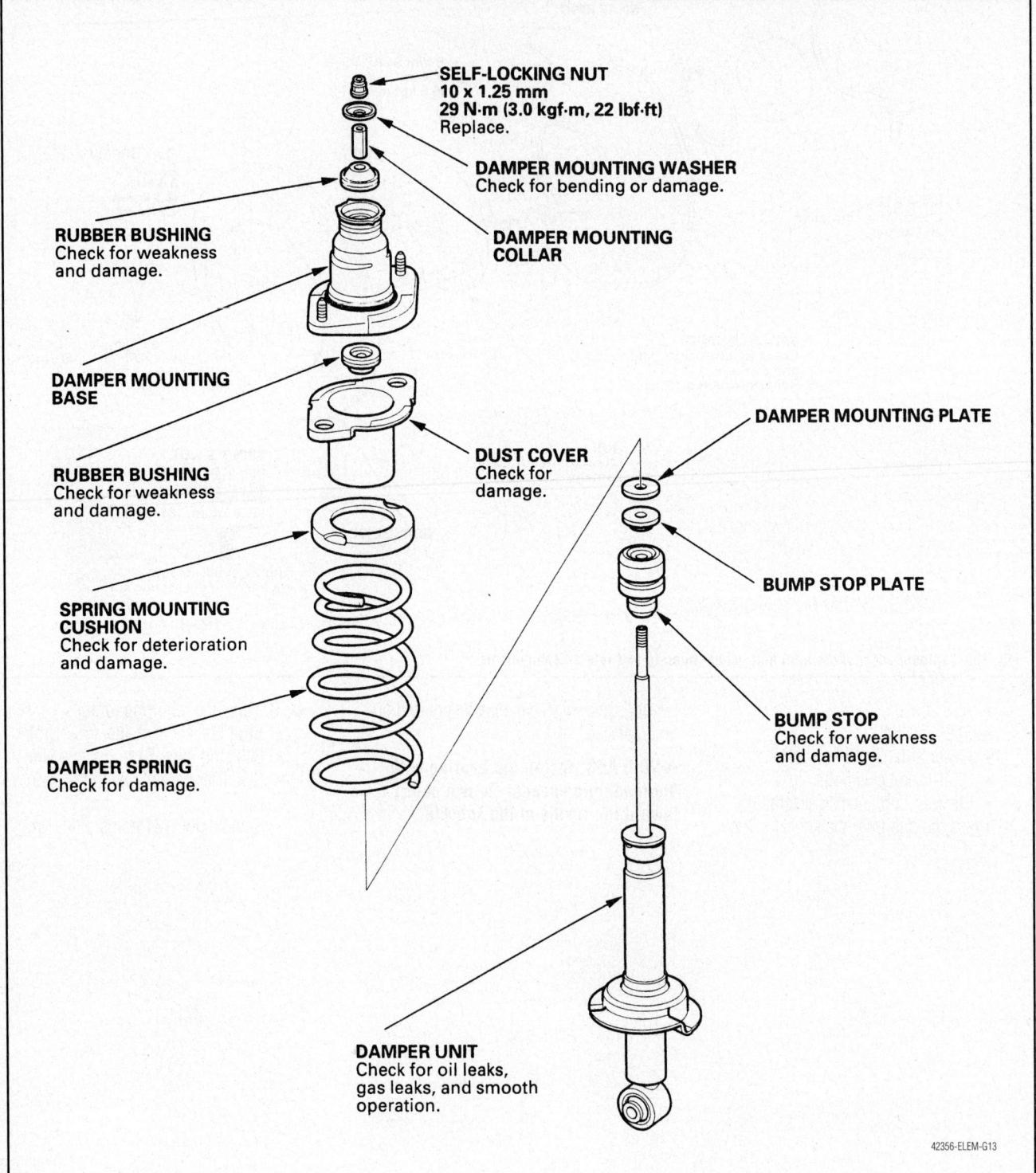

SELF-LOCKING NUT
10 x 1.25 mm
29 N·m (3.0 kgf·m, 22 lbf·ft)
Replace.

DAMPER MOUNTING WASHER
Check for bending or damage.

DAMPER MOUNTING COLLAR

RUBBER BUSHING
Check for weakness and damage.

DAMPER MOUNTING BASE

RUBBER BUSHING
Check for weakness and damage.

DUST COVER
Check for damage.

SPRING MOUNTING CUSHION
Check for deterioration and damage.

DAMPER SPRING
Check for damage.

DAMPER MOUNTING PLATE

BUMP STOP PLATE

BUMP STOP
Check for weakness and damage.

DAMPER UNIT
Check for oil leaks, gas leaks, and smooth operation.

42356-ELEM-G13

Fig. 109 Exploded view of the strut (damper and spring) assembly

- Upper strut mount, spring seat and related components
- Coil spring from the strut spring compressor

To install:

➡**Use a new self-locking nut.**

4. Compress the spring and position the strut so that the end of the spring aligns with the notch in the spring seat.
5. Install or connect the following:

- Upper strut mounting components and tighten the nut to 22 ft. lbs. (29 Nm).
- Strut to the vehicle

6. Check the wheel alignment and adjust as necessary.

STRUT/DAMPER

REMOVAL & INSTALLATION

See Figure 110.

1. Before servicing the vehicle, refer to the precautions section.
2. Support the vehicle under the lower control arm.
3. Remove or disconnect the following:

- Rear wheel
- Flange bolt from the bottom of the damper (strut)
- Evaporative emission (EVAP) canister bolts, and loosen the EVAP canister mounting (left side only)
- Interior access panel, if necessary

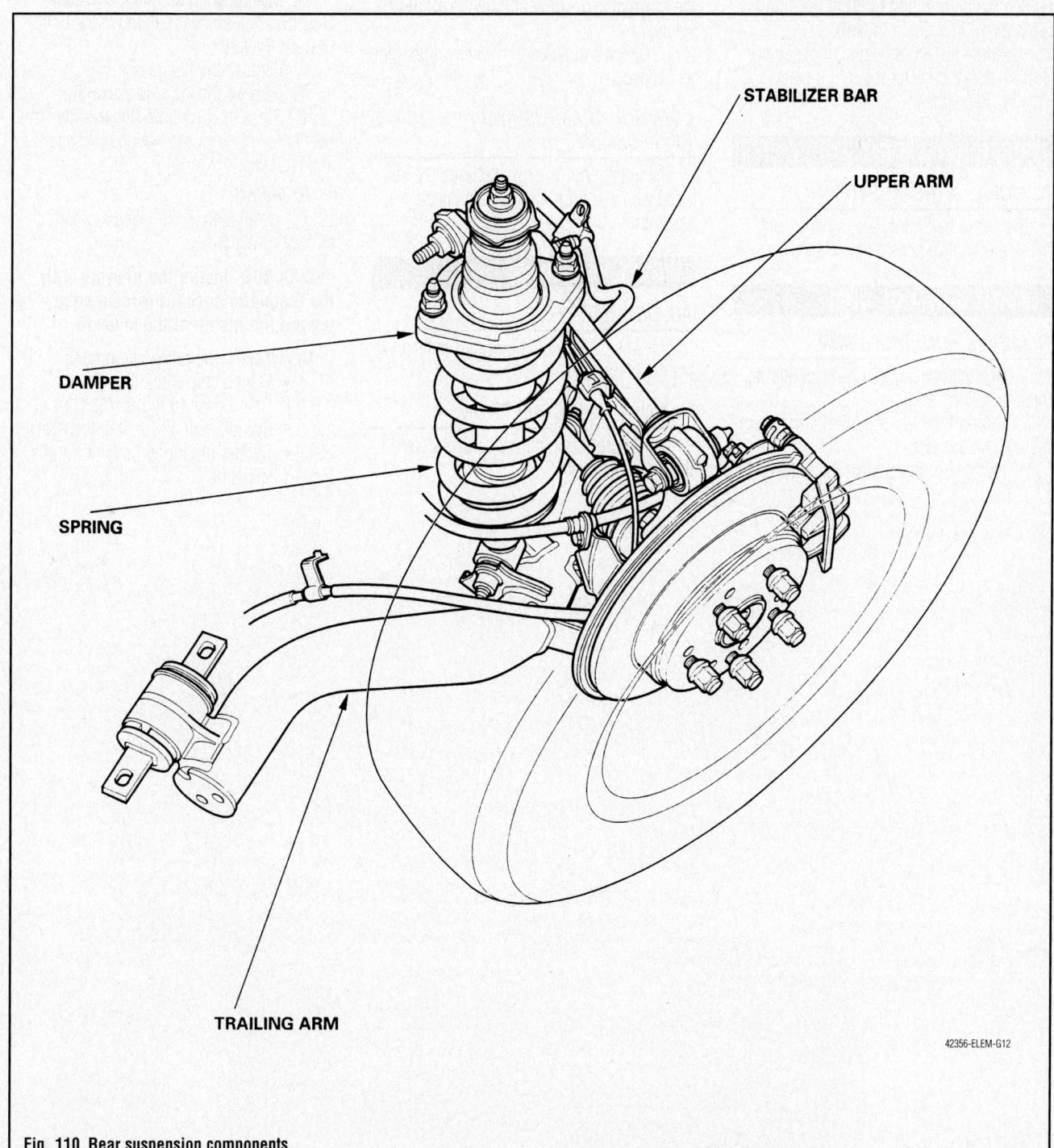

Fig. 110 Rear suspension components

42356-ELEM-G12

- Flange nuts from the top of the damper in the cargo area
- Strut

To install:

4. Install or connect the following:
  - Strut. Position the damper mounting base so the indent mark is toward the inside of the vehicle,
  - Upper flange nuts, hand-tight only
  - Bottom flange bolt, hand-tight only

5. With the suspension raised with a jack to load it with the vehicles weight, tighten the bottom bolt to 69 ft. lbs. and the top nuts to 54 ft. lbs. (74 Nm).
  - Interior access panel, if necessary
  - EVAP canister mounting bolts
  - Rear wheel

UPPER BALL JOINT

REMOVAL & INSTALLATION

The upper ball joints are replaced with the upper control arms as an assembly.

UPPER CONTROL ARM

REMOVAL & INSTALLATION

1. Before servicing the vehicle, refer to the precautions section.
2. Support the lower control arm assembly with a floor jack.
3. Remove or disconnect the following:

- Wheel speed sensor harness bracket, if equipped
- Flange bolts and the control arm

To install:

4. Install all suspension components and fasteners and hand tighten them. Place a jack under the trailing arm, raise the suspension with the jack and load the jack with the vehicle weight.
5. Tighten the upper control arm flange bolts to 69 ft. lbs. (93 Nm).
6. Clean the wheel, mating surface of the brake disc or drum and inside of the wheel.
7. Check and adjust the wheel alignment as needed.

CONTROL ARM BUSHING REPLACEMENT

The upper control arm bushings are serviced with the upper control arm as an assembly.

WHEEL BEARINGS

ADJUSTMENT

The wheel bearings are sealed units and are not adjustable.

REMOVAL & INSTALLATION

1. Before servicing the vehicle, refer to the precautions section.

2. Remove or disconnect the following:
  - Rear wheel
  - Brake caliper
  - Rotor
  - Spindle nut
  - Axle shaft (4wd)
  - Parking brake shoes
  - Parking brake cable
  - Wheel sensor, if equipped
3. Support the trailing arm.
4. Remove or disconnect the following:
  - Upper arm from the knuckle
5. Matchmark the trailing arm cam adjusting bolt and cam. Remove the bolt. Discard the nut.
6. Remove the flange bolt.
7. Remove the knuckle assembly.
8. Press the hub from the knuckle. The bearings and races can now be pressed out and replaced.

To install:

9. Installation is the reverse of the removal procedure.

➡ With ABS, install the bearing with the magnetic encoder (brown color) toward the inside of the knuckle.

10. Observe the following torques:
  - Flange bolt: 69 ft. lbs. (93 Nm)
  - Cam bolts: 43 ft. lbs. (59 Nm)
  - Spindle nut: 134 ft. lbs. (181 Nm)
  - Caliper mounting bolts: 41 ft. lbs. (55 Nm)

HONDA

Fit

7

SPECIFICATIONS AND MAINTENANCE CHARTS

ENGINE AND VEHICLE IDENTIFICATION

		Engine					Model Year	
Code ①	Liters (cc)	Cu. In.	Cyl.	Fuel Sys.	Engine Type	Eng. Mfg.	Code ②	Year
GD3	1.5 (1496)	91.00	I4	MPFI	SOHC	Honda	7	2007
							8	2008

MPFI: Multi-Point Fuel Injection

DOHC: Double Overhead Camshafts

① 4th-6th digits of VIN

② 10th digit of VIN

22140_HFIT_C0001

GENERAL ENGINE SPECIFICATIONS
All measurements are given in inches.

Year	Model	Engine Displacement Liters	Engine Series VIN	Net Horsepower @ rpm	Net Torque @ rpm (ft. lbs.)	Bore x Stroke (in.)	Com- pression Ratio	Oil Pressure @ rpm
2007	Fit	1.5	CD3	109@5800	105@4800	3.11 x 3.52	10.4:1	15.6@Idle
2008	Fit	1.5	CD3	109@5800	105@4800	3.11 x 3.52	10.4:1	15.6@Idle

NA: Not Available

22140_HFIT_C0002

GASOLINE ENGINE TUNE-UP SPECIFICATIONS

Year	Engine Displacement Liters	Engine VIN	Spark Plug Gap (in.)	Ignition Timing (deg.) MT	Ignition Timing (deg.) AT	Fuel Pump (psi)	Idle Speed (rpm) MT	Idle Speed (rpm) AT	Valve Clearance In.	Valve Clearance Ex.
2007	1.5	CD3	0.042-0.051	①	①	47-54	②	②	HYD	HYD
2008	1.5	CD3	0.042-0.051	①	①	47-54	②	②	HYD	HYD

NOTE: The Vehicle Emission Control Information label reflects specification changes made during production.

Follow the figures on the label if they differ from those in this chart.

HYD: Hydraulic

① Ignition timing is preset and cannot be adjusted

② Idle speed is maintained by the Electronic Control Module (ECM)

22140_HFIT_C0003

CAPACITIES

Year	Model	Engine Displacement Liters	Engine VIN	Engine Oil with Filter (qts.)	Transmission (pts.) Manual	Transmission (pts.) Auto. ①	Fuel Tank (gal.)	Cooling System (qts.)
2007	Fit	1.5	CD3	3.8	3.4	12.6	10.8	4
2008	Fit	1.5	CD3	3.8	3.4	12.6	10.8	4

NOTE: All capacities are approximate. Add fluid gradually and check to be sure a proper fluid level is obtained.

① Drain and refill

22140_HFIT_C0004

FLUID SPECIFICATIONS

Year	Model	Engine Displacement Liters	Engine ID/VIN	Engine Oil	Auto. Trans.	Manual Trans.	Power Steering Fluid	Brake Master Cylinder
2007	Fit	1.5	CD3	5W-20	ATF-Z1	①	—	②
2008	Fit	1.5	CD3	5W-20	ATF-Z1	①	—	②

DOT: Department Of Transportation

① Honda Manual Transmission Fluid (MTF): P/N 08798-9031

② DOT 3

22140_HFIT_C0005

VALVE SPECIFICATIONS

Year	Engine Displacement Liters	Engine VIN	Seat Angle (deg.)	Face Angle (deg.)	Spring Test Pressure (lbs. @ in.)	Spring Installed Height (in.)	Stem-to-Guide Clearance (in.) Intake	Stem-to-Guide Clearance (in.) Exhaust	Stem Diameter (in.) Intake	Stem Diameter (in.) Exhaust
2007	1.5	CD3	45	45	—	①	0.0008-0.0020	0.0020-0.0031	0.2157-0.2161	0.2146-0.2150
2008	1.5	CD3	45	45	—	①	0.0008-0.0020	0.0020-0.0031	0.2157-0.2161	0.2146-0.2150

① Free length - intake: 1.189 in.; exhaust: 2.259in.

22140_HFIT_C0006

CAMSHAFT AND BEARING SPECIFICATIONS CHART

All measurements are given in inches.

Year	Engine Displ. Liters	Engine ID/VIN	Journal Dia.	Brg. Oil Clearance	Shaft End-play	Runout	Journal Bore	Lobe Height Intake	Exhaust
2007	1.5	CD3	NA	0.0018-0.0033	0.002-0.0100	0.0010	NA	①	①
2008	1.5	CD3	NA	0.0018-0.0033	0.002-0.0100	0.0010	NA	①	①

NA: Not Available

① Intake Primary: 1.39291 inch

Intake Secondary: 1.20193 inch

Exhaust: 1.39321 inch

22140_HFIT_C0007

CRANKSHAFT AND CONNECTING ROD SPECIFICATIONS

All measurements are given in inches.

Year	Engine Displacement Liters	Engine VIN	Crankshaft Main Brg. Journal Dia.	Main Brg. Oil Clearance	Shaft End-play	Thrust on No.	Connecting Rod Journal Diameter	Oil Clearance	Side Clearance
2007	1.5	CD3	1.9676-1.9685	0.0020	0.018	—	1.5739-1.5748	0.0008-0.0015	0.006-0.0120
2008	1.5	CD3	1.9676-1.9685	0.0020	0.018	—	1.5739-1.5748	0.0008-0.0015	0.006-0.0120

22140_HFIT_C0008

PISTON AND RING SPECIFICATIONS

All measurements are given in inches.

Year	Engine Displ. Liters	Engine VIN	Piston Clearance	Ring Gap Top Compression	Bottom Compression	Oil Control	Ring Side Clearance Top Compression	Bottom Compression	Oil Control
2007	1.5	CD3	0.0004-0.0016	0.006-0.0120	0.014-0.0200	0.008-0.0280	0.0026-0.0035	0.0012-0.0022	NA
2008	1.5	CD3	0.0004-0.0016	0.006-0.0120	0.014-0.0200	0.008-0.0280	0.0026-0.0035	0.0012-0.0022	NA

NA: Not Available

22140_HFIT_C0009

TORQUE SPECIFICATIONS
All readings in ft. lbs.

Year	Engine Displacement Liters	Engine VIN	Cylinder Head Bolts	Main Bearing Bolts	Rod Bearing Bolts	Crankshaft Damper Bolts	Flywheel Bolts	Manifold		Spark Plugs	Oil Pan Drain Plug
								Intake	Exhaust		
2007	1.5	CD3	①	②	③	④	NA	17	33	13	NA
2008	1.5	CD3	①	②	③	④	NA	17	33	13	NA

NA: Not Available

① Step 1: 22 ft. lbs.
　Step 2: Plus 130 degrees

② Step 1: 18 ft. lbs.
　Step 2: Plus 40 degrees

③ Step 1: 7.2 ft. lbs.
　Step 2: Plus 90 degrees

④ Step 1: New bolt: 130 ft. lbs.
　Step 2: 27 ft. lbs.

22140_HFIT_C0010

WHEEL ALIGNMENT

Year	Model		Caster		Camber		Toe-in (mm)
			Range (+/-Deg.)	Preferred Setting (Deg.)	Range (+/-Deg.)	Preferred Setting (Deg.)	
2007	Fit	Front	3.75 +/- 1.0	0.0	0.0 +/- 1.0	0.0	0.0 +/- 3.0
		Rear	—	—	-1.5 +/- 1.0	—	2.5 +/- 2.5
2008	Fit	Front	3.75 +/- 1.0	0.0	0.0 +/- 1.0	0.0	0.0 +/- 3.0
		Rear	—	—	-1.5 +/- 1.0	—	2.5 +/- 2.5

22140_HFIT_C0011

TIRE, WHEEL AND BALL JOINT SPECIFICATIONS

| Year | Model | OEM Tires | | Tire Pressures (psi) | | Wheel Size | Ball Joint Inspection | Lug Nut Torque (ft. lbs.) |
		Standard	Optional	Front	Rear			
2007	Fit	P175/65R14	P195/55R15	①	①	N/A	②	80
2008	Fit	P175/65R14	P195/55R15	①	①	N/A	②	80

① Refer to placard on vehicle for proper inflation pressure.

② Replace if any measurable movement is found.

22140_HFIT_C0012

BRAKE SPECIFICATIONS
All measurements in inches unless noted

| Year | Model | | Brake Disc | | | Brake Drum Diameter | | | Minimum Lining Thickness | Brake Caliper | |
			Original Thickness	Minimum Thickness	Maximum Runout	Original Inside Diameter	Max. Wear Limit	Maximum Machine Diameter		Bracket Bolts (ft. lbs.)	Mounting Bolts (ft. lbs.)
2007	Fit	F	0.830	0.750	0.004	—	—	—	0.060	62-69	17
		R	—	—	—	7.870-7.874	7.91	7.91	0.080	—	—
2008	Fit	F	0.830	0.750	0.004	—	—	—	0.060	62-69	17
		R	—	—	—	7.870-7.874	7.91	7.91	0.080	—	—

22140_HFIT_C0013

SCHEDULED MAINTENANCE INTERVALS
HONDA - FIT

NOTE: HONDA FIT uses a Maintenance Service light system. There are few items associated with specific mileage intervals; most are based on SYMBOLS that appear with the Maintenance Service light; such as: If Service light shows symbol "A" at 15%, service is due soon; at 5%, service is due immediately; at 0%, service is past due.

22140_HFIT_C0014

PRECAUTIONS

Before servicing any vehicle, please be sure to read all of the following precautions, which deal with personal safety, prevention of component damage, and important points to take into consideration when servicing a motor vehicle:

• Never open, service or drain the radiator or cooling system when the engine is hot; serious burns can occur from the steam and hot coolant.

• Observe all applicable safety precautions when working around fuel. Whenever servicing the fuel system, always work in a well-ventilated area. Do not allow fuel spray or vapors to come in contact with a spark, open flame, or excessive heat (a hot drop light, for example). Keep a dry chemical fire extinguisher near the work area. Always keep fuel in a container specifically designed for fuel storage; also, always properly seal fuel containers to avoid the possibility of fire or explosion. Refer to the additional fuel system precautions later in this section.

• Fuel injection systems often remain pressurized, even after the engine has been turned **OFF**. The fuel system pressure must be relieved before disconnecting any fuel lines. Failure to do so may result in fire and/or personal injury.

• Brake fluid often contains polyglycol ethers and polyglycols. Avoid contact with the eyes and wash your hands thoroughly after handling brake fluid. If you do get brake fluid in your eyes, flush your eyes with clean, running water for 15 minutes. If eye irritation persists, or if you have taken brake fluid internally, IMMEDIATELY seek medical assistance.

• The EPA warns that prolonged contact with used engine oil may cause a number of skin disorders, including cancer. You should make every effort to minimize your exposure to used engine oil. Protective gloves should be worn when changing oil. Wash your hands and any other exposed skin areas as soon as possible after exposure to used engine oil. Soap and water, or waterless hand cleaner should be used.

• All new vehicles are now equipped with an air bag system, often referred to as a Supplemental Restraint System (SRS) or Supplemental Inflatable Restraint (SIR) system. The system must be disabled before performing service on or around system components, steering column, instrument panel components, wiring and sensors. Failure to follow safety and disabling procedures could result in accidental air bag deployment, possible personal injury and unnecessary system repairs.

• Always wear safety goggles when working with, or around, the air bag system. When carrying a non-deployed air bag, be sure the bag and trim cover are pointed away from your body. When placing a non-deployed air bag on a work surface, always face the bag and trim cover upward, away from the surface. This will reduce the motion of the module if it is accidentally deployed. Refer to the additional air bag system precautions later in this section.

• Clean, high quality brake fluid from a sealed container is essential to the safe and proper operation of the brake system. You should always buy the correct type of brake fluid for your vehicle. If the brake fluid becomes contaminated, completely flush the system with new fluid. Never reuse any brake fluid. Any brake fluid that is removed from the system should be discarded. Also, do not allow any brake fluid to come in contact with a painted surface; it will damage the paint.

• Never operate the engine without the proper amount and type of engine oil; doing so WILL result in severe engine damage.

• Timing belt maintenance is extremely important. Many models utilize an interference-type, non-freewheeling engine. If the timing belt breaks, the valves in the cylinder head may strike the pistons, causing potentially serious (also time-consuming and expensive) engine damage. Refer to the maintenance interval charts for the recommended replacement interval for the timing belt, and to the timing belt section for belt replacement and inspection.

• Disconnecting the negative battery cable on some vehicles may interfere with the functions of the on-board computer system(s) and may require the computer to undergo a relearning process once the negative battery cable is reconnected.

• When servicing drum brakes, only disassemble and assemble one side at a time, leaving the remaining side intact for reference.

• Only an MVAC-trained, EPA-certified automotive technician should service the air conditioning system or its components.

BRAKES

ANTI-LOCK BRAKE SYSTEM (ABS)

GENERAL INFORMATION

PRECAUTIONS

• Certain components within the ABS system are not intended to be serviced or repaired individually.

• Do not use rubber hoses or other parts not specifically specified for and ABS system. When using repair kits, replace all parts included in the kit. Partial or incorrect repair may lead to functional problems and require the replacement of components.

• Lubricate rubber parts with clean, fresh brake fluid to ease assembly. Do not use shop air to clean parts; damage to rubber components may result.

• Use only DOT 3 brake fluid from an unopened container.

• If any hydraulic component or line is removed or replaced, it may be necessary to bleed the entire system.

• A clean repair area is essential. Always clean the reservoir and cap thoroughly before removing the cap. The slightest amount of dirt in the fluid may plug an orifice and impair the system function. Perform repairs after components have been thoroughly cleaned; use only denatured alcohol to clean components. Do not allow ABS components to come into contact with any substance containing mineral oil; this includes used shop rags.

• The Anti-Lock control unit is a microprocessor similar to other computer units in the vehicle. Ensure that the ignition switch is **OFF** before removing or installing controller harnesses. Avoid static electricity discharge at or near the controller.

• If any arc welding is to be done on the vehicle, the control unit should be unplugged before welding operations begin.

BRAKES

BLEEDING THE BRAKE SYSTEM

BLEEDING PROCEDURE

BLEEDING PROCEDURE

✳✳ CAUTION

Do not reuse the drained fluid. Use only clean Honda DOT 3 Brake Fluid from an unopened container. Using a non-Honda brake fluid can cause corrosion and shorten the life of the system. Do not mix different brands of brake fluid; they may not be compatible. Make sure no dirt or other foreign matter is allowed to contaminate the brake fluid. Do not spill

brake fluid on the vehicle, it may damage the paint; if brake fluid does contact the paint, wash it off immediately with water.

1. Ensure the reservoir connected to the master cylinder is at the MAX (upper) level mark at the start of the bleeding procedure and checked after bleeding each brake system. Add fluid as required.

2. Have someone slowly pump the brake pedal several times, then apply steady pressure.

3. Start the bleeding at the driver's side of the front brake system.

4. Bleed the calipers or the wheel cylinders in the following sequence:
   - Left front
   - Right front
   - Right rear
   - Left rear

5. Attach a length of clear drain tube to the bleed screw, then, loosen the bleed screw to allow air to escape from the system. Then tighten the bleed screw securely.

6. Refill the master cylinder reservoir to the MAX (upper) level line.

7. Repeat the procedure for each brake circuit until no air bubbles are in the fluid.

BRAKES

FRONT DISC BRAKES

✳✳ CAUTION

Dust and dirt accumulating on brake parts during normal use may contain asbestos fibers from production or aftermarket brake linings. Breathing excessive concentrations of asbestos fibers can cause serious bodily harm. Exercise care when servicing brake parts. Do not sand or grind brake lining unless equipment used is designed to contain the dust residue. Do not clean brake parts with compressed air or by dry brushing. Cleaning should be done by dampening the brake components with a fine mist of water, then wiping the brake components clean with a dampened cloth. Dispose of cloth and all residue containing asbestos fibers in an impermeable container with the appropriate label. Follow practices prescribed by the Occupational Safety and Health Administration (OSHA) and the Environmental

Protection Agency (EPA) for the handling, processing, and disposing of dust or debris that may contain asbestos fibers.

BRAKE CALIPER

REMOVAL & INSTALLATION

1. Remove the wheel nuts and front wheel.

2. Remove the brake hose mounting bolt.

3. Remove the brake caliper bracket mounting bolts, and remove the caliper assembly from the knuckle.

4. Detach the brake caliper from the brake hose, if the caliper is being replaced.

5. If the caliper is not being replaced, it can be moved out of the way and suspended with wire, so the hose is not under strain.

✳✳ CAUTION

To prevent damage to the caliper assembly or brake hose, use a short

piece of wire to hang the caliper assembly from the undercarriage. Do not twist the brake hose excessively.

To install:

6. Position the caliper in place.

7. If removed, reattach the brake line to the caliper.

8. Install and tighten the caliper mounting bolts to 80 ft. lbs. (108 Nm).

9. Install the brake hose clip retaining bolt.

10. If the brake hose was removed from the caliper, the system, perform "Brake Bleeding" in this section.

DISC BRAKE PADS

REMOVAL & INSTALLATION

➡Manufacturer does not provide a separate procedure for disc brake pad removal and installation.

✳✳ CAUTION

Dust and dirt accumulating on brake parts during normal use may contain asbestos fibers from production or aftermarket brake linings. Breathing excessive concentrations of asbestos fibers can cause serious bodily harm. Exercise care when servicing brake parts. Do not sand or grind brake lining unless equipment used is designed to contain the dust residue. Do not clean brake parts with compressed air or by dry brushing. Cleaning should be done by dampening the brake components with a fine mist of water, then wiping the brake components clean with a dampened cloth. Dispose of cloth and all residue containing asbestos fibers in an impermeable container with the appropriate label. Follow practices prescribed by the Occupational Safety and Health Administration (OSHA) and the Environmental Protection Agency (EPA) for the handling, processing, and disposing of dust or debris that may contain asbestos fibers.

BRAKE DRUM

REMOVAL & INSTALLATION

See Figure 1.

✳✳ CAUTION

Keep any grease off the brake drum and brake shoes.

1. Raise the rear of the vehicle, and support it with safety stands in the proper locations.
2. Remove the rear wheel.
3. Remove the parking brake, and remove the brake drum from the hub bearing unit. If necessary, turn the adjuster bolt with a flat-tip screwdriver until the shoes become loose.
4. If the brake drum has clung to the hub bearing unit, thread two 8 x 1.25 mm bolts into the brake drum to push it away from the hub bearing unit. Turn each bolt 90 degrees at a time to prevent cocking the brake drum.

To install:

5. Install the brake drum in the reverse order of removal.
  a. After installation, press the brake pedal several times to make sure the

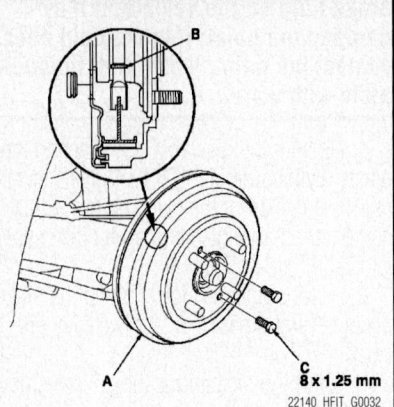

Fig. 1 Remove the parking brake, then the brake drum (A) from the hub bearing unit, turning the adjuster bolt (B) until the shoes become loose. If the brake drum has clung to the hub bearing unit, thread two 8 x 1.25 mm bolts (C) into the brake drum to push it away from the hub bearing unit

brakes work and self adjust the brake shoes.
  b. Before installing the brake drum, clean the mating surfaces of the rear hub and the inside of the brake drum.
  c. Clean the mating surfaces of the brake drum and the inside of the wheel, then install the rear wheel.

BRAKE SHOES

REMOVAL & INSTALLATION

See Figures 2 through 5.

1. Raise the rear of the vehicle, and support it with safety stands in the proper locations.
2. Remove the rear wheels.
3. Release the parking brake, and remove the brake drum.
4. Remove the tension pins by pushing respective retainer spring and turning the pin.
5. Remove the lower return spring, and remove the brake shoe assembly over the hub.
6. Remove the forward brake shoe by removing the upper return spring, and disassemble the brake shoe assembly.
7. Remove the rearward brake shoe by disconnecting the parking brake cable from the parking brake lever.
8. Remove the U-clip, wave washer, and pivot pin, and separate the parking brake lever from the brake shoe.

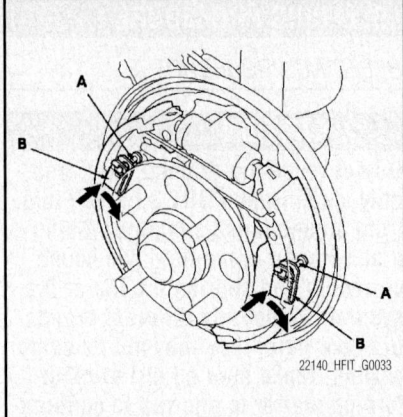

Fig. 2 Removing the tension pins (A) by pushing the retainer spring (B)

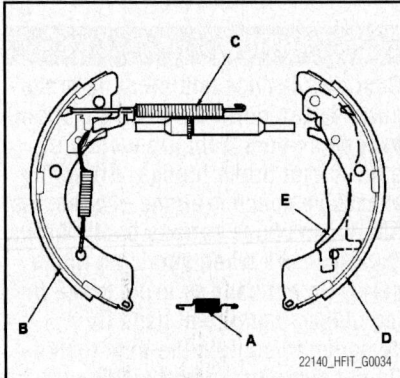

Fig. 3 Showing the lower return spring (A), forward brake shoe (B), upper return spring (C), and rear brake shoe (D), and parking brake lever (E) locations

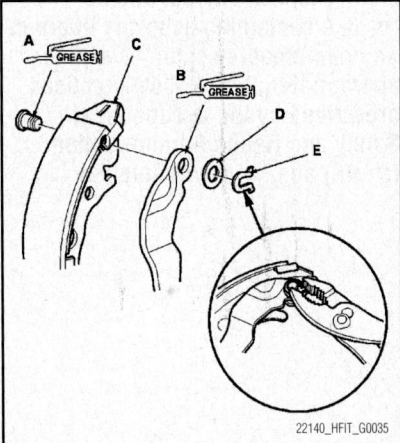

Fig. 4 Apply rubber grease to the sliding surface of the pivot pin (A) and parking brake lever (B) for the rearward brake shoe (C), then install the parking brake lever and the wave washer (D) on the pivot pin, and secure with a new U-clip (E)

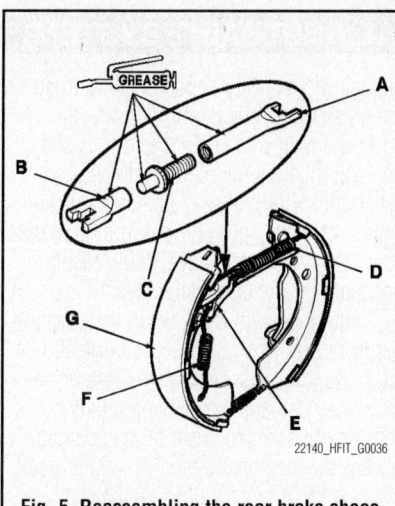

Fig. 5 Reassembling the rear brake shoes

To install:

9. Apply rubber grease to the sliding surface of the pivot pin and parking brake lever for the rearward brake shoe.

10. Install the parking brake lever and the wave washer on the pivot pin, and secure with a new U-clip.

➡**Pinch the U-clip securely to prevent the parking brake lever from coming out of the brake shoe.**

11. Connect the parking brake cable to the parking brake lever.

12. Apply a thin coat of rubber grease to the connecting rod ends and the sliding surfaces. Wipe off any excess. Keep grease off the brake linings.

13. Apply a thin coat of Molykote 44 MA grease to the shoe ends and the edge of the shoe surfaces that contact the backing plate. Wipe off any excess. Keep grease off the brake linings.

14. Install connecting rods A and B on the adjuster bolt (C).

➡**Clean the threaded portions of connecting rod A and the sliding surface of connecting rod B, then coat them with rubber grease.**

15. Shorten connecting rod A by fully turning the adjuster bolt.

16. Assemble the brake shoes, the upper return spring (D), and with the connecting rods the adjuster bolt on the backing plate, then install the self-adjuster lever (E) and

the self-adjuster spring (F) on the forward brake shoe (G).

17. Install the tension pins and the retainer springs by pushing in respective spring and turning each pin.

Install the lower return spring.

➡**Make sure the brake shoe positioning on the brake shoe bosses of the backing plate, and fitting the top of the brake shoes onto the wheel cylinder pistons.**

18. Before installing the brake drum, clean the mating surface of the rear hub and the inside of the brake drum.

19. Install the brake drum.

20. Install the rear wheels.

21. Press the brake pedal several times to make sure the brakes work and to set the self-adjusting brake.

➡**Engagement of the brakes may require a greater pedal stroke immediately after the brake shoes have been replaced as a set. Several applications of the brake pedal will restore the normal pedal stroke.**

22. Adjust the parking brake.

BRAKES

PARKING BRAKE SHOES

REMOVAL & INSTALLATION

See Figure 6.

1. Pull the parking brake lever with about 44 lbs. of force to fully apply the parking brake. The parking brake lever should be locked within 6–8 clicks. If the number of lever clicks is excessive, adjust the parking brake.

2. Remove the center console.

3. Release the parking brake lever fully.

4. Loosen the parking brake cable adjusting nut.

5. Press the brake pedal several times

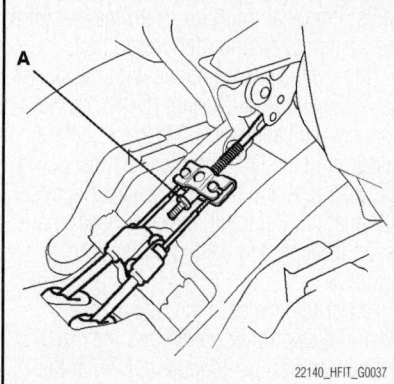

Fig. 6 Showing the parking brake cable adjusting nut (A)

PARKING BRAKE

to set the self-adjusting brake before adjusting the parking brake.

6. Pull the parking brake lever 1 click.

7. Tighten the parking brake cable adjusting nut until the parking brakes drag slightly when the rear wheels are turned.

8. Release the parking brake lever fully, and check that the parking brakes do not drag when the rear wheels are turned. Readjust if necessary.

9. Make sure the parking brakes are fully applied when the parking brake lever is pulled all the way.

10. Install the center console.

CHASSIS ELECTRICAL   AIR BAG (SUPPLEMENTAL RESTRAINT SYSTEM)

GENERAL INFORMATION

✶✶ CAUTION

These vehicles are equipped with an air bag system. The system must be disarmed before performing service on, or around, system components, the steering column, instrument panel components, wiring and sensors. Failure to follow the safety precautions and the disarming procedure could result in accidental air bag deployment, possible injury and unnecessary system repairs.

SERVICE PRECAUTIONS

Disconnect and isolate the battery negative cable before beginning any airbag system component diagnosis, testing, removal, or installation procedures. Allow system capacitor to discharge for two minutes before beginning any component service. This will disable the airbag system. Failure to disable the airbag system may result in accidental airbag deployment, personal injury, or death.

Do not place an intact undeployed airbag face down on a solid surface. The airbag will propel into the air if accidentally deployed and may result in personal injury or death.

When carrying or handling an undeployed airbag, the trim side (face) of the airbag should be pointing towards the body to minimize possibility of injury if accidental deployment occurs. Failure to do this may result in personal injury or death.

Replace airbag system components with OEM replacement parts. Substitute parts may appear interchangeable, but internal differences may result in inferior occupant protection. Failure to do so may result in occupant personal injury or death.

Wear safety glasses, rubber gloves, and long sleeved clothing when cleaning powder residue from vehicle after an airbag deployment. Powder residue emitted from a deployed airbag can cause skin irritation. Flush affected area with cool water if irritation is experienced. If nasal or throat irritation is experienced, exit the vehicle for fresh air until the irritation ceases. If irritation continues, see a physician.

Do not use a replacement airbag that is not in the original packaging. This may result in improper deployment, personal injury, or death.

The factory installed fasteners, screws and bolts used to fasten airbag components have a special coating and are specifically designed for the airbag system. Do not use substitute fasteners. Use only original equipment fasteners listed in the parts catalog when fastener replacement is required.

During, and following, any child restraint anchor service, due to impact event or vehicle repair, carefully inspect all mounting hardware, tether straps, and anchors for proper installation, operation, or damage. If a child restraint anchor is found damaged in any way, the anchor must be replaced. Failure to do this may result in personal injury or death.

Deployed and non-deployed airbags may or may not have live pyrotechnic material within the airbag inflator.

Do not dispose of driver/passenger/curtain airbags or seat belt tensioners unless you are sure of complete deployment. Refer to the Hazardous Substance Control System for proper disposal.

Dispose of deployed airbags and tensioners consistent with state, provincial, local, and federal regulations.

After any airbag component testing or service, do not connect the battery negative cable. Personal injury or death may result if the system test is not performed first.

If the vehicle is equipped with the Occupant Classification System (OCS), do not connect the battery negative cable before performing the OCS Verification Test using the scan tool and the appropriate diagnostic information. Personal injury or death may result if the system test is not performed properly.

Never replace both the Occupant Restraint Controller (ORC) and the Occupant Classification Module (OCM) at the same time. If both require replacement, replace one, then perform the Airbag System test before replacing the other.

Both the ORC and the OCM store Occupant Classification System (OCS) calibration data, which they transfer to one another when one of them is replaced. If both are replaced at the same time, an irreversible fault will be set in both modules and the OCS may malfunction and cause personal injury or death.

If equipped with OCS, the Seat Weight Sensor is a sensitive, calibrated unit and must be handled carefully. Do not drop or handle roughly. If dropped or damaged,

replace with another sensor. Failure to do so may result in occupant injury or death.

If equipped with OCS, the front passenger seat must be handled carefully as well. When removing the seat, be careful when setting on floor not to drop. If dropped, the sensor may be inoperative, could result in occupant injury, or possibly death.

If equipped with OCS, when the passenger front seat is on the floor, no one should sit in the front passenger seat. This uneven force may damage the sensing ability of the seat weight sensors. If sat on and damaged, the sensor may be inoperative, could result in occupant injury, or possibly death.

DISARMING THE SYSTEM

1. Before servicing the vehicle, refer to the Precautions Section.
2. Turn the ignition switch to the **LOCK** position.
3. Disconnect the negative battery cable.
4. Wait three minutes for the battery power to fully discharge from the system.

ARMING THE SYSTEM

1. Before servicing the vehicle, refer to the Precautions Section.
2. Connect the negative battery cable.
3. Turn the ignition switch **ON**.
4. Verify that the air bag indicator illuminates for 4–8 seconds, then goes off.

CABLE REEL CENTERING

See Figure 7.

Center the cable reel by first rotating the cable reel clockwise until it stops. Then rotate it counterclockwise (about three turns) until the arrow mark on the cable reel label points straight up.

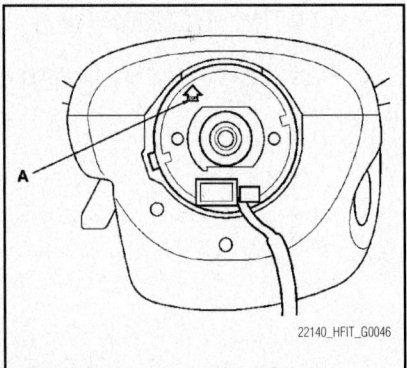

22140_HFIT_G0046

Fig. 7 Showing the cable reel and arrow mark

DRIVETRAIN

AUTOMATIC TRANSAXLE ASSEMBLY

REMOVAL & INSTALLATION

See Figures 8 through 14.

1. Make sure you have anti-theft code for the audio system.

2. Remove the air cleaner assembly.

3. Disconnect the negative cable from the battery, then disconnect the positive cable.

4. Remove the battery hold-down bracket, and remove the battery and the battery tray.

5. Remove the harness clamps from the battery base, and then remove the battery base.

6. Raise the vehicle on a lift, and make sure it is securely supported. Remove the front wheels.

7. Remove the splash shield.

8. Remove the drain plug, and drain the automatic transaxle fluid (ATF).

9. Reinstall the drain plug with a new sealing washer.

10. Disconnect the A/T clutch pressure control solenoid valve A connector, the solenoid valve B connector, and the solenoid valve C connector.

11. Remove the harness clamp (D).

12. Remove the mounting bolts securing the harness cover (E), and remove the clamp (F).

13. Disconnect the transaxle range switch harness connector (A), the 2nd clutch transaxle fluid pressure switch connector (B), the input shaft (mainshaft) speed sensor connector (C), and the output shaft (countershaft) speed sensor connector (D).

14. Remove the A/F sensor connector (E) from the clamp bracket.

15. Remove the harness clamp (F).

16. Disconnect the shift solenoid harness connector and the 3rd clutch transaxle fluid pressure switch connector.

17. Remove the harness clamps.

18. Disconnect the ATF cooler hoses. Turn the end of the ATF cooler hoses up to prevent ATF from flowing out, then plug the hoses and the lines.

19. Remove the air cleaner bracket.

20. Install a universal lifting eyelet, with a 10 mm (0.4 in.) spacer, to the engine.

21. Install a suitable engine support and lifting device, and remove the slack to support the engine.

22. Remove the front transaxle housing mounting bolts.

23. Remove the spindle nut, and tap the halfshaft inward with a plastic hammer to allow the removal tool to fit on the lower arm ball joint.

24. Insert a 5 mm Allen wrench in the end of the ball joint pin, and remove the nuts, then separate the front stabilizer link from the stabilizer.

25. Remove the cotter pins and the ball joint nuts from the tie-rod ends, then separate the steering tie-rod ends from the knuckles.

26. Remove the spring clips and the castle nuts, then separate the lower arms from the knuckles.

27. Remove the shift cable cover, then remove the shift cable holder.

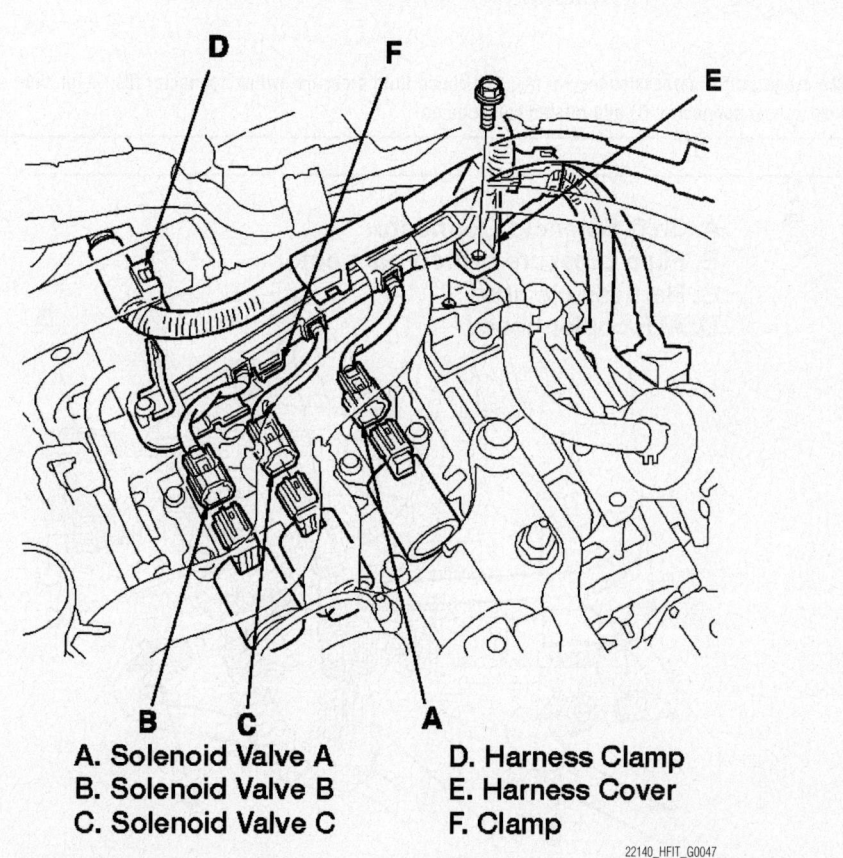

A. Solenoid Valve A
B. Solenoid Valve B
C. Solenoid Valve C
D. Harness Clamp
E. Harness Cover
F. Clamp

22140_HFIT_G0047

Fig. 8 Disconnect these electrical connectors and related components

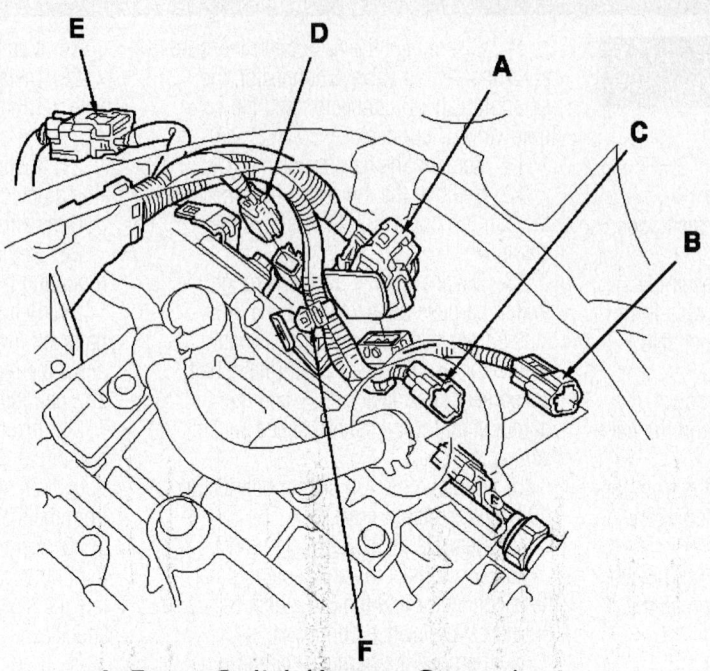

A. Range Switch Harness Connector
B. Trans. Fluid Pressure Switch Connector
C. Input Shaft Speed Sensor Connector
D. Output Shaft Speed Sensor Connector
E. A/F Sensor Connector
F. Clamp

22140_HFIT_G0048

Fig. 9 Disconnect the range switch harness connector (A), 2nd clutch fluid pressure switch connector (B), input shaft speed sensor connector (C), and output shaft speed sensor connector (D) and related components

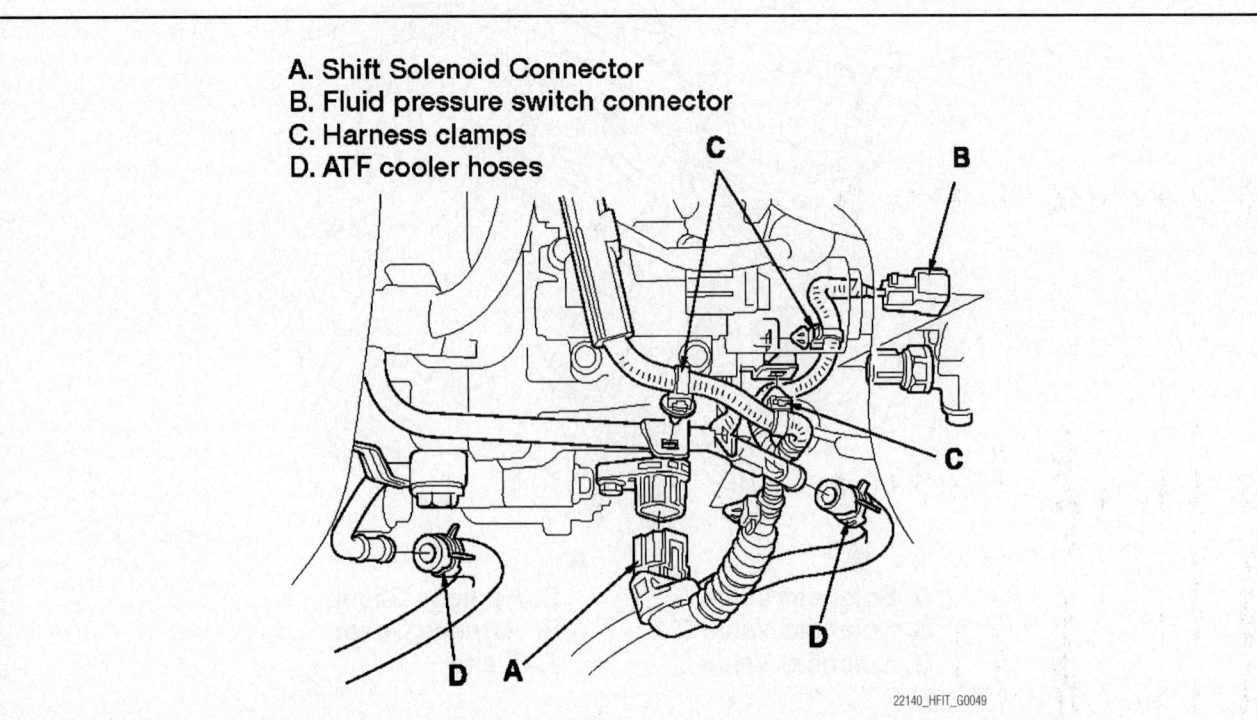

A. Shift Solenoid Connector
B. Fluid pressure switch connector
C. Harness clamps
D. ATF cooler hoses

22140_HFIT_G0049

Fig. 10 Showing the shift solenoid connector (A), fluid pressure switch connector (B), harness clamps (C), ATF cooler hoses (D)

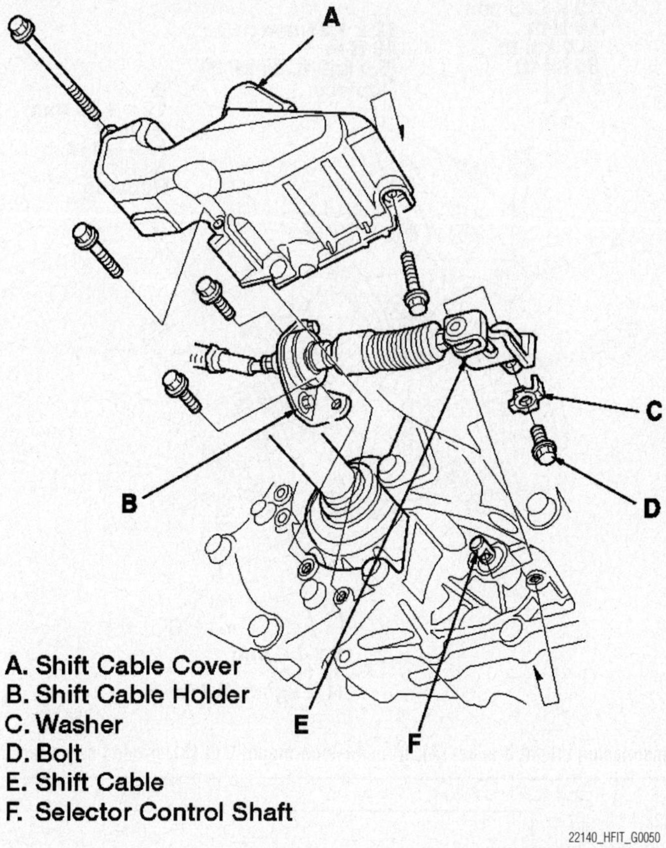

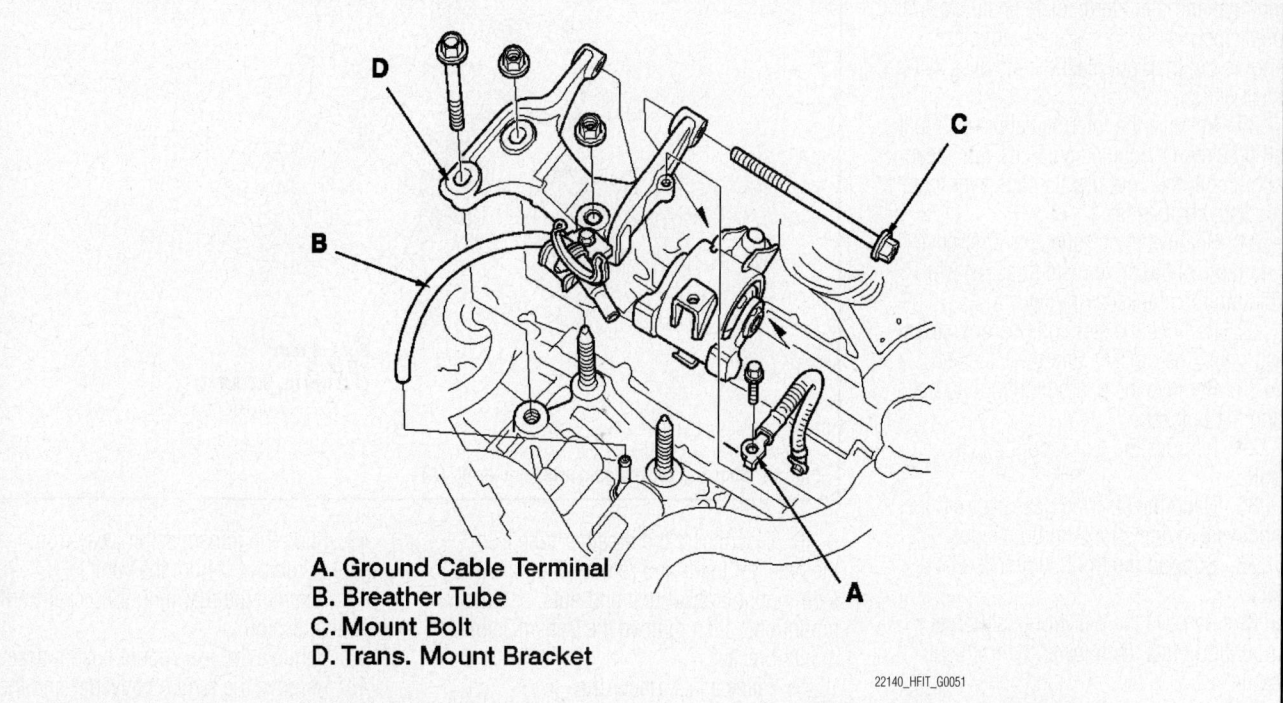

A. Shift Cable Cover
B. Shift Cable Holder
C. Washer
D. Bolt
E. Shift Cable
F. Selector Control Shaft

22140_HFIT_G0050

Fig. 11 Remove the shift cable cover (A), shift cable holder (B), lock tab of the lock washer (C), and lock bolt (D) and lock washer, then separate the shift cable (E) from the selector control shaft (F)

A. Ground Cable Terminal
B. Breather Tube
C. Mount Bolt
D. Trans. Mount Bracket

22140_HFIT_G0051

Fig. 12 Disconnect the breather tube (B) from the breather joint, and remove the transmission mount bracket bolt and nuts, and the mount bolt (C), then remove the transmission mount bracket (D).

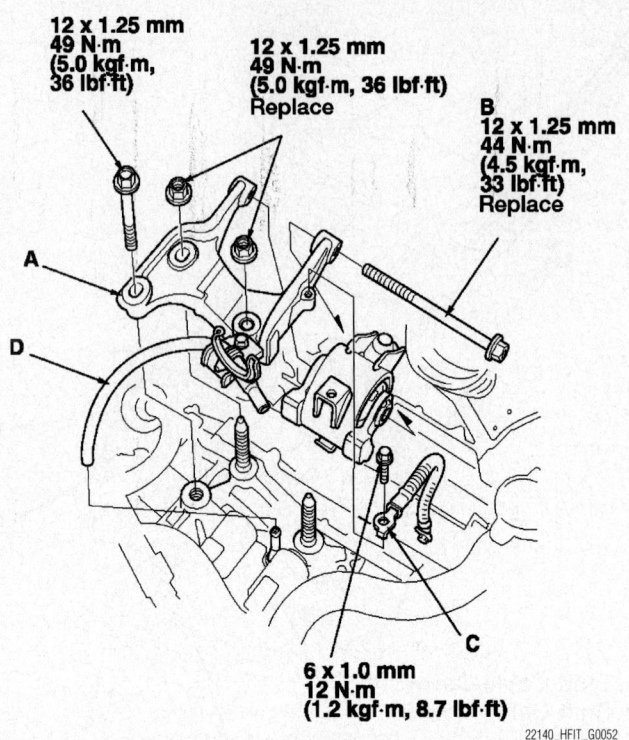

12 x 1.25 mm
49 N·m
(5.0 kgf·m,
36 lbf·ft)

12 x 1.25 mm
49 N·m
(5.0 kgf·m, 36 lbf·ft)
Replace

B
12 x 1.25 mm
44 N·m
(4.5 kgf·m,
33 lbf·ft)
Replace

A

D

C

6 x 1.0 mm
12 N·m
(1.2 kgf·m, 8.7 lbf·ft)

22140_HFIT_G0052

Fig. 13 Showing installation for transmission mount bracket (A), transmission mount bolt (B), ground cable terminal (C) and the breather tube (D).

28. Pry up the lock tab of the lock washer, and remove the lock bolt and the lock washer, then separate the shift cable from the selector control shaft. Do not bend the shift cable excessively.

29. Remove the halfshafts from the differential and the intermediate shaft. Coat all precision machined surfaces with clean engine oil, then put plastic bags over halfshaft ends.

30. Remove the intermediate shaft. Coat all precision machined surfaces with clean engine oil, then put plastic bags over intermediate shaft ends.

31. Remove the torque converter cover, and remove the drive plate bolts (8) while rotating the crankshaft pulley.

32. Remove the steering gearbox mounting bolts, the bracket, and the stiffener.

33. Remove the gearbox mounting bolts and the bracket.

34. Remove the rear mount bracket bolts.

35. Hang the steering gearbox from the body with nylon straps on both sides.

36. Support the front subframe with a jack.

37. Remove the four front subframe mounting bolts, then lower the front subframe.

38. Remove the headlight harness cover.

39. Remove the transmission ground cable terminal (A).

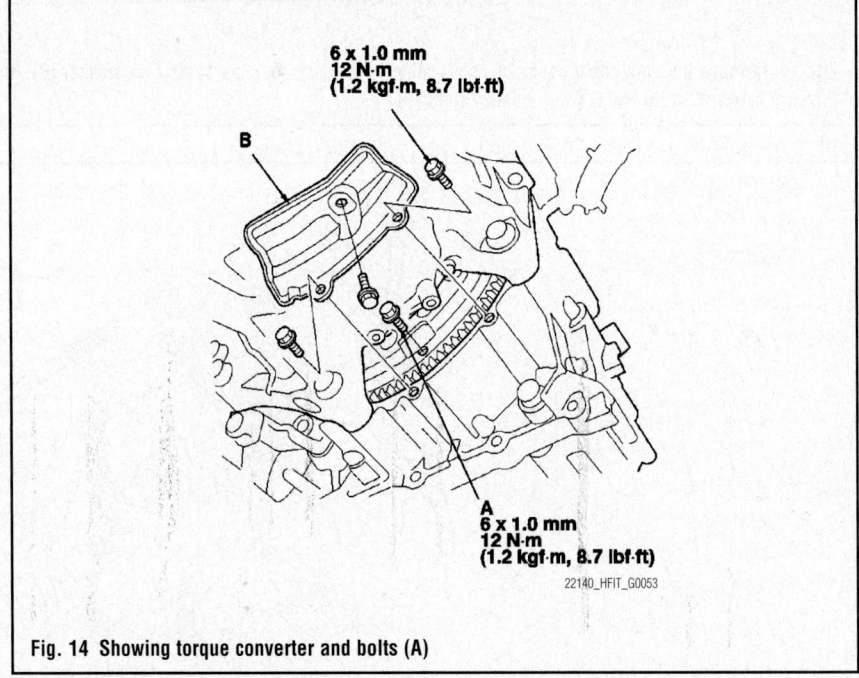

6 x 1.0 mm
12 N·m
(1.2 kgf·m, 8.7 lbf·ft)

B

A
6 x 1.0 mm
12 N·m
(1.2 kgf·m, 8.7 lbf·ft)

22140_HFIT_G0053

Fig. 14 Showing torque converter and bolts (A)

40. Disconnect the breather tube from the breather joint, and remove the transmission mount bracket bolt and nuts, and the mount bolt, then remove the transmission mount bracket.

41. Place a jack under the transmission.

42. Remove the rear and lower transmission housing mounting bolts.

43. Slide the transmission away from the engine to remove it from the vehicle.

44. Remove the rear mount/bracket from the transmission.

45. Remove the rear mount base bracket.

46. Remove the torque converter and the dowel pins.

47. Inspect the drive plate, and replace it if it is damaged.

To install:

48. Install the torque converter on the mainshaft with a new O-ring.

49. Install the 14 x 20 mm dowel pins in the torque converter housing.

50. Install the rear mount base bracket. Tighten the bracket bolts to 47 ft. lbs. (64 Nm).

51. Install the rear mount/bracket. Tighten the bracket bolts to 43 ft. lbs. (59 Nm).

52. Place the transmission on the jack, and raise the transmission to the engine level.

53. Install the rear and lower transmission housing mounting bolts. Tighten the bolts to 47 ft. lbs. (64 Nm).

54. Remove the jack from under the transmission.

55. Install the transmission mount bracket, and install the transmission mount bolt loosely.

56. Install and tighten the bracket bolt and nuts to 36 ft. lbs. (49 Nm), then tighten the mount bolt to 33 ft. lbs. (44 Nm).

57. Install the transmission ground cable terminal and the breather tube.

58. Install the headlight harness cover.

59. Support the front subframe with a jack, and lift it up to the body.

60. Install the front subframe on the body, and install the steering gearbox on the subframe, then tighten the new mounting bolts to 69 ft. lbs. (93 Nm).

61. Install the rear mount bracket bolts to 43 ft. lbs. (59 Nm).

62. Install the steering gearbox mounting bolts with the bracket and the stiffener. Tighten the two longer bolts to 38 ft. lbs. (52 Nm) and the two shorter bolts to 36 ft. lbs. (49 Nm).

63. Install the steering gearbox mounting bolts with its bracket. Tighten the bolts to 33 ft. lbs. (44 Nm).

64. Attach the torque converter to the drive plate with eight bolts. Rotate the crankshaft pulley as necessary to tighten the bolts to 1/2 of the specified torque, then to the final torque, in a crisscross pattern.

65. After tightening the last bolt, check that the crankshaft rotates freely. Then install the torque converter cover (B).

66. Install the new set ring on the intermediate shaft.

67. Clean the areas where the intermediate shaft contact the transmission (differential) with the solvent, and dry with the compressed air. Apply ATF to the intermediate shaft splines, then install the intermediate shaft. Be sure not to allow dust or other foreign particles to enter the transmission.

68. Apply grease to the contact area between the halfshaft and the wheel bearing.

69. Install new set ring on the halfshaft.

70. Clean the areas where the halfshaft contacts the transmission (differential) with the solvent, and dry with compressed air. Insert the halfshaft into the differential and intermediate shaft until the set ring locks in the groove.

71. Connect the lower arm ball joints to the knuckles, and install the castle nuts. Tighten to 36–43 ft. lbs. (49–59 Nm), then secure them with the spring clips.

72. Connect the tie-rod end ball joints to the knuckles, and install the ball joint nuts. Tighten the nuts to 32 ft. lbs. (43 Nm), then secure them with new cotter pins.

73. Install the stabilizer link to the front stabilizer. Insert a 5 mm Allen wrench in the ends of the ball joint pins, and tighten the nuts to 22 ft. lbs. (29 Nm).

74. Install the halfshaft through the hub, and install the new spindle nuts on the halfshafts. Stake the spindle nuts into the halfshafts.

75. Install the control lever over the selector control shaft. Secure the control lever with the new lock washer and the lock bolt, then bend the lock tab of the lock washer against the bolt head.

76. Install the shift cable holder, then install the shift cable cover.

77. Install the splash shield.

78. Install the front wheels.

79. Install the front transmission housing mounting bolts. Tighten the bolts to 47 ft. lbs. (64 Nm).

80. Remove the engine support hanger and the universal eyelet.

81. Install the air cleaner bracket.

82. Install the ATF cooler hoses.

83. Connect the shift solenoid harness connector and the 3rd clutch transmission fluid pressure switch connector.

84. Install the harness clamps

85. Connect the transmission range switch harness connector, the 2nd clutch transmission fluid pressure switch connector, the input shaft (mainshaft) speed sensor connector, and the output shaft (countershaft) speed sensor connector.

86. Install the A/F sensor connector to the clamp bracket.

87. Install the harness clamp.

88. Connect the A/T clutch pressure control solenoid valve A connector, the solenoid valve B connector, and the solenoid valve C connector.

89. Install the clamp, then install the mounting bolts securing the harness cover.

90. Install the harness clamp.

91. Install the battery base, then install the harness clamps to the battery base.

92. Install the battery tray and the battery, then secure the battery with its holddown bracket.

93. Refill the transmission with ATF.

94. Connect the battery terminals, and apply grease around the battery terminals to prevent corrosion.

95. Install the air cleaner assembly.

96. Set the parking brake. Start the engine, and shift the transmission through all positions three times.

97. Check the shift lever operation, the A/T gear position indicator operation, and the shift cable adjustment.

98. Check and adjust the front wheel alignment.

99. Start the engine in P or N, and warm it up to normal operating temperature (the radiator fan comes on).

100. Turn off the engine, and check the ATF level.

101. Enter the anti-theft code for the audio system. Set the clock.

102. Road-test the vehicle.

MANUAL TRANSAXLE ASSEMBLY

REMOVAL & INSTALLATION

See Figures 15 through 17.

1. Make sure you have the anti-theft codes for the audio system.

2. Disconnect the negative cable from the battery, then disconnect the positive cable. Remove the battery.

3. Remove the air cleaner assembly.

4. Remove the harness clamps and the battery base.

5. Carefully remove the slave cylinder to avoid bending the clutch line. Do not press the clutch pedal once the slave cylinder, boot and clamp have been removed. Remove the breather tube from the clamp.

6. Disconnect the back-up light switch 2-pin connector, then remove the harness clamps.

7. Remove the shift cable bracket, then disconnect the shift cables from the top of the transmission housing. Carefully remove both cables and the bracket together to avoid bending the cables.

8. Disconnect the vehicle speed sensor (VSS) connector.

9. Attach the universal eyelet to the air cleaner stay.

10. Remove the one upper transmission mounting bolt and lower starter motor mounting bolt.

11. Install a proper engine hanger assembly. Lift and support the engine and transaxle assembly.

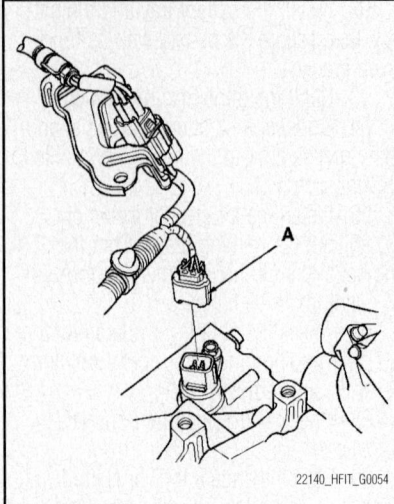

Fig. 15 Showing the Vehicle Speed Sensor (A)

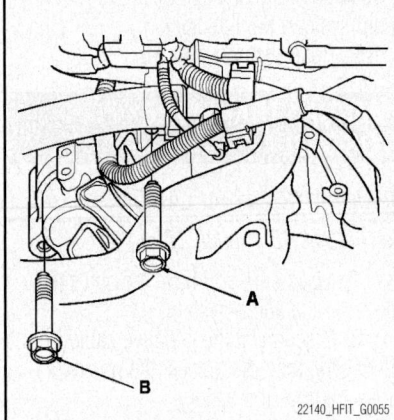

Fig. 16 Remove the one upper transmission mounting bolt (A) and lower starter motor mounting bolt (B)

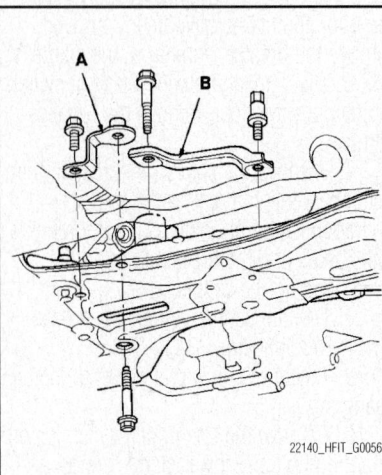

Fig. 17 Remove the steering gearbox mounting bolts, bracket (A), and stiffener (B)

12. Raise the vehicle on a lift, and make sure it is securely supported.

13. Drain the transmission fluid Install the drain plug with a new washer.

14. Remove the splash shield.

15. Remove the halfshafts. See "Driveshafts" in this article.

16. Remove the clutch cover.

17. Remove the steering gearbox mounting bolts, bracket, and stiffener.

18. Remove the bolts and the power steering gearbox bracket.

19. Attach a nylon strap or rope to the body and around the steering gearbox to support it on both sides.

CLUTCH

REMOVAL & INSTALLATION

See Figures 18 and 19.

1. Remove the manual transaxle.

2. Check the height of the diaphragm spring fingers using the dial indicator. If the height is more than the service limit of 0.04 in. (1.0 mm), replace the pressure plate.

3. Install the clutch alignment shaft, handle, and the ring gear holder.

4. To prevent warping, unscrew the pressure plate mounting bolts in a crisscross pattern in several steps, then remove the pressure plate.

5. Inspect the fingers of the diaphragm spring for wear at the release bearing contact area.

6. Inspect the pressure plate surface for wear, cracks, and burning.

7. Use a straight edge and feeler gauge to inspect the pressure plate for warpage. Measure across the pressure plate. If the

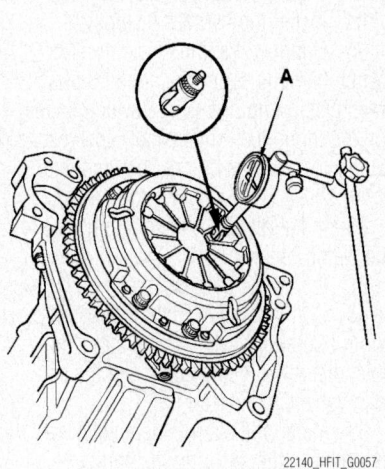

Fig. 18 Check the height of the diaphragm spring fingers using the dial indicator (A)

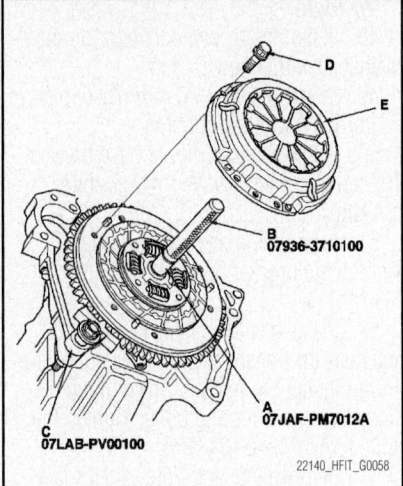

Fig. 19 Showing the clutch alignment shaft (A), handle (B), ring gear holder (C), pressure plate mounting bolts (D), and pressure plate (E)

warpage is more than 0.006 in. (0.15 mm) service limit, replace the pressure plate.

8. Remove the clutch disc, the clutch alignment shaft, and handle.

9. Inspect the lining of the clutch disc for signs of slipping or oil. If the clutch disc looks burnt or is soaked with oil, replace it. Find and repair the source of the oil leak if the clutch disc is soaked.

10. Measure the clutch disc thickness. If the thickness is less than 0.20 in. (5.0 mm) service limit, replace the clutch disc.

11. Measure the rivet depth from the clutch disc lining surface to the rivets on both sides. If the rivet depth is less than 0.08 in. (0.2 mm) service limit, replace the clutch disc.

➡️**At this point, the flywheel may also be removed and inspected. See "Flywheel" in this section.**

To install:

12. Temporarily install the clutch disc onto the splines of the transmission mainshaft. Make sure the clutch disc slides freely on the mainshaft.

13. Apply a light coat of super high temp urea grease (P/N 08798-9002) to the crankshaft pilot bearing.

14. Apply super-high temp urea grease to the splines of the clutch disc, then install the clutch disc using the clutch alignment shaft, and handle.

15. Install the pressure plate and the mounting bolts finger-tight.

16. Torque the mounting bolts in a crisscross pattern. Tighten the bolts, to 19 ft. lbs. (25 Nm), in several steps to prevent warping the diaphragm spring.

17. Remove the clutch alignment tool set and ring gear holder.
18. Make sure the diaphragm spring fingers are all the same height.
19. Install the transaxle.

FRONT HALFSHAFT

REMOVAL & INSTALLATION

With Automatic Transaxle

See Figure 20.

1. Raise the vehicle on a lift.
2. Remove the front wheels.
3. Lift up the locking tab on the spindle nut, then remove the nut.
4. Remove the splash shield.
5. Drain the transmission fluid Reinstall the drain plug using a new washer.
6. Separate the front stabilizer link.
7. Remove the lock pin from the lower arm ball joint castle nut and remove the nut, then separate the ball joint from the lower arm using the ball joint thread protector and remover.

❄ CAUTION

Be careful not to damage the ball joint boot when installing the remover. Do not force or hammer on the lower arm, or pry between the lower arm and the knuckle. You could damage the ball joint.

8. Remove the halfshaft outboard joint from the front wheel hub using a plastic hammer.
9. Pull the knuckle outward, and separate the halfshaft outboard joint from front wheel hub.
10. On the left halfshaft, pry the inboard joint from the transmission housing with a prybar. Remove the halfshaft as an assembly.

➡**Do not pull by the halfshaft or the inboard joint may come apart. Pull the halfshaft straight out to avoid damaging the oil seal.**

11. On the right halfshaft, drive the inboard joint off of the intermediate shaft using a drift and a hammer. Remove the halfshaft as an assembly.

➡**Do not pull by the halfshaft, or the inboard joint may come apart. Pull the halfshaft straight out to avoid damaging the oil seal.**

12. Inspect the halfshaft assembly. See "Inspection" below.

➡**Before starting installation, make sure the mating surfaces of the joint and the splined section are not dusty or dirty.**

13. Apply a small amount of Moly 60 paste (P/N 08734-0001) to the contact area of the outboard joint and the front wheel bearing.

➡**The paste helps to prevent noise and vibration.**

14. Install a new set ring onto the set ring groove of the halfshaft.
15. Apply super high-temp urea grease (P/N 08798-9002) to the whole splined surface of the right halfshaft. After applying grease, remove the grease from the splined grooves at intervals of 2-3 splines and from the set ring groove so that air can bleed from the intermediate shaft.
16. Clean the areas where the halfshaft contacts the differential thoroughly with solvent or brake cleaner, and dry with compressed air.

➡**Do not wash the rubber parts with solvent.**

17. Insert the inboard end of the halfshaft into the differential or the intermediate shaft until the set ring locks in the groove.

➡**Insert the halfshaft horizontally to prevent damaging the differential oil seal.**

18. Install the outboard joint into the front wheel hub.
19. Wipe off any grease contamination from the ball joint tapered section and threads, then install the knuckle onto the

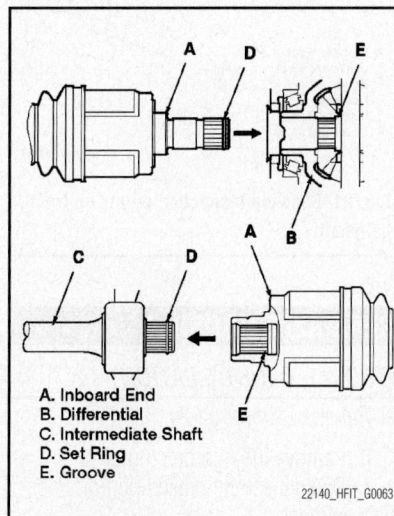

A. Inboard End
B. Differential
C. Intermediate Shaft
D. Set Ring
E. Groove

22140_HFIT_G0063

Fig. 20 Insert the inboard end (A) of the halfshaft into the differential (B) or the intermediate shaft (C) until the set ring (D) locks in the groove (E)

lower arm. Be careful not to damage the ball joint boot. Wipe off the grease before tightening the nut at the ball joint. Torque the new castle nut to the 36 ft. lbs. (49 Nm), then tighten it only far enough to align the slot with the ball joint pin hole.

➡**Make sure the ball joint boot is not damaged or cracked. Do not align the nut by loosening it.**

20. Install the lock pin into the ball joint pin hole.
21. Connect the front stabilizer link.
22. Install a new spindle nut, then tighten the nut to 134 ft. lbs. (181 Nm). After tightening, use a drift to stake the spindle nut shoulder against the halfshaft.
23. Clean the mating surfaces of the brake disc and the front wheel, then install the front wheel.
24. Turn the front wheel by hand, and make sure there is no inference between the halfshaft and surrounding parts.
25. Refill the transmission with the recommended transmission fluid.
26. Install the splash shield.
27. Lower the vehicle on the lift.
28. Check the wheel alignment, and adjust it if necessary.
29. Test-drive the vehicle.

With Manual Transaxle

1. Raise the vehicle on a lift.
2. Remove the front wheels.
3. Lift up the locking tab on the spindle nut, then remove the nut.
4. Remove the splash shield.
5. Drain the transmission fluid Reinstall the drain plug using a new washer.
6. Separate the front stabilizer link.
7. Remove the lock pin from the lower arm ball joint castle nut and remove the nut, then separate the ball joint from the lower arm using the ball joint thread protector and remover.

❄ CAUTION

Be careful not to damage the ball joint boot when installing the remover. Do not force or hammer on the lower arm, or pry between the lower arm and the knuckle. You could damage the ball joint.

8. Remove the halfshaft heat shield bolts and the heat shield.
9. Remove the halfshaft outboard joint from the front wheel hub using a plastic hammer.
10. Pull the knuckle outward, and separate the halfshaft outboard joint from the front wheel hub.

11. Pry the inboard joint from the transmission housing with a prybar. Remove the halfshaft as an assembly.

To install:

➡ **Before starting installation, make sure the mating surfaces of the joint and the splined section are clean.**

12. Apply a small amount of Moly 60 paste (P/N 08734-0001) to the contact area of the outboard joint and the front wheel bearing (the paste helps to prevent noise and vibration).

13. Install a new set ring onto the set ring groove of the halfshaft.

14. Clean the areas where the halfshaft contacts the differential thoroughly with solvent or brake cleaner, and dry with compressed air.

❋❋ CAUTION

Do not wash the rubber parts with solvent.

15. Insert the inboard end of the halfshaft into the differential until the new set ring locks in the groove.

❋❋ CAUTION

Insert the halfshaft horizontally to prevent damaging the differential oil seal.

16. Install the outboard joint into the front wheel hub.

17. Wipe off any grease contamination from the ball joint tapered section and threads, then install the knuckle onto the lower arm. Be careful not to damage the ball joint boot.

18. Wipe off the grease before tightening the nut at the ball joint. Torque the new castle nut to 36 ft. lbs. (49 Nm), then tighten it only far enough to align the slot with the ball joint pin hole.

➡ **Make sure the ball joint boot is not damaged or cracked. Do not align the nut by loosening it.**

19. Install the lock pin into the pin hole.

20. Install the heat shield. Tighten the 2 upper bolts and then the lower bolt to 7.2 ft. lbs. (9.8 Nm).

21. Connect the front stabilizer link.

22. Install a new spindle nut, then tighten the nut to 134 ft. lbs. (181 Nm). After tightening, use a drift to stake the spindle nut shoulder against the halfshaft.

23. Clean the mating surfaces of the brake disc and the front wheel, then install the front wheel.

24. Turn the front wheel by hand, and make sure there is no inference between the halfshaft and surrounding parts.

25. Refill the transmission with the recommended transmission fluid.

26. Install the splash shield.

27. Lower the vehicle on the lift.

28. Check the wheel alignment, and adjust it if necessary.

29. Test-drive the vehicle.

INSPECTION

See Figure 21.

1. Check the inboard boot (A) and the outboard boot (B) for cracks, damage, leaking grease, and loose boot bands (C). If any damage is found, replace the boot and boot bands.

2. Check the halfshaft (D) for cracks and damage. If any damage is found, replace the halfshaft.

3. Check the inboard joint (E) and the outboard joint (F) for cracks and damage. If any damage is found, replace the inboard joint or the outboard joint as an assembly.

4. Hold the inboard joint and turn the front wheel by hand, then make sure the joint is not excessively loose. If necessary, replace the inboard joint or the outboard joint as an assembly.

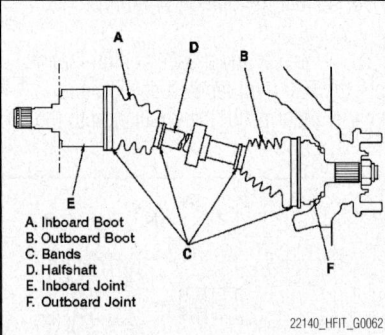

A. Inboard Boot
B. Outboard Boot
C. Bands
D. Halfshaft
E. Inboard Joint
F. Outboard Joint

22140_HFIT_G0062

Fig. 21 Showing inspection points on front halfshaft

REAR AXLE HOUSING

REMOVAL & INSTALLATION

See Figures 22 through 24.

1. Remove the hub bearing unit.

2. Disconnect the brake line from the wheel cylinder.

3. Remove the parking brake cable clamp from the axle housing, then remove the rear brake assembly from the spindle.

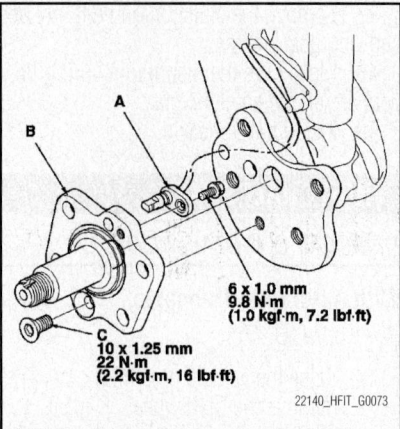

6 x 1.0 mm
9.8 N·m
(1.0 kgf·m, 7.2 lbf·ft)

10 x 1.25 mm
22 N·m
(2.2 kgf·m, 16 lbf·ft)

22140_HFIT_G0073

Fig. 22 Remove the wheel sensor (A), spindle (B) and flat-head screw (C) from the axle beam

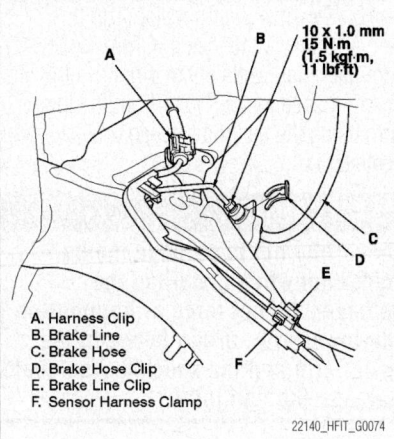

10 x 1.0 mm
15 N·m
(1.5 kgf·m, 11 lbf·ft)

A. Harness Clip
B. Brake Line
C. Brake Hose
D. Brake Hose Clip
E. Brake Line Clip
F. Sensor Harness Clamp

22140_HFIT_G0074

Fig. 23 Remove wheel sensor harness clip from the rear axle beam, brake line from the brake hose, and brake hose on the rear axle beam by removing the brake hose clip

❋❋ CAUTION

The parking brake cable must not be bent or distorted. This will lead to stiff operation and premature cable failure.

4. Remove the wheel sensor, and the spindle from the axle beam.

➡ **During installation, install a new flat-head screw.**

5. Remove the wheel sensor harness clip from the rear axle beam. Do not disconnect the wheel sensor connector.

6. Disconnect the brake line from the brake hose, and remove the brake hose on the rear axle beam by removing the brake hose clip.

➡ **Do not spill brake fluid on the vehicle; it may damage the paint;**

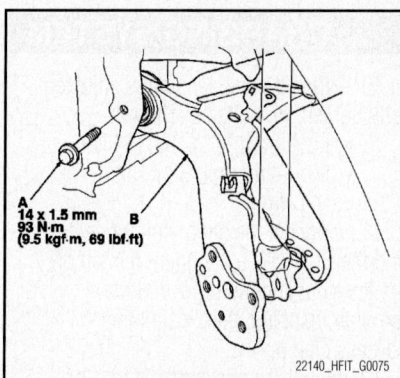

Fig. 24 Remove the axle beam mounting bolts (A) on the both sides, then lower the jack slowly, then remove the axle beam (B)

if brake fluid gets on the paint, wash it off immediately with water.

7. Remove the brake line clip, and wheel sensor harness clamp from the axle beam.

8. Remove the exhaust pipe and muffler.

9. Remove the fuel fill pipe as follows:
   a. Relieve the fuel pressure.
   b. Drain the fuel tank.
   c. Remove the fuel fill cap.
   d. Remove the fuel fill pipe cover.
   e. Remove the tank mount bracket.
   f. Disconnect the fuel fill hose and fuel suction tube from the pipes.
   g. Remove the fuel fill pipe.

10. Place a floor jack under the center of the axle beam, and support it by raising the floor jack.

11. Remove the axle beam mounting bolts on the both sides.

12. Lower the jack slowly, then remove the axle beam.

To install:

13. Install the axle beam in the reverse order of removal, and note these items:
   a. First install all the rear suspension components and lightly tighten bolts, then place a jack under the axle beam, and raise the suspension to load the vehicle weight before fully tightening bolts to the specified torque values.
   b. Before installing the brake drum, clean the mating surface of the hub bearing unit and the inside of the brake drum.
   c. Before installing the wheel, clean the mating surface of the brake drum and the inside of the wheel.
   d. After installing, fill the reservoir with new brake fluid, and bleed the brake system.
   e. Check the brake hose and line joint for leaks, and tighten if necessary.
   f. Check the brake hoses for interference and twisting.
   g. Check the wheel alignment, and adjust it if necessary.

➡**During installation, install a new axle beam mounting bolts, new flange bolts and a new brake hose clip.**

ENGINE COOLING

THERMOSTAT

REMOVAL & INSTALLATION

1. Drain the engine coolant.
2. Remove the air cleaner assembly.
3. Disconnect the radiator hoses.
4. Remove the harness holder, the ground cable, harness clamps, the water bypass hose and the engine junction connector clamp from near the thermostat housing.
5. Remove the thermostat housing, then remove the thermostat.

To install:

6. Position the new thermostat and gasket into place.
7. Remove old liquid gasket from the thermostat housing mating surfaces, bolts and bolt holes.
8. Apply a 1.5 mm wide bead of liquid gasket along the edges of the "rectangular" coolant bypass openings on the housing. Install the component within 5 minutes of applying the liquid gasket. If too much time has passed after applying the liquid gasket, remove the old liquid gasket and residue, then reapply new liquid gasket.
9. Install and tighten the thermostat housing bolts to 8.8 ft. lbs. (12 Nm).

WATER PUMP

REMOVAL & INSTALLATION

See Figure 25.

1. Drain the engine coolant.
2. Remove the drive belt by loosening the lock bolt, the mounting bolt, and the adjusting bolt, then remove the drive belt.
3. Remove the water pump pulley and idler pulley.
4. Remove the five bolts securing the water pump, then remove the water pump.
5. Inspect and clean the O-ring groove and mating surface with the engine block.
6. Install the water pump, with a new O-ring, in the reverse order of removal. Tighten the pump bolts to 8.8 ft. lbs. (12 Nm).
7. Clean up any spilled engine coolant.

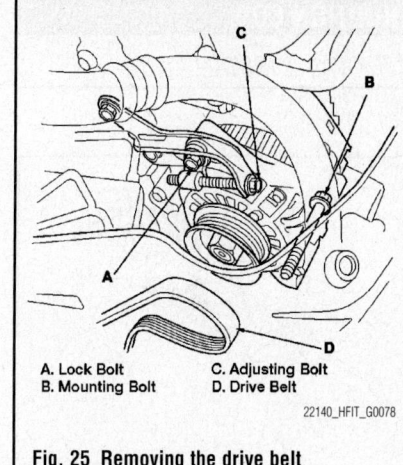

A. Lock Bolt   C. Adjusting Bolt
B. Mounting Bolt   D. Drive Belt

Fig. 25 Removing the drive belt

8. Install the water pump pulley. Tighten the bolts to 10 ft. lbs. (13 Nm)
9. Install the idler pulley. Tighten the bolt to 17 ft. lbs. (24 Nm).
10. Install the drive belt and adjust it.
11. Refill the radiator with engine coolant, and bleed air from the cooling system with the heater valve open.

ENGINE ELECTRICAL

ALTERNATOR

REMOVAL & INSTALLATION

See Figure 25.

1. Make sure you have the anti-theft code for the audio system, and that the ignition switch is in LOCK (0).
2. Disconnect the negative cable from the battery.
3. Remove the drive belt by loosening the lock bolt, the mounting bolt, and the adjusting bolt, then remove the drive belt.
4. Remove the intake manifold. See "Intake Manifold" section.
5. Disconnect the alternator connector, the BLK wire and the harness clamp from the alternator.
6. Remove the lock bolt and the mounting bolt, then remove the alternator.

To install:

7. Install the alternator, then loosely install the lock bolt and the mounting bolt.
8. Install the drive belt.
9. Adjust the drive belt tension as follows:

   a. Loosen the lock bolt and the mounting bolt.

   b. Turn the adjusting bolt to obtain the proper belt tension, then retighten the lock bolt and the mounting bolt.

   c. Recheck the belt tension. It should be 99–132 lbs. of pressure.

CHARGING SYSTEM

10. When the belt tension is adjusted, fully tighten the bolts as follows:
    - Lock bolt: 17 ft. lbs. (24 Nm)
    - Mounting bolt: 33 ft. lbs. (44 Nm)
11. Connect the alternator connector and the BLK wire by aligning the tab on the terminal and the groove of the terminal insulator, then install the harness clamp.
12. Install the intake manifold. See "Intake Manifold" section.
13. Connect the negative cable to the battery.
14. Enter the anti-theft code for the audio system.
15. Reset the clock.

ENGINE ELECTRICAL

FIRING ORDER

Firing order is: 1–3–4–2

IGNITION COIL

REMOVAL & INSTALLATION

See Figure 26.

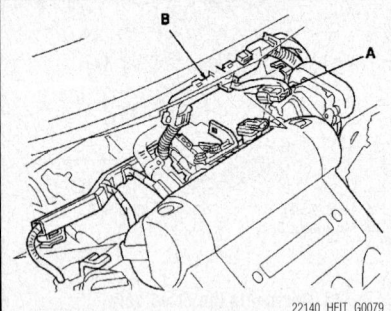

Fig. 26 Remove the MAP sensor connector (A), then remove the harness holder (B) from the brackets

22140_HFIT_G0079

1. Remove the MAP sensor connector, then remove the harness holder from the brackets.
2. Disconnect the ignition coil connector, then remove the ignition coils.
3. Install the ignition coils in the reverse order of removal.

IGNITION TIMING

ADJUSTMENT

1. Connect a scan tool to the data link connector (DLC).
2. Turn the ignition switch ON (II).
3. Make sure the scan tool communicates with the vehicle and the engine control module (ECM)/powertrain control module (PCM). If it doesn't, troubleshoot the DLC circuit.
4. Start the engine. Hold the engine speed at 3,000 rpm with no load (M/T in Neutral, or A/T in P or N) until the radiator fan comes on, then let it idle.

IGNITION SYSTEM

5. Check the idle speed. See "Idle Speed" under "FUEL SYSTEMS" section.
6. Jump the SCS line with the scan tool.
7. Connect the timing light to the No. 1 ignition coil harness.
8. Aim the light toward the pointer (A) on the cam chain case. Check the ignition timing under a no load condition (headlights, blower fan, rear window defogger, and air conditioner are turned off).
9. Ignition timing should be 8°+ / − 2° BTDC
10. If the ignition timing differs from the specification, check the cam timing. If the cam timing is OK, update the ECM/PCM if it does not have the latest software, or substitute a known-good ECM/PCM, then recheck.
11. If the system works properly, and the ECM/PCM was substituted, replace the original ECM/PCM.
12. Disconnect the scan tool and the timing light.

ENGINE ELECTRICAL

STARTER

REMOVAL & INSTALLATION

1. Make sure you have the anti-theft code for the audio system, and that the ignition switch is in LOCK (0).
2. Disconnect the negative cable from the battery.
3. Remove the dipstick and the dipstick tube.
4. Raise the vehicle on the lift, and make sure it is supported securely.
5. Remove the splash shield.
6. Disconnect the starter cable from the

B terminal, then disconnect the connector from the S terminal.
7. Remove the water hose clamp.
8. Remove the upper mounting bolt and the lower mounting bolt, then remove the starter.

To install:

9. Install the starter, then loosely install the upper mounting bolt and the lower mounting bolt.
10. Fully tighten the upper mounting bolt to 33 ft. lbs. (44 Nm). Tighten the lower mounting bolt to 47 ft. lbs. (64 Nm).
11. Install the water hose clamp.

STARTING SYSTEM

12. Install the starter cable by aligning the tab on the terminal and the groove of the terminal holder, then install the connector.
13. Install the splash shield.
14. Lower the vehicle on the lift.
15. Install the dipstick tube and the dipstick with new O-ring.
16. Connect the negative cable to the battery.
17. Start the engine to make sure the starter works properly.
18. Enter the anti-theft code for the audio system.
19. Reset the clock.

ENGINE MECHANICAL

➥Disconnecting the negative battery cable may interfere with the functions of the on board computer systems and may require the computer to undergo a relearning process, once the negative battery cable is reconnected.

ACCESSORY DRIVE BELTS

ACCESSORY BELT ROUTING

See Figure 27.

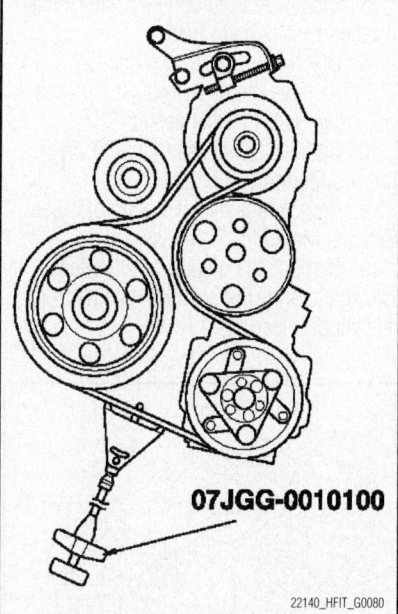

07JGG-0010100

22140_HFIT_G0080

Fig. 27 Showing the accessory drive belt routing (with belt tension tool attached)

INSPECTION

1. Attach the belt tension gauge to the belt and measure the tension. Tension should be 99–132 lbs.

2. Follow the gauge manufacturer's instructions. If the belt is worn or damaged, replace it. If the belt needs adjustment, go to see "Adjustment" below.

ADJUSTMENT

See Figure 28.

1. Loosen the lock bolt and the mounting bolt.
2. Turn the adjusting bolt to obtain the proper belt tension, then retighten the lock bolt and the mounting bolt.
3. Recheck the belt tension with the belt tension tool.

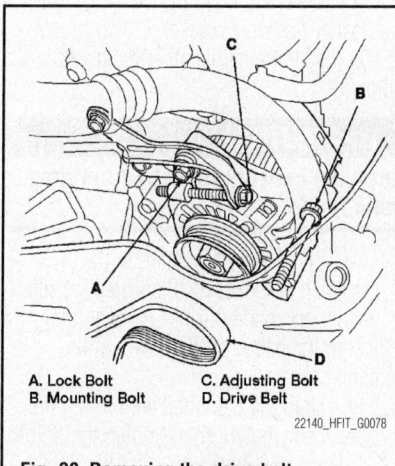

A. Lock Bolt
B. Mounting Bolt
C. Adjusting Bolt
D. Drive Belt

22140_HFIT_G0078

Fig. 28 Removing the drive belt

REMOVAL & INSTALLATION

1. Loosen the lock bolt, the mounting bolt, and the adjusting bolt, then remove the drive belt.
2. Install the drive belt in the reverse order of removal.
3. Adjust the drive belt tension. See "Adjustment."

CAMSHAFT AND VALVE LIFTERS

REMOVAL & INSTALLATION

See Figures 29 and 30.

1. Remove the air cleaner assembly.
2. Remove the camshaft sprocket as follows:
   a. Remove the cylinder head cover.
   b. Make a reference mark in one position across the camshaft sprocket and cam chain.
   c. Apply new engine oil to the slider surface of the cam chain tensioner slider through the oil return hole in the cylinder head.
   d. Remove the cylinder head plug.
   e. Hold the crankshaft pulley and set the socket wrench on the camshaft sprocket bolt.
   f. Remove the maintenance bolt, and turn the camshaft clockwise to compress the cam chain tensioner, then install the 6 x 1.0 mm bolt in the bolt hole on the engine block through the maintenance hole and cam chain tensioner.

✳✳ CAUTION

Turning torque should not exceed 41 ft. lbs. (56 Nm), when turning the camshaft. Do not turn the camshaft counterclockwise.

   g. Hold the camshaft with a 27 mm open-end wrench, then remove the camshaft sprocket.
3. Remove the rocker arm assembly. See "Rocker Arm" in this section.
4. Remove the air cleaner housing bracket, ground cable and harness clamps,

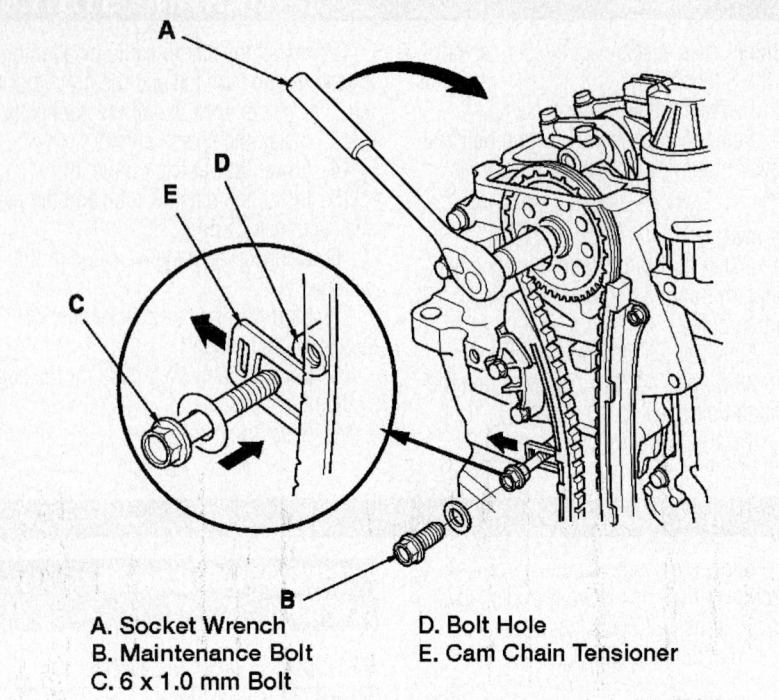

A. Socket Wrench
B. Maintenance Bolt
C. 6 x 1.0 mm Bolt
D. Bolt Hole
E. Cam Chain Tensioner

22140_HFIT_G0081

Fig. 29 Remove the maintenance bolt, turn the camshaft clockwise to compress the cam chain tensioner, then install a 6 x 1.0 mm bolt in the bolt hole on the engine block through the maintenance hole and cam chain tensioner

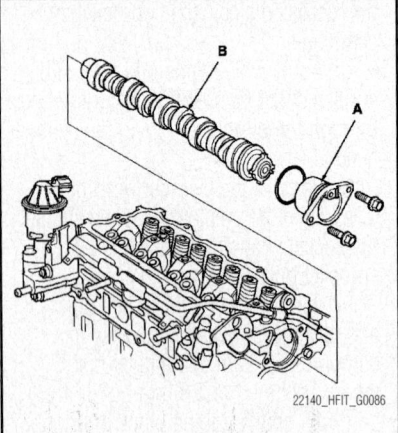

22140_HFIT_G0086

Fig. 30 Remove the camshaft thrust cover (A), then pull out the camshaft (B)

then remove the harness holder from the bracket.

5. Disconnect the camshaft position (CMP) sensor connector, then remove the CMP sensor.

6. Remove the camshaft thrust cover, then pull out the camshaft.

To install:

7. Install the camshaft into the cylinder head, then install the camshaft thrust cover with new O-ring. Tighten the bolts to 7.2 ft. lbs. (9.8 Nm).

8. Install the camshaft position (CMP) sensor with new O-ring, then connect the CMP sensor connector.

9. Install the harness holder, then install harness clamps, ground cable and air cleaner housing bracket.

10. Install the camshaft sprocket as follows:

❊❊ CAUTION

Keep the cam chain away from magnetic fields.

a. Install the cam chain to the camshaft sprocket by alignment the reference mark made during removal, then install the camshaft sprocket on the camshaft.

b. Hold the camshaft with a 27 mm open-end wrench, then tighten the bolt to 41 ft. lbs. (56 Nm).

c. Apply new engine oil to the slider surface of the cam chain tensioner slider through the oil return hole in the cylinder head.

d. Hold the crankshaft pulley and set the socket wrench on the camshaft sprocket bolt.

e. Turn the camshaft clockwise to compress the cam chain tensioner, then remove the 6 x 1.0 mm bolt.

❊❊ CAUTION

Turning torque should not exceed 41 ft. lbs. (56 Nm), when turning the camshaft. Do not turn the camshaft counterclockwise.

f. Install the maintenance bolt with a new washer to 14 ft. lbs. (20 Nm).

g. Install the new cylinder head plug.

h. Install the cylinder head cover.

11. Install the rocker arm assembly. See "Rocker Arm" in this section.

12. Install the air cleaner assembly.

CYLINDER HEAD

REMOVAL & INSTALLATION

See Figures 31 through 39.

➡ Mark all wiring and hoses to avoid misconnection. Also, be sure that they do not contact other wiring or hoses, or interfere with other parts. Keep the cam chain away from magnetic fields.

1. Relieve the fuel pressure.

2. Drain the engine coolant.

3. Remove the air cleaner assembly.

4. Disconnect the evaporative emission (EVAP) canister purge valve connector, throttle body connector, water bypass hoses and purge hose.

5. Disconnect the upper radiator hose and the lower radiator hose.

6. Disconnect the heater hoses.

7. Remove the drive belt. See "Accessory Drive Belts" in this section.

8. Remove the intake manifold. See "Intake Manifold" in this section.

9. Remove the exhaust manifold. See "Exhaust Manifold" in this section.

10. Remove the engine wire harness connectors and wire harness clamps from the cylinder head.

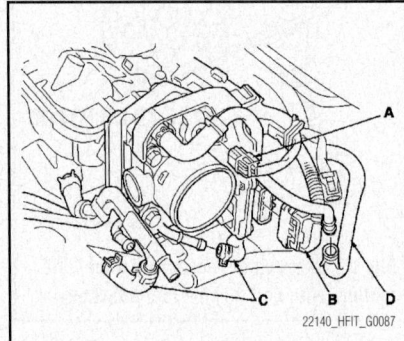

22140_HFIT_G0087

Fig. 31 Disconnect the evaporative emission (EVAP) canister purge valve connector (A), throttle body connector (B), water bypass hoses (C) and purge hose (D)

11. Remove or disconnect the following:
- Rocker arm oil control solenoid connector
- Rocker arm oil pressure switch connector
- Exhaust gas recirculation (EGR) valve/EGR valve position sensor connector
- Four injector connectors
- Engine coolant temperature (ECT) sensor 1 connector
- Camshaft position (CMP) sensor connector

12. Remove the fuel rail. See "FUEL SYSTEMS" section.

13. Remove the harness holder, ground cable, harness clamps, water bypass hose and engine junction connector clamp.

14. Disconnect the connecting pipe and water bypass hose from the side of the engine block.

15. Remove the cylinder head cover.

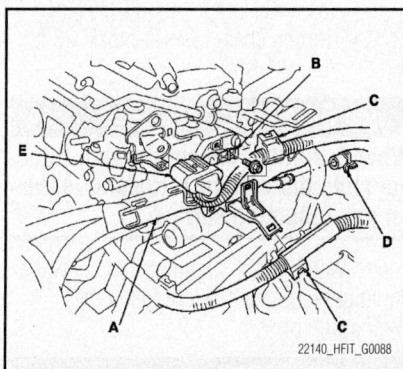

Fig. 32 Remove the harness holder (A), ground cable (B), harness clamps (C), water bypass hose (D) and engine junction connector clamp (E)

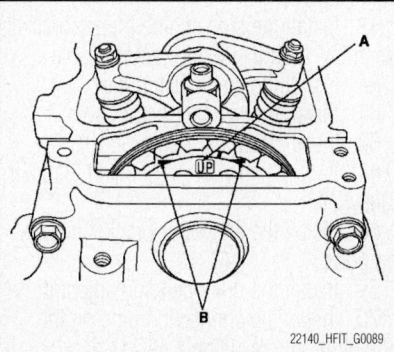

Fig. 33 Set the No. 1 piston at top dead center (TDC) so the "UP" mark (A) on the camshaft sprocket should be at the top, and the TDC grooves (B) on the camshaft sprocket should line up with the top edge of tho head

16. Remove the alternator bracket mounting bolts and loosen the alternator mounting bolt.

17. If equipped with A/C, remove the idler pulley.

18. Set the No. 1 piston at top dead center (TDC). The "UP" mark on the camshaft sprocket should be at the top, and the TDC grooves on the camshaft sprocket should line up with the top edge of the head.

19. Check that the TDC mark lines up with the pointer.

20. Remove the water pump pulley.

21. Remove the crankshaft pulley.

22. Disconnect the crankshaft position (CKP) sensor connector, then remove the harness clamps.

23. Support the engine with a wood block and a jack under the oil pan.

✳✳ CAUTION

Do not place the jack in the center of the oil pan to avoid damaging the oil pan.

24. Remove the ground cable, then remove the side engine mount/bracket assembly.

25. Keep the alternator away from the chain case. Remove the chain case.

26. Make a reference mark, at any position, across the camshaft sprocket and cam chain.

27. Loosely install the crankshaft pulley.

28. Apply new engine oil to the slider surface of the cam chain tensioner slider through the oil return hole in the cylinder head.

29. Hold the crankshaft pulley and set the socket wrench on the camshaft sprocket bolt.

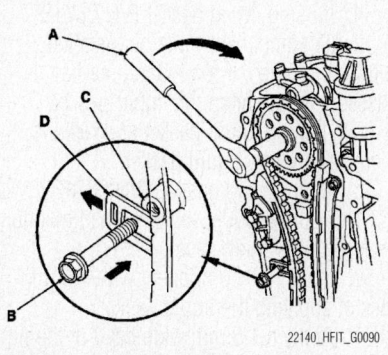

Fig. 34 Turn the camshaft clockwise with a wrench (A) to compress the cam chain tensioner, then install the 6 x 1.0 mm bolt (B) in the bolt hole (C) on the engine block through the cam chain tensioner (D)

30. Turn the camshaft clockwise to compress the cam chain tensioner, then install the 6 x 1.0 mm bolt in the bolt hole on the engine block through the cam chain tensioner.

✳✳ CAUTION

Turning torque should not exceed 41 ft. lbs. (56 Nm), when turning the camshaft. Do not turn the camshaft counterclockwise.

31. Hold the camshaft with a 27 mm open-end wrench, then remove the camshaft sprocket.

32. Remove the bolt securing the cam chain guide.

33. Remove the cylinder head bolts. To prevent warpage, loosen the bolts in sequence 1/3 turn at a time; repeat the sequence until all bolts are loosened.

34. Remove the cylinder head.

To install:

35. Clean the cylinder head and block surface.

36. Install the new cylinder head gasket and dowel pins on the engine block. Always use a new cylinder head gasket.

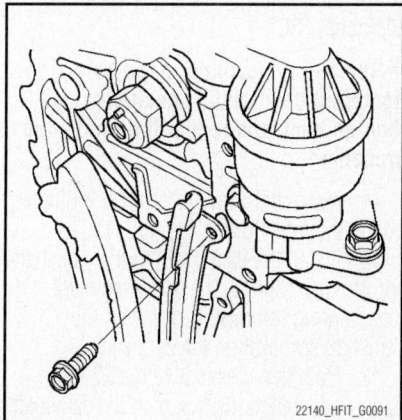

Fig. 35 Showing the location of the cam chain guide securing bolt

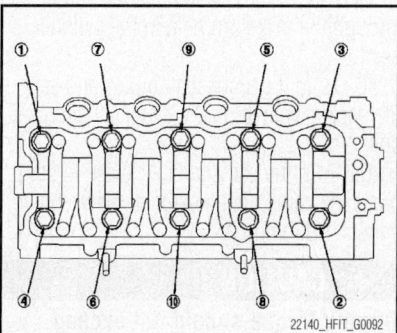

Fig. 36 Cylinder head bolt loosening sequence

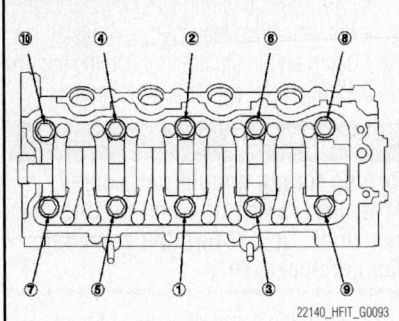

Fig. 37 Cylinder head bolt tightening sequence

37. Check that the crankshaft keyway is facing up.

38. Install the cylinder head on the engine block.

39. Using new cylinder head bolts, tighten, in sequence, to 22 ft. lbs. (29 Nm). Use a beam-type torque wrench. When using a preset-type torque wrench, be sure to tighten slowly and do not overtighten. If a bolt makes any noise while you are torquing it, loosen the bolt and retighten it from the first step.

40. Tighten all cylinder head bolts an additional 130°.

➡**Remove the cylinder head bolt if tightened beyond the specified angle, then go back and repeat the tightening procedure.**

41. Install the cam chain guide mounting bolt and tighten to 9 ft. lbs. (12 Nm).

42. Install the cam chain to the camshaft sprocket by aligning the reference mark made during removal, then install the camshaft sprocket on the camshaft.

43. Hold the camshaft with a 27 mm open-end wrench, then tighten the camshaft bolt to 41 ft. lbs. (56 Nm).

44. Loosely install the crankshaft pulley.

45. Apply new engine oil to the slider surface of the cam chain tensioner slider through the oil return hole in the cylinder head.

46. Hold the crankshaft pulley and set the socket wrench on the camshaft sprocket bolt.

47. Turn the camshaft clockwise to compress the cam chain tensioner, then remove the 6 x 1.0 mm bolt.

✳✳ **CAUTION**

Turning torque should not exceed 41 ft. lbs. (56 Nm), when turning the camshaft. Do not turn the camshaft counterclockwise.

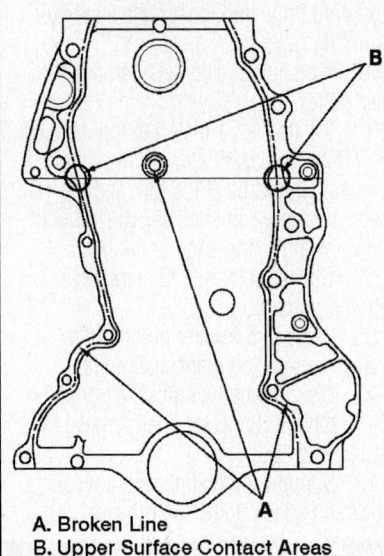

A. Broken Line
B. Upper Surface Contact Areas

Fig. 38 Showing locations for application of liquid gasket to timing chain cover

48. Check the chain case oil seal for damage If the oil seal is damaged, replace the chain case oil seal.

49. Remove old liquid gasket from the chain case mating surfaces, bolts, and bolt holes.

50. Clean and dry the chain case mating surfaces.

51. Apply liquid gasket to the cylinder head and the engine block mating surface of the chain case. Install the component within 5 minutes of applying the liquid gasket.

52. Apply a 1.5 mm wide bead of the liquid gasket along the broken line.

53. Apply a 3.0 mm wide bead of the liquid gasket to the engine block upper surface contact areas on the chain case.

54. If using liquid gasket P/N 08718-0012, the component must be installed within 4 minutes. If too much time has passed after applying the liquid gasket, remove the old liquid gasket and residue, then reapply new liquid gasket.

55. Apply liquid gasket, P/N 08718-0001 or 08718-0009, to the oil pan mating surface of the chain case.

56. Install the component within 5 minutes of applying the liquid gasket.

57. Apply a 1.5 mm wide bead of the liquid gasket along the broken line.

58. Apply a 5.0 mm wide bead of the liquid gasket to the shaded area.

59. If using liquid gasket P/N 08718-0012, the component must be installed within 4 minutes. If too much time has

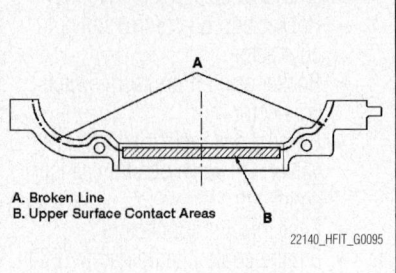

A. Broken Line
B. Upper Surface Contact Areas

Fig. 39 Showing locations for application of liquid gasket to timing chain cover at oil pan mating surface

passed after applying the liquid gasket, remove the old liquid gasket and residue, then reapply new liquid gasket.

60. Set the edge of the chain case to the edge of the oil pan, then install the chain case on the engine block. Tighten the bolts as follows:

- Timing chain case-to-oil pan: 9 ft. lbs. (12 Nm)
- Timing chain case-to-block: 23 ft. lbs. (31 Nm)

✳✳ **CAUTION**

When installing the chain case, do not slide the bottom surface onto the oil pan mounting surface.

61. Wait at least 30 minutes to allow the liquid gasket to cure before filling the engine with oil.

✳✳ **CAUTION**

Do not run the engine for at least 3 hours after installing the chain case.

62. Install the harness clamps, and connect the crankshaft position (CKP) sensor connector.

63. Install the side engine mount/bracket assembly, then install the ground cable.

64. Install the crankshaft pulley.

65. Install the cylinder head cover.

66. Install the water pump pulley.

67. If equipped with A/C, install the idler pulley.

68. Install the alternator bracket mounting bolts.

69. Install the drive belt and adjust it.

70. Install the connecting pipe on the side of the block, using a new O-ring.

71. Install the water bypass hose.

72. Install the harness holder, ground cable, harness clamps, water bypass hose and engine junction connector clamp.

73. Install the fuel rail.

74. Install the exhaust manifold.

75. Install the intake manifold.
76. Install the heater hoses.
77. Install the upper radiator hose and the lower radiator hose.
78. Install the purge hose, the water bypass hoses, the throttle body connector and the evaporative emission (EVAP) canister purge valve connector.
79. Install the air cleaner assembly.
80. Adjust the valve clearance.
81. After installation, check that all tubes, hoses and connectors are installed correctly:
- Rocker arm oil control solenoid connector
- Rocker arm oil pressure switch connector
- Exhaust gas recirculation (EGR) valve/EGR valve position sensor connector
- Four injector connectors
- Engine coolant temperature (ETC) sensor 1 connector
- Camshaft position (CMP) sensor connector

82. Inspect for fuel leaks. Turn the ignition switch to ON (II) (do not operate the starter) so the fuel pump runs for about 2 seconds and pressurizes the fuel line. Repeat this operation three times, then check for fuel leakage at any point in the fuel line.
83. Refill the radiator with engine coolant, and bleed air from the cooling system with the heater valve open.
84. Perform the crankshaft position (CKP) pattern clear/CKP pattern learn procedure as follows:
a. Start the engine. Hold the engine speed at 3,000 rpm without load (in Park or neutral) until the radiator fan comes on.
b. Test-drive the vehicle on a level road: Decelerate (with the throttle fully closed) from an engine speed of 2,500 rpm down to 1,000 rpm with the A/T in 2 position, or M/T in 1st gear.
c. Repeat the previous step several times.
d. Turn the ignition switch OFF.
e. Turn the ignition switch ON (II), and wait 30 seconds.
85. Inspect the idle speed. See "FUEL SYSTEMS" section.
86. Inspect the ignition timing. See "ENGINE ELECTRICAL" section.

ENGINE ASSEMBLY

REMOVAL & INSTALLATION

See Figures 40 through 50.

✳✳ CAUTION

To avoid damaging wire and terminals, unplug the wiring connectors carefully while holding the connector portion.

➡**During removal, mark all wiring and hoses to avoid misconnection. Also, be sure that they do not contact other wiring or hoses, or interfere with other parts.**

1. Make sure you have the anti-theft code for the audio system.
2. Secure the hood in the wide open position (support rod in the lower hole).
3. Relieve the fuel pressure.
4. Disconnect the negative cable from the battery, then disconnect the positive cable.
5. Remove the battery.
6. Remove the air cleaner assembly.
7. Remove the harness clamps, then remove the battery base.
8. Remove the passenger's dashboard undercover, then disconnect the engine control module (ECM)/powertrain control module (PCM) connectors, dashboard wire harness connector, and throttle actuator control module connector.
9. Loosen the blower unit mounting nuts (one inside the glove box opening and two at lower edge of housing).
10. In the engine compartment, remove the wiring harness clamps and the grommet from the bulkhead. Pull the engine wire harness through the bulkhead.
11. Disconnect the starter cable and the engine compartment wire harness connector.

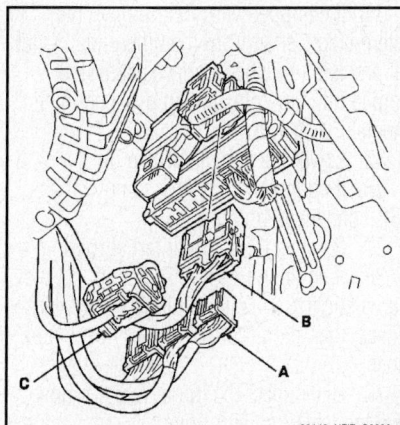

Fig. 40 Showing the ECM/PCM connector (A), dashboard wire harness connector (B) and throttle actuator control module connector (C)

12. Remove the battery cable from the under-hood fuse/relay box, then remove the cable clamp.
13. Remove the engine cover.
14. Disconnect the brake booster vacuum hose and the manifold absolute pressure (MAP) sensor connector, then remove the harness holder.
15. Disconnect the evaporative emission (EVAP) canister purge valve connector, the throttle body connector, the water bypass hoses, and the EVAP purge hose.
16. Remove the quick-connect fitting cover, then disconnect the fuel feed hose.
17. On M/T models, perform the following:
a. Remove the shift cables from the engine connections. Take care not to bend the cables when removing them.
b. Remove the breather tube from the clamp, the clutch slave cylinder and the clutch line bracket mounting bolt. Do not operate the clutch pedal once the slave cylinder has been removed.

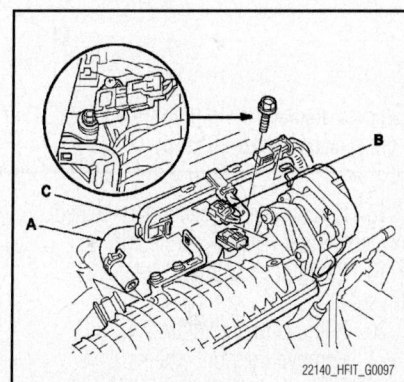

Fig. 41 Disconnect the brake booster vacuum hose (A) and the manifold absolute pressure (MAP) sensor connector (B), then remove the harness holder (C)

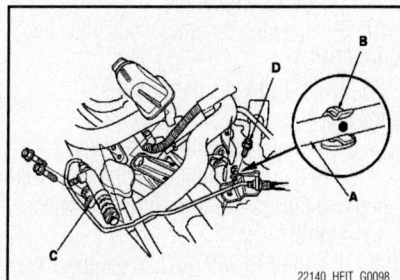

Fig. 42 Remove the breather tube (A) from the clamp (B), the clutch slave cylinder (C) and the clutch line bracket mounting bolt (D)—models with manual transmission

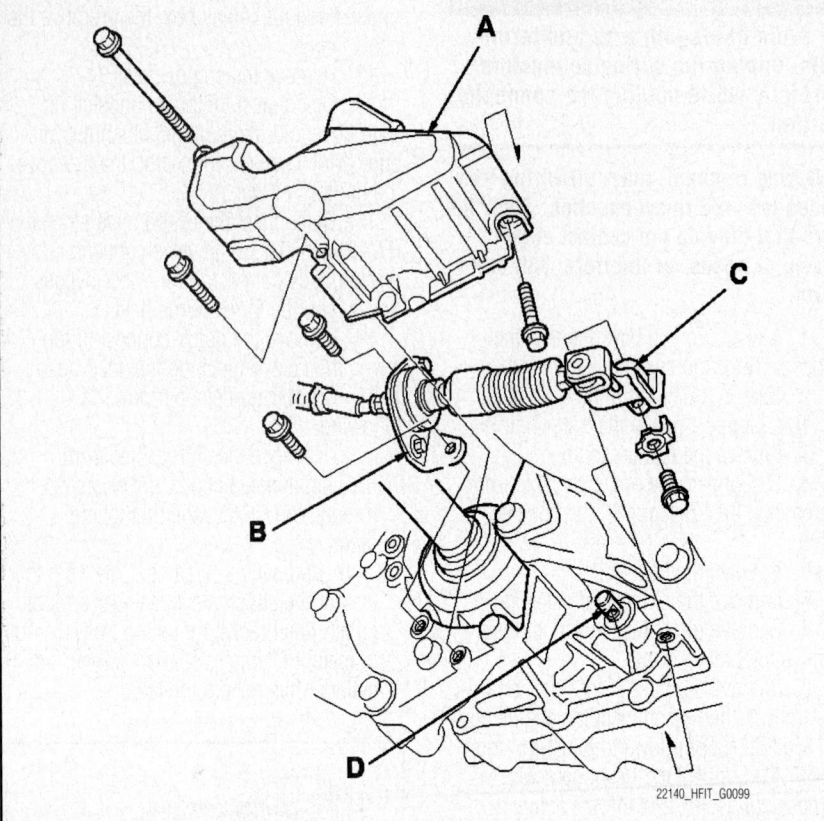

Fig. 43 Remove the shift cable cover (A), shift cable holder (B), and the control lever (C) from the selector control shaft (D)

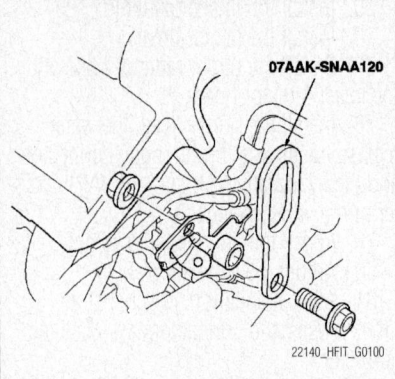

Fig. 44 Install the first universal eyelet to the air cleaner housing bracket

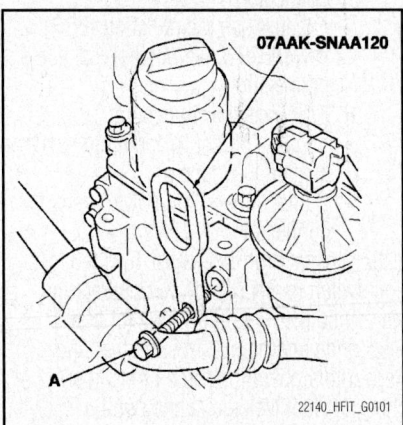

Fig. 45 Remove the bolt from the chain case cover, then install the second universal eyelet to the engine

18. Disconnect the secondary heated oxygen sensor (HO2S) connector.

19. If equipped with A/C, remove the drive belt.

20. Remove the radiator cap.

21. Remove the front wheels.

22. Raise the vehicle on the lift to full height.

23. Remove the splash shield.

24. Loosen the drain plug in the radiator to drain the engine coolant.

25. Drain the transmission fluid:

26. Drain the engine oil.

27. Remove the three way catalytic converter (TWC)

28. On A/T models perform the following:

   a. Remove the shift cable cover.

   b. Remove the shift cable holder, then remove the control lever from the selector control shaft.

29. If equipped with A/C, disconnect the A/C compressor clutch connector, then remove the A/C compressor without disconnecting the A/C hoses.

30. Disconnect the stabilizer links from the stabilizer bar. See "FRONT SUSPENSION" section.

31. Disconnect the suspension lower arm ball joints. See "FRONT SUSPENSION" section.

32. Remove the halfshafts. See "Halfshafts" under "AUTOMATIC TRANSAXLE" or "MANUAL TRANSAXLE", as applicable.

33. Coat all halfshaft precision finished surfaces with new engine oil. Tie plastic bags over the halfshaft ends.

34. If equipped with A/T, remove the intermediate shaft from the differential. Coat all precision finished surfaces with new engine oil. Tie a plastic bag over the intermediate shaft end.

35. Lower the vehicle on the lift.

36. If equipped with A/C, remove the A/C condenser fan shroud.

37. Remove the radiator fan shroud.

38. If equipped with A/T, disconnect the automatic transmission fluid (ATF) cooler hoses, then plug the ATF cooler hoses and lines.

39. Disconnect the upper radiator hose and the lower radiator hose.

40. Disconnect the heater hoses.

41. Install the first universal eyelet (07AAK-SNAA120) to the air cleaner housing bracket.

42. Install a proper engine hanger assembly to fully support the engine, using the universal eyelet on the air clean housing bracket.

43. Raise the vehicle on the lift.

44. Remove the steering gearbox mounting bolts, then remove the steering gearbox bracket.

45. Remove the steering gearbox mounting bolts, then remove the steering gearbox stiffeners.

46. Remove the rear engine mount bolts.

47. On M/T models, remove the left front transmission mount/bracket assembly.

48. Attach a nylon strap or rope to the body and around the steering gearbox to support it on both sides.

49. Support the front subframe with a jack.

50. Remove the front subframe.

51. Lower the vehicle on the lift.

52. Remove the bolt from the chain case cover, then install the second universal eyelet to the engine.

53. Attach the chain hoist to the engine.

54. Remove the engine support hanger.

55. Remove the headlight harness cover.

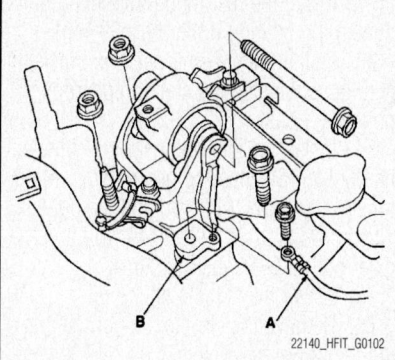

Fig. 46 Remove the ground cable (A), then remove the transmission mount bracket (B)

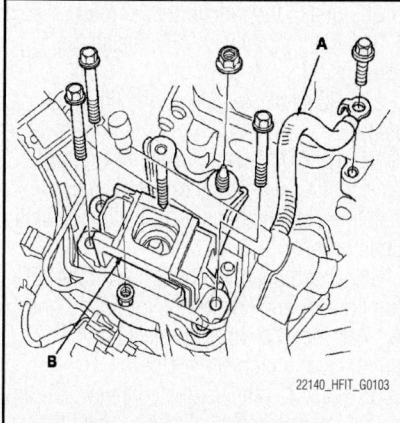

Fig. 47 Remove the ground cable (A), then remove the side engine mount/bracket assembly (B)

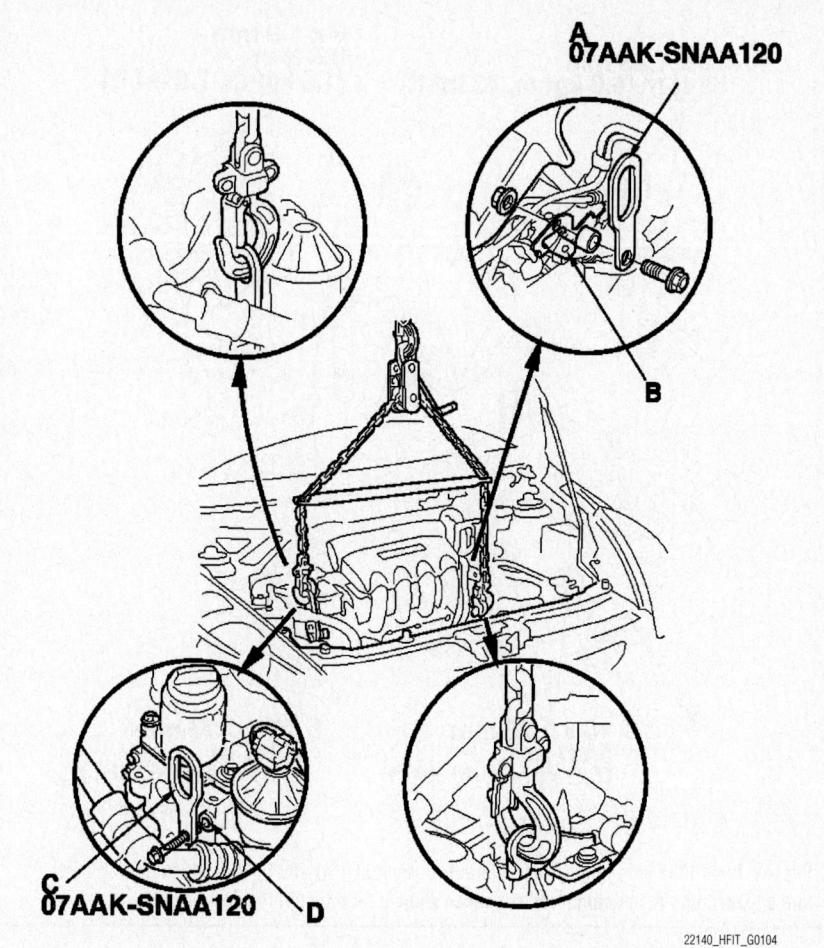

Fig. 48 Install the first universal eyelet (A) to the air cleaner housing bracket (B), then install the second universal eyelet (C) to the chain case cover (D)

56. Remove the ground cable, then remove the transmission mount bracket.

57. Remove the ground cable, then remove the side engine mount/bracket assembly.

58. Check that the engine/transmission is completely free of vacuum hoses, fuel and coolant hoses, and electrical wiring.

59. Slowly lower the engine/transmission about 150 mm (6 in.). Check once again that all hoses and electrical wiring are disconnected and free from the engine/transmission, then lower it all the way.

60. Remove the chain hoist from the engine/transmission.

61. Raise the vehicle.

➡Carefully position the A/C compressor and steering gearbox.

62. Remove the engine/transmission from under the vehicle.

To install:

63. Raise the vehicle on the lift.

64. Position the engine/transmission under the vehicle. Be sure that they are prop-

erly aligned. Carefully lower the vehicle until the engine and transmission are properly positioned in the engine compartment. Make sure the vehicle is not resting on any part of the engine or transmission. Lift and support the engine with a chain hoist and carefully raise the engine/transmission into place.

65. Reinstall the mounting bolts/support nuts in the sequence given in the following steps. Failure to follow this sequence may cause excessive noise and vibration, and reduce engine mount life:

a. Install the first universal eyelet to the air cleaner housing bracket, then install the second universal eyelet to the chain case cover.

b. Install the side engine mount/bracket assembly, then tighten the mounting bolts and support nuts in the numbered sequence shown.

c. Install the ground cable.

d. Install the transmission mount bracket, then tighten the support bolt/nuts and mounting bolt in the numbered sequence shown.

e. Install the ground cable.

66. Install the headlight harness cover.

67. Install the engine hanger assembly, as used for removal. Carefully position the engine support hanger on the vehicle, and attach the hook to the slotted hole in the universal eyelet. Tighten the wing nut by hand to lift and support the engine/transmission.

68. Remove the chain hoist, then raise the vehicle on the lift to full height.

69. Using the jack, raise the front subframe up to body.

70. Install the front subframe. Tighten the bolts to 69 ft. lbs. (93 Nm).

➡When installing, align both installation reference holes in the subframe with both reference holes in the body using a screwdriver or tapered punch as a guide.

71. On M/T models. install the left front transmission mount/bracket assembly. Tighten the top bolts to 33 ft. lbs. (44 Nm) and the lower bolts to 29 ft. lbs. (33 Nm).

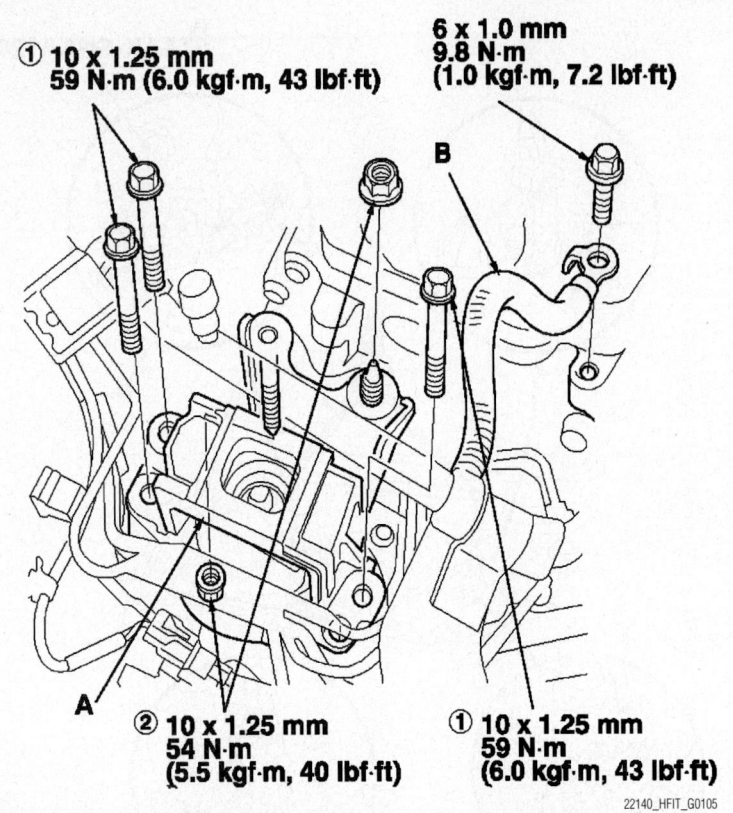

① 10 x 1.25 mm
59 N·m (6.0 kgf·m, 43 lbf·ft)

6 x 1.0 mm
9.8 N·m
(1.0 kgf·m, 7.2 lbf·ft)

B

A

② 10 x 1.25 mm
54 N·m
(5.5 kgf·m, 40 lbf·ft)

① 10 x 1.25 mm
59 N·m
(6.0 kgf·m, 43 lbf·ft)

22140_HFIT_G0105

Fig. 49 Install the side engine mount/bracket assembly (A) and tighten the mounting bolts and support nuts in the numbered sequence shown, then install the ground cable (B)

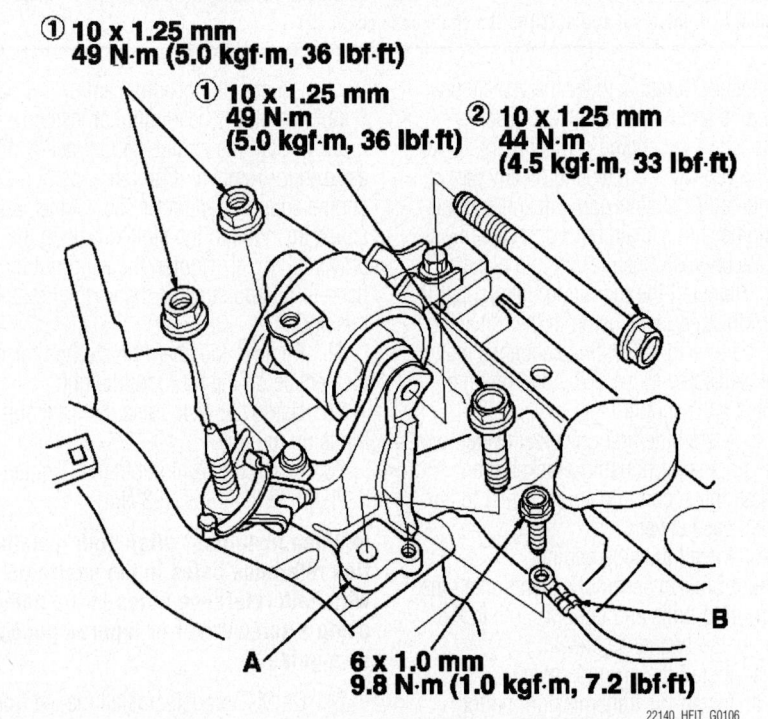

① 10 x 1.25 mm
49 N·m (5.0 kgf·m, 36 lbf·ft)

① 10 x 1.25 mm
49 N·m
(5.0 kgf·m, 36 lbf·ft)

② 10 x 1.25 mm
44 N·m
(4.5 kgf·m, 33 lbf·ft)

B

A

6 x 1.0 mm
9.8 N·m (1.0 kgf·m, 7.2 lbf·ft)

22140_HFIT_G0106

Fig. 50 Install the transmission mount bracket (A) and tighten the support bolt/nuts and mounting bolt in the numbered sequence shown, then, install the ground cable (B)

72. Install the rear mount mounting bolts. Tighten the bolts to 43 ft. lbs. (59 Nm).

73. Install the steering gearbox bracket, then loosely install the steering gearbox mounting bolts.

74. Install the steering gearbox stiffener, then tighten the steering gearbox mounting bolts to 33 ft. lbs. (44 Nm) and the upper stiffener bolts to 36 ft. lbs. (49 Nm) and the lower bolt to 38 ft. lbs. (52 Nm).

75. Lower the vehicle on the lift.

76. Remove the engine support hanger and the universal eyelets.

77. Install the heater hoses.

78. Install the upper radiator hose and the lower radiator hose.

79. On A/T models, install the automatic transmission fluid (ATF) cooler hoses.

80. Install the radiator fan shroud.

81. If equipped with A/C, install the A/C condenser fan shroud.

82. Raise the vehicle on the lift to full height.

83. On A/T models, install the control lever to the selector control shaft, then install the shift cable holder. Install the shift cable cover.

84. If equipped with A/C, install the A/C compressor and tighten the mounting bolts to 17 ft. lbs. (24 Nm). Connect the A/C compressor clutch connector.

85. On A/T models, install a new set ring on the intermediate shaft, then install the intermediate shaft.

86. Install a new set ring on the end of each halfshaft, then install the halfshafts. Make sure each ring "clicks" into place in the differential. See "AUTOMATIC TRANSAXLE" or "MANUAL TRANSAXLE" section, as applicable.

87. Connect the suspension lower arm ball joints and connect the stabilizer links to the stabilizer bar. See "FRONT SUSPENSION" section.

88. Install the three way catalytic converter (TWC). Use new gaskets at the flange connections. Install new self-locking nuts at the front flange and tighten to 25 ft. lbs. (33 Nm). Install and new bolts at the rear flange and tighten to 16 ft. lbs. (22 Nm).

➡**Tighten the bolts in steps, alternating side-to-side.**

89. Install the splash shield.

90. Lower the vehicle on the lift.

91. Install the front wheels.

92. Connect the secondary heated oxygen sensor (HO2S) connector.

93. On M/T models, perform the following:
   a. Reconnect the shift cables, using the plastic washers, washers, and new cotter pins.

b. Install the clutch slave cylinder and clutch line bracket mounting bolt. Tighten the slave cylinder mounting bolts to 16 ft. lbs. (22 Nm).

94. Align the reference mark made during removal on the breather tube with the clamp and install the tube in the clamp.

95. Connect the fuel feed hose, then install the quick-connect fitting cover.

96. Install the evaporative emission (EVAP) canister purge valve connector, the throttle body connector, the water bypass hoses and the purge hose.

97. Install the brake booster vacuum hose, the harness holder and the manifold absolute pressure (MAP) sensor connector.

98. Install the engine cover.

99. Install the starter cable and the engine compartment wire harness connector.

100. Install the battery cable to the under-hood fuse/relay box, then install the cable clamp.

101. Push the engine control module (ECM)/powertrain control module (PCM) connectors through the bulkhead, then install the bulkhead grommet. Install the harness clamps.

102. Tighten the blower unit mounting nuts to 7.2 ft. lbs. (9.8 Nm).

103. Connect the engine control module (ECM)/powertrain control module (PCM) connectors, dashboard wire harness connector, and the throttle actuator control module connector.

104. Install the passenger's dashboard undercover.

105. Install the battery base, then install the harness clamps.

106. Install the air cleaner assembly.

107. Install the battery. Clean the battery posts and cable terminals, then assemble them, and apply grease to prevent corrosion.

108. If equipped with A/C, install the drive belt and adjust it.

109. Refill the engine with engine oil.

110. Refill the transmission with fluid.

111. Inspect for fuel leaks. Turn the ignition switch ON (II) (do not operate the starter) so the fuel pump runs for about 2 seconds and pressurizes the fuel line. Repeat this operation three times, then check for fuel leakage at any point in the fuel line.

112. Refill the radiator with engine coolant, and bleed air from the cooling system with the heater valve open.

113. Perform the crankshaft position (CKP) pattern clear/CKP pattern learn procedure as follows:

a. Start the engine. Hold the engine speed at 3,000 rpm without load (in Park or neutral) until the radiator fan comes on.

b. Test-drive the vehicle on a level road: Decelerate (with the throttle fully closed) from an engine speed of 2,500 rpm down to 1,000 rpm with the A/T in 2 position, or M/T in 1st gear. Repeat this step several times.

c. Turn the ignition switch OFF.

d. Turn the ignition switch ON (II), and wait 30 seconds.

114. Inspect the idle speed.

115. Inspect the ignition timing.

116. Check the wheel alignment.

117. Enter the anti-theft code for the audio system.

118. Set the clock.

EXHAUST MANIFOLD

REMOVAL & INSTALLATION

See Figure 51.

1. Remove the cover and exhaust manifold bracket, then remove the exhaust manifold.

2. Install the exhaust manifold and tighten the bolts/nuts in a crisscross pattern in three steps, beginning with the inner bolt.

3. Install the other parts in the reverse order of removal.

INTAKE MANIFOLD

REMOVAL & INSTALLATION

See Figures 52 and 53.

1. Remove the engine cover.

2. Remove the air cleaner assembly.

3. Disconnect the brake booster vacuum hose, harness holder, the manifold absolute pressure (MAP) sensor connector and the bolt.

4. Remove the four coolant housing bolts as shown.

5. Remove the harness holder, the PCV hose, and the dipstick.

6. Remove the intake manifold and the exhaust gas recirculation (EGR) plate gasket.

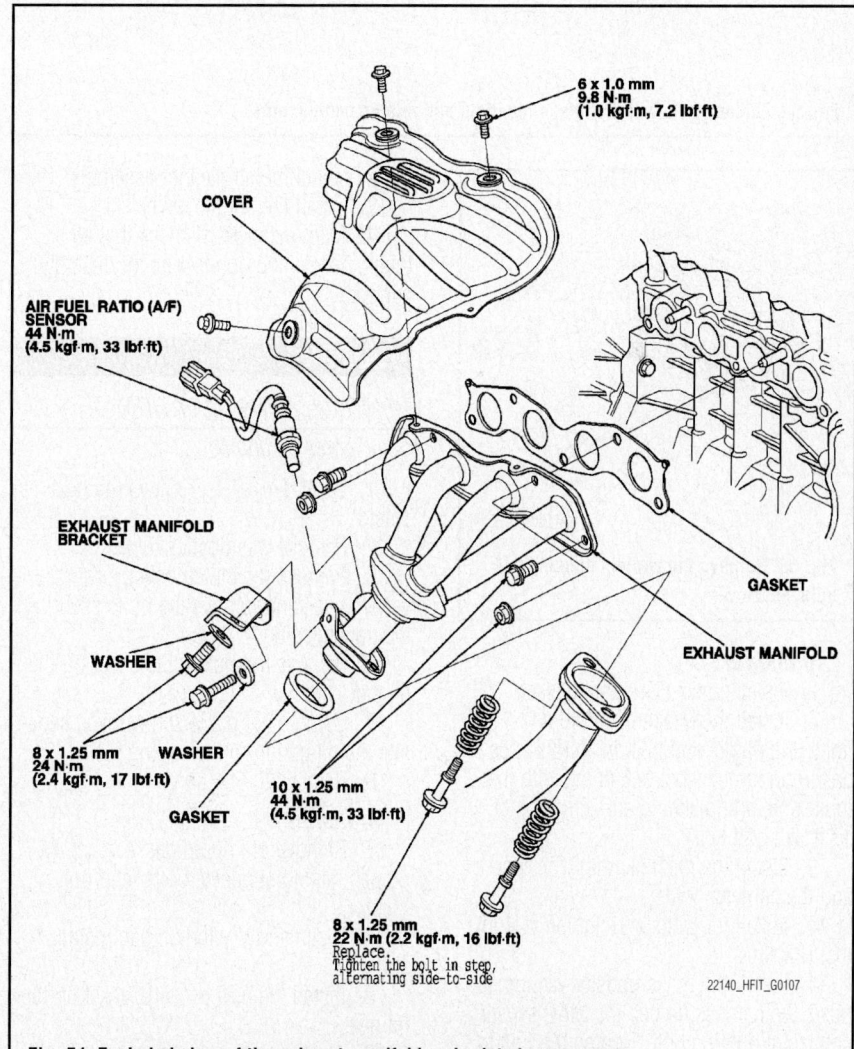

Fig. 51 Exploded view of the exhaust manifold and related components

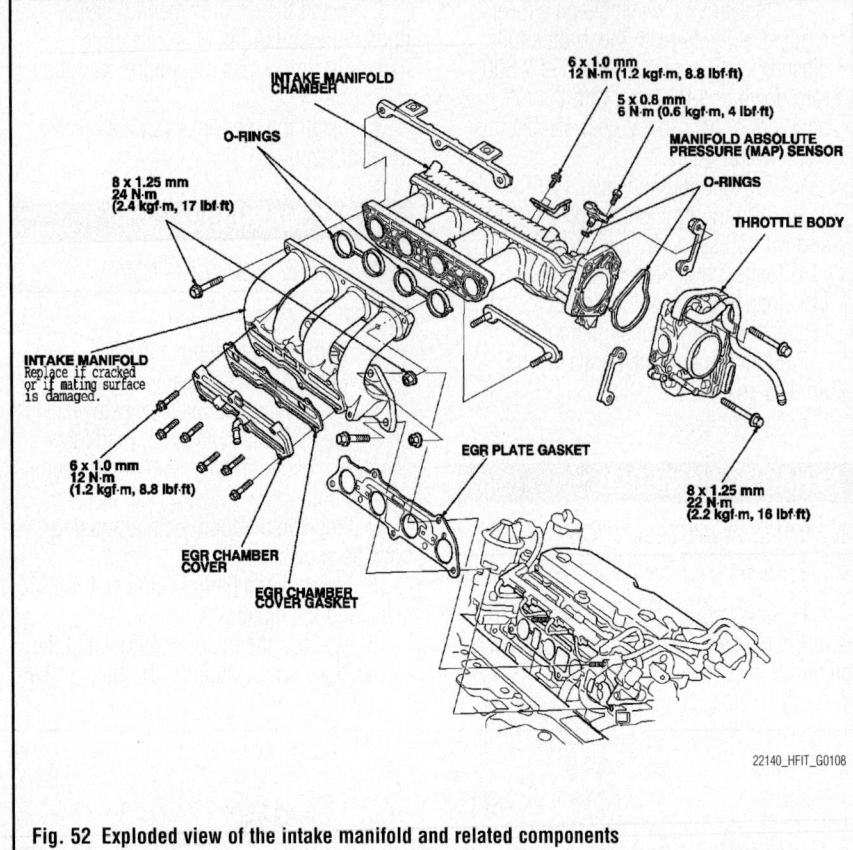

Fig. 52 Exploded view of the intake manifold and related components

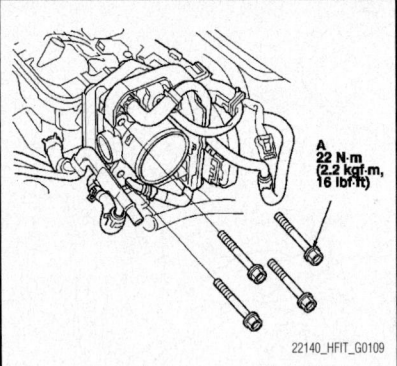

Fig. 53 Remove the coolant housing four bolts as shown

To install:

7. Install a new EGR plate gasket.

8. Install the intake manifold and tighten the bolts and nuts in a crisscross pattern in three steps, beginning with the inner bolt, and reaching a final torque of 17 ft. lbs. (24 Nm).

9. Install the dipstick, the PCV hose and the harness holder.

10. Install the bolts and tighten to 16 ft. lbs. (22 Nm).

11. Install the brake booster vacuum hose, the harness holder, the MAP sensor connector and the bolt. Tighten the bolt to 9 ft. lbs. (12 Nm).

12. Install the air cleaner assembly.

13. Install the engine cover.

14. After installation, check that all tubes, hoses, and connectors are installed correctly.

OIL PAN

REMOVAL & INSTALLATION

See Figures 54 and 55.

1. On M/T models, remove the heat shield.

2. Remove the dipstick tube.

3. Remove the clutch cover/torque converter cover, and remove the bolts securing the transmission.

4. Remove the bolts/nuts securing the oil pan.

5. Using a flat blade screwdriver, separate the oil pan from the engine block.

6. Remove the oil pan.

To install:

7. Remove old liquid gasket from the oil pan mating surfaces, bolts, and bolt holes.

8. Clean and dry the oil pan mating surfaces.

9. Install the new oil pan gasket on the oil pan.

10. Apply liquid gasket to the engine

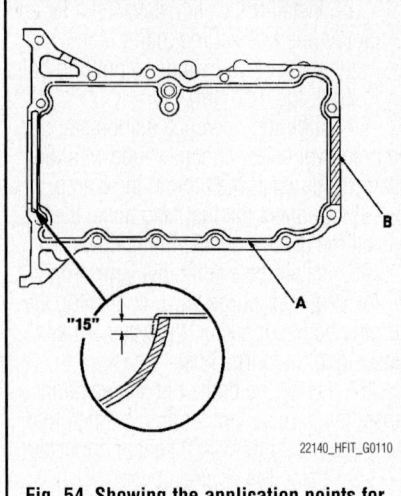

Fig. 54 Showing the application points for liquid gasket

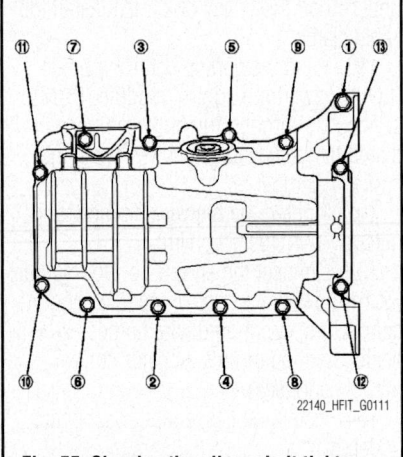

Fig. 55 Showing the oil pan bolt tightening sequence

block mating surface of the oil pan. Install the component within 5 minutes of applying the liquid gasket.

a. Apply a 1.5 mm wide bead of the liquid gasket along the broken line (A).

b. Apply a 5.0 mm wide bead of the liquid gasket to the shaded area (B).

➡If too much time has passed after applying the liquid gasket, remove the old liquid gasket and residue, then reapply new liquid gasket.

11. Install the oil pan with the dowel pin and a new O-ring.

12. Tighten the bolts in three steps:
- First step: 17 ft. lbs. (24 Nm)
- Second step: 8.7 ft. lbs. (12 Nm)
- Third step: 8.7 ft. lbs. (12 Nm)

➡Wait at least 30 minutes to allow liquid gasket to cure before filling the engine with oil. Do not run the engine for at least 3 hours after installing the oil pan.

13. Install the clutch cover/torque converter cover, and tighten the bolts securing the transmission and tighten to 47 ft. lbs. (64 Nm).

14. Install the dipstick tube (A) with a new O-ring (B).

15. On M/T models, install the heat shield.

OIL PUMP

REMOVAL & INSTALLATION

See Figure 56.

1. Remove the cam chain.
2. Remove the oil screen, then remove the oil pump.
3. Install in reverse of the removal procedure.

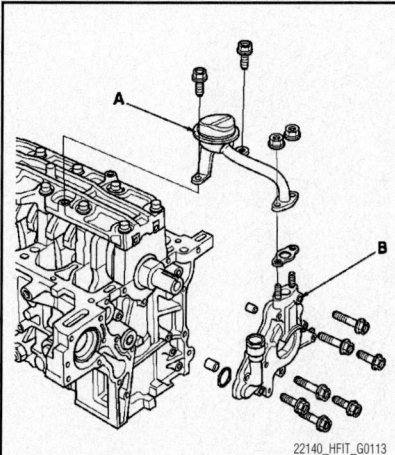

Fig. 56 Remove the oil screen (A), then remove the oil pump (B)

To install:

4. Install the dowel pins and new O-ring on the oil pump, then align the inner rotor with crankshaft, and install the oil pump.
5. Install the oil screen with new gasket.
6. Install the cam chain.

PISTON AND RING

POSITIONING

See Figure 57.

REAR MAIN SEAL

REMOVAL & INSTALLATION

With Automatic Transaxle

1. Remove the mainshaft bearing and the oil seal using the adjustable bearing puller and a slide hammer.
2. Install the new mainshaft bearing until

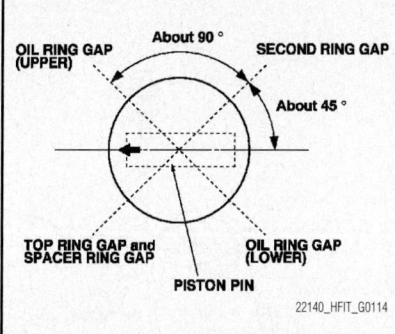

Fig. 57 Showing proper position of piston rings

it bottoms in the torque converter housing using the driver and the attachment (62 x 68 mm).

3. Install the new oil seal flush with the housing using the driver and the attachment (72 x 75 mm).

With Manual Transaxle

1. Remove the ball bearing from the clutch housing, using an adjustable bearing puller and a slide hammer.
2. Remove the oil seal from the transmission side.
3. Drive in the new oil seal from the transmission side using the driver and 37 x 40 mm attachment.
4. Drive in the new ball bearing from the transmission side using the driver and 52 x 55 mm attachment.

ROCKER ARM

REMOVAL & INSTALLATION

See Figures 58 through 60.

1. Remove the cylinder head cover.
2. Loosen the rocker arm adjusting screws.
3. Bundle the intake rocker arms with rubber bands to keep them together as a set. Unscrew the rocker shaft mounting

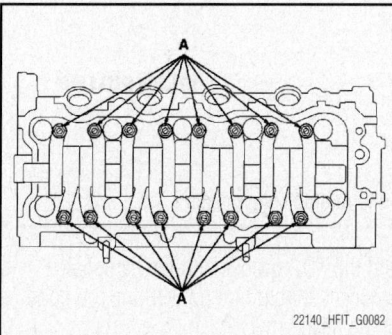

Fig. 58 Showing the rocker arm adjusting screws (A)

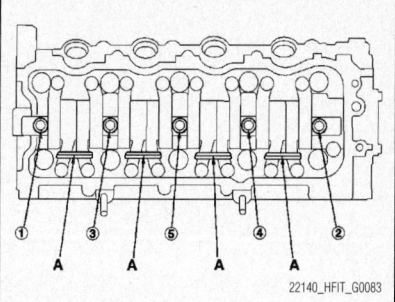

Fig. 59 Bundle the intake rocker arms with rubber bands (A) to keep them together as a set, then unscrew the rocker shaft mounting bolts two turns at a time, in sequence shown

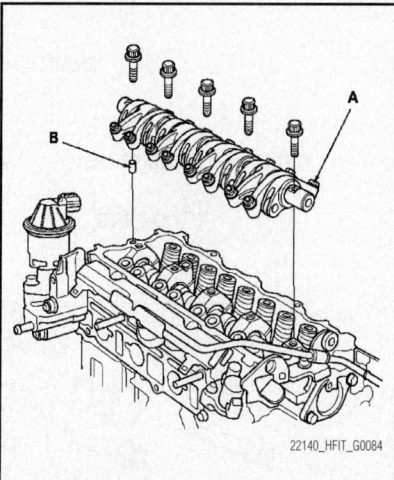

Fig. 60 Removing the dowel pin (B), then the rocker arm assembly (A)

bolts two turns at a time, in sequence shown.

4. Remove the rocker shaft mounting bolts, then remove the rocker arm assembly.
5. When disassembling the rocker arms, remove the dowel pin, then remove the rocker arm from cam chain side on the rocker shaft.

TIMING CHAIN, SPROCKETS, FRONT COVER AND SEAL

REMOVAL & INSTALLATION

See Figures 61 through 71.

➡Keep the cam (timing) chain away from magnetic fields.

1. Remove the front wheels.
2. Remove the splash shield.
3. Remove the drive belt.
4. Remove the alternator bracket mounting bolts and loosen the alternator mounting bolt.

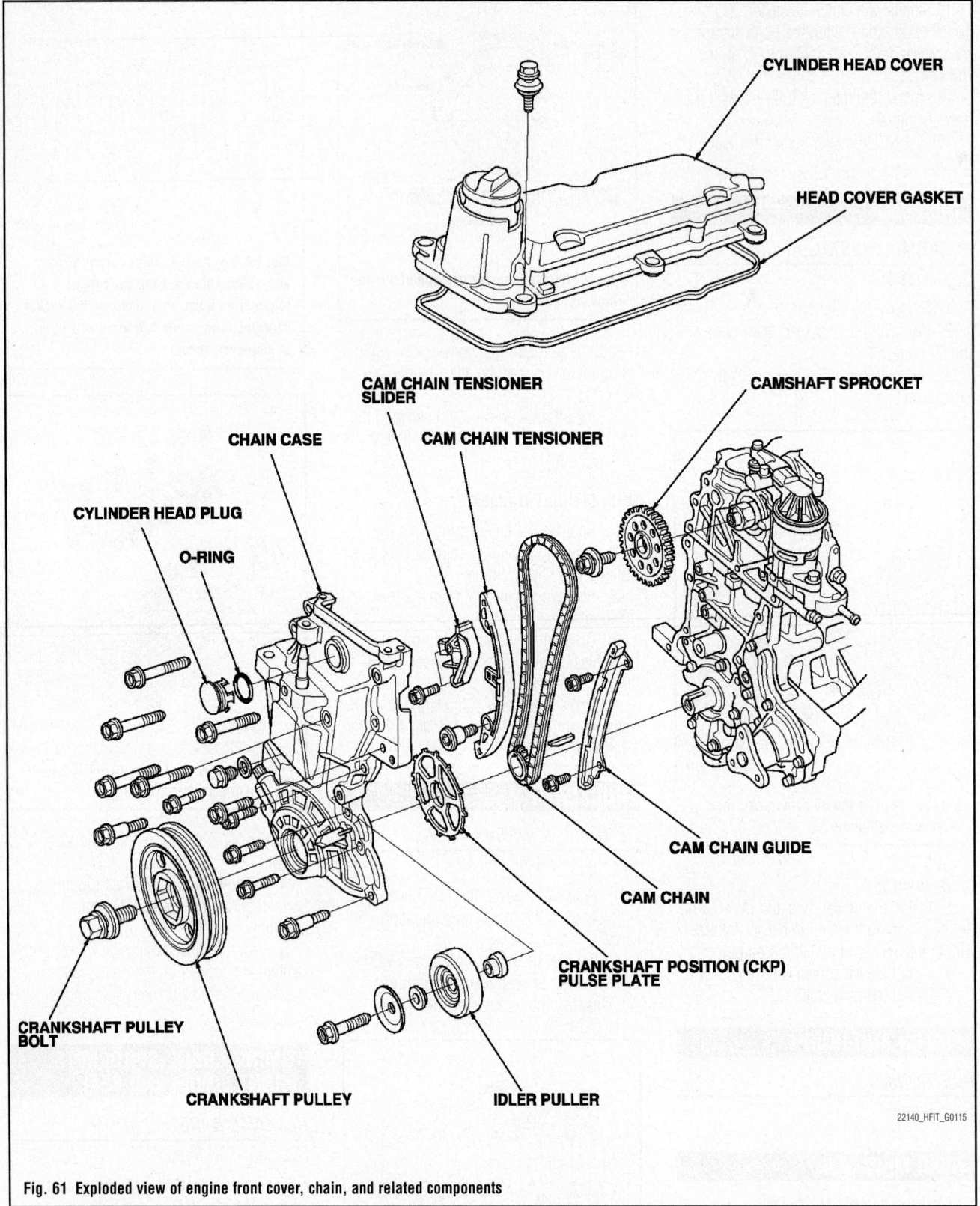

CYLINDER HEAD COVER

HEAD COVER GASKET

CAMSHAFT SPROCKET

CAM CHAIN TENSIONER SLIDER

CHAIN CASE

CAM CHAIN TENSIONER

CYLINDER HEAD PLUG

O-RING

CAM CHAIN GUIDE

CAM CHAIN

CRANKSHAFT POSITION (CKP) PULSE PLATE

CRANKSHAFT PULLEY BOLT

CRANKSHAFT PULLEY

IDLER PULLER

22140_HFIT_G0115

Fig. 61 Exploded view of engine front cover, chain, and related components

5. If equipped with A/C, remove the idler pulley.

6. Remove the cylinder head cover.

7. Set the No. 1 piston at top dead center (TDC). The "UP" mark on the camshaft sprocket should be at the top, and the TDC grooves on the camshaft sprocket should line up with the top edge of the head.

8. Check that the TDC mark on the crankshaft pulley lines up with the pointer.

9. Remove the water pump pulley.

10. Remove the crankshaft pulley.

11. Remove the oil pan. See "Oil Pan" in this section.

12. Disconnect the crankshaft position (CKP) sensor connector, then remove the harness clamps.

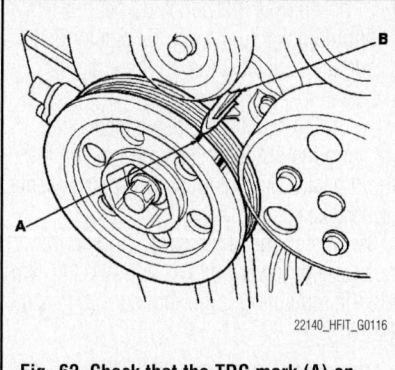

Fig. 62 Check that the TDC mark (A) on
the crankshaft pulley lines up with the
pointer (B)

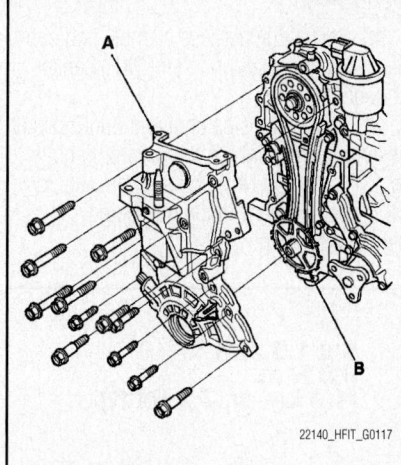

Fig. 63 Remove the chain case (A), then
remove the CKP pulse plate (B)

13. Support the engine with a jack and a
wood block under the engine block.

✳✳ CAUTION

**Do not hit the oil strainer when plac-
ing the jack on the edge of the engine
block.**

14. Remove the ground cable, then
remove the side engine mount/bracket
assembly.
15. Position the alternator away from the
chain case.
16. Remove the chain case (A), then
remove the CKP pulse plate.
17. Measure the cam chain separation.
If the distance is less than the service limit
of 0.59 in. (15 mm), replace the cam chain
and cam chain tensioner.
18. Apply new engine oil to the
sliding surface of the cam chain tensioner
slider.
19. Hold the cam chain tensioner slider
with the screwdriver, then remove the bolt,
and loosen the bolt.

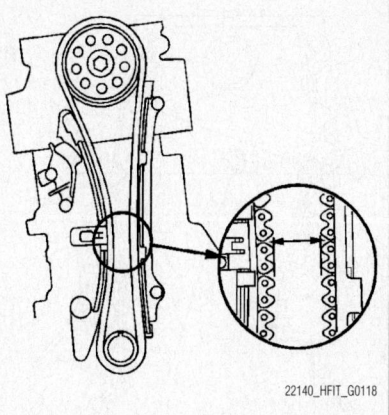

Fig. 64 Measure the cam chain separation
to be sure it is within service limits

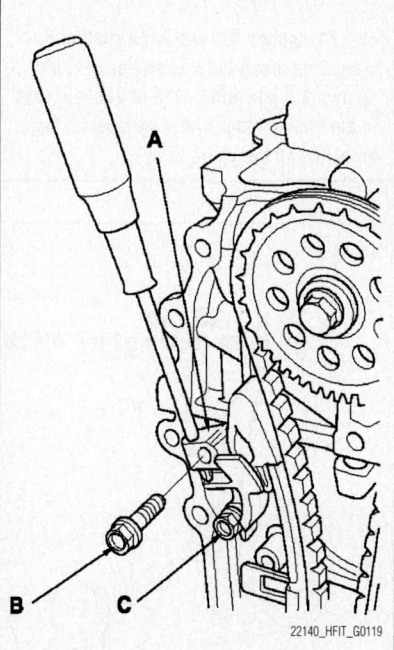

Fig. 65 Apply new engine oil to the sliding surface of the cam chain tensioner
slider (A), then hold the cam chain tensioner slider with the screwdriver, remove
the bolt (B), and loosen the bolt (C)

20. Remove the cam chain tensioner
slider.
21. Remove the cam chain tensioner and
the cam chain guide.
22. Remove the cam chain.

To install:

➡**Keep the cam chain away from
magnetic fields.**

23. Set the crankshaft to top dead center
(TDC). Align the TDC mark on the crankshaft sprocket with the pointer on the oil
pump.

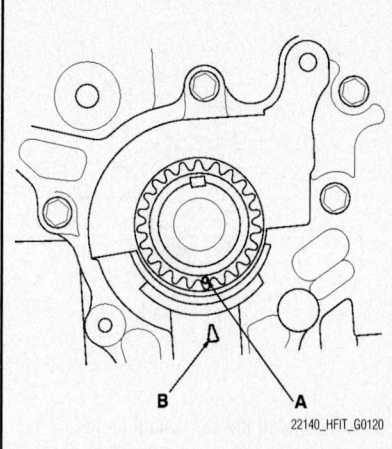

Fig. 66 Check that the TDC mark (A) on
the crankshaft pulley lines up with the
pointer (B) on the oil pump

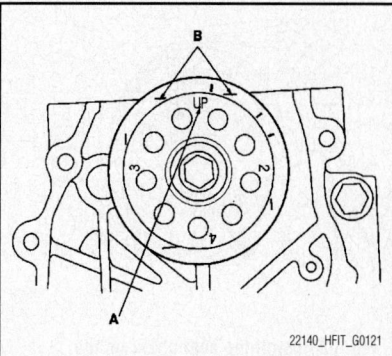

Fig. 67 Set the No. 1 piston at TDC: the
"UP" mark (A) on the camshaft sprocket
should be at the top, and the TDC grooves
(B) on the camshaft sprocket should line
up with the top edge of the head

24. Set the No. 1 piston at TDC. The
"UP" mark on the camshaft sprocket should
be at the top, and the TDC grooves on the
camshaft sprocket should line up with the
top edge of the head.
25. Install the cam chain on the crankshaft sprocket with the colored piece
aligned with the TDC mark on the crankshaft sprocket.
26. Install the cam chain on the
camshaft sprocket with the pointers
aligned with the three colored pieces as
shown.
27. Apply new engine oil to the
threads of the cam chain tensioner
mounting bolt. Tighten the bolt to
16 ft. lbs. (22 Nm).
28. Install the cam chain tensioner and
cam chain guide. Tighten the mounting
bolts to 9 ft. lbs. (12 Nm).
29. Install the cam chain tensioner
slider, and tighten the bolt loosely.

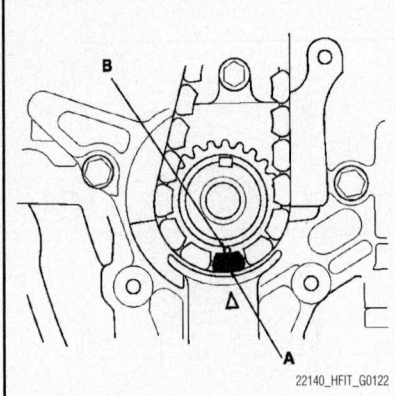

Fig. 68 Install the cam chain on the crankshaft sprocket with the colored piece (A) aligned with the TDC mark (B) on the crankshaft sprocket

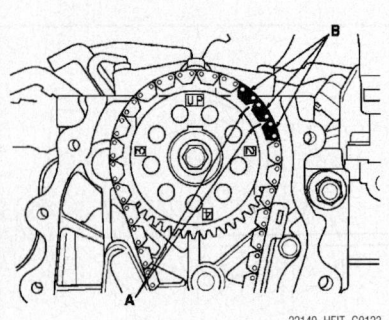

Fig. 69 Install the cam chain on the camshaft sprocket with the pointers (A) aligned with the three colored pieces (B) as shown

30. Apply new engine oil to the sliding surface of the cam chain tensioner slider.

31. Rotate the cam chain tensioner slider clockwise to compress the cam chain tensioner, and install the remaining bolt, then tighten the bolts to 9 ft. lbs. (12 Nm).

32. Check the chain case oil seal for damage If the oil seal is damaged, replace the chain case oil seal.

33. Remove the old liquid gasket from the chain case mating surfaces, the bolts, and the bolt holes.

34. Clean and dry the chain case mating surfaces.

35. Install the CKP pulse plate on the crankshaft.

36. Apply liquid gasket evenly to the engine block mating surface of the chain case.

37. Install the component within 5 minutes of applying the liquid gasket.

   a. Apply a 1.5 mm wide bead of liquid gasket along the broken lines.

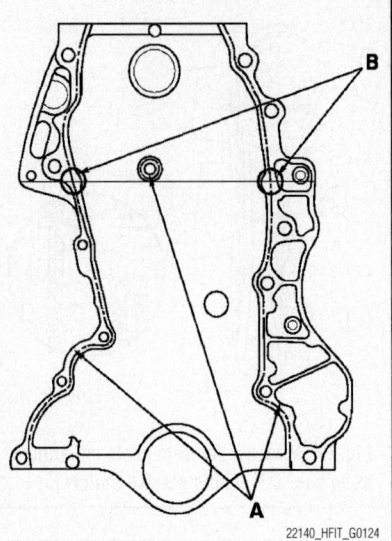

Fig. 70 Apply a 1.5 mm wide bead of liquid gasket along the broken lines (A) and apply a 3.0 mm wide bead of liquid gasket to the engine block upper surface contact areas (B) on the chain case

   b. Apply a 3.0 mm wide bead of liquid gasket to the engine block upper surface contact areas on the chain case.

38. If too much time has passed after applying the liquid gasket, remove the old liquid gasket and residue, then reapply new liquid gasket.

39. Install the chain case. Tighten the four smaller lower bolts to 9 ft. lbs. (12 Nm) and the remaining larger bolts to 23 ft. lbs. (31 Nm).

➡ **Wait at least 30 minutes to allow liquid gasket to cure before filling the engine with oil. Do not run the engine for at least 3 hours after installing the chain case.**

40. Install the harness clamps, and connect the crankshaft position (CKP) sensor connector.

41. Install the side engine mount/bracket assembly, then tighten the mounting bolts to 43 ft. lbs. (59 Nm) and support nuts to 40 ft. lbs. (54 Nm) in the numbered sequence shown.

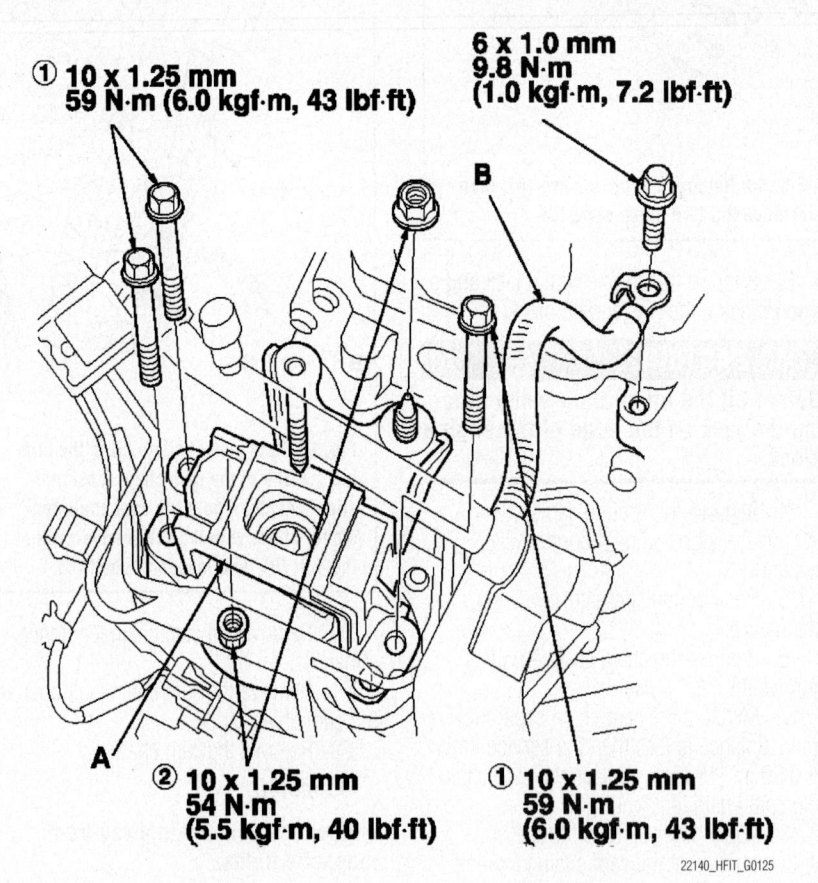

Fig. 71 Install the side engine mount/bracket assembly (A), then tighten the mounting bolts to 43 ft. lbs. (59 Nm) and support nuts to 40 ft. lbs. (54 Nm) in the numbered sequence shown

42. Install the ground cable.
43. Install the oil pan.
44. Install the crankshaft pulley as follows (do not use an impact wrench):

a. Hold the pulley with the holder handle and pulley holder attachment, then tighten the bolt to 27 ft. lbs. (37 Nm) with a torque wrench and a heavy duty 19 mm socket. If the pulley bolt or crankshaft are new, tighten the bolt to 130 ft. lbs. (177 Nm), then remove the bolt and tighten it to 27 ft. lbs. (37 Nm).

b. Mark the bolt head and the crankshaft pulley with a reference line, then tighten the bolt an additional 90° (the mark on the bolt head should line up with the mark on the crankshaft pulley).

45. Install the cylinder head cover.
46. Install the water pump pulley.
47. If equipped with A/C, install the idler pulley. Tighten the bolt to 17 ft. lbs. (24 Nm).
48. Install the alternator bracket mounting bolts. Tighten the bolt to 17 ft. lbs. (24 Nm).
49. Install the drive belt and adjust it.
50. Install the splash shield.
51. Install the front wheels.
52. Perform the crankshaft position (CKP) pattern clear/CKP pattern learn procedure as follows:

a. Start the engine. Hold the engine speed at 3,000 rpm without load (in Park or neutral) until the radiator fan comes on.

b. Test-drive the vehicle on a level road: Decelerate (with the throttle fully closed) from an engine speed of 2,500 rpm down to 1,000 rpm with the A/T in 2 position, or M/T in 1st gear. Repeat this step several times.

c. Turn the ignition switch OFF.
d. Turn the ignition switch ON (II), and wait 30 seconds.

VALVE LASH

ADJUSTMENT

See Figures 72 and 73.

➡**Valves should be adjusted only when the cylinder head temperature is less than 100°F (38°C).**

1. Remove the cylinder head cover.
2. Set the No. 1 piston at top dead center (TDC). The "UP" mark on the camshaft sprocket should be at the top, and the TDC grooves on the camshaft sprocket should line up with the top edge of the head.
3. Select the correct thickness feeler gauge for the valves to check.
4. Proper valve clearance should be:
   - Intake: 0.006–0.007 in. (0.15–0.19 mm)

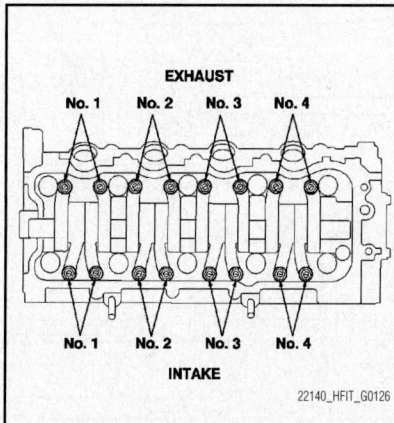

Fig. 72 Showing valve positions for reference during adjustment

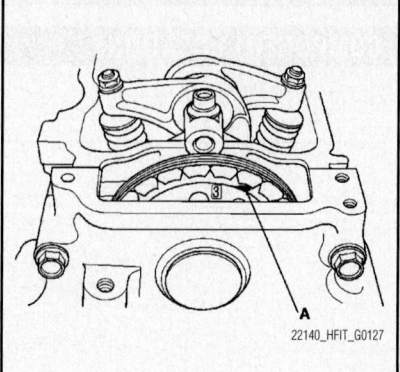

Fig. 73 Rotate the crankshaft clockwise. Align the No. 3 piston TDC groove (A) on the camshaft sprocket with the top edge of the head

- Exhaust: 0.010–0.012 in. (0.26–0.30 mm)

5. Insert the feeler gauge between the adjusting screw and the end of the valve stem and slide it back and forth; a slight amount of drag should be felt.
6. If too much or too little drag is present, loosen the locknut, and turn the adjusting screw until the drag on the feeler gauge is correct.
7. Tighten the locknut to 10 ft. lbs. (14 Nm) and recheck the clearance. Repeat the adjustment if necessary.
8. Rotate the crankshaft clockwise. Align the No. 3 piston TDC groove on the camshaft sprocket with the top edge of the head.
9. Check and if necessary, adjust the valve clearance on No. 3 cylinder.
10. Repeat this procedure for cylinder No. 4 and then for cylinder No. 2.

ENGINE PERFORMANCE & EMISSION CONTROL

COMPONENT LOCATIONS

See Figures 74 through 76.

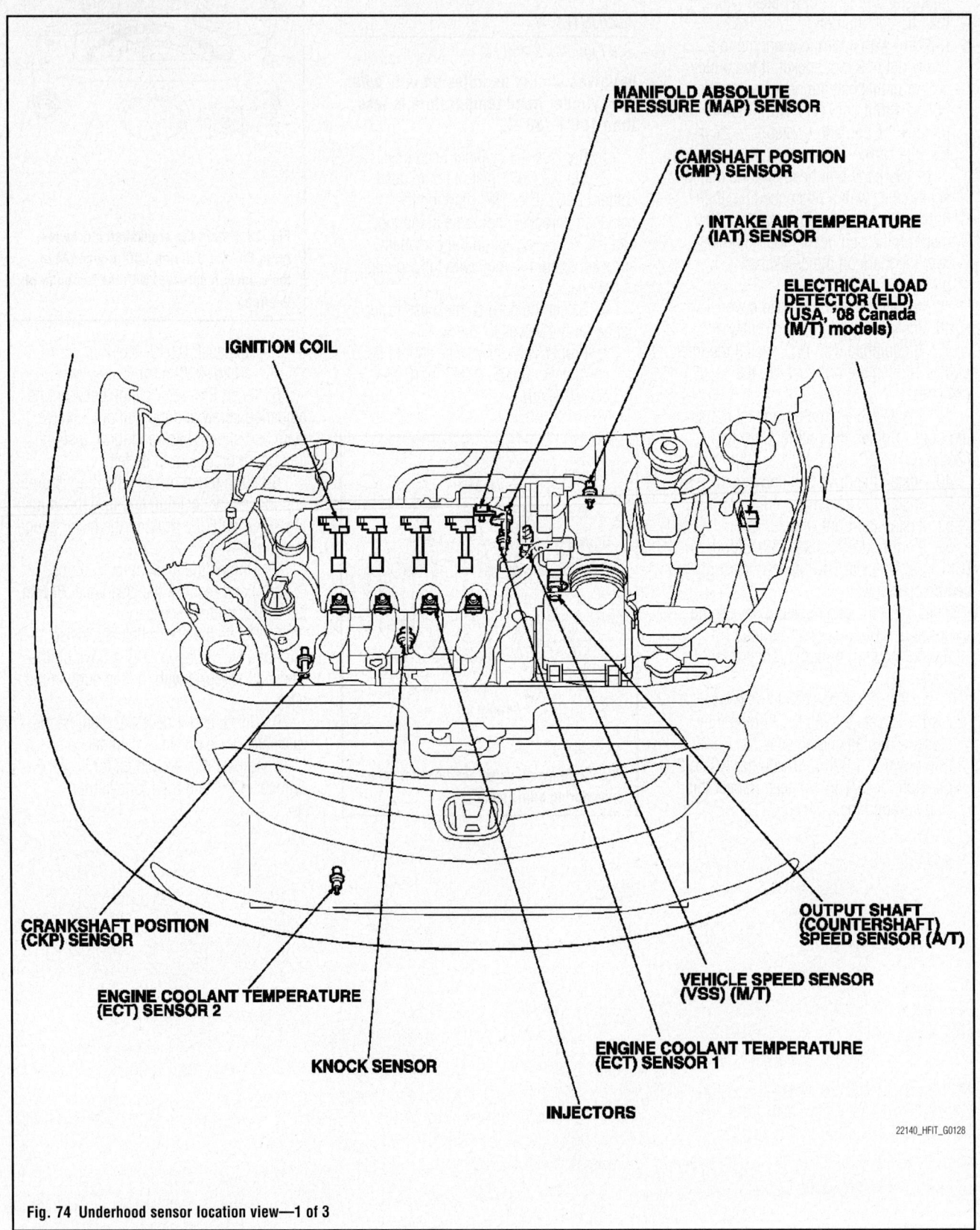

MANIFOLD ABSOLUTE
PRESSURE (MAP) SENSOR

CAMSHAFT POSITION
(CMP) SENSOR

INTAKE AIR TEMPERATURE
(IAT) SENSOR

ELECTRICAL LOAD
DETECTOR (ELD)
(USA, '08 Canada
(M/T) models)

IGNITION COIL

CRANKSHAFT POSITION
(CKP) SENSOR

OUTPUT SHAFT
(COUNTERSHAFT)
SPEED SENSOR (A/T)

VEHICLE SPEED SENSOR
(VSS) (M/T)

ENGINE COOLANT TEMPERATURE
(ECT) SENSOR 2

KNOCK SENSOR

ENGINE COOLANT TEMPERATURE
(ECT) SENSOR 1

INJECTORS

22140_HFIT_G0128

Fig. 74 Underhood sensor location view—1 of 3

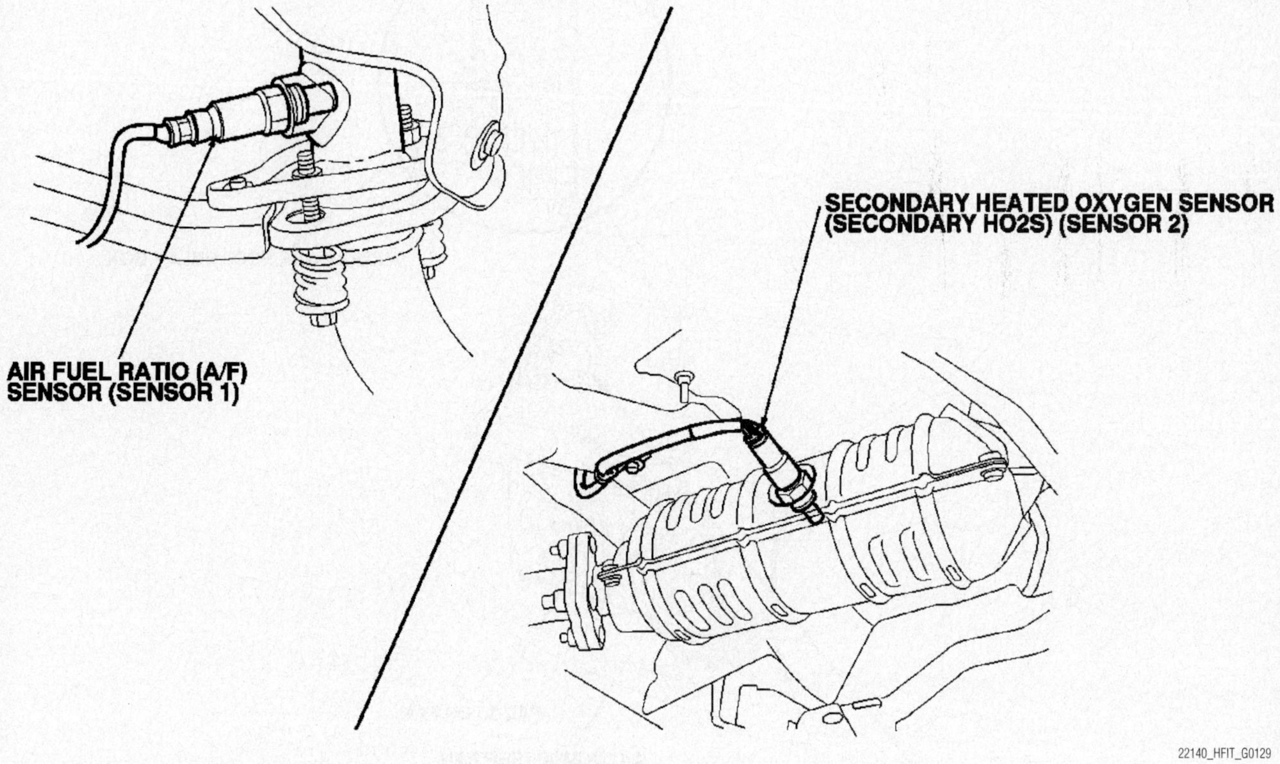

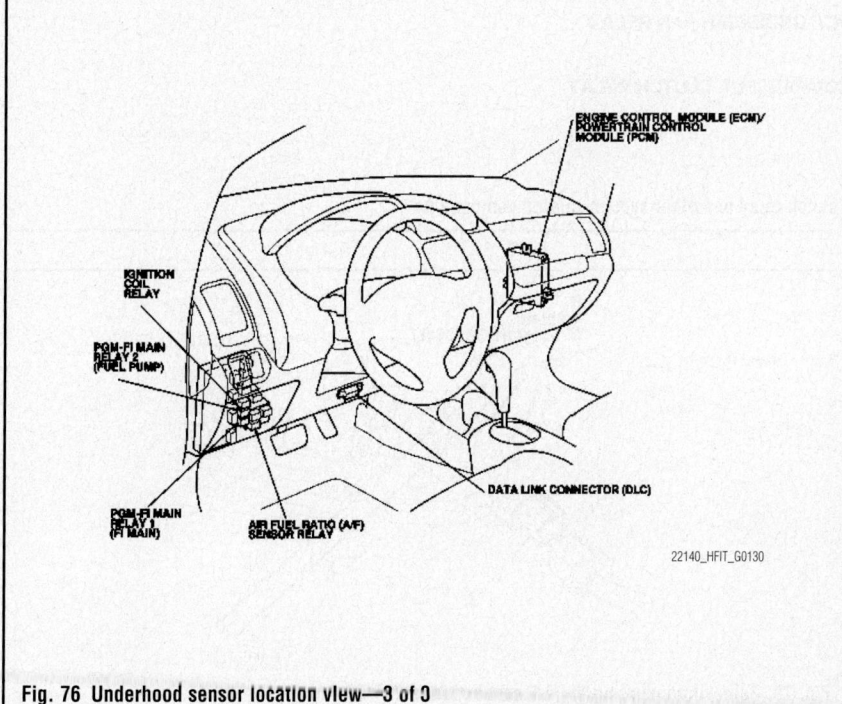

AIR FUEL RATIO (A/F)
SENSOR (SENSOR 1)

SECONDARY HEATED OXYGEN SENSOR
(SECONDARY HO2S) (SENSOR 2)

22140_HFIT_G0129

Fig. 75 Underhood sensor location view—2 of 3

ENGINE CONTROL MODULE (ECM)/
POWERTRAIN CONTROL
MODULE (PCM)

IGNITION
COIL
RELAY

PGM-FI MAIN
RELAY 2
(FUEL PUMP)

PGM-FI MAIN
RELAY 1
(FI MAIN)

AIR FUEL RATIO (A/F)
SENSOR RELAY

DATA LINK CONNECTOR (DLC)

22140_HFIT_G0130

Fig. 76 Underhood sensor location view—3 of 3

A/C COMPRESSOR CLUTCH RELAY

LOCATION
See Figure 77.

The relay is located in the underhood relay compartment as shown.

REMOVAL & INSTALLATION

1. The A/C compressor clutch relay is removed and installed in the underhood relay compartment.
2. Inspect the pins for any signs of damage or corrosion.
3. Replace the relay as needed.

AIR/FUEL RATIO SENSOR

LOCATION

➡See "Component Locations" illustration at the start of this section.

The A/F sensor is located in the exhaust manifold, just forward of the flange that connects the catalytic converter.

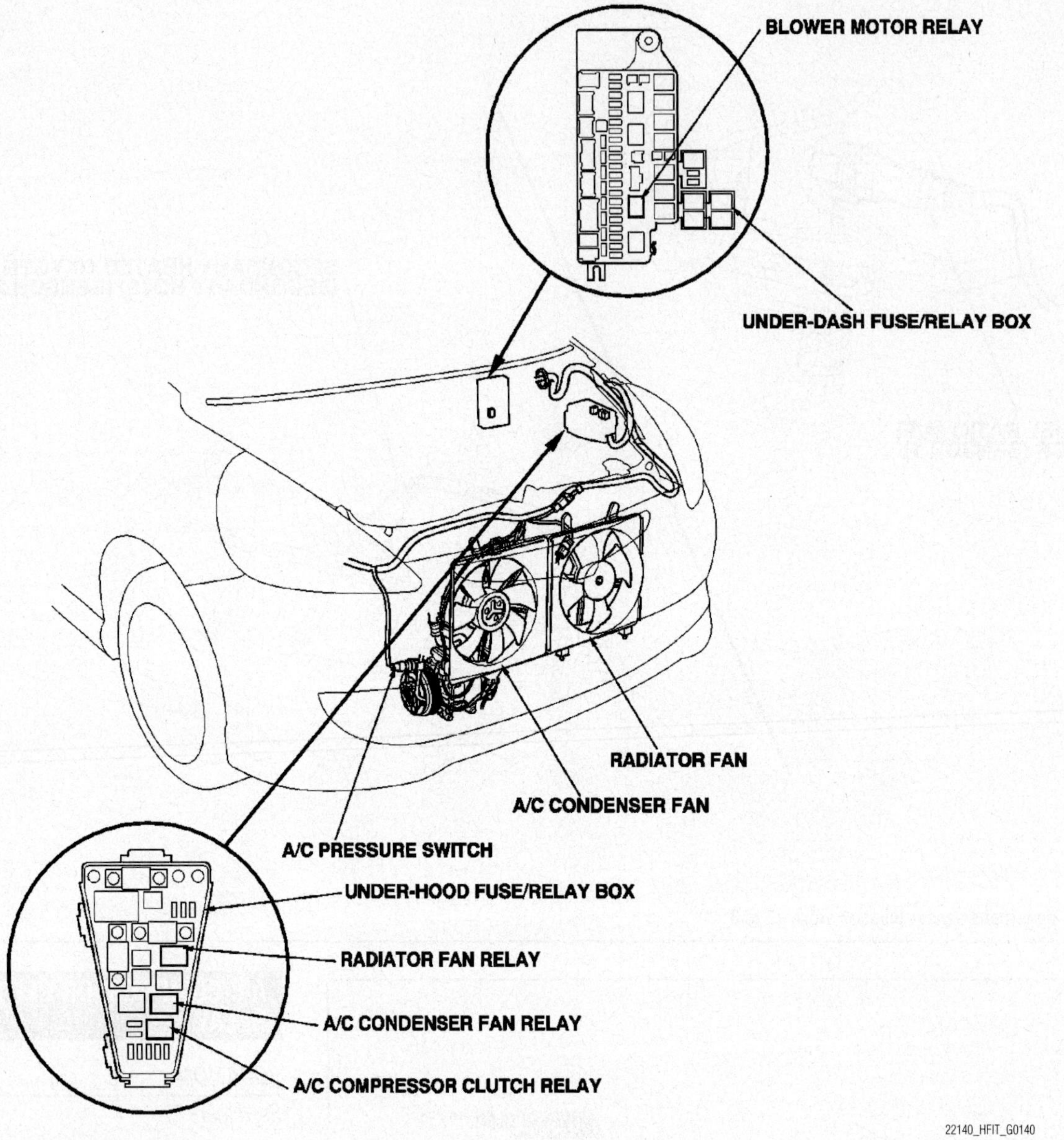

BLOWER MOTOR RELAY

UNDER-DASH FUSE/RELAY BOX

RADIATOR FAN

A/C CONDENSER FAN

A/C PRESSURE SWITCH

UNDER-HOOD FUSE/RELAY BOX

RADIATOR FAN RELAY

A/C CONDENSER FAN RELAY

A/C COMPRESSOR CLUTCH RELAY

22140_HFIT_G0140

Fig. 77 Showing the location of the A/C compressor clutch relay and other system-related components

REMOVAL & INSTALLATION

See Figure 78.

1. Disconnect the A/F sensor connector, then remove the A/F sensor.

2. Install the parts in the reverse order of removal.

3. Tighten A/F sensor to 33 ft. lbs. (44 Nm)

ACCELERATOR PEDAL POSITION (APP) SENSOR

LOCATION

See Figure 79.

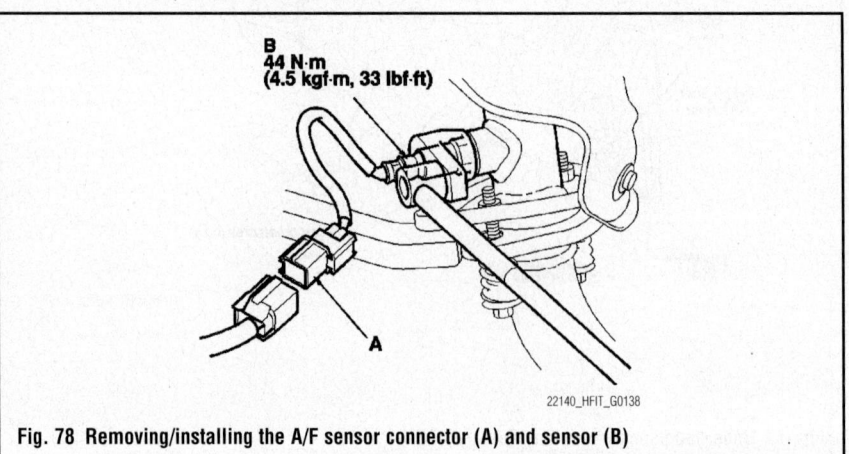

B
44 N·m
(4.5 kgf·m, 33 lbf·ft)

A

22140_HFIT_G0138

Fig. 78 Removing/installing the A/F sensor connector (A) and sensor (B)

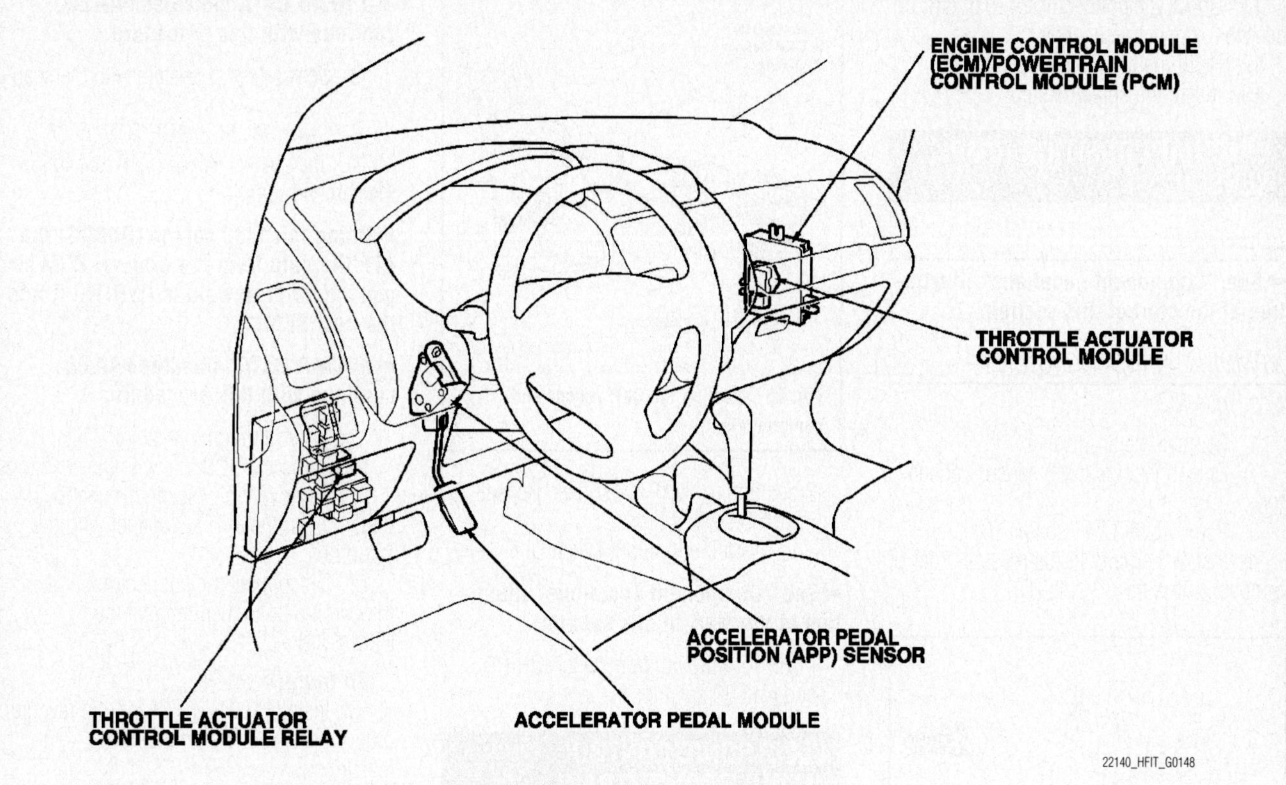

Fig. 79 Showing the location of the APP sensor and related system components

22140_HFIT_G0148

REMOVAL & INSTALLATION

See Figure 80.

1. Disconnect the accelerator pedal module connector.

2. Remove the accelerator pedal module.

➡ **The APP sensor is not available separately. Do not disassemble the accelerator pedal module.**

3. Install the parts in the reverse order of removal.

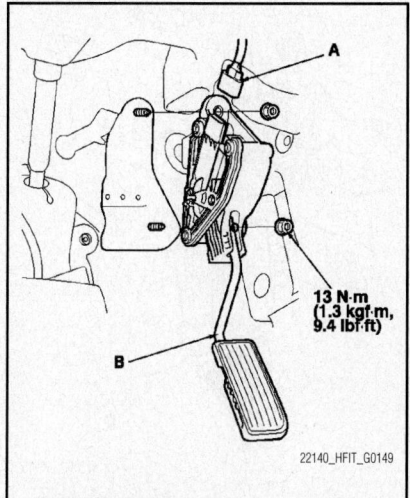

22140_HFIT_G0149

Fig. 80 Removing the accelerator pedal module (A, B), with APP sensor

AUTOMATIC TRANSMISSION FLUID (ATF) TEMPERATURE SENSOR

LOCATION

The ATF temperature sensor is located in the side of the automatic transmission housing.

REMOVAL & INSTALLATION

See Figure 81.

1. Raise the vehicle on a lift, or apply the parking brake, block both rear wheels, and raise the front of the vehicle. Make sure it is securely supported.

2. Remove the splash shield.

3. Remove the drain plug, and drain the automatic transmission fluid (ATF).

4. Reinstall the drain plug with a new sealing washer.

5. Remove the shift solenoid valve cover, dowel pins, and gasket.

6. Disconnect the connectors, remove the shift solenoid harness connector, and replace it.

7. Install the new O-ring (F) on the new shift solenoid harness connector, and install the connector in the transmission housing.

8. Connect the WHT and ORN wires connector to shift solenoid valve B. The ATF temperature sensor is assembled in the WHT wires connector.

9. Connect the harness terminals to the solenoids:

- BLU wire connector to shift solenoid valve A.
- GRN wire connector to shift solenoid valve C.
- YEL wire connector to shift solenoid valve D.

10. Install the shift solenoid valve cover, dowel pins, and a new gasket.

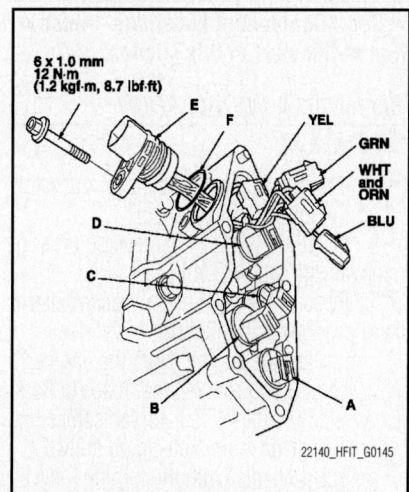

22140_HFIT_G0145

Fig. 81 Removing/installing the ATF temperature sensor

11. Check the connector for rust, dirt, or oil, then connect it securely.
12. Refill the transmission with ATF.
13. Install the splash shield.

CAMSHAFT POSITION (CMP) SENSOR

LOCATION

➡See "Component Locations" illustration at the start of this section.

REMOVAL & INSTALLATION
See Figure 82.

1. Remove the air cleaner.
2. Disconnect the CMP sensor connector (A).
3. Remove the CMP sensor (B).
4. Install the parts in the reverse order of removal with a new O-ring (C).

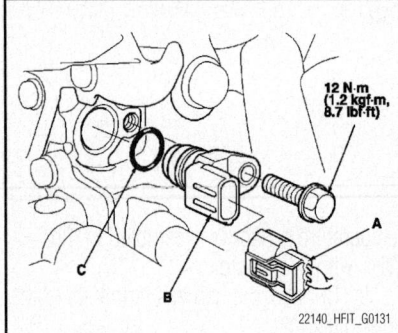

Fig. 82 Showing the CMP sensor and components

CRANKSHAFT POSITION (CKP) SENSOR

LOCATION

➡See "Component Locations" illustration at the start of this section.

REMOVAL & INSTALLATION
See Figure 83.

1. Disconnect the CKP sensor connector.
2. Remove the CKP sensor.
3. Install the parts in the reverse order of removal with a new O-ring.
4. Perform the CKP pattern clear/pattern learn procedure as follows:
   a. Start the engine. Hold the engine speed at 3,000 rpm without load (in Park or neutral) until the radiator fan comes on.
   b. Test-drive the vehicle on a level road: Decelerate (with the throttle fully closed) from an engine speed of 2,500 rpm down to 1,000 rpm with the A/T in

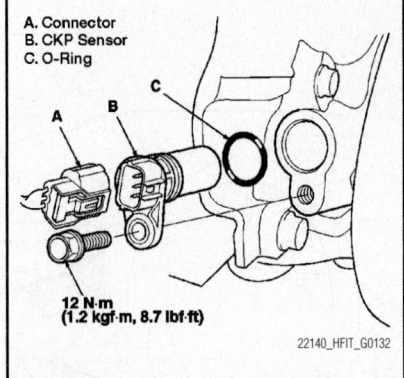

A. Connector
B. CKP Sensor
C. O-Ring

12 N·m
(1.2 kgf·m, 8.7 lbf·ft)

Fig. 83 Showing the CKP sensor and components

2 position, or M/T in 1st gear. Repeat this step several times.
   c. Turn the ignition switch OFF.

➡See "Component Locations" illustration at the start of this section.

   d. Turn the ignition switch ON (II), and wait 30 seconds.

ELECTRONIC CONTROL MODULE/POWERTRAIN CONTROL MODULE (ECM/PCM)

LOCATION

➡See "Component Locations" illustration at the start of this section.

REMOVAL & INSTALLATION
See Figure 84.

1. Connect a scan tool with proper software to the data link connector (DLC) located under the driver's side of the dashboard.
2. Turn the ignition switch ON (II).
3. Make sure the scan tool communicates with the ECM/PCM and other vehicle systems. If it doesn't, perform additional DLC circuit troubleshooting.
4. Select the PGM-FI system with the scan tool.
5. Select the INSPECTION MENU with the scan tool.
6. Select the ETCS TEST, then select the TP POSITION CHECK, and follow the screen prompts.

➡**If the TP POSITION CHECK indicates FAILED, continue with this procedure.**

7. Select the REPLACE ECM/PCM MENU, then select READ DATA and follow the screen prompts.

➡**Doing this step copies (READS) the engine oil life data from the original ECM/PCM so you can later download (WRITE) it into the new ECM/PCM.**

➡**If READ DATA indicates FAILED, continue with this procedure.**

8. Select the A/T system with the scan tool.
9. Select the REPLACE TCM/PCM MENU, then select READ DATA and follow the screen prompts.

➡**Doing this step copies (READS) the ATF life data from the original PCM so you can later download (WRITE) it into the new PCM.**

➡**If READ DATA indicates FAILED, continue with this procedure.**

   a. Turn the ignition switch OFF.
10. Jump the SCS line with the scan tool.
11. Remove the passenger's dashboard under cover, the side kick panel, and the glove box.
12. Disconnect the connectors, and remove the bolts then remove the ECM/PCM.

   To install:
13. Install the ECM/PCM in the reverse order of removal, then perform the following:
   a. Turn the ignition switch ON (II).
   b. Manually input the VIN to the ECM/PCM with the scan tool.

➡**DTC P0630 "VIN Not Programmed or Mismatch" may be stored because the VIN has not been programmed into the ECM/PCM; ignore it, and continue this procedure.**

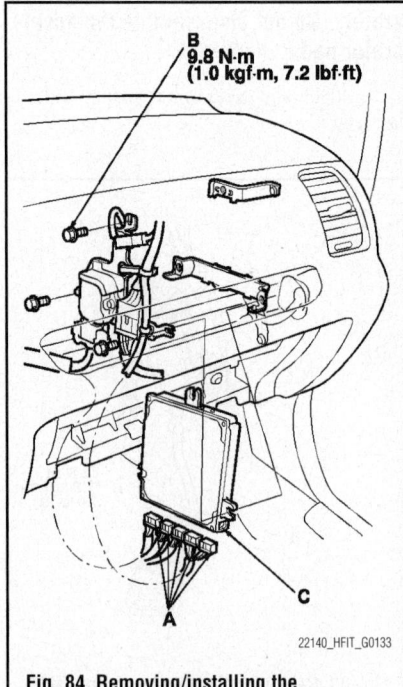

B
9.8 N·m
(1.0 kgf·m, 7.2 lbf·ft)

Fig. 84 Removing/installing the ECM/PCM—all transmissions

c. If the READ DATA (engine oil life) failed in step 7, go to step 20. Otherwise, go to step 18.

d. Select the PGM-FI system with the scan tool.

e. Select the REPLACE ECM/PCM MENU, then select WRITE DATA and follow the screen prompts.

➡**If the WRITE DATA indicates FAILED, continue with this procedure.**

f. If the READ DATA (ATF life) failed in step 9, go to step 23. Otherwise go to step 21.

g. Select the A/T SYSTEM with the scan tool.

h. Select the REPLACE TCM/PCM MENU, then select WRITE DATA and follow the screen prompts.

➡**If the WRITE DATA indicates FAILED, continue with this procedure.**

i. Select IMMOBI system with the scan tool.

j. Enter the immobilizer code with the ECM/PCM replacement procedure in the scan tool; it allows you to start the engine.

k. If the TP POSITION CHECK failed in step 6 clean the throttle body, then go to step 26.

l. If the READ DATA failed in step 7 or the WRITE DATA failed in step 19, replace the engine oil and engine oil filter, then go to step 27.

m. If the READ DATA failed in step 9 or the WRITE DATA failed in step 22, replace the ATF, then go to step 28.

n. Select PGM-FI system, and reset the ECM/PCM with the scan tool.

o. Update the ECM/PCM if it does not have the latest software.

p. Perform the ECM/PCM idle learn procedure as follows:

➡**The idle learn procedure must be done so the ECM/PCM can learn the engine idle characteristics.**

q. Do the idle learn procedure whenever you do any of these actions:
- Replace the ECM/PCM.
- Reset the ECM/PCM.
- Update the ECM/PCM.
- Replace or clean the throttle body.
- When the engine or transmission is disassembled.

➡**Clearing the DTCs with the scan tool does not require you to do the idle learn procedure.**

r. Make sure all electrical items (A/C, audio, lights, etc.) are off.

s. Reset the ECM/PCM with the scan tool.

t. Turn the ignition switch ON (II), and wait 2 seconds.

u. Start the engine. Hold the engine speed at 3,000 rpm without load (in Park or neutral) until the radiator fan comes on, or until the engine coolant temperature reaches 194°F (90°C).

v. Let the engine idle for about 5 minutes with the throttle fully closed.

➡**If the radiator fan comes on, do not include its running time in the 5 minutes**

14. Perform the CKP pattern learn procedure as follows:

a. Start the engine. Hold the engine speed at 3,000 rpm without load (in Park or neutral) until the radiator fan comes on.

b. Test-drive the vehicle on a level road: Decelerate (with the throttle fully closed) from an engine speed of 2,500 rpm down to 1,000 rpm with the A/T in 2 position, or M/T in 1st gear. Repeat this step several times.

c. Turn the ignition switch OFF.

d. Turn the ignition switch ON (II), and wait 30 seconds.

ENGINE COOLANT TEMPERATURE (ECT) SENSOR

LOCATION

The ECT sensors 1 and 2 are threaded into the engine block as shown in the illustrations below.

REMOVAL & INSTALLATION

ECT Sensor 1

See Figure 85.

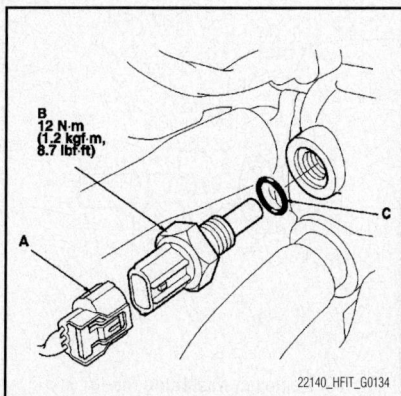

Fig. 85 Removing/installing the ECT sensor—sensor 1

1. Drain the engine coolant.
2. Remove the air cleaner.
3. Disconnect the ECT sensor 1 connector.
4. Remove the ECT sensor 1.
5. Install the parts in the reverse order of removal with a new O-ring.
6. Tighten the sensor to 9 ft. lbs. (12 Nm).
7. Refill the radiator with engine coolant.

ECT Sensor 2

See Figure 86.

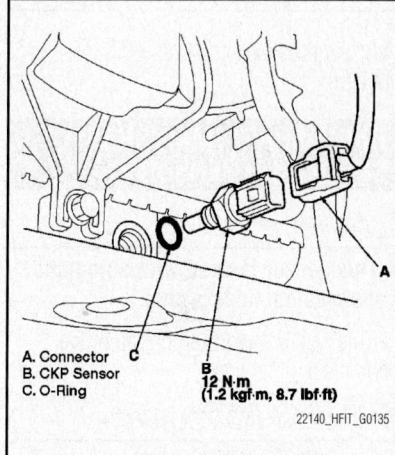

A. Connector
B. CKP Sensor
C. O-Ring
12 N·m
(1.2 kgf·m, 8.7 lbf·ft)

Fig. 86 Removing/installing the ECT sensor—sensor 2

1. Drain the engine coolant.
2. Remove the splash shield.
3. Disconnect the ECT sensor 2 connector.
4. Remove ECT sensor 2.
5. Install the parts in the reverse order of removal with a new O-ring.
6. Tighten the ECT sensor 2 to 9 ft. lbs. (12 Nm).
7. Refill the radiator with engine coolant.

HEATED OXYGEN (HO2S) SENSOR

LOCATION

See Figure 87.

The secondary HO2S is located on the TWC.

REMOVAL & INSTALLATION

1. Remove the center console, and the rear heater upper duct.
2. Disconnect the secondary HO2S connector.
3. Remove the secondary HO2S.
4. Install the parts in the reverse order of removal.

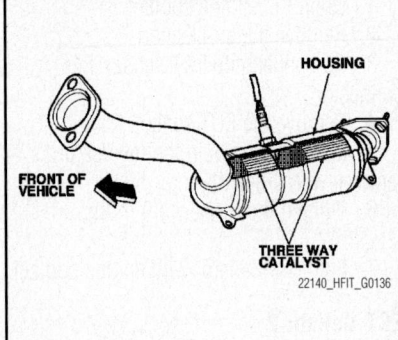

Fig. 87 Showing the location of the secondary HO2S sensor

5. Tighten the sensor to 33 ft. lbs. (44 Nm).

INPUT SHAFT (MAINSHAFT) SPEED (ISS) SENSOR

LOCATION

➡**This sensor is used with automatic transmission models only.**

This sensor is located on top of the transmission housing.

REMOVAL & INSTALLATION

See Figure 88.

1. Make sure you have anti-theft code for the audio system.
2. Remove the air cleaner assembly.
3. Disconnect the negative cable from the battery, then disconnect the positive cable.
4. Remove the battery hold-down bracket, and remove the battery and the battery tray.
5. Remove the harness clamps from the battery base, then remove the battery base.
6. Remove the ATF filter holder, then move the ATF filter aside, leaving it connected to the ATF hoses.
7. Disconnect the input shaft (mainshaft) speed sensor connector.
8. Remove the input shaft (mainshaft) speed sensor.

To install:

9. Install the new O-ring on the new input shaft (mainshaft) speed sensor, then install the input shaft (mainshaft) speed sensor in the transmission housing.
10. Check the connector for rust, dirt, or oil, then connect the connector securely.
11. Install the ATF filter.
12. Install the battery base, then install the harness clamps.

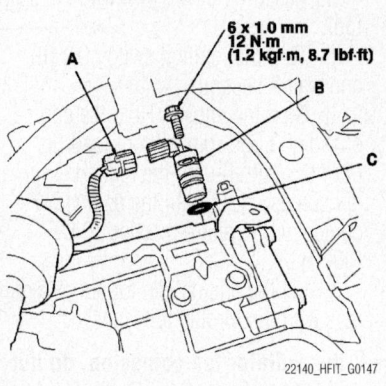

Fig. 88 Removing/installing the ISS sensor connector (A) and sensor (B) and O-ring (C)

13. Install the battery tray and the battery, then secure the battery with its hold-down bracket.
14. Connect the battery terminals, and apply grease around the battery terminals.
15. Install the air cleaner assembly.
16. Enter the anti-theft code for the audio system. Set the clock.

INTAKE AIR TEMPERATURE (IAT) SENSOR

LOCATION

➡**See "Component Locations" illustration at the start of this section.**

The IAT sensor is located in a grommet in the side of the air cleaner housing assembly.

REMOVAL & INSTALLATION

See Figure 89.

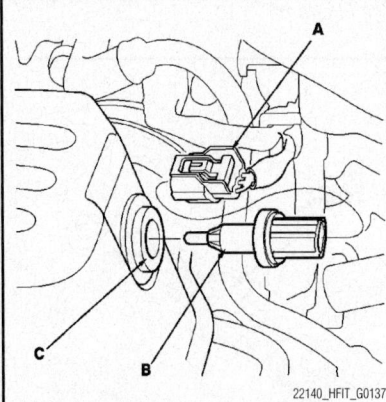

Fig. 89 Removing/installing the IAT sensor (B), connector (A) from the air cleaner housing grommet (C)

Disconnect the IAT sensor connector. Pull the IAT sensor (B) out of its grommet.

KNOCK SENSOR (KS)

LOCATION

➡**See "Component Locations" illustrations at the start of this section.**

REMOVAL & INSTALLATION

See Figure 90.

1. Remove the intake manifold. See "Intake Manifold" under "ENGINE MECHANICAL" section.
2. Disconnect the knock sensor connector.
3. Remove the knock sensor.
4. Install the parts in the reverse order of removal.
5. Tighten the knock sensor to 23 ft. lbs. (31 Nm).

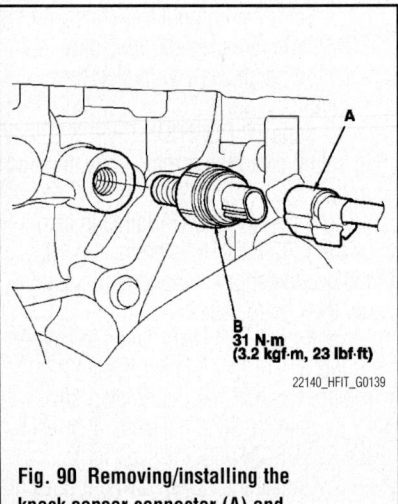

Fig. 90 Removing/installing the knock sensor connector (A) and sensor (B)

MANIFOLD ABSOLUTE PRESSURE (MAP) SENSOR

LOCATION

The MAP sensor is located on top of the intake manifold.

REMOVAL & INSTALLATION

See Figure 91.

1. Disconnect the MAP sensor connector.
2. Remove the screw.
3. Remove the MAP sensor.
4. Install the parts in the reverse order of removal with a new O-ring.

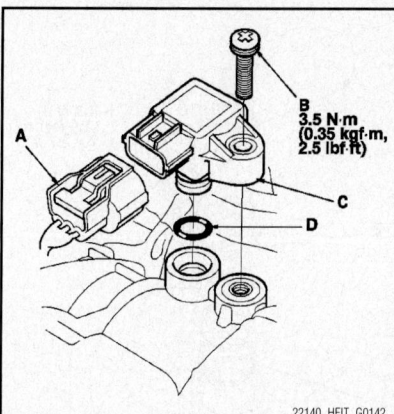

**Fig. 91 Removing/installing the MAP
sensor connector (A), screw (B) and
sensor (C) and O-ring (D)**

OUTPUT SHAFT SPEED (OSS) SENSOR

LOCATION

➡**This sensor is used on automatic transmission models only.**

The OSS sensor is located near the rear of the transmission housing, next to the bulkhead.

REMOVAL & INSTALLATION

See Figure 92.

1. Remove the air cleaner assembly.
2. Disconnect the output shaft (countershaft) speed sensor connector.
3. Remove the output shaft (countershaft) speed sensor.
4. Install the new O-ring on the new output shaft (countershaft) speed sensor, then install the output shaft (countershaft) speed sensor in the transmission housing. Tighten the sensor to 9 ft. lbs. (12 Nm).

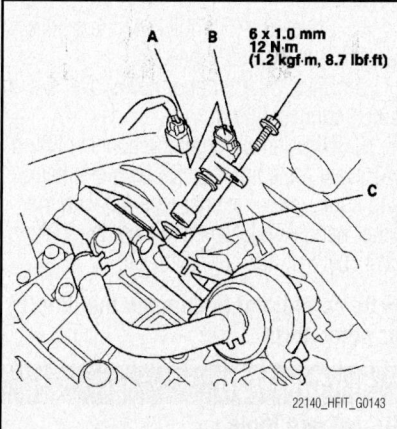

**Fig. 92 Removing/installing the OSS
sensor connector (A), sensor (B) and
O-ring (C)**

5. Check the connector for rust, dirt, or oil, then connect the connector securely.
6. Install the air cleaner assembly.

THROTTLE ACTUATOR CONTROL MODULE

LOCATION

The control module is located under the instrument panel on the passenger's side of the vehicle.

REMOVAL & INSTALLATION

See Figure 93.

1. Remove the passenger's dashboard under cover, the side kick panel, and the glove box.
2. Push the tab, and disconnect the throttle actuator control module connector.
3. Remove the bolts, and the throttle actuator control module.
4. Install the parts in the reverse order of removal.

THROTTLE POSITION SENSOR (TPS)

LOCATION

See Figure 94.

REMOVAL & INSTALLATION

➡**Manufacturer does not provide a specific procedure for this component.**

VEHICLE SPEED SENSOR (VSS)

LOCATION

➡**This sensor is used with manual transmission models only.**

REMOVAL & INSTALLATION

See Figure 95.

1. Remove the air cleaner.
2. Disconnect the VSS connector.
3. Remove the VSS.
4. Install the parts in the reverse order removal with new O-ring.

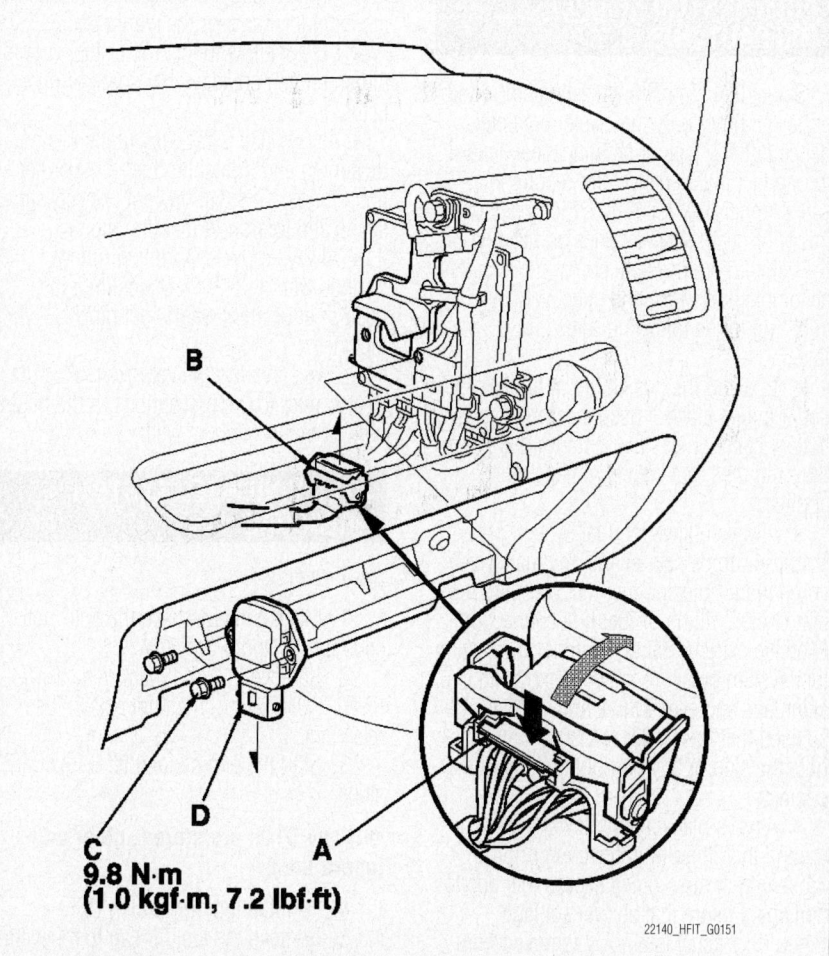

**Fig. 93 Removing/installing the throttle actuator control module by removing the tab (A),
connector (B), bolts (C) and module (D)**

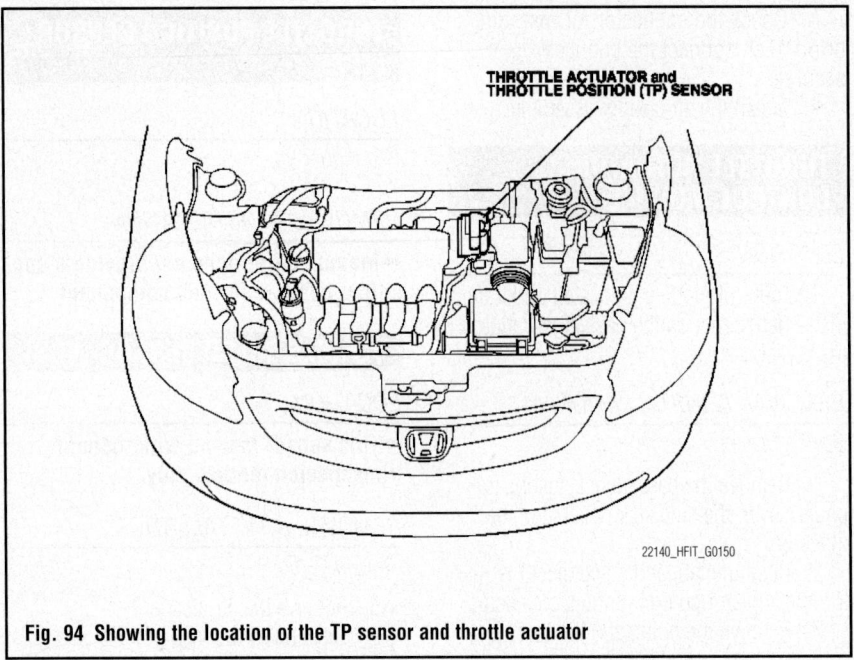

THROTTLE ACTUATOR and
THROTTLE POSITION (TP) SENSOR

22140_HFIT_G0150

Fig. 94 Showing the location of the TP sensor and throttle actuator

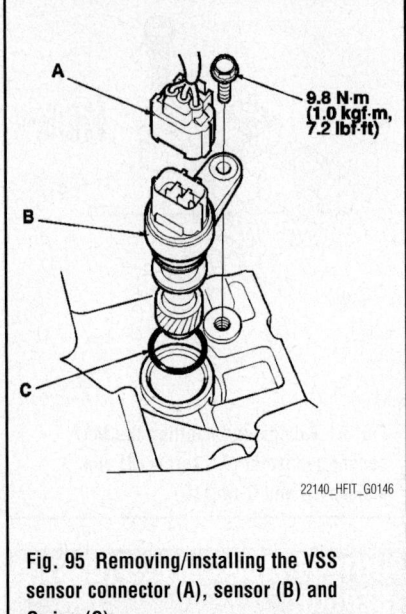

A

9.8 N·m
(1.0 kgf·m,
7.2 lbf·ft)

B

C

22140_HFIT_G0146

Fig. 95 Removing/installing the VSS sensor connector (A), sensor (B) and O-ring (C)

FUEL | GASOLINE FUEL INJECTION SYSTEM

FUEL SYSTEM SERVICE PRECAUTIONS

Safety is the most important factor when performing not only fuel system maintenance but any type of maintenance. Failure to conduct maintenance and repairs in a safe manner may result in serious personal injury or death. Maintenance and testing of the vehicle's fuel system components can be accomplished safely and effectively by adhering to the following rules and guidelines.

• To avoid the possibility of fire and personal injury, always disconnect the negative battery cable unless the repair or test procedure requires that battery voltage be applied.

• Always relieve the fuel system pressure prior to disconnecting any fuel system component (injector, fuel rail, pressure regulator, etc.), fitting or fuel line connection. Exercise extreme caution whenever relieving fuel system pressure to avoid exposing skin, face and eyes to fuel spray. Please be advised that fuel under pressure may penetrate the skin or any part of the body that it contacts.

• Always place a shop towel or cloth around the fitting or connection prior to loosening to absorb any excess fuel due to spillage. Ensure that all fuel spillage (should it occur) is quickly removed from engine surfaces. Ensure that all fuel soaked cloths or towels are deposited into a suitable waste container.

• Always keep a dry chemical (Class B) fire extinguisher near the work area.

• Do not allow fuel spray or fuel vapors to come into contact with a spark or open flame.

• Always use a back-up wrench when loosening and tightening fuel line connection fittings. This will prevent unnecessary stress and torsion to fuel line piping.

• Always replace worn fuel fitting O-rings with new Do not substitute fuel hose or equivalent where fuel pipe is installed.

Before servicing the vehicle, make sure to also refer to the precautions in the beginning of this section as well.

RELIEVING FUEL SYSTEM PRESSURE

See Figure 96.

1. Make sure you have the anti-theft code for the audio system.
2. Remove PGM-FI main relay 2 (FUEL PUMP) from the under-dash fuse/relay box.
3. Start the engine, and let it idle until it stalls.

➡**If any DTCs are stored, clear and ignore them.**

4. Turn the ignition switch OFF.
5. Remove the fuel fill cap to relieve the pressure in the fuel tank.
6. Disconnect the negative cable from the battery.

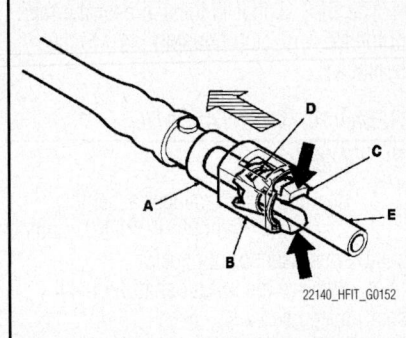

D
C
A
E
B

22140_HFIT_G0152

Fig. 96 Disconnecting the quick-connect fitting to relieve fuel system pressure

7. Remove the quick-connect fitting cover on the fuel line connection near the bulkhead.
8. Check the fuel quick-connect fitting for dirt, and clean it if needed.
9. Place a rag or shop towel over the quick-connect fitting.
10. Disconnect the quick-connect fitting (A): Hold the connector (B) with one hand, and squeeze the retainer tabs (C) with the other hand to release them from the locking tabs (D). Pull the connector off.

➡**Be careful not to damage the line (E) or other parts.**

❋❋ CAUTION
Do not use tools.

11. If the connector does not move, keep the retainer tabs pressed down, and

alternately pull and push the connector until it comes off easily.

12. Do not remove the retainer from the line; once removed, the retainer must be replaced with a new one.

13. After disconnecting the quick-connect fitting, check it for dirt or damage.

14. Reconnect the negative cable to the battery, then do this:

15. Enter the anti-theft code for the audio system.

16. Set the clock.

FUEL FILTER

REMOVAL & INSTALLATION
See Figure 97.

➡The fuel filter should be replaced whenever the fuel pressure drops below the specified value, after making sure that the fuel pump and the fuel pressure regulator are OK.

1. Remove the fuel tank pump/sending unit. See "Fuel Pump" in this section.
2. Remove the fuel filter set (A).

To install:
3. Install a new fuel filter set.
4. Check these items before installing the fuel tank unit:

a. When connecting the wire harness, make sure the connection is secure and the connectors are firmly locked into place.

b. When installing the fuel gauge sending unit, make sure the connection is secure and the connector is firmly locked into place. Be careful not to bend or twist it excessively.

5. Install the parts in the reverse order of removal with new O-rings and new base gasket. When installing the fuel tank unit, align the marks on the unit and the fuel tank.

6. Install the fuel pump assembly into the fuel tank. See "Fuel Pump" in this section.

➡Coat the O-rings with clean engine oil.

FUEL INJECTORS

REMOVAL & INSTALLATION
See Figures 98 and 99.

1. Relieve the fuel pressure. See procedure in this section.
2. Remove the air cleaner.
3. Remove the intake manifold. See "Intake Manifold" in "ENGINE MECHANICAL" section.
4. Disconnect the connectors from the injectors, and remove the wire harness from the harness clips from the fuel rail.
5. Disconnect the quick-connect fittings.
6. Remove the fuel rail bracket.
7. Remove the fuel rail mounting nuts from the fuel rail.
8. Remove the injector clips from the injectors.
9. Remove the injectors from the fuel rail.

To install:
10. Coat the new O-rings with clean engine oil, then insert the injectors into the fuel rail.
11. Install the injector clips.

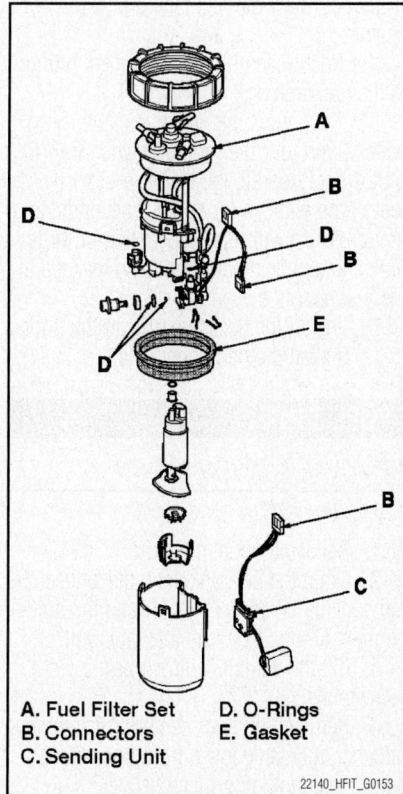

A. Fuel Filter Set   D. O-Rings
B. Connectors   E. Gasket
C. Sending Unit

22140_HFIT_G0153

Fig. 97 Exploded view of the fuel pump assembly, showing the fuel filter location

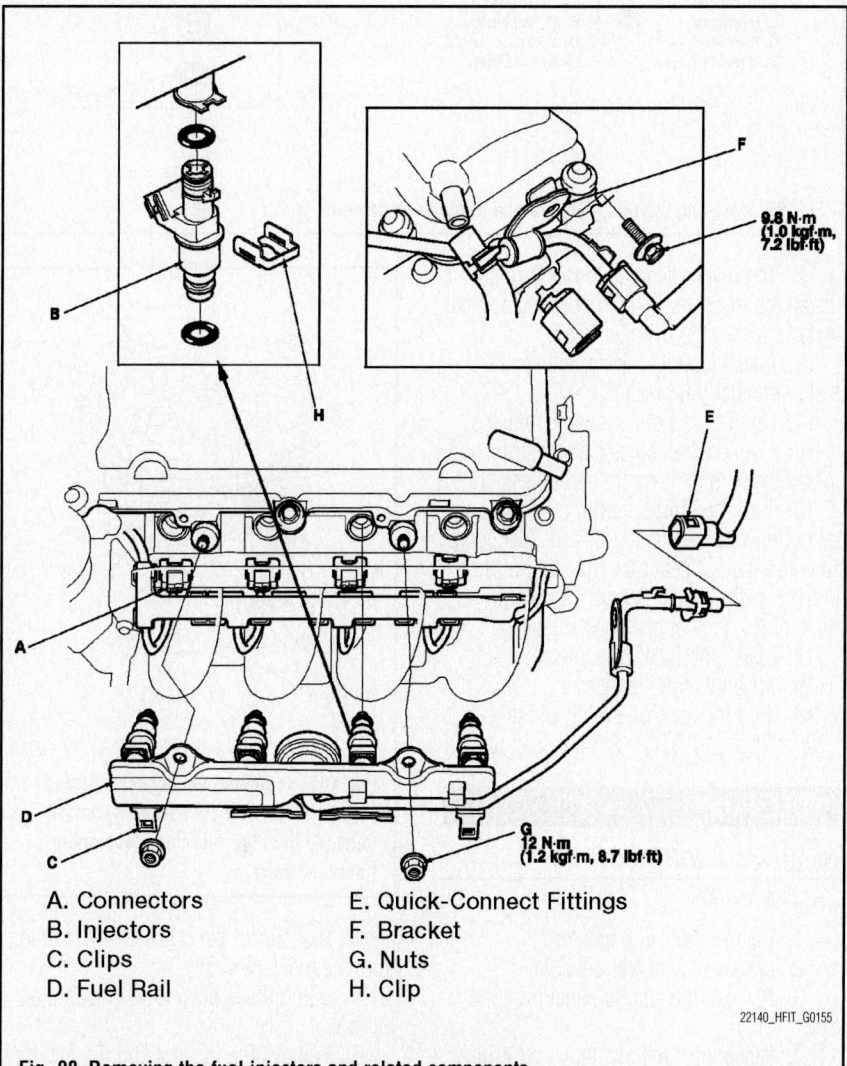

A. Connectors   E. Quick-Connect Fittings
B. Injectors   F. Bracket
C. Clips   G. Nuts
D. Fuel Rail   H. Clip

22140_HFIT_G0155

Fig. 98 Removing the fuel injectors and related components

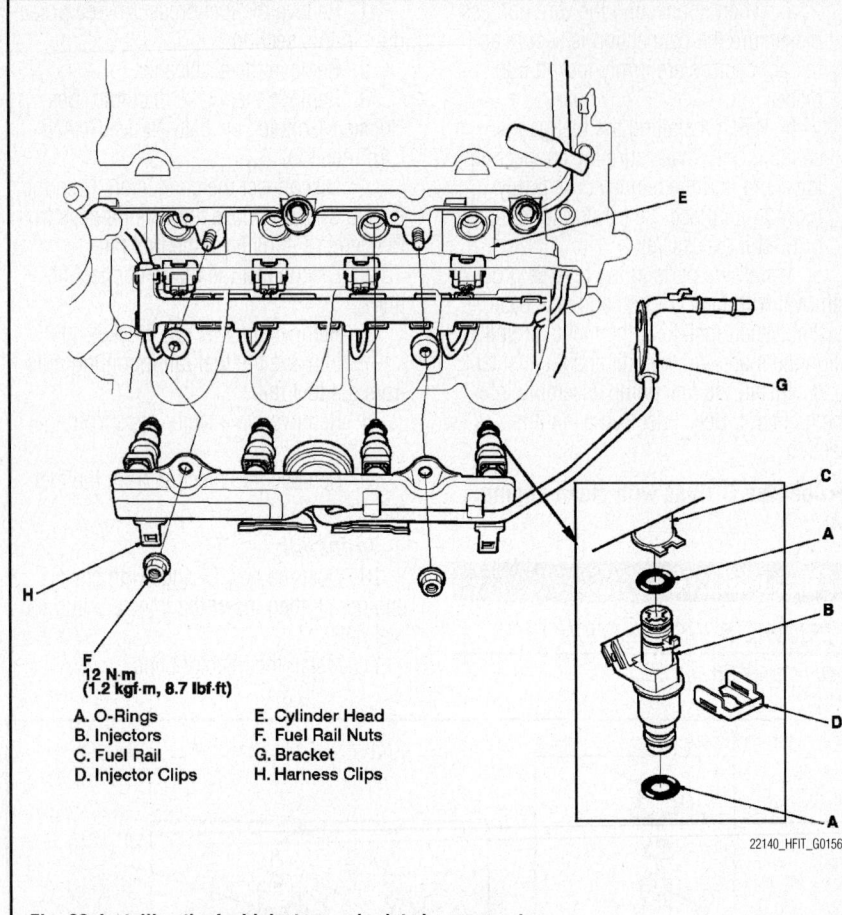

F
12 N·m
(1.2 kgf·m, 8.7 lbf·ft)

A. O-Rings   E. Cylinder Head
B. Injectors   F. Fuel Rail Nuts
C. Fuel Rail   G. Bracket
D. Injector Clips   H. Harness Clips

22140_HFIT_G0156

Fig. 99 Installing the fuel injectors and related components

12. To prevent damage to the O-rings, install the injectors in the fuel rail first, then install them in the cylinder head.

13. Install the fuel rail mounting nuts. Install the fuel rail bracket.

14. Connect the quick-connect fittings.

15. Connect the injector connectors, and install the wire harness to the harness clips.

16. Turn the ignition switch ON (II), but do not operate the starter. After the fuel pump runs for about 2 seconds, the fuel pressure in the fuel line rises. Repeat this two or three times, then check for fuel leakage.

17. Install the intake manifold.

18. Install the air cleaner.

19. Turn the ignition switch ON (II).

20. Clear any DTCs with the scan tool.

FUEL PUMP

REMOVAL & INSTALLATION

See Figure 100.

1. Relieve the fuel pressure.

2. Remove the center console.

3. Remove the access panel from the floor.

4. Disconnect the fuel tank unit 4-pin connector.

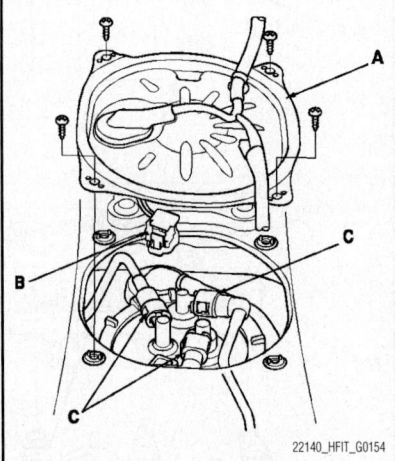

22140_HFIT_G0154

Fig. 100 Removing the access panel (A) and disconnecting connector (B) and fittings (C) from the fuel pump assembly (fuel tank unit)

5. Disconnect the quick-connect fittings from the fuel tank unit.

6. Using the special tool, loosen the locknut.

7. Remove the locknut and the fuel tank unit (fuel pump assembly).

8. Disassembly, if needed, to replace fuel filter or other components.

To install:

9. Temporarily attach a new base gasket to the fuel tank unit, then insert the fuel tank unit partially into the fuel tank.

> ❉❉ **CAUTION**
>
> **Be careful not to damage the new base gasket. Be careful not to bend the fuel gauge sending unit. Do not coat the base gasket with oil.**

10. Transfer the base gasket from the fuel tank unit to the fuel tank.

11. Align the marks on the fuel tank and the fuel tank unit, then insert the fuel tank unit until it sits on the base gasket.

> ❉❉ **CAUTION**
>
> **To prevent a fuel leak, check the base gasket, visually or by hand, to make sure it is not pinched.**

12. Using the tool, tighten the new locknut to 69 ft. lbs. (93 Nm).

13. After tightening, make sure the marks are still aligned.

14. After installation, check the base gasket visually or by hand to be sure it is not pinched.

15. Connect the fuel tank unit 4-pin connector.

16. Reconnect the quick-connect fitting to the fuel tank unit.

17. Reconnect the negative cable to the battery, and turn the ignition switch ON (II) (but do not operate the starter motor). The fuel pump will run for about 2 seconds, and fuel pressure will rise. Repeat two or three times, and check that there is no leakage in the fuel supply system.

18. Install the access panel on the floor.

19. Install the center console.

FUEL TANK

REMOVAL & INSTALLATION

See Figure 101.

1. Relieve the fuel pressure.

2. Drain the fuel tank, then disconnect the fuel tank unit 4-pin connector and the quick-connect fittings from the fuel tank unit.

3. Lift the vehicle, and support it with jack stands.

4. Remove the floor center crossmember stiffener, fuel tank cover, front floor cross beam, and tank mount bracket.

5. Disconnect the fuel fill hose, fuel suction tube, and fuel tank vapor control valve hose.

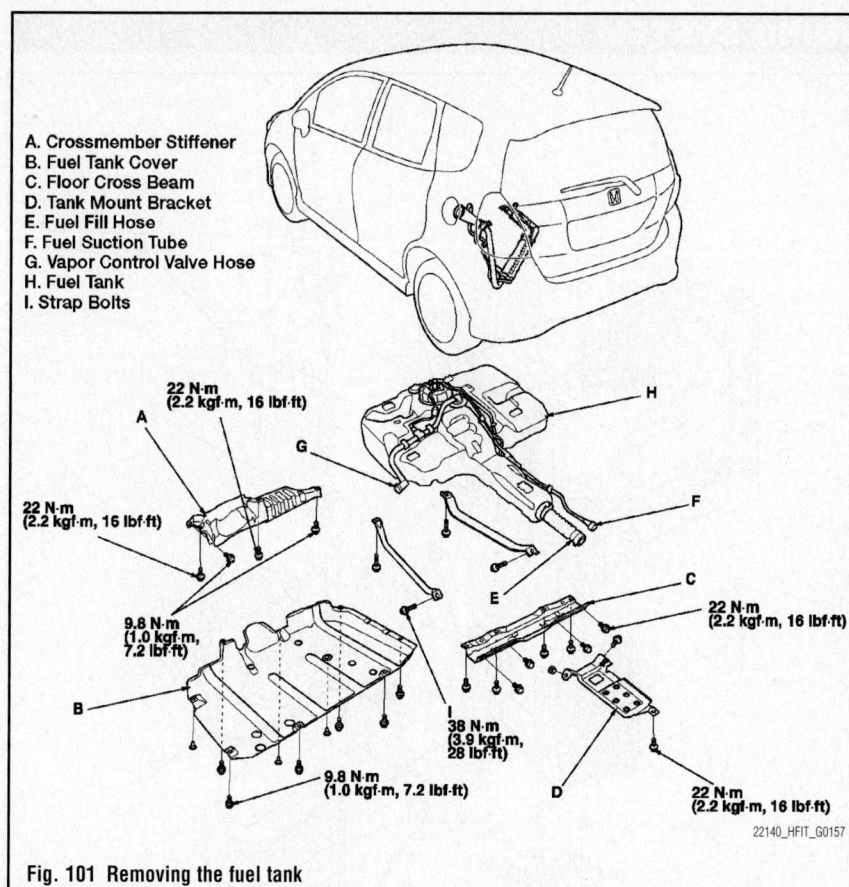

A. Crossmember Stiffener
B. Fuel Tank Cover
C. Floor Cross Beam
D. Tank Mount Bracket
E. Fuel Fill Hose
F. Fuel Suction Tube
G. Vapor Control Valve Hose
H. Fuel Tank
I. Strap Bolts

22 N·m (2.2 kgf·m, 16 lbf·ft)
22 N·m (2.2 kgf·m, 16 lbf·ft)
9.8 N·m (1.0 kgf·m, 7.2 lbf·ft)
9.8 N·m (1.0 kgf·m, 7.2 lbf·ft)
38 N·m (3.9 kgf·m, 28 lbf·ft)
22 N·m (2.2 kgf·m, 16 lbf·ft)
22 N·m (2.2 kgf·m, 16 lbf·ft)

22140_HFIT_G0157

Fig. 101 Removing the fuel tank

6. Place a jack or other support under the fuel tank, then remove the strap bolts.

7. Remove the fuel tank.

8. Install the parts in the reverse order of removal. Note the tightening specifications as shown.

THROTTLE BODY

REMOVAL & INSTALLATION

See Figure 102.

※ WARNING

Do not insert fingers into the installed throttle body when turning the ignition switch ON (II) or while the ignition switch is ON (II). Serious injury to fingers will result if the throttle valve is activated.

➡ If replacing or cleaning the throttle body, start at step 1. If removing the throttle body, start at step 5.

1. Connect the HDS while the engine is stopped.

2. Select the INSPECTION MENU on the scan tool.

3. Do the TP POSITION CHECK in the ETCS TEST.

4. Turn the ignition switch OFF.

5. Remove the air cleaner.

6. Disconnect the throttle body connector and the EVAP canister purge valve connector.

7. Disconnect and plug the water bypass hoses.

8. Disconnect the vacuum hose.

9. Remove the throttle body.

10. Install the parts in the reverse order of removal with a new gasket.

11. Tighten the retaining bolts to 17 ft. lbs. (24 Nm).

12. Perform the ECM/PCM idle learn procedure after replacing throttle body as follows:

a. Make sure all electrical items (A/C, audio, lights, etc.) are off.

b. Reset the ECM/PCM with the scan tool.

c. Turn the ignition switch ON (II), and wait 2 seconds.

d. Start the engine. Hold the engine speed at 3,000 rpm without load (in Park or neutral) until the radiator fan comes on, or until the engine coolant temperature reaches 194°F (90°C).

e. Let the engine idle for about 5 minutes with the throttle fully closed.

➡ If the radiator fan comes on, do not include its running time in the 5 minutes.

13. Refill the radiator with engine coolant.

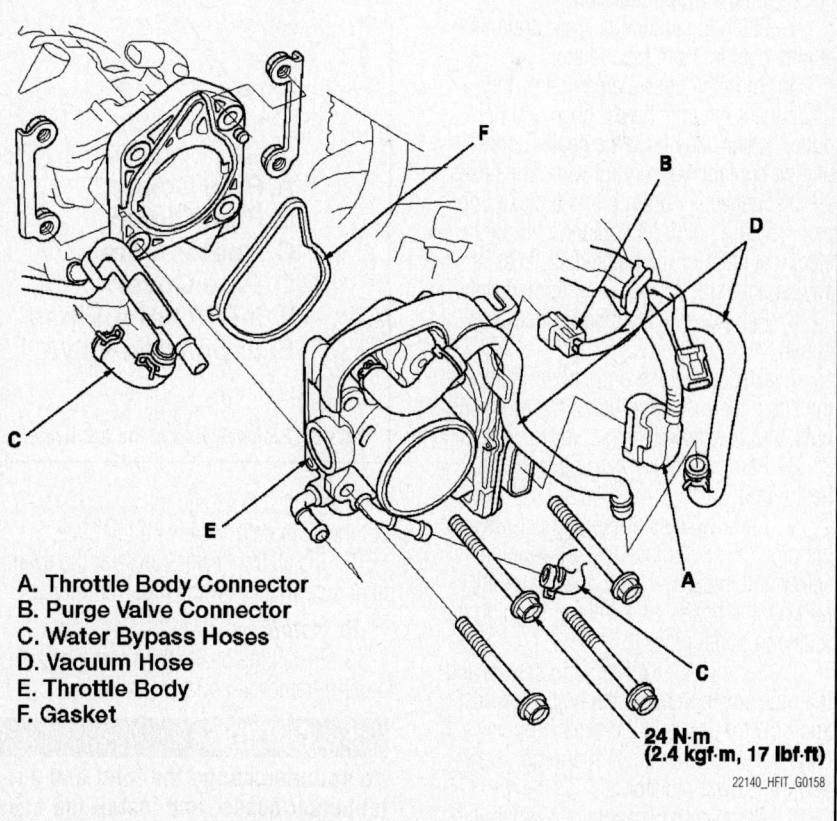

A. Throttle Body Connector
B. Purge Valve Connector
C. Water Bypass Hoses
D. Vacuum Hose
E. Throttle Body
F. Gasket

24 N·m (2.4 kgf·m, 17 lbf·ft)

22140_HFIT_G0158

Fig. 102 Removing the throttle body

HEATING & AIR CONDITIONING SYSTEM

BLOWER MOTOR

REMOVAL & INSTALLATION

1. Disconnect the negative battery cable.
2. Disconnect the blower motor electrical connector.
3. Remove the blower motor retaining screws.
4. Remove the blower motor.
5. Installation is the reverse of the removal procedure.

HEATER CORE

REMOVAL & INSTALLATION

See Figure 103.

❄ WARNING

SRS components are located in this area. Review the SRS component locations and precautions and procedures in the SRS section before doing repairs or service.

1. Make sure you have the anti-theft codes for the audio system.
2. Disconnect the negative cable from the battery.
3. Disconnect the suction and receiver lines from the evaporator core.
4. When the engine is cool, drain the engine coolant from the radiator.
5. Slide the hose clamps back, then disconnect the inlet heater hose and the outlet heater hose from the heater core. Engine coolant will run out when the hoses are disconnected; drain it into a clean drip pan. Be sure not to let coolant spill on the electrical parts or the painted surfaces. If any coolant spills, rinse it off immediately.
6. Remove the dashboard (instrument panel).
7. Disconnect the drain hose. Remove the clips, the mounting bolts, the mounting nuts, and the blower-heater unit.
8. Remove the blower unit from the heater unit.
9. Remove the self-tapping screws and the pipe cover. Remove the self-tapping screw and the pipe clamp, then carefully pull out the heater core without bending the inlet and outlet pipes.
10. Remove the self-tapping screws and the pipe cover. Remove the bolts, the inlet and outlet pipes, and the expansion valve. If necessary, remove the self-tapping screws and the blower resistor.
11. Remove the evaporator temperature sensor connector, then remove the

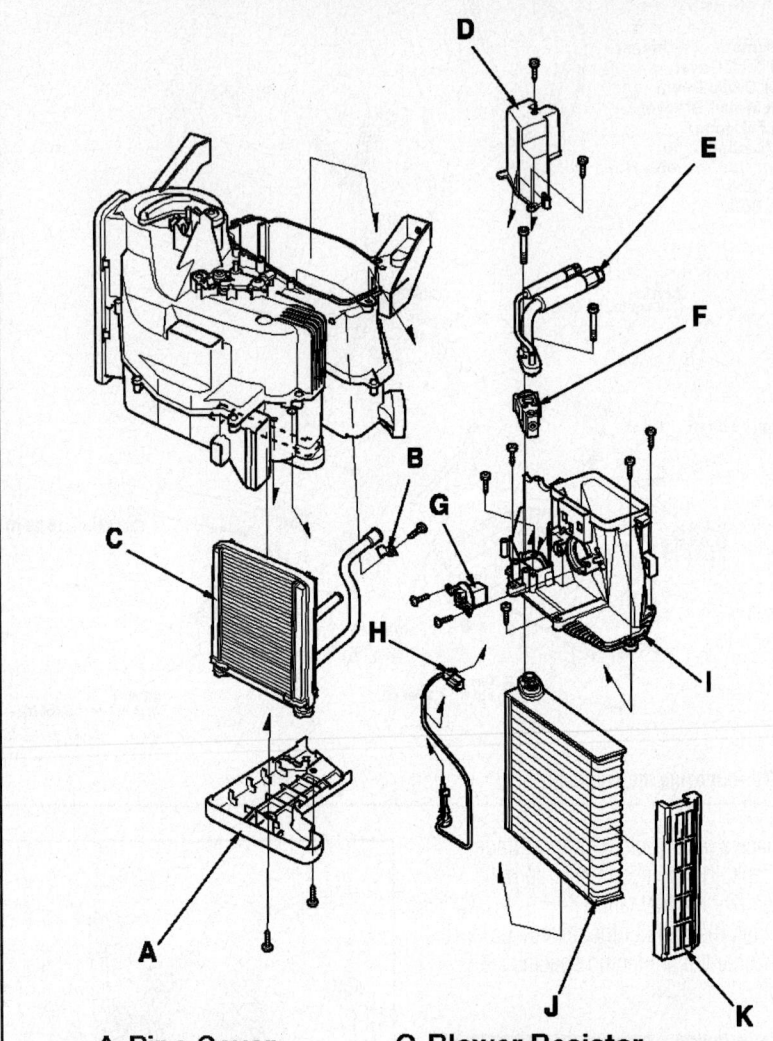

A. Pipe Cover
B. Pipe Clamp
C. Heater Core
D. Pipe Cover
E. Inlet/Outlet Pipes
F. Expansion Valve
G. Blower Resistor
H. Evaporator Temperature Sensor Connector
I. Duct
J. Evaporator Core
K. Plate

22140_HFIT_G0171

Fig. 103 Exploded view of the A/C-heater assembly, showing the heater core

self-tapping screws and the duct (I). Pull out the evaporator core with the plate.
12. Install the heater core and the evaporator core in the reverse order of removal.

To install:

13. Install the heater unit in the reverse order of removal, and note these items:

❄ CAUTION

Do not interchange the inlet and outlet heater hoses, and install the hose clamps securely.

14. With air conditioning; be sure to connect the drain hose securely.
15. Adjust the air mix control cable, the mode control cable, and the recirculation control cable.
16. Enter the anti-theft codes for the audio system.
17. Set the clock.
18. Refill the cooling system with engine coolant.
19. Make sure that there is no coolant leakage.
20. Make sure that there is no air leakage.

STEERING

ELECTRONIC POWER STEERING (EPS) CONTROL UNIT

REMOVAL & INSTALLATION

See Figure 104.

1. Make sure you have the anti-theft code for the audio system.
2. Make sure the ignition switch is OFF, then disconnect the negative cable from the battery.
3. Remove the driver's dashboard undercover.
4. Remove the under-dash fuse/relay box.
5. Disconnect EPS control unit 2-pin connector A, 2-pin connector B, 2-pin connector C, and 28-pin connector D.
6. Remove the nuts (E) from the EPS control unit (F).
7. Remove the EPS control unit.

To install:

8. Install the EPS control unit in the reverse order of removal.
9. Tighten the control unit retaining bolts to 7 ft. lbs. (10 Nm).

➡The EPS control unit has three 2-pin connectors; make sure they are correctly reconnected. The locations of the three 2-pin connectors are the top: connector C (RED, WHT/RED wires), the middle: connector B (GRN wire), the lower: connector A (BLK, WHT/BLU wires).

10. Reconnect the negative cable to the battery, and do the following items:
11. Enter the anti-theft code for the audio system.
12. Set the clock.
13. If the EPS control unit is replaced, the EPS control unit must memorize the torque sensor neutral position.
14. After installation, start the engine, allow it to idle, and turn the steering wheel from lock-to-lock several times. Check that the EPS indicator does not come on.

ELECTRONIC POWER STEERING (EPS) MOTOR

REMOVAL & INSTALLATION

See Figures 105 and 106.

➡Do not allow dust, dirt, or other foreign materials to enter the steering gearbox.

1. Remove the steering gearbox. See "Power Rack & Pinion Steering Gear" below.

2. Remove the connector bracket (A).
3. Remove the EPS motor angle sensor 6-pin connector (B), the EPS motor 1-pin connector (C), the EPS motor 2-pin connector (D), and the clip (E) from the connector bracket.
4. Remove the heat shield and the EPS motor from the steering gearbox, then remove the O-ring and discard it.

To install:

5. Clean the mating surface of the EPS motor and the steering gearbox.
6. Apply a thin coat of silicone grease to the new O-ring, and carefully fit it on the EPS motor.
7. Apply steering gear grease into the EPS motor shaft.
8. Install the EPS motor on the steering gearbox by engaging the EPS motor shaft and the worm shaft.
9. Before tightening the bolts, turn the EPS motor two or three times to the right and left about 45 degrees. Make sure the EPS motor is evenly seated on the steering gearbox, and that the O-ring is not pinched between the mating surfaces.
10. Install the heat shield, and in two or more steps, alternately tighten the three bolts on the steering gearbox to 14 ft. lbs. (20 Nm).

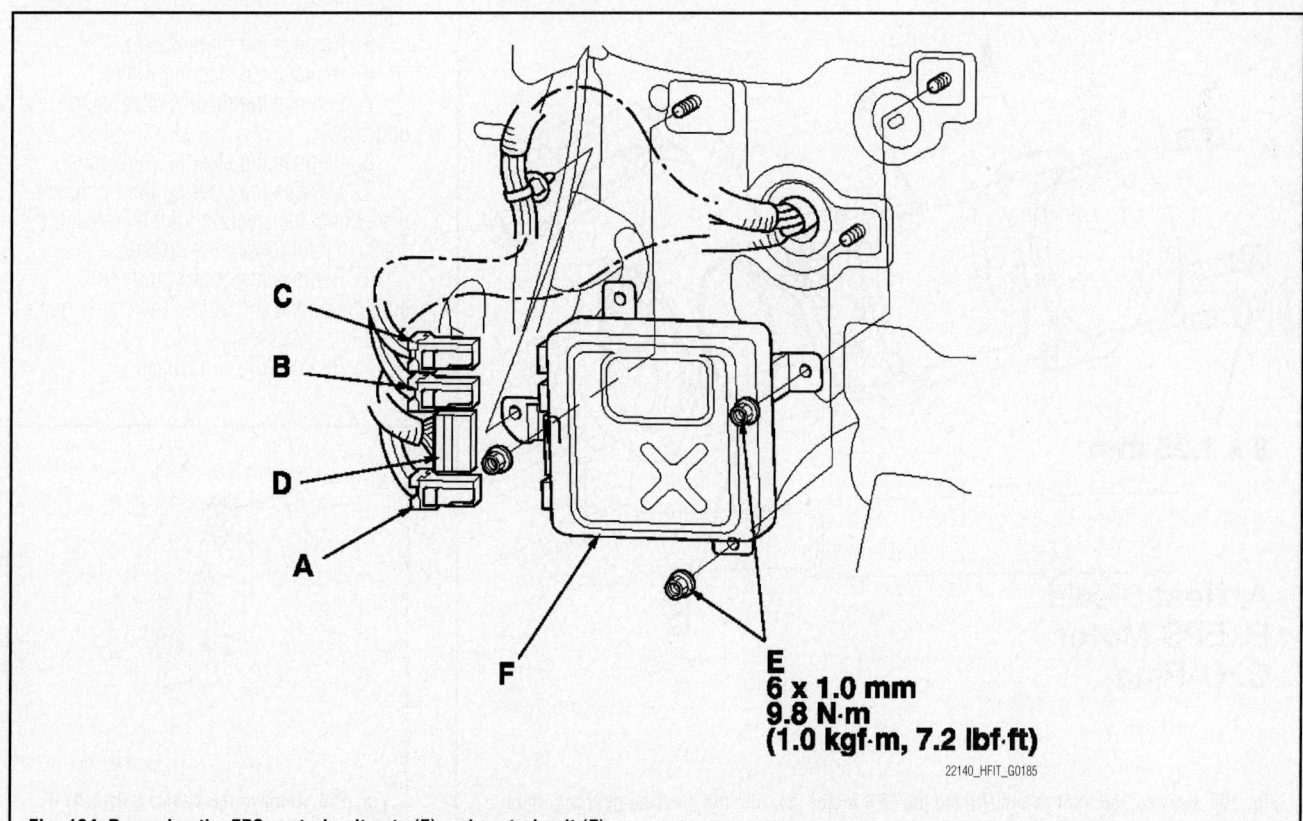

E
6 x 1.0 mm
9.8 N·m
(1.0 kgf·m, 7.2 lbf·ft)

22140_HFIT_G0185

Fig. 104 Removing the EPS control unit nuts (E) and control unit (F)

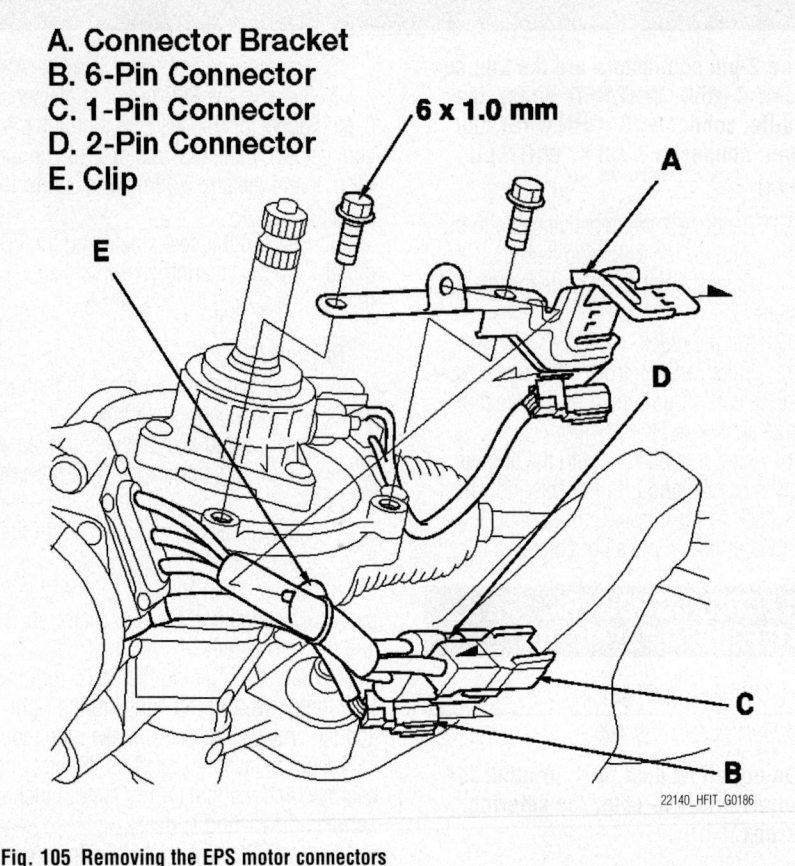

A. Connector Bracket
B. 6-Pin Connector
C. 1-Pin Connector
D. 2-Pin Connector
E. Clip

6 x 1.0 mm

22140_HFIT_G0186

Fig. 105 Removing the EPS motor connectors

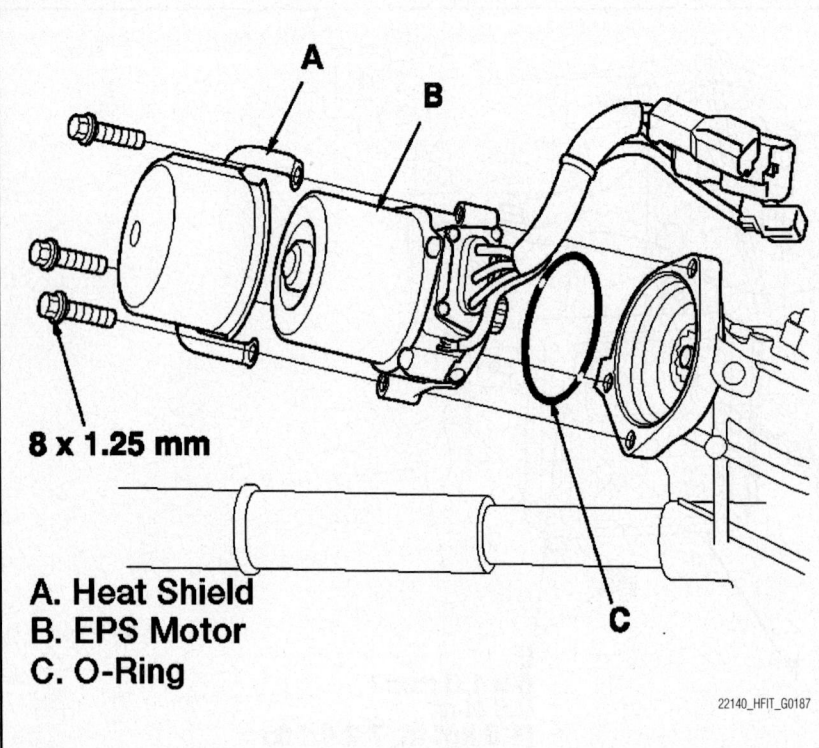

8 x 1.25 mm

A. Heat Shield
B. EPS Motor
C. O-Ring

22140_HFIT_G0187

Fig. 106 Remove the heat shield (A) and the EPS motor (B) from the steering gearbox, then remove the O-ring (C) and discard it

11. Install the EPS motor angle sensor 6-pin connector, the EPS motor 1-pin connector, the EPS motor 2-pin connector, and the clip to the connector bracket.

12. Install the connector bracket.

13. Finish the installation, make sure the EPS motor and the EPS wires are not caught or pinched by any parts.

14. Install the steering gearbox. See "Power Rack & Pinion Steering Gear" below.

POWER RACK & PINION STEERING GEAR

REMOVAL & INSTALLATION

See Figures 107 through 113.

1. Note these items during removal:

a. Using solvent and a brush, wash any oil and dirt off the end of the steering gearbox. Avoid any electrical parts. Blow dry with compressed air.

b. Be sure to remove the steering wheel before disconnecting the steering joint. Damage to the cable reel can occur.

2. Make sure you have the anti-theft code for the audio system.

3. Make sure the ignition switch is OFF, then disconnect the negative cable from the battery.

4. Raise the front of the vehicle, and support it with safety stands in the proper locations.

5. Remove the front wheels.

6. Remove the steering wheel.

7. Remove the driver's dashboard undercover.

8. Remove the steering joint cover.

9. Remove the steering joint bolt, and disconnect the steering joint by moving the steering joint toward the column.

10. Remove the center guide (if equipped), and discard it. The center guide is for factory assembly only.

11. Remove the air cleaner.

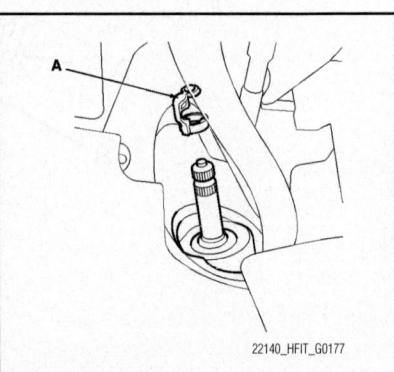

22140_HFIT_G0177

Fig. 107 Remove the center guide (A) if equipped

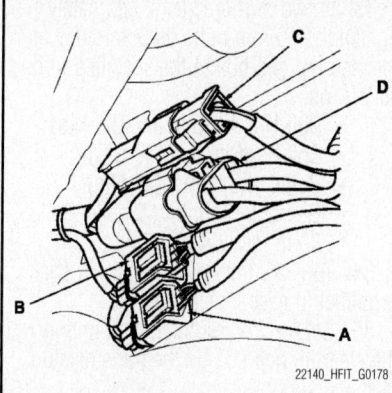

Fig. 108 Disconnect the torque sensor 4-pin connector (A), EPS motor angle sensor 6-pin connector (B), EPS motor 1-pin connector (C), and EPS motor 2-pin connector (D) from the driver's side of the steering gearbox

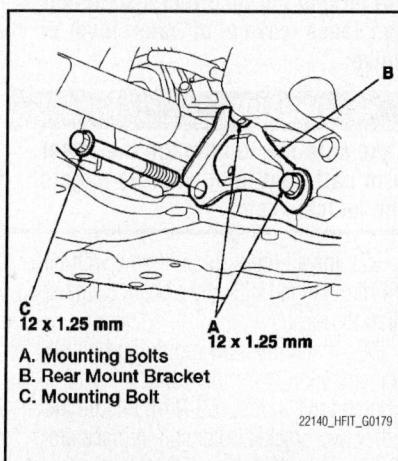

C
12 x 1.25 mm
A
12 x 1.25 mm
A. Mounting Bolts
B. Rear Mount Bracket
C. Mounting Bolt

22140_HFIT_G0179

Fig. 109 Removing transaxle rear mount bracket—A/T model shown; M/T similar

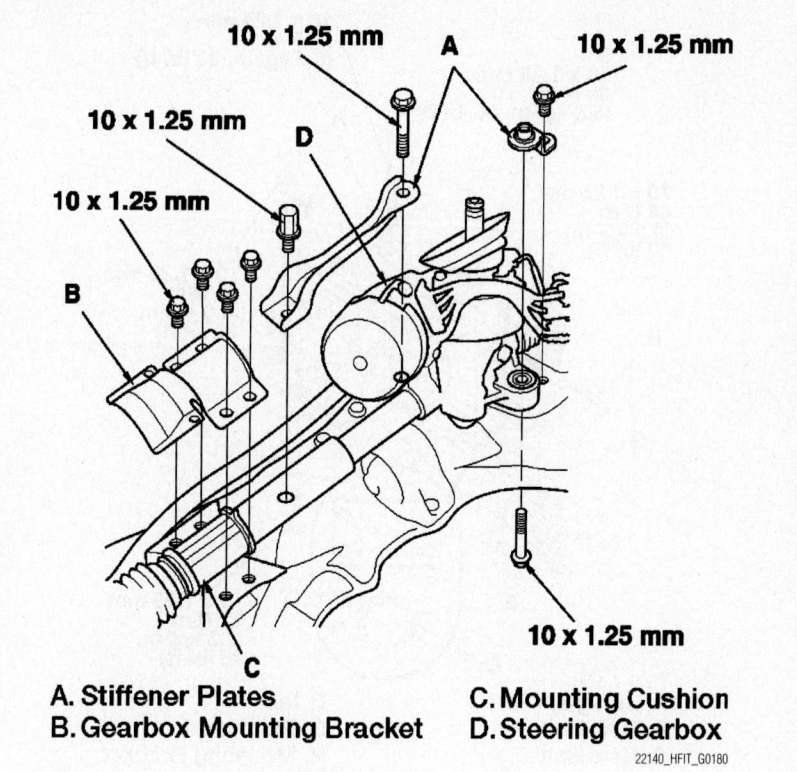

10 x 1.25 mm A 10 x 1.25 mm
10 x 1.25 mm D
10 x 1.25 mm
B
C
10 x 1.25 mm

A. Stiffener Plates   C. Mounting Cushion
B. Gearbox Mounting Bracket   D. Steering Gearbox

22140_HFIT_G0180

Fig. 110 Removing the rack and pinion steering gearbox

12. Install a proper engine lift and support assembly, and hold the engine weight.

13. Remove the cotter pin from the tie rod end ball joint nut, and loosen the nut.

14. Separate the tie-rod ball joint and the knuckle using the ball joint remover.

15. Remove the clip from the lower arm ball joint castle nut, and remove the nut.

16. Separate the lower arm ball joint and the knuckle using the ball joint remover and the ball joint thread protector.

17. Remove the stabilizer link from the stabilizer bar.

18. Disconnect the torque sensor 4-pin connector, EPS motor angle sensor 6-pin connector, EPS motor 1-pin connector, and EPS motor 2-pin connector from the driver's side of the steering gearbox. Wrap the connectors with vinyl tape to avoid the contamination from grease or water.

19. Remove the front splash shield.

20. On A/T models, loosen the mounting bolts of the transmission rear mount bracket on the front subframe.

21. Remove the mounting bolt of the transmission rear mount bracket on the front subframe, and discard it.

22. Attach a wood block to the lower center of the front subframe. Attach a jack to the center of the wood block and support the front subframe securely by raising the jack.

23. Remove the front subframe mounting bolts, and discard them.

24. Lower the front subframe and steering gearbox as an assembly by lowering the jack slowly.

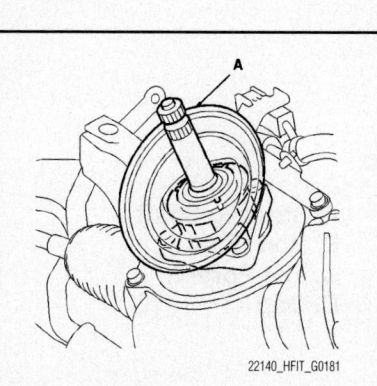

22140_HFIT_G0181

Fig. 111 Remove the pinion shaft grommet (A) from the top of the torque sensor

25. Remove the 10 mm flange bolts, then remove the stiffener plates, the gearbox mounting bracket and the mounting cushion.

26. Remove the steering gearbox from the front subframe.

27. Remove the pinion shaft grommet from the top of the torque sensor.

To install:

28. Install the pinion shaft grommet. Align the slot in the pinion shaft grommet with the lug portion on the torque sensor. The grommet must not have a gap at the mating surface of the grommet and the torque sensor.

29. Place the steering gearbox on the front subframe.

30. Loosely install the stiffener plates, and the gearbox mounting bolts on the driver's side of the steering gearbox.

➡**Use new mounting bolt and new bolt.**

31. Position the cutout on the mounting cushion as shown, and securely install it on the gearbox.

➡**Make sure to align the tab on the mounting cushion with the paint mark on the steering gearbox.**

32. Install the gearbox mounting bracket over the mounting cushion, and loosely install the 10 mm bolts.

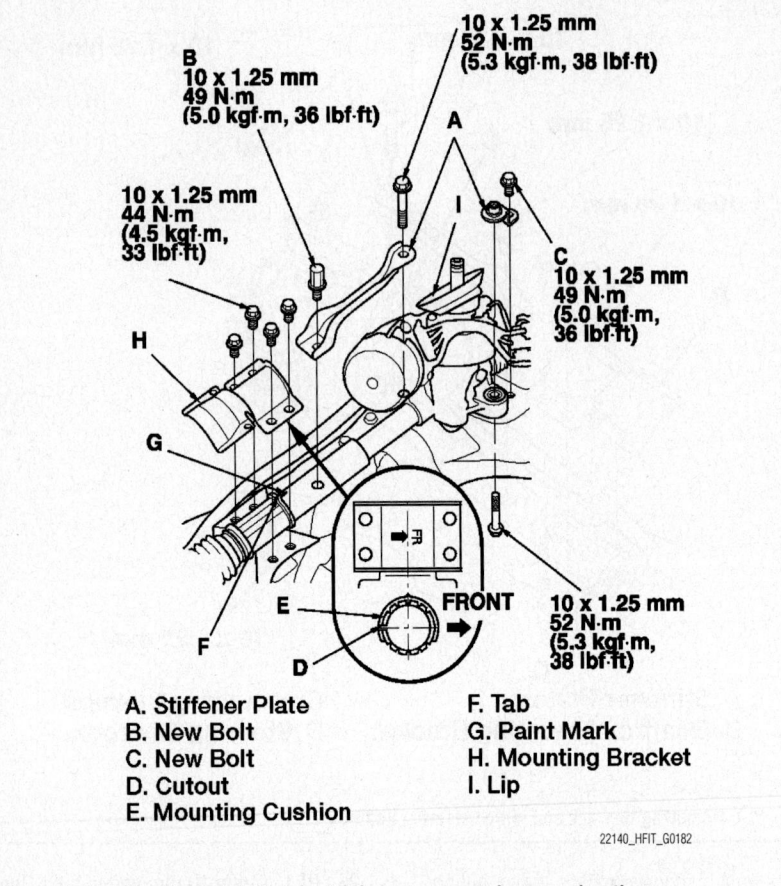

A. Stiffener Plate
B. New Bolt
C. New Bolt
D. Cutout
E. Mounting Cushion
F. Tab
G. Paint Mark
H. Mounting Bracket
I. Lip

22140_HFIT_G0182

Fig. 112 Installing the gearbox and retaining components in proper locations

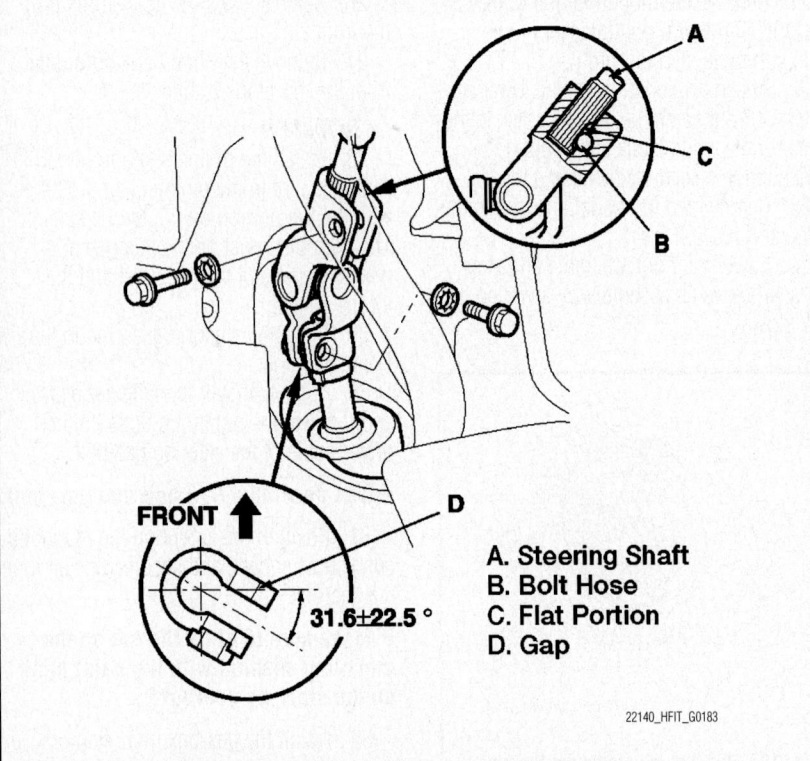

A. Steering Shaft
B. Bolt Hose
C. Flat Portion
D. Gap

22140_HFIT_G0183

Fig. 113 Installing the upper end of the steering joint to the pinion shaft

33. In two or more steps, alternately tighten the 10 mm bolts on both sides of the steering gearbox to the specified torque as follows:
- Bolt/Nut A: 38 ft. lbs. (52 Nm)
- Bolt B: 36 ft. lbs. (49 Nm)
- Bolt C: 36 ft. lbs. (49 Nm)
- Gearbox Mounting Bracket Bolts: 33 ft. lbs. (44 Nm)

34. Turn out the lip of the pinion shaft grommet to ease installation.

35. Set the front subframe mounting with the steering gearbox on the transmission jack and support it.

36. Carefully raise the front subframe with the jack, and pass the pinion shaft into the cabin. Return the lip of the pinion shaft grommet.

➡ Be sure that the pinion shaft grommet is securely in place. Check whether the pinion shaft grommet is not turning up. Incorrect installation can cause leakage of water, mud, or noise.

❊❊ CAUTION

Take care not to damage the lower arm ball joint boot with the edge of the knuckle, etc.

37. Install the front subframe with new 14 mm flange bolts and tighten to the specified torque.

38. Install the new mounting bolt on the transmission rear mount bracket, and tighten to 51 ft. lbs. (69 Nm). Ensure the other two bracket bolts still in place are tightened to 43 ft. lbs. (59 Nm).

39. Install the front splash shield.

40. Remove the vinyl tape, then firmly connect the connectors as marked during removal.

41. Connect the stabilizer links to the stabilizer bar.

42. Wipe off any grease contamination from the tapered section and threads of the lower arm ball joint. Then reconnect the lower arm to the knuckle. Install a new 12 mm castle nut and tighten it to 36 ft. lbs. (49 Nm), plus just enough additional to align the hole for the cotter pin. Do not align the castle nut by loosening.

➡ Insert the cotter pin from the inside to the outside of the vehicle.

43. Wipe off any grease contamination from the tapered section and the threads of the ball joint.

44. Reconnect the tie-rod ends to the steering knuckles. Install the 10 mm nut, and tighten to 32 ft. lbs. (43 Nm).

45. Install a new cotter pin.

46. Remove the engine support hanger and the universal eyelet.

47. Install the air cleaner.

48. Center the steering rack within its stroke.

49. Adjust the steering column to the full tilt up position.

50. Insert the upper end of the steering joint onto the steering shaft (line up the bolt hole with the flat portion on the shaft), and loosely install the upper joint bolt.

51. Slip the lower end of the steering joint onto the pinion shaft taking care to align the gap within the angle shown.

52. Align the bolt hole on the steering joint with the groove around the pinion shaft, then loosely install the lower joint bolt.

53. Pull on the steering joint to make sure that the steering joint is fully seated, then tighten the lower joint bolt, then the upper joint bolt to 21 ft. lbs. (28 Nm).

54. Install the steering joint cover.

55. Install the driver's dashboard undercover.

56. Install the front wheels, then set the wheels in the straight ahead position.

57. Install the steering wheel.

58. Reconnect the negative cable to the battery, and do these items:

a. Turn the ignition switch ON (II); the SRS indicator should come on for about 6 seconds and then go off.

b. Enter the anti-theft code for the audio system.

c. Set the clock.

d. Verify cruise control and turn signal switch operation.

59. After installation, do these checks:

a. Check the steering wheel spoke angle. If steering spoke angles to the right and left are not equal (steering wheel and rack are not centered), correct the engagement of the joint/pinion shaft serrations.

b. Set the steering column to the center tilt position, then perform an toe inspection.

c. Make sure the steering wheel is centered.

d. Perform the memorizing of the torque sensor neutral position procedure as follows:

➡**The torque sensor neutral position must be memorized whenever the gearbox, the motor, or the EPS control unit**

is replaced, and after doing a wheel alignment. Note that the torque sensor neutral position is not affected when erasing the DTC.

➡**The torque sensor is sensitive to temperature. This procedure should be performed within the range of 68±18°F (20±10°C).**

e. With the ignition switch OFF, connect the scan tool to the data link connector (DLC) located under the driver's side of the dashboard.

f. Turn the ignition switch ON (II).

g. Make sure the HDS communicates with the vehicle and the EPS control unit. If it doesn't, troubleshoot the DLC circuit.

h. Select MISCELLANEOUS TEST in the EPS MENU with the HDS.

i. Select TORQUE SENSOR LEARN, and follow the screen prompts.

➡**See the HDS Help menu for specific instructions.**

j. Turn the ignition switch OFF.

k. Start the engine, allow it to idle, and turn the steering wheel from lock-to-lock several times. Check that the EPS indicator does not come on.

SUSPENSION

FRONT SUSPENSION

LOWER CONTROL ARM

REMOVAL & INSTALLATION

See Figure 114.

1. Raise the front of the vehicle, and support it with safety stands in the proper locations.

2. Remove the front wheel.

3. Remove the front splash shield.

4. Remove the lock pin from the lower arm ball joint, then remove the nut.

➡**During installation, install the new lock pin after tightening the new castle nut.**

5. Disconnect the lower ball joint from the knuckle using the ball joint remover.

❊ CAUTION

When removing the lower ball joint, do not pull the driveshaft end outward. The inner driveshaft joint may come apart.

6. Remove the lower arm mounting bolts, then remove the lower arm from the front subframe.

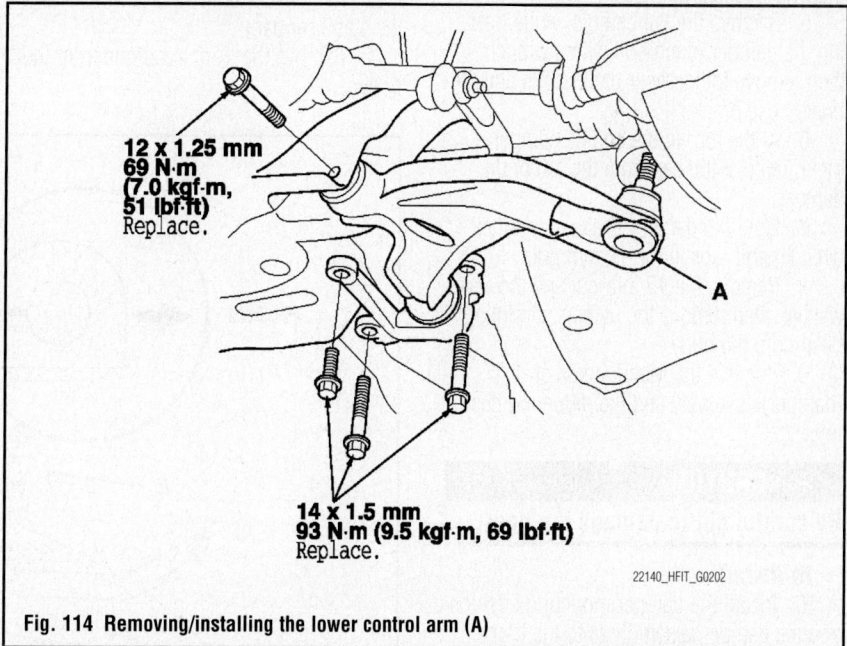

Fig. 114 Removing/installing the lower control arm (A)

➡**During installation, install new mounting bolts.**

7. Install the lower arm in the reverse order of removal, and note these items:

a. Be careful not to damage the ball joint boot when installing the knuckle.

b. Before connecting the lower ball joint to the knuckle, degrease the threaded section and tapered portion of the ball joint pin, the lower arm connecting hole, the threaded section and mating surface of the castle nut.

c. First install all the components and lightly tighten the bolts and nuts, then raise the suspension to load it with the vehicle's weight before fully tightening to the specified torque values.

d. Tighten the three lower arm mounting bolts to 69 ft. lbs. (93 Nm). Tighten the top mounting bolts to 51 ft. lbs. (69 Nm).

8. Torque the castle nut to 36 ft. lbs. (49 Nm), then tighten it only far enough to align the slot with the ball joint pin hole. Do not align the castle nut by loosening it.

9. Before installing the wheel, clean the mating surface of the brake disc and the inside of the wheel.

10. Check the wheel alignment, and adjust it if necessary.

MACPHERSON STRUT (DAMPER)

REMOVAL & INSTALLATION

1. Raise the front of the vehicle, and support it with safety stands in the proper locations.

2. Remove the front wheel.

3. Disconnect the stabilizer link from the damper.

4. Remove the wheel sensor harness clips from the damper. Do not disconnect the wheel sensor connector.

5. Remove the flange nuts, while holding the damper pinch bolt with a wrench, then remove the damper pinch bolts and the brake hose bracket.

6. At the top, in the engine compartment, remove the cap from the top of the damper.

7. Hold the damper shaft using a hex wrench, and loosen the 12 mm nut.

8. Remove the 12 mm nut and the wave washer, then remove the damper mounting base from the body.

9. Remove the MacPherson strut (damper) assembly and the mounting cushion.

❄❄ CAUTION

Be careful not to damage the body.

To install:

10. Install the damper mounting cushion and the damper assembly onto the frame.

11. Install the damper mounting base and the wave washer, then loosely install the 12 mm nut.

12. Loosely install new damper pinch bolts and new flange nuts with the brake hose bracket to the damper.

13. Align the pin of the brake hose bracket to the positioning hole on the damper.

14. Install the wheel sensor harness clip.

15. Raise the front suspension with a floor jack to load the suspension with the vehicle's weight.

16. Tighten the damper pinch bolts and new flange nuts to the specified torque value while holding the damper pinch bolt with a wrench.

17. Install the stabilizer link to the damper and tighten to 76 ft. lbs. (103 Nm).

18. Hold the damper shaft, and with a hex wrench, tighten the 12 mm nut to 33 ft. lbs. (44 Nm).

19. Install the cap to the top of the damper.

20. Clean the mating surface of the brake disc and the inside of the wheel, then install the front wheels.

21. Check the wheel alignment, and adjust it if necessary.

STABILIZER BAR

REMOVAL & INSTALLATION

See Figure 115.

1. Raise the front of the vehicle, and support it with safety stands in the proper locations.

2. Remove the front wheels.

3. Disconnect both stabilizer links from the stabilizer bar.

4. Remove the front subframe from the body.

5. Remove the flange bolts and the bushing holders, then remove the bushings and the stabilizer bar (D) from the front subframe.

To install:

6. Install the stabilizer bar in the reverse order of removal, and note these items:

a. Note the right and left direction of the stabilizer bar.

b. Tighten the stabilizer bar bracket bolts to 25 ft. lbs. (34 Nm).

c. Align the paint marks (F) on the stabilizer bar with the sides of the bushings.

d. Note the fore/aft direction of the bushing.

e. Refer to "Stabilizer Link" to connect the stabilizer bar to the links.

f. Check the wheel alignment, and adjust it if necessary.

STABILIZER LINK

REMOVAL & INSTALLATION

See Figure 116.

1. Raise the front of the vehicle, and support it with safety stands in the proper locations.

2. Remove the front wheel.

3. Remove the self-locking nut and the flange nut while holding the respective joint pin with a hex wrench, then remove the stabilizer link.

4. Install the stabilizer link on the stabilizer bar and damper with the joint pins set at the center of their range of movement.

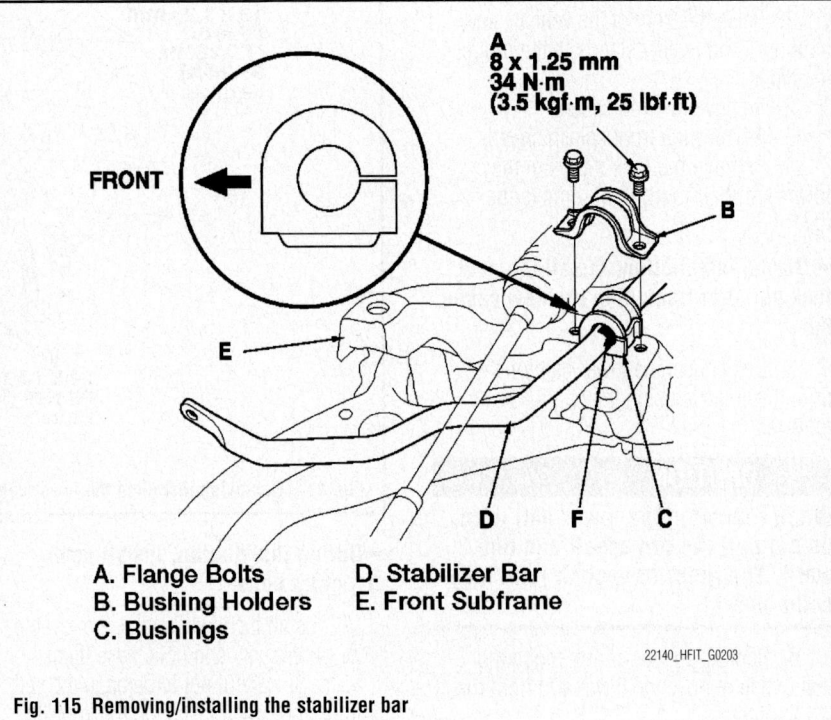

A
8 x 1.25 mm
34 N·m
(3.5 kgf·m, 25 lbf·ft)

FRONT

A. Flange Bolts
B. Bushing Holders
C. Bushings
D. Stabilizer Bar
E. Front Subframe

22140_HFIT_G0203

Fig. 115 Removing/installing the stabilizer bar

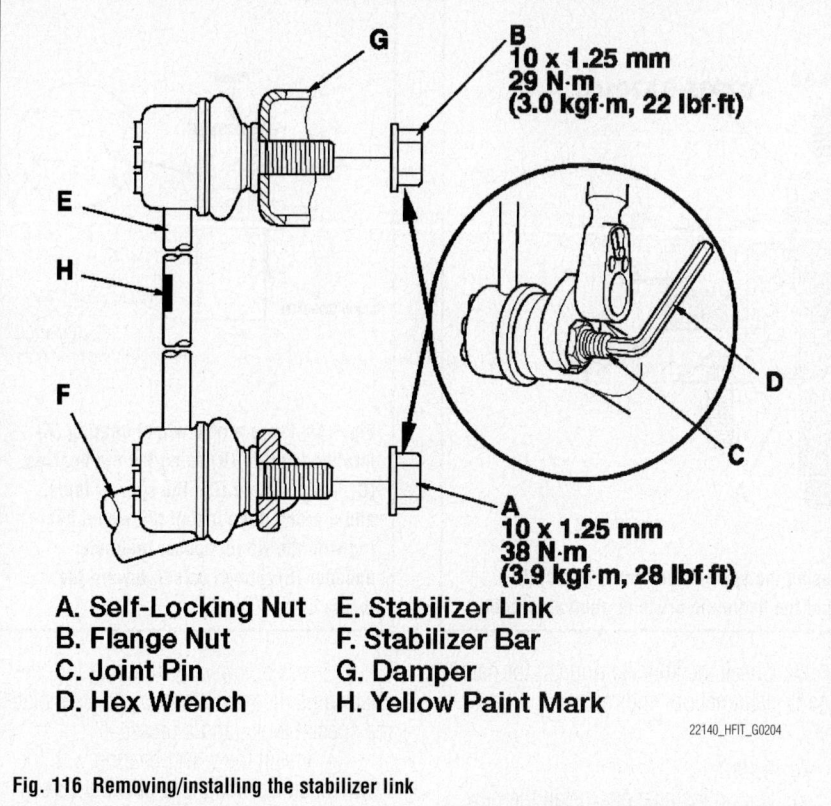

A. Self-Locking Nut
B. Flange Nut
C. Joint Pin
D. Hex Wrench
E. Stabilizer Link
F. Stabilizer Bar
G. Damper
H. Yellow Paint Mark

G

B
10 x 1.25 mm
29 N·m
(3.0 kgf·m, 22 lbf·ft)

A
10 x 1.25 mm
38 N·m
(3.9 kgf·m, 28 lbf·ft)

22140_HFIT_G0204

Fig. 116 Removing/installing the stabilizer link

➡**The left stabilizer link has a yellow paint mark, while the right stabilizer link has a white paint mark.**

5. Install a new self-locking nut and a new flange nut, and lightly tighten them.

6. Tighten the lower self-locking nut to 28 ft. lbs. (38 Nm) and the upper flange nut to 22 ft. lbs. (29 Nm), while holding the respective joint pin with a hex wrench.

7. Reinstall all removed parts, and test-drive the vehicle.

STEERING KNUCKLE/HUB ASSEMBLY

REMOVAL & INSTALLATION

1. Raise the front of the vehicle, and support it with safety stands in the proper locations.

2. Remove the wheel nuts and front wheel.

3. Remove the brake hose mounting bolt.

4. Remove the brake caliper bracket mounting bolts, and remove the caliper assembly from the knuckle. To prevent damage to the caliper assembly or brake hose, use a short piece of wire to hang the caliper assembly from the undercarriage. Do not twist the brake hose excessively.

5. Raise the stake, then remove the spindle nut.

6. Remove the brake disc. See "BRAKES" section.

7. Remove the wheel sensor from the knuckle, but DO NOT disconnect the wheel sensor connector.

8. Remove the cotter pin from the tie-rod end ball joint, then remove the nut.

9. Disconnect the tie-rod ball joint from the knuckle using the ball joint remover.

10. Remove the lock pin from the lower arm ball joint, then remove the nut.

11. Disconnect the lower ball joint from the knuckle using the ball joint remover.

12. Remove the damper pinch bolts and flange nuts while holding the damper pinch bolt with a wrench.

13. Remove the driveshaft outboard joint from the knuckle by tapping the driveshaft end with a soft face hammer while drawing the hub outward, then remove the knuckle.

✳✳ CAUTION

Do not pull the driveshaft end outward. The inner driveshaft joint may come apart.

To install:

14. During installation, apply grease to the mating surface of the wheel bearing and driveshaft outboard joint.

15. Install the knuckle/hub in the reverse order of removal, and note these items:

a. Be careful not to damage the ball joint boot when installing the knuckle.

b. Before connecting the lower ball joint to the knuckle, degrease the threaded section and tapered portion of the ball joint pin, the lower arm connecting hole, the threaded section and mating surface of the castle nut.

c. First install all the components and lightly tighten the bolts and nuts, then raise the suspension to load it with the vehicle's weight before fully tightening to the specified torque values.

d. Use a new spindle nut on reassembly.

e. Before installing the spindle nut, apply a small amount of engine oil to the seating surface of the nut.

f. After tightening, use a drift to stake the spindle nut shoulder against the driveshaft.

g. Before installing the brake disc, clean the mating surface of the front hub and the inside of the brake disc.

h. Tighten all mounting hardware to the specified torque values as follows:
- Damper pinch bolt nuts: 76 ft. lbs. (103 Nm)
- Lower ball joint nut: 36 ft. lbs. (43 Nm), then additional to align cotter pin hole
- Tie rod ball joint nut: 36 ft. lbs. (43 Nm), then additional to align cotter pin hole
- Wheel sensor bolt: 7 ft. lbs. (10 Nm)
- Wheel spindle nut: 134 ft. lbs. (181 Nm)
- Caliper bolts: 80 ft. lbs. (108 Nm)
- Brake hose retaining bolt: 16 ft. lbs. (22 Nm)

i. Before installing the wheel, clean the mating surface of the brake disc and the inside of the wheel.

16. Check the wheel alignment, and adjust it if necessary.

WHEEL BEARINGS

REMOVAL & INSTALLATION

See Figures 117 through 119.

1. Remove the front hub and knuckle assembly. See "Steering Knuckle/Hub Assembly" in this section.

2. Separate the hub from the knuckle using the special tool and a hydraulic press. Hold the knuckle with the attachment of the hydraulic press or equivalent tool. Be careful not to deform the splash guard. Hold onto the hub to keep it from falling when pressed clear.

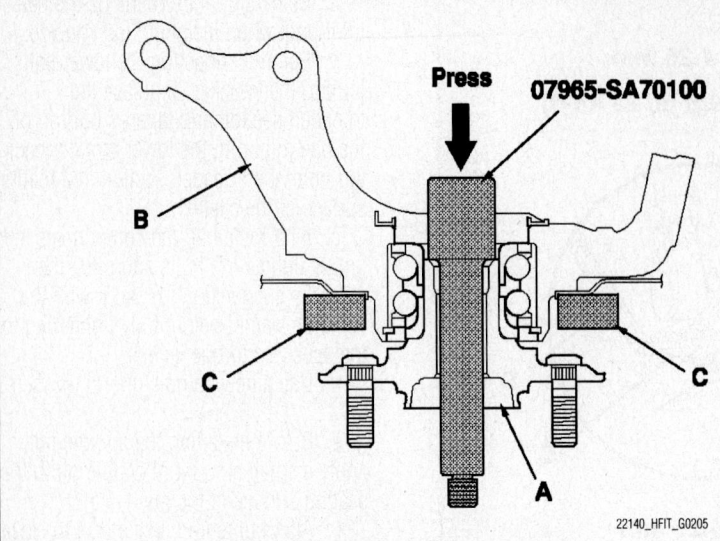

Fig. 117 Separate the hub (A) from the knuckle (B) using the special tool and a hydraulic press, then hold the knuckle with the attachment (C) of the hydraulic press or equivalent tool

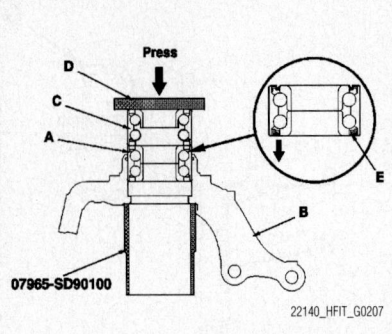

Fig. 119 Press a new wheel bearing (A) into the knuckle (B) using the old bearing (C), a steel plate (D), the special tools, and a press, then install the wheel bearing with the wheel sensor magnetic encoder (E) (brown color), toward the inside of the knuckle

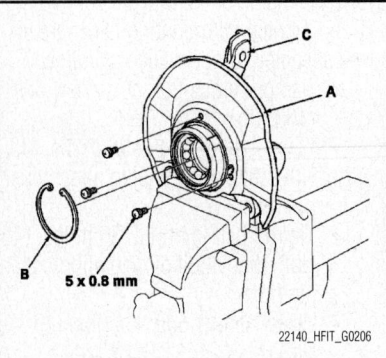

Fig. 118 Remove the splash guard (A) and the snap ring (B) from the knuckle (C)

3. Press the wheel bearing inner race off of the hub, using the special tool, a bearing separator, and a press.

4. Remove the splash guard and the snap ring from the knuckle.

5. Check the knuckle ring (A) for damage or deformation, and replace it if necessary.

To install:

6. During installation, install the new knuckle ring.

7. Before installing the knuckle ring, clean the mating surface of the knuckle.

8. Make sure to align the cutout portion on the knuckle ring with the cutout portion on the knuckle when installing it.

9. Be careful not to damage or deform the ring when installing it.

10. Make sure there is no gap between the mating surfaces of the knuckle ring and the knuckle.

11. Press the wheel bearing out of the knuckle, using the special tools and a press.

12. Wash the knuckle and hub thoroughly in high flash point solvent before reassembly.

13. Press a new wheel bearing into the knuckle using the old bearing, a steel plate, the special tools, and a press.

a. Install the wheel bearing with the wheel sensor magnetic encoder (brown color), toward the inside of the knuckle.

14. Remove any oil, grease, dust, metal debris, and other foreign material from the encoder surface.

15. Keep all magnetic tools away from the encoder surface.

16. Be careful not to damage the magnetic encoder surface when you insert the wheel bearing.

17. Install the snap ring securely in the knuckle.

18. Install the splash guard, and tighten the screws to 4 ft. lbs. (6 Nm).

19. Install the hub onto the knuckle, using the install tools and hydraulic press. Be careful not to distort the splash guard.

SUSPENSION

See Figure 120.

COIL SPRING

REMOVAL & INSTALLATION

See Figures 121 and 122.

1. Raise the rear of the vehicle, and support it with safety stands in the proper locations.

2. Remove the rear wheels.

3. Position the floor jack under the axle beam. Raise the floor jack until the rear suspension begins to compress.

4. Remove the flange bolts from the left and right sides of the rear damper.

5. Lower the jack and the axle beam, then remove the spring, the upper spring cushion, and the lower spring cushion.

To install:

6. Install the upper spring cushion and the lower spring cushion to the spring.

➡**Securely set the rear spring in the groove of the upper spring cushion.**

7. Align the upper end of the spring with the stepped part of the upper spring cushion.

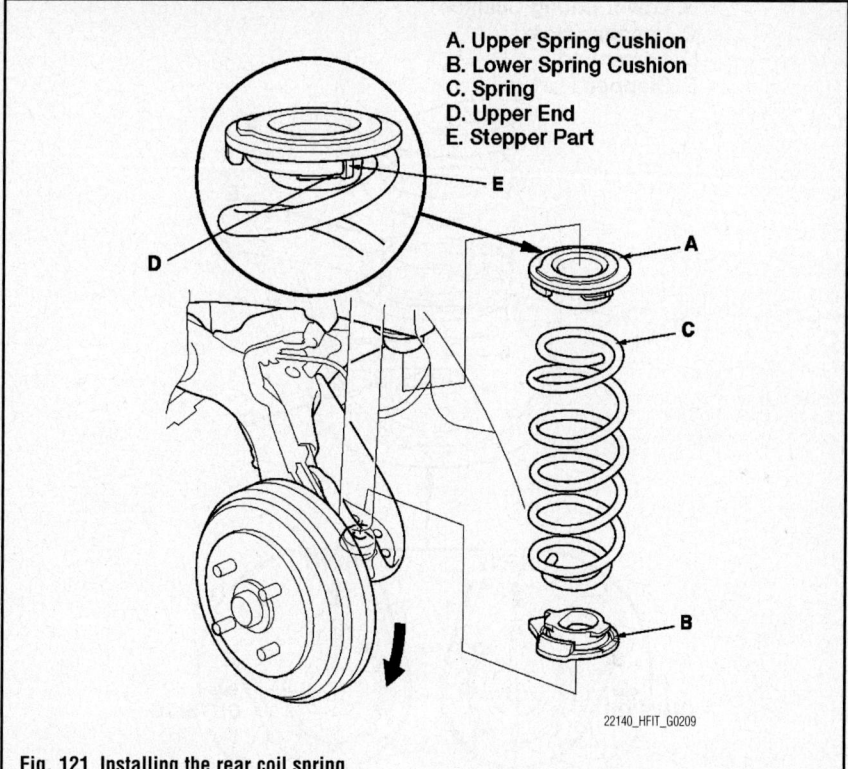

A. Upper Spring Cushion
B. Lower Spring Cushion
C. Spring
D. Upper End
E. Stepper Part

22140_HFIT_G0209

Fig. 121 Installing the rear coil spring

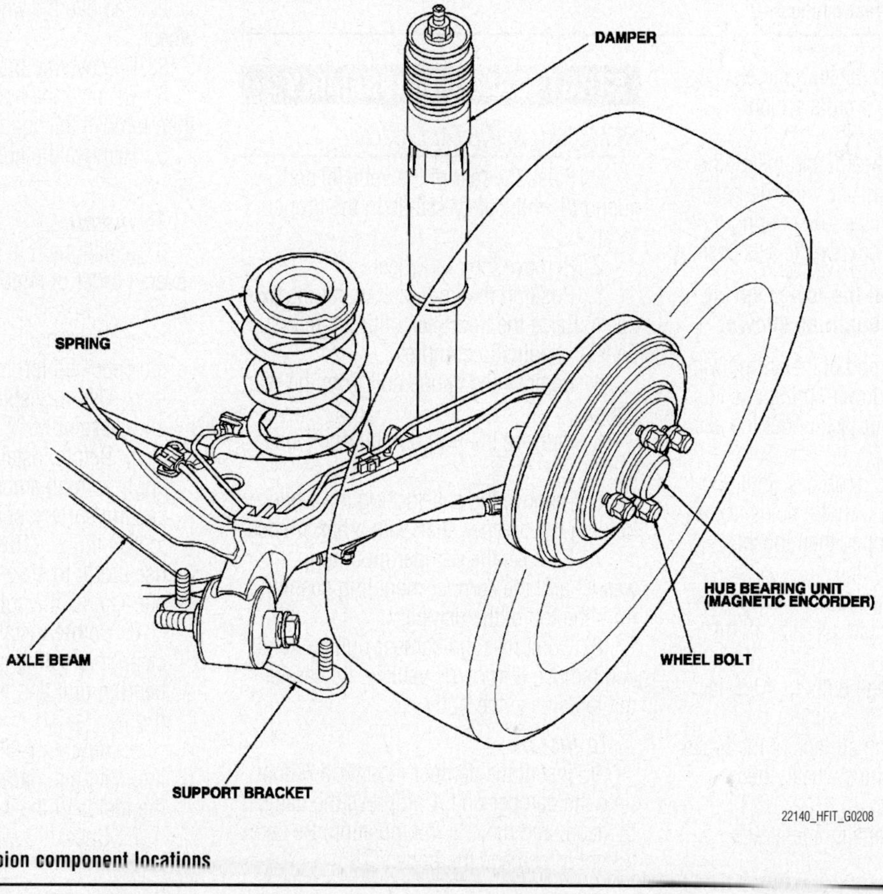

DAMPER

SPRING

HUB BEARING UNIT
(MAGNETIC ENCORDER)

AXLE BEAM

WHEEL BOLT

SUPPORT BRACKET

22140_HFIT_G0208

Fig. 120 Rear suspension component locations

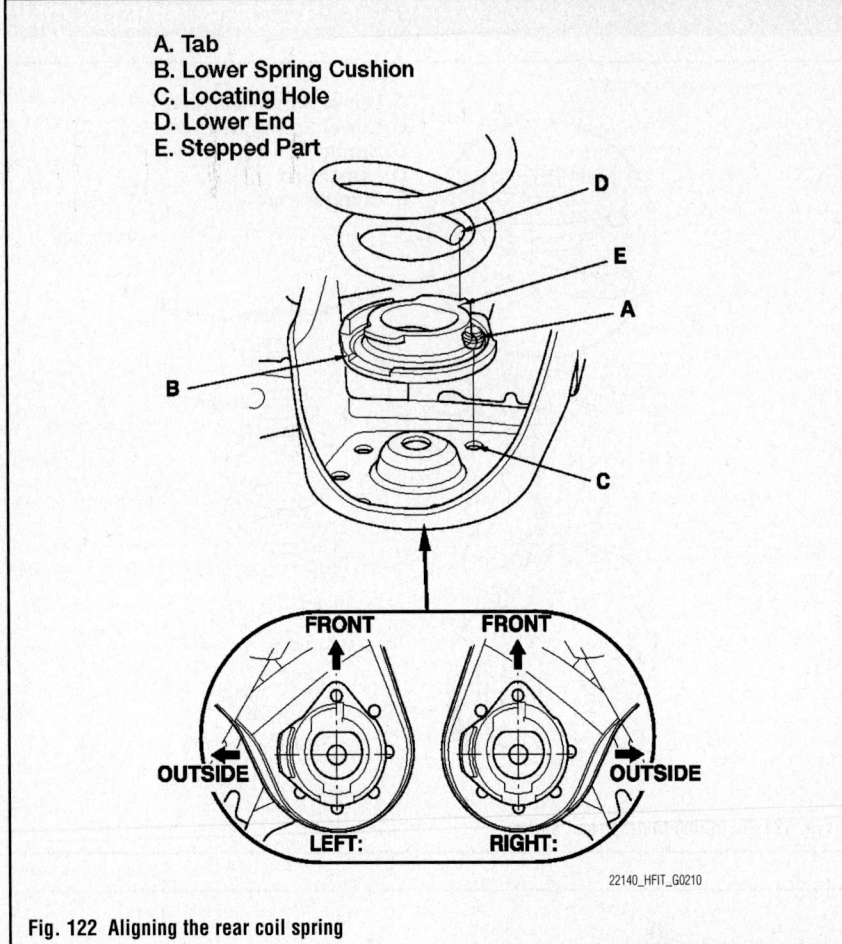

A. Tab
B. Lower Spring Cushion
C. Locating Hole
D. Lower End
E. Stepped Part

22140_HFIT_G0210

Fig. 122 Aligning the rear coil spring

8. After installation, be sure the stepped part and the other rubber parts are not pinched with the spring.

9. Lower the axle beam, and install the spring between the body and the axle beam.

10. Install the tab on the lower spring cushion in the locating hole on the axle beam.

➡**Make sure position the lower spring cushion on the axle beam as shown.**

11. Align the lower end of the spring with the stepped part of the lower spring cushion.

12. Position the floor jack under the axle beam.

13. Slowly raise the floor jack until you can align the bolt hole with the holes in the axle beam and the damper, then loosely install the new flange bolts.

14. Raise the rear suspension with the floor jack until the vehicle just lifts off the safety stands.

15. Tighten the flange bolts to 40 ft. lbs. (54 Nm).

16. Clean the mating surface of the brake drum and the inside of the wheel, then install the rear wheels.

17. Check the wheel alignment, and adjust it if necessary.

STRUT (DAMPER) ASSEMBLY

REMOVAL & INSTALLATION

1. Raise the rear of the vehicle, and support it with safety stands in the proper locations.

2. Remove the rear wheel.

3. Position the floor jack under the axle beam. Raise the floor jack until the rear suspension begins to compress.

4. Remove the flange bolt from the rear damper.

5. Remove the lid from the trunk side trim.

6. Remove the self-locking nut while holding the damper shaft with a hex wrench.

7. Remove the damper mounting washer and the damper mounting cushion from the top of the damper.

8. Compress the damper unit by hand, and remove it from the vehicle. Be careful not to damage the body.

To install:

9. Install the damper mounting rubber onto the damper unit. Compress the damper by hand, and move it into position. Be careful not to damage the body.

10. Position the floor jack under the axle beam.

11. Slowly raise the floor jack until you can align the bolt hole with the holes in the axle beam and the damper, then loosely install the new flange bolts.

12. Raise the rear suspension with the floor jack until the vehicle just lifts off the safety stands, then tighten the flange bolt to 40 ft. lbs. (54 Nm).

13. Install the damper mounting cushion, the damper mounting washer, and a new self-locking nut on the damper shaft.

14. Tighten the self-locking nut to 22 ft. lbs. (29 Nm), while holding the damper shaft with a hex wrench.

15. Install the lid on the trunk side trim.

16. Clean the mating surface of the brake drum and the inside of the wheel, then install the rear wheels.

17. Check the wheel alignment, and adjust it if necessary.

WHEEL BEARINGS

REMOVAL & INSTALLATION

1. Raise the rear of the vehicle, and support it with safety stands in the proper locations.

2. Remove the wheel nuts and the rear wheel.

3. Remove the brake drum.

4. Remove the hub cap. Raise the stake, then remove the spindle nut.

5. Remove the hub bearing unit from the spindle.

To install:

6. Install the hub bearing unit in the reverse order of removal, and note these items:

   a. Tighten all mounting hardware to the specified torque values.

   b. Use new spindle nut and hub cap on reassembly.

   c. Before installing the spindle nut, apply a small amount of engine oil to the seating surface of the nut. Tighten the nut to 134 ft. lbs. (181 Nm). After tightening, use a drift to stake the spindle nut shoulder against the spindle.

   d. Before installing the brake drum, clean the mating surface of the hub bearing unit and the inside of the brake drum.

   e. Before installing the wheel, clean the mating surface of the brake drum and the inside of the wheel.

   f. Check the wheel alignment.

HONDA

Odyssey

8

SPECIFICATIONS AND MAINTENANCE CHARTS

ENGINE AND VEHICLE IDENTIFICATION CHART

| | Engine Code | | | | | | Model Year | |
Code	Liters (cc)	Cu. In.	Cyl.	Fuel Sys.	Engine Type	Eng. Mfg.	Code ①	Year
J35A6	3.5 (3471)	212	6	SMFI	SOHC	Honda	7	2007
J35A7 ②	3.5 (3471)	212	6	SMFI	SOHC	Honda	8	2008

SOHC: Single Overhead Cam

SMFI: Sequential Multi-port Fuel Injection

① 10th position of VIN

② Variable cylinder management

22140_ODYS_C0001

GENERAL ENGINE SPECIFICATIONS

Year	Model	Engine Displacement Liters	Engine ID	Net Horsepower @ rpm	Net Torque @ rpm (ft. lbs.)	Bore x Stroke (in.)	Compression Ratio	Oil Pressure @ rpm
2007	Odyssey ①	3.5	J35A6	244@5750	240@5000	3.50x3.66	10.0:1	71@3000
	Odyssey ②	3.5	J35A7	244@5750	240@4500	3.50x3.66	10.0:1	71@3000
2008	Odyssey ①	3.5	J35A6	241@5700	242@4900	3.50x3.66	10.0:1	71@3000
	Odyssey ②	3.5	J35A7	244@5750	240@5000	3.50x3.66	10.5:1	71@3000

① LX and EX models

② EX-L and EX-L Touring models

22140_ODYS_C0002

ENGINE TUNE-UP SPECIFICATIONS

Year	Engine Displacement Liters	Engine ID	Spark Plug Gap (in.)	Ignition Timing (deg.) MT	AT	Fuel Pump (psi)	Idle Speed (rpm) MT	AT	Valve Clearance (in.) In.	Ex.
2007	3.5	J35A6 J35A7	0.039-0.043	—	8-12B	55-63	—	600-700	0.008-0.009	0.011-0.013
2008	3.5	J35A6 J35A7	0.039-0.043	—	8-12B	55-63	—	600-700	0.008-0.009	0.011-0.013

NOTE: The Vehicle Emission Control Information label often reflects changes made during production and must be used if they differ from this chart.

NOTE: The fuel pressure readings are given with the vacuum hose connected to the regulator and the engine running

B: Before top dead center

22140_ODYS_C0003

CAPACITIES

Year	Model	Engine Displacement Liters	Engine ID	Engine Oil with Filter (qts.)	Transmission (qts.)		Fuel Tank (gal.)	Cooling System (qts.)
					5-Spd	Auto.		
2007	Odyssey	3.5	J35A6/J35A7	4.5	—	6.6	21.0	7.4
2008	Odyssey	3.5	J35A6/J35A7	4.5	—	6.6	21.0	7.4

NOTE: All capacities are approximate. Add fluid gradually and check to be sure a proper fluid level is obtained.

22140_ODYS_C0004

FLUID SPECIFICATIONS

Year	Model	Engine Displ. Liters	Engine Oil	Man. Trans.	Auto. Trans.	Drive Axle		Transfer Case	Power Steering Fluid	Brake Master Cylinder	Cooling System
						Front	Rear				
2007	Odyssey	3.5	5W-20 Honda	N/A	Honda ATF-Z1	—	—	—	Honda PS Fluid	Honda DOT 3	①
2008	Odyssey	3.5	5W-20 Honda	N/A	Honda ATF-Z1	—	—	—	Honda PS Fluid	Honda DOT 3	①

DOT: Department Of Transpotation

① Honda Long Life Antifreeze/Coolant-Type2

22140_ODYS_C0005

VALVE SPECIFICATIONS

Year	Engine Displacement Liters	Engine ID	Seat Angle (deg.)	Face Angle (deg.)	Spring Test Pressure (lbs. @ in.)	Spring Installed Height (in.)	Stem-to-Guide Clearance (in.)		Stem Diameter (in.)	
							Intake	Exhaust	Intake	Exhaust
2007	3.5	J35A6	NA	NA	NA	①	0.0020-0.0040	0.0040-0.0060	0.2159-0.2163	0.2146-0.2150
		J35A7	NA	NA	NA	②	0.0020-0.0040	0.0040-0.0060	0.2159-0.2163	0.2146-0.2150
2008	3.5	J35A6	NA	NA	NA	①	0.0020-0.0040	0.0040-0.0060	0.2159-0.2163	0.2146-0.2150
		J35A7	NA	NA	NA	③	0.0020-0.0040	0.0040-0.0060	0.2159-0.2163	0.2146-0.2150

NA: Not Available

① Valve spring free length:
  Intake: 2.029 in.
  Exhaust: 2.010 in.

② Valve spring free length:
  Intake: 2.069 in.
  Exhaust: 2.069 in.

③ Valve spring free length:
  Intake: 2.027 in.
  Exhaust: 2.010 in.

22140_ODYS_C0007

CAMSHAFT AND BEARING SPECIFICATIONS

All measurements are given in inches.

Year	Engine Displacement Liters	Engine VIN	Journal Diameter	Brg. Oil Clearance	Shaft End-play	Runout	Journal Bore	Lobe Height Intake	Lobe Height Exhaust
2007	3.5	J35A6	NA	0.0020-0.0035	0.0020-0.0080	0.0010	NA	①	1.4302
	3.5	J35A7	NA	0.0020-0.0035	0.0020-0.0080	0.0010	NA	②	③
2008	3.5	J35A6	NA	0.0020-0.0035	0.0020-0.0080	0.0010	NA	①	1.4302
	3.5	J35A7	NA	0.0020-0.0035	0.0020-0.0080	0.0010	NA	④	⑤

NA: Information not available
① Primary: 1.3796 in.
   Mid: 1.4348 in.
   Secondary: 1.3891 in.
② Front: 1.4232 in.
   Rear: 1.3939 in.

③ Front: 1.4120 in.
   Rear: 1.4581 in.
④ Intake Nos. 1-4: 1.3843 in.
   Intake Nos. 5-6: 1.3841 in.
⑤ Intake Nos. 1-4: 1.4385 in.
   Intake Nos. 5-6: 1.4378 in.

22140_ODYS_C0006

CRANKSHAFT AND CONNECTING ROD SPECIFICATIONS

All measurements are given in inches

Year	Engine Displacement Liters	Engine ID	Crankshaft Main Brg. Journal Dia.	Crankshaft Main Brg. Oil Clearance	Crankshaft Shaft End-play	Crankshaft Thrust on No.	Connecting Rod Journal Diameter	Connecting Rod Oil Clearance	Connecting Rod Side Clearance
2007	3.5	J35A6 J35A7	2.8337-2.8346	0.0008-0.0017	0.0040-0.0140	3	2.1644-2.1654	0.0008-0.0017	0.0060-0.0140
2008	3.5	J35A6 J35A7	2.8337-2.8346	0.0007-0.0017	0.0040-0.0140	3	2.1644-2.1654	0.0008-0.0017	0.0060-0.0140

22140_ODYS_C0008

PISTON AND RING SPECIFICATIONS

All measurements are given in inches

Year	Engine Displacement Liters	Engine ID	Piston Clearance	Ring Gap Top Compression	Ring Gap Bottom Compression	Ring Gap Oil Control	Ring Side Clearance Top Compression	Ring Side Clearance Bottom Compression	Ring Side Clearance Oil Control
2007	3.5	J35A6 J35A7	0.0006-0.0016	0.0080-0.0140	0.0160-0.0220	0.0080-0.0280	0.0022-0.0031	0.0012-0.0022	NA
2008	3.5	J35A6 J35A7	0.0006-0.0016	0.0080-0.0140	0.0160-0.0220	0.0080-0.0280	0.0022-0.0031	0.0012-0.0022	NA

NA: Not Applicable

22140_ODYS_C0009

TORQUE SPECIFICATIONS
All readings in ft. lbs.

Year	Engine Displacement Liters	Engine ID	Cylinder Head Bolts	Main Bearing Bolts	Rod Bearing Bolts	Crankshaft Damper Bolts	Flywheel Bolts	Manifold Intake	Manifold Exhaust	Spark Plugs	Oil Pan Drain Plug
2007	3.5	J35A6	①	②	③	④	54	16	23	13	29
		J35A7	①	②	③	④	54	16	23	13	29
2008	3.5	J35A6	①	②	③	④	54	16	23	13	29
		J35A7	①	②	③	④	54	16	23	13	29

NOTE: Dip main bearing bolts and crankshaft damper bolt in clean engine oil prior to tightening.

① 12-point head bolts
  Step 1: Torque all bolts to 22 ft. lbs.
  Step 2: Torque all bolts an additional 90 deg.
  Step 3: Torque all bolts an additional 90 deg.
  New Bolt Only: an additional 90 deg.

② Cap bolts: 54 ft. lbs.
  Side bolts: 36 ft. lbs.
③ Step 1: 14 ft. lbs.
  Step 2: 90 degrees

④ Step 1: 47 ft. lbs.
  Step 2: plus 60 degrees

22140_ODYS_C0010

WHEEL ALIGNMENT

Year	Model		Caster Range (+/-Deg.)	Caster Preferred Setting (Deg.)	Camber Range (+/-Deg.)	Camber Preferred Setting (Deg.)	Toe-in (in.)
2007	Odyssey	F	1.00	+2.53	0.50	0	0+/-0.08
		R	—	—	0.75	-0.50	0.08+/-0.08
2008	Odyssey	F	1.00	+2.53	0.50	0	0+/-0.08
		R	—	—	0.75	-0.50	0.08+/-0.08

22140_ODYS_C0011

TIRE, WHEEL AND BALL JOINT SPECIFICATIONS

| Year | Model | OEM Tires | | Tire Pressures (psi) | | Wheel Size | Ball Joint Inspection | Lug Nut Torque (ft. lbs.) |
		Standard	Optional	Front	Rear			
2007	Except Touring	P235/65R16	None	①	①	7.0	NS	94
	Touring	P235-710R460A	None	①	①	7.0	NS	94
2008	Except Touring	P235/65R16	None	①	①	7.0	NS	94
	Touring	P235-710R460A	None	①	①	7.0	NS	94

OEM: Original Equipment Manufacturer

PSI: Pounds Per Square Inch

STD: Standard

OPT: Optional

NS: Not specified by manufacturer

① See placard on vehicle

22140_ODYS_C0012

BRAKE SPECIFICATIONS

All measurements in inches unless noted

| Year | Model | | Brake Disc | | | Parking Brake Brake Drum Diameter | | | Minimum Lining Thickness | | Brake Caliper | |
			Original Thickness	Minimum Thickness	Maximum Runout	Original Inside Diameter	Max. Wear Limit	Maximum Machine Diameter	Front	Rear	Bracket Bolts (ft. lbs.)	Mounting Bolts (ft. lbs.)
2007	Odyssey	F	1.100	1.020	0.004	—	—	—	0.060	—	101	37
		R	0.440	0.350	0.004	—	—	—	—	0.060	65	16
2008	Odyssey	F	1.100	1.020	0.004	—	—	—	0.060	—	101	37
		R	0.440	0.350	0.004	—	—	—	—	0.060	65	16

F: Front

R: Rear

① Parking brake shoe

22140_ODYS_C0013

MAINTENANCE MINDER SCHEDULE
Honda Odyssey

The Ridgeline displays engine oil life and maintenance service items in the information display to indicate when to perform maintenance service. If the engine oil life is 15% or less, based on the onboard computer's caluculations, you will see SERVICE DUE SOON in the information display every time the ignition key is turned to ON. The maintenance minder indicator will also come on and the maintenance code(s) for other scheduled maintenance items needing service will be displayed below the message.

Symbol	Item	Service
A	Engine oil ①	Change
B	Engine oil and filter	Change
	Tires	Rotate
	Brakes	Inspect
	Parking brake adjustment	Check
	Steering gear and linkage	Inspect
	Suspension components	Inspect
	Driveshaft boots	Inspect
	Brake hoses and lines	Inspect
	All fluid levels and condition	Inspect
	Exhaust system	Inspect
	Fuel lines and connections	Inspect
1	Tires	Rotate
2	Engine air filter ②	Replace
	Dust and pollen filter ③	Replace
	Accessory drive belt	Inspect
3	Transmission fluid ④	Replace
4	Spark plugs	Replace
	Timing belt ⑤	Replace
	Water pump	Inspect
	Valve clearance ⑥	Inspect
5	Engine coolant	Replace

① If the message SERVICE DUE NOW does not appear more than 12 months after the display is reset, change every year.

② If driven in dusty conditions, replace every 15,000 miles.

③ If driven in urban areas that have a high concentration of soot from industry and diesel, replace every 15,000 miles

④ If regularly driven in mountainous areas at very low speed or trailer towing, change the fluid every 30,000 miles.

⑤ If driven regularly in temperatures over 110 deg.F or below -20 deg.F, or towing a trailer, replace every 60,000 miles.

⑥ Adjust if necessary.

Additionally, replace the brake fluid every 3 years, and inspect the idle speed every 160,000 miles.

To reset the Engine Oil Life Display on LX, EX and EX-L models:

1. Turn the ignition switch to ON.
2. Press the SELECT/RESET button repeatedly until the engine oil life display or the service message is displayed.
3. Press the SELECT/RESET button for about 10 seconds. The engine oil life idicator and the codes will blink.
4. Press the SELECT/RESET knob for more than 5 seconds. The codes will disappear and the indicator will reset to 100.

To reset the Engine Oil Life Display on Touring models:

1. Turn the ignition switch to ON.
2. Press the SEL/RESET button on the steering wheel until the engine oil life is displayed.
3. Press the SEL/RESET button on the steering wheel for 10 seconds. The display will change to CUSTOM SETUP mode.
4. Press the SEL/RESET button on the steering wheel. The codes will disappear and the indicator will reset to 100.

22140_ODYS_C0014

PRECAUTIONS

Before servicing any vehicle, please be sure to read all of the following precautions, which deal with personal safety, prevention of component damage, and important points to take into consideration when servicing a motor vehicle:

• Never open, service or drain the radiator or cooling system when the engine is hot; serious burns can occur from the steam and hot coolant.

• Observe all applicable safety precautions when working around fuel. Whenever servicing the fuel system, always work in a well-ventilated area. Do not allow fuel spray or vapors to come in contact with a spark, open flame, or excessive heat (a hot drop light, for example). Keep a dry chemical fire extinguisher near the work area. Always keep fuel in a container specifically designed for fuel storage; also, always properly seal fuel containers to avoid the possibility of fire or explosion. Refer to the additional fuel system precautions later in this section.

• Fuel injection systems often remain pressurized, even after the engine has been turned **OFF**. The fuel system pressure must be relieved before disconnecting any fuel lines. Failure to do so may result in fire and/or personal injury.

• Brake fluid often contains polyglycol ethers and polyglycols. Avoid contact with the eyes and wash your hands thoroughly after handling brake fluid. If you do get brake fluid in your eyes, flush your eyes with clean, running water for 15 minutes. If eye irritation persists, or if you have taken brake fluid internally, IMMEDIATELY seek medical assistance.

• The EPA warns that prolonged contact with used engine oil may cause a number of skin disorders, including cancer. You should make every effort to minimize your exposure to used engine oil. Protective gloves should be worn when changing oil. Wash your hands and any other exposed skin areas as soon as possible after exposure to used engine oil. Soap and water, or waterless hand cleaner should be used.

• All new vehicles are now equipped with an air bag system, often referred to as a Supplemental Restraint System (SRS) or Supplemental Inflatable Restraint (SIR) system. The system must be disabled before performing service on or around system components, steering column, instrument panel components, wiring and sensors. Failure to follow safety and disabling procedures could result in accidental air bag deployment, possible personal injury and unnecessary system repairs.

• Always wear safety goggles when working with, or around, the air bag system. When carrying a non-deployed air bag, be sure the bag and trim cover are pointed away from your body. When placing a non-deployed air bag on a work surface, always face the bag and trim cover upward, away from the surface. This will reduce the motion of the module if it is accidentally deployed. Refer to the additional air bag system precautions later in this section.

• Clean, high quality brake fluid from a sealed container is essential to the safe and proper operation of the brake system. You should always buy the correct type of brake fluid for your vehicle. If the brake fluid becomes contaminated, completely flush the system with new fluid. Never reuse any brake fluid. Any brake fluid that is removed from the system should be discarded. Also, do not allow any brake fluid to come in contact with a painted surface; it will damage the paint.

• Never operate the engine without the proper amount and type of engine oil; doing so WILL result in severe engine damage.

• Timing belt maintenance is extremely important. Many models utilize an interference-type, non-freewheeling engine. If the timing belt breaks, the valves in the cylinder head may strike the pistons, causing potentially serious (also time-consuming and expensive) engine damage. Refer to the maintenance interval charts for the recommended replacement interval for the timing belt, and to the timing belt section for belt replacement and inspection.

• Disconnecting the negative battery cable on some vehicles may interfere with the functions of the on-board computer system(s) and may require the computer to undergo a relearning process once the negative battery cable is reconnected.

• When servicing drum brakes, only disassemble and assemble one side at a time, leaving the remaining side intact for reference.

• Only an MVAC-trained, EPA-certified automotive technician should service the air conditioning system or its components.

BRAKES

GENERAL INFORMATION

PRECAUTIONS

• Certain components within the ABS system are not intended to be serviced or repaired individually.

• Do not use rubber hoses or other parts not specifically specified for and ABS system. When using repair kits, replace all parts included in the kit. Partial or incorrect repair may lead to functional problems and require the replacement of components.

• Lubricate rubber parts with clean, fresh brake fluid to ease assembly. Do not use shop air to clean parts; damage to rubber components may result.

• Use only DOT 3 brake fluid from an unopened container.

• If any hydraulic component or line is removed or replaced, it may be necessary to bleed the entire system.

• A clean repair area is essential. Always clean the reservoir and cap thoroughly before removing the cap. The slightest amount of dirt in the fluid may plug an orifice and impair the system function. Perform repairs after components have been thoroughly cleaned; use only denatured alcohol to clean components. Do not allow ABS components to come into contact with any substance containing mineral oil; this includes used shop rags.

ANTI-LOCK BRAKE SYSTEM (ABS)

• The Anti-Lock control unit is a microprocessor similar to other computer units in the vehicle. Ensure that the ignition switch is **OFF** before removing or installing controller harnesses. Avoid static electricity discharge at or near the controller.

• If any arc welding is to be done on the vehicle, the control unit should be unplugged before welding operations begin.

BRAKES

BLEEDING PROCEDURE

BLEEDING PROCEDURE

See Figures 1 through 3.

➡ **Do not reuse the drained fluid. Use only clean Honda DOT 3 Brake Fluid**

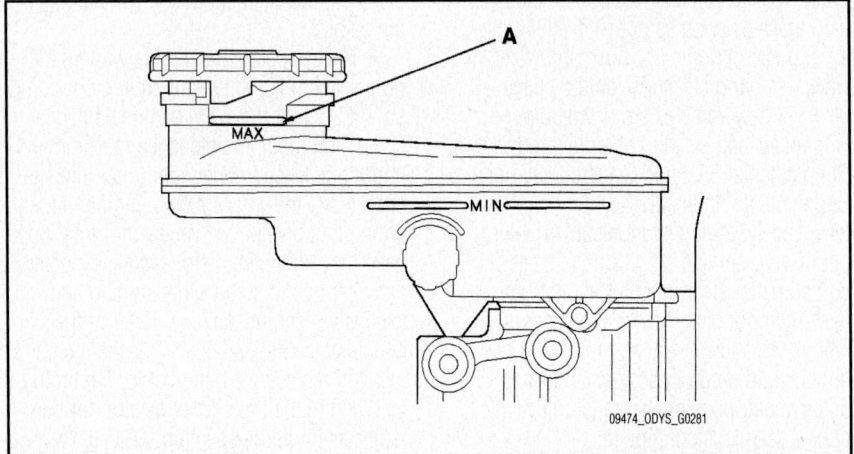

Fig. 1 The reservoir on the master cylinder must be at the MAX (upper) level mark at the start of the bleeding procedure

from an unopened container. Using a non-Honda brake fluid can cause corrosion and shorten the life of the system.

Do not mix different brands of brake fluid; they may not be compatible.

Make sure no dirt or other foreign matter is allowed to contaminate the brake fluid.

Do not spill brake fluid on the vehicle, it may damage the paint; if brake fluid does contact the paint, wash it off immediately with water.

The reservoir on the master cylinder must be at the MAX (upper) level mark at the start of the bleeding procedure and checked after bleeding each brake caliper. Add fluid as required.

1. Make sure the brake fluid level in the reservoir is at the MAX (upper) level line (A).

2. Attach a length of clear drain tube to the bleed screw.

3. Have someone slowly pump the brake pedal several times, then apply steady pressure.

4. Starting at the left-front, loosen the brake bleed screw to allow air to escape from the system. Then tighten the bleed screw securely.

5. Repeat the procedure for each caliper until no air bubbles are in the fluid. Bleed the calipers in the sequence shown.

6. Refill the master cylinder reservoir to the MAX (upper) level line.

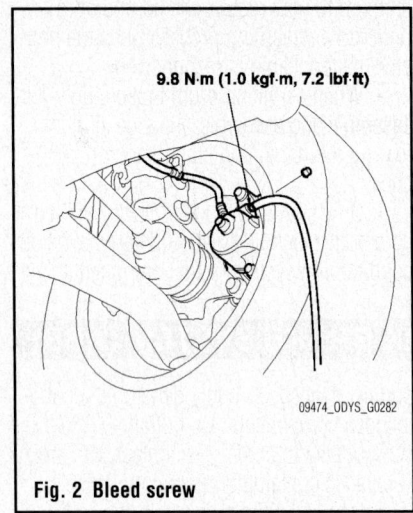

Fig. 2 Bleed screw

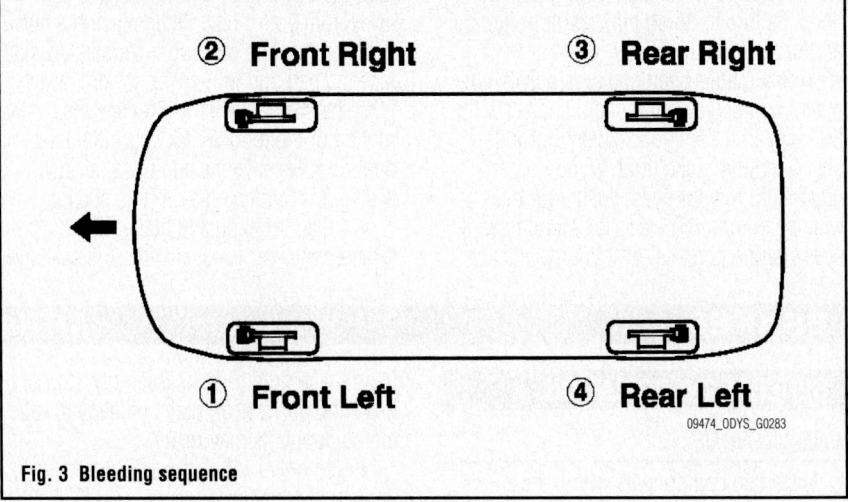

Fig. 3 Bleeding sequence

BRAKES   FRONT DISC BRAKES

❊❊ CAUTION

Dust and dirt accumulating on brake parts during normal use may contain asbestos fibers from production or aftermarket brake linings. Breathing excessive concentrations of asbestos fibers can cause serious bodily harm. Exercise care when servicing brake parts. Do not sand or grind brake lining unless equipment used is designed to contain the dust residue. Do not clean brake parts with compressed air or by dry brushing. Cleaning should be done by dampening the brake components with a fine mist of water, then wiping the brake components clean with a dampened cloth. Dispose of cloth and all residue containing asbestos fibers in an impermeable container with the appropriate label. Follow practices prescribed by the Occupational Safety and Health Administration (OSHA) and the Environmental Protection Agency (EPA) for the handling, processing, and disposing of dust or debris that may contain asbestos fibers.

BRAKE CALIPER

REMOVAL & INSTALLATION
See Figures 4 through 7.

❊❊ CAUTION

Dust and dirt accumulating on brake parts during normal use may contain

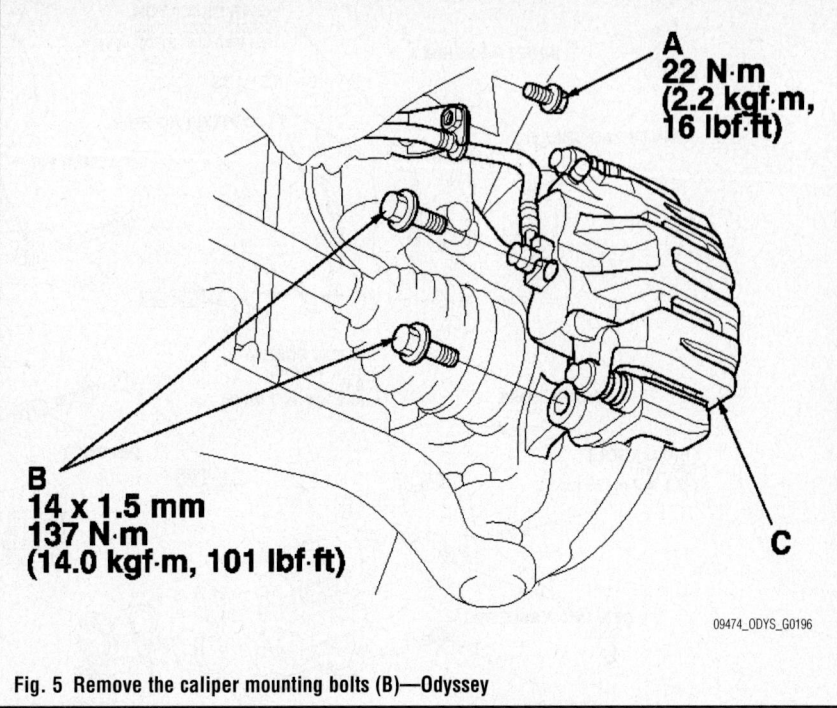

Fig. 5 Remove the caliper mounting bolts (B)—Odyssey

asbestos fibers from production or aftermarket brake linings. Breathing excessive concentrations of asbestos fibers can cause serious bodily harm. Exercise care when servicing brake parts. Do not sand or grind brake lining unless equipment used is designed to contain the dust residue. Do not clean brake parts with compressed air or by dry brushing. Cleaning should be done by dampening the brake components with a fine mist of water, then wiping the brake

components clean with a dampened cloth. Dispose of cloth and all residue containing asbestos fibers in an impermeable container with the appropriate label. Follow practices prescribed by the Occupational Safety and Health Administration (OSHA) and the Environmental Protection Agency (EPA) for the handling, processing, and disposing of dust or debris that may contain asbestos fibers.

1. Raise the front of the vehicle, and support it with safety stands in the proper locations.
2. Remove the wheel nuts (A) and front wheel (B).
3. Remove the brake hose mounting bolt (A).
4. Remove the brake hose-to-caliper union bolt and discard the washers. Plug the brake hose.
5. Remove the caliper mounting bolts (B).
6. Installation is the reverse of removal. Install the brake hose (A) on the damper bracket with flange bolt (B) first, then connect the brake hose to the caliper with the banjo bolt (C) and new sealing washers (D). Observe the torques given in the accompanying illustrations.
7. Bleed the brakes.
8. Press the brake pedal several times to make sure the brakes work.

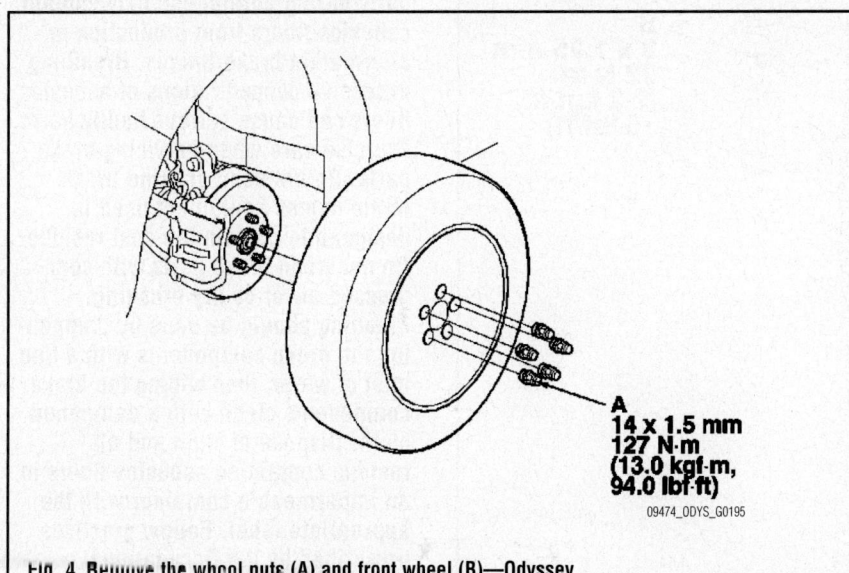

Fig. 4 Remove the wheel nuts (A) and front wheel (B)—Odyssey

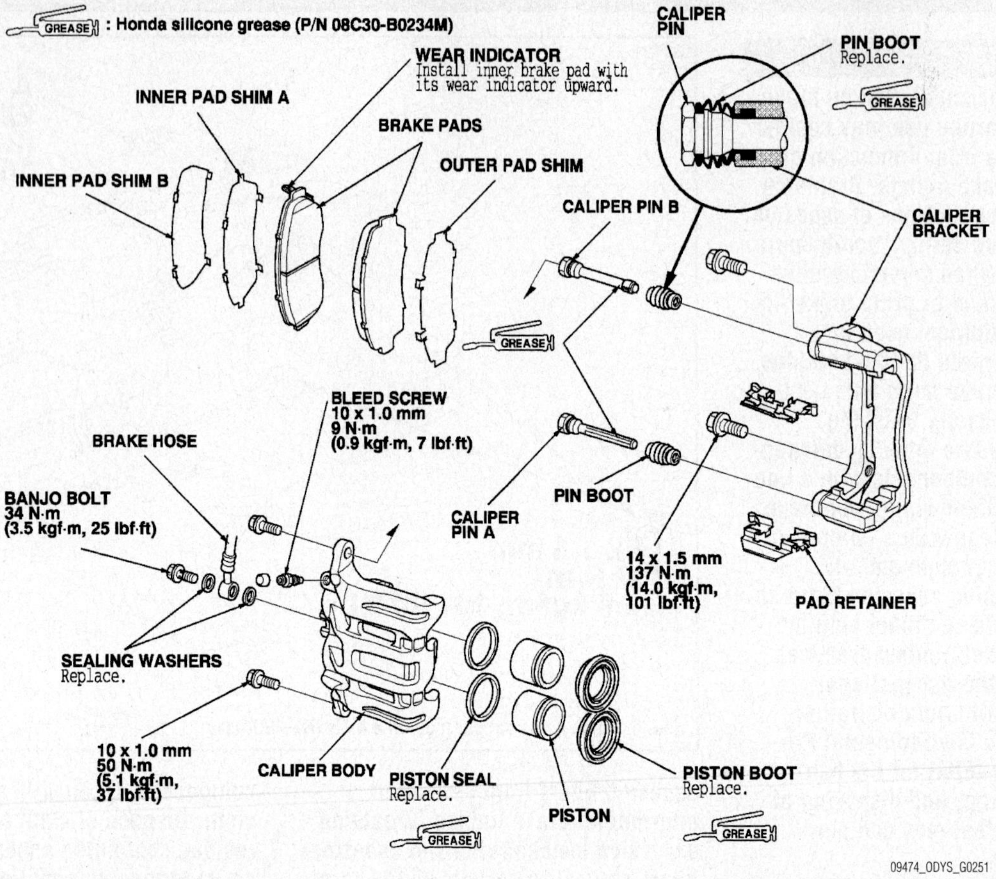

GREASE : Honda silicone grease (P/N 08C30-B0234M)

INNER PAD SHIM A

INNER PAD SHIM B

WEAR INDICATOR
Install inner brake pad with
its wear indicator upward.

BRAKE PADS

OUTER PAD SHIM

CALIPER PIN B

CALIPER
PIN

PIN BOOT
Replace.

GREASE

CALIPER
BRACKET

GREASE

BLEED SCREW
10 x 1.0 mm
9 N·m
(0.9 kgf·m, 7 lbf·ft)

BRAKE HOSE

PIN BOOT

BANJO BOLT
34 N·m
(3.5 kgf·m, 25 lbf·ft)

CALIPER
PIN A

14 x 1.5 mm
137 N·m
(14.0 kgf·m,
101 lbf·ft)

PAD RETAINER

SEALING WASHERS
Replace.

10 x 1.0 mm
50 N·m
(5.1 kgf·m,
37 lbf·ft)

CALIPER BODY

PISTON SEAL
Replace.

PISTON

PISTON BOOT
Replace.

GREASE

GREASE

09474_ODYS_G0251

Fig. 6 Exploded view of the front caliper—Odyssey

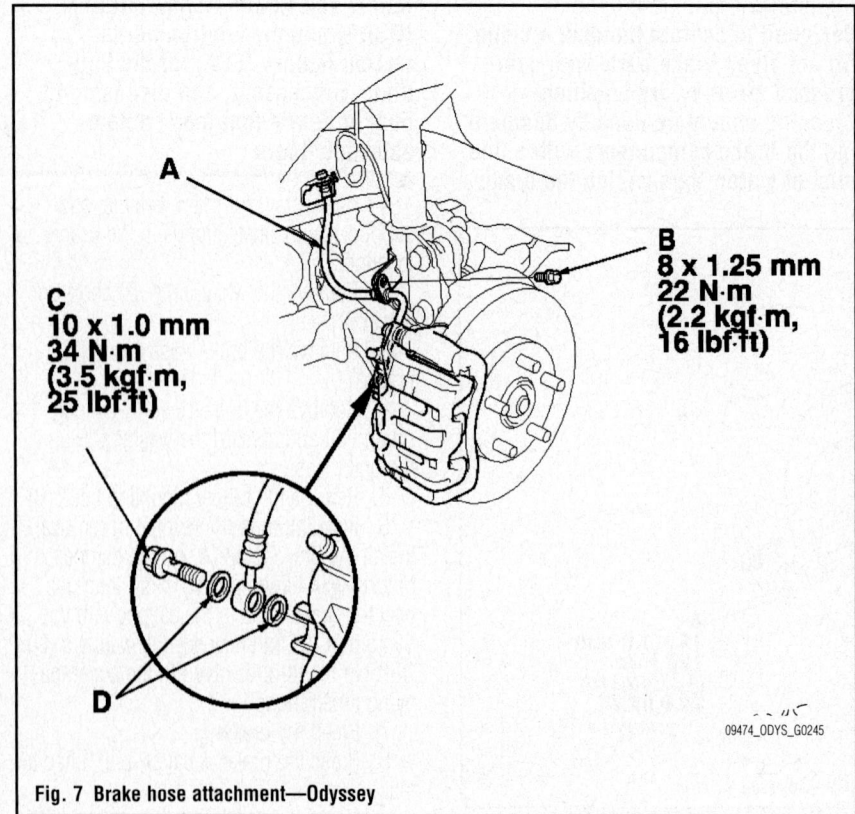

A

B
8 x 1.25 mm
22 N·m
(2.2 kgf·m,
16 lbf·ft)

C
10 x 1.0 mm
34 N·m
(3.5 kgf·m,
25 lbf·ft)

D

09474_ODYS_G0245

Fig. 7 Brake hose attachment—Odyssey

DISC BRAKE PADS

REMOVAL & INSTALLATION

See Figures 8 through 12.

❊❊ CAUTION

Dust and dirt accumulating on brake parts during normal use may contain asbestos fibers from production or aftermarket brake linings. Breathing excessive concentrations of asbestos fibers can cause serious bodily harm. Exercise care when servicing brake parts. Do not sand or grind brake lining unless equipment used is designed to contain the dust residue. Do not clean brake parts with compressed air or by dry brushing. Cleaning should be done by dampening the brake components with a fine mist of water, then wiping the brake components clean with a dampened cloth. Dispose of cloth and all residue containing asbestos fibers in an impermeable container with the appropriate label. Follow practices prescribed by the Occupational Safety and Health Administration

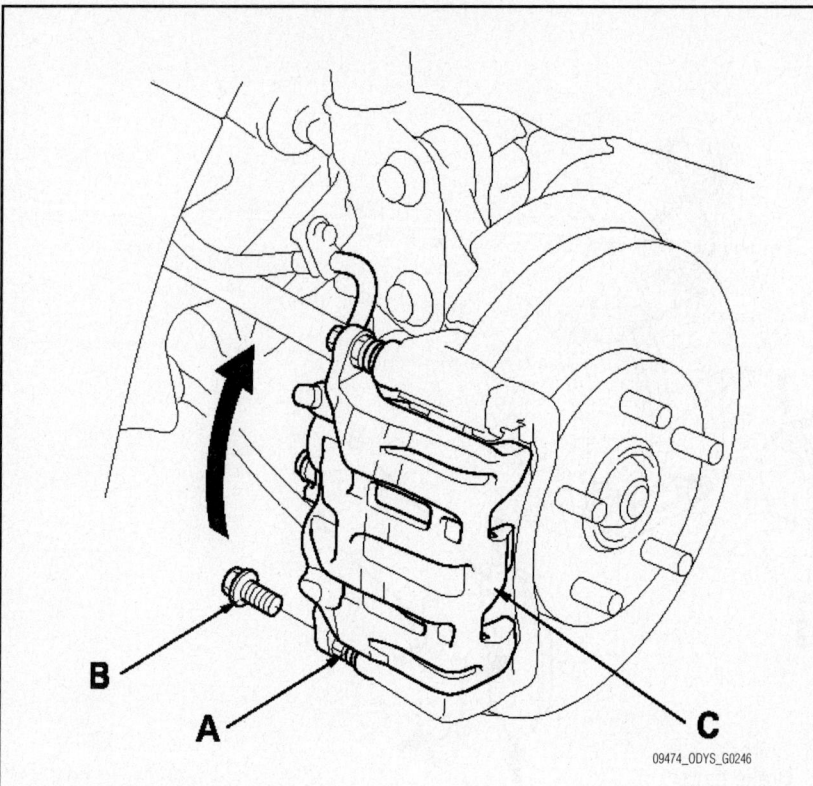

09474_ODYS_G0246

Fig. 8 While holding the caliper pin (A) with a wrench, remove the flange bolt (B). Then pivot the caliper (C) up out of the way—Odyssey

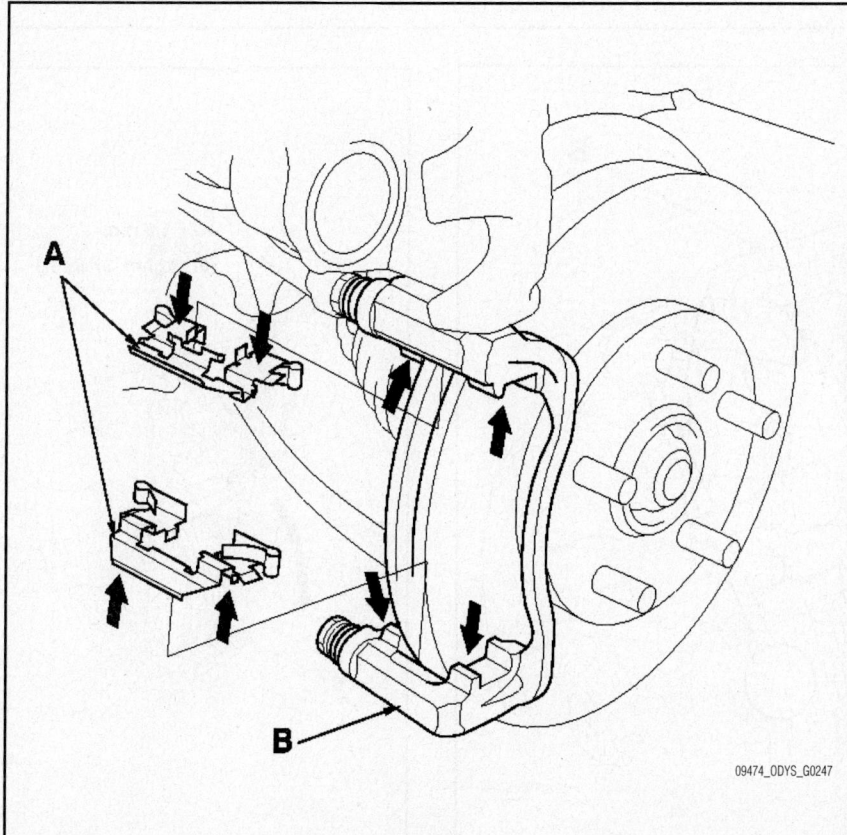

09474_ODYS_G0247

Fig. 9 Remove the pad retainers (A) from the caliper bracket (B)—Odyssey

(OSHA) and the Environmental Protection Agency (EPA) for the handling, processing, and disposing of dust or debris that may contain asbestos fibers.

1. Raise the front of the vehicle, and support it with safety stands in the proper locations.
2. Remove the front wheels.
3. While holding the caliper pin with a wrench, remove the flange bolt. Be careful not to damage the pin boot. Then pivot the caliper up out of the way. Check the hose and pin boots for damage and deterioration.
4. Remove the pad shims and brake pads.
5. Remove the pad retainers (A) from the caliper bracket (B).
6. Clean the caliper thoroughly; remove any rust, and check for grooves and cracks.
7. Check the brake disc for damage and cracks.

To install:

8. Apply a thin coat of M-77 assembly paste (P/N 08798-9010) to the retainers on their mating surfaces (indicated by the arrows) against the caliper bracket.
9. Install the pad retainers. Wipe excess paste off the retainers. Keep paste off the discs and pads.
10. Apply a thin coat of M-77 assembly paste (P/N 08798-9010) to the pad side of the shims, the back of the brake pads and the other areas indicated by the arrows. Wipe excess assembly paste off the shims and brake pads. Contaminated brake discs or brake pads reduce stopping ability. Keep grease and assembly paste off the brake discs and brake pads.
11. Install the brake pads and pad shims correctly. Install the brake pad with the wear indicator on the inside. If you are reusing the brake pads, always reinstall the brake pads in their original positions to prevent a loss of braking efficiency.
12. Mount the special tool on the caliper.
13. Press in the piston with the special tool so the caliper will fit over the brake pads. Make sure the piston boot is in position to prevent damaging it when pivoting the caliper down.

➡Be careful when pressing in the piston, brake fluid might overflow from the master cylinder's reservoir.

14. Remove the special tool.
15. Pivot the caliper down into position. Install the flange bolt (A), and

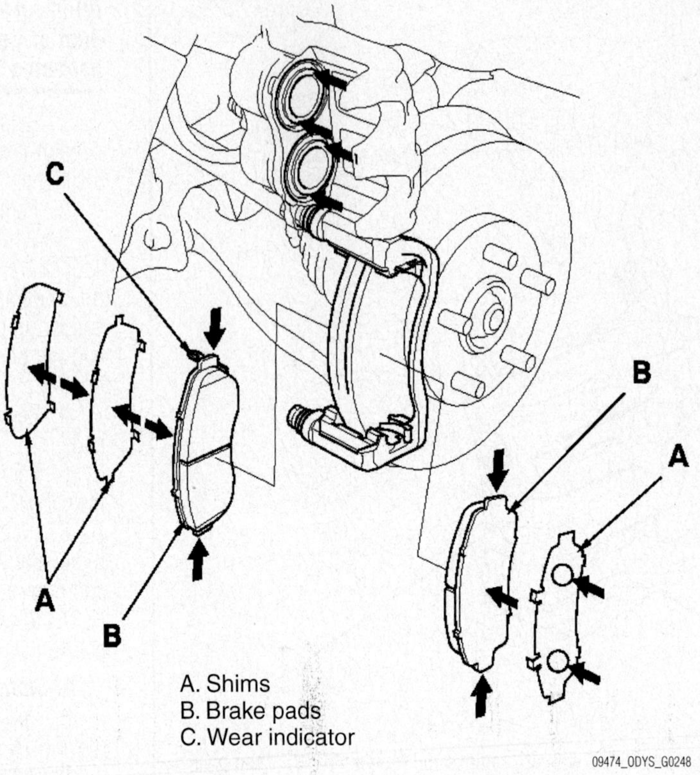

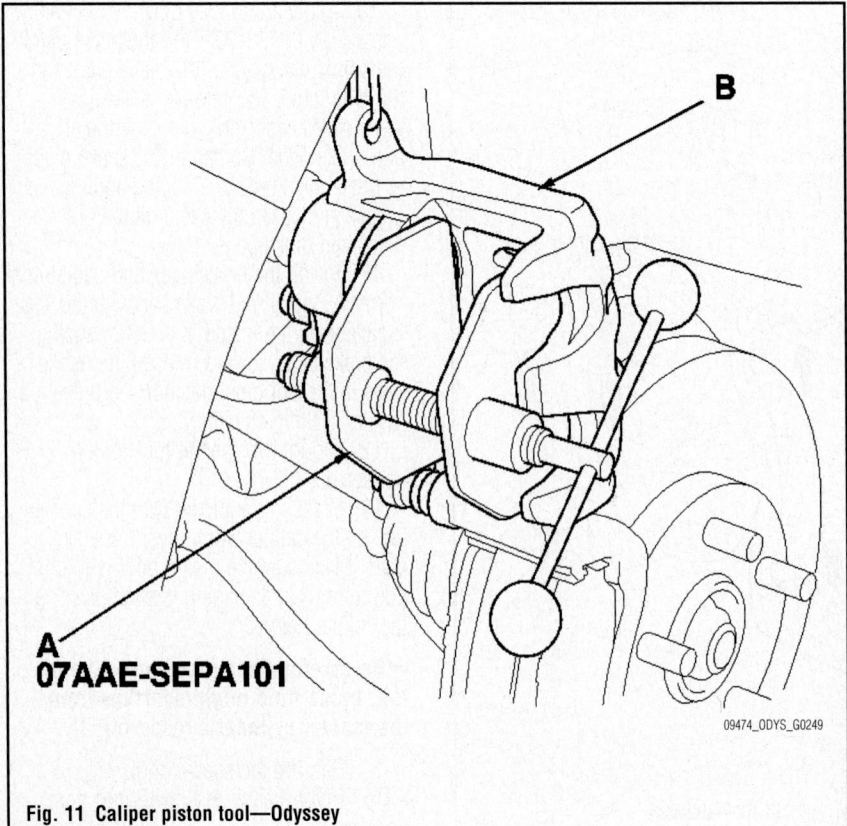

C
B
A
A
B
A. Shims
B. Brake pads
C. Wear indicator
09474_ODYS_G0248

Fig. 10 Front brake pads and shims—Odyssey

B
A
07AAE-SEPA101
09474_ODYS_G0249

Fig. 11 Caliper piston tool—Odyssey

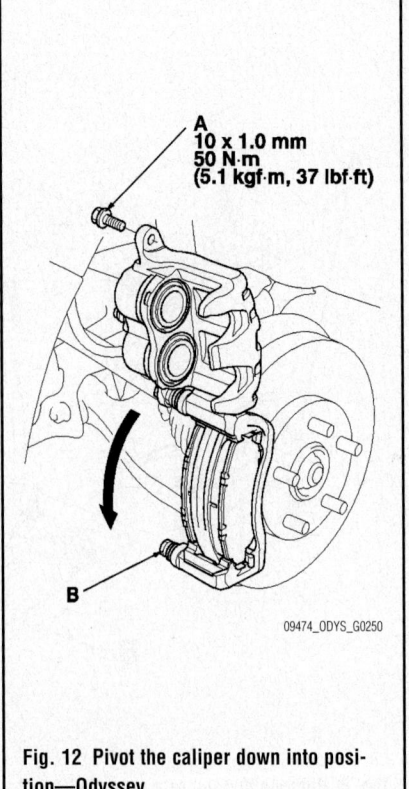

A
10 x 1.0 mm
50 N·m
(5.1 kgf·m, 37 lbf·ft)
B
09474_ODYS_G0250

Fig. 12 Pivot the caliper down into position—Odyssey

torque it to the specified torque while holding the caliper pin (B) with a wrench. Be careful not to damage the pin boot.

16. Press the brake pedal several times to make sure the brakes work.

➡Engagement of the brakes may require a greater pedal stroke immediately after the brake pads have been replaced. Several applications of the brake pedal will restore the normal pedal stroke.

17. After installation, check for leaks at hose and line joints or connections, and retighten if necessary.

18. Install the front wheels, then test-drive the vehicle.

BRAKES   REAR DISC BRAKES

✳✳ CAUTION

Dust and dirt accumulating on brake parts during normal use may contain asbestos fibers from production or aftermarket brake linings. Breathing excessive concentrations of asbestos fibers can cause serious bodily harm. Exercise care when servicing brake parts. Do not sand or grind brake lining unless equipment used is designed to contain the dust residue. Do not clean brake parts with compressed air or by dry brushing. Cleaning should be done by dampening the brake components with a fine mist of water, then wiping the brake components clean with a dampened cloth. Dispose of cloth and all residue containing asbestos fibers in an impermeable container with the appropriate label. Follow practices

prescribed by the Occupational Safety and Health Administration (OSHA) and the Environmental Protection Agency (EPA) for the handling, processing, and disposing of dust or debris that may contain asbestos fibers.

BRAKE CALIPER

REMOVAL & INSTALLATION
See Figure 13.

✳✳ CAUTION

Dust and dirt accumulating on brake parts during normal use may contain asbestos fibers from production or aftermarket brake linings. Breathing excessive concentrations of asbestos fibers can cause serious bodily harm. Exercise care when servicing brake

parts. Do not sand or grind brake lining unless equipment used is designed to contain the dust residue. Do not clean brake parts with compressed air or by dry brushing. Cleaning should be done by dampening the brake components with a fine mist of water, then wiping the brake components clean with a dampened cloth. Dispose of cloth and all residue containing asbestos fibers in an impermeable container with the appropriate label. Follow practices prescribed by the Occupational Safety and Health Administration (OSHA) and the Environmental Protection Agency (EPA) for the handling, processing, and disposing of dust or debris that may contain asbestos fibers.

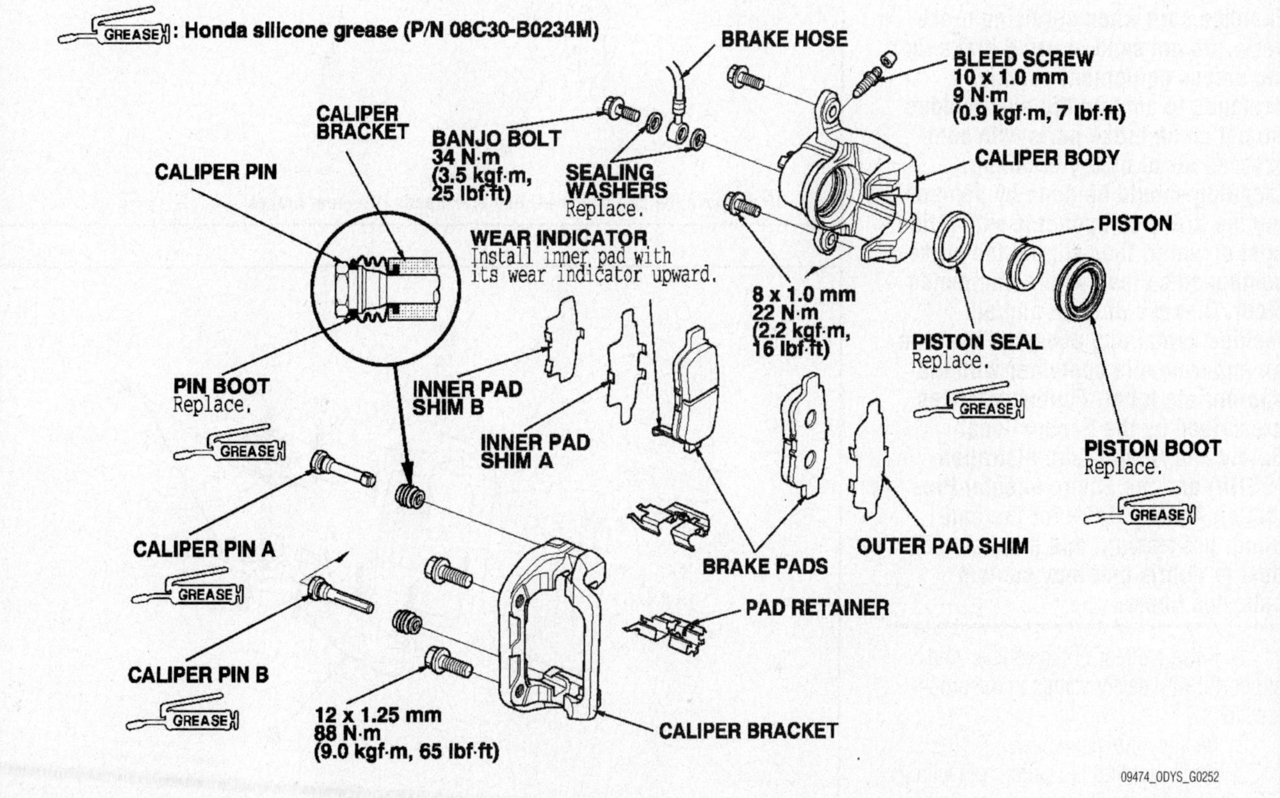

Fig. 13 Exploded view of the rear caliper—Odyssey

1. Raise the front of the vehicle, and support it with safety stands in the proper locations.

2. Remove the wheel nuts and front wheel.

3. Remove the brake hose mounting bolt.

4. Remove the brake hose-to-caliper union bolt and discard the washers. Plug the brake hose.

5. Remove the caliper mounting bolts.

6. Installation is the reverse of removal. Install the brake hose on the damper bracket with flange bolt first, then connect the brake hose to the caliper with the banjo bolt and new sealing washers. Observe the torques given in the accompanying illustrations.

7. Bleed the brakes.

8. Press the brake pedal several times to make sure the brakes work.

DISC BRAKE PADS

REMOVAL & INSTALLATION

See Figures 14 through 17.

✳✳ CAUTION

Dust and dirt accumulating on brake parts during normal use may contain asbestos fibers from production or aftermarket brake linings. Breathing excessive concentrations of asbestos fibers can cause serious bodily harm. Exercise care when servicing brake parts. Do not sand or grind brake lining unless equipment used is designed to contain the dust residue. Do not clean brake parts with compressed air or by dry brushing. Cleaning should be done by dampening the brake components with a fine mist of water, then wiping the brake components clean with a dampened cloth. Dispose of cloth and all residue containing asbestos fibers in an impermeable container with the appropriate label. Follow practices prescribed by the Occupational Safety and Health Administration (OSHA) and the Environmental Protection Agency (EPA) for the handling, processing, and disposing of dust or debris that may contain asbestos fibers.

1. Raise the rear of the vehicle, and support it with safety stands in the proper locations.

2. Remove the rear wheels.

3. While holding the caliper pin with a wrench, remove the flange bolt (B). Be careful not to damage the pin boot. Then pivot

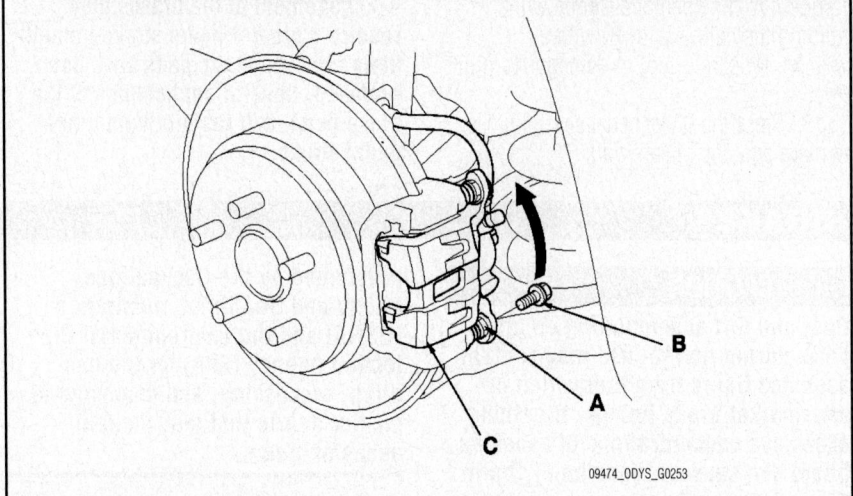

Fig. 14 While holding the caliper pin (A) with a wrench, remove the flange bolt (B). Then pivot the caliper (C) up out of the way—rear brakes

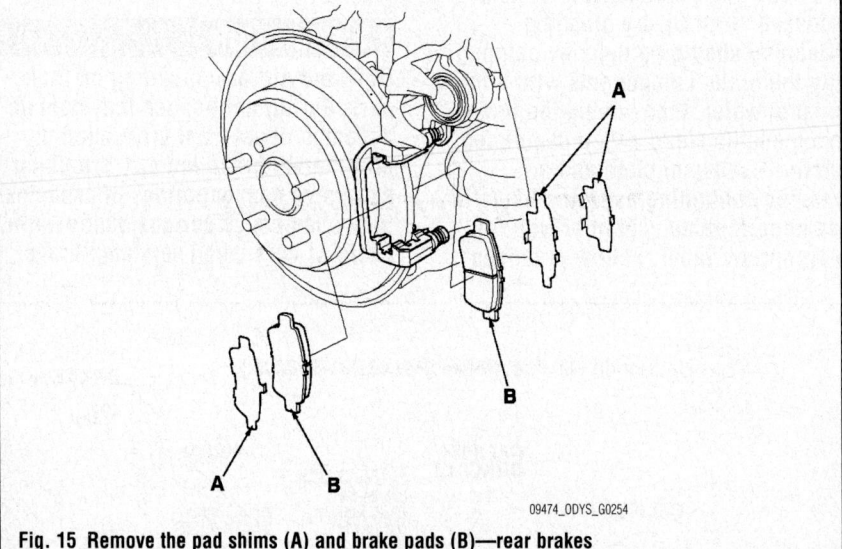

Fig. 15 Remove the pad shims (A) and brake pads (B)—rear brakes

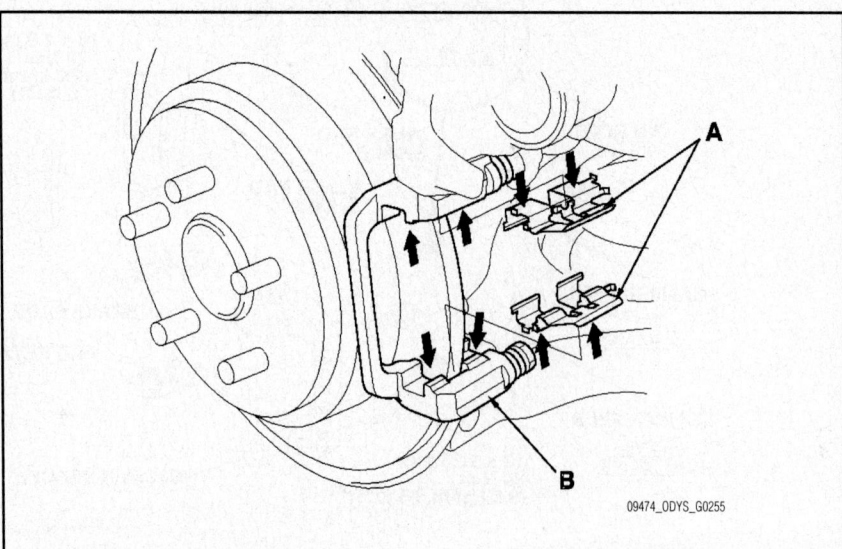

Fig. 16 Remove the pad retainers (A) from the caliper bracket (B)—rear brakes

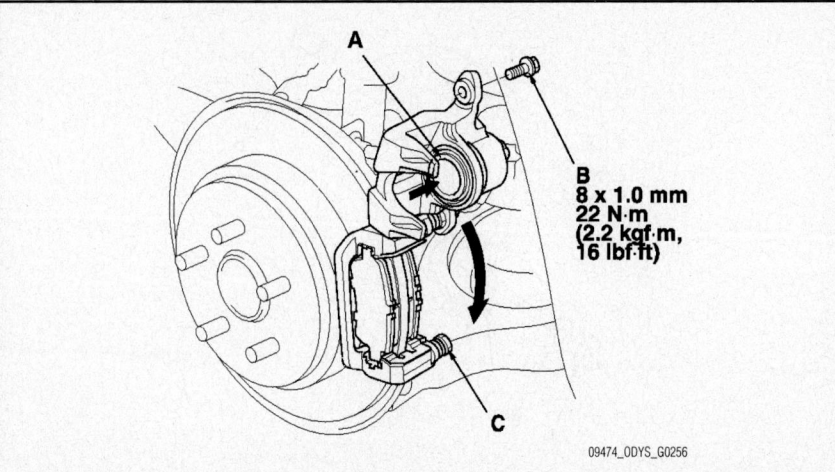

Fig. 17 Push in the piston (A) so the caliper will fit over the brake pads. Install the flange bolt (B), and torque it to the specified torque while holding the caliper pin (C) with a wrench—rear brakes

the caliper up out of the way. Check the hose and pin boots for damage and deterioration.

4. Remove the pad shims (A) and brake pads (B).

5. Remove the pad retainers (A) from the caliper bracket (B).

To install:

6. Clean the caliper bracket thoroughly; remove any rust, and check for grooves and cracks.

7. Check the brake disc for damage and cracks.

8. Apply a thin coat of M-77 assembly paste (P/N 08798-9010) to the retainers on their mating surfaces (indicated by the arrow) against the caliper bracket.

9. Install the pad retainers. Wipe excess assembly paste off the retainers. Keep any assembly paste off the discs and pads.

10. Apply a thin coat of M-77 assembly paste (P/N 08798-9010) to the pad side of the shims, the back of the brake pads, and the other areas indicated by the arrows. Wipe excess assembly paste off the pad shims and brake pads. Contaminated brake discs or pads reduce stopping ability. Keep assembly paste off the brake discs and pads.

11. Install the brake pads and pad shims correctly. Install the brake pad with the wear indicator on the inside. If you are reusing the brake pads, always reinstall the brake pads in their original positions to prevent a momentary loss of braking efficiency.

12. Push in the piston (A) so the caliper will fit over the brake pads. Make sure the piston boot is in position to prevent damaging it when pivoting the caliper down.

13. Pivot the caliper down into position. Install the flange bolt, and torque it to the

specified torque while holding the caliper pin with a wrench. Be careful not to damage the pin boot.

14. Press the brake pedal several times to make sure the brakes work.

➡**Brake engagement may require a greater pedal stroke immediately after the brake pads have been replaced. Several applications of the brake pedal will restore the normal pedal stroke.**

15. After installation, check for leaks at hose and line joints or connections, and retighten if necessary.

PARKING BRAKE SHOES

REMOVAL & INSTALLATION

See Figures 18 through 26.

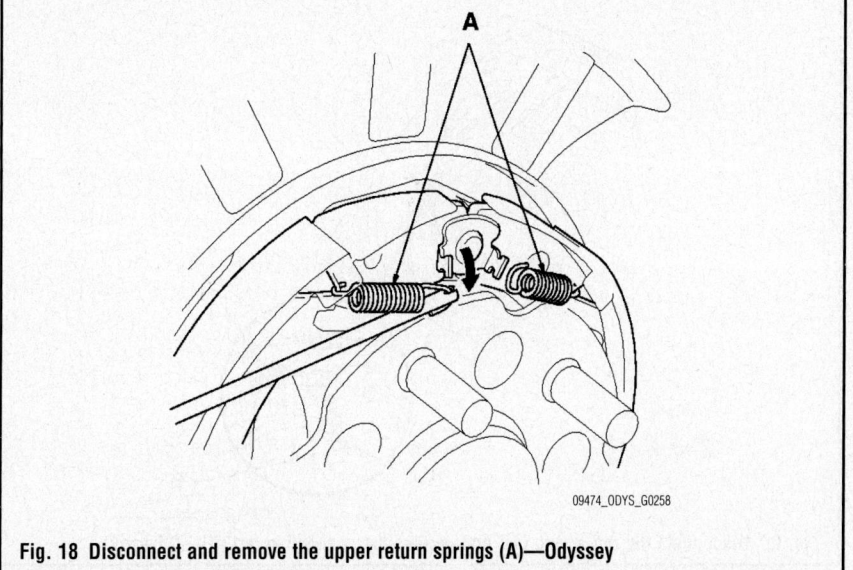

Fig. 18 Disconnect and remove the upper return springs (A)—Odyssey

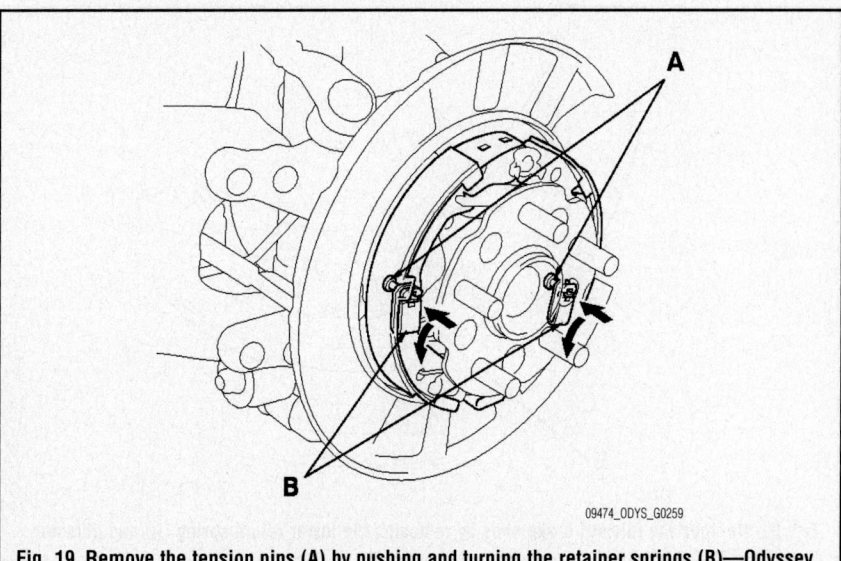

Fig. 19 Remove the tension pins (A) by pushing and turning the retainer springs (B)—Odyssey

1. Raise the rear of the vehicle, and support it with safety stands in the proper locations.

2. Remove the rear wheels.

3. Release the parking brake, and remove the rear brake caliper and brake disc.

4. Disconnect and remove the upper return springs.

5. Remove the tension pins (A) by pushing and turning the retainer springs (B).

6. Disconnect the rod spring (A), and remove the connecting rod (B).

7. Lower the parking brake shoe assembly.

8. Remove the forward brake shoe by removing the lower return spring (A) and adjuster assembly (B).

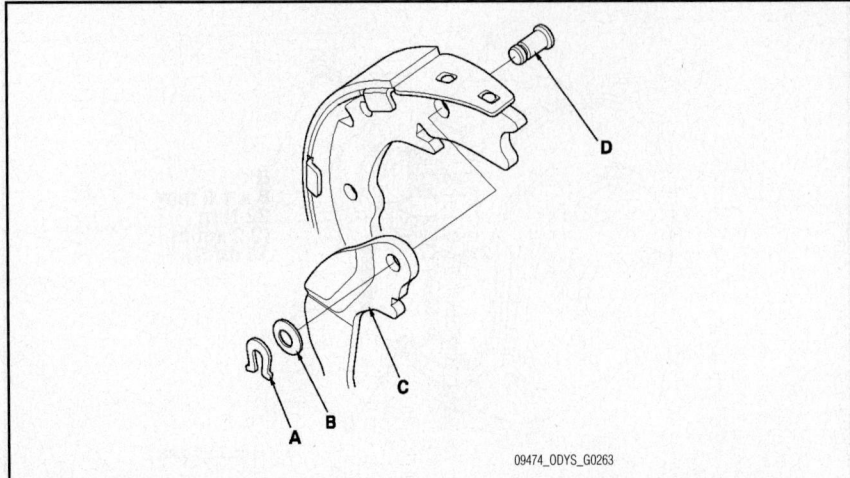

09474_ODYS_G0263

Fig. 22 Remove the U-clip (A), wave washer (B), parking brake lever (C), and pivot pin (D) from the brake shoe—Odyssey

9. Remove the rearward brake shoe by disconnecting the parking brake cable from the parking brake lever.

10. Remove the U-clip (A), wave washer (B), parking brake lever (C), and pivot pin (D) from the brake shoe.

To assemble:

11. Apply Molykote 44 MA grease to the sliding surface of the pivot pin and insert the pin into the brake shoe from the rear side.

12. Install the parking brake lever and wave washer on the pivot pin, and secure with a new U-clip.

   a. Install the wave washer with its convex side facing out.

   b. Pinch the U-clip securely to prevent the pivot pin from coming out of the brake shoe.

13. Connect the parking brake cable (A) to the parking brake lever (B). Apply silicone grease to the cable contact surface (C) on the backing plate.

14. Apply Molykote 44 MA grease to the sliding surfaces, the inner edges of the parking brake shoes, and the pivot of the parking brake lever as shown. Wipe off any excess. Keep grease off the brake linings.

15. Clean the threaded portions of clevis A, and coat the threads of clevis A with grease. Clean the sliding surface of clevis B, and coat the sliding surface of clevis B with grease. Thread clevis A all the way into the adjuster (C). Install clevis B.

16. Reinstall the brake shoe adjuster assembly (D), and hook the lower return spring (E) on the parking brake shoes.

17. Install the rod spring (A) to the connecting rod (B) first with the spring

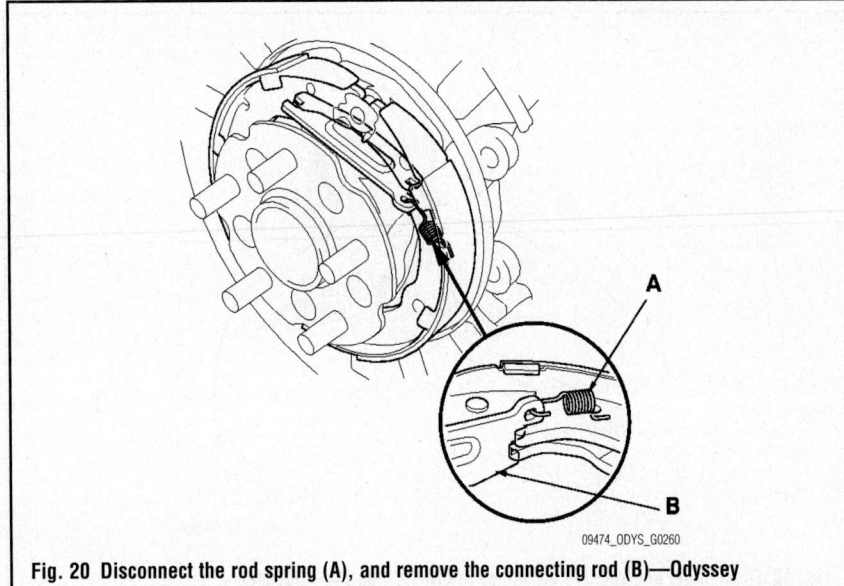

09474_ODYS_G0260

Fig. 20 Disconnect the rod spring (A), and remove the connecting rod (B)—Odyssey

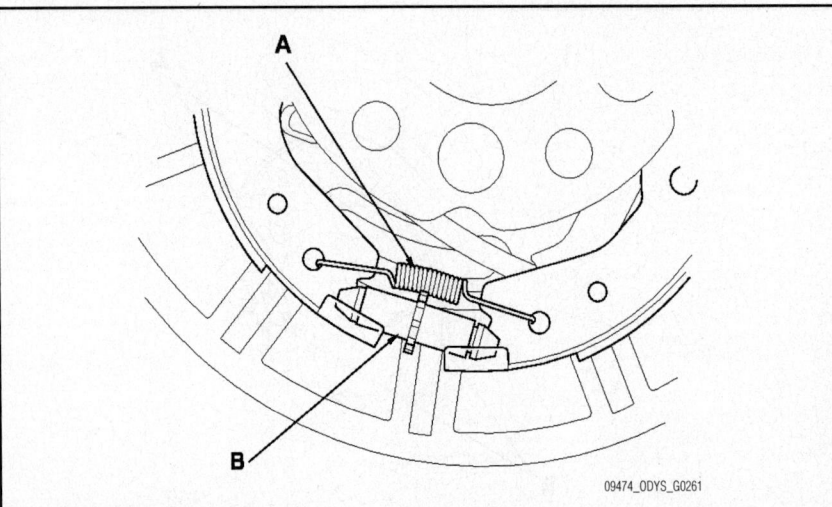

09474_ODYS_G0261

Fig. 21 Remove the forward brake shoe by removing the lower return spring (A) and adjuster assembly (B)—Odyssey

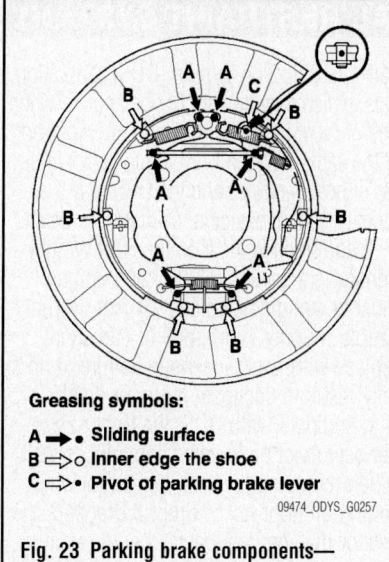

Greasing symbols:

A ➤● Sliding surface
B ⇨○ Inner edge the shoe
C ⇨● Pivot of parking brake lever

09474_ODYS_G0257

Fig. 23 Parking brake components—Odyssey

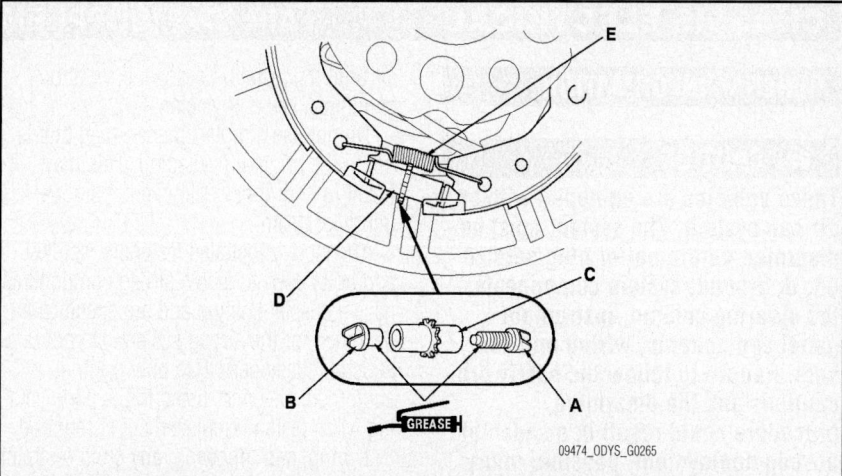

09474_ODYS_G0265

Fig. 25 Clean the threaded portions of clevis A, and coat the threads of clevis A with grease. Clean the sliding surface of clevis B, and coat the sliding surface of clevis B with grease. Thread clevis A all the way into the adjuster (C). Install clevis B. Reinstall the brake shoe adjuster assembly (D), and hook the lower return spring (E) on the parking brake shoes—Odyssey

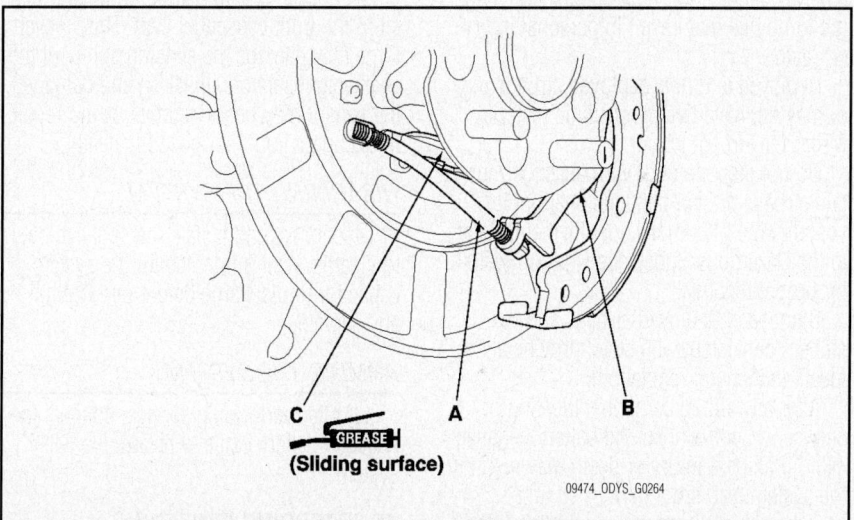

GREASE
(Sliding surface)

09474_ODYS_G0264

Fig. 24 Connect the parking brake cable (A) to the parking brake lever (B). Apply silicone grease to the cable contact surface (C) on the backing plate—Odyssey

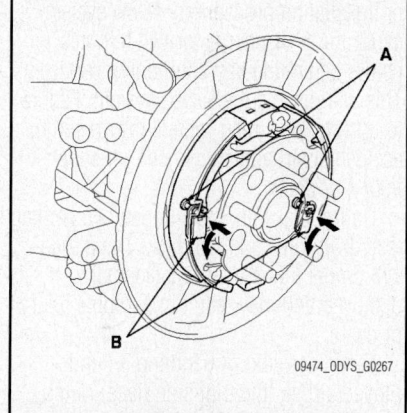

09474_ODYS_G0267

Fig. 26 Install the tension pins (A), and retainer springs (B). Make sure the tension pin does not contact to the parking brake lever—Odyssey

end (C) pointing downward. Then install the connecting rod on the parking brake shoes.

18. Install the tension pins (A), and retainer springs (B). Make sure the tension pin does not contact to the parking brake lever.

19. Install the upper return springs.

20. Install the brake disc/drum and rear brake caliper.

21. Adjust the parking brake.

CHASSIS ELECTRICAL · AIR BAG (SUPPLEMENTAL RESTRAINT SYSTEM)

GENERAL INFORMATION

❋❋ CAUTION

These vehicles are equipped with an air bag system. The system must be disarmed before performing service on, or around, system components, the steering column, instrument panel components, wiring and sensors. Failure to follow the safety precautions and the disarming procedure could result in accidental air bag deployment, possible injury and unnecessary system repairs.

SERVICE PRECAUTIONS

Disconnect and isolate the battery negative cable before beginning any airbag system component diagnosis, testing, removal, or installation procedures. Allow system capacitor to discharge for two minutes before beginning any component service. This will disable the airbag system. Failure to disable the airbag system may result in accidental airbag deployment, personal injury, or death.

Do not place an intact undeployed airbag face down on a solid surface. The airbag will propel into the air if accidentally deployed and may result in personal injury or death.

When carrying or handling an undeployed airbag, the trim side (face) of the airbag should be pointing towards the body to minimize possibility of injury if accidental deployment occurs. Failure to do this may result in personal injury or death.

Replace airbag system components with OEM replacement parts. Substitute parts may appear interchangeable, but internal differences may result in inferior occupant protection. Failure to do so may result in occupant personal injury or death.

Wear safety glasses, rubber gloves, and long sleeved clothing when cleaning powder residue from vehicle after an airbag deployment. Powder residue emitted from a deployed airbag can cause skin irritation. Flush affected area with cool water if irritation is experienced. If nasal or throat irritation is experienced, exit the vehicle for fresh air until the irritation ceases. If irritation continues, see a physician.

Do not use a replacement airbag that is not in the original packaging. This may result in improper deployment, personal injury, or death.

The factory installed fasteners, screws and bolts used to fasten airbag components have a special coating and are specifically designed for the airbag system. Do not use substitute fasteners. Use only original equipment fasteners listed in the parts catalog when fastener replacement is required.

During, and following, any child restraint anchor service, due to impact event or vehicle repair, carefully inspect all mounting hardware, tether straps, and anchors for proper installation, operation, or damage. If a child restraint anchor is found damaged in any way, the anchor must be replaced. Failure to do this may result in personal injury or death.

Deployed and non-deployed airbags may or may not have live pyrotechnic material within the airbag inflator.

Do not dispose of driver/passenger/curtain airbags or seat belt tensioners unless you are sure of complete deployment. Refer to the Hazardous Substance Control System for proper disposal.

Dispose of deployed airbags and tensioners consistent with state, provincial, local, and federal regulations.

After any airbag component testing or service, do not connect the battery negative cable. Personal injury or death may result if the system test is not performed first.

If the vehicle is equipped with the Occupant Classification System (OCS), do not connect the battery negative cable before performing the OCS Verification Test using the scan tool and the appropriate diagnostic information. Personal injury or death may result if the system test is not performed properly.

Never replace both the Occupant Restraint Controller (ORC) and the Occupant Classification Module (OCM) at the same time. If both require replacement, replace one, then perform the Airbag System test before replacing the other.

Both the ORC and the OCM store Occupant Classification System (OCS) calibration data, which they transfer to one another when one of them is replaced. If both are replaced at the same time, an irreversible fault will be set in both modules and the OCS may malfunction and cause personal injury or death.

If equipped with OCS, the Seat Weight Sensor is a sensitive, calibrated unit and must be handled carefully. Do not drop or handle roughly. If dropped or damaged, replace with another sensor. Failure to do so may result in occupant injury or death.

If equipped with OCS, the front passenger seat must be handled carefully as well. When removing the seat, be careful when setting on floor not to drop. If dropped, the sensor may be inoperative, could result in occupant injury, or possibly death.

If equipped with OCS, when the passenger front seat is on the floor, no one should sit in the front passenger seat. This uneven force may damage the sensing ability of the seat weight sensors. If sat on and damaged, the sensor may be inoperative, could result in occupant injury, or possibly death.

DISARMING THE SYSTEM

Disconnect and isolate the negative battery cable. Wait 3 minutes for the system capacitor to discharge before performing any service.

ARMING THE SYSTEM

1. After performing service, connect the negative battery cable to re-arm the SRS system.

CLOCKSPRING CENTERING

1. First rotate the cable reel clockwise until it stops.
2. Then rotate it counterclockwise (three full turns) until the arrow mark on the cable reel label points straight up.
3. Position the two tabs of the turn signal cancelling sleeve as shown, and install the steering wheel on to the steering column shaft, making sure the steering wheel hub engages the pins of the cable reel and tabs of the turn signal cancelling sleeve. Do not tap on the steering wheel or steering column shaft when installing the steering wheel.

DRIVETRAIN

AUTOMATIC TRANSAXLE ASSEMBLY

REMOVAL & INSTALLATION

See Figures 27 through 36.

1. Make sure you have the anti-theft codes for the radio and navigation system, then write down the audio presets.

2. Lift the vehicle up on a hoist, and make sure it is securely supported.

3. Remove the front inner fender and splash shield.

4. Remove the drain plug (A), and drain the automatic transmission fluid (ATF).

5. Reinstall the drain plug with a new sealing washer.

6. Remove the front grille cover.

7. Disconnect the negative cable from the battery first, then disconnect the positive cable.

8. Remove the battery hold-down bracket, battery, and battery tray.

9. Remove the intake air duct and air cleaner housing.

10. Remove the dipstick.

11. Remove the starter. For additional information, refer to the following section, "Starter, Removal & installation."

12. Remove the nuts securing the shift cable bracket.

13. Remove the spring clip and the control pin, then separate the shift cable end from the control lever.

14. Disconnect the shift solenoid wire harness connector and the S terminal connector.

15. Remove the bolt securing the harness cover from the harness cover/clamp bracket.

16. Disconnect the 4th clutch transmission fluid pressure switch connector.

17. Disconnect the A/T clutch pressure control solenoid valve A connector and the A/T clutch pressure control solenoid valve B connector.

18. Disconnect the A/T clutch pressure control solenoid valve C connector.

19. Remove the bolts securing the harness cover and the harness cover/clamp bracket.

20. Disconnect the transmission range switch connector, and remove the harness clamp from the clamp bracket.

21. Remove the transmission ground cable.

22. Disconnect the ATF cooler hoses from the ATF cooler lines. Turn the ends of the ATF cooler hoses up to prevent ATF from flowing out, and plug the ATF cooler

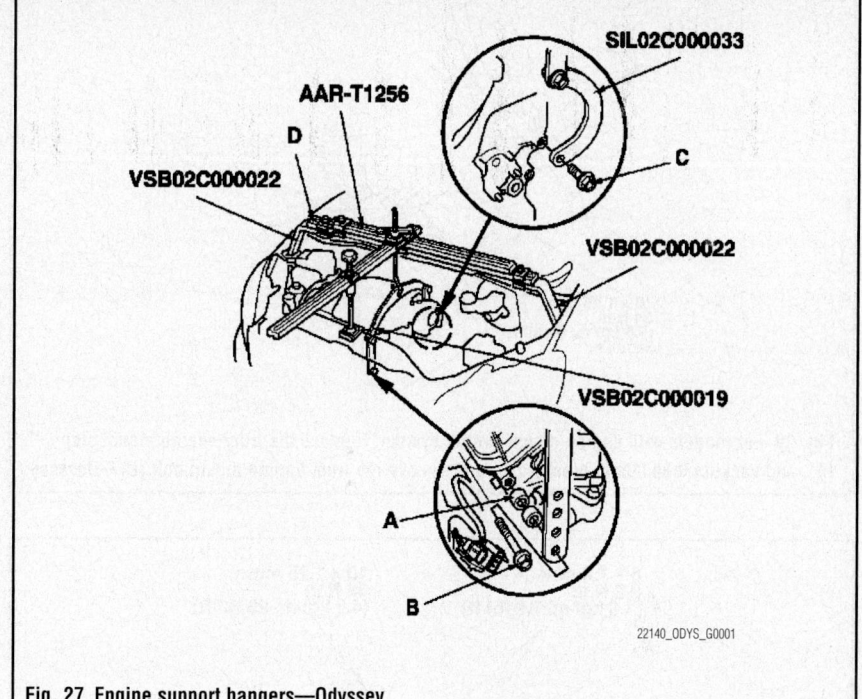

Fig. 27 Engine support hangers—Odyssey

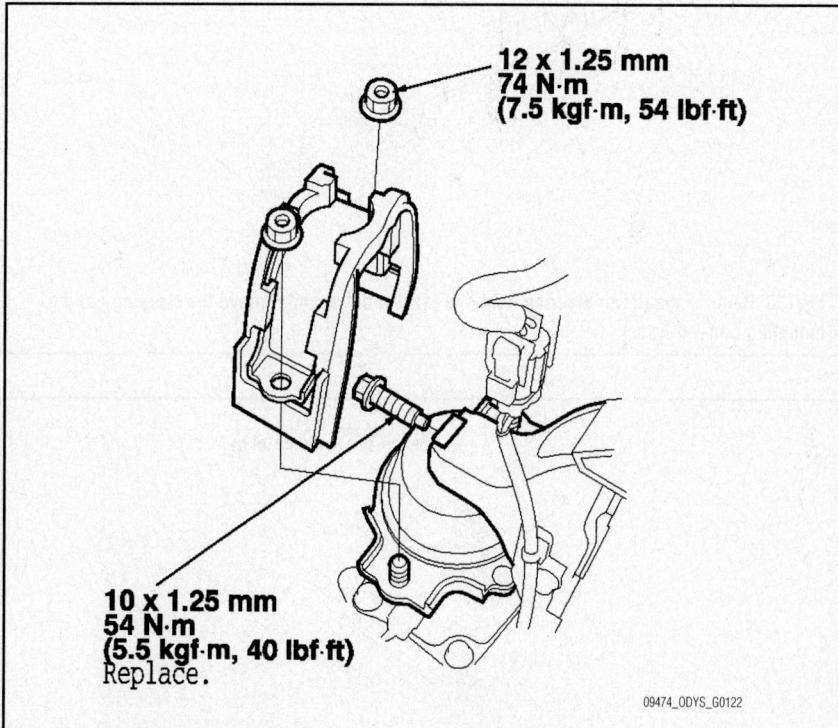

Fig. 28 For models with active engine mount control system: Remove the front engine mount stop (A), and remove the front engine mount bolt (B)—Odyssey

hoses and lines. Check for any signs of leakage at the hose joints.

23. Disconnect the input shaft (mainshaft) speed sensor connector, the output

shaft (countershaft) speed sensor connector, the ATF temperature sensor connector, and the 3rd clutch transmission fluid pressure switch connector.

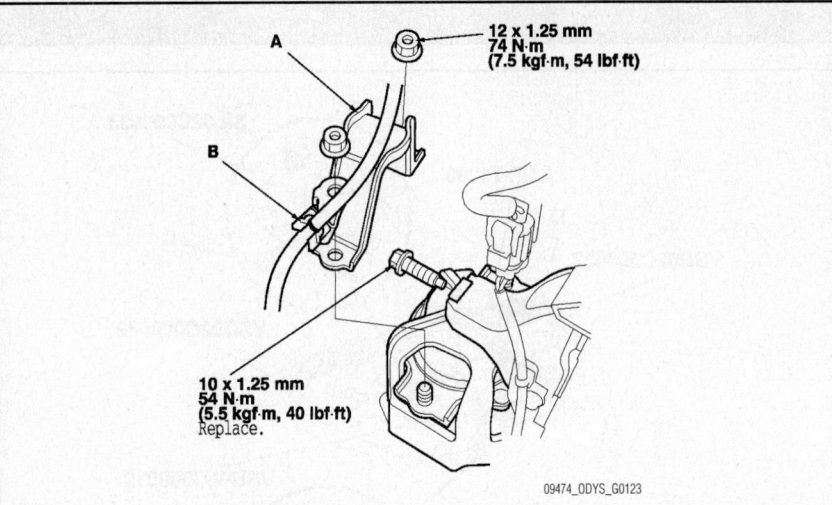

12 x 1.25 mm
74 N·m
(7.5 kgf·m, 54 lbf·ft)

A
B

10 x 1.25 mm
54 N·m
(5.5 kgf·m, 40 lbf·ft)
Replace.

09474_ODYS_G0123

Fig. 29 For models with engine mount control system: Remove the front engine mount stop (A), and vacuum tube clamp bracket (B), and remove the front engine mount bolt (C)—Odyssey

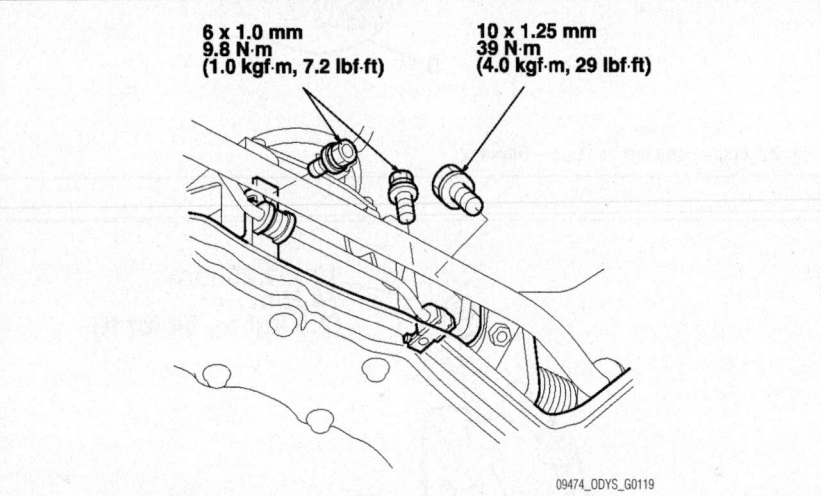

6 x 1.0 mm
9.8 N·m
(1.0 kgf·m, 7.2 lbf·ft)

10 x 1.25 mm
39 N·m
(4.0 kgf·m, 29 lbf·ft)

09474_ODYS_G0119

Fig. 30 Remove the power steering fluid line bracket bolts, and remove the steering gearbox mounting bolt—Odyssey

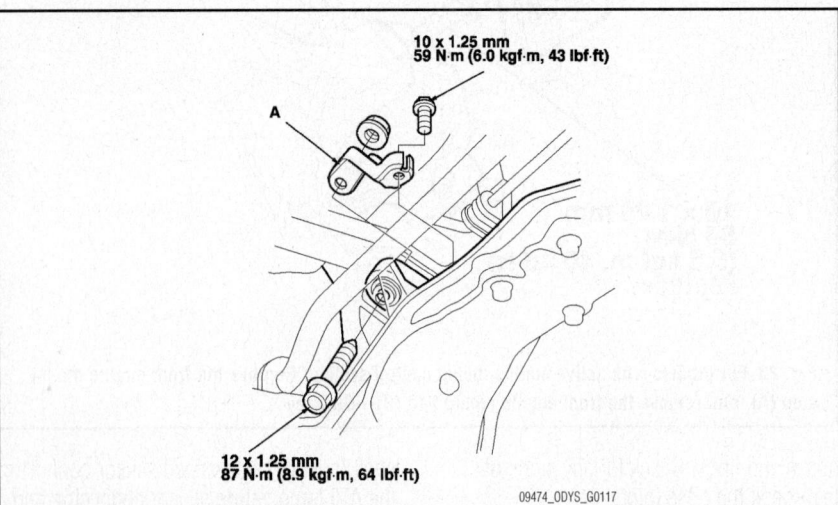

10 x 1.25 mm
59 N·m (6.0 kgf·m, 43 lbf·ft)

A

12 x 1.25 mm
87 N·m (8.9 kgf·m, 64 lbf·ft)

09474_ODYS_G0117

Fig. 31 Remove the steering gearbox mounting bolt and nut, and remove the steering gearbox stiffener (A)—Odyssey

24. Disconnect the 2nd clutch transmission fluid pressure switch connector.

25. Remove the front A/F sensor connector bracket from the engine front cylinder head; use the bracket bolt hole to attach the engine hanger balancer bar front arm.

26. Remove the rear A/F sensor connector bracket from the engine rear cylinder head; use the bracket bolt hole to attach the engine hanger balancer bar rear arm.

27. Lift and support the engine with the engine support hanger and the engine balancer bar (VSB02C000019). Attach the front arm to the front cylinder head with spacers (A) and the 10 x 1.25 mm bolt (B). Attach the rear arm (SIL02C000033) to the rear cylinder head with the 8 x 1.25 mm bolt (C).

28. Install the engine hanger adapters (VSB02C000022) on the end of the engine support beams (D), then install the engine support hanger (AAR-T1256) with the adapters to the vehicle.

29. Remove the upper transmission housing mounting bolts.

30. For models with active engine mount control system: Remove the front engine mount stop (A), and remove the front engine mount bolt (B).

31. For models with engine mount control system: Remove the front engine mount stop (A), and vacuum tube clamp bracket (B), and remove the front engine mount bolt (C).

32. Remove the exhaust pipe A and its mount.

33. Remove the lock pins and castle nuts, and separate the lower arms from the knuckles. See the procedures in the Steering and Suspension parts.

34. Insert a 6 mm Allen wrench in the top of the joint pin, remove the nuts, then separate the stabilizer links.

35. Remove the torque converter cover, and remove the drive plate bolts while rotating the crankshaft pulley.

36. Remove the engine-to-torque converter housing mounting bolts.

37. Remove the power steering fluid line bracket bolts, and remove the steering gearbox mounting bolt.

38. Remove the steering gearbox mounting bolt and nut, and remove the steering gearbox stiffener (A).

39. Remove the heat shield (A), and remove the steering gearbox mounting bolt (B).

40. Remove the steering gearbox mounting bolt and nut, and remove the steering gearbox stiffener (C).

41. Remove the rear mount mounting bolts.

42. Unclamp the power steering fluid line clamps on the front subframe.

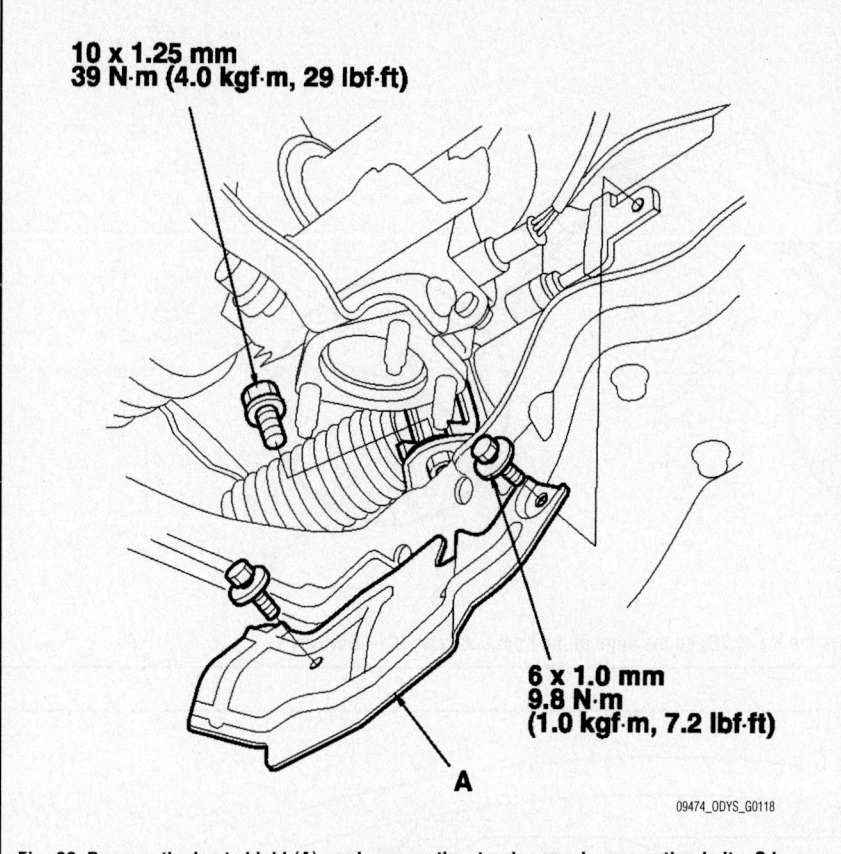

**10 x 1.25 mm
39 N·m (4.0 kgf·m, 29 lbf·ft)**

**6 x 1.0 mm
9.8 N·m
(1.0 kgf·m, 7.2 lbf·ft)**

A

09474_ODYS_G0118

Fig. 32 Remove the heat shield (A), and remove the steering gearbox mounting bolt—Odyssey

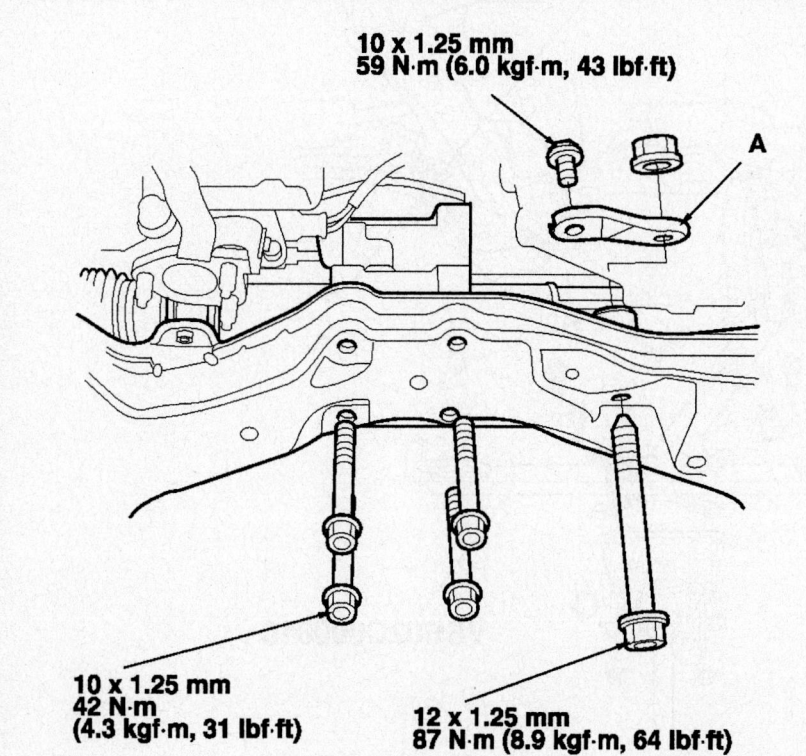

**10 x 1.25 mm
59 N·m (6.0 kgf·m, 43 lbf·ft)**

A

**10 x 1.25 mm
42 N·m
(4.3 kgf·m, 31 lbf·ft)**

**12 x 1.25 mm
87 N·m (8.9 kgf·m, 64 lbf·ft)**

09474_ODYS_G0116

Fig. 33 Remove the steering gearbox mounting bolt and nut, and remove the steering gearbox stiffener (C). Remove the rear mount mounting bolts—Odyssey

43. Remove the transmission lower mount nuts.

44. For models with active engine mount control system: Disconnect engine mount control solenoid valve connector at the front engine mount.

45. For models with engine mount control system: Remove the vacuum tube joint from its clamp from the front engine mount, disconnect the vacuum tube joint, then reinstall the joint in the clamp.

46. Make reference marks (A) on the body across the marks (B) on the edge of the front subframe (C).

47. Attach the front subframe adapter (VSB02C000016) to the front subframe by looping the belt over the front of the subframe, then securing the belt with its stop.

48. Raise the jack and line up the slots in the arms with the bolt holes on the corner of the jack base, then attach them with bolts.

49. Remove the four bolts securing the front stiffeners, two bolts securing the rear stiffeners, subframe mounting bolts, and remove the front and rear stiffeners.

50. Lower the front subframe with the jack.

51. Remove the transmission lower mounts.

52. Remove the left driveshaft from the differential and the right driveshaft from the intermediate shaft.

53. Move the left driveshaft to the front side. Coat all precision finished surfaces with clean engine oil, then put plastic bags over the driveshaft ends.

54. Remove the rear warm-up three-way catalytic converter (WU-TWC) stay (A) and heat shield (B).

55. Remove the intermediate shaft. Coat all precision finished surfaces with clean engine oil, then put plastic bags over the intermediate shaft ends.

56. Remove the lower transmission housing mounting bolts.

57. Remove the front mount bracket.

58. Remove the transmission housing mounting bolt.

59. Place a suitable jack under the transmission.

60. Remove the rear transmission housing mounting bolts.

61. Slide the transmission away from the engine to remove it from the vehicle.

To install:

62. Place the transmission on the jack, and raise it to engine level.

63. Attach the transmission to the engine, and install the lower transmission housing mounting bolts to 47 ft. lbs. (64 Nm).

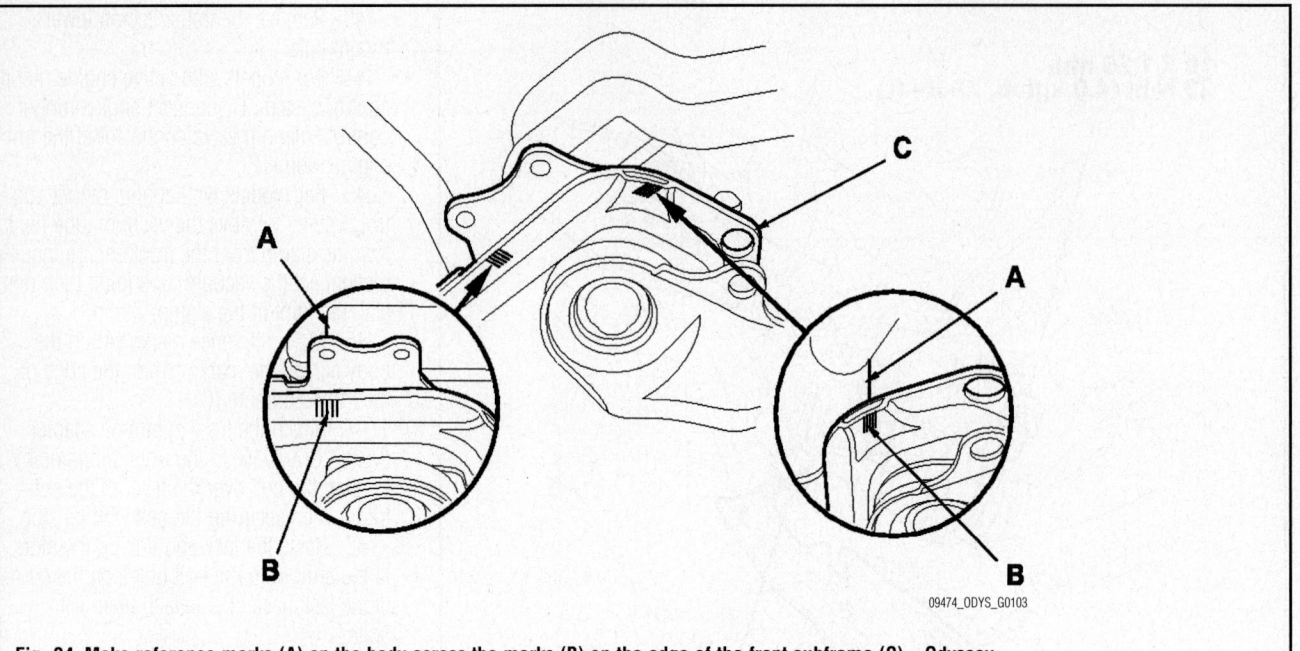

C
A
A
B
B
09474_ODYS_G0103

Fig. 34 Make reference marks (A) on the body across the marks (B) on the edge of the front subframe (C)—Odyssey

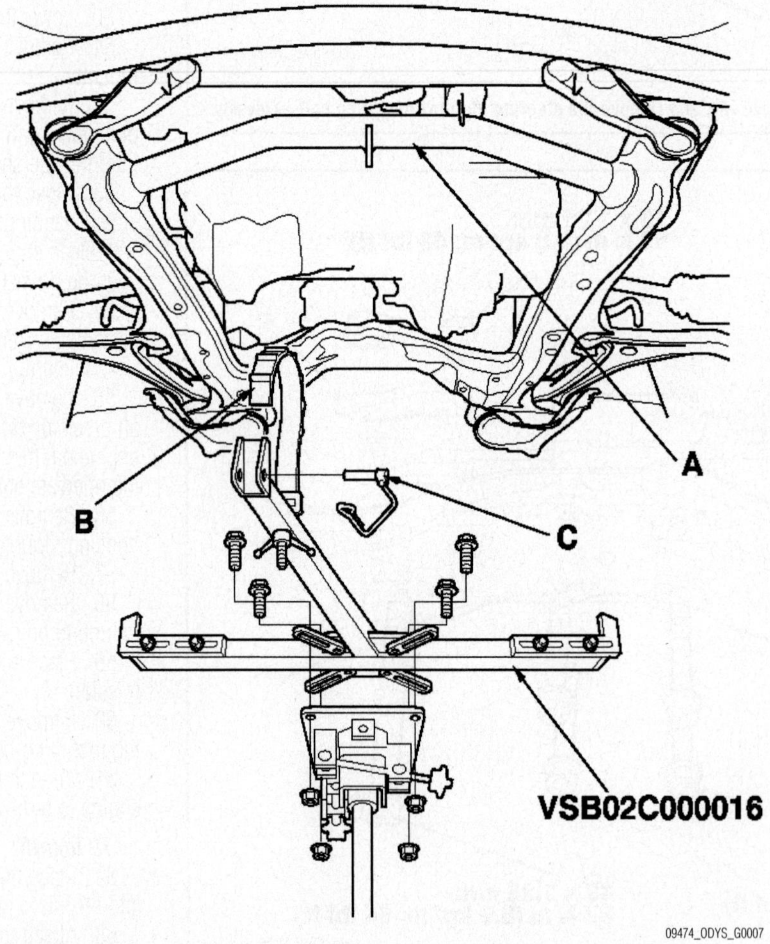

A
B
C
VSB02C000016
09474_ODYS_G0007

Fig. 35 Attach the front subframe adapter (VSB02C000016) to the front subframe (A) by looping the belt (B) over the front of the subframe, then securing the belt with its pin (C)—Odyssey

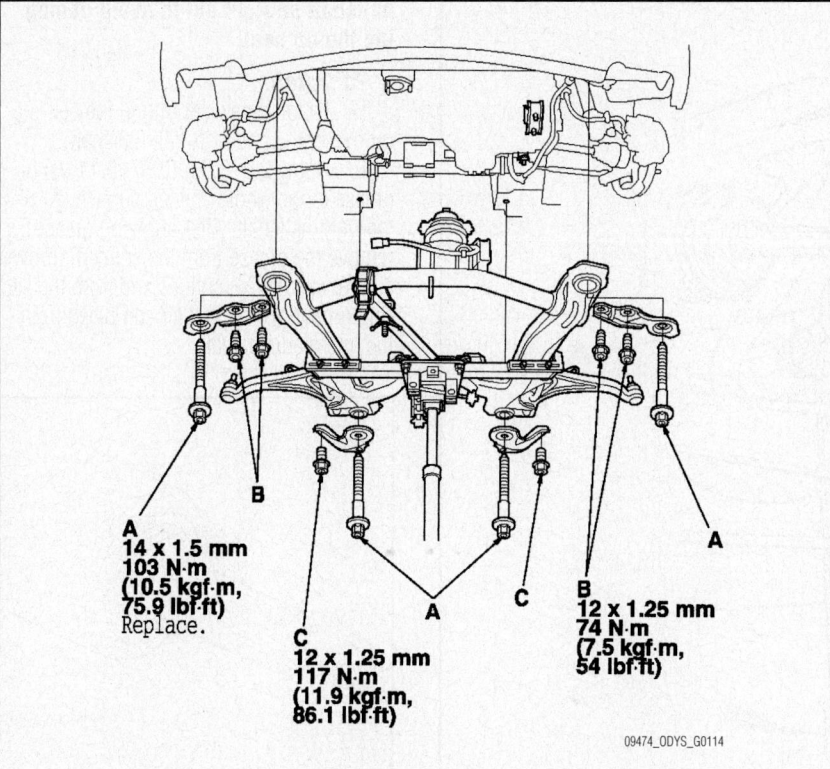

Fig. 36 Remove the four bolts securing the front stiffeners, two bolts securing the rear stiffeners, subframe mounting bolts, and remove the front and rear stiffeners—Odyssey

7. Remove the lock pin from the lower arm ball joint castle nut, and remove the nut, then separate the ball joint from the lower arm with the ball joint thread protector and remover.

➡**To avoid damaging the ball joint, install the ball joint thread protector onto the threads of the ball joint. Be careful not to damage the ball joint boot when installing the remover.**

8. Pull the knuckle outward, and remove the outboard joint from the front wheel hub using a plastic hammer.

9. Remove exhaust pipe A.

10. Left halfshaft: Pry the inboard joint (A) from the transmission housing with a prybar. Remove the halfshaft as an assembly.

➡**Do not pull on the halfshaft (B) or the inboard joint may come apart. Pull the halfshaft straight out to avoid damaging the oil seal.**

11. Right halfshaft: Drive the inboard joint (A) off of the intermediate shaft with a drift and hammer. Remove the halfshaft as an assembly.

➡**Do not pull on the halfshaft (B) or the inboard joint may come apart. Pull the**

64. Install the rear transmission housing mounting bolts to 47 ft. lbs. (64 Nm).

65. Remove the jack from the transmission.

66. Install the transmission housing mounting bolt and tighten to 47 ft. lbs. (64 Nm).

67. Install the front mount bracket with new mounting bolts and tighten to 40 ft. lbs. (54 Nm).

68. The remainder of the installation is the reverse order of removal.

FRONT HALFSHAFT

REMOVAL & INSTALLATION

See Figures 37 through 45.

1. Loosen the wheel nuts slightly.

2. Raise the front of the vehicle, and support it with safety stands in the proper locations.

3. Remove the wheel nuts and front wheels.

4. Lift up the locking tab on the spindle nut, then remove the nut.

5. Drain the automatic transmission fluid Reinstall the drain plug using a new washer.

6. Separate the front stabilizer link from the damper. See the procedure under Front Suspension.

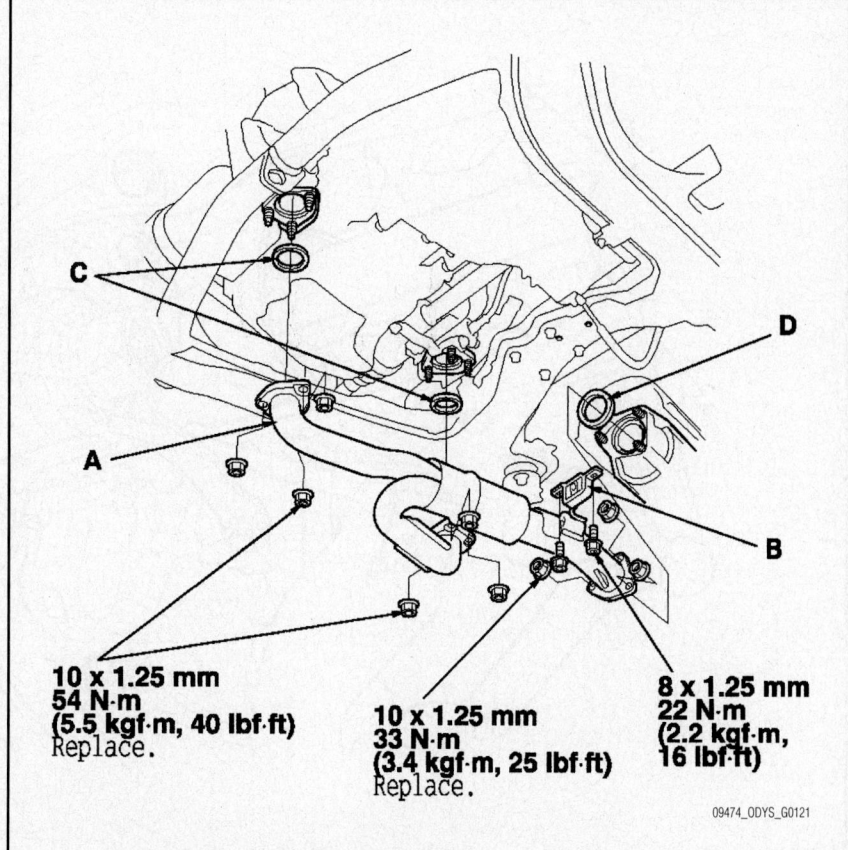

Fig. 37 Remove exhaust pipe A—Odyssey

halfshaft straight out to avoid damaging the oil seal.

To install:

12. Install a new set ring in the set ring groove of the halfshaft (left halfshaft).

13. Apply 2.0–3.0 g (0.07–0.11 oz) of grease to the whole splined surface (A) of the right halfshaft. After applying grease, remove the grease from the splined grooves at intervals of 2–3 splines and from the set ring groove (B) so that air can bleed from the intermediate shaft.

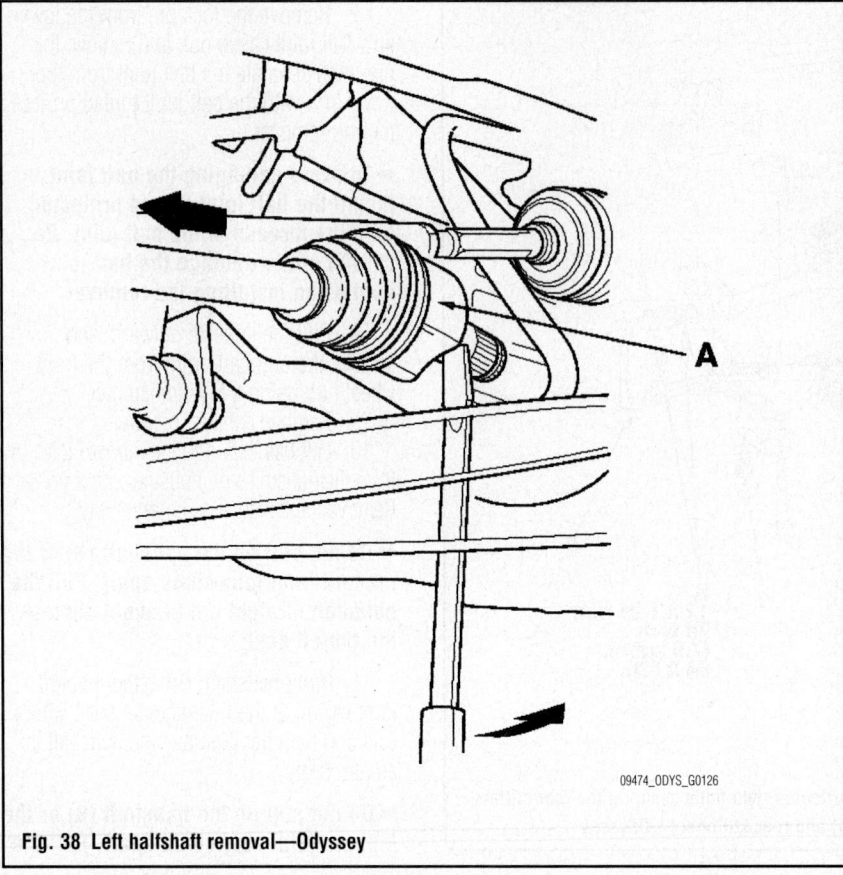

Fig. 38 Left halfshaft removal—Odyssey

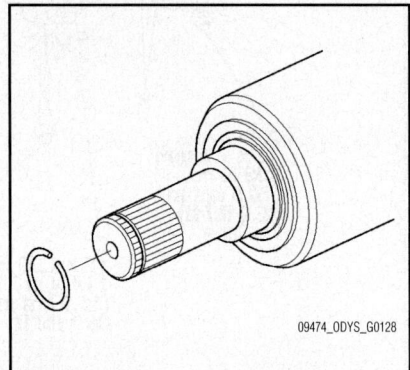

Fig. 40 Install a new set ring in the set ring groove of the halfshaft (left halfshaft)—Odyssey

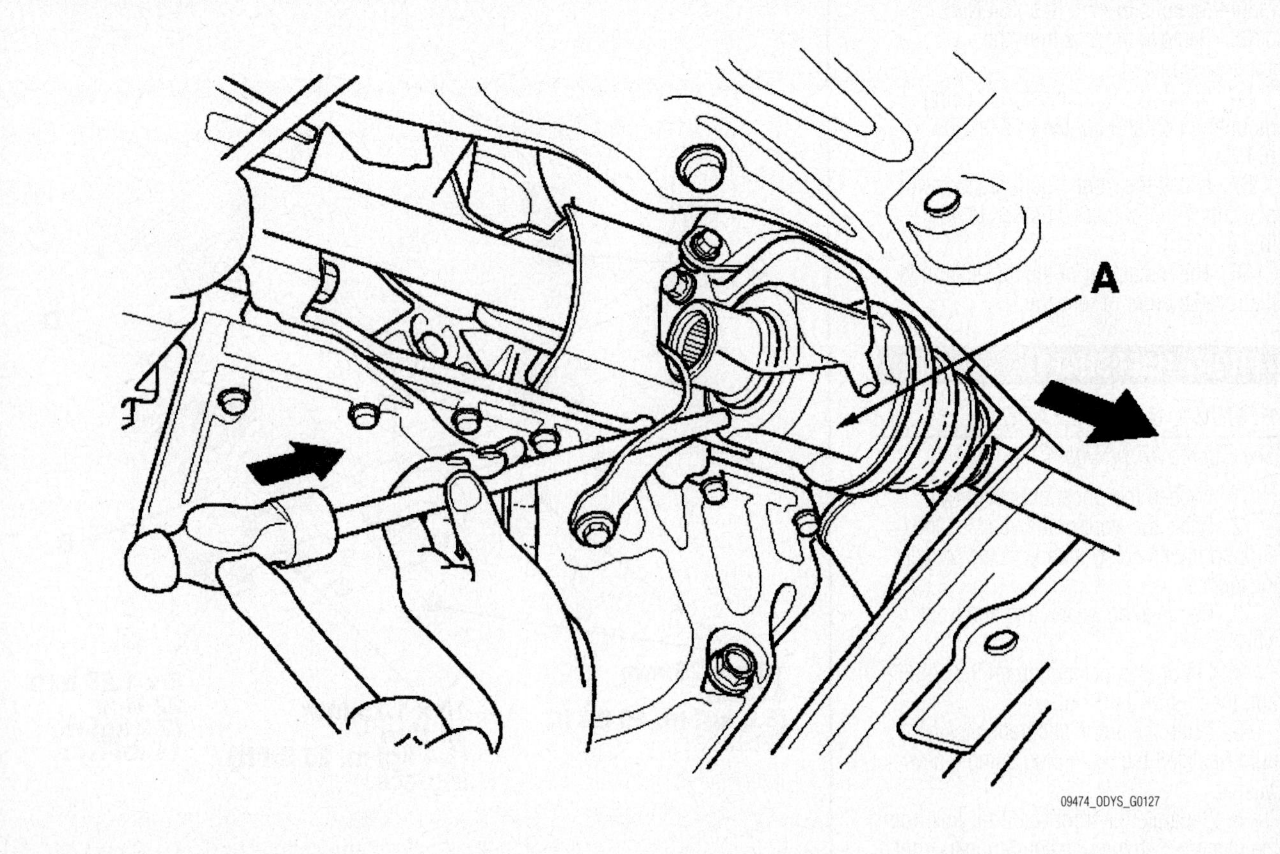

Fig. 39 Right halfshaft removal—Odyssey

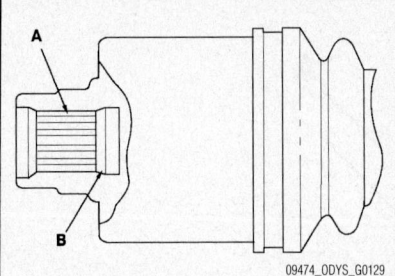

Fig. 41 Apply 2.0–3.0 g (0.07–0.11 oz) of grease to the whole splined surface (A) of the right halfshaft. After applying grease, remove the grease from the splined grooves at intervals of 2–3 splines and from the set ring groove (B) so that air can bleed from the intermediate shaft.— Odyssey

14. Clean the areas where the halfshaft contacts the differential thoroughly with brake cleaner, and dry with compressed air. Do not wash the rubber parts with solvent. Insert the inboard end (A) of the halfshaft into the differential (B) or intermediate shaft (C) until the new set ring (D) locks in the groove (E).

15. Install the outboard joint into the front hub.

16. Install exhaust pipe A.

17. Clean off any grease contamination from the ball joint tapered section and threads, then install the knuckle (A) onto the lower arm (B). Torque the new castle nut (C) to the lower torque specification, then tighten it only far enough to align the slot with the ball joint pin hole. Do not align the nut by loosening it.

➡**Make sure the ball joint boot is not damaged or cracked.**

18. Install the new lock pin (D) into the ball joint pin hole.

19. Connect the front stabilizer link (A) to the damper. Hold the stabilizer link ball joint pin (B) with a hex wrench (C), and tighten the new flange nut (D).

20. Install a new spindle nut (A), then tighten the nut. After tightening, use a drift to stake the spindle nut shoulder (B) against the halfshaft.

21. Clean the mating surfaces of the brake disc and the front wheel, then install the front wheel with the wheel nuts.

22. Turn the front wheel by hand, and make sure there is no interference between the halfshaft and surrounding parts.

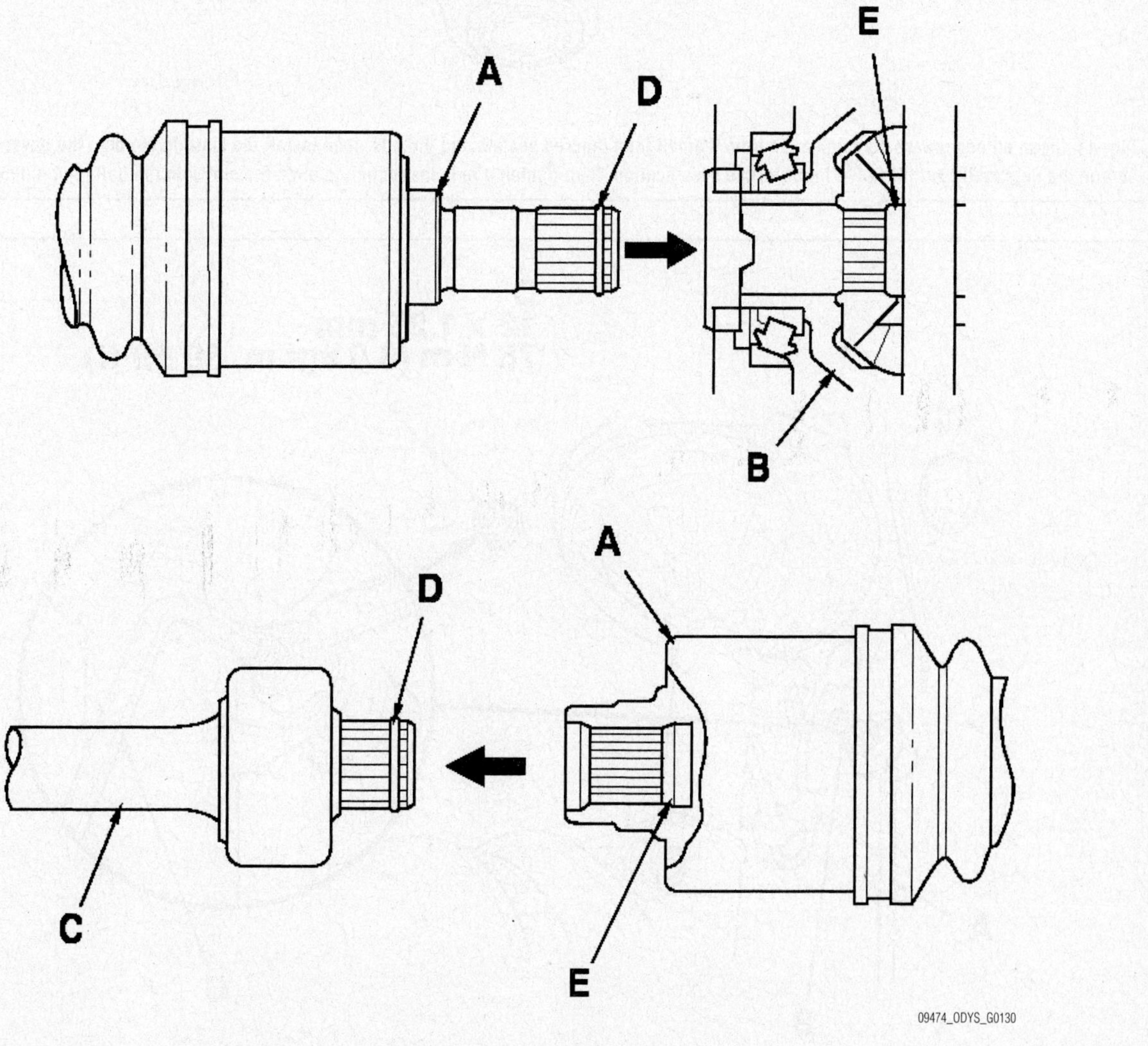

Fig. 42 Clean the areas where the halfshaft contacts the differential thoroughly with brake cleaner, and dry with compressed air. Do not wash the rubber parts with solvent. Insert the inboard end (A) of the halfshaft into the differential (B) or intermediate shaft (C) until the new set ring (D) locks in the groove (E)—Odyssey

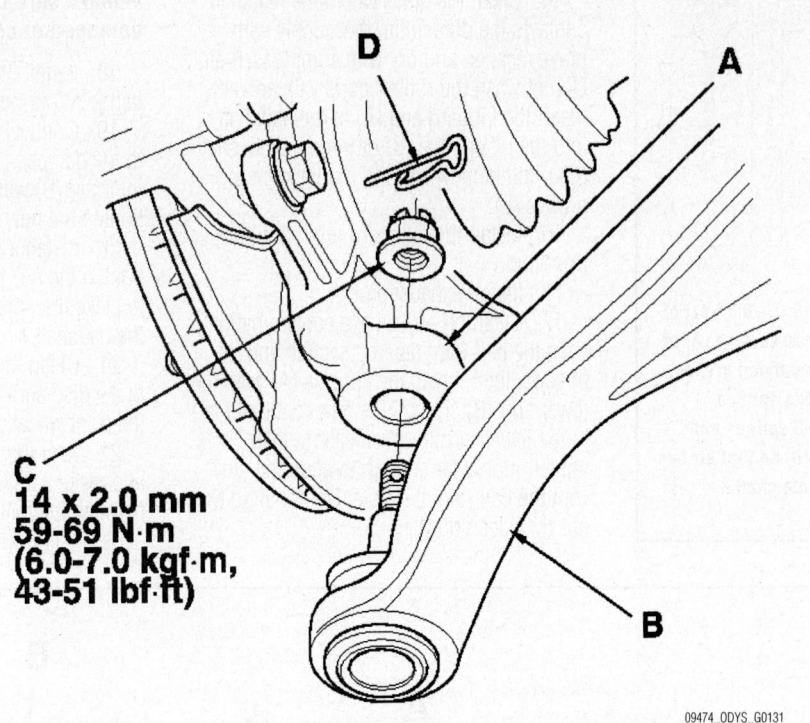

C
**14 x 2.0 mm
59-69 N·m
(6.0-7.0 kgf·m,
43-51 lbf·ft)**

09474_ODYS_G0131

Fig. 43 Clean off any grease contamination from the ball joint tapered section and threads, then install the knuckle (A) onto the lower arm (B). Torque the new castle nut (C) to the lower torque specification, then tighten it only far enough to align the slot with the ball joint pin hole—Odyssey

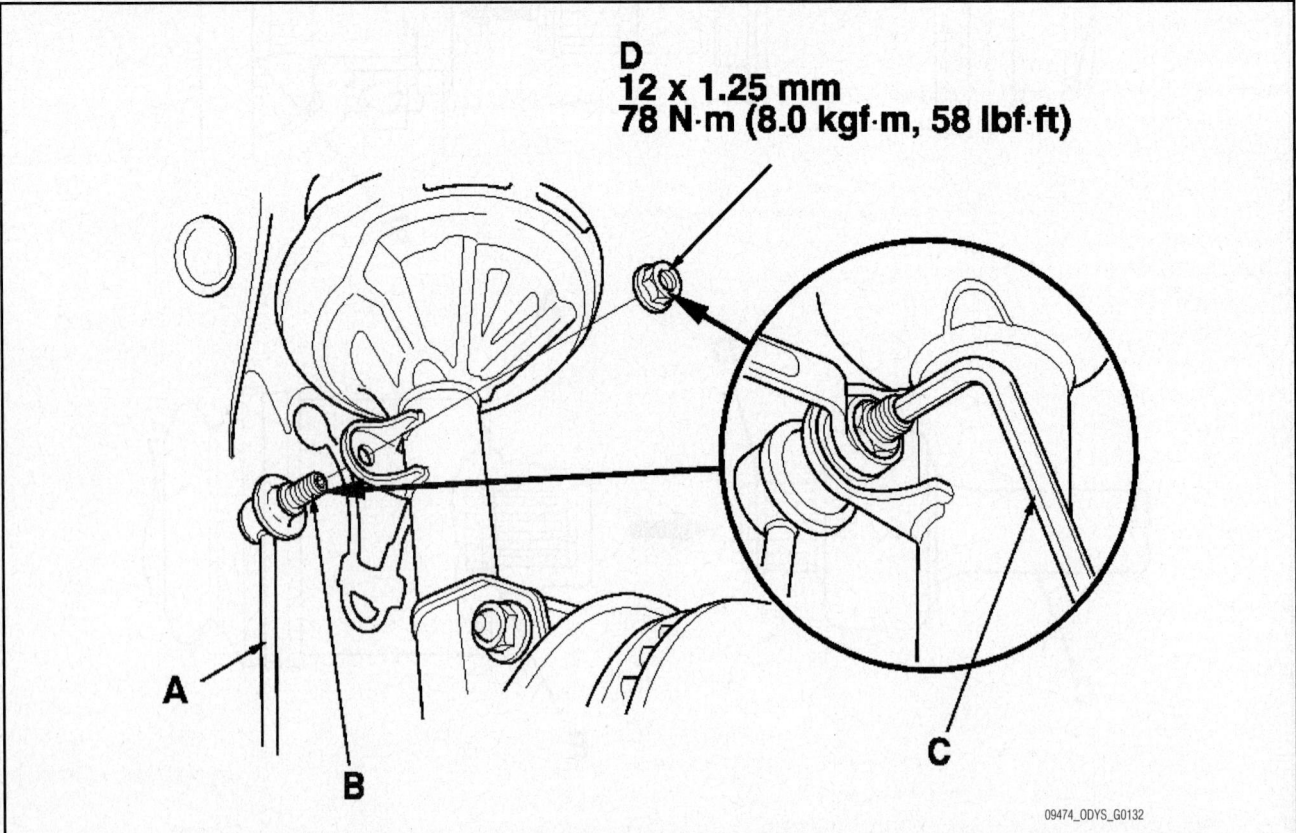

D
**12 x 1.25 mm
78 N·m (8.0 kgf·m, 58 lbf·ft)**

09474_ODYS_G0132

Fig. 44 Connect the front stabilizer link (A) to the damper. Hold the stabilizer link ball joint pin (B) with a hex wrench (C), and tighten the new flange nut (D)—Odyssey

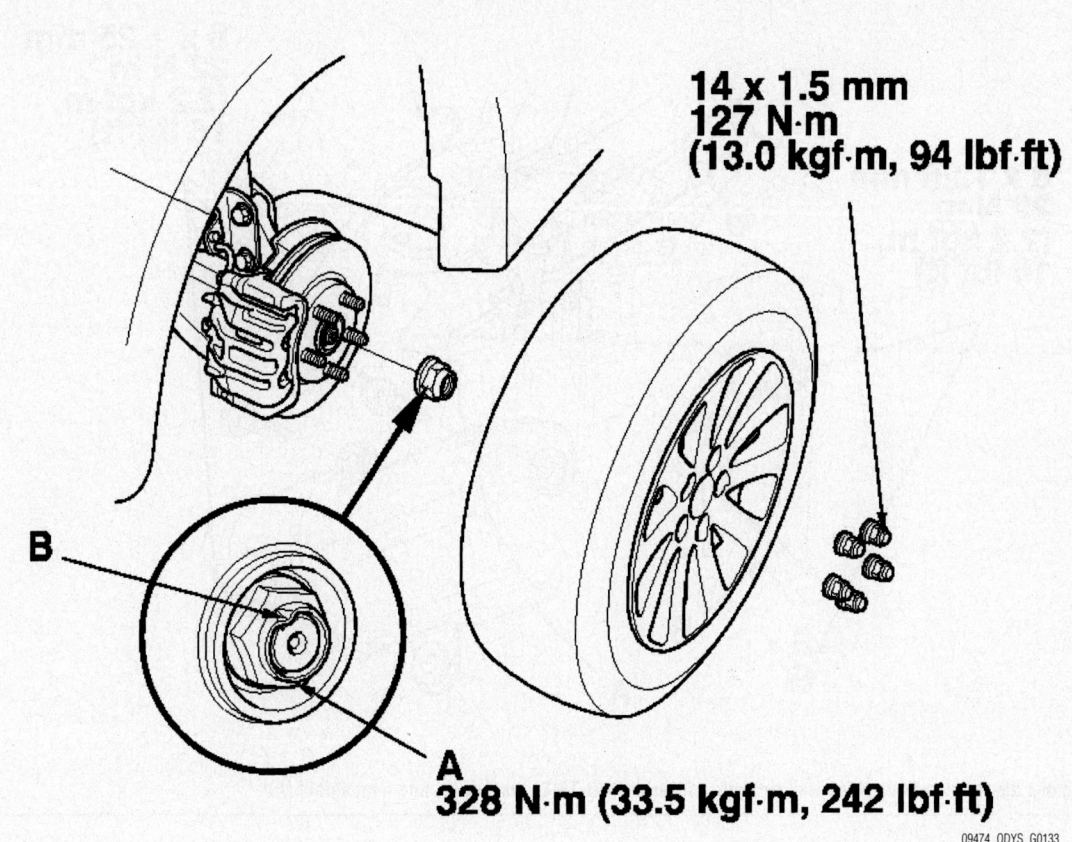

14 x 1.5 mm
127 N·m
(13.0 kgf·m, 94 lbf·ft)

B

A
328 N·m (33.5 kgf·m, 242 lbf·ft)

09474_ODYS_G0133

Fig. 45 Install a new spindle nut (A), then tighten the nut. After tightening, use a drift to stake the spindle nut shoulder (B) against the halfshaft—Odyssey

23. Refill the automatic transmission with recommended transmission fluid.

24. Check the front wheel alignment, and adjust it if necessary.

INTERMEDIATE SHAFT

REMOVAL & INSTALLATION

See Figures 46 through 48.

1. Drain the automatic transmission fluid Reinstall the drain plug, using a new washer.

2. Remove the right driveshaft.

3. Remove the rear warm-up three-way catalytic converter (WU-TWC) bracket (B) and heat shield (A).

4. Remove the flange bolt (A) and two dowel bolts (B).

5. Remove the intermediate shaft (A) from the differential. Hold the intermediate shaft horizontally until it is clear of the differential to prevent damage to the differential oil seal (B).

To install:

6. Use brake cleaner to thoroughly clean the areas where the intermediate shaft (A) contacts the transmission (differential), and dry them with compressed air. Do not wash the rubber parts with solvent. Insert the intermediate shaft assembly into the differential. Hold the intermediate shaft horizontally to prevent damage to the differential oil seal (B).

7. Install the flange bolt (A) and two dowel bolts (B).

8. Install the heat shield and rear warm-up three-way catalytic converter (WU-TWC) bracket.

9. Install the right driveshaft.

10. Refill the automatic transmission with the recommended transmission fluid.

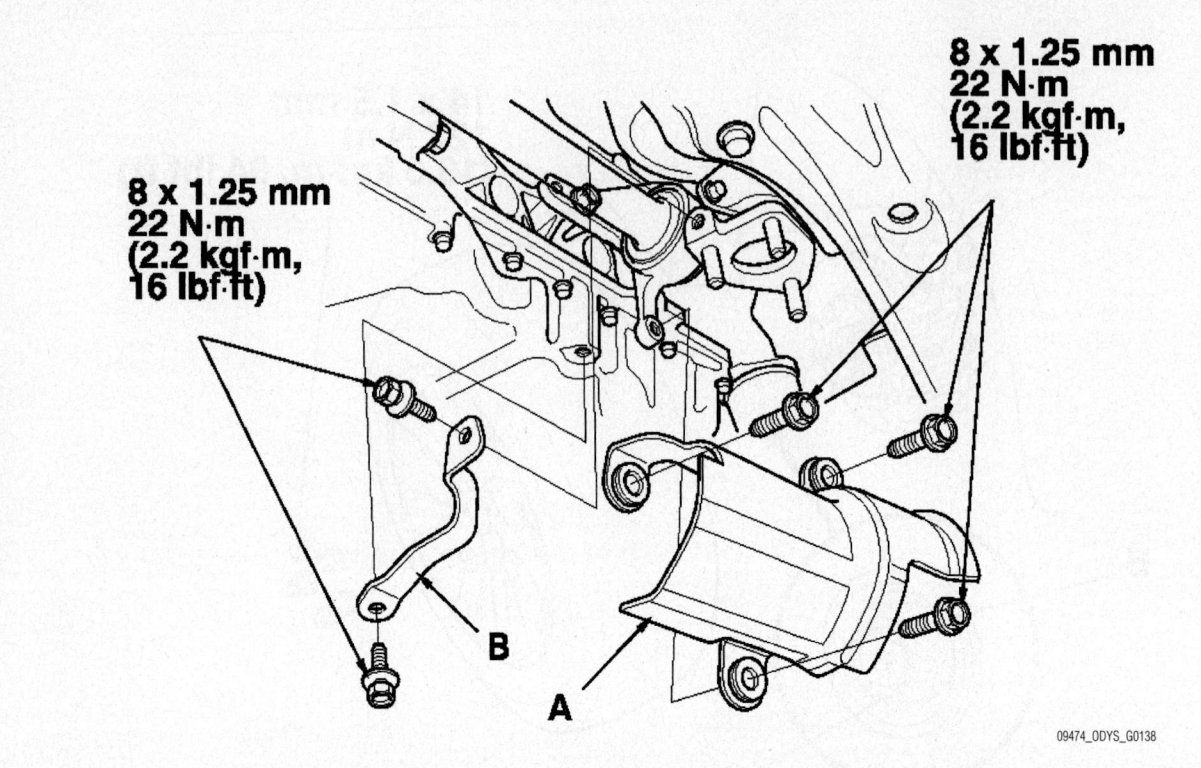

8 x 1.25 mm
22 N·m
(2.2 kgf·m,
16 lbf·ft)

8 x 1.25 mm
22 N·m
(2.2 kgf·m,
16 lbf·ft)

B

A

09474_ODYS_G0138

Fig. 46 Remove the rear warm-up three-way catalytic converter (WU-TWC) bracket (B) and heat shield (A)

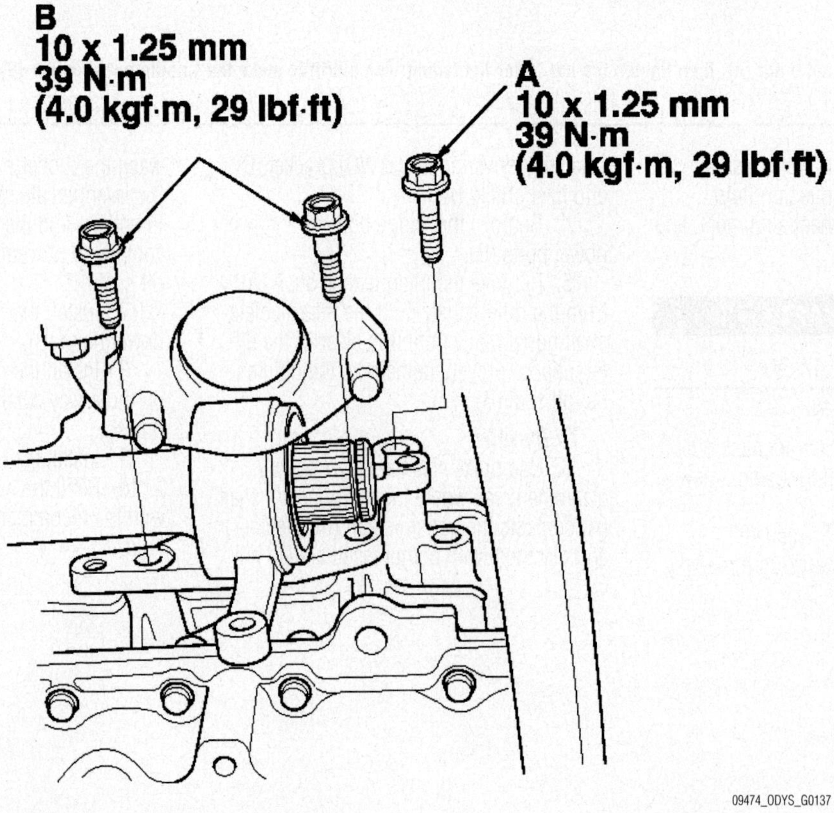

B
10 x 1.25 mm
39 N·m
(4.0 kgf·m, 29 lbf·ft)

A
10 x 1.25 mm
39 N·m
(4.0 kgf·m, 29 lbf·ft)

09474_ODYS_G0137

Fig. 47 Remove the flange bolt (A) and two dowel bolts (B)

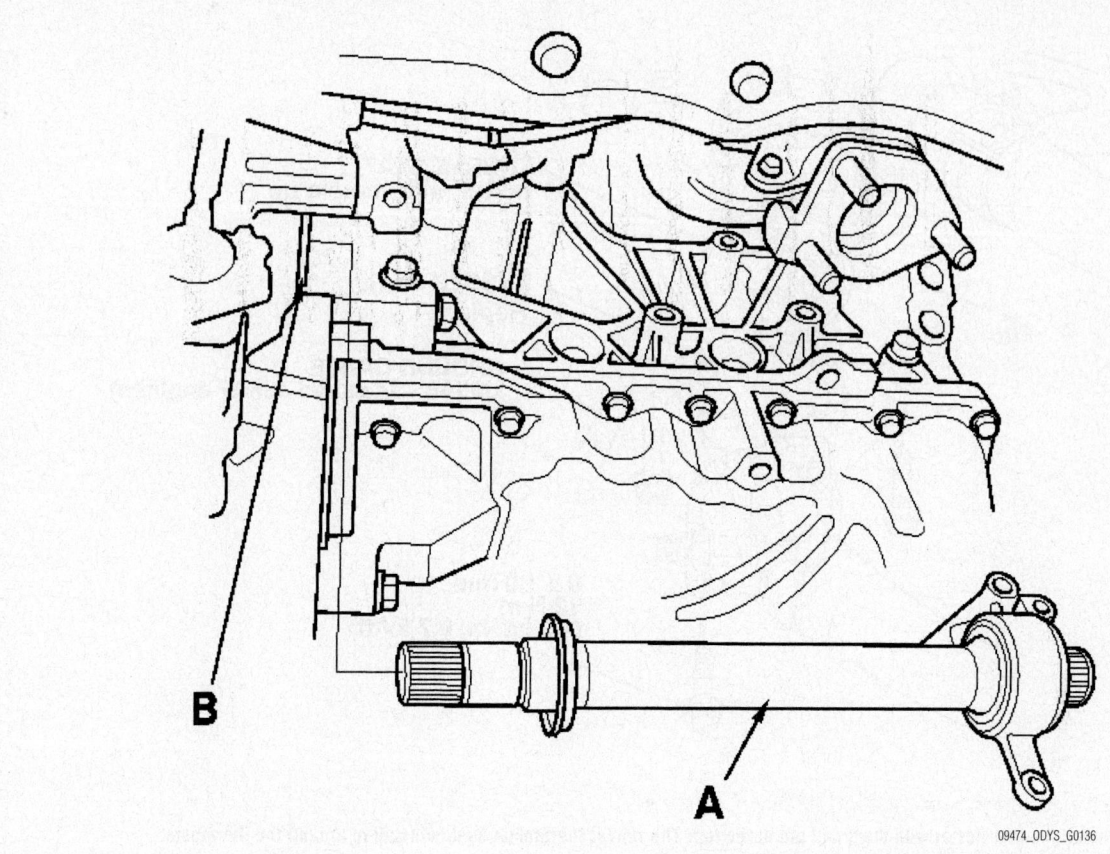

Fig. 48 Remove the intermediate shaft (A) from the differential. Hold the intermediate shaft horizontally until it is clear of the differential to prevent damage to the differential oil seal (B)

ENGINE COOLING

THERMOSTAT

REMOVAL & INSTALLATION
See Figure 49.

✻✻ CAUTION

Never open, service or drain the radiator or cooling system when hot; serious burns can occur from the steam and hot coolant. Also, when draining engine coolant, keep in mind that cats and dogs are attracted to ethylene glycol antifreeze and could drink any that is left in an uncovered container or in puddles on the ground. This will prove fatal in sufficient quantities. Always drain coolant into a sealable container. Coolant should be reused unless it is contaminated or is several years old.

1. Note the radio security code and station presets.
2. Disconnect the negative battery cable.

3. Drain the engine coolant into a sealable container.
4. Remove the fasteners from the thermostat housing, then remove the thermostat.

To install:

5. Install the thermostat using a new seal. If the thermostat has a small bleed hole, make sure the bleed hole is on the top.
6. Apply an anti-seize compound to the threads of the fasteners.
7. Reassemble in the reverse order of disassembly.
8. Set the heater to the full hot position. Set the heater to the full hot position.
9. Top off the cooling system and overflow reservoir with a 50/50 mixture of a recommended antifreeze and water solution and bleed the system to remove any air pockets as necessary. Simultaneously squeeze the upper and lower radiator hoses to help push any captured air pockets out of the system.

10. Inspect all coolant hoses and fittings to make sure they are properly installed and if previously opened, close the bleed valve.
11. Connect the negative battery cable.
12. Install the radiator cap loosely and start the engine. Allow the engine to run until the cooling fan has cycled two times, then turn the engine **OFF** and top off the cooling system as necessary.
13. Install the radiator cap and inspect for leaks.
14. Enter the radio security code.

✻✻ WARNING

The manufacturer does not recommend using a coolant concentration of greater than 60% antifreeze.

➡**When mixing a 50/50 solution of antifreeze and water, using distilled water may help to keep the cooling system from building up mineral deposits and internal blockage.**

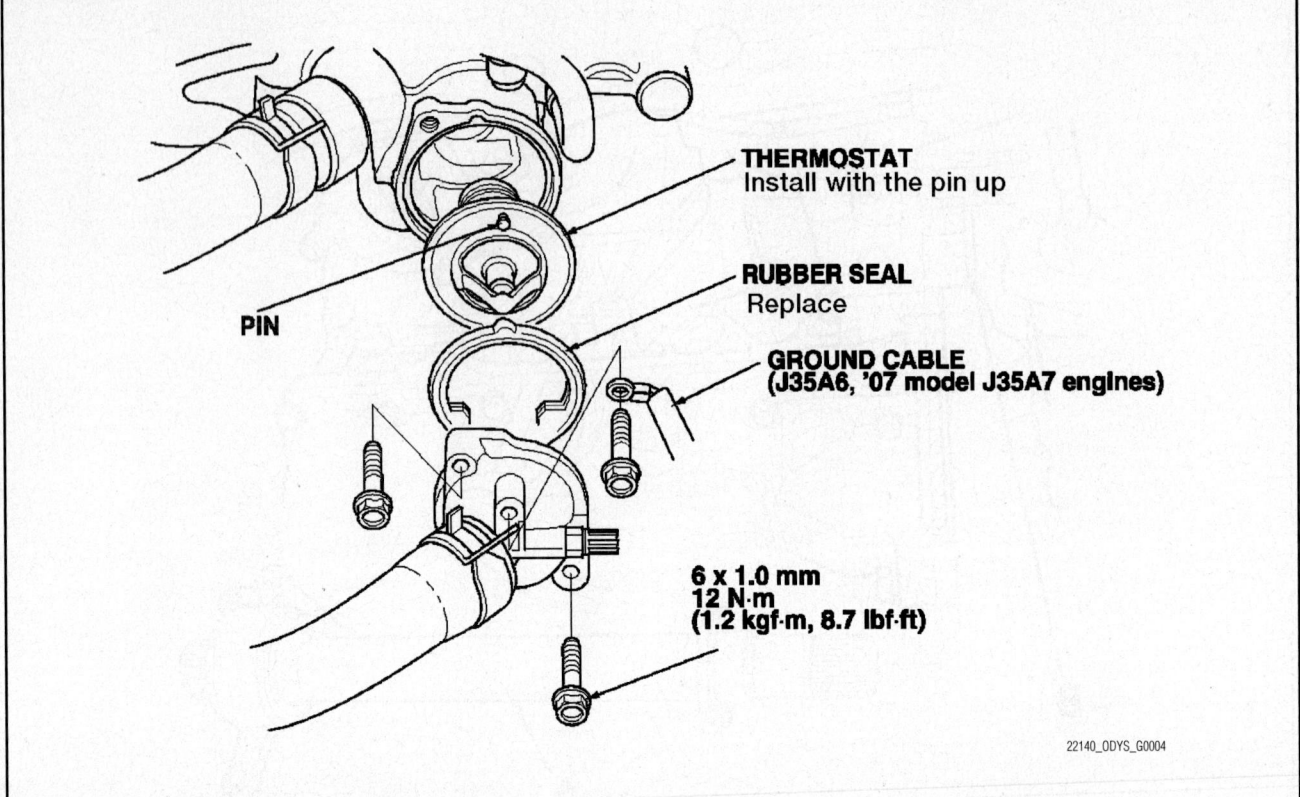

THERMOSTAT
Install with the pin up

RUBBER SEAL
Replace

GROUND CABLE
(J35A6, '07 model J35A7 engines)

PIN

6 x 1.0 mm
12 N·m
(1.2 kgf·m, 8.7 lbf·ft)

22140_ODYS_G0004

Fig. 49 Always install the thermostat with the small pin at the top. The rubber thermostat seal is installed around the thermostat

WATER PUMP

REMOVAL & INSTALLATION

See Figure 50.

1. Before servicing the vehicle, refer to the precautions in the beginning of this section.
2. Drain the cooling system.
3. Remove or disconnect the following:
   - Negative battery cable
   - Accessory drive belts
   - Front cover
   - Timing belt. Refer to the timing belt procedure.
   - Timing belt tensioner
   - Water pump

To install:

4. Install or connect the following:
   - Water pump. Use a new O-ring seal and tighten the bolts to 105 inch lbs. (12 Nm).
   - Timing belt tensioner
   - Timing belt
   - Front cover
   - Accessory drive belts
   - Negative battery cable
5. Refill the cooling system to the correct level.
6. Start the engine and check for leaks.

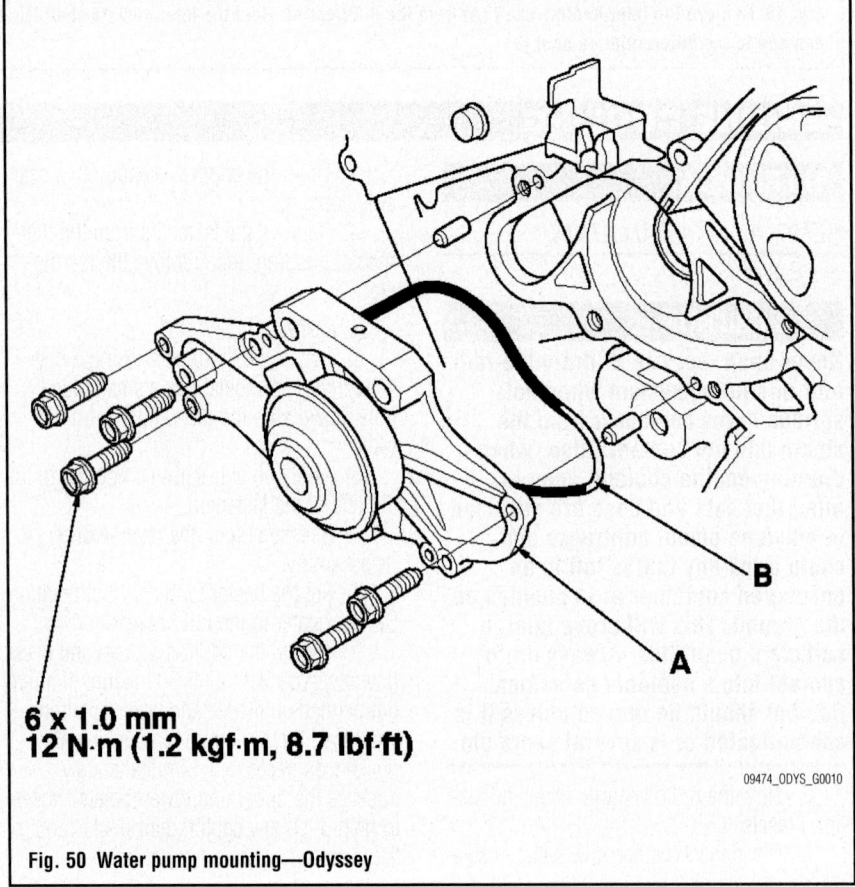

6 x 1.0 mm
12 N·m (1.2 kgf·m, 8.7 lbf·ft)

B

A

09474_ODYS_G0010

Fig. 50 Water pump mounting—Odyssey

ENGINE ELECTRICAL

CHARGING SYSTEM

ALTERNATOR

REMOVAL & INSTALLATION

See Figure 51.

1. Before servicing the vehicle, refer to the precautions in the beginning of this section.
2. Remove or disconnect the following:
   - Negative battery cable
   - Accessory drive belt
   - Alternator wiring harness connectors
   - Alternator mounting bolts
   - Wiring harness clamp
   - Alternator

To install:

3. Install or connect the following:
   - Alternator
   - Wiring harness clamp. Tighten the bolt to 105 inch lbs. (12 Nm).
   - Alternator mounting bolts. Tighten the 10mm bolt to 33 ft. lbs. (44 Nm) and the 8mm bolt to 16 ft. lbs. (22 Nm).
   - Alternator wiring harness connectors. Tighten the battery terminal nut to 105 inch lbs. (12 Nm).
   - Accessory drive belt
   - Negative battery cable

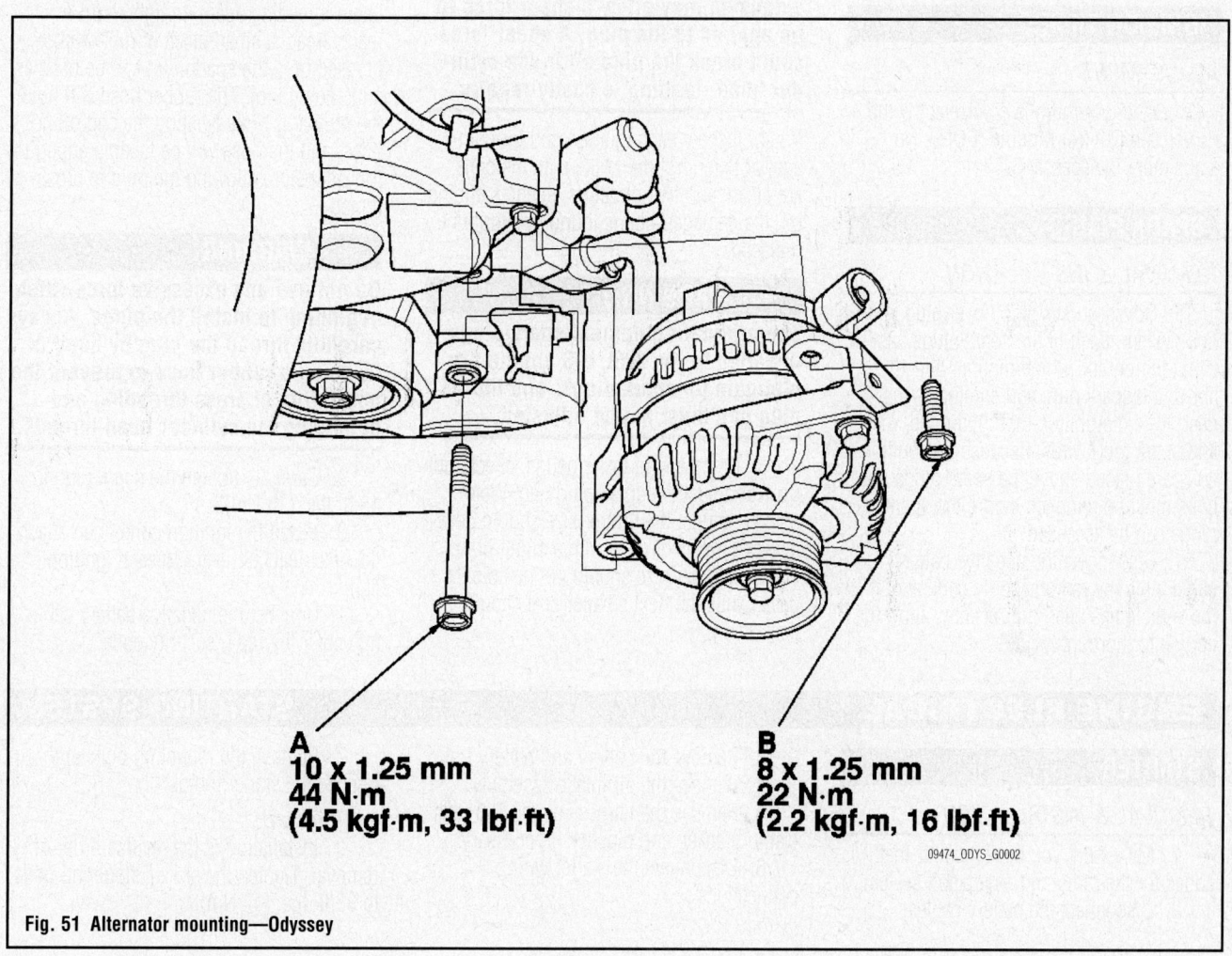

A
10 x 1.25 mm
44 N·m
(4.5 kgf·m, 33 lbf·ft)

B
8 x 1.25 mm
22 N·m
(2.2 kgf·m, 16 lbf·ft)

09474_ODYS_G0002

Fig. 51 Alternator mounting—Odyssey

ENGINE ELECTRICAL   IGNITION SYSTEM

IGNITION COIL

REMOVAL & INSTALLATION

1. Disconnect the negative battery cable.
2. Remove the engine appearance cover.
3. Disconnect the ignition coil connectors, then remove the ignition coils.
4. Installation is the reverse order of removal. Tighten the coil mounting bolts to 105 inch lbs. (12 Nm).

IGNITION TIMING

ADJUSTMENT

The ignition timing is controlled by the Powertrain Control Module (PCM). No adjustment is necessary.

SPARK PLUGS

REMOVAL & INSTALLATION

The Odyssey uses a coil over plug ignition system. Each of the spark plugs has its own ignition coil which mounts directly above the spark plug and eliminates the need for a distributor, distributor cap, rotor and spark plug wires. Because the ignition coils are placed above the spark plugs, the coils must be removed before the spark plugs can be accessed.

1. Disconnect the negative battery cable, note the radio security code and, if the vehicle has been run recently, allow the engine to thoroughly cool.

2. Remove the ignition coil for the spark plug that needs to be removed.

3. Using a spark plug socket equipped with a rubber insert to properly hold the plug, turn the spark plug counterclockwise to loosen and remove the spark plug from the threaded hole in the cylinder head.

✳✳ WARNING

Avoid using a flexible extension on the spark plug socket. A flexible extension may allow a shear force to be applied to the plug. A shear force could break the plug off in the cylinder head, leading to costly repairs.

4. Inspect each ignition coil for damage, or deterioration. Make sure the coils are clean, and free of debris, such as engine oil. If a damaged coil is found, it should be replaced.

✳✳ WARNING

The original equipment spark plugs installed in the 3.5L V-6 engine are platinum tip spark plugs, and the plug gap must not be adjusted.

5. Using a wire feeler gauge, check, but **do not** adjust the spark plug gap. When using a gauge, the proper size should pass between the electrodes with a slight drag. The next larger size should not be able to pass, while the next smaller size should pass freely.

6. The spark plug gap should be 0.039–0.043 inches (1.0–1.1mm). If the spark plug gap exceeds 0.051inches (1.3mm) replace the spark plug.

To install:

7. Apply a light coating of an anti-seize compound to the spark plug threads.
8. Carefully thread the plug into the threaded spark plug hole by hand. If resistance is felt before the plug is almost completely threaded, back the plug out and begin threading again. In tight, hard to reach areas, a small piece of rubber hose pressed onto the spark plug can be used as a threading tool. The rubber hose will hold the plug and while twisting the end of the hose, and the hose will be flexible enough to twist before allowing the plug to cross thread.

✳✳ WARNING

Do not use any excessive force when beginning to install the plugs. Always carefully thread the plug by hand or by using a rubber hose to prevent the possibility of cross threading and damaging the cylinder head threads.

9. Carefully tighten the spark plug to 13 ft. lbs. (18 Nm).
10. Install the ignition coils, then attach the electrical connector to each ignition coil.
11. Connect the negative battery cable and enter the radio security code.

ENGINE ELECTRICAL   STARTING SYSTEM

STARTER

REMOVAL & INSTALLATION

1. Make sure you have the anti-theft codes for the radio and navigation system.
2. Disconnect the battery cables.

3. Remove the battery and battery tray.
4. Remove the air intake assembly.
5. Remove the harness clamp, harness clamp bracket, and dipstick if necessary.
6. Disconnect the starter wiring connections.

7. Remove the mounting bolts and remove the starter motor.

To install:

8. Installation is the reverse order of removal. Tighten the starter mounting bolts to 33 ft. lbs. (44 Nm).

ENGINE MECHANICAL

➡Disconnecting the negative battery cable may interfere with the functions of the on board computer systems and may require the computer to undergo a relearning process, once the negative battery cable is reconnected.

ACCESSORY DRIVE BELTS

ACCESSORY BELT ROUTING

See Figure 52.

INSPECTION

J35A6 Engine

See Figure 53.

1. Inspect the belt for cracks and damage. If the belt is cracked or damaged, replace it.

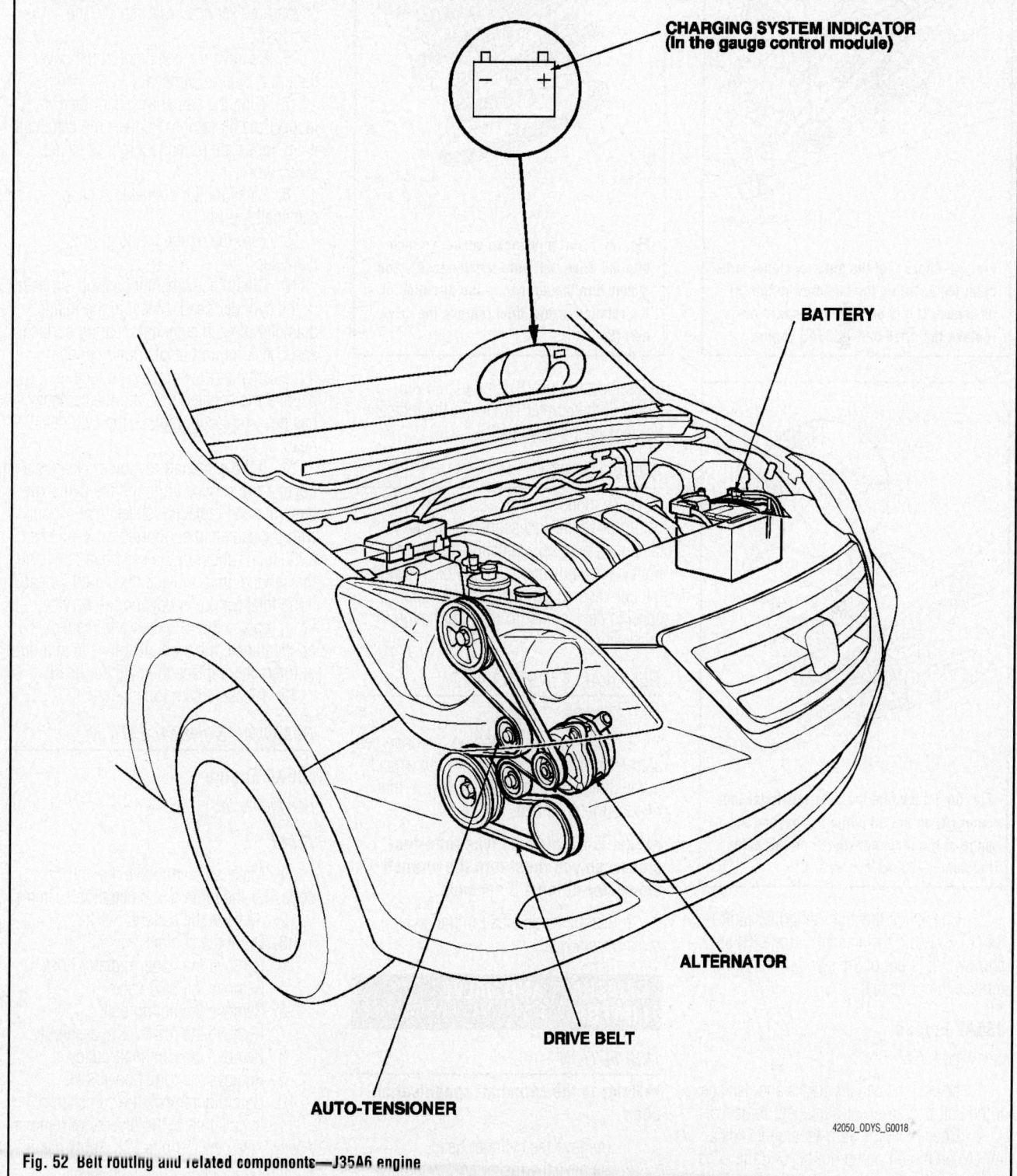

CHARGING SYSTEM INDICATOR
(in the gauge control module)

BATTERY

ALTERNATOR

DRIVE BELT

AUTO-TENSIONER

42050_ODYS_G0018

Fig. 52 Belt routing and related components—J35A6 engine

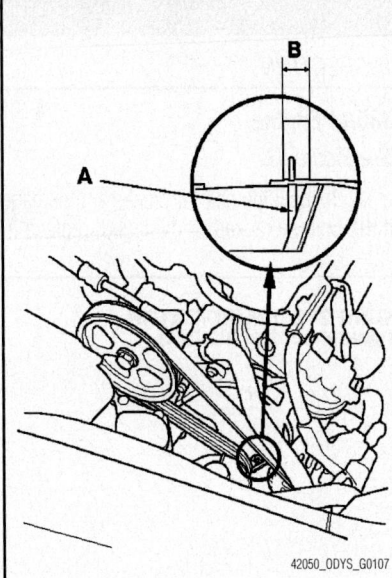

Fig. 53 Check that the auto-tensioner indicator (A) is within the standard range (B) as shown. If it is out of the standard range, replace the drive belt—J35A6 engine

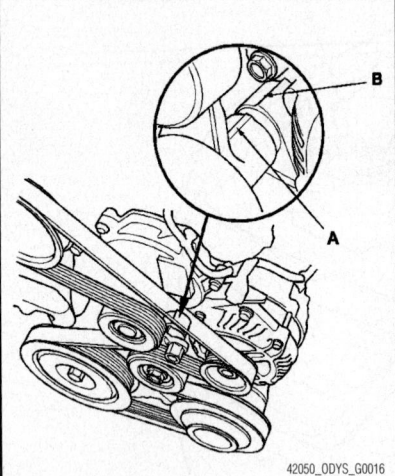

Fig. 54 Check that the auto-tensioner indicator (A) on the oil pump not beyond the edge of the indicator rib (B) on the auto-tensioner—J35A7 engine

2. Check that the auto-tensioner indicator (A) is within the standard range (B) as shown. If it is out of the standard range, replace the drive belt.

J35A7 Engine

See Figure 54.

1. Inspect the belt for cracks and damage. If the belt is cracked or damaged, replace it.
2. Check that the auto-tensioner indicator (A) on the oil pump not beyond the edge

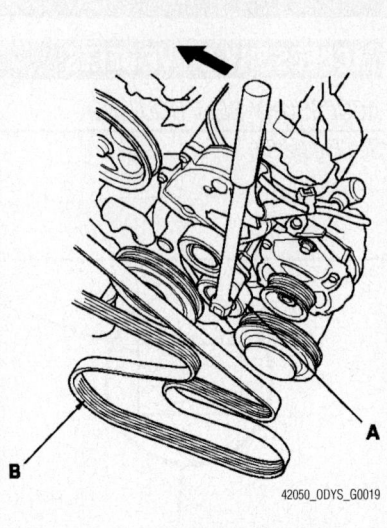

Fig. 55 Insert a hexagon socket wrench into the drive belt auto-tensioner (A), and slowly turn the wrench in the direction of the rotation arrow, then remove the drive belt (B)

of the indicator rib (B) on the auto-tensioner. If the pointer is beyond the indicator rib, replace the drive belt.

ADJUSTMENT

These models utilize a single serpentine belt to drive the accessories. The belt is tensioned by a self adjusting tensioning pulley. No adjustment is possible, however the belt tension can be checked. For specific details refer to the information in this section.

REMOVAL & INSTALLATION

See Figure 55.

1. Set a socket wrench on the drive belt auto-tensioner, and slowly turn the wrench in the direction of the rotation arrow, then remove the drive belt.

➡**This is a hydraulic type auto-tensioner, so you must turn the wrench slowly for at least 3 seconds.**

2. Install the new belt in the reverse order of removal.

CAMSHAFT AND VALVE LIFTERS

INSPECTION

➡**Refer to the camshaft specification chart.**

1. Remove the cylinder head.
2. Remove the rocker arm assembly.

3. Put the rocker shafts on the cylinder head, then tighten the bolts to 17ft. lbs. (24 Nm).
4. Seat the camshaft by pushing it toward the rear of the cylinder head.
5. Zero the dial indicator against the end of the camshaft. Push the camshaft back and forth and read the end play. If the end play is beyond the service limit, replace the thrust cover and recheck. If it is still beyond the service limit, replace the camshaft.
6. Remove the camshaft thrust cover, then pull out the camshaft.
7. Wipe the camshaft clean, then inspect the lift ramps. Replace the camshaft if any lobes are pitted, scored, or excessively worn.
8. Measure the diameter of each camshaft journal.
9. Zero the gauge to the journal diameter.
10. Clean the camshaft bearing surfaces in the cylinder head. Measure the inside diameter of each camshaft bearing surface, and check for an out-of-round condition. If the camshaft-to-holder clearance is beyond the service limit and the camshaft has been replaced, replace the cylinder head.
11. If the camshaft-to-holder clearance is beyond the service limit and the camshaft has not been replaced, check total runout with the camshaft supported on V-blocks. If the total runout of the camshaft is within the service limit, replace the cylinder head. If the total runout is beyond the service limit, replace the camshaft and recheck the oil clearance. If the oil clearance is still out of tolerance, replace the cylinder head.
12. Measure cam lobe height.

REMOVAL & INSTALLATION

J35A6 Engine

See Figure 56.

Front

1. Make sure you have the anti-theft codes for the radio and navigation systems.
2. Remove the battery.
3. Drain the coolant.
4. Remove the upper radiator hose.
5. Remove the EGR valve.
6. Remove the timing belt.
7. Remove the rocker arm assembly.
8. Remove the camshaft pulley.
9. Remove the thrust cover (A).
10. Carefully remove the camshaft (B).
11. Installation is the reverse of removal. Always use new O-rings (C). Apply clean engine oil to the journals and lobes.

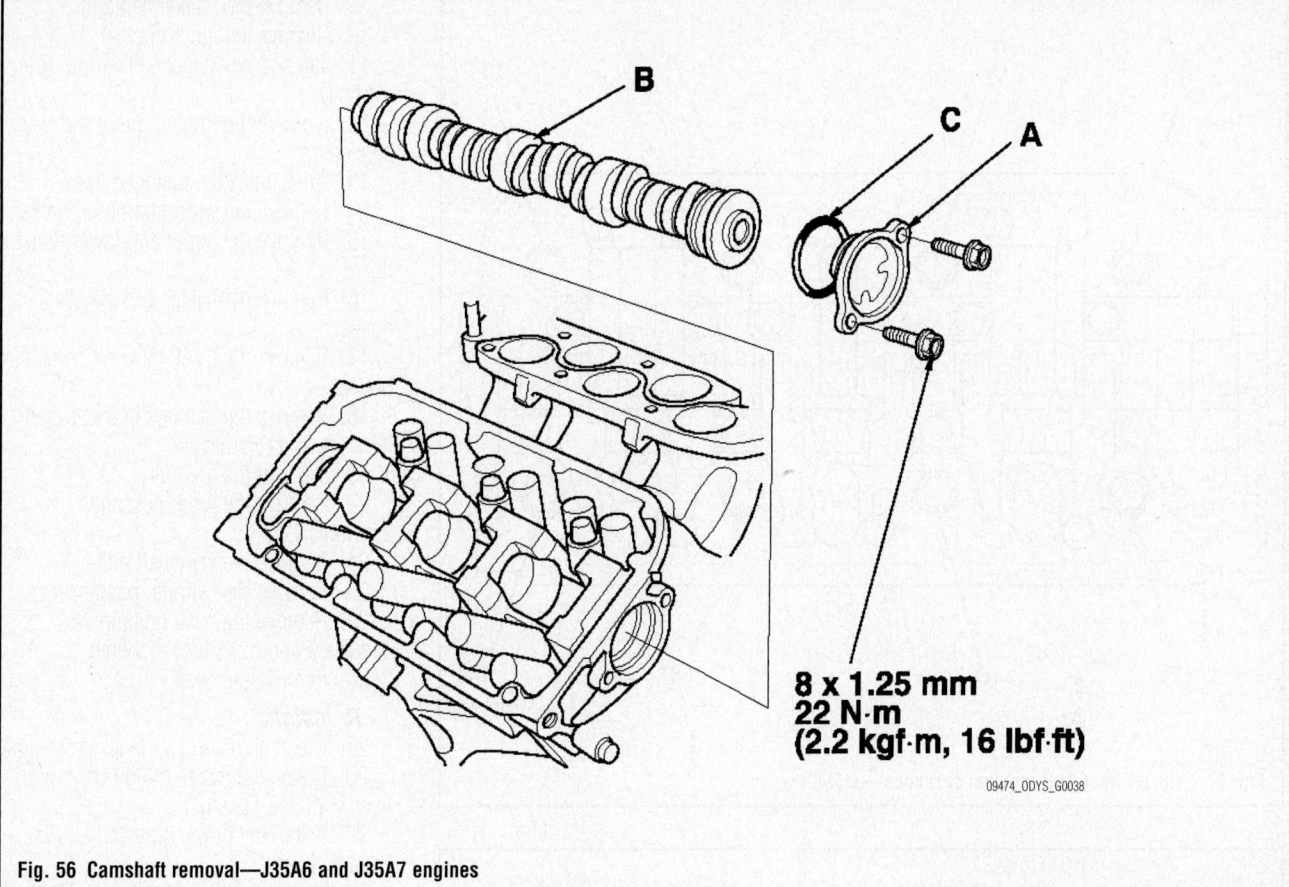

8 x 1.25 mm
22 N·m
(2.2 kgf·m, 16 lbf·ft)

09474_ODYS_G0038

Fig. 56 Camshaft removal—J35A6 and J35A7 engines

Rear

1. Relieve the fuel system pressure.
2. Remove the air cleaner assembly.
3. Remove the intake manifold.
4. Disconnect the fuel feed line.
5. Disconnect the brake lines at the master cylinder. Plug the openings.
6. Remove the timing belt.
7. Remove the rocker arm assembly.
8. Remove the camshaft pulley.
9. Remove the EVAP canister hose joint and bracket.
10. Remove the thrust cover (A) and camshaft (B).
11. Installation is the reverse of removal. Always use new O-rings (C). Apply clean engine oil to the journals and lobes. Bleed the brakes. Perform the CKP pattern clear/learn procedure.

J35A7 Engine

See Figure 56.

Front

1. Make sure you have the anti-theft codes for the radio and navigation systems.
2. Remove the battery.
3. Drain the coolant.
4. Remove the upper radiator hose.

5. Remove the EGR valve.
6. Remove the timing belt.
7. Remove the rocker arm assembly.
8. Remove the camshaft pulley.
9. Remove the thrust cover (A).
10. Carefully remove the camshaft (B).
11. Installation is the reverse of removal. Always use new O-rings (C). Apply clean engine oil to the journals and lobes.

Rear

1. Relieve the fuel system pressure.
2. Remove the air cleaner assembly.
3. Remove the intake manifold.
4. Disconnect the fuel feed line.
5. Disconnect the brake lines at the master cylinder. Plug the openings.
6. Remove the timing belt.
7. Remove the rocker arm assembly.
8. Remove the camshaft pulley.
9. Remove the thrust cover (A) and camshaft (B).
10. Installation is the reverse of removal. Always use new O-rings (C). Apply clean engine oil to the journals and lobes. Remove all traces of the old liquid gasket and apply new liquid gasket. Seat the rocker shaft assembly within 4 minutes of applying the liquid gasket. Bleed the brakes. Perform the CKP pattern clear/learn procedure.

CRANKSHAFT FRONT SEAL

REMOVAL & INSTALLATION

1. Remove the Crankshaft Position (CKP) sensor.
2. Remove the timing belt.
3. Remove the timing belt drive sprocket.
4. Remove the oil seal.

To install:

5. Dry and clean the oil seal housing.
6. Apply a light coating of multi-purpose grease to the crankshaft and oil seal lip.
7. Using a seal driver, install the seal until it bottoms.
8. The remainder of installation is the reverse of removal.

CYLINDER HEAD

REMOVAL & INSTALLATION

J35A6 Engine

See Figures 57 and 58.

→The engine coolant temperature must be below 100°F (38°C) before removing the head bolts.

1. Make sure that you have the anti-theft codes for the radio and navigation system.

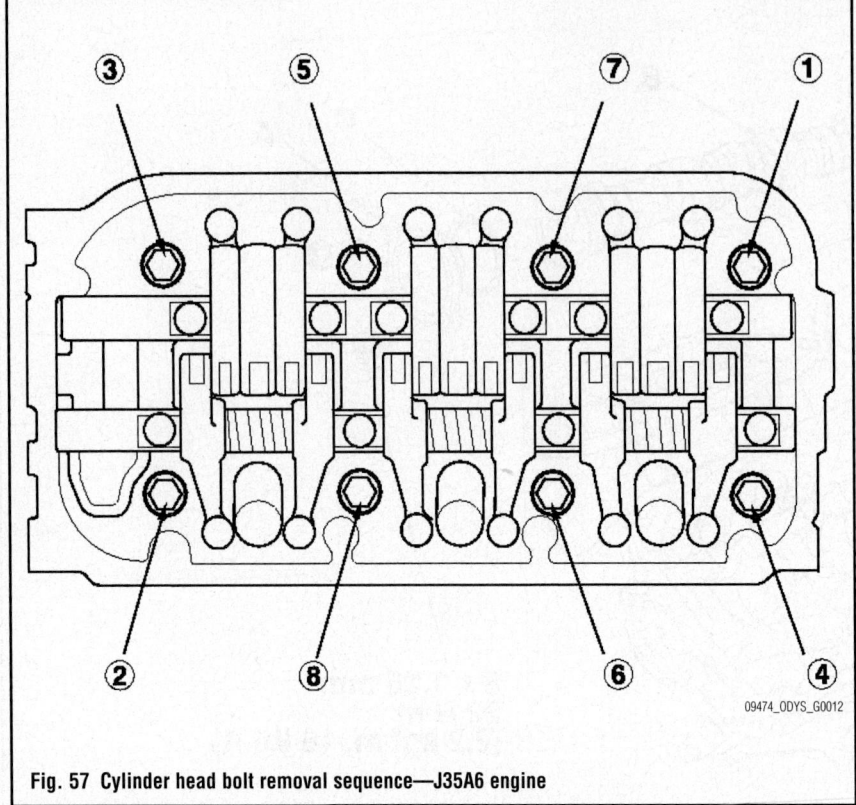

Fig. 57 Cylinder head bolt removal sequence—J35A6 engine

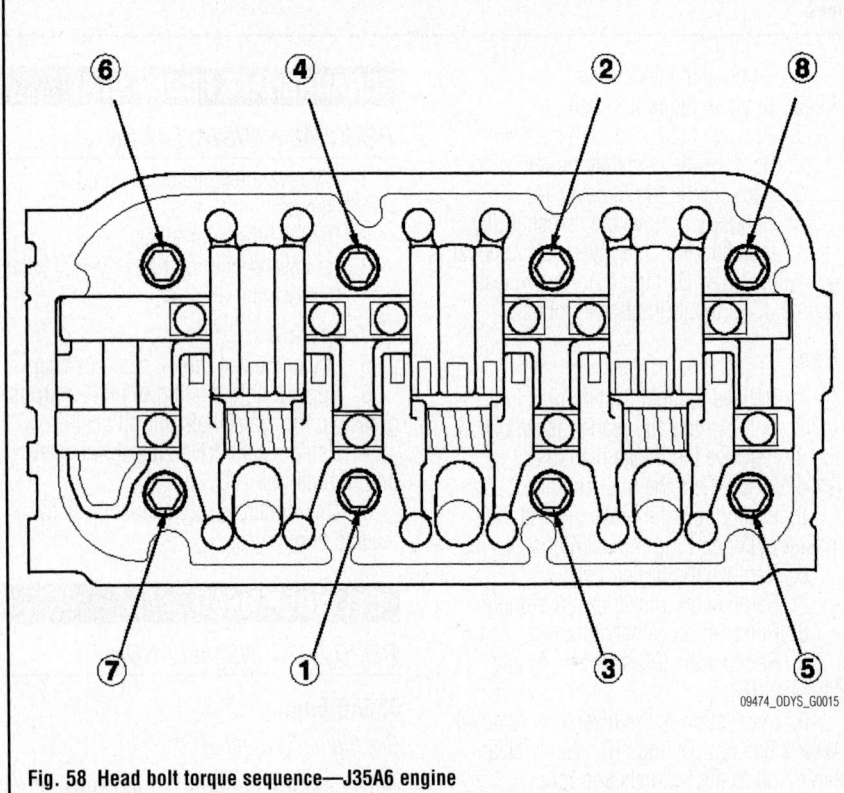

Fig. 58 Head bolt torque sequence—J35A6 engine

9. Remove the intake manifold.
10. Remove the ignition coils.
11. Tag and disconnect all wiring from the head.
12. Remove the front and rear warm-up converters.
13. Disconnect the fuel feed line.
14. Remove the wiring harness bracket.
15. Remove the upper and lower radiator hoses.
16. Remove the heater and bypass hoses.
17. Remove the EVAP canister hose joint bracket.
18. Remove the harness brackets from the front and rear heads.
19. Remove the fuel rail.
20. Remove the water passage assembly.
21. Remove the camshaft pulley.
22. Remove the cylinder head cover.
23. Remove the head bolts in the sequence shown ⅓ turn at a time.
24. Remove the head.

To install:
25. Clean all mating surfaces thoroughly.
26. Clean and install the oil control orifices with new O-rings.
27. Install the dowel pins and a new head gasket.
28. Set the crankshaft pulley to TDC by aligning the TDC mark (A) with the pointer (B). See the Timing Belt Procedure.
29. Set the camshaft pulley(s) to TDC by aligning the mark (A) with the pointer (B). See the Timing Belt Procedure.
30. Coat the threads and flanges of head bolts with clean engine oil.

✳✳ WARNING

There are 2 types of head bolts in service, 6-point and 12-point. Do not mix them on the same head.

➡**There are 2 different tightening methods based on which type of head bolt is used.**

✳✳ WARNING

When tightening the bolts, tighten them slowly. If any bolt makes a noise while tightening, loosen the bolt and retighten from Step 1.

 a. Step 1: Torque all bolts in sequence to 22 ft. lbs.
 b. Step 2: Torque all bolts in sequence an additional 90°
 c. Step 3: Torque all bolts in sequence an additional 90°
 • If new bolts are used, torque them in sequence an additional 90°

 2. Relieve the fuel system pressure.
 3. Disconnect the battery ground.
 4. Remove the accessory drive belt.
 5. Drain the coolant.

 6. Remove the power steering pump and the hose bracket.
 7. Remove the alternator.
 8. Remove the timing belt.

31. The remainder of installation is the reverse of removal. Note the following torques:
- Water passage 8mm bolts: 16 ft. lbs. (22 Nm)
- Water passage 6mm bolts: 105 inch lbs. (12 Nm)
- Front head bracket 10mm bolt: 33 ft. lbs. (44 Nm)
- Rear head bracket 8mm bolt: 16 ft. lbs. (22 Nm)
- EVAP canister joint bracket: 105 inch lbs. (12 Nm)
- Power steering pump bolts: 16 ft. lbs. (22 Nm)

J35A7 Engine

See Figures 59 through 62.

➡The engine coolant temperature must be below 100°F (38°C) before removing the head bolts.

1. Make sure that you have the anti-theft codes for the radio and navigation system.
2. Relieve the fuel system pressure.
3. Disconnect the battery ground.
4. Remove the accessory drive belt.
5. Drain the coolant.
6. Remove the power steering pump and the hose bracket.
7. Remove the alternator.
8. Remove the timing belt.
9. Remove the intake manifold.
10. Remove the ignition coils.
11. Tag and disconnect all wiring from the head.
12. Remove the front and rear warm-up converters.
13. Disconnect the fuel feed line.
14. Remove the wiring harness bracket.
15. Remove the upper and lower radiator hoses.
16. Remove the heater and bypass hoses.
17. Remove the EVAP canister hose joint bracket.
18. Remove the harness brackets from the front and rear heads.
19. Remove the fuel rail.
20. Remove the water passage assembly.
21. Remove the camshaft pulley.
22. Remove the cylinder head cover.
23. Remove the head bolts in the sequence shown ⅓ turn at a time.
24. Remove the head.

To install:

25. Clean all mating surfaces thoroughly.
26. Clean and install the oil control orifices with new O-rings.
27. Install the dowel pins and a new head gasket.

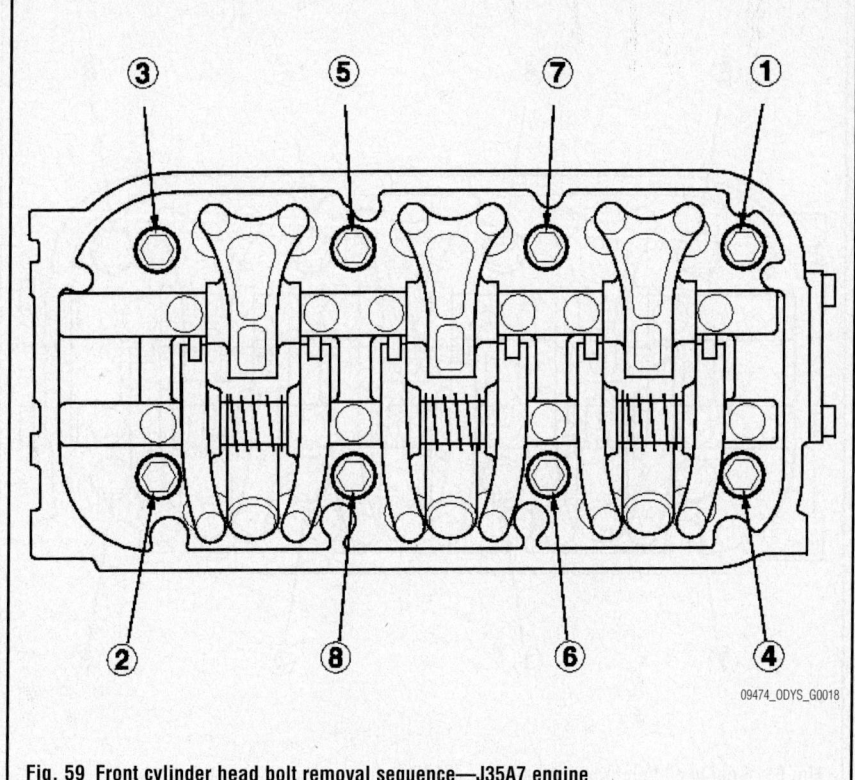

Fig. 59 Front cylinder head bolt removal sequence—J35A7 engine

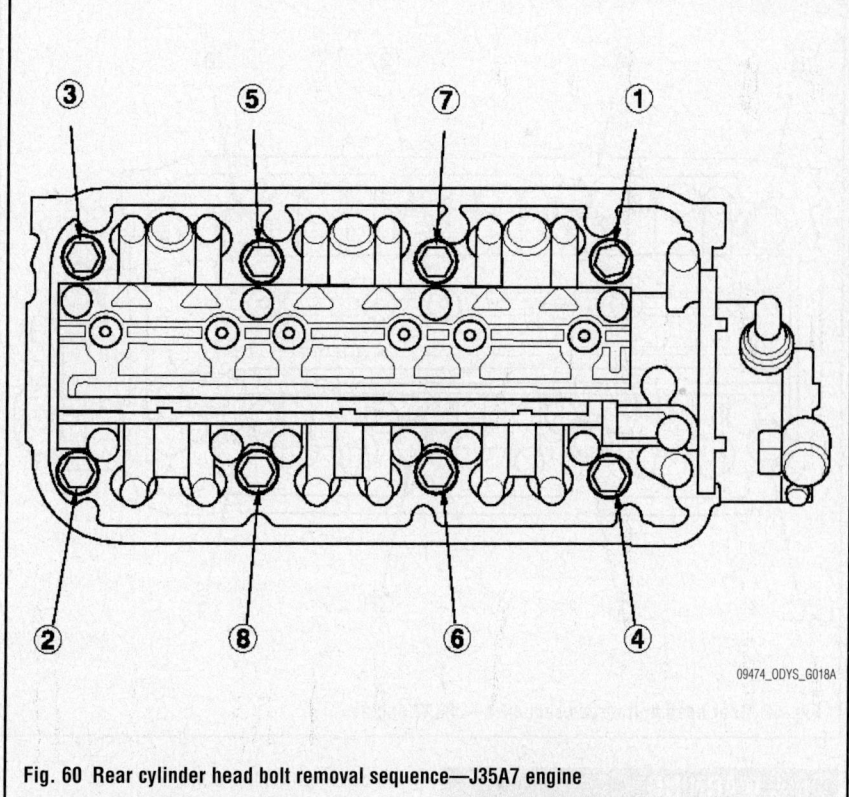

Fig. 60 Rear cylinder head bolt removal sequence—J35A7 engine

28. Set the crankshaft pulley to TDC by aligning the TDC mark (A) with the pointer (B). See the Timing Belt Procedure.
29. Set the camshaft pulley(s) to TDC by aligning the mark (A) with the pointer (B). See the Timing Belt Procedure.
30. Coat the threads and flanges of head bolts with clean engine oil.

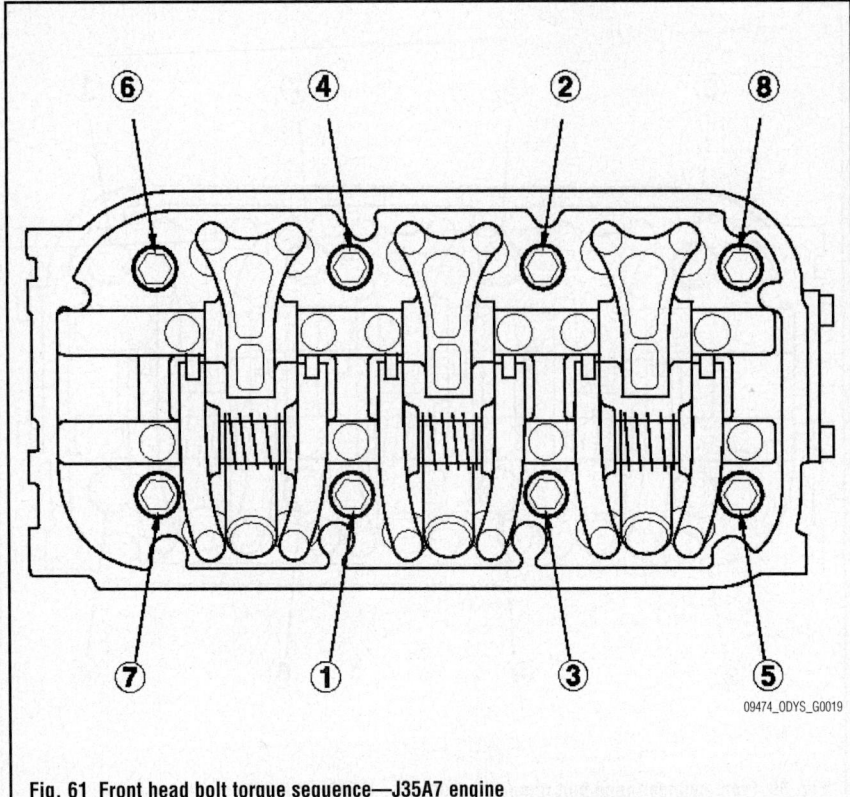

Fig. 61 Front head bolt torque sequence—J35A7 engine

09474_ODYS_G0019

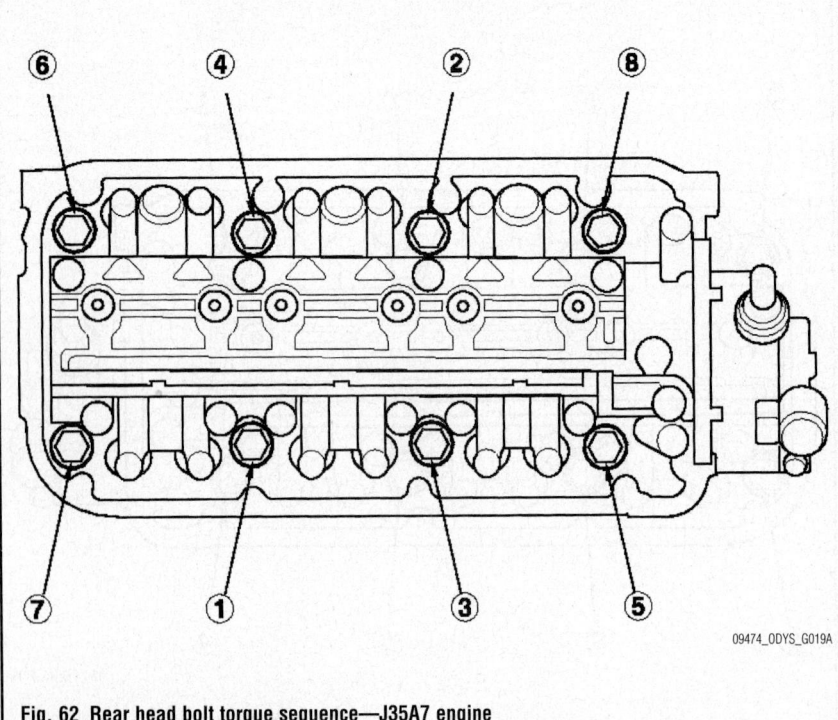

Fig. 62 Rear head bolt torque sequence—J35A7 engine

09474_ODYS_G019A

✳✳ WARNING

When tightening the bolts, tighten them slowly. If any bolt makes a noise while tightening, loosen the bolt and retighten from Step 1.

31. Tighten the bolts, in sequence, in 3 steps. Perform each step twice.

a. Step 1: 22 ft. lbs. (29 Nm)
b. Step 2: Torque all bolts in sequence an additional 90°
c. Step 3: Torque all bolts in sequence an additional 90°

32. The remainder of installation is the reverse of removal. Note the following torques:

- Water passage 8mm bolts: 16 ft. lbs. (22 Nm)
- Water passage 6mm bolts: 105 inch lbs. (12 Nm)
- Front head bracket 10mm bolt: 33 ft. lbs. (44 Nm)
- Rear head bracket 8mm bolt: 16 ft. lbs. (22 Nm)
- EVAP canister joint bracket: 105 inch lbs. (12 Nm)
- Power steering pump bolts: 16 ft. lbs. (22 Nm)

ENGINE ASSEMBLY

REMOVAL & INSTALLATION

See Figures 63 through 69.

1. Make sure that you have the anti-theft codes for the radio and navigation system.
2. Drain the power steering fluid.
3. Relive the fuel system pressure.
4. Disconnect the battery.
5. Remove the grille.
6. Remove the battery.
7. Remove the intake manifold cover.
8. Remove the air cleaner housing and related tubes.
9. Remove the air intake duct.
10. Remove the battery tray.
11. Disconnect the starter cables.
12. Disconnect the fuel supply line.
13. Tag and disconnect all wires and hoses connected to the engine.
14. Remove the driver's side console lower panel and pull back the carpet.
15. Remove the steering joint cover.
16. Lock the steering wheel and match-mark the steering shaft joint and the steering gear shaft.
17. Remove the bolt and disconnect the steering joint from the shaft.

➡**Do not turn the steering wheel with the joint disconnected.**

18. Remove the PCM cover.
19. Remove the accessory drive belt.
20. Disconnect the power steering outlet hose at the pump.
21. Remove the power steering reservoir and disconnect the return hose.
22. Remove the radiator cap.
23. Raise the vehicle on a hoist.
24. Remove the front wheels.
25. Remove the splash shield.
26. Drain the cooling system.

27. Drain the transmission.
28. Drain the engine oil.
29. Remove the front exhaust pipe.
30. Disconnect the stabilizer links.
31. Disconnect the tie rod ends from the knuckles.
32. Disconnect the lower control arms from the knuckles.
33. Remove the halfshafts.
34. Disconnect the shift cable and remove the control lever.
35. Disconnect the power steering hoses from the gear.
36. Disconnect the power steering pressure switch.
37. Remove the nuts from the transmission lower front and rear mounts.
38. Lower the vehicle.
39. Disconnect the cooling fan wiring.
40. Remove the coolant reservoir.
41. Remove the fan and shroud.
42. Remove the upper and lower radiator hoses.
43. Remove the heater hoses.

44. Remove the transmission cooler hoses.
45. Remove the radiator.
46. Remove the A/C compressor and secure it aside without disconnecting the hoses.
47. Install engine hanger adapters VSB02C000022 on the engine support beams (A). Then, install the engine support hanger AAR-T-12566 with the adapters.
48. Lift and support the engine with the support hanger and the balance bar VSB02C000019. Attach the front arm to the front cylinder head with a spacer (B) and the 10×1.25mm bolt (C). Attach the rear arm to the rear head with the 8×1.25mm bolt (D).
49. On the J35A6 engine, remove the vacuum hose (A) from the front engine mount stop (B), then remove the front engine mount bolt (C).
50. On the J35A7 engine, disconnect the front active control engine mount actuator connector (A) and clamp (B), then remove

the front engine mount stop (C) and the bolt (D).
51. On the J35A6 engine, remove the heat shield and the rear engine mount bolts.
52. On the J35A7 engine, disconnect the rear ACM connector (A), remove the rear engine mount stop (B) and the rear engine mount bolt (C).
53. Raise the vehicle on the hoist.
54. Matchmark the subframe and body.
55. Attach the subframe holding fixture (A) by hanging the belt (B) over the front of the subframe, then secure the belt with the retaining pin (C).
56. Raise the jack and install the 4 bolts at the corners.
57. Remove the six 12×1.25mm bolts (A) securing the subframe stiffeners (B), the 4 subframe mounting bolts (C) and the stiffeners. Lower the subframe (D).
58. Lower the vehicle.
59. Attach a chain hoist to the engine hook and the transmission hook. Lift the engine/transmission assembly until it is

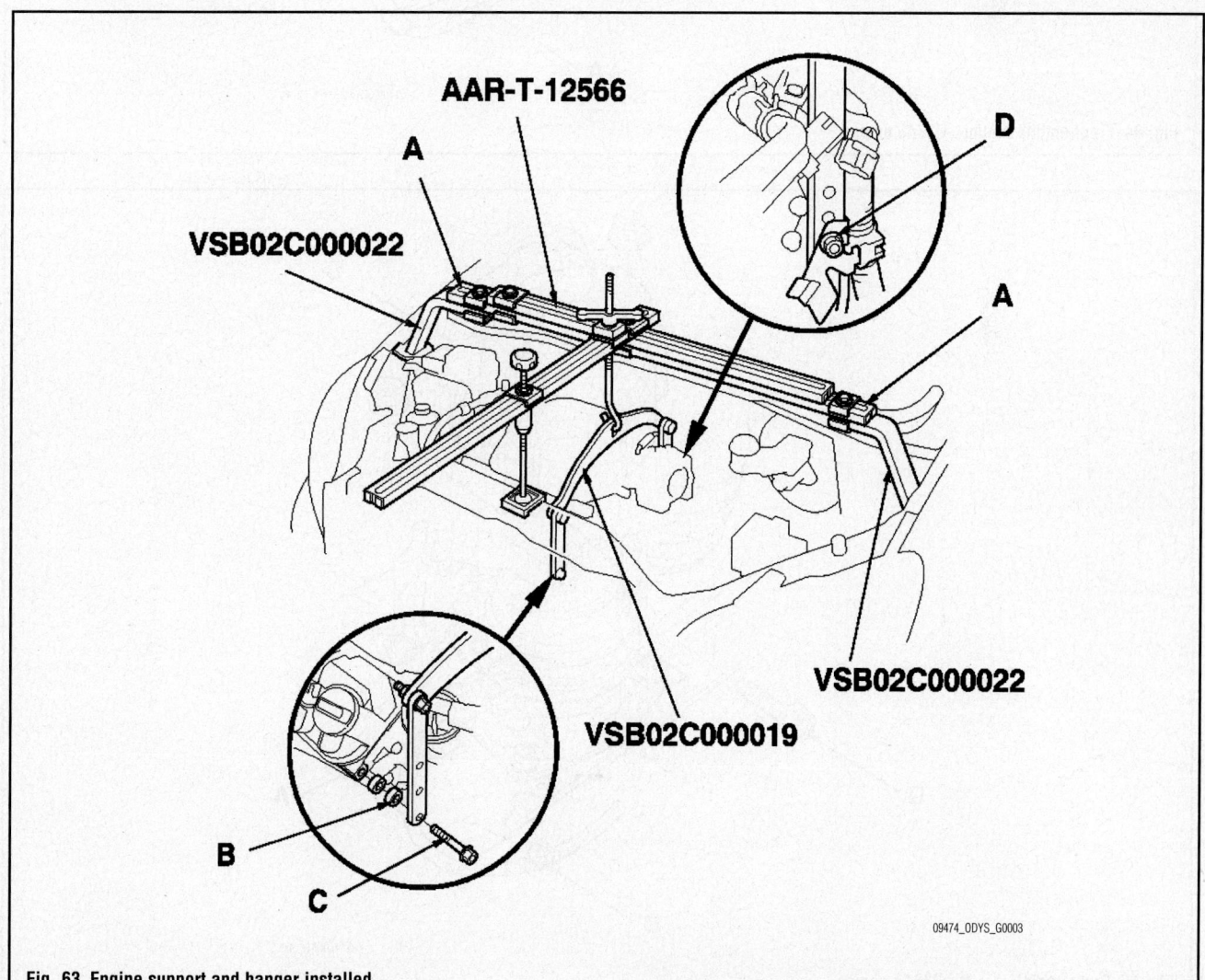

09474_ODYS_G0003

Fig. 63 Engine support and hanger installed

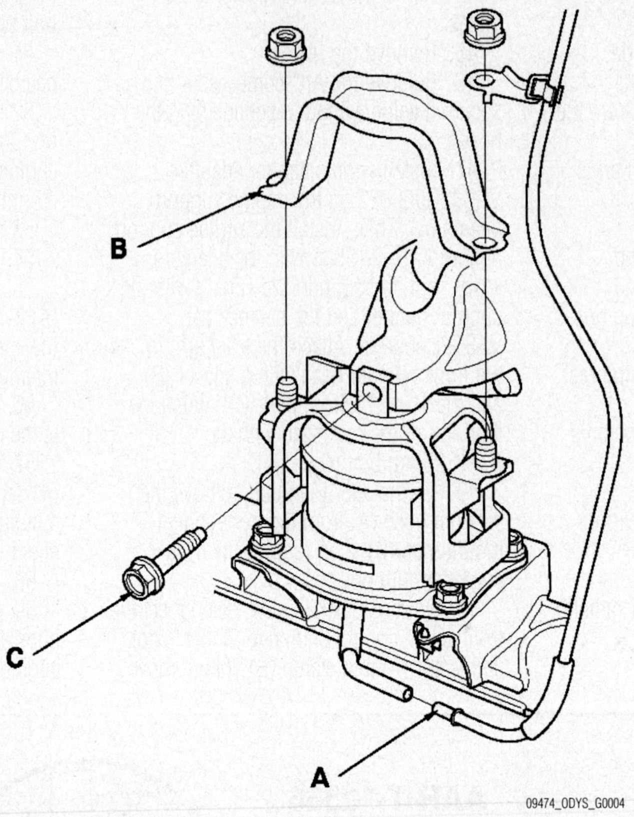

Fig. 64 Front engine mount—J35A6 engine

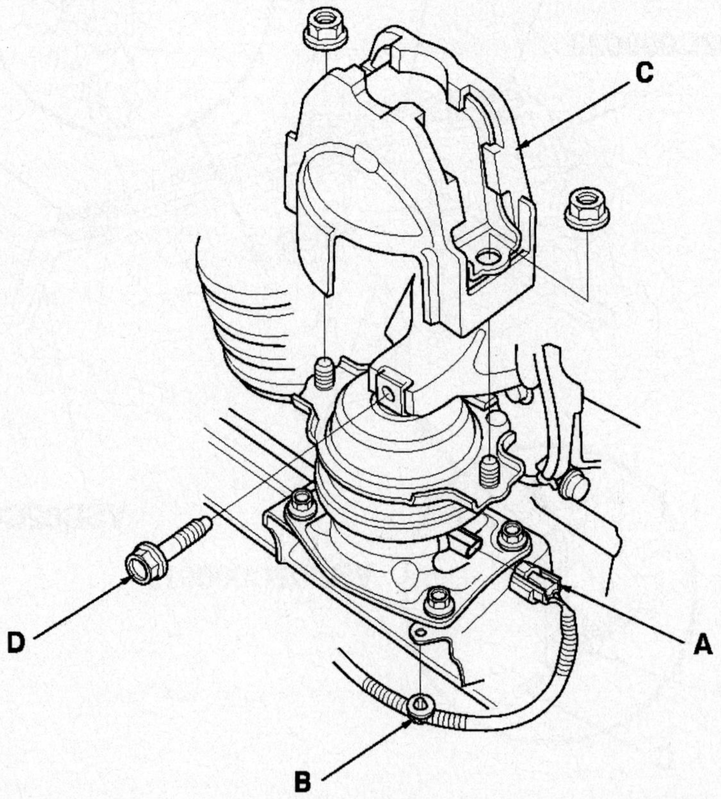

Fig. 65 Front engine mount—J35A7 engine

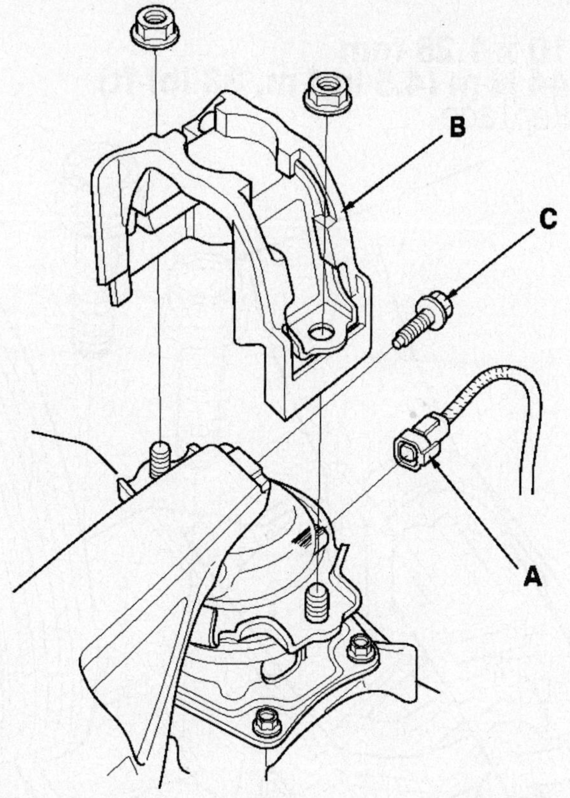

09474_ODYS_G0006

Fig. 66 Rear engine mount—J35A7 engine

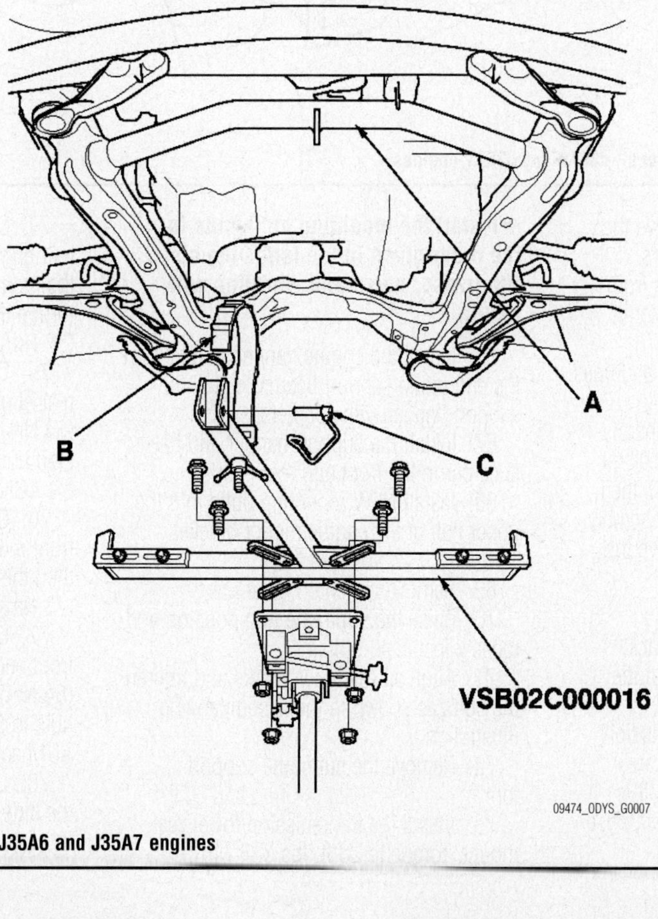

VSB02C000016

09474_ODYS_G0007

Fig. 67 Subframe holding fixture—J35A6 and J35A7 engines

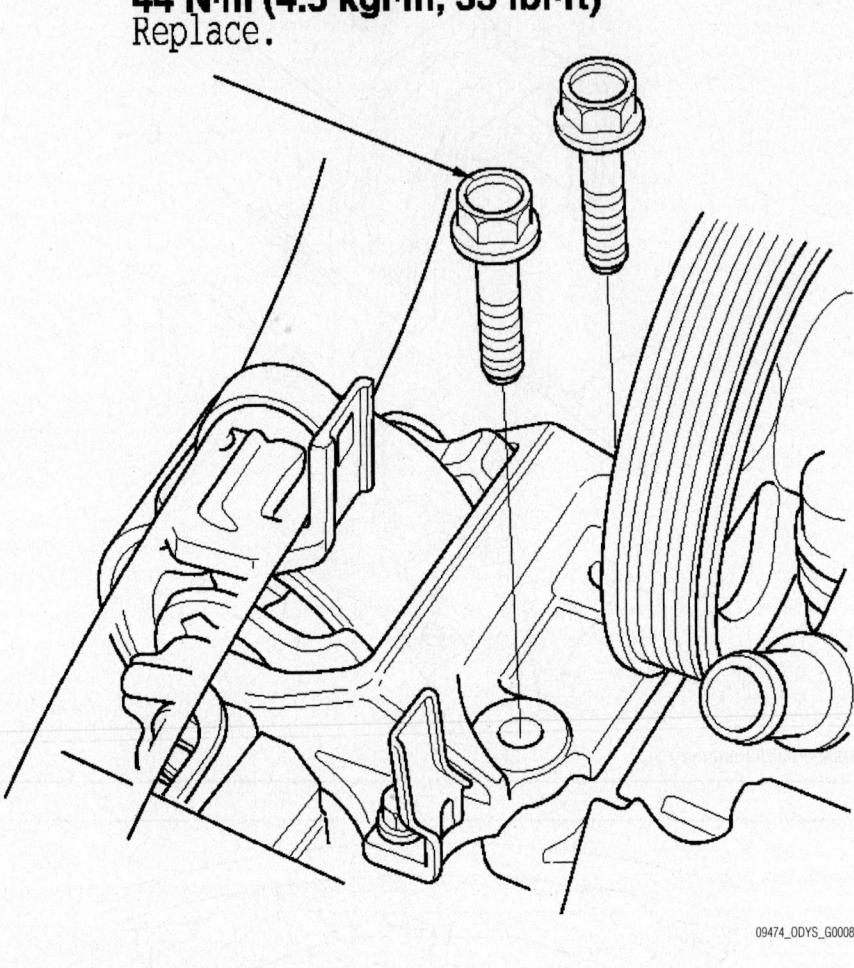

10 x 1.25 mm
44 N·m (4.5 kgf·m, 33 lbf·ft)
Replace.

09474_ODYS_G0008

Fig. 68 Engine mount bracket torques—J35A6 and J35A7 engines

supported by the hoist, then, remove the engine support hanger and adapters.

60. Remove the mounting bolts from the upper half of the side engine mount bracket.

61. Check that all hoses, wires and lines are free from the engine.

62. Slowly lower the engine/transmission assembly about 6 inches. Check for anything that might hinder removal, then remove the engine/transmission assembly.

63. Raise the vehicle and remove the unit.

To install:

64. Install the engine mount brackets and tighten to the torques shown in the accompanying illustration.

65. Position the engine/transmission under the vehicle. Make sure they are aligned and lower the vehicle until it is properly positioned. Using the hoist, raise the assembly into place.

➡**Install the mounting bolts/nuts in the order given in the following steps. Otherwise, noise and vibration will result.**

66. Install the engine hanger adapters on the end of the support beams. Install the support hanger with adapters.

67. Install the support hanger and balance bar to the front and rear heads.

68. Install NEW mounting bolts into the upper half of the engine mount bracket. Torque to 33 ft. lbs. (44 Nm).

69. Remove the chain hoist.

70. Raise the subframe into position and loosely install all bolts.

71. Align all reference marks and tighten the bolts as shown in the accompanying illustration.

72. Remove the subframe support tool.

73. Install the transmission lower rear mount. Torque to 28 ft. lbs. (38 Nm).

74. On the J35A6 engine, tighten the rear engine bolts and heat shield. Torque the mount bolts to 31 ft. lbs. (42 Nm); torque the heat shield bolts to 43 ft. lbs. (58 Nm).

75. On the J35A7 engine, torque the rear engine mount bolt to 40 ft. lbs. (54 Nm), then torque the rear engine stop nuts to 54 ft. lbs. (73 Nm). Connect the ACM.

76. On the J35A6 engine, tighten the front mount bolt to 40 ft. lbs. (54 Nm), then the front mount stop nuts to 54 ft. lbs. (73 Nm). Connect the vacuum hose.

77. On the J35A7 engine, tighten the front engine mount bolt to 40 ft. lbs. (54 Nm), then tighten the front mount stop nuts to 54 ft. lbs. (73 Nm). Connect the ACM actuator and install the harness clamp.

78. Loosen the bolts at the upper half of the side mount and retighten them to 33 ft. lbs. (44 Nm).

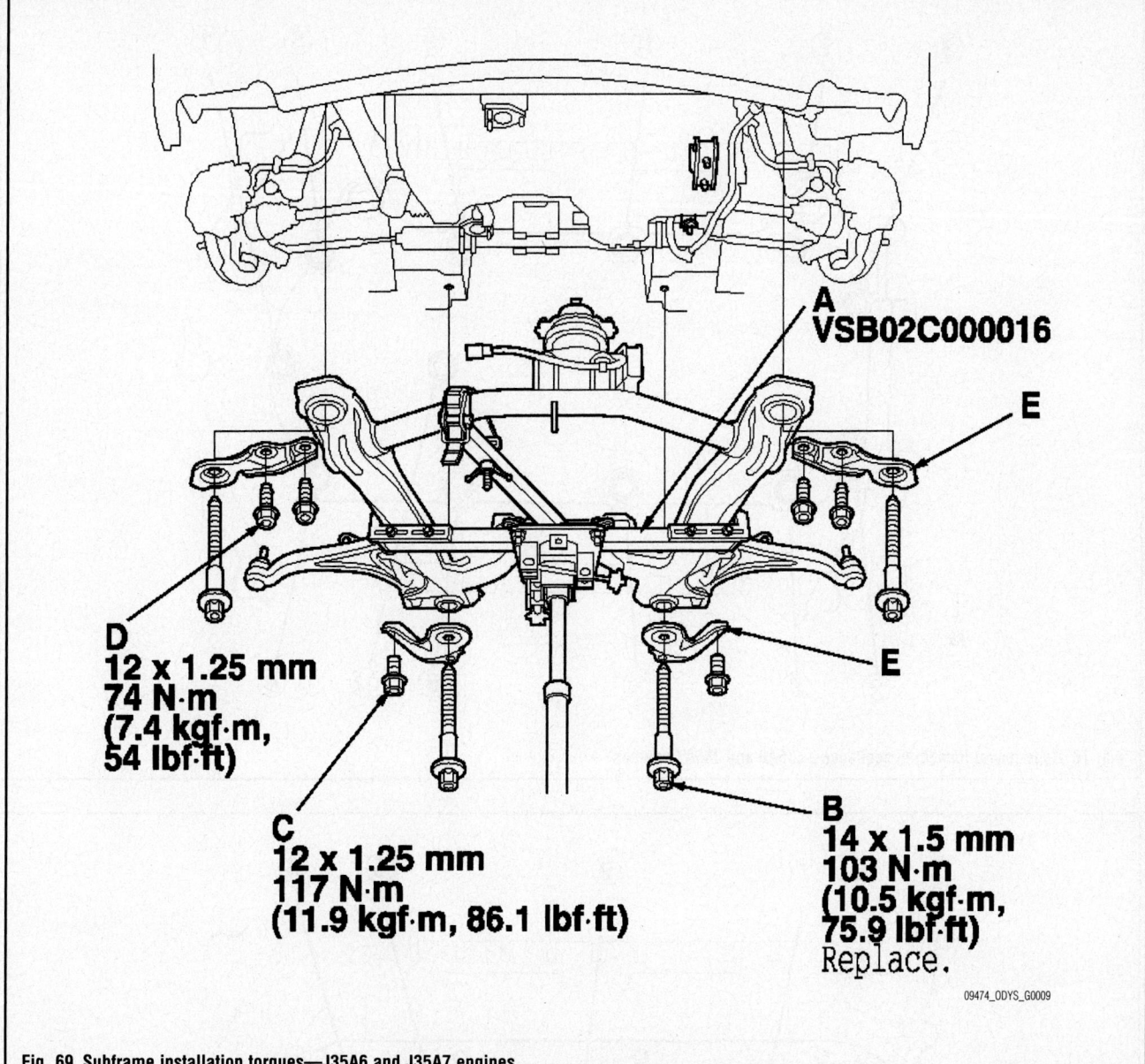

A
VSB02C000016

E

D
**12 x 1.25 mm
74 N·m
(7.4 kgf·m,
54 lbf·ft)**

E

C
**12 x 1.25 mm
117 N·m
(11.9 kgf·m, 86.1 lbf·ft)**

B
**14 x 1.5 mm
103 N·m
(10.5 kgf·m,
75.9 lbf·ft)**
Replace.

09474_ODYS_G0009

Fig. 69 Subframe installation torques—J35A6 and J35A7 engines

79. Remove the engine support pieces.
80. Install the A/C compressor. Torque to 16 ft. lbs. (22 Nm).
81. Install the radiator.
82. The remainder of installation is the reverse of removal. Observe the following torques:
- Exhaust pipe-to-manifold: 40 ft. lbs. (54 Nm)
- Exhaust pipe hanger nut: 16 ft. lbs. (22 Nm)
- Exhaust pipe-to-converter: 25 ft. lbs. (33 Nm)
- Steering joint bolt: 21 ft. lbs. (28 Nm)
- Battery tray bolts: 16 ft. lbs. (22 Nm)

83. Refill all fluids and check for leaks and proper operation.

EXHAUST MANIFOLD

REMOVAL & INSTALLATION

1. Remove the engine cover.
2. Remove the front A/F sensor (Sensor 1) and the front secondary HO2S (Sensor 2).
3. Remove the exhaust pipe A mounting nuts (front WU-TWC side).
4. Remove the A/C condenser fan assemblies.
5. Carefully remove the front WU-TWC.

To install:

6. Carefully install the front WU-TWC with a new gasket and new self-locking nuts. Tighten the nuts in a crisscross pattern in two or three steps to 23 ft. lbs. (31 Nm).

7. The remainder of the installation is the reverse order of removal.

INTAKE MANIFOLD

REMOVAL & INSTALLATION

See Figures 70 through 75.

1. Remove the intake manifold cover.
2. Remove the air inlet duct.
3. Remove the PCV hose and brake booster hose.
4. Remove the EVAP canister hose and water bypass hoses. Plug the bypass hose.
5. Tag and disconnect all wiring connected to the manifold.
6. Remove the upper cover mounting bolts and nuts, in 2 equal steps, in the sequence shown. Remove the cover.

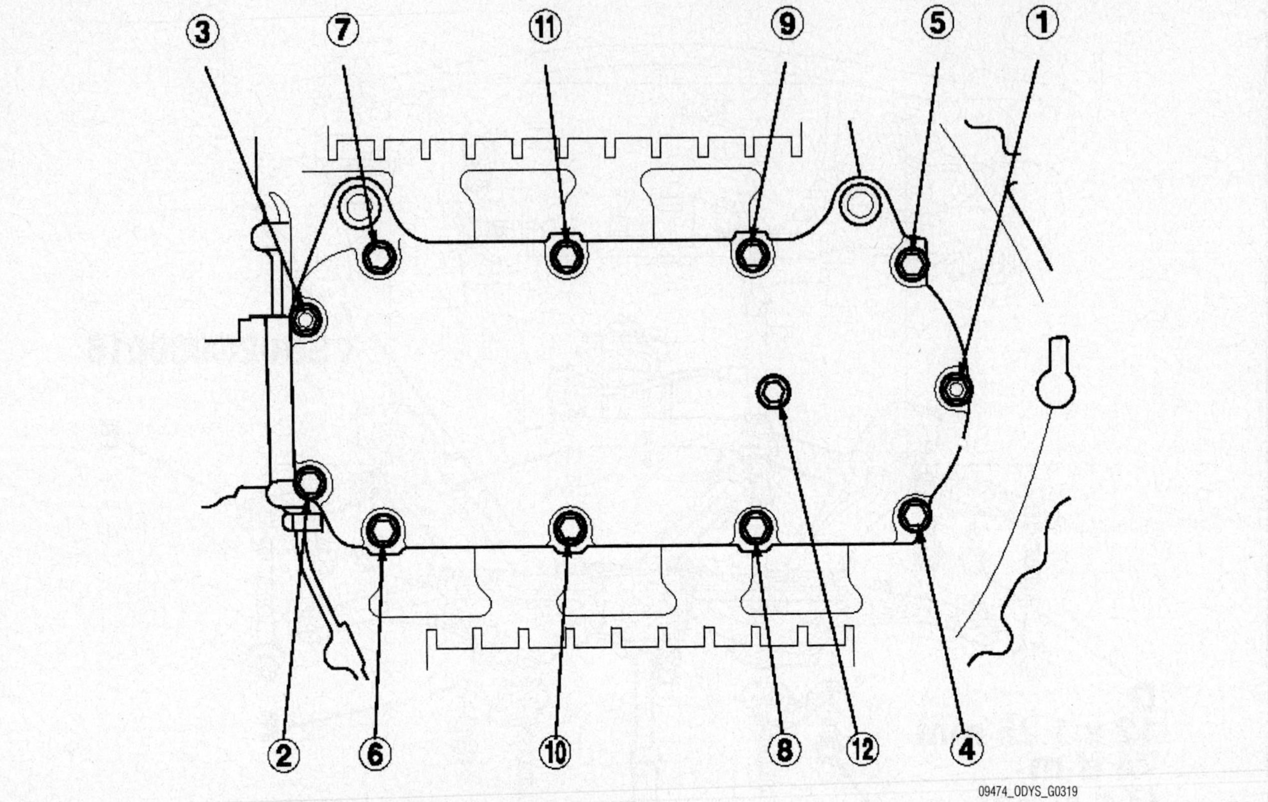

Fig. 70 Upper cover loosening sequence—J35A6 and J35A7 engines

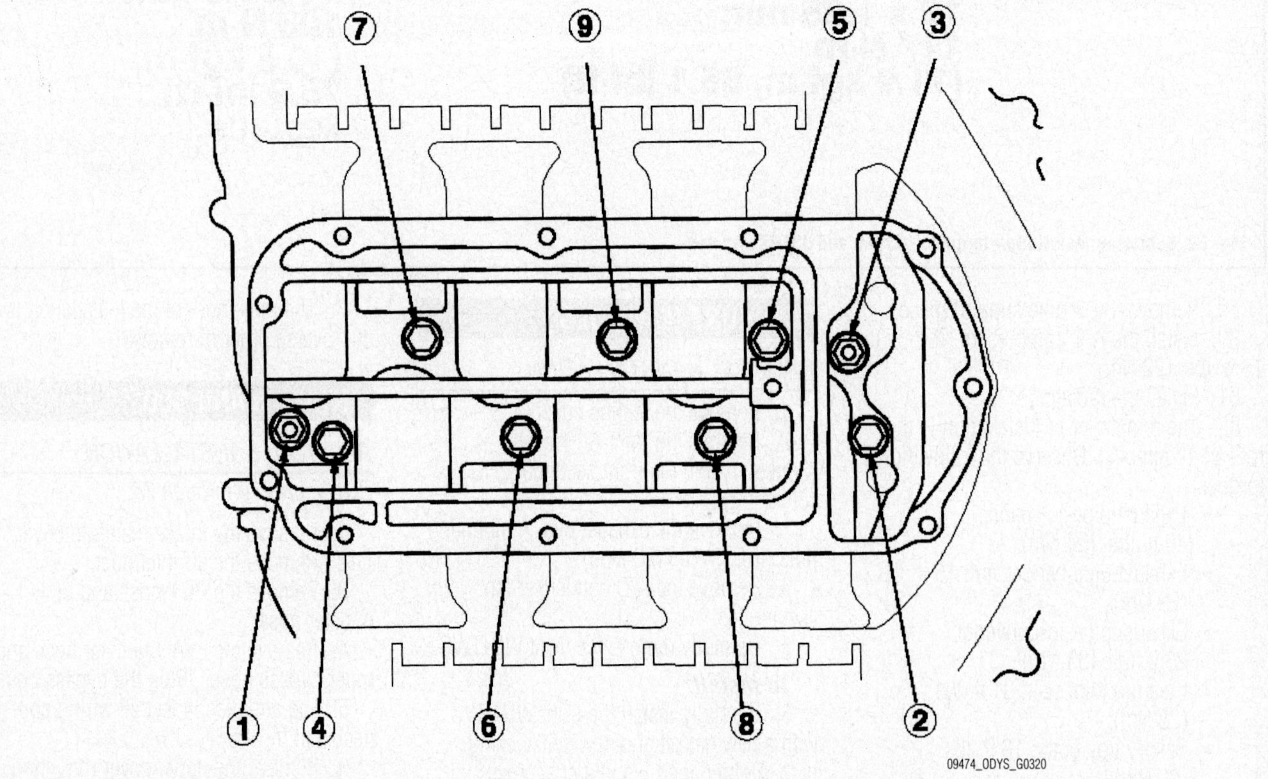

Fig. 71 Intake manifold loosening sequence—J35A6 and J35A7 engines

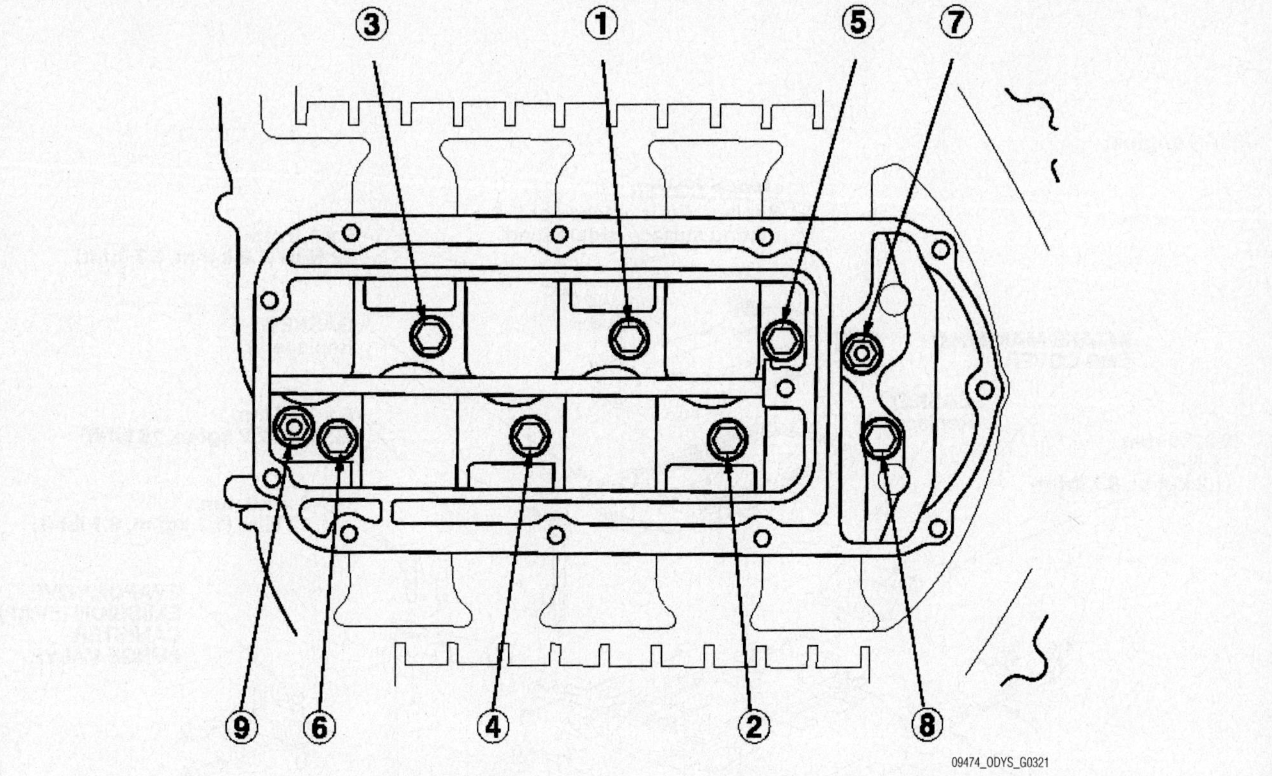
③
①
⑤
⑦
⑨
⑥
④
②
⑧
09474_ODYS_G0321

Fig. 72 Intake manifold torque sequence—J35A6 and J35A7 engines

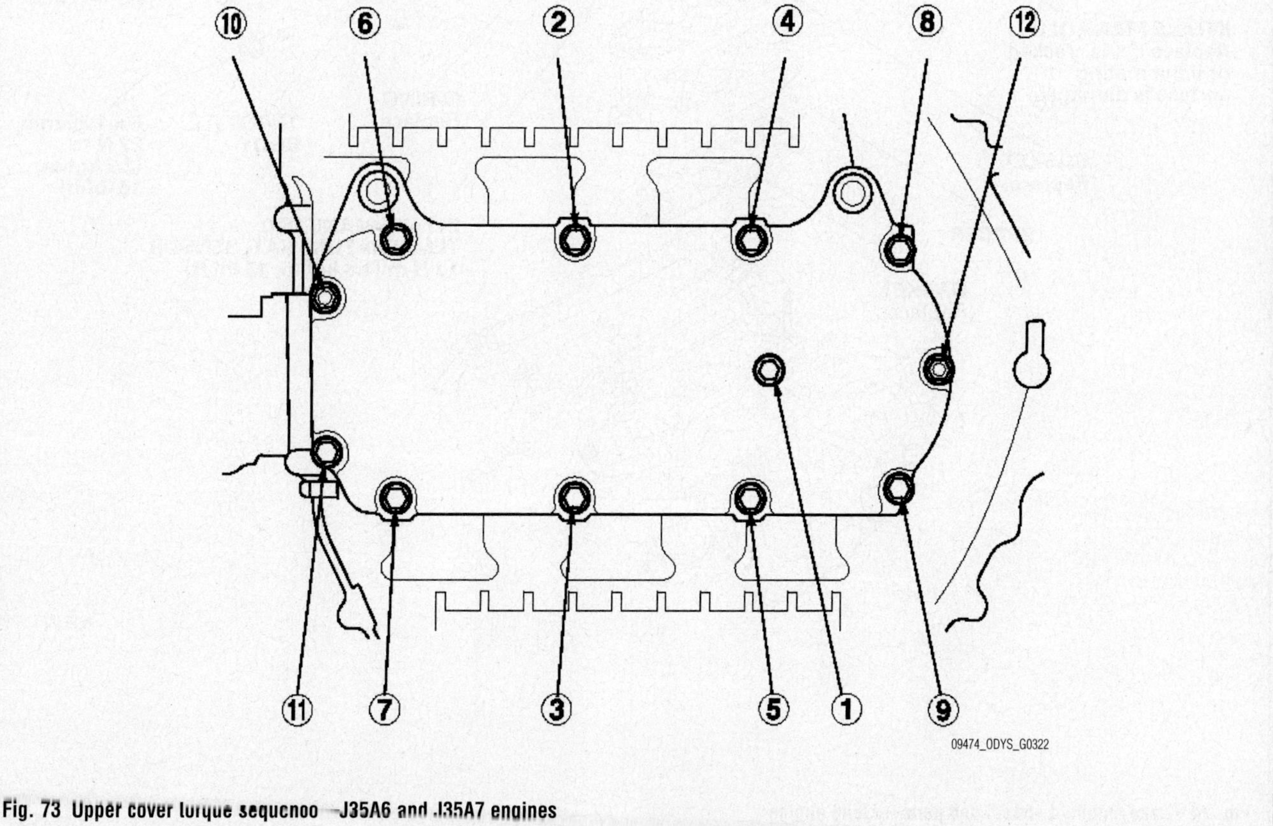
⑩
⑥
②
④
⑧
⑫
⑪
⑦
③
⑤
①
⑨
09474_ODYS_G0322

Fig. 73 Upper cover torque sequence—J35A6 and J35A7 engines

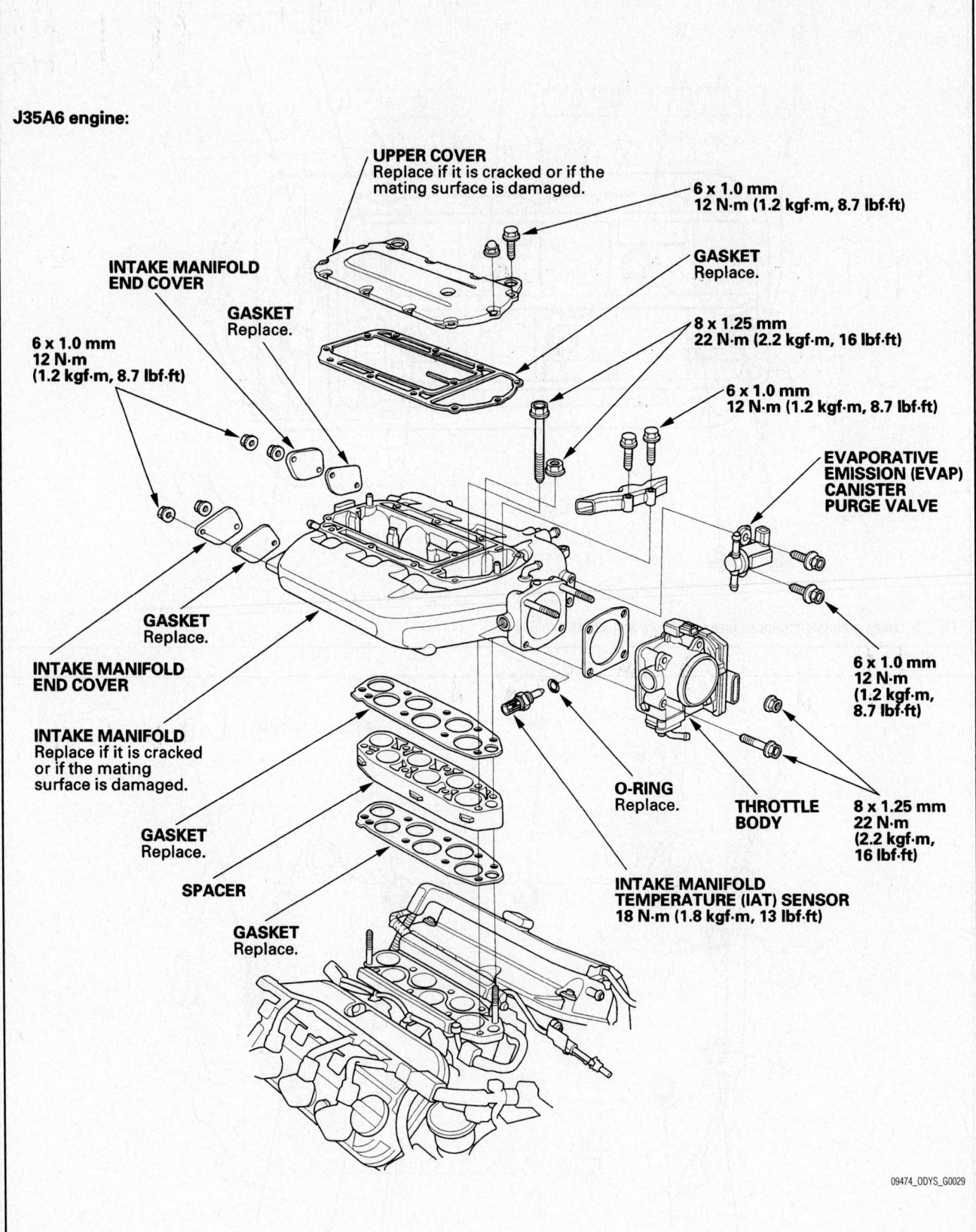

J35A6 engine:

UPPER COVER
Replace if it is cracked or if the
mating surface is damaged.

6 x 1.0 mm
12 N·m (1.2 kgf·m, 8.7 lbf·ft)

INTAKE MANIFOLD
END COVER

GASKET
Replace.

GASKET
Replace.

8 x 1.25 mm
22 N·m (2.2 kgf·m, 16 lbf·ft)

6 x 1.0 mm
12 N·m
(1.2 kgf·m, 8.7 lbf·ft)

6 x 1.0 mm
12 N·m (1.2 kgf·m, 8.7 lbf·ft)

EVAPORATIVE
EMISSION (EVAP)
CANISTER
PURGE VALVE

GASKET
Replace.

INTAKE MANIFOLD
END COVER

INTAKE MANIFOLD
Replace if it is cracked
or if the mating
surface is damaged.

6 x 1.0 mm
12 N·m
(1.2 kgf·m,
8.7 lbf·ft)

GASKET
Replace.

SPACER

O-RING
Replace.

THROTTLE
BODY

8 x 1.25 mm
22 N·m
(2.2 kgf·m,
16 lbf·ft)

GASKET
Replace.

INTAKE MANIFOLD
TEMPERATURE (IAT) SENSOR
18 N·m (1.8 kgf·m, 13 lbf·ft)

09474_ODYS_G0029

Fig. 74 Intake manifold and related parts—J35A6 engine

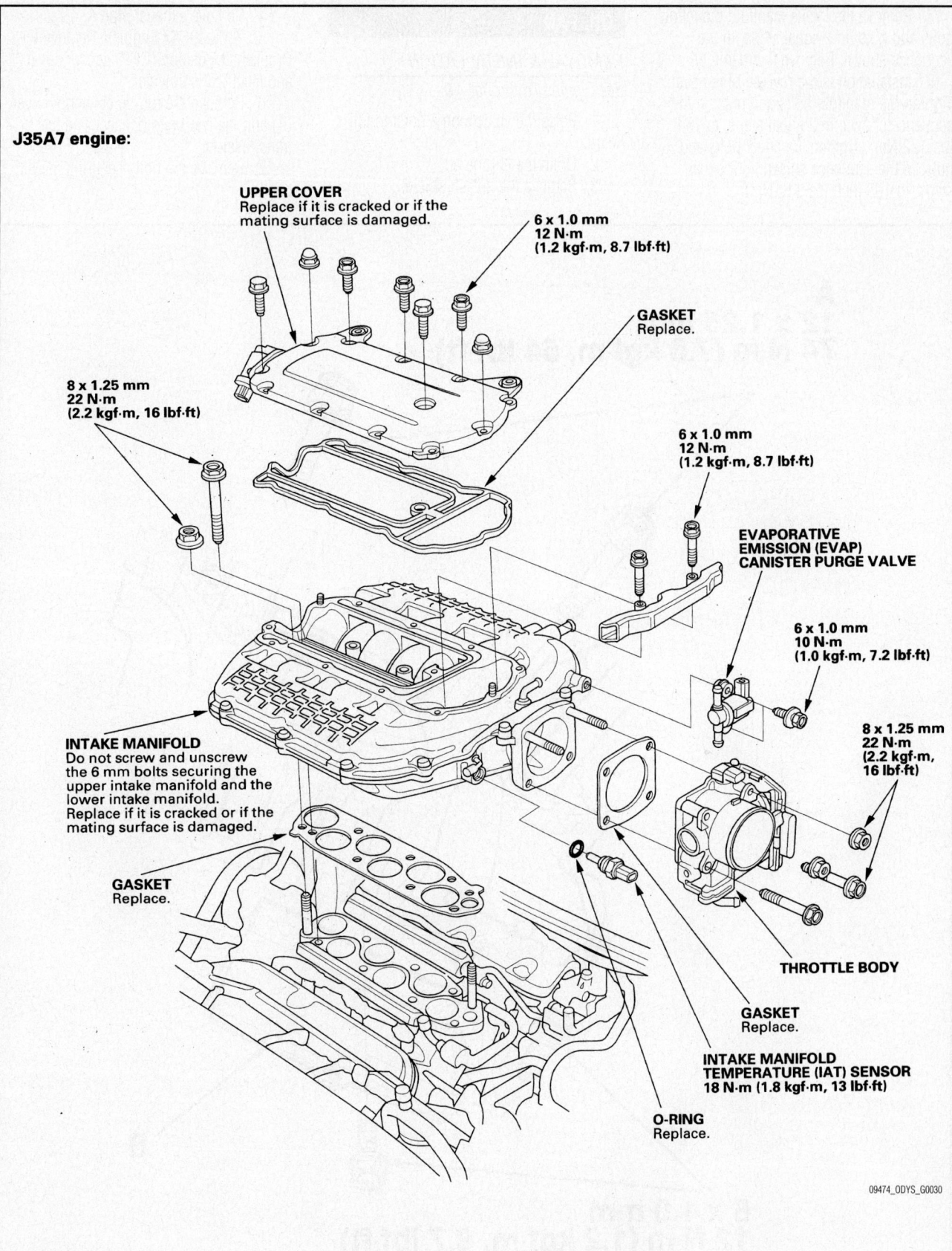

J35A7 engine:

UPPER COVER
Replace if it is cracked or if the
mating surface is damaged.

6 x 1.0 mm
12 N·m
(1.2 kgf·m, 8.7 lbf·ft)

GASKET
Replace.

8 x 1.25 mm
22 N·m
(2.2 kgf·m, 16 lbf·ft)

6 x 1.0 mm
12 N·m
(1.2 kgf·m, 8.7 lbf·ft)

**EVAPORATIVE
EMISSION (EVAP)
CANISTER PURGE VALVE**

6 x 1.0 mm
10 N·m
(1.0 kgf·m, 7.2 lbf·ft)

8 x 1.25 mm
22 N·m
(2.2 kgf·m,
16 lbf·ft)

INTAKE MANIFOLD
Do not screw and unscrew
the 6 mm bolts securing the
upper intake manifold and the
lower intake manifold.
Replace if it is cracked or if the
mating surface is damaged.

GASKET
Replace.

THROTTLE BODY

GASKET
Replace.

**INTAKE MANIFOLD
TEMPERATURE (IAT) SENSOR**
18 N·m (1.8 kgf·m, 13 lbf·ft)

O-RING
Replace.

09474_ODYS_G0030

Fig. 75 Intake manifold and related parts—J35A7 engine

7. Remove the intake manifold mounting bolts and nuts, in 2 equal steps, in the sequence shown. Remove the manifold.

8. Installation is the reverse of removal. Tighten the manifold bolts and nuts, in the sequence shown, in 2 equal steps, to 16 ft. lbs. (22 Nm). Tighten the cover bolts and nuts, in the sequence shown, in 2 equal steps, to 105 inch lbs. (12 Nm).

OIL PAN

REMOVAL & INSTALLATION

See Figures 76 through 79.

1. Raise the vehicle on a hoist to full height.
2. Drain the engine oil.
3. Remove the splash shield.

4. Remove exhaust pipe A.
5. 2008 J35A7 Engines: Remove the Crankshaft Position (CKP) sensor cover and electrical connector.
6. Remove the torque converter cover (A) and the two bolts (B) securing the transmission.
7. Remove the bolts securing the oil pan.

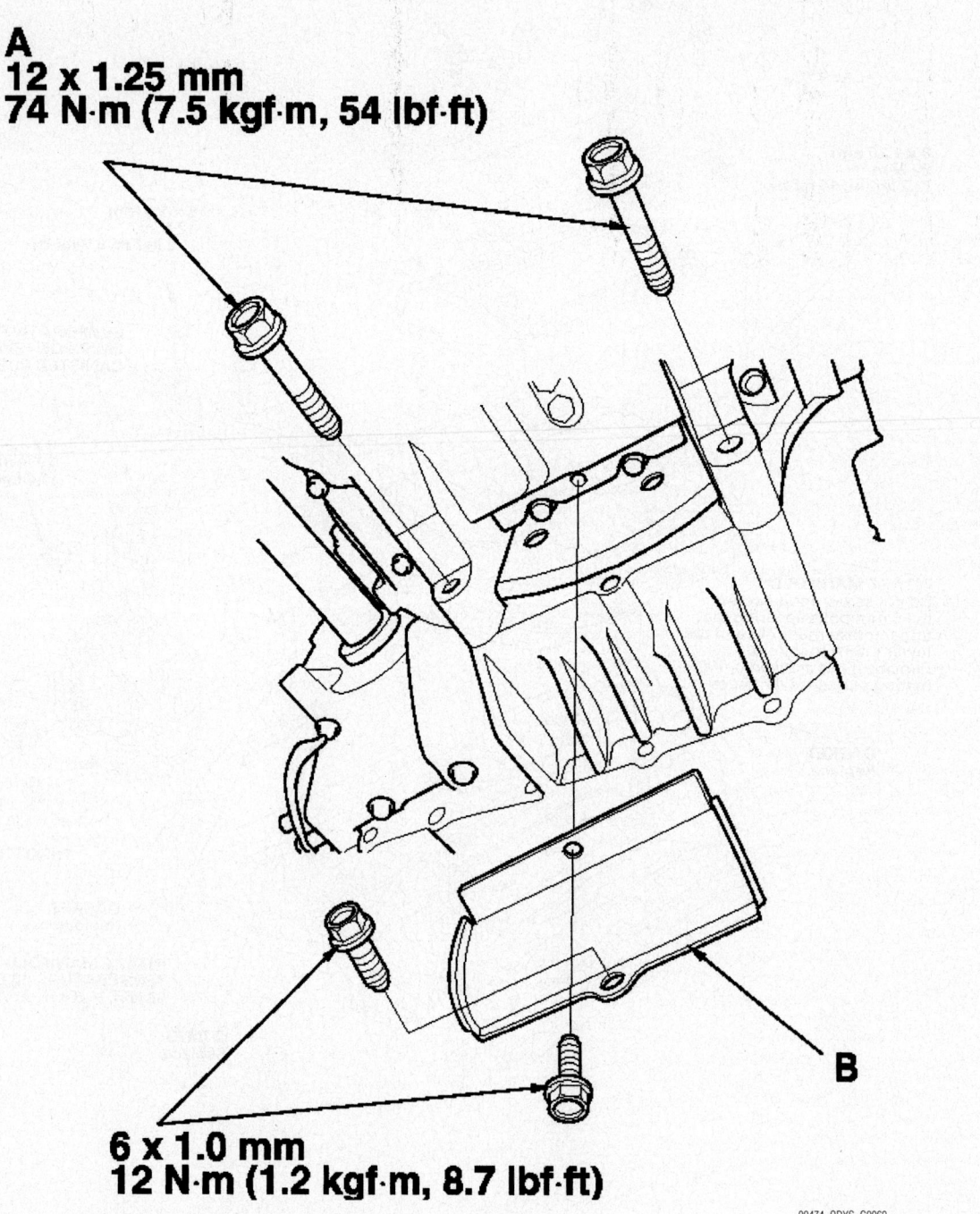

A
12 x 1.25 mm
74 N·m (7.5 kgf·m, 54 lbf·ft)

B

6 x 1.0 mm
12 N·m (1.2 kgf·m, 8.7 lbf·ft)

09474_ODYS_G0069

Fig. 76 Remove the torque converter cover (A) and the two bolts (B) securing the transmission

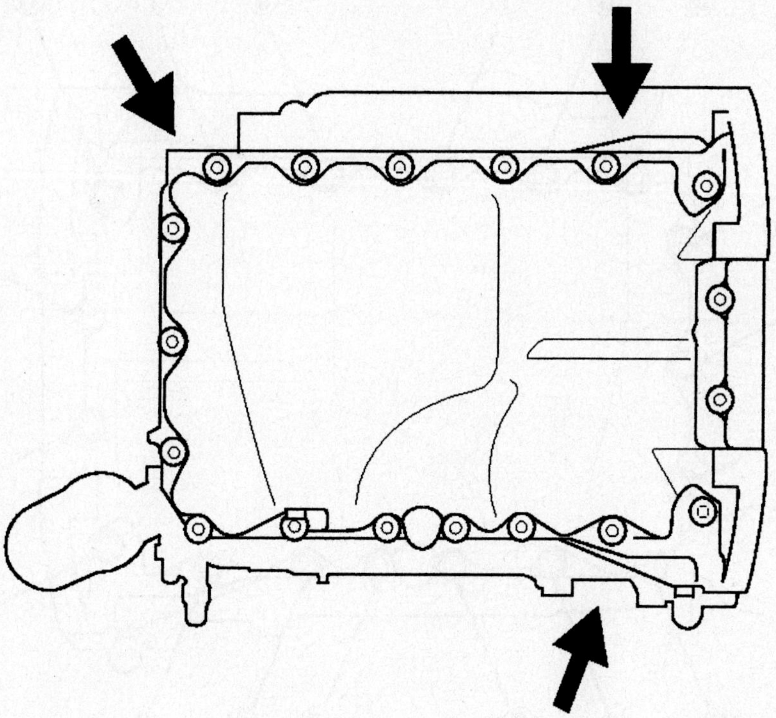

09474_ODYS_G0070

Fig. 77 Using a flat blade screwdriver, separate the oil pan from the block in the places shown

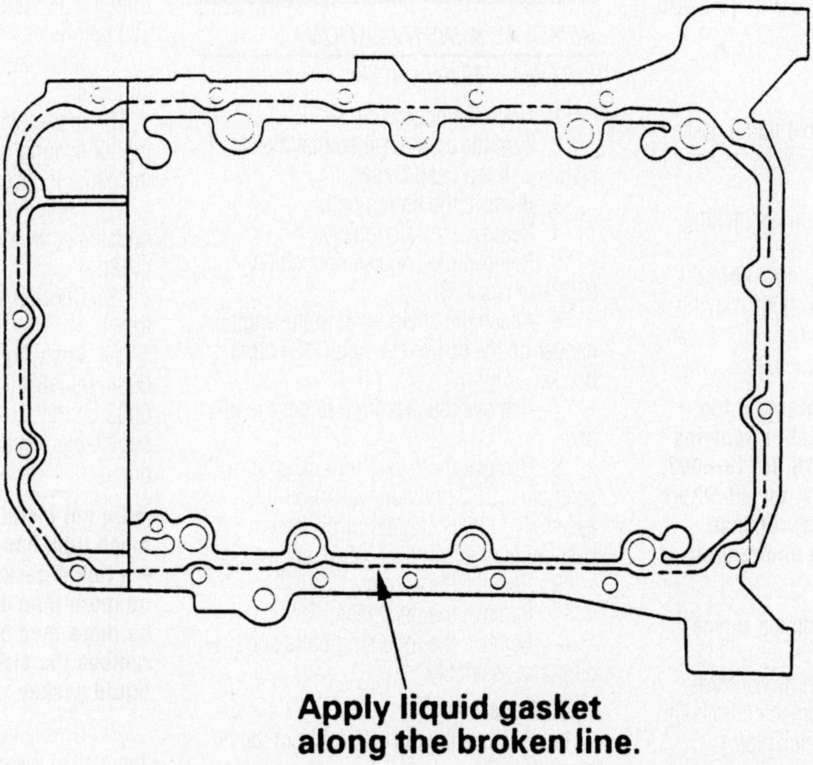

Apply liquid gasket along the broken line.

9302MG75

Fig. 78 Apply liquid gasket to the inner threads of the bolt holes and the engine block along the area indicated by the broken line

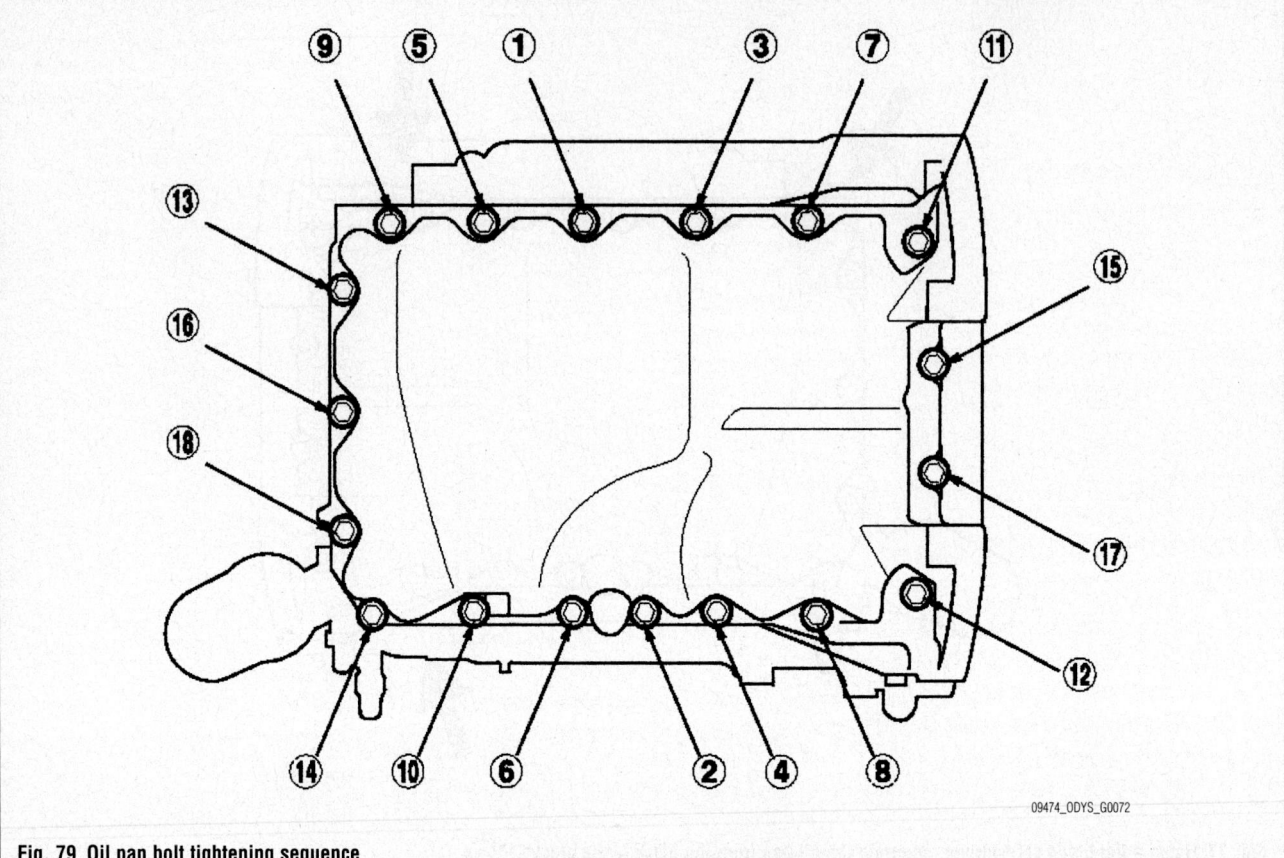

Fig. 79 Oil pan bolt tightening sequence

8. Using a flat blade screwdriver, separate the oil pan from the block in the places shown.

9. Remove the oil pan.

To install:

10. Remove all of the old liquid gasket from the oil pan mating surfaces, bolts, and bolt holes.

11. Clean and dry the oil pan mating surfaces.

12. Apply liquid gasket, P/N 08717-0004, 08718-0001, 08718-0002, 08718-0003, or 08718-009, evenly to the oil pan mating surface of the engine block.

➡**Do not install components if too much time has passed after applying the liquid gasket (for P/N 08718-0002, no more than 4 minutes, for all others, no more than 5 minutes). Instead, remove the old residue and reapply liquid gasket.**

13. Install the oil pan on the engine block.

14. Tighten the bolts in two or three steps. In the final step, tighten all bolts, in sequence, to 12 Nm (104 inch lbs.).

➡**After assembly, wait at least 30 minutes before filling the engine with oil.**

OIL PUMP

REMOVAL & INSTALLATION

See Figures 80 through 82.

1. Drain the engine oil.

2. Turn the crankshaft so that the No. 1 piston is at top dead center:

3. Remove the timing belt:

4. Remove the idler pulley.

5. Remove the crankshaft position (CKP) sensor A/B.

6. Attach the chain hoist to the engine hanger on the power steering (P/S) pump bracket.

7. Remove the jack from under the oil pan.

8. Remove the rocker arm oil control solenoid (VTEC solenoid valve)/oil filter assembly (J35A6 engine) or oil filter base/oil filter assembly (J35A7 engine).

9. Remove the oil pan.

10. Remove the oil screen.

11. Remove the mounting bolts and the oil pump assembly.

To install:

12. Remove the old oil seal from the oil pump.

13. Gently tap in the new oil seal until the oil seal driver bottoms on the pump.

14. Remove all of the old liquid gasket from the oil pump mating surfaces, bolts, and bolt holes.

15. Clean and dry the oil pump mating surfaces.

16. Inspect both rotors and pump housing for scoring or other damage. Replace the parts, if necessary.

17. Apply liquid thread lock to the pump housing screws, then install the oil pump cover.

18. Check that the oil pump turns freely.

19. Apply liquid gasket, P/N 08717-0004, 08718-0001, 08718-0002, 08718-0003, or 08718-0009, evenly to the engine block mating surface of the oil pump.

➡**Do not install components if too much time has passed after applying the liquid gasket (for P/N 08718-0002, no more than 4 minutes, for all others, no more than 5 minutes). Instead, remove the old residue and reapply liquid gasket.**

20. Grease the lip of the oil seal, and apply oil to the new O-ring (A).

21. Install the dowel pins (B), then align the inner rotor with the crankshaft, and install the oil pump (C).

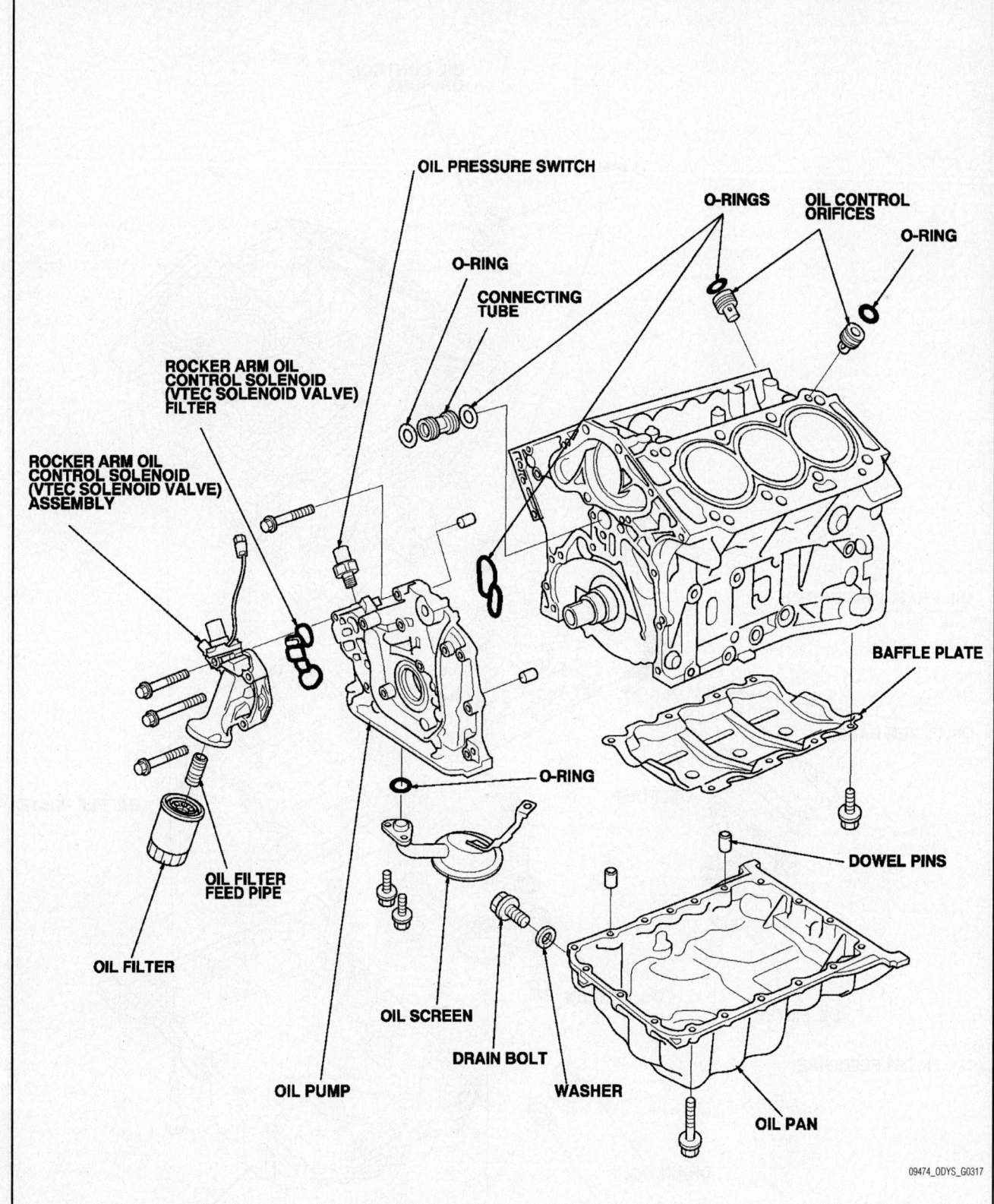

OIL PRESSURE SWITCH

O-RINGS

OIL CONTROL ORIFICES

O-RING

O-RING

CONNECTING TUBE

ROCKER ARM OIL CONTROL SOLENOID (VTEC SOLENOID VALVE) FILTER

ROCKER ARM OIL CONTROL SOLENOID (VTEC SOLENOID VALVE) ASSEMBLY

BAFFLE PLATE

O-RING

DOWEL PINS

OIL FILTER FEED PIPE

OIL FILTER

OIL SCREEN

DRAIN BOLT

OIL PUMP

WASHER

OIL PAN

09474_ODYS_G0317

Fig. 80 Oil pump and related parts—J35A6 engine

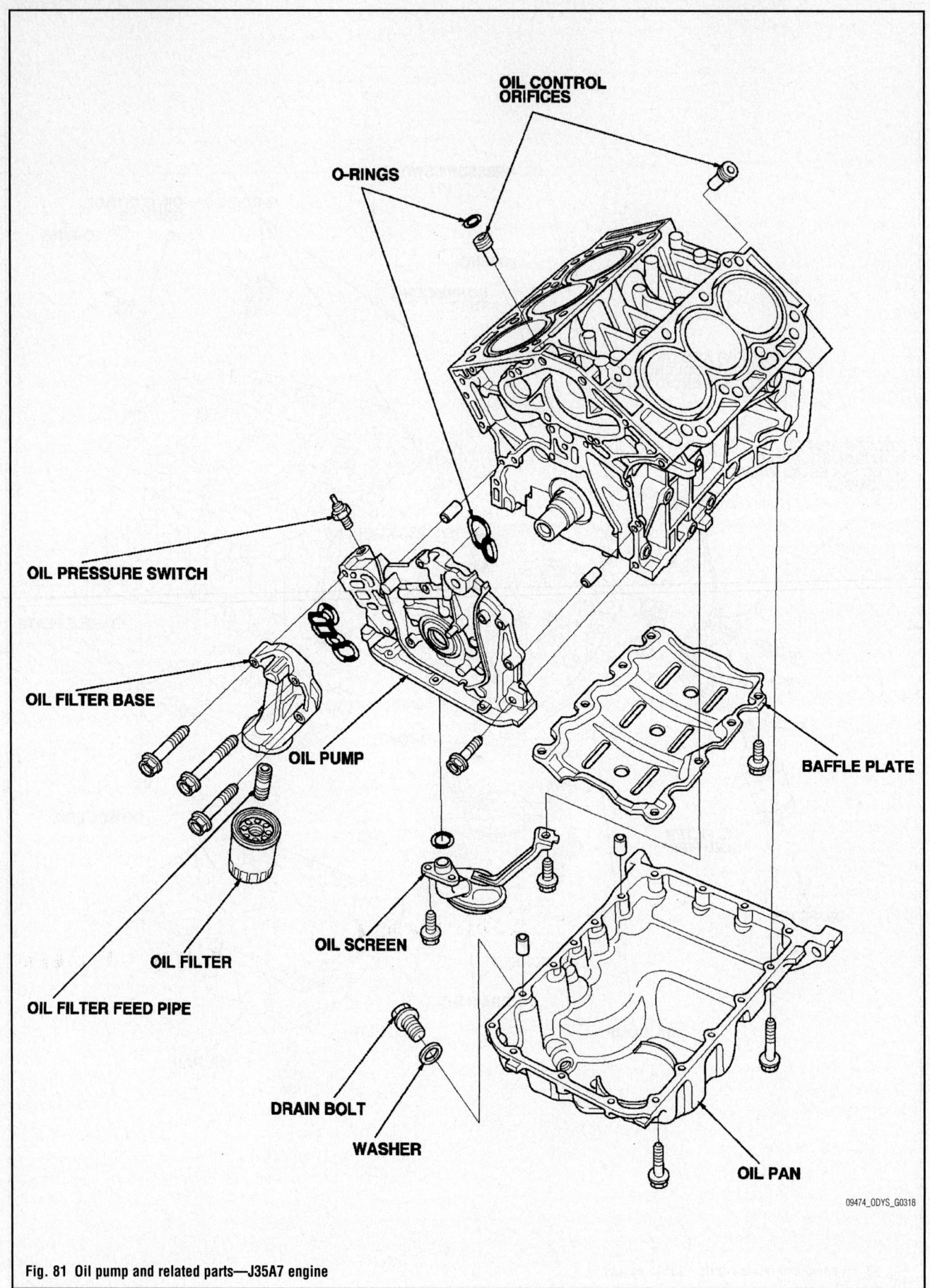

OIL CONTROL ORIFICES

O-RINGS

OIL PRESSURE SWITCH

OIL FILTER BASE

OIL PUMP

BAFFLE PLATE

OIL FILTER

OIL SCREEN

OIL FILTER FEED PIPE

DRAIN BOLT

WASHER

OIL PAN

09474_ODYS_G0318

Fig. 81 Oil pump and related parts—J35A7 engine

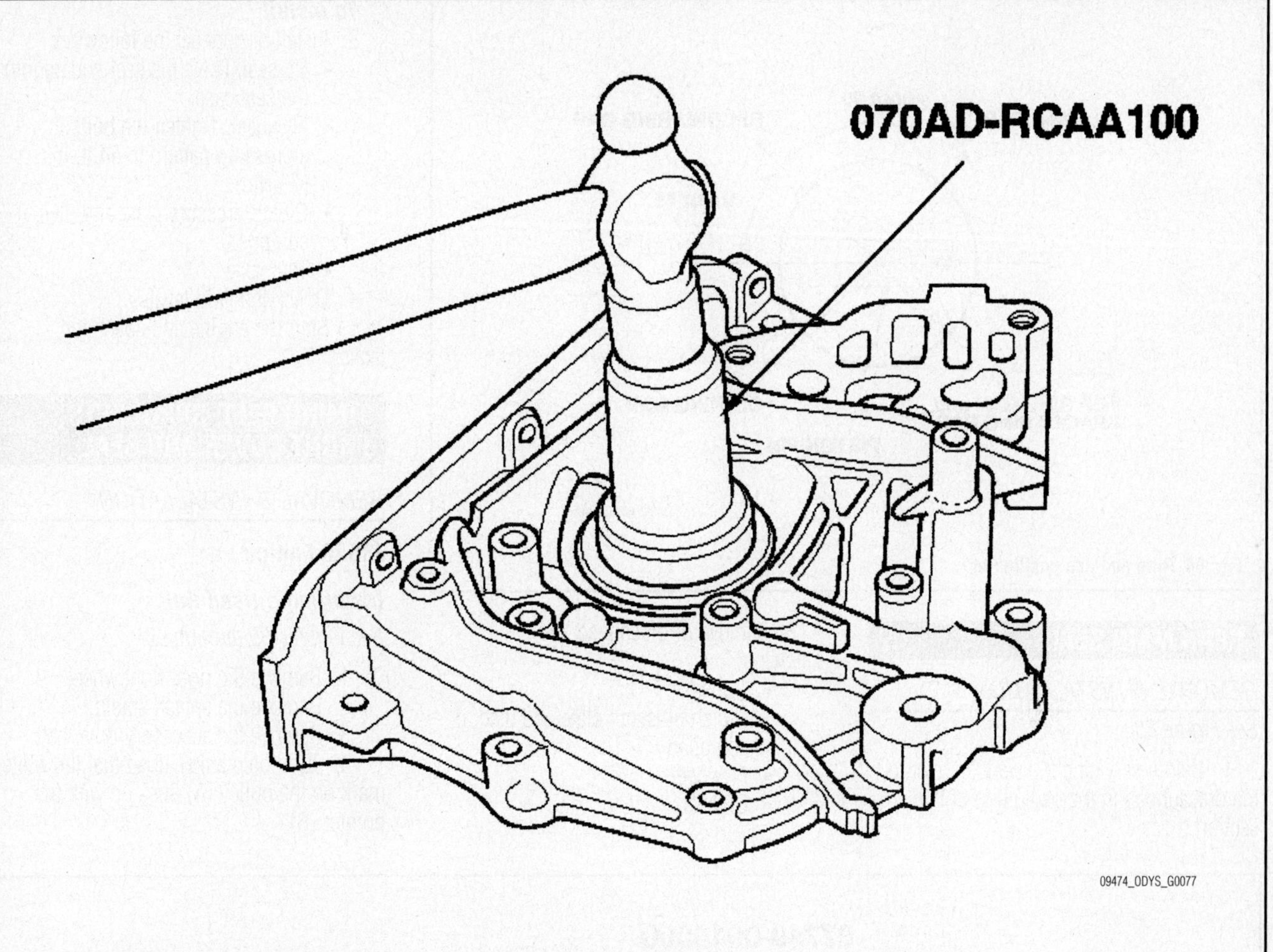

070AD-RCAA100
09474_ODYS_G0077

Fig. 82 Gently tap in the new oil seal until the oil seal driver bottoms on the pump

Clean the excess grease off the crankshaft, and check the seal for distortion.

22. Install the oil screen with new O-ring (E).
23. Install the rocker arm oil control solenoid (VTEC solenoid valve)/oil filter assembly, with a new rocker arm oil control solenoid filter (VTEC solenoid valve filter) (J35A6 engine), or oil filter base/oil filter assembly, with a new O-ring.
24. Install the oil pan.
25. Install the crankshaft position (CKP) sensor A/B.
26. Install the idler pulley.
27. Install the timing belt:
28. Remove the engine hanger.
29. After assembly, wait at least 30 minutes before filling the engine with oil.

PISTON AND RING

POSITIONING

See Figures 83 and 84.

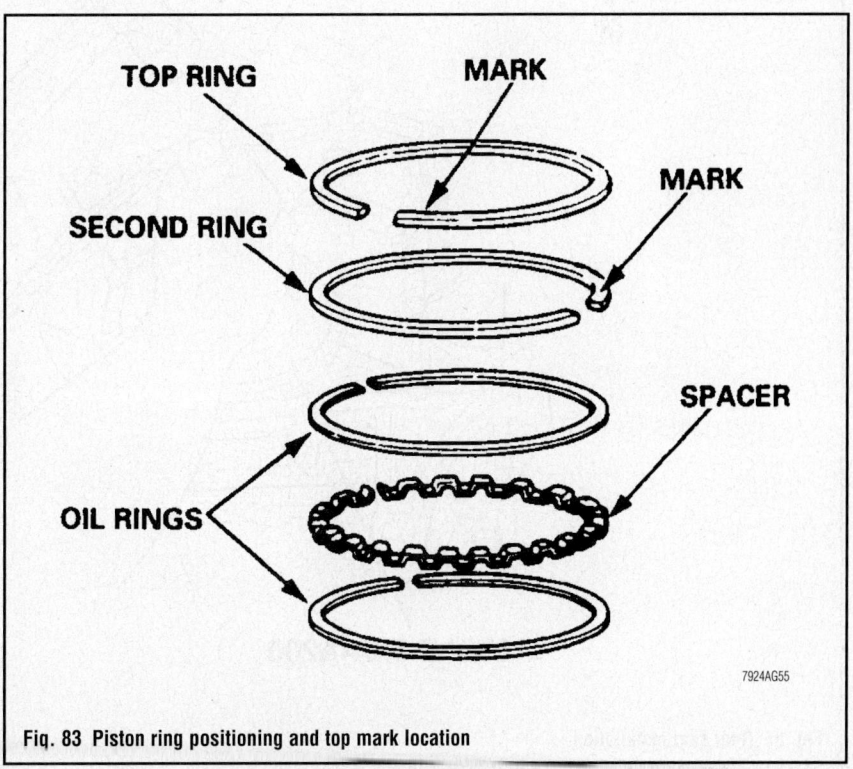

TOP RING
MARK
MARK
SECOND RING
SPACER
OIL RINGS
7924AG55

Fig. 83 Piston ring positioning and top mark location

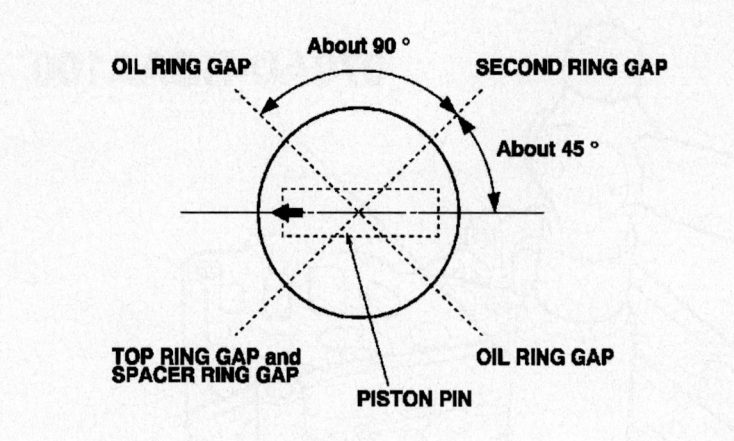

Fig. 84 Ring end gap positioning

REAR MAIN SEAL

REMOVAL & INSTALLATION

See Figure 85.

1. Before servicing the vehicle, refer to the precautions in the beginning of this section.

2. Remove or disconnect the following:
   • Transaxle
   • Clutch pressure plate and disc, if equipped
   • Flywheel
   • Oil seal

To install:

3. Install or connect the following:
   • Oil seal. Drive the seal square into the seal case.
   • Flywheel. Tighten the bolts in a crossing pattern to 54 ft. lbs. (73 Nm).
   • Clutch pressure plate and disc, if equipped
   • Transaxle

4. Check the fluid levels.

5. Start the engine and check for leaks.

TIMING BELT, SPROCKETS, FRONT COVER AND SEAL

REMOVAL & INSTALLATION

J35A6 Engine

Installing a Used Belt

See Figures 86 through 92.

1. Remove the right front wheel.
2. Remove the splash shield.
3. Remove the accessory drive belt.
4. Turn the crankshaft so that the white mark on the pulley (A) lines up with the pointer (B).

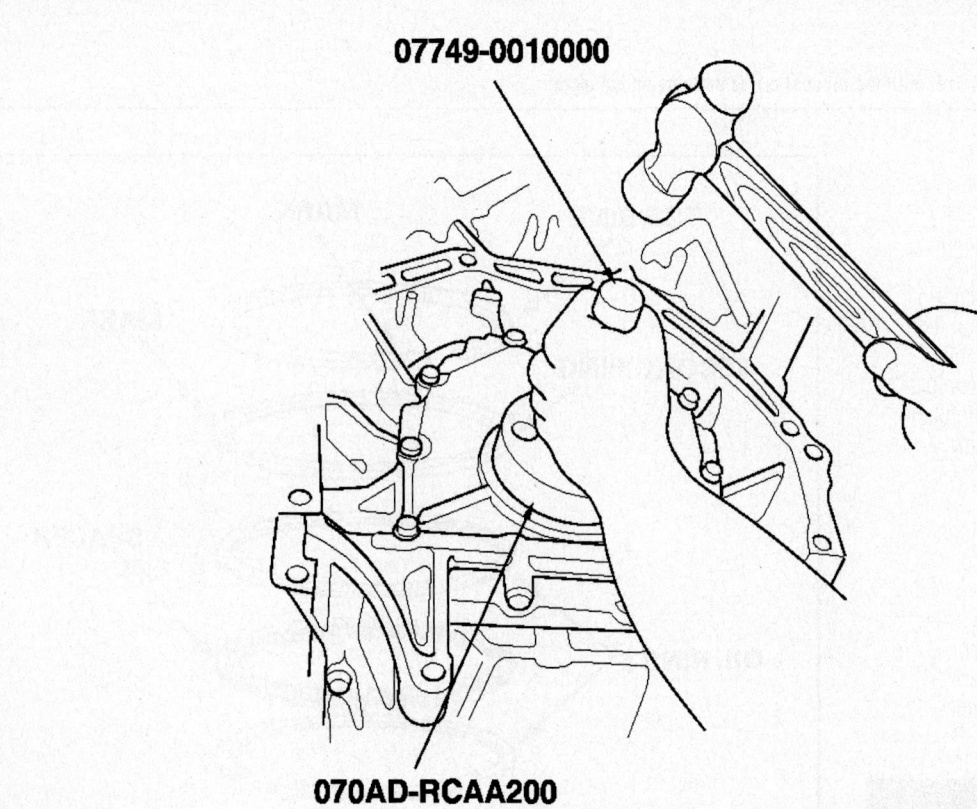

Fig. 85 Rear seal installation

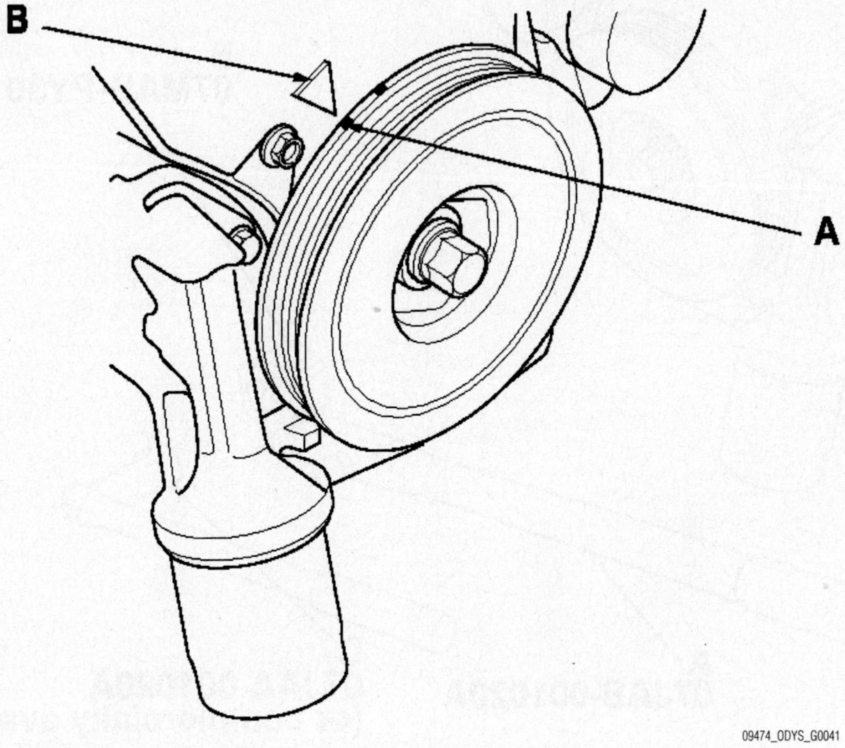
B
A
09474_ODYS_G0041

Fig. 86 Crankshaft timing marks aligned—J35A6 engine

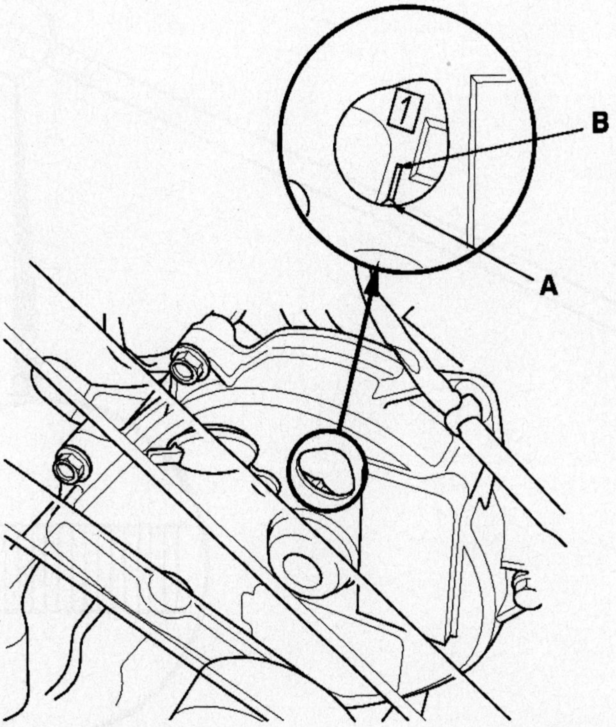
1
B
A
09474_ODYS_G0042

Fig. 87 Front camshaft timing marks aligned—J35A6 engine

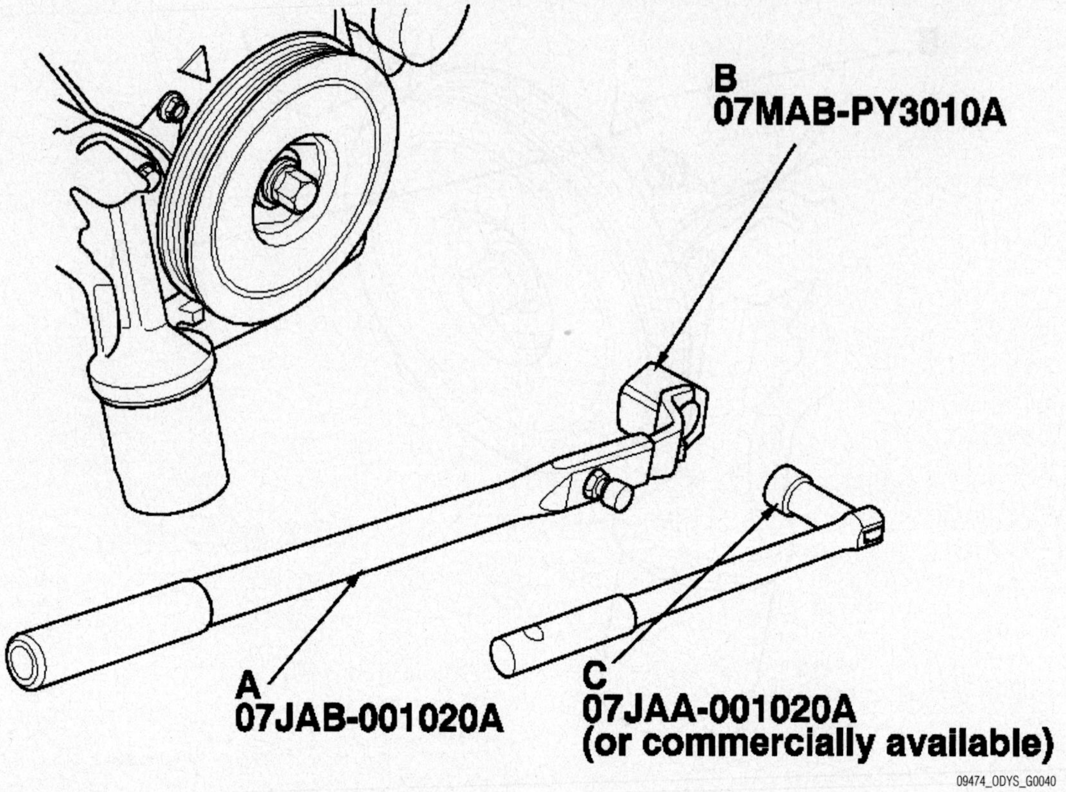

B
07MAB-PY3010A

A
07JAB-001020A

C
07JAA-001020A
(or commercially available)

09474_ODYS_G0040

Fig. 88 Damper bolt removal tools

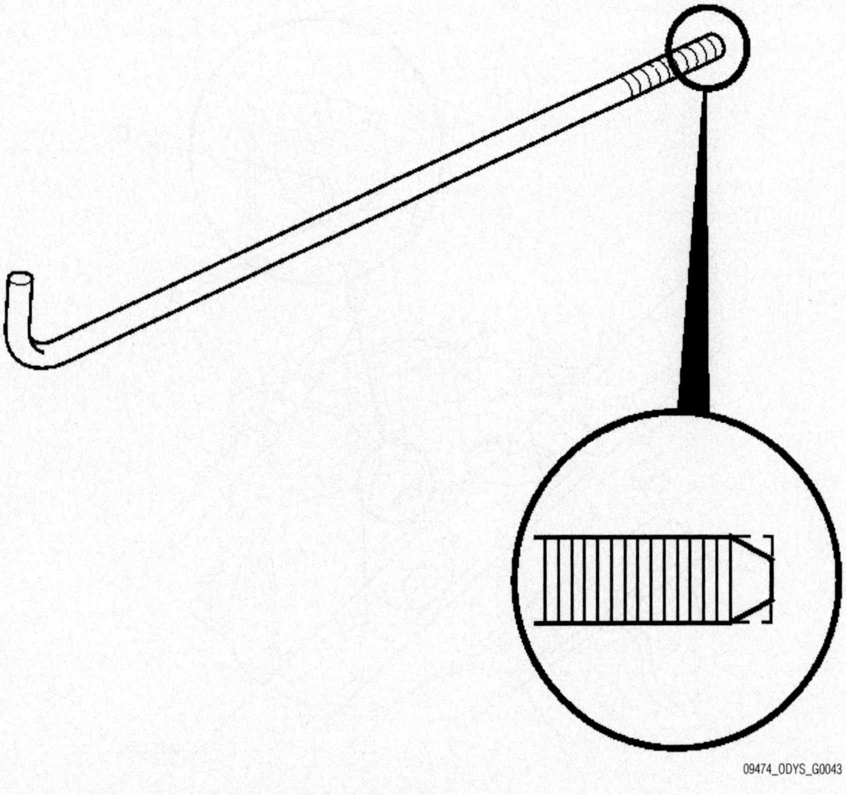

09474_ODYS_G0043

Fig. 89 Remove one of the battery hold-down clamp bolts and grind the end as shown

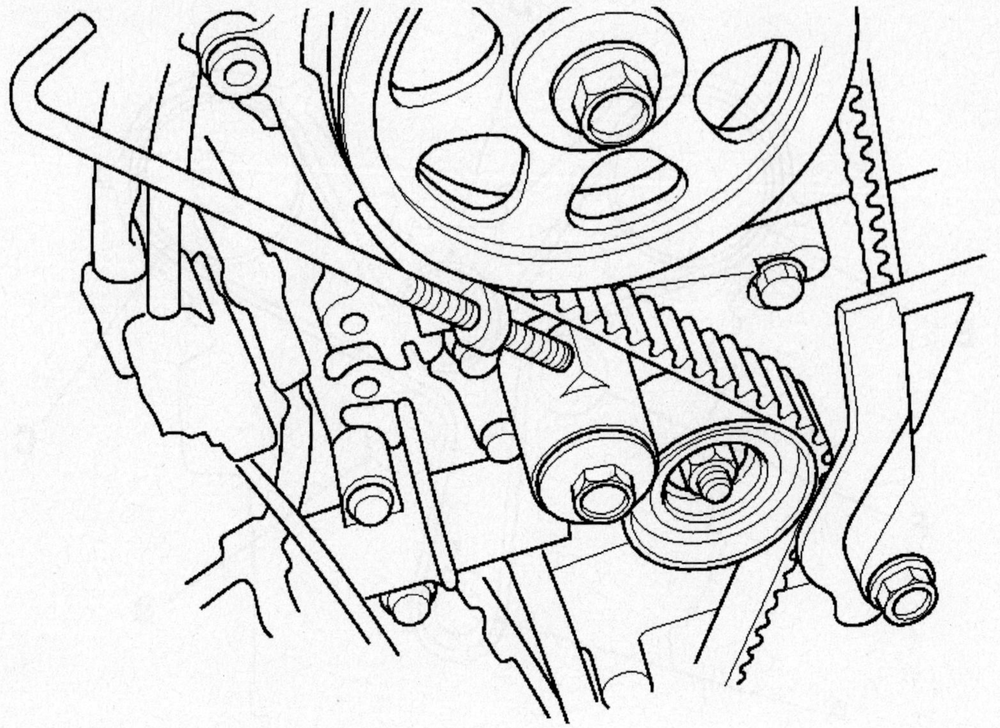

09474_ODYS_G0044

Fig. 90 Battery clamp bolt screwed into position—J35A6 engine

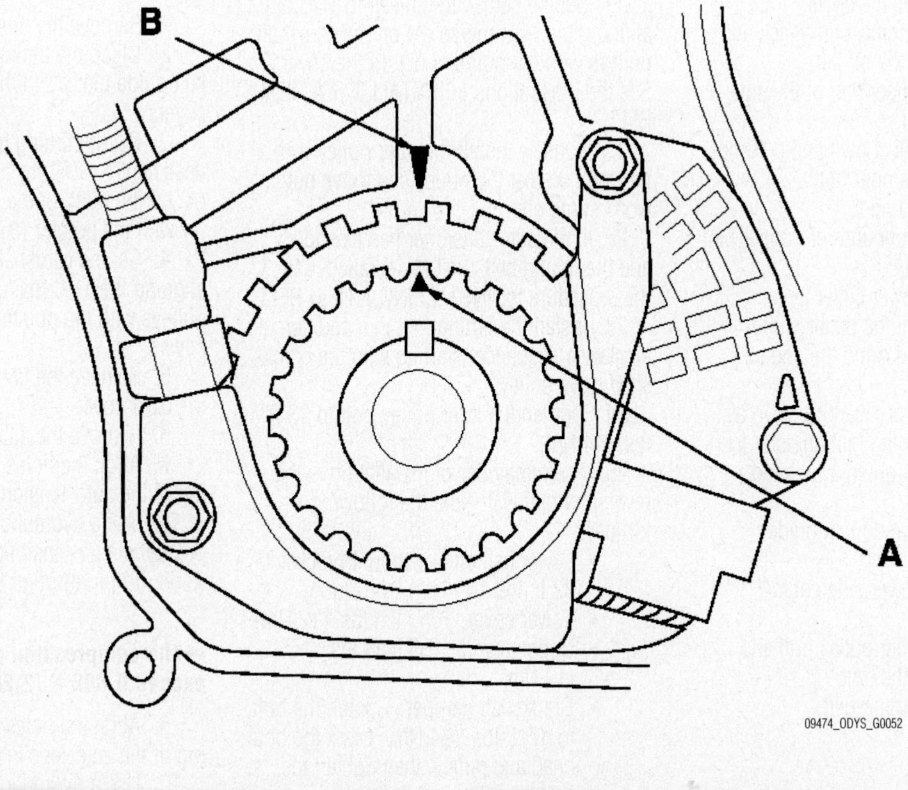

B

A

09474_ODYS_G0052

Fig. 91 Set the crankshaft sprocket to TDC by aligning tho TDC mark (A) with the pointer (B) on the oil pump—J35A6 engine

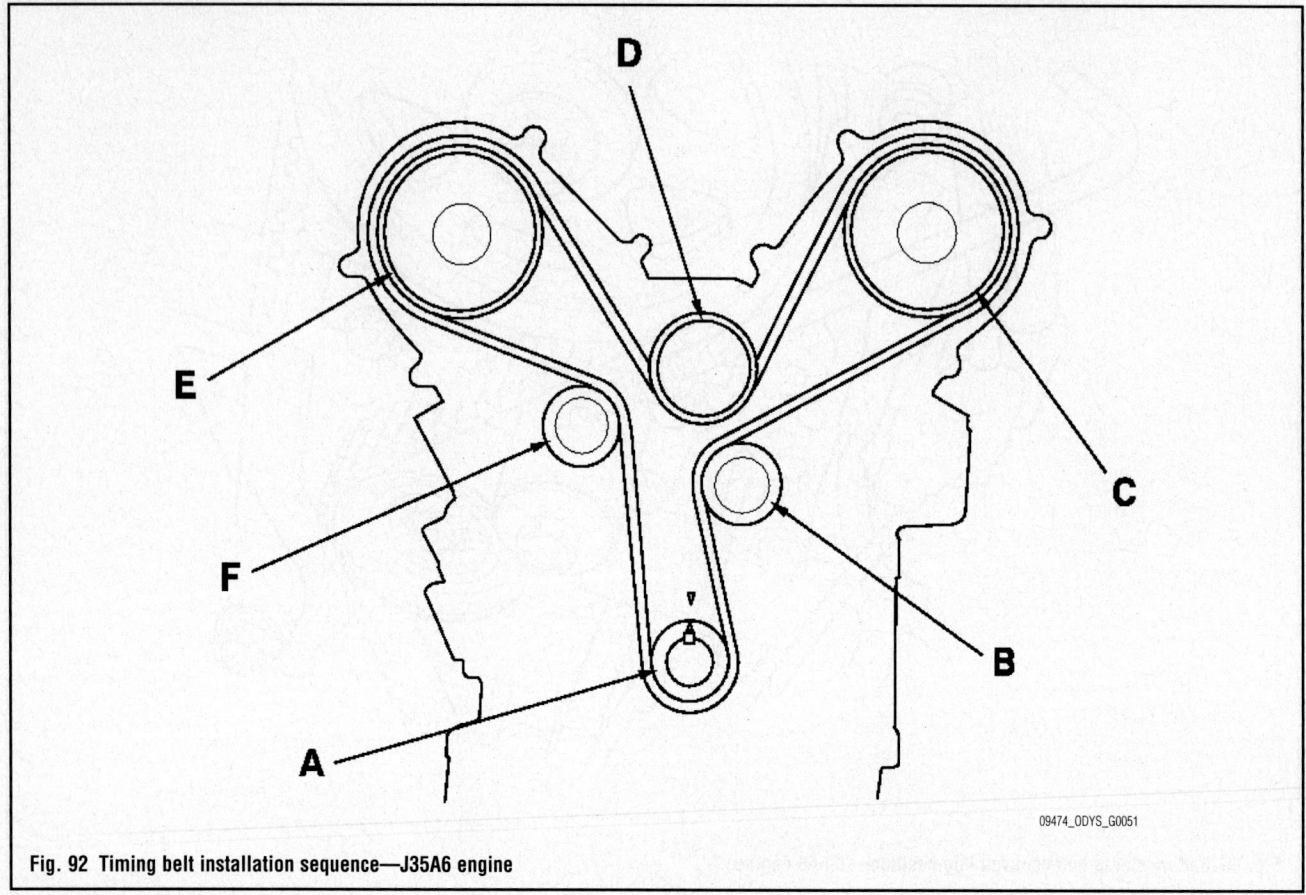

Fig. 92 Timing belt installation sequence—J35A6 engine

5. Check that the TDC timing mark (A) on the front camshaft sprocket is aligned with the pointer (B) on the cover.

6. Support the engine with a jack and a block of wood under the oil pan.

7. Remove the upper half of the side engine mount bracket.

8. Using the tools shown and a 19mm socket, remove the damper bolt.

9. Remove the damper.

10. Remove the front and rear upper belt covers.

11. Remove the lower cover.

12. Remove one of the battery hold-down clamp bolts and grind the end as shown.

13. Screw the battery clamp bolt in as shown to hold the timing belt adjuster in its current position. Tighten it by hand. Do not use a wrench.

14. Remove the timing belt guide plate.

15. Remove the lower side engine mount bracket.

16. Remove the idler pulley bolt and idler pulley. Discard the bolt.

17. Remove the timing belt.

To install:

18. Clean all parts.

19. Set the crankshaft sprocket to TDC

by aligning the TDC mark (A) with the pointer (B) on the oil pump.

20. Set the camshaft pulleys to TDC by aligning the TDC marks (A) on the camshaft pulleys with the pointers (B) on the covers. See the illustrations in INSTALLING A NEW BELT.

21. Loosely install the idler pulley with a new bolt so that the pulley can move but won't come off.

22. If the auto-tensioner has extended, and the timing belt can't be installed, see the procedure for INSTALLING A NEW BELT.

23. Install the timing belt in a counter-clockwise sequence, starting with the crank-shaft drive pulley.

24. Tighten the idler pulley bolt to 33 ft. lbs. (44 Nm).

25. The remainder of installation is the reverse of removal. Note the following torques:

- Lower half engine mount bracket: 33 ft. lbs. (44 Nm)
- Lower cover: 105 inch lbs. (12 Nm)
- Upper covers: 105 inch lbs. (12 Nm)
- Crankshaft damper: Tighten the bolt to 47 ft. lbs. (64 Nm). Mark the bolt head and pulley, then tighten the bolt an additional 60°.

Installing a New Belt

See Figures 93 through 107.

1. Remove the timing belt.

2. Clean the timing belt pulleys, timing belt guide plate, and the upper and lower covers.

3. Set the timing belt drive pulley to top dead center (TDC) by aligning the TDC mark (A) on the tooth of the timing belt drive pulley with the pointer (B) on the oil pump.

4. Set the camshaft pulleys to TDC by aligning the TDC marks (A) on the camshaft pulleys with the pointers (B) on the back covers.

5. Remove the battery clamp bolt from the back cover.

6. Remove the auto-tensioner.

7. Align the holes on the rod and housing of the auto-tensioner.

8. Use a hydraulic press to slowly compress the auto-tensioner. Insert a 2.0 mm (0.08 in.) pin through the housing and the rod.

➡**The compression pressure should not exceed 9,800 N (2,200 lbs.).**

9. Align the holes on the rod and housing of the auto-tensioner.

10. Use a hydraulic press to slowly compress the auto-tensioner. Insert a 2.0 mm

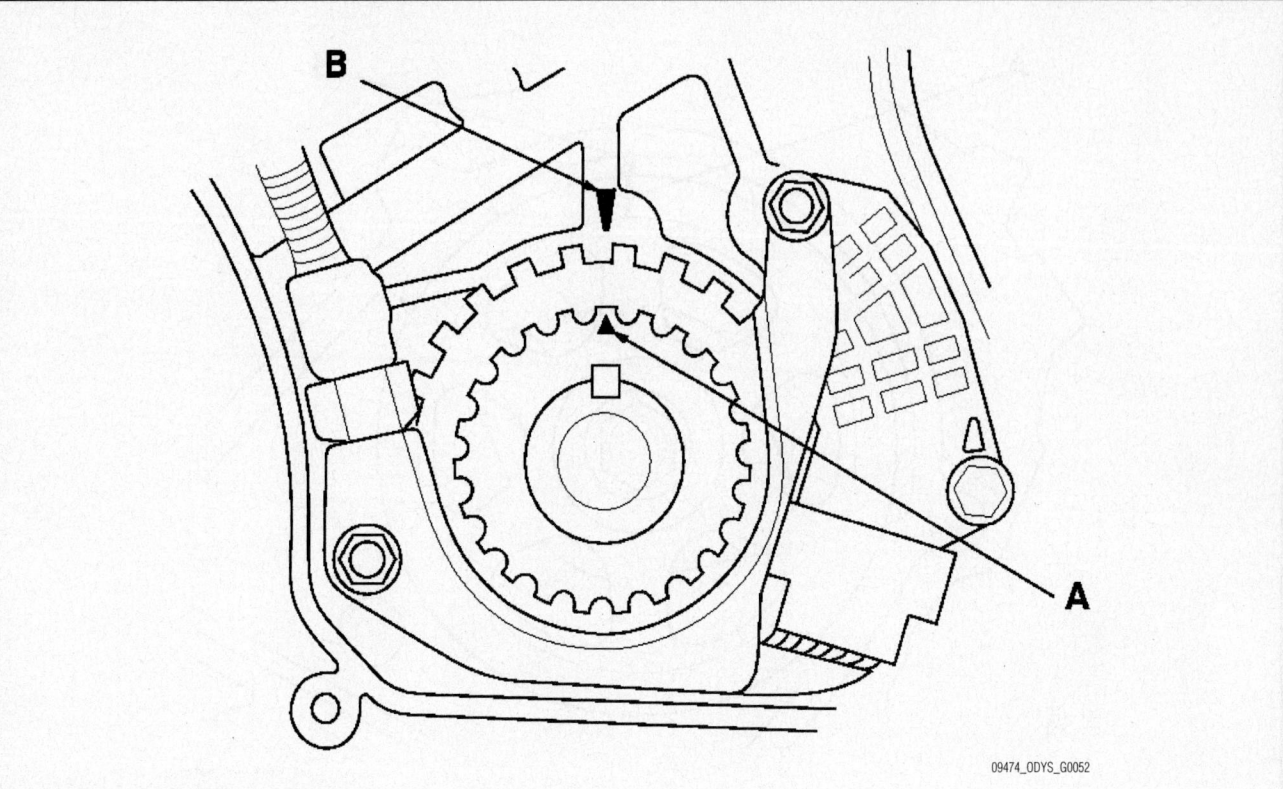

09474_ODYS_G0052

Fig. 93 Set the timing belt drive pulley to top dead center (TDC) by aligning the TDC mark (A) on the tooth of the timing belt drive pulley with the pointer (B) on the oil pump—J35A6 engine

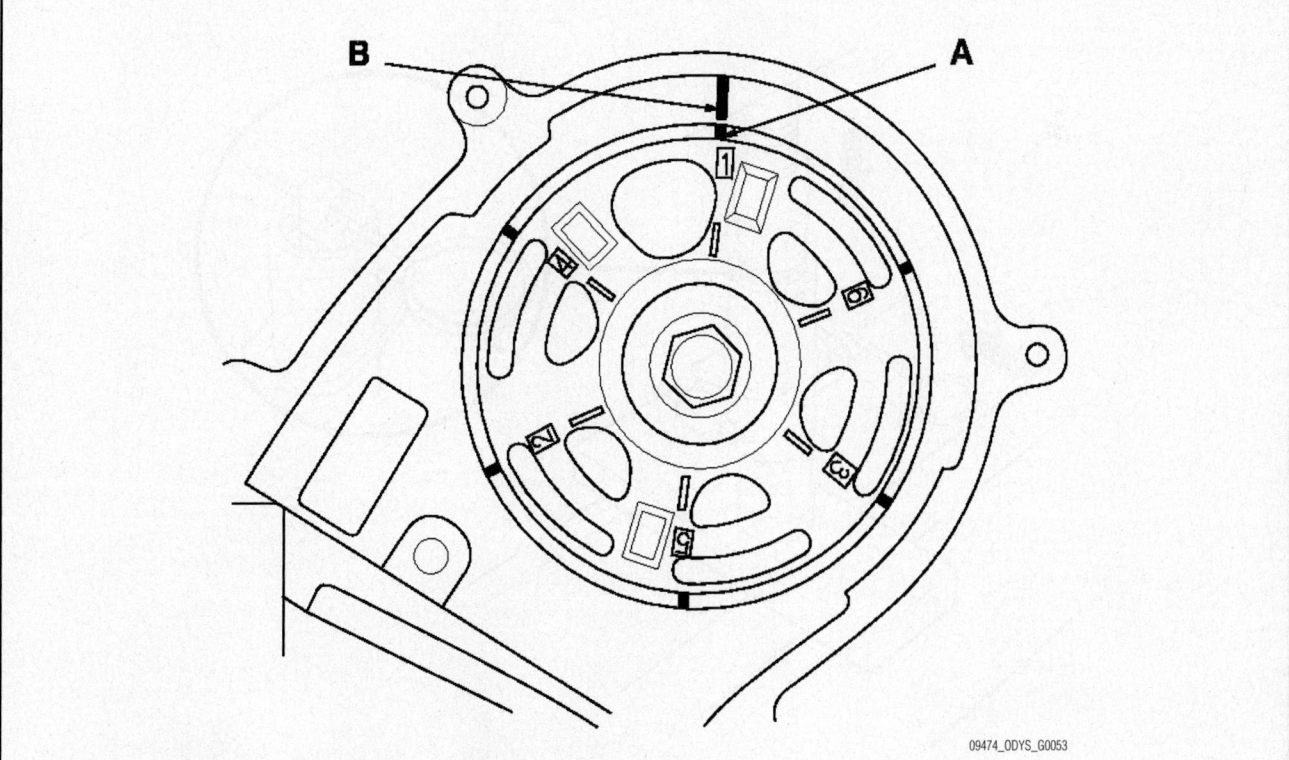

09474_ODYS_G0053

Fig. 94 Set the left camshaft pulley to TDC by aligning the TDC marks (A) on the camshaft pulleys with the pointers (B) on the back covers—J35A6 engine

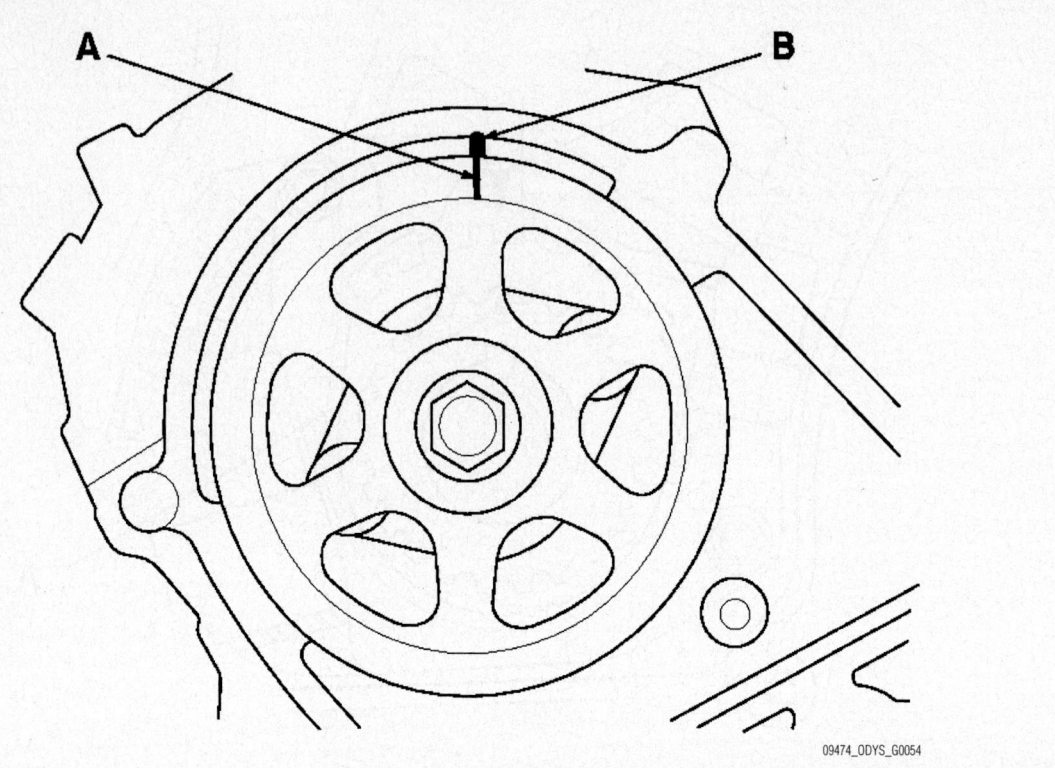

Fig. 95 Set the left camshaft pulley to TDC by aligning the TDC marks (A) on the camshaft pulleys with the pointers (B) on the back covers—J35A6 engine

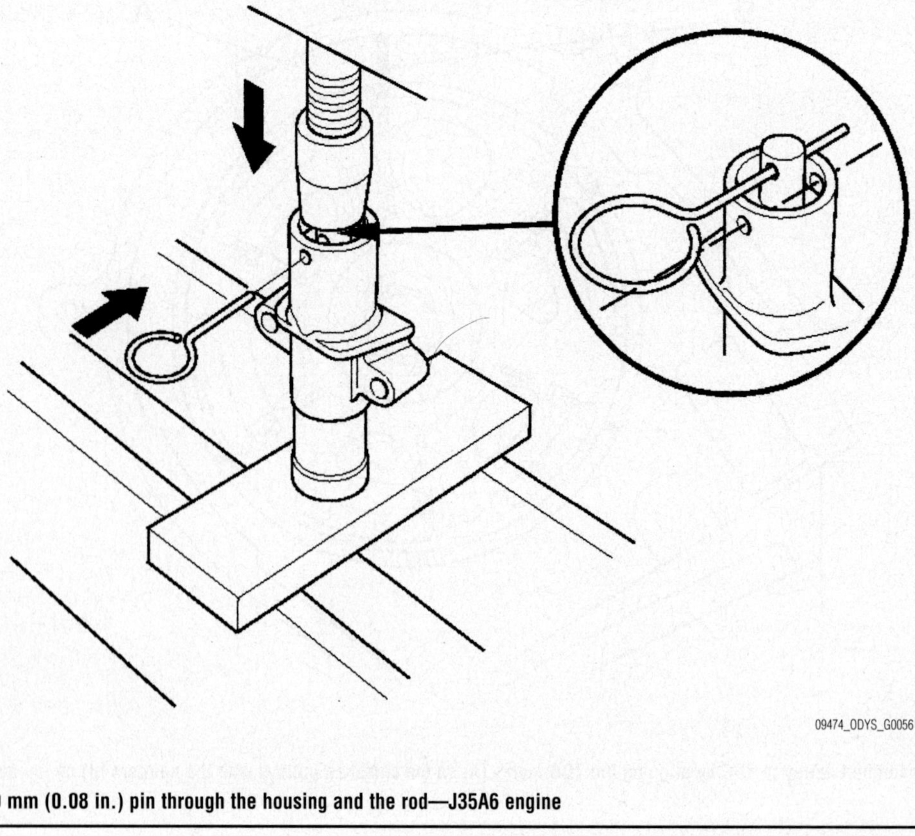

Fig. 96 Insert a 2.0 mm (0.08 in.) pin through the housing and the rod—J35A6 engine

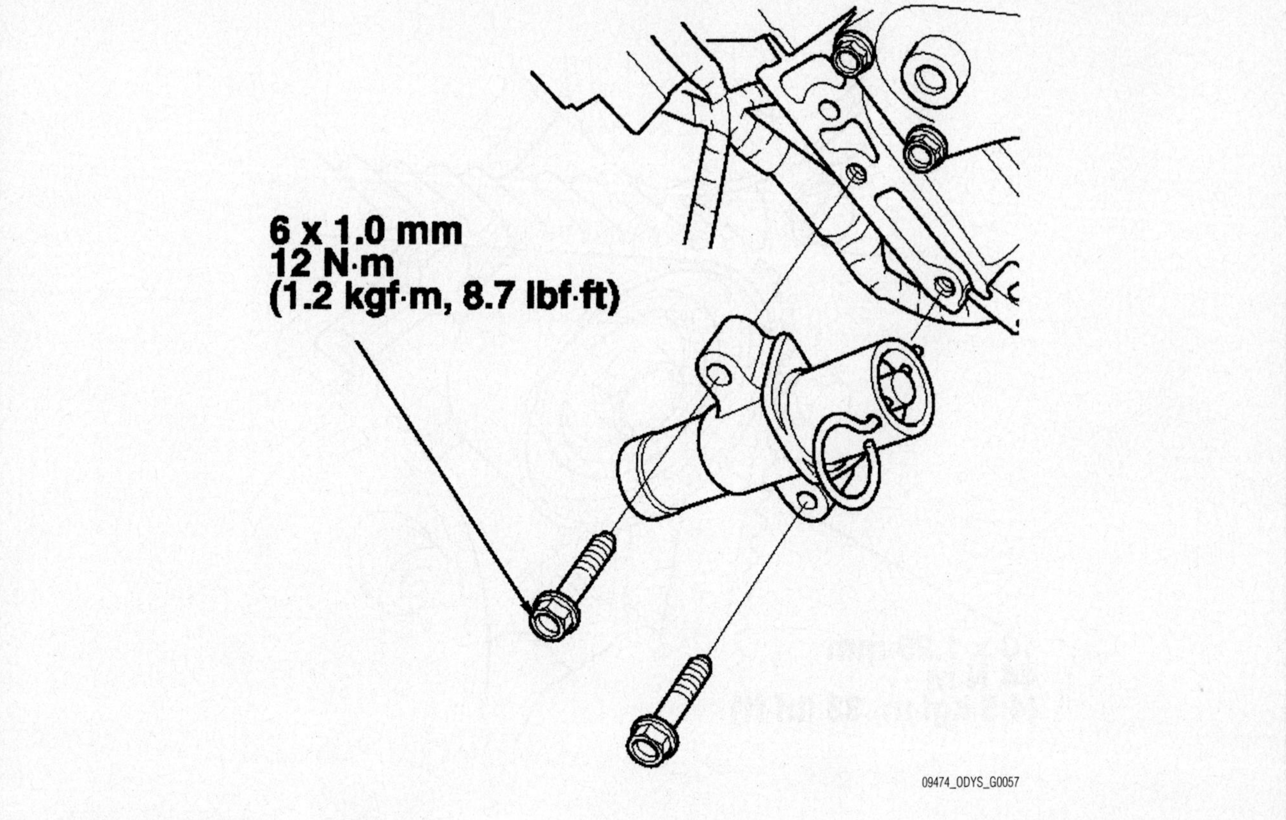

6 x 1.0 mm
12 N·m
(1.2 kgf·m, 8.7 lbf·ft)

09474_ODYS_G0057

Fig. 97 Install the auto-tensioner—J35A6 engine

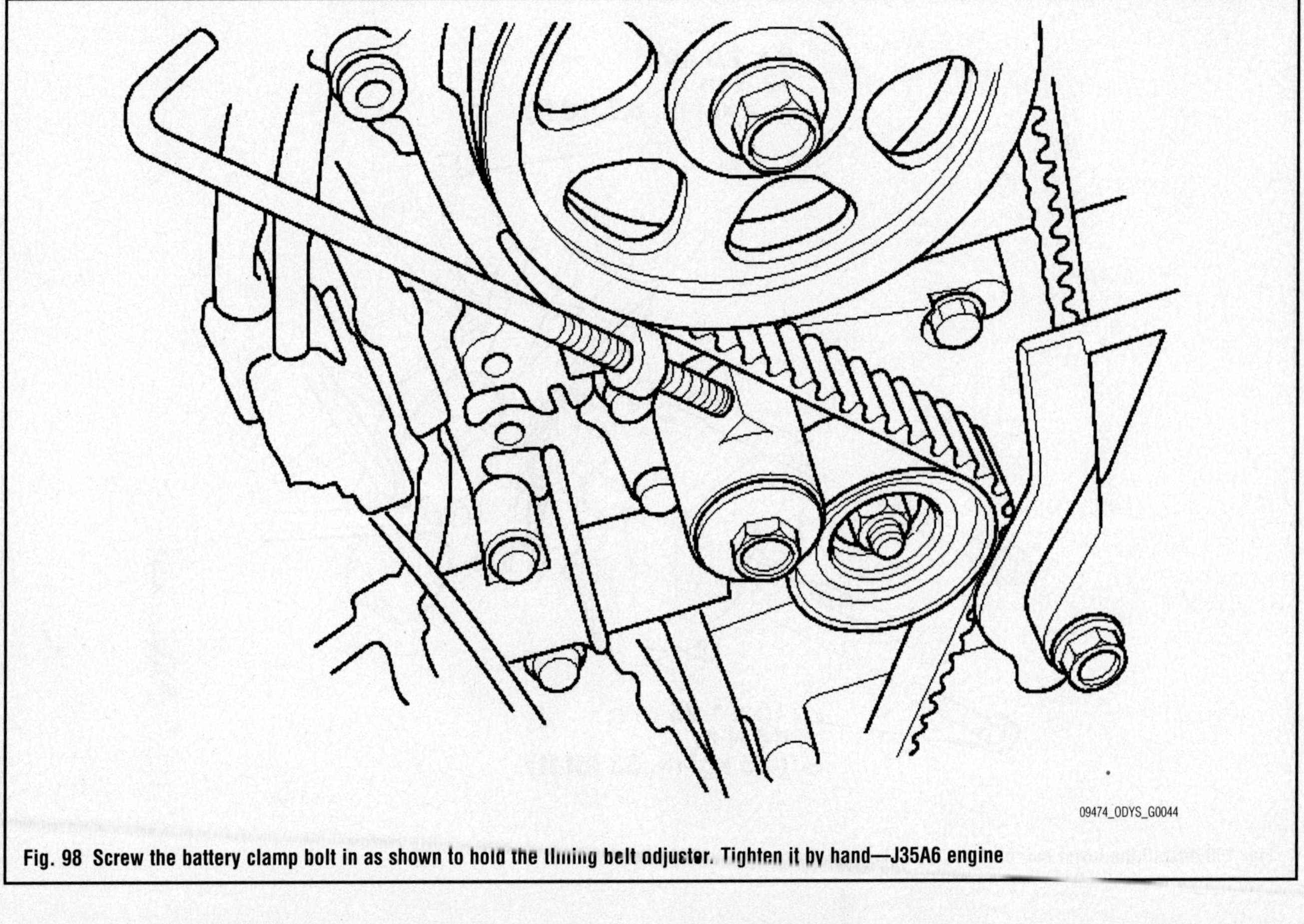

09474_ODYS_G0044

Fig. 98 Screw the battery clamp bolt in as shown to hold the timing belt adjuster. Tighten it by hand—J35A6 engine

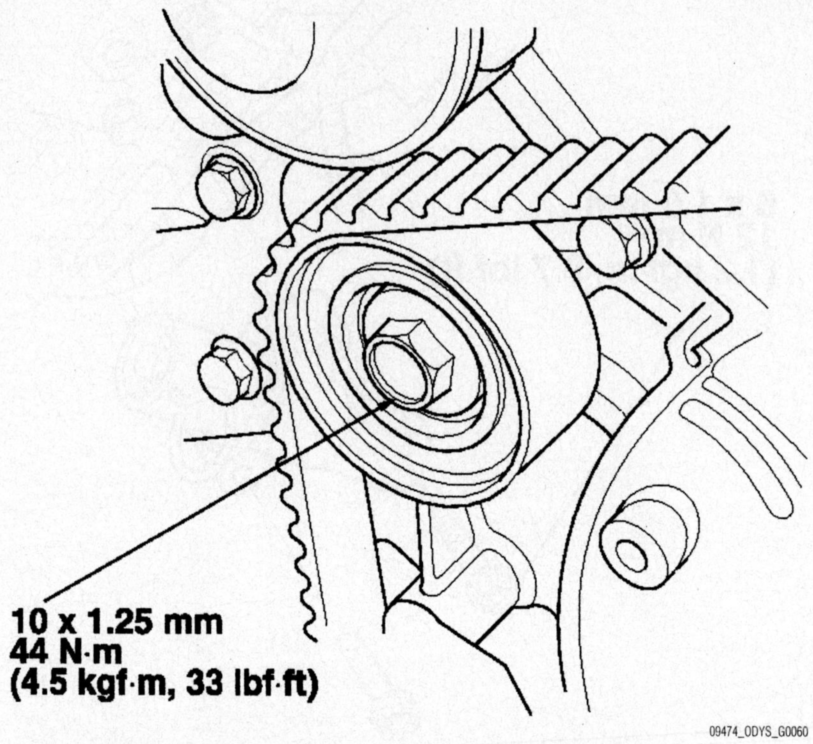

**10 x 1.25 mm
44 N·m
(4.5 kgf·m, 33 lbf·ft)**

09474_ODYS_G0060

Fig. 99 Tighten the idler pulley bolt—J35A6 engine

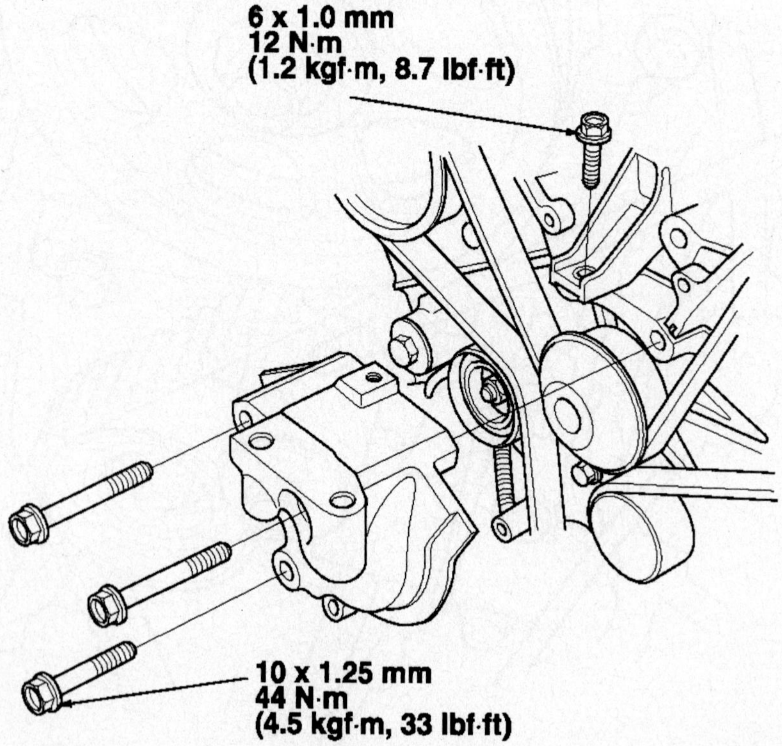

**6 x 1.0 mm
12 N·m
(1.2 kgf·m, 8.7 lbf·ft)**

**10 x 1.25 mm
44 N·m
(4.5 kgf·m, 33 lbf·ft)**

09474_ODYS_G0061

Fig. 100 Install the lower half of the side engine mount bracket—J35A6 engine

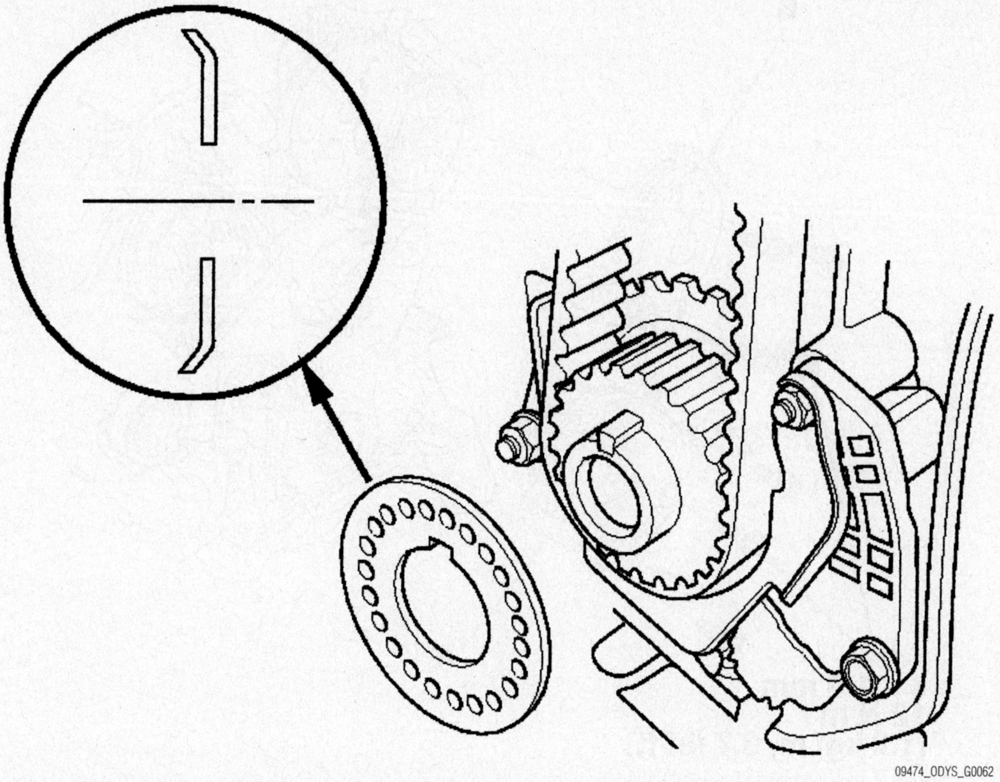

09474_ODYS_G0062

Fig. 101 Install the timing belt guide plate as shown—J35A6 engine

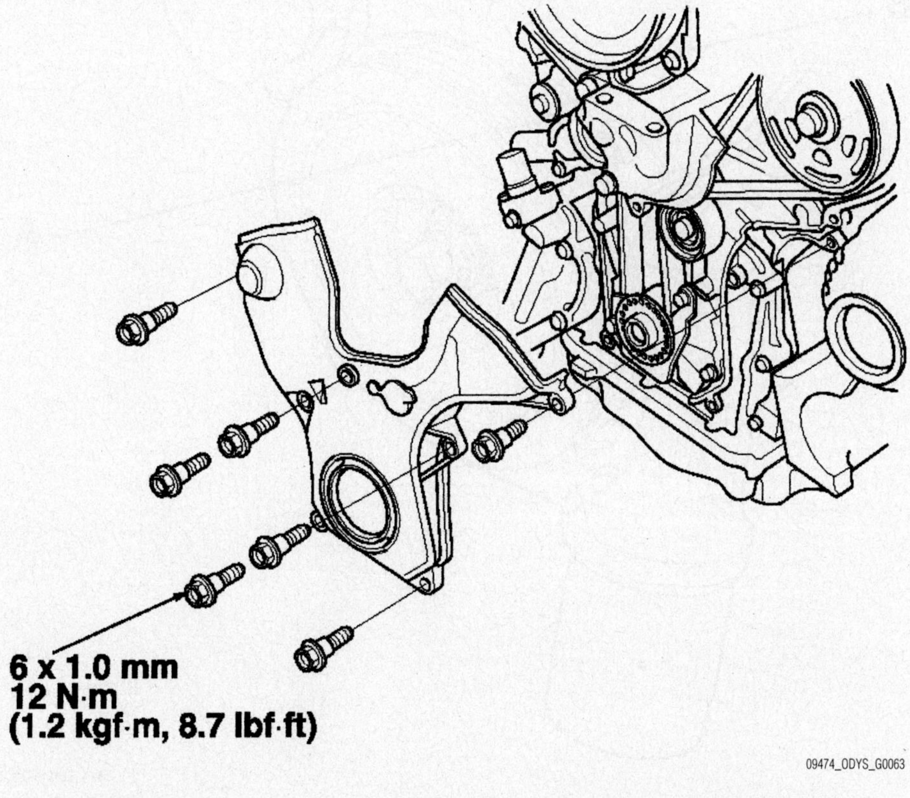

6 x 1.0 mm
12 N·m
(1.2 kgf·m, 8.7 lbf·ft)

09474_ODYS_G0063

Fig. 102 Install the lower cover—J35A6 engine

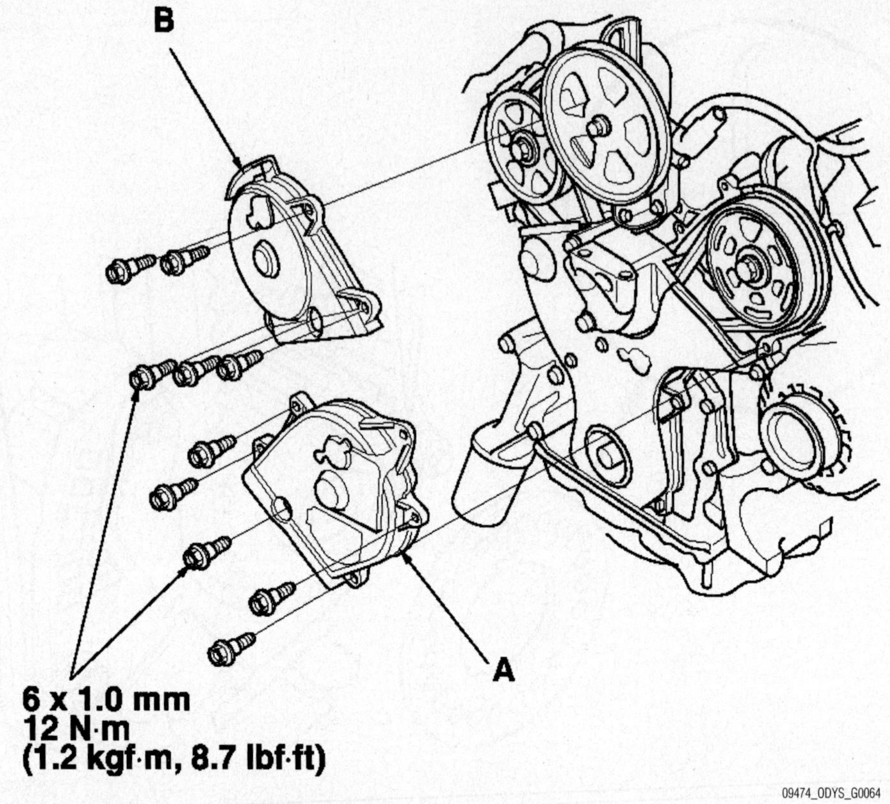

**6 x 1.0 mm
12 N·m
(1.2 kgf·m, 8.7 lbf·ft)**

09474_ODYS_G0064

Fig. 103 Install the front upper cover (A) and rear upper cover (B)—J35A6 engine

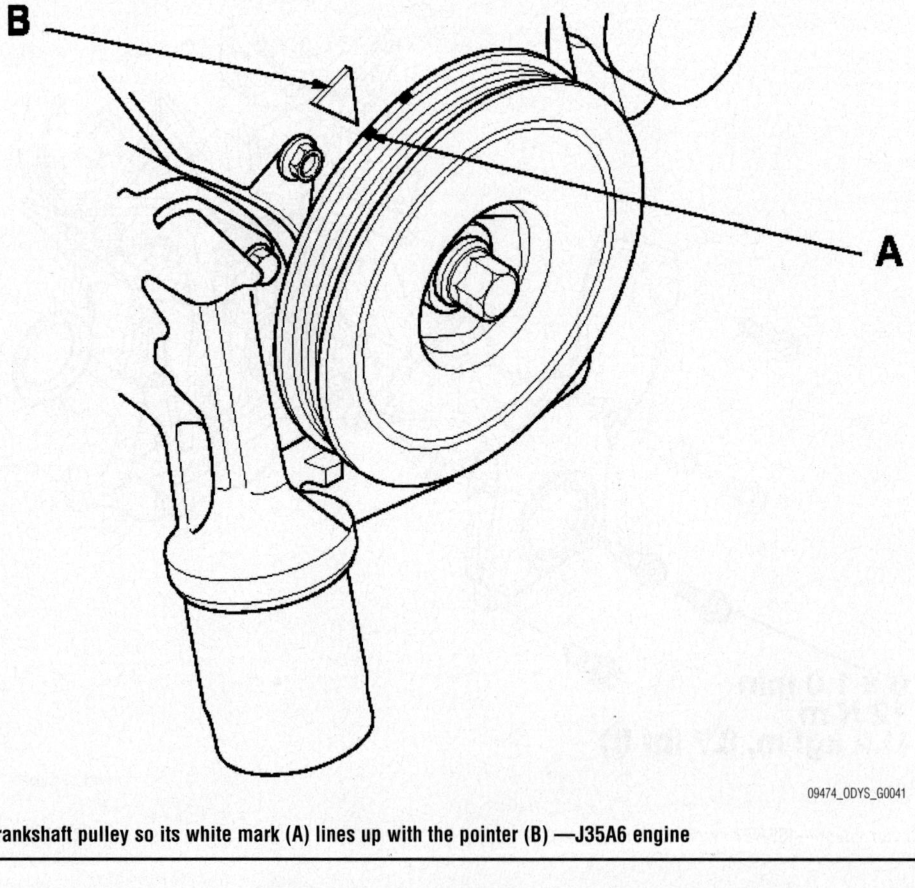

09474_ODYS_G0041

Fig. 104 Turn the crankshaft pulley so its white mark (A) lines up with the pointer (B) —J35A6 engine

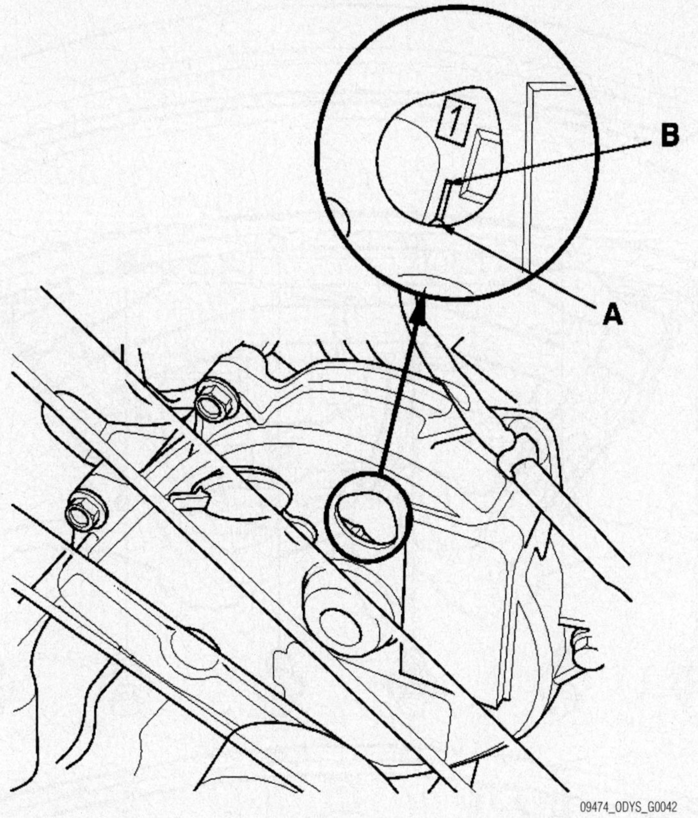

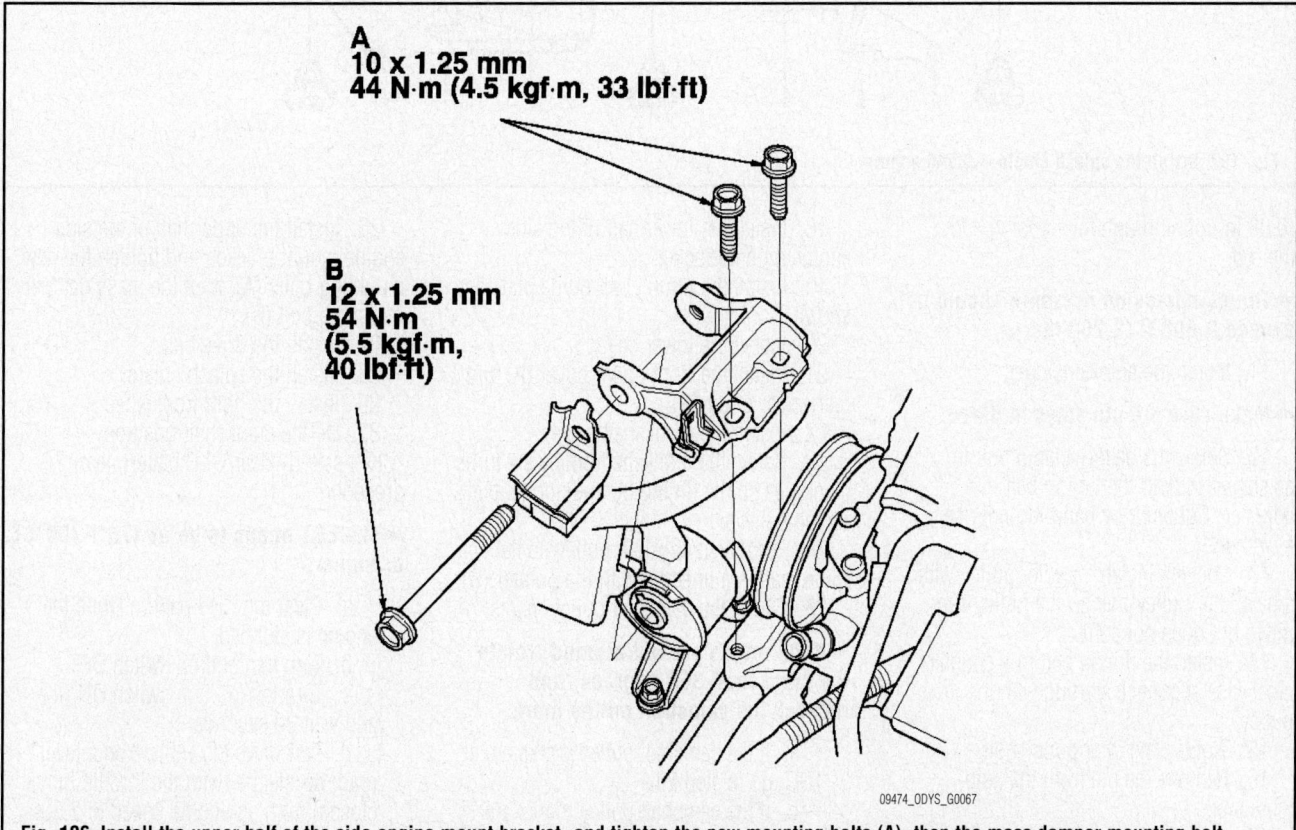

09474_ODYS_G0042

Fig. 105 Check the camshaft pulley marks—J35A6 engine

A
10 x 1.25 mm
44 N·m (4.5 kgf·m, 33 lbf·ft)

B
12 x 1.25 mm
54 N·m
(5.5 kgf·m,
40 lbf·ft)

09474_ODYS_G0067

Fig. 106 Install the upper half of the side engine mount bracket, and tighten the new mounting bolts (A), then the mass damper mounting bolt
(B) —J35A6 engine

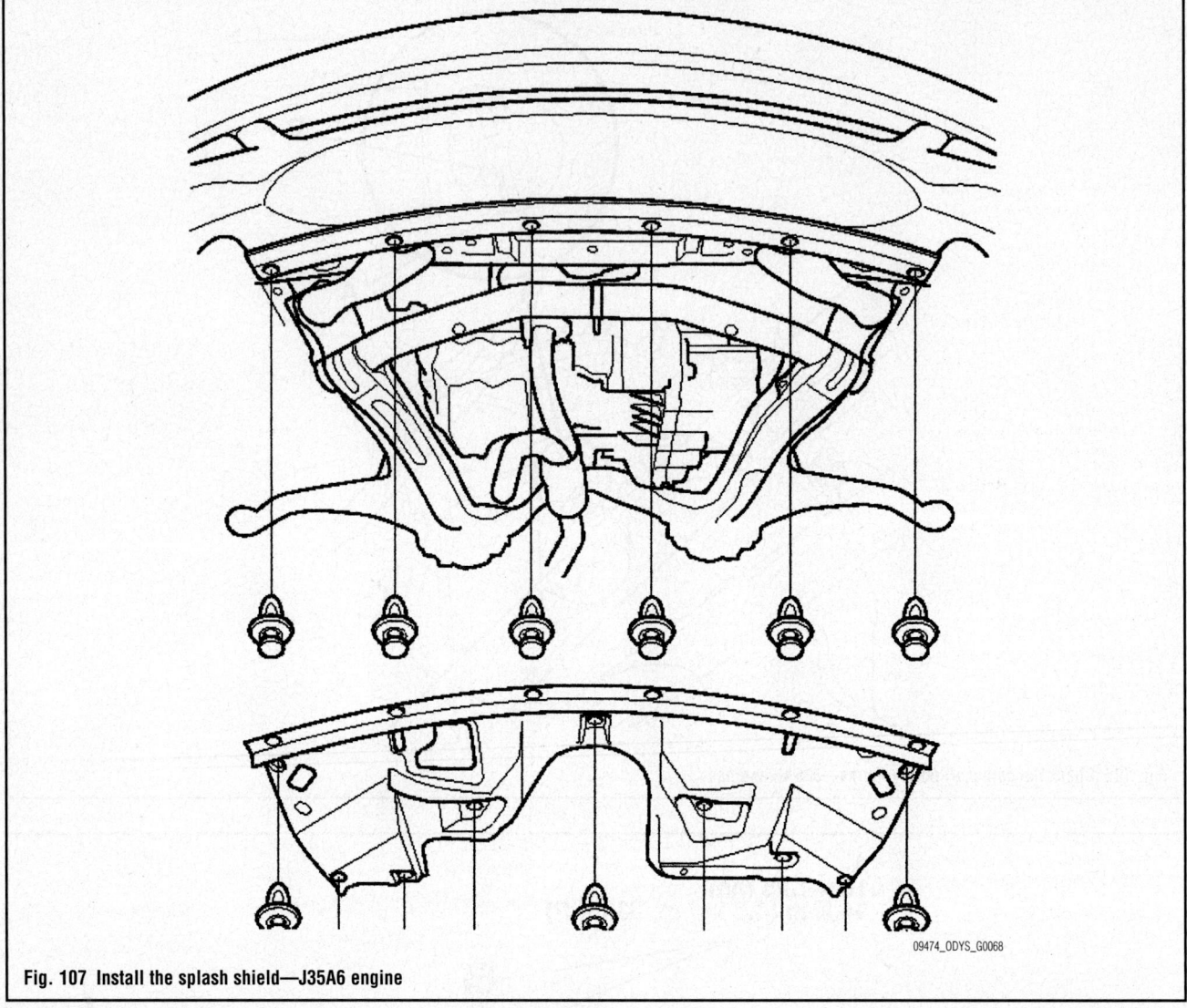

Fig. 107 Install the splash shield—J35A6 engine

(0.08 in.) pin through the housing and the rod.

➡**The compression pressure should not exceed 9,800 N (2,200 lbs.).**

11. Install the auto-tensioner.

➡**Make sure the pin stays in place.**

12. Screw the battery clamp bolt in as shown to hold the timing belt adjuster. Tighten it by hand; do not use a wrench.

13. Loosely install the idler pulley with a new idler pulley bolt so the pulley can move but does not come off.

14. Install the timing belt in a counter-clockwise sequence starting with the drive pulley.

15. Tighten the idler pulley bolt.

16. Remove the pin from the auto-tensioner.

17. Remove the battery clamp bolt from the back cover.

18. Install the lower half of the side engine mount bracket.

19. Install the timing belt guide plate as shown.

20. Install the lower cover.

21. Install the front upper cover (A) and rear upper cover (B).

22. Install the crankshaft pulley.

23. Rotate the crankshaft pulley six turns clockwise so the timing belt positions itself on the pulleys.

24. Turn the crankshaft pulley so its white mark (A) lines up with the pointer (B).

25. Check the camshaft pulley marks.

➡**If the marks are not aligned, rotate the crankshaft 360 degrees, and recheck the camshaft pulley mark.**

　　a. If the camshaft pulley marks are at TDC, go to step 24.

　　b. If the camshaft pulley marks are not at TDC, remove the timing belt and repeat step 3 through 22.

26. Install the upper half of the side engine mount bracket, and tighten the new mounting bolts (A), then the mass damper mounting bolt (B).

27. Install the drive belt.

28. Install the splash shield.

29. Install the right front wheel.

30. Do the crankshaft position (CKP) pattern clear/CKP pattern learn procedure.

➡**The ECT needs to be at 176°F (80°C) or higher.**

　　a. Clear the CKP pattern while the engine is stopped.

　　b. Turn the ignition switch OFF.

　　c. Turn the ignition switch ON (II), and wait 30 seconds.

　　d. Test-drive the vehicle on a level road: decelerate (with the throttle fully closed) from an engine speed of 2,500 RPM to 1,000 RPM with the A/T in 2 position.

e. Stop the vehicle, but keep the engine running.

f. Check PULSER F/B LEARN in the DATA LIST with the HDS. If it is NOT COMPLETED, go to step 4. If it is COMPLETED, go to step 7.

g. Test-drive the vehicle on a level road: decelerate (with the throttle fully closed) from an engine speed of 5,000 RPM to 3,000 RPM with the A/T in 2 position.

h. Stop the vehicle, but keep the engine running.

i. Check the PULSER F/B LEARN (HIGH RPM) in the DATA LIST with the HDS. If it is NOT COMPLETED, go to step 7. If it is COMPLETED, go to step 10.

j. Turn the ignition switch OFF.

k. Turn the ignition switch ON (II), and wait 30 seconds. The CKP learning procedure is complete.

J35A7 Engine

Installing a Used Belt

See Figures 108 through 115.

1. Remove the right front wheel.
2. Remove the splash shield.
3. Remove the accessory drive belt.

4. Turn the crankshaft so that the white mark on the pulley (A) lines up with the pointer (B).

5. Check that the TDC timing mark (A) on the front camshaft sprocket is aligned with the pointer (B) on the cover.

6. Support the engine with a jack and a block of wood under the oil pan.

7. Remove the upper half of the side engine mount bracket.

8. Using the tools shown and a 19mm socket, remove the damper bolt.

9. Remove the damper.

10. Remove the front and rear upper belt covers.

11. Remove the lower cover.

12. Remove one of the battery hold-down clamp bolts and grind the end as shown.

13. Screw the battery clamp bolt in as shown to hold the timing belt adjuster in its current position. Tighten it by hand. Do not use a wrench.

14. Remove the timing belt guide plate.

15. Remove the lower side engine mount bracket.

16. Remove the idler pulley bolt and idler pulley. Discard the bolt.

17. Remove the timing belt.

To install:

18. Clean all parts.

19. Set the crankshaft sprocket to TDC by aligning the TDC mark (A) with the pointer (B) on the oil pump.

20. Set the camshaft pulleys to TDC by aligning the TDC marks (A) on the camshaft pulleys with the pointers (B) on the covers. See the illustrations in INSTALLING A NEW BELT.

21. Loosely install the idler pulley with a new bolt so that the pulley can move but won't come off.

22. If the auto-tensioner has extended, and the timing belt can't be installed, see the procedure for INSTALLING A NEW BELT.

23. Install the timing belt in a counter-clockwise sequence, starting with the crankshaft drive pulley.

24. Tighten the idler pulley bolt to 33 ft. lbs. (44 Nm).

25. The remainder of installation is the reverse of removal. Note the following torques:

- Lower half engine mount bracket: 33 ft. lbs. (44 Nm)
- Lower cover: 105 inch lbs. (12 Nm)
- Upper covers: 105 inch lbs. (12 Nm)

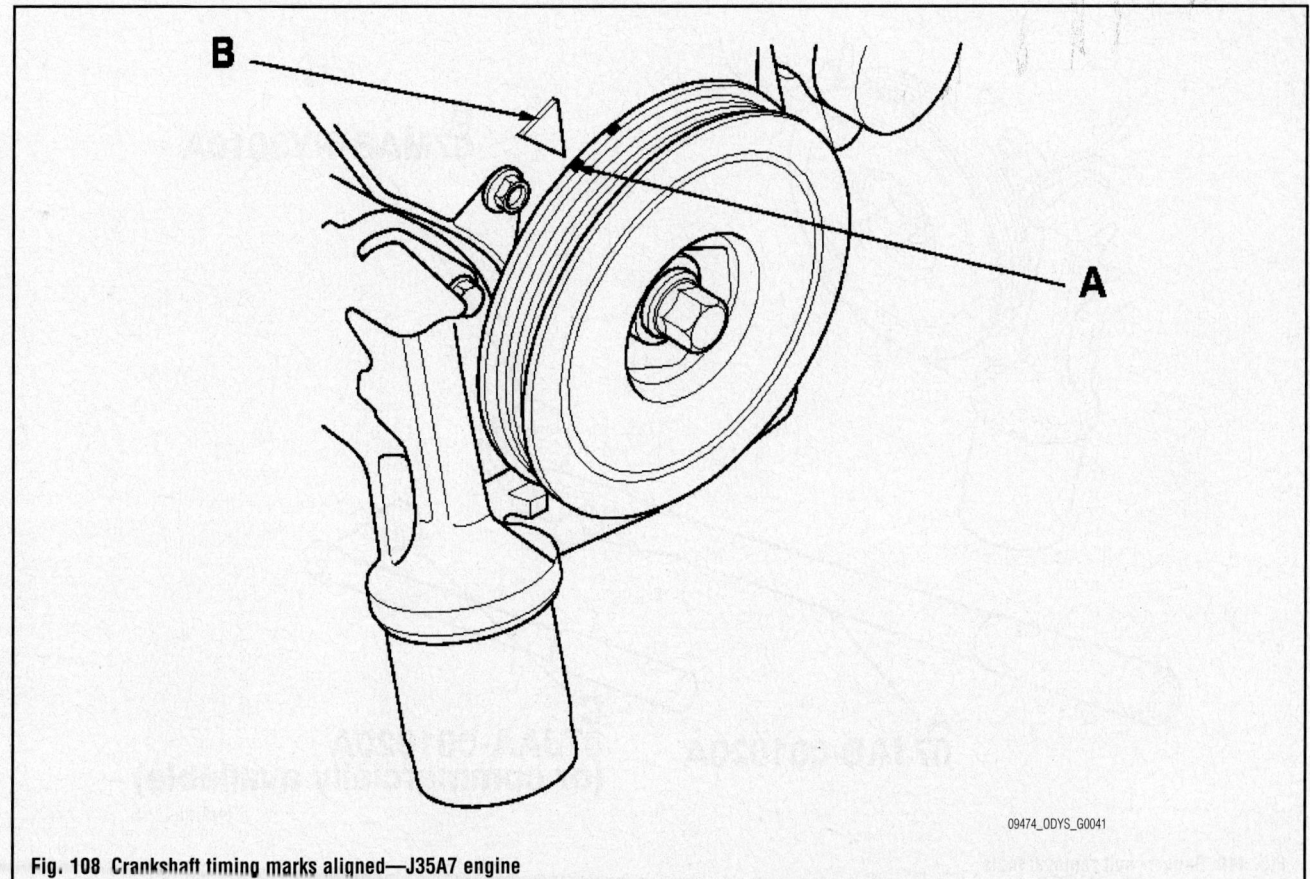

09474_ODYS_G0041

Fig. 108 Crankshaft timing marks aligned—J35A7 engine

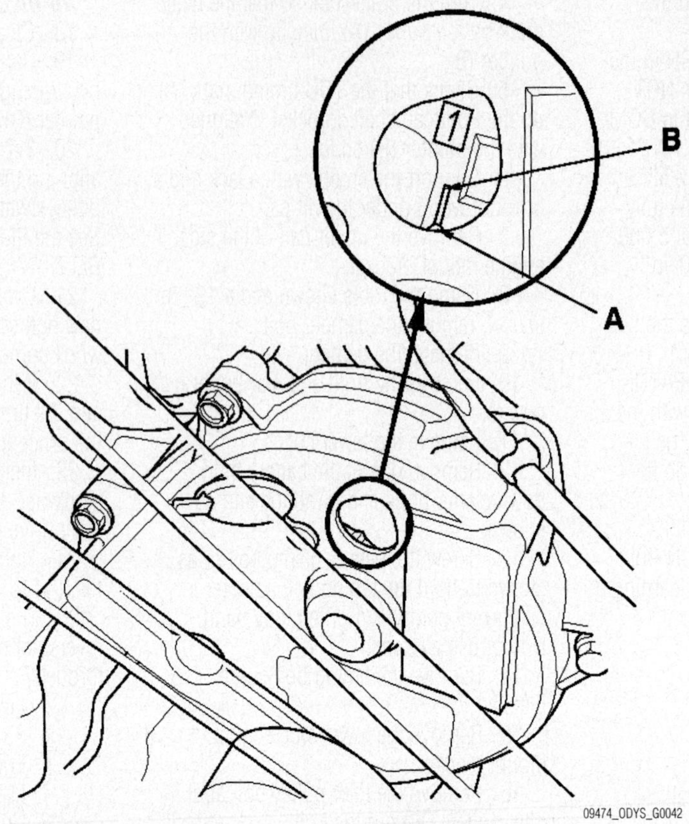

Fig. 109 Front camshaft timing marks aligned—J35A7 engine

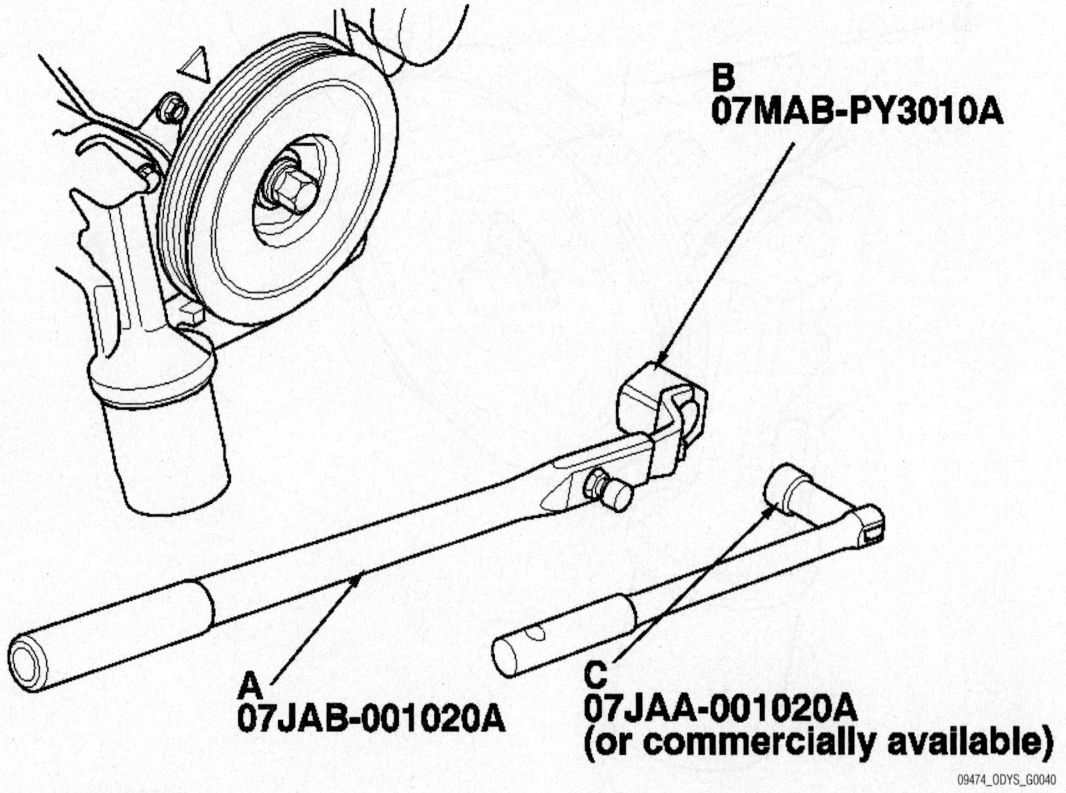

09474_ODYS_G0042

B
07MAB-PY3010A

A
07JAB-001020A

C
07JAA-001020A
(or commercially available)

09474_ODYS_G0040

Fig. 110 Damper bolt removal tools

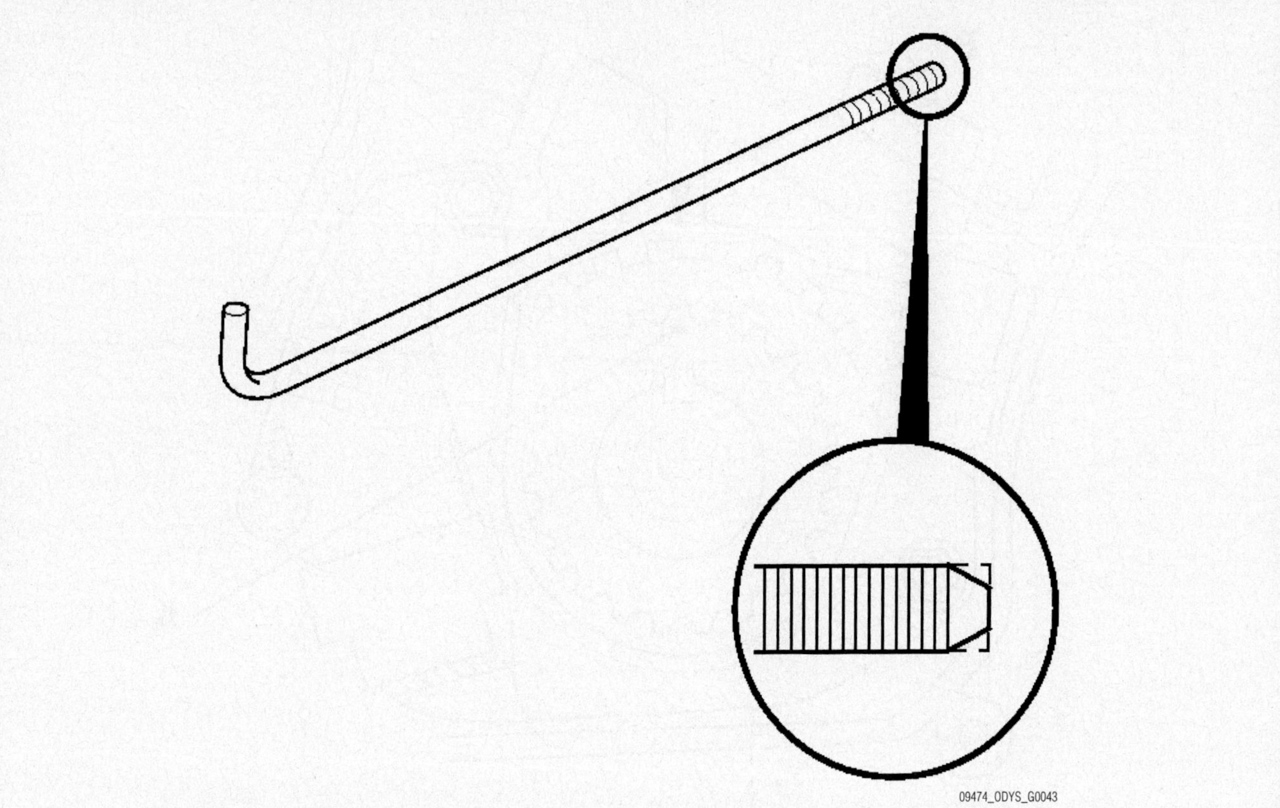

09474_ODYS_G0043

Fig. 111 Remove one of the battery hold-down clamp bolts and grind the end as shown

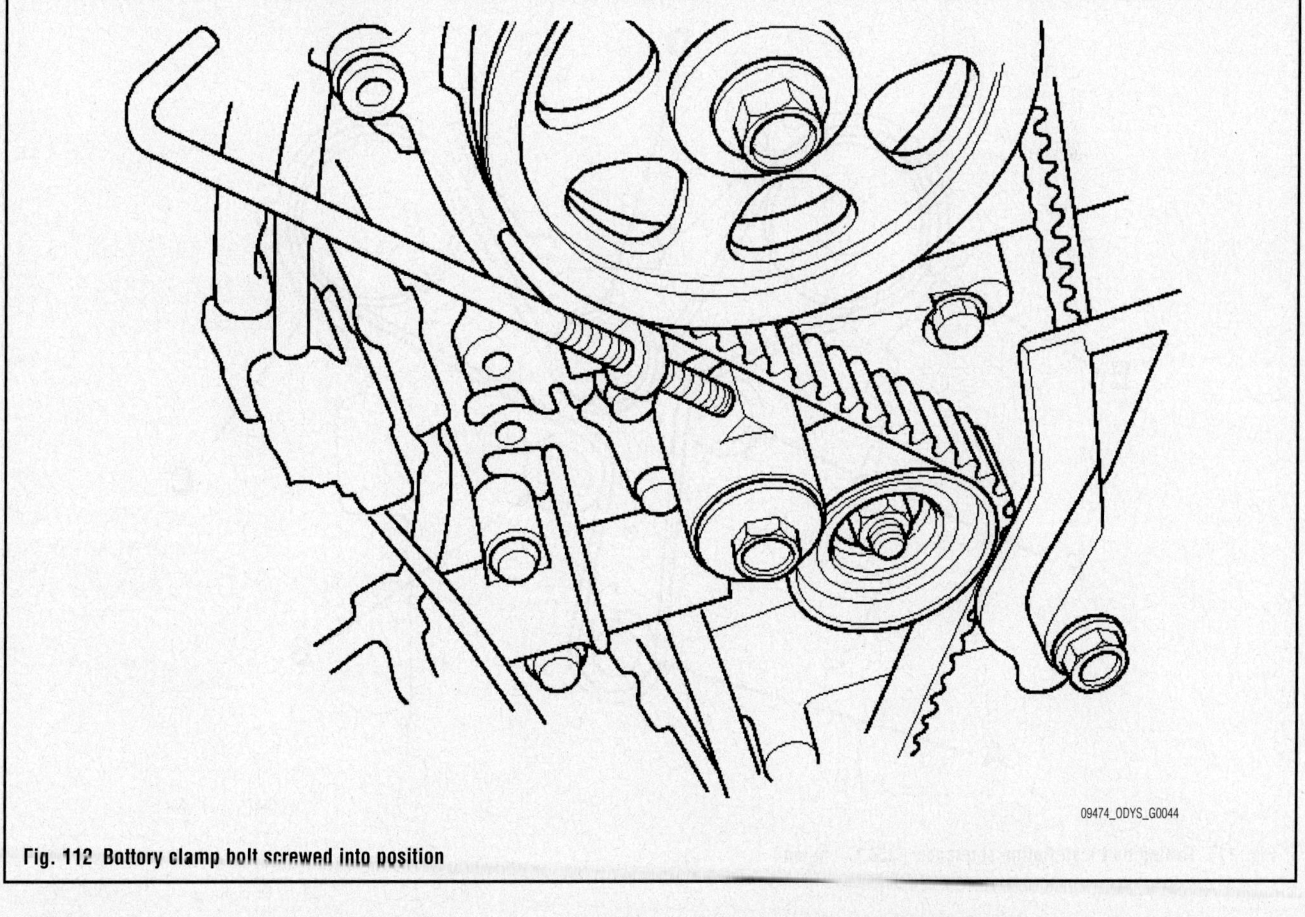

09474_ODYS_G0044

Fig. 112 Battery clamp bolt screwed into position

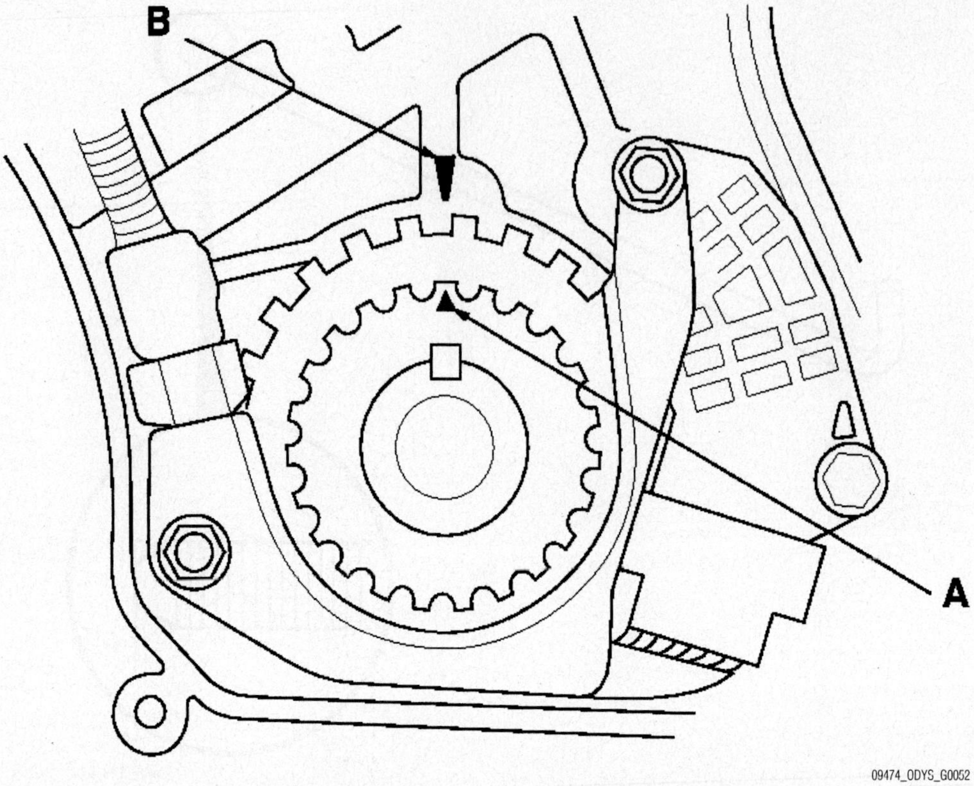
B
A
09474_ODYS_G0052

Fig. 113 Set the crankshaft sprocket to TDC by aligning the TDC mark (A) with the pointer (B) on the oil pump—J35A7 engine

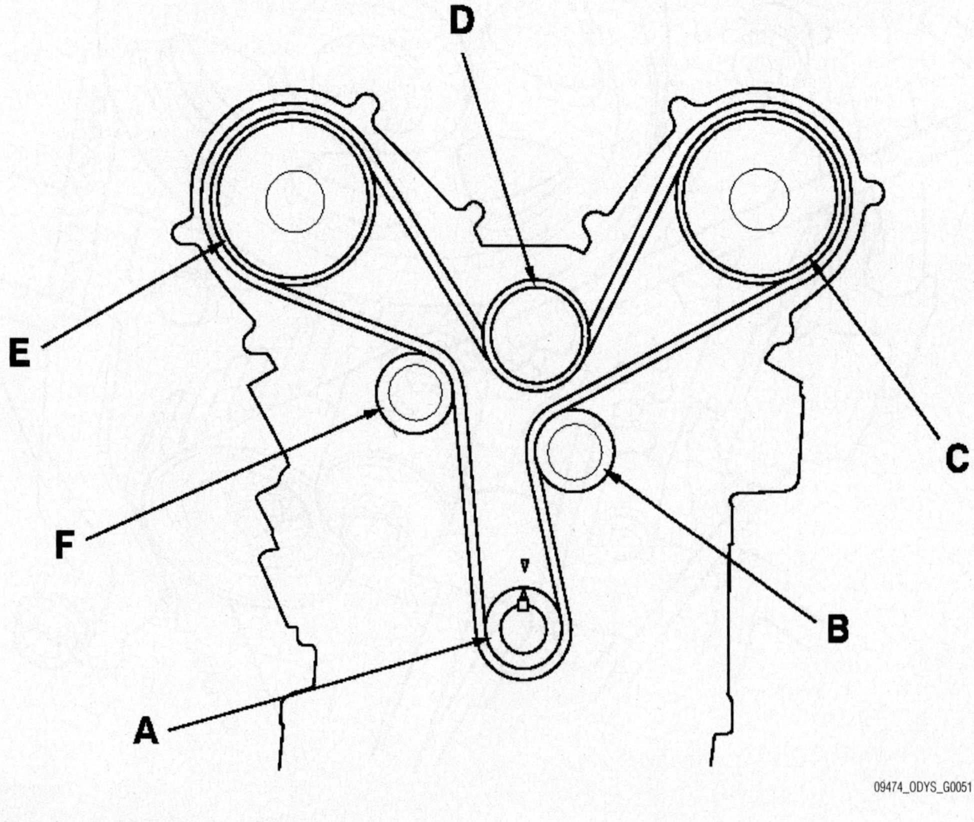
D
E
C
F
B
A
09474_ODYS_G0051

Fig. 114 Timing belt installation sequence—J35A7 engine

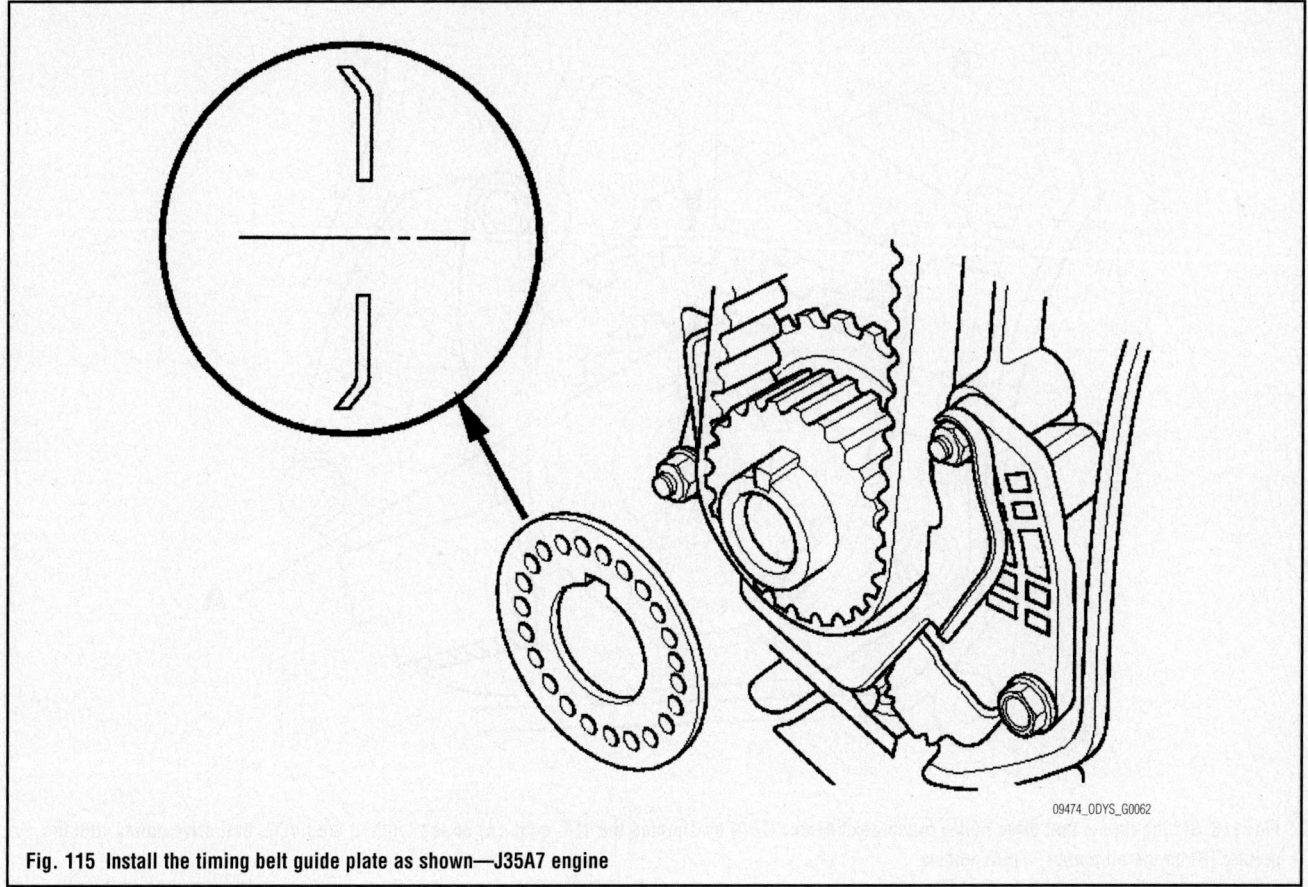

Fig. 115 Install the timing belt guide plate as shown—J35A7 engine

- Crankshaft damper: Tighten the bolt to 47 ft. lbs. (64 Nm). Mark the bolt head and pulley, then tighten the bolt an additional 60°.

26. Do the crankshaft position (CKP) pattern clear/CKP pattern learn procedure.

➡**The ECT needs to be at 176°F (80°C) or higher.**

a. Clear the CKP pattern while the engine is stopped.

b. Turn the ignition switch OFF.

c. Turn the ignition switch ON (II), and wait 30 seconds.

d. Test-drive the vehicle on a level road: decelerate (with the throttle fully closed) from an engine speed of 2,500 RPM to 1,000 RPM with the A/T in 2 position.

e. Stop the vehicle, but keep the engine running.

f. Check PULSER F/B LEARN in the DATA LIST with the HDS. If it is NOT COMPLETED, go to step 4. If it is COMPLETED, go to step 7.

g. Test-drive the vehicle on a level road: decelerate (with the throttle fully closed) from an engine speed of 5,000 RPM to 3,000 RPM with the A/T in 2 position.

h. Stop the vehicle, but keep the engine running.

i. Check the PULSER F/B LEARN (HIGH RPM) in the DATA LIST with the HDS. If it is NOT COMPLETED, go to step 7. If it is COMPLETED, go to step 10.

j. Turn the ignition switch OFF.

k. Turn the ignition switch ON (II), and wait 30 seconds. The CKP learning procedure is complete.

Installing a New Belt

See Figures 116 through 132.

1. Remove the timing belt.
2. Clean the timing belt pulleys, timing belt guide plate, and the upper and lower covers.
3. Set the timing belt drive pulley to top dead center (TDC) by aligning the TDC mark (A) on the tooth of the timing belt drive pulley with the pointer (B) on the oil pump.
4. Set the camshaft pulleys to TDC by aligning the TDC marks (A) on the camshaft pulleys with the pointers (B) on the back covers.
5. Remove the battery clamp bolt from the back cover.
6. Remove the auto-tensioner.

7. Align the holes on the rod and housing of the auto-tensioner.
8. Use a hydraulic press to slowly compress the auto-tensioner. Insert a 2.0 mm (0.08 in.) pin through the housing and the rod.

➡**The compression pressure should not exceed 9,800 N (2,200 lbs.).**

9. Align the holes on the rod and housing of the auto-tensioner.
10. Use a hydraulic press to slowly compress the auto-tensioner. Insert a 2.0 mm (0.08 in.) pin through the housing and the rod.

➡**The compression pressure should not exceed 9,800 N (2,200 lbs.).**

11. Install the auto-tensioner.

➡**Make sure the pin stays in place.**

12. Screw the battery clamp bolt in as shown to hold the timing belt adjuster. Tighten it by hand; do not use a wrench.
13. Loosely install the idler pulley with a new idler pulley bolt so the pulley can move but does not come off.
14. Install the timing belt in a counterclockwise sequence starting with the drive pulley.

- 1. Drive pulley (A)
- 2. Idler pulley (B)

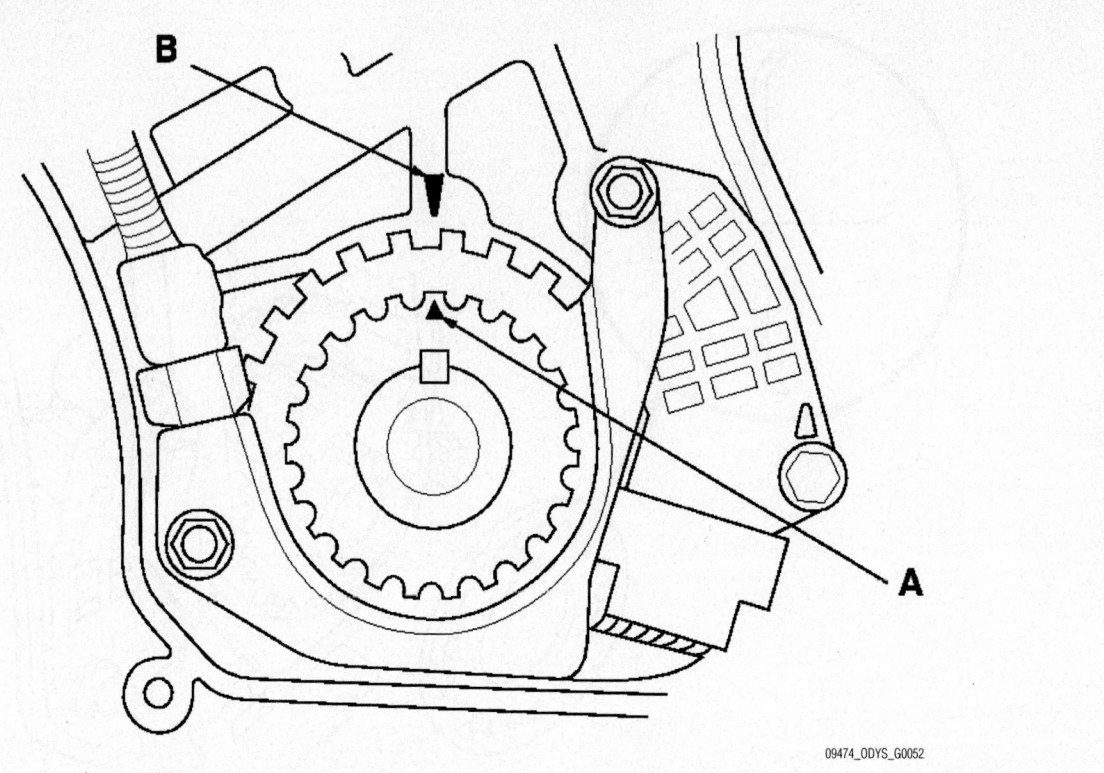

Fig. 116 Set the timing belt drive pulley to top dead center (TDC) by aligning the TDC mark (A) on the tooth of the timing belt drive pulley with the pointer (B) on the oil pump—J35A7 engine

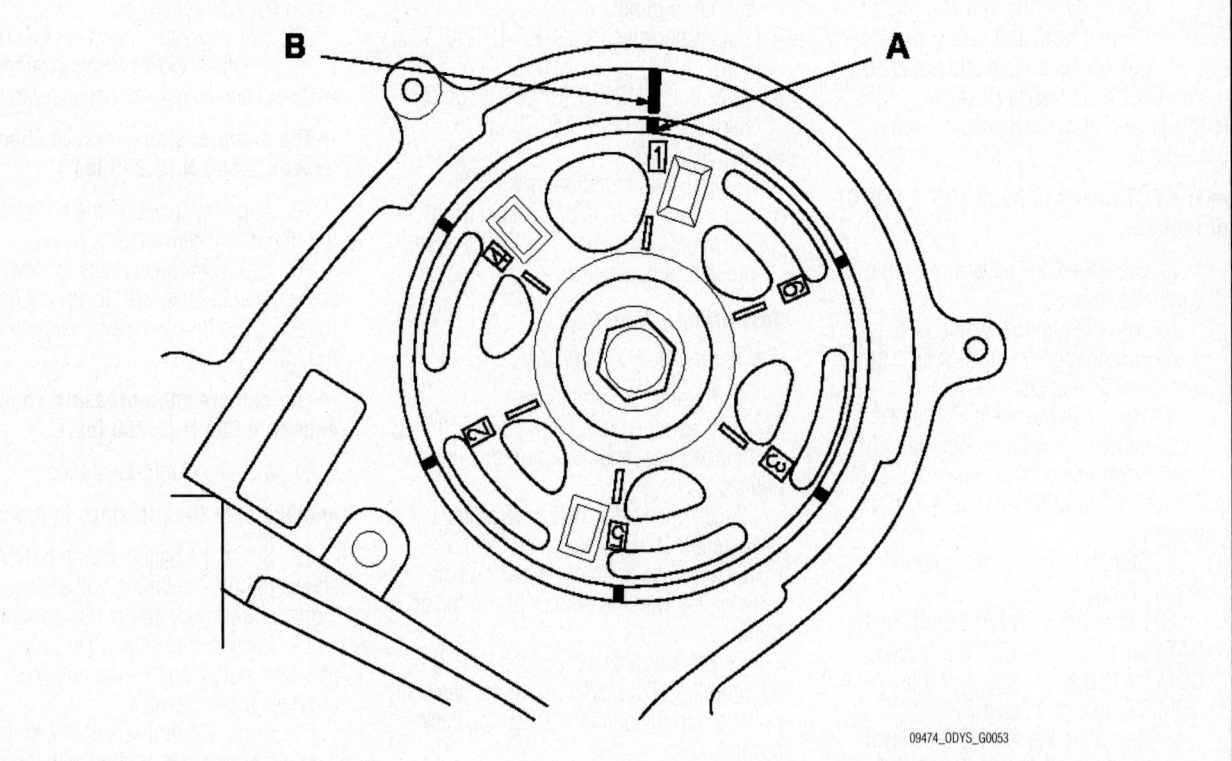

Fig. 117 Set the left camshaft pulley to TDC by aligning the TDC marks (A) on the camshaft pulleys with the pointers (B) on the back covers—J35A7 engine

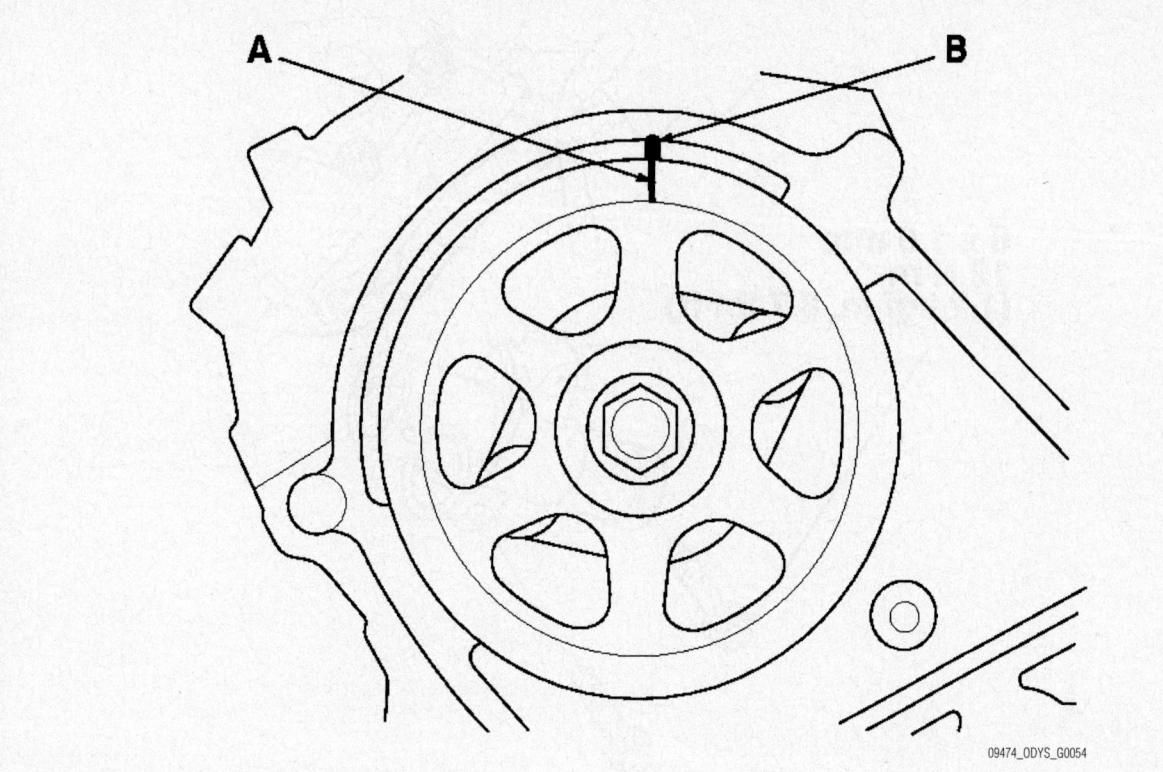

09474_ODYS_G0054

Fig. 118 Set the left camshaft pulley to TDC by aligning the TDC marks (A) on the camshaft pulleys with the pointers (B) on the back covers—J35A7 engine

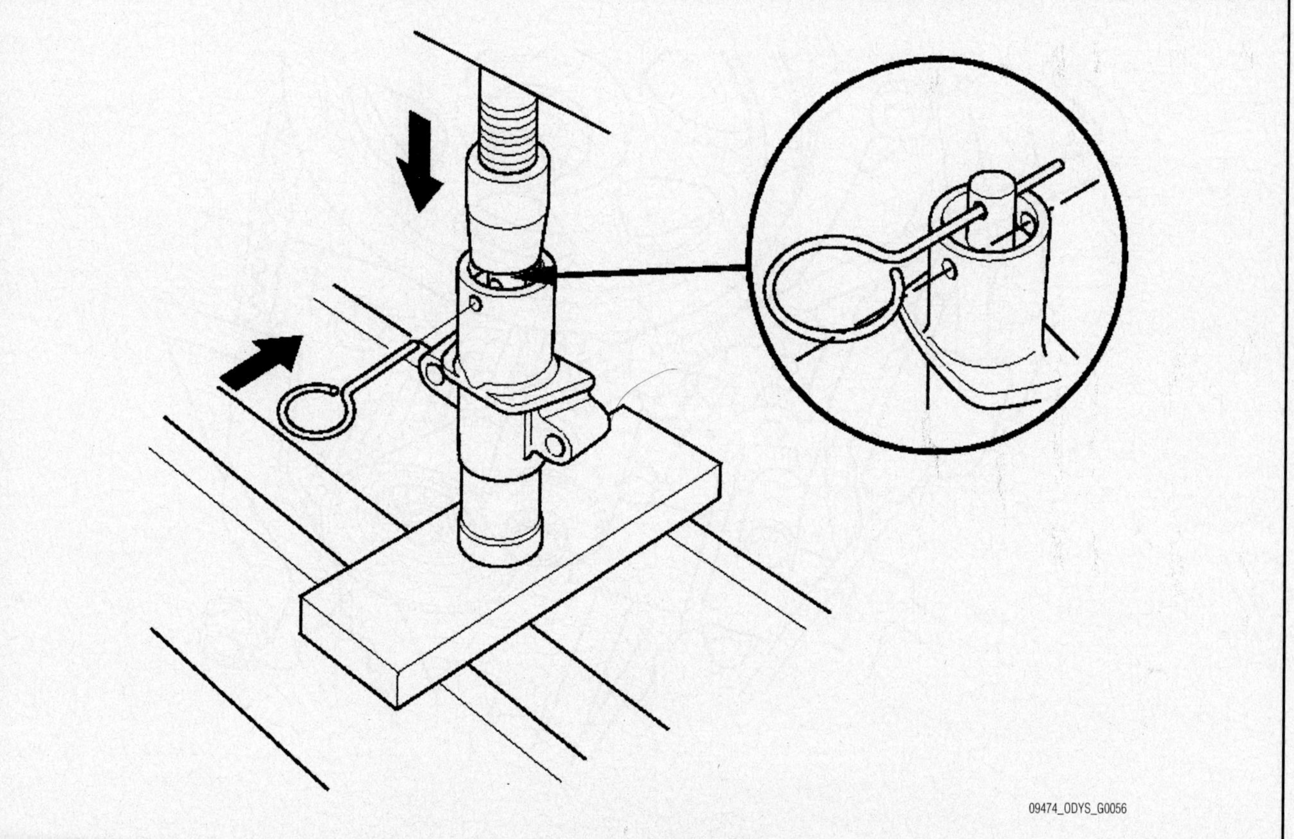

09474_ODYS_G0056

Fig. 119 Insert a 2.0 mm (0.08 in.) pin through the housing and the rod—J35A7 engine

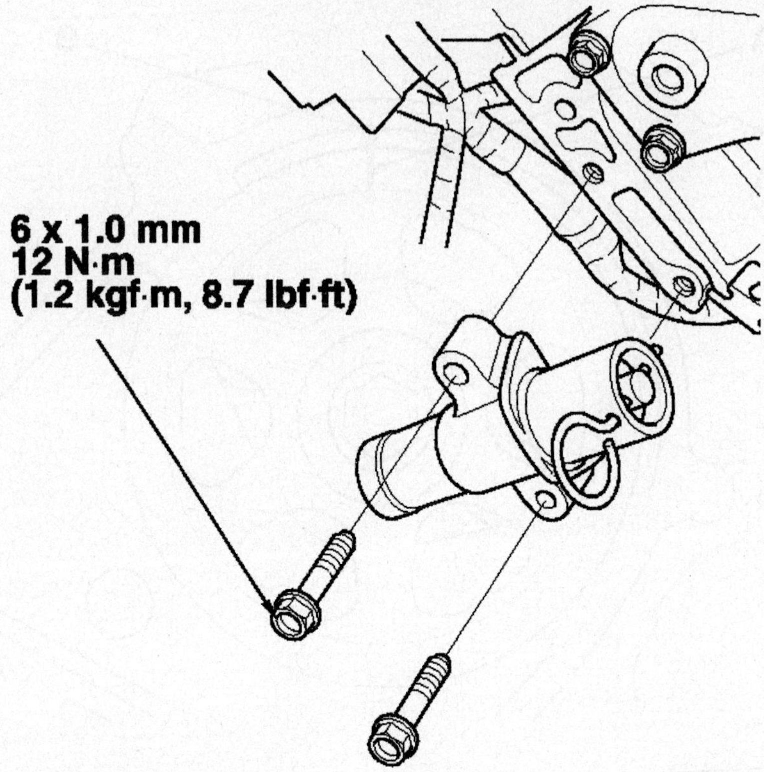

6 x 1.0 mm
12 N·m
(1.2 kgf·m, 8.7 lbf·ft)

09474_ODYS_G0057

Fig. 120 Install the auto-tensioner—J35A7 engine

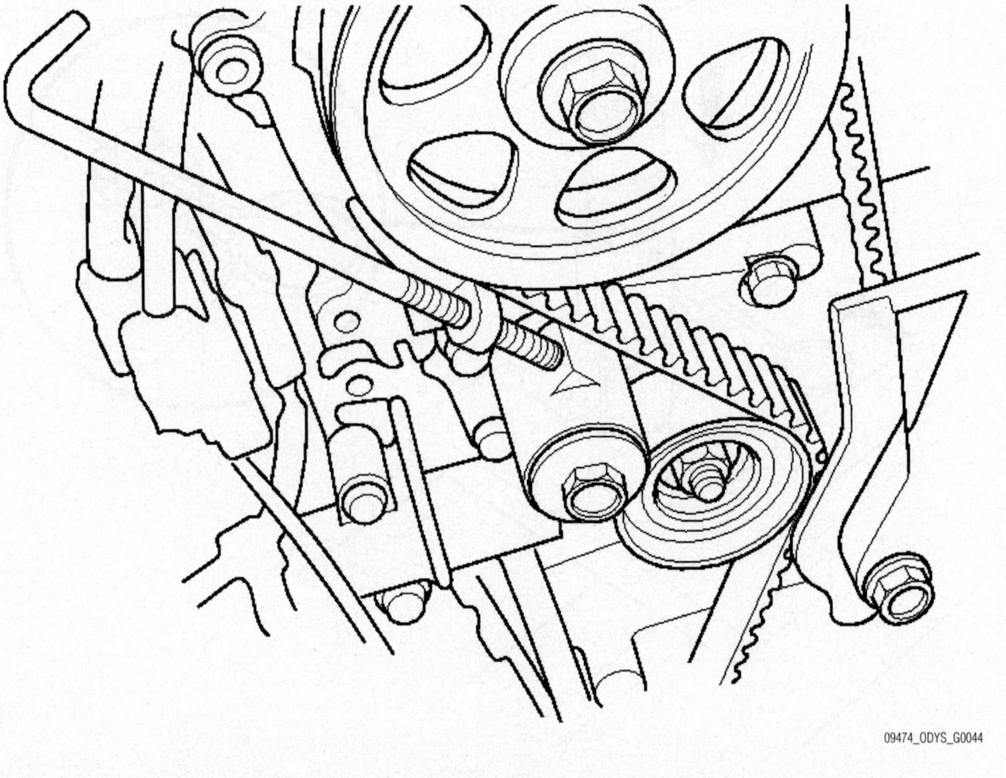

09474_ODYS_G0044

Fig. 121 Screw the battery clamp bolt in as shown to hold the timing belt adjuster. Tighten it by hand—J35A7 engine

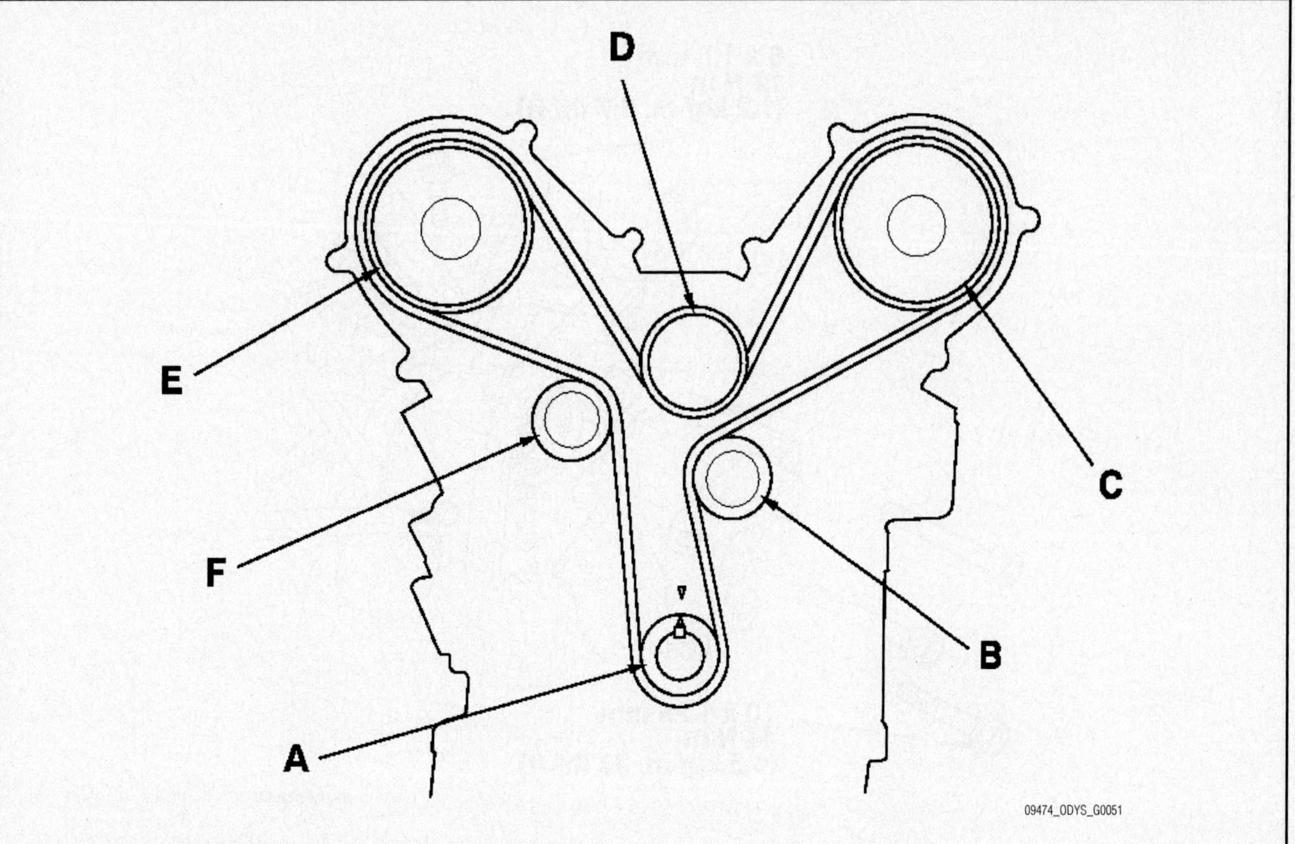
D
E
C
F
B
A
09474_ODYS_G0051

Fig. 122 Install the timing belt in a counterclockwise sequence starting with the drive pulley—J35A7 engine

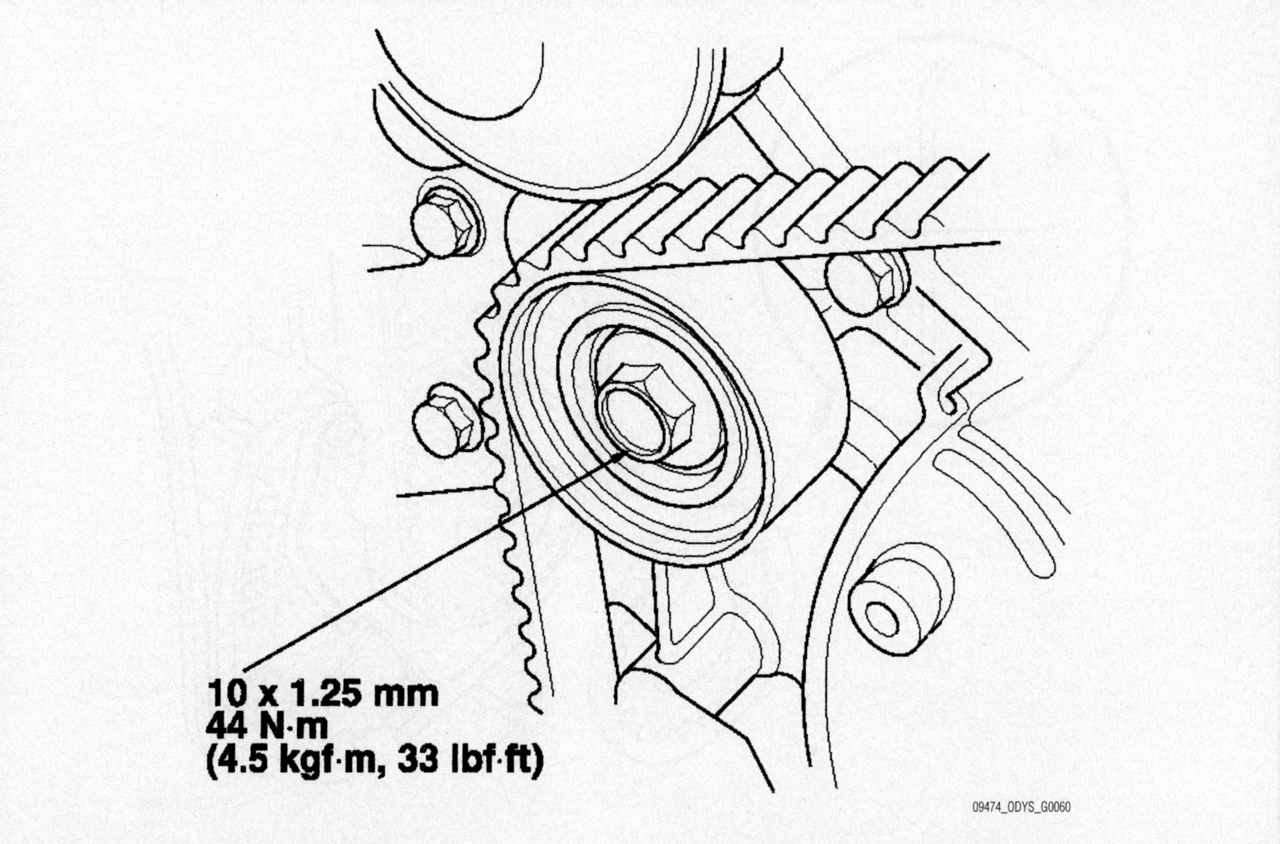
10 x 1.25 mm
44 N·m
(4.5 kgf·m, 33 lbf·ft)
09474_ODYS_G0060

Fig. 123 Tighten the idler pulley bolt—J35A7 engine

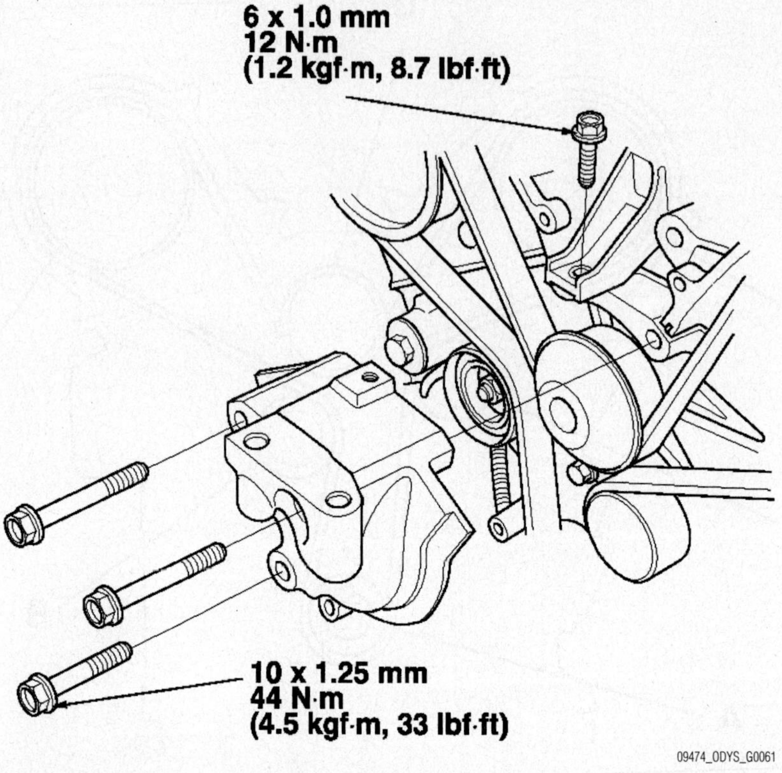

6 x 1.0 mm
12 N·m
(1.2 kgf·m, 8.7 lbf·ft)

10 x 1.25 mm
44 N·m
(4.5 kgf·m, 33 lbf·ft)

09474_ODYS_G0061

Fig. 124 Install the lower half of the side engine mount bracket—J35A7 engine

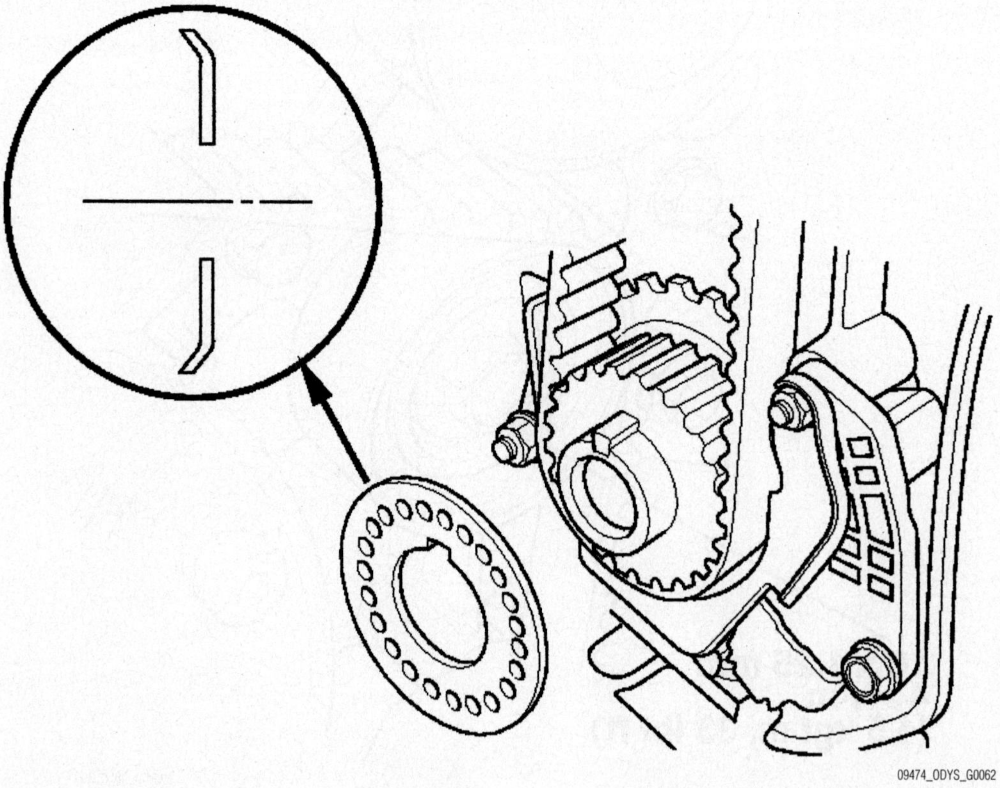

09474_ODYS_G0062

Fig. 125 Install the timing belt guide plate as shown—J35A7 engine

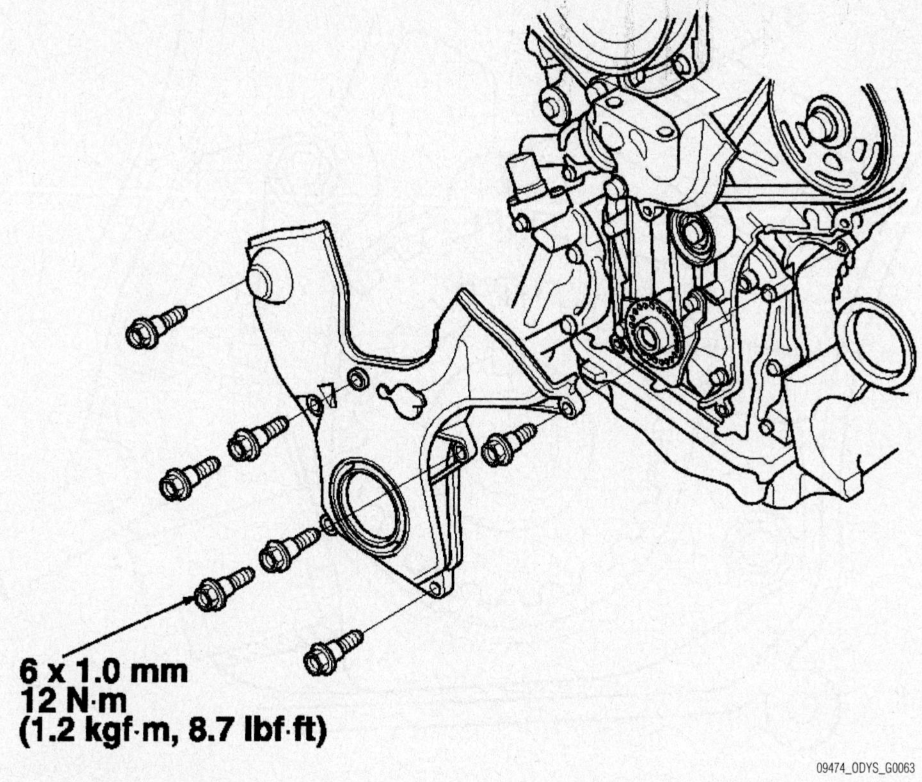

6 x 1.0 mm
12 N·m
(1.2 kgf·m, 8.7 lbf·ft)

09474_ODYS_G0063

Fig. 126 Install the lower cover—J35A7 engine

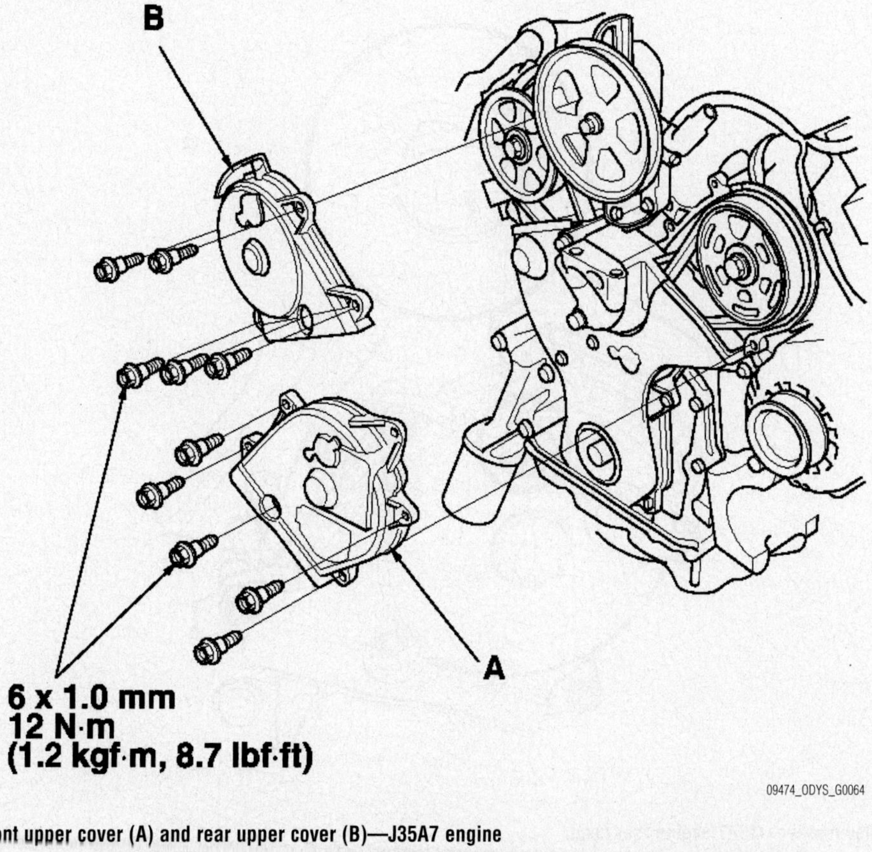

B

A

6 x 1.0 mm
12 N·m
(1.2 kgf·m, 8.7 lbf·ft)

09474_ODYS_G0064

Fig. 127 Install the front upper cover (A) and rear upper cover (B)—J35A7 engine

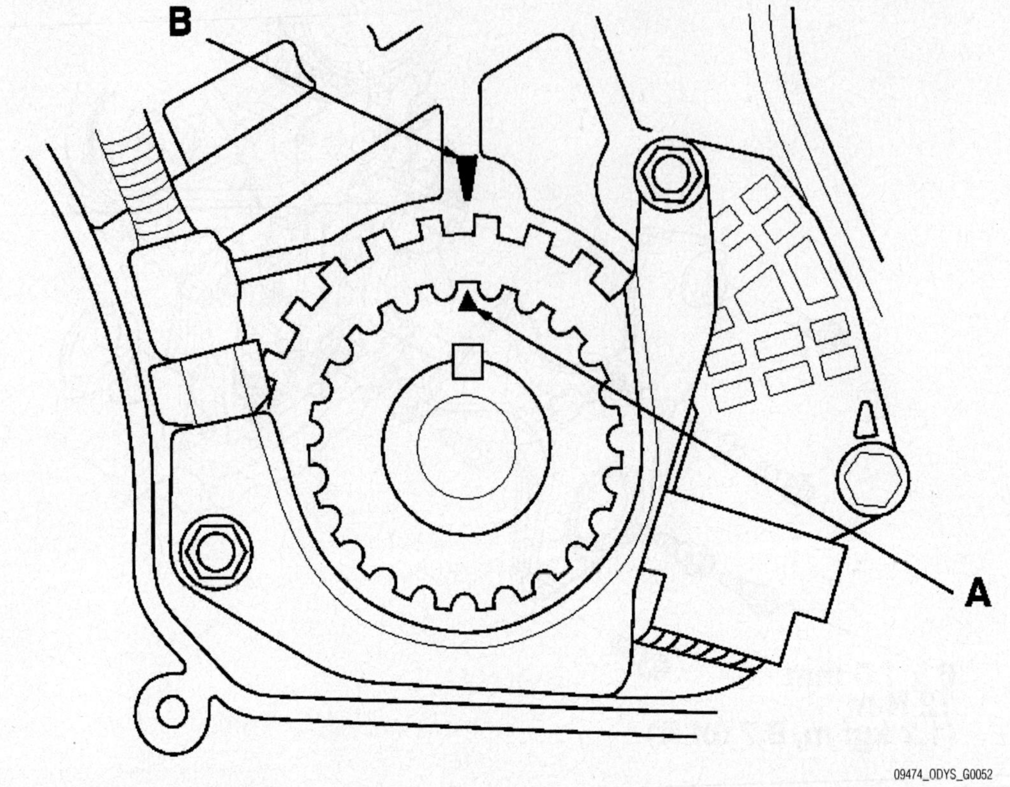

09474_ODYS_G0052

Fig. 128 Turn the crankshaft pulley so its white mark (A) lines up with the pointer (B) —J35A7 engine

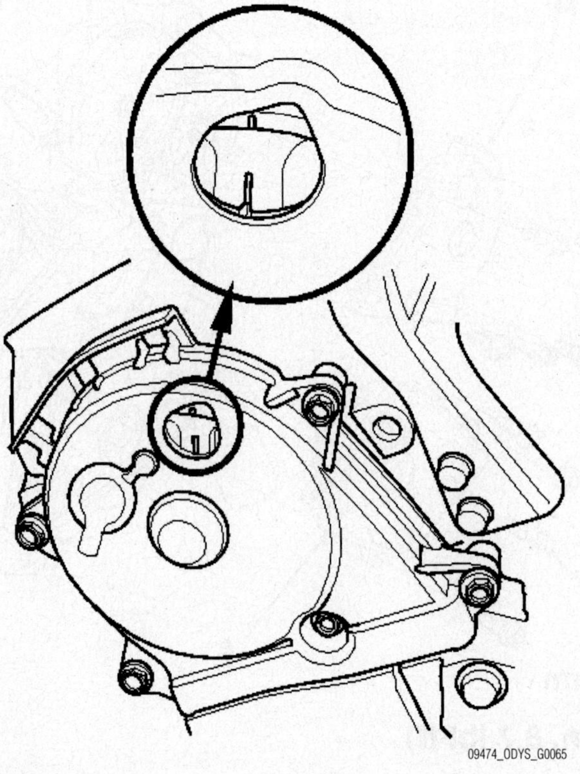

09474_ODYS_G0065

Fig. 129 Camshaft pulley marks—J35A7 engine rear head

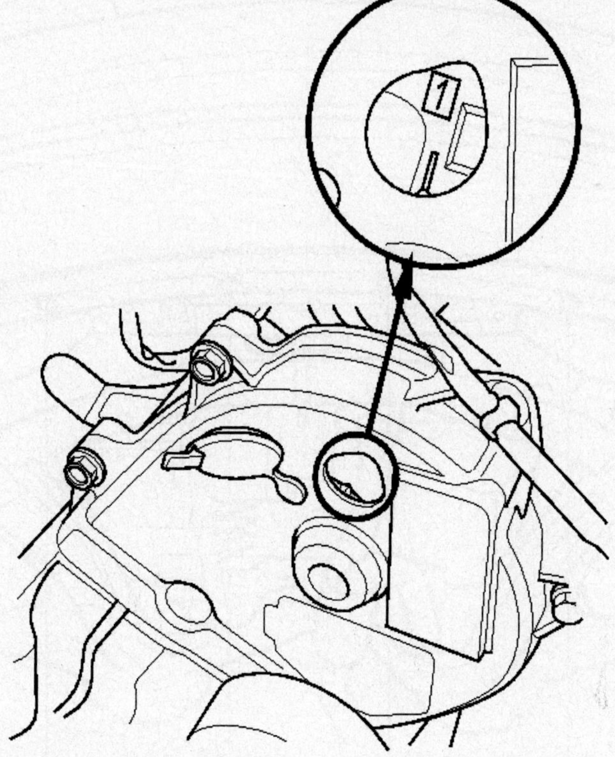

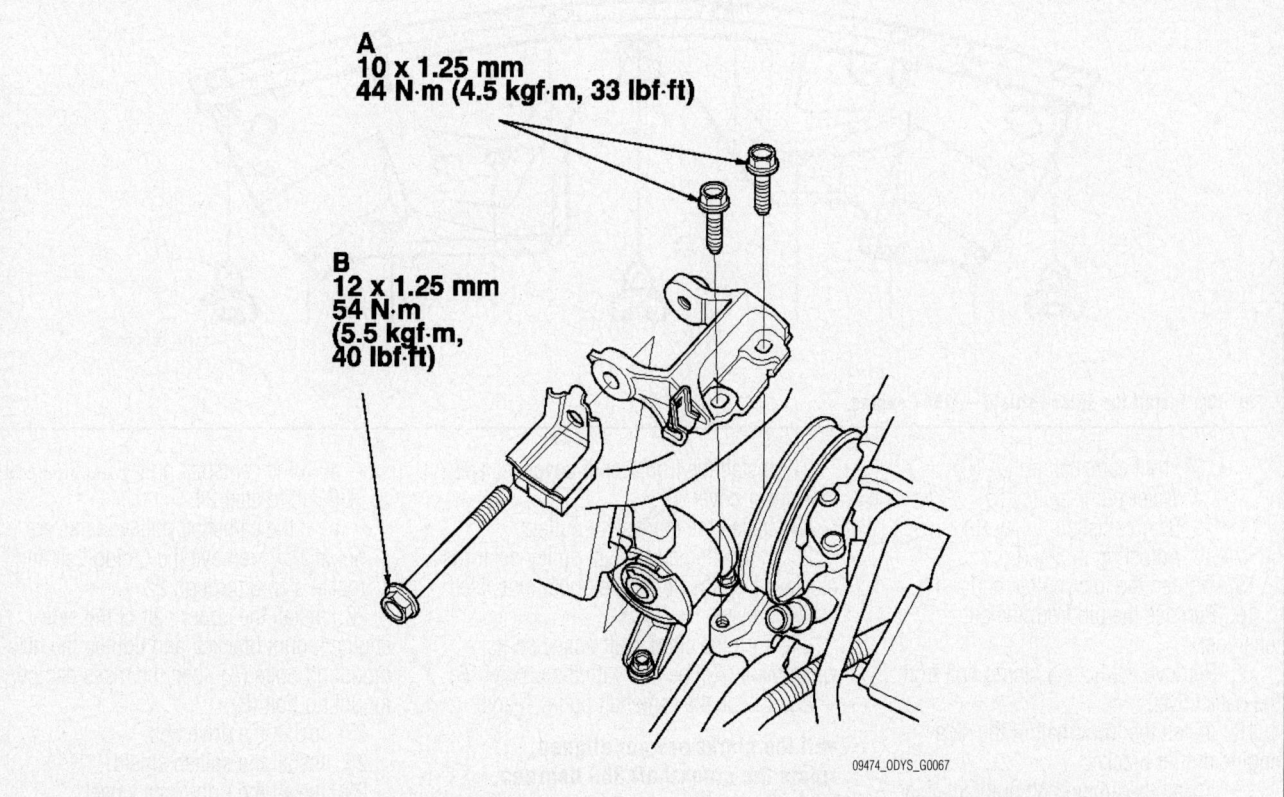

09474_ODYS_G0066

Fig. 130 Camshaft pulley marks—J35A7 engine front head

A
10 x 1.25 mm
44 N·m (4.5 kgf·m, 33 lbf·ft)

B
12 x 1.25 mm
54 N·m
(5.5 kgf·m,
40 lbf·ft)

09474_ODYS_G0067

Fig. 131 Install the upper half of the side engine mount bracket, and tighten the new mounting bolts (A), then the mass damper mounting bolt (B) —J35A7 engine

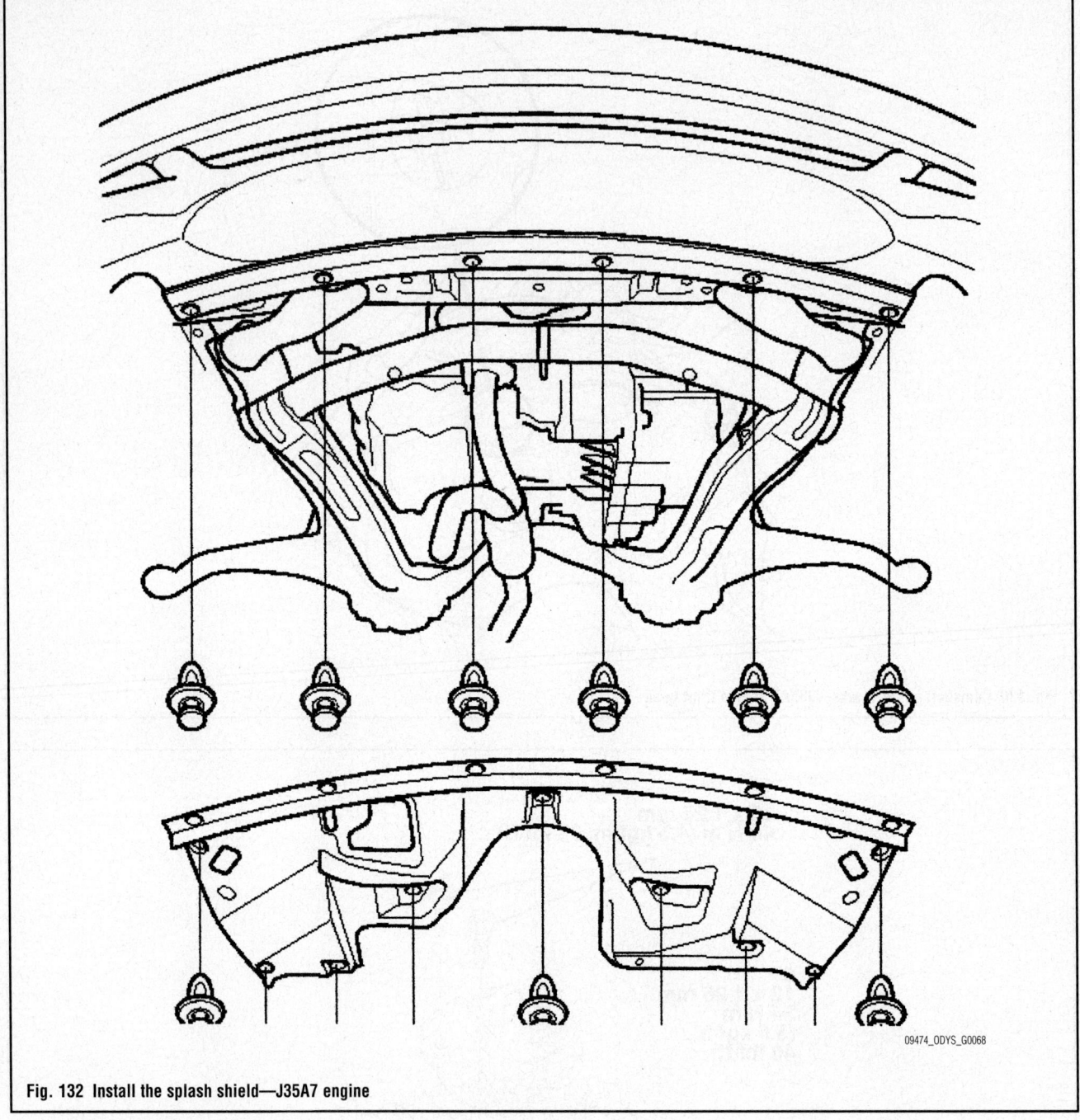

Fig. 132 Install the splash shield—J35A7 engine

- 3. Front camshaft pulley (C)
- 4. Water pump pulley (D)
- 5. Rear camshaft pulley (E)
- 6. Adjusting pulley (F)

15. Tighten the idler pulley bolt.
16. Remove the pin from the auto-tensioner.
17. Remove the battery clamp bolt from the back cover.
18. Install the lower half of the side engine mount bracket.
19. Install the timing belt guide plate as shown.
20. Install the lower cover.

21. Install the front upper cover (A) and rear upper cover (B).
22. Install the crankshaft pulley.
23. Rotate the crankshaft pulley six turns clockwise so the timing belt positions itself on the pulleys.
24. Turn the crankshaft pulley so its white mark (A) lines up with the pointer (B).
25. Check the camshaft pulley marks.

➡**If the marks are not aligned, rotate the crankshaft 360 degrees, and recheck the camshaft pulley mark.**

   a. If the camshaft pulley marks are at TDC, go to step 24.
   b. If the camshaft pulley marks are not at TDC, remove the timing belt and repeat step 3 through 22.
26. Install the upper half of the side engine mount bracket, and tighten the new mounting bolts (A), then the mass damper mounting bolt (B).
27. Install the drive belt.
28. Install the splash shield.
29. Install the right front wheel.
30. Do the crankshaft position (CKP) pattern clear/CKP pattern learn procedure.

➡**The ECT needs to be at 176°F (80°C) or higher.**

a. Clear the CKP pattern while the engine is stopped.

b. Turn the ignition switch OFF.

c. Turn the ignition switch ON (II), and wait 30 seconds.

d. Test-drive the vehicle on a level road: decelerate (with the throttle fully closed) from an engine speed of 2,500 RPM to 1,000 RPM with the A/T in 2 position.

e. Stop the vehicle, but keep the engine running.

f. Check PULSER F/B LEARN in the DATA LIST with the HDS. If it is NOT COMPLETED, go to step 4. If it is COMPLETED, go to step 7.

g. Test-drive the vehicle on a level road: decelerate (with the throttle fully closed) from an engine speed of 5,000 RPM to 3,000 RPM with the A/T in 2 position.

h. Stop the vehicle, but keep the engine running.

i. Check the PULSER F/B LEARN (HIGH RPM) in the DATA LIST with the HDS. If it is NOT COMPLETED, go to

step 7. If it is COMPLETED, go to step 10.

j. Turn the ignition switch OFF.

k. Turn the ignition switch ON (II), and wait 30 seconds. The CKP learning procedure is complete.

VALVE LASH

ADJUSTMENT

See Figures 133 through 137.

Adjust the valves only when the cylinder head temperature is less than 100°F (38°C).

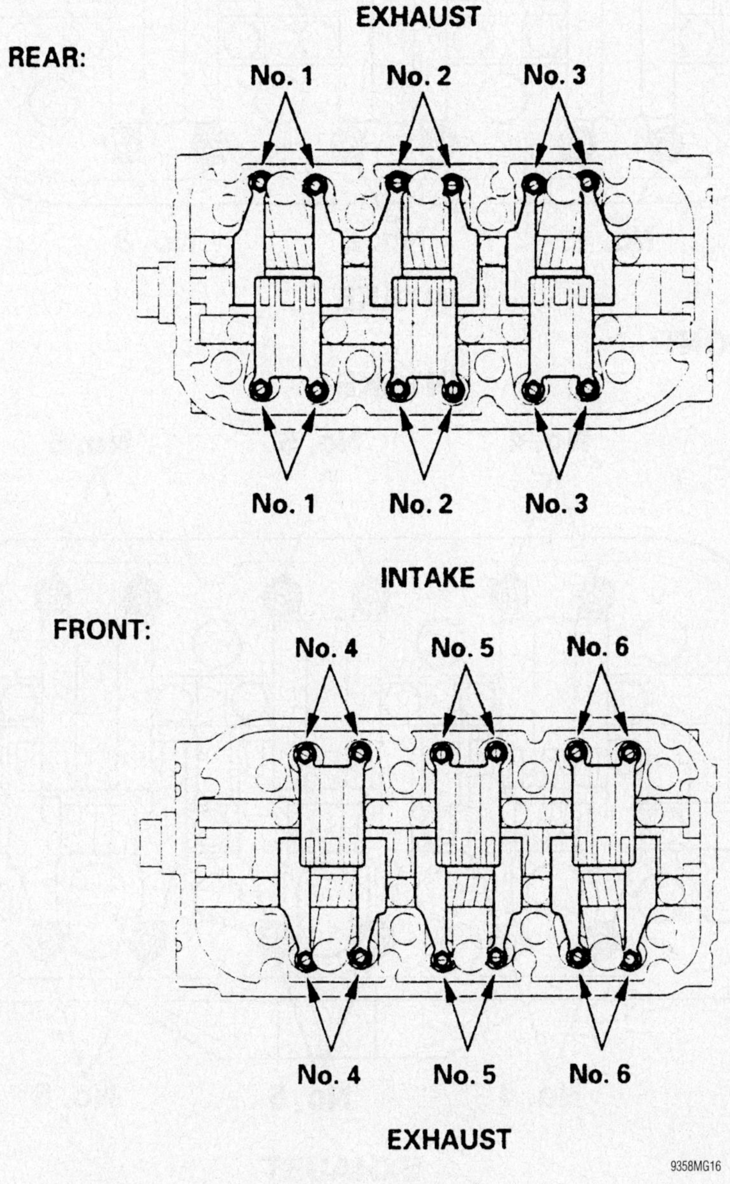

Fig. 133 Adjusting screw locations—J35A0 engine

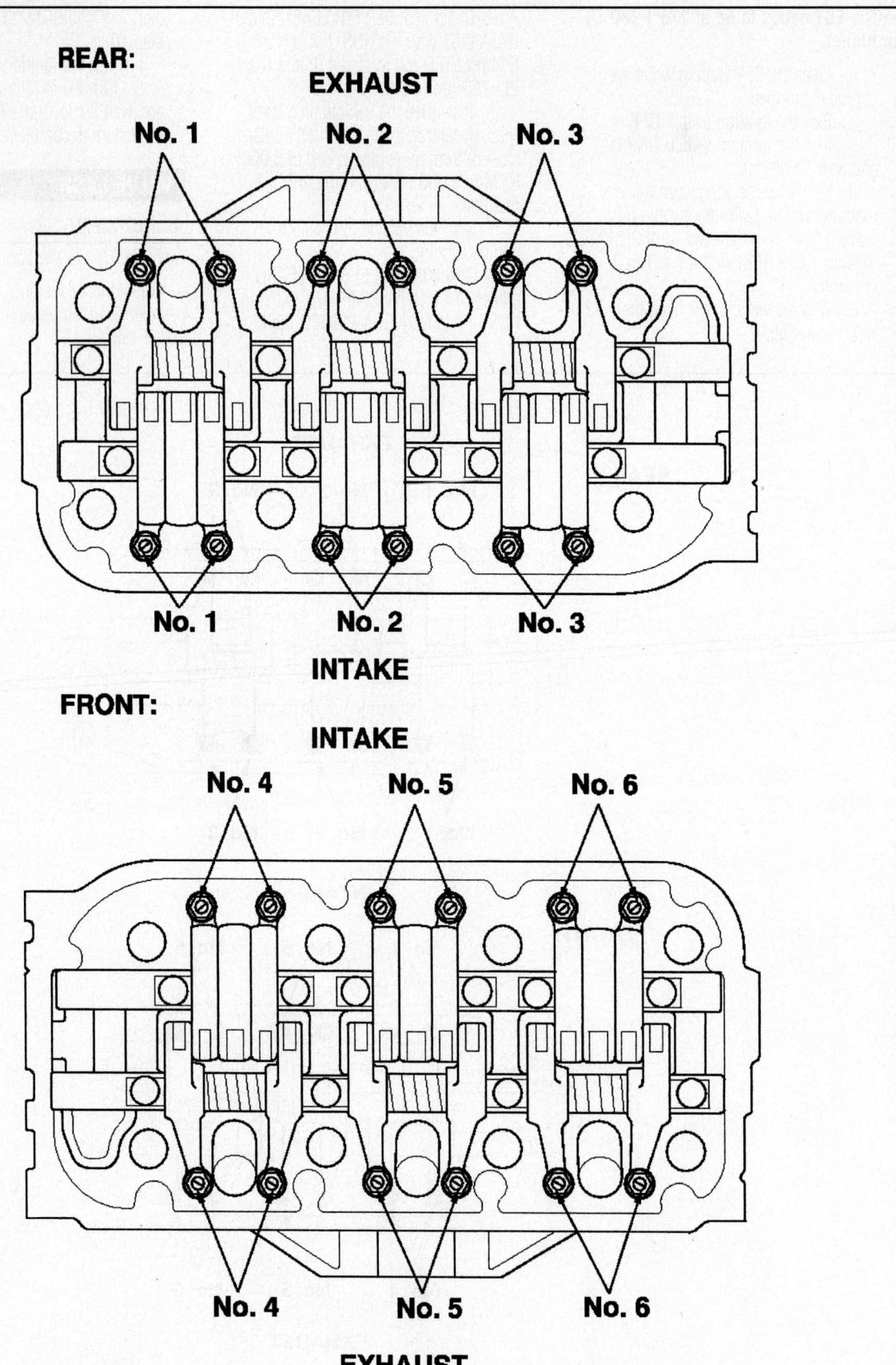

REAR:

EXHAUST

No. 1   No. 2   No. 3

No. 1   No. 2   No. 3

INTAKE

FRONT:

INTAKE

No. 4   No. 5   No. 6

No. 4   No. 5   No. 6

EXHAUST

09474_ODYS_G0027

Fig. 134 Adjusting screw locations—J35A7 engines

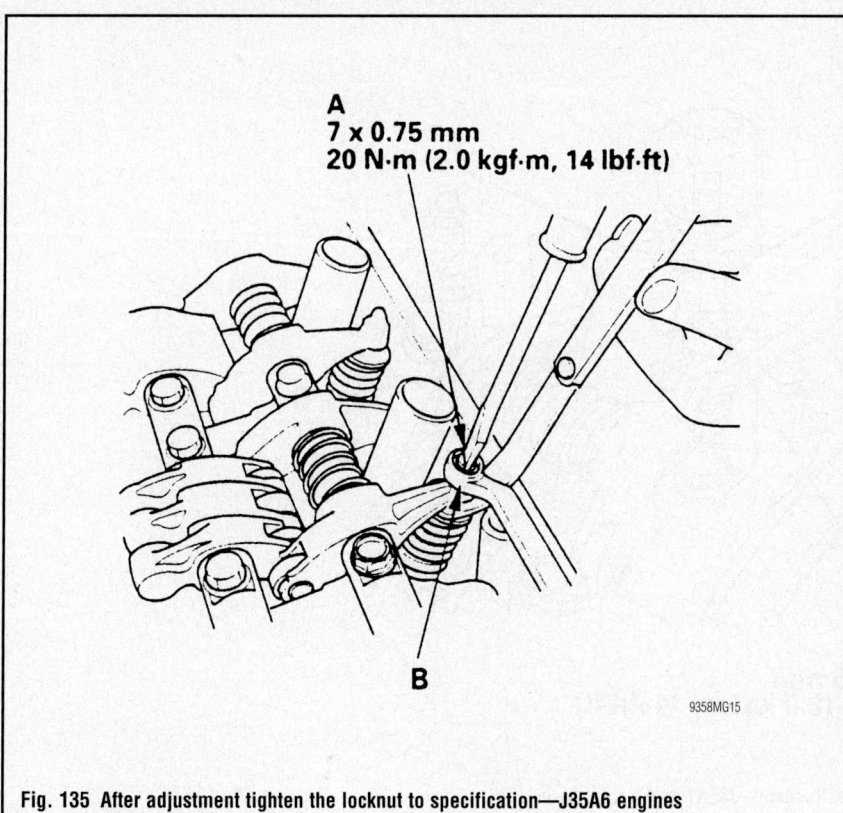

A
7 x 0.75 mm
20 N·m (2.0 kgf·m, 14 lbf·ft)

B

9358MG15

Fig. 135 After adjustment tighten the locknut to specification—J35A6 engines

1. Before servicing the vehicle, refer to the precautions in the beginning of this section.
2. Remove or disconnect the following:
   - Negative battery cable
   - Air intake tube
   - Intake manifold
   - Valve cover
3. Rotate the crankshaft so that the valves to be adjusted are closed and the rocker arm is contacting the camshaft lobe base circle.
4. Measure the valve clearance. If adjustment is necessary, loosen the locknut and turn the adjusting screw as necessary to achieve the correct valve clearance.
5. The correct valve clearance is:
   - Intake valves: 0.008–0.009 inches (0.20–0.24 mm)
   - Exhaust valves: 0.011–0.013 inches (0.28–0.32 mm)
6. After adjustment, tighten the locknuts to 14 ft. lbs. (20 Nm).
7. Install or connect the following:
   - Valve cover
   - Intake manifold
   - Air intake tube
   - Negative battery cable
8. Start the engine and check for proper operation.

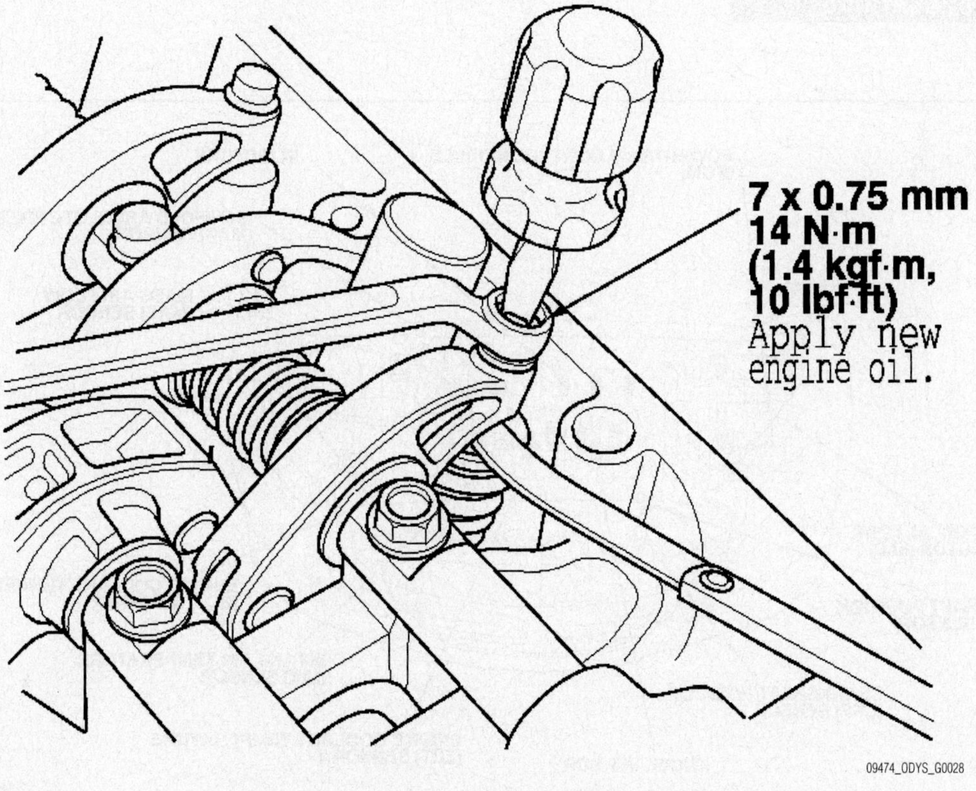

7 x 0.75 mm
14 N·m
(1.4 kgf·m,
10 lbf·ft)
Apply new
engine oil.

09474_ODYS_G0028

Fig. 136 After adjustment tighten the locknut to specification—J35A7 engine front head

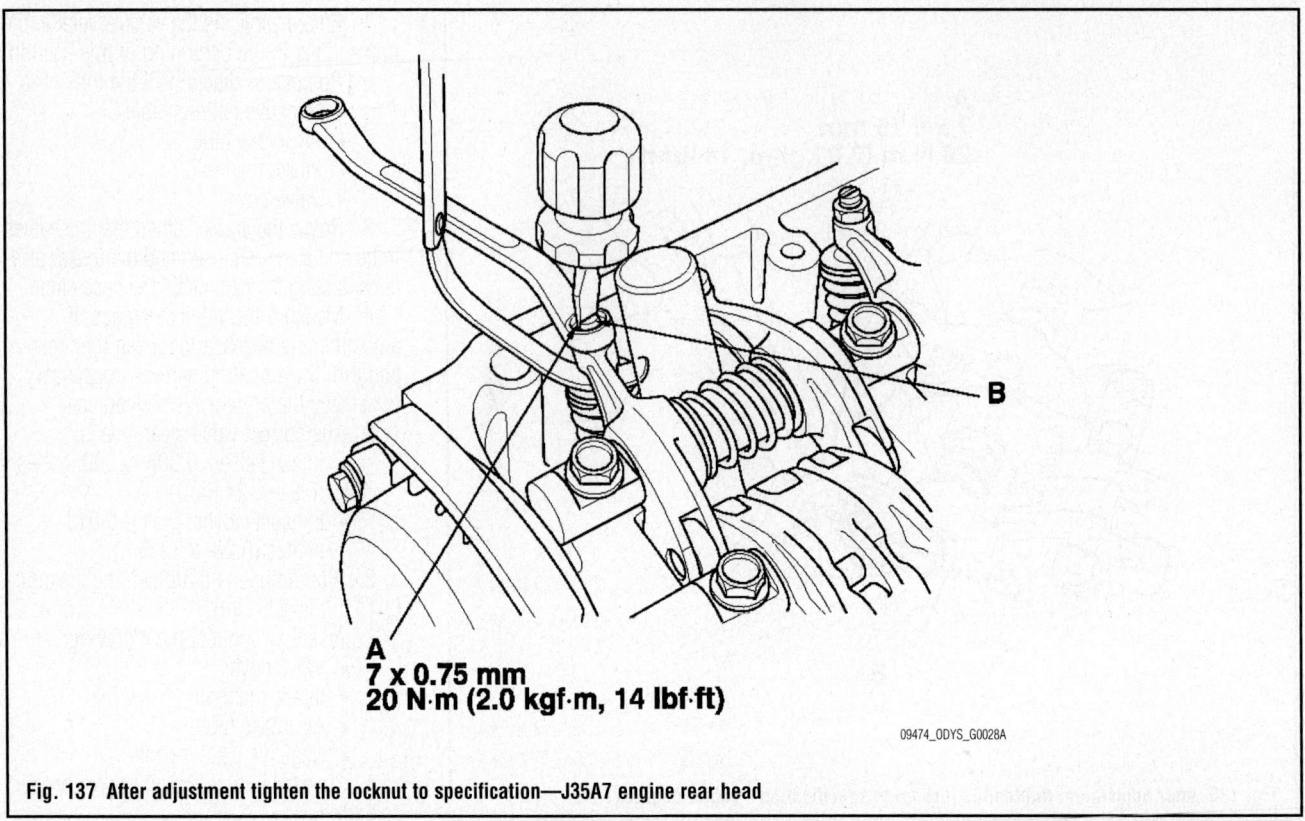

A
7 x 0.75 mm
20 N·m (2.0 kgf·m, 14 lbf·ft)

09474_ODYS_G0028A

Fig. 137 After adjustment tighten the locknut to specification—J35A7 engine rear head

ENGINE PERFORMANCE & EMISSION CONTROL

COMPONENT LOCATIONS

See Figure 138.

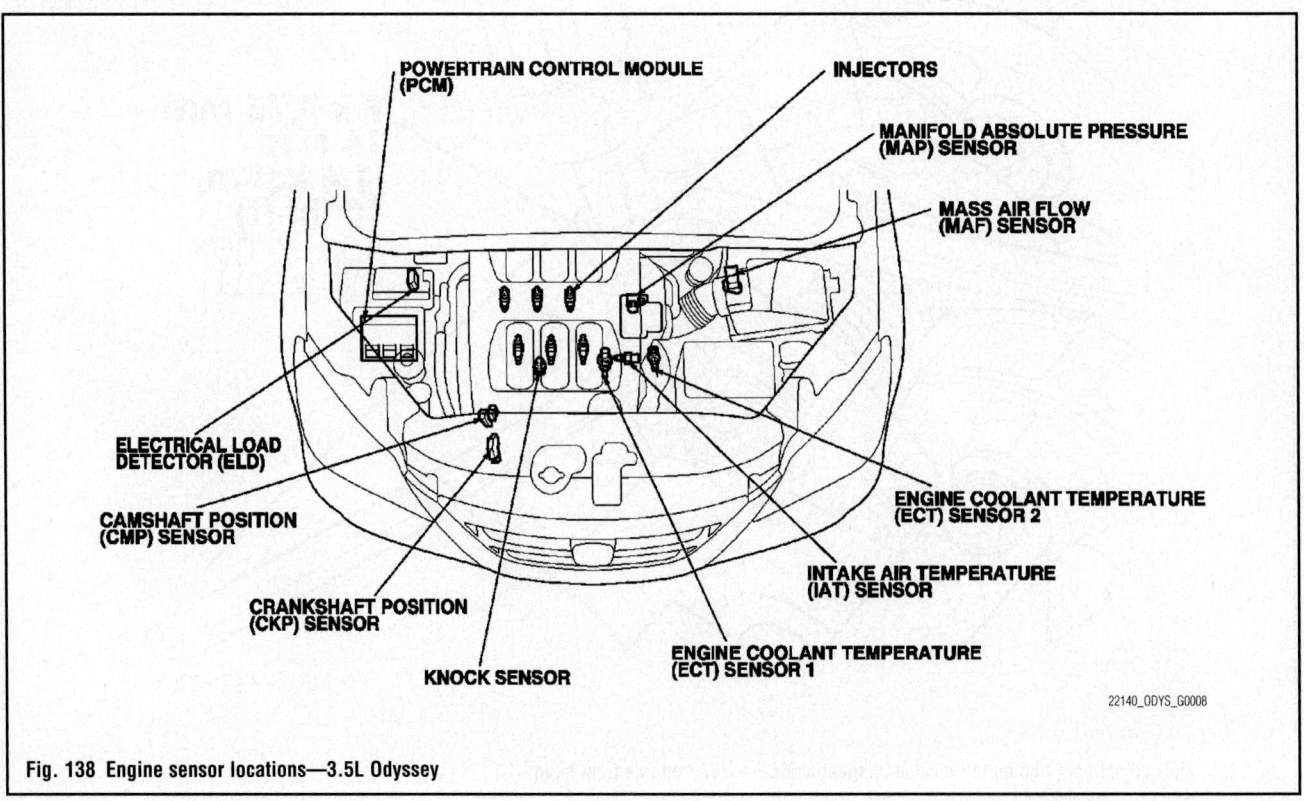

22140_ODYS_G0008

Fig. 138 Engine sensor locations—3.5L Odyssey

CAMSHAFT POSITION (CMP) SENSOR

LOCATION

The CMP sensor is located in the back cover behind the front camshaft pulley.

REMOVAL & INSTALLATION

1. Remove the timing belt.
2. Remove the front camshaft pulley (CMP sensor pulse plate).
3. Disconnect the CMP sensor 3P connector, then remove the back cover.
4. Remove the CMP sensor from the back cover.
5. Installation is the reverse order of removal.

CRANKSHAFT POSITION (CKP) SENSOR

LOCATION

The CKP sensor is mounting next to the crankshaft pulley behind the front cover.

REMOVAL & INSTALLATION

See Figure 139.

1. Move the auto-tensioner to remove tension from the drive belt, then remove the belt.
2. Remove the crankshaft pulley.
3. Remove the upper and lower front covers from the engine.
4. Remove the CKP sensor from the oil pump.

5. Installation is the reverse order of removal.

ENGINE COOLANT TEMPERATURE (ECT) SENSOR

LOCATION

There are two ECT sensors located below the engine appearance cover.

REMOVAL & INSTALLATION

Sensor 1

1. Drain the engine coolant.
2. Remove the engine cover.
3. Disconnect the ECT sensor 1 2P connector.
4. Remove the ECT sensor 1.
5. Install the parts in the reverse order of removal with a new O-ring (C), then refill the radiator with engine coolant, and bleed air from the cooling system with the heater valve open.

Sensor 2

1. Drain the engine coolant.
2. Remove the engine cover.
3. Disconnect the ECT sensor 2 2P connector.
4. Remove the ECT sensor 2.
5. Install the parts in the reverse order of removal with a new O-ring (C), then refill the radiator with engine coolant, and bleed air from the cooling system with the heater valve open.

HEATED OXYGEN (HO2S) SENSOR

LOCATION

There are two HO2S sensors, one located in the exhaust down pipe and one behind the catalytic converter.

REMOVAL & INSTALLATION

1. Disconnect the front secondary HO2S 4P connector, then remove the HO2S (front or rear).
2. Install the parts in the reverse order of removal.

INTAKE AIR TEMPERATURE (IAT) SENSOR

LOCATION

The IAT sensor is located in the air intake assembly.

REMOVAL & INSTALLATION

1. Disconnect the sensor wiring harness.
2. Remove the screws.
3. Remove the IAT sensor.
4. Install the parts in the reverse order of removal.

KNOCK SENSOR (KS)

LOCATION

The knock sensor is located in the cylinder head underneath the intake manifold.

REMOVAL & INSTALLATION

1. Remove the intake manifold.
2. Remove the injector rails and the injector base.
3. Disconnect the knock sensor 1P connector, then remove the knock sensor.
4. Install the parts in the reverse order of removal.

MANIFOLD ABSOLUTE PRESSURE (MAP) SENSOR

LOCATION

The MAP sensor is located on the throttle bottle assembly.

REMOVAL & INSTALLATION

1. Remove the engine appearance cover.
2. Disconnect the MAP sensor 3P connector.
3. Remove the screw.

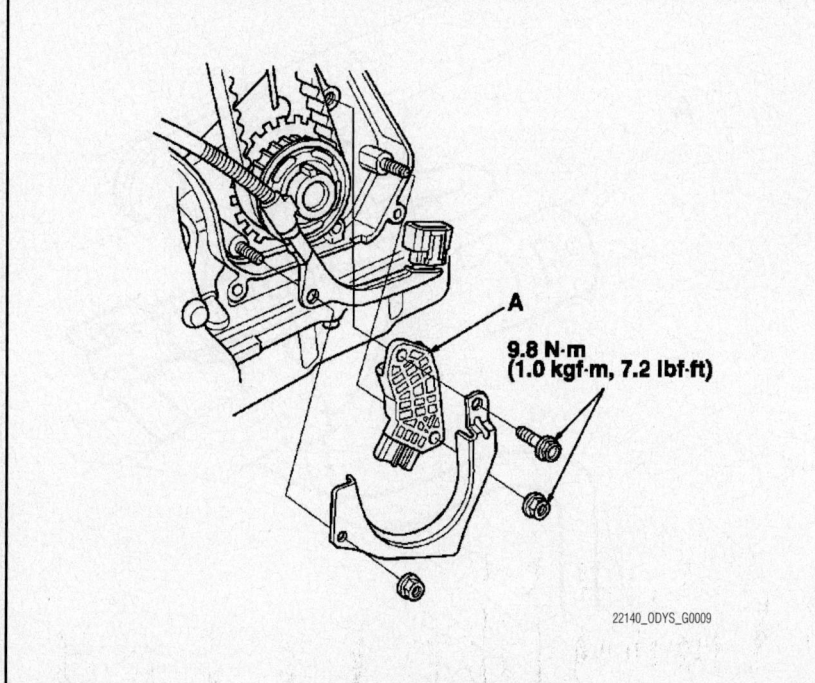

**9.8 N·m
(1.0 kgf·m, 7.2 lbf·ft)**

A

22140_ODYS_G0009

Fig. 130 Exploded view of the CKP sensor (A) mounting—Odyssey

4. Remove the MAP sensor.

5. Install the parts in the reverse order of removal with a new O-ring.

MASS AIR FLOW (MAF) SENSOR

LOCATION

The MAF sensor is located on the air intake tube of the air intake assembly.

REMOVAL & INSTALLATION

1. Disconnect the MAF sensor 5P connector.

2. Remove the screws.

3. Remove the MAF sensor.

4. Install the parts in the reverse order of removal.

THROTTLE POSITION SENSOR (TPS)

LOCATION

The TPS is part of the throttle actuator assembly and an integral part of the throttle body.

REMOVAL & INSTALLATION

For additional information, refer to the following section, "Throttle Body, Removal & Installation."

VEHICLE SPEED SENSOR (VSS)

LOCATION

The VSS is located on the transaxle assembly.

REMOVAL & INSTALLATION

1. Remove the engine splash shield.

2. Disconnect the output shaft (countershaft) speed sensor connector, and remove the output shaft (countershaft) speed sensor.

3. Install a new O-ring on a new output shaft (countershaft) speed sensor and apply clean ATF to the O-ring, then install the output shaft (countershaft) speed sensor in the transmission housing.

4. Check the connector for rust, dirt, or oil, then connect the connector securely.

5. Install the engine splash shield.

FUEL
GASOLINE FUEL INJECTION SYSTEM

FUEL SYSTEM SERVICE PRECAUTIONS

Safety is the most important factor when performing not only fuel system maintenance but any type of maintenance. Failure to conduct maintenance and repairs in a safe manner may result in serious personal injury or death. Maintenance and testing of the vehicle's fuel system components can be accomplished safely and effectively by adhering to the following rules and guidelines.

• To avoid the possibility of fire and personal injury, always disconnect the negative battery cable unless the repair or test procedure requires that battery voltage be applied.

• Always relieve the fuel system pressure prior to disconnecting any fuel system component (injector, fuel rail, pressure regulator, etc.), fitting or fuel line connection. Exercise extreme caution whenever relieving fuel system pressure to avoid exposing skin, face and eyes to fuel spray. Please be advised that fuel under pressure may penetrate the skin or any part of the body that it contacts.

• Always place a shop towel or cloth around the fitting or connection prior to loosening to absorb any excess fuel due to spillage. Ensure that all fuel spillage (should it occur) is quickly removed from engine surfaces. Ensure that all fuel soaked cloths or towels are deposited into a suitable waste container.

• Always keep a dry chemical (Class B) fire extinguisher near the work area.

• Do not allow fuel spray or fuel vapors to come into contact with a spark or open flame.

• Always use a back-up wrench when loosening and tightening fuel line connection fittings. This will prevent unnecessary stress and torsion to fuel line piping.

• Always replace worn fuel fitting O-rings with new Do not substitute fuel hose or equivalent where fuel pipe is installed.

Before servicing the vehicle, make sure to also refer to the precautions in the beginning of this section as well.

RELIEVING FUEL SYSTEM PRESSURE

Before disconnecting fuel lines or hoses, relieve pressure from the system by stopping the fuel pump and then disconnecting the fuel tube/quick connect fitting in the engine compartment.

With the HDS

See Figures 140 and 141.

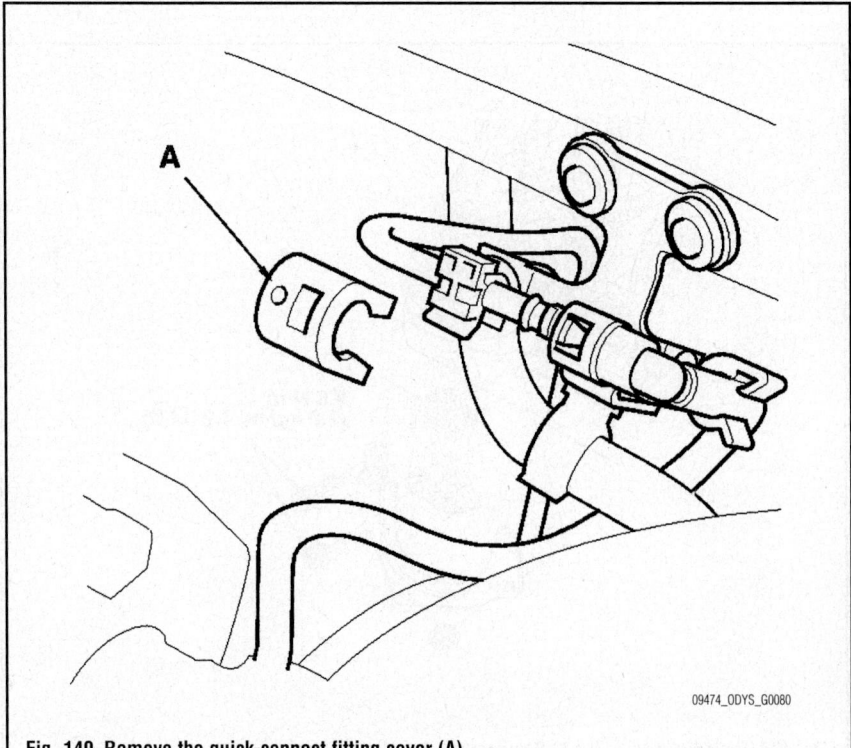

09474_ODYS_G0080

Fig. 140 Remove the quick-connect fitting cover (A)

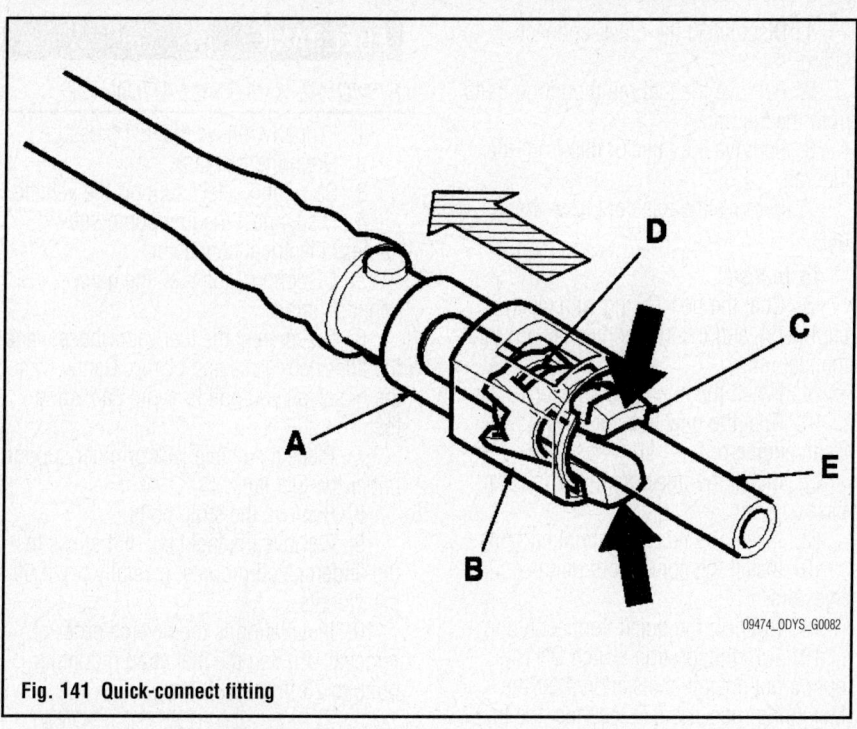

Fig. 141 Quick-connect fitting

1. Make sure you have the anti-theft codes for the radio and the navigation system (if equipped) then write down the radio station and XM radio channel presets.

2. Remove the fuel fill cap, and relieve the pressure in the fuel tank.

3. Turn the ignition switch ON (II).

4. From the INSPECTION MENU of the HDS, select Fuel Pump OFF, then start the engine and let it idle until it stalls.

5. Turn the ignition switch OFF.

➡**Do not allow the engine to idle above 1,000 RPM or the PCM will continue to operate the fuel pump. A DTC or a Temporary DTC may be set during this procedure. Check for DTCs, and clear them as needed.**

6. Turn the ignition switch OFF.

7. Disconnect the negative cable from the battery.

8. Remove the quick-connect fitting cover (A).

9. Check the fuel quick-connect fitting for dirt, and clean it if needed

10. Place a rag or shop towel over the quick-connect fitting (A).

11. Disconnect the quick-connect fitting (A): Hold the connector (B) with one hand, and squeeze the retainer tabs (C) with the other hand to release them from the locking tabs (D). Pull the connector off.

➡**Be careful not to damage the line (E) or other parts. Do not use tools. If the connector does not move, keep the retainer tabs pressed down, and alternately pull and push the connector**

until it comes off easily. Do not remove the retainer from the line; once removed, the retainer must be replaced with a new one.

12. After disconnecting the quick-connect fitting, check it for dirt or damage.

Reconnect the negative cable to the battery. Enter the anti-theft codes for the radio and the navigation system, then enter the customer's radio station and XM radio channel presets. Set the clock.

Without the HDS

See Figure 142.

1. Make sure you have the anti-theft codes for the radio and the navigation system (if equipped) then write down the radio station and XM radio channel presets.

2. Remove the left kick panel, then remove PGM-FI main relay 2 (FUEL PUMP) (A) from the driver's under-dash fuse/relay box.

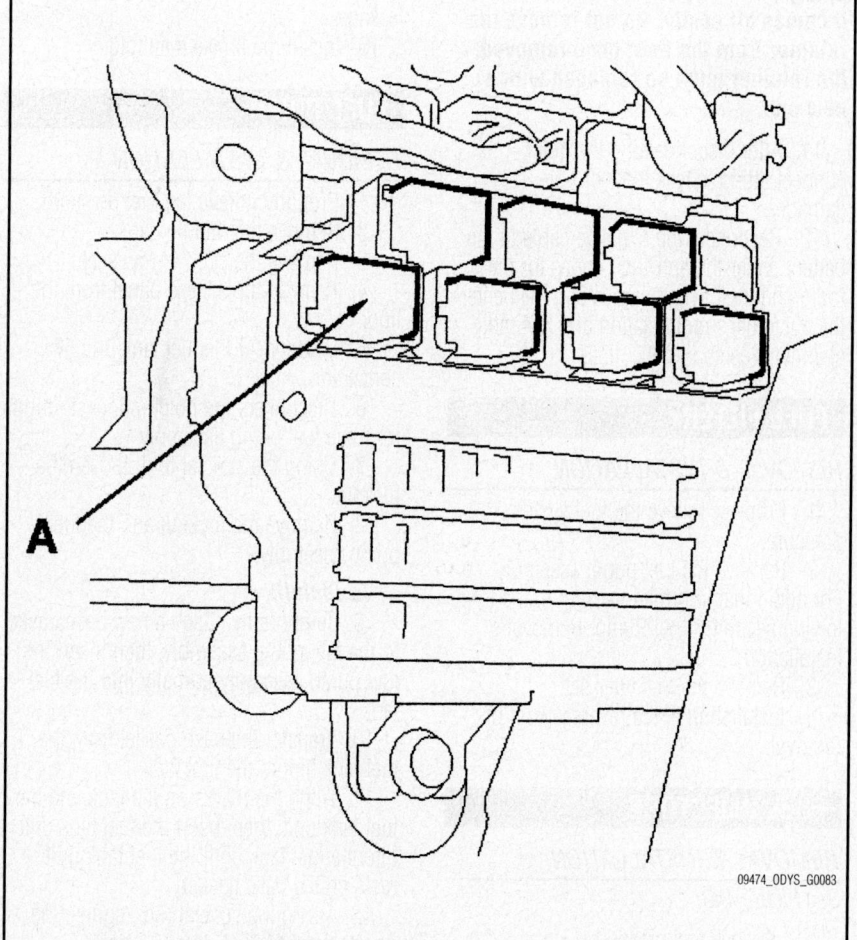

Fig. 142 Remove the left kick panel, then remove PGM-FI main relay 2 (FUEL PUMP) (A) from the driver's under-dash fuse/relay box

3. Start the engine, and let it idle until it stalls.

➡**If any DTCs are stored, clear and ignore them.**

4. Turn the ignition switch OFF.
5. Remove the fuel fill cap.
6. Disconnect the negative cable from the battery.
7. Remove the quick-connect fitting cover (A).
8. Check the fuel quick-connect fitting for dirt, and clean it if needed.
9. Place a rag or shop towel over the quick-connect fitting (A).
10. Disconnect the quick-connect fitting (A): Hold the connector (B) with one hand, and squeeze the retainer tabs (C) with the other hand to release them from the locking tabs (D). Pull the connector off.

➡**Be careful not to damage the line (E) or other parts. Do not use tools. If the connector does not move, keep the retainer tabs pressed down, and alternately pull and push the connector until it comes off easily. Do not remove the retainer from the line; once removed, the retainer must be replaced with a new one.**

11. After disconnecting the quick-connect fitting, check it for dirt or damage.
12. Reconnect the negative cable to the battery. Enter the anti-theft codes for the radio and the navigation system, then enter the customer's radio station and XM radio channel presets. Set the clock.

FUEL FILTER

REMOVAL & INSTALLATION

1. Properly relieve the fuel system pressure.
2. Remove the fuel pump assembly. For additional information, refer to the following section, "Fuel Pump, Removal & Installation."
3. Remove the fuel filter set.
4. Installation is the reverse order of removal.

FUEL INJECTORS

REMOVAL & INSTALLATION

See Figure 143.

1. Relieve fuel pressure.
2. Remove the intake manifold.
3. Disconnect the connectors from the injectors.
4. Disconnect the quick-connect fitting.
5. Remove the fuel rail mounting bolts from the fuel rail.
6. Remove the injector clip from the fuel rail.
7. Remove the injectors from the rails.

To install:

8. Coat the new O-ring with clean engine oil, and insert the injectors (B) into the fuel rail.
9. Install the injector clip.
10. Coat the new injector O-ring with clean engine oil.
11. Install the injectors in the injector base.
12. Install the fuel rail mounting bolts.
13. Install the connectors on the injectors.
14. Connect the quick-connect fitting.
15. Turn the ignition switch ON (II), but do not operate the starter. After the fuel pump runs about 2 seconds, the fuel pressure in the fuel line rises. Repeat this two or three times, then check for fuel leakage.
16. Install the intake manifold.

FUEL PUMP

REMOVAL & INSTALLATION

1. Properly relieve the fuel pressure.
2. Remove the fuel fill cap.
3. Remove the second row seat.
4. Remove the access panel from the floor.
5. Disconnect the fuel tank unit 5P connector.
6. Disconnect the quick-connect fittings from the fuel pump assembly.
7. Using the special tool, loosen the locknut.
8. Remove the locknut and the fuel pump assembly.

To install:

9. Temporarily attach a new base gasket to the fuel pump assembly, then insert the fuel pump assembly partially into the fuel tank.
10. Transfer the base gasket from the fuel tank unit to the fuel tank.
11. Align the marks on the tank and the fuel tank unit, then insert the fuel tank unit into the fuel tank until the fuel tank unit rests on the base gasket.
12. Using the special tool, tighten the new locknut to 69 ft. lbs. (93 Nm).
13. Connect the electrical connector.
14. The remainder of the installation is the reverse order of removal.

FUEL TANK

REMOVAL & INSTALLATION

1. Properly relieve the fuel pressure.
2. Drain the fuel tank.
3. Raise and safely support the vehicle.
4. Disconnect the fuel pump sub-harness electrical connector.
5. Disconnect the fuel line quick-connect fitting.
6. Disconnect the fuel vapor hoses, and the filler neck hose and clamp. Gently twist the hoses as you pull to avoid damaging them.
7. Place a suitable jack or other support under the fuel tank.
8. Remove the strap bolts.
9. Remove the fuel tank. If it sticks to the undercoated mounts, carefully pry it off the mounts.
10. Installation is the reverse order of removal. Tighten the fuel strap mounting bolts to 28 ft. lbs. (38 Nm).

IDLE SPEED

ADJUSTMENT

The idle speed is controlled by the Powertrain Control Module (PCM). No adjustment is necessary or possible.

THROTTLE BODY

REMOVAL & INSTALLATION

See Figure 144.

⁂ CAUTION

Do not insert your fingers into the installed throttle body when you turn the ignition switch ON (II) or while the ignition switch is ON (II). If you do, you will seriously injure your fingers. If the throttle valve is activated.

➡**If you are replacing the throttle body, start at step 1. If you are removing the throttle body, start at step 4. This procedure requires the use of the HDS or equivalent scan tool.**

1. Connect the HDS while the engine is stopped.
2. Select the INSPECTION MENU with the HDS.
3. Do the TP LEARNING CHECK in the ETCS TEST.
4. Disconnect the MAP sensor connector.
5. Remove the intake air duct.
6. Disconnect the throttle body connector.

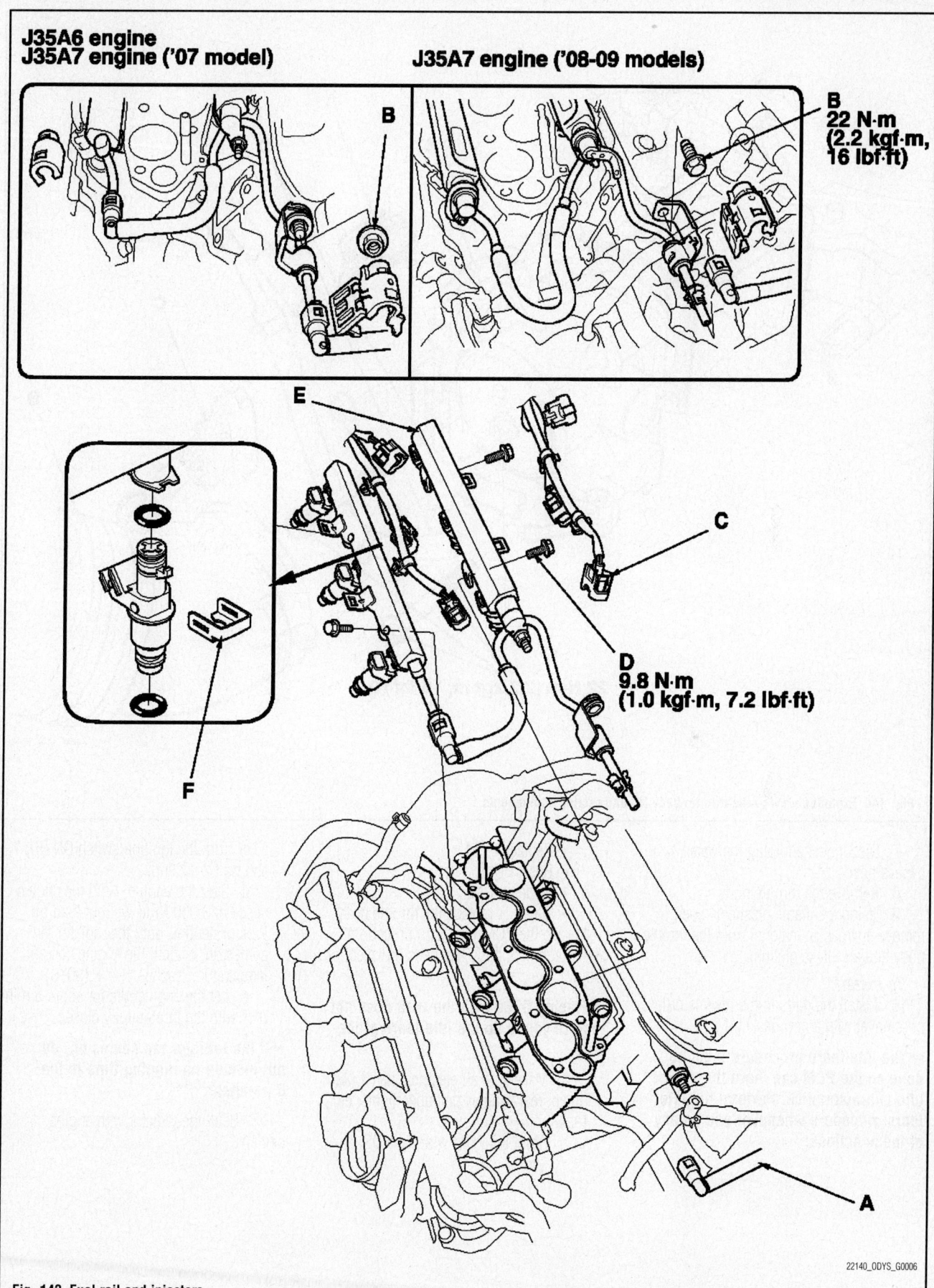

J35A6 engine
J35A7 engine ('07 model)

J35A7 engine ('08-09 models)

B

B
22 N·m
(2.2 kgf·m,
16 lbf·ft)

E

C

D
9.8 N·m
(1.0 kgf·m, 7.2 lbf·ft)

F

A

22140_ODYS_G0006

Fig. 143 Fuel rail and injectors

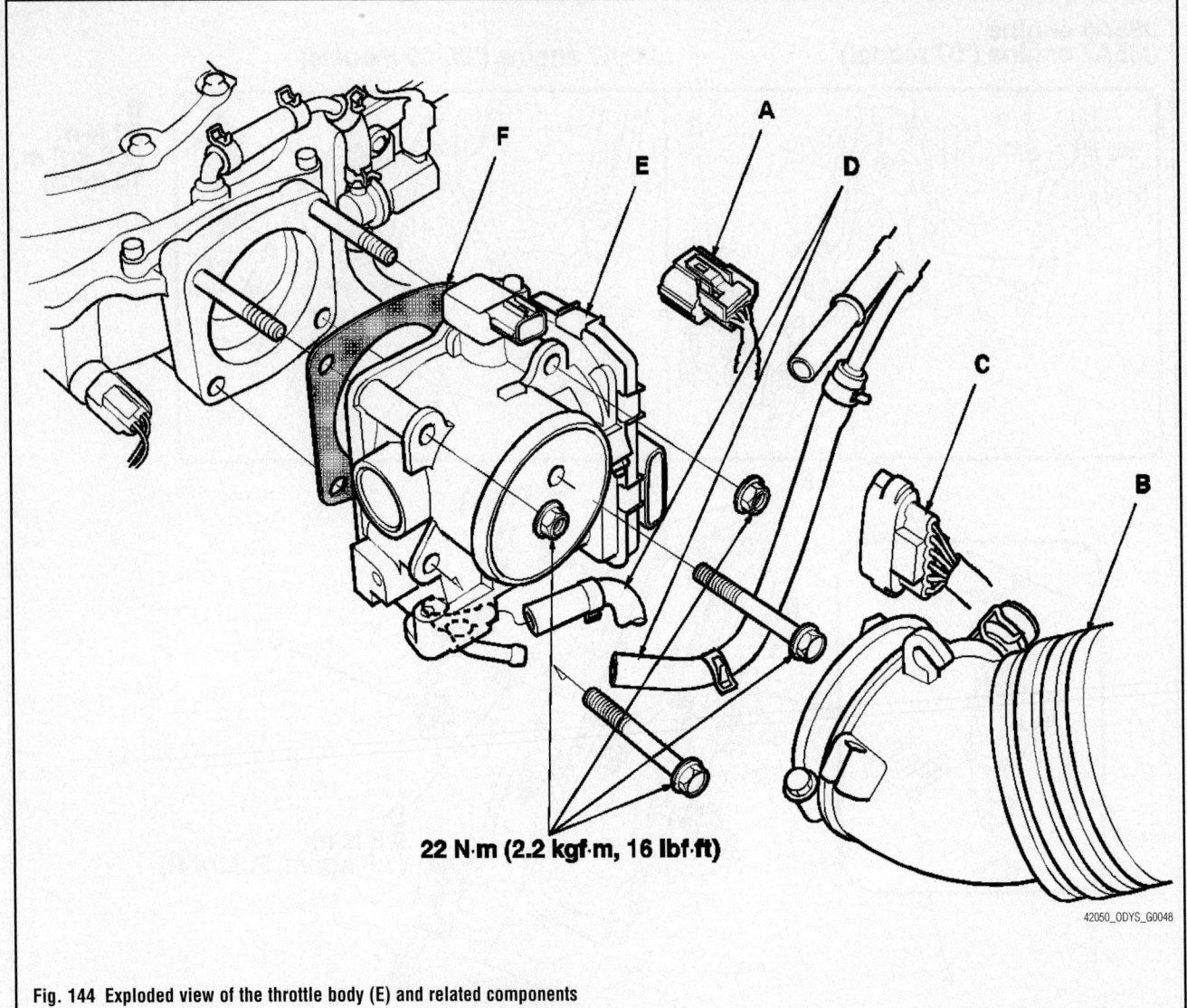

Fig. 144 Exploded view of the throttle body (E) and related components

22 N·m (2.2 kgf·m, 16 lbf·ft)

42050_ODYS_G0048

7. Disconnect and plug the water bypass hoses.

8. Remove the throttle body.

9. Using a suitable plastic scraper, remove any gasket material from the throttle body and air intake plenum.

To install:

10. Install the parts in the reverse order of removal with a new gasket (F).

➡**The idle learn procedure must be done so the PCM can learn the engine idle characteristics. Perform the idle learn procedure whenever you do any of these actions:**

- Replace PCM
- Reset PCM
- Update PCM
- Clean or replace the throttle body

11. Do the PCM idle learn procedure after the throttle body has been replaced, as follows:

➡**Erasing DTCs with the HDS does not require you to do the idle learn procedure.**

a. Make sure all electrical items (A/C, audio, rear window defogger, lights, etc.) are off.

b. Reset the PCM with the HDS.

c. Turn the ignition switch ON (II), and wait 2 seconds.

d. Start the engine. Hold the engine speed at 3,000 RPM without load (in Park or neutral) until the radiator fan comes on, or until the engine coolant temperature reaches 194°F (90°C).

e. Let the engine idle for about 5 minutes with the throttle fully closed.

➡**If the radiator fan comes on, do not include its running time in the 5 minutes.**

12. Refill the radiator with engine coolant.

HEATING & AIR CONDITIONING SYSTEM

BLOWER MOTOR

REMOVAL & INSTALLATION

The blower motor is located in the front passenger's right side foot well area.

1. If additional access is needed, remove the front passenger's right side lower kick panel.

2. Detach the electrical connectors from the blower motor.

3. Remove the fasteners on the blower motor mounting flange.

4. Remove the blower motor downward from the blower unit.

5. Installation is the reverse of the removal procedure.

HEATER CORE

REMOVAL & INSTALLATION

See Figures 145 through 161.

1. Make sure you have the anti-theft codes for the radio and the navigation system, then write down the XM radio channel presets.

2. Make sure the ignition is OFF, then disconnect the negative cable from the battery. Wait at least 3 minutes before proceeding.

3. Recover the refrigerant with a recovery/recycling/charging station.

4. Remove the intake manifold cover. See Intake Manifold Removal and Installation.

5. Remove the bolts and nut, then disconnect the suction line (A) and the receiver line (B) from the evaporator core.

6. From under the hood, open the cable clamp (A), then disconnect the heater valve cable (B) from the heater valve arm (C). Turn the heater valve arm to the fully opened position as shown.

7. When the engine is cool, drain the engine coolant from the radiator.

8. Slide the hose clamps (A) back. Remove the nut and the water valve (B), then disconnect the inlet heater hose (C) and the outlet heater hose (D) from the heater unit. Engine coolant will run out when the hoses are disconnected; drain it into a clean drip pan. Be sure not to let coolant spill on the electrical parts or the painted surfaces. If any coolant spills, rinse it off immediately.

9. Remove the mounting nut from the heater unit. Take care not to damage or bend the fuel lines and the brake lines, etc.

10. From under the dash, remove the clip (A) and detach the clip (B), release the hook (C) and stud (D), then remove the driver's heater lower cover (E) and passenger's heater lower cover (F).

11. Open the beverage holder (A), and remove the screws.

12. Remove the dashboard center lower cover (A).

a. Pull out the cover to release the clips (B, C).

b. Disconnect the seat heater connectors (D) and accessory socket connectors (E).

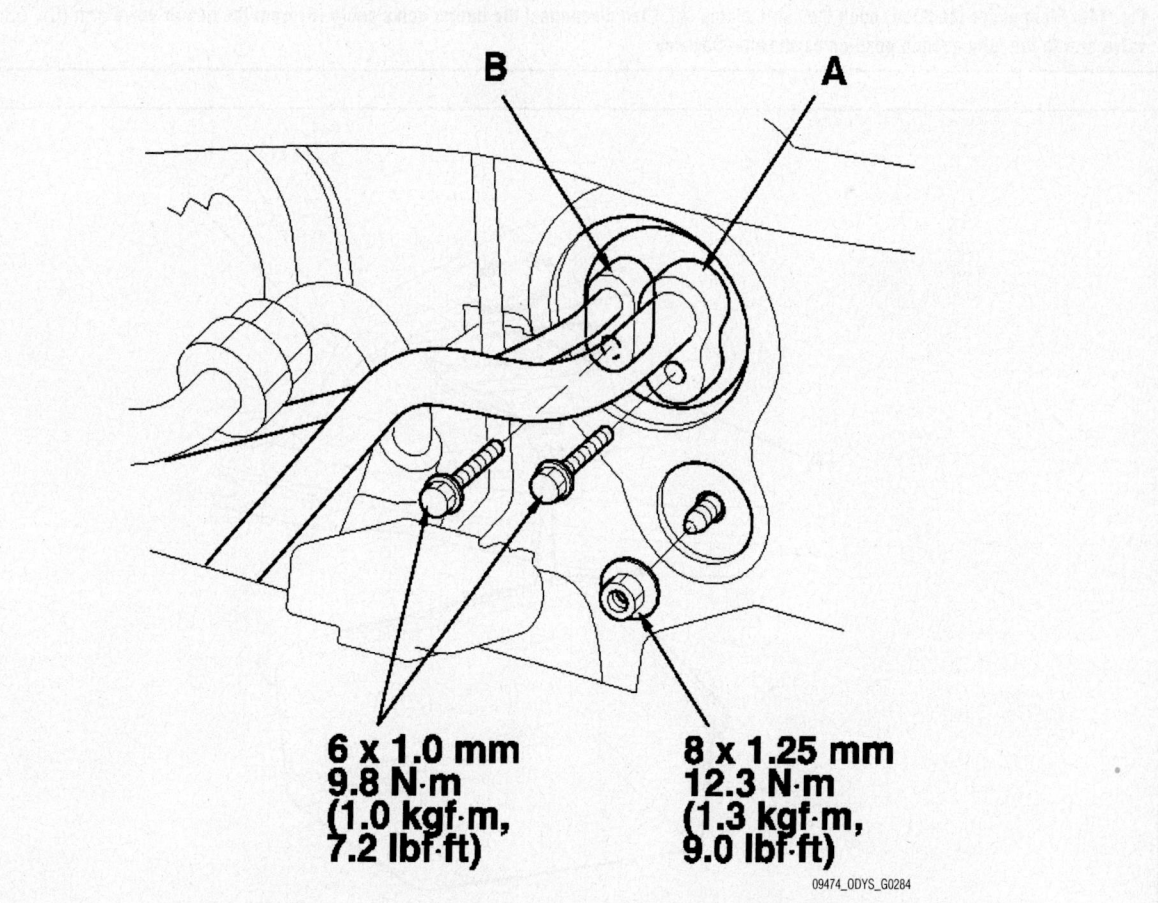

6 x 1.0 mm
9.8 N·m
(1.0 kgf·m,
7.2 lbf·ft)

8 x 1.25 mm
12.3 N·m
(1.3 kgf·m,
9.0 lbf·ft)

09474_ODYS_G0284

Fig. 145 Remove the bolts and nut, then disconnect the suction line (A) and the receiver line (B) from the evaporator core—Odyssey

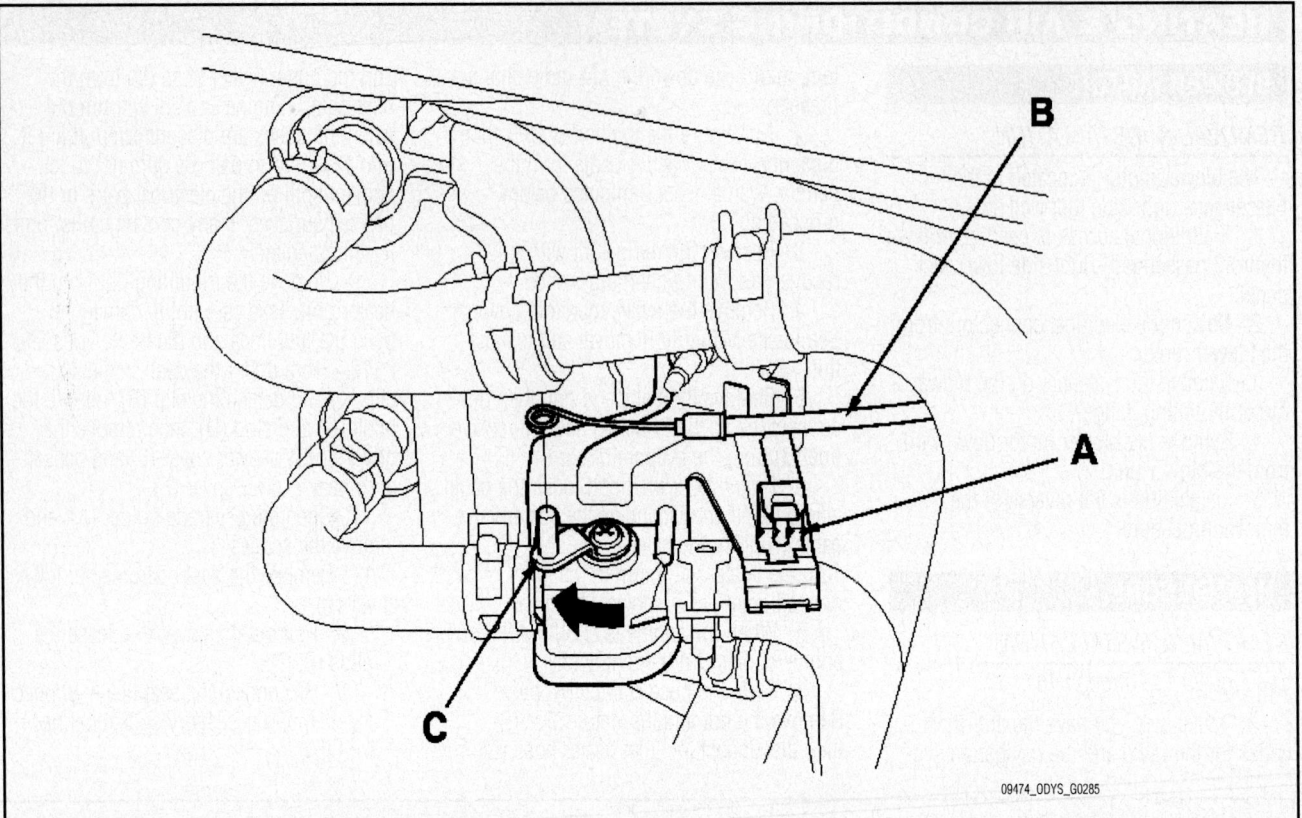

09474_ODYS_G0285

Fig. 146 From under the hood, open the cable clamp (A), then disconnect the heater valve cable (B) from the heater valve arm (C). Turn the heater valve arm to the fully opened position as shown—Odyssey

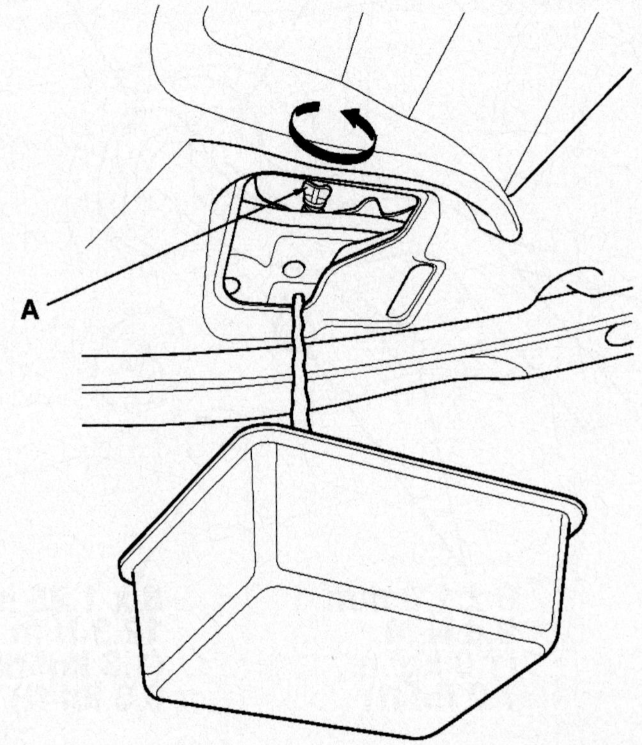

09474_ODYS_G0297

Fig. 147 Drain the engine coolant from the radiator

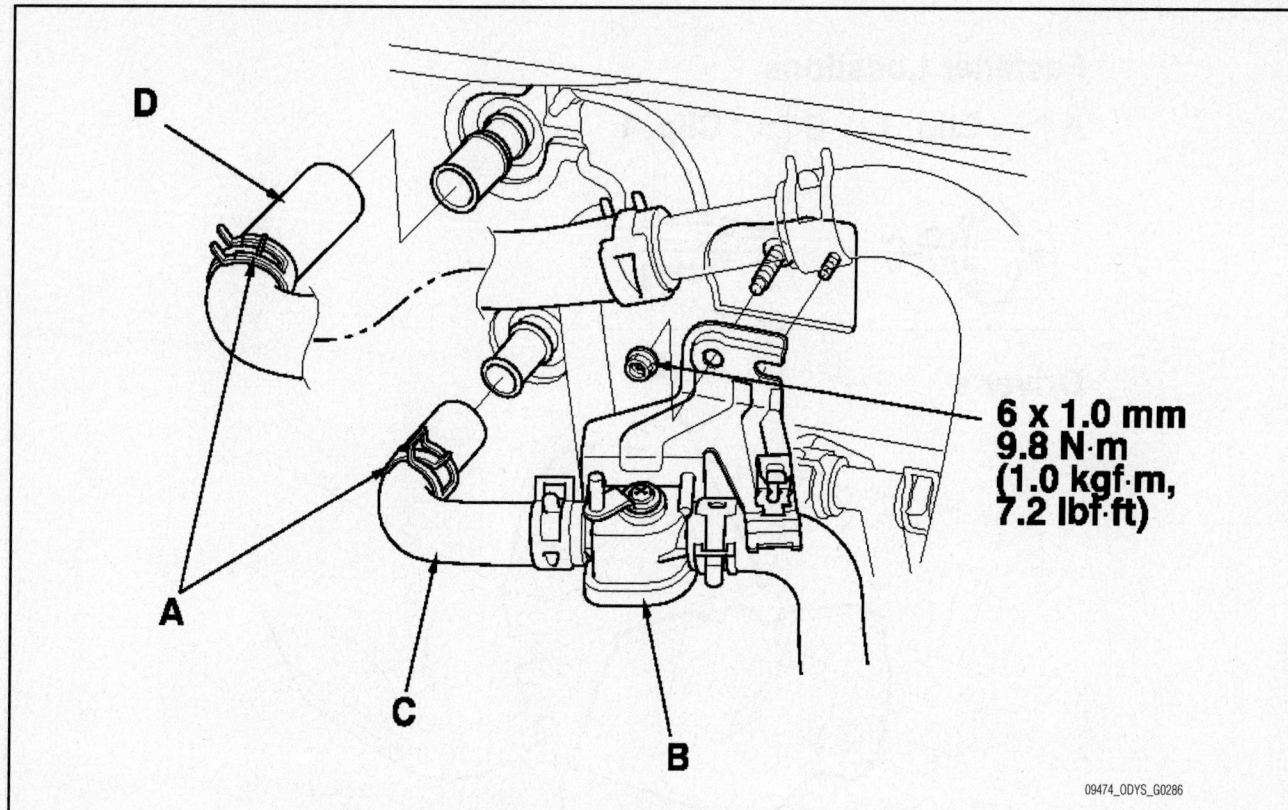

6 x 1.0 mm
9.8 N·m
(1.0 kgf·m,
7.2 lbf·ft)

09474_ODYS_G0286

Fig. 148 Slide the hose clamps (A) back. Remove the nut and the water valve (B), then disconnect the inlet heater hose (C) and the outlet heater hose (D) from the heater unit—Odyssey

8 x 1.25 mm
12.3 N·m
(1.3 kgf·m, 9.0 lbf·ft)

09474_ODYS_G0287

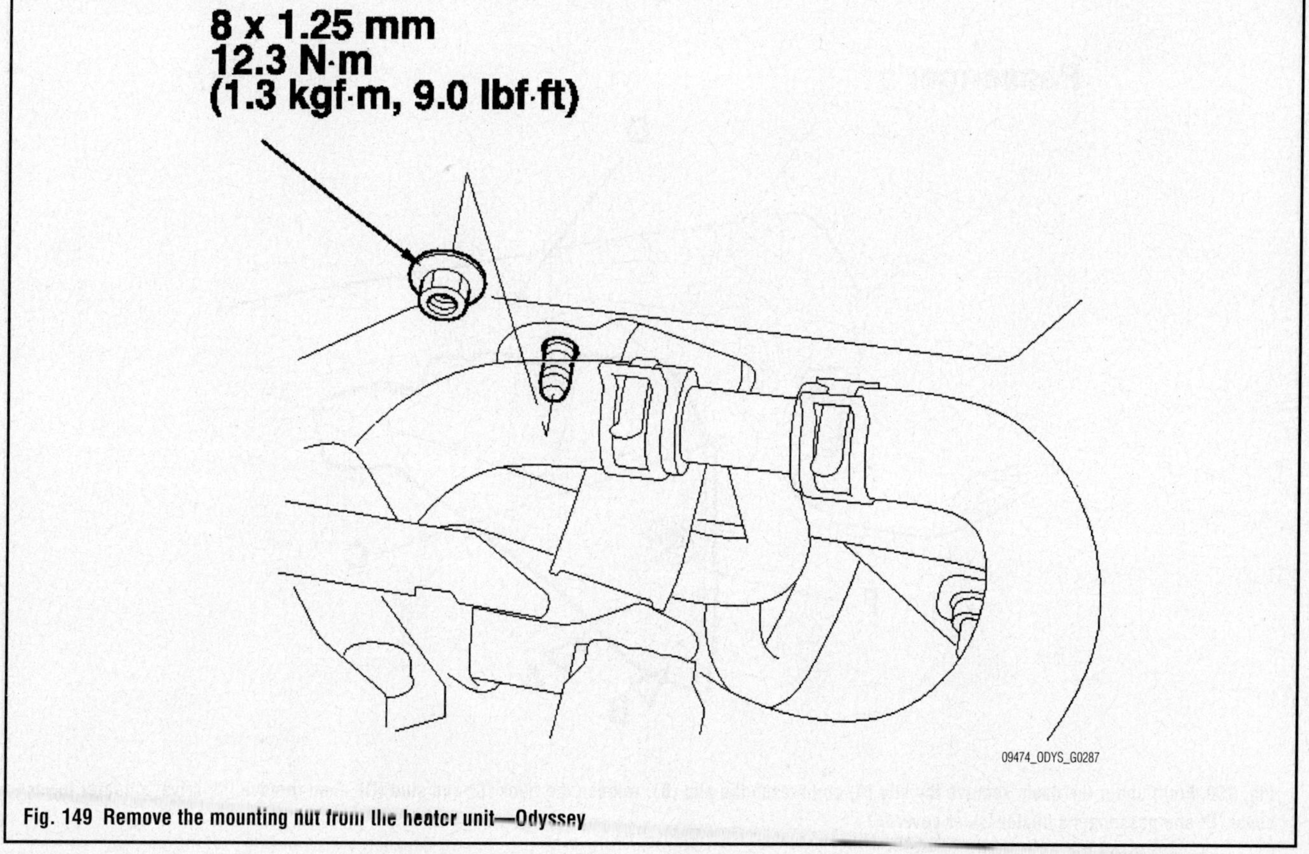

Fig. 149 Remove the mounting nut from the heater unit—Odyssey

Fastener Locations

A ▷ : Clip, 1   B ▷ : Clip, 1

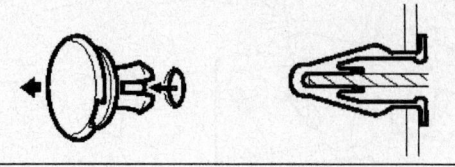

Driver's

Passenger's

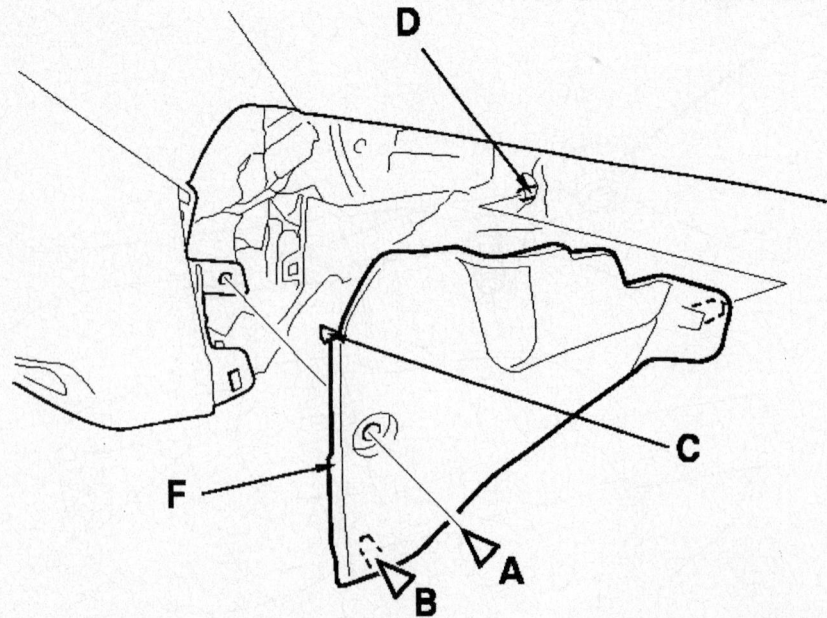

09474_ODYS_G0288

Fig. 150 From under the dash, remove the clip (A) and detach the clip (B), release the hook (C) and stud (D), then remove the driver's heater lower cover (E) and passenger's heater lower cover (F)

Fastener Locations

▶ : Screw, 4

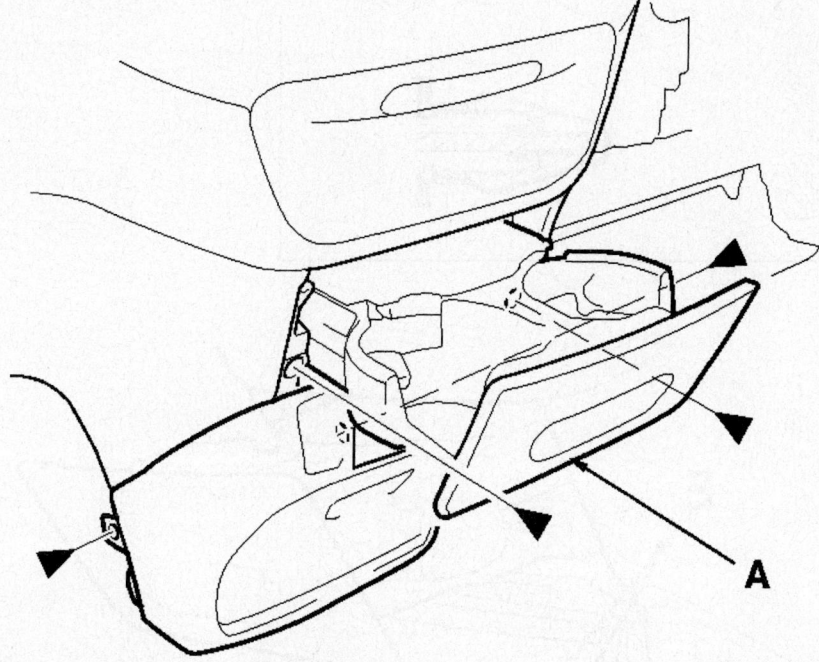

09474_ODYS_G0289

Fig. 151 Open the beverage holder (A), and remove the screws—Odyssey

13. Remove the screws, then remove the driver's A/C duct (A) from the dashboard (B).

14. Remove the screws, then remove the passenger's A/C duct (A) from the dashboard (B).

15. Remove the screws, then remove the center A/C duct from the dashboard.

16. Remove the screws, then remove the defogger duct (A) from the dashboard (B).

17. Disconnect the connectors (A) from the blower motor and the power transistor, then remove the wire harness clips (B).

18. Disconnect the connectors (A) from the mode control motor and the recirculation control motor, then remove the wire harness clips (B) and the connector clip (C).

19. Disconnect the connectors from the evaporator sensor and the air mix control motor, then remove the wire harness clips and the wire harness.

20. Remove the drain hose (A), then remove the nuts and the blower-heater unit (B).

21. Remove the self-tapping screws, the joint duct (A), and seal (B). Remove the self-tapping screws, then remove the passenger's heater outlet (C), and the heater core cover (D). Remove the self-tapping screws, the heater pipe brackets (E), the grommets (F), and carefully pull out the heater core (G) so you don't bend the inlet and outlet pipes.

22. Install the heater core in the reverse order of removal.

23. Install the heater unit in the reverse order of removal, and note these items:

- Do not interchange the inlet and outlet heater hoses, and install the hose clamps securely.
- Refill the cooling system with engine coolant.
- Make sure that there is no coolant leakage.
- Make sure that there is no air leakage.
- If you're installing a new evaporator core, add refrigerant oil (DENSO ND-OIL 8).

- Replace the O-rings with new ones at each fitting, and apply a thin coat of refrigerant oil before installing them. Be sure to use the correct O-rings for HFC-134a (R-134a) to avoid leakage.
- Immediately after using the oil, reinstall the cap on the container, and seal it to avoid moisture absorption.
- Do not spill the refrigerant oil on the vehicle; it may damage the paint; if the refrigerant oil contacts the paint, wash it off immediately.
- Make sure that there is no air leakage.
- Charge the system.

24. Adjust the heater valve cable.

a. Set the temperature control dial to Max Cool (Lo) with the ignition switch ON (II).

b. Attach the heater valve cable (B) to the air mix control linkage (C) as shown. Hold the end of the heater valve cable housing against the stop (D), then snap

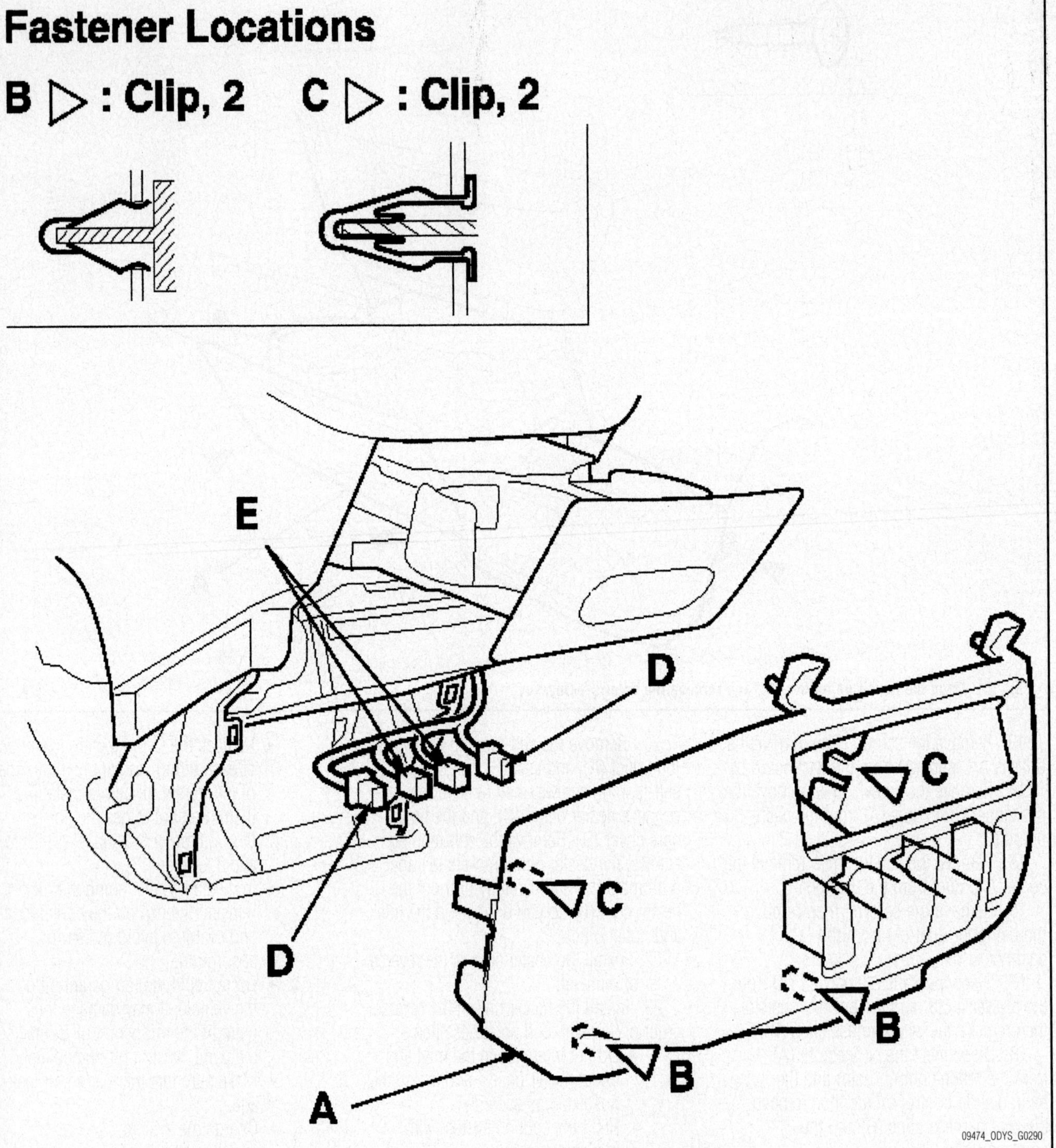

Fastener Locations
B ▷ : Clip, 2
C ▷ : Clip, 2
E
D
C
C
D
B
B
A
09474_ODYS_G0290

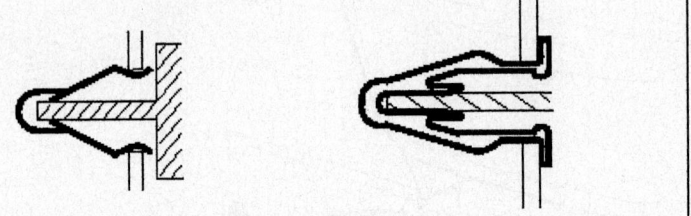

Fig. 152 Remove the dashboard center lower cover (A). Pull out the cover to release the clips (B, C). Disconnect the seat heater connectors (D) and accessory socket connectors (E)—Odyssey

Fastener Locations

▶ : Screw, 5

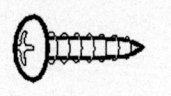

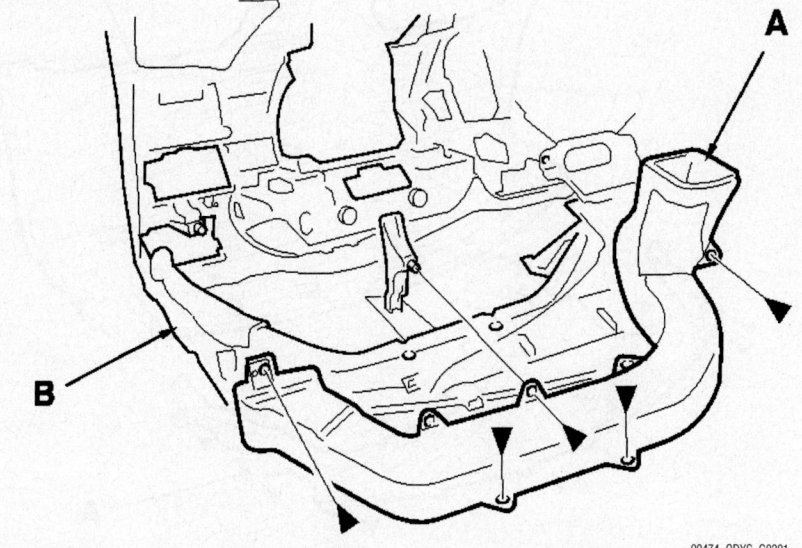

Fig. 153 Remove the screws, then remove the driver's A/C duct (A) from the dashboard (B)—Odyssey

09474_ODYS_G0291

Fastener Locations

▶ : Screw, 6

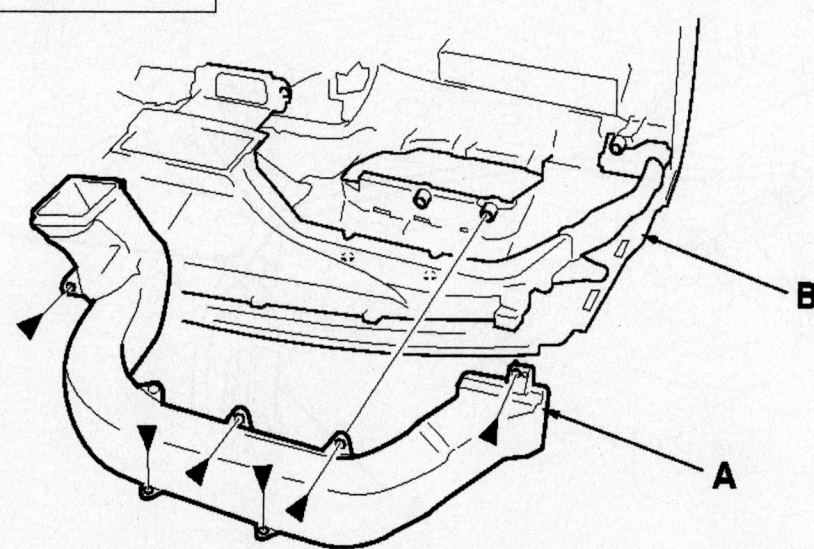

09474_ODYS_G0292

Fig. 154 Remove the screws, then remove the passenger's A/C duct (A) from the dashboard (B)—Odyssey

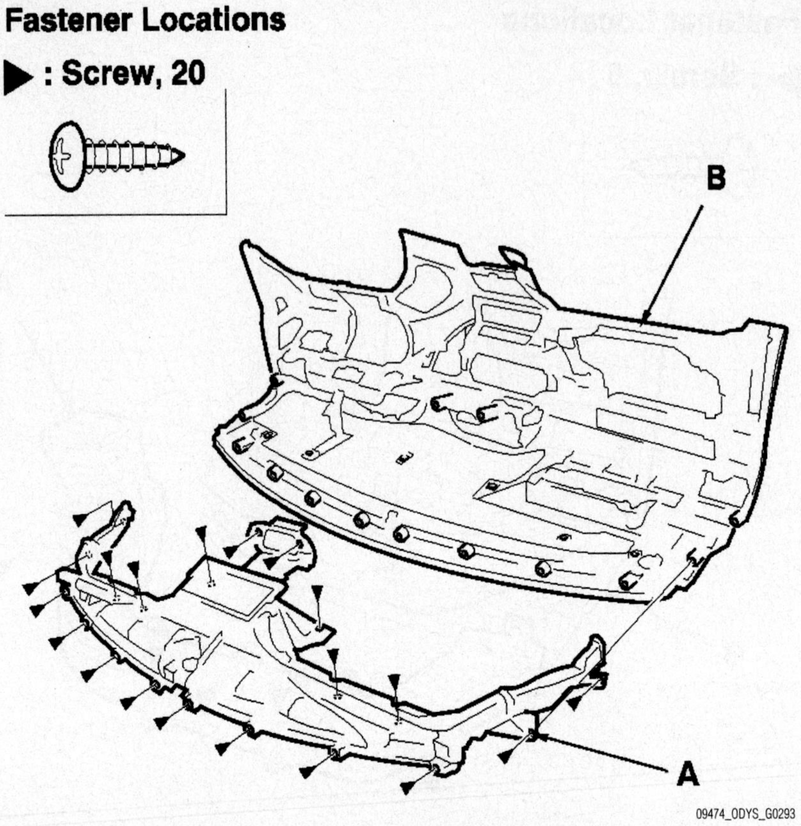
Fastener Locations
▶ : Screw, 20
B
A
09474_ODYS_G0293

Fig. 155 Remove the screws, then remove the defogger duct (A) from the dashboard (B)—Odyssey

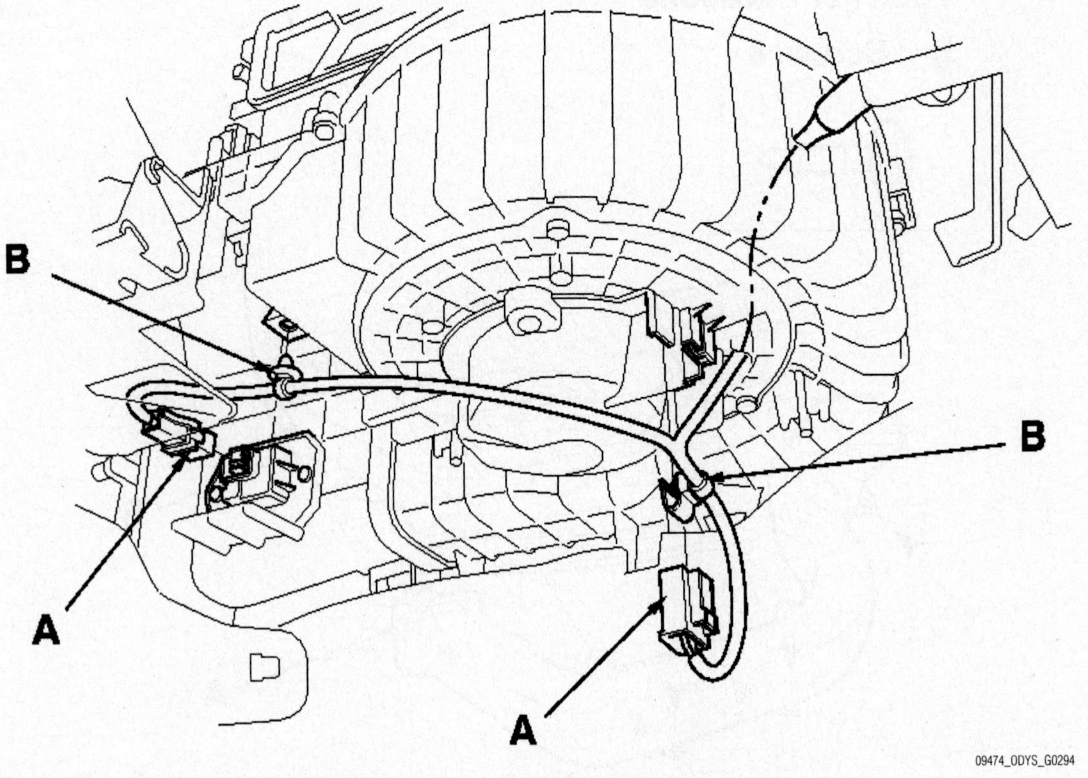
B
B
A
A
09474_ODYS_G0294

Fig. 156 Disconnect the connectors (A) from the blower motor and the power transistor, then remove the wire harness clips (B)—Odyssey

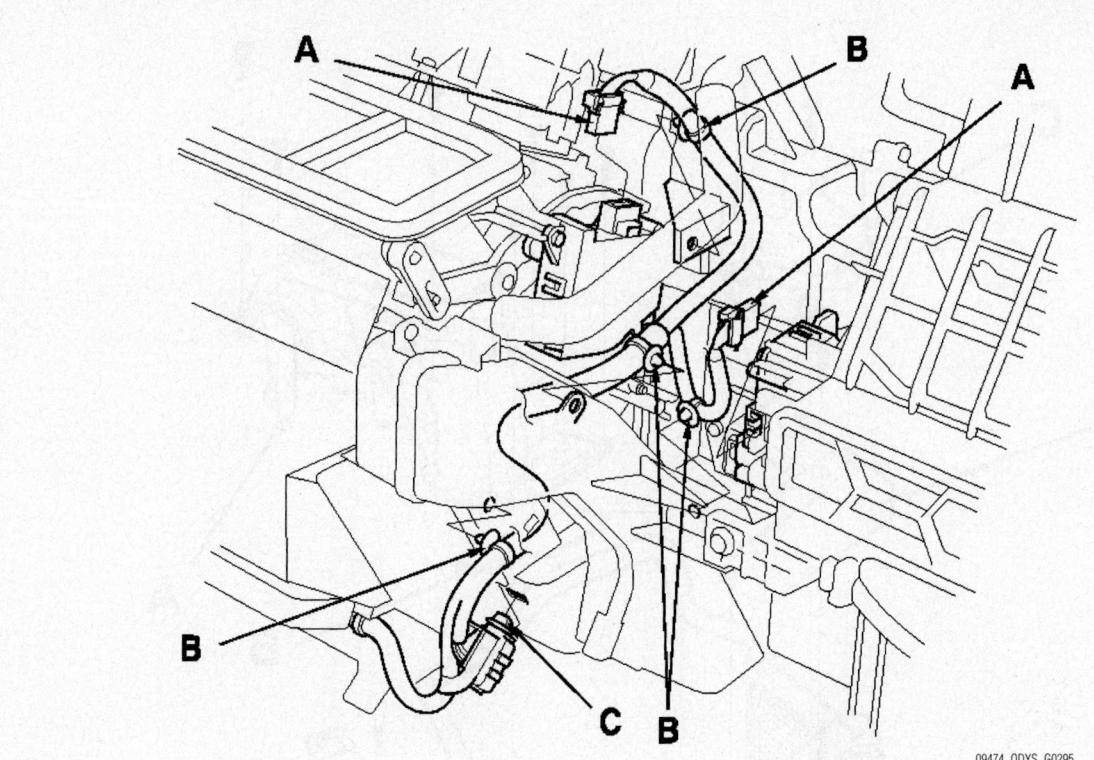

09474_ODYS_G0295

Fig. 157 Disconnect the connectors (A) from the mode control motor and the recirculation control motor, then remove the wire harness clips (B) and the connector clip (C)—Odyssey

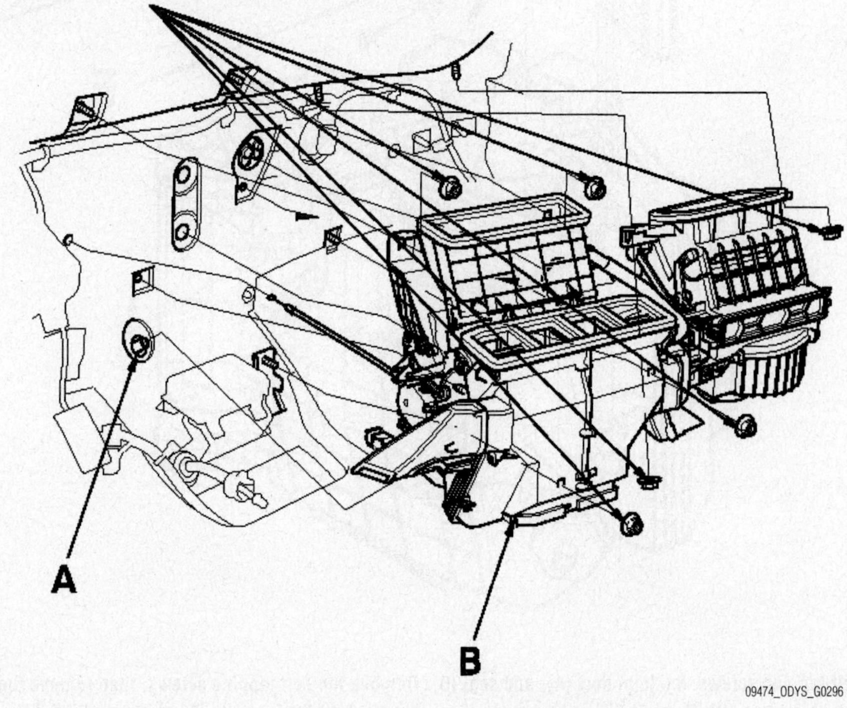

6 x 1.0 mm
9.8 N·m (1.0 kgf·m, 7.2 lbf·ft)

09474_ODYS_G0296

Fig. 158 Remove the drain hose (A), then remove the nuts and the blower-heater unit (B)—Odyssey

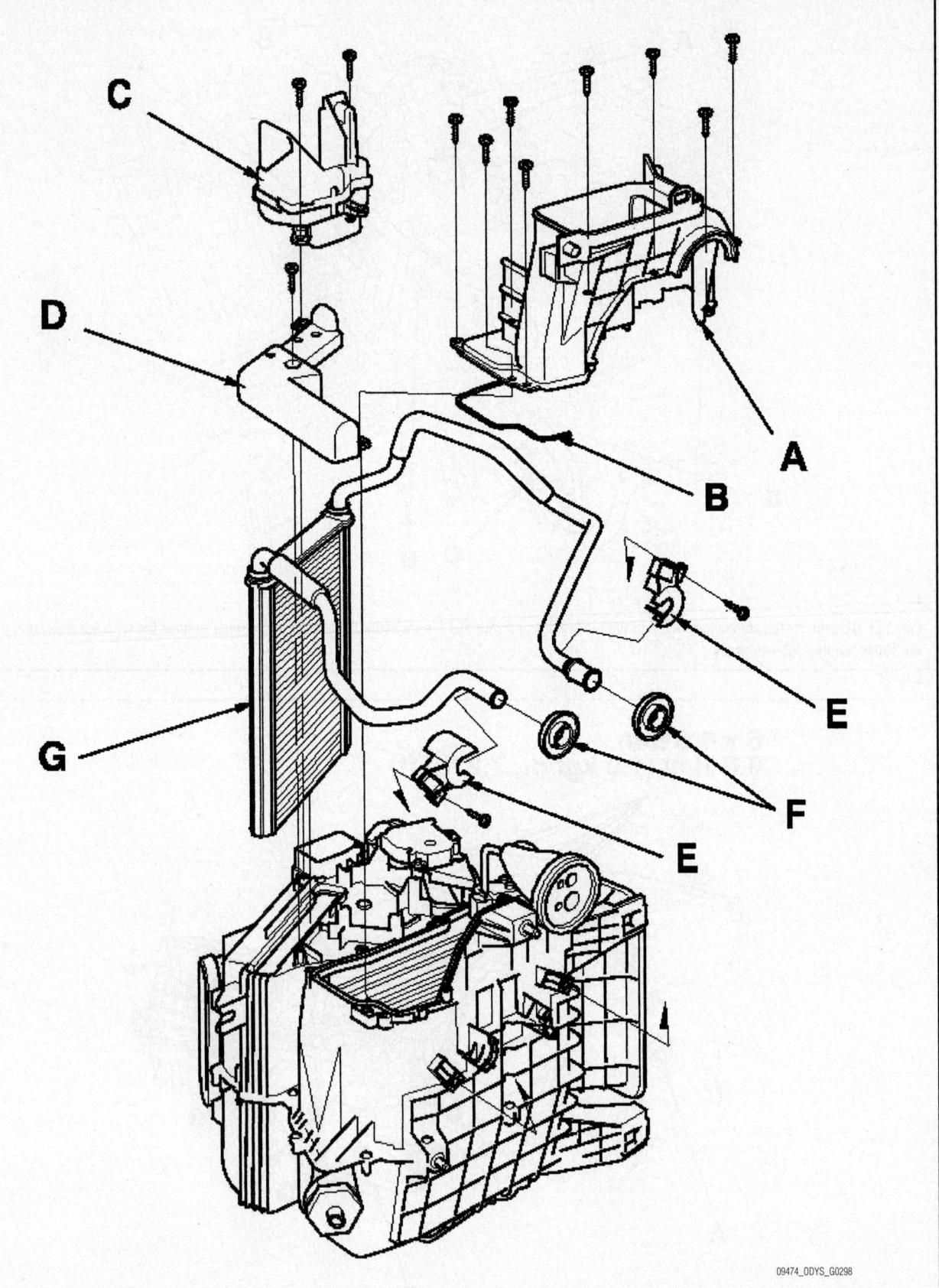

Fig. 159 Remove the self-tapping screws, the joint duct (A), and seal (B). Remove the self-tapping screws, then remove the passenger's heater outlet (C), and the heater core cover (D). Remove the self-tapping screws, the heater pipe brackets (E), the grommets (F), and carefully pull out the heater core (G)

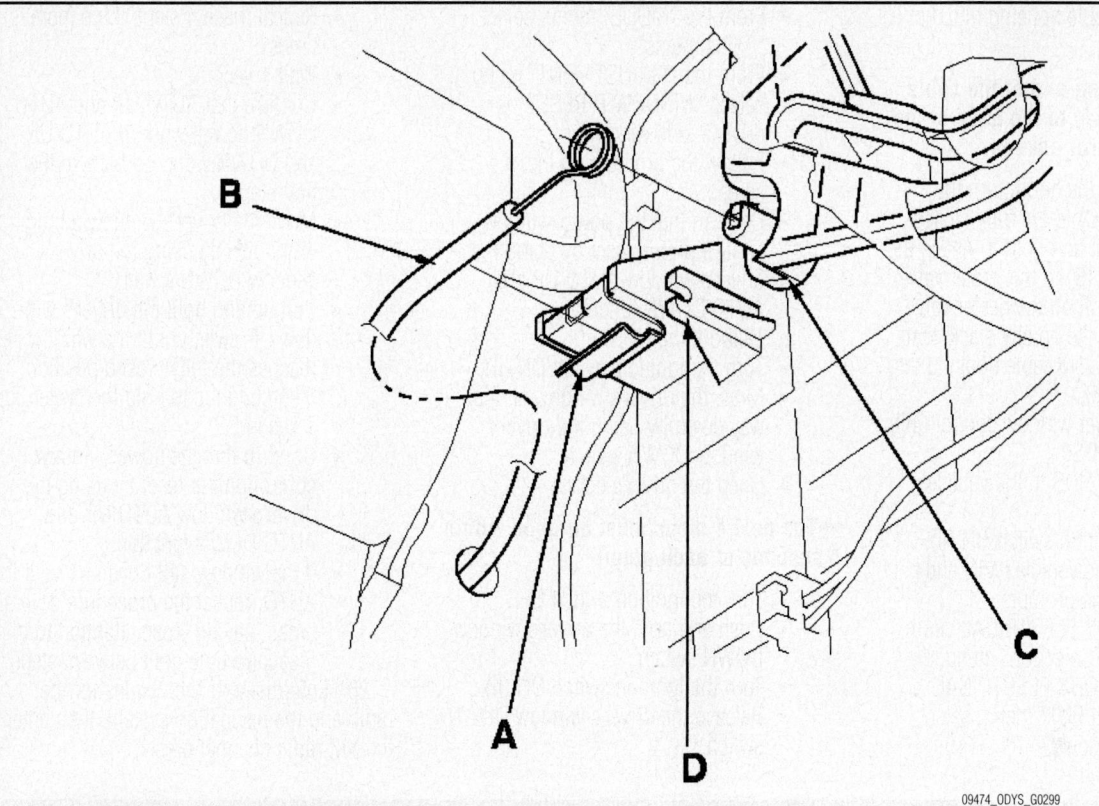

Fig. 160 Attach the heater valve cable (B) to the air mix control linkage (C) as shown. Hold the end of the heater valve cable housing against the stop (D), then snap the heater valve cable housing into the cable clamp (A).

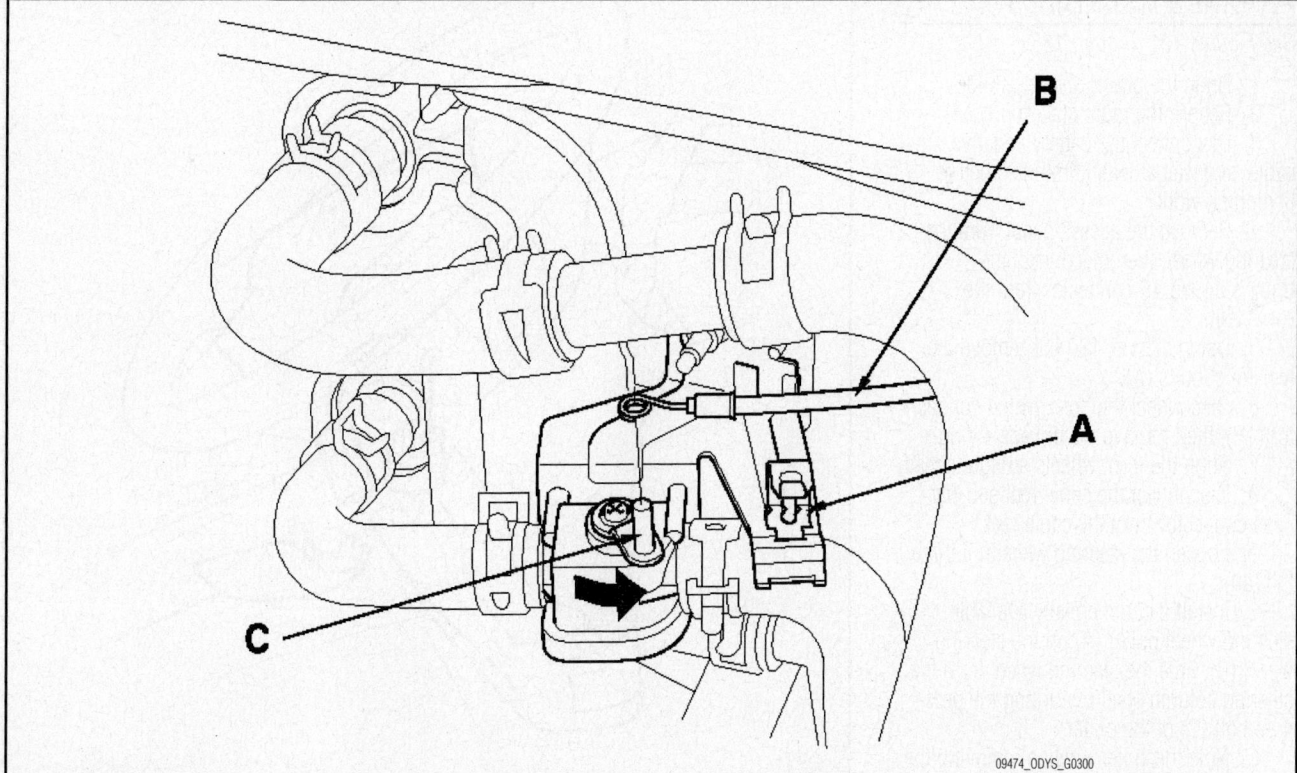

Fig. 161 From under the hood, turn the heater valve arm (C) to the fully closed position as shown, and hold it. Attach the heater valve cable (B) to the heater valve arm, and gently pull on the heater valve cable housing to take up any slack, then install the heater valve cable housing into the oablo clamp (A)

the heater valve cable housing into the cable clamp (A).

➡Make sure the ring-end of the cable is pushed all the way to the base of the pin on air mix control linkage.

c. From under the hood, turn the heater valve arm (C) to the fully closed position as shown, and hold it. Attach the heater valve cable (B) to the heater valve arm, and gently pull on the heater valve cable housing to take up any slack, then install the heater valve cable housing into the cable clamp (A).

25. Reset the power window control unit:
a. Using the HDS:
- Connect the HDS to the vehicle's DLC.
- Turn the ignition switch ON (II), then enter the vehicle's VIN and mileage at the prompts.
- Select "BODY ELECTRICAL" from the "System Selection" menu.
- From the "BODY ELECTRICAL SYSTEM SELECT" menu, select "Power Windows".

- From the "MODE" menu, select "Adjustments".
- From the "ADJUSTMENT" menu, select "WINDOW P RESET" for driver's side window.
- Follow the prompts on the screen.
- Confirm that the power window control unit is reset by using the driver's window AUTO UP and AUTO DOWN function.

b. Without the HDS:
- Turn the ignition switch ON (II).
- Move the driver's window all the way down by using the driver's window DOWN switch.
- Open the driver's door.

➡The next 4 steps must be done within 5 seconds of each other.

- Turn the ignition switch OFF.
- Push and hold the driver's window DOWN switch.
- Turn the ignition switch ON (II).
- Release the driver's window DOWN switch.

- Repeat these 4 steps three more times.
- Wait 1 second.
- Confirm that AUTO UP and AUTO DOWN do not work. If AUTO UP and DOWN work, go back to the first step.
- Move the driver's window all the way down by using the driver's window DOWN switch.
- Pull up and hold the driver's window UP switch until the window reaches the fully closed position, then continue to hold the switch for 1 second.
- Confirm that the power window control unit is reset by using the driver's window AUTO UP and AUTO DOWN function.
- If the window still does not work in AUTO, repeat the procedure several times, paying close attention to the 5 second time limit between steps.

26. Enter the anti-theft codes for the radio and the navigation system, then enter the XM radio channel presets.

STEERING

POWER STEERING GEAR

REMOVAL & INSTALLATION

See Figures 162 through 188.

1. Drain the power steering fluid.
2. Record the radio station preset.
3. Disconnect the battery negative cable, and wait at least 3 minutes before beginning work.
4. Remove the access panel from the steering wheel, then disconnect the driver's airbag 4P connector from the cable reel.
5. Using a Torx® T30 bit, remove the two Torx® bolts (A).
6. Disconnect the horn switch connector (1P), then remove the driver's airbag.
7. Align the front wheels straight ahead.
8. Disconnect the cable reel sub-harness connector from the cable reel.
9. Loosen the steering wheel nut three full turns.
10. Install a commercially available steering wheel puller (A) on the steering wheel (B). Free the steering wheel from the steering column shaft by turning the pressure bolt (C) of the puller.
11. Note these items when removing the steering wheel:
a. Do not tap on the steering wheel or the steering column shaft when removing the steering wheel.

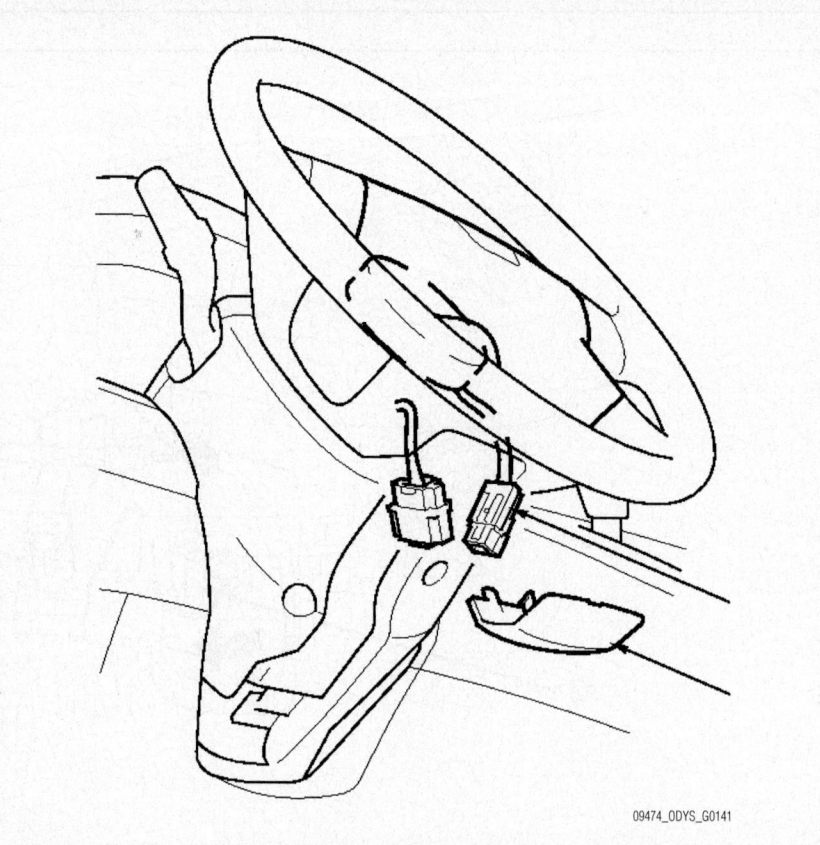

09474_ODYS_G0141

Fig. 162 Remove the access panel from the steering wheel, then disconnect the driver's airbag 4P connector from the cable reel—Odyssey

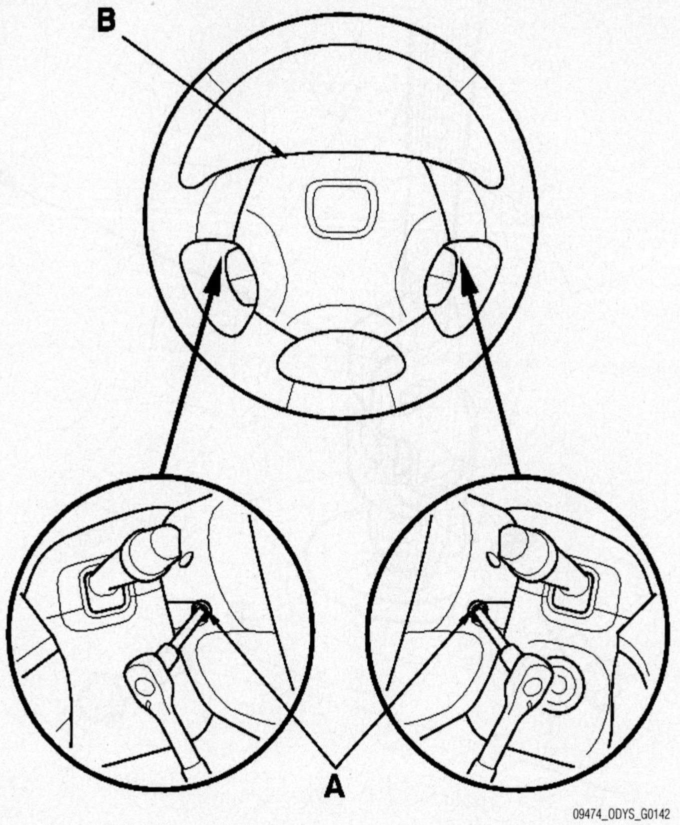

Fig. 163 Using a Torx® T30 bit, remove the two Torx® bolts (A) —Odyssey

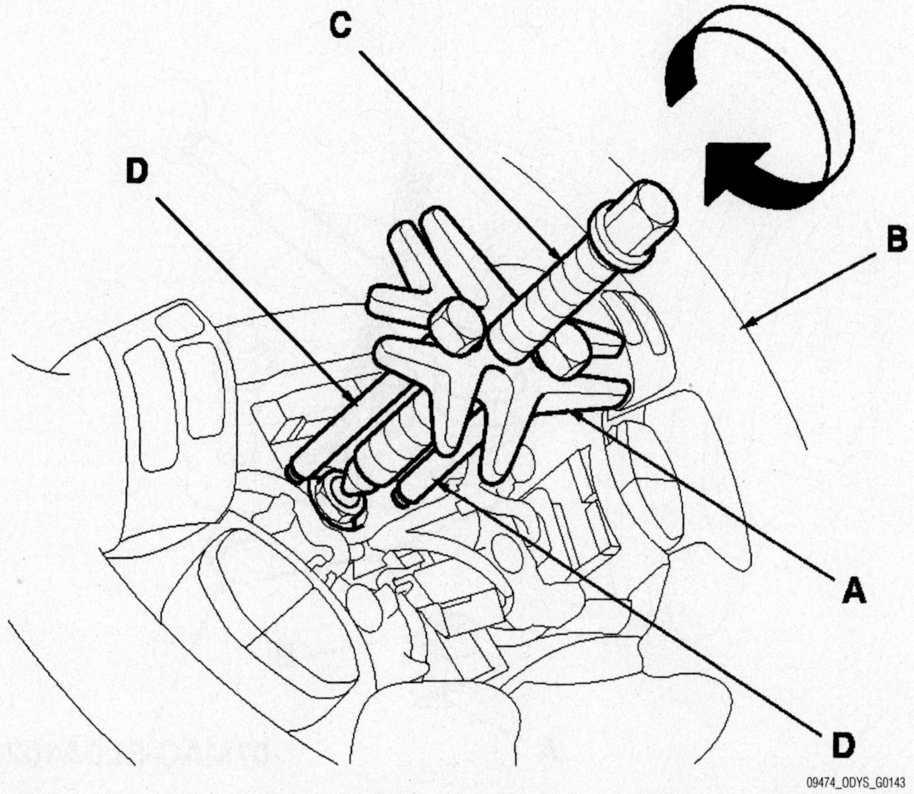

09474_ODYS_G0142

Fig. 164 Steering wheel puller

09474_ODYS_G0143

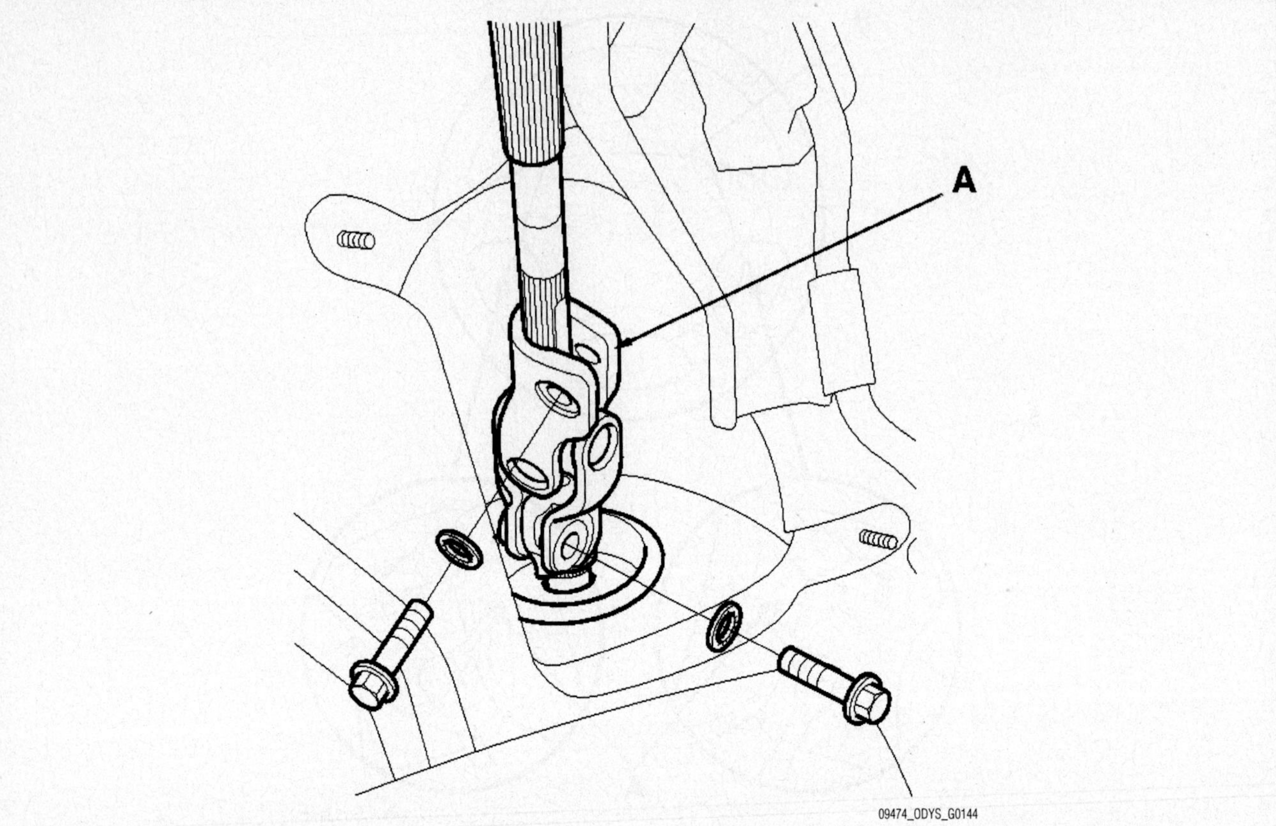

09474_ODYS_G0144

Fig. 165 Remove the steering joint bolts, disconnect the steering joint by moving the steering joint (A) toward the column—Odyssey

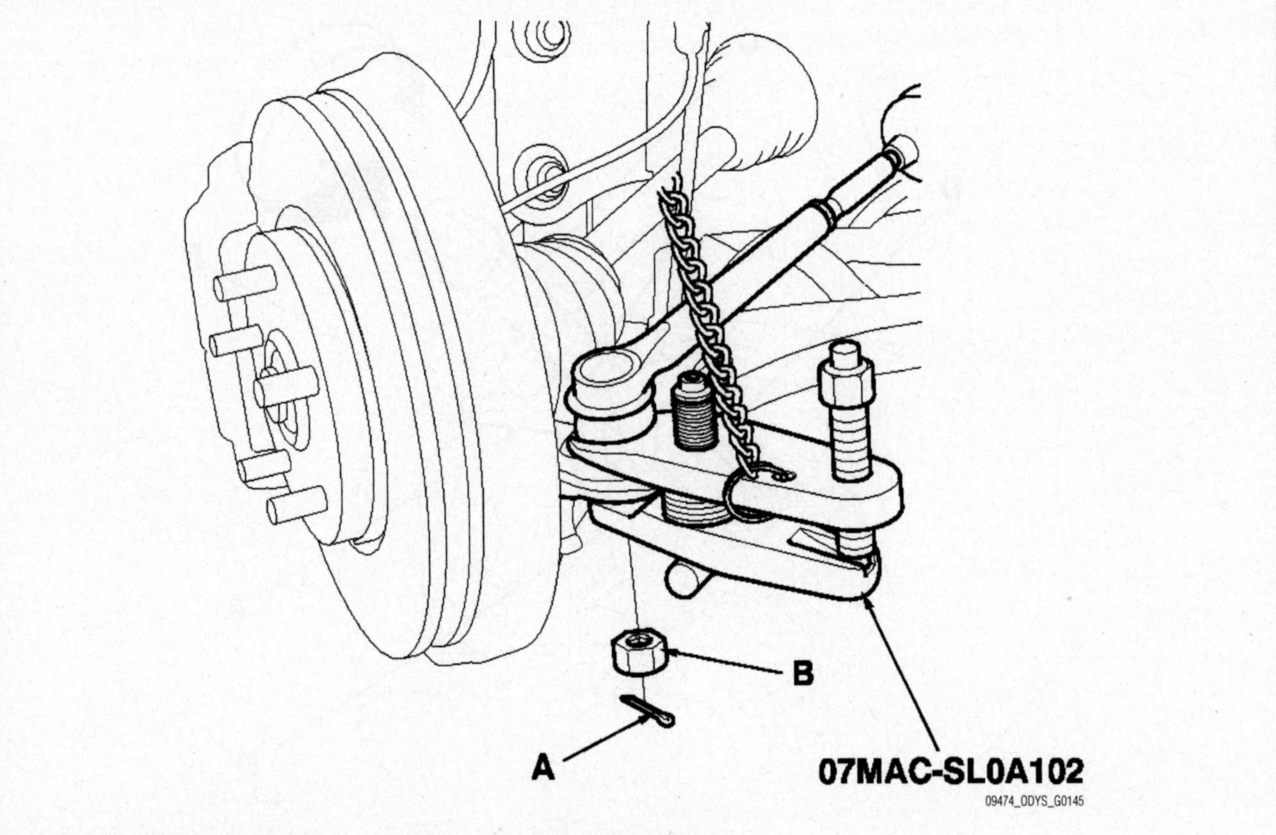

07MAC-SL0A102

09474_ODYS_G0145

Fig. 166 Remove the cotter pin (A) from the 12 mm nut (B), and loosen the nut—Odyssey

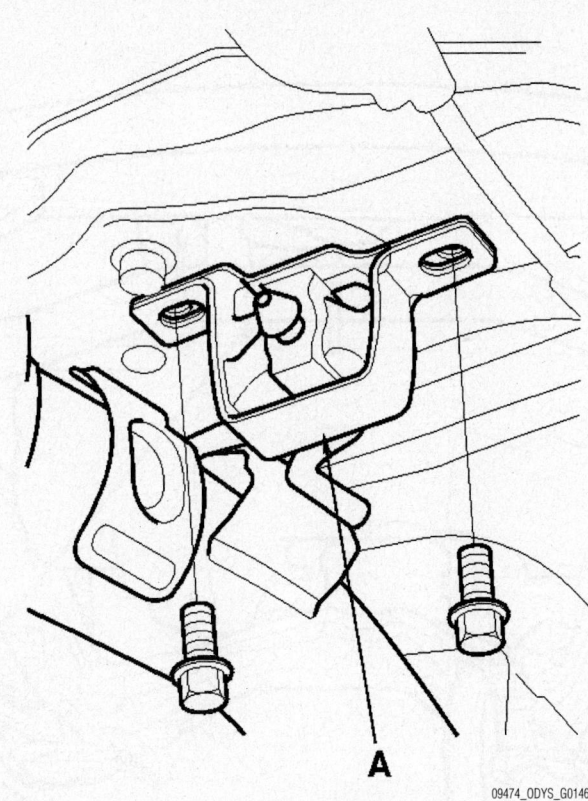

Fig. 167 Remove the 10 mm flange bolts of the exhaust rubber mount (A)—Odyssey

b. If you thread the puller bolts (D) into the wheel hub more than five threads, the bolts will hit the cable reel and damage it. To prevent this, install a pair of jam nuts five threads up on each puller bolt.

12. Remove the steering wheel puller, then remove the steering wheel nut and steering wheel from the steering column.

13. Remove the steering joint cover.

14. Remove the steering joint bolts, disconnect the steering joint by moving the steering joint (A) toward the column.

15. Remove the cotter pin (A) from the 12 mm nut (B), and loosen the nut.

16. Separate the tie-rod ball joint and knuckle using the special tool.

17. Remove the 10 mm flange bolts of the exhaust rubber mount (A).

18. Remove the three self-locking nuts, and disconnect the three way catalytic converter (TWC) from the muffler.

19. Remove exhaust pipe A.

20. Disconnect the power steering pressure (PSP) switch connector (A).

21. Remove the front splash shield (A).

22. Attach the special tool to the front subframe (A) by hanging the hook of the special tool over the front of the subframe, then tighten the special tool screw.

23. Raise the jack (B) and line up the slots in the arms with the bolt holes on the corner of the jack base, then attach them with bolts securely.

24. Loosen the front subframe front bracket (A) mounting bolts on the right and left of the vehicle so they are about 30 mm (1³⁄₁₆ in.) from the mounting surface.

25. Support the front subframe securely by raising the special tool, then remove the two 12 mm flange bolts (A).

26. Remove the two 14 mm special bolts (B) and front subframe rear brackets (C) from the front subframe.

27. Lower the jack supporting the front subframe with the special tool slowly until the front subframe has dropped about 50 mm (1¹⁵⁄₁₆ in.).

28. Remove the P/S line mounting brackets from the front subframe and gearbox mounting bracket.

29. Loosen the 16 mm flare nut, and disconnect the feed line from the gear box.

30. Loosen the adjustable hose clamp and disconnect the return hose.

31. Remove the P/S heat baffle plate.

32. Remove the two 10 mm flange bolts from the right side of the steering gearbox, then remove the mounting bracket (A) and cushion (B).

33. Remove the two 10 mm flange bolts and nuts from the left side of the gearbox.

34. Loosen the steering gearbox bracket mounting bolts (A).

35. EX-L, EX-L Touring models:

a. Remove the rear mount stop (A), then remove the rear mount bolt (B).

b. Remove the rear mount (A) from the base bracket (B), and disconnect the connector (C).

c. Remove the base bracket from the front subframe.

36. VAN, LX, EX models

a. Remove the rear mount bracket bolt (A).

b. Remove the rear mount bracket from the front subframe.

37. Disconnect the return hose clip.

38. Move the steering gearbox to the driver's side, and rotate it so the pinion shaft points toward the front of the vehicle.

39. Carefully move the steering gearbox as an assembly toward the left side of the vehicle until the pinion shaft clears the wheel well opening. Be careful not to damage the brake lines with the pinion shaft.

40. Remove the steering gearbox through the wheel well opening on the driver's side.

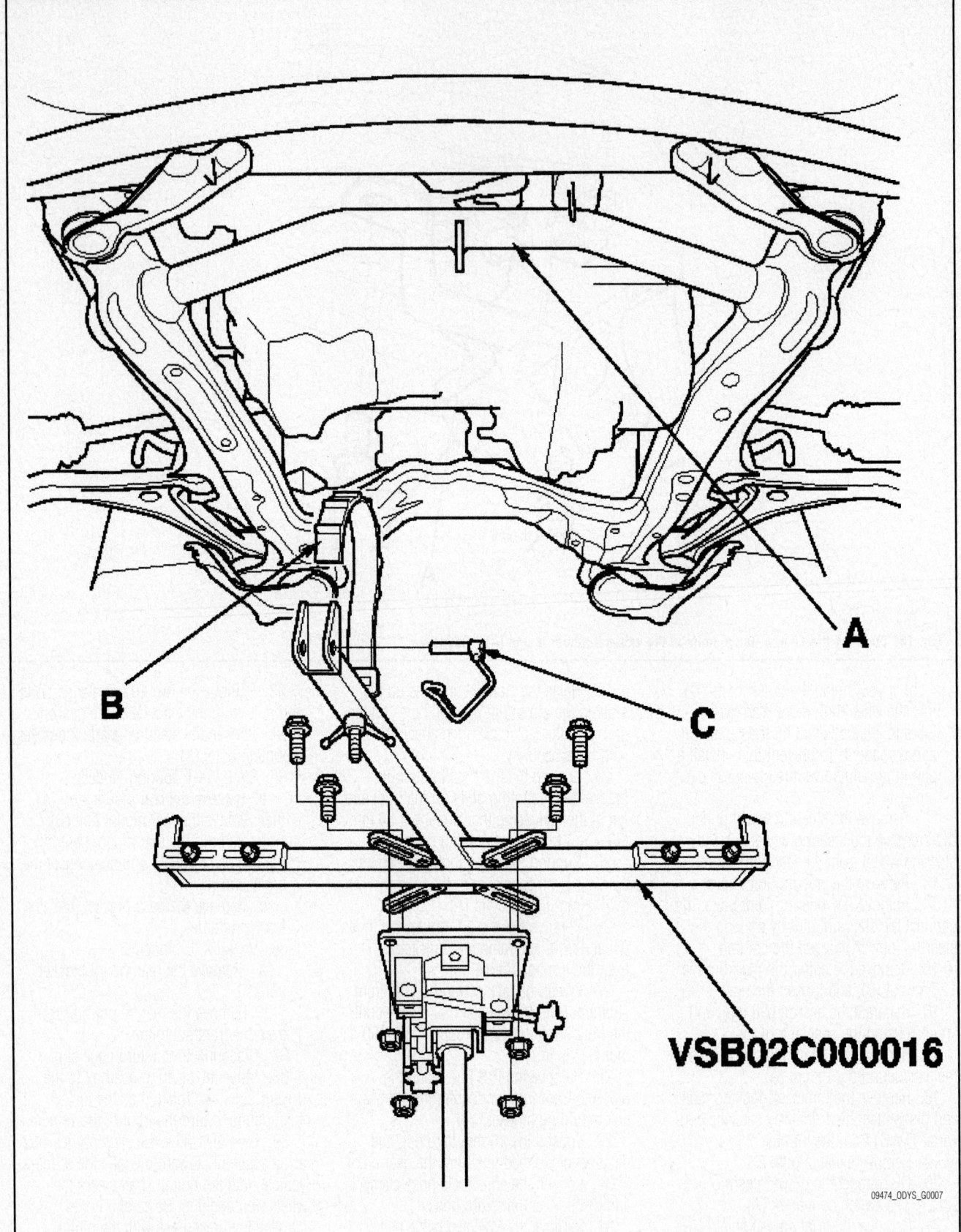

09474_ODYS_G0007

VSB02C000016

Fig. 168 Attach the special tool to the front subframe (A) by hanging the belt (B) of the special tool over the front of the subframe, secure it with the pin (C), then tighten the special tool screw—Odyssey

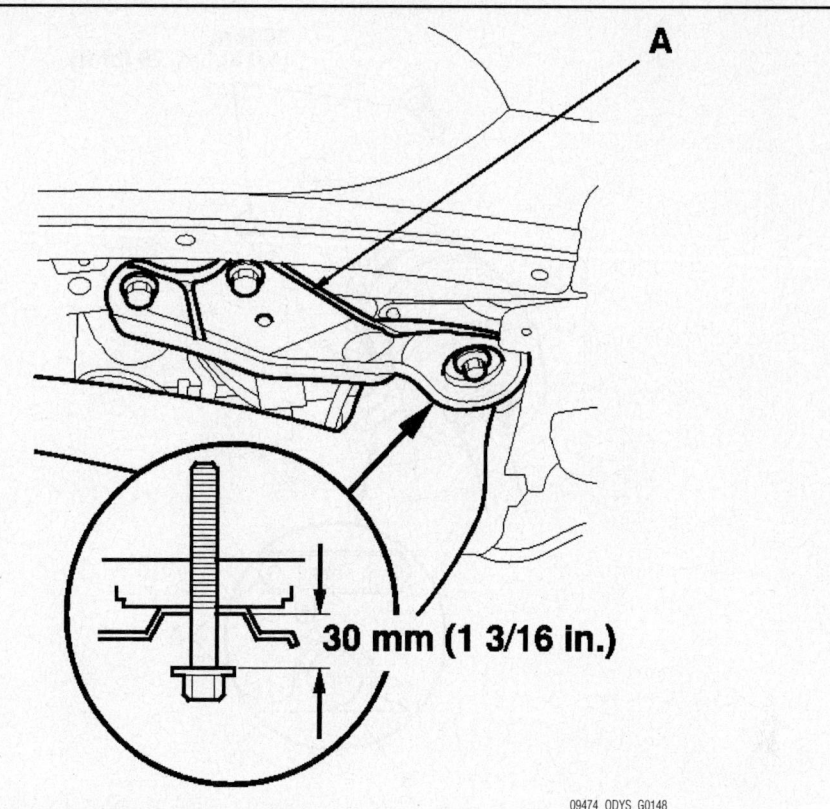

09474_ODYS_G0148

Fig. 169 Loosen the front subframe front bracket (A) mounting bolts on the right and left of the vehicle so they are about 30 mm (1³⁄₁₆ in.) from the mounting surface—Odyssey

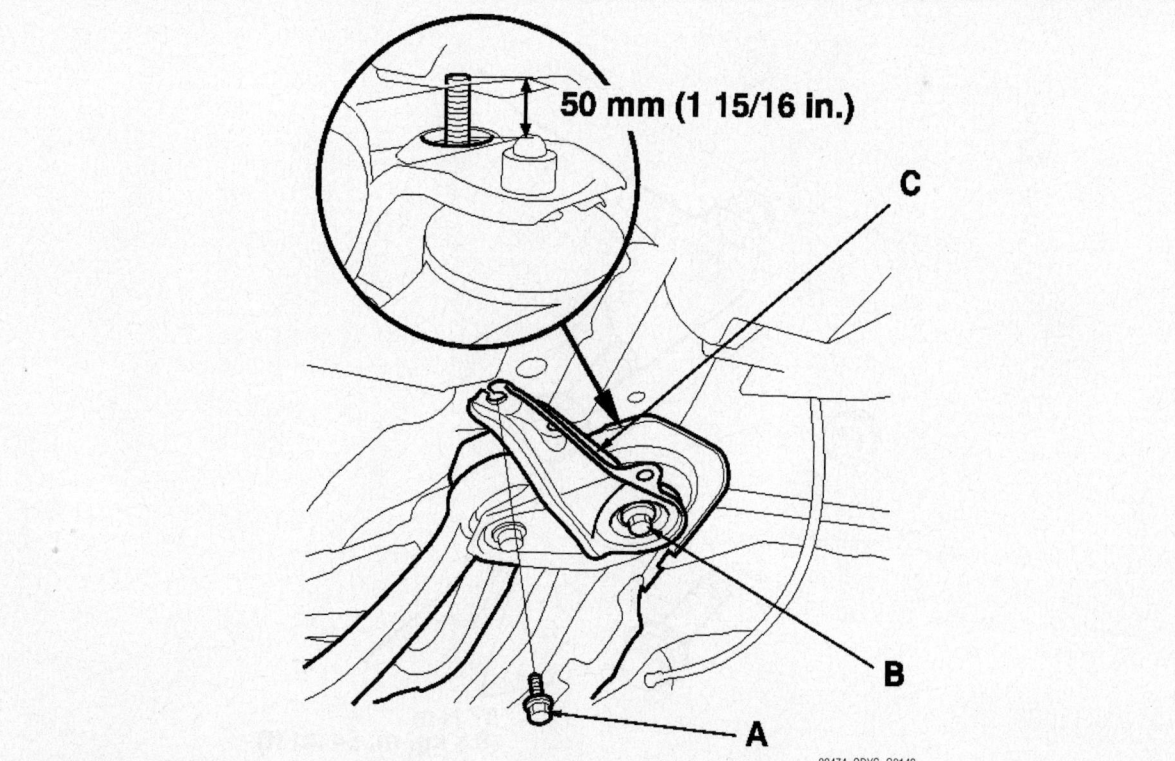

09474_ODYS_G0149

Fig. 170 Support the front subframe securely by raising the special tool, then remove the two 12 mm flange bolts (A). Remove the two 14 mm special bolts (B) and front subframe rear brackets (C) from the front subframe. Lower the jack supporting the front subframe with the special tool slowly until the front subframe has dropped about 50 mm (1¹⁵⁄₁₆ in.)—Odyssey

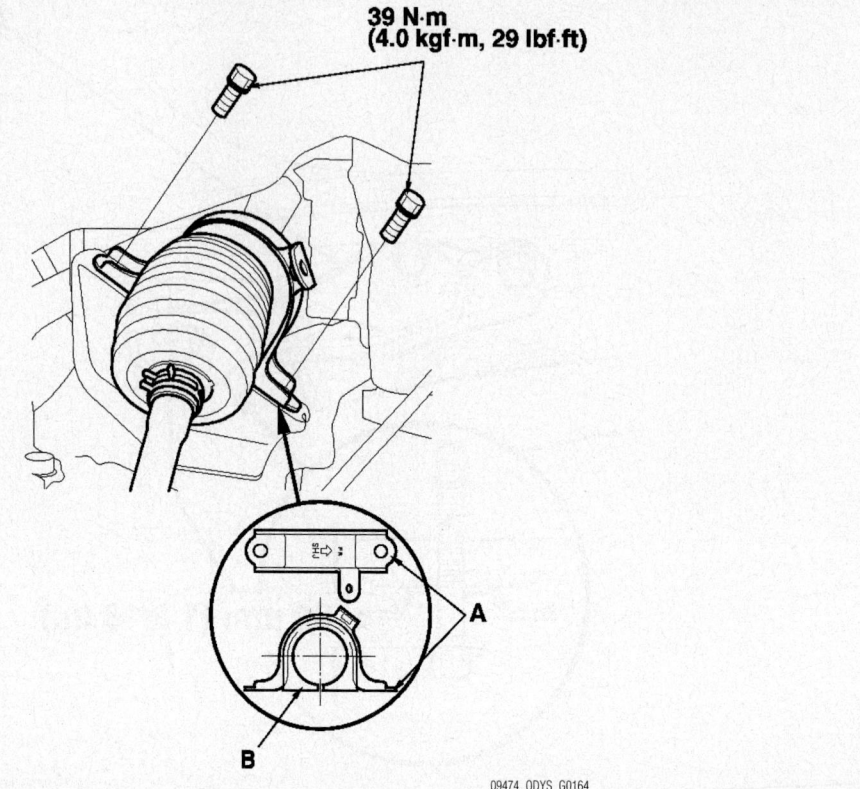

**39 N·m
(4.0 kgf·m, 29 lbf·ft)**

A

B

09474_ODYS_G0164

Fig. 171 Remove the two 10 mm flange bolts from the right side of the steering gearbox, then remove the mounting bracket (A) and cushion (B)—Odyssey

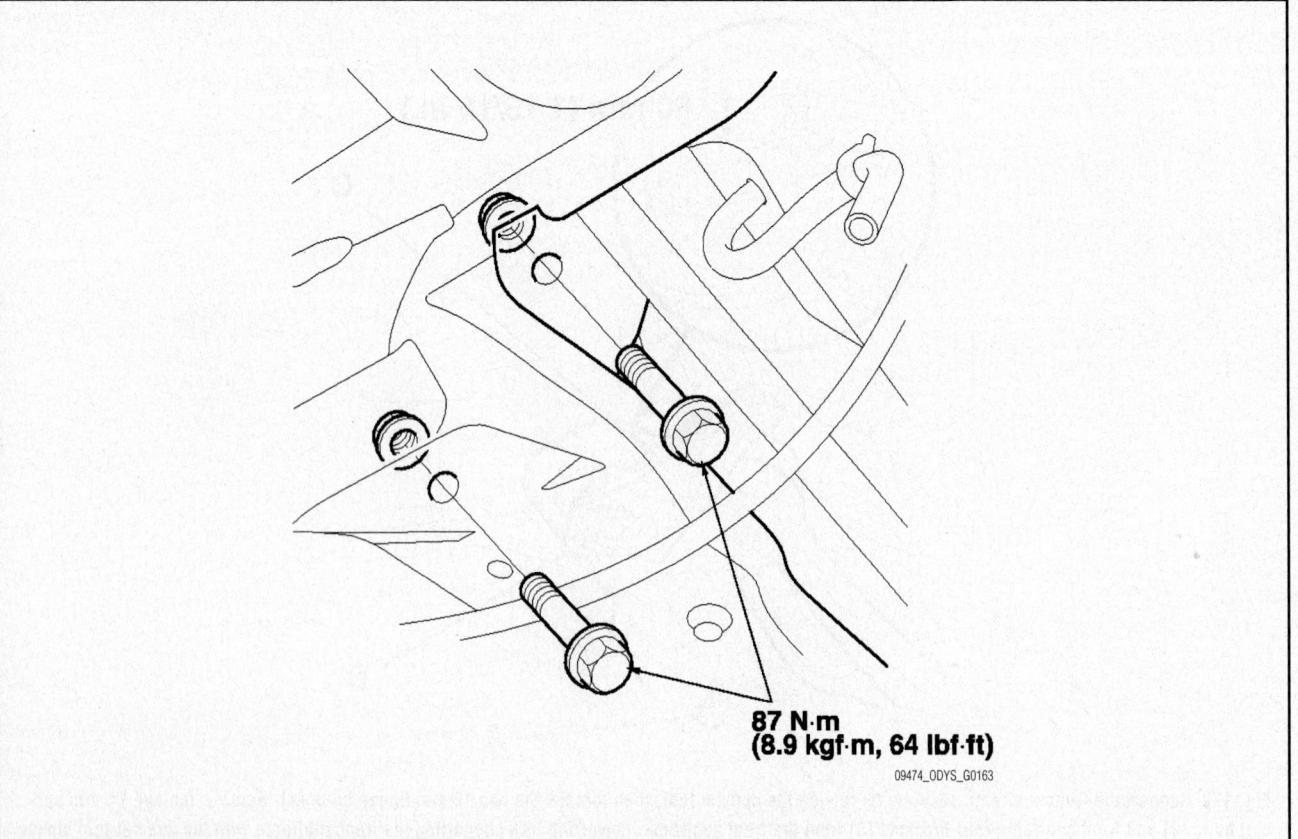

**87 N·m
(8.9 kgf·m, 64 lbf·ft)**

09474_ODYS_G0163

Fig. 172 Remove the two 10 mm flange bolts and nuts from the left side of the gearbox—Odyssey

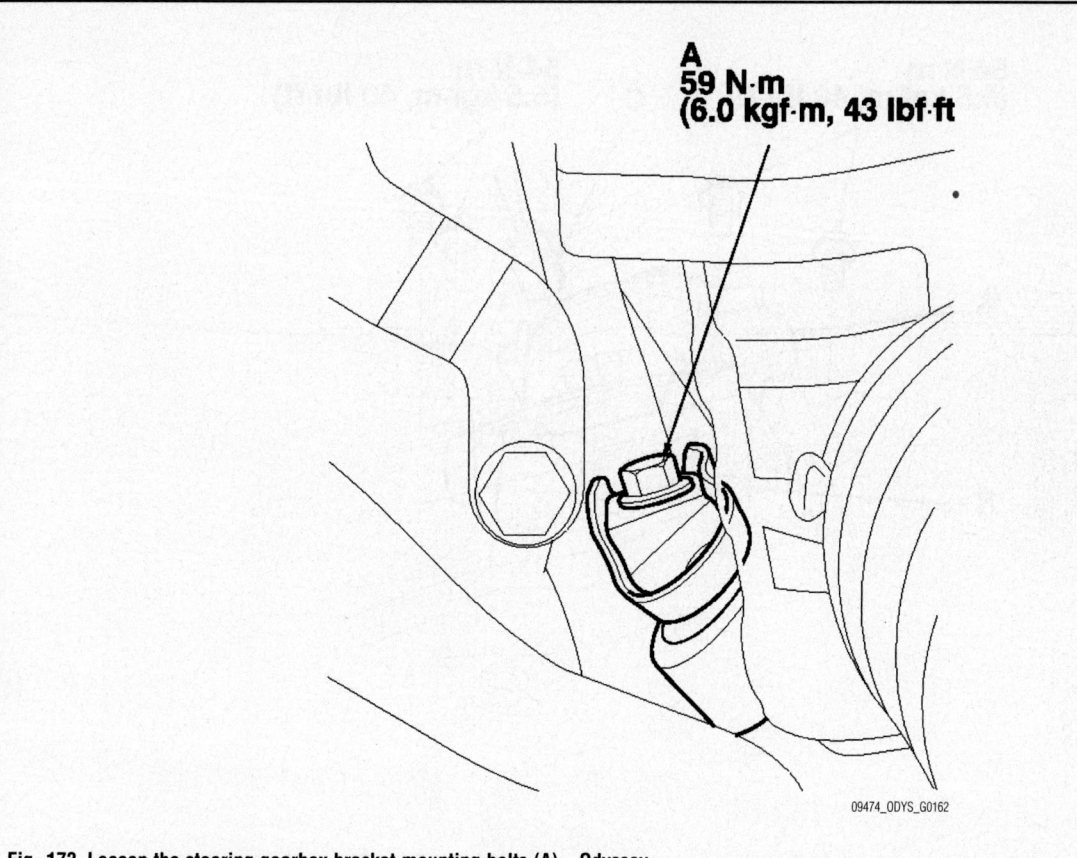

**A
59 N·m
(6.0 kgf·m, 43 lbf·ft)**

09474_ODYS_G0162

Fig. 173 Loosen the steering gearbox bracket mounting bolts (A)—Odyssey

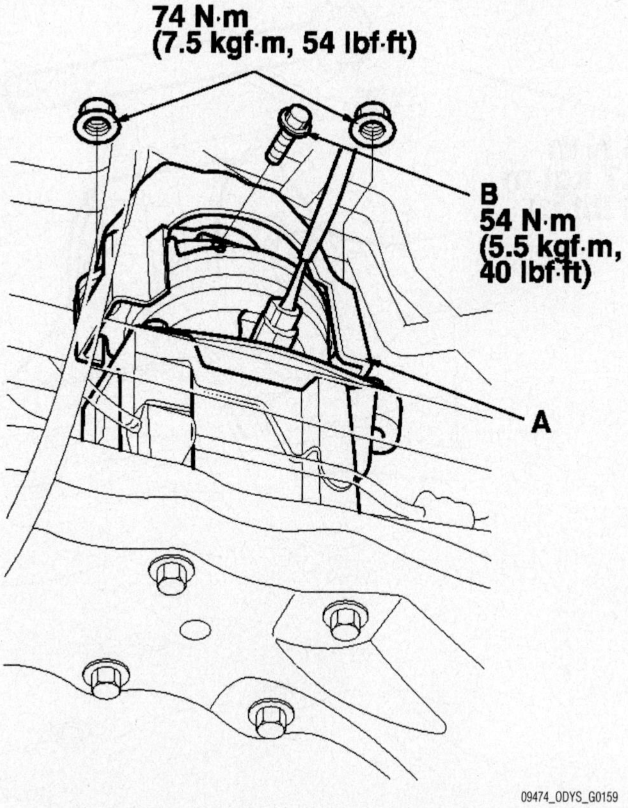

**74 N·m
(7.5 kgf·m, 54 lbf·ft)**

**B
54 N·m
(5.5 kgf·m,
40 lbf·ft)**

A

09474_ODYS_G0159

Fig. 174 Remove the rear mount stop (A), then remove the rear mount bolt (R)—EX-L, EX-L Touring models

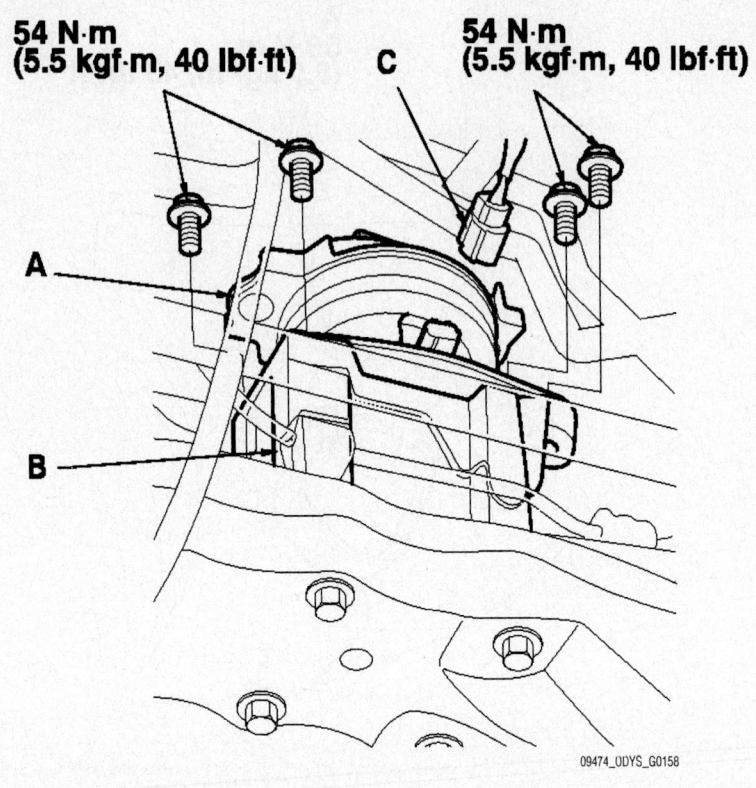

54 N·m (5.5 kgf·m, 40 lbf·ft)

C

54 N·m (5.5 kgf·m, 40 lbf·ft)

A

B

09474_ODYS_G0158

Fig. 175 Remove the rear mount (A) from the base bracket (B), and disconnect the connector (C)—EX-L, EX-L Touring models

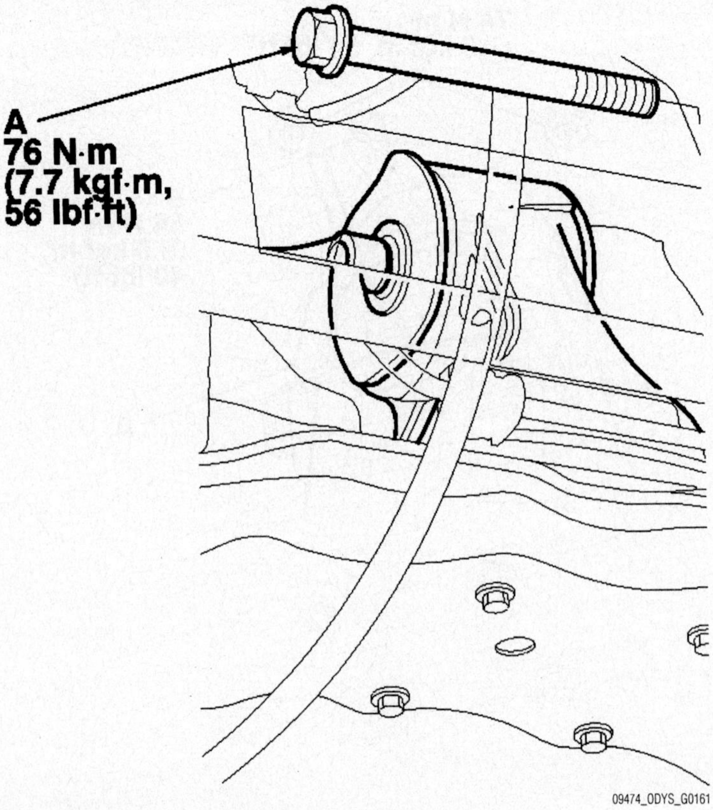

A
76 N·m (7.7 kgf·m, 56 lbf·ft)

09474_ODYS_G0161

Fig. 176 Remove the rear mount bracket bolt (A)—LX, EX models

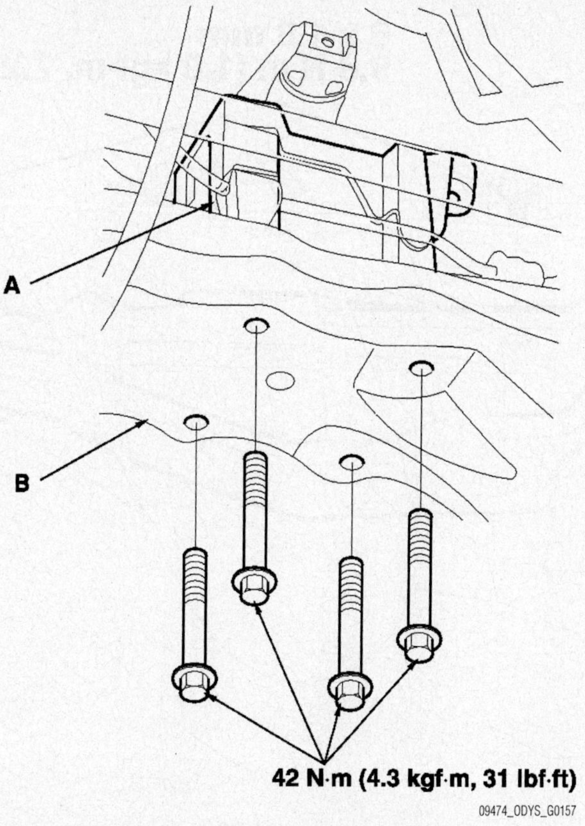

42 N·m (4.3 kgf·m, 31 lbf·ft)

09474_ODYS_G0157

Fig. 177 Install the base bracket (A) on the front subframe (B)— EX-L, EX-L Touring models

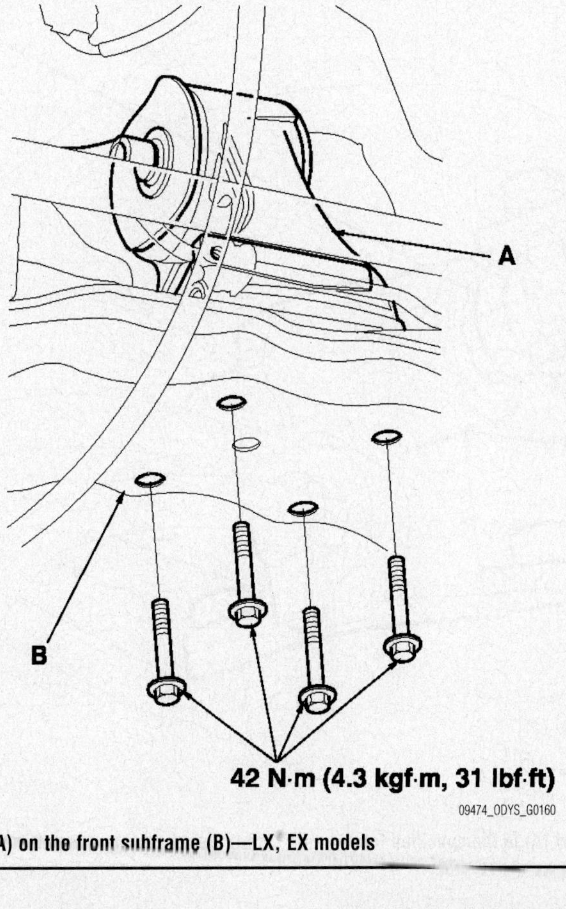

42 N·m (4.3 kgf·m, 31 lbf·ft)

09474_ODYS_G0160

Fig. 178 Install the rear mount bracket (A) on the front subframe (B)—LX, EX models

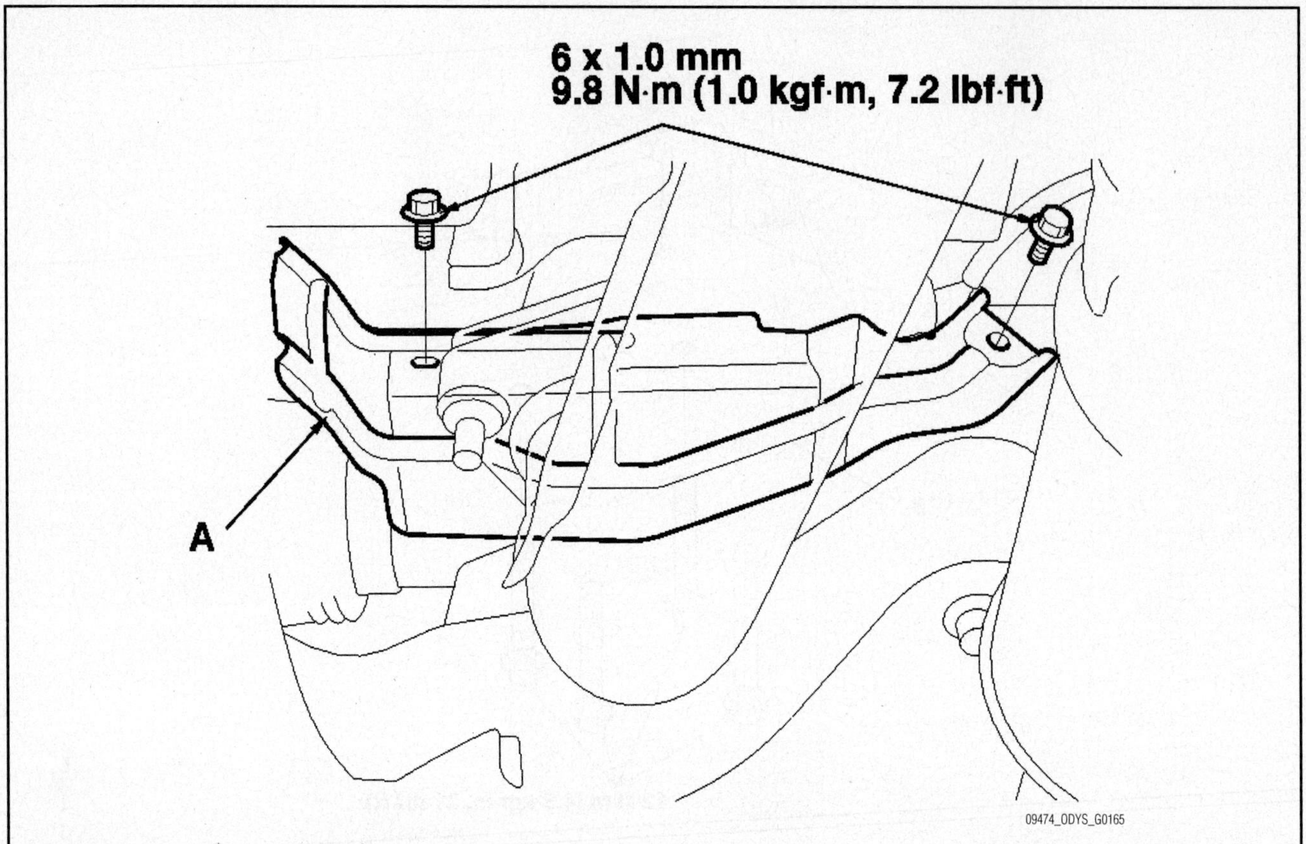

6 x 1.0 mm
9.8 N·m (1.0 kgf·m, 7.2 lbf·ft)

A

09474_ODYS_G0165

Fig. 179 Install the P/S heat baffle plate (A)—Odyssey

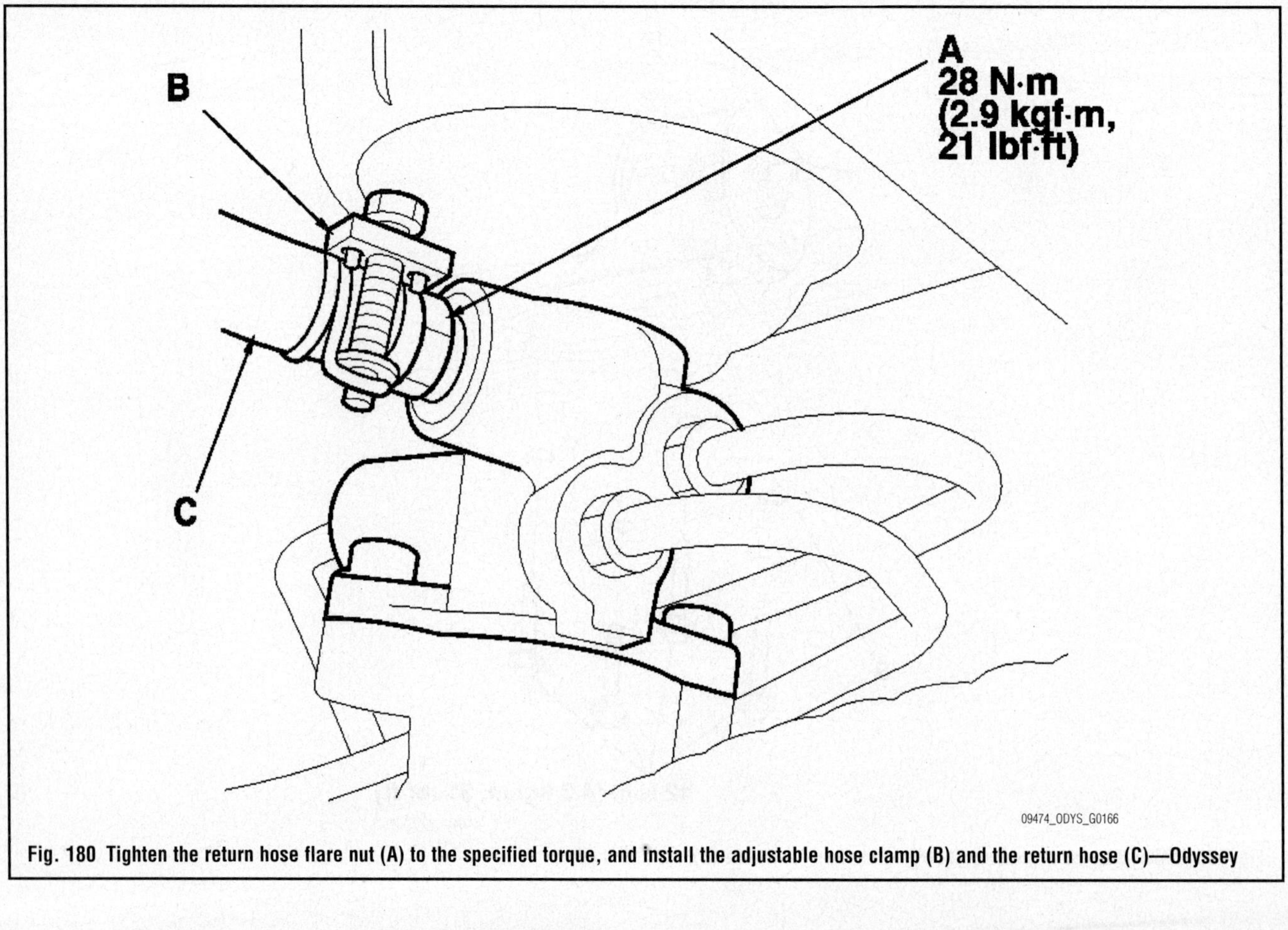

B

A
28 N·m
(2.9 kgf·m,
21 lbf·ft)

C

09474_ODYS_G0166

Fig. 180 Tighten the return hose flare nut (A) to the specified torque, and install the adjustable hose clamp (B) and the return hose (C)—Odyssey

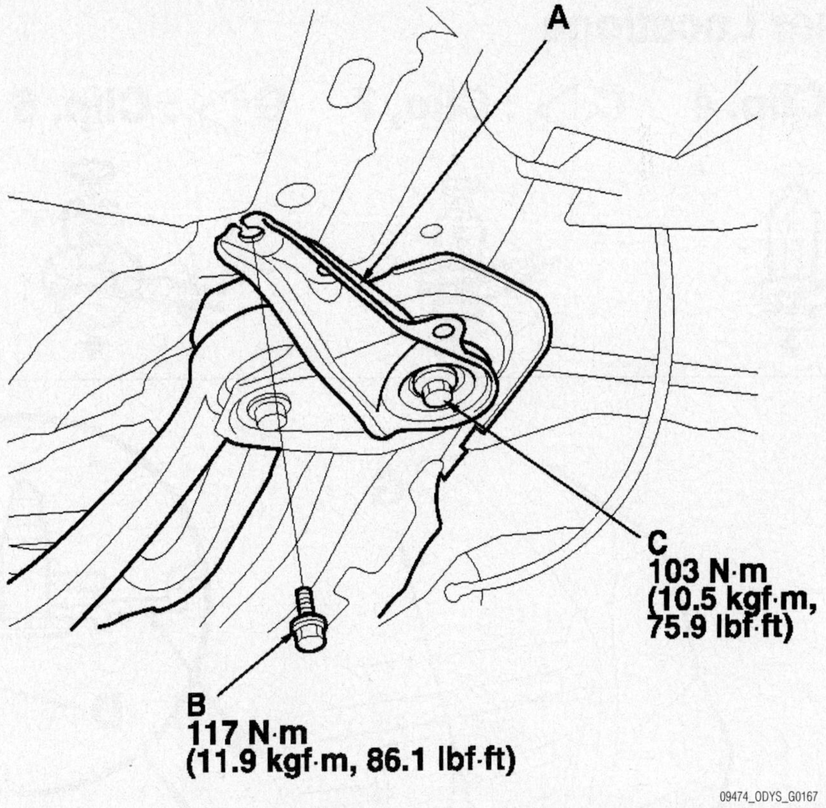

A

C
**103 N·m
(10.5 kgf·m,
75.9 lbf·ft)**

B
**117 N·m
(11.9 kgf·m, 86.1 lbf·ft)**

09474_ODYS_G0167

Fig. 181 Install the front subframe rear bracket (A) with 12 mm flange bolts (B) and 14 mm special bolts (C), and tighten to specified torque—Odyssey

A

B
**74 N·m
(7.5 kgf·m, 54 lbf·ft)**

C
**103 N·m
(10.5 kgf·m, 75.9 lbf·ft)**

09474_ODYS_G0168

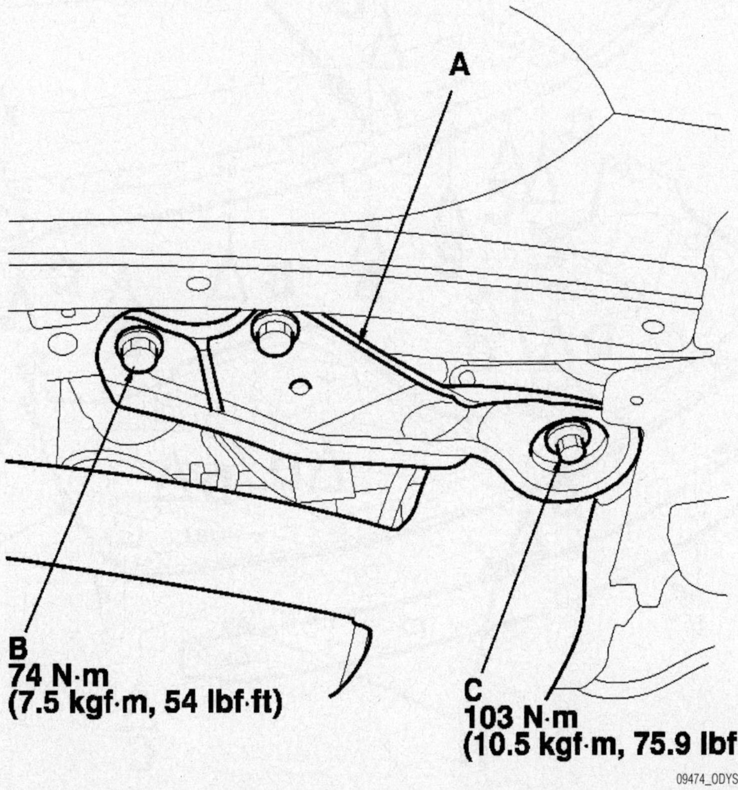

Fig. 182 Install the front subframe front bracket (A) with 12 mm flange bolts (B) and 14 mm special bolts (C), and tighten to specified torque—Odyssey

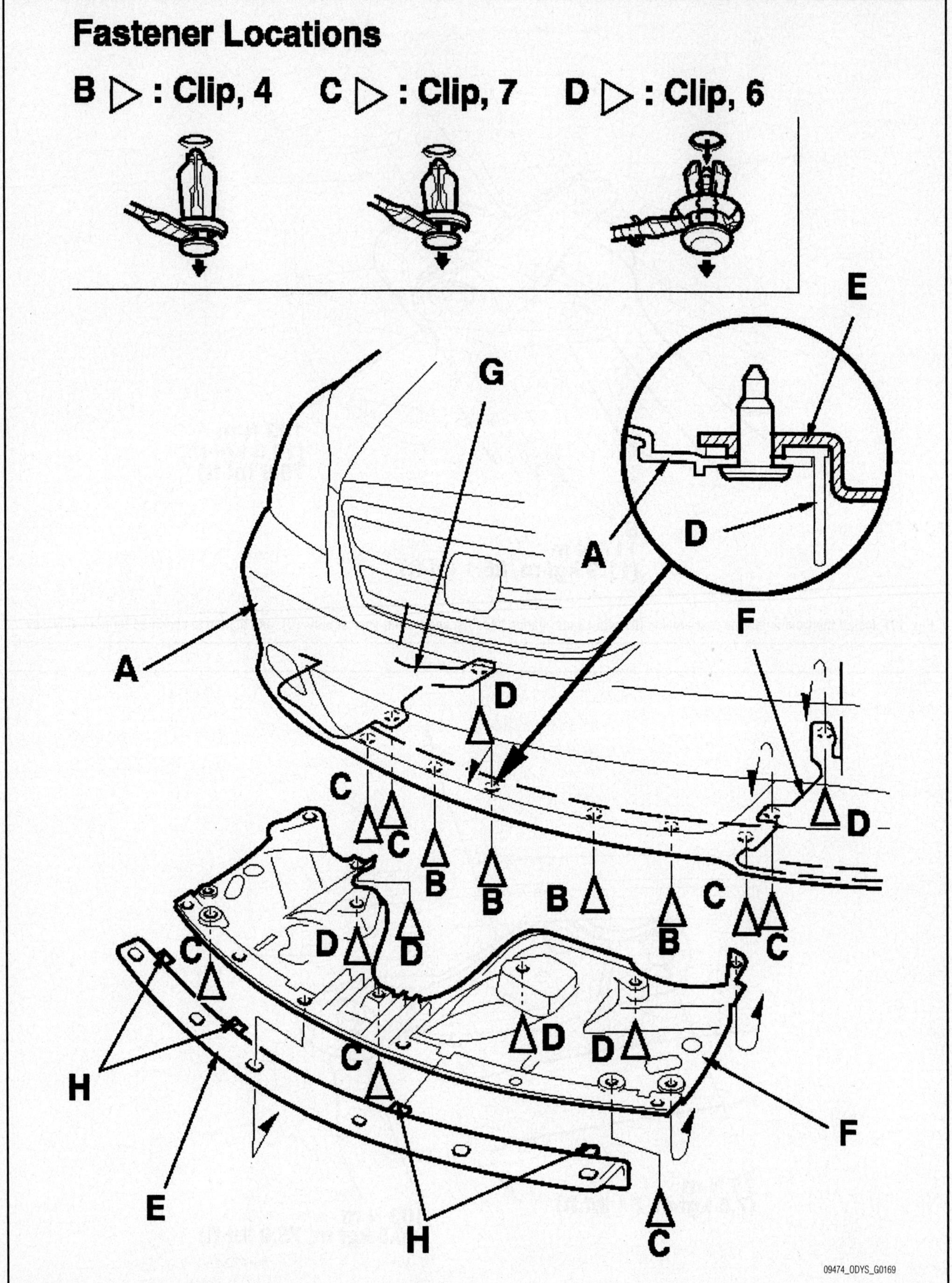

Fastener Locations

B ▷ : Clip, 4   C ▷ : Clip, 7   D ▷ : Clip, 6

E

G

A

D

A

D

F

E

A

C

C

B

D

D

D

B

B

B

C

C

F

H

C

D

D

F

H

E

C

H

C

09474_ODYS_G0169

Fig. 183 Front splash shield installation—Odyssey

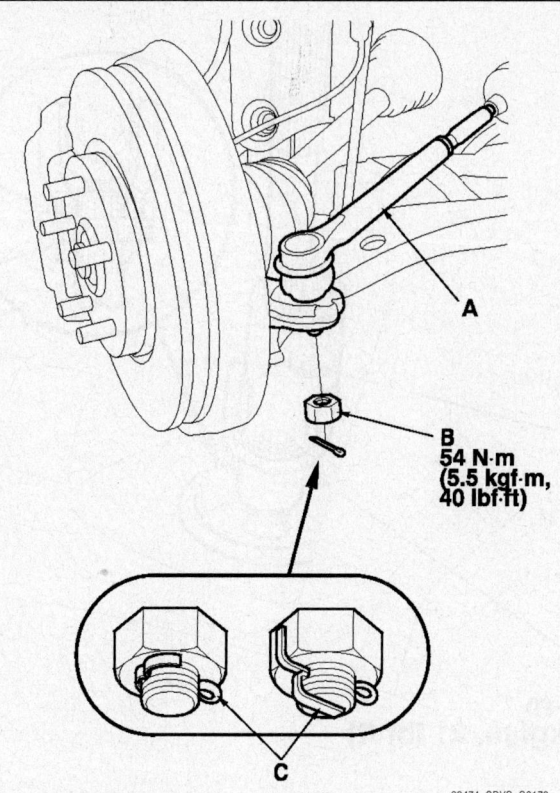

09474_ODYS_G0170

Fig. 184 Reconnect the tie-rod ends (A) to the steering knuckles. Install the 12 mm nut (B) and tighten it. Install a new cotter pin (C), and bend it as shown—Odyssey

B
54 N·m
(5.5 kgf·m,
40 lbf·ft)

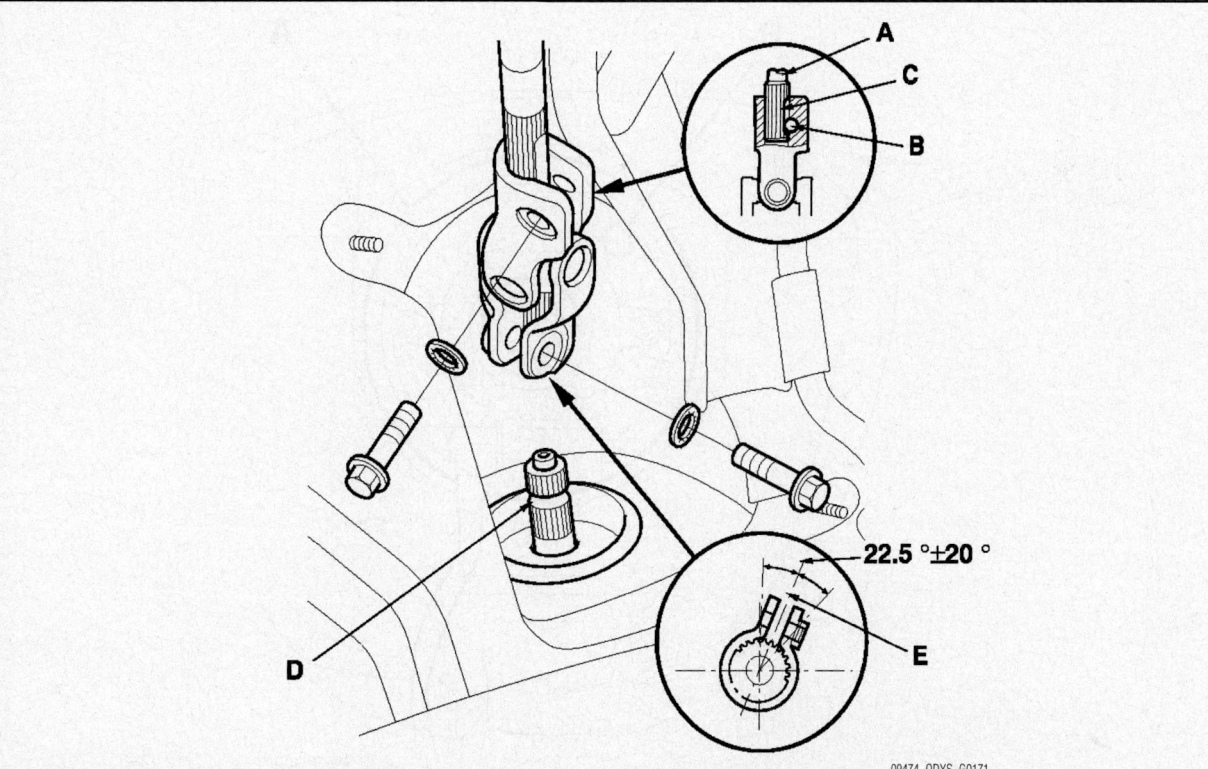

22.5 °±20 °

09474_ODYS_G0171

Fig. 185 Insert the upper end of the steering joint onto the steering shaft (A) (line up the bolt hole (B) with the flat portion (C) on the shaft), and loosely install the upper joint bolt. Slip the lower end of the steering joint onto the pinion shaft (D) taking care to align the gap (E) within the angle—Odyssey

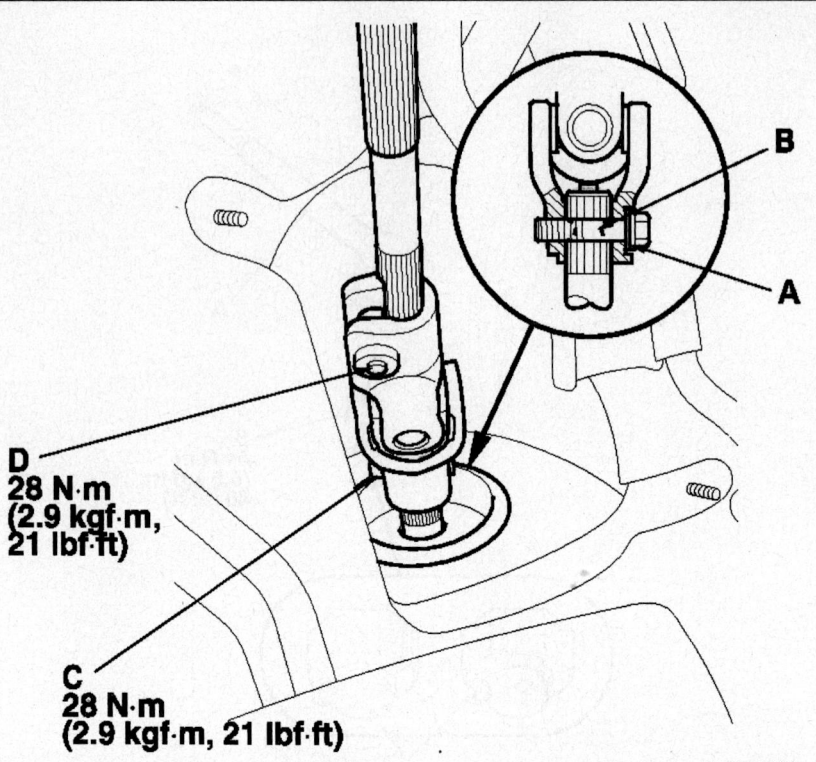

D
28 N·m
(2.9 kgf·m,
21 lbf·ft)

C
28 N·m
(2.9 kgf·m, 21 lbf·ft)

09474_ODYS_G0172

Fig. 186 Align the bolt hole (A) on the steering joint with the groove (B) around the pinion shaft, then loosely install the lower joint bolt (C). Pull on the steering joint to make sure that the steering joint is fully seated, then tighten the lower joint bolt to the specified torque. Tighten the upper joint bolt (D) to the specified torque—Odyssey

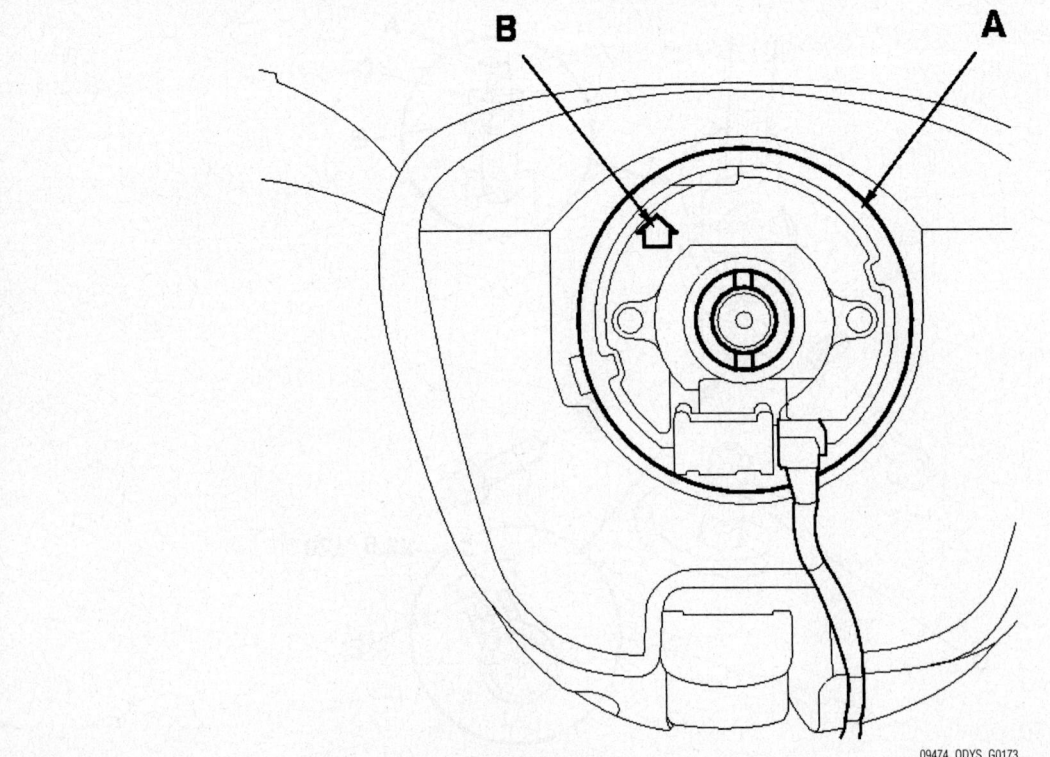

09474_ODYS_G0173

Fig. 187 Make sure the front wheels are aligned straight ahead, then center the cable reel (A). Do this by first rotating the cable reel clockwise until it stops. Then rotate it counterclockwise about three full turns. The arrow mark (B) on the cable reel label point should point straight up—Odyssey

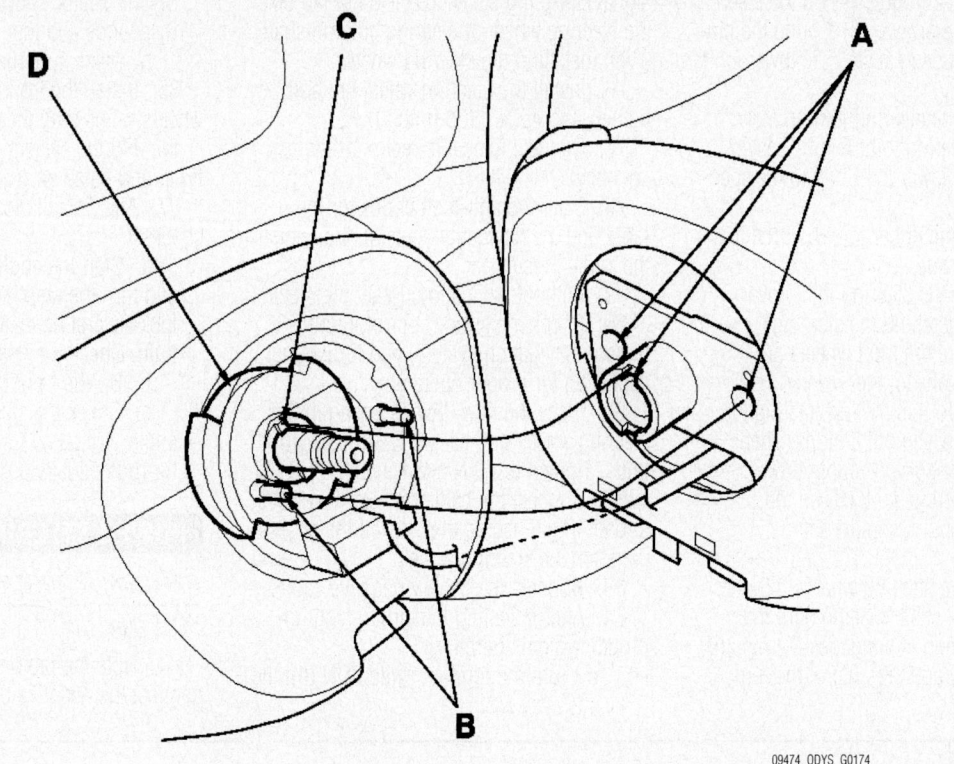

09474_ODYS_G0174

Fig. 188 Install the steering wheel on to the steering column shaft, making sure the steering wheel hub (A) engages the pins (B) of the cable reel and tabs (C) of the turn signal canceling sleeve (D). Do not tap on the steering wheel or steering column shaft when installing the steering wheel— Odyssey

41. Remove the pinion shaft grommet from the steering joint cover B.

42. After removing the steering gearbox, make sure that no power steering fluid gets on the gearbox mount cushions, gearbox housing, surface of the front subframe and stiffener. Wipe off any spilled fluid at once.

To install:

43. Install the pinion shaft grommet on the valve housing.

44. Slide the steering gearbox between the front subframe and body from the driver's side. Place the gearbox in position on the front suspension subframe.

45. Rotate the steering gearbox so the pinion shaft points upward.

46. Continue moving the gearbox toward the passenger's side until the steering gearbox is in position. Make sure the power steering return line and feed line are routed above the gearbox.

47. Connect the return hose clip.

48. EX-L, EX-L Touring models:
   a. Install the base bracket (A) on the front subframe (B).
   b. Install the rear mount (A) on the base bracket (B), and connect the connector (C).
   c. Install the rear mount stop (A), then install the rear mount bolt (B).

49. Van, LX, EX models:
   a. Install the rear mount bracket (A) on the front subframe (B).
   b. Install the rear mount bracket bolt (A).

50. Install the steering gearbox bracket mounting bolts (A).

51. Install the two 12 mm flange bolts and nuts on the left side of the gearbox.

52. Install the two 10 mm flange bolts on the right side of the steering gearbox, then install the mounting bracket and cushion.

53. Install the P/S heat baffle plate (A).

54. Tighten the return hose flare nut (A) to the specified torque, and install the adjustable hose clamp (B) and the return hose (C).

55. Tighten the feed line flare nut to 37 Nm (27 ft. lbs.).

56. Install the P/S line mounting brackets on the front subframe and gearbox mounting bracket.

57. Install the front subframe rear bracket (A) with 12 mm flange bolts (B) and 14 mm special bolts (C), and tighten to specified torque.

58. Install the front subframe front bracket (A) with 12 mm flange bolts (B) and

14 mm special bolts (C), and tighten to specified torque.

59. Install the front splash shield.

60. Connect the power steering pressure (PSP) switch connector.

61. Install exhaust pipe A using new gaskets and new self-locking nuts.

62. Connect the three way catalytic converter (TWC) (A) to the muffler (B).

63. Install the new 10 mm self-locking nuts, and tighten them to 33 Nm (25 ft. lbs.).

64. Install the exhaust rubber mount on the frame. Torque to 22 Nm (16 ft. lbs.).

65. Wipe off any grease contamination from the ball joint tapered section and threads. Reconnect the tie-rod ends (A) to the steering knuckles. Install the 12 mm nut (B) and tighten it.

66. Install a new cotter pin (C), and bend it as shown.

67. Center the steering rack within its stroke.

68. Insert the upper end of the steering joint onto the steering shaft (A) (line up the bolt hole (B) with the flat portion (C) on the shaft), and loosely install the upper joint bolt.

69. Slip the lower end of the steering joint onto the pinion shaft (D) taking care to align the gap (E) within the angle.

70. Align the bolt hole (A) on the steering joint with the groove (B) around the pinion shaft, then loosely install the lower joint bolt (C).

71. Pull on the steering joint to make sure that the steering joint is fully seated, then tighten the lower joint bolt to the specified torque.

72. Tighten the upper joint bolt (D) to the specified torque.

73. Reinstall the steering joint cover.

74. Before installing the steering wheel, make sure the front wheels are aligned straight ahead, then center the cable reel (A). Do this by first rotating the cable reel clockwise until it stops. Then rotate it counterclockwise about three full turns. The arrow mark (B) on the cable reel label point should point straight up.

75. Install the steering wheel on to the steering column shaft, making sure the steering wheel hub (A) engages the pins (B) of the cable reel and tabs (C) of the turn

signal canceling sleeve (D). Do not tap on the steering wheel or steering column shaft when installing the steering wheel.

76. Install the steering wheel nut and tighten it to 49 Nm (36 ft. lbs.).

77. Connect the cable reel sub-harness connector.

78. Enter the anti-theft codes for the radio and the navigation system, then enter the audio presets.

79. Without navigation: Reset the clock.

80. Check the cruise control, radio remote, navigation system, and turn signal canceling for proper operation.

81. Place the new driver's airbag in the steering wheel, and secure it with new Torx® bolts. Tighten to 10 Nm (84 inch lbs.).

82. Connect the cable reel to the driver's airbag 4P connector, then install the access panel on the steering wheel.

83. Connect the battery negative cable.

84. After installing the airbag, confirm proper system operation:

a. Turn the ignition switch ON (II); the

SRS indicator should come on for about 6 seconds and then go off.

b. Make sure the horn button works.

85. Install the front wheel, then set the wheels in the straight ahead position.

86. Fill the system with power steering fluid, and bleed air from the system.

87. After installation, do the following checks.

a. Start the engine, allow it to idle, and turn the steering wheel from lock-to-lock several times to warm up to the fluid. Check the gearbox for leaks.

b. Do the front toe inspection.

c. Check the steering wheel spoke angle. Adjust by turning the right and left tie-rods equally if necessary.

POWER STEERING PUMP

REMOVAL & INSTALLATION

See Figure 189.

1. Note the radio security code and disconnect the negative battery cable.

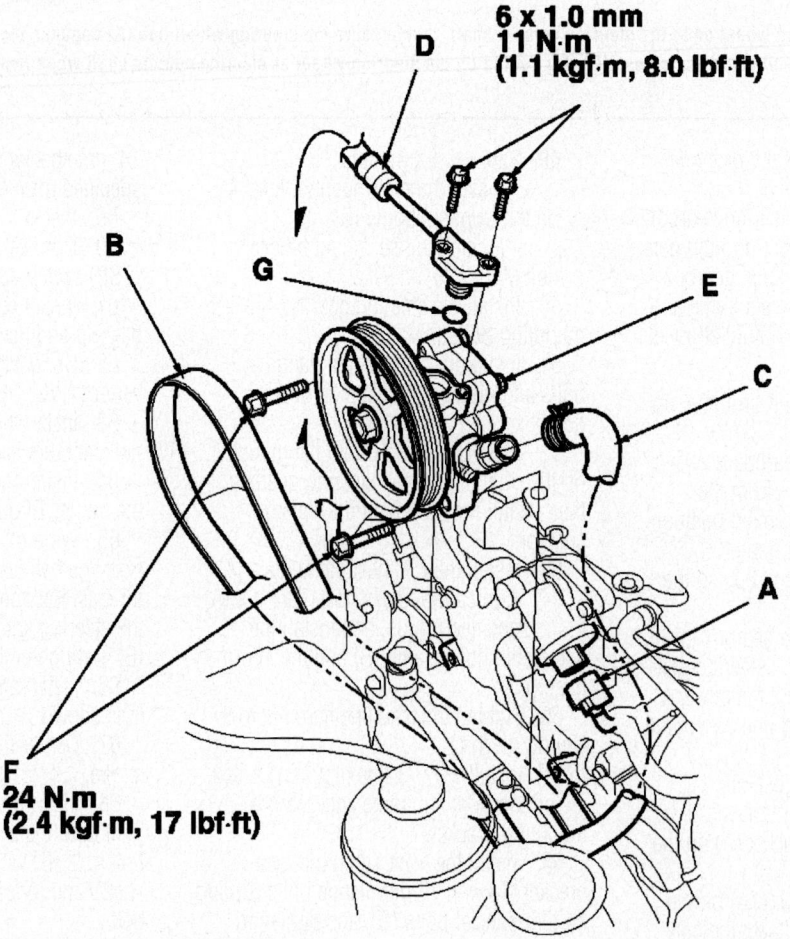

Fig. 189 Exploded view of the IMT actuator connector (A), drive belt (B). power steering pump inlet hose (C), outlet hose (D), power steering pump (E), mounting bolts (F) and nut (G)

2. Place a suitable drain pan under the vehicle.

3. Drain the power steering fluid from the reservoir.

4. J35A7 Engines: Disconnect the IMT actuator connector .

5. Remove the accessory drive belt. For additional information, refer to the following section, "Accessory Drive Belt, Removal & Installation."

6. Cover the components around the pump and drive belts with several shop towels to absorb any spilled fluid.

✳ WARNING

If any power steering fluid is spilled on a painted surface, wipe it off immediately.

7. Squeeze the power steering hose clamp at the pump, slide it up the hose and remove the hose from the pump.

8. Loosen the power steering pump outlet hose retaining bolts, and remove the hose fitting along with its O-ring from the pump.

✳✳ WARNING

Protect all open power steering lines and fittings from debris and contaminants, otherwise internal damage may occur.

9. Wrap a clean cloth around the outlet hose fitting, and the port on the pump and hold in place with duct electrical tape.

10. Plug and cap the reservoir hose and metal hose fitting on the pump.

✳✳ WARNING

Do not turn the steering wheel with the pump removed.

11. Remove the power steering pump retainers, then remove the pump.

To install:

12. Install the pump and loosely install the lock and pivot bolts.

13. Install a new O-ring on the power steering hose outlet fitting, lubricate the O-ring with a light coating of Honda power steering fluid and install the fitting.

14. The balance of the installation is the reverse of the removal procedure.

15. Make note of the following points:
- Tighten the pump fasteners to 16 ft. lbs. (22 Nm)
- Make sure all fasteners, hose clamps and fittings are properly installed and tightened.
- When topping off the system use only Genuine Honda Power Steering Fluid-V or S. Substituting another brand may damage internal components.
- Clean any spilled fluid before starting the vehicle.
- Bleed the system and top off as necessary.

BLEEDING

1. Fill the reservoir to the upper line with Genuine Honda Power Steering Fluid-V or S.

2. Run the engine at idle and turn the steering wheel lock-to-lock several times to bleed air from the system and fill the rack valve body.

3. Recheck the fluid level and add more if necessary. Don't overfill the reservoir.

4. Check the power steering system for leaks.

SUSPENSION

LOWER BALL JOINT

REMOVAL & INSTALLATION

See Figure 190.

1. Install a hex nut onto the threads of the ball joint. Make sure the nut is flush with the ball joint pin end to prevent damage to the threaded end of the ball joint pin.

2. Apply grease to the ball joint remover on the areas shown. This will ease

installation of the tool and prevent damage to the pressure bolt threads.

3. Loosen the pressure bolt, and install the ball joint remover as shown. Insert the jaws carefully, making sure not to damage the ball joint boot. Adjust the jaw spacing by turning the adjusting bolt.

4. After adjusting the adjusting bolt, make sure the head of the adjusting bolt is in the position shown to allow the jaw to pivot.

5. With a wrench, tighten the pressure bolt until the ball joint pin pops loose from the ball joint pin hole. If necessary, apply penetrating type lubricant to loosen the ball joint pin.

6. Remove the ball joint remover, then remove the nut from the end of the ball joint pin, and pull the ball joint out of the ball joint pin hole. Inspect the ball joint boot, and replace it if damaged.

LOWER CONTROL ARM

REMOVAL & INSTALLATION

See Figures 191 through 193.

1. Raise the front of the vehicle, and support it with safety stands in the proper locations.

2. Remove the front wheel.

FRONT SUSPENSION

3. Remove the flange nut (A) while holding the respective joint pin (B) with a hex wrench (C), then disconnect the stabilizer links from the strut (D).

4. Turn the stabilizer bar backward to gain easier access to the front side of the lower arm mounting bolt.

5. Remove the lock pin from the lower arm ball joint castle nut, and remove the nut.

6. Remove the lower ball joint from the knuckle using the special tools.

7. Remove the flange bolts (A), then remove the lower arm (B).

8. Install the lower arm in the reverse order of removal, and note these items:
- Be careful not to damage the ball joint boot when connecting the lower arm to the knuckle.
- Place a jack under the lower arm, and raise the lower arm to load the suspension with the vehicle's weight.
- Tighten all mounting hardware to the specified torque values.
- Torque the castle nut to the lower torque specification, then tighten it only far enough to align the slot with the hole in the stud. Do not align the castle nut by loosening it.

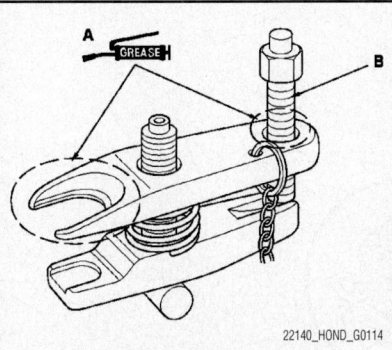

22140_HOND_G0114

Fig. 190 Apply grease to the ball joint remover on the areas shown (A). This will ease installation of the tool and prevent damage to the pressure bolt (B) threads.

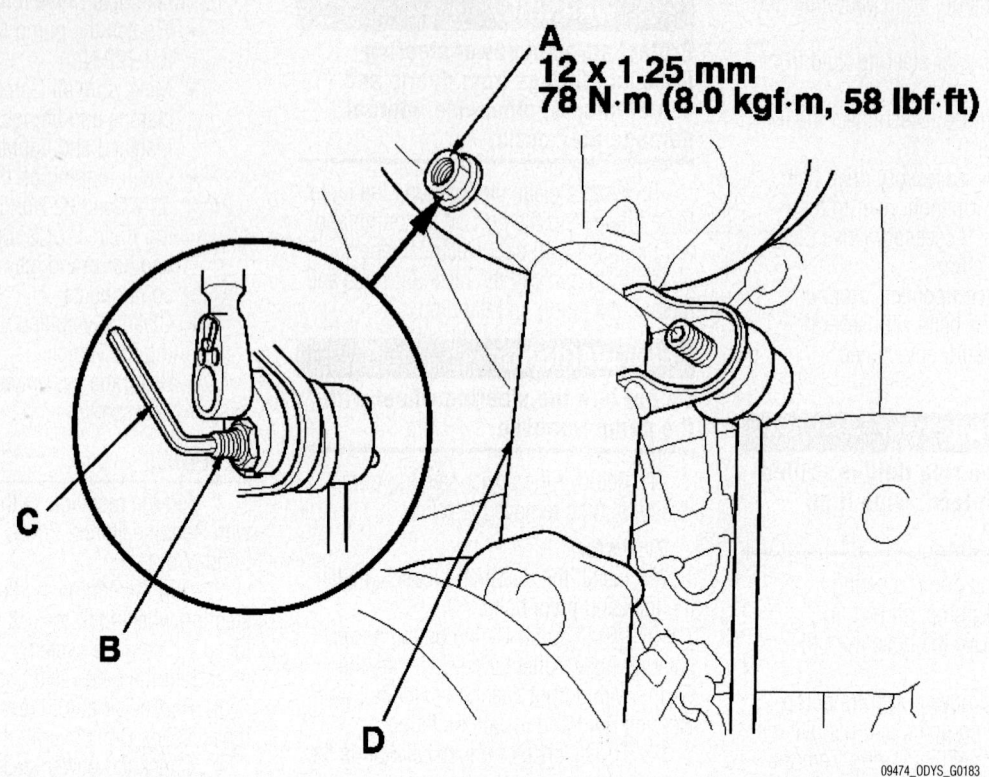

A
12 x 1.25 mm
78 N·m (8.0 kgf·m, 58 lbf·ft)

C

B

D

09474_ODYS_G0183

Fig. 191 Remove the flange nut (A) while holding the respective joint pin (B) with a hex wrench (C), then disconnect the stabilizer links from the strut (D)—Odyssey

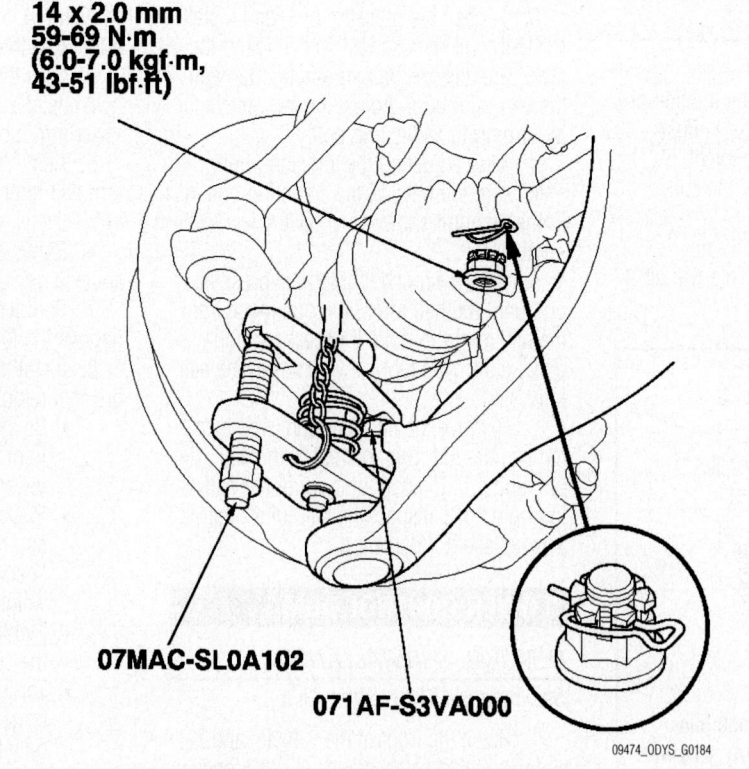

14 x 2.0 mm
59-69 N·m
(6.0-7.0 kgf·m,
43-51 lbf·ft)

07MAC-SL0A102

071AF-S3VA000

09474_ODYS_G0184

Fig. 192 Remove the lock pin from the lower arm ball joint castle nut, and remove the nut—Odyssey

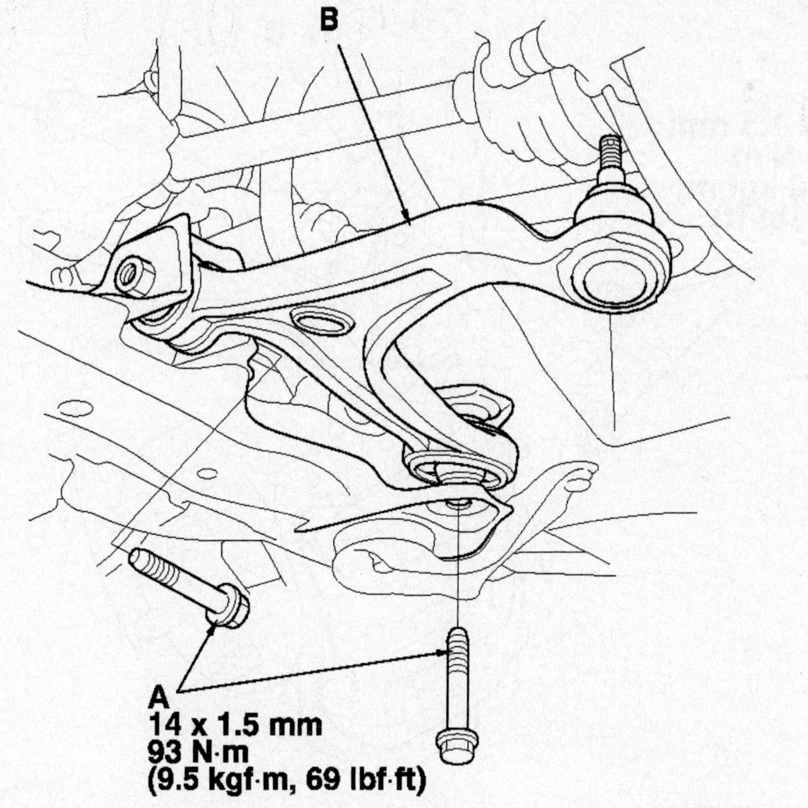

Fig. 193 Remove the flange bolts (A), then remove the lower arm (B)—Odyssey

A
14 x 1.5 mm
93 N·m
(9.5 kgf·m, 69 lbf·ft)

09474_ODYS_G0185

- Use a new lock pin on the castle nut.
- Connect the struts to the links.
- Before installing the wheel, clean the mating surface on the brake disc and the inside of the wheel.
- Check the front wheel alignment, and adjust it if necessary.

MACPHERSON STRUT

REMOVAL & INSTALLATION

See Figures 194 through 196.

➡When compressing the damper spring, use a commercially available strut spring compressor (Branick MST-580A or Model 7200 or equivalent), according to the manufacturer's instructions.

1. Raise the vehicle, and support it with safety stands in the proper locations.
2. Remove the front wheel.
3. Disconnect the stabilizer link from the damper.
4. Remove the wheel sensor harness and the brake hose bracket from the damper. Do not disconnect the wheel sensor connector.
5. Remove the flange nuts and damper pinch bolts from the damper.

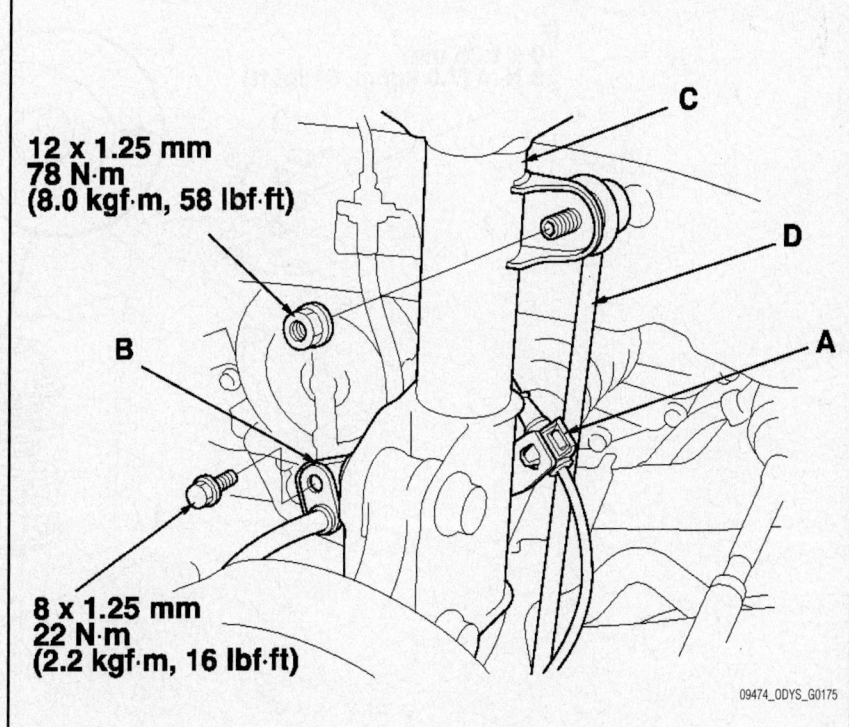

12 x 1.25 mm
78 N·m
(8.0 kgf·m, 58 lbf·ft)

8 x 1.25 mm
22 N·m
(2.2 kgf·m, 16 lbf·ft)

09474_ODYS_G0175

Fig. 194 Disconnect the stabilizer link (A) from the damper (B). Remove the wheel sensor harness (C) and the brake hose bracket (D) from the damper—Odyssey

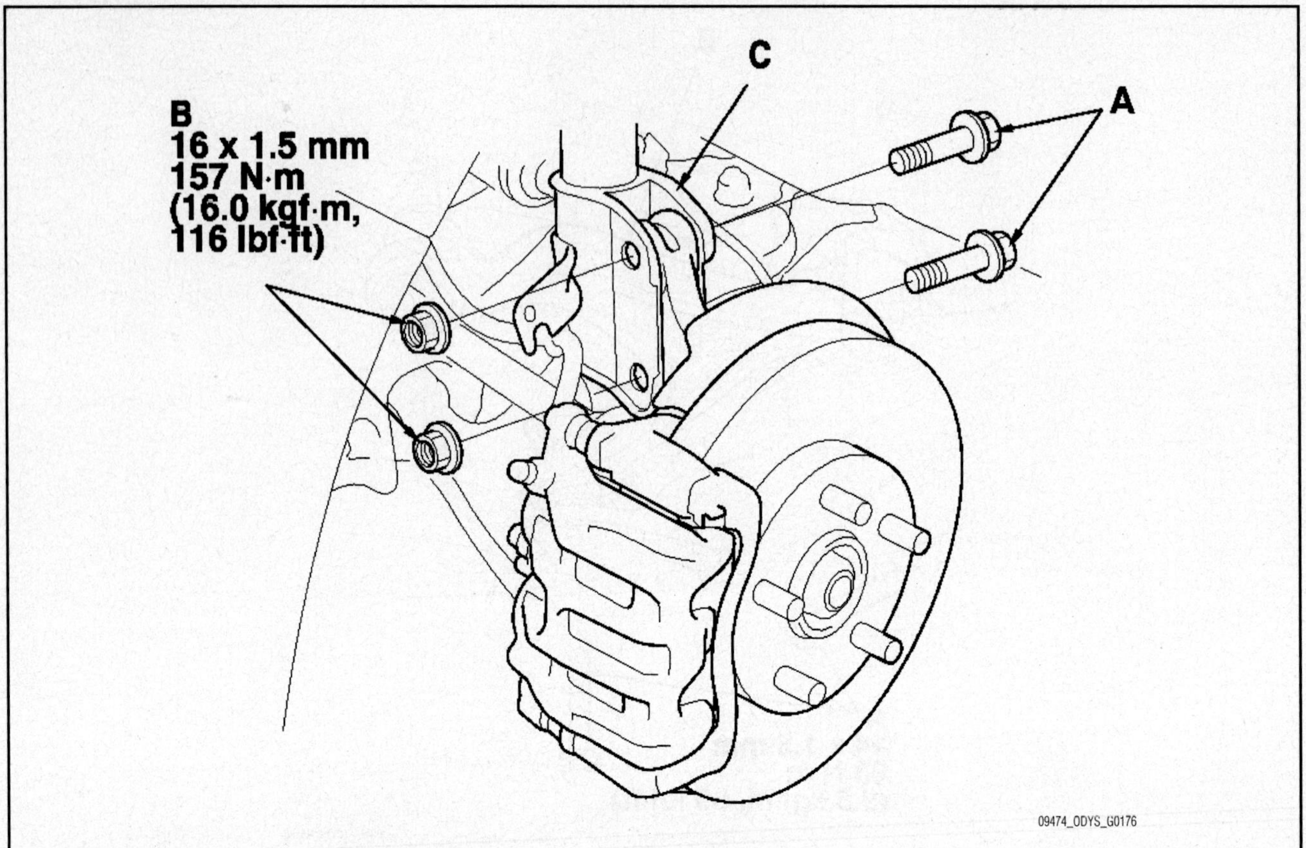

B
16 x 1.5 mm
157 N·m
(16.0 kgf·m,
116 lbf·ft)

C

A

09474_ODYS_G0176

Fig. 195 Remove the flange nuts (A) and damper pinch bolts (B) from the damper—Odyssey

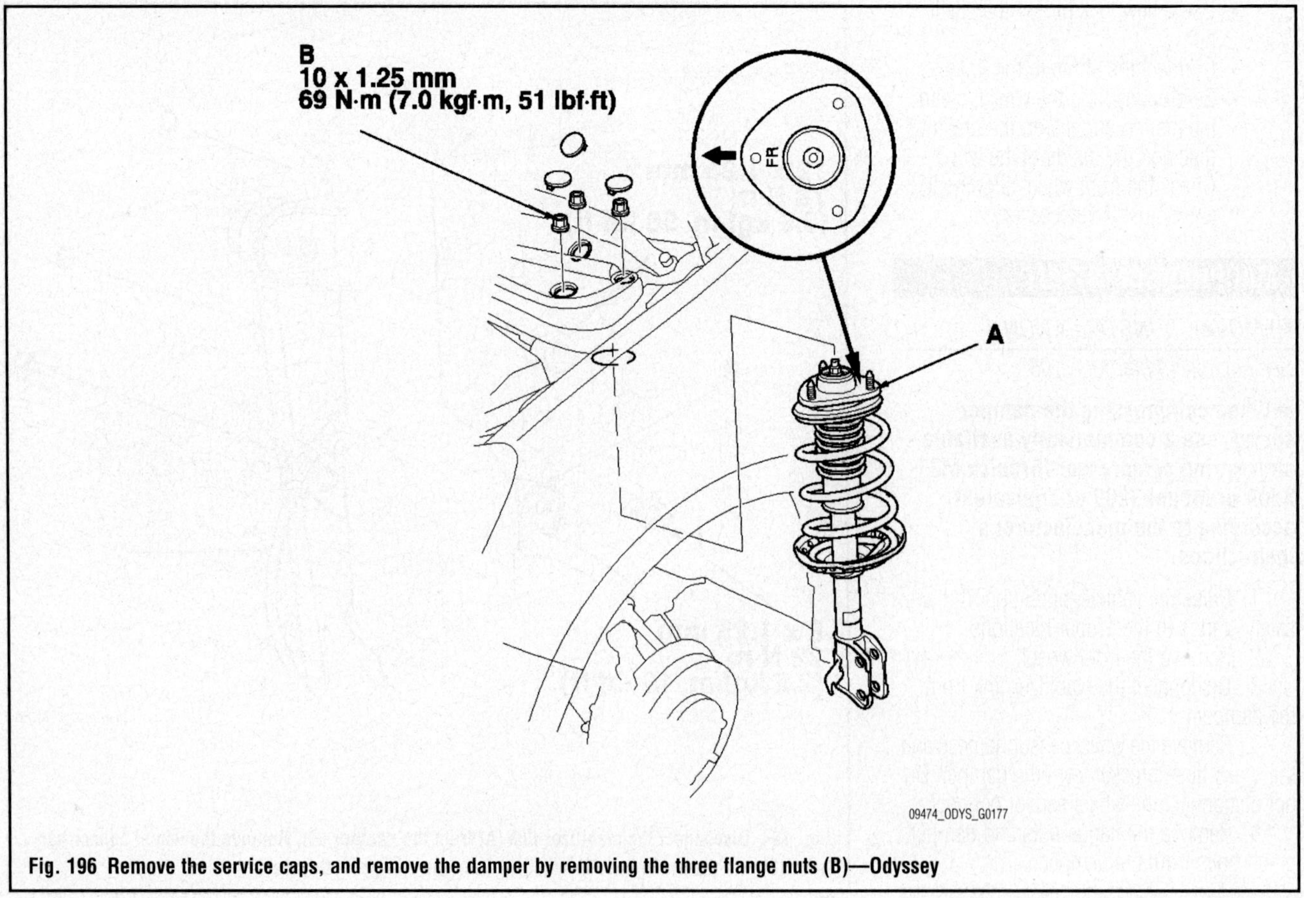

B
10 x 1.25 mm
69 N·m (7.0 kgf·m, 51 lbf·ft)

FR

A

09474_ODYS_G0177

Fig. 196 Remove the service caps, and remove the damper by removing the three flange nuts (B)—Odyssey

6. Remove the service caps, and remove the damper by removing the three flange nuts.

7. Remove the damper.

➡**Damper springs are different, left and right. Mark the springs L and R before you continue.**

To install:

8. Install the damper onto the frame, then loosely install the three flange nuts.

9. Loosely install the damper pinch bolts and flange nuts to the damper.

10. Install the wheel sensor harness and the brake hose bracket to the damper.

11. Loosely install the stabilizer link to the damper.

12. Raise the front suspension with a floor jack to load the suspension with the vehicle's weight.

13. Tighten the damper pinch bolts and flange nuts to the specified torque value.

14. Tighten the flange nuts on top of the damper to the specified torque value.

15. Tighten the stabilizer link nuts to the specified torque value.

16. Install the service caps.

17. Install the front wheel.

18. Check the front wheel alignment, and adjust it if necessary.

STABILIZER BAR

REMOVAL & INSTALLATION

See Figures 197 through 202.

Special tool required: front subframe adapter VSB02C000016

1. Raise the front of the vehicle, and support it with safety stands in the proper locations.

2. Remove the front wheels.

3. Disconnect the stabilizer links from the stabilizer bar on the right and left sides.

4. Remove the splash shield (A).

5. Attach the special tool to the front subframe (A) by hanging the belt (B) of the special tool over the front of the subframe, secure it with the pin (C), then tighten the special tool screw.

6. Raise the jack (B) and line up the slots in the arms with the bolt holes on the corner of the jack base, then attach them with bolts securely.

7. Loosen the front subframe front bracket (A) mounting bolts on the right and left of the vehicle so they are about 14mm (⁹⁄₁₆ in.) from the mounting surface.

8. Support the front subframe securely by raising the special tool, then remove the two 12mm flange bolts (A).

9. Loosen the two 14mm special bolts (B) so they are about 14mm (⁹⁄₁₆ in.) from the mounting surface.

10. Lower the jack supporting the front subframe with the special tool slowly until the front subframe has dropped about 14mm (⁹⁄₁₆ in.).

11. Remove the flange bolts (A) and bushing holders (B), then remove the bushings (C) and the stabilizer bar (D) from the front subframe (E).

12. Install the stabilizer bar in the reverse order of removal, and note these items:

- Note the right and left direction of the stabilizer bar.
- Align the paint marks (A) on the stabilizer bar with the sides of the bushings (B).
- Note the fore/aft direction of the bushing holders.
- Raise the front subframe up with the jack and special tool until it contacts the body frame, then tighten the mounting bolts to the specified torque.
- Refer to Stabilizer Link Removal/Installation to connect the stabilizer bar to the links.
- Do the front toe inspection, and adjust it if necessary.

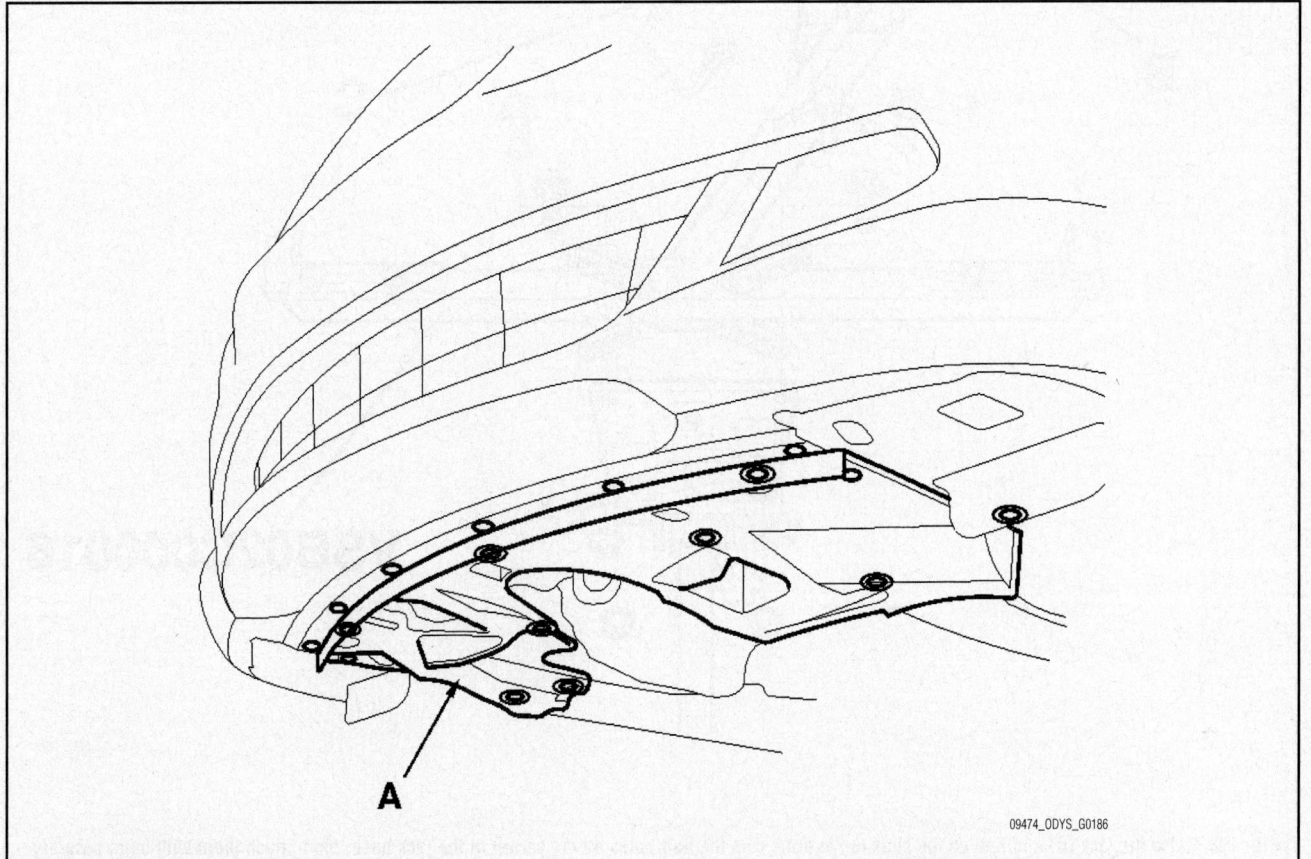

09474_ODYS_G0186

Fig. 197 Remove the splash shield (A)

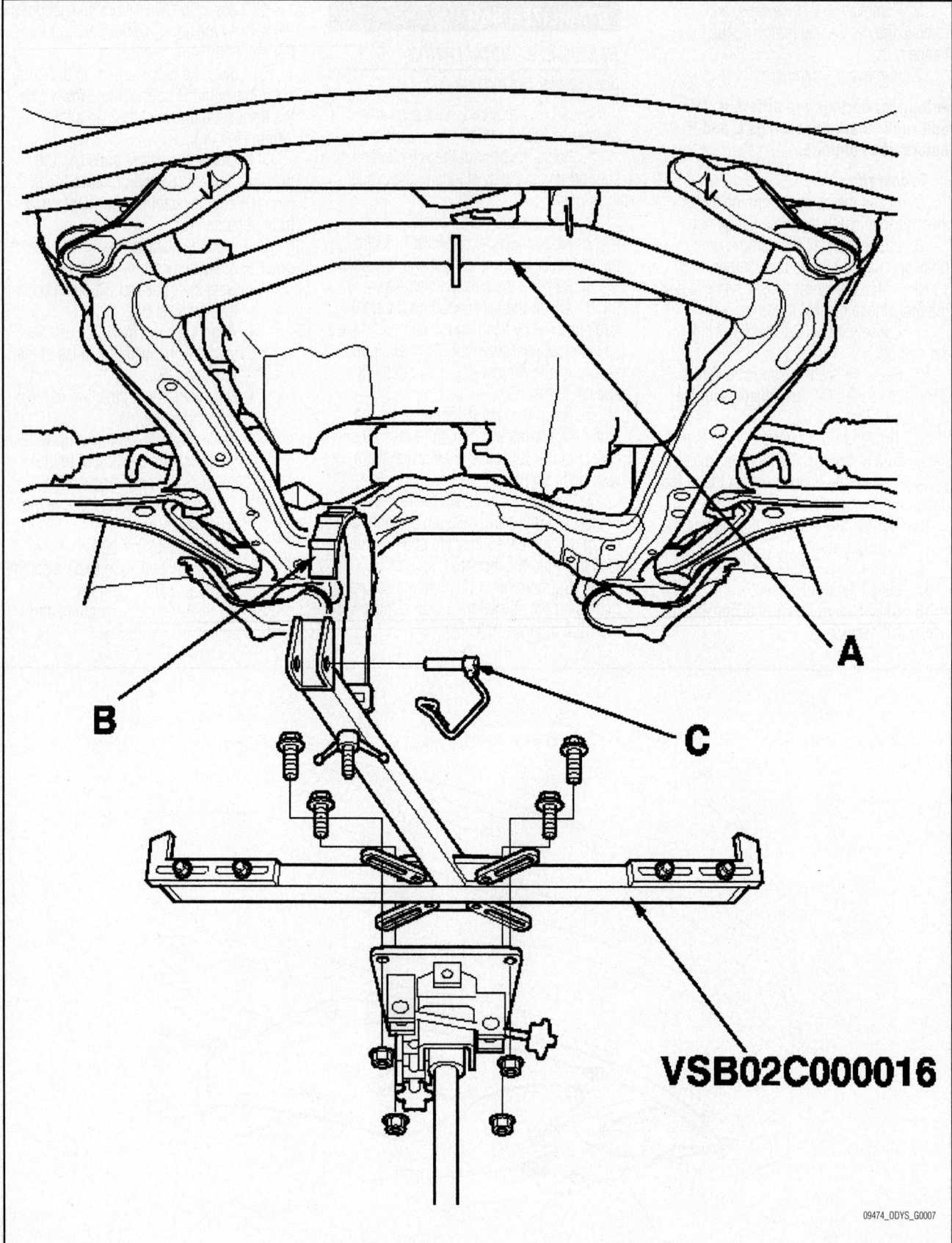

A

B

C

VSB02C000016

Fig. 198 Raise the jack (B) and line up the slots in the arms with the bolt holes on the corner of the jack base, then attach them with bolts securely.
Raise the jack (B) and line up the slots in the arms with the bolt holes on the corner of the jack base, then attach them with bolts securely

09474_ODYS_G0007

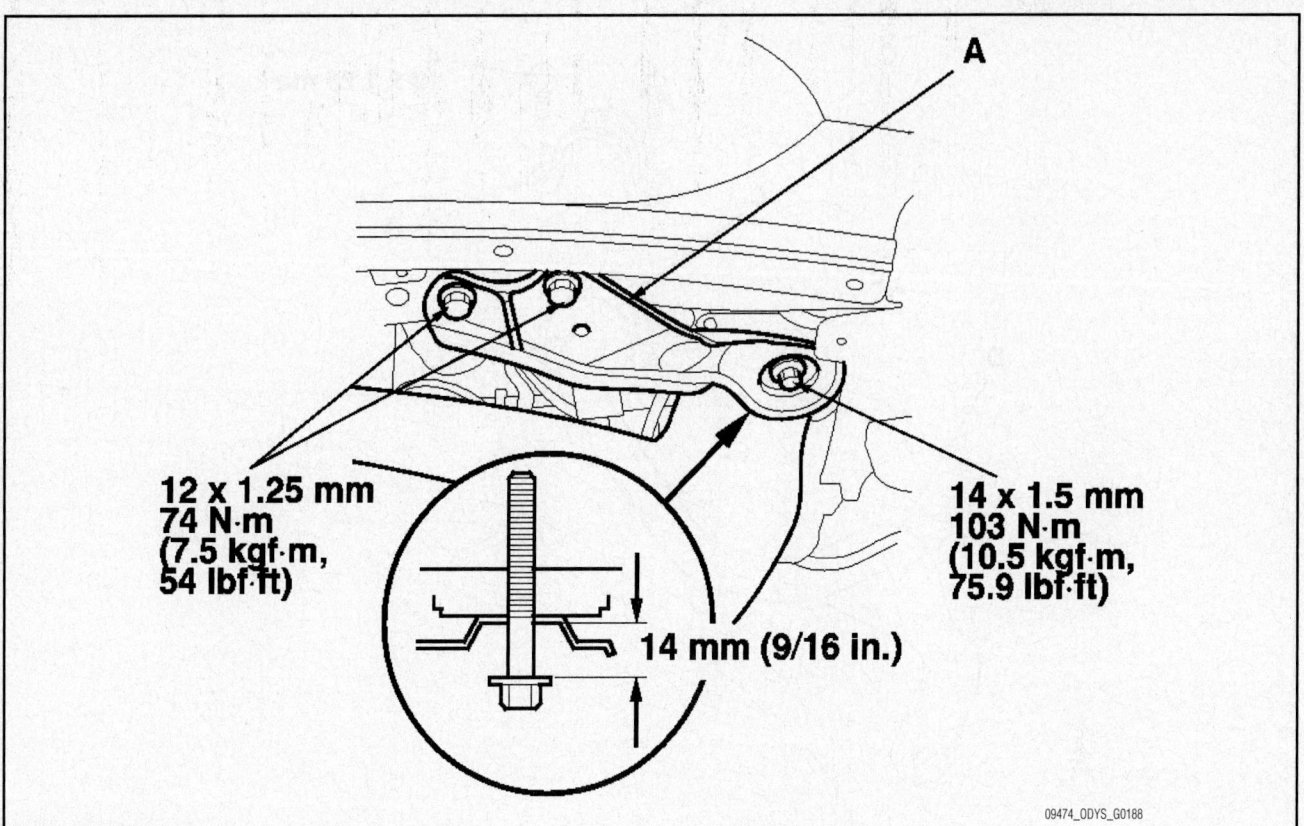

12 x 1.25 mm
74 N·m
(7.5 kgf·m,
54 lbf·ft)

14 x 1.5 mm
103 N·m
(10.5 kgf·m,
75.9 lbf·ft)

14 mm (9/16 in.)

09474_ODYS_G0188

Fig. 199 Loosen the front subframe front bracket (A) mounting bolts on the right and left of the vehicle so they are about 14mm (9/16 in.) from the mounting surface

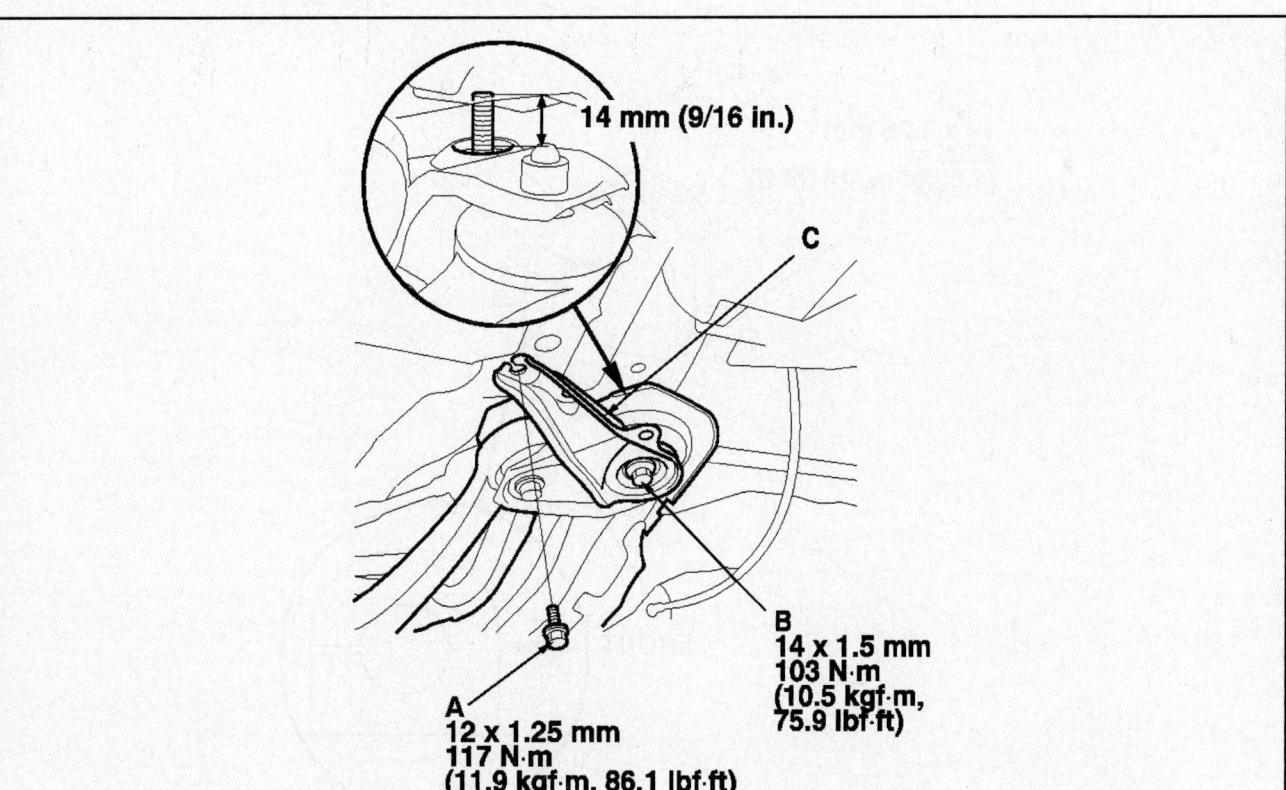

14 mm (9/16 in.)

C

B
14 x 1.5 mm
103 N·m
(10.5 kgf·m,
75.9 lbf·ft)

A
12 x 1.25 mm
117 N·m
(11.9 kgf·m, 86.1 lbf·ft)

09474_ODYS_G0189

Fig. 200 Support the front subframe securely by raising the special tool, then remove the two 12mm flange bolts (A). Loosen the two 14mm special bolts (B) so they are about 14mm (9/16 in.) from the mounting surface

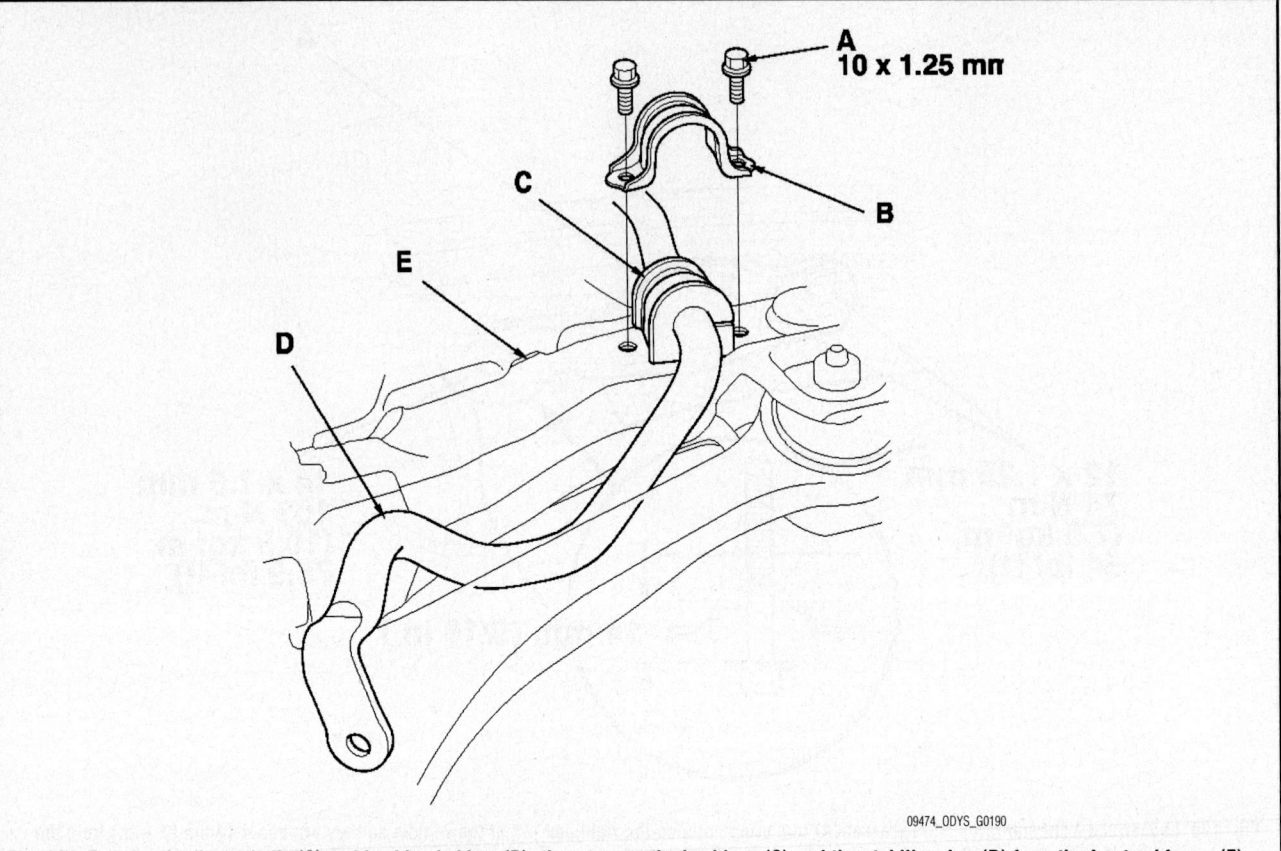

A
10 x 1.25 mm
B
C
E
D
09474_ODYS_G0190

Fig. 201 Remove the flange bolts (A) and bushing holders (B), then remove the bushings (C) and the stabilizer bar (D) from the front subframe (E)

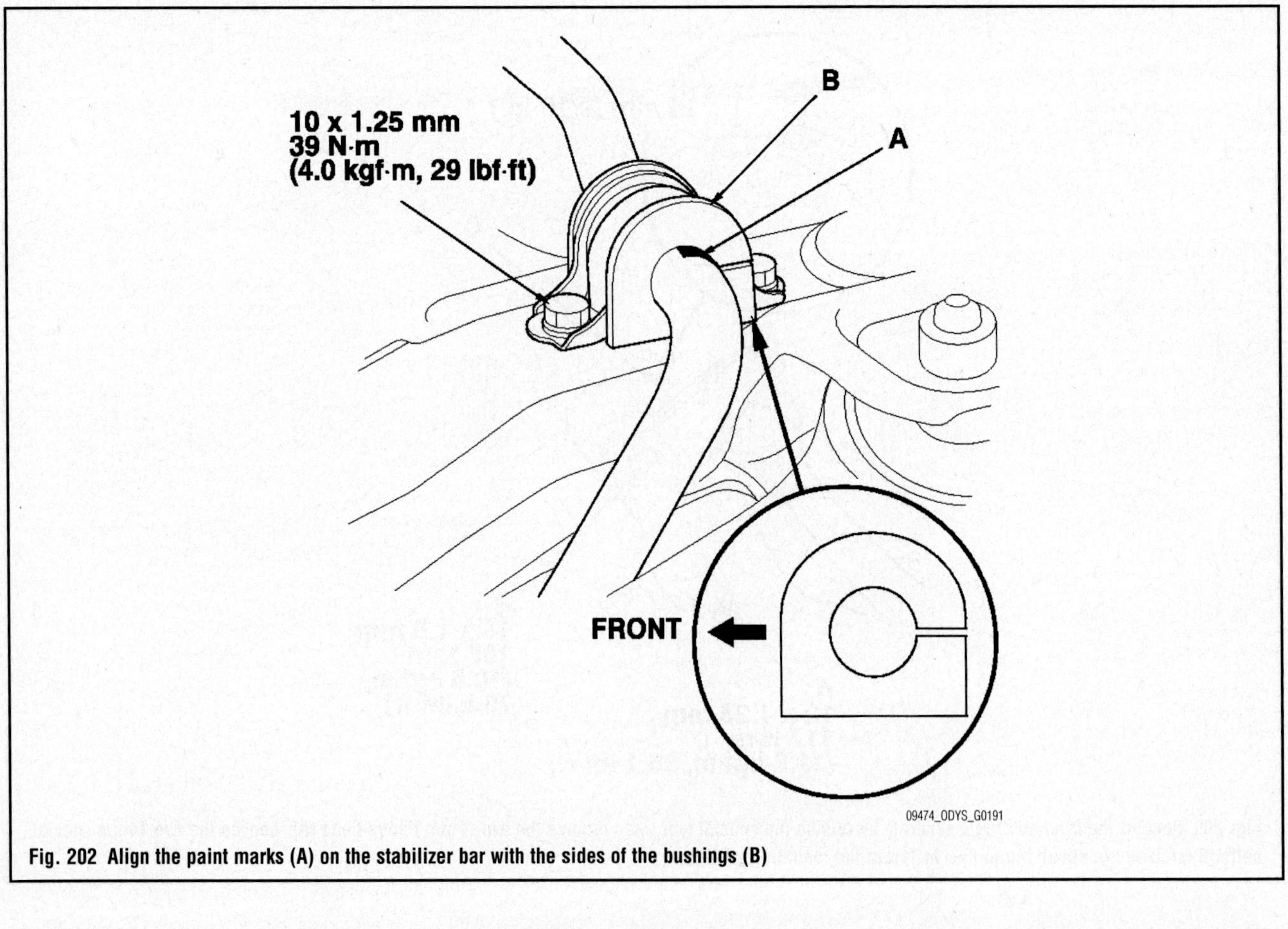

B
A
10 x 1.25 mm
39 N·m
(4.0 kgf·m, 29 lbf·ft)
FRONT
09474_ODYS_G0191

Fig. 202 Align the paint marks (A) on the stabilizer bar with the sides of the bushings (B)

STEERING KNUCKLE

REMOVAL & INSTALLATION

See Figures 203 through 208.

1. Raise the front of the vehicle, and support it with safety stands in the proper locations.

2. Remove the wheel nuts and front wheel.

3. Remove the brake hose mounting bolt (A).

4. Remove the caliper mounting bolts (B), and hang the caliper assembly (C) to one side. To prevent damage to the caliper assembly or brake hose, use a short piece of wire to hang the caliper from the undercarriage.

5. Remove the wheel sensor from the knuckle. Do not disconnect the wheel sensor connector.

6. Raise the stake, and then remove the spindle nut.

7. Remove the 6mm brake disc retaining screws (A).

8. Screw two 8 x 1.25mm bolts (B) into the brake disc to push it away from the hub. Turn each bolt two turns at a time to prevent cocking the disc excessively.

9. Remove the brake disc from the knuckle.

10. Check the front hub for damage and cracks.

11. Remove the cotter pin (A) from the tie-rod end ball joint, then loosen the nut (B).

12. Remove the tie-rod ball joint from the knuckle using the special tool.

13. Remove the lock pin (A) from the lower arm ball joint castle nut (B), and remove the nut.

14. Remove the lower ball joint from the knuckle using the special tools.

15. Remove the damper pinch bolts (A) and flange nuts (B) from the damper.

16. Pull the knuckle outward, and remove the driveshaft outboard joint (C) from the knuckle (D) by tapping the driveshaft end (E) with a plastic hammer, then remove the knuckle.

To install:

17. Install the knuckle in the reverse order of removal, and pay particular attention to the following items:

- Be careful not to damage the ball joint boot when installing the knuckle.
- First, install all the components and lightly tighten the bolts and nuts, then raise the suspension to load it with the vehicle's weight before fully tightening to the specified torque values. Do not place the jack against the ball joint pin of the knuckle.
- Tighten all mounting hardware to the specified torque values.
- Before connecting the ball joint to the knuckle, degrease the threaded section and tapered portion of the ball joint pin, the connecting hole, the threaded section and mating surface of the castle nut.
- Torque the castle nut to the lower torque specification, then tighten it only far enough to align the slot with the ball joint pin hole. Do not align the castle nut by loosening it.
- Use a new lock pin on the castle nut.
- Before installing the brake disc, clean the mating surface of the front hub and the inside of the brake disc.
- Use a new spindle nut on reassembly.
- Before installing the new spindle nut, apply a small amount of engine oil to the seating surface of the nut. After tightening, use a drift to stake the spindle nut shoulder against the driveshaft.

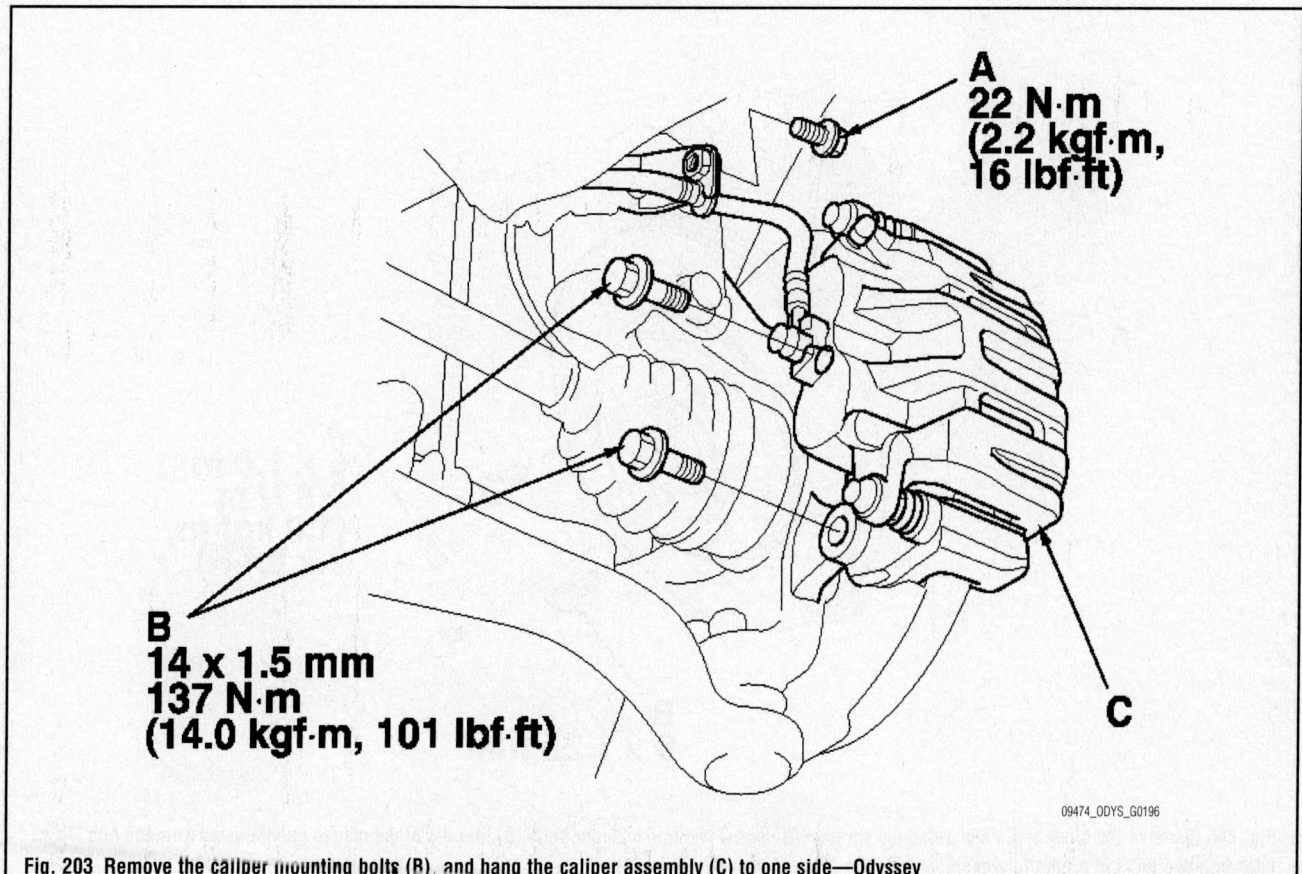

**A
22 N·m
(2.2 kgf·m,
16 lbf·ft)**

**B
14 x 1.5 mm
137 N·m
(14.0 kgf·m, 101 lbf·ft)**

C

09474_ODYS_G0196

Fig. 203 Remove the caliper mounting bolts (B), and hang the caliper assembly (C) to one side—Odyssey

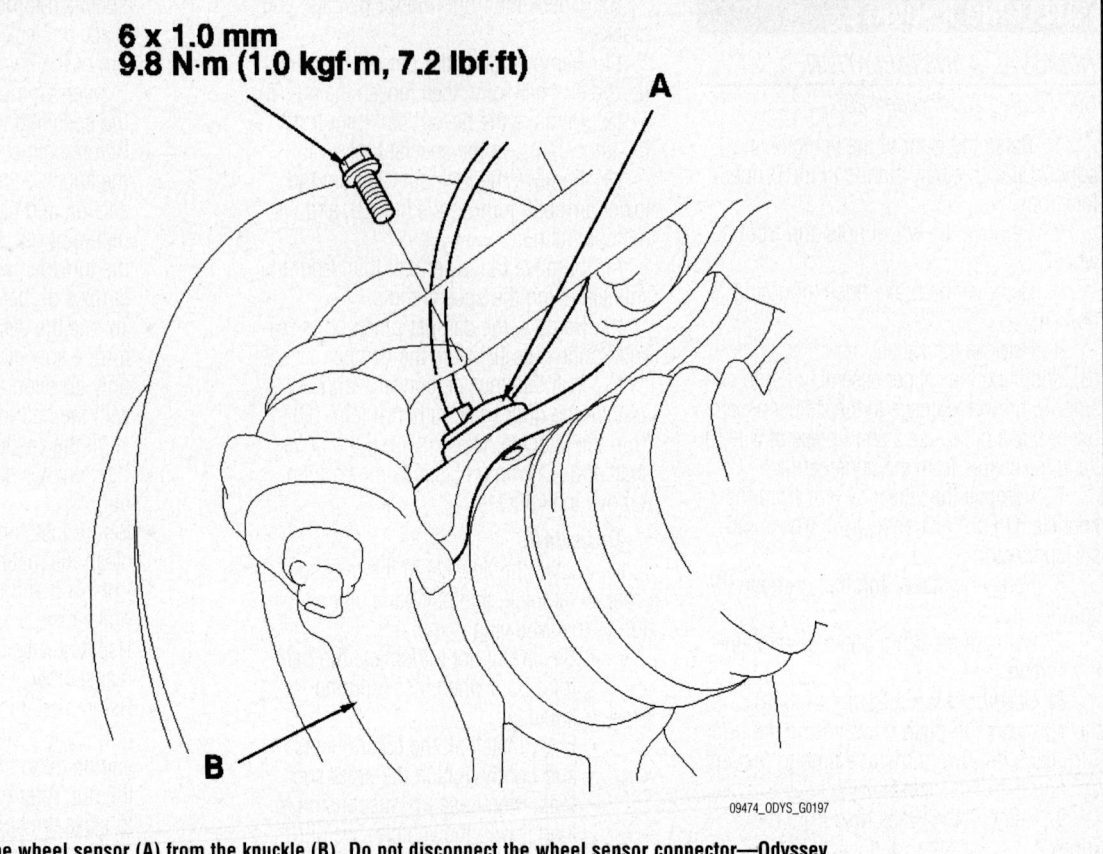

6 x 1.0 mm
9.8 N·m (1.0 kgf·m, 7.2 lbf·ft)

A

B

09474_ODYS_G0197

Fig. 204 Remove the wheel sensor (A) from the knuckle (B). Do not disconnect the wheel sensor connector—Odyssey

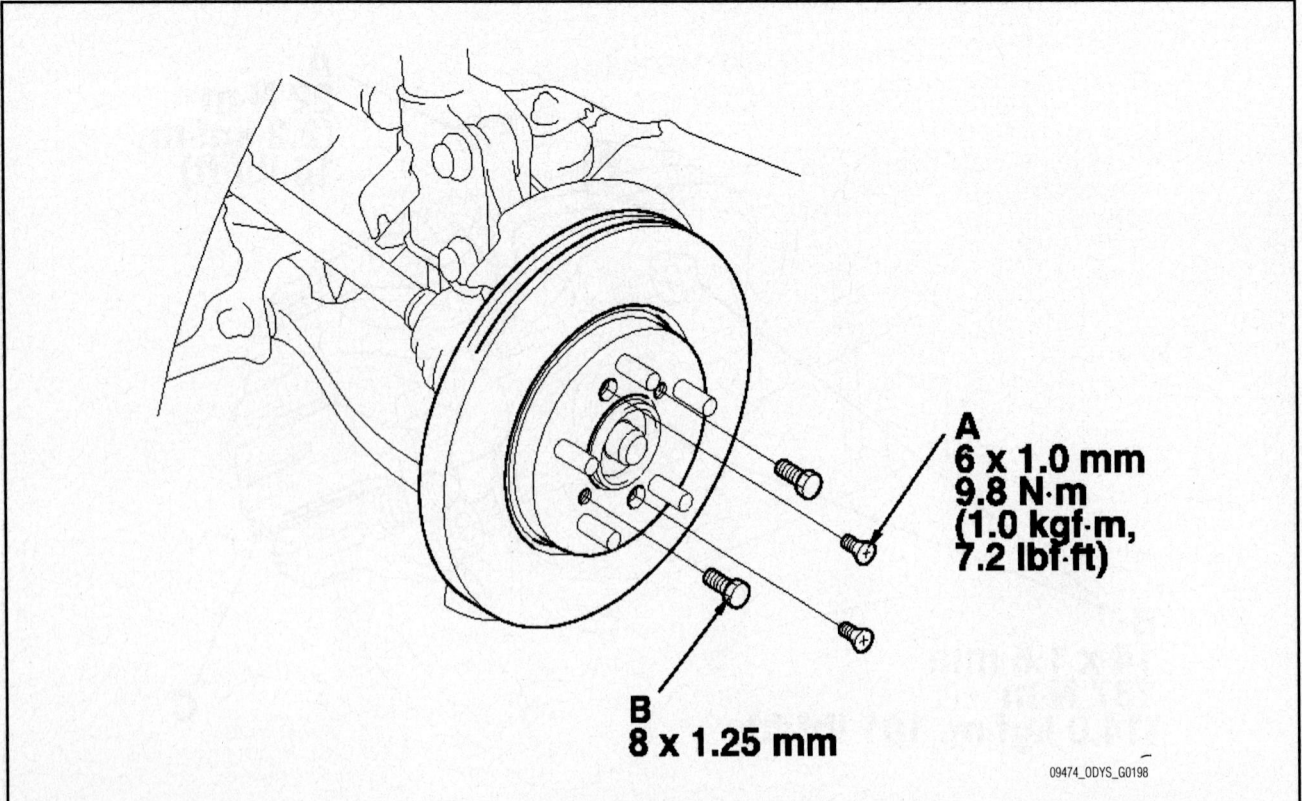

A
6 x 1.0 mm
9.8 N·m
(1.0 kgf·m,
7.2 lbf·ft)

B
8 x 1.25 mm

09474_ODYS_G0198

Fig. 205 Remove the 6mm brake disc retaining screws (A). Screw two 8 × 1.25mm bolts (B) into the brake disc to push it away from the hub. Turn each bolt two turns at a time to prevent cocking the disc excessively—Odyssey

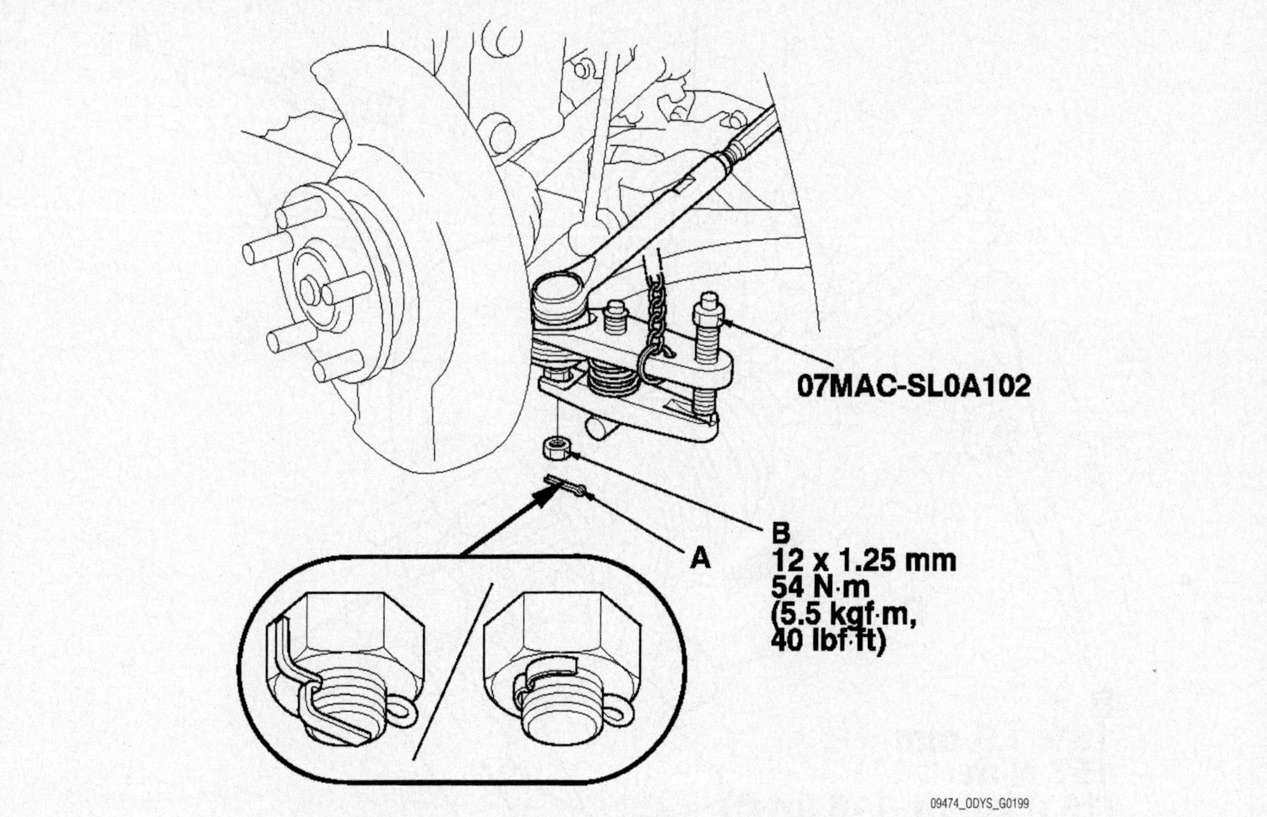

07MAC-SL0A102

B
12 x 1.25 mm
54 N·m
(5.5 kgf·m,
40 lbf·ft)

A

09474_ODYS_G0199

Fig. 206 Remove the cotter pin (A) from the tie-rod end ball joint, then loosen the nut (B)—Odyssey

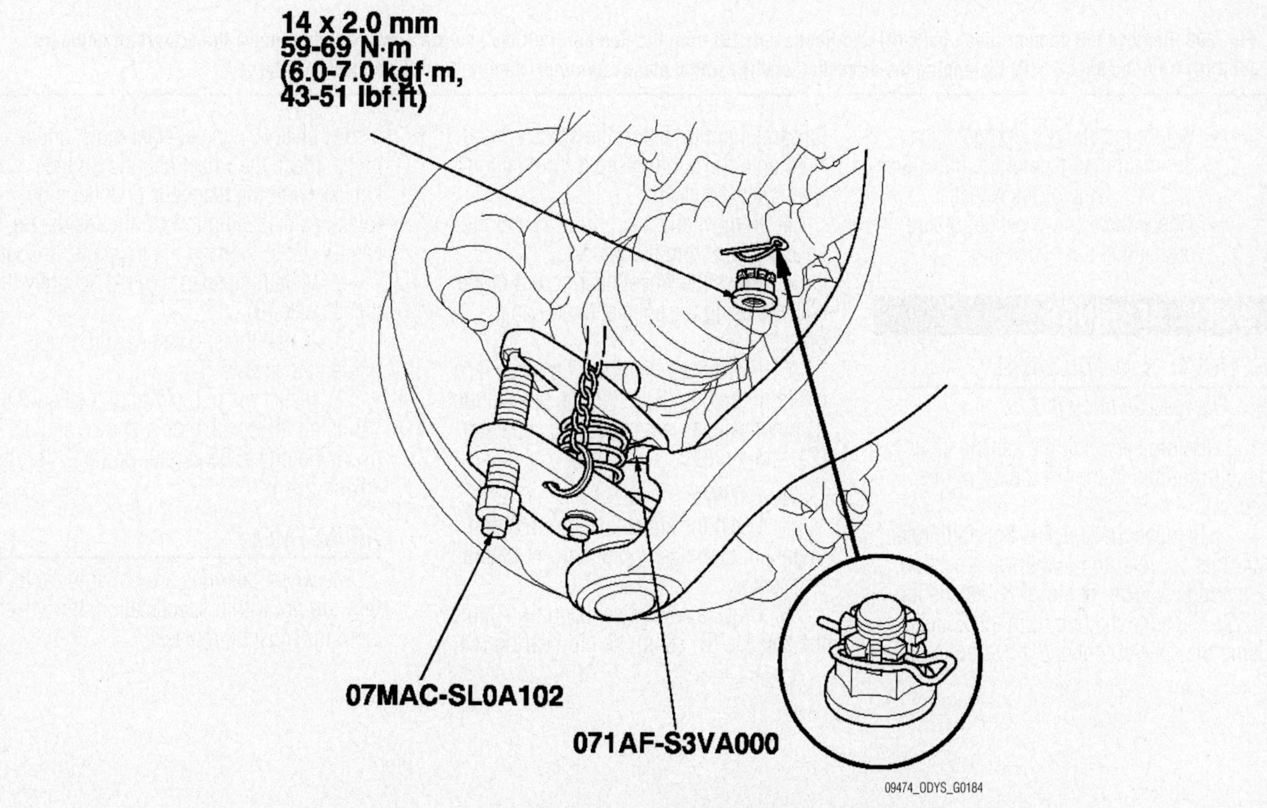

14 x 2.0 mm
59-69 N·m
(6.0-7.0 kgf·m,
43-51 lbf·ft)

07MAC-SL0A102

071AF-S3VA000

09474_ODYS_G0184

Fig. 207 Remove the lock pin (A) from the lower arm ball joint castle nut (B), and remove the nut—Odyssey

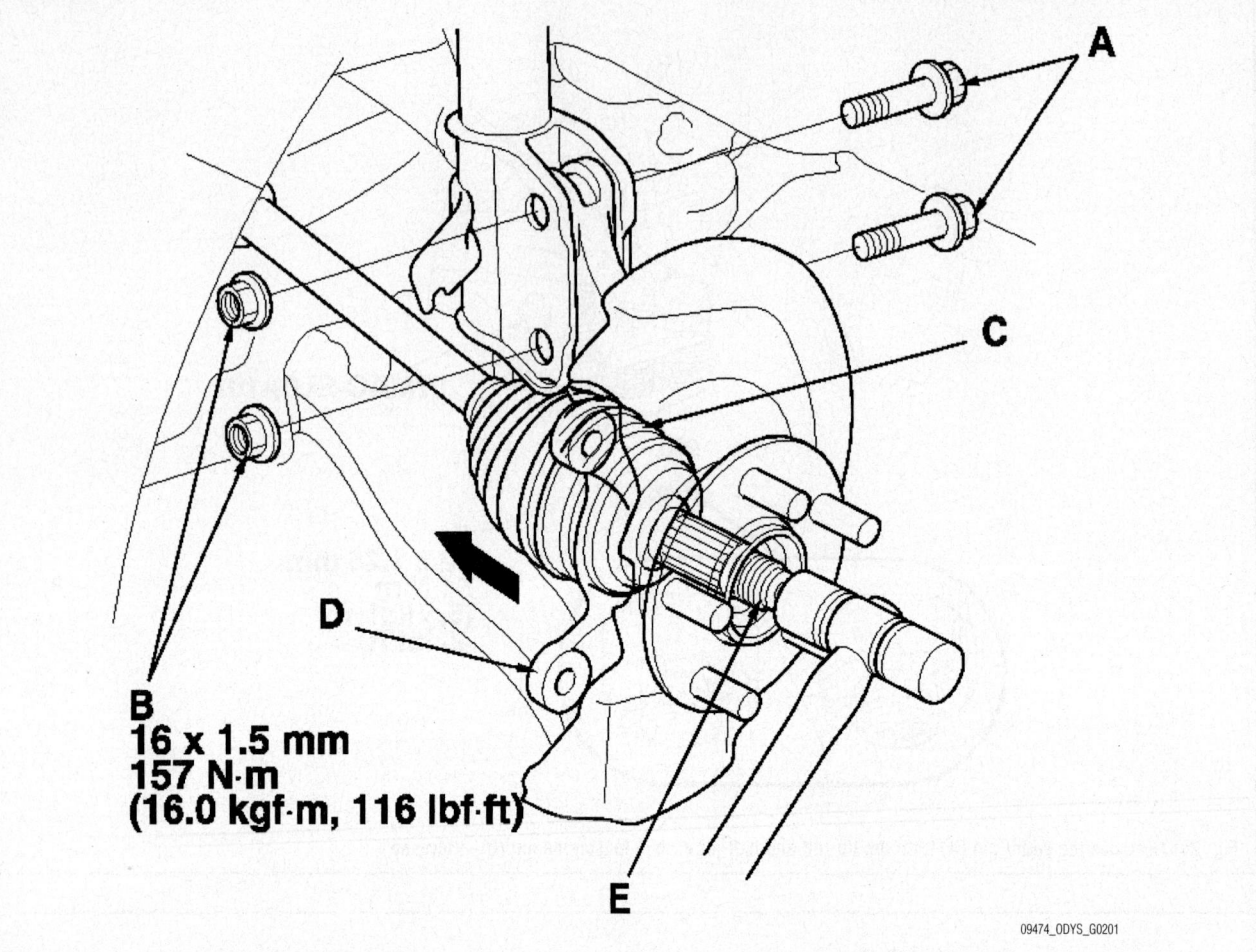

B
16 x 1.5 mm
157 N·m
(16.0 kgf·m, 116 lbf·ft)

09474_ODYS_G0201

Fig. 208 Remove the damper pinch bolts (A) and flange nuts (B) from the damper. Pull the knuckle outward, and remove the driveshaft outboard joint (C) from the knuckle (D) by tapping the driveshaft end (E) with a plastic hammer, then remove the knuckle—Odyssey

- Before installing the wheel, clean the mating surface of the brake disc and the inside of the wheel.
- Check the front wheel alignment, and adjust it if necessary.

WHEEL BEARINGS

REMOVAL & INSTALLATION

See Figures 209 through 215.

1. Before servicing the vehicle, refer to the precautions in the beginning of this section.

2. Remove the knuckle. For additional information, refer to the following section, "Steering Knuckle, Removal & Installation."

3. Separate the hub from the knuckle using the special tool and a hydraulic press.

Be careful not to distort the splash guard. Hold onto the hub to keep it from falling when pressed clear.

4. Remove the snap ring (A) and the splash guard from the knuckle.

5. Press the wheel bearing out of the knuckle using the special tools and a press.

6. Press the wheel bearing inner race (A) from the hub (B) using the special tool, commercially available bearing separator (C), and a press.

To install:

7. Wash the knuckle and hub thoroughly in high flash point solvent before reassembly.

8. Press a new wheel bearing (A) into the knuckle (B) using the old bearing (C),

a steel plate (D), the special tools, and a press. Place the wheel bearing on the knuckle with the magnetic encoder side (black color) facing toward the inside. Be careful not to damage the magnetic encoder.

9. Install the snap ring (A) securely in the knuckle (B).

10. Install the splash guard (C), and tighten the screws (D).

11. Install the hub (A) onto the knuckle (B) using the special tools shown and a hydraulic press. Be careful not to distort the splash guard (C).

ADJUSTMENT

The wheel bearings are not adjustable. If they are not within specification, the wheel bearings must be replaced.

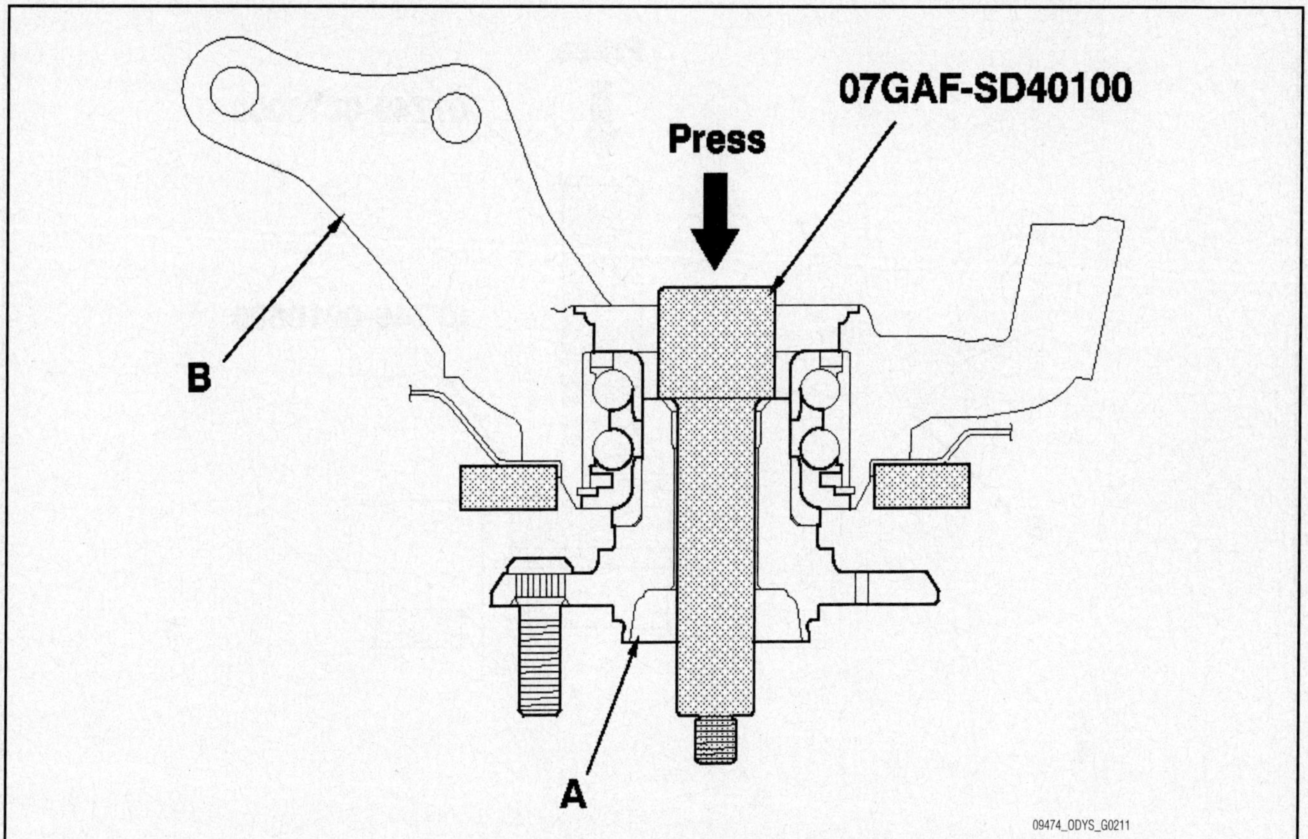

07GAF-SD40100

Press

B

A

09474_ODYS_G0211

Fig. 209 Separate the hub (A) from the knuckle (B) using the special tool and a hydraulic press—Odyssey

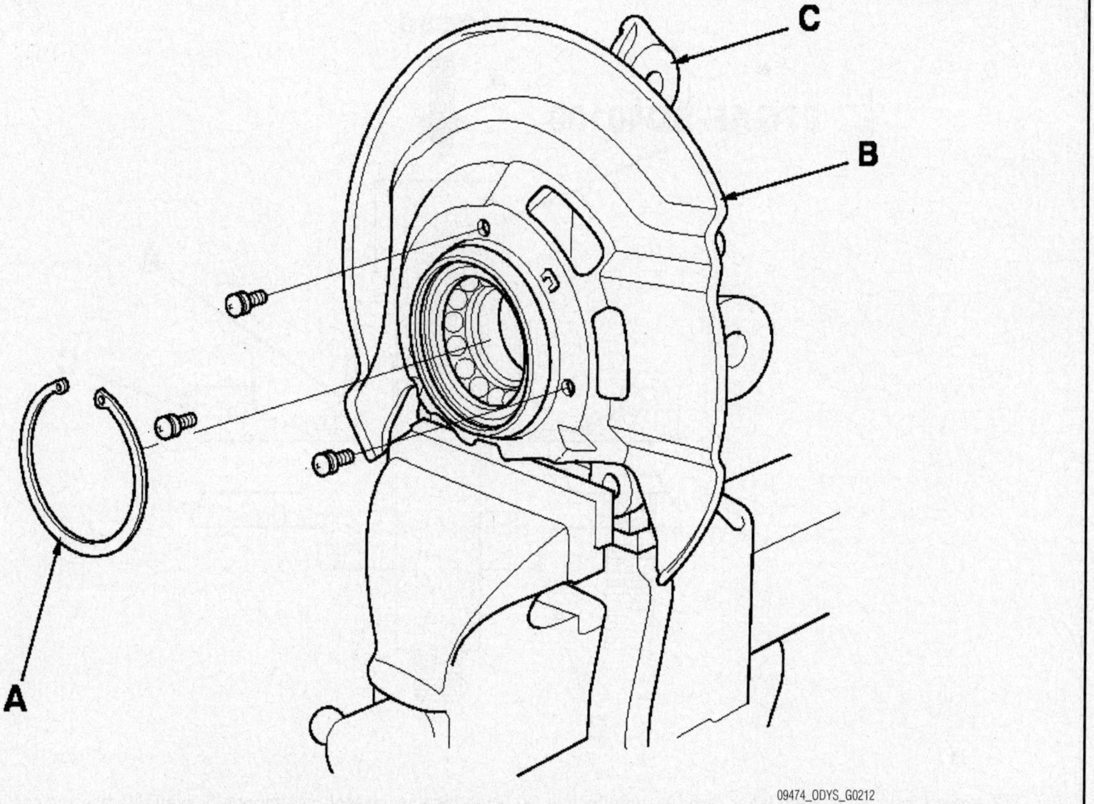

C

B

A

09474_ODYS_G0212

Fig. 210 Remove the snap ring (A) and the splash guard (B) from the knuckle (C)

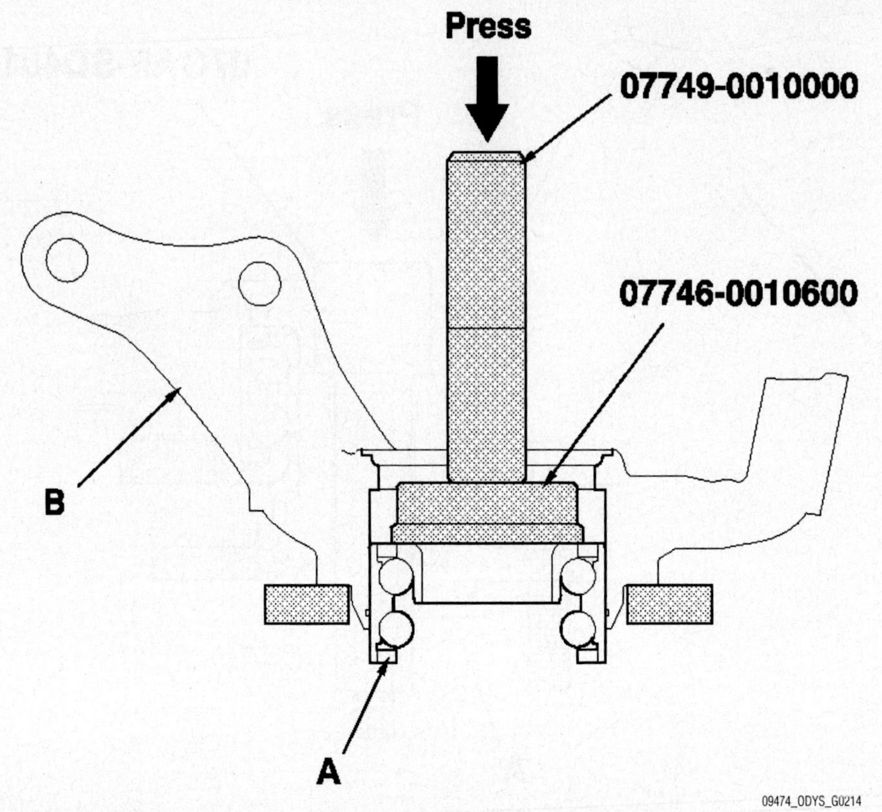

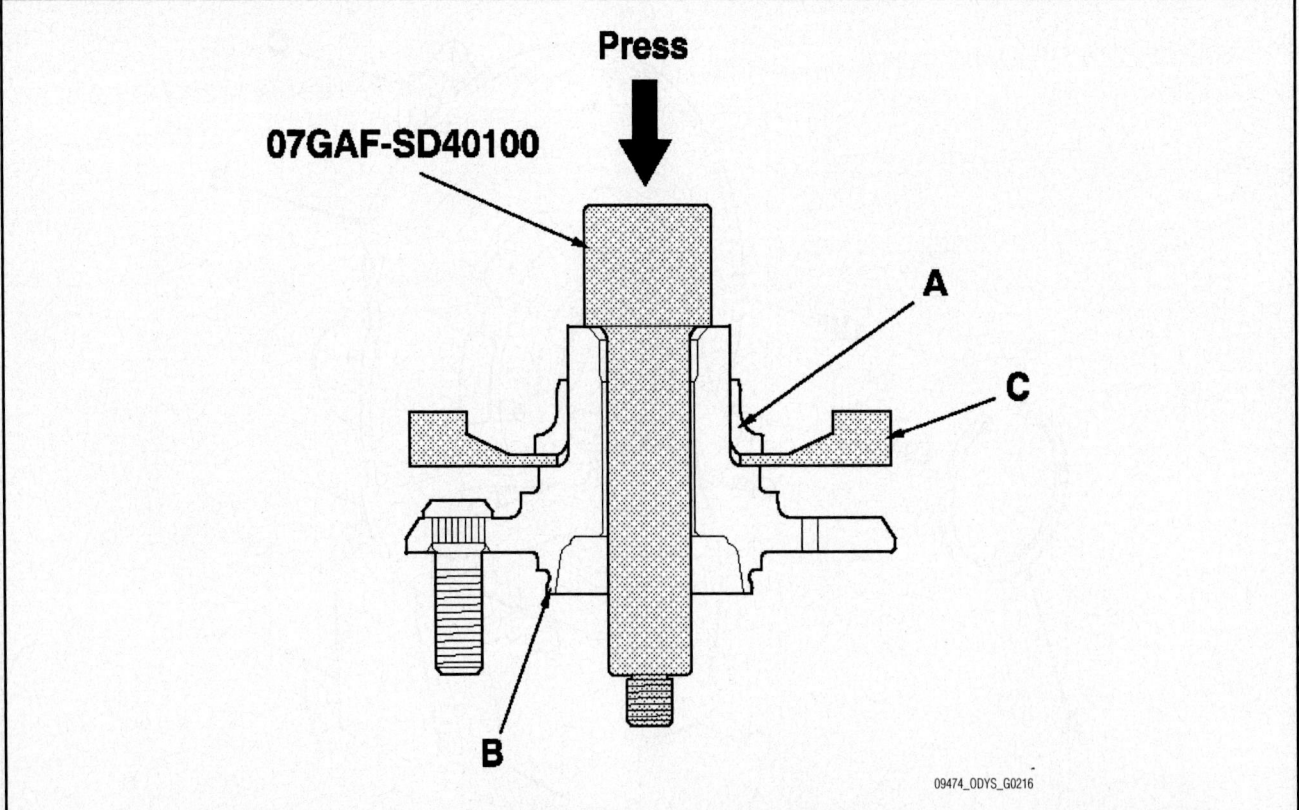

Press

07749-0010000

07746-0010600

B

A

09474_ODYS_G0214

Fig. 211 Press the wheel bearing (A) out of the knuckle (B) using the special tools and a press—Odyssey

Press

07GAF-SD40100

A

C

B

09474_ODYS_G0216

Fig. 212 Press the wheel bearing inner race (A) from the hub (B) using the special tool, commercially available bearing separator (C), and a press—Odyssey

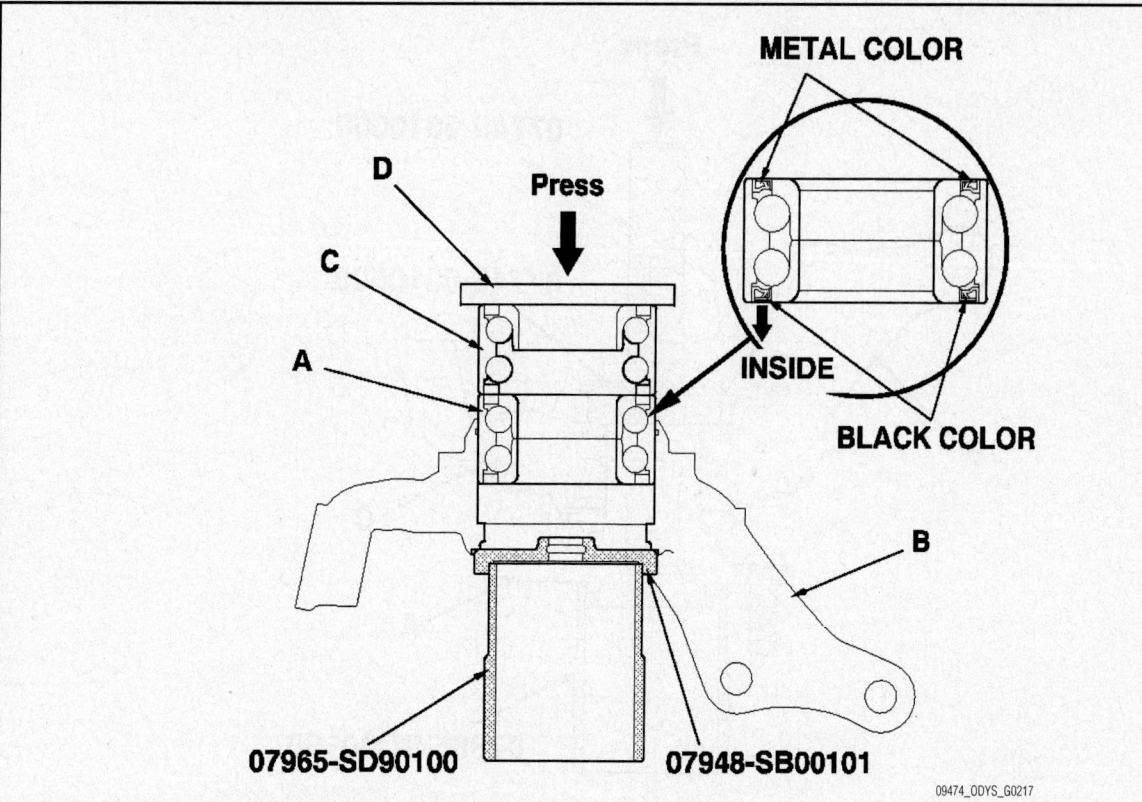

METAL COLOR

Press

D

C

A

INSIDE

BLACK COLOR

B

07965-SD90100

07948-SB00101

09474_ODYS_G0217

Fig. 213 Press a new wheel bearing (A) into the knuckle (B) using the old bearing (C), a steel plate (D), the special tools, and a press. Place the wheel bearing on the knuckle with the magnetic encoder side (black color) facing toward the inside

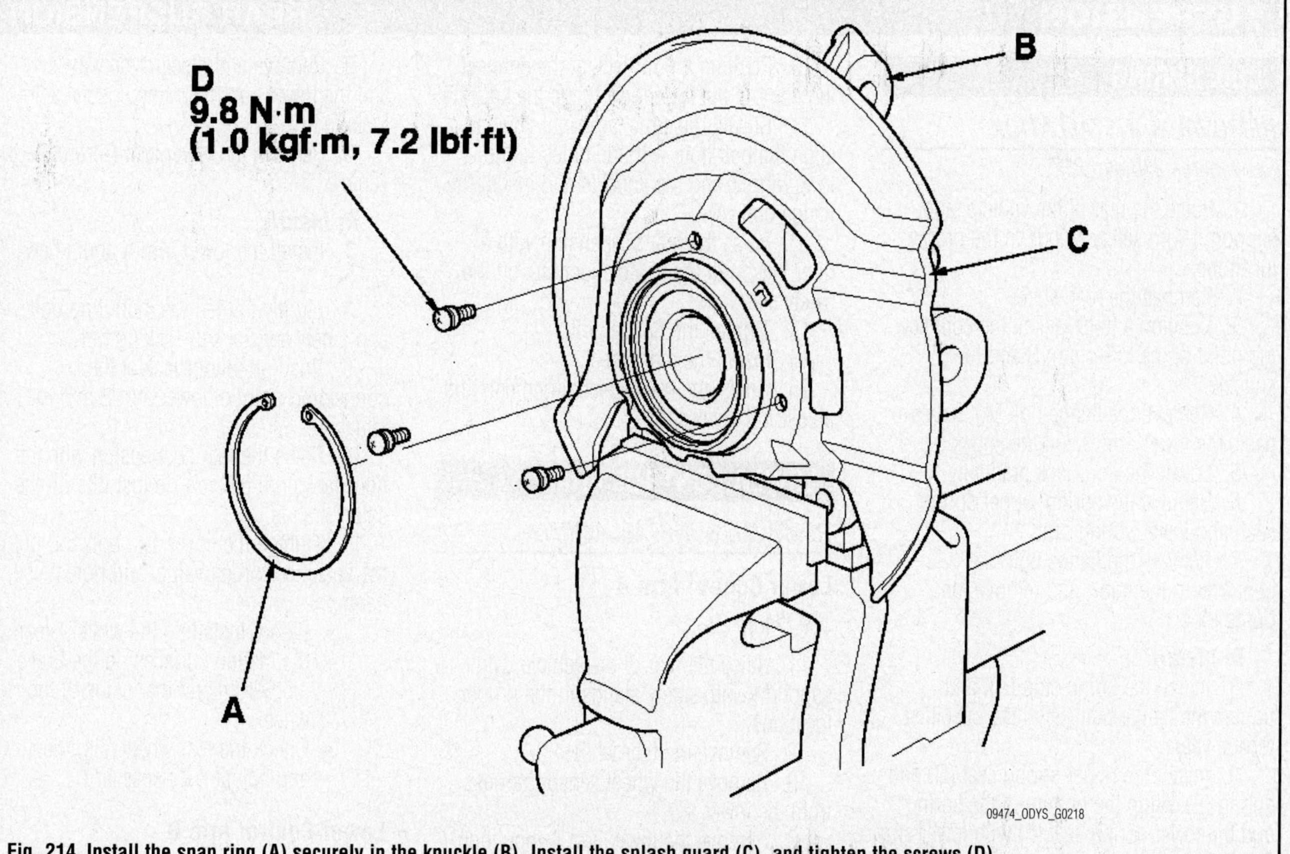

D
9.8 N·m
(1.0 kgf·m, 7.2 lbf·ft)

B

C

A

09474_ODYS_G0218

Fig. 214 Install the snap ring (A) securely in the knuckle (B). Install the splash guard (C), and tighten the screws (D)

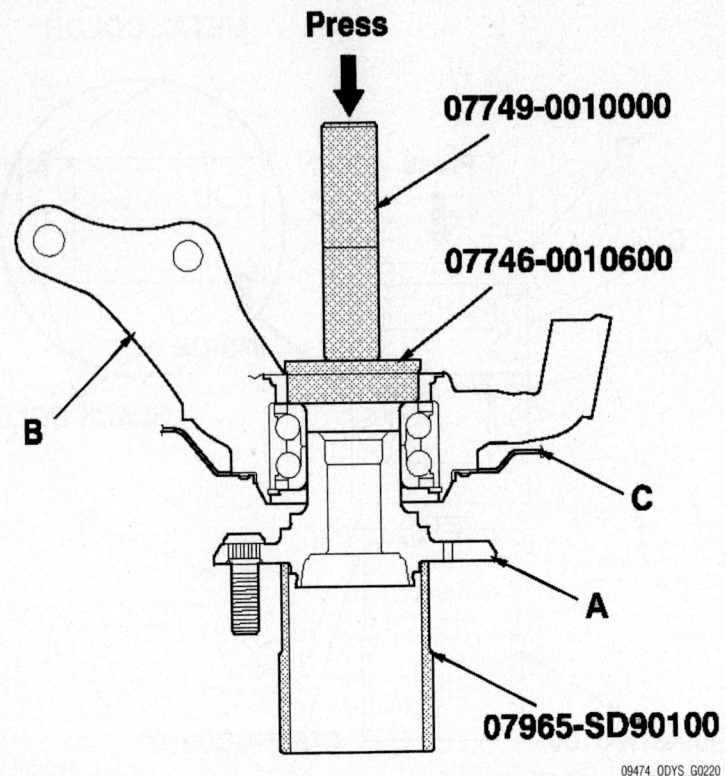

Press

07749-0010000

07746-0010600

B

C

A

07965-SD90100

09474_ODYS_G0220

Fig. 215 Install the hub (A) onto the knuckle (B) using the special tools shown and a hydraulic press. Be careful not to distort the splash guard (C)—Odyssey

SUSPENSION

REAR SUSPENSION

COIL SPRING

REMOVAL & INSTALLATION

See Figures 216 and 217.

1. Raise the rear of the vehicle, and support it with safety stands in the proper locations.
2. Remove the rear wheel.
3. Position a floor jack at the connecting point of the lower arm B and the knuckle.
4. Remove the flange bolt (A) that connects the lower arm B and the knuckle.
5. Lower the floor jack gradually.
6. Remove the spring, upper spring seat, and lower spring seat.
7. Remove the flange bolt that connects to the body, and remove the bump stop.

To install:

8. Install the bump stop (A), and tighten the flange bolt (C) to the specified torque value.
9. Install the upper spring seat (D) and spring (E). Align the bottom of the spring and the lower spring seat (F) with lower arm B as shown.

10. Position a floor jack at the connecting point of the lower arm B and the knuckle.
11. Slowly raise the jack until you can align the bolt hole with the holes in the lower arm B and the knuckle and install the flange bolt (A).
12. Raise the rear suspension with a floor jack until the vehicle just lifts off the safety stands.
13. Tighten the flange bolt.
14. Install the rear wheel.
15. Check the rear wheel alignment, and adjust it if necessary.

LOWER CONTROL ARM

REMOVAL & INSTALLATION

Lower Control Arm A

See Figure 218.

1. Raise the rear of the vehicle, and support it with safety stands in the proper locations.
2. Remove the rear wheel.
3. Remove the wheel sensor harness from the lower arm A.
4. Remove the lower arm A mounting bolt.

5. Remove and discard the washer and the lower arm mounting special self-locking nut.
6. Remove the lower arm A from the vehicle.

To install:

7. Install the lower arm A and a new washer.
8. Lightly tighten the mounting bolt and a new special self-locking nut.
9. Position a floor jack at the connecting point of lower arm B and the knuckle.
10. Raise the rear suspension with the floor jack until the vehicle just lifts off the safety stands.
11. Tighten the lower arm bolt and nut to the specified torque value, and note these items:

- Before installing the wheel, clean the mating surfaces on the brake disc/drum and the inside of the wheel.
- Check the rear wheel alignment, and adjust it if necessary

Lower Control Arm B

See Figure 219.

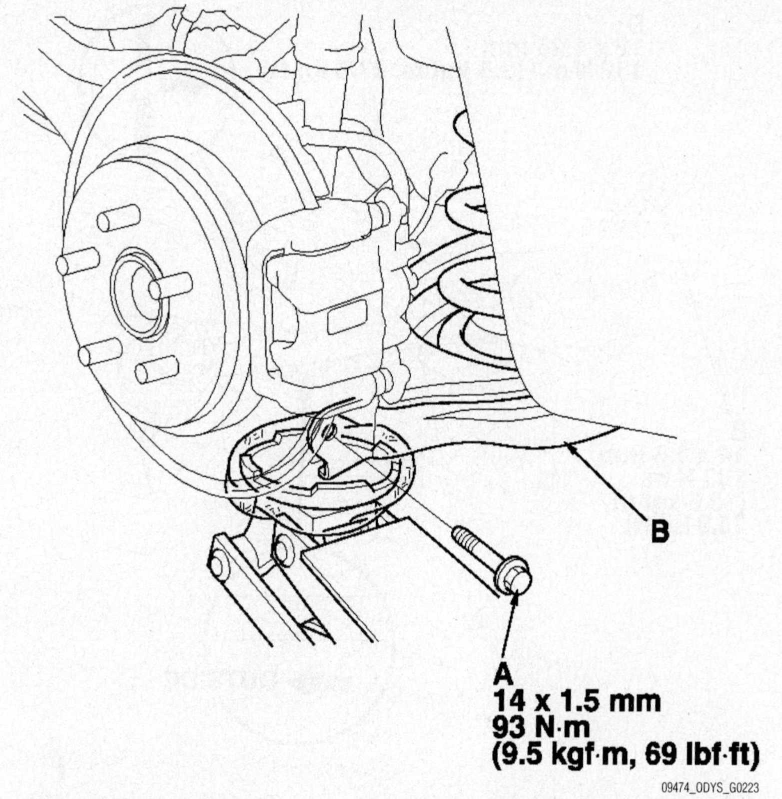

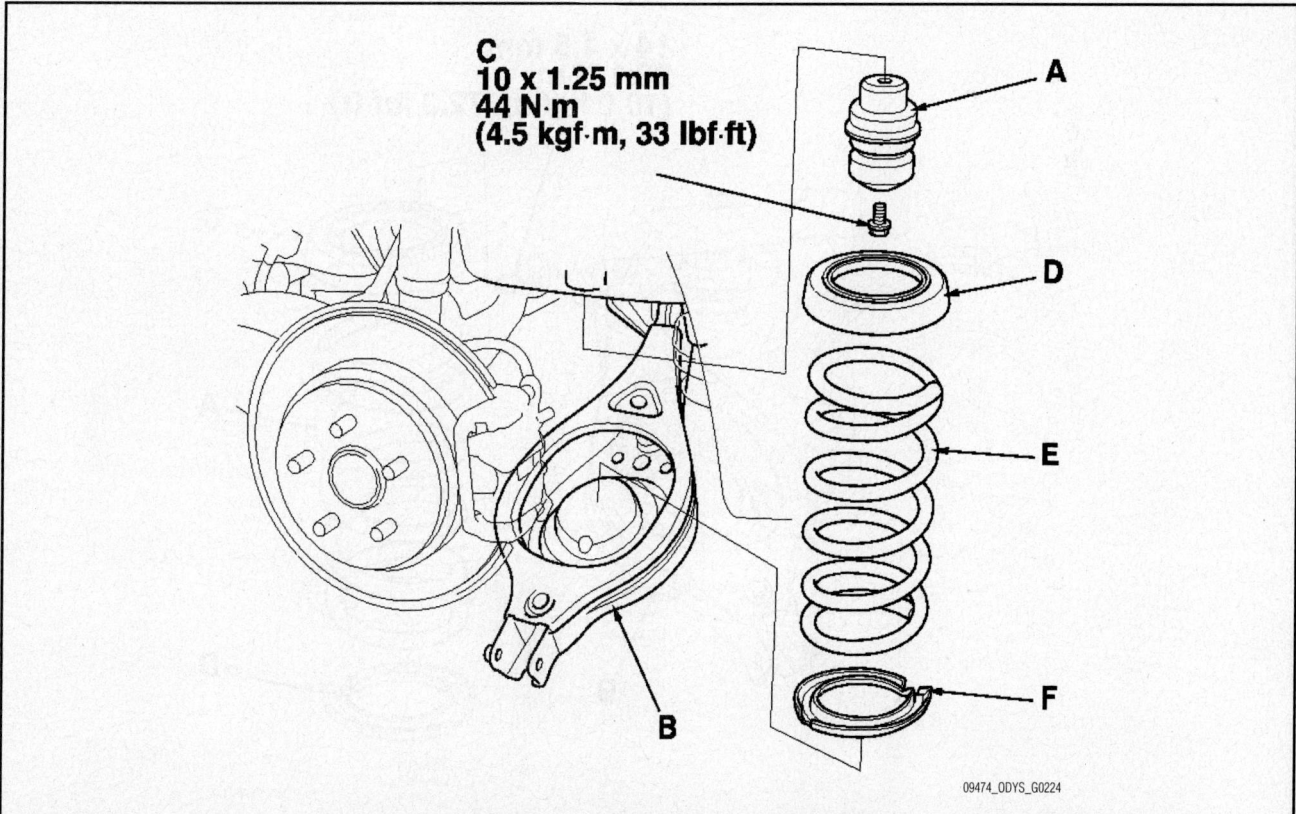

14 x 1.5 mm
93 N·m
(9.5 kgf·m, 69 lbf·ft)

09474_ODYS_G0223

Fig. 216 Remove the flange bolt (A) that connects the lower arm B and the knuckle—Odyssey

C
10 x 1.25 mm
44 N·m
(4.5 kgf·m, 33 lbf·ft)

A

D

E

F

B

09474_ODYS_G0224

Fig. 217 Install the bump stop (A), and tighten the flange bolt (C) to the specified torque value. Install the upper spring seat (D) and spring (E). Align the bottom of the spring and the lower spring seat (F) with lower arm B as shown—Odyssey

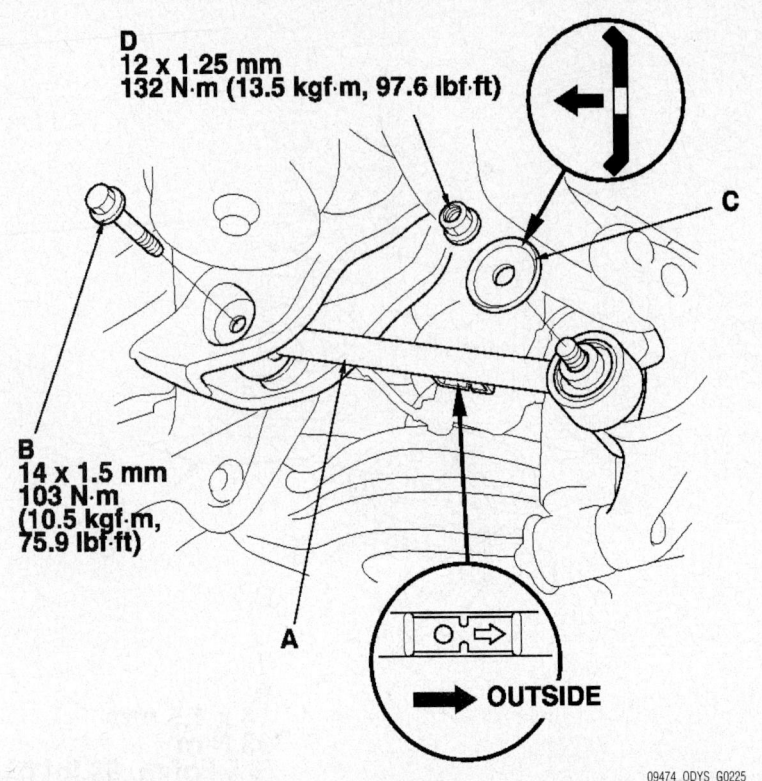

D
12 x 1.25 mm
132 N·m (13.5 kgf·m, 97.6 lbf·ft)

C

B
14 x 1.5 mm
103 N·m
(10.5 kgf·m,
75.9 lbf·ft)

A

OUTSIDE

09474_ODYS_G0225

Fig. 218 Remove the lower arm A mounting bolt (B). Remove and discard the washer (C) and the lower arm mounting special self-locking nut (D)—Odyssey

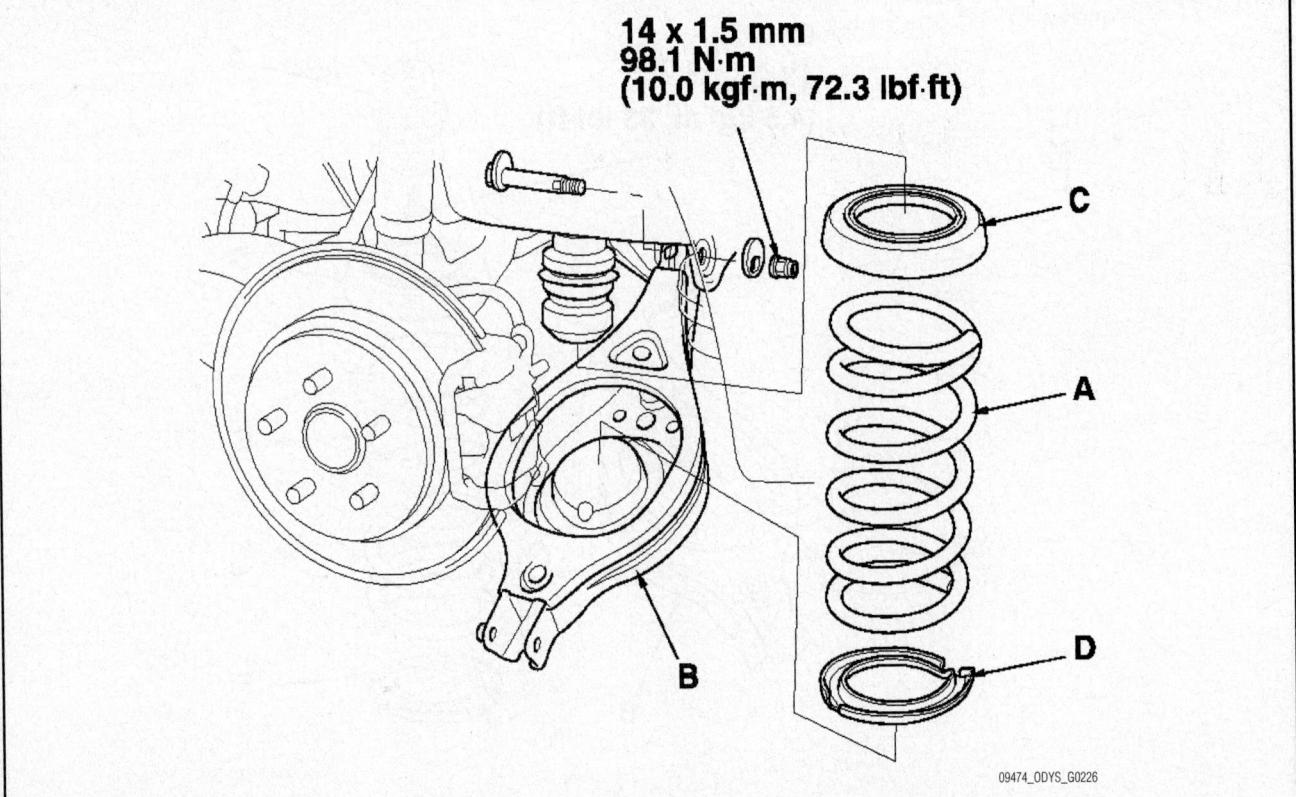

14 x 1.5 mm
98.1 N·m
(10.0 kgf·m, 72.3 lbf·ft)

C

A

B

D

09474_ODYS_G0226

Fig. 219 Install the spring (A), and upper spring seat (C). Align the bottom of the spring and the lower spring seat (D) with the lower arm B as shown—Odyssey

1. Raise the rear of the vehicle, and support it with safety stands in the proper locations.

2. Remove the rear wheel.

3. Position a floor jack at the connecting point of lower arm B and the knuckle.

4. Remove the flange bolt (A) that connects the lower arm B and the knuckle.

5. Lower the floor jack gradually.

6. Remove the spring, upper spring seat and lower spring seat.

7. Remove the self-locking nut and flange bolt.

To install:

8. Position the lower arm B, and loosely install the flange bolt and the self-locking nut.

9. Install the spring (A), and upper spring seat (C). Align the bottom of the spring and the lower spring seat (D) with the lower arm B as shown.

SHOCK ABSORBER

REMOVAL & INSTALLATION

See Figure 220.

1. Raise the rear of the vehicle, and support it with safety stands in the proper locations.

2. Remove the rear wheel.

3. Position a floor jack at the connecting point of the lower control arm and knuckle (A). Raise the floor jack until the suspension begins to compress.

4. Remove the flange bolt and nut from the body.

5. Remove the self-locking nut from the knuckle.

6. Compress the damper by hand, and remove it from the vehicle.

7. Compress the damper assembly by hand, and check for smooth operation through a full stroke, both compression and extension. The damper should move smoothly and constantly when compression is released. If it does not, the gas is leaking and the damper should be replaced.

8. Check for oil leaks, abnormal noises, or binding during these tests.

To install:

9. Lower the rear suspension. Compress the damper (A) by hand, and move it into position. Loosely install the flange nut (B), bolt (C), and new self-locking nut (D).

10. Raise the rear suspension with a floor jack until the vehicle just lifts off the safety stands.

11. Tighten the flange bolt and self-locking nut on the bottom of the damper to the specified torque value.

12. Install the rear wheel.

13. Check the rear wheel alignment, and adjust it if necessary.

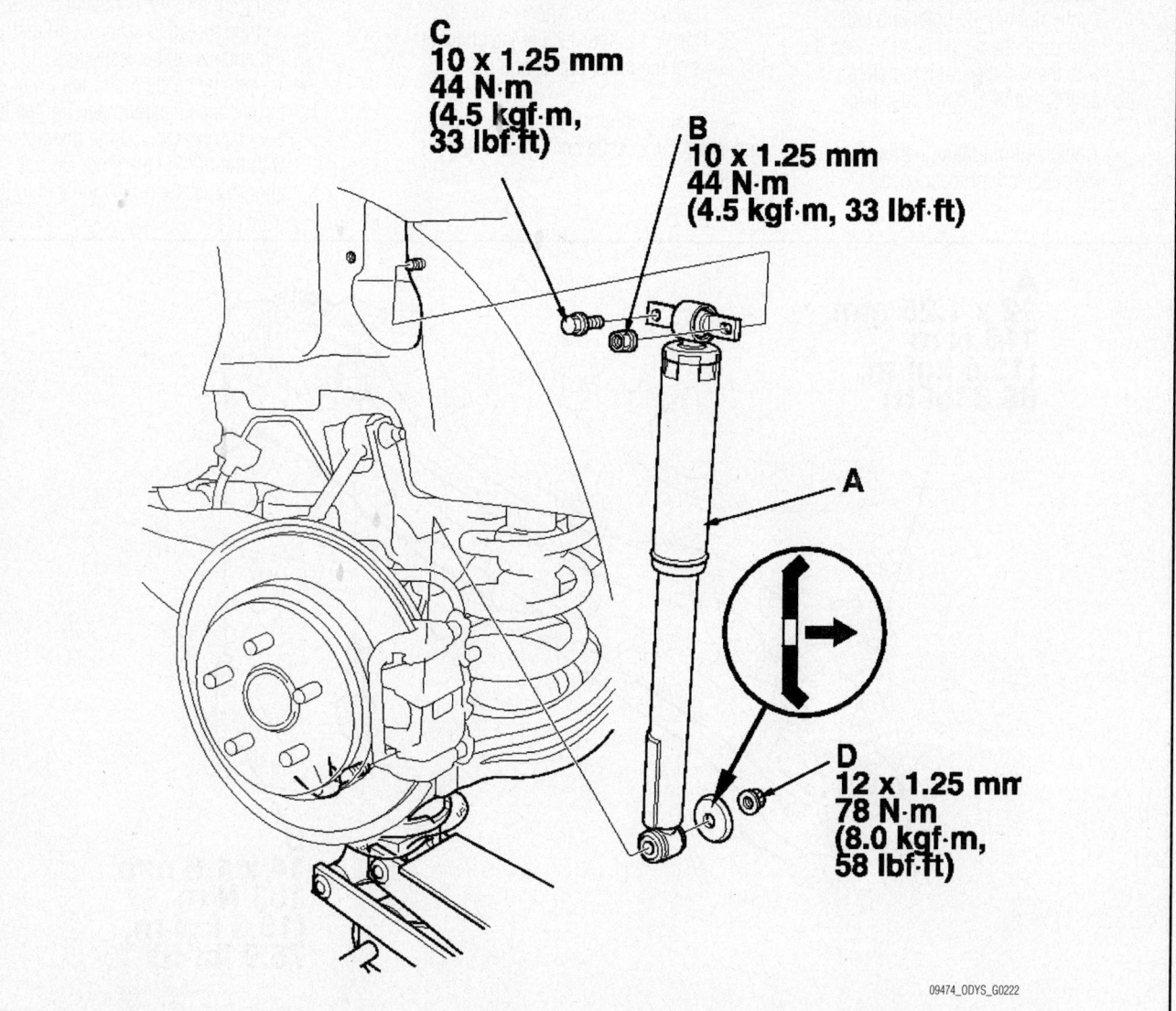

C
10 x 1.25 mm
44 N·m
(4.5 kgf·m, 33 lbf·ft)

B
10 x 1.25 mm
44 N·m
(4.5 kgf·m, 33 lbf·ft)

A

D
12 x 1.25 mm
78 N·m
(8.0 kgf·m, 58 lbf·ft)

09474_ODYS_G0222

Fig. 220 Rear shock absorber—Odyssey

TRAILING ARM

REMOVAL & INSTALLATION

See Figure 221.

1. Raise the rear of the vehicle, and support it with safety stands in the proper locations.
2. Remove the rear wheel.
3. Remove the brake disc.
4. Remove the parking brake cable from the trailing arm.
5. Remove the brake hose mounting nut from the trailing arm.
6. Position a floor jack at the connecting point of the lower arm and the knuckle.
7. Remove the self-locking nuts (A) and bolts (B) from the trailing arm.
8. Install the trailing arm in the reverse order of removal, and note these items:
   - Use a new self-locking nut on reassembly.
   - First install all the components and lightly tighten the bolts and nuts, then raise the suspension to load it with the vehicle's weight before fully tightening to the specified torque values.
   - Tighten all mounting hardware to the specified torque values.

- Check the brake hose and line joint for leaks, and tighten if necessary.
- Check the brake hose for interference and twisting.
- Before installing the wheel, clean the mating surfaces on the brake disc/drum and the inside of the wheel.
- Fill up the brake reservoir, and bleed the brake system.
- After installation, check for leaks at the line joint, and retighten if necessary. Check the rear wheel alignment, and adjust it if necessary.

UPPER CONTROL ARM

REMOVAL & INSTALLATION

See Figures 222 and 223.

1. Raise the rear of the vehicle, and support it with safety stands in the proper locations.
2. Remove the rear wheel.
3. Position a floor jack at the connecting point of the lower arm B and the knuckle.
4. Remove the lock pin (A) from the upper ball joint castle nut (B), and remove the nut.

5. Remove the upper ball joint from the knuckle using the special tool.
6. Remove the upper arm bolt (A). Remove the upper arm (B) from the vehicle.

To install:

7. Install the upper arm.
8. Lightly tighten the upper arm bolt and castle nut.
9. Position a floor jack at the connecting point of the lower arm B and the knuckle.
10. Raise the rear suspension with the floor jack until the vehicle just lifts off the safety stands.
11. Tighten the upper arm bolt to the specified torque value, and note these items:

- Be careful not to damage the ball joint boot when connecting the upper arm to the knuckle.
- Before connecting the ball joint to the knuckle, degrease the threaded section and tapered portion of the ball joint pin, the connecting hole, and the threaded section and mating surface of the castle nut.
- Torque the castle nut to the lower torque specification, then tighten it only far enough to align the slot with the hole in the stud. Do not align the castle nut by loosening it.

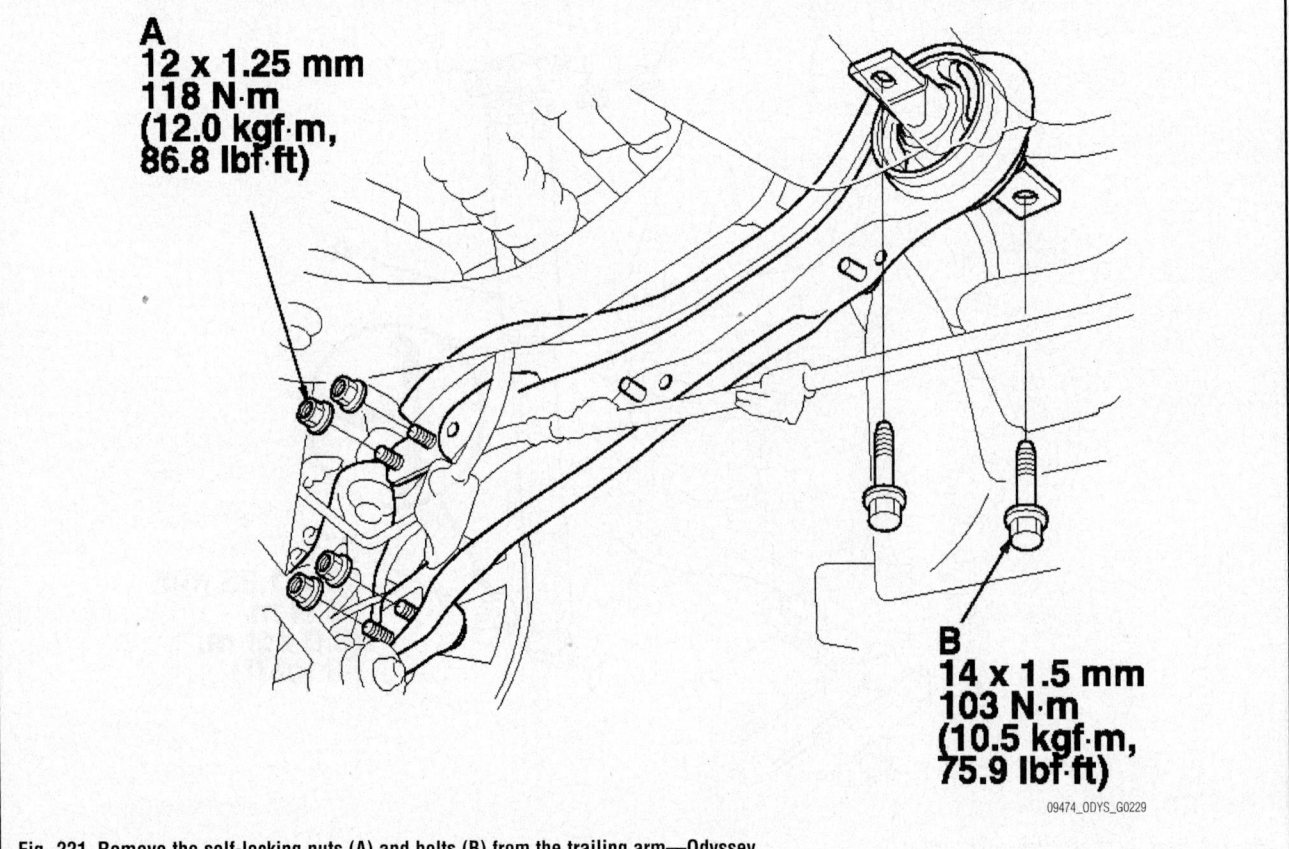

A
12 x 1.25 mm
118 N·m
(12.0 kgf·m,
86.8 lbf·ft)

B
14 x 1.5 mm
103 N·m
(10.5 kgf·m,
75.9 lbf·ft)

09474_ODYS_G0229

Fig. 221 Remove the self-locking nuts (A) and bolts (B) from the trailing arm—Odyssey

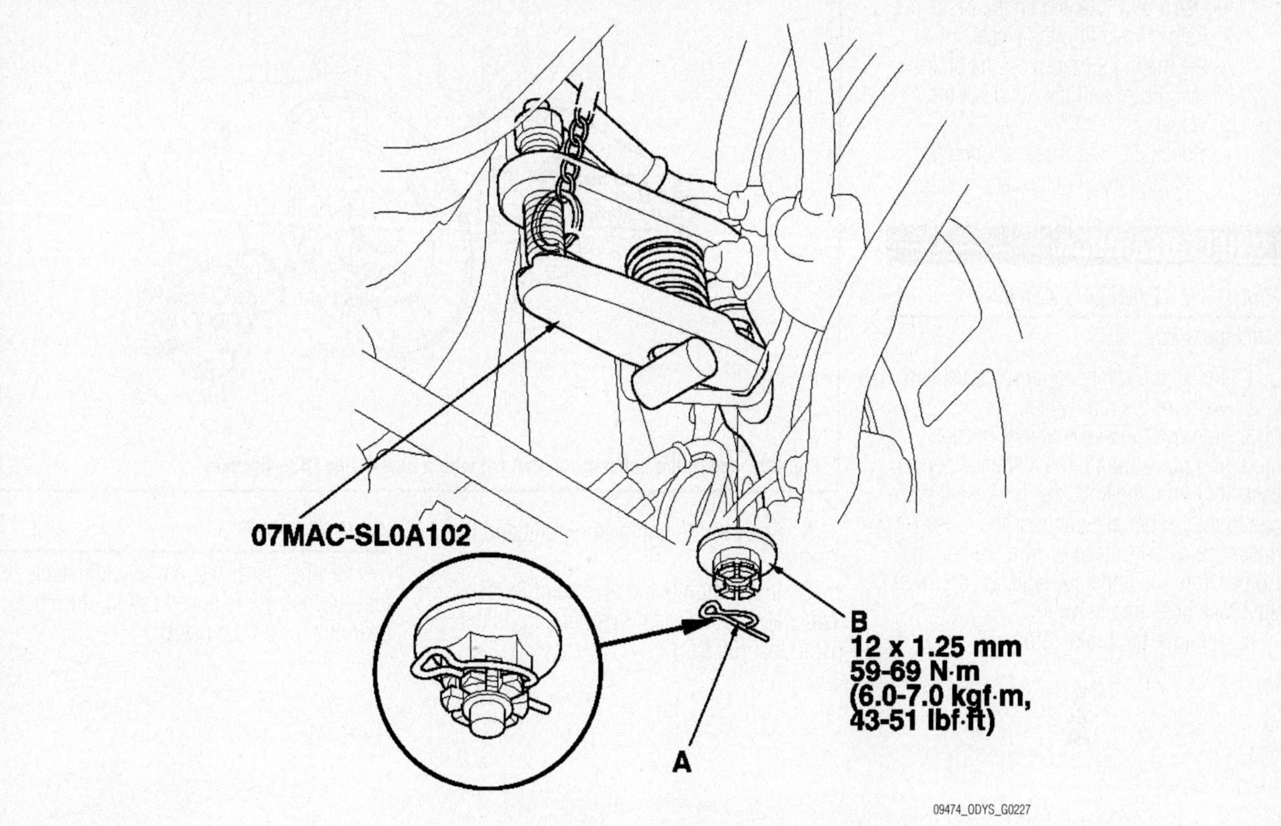

07MAC-SL0A102

B
12 x 1.25 mm
59-69 N·m
(6.0-7.0 kgf·m,
43-51 lbf·ft)

A

09474_ODYS_G0227

Fig. 222 Remove the lock pin (A) from the upper ball joint castle nut (B), and remove the nut—Odyssey

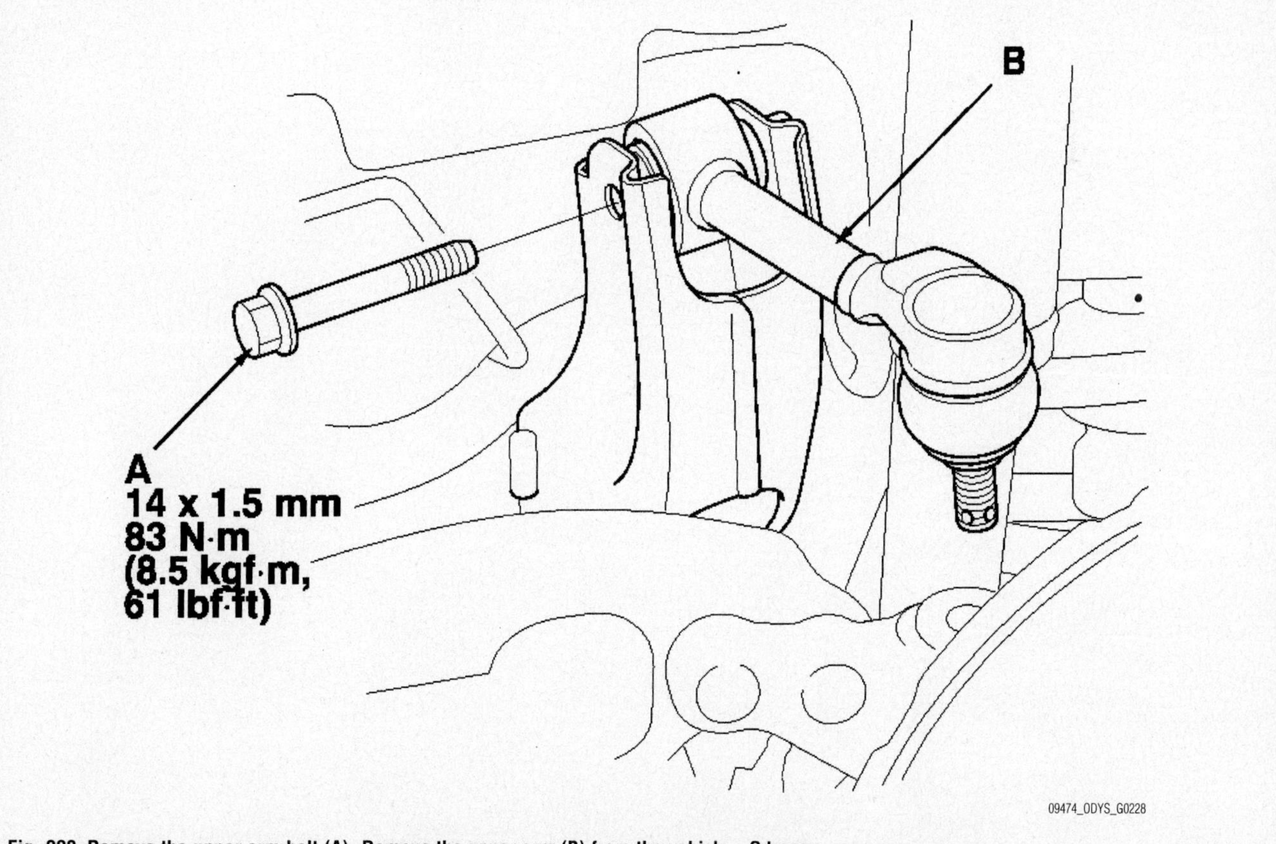

B

A
14 x 1.5 mm
83 N·m
(8.5 kgf·m,
61 lbf·ft)

09474_ODYS_G0228

Fig. 223 Remove the upper arm bolt (A). Remove the upper arm (B) from the vehicle—Odyssey

- Use a new lock pin on the castle nut.
- Before installing the wheel, clean the mating surfaces on the brake disc/drum and the inside of the wheel.
- Check the rear wheel alignment, and adjust it if necessary.

WHEEL BEARINGS

REMOVAL & INSTALLATION

See Figure 224.

1. Raise and safely support the vehicle.
2. Remove the rear wheel.
3. Remove the brake caliper bracket mounting bolts, then remove the caliper assembly from the knuckle. To prevent damage to the caliper assembly or the brake hose, use a short piece of wire to hang the caliper from the undercarriage. Do not twist the brake hose excessively.
4. Remove the brake rotor.

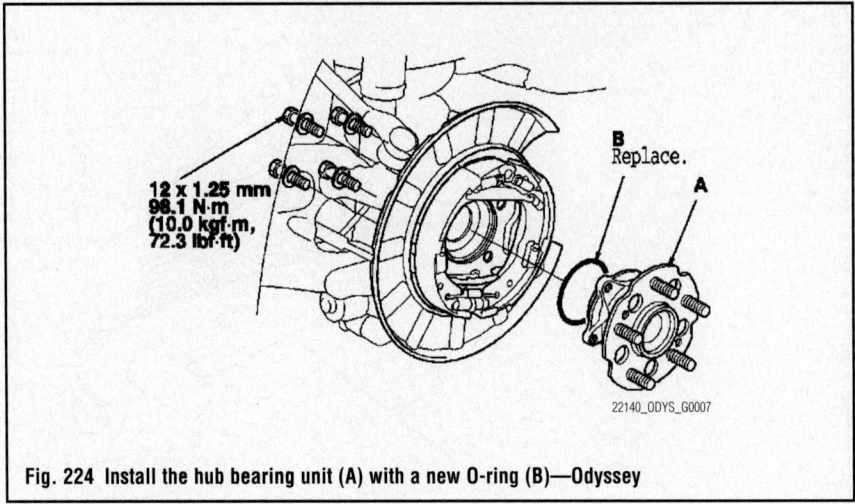

Fig. 224 Install the hub bearing unit (A) with a new O-ring (B)—Odyssey

5. Remove the hub bearing unit and the O-ring.
6. Installation is the reverse order of removal. Tighten the hub mounting bolts to 72 ft. lbs. (98 Nm).

ADJUSTMENT

The wheel bearings are not adjustable. If they are not within specification, the wheel bearings must be replaced.

HONDA

Pilot

9

SPECIFICATIONS AND MAINTENANCE CHARTS

ENGINE AND VEHICLE IDENTIFICATION CHART

		Engine Code						Model Year	
Code	Liters (cc)	Cu. In.	Cyl.	Fuel Sys.	Engine Type	Eng. Mfg.		Code ①	Year
J35A9	3.5 (3471)	212	6	SMFI	SOHC	Honda		6	2006
J35Z1 ②	3.5 (3471)	212	6	SMFI	SOHC	Honda		7	2007
								8	2008

SOHC: Single Overhead Cam

SMFI: Sequential Multi-port Fuel Injection

① 10th position of VIN

② Variable Cylinder Management

22140_PILO_C0001

GENERAL ENGINE SPECIFICATIONS

Year	Model	Engine Displacement Liters (VIN)	Net Horsepower @ rpm	Net Torque @ rpm (ft. lbs.)	Bore x Stroke (in.)	Com-pression Ratio	Oil Pressure @ rpm
2006	Pilot	3.5 (J35A9)	NA	NA	3.50x3.66	10.0:1	71@3000
	Pilot	3.5 (J35Z1)	244@5750	240@5400	3.50x3.66	10.0:1	71@3000
2007	Pilot	3.5 (J35A9)	NA	NA	3.50x3.66	10.0:1	71@3000
	Pilot	3.5 (J35Z1)	244@5750	240@5400	3.50x3.66	10.0:1	71@3000
2008	Pilot	3.5 (J35A9)	NA	NA	3.50x3.66	10.0:1	71@3000
	Pilot	3.5 (J35Z1)	244@5750	240@5400	3.50x3.66	10.0:1	71@3000

NA: Not Available

22140_PILO_C0002

ENGINE TUNE-UP SPECIFICATIONS

Year	Engine Displacement Liters (VIN)	Spark Plug Gap (in.)	Ignition Timing (deg.) MT	Ignition Timing (deg.) AT	Fuel Pump (psi)	Idle Speed (rpm) MT	Idle Speed (rpm) AT	Valve Clearance (in.) In.	Valve Clearance (in.) Ex.
2006	3.5 (J35A9)	0.039-0.043	—	8-12B	55-63	—	625-750	0.008-0.009	0.011-0.013
	3.5 (J35Z1)	0.039-0.043	—	8-12B	55-63	—	680-780	0.008-0.009	0.011-0.013
2007	3.5 (J35A9)	0.039-0.043	—	8-12B	55-63	—	625-750	0.008-0.009	0.011-0.013
	3.5 (J35Z1)	0.039-0.043	—	8-12B	55-63	—	680-780	0.008-0.009	0.011-0.013
2008	3.5 (J35A9)	0.039-0.043	—	8-12B	55-63	—	625-750	0.008-0.009	0.011-0.013
	3.5 (J35Z1)	0.039-0.043	—	8-12B	55-63	—	680-780	0.008-0.009	0.011-0.013

NOTE: The Vehicle Emission Control Information label often reflects changes made during production and must be used if they differ from this chart.

NOTE: Pressure with fuel pressure gauge connected

B: Before top dead center

22140_PILO_C0003

CAPACITIES

Year	Model	Engine Displacement Liters (VIN)	Engine Oil with Filter (qts.)	Transmission (pts.) 5-Spd	Transmission (pts.) Auto.	Transfer Case (pts.)	Drive Axle Front (pts.)	Drive Axle Rear (pts.)	Fuel Tank (gal.)	Cooling System (qts.)
2006	Pilot	3.5 (J35A9)	4.5	—	①	—	②	5.5	20.3	8.0
	Pilot	3.5 (J35Z1)	4.5	—	①	—	②	5.5	20.3	8.0
2007	Pilot	3.5 (J35A9)	4.5	—	①	—	②	5.5	20.3	8.0
	Pilot	3.5 (J35Z1)	4.5	—	①	—	②	5.5	20.3	8.0
2008	Pilot	3.5 (J35A9)	4.5	—	①	—	②	5.5	20.3	8.0
	Pilot	3.5 (J35Z1)	4.5	—	①	—	②	5.5	20.3	8.0

NOTE: All capacities are approximate. Add fluid gradually and check to be sure a proper fluid level is obtained.

① Fluid change: 3.5 quarts, 4WD. 4.0 quarts 2WD.

  Overhaul: 8.3 quarts, 4WD. 8.9 quarts 2WD.

② Fluid change: 0.45 quart

  Overhaul: 0.48 quart

22140_QX56_C0004

FLUID SPECIFICATIONS

Year	Model	Engine Displacement Liters (VIN)	Engine Oil	Auto. Trans.	Drive Axle	Power Steering Fluid	Brake Master Cylinder
2006	Pilot	3.5 (J35A9)	①	②	③	④	DOT 3
	Pilot	3.5 (J35Z1)	①	②	③	④	DOT 3
2007	Pilot	3.5 (J35A9)	①	②	③	④	DOT 3
	Pilot	3.5 (J35Z1)	①	②	③	④	DOT 3
2008	Pilot	3.5 (J35A9)	①	②	③	④	DOT 3
	Pilot	3.5 (J35Z1)	①	②	③	④	DOT 3

DOT: Department Of Transportation

Note: If specification disagrees with specification in owners manual, use specification in owners manaual

① Mobil 1 (P/N 5w-30-MB1-000) or equivalent that meets Honda HTO-06 standard

② Acura ATF-Z1 fluid

③ Transfer case: API classified GL4 or GL5 only. SAE 90 or SAE 80W-90 viscosity.

  Rear differential: Honda VTM-4 differential fluid

④ Honda power steering fluid

22140_PILO_C0014

VALVE SPECIFICATIONS

Year	Engine Displacement Liters (VIN)	Seat Angle (deg.)	Face Angle (deg.)	Spring Test Pressure (lbs. @ in.)	Spring Installed Height (in.)	Stem-to-Guide Clearance (in.)		Stem Diameter (in.)	
						Intake	Exhaust	Intake	Exhaust
2006	3.5 (J35A9)	NA	NA	NA	①	0.0008-0.0018	0.0022-0.0031	0.2159-0.2163	0.2146-0.2150
	3.5 (J35Z1)	NA	NA	NA	①	0.0008-0.0018	0.0022-0.0031	0.2156-0.2159	0.2156-0.2159
2007	3.5 (J35A4)	NA	NA	NA	①	0.0008-0.0018	0.0022-0.0031	0.2159-0.2163	0.2146-0.2150
	3.5 (J35A4)	NA	NA	NA	①	0.0008-0.0018	0.0022-0.0031	0.2159-0.2163	0.2146-0.2150
2008	3.5 (J35A4)	NA	NA	NA	①	0.0008-0.0018	0.0022-0.0031	0.2159-0.2163	0.2146-0.2150
	3.5 (J35A6)	NA	NA	NA	①	0.0008-0.0018	0.0022-0.0031	0.2159-0.2163	0.2146-0.2150

NA: Not Available

① Valve spring free length:

Intake: 2.029 in. for J35A9 engine. 2.111 in. (front side) for J35Z1 engine.

Exhaust: 2.010 in. for J35A9 engine. 2.069 in. (front side) for J35Z1 engine.

22140_PILO_C0005

CAMSHAFT SPECIFICATIONS
All measurements in inches unless noted

Year	Model	Engine Displacement Liters (VIN)	Journal Dia.	Brg. Oil Clearance	Shaft End-play	Circle Runout	Lobe Height	
							Intake	Exhaust
2006	Pilot	3.5 (J35A9)	NA	NA	0.0020-0.0080	NA	①	1.4302
	Pilot	3.5 (J35Z1)	NA	NA	0.0020-0.0080	NA	②	③
2007	Pilot	3.5 (J35A9)	NA	NA	0.0020-0.0080	NA	①	1.4302
	Pilot	3.5 (J35Z1)	NA	NA	0.0020-0.0080	NA	②	③
2008	Pilot	3.5 (J35A9)	NA	NA	0.0020-0.0080	NA	①	1.4302
	Pilot	3.5 (J35Z1)	NA	NA	0.0020-0.0080	NA	②	③

NA: Not Available

① Primary: 1.3796 in. Mid: 1.4348 in. Secondary: 1.3891 in.

② Front side: 1.4232 in. Rear side: 1.3891 in.

③ Front side: 1.4120 in. Rear side: 1.4374 in.

22140_PILO_C0006

CRANKSHAFT AND CONNECTING ROD SPECIFICATIONS

All measurements are given in inches

Year	Engine Displacement Liters (VIN)	Crankshaft				Connecting Rod		
		Main Brg. Journal Dia.	Main Brg. Oil Clearance	Shaft End-play	Thrust on No.	Journal Diameter	Oil Clearance	Side Clearance
2006	3.5 (J35A9)	2.8337-2.8346	0.0008-0.0017	0.0040-0.0140	3	2.1644-2.1654	0.0008-0.0017	0.0060-0.0140
	3.5 (J35Z1)	2.8337-2.8346	0.0008-0.0017	0.0040-0.0140	3	2.1644-2.1654	0.0008-0.0017	0.0060-0.0140
2007	3.5 (J35A9)	2.8337-2.8346	0.0008-0.0017	0.0040-0.0140	3	2.1644-2.1654	0.0008-0.0017	0.0060-0.0140
	3.5 (J35Z1)	2.8337-2.8346	0.0008-0.0017	0.0040-0.0140	3	2.1644-2.1654	0.0008-0.0017	0.0060-0.0140
2008	3.5 (J35A9)	2.8337-2.8346	0.0008-0.0017	0.0040-0.0140	3	2.1644-2.1654	0.0008-0.0017	0.0060-0.0140
	3.5 (J35Z1)	2.8337-2.8346	0.0008-0.0017	0.0040-0.0140	3	2.1644-2.1654	0.0008-0.0017	0.0060-0.0140

22140_PILO_C0007

PISTON AND RING SPECIFICATIONS

All measurements are given in inches

Year	Engine Displacement Liters (VIN)	Piston Clearance	Ring Gap			Ring Side Clearance		
			Top Compression	Bottom Compression	Oil Control	Top Compression	Bottom Compression	Oil Control
2006	3.5 (J35A9)	0.0006-0.0016	0.0080-0.0140	0.0160-0.0220	0.0080-0.0280	0.0022-0.0031	0.0012-0.0022	NA
	3.5 (J35Z1)	0.0006-0.0016	0.0080-0.0140	0.0160-0.0220	0.0080-0.0280	0.0022-0.0031	0.0012-0.0022	NA
2007	3.5 (J35A9)	0.0006-0.0016	0.0080-0.0140	0.0160-0.0220	0.0080-0.0280	0.0022-0.0031	0.0012-0.0022	NA
	3.5 (J35Z1)	0.0006-0.0016	0.0080-0.0140	0.0160-0.0220	0.0080-0.0280	0.0022-0.0031	0.0012-0.0022	NA
2008	3.5 (J35A9)	0.0006-0.0016	0.0080-0.0140	0.0160-0.0220	0.0080-0.0280	0.0022-0.0031	0.0012-0.0022	NA
	3.5 (J35Z1)	0.0006-0.0016	0.0080-0.0140	0.0160-0.0220	0.0080-0.0280	0.0022-0.0031	0.0012-0.0022	NA

NA: Not Available

22140_PILO_C0008

TORQUE SPECIFICATIONS
All readings in ft. lbs.

Year	Engine Displacement Liters (VIN)	Cylinder Head Bolts	Main Bearing Bolts	Rod Bearing Bolts	Crankshaft Damper Bolts	Flywheel Bolts	Manifold		Spark Plugs	Oil Pan Drain Plug
							Intake	Exhaust		
2006	3.5 (J35A9)	①	②	③	④	54	16	23	13	29
	3.5 (J35Z1)	①	②	③	④	54	16	23	13	29
2007	3.5 (J35A9)	①	②	③	④	54	16	23	13	29
	3.5 (J35Z1)	①	②	③	④	54	16	23	13	29
2008	3.5 (J35A9)	①	②	③	④	54	16	23	13	29
	3.5 (J35Z1)	①	②	③	④	54	16	23	13	29

NOTE: Dip main bearing bolts and crankshaft damper bolt in clean engine oil prior to tightening.

① Step 1: 29 ft. lbs.
Step 2: plus 90 degrees
Step 3: plus 90 degrees
Step 4: If using a new bolt, plus 90 degrees

② Cap bolts: 55 ft. lbs.
Side bolts: 36 ft. lbs.

③ Step 1: 15 ft. lbs.
Step 2: 90 degrees

④ Step 1: 48 ft. lbs.
Step 2: plus 60 degrees

22140_PILO_C0009

WHEEL ALIGNMENT

Year	Model		Caster		Camber		Toe-in (in.)
			Range (+/-Deg.)	Preferred Setting (Deg.)	Range (+/-Deg.)	Preferred Setting (Deg.)	
2006	Pilot	F	1.00	1.00	1.00	30	0+/-1/16
		R	—	—	0	30	0+/-1/16
2007	Pilot	F	1.00	1.00	1.00	30	0+/-1/16
		R	—	—	0	30	0+/-1/16
2008	Pilot	F	1.00	1.00	1.00	30	0+/-1/16
		R	—	—	0	30	0+/-1/16

22140_PILO_C0010

TIRE, WHEEL AND BALL JOINT SPECIFICATIONS

| Year | Model | OEM Tires | | Tire Pressures (psi) | | Wheel Size | Ball Joint Inspection | Lug Nut (ft. lbs.) |
		Standard	Optional	Front	Rear			
2006	Pilot	P235/65R16	None	32	32	R16	NS	80
2007	Pilot	P235/65R16	None	32	32	R16	NS	80
2008	Pilot	P235/65R16	None	32	32	R16	NS	80

OEM: Original Equipment Manufacturer

PSI: Pounds Per Square Inch

NS: Not specified by manufacturer

22140_PILO_C0011

BRAKE SPECIFICATIONS

All measurements in inches unless noted

| Year | Model | | Brake Disc | | | Brake Drum Diameter | | | Minimum Lining Thickness | | Brake Caliper | |
			Original Thickness	Minimum Thickness	Maximum Runout	Original Inside Diameter	Max. Wear Limit	Maximum Machine Diameter	Front	Rear	Bracket Bolts (ft. lbs.)	Mounting Bolts (ft. lbs.)
2006	Pilot	F	1.100	1.020	0.0016	—	—	—	0.060	—	80	27
		R	0.430	0.350	0.0016	—	—	—	—	0.060	41	27
2007	Pilot	F	1.100	1.020	0.004	—	—	—	0.060	—	80	27
		R	0.430	0.350	0.004	—	—	—	—	0.060	41	27
2008	Pilot	F	1.100	1.020	0.004	—	—	—	0.060	—	80	27
		R	0.430	0.350	0.004	—	—	—	—	0.060	41	27

F: Front

R: Rear

22140_PILO_C0012

SCHEDULED MAINTENANCE INTERVALS
Honda Pilot

TO BE SERVICED	OF SERVIC	VEHICLE MILEAGE INTERVAL (x1000)															
		7.5	15	22.5	30	37.5	45	52.5	60	67.5	75	82.5	90	97.5	105	112.5	120
Accessory drive belts	I & A				✓				✓				✓				✓
Air cleaner element	R				✓				✓				✓				✓
Brake fluid	R	Every 3 years															
Brake hoses & lines (incl. ABS)	I		✓		✓		✓		✓		✓		✓		✓		✓
Cooling system hoses & connections	I		✓		✓		✓		✓		✓		✓		✓		✓
Engine coolant ①	R					✓							✓				
Engine oil	R	✓	✓	✓	✓	✓	✓	✓	✓	✓	✓	✓	✓	✓	✓	✓	✓
Engine oil and coolant levels	I	Inspect at each fuel stop															
Engine oil filter	R		✓		✓		✓		✓		✓		✓		✓		✓
Exhaust system	I		✓		✓		✓		✓		✓		✓		✓		✓
Fluid levels and condition	I		✓		✓		✓		✓		✓		✓		✓		✓
Front and rear brakes	I		✓		✓		✓		✓		✓		✓		✓		✓
Fuel lines & connection	I		✓		✓		✓		✓		✓		✓		✓		✓
Halfshaft boots	I		✓		✓		✓		✓		✓		✓		✓		✓
Idle speed	I & A														✓		
Parking brake system	I & A		✓		✓		✓		✓		✓		✓		✓		✓
Rear differential fluid	R	✓			✓		✓		✓				✓				✓
Rotate and inspect tires	I	✓	✓	✓	✓	✓	✓	✓	✓	✓	✓	✓	✓	✓	✓	✓	✓
Spark plugs	R														✓		
Supplemental Restrain System	I	Inspect the SRS 10 years after production															
Suspension components	I		✓		✓		✓		✓		✓		✓		✓		✓
Tie rod ends, steering gear box & boots	I		✓		✓		✓		✓		✓		✓		✓		✓
Timing belt	R														✓		
Transmission fluid	R					✓						✓			✓		
Valve clearance	I	Adjust if valves are noisy															
Water pump	S/I														✓		

R: Replace   I: Inspect   A: Adjust

① Every 12,000 miles or 10 years, then every 60,000 miles or 5 years

FREQUENT OPERATION MAINTENANCE (SEVERE SERVICE)

If a vehicle is operated under any of the following conditions it is considered severe service:

- Towing a trailer or using a camper or car-top carrier.
- Repeated short trips of less than 5 miles in temperatures below freezing, or trips of less than 10 miles in any temperature.
- Extensive idling or low-speed driving for long distances as in heavy commercial use, such as delivery, taxi or police cars.
- Operating on rough, muddy or salt-covered roads.
- Operating on unpaved or dusty roads.
- Driving in extremely hot (over 90°) conditions.

Air cleaner element: replace every 15,000 miles

Engine oil and filter: replace every 3750 miles or 6 months, whichever occurs first.

Timing belt: replace every 60,000 miles if the vehicle is regularly driven in temperatures above 110°F or below -20°F, or if frequently towing a trailer.

Transmission fluid: replace every 30,000 miles.

Rear differential fluid: replace every 60,000 miles.

Front and rear brakes: inspect every 7500 miles or 6 months, whichever occurs first.

Locks and hinges: lubricate every 15,000 miles.

Tie rods, steering gear box, boots: inspect every 7500 miles or 6 months, whichever occurs first.

Suspension components: inspect every 7500 miles or 6 months, whichever occurs first.

Halfshaft boots: inspect every 7500 miles or 6 months, whichever occurs first.

PRECAUTIONS

Before servicing any vehicle, please be sure to read all of the following precautions, which deal with personal safety, prevention of component damage, and important points to take into consideration when servicing a motor vehicle:

• Never open, service or drain the radiator or cooling system when the engine is hot; serious burns can occur from the steam and hot coolant.

• Observe all applicable safety precautions when working around fuel. Whenever servicing the fuel system, always work in a well-ventilated area. Do not allow fuel spray or vapors to come in contact with a spark, open flame, or excessive heat (a hot drop light, for example). Keep a dry chemical fire extinguisher near the work area. Always keep fuel in a container specifically designed for fuel storage; also, always properly seal fuel containers to avoid the possibility of fire or explosion. Refer to the additional fuel system precautions later in this section.

• Fuel injection systems often remain pressurized, even after the engine has been turned **OFF**. The fuel system pressure must be relieved before disconnecting any fuel lines. Failure to do so may result in fire and/or personal injury.

• Brake fluid often contains polyglycol ethers and polyglycols. Avoid contact with the eyes and wash your hands thoroughly after handling brake fluid. If you do get brake fluid in your eyes, flush your eyes with clean, running water for 15 minutes. If eye irritation persists, or if you have taken brake fluid internally, IMMEDIATELY seek medical assistance.

• The EPA warns that prolonged contact with used engine oil may cause a number of skin disorders, including cancer. You should make every effort to minimize your exposure to used engine oil. Protective gloves should be worn when changing oil. Wash your hands and any other exposed skin areas as soon as possible after exposure to used engine oil. Soap and water, or waterless hand cleaner should be used.

• All new vehicles are now equipped with an air bag system, often referred to as a Supplemental Restraint System (SRS) or Supplemental Inflatable Restraint (SIR) system. The system must be disabled before performing service on or around system components, steering column, instrument panel components, wiring and sensors. Failure to follow safety and disabling procedures could result in accidental air bag deployment, possible personal injury and unnecessary system repairs.

• Always wear safety goggles when working with, or around, the air bag system. When carrying a non-deployed air bag, be sure the bag and trim cover are pointed away from your body. When placing a non-deployed air bag on a work surface, always face the bag and trim cover upward, away from the surface. This will reduce the motion of the module if it is accidentally deployed. Refer to the additional air bag system precautions later in this section.

• Clean, high quality brake fluid from a sealed container is essential to the safe and proper operation of the brake system. You should always buy the correct type of brake fluid for your vehicle. If the brake fluid becomes contaminated, completely flush the system with new fluid. Never reuse any brake fluid. Any brake fluid that is removed from the system should be discarded. Also, do not allow any brake fluid to come in contact with a painted surface; it will damage the paint.

• Never operate the engine without the proper amount and type of engine oil; doing so WILL result in severe engine damage.

• Timing belt maintenance is extremely important. Many models utilize an interference-type, non-freewheeling engine. If the timing belt breaks, the valves in the cylinder head may strike the pistons, causing potentially serious (also time-consuming and expensive) engine damage. Refer to the maintenance interval charts for the recommended replacement interval for the timing belt, and to the timing belt section for belt replacement and inspection.

• Disconnecting the negative battery cable on some vehicles may interfere with the functions of the on-board computer system(s) and may require the computer to undergo a relearning process once the negative battery cable is reconnected.

• When servicing drum brakes, only disassemble and assemble one side at a time, leaving the remaining side intact for reference.

• Only an MVAC-trained, EPA-certified automotive technician should service the air conditioning system or its components.

BRAKES

ANTI-LOCK BRAKE SYSTEM (ABS)

GENERAL INFORMATION

PRECAUTIONS

• Certain components within the ABS system are not intended to be serviced or repaired individually.

• Do not use rubber hoses or other parts not specifically specified for and ABS system. When using repair kits, replace all parts included in the kit. Partial or incorrect repair may lead to functional problems and require the replacement of components.

• Lubricate rubber parts with clean, fresh brake fluid to ease assembly. Do not use shop air to clean parts; damage to rubber components may result.

• Use only DOT 3 brake fluid from an unopened container.

• If any hydraulic component or line is removed or replaced, it may be necessary to bleed the entire system.

• A clean repair area is essential. Always clean the reservoir and cap thoroughly before removing the cap. The slightest amount of dirt in the fluid may plug an orifice and impair the system function. Perform repairs after components have been thoroughly cleaned; use only denatured alcohol to clean components. Do not allow ABS components to come into contact with any substance containing mineral oil; this includes used shop rags.

• The Anti-Lock control unit is a microprocessor similar to other computer units in the vehicle. Ensure that the ignition switch is **OFF** before removing or installing controller harnesses. Avoid static electricity discharge at or near the controller.

• If any arc welding is to be done on the vehicle, the control unit should be unplugged before welding operations begin.

BRAKES

BLEEDING PROCEDURE

BLEEDING PROCEDURE

See Figure 1.

➥Do not reuse the drained fluid. Use only clean Honda DOT 3 Brake Fluid from an unopened container. Using a non-Honda brake fluid can cause corrosion and shorten the life of the system.

❇ WARNING

Make sure no dirt or other foreign matter is allowed to contaminate the brake fluid.

❇ WARNING

Do not spill brake fluid on the vehicle, it may damage the paint; if brake fluid does contact the paint, wash it off immediately with water.

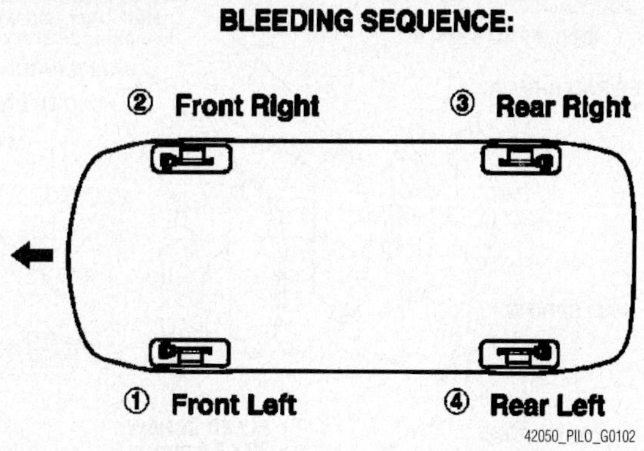

1. The reservoir on the master cylinder must be at the MAX (upper) level mark at the start of the bleeding procedure and

BLEEDING SEQUENCE:

② **Front Right**   ③ **Rear Right**

① **Front Left**   ④ **Rear Left**

42050_PILO_G0102

Fig. 1 Brake bleeding sequence

checked after bleeding each brake caliper. Add fluid as required.

2. Make sure the brake fluid level in the reservoir is at the MAX (upper) level line.

3. Slide a piece of clear plastic hose over the first bleed screw, and submerge the other end in a container of new brake fluid.

4. Have someone slowly pump the brake pedal several times, and then apply steady pressure.

5. Starting at the left-front, loosen the brake bleed screw to allow air to escape from the system. Then tighten the bleed screw securely.

6. Repeat the procedure for each wheel in the sequence shown following until air bubbles no longer appear in the fluid.

7. Refill the master cylinder reservoir to the MAX (upper) level line.

BRAKES

❇ CAUTION

Dust and dirt accumulating on brake parts during normal use may contain asbestos fibers from production or aftermarket brake linings. Breathing excessive concentrations of asbestos fibers can cause serious bodily harm. Exercise care when servicing brake parts. Do not sand or grind brake lining unless equipment used is designed to contain the dust residue. Do not clean brake parts with compressed air or by dry brushing. Cleaning should be done by dampening the brake components with a fine mist of water, then wiping the brake components clean with a dampened cloth. Dispose of cloth and all residue containing asbestos fibers in an impermeable container with the appropriate label. Follow practices prescribed by the Occupational Safety and Health Administration (OSHA) and the Environmental Protection Agency (EPA) for the handling, processing, and disposing of dust or debris that may contain asbestos fibers.

BRAKE CALIPER

REMOVAL & INSTALLATION

See Figure 2.

❇❇ CAUTION

Dust and dirt accumulating on brake parts during normal use may contain asbestos fibers from production or aftermarket brake linings. Breathing excessive concentrations of asbestos fibers can cause serious bodily harm. Exercise care when servicing brake parts. Do not sand or grind brake lining unless equipment used is designed to contain the dust residue. Do not clean brake parts with compressed air or by dry brushing. Cleaning should be done by dampening the brake components with a fine mist of water, then wiping the brake components clean with a dampened cloth. Dispose of cloth and all residue containing asbestos fibers in an impermeable container with the appropriate label. Follow practices prescribed by the Occupational Safety and Health Administration (OSHA) and the Environmental Pro-

tection Agency (EPA) for the handling, processing, and disposing of dust or debris that may contain asbestos fibers.

1. Before servicing the vehicle, refer to the precautions section.

2. Remove some fluid from the reservoir with a suction pump.

3. Remove or disconnect the following:
   - Front wheels
   - Banjo bolt and disconnect the brake hose from the caliper. Plug the hose to prevent fluid loss and contamination.
   - Mounting bolts and the caliper from its mounting bracket

To Install:

4. Install or connect the following:
   - Caliper over the pads and onto its mounting bracket. Torque both caliper bolts to 27 ft. lbs. (36 Nm).
   - Brake hose to the caliper using new sealing washers. Carefully torque the banjo bolt to 25 ft. lbs. (34 Nm).

5. Fill the reservoir with fluid and bleed the brakes.
   - Front wheels

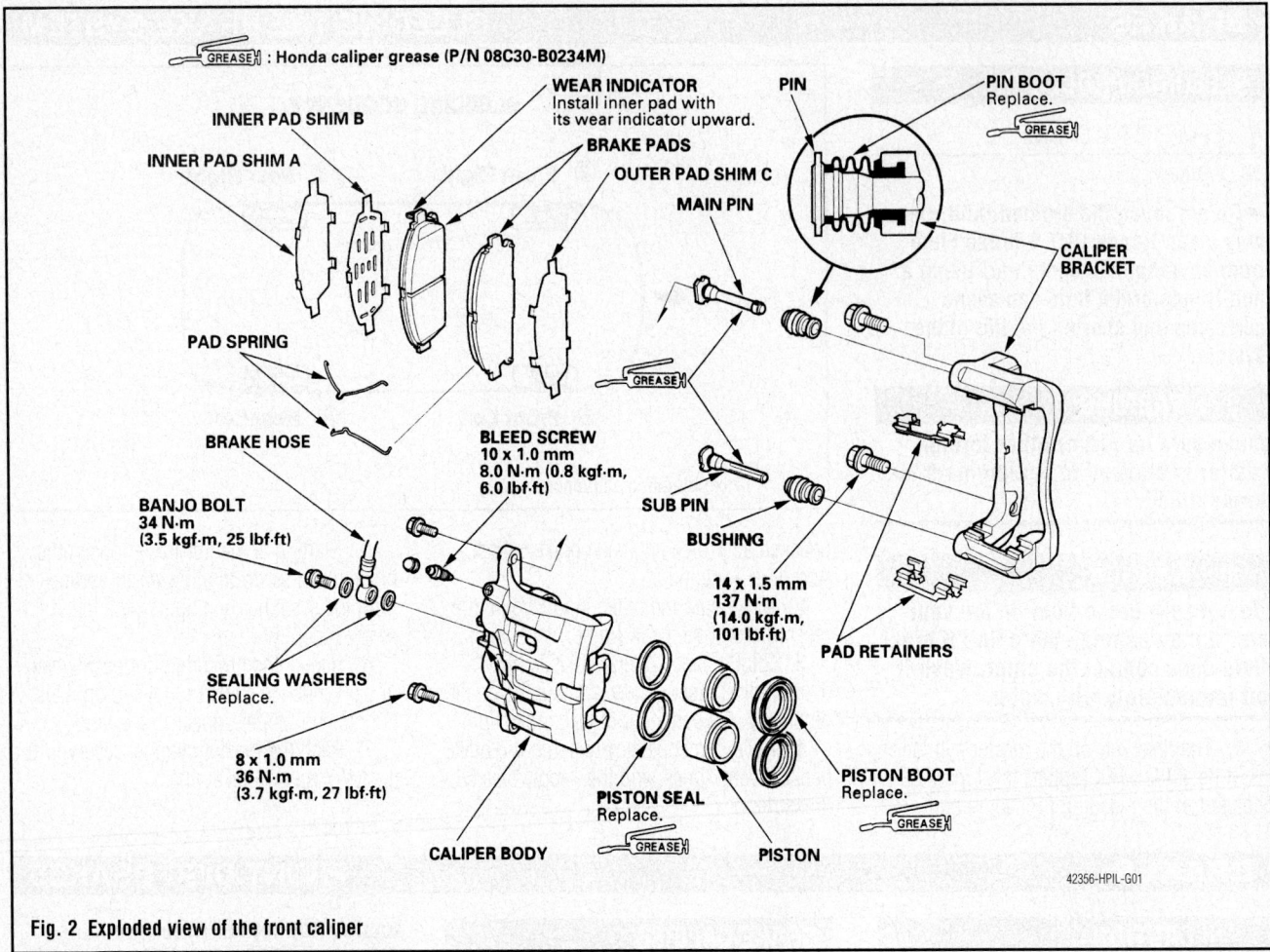

Fig. 2 Exploded view of the front caliper

DISC BRAKE PADS

REMOVAL & INSTALLATION

✳✳ CAUTION

Dust and dirt accumulating on brake parts during normal use may contain asbestos fibers from production or aftermarket brake linings. Breathing excessive concentrations of asbestos fibers can cause serious bodily harm. Exercise care when servicing brake parts. Do not sand or grind brake lining unless equipment used is designed to contain the dust residue. Do not clean brake parts with compressed air or by dry brushing. Cleaning should be done by dampening the brake components with a fine mist of water, then wiping the brake components clean with a dampened cloth. Dispose of cloth and all residue containing asbestos fibers in an impermeable container with the appropriate label. Follow practices prescribed by the Occupational Safety and Health Administration (OSHA) and the Environmental Protection Agency (EPA) for the handling, processing, and disposing of dust or debris that may contain asbestos fibers.

1. Before servicing the vehicle, refer to the precautions section.
2. Remove or disconnect the following:
   - Front wheels
3. Remove a small amount of brake fluid from the reservoir using a suction pump.
   - Brake hose clamp from the knuckle by unfastening the retaining bolts
   - Lower caliper retaining bolt and pivot the caliper upward, off of the pads
   - Pad springs while holding the pads
   - Pad shim and pad retainers
   - Disc brake pads from the caliper

To install:

4. Clean the caliper thoroughly; remove any rust from the lip of the disc or rotor. Check the brake rotor for grooves or cracks. If any heavy scoring is present, the rotor must be replaced.
5. Install or connect the following:
   - Pad retainers. Apply molybdenum brake grease to both surfaces of the shims and the back of the disc brake pads.
   - Pads and shims. The pad with the wear indicator goes in the inboard position.
   - Pad springs while holding the pads
6. Push in the caliper piston so the caliper will fit over the pads. This is most easily accomplished with a pad spreader or large C-clamp.
   - Caliper down into position and tighten the mounting bolt to 27 ft. lbs. (37 Nm)
   - Brake hose to the knuckle, if removed
   - Wheels
7. Add brake fluid to the master cylinder reservoir and install the cap.
8. Depress the brake pedal several times and make sure that the movement feels normal. The first brake pedal application may result in a very long pedal action due to the pistons being retracted. Always make several brake applications before starting the vehicle. Bleed the system if necessary.

BRAKES   **REAR DISC BRAKES**

✴✴ CAUTION

Dust and dirt accumulating on brake parts during normal use may contain asbestos fibers from production or aftermarket brake linings. Breathing excessive concentrations of asbestos fibers can cause serious bodily harm. Exercise care when servicing brake parts. Do not sand or grind brake lining unless equipment used is designed to contain the dust residue. Do not clean brake parts with compressed air or by dry brushing. Cleaning should be done by dampening the brake components with a fine mist of water, then wiping the brake components clean with a dampened cloth. Dispose of cloth and all residue containing asbestos fibers in an impermeable container with the appropriate label. Follow practices prescribed by the Occupational Safety and Health Administration (OSHA) and the Environmental Protection Agency (EPA) for the handling, processing, and disposing of

dust or debris that may contain asbestos fibers.

BRAKE CALIPER

REMOVAL & INSTALLATION
See Figure 3.

✴✴ CAUTION

Dust and dirt accumulating on brake parts during normal use may contain asbestos fibers from production or aftermarket brake linings. Breathing excessive concentrations of asbestos fibers can cause serious bodily harm. Exercise care when servicing brake parts. Do not sand or grind brake lining unless equipment used is designed to contain the dust residue. Do not clean brake parts with compressed air or by dry brushing. Cleaning should be done by dampening the brake components with a fine mist of water, then wiping the brake components clean with a dampened cloth. Dispose of cloth and all residue

containing asbestos fibers in an impermeable container with the appropriate label. Follow practices prescribed by the Occupational Safety and Health Administration (OSHA) and the Environmental Protection Agency (EPA) for the handling, processing, and disposing of dust or debris that may contain asbestos fibers.

1. Before servicing the vehicle, refer to the precautions section.
2. Remove some fluid from the reservoir with a suction pump.
3. Remove or disconnect the following:
   • Rear wheels
   • Banjo bolt and disconnect the brake hose from the caliper. Plug the hose to prevent fluid loss and contamination.
   • 2 caliper mounting bolts and the caliper from its mounting bracket

To Install:
4. Install or connect the following:
   • Caliper over the pads and onto its mounting bracket. Tighten the caliper bolts to 27 ft. lbs. (37 Nm).

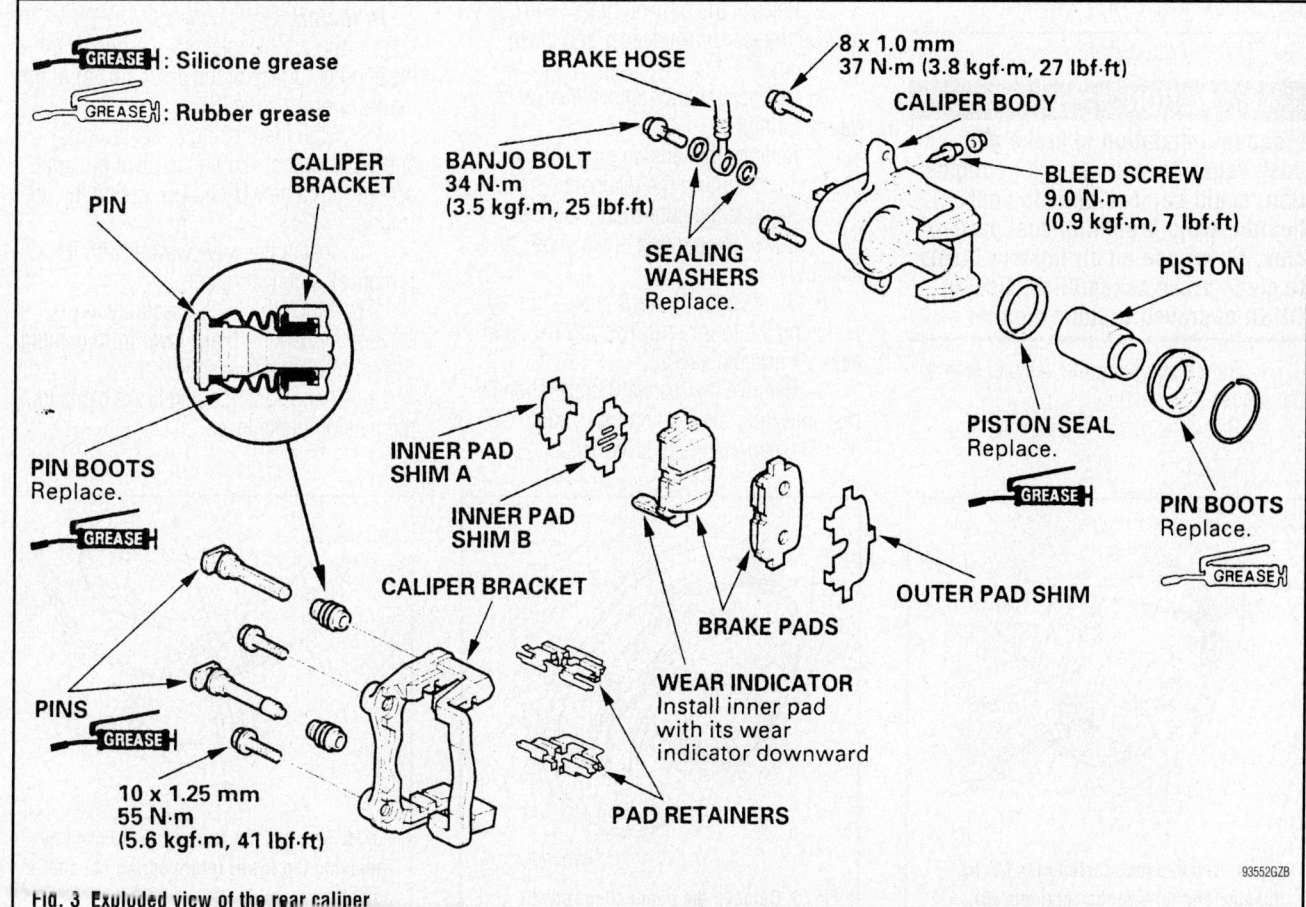

Fig. 3 Exploded view of the rear caliper

- Brake hose with new sealing washers. Tighten the banjo bolt to 25 ft. lbs. (34 Nm).

5. Fill the reservoir with fluid and bleed the brake system. Adjust the parking brake if necessary.

- Rear wheels

DISC BRAKE PADS

REMOVAL & INSTALLATION

⊕ CAUTION

Dust and dirt accumulating on brake parts during normal use may contain asbestos fibers from production or aftermarket brake linings. Breathing excessive concentrations of asbestos fibers can cause serious bodily harm. Exercise care when servicing brake parts. Do not sand or grind brake lining unless equipment used is designed to contain the dust residue. Do not clean brake parts with compressed air or by dry brushing.

Cleaning should be done by dampening the brake components with a fine mist of water, then wiping the brake components clean with a dampened cloth. Dispose of cloth and all residue containing asbestos fibers in an impermeable container with the appropriate label. Follow practices prescribed by the Occupational Safety and Health Administration (OSHA) and the Environmental Protection Agency (EPA) for the handling, processing, and disposing of dust or debris that may contain asbestos fibers.

1. Before servicing the vehicle, refer to the precautions section.
2. Remove a small amount of brake fluid from the reservoir using a suction pump.
3. Remove or disconnect the following:
   - Rear wheels
   - 2 caliper mounting bolts and the caliper from the bracket
   - Pads, shims, and pad retainers

To install:

4. Clean the caliper thoroughly; remove any dirt or dust. Check the brake rotor for grooves or cracks and machine or replace, as necessary.
5. Install or connect the following:
   - Pad retainers. Apply molybdenum brake grease to both surfaces of the shims and the back of the disc brake pads.
   - Pads and shims. The wear retainer on the inboard pad faces down.
6. Use a suitable tool to push caliper piston into its bore and enable the caliper to fit over the pads. Lubricate the piston boot with silicon grease. Avoid twisting the boot.
   - Brake caliper and tighten the mounting bolts to 27 ft. lbs. (37 Nm)
   - Rear wheels
7. Add brake fluid to the master cylinder reservoir. Depress the brake pedal several times to seat the pads. Bleed the brakes if necessary.

BRAKES

PARKING BRAKE

PARKING BRAKE SHOES

REMOVAL & INSTALLATION

See Figures 4 through 13.

⊕ CAUTION

Frequent inhalation of brake pad dust, regardless of material composition, could be hazardous to your health. Avoid breathing dust particles. Never use an air hose or brush to clean brake assemblies. Use an OHSA-approved vacuum cleaner.

1. Before servicing the vehicle, refer to the precautions section.
2. Raise the rear of the vehicle, and

make sure it is securely supported. Remove the rear wheels.
3. Release the parking brake, and remove the rear brake caliper and brake disc/drum.
4. Disconnect and remove the upper return springs.
5. Remove the tension pins (A) by pushing and turning the retainers (B).
6. Remove the connecting rod (A).
7. Lower the parking brake shoe assembly.
8. Remove the forward brake shoe by removing the lower return spring (A) and adjuster assembly (B).
9. Remove the rearward brake shoe by disconnecting the parking brake cable (A) from the parking brake lever (B).

10. Remove the U-clip, wave washer, and parking brake lever from the brake shoe.

To install:

11. Apply Molykote® 44 MA grease to the sliding surface of the pivot pin (A) of the rearward brake shoe (B).
12. Install the parking brake lever (C) and wave washer (D) on the pivot pin, and secure with a new U-clip (E), noting the following:
   a. Install the wave washer with its convex side facing out.
   b. Pinch the U-clip securely to prevent the parking brake lever from coming out from the brake shoe.
13. Connect the parking brake cable to the parking brake lever.

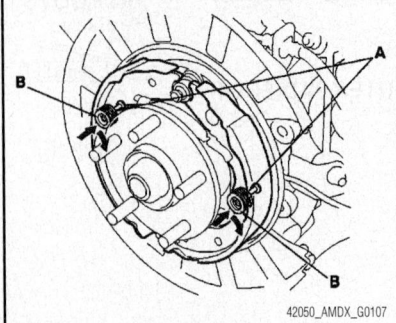

42050_AMDX_G0107

Fig. 4 Remove the tension pins (A) by pushing and turning the retainers (B)

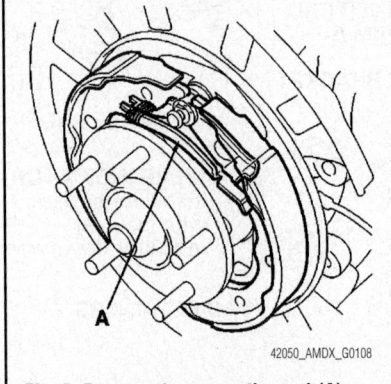

42050_AMDX_G0108

Fig. 5 Remove the connecting rod (A)

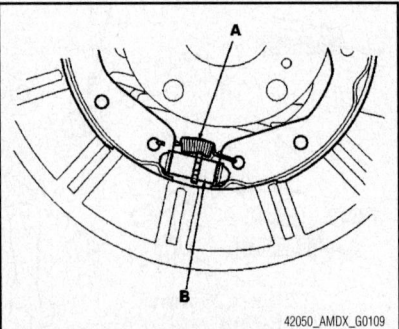

42050_AMDX_G0109

Fig. 6 Remove the forward brake shoe by removing the lower return spring (A) and adjuster assembly (B)

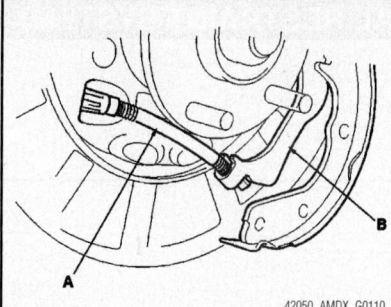

Fig. 7 Remove the rearward brake shoe by disconnecting the parking brake cable (A) from the parking brake lever (B)

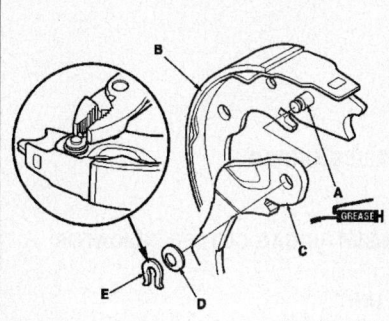

Fig. 8 Apply Molykote® 44 MA grease to the sliding surface of the pivot pin (A) of the rearward brake shoe (B). Install the parking brake lever (C) and wave washer (D) on the pivot pin, and secure with a new U-clip (E)

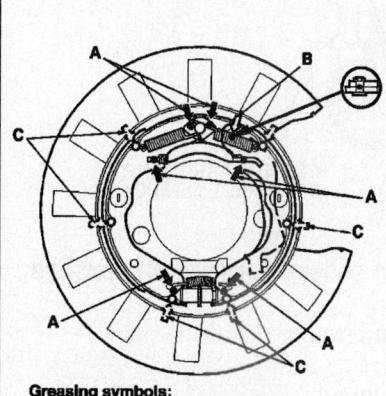

Greasing symbols:
➡● Brake shoe ends and connecting rod ends
▱○ Opposite edge of the shoe
⇨● Sliding surface

Fig. 9 Apply Molykote 44 MA grease to the shoe ends and connecting rod ends (A), sliding surfaces (B), and opposite edges of the parking brake shoe (C) as shown

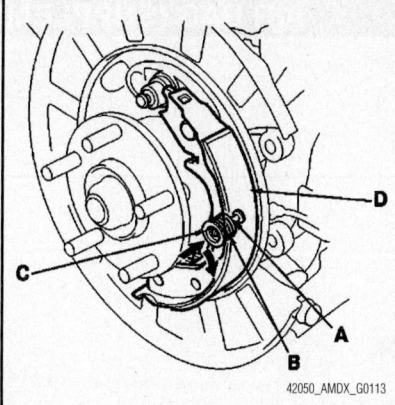

Fig. 10 Install the tension pin (A), retainer spring (B), and retainer (C) of the rearward brake shoe (D). Make sure the tension pin does not contact the parking brake lever

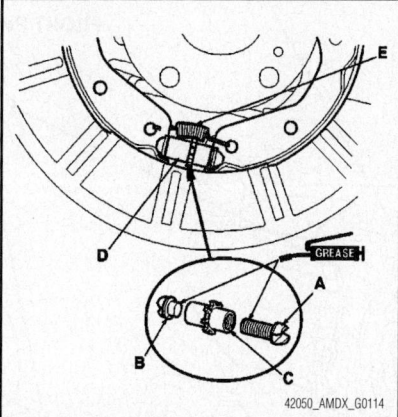

Fig. 11 View of clevis (A & B), adjuster (C), adjuster assembly (D) and lower return spring (E)

14. Apply Molykote® 44 MA grease to the shoe ends and connecting rod ends (A), sliding surfaces (B), and opposite edges of the parking brake shoe (C) as shown. Wipe off any excess. Keep grease off the brake linings.

15. Install the tension pin (A), retainer spring (B), and retainer (C) of the rearward brake shoe (D). Make sure the tension pin does not contact the parking brake lever.

16. Clean the threaded portions of clevis A, and coat the threads of clevis A with grease. Clean the sliding surface of clevis B, and coat the sliding surface of clevis B with grease. Install clevis A and B on the adjuster (C), and shorten clevis A by turning the adjuster.

17. Reinstall the brake shoe adjuster assembly (D), and hook the lower return spring (E) on the parking brake shoes.

18. Install the rod spring to the connecting rod first. Then install the connecting rod on the parking brake shoes.

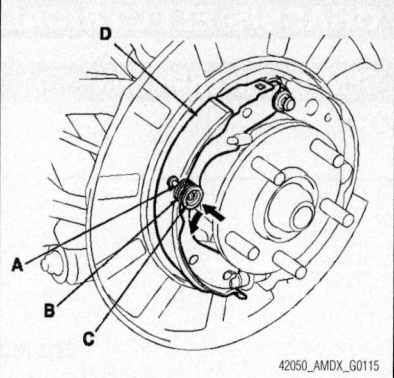

Fig. 12 Install the tension pin (A), retainer spring (B), and retainer (C) of the forward brake shoe (D)

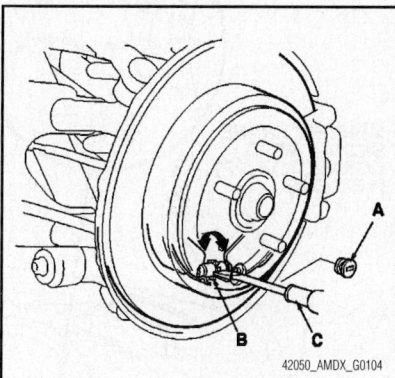

Fig. 13 Remove the access plug (A). Turn up the ratchet teeth (B) or the adjuster assembly with a suitable tool (C) until the shoes lock against the drum. Then back off ten clicks, and install the access plug.

19. Install the tension pin (A), retainer spring (B), and retainer (C) of the forward brake shoe (D).

20. Install the upper return springs.

21. Install the rear brake disc/drum and rear brake caliper.

22. Adjust the parking brake.

➡This procedure should be done when you replace the brake shoes.

23. Raise the rear of the vehicle, and make sure it is securely supported. Remove the rear wheels.

24. Release the parking brake, and back off the adjusting nut.

25. Remove the access plug (A).

26. Turn up the ratchet teeth (B) or the adjuster assembly with a flat-tip screwdriver (C) until the shoes lock against the drum. Then back off ten clicks, and install the access plug.

27. Do the minor adjustment procedure.

28. Install the rear wheels.

CHASSIS ELECTRICAL   AIR BAG (SUPPLEMENTAL RESTRAINT SYSTEM)

GENERAL INFORMATION

See Figures 14 and 15.

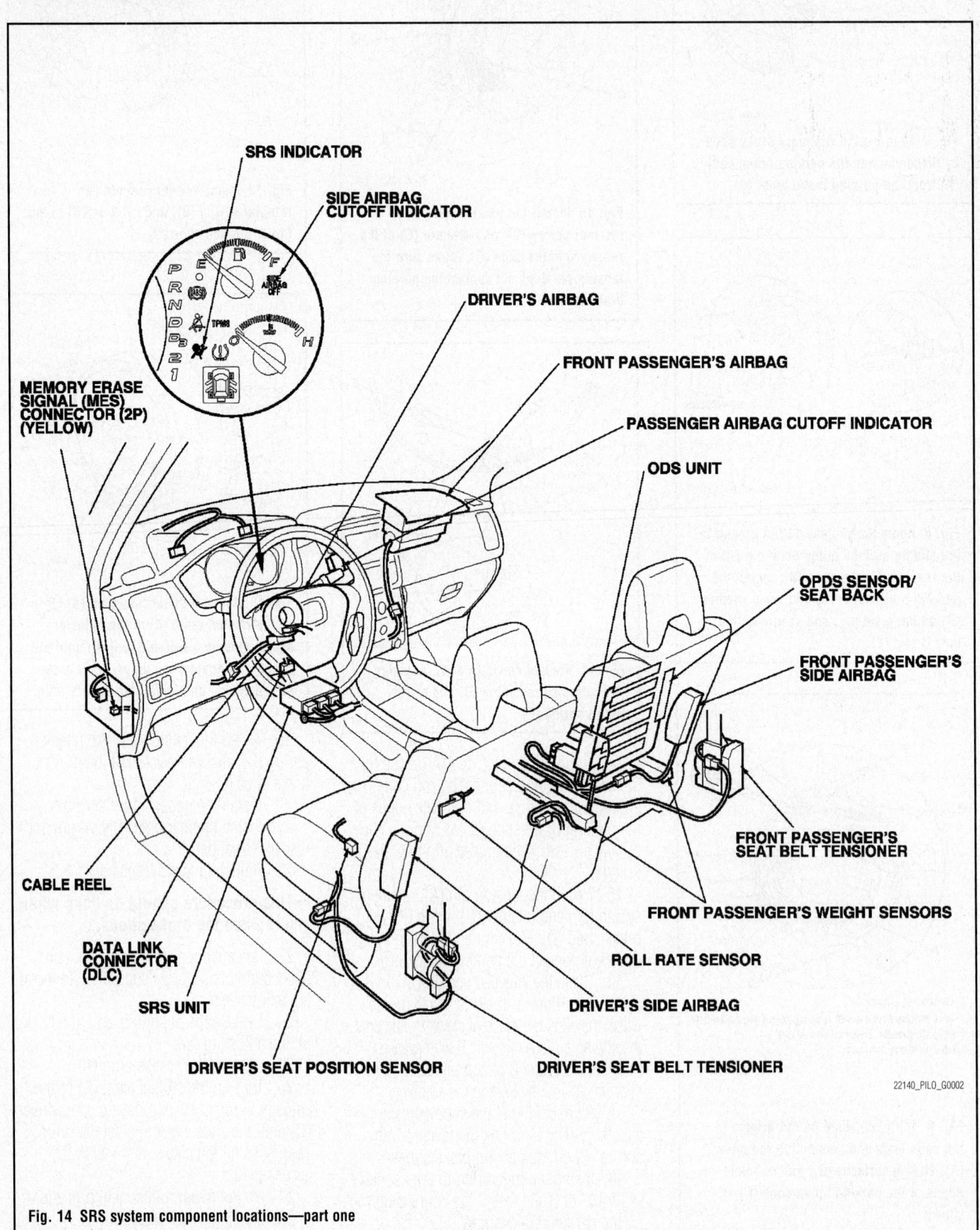

Fig. 14 SRS system component locations—part one

22140_PILO_G0002

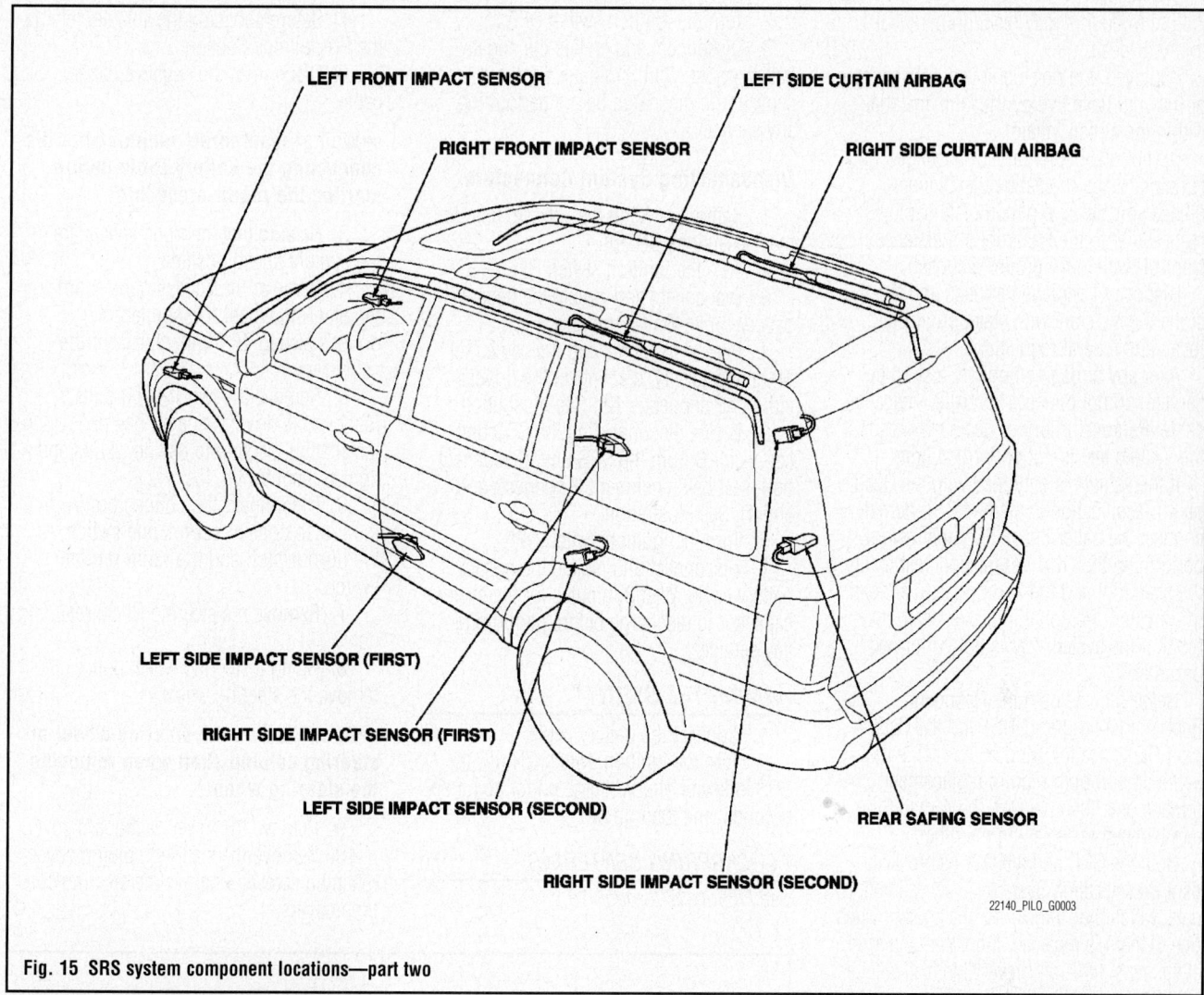

LEFT FRONT IMPACT SENSOR

RIGHT FRONT IMPACT SENSOR

LEFT SIDE CURTAIN AIRBAG

RIGHT SIDE CURTAIN AIRBAG

LEFT SIDE IMPACT SENSOR (FIRST)

RIGHT SIDE IMPACT SENSOR (FIRST)

LEFT SIDE IMPACT SENSOR (SECOND)

RIGHT SIDE IMPACT SENSOR (SECOND)

REAR SAFING SENSOR

22140_PILO_G0003

Fig. 15 SRS system component locations—part two

❊❊ CAUTION

These vehicles are equipped with an air bag system. The system must be disarmed before performing service on, or around, system components, the steering column, instrument panel components, wiring and sensors. Failure to follow the safety precautions and the disarming procedure could result in accidental air bag deployment, possible injury and unnecessary system repairs.

SERVICE PRECAUTIONS

Disconnect and isolate the battery negative cable before beginning any airbag system component diagnosis, testing, removal, or installation procedures. Allow system capacitor to discharge for two minutes before beginning any component service. This will disable the airbag system. Failure to disable the airbag system may result in

accidental airbag deployment, personal injury, or death.

Do not place an intact undeployed airbag face down on a solid surface. The airbag will propel into the air if accidentally deployed and may result in personal injury or death.

When carrying or handling an undeployed airbag, the trim side (face) of the airbag should be pointing towards the body to minimize possibility of injury if accidental deployment occurs. Failure to do this may result in personal injury or death.

Replace airbag system components with OEM replacement parts. Substitute parts may appear interchangeable, but internal differences may result in inferior occupant protection. Failure to do so may result in occupant personal injury or death.

Wear safety glasses, rubber gloves, and long sleeved clothing when cleaning powder residue from vehicle after an airbag deployment. Powder residue emitted from a deployed airbag can cause skin irritation.

Flush affected area with cool water if irritation is experienced. If nasal or throat irritation is experienced, exit the vehicle for fresh air until the irritation ceases. If irritation continues, see a physician.

Do not use a replacement airbag that is not in the original packaging. This may result in improper deployment, personal injury, or death.

The factory installed fasteners, screws and bolts used to fasten airbag components have a special coating and are specifically designed for the airbag system. Do not use substitute fasteners. Use only original equipment fasteners listed in the parts catalog when fastener replacement is required.

During, and following, any child restraint anchor service, due to impact event or vehicle repair, carefully inspect all mounting hardware, tether straps, and anchors for proper installation, operation, or damage. If a child restraint anchor is found damaged in any way, the anchor must be replaced.

Failure to do this may result in personal injury or death.

Deployed and non-deployed airbags may or may not have live pyrotechnic material within the airbag inflator.

Do not dispose of driver/passenger/ curtain airbags or seat belt tensioners unless you are sure of complete deployment. Refer to the Hazardous Substance Control System for proper disposal.

Dispose of deployed airbags and tensioners consistent with state, provincial, local, and federal regulations.

After any airbag component testing or service, do not connect the battery negative cable. Personal injury or death may result if the system test is not performed first.

If the vehicle is equipped with the Occupant Classification System (OCS), do not connect the battery negative cable before performing the OCS Verification Test using the scan tool and the appropriate diagnostic information. Personal injury or death may result if the system test is not performed properly.

Never replace both the Occupant Restraint Controller (ORC) and the Occupant Classification Module (OCM) at the same time. If both require replacement, replace one, then perform the Airbag System test before replacing the other.

Both the ORC and the OCM store Occupant Classification System (OCS) calibration data, which they transfer to one another when one of them is replaced. If both are replaced at the same time, an irreversible fault will be set in both modules and the OCS may malfunction and cause personal injury or death.

If equipped with OCS, the Seat Weight Sensor is a sensitive, calibrated unit and must be handled carefully. Do not drop or handle roughly. If dropped or damaged, replace with another sensor. Failure to do so may result in occupant injury or death.

If equipped with OCS, the front passenger seat must be handled carefully as well. When removing the seat, be careful when setting on floor not to drop. If dropped, the sensor may be inoperative, could result in occupant injury, or possibly death.

If equipped with OCS, when the passenger front seat is on the floor, no one should sit in the front passenger seat. This uneven force may damage the sensing ability of the seat weight sensors. If sat on and damaged, the sensor may be inoperative, could result in occupant injury, or possibly death.

DISARMING THE SYSTEM

1. Before servicing the vehicle, refer to the Precautions Section.

2. Turn the ignition switch OFF.
3. Disconnect and isolate the negative battery cable. Wait 3 minutes for the system capacitor to discharge before performing any service.

Disconnecting System Connectors

1. Before servicing the vehicle, refer to the Precautions Section.
2. Turn the ignition switch OFF.
3. Disconnect and isolate the negative battery cable. Wait 3 minutes.
4. Before disconnecting the cable reel 4P connector (1), disconnect the driver's airbag 4P connector (2). See illustration.
5. Before disconnecting the SRS unit connector B from the SRS unit, disconnect both seat belt tensioner 2P connectors (3 and 4). See illustration.
6. Turn the ignition switch OFF.
7. Disconnect and isolate the negative battery cable. Wait 3 minutes for the system capacitor to discharge before performing any service.

ARMING THE SYSTEM

1. Connect the battery cable.
2. Turn the ignition switch ON (II), the SRS indicator should come on for about six seconds, and then go off.

CLOCKSPRING CENTERING

See Figures 17 through 19.

1. Before servicing the vehicle, refer to the Precautions Section.
2. Disconnect the negative battery cable.

➡ Wait at least three minutes after disconnecting the battery cable before starting the repair procedure.

3. Be sure that the front wheels are in the straight ahead position.
4. Remove the access panel from the steering wheel. Disconnect the driver's airbag 4P connector from the cable reel.
5. Remove the two Torx bit bolts. Remove the driver's side air bag pad. Upon installation, be sure to use new bolts and tighten them to 7 ft. lbs.
6. Disconnect the connectors from the cruise control set/resume switch, the horn switch and the radio remote switch.
7. Remove the steering wheel retaining nut.
8. Using a steering wheel puller, remove the steering wheel.

➡ Do not tap on the steering wheel or steering column shaft when removing the steering wheel.

9. Remove the lower dashboard cover.
10. Remove the steering column cover retaining screws. Remove the steering column covers.

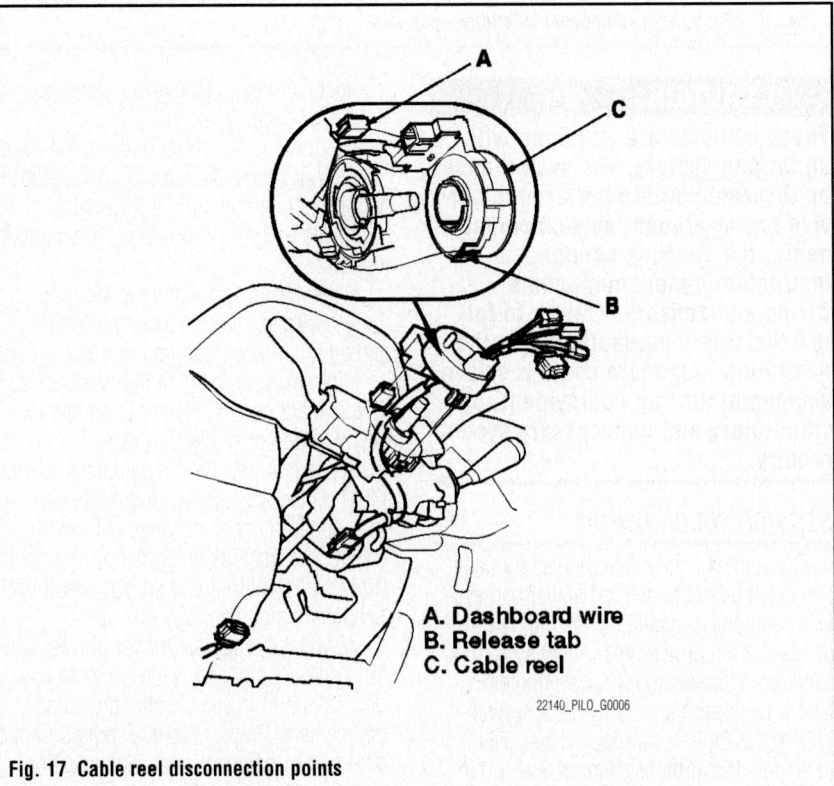

A. Dashboard wire
B. Release tab
C. Cable reel

22140_PILO_G0006

Fig. 17 Cable reel disconnection points

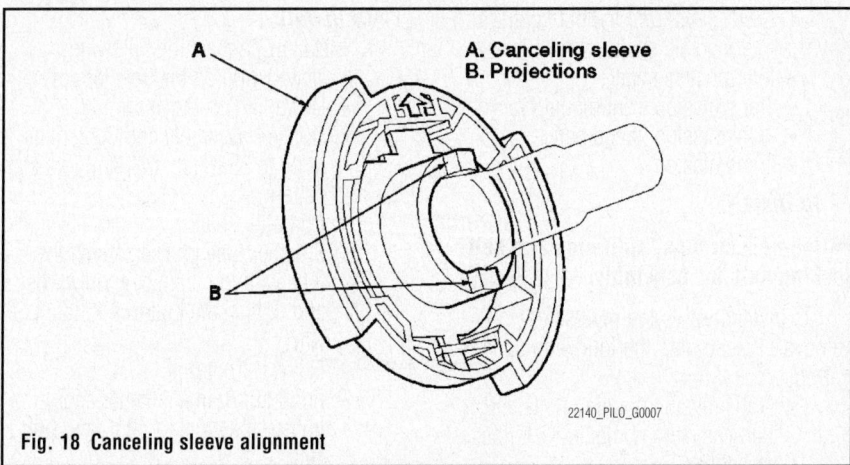

A. Canceling sleeve
B. Projections

Fig. 18 Canceling sleeve alignment

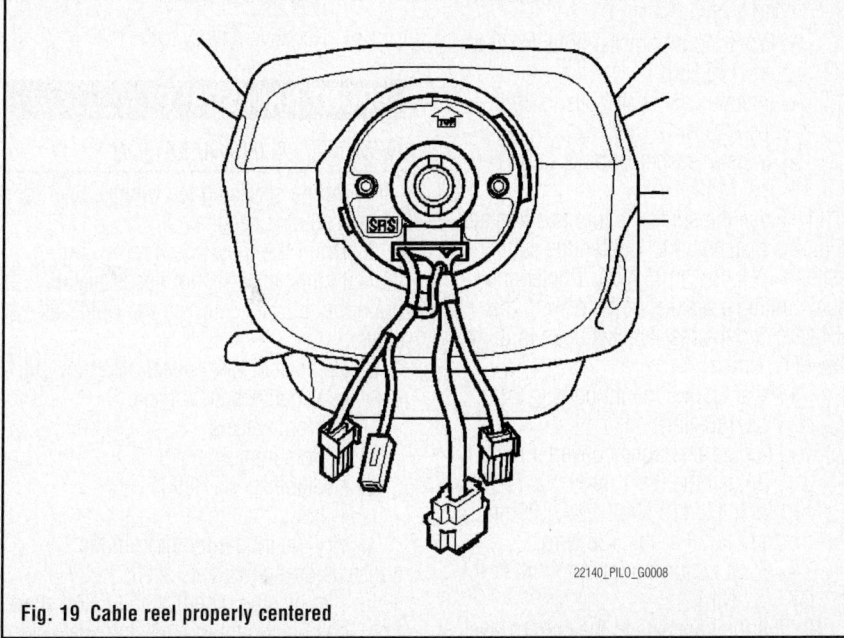

Fig. 19 Cable reel properly centered

11. Disconnect the dashboard wire harness connector from the cable reel connector.

12. Disconnect the dashboard wire harness connector and release the tab. Pull off the cable reel.

To install:

13. Be sure that the wheels are in the straight ahead position.

14. Set the turn signal canceling sleeve so the projections are aligned vertically. See illustration.

15. Carefully install the cable reel.

16. Install the steering column covers.

17. If necessary, center the cable. First rotate the cable reel clockwise until it stops. Then rotate it counterclockwise about 2.5 turns, until the arrow mark on the cable reel label points straight up.

➡**New cable reels come centered.**

18. Align the projections on the cable reel with the holes on the steering wheel. Install the steering wheel. Tighten the retaining nut to 36 ft. lbs.

19. Continue the installation in the reverse order of the removal procedure.

20. When installing the driver's side airbag, be sure to use new bolts and tighten them to 7 ft. lbs.

21. Connect the battery cable.

22. Turn the ignition switch ON (II), the SRS indicator should come on for about six seconds, and then go off.

23. After the SRS indicator has turned off, turn the steering wheel fully to the left and fully to the right to ensure that the SRS indicator light does not come on. Correct as required.

DRIVETRAIN

AUTOMATIC TRANSAXLE ASSEMBLY

REMOVAL & INSTALLATION

See Figures 20 and 21.

1. Before servicing the vehicle, refer to the precautions section.

2. Drain the transaxle.

3. Drain the power steering system.

4. Remove the engine appearance covers.

5. Remove the driver's side center console lower panel and pull back the cover to access steering joint cover.

6. Remove or disconnect the following:
   - Steering joint bolt
   - Steering joint from the steering gearbox pinion shaft
   - Battery
   - Battery tray
   - Intake manifold cover
   - Air intake assembly
   - Power steering pump hose and the clamp bolt
   - Splash shield
   - Transmission breather tube
   - Transaxle oil cooler lines
   - Cooler hose from the clamp on the starter, if equipped
   - Starter motor
   - Shift control solenoid valve connectors
   - Transaxle ground cable
   - Connector from the bracket and the connector
   - Harness clip from the brackets
   - Clutch pressure switch connectors
   - Joint connector and transmission range switch connector from the brackets
   - Countershaft speed sensor connector

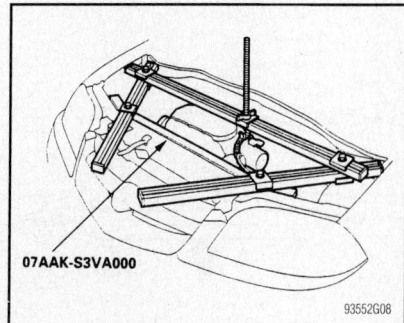

Fig. 20 Support the engine while removing the transaxle

- Heated Oxygen (HO₂S) sensor connectors
- Transmission housing mounting bolts
- Nut from the front mount and the ground cable from the engine
- Bulkhead cover, windshield wiper arms, cowl cover sealing and cover
- Install a support fixture to the engine lifting eyes.
- Front sub-frame stiffener
- Primary HO₂S sensor clamp bracket from the transmission and harness from the clamp
- Exhaust front pipe
- Lower control arms from the knuckle
- Stabilizer bar links
- Tie rod ends from the knuckle
- Left driveshaft from the differential
- Right driveshaft from the intermediate shaft
- Propeller shaft from the companion flange
- Shift cable cover and holder
- Shift control cable and lever

7. Install a 6 x 1 x 14mm bolt and nut on the cable cover, then reinstall the cable cover to the torque converter housing. If this is not done, the bolt head of the cable cover may prevent torque converter removal.

- Transfer assembly
- Engine-to-torque converter bolts
- Power steering pressure switch connection
- Power steering hose clamp, then the hose from the pipe at the subframe
- Transmission lower mount nuts

8. Matchmark the front subframe to the vehicle body.

- Rear mount bracket bolts

9. Support the sub-frame with a 4 x 4 x 50 inch piece of wood and a jack.

- Sub-frame
- Transaxle lower mounts

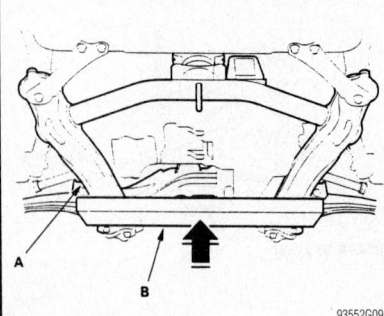

Fig. 21 Support the sub-frame with a 4 x 4 x 50 inch piece of wood and a jack

- Driveshafts from the differential and intermediate shaft
- Intermediate shaft
- Transmission front mount bracket
- Transmission flange bolts
- Transmission

To install:

➡**Use new circlips, split pins and self-locking nuts for assembly.**

10. Installation is the reverse of removal. Please note the following specifications:

- Transmission housing bolts and harness clamp bolts to 47 ft. lbs. (64 Nm)
- Transmission housing bolts to 40 ft. lbs. (54 Nm)
- Front mount bracket bolts to 28 ft. lbs. (38 Nm)
- Intermediate shaft bolts to 29 ft. lbs. (39 Nm)
- Transfer assembly bolts to 33 ft. lbs. (44 Nm)

11. Raise the subframe into position and align the matchmarks. Tighten the subframe bolts to 76 ft. lbs. (103 Nm). Tighten the front subframe bracket bolts to 54 ft. lbs. (74 Nm) and the rear bracket bolts to 86 ft. lbs. (117 Nm).

- Rear engine mount bolts to 28 ft. lbs. (38 Nm)
- Engine-to-torque converter bolts. Tighten the 6 x 1 mm bolts to 105 inch lbs. (12 Nm), 10 x 1.25mm bolt to 28 ft. lbs. (38 Nm).
- Front motor mount nut to 40 ft. lbs. (54 Nm)

12. Fill the transaxle to the correct level.
13. Start the engine and check for leaks.
14. Check the wheel alignment and adjust as necessary.

TRANSFER CASE ASSEMBLY

REMOVAL & INSTALLATION

1. Before servicing the vehicle, refer to the precautions section.
2. Drain the transmission fluid.
3. Remove or disconnect the following:

- Negative battery cable
- Heated Oxygen (HO₂S) sensor connectors
- Front sub-frame stiffener
- Exhaust front pipe
- Breather tube bracket bolt, then the tube from the breather pipe
- Propeller shaft from the transfer assembly
- Transfer assembly bolts and the assembly

To install:

4. Install or connect the following:

- New O-ring on the transfer cover
- Dowel pin on the assembly
- Transfer assembly and tighten the bolts to 33 ft. lbs. (44 Nm) in a star pattern
- Propeller shaft
- Breather tube bracket, attach the tube with the dot facing outwards and tighten the bolt to 9 ft. lbs. (12 Nm)
- Exhaust front pipe
- Front sub-frame stiffener and tighten the bolts to 40 ft. lbs. (54 Nm)
- Heated Oxygen (HO₂S) sensor connectors
- Negative battery cable

FRONT HALFSHAFT

REMOVAL & INSTALLATION

1. Before servicing the vehicle, refer to the precautions section.
2. Drain the transaxle if removing the left halfshaft. It is not necessary to drain the fluid if removing the right halfshaft.
3. Remove or disconnect the following:

- Negative battery cable
- Front wheels
- Spindle nut
- Stabilizer bar link
- Lower ball joint

4. Pry the inboard joint from the transaxle or intermediate shaft.
5. Remove the outer CV-joint stub shaft from the hub by tapping the stub shaft with a plastic hammer.

To install:

➡**Use new circlips, split pins and self-locking nuts for assembly.**

6. Install the outer CV-joint stub shaft into the hub.
7. Install the inner CV-joint to the transaxle or intermediate shaft until the circlip locks in the retaining groove.
8. Install or connect the following:

- Lower ball joint. Tighten the nut to 47 ft. lbs. (64 Nm).
- Stabilizer bar link. Tighten the nut to 58 ft. lbs. (78 Nm).
- Spindle nut. Tighten the nut to 210 ft. lbs.
- Front wheels
- Negative battery cable

9. Fill the transaxle to the correct level and check for leaks.

REAR HALFSHAFT

REMOVAL & INSTALLATION

4WD Vehicles

See Figures 22 through 24.

1. Before servicing the vehicle, refer to the precautions section.
2. Raise and support the vehicle safely.
3. Drain the differential fluid. Be sure to properly dispose of used fluid.
4. Remove the rear tire and wheel assemblies.
5. Lift the locking nut on the spindle nut. Remove and discard the spindle nut.
6. Safely support the lower control arm, using a jack at the connecting point of the control arm and knuckle. Remove the upper arm bolt.

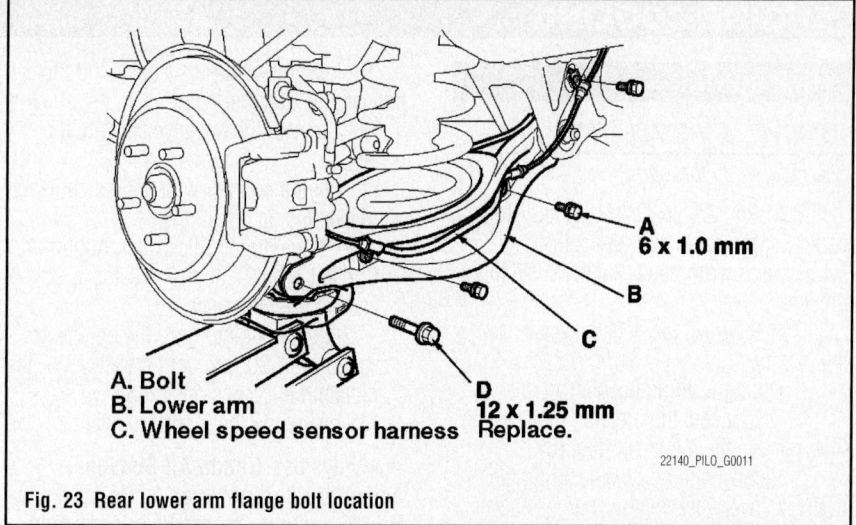

A. Bolt
B. Lower arm
C. Wheel speed sensor harness
A. 6 x 1.0 mm
B.
C.
D. 12 x 1.25 mm Replace.

22140_PILO_G0011

Fig. 23 Rear lower arm flange bolt location

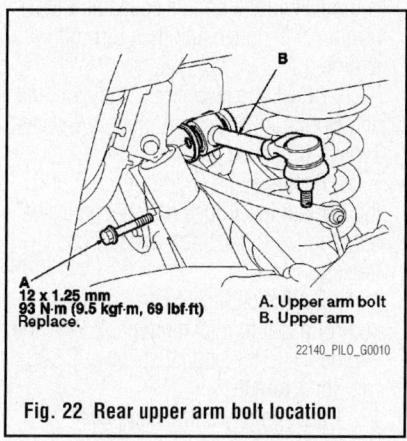

A. 12 x 1.25 mm
93 N·m (9.5 kgf·m, 69 lbf·ft)
Replace.

A. Upper arm bolt
B. Upper arm

22140_PILO_G0010

Fig. 22 Rear upper arm bolt location

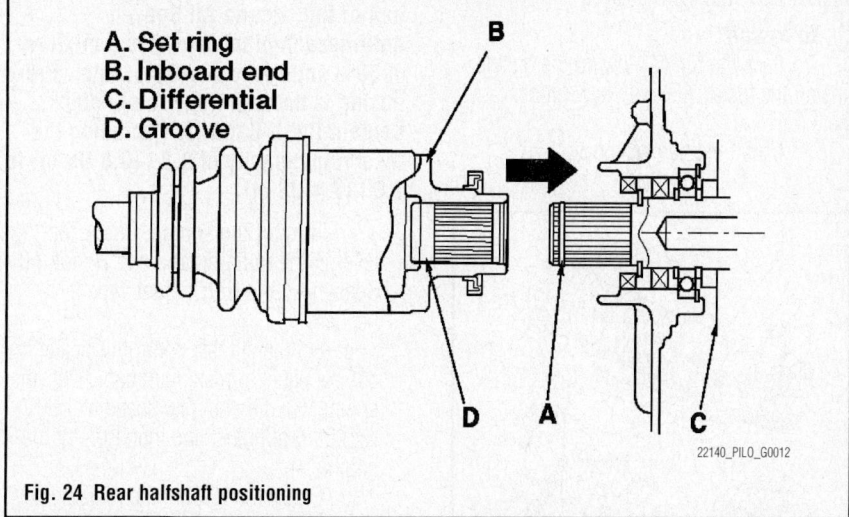

A. Set ring
B. Inboard end
C. Differential
D. Groove

22140_PILO_G0012

Fig. 24 Rear halfshaft positioning

7. Remove the rear shock absorber assembly.
8. Remove the lower arm flange bolt.
9. Remove the rear wheel speed sensor bracket.
10. Pull the knuckle outward and disconnect the rear halfshaft outboard joint from the rear wheel hub, using a plastic hammer.
11. Remove the rear halfshaft outboard joint.

To install:

➡ **Be sure that the mating surfaces of the joint and the splined section are free of dirt or dust.**

12. Apply 0.05—0.07 ounces of super high temperature grease, part number 08798-9002 or equivalent, to the whole splined surface of the assembly.
13. After application of the grease remove the grease from the splined groves at intervals of two or three splines and from the set ring groove so that air can bleed from the differential.
14. Install a new set ring in the set groove of the differential.

➡ **Clean the areas where the halfshaft contacts the differential thoroughly with solvent and dry with compressed air. Do not wash the rubber parts with solvent.**

15. Insert the inboard end of the halfshaft into the differential until the set ring locks in the groove. Install the driveshaft.
16. Continue the installation in the reverse order of the removal procedure.
17. Be sure to use new bolts and nuts, as required.
18. Tighten the new rear shock absorber bolt and nut to 47 ft. lbs.
19. Tighten the new spindle nut to 181 ft. lbs.
20. Refill the differential, using the proper grade and type fluid.
21. Check and adjust the rear wheel alignment, as required.

ENGINE COOLING

THERMOSTAT

REMOVAL & INSTALLATION

See Figures 25 through 27.

1. Make sure you have the anti-theft codes for the radio and navigation system, and write down the radio station presets.

2. Disconnect the negative cable from the battery.

3. Drain the engine coolant.

4. Disconnect the breather pipe (A), then remove the intake air duct (B).

5. Remove the bolt (A) securing the harness cover, then remove harness cover.

6. Remove the thermostat cover (A), then remove the thermostat (B).

To install:

7. Face the pin (C) toward the upside, install the thermostat with new rubber seal (D).

8. Install the harness cover.

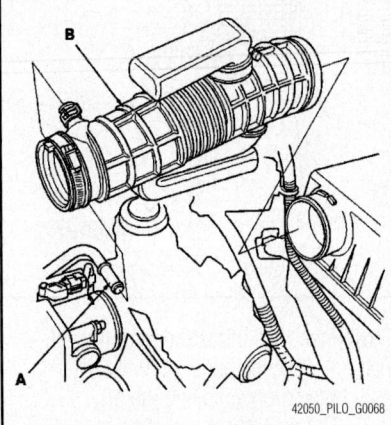

Fig. 25 Disconnect the breather pipe (A) and remove the intake air duct (B)

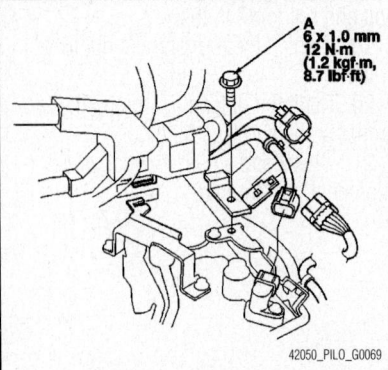

Fig. 26 Remove the bolt (A) securing the harness cover, then remove harness cover

9. Install the intake air duct, and then connect the breather pipe.

10. Connect the negative cable to the battery.

11. Fill the radiator with engine coolant, and bleed the air, as follows.

a. Pour Honda All Season Antifreeze/Coolant Type 2 into the radiator up to the base of the filler neck.

Start the engine. Hold the engine speed at 1,500 rpm until it warms up (the radiator fan comes on at least twice). Make sure the thermostat is open.

➡**Always use Honda All Season Antifreeze/Coolant Type 2. Using a non-Honda coolant can result in corrosion, causing the cooling system to malfunction or fail. Honda All Season Antifreeze/Coolant Type 2 is a mixture of 50% antifreeze and 50% water. Pre-mixing is not required. The Engine Coolant Refill Capacity (including the reservoir capacity of 0.8 l (0.8 US qt) is 7.0 l (7.4 US qt).**

b. Turn off the engine. Check the level in the radiator, and add Honda All Season Antifreeze/Coolant Type 2, if needed.

c. Set the climate control or heater control panel to maximum cool. Start the engine. Hold the engine speed at 1,500 rpm for 5 minutes, and then turn off the engine.

d. Check the level in the radiator, and add Honda All Season Antifreeze/Coolant Type 2, if needed.

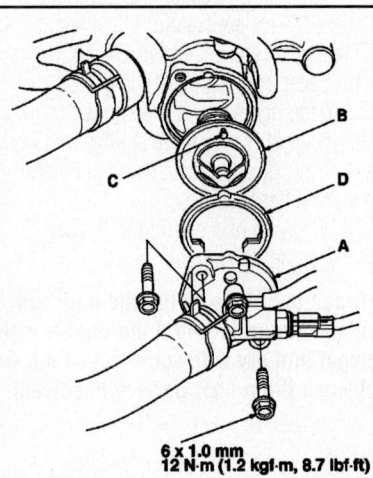

Fig. 27 View of the thermostat cover (A) and thermostat (B). When installing, make sure to face the pin (C) upward and use a new rubber seal (D)

e. Set the climate control or heater control panel to maximum heat. Start the engine. Hold the engine speed at 1,500 rpm for 5 minutes, and then turn off the engine.

f. Check the level in the radiator, and add Honda All Season Antifreeze/Coolant Type 2, if needed.

g. Set the climate control or heater control panel to maximum cool. Start the engine. Hold the engine speed at 1,500 rpm for 3 minutes, and then turn off the engine.

h. Check the level in the radiator, and add Honda All Season Antifreeze/Coolant Type 2, if needed.

i. Set the climate control or heater control panel to maximum heat. Start the engine. Hold the engine speed at 1,500 rpm for 3 minutes, and then turn off the engine.

j. Check the level in the radiator, and add Honda All Season Antifreeze/Coolant Type 2, if needed.

k. Repeat steps O. through R. until the coolant level does not change in the radiator, then install the radiator cap loosely.

l. Set the climate control or heater control panel to maximum cool. Start the engine. Hold the engine speed at 2,500 rpm for 1 minute.

12. Clean up any spilled engine coolant.

13. Perform the power window control unit reset procedure

a. Turn the ignition switch OFF, and then back ON (II).

b. Move the driver's window all the way down by holding the driver's switch firmly down to the second detent, when the window reaches the bottom, hold the driver's window switch in the AUTO DOWN position for 2 seconds.

c. Move the driver's window all the way up without stopping by holding the driver's switch firmly up to the second detent, when the window reaches the top, hold the driver's window switch in the AUTO UP position for 2 seconds.

d. If the window does not work in AUTO, reset the power window master switch according to the above procedures again.

14. Enter the anti-theft codes for the radio and the navigation system, and enter the radio station presets. Set the clock.

WATER PUMP

REMOVAL & INSTALLATION

See Figure 28.

1. Before servicing the vehicle, refer to the precautions section.
2. Drain the cooling system.
3. Remove or disconnect the following:
   - Negative battery cable
   - Accessory drive belts
   - Front cover
   - Timing belt
   - Timing belt tensioner
   - Water pump

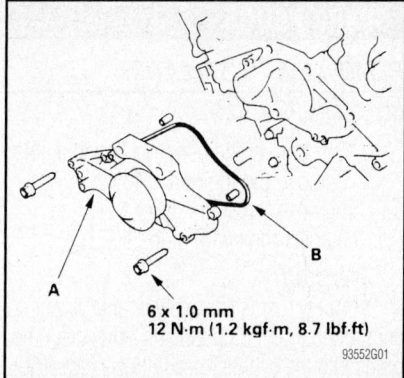

6 x 1.0 mm
12 N·m (1.2 kgf·m, 8.7 lbf·ft)

93552G01

Fig. 28 Exploded view of the water pump mounting

To install:

4. Install or connect the following:
   - Water pump. Use a new O-ring seal and tighten the bolts to 105 inch lbs. (12 Nm).
   - Timing belt tensioner
   - Timing belt
   - Front cover
   - Accessory drive belts
   - Negative battery cable
5. Fill the cooling system.
6. Start the engine and check for leaks.

ENGINE ELECTRICAL

ALTERNATOR

REMOVAL & INSTALLATION

See Figure 29.

1. Before servicing the vehicle, refer to the precautions section.
2. Remove or disconnect the following:
   - Negative battery cable
   - Accessory drive belt
   - Intake manifold cover
   - Ignition coil covers (J35A4 engine only)
   - Alternator wiring harness connectors
   - Alternator mounting bolts
   - Wiring harness clamp
   - Alternator

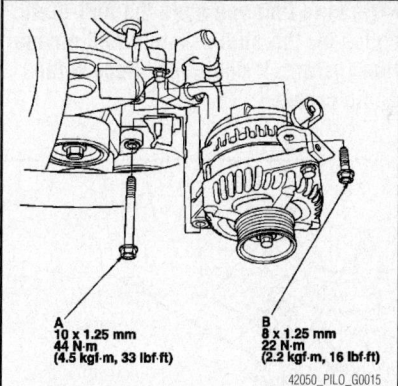

A
10 x 1.25 mm
44 N·m
(4.5 kgf·m, 33 lbf·ft)

B
8 x 1.25 mm
22 N·m
(2.2 kgf·m, 16 lbf·ft)

42050_PILO_G0015

Fig. 29 Exploded view of the alternator, mounting bolt (A) and bracket mounting bolt (B)

CHARGING SYSTEM

To install:

3. Install or connect the following:
   - Alternator
   - Wiring harness clamp. Tighten the bolt to 105 inch lbs. (12 Nm).
   - Alternator mounting bolts. Tighten the bolts to the specifications shown in the accompanying figures.
   - Alternator wiring harness connectors. Tighten the battery terminal nut to 105 inch lbs. (12 Nm).
   - Accessory drive belt
   - Intake manifold and ignition coil covers, as applicable
   - Negative battery cable

ENGINE ELECTRICAL

FIRING ORDER

See Figure 30.

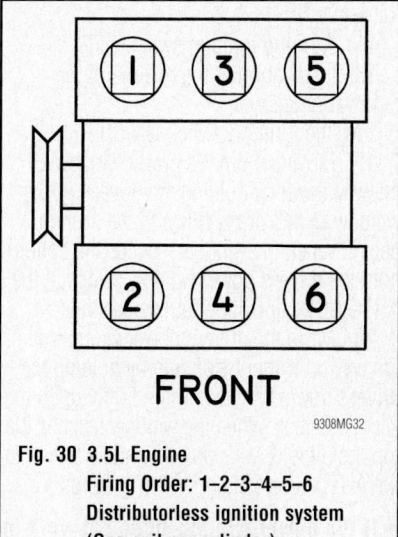

FRONT

9308MG32

**Fig. 30 3.5L Engine
Firing Order: 1–2–3–4–5–6
Distributorless ignition system
(One coil per cylinder)**

IGNITION COIL

REMOVAL & INSTALLATION

See Figures 31 and 32.

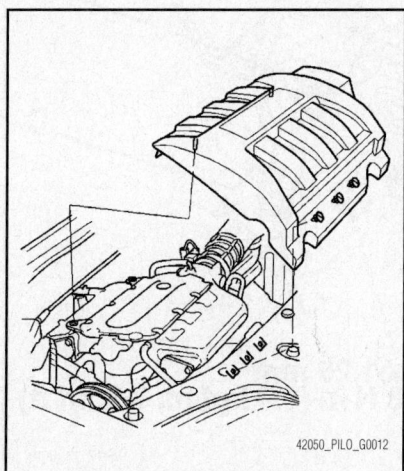

42050_PILO_G0012

Fig. 31 Intake manifold cover

IGNITION SYSTEM

1. Remove the intake manifold cover and ignition coil cover, if equipped.
2. Detach the connectors, if necessary, and remove the ignition coils.

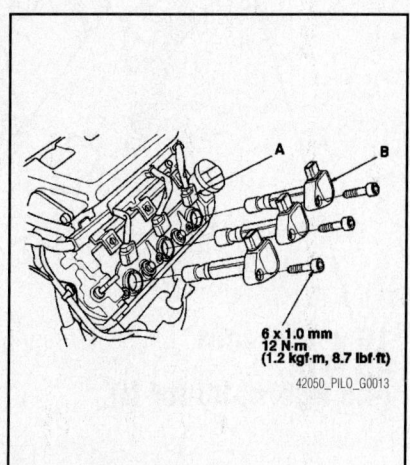

6 x 1.0 mm
12 N·m
(1.2 kgf·m, 8.7 lbf·ft)

42050_PILO_G0013

Fig. 32 View of the ignition coil connectors (A) and ignition coils (B)

To install:

3. Install the ignition coils in the reverse order of removal. Tighten the retainers to the specifications shown in the accompanying illustrations.

IGNITION TIMING

ADJUSTMENT

This vehicle is equipped with a Distributorless Ignition System (DIS). The ignition timing is controlled by the Powertrain Control module (PCM). No adjustment is necessary.

SPARK PLUGS

REMOVAL & INSTALLATION

See Figure 33.

1. Disconnect the negative battery cable.
2. Remove the ignition coil(s).
3. Remove the spark plug.
4. Inspect the spark plug.

To install:

5. Apply a small quantity of anti-seize compound to the plug threads, and screw the plugs into the cylinder head finger-tight. Then tighten the spark plugs to 13 ft. lbs. (18 Nm).
6. Install the ignition coil(s).

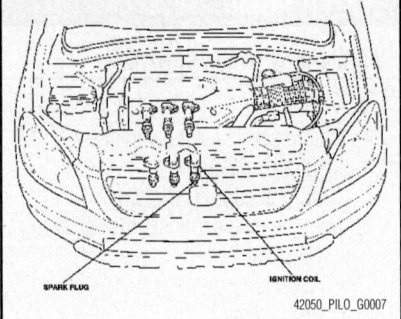

42050_PILO_G0007

Fig. 33 The spark plugs are located under the ignition coils

ENGINE ELECTRICAL

STARTER

REMOVAL & INSTALLATION

See Figure 34.

➡Be sure that you have the anti-theft codes for the audio system and navigation system, if equipped. Record the audio presets.

STARTING SYSTEM

1. Before servicing the vehicle, refer to the Precautions Section.
2. Turn the ignition switch OFF.
3. Disconnect and isolate the negative battery cable. Wait 3 minutes for the system capacitor to discharge before performing any service.
4. Disconnect the positive battery cable.
5. Remove the battery.
6. Remove the transaxle fluid dipstick.
7. Remove the harness clamp.
8. Disconnect the starter cable from the terminal (B) and the BLK wire from the solenoid terminal (S).
9. Remove the starter retaining bolts.
10. Remove the starter from the vehicle.

To install:

11. Installation is the reverse of the removal procedure.
12. Be sure to use a new gasket.
13. Start the engine to be sure that the starter performs properly.
14. Reset the power window control, as follows.
15. Turn the ignition switch OFF.
16. Installation is the reverse of the removal procedure.
Turn the ignition switch ON (II).
17. Move the driver's power window all the way down by holding the driver's power window switch firmly down to the second detent. When the window reaches the bottom, hold the driver's power window switch in the AUTO DOWN position for two seconds.
18. Move the driver's power window all the way up without stopping by holding the driver's power window switch firmly up to the second detent. When the window reaches the top, hold the driver's power window switch in the AUTO UP position for two seconds.

➡If the power window does not work in AUTO, repeat the procedure.

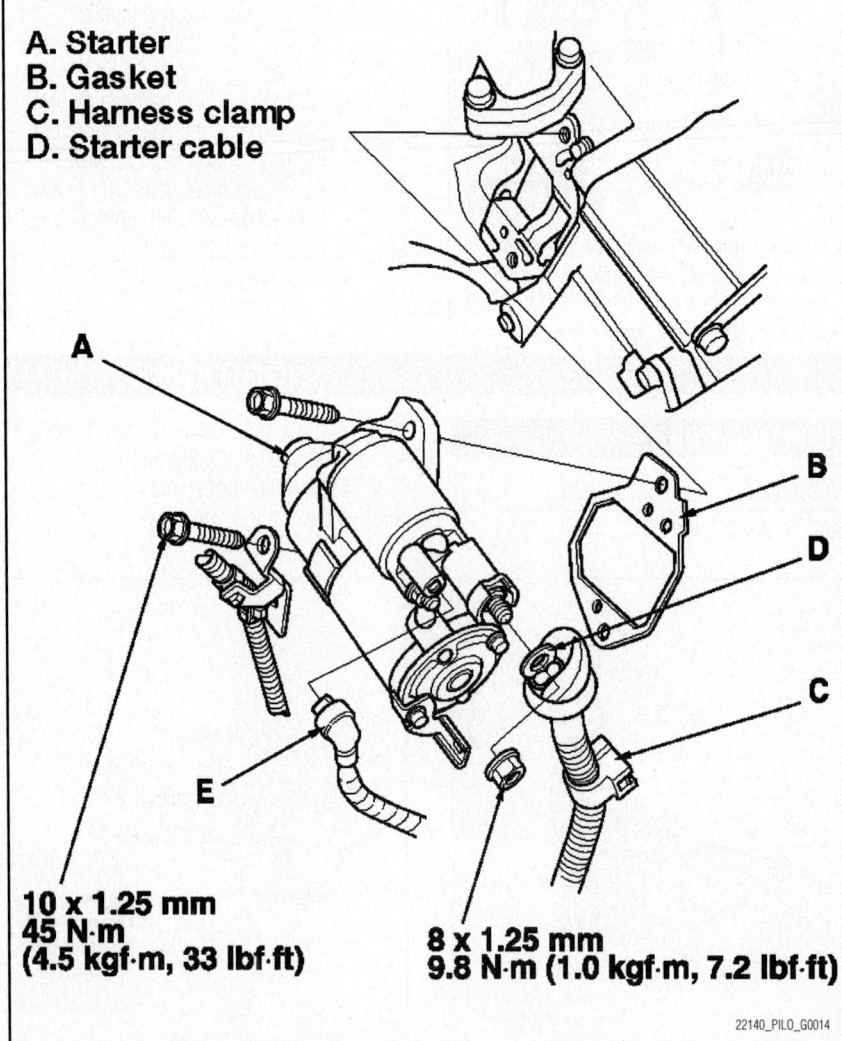

A. Starter
B. Gasket
C. Harness clamp
D. Starter cable

**10 x 1.25 mm
45 N·m
(4.5 kgf·m, 33 lbf·ft)**

**8 x 1.25 mm
9.8 N·m (1.0 kgf·m, 7.2 lbf·ft)**

22140_PILO_G0014

Fig. 34 Starter and related components

ENGINE MECHANICAL

➡Disconnecting the negative battery cable may interfere with the functions of the on board computer systems and may require the computer to undergo a relearning process, once the negative battery cable is reconnected.

➡The power window reset procedure must be performed when disconnecting the battery, removing the No. 1 (20amp) fuse in the passenger's under-dash fuse/relay box, disconnecting the 18P connector from the power window control unit, removing the window regulator, glass or glass run channel and

disconnecting the driver's door wire harness.

1. Turn the ignition switch OFF.
2. Installation is the reverse of the removal procedure.
Turn the ignition switch ON (II).
3. Move the driver's power window all the way down by holding the driver's power window switch firmly down to the second detent. When the window reaches the bottom, hold the driver's power window switch in the AUTO DOWN position for two seconds.
4. Move the driver's power window all

the way up without stopping by holding the driver's power window switch firmly up to the second detent. When the window reaches the top, hold the driver's power window switch in the AUTO UP position for two seconds.

➡If the power window does not work in AUTO, repeat the procedure.

ACCESSORY DRIVE BELTS

ACCESSORY BELT ROUTING

See Figure 35.

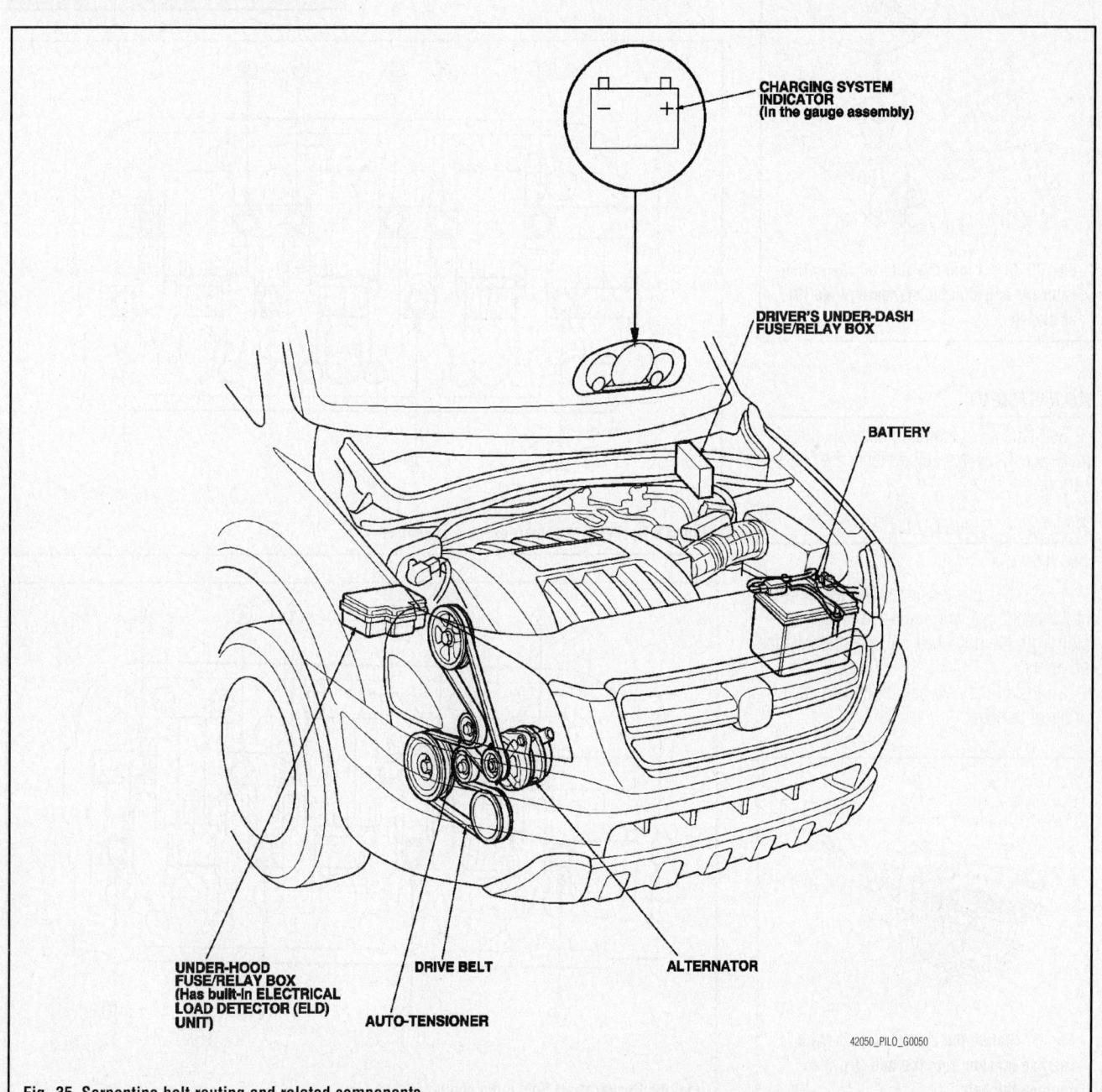

CHARGING SYSTEM
INDICATOR
(in the gauge assembly)

DRIVER'S UNDER-DASH
FUSE/RELAY BOX

BATTERY

UNDER-HOOD
FUSE/RELAY BOX
(Has built-in ELECTRICAL
LOAD DETECTOR (ELD)
UNIT)

AUTO-TENSIONER

DRIVE BELT

ALTERNATOR

42050_PILO_G0050

Fig. 35 Serpentine belt routing and related components

INSPECTION

See Figure 36.

1. Inspect the belt for cracks and damaged. If the belt is cracked or damaged, replace it.
2. Check that the auto-tensioner indicator (A) is within the standard range (B) as shown. If it is out of the standard range, replace the drive belt.

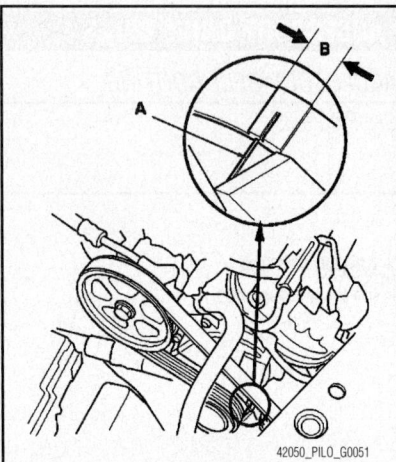

Fig. 36 Check that the auto-tensioner indicator (A) is within the standard range (B) as shown

ADJUSTMENT

Belt tension is maintained by an automatic tensioner. No adjustments are necessary or possible.

REMOVAL & INSTALLATION

See Figure 37.

1. Rotate auto-tensioner (A), as shown in the accompanying figure, to release tension from the drive belt (B), then remove the drive belt.
2. Install the new belt in the reverse order of removal.

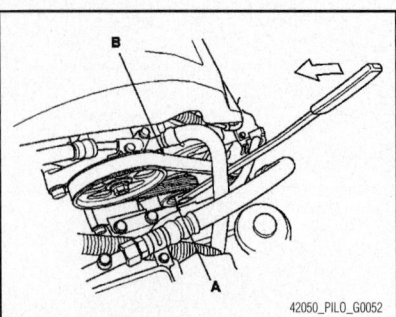

Fig. 37 Rotate the auto-tensioner (A) to release tension from the belt (B), then remove the belt

CAMSHAFT AND VALVE LIFTERS

REMOVAL & INSTALLATION

Front

See Figures 38 through 40.

1. Before servicing the vehicle, refer to the precautions section.
2. Remove or disconnect the following:
   - Negative and positive battery cables
   - Battery
3. Drain the coolant.
   - Exhaust Gas Recirculation (EGR) valve

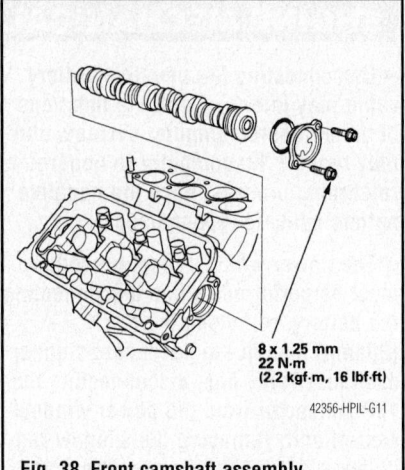

8 x 1.25 mm
22 N·m
(2.2 kgf·m, 16 lbf·ft)

42356-HPIL-G11

Fig. 38 Front camshaft assembly

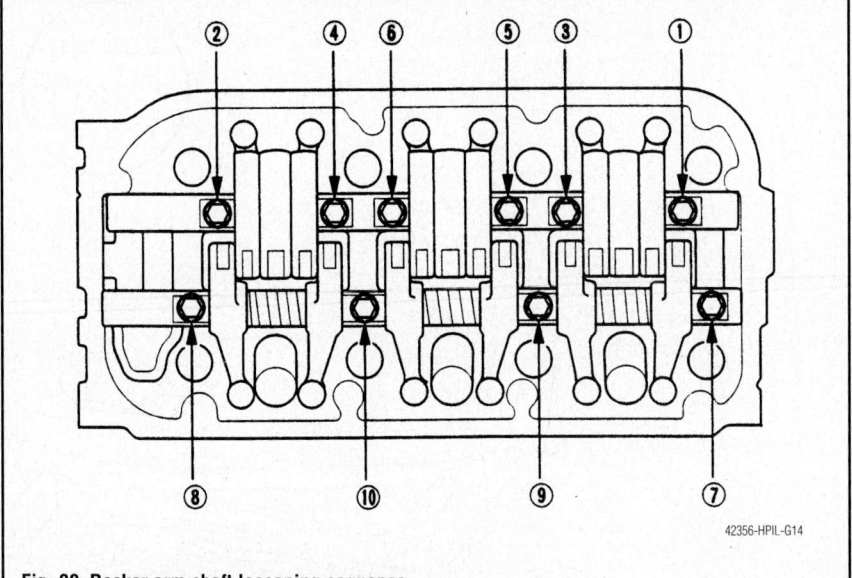

42356-HPIL-G14

Fig. 39 Rocker arm shaft loosening sequence

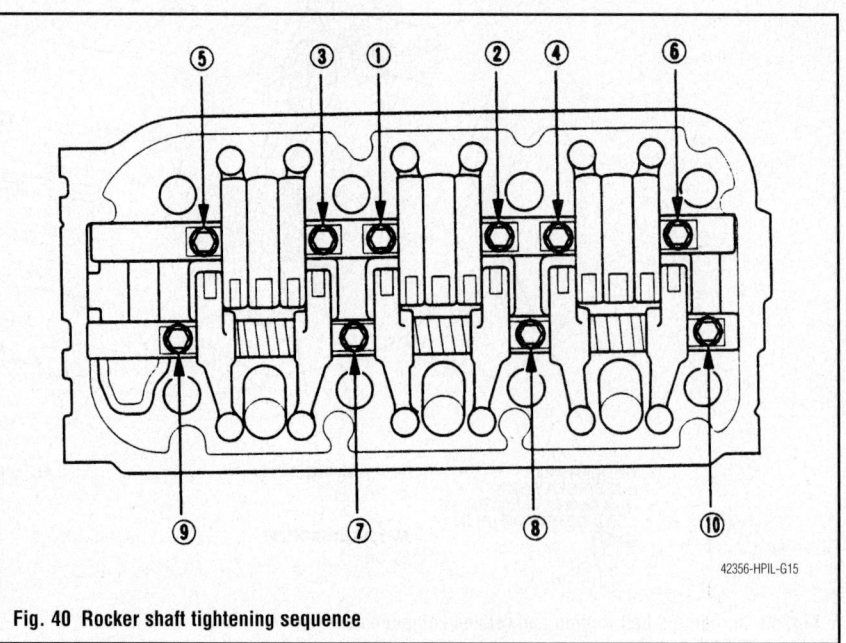

42356-HPIL-G15

Fig. 40 Rocker shaft tightening sequence

- Timing belt
- Rocker arm assembly
- Front camshaft pulley
- Thrust plate and camshaft

To install:

4. Install or connect the following:
- Camshaft using a new O-ring. Tighten the thrust plate to 16 ft. lbs. (22 Nm).
- Front camshaft pulley
- Rocker arm assembly
- Timing belt
- Exhaust Gas Recirculation (EGR) valve
- Battery
- Positive, then negative battery cables

5. Fill the cooling system.
- Camshaft

6. Start the engine and check for leaks.

Rear

See Figure 41.

1. Before servicing the vehicle, refer to the precautions section.
2. Drain the cooling system.
3. Relieve the fuel system pressure.
4. Remove or disconnect the following:
- Negative battery cable
- Under-hood fuse box
- Fuel feed hose
- Nuts securing the fuel line
- Brake lines from the master cylinder
- Timing belt
- Rocker arm assembly
- Rear camshaft pulley
- Thrust plate and camshaft

To install:

5. Install or connect the following:
- Camshaft using a new O-ring.

Tighten the thrust plate to 16 ft. lbs. (22 Nm).
- Rear camshaft pulley
- Rocker arm assembly
- Timing belt
- Brake lines to the master cylinder
- Nuts securing the fuel line
- Fuel feed hose
- Under-hood fuse box
- Negative battery cable

CRANKSHAFT FRONT SEAL

REMOVAL & INSTALLATION

See Figure 42.

1. Before servicing the vehicle, refer to the precautions section.
2. Remove or disconnect the following:
- Negative battery cable
- Accessory drive belts
- Side engine mount
- Valve cover
- Crankshaft pulley
- Front cover
- Balance shaft belt, if equipped
- Timing belt
- Top Dead Center (TDC) sensor, if equipped
- Crankshaft timing sprocket
- Front crankshaft seal

To install:

3. Lubricate the crankshaft seal lip with grease prior to installation.
4. Install the front crankshaft seal so that it is flush with the surface of the oil pump housing.
5. Install or connect the following:
- Crankshaft timing sprocket
- Top Dead Center (TDC) sensor, if equipped
- Timing belt
- Balance shaft belt, if equipped

- Front cover
- Crankshaft pulley. Tighten the bolt to 47 ft. lbs. (64 Nm) plus 60 degrees
- Valve cover
- Side engine mount
- Accessory drive belts
- Negative battery cable

6. Check the engine oil level and add if necessary.

7. Start the engine and check for leaks.

CYLINDER HEAD

REMOVAL & INSTALLATION

See Figures 43 through 46.

1. Before servicing the vehicle, refer to the precautions section.
2. Drain the cooling system.
3. Relieve the fuel system pressure.
4. Remove or disconnect the following:
- Negative battery cable
- Drive belts
- Power steering and pump
- Power steering hose clamp

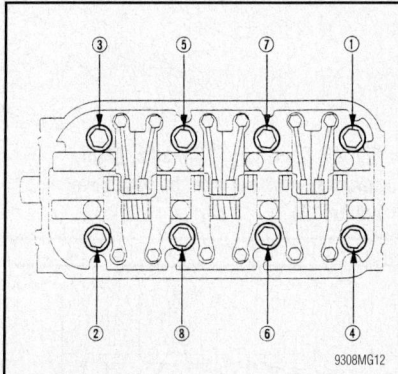

9308MG12

Fig. 43 Cylinder head bolt loosening sequence

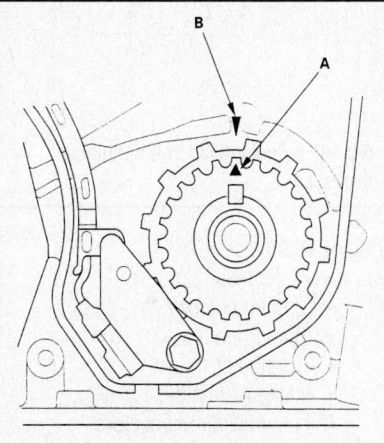

9302MG74

Fig. 44 Crankshaft timing belt sprocket TDC marks. Align sprocket mark (A) with pointer (B)

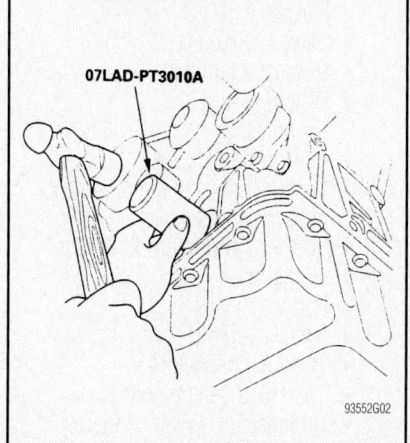

6 x 1.0 mm
12 N·m (1.2 kgf·m, 8.7 lbf·ft)

42356-HPIL-G12

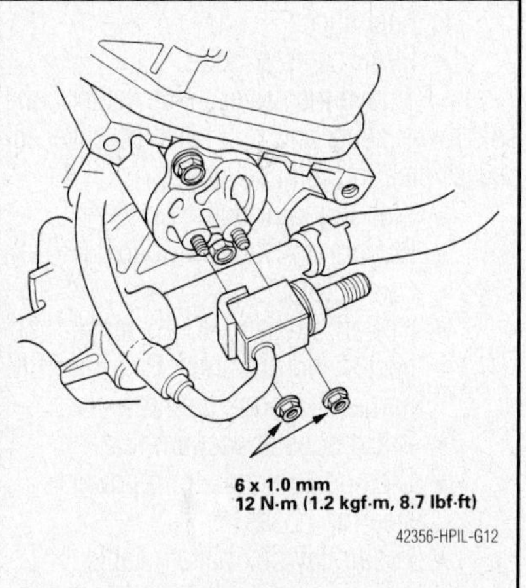

Fig. 41 Remove the nuts attaching the fuel line when removing the rear camshaft

07LAD-PT3010A

93552G02

Fig. 42 Front crankshaft seal installation

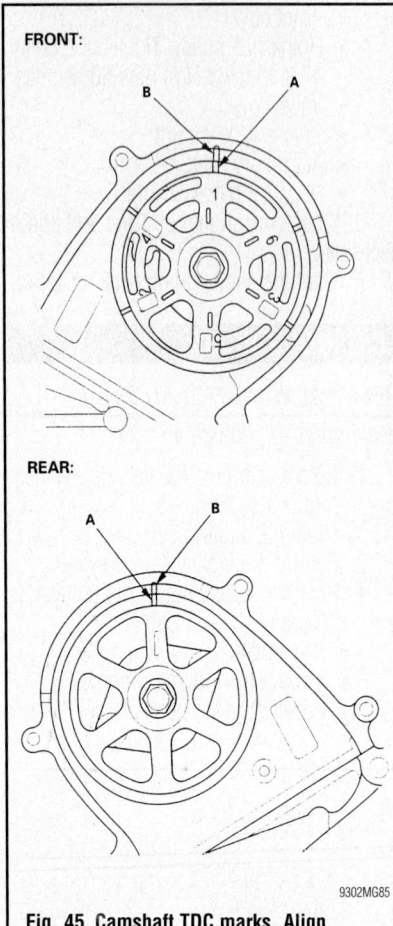

Fig. 45 Camshaft TDC marks. Align sprocket mark (A) with the back cover pointer (B)

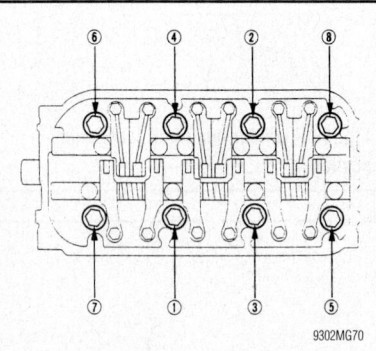

Fig. 46 Cylinder head bolt tightening sequence

- Alternator
- Intake manifold cover
- Ignition coil covers
- Ignition coils
- Timing belt
- Intake manifold
- Fuel injector connectors
- Engine Coolant Temperature (ECT) sensor connector
- Crankshaft Position (CKP) sensor connector

- Camshaft Position (CMP) sensor connector
- Exhaust Gas Recirculation (EGR) connector
- Radiator fan switch connectors
- Valve Lift Electronic Control (VTEC) solenoid valve connector and oil pressure switch connections
- Oil pressure switch connector
- Two Air/Fuel (A/F) connectors
- Two Heated Oxygen (HO$_2$S) sensor connectors
- Radiator hoses
- Heater hose
- Water bypass hose
- Harness bracket
- EVAP canister hose
- Fuel feed and return lines
- Fuel rails
- Exhaust manifolds
- Water passage
- Vacuum hoses from the intake air bypass control valve
- Ground cable
- Camshaft pulleys and back covers
- Valve covers

5. Loosen the cylinder head bolts in sequence and ⅓ turns until all bolts are loose.

6. Remove the cylinder head.

To install:

7. Align the crankshaft and camshaft sprocket TDC marks as shown.

8. Install the cylinder heads with new gaskets.

9. Apply clean engine oil to the cylinder head bolt threads and flanges.

10. Tighten the cylinder head bolts in sequence as follows:

   a. Step 1: 29 ft. lbs. (39 Nm).
   b. Step 2: 51 ft. lbs. (69 Nm).
   c. Step 3: 72 ft. lbs. (98 Nm).

11. Install or connect the following:

- Timing belt and adjust the valve clearance
- Valve covers
- Exhaust manifolds
- Water passage
- Fuel rails
- EVAP canister hose
- Vacuum hoses to the intake air bypass control valve
- Fuel feed and return lines
- Heater hose
- Radiator hoses
- Intake manifold
- Two A/F connectors
- Two HO$_2$S sensor connectors
- Oil pressure switch connector
- VTEC solenoid valve connector and oil pressure switch connections

- EGR connector
- CMP sensor connector
- CKP sensor connector
- Radiator fan switch connectors
- ECT sensor connector
- Fuel injector connectors
- Ignition coils
- Power steering pump and belt
- Power steering hose clamp
- Alternator
- Ground cable
- Ignition coil covers
- Intake manifold cover
- Alternator belt
- Negative battery cable

12. Fill the cooling system.

13. Start the engine and check for leaks.

ENGINE ASSEMBLY

REMOVAL & INSTALLATION

See Figure 47.

➡**The engine and transaxle are removed from the vehicle as a unit.**

1. Before servicing the vehicle, refer to the precautions section.

2. Drain the cooling system.

3. Drain the power steering system.

4. Drain the transaxle fluid.

5. Drain the engine oil.

6. Relieve fuel system pressure.

7. Remove or disconnect the following:

- Negative battery cable
- Battery
- Intake and ignition coil covers
- Air intake duct
- Left engine wire harness connectors
- Relay bracket
- Battery and tray
- Starter cable and harness clamp
- Accelerator cable
- Cruise control cable
- Fuel lines
- EVAP canister hose

8. Remove the driver's side center console lower panel and pull back the cover to access steering joint cover.

- Steering joint bolt
- Powertrain Control Module (PCM) connectors
- Heated Oxygen (HO$_2$S) sensor connector and grommet. Pull the PCM harness through the firewall.
- Brake booster vacuum line
- Clamps and clips from power steering hoses
- Fuse/relay box battery cable
- Accessory drive belts
- Front wheels

- Splash shield
- Front sub-frame stiffener
- Exhaust front pipe
- Propeller shaft
- Shift control cable
- Transfer assembly
- Ball joints
- Stabilizer bar links
- Halfshafts
- Power steering hose and pressure switch connector
- Transaxle lower front mount
- Transaxle lower rear mount
- A/C compressor
- Heater hoses
- Radiator hoses
- Ground cable
- Transaxle oil cooler lines
- Radiator

9. Attach a hoist to the engine lifting eyes and support the powertrain weight.
10. Remove or disconnect the following:
- Side engine mount bracket
- Front mount bracket support nut

11. Matchmark the front subframe to the mounting points.
12. Remove or disconnect the following:
- Front subframe
- All remaining hoses and electrical connections

13. Lower the powertrain away from the vehicle.

To install:
14. Raise the powertrain into position.
15. Installation is the reverse of removal but please note the following steps:
- A/C compressor bolts to 16 ft. lbs. (22 Nm)

- Front subframe. Use new bolts and tighten the 14mm bolts to 76 ft. lbs. (103 Nm). Tighten the front brace bolts to 54 ft. lbs. (74 Nm) and the rear brace bolts to 86 ft. lbs. (117 Nm).
- Transaxle lower front mount nuts to 28 ft. lbs. (38 Nm)
- Transaxle lower rear mount bolts to 28 ft. lbs. (38 Nm)
- Front mount bracket support nut to 40 ft. lbs. (54 Nm)
- Side engine mount bracket bolts to 33 ft. lbs. (44 Nm) and the through bolt to 40 ft. lbs. (54 Nm)

16. Fill the engine crankcase to the correct level.
17. Fill the transaxle to the correct level.
18. Fill the cooling system.
19. Fill the power steering system.
20. Start the engine and check for leaks.
21. Check the wheel alignment and adjust as necessary.

EXHAUST MANIFOLD

REMOVAL & INSTALLATION

1. Before servicing the vehicle, refer to the precautions section.
2. Remove or disconnect the following:
- Negative battery cable
- Exhaust manifold heat shield
- Heated Oxygen (HO$_2$S) sensor connector
- Exhaust front pipe
- Exhaust manifold bracket, if equipped
- Exhaust manifold

To install:
3. Install or connect the following:
- Exhaust manifold. Tighten the fasteners to 23 ft. lbs. (31 Nm).
- Exhaust manifold bracket, if equipped. Tighten the bolts to 33 ft. lbs. (44 Nm).
- Exhaust front pipe. Tighten the nuts to 40 ft. lbs. (55 Nm).
- Heated Oxygen (HO$_2$S) sensor connector
- Exhaust manifold heat shield
- Negative battery cable

INTAKE MANIFOLD

REMOVAL & INSTALLATION
See Figures 48 through 52.

1. Before servicing the vehicle, refer to the precautions section.
2. Remove or disconnect the following:
- Negative battery cable

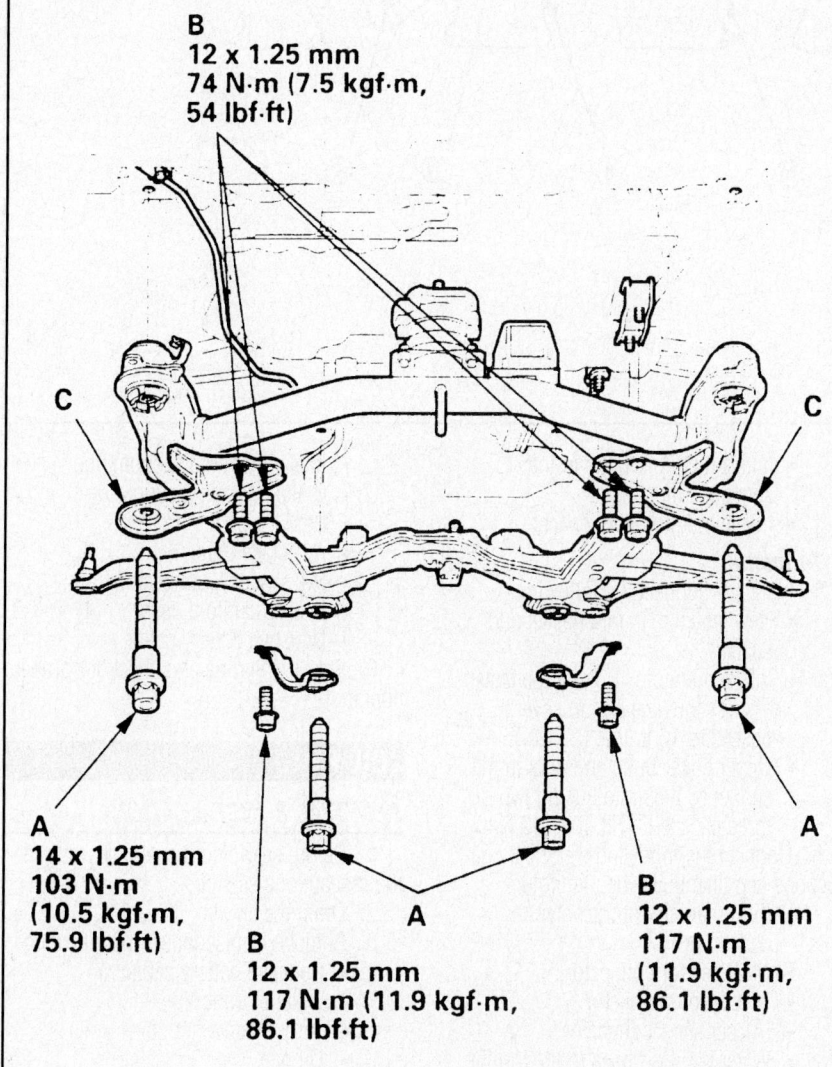

B
12 x 1.25 mm
74 N·m (7.5 kgf·m, 54 lbf·ft)

C

C

A
14 x 1.25 mm
103 N·m (10.5 kgf·m, 75.9 lbf·ft)

A

B
12 x 1.25 mm
117 N·m (11.9 kgf·m, 86.1 lbf·ft)

B
12 x 1.25 mm
117 N·m (11.9 kgf·m, 86.1 lbf·ft)

A

9302MG69

Fig. 47 Sub-frame fastener locations and tightening torque

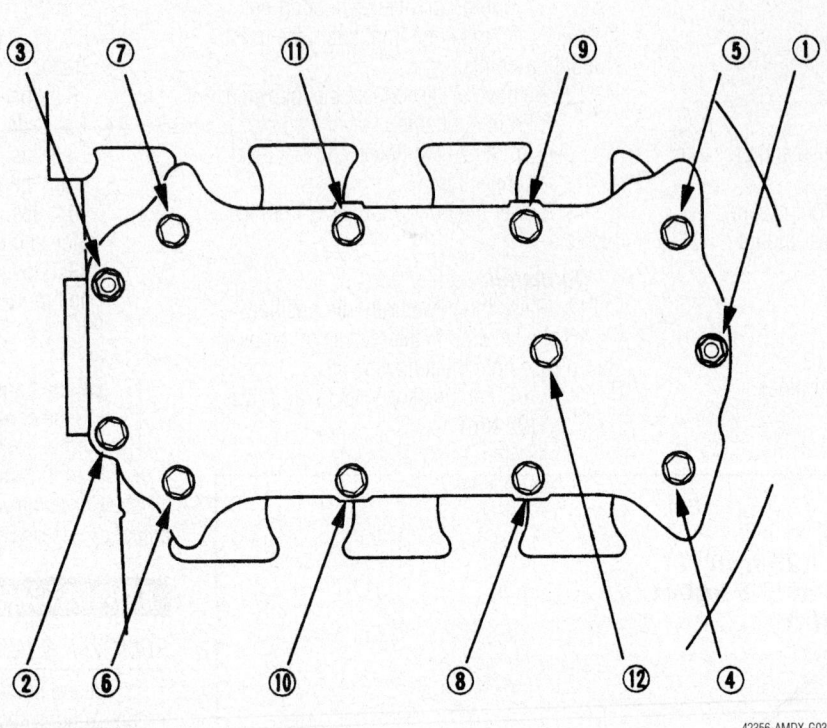

Fig. 48 Upper cover loosening sequence

42356-AMDX-G03

- Intake manifold cover
- Intake Air Temperature (IAT) sensor 2 connector
- Air intake tube
- Positive Crankcase Ventilation (PCV) hose
- Brake booster vacuum line
- Evaporative Emissions (EVAP) control canister hose and transmission breather hose clamp bracket
- Water bypass hoses from the throttle body and plug the hoses

3. Remove the following electrical connections and clamps from the manifold:
- Intake Air Temperature (IAT) sensor connector
- Throttle Position (TP) sensor connector
- Manifold Absolute Pressure (MAP) sensor connector
- EVAP canister purge valve connector
- Upper cover bolts and nuts in the sequence illustrated using two or three passes

- Intake manifold bolts in the sequence illustrated
- Intake manifold and spacer

To install:
4. Install or connect the following:
- New intake manifold gasket and spacer
- Intake manifold. Tighten the fasteners in sequence and in several passes to 16 ft. lbs. (22 Nm).
- Upper cover bolts and nuts in the sequence illustrated using two or three passes to 9 ft. lbs. (12 Nm)

5. Connect the following electrical connections and clamps to the manifold:
- EVAP canister purge valve connector
- MAP sensor connector
- TP sensor connector
- IAT sensor 1 connector
- Water bypass hoses to the throttle body
- EVAP control canister hose and clamp bracket

- Intake manifold vacuum line
- Brake booster vacuum line
- PCV hose
- Air intake tube
- IAT sensor 2 connector
- Intake manifold cover
- Negative battery cable

6. Start the engine and check for proper operation.

OIL PAN

REMOVAL & INSTALLATION

1. Before servicing the vehicle, refer to the precautions section.
2. Drain the engine oil.
3. Remove or disconnect the following:
- Negative battery cable
- Exhaust front pipe
- Torque converter cover
- Oil pan

To install:
4. Install the oil pan. Apply liquid gasket as shown.

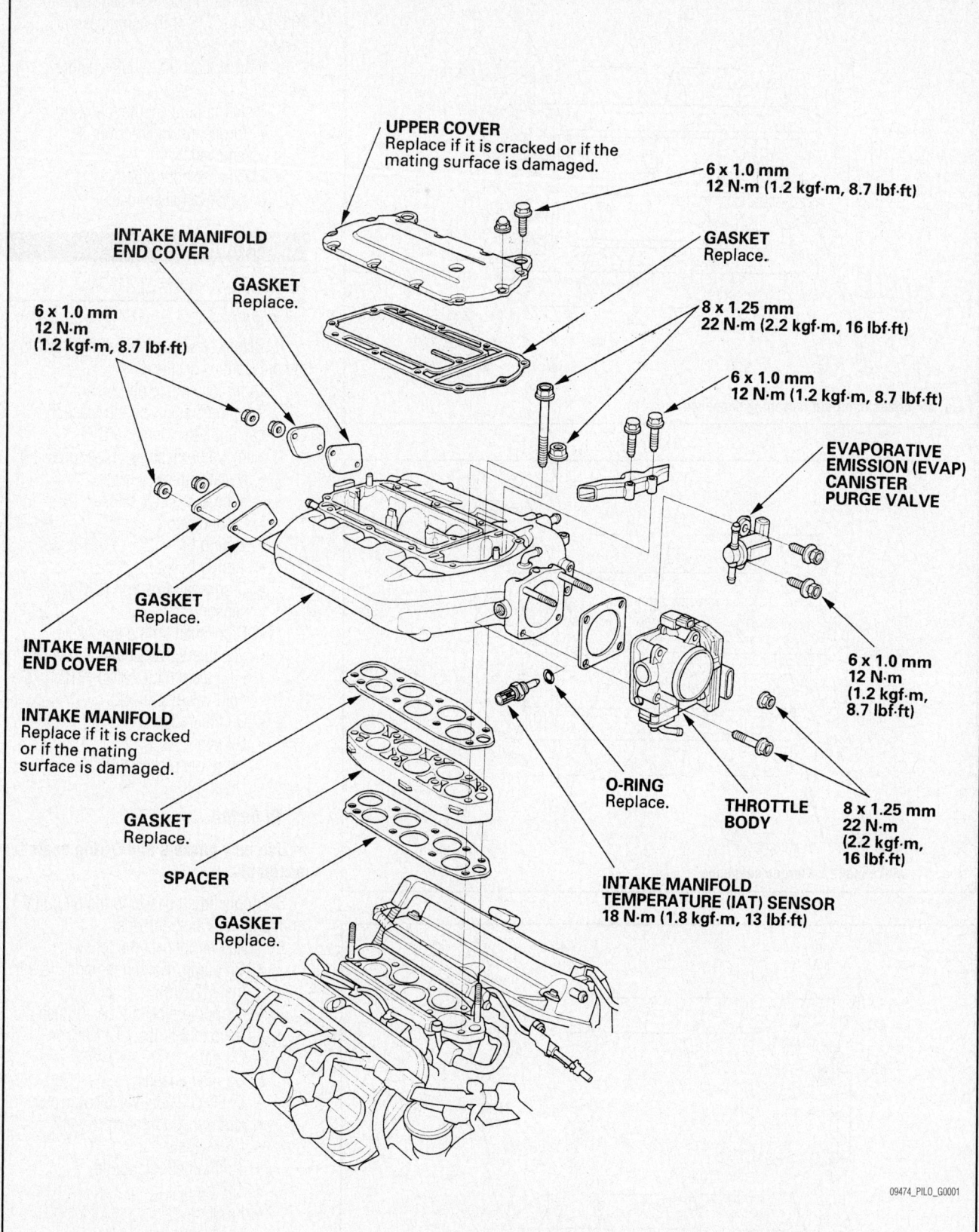

UPPER COVER
Replace if it is cracked or if the mating surface is damaged.

6 x 1.0 mm
12 N·m (1.2 kgf·m, 8.7 lbf·ft)

GASKET
Replace.

INTAKE MANIFOLD END COVER

GASKET
Replace.

8 x 1.25 mm
22 N·m (2.2 kgf·m, 16 lbf·ft)

6 x 1.0 mm
12 N·m
(1.2 kgf·m, 8.7 lbf·ft)

6 x 1.0 mm
12 N·m (1.2 kgf·m, 8.7 lbf·ft)

EVAPORATIVE EMISSION (EVAP) CANISTER PURGE VALVE

GASKET
Replace.

INTAKE MANIFOLD END COVER

INTAKE MANIFOLD
Replace if it is cracked or if the mating surface is damaged.

GASKET
Replace.

SPACER

GASKET
Replace.

O-RING
Replace.

THROTTLE BODY

6 x 1.0 mm
12 N·m
(1.2 kgf·m,
8.7 lbf·ft)

8 x 1.25 mm
22 N·m
(2.2 kgf·m,
16 lbf·ft)

INTAKE MANIFOLD TEMPERATURE (IAT) SENSOR
18 N·m (1.8 kgf·m, 13 lbf·ft)

09474_PILO_G0001

Fig. 40 Exploded view of the intake manifold

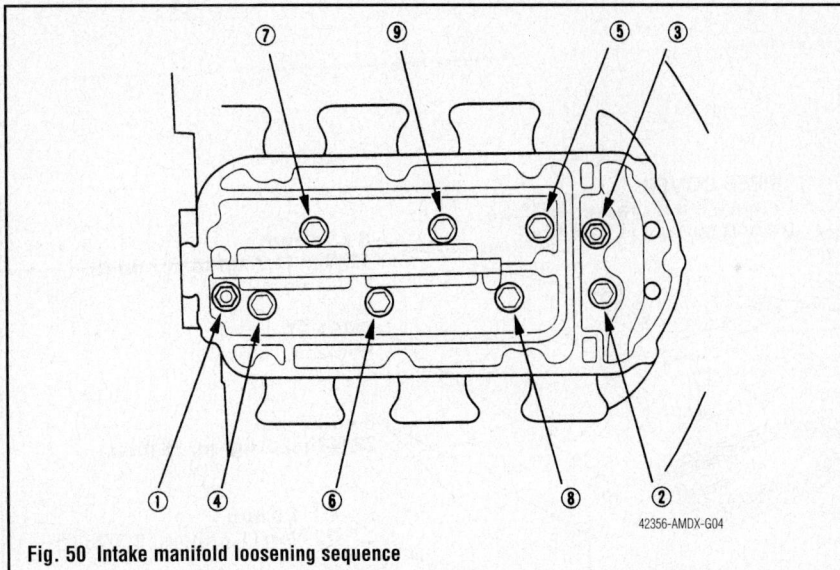

Fig. 50 Intake manifold loosening sequence

42356-AMDX-G04

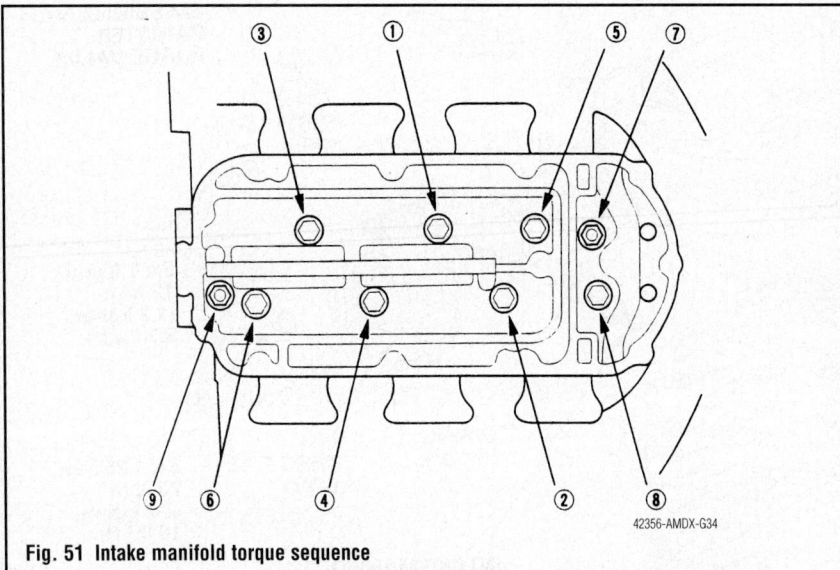

Fig. 51 Intake manifold torque sequence

42356-AMDX-G34

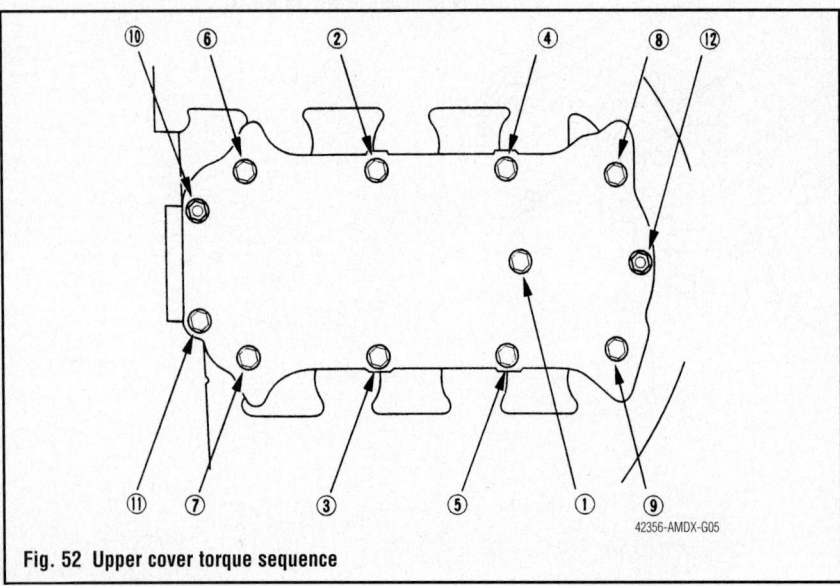

Fig. 52 Upper cover torque sequence

42356-AMDX-G05

5. Tighten the bolts in sequence to 105 inch lbs. (12 Nm) using several passes.

6. Wait at least 30 minutes before adding oil to the engine.

7. Install or connect the following:
- Torque converter cover, if removed
- Exhaust front pipe
- Negative battery cable

OIL PUMP

REMOVAL & INSTALLATION

See Figure 53.

1. Before servicing the vehicle, refer to the precautions section.

2. Drain the engine oil.

3. Turn the crankshaft and place the engine at Top Dead Center (TDC).

4. Remove or disconnect the following:
- Negative battery cable
- Accessory drive belts
- Front cover
- Timing belt
- Timing belt idler pulley
- Crankshaft Position (CKP) sensor
- Crankshaft timing sprocket
- Variable Valve Timing and Valve Lift Electronic Control (VTEC) solenoid valve connector
- Oil filter adapter
- Oil pan
- Oil pump pickup tube
- Oil pump

To install:

➡**Use new gaskets and O-ring seals for assembly.**

5. Apply liquid gasket to the oil pump and to the bolt hole threads.

6. Install or connect the following:
- Oil pump. Tighten the bolts to 9 ft. lbs. (12 Nm).
- Oil pump pickup tube. Tighten the bolts to 9 ft. lbs. (12 Nm).
- Oil pan
- Oil filter adapter
- VTEC solenoid valve connector
- Crankshaft timing sprocket
- CKP sensor
- Timing belt idler pulley
- Timing belt
- Front cover
- Accessory drive belts
- Negative battery cable

7. Fill the crankcase to the correct level.

8. Start the engine and check for leaks.

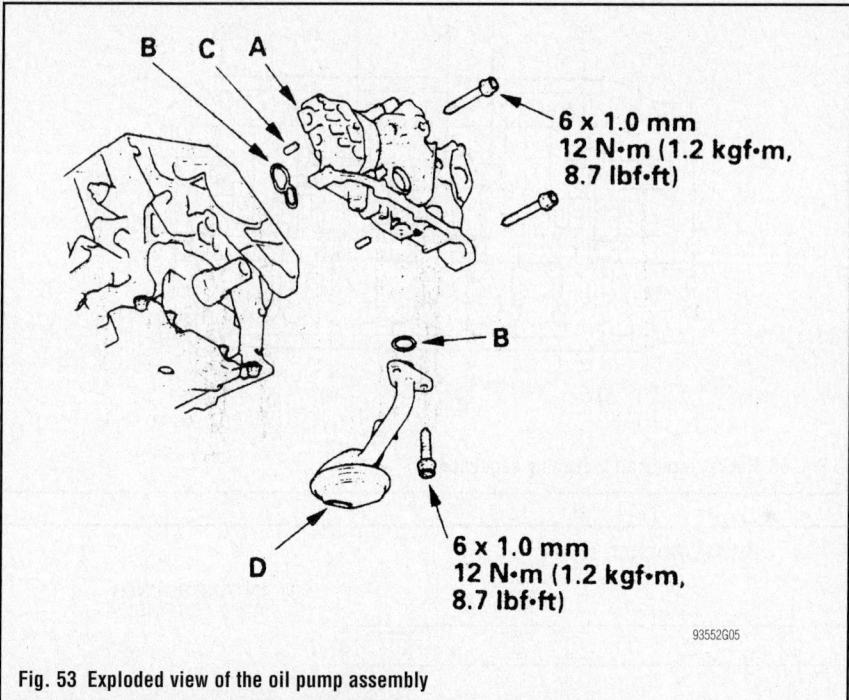

Fig. 53 Exploded view of the oil pump assembly

6 x 1.0 mm
12 N•m (1.2 kgf•m, 8.7 lbf•ft)

6 x 1.0 mm
12 N•m (1.2 kgf•m, 8.7 lbf•ft)

PISTON AND RING

POSITIONING

See Figures 54 and 55.

REAR MAIN SEAL

REMOVAL & INSTALLATION

1. Before servicing the vehicle, refer to the precautions section.
2. Remove or disconnect the following:
   - Transaxle
   - Flywheel
   - Oil seal

To install:

3. Install or connect the following:
   - Oil seal. Drive the seal square into the seal case.
   - Flywheel. Tighten the bolts in a crossing pattern to 54 ft. lbs. (73 Nm).
   - Transaxle
4. Check the fluid levels.
5. Start the engine and check for leaks.

ROCKER ARMS/SHAFTS

REMOVAL & INSTALLATION

See Figures 56 through 59.

1. Before servicing the vehicle, refer to the precautions section.
2. Remove or disconnect the following:
   - Negative battery cable
   - Intake manifold
   - Ignition coils
   - Valve cover
   - Rocker arm adjusting screws. Refer to the illustration for location.
3. Remove the rocker arm assembly as follows:
   a. Unscrew the rocker shaft bolts 2 turns at a time in a criss-cross pattern to avoid damaging the valves or rocker assembly.
   b. Do not remove the rocker shaft bolts. These bolts keep the springs and rocker arms on the shafts.
4. Loosen the valve adjuster locknuts and screws so that all valves are closed.
5. Remove the rocker arms and shafts from the vehicle as an assembly.

➡**Keep all valve train components in order for assembly.**

6. Remove the rocker arms and springs from the rocker arm shafts.

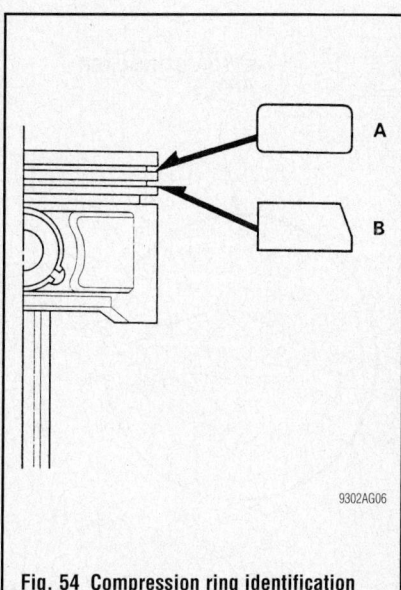

Fig. 54 Compression ring identification

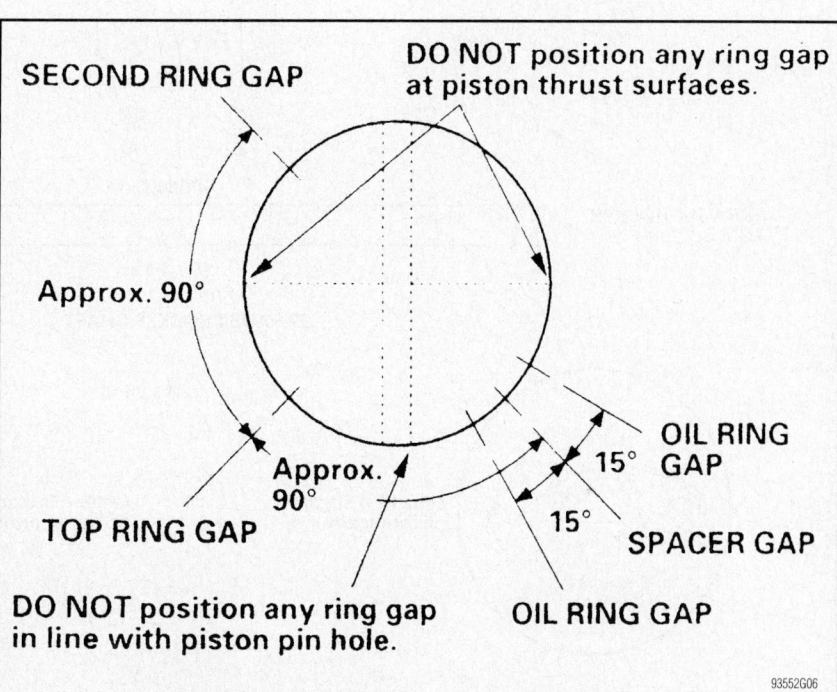

Fig. 55 Ring end gap positioning

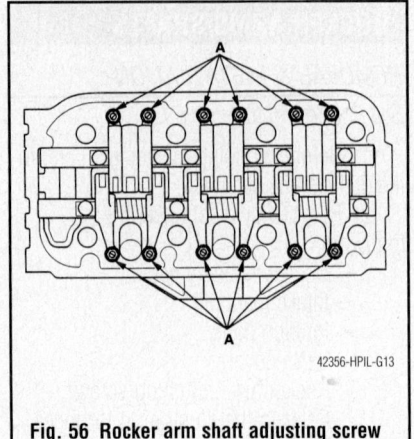

Fig. 56 Rocker arm shaft adjusting screw locations

42356-HPIL-G13

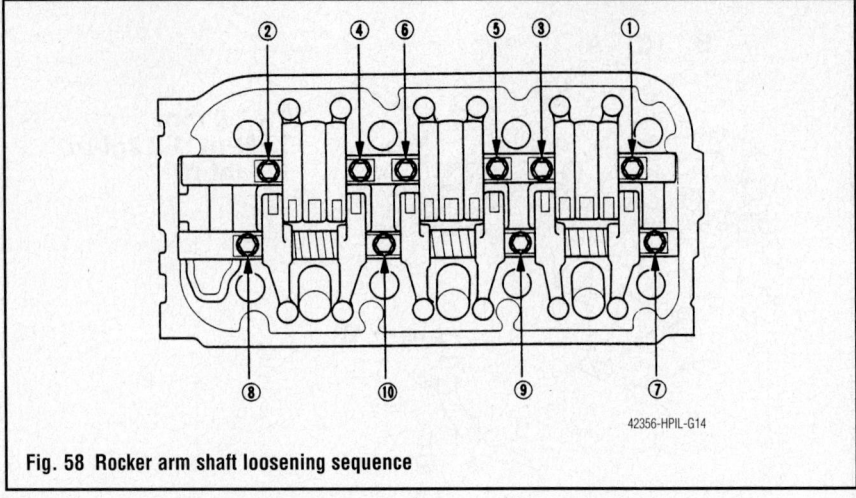

Fig. 58 Rocker arm shaft loosening sequence

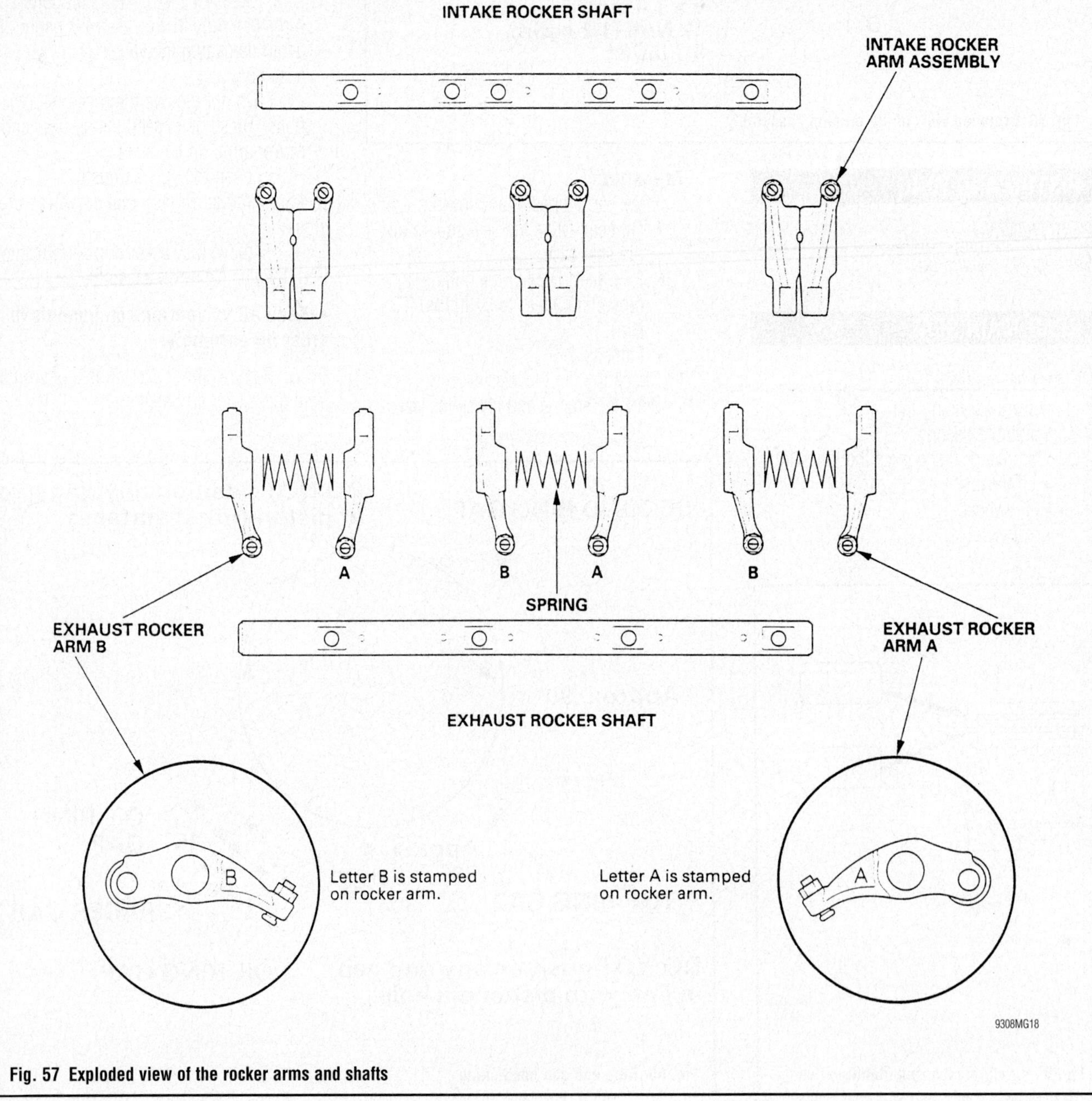

42356-HPIL-G14

INTAKE ROCKER SHAFT

INTAKE ROCKER ARM ASSEMBLY

EXHAUST ROCKER ARM B

A   B   A   B

SPRING

EXHAUST ROCKER ARM A

EXHAUST ROCKER SHAFT

Letter B is stamped on rocker arm.

Letter A is stamped on rocker arm.

9308MG18

Fig. 57 Exploded view of the rocker arms and shafts

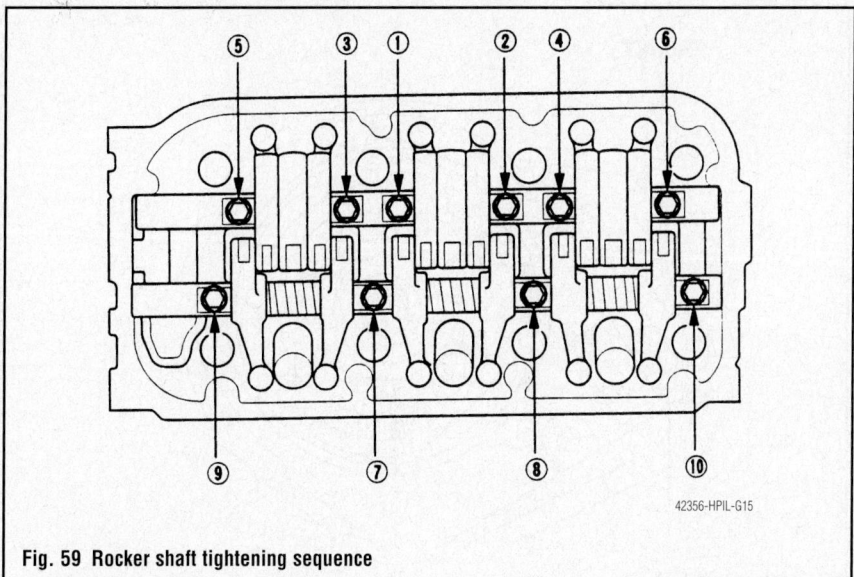

Fig. 59 Rocker shaft tightening sequence

To install:

7. Assemble the rocker arms and springs to the rocker arm shafts in their original positions.

8. Install the rocker arm assemblies. Tighten the bolts in sequence and in multiple passes to 17 ft. lbs. (24 Nm).

9. Adjust the valve clearance.

10. Install or connect the following:
- Valve covers
- Ignition coils and torque the retainers to 9 ft. lbs. (12 Nm)
- Ignition coil covers
- Intake manifold
- Negative battery cable

TIMING BELT AND SPROCKETS

REMOVAL & INSTALLATION

See Figures 60 through 71.

1. Before servicing the vehicle, refer to the precautions section.

2. Turn the crankshaft so the white mark aligns with the pointer.

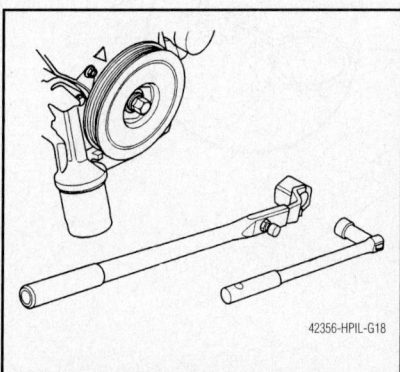

Fig. 60 Remove the crankshaft pulley using holder tool shown

3. Make sure the number 1 piston is at Top Dead center (TDC).

4. Remove or disconnect the following:
- Negative battery cable
- Wheels and splash shield

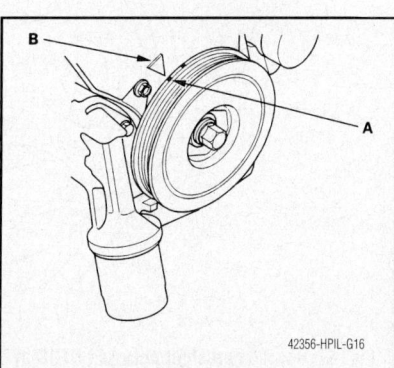

Fig. 61 Turn the crankshaft so the white mark (A) aligns with the pointer (B)

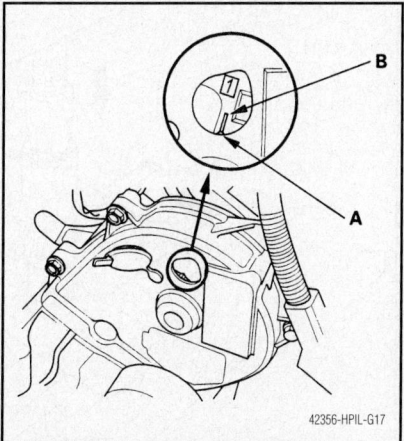

Fig. 62 Make sure the number 1 piston is at top dead center (A) on the front camshaft pulley and pointer (B)

- Drive belts

5. Support the engine with a block of wood and a jack under the oil pan.
- Upper side engine mount
- Dipstick tube

6. Remove the crankshaft pulley using holder tool shown in the accompanying illustration and a breaker and socket, loosen the 19mm bolt and remove the pulley.
- Front upper cover, rear upper cover and the lower cover
- One of the battery clamp bolts and grind the end as illustrated

7. Screw the battery clamp bolt as illustrated to hold the belt adjuster in position. Do not use a wrench, hand tighten only.
- Lower side engine mount
- Idler pulley bolt and the pulley
- Timing belt

To install:

8. If installing a new belt, perform the following steps:

a. Clean the pulleys, belt guide plate and the upper and lower covers.

b. Set the timing belt drive pulley to TDC by aligning the TDC mark on the

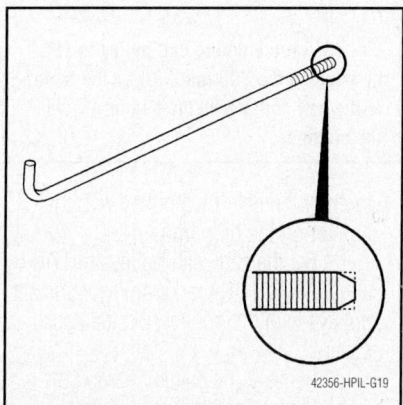

Fig. 63 Remove a battery clamp bolt and grind the end as shown

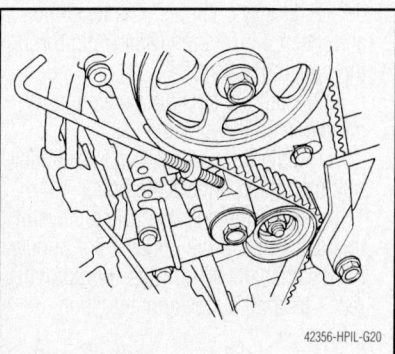

Fig. 64 Install the battery clamp bolt as shown to hold the belt adjuster in position

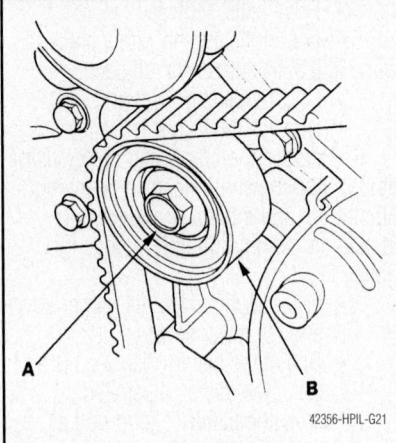

Fig. 65 Remove the idler pulley bolt (A), pulley (B) and the timing belt

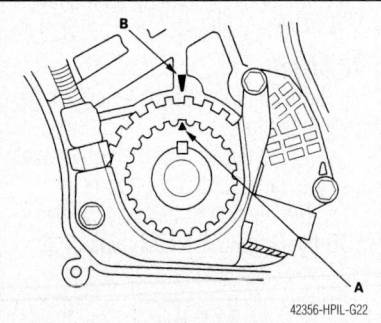

Fig. 66 Set the timing belt pulley to TDC by aligning the TDC mark (A) on the tooth of the belt pulley with the pointer (B) on the oil pump

tooth of the belt drive pulley with the pointer on the oil pump.

c. Set the camshaft pulleys to TDC by aligning the TDC marks on the camshaft pulleys with the pointers on the back covers.

d. Remove the battery clamp bolt.

e. Remove the belt tensioner.

f. Align the holes on the rod and housing of the tensioner.

g. Using a press or other suitable device, slowly compress the tensioner and insert a 0.08 inch (2mm) pin through the housing and rod.

h. Install the tensioner making sure the pin is still installed.

i. Apply thread locker to idler pulley bolt then hand tighten the bolt.

j. Install the belt over the pulleys in this sequence; drive pulley, idler pulley, front camshaft pulley, water pump pulley, rear camshaft pulley and adjusting pulley.

k. Tighten the idler pulley bolt to 33 ft. lbs. (44 Nm).

l. Remove the pin from the tensioner.

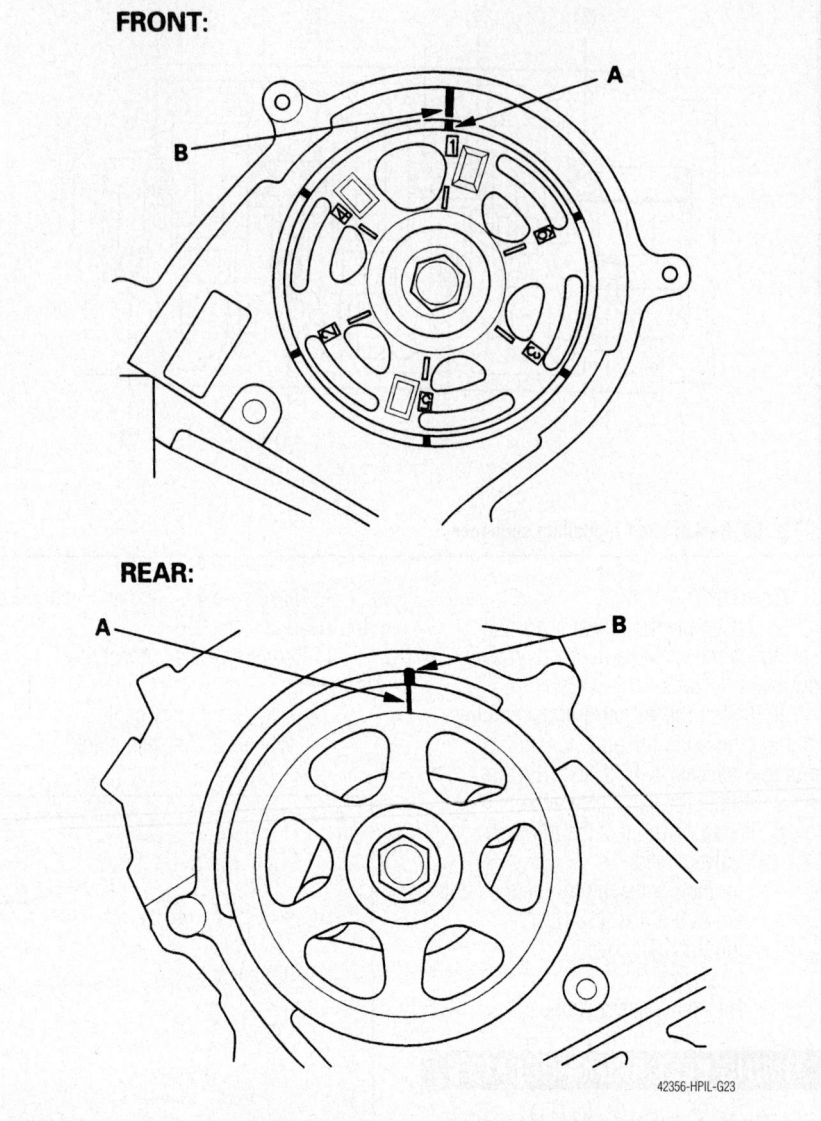

FRONT:

REAR:

Fig. 67 Set the camshaft pulleys to TDC by aligning the TDC marks (A) on the camshaft pulleys with the pointers (B) on the back covers

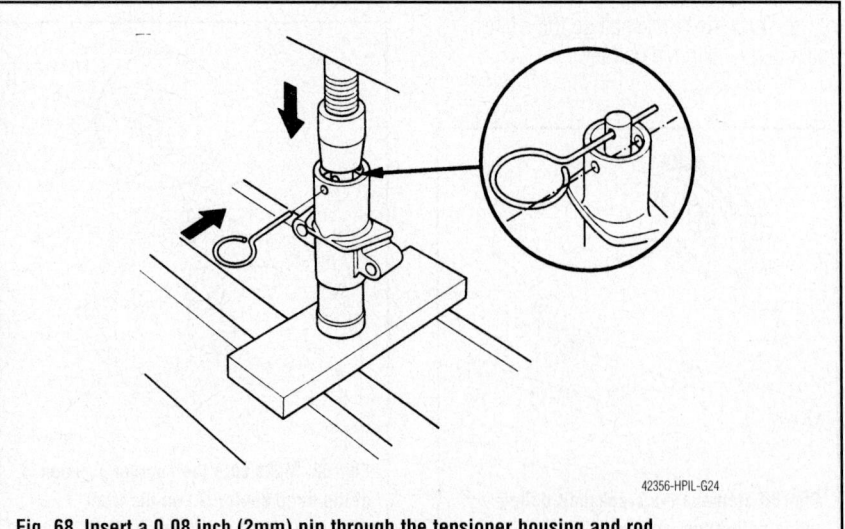

Fig. 68 Insert a 0.08 inch (2mm) pin through the tensioner housing and rod

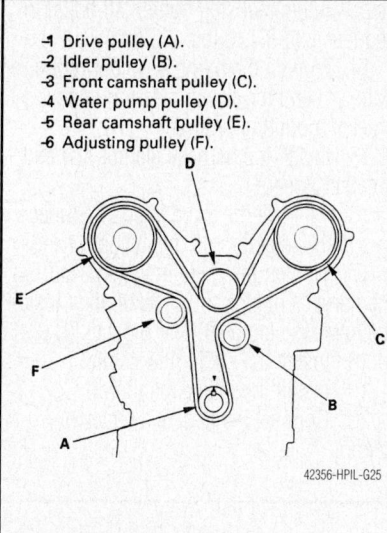

-1 Drive pulley (A).
-2 Idler pulley (B).
-3 Front camshaft pulley (C).
-4 Water pump pulley (D).
-5 Rear camshaft pulley (E).
-6 Adjusting pulley (F).

42356-HPIL-G25

Fig. 69 Route the belt as shown in the sequence listed

9. Install or connect the following:
   - Lower half of the side mount and tighten the 3 long bolts to 33 ft. lbs. (44 Nm) and the one short bolt to 9 ft. lbs. (12 Nm)
   - Timing belt guide plate as illustrated
   - Lower timing cover and tighten the bolts to 9 ft. lbs. (12 Nm)
   - Front and rear upper timing covers and tighten the bolts to 9 ft. lbs. (12 Nm)
   - Crankshaft pulley and tighten the bolts to 181 ft. lbs. (245 Nm), using the holding tool to prevent the unit from turning

10. Rotate the crankshaft pulley about 5 or 6 degrees clockwise so the belt positions itself on the pulleys.

11. Turn the crankshaft pulley so the white mark aligns with the pointer.

12. Check the camshaft pulley marks are aligned. If the marks are aligned, proceed to the next step. If the marks are not aligned, remove the timing belt and reinstall using the steps outlined before this step.

13. Remove or disconnect the following:
    - Drive belt
    - Upper side mount and tighten the bolts in the sequence illustrated to the specifications in the illustration

14. Using a suitable scan tool, perform the Powertain Control Module (PCM) reset and the Crankshaft position (CKP) pattern clear/learn procedures, following the scan tool manufactures instructions.

15. If installing the old belt, perform the following steps:
    a. Clean the pulleys, belt guide plate and the upper and lower covers.
    b. Set the timing belt drive pulley to TDC by aligning the TDC mark on the tooth of the belt drive pulley with the pointer on the oil pump.
    c. Set the camshaft pulleys to TDC by aligning the TDC marks on the camshaft pulleys with the pointers on the back covers.
    d. Apply thread locker to idler pulley bolt then hand tighten the bolt.

16. If the tensioner was extended and the belt cannot be installed, perform the steps above for the new belt installation.
    a. Install the belt over the pulleys in this sequence; drive pulley, idler pulley, front camshaft pulley, water pump pulley, rear camshaft pulley and adjusting pulley.
    b. Tighten the idler pulley bolt to 33 ft. lbs. (44 Nm).
    c. Remove the battery clamp bolt.

17. Install or connect the following:
    - Lower half of the side mount and tighten the 3 long bolts to 33 ft. lbs. (44 Nm) and the one short bolt to 9 ft. lbs. (12 Nm)

FRONT CAMSHAFT PULLEY:

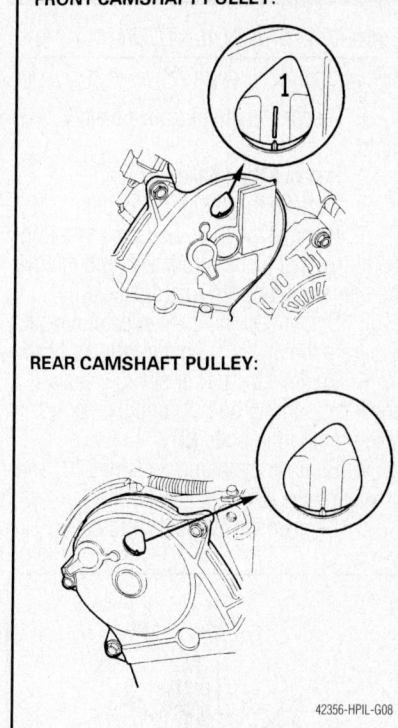

REAR CAMSHAFT PULLEY:

42356-HPIL-G08

Fig. 71 Check that the camshaft pulley marks are aligned as shown

- Timing belt guide plate as illustrated
- Lower timing cover and tighten the bolts to 9 ft. lbs. (12 Nm)
- Front and rear upper timing covers and tighten the bolts to 9 ft. lbs. (12 Nm)
- Crankshaft pulley and tighten the bolts to 181 ft. lbs. (245 Nm), using the holding tool to prevent the

18. Rotate the crankshaft pulley about 5 or 6 degrees clockwise so the belt positions itself on the pulleys.

19. Turn the crankshaft pulley so the white mark aligns with the pointer.

20. Check the camshaft pulley marks are aligned. If the marks are aligned, proceed to the next step. If the marks are not aligned, remove the timing belt and reinstall using the steps outlined before this step.

21. Install or connect the following:
    - Drive belt
    - Upper side mount and tighten the bolts in the sequence illustrated to the specifications in the illustration
    - Dipstick tube

22. Using a suitable scan tool, perform the Powertain Control Module (PCM) reset and the Crankshaft position (CKP) pattern clear/learn procedures, following the scan tool manufactures instructions.

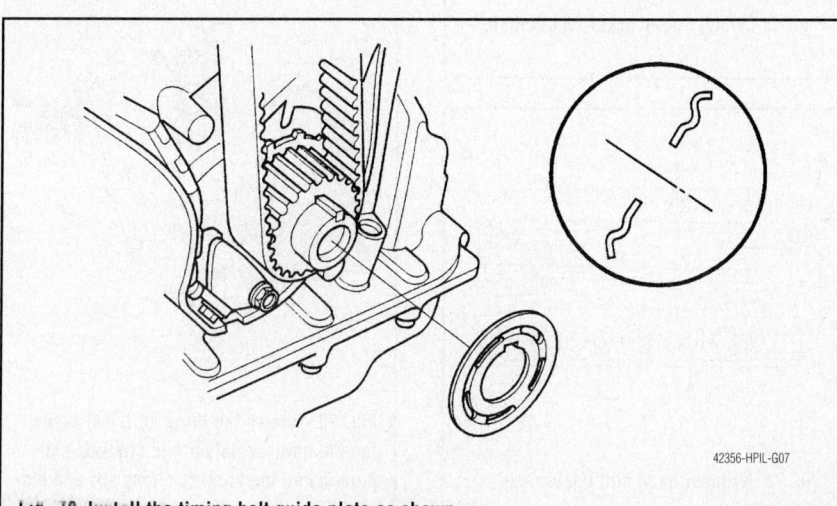

42356-HPIL-G07

Fig. 70 Install the timing belt guide plate as shown

VALVE COVERS

REMOVAL & INSTALLATION

See Figures 72 through 78.

1. Disconnect the negative battery cable.
2. Remove the intake manifold.
3. Remove the six ignition coils.
4. Remove the three bolts (A) securing the harness holder and bracket, and remove the harness clamp (B) and dipstick (C).
5. Disconnect the three injector connectors from the injectors on the cylinder head.
6. Remove the power steering hose bracket mounting bolt (A) and the harness holder mounting bolts (B).
7. Remove the harness clamps (C) and breather hose (D).
8. Remove the cylinder head cover.

9. Clean the head cover contacting surfaces with a shop towel.
10. Visually check the spark plug seals for damage. Replace if necessary.

To install:

11. Set the spark plug seals (A) on the spark plug tubes, and install the cylinder head cover (B).
12. Inspect the cover washer (C). Replace any washer that is damaged or deteriorated.
13. Tighten the bolts in two or three

steps. In the final step tighten all bolts, in sequence, to 8.7 ft. lbs. (12 Nm).
14. Tighten the harness holder mounting bolts (A) and the power steering hose bracket mounting bolt (B).
15. Install the harness clamps (C) and breather hose (D).
16. Connect the three injector connectors to the injectors on the cylinder head.
17. Tighten the three bolts (A) securing the harness holder and bracket, then install the harness clamp (B) and dipstick (C).
18. Install the six ignition coils.
19. Install the intake manifold.
20. Connect the negative battery cable.

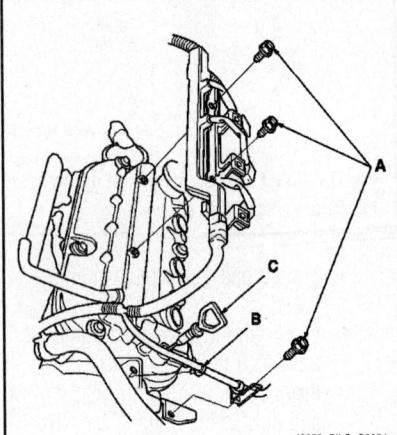

42050_PILO_G0054

Fig. 72 Remove the three bolts (A) securing the harness holder and bracket, and remove the harness clamp (B) and dipstick (C)

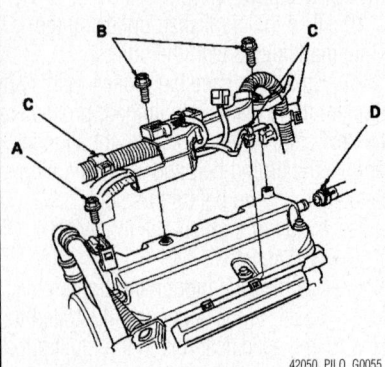

42050_PILO_G0055

Fig. 73 Remove the power steering hose bracket mounting bolt (A) and the harness holder mounting bolts (B). Then, remove the harness clamps (C) and breather hose (D)

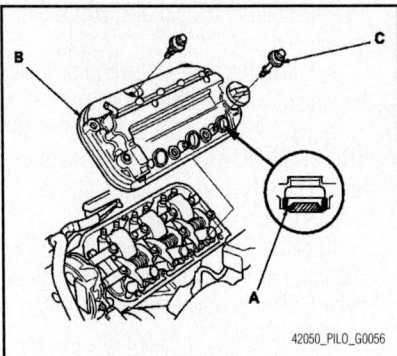

42050_PILO_G0056

Fig. 74 Cylinder head cover installation—front

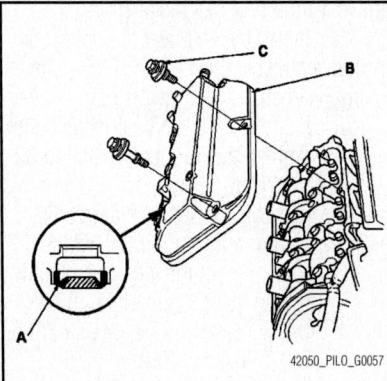

42050_PILO_G0057

Fig. 75 Cylinder head cover installation—rear

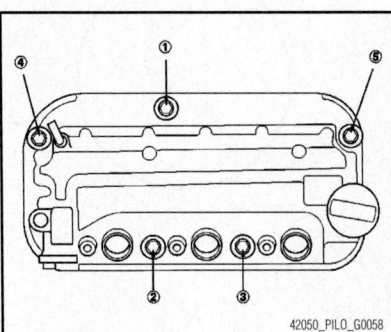

42050_PILO_G0058

Fig. 76 Cylinder head bolt tightening sequence

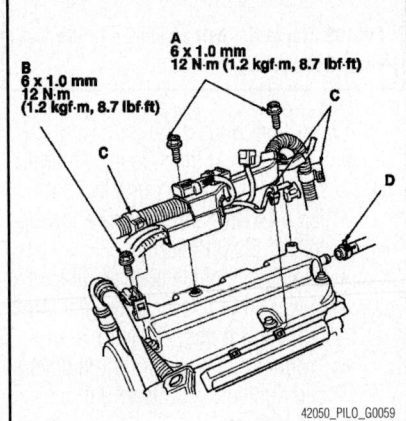

42050_PILO_G0059

Fig. 77 Tighten the harness holder mounting bolts (A) and the power steering hose bracket mounting bolt (B). Then, install the harness clamps (C) and breather hose (D)

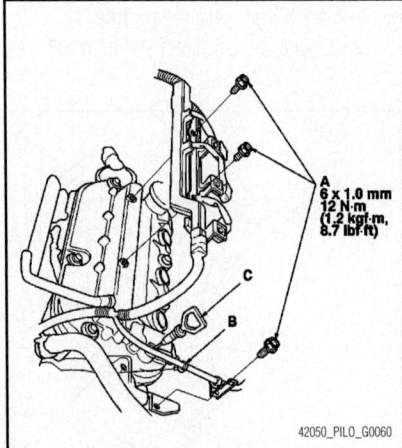

42050_PILO_G0060

Fig. 78 Tighten the three bolts (A) securing the harness holder and bracket, and then install the harness clamp (B) and dipstick (C)

VALVE LASH

ADJUSTMENT

See Figures 79 and 80.

Adjust the valves only when the cylinder head temperature is less than 100°F (38°C).

1. Before servicing the vehicle, refer to the precautions section.

2. Remove or disconnect the following:
   - Negative battery cable
   - Air intake tube
   - Intake manifold
   - Valve cover

3. Rotate the crankshaft so that the valves to be adjusted are closed and the rocker arm is contacting the camshaft lobe base circle.

4. Measure the valve clearance. If adjustment is necessary, loosen the locknut and turn the adjusting screw as necessary to achieve the correct valve clearance.

5. The correct valve clearance is:
   - Intake valves: 0.008–0.009 inches (0.20–0.24mm)
   - Exhaust valves: 0.011–0.013 inches (0.28–0.32mm)

6. After adjustment, tighten the locknuts to 14 ft. lbs. (20 Nm).

7. Install or connect the following:

   - Valve cover
   - Intake manifold
   - Air intake tube

   - Negative battery cable

8. Start the engine and check for proper operation.

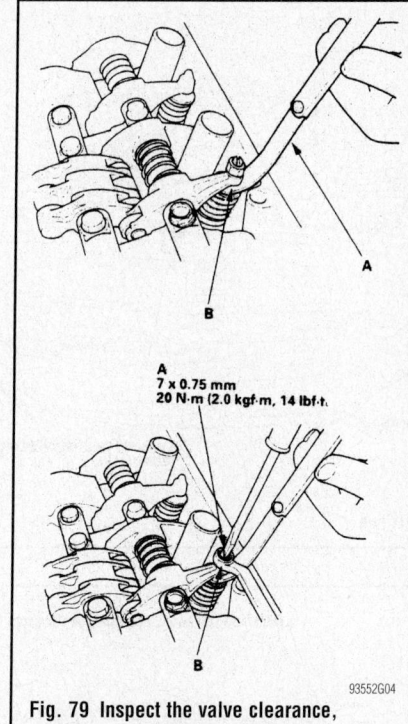
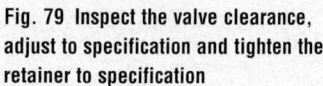

Fig. 79 Inspect the valve clearance, adjust to specification and tighten the retainer to specification

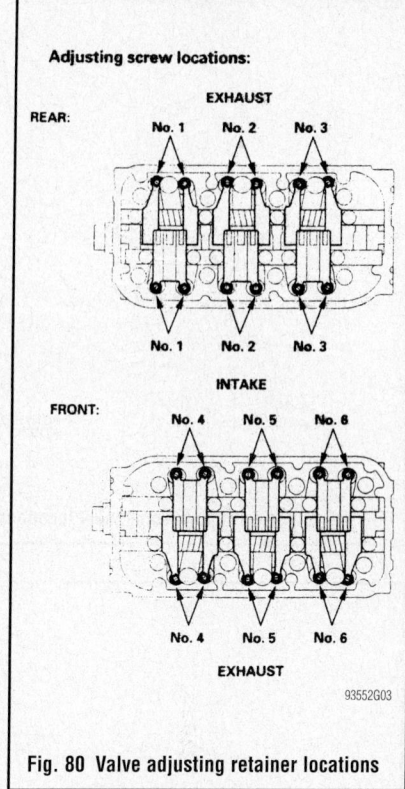

Fig. 80 Valve adjusting retainer locations

ENGINE PERFORMANCE & EMISSION CONTROL

COMPONENT LOCATIONS

See Figures 81 through 93.

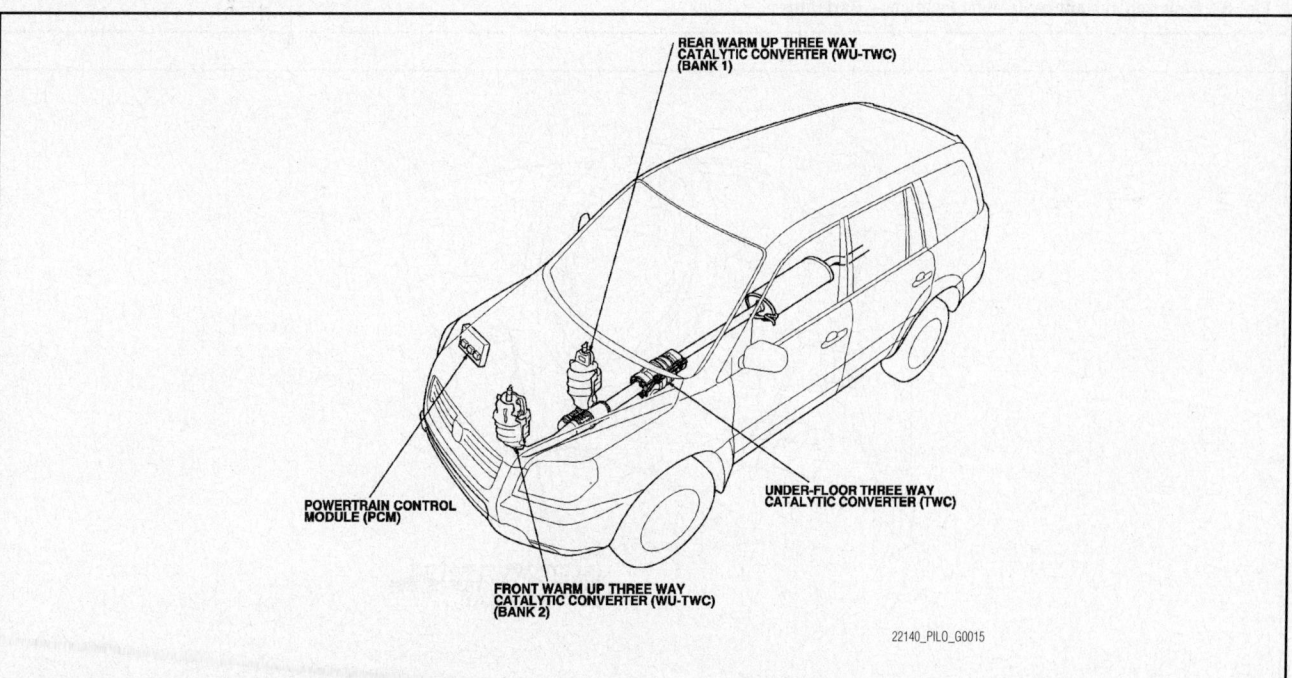

Fig. 81 Emission system component locations—part one

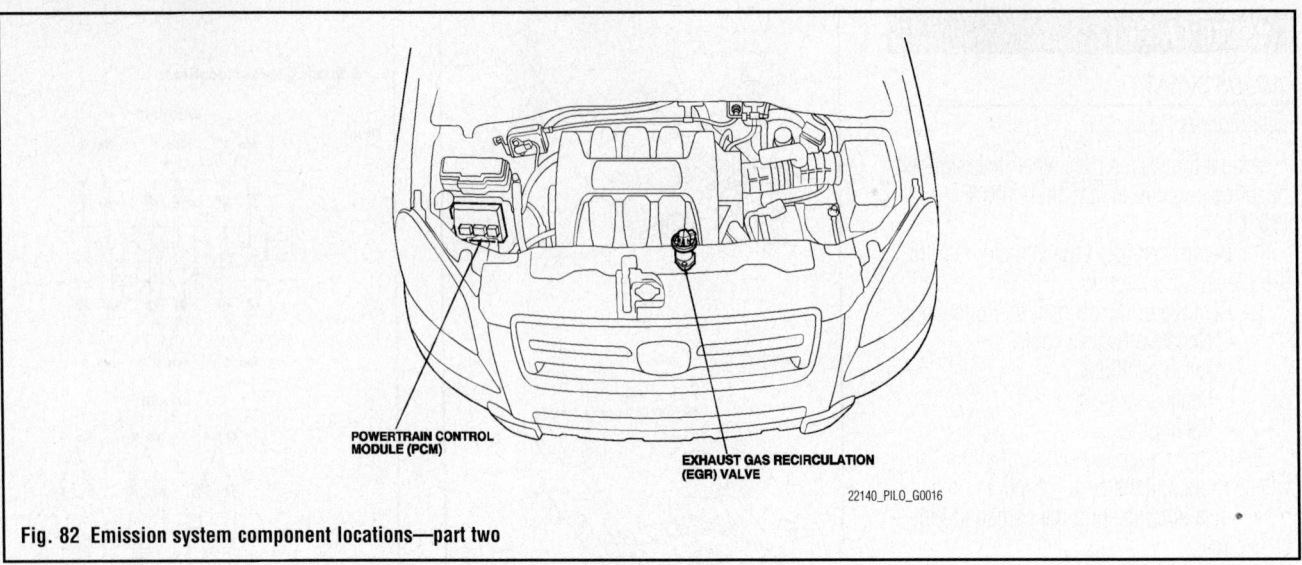

POWERTRAIN CONTROL MODULE (PCM)

EXHAUST GAS RECIRCULATION (EGR) VALVE

22140_PILO_G0016

Fig. 82 Emission system component locations—part two

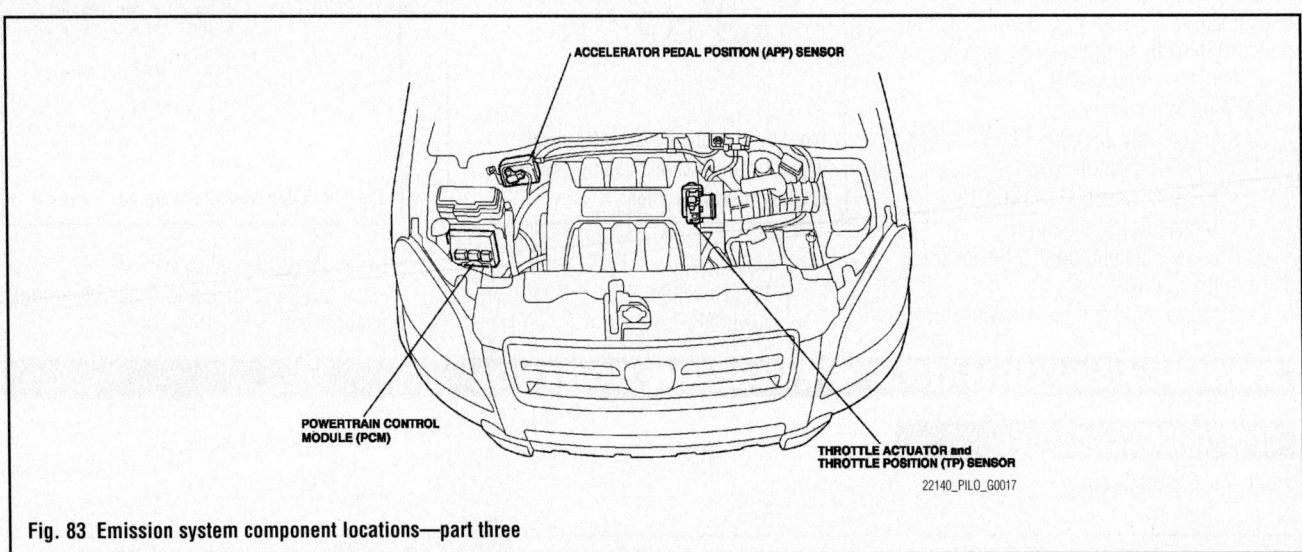

ACCELERATOR PEDAL POSITION (APP) SENSOR

POWERTRAIN CONTROL MODULE (PCM)

THROTTLE ACTUATOR and THROTTLE POSITION (TP) SENSOR

22140_PILO_G0017

Fig. 83 Emission system component locations—part three

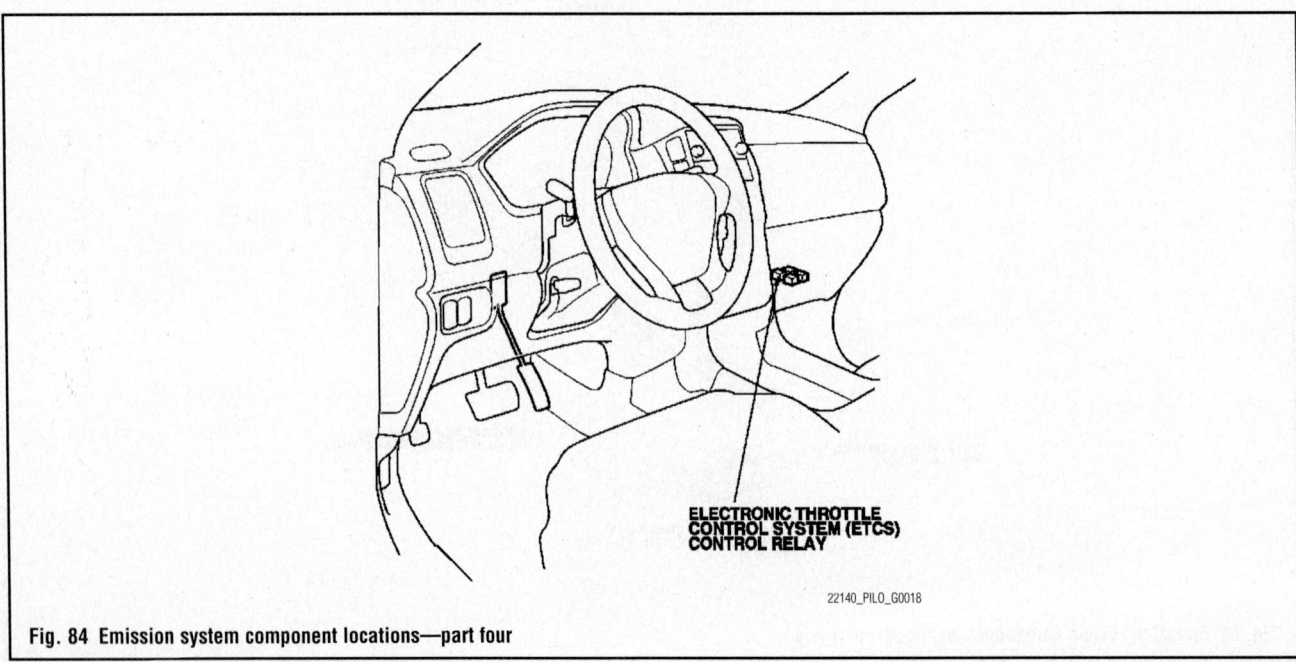

ELECTRONIC THROTTLE CONTROL SYSTEM (ETCS) CONTROL RELAY

22140_PILO_G0018

Fig. 84 Emission system component locations—part four

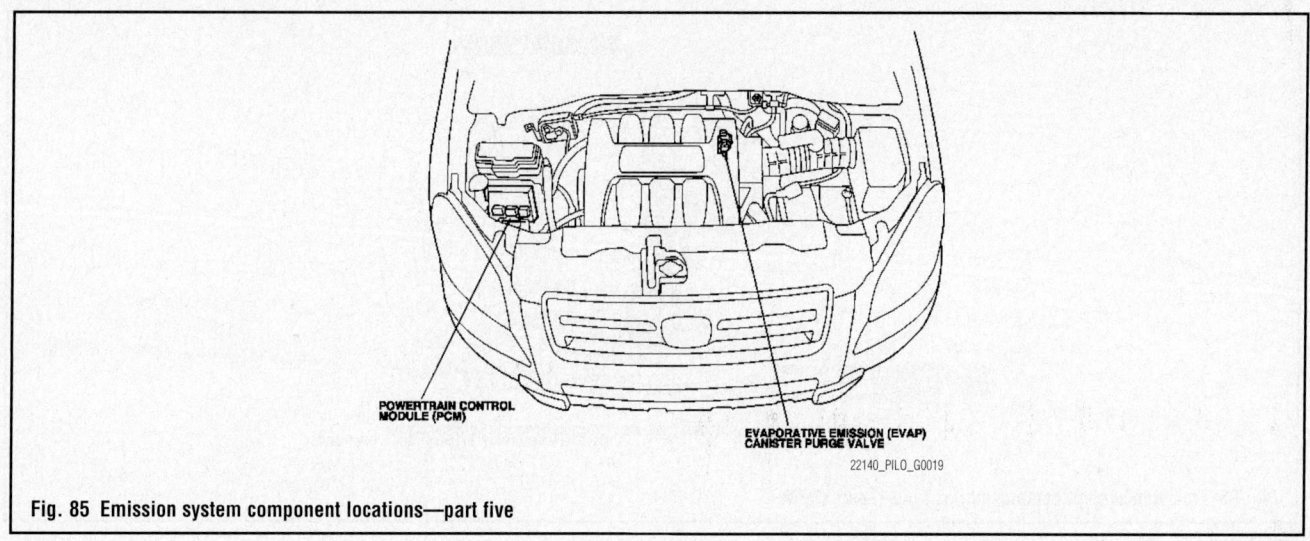

Fig. 85 Emission system component locations—part five

22140_PILO_G0019

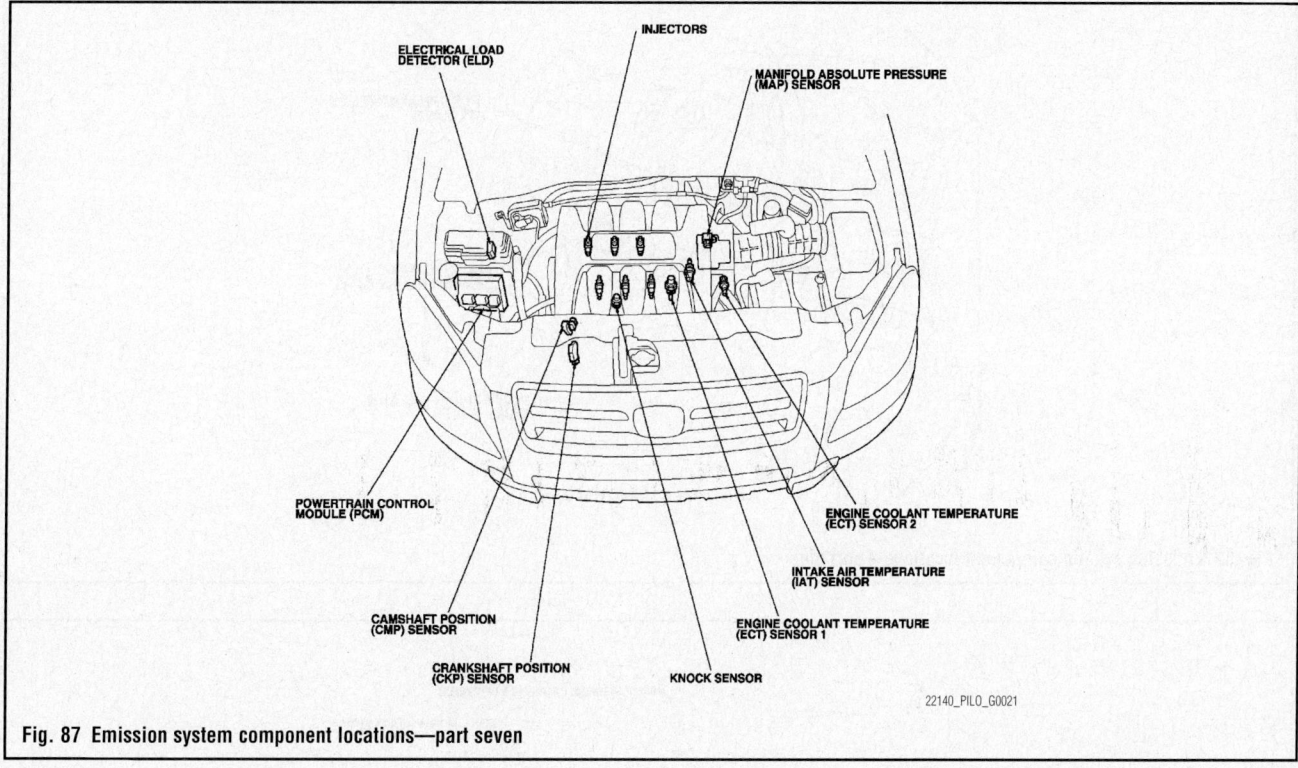

ELECTRICAL LOAD
DETECTOR (ELD)

INJECTORS

MANIFOLD ABSOLUTE PRESSURE
(MAP) SENSOR

POWERTRAIN CONTROL
MODULE (PCM)

ENGINE COOLANT TEMPERATURE
(ECT) SENSOR 2

INTAKE AIR TEMPERATURE
(IAT) SENSOR

CAMSHAFT POSITION
(CMP) SENSOR

ENGINE COOLANT TEMPERATURE
(ECT) SENSOR 1

CRANKSHAFT POSITION
(CKP) SENSOR

KNOCK SENSOR

22140_PILO_G0021

Fig. 87 Emission system component locations—part seven

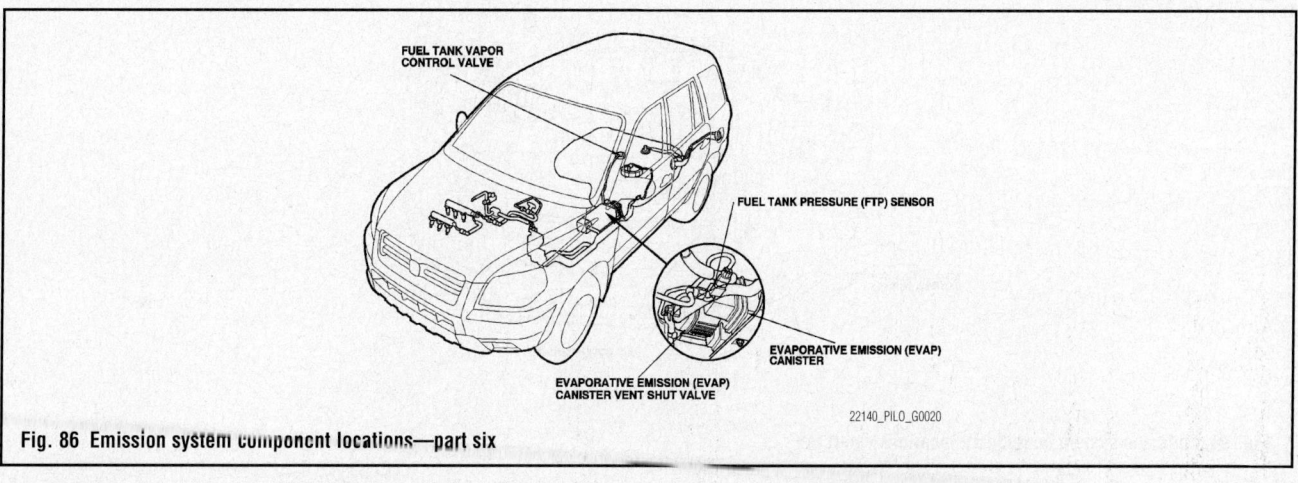

FUEL TANK VAPOR
CONTROL VALVE

FUEL TANK PRESSURE (FTP) SENSOR

EVAPORATIVE EMISSION (EVAP)
CANISTER

EVAPORATIVE EMISSION (EVAP)
CANISTER VENT SHUT VALVE

22140_PILO_G0020

Fig. 86 Emission system component locations—part six

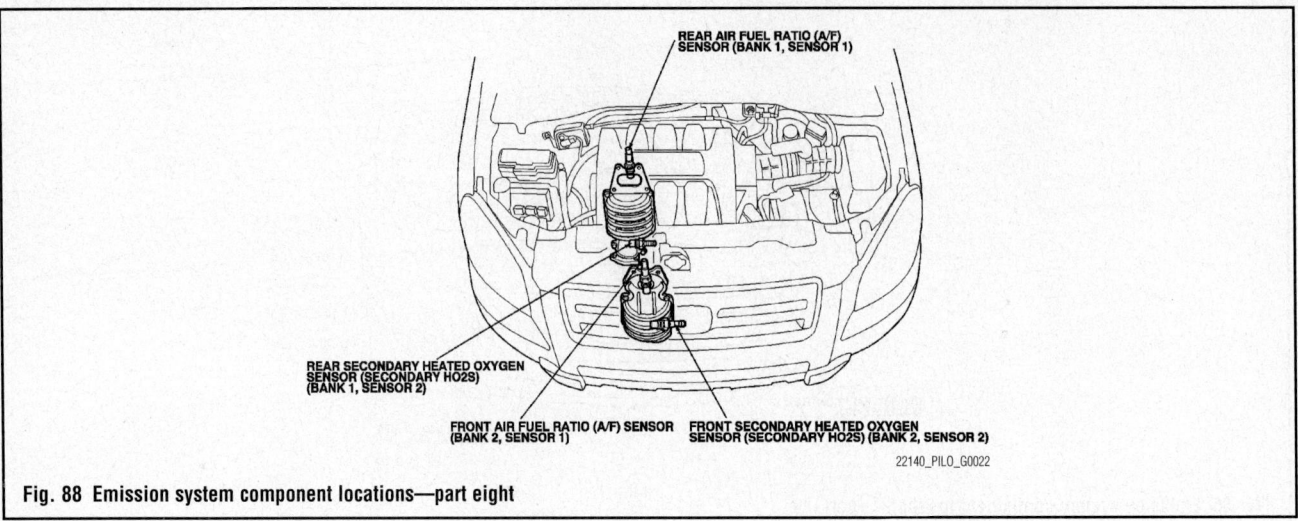

Fig. 88 Emission system component locations—part eight

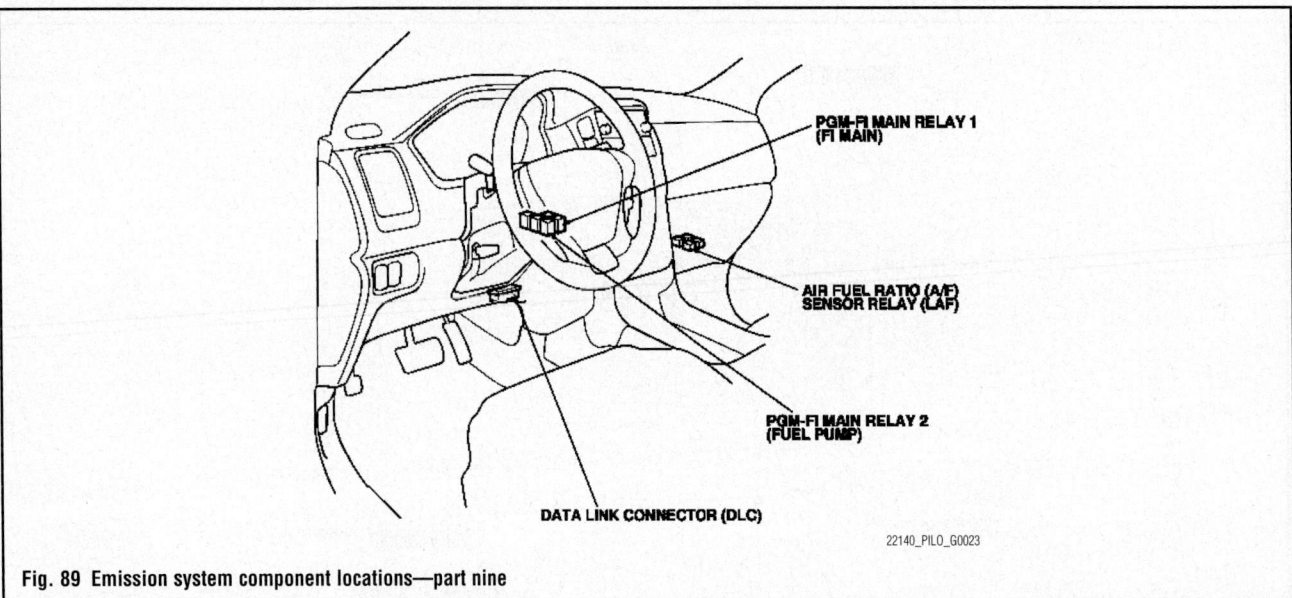

Fig. 89 Emission system component locations—part nine

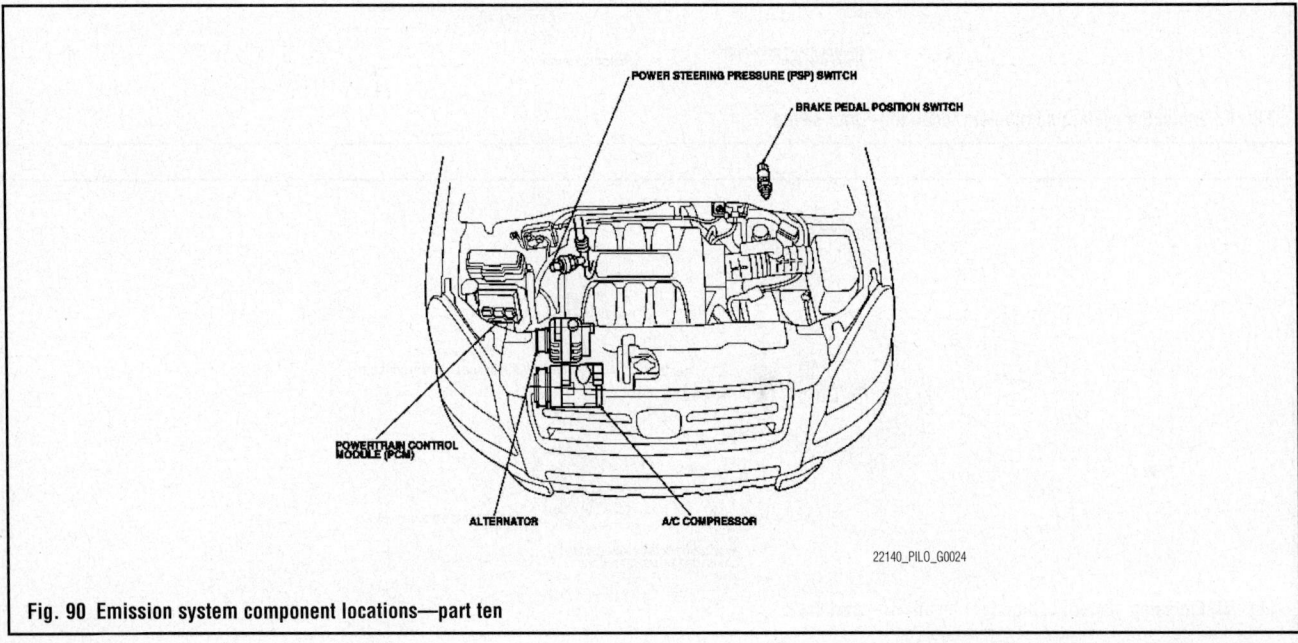

Fig. 90 Emission system component locations—part ten

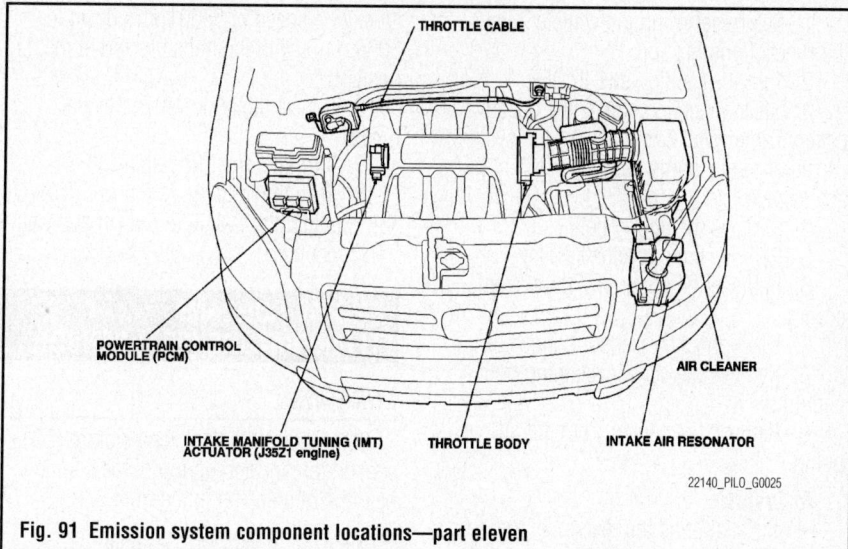

Fig. 91 Emission system component locations—part eleven

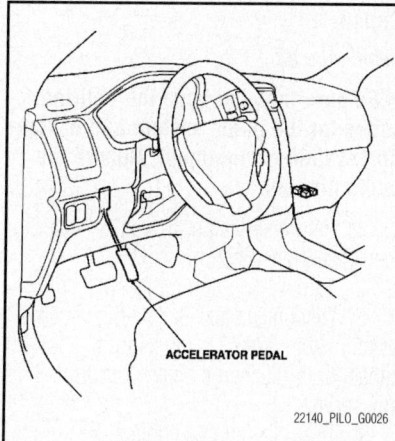

Fig. 92 Emission system component locations—part twelve

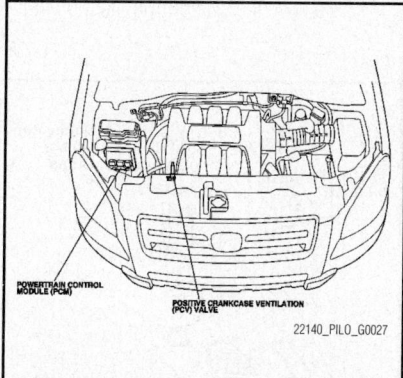

Fig. 93 Emission system component locations—part thirteen

AIR FUEL (A/F) SENSOR

LOCATION

Both Air Fuel (A/F) sensors are mounted on top of the exhaust manifolds.

REMOVAL & INSTALLATION

See Figures 94 and 95.

➡ **Be sure that you have the anti-theft codes for the audio system and navigation system, if equipped. Record the audio presets.**

1. Before servicing the vehicle, refer to the Precautions Section.
2. Turn the ignition switch OFF.
3. Disconnect and isolate the negative battery cable. Wait 3 minutes for the system capacitor to discharge before performing any service.
4. To remove the rear bank (bank one) sensor, remove the engine cover retaining screws. Remove the engine cover.
5. Disconnect the electrical connector.
6. Remove the component from its mounting, using a sensor removal tool.

To install:

7. Installation is the reverse of the removal procedure.
8. Tighten the sensor to 33 ft. lbs.

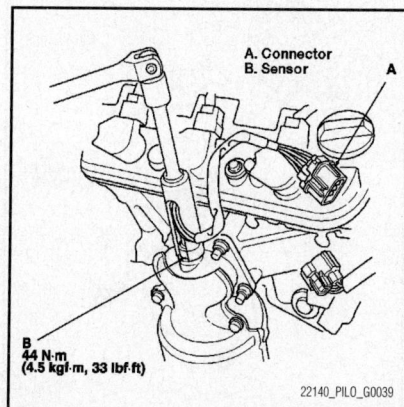

Fig. 94 Air fuel sensor and related components—front bank (bank two)

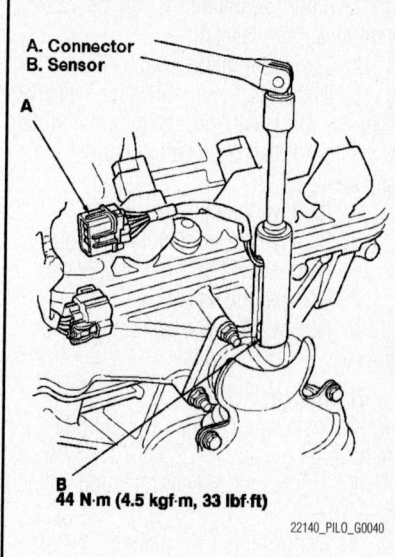

Fig. 95 Air fuel sensor and related components—rear bank (bank one)

CAMSHAFT POSITION (CMP) SENSOR

LOCATION

The Camshaft Position (CMP) Sensor is located on the back cover.

REMOVAL & INSTALLATION

See Figure 96.

➡ **Be sure that you have the anti-theft codes for the audio system and navigation system, if equipped. Record the audio presets.**

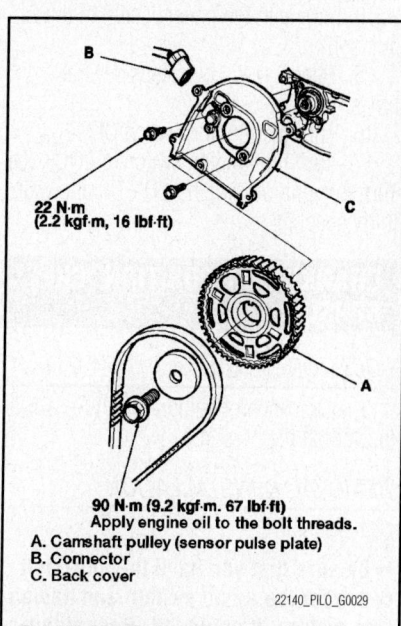

Fig. 96 Camshaft position sensor and related components

1. Before servicing the vehicle, refer to the Precautions Section.

2. Turn the ignition switch OFF.

3. Disconnect and isolate the negative battery cable. Wait 3 minutes for the system capacitor to discharge before performing any service.

4. Remove the timing belt.

5. Remove the front crankshaft pulley.

6. Disconnect the sensor connector.

7. Remove the back cover.

8. Remove the sensor from the back cover.

To install:

9. Installation is the reverse of the removal procedure.

10. Tighten the sensor retaining bolt to 3 ft. lbs.

11. Perform the CKP pattern clear/relearn procedure as follows.

➡**The Honda Diagnostic Service (HDS) tool should be used to perform this operation. Follow the directions using the HDS tool. If the tool is not available, precede as indicated below.**

12. Start the engine. Hold the engine speed at 3000 rpm's, without a load in either PARK or NEUTRAL until the radiator fan comes on.

13. Test drive the vehicle on a level road. Decelerate with the throttle fully closed from an engine speed of 2500 rpm's down to 1000 rpm's with the transmission in the 2 position.

14. Test drive the vehicle on a level road. Decelerate with the throttle fully closed from an engine speed of 5000 rpm's down to 3000 rpm's with the transmission in the 2 position.

15. Repeat the above steps several times.

16. Turn the ignition switch OFF.

17. Turn the ignition switch to LOCK (0). Turn the ignition switch to ON (II) and wait thirty seconds.

CRANKSHAFT POSITION (CKP) SENSOR

LOCATION

The Crankshaft Position (CKP) Sensor is located on the oil pump.

REMOVAL & INSTALLATION

See Figure 97.

➡**Be sure that you have the anti-theft codes for the audio system and navigation system, if equipped. Record the audio presets.**

1. Before servicing the vehicle, refer to the Precautions Section.

2. Turn the ignition switch OFF.

3. Disconnect and isolate the negative battery cable. Wait 3 minutes for the system capacitor to discharge before performing any service.

4. Remove the timing belt.

5. Remove the crankshaft pulley.

6. Remove the upper and lower front covers.

7. Disconnect the sensor connector.

8. Remove the sensor retaining bolt and nuts.

9. Remove the sensor from the oil pump.

To install:

10. Installation is the reverse of the removal procedure.

11. Tighten the sensor retaining nuts to 7 ft. lbs.

12. Perform the CKP pattern clear/relearn procedure as follows.

➡**The Honda Diagnostic Service (HDS) tool should be used to perform this operation. Follow the directions using the HDS tool. If the tool is not available, precede as indicated below.**

13. Start the engine. Hold the engine speed at 3000 rpm's, without a load in either PARK or NEUTRAL until the radiator fan comes on.

14. Test drive the vehicle on a level road. Decelerate with the throttle fully closed from an engine speed of 2500 rpm's down to 1000 rpm's with the transmission in the 2 position.

15. Test drive the vehicle on a level road. Decelerate with the throttle fully closed from an engine speed of 5000 rpm's down to 3000 rpm's with the transmission in the 2 position.

16. Repeat the above steps several times.

17. Turn the ignition switch OFF.

18. Turn the ignition switch to LOCK (0). Turn the ignition switch to ON (II) and wait thirty seconds.

ENGINE COOLANT TEMPERATURE (ECT) SENSOR

LOCATION

Both Engine Coolant Temperature (ECT) sensors are located on top of the engine, near the middle and in the front.

REMOVAL & INSTALLATION

Sensor 1

See Figure 98.

➡**Be sure that you have the anti-theft codes for the audio system and navigation system, if equipped. Record the audio presets.**

1. Before servicing the vehicle, refer to the Precautions Section.

2. Turn the ignition switch OFF.

3. Disconnect and isolate the negative battery cable. Wait 3 minutes for the system capacitor to discharge before performing any service.

4. Drain the engine coolant. Be sure to properly dispose of used coolant.

5. Remove the engine cover retaining screws. Remove the engine cover.

6. Disconnect the connector.

7. Remove the sensor from its mounting.

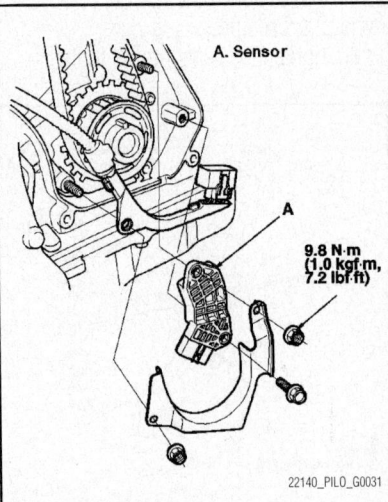

Fig. 97 Crankshaft position sensor and related components

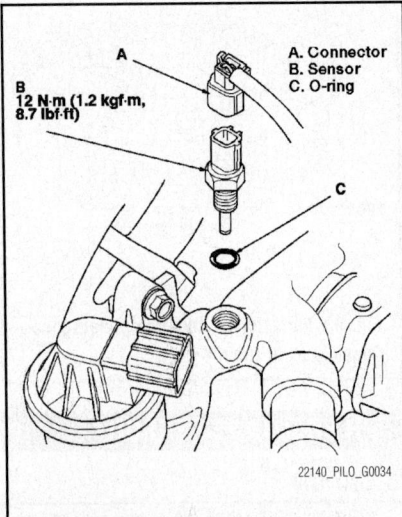

Fig. 98 Engine coolant temperature sensor 1 and related components

8. Discard the O-ring.

To install:

9. Installation is the reverse of the removal procedure.

10. Be sure to use a new O-ring.

11. Refill the radiator, using the proper grade and type engine coolant.

12. Be sure to bleed the cooling system with the heater valve open.

Sensor 2

See Figure 99.

➡ **Be sure that you have the anti-theft codes for the audio system and navigation system, if equipped. Record the audio presets.**

1. Before servicing the vehicle, refer to the Precautions Section.

2. Turn the ignition switch OFF.

3. Disconnect and isolate the negative battery cable. Wait 3 minutes for the system capacitor to discharge before performing any service.

4. Drain the engine coolant. Be sure to properly dispose of used coolant.

5. Remove the engine cover retaining screws. Remove the engine cover.

6. Disconnect the connector.

7. Remove the sensor from its mounting.

8. Discard the O-ring.

To install:

9. Installation is the reverse of the removal procedure.

10. Be sure to use a new O-ring.

11. Refill the radiator, using the proper grade and type engine coolant.

12. Be sure to bleed the cooling system with the heater valve open.

HEATED OXYGEN (HO2S) SENSOR

LOCATION

Both Heated Oxygen (HO2S) sensors are mounted at the bottom of the exhaust manifolds.

REMOVAL & INSTALLATION

See Figures 100 and 101.

➡ **Be sure that you have the anti-theft codes for the audio system and navigation system, if equipped. Record the audio presets.**

1. Before servicing the vehicle, refer to the Precautions Section.

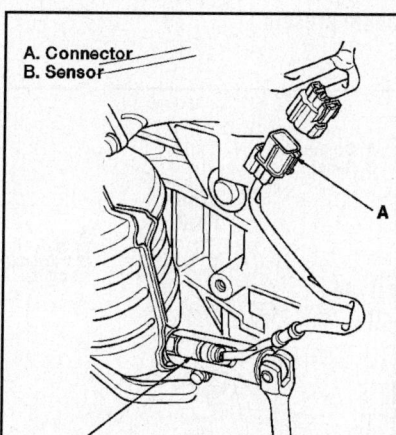

Fig. 100 Heated Oxygen sensor and related components—front bank (bank two)

2. Turn the ignition switch OFF.

3. Disconnect and isolate the negative battery cable. Wait 3 minutes for the system capacitor to discharge before performing any service.

4. To remove the front bank (bank two) sensor, remove the splash shield.

5. Disconnect the electrical connector.

6. Remove the component from its mounting, using a sensor removal tool.

To install:

7. Installation is the reverse of the removal procedure.

8. Tighten the sensor to 33 ft. lbs.

INTAKE AIR TEMPERATURE (IAT) SENSOR

LOCATION

The Intake Air Temperature (IAT) sensor is located on top of the engine, near the middle and in the front.

REMOVAL & INSTALLATION

See Figure 102.

➡ **Be sure that you have the anti-theft codes for the audio system and navigation system, if equipped. Record the audio presets.**

1. Before servicing the vehicle, refer to the Precautions Section.

2. Turn the ignition switch OFF.

3. Disconnect and isolate the negative battery cable. Wait 3 minutes for the system capacitor to discharge before performing any service.

4. Remove the engine cover.

5. Disconnect the electrical connector.

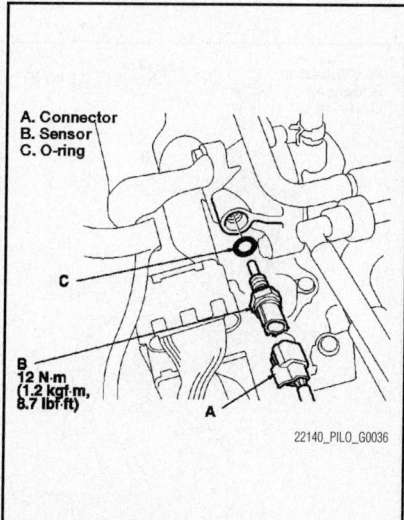

Fig. 99 Engine coolant temperature sensor 2 and related components

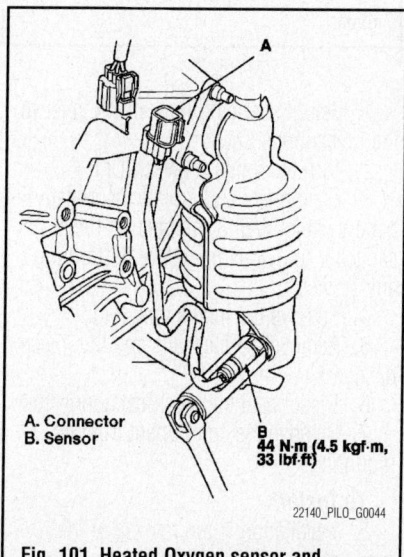

Fig. 101 Heated Oxygen sensor and related components—rear bank (bank one)

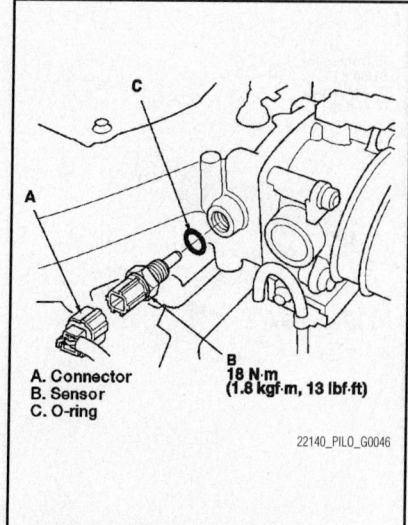

Fig. 102 Intake air temperature sensor and related components

6. Remove the component from its mounting.

7. Discard the O-ring.

To install:

8. Installation is the reverse of the removal procedure.

9. Be sure to use a new O-ring.

10. Tighten the sensor to 13 ft. lbs.

INTAKE MANIFOLD TUNING (IMT) SENSOR

LOCATION

The Intake manifold tuning (IMT) sensor is located on top of the engine, near the middle and near the front of the passenger's side of the vehicle.

REMOVAL & INSTALLATION

See Figure 103.

➡ **Be sure that you have the anti-theft codes for the audio system and navigation system, if equipped. Record the audio presets.**

1. Before servicing the vehicle, refer to the Precautions Section.

2. Turn the ignition switch OFF.

3. Disconnect and isolate the negative battery cable. Wait 3 minutes for the system capacitor to discharge before performing any service.

4. Disconnect the electrical connector.

5. Remove the retaining bolts.

6. Remove the component from its mounting.

7. Discard the O-ring.

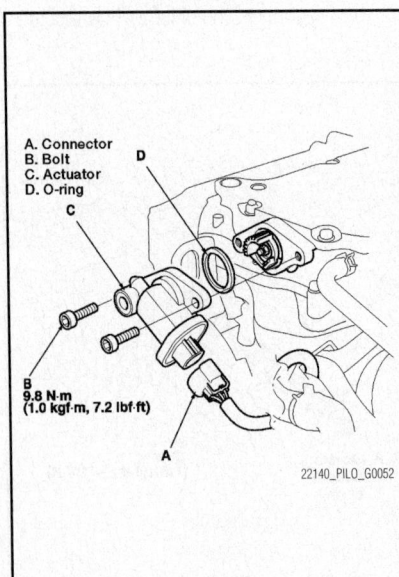

Fig. 103 Intake manifold tuning sensor and related components

To install:

8. Installation is the reverse of the removal procedure.

9. Be sure to use a new O-ring.

10. Tighten the sensor to 7 ft. lbs.

KNOCK SENSOR (KS)

LOCATION

The Knock (KS) sensor is located on top of the engine. It is visible once the intake manifold has been removed.

REMOVAL & INSTALLATION

See Figure 104.

➡ **Be sure that you have the anti-theft codes for the audio system and navigation system, if equipped. Record the audio presets.**

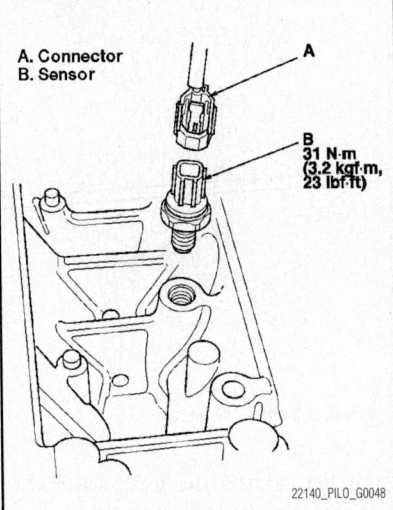

Fig. 104 Knock sensor and related components

1. Before servicing the vehicle, refer to the Precautions Section.

2. Turn the ignition switch OFF.

3. Disconnect and isolate the negative battery cable. Wait 3 minutes for the system capacitor to discharge before performing any service.

4. Remove the intake manifold.

5. Remove the fuel rails and the injector base.

6. Disconnect the electrical connector.

7. Remove the component from its mounting.

To install:

8. Installation is the reverse of the removal procedure.

9. Tighten the sensor to 23 ft. lbs.

MANIFOLD ABSOLUTE PRESSURE (MAP) SENSOR

LOCATION

The Manifold Absolute Pressure (MAP) sensor is located on top of the engine, near the middle and in the rear.

REMOVAL & INSTALLATION

See Figure 105.

➡ **Be sure that you have the anti-theft codes for the audio system and navigation system, if equipped. Record the audio presets.**

1. Before servicing the vehicle, refer to the Precautions Section.

2. Turn the ignition switch OFF.

3. Disconnect and isolate the negative battery cable. Wait 3 minutes for the system capacitor to discharge before performing any service.

4. Disconnect the electrical connector.

5. Remove the component from its mounting.

6. Discard the O-ring.

To install:

7. Installation is the reverse of the removal procedure.

8. Be sure to use a new O-ring.

9. Tighten the sensor to 3 ft. lbs.

POWERTRAIN CONTROL MODULE (PCM)

LOCATION

The Power Train Control Module (PCM) is located under the hood on the passenger's side of the vehicle, toward the front of the fender.

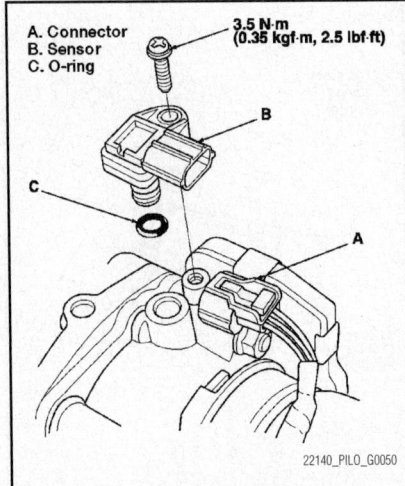

Fig. 105 Manifold absolute pressure sensor and related components

REMOVAL & INSTALLATION

See Figure 106.

➡**Be sure that you have the anti-theft codes for the audio system and navigation system, if equipped. Record the audio presets.**

1. Before servicing the vehicle, refer to the Precautions Section.
2. Turn the ignition switch OFF.
3. Disconnect and isolate the negative battery cable. Wait 3 minutes for the system capacitor to discharge before performing any service.

➡**Refer to the HDS tool instructions for the particular model you are repairing. Use those procedures to position the PCM for replacement. Diagnostics for engine oil replacement, ATF fluid replacement, and throttle body cleaning will be addressed on both US and Canadian vehicles.**

4. Remove the cover.
5. Disconnect the electrical connectors.

➡**These connectors have symbols embossed in them for identification, see illustration.**

6. Remove the component retaining bolts.
7. Remove the component from its mounting.
8. Remove the cover and the bracket from the PCM.

To install:

9. Installation is the reverse of the removal procedure.
10. Tighten the retaining bolt to 7 ft. lbs.

➡**Refer to the HDS tool instructions for the particular model you are repairing. Use those procedures to check the PCM**

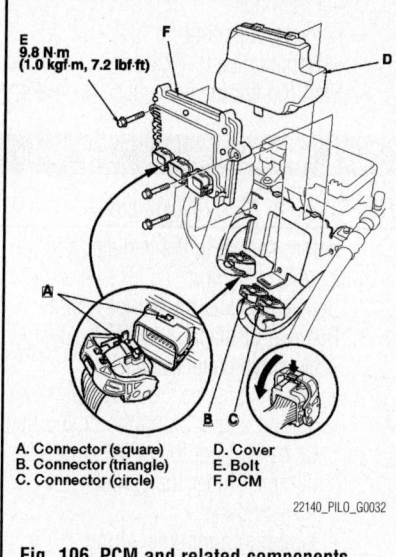

E
9.8 N·m
(1.0 kgf·m, 7.2 lbf·ft)

F

D

A

B C

A. Connector (square)
B. Connector (triangle)
C. Connector (circle)
D. Cover
E. Bolt
F. PCM

22140_PILO_G0032

Fig. 106 PCM and related components

after replacement. Diagnostics for engine oil replacement, ATF fluid replacement, and throttle body cleaning will be addressed on both US and Canadian vehicles.

➡**Update the PCM if it does not have the latest software.**

➡**The PCM relearn procedure must be performed whenever you replace the PCM, Reset the PCM, update the PCM, remove the engine or replace/clean the throttle body. Erasing DTC's with the HDS tool does not require you to perform the relearn procedure.**

11. Reset the PCM with the Honda Diagnostic System (HDS) tool while the engine is stopped. Turn the ignition switch to LOCK (0). Turn the ignition switch to ON (II), and wait thirty seconds. Turn the ignition switch

to LOCK (0) and disconnect the tool from the DLC.

12. Make sure that all electrical items are turned off.
13. Reset the PCM with the HDS tool.
14. Turn the ignition switch ON (II) and wait two seconds.
15: Start the engine. Hold the engine speed at 3000 rpm's, without a load in either PARK or NEUTRAL until the radiator fan comes on or until the coolant temperature reaches 194 degrees F.
16. Let the engine idle for about five minutes with the throttle fully closed

➡**If the radiator fan comes on, do not include its running time in the five minutes.**

17. Perform the CKP pattern clear/relearn procedure as follows.

➡**The Honda Diagnostic Service (HDS) tool should be used to perform this operation. Follow the directions using the HDS tool. If the tool is not available, precede as indicated below.**

18. Start the engine. Hold the engine speed at 3000 rpm's, without a load in either PARK or NEUTRAL until the radiator fan comes on.
19. Test drive the vehicle on a level road. Decelerate with the throttle fully closed from an engine speed of 2500 rpm's down to 1000 rpm's with the transmission in the 2 position.
20. Test drive the vehicle on a level road. Decelerate with the throttle fully closed from an engine speed of 5000 rpm's down to 3000 rpm's with the transmission in the 2 position.
21. Repeat the above steps several times.
22. Turn the ignition switch OFF.
23. Turn the ignition switch to LOCK (0). Turn the ignition switch to ON (II) and wait thirty seconds.

FUEL

GASOLINE FUEL INJECTION SYSTEM

FUEL SYSTEM SERVICE PRECAUTIONS

Safety is the most important factor when performing not only fuel system maintenance but any type of maintenance. Failure to conduct maintenance and repairs in a safe manner may result in serious personal injury or death. Maintenance and testing of the vehicle's fuel system components can be accomplished safely and effectively by adhering to the following rules and guidelines.

• To avoid the possibility of fire and personal injury, always disconnect the negative battery cable unless the repair or test procedure requires that battery voltage be applied.

• Always relieve the fuel system pressure prior to disconnecting any fuel system component (injector, fuel rail, pressure regulator, etc.), fitting or fuel line connection. Exercise extreme caution whenever relieving fuel system pressure to avoid exposing skin, face and eyes to fuel spray. Please be advised that fuel under pressure may penetrate the skin or any part of the body that it contacts.

• Always place a shop towel or cloth around the fitting or connection prior to loosening to absorb any excess fuel due to spillage. Ensure that all fuel spillage (should it occur) is quickly removed from engine surfaces. Ensure that all fuel soaked cloths or towels are deposited into a suitable waste container.

• Always keep a dry chemical (Class B) fire extinguisher near the work area.

• Do not allow fuel spray or fuel vapors to come into contact with a spark or open flame.

• Always use a back-up wrench when loosening and tightening fuel line connection fittings. This will prevent unnecessary stress and torsion to fuel line piping.

• Always replace worn fuel fitting O-rings with new Do not substitute fuel hose or equivalent where fuel pipe is installed.

Before servicing the vehicle, make sure to also refer to the precautions in the beginning of this section as well.

RELIEVING FUEL SYSTEM PRESSURE

1. Before servicing the vehicle, refer to the precautions section.
2. Remove the driver's side dashboard lower cover and disconnect the PGM-FI main relay.
3. Remove the fuel filler cap.
4. Start the engine and let it stall.

➡**A temporary code may be set during this procedure and the codes must be cleared after repairs are completed.**

5. Turn the ignition off.
6. Disconnect the negative battery cable.
7. Remove the quick connect fitting cover.
8. Place a shop towel over the quick connect fitting and disconnect the fitting.
9. After the pressure is release, reconnect the fitting and install the cover.
10. After repairs are complete install the relay and connect the negative battery cable.

FUEL FILTER

REMOVAL & INSTALLATION

The fuel filter is part of the fuel tank unit and is serviced along with that component.

FUEL INJECTORS

REMOVAL & INSTALLATION

1. Before servicing the vehicle, refer to the precautions section.
2. Relieve the fuel system pressure.
3. Remove or disconnect the following:
   - Negative battery cable
   - Intake manifold
   - Fuel lines
   - Fuel injector connectors
   - Fuel pressure regulator vacuum line
   - Fuel supply manifold
4. Separate the fuel injectors from the fuel supply manifold.

To install:

5. Install the fuel injectors to the fuel supply manifold with new cushion rings and O-rings.
6. Install new seal rings to the intake manifold.
7. Install or connect the following:
   - Fuel supply manifold and injector assembly. Tighten the bolts to 86 inch lbs. (10 Nm).
   - Fuel pressure regulator vacuum line
   - Fuel injector connectors

- Fuel lines
- Intake manifold
- Negative battery cable
8. Start the engine and check for leaks.

FUEL PUMP

REMOVAL & INSTALLATION

1. Before servicing the vehicle, refer to the precautions section.
2. Relieve the fuel system pressure.
3. Remove or disconnect the following:
   - Negative battery cable
   - Driver's side second row seat and cut the carpet along the dotted line. Be careful not to cut the wiring harness under the carpet.
   - Access panel
   - Fuel pump module wiring connector

- Fuel supply and return lines
- Fuel pump locknut
- Fuel pump module

To install:

4. Install or connect the following:
   - Fuel pump module. Use a new seal and align the matchmarks.
   - Fuel pump locknut
   - Fuel supply and return lines
   - Fuel pump module wiring connector
   - Access panel
   - Carpet and seat
   - Negative battery cable
5. Start the engine and check for leaks.

FUEL TANK

REMOVAL & INSTALLATION

See Figure 107.

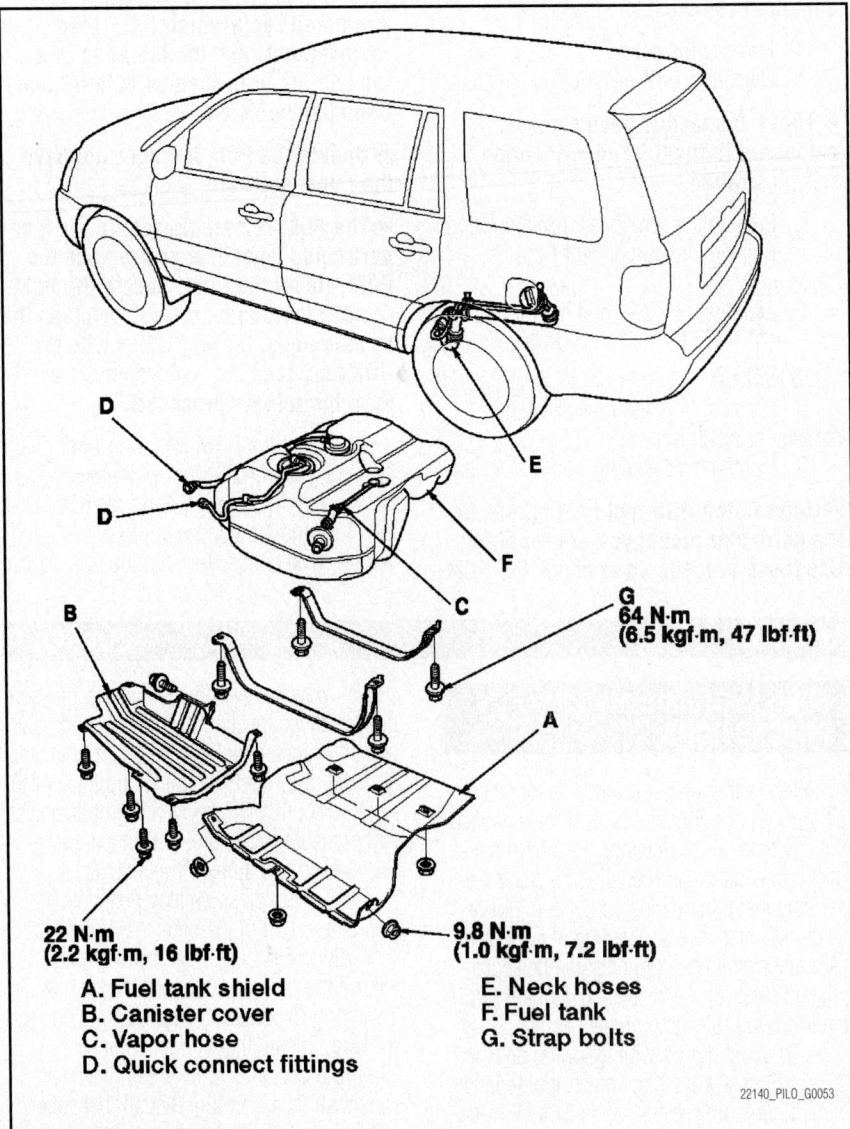

G
64 N·m
(6.5 kgf·m, 47 lbf·ft)

22 N·m
(2.2 kgf·m, 16 lbf·ft)

9.8 N·m
(1.0 kgf·m, 7.2 lbf·ft)

A. Fuel tank shield
B. Canister cover
C. Vapor hose
D. Quick connect fittings

E. Neck hoses
F. Fuel tank
G. Strap bolts

22140_PILO_G0053

Fig. 107 Fuel tank and related components

➡Be sure that you have the anti-theft codes for the audio system and navigation system, if equipped. Record the audio presets.

1. Before servicing the vehicle, refer to the Precautions Section.

2. Turn the ignition switch OFF.

3. Disconnect and isolate the negative battery cable. Wait 3 minutes for the system capacitor to discharge before performing any service.

4. Properly relieve the fuel system. Remove the fuel cap.

`5. Drain the fuel tank. Be sure to properly dispose of the fuel in the tank using an approved container.

6. Raise and support the vehicle safely.

7. Remove the exhaust muffler.

8. On 4WD vehicles, matchmark the driveshaft. Remove the driveshaft.

9. Remove the fuel tank shield. Remove the canister cover.

10. Disconnect the vapor hose and the quick connect fittings.

11. Disconnect the filler neck hoses. Slide back the clamps and then twist the hoses as you pull to avoid damaging them.

12. Position a suitable jack, or other type of support, under the fuel tank.

13. Remove the strap bolts.

14. Remove the fuel tank. If it sticks to the undercoated mounts, carefully pry it off of its mounts.

To install:

15. Installation is the reverse of the removal procedure.

16. Tighten the strap retaining bolts to 47 ft. lbs.

17. Start the engine and check for leaks, correct as required.

IDLE SPEED

ADJUSTMENT

The idle speed is controlled by the PCM and is not adjustable.

THROTTLE BODY

REMOVAL & INSTALLATION

See Figure 108.

1. Disconnect the negative battery cable.

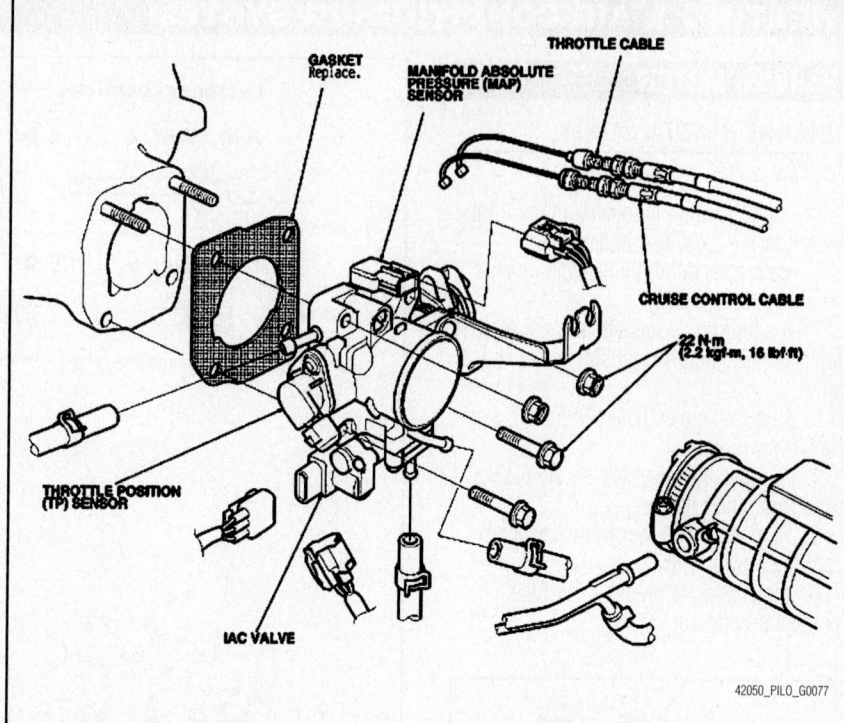

Fig. 108 Exploded view of the throttle body and related components—typical

2. Disconnect the Manifold Absolute Pressure (MAP) sensor connector.

3. Remove the intake air duct.

4. Disconnect the throttle body connector.

5. Disconnect the water bypass hoses, and plug the water bypass hoses.

6. Remove the throttle body.

7. Clean the throttle body and intake manifold surface.

To install:

8. Install the parts in the reverse order of removal with a new gasket.

➡For the PCM to learn the engine idle characteristics, the idle learn procedure must be done. Do the idle learn procedure whenever you do any of these actions:

- Disconnect the battery.
- Replace or clean the throttle body.
- Replace the Idle Air Control (IAC) valve.
- Replace the PCM.
- Reset the PCM.

9. Perform the idle learn procedure, as follows:

a. Make sure all electrical items (A/C, audio, rear window defogger, lights, etc.) are off.

b. Start the engine, and hold it at 3,000 rpm with no load (in Park or neutral) until the radiator fan comes on, or until the engine coolant temperature reaches 194°F (90°C).

c. Let the engine idle for a minimum of 5 minutes with no load.

➡If the radiator fan cycles during this time, do not include its running time in the 5 minutes.

10. Refill the radiator with engine coolant.

11. Connect the HDS while the engine stopped.

12. Select the ETC (TAC) test from the INSPECTION MENU with the HDS.

13. Do the ETCS TEST. If test results indicates the system is normal, the repair is complete. If any temporary DTCs or DTCs are indicated, go to the indicated DTC's troubleshooting.

HEATING & AIR CONDITIONING SYSTEM

BLOWER MOTOR

REMOVAL & INSTALLATION

See Figure 109.

The blower motor is located in the front passenger's right side foot well area.

1. Before servicing the vehicle, refer to the precautions section.

2. If additional access is needed, remove the front passenger's right side lower kick panel.

3. Detach the electrical connectors from the blower motor.

4. Remove the fasteners on the blower motor mounting flange.

5. Remove the blower motor downward from the blower unit.

6. Installation is the reverse of the removal procedure.

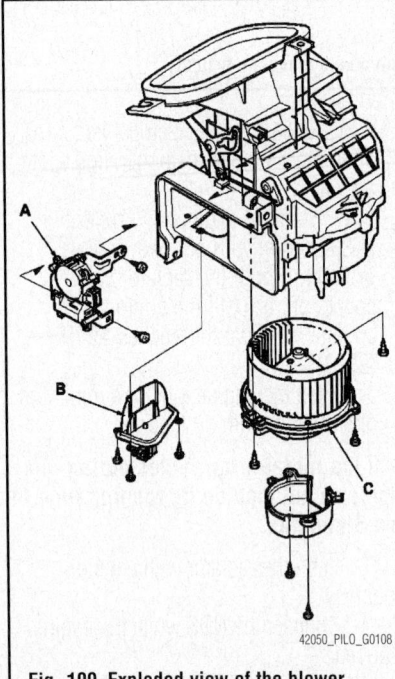

Fig. 109 Exploded view of the blower motor (C) and related components

HEATER CORE

REMOVAL & INSTALLATION

See Figures 110 through 114.

1. Before servicing the vehicle, refer to the precautions section.

2. Drain the cooling system.

3. Remove or disconnect the following:
   - Negative battery cable

4. Recover the refrigerant using approved equipment.

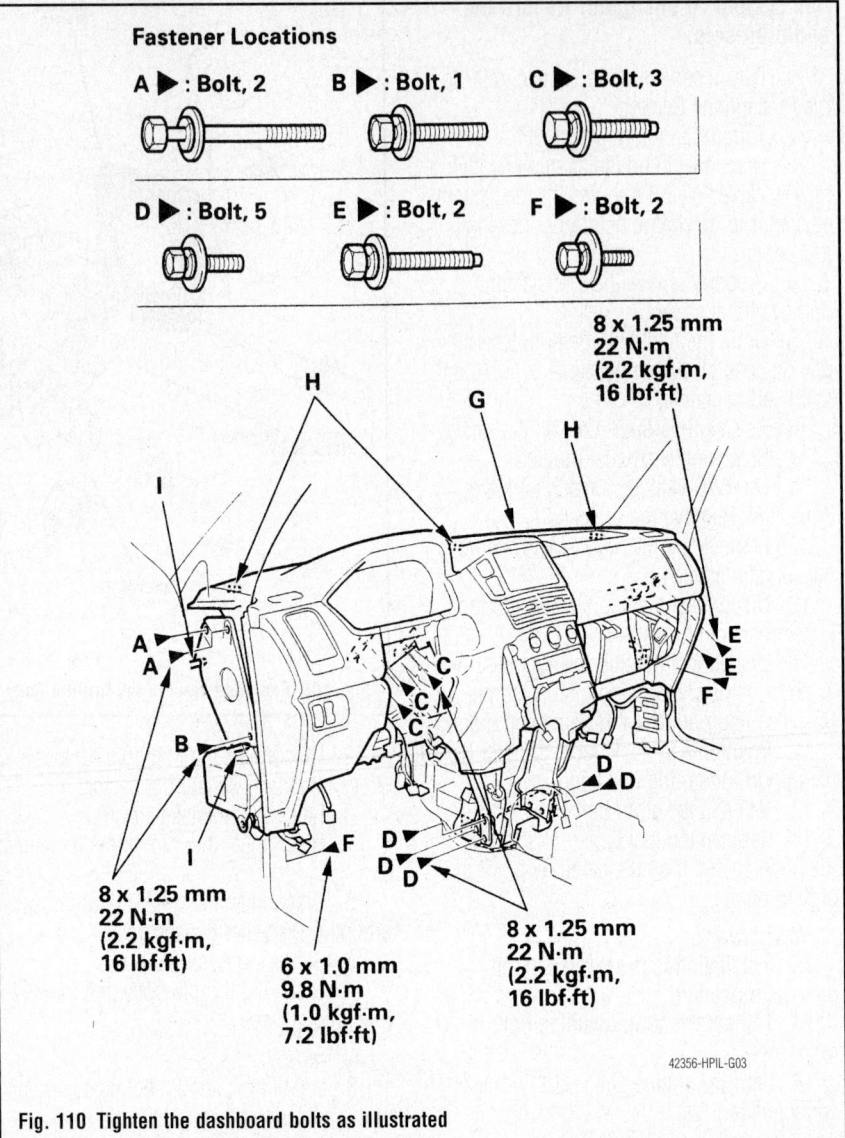

Fastener Locations

A ▶ : Bolt, 2   B ▶ : Bolt, 1   C ▶ : Bolt, 3

D ▶ : Bolt, 5   E ▶ : Bolt, 2   F ▶ : Bolt, 2

8 x 1.25 mm
22 N·m
(2.2 kgf·m,
16 lbf·ft)

8 x 1.25 mm
22 N·m
(2.2 kgf·m,
16 lbf·ft)

6 x 1.0 mm
9.8 N·m
(1.0 kgf·m,
7.2 lbf·ft)

8 x 1.25 mm
22 N·m
(2.2 kgf·m,
16 lbf·ft)

42356-HPIL-G03

Fig. 110 Tighten the dashboard bolts as illustrated

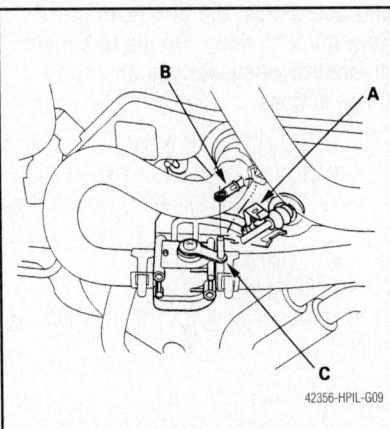

42356-HPIL-G09

Fig. 111 In the engine compartment, open the cable clamp (A), then disconnect the heater valve cable (B) from the valve arm (C)

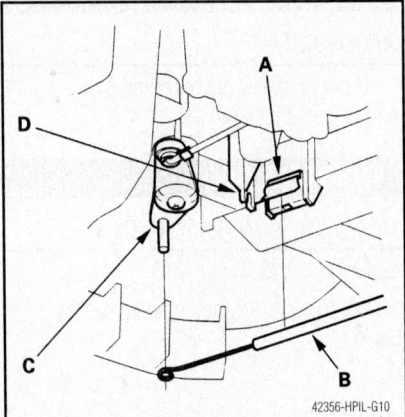

42356-HPIL-G10

Fig. 112 Under the dashboard, disconnect the valve cable housing from the clamp (A) and the cable (B) from the mix control linkage (C)

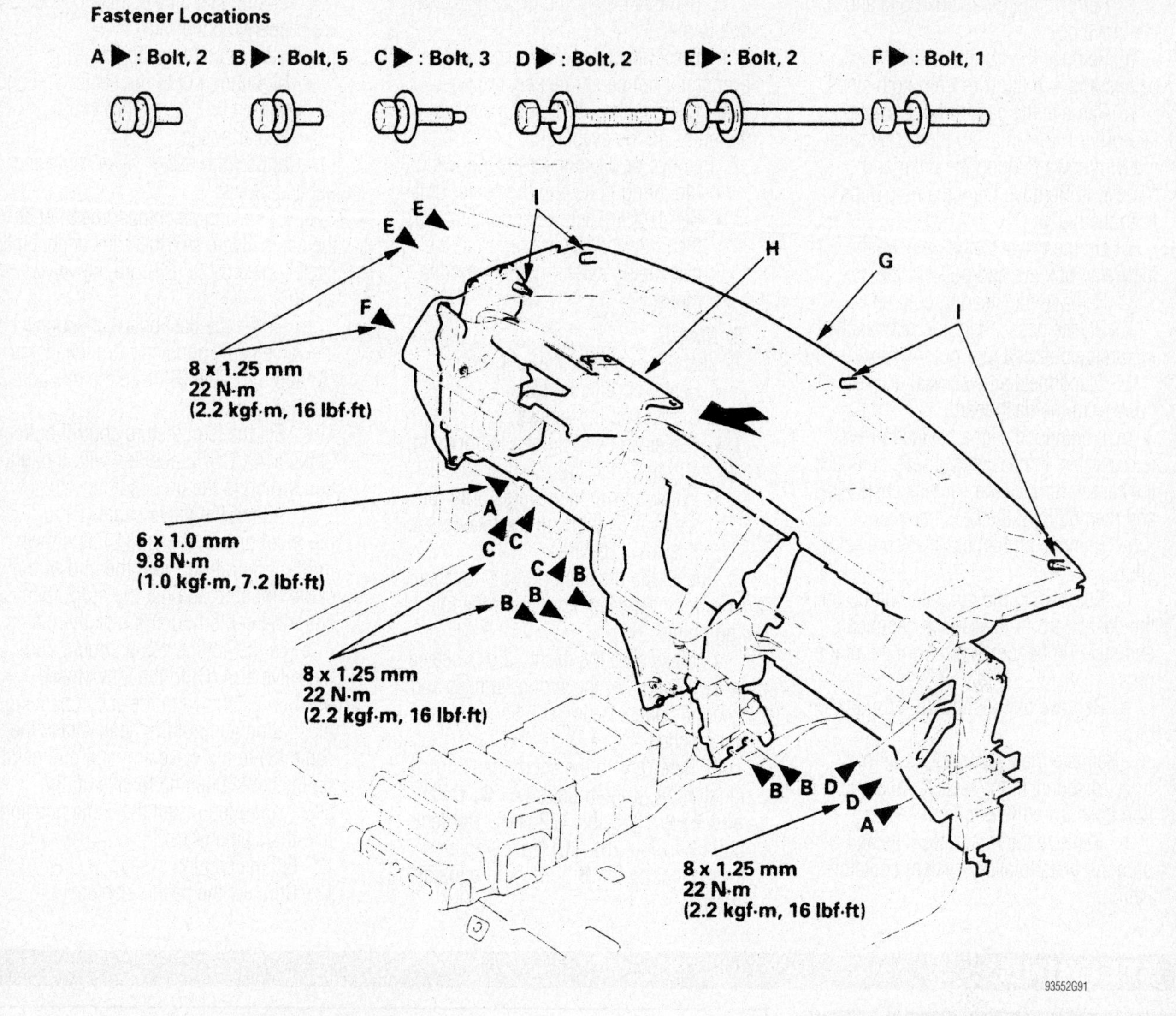

Fastener Locations

A ▶ : Bolt, 2   B ▶ : Bolt, 5   C ▶ : Bolt, 3   D ▶ : Bolt, 2   E ▶ : Bolt, 2   F ▶ : Bolt, 1

8 x 1.25 mm
22 N·m
(2.2 kgf·m, 16 lbf·ft)

6 x 1.0 mm
9.8 N·m
(1.0 kgf·m, 7.2 lbf·ft)

8 x 1.25 mm
22 N·m
(2.2 kgf·m, 16 lbf·ft)

8 x 1.25 mm
22 N·m
(2.2 kgf·m, 16 lbf·ft)

93552G91

Fig. 113 Exploded view of the dashboard mounting

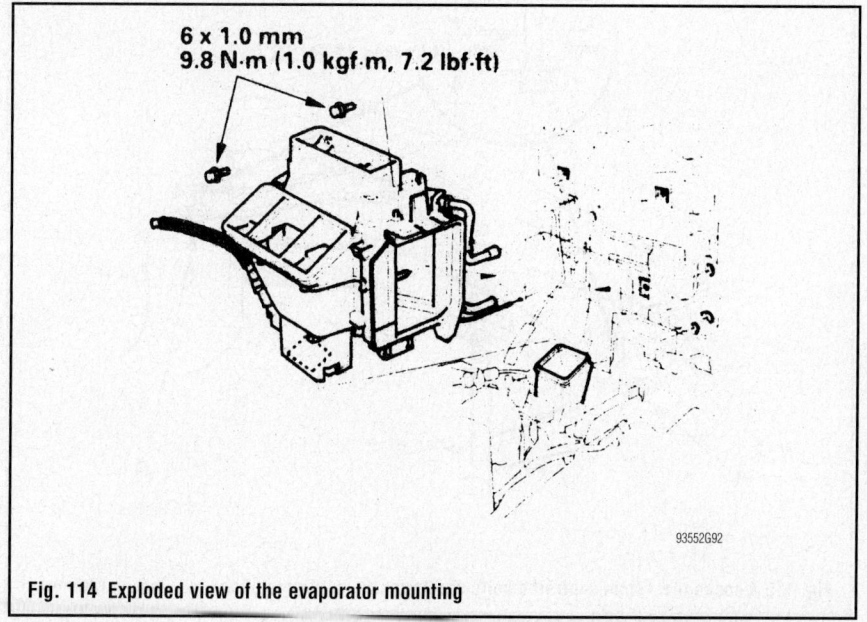

6 x 1.0 mm
9.8 N·m (1.0 kgf·m, 7.2 lbf·ft)

93552G92

Fig. 114 Exploded view of the evaporator mounting

- Heater valve cable from the valve arm. Turn the valve arm to the fully opened position.
- Heater hoses from the heater unit
- Mounting nut from the heater unit. Be careful not to bend or damage fuel or brake lines.

5. Remove the dashboard as follows:

a. Remove the center console by unlatching the clips.

b. Remove the dashboard lower cover screw, gently pull down on the cover to disengage the clips and disconnect the electrical connections.

c. Remove the dashboard side cover by gently pulling and turning to unfasten the clips.

d. Remove the right kick panel.

e. While holding the glove box, remove the box stop from each side, then disconnect the lock from the damper.

f. Remove the glove box bolts and the glove box.

g. Remove the front door trim, kick panels and A-pillar trim from both sides.

h. Remove the cap from the front pillar corner trim. Unfasten the screw, slide the trim upward along the pillar and remove it. Remove the remaining clips from the body.

i. On the driver's side, remove the fuel/relay box nut and pull out the box.

j. Remove the steering column

k. On the passenger side remove the fuse/relay bolt and pull out the box.

l. Disconnect all electrical connections from the dashboard.

m. If equipped with a navigation system, remove the passenger seat, pull back the carpet, remove the harness cushions and then pull out the GPS harness.

n. Remove all harness and connector clips.

o. Remove all the bolts and lift up on the dashboard to release the dashboard and steering hanger beam from the guide pins.

p. Remove the dashboard through the door.

6. Remove the evaporator as follows:

a. Disconnect the receiver and suction lines from the evaporator.

b. Remove the mounting nuts and plug the lines to avoid system contamination.

c. Remove the plastic brace and glove box frame.

d. Disconnect the wire harness and evaporator temperature sensor connector.

e. Remove the self-tapping screws, the nuts and the evaporator.

7. Remove or disconnect the following:
- Mounting bolts and the heater unit
- Self-tapping screws and the clamp, then pull the heater core from the case being careful not to bend the pipes

To install:

8. Install or connect the following:
- Heater core in the case
- Clamp and the screws
- Heater unit and tighten the bolts to 7 ft. lbs. (10 Nm)
- Evaporator in the reverse order of removal. Tighten all the retainers to 7 ft. lbs. (10 Nm) .

9. Install the dashboard in the reverse order of removal keeping in mind the following points:

a. Make sure the dashboard is seated properly and that the wiring harness and steering hanger beam wire harness are not pinched.

b. Referring to the accompanying illustration, tighten bolts **(A, B, C, D and E)** to 16 ft. lbs. (22 Nm). Tighten bolts **F** to 7 ft. lbs. (10 Nm). Apply thread lock to the **B** bolts before installation.

c. Ensure that all electrical connectors are properly connected.

10. Install or connect the following:
- Mounting nut to the heater unit and tighten to 7 ft. lbs. (10 Nm)
- Heater hoses

11. Connect the heater valve cable and adjust as follows:

a. In the engine compartment, open the cable clamp (A), then disconnect the heater valve cable (B) from the valve arm (C).

b. Under the dashboard, disconnect the valve cable housing from the clamp (A) and the cable (B) from the mix control linkage (C).

c. Set the temperature control button to the MAX COOL position with the ignition switch in the on position.

d. Attach the valve cable (B) to the mix control linkage (C) as shown in the illustration, hold the end of the cable housing against the stop, then snap the cable housing into the clamp.

e. In the engine compartment, turn the valve arm (C) to the fully closed position as shown in the accompanying illustration and hold it there. Attach the cable (B) to the valve arm and pull gently on the cable housing to take up the slack, and then install the cable housing into the clamp (A).

12. Fill the cooling system

13. Connect the battery cable.

STEERING

POWER STEERING GEAR

REMOVAL & INSTALLATION

See Figures 115 through 119.

WARNING

Do not permit the steering wheel to turn whenever the steering gear is disconnected from the steering column. Damage to the air bag wiring can result.

1. Before servicing the vehicle, refer to the precautions section.

2. Center the steering wheel and lock it in position.

3. Attach a support fixture to the engine lifting eyes.

4. Remove or disconnect the following:
- Negative battery cable
- Air bag and steering wheel
- Steering joint cover
- Steering flexible joint

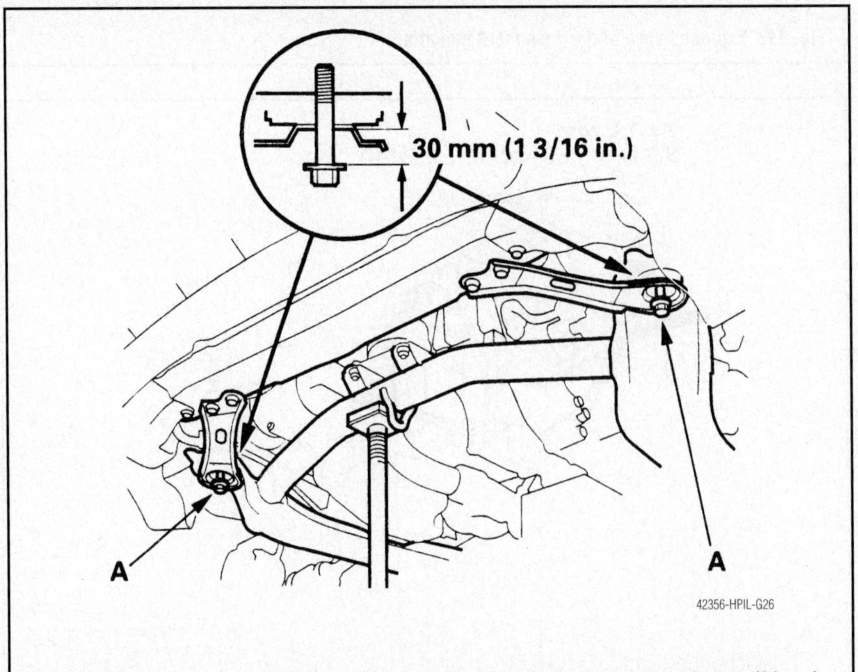
30 mm (1 3/16 in.)
A
A
42356-HPIL-G26

Fig. 115 Loosen the 14mm subframe bolts and lower the subframe about 1³⁄₁₆ inches (30mm)

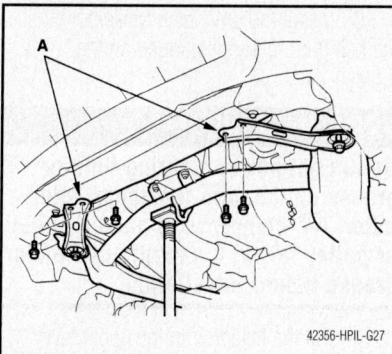

Fig. 116 Remove the Four 12mm stiffener plate bolts

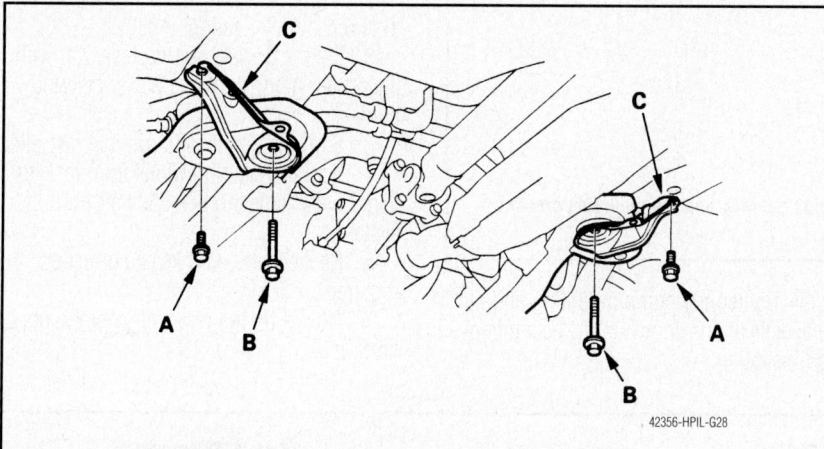

Fig. 117 Remove the Two 12mm bolts (A), then the 14mm bolts (B) and the rear stiffener plates (C) from the sub-frame

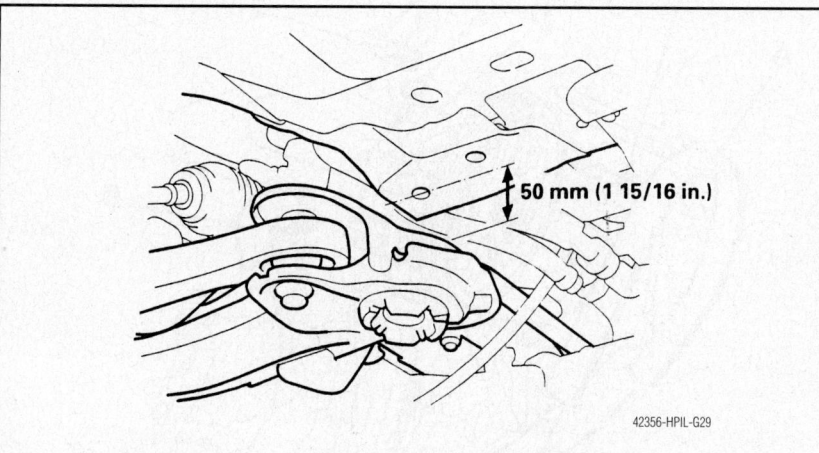

Fig. 118 Lower the transmission jack until the front subframe has dropped about 1 $^{15}/_{16}$ inch (50mm)

- Power steering fluid lines
- 10mm bolt on the engine side mount bracket
- Front wheels
- Outer tie rod ends
- Sub-frame stiffener
- Heated Oxygen (HO$_2$S) sensor connoctors

- 3 way catalytic converter from the mufflers
- Flange bolts from the exhaust rubber mount
- Power steering pressure switch connector
- Propeller shaft protector
- Splash shield

5. Support the front subframe with a jack and support the transmission with a second jack.

6. Loosen the 14mm subframe bolts.

7. Lower the subframe about 1 $^{3}/_{16}$ inches (30mm).

8. Remove or disconnect the following:
- Two 12mm and two 14 stiffener plate bolts

9. Support the transfer case by raising the transmission jack and remove the two 12mm bolts.
- Two 14mm bolts and the rear stiffener plats from the sub-frame

10. Lower the transmission jack until the front subframe has dropped about 1 $^{15}/_{16}$ inch (50mm).
- Power steering line brackets
- Feed line
- Return hose
- Two 10mm bolts from the right side gearbox
- Mounting bracket and cushion
- Two 10mm bolts from the left side gearbox

11. Lower the transmission jack until the front subframe has dropped about 3 $^{15}/_{16}$ inch (100mm).
- Gearbox stiffener bracket

12. Slide the gearbox between the body and front sub-frame towards the left and from the vehicle.

To install:

13. Position the steering gear in the vehicle.

14. Install or connect the following:
- Left steering gear mounting bolts. Tighten the bolts to 43 ft. lbs. (58 Nm).
- Right steering gear mounting bracket. Tighten the bolts to 29 ft. lbs. (39 Nm).
- Return hose
- Feed line
- Power steering line mounting brackets and tighten the bolts to 7 ft. lbs. (10 Nm)

15. Raise the subframe into position. Tighten the 14mm bolts to 76 ft. lbs. (103 Nm) and the 12mm bolts to 86 ft. lbs. (117 Nm).

16. Install or connect the following:
- Front stiffener plates. Tighten the 14mm bolts to 76 ft. lbs. (103 Nm) and the 12mm bolts to 54 ft. lbs. (74 Nm).
- Splash shield
- Propeller shaft protector
- Power steering pressure switch
- 3 way catalytic converter and mufflers. Tighten the nuts to 25 ft. lbs. (33 Nm)
- Rubber exhaust mount and tighten the bolts to 28 ft. lbs. (38 Nm)
- HO$_2$S sensor connectors
- Sub-frame stiffener plate
- 10mm flange bolts on the engine side mount bracket to 33 ft. lbs. (44 Nm)
- Power steering hoses
- Outer tie rod ends
- Front wheels. Position the wheels straight-ahead.
- Steering flexible joint. Tighten the pinch bolts to 16 ft. lbs. (22 Nm)
- Steering joint cover

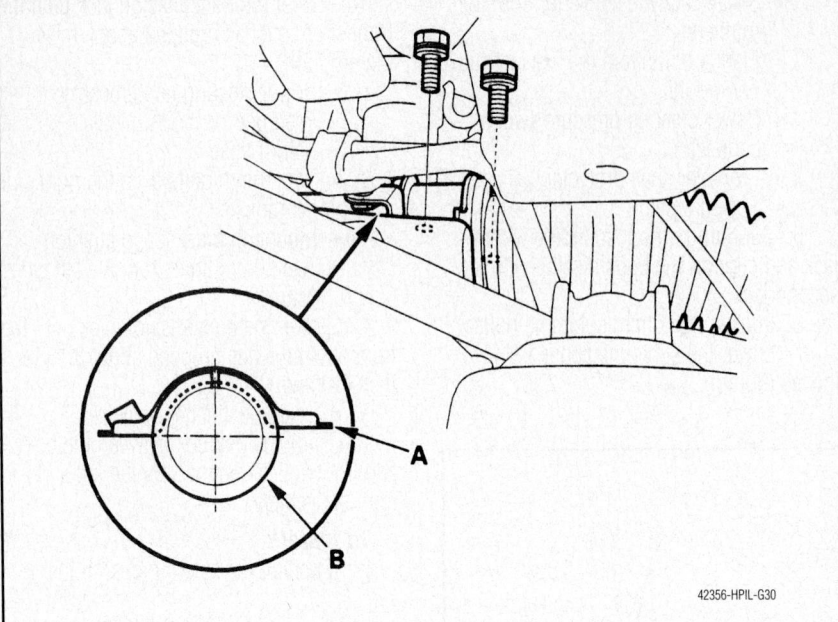

42356-HPIL-G30

Fig. 119 Remove the two 10mm bolts from the right side gearbox and the mounting bracket and cushion

• Negative battery cable

17. Fill the power steering system.

18. Check the wheel alignment and adjust as necessary.

POWER STEERING PUMP

REMOVAL & INSTALLATION

See Figure 120.

1. Place a suitable container under the vehicle.

2. Drain the power steering fluid from the reservoir.

3. Remove the drive belt (A) from the pump pulley.

4. Cover the auto-tensioner, alternator, and A/C compressor with several shop towels to protect them from spilled power steering fluid. Disconnect the pump inlet hose (B) and pump outlet hose (C) from the pump (D), and plug them. Take care not to spill the fluid on the body or parts. Wipe off any spilled fluid at once. Do not turn the steering wheel with the pump removed.

5. Remove the pump mounting bolts (E).

6. Cover the opening of the pump with a piece of tape to prevent foreign material from entering the pump.

To install:

7. Connect the pump inlet hose and pump outlet hose onto the new pump with new O-ring (F).

8. Loosely install the pump in the pump bracket with the mounting bolts, then tighten the pump fittings securely.

9. Tighten the pump mounting bolts to the specifications shown in the accompanying illustration.

10. Install the drive belt. Make sure that the belt is properly positioned on the pulleys.

✳✳ WARNING

Do not get power steering fluid or grease on the auto-tensioner, alternator, A/C compressor, and drive belt or pulley faces. Clean off any fluid or grease before installation.

11. Fill the reservoir to the upper level line.

BLEEDING

1. Before servicing the vehicle, refer to the Precautions Section.

2. Fill the power steering reservoir with the proper grade and type power steering fluid.

3. Start the engine and run it at fast idle.

4. Turn the steering wheel from lock-to-lock several times to bleed the air from the system.

5. Recheck the fluid level, correct as required.

6. Do not fill the reservoir past the upper level line.

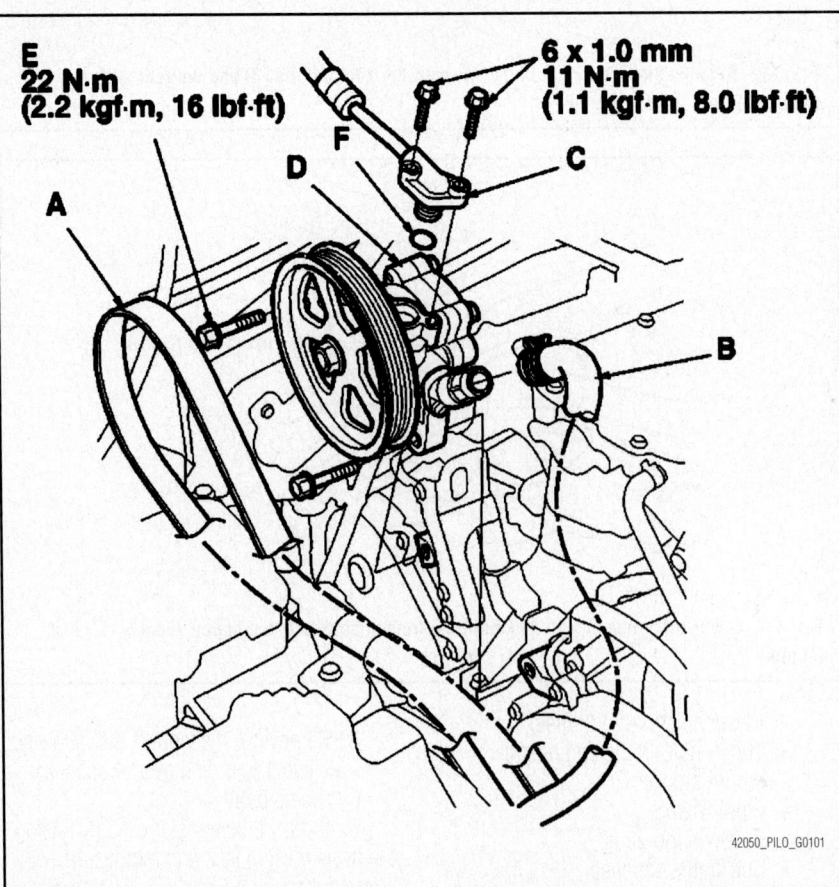

E
22 N·m
(2.2 kgf·m, 16 lbf·ft)

6 x 1.0 mm
11 N·m
(1.1 kgf·m, 8.0 lbf·ft)

42050_PILO_G0101

Fig. 120 Power steering pump removal and installation and tightening specifications

SUSPENSION

COIL SPRING

REMOVAL & INSTALLATION

1. Before servicing the vehicle, refer to the precautions section.

2. Remove the strut from the vehicle and install in a strut spring compressor. Compress the spring until the end of the spring comes away from the spring seat.

3. Remove the upper strut mount, spring seat and related components.

4. Remove the coil spring from the strut spring compressor.

To install:

➡**Use a new self-locking nut.**

5. Compress the spring and position the strut so that the end of the spring aligns with the notch in the spring seat.

6. Install the upper strut mounting components and tighten the nut to 33 ft. lbs. (44 Nm).

7. Install the strut to the vehicle.

8. Check the wheel alignment and adjust as necessary.

LOWER BALL JOINT

REMOVAL & INSTALLATION

The lower ball joints are replaced with the control arms as an assembly.

LOWER CONTROL ARM

REMOVAL & INSTALLATION

1. Before servicing the vehicle, refer to the precautions section.

2. Remove or disconnect the following:
   - Front wheel
   - Lower ball joint
   - Front inner flange bolt
   - Rear inner flange bolt
   - Lower control arm

To install:

➡**Use a new split pin for assembly.**

3. Install or connect the following:
   - Lower control arm. Tighten the inner flange bolts to 69 ft. lbs. (93 Nm).
   - Lower ball joint. Tighten the nut to 43–51 ft. lbs. (59–69 Nm).
   - Front wheel

4. Check the wheel alignment and adjust as necessary.

MACPHERSON STRUT

REMOVAL & INSTALLATION

See Figure 121.

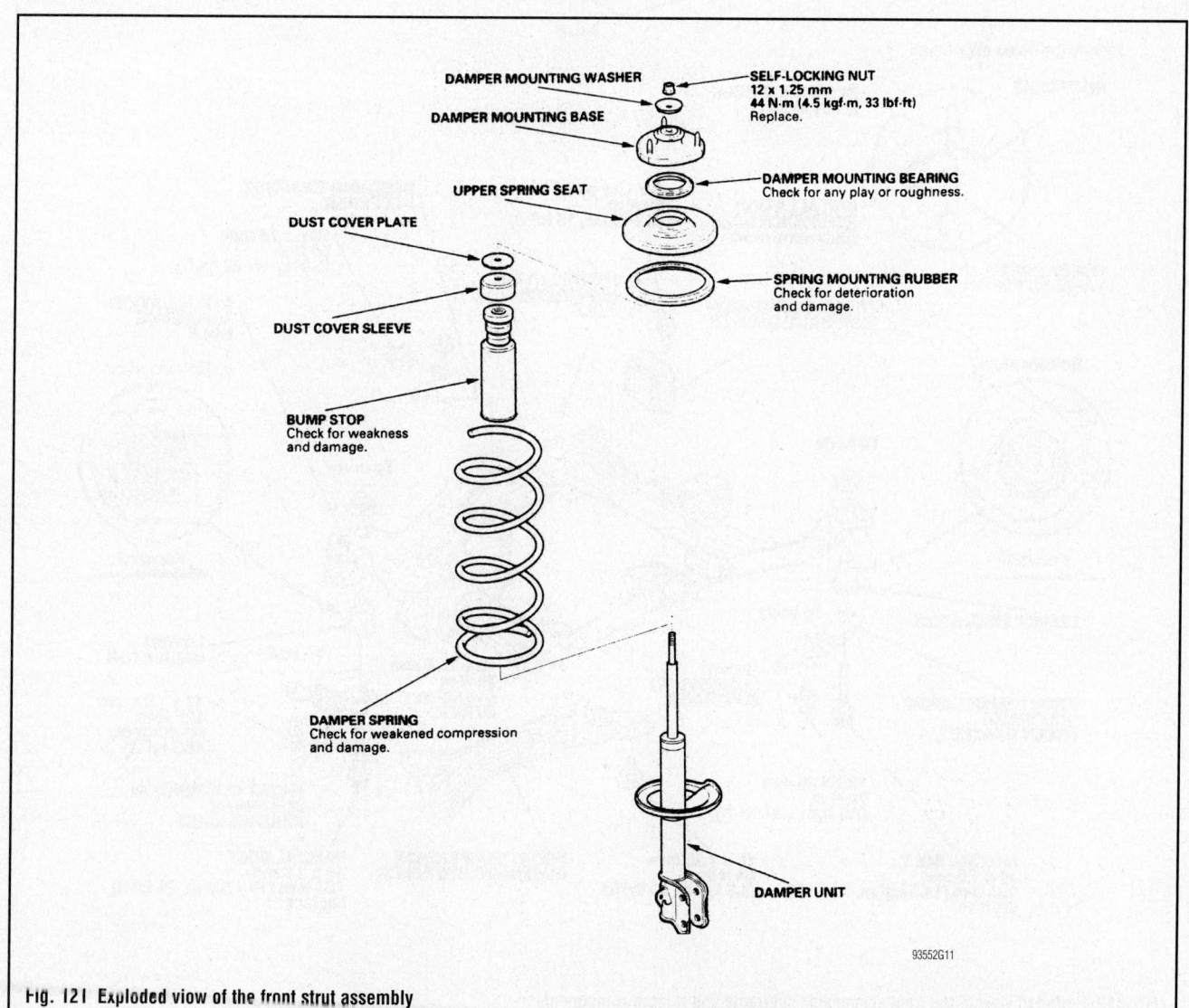

Fig. 121 Exploded view of the front strut assembly

1. Before servicing the vehicle, refer to the precautions section.
2. Remove or disconnect the following:
   - Front wheel
   - Wheel speed sensor wiring bracket
   - Brake hose bracket
   - Stabilizer bar link
   - Strut pinch bolts
   - Upper mount nuts
   - Strut

To install:
3. Install or connect the following:

- Strut. Tighten the upper mount nuts to 43 ft. lbs. (59 Nm).
- Strut pinch bolts. Tighten the nuts to 116 ft. lbs. (157 Nm).
- Stabilizer bar link. Tighten the nut to 58 ft. lbs. (78 Nm).
- Brake hose bracket
- Wheel speed sensor wiring bracket
- Front wheel

4. Check the wheel alignment and adjust as necessary.

STABILIZER BAR

REMOVAL & INSTALLATION
See Figures 122 through 124.

1. Raise and safely support the front of the vehicle
2. Disconnect the stabilizer links from the stabilizer bar on the right and left sides. Refer to the Stabilizer Link procedure in this section.
3. Remove the front suspension subframe

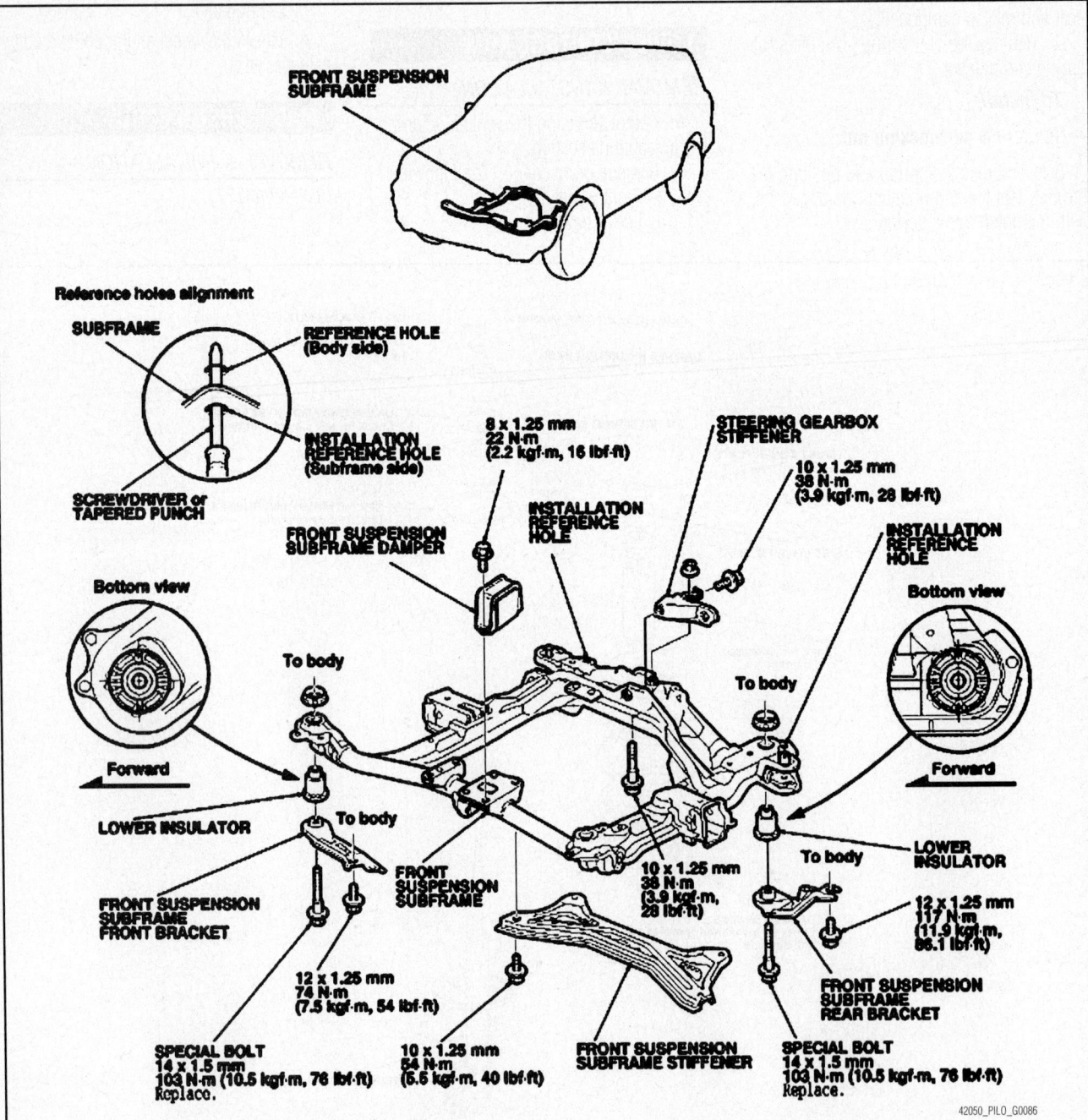

Fig. 122 Exploded view of the front suspension subframe and related components

42050_PILO_G0086

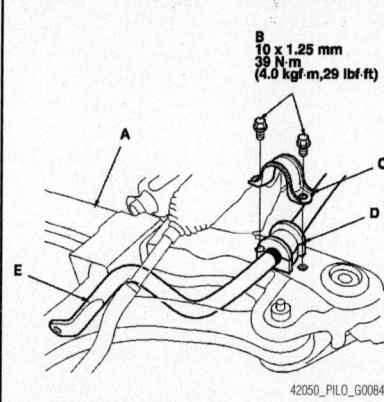

Fig. 123 Remove the front suspension subframe (A), then remove the flange bolts (B) and bushing holders (C), then remove the bushings (D) and the stabilizer bar (E)

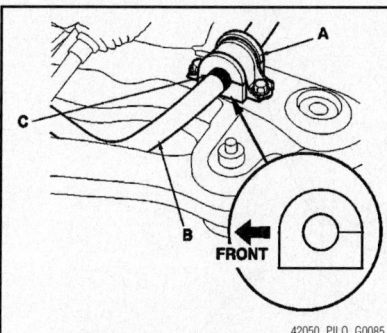

Fig. 124 When installing the stabilizer bar, do not set the bushing (A) on the bent or curved part of the stabilizer bar (B). Also, make sure to align the ends of the paint marks (C) on the stabilizer bar with each end of the bushings

from the body. Refer to the accompanying illustration.

4. Remove the flange bolts (B) and bushing holders (C), then remove the bushings (D) and the stabilizer bar (E).

To install:

5. Install the stabilizer bar in the reverse order of removal, and note these items:

 a. Note the right and left direction of the stabilizer bar.

 b. Do not set the bushing (A) on the bent or curved part of the stabilizer bar (B).

 c. Align the ends of the paint marks (C) on the stabilizer bar with each end of the bushings.

 d. Note the fore/aft direction of the bushing holders.

 e. Refer to the Stabilizer Link Procedure in this section to connect the stabilizer bar to the links.

STABILIZER LINK

REMOVAL & INSTALLATION

See Figures 125 through 127.

1. Raise and safely support the front of the vehicle
2. Remove the front wheels.
3. Remove the flange nuts (A) while holding the respective joint pins (B) with a hex wrench (C), and remove the stabilizer link (D).

To install:

4. Install the stabilizer link (A) on the stabilizer bar (B) and damper (C) with the joint pins (D) set at the center of each moving range.
5. Install the flange nuts, and lightly tighten them.
6. Place a jack under the lower arm, and raise the suspension to load it with the vehicle's weight.

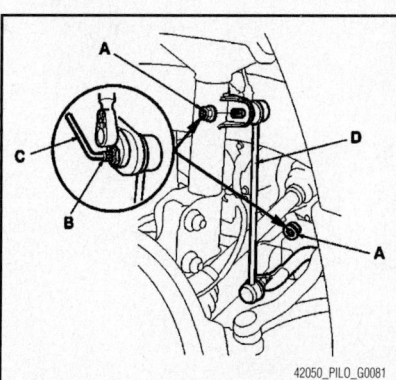

Fig. 125 Remove the flange nuts (A) while holding the respective joint pins (B) with a hex wrench (C), and remove the stabilizer link (D)

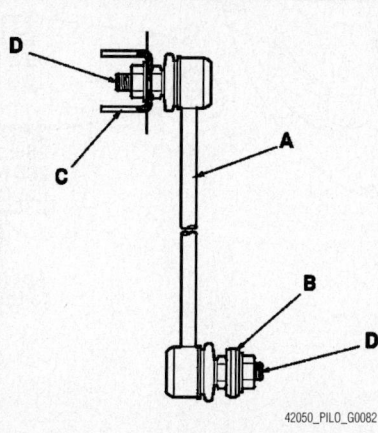

Fig. 126 Install the stabilizer link (A) on the stabilizer bar (B) and damper (C) with the joint pins (D) set at the center of each moving range

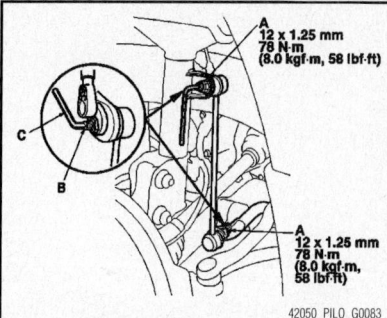

Fig. 127 Tighten the flange nuts (A) to the torque specifications shown here, while holding the respective joint pins (B) with a hex wrench (C)

7. Tighten the flange nuts (A) to the torque specifications shown in the accompanying illustration, while holding the respective joint pins (B) with a hex wrench (C).

STEERING KNUCKLE

REMOVAL & INSTALLATION

The steering knuckle is removed as part of the Wheel Bearing removal & Installation. Please refer to that procedure in this section.

WHEEL BEARINGS

REMOVAL & INSTALLATION

See Figures 128 through 131.

1. Before servicing the vehicle, refer to the precautions section.
2. Remove or disconnect the following:
   - Front wheel
   - Brake hose mounting bolt
   - Brake caliper
   - Wheel speed sensor
   - Spindle nut
   - Brake rotor
   - Outer tie rod end
   - Lower ball joint
   - Steering knuckle
3. Press the hub out of the wheel bearing.
   - Splash guard
   - Snapring and press the wheel bearing out of the steering knuckle
4. If necessary, press the inner bearing race off of the hub.

To install:

➡️Use a new ball joint nut, split pin, snapring and spindle nut for assembly.

5. Press the bearing into the steering knuckle and install the snapring.
6. Install the splash guard.
7. Press the hub into the bearing.

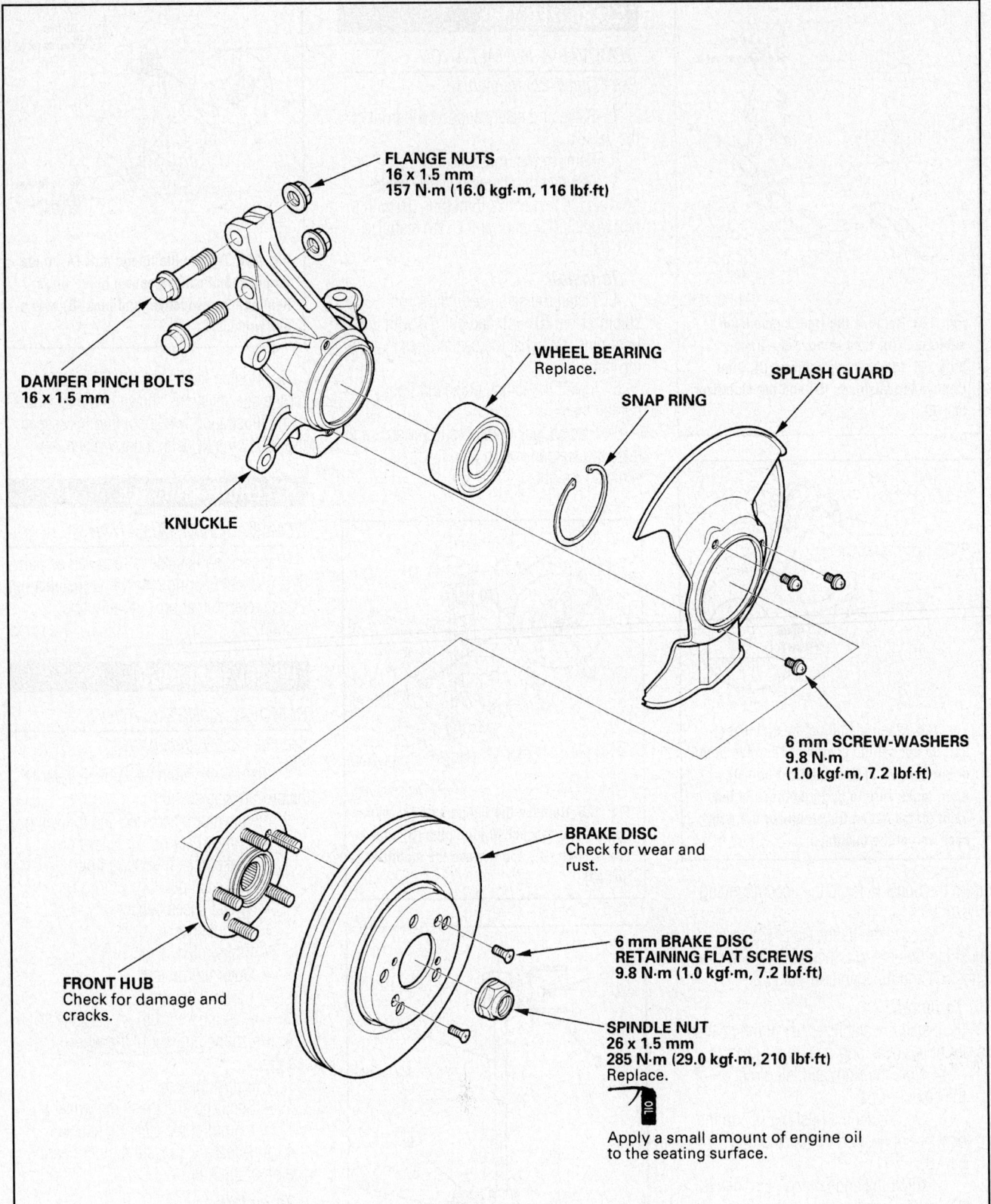

FLANGE NUTS
16 x 1.5 mm
157 N·m (16.0 kgf·m, 116 lbf·ft)

DAMPER PINCH BOLTS
16 x 1.5 mm

KNUCKLE

WHEEL BEARING
Replace.

SNAP RING

SPLASH GUARD

6 mm SCREW-WASHERS
9.8 N·m
(1.0 kgf·m, 7.2 lbf·ft)

BRAKE DISC
Check for wear and
rust.

FRONT HUB
Check for damage and
cracks.

**6 mm BRAKE DISC
RETAINING FLAT SCREWS**
9.8 N·m (1.0 kgf·m, 7.2 lbf·ft)

SPINDLE NUT
26 x 1.5 mm
285 N·m (29.0 kgf·m, 210 lbf·ft)
Replace.

OIL

Apply a small amount of engine oil
to the seating surface.

42356-HPIL-G02

Fig. 128 Front wheel bearing assembly

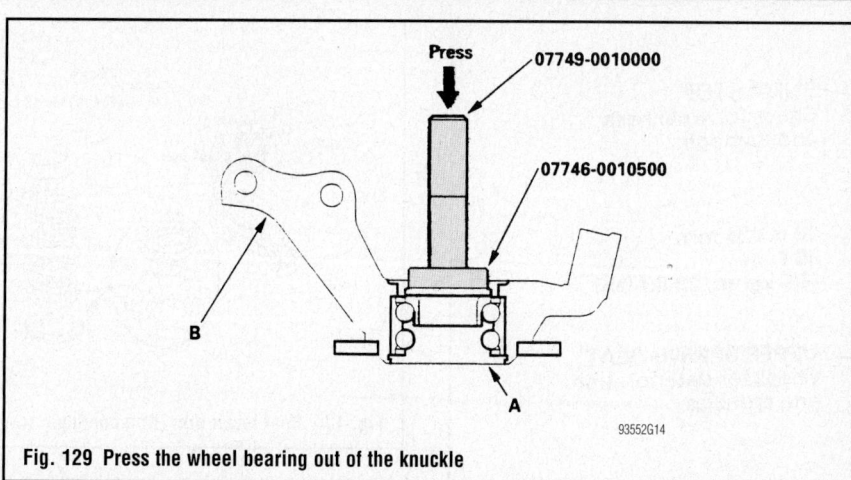

Fig. 129 Press the wheel bearing out of the knuckle

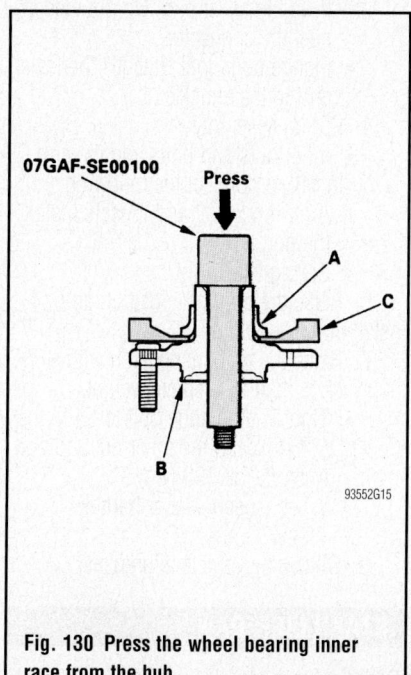

Fig. 130 Press the wheel bearing inner race from the hub

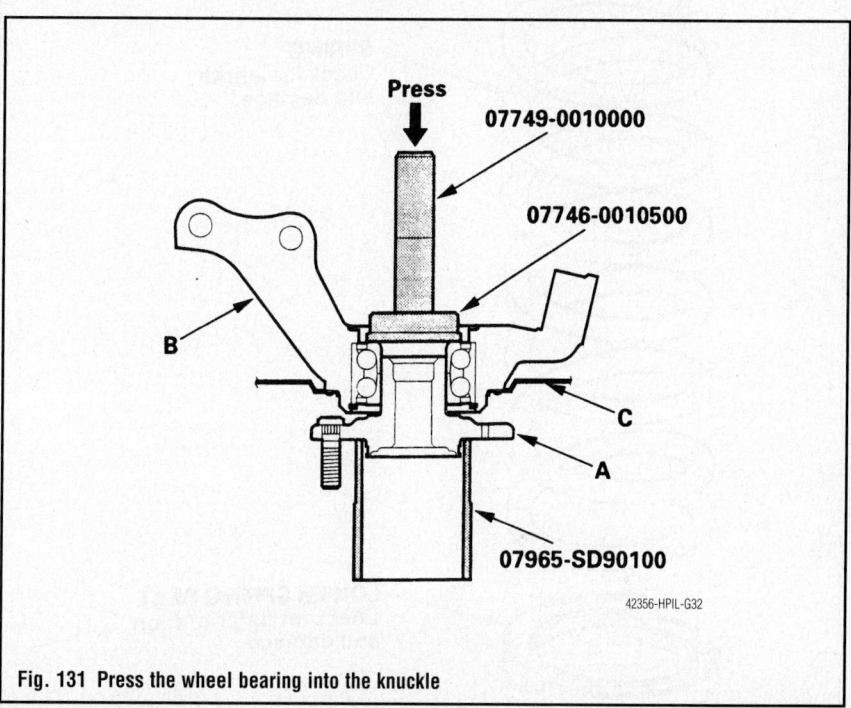

Fig. 131 Press the wheel bearing into the knuckle

8. Install or connect the following:
   - Steering knuckle. Tighten the ball joint nut to 43–51 ft. lbs. (59–69 Nm) and the damper flange bolts to 116 ft. lbs. (157 Nm).
   - Outer tie rod end. Tighten the nut to 40 ft. lbs. (54 Nm).
   - Wheel speed sensor, if equipped
   - Brake caliper and rotor
   - Brake hose
   - Spindle nut. Tighten the nut to 210 ft. lbs. (285 Nm).
   - Front wheel
9. Check the wheel alignment and adjust as necessary.

SUSPENSION

COIL SPRING

REMOVAL & INSTALLATION

See Figure 132.

1. Before servicing the vehicle, refer to the precautions section.
2. Support the vehicle under the lower control arm.
3. Remove or disconnect the following:
   - Rear wheel
   - Stabilizer link from the lower arm
   - Wheel speed sensor wiring harness from the lower arm. Do not disconnect the connector.
   - Upper shock absorber flange bolt
   - Lower control arm bolts

4. Lower the floor jack and remove the coil spring and spring seats.

To install:

➡ **Use new self-locking nuts for assembly.**

5. Place the coil spring and spring seats on the lower control arm and raise into position. Tighten the inboard bolt to 61 ft. lbs. and the outer bolt to 54 ft. lbs. (74 Nm).
6. Install or connect the following:
   - Rear wheel

LOWER CONTROL ARM

REMOVAL & INSTALLATION

Lower Arm (A)

See Figure 133.

REAR SUSPENSION

1. Before servicing the vehicle, refer to the precautions section.
2. Remove or disconnect the following:
   - Lower arm mounting bolt and nut
   - Lower arm
3. Installation is the reverse of removal. Tighten the bolt to 105 ft. lbs. (142 Nm) and the nut to 47 ft. lbs. (64 Nm).

Lower Arm (B)

See Figure 134.

1. Before servicing the vehicle, refer to the precautions section.
2. Support the control arm with a jack.
3. Remove or disconnect the following:
   - Wheel
   - Stabilizer link from the lower arm

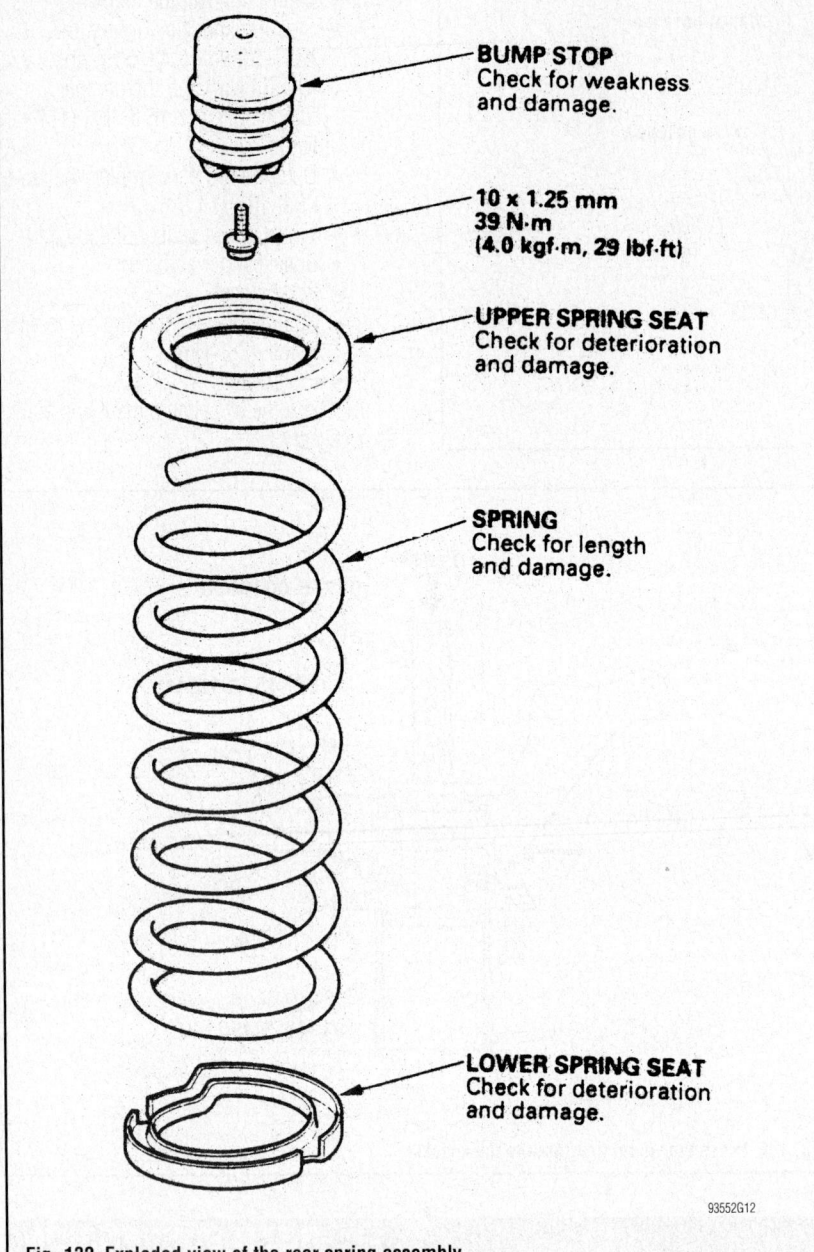

Fig. 132 Exploded view of the rear spring assembly

BUMP STOP
Check for weakness and damage.

10 x 1.25 mm
39 N·m
(4.0 kgf·m, 29 lbf·ft)

UPPER SPRING SEAT
Check for deterioration and damage.

SPRING
Check for length and damage.

LOWER SPRING SEAT
Check for deterioration and damage.

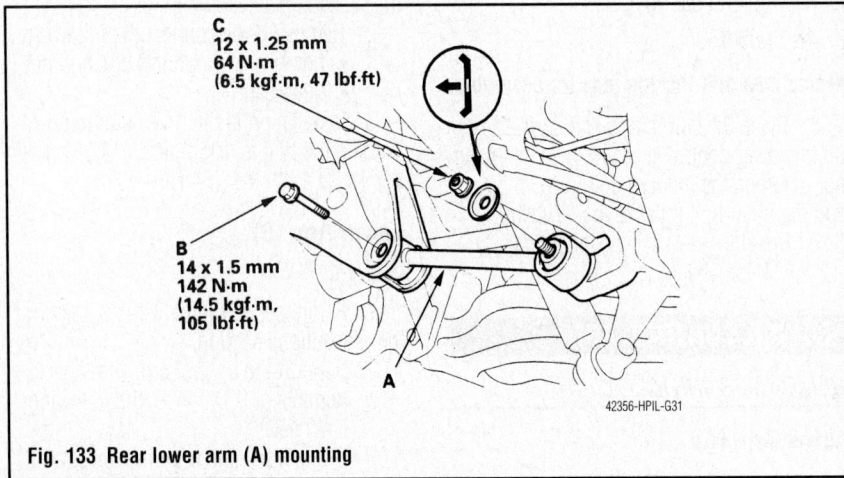

Fig. 133 Rear lower arm (A) mounting

C
12 x 1.25 mm
64 N·m
(6.5 kgf·m, 47 lbf·ft)

B
14 x 1.5 mm
142 N·m
(14.5 kgf·m, 105 lbf·ft)

A

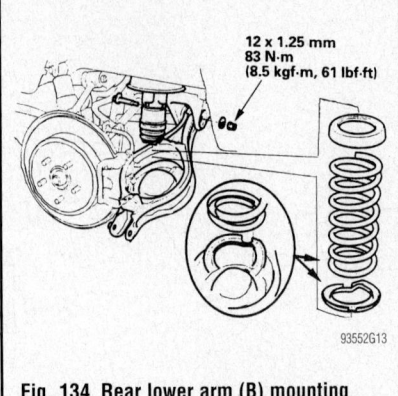

12 x 1.25 mm
83 N·m
(8.5 kgf·m, 61 lbf·ft)

Fig. 134 Rear lower arm (B) mounting

- Wheel speed sensor wiring harness from the lower arm. Do not disconnect the connector.
- Flange bolts that attaches the lower arm to the knuckle

4. Spring assembly
- Inner nuts and bolts and the arm
5. Install or connect the following:
- Arm, inner bolt and loosely install the nut
- Spring assembly
6. Raise the arm into position and install the flange bolt.
7. Raise the rear suspension with a floor jack to load the vehicle weight.
- Tighten the flange bolt to 54 ft. lbs. (74 Nm) and the inner nut and bolt to 61 ft. lbs. (83 Nm).
- Wheel speed sensor harness
- Wheel
8. Check the vehicle alignment.

STABILIZER BAR

REMOVAL & INSTALLATION

See Figures 135 and 136.

➡**Be sure that you have the anti-theft codes for the audio system and navigation system, if equipped. Record the audio presets.**

1. Before servicing the vehicle, refer to the Precautions Section.
2. Turn the ignition switch OFF.
3. Disconnect and isolate the negative battery cable. Wait 3 minutes for the system capacitor to discharge before performing any service.
4. Raise and support the vehicle safely. Remove the tire and wheel assemblies.
5. Disconnect both stabilizer links from the stabilizer bar.
6. Remove the rear suspension subframe.
7. Remove the flange bolts and the bushing holders.

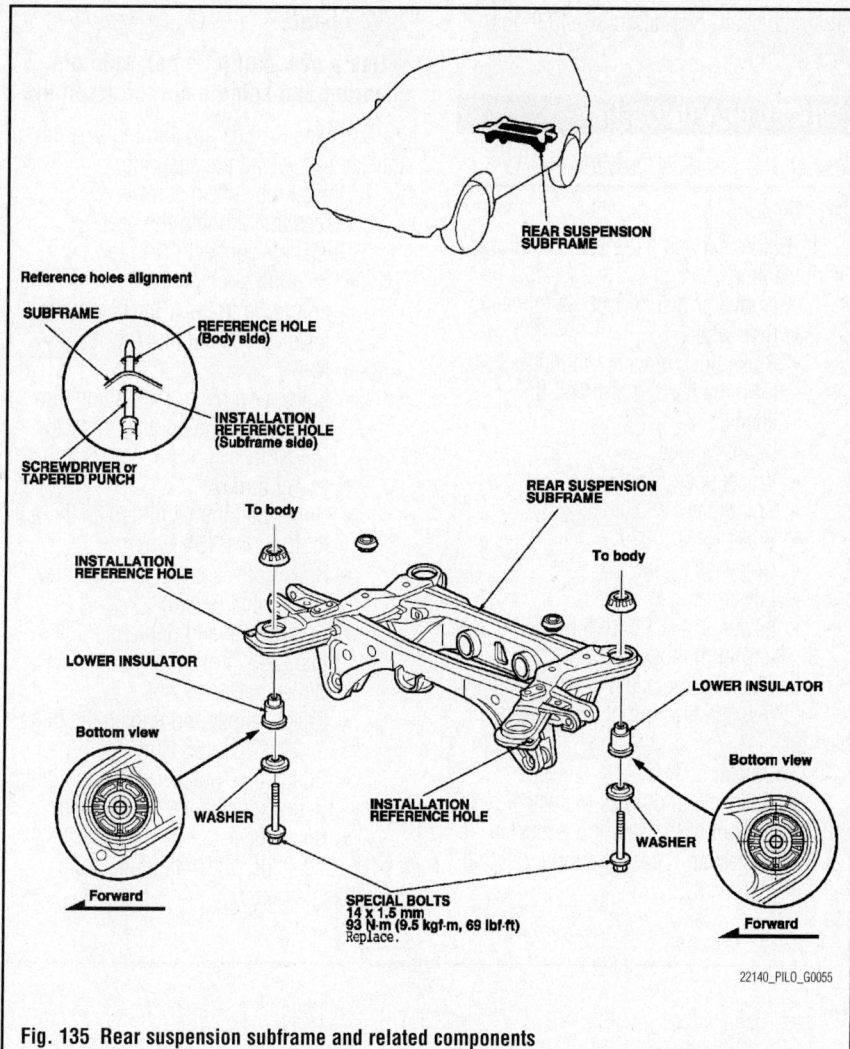

Fig. 135 Rear suspension subframe and related components

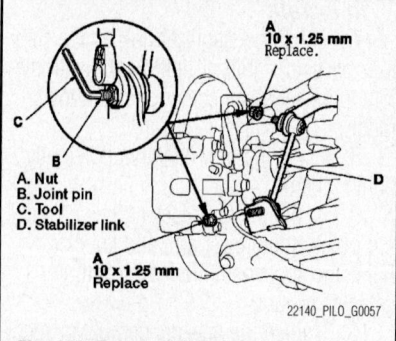

Fig. 137 Rear stabilizer bar link and related components

A. Stabilizer link
B. Stabilizer bar
C. Stabilizer link bracket
D. Joint pins

Fig. 138 Proper rear stabilizer bar link installation points

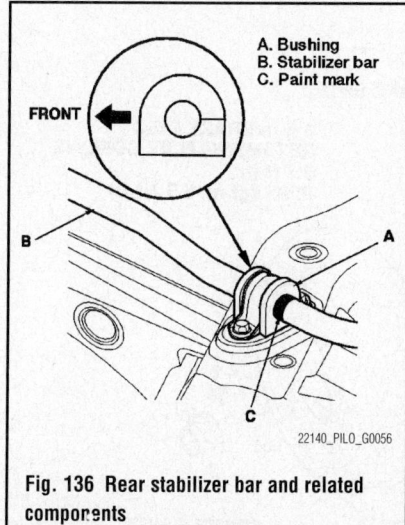

Fig. 136 Rear stabilizer bar and related components

8. Remove the stabilizer bar from its mounting.

To install:

9. Installation is the reverse of the removal procedure.

10. Tighten the stabilizer bar bushing bolts to 16 ft. lbs.

➡ Note the right and left direction of the stabilizer bar. The bar has a paint mark on the center facing downward. Do not set the bushing on the bent or curved part of the stabilizer bar.

11. Check and adjust the rear alignment, as required.

STABILIZER LINK

REMOVAL & INSTALLATION
See Figures 137 through 139.

➡ Be sure that you have the anti-theft codes for the audio system and navigation system, if equipped. Record the audio presets.

1. Before servicing the vehicle, refer to the Precautions Section.
2. Turn the ignition switch OFF.
3. Disconnect and isolate the negative battery cable. Wait 3 minutes for the system

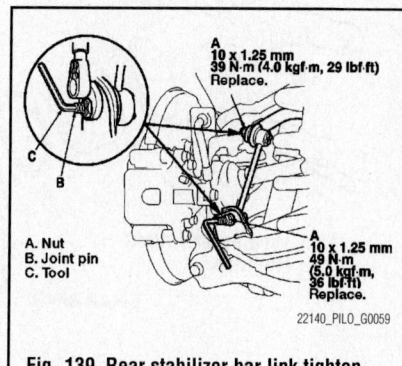

Fig. 139 Rear stabilizer bar link tightening specifications

capacitor to discharge before performing any service.

4. Raise and support the vehicle safely. Remove the tire and wheel assemblies.
5. Remove the self locking nut and the flange nut while holding the respective joint pin with the hex wrench.
6. Remove the stabilizer link from the vehicle.

To install:

7. Install the stabilizer link on the stabilizer bar and stabilizer link bracket with the joint pins set at the center of each moving range.

8. Install the new flange nuts. Lightly tighten them.

9. Position a floor jack at the connecting point of the lower arm and knuckle. Raise the suspension to load it with the vehicles weight.

10. Tighten the new flange nuts to specification. See illustration.

11. Continue the installation in the reverse order of the removal procedure.

12. Check and adjust the rear alignment, as required.

UPPER CONTROL ARM

REMOVAL & INSTALLATION

1. Before servicing the vehicle, refer to the precautions section.

2. Support the control arm at the knuckle.

3. Remove or disconnect the following:
- Wheel
- Upper ball joint from the knuckle
- Upper arm bolt and the arm

4. Installation is the reverse of removal. Tighten the arm bolt to 69 ft. lbs. (93 Nm).

Tighten the ball joint nut to 36–43 ft. lbs. (49–59 Nm).

WHEEL BEARINGS

REMOVAL & INSTALLATION

See Figure 140.

1. Before servicing the vehicle, refer to the precautions section.

2. Remove or disconnect the following:
- Rear wheel
- Brake hose bracket mounting bolts from the trailing arm and the knuckle
- Brake caliper
- Wheel speed sensor
- Spindle nut
- Brake rotor
- Upper ball joint
- Lower arm (A)
- Lower arm (B) from the trailing arm

3. Support the lower arm (B)
- Steering knuckle

4. Press the hub out of the wheel bearing.
- Splash guard
- Snapring and press the wheel bearing out of the steering knuckle

5. If necessary, press the inner bearing race off of the hub.

To install:

→ Use a new ball joint nut, split pin, snapring and spindle nut for assembly.

6. Press the bearing into the steering knuckle and install the snapring.

7. Install the splash guard.

8. Press the hub into the bearing.

9. Install or connect the following:
- Steering knuckle. Tighten the flange bolt to 54 ft. lbs. (74 Nm) and the lower shock nut to 47 ft. lbs. (64 Nm)
- Lower arm (B) to the trailing arm and tighten the bolts to 47 ft. lbs. (64 Nm)
- Lower arm (A)
- Upper ball joint and tighten the nut to 40 ft. lbs. (54 Nm)
- Brake rotor and tighten the screws to 7 ft. lbs. (9 Nm)
- Spindle nut and tighten to 181 ft, lbs. (245 Nm)
- Wheel speed sensor
- Brake caliper and tighten the bolts to 41 ft. lbs. (55 Nm)
- Brake hose bracket mounting bolts to the knuckle and trailing arm
- Rear wheel

10. Check the wheel alignment and adjust as necessary.

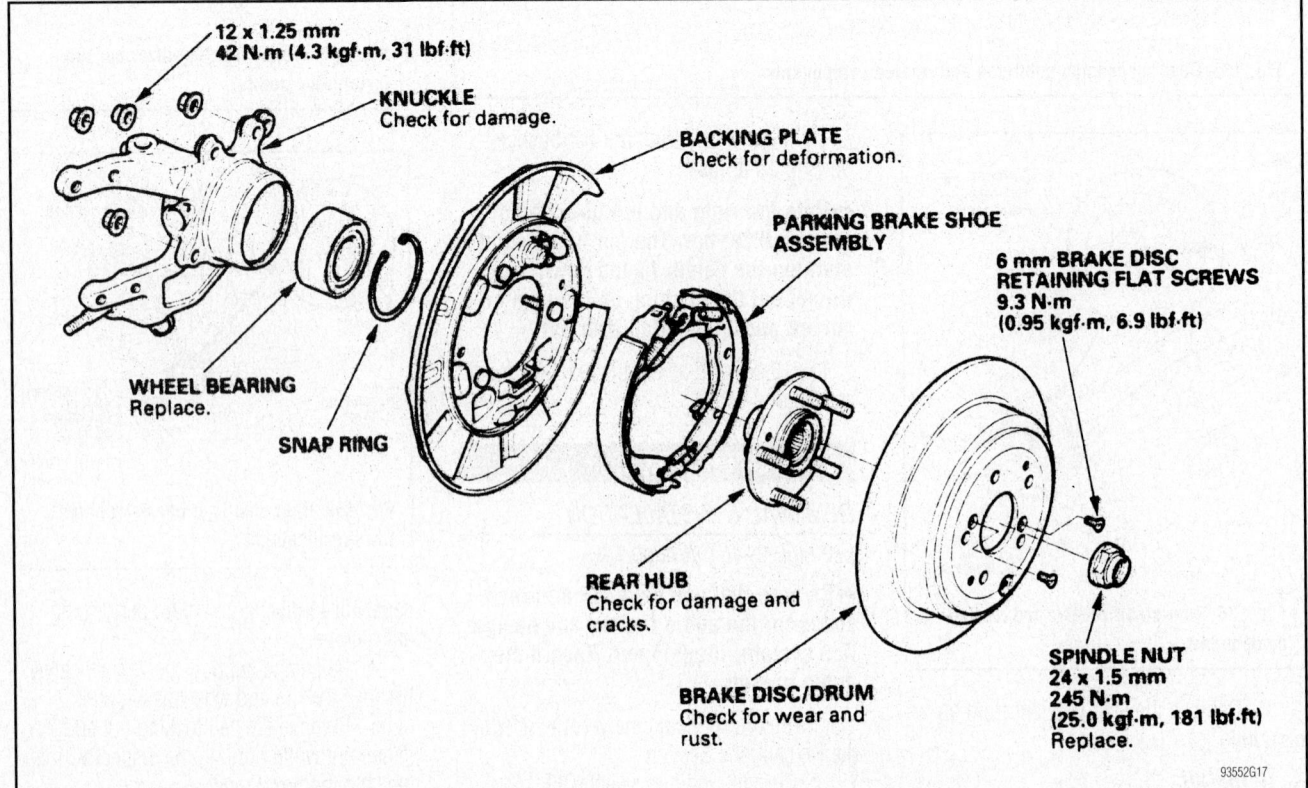

Fig. 140 Exploded view of the rear wheel bearing assembly

12 x 1.25 mm
42 N·m (4.3 kgf·m, 31 lbf·ft)

KNUCKLE
Check for damage.

BACKING PLATE
Check for deformation.

PARKING BRAKE SHOE ASSEMBLY

6 mm **BRAKE DISC RETAINING FLAT SCREWS**
9.3 N·m
(0.95 kgf·m, 6.9 lbf·ft)

WHEEL BEARING
Replace.

SNAP RING

REAR HUB
Check for damage and cracks.

BRAKE DISC/DRUM
Check for wear and rust.

SPINDLE NUT
24 x 1.5 mm
245 N·m
(25.0 kgf·m, 181 lbf·ft)
Replace.

93552G17

HONDA

Ridgeline

10

SPECIFICATIONS AND MAINTENANCE CHARTS

ENGINE AND VEHICLE IDENTIFICATION CHART

		Engine Code					Model Year	
Code	Liters (cc)	Cu. In.	Cyl.	Fuel Sys.	Engine Type	Eng. Mfg.	Code ①	Year
J35A9	3.5 (3471)	222	6	SMFI	SOHC	Honda	6	2006
							7	2007
							8	2008

SOHC: Single Overhead Cam

SMFI: Sequential Multi-port Fuel Injection

① 10th position of VIN

22140_RIDG_C0001

GENERAL ENGINE SPECIFICATIONS

Year	Model	Engine Displacement Liters	Engine ID	Net Horsepower @ rpm	Net Torque @ rpm (ft. lbs.)	Bore x Stroke (in.)	Com-pression Ratio	Oil Pressure @ rpm
2006	Ridgeline	3.5	J35A9	255@5750	252@4500	3.50x3.66	10:01	71@3000 ①
2007	Ridgeline	3.5	J35A9	247@5750	245@4500	3.50x3.66	10:01	71@3000 ①
2008	Ridgeline	3.5	J35A9	247@5750	245@4500	3.50x3.66	10:01	71@3000 ①

① At idle: 10 psi

22140_RIDG_C0002

ENGINE TUNE-UP SPECIFICATIONS

Year	Engine Displacement Liters	Engine ID	Spark Plug Gap (in.)	Ignition Timing (deg. BTDC)	Fuel Pump (psi)	Idle Speed (rpm)	Valve Clearance (in.) In.	Ex.
2006	3.5	J35A9	0.039-0.043	8-12	57-64	680-780	0.008-0.009	0.011-0.013
2007	3.5	J35A9	0.039-0.043	8-12	57-64	680-780	0.008-0.009	0.011-0.013
2008	3.5	J35A9	0.039-0.043	8-12	57-64	680-780	0.008-0.009	0.011-0.013

NOTE: The Vehicle Emission Control Information label often reflects changes made during production and must be used if they differ from this chart.

BTDC: Before top dead center

22140_RIDG_C0003

CAPACITIES

Year	Model	Engine Displacement Liters	Engine ID	Engine Oil with Filter (qts.)	Transmission (pts.) *	Transfer Case (pts.)	Rear Drive Axle (pts.)	Fuel Tank (gal.)	Cooling System (qts.)
2006	Ridgeline	3.5	J35A9	4.5	6.6	0.9	5.58	22.0	①
2007	Ridgeline	3.5	J35A9	4.5	6.6	0.9	5.58	22.0	①
2008	Ridgeline	3.5	J35A9	4.5	6.6	0.9	5.58	22.0	①

NOTE: All capacities are approximate. Add fluid gradually and check to be sure a proper fluid level is obtained.

* Fluid change

① Total capacity: 8.56 qts.

   At coolant change: 6.26 qts.

22140_RIDG_C0004

FLUID SPECIFICATIONS

Year	Model	Engine Displacement Liters	Engine Oil	Auto. Trans.	Drive Axle	Power Steering Fluid	Brake Master Cylinder
2006	Ridgeline	3.5	①	Honda ATF-Z1	Honda VTM-4	Honda P/S Fluid	DOT 3
2007	Ridgeline	3.5	①	Honda ATF-Z1	Honda VTM-4	Honda P/S Fluid	DOT 3
2008	Ridgeline	3.5	①	Honda ATF-Z1	Honda VTM-4	Honda P/S Fluid	DOT 3

DOT: Department Of Transportation

① See oil filler cap.

22140_RIDG_C0013

VALVE SPECIFICATIONS

Year	Engine Displacement Liters	Engine ID	Seat Angle (deg.)	Face Angle (deg.)	Spring Test Pressure (lbs. @ in.)	Spring Free Length (in.)	Stem-to-Guide Clearance (in.) Intake	Stem-to-Guide Clearance (in.) Exhaust	Stem Diameter (in.) Intake	Stem Diameter (in.) Exhaust
2006	3.5	J35A9	45	45	NA	①	0.0008-0.0018	0.0022-0.0031	0.2159-0.2163	0.2146-0.2150
2007	3.5	J35A9	45	45	NA	①	0.0008-0.0018	0.0022-0.0031	0.2159-0.2163	0.2146-0.2150
2008	3.5	J35A9	45	45	NA	①	0.0008-0.0018	0.0022-0.0031	0.2159-0.2163	0.2146-0.2150

NA: Information not available

① Intake: 1.9713 in.

   Exhaust: 2.1060 in.

22140_RIDG_C0005

CAMSHAFT AND BEARING SPECIFICATIONS

All measurements are given in inches.

Year	Engine Displacement Liters	Engine ID	Journal Diameter	Brg. Oil Clearance	Shaft End-play	Runout	Journal Bore	Lobe Lift	
								Intake	Exhaust
2006	3.5	J35A9	NA	0.0002-0.0035	0.0002-0.0008	0.001	NA	①	1.4302
2007	3.5	J35A9	NA	0.0002-0.0035	0.0002-0.0008	0.001	NA	①	1.4302
2008	3.5	J35A9	NA	0.0002-0.0035	0.0002-0.0008	0.001	NA	①	1.4302

NA: Information not available

① Primary: 1.3796

Mid: 1.4348

Secondary: 1.3891

22140_RIDG_C0014

CRANKSHAFT AND CONNECTING ROD SPECIFICATIONS

All measurements are given in inches

Year	Engine Displacement Liters	Engine ID	Crankshaft				Connecting Rod		
			Main Brg. Journal Dia.	Main Brg. Oil Clearance	Shaft End-play	Thrust on No.	Journal Diameter	Oil Clearance	Side Clearance
2006	3.5	J35A9	2.8337-2.8346	0.0008-0.0017	0.0040-0.0140	3	2.1644-2.1654	0.0008-0.0017	0.0060-0.0140
2007	3.5	J35A9	2.8337-2.8346	0.0008-0.0017	0.0040-0.0140	3	2.1644-2.1654	0.0008-0.0017	0.0060-0.0140
2008	3.5	J35A9	2.8337-2.8346	0.0008-0.0017	0.0040-0.0140	3	2.1644-2.1654	0.0008-0.0017	0.0060-0.0140

NA: Information not available

22140_RIDG_C0006

PISTON AND RING SPECIFICATIONS

All measurements are given in inches

Year	Engine Displacement Liters	Engine ID	Piston Clearance	Ring Gap			Ring Side Clearance		
				Top Compression	Bottom Compression	Oil Control	Top Compression	Bottom Compression	Oil Control
2006	3.5	J35A9	0.0006-0.0016	0.0080-0.0140	0.0160-0.0220	0.0080-0.0280	0.0022-0.0031	0.0012-0.0022	snug
2007	3.5	J35A9	0.0006-0.0016	0.0080-0.0140	0.0160-0.0220	0.0080-0.0280	0.0022-0.0031	0.0012-0.0022	snug
2008	3.5	J35A9	0.0006-0.0016	0.0080-0.0140	0.0160-0.0220	0.0080-0.0280	0.0022-0.0031	0.0012-0.0022	snug

22140_RIDG_C0007

TORQUE SPECIFICATIONS
All readings in ft. lbs.

Year	Engine Displacement Liters	Engine ID	Cylinder Head Bolts	Main Bearing Bolts	Rod Bearing Bolts	Crankshaft Damper Bolts	Flywheel Bolts	Manifold Intake	Manifold Exhaust	Spark Plugs	Oil Pan Drain Plug
2006	3.5	J35A9	①	②	③	④	54	16	23	13	29
2007	3.5	J35A9	①	②	③	④	54	16	23	13	29
2008	3.5	J35A9	①	②	③	④	54	16	23	13	29

① Step 1: 29 ft. lbs.
Step 2: 51 ft. lbs.
Step 3: 72 ft. lbs.

② Cap bolts: 54 ft. lbs.
Side bolts: 36 ft. lbs.

③ Step 1: 14 ft. lbs.
Step 2: plus 90 degrees

④ Step 1: 47 ft. lbs.
Step 2: plus 60 degrees

22140_RIDG_C0008

WHEEL ALIGNMENT

Year	Model		Caster Range (+/-Deg.)	Caster Preferred Setting (Deg.)	Camber Range (+/-Deg.)	Camber Preferred Setting (Deg.)	Toe-in (in.)
2006	Ridgeline	F	1.00	+1.88	1.00	-0.50	0 +/- 1/16
		R	—	—	0.75	-0.50	0 +/- 1/16
2007	Ridgeline	F	1.00	+1.88	1.00	-0.50	0 +/- 1/16
		R	—	—	0.75	-0.50	0 +/- 1/16
2008	Ridgeline	F	1.00	+1.88	1.00	-0.50	0 +/- 1/16
		R	—	—	0.75	-0.50	0 +/- 1/16

22140_RIDG_C0010

TIRE, WHEEL AND BALL JOINT SPECIFICATIONS

Year	Model	OEM Tires		Tire Pressures (psi)		Wheel Size	Ball Joint Inspection	Lug Nut Torque (ft. lbs.)
		Standard	Optional	Front	Rear			
2006	Ridgeline	P245/65R17	None	32	32	7.5	NA	94
2007	Ridgeline	P245/65R17	None	32	32	7.5	NA	94
2008	Ridgeline	P245/65R17	None	32	32	7.5	NA	94

NA: Information not available

OEM: Original Equipment Manufacturer

PSI: Pounds Per Square Inch

22140_RIDG_C0009

BRAKE SPECIFICATIONS
All measurements in inches unless noted

Year	Model		Brake Disc			Minimum Lining Thickness	Brake Caliper	
			Original Thickness	Minimum Thickness	Maximum Runout		Bracket Bolts (ft. lbs.)	Mounting Bolts (ft. lbs.)
2006	Ridgeline	F	1.105	1.020	0.0006	0.040	101	53
		R	0.435	0.350	0.0006	0.040	80	16
2007	Ridgeline	F	1.105	1.020	0.0006	0.040	101	53
		R	0.435	0.350	0.0006	0.040	80	16
2008	Ridgeline	F	1.105	1.020	0.0006	0.040	101	53
		R	0.435	0.350	0.0006	0.040	80	16

F: Front

R: Rear

22140_RIDG_C0011

MAINTENANCE MINDER SCHEDULE
Honda Ridgeline

The Ridgeline displays engine oil life and maintenance service items in the information display to indicate when to perform maintenance service. If the engine oil life is 15% or less, based on the onboard computer's caluculations, you will see SERVICE DUE SOON in the information display every time the ignition key is turned to ON. The maintenance minder indicator will also come on and the maintenance code(s) for other scheduled maintenance items needing service will be displayed below the message.

Symbol	Item	Service
A	Engine oil ①	Change
B	Engine oil and filter	Change
	Tires	Rotate
	Brakes	Inspect
	Parking brake adjustment	Check
	Steering gear and linkage	Inspect
	Suspension components	Inspect
	Driveshaft boots	Inspect
	Brake hoses and lines	Inspect
	Exhaust system	Inspect
	Fuel lines and connections	Inspect
1	Tires	Rotate
2	Engine air filter ②	Replace
	Dust and pollen filter ③	Replace
	Accessory drive belt	Inspect
3	Transmission fluid ④	Replace
	Transfer case fluid ④	Replace
4	Spark plugs	Replace
	Timing belt ⑤	Replace
	Water pump	Inspect
	Valve clearance ⑥	Inspect
5	Engine coolant	Replace
6	VTM-4 rear differential fluid	Replace

① If the message SERVICE DUE NOW does not appear more than 12 months after the display is reset, change every year.

② If driven in dusty conditions, replace every 15,000 miles.

③ If driven in urban areas that have a high concentration of soot from industry and diesel, replace every 15,000 miles

④ If regularly driven in mountainous areas at very low speed or trailer towing, change the fluid every 30,000 miles.

⑤ If driven regularly in temperatures over 110 deg.F or below -20 deg.F, or towing a trailer, replace every 60,000 miles.

⑥ Adjust if necessary.

Additionally, replace the brake fluid every 3 years, and inspect the idle speed every 160,000 miles.

To reset the Engine Oil Life Display:

1. Turn the ignition switch to ON.

2. Press the SELECT button repeatedly until the engine oil life display or the service message is displayed.

3. Press the RESET button for about 10 seconds. You will see a MAINT RESET message.

4. Select the appropriate answer, MAINT RESET >N (NO) or MAINT RESET > y (YES) by pressing the SELECT button repeatedly. >N or >Y is displayed on the outside temperature >N or >Y is displayed on the outside temperature display.

5. Select the MAINT RESET > Y (YES), and press and hold the RESET button again to reset the engine oil life to 100%.

22140_RIDG_C0012

PRECAUTIONS

Before servicing any vehicle, please be sure to read all of the following precautions, which deal with personal safety, prevention of component damage, and important points to take into consideration when servicing a motor vehicle:

• Never open, service or drain the radiator or cooling system when the engine is hot; serious burns can occur from the steam and hot coolant.

• Observe all applicable safety precautions when working around fuel. Whenever servicing the fuel system, always work in a well-ventilated area. Do not allow fuel spray or vapors to come in contact with a spark, open flame, or excessive heat (a hot drop light, for example). Keep a dry chemical fire extinguisher near the work area. Always keep fuel in a container specifically designed for fuel storage; also, always properly seal fuel containers to avoid the possibility of fire or explosion. Refer to the additional fuel system precautions later in this section.

• Fuel injection systems often remain pressurized, even after the engine has been turned **OFF**. The fuel system pressure must be relieved before disconnecting any fuel lines. Failure to do so may result in fire and/or personal injury.

• Brake fluid often contains polyglycol ethers and polyglycols. Avoid contact with the eyes and wash your hands thoroughly after handling brake fluid. If you do get brake fluid in your eyes, flush your eyes with clean, running water for 15 minutes. If eye irritation persists, or if you have taken brake fluid internally, IMMEDIATELY seek medical assistance.

• The EPA warns that prolonged contact with used engine oil may cause a number of skin disorders, including cancer. You should make every effort to minimize your exposure to used engine oil. Protective gloves should be worn when changing oil. Wash your hands and any other exposed skin areas as soon as possible after exposure to used engine oil. Soap and water, or waterless hand cleaner should be used.

• All new vehicles are now equipped with an air bag system, often referred to as a Supplemental Restraint System (SRS) or Supplemental Inflatable Restraint (SIR) system. The system must be disabled before performing service on or around system components, steering column, instrument panel components, wiring and sensors. Failure to follow safety and disabling procedures could result in accidental air bag deployment, possible personal injury and unnecessary system repairs.

• Always wear safety goggles when working with, or around, the air bag system. When carrying a non-deployed air bag, be sure the bag and trim cover are pointed away from your body. When placing a non-deployed air bag on a work surface, always face the bag and trim cover upward, away from the surface. This will reduce the motion of the module if it is accidentally deployed. Refer to the additional air bag system precautions later in this section.

• Clean, high quality brake fluid from a sealed container is essential to the safe and proper operation of the brake system. You should always buy the correct type of brake fluid for your vehicle. If the brake fluid becomes contaminated, completely flush the system with new fluid. Never reuse any brake fluid. Any brake fluid that is removed from the system should be discarded. Also, do not allow any brake fluid to come in contact with a painted surface; it will damage the paint.

• Never operate the engine without the proper amount and type of engine oil; doing so WILL result in severe engine damage.

• Timing belt maintenance is extremely important. Many models utilize an interference-type, non-freewheeling engine. If the timing belt breaks, the valves in the cylinder head may strike the pistons, causing potentially serious (also time-consuming and expensive) engine damage. Refer to the maintenance interval charts for the recommended replacement interval for the timing belt, and to the timing belt section for belt replacement and inspection.

• Disconnecting the negative battery cable on some vehicles may interfere with the functions of the on-board computer system(s) and may require the computer to undergo a relearning process once the negative battery cable is reconnected.

• When servicing drum brakes, only disassemble and assemble one side at a time, leaving the remaining side intact for reference.

• Only an MVAC-trained, EPA-certified automotive technician should service the air conditioning system or its components.

BRAKES

GENERAL INFORMATION

PRECAUTIONS

• Certain components within the ABS system are not intended to be serviced or repaired individually.

• Do not use rubber hoses or other parts not specifically specified for and ABS system. When using repair kits, replace all parts included in the kit. Partial or incorrect repair may lead to functional problems and require the replacement of components.

• Lubricate rubber parts with clean, fresh brake fluid to ease assembly. Do not use shop air to clean parts; damage to rubber components may result.

• Use only DOT 3 brake fluid from an unopened container.

• If any hydraulic component or line is removed or replaced, it may be necessary to bleed the entire system.

• A clean repair area is essential. Always clean the reservoir and cap thoroughly before removing the cap. The slightest amount of dirt in the fluid may plug an orifice and impair the system function. Perform repairs after components have been

ANTI-LOCK BRAKE SYSTEM (ABS)

thoroughly cleaned; use only denatured alcohol to clean components. Do not allow ABS components to come into contact with any substance containing mineral oil; this includes used shop rags.

• The Anti-Lock control unit is a microprocessor similar to other computer units in the vehicle. Ensure that the ignition switch is **OFF** before removing or installing controller harnesses. Avoid static electricity discharge at or near the controller.

• If any arc welding is to be done on the vehicle, the control unit should be unplugged before welding operations begin.

BRAKES
BLEEDING THE BRAKE SYSTEM

BLEEDING PROCEDURE

BLEEDING PROCEDURE

See Figure 1.

When bleeding the brake system, observe the following:

• Do not reuse the drained fluid. Use only clean Honda DOT 3 Brake Fluid from an unopened container. Using a non-Honda brake fluid can cause corrosion and shorten the life of the system. Do not mix different brands of brake fluid; they may not be compatible.

• Make sure no dirt or other foreign matter is allowed to contaminate the brake fluid.

• Do not spill brake fluid on the vehicle, it may damage the paint; if brake fluid does

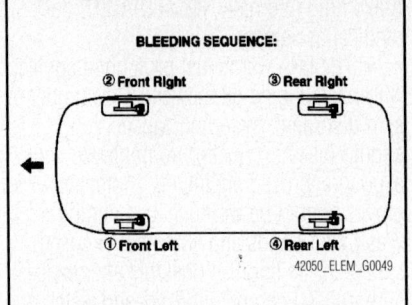

BLEEDING SEQUENCE:

② Front Right   ③ Rear Right

① Front Left   ④ Rear Left

42050_ELEM_G0049

Fig. 1 Proper brake bleeding sequence

contact the paint, wash it off immediately with water.

1. The reservoir on the master cylinder must be at the MAX (upper) level mark at the start of the bleeding procedure and

checked after bleeding each brake caliper. Add fluid as required.

2. Make sure the brake fluid level in the reservoir is at the MAX (upper) level line.

3. Slide a piece of clear plastic hose over the bleed screw, and submerge the other end in a container of new brake fluid.

4. Have someone slowly pump the brake pedal several times, then apply steady pressure.

5. Loosen the left-front brake bleed screw to allow air to escape from the system. Then tighten the bleed screw securely.

6. Repeat the procedure for each caliper until no air bubbles are in the fluid. Bleed the calipers in the sequence shown.

7. Refill the master cylinder reservoir to the MAX (upper) level line.

BRAKES
FRONT DISC BRAKES

❊ CAUTION

Dust and dirt accumulating on brake parts during normal use may contain asbestos fibers from production or aftermarket brake linings. Breathing excessive concentrations of asbestos fibers can cause serious bodily harm. Exercise care when servicing brake parts. Do not sand or grind brake lining unless equipment used is designed to contain the dust residue.

Do not clean brake parts with compressed air or by dry brushing. Cleaning should be done by dampening the brake components with a fine mist of water, then wiping the brake components clean with a dampened cloth. Dispose of cloth and all residue containing asbestos fibers in an impermeable container with the appropriate label. Follow practices prescribed by the Occupational

Safety and Health Administration (OSHA) and the Environmental Protection Agency (EPA) for the handling, processing, and disposing of dust or debris that may contain asbestos fibers.

BRAKE CALIPER

REMOVAL & INSTALLATION

See Figure 2.

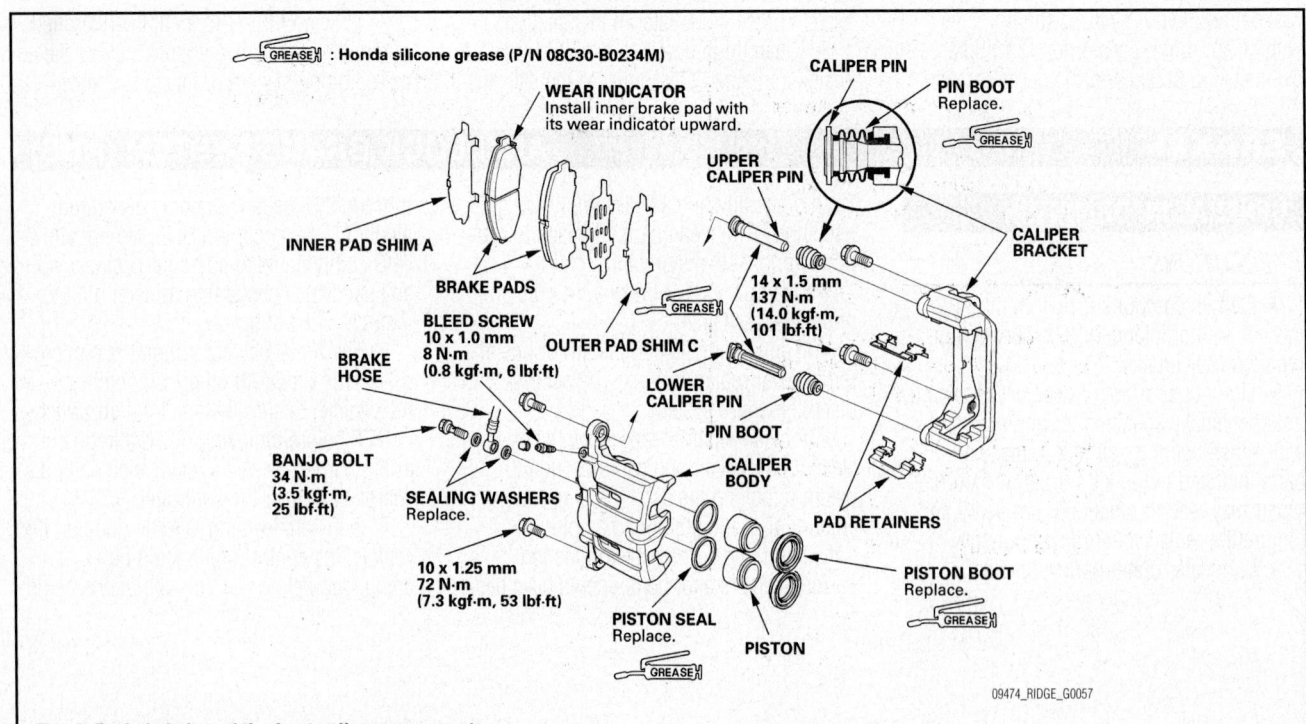

Fig. 2 Exploded view of the front caliper components

1. Before servicing the vehicle, refer to the precautions.

2. Remove some fluid from the reservoir with a suction pump.

3. Remove the front wheels.

4. Remove the banjo bolt and disconnect the brake hose from the caliper. Plug the hose to prevent fluid loss and contamination.

5. Remove the caliper pin bolts and the caliper from its mounting bracket.

To install:

6. Install the caliper over the pads and onto its mounting bracket. Torque both caliper pin bolts to 53 ft. lbs. (72 Nm).

7. Install the brake hose to the caliper using new sealing washers. Carefully torque the banjo bolt to 25 ft. lbs. (34 Nm).

8. Fill the reservoir with fluid and bleed the brakes.

9. Install the front wheels

DISC BRAKE PADS

REMOVAL & INSTALLATION

See Figures 3 and 4.

1. Before servicing the vehicle, refer to the precautions.

2. Remove a small amount of brake fluid from the reservoir using a suction pump.

3. Remove the front wheels.

4. Remove the lower caliper retaining bolt and pivot the caliper upward, off of the pads.

5. Remove the pad springs while holding the pads.

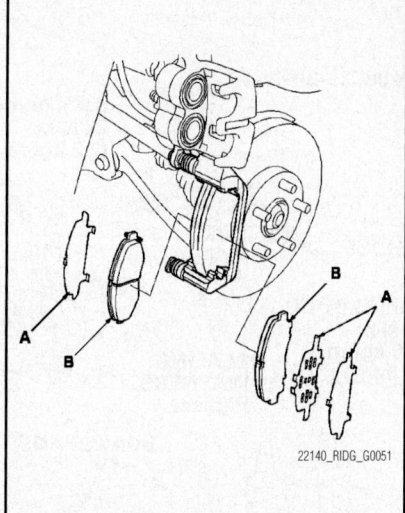

Fig. 3 Front pad shims (A) and the brake pads (B)

6. Remove the pad shim and pad retainers.

7. Remove the disc brake pads from the caliper.

To install:

8. Clean the caliper thoroughly; remove any rust from the lip of the rotor. Check the brake rotor for grooves or cracks. If any heavy scoring is present, the rotor must be replaced.

9. Install the pad retainers. Apply molybdenum brake grease to both surfaces of the shims and the back of the disc brake pads.

10. Install the pads and shims. The pad with the wear indicator goes in the inboard position.

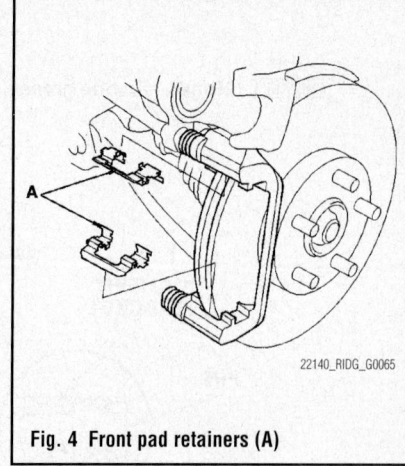

Fig. 4 Front pad retainers (A)

11. Install the pad springs while holding the pads.

→Push in the caliper piston so the caliper will fit over the pads. This is most easily accomplished with a pad spreader or large C-clamp.

12. Install the caliper down into position and tighten the mounting bolt to 53 ft. lbs. (72 Nm).

13. Install the wheels.

14. Add brake fluid to the master cylinder reservoir and install the cap.

15. Depress the brake pedal several times and make sure that the movement feels normal. The first brake pedal application may result in a very long pedal action due to the pistons being retracted. Always make several brake applications before starting the vehicle. Bleed the system if necessary.

BRAKES

✳✳ CAUTION

Dust and dirt accumulating on brake parts during normal use may contain asbestos fibers from production or aftermarket brake linings. Breathing excessive concentrations of asbestos fibers can cause serious bodily harm. Exercise care when servicing brake parts. Do not sand or grind brake lining unless equipment used is designed to contain the dust residue. Do not clean brake parts with compressed air or by dry brushing. Cleaning should be done by dampening the brake components with a fine mist of water, then wiping the brake components clean with a dampened cloth. Dispose of cloth and all residue containing asbestos fibers in

an impermeable container with the appropriate label. Follow practices prescribed by the Occupational Safety and Health Administration (OSHA) and the Environmental Protection Agency (EPA) for the handling, processing, and disposing of dust or debris that may contain asbestos fibers.

BRAKE CALIPER

REMOVAL & INSTALLATION

See Figure 5.

1. Before servicing the vehicle, refer to the precautions.

2. Remove some fluid from the reservoir with a suction pump.

REAR DISC BRAKES

3. Remove the rear wheels.

4. Remove the banjo bolt and disconnect the brake hose from the caliper. Plug the hose to prevent fluid loss and contamination.

5. Remove the caliper mounting bolts and the caliper from its mounting bracket.

To Install:

6. Install the caliper over the pads and onto its mounting bracket. Tighten the caliper bolts to 80 ft. lbs. (108 Nm).

7. Install the brake hose with new sealing washers. Tighten the banjo bolt to 25 ft. lbs. (34 Nm).

8. Fill the reservoir with fluid and bleed the brake system. Adjust the parking brake if necessary.

9. Install the rear wheels.

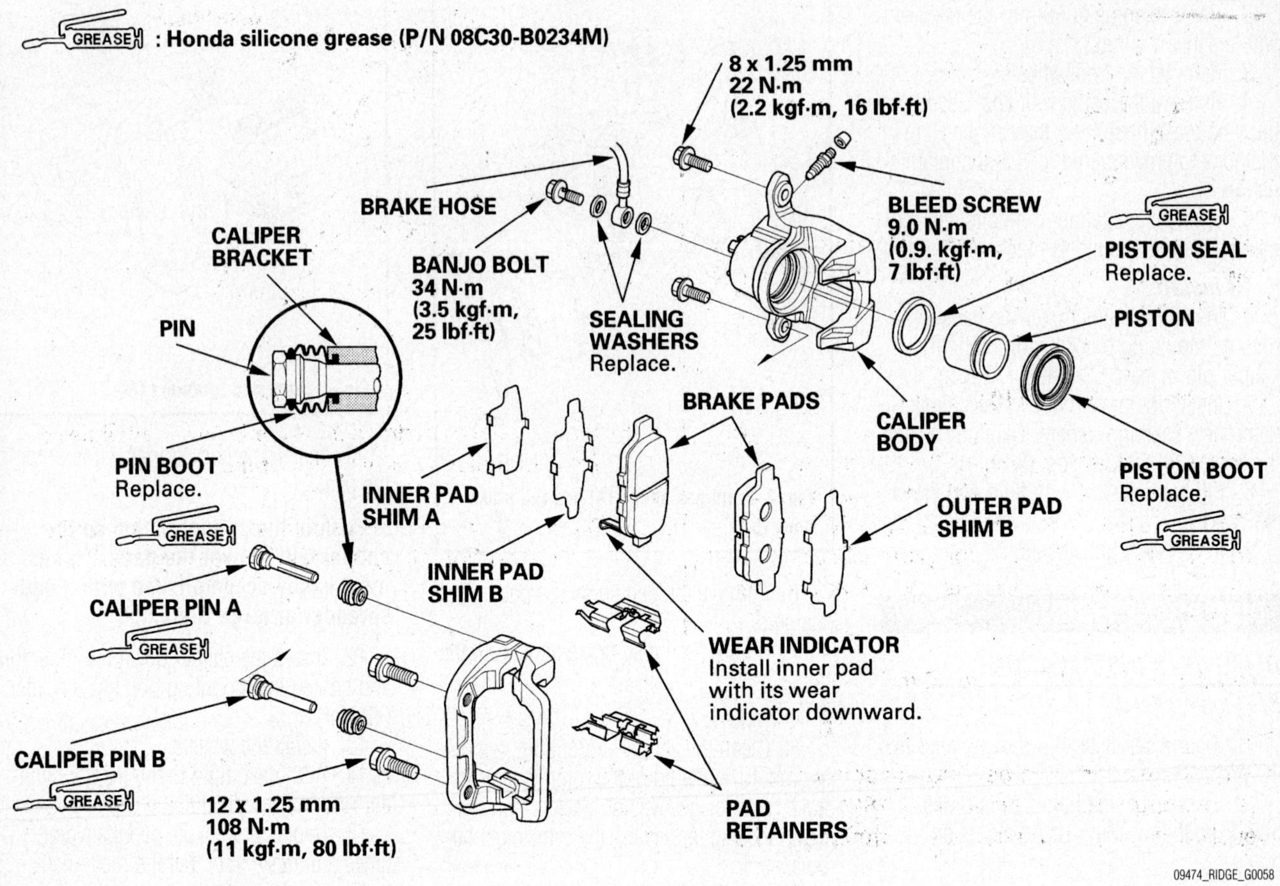

GREASE : Honda silicone grease (P/N 08C30-B0234M)

8 x 1.25 mm
22 N·m
(2.2 kgf·m, 16 lbf·ft)

BRAKE HOSE

BANJO BOLT
34 N·m
(3.5 kgf·m,
25 lbf·ft)

BLEED SCREW
9.0 N·m
(0.9. kgf·m,
7 lbf·ft)

GREASE

PISTON SEAL
Replace.

PISTON

CALIPER
BRACKET

PIN

SEALING
WASHERS
Replace.

BRAKE PADS

CALIPER
BODY

PIN BOOT
Replace.

INNER PAD
SHIM A

OUTER PAD
SHIM B

PISTON BOOT
Replace.

GREASE

GREASE

CALIPER PIN A

INNER PAD
SHIM B

GREASE

WEAR INDICATOR
Install inner pad
with its wear
indicator downward.

CALIPER PIN B

GREASE

12 x 1.25 mm
108 N·m
(11 kgf·m, 80 lbf·ft)

PAD
RETAINERS

09474_RIDGE_G0058

Fig. 5 Exploded view of the rear caliper components

DISC BRAKE PADS

REMOVAL & INSTALLATION

See Figure 6.

1. Remove a small amount of brake fluid from the reservoir using a suction pump.

2. Remove the lower caliper pin bolt and pivot the caliper upward.

3. Remove the pads, shims, and pad retainers.

To install:

4. Clean the caliper thoroughly; remove any dirt or dust. Check the brake rotor for grooves or cracks and machine or replace, as necessary.

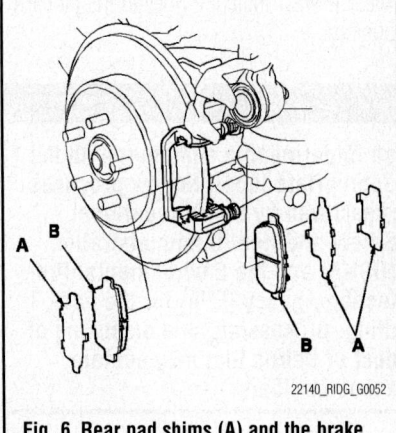

22140_RIDG_G0052

Fig. 6 Rear pad shims (A) and the brake pads (B)

5. Install the pad retainers. Apply molybdenum brake grease to both surfaces of the shims and the back of the disc brake pads.

6. Install the pads and shims. The wear retainer on the inboard pad faces down.

7. Use a suitable tool to push caliper piston into its bore and enable the caliper to fit over the pads. Lubricate the piston boot with silicon grease. Avoid twisting the boot.

8. Rotate the caliper down and tighten the mounting bolts to 80 ft. lbs. (108 Nm)

9. Install the rear wheels

10. Add brake fluid to the master cylinder reservoir. Depress the brake pedal several times to seat the pads. Bleed the brakes if necessary.

PARKING BRAKE SHOES

REMOVAL & INSTALLATION

See Figure 7.

1. Before servicing the vehicle, refer to the precautions.
2. Remove the rear wheels.
3. Release the parking brake tension.
4. Remove the caliper and support it out of the way.
5. Remove the rotor/drum assembly.
6. Remove the upper return springs.
7. Remove the hold-down pins and retainers.
8. Remove the connecting rod.
9. Remove the lower return spring and adjuster.
10. Remove the forward shoe.
11. Disconnect the cable and remove the rear shoe.

To install:

12. Installation is the reverse of removal.
13. Clean the backing plate thoroughly and apply a suitable brake grease to all mounting points and the adjuster threads.
14. Adjust the parking brake shoes.
15. While driving the vehicle safely, pull the parking brake release lever.
16. Press the parking brake pedal 2–4 clicks.
17. Drive the vehicle for one-quarter mile at no more that 30 mph.

18. Stop the vehicle and release the parking brake for 10 minutes to allow the drums to cool.
19. Repeat the procedure 3 more times.
20. Recheck the adjustment.

ADJUSTMENT

See Figures 8 and 9.

1. Before servicing the vehicle, refer to the precautions.
2. With the parking brake released, back off the adjusting nut at the pedal.
3. Remove the access plug (A) from the drum.

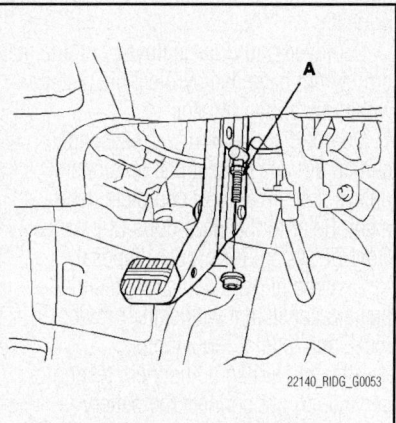

Fig. 8 With the parking brake released, back off the adjusting nut (A) at the pedal

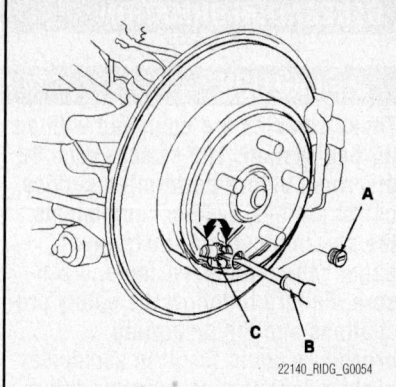

Fig. 9 Remove the access plug (A) from the drum. Using an adjusting spoon (B), turn up the ratchet teeth on the adjuster (C) until the shoes lock the drum. Then, back off 10 clicks and install the plug

4. Using an adjusting spoon, turn up the ratchet teeth on the adjuster until the shoes lock the drum. Then, back off 10 clicks and install the plug.
5. Press the parking brake pedal with about 66 lbs. of force. The pedal should travel 10–12 clicks. If it travels more than that:
6. Turn the adjusting nut at the pedal until the brake shoes drag slightly when the rear wheels are turned.
7. Back off the adjusting nut in half-turn increments until the proper pressure gives the proper number of clicks.

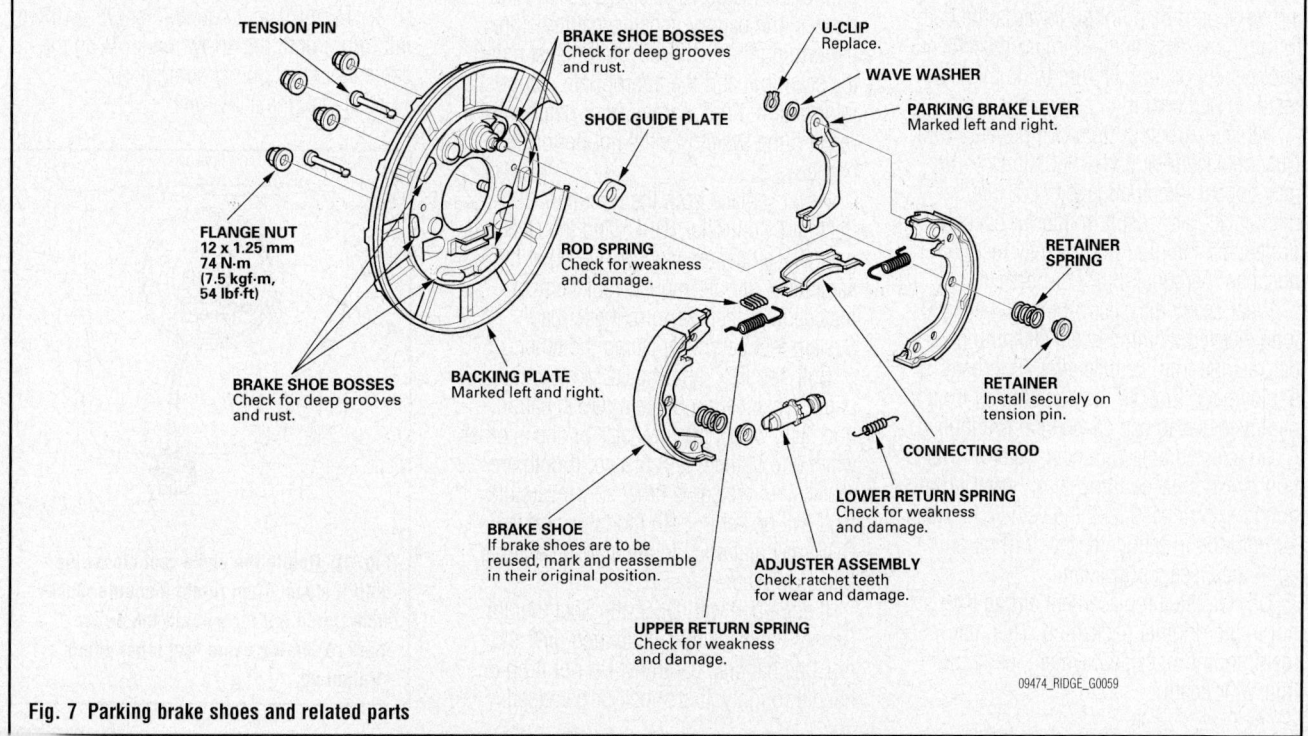

Fig. 7 Parking brake shoes and related parts

CHASSIS ELECTRICAL   **AIR BAG (SUPPLEMENTAL RESTRAINT SYSTEM)**

GENERAL INFORMATION

✳✳ CAUTION

These vehicles are equipped with an air bag system. The system must be disarmed before performing service on, or around, system components, the steering column, instrument panel components, wiring and sensors. Failure to follow the safety precautions and the disarming procedure could result in accidental air bag deployment, possible injury and unnecessary system repairs.

SERVICE PRECAUTIONS

Disconnect and isolate the battery negative cable before beginning any airbag system component diagnosis, testing, removal, or installation procedures. Allow system capacitor to discharge for two minutes before beginning any component service. This will disable the airbag system. Failure to disable the airbag system may result in accidental airbag deployment, personal injury, or death.

Do not place an intact undeployed airbag face down on a solid surface. The airbag will propel into the air if accidentally deployed and may result in personal injury or death.

When carrying or handling an undeployed airbag, the trim side (face) of the airbag should be pointing towards the body to minimize possibility of injury if accidental deployment occurs. Failure to do this may result in personal injury or death.

Replace airbag system components with OEM replacement parts. Substitute parts may appear interchangeable, but internal differences may result in inferior occupant protection. Failure to do so may result in occupant personal injury or death.

Wear safety glasses, rubber gloves, and long sleeved clothing when cleaning powder residue from vehicle after an airbag deployment. Powder residue emitted from a deployed airbag can cause skin irritation. Flush affected area with cool water if irritation is experienced. If nasal or throat irritation is experienced, exit the vehicle for fresh air until the irritation ceases. If irritation continues, see a physician.

Do not use a replacement airbag that is not in the original packaging. This may result in improper deployment, personal injury, or death.

The factory installed fasteners, screws and bolts used to fasten airbag components have a special coating and are specifically designed for the airbag system. Do not use substitute fasteners. Use only original equipment fasteners listed in the parts catalog when fastener replacement is required.

During, and following, any child restraint anchor service, due to impact event or vehicle repair, carefully inspect all mounting hardware, tether straps, and anchors for proper installation, operation, or damage. If a child restraint anchor is found damaged in any way, the anchor must be replaced. Failure to do this may result in personal injury or death.

Deployed and non-deployed airbags may or may not have live pyrotechnic material within the airbag inflator.

Do not dispose of driver/passenger/curtain airbags or seat belt tensioners unless you are sure of complete deployment. Refer to the Hazardous Substance Control System for proper disposal.

Dispose of deployed airbags and tensioners consistent with state, provincial, local, and federal regulations.

After any airbag component testing or service, do not connect the battery negative cable. Personal injury or death may result if the system test is not performed first.

If the vehicle is equipped with the Occupant Classification System (OCS), do not connect the battery negative cable before performing the OCS Verification Test using the scan tool and the appropriate diagnostic information. Personal injury or death may result if the system test is not performed properly.

Never replace both the Occupant Restraint Controller (ORC) and the Occupant Classification Module (OCM) at the same time. If both require replacement, replace one, then perform the Airbag System test before replacing the other.

Both the ORC and the OCM store Occupant Classification System (OCS) calibration data, which they transfer to one another when one of them is replaced. If both are replaced at the same time, an irreversible fault will be set in both modules and the OCS may malfunction and cause personal injury or death.

If equipped with OCS, the Seat Weight Sensor is a sensitive, calibrated unit and must be handled carefully. Do not drop or handle roughly. If dropped or damaged,

replace with another sensor. Failure to do so may result in occupant injury or death.

If equipped with OCS, the front passenger seat must be handled carefully as well. When removing the seat, be careful when setting on floor not to drop. If dropped, the sensor may be inoperative, could result in occupant injury, or possibly death.

If equipped with OCS, when the passenger front seat is on the floor, no one should sit in the front passenger seat. This uneven force may damage the sensing ability of the seat weight sensors. If sat on and damaged, the sensor may be inoperative, could result in occupant injury, or possibly death.

DISARMING THE SYSTEM

Disconnect and isolate the negative battery cable. Wait 3 minutes for the system capacitor to discharge before performing any service.

ARMING THE SYSTEM

To arm the system, connect the negative battery cable.

CLOCKSPRING CENTERING

See Figure 10.

1. Only used cable reel need be centered, new replacement cable reels come centered.

2. Rotate the cable reel clockwise until it stops.

3. Then rotate it counterclockwise (three full turns) until the arrow mark (A) on the cable reel label points straight up.

4. Install the cable reel.

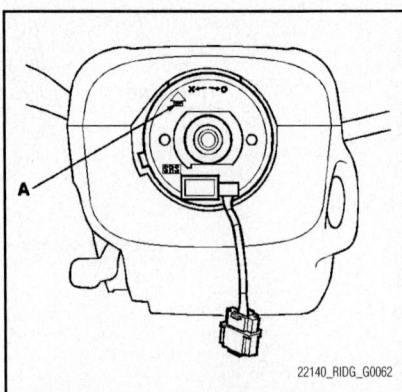

22140_RIDG_G0062

Fig. 10 Rotate the cable reel clockwise until it stops. Then rotate it counterclockwise (three full turns) until the arrow mark (A) on the cable reel label points straight up

DRIVETRAIN

AUTOMATIC TRANSMISSION ASSEMBLY

REMOVAL & INSTALLATION

See Figures 11 through 16.

1. Before servicing the vehicle, refer to the precautions.
2. Make sure you have the radio and navigation system anti-theft codes.
3. Set the wheels in the straight-ahead position and lock the steering.
4. Drain the power steering reservoir.
5. Make sure that the ignition switch is in the **OFF** position.
6. Disconnect the negative, then the positive battery cables.
7. Remove the battery and battery tray.
8. Remove the intake manifold cover.
9. Remove the intake air duct, resonator and air cleaner housing.
10. Remove the battery base and bracket.
11. Remove the front bulkhead cover.
12. Raise and support the vehicle.
13. Remove the splash shield.
14. Drain the transaxle fluid. When the fluid is drained, install the drain plug and new sealing washer. Torque the plug to 36 ft. lbs. (49 Nm).
15. Matchmark and disconnect the steering coupler.
16. Disconnect the power steering outlet line at the pump and remove the hose clamp bolt.
17. Disconnect the transaxle breather hose at the transaxle.
18. Disconnect the solenoid wiring.
19. Disconnect the starter wiring at the starter.
20. Remove the dipstick.
21. Remove the starter.
22. Remove the shift cable bracket nuts.
23. Disconnect the shift cable from the control lever.
24. Disconnect all wiring from the transaxle.
25. Disconnect the cooler hoses.
26. Disconnect the transfer case breather.
27. Remove the connector bracket from the front cylinder head and use that bolt hole to attach the balancer bar front arm.
28. Remove the harness clamp bracket from the rear head and use that bolt hole to attach the balancer bar rear arm.
29. Remove the caps from the front strut mounting nuts. Position the engine hanger adapters (VSB02C000024) with the FRONT mark facing forward, over the flange nuts.

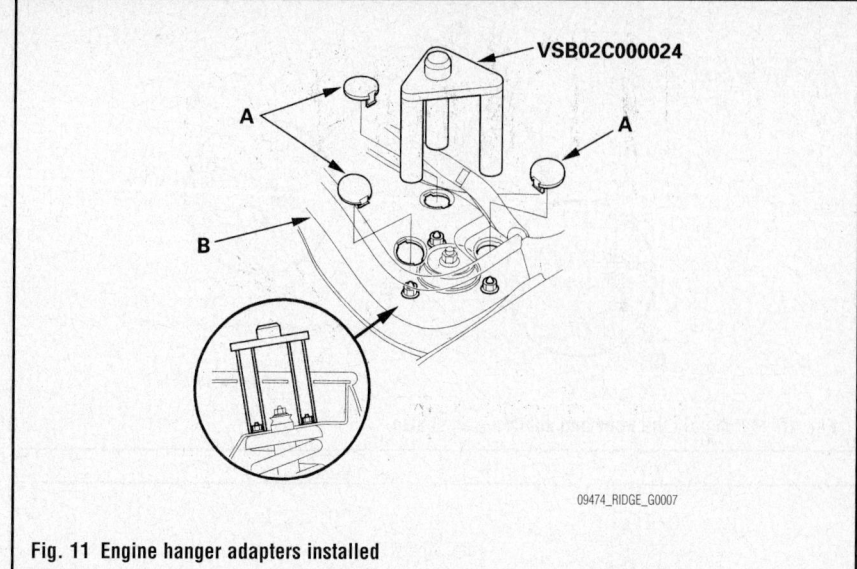

Fig. 11 Engine hanger adapters installed

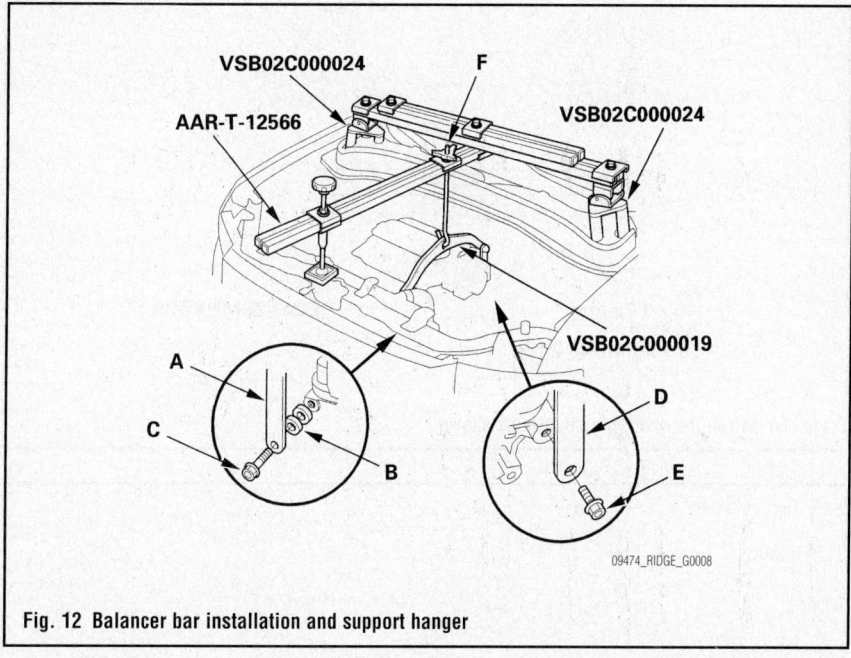

Fig. 12 Balancer bar installation and support hanger

30. Install the balancer bar (VSB02C000019):
    a. Attach the front arm (A) to the front head with spacer (B) and a 10 × 1.25mm bolt (C)
    b. Attach the rear arm (D) to the rear head with an 8 × 1.25mm bolt (E).
31. Install the support hanger (AAR-T-12566):
    a. Attach the hook to the slotted hole in the balancer bar.
    b. Tighten the wing nut (F) by hand to lift and support the engine/transaxle assembly.
32. Remove and discard the front engine mount nut.
33. Remove the subframe stiffener.
34. Remove the exhaust pipe and bracket.
35. Separate the lower arms from the knuckles.
36. Separate the stabilizer bar links from the knuckles.
37. Separate the tie rod ends from the knuckles.
38. Disconnect the transfer case breather.
39. Matchmark and disconnect the driveshaft from the transfer case.
40. Remove the transfer case.
41. Remove the torque converter cover.
42. Remove the 8 torque converter-to-flexplate bolts.

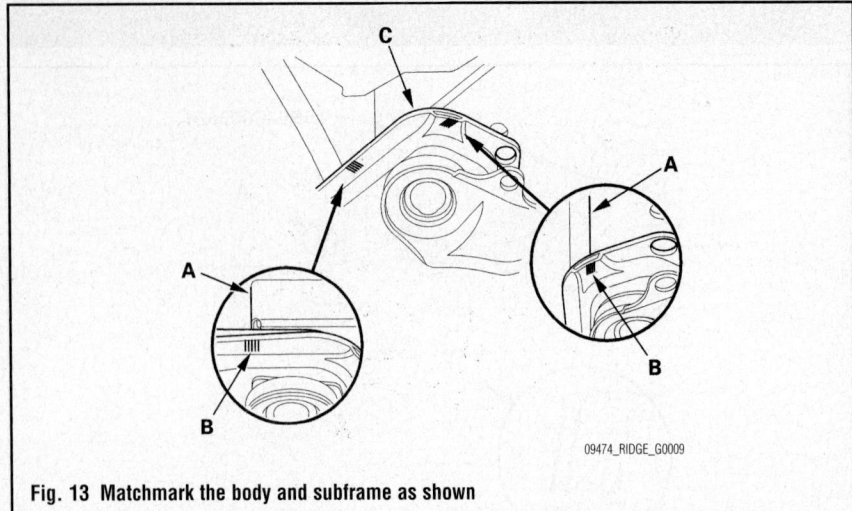

Fig. 13 Matchmark the body and subframe as shown

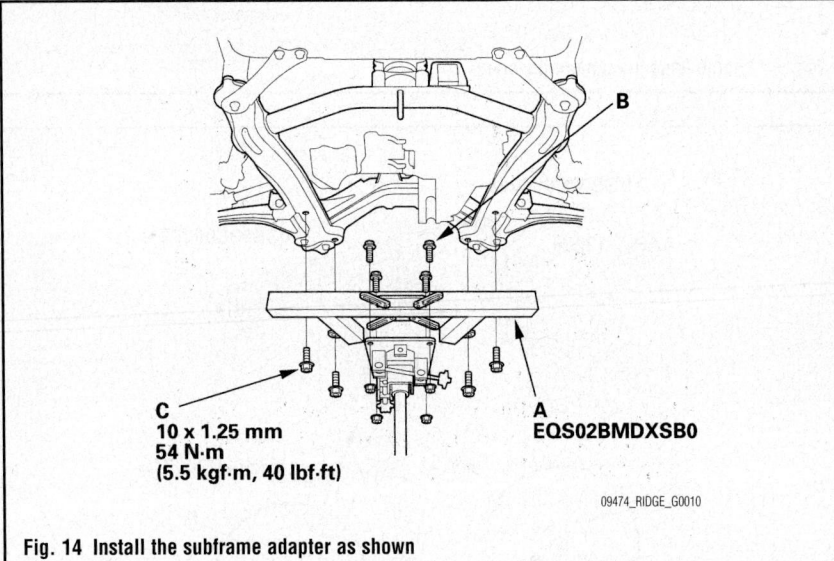

C
10 x 1.25 mm
54 N·m
(5.5 kgf·m, 40 lbf·ft)

A
EQS02BMDXSB0

Fig. 14 Install the subframe adapter as shown

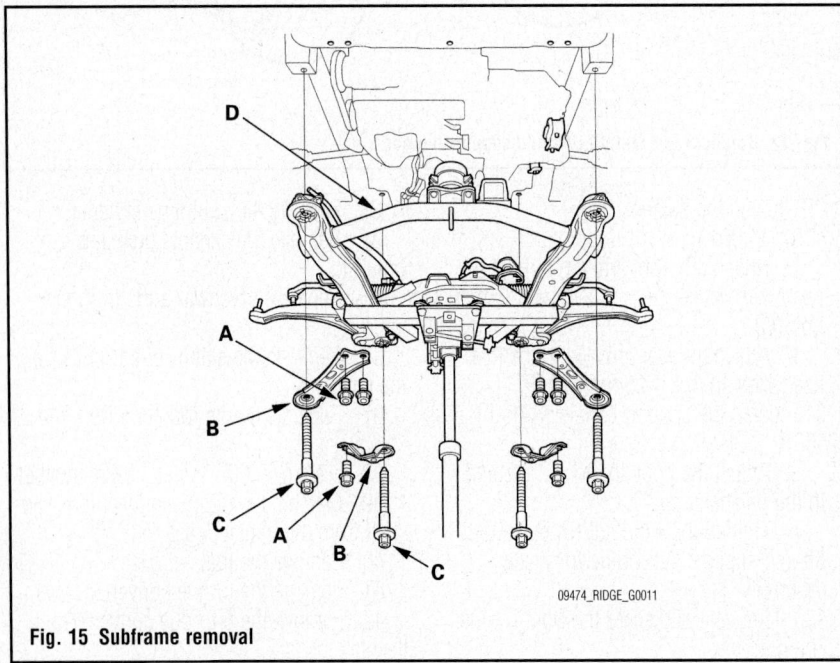

Fig. 15 Subframe removal

43. Disconnect the power steering pressure switch.

44. Remove the power steering hose clamp from the subframe.

45. Remove the 4 transaxle lower mount nuts.

46. Matchmark the body and subframe as shown.

47. Install the subframe adapter (A) as shown.

48. Remove the six 12 × 1.25mm bolts (A) securing the subframe stiffeners (B), the 4 subframe mounting bolts (C) and the stiffeners. Then, lower the subframe (D).

49. Remove the 3 rear mount bracket bolts.

50. Remove the transaxle lower mounts.

51. Remove the halfshafts.

52. Remove the exhaust manifold bracket and heat shield.

53. Remove the intermediate shaft.

54. Remove the front mount bracket.

55. Remove the transaxle-to-engine bolts.

56. lower the transaxle by loosening the wing nut of the support and tile the engine just enough for the transaxle to clear the frame.

57. Slide a jack under the transaxle and roll it out.

To install:

58. Make sure the dowel pins are installed in the converter housing.

59. Install the transmission lower mounts. Torque to 33 ft. lbs. (44 Nm).

60. Mate the transaxle to the engine. Install the bolts and torque to 47 ft. lbs. (64 Nm).

61. Install the front bracket using new bolts. Torque to 40 ft. lbs. (54 Nm).

62. Install the intermediate shaft.

63. Install the halfshafts.

64. Install the subframe, aligning the matchmarks and tightening the bolts as shown.

65. Install the rear mount bracket bolts. Torque to 28 ft. lbs. (38 Nm).

66. The remainder of installation is the reverse of removal. Observe the following torques:

- Lower mount nuts: 33 ft. lbs. (44 Nm)
- Torque converter bolts, in 2 equal steps, alternating the tightening sequence, to 104 inch lbs. (12 Nm).
- Torque converter cover: 104 inch lbs. (12 Nm)
- Transfer case: 38 ft. lbs. (51 Nm)
- Driveshaft-to-transfer case: 53 ft. lbs. (72 Nm)

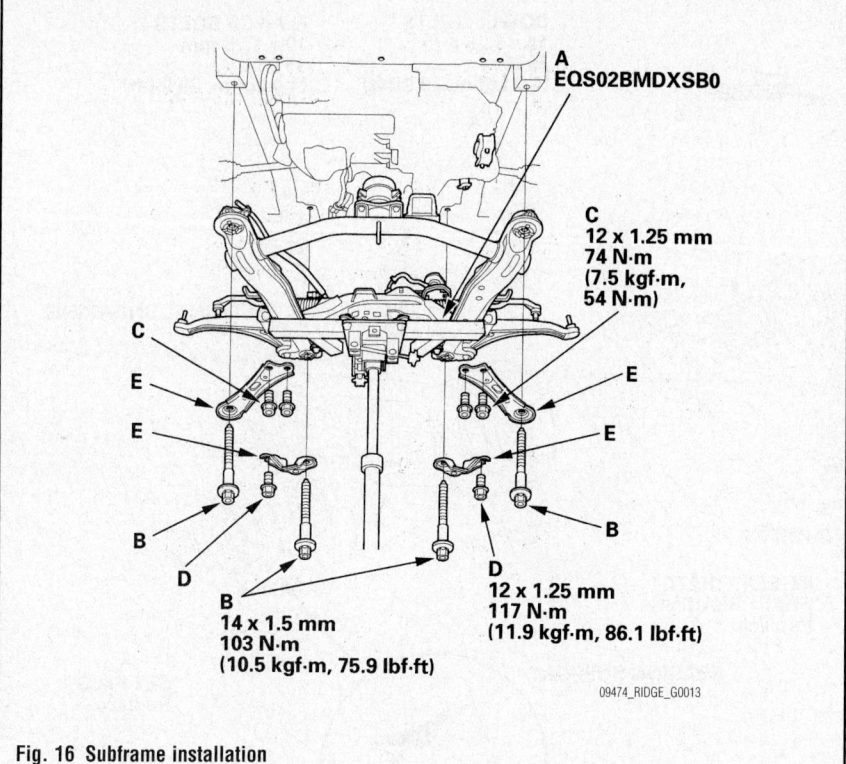

Fig. 16 Subframe installation

A
EQS02BMDXSB0

C
12 x 1.25 mm
74 N·m
(7.5 kgf·m,
54 N·m)

D
12 x 1.25 mm
117 N·m
(11.9 kgf·m, 86.1 lbf·ft)

B
14 x 1.5 mm
103 N·m
(10.5 kgf·m, 75.9 lbf·ft)

09474_RIDGE_G0013

- Tie rod end ball stud nut: 40 ft. lbs. (54 Nm)
- Lower ball joint stud nut: 65–72 ft. lbs. (88–98 Nm)
- Stabilizer link nut: 58 ft. lbs. (78 Nm)
- Exhaust pipe-to-converter: 25 ft. lbs. (33 Nm)
- Exhaust pipe hanger: 16 ft. lbs. (22 Nm)
- Exhaust pipe-to-manifold: 40 ft. lbs. (54 Nm)
- Stiffener (new bolts): 40 ft. lbs. (54 Nm)
- Front mount nut (new): 40 ft. lbs. (54 Nm)
- Connector bracket-to-front head: 33 ft. lbs. (44 Nm)
- Harness bracket-to-rear head: 20 ft. lbs. (26 Nm)

DRIVESHAFT

REMOVAL & INSTALLATION

See Figures 17 through 19.

1. Before servicing the vehicle, refer to the precautions.
2. Raise and support the vehicle.
3. Remove the front and rear driveshaft protector loops.
4. Matchmark the front driveshaft and transfer case flange.
5. Remove the bolts.

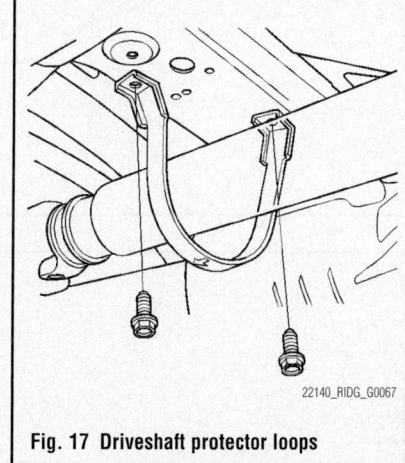

Fig. 17 Driveshaft protector loops

22140_RIDG_G0067

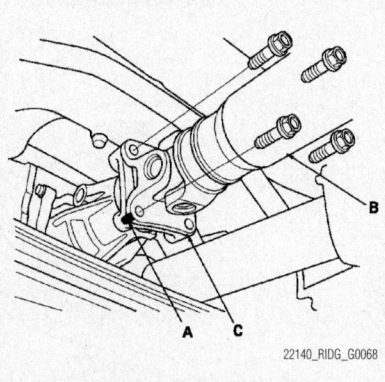

Fig. 18 Matchmark the driveshaft to the transfer case and differential flanges

22140_RIDG_G0068

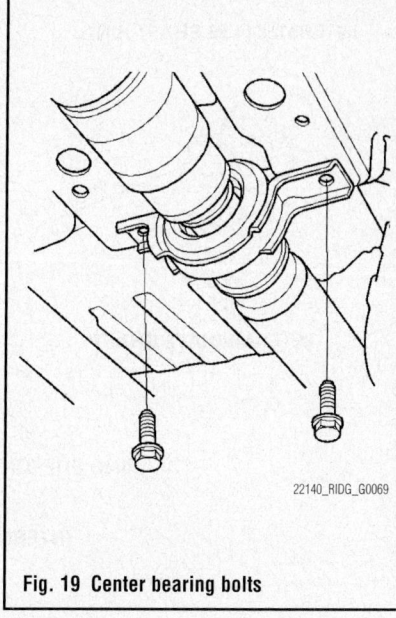

Fig. 19 Center bearing bolts

22140_RIDG_G0069

6. Remove the center bearing bolts.
7. Matchmark the rear driveshaft and differential flange.
8. Remove the bolts and remove the driveshaft assembly.

To install:

9. Installation is the reverse of removal, noting the following:
   a. Align all matchmarks.
   b. Torque the flange bolts to 53 ft. lbs. (72 Nm).
   c. Torque the center bearing bolts to 29 ft. lbs. (39 Nm).
   d. Torque the loop bolts to 16 ft. lbs. (22 Nm).

INTERMEDIATE SHAFT

REMOVAL & INSTALLATION

See Figure 20.

1. Before servicing the vehicle, refer to the precautions.
2. Drain the transaxle.
3. Remove the right halfshaft.
4. Remove the subframe stiffener.
5. Remove the exhaust pipe and bracket.
6. Remove the heat shield.
7. Remove the intermediate shaft from the differential.

➡Hold the intermediate shaft horizontally to avoid damage to the seal.

To install:

8. Installation is the reverse of removal, observing the following torques:
   a. Torque the heat shield bolts to 29 ft. lbs. (39 Nm).

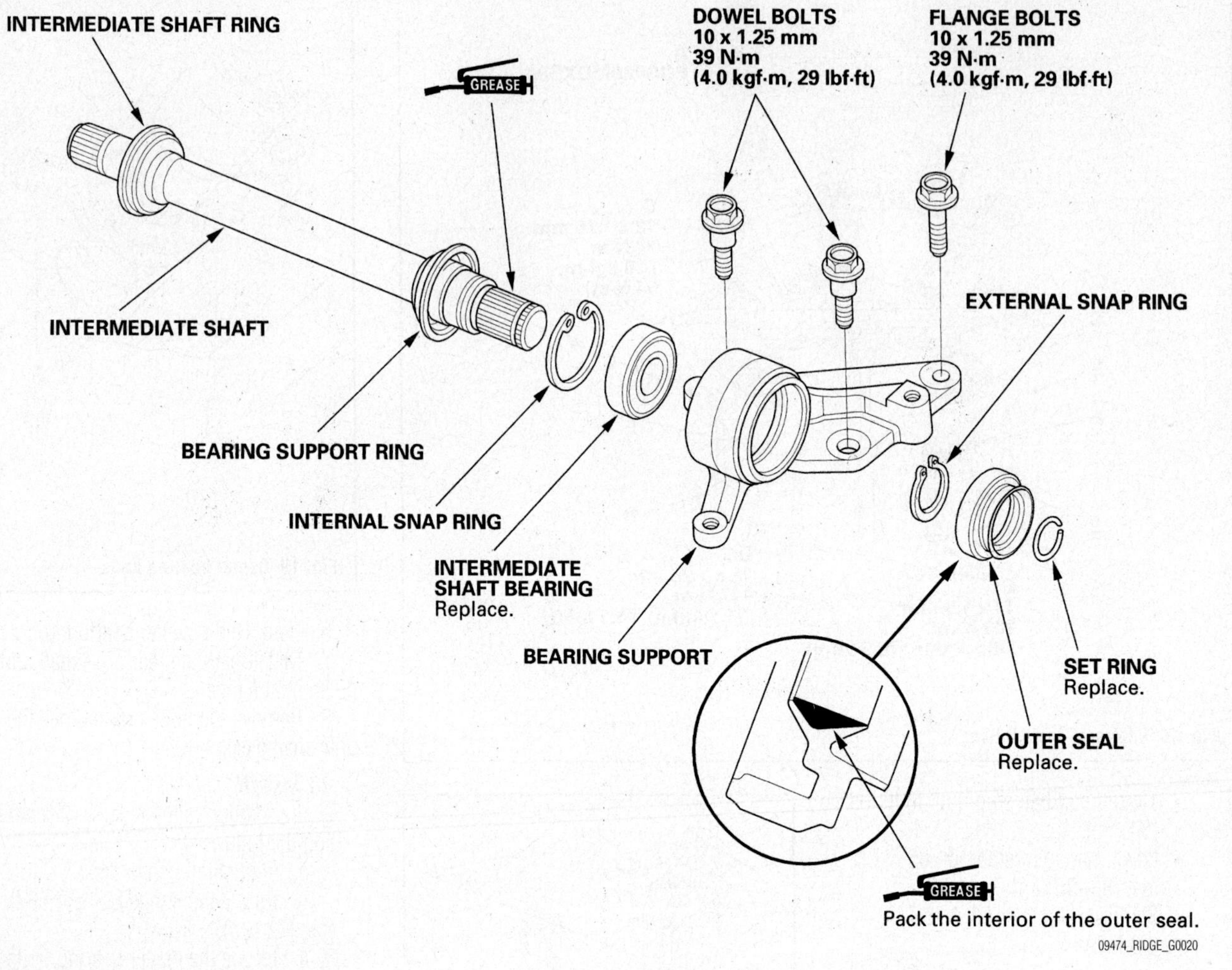

INTERMEDIATE SHAFT RING

GREASE

DOWEL BOLTS
10 x 1.25 mm
39 N·m
(4.0 kgf·m, 29 lbf·ft)

FLANGE BOLTS
10 x 1.25 mm
39 N·m
(4.0 kgf·m, 29 lbf·ft)

INTERMEDIATE SHAFT

EXTERNAL SNAP RING

BEARING SUPPORT RING

INTERNAL SNAP RING

INTERMEDIATE
SHAFT BEARING
Replace.

BEARING SUPPORT

SET RING
Replace.

OUTER SEAL
Replace.

GREASE

Pack the interior of the outer seal.

09474_RIDGE_G0020

Fig. 20 Intermediate shaft exploded view

TRANSFER CASE ASSEMBLY

REMOVAL & INSTALLATION

See Figure 21.

1. Before servicing the vehicle, refer to the precautions.

2. Raise and support the vehicle.

3. Place the transaxle in **N**.

4. Drain the transaxle. When the fluid is drained, install the drain plug, using a new washer and torque to 36 ft. lbs. (49 Nm).

5. Remove the subframe stiffener.

6. Remove the exhaust pipe and bracket.

7. Disconnect the transfer case breather.

8. Matchmark the driveshaft and disconnect it from the transfer case.

9. Remove the bolts and remove the transfer case.

10. Installation is the reverse of removal. Observe the following torques:

- Transfer case bolts: 38 ft. lbs. (51 Nm)

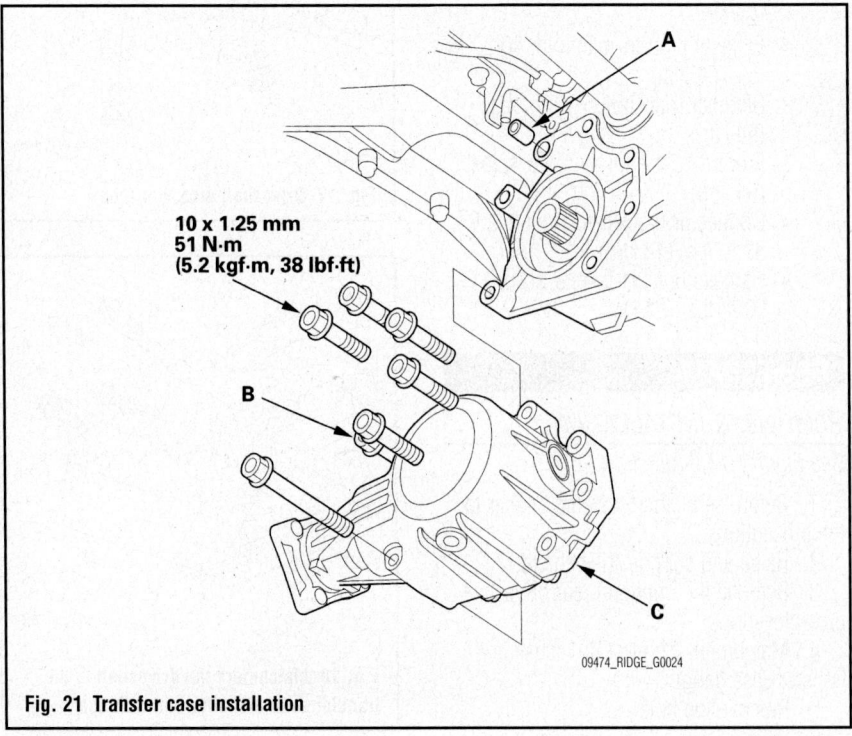

A

10 x 1.25 mm
51 N·m
(5.2 kgf·m, 38 lbf·ft)

B

C

09474_RIDGE_G0024

Fig. 21 Transfer case installation

- Driveshaft bolts: 53 ft. lbs. (72 Nm)
- Exhaust pipe-to-converter: 25 ft. lbs. (33 Nm)
- Exhaust pipe hanger: 16 ft. lbs. (22 Nm)
- Exhaust pipe-to-manifold: 40 ft. lbs. (54 Nm)

- Stiffener (new bolts): 40 ft. lbs. (54 Nm)

FRONT HALFSHAFT

REMOVAL & INSTALLATION

See Figure 22.

1. Before servicing the vehicle, refer to the precautions.
2. Drain the transaxle if removing the left halfshaft. If is not necessary to drain the fluid if removing the right halfshaft.
3. Connect the negative battery cable.

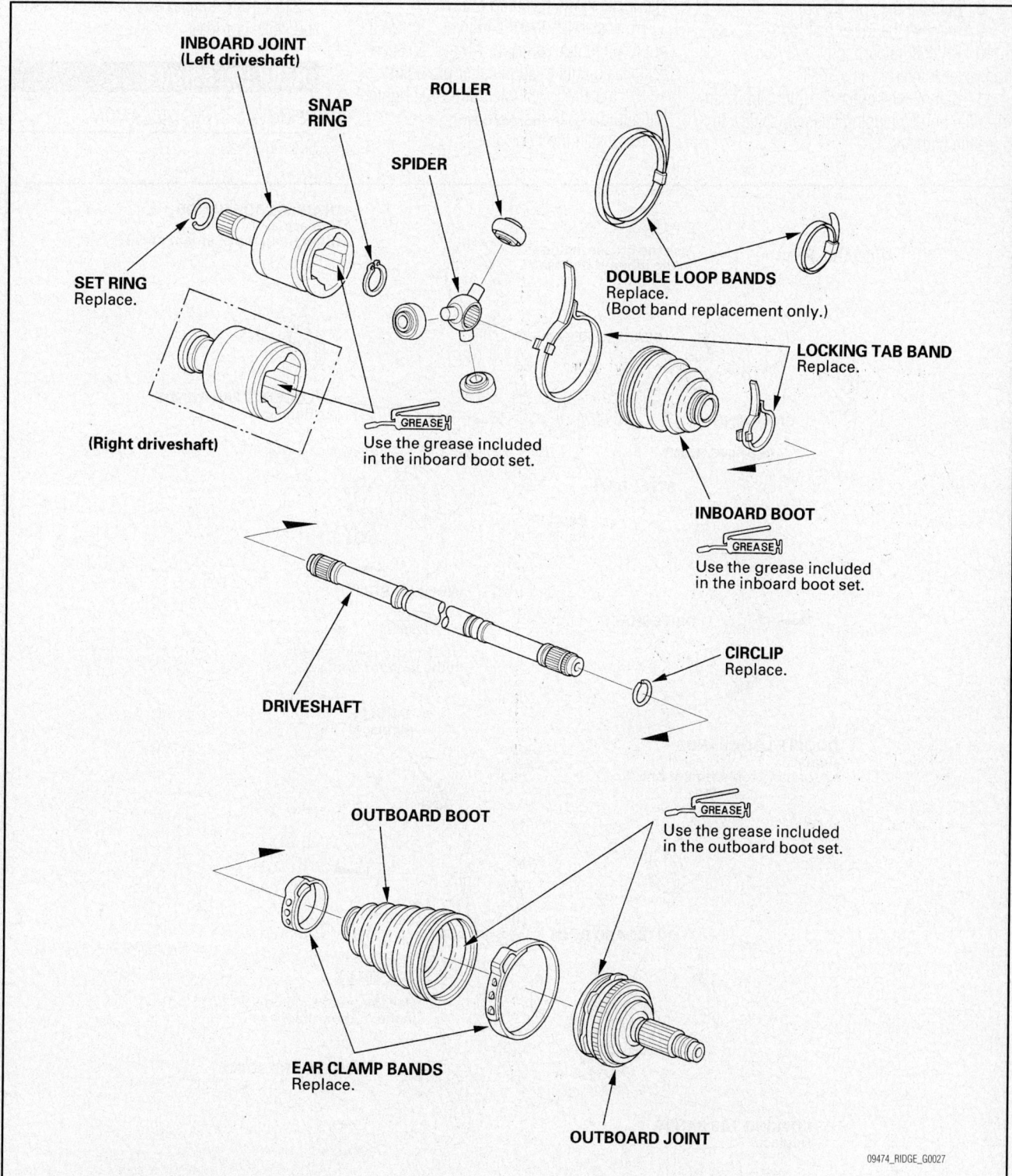

INBOARD JOINT
(Left driveshaft)

SNAP RING

SET RING
Replace.

(Right driveshaft)

SPIDER

ROLLER

GREASE
Use the grease included in the inboard boot set.

DOUBLE LOOP BANDS
Replace.
(Boot band replacement only.)

LOCKING TAB BAND
Replace.

INBOARD BOOT

GREASE
Use the grease included in the inboard boot set.

CIRCLIP
Replace.

DRIVESHAFT

OUTBOARD BOOT

GREASE
Use the grease included in the outboard boot set.

EAR CLAMP BANDS
Replace.

OUTBOARD JOINT

09474_RIDGE_G0027

Fig. 22 Front halfshaft exploded view

4. Enter the anti-theft codes for the audio system and the navigation system (if equipped).

5. Set the clock on vehicles without navigation.

6. Perform the power window control unit reset procedure.

7. Remove the front wheels.

8. Remove the spindle nut.

9. Remove the lower ball joint.

10. Pry the inboard joint from the transaxle or intermediate shaft.

11. Remove the outer CV-joint stub shaft from the hub by tapping the stub shaft with a plastic hammer.

To install:

➡ **Use new circlips, split pins and self-locking nuts for assembly.**

12. Install the outer CV-joint stub shaft into the hub.

13. Install the inner CV-joint to the transaxle or intermediate shaft until the circlip locks in the retaining groove.

14. Install the lower ball joint. Tighten the nut to 69 ft. lbs. (93 Nm). Always advance castellated nuts to align cotter pin holes.

15. Install the spindle nut (new). Tighten the nut to 242 ft. lbs. (328 Nm).

16. Install the front wheels.

17. Connect the negative battery cable.

18. Enter the anti-theft codes for the audio system and the navigation system (if equipped).

19. Set the clock on vehicles without navigation.

20. Perform the power window control unit reset procedure.

21. Fill the transaxle to the correct level and check for leaks.

REAR HALFSHAFT

REMOVAL & INSTALLATION

See Figure 23.

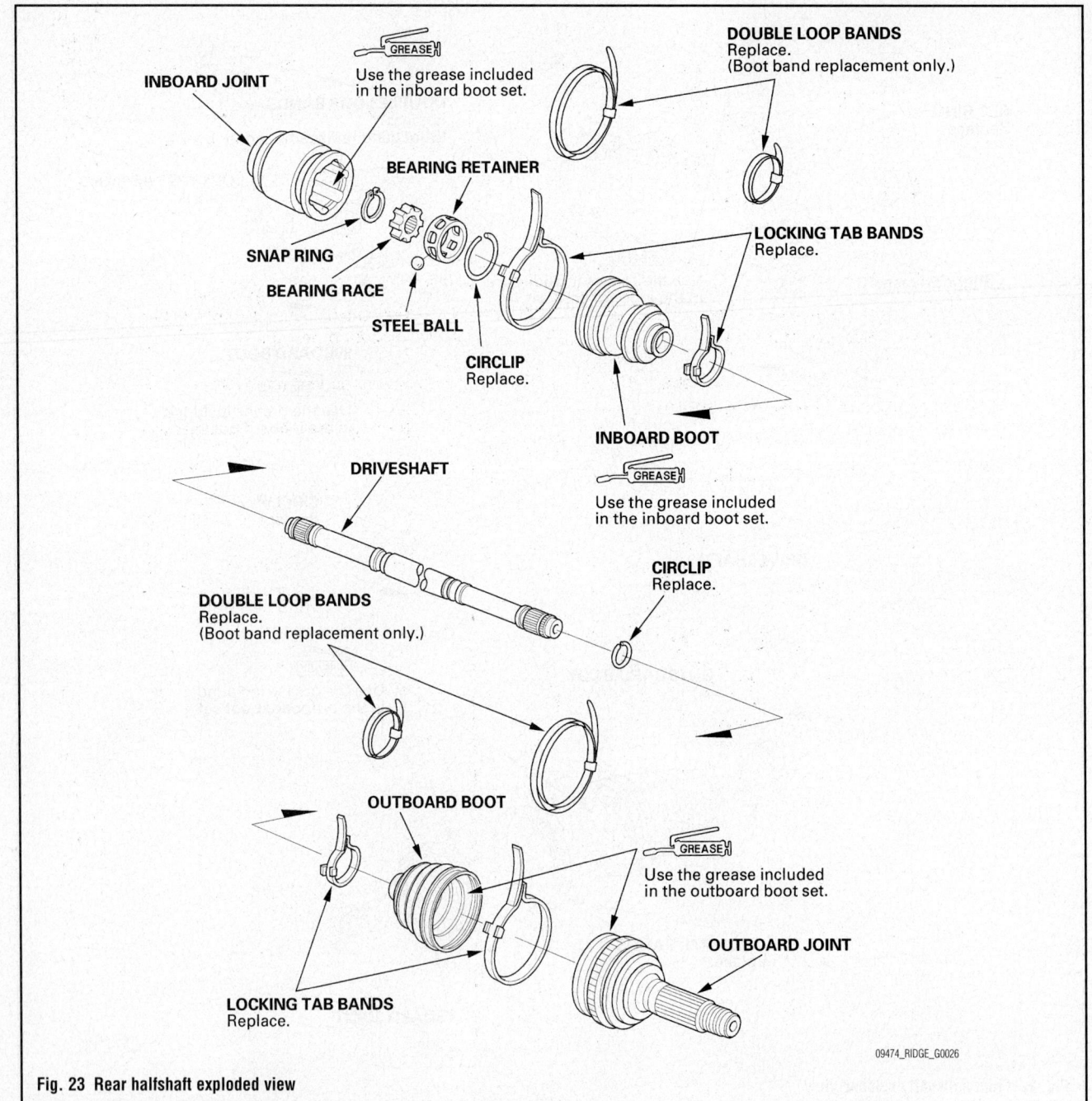

Fig. 23 Rear halfshaft exploded view

09474_RIDGE_G0026

1. Before servicing the vehicle, refer to the precautions.

2. Raise and support the rear of the vehicle.

3. Drain the differential.

4. Remove the wheels.

5. Remove and discard the spindle nut.

6. Remove the VSA rear wheel sensor.

7. Separate the upper ball joint from the upper arm.

8. Remove the lower track rod.

9. Separate the lower arm from the knuckle.

10. Pull the knuckle outwards and separate the halfshaft from the hub.

11. Pry the halfshaft from the differential.

12. Installation is the reverse of removal.

  a. Use new snaprings and cotter pins.

  b. Always advance castellated nuts to align cotter pin holes.

  c. Observe the following torques:
- Lower arm-to-knuckle bolt: 105 ft. lbs. (142 Nm)
- Track rod inner end: 69 ft. lbs. (93 Nm)
- Track rod outer end: 74 ft. lbs. (101 Nm)
- Upper arm ball stud nut: 40 ft. lbs. (54 Nm)
- Halfshaft end nut: 181 ft. lbs. (245 Nm)

DIFFERENTIAL ASSEMBLY

REMOVAL & INSTALLATION

See Figure 24.

1. Before servicing the vehicle, refer to the precautions.

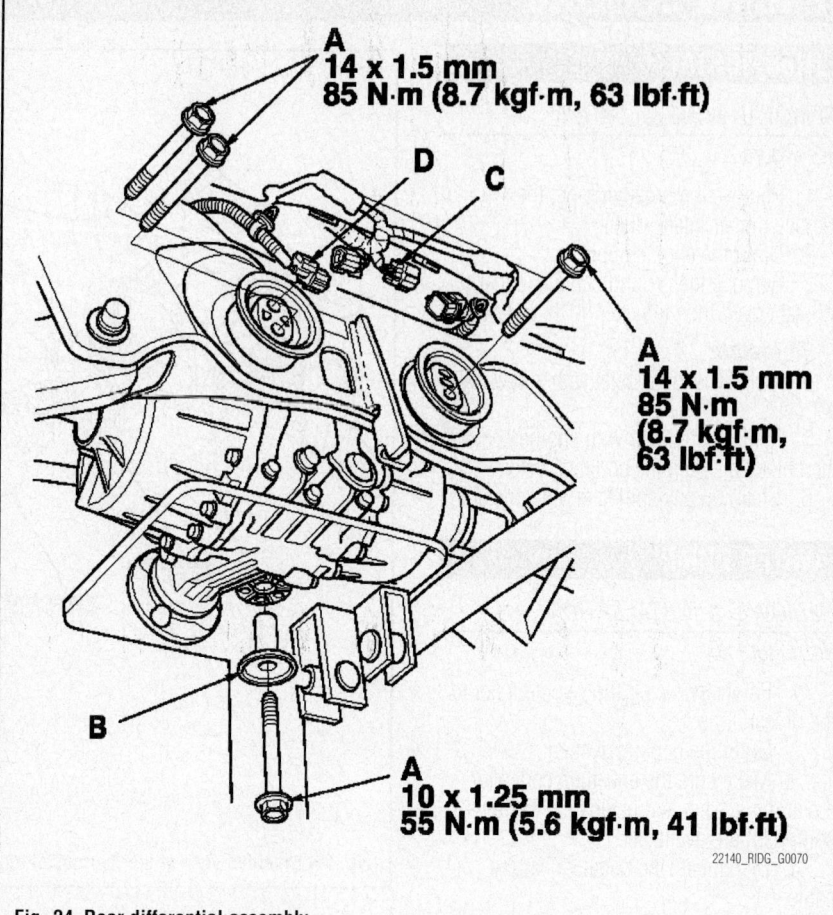

A
14 x 1.5 mm
85 N·m (8.7 kgf·m, 63 lbf·ft)

A
14 x 1.5 mm
85 N·m (8.7 kgf·m, 63 lbf·ft)

A
10 x 1.25 mm
55 N·m (5.6 kgf·m, 41 lbf·ft)

22140_RIDG_G0070

Fig. 24 Rear differential assembly

2. Matchmark and disconnect the driveshaft.

3. Remove the halfshafts.

4. Place a transmission jack under the rear differential (A).

5. Disconnect the 6P (B) and 2P (C) connectors, then remove the mounting bolts (D) and washer (E).

6. Lower the unit slightly and disconnect the breather hose.

7. Remove the differential.

To install:

8. Installation is the reverse of removal, noting the following:

  a. Tighten the horizontal mounting bolts to 63 ft. lbs. (85 Nm).

  b. Tighten the vertical mounting bolts to 41 ft. lbs. (55 Nm).

ENGINE COOLING

THERMOSTAT

REMOVAL & INSTALLATION

See Figure 25.

1. Remove the breather pipe, then remove the air intake duct.
2. Drain the engine coolant.
3. Remove the ground cable and thermostat cover, then remove the thermostat.

To install:

4. Install the thermostat with a new rubber seal.
5. Refill the radiator with engine coolant, then bleed air from the cooling system.
6. Clean up any spilled engine coolant.

WATER PUMP

REMOVAL & INSTALLATION

See Figure 26.

1. Before servicing the vehicle, refer to the precautions.
2. Drain the cooling system.
3. Make sure the anti-theft codes for the audio system and navigation system (if equipped) are available.
4. Disconnect the negative battery cable.
5. Remove the accessory drive belts.
6. Remove the front cover.
7. Remove the timing belt.
8. Remove the timing belt tensioner.
9. Remove the water pump.

To install:

10. Install the water pump. Use a new O-ring seal and tighten the bolts to 105 inch lbs. (12 Nm).
11. Install the timing belt tensioner.
12. Install the timing belt.
13. Install the front cover.
14. Install the accessory drive belts
15. Connect the negative battery cable.
16. Enter the anti-theft codes for the audio system and the navigation system (if equipped).
17. Set the clock on vehicles without navigation.
18. Perform the power window control unit reset procedure.
19. Fill and bleed the cooling system.
20. Start the engine and check for leaks.

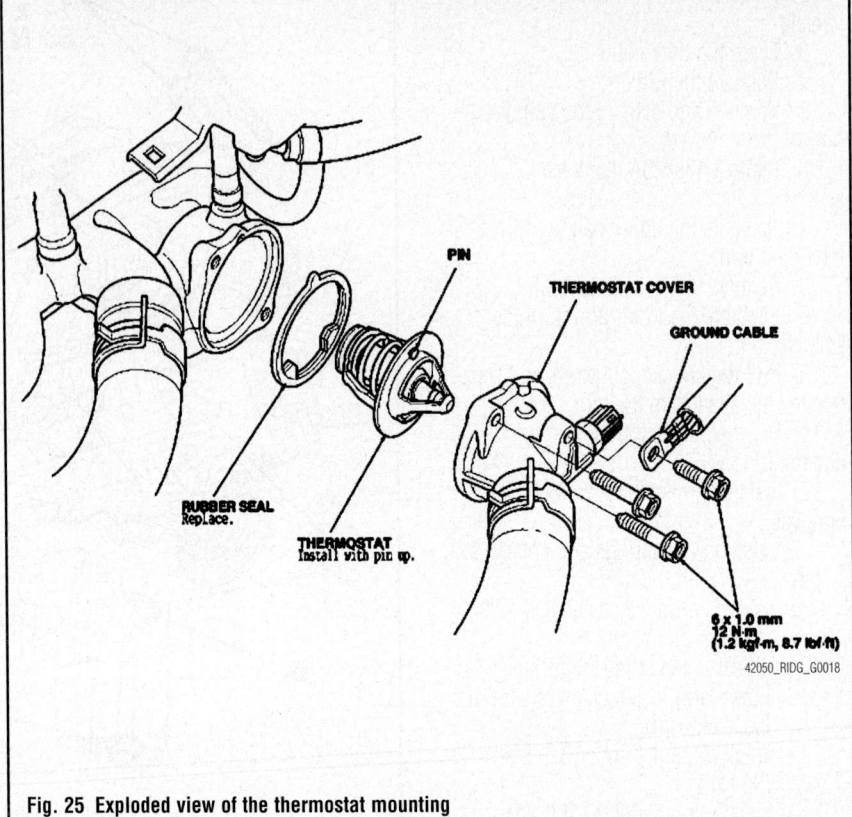

Fig. 25 Exploded view of the thermostat mounting

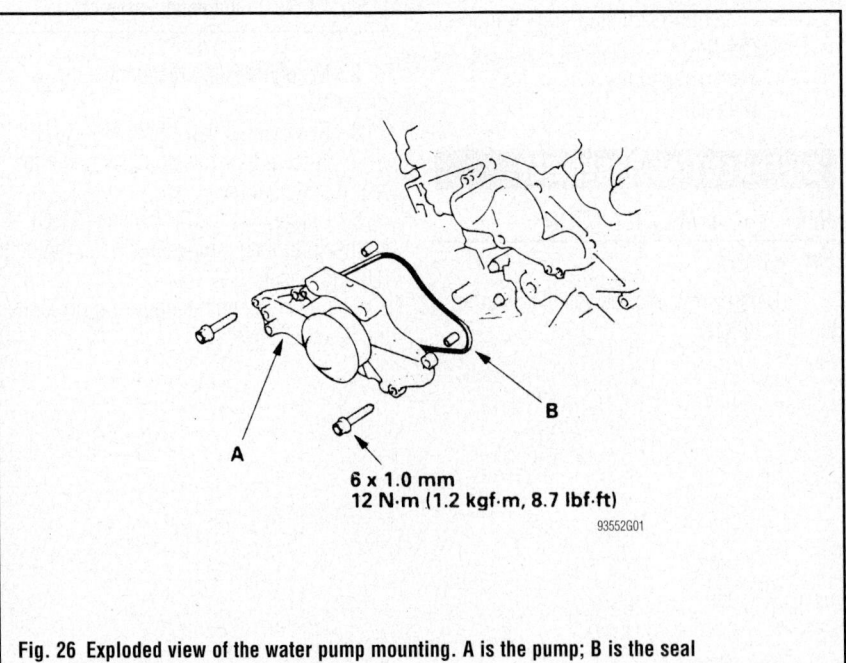

Fig. 26 Exploded view of the water pump mounting. A is the pump; B is the seal

ENGINE ELECTRICAL

CHARGING SYSTEM

ALTERNATOR

REMOVAL & INSTALLATION

See Figure 27.

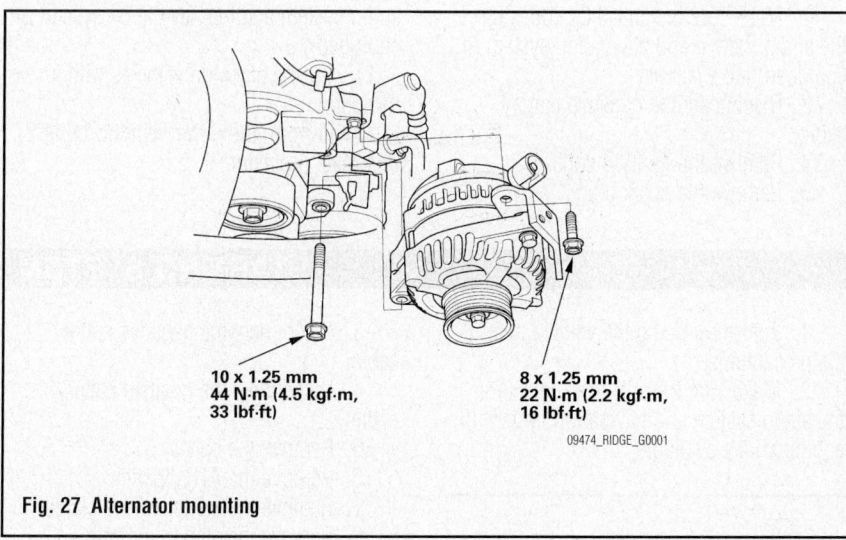

10 x 1.25 mm
44 N·m (4.5 kgf·m,
33 lbf·ft)

8 x 1.25 mm
22 N·m (2.2 kgf·m,
16 lbf·ft)

09474_RIDGE_G0001

Fig. 27 Alternator mounting

1. Before servicing the vehicle, refer to the precautions.
2. Make sure you have the anti-theft codes for the radio and navigation systems.

3. Make sure the anti-theft codes for the audio system and navigation system (if equipped) are available.
4. Disconnect the negative battery cable.

5. Disconnect the positive battery cable.
6. Remove the intake manifold cover.

7. Remove the accessory drive belt.
8. Remove the alternator wiring harness connectors.
9. Remove the alternator mounting bolts.
10. Remove the wiring harness clamp.
11. Remove the alternator.

To install:
12. Install the alternator.
13. Install the wiring harness clamp. Tighten the bolt to 105 inch lbs. (12 Nm).
14. Install the alternator mounting bolts. Tighten the 10mm bolt to 33 ft. lbs. (44 Nm) and the 8mm bolt to 16 ft. lbs. (22 Nm).
15. Install the alternator wiring harness connectors. Tighten the battery terminal nut to 105 inch lbs. (12 Nm).
16. Install the accessory drive belt.
17. Install the intake manifold cover.
18. Connect the positive cable.
19. Connect the negative battery cable.
20. Enter the anti-theft codes for the audio system and the navigation system (if equipped).
21. Set the clock on vehicles without navigation.
22. Perform the power window control unit reset procedure.

ENGINE ELECTRICAL

IGNITION SYSTEM

FIRING ORDER

See Figure 28.

```
      ┌─────────────────┐
      │  1    3    5     │
  ◄═══┤                 │
      │  2    4    6     │
      └─────────────────┘
         FRONT
```

9308MG32

Fig. 28 3.5L Engine
      Firing Order: 1–4–2–5–3–6
      Distributorless ignition system

IGNITION COIL PACK

REMOVAL & INSTALLATION

See Figure 29.

1. Before servicing the vehicle, refer to the precautions.
2. Make sure the anti-theft codes for the audio system and navigation system (if equipped) are available.

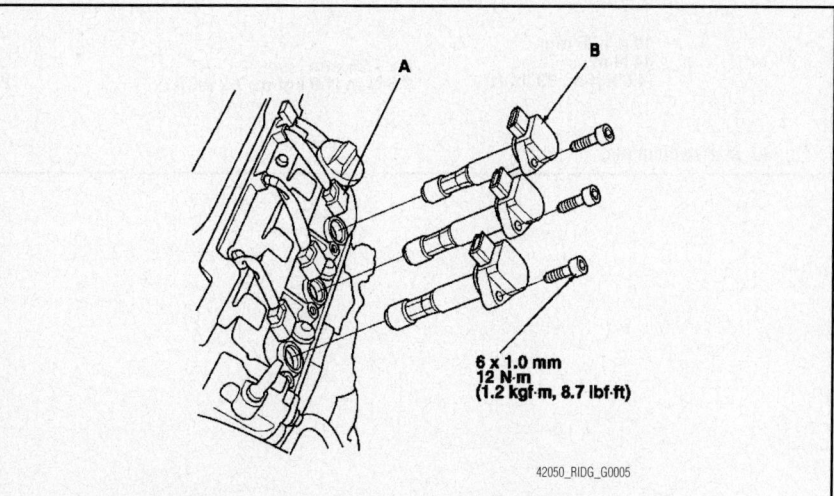

A
B

6 x 1.0 mm
12 N·m
(1.2 kgf·m, 8.7 lbf·ft)

42050_RIDG_G0005

Fig. 29 View of the ignition coil connectors (A) and ignition coils (B)—front bank shown, rear bank similar

3. Disconnect the negative battery cable.
4. Remove the intake manifold cover.
5. Disconnect the ignition coil connectors, then remove the ignition coils.

To install:

6. Install the ignition coils in the reverse order of removal, noting the following:

    a. Tighten the coil retaining bolts to 9 ft. lbs. (12 Nm).

IGNITION TIMING

ADJUSTMENT

The ignition timing is controlled by the Powertrain Control Module (PCM). No adjustment is necessary or possible.

SPARK PLUGS

REMOVAL & INSTALLATION

1. Before servicing the vehicle, refer to the precautions.
2. Make sure the anti-theft codes for the audio system and navigation system (if equipped) are available.
3. Disconnect the negative battery cable.
4. Remove the ignition coils.
5. Remove the spark plug.
6. Inspect the spark plug.

To install:

7. Install the spark plug and tighten to 13 ft. lbs. (18 Nm).
8. Install the ignition coils.
9. Connect the negative battery cable.
10. Enter the anti-theft codes for the audio system and the navigation system (if equipped).
11. Set the clock on vehicles without navigation.
12. Perform the power window control unit reset procedure.

ENGINE ELECTRICAL

STARTER

REMOVAL & INSTALLATION

See Figure 30.

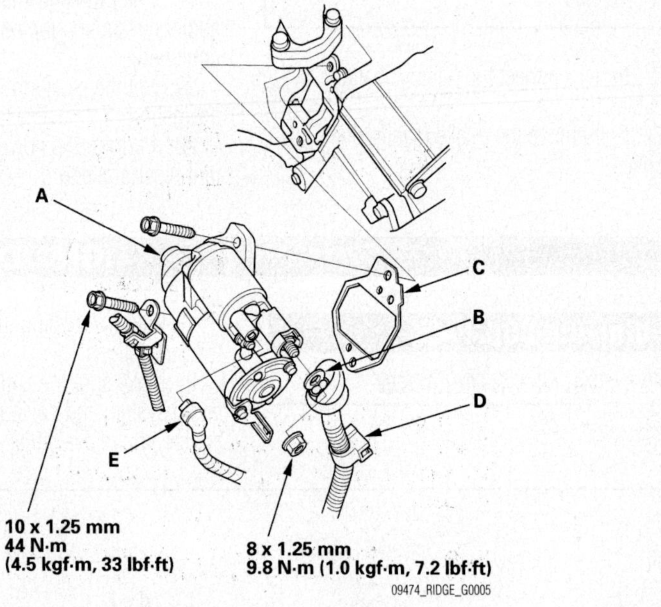

1. Before servicing the vehicle, refer to the precautions.
2. Make sure the anti-theft codes for the audio system and navigation system (if equipped) are available.

STARTING SYSTEM

3. Disconnect the negative battery cable.
4. Disconnect the positive battery cable.
5. Remove the battery.
6. Remove the ATF dipstick.
7. Remove the starter wiring clamp.
8. Remove the starter wiring harness connectors.
9. Remove the starter motor bolts.

To install:

10. Install the starter motor. Tighten the bolts to 33 ft. lbs. (44 Nm).
11. Install the starter wiring harness connectors. Tighten the battery cable nut to 79 inch lbs. (9 Nm).
12. Install the ATF dipstick.
13. Install the battery.
14. Connect the positive battery cable.
15. Connect the negative battery cable.
16. Enter the anti-theft codes for the audio system and the navigation system (if equipped).
17. Set the clock on vehicles without navigation.
18. Perform the power window control unit reset procedure.

10 x 1.25 mm
44 N·m
(4.5 kgf·m, 33 lbf·ft)

8 x 1.25 mm
9.8 N·m (1.0 kgf·m, 7.2 lbf·ft)

09474_RIDGE_G0005

Fig. 30 Starter mounting

ENGINE MECHANICAL

→Disconnecting the negative battery cable may interfere with the functions of the on board computer systems and may require the computer to undergo a relearning process, once the negative battery cable is reconnected.

ACCESSORY DRIVE BELTS

ACCESSORY BELT ROUTING

See Figure 31.

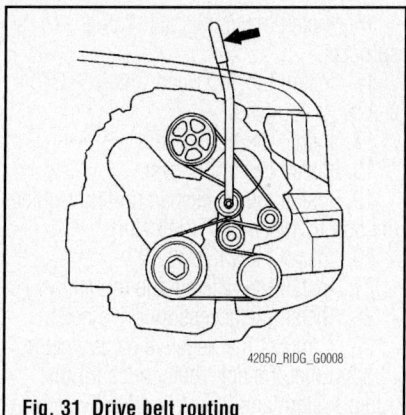

Fig. 31 Drive belt routing

INSPECTION

See Figure 32.

1. Inspect the belt for cracks or damage. If the belt is cracked or damaged, replace it.
2. Check that the auto-tensioner indicator (A) is within the standard range (B) as shown. If it is out of the standard range, replace the drive belt.

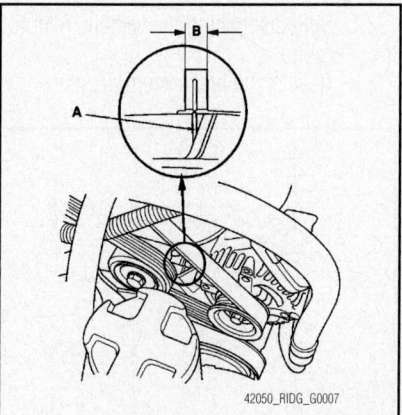

Fig. 32 Check that the auto-tensioner indicator (A) is within the standard range (B) as shown. If it is out of the standard range, replace the drive belt

ADJUSTMENT

Belt tension is maintained by an automatic tensioner. No adjustments are necessary or possible.

REMOVAL & INSTALLATION

See Figure 33.

1. Move the auto-tensioner (A) to relieve tension from the drive belt, then remove the drive belt.
2. Install the new belt in the reverse order of removal.

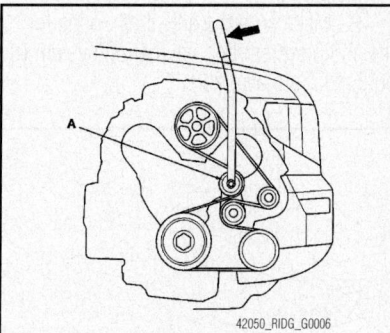

Fig. 33 Move the auto-tensioner (A) to relieve tension from the drive belt, then remove the drive belt

CAMSHAFT AND VALVE LIFTERS

REMOVAL & INSTALLATION

Front

See Figure 34.

1. Before servicing the vehicle, refer to the precautions.
2. Make sure the anti-theft codes for the audio system and navigation system (if equipped) are available.
3. Disconnect the negative battery cable.
4. Disconnect the positive battery cable.
5. Remove the battery and battery box.
6. Drain the coolant.
7. Remove the upper radiator hose.
8. Remove the exhaust Gas Recirculation (EGR) valve.
9. Remove the timing belt.
10. Remove the rocker arm assembly.
11. Remove the front camshaft pulley.
12. Remove the thrust plate and camshaft.

To install:

13. Install the camshaft using a new O-ring. Tighten the thrust plate to 16 ft. lbs. (22 Nm).
14. Install the front camshaft pulley.
15. Install the rocker arm assembly.
16. Install the timing belt.
17. Install the exhaust gas recirculation (EGR) valve.
18. Install the battery.
19. Fill the cooling system.
20. Connect the positive battery cable.
21. Connect the negative battery cable.
22. Enter the anti-theft codes for the audio system and the navigation system (if equipped).
23. Set the clock on vehicles without navigation.

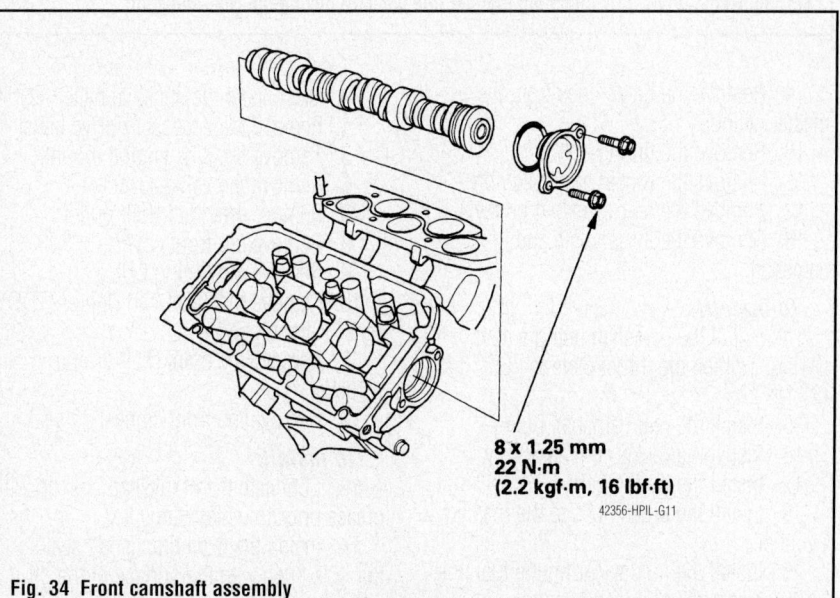

8 x 1.25 mm
22 N·m
(2.2 kgf·m, 16 lbf·ft)

Fig. 34 Front camshaft assembly

24. Perform the power window control unit reset procedure.

25. Start the engine and check for leaks.

Rear

See Figure 35.

1. Before servicing the vehicle, refer to the precautions.

2. Drain the cooling system.

3. Relieve the fuel system pressure.

4. Make sure the anti-theft codes for the audio system and navigation system (if equipped) are available.

5. Disconnect the negative battery cable.

6. Remove the under-hood fuse box.

7. Remove the fuel feed hose.

8. Remove the nuts securing the fuel line.

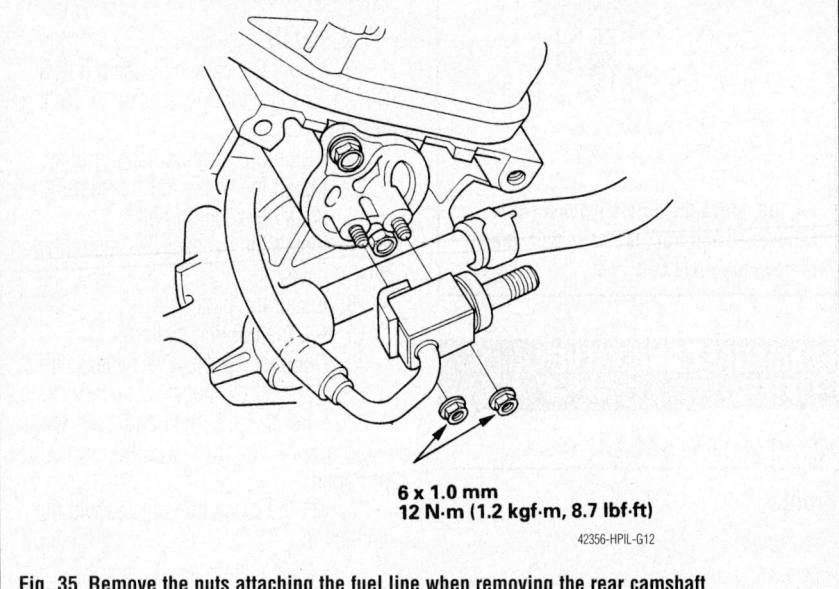

6 x 1.0 mm
12 N·m (1.2 kgf·m, 8.7 lbf·ft)
42356-HPIL-G12

Fig. 35 Remove the nuts attaching the fuel line when removing the rear camshaft

9. Remove the brake lines from the master cylinder.

10. Remove the timing belt.

11. Remove the rocker arm assembly.

12. Remove the rear camshaft pulley.

13. Remove the thrust plate and camshaft.

To install:

14. Install the camshaft using a new O-ring. Tighten the thrust plate to 16 ft. lbs. (22 Nm).

15. Install the rear camshaft pulley.

16. Install the rocker arm assembly.

17. Install the timing belt.

18. Install the brake lines to the master cylinder.

19. Install the nuts securing the fuel line.

20. Install the fuel feed hose.

21. Install the under-hood fuse box.

22. Connect the negative battery cable.

23. Enter the anti-theft codes for the audio system and the navigation system (if equipped).

24. Set the clock on vehicles without navigation.

25. Perform the power window control unit reset procedure.

CRANKSHAFT FRONT SEAL

REMOVAL & INSTALLATION

See Figure 36.

1. Before servicing the vehicle, refer to the precautions.

2. Make sure the anti-theft codes for the audio system and navigation system (if equipped) are available.

3. Disconnect the negative battery cable.

4. Remove the accessory drive belts.

5. Remove the side engine mount.

6. Remove the valve cover.

7. Remove the crankshaft pulley.

8. Remove the front cover.

9. Remove the timing belt.

10. Remove the Top Dead Center (TDC) sensor, if equipped.

11. Remove the crankshaft timing sprocket.

12. Remove the front crankshaft seal.

To install:

13. Lubricate the crankshaft seal lip with grease prior to installation.

14. Install the front crankshaft seal so that it is flush with the surface of the oil pump housing.

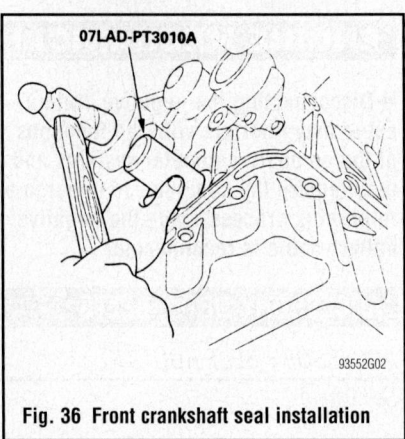

07LAD-PT3010A

93552G02

Fig. 36 Front crankshaft seal installation

15. Install the crankshaft timing sprocket.

16. Install the Top Dead Center (TDC) sensor, if equipped.

17. Install the timing belt.

18. Install the front cover.

19. Install the crankshaft pulley. Tighten the bolt to 181 ft. lbs. (245 Nm).

20. Install the valve cover.

21. Install the side engine mount.

22. Install the accessory drive belts.

23. Connect the negative battery cable.

24. Enter the anti-theft codes for the audio system and the navigation system (if equipped).

25. Set the clock on vehicles without navigation.

26. Perform the power window control unit reset procedure.

27. Check the engine oil level and add if necessary.

28. Start the engine and check for leaks.

CYLINDER HEAD

REMOVAL & INSTALLATION

See Figures 37 through 40.

1. Before servicing the vehicle, refer to the precautions.

2. Relieve the fuel system pressure.

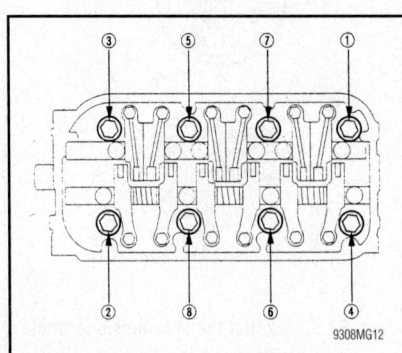

9308MG12

Fig. 37 Cylinder head bolt loosening sequence

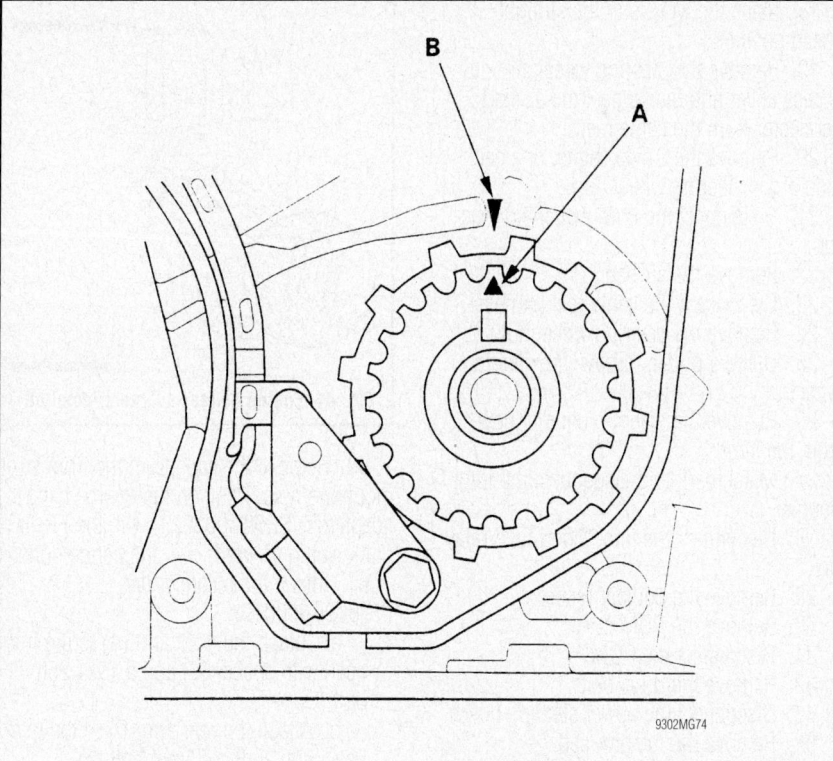

Fig. 38 Crankshaft timing belt sprocket TDC marks. Align sprocket mark (A) with pointer (B).

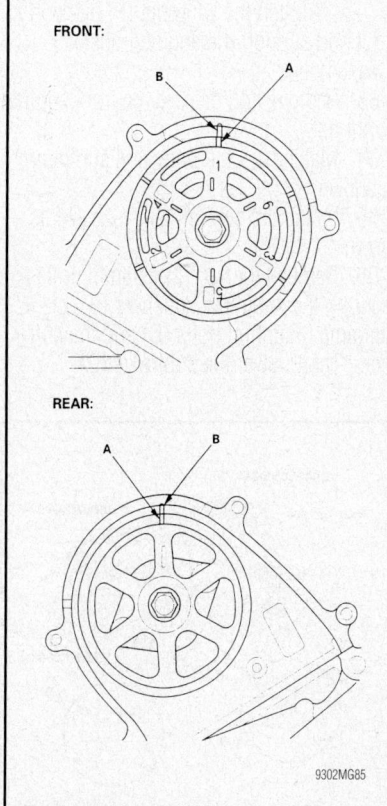

Fig. 39 Camshaft TDC marks. Align sprocket mark (A) with the back cover pointer (B).

3. Make sure the anti-theft codes for the audio system and navigation system (if equipped) are available.

4. Disconnect the negative battery cable.

5. Drain the cooling system.

6. Remove the alternator belt.

7. Remove the intake manifold cover.

8. Remove the ignition coil covers.

9. Remove the power steering belt and pump.

10. Remove the power steering hose clamp.

11. Remove the alternator.

12. Remove the fuel feed and return lines.

13. Remove the EVAP canister hose.

14. Remove the intake manifold.

15. Remove the ignition coils.

16. Remove the timing belt.

17. Remove the fuel injector connectors.

18. Disconnect the Engine Coolant Temperature (ECT) sensor connector

19. Remove the radiator fan switch connectors.

20. Disconnect the Crankshaft Position (CKP) sensor connector

21. Disconnect the Camshaft Position (CMP) sensor connector

22. Disconnect the Exhaust Gas Recirculation (EGR) connector

23. Disconnect the Valve Lift Electronic Control (VTEC) solenoid valve connector and oil pressure switch connections.

24. Remove the oil pressure switch connector.

25. Remove the vacuum hoses from the intake air bypass control valve.

26. Remove the fuel rails.

27. Remove the heater hose.

28. Remove the radiator hoses.

29. Remove the ground cable.

30. Remove the exhaust manifolds.

31. Remove the water passage.

32. Remove the camshaft pulleys and back covers.

33. Remove the valve covers.

34. Loosen the cylinder head bolts in sequence and 1/3 turns until all bolts are loose.

35. Remove the cylinder head.

To install:

36. Align the crankshaft and camshaft sprocket TDC marks as shown.

37. Install the cylinder heads with new gaskets.

38. Apply clean engine oil to the cylinder head bolt threads and flanges.

39. Tighten the cylinder head bolts in sequence as follows:
- Step 1: 29 ft. lbs. (39 Nm)
- Step 2: 51 ft. lbs. (69 Nm)
- Step 3: 72 ft. lbs. (98 Nm)

40. Install the timing belt and adjust the valve clearance.

41. Install the valve covers.

42. Install the exhaust manifolds.

43. Install the water passage.

44. Install the fuel rails.

45. Install the vacuum hoses to the intake air bypass control valve.

46. Install the radiator hoses.

47. Install the heater hose.

48. Install the oil pressure switch connector.

49. Disconnect the VTEC solenoid valve connector and oil pressure switch connections.

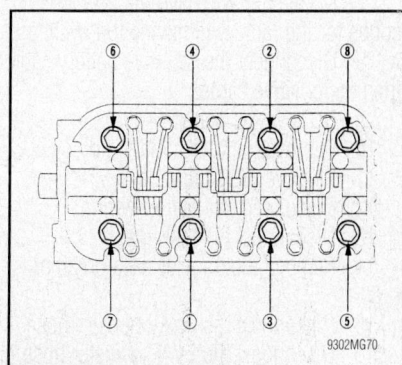

Fig. 40 Cylinder head bolt tightening sequence

50. Disconnect the EGR connector.
51. Disconnect the CMP sensor connector.
52. Disconnect the CKP sensor connector.
53. Disconnect the radiator fan switch connectors.
54. Disconnect the ECT sensor connector.
55. Disconnect the fuel injector connectors.
56. Install the ignition coils.
57. Install the intake manifold.
58. Install the EVAP canister hose.
59. Install the fuel feed and return lines.
60. Install the alternator.
61. Install the power steering hose clamp.
62. Install the power steering pump and belt.
63. Install the ground cable.
64. Install the ignition coil covers.
65. Install the intake manifold cover.
66. Install the alternator belt.
67. Fill the cooling system.
68. Connect the negative battery cable.
69. Enter the anti-theft codes for the audio system and the navigation system (if equipped).
70. Set the clock on vehicles without navigation.
71. Perform the power window control unit reset procedure.
72. Start the engine and check for leaks.

ENGINE ASSEMBLY

REMOVAL & INSTALLATION

See Figures 41 through 47.

1. Before servicing the vehicle, refer to the precautions.
2. Relieve the fuel system pressure.
3. Place the vehicle on a hoist.
4. Drain the power steering fluid.
5. Remove the air intake duct.
6. Remove the bulkhead cover.
7. Make sure you have the anti-theft codes for the radio and navigation systems.
8. Disconnect the negative battery cable then the positive cable.
9. Remove the battery.
10. Remove the intake manifold cover.
11. Remove the battery box.
12. Disconnect the starter wiring.
13. Disconnect the fuel line.
14. Remove the transfer case breather box.
15. Disconnect the brake booster hose.
16. Disconnect the EVAP canister hose.
17. Disconnect the battery ground cable and wait at least 3 minutes.

18. Place the wheels in the straight-ahead position.
19. Remove the steering wheel air bag access cover and disconnect the air bag connector from the cable reel.
20. Remove the 2 Torx®bolts, one each side of the steering wheel hub.
21. Disconnect the horn switch connector.
22. Remove the air bag.
23. Disconnect the cable reel harness.
24. Remove the steering wheel nut.
25. Using a puller, remove the steering wheel.
26. Remove the steering coupler cover from the floor.
27. Matchmark and disconnect the joint coupler.
28. Disconnect the underhood fuse/relay box.
29. Remove the coolant reservoir.
30. Remove the PCM cover.
31. Disconnect the PCM.
32. Remove the drive belt.
33. Disconnect the power steering hoses.
34. Remove the radiator cap.
35. Raise the hoist full height.
36. Remove the front wheels.
37. Remove the splash shield.
38. Drain the coolant.
39. Drain the transaxle.
40. Drain the engine oil.
41. Remove the front subframe stiffener.
42. Remove the front exhaust pipe.
43. Disconnect the stabilizer bar links.
44. Disconnect the tie rod ends from the knuckles.
45. Disconnect the knuckles from the lower arms.
46. Remove the halfshafts.
47. Remove the transfer case.
48. Disconnect the fluid lines from the steering gear.
49. Disconnect the power steering pressure switch.
50. Remove the transaxle lower front mount bolts.
51. Remove the rear mount bolts.
52. Remove the A/C compressor and set it aside without disconnecting the hoses.
53. Lower the vehicle.
54. Remove the heater hoses.
55. Remove the radiator hoses.
56. Disconnect the transaxle cooler lines.
57. Remove the radiator.
58. Remove the connector bracket from the front cylinder head and use that bolt hole to attach the balancer bar front arm.
59. Remove the harness clamp bracket from the rear head and use that bolt hole to attach the balancer bar rear arm.

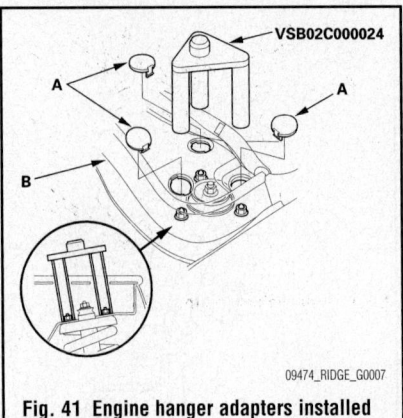

Fig. 41 Engine hanger adapters installed

60. Remove the caps from the front strut mounting nuts. Position the engine hanger adapters (VSB02C000024) with the FRONT mark facing forward, over the flange nuts.
61. Install the balancer bar (VSB02C000019):
   a. Attach the front arm (A) to the front head with spacer (B) and a 10x1.25mm bolt (C)
   b. Attach the rear arm (D) to the rear head with an 8x1.25mm bolt (E).
62. Install the support hanger (AAR-T-12566):
   a. Attach the hook to the slotted hole in the balancer bar.
   b. Tighten the wing nut (F) by hand to lift and support the engine/transaxle assembly.
63. Remove and discard the front engine mount nut.
64. Matchmark the body and subframe as shown.
65. Install the subframe adapter (A) as shown.
66. Remove the six 12x1.25mm bolts (A) securing the subframe stiffeners (B), the 4 subframe mounting bolts (C) and the stiffeners. Then, lower the subframe (D).

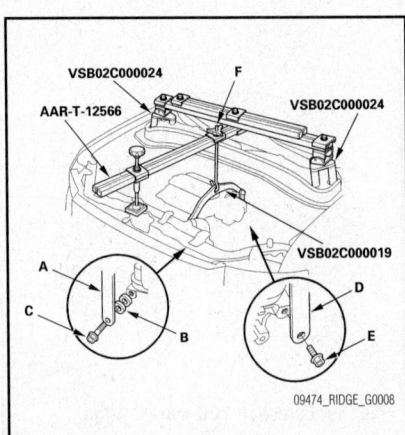

Fig. 42 Balancer bar installation and support hanger

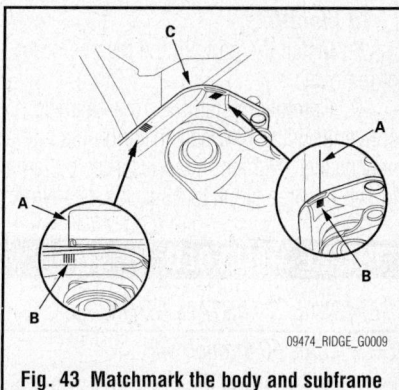

Fig. 43 Matchmark the body and subframe as shown

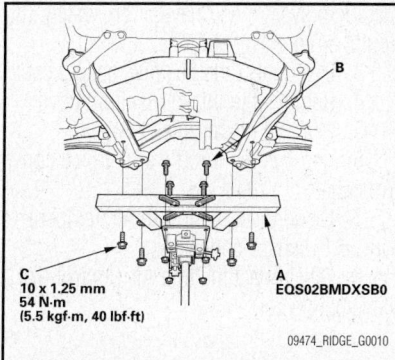

C
10 x 1.25 mm
54 N·m
(5.5 kgf·m, 40 lbf·ft)

Fig. 44 Install the subframe adapter as shown

67. Lower the vehicle and attach an engine crane as shown. Lift the engine until the crane has the weight and remove the support hanger.

68. Remove the upper side engine mount bolts.

69. Slowly lower the engine about 6 inches, make sure everything is clear, then lower it all the way.

70. Remove the crane.

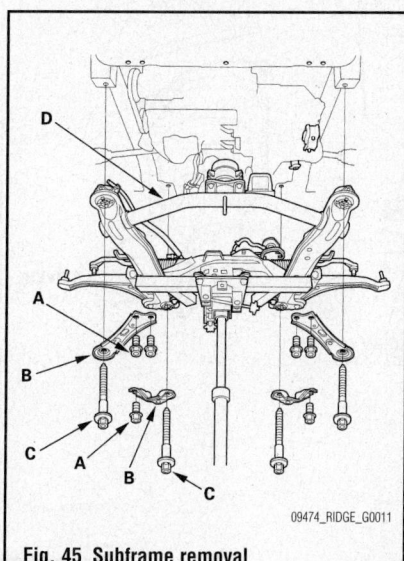

Fig. 45 Subframe removal

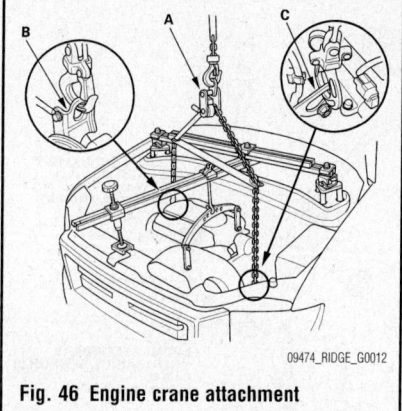

Fig. 46 Engine crane attachment

71. Raise the vehicle and remove the engine assembly.

To install:

72. Installation is the reverse of removal. Observe the following torques:
- Engine mount bracket: 33 ft. lbs. (44 Nm)
- Rear engine mount bracket: 65 ft. lbs. (88 Nm)
- Front engine mount bracket: 28 ft. lbs. (38 Nm)
- A/C compressor bracket: 33 ft. lbs. (44 Nm)
- P/S pump bracket: 16 ft. lbs. (22 Nm)
- Upper half engine mount bracket bolts: 33 ft. lbs. (44 Nm)

- Subframe bolts: as shown
- Transaxle lower front and rear lower mount nuts: 33 ft. lbs. (44 Nm)
- Rear mount bolts: 28 ft. lbs. (38 Nm)
- NEW front mount nut: 40 ft. lbs. (54 Nm)
- Exhaust pipe to crossover pipe: 40 ft. lbs. (54 Nm)
- Exhaust pipe hanger: 16 ft. lbs. (22 Nm)
- Exhaust pipe to converter: 25 ft. lbs. (33 Nm)
- Subframe stiffener: 40 ft. lbs. (54 Nm)
- Connector bracket to front head: 33 ft. lbs. (44 Nm)
- Harness clamp to rear head: 16 ft. lbs. (22 Nm)
- Steering shaft coupler: 16 ft. lbs. (22 Nm)
- Steering wheel nut to 36 ft. lbs. (49 Nm)

EXHAUST MANIFOLD

REMOVAL & INSTALLATION

Front

See Figure 48.

➡This engine does not use traditional exhaust manifolds. The catalytic converters are connected directly to the cylinder head.

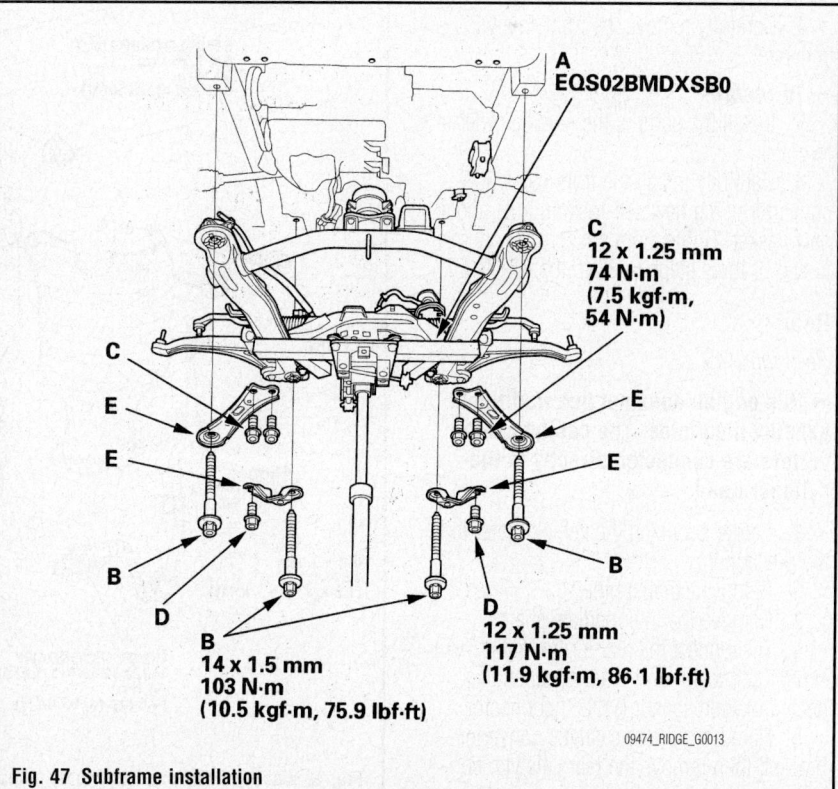

A
EQS02BMDXSB0

C
12 x 1.25 mm
74 N·m
(7.5 kgf·m, 54 N·m)

B
14 x 1.5 mm
103 N·m
(10.5 kgf·m, 75.9 lbf·ft)

D
12 x 1.25 mm
117 N·m
(11.9 kgf·m, 86.1 lbf·ft)

Fig. 47 Subframe installation

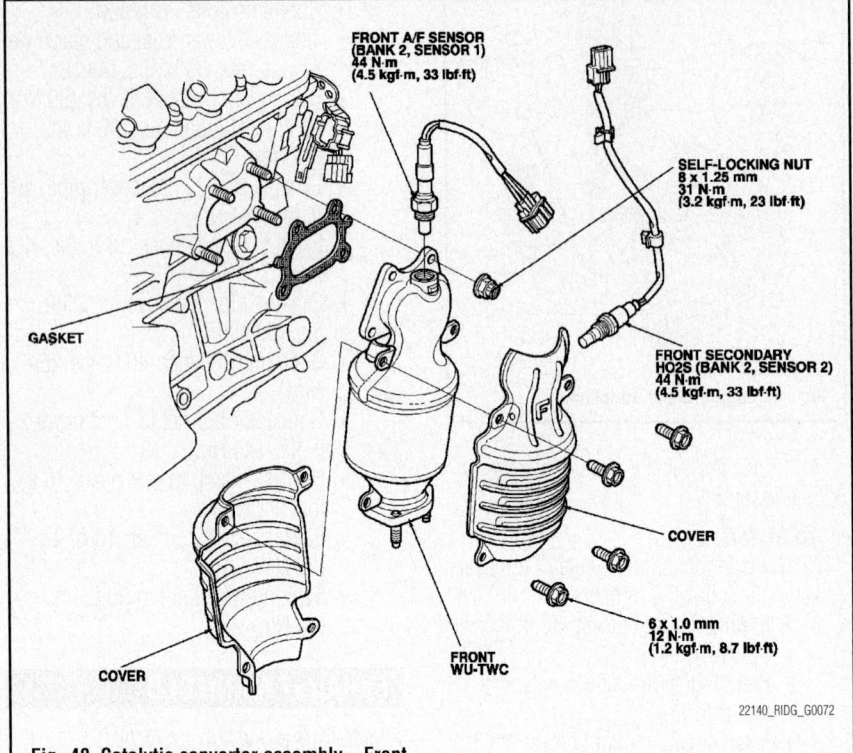

Fig. 48 Catalytic converter assembly—Front

1. Before servicing the vehicle, refer to the precautions.
2. Remove the radiator and A/C condenser fan assemblies.
3. Disconnect the front air fuel ratio (A/F) sensor connector and front secondary heated oxygen sensor (HO2S) connector.
4. Carefully remove the front catalytic converter.

To install:

5. Install the parts in the reverse order of removal.
6. Carefully install the front catalytic converter using new self-locking nuts and a new gasket. Tighten in a crisscross pattern in two or three steps to 23 ft. lbs. (31 Nm).

Rear

See Figure 49.

➠**This engine does not use traditional exhaust manifolds. The catalytic converters are connected directly to the cylinder head.**

1. Before servicing the vehicle, refer to the precautions.
2. Remove exhaust pipe.
3. Remove the intermediate shaft.
4. Disconnect the rear air fuel ratio (A/F) sensor connector and the rear secondary heated oxygen sensor (HO2S) connector.
5. Remove the rear catalytic converter bracket, then remove the rear catalytic converter.

To install:

6. Install the parts in the reverse order of removal.
7. Carefully install the front catalytic converter using new self-locking nuts and a new gasket. Tighten in a crisscross pattern in two or three steps to 23 ft. lbs. (31 Nm).

INTAKE MANIFOLD

REMOVAL & INSTALLATION

See Figures 50 through 54.

1. Before servicing the vehicle, refer to the precautions.
2. Make sure the anti-theft codes for the audio system and navigation system (if equipped) are available.
3. Disconnect the negative battery cable.
4. Remove the intake manifold cover.
5. Remove the air intake tube.
6. Remove the throttle and cruise control cables.
7. Disconnect the Intake Air Temperature (IAT) sensor connector.
8. Disconnect the Idle Air Control (IAC) valve connector.

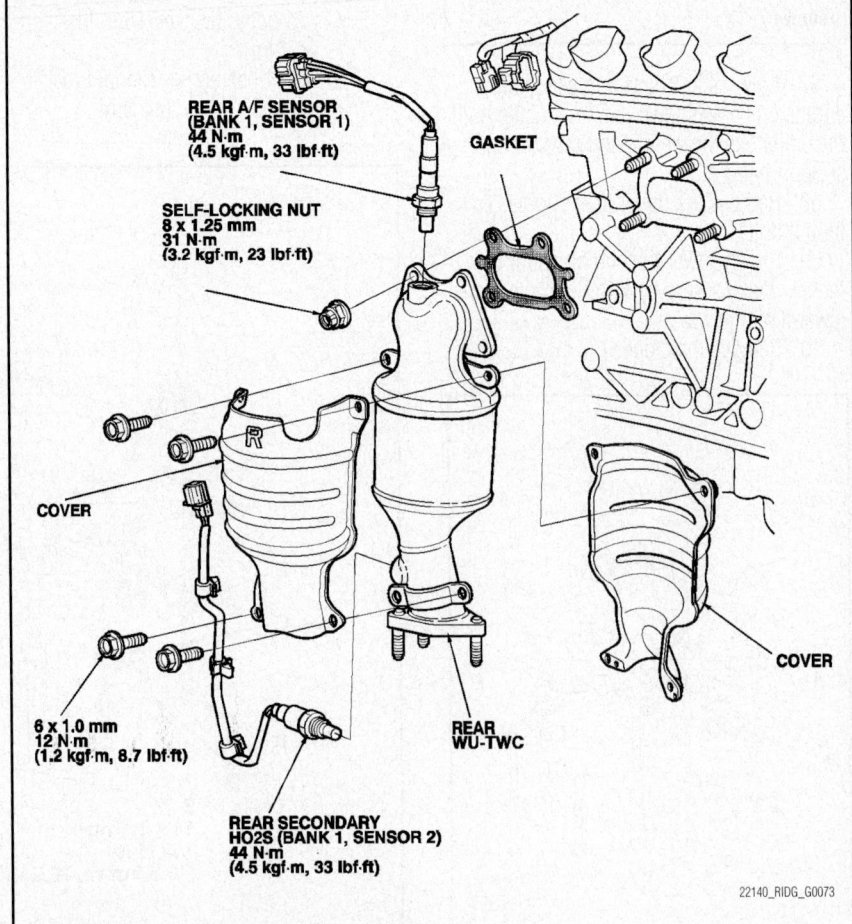

Fig. 49 Catalytic converter assembly—Rear

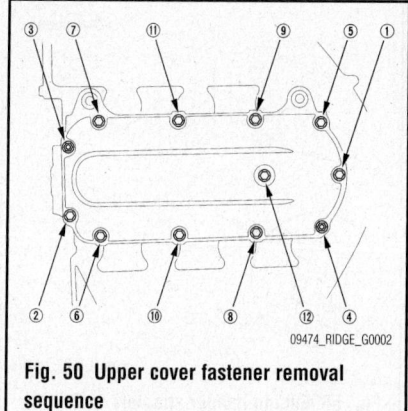

Fig. 50 Upper cover fastener removal sequence

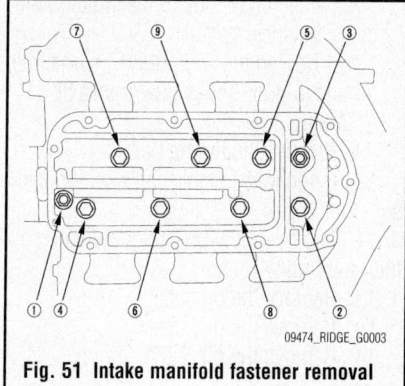

Fig. 51 Intake manifold fastener removal sequence

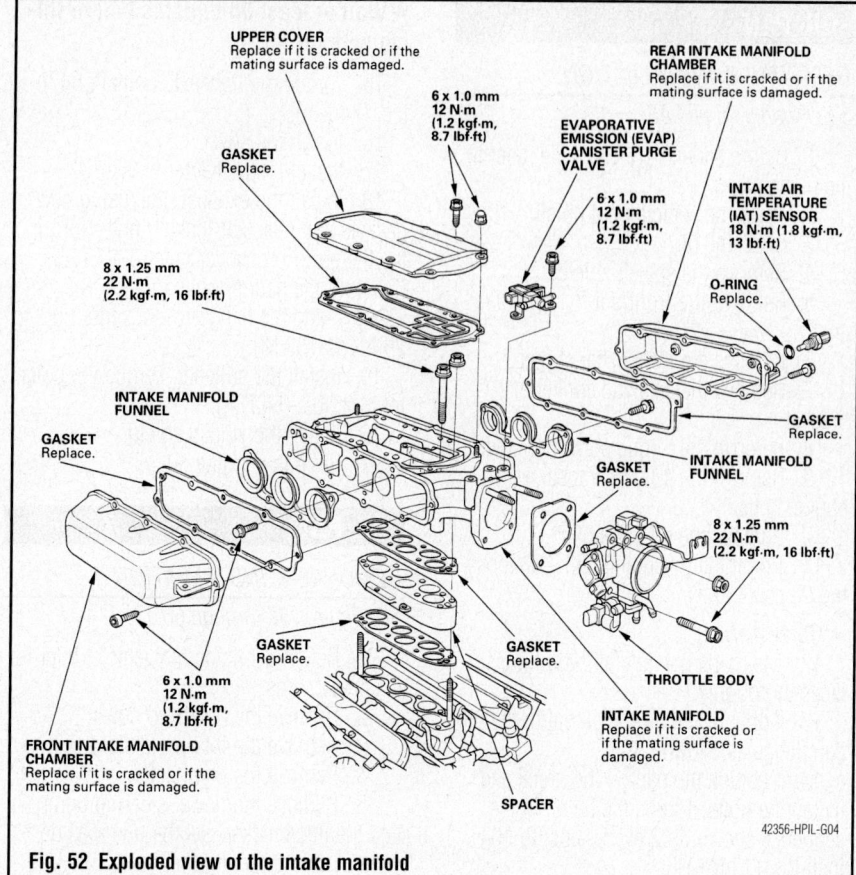

Fig. 52 Exploded view of the intake manifold

9. Disconnect the Throttle Position (TP) sensor connector.

10. Disconnect the Manifold Absolute Pressure (MAP) sensor connector.

11. Disconnect the Evaporative Emissions (EVAP) control canister purge valve connector.

12. Disconnect the brake booster vacuum line.

13. Disconnect the Positive Crankcase Ventilation (PCV) hose.

14. Disconnect the water bypass hoses from the throttle body and plug the hoses.

15. Disconnect the EVAP control canister hose .

16. Remove the upper cover bolts and nuts in the sequence illustrated using two or three passes.

17. Remove the intake manifold bolts in sequence using two or three passes.

18. Remove the intake manifold and spacer.

To install:

19. Install the New intake manifold gasket and spacer.

20. Install the intake manifold. Tighten the fasteners in sequence and in several passes to 16 ft. lbs. (22 Nm).

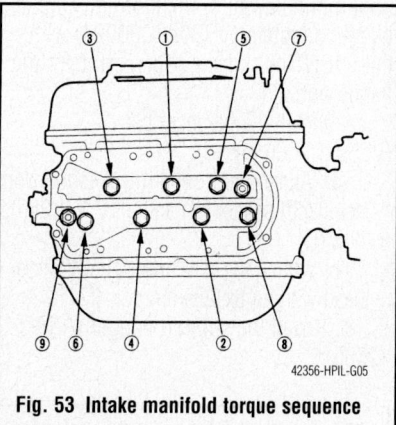

Fig. 53 Intake manifold torque sequence

21. Install the upper cover bolts and nuts in the sequence illustrated using two or three passes to 9 ft. lbs. (12 Nm).

22. Install the EVAP control canister hose .

23. Install the water bypass hoses to the throttle body.

24. Install the brake booster vacuum line.

25. Install the PCV hose.

26. Connect the EVAP control canister purge valve connector.

27. Connect the MAP sensor connector.

28. Connect the TP sensor connector.

29. Connect the IAC valve connector.

30. Connect the IAT sensor connector.

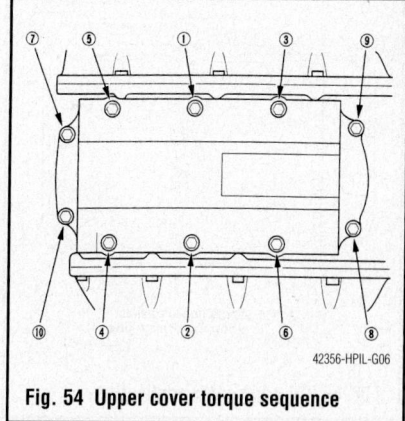

Fig. 54 Upper cover torque sequence

31. Install the throttle and cruise control cables.

32. Install the air intake tube.

33. Install the intake manifold cover.

34. Connect the negative battery cable.

35. Enter the anti-theft codes for the audio system and the navigation system (if equipped).

36. Set the clock on vehicles without navigation.

37. Perform the power window control unit reset procedure.

38. Start the engine and check for proper operation.

OIL PAN

REMOVAL & INSTALLATION

See Figures 55 and 56.

1. Before servicing the vehicle, refer to the precautions.
2. Raise the vehicle on a hoist.
3. Drain the oil.
4. Remove the splash shield.
5. Remove the front subframe stiffener.
6. Remove the front exhaust pipe.
7. Remove the catalytic converter bracket.
8. Remove the torque converter cover.
9. Remove the 4 lower transaxle-to-engine bolts.
10. Remove the oil pan bolts.
11. Pry at the pry-points to break loose the oil pan.

To install:

12. Clean the pan and all mounting surfaces thoroughly.
13. Apply RTV gasket material to the oil pan flange as shown.
14. Position the pan on the block and install the bolts. Torque the bolts, in the sequence shown, in 2 even steps, to 104 inch lbs. (12 Nm).

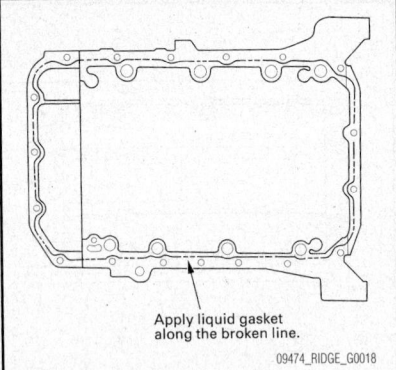

Apply liquid gasket along the broken line.

09474_RIDGE_G0018

Fig. 55 Oil pan sealer application

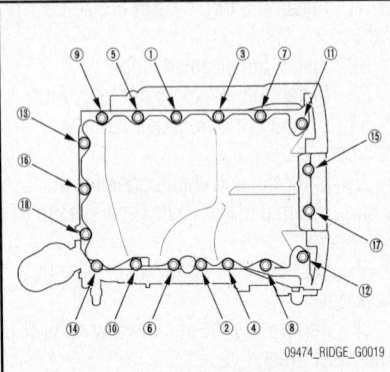

09474_RIDGE_G0019

Fig. 56 Oil pan bolt torque sequence

➡ Wait at least 30 minutes before filling with oil.

15. Torque the transaxle bolts to 54 ft. lbs. (74 Nm).
16. Install the cover.
17. Install the converter bracket.
18. Install the exhaust pipe using new gaskets and new self-locking nuts. Torque the flange-to-crossover nuts to 40 ft. lbs. (54 Nm), the hanger bolts to 16 ft. lbs. (22 Nm) and the flange-to-converter nuts to 25 ft. lbs. (33 Nm).
19. Install the stiffener. Torque the bolts to 40 ft. lbs. (54 Nm).
20. Install the splash shield.
21. Refill the engine oil.

OIL PUMP

REMOVAL & INSTALLATION

See Figures 57 through 60.

1. Before servicing the vehicle, refer to the precautions.
2. Remove the bulkhead cover.
3. Remove the intake manifold cover.
4. Remove the drive belt.
5. Remove the power steering pump and the line bracket. Set the pump aside without disconnecting the lines.
6. Remove the caps from the front strut mounting nuts. Position the engine hanger adapters (VSB02C000024) with the FRONT mark facing forward, over the flange nuts.
7. Install the balancer bar (VSB02C000019):
   a. Attach the front arm (A) to the front head with spacer (B) and a 10 × 1.25mm bolt (C)
   b. Attach the rear arm (D) to the rear head with an 8x1.25mm bolt (E).
8. Install the support hanger (AAR-T-12566):

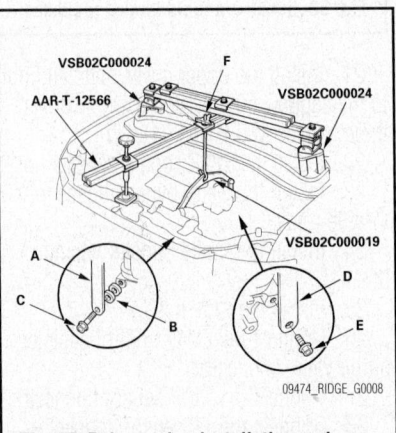

VSB02C000024 F
AAR-T-12566 VSB02C000024

VSB02C000019

A D
C B E

09474_RIDGE_G0008

Fig. 57 Balancer bar installation and support hanger

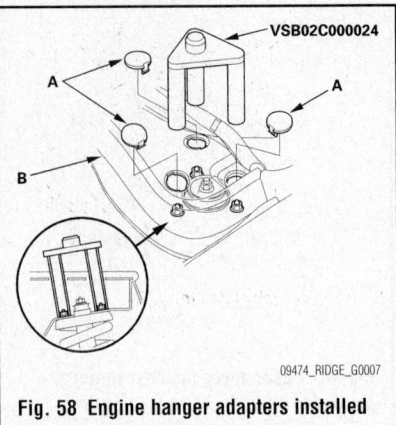

VSB02C000024
A A
B

09474_RIDGE_G0007

Fig. 58 Engine hanger adapters installed

   a. Attach the hook to the slotted hole in the balancer bar.
   b. Tighten the wing nut (F) by hand to lift and support the engine/transaxle assembly.
9. Remove the timing belt.
10. Remove the crankshaft position sensor.
11. Remove the VTEC solenoid valve/oil filter assembly.
12. Remove the oil pan.
13. Remove the oil screen.
14. Remove the oil pump.

To install:

15. Discard the oil seal.
16. Tap a new seal into place until it bottoms.
17. Thoroughly clean and dry all gasket surfaces.
18. Apply a bead of RTV gasket material to the pump mating surface of the block.

➡ The oil pump must be installed within 4 minutes of applying the gasket material.

19. Apply all-purpose grease to the seal lip and clean engine oil to the new O-ring (A).

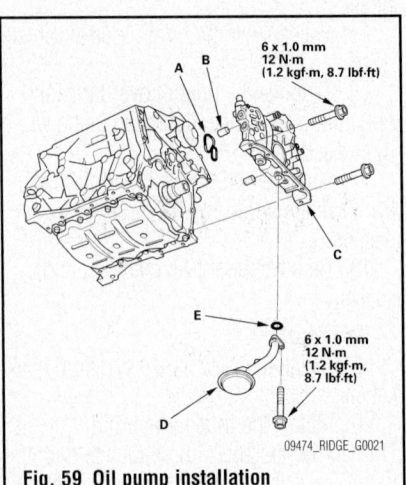

6 x 1.0 mm
12 N·m
(1.2 kgf·m, 8.7 lbf·ft)
A B

C

E

6 x 1.0 mm
12 N·m
(1.2 kgf·m, 8.7 lbf·ft)
D

09474_RIDGE_G0021

Fig. 59 Oil pump installation

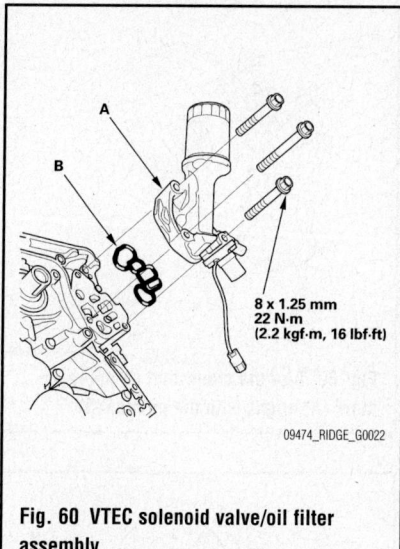

Fig. 60 VTEC solenoid valve/oil filter assembly

20. Install the dowel pins (B) then align the inner rotor with the crankshaft and install the oil pump (C).
21. Install the screen (D) with a new O-ring (E).
22. Install the VTEC solenoid valve/oil filter assembly, using a new filter.
23. Install the oil pan.
24. Install the CKP.
25. Install the timing belt.
26. Remove the engine hanger.
27. Install the caps on the strut.
28. Install the power steering pump. Torque the mounting bolts to 16 ft. lbs. (22 Nm).
29. Install the drive belt.
30. Install the bulkhead cover.

MAIN BEARING TORQUE SEQUENCE

See Figure 61.

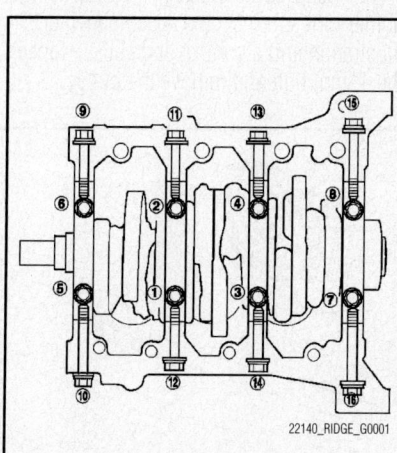

Fig. 61 Main bearing torque sequence—3.5L engine

PISTON AND RING

POSITIONING

See Figures 62 and 63.

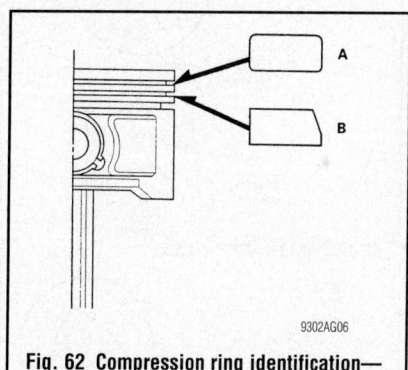

Fig. 62 Compression ring identification—3.5L engine

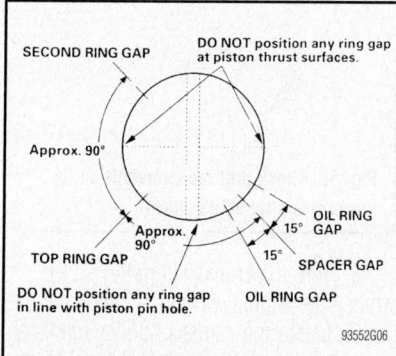

Fig. 63 Ring end gap positioning—3.5L engine

REAR MAIN SEAL

REMOVAL & INSTALLATION

See Figure 64.

➡**Oil seal Driver 07749-0010000 and Driver attachment, 106 mm 070AD-RCA0200 are required to perform this procedure.**

1. Before servicing the vehicle, refer to the precautions.
2. Remove the transmission and the drive plate.
3. Remove the transmission end crankshaft oil seal.

To install:
4. Clean and dry the crankshaft oil seal housing.
5. Apply a light coat of multipurpose grease to the crankshaft and to the lip of the seal.
6. Using the special tools, drive in the crankshaft oil seal until the driver attachment bottoms against the engine block end cover. Align the hole in the

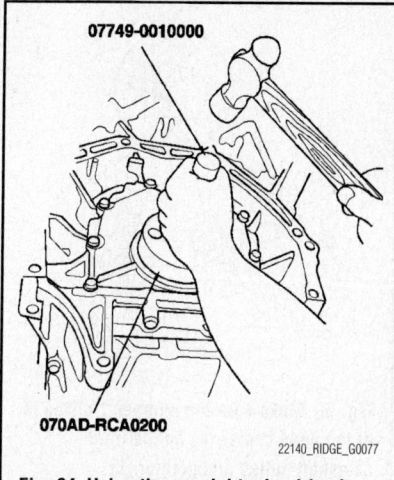

Fig. 64 Using the special tools, drive in the crankshaft oil seal until the driver attachment bottoms against the engine block end cover

driver attachment with the pin on the crankshaft.
7. Clean any excess grease off the crankshaft, and check that the oil seal lip is not distorted.
8. Install the drive plate, and the transmission.
9. Check the fluid levels.
10. Start the engine and check for leaks.

TIMING BELT FRONT COVER

REMOVAL & INSTALLATION

See Figures 65 through 68.

1. Before servicing the vehicle, refer to the precautions.
2. Turn the crankshaft so the white mark aligns with the pointer.
3. Make sure the number 1 piston is at Top Dead center (TDC).
4. Make sure the anti-theft codes for the audio system and navigation system (if equipped) are available.

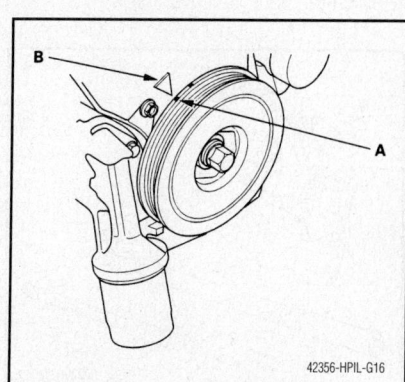

Fig. 65 Turn the crankshaft so the white mark (A) aligns with the pointer (B)

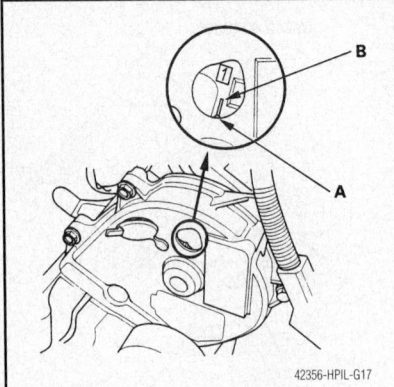

Fig. 66 Make sure the number 1 piston is at top dead center (A) on the front camshaft pulley and pointer (B)

5. Disconnect the negative battery cable
6. Remove the wheels and engine splash shield.
7. Remove the drive belts.
8. Support the engine with a block of wood and a jack under the oil pan.
9. Remove the ground cable bracket.
10. Remove the upper side engine mount.
11. Remove the dipstick tube.
12. Remove the crankshaft pulley using holder tool shown in the accompanying illustration and a breaker and socket, loosen the 19mm bolt and remove the pulley.
13. Remove the front upper cover, rear upper cover and the lower cover.

To install:

14. Install the lower timing cover and tighten the bolts to 9 ft. lbs. (12 Nm).
15. Install the front and rear upper timing covers and tighten the bolts to 9 ft. lbs. (12 Nm).
16. Install the crankshaft pulley and tighten the bolts to 181 ft. lbs. (245 Nm), using the holding tool to prevent the unit from turning.
17. Rotate the crankshaft pulley about 5 or 6 degrees clockwise so the belt positions itself on the pulleys.

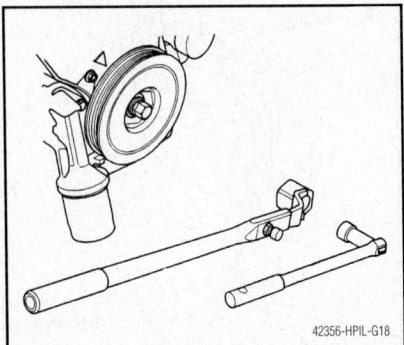

Fig. 67 Remove the crankshaft pulley using holder tool shown

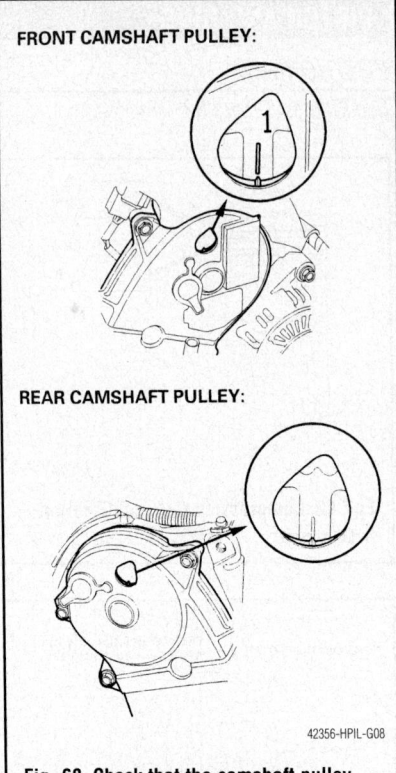

FRONT CAMSHAFT PULLEY:

REAR CAMSHAFT PULLEY:

Fig. 68 Check that the camshaft pulley marks are aligned as shown

18. Turn the crankshaft pulley so the white mark aligns with the pointer.
19. Check the camshaft pulley marks are aligned. If the marks are aligned, proceed to the next step. If the marks are not aligned, remove the timing belt and reinstall using the steps outlined before this step.
20. Install the drive belt.
21. Install the upper side mount and tighten the bolts in the sequence illustrated to the specifications in the illustration.
22. Using a suitable scan tool, perform the Powertrain Control Module (PCM) reset and the Crankshaft position (CKP) pattern clear/learn procedures, following the scan tool manufactures instructions.

TIMING BELT AND SPROCKETS

REMOVAL & INSTALLATION

See Figures 69 through 80.

1. Before servicing the vehicle, refer to the precautions.
2. Turn the crankshaft so the white mark aligns with the pointer.
3. Make sure the number 1 piston is at Top Dead center (TDC).
4. Make sure the anti-theft codes for the audio system and navigation system (if equipped) are available.
5. Disconnect the negative battery cable

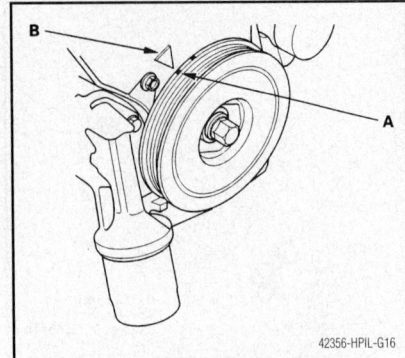

Fig. 69 Turn the crankshaft so the white mark (A) aligns with the pointer (B)

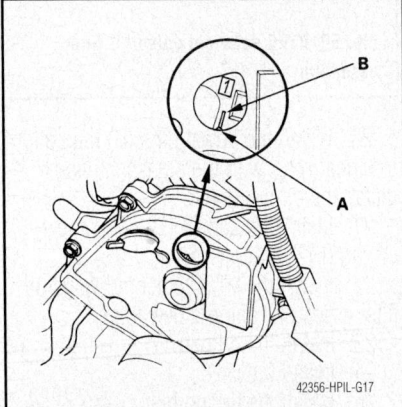

Fig. 70 Make sure the number 1 piston is at top dead center (A) on the front camshaft pulley and pointer (B)

6. Remove the wheels and engine splash shield.
7. Remove the drive belts.
8. Support the engine with a block of wood and a jack under the oil pan.
9. Remove the ground cable bracket.
10. Remove the upper side engine mount.
11. Remove the dipstick tube.
12. Remove the crankshaft pulley using holder tool shown in the accompanying illustration and a breaker and socket, loosen the 19mm bolt and remove the pulley.

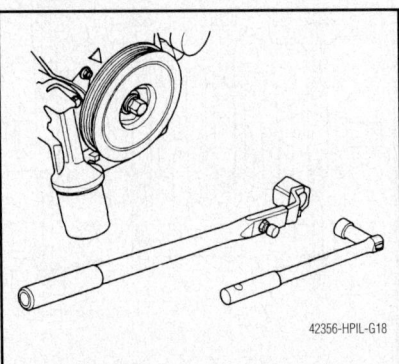

Fig. 71 Remove the crankshaft pulley using holder tool shown

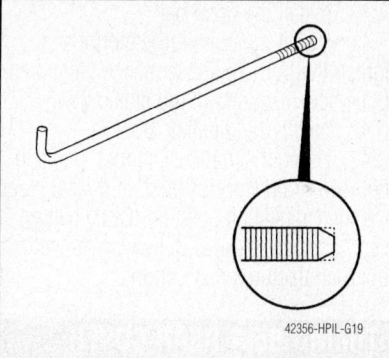

Fig. 72 Remove a battery clamp bolt and grind the end as shown

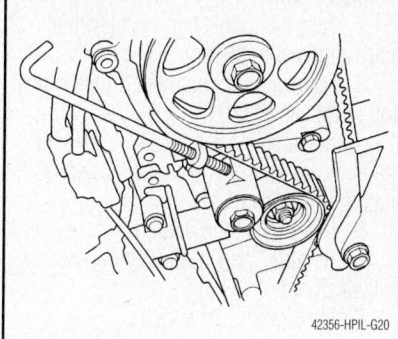

Fig. 73 Install the battery clamp bolt as shown to hold the belt adjuster in position

13. Remove the front upper cover, rear upper cover and the lower cover.

14. Remove the one of the battery clamp bolts and grind the end as shown in the illustration.

15. Screw the battery clamp bolt as shown in the illustration to hold the belt adjuster in position. Do not use a wrench, hand tighten only.

16. Remove the lower side engine mount.

17. Remove the idler pulley bolt and the pulley.

• Timing belt

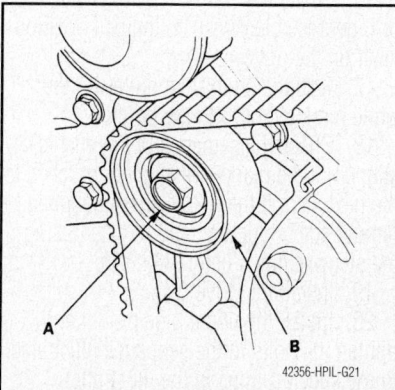

42356-HPIL-G21

Fig. 74 Remove the idler pulley bolt (A), pulley (B) and the timing belt

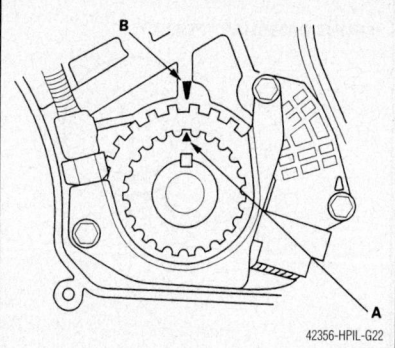

Fig. 75 Set the timing belt pulley to TDC by aligning the TDC mark (A) on the tooth of the belt pulley with the pointer (B) on the oil pump

To install:

18. If installing a new belt, perform the following steps:

a. Clean the pulleys, belt guide plate and the upper and lower covers.

b. Set the timing belt drive pulley to TDC by aligning the TDC mark on the tooth of the belt drive pulley with the pointer on the oil pump.

c. Set the camshaft pulleys to TDC by aligning the TDC marks on the camshaft pulleys with the pointers on the back covers.

d. Remove the battery clamp bolt.

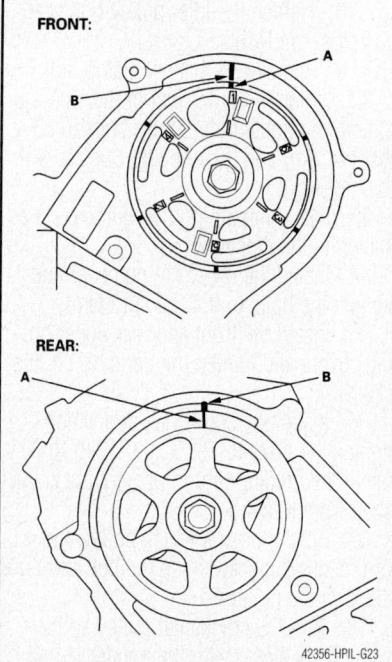

Fig. 76 Set the camshaft pulleys to TDC by aligning the TDC marks (A) on the camshaft pulleys with the pointers (B) on the back covers

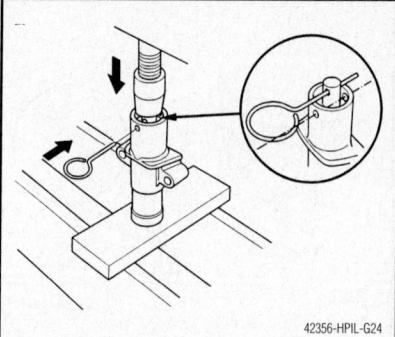

Fig. 77 Insert a 0.08 inch (2mm) pin through the tensioner housing and rod

e. Remove the belt tensioner.

f. Align the holes on the rod and housing of the tensioner.

g. Using a press or other suitable device, slowly compress the tensioner and insert a 0.08 inch (2mm) pin through the housing and rod.

h. Install the tensioner making sure the pin is still installed.

i. Apply thread locker to idler pulley bolt then hand tighten the bolt.

j. Install the belt over the pulleys in this sequence; drive pulley, idler pulley, front camshaft pulley, water pump pulley, rear camshaft pulley and adjusting pulley.

k. Tighten the idler pulley bolt to 33 ft. lbs. (44 Nm).

l. Remove the pin from the tensioner.

19. Install the lower half of the side mount and tighten the 3 long bolts to 33 ft. lbs. (44 Nm) and the one short bolt to 9 ft. lbs. (12 Nm).

20. Install the timing belt guide plate as shown in the illustration.

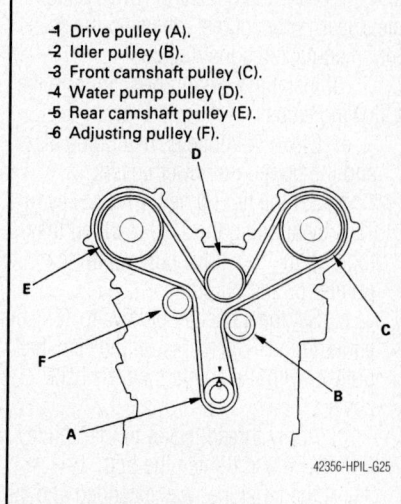

1 Drive pulley (A).
2 Idler pulley (B).
3 Front camshaft pulley (C).
4 Water pump pulley (D).
5 Rear camshaft pulley (E).
6 Adjusting pulley (F).

Fig. 78 Route the belt as shown in the sequence listed

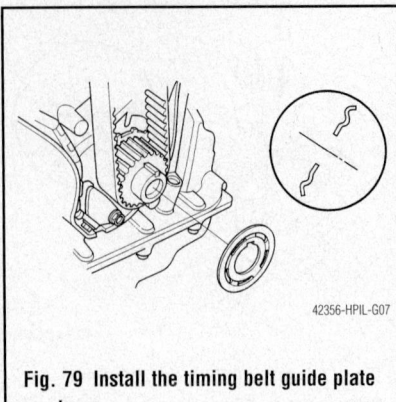

Fig. 79 Install the timing belt guide plate as shown

21. Install the lower timing cover and tighten the bolts to 9 ft. lbs. (12 Nm).

22. Install the front and rear upper timing covers and tighten the bolts to 9 ft. lbs. (12 Nm).

23. Install the crankshaft pulley and tighten the bolts to 181 ft. lbs. (245 Nm), using the holding tool to prevent the unit from turning.

24. Rotate the crankshaft pulley about 5 or 6 degrees clockwise so the belt positions itself on the pulleys.

25. Turn the crankshaft pulley so the white mark aligns with the pointer.

26. Check the camshaft pulley marks are aligned. If the marks are aligned, proceed to the next step. If the marks are not aligned, remove the timing belt and reinstall using the steps outlined before this step.

27. Install the drive belt.

28. Install the upper side mount and tighten the bolts in the sequence illustrated to the specifications in the illustration.

29. Using a suitable scan tool, perform the Powertrain Control Module (PCM) reset and the Crankshaft position (CKP) pattern clear/learn procedures, following the scan tool manufactures instructions.

30. If installing the old belt, perform the following steps:

    a. Clean the pulleys, belt guide plate and the upper and lower covers.

    b. Set the timing belt drive pulley to TDC by aligning the TDC mark on the tooth of the belt drive pulley with the pointer on the oil pump.

    c. Set the camshaft pulleys to TDC by aligning the TDC marks on the camshaft pulleys with the pointers on the back covers.

    d. Apply thread locker to idler pulley bolt then hand tighten the bolt.

31. If the tensioner was extended and the belt cannot be installed, perform the steps above for the new belt installation.

FRONT CAMSHAFT PULLEY:

REAR CAMSHAFT PULLEY:

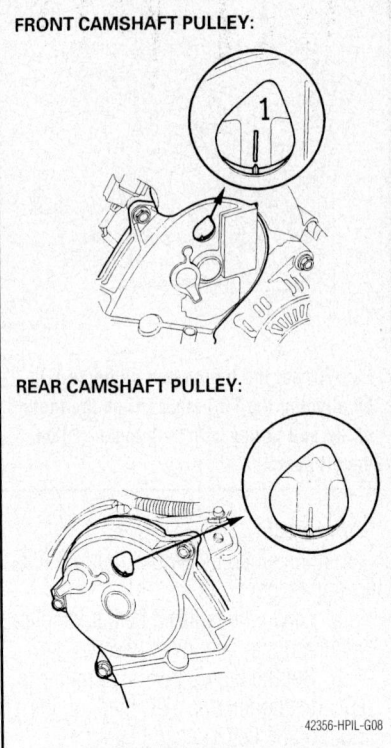

Fig. 80 Check that the camshaft pulley marks are aligned as shown

    a. Install the belt over the pulleys in this sequence; drive pulley, idler pulley, front camshaft pulley, water pump pulley, rear camshaft pulley and adjusting pulley.

    b. Tighten the idler pulley bolt to 33 ft. lbs. (44 Nm).

    c. Remove the battery clamp bolt.

32. Install the lower half of the side mount and tighten the 3 long bolts to 33 ft. lbs. (44 Nm) and the one short bolt to 9 ft. lbs. (12 Nm).

33. Install the timing belt guide plate as shown in the illustration.

34. Install the lower timing cover and tighten the bolts to 9 ft. lbs. (12 Nm).

35. Install the front and rear upper timing covers and tighten the bolts to 9 ft. lbs. (12 Nm).

36. Install the crankshaft pulley and tighten the bolts to 181 ft. lbs. (245 Nm), using the holding tool to prevent the crankshaft from rotating.

37. Rotate the crankshaft pulley about 5 or 6 degrees clockwise so the belt positions itself on the pulleys.

38. Turn the crankshaft pulley so the white mark aligns with the pointer.

39. Check the camshaft pulley marks are aligned. If the marks are aligned, proceed to the next step. If the marks are not aligned, remove the timing belt and reinstall using the steps outlined before this step.

40. Install the drive belt.

41. Install the upper side mount and tighten the bolts in the sequence illustrated to the specifications in the illustration.

42. Install the dipstick tube.

43. Using a suitable scan tool, perform the Powertrain Control Module (PCM) reset and the Crankshaft position (CKP) pattern clear/learn procedures, following the scan tool manufactures instructions.

TIMING BELT REAR COVER

REMOVAL & INSTALLATION

1. Before servicing the vehicle, refer to the precautions.

2. Turn the crankshaft so the white mark aligns with the pointer.

3. Make sure the number 1 piston is at Top Dead center (TDC).

4. Make sure the anti-theft codes for the audio system and navigation system (if equipped) are available.

5. Disconnect the negative battery cable

6. Remove the wheels and engine splash shield.

7. Remove the drive belts.

8. Support the engine with a block of wood and a jack under the oil pan.

9. Remove the ground cable bracket.

10. Remove the upper side engine mount.

11. Remove the dipstick tube.

12. Remove the crankshaft pulley using holder tool shown in the accompanying illustration and a breaker and socket, loosen the 19mm bolt and remove the pulley.

13. Remove the rear upper cover.

To install:

14. Install the rear upper timing covers and tighten the bolts to 9 ft. lbs. (12 Nm).

15. Install the crankshaft pulley and tighten the bolts to 181 ft. lbs. (245 Nm), using the holding tool to prevent the unit from turning.

16. Rotate the crankshaft pulley about 5 or 6 degrees clockwise so the belt positions itself on the pulleys.

17. Turn the crankshaft pulley so the white mark aligns with the pointer.

18. Check the camshaft pulley marks are aligned. If the marks are aligned, proceed to the next step. If the marks are not aligned, remove the timing belt and reinstall using the steps outlined before this step.

19. Install the drive belt.

20. Install the upper side mount and tighten the bolts in the sequence illustrated to the specifications in the illustration.

21. Using a suitable scan tool, perform the Powertrain Control Module (PCM) reset

and the Crankshaft position (CKP) pattern clear/learn procedures, following the scan tool manufactures instructions.

VALVE LASH

ADJUSTMENT

See Figures 81 and 82.

Adjust the valves only when the cylinder head temperature is less than 100°F (38°C).

1. Before servicing the vehicle, refer to the precautions.

2. Make sure the anti-theft codes for the audio system and navigation system (if equipped) are available.

3. Disconnect the negative battery cable.

4. Remove the air intake tube.

5. Remove the intake manifold.

6. Remove the valve cover.

7. Rotate the crankshaft so that the valves to be adjusted are closed and the rocker arm is contacting the camshaft lobe base circle.

8. Measure the valve clearance. If adjustment is necessary, loosen the locknut and turn the adjusting screw as necessary to achieve the correct valve clearance.

9. After adjustment, tighten the locknuts to 14 ft. lbs. (20 Nm).

10. Install the valve cover.

11. Install the intake manifold.

12. Install the air intake tube.

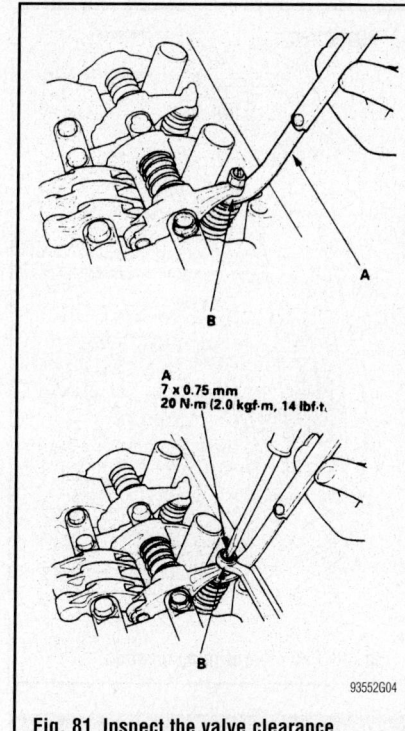

7 x 0.75 mm
20 N·m (2.0 kgf·m, 14 lbf·ft.)

Fig. 81 Inspect the valve clearance, adjust to specification and tighten the retainer to specification

13. Connect the negative battery cable.

14. Enter the anti-theft codes for the audio system and the navigation system (if equipped).

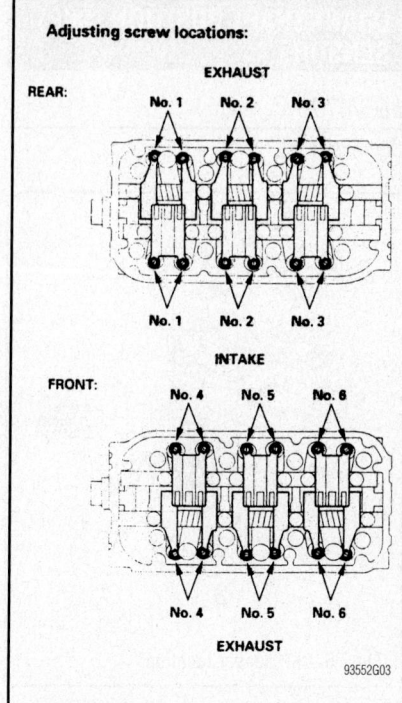

Adjusting screw locations:

Fig. 82 Valve adjusting retainer locations

15. Set the clock on vehicles without navigation.

16. Perform the power window control unit reset procedure.

17. Start the engine and check for proper operation.

ENGINE PERFORMANCE & EMISSION CONTROL

ACCELERATOR PEDAL POSITION (APP) SENSOR

LOCATION
See Figure 83.

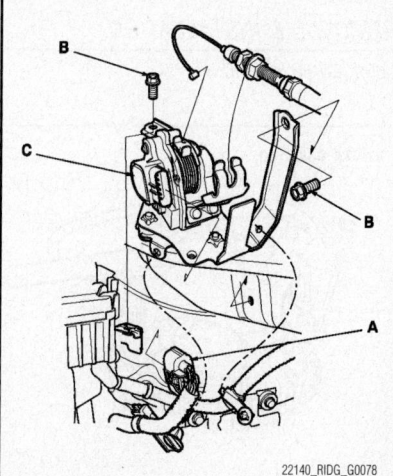

Fig. 83 APP sensor (C), attaching bolt (B) and connector (A)

REMOVAL & INSTALLATION

1. Remove the throttle cable.

2. Disconnect the accelerator pedal position (APP) sensor 6P connector (A).

3. Remove the bolts (B) and the APP sensor (C).

To install:

4. Install the parts in the reverse order of removal.

CAMSHAFT POSITION (CMP) SENSOR

LOCATION
See Figure 84.

REMOVAL & INSTALLATION

1. Before servicing the vehicle, refer to the precautions.

2. Remove the timing belt.

3. Remove the front camshaft pulley.

4. Disconnect the CMP sensor connector, then remove the back cover.

5. Remove the CMP sensor from the back cover.

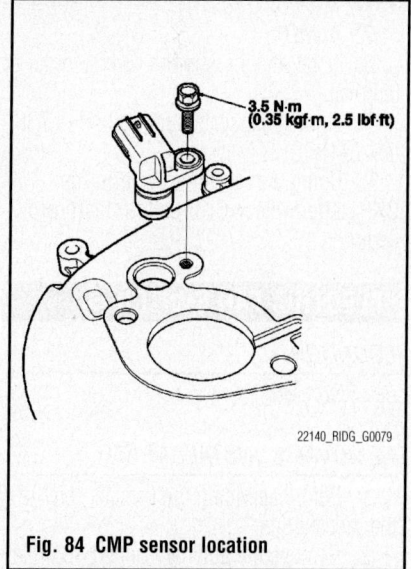

3.5 N·m
(0.35 kgf·m, 2.5 lbf·ft)

Fig. 84 CMP sensor location

To install:

6. Install the parts in the reverse order of removal.

7. Using a scan tool, perform the CKP pattern clear/CKP pattern learn procedure.

CRANKSHAFT POSITION (CKP) SENSOR

LOCATION

See Figure 85.

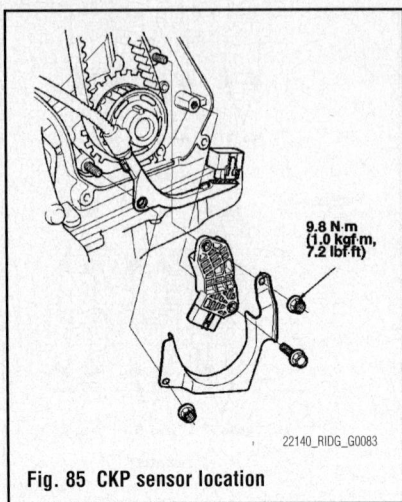

Fig. 85 CKP sensor location

REMOVAL & INSTALLATION

1. Before servicing the vehicle, refer to the precautions.
2. Move the auto-tensioner to remove tension from the drive belt, then remove the belt.
3. Remove the crankshaft pulley.
4. Remove the upper and lower front covers from the engine.
5. Remove the CKP sensor from the oil pump.

To install:

6. Install the parts in the reverse order of removal.
7. Tighten sensor mounting bolt to 7 ft. lbs. (9 Nm).
8. Using a scan tool, perform the CKP pattern clear/CKP pattern learn procedure.

ELECTRIC FAN SWITCH

LOCATION

See Figure 86.

REMOVAL & INSTALLATION

1. Before servicing the vehicle, refer to the precautions.
2. Remove the auxiliary underhood relay box cover.
3. Remove the radiator fan control relay.

To install:

4. Install the parts in the reverse order of removal.

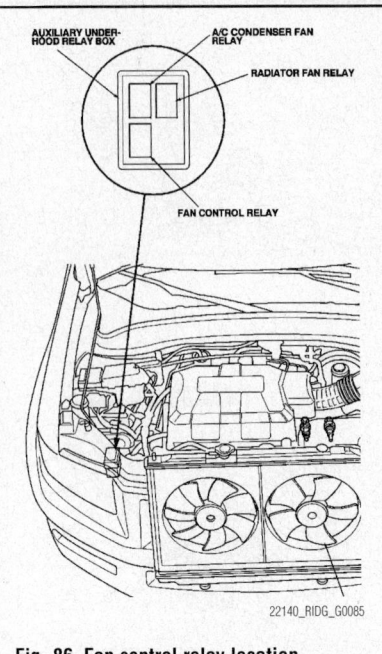

Fig. 86 Fan control relay location

ENGINE COOLANT TEMPERATURE (ECT) SENSOR

LOCATION

See Figures 87 and 88.

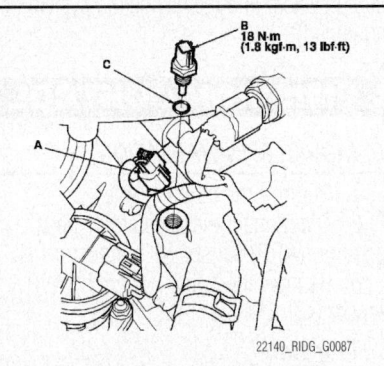

Fig. 87 ECT sensor 1 (B), O-ring (C) and connector (A) location

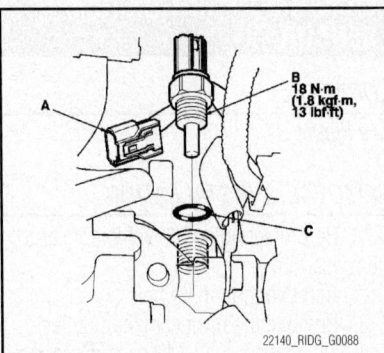

Fig. 88 ECT sensor 2 (B), O-ring (C) and connector (A) location

REMOVAL & INSTALLATION

1. Before servicing the vehicle, refer to the precautions.
2. Drain the engine coolant.
3. Remove the intake manifold cover.
4. Disconnect the ECT sensor 2P connector.
5. Remove ECT sensor.

To install:

6. Install the parts in the reverse order of removal with a new O-ring.
7. Tighten to 13 ft. lbs. (18 Nm).
8. Refill the radiator with engine coolant.

FUEL LEVEL SENDING UNIT

LOCATION

See Figure 89.

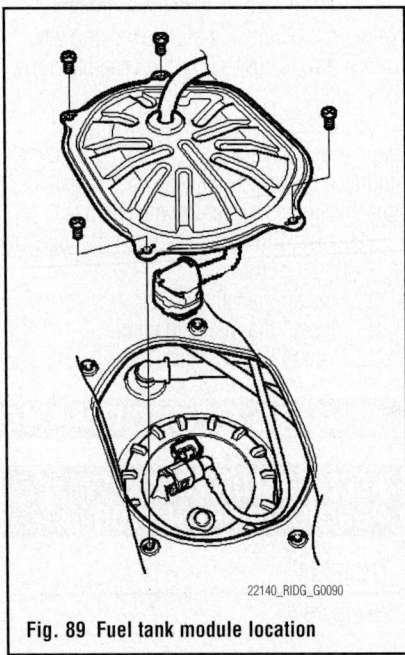

Fig. 89 Fuel tank module location

REMOVAL & INSTALLATION

See Figures 90 and 91.

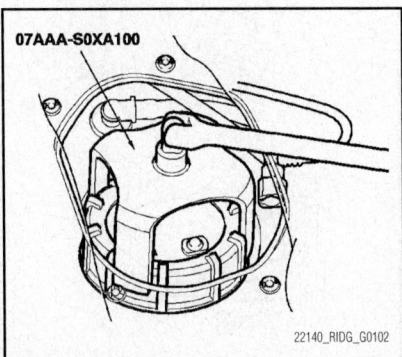

Fig. 90 Fuel tank module lockring special tool

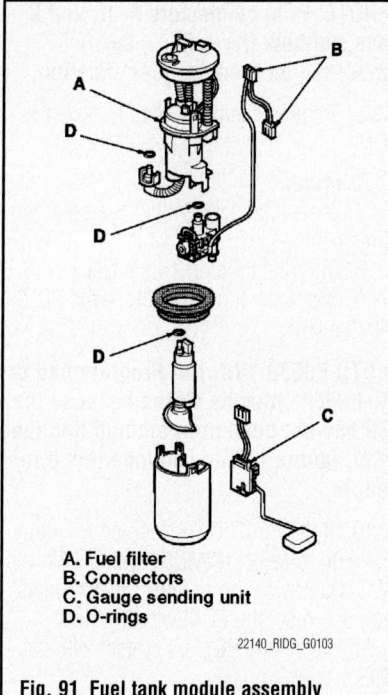

A. Fuel filter
B. Connectors
C. Gauge sending unit
D. O-rings

22140_RIDG_G0103

Fig. 91 Fuel tank module assembly

1. Before servicing the vehicle, refer to the precautions.
2. Relieve the fuel system pressure.
3. Remove the rear back seat.
4. Remove the access panel on the floor.
5. Remove the fuel tank module lock-ring using the special tool.
6. Remove the fuel level sending unit.

To install:

7. Check these items before installing the fuel tank unit:

   a. When connecting the wire harness, make sure the connection is secure and the connectors are firmly locked into place.

   b. When installing the fuel gauge sending unit, make sure the connection is secure and the connector is firmly locked into place. Be careful not to bend or twist it excessively.

8. Install the parts in the reverse order of removal with new O-rings.

➡ **Coat the O-rings with clean engine oil.**

9. When installing the fuel tank unit, align the marks on the unit and the fuel tank.
10. Torque the lockring to 69 ft. lbs. (93 Nm).

HEATED OXYGEN (HO2S) SENSOR

LOCATION
See Figures 92 and 93.

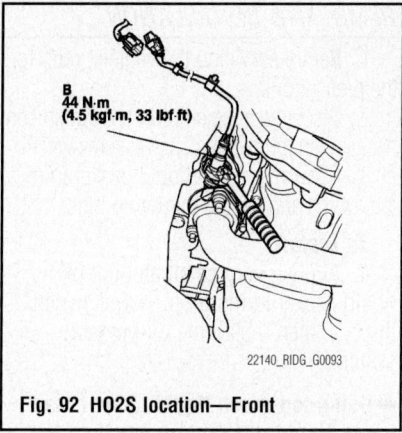

B
44 N·m
(4.5 kgf·m, 33 lbf·ft)

22140_RIDG_G0093

Fig. 92 HO2S location—Front

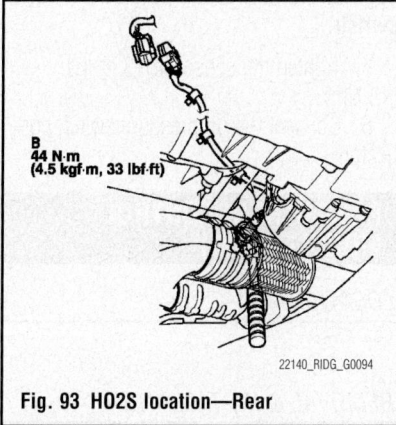

B
44 N·m
(4.5 kgf·m, 33 lbf·ft)

22140_RIDG_G0094

Fig. 93 HO2S location—Rear

REMOVAL & INSTALLATION

➡ **O2 sensor socket wrench, Snap-on YA8875, SP Tools 93750, or equivalent is needed to perform this procedure.**

1. Before servicing the vehicle, refer to the precautions.
2. Disconnect the oxygen sensor connector.
3. Remove the oxygen sensor using the special tool.

To install:

4. Install the parts in the reverse order of removal.
5. Tighten to 33 ft. lbs. (44 Nm).

INTAKE AIR TEMPERATURE (IAT) SENSOR

LOCATION
See Figure 94.

REMOVAL & INSTALLATION

1. Before servicing the vehicle, refer to the precautions.
2. Remove the intake manifold cover.
3. Disconnect the IAT sensor 2P connector (A).
4. Remove the IAT sensor (B).

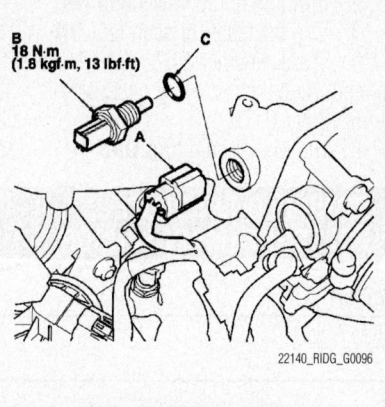

B
18 N·m
(1.8 kgf·m, 13 lbf·ft)
C

22140_RIDG_G0096

Fig. 94 IAT sensor (B), O-ring (C) and connector (A) location

To install:

5. Install the parts in the reverse order of removal with a new O-ring (C).
6. Tighten to 13 ft. lbs. (18 Nm).

KNOCK SENSOR (KS)

LOCATION
See Figure 95.

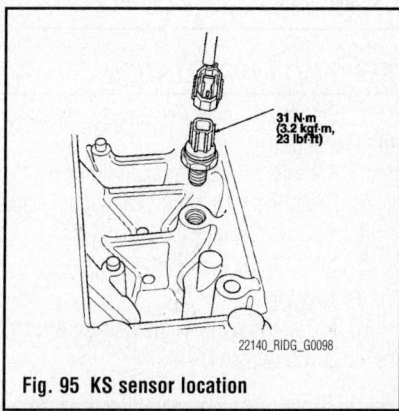

31 N·m
(3.2 kgf·m, 23 lbf·ft)

22140_RIDG_G0098

Fig. 95 KS sensor location

REMOVAL & INSTALLATION

1. Before servicing the vehicle, refer to the precautions.
2. Remove the intake manifold.
3. Remove the injector rails and the base.
4. Disconnect the knock sensor 1P connector (A), then remove the knock sensor (B).

To install:

5. Install the parts in the reverse order of removal.
6. Tighten to 23 ft. lbs. (31 Nm).

MALFUNCTION INDICATOR LIGHT (MIL)

RESET PROCEDURES

1. Connect an HDS or equivalent OBD II scan tool to the diagnostic connector.

2. Turn the ignition switch to **ON**.

3. Turn the tester or scan tool **ON**.

4. Check whether any DTCs have been stored. Note them down if necessary.

5. Clear DTCs.

6. The MIL should turn **OFF**.

MANIFOLD ABSOLUTE PRESSURE (MAP) SENSOR

LOCATION

See Figure 96.

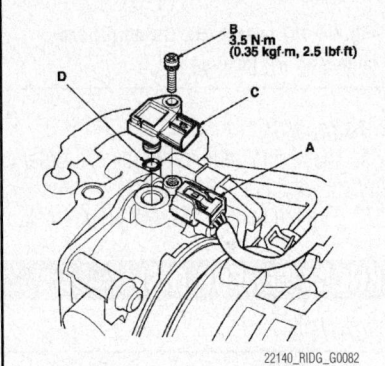

Fig. 96 MAP sensor (C), connector (A), screw location

REMOVAL & INSTALLATION

1. Before servicing the vehicle, refer to the precautions.

2. Remove the intake manifold cover.

3. Disconnect the MAP sensor connector.

4. Remove the screw.

5. Remove the MAP sensor.

To install:

6. Install the parts in the reverse order of removal with a new O-ring.

OIL PRESSURE SENSOR

LOCATION

See Figure 97.

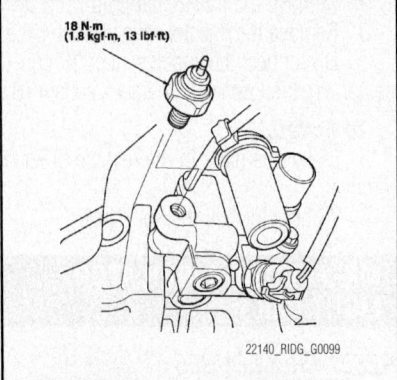

Fig. 97 Oil pressure sensor (B), positive terminal (C) and connector (A) location

REMOVAL & INSTALLATION

1. Before servicing the vehicle, refer to the precautions.

2. Disconnect the oil pressure switch connector, then remove the oil pressure switch.

3. Remove any old liquid gasket from the switch and switch mounting hole.

To install:

4. Apply a very small amount of liquid gasket to the oil pressure switch threads, then install the oil pressure switch.

➡ **Using too much liquid gasket may cause liquid gasket to enter the oil passage or the end of the new oil pressure switch.**

5. Tighten the sensor to 13 ft. lbs. (18 Nm).

6. Connect the oil pressure switch connector.

POWERTRAIN CONTROL MODULE (PCM)

LOCATION

See Figure 98.

REMOVAL & INSTALLATION

1. Before servicing the vehicle, refer to the precautions.

2. Turn the ignition switch **OFF**.

3. Jump the SCS line with the HDS.

4. Remove the cover.

5. Disconnect PCM connectors A, B, and C.

➡ **NOTE: PCM connectors A, B, and C have symbols (A=□, B=□, C=○) embossed on them for identification.**

6. Remove the bolts, then remove the PCM.

To install:

7. Install the parts in the reverse order of removal.

8. Turn the ignition switch **ON**.

9. Manually input the VIN to the PCM with the HDS.

➡ **DTC P0630 "VIN Not Programmed or Mismatch" may be stored because the VIN has not been programmed into the PCM, ignore it, and continue this procedure.**

10. If the READ DATA (engine oil life) failed go to select IMMOBI system with the HDS. Otherwise, go to Select PGM-FI system, and reset the PCM with the HDS.

11. Select the PGM-FI system with the HDS.

12. Select the REPLACE PCM MENU, then select WRITE DATA and follow the screen prompts.

➡ **If the WRITE DATA indicates FAILED, continue this procedure.**

13. Select IMMOBI system with the HDS.

14. Enter the immobilizer code with the PCM replacement procedure in the HDS, it allows you to start the engine.

15. If the TP POSITION CHECK failed, clean the throttle body, then go to the next step.

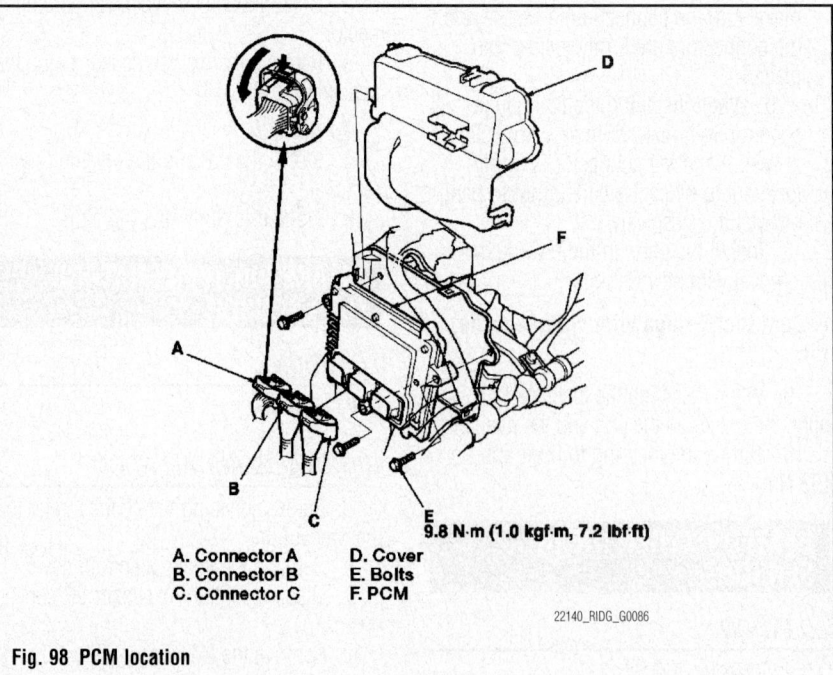

A. Connector A   D. Cover
B. Connector B   E. Bolts
C. Connector C   F. PCM

Fig. 98 PCM location

16. If the READ DATA failed or the WRITE DATA failed, replace the engine oil and engine oil filter, then go to the next step.

17. Select PGM-FI system, and reset the PCM with the HDS.

18. Update the PCM if it does not have the latest software.

19. Do the PCM idle learn procedure.

20. Do the CKP pattern learn procedure.

THROTTLE POSITION SENSOR (TPS)

LOCATION

See Figure 99.

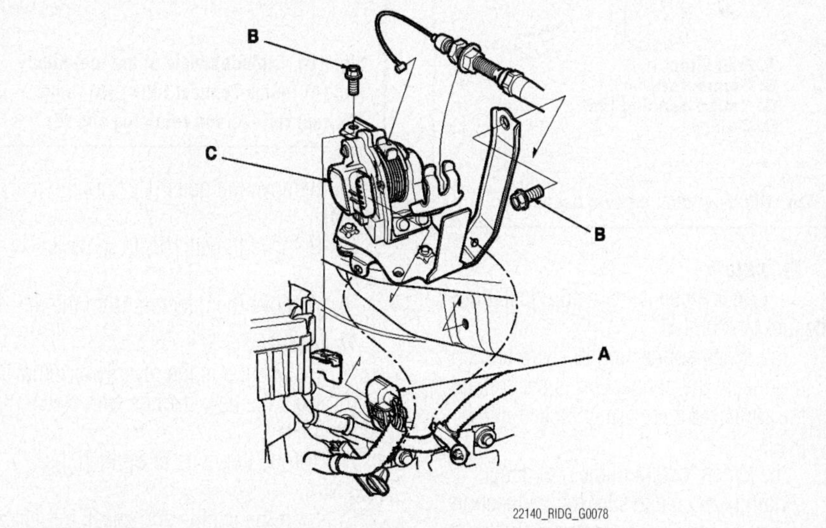

REMOVAL & INSTALLATION

1. Before servicing the vehicle, refer to the precautions.
2. Remove the throttle cable.
3. Disconnect the TPS sensor 6P connector.
4. Remove the bolts and the TPS sensor.

To install:

5. Install the parts in the reverse order of removal.

VEHICLE SPEED SENSOR (VSS)

LOCATION

See Figure 100.

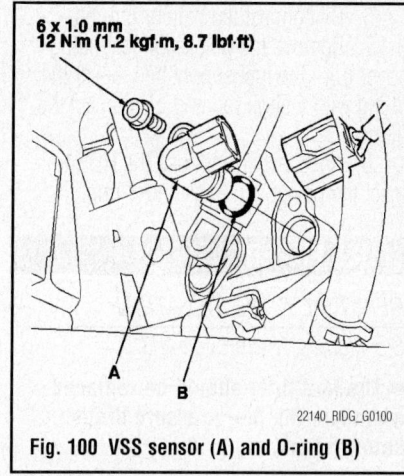

Fig. 100 VSS sensor (A) and O-ring (B)

REMOVAL & INSTALLATION

1. Before servicing the vehicle, refer to the precautions.
2. Remove the splash shield.
3. Disconnect the output shaft (countershaft) speed sensor connector, and remove the output shaft (countershaft) speed sensor (A).

To install:

4. Install a new O-ring (B) on the new output shaft (countershaft) speed sensor, then install the output shaft (countershaft) speed sensor in the transmission housing.
5. Tighten the sensor to 9 ft. lbs. (12 Nm).
6. Check the connector for rust, dirt, or oil, then connect the connector securely.
7. Install the splash shield.

Fig. 99 TPS sensor (C), attaching bolt (B) and connector (A)

FUEL   GASOLINE FUEL INJECTION SYSTEM

FUEL SYSTEM SERVICE PRECAUTIONS

Safety is the most important factor when performing not only fuel system maintenance but any type of maintenance. Failure to conduct maintenance and repairs in a safe manner may result in serious personal injury or death. Maintenance and testing of the vehicle's fuel system components can be accomplished safely and effectively by adhering to the following rules and guidelines.

• To avoid the possibility of fire and personal injury, always disconnect the negative battery cable unless the repair or test procedure requires that battery voltage be applied.

• Always relieve the fuel system pressure prior to disconnecting any fuel system component (injector, fuel rail, pressure regulator,

etc.), fitting or fuel line connection. Exercise extreme caution whenever relieving fuel system pressure to avoid exposing skin, face and eyes to fuel spray. Please be advised that fuel under pressure may penetrate the skin or any part of the body that it contacts.

• Always place a shop towel or cloth around the fitting or connection prior to loosening to absorb any excess fuel due to spillage. Ensure that all fuel spillage (should it occur) is quickly removed from engine surfaces. Ensure that all fuel soaked cloths or towels are deposited into a suitable waste container.

• Always keep a dry chemical (Class B) fire extinguisher near the work area.

• Do not allow fuel spray or fuel vapors to come into contact with a spark or open flame.

• Always use a back-up wrench when loosening and tightening fuel line connection

fittings. This will prevent unnecessary stress and torsion to fuel line piping.

• Always replace worn fuel fitting O-rings with new Do not substitute fuel hose or equivalent where fuel pipe is installed.

Before servicing the vehicle, make sure to also refer to the precautions in the beginning of this section as well.

RELIEVING FUEL SYSTEM PRESSURE

1. Before servicing the vehicle, refer to the precautions.
2. Remove the left kick panel, then remove the PGM-FI main relay 2 (FUEL PUMP).
3. Start the engine and let it idle until it stalls.
4. Turn the ignition off.
5. Remove the fuel filler cap.

6. Disconnect the battery ground.

7. Remove the quick-connect fitting cover from the fuel supply line, cover the fitting with a shop rag and disconnect the fitting.

8. After recharging, use the HDS to reset the power window control unit.

FUEL FILTER

REMOVAL & INSTALLATION

See Figures 101 through 103.

➡**The fuel filter should be replaced whenever the fuel pressure drops below 57 psi.**

1. Before servicing the vehicle, refer to the precautions.

2. Relieve the fuel system pressure.

3. Remove the rear back seat.

4. Remove the access panel on the floor.

5. Remove the fuel tank module lockring using the special tool.

6. Remove the fuel filter.

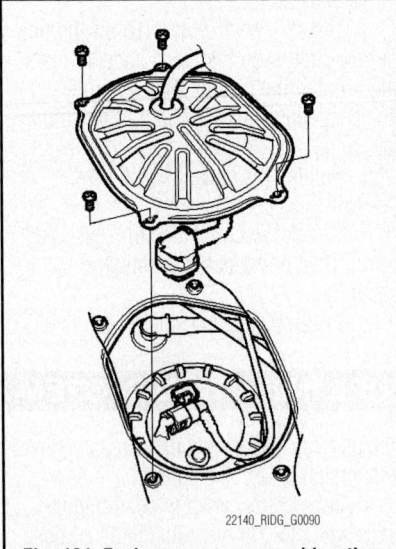

22140_RIDG_G0090

Fig. 101 Fuel pump access panel location

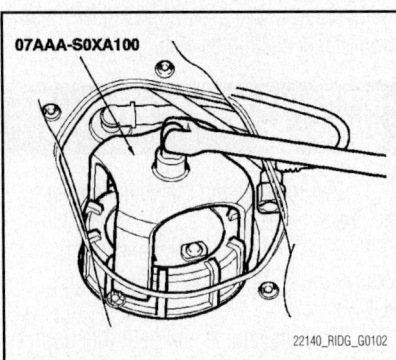

07AAA-S0XA100

22140_RIDG_G0102

Fig. 102 Fuel tank module lockring special tool

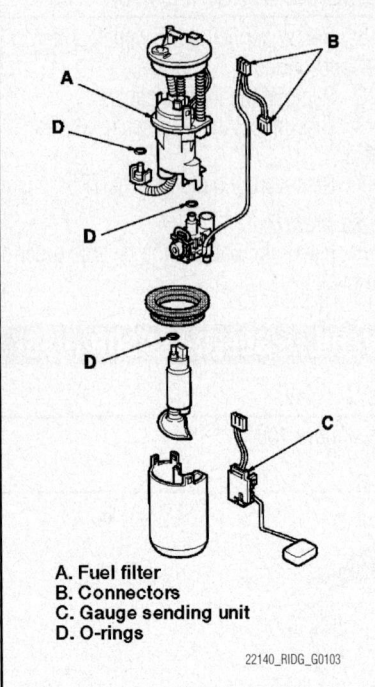

A. Fuel filter
B. Connectors
C. Gauge sending unit
D. O-rings

22140_RIDG_G0103

Fig. 103 Fuel tank module assembly

To install:

7. Check these items before installing the fuel tank unit:

a. When connecting the wire harness, make sure the connection is secure and the connectors are firmly locked into place.

b. When installing the fuel gauge sending unit, make sure the connection is secure and the connector is firmly locked into place. Be careful not to bend or twist it excessively.

8. Install the parts in the reverse order of removal with new O-rings.

➡**Coat the O-rings with clean engine oil.**

9. When installing the fuel tank unit, align the marks on the unit and the fuel tank.

10. Torque the lockring to 69 ft. lbs. (93 Nm).

FUEL RAIL & INJECTORS

REMOVAL & INSTALLATION

See Figure 104.

1. Before servicing the vehicle, refer to the precautions.

2. Relieve the fuel system pressure.

3. Remove the intake manifold.

4. Disconnect the wiring from the injectors (A).

5. Disconnect the quick-connect fitting (B).

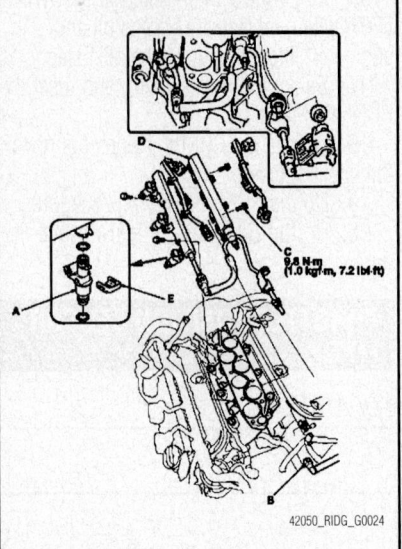

42050_RIDG_G0024

Fig. 104 Exploded view of the fuel injectors (A), quick-connect fitting (B), bolts, (C), fuel rail (D) and retaining clip (E).

6. Remove the bolts (C) from the fuel rail (D).

7. Remove the clip (E) from the fuel rail.

8. Remove the injectors from the rail

To install:

9. Installation is the reverse of removal.

10. Coat the new O-rings with clean engine oil.

11. Torque the bolts to 89 inch lbs. (9.8 Nm).

12. Start the engine and check for leaks.

FUEL PUMP

REMOVAL & INSTALLATION

See Figures 105 through 107.

1. Before servicing the vehicle, refer to the precautions.

2. Relieve the fuel system pressure.

3. Remove the rear back seat.

4. Remove the access panel on the floor.

5. Remove the fuel tank module lockring using the special tool.

6. Disassemble the fuel tank module and remove the fuel pump.

To install:

7. Check these items before installing the fuel tank unit:

a. When connecting the wire harness, make sure the connection is secure and the connectors are firmly locked into place.

b. When installing the fuel gauge sending unit, make sure the connection is secure and the connector is firmly

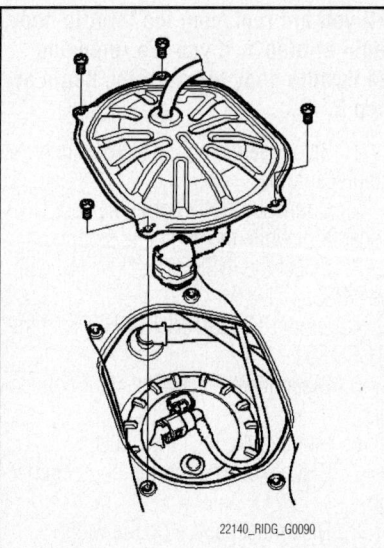

Fig. 105 Fuel pump access panel location

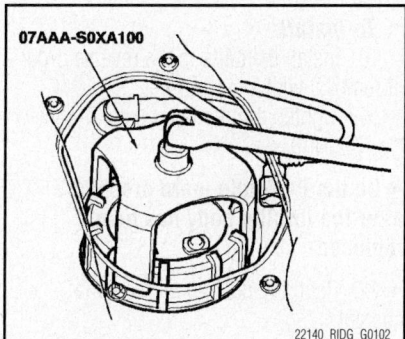

Fig. 106 Fuel tank module lockring special tool

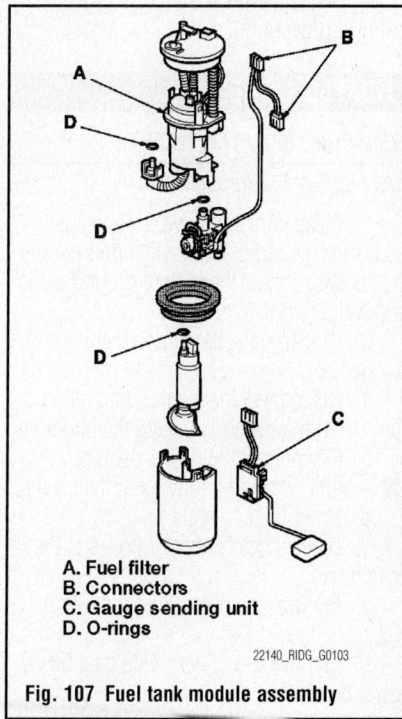

A. Fuel filter
B. Connectors
C. Gauge sending unit
D. O-rings

Fig. 107 Fuel tank module assembly

locked into place. Be careful not to bend or twist it excessively.

8. Install the parts in the reverse order of removal with new O-rings.

➡ **Coat the O-rings with clean engine oil.**

9. When installing the fuel tank unit, align the marks on the unit and the fuel tank.

10. Torque the lockring to 69 ft. lbs. (93 Nm).

FUEL TANK

REMOVAL & INSTALLATION

See Figure 108.

1. Before servicing the vehicle, refer to the precautions.
2. Drain the fuel tank.
3. Raise and safely support the vehicle securely on jackstands.
4. Remove the exhaust pipe.
5. Remove the propeller shaft, and support it with jackstands.
6. Remove the fuel tank protector (A).
7. Loosen the clamp (B), and disconnect the tube (C).

8. Open the clamp (D).
9. Disconnect the hoses (E). Slide back the clamps, then twist the hoses as you pull to avoid damaging them.
10. Place a jack or other support under the tank (F).
11. Remove the strap bolts and the straps (G).
12. Remove the fuel tank.

To install:

13. Install the parts in the reverse order of removal.
14. Tighten tank strap bolts to 47 ft. lbs. (64 Nm).
15. Tighten tank protector nuts to 7 ft. lbs. (10 Nm).

IDLE SPEED

ADJUSTMENT

Idle speed is maintained by the Powertrain Control Module (PCM). No adjustment is necessary or possible.

THROTTLE BODY

REMOVAL & INSTALLATION

See Figure 109.

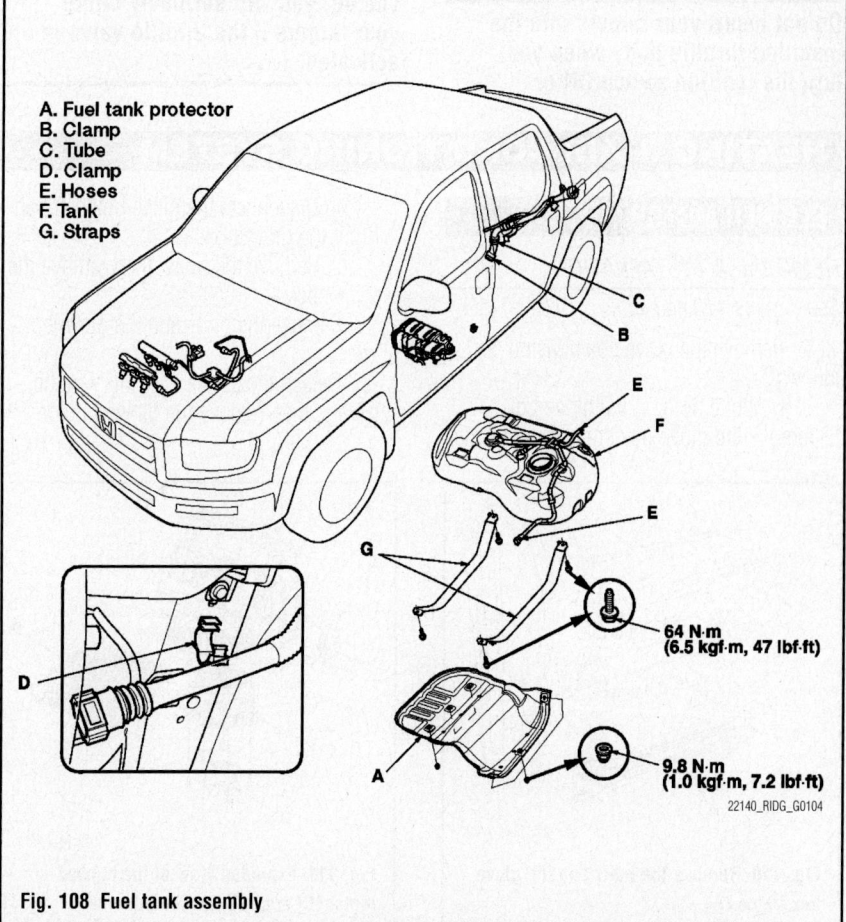

A. Fuel tank protector
B. Clamp
C. Tube
D. Clamp
E. Hoses
F. Tank
G. Straps

64 N·m
(6.5 kgf·m, 47 lbf·ft)

9.8 N·m
(1.0 kgf·m, 7.2 lbf·ft)

Fig. 108 Fuel tank assembly

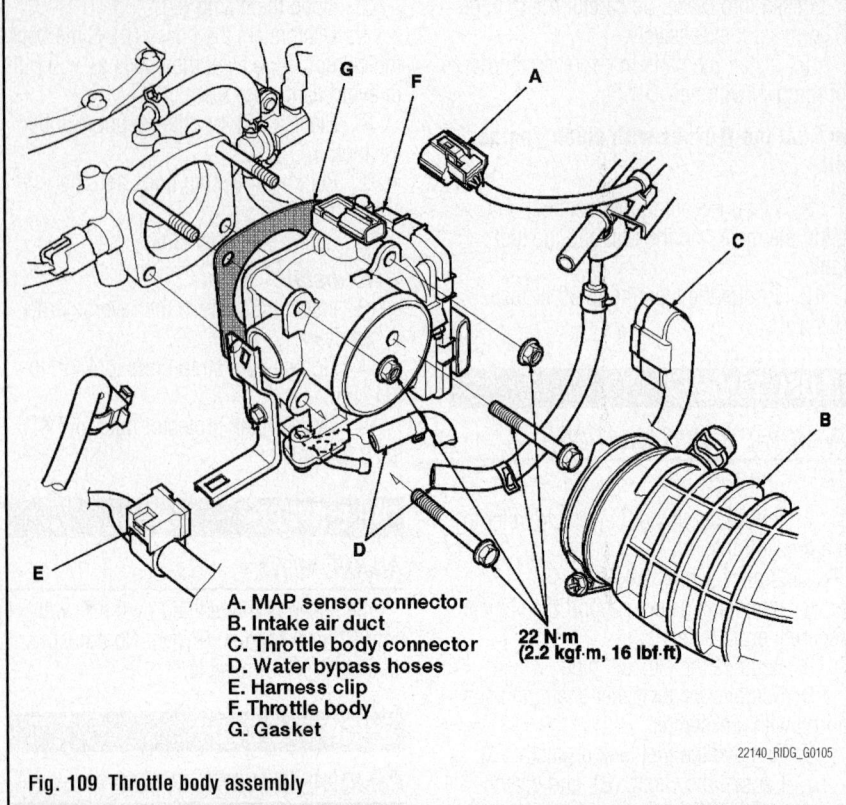

A. MAP sensor connector
B. Intake air duct
C. Throttle body connector
D. Water bypass hoses
E. Harness clip
F. Throttle body
G. Gasket

22 N·m
(2.2 kgf·m, 16 lbf·ft)

22140_RIDG_G0105

Fig. 109 Throttle body assembly

✳✳ CAUTION

Do not insert your fingers into the installed throttle body when you turn the ignition switch ON or while the ignition switch is ON. If you do, you will seriously injure your fingers if the throttle valve is activated.

➥If you are replacing the throttle body, begin at Step 1. If you are removing the throttle body temporarily, begin at Step 5.

1. Before servicing the vehicle, refer to the precautions.

2. Connect the HDS or equivalent OBD II scan tool while the engine is stopped.

3. Select the INSPECTION MENU with the HDS.

4. Do the TP LEARNING CHECK in the ETCS TEST.

5. Disconnect the MAP sensor connector.

6. Remove the intake air duct.

7. Disconnect the throttle body connector.

8. Disconnect and plug the water bypass hoses.

9. Remove the harness clip and the throttle body.

To install:

10. Install the parts in the reverse order of removal with a new gasket.

11. Tighten throttle body nuts to 16 ft. lbs. (22 Nm).

➥Do the PCM idle learn procedure after the throttle body has been replaced.

12. Refill the radiator with engine coolant.

HEATING & AIR CONDITIONING SYSTEM

BLOWER MOTOR

REMOVAL & INSTALLATION

See Figures 110 and 111.

1. Remove the glove box housing, as follows:

   a. While holding the glove box, remove the glove box stop on each side.

   b. Disconnect the glove box damper from the glove box.

   c. Remove the bolts, then remove the glove box.

2. Remove the bolts and the glove box frame.

3. Detach the connector, unfasten the retainers, then remove the blower motor.

4. Installation is the reverse of the removal procedure.

HEATER CORE

REMOVAL & INSTALLATION

See Figures 112 through 114.

1. Make sure you have the anti-theft codes for the radio and navigation systems.

2. Disconnect the battery ground cable and wait at least 3 minutes.

3. Properly discharge and recover the refrigerant.

4. Disconnect the suction and receiver lines from the evaporator. Cap the openings.

5. Disconnect the heater coolant valve cable at the valve and fully open the valve.

6. Drain the coolant.

7. Disconnect the heater hoses at the core tubes.

8. Remove the heater unit mounting nut.

9. Remove the driver's side dashboard lower cover.

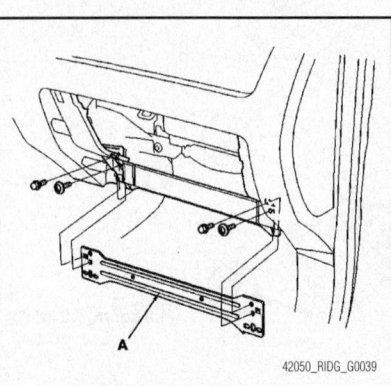

42050_RIDG_G0039

Fig. 110 Remove the bolts and the glove box frame (A)

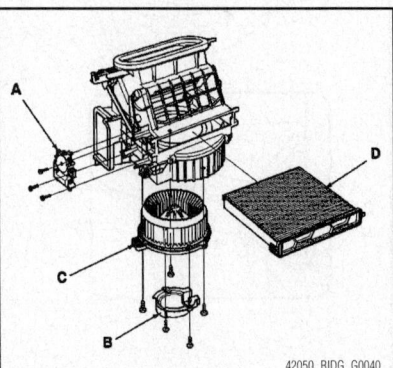

42050_RIDG_G0040

Fig. 111 Exploded view of the blower motor (C) and related components

10. Remove the dashboard center lower cover.

11. Remove the center console.

12. Remove the glove box.

13. Remove the driver's side dashboard side cover.

14. Remove the both kick panels.

15. Remove the both A-pillar trim panels.

16. Place the wheels in the straight-ahead position.

17. Remove the steering wheel air bag access cover and disconnect the air bag connector from the cable reel.

18. Remove the 2 Torx® bolts, one each side of the steering wheel hub.

19. Disconnect the horn switch connector.

20. Remove the air bag.

21. Disconnect the cable reel harness.

22. Remove the steering wheel nut.

23. Using a puller, remove the steering wheel.

24. Remove the steering coupler cover from the floor.

25. Set the column shaft to the neutral position by raising the column to the upper-most position, then lower it 8mm (0.31 in.). Tighten the tilt lever.

26. Remove the column covers.

27. Move the shift lever to N and remove the shift cable from the column.

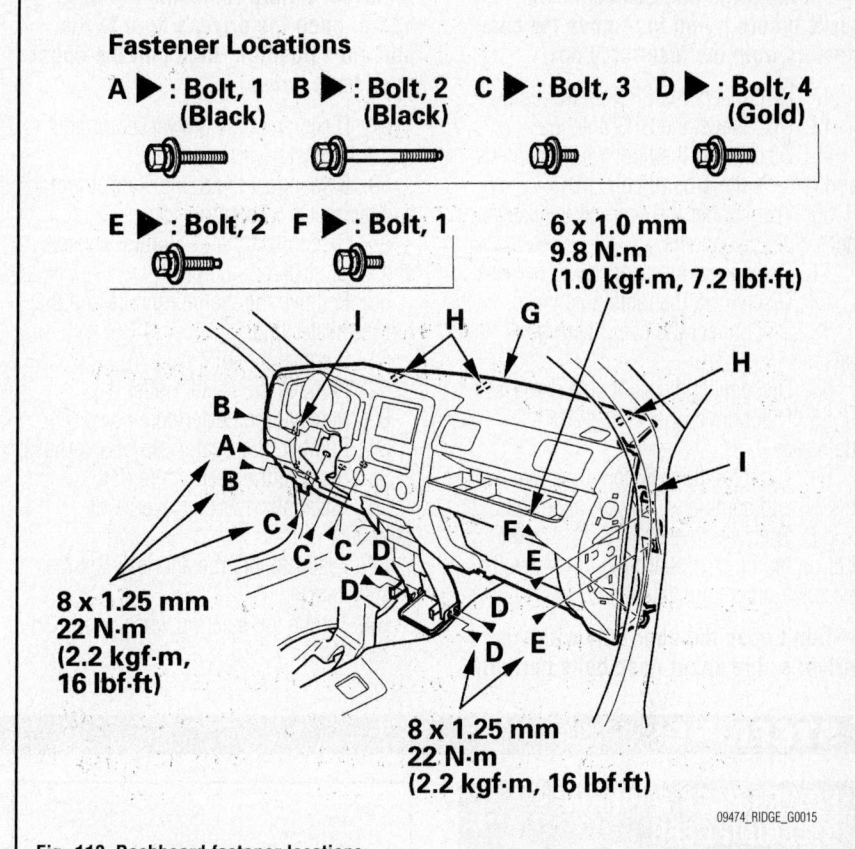

Fig. 113 Dashboard fastener locations

28. Remove the combination switch.

29. Disconnect the ignition switch.

30. Disconnect the immobilizer receiver unit, the park pin switch and the shift lock solenoid.

31. Matchmark and disconnect the steering joint from the column shaft.

32. Remove the attaching bolts and nuts and remove the steering column.

33. Disconnect the shift cable.

34. Remove the parking brake release lever bolt.

35. Remove the passenger's side dashboard side cover.

36. Remove the right speaker grille.

37. Remove the console front bracket.

38. Remove the rear vent duct.

39. Remove the rear heater duct.

40. Disconnect the cabin wiring harness.

41. Disconnect the interior wiring harness.

42. Disconnect the left side wiring harness.

43. Disconnect the parking brake switch.

44. Disconnect the brake switch.

45. Disconnect the dashboard wiring harness.

46. Disconnect the under-dash fuse/relay box.

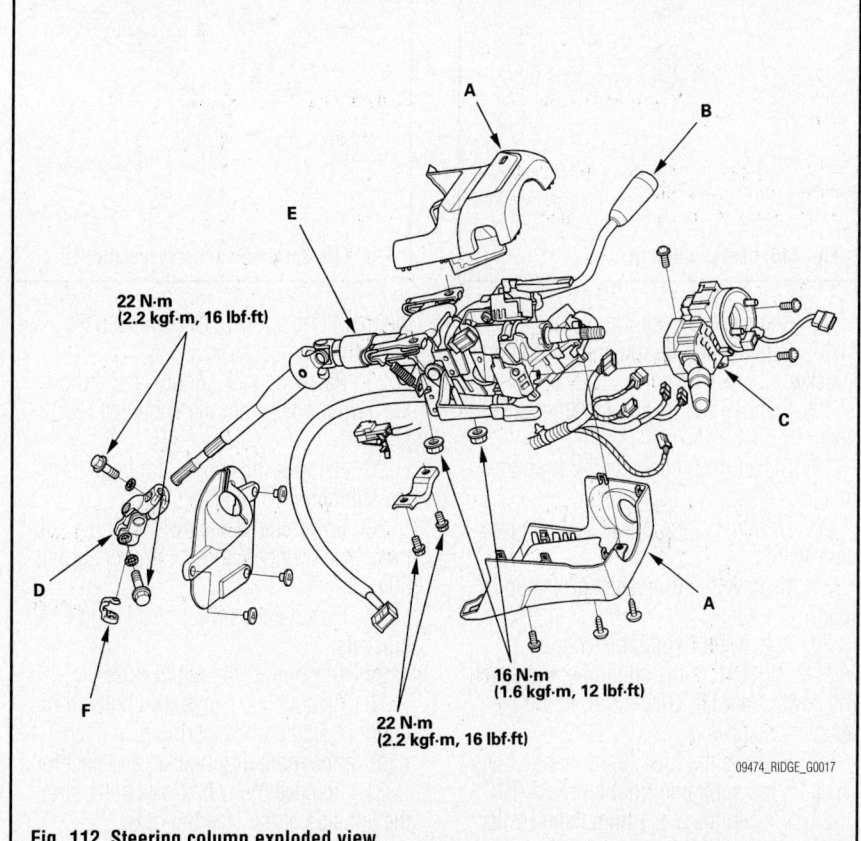

Fig. 112 Steering column exploded view

➡Lift the large harness connector locks before trying to remove the connectors from the fuse/relay box.

47. Disconnect the SRS control unit.
48. Disconnect the GPS antenna.
49. Disconnect the floor wiring harness, and remove the ground bolt.
50. From under the dash on the passenger's side, disconnect:
51. Disconnect the door wiring harness.
52. Disconnect the radio antenna.
53. Disconnect the cabin wiring harness.
54. Disconnect the right side harness.
55. Disconnect the front glass defogger.
56. Open the driver's door. See the illustration and completely loosen upper bolts A and B. Then, remove bolts C, D, E and F. Lift up on the dashboard (G) to release it from the guide pins (H&I).

➡Don't open the door fully with the driver's side upper dash bolts partially removed. Before removing the dashboard, open the driver's door to the half open position, then pull the upper dash bolts outward.

57. Disconnect the blower motor and power transistor wiring.
58. Disconnect the mode control motor and recirculation control motor.
59. Disconnect the evaporator sensor and air control mix motor.
60. Remove the mounting nuts and the blower/heater unit.
61. Remove the joint duct (A).
62. Remove the heater outlet (C).
63. Remove the heater core cover (D).
64. Remove the heater pip brackets (E).
65. Remove the heater core (G).
66. Installation is the reverse of removal.
67. Evacuate, charge and leak test the A/C system.
68. Torque the steering wheel nut to 36 ft. lbs. (49 Nm).

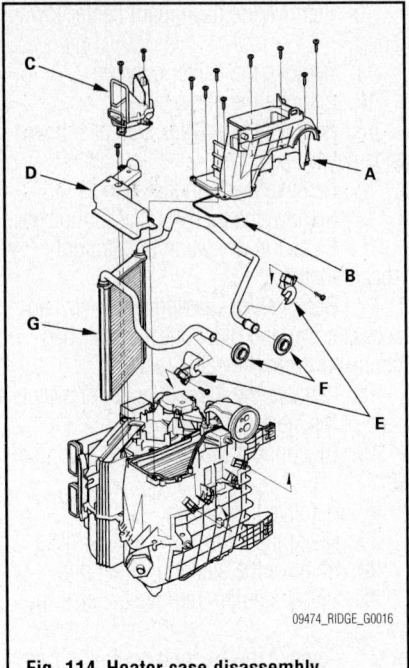

Fig. 114 Heater case disassembly

STEERING

POWER RACK & PINION STEERING GEAR

REMOVAL & INSTALLATION

See Figures 115 through 118.

1. Before servicing the vehicle, refer to the precautions.
2. Drain the power steering fluid.
3. Disconnect the battery ground cable and wait at least 3 minutes.
4. Place the wheels in the straight-ahead position.
5. Remove the steering wheel air bag access cover and disconnect the air bag connector from the cable reel.
6. Remove the 2 Torx® bolts, one each side of the steering wheel hub.
7. Disconnect the horn switch connector.
8. Remove the air bag.
9. Disconnect the cable reel harness.
10. Remove the steering wheel nut.
11. Using a puller, remove the steering wheel.
12. Remove the steering coupler cover from the floor.
13. Matchmark and disconnect the joint coupler. If equipped with a center guide, discard it. It's for factory assembly only.
14. Disconnect the power steering outlet hose from the pump and remove the clamp.
15. Remove the 10mm flange bolts from the engine side mount bracket.

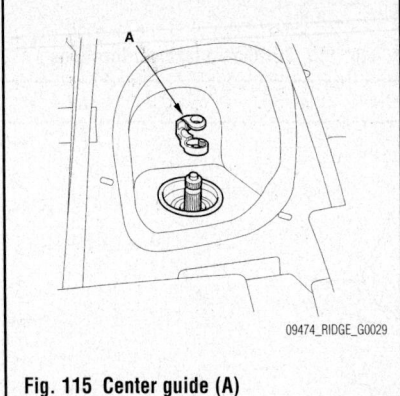

Fig. 115 Center guide (A)

16. Remove the front wheels.
17. Separate the tie rod ends from the knuckle.
18. Remove the front subframe stiffener.
19. Disconnect the converter from the muffler.
20. Disconnect the power steering pressure switch.
21. Remove the driveshaft protector loop.
22. Attach front subframe adapter VSB02C000016 to the subframe and attach the power train lift, OTC-1585, to the adapter as shown.
23. Remove the four 12mm flange bolts (A) from the subframe front brackets (B).
24. Loosen the two 14mm flange bolts (C) so they are ¹³⁄₁₆ in. (about 30mm) from

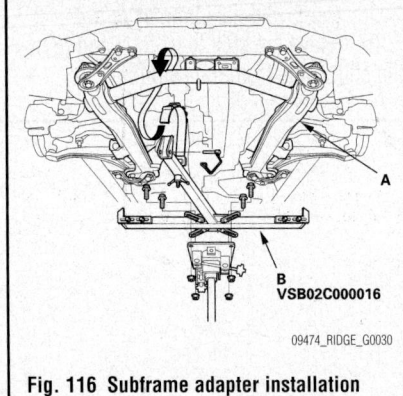

Fig. 116 Subframe adapter installation

the mounting surface. Don't loosen the more than is necessary.
25. Raise the jack slightly and remove the 12mm bolts from the subframe rear brackets.
26. Remove the two 14mm bolts from the subframe rear brackets.
27. Lower the jack slowly until the subframe has dropped about 1¹⁵⁄₁₆ in. (about 50mm).
28. Remove the power steering line brackets.
29. Disconnect the return hose.
30. Remove the two 10mm bolts from the right side of the steering gear, then remove the mounting bracket and cushion.
31. Remove the two 10mm bolts from the left side of the steering gear.
32. Lower the jack another 2 inches.

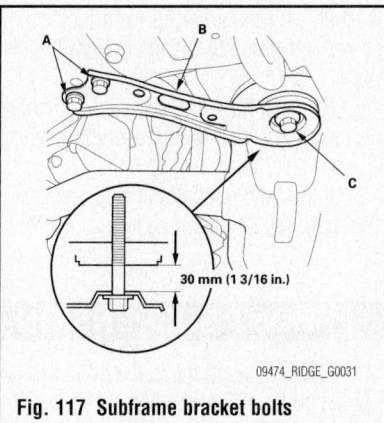

Fig. 117 Subframe bracket bolts

33. Remove the steering gear stiffener from the left side of the subframe.

34. Disconnect the feed line from the gear.

35. Slide the gear towards the left side and out.

36. Installation is the reverse of removal. Observe the following torques:

- Feed line to gear: 31 ft. lbs. (42 Nm)
- Steering gear stiffener bolts: 28 ft. lbs. (38 Nm)
- Left side gear mounting bolts (loose until right side bolts are installed) then: 43 ft. lbs. (58 Nm)
- Right side gear mounting bolts: 29 ft. lbs. (39 Nm)
- Rear subframe brackets 12mm bolts: 86 ft. lbs. (117 Nm)
- Rear subframe brackets 14mm bolts: 76 ft. lbs. (103 Nm)
- Front subframe brackets 12mm bolts: 54 ft. lbs. (74 Nm)
- Front subframe brackets 14mm bolts: 76 ft. lbs. (103 Nm)
- Converter-to-muffler: 25 ft. lbs. (33 Nm)

- Subframe stiffener: 40 ft. lbs. (54 Nm)
- Engine side mount bolts: 33 ft. lbs. (44 Nm)
- Tie rod nuts: 40 ft. lbs. (54 Nm)
- Coupler bolt: 16 ft. lbs. (22 Nm)
- Steering wheel nut to 36 ft. lbs. (49 Nm)

37. Fill the system with fluid and bleed the air.

38. Reset the alignment.

POWER STEERING PUMP

REMOVAL & INSTALLATION

See Figure 119.

1. Before servicing the vehicle, refer to the precautions.

2. Place a suitable container under the vehicle.

3. Drain the power steering fluid from the reservoir.

4. Remove the engine cover.

5. Remove the drive belt (A) from the pump pulley, as outlined in the Engine Mechanical Section.

6. Cover the auto-tensioner, alternator, and A/C compressor with several shop towels to protect them from spilled power steering fluid. Disconnect the pump inlet hose (B) and pump outlet hose (C) from the pump (D), and plug them. Take care not to spill the fluid on the body or parts. Wipe off any spilled fluid at once. Do not turn the steering wheel with the pump removed.

7. Remove the pump mounting bolts (E).

8. Cover the opening of the pump with a piece of tape to prevent foreign material from entering the pump.

To install:

9. Connect the pump inlet hose and pump outlet hose onto the new pump with new O-ring (F).

10. Loosely install the pump in the pump bracket with the mounting bolts, then tighten the pump fittings securely.

11. Tighten the pump mounting bolts to 16 ft. lbs. (22 Nm).

12. Install the drive belt. Make sure that the belt is properly positioned on the pulleys.

✶✶ WARNING

Do not get power steering fluid or grease on the auto-tensioner, alternator, A/C compressor, and drive belt or pulley faces. Clean off any fluid or grease before installation.

13. Fill the reservoir to the upper level line and bleed the system, as outlined in this section.

BLEEDING

See Figure 120.

Check the power steering fluid reservoir at regular intervals, and add the recommended fluid as necessary. Always use Honda Power Steering Fluid. Using any other type of power steering fluid or automatic transmission fluid can cause increased wear and poor steering in cold weather.

➡ **If the fluid is contaminated, the screen in the reservoir may be partially blocked. Replace the reservoir if necessary.**

The power steering fluid system capacity is 1.22 qt. at disassembly and reservoir capacity is 0.36 qt.

1. Pull the cover (A) up, raise the reservoir, then disconnect the return hose (B) to drain the reservoir. Take care not to spill the fluid on the body and parts. Wipe off any spilled fluid at once.

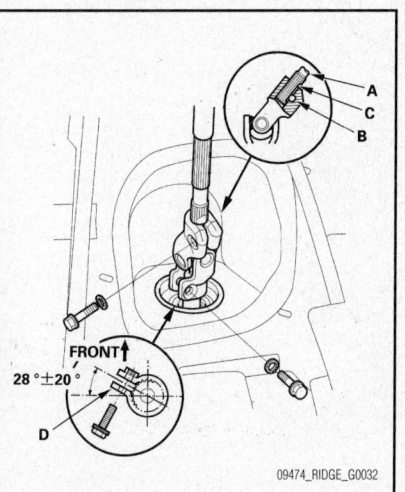

Fig. 118 Coupler alignment. (A) steering shaft, (B) bolt hole, (C) flat, (D) gap

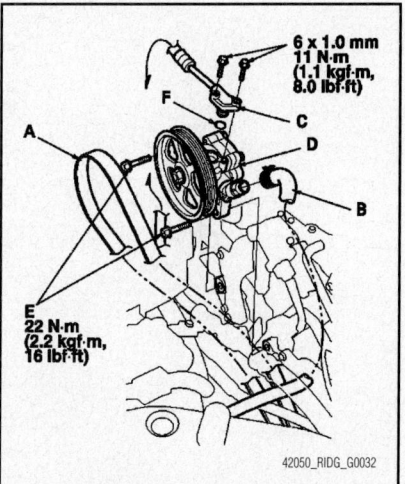

Fig. 119 Exploded view of the power steering pump (D) and related components

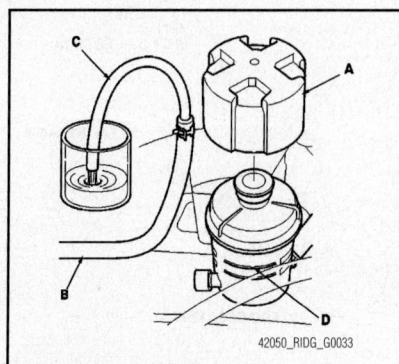

Fig. 120 Power steering filling and bleeding

➡Check the reservoir screen for any debris. If the reservoir screen is clogged, replace the reservoir.

2. Connect a hose (C) of suitable diameter to the disconnected return hose, and put the hose end in a suitable container.

3. Start the engine, let it run at idle, and turn the steering wheel from lock-to-lock several times. When fluid stops running out of the hose, shut off the engine. Discard the fluid.

4. Reinstall the return hose on the reservoir.

5. Fill the reservoir to the upper level line (D).

6. Start the engine and run it at fast idle, then turn the steering from lock-to-lock several times to bleed air from the system.

7. Recheck the fluid level and add some if necessary. Do not fill the reservoir beyond the upper level line.

8. If the fluid is contaminated, dark, or discolored, repeat the procedure as necessary.

SUSPENSION

LOWER BALL JOINT

REMOVAL & INSTALLATION

The lower ball joints are an integral part of the lower control arms and not serviced separately.

LOWER CONTROL ARM

REMOVAL AND & INSTALLATION

See Figures 121 and 122.

➡The control arm bushings are an integral part of the lower control arms and not serviced separately.

1. Before servicing the vehicle, refer to the precautions.

2. Remove the front wheels.

3. Disconnect the stabilizer link from the strut.

4. Remove and discard the lower ball joint nut.

5. Separate the lower ball stud from the knuckle.

6. Remove and discard the lower arm horizontal and vertical bolts.

7. Remove the lower arm.

To install:

8. Installation is the reverse of removal, noting the following:

9. Use new bolts and nuts.

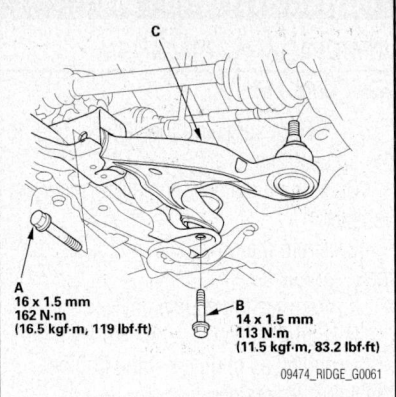

Fig. 122 Lower control arm mounting. (A) horizontal bolt, (B) vertical bolt, (C) control arm.

10. Install all fasteners loosely, then, final torque all fasteners with the suspension loaded (weight of the car).

FRONT SUSPENSION

11. Never back off the ball stud nut to align the cotter pin holes. Always tighten it to align.

12. Observe the following torque specifications:

- Vertical lower arm bolt: 83 ft. lbs. (113 Nm)
- Horizontal lower arm bolt: 119 ft. lbs. (162 Nm)
- Ball stud nut: 69 ft. lbs. (93 Nm)
- Stabilizer link nut: 58 ft. lbs. (78 Nm)

MACPHERSON STRUT

REMOVAL & INSTALLATION

See Figure 123.

1. Before servicing the vehicle, refer to the precautions.

2. Remove the front wheel.

3. Remove the stabilizer bar link.

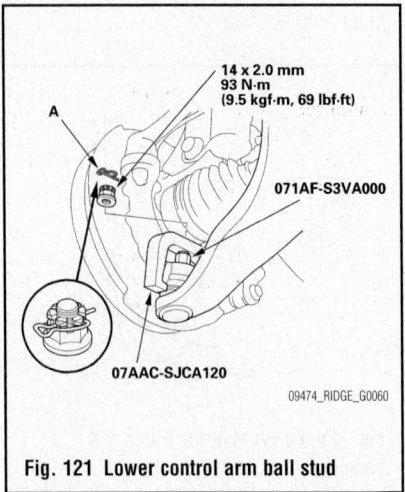

Fig. 121 Lower control arm ball stud

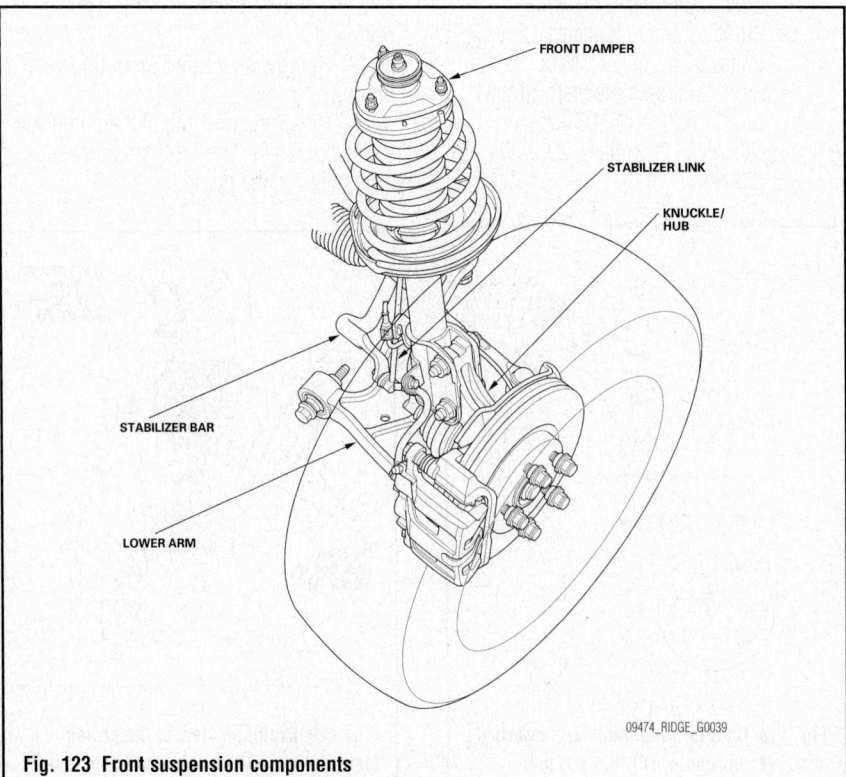

Fig. 123 Front suspension components

4. Remove the wheel speed sensor wiring bracket.

5. Remove the brake hose bracket.

6. Remove the strut pinch bolts/nuts.

7. Remove the upper mount nuts.

8. Remove the strut.

➥**The left and right struts are not interchangeable.**

To install:

9. Install the strut. Tighten the upper mount nuts to 43 ft. lbs. (59 Nm).

10. Install the strut pinch bolts. Tighten the nuts to 156 ft. lbs. (211 Nm).

11. Install the stabilizer bar link. Tighten the nut to 58 ft. lbs. (78 Nm).

12. Install the brake hose bracket.

13. Install the wheel speed sensor wiring bracket.

14. Install the front wheel.

15. Check the wheel alignment and adjust as necessary.

OVERHAUL

See Figures 124 through 127.

1. Before servicing the vehicle, refer to the precautions.

2. Remove the strut from the vehicle and install in a strut spring compressor. Compress the spring until the end of the spring comes away from the spring seat.

3. Remove the upper strut mount, spring seat and related components.

4. Remove the coil spring from the strut spring compressor.

To assemble:

➥**Use a new self-locking nut.**

5. Install the cushion (A) on the upper spring seat (B) by aligning the tab (C) with the notch (D) in the seat.

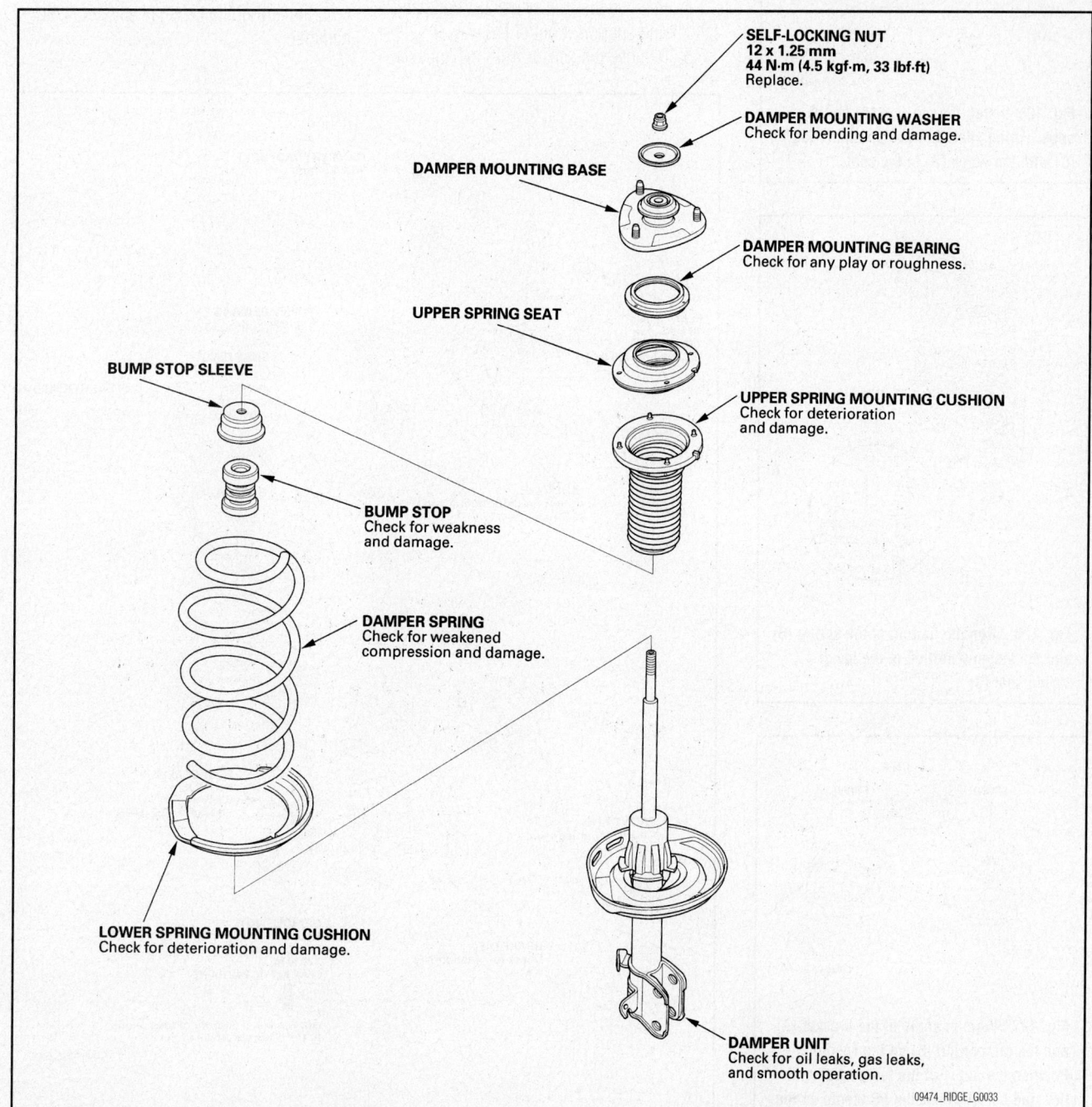

SELF-LOCKING NUT
12 x 1.25 mm
44 N·m (4.5 kgf·m, 33 lbf·ft)
Replace.

DAMPER MOUNTING WASHER
Check for bending and damage.

DAMPER MOUNTING BASE

DAMPER MOUNTING BEARING
Check for any play or roughness.

UPPER SPRING SEAT

UPPER SPRING MOUNTING CUSHION
Check for deterioration and damage.

BUMP STOP SLEEVE

BUMP STOP
Check for weakness and damage.

DAMPER SPRING
Check for weakened compression and damage.

LOWER SPRING MOUNTING CUSHION
Check for deterioration and damage.

DAMPER UNIT
Check for oil leaks, gas leaks, and smooth operation.

09474_RIDGE_G0033

Fig. 124 Exploded view of the front strut assembly

6. Compress the spring.

7. Align an angle of the bracket (A) and the tab portion (B) on the cushion. Position the angle of the tab portion and the stud

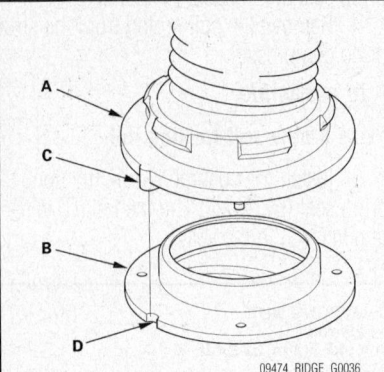

Fig. 125 Install the cushion (A) on the upper spring seat (B) by aligning the tab (C) with the notch (D) in the seat

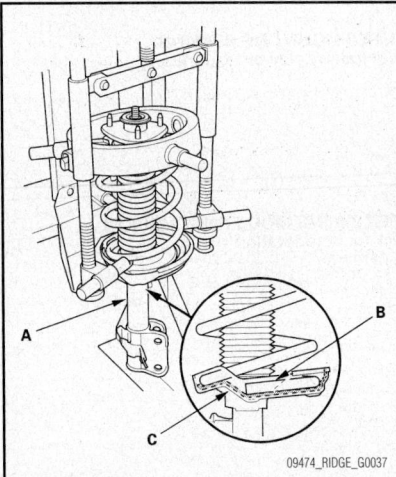

Fig. 126 Align the bottom of the spring (B) and the stepped portion of the lower spring seat (C)

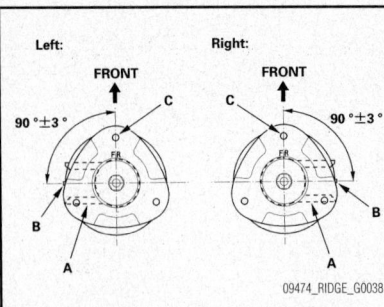

Fig. 127 Align an angle of the bracket (A) and the tab portion (B) on the cushion. Position the angle of the tab portion and the stud bolt (C) near the FR stamp on the upper spring seat.

bolt (C) near the **FR** stamp on the upper spring seat.

8. Install the upper strut mounting components and tighten the nut to 33 ft. lbs. (44 Nm).

STEERING KNUCKLE

REMOVAL & INSTALLATION

See Figure 128.

1. Before servicing the vehicle, refer to the precautions.

2. Remove the wheel.

3. Remove the brake hose bracket bolt.

4. Remove the caliper bracket bolts and caliper and support it out of the way.

5. Remove the sensor from the knuckle.

6. Unstake and remove the halfshaft end nut.

7. Remove the brake rotor.

➡ **It may be necessary to push off the rotor by using two 8 ×1.25mm bolts in the holes provided.**

8. Remove the nut and disconnect the tie rod end from the knuckle.

9. Remove the ball stud nut and separate the lower control arm from the knuckle.

10. Remove the strut-to-knuckle bolts/nuts.

11. Pull the knuckle outwards while tapping the end of the halfshaft with a plastic hammer.

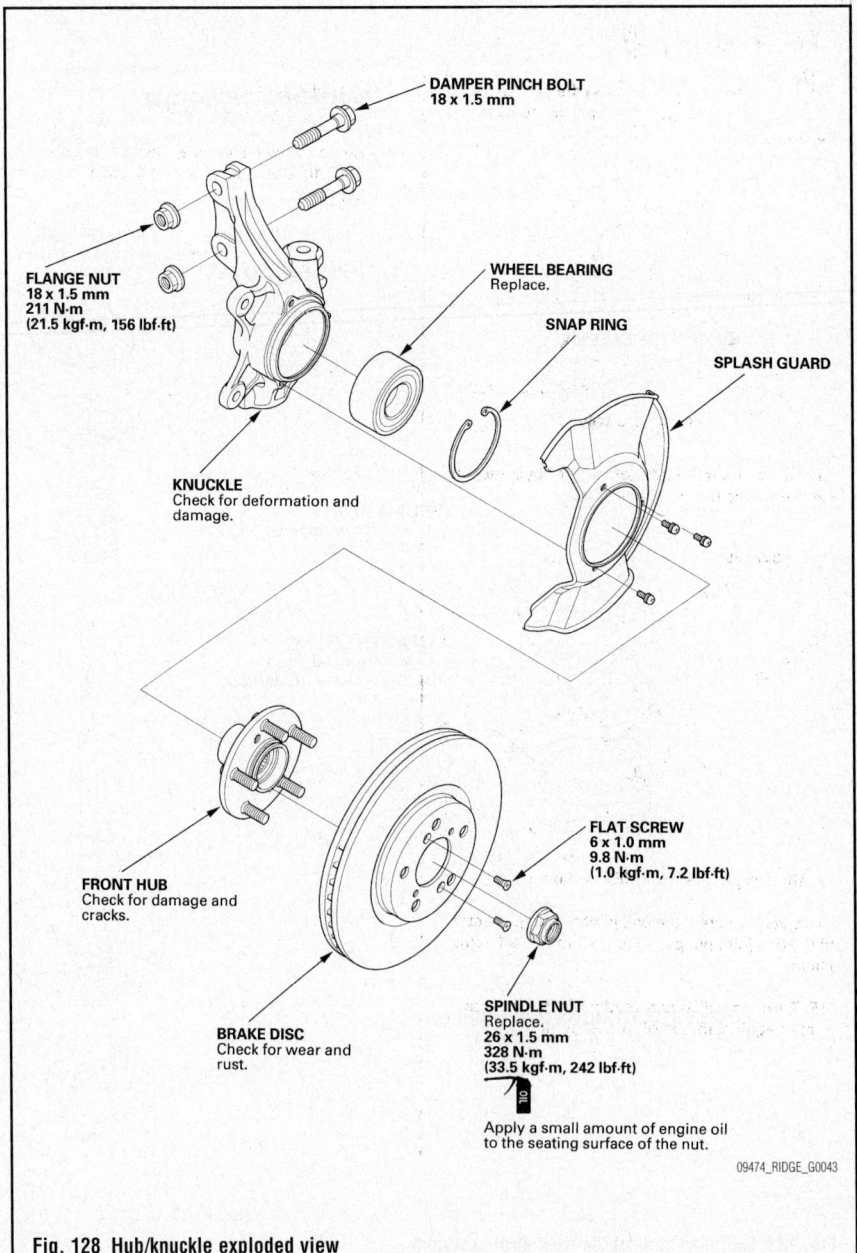

Fig. 128 Hub/knuckle exploded view

12. Installation is the reverse of removal, noting the following:

13. Use new bolts and nuts.

14. Install all fasteners loosely, then, final torque all fasteners with the suspension loaded (weight of the car). Never back off the ball stud nut to align the cotter pin holes.

15. Always tighten it to align.

16. Observe the following torque specifications:

- Strut-to-knuckle: 156 ft. lbs. (211 Nm)
- Lower arm ball stud nut: 69 ft. lbs. (93 Nm)
- Tie rod end nut: 40 ft. lbs. (54 Nm)
- Halfshaft end nut: 242 ft. lbs. (328 Nm)
- Caliper bracket bolts: 101 ft. lbs. (137 Nm)

STABILIZER BAR

REMOVAL & INSTALLATION

See Figures 129 through 133.

1. Before servicing the vehicle, refer to the precautions.

2. Remove the front wheels.

3. Remove the caps from the front strut mounting nuts. Position the engine hanger adapters (VSB02C000024) with the FRONT mark facing forward, over the flange nuts.

4. Install the balancer bar (VSB02C000019):

    a. Attach the front arm (A) to the front head with spacer (B) and a 10 × 1.25mm bolt (C)

    b. Attach the rear arm (D) to the rear head with an 8 × 1.25mm bolt (E).

5. Install the support hanger (AAR-T-12566):

    a. Attach the hook to the slotted hole in the balancer bar.

    b. Tighten the wing nut (F) by hand to lift and support the engine/transaxle assembly.

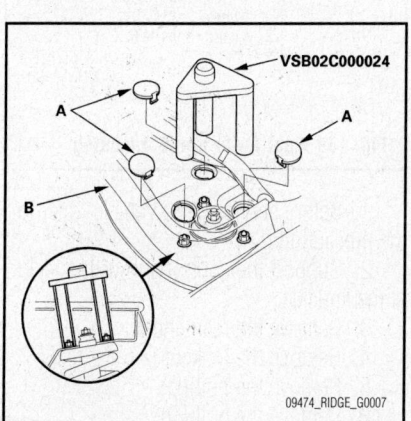

Fig. 129 Engine hanger adapters installed

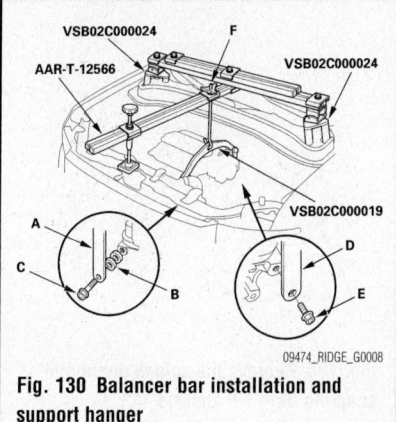

Fig. 130 Balancer bar installation and support hanger

6. Disconnect the stabilizer links from the stabilizer bar.

7. Attach front subframe adapter VSB02C000016 to the subframe and attach the power train lift, OTC-1585, to the adapter as shown.

8. Remove the four 12mm flange bolts (A) from the subframe front brackets (B).

9. Loosen the two 14mm flange bolts (C) so they are ⁹⁄₁₆ in. (about 14mm) from the mounting surface. Don't loosen the more than is necessary.

10. Raise the jack slightly and remove the 12mm bolts from the subframe rear brackets.

11. Remove the two 14mm bolts from the subframe rear brackets.

12. Lower the jack slowly until the subframe has dropped about ⁹⁄₁₆ in. (about 14mm).

13. Matchmark the bushings and bar.

14. Remove the stabilizer bar brackets and bar.

To install:

15. Installation is the reverse of removal, noting the following:

16. Align the subframe and perform a front end alignment.

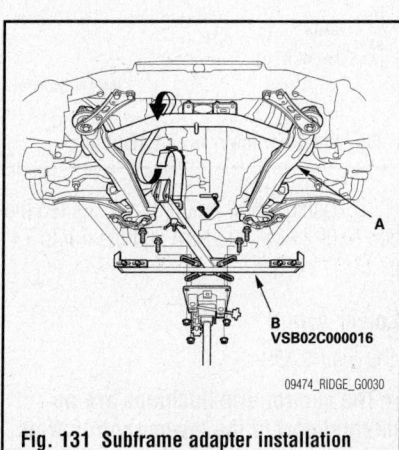

Fig. 131 Subframe adapter installation

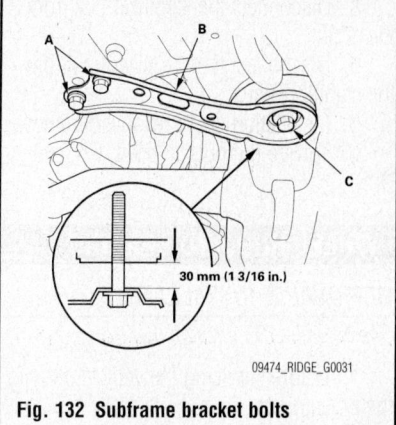

Fig. 132 Subframe bracket bolts

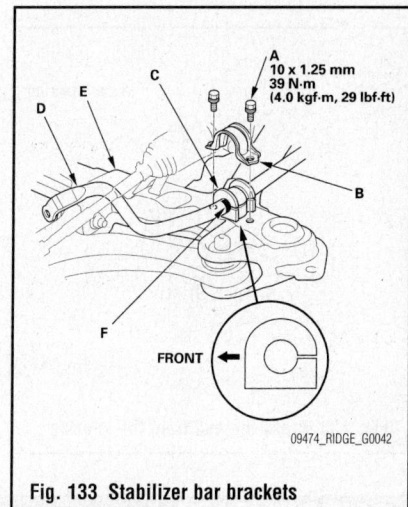

Fig. 133 Stabilizer bar brackets

17. Torque the bracket bolts to 29 ft. lbs. (39 Nm).

Stabilizer Link

See Figure 134.

1. Before servicing the vehicle, refer to the precautions.

2. Remove the front wheels.

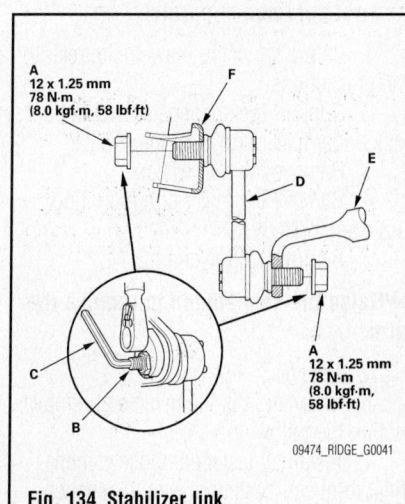

Fig. 134 Stabilizer link

3. Disconnect the stabilizer link from the strut.

4. Disconnect the stabilizer link from the stabilizer bar.

5. Installation is the reverse of removal.

6. Torque the nuts to 58 ft. lbs. (78 Nm).

WHEEL HUB AND BEARING

REMOVAL & INSTALLATION

See Figures 135 through 137.

1. Before servicing the vehicle, refer to the precautions.

2. Remove the knuckle.

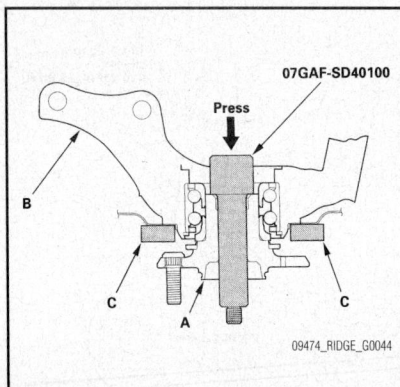

Fig. 135 Press the hub from the knuckle

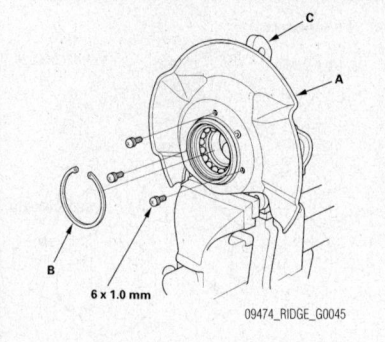

Fig. 136 Remove the splash guard and snapring from the knuckle

3. Place the knuckle in a press. Separate the hub (A) from the knuckle (B). Hold the knuckle with the press attachment (C).

4. Press the inner race from the hub.

5. Remove the splash guard and snapring from the knuckle.

6. Press the wheel bearing from the knuckle.

7. Wash the knuckle and hub thoroughly with a safe solvent.

To install:

8. Press a new bearing (A) into the knuckle (B) using the old bearing (C), a steel plate (D) and the special tools shown.

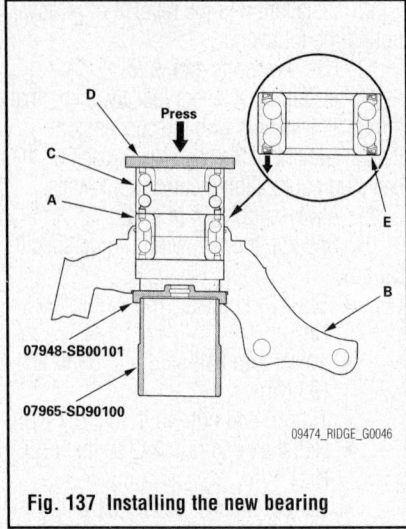

Fig. 137 Installing the new bearing

➡Install the bearing with the sensor magnetic encoded (E), brown color, towards the outside of the knuckle.

❊❊ WARNING

Take care to avoid damaging the encoder surface. Keep all magnetic tools away from the encoder.

9. The remainder of installation is the reverse of removal.

10. Torque the splash shield bolts to 86 inch lbs. (9.8 Nm).

SUSPENSION

CONTROL ARMS/LINKS

REMOVAL & INSTALLATION

Lower Arm A

See Figure 138.

➡The control arm bushings are an integral part of the lower control arms and not serviced separately.

1. Before servicing the vehicle, refer to the precautions.

2. Support the suspension with a floor jack under lower arm B, at the knuckle.

3. Remove the rear wheel.

4. Remove the lower arm mounting bolts and nuts.

5. Remove the lower arm.

➡Raise the suspension to remove the arm.

To install:

6. Installation is the reverse of removal, noting the following:

7. Install all fasteners loosely, then, final torque all fasteners with the suspension loaded (weight of the car).

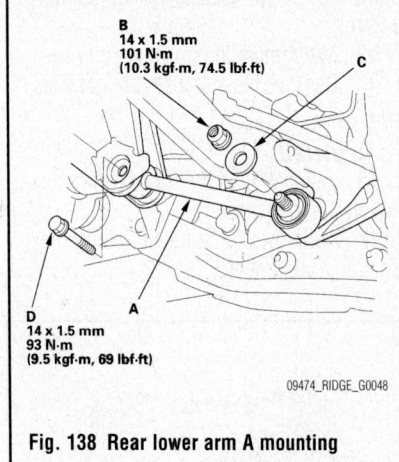

Fig. 138 Rear lower arm A mounting

8. Using a new bolt and nut, tighten the bolt to 69 ft. lbs. (93 Nm) and the nut to 74 ft. lbs. (101 Nm).

Lower Arm B

See Figure 139.

➡The control arm bushings are an integral part of the lower control arms and not serviced separately.

REAR SUSPENSION

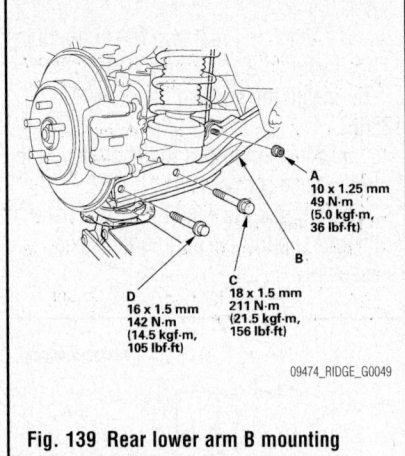

Fig. 139 Rear lower arm B mounting

1. Before servicing the vehicle, refer to the precautions.

2. Support the control arm with a jack at the knuckle.

3. Remove the rear wheel.

4. Remove the locknut (A).

5. Remove the bolt (C).

6. Remove the bolt (D).

7. Gradually lower the arm.

8. Matchmark the adjusting cam.

9. Remove the locknut, adjusting bolt and adjusting cam at the frame.

10. Remove the spring assembly.

11. Remove the inner nuts and bolts and the arm.

To install:

12. Installation is the reverse of removal.

13. Install all fasteners loosely, then, final torque all fasteners with the suspension loaded (weight of the car).

14. Using new bolts and nuts, observe the following torques:
   - Adjusting cam nut: 61 ft. lbs. (83 Nm)
   - Nut A: 36 ft. lbs. (49 Nm)
   - Bolt C: 156 ft. lbs. (211 Nm)
   - Bolt D: 105 ft. lbs. (142 Nm)

Trailing Arm

See Figure 140.

1. Before servicing the vehicle, refer to the precautions.

2. Support control arm B with a jack at the knuckle.

3. Remove the caliper and support it out of the way.

4. Remove the rotor.

5. Disconnect the brake hose from the brake pipe and discard the clip. Plug the lines.

6. Remove the parking brake cable from the training arm.

7. Remove the trailing arm-to-knuckle bolts.

8. Remove the trailing arm mounting bolts.

To install:

9. Installation is the reverse of removal.

10. Install all fasteners loosely, then, final torque all fasteners with the suspension loaded (weight of the car).

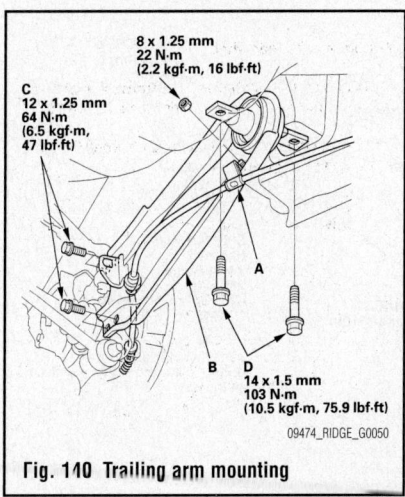

Fig. 140 Trailing arm mounting

11. Using new bolts and nuts and a new brake line clip, observe the following torque specifications:
   - Trailing arm-to-knuckle bolts: 47 ft. lbs. (64 Nm)
   - Trailing arm mounting bolts: 76 ft. lbs. (103 Nm).

12. Check the alignment.

13. Bleed the brakes.

Upper Control Arm

1. Before servicing the vehicle, refer to the precautions.

2. Support the control arm at the knuckle.

3. Remove the wheel.

4. Remove the upper ball joint from the knuckle.

5. Remove the upper arm bolt and the arm.

To install:

6. Installation is the reverse of removal.

7. Install all fasteners loosely, then, final torque all fasteners with the suspension loaded (weight of the car).

8. Using new bolts and nuts, and cotter pin, observe the following torque specifications:
   - Upper control arm bolt: 69 ft. lbs. (94 Nm)
   - Ball joint nut: 36–43 ft. lbs. (49–59 Nm).

MACPHERSON STRUTS

REMOVAL & INSTALLATION

See Figure 141.

1. Before servicing the vehicle, refer to the precautions.

2. Raise and support the rear end.

3. Remove the wheel(s).

4. Remove lower arm B.

5. Remove the 3 bolts from the top of the strut.

6. Installation is the reverse of removal.

7. Torque the upper bolts to 25 ft. lbs. (34 Nm).

OVERHAUL

See Figures 142 and 143.

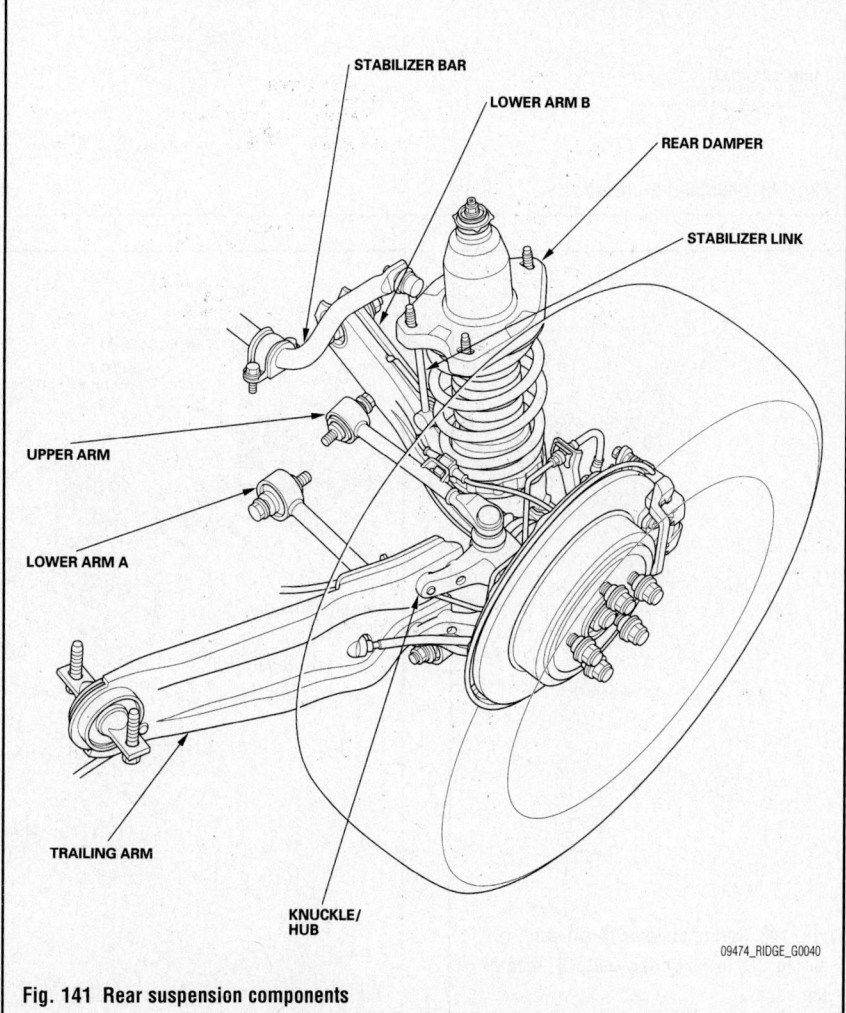

Fig. 141 Rear suspension components

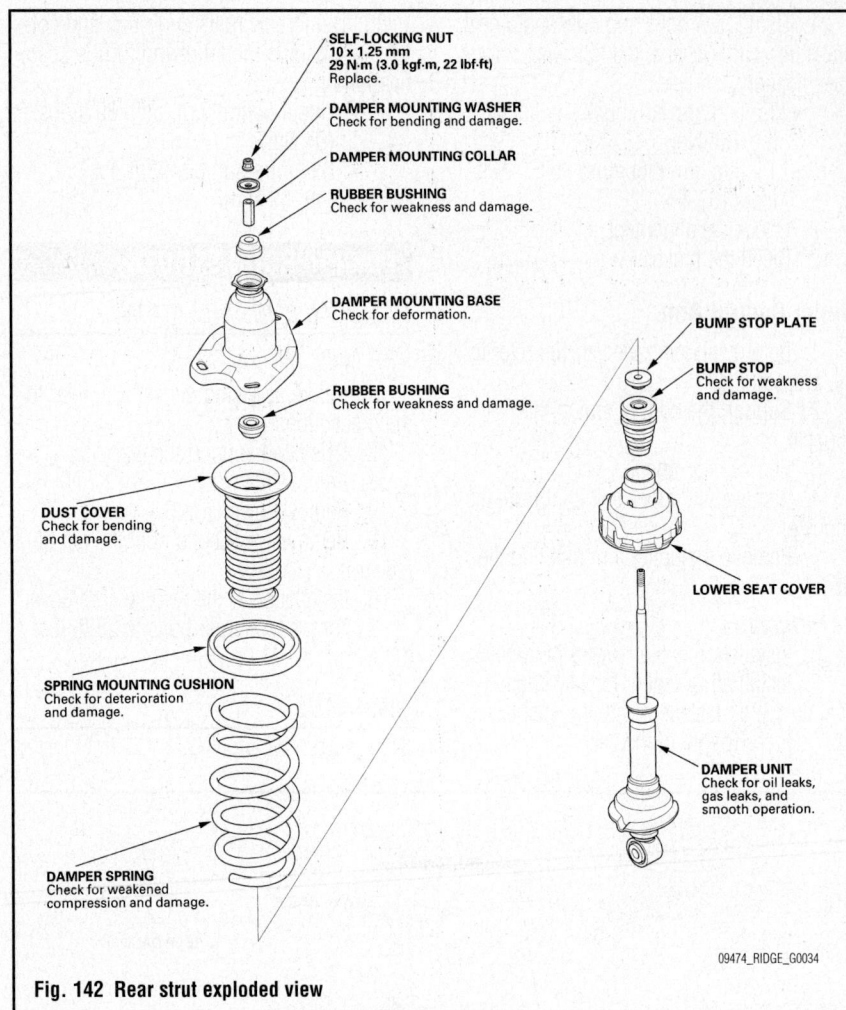

Fig. 142 Rear strut exploded view

1. Before servicing the vehicle, refer to the precautions.

2. Place the strut in a spring compressor.

3. Compress the spring and remove the shaft nut.

4. Release the spring pressure

To assemble:

5. Compress the spring.

6. Assemble all parts except the shaft washer and nut.

7. Align the spring as shown.

8. Install the washer and NEW nut. Torque to 22 ft. lbs. (29 Nm).

REAR KNUCKLE

REMOVAL & INSTALLATION

See Figure 144.

1. Before servicing the vehicle, refer to the precautions.

2. Remove the wheel.

3. Remove and discard the brake hose clip.

4. Remove the brake hose bracket bolts.

5. Remove the caliper bracket mounting bolts and support the caliper out of the way.

6. Remove the sensor from the knuckle. Don't disconnect it.

7. Remove the halfshaft end nut.

8. Remove the rotor.

9. Remove the parking brake shoes and cable.

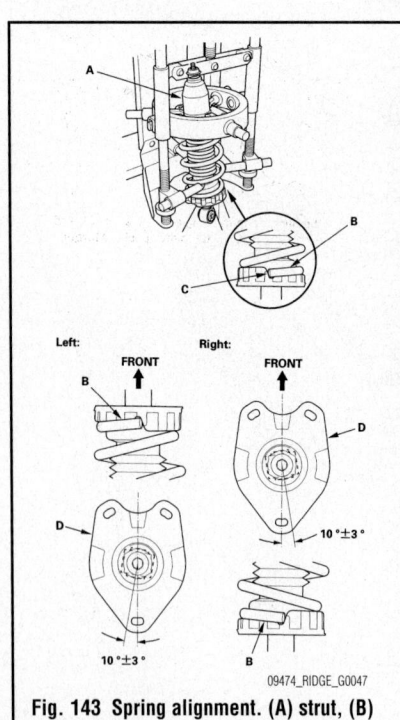

Fig. 143 Spring alignment. (A) strut, (B) spring, (C) lower spring seat, (D) mounting base.

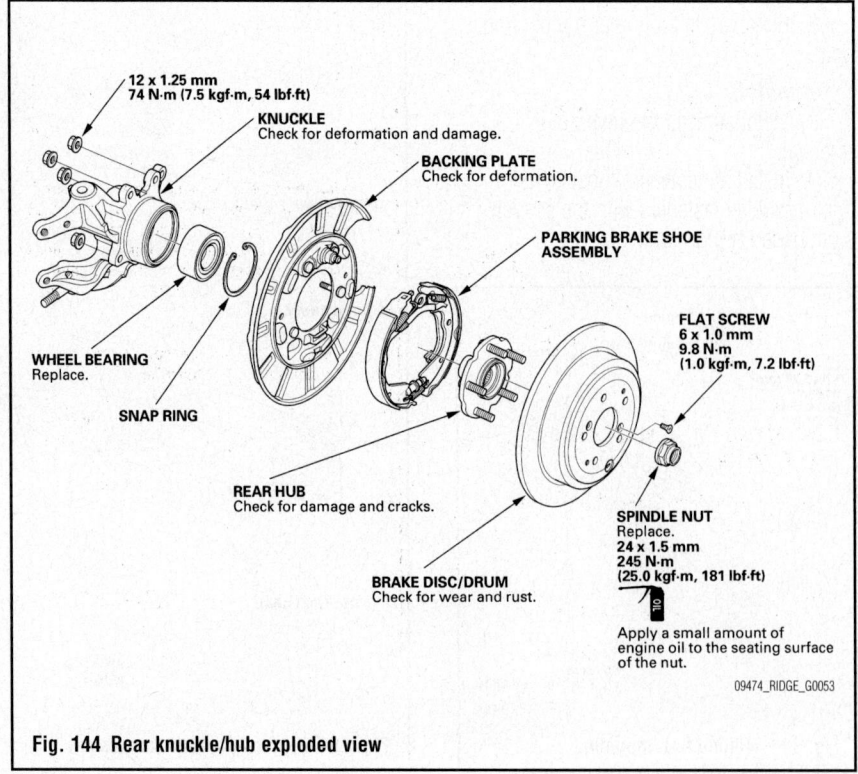

Fig. 144 Rear knuckle/hub exploded view

10. Disconnect the upper arm from the knuckle.

11. Remove lower arm A.

12. Support the lower arm B with a floor jack.

13. Disconnect lower arm B from the knuckle.

14. Pull outward on the knuckle while tapping the end of the halfshaft with a plastic hammer.

15. Installation is the reverse of removal.

16. Install all fasteners loosely, then, final torque all fasteners with the suspension loaded (weight of the car).

17. Use new bolts and nuts and a new brake line clip. See the relevant procedures for torque values.

REMOVAL & INSTALLATION

See Figure 145.

1. Before servicing the vehicle, refer to the precautions.

2. Remove the wheel.

3. Remove the link-to-bar nuts.

4. Remove the rear subframe.

Rear Subframe Torque

NOTE:
- When installing, align both installation reference holes in the subframe with the reference holes in the body using a screwdriver or tapered punch as a guide.
- After removing the subframe mounting bolts, be sure to replace them with new ones.

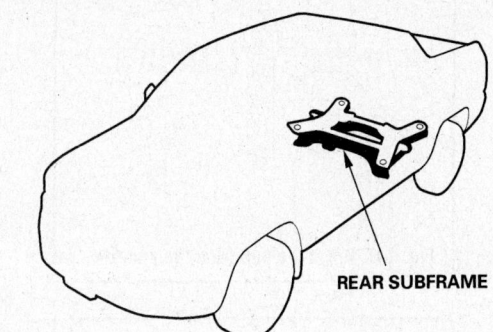

REAR SUBFRAME

Reference hole alignment

SUBFRAME

REFERENCE HOLE (Body side)

INSTALLATION REFERENCE HOLE (Subframe side)

SCREWDRIVER or TAPERED PUNCH

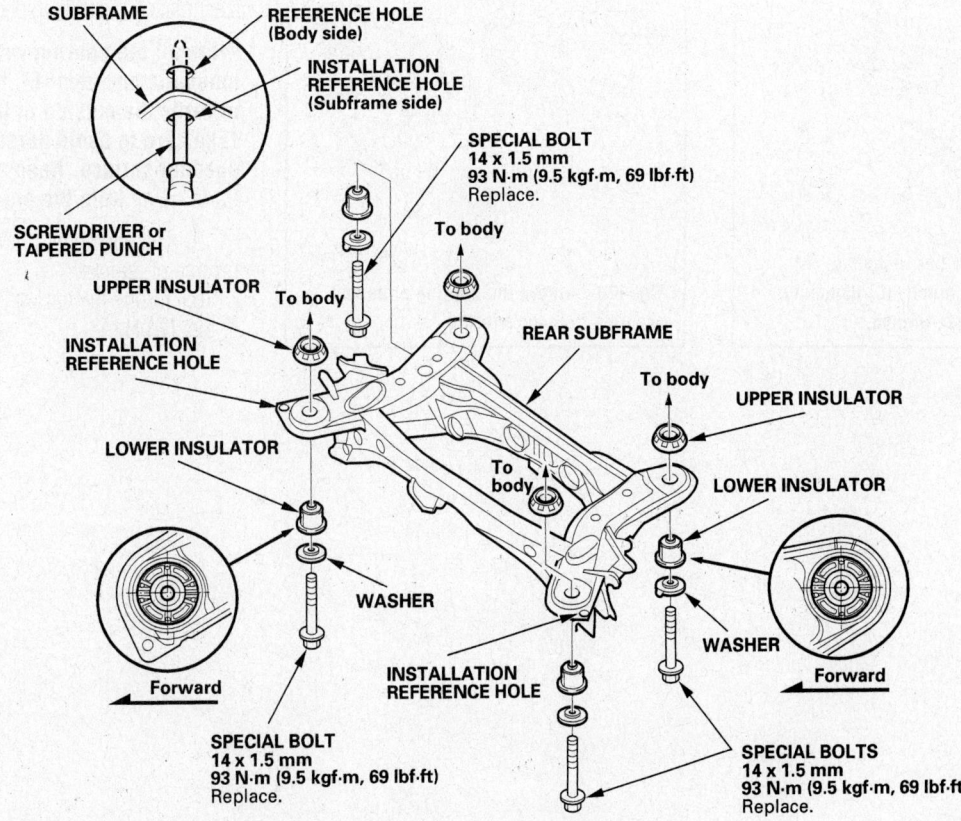

SPECIAL BOLT
14 x 1.5 mm
93 N·m (9.5 kgf·m, 69 lbf·ft)
Replace.

To body

UPPER INSULATOR

To body

INSTALLATION REFERENCE HOLE

REAR SUBFRAME

To body

UPPER INSULATOR

LOWER INSULATOR

To body

LOWER INSULATOR

WASHER

Forward

WASHER

Forward

INSTALLATION REFERENCE HOLE

SPECIAL BOLT
14 x 1.5 mm
93 N·m (9.5 kgf·m, 69 lbf·ft)
Replace.

SPECIAL BOLTS
14 x 1.5 mm
93 N·m (9.5 kgf·m, 69 lbf·ft)
Replace.

09474_RIDGE_G0062

Fig. 145 Rear subframe mounting

5. Remove the stabilizer bar bracket bolts.

6. Matchmark the bushings, and remove the bar.

To install:

7. Installation is the reverse of removal.

8. Torque the bracket bolts to 16 ft. lbs. (22 Nm).

Stabilizer Link

See Figure 146.

1. Before servicing the vehicle, refer to the precautions.

2. Remove the wheel.

3. Remove the link nuts.

➡**The left and right links aren't interchangeable. The left link has a yellow paint mark; the right has a white paint mark.**

4. Installation is the reverse of removal.

5. Install the link on the bar with the joint pins set at the center of their movement.

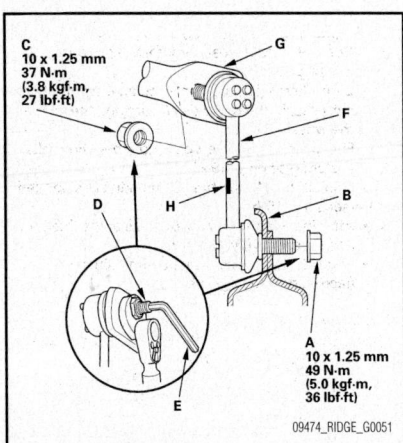

Fig. 146 Stabilizer link mounting. (A) locknut, (B) lower arm B, (C) flange nut, (D) joint pin, (E) hex wrench.

6. Using new nuts, torque the link-to-bar nut to 27 ft. lbs. (37 Nm); the link-to-lower arm B nut to 36 ft. lbs. (49 Nm).

WHEEL HUB AND BEARING

REMOVAL & INSTALLATION

See Figures 147 through 149.

1. Before servicing the vehicle, refer to the precautions.

2. Remove the knuckle.

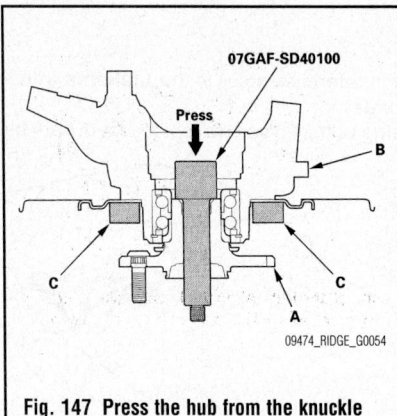

Fig. 147 Press the hub from the knuckle

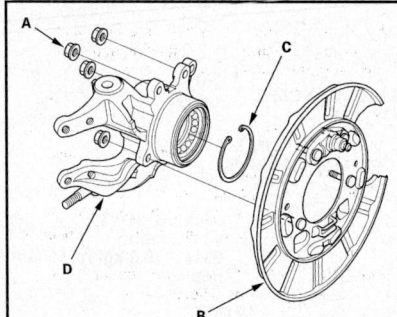

Fig. 148 Remove the backing plate and snapring from the knuckle

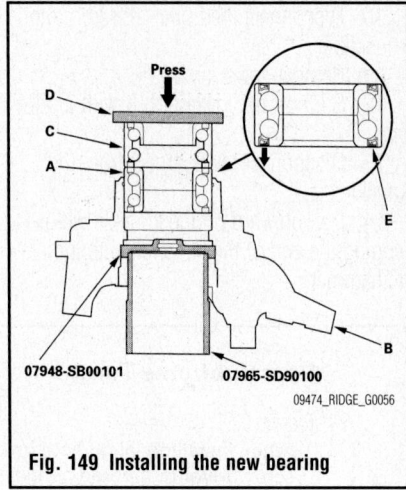

Fig. 149 Installing the new bearing

3. Place the knuckle in a press. Separate the hub (A) from the knuckle (B). Hold the knuckle with the press attachment (C).

4. Press the inner race from the hub.

5. Remove the backing plate and snapring from the knuckle.

6. Press the wheel bearing from the knuckle.

7. Wash the knuckle and hub thoroughly with a safe solvent.

To install:

8. Press a new bearing (A) into the knuckle (B) using the old bearing (C), a steel plate (D) and the special tools shown.

➡**Install the bearing with the sensor magnetic encoded (E), brown color, towards the outside of the knuckle. Take care to avoid damaging the encoder surface. Keep all magnetic tools away from the encoder.**

9. The remainder of installation is the reverse of removal.

10. Torque the backing plate bolts to 54 ft. lbs. (74 Nm).

ACURA

Diagnostic Trouble Codes

11

DIAGNOSTIC TROUBLE CODES

OBD II VEHICLE APPLICATIONS

ACURA

MDX
2007–2008
• 3.7L V6 . . . . J37A1
RDX
2007–2008
• 2.3L I4 . . . . K23A1

RL
2007–2008
• 3.5L V6 . . . . J35A8
TL
2007–2008
• 3.2L V6 . . . . J32A3
• 3.5L V6 . . . . J35A8
TSX
2007–2008
• 2.4L I4 . . . . K24A2

OBD II Trouble Code List (P0xxx Codes)

DTC	Trouble Code Title, Conditions & Possible Causes
DTC: P0010 **1T CCM, MIL: Yes** **Years:** 2007, 2008 **Models:** RDX, TSX **Engines:** All **Transmissions:** All	**Variable Valve Timing Control Oil Control Solenoid Valve Circuit Malfunction** Key on or engine running; and the PCM detected an unexpected voltage condition on the VVT Oil Control Solenoid control signal. **Possible Causes:** • VVT oil control solenoid control circuit is open • VVT oil control solenoid control circuit is shorted to ground • VVT oil control solenoid control circuit is damaged or has failed • PCM has failed
DTC: P0011 **2T CCM, MIL: Yes** **Years:** 2007, 2008 **Models:** RDX, TSX **Engines:** All **Transmissions:** All	**Variable Valve Timing Control System Malfunction** Engine started, vehicle driven through a hard acceleration period, then returned back to idle speed, and the PCM detected a problem in the VVT System operation. **Possible Causes:** • VVT oil control solenoid valve is stuck • VVT oil control solenoid is damaged or has failed • PCM has failed
DTC: P0107 **1T CCM, MIL: Yes** **Years:** 2007, 2008 **Models:** MDX, MDX, RDX, RL, TL, TSX **Engines:** All **Transmissions:** All	**Manifold Air Pressure Sensor Circuit Low Input** Engine started, engine running in closed loop at idle speed, and the PCM detected the MAP sensor indicated a value of 0.0" Hg during the CCM test. **Note: The key on, engine off MAP sensor input should be near 2.9v.** **Possible Causes:** • MAP sensor 5-volt power circuit open or shorted to ground • MAP Sensor signal circuit is shorted to ground • MAP Sensor is damaged or has failed • PCM has failed
DTC: P0108 **1T CCM, MIL: Yes** **Years:** 2007, 2008 **Models:** MDX, MDX, RDX, RL, TL, TSX **Engines:** All **Transmissions:** All	**Manifold Air Pressure Sensor Circuit High Input** Engine started, engine running in closed loop at idle speed, and the PCM detected the MAP sensor indicated a value of 29.9" Hg during the CCM test. **Note: The key on, engine off MAP sensor input should be near 2.9v.** **Possible Causes:** • MAP sensor signal circuit is open, or the ground circuit is open • MAP sensor signal circuit shorted to 5v VREF or system power • MAP sensor is damaged (due to an open circuit) or has failed • PCM has failed
DTC: P0112 **1T CCM, MIL: Yes** **Years:** 2007, 2008 **Models:** MDX, RDX, RL, TL, TSX **Engines:** All **Transmissions:** All	**Intake Air Temperature Sensor Circuit Low Input** Key on or engine running; and the PCM detected the Intake Air Temperature (IAT) sensor signal indicated less than 0.1v (Scan Tool reads 302 degrees F) during the CCM test. **Possible Causes:** • IAT sensor signal shorted to chassis ground • IAT sensor signal shorted to sensor ground circuit • IAT sensor has an internal failure (it is shorted) or has failed • PCM has failed
DTC: P0113 **1T CCM, MIL: Yes** **Years:** 2007, 2008 **Models:** MDX, RDX, RL, TL, TSX **Engines:** All **Transmissions:** All	**Intake Air Temperature Sensor Circuit High Input** Key on or engine running; and the PCM detected the Intake Air Temperature (IAT) sensor signal indicated more than 4.90v (Scan Tool reads −4 degrees F) during the CCM test. **Possible Causes:** • IAT sensor signal shorted to VREF or system power • IAT sensor signal circuit is open • IAT sensor ground circuit is open • Sensor has an internal failure (it is open) • PCM has failed
DTC: P0116 **1T CCM, MIL: Yes** **Years:** 2007, 2008 **Models:** MDX, RDX, RL, TL, TSX **Engines:** All **Transmissions:** All	**Engine Coolant Temperature Sensor Range/Performance** DTC P0116 and P0117 not set, Engine started, and the PCM detected too much change in the ECT sensor signal in too short a period of time during the CCM Rationality test. **Note: The ECT sensor should read 0.47v-0.78v at hot idle speed.** **Possible Causes:** • ECT sensor ground circuit is open (fault may be intermittent) • ECT sensor signal circuit is open (fault may be intermittent) • ECT sensor is damaged or has failed (fault may be intermittent) • PCM has failed

DTC	Trouble Code Title, Conditions & Possible Causes
DTC: P0117 **1T CCM, MIL: Yes** **Years:** 2007, 2008 **Models:** MDX, RDX, RL, TL, TSX **Engines:** All **Transmissions:** All	**Engine Coolant Temperature Sensor Circuit Low Input** Key on or engine running; and the PCM detected the Engine Coolant Temperature (ECT) sensor indicated less than 0.10v (Scan Tool reads more than 302 degrees F) during the CCM test. **Note: The ECT sensor should read 0.47v-0.78v at hot idle speed.** **Possible Causes:** • ECT sensor signal is shorted to chassis ground • ECT sensor signal is shorted to sensor ground circuit • ECT sensor is damaged or has failed (it is shorted internally) • PCM has failed
DTC: P0118 **1T CCM, MIL: Yes** **Years:** 2007, 2008 **Models:** MDX, RDX, RL, TL, TSX **Engines:** All **Transmissions:** All	**Engine Coolant Temperature Sensor Circuit High Input** Key on or engine running; and the PCM detected the Engine Coolant Temperature (ECT) sensor indicated more than 4.90v (Scan Tool reads less than −4 degrees F) during the CCM test. **Note: The ECT sensor should read 0.47v-0.78v at hot idle speed.** **Possible Causes:** • ECT sensor signal circuit is open between the sensor and PCM • ECT sensor signal circuit is shorted to VREF or system power • ECT sensor signal circuit is open between sensor and the PCM • ECT sensor is damaged or has failed (it is open internally) • PCM has failed
DTC: P0122 **1T CCM, MIL: Yes** **Years:** 2007, 2008 **Models:** MDX, RDX, RL, TL, TSX **Engines:** All **Transmissions:** All	**Throttle Position Sensor Circuit Low Input** Engine started, engine running in closed loop at idle speed, and the PCM detected the TP signal indicated less than 0.16v (Scan Tool reads less than 10 percent) during the CCM test. **Possible Causes:** • TP sensor signal circuit is shorted to ground • TP sensor VREF circuit is open • TP sensor VREF circuit is shorted to ground • TP sensor is damaged (it may be shorted internally) • PCM has failed
DTC: P0123 **1T CCM, MIL: Yes** **Years:** 2007, 2008 **Models:** MDX, RDX, RL, TL, TSX **Engines:** All **Transmissions:** All	**Throttle Position Sensor Circuit High Input** Engine started, engine running under a momentary WOT condition, and the PCM detected the TP signal was more than 4.60v (Scan Tool reads more than 90 percent) during the CCM test. **Possible Causes:** • TP sensor signal circuit is shorted to VREF • TP sensor signal circuit is shorted to system power • TP sensor ground circuit is open between sensor and the PCM • TP sensor is damaged or has failed • PCM has failed
DTC: P0128 **2T ECT, MIL: Yes** **Years:** 2007, 2008 **Models:** MDX, RDX, RL, TL, TSX **Engines:** All **Transmissions:** All	**Cooling System Malfunction** DTC P0107, P0108, P0112, P0113, P0116, P0117, P0118, P0335, P0336, P0300, P0301-P0306, P0335, P0336, P0401, P0505, P1106, P0117, P1108, P1129, P1259 and P1519, engine started, engine running at road load for 10 minutes, and the PCM detected the ECT sensor input did not reach the correct closed loop value. **Possible Causes:** • Inspect for low coolant level or for an incorrect coolant mixture • Check the operation of the thermostat (it may be stuck open) • TSB 02-009 (3/11/02) contains a repair procedure for this code
DTC: P0131 **2T CCM, MIL: Yes** **Years:** 2007, 2008 **Models:** TL **Engines:** All **Transmissions:** All	**HO2S-11 (Bank 1 Sensor 1) Circuit Low Input** Engine started, vehicle driven in 4th gear at cruise speed, and the PCM detected the HO2S signal was fixed at less than 0.50v. **Possible Causes:** • HO2S signal circuit is open or it is shorted to ground • HO2S may be contaminated or may have failed • Fuel supply system is too lean (fuel filter is clogged or dirty) • PCM has failed
DTC: P0131 **1T CCM, MIL: Yes** **Years:** 2007, 2008 **Models:** MDX, RDX, RL, TL, TSX **Engines:** All **Transmissions:** All	**HO2S-11 (Bank 1 Sensor 1) Circuit Low Input** Engine started, vehicle driven in closed loop in 4th gear at cruise speed, and the PCM detected the HO2S signal was fixed at less than 0.50v (the actual value is stored in the PCM memory). **Possible Causes:** • HO2S signal circuit is open or it is shorted to ground • HO2S may be contaminated or may have failed • Fuel supply system is too lean (fuel filter is clogged or dirty) • PCM has failed

DTC	Trouble Code Title, Conditions & Possible Causes
DTC: P0132 **1T CCM, MIL: Yes** **Years:** 2007, 2008 **Models:** MDX, RDX, RL, TL, TSX **Engines:** All **Transmissions:** All	**HO2S-11 (Bank 1 Sensor 1) Circuit High Input** Engine started, vehicle driven in D4 or D5 position in closed loop at cruise speed, and the PCM detected the HO2S signal was fixed at more than 0.90v (the actual value is stored in the PCM memory). **Possible Causes:** • HO2S signal tracking (wet/oily) in connector causing a short between the signal circuit and heater power circuit • HO2S signal circuit is open, or the ground circuit is open • HO2S heater supply circuit is open • PCM has failed
DTC: P0133 **2T O2, MIL: Yes** **Years:** 2007, 2008 **Models:** MDX, RDX, RL, TL, TSX **Engines:** All **Transmissions:** All	**HO2S-11 (Bank 1 Sensor 1) Circuit Slow Response** Engine started, vehicle driven in closed loop in 4th or 6th gear at over 55 mph at steady speed, and the PCM detected the HO2S response time to switch between 300-600 mv was too slow, or that the rich to lean or lean to rich switch time was too slow. **Possible Causes:** • Exhaust leak present in the exhaust manifold or exhaust pipes • O2S element fuel contamination • O2S element has deteriorated • PCM has failed
DTC: P0134 **2T O2, MIL: Yes** **Years:** 2007, 2008 **Models:** RSX **Engines:** All **Transmissions:** All	**Air Fuel Sensor (Bank 1 Sensor 1) Circuit Stuck Lean** Engine started, vehicle driven in closed loop in 4th or 6th gear at over 55 mph at steady speed, and the PCM detected the AFS1 signal indicated that it was stuck in a "lean" air/fuel ratio condition. **Possible Causes:** • Exhaust leak present in the exhaust manifold or exhaust pipes • AFS1 element is loose at its mounting location • AFS1 element is contaminated or deteriorated, or it has failed • PCM has failed
DTC: P0135 **1T O2, MIL: Yes** **Years:** 2007, 2008 **Models:** TL **Engines:** All **Transmissions:** All	**HO2S-11 (Bank 1 Sensor 1) Heater Circuit Malfunction** Engine started, engine runtime over 80 seconds, and the PCM detected an unexpected voltage condition on the HO2S heater circuit during the CCM test. **Possible Causes:** • Main relay output (power) circuit to the heater is open • O2S heater ground circuit is open • O2S heater element has high resistance or an open circuit • O2S heater element has a shorted condition • PCM has failed
DTC: P0135 **1T O2, MIL: Yes** **Years:** 2007, 2008 **Models:** MDX, RDX, RL, TL, TSX **Engines:** All **Transmissions:** All	**HO2S-11 (Bank 1 Sensor 1) Heater Circuit Malfunction** Engine started, engine runtime over 80 seconds, and the PCM detected an unexpected voltage condition on the HO2S heater circuit during the CCM test. **Possible Causes:** • Main relay output (power) circuit to the heater is open • O2S heater ground circuit is open • O2S heater element has high resistance or an open circuit • O2S heater element has a shorted condition • PCM has failed
DTC: P0137 **1T CCM, MIL: Yes** **Years:** 2007, 2008 **Models:** MDX, RDX, RL, TL, TSX **Engines:** All **Transmissions:** All	**HO2S-12 (Bank 1 Sensor 2) Circuit Low Input** Engine started, vehicle driven in 4th gear at cruise speed, and the PCM detected the HO2S signal was fixed at less than 0.30v. **Note: The actual value where the code sets is in the PCM memory.** **Possible Causes:** • HO2S signal circuit is open • HO2S signal circuit is shorted to ground • HO2S ground circuit is open • HO2S may be contaminated or may have failed • PCM has failed
DTC: P0138 **1T CCM, MIL: Yes** **Years:** 2007, 2008 **Models:** MDX, RDX, RL, TL, TSX **Engines:** All **Transmissions:** All	**HO2S-12 (Bank 1 Sensor 2) Circuit High Input** Engine started, vehicle driven in closed loop in 4th gear at cruise speed, and the PCM detected the HO2S signal was fixed at more than 0.60v. **Note: The actual value where the code sets is in the PCM memory.** **Possible Causes:** • HO2S signal tracking (wet/oily) in connector causing a short between the signal circuit and heater power circuit • HO2S signal circuit is shorted to system power • HO2S heater supply circuit is open • PCM has failed

DTC	Trouble Code Title, Conditions & Possible Causes
DTC: P0139 **2T O2, MIL: Yes** **Years:** 2007, 2008 **Models:** MDX, RDX, RL, TL, TSX **Engines:** All **Transmissions:** All	**HO2S-12 (Bank 1 Sensor 2) Circuit Slow Response** Engine started, vehicle driven in closed loop in 4th or 6th gear at over 55 mph at steady speed, and the PCM detected the HO2S response time to switch between 300-600 mv was too slow, or that the rich to lean or lean to rich switch time was too slow. **Possible Causes:** • Exhaust leak present in the exhaust manifold or exhaust pipes • HO2S element fuel contamination • HO2S element has deteriorated • PCM has failed
DTC: P0141 **2T O2, MIL: Yes** **Years:** 2007, 2008 **Models:** MDX, RDX, RL, TL, TSX **Engines:** All **Transmissions:** All	**HO2S-12 (Bank 1 Sensor 2) Heater Circuit Malfunction** Engine started, engine runtime over 80 seconds, and the PCM detected an unexpected voltage condition on the HO2S heater circuit during the CCM test. **Possible Causes:** • Main relay output (power) circuit to the heater is open • O2S heater ground circuit is open • O2S heater element has high resistance or an open circuit • O2S heater element has a shorted condition • PCM has failed
DTC: P0153 **2T O2S1, MIL: YES** **Years:** 2007, 2008 **Models:** RL, TL **Engines:** All **Transmissions:** All	**Air Fuel Sensor (Bank 2 Sensor 1) Circuit Slow Response** Engine started, ECT sensor signal more than 158 degrees F, vehicle driven at an engine speed from 1200-2250 rpm at a speed over 48 mph, and the PCM detected the AFS2 signal response time was too slow. **Possible Causes:** • Exhaust leak present in the exhaust manifold or exhaust pipes • AFS2 element is loose at its mounting location • AFS2 element is contaminated or deteriorated, or it has failed • PCM has failed
DTC: P0154 **2T O2S HTR1, MIL: YES** **Years:** 2007, 2008 **Models:** RL, TL **Engines:** All **Transmissions:** All	**Air Fuel Sensor (Bank 2 Sensor 1) Heater Malfunction** Engine started, ECT sensor signal more than 158 degrees F, and the PCM detected a problem in the AFS2 Heater circuit operation in the test. **Possible Causes:** • AFS2 heater power circuit is open • AFS2 heater control circuit is open or shorted to power • AFS2 (heater portion) is damaged or has failed • PCM has failed
DTC: P0155 **2T O2S HTR1, MIL: YES** **Years:** 2007, 2008 **Models:** RL, TL **Engines:** All **Transmissions:** All	**Air Fuel Sensor (Bank1 Sensor 1) Heater Circuit Malfunction** Engine started, ECT sensor input more than 158 degrees F, and the PCM detected a problem in the AFS1 Heater circuit operation in the test. **Possible Causes:** • AFS1 heater power circuit is open (check the LAFHT 15A fuse) • AFS1 heater ground circuit is open • AFS1 heater element has high resistance or an open circuit • AFS1 heater element has a shorted condition • PCM has failed
DTC: P0157 **1T CCM, MIL: YES** **Years:** 2007, 2008 **Models:** RL, TL **Engines:** All **Transmissions:** All	**HO2S-22 (Bank 2 Sensor 2) Circuit Low Input** Engine started, vehicle driven in closed loop in D4/D5 position at cruise speed, and the PCM detected the HO2S signal was fixed at less than 0.30v (actual value where the code sets is in its memory). **Possible Causes:** • HO2S signal circuit is open or shorted to ground • HO2S ground circuit is open • HO2S may be contaminated or may have failed • PCM has failed
DTC: P0158 **1T CCM, MIL: YES** **Years:** 2007, 2008 **Models:** RL, TL **Engines:** All **Transmissions:** All	**HO2S-22 (Bank 2 Sensor 2) Circuit High Input** Engine started, vehicle driven in closed loop in D4/D5 position at cruise speed, and the PCM detected the HO2S signal was fixed at more than 0.60v **Note: The actual value where the code sets is in the PCM memory.** **Possible Causes:** • HO2S signal tracking (wet/oily) in connector causing a short between the signal circuit and heater power circuit • HO2S signal circuit is open, or the ground circuit is open • HO2S heater supply circuit is open • PCM has failed

DTC	Trouble Code Title, Conditions & Possible Causes
DTC: P0159 **1T O2S1, MIL: YES** **Years:** 2007, 2008 **Models:** RL, TL **Engines:** All **Transmissions:** All	**HO2S-22 (Bank 2 Sensor 2) Circuit Slow Response** Engine started, vehicle driven in closed loop at over 55 mph at cruise speed, and the PCM detected the HO2S response time to switch from rich to lean or from lean to rich switch time was too slow. **Possible Causes:** • Exhaust leak present in the exhaust manifold or exhaust pipes • HO2S element fuel contamination • HO2S element has deteriorated • PCM has failed
DTC: P0161 **1T O2S HTR1, MIL: YES** **Years:** 2007, 2008 **Models:** RL, TL **Engines:** All **Transmissions:** All	**HO2S-22 (Bank 2 Sensor 2) Heater Circuit Malfunction** Engine started, engine runtime 80 seconds, and the PCM detected an unexpected voltage condition on the HO2S heater circuit. **Possible Causes:** • Main relay power circuit to the heater is open (intermittent) • HO2S heater ground circuit is open • HO2S heater element has high resistance or has failed • PCM has failed
DTC: P0171 **2T FUEL, MIL: Yes** **Years:** 2007, 2008 **Models:** MDX, RDX, RL, TL, TSX **Engines:** All **Transmissions:** All	**Fuel System Too Lean (Bank 1)** DTC P0107, P0108, P0135, P0137, P0138, P0141, P1128, P1129 and P1259 not set, engine running in closed loop, and the PCM detected the LONGFT value exceeded the calibrated lean limit value. **Possible Causes:** • Air leaks in intake manifold, exhaust pipes or exhaust manifold • One or more injectors restricted or pressure regulator has failed • Air is being drawn in from leaks in gaskets or other seals • O2S element is deteriorated or has failed • A "fuel control" sensor is out of calibration (ECT, IAT or MAP) • PCM has failed
DTC: P0172 **2T FUEL, MIL: Yes** **Years:** 2007, 2008 **Models:** MDX, RDX, RL, TL, TSX **Engines:** All **Transmissions:** All	**Fuel System Too Rich (Bank 1)** DTC P0107, P0108, P0135, P0137, P0138, P0141, P1128, P1129 and P1259 not set, engine running in closed loop, and the PCM detected the LONGFT value exceeded the calibrated rich limit. **Note: A high MAP sensor signal at idle can cause this code to set.** **Possible Causes:** • Base engine fault (i.e., cam timing incorrect, oil level too high) • EVAP vapor recovery system has failed (pulling vacuum) • HO2S element may be contaminated with water or alcohol • Leaking/contaminated fuel injector(s) or fuel pressure regulator • Partial engine misfire condition is present
DTC: P0174 **2T FUEL, MIL: YES** **Years:** 2007, 2008 **Models:** MDX, RDX, RL, TL, TSX **Engines:** All **Transmissions:** All	**Fuel System Too Lean (Bank 2)** DTC P0107, P0108, P0133-P0141, P0153-P0161, P0401-P0406, P2251-P2255, P2A00, P2A03 and P2279 not set, engine running in closed loop, and the PCM detected a very lean Air/Fuel condition. **Possible Causes:** • Air leaks in intake manifold, exhaust pipes or exhaust manifold • One or more injectors restricted or pressure regulator has failed • Air is being drawn in from leaks in gaskets or other seals • Fuel control sensor is out of calibration (ECT, IAT or MAP)
DTC: P0175 **2T FUEL, MIL: YES** **Years:** 2007, 2008 **Models:** MDX, RDX, RL, TL, TSX **Engines:** All **Transmissions:** All	**Fuel System Too Rich (Bank 2)** DTC P0107, P0108, P0135, P0137, P0138, P0141, P1128, P1129 and P1259 not set, engine running in closed loop, and the PCM detected the LONGFT value exceeded the calibrated rich limit. **Note: A high MAP sensor signal at idle can cause this code to set.** **Possible Causes:** • Base engine fault (i.e., cam timing incorrect, oil level too high) • EVAP vapor recovery system has failed (pulling vacuum) • HO2S element may be contaminated with water or alcohol • Leaking/contaminated fuel injector(s) or fuel pressure regulator
DTC: P0222 **1T CCM, MIL: Yes** **Years:** 2007, 2008 **Models:** MDX, RDX, RL, TL, TSX **Engines:** All **Transmissions:** All	**Throttle Position Sensor 2 Circuit Low Input** Engine started, engine running in closed loop conditions, and the PCM detected the closed throttle TP2 signal was less than 0.16v (Scan Tool reads less than 10 percent open) during the CCM test. **Possible Causes:** • TP2 sensor signal circuit is shorted to ground • TP2 sensor VREF circuit is open or shorted to ground • TP2 sensor is damaged (it may be shorted internally) • PCM has failed

DTC	Trouble Code Title, Conditions & Possible Causes
DTC: P0223 **1T CCM, MIL: Yes** **Years:** 2007, 2008 **Models:** MDX, RDX, RL, TL, TSX **Engines:** All **Transmissions:** All	**Throttle Position Sensor 2 Circuit High Input** Engine started, and after a momentary WOT condition, the PCM detected the TP signal indicated over 4.60v (Scan Tool over 90 percent). **Possible Causes:** • TP2 sensor signal circuit is shorted to VREF or system power • TP2 sensor ground circuit is open between sensor and PCM • TP2 sensor is damaged or has failed • PCM has failed
DTC: P0300 **2T MISFIRE, MIL: Yes** **Years:** 2007, 2008 **Models:** MDX, RDX, RL, TL, TSX **Engines:** All **Transmissions:** All	**Multiple Misfire Detected** DTC P0101-P0103, P0171, P0172, P0401, P1102, P1103, P1361, P1362, P1381, P1382, P1491, P1498 and P1508 not set, engine running under positive torque conditions, and the PCM detected a misfire in only one cylinder (MIL will flash if the misfire is severe). **Possible Causes:** • Base engine mechanical fault affecting more than 1 cylinder • CKP or CMP sensor problem affecting more than one cylinder • Fuel system problem affecting more than one cylinder • Ignition system problem affecting more than one cylinder
DTC: P0301 **2T MISFIRE, MIL: Yes** **Years:** 2007, 2008 **Models:** MDX, RDX, RL, TL, TSX **Engines:** All **Transmissions:** All	**Cylinder 1 Misfire Detected** DTC P0107, P0108, P0131, P0132, P0171, P0172, P1128, P0335, P0336, P0505, P1128, P1129, P1259, P1361, P1362, P1366, P1367 and P1519 not set, engine running under positive torque conditions, and the PCM detected a misfire condition in one cylinder. **Note: If the misfire is severe, the MIL will flash on/off on the 1st trip!** **Possible Causes:** • Air leak in the intake manifold, or in the EGR or PCM system • Base engine (mechanical) problem affecting only one cylinder • Fuel system problem affecting only one cylinder • Ignition problem (i.e., coil or spark plug) affecting one cylinder
DTC: P0302 **2T MISFIRE, MIL: Yes** **Years:** 2007, 2008 **Models:** MDX, RDX, RL, TL, TSX **Engines:** All **Transmissions:** All	**Cylinder 2 Misfire Detected** DTC P0107, P0108, P0131, P0132, P0171, P0172, P1128, P0335, P0336, P0505, P1128, P1129, P1259, P1361, P1362, P1366, P1367 and P1519 not set, engine running under positive torque conditions, and the PCM detected a misfire condition in one cylinder. **Note: If the misfire is severe, the MIL will flash on/off on the 1st trip!** **Possible Causes:** • Air leak in the intake manifold, or in the EGR or PCM system • Base engine (mechanical) problem affecting only one cylinder • Fuel system problem affecting only one cylinder • Ignition problem (i.e., coil or spark plug) affecting one cylinder
DTC: P0303 **2T MISFIRE, MIL: Yes** **Years:** 2007, 2008 **Models:** MDX, RDX, RL, TL, TSX **Engines:** All **Transmissions:** All	**Cylinder 3 Misfire Detected** DTC P0107, P0108, P0131, P0132, P0171, P0172, P1128, P0335, P0336, P0505, P1128, P1129, P1259, P1361, P1362, P1366, P1367 and P1519 not set, engine running under positive torque conditions, and the PCM detected a misfire condition in one cylinder. **Note: If the misfire is severe, the MIL will flash on/off on the 1st trip!** **Possible Causes:** • Air leak in the intake manifold, or in the EGR or PCM system • Base engine (mechanical) problem affecting only one cylinder • Fuel system problem affecting only one cylinder • Ignition problem (i.e., coil or spark plug) affecting one cylinder
DTC: P0304 **2T MISFIRE, MIL: Yes** **Years:** 2007, 2008 **Models:** MDX, RDX, RL, TL, TSX **Engines:** All **Transmissions:** All	**Cylinder 4 Misfire Detected** DTC P0107, P0108, P0131, P0132, P0171, P0172, P1128, P0335, P0336, P0505, P1128, P1129, P1259, P1361, P1362, P1366, P1367 and P1519 not set, engine running under positive torque conditions, and the PCM detected a misfire condition in one cylinder. **Note: If the misfire is severe, the MIL will flash on/off on the 1st trip!** **Possible Causes:** • Air leak in the intake manifold, or in the EGR or PCM system • Base engine (mechanical) problem affecting only one cylinder • Fuel system problem affecting only one cylinder • Ignition problem (i.e., coil or spark plug) affecting one cylinder

DTC	Trouble Code Title, Conditions & Possible Causes
DTC: P0305 **2T MISFIRE, MIL: Yes** **Years:** 2007, 2008 **Models:** TL, RL **Engines:** All **Transmissions:** All	**Cylinder 5 Misfire Detected** DTC P0107, P0108, P0131, P0132, P0171, P0172, P1128, P0335, P0336, P0505, P1128, P1129, P1259, P1361, P1362, P1366, P1367 and P1519 not set, engine running under positive torque conditions, and the PCM detected a misfire condition in one cylinder. **Note: If the misfire is severe, the MIL will flash on/off on the 1st trip!** **Possible Causes:** • Base engine (mechanical) problem affecting only one cylinder • Fuel system problem affecting only one cylinder • Ignition problem (i.e., coil or spark plug) affecting one cylinder
DTC: P0306 **2T MISFIRE, MIL: Yes** **Years:** 2007, 2008 **Models:** TL, RL **Engines:** All **Transmissions:** All	**Cylinder 6 Misfire Detected** DTC P0107, P0108, P0131, P0132, P0171, P0172, P1128, P0335, P0336, P0505, P1128, P1129, P1259, P1361, P1362, P1366, P1367 and P1519 not set, engine running under positive torque conditions, and the PCM detected a misfire condition in one cylinder. **Note: If the misfire is severe, the MIL will flash on/off on the 1st trip!** **Possible Causes:** • Air leak in the intake manifold, or in the EGR or PCM system • Base engine (mechanical) problem affecting only one cylinder • Fuel system problem affecting only one cylinder • Ignition problem (i.e., coil or spark plug) affecting one cylinder
DTC: P0325 **1T CCM, MIL: Yes** **Years:** 2007, 2008 **Models:** MDX, RDX, RL, TL, TSX **Engines:** All **Transmissions:** All	**Knock Sensor 1 Circuit Malfunction** Engine started, engine running for 1 minute, and the PCM detected an unexpected voltage condition on the Bank 1 Knock sensor circuit. **Possible Causes:** • Knock sensor signal circuit is open (rear bank of engine) • Knock sensor signal circuit is grounded (rear bank of engine) • Knock sensor not tightened properly • Knock sensor damaged or has failed (it may be open internally) • PCM has failed
DTC: P0330 **1T CCM, MIL: Yes** **Years:** 2007, 2008 **Models:** RL **Engines:** All **Transmissions:** All	**Knock Sensor 2 Circuit Malfunction** Engine started, engine running for 1 minute, and the PCM detected an unexpected voltage condition on the Bank2 Knock Sensor circuit. **Possible Causes:** • Knock sensor signal circuit is open (front bank of engine) • Knock sensor signal circuit is grounded (front bank of engine) • Knock sensor not tightened properly • Knock sensor damaged or has failed (it may be open internally) • PCM has failed
DTC: P0335 **1T CCM, MIL: Yes** **Years:** 2007, 2008 **Models:** MDX, RDX, RL, TL, TSX **Engines:** All **Transmissions:** All	**CKP Sensor 1 Circuit Malfunction (No Signal)** Engine cranking or running; and the PCM did not detect any signals from the Crankshaft Position (CKP) Sensor 1 during the test. **Note: The engine will crank for a longer period of time, may buck or jerk, but it will start and run without the CKP sensor signal present.** **Possible Causes:** • CKP1 sensor signal circuit is open • CKP1 sensor is shorted to ground • CKP1 sensor is damaged or has failed • PCM has failed
DTC: P0336 **1T CCM, MIL: Yes** **Years:** 2007, 2008 **Models:** MDX, RDX, RL, TL, TSX **Engines:** All **Transmissions:** All	**CKP Sensor 1 Circuit Intermittent Signal** Engine started, and the PCM detected the CKP Sensor 1 signal was interrupted during the CCM test. **Note: This trouble code is usually caused by an intermittent fault.** **Possible Causes:** • CKP1 sensor signal circuit is open (intermittent) • CKP1 sensor is shorted to ground (intermittent) • CKP1 sensor is damaged or has failed (intermittent fault) • PCM has failed

DTC	Trouble Code Title, Conditions & Possible Causes
DTC: P0339 **1T CCM, MIL: Yes** **Years:** 2007, 2008 **Models:** MDX, RDX, RL, TL, TSX **Engines:** All **Transmissions:** All	**CKP Sensor 'A' Circuit Intermittent Signal** Engine started, and the PCM detected an unexpected interruption in the CKP Sensor 'A' signal during testing. **Possible Causes:** • CKP 'A' sensor signal circuit is open (intermittent) • CKP 'A' sensor is shorted to ground (intermittent) • CKP 'A' sensor is damaged or has failed (intermittent fault) • PCM has failed
DTC: P0340 **1T CCM, MIL: Yes** **Years:** 2007, 2008 **Models:** MDX, RDX, RL, TL, TSX **Engines:** All **Transmissions:** All	**Camshaft Position Sensor Circuit No Signal** Engine started, CKP sensor signals received, and the PCM detected an intermittent loss of the CMP sensor signal. **Possible Causes:** • CMP sensor signal circuit is open (intermittent fault) • CMP sensor signal circuit is shorted to ground (Intermittent) • CMP sensor is damaged or has failed (intermittent fault) • PCM has failed
DTC: P0344 **1T CCM, MIL: Yes** **Years:** 2007, 2008 **Models:** MDX, RDX, RL, TL, TSX **Engines:** All **Transmissions:** All	**Camshaft Position Sensor Circuit Intermittent Signal** Engine started, and the PCM detected an unexpected interruption of the CMP sensor signal during testing. **Possible Causes:** • CMP Sensor 1 is damaged or has failed • CMP Sensor 1 signal circuit is open or shorted to ground • Engine timing belt has a skipped teeth condition • PCM has failed
DTC: P0385 **1T CCM, MIL: Yes** **Years:** 2007, 2008 **Models:** MDX, RDX, RL, TL, TSX **Engines:** All **Transmissions:** All	**Camshaft Position Sensor 'B' Circuit Malfunction** Engine started, CKP sensor signals received, and the PCM detected a loss of the CMP sensor 'B' signal during the CCM test. **Possible Causes:** • CMP sensor signal 'B' circuit is open • CMP sensor signal 'B' circuit is shorted to ground • CMP sensor 'B' is damaged or has failed • PCM has failed
DTC: P0389 **1T CCM, MIL: Yes** **Years:** 2007, 2008 **Models:** MDX, RDX, RL, TL, TSX **Engines:** All **Transmissions:** All	**Camshaft Position Sensor 'B' Circuit Intermittent Signal** Engine started, CKP sensor signals received, and the PCM detected an intermittent loss of the CMP Sensor 'B' signal. **Possible Causes:** • CMP sensor signal 'B' circuit is open (intermittent fault) • CMP sensor signal 'B' circuit is shorted to ground (Intermittent) • CMP sensor 'B' is damaged or has failed (intermittent fault) • PCM has failed
DTC: P0401 **2T EGR, MIL: Yes** **Years:** 2007, 2008 **Models:** MDX, RDX, RL, TL, TSX **Engines:** All **Transmissions:** All	**EGR System Insufficient Flow Detected** Cold engine startup (ECT sensor less than 76 degrees F at startup), engine running in closed loop at 40-55 mph for 5 minutes in high gear followed by a deceleration period back to 35 mph for 5 seconds with the throttle closed, and the PCM detected a signal from the EGR position sensor that indicated insufficient EGR flow during the test. **Possible Causes:** • EGR valve source vacuum supply line open or restricted • EGR exhaust manifold passages are clogged or restricted • EGR valve assembly or solenoid valve damaged or has failed • PCM has failed
DTC: P0403 **2T CCM, MIL: Yes** **Years:** 2007, 2008 **Models:** MDX, RDX, RL, TL, TSX **Engines:** All **Transmissions:** All	**EGR Solenoid Control Circuit Malfunction** Key on or engine running; and the PCM detected an unexpected voltage condition on the EGR solenoid after it was cycled from "on" to "off". The solenoid resistance is 6.3-6.7 ohms. **Possible Causes:** • EGR solenoid control circuit open or shorted to ground • EGR solenoid control circuit shorted to VREF or system power • EGR solenoid is damaged or has failed • PCM has failed (the EGR solenoid driver may be open/shorted)

DTC	Trouble Code Title, Conditions & Possible Causes

DTC: P0404
2T CCM, MIL: Yes
Years: 2007, 2008
Models: MDX, RDX, RL, TL, TSX
Engines: All
Transmissions: All
	EGR Solenoid Control Circuit Range/Performance
	Engine started, engine running at cruise speed in closed loop, and the PCM did not detect the correct amount of change in the EGR valve position sensor signal after the valve was cycled "on" to "off".
	Possible Causes:
	  • EGR valve source vacuum supply line clogged (remove EGR valve plate to gain access to the manifold port that is clogged)
	  • EGR valve or exhaust manifold passages are restricted
	  • EGR solenoid connector is disconnected or damaged

DTC: P0406
2T CCM, MIL: Yes
Years: 2007, 2008
Models: MDX, RDX, RL, TL, TSX
Engines: All
Transmissions: All
	EGR Valve Position Sensor Circuit High Input
	Key on or engine running; and the PCM detected an unexpected "high" voltage condition (more than 4.88v) on the EGR valve position sensor signal circuit during the CCM test.
	Possible Causes:
	  • EGR valve position sensor circuit is open
	  • EGR valve position sensor circuit is shorted to VREF or power
	  • EGR valve position sensor is damaged or has failed
	  • PCM has failed

DTC: P0420
3T CAT, MIL: Yes
Years: 2007, 2008
Models: MDX, RDX, RL, TL, TSX
Engines: All
Transmissions: All
	Catalyst Efficiency Below Thresholds (Bank 1)
	DTC P0137, P0138 and P0141 not set, engine running in closed loop at 40-55 mph for 2 minutes, followed by a deceleration period to 35 mph at closed throttle, and the PCM detected excessive activity in the Catalyst oxygen sensor (Bank 1) in the Catalyst Monitor test.
	Possible Causes:
	  • Air leaks in at the exhaust manifold or exhaust pipes
	  • Catalytic converter damaged or has failed (deteriorated)
	  • Front HO2S is more aged than the rear HO2S (HO2S is lazy)
	  • PCM has failed

DTC: P0430
3T CAT, MIL: Yes
Years: 2007, 2008
Models: MDX, RDX, RL, TL, TSX
Engines: All
Transmissions: All
	Catalyst Efficiency Below Thresholds (Bank 2)
	DTC P0157, P0158 and P0161 not set, engine started, ECT sensor signal more than 158 degrees F, vehicle driven in closed loop at 50-55 mph for 2 minutes, and the PCM detected excessive activity in the rear HO2S-22 during the Catalyst Monitor test.
	Possible Causes:
	  • Air leaks in at the exhaust manifold or exhaust pipes
	  • Catalytic converter is damaged or has failed (deteriorated)
	  • Front HO2S is more aged than the rear HO2S (HO2S is lazy)
	  • Fuel quality is poor or fuel is contaminated
	  • PCM has failed

DTC: P0443
1T CCM, MIL: Yes
Years: 2007, 2008
Models: MDX, RDX, RL, TL, TSX
Engines: All
Transmissions: All
	EVAP Canister Purge Valve Circuit Malfunction
	Engine started, and PCM detected an unexpected voltage condition on the EVAP canister purge solenoid after it was cycled from "on" to "off". The solenoid resistance is 26-28 ohms.
	Possible Causes:
	  • EGR solenoid control circuit open or shorted to ground
	  • EGR solenoid control circuit shorted to VREF or system power
	  • EGR solenoid is damaged or has failed
	  • PCM has failed (the EGR solenoid driver may be open/shorted)

DTC: P0451
1T CCM, MIL: Yes
Years: 2007, 2008
Models: MDX, RDX, RL, TL, TSX
Engines: All
Transmissions: All
	Fuel Tank Pressure Sensor Range/Performance
	Engine started, and the PCM detected the fuel tank pressure (FTP) sensor signal was less than the allowable range stored in the PCM memory (a calibrated range adjusted to current conditions).
	Note: The FTP sensor PID should be near 2.5v with the fuel cap off.
	Possible Causes:
	  • FTP sensor vacuum lines loose, damaged or disconnected
	  • Fuel tank pressure sensor is damaged or has failed
	  • PCM has failed

DTC: P0453
1T CCM, MIL: Yes
Years: 2007, 2008
Models: MDX, RDX, RL, TL, TSX
Engines: All
Transmissions: All
	Fuel Tank Pressure Sensor Circuit High Input
	Key on or engine running; and the PCM detected the fuel tank pressure (FTP) sensor signal was more than 4.90v during the test.
	Note: The FTP sensor PID should be near 2.5v with the fuel cap off.
	Possible Causes:
	  • FTP sensor signal circuit is shorted to VREF or power (B+)
	  • FTP sensor ground circuit is open
	  • Fuel tank pressure sensor is damaged or has failed
	  • PCM has failed

DTC	Trouble Code Title, Conditions & Possible Causes
DTC: P0455 **2T EVAP, MIL: Yes** **Years:** 2007, 2008 **Models:** MDX, RDX, RL, TL, TSX **Engines:** All **Transmissions:** All	**EVAP System Very Large Leak (0.080") Detected** IAT sensor signal from 32-86 degrees F at startup, ECT signal over 154 degrees F during testing, engine runtime over 10 minutes, TP sensor signal from 1-4v, vehicle driven to a speed of over 20 mph at an engine speed over 1200 rpm, then with both the purge and vent solenoids closed, the PCM detected a large amount of change in the fuel tank pressure during the EVAP leak test. **Possible Causes:** • Fuel tank cap damaged, loose or the wrong part number • Fuel tank leaks at the fuel fill pipe or at the fuel tank seals • Fuel vapor control valve is damaged or has failed • Fuel tank vapor recirculation valve or vapor tube is damaged • Fuel tank vapor control vent tube is damaged or has failed
DTC: P0456 **2T EVAP, MIL: Yes** **Years:** 2007, 2008 **Models:** MDX, RDX, RL, TL, TSX **Engines:** All **Transmissions:** All	**EVAP System Very Small Leak (0.020") Detected** IAT sensor signal from 32-86 degrees F at startup, vehicle driven at over 5 mph for over 2 minutes, then with ECT sensor signal more than 154 degrees F and the EVAP Control and Vent solenoids enabled, the PCM detected the fuel tank pressure was incorrect due to a leak in the fuel tank area during the EVAP Monitor Leak Test. **Possible Causes:** • Fuel tank cap damaged, loose or the wrong part number • Fuel tank leaks at the fuel fill pipe or at the fuel tank seals • Fuel vapor control valve is damaged or has failed • Fuel tank vapor recirculation valve or vapor tube is damaged • Fuel tank vapor control vent tube is damaged or has failed
DTC: P0457 **2T EVAP, MIL: Yes** **Years:** 2007, 2008 **Models:** MDX, RDX, RL, TL, TSX **Engines:** All **Transmissions:** All	**EVAP System Leak Detected (Fuel Cap Loose Or Off)** Engine started, engine running at a speed of more than 5 mph for over 1 minute, and the PCM detected a fuel tank pressure value that indicated a leak present the fuel tank area due to a missing fuel cap. **Possible Causes:** • Fuel tank cap damaged, loose or the wrong part number • Fuel tank leaks at the fuel fill pipe or at the fuel tank seals • Fuel vapor control valve is damaged or has failed • Fuel tank vapor recirculation valve or vapor tube is damaged • Fuel tank vapor control vent tube is damaged or has failed
DTC: P0461 **2T CCM, MIL: Yes** **Years:** 2007, 2008 **Models:** MDX, RDX, RL, TL, TSX **Engines:** All **Transmissions:** All	**Fuel Level Sensor Circuit Range/Performance** Key on or engine running; and the PCM detected an unexpected high or low voltage condition on the Fuel Gauge Unit circuit during the CCM test. **Possible Causes:** • Fuel gauge circuit is open, shorted to ground or to power • Fuel gauge unit is damaged or has failed • PCM has failed
DTC: P0462 **1T CCM, MIL: Yes** **Years:** 2007, 2008 **Models:** MDX, RDX, RL, TL, TSX **Engines:** All **Transmissions:** All	**Fuel Level Sensor Circuit Low Input** Key on or engine running; and the PCM detected an unexpected low voltage condition (less than 0.05v) on the Fuel Gauge Unit circuit. **Possible Causes:** • Fuel gauge circuit is open, shorted to ground • Fuel gauge unit is damaged or has failed • PCM has failed
DTC: P0463 **1T CCM, MIL: Yes** **Years:** 2007, 2008 **Models:** MDX, RDX, RL, TL, TSX **Engines:** All **Transmissions:** All	**Fuel Level Sensor Circuit High Input** Key on or engine running; and PCM detected an unexpected high voltage condition (less than 0.05v) on the Fuel Gauge Unit circuit. **Possible Causes:** • Fuel gauge circuit is shorted to power • Fuel gauge unit is damaged or has failed • PCM has failed
DTC: P0496 **2T EVAP, MIL: Yes** **Years:** 2007, 2008 **Models:** MDX, RDX, RL, TL, TSX **Engines:** All **Transmissions:** All	**EVAP Canister System High Purge Flow** IAT sensor signal from 32-86 degrees F at startup, vehicle driven at over 5 mph for over 2 minutes, then with the EVAP canister purge solenoid "on", the PCM detected the fuel tank pressure was incorrect due to a high purge flow condition. The EVAP Function test can also be used. **Possible Causes:** • EVAP canister purge valve has a poor connection • EVAP canister purge valve is damaged or has failed • EVAP canister vent valve has a poor connection • Fuel tank pressure sensor has a poor connection

DTC	Trouble Code Title, Conditions & Possible Causes
DTC: P0497 **2T EVAP, MIL: Yes** **Years:** 2007, 2008 **Models:** MDX, RDX, RL, TL, TSX **Engines:** All **Transmissions:** All	**EVAP Canister System Low Purge Flow** IAT sensor signal from 32-86 degrees F at engine startup, vehicle driven at over 5 mph for over 2 minutes, then with the EVAP Control solenoid enabled, the PCM detected the fuel tank pressure was incorrect due to a low purge flow condition. The EVAP Function test can be used. **Possible Causes:** • Check for leaks or restrictions in the canister purge vacuum line between the canister purge valve and the in take manifold • Check the vacuum line from canister to the solenoid for leaks • EVAP canister purge valve is damaged or has failed • Check the FTP, purge solenoid and vent solenoid connections
DTC: P0498 **1T CCM, MIL: Yes** **Years:** 2007, 2008 **Models:** MDX, RDX, RL, TL, TSX **Engines:** All **Transmissions:** All	**EVAP Canister Vent Shut Valve Control Circuit Low Voltage** Key on or engine running; and the PCM detected an unexpected low voltage condition on the EVAP canister vent shut valve control circuit during the CCM test. The solenoid resistance is 25-30 ohms at 68 degrees F. **Possible Causes:** • EVAP canister vent shut valve control circuit shorted to ground • EVAP canister vent shut valve power circuit is open (test fuse) • EVAP canister vent control circuit is open • EVAP canister vent control circuit is damaged or has failed • PCM has failed
DTC: P0499 **1T CCM, MIL: Yes** **Years:** 2007, 2008 **Models:** MDX, RDX, RL, TL, TSX **Engines:** All **Transmissions:** All	**EVAP Canister Vent Shut Valve Control Circuit High Voltage** Key on or engine running; and the PCM detected an unexpected high voltage condition on the EVAP canister vent shut valve control circuit during testing. The solenoid resistance is 25-30 ohms at 68 degrees F. **Possible Causes:** • Check for loose connections at the EVAP canister vent shut valve (i.e., check the control and ground circuit connections) • Check for a loose EVAP vent shut valve connection at the PCM • PCM has failed
DTC: P0500 **1T CCM, MIL: Yes** **Years:** 2007, 2008 **Models:** MDX, RDX, RL, TL, TSX **Engines:** All **Transmissions:** All	**Vehicle Speed Sensor Circuit Low Input** Engine started, vehicle driven at cruise speed, followed by a deceleration period with the throttle closed, and the PCM did not detect any VSS signals during the CCM test. **Possible Causes:** • VSS signal circuit is open or shorted to ground • VSS signal circuit is shorted to VREF or system power (B+) • VSS is damaged or has failed
DTC: P0505 **1T CCM, MIL: Yes** **Years:** 2007, 2008 **Models:** MDX, RDX, RL, TL, TSX **Engines:** All **Transmissions:** All	**Idle Speed Control System** DTC P1519 not set, engine started, engine running at hot idle, and the PCM detected the Actual idle speed and the Target idle speed were too far apart during the CCM Rationality test. **Possible Causes:** • IAC valve circuit open, shorted to ground or to power (B+) • IAC valve is damaged or has failed • Fast idle thermo valve is damaged or has failed (some models) • Throttle body is dirty or full of sludge (clean and then retest) • PCM has failed • TSB 03-006 (3/17/03) contains a repair procedure for this code
DTC: P0506 **1T CCM, MIL: Yes** **Years:** 2007, 2008 **Models:** MDX, RDX, RL, TL, TSX **Engines:** All **Transmissions:** All	**Idle Speed Control System Lower Than Expected** Engine started, ECT sensor input over 158 degrees F, vehicle not moving, Fuel Trim from 0.73 to 1.47, and the PCM detected the Actual idle speed was more than 100 rpm lower than the Target idle speed. **Possible Causes:** • IAC valve may be damaged or have failed • Throttle body bore/plate dirty or full of sludge (clean and retest) • Throttle body is damaged • PCM is damaged or has failed
DTC: P0507 **1T CCM, MIL: Yes** **Years:** 2007, 2008 **Models:** MDX, RDX, RL, TL, TSX **Engines:** All **Transmissions:** All	**Idle Speed Control System Higher Than Expected** Engine started, ECT sensor input over 158 degrees F, vehicle not moving, Fuel Trim from 0.73 to 1.47, and the PCM detected the Actual idle speed was more than 100 rpm higher than the Target idle speed. **Possible Causes:** • IAC valve may be damaged or have failed • Inspect for any air intake system leaks in the engine or hoses • Throttle body bore/plate dirty or full of sludge (clean and retest) • PCM is damaged or has failed

DTC	Trouble Code Title, Conditions & Possible Causes
DTC: P0560 **1T CCM, MIL: Yes** **Years:** 2007, 2008 **Models:** RL **Engines:** All **Transmissions:** All	**PCM Backup Circuit Low Voltage** Key on or engine running; and the PCM detected a low voltage condition on the PCM Backup circuit. **Note: This circuit is connected to the Backup/Radio 7.5 amp fuse.** **Possible Causes:** • PCM backup circuit is open or shorted to ground • PCM backup circuit has high resistance • PCM has failed
DTC: P0562 **1T CCM, MIL: No** **Years:** 2007, 2008 **Models:** MDX, RDX, RL, TL, TSX **Engines:** All **Transmissions:** All	**System Voltage Low Input** Engine started, and the PCM detected the system voltage was less than 11.5v for 15 minutes during the CCM test. **Note: For additional help with this code, view the Failure Records.** **Possible Causes:** • Check the drive belt for excessive wear and the proper tension • Check for high resistance at the battery connections or at the starter solenoid connection that connects to PCM power circuit • Check the generator output and the battery condition
DTC: P0563 **1T CCM, MIL: Yes** **Years:** 2007, 2008 **Models:** MDX, RDX, RL, TL, TSX **Engines:** All **Transmissions:** All	**PCM Power Source Circuit Unexpected Voltage** Key on or engine running; and the PCM detected an unexpected loss of voltage condition on the PCM power source circuit during the CCM test. **Possible Causes:** • PCM power source circuit is open or shorted to ground • PCM power source circuit has high resistance • PCM has failed
DTC: P0600 **1T CCM, MIL: Yes** **Years:** 2007, 2008 **Models:** MDX, RDX, RL, TL, TSX **Engines:** All **Transmissions:** All	**Serial Communication Link Circuit Malfunction** Key on or engine running; and the PCM detected an unexpected voltage condition on the Serial Communication Link circuit. **Possible Causes:** • PCM serial communication link circuit is open • PCM serial communication link circuit shorted to ground • PCM serial communication link circuit shorted to system power • PCM has failed
DTC: P0601 **1T CCM, MIL: No** **Years:** 2007, 2008 **Models:** MDX, RDX, RL, TL, TSX **Engines:** All **Transmissions:** All	**PCM Internal Check Sum Error** Key on, and the PCM detected a check sum error had occurred. **Note: For additional help with this code, view the Failure Records.** **Possible Causes:** • The contents of the EEPROM have changed • PCM must be replaced to repair this trouble code
DTC: P0603 **1T CCM, MIL: Yes** **Years:** 2007, 2008 **Models:** MDX, RDX, RL, TL, TSX **Engines:** All **Transmissions:** All	**PCM Internal Module Keep Alive Memory Error** Key on or engine running; and the PCM detected a problem in the Keep Alive Memory (KAM) portion of its internal interface. **Possible Causes:** • Clear codes and recheck for this trouble code. If it resets, update the PCM to the latest software. Then recheck for the same trouble code. If it resets, substitute a known good PCM to determine if the PCM has failed and is causing the code.
DTC: P0606 **1T CCM, MIL: Yes** **Years:** 2007, 2008 **Models:** MDX, RDX, RL, TL, TSX **Engines:** All **Transmissions:** All	**PCM Processor Error** Key on or engine running; and the PCM detected a problem in its internal processor. **Possible Causes:** • Clear codes and recheck for this trouble code. If it resets, update the PCM to the latest software. Then recheck for the same trouble code. If it resets, substitute a known good PCM to determine if the PCM has failed and is causing the code.
DTC: P0627 **1T CCM, MIL: Yes** **Years:** 2007, 2008 **Models:** MDX, RDX, RL, TL, TSX **Engines:** All **Transmissions:** All	**PGM-FI Main Relay 2 (Fuel Pump) Circuit Malfunction** Key on or engine running; and the PCM detected an unexpected voltage condition on the PGM-FI main relay control circuit. **Possible Causes:** • PGM-FI main relay control circuit is open or shorted to ground • PGM-FI main relay power circuit is open (check the fuse) • PGM-FI main relay is damaged or has failed • PCM is damaged or has failed

DTC	Trouble Code Title, Conditions & Possible Causes
DTC: P0641 **1T CCM, MIL: Yes** **Years:** 2007, 2008 **Models:** MDX, RDX, RL, TL, TSX **Engines:** All **Transmissions:** All	**PCM Sensor Reference Voltage 'A' Circuit Malfunction** Key on or engine running; and the PCM detected an unexpected low voltage condition on the Sensor Reference Voltage 'A' circuit. **Possible Causes:** • One or more of these related sensors or circuits (APP1, APP2, MAP, or Counter Shaft Speed sensor) circuits might be shorted to ground. Disconnect one sensor at a time, clear the codes and then recheck for the same trouble code to determine which sensor is causing this trouble code to set. • PCM is damaged or has failed
DTC: P0651 **1T CCM, MIL: Yes** **Years:** 2007, 2008 **Models:** MDX, RDX, RL, TL, TSX **Engines:** All **Transmissions:** All	**PCM Sensor Reference Voltage 'B' Circuit Malfunction** Key on or engine running; and the PCM detected an unexpected low voltage condition on the Sensor Reference Voltage 'B' circuit. **Possible Causes:** • One or more of these related sensors or circuits (EGR Valve Position, Mainshaft Speed, FTP or the TAC module) circuits might be shorted to ground. Disconnect one sensor at a time, clear the codes and then recheck for the same trouble code to determine which sensor is causing this trouble code to set. • PCM is damaged or has failed
DTC: P0657 **1T CCM, MIL: Yes** **Years:** 2007, 2008 **Models:** MDX, RDX, RL, TL, TSX **Engines:** All **Transmissions:** All	**Air Fuel Sensor Power Relay Circuit Malfunction** Key on or engine running; and the PCM detected an unexpected voltage condition on the Air Fuel Sensor power relay control circuit. **Possible Causes:** • AFS power relay control circuit is open or shorted to ground • AFS relay power circuit is open (check the No. 9 LAF HT fuse) • AFS power relay is damaged or has failed • PCM is damaged or has failed
DTC: P0661 **1T CCM, MIL: Yes** **Years:** 2007, 2008 **Models:** MDX, RDX, RL, TL, TSX **Engines:** All **Transmissions:** All	**Intake Manifold Runner Control Valve Position Sensor Circuit Low Input** Key on or engine running; and the PCM detected an unexpected low voltage condition on the IMRC Valve Position Sensor circuit. **Possible Causes:** • IMRC valve position sensor circuit shorted to sensor ground • IMRC valve position sensor circuit is shorted to chassis ground • PCM has failed
DTC: P0662 **1T CCM, MIL: Yes** **Years:** 2007, 2008 **Models:** MDX, RDX, RL, TL, TSX **Engines:** All **Transmissions:** All	**Intake Manifold Runner Control Valve Position Sensor Circuit High Input** Key on or engine running; and the PCM detected an unexpected high voltage condition on the IMRC Valve Position Sensor circuit. **Possible Causes:** • IMRC valve position ground circuit is open • IMRC valve position sensor circuit shorted to VREF • IMRC valve position sensor circuit is shorted to system power • PCM has failed
DTC: P0685 **1T CCM, MIL: Yes** **Years:** 2007, 2008 **Models:** MDX, RDX, RL, TL, TSX **Engines:** All **Transmissions:** All	**PCM Power Control Circuit Malfunction** Key on or engine running; and the PCM detected an unexpected voltage condition on the Air Fuel Sensor power relay control circuit. **Possible Causes:** • Clear codes and recheck for this trouble code. If it resets, update the PCM to the latest software. Then recheck for the same trouble code. If it resets, substitute a known good PCM to determine if the PCM has failed and is causing the code.
DTC: P0700 **1T CCM, MIL: Yes** **Years:** 2007, 2008 **Models:** MDX, RDX, RL, TL, TSX **Engines:** All **Transmissions:** All	**Automatic Transaxle System Malfunction** Engine started, vehicle driven to over 30 mph for several minutes, and the PCM detected a fault in the Automatic Transaxle system. **Note: DTC P0700 sets along with several other TCM trouble codes.** **Possible Causes:** • Check for other A/T related trouble codes, and then refer to the Possible Causes for these trouble codes for more information.
DTC: P0705 **1T CCM, MIL: No** **Years:** 2007, 2008 **Models:** MDX, RDX, RL, TL, TSX **Engines:** All **Transmissions:** All	**Transmission Range Switch Illegal Position Malfunction** Engine started, then driven to a speed of over 8 mph, and the PCM detected "illegal" TR Range or Mode switch signals for 5 seconds. **Note: For additional help with this code, view the Failure Records.** **Possible Causes:** • TR range switch signal is open • TR range switch signal shorted to another switch position signal • TR range switch is damaged or has failed • PCM has failed

DTC	Trouble Code Title, Conditions & Possible Causes
DTC: P0706 **1T CCM, MIL: No** **Years:** 2007, 2008 **Models:** MDX, RDX, RL, TL, TSX **Engines:** All **Transmissions:** All	**Transmission Range Switch Circuit Performance** DTC P0122, P0123, P0722 and P0723 not set, engine started, then driven with the output speed over 3200 rpm, and the PCM detected the TR Switch indicated Reverse position, or with the Output speed under 3000 rpm and the throttle angle over 20 percent, it detected the TR switch indicated Park or Neutral position for 4 seconds in the test. **Note: For additional help with this code, view the Failure Records.** **Possible Causes:** • TR switch signal is open • TR switch signal shorted to another switch position signal • TR switch is damaged or has failed • PCM has failed
DTC: P0710 **1T CCM, MIL: Yes** **Years:** 2007, 2008 **Models:** MDX, RDX, RL, TL, TSX **Engines:** All **Transmissions:** All	**Automatic Transmission Fluid Circuit Malfunction** Key on or engine running; and the PCM detected an unexpected voltage condition on the ATF sensor signal circuit during the test. **Possible Causes:** • ATF sensor signal is open between the sensor and the PCM • ATF sensor signal is shorted to sensor or chassis ground • ATF sensor connector is disconnected or damaged • ATF sensor is damaged or has failed • PCM has failed
DTC: P0711 **1T CCM, MIL: No** **Years:** 2007, 2008 **Models:** MDX, RDX, RL, TL, TSX **Engines:** All **Transmissions:** All	**Transmission Fluid Temperature Sensor Performance** DTC P0722, P0723 and P1870 not set, engine started, system voltage from 11-16v, TFT sensor from −40 degrees F to 69.8 degrees F at startup, ECT sensor more than 150 degrees F and has changed more than 90 degrees F since startup, vehicle speed over 5 mph with the TCC slip speed over 120 rpm for 410 seconds, and the PCM detected the TFT sensor changed less than 2 counts since startup, or that its delta change was over 36 degrees F at least 14 times during a 7 second period. **Possible Causes:** • TFT signal or ground circuit has a high resistance condition • TFT sensor is out-of-calibration (it may be skewed) • TFT sensor is damaged or has failed • PCM has failed
DTC: P0712 **1T CCM, MIL: No** **Years:** 2007, 2008 **Models:** MDX, RDX, RL, TL, TSX **Engines:** All **Transmissions:** All	**Transmission Fluid Temperature Sensor Low Input** Engine started, system voltage from 11-16v, and the PCM detected the TFT sensor was less than 0.40v for 20 seconds during the test. **Note: For additional help with this code, view the Failure Records.** **Possible Causes:** • TFT sensor signal circuit is shorted to sensor ground • TFT sensor signal circuit is shorted to chassis ground • TFT sensor is damaged (it may be shorted internally) • PCM has failed
DTC: P0713 **1T CCM, MIL: No** **Years:** 2007, 2008 **Models:** MDX, RDX, RL, TL, TSX **Engines:** All **Transmissions:** All	**Transmission Fluid Temperature Sensor High Input** Engine started, system voltage from 11-16v, and the PCM detected the TFT sensor signal was more than 4.92v for 409 seconds. **Possible Causes:** • TFT sensor signal circuit is open between the sensor and PCM • TFT sensor ground circuit is open between sensor and ground • TFT sensor signal circuit is shorted to VREF or system power • TFT sensor is damaged (it may be open internally) • PCM has failed
DTC: P0715 **1T CCM, MIL: Yes** **Years:** 2007, 2008 **Models:** MDX, RDX, RL, TL, TSX **Engines:** All **Transmissions:** All	**A/T Mainshaft Speed Sensor Circuit Malfunction** Engine running with VSS inputs received, and the PCM detected an unexpected voltage condition on the Mainshaft speed sensor circuit. **Possible Causes:** • Mainshaft speed sensor circuit is open or shorted to ground • Mainshaft speed sensor circuit is shorted to VREF or power • Mainshaft speed sensor connector is disconnected or damaged • Mainshaft speed sensor is damaged or has failed • PCM has failed

DTC	Trouble Code Title, Conditions & Possible Causes
DTC: P0720 **1T CCM, MIL: Yes** **Years:** 2007, 2008 **Models:** MDX, RDX, RL, TL, TSX **Engines:** All **Transmissions:** All	**A/T Countershaft Speed Sensor Circuit Malfunction** Engine running with VSS inputs received, and the PCM detected an unexpected voltage on the Countershaft Speed Sensor circuit. **Possible Causes:** • Countershaft speed sensor circuit is open or shorted to ground • Countershaft speed sensor circuit is shorted to VREF or power • Countershaft speed sensor connector is disconnected or damaged • Countershaft speed sensor is damaged or has failed • PCM has failed
DTC: P0723 **1T CCM, MIL: No** **Years:** 2007, 2008 **Models:** MDX, RDX, RL, TL, TSX **Engines:** All **Transmissions:** All	**Output Speed Sensor Circuit Malfunction** Engine started, TR switch indicating other than Park or Neutral position, throttle angle over 10 percent, engine vacuum from 0-70 kPa, engine speed from 3000-5000 rpm, and the PCM detected an interruption in the Output Speed Sensor (OSS) circuit in the test. **Possible Causes:** • OSS (+) signal circuit open or shorted to ground (intermittent) • OSS (−) signal circuit open or shorted to ground (intermittent) • OSS is damaged or has failed (an intermittent fault) • PCM has failed
DTC: P0730 **1T CCM, MIL: Yes** **Years:** 2007, 2008 **Models:** MDX, RDX, RL, TL, TSX **Engines:** All **Transmissions:** A/T	**Automatic Transaxle Shift Control System** No other A/T trouble codes set, engine started, vehicle driven at cruise speed with VSS inputs received, and the PCM detected the lockup clutch did not lock or unlock correctly during the CCM test. **Possible Causes:** • One or more A/T clutch pressure control switches is damaged • Refer to the repair instructions in a transmission repair manual or the information in other electronic media to repair this code.
DTC: P0740 **1T CCM, MIL: Yes** **Years:** 2007, 2008 **Models:** MDX, RDX, RL, TL, TSX **Engines:** All **Transmissions:** A/T	**Automatic Transaxle Lockup Clutch System Malfunction** No other A/T trouble codes set, engine started, vehicle driven at cruise speed with VSS inputs received, and the PCM detected the lockup clutch did not engage or disengage correctly during the test. **Possible Causes:** • Possible problem in the transmission or torque converter • Refer to the repair instructions in a transmission repair manual or the information in other electronic media to repair this code.
DTC: P0751 **1T CCM, MIL: No** **Years:** 2007, 2008 **Models:** MDX, RDX, RL, TL, TSX **Engines:** All **Transmissions:** All	**Shift Solenoid 'A' Performance Without Input Speed** DTC P0122, P0123, P0722, P0723, P0742, P0753, P0758 and P1860 not set, vehicle driven in D4 Gear at over 6.25 mph, TFT sensor signal at 68-257 degrees F, then during a 1-2 Shift, TP angle at 10-60 percent (± 3 percent), VSS at 11-31 mph, the PCM detected the engine speed in 2nd Gear was 100 rpm more than it was in 1st Gear (1); or during a 2-3 Shift, TP angle at 13-60 percent (± 5 percent), VSS at 20-45 mph, the engine speed in 3rd Gear was 64 rpm less than it was in 2nd Gear (2); or during a 3-4 Shift, TP at 7-60 percent (± 5 percent), VSS at 25-87 mph, the engine speed in 4th gear was 60 rpm more than it was in 3rd Gear (3); or while in 4th Gear, TP angle at 13-60 percent (± 5 percent), speed ratio at 0.85 to 1.2, the TCC slip speed was 100-2000 rpm for 3 seconds (4); or while in 4th Gear with TCC "on", speed ratio at 0.5-0.85, the TCC slip speed was −50 to +500 for 3 seconds (5). **Note: This code is set if the conditions in (1), (2), (3) or (4) are met, or if the conditions in (1), (2), (3) or (5) are met twice in a row.** **Possible Causes:** • Shift solenoid 'A' is damaged or has failed mechanically (on) • Other internal transmission concerns can cause this problem
DTC: P0753 **1T CCM, MIL: Yes** **Years:** 2007, 2008 **Models:** MDX, RDX, RL, TL, TSX **Engines:** All **Transmissions:** A/T	**A/T Lockup Solenoid 'A' Circuit Malfunction** Engine started, engine running at cruise speed with VSS inputs, and the PCM detected an unexpected voltage condition on the Solenoid Valve 'A' circuit during the CCM test. **Possible Causes:** • A/T Solenoid 'A' control circuit is open or shorted to ground • A/T Solenoid 'A' control circuit is shorted to system power • A/T Solenoid 'A' connector is disconnected or damaged • A/T Solenoid 'A' is damaged or has failed • TCM or PCM has failed

DTC	Trouble Code Title, Conditions & Possible Causes
DTC: P0758 **1T CCM, MIL: Yes** **Years:** 2007, 2008 **Models:** MDX, RDX, RL, TL, TSX **Engines:** All **Transmissions:** A/T	**A/T Lockup Solenoid 'B' Circuit Malfunction** Engine started, engine running at cruise speed, and the PCM detected an unexpected voltage condition on the A/t Solenoid Valve 'B' circuit during the CCM test. **Possible Causes:** • A/T Solenoid 'B' circuit is open, shorted to ground or to power • A/T Solenoid 'B' control circuit is shorted to system power • A/T Solenoid 'B' connector is disconnected or damaged • A/T Solenoid 'B' is damaged or has failed • TCM or PCM has failed
DTC: P0773 **1T CCM, MIL: Yes** **Years:** 2007, 2008 **Models:** MDX, RDX, RL, TL, TSX **Engines:** All **Transmissions:** All	**A/T Lockup Solenoid 'E' Circuit Malfunction** Engine started, engine running at cruise speed, and the PCM detected an unexpected voltage condition on the A/T Solenoid Valve 'E' circuit during the CCM test. **Possible Causes:** • A/T Solenoid 'E' circuit is open, shorted to ground or to power • A/T Solenoid 'E' connector is disconnected or damaged • A/T Solenoid 'E' is damaged or has failed • TCM or PCM has failed
DTC: P0775 **1T CCM, MIL: Yes** **Years:** 2007, 2008 **Models:** MDX, RDX, RL, TL, TSX **Engines:** All **Transmissions:** All	**A/T Clutch Pressure Control Solenoid 'B' Circuit Malfunction** Engine started, engine running at 25-40 mph for 10 seconds followed by a deceleration to a complete stop, and the PCM detected invalid operation of the clutch pressure control solenoid 'B'. **Possible Causes:** • Clutch pressure control solenoid 'B' is damaged or has failed • Automatic transmission may also me damaged or have failed
DTC: P0778 **1T CCM, MIL: Yes** **Years:** 2007, 2008 **Models:** MDX, RDX, RL, TL, TSX **Engines:** All **Transmissions:** All	**A/T Clutch Pressure Control Solenoid 'B' Circuit Malfunction** Engine started, engine running at 25-40 mph for 10 seconds followed by a deceleration to a complete stop, and the PCM detected an unexpected voltage condition on the A/T Solenoid Valve 'B' circuit during the CCM test. **Possible Causes:** • Clutch pressure control solenoid 'B' circuit is open or shorted • Clutch pressure control solenoid 'B' is damaged or has failed • TCM or PCM has failed
DTC: P0780 **1T CCM, MIL: Yes** **Years:** 2007, 2008 **Models:** MDX, RDX, RL, TL, TSX **Engines:** All **Transmissions:** All	**Automatic Transaxle System Malfunction** Engine started, vehicle driven at cruise speed for several minutes, and the PCM detected an Automatic Transaxle fault. **Note: This trouble code sets with along with several TCM related trouble codes.** **Possible Causes:** • Refer to the repair instructions in a transmission repair manual or the information in other electronic media to repair this code.
DTC: P0795 **1T CCM, MIL: Yes** **Years:** 2007, 2008 **Models:** MDX, RDX, RL, TL, TSX **Engines:** All **Transmissions:** All	**A/T Clutch Pressure Control Solenoid 'C' Circuit Malfunction** Engine started, vehicle driven at cruise speed, and the PCM detected an unexpected voltage condition on the A/T Clutch Pressure Control solenoid valve 'C' circuit during the CCM test. **Possible Causes:** • Clutch pressure control solenoid 'C' circuit is open • Clutch pressure control solenoid 'C' connector is disconnected • Clutch pressure control solenoid 'C" is damaged or has failed • Clear codes and recheck for this trouble code. If it resets, update the PCM to the latest software. Then recheck for the same trouble code. If it resets, substitute a known good PCM to determine if the PCM has failed and is causing the code.
DTC: P0798 **1T CCM, MIL: Yes** **Years:** 2007, 2008 **Models:** MDX, RDX, RL, TL, TSX **Engines:** All **Transmissions:** All	**A/T Clutch Pressure Control Solenoid 'C' Circuit Malfunction** Engine started, vehicle driven at cruise speed, and the PCM detected an unexpected voltage condition on the A/T Clutch Pressure Control solenoid valve 'C' circuit during the CCM test. **Possible Causes:** • Clutch pressure control solenoid 'C' circuit is shorted to ground • Clutch pressure control solenoid 'C" is damaged or has failed • Clear codes and recheck for this trouble code. If it resets, update the PCM to the latest software. Then recheck for the same trouble code. If it resets, substitute a known good PCM to determine if the PCM has failed and is causing the code.

DTC	Trouble Code Title, Conditions & Possible Causes
DTC: P0840 **1T CCM, MIL: Yes** **Years:** 2007, 2008 **Models:** MDX, RDX, RL, TL, TSX **Engines:** All **Transmissions:** All	**A/T Clutch Pressure Switch No. 2 Circuit Malfunction** Engine started, vehicle driven at cruise speed, and the PCM detected an unexpected voltage condition on the A/T Clutch pressure switch No. 2 circuit during the CCM test period. **Possible Causes:** • Clutch Pressure Switch No. 2 circuit is open • Clutch Pressure Switch No. 2 circuit is shorted to ground • Clutch Pressure Switch connector is disconnected or damaged • Clear codes and recheck for this trouble code. If it resets, update the PCM to the latest software. Then recheck for the same trouble code. If it resets, substitute a known good PCM to determine if the PCM has failed and is causing the code.
DTC: P0845 **1T CCM, MIL: Yes** **Years:** 2007, 2008 **Models:** MDX, RDX, RL, TL, TSX **Engines:** All **Transmissions:** All	**A/T Clutch Pressure Switch No. 3 Circuit Malfunction** Engine started, vehicle driven at cruise speed, and the PCM detected an unexpected voltage condition on the A/T Clutch pressure switch No. 3 circuit during the CCM test period. **Possible Causes:** • Clutch Pressure Switch No. 3 circuit is open • Clutch Pressure Switch No. 3 circuit is shorted to ground • Clutch Pressure Switch connector is disconnected or damaged • Clear codes and recheck for this trouble code. If it resets, update the PCM to the latest software. Then recheck for the same trouble code. If it resets, substitute a known good PCM to determine if the PCM has failed and is causing the code.
DTC: P1077 **1T CCM, MIL: Yes** **Years:** 2007, 2008 **Models:** MDX, RDX, RL, TL, TSX **Engines:** All **Transmissions:** All	**Intake Manifold Runner Control System (Low RPM) Malfunction** Engine started, vehicle driven at cruise speed and then back to idle speed, and the PCM detected a problem in the IMRC system at low engine speed during the CCM test. If P0651 is also set, repair it first. **Possible Causes:** • IMRC component failure affecting low engine speed operation • IMRC solenoid is damaged or stuck at low engine speed • PCM has failed

OBD II Trouble Code List (P1xxx Codes)

DTC	Trouble Code Title, Conditions & Possible Causes
DTC: P1078 **1T CCM, MIL : Yes** **Years:** 2007, 2008 **Models:** MDX, RDX, RL, TL, TSX **Engines:** All **Transmissions:** All	**Intake Manifold Runner Control System (High RPM) Malfunction** Engine started, vehicle driven at cruise speed and then back to idle speed, and the PCM detected a problem in the IMRC system at high engine speed during the CCM test. **Possible Causes:** • IMRC component failure affecting high engine speed operation • IMRC solenoid is damaged or stuck at high engine speed • PCM has failed
DTC: P1106 **1T CCM, MIL : Yes** **Years:** 2007, 2008 **Models:** MDX, RDX, RL, TL, TSX **Engines:** All **Transmissions:** All	**BARO Pressure Sensor Circuit Range/Performance** Engine started, engine running in 4th gear and accelerated under wide-open-throttle conditions, and the PCM detected the BARO sensor signal did not change enough under these test conditions. **Possible Causes:** • BARO sensor signal circuit is open or shorted to ground • BARO sensor ground circuit has high resistance • BARO sensor is damaged or it may be out of calibration • PCM has failed
DTC: P1107 **1T CCM, MIL : Yes** **Years:** 2007, 2008 **Models:** MDX, RDX, RL, TL, TSX **Engines:** All **Transmissions:** All	**BARO Pressure Sensor Circuit Low Input** Key on or engine running; and the PCM detected an unexpected "low" voltage on the BARO sensor circuit during the CCM test. **Possible Causes:** • BARO sensor signal circuit is shorted to signal ground • BARO sensor signal circuit is shorted to chassis ground • BARO sensor is damaged (it may be shorted internally) • BARO sensor signal circuit to the TCM is open or grounded • TCM or the PCM has failed
DTC: P1108 **1T CCM, MIL : Yes** **Years:** 2007, 2008 **Models:** MDX, RDX, RL, TL, TSX **Engines:** All **Transmissions:** All	**BARO Pressure Sensor Circuit High Input** Key on or engine running; and the PCM detected an unexpected "high" voltage on the BARO sensor circuit during the CCM test. **Possible Causes:** • BARO sensor signal circuit shorted to VREF • BARO sensor signal circuit is shorted to system power (B+) • BARO sensor is damaged (it may be open internally) • BARO sensor signal circuit to the TCM is shorted to power • TCM or the PCM has failed

DTC	Trouble Code Title, Conditions & Possible Causes
DTC: P1121 **1T CCM, MIL : Yes** **Years:** 2007, 2008 **Models:** TL **Engines:** All **Transmissions:** All	**TP Sensor Signal Lower Than Expected** Engine started, and the PCM detected the TP sensor was lower than the expected value with the throttle wide open (i.e., the Scan Tool reads less than 14.1 percent under these conditions) during the CCM Rationality Test. **Possible Causes:** • TP sensor signal circuit is shorted to ground between the PCM and the TCM • TP sensor is damaged or has failed • PCM has failed
DTC: P1122 **1T CCM, MIL : Yes** **Years:** 2007, 2008 **Models:** MDX, RDX, RL, TL, TSX **Engines:** All **Transmissions:** All	**TP Sensor Signal Higher Than Expected** Engine started, and the PCM detected the TP sensor was more than the expected value with the throttle wide open (i.e., the Scan Tool reads more than 16.5 percent under these conditions) during the CCM Rationality test. **Possible Causes:** • TP sensor signal circuit is open or shorted to VREF between the PCM and the TCM • TP sensor is damaged or has failed • PCM has failed
DTC: P1128 **1T CCM, MIL : Yes** **Years:** 2007, 2008 **Models:** MDX, RDX, RL, TL, TSX **Engines:** All **Transmissions:** All	**MAP Sensor Signal Lower Than Expected** Engine started, engine running at cruise speed and then back to idle speed, and the PCM detected a MAP sensor signal was less than the expected value during the CCM Rationality Test. **Possible Causes:** • MAP sensor signal circuit shorted to ground (intermittent fault) • MAP sensor vacuum line bent or plugged at intake manifold • MAP sensor is damaged or it is out-of-calibration • PCM has failed
DTC: P1129 **1T CCM, MIL : Yes** **Years:** 2007, 2008 **Models:** TL **Engines:** All **Transmissions:** All	**MAP Sensor Value Higher Than Expected** Engine started, engine running at cruise speed and then back to idle speed, and the PCM detected a MAP sensor signal was higher than the expected value during the CCM Rationality Test. **Possible Causes:** • Check for signs of a vacuum leak at the PCV valve or hose, the engine mount vacuum hose, brake booster and throttle body • MAP sensor signal circuit shorted to VREF (intermittent fault) • MAP sensor is damaged or it is out-of-calibration • PCM has failed
DTC: P1162 **1T CCM, MIL : Yes** **Years:** 2007, 2008 **Models:** MDX, RDX, RL, TL, TSX **Engines:** All **Transmissions:** All	**Air Fuel Sensor (Bank 1 Sensor 1) Circuit Malfunction** DTC P0131, P0132, P0133, P1163 not set, vehicle driven while in closed loop at over 55 mph in D4 for 1-2 minutes, and the PCM detected an unexpected voltage condition on the A/F circuit during the HO2S Monitor test. **Possible Causes:** • A/F sensor may be contaminated or may have failed • Fuel supply system is too lean (exhaust leaks in front of HO2S) • Fuel supply system is too rich (fuel filter is clogged or dirty) • PCM has failed
DTC: P1163 **1T CCM, MIL : Yes** **Years:** 2007, 2008 **Models:** MDX, RDX, RL, TL, TSX **Engines:** All **Transmissions:** All	**Air Fuel Sensor (Bank 1 Sensor 1) Slow Response** Engine at idle speed, then accelerated to 55 mph for 5 seconds, then back to idle speed for 5 seconds, and the PCM detected the A/F sensor response time was too slow, or the rich-lean or lean-rich response switch rate was too slow during the HO2S Monitor test. **Possible Causes:** • A/F sensor may be contaminated or may have failed • Fuel supply system is too lean (exhaust leaks in front of LAF) • Fuel supply system is too rich (fuel filter is clogged or dirty) • PCM has failed • TSB 01-038 (10/29/01) contains a repair procedure for the code
DTC: P1164 **1T CCM, MIL : Yes** **Years:** 2007, 2008 **Models:** MDX, RDX, RL, TL, TSX **Engines:** All **Transmissions:** All	**Air Fuel Sensor (Bank 1 Sensor 1) Range/Performance** Engine speed over 1500 rpm in 4th gear in closed loop, then a quick acceleration to WOT, followed by a 5 second deceleration period with the throttle closed, and the PCM detected the A/F sensor response time or the R-L or L-R switch rate was too slow. **Possible Causes:** • A/F sensor may be contaminated or may have failed • Fuel supply system too lean (exhaust leaks in front of the LAF) • Fuel supply system too rich (the fuel filter may be very dirty) • PCM has failed

DTC	Trouble Code Title, Conditions & Possible Causes
DTC: P1166 **1T O2, MIL: Yes** **Years:** 2007, 2008 **Models:** MDX, RDX, RL, TL, TSX **Engines:** All **Transmissions:** All	**Air Fuel Sensor (Bank 1 Sensor 1) Heater Circuit Malfunction** Engine runtime more than 80 seconds; and the PCM detected an unexpected voltage condition on the A/F sensor heater circuit. **Possible Causes:** • A/F sensor heater circuit is open or shorted to ground • A/F sensor heater is damaged or has failed • PCM has failed
DTC: P1167 **1T O2, MIL: Yes** **Years:** 2007, 2008 **Models:** MDX, RDX, RL, TL, TSX **Engines:** All **Transmissions:** All	**Air Fuel Sensor (Bank 1 Sensor 1) Heater Circuit Malfunction** Engine runtime over 80 seconds, and the PCM detected an unexpected voltage condition on the A/F sensor Heater circuit. **Possible Causes:** • A/F sensor heater power supply circuit is open (check the fuse) • A/F sensor heater circuit is shorted to system power (B+) • A/F sensor heater is damaged or has failed • PCM has failed
DTC: P1201 **2T MISFIRE, MIL : Yes** **Years:** 2007, 2008 **Models:** RL **Engines:** All **Transmissions:** All	**Cylinder 1 Misfire Detected** Engine runtime 10 seconds, engine under positive load conditions, and the PCM detected a random misfire condition in one cylinder during the 200 (Catalyst) or 1000 revolution (Emission) test range. **Note: If the misfire is severe, the MIL will flash on/off on the 1st trip!** **Possible Causes:** • Air leaks (intake) or exhaust leaks affecting only one cylinder • Base engine (compression) problem affecting only one cylinder • Fuel metering (fuel injector dirty) problem affecting one cylinder • Ignition system (spark plug or wire) fault affecting one cylinder
DTC: P1202 **2T MISFIRE, MIL : Yes** **Years:** 2007, 2008 **Models:** RL **Engines:** All **Transmissions:** All	**Cylinder 2 Misfire Detected** Engine runtime 10 seconds, engine under positive load conditions, and the PCM detected a random misfire condition in one cylinder during the 200 (Catalyst) or 1000 revolution (Emission) test range. **Note: If the misfire is severe, the MIL will flash on/off on the 1st trip!** **Possible Causes:** • Air leaks (intake) or exhaust leaks affecting only one cylinder • Base engine (compression) problem affecting only one cylinder • Fuel metering (fuel injector dirty) problem affecting one cylinder • Ignition system (spark plug or wire) fault affecting one cylinder
DTC: P1204 **2T MISFIRE, MIL : Yes** **Years:** 2007, 2008 **Models:** RL **Engines:** All **Transmissions:** All	**Cylinder 4 Misfire Detected** Engine runtime 10 seconds, engine under positive load conditions, and the PCM detected a random misfire condition in one cylinder during the 200 (Catalyst) or 1000 revolution (Emission) test range. **Note: If the misfire is severe, the MIL will flash on/off on the 1st trip!** **Possible Causes:** • Air leaks (intake) or exhaust leaks affecting only one cylinder • Base engine (compression) problem affecting only one cylinder • Fuel metering (fuel injector dirty) problem affecting one cylinder • Ignition system (spark plug or wire) fault affecting one cylinder
DTC: P1205 **2T MISFIRE, MIL : Yes** **Years:** 2007, 2008 **Models:** RL **Engines:** All **Transmissions:** All	**Cylinder 5 Misfire Detected** Engine runtime 10 seconds, engine under positive load conditions, and the PCM detected a random misfire condition in one cylinder during the 200 (Catalyst) or 1000 revolution (Emission) test range. **Note: If the misfire is severe, the MIL will flash on/off on the 1st trip!** **Possible Causes:** • Air leaks (intake) or exhaust leaks affecting only one cylinder • Base engine (compression) problem affecting only one cylinder • Fuel metering (fuel injector dirty) problem affecting one cylinder • Ignition system (spark plug or wire) fault affecting one cylinder
DTC: P1206 **2T MISFIRE, MIL : Yes** **Years:** 2007, 2008 **Models:** RL **Engines:** All **Transmissions:** All	**Cylinder 6 Misfire Detected** Engine runtime 10 seconds, engine under positive load conditions, and the PCM detected a random misfire condition in one cylinder during the 200 (Catalyst) or 1000 revolution (Emission) test range. **Note: If the misfire is severe, the MIL will flash on/off on the 1st trip!** **Possible Causes:** • Air leaks (intake) or exhaust leaks affecting only one cylinder • Base engine (compression) problem affecting only one cylinder • Fuel metering (fuel injector dirty) problem affecting one cylinder • Ignition system (spark plug or wire) fault affecting one cylinder

DTC	Trouble Code Title, Conditions & Possible Causes
DTC: P1259 **1T CCM, MIL : Yes** **Years:** 2007, 2008 **Models:** MDX, RDX, RL, TL, TSX **Engines:** All **Transmissions:** All	**VTEC System Malfunction** Engine running under hard acceleration, than back to road load, and the PCM detected a fault in the operation of the VTEC system. **Possible Causes:** • VTEC solenoid control circuit is open or shorted to ground • VTEC solenoid is damaged or has failed • VTEC pressure switch circuit is open or shorted to ground • VTEC pressure switch is damaged or has failed • PCM has failed
DTC: P1297 **1T CCM, MIL : Yes** **Years:** 2007, 2008 **Models:** MDX, RDX, RL, TL, TSX **Engines:** All **Transmissions:** All	**Electrical Load Detector Circuit Low Input** Engine running at hot idle speed or at cruise speed, headlights "on", and the PCM detected the ELD signal was less than a stored value. **Possible Causes:** • ELD sensor signal circuit is open or shorted to ground • ELD sensor power circuit is open or shorted to ground • ELD sensor is damaged or has failed • PCM has failed
DTC: P1298 **1T CCM, MIL : Yes** **Years:** 2007, 2008 **Models:** MDX, RDX, RL, TL, TSX **Engines:** All **Transmissions:** All	**Electrical Load Detector Circuit High Input** Engine running at hot idle speed or at cruise speed, headlights "on", and the PCM detected the ELD signal was more than a stored value. **Possible Causes:** • ELD sensor signal circuit is shorted to VREF • ELD sensor signal circuit is shorted to system power (B+) • ELD sensor is damaged or has failed • PCM has failed
DTC: P1300 **1T CCM, MIL : Yes** **Years:** 2007, 2008 **Models:** RL **Engines:** All **Transmissions:** All	**Random Misfire Detected** DTC P0107, P0108, P0131, P0132, P0171, P0172, P1128, P0335, P0336, P0505, P1128, P1129, P1259, P1361, P1362, P1366, P1367 and P1519 not set, engine running under positive torque conditions, and the PCM detected a misfire in 2 or more cylinders. **Note: If the misfire is severe, the MIL will flash on/off on the 1st trip!** **Possible Causes:** • CKP or CMP sensor signal erratic or intermittent • Fuel system problem affecting more than one cylinder • Ignition system problem affecting more than one cylinder • Base engine mechanical fault affecting more than one cylinder
DTC: P1301 **1T MISFIRE, MIL : Yes** **Years:** 2007, 2008 **Models:** RL **Engines:** All **Transmissions:** All	**Cylinder 1 Misfire Detected** DTC P0107, P0108, P0131, P0132, P0171, P0172, P1128, P0335, P0336, P0505, P1128, P1129, P1259, P1361, P1362, P1366, P1367 and P1519 not set, engine running under positive torque conditions, and the PCM detected a misfire condition in one cylinder. **Note: If the misfire is severe, the MIL will flash on/off on the 1st trip!** **Possible Causes:** • Base engine (mechanical) problem affecting only one cylinder • Fuel system problem affecting only one cylinder • Ignition system problem affecting one cylinder
DTC: P1302 **1T MISFIRE, MIL : Yes** **Years:** 2007, 2008 **Models:** RL **Engines:** All **Transmissions:** All	**Cylinder 2 Misfire Detected** DTC P0107, P0108, P0131, P0132, P0171, P0172, P1128, P0335, P0336, P0505, P1128, P1129, P1259, P1361, P1362, P1366, P1367 and P1519 not set, engine running under positive torque conditions, and the PCM detected a misfire condition in one cylinder. **Note: If the misfire is severe, the MIL will flash on/off on the 1st trip!** **Possible Causes:** • Base engine (mechanical) problem affecting only one cylinder • Fuel system problem affecting only one cylinder • Ignition system problem affecting one cylinder
DTC: P1303 **1T MISFIRE, MIL : Yes** **Years:** 2007, 2008 **Models:** RL **Engines:** All **Transmissions:** All	**Cylinder 3 Misfire Detected** DTC P0107, P0108, P0131, P0132, P0171, P0172, P1128, P0335, P0336, P0505, P1128, P1129, P1259, P1361, P1362, P1366, P1367 and P1519 not set, engine running under positive torque conditions, and the PCM detected a misfire condition in one cylinder. **Note: If the misfire is severe, the MIL will flash on/off on the 1st trip!** **Possible Causes:** • Base engine (mechanical) problem affecting only one cylinder • Fuel system problem affecting only one cylinder • Ignition system problem affecting one cylinder

DTC	Trouble Code Title, Conditions & Possible Causes
DTC: P1304 **1T MISFIRE, MIL : Yes** **Years:** 2007, 2008 **Models:** RL **Engines:** All **Transmissions:** All	**Cylinder 4 Misfire Detected** DTC P0107, P0108, P0131, P0132, P0171, P0172, P1128, P0335, P0336, P0505, P1128, P1129, P1259, P1361, P1362, P1366, P1367 and P1519 not set, engine running under positive torque conditions, and the PCM detected a misfire condition in one cylinder. **Note: If the misfire is severe, the MIL will flash on/off on the 1st trip!** **Possible Causes:** • Base engine (mechanical) problem affecting only one cylinder • Fuel system problem affecting only one cylinder • Ignition system problem affecting one cylinder
DTC: P1305 **1T MISFIRE, MIL : Yes** **Years:** 2007, 2008 **Models:** RL **Engines:** All **Transmissions:** All	**Cylinder 5 Misfire Detected** DTC P0107, P0108, P0131, P0132, P0171, P0172, P1128, P0335, P0336, P0505, P1128, P1129, P1259, P1361, P1362, P1366, P1367 and P1519 not set, engine running under positive torque conditions, and the PCM detected a misfire condition in one cylinder. **Note: If the misfire is severe, the MIL will flash on/off on the 1st trip!** **Possible Causes:** • Base engine (mechanical) problem affecting only one cylinder • Fuel system problem affecting only one cylinder • Ignition system problem affecting one cylinder
DTC: P1306 **1T MISFIRE, MIL : Yes** **Years:** 2007, 2008 **Models:** RL **Engines:** All **Transmissions:** All	**Cylinder 6 Misfire Detected** DTC P0107, P0108, P0131, P0132, P0171, P0172, P1128, P0335, P0336, P0505, P1128, P1129, P1259, P1361, P1362, P1366, P1367 and P1519 not set, engine running under positive torque conditions, and the PCM detected a misfire condition in one cylinder. **Note: If the misfire is severe, the MIL will flash on/off on the 1st trip!** **Possible Causes:** • Base engine (mechanical) problem affecting only one cylinder • Fuel system problem affecting only one cylinder • Ignition system problem affecting one cylinder
DTC: P1316 **1T CCM, MIL : Yes** **Years:** 2007, 2008 **Models:** RL **Engines:** All **Transmissions:** All	**Spark Plug Detection Module Circuit Fault (Bank 2)** Engine started, and the PCM detected an unexpected condition on the Spark Plug Detection Module circuit for Cylinder Bank 2. **Possible Causes:** • Spark Plug Detection Module circuit open or shorted to ground • Spark Plug Detection Module circuit is shorted to system power • Spark Plug Detection Module is damaged or has failed • PCM has failed
DTC: P1317 **1T CCM, MIL : Yes** **Years:** 2007, 2008 **Models:** RL **Engines:** All **Transmissions:** All	**Spark Plug Detection Module Circuit Fault (Bank 1)** Engine started, and the PCM detected an unexpected condition on the Spark Plug Detection Module circuit for Cylinder Bank 1. **Possible Causes:** • Spark Plug Detection Module circuit open or shorted to ground • Spark Plug Detection Module circuit is shorted to system power • Spark Plug Detection Module is damaged or has failed • PCM has failed
DTC: P1318 **1T CCM, MIL : Yes** **Years:** 2007, 2008 **Models:** RL **Engines:** All **Transmissions:** All	**Spark Plug Detection Module Reset Circuit Fault (Bank 2)** Engine started, and the PCM detected an unexpected condition on the Spark Plug Detection Module Reset circuit for Cylinder Bank 2. **Possible Causes:** • Spark Plug Detection Module circuit open or shorted to ground • Spark Plug Detection Module circuit is shorted to system power • Spark Plug Detection Module is damaged or has failed • PCM has failed
DTC: P1319 **1T CCM, MIL : Yes** **Years:** 2007, 2008 **Models:** RL **Engines:** All **Transmissions:** All	**Spark Plug Detection Module Reset Circuit Fault (Bank 1)** Engine started, and the PCM detected an unexpected condition on the Spark Plug Detection Module Reset circuit for Cylinder Bank 1. **Possible Causes:** • Spark Plug Detection Module circuit open or shorted to ground • Spark Plug Detection Module circuit is shorted to system power • Spark Plug Detection Module is damaged or has failed • PCM has failed

DTC	Trouble Code Title, Conditions & Possible Causes
DTC: P1336 **1T CCM, MIL : Yes** **Years:** 2007, 2008 **Models:** RL **Engines:** All **Transmissions:** All	**Crankshaft Position Sensor 2 Circuit Malfunction** Engine started, and the PCM detected an unexpected (intermittent) interruption of the Crankshaft position (CKP) 'B' sensor signal. **Possible Causes:** • CSF2 signal circuit is open, short to ground or to system power • CSF2 is damaged or has failed • CSF2 pickup assembly or its pulse rotor is damaged • PCM has failed
DTC: P1337 **1T CCM, MIL : Yes** **Years:** 2007, 2008 **Models:** RL **Engines:** All **Transmissions:** All	**Crankshaft Position Sensor 2 Circuit No Signal** Engine cranking or running; and the PCM did not detect any signals from the Crankshaft Position (CKP) 2 sensor during the CCM test. **Possible Causes:** • CKP Sensor 2 signal circuit is open or shorted to ground • CKP Sensor 2 signal circuit is shorted to VREF or power • CKP Sensor 2 is damaged or has failed • CKP Sensor 2 pickup assembly or its pulse rotor is damaged • PCM has failed
DTC: P1361 **1T CCM, MIL : Yes** **Years:** 2007, 2008 **Models:** MDX, RDX, RL, TL, TSX **Engines:** All **Transmissions:** All	**Top Dead Center Sensor Circuit Intermittent Signal** Engine started, and the PCM detected an unexpected (intermittent) interruption of the Top Dead Center (TDC) sensor signal in the test. **Possible Causes:** • TDC signal circuit is open (intermittent fault) • TDC signal circuit is shorted to ground (intermittent fault) • TDC pickup assembly or its pulse rotor is damaged (intermittent fault) • TDC is damaged or has failed (intermittent fault) • PCM has failed
DTC: P1362 **1T CCM, MIL : Yes** **Years:** 2007, 2008 **Models:** MDX, RDX, RL, TL, TSX **Engines:** All **Transmissions:** All	**Top Dead Center 1 Sensor Circuit No Signal** Engine cranking or running; and the PCM did not receive any signals from the Top Dead Center 1(TDC1) sensor during the CCM test. **Note: The engine will start and run without the TDC sensor 1 signal.** **Possible Causes:** • TDC1 signal circuit is open or shorted to ground • TDC1 pickup assembly or its pulse rotor is damaged • TDC1 is damaged or has failed • PCM has failed
DTC: P1366 **1T CCM, MIL : Yes** **Years:** 2007, 2008 **Models:** MDX, RDX, RL, TL, TSX **Engines:** All **Transmissions:** All	**Top Dead Center Sensor 2 Circuit Intermittent Signal** Engine started, and the PCM detected an unexpected (intermittent) interruption of the Top Dead Center 2 (TDC2) sensor signal. **Possible Causes:** • TDC2 signal circuit is open (intermittent fault) • TDC2 signal circuit is shorted to ground (intermittent fault) • TDC2 pickup assembly or its pulse rotor is damaged (intermittent fault) • TDC2 is damaged or has failed (intermittent fault) • PCM has failed
DTC: P1367 **1T CCM, MIL : Yes** **Years:** 2007, 2008 **Models:** MDX, RDX, RL, TL, TSX **Engines:** All **Transmissions:** All	**Top Dead Center Sensor 2 No Signal** Engine cranking or running; and the PCM did not detect any signals from the Top Dead Center 2 (TDC2) sensor during the CCM test. **Note: The engine will start and run without the TDC sensor 2 signal.** **Possible Causes:** • TDC2 signal circuit is open or shorted to ground • TDC2 pickup assembly or its pulse rotor is damaged • TDC2 is damaged or has failed • PCM has failed
DTC: P1381 **1T CCM, MIL : Yes** **Years:** 2007, 2008 **Models:** RL **Engines:** All **Transmissions:** All	**Camshaft Position Sensor 1 No Signal** Engine cranking or running; and the PCM did not detect any signals from the Camshaft Position (CMP) sensor 1 during the CCM test. **Note: The engine will start and run without the CMP sensor 1 signal.** **Possible Causes:** • CMP signal circuit is open or shorted to ground • CMP pickup assembly or CMP sensor is damaged or has failed • PCM has failed

DTC	Trouble Code Title, Conditions & Possible Causes
DTC: P1382 **1T CCM, MIL : Yes** **Years:** 2007, 2008 **Models:** RL **Engines:** All **Transmissions:** All	**Camshaft Position Sensor 1 No Signal** Engine cranking or running; and the PCM did not detect any signals from the Camshaft Position 1 (CMP) sensor during the CCM test. **Note: The engine will start and run without the CMP sensor signal.** **Possible Causes:** • CMP1 signal circuit is open or shorted to ground • CMP1 pickup unit or the CMP sensor is damaged or has failed • PCM has failed
DTC: P1386 **1T CCM, MIL : Yes** **Years:** 2007, 2008 **Models:** RL **Engines:** All **Transmissions:** All	**Camshaft Position Sensor 2 Circuit Malfunction** Engine running and the PCM detected an unexpected or intermittent interruption of the Camshaft Position 2 (CMP) sensor signal. **Possible Causes:** • CMP2 signal circuit is open or shorted to ground • CMP2 signal circuit is shorted to VREF or system power • CMP2 pickup unit or the CMP sensor is damaged or has failed • PCM has failed
DTC: P1387 **1T CCM, MIL : Yes** **Years:** 2007, 2008 **Models:** RL **Engines:** All **Transmissions:** All	**Camshaft Position Sensor 2 No Signal** Engine cranking or running; and the PCM did not detect any signals from the Camshaft Position 2 (CMP) sensor during the CCM test. **Note: The engine will start and run without the CMP sensor signal.** **Possible Causes:** • CMP2 signal circuit is open or shorted to ground • CMP2 pickup unit is damaged or has failed • CMP2 sensor is damaged or has failed • PCM has failed
DTC: P1450 **1T CCM, MIL : Yes** **Years:** 2007, 2008 **Models:** MDX, RDX, RL, TL, TSX **Engines:** All **Transmissions:** All	**EVAP Two-Way Valve Bypass Valve Control Circuit Low Input** Key on or engine running; and the PCM detected an unexpected low voltage condition on the EVAP two-way bypass valve control circuit. **Possible Causes:** • 2-Way bypass valve control circuit is open • 2-Way bypass valve control power circuit is open (test the fuse) • 2-Way bypass control valve is damaged or has failed • PCM is damaged or has failed
DTC: P1451 **1T CCM, MIL : Yes** **Years:** 2007, 2008 **Models:** MDX, RDX, RL, TL, TSX **Engines:** All **Transmissions:** All	**EVAP Two-Way Valve Bypass Valve Control Circuit High Input** Key on or engine running; and the PCM detected an unexpected high voltage condition on the two-way bypass valve control circuit. **Possible Causes:** • 2-Way bypass valve control circuit is shorted to system power • 2-Way bypass control valve is damaged or has failed • PCM is damaged or has failed
DTC: P1454 **1T CCM, MIL : Yes** **Years:** 2007, 2008 **Models:** MDX, RDX, RL, TL, TSX **Engines:** All **Transmissions:** All	**Fuel Tank Pressure Sensor Circuit Range/Performance** Key on or engine running; and the PCM detected an unexpected voltage condition on the fuel tank pressure sensor signal circuit. **Possible Causes:** • FTP sensor circuit is open or shorted (intermittent fault) • FTP sensor air tube is clogged with debris or dirt • Fuel tank pressure sensor is damaged or has failed • PCM is damaged or has failed
DTC: P1456 **2T EVAP, MIL: Yes** **Years:** 2007, 2008 **Models:** MDX, RDX, RL, TL, TSX **Engines:** All **Transmissions:** All	**EVAP System Leak Detected (Fuel Tank Area)** IAT sensor signal from 32-86 degrees F at startup, vehicle driven to a speed over 5 mph for 2-3 minutes, ECT sensor signal over 154 degrees F, then with the EVAP Control and Vent solenoids "on", the PCM detected an incorrect fuel tank pressure value due to a leak in the fuel tank area. **Possible Causes:** • Fuel tank cap damaged, loose or the wrong part number • Fuel tank leaks at the fuel fill pipe or at the fuel tank seals • Fuel vapor control valve is damaged or has failed • Fuel tank vapor recirculation valve or vapor tube is damaged • Fuel tank vapor control vent tube is damaged or has failed • TSB 02-005 (2/18/02) contains a repair procedure for this code

DTC	Trouble Code Title, Conditions & Possible Causes
DTC: P1456 **2T EVAP, MIL: Yes** **Years:** 2007, 2008 **Models:** RL **Engines:** All **Transmissions:** All	**EVAP System Leak Detected (Fuel Tank Area)** IAT sensor signal from 32-86 degrees F at startup, vehicle driven at over 5 mph for over 2 minutes, ECT sensor signal over 154 degrees F, the with the EVAP Control and Vent solenoids "on", the PCM detected the fuel tank pressure was incorrect due to a leak in the fuel tank area. **Possible Causes:** • Fuel tank cap damaged, loose or the wrong part number • Fuel tank leaks at the fuel fill pipe or at the fuel tank seals • Fuel vapor control valve is damaged or has failed • Fuel tank vapor recirculation valve or vapor tube is damaged • Fuel tank vapor control vent tube is damaged or has failed
DTC: P1457 **2T EVAP, MIL: Yes** **Years:** 2007, 2008 **Models:** MDX, RDX, RL, TL, TSX **Engines:** All **Transmissions:** All	**EVAP System Leak Detected (Canister Area)** Cold startup completed (IAT sensor signal from 32-86 degrees F at engine startup), vehicle driven at over 5 mph for over 2 minutes, then with ECT sensor signal more than 154 degrees F and the EVAP Control and Vent solenoids enabled, and the PCM detected an invalid fuel tank pressure value due to a leak in the canister area in the Leak Test. **Possible Causes:** • EVAP canister is leaking, damaged or full of water • EVAP canister purge line is loose, damaged or blocked • EVAP two-way valve or ORVR vent shut valve is damaged • EVAP fuel tank vapor control valve is damaged or has failed • PCM has failed • TSB 03-001 (1/27/03) contains a repair procedure for this code
DTC: P1457 **2T EVAP, MIL: Yes** **Years:** 2007, 2008 **Models:** RL **Engines:** All **Transmissions:** All	**EVAP System Leak Detected (Canister Area)** Cold startup completed (IAT sensor signal from 32-86 degrees F at engine startup), vehicle driven at over 5 mph for over 2 minutes, then with ECT sensor signal more than 154 degrees F and the EVAP Control and Vent solenoids enabled, the PCM detected the fuel tank pressure was incorrect due to a leak in the canister area during the Leak Test. **Possible Causes:** • EVAP canister is leaking, damaged or full of water • EVAP canister purge line is loose, damaged or blocked • EVAP two-way valve or ORVR vent shut valve is damaged • EVAP fuel tank vapor control valve is damaged or has failed • PCM has failed • TSB 03-001 (1/27/03) contains a repair procedure for this code
DTC: P1460 **1T CCM, MIL : Yes** **Years:** 2007, 2008 **Models:** MDX, RDX, RL, TL, TSX **Engines:** All **Transmissions:** All	**Fuel Level Sensor Power Supply Circuit Malfunction** Key on or engine running; and the PCM detected an unexpected voltage condition on the fuel level sensor power supply circuit. **Possible Causes:** • Fuel level sensor power circuit is open or shorted to ground • Fuel level sensor circuit to fuel gauge assembly is damaged • Fuel level sensor is damaged or has failed • PCM is damaged or has failed
DTC: P1486 **2T ECT, MIL: Yes** **Years:** 2007, 2008 **Models:** MDX, RDX, RL, TL, TSX **Engines:** All **Transmissions:** All	**Cooling System Malfunction** DTC P0107, P0108, P0112, P0113, P0116, P0117, P0118, P0300, P0301, P0302, P0303, P0304, P0305, P0306, P0335, P0336, P0401, P0500, P0505, P1106, P1107, P1108, P1128, P1129, P1253, P1257, P1258, P1259, P1359, P1399, P1491, P1498 and P1519 not set, vehicle driven for over 10 minutes, and the PCM detected the ECT signal did not reach the correct closed loop value. **Note: This trouble code can set if the engine remains under hot idle conditions with the hood open for an extended period of time.** **Possible Causes:** • ECT sensor is out of calibration • Check for low coolant level or incorrect coolant mixture • Cooling system component failure (thermostat stuck open) • TSB 01-016 (9/10/01) contains a repair procedure for this code
DTC: P1486 **2T ECT, MIL: Yes** **Years:** 2007, 2008 **Models:** RL **Engines:** All **Transmissions:** All	**Cooling System Malfunction** DTC P0107, P0108, P0112, P0113, P0116, P0117, P0118, P0300, P0301, P0302, P0303, P0304, P0305, P0306, P0335, P0336, P0401, P0500, P0505, P1106, P1107, P1108, P1128, P1129, P1253, P1257, P1258, P1259, P1359, P1399, P1491, P1498 and P1519 not set, vehicle driven for over 10 minutes, and the PCM detected the ECT signal did not reach the correct closed loop value. **Possible Causes:** • ECT sensor is out of calibration • Check for low coolant level or incorrect coolant mixture • Cooling system component failure (thermostat stuck open)

DTC	Trouble Code Title, Conditions & Possible Causes
DTC: P1491 **2T EGR, MIL: Yes** **Years:** 2007, 2008 **Models:** MDX, RDX, RL, TL, TSX **Engines:** All **Transmissions:** All	**EGR Valve Lift Sensor Insufficient Flow Detected** Engine started, vehicle driven while in closed loop at 1700-2500 rpm for at least 10 minutes, and the PCM detected the EGR valve lift sensor (EGRV) signal indicated insufficient EGR flow during the EGR Monitor test. **Possible Causes:** • EGR valve lift sensor is stuck, damaged or has failed • EGR control solenoid circuit is open or shorted to ground • EGR control solenoid valve is damaged or has failed • PCM has failed
DTC: P1498 **1T CCM, MIL : Yes** **Years:** 2007, 2008 **Models:** MDX, RDX, RL, TL, TSX **Engines:** All **Transmissions:** All	**EGR Valve Lift Sensor Circuit High Input** Key on or engine running; and the PCM detected the EGR Valve Lift sensor signal was more than an allowable range stored in memory. **Possible Causes:** • EGR valve lift sensor circuit is open or shorted to power • EGR valve lift sensor is shorted to VREF or system power (B+) • EGR valve lift sensor is stuck, damaged or has failed • PCM has failed
DTC: P1519 **1T CCM, MIL : Yes** **Years:** 2007, 2008 **Models:** MDX, RDX, RL, TL, TSX **Engines:** All **Transmissions:** All	**Idle Air Control Valve Circuit Malfunction** Key on or engine running; and the PCM detected an unexpected voltage condition on the Idle Air Control (IAC) valve control circuit. **Possible Causes:** • IAC valve circuit is open, shorted to ground or to system power • IAC valve power circuit is open or shorted to ground • IAC valve is damaged or has failed • PCM is damaged • TSB 03-006 (3/17/03) contains a repair procedure for this code
DTC: P1519 **1T CCM, MIL : Yes** **Years:** 2007, 2008 **Models:** TL **Engines:** All **Transmissions:** All	**Idle Air Control Valve Circuit Malfunction** Key on or engine running; and the PCM detected an unexpected voltage condition on the Idle Air Control (IAC) valve control circuit. **Possible Causes:** • IAC valve circuit is open, shorted to ground or to system power • IAC valve power circuit is open or shorted to ground • IAC valve is damaged or has failed • PCM is damaged • TSB 03-006 (3/17/03) contains a repair procedure for this code
DTC: P1607 **1T CCM, MIL : Yes** **Years:** 2007, 2008 **Models:** MDX, RDX, RL, TL, TSX **Engines:** All **Transmissions:** All	**PCM Internal Circuit Malfunction** Key on, and the PCM detected a fault in one of its internal circuits. **Note: This trouble code indicates an internal failure in the PCM. The OEM repair procedure recommends replacing the original PCM with a "known good" PCM and then verify the code does not reset.** **Possible Causes:** • PCM is damaged or has failed
DTC: P1656 **1T CCM, MIL : Yes** **Years:** 2007, 2008 **Models:** MDX, RDX, RL, TL, TSX **Engines:** All **Transmissions:** All	**PCM To VSA Unit Signal Line Circuit Malfunction** Engine started, and the PCM detected an unexpected voltage condition on the PCM to VSA circuit during the CCM test. **Possible Causes:** • SEFA or SEAF signal line is open or shorted to ground • SEFA or SEAF signal line is shorted to system power • TCM or PCM has failed
DTC: P1676 **1T CCM, MIL : Yes** **Years:** 2007, 2008 **Models:** TL **Engines:** All **Transmissions:** All	**A/T (TCM) FPTDR Signal Line Circuit Malfunction** Key on or engine running; and the PCM detected an unexpected voltage condition on the TCM FPTDR Line circuit during the test. **Possible Causes:** • A/T FTPDR data line is open or shorted to ground • A/T FTPDR data line is shorted to VREF or system power • TCM or PCM has failed
DTC: P1678 **1T CCM, MIL : Yes** **Years:** 2007, 2008 **Models:** TL **Engines:** All **Transmissions:** All	**A/T (TCM) FPTDR Signal Line Circuit Malfunction** Key on or engine running; and the PCM detected an unexpected (intermittent) interruption of the TCM FPTDR Line signal in the test. **Possible Causes:** • A/T FTPDR data line is open (intermittent fault) • A/T FTPDR data line is shorted to ground (intermittent fault) • A/T FTPDR data line is shorted to VREF or power (intermittent) • TCM or PCM has failed

DTC	Trouble Code Title, Conditions & Possible Causes

DTC: P1683
1T CCM, MIL : Yes
Years: 2007, 2008
Models: MDX, RDX, RL, TL, TSX
Engines: All
Transmissions: All
	Throttle Valve Default Position Spring Range/Performance
	Key on or engine running; and the PCM detected an unexpected voltage condition on the throttle valve position during the CCM test. Do not check the throttle valve position spring action with the key on.
	Possible Causes:
	   • Check for a binding throttle valve position spring
	   • Check for dirt and sludge on the throttle body causing the throttle valve default position to indicate an incorrect value

DTC: P1684
1T CCM, MIL : Yes
Years: 2007, 2008
Models: MDX, RDX, RL, TL, TSX
Engines: All
Transmissions: All
	Throttle Valve Default Position Spring Range/Performance
	Key on or engine running; and the PCM detected an unexpected voltage condition on the throttle valve position during the CCM test. Do not check the throttle valve position spring action with the key on.
	Possible Causes:
	   • Check for a binding throttle valve position spring
	   • Check for dirt and sludge on the throttle body causing the throttle valve default position to indicate an incorrect value

DTC: P1696
1T CCM, MIL : Yes
Years: 2007, 2008
Models: RL
Engines: All
Transmissions: A/T
	TCFC Data Line Signal Low Input
	Key on or engine running; and the PCM detected an unexpected "low" input on the TCFC Data Line circuit during the CCM test.
	Note: This code applies to models with a Traction Control System.
	Possible Causes:
	   • A/T TCFC data line is shorted to sensor ground
	   • A/T TCFC data line is shorted to chassis ground
	   • TCM or PCM has failed

DTC: P1697
1T CCM, MIL : Yes
Years: 2007, 2008
Models: RL
Engines: All
Transmissions: A/T
	TUFT Data Line Signal High Input
	Key on or engine running; and the PCM detected an unexpected "high" input on the TUFT Data Line circuit.
	Note: This code applies to models with a Traction Control System.
	Possible Causes:
	   • A/T TUFT data line is shorted to VREF (voltage reference)
	   • A/T TUFT data line is shorted to system power (B+)
	   • TCM or PCM has failed

DTC: P1705
1T CCM, MIL : Yes
Years: 2007, 2008
Models: RL
Engines: All
Transmissions: A/T
	A/T Transmission Range Switch Low Input
	Engine started, vehicle driven and the PCM detected two gear position switch inputs simultaneously (two input at the same time).
	Note: The D4 lamp on the dash will flash 5 times if this code is set.
	Possible Causes:
	   • A/T gear position switch signal circuit is shorted to ground
	   • A/T gear position switch signal circuit is shorted to another wire
	   • A/T gear position switch is damaged or has failed
	   • TCM or PCM has failed

DTC: P1705
1T CCM, MIL : Yes
Years: 2007, 2008
Models: TL
Engines: All
Transmissions: A/T
	A/T Transmission Range Switch Low Input
	Engine started, vehicle driven and the PCM detected two gear position switch inputs simultaneously (two input at the same time).
	Note: The D4 or D5 lamp on the dash will flash when this code sets.
	Possible Causes:
	   • A/T gear position switch signal circuit is shorted to ground
	   • A/T gear position switch signal circuit is shorted to another wire
	   • A/T gear position switch is damaged or has failed
	   • TCM or PCM has failed

DTC: P1706
1T CCM, MIL : Yes
Years: 2007, 2008
Models: MDX, RDX, RL, TL, TSX
Engines: All
Transmissions: All
	A/T Transmission Range Switch Circuit No Gear Inputs
	Engine started, vehicle driven and the PCM detected an unexpected high voltage condition on the Gear Position Switch circuit.
	Note: The D4 or D5 indicator on the dash will flash 6 times if this code is set.
	Possible Causes:
	   • A/T gear position switch signal circuit is shorted to ground
	   • A/T gear position switch signal circuit is shorted to another wire
	   • A/T gear position switch is damaged or has failed
	   • TCM or PCM has failed

DTC: P1709
1T CCM, MIL : Yes
Years: 2007, 2008
Models: MDX, RDX, RL, TL, TSX
Engines: All
Transmissions: All
	A/T Transmission Gear Switch Circuit Malfunction
	Key on or engine running; and the PCM detected more than one signal on the Position Switch circuit at the same time during the test.
	Possible Causes:
	   • A/T gear position switch signal circuit is shorted to another wire inside the switch or in the wiring harness (i.e., the P/N signal is present along with a Drive, 1st, 2nd, 3rd, 4th or Reverse signal)
	   • A/T gear position switch is damaged or has failed
	   • TCM or PCM has failed

DTC	Trouble Code Title, Conditions & Possible Causes
DTC: P1710 **1T CCM, MIL: No** **Years:** 2007, 2008 **Models:** MDX, RDX, RL, TL, TSX **Engines:** All **Transmissions:** All	**A/T First Hold Switch Circuit Malfunction** Key on or engine running; and the PCM detected an unexpected voltage condition on the First Hold switch circuit during the test. **Possible Causes:** • First hold witch signal circuit is open • First hold switch is shorted to ground • First hold switch ground circuit is open • First hold switch is shorted to VREF or system power • First hold switch is damaged or has failed • TCM or PCM has failed
DTC: P1717 **1T CCM, MIL : Yes** **Years:** 2007, 2008 **Models:** MDX, RDX, RL, TL, TSX **Engines:** All **Transmissions:** All	**A/T Transmission Range Switch Circuit Malfunction** P1705 and P1706 not set, engine started, vehicle driven, and the PCM detected an unexpected voltage condition on the Transmission Range Switch ATP RVS circuit during the CCM test. **Possible Causes:** • Transmission range switch circuit is open or shorted to ground • Transmission range switch circuit is shorted to VREF or power • Transmission range switch connector is disconnected • TCM or PCM has failed
DTC: P1739 **1T CCM, MIL: No** **Years:** 2007, 2008 **Models:** MDX, RDX, RL, TL, TSX **Engines:** All **Transmissions:** All	**A/T 3rd Pressure Switch Circuit Malfunction** Engine running in gear, and the PCM detected an unexpected voltage condition on the 3rd pressure switch circuit during the test. **Possible Causes:** • A/T 3rd pressure switch signal circuit is open • A/T 3rd pressure switch signal circuit is shorted to ground • A/T 3rd pressure switch signal circuit shorted to system power • A/T 3rd pressure switch is damaged or has failed • TCM or PCM has failed
DTC: P1740 **1T CCM, MIL: No** **Years:** 2007, 2008 **Models:** MDX, RDX, RL, TL, TSX **Engines:** All **Transmissions:** All	**A/T 4th Pressure Switch Circuit Malfunction** Engine running in gear, and the PCM detected an unexpected voltage condition on the 4th pressure switch circuit during the test. **Possible Causes:** • A/T 4th pressure switch signal circuit is open • A/T 4th pressure switch signal circuit is shorted to ground • A/T 4th pressure switch signal circuit shorted to system power • A/T 4th pressure switch is damaged or has failed • TCM or PCM has failed
DTC: P1750 **1T CCM, MIL : Yes** **Years:** 2007, 2008 **Models:** MDX, RDX, RL, TL, TSX **Engines:** All **Transmissions:** All	**A/T Hydraulic Pressure Control Malfunction (Mechanical)** Engine started, engine running in gear, and the PCM detected a mechanical fault existed in the hydraulic portion of the A/T (as it relates to Shift Solenoids 'A' and 'B') during the CCM test. **Note: The D4 lamp on the dash will flash when this code sets.** **Possible Causes:** • Shift Solenoid 'A' is damaged or has failed • Shift Solenoid 'B' is damaged or has failed • The transmission is damaged or has failed
DTC: P1751 **1T CCM, MIL : Yes** **Years:** 2007, 2008 **Models:** MDX, RDX, RL, TL, TSX **Engines:** All **Transmissions:** All	**A/T Hydraulic Pressure Control Malfunction (Mechanical)** Engine started, engine running in gear, and the PCM detected a mechanical fault existed in the hydraulic portion of the A/T (as it relates to Clutch Pressure Control Solenoids 'A' and 'B') during the CCM test. **Note: The D4 lamp on the dash will flash once when this code is set.** **Possible Causes:** • Clutch Pressure Control Solenoid 'A' is damaged or has failed • Clutch Pressure Control Solenoid 'B' is damaged or has failed • Shift Solenoid 'B' is damaged or has failed • The transmission is damaged or has failed
DTC: P1753 **1T CCM, MIL : Yes** **Years:** 2007, 2008 **Models:** RL **Engines:** All **Transmissions:** A/T	**A/T Lockup Solenoid Valve 'A' Circuit Malfunction** Vehicle driven in 1st, 2nd, 3rd and 4th gears, and the PCM detected an unexpected voltage condition on the Solenoid 'A' control circuit. **Note: The D4 lamp on the dash will flash once when this code is set.** **Possible Causes:** • Lockup solenoid 'A' control circuit is open or shorted to ground • Lockup solenoid 'A' circuit is shorted to system power • Lockup solenoid 'A' connector is disconnected or damaged • Lockup solenoid 'A' is damaged or has failed • TCM or PCM has failed

DTC	Trouble Code Title, Conditions & Possible Causes
DTC: P1758 **1T CCM, MIL : Yes** **Years:** 2007, 2008 **Models:** MDX, RDX, RL, TL, TSX **Engines:** All **Transmissions:** All	**A/T Lockup Solenoid Valve 'B' Circuit Malfunction** Vehicle driven in 1st, 2nd, 3rd and 4th gears, and the PCM detected an unexpected voltage condition on the Solenoid 'B' circuit during the CCM test. **Note: The D4 lamp on the dash will flash twice when this code is set.** **Possible Causes:** • Lockup solenoid 'B' control circuit is open or shorted to ground • Lockup solenoid 'B' circuit is shorted to system power • Lockup solenoid 'B' connector is disconnected or damaged • Lockup solenoid 'B' is damaged or has failed • TCM or PCM has failed
DTC: P1768 **1T CCM, MIL : Yes** **Years:** 2007, 2008 **Models:** MDX, RDX, RL, TL, TSX **Engines:** All **Transmissions:** All	**Automatic Transaxle Linear Solenoid Circuit Malfunction** Key on or engine running; and the PCM detected an unexpected voltage condition on the A/T Linear Solenoid circuit to the TCM. **Note: The D4 lamp on the dash will flash 16 times if this code is set.** **Possible Causes:** • A/T linear solenoid control circuit to the TCM is open • A/T linear solenoid circuit to the TCM is shorted to ground • A/T linear solenoid connector is disconnected or damaged • A/T linear solenoid circuit to the TCM is shorted to VREF • TCM or PCM has failed
DTC: P1773 **1T CCM, MIL : Yes** **Years:** 2007, 2008 **Models:** TL **Engines:** All **Transmissions:** All	**A/T Lockup Solenoid Valve 'B' Circuit Malfunction** Vehicle driven in 1st, 2nd, 3rd and 4th gears, and the PCM detected an unexpected voltage condition on the Solenoid 'B' control circuit. **Note: The D4 lamp on the dash will blink when this code is set.** **Possible Causes:** • Lockup solenoid 'B' control circuit is open or shorted to ground • Lockup solenoid 'B' circuit is shorted to system power • Lockup solenoid 'B' is damaged or has failed • TCM or PCM has failed
DTC: P1778 **1T CCM, MIL : Yes** **Years:** 2007, 2008 **Models:** MDX, RDX, RL, TL, TSX **Engines:** All **Transmissions:** All	**A/T Lockup Solenoid Valve 'C' Circuit Malfunction** Vehicle driven in 1st, 2nd, 3rd and 4th gears, and the PCM detected an unexpected voltage condition on the Solenoid 'C' control circuit. **Note: The D4 lamp on the dash will blink when this code is set.** **Possible Causes:** • Lockup solenoid 'C' control circuit is open or shorted to ground • Lockup solenoid 'C' circuit is shorted to system power • Lockup solenoid 'C' is damaged or has failed • TCM or PCM has failed
DTC: P1791 **1T CCM, MIL : Yes** **Years:** 2007, 2008 **Models:** RL, TL **Engines:** All **Transmissions:** A/T	**A/T Vehicle Speed Sensor Circuit Malfunction** Vehicle driven for 15 seconds, and the PCM detected an unexpected voltage condition on the vehicle speed signal circuit to the TCM. **Note: The lockup clutch does not engage when this code is set.** **Note: The D4 lamp on the dash will flash 4 times if this code is set.** **Possible Causes:** • VSS signal circuit to the TCM is open • VSS signal circuit is shorted to ground • VSS signal circuit to the TCM is shorted to VREF • VSS connector is disconnected or damaged • TCM or PCM has failed

OBD II Trouble Code List (P2xxx Codes)

DTC	Trouble Code Title, Conditions & Possible Causes
DTC: P2101 **1T CCM, MIL: Yes** **Years:** 2007, 2008 **Models:** MDX, RDX, RL, TL, TSX **Engines:** All **Transmissions:** All	**Throttle Actuator System Malfunction** Engine started, engine running (at the speed indicated in the trouble code Freeze Frame data), and the PCM detected a signal from the TAC module indicating a problem present in the TAC system. **Possible Causes:** • Check for loose connections at TAC module (intermittent fault) • Key off - check the throttle blade for free movement (be careful not to pinch your finger while performing this inspection step) Clean the throttle body as needed and recheck for the code • Update the PCM to the latest software. If the problem is still present, substitute a known good PCM and then retest. If the problem goes away, the original PCM has failed.

DTC	Trouble Code Title, Conditions & Possible Causes
DTC: P2108 **1T CCM, MIL: Yes** **Years:** 2007, 2008 **Models:** MDX, RDX, RL, TL, TSX **Engines:** All **Transmissions:** All	**Throttle Actuator Control Module Malfunction** Key on or engine running; and the PCM detected a signal from the TAC module indicating a problem present in the TAC module. **Possible Causes:** • Check for loose connections at TAC module (intermittent fault) • Check for loose connections at the PCM (intermittent fault) • Substitute a known good throttle body assembly and retest. If the problem goes away, the original throttle body has failed.
DTC: P2118 **1T CCM, MIL: Yes** **Years:** 2007, 2008 **Models:** MDX, RDX, RL, TL, TSX **Engines:** All **Transmissions:** All	**Throttle Actuator Motor Current Range/Performance** Engine started, vehicle driven with several changes in the throttle position angle, and the PCM detected a signal from the TAC module indicating a TAC motor current problem had occurred. **Possible Causes:** • Check for loose connections at TAC module (intermittent fault) • Substitute a known good throttle body assembly and retest. If the problem goes away, the original throttle body has failed. • Update the PCM to the latest software. If the problem is still present, substitute a known good PCM and then retest. If the problem goes away, the original PCM has failed.
DTC: P2122 **1T CCM, MIL: Yes** **Years:** 2007, 2008 **Models:** MDX, RDX, RL, TL, TSX **Engines:** All **Transmissions:** All	**Accelerator Pedal Position Sensor 1 Circuit Low Input** Key on or engine running; and the PCM detected an unexpected low voltage condition (less than 0.1v) on the APP1 sensor signal circuit. If P0641 (low VREF) is set along with P2127, repair the P0641 first. **Possible Causes:** • Check for loose connections at APP1 sensor (intermittent fault) • APP1 Sensor VREF circuit is open • APP1 Sensor signal circuit is shorted to ground • APP1 Sensor is damaged or has failed. • Update the PCM to the latest software. If the problem is still present, substitute a known good PCM and then retest. If the problem goes away, the original PCM has failed.
DTC: P2123 **1T CCM, MIL: Yes** **Years:** 2007, 2008 **Models:** MDX, RDX, RL, TL, TSX **Engines:** All **Transmissions:** All	**Accelerator Pedal Position Sensor 1 Circuit High Input** Key on or engine running; and the PCM detected an unexpected high voltage condition (more than 4.85v) on the APP1 sensor signal circuit. **Possible Causes:** • Check for loose connections at APP1 sensor (intermittent fault) • APP1 Sensor signal circuit is shorted to the VREF circuit • APP1 Sensor signal circuit is open between sensor and PCM • APP1 Sensor is damaged or has failed • Update the PCM to the latest software. If the problem is still present, substitute a known good PCM and then retest. If the problem goes away, the original PCM has failed.
DTC: P2127 **1T CCM, MIL: Yes** **Years:** 2007, 2008 **Models:** MDX, RDX, RL, TL, TSX **Engines:** All **Transmissions:** All	**Accelerator Pedal Position Sensor 2 Circuit Low Input** Key on or engine running; and the PCM detected an unexpected low voltage condition (less than 0.1v) on the APP2 sensor signal circuit. If P0641 (low VREF) is set along with P2127, repair the P0641 first. **Possible Causes:** • Check for loose connections at APP2 sensor (intermittent fault) • APP2 Sensor VREF circuit is open • APP2 Sensor signal circuit is shorted to ground • APP2 Sensor is damaged or has failed. • Update the PCM to the latest software. If the problem is still present, substitute a known good PCM and then retest. If the problem goes away, the original PCM has failed.
DTC: P2128 **1T CCM, MIL: Yes** **Years:** 2007, 2008 **Models:** MDX, RDX, RL, TL, TSX **Engines:** All **Transmissions:** All	**Accelerator Pedal Position Sensor 2 Circuit High Input** Key on or engine running; and the PCM detected an unexpected high voltage condition (more than 4.80v) on the APP2 sensor signal circuit. **Possible Causes:** • Check for loose connections at APP2 sensor (intermittent fault) • APP2 Sensor signal circuit is shorted to the VREF circuit • APP2 Sensor signal circuit is open between sensor and PCM • APP2 Sensor is damaged or has failed • Update the PCM to the latest software. If the problem is still present, substitute a known good PCM and then retest. If the problem goes away, the original PCM has failed.
DTC: P2135 **1T CCM, MIL: Yes** **Years:** 2007, 2008 **Models:** MDX, RDX, RL, TL, TSX **Engines:** All **Transmissions:** All	**Throttle Position Sensor 1/2 Incorrect Voltage Correlation** Key on or engine running; and the PCM detected an incorrect voltage correlation on the TP1/2 sensor circuit during the CCM test. **Possible Causes:** • Check for loose connections at TAC module (intermittent fault) • Clear the codes, and then use the Scan Tool ETCS test to check for any related trouble codes. If any codes are present, repair these trouble codes and then retest for the original code.

DTC	Trouble Code Title, Conditions & Possible Causes
DTC: P2138 **1T CCM, MIL: Yes** **Years:** 2007, 2008 **Models:** MDX, RDX, RL, TL, TSX **Engines:** All **Transmissions:** All	**Accelerator Pedal Position Sensor 1/2 Incorrect Voltage Correlation** Key on or engine running; and the PCM detected an incorrect voltage correlation on the APP1/2 sensor circuit in the CCM test. **Possible Causes:** • Check for loose connections at APP sensor (intermittent fault) • Clear the codes, and then use the Scan Tool ETCS test to check for any related trouble codes. If any codes are present, repair these trouble codes and then retest for the original code. • Update the PCM to the latest software. If the problem is still present, substitute a known good PCM and then retest. If the problem goes away, the original PCM has failed.
DTC: P2176 **1T CCM, MIL: Yes** **Years:** 2007, 2008 **Models:** MDX, RDX, RL, TL, TSX **Engines:** All **Transmissions:** All	**Throttle Actuator Control System Idle Position Not Learned** Key on or engine running; and the PCM detected an incorrect voltage correlation on the APP1/2 sensor circuit in the CCM test. **Possible Causes:** • Check for loose connections at TAC module (intermittent fault) • Clear the codes, and then use the Scan Tool ETCS test to check the throttle valve movement with the key on, engine off and the intake air duct removed (keep your fingers clear during the inspection step). If the throttle valve does not operate to its fully closed position, substitute a known good throttle assembly and retest for the same code. If okay, clean the throttle body. • Update the PCM to the latest software. If the problem is still present, substitute a known good PCM and then retest. If the problem goes away, the original PCM has failed.
DTC: P2195 **1T CCM, MIL: Yes** **Years:** 2007, 2008 **Models:** MDX, RDX, RL, TL, TSX **Engines:** All **Transmissions:** All	**Air Fuel Sensor 1 (Bank 1 Sensor 1) Signal Stuck Lean** Engine started, vehicle driven for several minutes at various cruise speeds, and the PCM detected the AFS1 signal was stuck "lean". **Possible Causes:** • Check for a loose AFS1 assembly (not tightened) at the exhaust pipe connection. Clear the codes and retest the code. • If DTC P2195 is present, replace the AFS1 and then retest. • If this code is an intermittent problem, check for loose connections at the AFS1 connector.
DTC: P2197 **1T CCM, MIL: Yes** **Years:** 2007, 2008 **Models:** MDX, RDX, RL, TL, TSX **Engines:** All **Transmissions:** All	**Air Fuel Sensor 2 (Bank 2 Sensor 1) Signal Stuck Lean** Engine started, vehicle driven for several minutes at various cruise speeds, and the PCM detected the AFS2 signal was stuck "lean". **Possible Causes:** • Check for a loose AFS2 assembly (not tightened) at the exhaust pipe connection. Clear the codes and retest the code. • If DTC P2197 is present, replace the AFS1 and then retest. • If this code is an intermittent problem, check for loose connections at the AFS2 connector.
DTC: P2199 **2T CCM, MIL: Yes** **Years:** 2007, 2008 **Models:** MDX, RDX, RL, TL, TSX **Engines:** All **Transmissions:** All	**Intake Air Temperature Sensor 1/2 Voltage Correlation** Engine started cold (both IAT sensors less than 77 degrees F), and the PCM detected the voltage correlation difference between the IAT Sensor 1 and IAT Sensor 2 was more than 45 degrees F during the CCM test period. **Possible Causes:** • Check for loose connections at the IAT 1/2 (intermittent fault) • Let the vehicle sit for at least 8 hours. If the code resets, replace the IAT sensor that has the largest difference from the room temperature (after testing at KOEO with a cold engine).
DTC: P2227 **1T CCM, MIL: Yes** **Years:** 2007, 2008 **Models:** MDX, RDX, RL, TL, TSX **Engines:** All **Transmissions:** All	**BARO Sensor Circuit Range/Performance** DTC P0107, P0108, P1128 and P1129 not set, ECT sensor input over 158 degrees F, vehicle driven in Drive at a throttle angle of 12-20 degrees, and the PCM detected a problem in the BARO sensor. **Possible Causes:** • Check for loose connections at the PCM (intermittent fault) • Update the PCM to the latest software. If the problem is still present, substitute a known good PCM and then retest. If the problem goes away, the original PCM has failed.
DTC: P2228 **1T CCM, MIL: Yes** **Years:** 2007, 2008 **Models:** MDX, RDX, RL, TL, TSX **Engines:** All **Transmissions:** All	**BARO Sensor Circuit Low Input** Key on or engine running; and the PCM detected an unexpected low voltage (less than 1.58v [53 kPa]) on the BARO sensor signal circuit. **Possible Causes:** • Check for loose connections at the PCM (intermittent fault) • Update the PCM to the latest software. If the problem is still present, substitute a known good PCM and then retest. If the problem goes away, the original PCM has failed.
DTC: P2229 **1T CCM, MIL: Yes** **Years:** 2007, 2008 **Models:** MDX, RDX, RL, TL, TSX **Engines:** All **Transmissions:** All	**BARO Sensor Circuit High Input** Key on or engine running; and the PCM detected an unexpected high voltage (more than 4.50v [160 kPa]) on the BARO sensor signal circuit. **Possible Causes:** • Check for loose connections at the PCM (intermittent fault) • Update the PCM to the latest software. If the problem is still present, substitute a known good PCM and then retest. If the problem goes away, the original PCM has failed.

DTC	Trouble Code Title, Conditions & Possible Causes
DTC: P2237 **1T CCM, MIL: Yes** **Years:** 2007, 2008 **Models:** MDX, RDX, RL, TL, TSX **Engines:** All **Transmissions:** All	**Air Fuel Sensor 1 (Bank 1 Sensor 1) I/P Line High Voltage** DTC P0134 and P2237 not set, engine started and engine running, and the PCM detected an unexpected high voltage condition on the AFS1 I/P circuit during the CCM test. **Possible Causes:** • Check for loose connections at the AFS1 unit (intermittent fault) • AFS1 ground circuit is open • AFS1 assembly is damaged or has failed • Update the PCM to the latest software. If the problem is still present, substitute a known good PCM and then retest. If the problem goes away, the original PCM has failed.
DTC: P2238 **1T CCM, MIL: Yes** **Years:** 2007, 2008 **Models:** MDX, RDX, RL, TL, TSX **Engines:** All **Transmissions:** All	**Air Fuel Sensor 1 (Bank 1 Sensor 1) I/P Line Low Voltage** DTC P0134 and P2237 not set, engine started and engine running, and the PCM detected an unexpected low voltage condition on the AFS1 I/P circuit during the CCM test. **Possible Causes:** • Check for loose connections at the AFS1 unit (intermittent fault) • AFS1 signal circuit is shorted to ground • AFS1 assembly is damaged or has failed • Update the PCM to the latest software. If the problem is still present, substitute a known good PCM and then retest. If the problem goes away, the original PCM has failed.
DTC: P2240 **1T CCM, MIL: Yes** **Years:** 2007, 2008 **Models:** MDX, RDX, RL, TL, TSX **Engines:** All **Transmissions:** All	**Air Fuel Sensor 2 (Bank 2 Sensor 1) I/P Line High Voltage** DTC P0154 and P2240 not set, engine started and engine running, and the PCM detected an unexpected high voltage condition on the AFS2 I/P circuit during the CCM test. **Possible Causes:** • Check for loose connections at the AFS2 unit (intermittent fault) • AFS2 ground circuit is open • AFS2 assembly is damaged or has failed • Update the PCM to the latest software. If the problem is still present, substitute a known good PCM and then retest. If the problem goes away, the original PCM has failed.
DTC: P2241 **1T CCM, MIL: Yes** **Years:** 2007, 2008 **Models:** MDX, RDX, RL, TL, TSX **Engines:** All **Transmissions:** All	**Air Fuel Sensor 2 (Bank 2 Sensor 1) I/P Line Low Voltage** DTC P0154 and P2240 not set, engine started and engine running, and the PCM detected an unexpected high voltage condition on the AFS2 I/P circuit during the CCM test. **Possible Causes:** • Check for loose connections at the AFS2 unit (intermittent fault) • AFS2 ground circuit is open • AFS2 assembly is damaged or has failed • Update the PCM to the latest software. If the problem is still present, substitute a known good PCM and then retest. If the problem goes away, the original PCM has failed.
DTC: P2243 **1T CCM, MIL: Yes** **Years:** 2007, 2008 **Models:** MDX, RDX, RL, TL, TSX **Engines:** All **Transmissions:** All	**Air Fuel Sensor 1 (Bank 1 Sensor 1) VCENT Line High Voltage** DTC P2243 not set, engine started and engine running, and the PCM detected an unexpected high voltage condition on the AFS1 VCENT circuit during the CCM test. **Possible Causes:** • Check for loose connections at the AFS1 unit (intermittent fault) • AFS1 VCENT ground circuit is open (Circuit A54) • AFS1 assembly is damaged or has failed • Update the PCM to the latest software. If the problem is still present, substitute a known good PCM and then retest. If the problem goes away, the original PCM has failed.
DTC: P2245 **1T CCM, MIL: Yes** **Years:** 2007, 2008 **Models:** MDX, RDX, RL, TL, TSX **Engines:** All **Transmissions:** All	**Air Fuel Sensor 1 (Bank 1 Sensor 1) VCENT Line Low Voltage** Engine started, and the PCM detected an unexpected low voltage condition on the AFS1 VCENT circuit during the CCM test. **Possible Causes:** • Check for loose connections at the AFS1 unit (intermittent fault) • AFS1 VCENT signal circuit is shorted to ground • AFS1 assembly is damaged or has failed • Update the PCM to the latest software. If the problem is still present, substitute a known good PCM and then retest. If the problem goes away, the original PCM has failed.

DTC	Trouble Code Title, Conditions & Possible Causes
DTC: P2247 **1T CCM, MIL: Yes** **Years:** 2007, 2008 **Models:** MDX, RDX, RL, TL, TSX **Engines:** All **Transmissions:** All	**Air Fuel Sensor 2 (Bank 2 Sensor 1) VCENT Line High Voltage** DTC P2247 not set, engine started and engine running, and the PCM detected an unexpected high voltage condition on the AFS2 VCENT circuit during the CCM test. **Possible Causes:** • Check for loose connections at the AFS2 unit (intermittent fault) • AFS1 VCENT ground circuit is open (Circuit A50) • AFS1 assembly is damaged or has failed • Update the PCM to the latest software. If the problem is still present, substitute a known good PCM and then retest. If the problem goes away, the original PCM has failed.
DTC: P2249 **1T CCM, MIL: Yes** **Years:** 2007, 2008 **Models:** MDX, RDX, RL, TL, TSX **Engines:** All **Transmissions:** All	**Air Fuel Sensor 2 (Bank 2 Sensor 1) VCENT Line Low Voltage** Engine started, and the PCM detected an unexpected low voltage condition on the AFS2 VCENT circuit during the CCM test. **Possible Causes:** • Check for loose connections at the AFS2 unit (intermittent fault) • AFS2 VCENT signal circuit is shorted to ground • AFS2 assembly is damaged or has failed • Update the PCM to the latest software. If the problem is still present, substitute a known good PCM and then retest. If the problem goes away, the original PCM has failed.
DTC: P2251 **1T CCM, MIL: Yes** **Years:** 2007, 2008 **Models:** MDX, RDX, RL, TL, TSX **Engines:** All **Transmissions:** All	**Air Fuel Sensor 1 (Bank 1 Sensor 1) VS Line High Voltage** Engine started, and the PCM detected an unexpected high voltage condition on the AFS1 VS circuit during the CCM test. **Possible Causes:** • Check for loose connections at the AFS1 unit (intermittent fault) • AFS1 VS circuit is open (Circuit A53) • AFS1 assembly is damaged or has failed • Update the PCM to the latest software. If the problem is still present, substitute a known good PCM and then retest. If the problem goes away, the original PCM has failed.
DTC: P2252 **1T CCM, MIL: Yes** **Years:** 2007, 2008 **Models:** MDX, RDX, RL, TL, TSX **Engines:** All **Transmissions:** All	**Air Fuel Sensor 1 (Bank 1 Sensor 1) VS Line Low Voltage** Engine started, and the PCM detected an unexpected low voltage condition on the AFS1 VS circuit during the CCM test. **Possible Causes:** • Check for loose connections at the AFS1 unit (intermittent fault) • AFS1 VS circuit is shorted to ground • AFS1 assembly is damaged or has failed • Update the PCM to the latest software. If the problem is still present, substitute a known good PCM and then retest. If the problem goes away, the original PCM has failed.
DTC: P2254 **1T CCM, MIL: Yes** **Years:** 2007, 2008 **Models:** MDX, RDX, RL, TL, TSX **Engines:** All **Transmissions:** All	**Air Fuel Sensor 2 (Bank 2 Sensor 1) VS Line High Voltage** Engine started, and the PCM detected an unexpected high voltage condition on the AFS2 VS circuit during the CCM test. **Possible Causes:** • Check for loose connections at the AFS1 unit (intermittent fault) • AFS2 VS circuit is open (Circuit A49) • AFS2 assembly is damaged or has failed • Update the PCM to the latest software. If the problem is still present, substitute a known good PCM and then retest. If the problem goes away, the original PCM has failed.
DTC: P2255 **1T CCM, MIL: Yes** **Years:** 2007, 2008 **Models:** MDX, RDX, RL, TL, TSX **Engines:** All **Transmissions:** All	**Air Fuel Sensor 2 (Bank 2 Sensor 1) VS Line Low Voltage** Engine started, and the PCM detected an unexpected low voltage condition on the AFS2 VS circuit during the CCM test. **Possible Causes:** • Check for loose connections at the AFS2 unit (intermittent fault) • AFS2 VS circuit is shorted to ground • AFS2 assembly is damaged or has failed • Update the PCM to the latest software. If the problem is still present, substitute a known good PCM and then retest. If the problem goes away, the original PCM has failed.
DTC: P2279 **1T CCM, MIL: Yes** **Years:** 2007, 2008 **Models:** MDX, RDX, RL, TL, TSX **Engines:** All **Transmissions:** All	**Intake Air System Leak** DTC P0443 not set, engine started and engine running, and the PCM detected a problem in the intake Air System (a vacuum leak). **Possible Causes:** • Check for a vacuum leak at the PCV valve or vacuum hose • Check for a vacuum leak at EVAP purge valve vacuum hose • Check for a vacuum leak the throttle body assembly • Check for a vacuum leak at the brake booster hose • Check for a vacuum leak at the engine mount control solenoid vacuum hose • Check the camshaft timing to determine if it is correct

DTC	Trouble Code Title, Conditions & Possible Causes
DTC: P2413 **2T CCM, MIL: Yes** **Years:** 2007, 2008 **Models:** MDX, RDX, RL, TL, TSX **Engines:** All **Transmissions:** All	**EGR System Malfunction** DTC P0651 not set, engine started and engine running, and the PCM detected a problem in the intake Air System (a vacuum leak). **Possible Causes:** • Check for loose connections at EGR valve (intermittent fault) • EGR valve position sensor ground circuit is open • EGR valve position sensor VREF circuit is open or shorted • EGR valve position sensor signal circuit is open or shorted • EGR valve position sensor is damaged or has failed
DTC: P2422 **2T CCM, MIL: Yes** **Years:** 2007, 2008 **Models:** MDX, RDX, RL, TL, TSX **Engines:** All **Transmissions:** All	**EVAP Canister Vent Shut Valve Close Malfunction** DTC P0651 not set, engine started and engine running, and the PCM detected a problem in the intake Air System (a vacuum leak). **Possible Causes:** • Check for loose connections at EVAP canister vent shut valve (intermittent fault) • EVAP canister is clogged or restricted • EVAP canister vent shut valve is damaged or has failed
DTC: P2552 **2T CCM, MIL: Yes** **Years:** 2007, 2008 **Models:** MDX, RDX, RL, TL, TSX **Engines:** All **Transmissions:** All	**Throttle Actuator Control Module Relay Malfunction** Key on or engine running; and the PCM detected an unexpected voltage condition on the TAC module relay control circuit. **Possible Causes:** • Check for loose connections at the TAC module relay (intermittent fault) • TAC module relay control circuit is shorted to power • Throttle body assembly is damaged or has failed
DTC: P2610 **1T CCM, MIL: Yes** **Years:** 2007, 2008 **Models:** MDX, RDX, RL, TL, TSX **Engines:** All **Transmissions:** All	**24 Hour Time Malfunction** Key on or engine running; and the PCM detected an unexpected voltage condition on the TAC module relay control circuit. **Possible Causes:** • Clear codes and retest. If the code does not reset, the problem was an intermittent fault and is not present at this time. • Update the PCM to the latest software. If the problem is still present, substitute a known good PCM and then retest. If the problem goes away, the original PCM has failed.
DTC: P2627 **1T CCM, MIL: Yes** **Years:** 2007, 2008 **Models:** MDX, RDX, RL, TL, TSX **Engines:** All **Transmissions:** All	**Air Fuel Sensor 1 (Bank 1 Sensor 1) Label Circuit Low Input** Engine started, vehicle driven at cruise speed for 2-3 minutes in closed loop, and the PCM detected an unexpected low voltage condition on the AFS1 Label circuit during the CCM test. **Possible Causes:** • Clear the codes and recheck for DTC P2627. If it does, check for a short to ground condition on Circuit A56. • Check for loose connections at the AFS1 (intermittent fault) • If DTC P0107, P0112, P0117, P0452, P0700, P2122, P2128 or P02630 are set, repair these trouble codes first and then retest to determine if DTC P2627 is still present.
DTC: P2628 **1T CCM, MIL: Yes** **Years:** 2007, 2008 **Models:** MDX, RDX, RL, TL, TSX **Engines:** All **Transmissions:** All	**Air Fuel Sensor 1 (Bank 1 Sensor 1) Label Circuit High Input** Engine started, vehicle driven at cruise speed for 2-3 minutes in closed loop, and the PCM detected an unexpected high voltage condition on the AFS1 Label circuit during the CCM test. **Possible Causes:** • Check for loose connections at the AFS1 (intermittent fault) • Check for an open circuit condition on Circuit A56 • AFS2 assembly is damaged or has failed
DTC: P2630 **1T CCM, MIL: Yes** **Years:** 2007, 2008 **Models:** MDX, RDX, RL, TL, TSX **Engines:** All **Transmissions:** All	**Air Fuel Sensor 2 (Bank 2 Sensor 1) Label Circuit Low Input** Engine started, vehicle driven at cruise speed for 2-3 minutes in closed loop, and the PCM detected an unexpected low voltage condition on the AFS2 Label circuit during the CCM test. **Possible Causes:** • Clear the codes and recheck for DTC P2630. If it does, check for a short to ground condition on Circuit A52. • Check for loose connections at the AFS1 (intermittent fault) • If DTC P0107, P0112, P0117, P0452, P0700, P2122, P2127 or P02628 are set, repair these trouble codes first and then retest to determine if DTC P2630 is still present.
DTC: P2631 **1T CCM, MIL: Yes** **Years:** 2007, 2008 **Models:** MDX, RDX, RL, TL, TSX **Engines:** All **Transmissions:** All	**Air Fuel Sensor 2 (Bank 2 Sensor 1) Label Circuit High Input** Engine started, vehicle driven at cruise speed for 2-3 minutes in closed loop, and the PCM detected an unexpected high voltage condition on the AFS2 Label circuit during the CCM test. **Possible Causes:** • Check for loose connections at the AFS2 (intermittent fault) • Check for an open circuit condition on Circuit A52 • AFS2 assembly is damaged or has failed

DTC	Trouble Code Title, Conditions & Possible Causes
DTC: P2646 **1T CCM, MIL: Yes** **Years:** 2007, 2008 **Models:** MDX, RDX, RL, TL, TSX **Engines:** All **Transmissions:** All	**VTEC Oil Pressure Switch Circuit Low Input** Engine started, and the PCM detected an unexpected low voltage condition on the VTEC oil pressure switch circuit in the CCM test. **Possible Causes:** • Check for loose connections at VTEC switch (intermittent fault) • VTEC oil pressure switch circuit is shorted to ground • VTEC oil pressure switch is damaged or has failed • PCM has failed (substitute known good PCM and then retest)
DTC: P2647 **1T CCM, MIL: Yes** **Years:** 2007, 2008 **Models:** MDX, RDX, RL, TL, TSX **Engines:** All **Transmissions:** All	**VTEC Oil Pressure Switch Circuit High Input** Engine started, and the PCM detected an unexpected high voltage condition on the VTEC oil pressure switch circuit in the CCM test. **Possible Causes:** • Check for loose connections at VTEC switch (intermittent fault) • VTEC oil pressure switch circuit is shorted to ground • VTEC oil pressure switch is damaged or has failed • PCM has failed (substitute known good PCM and then retest)
DTC: P2648 **1T CCM, MIL: Yes** **Years:** 2007, 2008 **Models:** MDX, RDX, RL, TL, TSX **Engines:** All **Transmissions:** All	**VTEC Solenoid Valve Circuit Low Input** Key on or engine running; and the PCM detected an unexpected low voltage condition on the VTEC solenoid valve circuit during the test. The VTEC solenoid resistance at 68 degrees F is 14-30 ohms. **Possible Causes:** • Check for loose connections at VTEC solenoid valve (intermittent fault) • VTEC solenoid valve control circuit is shorted to ground • VTEC solenoid valve is damaged or has failed • PCM has failed (substitute known good PCM and then retest)
DTC: P2649 **1T CCM, MIL: Yes** **Years:** 2007, 2008 **Models:** MDX, RDX, RL, TL, TSX **Engines:** All **Transmissions:** All	**VTEC Solenoid Valve Circuit High Input** Key on or engine running; and the PCM detected an unexpected high voltage condition on the VTEC solenoid valve circuit in the test. The VTEC solenoid resistance at 68 degrees F is 14-30 ohms. **Possible Causes:** • Check for loose connections at VTEC solenoid valve (intermittent fault) • VTEC solenoid valve control circuit is open • VTEC solenoid valve is damaged or has failed • PCM has failed (substitute known good PCM and then retest)
DTC: P2A00 **1T CCM, MIL: Yes** **Years:** 2007, 2008 **Models:** MDX, RDX, RL, TL, TSX **Engines:** All **Transmissions:** All	**Air Fuel Sensor 1 (Bank 1 Sensor 1) Range/Performance** Engine started, vehicle driven at over 45 mph for 2-3 minutes followed by a deceleration period of 5 seconds with the throttle closed, and the PCM detected a problem in the AFS1 signal (it was not operating with a range of 0.8-1.2v during the test period). **Possible Causes:** • Check for loose connections at the AFS1 unit (intermittent fault) • AFS1 assembly is contaminated or has failed
DTC: P2A02 **1T CCM, MIL: Yes** **Years:** 2007, 2008 **Models:** MDX, RDX, RL, TL, TSX **Engines:** All **Transmissions:** All	**Air Fuel Sensor 2 (Bank 2 Sensor 1) Range/Performance** Engine started, vehicle driven at over 45 mph for 2-3 minutes followed by a deceleration period of 5 seconds with the throttle closed, and the PCM detected a problem in the AFS2 signal (it was not operating with a range of 0.8-1.2v during the test period). **Possible Causes:** • Check for loose connections at the AFS1 unit (intermittent fault) • AFS2 assembly is contaminated or has failed

OBD II Trouble Code List (U0xxx Codes)

DTC	Trouble Code Title, Conditions & Possible Causes
DTC: U0073 **1T CCM, MIL: Yes** **Years:** 2007, 2008 **Models:** MDX, RDX, RL, TL, TSX **Engines:** All **Transmissions:** All	**FCAN Malfunction (BUS Off)** DTC U0122 and U0155 not set; and a message from a learned ID number was not detected for the five seconds from the Vehicle Navigation Module (VTM) or the VSA control module. **Possible Causes:** • Check for loose connections at the VSA control unit (intermittent fault) • Communication is open, shorted to ground or to B+ • VSA has failed (substitute known good VSA and then retest)

DTC	Trouble Code Title, Conditions & Possible Causes
DTC: U0107 **1T CCM, MIL: Yes** **Years:** 2007, 2008 **Models:** MDX, RDX, RL, TL, TSX **Engines:** All **Transmissions:** All	**Lost Communication With TAC Module** DTC U0122 and U0155 not set; and the PCM detected a problem in the FCAN communication circuit to the Vehicle Navigation Module (VTM) or the VSA control module. **Possible Causes** • Check for loose connections at the TAC control unit (intermittent fault) • FCAN communication circuit to the TAC is open, shorted to ground or B+ • Throttle body assembly is damaged or has failed
DTC: U0114 **1T CCM, MIL: Yes** **Years:** 2007, 2008 **Models:** MDX, RDX, RL, TL, TSX **Engines:** All **Transmissions:** All	**FCAN Malfunction (VTM-4 Control Module To PCM)** DTC U0073 not set; and the PCM detected a problem in the FCAN communication circuit to the Vehicle Navigation Module (VTM) or the VSA control module. **Possible Causes:** • Check for loose connections at VTM-4 unit (intermittent fault) • Check for any Body Control Module trouble codes • FCAN communication circuit to the VTM-4 is open, shorted to ground or B+
DTC: U0122 **1T CCM, MIL: Yes** **Years:** 2007, 2008 **Models:** MDX, RDX, RL, TL, TSX **Engines:** All **Transmissions:** All	**FCAN Malfunction (VAS To PCM)** DTC U0155 not set; and the PCM detected a problem in the FCAN communication circuit to the Fuel Control Module (FCM). **Possible Causes:** • Check for loose connections at the VSA control unit (intermittent fault) • FCAN communication circuit to the VSA is open, shorted to ground or B+ • VSA has failed (substitute known good VSA and then retest)

HONDA

Diagnostic Trouble Codes

DIAGNOSTIC TROUBLE CODES

OBD II VEHICLE APPLICATIONS

HONDA

Accord
2007–2008
- 2.4L I4 .... K24A8
- 3.0L V6................... J30A5

Civic
2007–2008
- 1.8L I4 .... R18A1
- 2.0L I4 .... K20Z3

CR-V
2007–2008
- 2.4L I4 .... K24Z1

Element
2007–2008
- 2.4L I4 .... K24A4

Fit
2007–2008
- 1.5L I4 ....CD3

Odyssey
2007–2008
- 3.5L V6............. J35A6, J35A7

Pilot
2007–2008
- 3.5L V6............. J35A9, J3521

Ridgeline
2007–2008
- 3.5L V6................... J35A9

S2000
2007–2008
- 2.2L I4 .... F22C1

OBD II Trouble Code List (P0xxx Codes)

DTC	Trouble Code Title, Conditions & Possible Causes
DTC: P0010 **2T CCM, MIL: Yes** **Years:** 2007, 2008 **Models:** Accord, Civic, CR-V, Element, Fit, Odyssey, Pilot, Ridgeline, S2000 **Engines:** All **Transmissions:** All	**"A" Camshaft Position Actuator Circuit (Bank 1) Conditions:** Key on or engine running; and the ECM detected an unexpected high voltage or low voltage condition on the camshaft position sensor. The relative position between the camshaft and crankshaft needs to be optimal so the engine has better torque, fuel economy and emissions. **Note: The camshaft adjustment is load- and RPM dependant. The electrical camshaft adjustment valve 1 switches oil pressure onto camshaft adjuster (mechanical adjustment mechanism), which adjusts the camshaft.** **Possible Causes:** • Fuel pump has failed • Actuator circuit is open • ECM has failed • Battery voltage below 11.5 volts • Position actuator circuit may short to B+ or Ground
DTC: P0011 **2T CCM, MIL: Yes** **Years:** 2007, 2008 **Models:** Accord, Civic, CR-V, Element, Fit, Odyssey, Pilot, Ridgeline, S2000 **Engines:** All **Transmissions:** All	**"A" Camshaft Position Timing Over-Advanced (Bank 1) Conditions:** Engine started and driven at an engine speed of more than 400rpm; and the ECM detected the camshaft timing exceeded the maximum calibrated advance value, or the camshaft remained in an advanced position during the CCM test. The valve timing did not change from the current valve timing or it remained fixed during the testing. **Note: The camshaft adjustment is load- and RPM dependant. The electrical camshaft adjustment valve 1 switches oil pressure onto camshaft adjuster (mechanical adjustment mechanism), which adjusts the camshaft.** **Possible Causes:** • Fuel pump has failed • CPS circuit is open, shorted to ground or shorted to power • ECM has failed • Battery voltage below 11.5 volts • Position actuator circuit may short to B+ or Ground • Camshaft timing improperly set, or continuous oil flow to the VCT piston chamber • Camshaft advance mechanism (the VCT unit) is sticking or binding mechanically • VCT solenoid valve is stuck in open position
DTC: P0101 **2T CCM, MIL: Yes** **Years:** 2007, 2008 **Models:** Accord, Civic, CR-V, Element, Fit, Odyssey, Pilot, Ridgeline, S2000 **Engines:** All **Transmissions:** All	**Mass or Volume Air Flow Circuit Range/Performance Conditions** Engine running, with the system voltage more than 11.0v, and the temperature must be at least 185-degrees (F) and all electrical equipment (A/C, lights, etc) must be off. The ECM has detected that the MAF signal was out of a calculated range with the engine (or undetectable) for a certain period of time. **Possible Causes:** • Mass air flow (MAF) sensor has failed or is damaged • ECM has failed • Signal and ground wires of Mass Air Flow (MAF) sensor has short circuited
DTC: P0102 **2T CCM, MIL: Yes** **Years:** 2007, 2008 **Models:** Accord, Civic, CR-V, Element, Fit, Odyssey, Pilot, Ridgeline, S2000 **Engines:** All **Transmissions:** All	**MAF Sensor Circuit Low Input Conditions:** Key on, engine started, and the ECM detected the MAF sensor signal was less than the minimum calibrated value. The engine temperature must beat least 185-degrees (F) and all electrical equipment (A/C, lights, etc) must be off. The ECM has detected that the MAF signal was less than the required minimum. **Possible Causes:** • Check for leaks between MAF sensor and throttle valve control module • Voltage supply faulty. • Sensor power circuit open from fuel pump relay to MAF sensor • Sensor signal circuit open (may be disconnected) from ECM and MAF • Faulty ground cable resistance between connector terminal 1 and Ground • MAF Sensor malfunction
DTC: P0103 **2T CCM, MIL: Yes** **Years:** 2007, 2008 **Models:** Accord, Civic, CR-V, Element, Fit, Odyssey, Pilot, Ridgeline, S2000 **Engines:** All **Transmissions:** All	**MAF Sensor Circuit High Input Conditions:** Key on, engine started, and the ECM detected the MAF sensor signal was more than the minimum calibrated value. The engine temperature must beat least 185-degrees (F) and all electrical equipment (A/C, lights, etc) must be off. The ECM has detected that the MAF signal was more than the required minimum. **Possible Causes:** • Check for leaks between MAF sensor and throttle valve control module • Voltage supply faulty. • Sensor power circuit open from fuel pump relay to MAF sensor • Sensor signal circuit open (may be disconnected) from ECM and MAF • Faulty ground cable resistance between connector terminal 1 and Ground • MAF Sensor malfunction

DTC	Trouble Code Title, Conditions & Possible Causes

DTC: P0106
2T CCM, MIL: Yes
Years: 2007, 2008
Models: Accord, Civic, CR-V, Element, Fit, Odyssey, Pilot, Ridgeline, S2000
Engines: All
Transmissions: All
	Manifold Air Pressure Sensor Range/Performance
	Engine started, engine runtime over 1 second, and the PCM detected the MAP sensor was more 11.8" Hg during the test period.
	Possible Causes:
	    • MAP sensor source vacuum line is leaking or disconnected
	    • MAP sensor source vacuum line is plugged at intake manifold
	    • MAP sensor is damaged, out-of-calibration or has failed
	    • PCM has failed

DTC: P0107
1T CCM, MIL: Yes
Years: 2007, 2008
Models: Accord, Civic, CR-V, Element, Fit, Odyssey, Pilot, Ridgeline, S2000
Engines: All
Transmissions: All
	Manifold Air Pressure Sensor Circuit Low Input
	Engine started, engine running in closed loop, and the PCM detected the MAP sensor was near 0.0" Hg during the CCM test.
	Note: The key on, engine off MAP sensor input should be near 2.9v.
	Possible Causes:
	    • MAP sensor 5-volt power circuit open or shorted to ground
	    • MAP Sensor signal circuit is shorted to ground
	    • MAP Sensor is damaged or has failed
	    • PCM has failed

DTC: P0108
1T CCM, MIL: Yes
Years: 2007, 2008
Models: Accord, Civic, CR-V, Element, Fit, Odyssey, Pilot, Ridgeline, S2000
Engines: All
Transmissions: All
	Manifold Air Pressure Sensor Circuit High Input
	Engine started, engine running in closed loop, and the PCM detected the MAP sensor was near 29.9" Hg during the test period.
	Note: The key on, engine off MAP sensor input should be near 2.9v.
	Possible Causes:
	    • MAP sensor signal circuit is open, or the ground circuit is open
	    • MAP sensor signal circuit shorted to 5v VREF or system power
	    • MAP sensor is damaged (due to an open circuit) or has failed
	    • PCM has failed

DTC: P0111
2T CCM, MIL: Yes
Years: 2007, 2008
Models: Accord, Civic, CR-V, Element, Fit, Odyssey, Pilot, Ridgeline, S2000
Engines: All
Transmissions: All
	Intake Air Temperature Sensor Range/Performance
	Engine started, engine runtime over 10 minutes, and the PCM detected the Intake Air Temperature (IAT) sensor signal changed too much in too short a period of time during the CCM test.
	Possible Causes:
	    • IAT sensor signal value less than 1.0v
	    • IAT sensor ground circuit has high resistance
	    • IAT sensor signal circuit has high resistance
	    • IAT sensor is damaged or has failed
	    • PCM has failed

DTC: P0112
1T CCM, MIL: Yes
Years: 2007, 2008
Models: Accord, Civic, CR-V, Element, Fit, Odyssey, Pilot, Ridgeline, S2000
Engines: All
Transmissions: All
	Intake Air Temperature Sensor Circuit Low Input
	Key on or engine running, and the PCM detected the IAT sensor signal indicated less than 0.1v (Scan Tool reads over 302 degrees F).
	Possible Causes:
	    • IAT sensor signal shorted to chassis ground
	    • IAT sensor signal shorted to sensor ground circuit
	    • IAT sensor has an internal failure (it is shorted) or has failed
	    • PCM has failed

DTC: P0113
1T CCM, MIL: Yes
Years: 2007, 2008
Models: Accord, Civic, CR-V, Element, Fit, Odyssey, Pilot, Ridgeline, S2000
Engines: All
Transmissions: All
	Intake Air Temperature Sensor Circuit High Input
	Key on or engine running, and the PCM detected the IAT sensor signal indicated more than 4.90 (Scan Tool reads less than —4 degrees F).
	Possible Causes:
	    • IAT sensor signal shorted to VREF or system power
	    • IAT sensor signal circuit is open
	    • IAT sensor ground circuit is open
	    • Sensor has an internal failure (it is open)
	    • PCM has failed

DTC: P0116
1T CCM, MIL: Yes
Years: 2007, 2008
Models: Accord, Civic, CR-V, Element, Fit, Odyssey, Pilot, Ridgeline, S2000
Engines: All
Transmissions: All
	Engine Coolant Temperature Sensor Range/Performance
	Engine running, and the PCM detected the Engine Coolant Temperature (ECT) sensor signal changed too much too quickly.
	Note: The ECT sensor should read 0.47v-0.78v at hot idle speed.
	Possible Causes:
	    • ECT sensor signal circuit is open (an intermittent fault)
	    • ECT sensor signal circuit is shorted to ground (intermittent)
	    • ECT sensor is damaged or has failed
	    • PCM has failed

DTC	Trouble Code Title, Conditions & Possible Causes
DTC: P0117 **1T CCM, MIL: Yes** **Years:** 2007, 2008 **Models:** Accord, Civic, CR-V, Element, Fit, Odyssey, Pilot, Ridgeline, S2000 **Engines:** All **Transmissions:** All	**Engine Coolant Temperature Sensor Circuit Low Input** Key on or engine running, and the PCM detected the Engine Coolant Temperature (ECT) sensor signal indicated more than 302 degrees F (0.1v). **Note: The normal range of the ECT sensor is from 0.47v to 0.78v.** **Possible Causes:** • ECT sensor signal shorted to chassis ground • ECT sensor signal shorted to sensor ground circuit • ECT sensor has an internal failure (it is shorted) or has failed • PCM has failed
DTC: P0118 **1T CCM, MIL: Yes** **Years:** 2007, 2008 **Models:** Accord, Civic, CR-V, Element, Fit, Odyssey, Pilot, Ridgeline, S2000 **Engines:** All **Transmissions:** All	**Engine Coolant Temperature Sensor Circuit High Input** Key on or engine running, and the PCM detected the Engine Coolant Temperature (ECT) sensor signal indicated less than –4 degrees F (4.9v). **Note: The normal range of the ECT sensor is from 0.47v to 0.78v.** **Possible Causes:** • IAT sensor signal shorted to VREF or system power • IAT sensor signal circuit is open • IAT sensor ground circuit is open • Sensor has an internal failure (it is open) • PCM has failed
DTC: P0122 **1T CCM, MIL: Yes** **Years:** 2007, 2008 **Models:** Accord, Civic, CR-V, Element, Fit, Odyssey, Pilot, Ridgeline, S2000 **Engines:** All **Transmissions:** All	**Throttle Position Sensor Circuit Low Input** Engine running in closed loop conditions, and the PCM detected the closed throttle TP signal was less than 0.16v (less than 10 percent open). **Possible Causes:** • TP sensor signal circuit is shorted to ground • TP sensor VREF circuit is open • TP sensor VREF circuit is shorted to ground • TP sensor is damaged (it may be shorted internally) • PCM has failed
DTC: P0123 **1T CCM, MIL: Yes** **Years:** 2007, 2008 **Models:** Accord, Civic, CR-V, Element, Fit, Odyssey, Pilot, Ridgeline, S2000 **Engines:** All **Transmissions:** All	**Throttle Position Sensor Circuit High Input** Engine running in closed loop conditions, and the PCM detected the wide-open-throttle TP signal was more than 4.60v (more than 90 percent). **Possible Causes:** • TP sensor signal circuit is shorted to VREF • TP sensor signal circuit is shorted to system power • TP sensor ground circuit is open between sensor and the PCM • TP sensor is damaged or has failed • PCM has failed
DTC: P0128 **2T ECT, MIL: Yes** **Years:** 2007, 2008 **Models:** Accord, Civic, CR-V, Element, Fit, Odyssey, Pilot, Ridgeline, S2000 **Engines:** All **Transmissions:** All	**Thermostat Range/Performance** DTC P0107, P0108, P0112, P0113, P0116-118, P0335, P0336, P0300-P0306, P0401, P0505, P1106-P1108, P1259 and P1519, engine running at road load for 10 minutes, and the PCM detected the ECT sensor input did not reach the correct closed loop value. **Note: It is possible for this code to set if the engine is left running while the hood is open for an extended period in a warm climate.** **Possible Causes:** • Inspect for low coolant level or for an incorrect coolant mixture • Check the operation of the thermostat (it may be stuck open) • TSB 01-064 (9/11/01) contains a repair procedure for this code
DTC: P0131 **1T CCM, MIL: Yes** **Years:** 2007, 2008 **Models:** Accord, Civic, CR-V, Element, Fit, Odyssey, Pilot, Ridgeline, S2000 **Engines:** All **Transmissions:** All	**HO2S-11 (Bank 1 Sensor 1) Circuit Low Input** Engine running in closed loop in D4 position at cruise speed, and the PCM detected the HO2S signal was fixed at less than 0.50v. **Note: The actual value to set the code is stored in the PCM memory.** **Possible Causes:** • HO2S signal circuit is open or it is shorted to ground • HO2S may be contaminated or may have failed • Fuel supply system is too lean (fuel filter is clogged or dirty) • PCM has failed

DTC	Trouble Code Title, Conditions & Possible Causes
DTC: P0132 **1T CCM, MIL: Yes** **Years:** 2007, 2008 **Models:** Accord, Civic, CR-V, Element, Fit, Odyssey, Pilot, Ridgeline, S2000 **Engines:** All **Transmissions:** All	**HO2S-11 (Bank 1 Sensor 1) Circuit High Input** Engine running in closed loop in D4 position at cruise speed, and the PCM detected the HO2S signal was fixed at more than 0.90v **Note: The actual value to set the code is stored in the PCM memory.** **Possible Causes:** • HO2S signal tracking (wet/oily) in connector causing a short between the signal circuit and heater power circuit • HO2S signal circuit is open, or the ground circuit is open • HO2S heater supply circuit is open • PCM has failed
DTC: P0133 **1T O2S1, MIL: Yes** **Years:** 2007, 2008 **Models:** Accord, Civic, CR-V, Element, Fit, Odyssey, Pilot, Ridgeline, S2000 **Engines:** All **Transmissions:** All	**HO2S-11 (Bank 1 Sensor 1) Circuit Slow Response** Engine running in closed loop in D4 position at over 55 mph at steady speed, and the PCM detected the HO2S response time to switch between 300-600 mv was too slow, or that the rich to lean or lean to rich switch time was too slow. **Possible Causes:** • Exhaust leak present in the exhaust manifold or exhaust pipes • O2S element fuel contamination • O2S element has deteriorated • PCM has failed
DTC: P0135 **1T O2S HTR1, MIL: Yes** **Years:** 2007, 2008 **Models:** Accord, Civic, CR-V, Element, Fit, Odyssey, Pilot, Ridgeline, S2000 **Engines:** All **Transmissions:** All	**HO2S-11 (Bank 1 Sensor 1) Heater Circuit Malfunction** Engine runtime over 80 seconds, and the PCM detected an incorrect signal value at the HO2S heater circuit during the test period. **Possible Causes:** • Main relay output (power) circuit to the heater is open • O2S heater ground circuit is open • O2S heater element has high resistance • O2S heater element has an open condition • O2S heater element has a shorted condition • PCM has failed
DTC: P0137 **1T CCM, MIL: Yes** **Years:** 2007, 2008 **Models:** Accord, Civic, CR-V, Element, Fit, Odyssey, Pilot, Ridgeline, S2000 **Engines:** All **Transmissions:** All	**HO2S-12 (Bank 1 Sensor 2) Circuit Low Input** Engine running in closed loop in D4 position at cruise speed, and the PCM detected the HO2S signal was fixed at less than 0.30v. **Note: The actual value where the code sets is in the PCM memory.** **Possible Causes:** • HO2S signal circuit is open • HO2S signal circuit is shorted to ground • HO2S ground circuit is open • HO2S may be contaminated or may have failed • PCM has failed
DTC: P0138 **1T CCM, MIL: Yes** **Years:** 2007, 2008 **Models:** Accord, Civic, CR-V, Element, Fit, Odyssey, Pilot, Ridgeline, S2000 **Engines:** All **Transmissions:** All	**HO2S-12 (Bank 1 Sensor 2) Circuit High Input** Engine running in closed loop in 4th or 6th gear at cruise speed, and the PCM detected the HO2S signal was fixed at more than 0.60v. **Note: The actual value where the code sets is in the PCM memory.** **Possible Causes:** • HO2S signal tracking (wet/oily) in connector causing a short between the signal circuit and heater power circuit • HO2S signal circuit is open, or the ground circuit is open • HO2S heater supply circuit is open • PCM has failed
DTC: P0139 **2T O2S1, MIL: Yes** **Years:** 2007, 2008 **Models:** Accord, Civic, CR-V, Element, Fit, Odyssey, Pilot, Ridgeline, S2000 **Engines:** All **Transmissions:** All	**HO2S-12 (Bank 1 Sensor 2) Circuit Slow Response** Engine running in closed loop in D4 position at over 55 mph at steady speed, and the PCM detected the HO2S response time to switch between 300-600 mv was too slow, or that the rich to lean or lean to rich switch time was too slow. **Possible Causes:** • Exhaust leak present in the exhaust manifold or exhaust pipes • HO2S element fuel contamination • HO2S element has deteriorated • PCM has failed
DTC: P0141 **1T O2S HTR1, MIL: Yes** **Years:** 2007, 2008 **Models:** Accord, Civic, CR-V, Element, Fit, Odyssey, Pilot, Ridgeline, S2000 **Engines:** All **Transmissions:** All	**HO2S-12 (Bank 1 Sensor 2) Heater Circuit Malfunction** Engine runtime over 80 seconds, and the PCM detected an incorrect signal value at the HO2S heater circuit during the test period. **Possible Causes:** • Main relay output (power) circuit to the heater is open • HO2S heater ground circuit is open • HO2S heater element has high resistance • HO2S heater element has an open condition • HO2S heater element has a shorted condition • PCM has failed

DTC	Trouble Code Title, Conditions & Possible Causes
DTC: P0151 **1T CCM, MIL: Yes** **Years:** 2007, 2008 **Models:** Accord, Odyssey **Engines:** All **Transmissions:** All	**HO2S-21 (Bank 2 Sensor 1) Circuit Low Input** Engine running in closed loop in D4 position at cruise speed, and the PCM detected the HO2S signal was fixed at less than 0.10v. **Note: The actual value where the code sets is in the PCM memory.** **Possible Causes:** • HO2S signal circuit is open • HO2S signal circuit is shorted to ground • HO2S ground circuit is open • HO2S may be contaminated or may have failed • PCM has failed
DTC: P0152 **1T CCM, MIL: Yes** **Years:** 2007, 2008 **Models:** Accord, Odyssey **Engines:** All **Transmissions:** All	**HO2S-21 (Bank 2 Sensor 1) Circuit High Input** Engine running in closed loop in D4 position at cruise speed, and the PCM detected the HO2S signal was fixed at more than 0.90v **Note: The actual value where the code sets is in the PCM memory.** **Possible Causes:** • HO2S signal tracking (wet/oily) in connector causing a short between the signal circuit and heater power circuit • HO2S signal circuit is open, or the ground circuit is open • HO2S heater supply circuit is open • PCM has failed
DTC: P0153 **2T O2S1, MIL: Yes** **Years:** 2007, 2008 **Models:** Accord, Odyssey **Engines:** All **Transmissions:** All	**HO2S-21 (Bank 2 Sensor 1) Circuit Slow Response** Engine running in closed loop in D4 position at over 55 mph at steady speed, and the PCM detected the HO2S response time to switch between 300-600 mv was too slow, or that the rich to lean or lean to rich switch time was too slow. **Possible Causes:** • Exhaust leak present in the exhaust manifold or exhaust pipes • HO2S element fuel contamination • HO2S element has deteriorated • PCM has failed
DTC: P0155 **1T O2S HTR1, MIL: Yes** **Years:** 2007, 2008 **Models:** Accord, Odyssey **Engines:** All **Transmissions:** All	**HO2S-21 (Bank 2 Sensor 1) Heater Circuit Malfunction** Engine runtime over 80 seconds, and the PCM detected an incorrect signal value at the HO2S heater circuit during the test period. **Possible Causes:** • Main relay output (power) circuit to the heater is open • O2S heater ground circuit is open • O2S heater element has high resistance • O2S heater element has an open condition • O2S heater element has a shorted condition • PCM has failed
DTC: P0157 **1T CCM, MIL: Yes** **Years:** 2007, 2008 **Models:** Accord, Odyssey **Engines:** All **Transmissions:** All	**HO2S-22 (Bank 2 Sensor 2) Circuit Low Input** Engine running in closed loop in D4 position at cruise speed, and the PCM detected the HO2S signal was fixed at less than 0.30v. **Note: The actual value where the code sets is in the PCM memory.** **Possible Causes:** • HO2S signal circuit is open • HO2S signal circuit is shorted to ground • HO2S ground circuit is open • HO2S may be contaminated or may have failed • PCM has failed
DTC: P0158 **1T CCM, MIL: Yes** **Years:** 2007, 2008 **Models:** Accord, Odyssey **Engines:** All **Transmissions:** All	**HO2S-22 (Bank 2 Sensor 2) Circuit High Input** Engine running in closed loop in D4 position at cruise speed, and the PCM detected the HO2S signal was fixed at more than 0.60v **Note: The actual value where the code sets is in the PCM memory.** **Possible Causes:** • HO2S signal tracking (wet/oily) in connector causing a short between the signal circuit and heater power circuit • HO2S signal circuit is open, or the ground circuit is open • HO2S heater supply circuit is open • PCM has failed
DTC: P0159 **2T O2S1, MIL: Yes** **Years:** 2007, 2008 **Models:** Accord, Odyssey **Engines:** All **Transmissions:** All	**HO2S-22 (Bank 2 Sensor 2) Circuit Slow Response** Engine running in closed loop in D4 position at over 55 mph at steady speed, and the PCM detected the HO2S response time to switch between 300-600 mv was too slow, or that the rich to lean or lean to rich switch time was too slow. **Possible Causes:** • Exhaust leak present in the exhaust manifold or exhaust pipes • O2S element fuel contamination • O2S element has deteriorated • PCM has failed

DTC	Trouble Code Title, Conditions & Possible Causes
DTC: P0161 **1T O2S HTR1, MIL: Yes** **Years:** 2007, 2008 **Models:** Accord, Odyssey **Engines:** All **Transmissions:** All	**HO2S-22 (Bank 2 Sensor 2) Heater Circuit Malfunction** Engine runtime over 80 seconds, and the PCM detected an incorrect signal value at the HO2S heater circuit during the test period. **Possible Causes:** • Main relay output (power) circuit to the heater is open • O2S heater ground circuit is open • O2S heater element has high resistance • O2S heater element has an open condition • O2S heater element has a shorted condition • PCM has failed
DTC: P0171 **2T FUEL, MIL: Yes** **Years:** 2007, 2008 **Models:** Accord, Civic, CR-V, Element, Fit, Odyssey, Pilot, Ridgeline, S2000 **Engines:** All **Transmissions:** All	**Fuel System Too Lean (Bank 1)** DTC P0107, P0108, P0135, P0137, P0138, P0141, P1128, P1129 and P1259 not set, engine running in closed loop, and the PCM detected the LONGFT value exceeded the calibrated lean limit value **Possible Causes:** • Air leaks in intake manifold, exhaust pipes or exhaust manifold • One or more injectors restricted or pressure regulator has failed • Air is being drawn in from leaks in gaskets or other seals • O2S element is deteriorated or has failed • A "fuel control" sensor is out of calibration (ECT, IAT or MAP) • PCM has failed
DTC: P0172 **2T FUEL, MIL: Yes** **Years:** 2007, 2008 **Models:** Accord, Civic, CR-V, Element, Fit, Odyssey, Pilot, Ridgeline, S2000 **Engines:** All **Transmissions:** All	**Fuel System Too Rich (Bank 1)** DTC P0107, P0108, P0135, P0137, P0138, P0141, P1128, P1129 and P1259 not set, engine running in closed loop, and the PCM detected the LONGFT value exceeded the calibrated rich limit. **Note: A high MAP sensor signal at idle can cause this code to set.** **Possible Causes:** • Leaking/contaminated fuel injector(s) or fuel pressure regulator • HO2S element may be contaminated with water or alcohol • EVAP vapor recovery system has failed (pulling vacuum) • Base engine fault (i.e., cam timing incorrect, engine oil too high
DTC: P0174 **2T FUEL, MIL: Yes** **Years:** 2007, 2008 **Models:** Accord, Odyssey **Engines:** All **Transmissions:** All	**Fuel System Too Lean (Bank 2)** DTC P0107, P0108, P0135, P0137, P0138, P0141, P1128, P1129 and P1259 not set, engine running in closed loop, and the PCM detected the LONGFT value exceeded the calibrated lean limit value **Possible Causes:** • Air leaks in intake manifold, exhaust pipes or exhaust manifold • One or more injectors restricted or pressure regulator has failed • Air is being drawn in from leaks in gaskets or other seals • O2S element is deteriorated or has failed • A "fuel control" sensor is out of calibration (ECT, IAT or MAP)
DTC: P0175 **2T FUEL, MIL: Yes** **Years:** 2007, 2008 **Models:** Accord, Odyssey **Engines:** All **Transmissions:** All	**Fuel System Too Rich (Bank 2)** DTC P0107, P0108, P0135, P0137, P0138, P0141, P1128, P1129 and P1259 not set, engine running in closed loop, and the PCM detected the LONGFT value exceeded the calibrated rich limit. **Note: A high MAP sensor signal at idle can cause this code to set.** **Possible Causes:** • Leaking/contaminated fuel injector(s) or fuel pressure regulator • HO2S element may be contaminated with water or alcohol • EVAP vapor recovery system has failed (pulling vacuum) • A "fuel control" sensor is out of calibration (ECT, IAT or MAP) • Base engine fault (i.e., cam timing incorrect, engine oil too high
DTC: P0191 **1T CCM, MIL: Yes** **Years:** 2007, 2008 **Models:** Civic **Engines:** 1.7L VIN EN2 **Transmissions:** All	**CNG Fuel Pressure Sensor Range/Performance** Engine running at hot idle speed, and the PCM detected the Fuel Pressure sensor signal indicated a fuel pressure out of specification. **Note: The fuel pressure sensor PID reading a hot idle is 2.33-2.93v.** **Possible Causes:** • Fuel leaks somewhere in the system • Fuel pressure regulator is damaged or has failed • Fuel pressure sensor is out of calibration or has failed • PCM has failed

DTC	Trouble Code Title, Conditions & Possible Causes
DTC: P0192 **1T CCM, MIL: Yes** **Years:** 2007, 2008 **Models:** Civic **Engines:** 1.7L VIN EN2 **Transmissions:** All	**CNG Fuel Pressure Sensor Circuit Low Input** Key on or engine running, and the PCM detected an unexpected low voltage condition on the Fuel Pressure sensor signal circuit. **Note: The fuel pressure sensor PID reading a hot idle is 2.33-2.93v.** **Possible Causes:** • Fuel pressure sensor signal circuit is shorted to ground • Fuel pressure sensor is damaged or has failed • PCM has failed
DTC: P0193 **1T CCM, MIL: Yes** **Years:** 2007, 2008 **Models:** Civic **Engines:** 1.7L VIN EN2 **Transmissions:** All	**CNG Fuel Pressure Sensor Circuit High Input** Key on or engine running, and the PCM detected an unexpected high voltage condition on the Fuel Pressure sensor signal circuit. **Note: The fuel pressure sensor PID reading a hot idle is 2.33-2.93v.** **Possible Causes:** • Fuel pressure sensor signal circuit is open • Fuel pressure sensor signal circuit is shorted to VREF • Fuel pressure sensor is damaged or has failed • PCM has failed
DTC: P0222 **2T, MIL: Yes** **Years:** 2007, 2008 **Models:** Accord, Civic, CR-V, Element, Fit, Odyssey, Pilot, Ridgeline, S2000 **Engines:** All **Transmissions:** All	**Throttle Position Sensor 'B' Circuit Low Input Conditions:** Engine started, battery voltage at least 11.5v, all electrical components off, ground connections between engine and chassis well connected, coolant temperature at least 80-degrees Celicius and the throttle valve must not be damaged or dirty; and the ECM detected the TP Sensor 'B' circuit was out of its normal operating range during a condition with the throttle wide open, or with it completely closed. The throttle valve activation occurs via an electric motor (throttle drive) in the throttle valve control module. It is activated by the ECM according to specifications of the two sensors, Throttle Position Sensor and Accelerator Pedal Position Sensor 2. Slowly depress accelerator pedal up to Wide Open Throttle (WOT) stop while observing the percentage display on the PID data function of the scan tool. The percentage display must increase uniformly. **Possible Causes:** • ETC TP Sensor 'B' connector is damaged or shorted • ETC TP Sensor 'B' signal circuit is shorted to ground • ETC TP Sensor 'B' is damaged or it has failed • ECM has failed
DTC: P0223 **2T, MIL: Yes** **Years:** 2007, 2008 **Models:** Accord, Civic, CR-V, Element, Fit, Odyssey, Pilot, Ridgeline, S2000 **Engines:** All **Transmissions:** All	**Throttle Position Sensor 'B' Circuit High Input Conditions:** Engine started, battery voltage at least 11.5v, all electrical components off, ground connections between engine and chassis well connected, coolant temperature at least 80-degrees Celicius and the throttle valve must not be damaged or dirty; and the ECM detected the TP Sensor 'B' circuit was out of its normal operating range during a condition with the throttle wide open, or with it completely closed. The throttle valve activation occurs via an electric motor (throttle drive) in the throttle valve control module. It is activated by the ECM according to specifications of the two sensors, Throttle Position Sensor and Accelerator Pedal Position Sensor 2. Slowly depress accelerator pedal up to Wide Open Throttle (WOT) stop while observing the percentage display on the PID data function of the scan tool. The percentage display must increase uniformly. **Possible Causes:** • ETC TP Sensor 'B' connector is damaged or open • ETC TP Sensor 'B' signal circuit is open • ETC TP Sensor 'B' signal circuit is shorted to VREF (5v) • ETC TP Sensor 'B' is damaged or it has failed
DTC: P0300 **2T MISFIRE, MIL: Yes** **Years:** 2007, 2008 **Models:** Accord, Civic, CR-V, Element, Fit, Odyssey, Pilot, Ridgeline, S2000 **Engines:** All **Transmissions:** All	**Multiple Misfire Detected** DTC P0107, P0108, P0131, P0132, P0171, P0172, P1128, P0335, P0336, P0505, P1128, P1129, P1259, P1361, P1362, P1366, P1367 and P1519 not set, engine running under positive torque conditions, and the PCM detected a misfire in 2 or more cylinders. **Note: If the misfire is severe, the MIL will flash on/off on the 1st trip!** **Possible Causes:** • CKP or CMP sensor problem affecting more than one cylinder • Fuel system problem affecting more than one cylinder • Ignition system problem affecting more than one cylinder • Base engine mechanical fault affecting more than 1 cylinder
DTC: P0301 **2T MISFIRE, MIL: Yes** **Years:** 2007, 2008 **Models:** Accord, Civic, CR-V, Element, Fit, Odyssey, Pilot, Ridgeline, S2000 **Engines:** All **Transmissions:** All	**Cylinder 1 Misfire Detected** DTC P0107, P0108, P0131, P0132, P0171, P0172, P1128, P0335, P0336, P0505, P1128, P1129, P1259, P1361, P1362, P1366, P1367 and P1519 not set, engine running under positive torque conditions, and the PCM detected a misfire condition in one cylinder. **Note: If the misfire is severe, the MIL will flash on/off on the 1st trip!** **Possible Causes:** • Fuel system problem affecting only Cylinder 1 • Ignition system problem affecting Cylinder 1 • Base engine (mechanical) problem affecting only Cylinder 1

DTC	Trouble Code Title, Conditions & Possible Causes
DTC: P0302 **2T MISFIRE, MIL: Yes** **Years:** 2007, 2008 **Models:** Accord, Civic, CR-V, Element, Fit, Odyssey, Pilot, Ridgeline, S2000 **Engines:** All **Transmissions:** All	**Cylinder 2 Misfire Detected** DTC P0107, P0108, P0131, P0132, P0171, P0172, P1128, P0335, P0336, P0505, P1128, P1129, P1259, P1361, P1362, P1366, P1367 and P1519 not set, engine running under positive torque conditions, and the PCM detected a misfire condition in one cylinder. **Note: If the misfire is severe, the MIL will flash on/off on the 1st trip!** **Possible Causes:** • Fuel system problem affecting only Cylinder 2 • Ignition system problem affecting Cylinder 2 • Base engine (mechanical) problem affecting only Cylinder 2
DTC: P0303 **2T MISFIRE, MIL: Yes** **Years:** 2007, 2008 **Models:** Accord, Civic, CR-V, Element, Fit, Odyssey, Pilot, Ridgeline, S2000 **Engines:** All **Transmissions:** All	**Cylinder 3 Misfire Detected** DTC P0107, P0108, P0131, P0132, P0171, P0172, P1128, P0335, P0336, P0505, P1128, P1129, P1259, P1361, P1362, P1366, P1367 and P1519 not set, engine running under positive torque conditions, and the PCM detected a misfire condition in one cylinder. **Note: If the misfire is severe, the MIL will flash on/off on the 1st trip!** **Possible Causes:** • Fuel system problem affecting only Cylinder 3 • Ignition system problem affecting Cylinder 3 • Base engine (mechanical) problem affecting only Cylinder 3
DTC: P0304 **2T MISFIRE, MIL: Yes** **Years:** 2007, 2008 **Models:** Accord, Civic, CR-V, Element, Fit, Odyssey, Pilot, Ridgeline, S2000 **Engines:** All **Transmissions:** All	**Cylinder 4 Misfire Detected** DTC P0107, P0108, P0131, P0132, P0171, P0172, P1128, P0335, P0336, P0505, P1128, P1129, P1259, P1361, P1362, P1366, P1367 and P1519 not set, engine running under positive torque conditions, and the PCM detected a misfire condition in one cylinder. **Note: If the misfire is severe, the MIL will flash on/off on the 1st trip!** **Possible Causes:** • Fuel system problem affecting only Cylinder 4 • Ignition system problem affecting Cylinder 4 • Base engine (mechanical) problem affecting only Cylinder 4
DTC: P0305 **2T MISFIRE, MIL: Yes** **Years:** 2007, 2008 **Models:** Accord, Odyssey, Pilot, Ridgeline **Engines:** All **Transmissions:** All	**Cylinder 5 Misfire Detected** DTC P0107, P0108, P0131, P0132, P0171, P0172, P1128, P0335, P0336, P0505, P1128, P1129, P1259, P1361, P1362, P1366, P1367 and P1519 not set, engine running under positive torque conditions, and the PCM detected a misfire condition in one cylinder. **Note: If the misfire is severe, the MIL will flash on/off on the 1st trip!** **Possible Causes:** • Fuel system problem affecting only Cylinder 5 • Ignition system problem affecting Cylinder 5 • Base engine (mechanical) problem affecting only Cylinder 5
DTC: P0306 **2T MISFIRE, MIL: Yes** **Years:** 2007, 2008 **Models:** Accord, Odyssey, Pilot, Ridgeline **Engines:** All **Transmissions:** All	**Cylinder 6 Misfire Detected** DTC P0107, P0108, P0131, P0132, P0171, P0172, P1128, P0335, P0336, P0505, P1128, P1129, P1259, P1361, P1362, P1366, P1367 and P1519 not set, engine running under positive torque conditions, and the PCM detected a misfire condition in one cylinder. **Note: If the misfire is severe, the MIL will flash on/off on the 1st trip!** **Possible Causes:** • Fuel system problem affecting only Cylinder 6 • Ignition system problem affecting Cylinder 6 • Base engine (mechanical) problem affecting only Cylinder 6
DTC: P0325 **2T CCM, MIL: Yes** **Years:** 2007, 2008 **Models:** Accord, Civic, CR-V, Element, Fit, Odyssey, Pilot, Ridgeline, S2000 **Engines:** All **Transmissions:** All	**Knock Sensor (Rear) Circuit Malfunction** Engine running for over 1 minute, and the PCM detected an incorrect signal at the rear Knock Sensor (KS) circuit during the test. **Possible Causes:** • Knock sensor signal circuit is open (rear bank of engine) • Knock sensor signal circuit is grounded (rear bank of engine) • Knock sensor not tightened properly • Knock sensor damaged or has failed (it may be open internally) • PCM has failed

DTC	Trouble Code Title, Conditions & Possible Causes
DTC: P0330 **2T CCM, MIL: Yes** **Years:** 2007, 2008 **Models:** Accord, Civic, CR-V, Element, Fit, Odyssey, Pilot, Ridgeline, S2000 **Engines:** All **Transmissions:** All	**Knock Sensor (Front) Circuit Malfunction** Engine running for over 1 minute, and the PCM detected an incorrect signal at the front Knock Sensor (KS) circuit during the test. **Possible Causes:** • Knock sensor signal circuit is open (front bank of engine) • Knock sensor signal circuit is grounded (front bank of engine) • Knock sensor not tightened properly • Knock sensor damaged or has failed (it may be open internally) • PCM has failed
DTC: P0335 **2T CCM, MIL: Yes** **Years:** 2007, 2008 **Models:** Accord, Civic, CR-V, Element, Fit, Odyssey, Pilot, Ridgeline, S2000 **Engines:** All **Transmissions:** All	**CKP Sensor 'A' Circuit Malfunction (No Signal)** Engine running, and the PCM did not detect any signals from the Crankshaft Position (CKP) Sensor 'A' during the test period. **Note: The engine will crank for a longer period of time, may buck or jerk, but it will start and run without the CKP sensor signal present.** **Possible Causes:** • CKP Sensor 'A' signal circuit is open or shorted to ground • CKP Sensor 'A' signal circuit shorted to VREF or system power • CKP Sensor 'A' is damaged or has failed • PCM has failed
DTC: P0336 **2T CCM, MIL: Yes** **Years:** 2007, 2008 **Models:** Accord, Civic, CR-V, Element, Fit, Odyssey, Pilot, Ridgeline, S2000 **Engines:** All **Transmissions:** All	**CKP Sensor 'A' Circuit Range/Performance** Engine running, and the PCM detected the Crankshaft Position (CKP) Sensor 'A' signal was missing for a short period of time. **Note: This trouble code is usually caused by an intermittent fault.** **Possible Causes:** • CKP Sensor 'A' signal circuit is open or shorted to ground • CKP Sensor 'A' signal circuit shorted to VREF or system power • CKP Sensor 'A' is damaged or has failed • PCM has failed
DTC: P0340 **2T, MIL: Yes** **Years:** 2007, 2008 **Models:** Accord, Civic, CR-V, Element, Fit, Odyssey, Pilot, Ridgeline, S2000 **Engines:** All **Transmissions:** All	**Camshaft Position Sensor Circuit Malfunction Conditions:** Engine started, battery voltage must be at least 11.5v, all electrical components must be off, parking brake must be engaged (to keep daytime driving lights off), automatic transmission selector must be in park and the ground between the engine and the chassis must be well connected. The ECM detected the CMP sensor signal was missing or it was erratic. **Possible Causes:** • CMP sensor circuit is open or shorted to ground • CMP sensor circuit is shorted to power • CMP sensor ground (return) circuit is open • CMP sensor installation incorrect (Hall-effect type) • CMP sensor is damaged or CMP sensor shielding damaged • CMP sensor has failed • ECM has failed
DTC: P0341 **2T, MIL: Yes** **Years:** 2007, 2008 **Models:** Accord, Civic, CR-V, Element, Fit, Odyssey, Pilot, Ridgeline, S2000 **Engines:** All **Transmissions:** All	**Camshaft Position Sensor Circ Range/Performance Conditions:** Engine started, battery voltage must be at least 11.5v, all electrical components must be off, parking brake must be engaged (to keep daytime driving lights off), automatic transmission selector must be in park and the ground between the engine and the chassis must be well connected. The ECM detected the CMP sensor signal was implausible. **Possible Causes:** • CMP sensor circuit is open or shorted to ground • CMP sensor circuit is shorted to power • CMP sensor ground (return) circuit is open • CMP sensor installation incorrect (Hall-effect type) • CMP sensor is damaged or CMP sensor shielding damaged • ECM has failed
DTC: P0401 **2T EGR1, MIL: Yes** **Years:** 2007, 2008 **Models:** Accord, Civic, CR-V, Element, Fit, Odyssey, Pilot, Ridgeline, S2000 **Engines:** All **Transmissions:** All	**EGR System Insufficient Flow Detected** Cold engine startup requirement met (ECT sensor less than 76 degrees F at startup), engine running in closed loop at 40-55 mph for 2 minutes in Drive (D4), followed by a deceleration period back to 35 mph with the throttle closed, and the PCM detected a signal from the EGR position sensor that indicated insufficient EGR flow during the test. **Possible Causes:** • EGR valve source vacuum supply line open or restricted • EGR intake or exhaust manifold passages are restricted • EGR valve assembly or solenoid valve damaged or has failed • EGR constant vacuum control (CVC) valve is dirty or damaged • PCM has failed

DTC	Trouble Code Title, Conditions & Possible Causes
DTC: P0410 **2T AIR, MIL: Yes** **Years:** 2007, 2008 **Models:** S2000 **Engines:** All **Transmissions:** All	**Secondary Air Pump Circuit Malfunction** Cold startup requirement met (ECT sensor from 32-158 degrees F at engine startup), engine running, and the PCM detected an unexpected voltage condition on the Air Pump control circuit during the test. **Possible Causes:** • Secondary air pump control circuit is open or shorted to ground • Secondary air pump power (B+) circuit is open • Secondary air pump solenoid is damaged or has failed • PCM has failed
DTC: P0411 **2T AIR, MIL: Yes** **Years:** 2007, 2008 **Models:** S2000 **Engines:** All **Transmissions:** All	**Secondary Air System Incorrect Flow** Cold startup requirement met (ECT sensor from 32-158 degrees F at engine startup), engine running at idle speed, and the PCM detected an incorrect amount of airflow from the Air Injection system in the test. **Possible Causes:** • AIR solenoid air injection valve is damaged or has failed • AIR solenoid source vacuum hoses loose or disconnected • AIR solenoid air injection tube is restricted or clogged • PCM has failed
DTC: P0420 **2T CAT1, MIL: Yes** **Years:** 2007, 2008 **Models:** Accord, Civic, CR-V, Element, Fit, Odyssey, Pilot, Ridgeline, S2000 **Engines:** All **Transmissions:** All	**Catalyst Efficiency Below Threshold (Bank 1)** DTC P0137, P0138 and P0141 not set, engine running in closed loop at 40-55 mph for 2 minutes, followed by a deceleration period to 35 mph at closed throttle, and the PCM detected excessive activity in the Catalyst oxygen sensor (rear HO2S) during the test period. **Possible Causes:** • Air leaks at the exhaust manifold or in the exhaust pipes • Catalytic converter damaged or has failed (deteriorated) • Front HO2S older (aged) than the rear HO2S (HO2S is lazy) • Front HO2S or rear HO2S is contaminated with fuel or moisture
DTC: P0441 **2T EVADTC: P1, MIL: Yes** **Years:** 2007, 2008 **Models:** Accord, Civic, CR-V, Element, Fit, Odyssey, Pilot, Ridgeline, S2000 **Engines:** All **Transmissions:** All	**EVAP System Incorrect Purge Flow** Cold engine startup (ECT sensor signal less than 154 degrees F and IAT sensor signal more than 14 degrees F at startup), engine runtime from 3-5 minutes, followed by an acceleration period to 50-60 mph at steady throttle, and the PCM did not detect enough purge flow through the EVAP system during the EVAP Monitor flow test. **Possible Causes:** • EVAP purge control diaphragm valve hose loose/disconnected • EVAP purge cutoff or purge control solenoid valve is damaged • EVAP purge control diaphragm valve is damaged or has failed • PCM has failed
DTC: P0442 **2T CCM, MIL: Yes** **Years:** 2007, 2008 **Models:** Accord, Civic, CR-V, Element, Fit, Odyssey, Pilot, Ridgeline, S2000 **Engines:** All **Transmissions:** All	**EVAP Control System Small Leak Detected Conditions:** Engine started, battery voltage must be at least 11.5v, all electrical components must be off, parking brake must be engaged (to keep daytime driving lights off), automatic transmission selector must be in park, the exhaust system must be properly sealed between the catalytic converter and the cylinder head, coolant temperature must be at least 80 degrees Celsius and oxygen sensor heaters for oxygen sensors before the catalytic converter must be functioning properly and the ground between the engine and the chassis must be well connected. The ECM detected a leak in the EVAP system as small as 0.040" during the EVAP Monitor Test. **Possible Causes:** • Aftermarket EVAP parts that do not conform to specifications • CV solenoid remains partially open when commanded to close • EVAP component seals leaking (i.e., leaks in the Purge valve, fuel tank pressure sensor, canister vent solenoid, fuel vapor control valve tube assembly or fuel vapor vent valve). • Fuel filler cap damaged, cross-threaded or loosely installed • Loose fuel vapor hose/tube connections to EVAP components • Small holes or cuts in fuel vapor hoses or EVAP canister tubes
DTC: P0443 **2T CCM, MIL: Yes** **Years:** 2007, 2008 **Models:** Accord, Civic, CR-V, Element, Fit, Odyssey, Pilot, Ridgeline, S2000 **Engines:** All **Transmissions:** All	**EVAP Vapor Management Valve Circuit Malfunction Conditions:** Engine started, battery voltage must be at least 11.5v, all electrical components must be off, parking brake must be engaged (to keep daytime driving lights off), automatic transmission selector must be in park, the exhaust system must be properly sealed between the catalytic converter and the cylinder head, coolant temperature must be at least 80 degrees Celsius and oxygen sensor heaters for oxygen sensors before the catalytic converter must be functioning properly and the ground between the engine and the chassis must be well connected. The ECM detected an unexpected high or low voltage condition on the Vapor Management Valve (VMV) circuit when the device was cycled On/Off during testing. **Possible Causes:** • EVAP power supply circuit is open • EVAP solenoid control circuit is open or shorted to ground • EVAP solenoid control circuit is shorted to power (B+) • EVAP solenoid valve is damaged or it has failed • ECM has failed

DTC	Trouble Code Title, Conditions & Possible Causes
DTC: P0451 **2T CCM, MIL: Yes** **Years:** 2007, 2008 **Models:** Accord, Civic, CR-V, Element, Fit, Odyssey, Pilot, Ridgeline, S2000 **Engines:** All **Transmissions:** All	**Fuel Tank Pressure Sensor Range/Performance** Engine running, and the PCM detected the fuel tank pressure (FTP) sensor signal was less than the allowable range stored in the PCM memory (a calibrated range adjusted to current conditions). **Note: The FTP sensor PID should be near 2.5v with the fuel cap off.** **Possible Causes:** • FTP sensor vacuum lines loose, damaged or disconnected • Fuel tank pressure sensor is damaged or has failed • PCM has failed
DTC: P0452 **1T CCM, MIL: Yes** **Years:** 2007, 2008 **Models:** Accord, Civic, CR-V, Element, Fit, Odyssey, Pilot, Ridgeline, S2000 **Engines:** All **Transmissions:** All	**Fuel Tank Pressure Sensor Circuit Low Input** Key on or engine running, and the PCM detected the fuel tank pressure (FTP) sensor signal was less than 0.16v during the test. **Note: The FTP sensor PID should be near 2.5v with the fuel cap off.** **Possible Causes:** • FTP sensor signal circuit is shorted to ground • FTP sensor vacuum lines loose, damaged or disconnected • Fuel tank pressure sensor is damaged or has failed • PCM has failed
DTC: P0453 **1T CCM, MIL: Yes** **Years:** 2007, 2008 **Models:** Accord, Civic, CR-V, Element, Fit, Odyssey, Pilot, Ridgeline, S2000 **Engines:** All **Transmissions:** All	**Fuel Tank Pressure Sensor Circuit High Input** Key on or engine running, and the PCM detected the fuel tank pressure (FTP) sensor signal was more than 4.90v during the test. **Note: The FTP sensor PID should be near 2.5v with the fuel cap off.** **Possible Causes:** • FTP sensor signal circuit is shorted to VREF or power (B+) • FTP sensor ground circuit is open • FTP sensor vacuum lines loose, damaged or disconnected • Fuel tank pressure sensor is damaged or has failed • PCM has failed
DTC: P0455 **2T, MIL: Yes** **Years:** 2007, 2008 **Models:** Accord, Civic, CR-V, Element, Fit, Odyssey, Pilot, Ridgeline, S2000 **Engines:** All **Transmissions:** All	**EVAP Control System Large Leak Detected Conditions:** Engine started, battery voltage must be at least 11.5v, all electrical components must be off, parking brake must be engaged (to keep daytime driving lights off), automatic transmission selector must be in park, the exhaust system must be properly sealed between the catalytic converter and the cylinder head, coolant temperature must be at least 80 degrees Celsius and oxygen sensor heaters for oxygen sensors before the catalytic converter must be functioning properly and the ground between the engine and the chassis must be well connected. The ECM detected multiple small fuel vapor leaks; or it detected a large leak in the system during the leak test. **Possible Causes:** • Aftermarket EVAP hardware non-conforming to specifications • EVAP canister tube, EVAP canister purge outlet tube or EVAP return tube disconnected or cracked, or canister is damaged • EVAP canister purge valve stuck closed, or canister damaged • Fuel filler cap missing, loose (not tightened) or the wrong part • Loose fuel vapor hose/tube connections to EVAP components • Canister vent (CV) solenoid stuck open • Fuel tank pressure (FTP) sensor has failed mechanically
DTC: P0456 **2T, MIL: Yes** **Years:** 2007, 2008 **Models:** Accord, Civic, CR-V, Element, Fit, Odyssey, Pilot, Ridgeline, S2000 **Engines:** All **Transmissions:** All	**EVAP Control System Small Leak Detected Conditions:** Engine started, battery voltage must be at least 11.5v, all electrical components must be off, parking brake must be engaged (to keep daytime driving lights off), automatic transmission selector must be in park, the exhaust system must be properly sealed between the catalytic converter and the cylinder head, coolant temperature must be at least 80 degrees Celsius and oxygen sensor heaters for oxygen sensors before the catalytic converter must be functioning properly and the ground between the engine and the chassis must be well connected. The ECM detected multiple small fuel vapor leaks; or it detected a large leak in the system during the leak test. **Possible Causes:** • Aftermarket EVAP hardware non-conforming to specifications • EVAP canister tube, EVAP canister purge outlet tube or EVAP return tube disconnected or cracked, or canister is damaged • EVAP canister purge valve stuck closed, or canister damaged • Fuel filler cap missing, loose (not tightened) or the wrong part • Loose fuel vapor hose/tube connections to EVAP components • Canister vent (CV) solenoid stuck open • Fuel tank pressure (FTP) sensor has failed mechanically

DTC	Trouble Code Title, Conditions & Possible Causes
DTC: P0500 **1T CCM, MIL: Yes** **Years:** 2007, 2008 **Models:** Accord, Civic, CR-V, Element, Fit, Odyssey, Pilot, Ridgeline, S2000 **Engines:** All **Transmissions:** All	**Vehicle Speed Sensor Circuit Low Input** Engine running, then the vehicle was accelerated to 4000 rpm in 2nd gear, followed by a deceleration period to 1500 rpm at closed throttle and the PCM did not detect any VSS signal during the CCM test. **Note: The VSS signal should pulse from 0-5v as the vehicle moves.** **Possible Causes:** • VSS signal circuit is open • VSS signal circuit is shorted to ground • VSS signal circuit is shorted to VREF or system power (B+) • VSS is damaged or has failed
DTC: P0501 **2T CCM, MIL: Yes** **Years:** 2007, 2008 **Models:** Accord, Civic, Odyssey, S2000 **Engines:** All **Transmissions:** All	**Vehicle Speed Sensor Circuit Performance** Engine running at Cruise speed under road load conditions, and the PCM detected the VSS signal was erratic or too low during the test. **Note: The VSS signal should pulse from 0-5v as the vehicle moves.** **Possible Causes:** • VSS signal circuit is open • VSS signal circuit is shorted to ground • VSS signal circuit is shorted to VREF or system power (B+) • VSS is damaged or has failed
DTC: P0505 **1T CCM, MIL: Yes** **Years:** 2007, 2008 **Models:** Accord, Civic, CR-V, Element, Fit, Odyssey, Pilot, Ridgeline, S2000 **Engines:** All **Transmissions:** All	**Idle Speed Control System** DTC P1519 not set, engine running at hot idle speed, and the PCM detected the Actual and Target idle speed values were too far apart. **Possible Causes:** • IAC valve circuit open, shorted to ground or to power (B+) • IAC valve is damaged or has failed • Fast idle thermo valve is damaged or has failed (some models) • Throttle body is dirty or full of sludge • PCM has failed
DTC: P0560 **1T CCM, MIL: Yes** **Years:** 2007, 2008 **Models:** Accord, Civic, CR-V, Odyssey **Engines:** All **Transmissions:** All	**PCM Backup Circuit Low Voltage** Key on or engine running, and the PCM detected a low voltage condition on the PCM Backup circuit. **Note: This circuit is connected to the Backup/Radio 7.5 amp fuse.** **Possible Causes:** • PCM backup circuit is open • PCM backup circuit is shorted to ground • PCM backup circuit has high resistance • PCM has failed
DTC: P0700 **1T CCM, MIL: Yes** **Years:** 2007, 2008 **Models:** Accord, Civic, CR-V, Odyssey **Engines:** All **Transmissions:** All	**Automatic Transaxle** Engine running and the PCM detected an Automatic Transaxle fault. **Note: DTC P0700 sets along with several other TCM trouble codes.** **Possible Causes:** • Check for other A/T related trouble codes, and then refer to the Possible Causes for these trouble codes for more information.
DTC: P0715 **1T CCM, MIL: Yes** **Years:** 2007, 2008 **Models:** Accord, Civic, CR-V, Odyssey **Engines:** All **Transmissions:** All	**TCM A/T Mainshaft Speed Sensor Circuit Malfunction** Engine running with VSS inputs received, and the PCM detected an unexpected voltage condition on the Mainshaft speed sensor circuit. **Possible Causes:** • Mainshaft speed sensor circuit is open or shorted to ground • Mainshaft speed sensor circuit is shorted to VREF or power • Mainshaft speed sensor is damaged or has failed • PCM has failed
DTC: P0720 **1T CCM, MIL: Yes** **Years:** 2007, 2008 **Models:** Accord, Civic, CR-V, Odyssey **Engines:** All **Transmissions:** All	**TCM A/T Countershaft Speed Sensor Circuit Malfunction** Engine running with VSS inputs received, and the PCM detected an unexpected voltage on the Countershaft Speed Sensor circuit. **Possible Causes:** • Countershaft speed sensor circuit is open or shorted to ground • Countershaft speed sensor circuit is shorted to VREF or power • Countershaft speed sensor is damaged or has failed • PCM has failed

DTC	Trouble Code Title, Conditions & Possible Causes
DTC: P0725 **1T CCM, MIL: Yes** **Years:** 2007, 2008 **Models:** Accord, Civic, CR-V, Odyssey **Engines:** All **Transmissions:** All	**Automatic Transaxle** Engine running and the PCM detected an Automatic Transaxle fault. **Note: This trouble code sets along with several other Automatic Transaxle related trouble codes.** **Possible Causes:** • Check for other A/T related trouble codes, and then refer to the Possible Causes for these trouble codes for more information.
DTC: P0730 **1T CCM, MIL: Yes** **Years:** 2007, 2008 **Models:** Accord, Civic, CR-V, Odyssey **Engines:** All **Transmissions:** All	**TCM A/T Shift Control System** No other A/T trouble codes set, engine running at cruise speed with VSS inputs received, and the PCM detected the lockup clutch did not lock or unlock correctly. **Possible Causes:** • Refer to the repair instructions in a transmission repair manual or the information in other electronic media to repair this code.
DTC: P0740 **1T CCM, MIL: Yes** **Years:** 2007, 2008 **Models:** Accord, Civic, CR-V, Element, Fit, Odyssey, Pilot, Ridgeline, S2000 **Engines:** All **Transmissions:** All	**TCM A/T Lockup Clutch System** No other A/T trouble codes set, engine running at cruise speed with VSS inputs received, and the PCM detected the lockup clutch did not engage or disengage correctly. **Possible Causes:** • Refer to the repair instructions in a transmission repair manual or the information in other electronic media to repair this code.
DTC: P0753 **1T CCM, MIL: Yes** **Years:** 2007, 2008 **Models:** Accord, Civic, CR-V, Odyssey **Engines:** All **Transmissions:** All	**TCM A/T Lockup Solenoid 'A' Circuit Malfunction** Engine running with VSS inputs, and the PCM detected an unexpected voltage condition on the Solenoid Valve 'A' circuit. **Possible Causes:** • TCM Solenoid 'A' control circuit is open or shorted to ground • TCM Solenoid 'A' control circuit is shorted to system power • TCM Solenoid 'A' is damaged or has failed • TCM or PCM has failed
DTC: P0758 **1T CCM, MIL: Yes** **Years:** 2007, 2008 **Models:** Accord, Civic, CR-V, Odyssey **Engines:** All **Transmissions:** All	**TCM A/T Lockup Solenoid 'B' Circuit Malfunction** Engine running with VSS inputs, and the PCM detected an unexpected voltage condition on the Solenoid Valve 'B' circuit. **Possible Causes:** • TCM Solenoid 'B' control circuit is open or shorted to ground • TCM Solenoid 'B' control circuit is shorted to system power • TCM Solenoid 'B' is damaged or has failed • TCM or PCM has failed
DTC: P0763 **1T CCM, MIL: Yes** **Years:** 2007, 2008 **Models:** Accord, Civic, CR-V, Odyssey **Engines:** All **Transmissions:** All	**TCM A/T Control Unit or Related Circuit Malfunction** Engine running with VSS inputs received, and the PCM detected a fault in the TCM A/T Control Unit or one of its related circuits. **Possible Causes:** • Refer to the repair instructions in a transmission repair manual or the information in other electronic media to repair this code.
DTC: P0780 **1T CCM, MIL: Yes** **Years:** 2007, 2008 **Models:** Accord, Civic, CR-V, Odyssey **Engines:** All **Transmissions:** All	**Automatic Transaxle Malfunction** Engine running and the PCM detected an Automatic Transaxle fault. **Note: This trouble code (P0780) sets with along with several TCM related trouble codes.** **Possible Causes:** • Refer to the repair instructions in a transmission repair manual or the information in other electronic media to repair this code.

OBD II Trouble Code List (P1xxx Codes)

DTC	Trouble Code Title, Conditions & Possible Causes
DTC: P1106 **1T CCM, MIL: Yes** **Years:** 2007, 2008 **Models:** Accord, Civic, CR-V, Element, Fit, Odyssey, Pilot, Ridgeline, S2000 **Engines:** All **Transmissions:** All	**BARO Pressure Sensor Performance** Engine running in 4th gear, then accelerated to WOT, and the PCM detected the BARO sensor input did not change sufficiently within a specified period of time. **Possible Causes:** • BARO sensor signal circuit is open or shorted to ground • BARO sensor ground circuit has high resistance • BARO sensor is damaged or it may be out of calibration • PCM has failed
DTC: P1107 **1T CCM, MIL: Yes** **Years:** 2007, 2008 **Models:** Accord, Civic, CR-V, Element, Fit, Odyssey, Pilot, Ridgeline, S2000 **Engines:** All **Transmissions:** All	**BARO Pressure Sensor Circuit Low Input** Key on or engine running, and the PCM detected the BARO sensor signal was less than a value in stored in backup memory. **Possible Causes:** • BARO sensor signal circuit is shorted to signal ground • BARO sensor signal circuit is shorted to chassis ground • BARO sensor is damaged (it may be shorted internally) • BARO sensor signal circuit to the TCM is open or grounded • TCM or the PCM has failed
DTC: P1108 **1T CCM, MIL: Yes** **Years:** 2007, 2008 **Models:** Accord, Civic, CR-V, Element, Fit, Odyssey, Pilot, Ridgeline, S2000 **Engines:** All **Transmissions:** All	**BARO Pressure Sensor Circuit High Input** Key on or engine running, and the PCM detected the BARO sensor signal was more than a value in stored in backup memory. **Possible Causes:** • BARO sensor signal circuit shorted to VREF • BARO sensor signal circuit is shorted to system power (B+) • BARO sensor is damaged (it may be open internally) • BARO sensor signal circuit to the TCM is shorted to power • TCM or the PCM has failed
DTC: P1121 **1T CCM, MIL: Yes** **Years:** 2007, 2008 **Models:** Accord, Civic, CR-V, Element, Fit, Odyssey, Pilot, Ridgeline, S2000 **Engines:** All **Transmissions:** All	**TP Sensor Input Lower Than Expected** Engine running, and the PCM detected the TP sensor input was lower than an expected value with the throttle wide open (<13.7 percent). **Note: This trouble code sets if this circuit fails the rationality test.** **Possible Causes:** • Throttle plate is dirty, clogged, or it is binding • TP sensor circuit open or shorted to ground between the PCM and the TCM • TP sensor is damaged or has failed • PCM has failed
DTC: P1122 **1T CCM, MIL: Yes** **Years:** 2007, 2008 **Models:** Accord, Civic, CR-V, Element, Fit, Odyssey, Pilot, Ridgeline, S2000 **Engines:** All **Transmissions:** All	**TP Sensor Input Higher Than Expected** Engine running, and the PCM detected that the TP sensor input higher than the expected value with the throttle closed (>16.9 percent). **Note: This trouble code sets if this circuit fails the rationality test.** **Possible Causes:** • Throttle plate is dirty, clogged, or it is binding • TP sensor signal circuit shorted to VREF or it is open between the PCM and the TCM • TP sensor is damaged or has failed • PCM has failed
DTC: P1128 **1T CCM, MIL: Yes** **Years:** 2007, 2008 **Models:** Accord, Civic, CR-V, Element, Fit, Odyssey, Pilot, Ridgeline, S2000 **Engines:** All **Transmissions:** All	**MAP Sensor Value Less Than Expected** Engine running at cruise speed, then back to idle speed, and the PCM detected a MAP sensor signal lower than the expected value. **Note: This trouble code sets if this circuit fails the rationality test.** **Possible Causes:** • MAP sensor signal circuit shorted to ground (intermittent fault) • MAP sensor vacuum line bent or plugged at intake manifold • MAP sensor is damaged or it is out-of-calibration • PCM has failed
DTC: P1129 **1T CCM, MIL: Yes** **Years:** 2007, 2008 **Models:** Accord, Civic, CR-V, Element, Fit, Odyssey, Pilot, Ridgeline, S2000 **Engines:** All **Transmissions:** All	**MAP Sensor Value Higher Than Expected** Engine running at cruise speed, then back to idle speed, and the PCM detected a MAP sensor signal higher than the expected value. **Note: This trouble code sets if this circuit fails the rationality test.** **Possible Causes:** • MAP sensor signal circuit shorted to VREF (intermittent fault) • MAP sensor ground circuit has high resistance • MAP sensor is damaged or it is out-of-calibration • PCM has failed

DTC	Trouble Code Title, Conditions & Possible Causes
DTC: P1149 **1T O2S1, MIL: Yes** **Years:** 2007, 2008 **Models:** Accord, Civic, CR-V, Odyssey **Engines:** All **Transmissions:** All	**HO2S-11 (Bank 1 Sensor 1) Performance** Vehicle driven at 55 mph in closed loop at a steady throttle, then back to idle speed, and the PCM detected the front HO2S response time was too slow between 300-600 mv, or it detected the rich to lean or lean to rich switch rate was too slow. **Possible Causes:** • Exhaust leak present in the exhaust manifold or exhaust pipes • O2S element fuel contamination • O2S element has deteriorated • PCM has failed
DTC: P1149 **1T O2S1, MIL: Yes** **Years:** 2007, 2008 **Models:** Accord, Civic, CR-V, Odyssey **Engines:** All **Transmissions:** All	**HO2S-11 (Bank 1 Sensor 1) Performance** Vehicle driven at 55 mph in closed loop at a steady throttle in Drive, then back to idle speed, and the PCM detected the front HO2S response time was slow between 300-600 mv, or it detected the rich to lean or lean to rich switch rate was too slow. **Possible Causes:** • Exhaust leak present in the exhaust manifold or exhaust pipes • O2S element fuel contamination • O2S element has deteriorated • PCM has failed
DTC: P1162 **1T CCM, MIL: Yes** **Years:** 2007, 2008 **Models:** Accord **Engines:** All **Transmissions:** All	**Lean A/F Sensor (Bank 1 Sensor 1) Circuit Malfunction** DTC P0131, P0132, P0133, P1163 not set, vehicle driven while in closed loop at over 55 mph in D4 for 1-2 minutes, and the PCM detected an unexpected voltage condition on the LAF circuit during the HO2S Monitor test. **Possible Causes:** • HO2S may be contaminated or may have failed • Fuel supply system is too lean (exhaust leaks in front of HO2S) • Fuel supply system is too rich (fuel filter is clogged or dirty) • PCM has failed
DTC: P1162 **1T CCM, MIL: Yes** **Years:** 2007, 2008 **Models:** Accord **Engines:** All **Transmissions:** All	**Lean A/F Sensor (Bank 1 Sensor 1) Circuit Malfunction** DTC P0131, P0132, P0133, P1163 not set, vehicle driven while in closed loop at over 55 mph for 1-2 minutes in 5th gear, and the PCM detected an unexpected voltage condition on the front Lean Air Fuel Sensor (LAF) circuit during the HO2S Monitor test. **Possible Causes:** • LAF sensor may be contaminated or may have failed • Fuel supply system is too lean (exhaust leaks in front of LAF) • Fuel supply system is too rich (fuel filter is clogged or dirty) • PCM has failed
DTC: P1163 **2T O2S1, MIL: Yes** **Years:** 2007, 2008 **Models:** Accord, Civic, CR-V, Element, Fit, Odyssey, Pilot, Ridgeline, S2000 **Engines:** All **Transmissions:** All	**HO2S (Bank 1 Sensor 1) Signal Slow Response** Engine at idle speed, then accelerated to over 55 mph for 5 seconds at a steady throttle, then back to idle speed for 5 seconds, and the PCM detected the front HO2S response time was too slow, or that the R-L or L-R switch rate was too slow during the HO2S test period. **Possible Causes:** • HO2S may be contaminated or may have failed • Fuel supply system is too lean (exhaust leaks in front of LAF) • Fuel supply system is too rich (fuel filter is clogged or dirty) • PCM has failed
DTC: P1164 **2T O2S1, MIL: Yes** **Years:** 2007, 2008 **Models:** Accord, Civic, CR-V, Element, Fit, Odyssey, Pilot, Ridgeline, S2000 **Engines:** All **Transmissions:** All	**HO2S-11 (Bank 1 Sensor 1) Signal Range/Performance** Vehicle driven to over 1500 rpm in high gear in closed loop, then a quick acceleration to WOT, followed by a 5 second deceleration period with the throttle closed, and the PCM detected the front HO2S response time or the R-L or L-R switch rate was too slow. **Possible Causes:** • HO2S may be contaminated or may have failed • Fuel supply system is too lean (exhaust leaks in front of LAF) • Fuel supply system is too rich (fuel filter is clogged or dirty) • PCM has failed
DTC: P1165 **2T O2S1, MIL: Yes** **Years:** 2007, 2008 **Models:** Accord, Civic, CR-V, Element, Fit, Odyssey, Pilot, Ridgeline, S2000 **Engines:** All **Transmissions:** All	**HO2S-11 (Bank 1 Sensor 1) Signal Range/Performance** Vehicle driven at an engine speed of 1500-2500 rpm at 45-60 mph for 2-3 minutes, followed by a deceleration period of 3 seconds back to idle speed with the throttle closed, and the PCM detected the HO2S signal was too high or too low during the HO2S Monitor test. **Possible Causes:** • HO2S signal circuit is open or grounded (intermittent fault) • HO2S signal circuit is shorted to heater power (intermittent) • HO2S may be contaminated or may have failed • PCM has failed

DTC	Trouble Code Title, Conditions & Possible Causes
DTC: P1166 **1T O2S HTR1, MIL: Yes** **Years:** 2007, 2008 **Models:** Accord, Civic, CR-V, Element, Fit, Odyssey, Pilot, Ridgeline, S2000 **Engines:** All **Transmissions:** All	**Lean A/F-11 (Bank 1 Sensor 1) Heater Circuit Malfunction** Engine runtime over 80 seconds, and the PCM detected an unexpected voltage condition on the LAF sensor heater circuit. **Possible Causes:** • LAF sensor heater circuit is open or shorted to ground • LAF sensor heater is damaged or has failed • PCM has failed
DTC: P1167 **1T O2S HTR1, MIL: Yes** **Years:** 2007, 2008 **Models:** Accord, Civic, CR-V, Element, Fit, Odyssey, Pilot, Ridgeline, S2000 **Engines:** All **Transmissions:** All	**Lean A/F-11 (Bank 1 Sensor 1) Heater Circuit Malfunction** Engine runtime over 80 seconds, and the PCM detected an unexpected voltage condition on the LAF sensor Heater circuit. **Possible Causes:** • LAF sensor heater power supply circuit is open (check the fuse) • LAF sensor heater circuit is shorted to system power (B+) • LAF sensor heater is damaged or has failed • PCM has failed
DTC: P1168 **1T CCM, MIL: Yes** **Years:** 2007, 2008 **Models:** Accord, Civic, CR-V, Element, Fit, Odyssey, Pilot, Ridgeline, S2000 **Engines:** All **Transmissions:** All	**HO2S-11 (Bank 1 Sensor 1) Label Circuit Low Input** Engine runtime over 2 minutes at idle speed, and the PCM detected the front HO2S signal remained at less than the low threshold limit for too long a period of time during the HO2S Monitor test period. **Possible Causes:** • HO2S signal circuit is open or it is shorted to ground • HO2S is damaged or has failed (it may be contaminated) • PCM has failed
DTC: P1169 **2T CCM, MIL: Yes** **Years:** 2007, 2008 **Models:** Accord, Civic, CR-V, Element, Fit, Odyssey, Pilot, Ridgeline, S2000 **Engines:** All **Transmissions:** All	**HO2S-11 (Bank 1 Sensor 1) Label Circuit High Input** Engine running for 2 minutes, and the PCM detected the front HO2S signal remained at more than the high threshold limit for too long a period of time during the HO2S Monitor test period. **Possible Causes:** • HO2S signal circuit is shorted to system power (B+) • HO2S is damaged or has failed (it may be contaminated) • PCM has failed
DTC: P1182 **1T CCM, MIL: Yes** **Years:** 2007, 2008 **Models:** Accord, Civic, CR-V, Element, Fit, Odyssey, Pilot, Ridgeline, S2000 **Engines:** All **Transmissions:** All	**CNG Fuel Temperature Sensor Circuit Low Input** Key on or engine running, and the PCM detected an unexpected low voltage condition on the Fuel Temperature sensor signal circuit. **Possible Causes:** • Fuel temperature sensor signal circuit is shorted to ground • Fuel temperature sensor is damaged or has failed • PCM has failed
DTC: P1183 **1T CCM, MIL: Yes** **Years:** 2007, 2008 **Models:** Accord, Civic, CR-V, Element, Fit, Odyssey, Pilot, Ridgeline, S2000 **Engines:** All **Transmissions:** All	**CNG Fuel Temperature Sensor Circuit High Input** Key on or engine running, and the PCM detected an unexpected high voltage condition on the Fuel Temperature sensor signal circuit. **Possible Causes:** • Fuel temperature sensor signal circuit is open • Fuel temperature sensor signal is shorted to VREF • Fuel temperature sensor is damaged or has failed • PCM has failed
DTC: P1187 **1T CCM, MIL: Yes** **Years:** 2007, 2008 **Models:** Accord, Civic, CR-V, Element, Fit, Odyssey, Pilot, Ridgeline, S2000 **Engines:** All **Transmissions:** All	**CNG Fuel Tank Temperature Sensor Circuit Low Input** Key on or engine running, and the PCM detected an unexpected low voltage condition on the Fuel Tank Temperature sensor signal circuit. **Possible Causes:** • Fuel tank temperature sensor signal circuit is shorted to ground • Fuel tank temperature sensor is damaged or has failed • PCM has failed

DTC	Trouble Code Title, Conditions & Possible Causes
DTC: P1188 **1T CCM, MIL: Yes** **Years:** 2007, 2008 **Models:** Accord, Civic, CR-V, Element, Fit, Odyssey, Pilot, Ridgeline, S2000 **Engines:** All **Transmissions:** All	**CNG Fuel Tank Temperature Sensor Circuit High Input** Key on or engine running, and the PCM detected an unexpected high voltage condition on the Fuel Tank Temperature sensor signal circuit. **Possible Causes:** • Fuel tank temperature sensor signal circuit is open • Fuel tank temperature sensor signal circuit is shorted to ground • Fuel tank temperature sensor is damaged or has failed • PCM has failed
DTC: P1192 **1T CCM, MIL: Yes** **Years:** 2007, 2008 **Models:** Accord, Civic, CR-V, Element, Fit, Odyssey, Pilot, Ridgeline, S2000 **Engines:** All **Transmissions:** All	**CNG Fuel Tank Pressure Sensor Circuit Low Input** Key on or engine running, and the PCM detected an unexpected low voltage condition on the Fuel Tank Pressure sensor signal circuit. **Possible Causes:** • Fuel tank pressure sensor signal circuit is shorted to ground • Fuel tank pressure sensor is damaged or has failed • PCM has failed
DTC: P1193 **1T CCM, MIL: Yes** **Years:** 2007, 2008 **Models:** Accord, Civic, CR-V, Element, Fit, Odyssey, Pilot, Ridgeline, S2000 **Engines:** All **Transmissions:** All	**CNG Fuel Tank Pressure Sensor Circuit High Input** Key on or engine running, and the PCM detected an unexpected high voltage condition on the Fuel Tank Pressure sensor signal circuit. **Possible Causes:** • Fuel tank pressure sensor signal circuit is open • Fuel tank pressure sensor signal circuit is shorted to VREF • Fuel tank pressure sensor is damaged or has failed • PCM has failed
DTC: P1259 **1T CCM, MIL: Yes** **Years:** 2007, 2008 **Models:** Accord, Civic, CR-V, Element, Fit, Odyssey, Pilot, Ridgeline, S2000 **Engines:** All **Transmissions:** All	**VTEC System Malfunction (Bank 1)** Engine running in closed loop, then accelerated in 1st gear to over 6000 rpm for 2 seconds, and the PCM detected a fault in the VTEC solenoid or the VTEC switch. **Possible Causes:** • VTEC solenoid is damaged or has failed • VTEC switch is damaged or has failed • PCM has failed
DTC: P1297 **1T CCM, MIL: Yes** **Years:** 2007, 2008 **Models:** Accord, Civic, CR-V, Element, Fit, Odyssey, Pilot, Ridgeline, S2000 **Engines:** All **Transmissions:** All	**Electrical Load Detector Circuit Low Input** Engine running at hot idle speed or at cruise speed, headlights "on", and the PCM detected the ELD signal was less than a stored value. **Possible Causes:** • ELD sensor signal circuit is open or shorted to ground • ELD sensor power circuit is open or shorted to ground • ELD sensor is damaged or has failed • PCM has failed
DTC: P1298 **1T CCM, MIL: Yes** **Years:** 2007, 2008 **Models:** Accord, Civic, CR-V, Element, Fit, Odyssey, Pilot, Ridgeline, S2000 **Engines:** All **Transmissions:** All	**Electrical Load Detector Circuit High Input** Engine running at hot idle speed or at cruise speed, headlights "on", and the PCM detected the ELD signal was more than a stored value. **Possible Causes:** • ELD sensor signal circuit is shorted to VREF • ELD sensor signal circuit is shorted to system power (B+) • ELD sensor is damaged or has failed • PCM has failed
DTC: P1324 **1T CCM, MIL: Yes** **Years:** 2007, 2008 **Models:** Accord, Civic, CR-V, Element, Fit, Odyssey, Pilot, Ridgeline, S2000 **Engines:** All **Transmissions:** All	**Knock Sensor Power Source Circuit Low Input** Key on or engine running, and the PCM detected a low input on the Knock Sensor Power Source circuit during the CCM test. **Possible Causes:** • Knock sensor power source circuit is open • Knock sensor power source circuit is shorted to ground • PCM has failed

DTC	Trouble Code Title, Conditions & Possible Causes
DTC: P1336 **1T CCM, MIL: Yes** **Years:** 2007, 2008 **Models:** Accord, Civic, CR-V, Element, Fit, Odyssey, Pilot, Ridgeline, S2000 **Engines:** All **Transmissions:** All	**Engine Speed Fluctuation Sensor Circuit Malfunction** Engine running, and PCM detected an unexpected or intermittent interruption of the engine speed fluctuation (ESF) sensor signal. **Possible Causes:** • ESF signal circuit is open or shorted to ground • ESF signal circuit is shorted to VREF or system power (B+) • ESF is damaged or has failed • ESF pickup assembly or its pulse rotor is damaged • PCM has failed
DTC: P1337 **1T CCM, MIL: Yes** **Years:** 2007, 2008 **Models:** Accord, Civic, CR-V, Element, Fit, Odyssey, Pilot, Ridgeline, S2000 **Engines:** All **Transmissions:** All	**Crankshaft Speed Fluctuation Sensor No Signal** Engine running, and the PCM detected that it did not receive any signals from the Crankshaft Speed Fluctuation (CSF) sensor. **Possible Causes:** • CSF signal circuit is open or shorted to ground • CSF signal circuit is shorted to VREF or system power (B+) • CSF is damaged or has failed • CSF pickup assembly or its pulse rotor is damaged • PCM has failed
DTC: P1359 **1T CCM, MIL: Yes** **Years:** 2007, 2008 **Models:** Accord, CR-V **Engines:** All **Transmissions:** All	**CKP/TDC Sensor Circuit Malfunction** Engine running, and the PCM detected an unexpected voltage condition on the CKP/TDC sensor circuit during the CCM test. **Possible Causes:** • TDC signal circuit is open or shorted to ground • TDC signal circuit is shorted to VREF or system power (B+) • TDC pickup assembly or its pulse rotor is damaged • PCM has failed
DTC: P1361 **1T CCM, MIL: Yes** **Years:** 2007, 2008 **Models:** Accord, Civic, CR-V, Element, Fit, Odyssey, Pilot, Ridgeline, S2000 **Engines:** All **Transmissions:** All	**Top Dead Center Sensor 1 Circuit Intermittent Signal** Engine running, and the PCM detected an unexpected or intermittent interruption of the Top Dead Center 1 (TDC1) sensor signal. **Possible Causes:** • TDC1 signal circuit is open or shorted to ground • TDC1 signal circuit is shorted to VREF or system power • TDC1 pickup assembly or its pulse rotor is damaged • TDC1 is damaged or has failed • PCM has failed
DTC: P1362 **1T CCM, MIL: Yes** **Years:** 2007, 2008 **Models:** Accord, Civic, CR-V, Element, Fit, Odyssey, Pilot, Ridgeline, S2000 **Engines:** All **Transmissions:** All	**Top Dead Center Sensor 1 No Signal** Engine cranking or running, and the PCM did not receive any signals from the Top Dead Center 1 (TDC1) sensor during the CCM test. **Note: The engine will start and run without the TDC sensor 1 signal.** **Possible Causes:** • TDC1 signal circuit is open or shorted to ground • TDC1 pickup assembly or its pulse rotor is damaged • TDC1 is damaged or has failed • PCM has failed
DTC: P1366 **1T CCM, MIL: Yes** **Years:** 2007, 2008 **Models:** Accord, Civic, CR-V, Element, Fit, Odyssey, Pilot, Ridgeline, S2000 **Engines:** All **Transmissions:** All	**Top Dead Center Sensor 2 Circuit Intermittent Signal** Engine running and the PCM detected an unexpected or intermittent interruption of the Top Dead Center Sensor 2 (TDC2) signal. **Possible Causes:** • TDC2 signal circuit is open or shorted to ground • TDC2 signal circuit is shorted to VREF or system power • TDC2 pickup assembly or its pulse rotor is damaged • TDC2 is damaged or has failed • PCM has failed
DTC: P1366 **1T CCM, MIL: Yes** **Years:** 2007, 2008 **Models:** Accord, Civic, CR-V, Element, Fit, Odyssey, Pilot, Ridgeline, S2000 **Engines:** All **Transmissions:** All	**Top Dead Center Sensor 2 Circuit Intermittent Signal** Engine running and the PCM detected an unexpected or intermittent interruption of the Top Dead Center Sensor 2 (TDC2) signal. **Possible Causes:** • TDC2 signal circuit is open or shorted to ground • TDC2 signal circuit is shorted to VREF or system power • TDC2 pickup assembly or its pulse rotor is damaged • TDC2 is damaged or has failed • PCM has failed

DTC	Trouble Code Title, Conditions & Possible Causes
DTC: P1367 **1T CCM, MIL: Yes** **Years:** 2007, 2008 **Models:** Accord, Civic, CR-V, Element, Fit, Odyssey, Pilot, Ridgeline, S2000 **Engines:** All **Transmissions:** All	**Top Dead Sensor 2 No Signals** Engine cranking or running, and the PCM did not detect any signals from the Top Dead Center Sensor 2 (TDC2) during the CCM test. **Note: The engine will start and run without the TDC sensor 2 signal.** **Possible Causes:** • TDC2 signal circuit is open or shorted to ground • TDC2 pickup assembly or its pulse rotor is damaged • TDC2 is damaged or has failed • PCM has failed
DTC: P1367 **1T CCM, MIL: Yes** **Years:** 2007, 2008 **Models:** Accord, Civic, CR-V, Element, Fit, Odyssey, Pilot, Ridgeline, S2000 **Engines:** All **Transmissions:** All	**Top Dead Sensor 2 No Signals** Engine cranking or running, and the PCM did not detect any signals from the Top Dead Center Sensor 2 (TDC2) during the CCM test. **Note: The engine will start and run without the TDC sensor 2 signal.** **Possible Causes:** • TDC2 signal circuit is open or shorted to ground • TDC2 pickup assembly or its pulse rotor is damaged • TDC2 is damaged or has failed • PCM has failed
DTC: P1381 **1T CCM, MIL: Yes** **Years:** 2007, 2008 **Models:** Accord **Engines:** All **Transmissions:** All	**Camshaft Position Sensor 1 Circuit Malfunction** Engine running and the PCM detected an unexpected or intermittent interruption of the Camshaft Position (CMP) sensor 1 signal. **Possible Causes:** • CMP signal circuit is open or shorted to ground • CMP signal circuit is shorted to VREF or system power • CMP pickup assembly or CMP sensor is damaged or has failed • PCM has failed
DTC: P1381 **1T CCM, MIL: Yes** **Years:** 2007, 2008 **Models:** Accord, Civic, CR-V, Element, Fit, Odyssey, Pilot, Ridgeline, S2000 **Engines:** All **Transmissions:** All	**Camshaft Position Sensor 1 Circuit Malfunction** Engine running and the PCM detected an unexpected or intermittent interruption of the Camshaft Position (CMP) sensor 'A' signal. **Possible Causes:** • CMP signal circuit is open or shorted to ground • CMP signal circuit is shorted to VREF or system power • CMP pickup assembly or CMP sensor is damaged or has failed • PCM has failed
DTC: P1381 **1T CCM, MIL: Yes** **Years:** 2007, 2008 **Models:** Accord, Civic, CR-V, Element, Fit, Odyssey, Pilot, Ridgeline, S2000 **Engines:** All **Transmissions:** All	**Camshaft Position Sensor 1 Circuit Malfunction** Engine running and the PCM detected an unexpected or intermittent interruption of the Camshaft Position (CMP) sensor 1 signal. **Possible Causes:** • CMP signal circuit is open or shorted to ground • CMP signal circuit is shorted to VREF or system power • CMP pickup assembly or CMP sensor is damaged or has failed • PCM has failed
DTC: P1381 **1T CCM, MIL: Yes** **Years:** 2007, 2008 **Models:** CR-V **Engines:** All **Transmissions:** All	**Camshaft Position Sensor 1 Circuit Malfunction** Engine running and the PCM detected an unexpected or intermittent interruption of the Camshaft Position (CMP) sensor 1 signal. **Possible Causes:** • CMP signal circuit is open or shorted to ground • CMP signal circuit is shorted to VREF or system power • CMP pickup assembly or CMP sensor is damaged or has failed • PCM has failed
DTC: P1382 **1T CCM, MIL: Yes** **Years:** 2007, 2008 **Models:** Accord **Engines:** All **Transmissions:** All	**Camshaft Position Sensor 1 No Signal** Engine cranking or running, and the PCM did not detect any signals from the Camshaft Position (CMP) sensor 1 during the CCM test. **Note: The engine will start and run without the CMP sensor 1 signal.** **Possible Causes:** • CMP signal circuit is open or shorted to ground • CMP pickup assembly or CMP sensor is damaged or has failed • PCM has failed

DTC	Trouble Code Title, Conditions & Possible Causes

DTC: P1382
1T CCM, MIL: Yes
Years: 2007, 2008
Models: Accord, Civic, CR-V, Element, Fit, Odyssey, Pilot, Ridgeline, S2000
Engines: All
Transmissions: All
	Camshaft Position Sensor 1 No Signal
	Engine cranking or running, and the PCM did not detect any signals from the Camshaft Position (CMP) sensor 1 during the CCM test.
	Note: The engine will start and run without the CMP sensor 1 signal.
	Possible Causes:
	  • CMP signal circuit is open or shorted to ground
	  • CMP pickup assembly or CMP sensor is damaged or has failed
	  • PCM has failed

DTC: P1382
1T CCM, MIL: Yes
Years: 2007, 2008
Models: Accord, Civic, CR-V, Element, Fit, Odyssey, Pilot, Ridgeline, S2000
Engines: All
Transmissions: All
	Camshaft Position Sensor 1 No Signal
	Engine cranking or running, and the PCM did not detect any signals from the Camshaft Position (CMP) sensor 1 during the CCM test.
	Note: The engine will start and run without the CMP sensor 1 signal.
	Possible Causes:
	  • CMP signal circuit is open or shorted to ground
	  • CMP pickup assembly or CMP sensor is damaged or has failed
	  • PCM has failed

DTC: P1382
1T CCM, MIL: Yes
Years: 2007, 2008
Models: CR-V
Engines: All
Transmissions: All
	Camshaft Position Sensor 1 No Signal
	Engine cranking or running, and the PCM did not detect any signals from the Camshaft Position (CMP) sensor 1 during the CCM test.
	Note: The engine will start and run without the CMP sensor 1 signal.
	Possible Causes:
	  • CMP signal circuit is open or shorted to ground
	  • CMP pickup assembly or CMP sensor is damaged or has failed
	  • PCM has failed

DTC: P1410
2T AIR, MIL: Yes
Years: 2007, 2008
Models: S2000
Engines: All
Transmissions: All
	Air Pump System Malfunction
	Cold startup completed (ECT sensor from 32-158 degrees F at startup), and then with engine running for 10 seconds at idle speed under no load conditions after startup, the PCM detected the amount of airflow was incorrect with the Air System air pump commanded "on".
	Possible Causes:
	  • Secondary AIR system component problem (check the air pump, air injection hoses, or AIR relay or AIR solenoid
	  • PCM has failed

DTC: P1415
1T CCM, MIL: Yes
Years: 2007, 2008
Models: S2000
Engines: All
Transmissions: All
	AIR Pump Electric Current Sensor Signal Low Input
	Key on or engine running, and the PCM detected a low voltage condition on the Secondary Air Pump control circuit during the test.
	Possible Causes:
	  • AIR pump electric current sensor circuit is open
	  • AIR pump electric current sensor circuit is shorted to ground
	  • AIR pump electric current sensor power circuit is open
	  • PCM has failed

DTC: P1416
2T CCM, MIL: Yes
Years: 2007, 2008
Models: S2000
Engines: All
Transmissions: All
	AIR Pump Electric Current Sensor Signal High Input
	Key on or engine running, and the PCM detected a high voltage condition on the Secondary Air Pump control circuit during the test.
	Possible Causes:
	  • Air pump electric current sensor power circuit is open
	  • Air pump electric current sensor circuit shorted to system power
	  • PCM has failed

DTC: P1456
2T EVADTC: P1, MIL: Yes
Years: 2007, 2008
Models: Accord, Civic, CR-V, Element, Fit, Odyssey, Pilot, Ridgeline, S2000
Engines: All
Transmissions: All
	EVAP System Leak Detected (Fuel Tank Area)
	Cold startup completed (IAT sensor signal from 32-86 degrees F at engine startup), vehicle driven at over 5 mph for over 2 minutes, then with ECT sensor signal more than 154 degrees F and the EVAP Control and Vent solenoids enabled, the PCM detected the fuel tank pressure was incorrect due to a leak in the fuel tank area during the Leak Test.
	Possible Causes:
	  • Fuel tank cap damaged, loose or the wrong part number
	  • Fuel tank leaks at the fuel fill pipe or at the fuel tank seals
	  • Fuel vapor control valve is damaged or has failed
	  • Fuel tank vapor recirculation valve or vapor tube is damaged
	  • Fuel tank vapor control vent tube is damaged or has failed

DTC	Trouble Code Title, Conditions & Possible Causes
DTC: P1457 **2T EVADTC: P1, MIL: Yes** **Years:** 2007, 2008 **Models:** Accord, Civic, CR-V, Element, Fit, Odyssey, Pilot, Ridgeline, S2000 **Engines:** All **Transmissions:** All	**EVAP System Leak Detected (Canister Area)** Cold startup completed (IAT sensor signal from 32-86 degrees F at engine startup), vehicle driven at over 5 mph for over 2 minutes, then with ECT sensor signal more than 154 degrees F and the EVAP Control and Vent solenoids enabled, the PCM detected the fuel tank pressure was incorrect due to a leak in the canister area during the Leak Test. **Possible Causes:** • EVAP canister is leaking, damaged or full of water • EVAP canister purge line is loose, damaged or blocked • EVAP two-way valve or ORVR vent shut valve is damaged • EVAP fuel tank vapor control valve is damaged or has failed • PCM has failed
DTC: P1486 **2T ECT, MIL: Yes** **Years:** 2007, 2008 **Models:** Accord, Civic, CR-V, Element, Fit, Odyssey, Pilot, Ridgeline, S2000 **Engines:** All **Transmissions:** All	**Thermostat Range/Performance Malfunction** DTC P0107, P0108, P0112, P0113, P0116, P0117, P0118, P0300, P0301, P0302, P0303, P0304, P0305, P0306, P0335, P0336, P0401, P0500, P0505, P1106, P1107, P1108, P1128, P1129, P1253, P1257, P1258, P1259, P1359, P1399, P1491, P1498 and P1519 not set, vehicle driven for over 10 minutes, and the PCM detected the ECT signal did not reach the correct closed loop value. **Note: This trouble code can set if the engine remains under hot idle conditions with the hood open for an extended period of time.** **Possible Causes:** • ECT sensor is out of calibration • Check for low coolant level or incorrect coolant mixture • Cooling system component failure (thermostat is stuck open) • TSB 01-064 (9/11/01) contains a repair procedure for this code
DTC: P1486 **2T ECT, MIL: Yes** **Years:** 2007, 2008 **Models:** Insight **Engines:** All **Transmissions:** All	**Thermostat Range/Performance Malfunction** DTC P0107, P0108, P0112, P0113, P0116, P0117, P0118, P0300, P0301, P0302, P0303, P0304, P0305, P0306, P0335, P0336, P0401, P0500, P0505, P1106, P1107, P1108, P1128, P1129, P1253, P1257, P1258, P1259, P1359, P1399, P1491, P1498 and P1519 not set, vehicle driven for over 10 minutes, and the PCM detected the ECT signal did not reach the correct closed loop value. **Note: This trouble code can set if the engine remains under hot idle conditions with the hood open for an extended period of time.** **Possible Causes:** • ECT sensor is out of calibration • Check for low coolant level or incorrect coolant mixture • Cooling system component failure (thermostat is stuck open)
DTC: P1491 **2T EGR1, MIL: Yes** **Years:** 2007, 2008 **Models:** Accord, Insight, Odyssey **Engines:** All **Transmissions:** All	**EGR Valve Lift Sensor Insufficient Flow Detected** Vehicle driven in closed loop at 1700-2500 rpm for over 10 minutes, and the PCM detected the EGR valve lift sensor (EGRV) signal indicated insufficient EGR flow during the EGR Monitor test. **Possible Causes:** • EGR valve lift sensor is stuck, damaged or has failed • EGR control solenoid circuit is open or shorted to ground • EGR control solenoid valve is damaged or has failed • PCM has failed
DTC: P1491 **2T EGR1, MIL: Yes** **Years:** 2007, 2008 **Models:** Civic **Engines:** All **Transmissions:** All	**EGR Valve Lift Sensor Insufficient Flow Detected** Vehicle driven in closed loop at 1700-2500 rpm for over 10 minutes, and the PCM detected the EGR valve lift sensor (EGRV) signal indicated insufficient EGR flow during the EGR Monitor test. **Possible Causes:** • EGR valve lift sensor is stuck, damaged or has failed • EGR control solenoid circuit is open or shorted to ground • EGR control solenoid valve is damaged or has failed • PCM has failed
DTC: P1498 **1T CCM, MIL: Yes** **Years:** 2007, 2008 **Models:** Accord, Insight, Odyssey **Engines:** All **Transmissions:** All	**EGR Valve Lift Sensor High Input** Key on or engine running, and the PCM detected the EGR Valve Lift sensor signal was more than an allowable range stored in memory. **Possible Causes:** • EGR valve lift sensor circuit is open or shorted to power • EGR valve lift sensor is shorted to VREF or system power (B+) • EGR valve lift sensor is stuck, damaged or has failed • PCM has failed
DTC: P1498 **1T CCM, MIL: Yes** **Years:** 2007, 2008 **Models:** Civic **Engines:** All **Transmissions:** All	**EGR Valve Lift Sensor High Input** Key on or engine running, and the PCM detected the EGR Valve Lift sensor signal was more than an allowable range in memory. **Possible Causes:** • EGR valve lift sensor circuit is open or shorted to power • EGR valve lift sensor is shorted to VREF or system power (B+) • EGR valve lift sensor is stuck, damaged or has failed • PCM has failed

DTC	Trouble Code Title, Conditions & Possible Causes
DTC: P1508 **1T CCM, MIL: Yes** **Years:** 2007, 2008 **Models:** CR-V **Engines:** All **Transmissions:** All	**Idle Air Control System Low RPM** Engine running at idle speed while in closed loop, and the PCM detected the Actual idle speed was more than 100 rpm below the Target idle speed during the CCM test. **Possible Causes:** • Air inlet dirty, restricted or the IAC valve is stuck partially closed • IAC valve is damaged or has failed • The throttle plate is carbon fouled (it may need to be cleaned) • PCM has failed
DTC: P1509 **1T CCM, MIL: Yes** **Years:** 2007, 2008 **Models:** CR-V **Engines:** All **Transmissions:** All	**Idle Air Control System High RPM** Engine running at idle speed while in closed loop, and the PCM detected the Actual idle speed was more than 100 rpm below the Target idle speed during the CCM test. **Possible Causes:** • Air inlet dirty, restricted or the IAC valve is stuck partially closed • IAC valve is damaged or has failed • The throttle plate is carbon fouled (it may need to be cleaned) • PCM has failed
DTC: P1519 **1T CCM, MIL: Yes** **Years:** 2007, 2008 **Models:** Accord, Insight, Odyssey, S2000 **Engines:** All **Transmissions:** All	**Idle Air Control Valve Circuit Malfunction** Key on or engine running, and the PCM detected an unexpected voltage condition on the Idle Air Control (IAC) valve control circuit. **Possible Causes:** • IAC valve control circuit is open or shorted to power • IAC valve control circuit is shorted to system power (B+) • IAC valve power circuit is open or shorted to ground • IAC valve is damaged or has failed • PCM is damaged
DTC: P1519 **1T CCM, MIL: Yes** **Years:** 2007, 2008 **Models:** Civic **Engines:** All **Transmissions:** All	**Idle Air Control Valve Circuit Malfunction** Key on or engine running, and the PCM detected an unexpected voltage condition on the Idle Air Control (IAC) valve control circuit. **Possible Causes:** • IAC valve control circuit is open or shorted to power • IAC valve control circuit is shorted to system power (B+) • IAC valve power circuit is open or shorted to ground • IAC valve is damaged or has failed • PCM is damaged
DTC: P1522 **1T CCM, MIL: Yes** **Years:** 2007, 2008 **Models:** Insight **Engines:** All **Transmissions:** All	**Master Power Vacuum Sensor Low Input** Engine running at hot idle speed, and the PCM detected the Master Power Vacuum Sensor signal remained in a low state. **Possible Causes:** • Master power vacuum sensor power circuit is open • Master power vacuum sensor ground circuit open • Master power vacuum sensor signal circuit shorted to ground • Master power vacuum sensor is damaged or has failed • PCM has failed
DTC: P1523 **1T CCM, MIL: Yes** **Years:** 2007, 2008 **Models:** Insight **Engines:** All **Transmissions:** All	**Master Power Vacuum Sensor High Input** Engine running at hot idle speed, and the PCM detected the Master Power Vacuum Sensor signal remained in a high state. **Possible Causes:** • Master power vacuum sensor signal circuit shorted to VREF • Master power vacuum sensor signal circuit shorted to power • Master power vacuum sensor is damaged or has failed • PCM has failed
DTC: P1541 **1T CCM, MIL: Yes** **Years:** 2007, 2008 **Models:** Insight **Engines:** All **Transmissions:** All	**HTRS Passenger Heater Standby Circuit Low Input** Key on or engine running, and the PCM detected the HTRS Passenger Compartment Heater Standby signal was in a low state. **Possible Causes:** • HTRS passenger heater standby circuit is open • HTRS passenger heater standby circuit is shorted to ground • HTRS passenger heater is damaged or has failed • PCM has failed

DTC	Trouble Code Title, Conditions & Possible Causes
DTC: P1542 **1T CCM, MIL: Yes** **Years:** 2007, 2008 **Models:** Insight **Engines:** All **Transmissions:** All	**HTRS Passenger Heater Standby Circuit High Input** Key on or engine running, and the PCM detected the HTRS Passenger Compartment Heater Standby signal was in a high state. **Possible Causes:** • HTRS passenger heater standby circuit is shorted to VREF • HTRS passenger heater standby circuit is shorted to power • HTRS passenger heater is damaged or has failed • PCM has failed
DTC: P1607 **1T DTC: PCM, MIL: Yes** **Years:** 2007, 2008 **Models:** Accord, Civic, CR-V, Element, Fit, Odyssey, Pilot, Ridgeline, S2000 **Engines:** All **Transmissions:** All	**PCM Internal Circuit 'A' Malfunction** Key on, and the PCM detected an Internal Fault 'A' condition. **Note: This trouble code indicates an internal failure in the PCM. The OEM repair procedure recommends replacing the original PCM with a "known good" PCM and then verify the code does not reset.** **Possible Causes:** • PCM is damaged or has failed
DTC: P1640 **1T CCM, MIL: Yes** **Years:** 2007, 2008 **Models:** Insight **Engines:** All **Transmissions:** All	**ACTRQ Motor Torque Signal Circuit Low Input** Key on or engine running, and the PCM detected the ACTRQ Motor Torque Signal circuit remained in a low state. **Possible Causes:** • ACTRQ motor torque signal circuit is open • ACTRQ motor torque signal circuit is shorted to ground • ACTRQ motor is damaged or has failed • PCM has failed
DTC: P1641 **1T CCM, MIL: Yes** **Years:** 2007, 2008 **Models:** Insight **Engines:** All **Transmissions:** All	**ACTRQ Motor Torque Signal Circuit High Input** Key on or engine running, and the PCM detected the ACTRQ Motor Torque Signal circuit remained in a high state. **Possible Causes:** • ACTRQ motor torque signal circuit is shorted to VREF • ACTRQ motor torque signal circuit is shorted to power • ACTRQ motor is damaged or has failed • PCM has failed
DTC: P1642 **1T CCM, MIL: Yes** **Years:** 2007, 2008 **Models:** Insight **Engines:** All **Transmissions:** All	**QBATT Battery Signal Circuit Low Input** Key on or engine running, and the PCM detected that the QBATT Battery Signal circuit remained in a low state. **Possible Causes:** • QBATT battery signal circuit is open • QBATT battery signal circuit is shorted to ground • QBATT battery is damaged or has failed • PCM has failed
DTC: P1643 **1T CCM, MIL: Yes** **Years:** 2007, 2008 **Models:** Insight **Engines:** All **Transmissions:** All	**QBATT Battery Signal Circuit High Input** Key on or engine running, and the PCM detected the QBATT Battery Signal circuit remained in a high state. **Possible Causes:** • QBATT battery signal circuit is shorted to VREF • QBATT battery signal circuit is shorted to power • QBATT battery is damaged or has failed • PCM has failed
DTC: P1644 **1T CCM, MIL: Yes** **Years:** 2007, 2008 **Models:** Insight **Engines:** All **Transmissions:** All	**MOTFSA Signal Circuit Malfunction** Key on or engine running, and the PCM detected an unexpected voltage condition on the MOTFSA signal circuit during the CCM test. **Possible Causes:** • MOTFSA signal circuit is shorted to VREF or system power • MOTFSA signal circuit is open or shorted to ground • MOTFSA is damaged or has failed • PCM has failed
DTC: P1645 **1T CCM, MIL: Yes** **Years:** 2007, 2008 **Models:** Insight **Engines:** All **Transmissions:** All	**MOTFSA Signal Circuit Malfunction** Key on or engine running, and the PCM detected an unexpected voltage condition on the MOTFSA signal circuit during the CCM test. **Possible Causes:** • MOTFSA signal circuit is shorted to VREF or system power • MOTFSA signal circuit is open or shorted to ground • MOTFSA is damaged or has failed • PCM has failed

DTC	Trouble Code Title, Conditions & Possible Causes
DTC: P1646 **1T CCM, MIL: Yes** **Years:** 2007, 2008 **Models:** Insight **Engines:** All **Transmissions:** All	**MOTFSA Signal Circuit Malfunction** Key on or engine running, and the PCM detected an unexpected voltage condition on the MOTFSA signal circuit during the CCM test. **Possible Causes:** • MOTFSA signal circuit is shorted to VREF or system power • MOTFSA signal circuit is open or shorted to ground • MOTFSA is damaged or has failed • PCM has failed
DTC: P1655 **1T CCM, MIL: Yes** **Years:** 2007, 2008 **Models:** Civic **Engines:** 1.6L VIN EJ7 **Transmissions:** All	**TMA or TMB Signal Line Circuit Malfunction** Engine running, and the PCM detected an unexpected voltage condition on the TMA or TMB circuit during the CCM test. **Note: This trouble code is for vehicles with a CVT style transaxle.** **Possible Causes:** • TMA or TMB signal line is open or shorted to ground • TMA or TMB signal line is shorted to VREF or system power • TCM or PCM has failed
DTC: P1671 **1T CCM, MIL: Yes** **Years:** 2007, 2008 **Models:** Accord, Civic, CR-V, Odyssey **Engines:** All **Transmissions:** All	**TCM A/T FI Data Line No Signal** Key on or engine running, and the PCM detected a fault in the TCM FI Data Line circuit during the CCM test. **Possible Causes:** • A/T FI data line is open or shorted to ground • A/T FI data line is shorted to VREF or system power • TCM or PCM has failed
DTC: P1672 **1T CCM, MIL: Yes** **Years:** 2007, 2008 **Models:** Accord, Civic, CR-V, Odyssey **Engines:** All **Transmissions:** All	**TCM A/T FI Data Line Circuit Malfunction** Key on or engine running, and the PCM detected an unexpected voltage condition on the TCM FI Data Line circuit during the test. **Possible Causes:** • A/T FI data line is open or shorted to ground • A/T FI data line is shorted to VREF or system power • TCM or PCM has failed
DTC: P1676 **1T CCM, MIL: Yes** **Years:** 2007, 2008 **Models:** Accord, Civic, CR-V, Odyssey **Engines:** All **Transmissions:** All	**TCM A/T FI Data Line Circuit Malfunction** Key on or engine running, and the PCM detected an unexpected voltage condition on the TCM FI Data Line circuit during the test. **Possible Causes:** • A/T FI data line is open or shorted to ground • A/T FI data line is shorted to VREF or system power • TCM or PCM has failed
DTC: P1677 **1T CCM, MIL: Yes** **Years:** 2007, 2008 **Models:** Accord, Civic, CR-V, Odyssey **Engines:** All **Transmissions:** All	**TCM A/T FI Data Line Circuit Malfunction** Key on or engine running, and the PCM detected an unexpected voltage condition on the TCM FI Data Line circuit during the test. **Possible Causes:** • A/T FI data line is open or shorted to ground • A/T FI data line is shorted to VREF or system power • TCM or PCM has failed
DTC: P1678 **1T CCM, MIL: Yes** **Years:** 2007, 2008 **Models:** Accord, Civic, CR-V, Odyssey **Engines:** All **Transmissions:** All	**TCM A/T FPTDR Signal Line Circuit Malfunction** Key on or engine running, and the PCM detected an unexpected voltage condition on the TCM FPTDR Line circuit during the test. **Possible Causes:** • A/T FTPDR data line is open or shorted to ground • A/T FTPDR data line is shorted to VREF or system power • TCM or PCM has failed
DTC: P1705 **1T CCM, MIL: Yes** **Years:** 2007, 2008 **Models:** Accord, Civic, CR-V, Odyssey **Engines:** All **Transmissions:** All	**TCM A/T Gear Position Switch Low Input** Engine running, and the PCM detected a "low input" condition in the Gear Position Switch. **Note: The transaxle has no lockup function when this code is set.** **Possible Causes:** • A/T gear position switch signal circuit is shorted to ground • A/T gear position switch signal circuit is shorted to another wire • A/T gear position switch is damaged or has failed

DTC	Trouble Code Title, Conditions & Possible Causes
DTC: P1706 **1T CCM, MIL: Yes** **Years:** 2007, 2008 **Models:** Accord, Civic, CR-V, Odyssey **Engines:** All **Transmissions:** All	**TCM A/T Gear Position Switch High Input** Key on or engine running, and the PCM detected a "high input" condition in the Gear Position Switch. **Note: The transaxle has no lockup function when this code is set.** **Possible Causes:** • A/T gear position switch signal circuit is open • A/T gear position switch is damaged or has failed
DTC: P1709 **1T CCM, MIL: Yes** **Years:** 2007, 2008 **Models:** Odyssey **Engines:** All **Transmissions:** All	**TCM A/T Mode Switch Circuit Malfunction** Key on or engine running, and the PCM detected an unexpected voltage condition on the A/T Mode switch circuit during the test. **Possible Causes:** • A/T mode switch signal circuit is open • A/T mode switch signal circuit is shorted to ground • A/T mode switch signal circuit is shorted to system power • A/T mode switch is damaged or has failed • TCM or PCM has failed
DTC: P1738 **1T CCM, MIL: No** **Years:** 2007, 2008 **Models:** Accord, Odyssey **Engines:** All **Transmissions:** All	**TCM A/T 2nd Pressure Switch Circuit Malfunction** Engine running in gear, and the PCM detected an unexpected voltage condition on the 2nd pressure switch circuit during the test. **Possible Causes:** • A/T 2nd pressure switch signal circuit is open • A/T 2nd pressure switch signal circuit is shorted to ground • A/T 2nd pressure switch signal circuit shorted to system power • A/T 2nd pressure switch is damaged or has failed • TCM or PCM has failed
DTC: P1739 **1T CCM, MIL: No** **Years:** 2007, 2008 **Models:** Accord, Odyssey **Engines:** All **Transmissions:** All	**TCM A/T 3rd Pressure Switch Circuit Malfunction** Engine running in gear, and the PCM detected an unexpected voltage condition on the 3rd pressure switch circuit during the test. **Possible Causes:** • A/T 3rd pressure switch signal circuit is open • A/T 3rd pressure switch signal circuit is shorted to ground • A/T 3rd pressure switch signal circuit shorted to system power • A/T 3rd pressure switch is damaged or has failed • TCM or PCM has failed
DTC: P1750 **1T CCM, MIL: Yes** **Years:** 2007, 2008 **Models:** Accord **Engines:** All **Transmissions:** All	**TCM A/T System Mechanical Malfunction** Vehicle driven through 1st, 2nd, 3rd and 4th gears, and the PCM detected a mechanical problem in the A/T Control system. **Possible Causes:** • Clutch pressure control solenoid valve 'A' is damaged or failed • Clutch pressure control solenoid valve 'B' is damaged or failed • A/T hydraulic control system mechanical problem • TCM or PCM has failed
DTC: P1751 **1T CCM, MIL: Yes** **Years:** 2007, 2008 **Models:** Accord **Engines:** All **Transmissions:** All	**TCM A/T System Mechanical Malfunction** Vehicle driven through 1st, 2nd, 3rd and 4th gears, and the PCM detected a mechanical problem in the A/T Control system. **Possible Causes:** • Shift solenoid valve 'B' is damaged or has failed • Clutch pressure control solenoid valve 'A' is damaged or failed • Clutch pressure control solenoid valve 'B' is damaged or failed • A/T hydraulic control system mechanical problem • TCM or PCM has failed
DTC: P1753 **1T CCM, MIL: Yes** **Years:** 2007, 2008 **Models:** Accord, Civic, CR-V, Odyssey **Engines:** All **Transmissions:** All	**TCM A/T Lockup Solenoid Valve 'A' Circuit Malfunction** Vehicle driven in 1st, 2nd, 3rd and 4th gears, and the PCM detected an unexpected voltage condition on the Solenoid 'A' circuit during the CCM test. **Note: The D4 lamp on the dash will blink when this code is set.** **Possible Causes:** • Lockup solenoid 'A' control circuit is open or shorted to ground • Lockup solenoid 'A' circuit is shorted to system power • Lockup solenoid 'A' is damaged or has failed • TCM or PCM has failed

DTC	Trouble Code Title, Conditions & Possible Causes
DTC: P1758 **1T CCM, MIL: Yes** **Years:** 2007, 2008 **Models:** Accord, Civic, CR-V, Odyssey **Engines:** All **Transmissions:** All	**TCM A/T Lockup Solenoid Valve 'B' Circuit Malfunction** Vehicle driven in 1st, 2nd, 3rd and 4th gears, and the PCM detected an unexpected voltage condition on the Solenoid 'B' circuit during the CCM test. **Note: The D4 lamp on the dash will blink when this code is set.** **Possible Causes:** • Lockup solenoid 'B' control circuit is open or shorted to ground • Lockup solenoid 'B' circuit is shorted to system power • Lockup solenoid 'B' is damaged or has failed • TCM or PCM has failed
DTC: P1768 **1T CCM, MIL: Yes** **Years:** 2007, 2008 **Models:** Accord, Civic, CR-V, Odyssey **Engines:** All **Transmissions:** All	**TCM A/T Clutch Pressure Solenoid 'A' Circuit Malfunction** Vehicle driven in 1st, 2nd, 3rd and 4th gears, and the PCM detected an unexpected voltage condition on the Clutch Pressure Solenoid 'A' during the CCM test. **Note: The D4 lamp on the dash will blink when this code is set.** **Possible Causes:** • Clutch pressure solenoid 'A' circuit is open or shorted to ground • Clutch solenoid 'A' control circuit is shorted to system power • Clutch solenoid valve 'A' is damaged or has failed • TCM or PCM has failed
DTC: P1773 **1T CCM, MIL: Yes** **Years:** 2007, 2008 **Models:** Accord, Civic, CR-V, Odyssey **Engines:** All **Transmissions:** All	**TCM A/T Clutch Pressure Solenoid 'B' Circuit Malfunction** Vehicle driven in 1st, 2nd, 3rd and 4th gears, and the PCM detected an unexpected voltage condition on the Clutch Pressure Solenoid 'B' during the CCM test. **Note: The D4 lamp on the dash will blink when this code is set.** **Possible Causes:** • Clutch pressure solenoid 'B' circuit is open or shorted to ground • Clutch solenoid 'B' control circuit is shorted to system power • Clutch solenoid valve 'B' is damaged or has failed • TCM or PCM has failed
DTC: P1785 **1T CCM, MIL: Yes** **Years:** 2007, 2008 **Models:** Accord, Civic **Engines:** All **Transmissions:** All	**TCM A/T System Malfunction** Vehicle driven in 1st, 2nd, 3rd and 4th gears, and the PCM detected a fault somewhere in the A/T system during the CCM test. **Note: The D4 lamp on the dash will blink when this code is set.** **Possible Causes:** • A/T hydraulic system problem in the transmission • TCM or PCM has failed
DTC: P1882 **1T CCM, MIL: Yes** **Years:** 2007, 2008 **Models:** Civic **Engines:** 1.6L VIN EJ7 **Transmissions:** All	**A/T Inhibitor Solenoid Circuit Malfunction** Engine running, and the PCM detected an unexpected voltage condition on the A/T Inhibitor solenoid circuit during the CCM test. **Note: This trouble code is for vehicles with a CVT style transaxle.** **Possible Causes:** • A/T inhibitor solenoid control circuit open or shorted to ground • A/T inhibitor solenoid is shorted to VREF or system power • A/T inhibitor solenoid is damaged or has failed • TCM or PCM has failed
DTC: P1885 **1T CCM, MIL: Yes** **Years:** 2007, 2008 **Models:** Civic **Engines:** 1.6L VIN EJ7 **Transmissions:** All	**A/T Drive Pulley Speed Sensor Circuit Malfunction** Engine running, and the PCM detected an unexpected voltage condition on the A/T Drive Pulley Speed sensor circuit in the test. The drive pulley sensor resistance is f360-600 ohms at 68 degrees F. **Note: This trouble code is for vehicles with a CVT style transaxle.** **Possible Causes:** • A/T drive pulley speed sensor circuit open or shorted to ground • A/T drive pulley speed sensor is shorted to VREF • A/T drive pulley speed sensor is damaged or has failed • TCM or PCM has failed
DTC: P1886 **1T CCM, MIL: Yes** **Years:** 2007, 2008 **Models:** Civic **Engines:** 1.6L VIN EJ7 **Transmissions:** All	**A/T Driven Pulley Speed Sensor Circuit Malfunction** Engine running, and the PCM detected an unexpected voltage condition on the A/T Driven Pulley Speed sensor circuit in the test. The drive pulley sensor resistance is f360-600 ohms at 68 degrees F. **Note: This trouble code is for vehicles with a CVT style transaxle.** **Possible Causes:** • A/T driven pulley speed sensor circuit is open • A/T driven pulley speed sensor circuit is shorted to ground • A/T driven pulley speed sensor is shorted to VREF • A/T driven pulley speed sensor is damaged or has failed • TCM or PCM has failed

DTC	Trouble Code Title, Conditions & Possible Causes
DTC: P1888 **1T CCM, MIL: Yes** **Years:** 2007, 2008 **Models:** Civic **Engines:** 1.6L VIN EJ7 **Transmissions:** All	**A/T Second Gear Shaft Speed Sensor Circuit Malfunction** Engine running, and the PCM detected an unexpected voltage condition on the A/T 2nd Gear Shaft Speed sensor circuit in the test. **Note: This trouble code is for vehicles with a CVT style transaxle.** **Possible Causes:** • A/T 2nd gear shaft speed sensor circuit is open • A/T 2nd gear shaft speed sensor circuit is shorted to ground • A/T 2nd gear shaft speed sensor is shorted to VREF • A/T 2nd gear shaft speed sensor is damaged or has failed • TCM or PCM has failed
DTC: P1890 **1T CCM, MIL: Yes** **Years:** 2007, 2008 **Models:** Civic **Engines:** 1.6L VIN EJ7 **Transmissions:** All	**A/T Shift Control System Malfunction** Vehicle driven through several gears, and the PCM detected a problem somewhere in the A/T Shift Control system during the test. **Note: This trouble code is for vehicles with a CVT style transaxle.** **Possible Causes:** • A/T shift control system is damaged • A/T shift control system has failed • TCM or PCM has failed
DTC: P1891 **1T CCM, MIL: Yes** **Years:** 2007, 2008 **Models:** Civic **Engines:** 1.6L VIN EJ7 **Transmissions:** All	**A/T Start Clutch Control System Malfunction** Vehicle driven through several gears, and the PCM detected a problem somewhere in the Start Clutch Control system in the test. **Note: This trouble code is for vehicles with a CVT style transaxle.** **Possible Causes:** • A/T start clutch control system is damaged • A/T start clutch control system has failed • TCM or PCM has failed

OBD II Trouble Code List (P2xxx Codes)

DTC	Trouble Code Title, Conditions & Possible Causes
DTC: P2101 **2T, MIL: Yes** **Years:** 2007, 2008 **Models:** Accord, Civic, CR-V, Element, Fit, Odyssey, Pilot, Ridgeline, S2000 **Engines:** All **Transmissions:** All	**Throttle Actuator Control Motor Range/Performance Conditions:** Engine started, battery voltage must be at least 11.5v, all electrical components must be off, parking brake must be engaged (to keep daytime driving lights off), automatic transmission selector must be in park, the exhaust system must be properly sealed between the catalytic converter and the cylinder head, coolant temperature must be at least 80 degrees Celsius. The ECM detected an unexpected low or high voltage condition on the Throttle Actuator Control Motor (TACM) circuit during the CCM test. **Note: The throttle valve activation occurs via an electric motor (throttle drive) in the throttle valve control module. It is activated by the Engine Control Module (ECM) according to specifications of the two sensors, Throttle Position (TP) Sensor and Sender 2 for accelerator pedal position.** **Possible Causes:** • TACM wiring harness connector is damaged or open • TACM wiring may be crossed in the wire harness assembly • TACM (motor) circuit is open, or TACM assembly is damaged (possible open circuit) • TACM or the Throttle Valve is dirty • Throttle Position sensor has failed • ECM has failed
DTC: P2106 **Years:** 2007, 2008 **Models:** Accord, Civic, CR-V, Element, Fit, Odyssey, Pilot, Ridgeline, S2000 **Engines:** All **Transmissions:** All	**Throttle Actuator Control System – Forced Limited Power Conditions** Engine started, battery voltage must be at least 11.5v, all electrical components must be off, parking brake must be engaged (to keep daytime driving lights off), automatic transmission selector must be in park, the exhaust system must be properly sealed between the catalytic converter and the cylinder head, coolant temperature must be at least 80 degrees Celsius. The ECM detected an unexpected low or high voltage condition on the Throttle Actuator Control Motor (TACM) circuit during the CCM test. **Note: The throttle valve activation occurs via an electric motor (throttle drive) in the throttle valve control module. It is activated by the Engine Control Module (ECM) according to specifications of the two sensors, Throttle Position (TP) Sensor and Sender 2 for accelerator pedal position.** **Possible Causes:** • TACM wiring harness connector is damaged or open • TACM wiring may be crossed in the wire harness assembly • TACM (motor) circuit is open, or TACM assembly is damaged (possible open circuit) • TACM or the Throttle Valve is dirty • Throttle Position sensor has failed • ECM has failed

DTC	Trouble Code Title, Conditions & Possible Causes
DTC: P2108 **2T ECM, MIL: No** **Years:** 2007, 2008 **Models:** Accord, Civic, CR-V, Element, Fit, Odyssey, Pilot, Ridgeline, S2000 **Engines:** All **Transmissions:** All	**Throttle Actuator Control Motor Performance Conditions:** Engine started, battery voltage must be at least 11.5v, all electrical components must be off, parking brake must be engaged (to keep daytime driving lights off), automatic transmission selector must be in park, the exhaust system must be properly sealed between the catalytic converter and the cylinder head, coolant temperature must be at least 80 degrees Celsius. The ECM detected an unexpected low or high voltage condition on the Throttle Actuator Control Motor (TACM) circuit during the CCM test. **Note: The throttle valve activation occurs via an electric motor (throttle drive) in the throttle valve control module. It is activated by the Engine Control Module (ECM) according to specifications of the two sensors, Throttle Position (TP) Sensor and Sender 2 for accelerator pedal position.** **Possible Causes:** • TACM wiring harness connector is damaged or open • TACM wiring may be crossed in the wire harness assembly • TACM (motor) circuit is open, or TACM assembly is damaged (possible open circuit) • TACM or the Throttle Valve is dirty • Throttle Position sensor has failed • ECM has failed
DTC: P2110 **2T ECM, MIL: No** **Years:** 2007, 2008 **Models:** Accord, Civic, CR-V, Element, Fit, Odyssey, Pilot, Ridgeline, S2000 **Engines:** All **Transmissions:** All	**Throttle Actuator Control System – Forced Limited RPM Conditions:** Engine started, battery voltage must be at least 11.5v, all electrical components must be off, parking brake must be engaged (to keep daytime driving lights off), automatic transmission selector must be in park, the exhaust system must be properly sealed between the catalytic converter and the cylinder head, coolant temperature must be at least 80 degrees Celsius. The ECM detected an unexpected low or high voltage condition on the Throttle Actuator Control Motor (TACM) circuit during the CCM test. **Note: The throttle valve activation occurs via an electric motor (throttle drive) in the throttle valve control module. It is activated by the Engine Control Module (ECM) according to specifications of the two sensors, Throttle Position (TP) Sensor and Sender 2 for accelerator pedal position.** **Possible Causes:** • TACM wiring harness connector is damaged or open • TACM wiring may be crossed in the wire harness assembly • TACM (motor) circuit is open, or TACM assembly is damaged (possible open circuit) • TACM or the Throttle Valve is dirty • Throttle Position sensor has failed • ECM has failed
DTC: P2122 **2T, MIL: Yes** **Years:** 2007, 2008 **Models:** Accord, Civic, CR-V, Element, Fit, Odyssey, Pilot, Ridgeline, S2000 **Engines:** All **Transmissions:** All	**Accelerator Pedal Position Sensor 'D' Circuit Low Input Conditions:** Engine started, battery voltage at least 11.5v, all electrical components off, ground connections between engine and chassis well connected, the ECM detected that the accelerator pedal position sensor signal was outside the parameters to function normally. **Note: Both the Throttle Position (TP) Sensor and Accelerator Pedal Position Sensor are located at the accelerator pedal module and communicate the driver's intentions to the ECM completely independently of each other. Both sensors are stored in one housing.** **Possible Causes:** • Ground between engine and chassis may be broken • Throttle position sensor may have failed • Accelerator Pedal Position Sensor has failed • Throttle position sensor wiring may have shorted • Throttle position sensor has failed • Faulty voltage supply • ECM has failed
DTC: P2123 **2T, MIL: Yes** **Years:** 2007, 2008 **Models:** Accord, Civic, CR-V, Element, Fit, Odyssey, Pilot, Ridgeline, S2000 **Engines:** All **Transmissions:** All	**Accelerator Pedal Position Sensor 'D' Circuit High Input Conditions:** Engine started, battery voltage at least 11.5v, all electrical components off, ground connections between engine and chassis well connected, the ECM detected that the accelerator pedal position sensor signal was outside the parameters to function normally. **Note: Both the Throttle Position (TP) Sensor and Accelerator Pedal Position Sensor are located at the accelerator pedal module and communicate the driver's intentions to the ECM completely independently of each other. Both sensors are stored in one housing.** **Possible Causes:** • Ground between engine and chassis may be broken • Throttle position sensor may have failed • Accelerator Pedal Position Sensor has failed • Throttle position sensor wiring may have shorted • Throttle position sensor has failed • Faulty voltage supply • ECM has failed

DTC	Trouble Code Title, Conditions & Possible Causes
DTC: P2127 **2T, MIL: Yes** **Years:** 2007, 2008 **Models:** Accord, Civic, CR-V, Element, Fit, Odyssey, Pilot, Ridgeline, S2000 **Engines:** All **Transmissions:** All	**Accelerator Pedal Position Sensor 'E' Circuit Low Input Conditions:** Engine started, battery voltage at least 11.5v, all electrical components off, ground connections between engine and chassis well connected, the ECM detected that the accelerator pedal position sensor signal was outside the parameters to function normally. **Note: Both the Throttle Position (TP) Sensor and Accelerator Pedal Position Sensor are located at the accelerator pedal module and communicate the driver's intentions to the ECM completely independently of each other. Both sensors are stored in one housing.** **Possible Causes:** • Ground between engine and chassis may be broken • Throttle position sensor may have failed • Accelerator Pedal Position Sensor has failed • Throttle position sensor wiring may have shorted • Throttle position sensor has failed • Faulty voltage supply • ECM has failed
DTC: P2128 **2T, MIL: Yes** **Years:** 2007, 2008 **Models:** Accord, Civic, CR-V, Element, Fit, Odyssey, Pilot, Ridgeline, S2000 **Engines:** All **Transmissions:** All	**Accelerator Pedal Position Sensor 'E' Circuit High Input Conditions:** Engine started, battery voltage at least 11.5v, all electrical components off, ground connections between engine and chassis well connected, the ECM detected that the accelerator pedal position sensor signal was outside the parameters to function normally. **Note: Both the Throttle Position (TP) Sensor and Accelerator Pedal Position Sensor are located at the accelerator pedal module and communicate the driver's intentions to the ECM completely independently of each other. Both sensors are stored in one housing.** **Possible Causes:** • Ground between engine and chassis may be broken • Throttle position sensor may have failed • Accelerator Pedal Position Sensor has failed • Throttle position sensor wiring may have shorted • Throttle position sensor has failed • Faulty voltage supply • ECM has failed
DTC: P2133 **2T, MIL: Yes** **Years:** 2007, 2008 **Models:** Accord, Civic, CR-V, Element, Fit, Odyssey, Pilot, Ridgeline, S2000 **Engines:** All **Transmissions:** All	**Accelerator Pedal Position Sensor 'F' Circuit High Input Conditions:** Engine started, battery voltage at least 11.5v, all electrical components off, ground connections between engine and chassis well connected, the ECM detected that the accelerator pedal position sensor signal was outside the parameters to function normally. **Note: Both the Throttle Position (TP) Sensor and Accelerator Pedal Position Sensor are located at the accelerator pedal module and communicate the driver's intentions to the ECM completely independently of each other. Both sensors are stored in one housing.** **Possible Causes:** • Ground between engine and chassis may be broken • Throttle position sensor may have failed • Accelerator Pedal Position Sensor has failed • Throttle position sensor wiring may have shorted • Throttle position sensor has failed • Faulty voltage supply • ECM has failed
DTC: P2138 **2T, MIL: Yes** **Years:** 2007, 2008 **Models:** Accord, Civic, CR-V, Element, Fit, Odyssey, Pilot, Ridgeline, S2000 **Engines:** All **Transmissions:** All	**Throttle Position Sensor D/E Voltage Correlation Conditions:** Engine started, battery voltage must be at least 11.5v, all electrical components must be off, parking brake must be engaged (to keep daytime driving lights off), automatic transmission selector must be in park; and the ECM detected the Throttle Position 'D' (TPD) and Throttle Position 'B' (TPE) sensors disagreed, or that the TPD sensor should not be in its detected position, or that the TPE sensor should not be in its detected position during testing. **Note: Both the Throttle Position (TP) Sensor and Accelerator Pedal Position Sensor are located at the accelerator pedal module and communicate the driver's intentions to the ECM completely independently of each other. Both sensors are stored in one housing.** **Possible Causes:** • ETC TP sensor connector is damaged or shorted • ETC TP sensor circuits shorted together in the wire harness • ETC TP sensor signal circuit is shorted to VREF (5v) • ETC TP sensor is damaged or the ECM has failed

DTC	Trouble Code Title, Conditions & Possible Causes
DTC: P2146 **Years:** 2007, 2008 **Models:** Accord, Civic, CR-V, Element, Fit, Odyssey, Pilot, Ridgeline, S2000 **Engines:** All **Transmissions:** All	**Fuel Injector Group "A" Supply Voltage Circuit/Open Conditions:** Engine started, battery voltage must be at least 11.5v, all electrical components must be off, the ground between the engine and the chassis must be well connected, the exhaust system must be properly sealed between the catalytic converter and the cylinder head, and the oxygen sensor heater for oxygen sensor before the catalytic converter must be properly functioning. The ECM detected the fuel injector supply voltage circuit was outside the normal range during the test period. **Note: For resistance testing of sensor heating, oxygen sensor should be cooled to ambient temperature. High temperatures at oxygen sensor may lead to inaccurate measurements.** **Possible Causes:** • Oxygen sensor (before catalytic converter) is faulty • Oxygen sensor (behind catalytic converter) is faulty • Oxygen sensor heater (before catalytic converter) is faulty • Oxygen sensor heater (behind catalytic converter) is faulty • Check circuits for shorts to each other, ground or power • Fuel Injector(s) may have failed • ECM has failed
DTC: P2149 **2T, MIL: Yes** **Years:** 2007, 2008 **Models:** Accord, Civic, CR-V, Element, Fit, Odyssey, Pilot, Ridgeline, S2000 **Engines:** All **Transmissions:** All	**Fuel Injector Group "B" Supply Voltage Circuit/Open Conditions:** Engine started, battery voltage must be at least 11.5v, all electrical components must be off, the ground between the engine and the chassis must be well connected, the exhaust system must be properly sealed between the catalytic converter and the cylinder head, and the oxygen sensor heater for oxygen sensor before the catalytic converter must be properly functioning. The ECM detected the fuel injector supply voltage circuit was outside the normal range during the test period. **Note: For resistance testing of sensor heating, oxygen sensor should be cooled to ambient temperature. High temperatures at oxygen sensor may lead to inaccurate measurements.** **Possible Causes:** • Oxygen sensor (before catalytic converter) is faulty • Oxygen sensor (behind catalytic converter) is faulty • Oxygen sensor heater (before catalytic converter) is faulty • Oxygen sensor heater (behind catalytic converter) is faulty • Check circuits for shorts to each other, ground or power • Fuel Injector(s) may have failed • ECM has failed
DTC: P2177 **2T, MIL: Yes** **Years:** 2007, 2008 **Models:** Accord, Civic, CR-V, Element, Fit, Odyssey, Pilot, Ridgeline, S2000 **Engines:** All **Transmissions:** All	**System Too Lean Off Idle Bank 1 Conditions:** Engine started, battery voltage must be at least 11.5v, all electrical components must be off, the ground between the engine and the chassis must be well connected, the exhaust system must be properly sealed between the catalytic converter and the cylinder head, and the oxygen sensor heater for oxygen sensor before the catalytic converter must be properly functioning. The ECM detected the system indicated a lean signal, or it could no longer control bank 1 because it was at its lean limit. **Possible Causes:** • Intake Manifold Runner Position Sensor has failed • Intake system has leaks (false air) • Motor for intake flap is faulty • Oxygen sensor (before catalytic converter) is faulty • Oxygen sensor (behind catalytic converter) is faulty • Oxygen sensor heater (before catalytic converter) is faulty • Oxygen sensor heater (behind catalytic converter) is faulty • Check circuits for shorts to each other, ground or power • Fuel Injector(s) may have failed • ECM has failed
DTC: P2178 **2T, MIL: Yes** **Years:** 2007, 2008 **Models:** Accord, Civic, CR-V, Element, Fit, Odyssey, Pilot, Ridgeline, S2000 **Engines:** All **Transmissions:** All	**System Too Rich Off Idle Bank 1 Conditions:** Engine started, battery voltage must be at least 11.5v, all electrical components must be off, the ground between the engine and the chassis must be well connected, the exhaust system must be properly sealed between the catalytic converter and the cylinder head, and the oxygen sensor heater for oxygen sensor before the catalytic converter must be properly functioning. The ECM detected the system indicated a rich signal, or it could no longer control bank 1 because it was at its rich limit. **Possible Causes:** • Intake Manifold Runner Position Sensor has failed • Intake system has leaks (false air) • Motor for intake flap is faulty • Oxygen sensor (before catalytic converter) is faulty • Oxygen sensor (behind catalytic converter) is faulty • Oxygen sensor heater (before catalytic converter) is faulty • Oxygen sensor heater (behind catalytic converter) is faulty • Check circuits for shorts to each other, ground or power • Fuel Injector(s) may have failed • ECM has failed

DTC	Trouble Code Title, Conditions & Possible Causes
DTC: P2181 **2T, MIL: Yes** **Years:** 2007, 2008 **Models:** Accord, Civic, CR-V, Element, Fit, Odyssey, Pilot, Ridgeline, S2000 **Engines:** All **Transmissions:** All	**Cooling System Performance Malfunction Conditions:** Key on, engine cold; and the Engine Coolant Temperature (ECM) detected the ECT sensor signal was more or less than the self-test limits or has failed to gain a signal. This is a thermistor-type sensor with a variable resistance that changes when exposed to different temperatures **Possible Causes:** • ECT sensor has failed • ECT Sensor (on Radiator) has failed • ECT sensor signal circuit is open (inspect wiring & connector) • ECT sensor signal circuit is shorted • Cooling system malfunction, or the thermostat is stuck • Engine not operating at normal operating temperature • EOT sensor is damaged or it has failed
DTC: P2184 **Years:** 2007, 2008 **Models:** Accord, Civic, CR-V, Element, Fit, Odyssey, Pilot, Ridgeline, S2000 **Engines:** All **Transmissions:** All	**Engine Coolant Temperature Sensor 2 Circuit Low Conditions:** Key on or engine running; and the Engine Coolant Temperature (ECM) detected the ECT sensor signal was less than the self-test minimum. This is a thermistor-type sensor with a variable resistance that changes when exposed to different temperatures **Possible Causes:** • ECT sensor has failed • ECT Sensor (on Radiator) has failed • ECT sensor signal circuit is open (inspect wiring & connector) • ECT sensor signal circuit is shorted • Cooling system malfunction, or the thermostat is stuck • Engine not operating at normal operating temperature • EOT sensor is damaged or it has failed
DTC: P2185 **Years:** 2007, 2008 **Models:** Accord, Civic, CR-V, Element, Fit, Odyssey, Pilot, Ridgeline, S2000 **Engines:** All **Transmissions:** All	**Engine Coolant Temperature Sensor 2 Circuit High Conditions:** Key on or engine running; and the Engine Coolant Temperature (ECM) detected the ECT sensor signal was more than the self-test maximum. This is a thermistor-type sensor with a variable resistance that changes when exposed to different temperatures **Possible Causes:** • ECT sensor has failed • ECT Sensor (on Radiator) has failed • ECT sensor signal circuit is open (inspect wiring & connector) • ECT sensor signal circuit is shorted • Cooling system malfunction, or the thermostat is stuck • Engine not operating at normal operating temperature • EOT sensor is damaged or it has failed
DTC: P2187 **2T, MIL: Yes** **Years:** 2007, 2008 **Models:** Accord, Civic, CR-V, Element, Fit, Odyssey, Pilot, Ridgeline, S2000 **Engines:** All **Transmissions:** All	**System Too Lean On Idle Bank 1 Conditions:** Engine started, battery voltage must be at least 11.5v, all electrical components must be off, the ground between the engine and the chassis must be well connected, the exhaust system must be properly sealed between the catalytic converter and the cylinder head, and the oxygen sensor heater for oxygen sensor before the catalytic converter must be properly functioning. ECM detected the system indicated a lean signal, or it could no longer control bank 1 because it was at its lean limit. **Possible Causes:** • Evaporative Emission (EVAP) canister purge regulator valve is faulty • Exhaust system components are damaged • Fuel injectors are faulty • Fuel pressure regulator and residual pressure have failed • Fuel Pump (FP) in fuel tank is faulty • Intake system has leaks (false air) • Secondary Air Injection (AIR) system has an improper seal • Intake Manifold Runner Position Sensor has failed • Motor for intake flap is faulty • Oxygen sensor (before catalytic converter) is faulty • Oxygen sensor (behind catalytic converter) is faulty • Oxygen sensor heater (before catalytic converter) is faulty • Oxygen sensor heater (behind catalytic converter) is faulty • Check circuits for shorts to each other, ground or power • ECM has failed

DTC	Trouble Code Title, Conditions & Possible Causes
DTC: P2188 **2T, MIL:** Yes **Years:** 2007, 2008 **Models:** Accord, Civic, CR-V, Element, Fit, Odyssey, Pilot, Ridgeline, S2000 **Engines:** All **Transmissions:** All	**System Too Rich On Idle Bank 1 Conditions:** Engine started, battery voltage must be at least 11.5v, all electrical components must be off, the ground between the engine and the chassis must be well connected, the exhaust system must be properly sealed between the catalytic converter and the cylinder head, and the oxygen sensor heater for oxygen sensor before the catalytic converter must be properly functioning. ECM detected the system indicated a rich signal, or it could no longer control bank 1 because it was at its rich limit. **Possible Causes:** • Evaporative Emission (EVAP) canister purge regulator valve is faulty • Exhaust system components are damaged • Fuel injectors are faulty • Fuel pressure regulator and residual pressure have failed • Fuel Pump (FP) in fuel tank is faulty • Intake system has leaks (false air) • Secondary Air Injection (AIR) system has an improper seal • Intake Manifold Runner Position Sensor has failed • Motor for intake flap is faulty • Oxygen sensor (before catalytic converter) is faulty • Oxygen sensor (behind catalytic converter) is faulty • Oxygen sensor heater (before catalytic converter) is faulty • Oxygen sensor heater (behind catalytic converter) is faulty • Check circuits for shorts to each other, ground or power • ECM has failed
DTC: P2191 **2T, MIL:** Yes **Years:** 2007, 2008 **Models:** Accord, Civic, CR-V, Element, Fit, Odyssey, Pilot, Ridgeline, S2000 **Engines:** All **Transmissions:** All	**System Too Lean at Higher Load Bank 1 Conditions:** Engine started, battery voltage must be at least 11.5v, all electrical components must be off, the ground between the engine and the chassis must be well connected, the exhaust system must be properly sealed between the catalytic converter and the cylinder head, and the oxygen sensor heater for oxygen sensor before the catalytic converter must be properly functioning. ECM detected the system indicated a lean signal, or it could no longer control bank 1 because it was at its lean limit. **Possible Causes:** • Evaporative Emission (EVAP) canister purge regulator valve is faulty • Exhaust system components are damaged • Fuel injectors are faulty • Fuel pressure regulator and residual pressure have failed • Fuel Pump (FP) in fuel tank is faulty • Intake system has leaks (false air) • Secondary Air Injection (AIR) system has an improper seal • Intake Manifold Runner Position Sensor has failed • Motor for intake flap is faulty • Oxygen sensor (before catalytic converter) is faulty • Oxygen sensor (behind catalytic converter) is faulty • Oxygen sensor heater (before catalytic converter) is faulty • Oxygen sensor heater (behind catalytic converter) is faulty • Check circuits for shorts to each other, ground or power • ECM has failed
DTC: P2192 **2T, MIL:** Yes **Years:** 2007, 2008 **Models:** Accord, Civic, CR-V, Element, Fit, Odyssey, Pilot, Ridgeline, S2000 **Engines:** All **Transmissions:** All	**System Too Rich at Higher Load Bank 1 Conditions:** Engine started, battery voltage must be at least 11.5v, all electrical components must be off, the ground between the engine and the chassis must be well connected, the exhaust system must be properly sealed between the catalytic converter and the cylinder head, and the oxygen sensor heater for oxygen sensor before the catalytic converter must be properly functioning. ECM detected the system indicated a rich signal, or it could no longer control bank 1 because it was at its rich limit. **Possible Causes:** • Evaporative Emission (EVAP) canister purge regulator valve is faulty • Exhaust system components are damaged • Fuel injectors are faulty • Fuel pressure regulator and residual pressure have failed • Fuel Pump (FP) in fuel tank is faulty • Intake system has leaks (false air) • Secondary Air Injection (AIR) system has an improper seal • Intake Manifold Runner Position Sensor has failed • Motor for intake flap is faulty • Oxygen sensor (before catalytic converter) is faulty • Oxygen sensor (behind catalytic converter) is faulty • Oxygen sensor heater (before catalytic converter) is faulty • Oxygen sensor heater (behind catalytic converter) is faulty • Check circuits for shorts to each other, ground or power • ECM has failed

DTC	Trouble Code Title, Conditions & Possible Causes
DTC: P2195 **2T, MIL: Yes** **Years:** 2007, 2008 **Models:** Accord, Civic, CR-V, Element, Fit, Odyssey, Pilot, Ridgeline, S2000 **Engines:** All **Transmissions:** All	**O2 Sensor Signal Stuck Lean Bank 1 Sensor 1 Conditions:** Engine running in closed loop, and the ECM detected the O2S indicated a lean signal, or it could no longer control Fuel Trim because it was at lean limit. **Possible Causes:** • Engine oil level high • Camshaft timing error • Cylinder compression low • Exhaust leaks in front of O2S • EGR valve is stuck open • EGR gasket is leaking • EVR diaphragm is leaking • Damaged fuel pressure regulator or extremely low fuel pressure • O2S circuit is open or shorted in the wiring harness • Oxygen sensor (before catalytic converter) is faulty • Oxygen sensor (behind catalytic converter) is faulty • Oxygen sensor heater (before catalytic converter) is faulty • Oxygen sensor heater (behind catalytic converter) is faulty • Air leaks after the MAF sensor • PCV system leaks • Dip stick not seated properly
DTC: P2270 **2T, MIL: Yes** **Years:** 2007, 2008 **Models:** Accord, Civic, CR-V, Element, Fit, Odyssey, Pilot, Ridgeline, S2000 **Engines:** All **Transmissions:** All	**O2 Sensor Signal Stuck Lean Bank 1 Sensor 2 Conditions:** Engine started, battery voltage must be at least 11.5v, all electrical components must be off, parking brake must be engaged (to keep daytime driving lights off), automatic transmission selector must be in park. The ECM detected an unexpected voltage condition, or it detected an unexpected current draw in the heater circuit during the CCM test. **Note: Vehicle must be raised before connector for oxygen sensors is accessible.** **Possible Causes:** • Oxygen sensor (before catalytic converter) is faulty • Oxygen sensor heater (before catalytic converter) is faulty • Oxygen sensor heater (before catalytic converter) is faulty • Oxygen sensor heater (behind catalytic converter) is faulty • O2S circuit is open or shorted in the wiring harness • ECM has failed
DTC: P2271 **2T, MIL: Yes** **Years:** 2007, 2008 **Models:** Accord, Civic, CR-V, Element, Fit, Odyssey, Pilot, Ridgeline, S2000 **Engines:** All **Transmissions:** All	**O2 Sensor Signal Stuck Rich Bank 1 Sensor 2 Conditions:** Engine started, battery voltage must be at least 11.5v, all electrical components must be off, parking brake must be engaged (to keep daytime driving lights off), automatic transmission selector must be in park. The ECM detected an unexpected voltage condition, or it detected an unexpected current draw in the heater circuit during the CCM test. **Note: Vehicle must be raised before connector for oxygen sensors is accessible.** **Possible Causes:** • Oxygen sensor (before catalytic converter) is faulty • Oxygen sensor heater (before catalytic converter) is faulty • Oxygen sensor heater (before catalytic converter) is faulty • Oxygen sensor heater (behind catalytic converter) is faulty • O2S circuit is open or shorted in the wiring harness • ECM has failed

ISUZU

i-290 • i-370

SPECIFICATIONS AND MAINTENANCE CHARTS

ENGINE AND VEHICLE IDENTIFICATION

Code ①	Liters (cc)	Cu. In.	Cyl.	Fuel Sys.	Engine Type	Eng. Mfg.	Code ②	Year
				Engine			Model Year	
9	2.9 (2900)	178	4	SFI	DOHC	CPC	7	2007
E	3.7 (3700)	223	5	SFI	DOHC	CPC	8	2008

CPC: Chevrolet/Pontiac/Canada

SFI: Sequential Fuel Injection

① 8th position of VIN

② 10th position of VIN

22140_I290_C0001

GENERAL ENGINE SPECIFICATIONS

All measurements are given in inches.

Year	Model	Engine Displacement Liters	Engine Series VIN	Net Horsepower @ rpm	Net Torque @ rpm (ft. lbs.)	Bore x Stroke (in.)	Compression Ratio	Oil Pressure @ rpm
2007	i290	2.9	9	185@5600	190@2800	3.76x4.02	10.3:1	12@1200
	i370	3.7	E	242@5600	242@4600	3.76x4.02	10.3:1	12@1200
2008	i290	2.9	9	185@5600	190@2800	3.76x4.02	10.3:1	12@1200
	i370	3.7	E	242@5600	242@4600	3.76x4.02	10.3:1	12@1200

22140_I290_C0002

GASOLINE ENGINE TUNE-UP SPECIFICATIONS

Year	Engine Displacement Liters	Engine VIN	Spark Plug Gap (in.)	Ignition Timing (deg.) MT	Ignition Timing (deg.) AT	Fuel Pump (psi)	Idle Speed (rpm) MT	Idle Speed (rpm) AT	Valve Clearance In.	Valve Clearance Ex.
2007	2.9	9	0.042	①	①	50-57	②	②	HYD	HYD
	3.7	E	0.042	①	①	50-57	②	②	HYD	HYD
2008	2.9	9	0.042	①	①	50-57	②	②	HYD	HYD
	3.7	E	0.042	①	①	50-57	②	②	HYD	HYD

NOTE: The Vehicle Emission Control Information label often reflects specification changes made during production.

The label figures must be used if they differ from those in this chart.

HYD: Hydraulic

① Ignition timing is preset and cannot be adjusted

② Idle speed is maintained by the PCM

22140_I290_C0003

CAPACITIES

Year	Model	Engine Displacement Liters	Engine VIN	Engine Oil with Filter (qts.)	Transmission (pts.) 5-Spd	Transmission (pts.) Auto.	Transfer Case (pts.)	Drive Axle Front (pts.)	Drive Axle Rear (pts.)	Fuel Tank (gal.)	Cooling System (qts.)
2007	i290	2.9	9	5	①	②	—	3.2	3.4-3.8	19.0	10.4
	i370	3.7	E	6	①	②	—	3.2	3.4-3.8	19.0	10.6
2008	i290	2.9	9	5	①	②	—	3.2	3.4-3.8	19.0	10.4
	i370	3.7	E	6	①	②	—	3.2	3.4-3.8	19.0	10.6

NOTE: All capacities are approximate. Add fluid gradually and check to be sure a proper fluid level is obtained.

① RWD: 4.6
   4WD: 4.8

② w/245mm Torque Converter (Dry Fill) 19.8 qts./39.6 pts.
   w/258mm Torque Converter (Dry Fill) 20.3 qts./40.6 pts.

22140_I290_C0004

FLUID SPECIFICATIONS

Year	Model	Engine Displacement Liters	Engine ID/VIN	Engine Oil	Auto. Trans. ①	Drive Axle	Power Steering Fluid	Brake Master Cylinder
2007	i290	2.9	9	5W-30	Dexron VI	75W-90	GM Part No. 89021184	DOT 3
	i370	3.7	E	5W-30	Dexron VI	75W-90	GM Part No. 89021184	DOT 3
2008	i290	2.9	9	5W-30	Dexron VI	75W-90	GM Part No. 89021184	DOT 3
	i370	3.7	E	5W-30	Dexron VI	75W-90	GM Part No. 89021184	DOT 3

DOT: Department Of Transpotation

① Type 9601 may be substituted

22140_I290_C0015

VALVE SPECIFICATIONS

Year	Engine Displacement Liters	Engine VIN	Seat Angle (deg.)	Face Angle (deg.)	Spring Test Pressure (lbs. @ in.)	Spring Installed Height (in.)	Stem-to-Guide Clearance (in.) Intake	Stem-to-Guide Clearance (in.) Exhaust	Stem Diameter (in.) Intake	Stem Diameter (in.) Exhaust
2007	2.9	9	NA	NA	130-142 @0.965	1.379	0.0011-0.0025	0.0015-0.0030	NA	NA
	3.7	E	NA	NA	130-142 @0.965	1.701	0.0011-0.0025	0.0015-0.0030	NA	NA
2008	2.9	9	NA	NA	130-142 @0.965	1.379	0.0011-0.0025	0.0015-0.0030	NA	NA
	3.7	E	NA	NA	130-142 @0.965	1.701	0.0011-0.0025	0.0015-0.0030	NA	NA

NA: Not Available

22140_I290_C0005

CAMSHAFT AND BEARING SPECIFICATIONS CHART

All measurements are given in inches.

Year	Engine Displ. Liters	Engine ID/VIN	Journal Dia.	Brg. Oil Clearance	Shaft End-play	Runout	Journal Bore	Lobe Height Intake	Lobe Height Exhaust
2007	2.9	9	①	NA	②	NA	0.0015-0.0033	NA	NA
	3.7	E	①	NA	②	NA	0.0015-0.0033	NA	NA
2008	2.9	9	①	NA	②	NA	0.0015-0.0033	NA	NA
	3.7	E	①	NA	②	NA	0.0015-0.0033	NA	NA

NA: Not Available

① All intake and exhaust no's. 2 through 7: 1.0612 - 1.0622 in.
    Exhaust no.1: 1.794 - 1.1804 in.

② Exhaust: 0.0017 - 0.0084 in.
    Intake: 0.0020 - 0.0079 in.

22140_I290_C0016

CRANKSHAFT AND CONNECTING ROD SPECIFICATIONS

All measurements are given in inches.

Year	Engine Displacement Liters	Engine VIN	Crankshaft Main Brg. Journal Dia.	Crankshaft Main Brg. Oil Clearance	Crankshaft Shaft End-play	Crankshaft Thrust on No.	Connecting Rod Journal Diameter	Connecting Rod Oil Clearance	Connecting Rod Side Clearance
2007	2.9	9	2.7567-2.7574	0.0004-0.0025	0.0044-0.0153	3	2.3749-2.3755	0.0008-0.0025	0.0019-0.0137
	3.7	E	2.7567-2.7574	0.0004-0.0025	0.0044-0.0153	4	2.3749-2.3755	0.0008-0.0025	0.0019-0.0137
2008	2.9	9	2.7567-2.7574	0.0004-0.0025	0.0044-0.0153	3	2.3749-2.3755	0.0008-0.0025	0.0019-0.0137
	3.7	E	2.7567-2.7574	0.0004-0.0025	0.0044-0.0153	4	2.3749-2.3755	0.0008-0.0025	0.0019-0.0137

22140_I290_C0006

PISTON AND RING SPECIFICATIONS

All measurements are given in inches.

Year	Engine Displ. Liters	Engine VIN	Piston Clearance	Ring Gap Top Compression	Ring Gap Bottom Compression	Ring Gap Oil Control	Ring Side Clearance Top Compression	Ring Side Clearance Bottom Compression	Ring Side Clearance Oil Control
2007	2.9	9	0.0006-0.0014	0.00787-0.0157	0.0142-0.0201	0.0098-0.0299	0.0017-0.0037	0.0021-0.0037	0.0023-0.0085
	3.7	E	0.0004-0.0017	0.0079-0.0157	0.0142-0.0201	0.0098-0.0299	0.0017-0.0037	0.0021-0.0037	0.0023-0.0085
2008	2.9	9	0.0006-0.0014	0.00787-0.0157	0.0142-0.0201	0.0098-0.0299	0.0017-0.0037	0.0021-0.0037	0.0023-0.0085
	3.7	E	0.0004-0.0017	0.0079-0.0157	0.0142-0.0201	0.0098-0.0299	0.0017-0.0037	0.0021-0.0037	0.0023-0.0085

22140_I290_C0007

TORQUE SPECIFICATIONS
All readings in ft. lbs.

Year	Engine Displacement Liters	Engine VIN	Cylinder Head Bolts	Main Bearing Bolts	Rod Bearing Bolts	Crankshaft Damper Bolts	Flywheel Bolts	Manifold Intake	Manifold Exhaust	Spark Plugs	Oil Pan Drain Plug
2007	2.9	9	①	②	③	④	⑤	⑥	⑦	13	19
	3.7	E	①	②	③	④	⑤	⑥	⑦	13	19
2008	2.9	9	①	②	③	④	⑤	⑥	⑦	13	19
	3.7	E	①	②	③	④	⑤	⑥	⑦	13	19

① 1st pass: 22 ft. lbs.
  2nd pass: Plus 155 degrees
  Short end bolt
  1st pass: 62 inch lbs.
  2nd pass: plus 60 degrees
  Long end bolt
  1st pass: 62 inch lbs.
  2nd pass: plus 120 degrees
② 1st pass: 18 ft. lbs.
  2nd pass: plus 180 degrees

③ 1st pass: 18 ft. lbs.
  2nd pass: plus 110 degrees
④ 1st pass: 110 ft. lbs.
  2nd pass: plus 180 degrees
⑤ 1st pass: 30 ft. lbs.
  2nd pass: plus 45 degrees
⑥ 89 inch lbs.
⑦ 1st pass: 15 ft. lbs.
  Repeat twice more in sequence

22140_I290_C0008

WHEEL ALIGNMENT

Year	Model		Caster Range (+/-Deg.)	Caster Preferred Setting (Deg.)	Camber Range (+/-Deg.)	Camber Preferred Setting (Deg.)	Toe-in (Deg.)
2007	2WD	Left	1.0	+4.3	0.60	0	0+/-0.20
		Right	1.0	+4.5	—	—	0+/-0.20
	4WD	Left	1.0	+3.8	0.60	0	0+/-0.20
		Right	1.0	+4.0	—	—	0+/-0.20
2008	2WD	Left	1.0	+4.3	0.60	0	0+/-0.20
		Right	1.0	+4.5	—	—	0+/-0.20
	4WD	Left	1.0	+3.8	0.60	0	0+/-0.20
		Right	1.0	+4.0	—	—	0+/-0.20

22140_I290_C0009

TIRE, WHEEL AND BALL JOINT SPECIFICATIONS

| Year | Model | OEM Tires | | Tire Pressures (psi) | | Wheel Size | Ball Joint Inspection | Lug Nut Torque (ft. lbs.) |
		Standard	Optional	Front	Rear			
2007	2WD ①	P205/75R15	P225/75R15	②	②	15x6 15x6.5	NA	103
	4WD ③	P235/75R15	None	②	②	15x6 15x6.5	NA	103
2008	Z85 2WD ②	P205/75R15	P225/75R15	②	②	15x6 15x6.5	NA	103
	Z85 4WD ③	P235/75R15	None	②	②	15x6 15x6.5	NA	103

NA: Not Available

OEM: Original Equipment Manufacturer

OPT: Optional

PSI: Pounds Per Square Inch

STD: Standard

① Compact Spare Standard, P225/75R15 Optional

② Refer to placard on vehicle for proper inflation pressure

③ Compact Spare Standard, P235/75R15 Optional

22140_I290_C0010

BRAKE SPECIFICATIONS
All measurements in inches unless noted

| Year | Model | | Brake Disc | | | Brake Drum Diameter | | | Minimum Lining Thickness | Brake Caliper | |
			Original Thickness	Minimum Thickness	Maximum Runout	Original Inside Diameter	Max. Wear Limit	Maximum Machine Diameter		Bracket Bolts (ft. lbs.)	Mounting Bolts (ft. lbs.)
2007	i290	F	1.060	1.000	0.002	—	—	—	0.070	129	29
		R	—	—	—	—	—	11.673	0.030	—	—
	i370	F	1.060	1.000	0.002	—	—	—	0.070	129	29
		R	—	—	—	—	—	11.673	0.030	—	—
2008	i290	F	1.060	1.000	0.002	—	—	—	0.070	129	29
		R	—	—	—	—	—	11.673	0.030	—	—
	i370	F	1.060	1.000	0.002	—	—	—	0.070	129	29
		R	—	—	—	—	—	11.673	0.030	—	—

22140_I290_C0011

SCHEDULED MAINTENANCE INTERVALS
2007-08 ISUZU i290 & i370

When the CHANGE ENGINE OIL light appears, certain services and inspections are required.

Required services are described as Maintenance I and Maintenance II.

The first service on a vehicle should be Maintenance I, and the second service should be Maintenance II. Alternate between the 2 thereafter. However, in some cases, Maintenance II may be required more often.

Maintenance I: Use Maintenance I if the CHANGE ENGINE OIL light comes on within 10 months since vehicle was purchased or, if Maintenance II was performed.

Maintenance II: Use Maintenance II if the previous service performed was Maintenance I.

Always use Maintenance II whenever the CHANGE ENGINE OIL light comes on 10 months or more since the last service, or, if the CHANGE ENGINEOIL light has not come on at all for one year.

Service	Maintenance I	Maintenance II
Change oil and oil filter, then Reset Oil Life Monitor ①	✓	✓
Visually inspect vehicle for any leaks or damage	✓	✓
Inspect engine air filter and replace as necessary		✓
Rotate tires, check for unusual wear and reset tire pressures ②	✓	✓
Inspect brake system	✓	✓
Check engine coolant and add fluid as needed	✓	✓
See additional required services	✓	✓
Inspect suspension and steering components		✓
Inspect engine cooling system		✓
Inspect wiper blades and windshield washer fluid level		✓
Inspect restraint system components		✓
Lubricate body components		✓
Check transmission and transfer case fluid level and add fluid as required		✓

① Mileage interval varies based on your driving habits.

② See Placard on Vehicle for Proper Inflation Pressure

Engine Oil Life System Reset Instructions

1). Begin with the engine off and the key in the "Lock" position.

2). Turn key to the "On" position

3). Press and release the stem located at the lower center of the instrument cluster until the "Oil Life" message is displayed.

4). Wait for the "Oil Life" and "Reset" message appear, then press and hold the stem down until several beeps are heard.

5). Turn the key to the"Lock" position.

22140_I290_C0012

ADDITIONAL SERVICE REQUIREMENTS
2007-2008 ISUZU i290 & i370

TO BE SERVICED	TYPE OF SERVICE	VEHICLE MILEAGE INTERVAL (x1000)					
		25	50	75	100	125	150
Fuel system	I	✓	✓	✓	✓	✓	✓
Exhaust system	I	✓	✓	✓	✓	✓	✓
Air filter	R		✓		✓		✓
Automatic transmission fluid and filter (Severe Service)	S		✓		✓		✓
Automatic transmission fluid and filter (Normal Service)	S				✓		
Spark plugs	R				✓		
Cooling system	S						✓
Accessory drive belt	I						✓

22140_I290_C0013

PRECAUTIONS

Before servicing any vehicle, please be sure to read all of the following precautions, which deal with personal safety, prevention of component damage, and important points to take into consideration when servicing a motor vehicle:

• Never open, service or drain the radiator or cooling system when the engine is hot; serious burns can occur from the steam and hot coolant.

• Observe all applicable safety precautions when working around fuel. Whenever servicing the fuel system, always work in a well-ventilated area. Do not allow fuel spray or vapors to come in contact with a spark, open flame, or excessive heat (a hot drop light, for example). Keep a dry chemical fire extinguisher near the work area. Always keep fuel in a container specifically designed for fuel storage; also, always properly seal fuel containers to avoid the possibility of fire or explosion. Refer to the additional fuel system precautions later in this section.

• Fuel injection systems often remain pressurized, even after the engine has been turned **OFF**. The fuel system pressure must be relieved before disconnecting any fuel lines. Failure to do so may result in fire and/or personal injury.

• Brake fluid often contains polyglycol ethers and polyglycols. Avoid contact with the eyes and wash your hands thoroughly after handling brake fluid. If you do get brake fluid in your eyes, flush your eyes with clean, running water for 15 minutes. If eye irritation persists, or if you have taken brake fluid internally, IMMEDIATELY seek medical assistance.

• The EPA warns that prolonged contact with used engine oil may cause a number of skin disorders, including cancer. You should make every effort to minimize your exposure to used engine oil. Protective gloves should be worn when changing oil. Wash your hands and any other exposed skin areas as soon as possible after exposure to used engine oil. Soap and water, or waterless hand cleaner should be used.

• All new vehicles are now equipped with an air bag system, often referred to as a Supplemental Restraint System (SRS) or Supplemental Inflatable Restraint (SIR) system. The system must be disabled before performing service on or around system components, steering column, instrument panel components, wiring and sensors. Failure to follow safety and disabling procedures could result in accidental air bag deployment, possible personal injury and unnecessary system repairs.

• Always wear safety goggles when working with, or around, the air bag system. When carrying a non-deployed air bag, be sure the bag and trim cover are pointed away from your body. When placing a non-deployed air bag on a work surface, always face the bag and trim cover upward, away from the surface. This will reduce the motion of the module if it is accidentally deployed. Refer to the additional air bag system precautions later in this section.

• Clean, high quality brake fluid from a sealed container is essential to the safe and proper operation of the brake system. You should always buy the correct type of brake fluid for your vehicle. If the brake fluid becomes contaminated, completely flush the system with new fluid. Never reuse any brake fluid. Any brake fluid that is removed from the system should be discarded. Also, do not allow any brake fluid to come in contact with a painted surface; it will damage the paint.

• Never operate the engine without the proper amount and type of engine oil; doing so WILL result in severe engine damage.

• Timing belt maintenance is extremely important. Many models utilize an interference-type, non-freewheeling engine. If the timing belt breaks, the valves in the cylinder head may strike the pistons, causing potentially serious (also time-consuming and expensive) engine damage. Refer to the maintenance interval charts for the recommended replacement interval for the timing belt, and to the timing belt section for belt replacement and inspection.

• Disconnecting the negative battery cable on some vehicles may interfere with the functions of the on-board computer system(s) and may require the computer to undergo a relearning process once the negative battery cable is reconnected.

• When servicing drum brakes, only disassemble and assemble one side at a time, leaving the remaining side intact for reference.

• Only an MVAC-trained, EPA-certified automotive technician should service the air conditioning system or its components.

BRAKES

GENERAL INFORMATION

PRECAUTIONS

• Certain components within the ABS system are not intended to be serviced or repaired individually.

• Do not use rubber hoses or other parts not specifically specified for and ABS system. When using repair kits, replace all parts included in the kit. Partial or incorrect repair may lead to functional problems and require the replacement of components.

• Lubricate rubber parts with clean, fresh brake fluid to ease assembly. Do not use shop air to clean parts; damage to rubber components may result.

• Use only DOT 3 brake fluid from an unopened container.

• If any hydraulic component or line is removed or replaced, it may be necessary to bleed the entire system.

• A clean repair area is essential. Always clean the reservoir and cap thoroughly before removing the cap. The slightest amount of dirt in the fluid may plug an orifice and impair the system function. Perform repairs after components have been thoroughly cleaned; use only denatured alcohol

ANTI-LOCK BRAKE SYSTEM (ABS)

to clean components. Do not allow ABS components to come into contact with any substance containing mineral oil; this includes used shop rags.

• The Anti-Lock control unit is a microprocessor similar to other computer units in the vehicle. Ensure that the ignition switch is **OFF** before removing or installing controller harnesses. Avoid static electricity discharge at or near the controller.

• If any arc welding is to be done on the vehicle, the control unit should be unplugged before welding operations begin.

BRAKES   BLEEDING THE BRAKE SYSTEM

BLEEDING PROCEDURE

When bleeding the brake system, bleed one brake cylinder at a time, beginning at the cylinder with the longest hydraulic line (farthest from the master cylinder) first. ALWAYS Keep the master cylinder reservoir filled with brake fluid during the bleeding operation. Never use brake fluid that has been drained from the hydraulic system, no matter how clean it is.

The primary and secondary hydraulic brake systems are separate and are bled independently. During the bleeding operation, do not allow the reservoir to run dry. Keep the master cylinder reservoir filled with brake fluid.

1. Clean all dirt from around the master cylinder fill cap, remove the cap and fill the master cylinder with brake fluid until the level is within ¼ in. (6mm) of the top edge of the reservoir.

2. Clean the bleeder screws at all 4 wheels. The bleeder screws are located on the back of the brake backing plate (drum brakes) and on the top of the brake calipers (disc brakes).

3. Attach a length of rubber hose over the bleeder screw and place the other end of the hose in a glass jar, submerged in brake fluid.

4. Open the bleeder screw ½–¾ turn. Have an assistant slowly depress the brake pedal.

5. Close the bleeder screw and tell your assistant to allow the brake pedal to return slowly. Continue this process to purge all air from the system.

6. When bubbles cease to appear at the end of the bleeder hose, close the bleeder screw and remove the hose.

7. Check the master cylinder fluid level and add fluid accordingly. Do this after bleeding each wheel.

✴✴ WARNING

Clean, high quality brake fluid is essential to the safe and proper operation of the brake system. You should always buy the highest quality brake fluid that is available. If the brake fluid becomes contaminated, drain and flush the system, then refill the master cylinder with new fluid. Never reuse any brake fluid. Any brake fluid that is removed from the system should be discarded. Also, do not allow any brake fluid to come in contact with a painted surface; it will damage the paint.

8. Repeat the bleeding operation at the remaining 3 wheels, ending with the one closet to the master cylinder.

9. Fill the master cylinder reservoir to the proper level.

BRAKES   FRONT DISC BRAKES

✴✴ CAUTION

Dust and dirt accumulating on brake parts during normal use may contain asbestos fibers from production or aftermarket brake linings. Breathing excessive concentrations of asbestos fibers can cause serious bodily harm. Exercise care when servicing brake parts. Do not sand or grind brake lining unless equipment used is designed to contain the dust residue. Do not clean brake parts with compressed air or by dry brushing. Cleaning should be done by dampening the brake components with a fine mist of water, then wiping the brake components clean with a dampened cloth. Dispose of cloth and all residue containing asbestos fibers in an impermeable container with the appropriate label. Follow practices prescribed by the Occupational Safety and Health Administration (OSHA) and the Environmental Protection Agency (EPA) for the handling, processing, and disposing of dust or debris that may contain asbestos fibers.

BRAKE CALIPER

REMOVAL & INSTALLATION

1. Before servicing the vehicle, refer to the Precautions Section.

2. If brake fluid level is midway between MAX and MIN level in the reservoir, no fluid needs to be removed. If brake fluid level is higher than midway, remove the brake fluid to the midway point.

3. Remove or disconnect the following:
- Tire and wheel assembly
- Brake caliper fluid line, then plug it
- Caliper slide pin bolts
- Caliper from the mounting bracket

To install:

4. Clean and lubricate the sleeves and bushings with silicon grease.

5. Install or connect the following:

- Caliper in mounting bracket
- Slide pin bolts. Tighten to 29 ft. lbs. (40 Nm).
- Fluid lines to the caliper using new Copper washers, and tighten to 29 ft. lbs. (40 Nm)
- Wheel and tire assembly

6. Refill the master cylinder to the correct level. Bleed the brake system.

DISC BRAKE PADS

REMOVAL & INSTALLATION

See Figure 1.

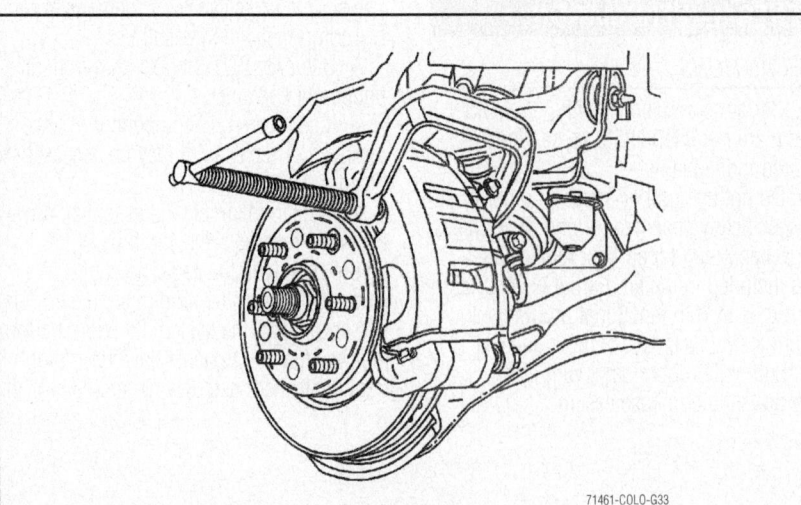

71461-COLO-G33

Fig. 1 Compressing the caliper piston with a C-clamp

1. Before servicing the vehicle, refer to the Precautions Section.

2. If brake fluid level is midway between MAX and MIN level in the reservoir, no fluid needs to be removed. If brake fluid level is higher than midway, remove the brake fluid to the midway point.

3. Remove or disconnect the following:

4. Place a C-clamp around the outer pad and caliper; tighten the C-clamp until the piston is fully compressed in the caliper.

- Remove top caliper retainer, and rotate caliper away from rotor
- Inboard pad and retaining clips
- Outboard pad from the caliper
- Shims

To install:

5. Install or connect the following:
- New retaining clips onto the inboard pad

- Shims
- Outboard pad into the caliper
- Inboard pad in the caliper
- Caliper in position over the rotor
- Caliper bolts and tighten to 29 ft. lbs. (40 Nm).
- Wheel and tire

6. Refill the master cylinder and pump pedal to attain full brake pedal before road-testing the vehicle.

BRAKES

❊❊ CAUTION

Dust and dirt accumulating on brake parts during normal use may contain asbestos fibers from production or aftermarket brake linings. Breathing excessive concentrations of asbestos fibers can cause serious bodily harm. Exercise care when servicing brake parts. Do not sand or grind brake lining unless equipment used is designed to contain the dust residue. Do not clean brake parts with compressed air or by dry brushing. Cleaning should be done by dampening the brake components with a fine mist of water, then wiping the brake components clean with a dampened cloth. Dispose of cloth and all residue containing asbestos fibers in an impermeable container with the appropriate label. Follow practices prescribed by the Occupational Safety and Health Administration (OSHA) and the Environmental Protection Agency (EPA) for the handling, processing, and disposing of dust or debris that may contain asbestos fibers.

BRAKE DRUM

REMOVAL & INSTALLATION

1. Before servicing the vehicle, refer to the Precautions Section.

2. Remove or disconnect the following:
- Wheel and tire assembly
- Retaining clip
- Brake drum. If the drum will not pull of the axle, use a rubber mallet and tap it around the edge.

To install:

3. Install or connect the following:
- Drum on the axle
- Retaining clip
- Wheel and tire assembly

4. Refill the master cylinder and pump pedal to attain full brake pedal before road-testing the vehicle.

BRAKE SHOES

REMOVAL & INSTALLATION

1. Before servicing the vehicle, refer to the Precautions Section.

2. Remove or disconnect the following:
- Wheel and tire assembly
- Brake drum
- Adjuster assembly

REAR DRUM BRAKES

- Retractor spring from secondary shoe
- Secondary shoe
- Retractor spring from primary shoe
- Primary shoe
- Return spring
- Depress lock tab on parking brake cable
- Hold lock tab and push parking brake cable forward
- Parking brake cable from lever

To install:

3. Lubricate the contact points on the backing plate with high temperature silicone grease.

4. Install or connect the following:
- Parking brake cable in the lever
- Primary shoe
- Retractor spring on primary shoe
- Secondary shoe
- Retractor spring on secondary shoe
- Adjuster assembly

5. Adjust the brake shoes so there is 0.030 inch (0.76 mm) clearance between the lining and the drum.

6. Install the brake drum.

7. Adjust the parking brake cable as necessary.

8. Install the wheel and tire assemblies.

9. Refill the master cylinder and pump pedal to attain full brake pedal before Road-testing the vehicle.

BRAKES

PARKING BRAKE

PARKING BRAKE SHOES

The rear drum brake shoes serve as the parking brakes. Refer to the procedures under Rear Drum Brakes.

CHASSIS ELECTRICAL

AIR BAG (SUPPLEMENTAL RESTRAINT SYSTEM)

GENERAL INFORMATION

☀ CAUTION

These vehicles are equipped with an air bag system. The system must be disarmed before performing service on, or around, system components, the steering column, instrument panel components, wiring and sensors. Failure to follow the safety precautions and the disarming procedure could result in accidental air bag deployment, possible injury and unnecessary system repairs.

SERVICE PRECAUTIONS

Disconnect and isolate the battery negative cable before beginning any airbag system component diagnosis, testing, removal, or installation procedures. Allow system capacitor to discharge for two minutes before beginning any component service. This will disable the airbag system. Failure to disable the airbag system may result in accidental airbag deployment, personal injury, or death.

Do not place an intact undeployed airbag face down on a solid surface. The airbag will propel into the air if accidentally deployed and may result in personal injury or death.

When carrying or handling an undeployed airbag, the trim side (face) of the airbag should be pointing towards the body to minimize possibility of injury if accidental deployment occurs. Failure to do this may result in personal injury or death.

Replace airbag system components with OEM replacement parts. Substitute parts may appear interchangeable, but internal differences may result in inferior occupant protection. Failure to do so may result in occupant personal injury or death.

Wear safety glasses, rubber gloves, and long sleeved clothing when cleaning powder residue from vehicle after an airbag deployment. Powder residue emitted from a deployed airbag can cause skin irritation. Flush affected area with cool water if irrita-

tion is experienced. If nasal or throat irritation is experienced, exit the vehicle for fresh air until the irritation ceases. If irritation continues, see a physician.

Do not use a replacement airbag that is not in the original packaging. This may result in improper deployment, personal injury, or death.

The factory installed fasteners, screws and bolts used to fasten airbag components have a special coating and are specifically designed for the airbag system. Do not use substitute fasteners. Use only original equipment fasteners listed in the parts catalog when fastener replacement is required.

During, and following, any child restraint anchor service, due to impact event or vehicle repair, carefully inspect all mounting hardware, tether straps, and anchors for proper installation, operation, or damage. If a child restraint anchor is found damaged in any way, the anchor must be replaced. Failure to do this may result in personal injury or death.

Deployed and non-deployed airbags may or may not have live pyrotechnic material within the airbag inflator.

Do not dispose of driver/passenger/curtain airbags or seat belt tensioners unless you are sure of complete deployment. Refer to the Hazardous Substance Control System for proper disposal.

Dispose of deployed airbags and tensioners consistent with state, provincial, local, and federal regulations.

After any airbag component testing or service, do not connect the battery negative cable. Personal injury or death may result if the system test is not performed first.

If the vehicle is equipped with the Occupant Classification System (OCS), do not connect the battery negative cable before performing the OCS Verification Test using the scan tool and the appropriate diagnostic information. Personal injury or death may result if the system test is not performed properly.

Never replace both the Occupant Restraint Controller (ORC) and the Occupant Classification Module (OCM) at the same time. If both require replacement,

replace one, then perform the Airbag System test before replacing the other.

Both the ORC and the OCM store Occupant Classification System (OCS) calibration data, which they transfer to one another when one of them is replaced. If both are replaced at the same time, an irreversible fault will be set in both modules and the OCS may malfunction and cause personal injury or death.

If equipped with OCS, the Seat Weight Sensor is a sensitive, calibrated unit and must be handled carefully. Do not drop or handle roughly. If dropped or damaged, replace with another sensor. Failure to do so may result in occupant injury or death.

If equipped with OCS, the front passenger seat must be handled carefully as well. When removing the seat, be careful when setting on floor not to drop. If dropped, the sensor may be inoperative, could result in occupant injury, or possibly death.

If equipped with OCS, when the passenger front seat is on the floor, no one should sit in the front passenger seat. This uneven force may damage the sensing ability of the seat weight sensors. If sat on and damaged, the sensor may be inoperative, could result in occupant injury, or possibly death.

DISARMING THE SYSTEM

1. Turn the steering wheel so that the vehicle's wheels are pointing straight ahead.
2. Turn the ignition switch to **LOCK**, remove the key, then disconnect the negative battery cable.
3. Remove the SIR fuse from the fuse block.
4. Remove the steering column filler panel or knee bolster.
5. Unplug the Connector Position Assurance (CPA) and yellow four way connector at the base of the steering column.
6. Remove the Connector Position Assurance (CPA) from the passenger yellow four way connector located behind the glove box.
7. Unplug the yellow four way connector located behind the glove box.
8. Connect the negative battery cable.

→With the AIR BAG fuse removed, the battery cable connected and the ignition in the ON position, the AIR BAG warning lamp will be ON. This is normal and does not indicate a system malfunction.

ARMING THE SYSTEM

1. Disconnect the negative battery cable.
2. Attach the yellow four way connector located behind the glove box.
3. Install the Connector Position Assurance (CPA) to the passenger yellow four way connector located behind the glove box.
4. Turn the ignition switch to **LOCK**, then remove the key.
5. Attach the four way connector at the base of the steering column and the Connector Position Assurance (CPA).
6. Install the steering column filler panel or knee bolster.
7. Install the AIR BAG fuse to the fuse block.
8. Connect the negative battery cable.
9. Staying away from the air bags, turn the ignition switch to **RUN** and make sure that the AIR BAG warning lamp flashes seven times and then shuts off. If the warning lamp does not shut off, make sure that the wiring is properly connected. If the light remains on, take the vehicle to a reputable repair facility for service.

STEERING WHEEL MODULE COIL CENTERING

→The new SIR coil assembly will be centered. Improper alignment of the SIR coil assembly may damage the unit, causing an inflatable restraint malfunction.

✳✳ CAUTION

If double wire harness strap is installed onto the wire harness assembly and column, you must reuse the holder for the wire straps during installation. Remove the wire harness strap(s) where necessary.

1. Verify the following conditions before centering the SIR coil:
   - The wheels on the vehicle are straight ahead
   - The block tooth (1) of the steering shaft assembly is in the 12 o'clock position
   - The ignition switch is in the LOCK position
2. If the front of the SIR coil has a centering window, and the back side includes a spring service lock, perform the following steps:
   - Hold the SIR coil with the face up
   - While depressing the spring service lock, rotate the coil hub clockwise until the coil ribbon stops
   - Rotate the coil hub slowly, counterclockwise, until the centering window appears yellow and both arrows line up
   - Release spring service lock between the locking tab. The SIR coil is now centered
   - Align the centered SIR coil with the horn tower and slide onto the steering shaft assembly
3. If the front of the SIR coil has a centering window and the back side includes NO spring service lock, perform the following steps:
   - Hold the SIR coil with the face up
   - Rotate the coil hub clockwise until the coil ribbon stops
   - Rotate the coil hub slowly, counterclockwise until the centering window appears yellow and both arrows line up. This is the CENTER position
   - While holding the coil hub in the CENTER position, align the SIR coil with the horn tower and slide onto the steering shaft assembly
4. If the front side of the SIR coil has NO centering window, but the back side includes a spring service lock, perform the following steps:
   - Hold the SIR coil with the back side up
   - While depressing the spring service lock, rotate the coil hub in the direction of the arrow until the coil ribbon stops
   - Still pressing the spring service lock, rotate the coil hub in the opposite direction 2½ revolutions
   - Release the spring service lock between locking tabs. The SIR coil is now centered
   - Align the centered SIR coil with the horn tower and slide onto the steering shaft assembly
5. If the front side of the SIR coil has NO centering window, and the back side includes NO spring service lock, perform the following steps:
   - Hold the SIR coil with the face up
   - Rotate the coil hub in the direction of the arrow until the coil ribbon stops
   - Rotate the coil hub, slowly, counterclockwise, for 2½ revolutions. This is the CENTER position
   - While maintaining the coil hub in the CENTER position, align the centered SIR coil with the horn tower and slide onto the steering shaft assembly
6. If double wire harness strap is installed onto the wire harness assembly and column, you must route the wires up against the steering column. One wire harness strap will surround one lead from the coil to the steering column. The other wire harness strap will surround all leads to the steering column.

DRIVETRAIN

AUTOMATIC TRANSMISSION ASSEMBLY

REMOVAL & INSTALLATION

See Figure 2.

1. Before servicing the vehicle, refer to the Precautions Section.
2. Drain the transmission fluid.
3. Remove or disconnect the following:
   - Negative battery cable
   - Dipstick and filler tube
   - Rear driveshaft
   - Front driveshaft, if equipped with 4WD
   - Transfer case, if equipped with 4WD
   - Range selector cable from selector lever
   - Transmission main harness connector
   - Engine wiring harness retainers
   - Park/neutral switch connector
   - Vent hose retainer
   - Fuel line bracket retainers and position fuel line aside
   - Transmission service access plug
   - 3 bolts securing the torque converter to the flywheel
   - Transmission cooler lines from the transmission. Plug the lines and the ports in the transmission.
4. Place a transmission jack under the transmission.

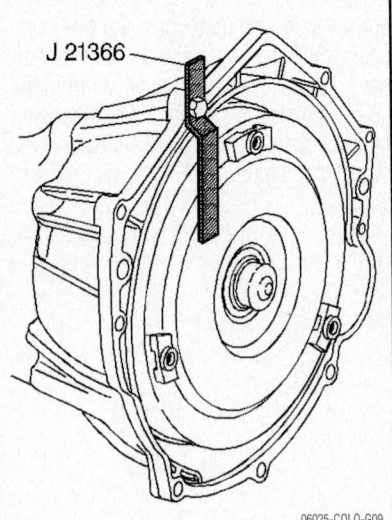

J 21366

06025-COLO-G09

Fig. 2 Install Special tool 21366 to secure the torque converter to the transmission—Automatic transmission

5. Remove the transmission crossmember.
6. Inspect for any other wiring, brackets etc. which may interfere with the removal of the transmission.
7. Remove the transmission from the engine by pulling the transmission rearward to disengage it from the locator dowel pins on the back of the block. Carefully lower the transmission from the vehicle. Use care that the torque converter does not fall out of the front of the transmission.

➡**Use converter holding strap tool No. J-21366, to secure the torque converter to the transmission during removal and installation procedures.**

To install:

Installation is the reverse of removal, but please note the following important steps.

8. Make sure the torque converter is fully seated in the pump drive. If not, the transmission will not fit tightly to the rear of the engine block.
9. Raise the transmission into position and remove the torque converter holding strap and carefully. Slide the transmission forward until the dowel pins are engaged.
10. The torque converter should be flush with the flywheel and turn freely by hand.
11. Install the transmission–to–engine bolts. Tighten the bolts to 37 ft. lbs. (50 Nm).
12. Tighten the torque converter-to-flywheel bolts to 44 ft. lbs. (60 Nm).
13. Refill the transmission with the proper amount and type of fluid.
14. Connect the negative battery cable. Start the vehicle and allow to warm while checking for leaks. Road test the vehicle to check for shift quality.

MANUAL TRANSMISSION ASSEMBLY

REMOVAL & INSTALLATION

1. Before servicing the vehicle, refer to the Precautions Section.
2. Remove or disconnect the following:
   - Negative battery cable
   - Shift lever housing and boot
   - Rear driveshaft
   - Front driveshaft, if equipped with 4WD
   - Transfer case and shift lever, if equipped with 4WD
   - All wiring harness that would interfere with transmission removal

3. Disconnect the hydraulic clutch quick-connect from the slave cylinder using special tool J–42371 to depress the white plastic sleeve on the quick connect to separate the clutch line end from the slave cylinder quick connect.
   - Engine wiring harness and fuel line retainers from transmission
4. Support the transmission with a transmission jack or equivalent.
5. Remove the transmission cross member.
6. Remove the transmission mounting bolts. Pull the transmission straight back on the clutch hub splines.
7. Lower the transmission using the transmission jack.

To install:

Installation is the reverse of removal, but please note the following important steps.

8. Place a THIN coat of high-temperature grease on the main drive gear (input shaft) splines.
9. Secure the transmission to the floor jack and raise the transmission into position.
10. Slowly insert the input shaft through the clutch. Rotate the output shaft slowly to engage the splines of the input shaft into the clutch while pushing the transmission forward into place. Do not force the transmission into position, the transmission should easily fall into place once everything is properly aligned.
11. Tighten the transmission mounting bolts to 37 ft. lbs. (50 Nm).
12. Tighten the transmission crossmember horizontal nuts to 37 ft. lbs. (50 Nm).
13. Tighten the transmission crossmember vertical bolts to 74 ft. lbs. (100 Nm).
14. Do not remove the transmission jack until the crossmember has been installed.
15. Check the transmission fluid level and replenish as necessary.

CLUTCH

REMOVAL & INSTALLATION

See Figures 3 and 4.

1. Before servicing the vehicle, refer to the Precautions Section.
2. Remove or disconnect the following:
   - Negative battery cable
   - Transmission
3. Install a clutch alignment tool or a

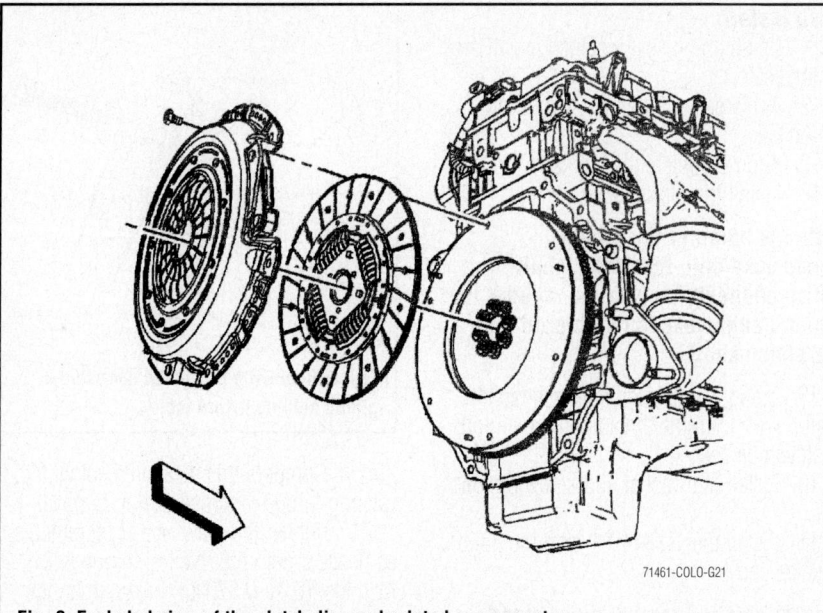

Fig. 3 Exploded view of the clutch disc and related components

71461-COLO-G21

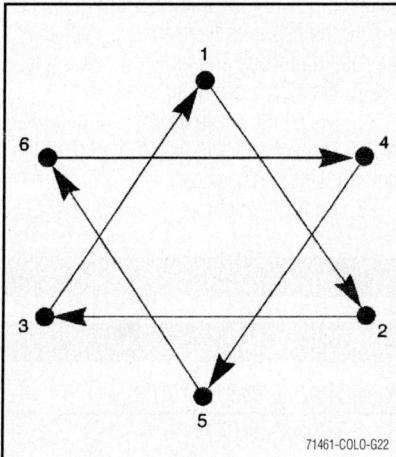

Fig. 4 Clutch pressure plate bolt tightening sequence

71461-COLO-G22

used transmission input shaft to support the clutch.

4. If the clutch assembly is going to be reused, mark the flywheel, clutch cover and a pressure plate lug for alignment when installing.

5. Remove or disconnect the following:
- Clutch cover bolts and washers
- Clutch cover assembly and the clutch plate
- Clutch alignment tool

6. Clean all parts and inspect for damage.

To install:

7. Install or connect the following:
- Clutch alignment tool, to support the clutch

- Clutch plate/clutch cover assembly to the flywheel. Tighten the bolts in sequence shown to 15 ft. lbs. (20 Nm).

8. Remove the clutch alignment tool.

9. Install or connect the following:
- Transmission
- Negative battery cable

TRANSFER CASE ASSEMBLY

REMOVAL & INSTALLATION

1. Before servicing the vehicle, refer to the Precautions Section.

2. Disconnect the negative battery cable.

3. Drain the transfer case fluid.

4. Support the transfer case.

5. Remove or disconnect the following:
- Front and rear driveshafts from the transfer case. Matchmark the shafts prior to removal.
- Vacuum lines and/or the electrical connectors, as equipped
- Transfer case encoder motor
- Engine wiring harness and position aside
- Transfer case

6. Remove all traces of old gasket material from the mating surfaces.

To install:

7. Install or connect the following:
- New gasket using sealer to hold it in position
- Transfer case. Torque the bolts to 37 ft. lbs. (50 Nm).
- Engine wiring harness
- Encoder motor

- Vacuum lines and/or electrical connections, as necessary
- Front and rear driveshafts by aligning the matchmarks

8. Refill the transfer case.

9. Connect the negative battery cable.

FRONT HALFSHAFT

REMOVAL & INSTALLATION

1. Before servicing the vehicle, refer to the Precautions Section.

2. Unlock the steering column so the steering linkage is free to move.

3. Remove or disconnect the following:
- Negative battery cable
- Front wheel
- Halfshaft nut and washer
- Steering knuckle assembly

4. Place a drift against the tripot housing and hammer the drift to release the shaft.

5. Remove the halfshaft from the vehicle.

To install:

6. Install the halfshaft. A snap or pop should be heard and felt when the shaft properly seats in the differential housing.

7. Install the steering knuckle assembly.

8. Install a new halfshaft nut and tighten the nut to 191 ft. lbs. (260 Nm).

9. Install the front wheel.

10. Lower the vehicle.

REAR AXLE SHAFT, BEARING & SEAL

REMOVAL & INSTALLATION

1. Before servicing the vehicle, refer to the Precautions Section.

2. Raise and support the vehicle.

3. Remove the rear wheel.

4. Remove the brake drum.

5. Remove the rear axle housing cover.

6. Remove the pinion shaft lock bolt.

7. Remove the pinion shaft.

8. Remove the c-lock from the rear axle.

9. Pull the axle shaft out of the housing.

10. Remove the axle shaft seal and bearing.

To install:

11. Press on a new bearing and install the seal.

12. Install the axle shaft.

13. Install the c-lock to the rear axle.

14. Install the pinion shaft.

15. Install the pinion shaft lock bolt and tighten to 18 ft. lbs. (25 Nm).

16. Install a new gasket and the rear axle housing cover.

REAR PINION SEAL

REMOVAL & INSTALLATION

See Figure 5.

1. Raise and support the vehicle.
2. Remove the rear brake drums.
3. Remove the propeller shaft.
4. Measure the amount of torque required to rotate the pinion. Use an inch-pound torque wrench. Record this measurement for reassembly.
5. Place an alignment mark between the pinion and the pinion yoke.
6. Install the flange and holding tool J-8614-01, and remove the pinion nut while holding the tool.
7. Remove the washer.
8. Install the J 8614-3 (2) and the J 8614-2 (3) into the as shown.
9. Remove the pinion yoke by turning the J 8614-3 (3) clockwise while holding the J-8614-01 (1). Use a container in order to retrieve the lubricant.
10. Remove the pinion oil seal. Use a suitable seal removal tool. Do not damage the housing.

To install:

11. Using an oil seal installer, install the pinion seal.
12. Apply sealant to the splines of the pinion yoke.
13. Align the reference marks.
14. Install the pinion yoke.

➡**Do not hammer the pinion flange/yoke onto the pinion shaft. Pinion components may be damaged if the pinion flange/yoke is hammered onto the pinion shaft.**

15. Using a soft-faced hammer, tap the pinion yoke until the threads on the pinion shaft can be seen.
16. Install the washer and a new pinion nut.
17. Install the J-8614-01 onto the pinion yoke as shown.

➡**If the rotating torque is exceeded, the pinion will have to be removed and a new collapsible spacer installed.**

18. Tighten the pinion nut while holding the tool.
19. Tighten the nut until the pinion end play is just taken up. Rotate the pinion while tightening the nut to seat the bearings.
20. Measure the rotating torque of the

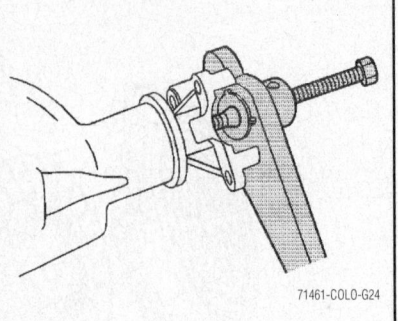

71461-COLO-G24

Fig. 5 Removing the pinion nut using a pinion holding fixture tool

pinion. Compare this measurement with the rotating torque recorded during removal.

21. Tighten the nut in small increments, as needed, until the rotating torque is 3-5 inch lbs. (0.40-0.57 Nm) greater than the rotating torque recorded during removal.
22. Once the specified torque is obtained, rotate the pinion several times to ensure the bearings have seated. Recheck the rotating torque and adjust if necessary.
23. Install the propeller shaft.
24. Install the brake drums.
25. Inspect and add axle lubricant to the axle housing, if necessary.
26. Lower the vehicle

ENGINE COOLING

THERMOSTAT

REMOVAL & INSTALLATION

1. Drain the cooling system.
2. Raise and support the vehicle only high enough to access the thermostat housing through the wheelhouse.
3. Remove the left wheelhouse liner.
4. Position the hose clamp pliers to the clamp to remove the radiator inlet hose from the thermostat housing.
5. Remove the thermostat housing bolts.
6. Remove the thermostat housing from the engine block.
7. Clean and inspect the thermostat housing.
8. Clean and inspect the sealing surface of the engine block.

To install:

9. Position the thermostat housing to the engine block.
10. Install the thermostat housing bolts and tighten to 89 inch lbs. (10 Nm).
11. Connect the radiator inlet hose to the thermostat housing.
12. Install the left wheelhouse liner.

13. Lower the vehicle.
14. Fill the cooling system.
15. Inspect all sealing surfaces for leaks after starting the engine.

WATER PUMP

REMOVAL & INSTALLATION

See Figures 6 and 7.

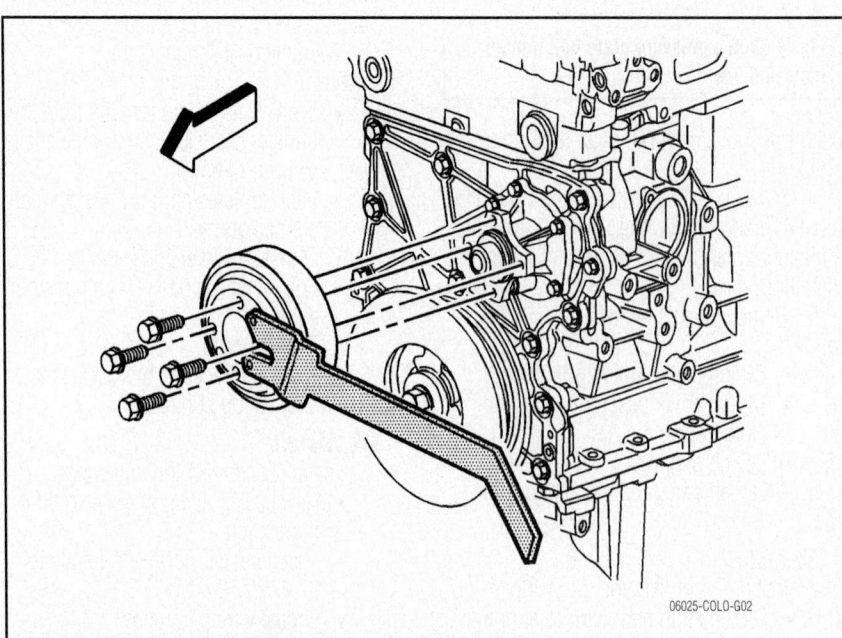

06025-COLO-G02

Fig. 6 Use Special Tool J-46406 to secure the pump pulley

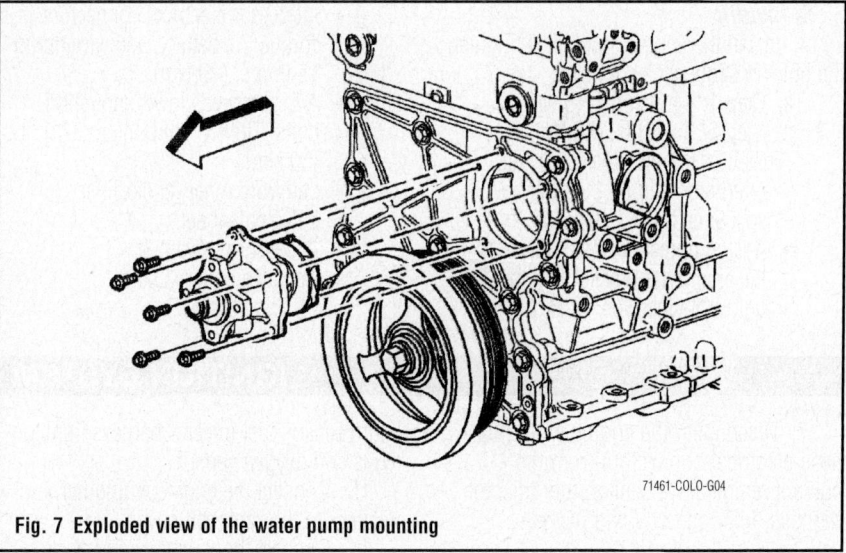

Fig. 7 Exploded view of the water pump mounting

1. Before servicing the vehicle, refer to the Precautions Section.
2. Disconnect the negative battery cable.

3. Drain the engine cooling system.
4. Remove fan.
5. Relieve the belt tension and remove the accessory drive belt.

6. Using Special Tool J-46406, secure the water pump pulley and remove the water pump pulley bolts.
7. Remove or disconnect the following:
   • Water pump pulley
   • Coolant hose(s) from the water pump
   • Water pump bolts
   • Water pump

To install:

8. Clean the gasket mounting surfaces.
9. Install or connect the following:
   • Water pump using a new gasket. Tighten the water pump bolts to 89 inch lbs. (10 Nm).
   • Coolant hose(s)
   • Water pump pulley and tighten the bolts to 18 ft. lbs. (25 Nm).
   • Drive belt and adjust the tension
   • Negative battery cable
10. Refill the engine cooling system.
11. Run the engine and check for leaks.

ENGINE ELECTRICAL   CHARGING SYSTEM

ALTERNATOR

REMOVAL & INSTALLATION

See Figures 8 and 9.

1. Before servicing the vehicle, refer to the Precautions Section.
2. Remove or disconnect the following:
   • Negative battery cable
   • Accessory drive belt
   • Left front wheel
   • Left front inner fender liner
   • A/C compressor electrical connector
   • Lower A/C compressor mounting bolts

➡The lower mounting bolts are removed to allow the engine lift

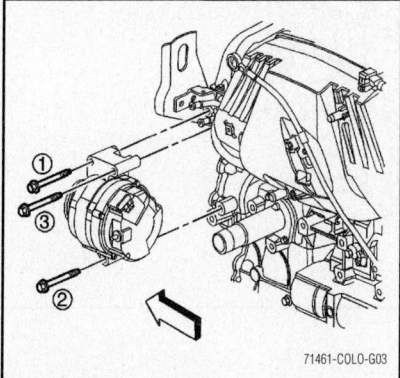

Fig. 8 Alternator mounting bolt tightening sequence

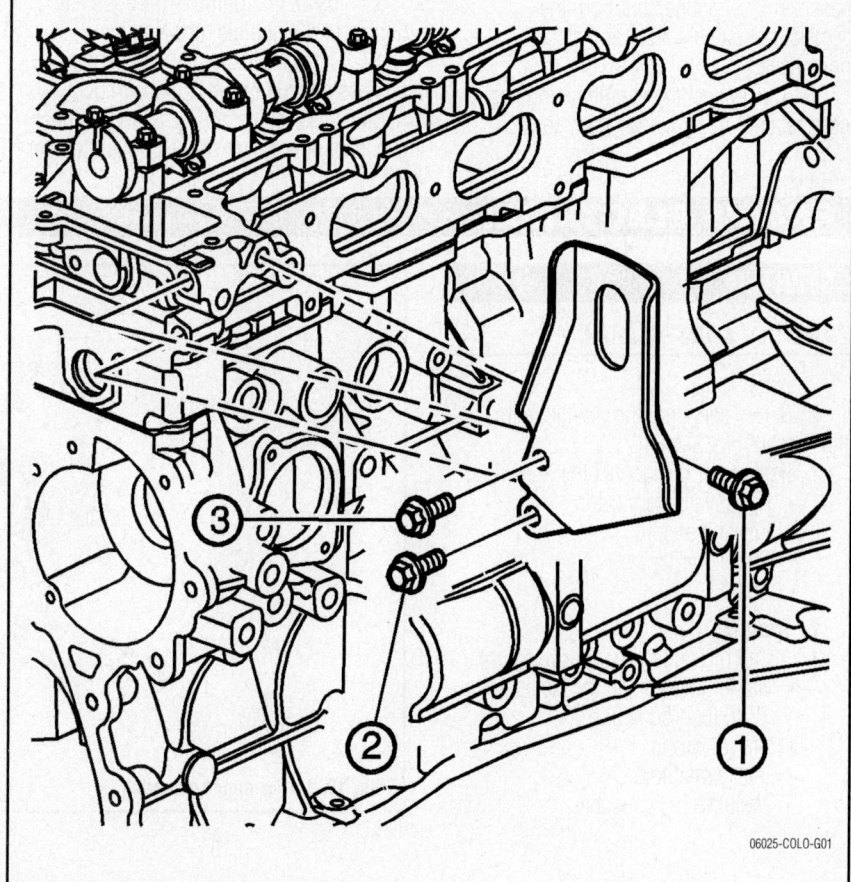

Fig. 9 Engine lift bracket torque sequence

bracket to be removed. Do not remove the upper A/C compressor mounting bolt.

- Alternator wiring
- Alternator mounting bolts
- A/C compressor hose bracket from the engine lift bracket
- Engine lift bracket
- Alternator

To install:

3. Install the engine lift bracket. Tighten the bolts in sequence as follows:
   a. Step 1: 44 inch lbs. (5 Nm)
   b. Step 2: 37 ft. lbs. (50 Nm)
4. Install or connect the following:
   - Alternator. Torque the bolts in sequence to 37 ft. lbs. (50 Nm).
   - A/C compressor hose bracket to the engine lift bracket. Tighten the bolt to 80 inch lbs. (9 Nm).

- Alternator electrical connectors. Torque the battery feed wire nut to 15 ft. lbs. (20 Nm).
- A/C compress lower mounting bolts. Tighten the bolts to 37 ft. lbs. (50 Nm).
- Left front inner fender liner
- Left front wheel
- Accessory drive belt
- Negative battery cable

ENGINE ELECTRICAL

IGNITION COIL

REMOVAL & INSTALLATION

1. Remove the air cleaner resonator and outlet duct.
2. Disconnect the engine wiring harness electrical connector from the oil pressure sensor.
3. Disconnect the engine wiring harness retainer from the oil filter adapter.
4. Disconnect the engine wiring harness retainers from the power steering pump.
5. Disconnect the engine wiring harness electrical connectors from the camshaft position sensor and actuator solenoid valve.
6. Disconnect the engine wiring harness retainer from the camshaft cover.

7. Disconnect the engine wiring harness electrical connectors from the coolant temperature sensor, fuel injector harness, ignition coils and oxygen sensor.
8. Carefully disengage the engine wiring harness conduit from the camshaft cover, and position aside.
9. Remove the ignition coil bolts.
10. Remove the ignition coils from the camshaft cover.

To install:

11. Install the ignition coils to the camshaft cover.
12. Install the ignition coil bolts and tighten to 89 inch lbs. (10 Nm).
13. Carefully install the engine wiring harness conduit to the camshaft cover.
14. Connect the engine wiring harness electrical connectors to the coolant temper-

IGNITION SYSTEM

ature sensor, fuel injector harness, ignition coils and oxygen sensor.

15. Connect the engine wiring harness retainer to the camshaft cover.
16. Connect the engine wiring harness electrical connectors to the camshaft position sensor and actuator solenoid valve.
17. Connect the engine wiring harness retainers to the power steering pump.
18. Connect the engine wiring harness retainer to the oil filter adapter.
19. Connect the engine wiring harness electrical connector to the oil pressure sensor.

IGNITION TIMING

ADJUSTMENT

Ignition timing is controlled by the PCM and is not adjustable.

ENGINE ELECTRICAL

STARTING SYSTEM

STARTER

REMOVAL & INSTALLATION

See Figure 10.

1. Before servicing the vehicle, refer to the Precautions Section.
2. Remove or disconnect the following:
   - Negative battery cable
   - Intake manifold
   - Starter wiring
   - Starter

To install:

3. Install or connect the following:
   - Starter and tighten the fasteners to 37 ft. lbs. (50 Nm)
   - Starter wiring
   - Intake manifold
   - Negative battery cable

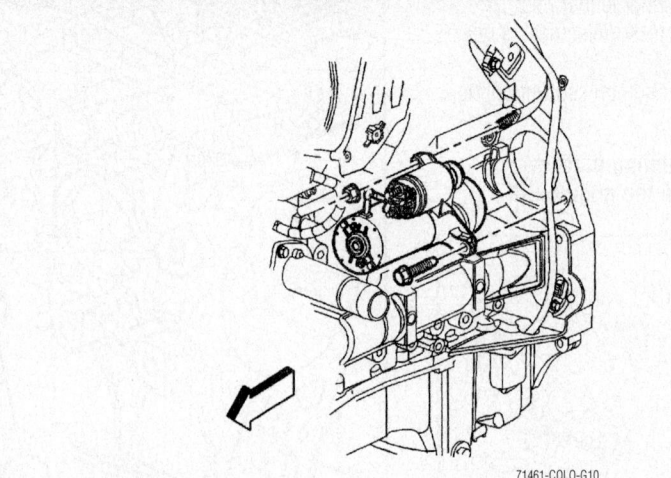

71461-COLO-G10

Fig. 10 Starter motor mounting

ENGINE MECHANICAL

➡️Disconnecting the negative battery cable may interfere with the functions of the on board computer systems and may require the computer to undergo a relearning process, once the negative battery cable is reconnected.

ACCESSORY DRIVE BELTS

ACCESSORY BELT ROUTING

See Figure 11.

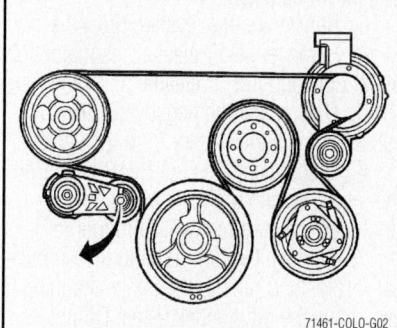

71461-COLO-G02

Fig. 11 Accessory drive belt routing—2.9L and 3.7L engines with A/C

INSPECTION

Inspect the drive belt for signs of glazing or cracking. A glazed belt will be perfectly smooth from slippage, while a good belt will have a slight texture of fabric visible. Cracks will usually start at the inner edge of the belt and run outward. All worn or damaged drive belts should be replaced immediately.

ADJUSTMENT

Drive belt tension is automatically adjusted by the drive belt tensioner.

REMOVAL & INSTALLATION

1. Install a 3/8 inch breaker bar into the drive belt tensioner and rotate the tensioner clockwise, enough to relieve the tension on the drive belt.
2. Slide the drive belt from the water pump pulley.
3. Allow the drive belt tensioner to return to the relaxed position.
4. Remove the drive belt from the remaining pulleys.

To install:
5. Route the drive belt over all the pulleys, excluding the water pump pulley.
6. Install the ⅜ inch breaker bar on the drive belt tensioner and rotate the tensioner clockwise.

7. Route the drive belt over the top of the water pump pulley.
8. Slowly release the tension to the drive belt tensioner.
9. Ensure the drive belt is properly aligned and seated into the grooves of the drive pulleys.
10. Inspect for proper installation of the drive belt on the pulleys.

CAMSHAFT AND VALVE LIFTERS

REMOVAL & INSTALLATION

See Figures 12 and 13.

1. Before servicing the vehicle, refer to the Precautions Section.
2. Properly relieve the fuel system pressure.
3. Disconnect the negative battery cable.
4. Drain the engine cooling system and the engine oil.
5. Remove or disconnect the following:
   - Intake manifold
   - Ignition coils
   - Coolant temperature sensor connector
   - Injector harness connector
   - Ignition coil connectors
   - Oxygen Sensor (O$_2$S) connector
   - Fuel pressure regulator screw
   - Camshaft cover
   - Both Camshaft Position (CMP) sensors
6. Rotate the crankshaft clockwise until the no. 1 cylinder is at TDC on the compression stroke. The word DELPHI on the camshaft position actuator will be parallel with the cylinder head surface.
7. Install camshaft locking tool J-44221 to the rear of the camshafts.
8. Remove the camshaft sprocket bolts and discard them.
9. Install tension tool J-44222 on the cylinder head and install the holding bolts in the camshaft sprocket bolt holes to lock the timing chain and sprockets in position.
10. Carefully slide the sprockets and timing chain onto the tension tool.
11. Remove the camshaft bearing caps.
12. Remove the Camshaft Locking Tool from the camshafts.
13. Remove the camshafts.
14. Remove the valve rocker arms.
15. Remove the valve lash adjusters.
16. Clean and inspect the valve rocker arms and valve lash adjusters.

To install:
17. Lubricate and fill the valve lash adjusters with engine oil.

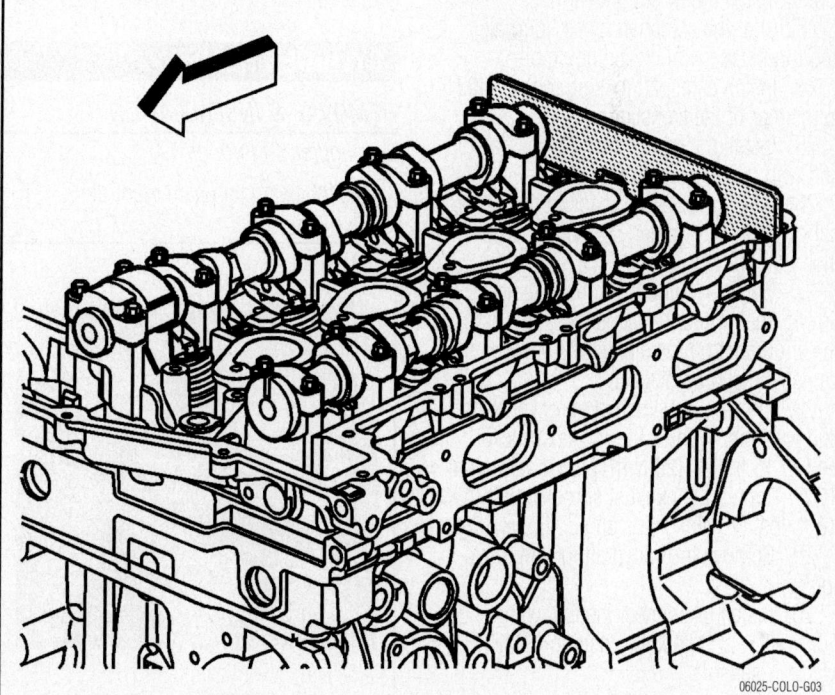

06025-COLO-G03

Fig. 12 Install the camshaft locking tool J-44221 to the rear of the camshafts

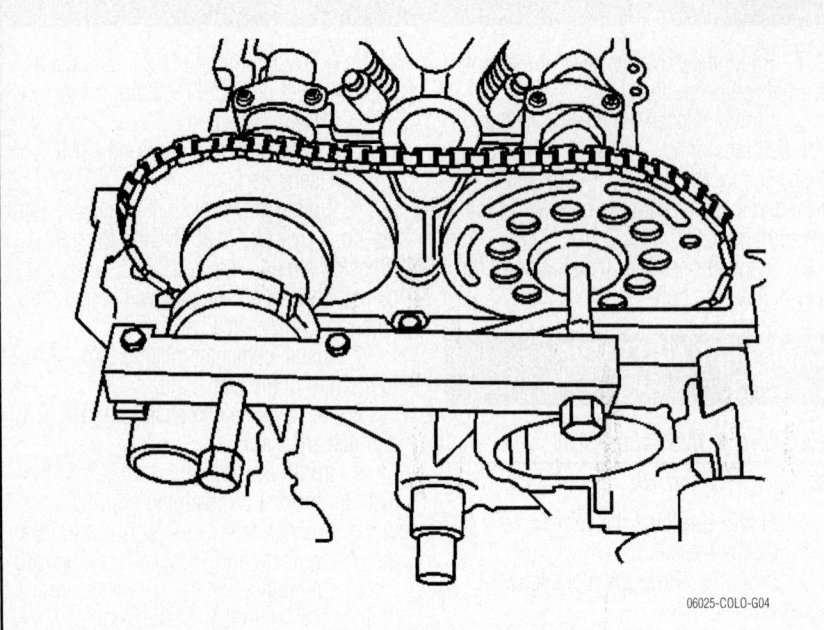

Fig. 13 Use Special Tool J-44222 camshaft sprocket holding tool to prevent the timing chain and sprockets from turning

18. Install the valve lash adjusters in their original locations.

19. Lubricate the entire valve rocker arm.

20. Install the valve rocker arms in their original locations.

21. Inspect the camshaft, journals and lobes for wear and replace, if necessary.

22. If removed, use the camshaft bearing tool to install a new set of bearings.

23. Coat the camshaft lobes, journals and thrust face with clean engine oil.

24. Install camshaft locking tool J-44221 to the rear of the camshafts.

25. Install the camshafts with the flats up and with cylinder no. 1 at TDC.

26. Install the bearing caps in their original position and tighten the bolts to 106 inch lbs. (12 Nm).

27. Carefully slide the sprockets and timing chain onto the camshafts, ensuring the alignment pins are engaged between the camshafts and sprockets.

28. Install new camshaft sprocket bolts and washers. Tighten the intake sprocket bolt to 15 ft. lbs. (20 Nm), plus it additional 100°. Tighten the exhaust sprocket bolt to 18 ft. lbs. (25 Nm), plus an additional 135°.

29. Remove the camshaft locking plate tool.

30. Install or connect the following:
- Both Camshaft Position (CMP) sensors
- Camshaft cover
- Fuel pressure regulator screw
- Oxygen Sensor (O2S) connector
- Ignition coil connectors
- Injector harness connector
- Coolant temperature sensor connector
- Ignition coils
- Intake manifold
- Negative battery cable

31. Refill the engine cooling system and engine oil.

CYLINDER HEAD

REMOVAL & INSTALLATION

See Figures 14 through 17.

The following tools are required:

Fig. 14 Installing –48464–Lower Timing Chain Tensioner Holding Tool

- EN–48464–Lower Timing Gear Tensioner Holding Tool
- EN–46547–Flywheel Holding Tool (LK5 or L52 Engine Only)
- J 44221–Camshaft Holding Tool
- J 45059–Torque Angle Meter
- TDC Indicator Tool (Some models may use a dial Indicator or graduation marks on the side of a sliding shaft for visual indication)

1. Before servicing the vehicle, refer to the Precautions Section.

2. Install protective covering to the front of the vehicle. This is to protect the grille from damage.

3. Relieve the fuel system pressure.

4. Remove the air cleaner resonator, the outlet duct and the air cleaner assembly.

5. Disconnect the negative battery cable. Remove the battery from the vehicle.

6. Disconnect the Fuel/EVAP lines from the intake manifold and move aside (includes fuel line removal from fuel rail).

7. Remove the bolt holding the oil indicator (dipstick) tube to the intake manifold and move the oil dipstick aside. Do not remove.

8. Drain and recycle the engine coolant.

9. Raise and safely support the vehicle securely.

10. Remove or disconnect the following:
- Drive belt
- Engine oil
- Left front tire and wheel
- Left front wheelhouse panel
- Fir tree wiring harness connectors from the engine wiring harness bracket from the left front wheelhouse opening
- From the left front wheelhouse, the engine wiring harness bracket from the engine and set aside. It may be necessary to loosen the fasteners on the frame rail (3.7L).
- Intake manifold bolts from the wheelhouse access (bolts stay with intake manifold). On two-wheel drive vehicles, the intake manifold bolts are removed from the top of the engine, not through the left front wheelhouse panel.

11. Lower the vehicle.

12. Remove or disconnect the following:
- PCV pipes from the cam cover and remove the intake manifold from the vehicle. PCM engine harness connectors from the PCM (LL8 Engine Only). PCM from the intake manifold bracket (3.7L)
- Generator output battery terminal nut
- Generator lead from the generator
- Generator electrical connector

Fig. 15 Placing the narrow ramp of the wedge tool so that it faces the timing chain

- Generator bolts and set the generator aside. The generator does not have to be removed from the vehicle. On a two-wheel drive vehicle, the generator must be removed from the vehicle.
- A/C pipe clamp from the engine lift hook bracket
- Engine lift hook bracket bolts and bracket from the vehicle
- P/S pump bolts and reposition the pump (3.7L)
- PCM engine wiring harness connector

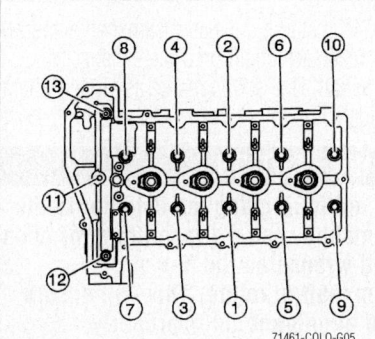

Fig. 16 Cylinder head bolt torque sequence—2.9Lengines

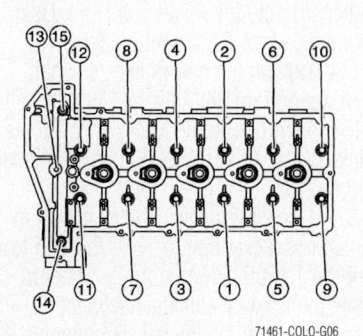

Fig. 17 Cylinder head bolt torque sequence—3.7L engines

- Engine Coolant Temperature Sensor wiring harness connector
- MAP Sensor wiring harness connector
- Ignition Coils wiring harness connector
- Harness clamps at power steering pump wiring harness connector
- Wiring Harness fastener at right front inner fender
- Throttle Body wiring harness connector
- Camshaft Sensors wiring harness connector
- Exhaust Camshaft Actuator wiring harness connector
- Fuel Injectors wiring harness connectors
- HO2S No. 1 wiring harness connectors
- Set aside the cross-vehicle engine wiring harness on the left side of the vehicle
- A.I.R. injection pipe block-off plate bolts from the cylinder head
- A.I.R. injection pipe block-off plate
- Bolts from the exhaust manifold heat shield
- Exhaust manifold back and away from the cylinder head
- Bolts to all ignition coil assemblies and remove all ignition coil assemblies from the cam cover
- All the spark plugs from the cylinder head
- Ground strap bolt (3.7L)
- Cam cover bolts
- Cam cover from the cylinder head
- Upper radiator hose and clamp from the cylinder head

✳✳ CAUTION

Before performing the TDC procedures, break loose both the exhaust and intake camshaft sprocket bolts. Use a 1 inch (25 mm) open end wrench on the camshaft hexes to hold the camshaft from turning. Do not remove the bolts.

13. Rotate the engine clockwise by hand to TDC on the compression stroke by using a piston TDC indicator tool and/or dial indicator in the number 1 cylinder. The TDC indicator tool graduation marks on the shaft should note top of the piston stroke. When the piston is at TDC, the flats at the rear of the camshafts will be facing up and flat when using a straight edge across the camshaft flats. There may be a variation of build and the straight edge may not perfectly lay flat across back of the camshafts.

✳✳ CAUTION

Do not rotate the engine counterclockwise to find TDC, as this will put slack in the camshaft actuator and the timing chain. If the engine is rotated counterclockwise, even a small amount, then the engine should be rotated at least 90 degrees and then brought back to TDC by rotating clockwise only.

✳✳ CAUTION

The word Delphi on the exhaust camshaft position actuator will be parallel with the cylinder head to cam cover mating surface. When the piston is at TDC, the flats at the rear of the camshafts will be facing up.

14. Use a metal scribe or equivalent to place a reference mark on the harmonic balancer to the front cover for alignment purposes.

15. Lower the vehicle.

✳✳ CAUTION

The camshaft holding tools must be installed on the camshafts to prevent camshaft rotation. When performing service to the valve train and/or timing components, valve spring pressure can cause the camshafts to rotate unexpectedly and can cause personal injury.

✳✳ CAUTION

If the timing is correct (TDC compression stroke number 1 cylinder), the camshaft flats will be in the up position.

16. Install the J 44221–Camshaft Holding Tool to the back of the camshafts.

17. Remove the upper timing chain guide from the cylinder head (not found on later vehicles).

✳✳ CAUTION

The exhaust camshaft actuator must be fully advanced before scribing your reference mark on both timing gear sprockets and the timing chain to mark location prior to disassembly. There may be some normal play in the exhaust camshaft actuator itself. To ensure camshaft actuator is fully advanced, do not rotate engine counterclockwise when finding TDC No. 1.

18. Clean the timing chain and gears with Brake Cleaner or suitable solvent. Use a metal scribe or equivalent to place a reference mark on both timing gear sprockets and the timing chain to mark the location prior to disassembly. It is recommended that a metal scribe or equivalent be in the twelve 'o–clock position.

> ❋❋ **CAUTION**
>
> **Do not use excessive force to seat the wedge tool. If excessive force is used, you may damage the timing chain tensioner or break the front cover bolt requiring complete disassembly of the front engine.**

19. Install EN–48464—Lower Timing Chain Tensioner Holding Tool. This wedge tool will hold the timing chain and tensioner in position. It is important to install the tool with the proper orientation and to ensure that it is seated square against the timing chain and against the timing cover center bolt. The narrow ramp of the wedge tool needs to be placed so that it faces the timing chain. The wedge tool should be lightly seated using a couple of very light taps with a small plastic or brass hammer.

20. Once the tool is correctly installed, unscrew the handle and remove the handle.

21. Remove (1 long and 2 short) cylinder head bolts next to the exhaust and intake timing chain tensioner shoes and discard the bolts.

> ❋❋ **CAUTION**
>
> **Use a 25 mm (1 inch) open end wrench on the camshaft hexes to hold the camshaft from turning. It is critical that the crankshaft does not move and is held at TDC when the intake and exhaust camshaft sprocket bolts are removed.**

> ❋❋ **CAUTION**
>
> **If the crankshaft is not held in place, the wedge tool could be dislodged. If the crankshaft moves, or if the tool is not seated properly allowing the timing chain tensioner to extend, the repair will have to be completed by removing the front cover to release the timing chain tensioner.**

22. Remove both upper cylinder head access hole plugs from the front of the cylinder head.

23. Remove both upper timing chain tensioner shoe bolts.

24. Remove the exhaust and the intake camshaft sprocket bolts. Discard the bolts.

25. Carefully remove the exhaust and intake camshaft sprockets from the exhaust and intake camshafts while still engaged with the timing chain. Refer to the above illustration. The above illustration shows the exhaust camshaft sprocket already removed.

26. Remove the sprockets from the chain. Tie a piece of mechanics wire on the timing chain and let it drop.

27. Remove the cylinder head bolts from the cylinder head. Before removing the cylinder head bolts, use a drift punch and hammer to shock the bolts. This will ensure that the cylinder head bolts will not strip out the threads in the engine block or break. Once the bolts are removed, discard them.

28. Remove the AIR pump bolt and fir tree fastener from the back of the cylinder head.

29. Remove the cylinder head from the vehicle.

To install:

> ❋❋ **CAUTION**
>
> **Install the cylinder head without the camshafts.**

30. Install the engine cylinder head to the engine block.

31. Install the AIR pump bolt and fir tree fastener from the back of the cylinder head.

32. Install new cylinder head bolts and tighten the bolts.

> ❋❋ **CAUTION**
>
> **The camshaft holding tools must be installed on the camshafts to prevent camshaft rotation. When performing service to the valve train and/or timing components, valve spring pressure can cause the camshafts to rotate unexpectedly and can cause personal injury.**

> ❋❋ **CAUTION**
>
> **Before installing the camshafts, Clean and inspect the gasket mounting surfaces.**

33. Install the camshafts with the flats up using the J 44221–Camshaft Holding Tool.

➡ **Tension must be always kept on the intake side of the timing chain to properly keep the engine in time. If the chain is loose the timing will be off, which may cause internal engine damage or set DTC P0017.**

➡ **Use the correct fastener in the correct location. Replacement fasteners**

must be the correct part number for that application. Fasteners requiring replacement or fasteners requiring the use of thread locking compound or sealant are identified in the service procedure. Do not use paints, lubricants, or corrosion inhibitors on fasteners or fastener joint surfaces unless specified. These coatings affect fastener torque and joint clamping force and may damage the fastener. Use the correct tightening sequence and specifications when installing fasteners in order to avoid damage to parts and systems.

➡ **The exhaust camshaft actuator must be fully advanced during installation. Engine damage may occur if the camshaft actuator is not fully advanced.**

> ❋❋ **CAUTION**
>
> **To aid in aligning the actuator to the camshaft, use a 1 inch (25 mm) open end wrench on the hex of the camshaft to rotate. This will ensure the alignment pin is properly engaged with the camshaft and hand tighten the new exhaust camshaft sprocket bolt.**

34. Install the exhaust camshaft actuator/sprocket and chain onto the exhaust camshaft. Use scribe marks as an alignment guide.

> ❋❋ **CAUTION**
>
> **To aid in aligning the actuator to the camshaft, use a 1 inch (25 mm) open end wrench on the hex of the camshaft to rotate. This will ensure the alignment pin is properly engaged with the camshaft and hand tighten the new exhaust camshaft sprocket bolt.**

35. Install the intake camshaft sprocket and chain onto the intake camshaft. Use scribe marks as alignment guide.

36. Tighten the new intake camshaft sprocket bolt to 15 ft. lbs. (20 Nm). Use the J 45059–Torque Angle Meter to rotate the intake camshaft sprocket bolt an additional 100 degrees.

37. Tighten the new exhaust camshaft actuator sprocket bolt to 18 ft. lbs. (25 Nm). Use the J 45059–Torque Angle Meter to rotate the exhaust camshaft actuator sprocket bolt an additional 135 degrees.

38. Install both upper timing chain tensioner shoe bolts to 18 ft. lbs. (25 Nm).

39. Install both upper cylinder head access hole plugs to the front of the cylinder head. Tighten the plugs to 44 inch lbs. (5 Nm).

40. Lift the vehicle and remove the EN–46547—Flywheel Holding Tool (LK5 or L52 Engine Only).

41. Lower the vehicle.

42. Remove the J 44221–Camshaft Holding Tool from the back of the camshafts.

➡ **Ensure that the wedge tool is removed from the engine prior to rotation. If the wedge tool is not removed, engine damage will result.**

43. Install the handle of the EN–48464—Lower Timing Chain Tensioner Holding Tool and remove the wedge portion of the tool from the engine.

✳✳ CAUTION

It is critical that the engine is at TDC and not a couple of degrees off. If in doubt, repeat this following step.

44. Rotate the engine clockwise by hand two complete revolutions to TDC No. 1 on the compression stroke. Refer to removal procedure. If you go past TDC, rotate the engine back approximately 45 degrees before TDC and then rotate clockwise up to TDC to ensure that the timing chain is tight (no slack) between the crank sprocket and the timing gears.

✳✳ CAUTION

Do not use the J 44221–Camshaft Holding Tool , installed to the back of the camshafts, as a method to verify timing.

45. Both intake and exhaust camshaft flats should be facing up and flat with the cylinder head. If the J 44221–Camshaft Holding Tool is used to verify cam timing, you could be off approximately one tooth and cause DTC P0017 to set. If a worn or new J 44221–Camshaft Holding Tool is used to verify timing, the timing will be off.

46. To verify timing, set a straight edge across the flats of the camshafts. Both camshaft flats should be flat. there may be some variation of build and the straight edge may not lay perfectly flat across back of the camshafts. If one or both camshaft flats are off, then the timing is off. Repeat step 15 and recheck. If the camshaft flats are still not flat, the camshaft timing will have to be reset. This may require removal and reinstallation of one or both camshaft sprockets. Refer to removal procedure.

47. Install (1 long and 2 short) cylinder head bolts next to the exhaust and intake timing chain tensioner shoes and tighten the bolts. Refer to Cylinder Head Replacement.

48. Position the upper timing chain guide to the cylinder head. Apply threadlocker (P/N 8902129), to the upper timing chain guide bolt threads.

49. Install the upper timing chain guide bolts and tighten to 89 inch lbs. (10 Nm).

50. Install the upper radiator hose and clamp to the cylinder head.

51. Clean and inspect the camshaft cover.

52. Install a new camshaft cover seal and new ignition control module seals to the cam cover. Position the camshaft cover to the cylinder head.

53. Install the camshaft cover bolts and tighten to 89 inch lbs. (10 Nm).

54. Check the gap on all of the spark plugs. The gap should be 0.042 inches (1.08 mm). Tighten all of the spark plugs to 13 ft lbs. (18 Nm).

55. Install the ignition coils into the camshaft cover.

56. Install the ignition coil bolts and tighten to 89 inch lbs. (10 Nm).

57. Reposition the exhaust manifold to the cylinder head and install the exhaust manifold bolts to the cylinder head.

58. Install a new A.I.R. injection gasket, then the cover and pipe studs to the cylinder head. Tighten the pipe studs to 18 ft. lbs. (25 Nm).

59. Install ground strap to cylinder head to exhaust manifold (3.7L).

60. Install the exhaust manifold heat shield to the exhaust manifold.

61. Apply anti-seize, GM P/N 12371386, to the exhaust manifold heat shield nuts.

62. Install the exhaust manifold heat shield nuts and tighten 89 inch lbs. (10 Nm).

63. Install the intake manifold to the cylinder head. Raise the vehicle and install the blind intake manifold bolts from the left front wheelhouse access.

64. Reposition the engine wiring harness bracket to the engine and harnesses. Install the engine wiring harness bracket bolts. Tighten the bracket bolts to 89 inch lbs. (10 Nm).

65. Install the left front wheelhouse panel. Install the left wheel and tire.

66. Refill the engine oil.

67. Install the lower radiator hose if removed.

68. Lower the vehicle.

69. Install the cross-vehicle wiring harness connectors to the following components:
- PCM

a. PCM engine harness connectors from PCM (3.7L)

b. PCM from intake manifold Bracket (3.7L)
- Map Sensor
- Ignition Coils
- Harness clamps at power steering pump
- Wiring harness fastener at right front inner fender
- Throttle Body
- Camshaft Sensors
- Exhaust Camshaft Actuator
- Fuel Injectors
- HO2S No. 1

70. Install the PCV pipes to the intake manifold.

71. Reposition the oil indicator (dipstick) tube and tighten the bolt to the intake manifold.

72. Reposition the Fuel/EVAP lines to the intake manifold retainer.

73. Install the following components.
- Power steering pump bolts
- Generator
- A/C compressor hose/pipe bracket clamp for the engine lift bracket.
- Drive Belt

74. Install the negative battery cable (3.7L).

75. Install the air induction assembly.

76. Refill with new engine oil.

77. Refill with new coolant.

78. Install the air cleaner resonator, the outlet duct and the air cleaner assembly.

79. Remove the fender covers.

80. Remove the protective covering from the front of the vehicle.

81. Install the scan tool and start the engine.

82. Check for DTC's

83. Road test the vehicle. DTC P0017 is a Type B diagnostic code. Three consecutive ignition key cycles must be performed during the road test with a minimum of a one minute run time between key cycles to verify that a DTC P0017 did not set.

ENGINE ASSEMBLY

REMOVAL & INSTALLATION

1. Disconnect the battery cables and properly relieve the fuel system pressure.

2. Drain the engine cooling system.

3. Drain the engine oil.

4. Remove or disconnect the following:
- Hood
- Battery box
- Radiator hoses
- Cooling fan

- Air cleaner assembly
- Engine lifting bracket
- Alternator

➡ **Reinstall the engine lift bracket after removing the alternator. Tighten the bolts to 44 inch lbs. (5 Nm) and then retighten to 37 ft. lbs. (50 Nm).**

- Washer reservoir bolts
- Engine wiring harness connector at PCM
- Wiring harness retainers at fender, power steering pump, throttle body and camshaft cover
- Oil pressure switch connector
- 4WD motor connector, if equipped
- Camshaft Position (CMP) sensor connector
- Exhaust camshaft actuator connector
- Coolant temperature sensor connector
- Injector harness connector
- Ignition coil connectors
- Oxygen Sensor (O2S) connector
- Wiring harness conduit at camshaft cover
- Automatic transmission filler tube, if equipped
- Air injection pipe cover
- Install engine lifting eye in air pipe cover location
- Heater hoses from heater core
- A/C suction hose bracket
- Power steering pump bolts and position pump aside
- Right engine mount-to-frame bracket bolt
- Wiring harness retainer from intake manifold and position aside
- Fuel lines from fuel rail
- Evaporative pipe at intake manifold
- Dipstick and tube
- Brake booster hose
- Manifold Absolute Pressure (MAP) sensor
- MAP wiring harness retainer
- Upper 2 engine wiring harness bracket bolts
- Raise and support vehicle
- Left front wheel
- Left front fender inner liner
- Wiring harness retainers from engine wiring bracket
- Engine wiring harness bracket
- A/C compressor mounting bolts and position compressor aside
- Starter wiring
- Negative battery cable from block
- EVAP canister connector
- Knock sensor connectors
- Heater outlet hose

- Crankcase Position (CKP) sensor connector
- Engine wiring ground leads from block
- Engine wiring harness retainer at oil pan rail and position harness aside
- Catalytic converter
- Exhaust donut gasket
- Automatic transmission oil cooler and fuel line brackets, if equipped
- Left engine mount-to-frame bracket bolt
- On 2WD drive models, front crossmember
- On 4WD models, differential carrier
- On automatic transmission, torque converter bolts after marking torque converter-to-flexplate location
- Leave 2 upper transmission-to-engine bolts, but remove all other bolts

5. Lower the vehicle and place a transmission jack under the transmission

6. Remove the remaining transmission mounting bolts

7. Install a suitable lifting device to the engine.

8. Remove the engine mount bolts and carefully lift the engine from the vehicle. Pause several times while lifting the engine to make sure no wires or hoses have become snagged.

To install:

9. Carefully lower the engine into the vehicle and align the engine dowels with the transmission.

10. Install the engine mount bolts and tighten the bolts to 37 ft. lbs. (50 Nm).

11. Lower the engine onto the engine mounts.

12. Remove the engine lifting device.

13. Raise and support the vehicle.

14. Install the transmission-to-engine bolts and tighten the bolts to 37 ft. lbs. (50 Nm).

15. Remove the transmission jack.

16. Aligning the torque converter to the flexplate, install the bolts and tighten to 44 ft. lbs. (60 Nm).

17. On 2WD, install the crossmember and tighten the bolts to 44 ft. lbs. (60 Nm).

18. On 4WD, install the differential carrier.

19. Install or connect the following:

- Automatic transmission oil cooler and fuel line brackets, if equipped
- Left engine mount-to-frame bracket bolt and tighten to 63 ft. lbs. (85 Nm).
- New exhaust donut gasket

- Catalytic converter and tighten the bolts to 37 ft. lbs. (50 Nm).
- Engine wiring harness retainers at oil pan rail
- Engine wiring ground leads to block
- CKP sensor connector
- Heater outlet hose
- Knock sensor connectors
- EVAP canister connector
- Negative battery cable to block
- Starter wiring
- A/C compressor mounting bolts tighten the bolts to 37 ft. lbs. (50 Nm)
- Engine wiring harness bracket
- Wiring harness retainers from engine wiring bracket
- Left front fender inner liner
- Left front wheel
- Lower the vehicle
- Upper 2 engine wiring harness bracket bolts
- MAP wiring harness retainer
- MAP sensor
- Brake booster hose
- Dipstick and tube
- Wiring harness retainer to intake manifold
- Evaporative pipe at intake manifold
- Fuel lines to fuel rail
- Right engine mount-to-frame bracket bolt and tighten bolt to 63 ft. lbs. (85 Nm)
- Power steering pump
- A/C suction hose bracket
- Heater hoses from heater core
- Remove engine lifting eye in air pipe cover location
- Air injection pipe cover using new gasket
- Automatic transmission filler tube
- Wiring harness conduit at camshaft cover
- Wiring harness retainers at fender, power steering pump, throttle body and camshaft cover
- Coolant temperature sensor connector
- Injector harness connector
- Ignition coil connectors
- Oxygen Sensor (O2S) connector
- Exhaust camshaft actuator connector
- CMP sensor connector
- 4WD motor connector, if equipped
- Oil pressure switch connector
- Engine wiring harness connector at PCM
- Washer reservoir bolts
- Alternator
- Engine lifting bracket

- Air cleaner
- Cooling fan
- Radiator hoses
- Battery box
- Hood

20. Check all powertrain fluid levels and add, as necessary. Be sure to properly fill the engine crankcase with clean engine oil.

21. Connect the battery cables and properly fill the engine cooling system.

22. Start and run the engine, then check for leaks.

EXHAUST MANIFOLD

REMOVAL & INSTALLATION

See Figures 18 and 19.

1. Before servicing the vehicle, refer to the Precautions Section.
2. Remove or disconnect the following:
   - Negative battery cable
   - Catalytic converter from the exhaust manifold
   - Exhaust seal
   - Air intake assembly
   - Transmission fill tube bracket nut, if equipped with automatic transmission
   - Heated Oxygen Sensor (HO2S) from exhaust manifold
   - Exhaust manifold heat shield
   - Exhaust manifold bolts
   - Exhaust manifold and gasket

To install:

3. Clean the exhaust manifold retainer threads and the gasket mating surfaces.
4. Coat the bolt threads with a suitable threadlock.
5. Install the exhaust manifold and tighten the bolts in sequence to 15 ft. lbs. (20 Nm), tighten the bolts again to 15 ft.

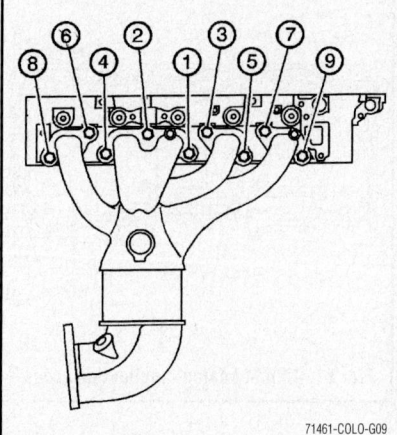

Fig. 19 Exhaust manifold tightening sequence—3.7L engines

lbs. (20 Nm), then tighten the bolts again to 15 ft. lbs. (20 Nm).

6. Install or connect the following:
   - Heat shield and tighten the nuts to 89 inch lbs. (10 Nm)
   - HO2S and tighten to 31 ft. lbs. (42 Nm)
   - Transmission fill tube bracket nut, if equipped with automatic transmission.
   - Air intake assembly
   - New exhaust seal to the exhaust manifold flange
   - Catalytic converter to the manifold. Tighten the nuts to 37 ft. lbs. (50 Nm).
   - Negative battery cable

INTAKE MANIFOLD

REMOVAL & INSTALLATION

See Figure 20.

1. Before servicing the vehicle, refer to the Precautions Section.
2. Remove or disconnect the following:
   - Negative battery cable
   - Air intake assembly
   - Throttle body
   - Battery and battery box
   - Dipstick and tube
   - Brake booster hose
   - Manifold Absolute Pressure (MAP) sensor connector and harness retainer
   - PCV tube
   - Alternator
   - Engine wiring harness retainer
   - Upper 2 engine wiring harness bracket bolts
   - Raise and support vehicle
   - Left front wheel
   - Left front fender inner liner
   - Wiring harness retainers for battery cable, engine and MAP sensor
   - Engine wiring harness bracket from the intake manifold
   - Intake manifold bolts
   - Lower the vehicle
   - Intake manifold and gasket

To install:

3. Install or connect the following:
   - Intake manifold with new gasket
   - Raise and support the vehicle
   - Tighten the intake manifold bolts,

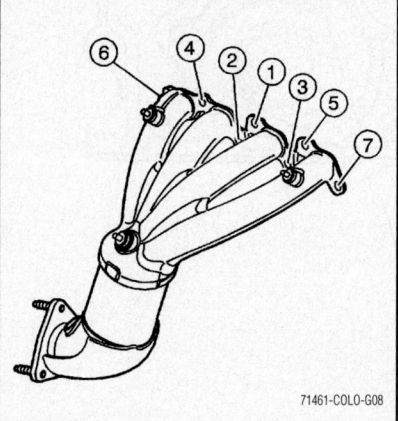

Fig. 18 Exhaust manifold tightening sequence—2.9L engines

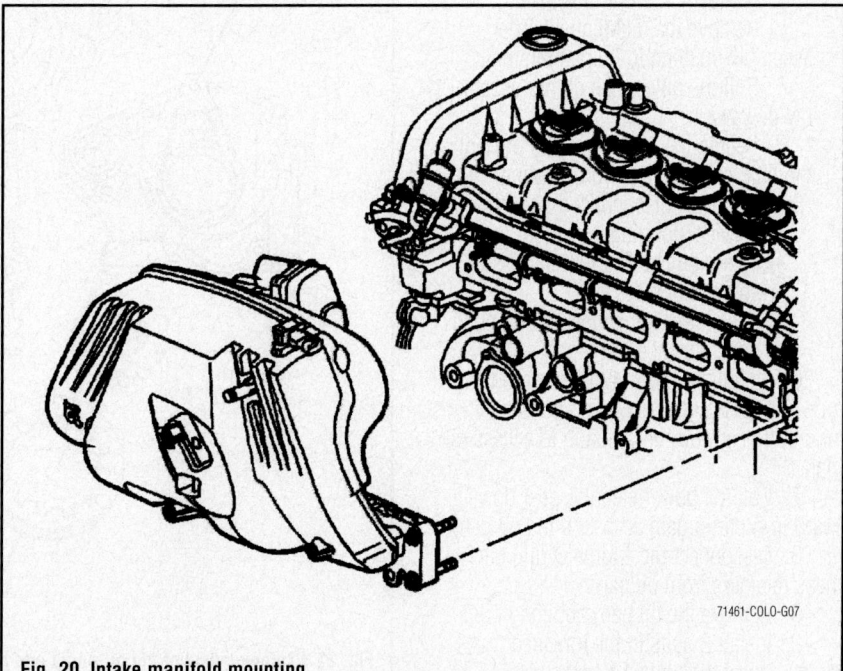

Fig. 20 Intake manifold mounting

working from the center outward, to 89 inch lbs. (10 Nm)

- Engine wiring harness bracket
- Wiring harness retainers for the battery cable, engine and MAP sensor
- Left front fender inner liner
- Left front wheel
- Lower the vehicle
- Upper 2 engine wiring harness bracket bolts
- Engine wiring harness retainer
- Alternator
- PCV tube
- MAP sensor connector and harness retainer
- Brake booster hose
- Dipstick and tube
- Throttle body
- Negative battery cable

4. Start the engine and check for leaks.

OIL PAN

REMOVAL & INSTALLATION

See Figure 21.

1. Before servicing the vehicle, refer to the Precautions Section.
2. Drain the engine oil.
3. Remove or disconnect the following:
- Oil dipstick tube
- Engine splash guard
- Right front halfshaft, if equipped with 3.7L Engines
- Power steering gear, if equipped with RWD

4. If equipped with 4WD:
   a. Remove the front driveshaft.
   b. Remove the differential carrier assembly bushing to frame bolts only.
   c. Pull the differential carrier assembly downward.
   d. Secure the pinion yoke to prevent the differential carrier from rotating.
5. Remove or disconnect the following:
- Service slot plug
- Fuel pipe bracket at transmission and position aside
- Four lower transmission mounting bolts attached to oil pan

6. If equipped with 4WD, remove the power steering gear mounting bolts only and pull gear down far enough to access oil pan.
7. Pull the power steering gear downward in order to gain access to the oil pan.
8. Disconnect the engine wiring harness retainers from oil pan
9. Remove the oil pan mounting bolts.
10. Install 2 bolts in the threaded holes at the rear of the oil pan to act as jack

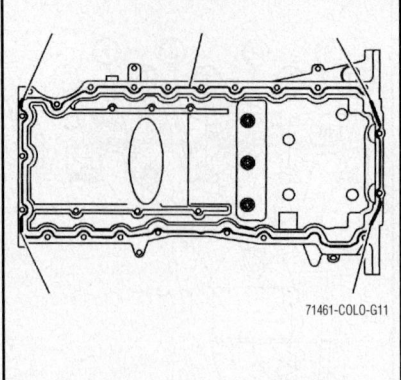

Fig. 21 Oil pan sealant application areas

screws. Tighten evenly to release the oil pan from the engine block.

11. Remove the oil pan and 2 bolts from the jack screw holes.
12. Clean and inspect the engine block sealing surface.

To install:

13. Apply a bead of sealant around the oil pan as shown.

→**Install the oil pan within 10 minutes of applying the sealant.**

14. Install the oil pan, making sure that the pan if positioned fully rearward against the transmission mounting surface.
15. Install the oil pan bolts and tighten the side bolts to 18 ft. lbs. (25 Nm). Tighten the end bolts to 89 inch lbs. (10 Nm).

16. Connect the engine wiring harness retainers to oil pan.
17. If equipped with 4WD, position the steering gear upward to the frame assembly and install the steering gear mounting bolts.
18. Install the four lower transmission mounting bolts and tighten to 37 ft. lbs. (50 Nm).
19. Install the nuts securing the fuel pipe bracket at transmission and tighten to 15 ft. lbs. (20 Nm).
20. Install the service slot plug.
21. If equipped with RWD, install the power steering gear.
22. If equipped with 4WD:
   a. Position the differential carrier assembly to the frame.
   b. Install the differential carrier assembly bushing to frame bolts. Tighten to 112 ft. lbs. (152 Nm).
   c. Install the front driveshaft.
23. Install the engine splash shield.
24. Install the oil dipstick tube. Tighten the bolt to 89 inch lbs. (10 Nm).
25. Fill the engine with oil to the correct level.
26. Start the engine and check for leaks.

OIL PUMP

REMOVAL & INSTALLATION

See Figure 22.

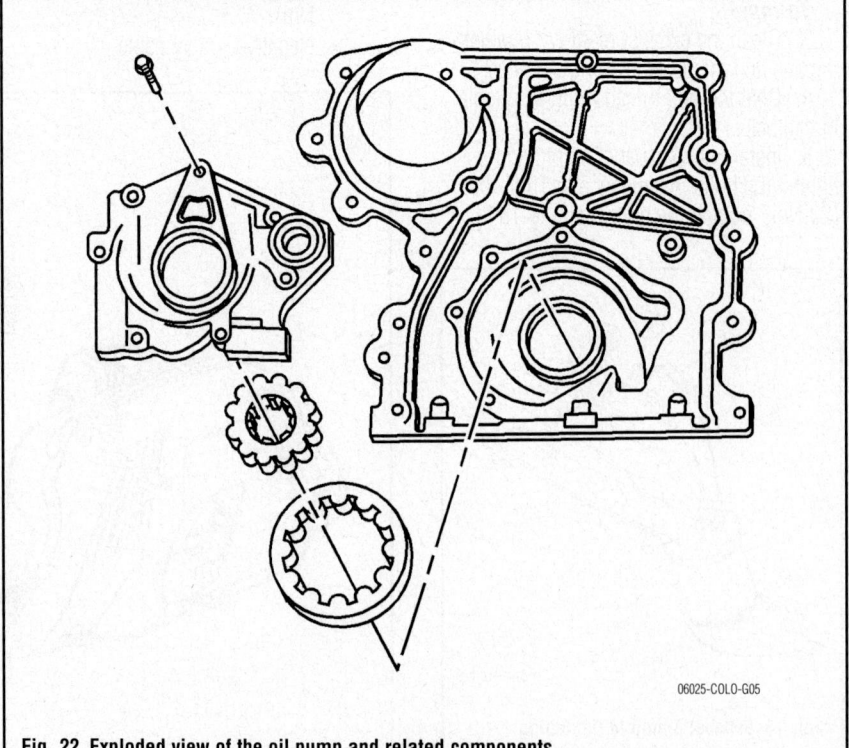

Fig. 22 Exploded view of the oil pump and related components

1. Before servicing the vehicle, refer to the Precautions Section.
2. Remove or disconnect the following:
   - Engine front cover
   - Oil pump cover bolts
   - Oil pump cover
3. Matchmark the inner and outer gears in relation to the oil pump housing.
4. Remove or disconnect the following:
   - Inner and outer oil pump gears
   - Oil pump pressure relief valve plug
   - Oil pump pressure relief valve and spring

To install:

5. Install or connect the following:
   - Oil pump pressure relief valve and spring
   - Oil pump pressure relief valve plug and tighten to 124 inch lbs. (14 Nm)
   - Oil pump outer and inner gears
   - Oil pump cover and tighten the bolts to 89 inch lbs. (10 Nm).
   - Engine front cover
6. Start the engine and check for leaks.

PISTON AND RING

POSITIONING

See Figures 23 and 24.

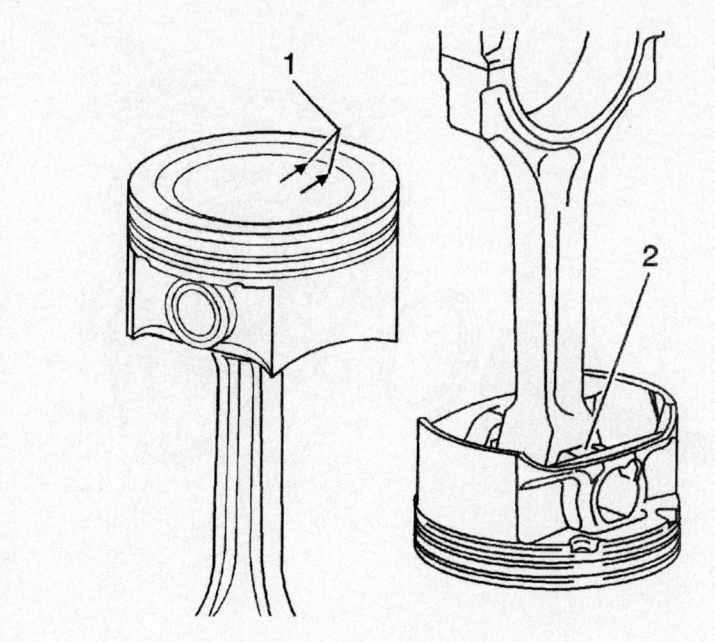

1 - Piston arrow face forward in block
2 - Flat casting surface faces forward

71461-COLO-G17

Fig. 24 Piston and connecting rod orientation

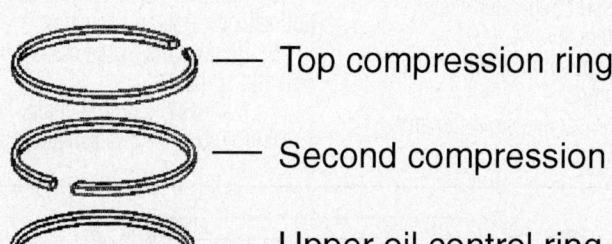
— Top compression ring

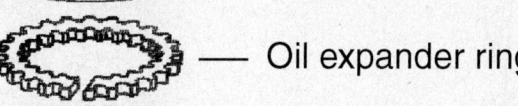
— Second compression ring

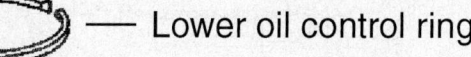
— Upper oil control ring

— Oil expander ring

— Lower oil control ring

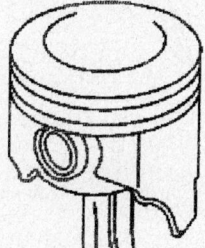

71461-COLO-G16

Fig. 23 Piston ring positioning

REAR MAIN SEAL

REMOVAL & INSTALLATION

See Figures 25 and 26.

1. Before servicing the vehicle, refer to the Precautions Section.
2. Remove or disconnect the following:
   - Negative battery cable
   - Transmission assembly and transfer case, if equipped
   - Clutch assembly, if equipped
   - Flywheel

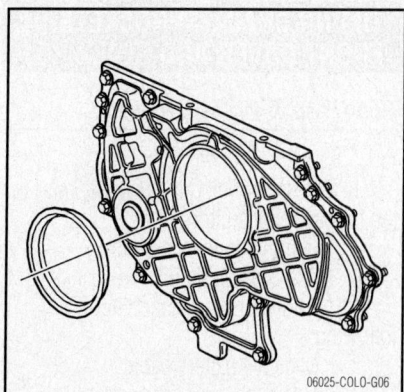

06025-COLO-G06

Fig. 25 Remove the seal from the oil seal housing.

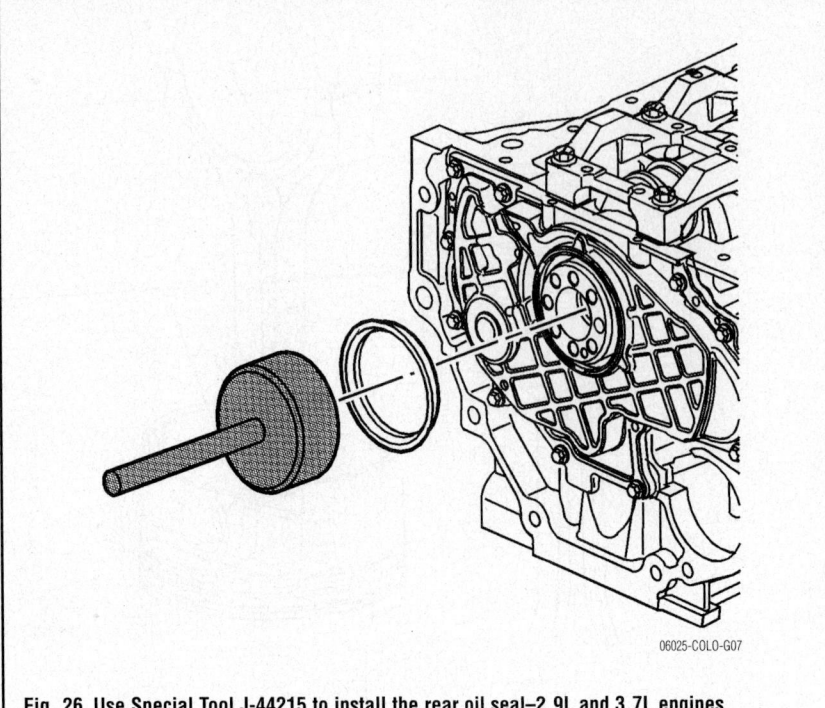

Fig. 26 Use Special Tool J-44215 to install the rear oil seal–2.9L and 3.7L engines

- Crankshaft seal by prying it from out oil seal housing

➡**Be careful not to damage the crankshaft seal surface with the prying tool.**

To install:

3. Install the new rear seal by lubricating it with engine oil and using a seal installer Special Tool J-44215. The spring side goes toward the engine and the seal will bottom out when installed fully.
4. Install or connect the following:
- Flywheel/clutch assembly or flexplate
- Transmission assembly and transfer case, if equipped
- Negative battery cable
5. Start the engine and check for leaks.

TIMING CHAIN, SPROCKETS, FRONT COVER AND SEAL

REMOVAL & INSTALLATION

See Figures 27 through 30.

1. Before servicing the vehicle, refer to the Precautions Section.
2. Drain the cooling system.
3. Drain the engine oil.
4. Remove or disconnect the following:
- Negative battery cable
- Number 1 cylinder spark plug
- Intake manifold
- Ignition coils
- Engine coolant temperature (ECT)

sensor electrical connector from camshaft cover
- Fuel injector connector from camshaft cover
- Heated oxygen sensor (HO2S) connector from camshaft cover
- Fuel pressure regulator screw
- Camshaft cover
- Camshaft position (CMP) sensor
- Water pump
5. Remove the service slot plug and

install Flywheel Holding Tool EN-46547 into the flywheel teeth.
6. Remove the crankshaft balancer bolt and discard.
7. Install Crankshaft end protector J-41816-2 into the end of the crankshaft and remove the crankshaft balancer using a 3-jaw puller.
8. Remove or disconnect the following:
- Drive belt tensioner
- Power steering pump
- Oil pan
- Oil pump pipe and screen assembly
- 7mm center front cover bolt
- Remaining engine front cover bolts
9. Install 2 bolts into the threaded holes to act as jack screws and tighten evenly to release the front cover.
10. Rotate the crankshaft clockwise until the no. 1 cylinder is at TDC on the compression stroke. The word DELPHI on the camshaft position actuator will be parallel with the cylinder head surface.
11. Install camshaft locking tool J-44221 to the rear of the camshafts.
12. Release the tension on the timing chain by moving the tensioner shoe in by hand.
13. Place a tee in the tensioner to hold the shoe in place.
14. Remove the top timing chain guide.
15. Remove the exhaust camshaft position actuator bolt and actuator.
16. Remove the intake camshaft sprocket bolt and sprocket.
17. Remove the timing chain.
18. Remove the crankshaft sprocket.

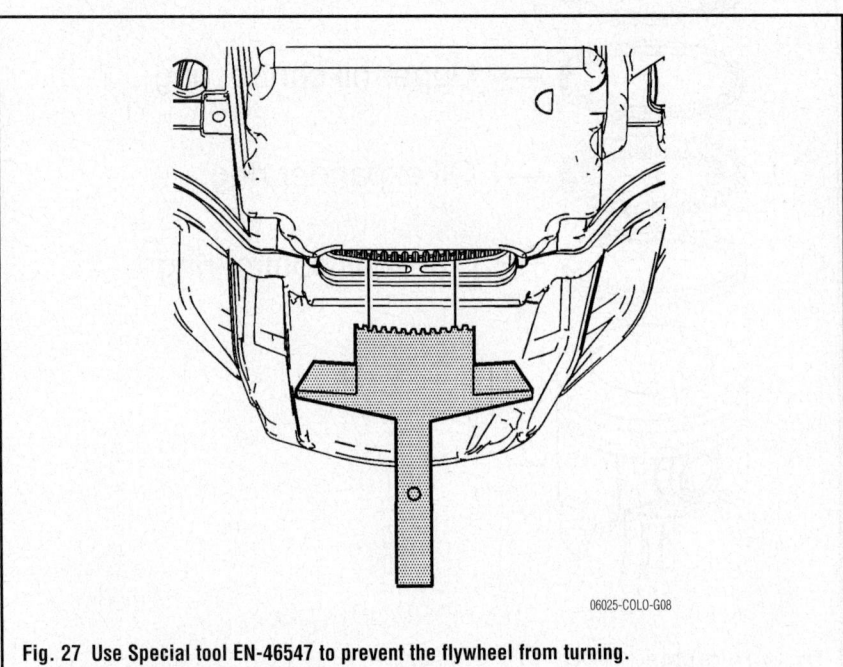

Fig. 27 Use Special tool EN-46547 to prevent the flywheel from turning.

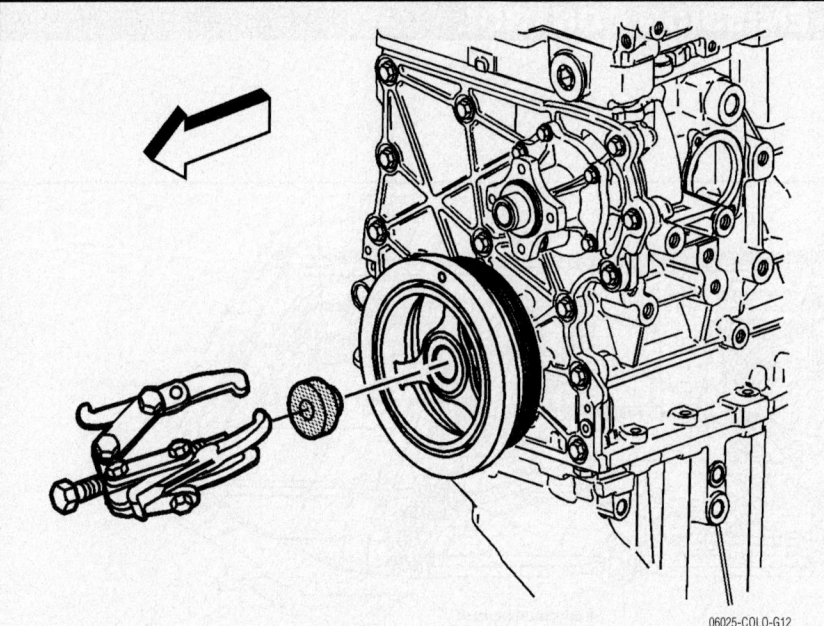

Fig. 28 Remove the crankshaft balancer using a suitable puller after installing End Protector J-41816-2.

06025-COLO-G12

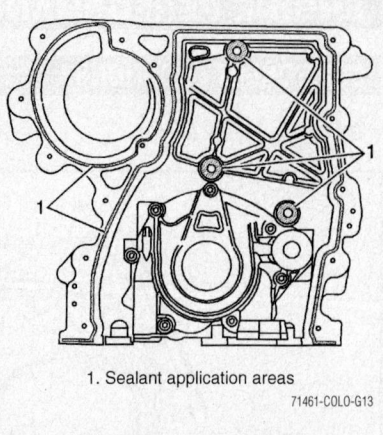

1. Sealant application areas

71461-COLO-G13

Fig. 30 Front cover sealant application

To install:

19. Install the crankshaft sprocket.

20. Install the intake camshaft sprocket on the timing chain and align the dark link on the timing chain with the timing mark on the intake sprocket as shown.

21. Feed the timing chain through the opening in the cylinder head.

22. Install the timing chain on the crankshaft sprocket and align the dark link of the timing chain with the timing mark on the crankshaft sprocket.

23. Install a new intake camshaft sprocket bolt and washer and tighten the bolt to 15 ft. lbs. (20 Nm), plus an additional 100°.

24. Install the exhaust camshaft actuator on the timing chain with the word DELPHI facing horizontal to the cylinder head surface and the dark link of the timing chain aligned with the timing mark on the camshaft actuator sprocket.

➡**Ensure the alignment pin is engaged between the camshaft and exhaust camshaft actuator sprocket.**

25. Install the exhaust camshaft actuator onto the exhaust camshaft.

➡**Rotate the camshaft actuator clockwise until it stops. This will fully advance the actuator. Engine damage may occur if the actuator is not fully advanced.**

26. Install a new exhaust camshaft sprocket bolt and washer and tighten the bolt to 18 ft. lbs. (25 Nm), plus an additional 135°.

27. Remove the tee in the timing chain tensioner to tension the timing chain.

28. Remove the camshaft locking tool from the camshafts.

29. The dark links should be aligned with the camshaft and crankshaft sprockets as shown.

30. Thread alignment pins into the engine block to aid front cover installation.

31. Apply sealant to the front cover surfaces as shown.

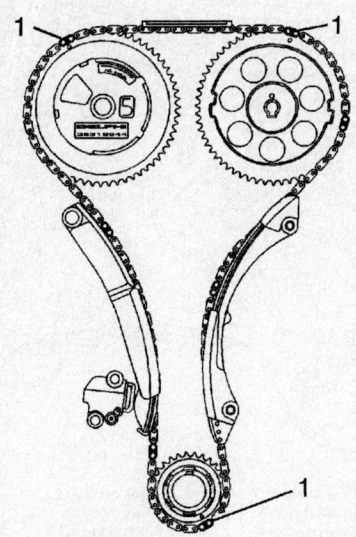

1. Timing chain dark link locations

71461-COLO-G12

Fig. 29 Aligning timing chain dark links with camshaft and crankshaft sprockets

32. Align the oil pump with the crankshaft sprocket splines.

33. Install the front cover over the alignment pins, and loosely install the front cover bolts.

34. Remove the alignment pins and install the remaining 2 bolts.

35. Tighten the front cover bolts to 89 inch lbs. (10 Nm).

36. Tighten the small center cover bolt to 71 inch lbs. (8 Nm).

37. Install clean engine oil to the outside diameter of the new front crankshaft oil seal.

38. Using a seal installer, install the front oil seal.

39. Install a new o-ring to the oil pump pipe screen assembly and install the oil pump pipe screen.

40. Apply sealant to the oil pump pipe bolt threads and tighten the bolts to 89 inch lbs. (10 Nm).

41. Install or connect the following:
- Oil pan
- Power steering pump
- Drive belt tensioner
- Crankshaft damper and tighten new bolt and tighten to 111 ft. lbs. (150 Nm), plus an additional 180°.

42. Remove the flywheel locking tool.

43. Install or connect the following:
- Flywheel access service plug
- Water pump
- Accessory drive belt
- Cooling fan
- Negative battery cable

44. Fill the engine with coolant and oil.

45. Start the engine and check for leaks.

VALVE LASH

ADJUSTMENT

Valve lash is maintained by the automatic lash adjusters.

ENGINE PERFORMANCE & EMISSION CONTROL

COMPONENT LOCATIONS

See Figures 31 through 36.

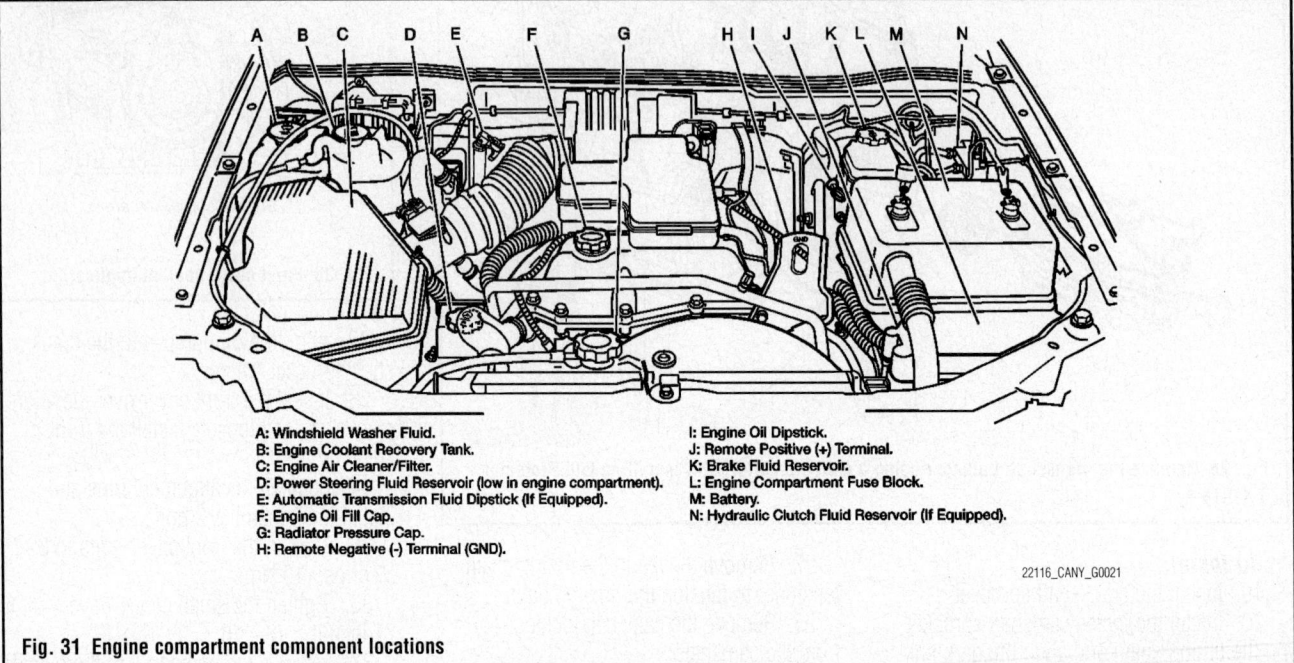

A: Windshield Washer Fluid.
B: Engine Coolant Recovery Tank.
C: Engine Air Cleaner/Filter.
D: Power Steering Fluid Reservoir (low in engine compartment).
E: Automatic Transmission Fluid Dipstick (If Equipped).
F: Engine Oil Fill Cap.
G: Radiator Pressure Cap.
H: Remote Negative (-) Terminal (GND).

I: Engine Oil Dipstick.
J: Remote Positive (+) Terminal.
K: Brake Fluid Reservoir.
L: Engine Compartment Fuse Block.
M: Battery.
N: Hydraulic Clutch Fluid Reservoir (If Equipped).

22116_CANY_G0021

Fig. 31 Engine compartment component locations

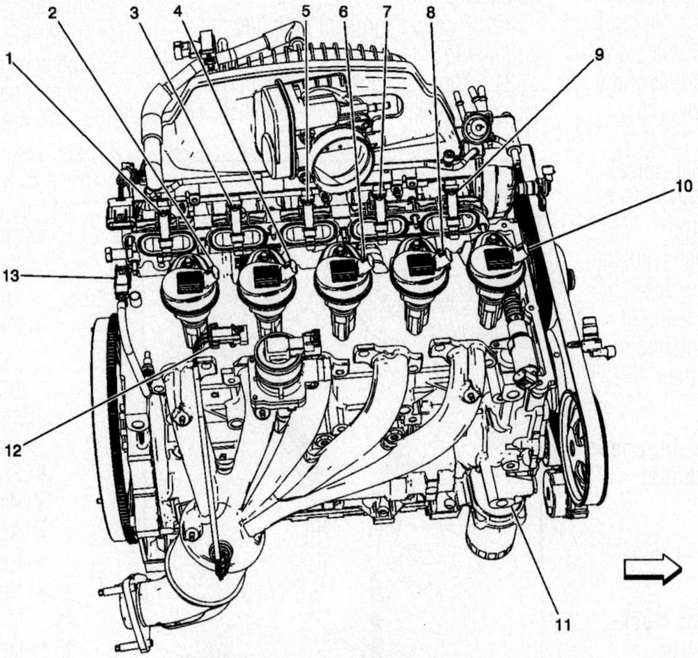

1. Fuel Injector 5
2. Ignition Coil 5
3. Fuel Injector 4
4. Ignition Coil 4
5. Fuel Injector 3
6. Ignition Coil 3
7. Fuel Injector 2

8. Ignition Coil 2
9. Fuel Injector 1
10. Ignition Coil 1
11. Engine Oil Pressure (EOP) Sensor
12. Heated Oxygen Sensor (HO2S) 1
13. Engine Coolant Temperature
   (ECT) Sensor Connector

22116_CANY_G0022

Fig. 32 Engine control component locations—1 of 4

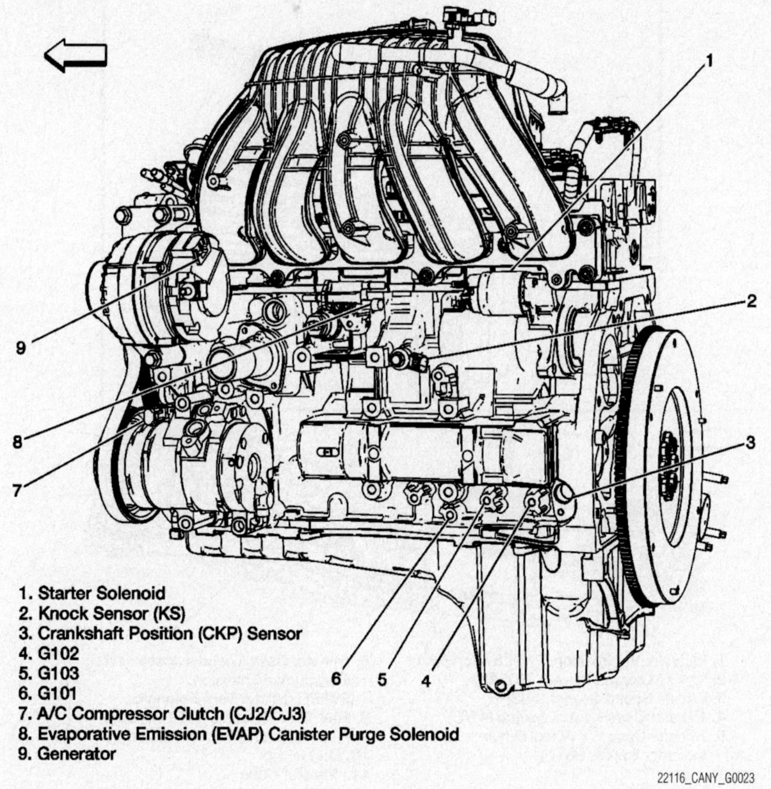

1. Starter Solenoid
2. Knock Sensor (KS)
3. Crankshaft Position (CKP) Sensor
4. G102
5. G103
6. G101
7. A/C Compressor Clutch (CJ2/CJ3)
8. Evaporative Emission (EVAP) Canister Purge Solenoid
9. Generator

22116_CANY_G0023

Fig. 33 Engine control component locations–2 of 4

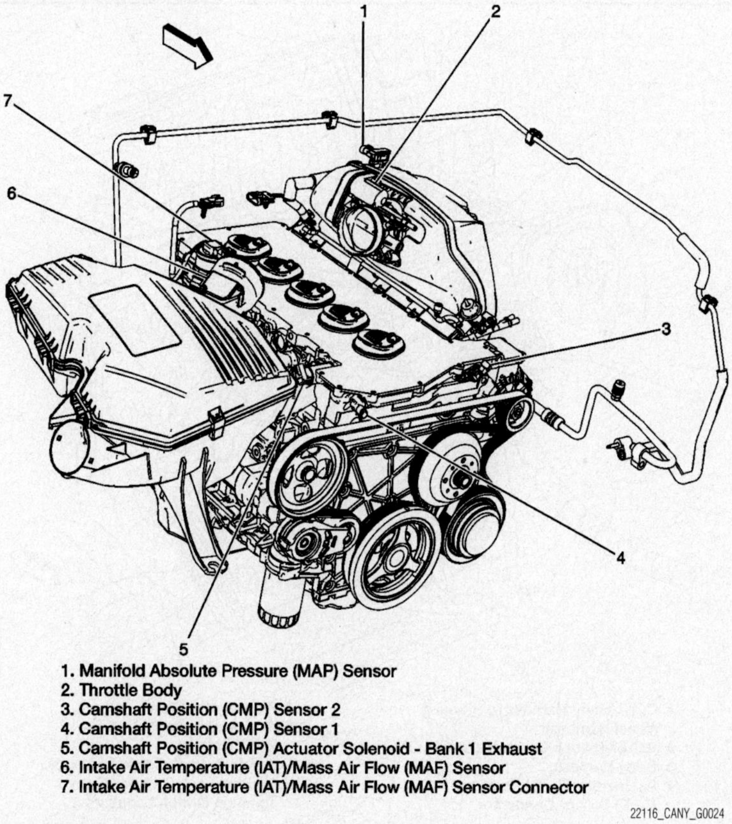

1. Manifold Absolute Pressure (MAP) Sensor
2. Throttle Body
3. Camshaft Position (CMP) Sensor 2
4. Camshaft Position (CMP) Sensor 1
5. Camshaft Position (CMP) Actuator Solenoid - Bank 1 Exhaust
6. Intake Air Temperature (IAT)/Mass Air Flow (MAF) Sensor
7. Intake Air Temperature (IAT)/Mass Air Flow (MAF) Sensor Connector

22116_CANY_G0024

Fig. 34 Engine control component locations–3 of 4

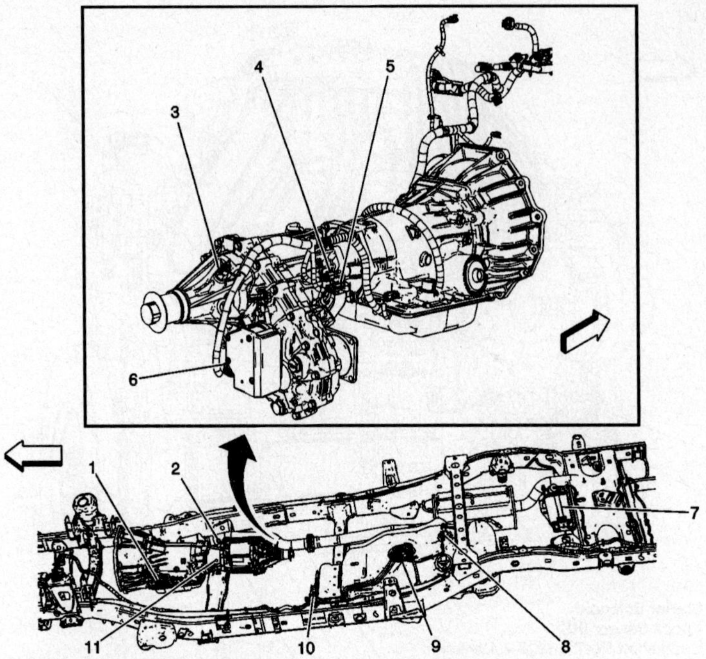

Fig. 35 Engine control component locations–4 of 4

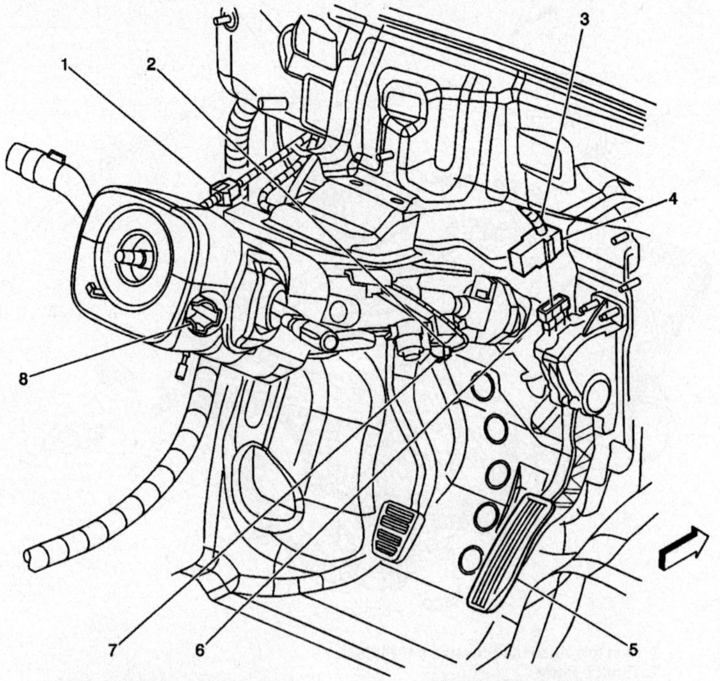

1. Park/Neutral Position (PNP) Switch (M30)
2. Heated Oxygen Sensor (HO2S) 2
3. Vehicle Speed Sensor (VSS)
4. Transfer Case Neutral Switch (4WD)
5. Transfer Case 2/4 Wheel Driver
   Indicator Switch (4WD)
6. Transfer Case Encoder Motor (4WD)
7. Evaporative Emission
   (EVAP) Canister Vent Solenoid
8. Fuel Tank Pressure (FTP) Sensor
9. Fuel Pump and Sender Assembly
10. Fuel Pump
11. Transfer Case

22116_CANY_G0025

Fig. 36 Left side of instrument panel

1. C275 (Body Harness to Steering
   Wheel Harness)
2. A/T Shift Lock Control Solenoid Connector
3. Body Harness
4. Accelerator Pedal Position
   (APP) Sensor Connector
5. Accelerator Pedal
6. Accelerator Pedal Position
   (APP) Sensor
7. A/T Shift Lock Control Solenoid
8. Ignition Lock Cylinder Case
   (Ignition Switch Location)

22116_CANY_G0026

ACCELERATOR PEDAL POSITION (APP) SENSOR

LOCATION

See Figure 37.

The APP sensor is located above the accelerator pedal arm.

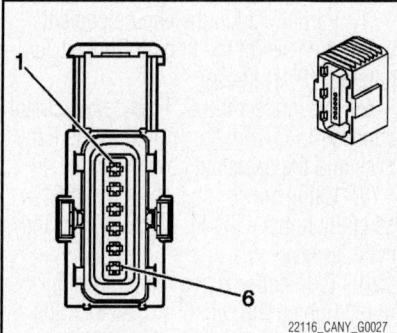

Fig. 37 Identifying accelerator pedal position sensor connector

REMOVAL & INSTALLATION

See Figure 38.

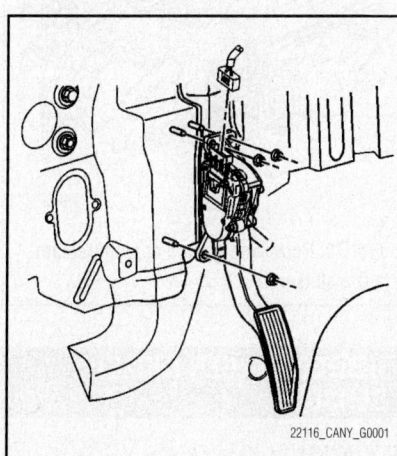

Fig. 38 Removing accelerator pedal position sensor and components

1. Disconnect the Accelerator Pedal Position (APP) sensor electrical connector.
2. Remove the APP sensor nuts.
3. Remove the APP sensor from the vehicle.
4. To install, reverse removal procedure. Inspect below the pedal for binding, to ensure full range of motion.

CAMSHAFT POSITION (CMP) SENSOR

LOCATION

See Figures 39 and 40.

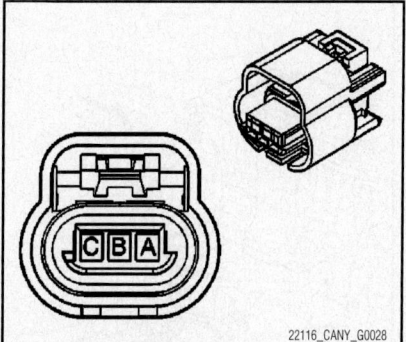

Fig. 39 Identifying camshaft position sensor 1 connector

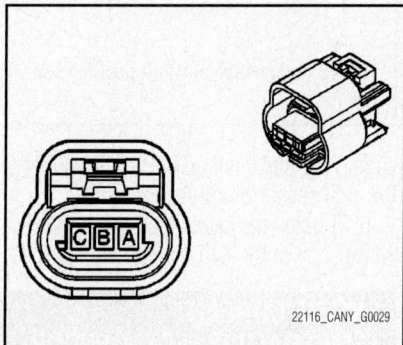

Fig. 40 Identifying camshaft position sensor 2 connector

Refer to illustrations under removal and installation for camshaft position sensor location.

REMOVAL & INSTALLATION

Exhaust Camshaft

See Figure 41.

1. Disconnect the engine wiring harness electrical connector from the Camshaft Position (CMP) sensor.
2. Remove the CMP sensor retaining bolt.

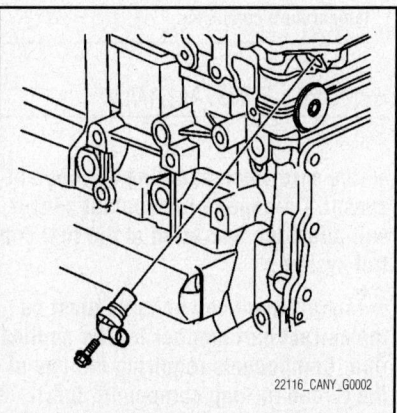

Fig. 41 Removing camshaft position sensor

3. Remove the CMP sensor from the cylinder head. Discard the O-ring seal.
4. To install, reverse removal procedure. Install a new O-ring seal to the CMP sensor. Lightly lubricate the O-ring seal with clean engine oil.
5. Tighten the camshaft position (CMP) sensor bolt to 89 inch lbs. (10 Nm).

Intake Camshaft

See Figures 42 and 43.

1. Disconnect the engine wiring harness electrical connector from the intake Camshaft Position (CMP) sensor.
2. Remove the intake CMP sensor retaining bolt.
3. Remove the intake CMP sensor from the cylinder head.
4. To install, reverse removal procedure. Install a new O-ring seal to the CMP sensor.

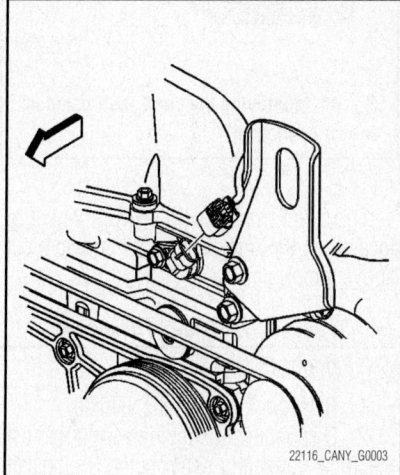

Fig. 42 Disconnecting CMP electrical connector—intake

Fig. 43 Removing camshaft position sensor

Lightly lubricate the O-ring seal with clean engine oil. Add sealer GM P/N 12346004 to the CMP sensor bolt threads.

5. Tighten the camshaft position (CMP) sensor bolt to 89 inch lbs. (10 Nm).

CRANKSHAFT POSITION (CKP) SENSOR

LOCATION

See Figure 44.

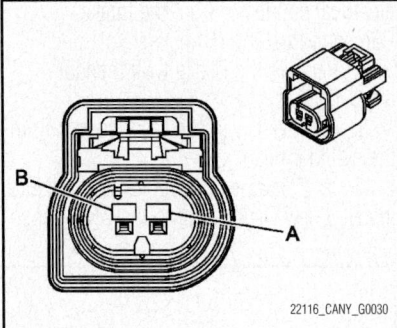

Fig. 44 Identifying the crankshaft position sensor connector

The Crankshaft Position (CKP) sensor is located on the right side of the engine block above the oil pan.

REMOVAL & INSTALLATION

See Figures 45 and 46.

1. Raise and support the vehicle.
2. Disconnect the engine wiring harness electrical connector (3) from the Crankshaft Position (CKP) sensor (1).
3. Remove the CKP sensor bolt.
4. Remove the CKP sensor from the engine block. Discard the O-ring seals.
5. To install, reverse removal procedure. Install new O-ring seals to the CKP sensor. Lightly lubricate the O-ring seals with clean

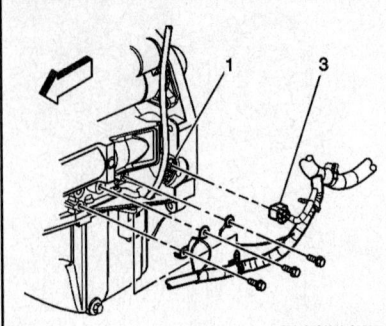

Fig. 45 Disconnecting CKP electrical connector

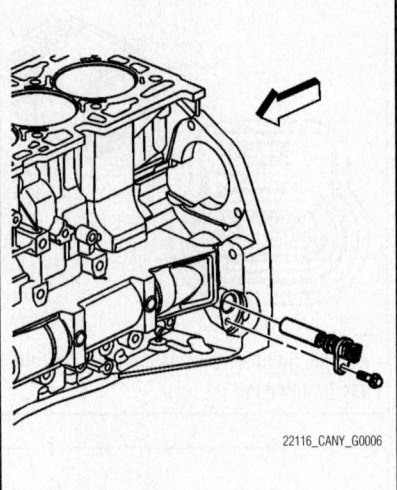

Fig. 46 Removing camshaft position sensor

engine oil. Add sealer GM P/N 12346004 to the CKP sensor bolt threads.

6. Tighten the camshaft position (CKP) sensor bolt to 89 inch lbs. (10 Nm).

ENGINE COOLANT TEMPERATURE (ECT) SENSOR

LOCATION

See Figure 47.

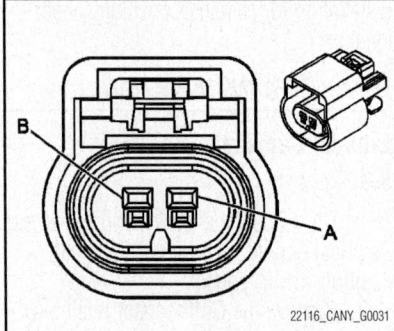

Fig. 47 Identifying the engine coolant temperature connector

REMOVAL & INSTALLATION

See Figure 48.

➡Use care when handling the coolant sensor. Damage to the coolant sensor will affect the operation of the fuel control system.

➡Replacement components must be the correct part number for the application. Components requiring the use of the thread locking compound, lubricants, corrosion inhibitors, or sealants are identified in the service procedure.

Some replacement components may come with these coatings already applied. Do not use these coatings on components unless specified. These coatings can affect the final torque, which may affect the operation of the component. Use the correct torque specification when installing components in order to avoid damage.

1. Partially drain the engine coolant below the level of the Engine Coolant Temperature (ECT) sensor.
2. Disconnect the ECT sensor electrical connector (1) from the engine wiring harness and the camshaft cover.
3. Using sensor socket (2) J-45861, carefully remove the ECT sensor from the cylinder head.
4. To install, reverse the removal procedure. Tighten the sensor to 124 inch lbs. (14 Nm). Refill the engine coolant.

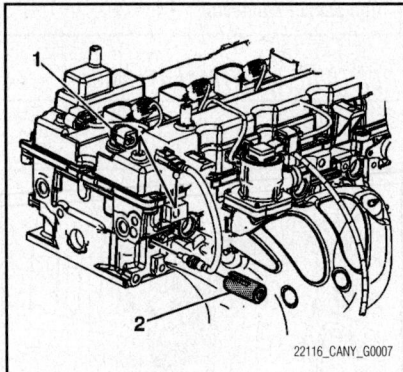

Fig. 48 Removing engine coolant temperature sensor

HEATED OXYGEN (HO2S) SENSOR

LOCATION

See Figures 49 and 50.

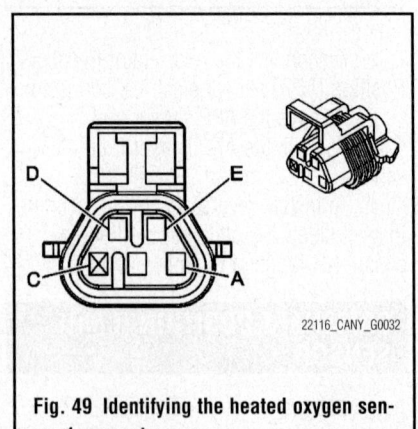

Fig. 49 Identifying the heated oxygen sensor 1 connector

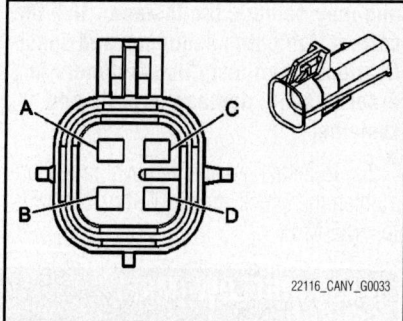

Fig. 50 Identifying the heated oxygen sensor 2 connector

The Heated Oxygen (HO2S) sensors are located on the exhaust manifold and on the exhaust pipe before the catalytic converter

REMOVAL & INSTALLATION

➡**When replacing the HO2S perform the following:**

- A code clear with a scan tool, regardless of whether or not a DTC is set
- HO2S heater resistance learn reset with a scan tool, where available

Perform the above in order to reset the HO2S resistance learned value and avoid possible HO2S failure.

➡**The oxygen sensor may be difficult to remove when the engine temperature is below 120°F (48°C). Excessive force may damage threads in the exhaust manifold or the exhaust pipe.**

Sensor 1

See Figures 51 and 52.

1. Remove the connector position assurance (CPA) retainer.
2. Disconnect the engine wiring harness electrical connector (1) from the Heated Oxygen sensor (HO2S) electrical connector.
3. Remove the HO2S sensor electrical connector from the camshaft cover.

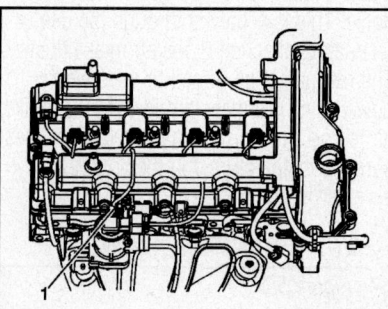

Fig. 51 Disconnecting the engine wiring harness electrical connector–sensor 1

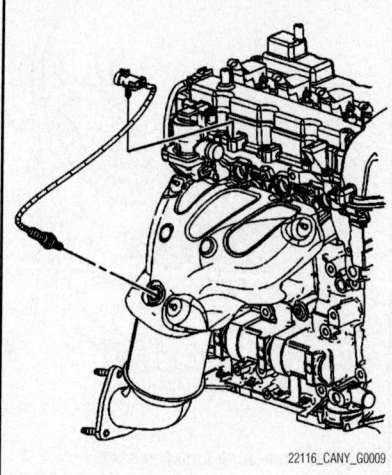

Fig. 52 Removing heated oxygen sensor (HO2S) sensor–sensor 1

4. Using the Heated Oxygen Sensor Wrench (J 39194-B) remove the HO2S from the exhaust manifold.

To install:

※※ CAUTION

Use special anti-seize compound on the HO2S threads. The compound consists of graphite suspended in fluid and glass beads. The graphite burns away, but the glass beads remain, making the sensor easier to remove. New service sensors already have the compound applied to the threads. If you remove an oxygen sensor and if for any reason you must install the same oxygen sensor, apply the anti-seize compound to the threads before reinstallation.

5. Coat the threads of the HO2S with the anti-seize compound P/N 5613695, or equivalent if necessary.
6. Thread the HO2S into the exhaust manifold by hand.

➡**Replacement components must be the correct part number for the application. Components requiring the use of the thread locking compound, lubricants, corrosion inhibitors, or sealants are identified in the service procedure. Some replacement components may come with these coatings already applied. Do not use these coatings on components unless specified. These coatings can affect the final torque, which may affect the operation of the component. Use the correct torque specification when installing components in order to avoid damage.**

7. Tighten the HO2S to 31ft. lbs. (42 Nm).
8. To complete installation, reverse remaining removal procedure.

Sensor 2

See Figures 53 and 54.

1. Raise and support the vehicle.
2. Disconnect the heated oxygen sensor (HO2S) electrical connector (3) from the engine wiring harness connector (1).
3. Using the Heated Oxygen Sensor Wrench (J 39194-B), remove the HO2S from the catalytic converter.

To install:

※※ CAUTION

Use special anti-seize compound on the HO2S threads. The compound consists of graphite suspended in fluid and glass beads. The graphite burns away, but the glass beads remain, making the sensor easier to remove. New service sensors already have the compound applied to the threads. If you remove an oxygen sensor and if for any reason you must install the same oxygen sensor, apply the anti-seize compound to the threads before reinstallation.

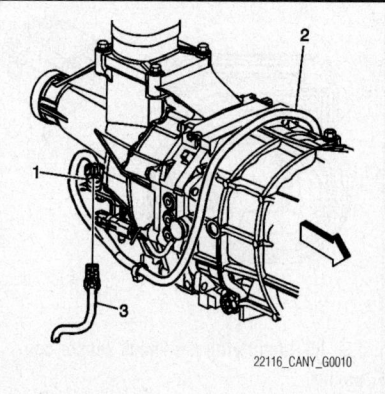

Fig. 53 Disconnecting the engine wiring harness electrical connector–sensor 2

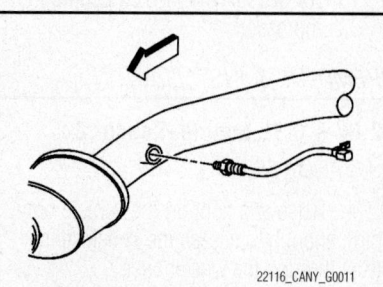

Fig. 54 Removing heated oxygen sensor (HO2S) sensor–sensor 2

4. Coat the threads of the HO2S with the anti-seize compound P/N 5613695, or equivalent if necessary.

5. Thread the HO2S into the exhaust manifold by hand.

➡Replacement components must be the correct part number for the application. Components requiring the use of the thread locking compound, lubricants, corrosion inhibitors, or sealants are identified in the service procedure. Some replacement components may come with these coatings already applied. Do not use these coatings on components unless specified. These coatings can affect the final torque, which may affect the operation of the component. Use the correct torque specification when installing components in order to avoid damage.

6. Tighten the HO2S to 31ft. lbs. (42 Nm).

7. To complete installation, reverse remaining removal procedure.

KNOCK SENSOR (KS)

LOCATION
See Figure 55.

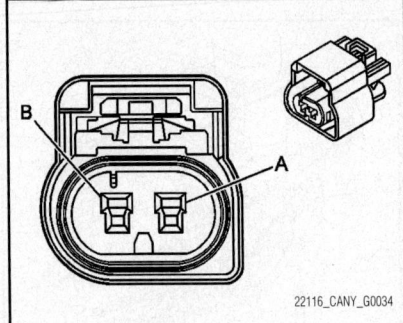

Fig. 55 Identifying the knock sensor connector

The Knock Sensor 2 is located on the side of the engine block behind the alternator. Knock sensor 1 is located behind the A/C compressor.

REMOVAL & INSTALLATION

2.9L & 3.7L Engine–Sensor 2
See Figure 56.

1. Raise and support the vehicle only high enough to access the knock sensor (KS) through the wheelhouse.

2. Remove the left wheelhouse liner.

3. Disconnect the engine wiring harness (2) electrical connector (3) from the KS (1).

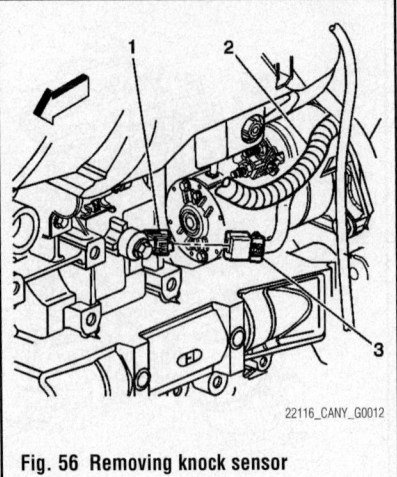

Fig. 56 Removing knock sensor

4. Remove the KS retaining bolt. Remove the KS (1) from the engine block.

➡Use the correct fastener in the correct location. Replacement fasteners must be the correct part number for that application. Fasteners requiring replacement or fasteners requiring the use of thread locking compound or sealant are identified in the service procedure. Do not use paints, lubricants, or corrosion inhibitors on fasteners or fastener joint surfaces unless specified. These coatings affect fastener torque and joint clamping force and may damage the fastener. Use the correct tightening sequence and specifications when installing fasteners in order to avoid damage to parts and systems.

5. To install, reverse removal procedure. Tighten the knock sensor (KS) bolt to 18 ft. lbs. (25 Nm).

3.7L Engine–Sensor 1

1. Raise and support the vehicle.

2. Disconnect the engine wiring harness electrical connector from the KS (2).

3. Remove the KS retaining bolt.

4. Remove the KS (2) from the engine block.

➡Use the correct fastener in the correct location. Replacement fasteners must be the correct part number for that application. Fasteners requiring replacement or fasteners requiring the use of thread locking compound or sealant are identified in the service procedure. Do not use paints, lubricants, or corrosion inhibitors on fasteners or fastener joint surfaces unless specified. These coatings affect fastener torque and joint clamping force

and may damage the fastener. Use the correct tightening sequence and specifications when installing fasteners in order to avoid damage to parts and systems.

5. To install, reverse removal procedure. Tighten the knock sensor (KS) bolt to 18 ft. lbs. (25 Nm).

MASS AIRFLOW (MAF) SENSOR

➡This is also known as the Intake Air Temperature (IAT) Sensor.

LOCATION
See Figure 57.

The Mass Airflow (MAF)/Intake Air Temperature (IAT) sensor is located on the air intake housing tube inlet.

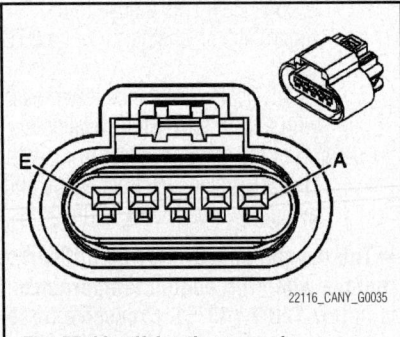

Fig. 57 Identifying the mass air flow/intake air temperature sensor connector

OPERATION

The primary function of the Air Intake System is to provide filtered air to the engine. The system uses a cleaner element mounted in a housing. The cleaner housing is remotely mounted and uses intake ducts to route the incoming air into the throttle body. The secondary function of the Air Intake System is to muffle air induction noise. This is achieved through the use of resonators attached to the air intake ducts. The resonators are tuned to the specific powertrain. The mass air flow (MAF)/intake air temperature (IAT) sensor is used to measure the temperature and the volume of the air entering the engine.

REMOVAL & INSTALLATION
See Figure 58.

❊❊ CAUTION

Handle the MAF sensor carefully. Do not drop the MAF sensor in order to

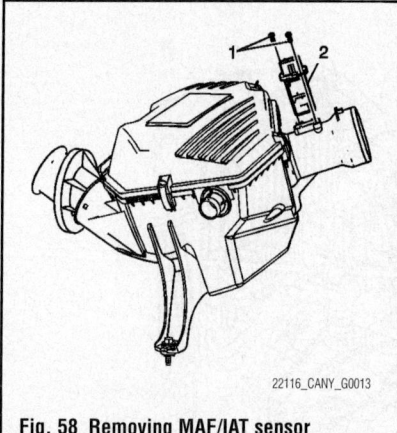

Fig. 58 Removing MAF/IAT sensor

prevent damage to the MAF sensor. Do not damage the screen located on the air inlet end of the MAF. Do not touch the sensing elements. Do not allow solvents and lubricants to come in contact with the sensing elements. Use a small amount of a soap based solution in order to aid in the installation.

1. Disconnect the electrical connector from the MAF/IAT sensor.
2. Remove the 2 screws (1) securing the MAF/IAT sensor to the air cleaner assembly.
3. Remove the MAF/IAT sensor (2) from the air cleaner assembly.
4. To install, reverse removal procedure.

MANIFOLD ABSOLUTE PRESSURE (MAP) SENSOR

LOCATION
See Figure 59.

Refer to illustration under removal and installation for Manifold Absolute Pressure (MAP) sensor location.

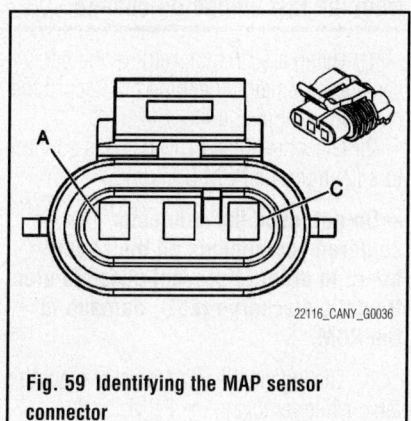

Fig. 59 Identifying the MAP sensor connector

REMOVAL & INSTALLATION
See Figure 60.

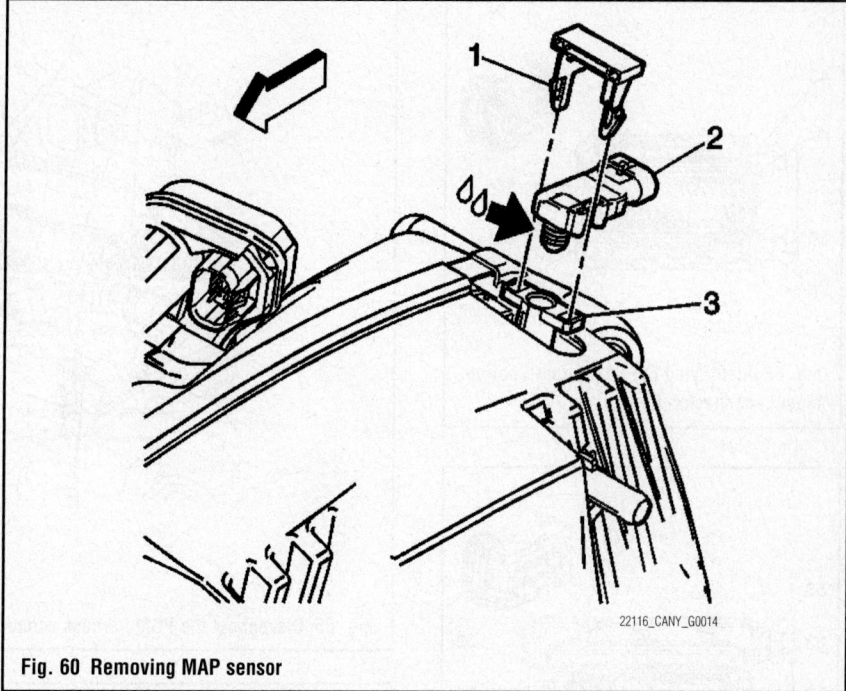

Fig. 60 Removing MAP sensor

1. Disconnect the MAP sensor electrical connector (3).
2. Press the retainer locking tabs inward, then pull the retainer (1) up to remove.
3. Remove the MAP sensor (2) from the intake manifold (3).
4. Inspect the MAP sensor seal for damage and replace as necessary.

5. To install, reverse removal procedure.

POWERTRAIN CONTROL MODULE (PCM)

LOCATION
See Figures 61 through 64.

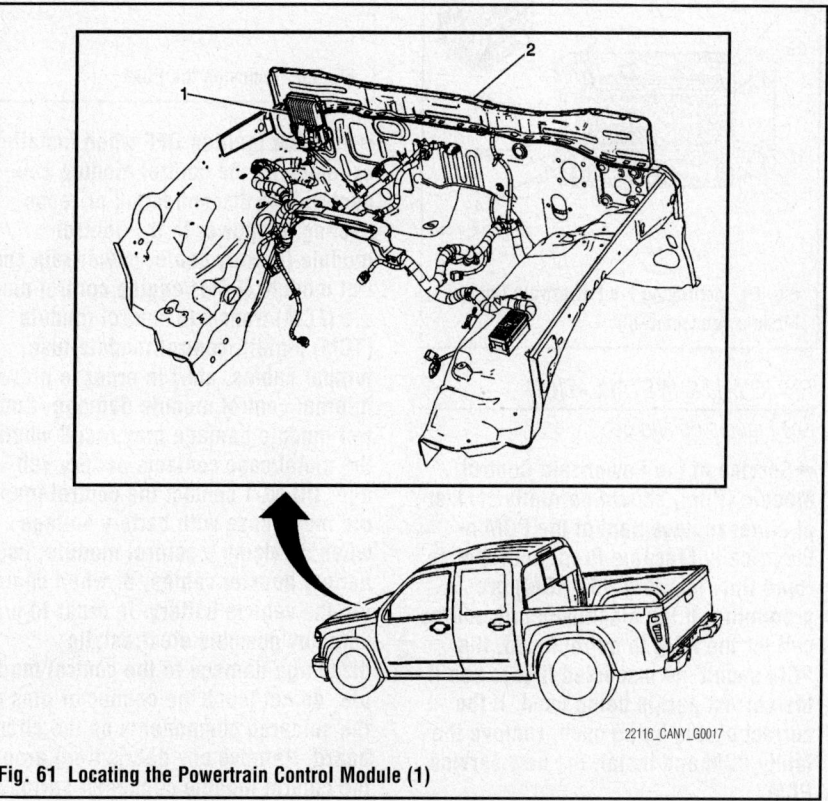

Fig. 61 Locating the Powertrain Control Module (1)

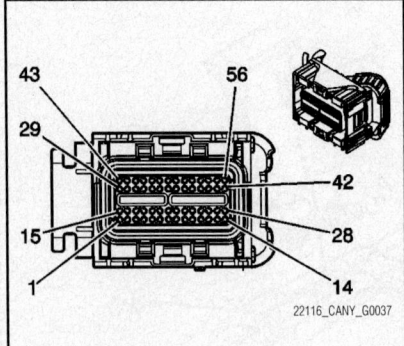

Fig. 62 Identifying the Powertrain Control Module connector—C1

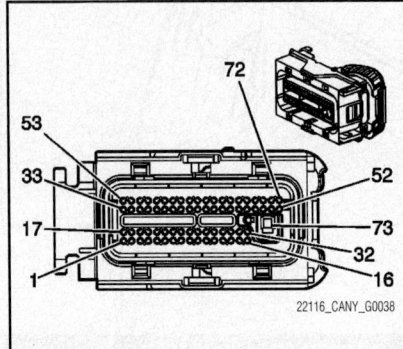

Fig. 63 Identifying the Powertrain Control Module connector—C2

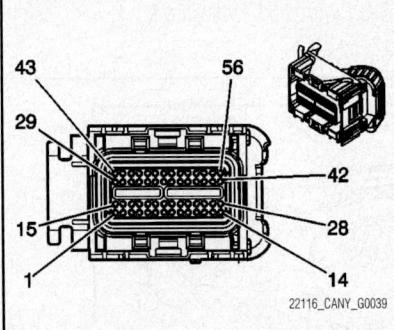

Fig. 64 Identifying the Powertrain Control Module connector—C3

REMOVAL & INSTALLATION

See Figures 65 and 66.

➡Service of the Powertrain Control Module (PCM) should normally consist of either replacement of the PCM or Electrically Erasable Programmable Read Only Memory (EEPROM) programming. If the diagnostic procedures call for the PCM to be replaced, the PCM should be inspected first to see if the correct part is being used. If the correct part is being used, remove the faulty PCM and install the new service PCM.

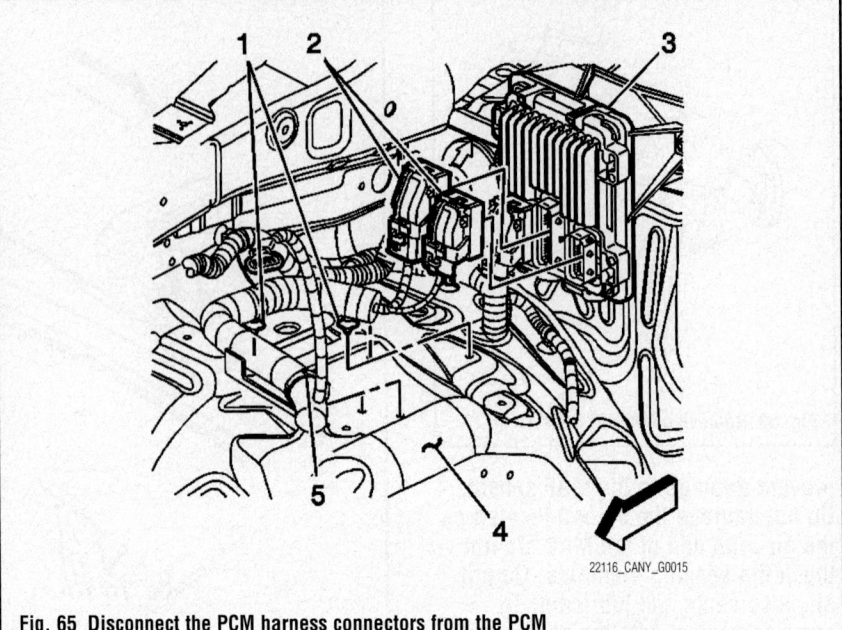

Fig. 65 Disconnect the PCM harness connectors from the PCM

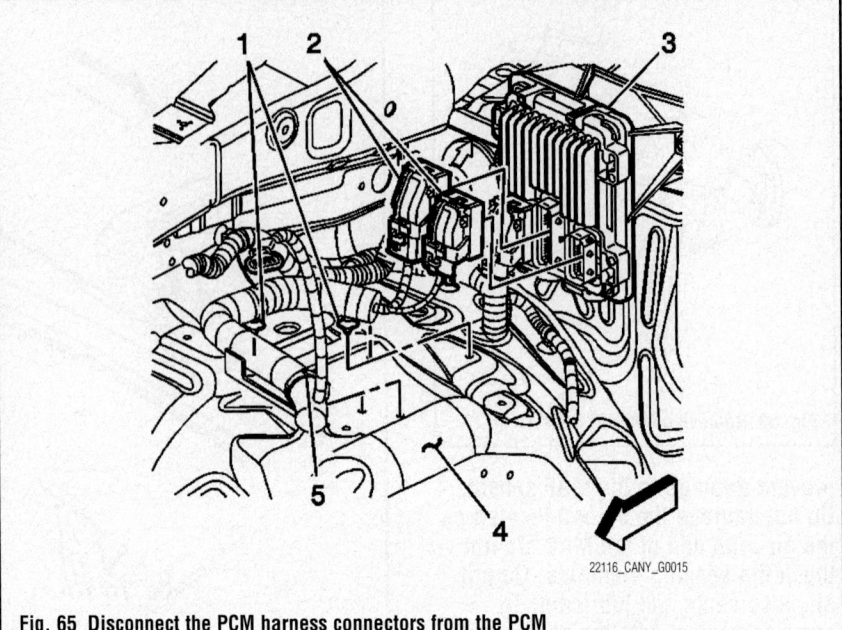

Fig. 66 Removing the PCM

➡Turn the ignition OFF when installing or removing the control module connectors and disconnecting or reconnecting the power to the control module (battery cable, powertrain control module (PCM)/engine control module (ECM)/transaxle control module (TCM) pigtail, control module fuse, jumper cables, etc.) in order to prevent internal control module damage. Control module damage may result when the metal case contacts battery voltage. DO NOT contact the control module metal case with battery voltage when servicing a control module, using battery booster cables, or when charging the vehicle battery. In order to prevent any possible electrostatic discharge damage to the control module, do not touch the connector pins or the soldered components on the circuit board. Remove any debris from around the control module connector surfaces

before servicing the control module. Inspect the control module connector gaskets when diagnosing or replacing the control module. Ensure that the gaskets are installed correctly. The gaskets prevent contaminant intrusion into the control module. The replacement control module must be programmed.

✳✳ CAUTION

It is necessary to record the remaining engine oil life. If the replacement module is not programmed with the remaining engine oil life, the engine oil life will default to 100 percent. If the replacement module is not programmed with the remaining engine oil life, the engine oil will need to be changed at 3,000 miles (5000 km) from the last engine oil change.

1. Using a scan tool, retrieve the percentage of remaining engine oil. Record the remaining engine oil life.
2. Disconnect the PCM harness connectors (2) from the PCM (3).

➡Do not touch the connector pins or soldered components on the circuit board in order to prevent possible electrostatic discharge (ESD) damage to the PCM.

3. Disengage the PCM bracket mounting tabs, while removing the PCM.
4. To install, reverse removal procedure. If a new PCM is being installed, the PCM must be programmed.

VEHICLE SPEED SENSOR (VSS)

LOCATION

See Figures 67 and 68.

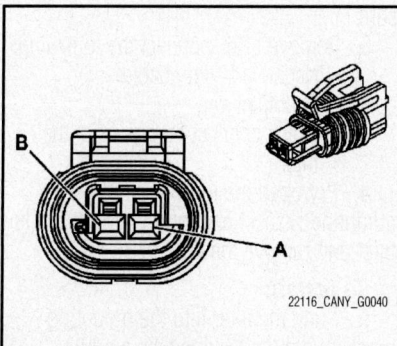

Fig. 67 Identifying the vehicle speed sensor connector—M/T

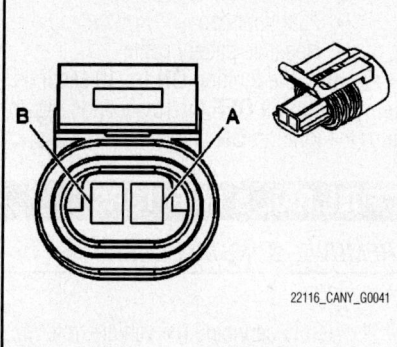

Fig. 68 Identifying the vehicle speed sensor connector—A/T

The Vehicle Speed Sensor (VSS) is located on the transmission housing. Refer to the illustrations under removal and installation.

REMOVAL & INSTALLATION

See Figure 69.

➡**Replacement components must be the correct part number for the applica-**

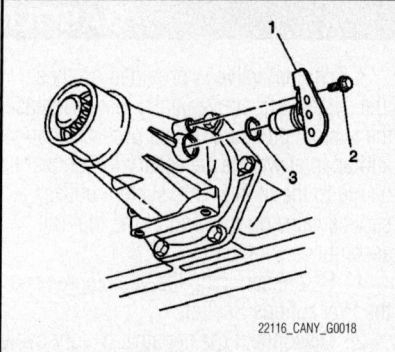

Fig. 69 Removing vehicle speed sensor—A/T

tion. Components requiring the use of the thread locking compound, lubricants, corrosion inhibitors, or sealants are identified in the service procedure. Some replacement components may come with these coatings already applied. Do not use these coatings on components unless specified. These coatings can affect the final torque, which may affect the operation of the component. Use the correct torque specification when installing components in order to avoid damage.

Automatic Transmission—4L60-E/4L65-E/4L70-E

1. Raise and support the vehicle.
2. Disconnect the wiring harness electrical connector from the vehicle speed sensor.
3. Remove the harness connector.
4. Remove the bolt (2). Remove the vehicle speed sensor (1). Remove the O-ring seal (3).
5. To install, reverse removal procedure. Install the O-ring seal on the vehicle speed sensor. Coat the O-ring seal with a thin film of transmission fluid.
6. Install the vehicle speed sensor into the transmission case. Install the bolt

and tighten the bolt to 97 inch lbs. (11 Nm).

Manual Transmission—Aisin AR5

See Figures 70 and 71.

1. Raise and support the vehicle.
2. Disconnect the vehicle speed sensor (VSS) electrical connector (2).
3. Remove the VSS with O-ring seal.
4. To install, reverser removal procedure. Install the VSS with O-ring seal. Tighten the vehicle speed sensor (VSS) to 13 ft. lbs. (17 Nm).

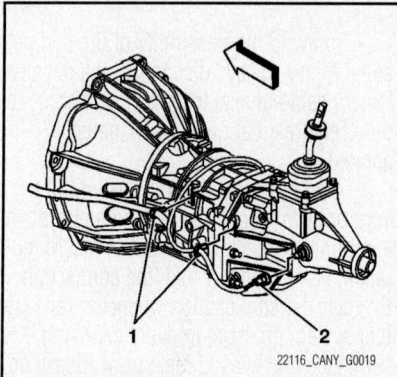

Fig. 70 Disconnecting the vehicle speed sensor electrical connector—M/T

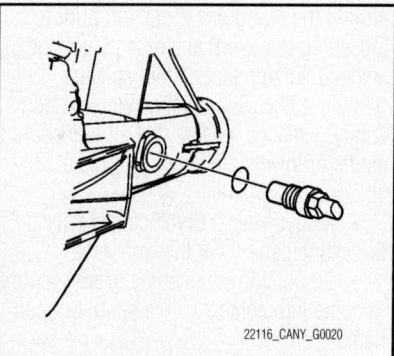

Fig. 71 Removing vehicle speed sensor—M/T

FUEL | **GASOLINE FUEL INJECTION SYSTEM**

FUEL SYSTEM SERVICE PRECAUTIONS

Safety is the most important factor when performing not only fuel system maintenance but any type of maintenance. Failure to conduct maintenance and repairs in a safe manner may result in serious personal injury or death. Maintenance and testing of the vehicle's fuel system components can be accomplished safely and effectively by adhering to the following rules and guidelines.

• To avoid the possibility of fire and personal injury, always disconnect the negative battery cable unless the repair or test procedure requires that battery voltage be applied.

• Always relieve the fuel system pressure prior to disconnecting any fuel system component (injector, fuel rail, pressure regulator, etc.), fitting or fuel line connection. Exercise extreme caution whenever relieving fuel system pressure to avoid exposing skin, face and eyes to fuel spray. Please be advised that fuel under pressure may penetrate the skin or any part of the body that it contacts.

• Always place a shop towel or cloth around the fitting or connection prior to loosening to absorb any excess fuel due to spillage. Ensure that all fuel spillage (should it occur) is quickly removed from engine surfaces. Ensure that all fuel soaked cloths or towels are deposited into a suitable waste container.

• Always keep a dry chemical (Class B) fire extinguisher near the work area.

• Do not allow fuel spray or fuel vapors to come into contact with a spark or open flame.

• Always use a back-up wrench when loosening and tightening fuel line connection fittings. This will prevent unnecessary stress and torsion to fuel line piping.

• Always replace worn fuel fitting O-rings with new Do not substitute fuel hose or equivalent where fuel pipe is installed.

Before servicing the vehicle, make sure to also refer to the precautions in the beginning of this section as well.

RELIEVING FUEL SYSTEM PRESSURE

The fuel systems operate under high fuel pressures. It is very important that the pressure be properly relieved prior to servicing the system or any of its components.

A Schrader valve is provided on these fuel systems to conveniently test or release the system pressure. A fuel pressure gauge and adapter will be necessary to connect the gauge to the fitting. This system utilizes a service valve on one end of the fuel rail assembly.

1. Before servicing the vehicle, refer to the Precautions Section.
2. Disconnect the negative battery cable to assure the prevention of fuel spillage if the ignition switch is accidentally turned **ON** while a fitting is still detached.
3. Loosen the fuel filler cap to release the fuel tank pressure.
4. Be sure the release valve on the fuel gauge is closed, then connect the fuel gauge to the pressure fitting located on the inlet fuel pipe fitting.

❋❋ CAUTION

When connecting the gauge to the fitting, be sure to wrap a rag around the fitting to avoid spillage. After repairs, place the rag in an approved container.

5. Install the bleed hose portion of the fuel gauge assembly into an approved container, then open the gauge release valve and bleed the fuel pressure from the system.
6. When the gauge is removed, be sure to open the bleed valve and drain all fuel from the gauge assembly.
7. When fuel service is finished, tighten the fuel filler cap and connect the negative battery cable.

FUEL FILTER

REMOVAL & INSTALLATION
See Figure 72.

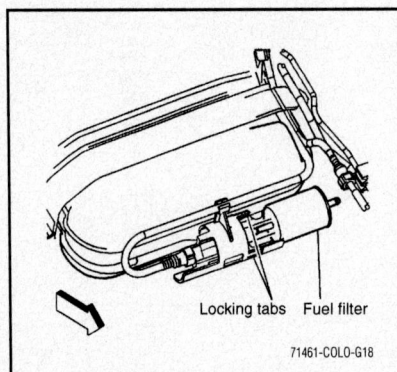

Locking tabs   Fuel filter

71461-COLO-G18

Fig. 72 Fuel filter mounting at the fuel tank

1. Before servicing the vehicle, refer to the Precautions Section.
2. Properly relieve the fuel system pressure.
3. Remove or disconnect the following:
   • Negative battery cable
   • Fuel filler cap
   • Quick connect fittings from the filter
4. Pry open the locking tabs of the mounting bracket enough to remove the fuel filter and remove the filter.

To install:

5. Slide the filter into the mounting bracket until the lacking tabs are fully engaged.
6. Install or connect the following:
   • Fuel quick disconnect fittings to the filter
   • Fuel filler cap
   • Negative battery cable
7. Turn the ignition **ON** for 10 seconds and then turn it **OFF** for 10 seconds. Again turn the ignition **ON** and check for leaks.

FUEL RAIL & INJECTORS

REMOVAL & INSTALLATION
See Figure 73.

1. Before servicing the vehicle, refer to the Precautions Section.
2. Relieve the fuel system pressure.
3. Remove or disconnect the following:
   • Negative battery cable
   • Fuel feed and return lines from fuel rail.

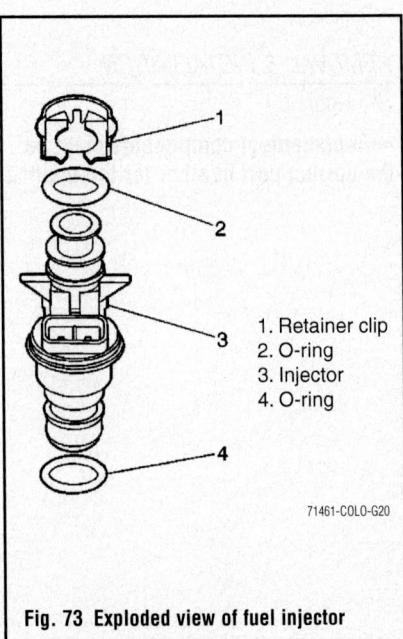

1. Retainer clip
2. O-ring
3. Injector
4. O-ring

71461-COLO-G20

Fig. 73 Exploded view of fuel injector

- Vent hoses from the air cleaner resonator and fuel pressure regulator
- EVAP purge hose
- Intake manifold
- Fuel injector harness from the engine wiring harness
- Fuel rail mounting bolts
- Fuel rail
- Fuel injector connector from the injector
- Injector retaining clip
- Fuel injector from the fuel rail

4. Discard the retainer clip and remove and discard the 2 O-rings from the injector.

To install:

5. Lubricate the new injector O-ring seats with engine oil.

6. Install or connect the following:
- O-rings and retainer clip on the injector
- Fuel injector into the fuel rail socket
- Fuel rail. Torque the bolts to 89 inch lbs. (10 Nm).
- Fuel injector harness
- Intake manifold
- EVAP purge hose
- Vent hoses to air cleaner resonator and fuel pressure regulator
- Fuel feed and return lines
- Negative battery cable

7. Turn the ignition **ON** for 10 seconds and then turn it **OFF** for 10 seconds. Again turn the ignition **ON** and check for leaks.

FUEL PUMP

REMOVAL & INSTALLATION

See Figure 74.

1. Before servicing the vehicle, refer to the Precautions Section.

2. Properly relieve the fuel system pressure.

3. Disconnect the negative battery cable.

4. Drain the fuel tank into an approved container.

5. Raise and support the rear of the vehicle.

6. Remove or disconnect the following:
- Left rear tire
- Left rear inner fender liner
- Fuel filler tube from the fuel tank
- EVAP hose from the filler vent tube
- Electrical connectors and wiring harness retainers from the fuel tank
- Fuel return line and fuel filter
- Upper fuel tank retaining strap

7. Support the fuel tank.

8. Remove the lower tank retaining strap and lower the fuel tank.

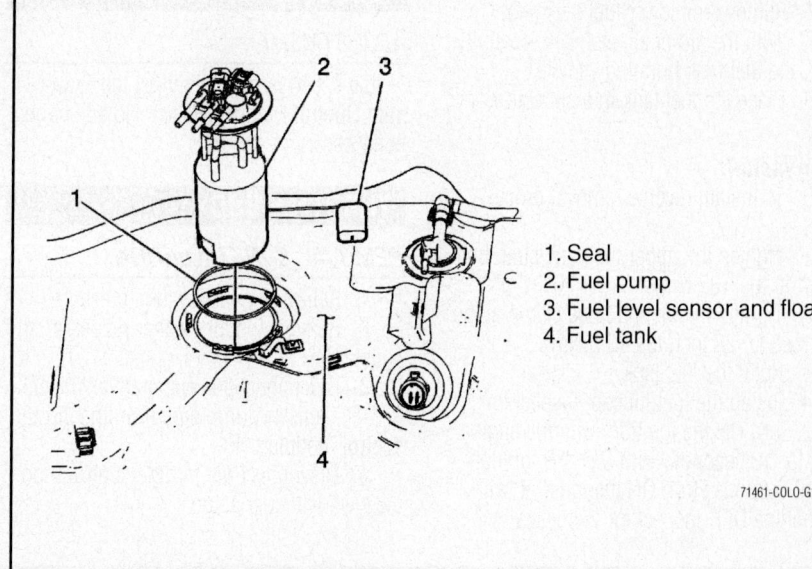

1. Seal
2. Fuel pump
3. Fuel level sensor and float
4. Fuel tank

71461-COLO-G19

Fig. 74 Fuel pump mounting in fuel tank

9. Disconnect the fuel lines from the fuel pump module.

10. Using tool J-39765, rotate the fuel pump cam locking ring counterclockwise and remove the ring.

11. Raise the fuel pump and tilt it back to allow the fuel level sensor and float to clear the opening.

12. Remove and discard the fuel pump seal.

To install:

13. Install a new seal on the fuel pump.

14. Tilt the fuel pump until the fuel level sensor and float can enter the opening.

15. Lower the fuel pump and align the tang on the pump with the notch in the opening.

16. Using tool J-39765, rotate the fuel pump cam locking ring clockwise until fully seated.

17. Reconnect the fuel lines to the fuel pump module.

18. Raise the fuel tank and install the lower tank retaining strap.

19. Install or connect the following:
- Upper fuel tank retaining strap and tighten both strap bolts to 24 ft. lbs. (32 Nm)
- Fuel return line and fuel filter
- Electrical connectors and wiring harness retainers to the fuel tank
- EVAP hose to the vent tube
- Fuel filler tube to the fuel tank
- Left rear inner fender liner
- Left rear tire

20. Lower the vehicle.

21. Fill the tank with gasoline.

22. Turn the ignition **ON** for 10 seconds

and then turn it **OFF** for 10 seconds. Again turn the ignition **ON** and check for leaks.

FUEL TANK

REMOVAL & INSTALLATION

1. Before servicing the vehicle, refer to the Precautions Section.

2. Relieve the fuel system pressure.

3. Drain the fuel tank.

4. Raise and support the vehicle.

5. Remove the left rear pickup box wheelhouse liner.

6. Loosen the fuel fill hose clamp (2) at the fuel tank.

7. Disconnect the fuel tank evaporative emission (EVAP) line (1) quick connect fitting from the fill tube vent line.

8. Separate the fuel fill hose from the fuel tank.

9. Disconnect the chassis wiring harness electrical connectors from the pressure sensor and the module.

10. Disengage the harness from the retainer on the fuel tank.

11. Raise the vehicle completely.

12. Disconnect and remove the middle EVAP vapor line (1) from the fuel tank (3) and the EVAP canister (5).

13. Disconnect the fuel feed line quick connect fitting from the fuel tank line.

➡**Do not bend the fuel tank straps. Bending the fuel tank straps may damage the straps.**

14. Remove the upper fuel tank strap bolt.

15. Remove the upper fuel tank strap.

16. Remove the lower fuel tank strap bolt.

17. Remove the lower fuel tank strap.

18. With the aid of an assistant, carefully lower the fuel tank from the vehicle.

19. Place the fuel tank in a suitable work area.

To install:

20. To install, reverse removal procedure.

21. Tighten the upper fuel tank strap bolt to: 24 ft. lbs. (32 Nm).

22. Tighten the fuel fill hose clamp at the fuel tank to: 22 inch lbs. (2.5 Nm).

23. Refill the fuel tank.

24. Install the fuel fill cap. Inspect for leaks. Turn ON the ignition, with the engine OFF for 10 seconds. Turn OFF the ignition for 10 seconds. Turn ON the ignition, with the engine OFF. Inspect for fuel leaks.

IDLE SPEED

ADJUSTMENT

Idle speed is maintained by the Powertrain Control Module (PCM). No adjustment is necessary or possible.

THROTTLE BODY

REMOVAL & INSTALLATION

1. Relieve the fuel system pressure.

2. Remove the air cleaner resonator and outlet duct.

3. Disconnect the evaporative emission (EVAP) canister purge pipe from the throttle control module.

4. Disconnect the throttle control module electrical connector.

5. Remove the throttle control module bolts.

6. Remove the throttle control module and the seal from the intake manifold.

7. Clean the gasket surface.

To install:

8. Insert a new seal into the intake manifold groove.

9. Position the throttle control module to the intake manifold.

10. Install the throttle control module bolts and tighten to 89 inch lbs. (10 Nm).

11. Connect the throttle control module electrical connector.

12. Connect the EVAP canister purge pipe to the throttle control module.

13. Install the air cleaner resonator and outlet duct.

HEATING & AIR CONDITIONING SYSTEM

BLOWER MOTOR

REMOVAL & INSTALLATION

1. Remove the right hinge pillar trim panel.

2. Remove the blower motor mounting screws.

3. Remove the blower motor cooling tube.

4. Disconnect the blower motor electrical connector.

5. Remove the blower motor.

To install:

6. Install the blower motor.

7. Connect the blower motor electrical connector.

8. Install the blower motor cooling tube.

9. Install the blower motor mounting screws.

10. Install the right hinge pillar trim panel.

HEATER CORE

REMOVAL & INSTALLATION

1. Before servicing the vehicle, refer to the Precautions Section.

2. Disable the air bag system, as outlined in the Chassis Electrical Section.

3. Disconnect the negative battery cable.

4. Drain the engine cooling system.

5. Discharge and recovery the A/C refrigerant using approved recycling equipment.

6. Remove or disconnect the following:
   - Glove box door
   - A pillar trim panels

- Door sill plates
- Hinge pillar trim panels
- Lower center instrument panel extension
- Accessory trim plate panel
- Radio
- A/C-heater control panel
- Center air outlets
- Left and right air outlets
- Knee bolster trim panel and bolster
- Instrument luster bezel
- Instrument cluster
- Headlight switch
- Daytime running light sensor
- Upper 3 instrument panel nuts
- Upper HVAC module screws
- Instrument panel screws at instrument cluster and glove box openings
- Hazard warning light connector
- Six passenger air bag fasteners and passenger air bag module
- Two screws in passenger air bag opening
- Open the left side compartment and remove the screw
- Screw at the back of the center storage compartment
- Grasp the lower edge of the center storage compartment and the right lower edge of the instrument panel and pull out to disengage the clips
- Partially pull out the instrument panel from the carrier
- Release all the wiring harness clips from the instrument panel

7. With the aid of an assistant, carefully remove the instrument panel.

8. Disconnect the heater hoses from the HVAC module.

9. Disconnect the A/C lines from the thermal expansion valve.

10. Disconnect all HVAC module electrical connectors.

11. Remove the HVAC module retaining screws from the firewall

12. Remove the HVAC module.

13. Remove the core pipe clamp screws and clamp.

14. Remove the heater core from the HVAC module.

To install:

15. Install or connect the following:
   - Heater core to the HVAC module
   - Heater core pipe clamp screws and clamp
   - HVAC module
   - HVAC module retaining screws at the firewall
   - HVAC module electrical connectors
   - A/C lines at the thermal expansion valve
   - Heater hoses to HVAC module
   - With the aid of an assistant, carefully install the instrument panel
   - Wiring harness clips to instrument panel
   - Push the lower edge of the center storage compartment and the right lower edge of the instrument panel in to engage the clips
   - Screw at the back of the center storage compartment
   - Screw inside the left side compartment

- Two screws in passenger air bag opening
- Hazard warning light connector
- Passenger air bag fasteners and passenger air bag module. Tighten the fasteners to 80 inch lbs. (9 Nm).
- Instrument panel screws at instrument cluster and glove box openings
- Upper HVAC module screws
- Upper 3 instrument panel nuts

- Daytime running light sensor
- Headlight switch
- Instrument cluster
- Instrument luster bezel
- Knee bolster trim panel and bolster
- Left and right air outlets
- Center air outlets
- A/C-heater control panel
- Radio
- Accessory trim plate panel
- Lower center instrument panel extension

- Hinge pillar trim panels
- Door sill plates
- A pillar trim panels
- Glove box door

16. Charge the A/C refrigerant using approved equipment.
17. Refill the cooling system.
   a. Enable the air bag system.
18. Connect the negative battery cable.
19. Run the engine to normal operating temperatures; then, check the climate control operation and check for leaks.

STEERING

POWER STEERING GEAR

REMOVAL & INSTALLATION

See Figure 75.

1. Before servicing the vehicle, refer to the Precautions Section.
2. Position a fluid catch pan under the power steering gear.
3. Place the front wheels in the straight ahead position.
4. Turn the ignition key to the LOCK position and remove the key.
5. Insert Steering Column Locking Pin J-42640 into the access hole in the lower steering column trim cover.
6. Raise and support the vehicle.
7. Remove or disconnect the following:
   - Front wheels
   - Engine skid plate, if equipped
   - Front axle housing, if equipped
   - Outer tie rod ends from steering knuckle
   - Feed and return fluid hoses from the steering gear. Immediately cap or plug all openings to prevent sys-

tem contamination or excessive fluid loss.
   - Intermediate shaft-to-lower steering column shaft pinch bolt
   - Lower shaft-to-steering gear pinch bolt
   - Power steering gear-to-frame bolts and washers
   - Front crossmember, if equipped with RWD only
   - Power steering gear

To install:

8. Install or connect the following:
   - Steering gear to the vehicle. Loosely install the mounting nuts, washers and bolts.
   - Crossmember, if equipped with RWD only. Tighten the mounting bolts to 44 ft. lbs. (60 Nm).
   - Steering gear mounting bolts. Tighten the vertical bolts to 96 ft. lbs. (130 Nm) and the horizontal bolts to 74 ft. lbs. (100 Nm).
   - Lower shaft-to-steering gear pinch bolt and tighten the bolt to 33 ft. lbs. (45 Nm).

   - Intermediate shaft-to-lower shaft pinch bolt, making sure the matchmarks line up. Tighten the bolt to 17 ft. lbs. (23 Nm).
   - Pressure and return hoses to the power steering gear. Tighten the hoses to 106 inch lbs. (12 Nm).
   - Outer tie rod ends
   - Front axle housing, if equipped
   - Engine undercover
   - Front wheels

9. Remove the steering column lock pin
10. Bleed the power steering system

POWER STEERING PUMP

REMOVAL & INSTALLATION

1. Remove the air cleaner assembly.
2. Remove the drive belt.
3. Remove the power steering pump pulley.
4. Disconnect the oil pressure sensor harness clip from the pump body.
5. Install a drain pan under the vehicle.
6. Disconnect the power steering pressure hoses from the power steering pump.
7. Remove the power steering pump mounting bolts.
8. Remove the power steering pump.

To install:

9. Install the power steering pump.
10. Install the power steering pump mounting bolts and tighten to 18 ft. lbs. (25 Nm).
11. Connect the power steering pressure hoses to the power steering pump and tighten to 18 ft. lbs. (25 Nm).
12. Remove the drain pan from under the vehicle.
13. Connect the oil pressure sensor harness clip to the pump body.
14. Install the power steering pump pulley.
15. Install the drive belt.
16. Install the air cleaner assembly.
17. Bleed the power steering system.

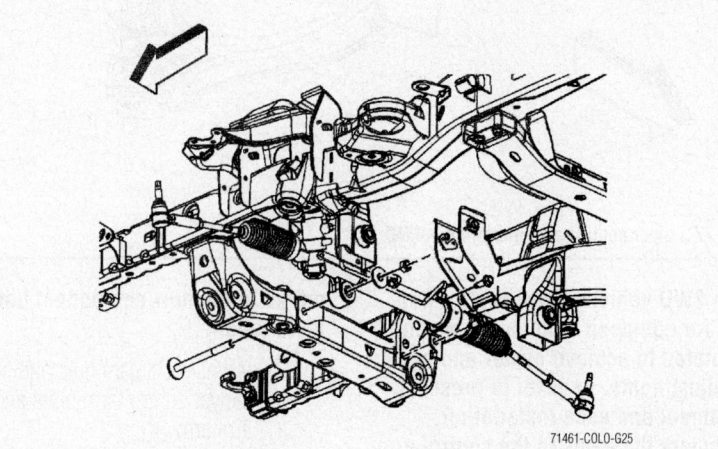

71461-COLO-G25

Fig. 75 Power steering gear mounting

COIL SPRING

REMOVAL & INSTALLATION

➡On 2WD models, the coil springs are removed with the front shock absorber/coil spring assembly. 4WD models do not use coil springs.

LOWER BALL JOINT

REMOVAL & INSTALLATION

2WD Models

1. Before servicing the vehicle, refer to the Precautions Section.
2. Remove the steering knuckle.
3. Remove the ball joint nuts and bolts.
4. Remove the ball joint.

To install:

5. Install or connect the following:
   - Ball joint nuts and bolts and tighten to 44 ft. lbs. (60 Nm)
   - Steering knuckle
6. Check and adjust the front end alignment, as necessary.

4WD Models

1. Before servicing the vehicle, refer to the Precautions Section.
2. Remove the steering knuckle.
3. Remove the ball joint nuts and bolts.
4. Remove the ball joint.

To install:

5. Install or connect the following:
   - Ball joint nuts and bolts and tighten to 47 ft. lbs. (64 Nm)
   - Steering knuckle
6. Check and adjust the front end alignment, as necessary.

LOWER CONTROL ARM

REMOVAL & INSTALLATION

2WD Models

See Figure 76.

1. Before servicing the vehicle, refer to the Precautions Section.
2. Raise and support the vehicle. Remove or disconnect the following:
   - Front wheel
   - Stabilizer bar links from control arm
   - Lower shock nut and through bolt
   - Lower ball joint stud from steering knuckle

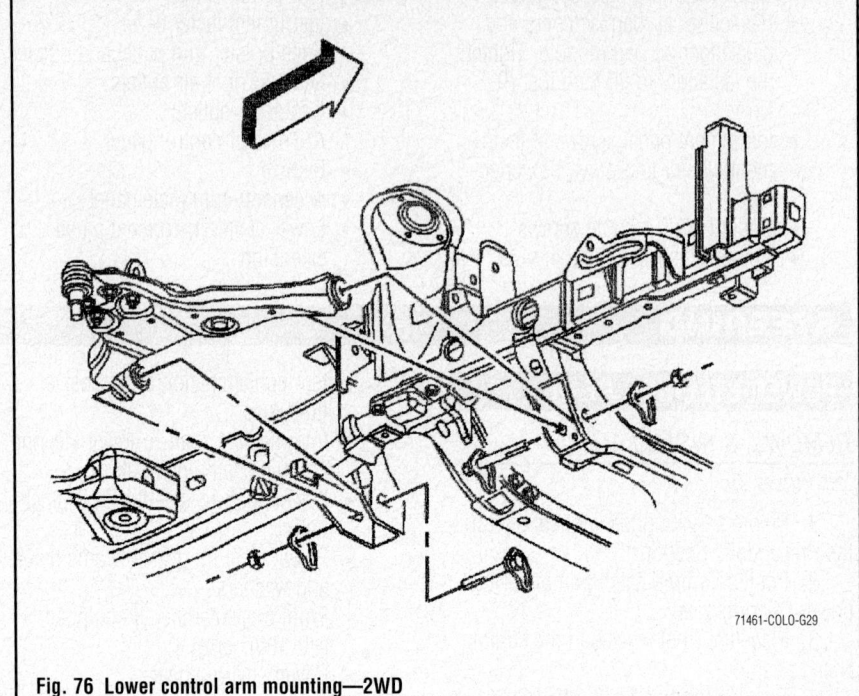

Fig. 76 Lower control arm mounting—2WD

71461-COLO-G29

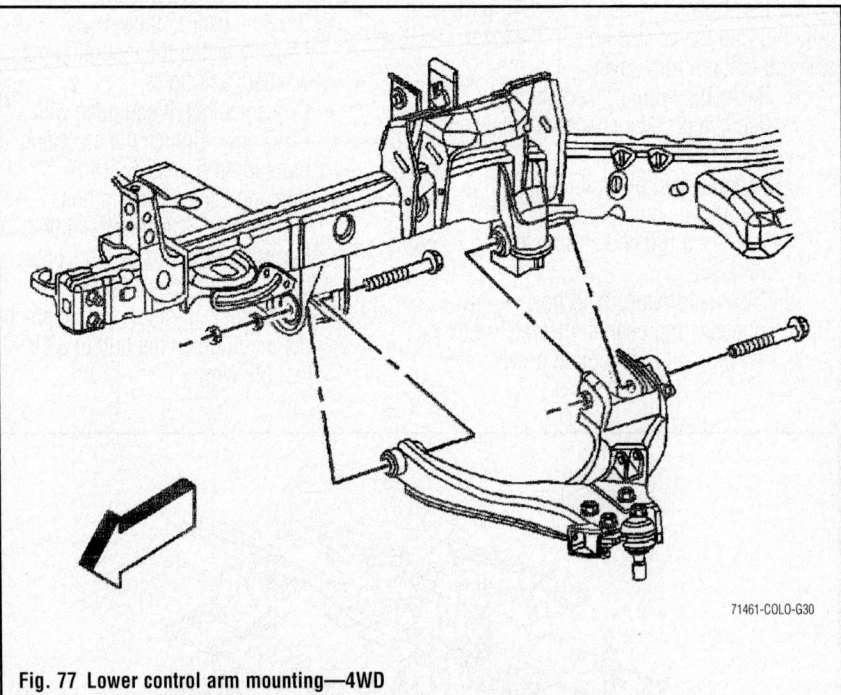

Fig. 77 Lower control arm mounting—4WD

71461-COLO-G30

➡The 2WD vehicle lower control arm bolts are equipped with cams, which are rotated to achieve caster and camber adjustments. In order to preserve adjustment and ease installation, matchmark the cams to the control arm before removal. If the control arm is being replaced, transfer the alignment marks to the new component before installation.

   - Adjustment cam nuts and cams
   - Lower control arm bolts and control arm

To install:

3. Install or connect the following:

➡ **The nuts must be tightened in sequence.**

- Lower control arm. Tighten the rear nuts first, then the front nuts to 114 ft. lbs. (155 Nm).
- Lower ball joint stud into the steering knuckle
- Ball joint-to-steering knuckle nut and tighten to 107 ft. lbs. (145 Nm).
- Lower shock nut and through bolt and tighten
- Stabilizer bar links
- Front wheel

4. Check the front wheel alignment.

4WD Models

See Figure 77.

1. Before servicing the vehicle, refer to the Precautions Section.
2. Raise and support the vehicle.
3. Remove or disconnect the following:
   - Front wheels
   - Steering knuckle
   - Stabilizer bar links from the control arm
   - Lower shock bolt and nut
   - Torsion bar, if necessary
   - Lower control arm

To install:

4. Install the control arm, bolts and nuts. Install the washer on the front bolt with the shoulder facing the control arm.

➡ **The nuts must be tightened in sequence.**

5. Tighten the rear nut first to 107 ft. lbs. (145 Nm), then tighten the front nut to 122 ft. lbs. (165 Nm).
6. Install or connect the following:
   - Steering knuckle
   - Torsion bar, if removed
   - Lower shock bolt and nut
   - Stabilizer bar links
   - Front wheels
7. Check and adjust the front end alignment, as necessary.

SHOCK ABSORBER

REMOVAL & INSTALLATION

2WD Models

1. Before servicing the vehicle, refer to the Precautions Section.
2. Remove or disconnect the following:
   - Upper shock mounting nuts
   - Wheel
   - Lower mounting bolt and nut

- Shock absorber/coil spring assembly

3. Place the shock absorber/coil spring assembly into a coil spring compressor and compress the spring.
4. Remove the upper shock retaining nut, bushings and washer.
5. Remove the shock absorber from the spring compressor.
6. If the coil spring is being replaced, remove the spring and mounting plate from the compressor, noting the mounting plate-to-spring orientation.

To install:

7. If the coil spring was replaced, align the spring to the mounting plate.
8. Align the centerline of the upper mounting studs with the centerline of the lower shock mount.

Install the mounting plate and spring into the compressor and compress the spring.

9. Install the shock absorber into the coil spring and compressor.
10. Install the washers, bushings and upper retaining nut. Tighten the retaining nut to 15 ft. lbs. (20 Nm).
11. Relax the tension on the spring compressor and remove the compressor.
12. Install the shock absorber/coil spring assembly to the lower mount. Tighten the lower mounting bolt and nut to 81 ft. lbs. (110 Nm).
13. Install the wheel and lower the vehicle.
14. Install the upper mounting nuts and tighten to 20 ft. lbs. (27 Nm).

4WD Models

1. Before servicing the vehicle, refer to the Precautions Section.
2. Place a jack or stand under the lower control arm.
3. Remove or disconnect the following:
   - Shock absorber upper nut and isolator
   - Lower nut/bolt
   - Shock absorber through the control arm

To install:

4. Install the shock through the lower control arm and insert it through the mounting hole in the upper spring pocket.
5. Install or connect the following:
   - Shock absorber to the lower control arm. Tighten the nuts/bolts to 52 ft. lbs. (70 Nm).
   - Upper isolator to the shock
6. Tighten the upper mounting nut to 18 ft. lbs. (25 Nm).

STABILIZER BAR

REMOVAL & INSTALLATION

1. Before servicing the vehicle, refer to the Precautions Section.
2. Raise and support the vehicle.
3. Remove or disconnect the following:
   - Front wheels
   - Stabilizer link nuts from control arm
   - Stabilizer links
   - Stabilizer bar insulator clamps
   - Stabilizer bar
   - Insulators

To install:

4. Install new insulators on the stabilizer bar.
5. Install or connect the following:
   - Stabilizer bar and insulator clamps. Tighten the bolts to 37 ft. lbs. (50 Nm).
6. Support the lower control arms at ride height.
7. Install the stabilizer links and tighten the nuts to 32 ft. lbs. (44 Nm).
8. Install the front wheels and lower the vehicle.

STEERING KNUCKLE

REMOVAL & INSTALLATION

2WD Models

1. Before servicing the vehicle, refer to the Precautions Section.
2. Raise and support the vehicle.
3. Place a jack stand under the lower control arm.
4. Remove or disconnect the following:
   - Wheel
   - Wheel hub/bearing assembly
   - Outer tie rod end
   - Upper and lower ball joint nuts and discard
   - Separate the ball joints from the steering knuckle
   - Steering knuckle

To install:

5. Clean the ball joints and tie rod ends. Inspect the tapered holes and mounting surfaces of the steering knuckle for damage or being out of round. Replace the knuckle if the holes are damaged.
6. Install the steering knuckle and connect the lower ball joint. Tighten the nut to 107 ft. lbs. (145 Nm).
7. Connect the upper ball joint to the knuckle and tighten the nut to 55 ft. lbs. (75 Nm).
8. Install or connect the following:

- Outer tie rod end
- Wheel hub/bearing assembly
- Wheel

9. Check the front wheel alignment.

4WD Models

1. Before servicing the vehicle, refer to the Precautions Section.
2. Raise and support the vehicle.
3. Place a jack stand under the lower control arm.
4. Remove or disconnect the following:
- Wheel
- Wheel hub nut and discard
- Brake caliper bracket
- Speed sensor harness from chassis harness and fender panel
- Outer tie rod end from steering knuckle
- Upper ball joint stud nuts
- Lower ball joint nut
- Separate ball joint from knuckle
- Steering knuckle

5. If the knuckle is being replaced, remove the wheel hub/bearing assembly.

To install:

6. Clean the ball joints and tie rod ends. Inspect the tapered holes and mounting surfaces of the steering knuckle for damage or being out of round. Replace the knuckle if the holes are damaged.
7. If removed, install the wheel hub/bearing assembly.
8. Install the steering knuckle and connect the lower ball joint. Tighten the nut to 107 ft. lbs. (145 Nm).
9. Connect the upper ball joint to the knuckle and tighten the nut to 55 ft. lbs. (75 Nm).
10. Install or connect the following:
- Outer tie rod end
- Speed sensor harness to chassis harness and fender panel
- Brake caliper bracket
- New wheel hub nut and tighten to 191 ft. lbs. (260 Nm).
- Wheel

11. Check the front wheel alignment.

UPPER BALL JOINT

REMOVAL & INSTALLATION

2WD Models

See Figure 78.

1. Before servicing the vehicle, refer to the Precautions Section.
2. Raise and support the vehicle.
3. Remove the front wheels.
4. Support the lower control arm with a suitable jack.

5. Disconnect the brake lines from the upper control arm.
6. Remove the wheel speed sensor bracket bolt and disconnect the sensor brackets.
7. Remove the upper ball joint nut.
8. Separate the ball joint from the knuckle using a separator.
9. Remove the 4 ball joint nuts and bolts from the control arm and discard them.
10. Remove the ball joint from the arm.

To install:

11. Install or connect the following:
- Ball joint in the upper control arm
- New ball joint retaining nuts and bolts. Tighten the ball joint retainers to 12 ft. lbs. (16 Nm).
- Ball joint to steering knuckle. Tighten the new nut to 55 ft. lbs. (75 Nm).
- Speed sensor bracket and tighten the nuts to 15 ft. lbs. (20 Nm).
- Brake hose to the control arm
- Remove the lower control arm jack
- Tire and wheel assembly

12. Check and adjust the front end alignment, as necessary.

4WD Models

1. Before servicing the vehicle, refer to the Precautions Section.
2. Raise and support the vehicle.
3. Remove the front wheels.
4. Support the lower control arm with a suitable jack.

5. Disconnect the brake lines from the upper control arm.
6. Remove the wheel speed sensor bracket bolt and disconnect the sensor brackets.
7. Remove the upper ball joint nut.
8. Disconnect the upper control arm from the ball stud by removing the retaining nuts.
9. Remove and discard the upper ball joint retaining nut.
10. Separate the ball joint from the knuckle using a separator.
11. Remove the ball joint from the arm.

To install:

12. Install or connect the following:
- Ball joint in the steering knuckle
- Ball joint retaining nut and tighten the nut to 55 ft. lbs. (75 Nm).
- Upper control arm to the ball stud and tighten the nut to 35 ft. lbs. (47 Nm).
- Wheel speed sensor bracket bolt and tighten to 15 ft. lbs. (20 Nm).
- Brake line to the upper control arm
- Front tires

13. Check and adjust the front end alignment, as necessary.

WHEEL BEARINGS & HUB

REMOVAL & INSTALLATION

See Figures 79 and 80.

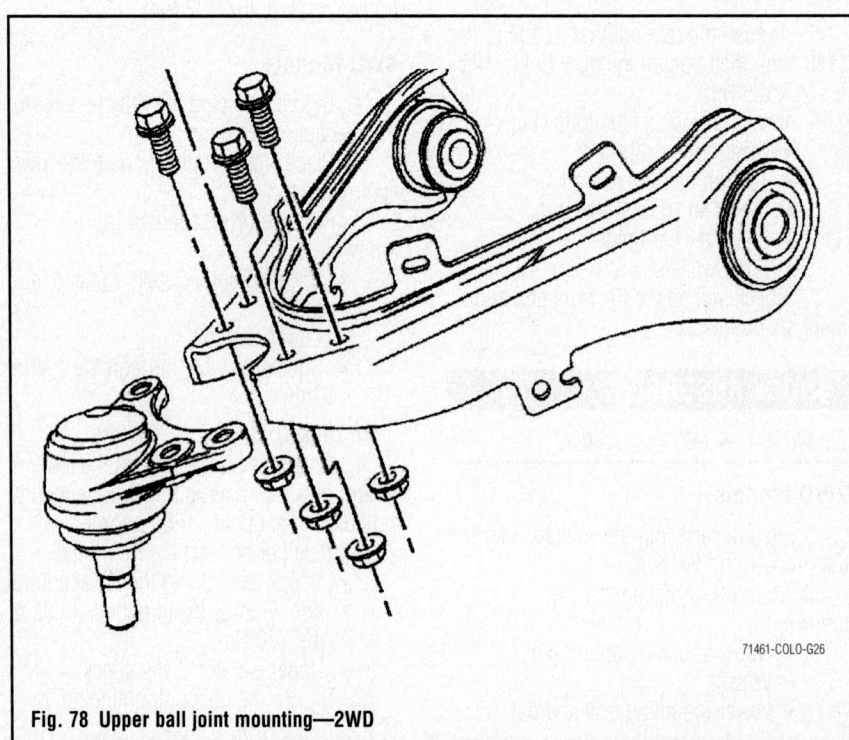

71461-COLO-G26

Fig. 78 Upper ball joint mounting—2WD

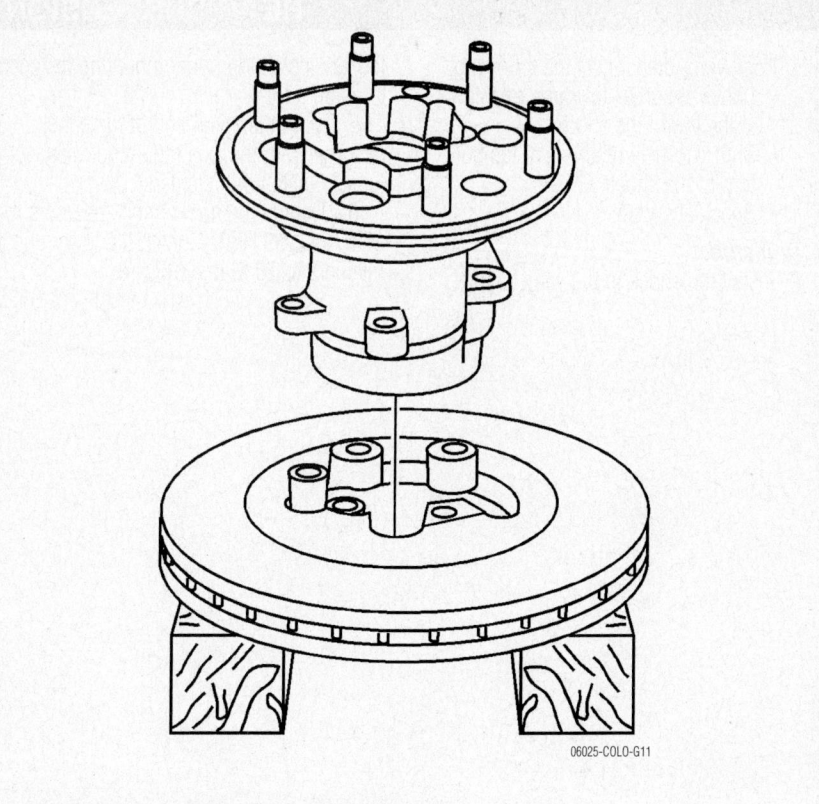

06025-COLO-G11

Fig. 79 Separating the wheel hub from the brake rotor

To install:

    7. Clean the contact surface between the wheel hub and brake rotor.

    8. Position the new hub/bearing assembly onto the brake rotor and tighten the bolts to 15 ft. lbs. (20 Nm) in a criss-cross pattern.

    9. Install or connect the following:

- Wheel hub/rotor assembly to backing plate
- On 4WD models, the steering knuckle
- Wheel hub/rotor to steering knuckle. Tighten the bolts to 92 ft. lbs. (125 Nm).
- Speed sensor electrical connector to body

    10. While holding the rotor from turning, tighten the rotor mounting bolts in the sequence shown to 88 ft. lbs. (120 Nm).

    11. Install or connect the following:

- Brake caliper
- Speed sensor bracket to control arm
- Wheel speed sensor harness to upper control arm
- Wheel

    1. Before servicing the vehicle, refer to the Precautions Section.

    2. Remove or disconnect the following:

- Wheel
- Brake caliper with the pads without disconnecting the brake line
- Wheel speed sensor harness from upper control arm
- Speed sensor bracket from control arm
- Speed sensor electrical connector from body

    3. Using white paint, mark the location of the speed sensor wiring harness to the steering knuckle for installation reference. Coil up the sensor wiring so it is out of the way of suspension components.

    4. On 4WD models, remove the steering knuckle.

    5. On all models, remove the wheel hub/brake rotor rear mounting bolts and remove the hub/rotor assembly. The backing plate will come off when the assembly is removed.

    6. Separate the brake rotor from the wheel hub by removing the 6 mounting bolts.

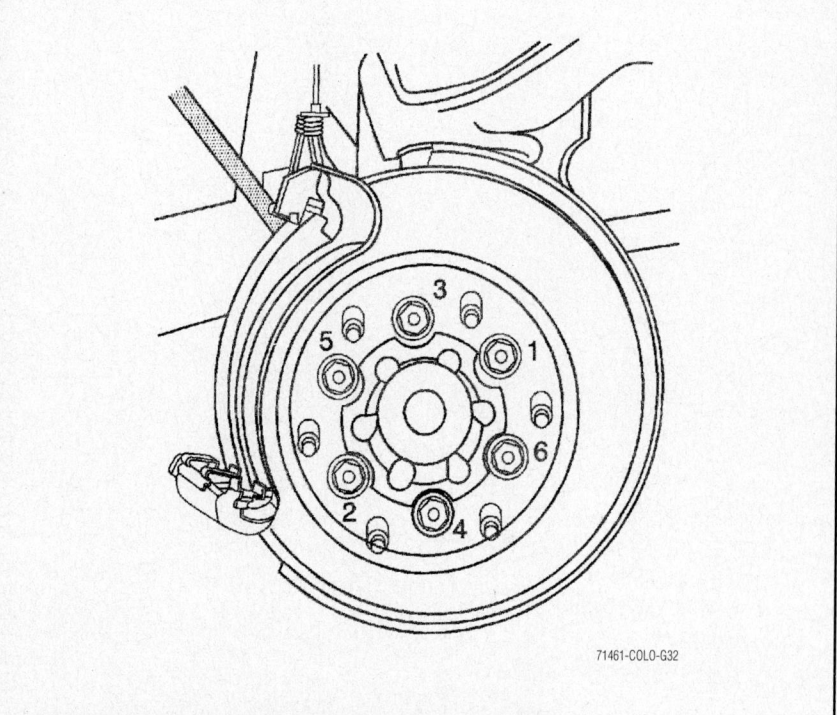

71461-COLO-G32

Fig. 80 Brake rotor/wheel hub mounting bolt tightening sequence

SHOCK ABSORBER

REMOVAL & INSTALLATION

1. Before servicing the vehicle, refer to the Precautions Section.

2. Support the rear axle assembly at ride height.

3. Remove or disconnect the following:
   - Shock absorber-to-frame retainers at the top of the shock
   - Shock-to-axle retainers at the bottom of the shock
   - Shock absorber

To install:

4. Install the shock in the vehicle and loosely install the upper mounting fasteners to retain it.

5. Align the lower-end of the shock absorber with the axle mounting, then loosely install the retainers.

6. Tighten the upper shock retainers to 26 ft. lbs. (35 Nm). Tighten the lower shock retainers to 70 ft. lbs. (95 Nm).

ISUZU

Ascender

14

SPECIFICATIONS AND MAINTENANCE CHARTS

ENGINE AND VEHICLE IDENTIFICATION

			Engine					Model Year	
Code ①	Liters (cc)	Cu. In.	Cyl.	Fuel Sys.	Engine Type	Eng. Mfg.	Code ②		Year
S	4.2 (4200)	256	6	MFI	DOHC	CPC	6		2006
P	5.3 (5326)	325	8	SFI	OHV	CPC	7		2007
M	5.3 (5326)	325	8	SFI	OHV	CPC	8		2008

CPC: Chevrolet/Pontiac/Canada

MFI: Multi-port Fuel Injection

SFI: Sequential Fuel Injection

① 8th position of VIN

② 10th position of VIN

22140_ASCE_C0001

GENERAL ENGINE SPECIFICATIONS

All measurements are given in inches.

Year	Model	Engine Displacement Liters	Engine Series (ID/VIN)	Net Horsepower @ rpm	Net Torque @ rpm (ft. lbs.)	Bore x Stroke (in.)	Compression Ratio	Oil Pressure @ rpm
2006	Ascender	4.2	S	291@6000	277@4800	3.66x4.02	10.3:1	12@1200
		5.3	M	300@5200	330@4000	3.78x3.62	9.95:1	18@2000
2007	Ascender	4.2	S	291@6000	277@4800	3.66x4.02	10.3:1	12@1200
2008	Ascender	4.2	S	291@6000	277@4800	3.66x4.02	10.3:1	12@1200

22140_ASCE_C0002

GASOLINE ENGINE TUNE-UP SPECIFICATIONS

Year	Engine Displacement Liters	Engine ID/VIN	Spark Plugs Gap (in.)	Ignition Timing (deg.) MT	Ignition Timing (deg.) AT	Fuel Pump (psi)	Idle Speed (rpm) MT	Idle Speed (rpm) AT	Valve Clearance In.	Valve Clearance Ex.
2006	4.2	S	0.042	—	①	50-57 ②	—	③	HYD	HYD
	5.3	M	0.040	—	①	50-60 ②	—	③	HYD	HYD
2007	4.2	S	0.042	—	①	50-57 ②	—	③	HYD	HYD
2008	4.2	S	0.042	—	①	50-57 ②	—	③	HYD	HYD

NOTE: The Vehicle Emission Control Information label often reflects specification changes made during production.

The label figures must be used if they differ from those in this chart.

HYD: Hydraulic

① Distributorless ignition, cannot be adjusted

② With key ON and engine OFF

③ Distributorless ignition, cannot be adjusted

22140_ASCE_C0003

CAPACITIES

Year	Model	Engine Displacement Liters	Engine ID/VIN	Engine Oil with Filter (qts.)	Transmission (pts.) 5-Spd	Transmission (pts.) Auto.	Transfer Case (pts.)	Drive Axle Front (pts.)	Drive Axle Rear (pts.)	Fuel Tank (gal.)	Cooling System (qts.)
2006	Ascender	4.2	S	7.0	—	10.0	4.0	1.7	3.6	①	②
		5.3	M	6.0	—	10.0	4.0	1.7	4.3	①	③
2007	Ascender	4.2	S	7.0	—	10.0	4.0	1.7	3.6	22	9.7
2007	Ascender	4.2	S	7.0	—	10.0	4.0	1.7	3.6	22	9.7

NOTE: All capacities are approximate. Add fluid gradually and check to be sure a proper fluid level is obtained.

① Short wheelbase: 22.0 gal.
Long wheelbase: 25.3 gal.

② Short wheelbase: 10.8 qts.
Long wheelbase: 13.8 qts.

22140_ASCE_C0004

FLUID SPECIFICATIONS

Year	Model	Engine Displacement Liters	Engine ID/VIN	Engine Oil	Auto. Trans.	Drive Axle	Transfer Case	Power Steering Fluid	Brake Master Cylinder
2006	Ascender	4.2	S	5W-30	Dexron-VI	75W-90	Auto-Trak II	GM PS Fluid	DOT 3
		5.3	M	5W-30	Dexron-VI	75W-90	Auto-Trak II	GM PS Fluid	DOT 3
2007	Ascender	4.2	S	5W-30	Dexron-VI	75W-90	Auto-Trak II	GM PS Fluid	DOT 3
2008	Ascender	4.2	S	5W-30	Dexron-VI	75W-90	Auto-Trak II	GM PS Fluid	DOT 3

DOT: Department Of Transpotation

22140_ASCE_C0014

VALVE SPECIFICATIONS

Year	Engine Displacement Liters	Engine ID/VIN	Seat Angle (deg.)	Face Angle (deg.)	Spring Test Pressure (lbs. @ in.)	Spring Installed Height (in.)	Stem-to-Guide Clearance (in.) Intake	Stem-to-Guide Clearance (in.) Exhaust	Stem Diameter (in.) Intake	Stem Diameter (in.) Exhaust
2006	4.2	S	NA	NA	130-142@1.26	NA	0.0011-0.0025	0.0015-0.0030	NA	NA
	5.3	M	46	45	220@1.32	1.80	0.0010-0.0026	0.0010-0.0026	0.313-0.314	0.313-0.314
2007	4.2	S	NA	NA	130-142@1.26	NA	0.0011-0.0025	0.0015-0.0030	NA	NA
2008	4.2	S	NA	NA	130-142@1.26	NA	0.0011-0.0025	0.0015-0.0030	NA	NA

NA: Not Available

22140_ASCE_C0005

CRANKSHAFT AND CONNECTING ROD SPECIFICATIONS

All measurements are given in inches.

| Year | Engine Displacement Liters | Engine ID/VIN | Crankshaft | | | | Connecting Rod | | |
			Main Brg. Journal Dia.	Main Brg. Oil Clearance	Shaft End-play	Thrust on No.	Journal Diameter	Oil Clearance	Side Clearance
2006	4.2	S	2.7567-2.7574	0.0004-0.0025	0.0044-0.0153	4	2.2337-2.2342	0.0008-0.0025	0.0019-0.0137
	5.3	M	2.5587-2.5593	0.0008-0.0021	0.0015-0.0078	4	2.0991-2.0999	0.0009-0.0025	0.0043-0.0200
2007	4.2	S	2.7567-2.7574	0.0004-0.0025	0.0044-0.0153	4	2.2337-2.2342	0.0008-0.0025	0.0019-0.0137
2007	4.2	S	2.7567-2.7574	0.0004-0.0025	0.0044-0.0153	4	2.2337-2.2342	0.0008-0.0025	0.0019-0.0137

22140_ASCE_C0006

PISTON AND RING SPECIFICATIONS

All measurements are given in inches.

| Year | Engine Displ. Liters | Engine ID/VIN | Piston Clearance | Ring Gap | | | Ring Side Clearance | | |
				Top Compression	Bottom Compression	Oil Control	Top Compression	Bottom Compression	Oil Control
2006	4.2	S	-0.0006 0.0014	0.0059-0.0118	0.0142-0.0201	0.0098-0.0299	0.0017-0.0037	0.0017-0.0037	0.0023-0.0085
	5.3	M	-0.0014 0.0006	0.0090-0.0173	0.0173-0.0275	0.0070-0.0295	0.0015-0.0033	0.0015-0.0031	0.0005-0.0078
2007	4.2	S	-0.0006 0.0014	0.0059-0.0118	0.0142-0.0201	0.0098-0.0299	0.0017-0.0037	0.0017-0.0037	0.0023-0.0085
2008	4.2	S	-0.0006 0.0014	0.0059-0.0118	0.0142-0.0201	0.0098-0.0299	0.0017-0.0037	0.0017-0.0037	0.0023-0.0085

22140_ASCE_C0007

TORQUE SPECIFICATIONS
All readings in ft. lbs.

Year	Engine Displacement Liters	Engine ID/VIN	Cylinder Head Bolts	Main Bearing Bolts	Rod Bearing Bolts	Crankshaft Damper Bolts	Flywheel Bolts	Manifold Intake *	Manifold Exhaust	Spark Plugs	Oil Pan Drain Plug
2006	4.2	S	①	②	③	④	⑤	⑥	⑦	13	19
	5.3	M	⑧	⑨	⑩	⑪	⑫	⑬	⑭	11	18
2007	4.2	S	①	②	③	④	⑤	⑥	⑦	13	19
2007	4.2	S	①	②	③	④	⑤	⑥	⑦	13	19

* NOTE: Applies to Lower Manifold only.

① Cylinder head bolts (14)
1st pass: 22 ft. lbs.
2nd pass: Plus 155 degrees
2 short end bolts: 62 INCH lbs.
2nd pass: plus 60 degrees
1 long end bolt: 62 INCH lbs.
2nd pass: plus 120 degrees

② 18 ft. lbs., plus 180 depress

③ 18 ft. lbs., plus 110 degrees

④ 110 ft. lbs., plus 180 degrees

⑤ 18 ft. lbs., plus 50 degrees

⑥ 89 inch lbs.

⑦ 1st pass: 15 ft. lbs.
2nd pass: 15 ft. lbs.
3rd pass: 15 ft. lbs.

⑧ M11 bolts: 22 ft. lbs.
2nd pass: Plus 90 degrees
3rd pass: Plus 70 degrees
M8 bolts: 22 ft. lbs.

⑨ Inner bolts:
1st pass: 15 ft. lbs.
Final pass: Plus 80 degrees
Outer bolts:
1st pass: 15 ft. lbs.
Final pass: Plus 51 degrees
M8 bolts: 18 ft. lbs.

⑩ 15 ft. lbs. plus 85 degrees

⑪ Installation pass: 240 ft. lbs. (discard bolt)
First pass: 37 ft. lbs. (new bolt)
Final pass: Plus 140 degrees

⑫ 1st pass: 15 ft. lbs.
2nd pass: 37 ft. lbs.
3rd pass: 74 ft. lbs.

⑬ 1st pass: 44 inch lbs.
2nd pass: 89 inch lbs.

⑭ 1st pass: 11 ft. lbs.
2nd pass: 18 ft. lbs.

22140_ASCE_C0008

WHEEL ALIGNMENT

Year	Model		Caster Range (+/-Deg.)	Caster Preferred Setting (Deg.)	Camber Range (+/-Deg.)	Camber Preferred Setting (Deg.)	Toe-in (in.)
2006	Ascender	①	0.60	+4.00	0.60	0.00	-0.10+/-0.20
		②	0.60	+4.25	0.60	0.00	-0.10+/-0.20
2007	Ascender	①	0.60	+4.00	0.60	0.00	-0.10+/-0.20
		②	0.60	+4.25	0.60	0.00	-0.10+/-0.20
2009	Ascender	①	0.60	+4.00	0.60	0.00	-0.10+/-0.20
		②	0.60	+4.25	0.60	0.00	-0.10+/-0.20

① With rear coil spring suspension.
② With rear air spring suspension.

22140_ASCE_C0009

TIRE, WHEEL AND BALL JOINT SPECIFICATIONS

Year	Model	OEM Tires		Tire Pressures (psi)		Wheel Size	Ball Joint Inspection	Lug Nut (ft. lbs.)
		Standard	Optional	Front	Rear			
2006	Ascender	P245/65R17	None	36	36	7-JJ	L ①	103
2007	Ascender	P245/65R17	None	36	36	7-JJ	L ①	103
2008	Ascender	P245/65R17	None	36	36	7-JJ	L ①	103

OEM: Original Equipment Manufacturer

PSI: Pounds Per Square Inch

L: Lower (ball joint)

① Do not lift truck. Inspect the boss into which the grease fitting is threaded. Replace if the boss is flush or receded below the surface of the ball joint

22140_ASCE_C0010

BRAKE SPECIFICATIONS

All measurements in inches unless noted

Year	Model	Front Brake Disc			Rear Brake Disc			Minimum Lining Thickness	Brake Caliper	
									Bracket Bolts (ft. lbs.)	Mounting Bolts (ft. lbs.)
		Original Thickness	Minimum Thickness	Maximum Runout	Original Thickness	Minimum Thickness	Maximum Runout			
2006	Ascender	1.140	1.080	0.002	0.787	0.728	0.002	NA	①	②
2007	Ascender	1.140	1.080	0.002	0.787	0.728	0.002	NA	①	②
2008	Ascender	1.140	1.080	0.002	0.787	0.728	0.002	NA	①	②

NA: Not Available

① Front: 118 ft. lbs.

   Rear: 148 ft. lbs.

② Front: 31 ft. lbs.

   Rear: 23 ft. lbs.

22140_ASCE_C0011

MAINTENANCE I AND II SERVICE SCHEDULES
2006-08 Ascender

When the CHANGE ENGINE OIL light appears, certain services and inspections are required.
Required services are described as Maintenance I and Maintenance II.
The first service on a vehicle should be Maintenance I, and the second service should be Maintenance II.
Alternate between the 2 thereafter. However, in some cases, Maintenance II may be required more often.
Maintenance I: Use Maintenance I if the CHANGE ENGINE OIL light comes on within 10 months
since vehicle was purchased or, if Maintenance II was performed.
Maintenance II: Use Maintenance II if the previous service performed was Maintenance I.
Always use Maintenance II whenever the CHANGE ENGINE OIL light comes on 10 months or more since the last
service, or, if the CHANGE ENGINE OIL light has not come on at all for one year.

Service	Maintenance I	Maintenance II
Change the engine oil and filter. Reset the oil life system.	✓	✓
Visually inspect the vehicle for leaks or damage. A fluid loss in the vehicle system could indicate a problem. Inspected, repair and add fluid to the system if necessary.	✓	✓
Inspect the engine air cleaner filter. If necessary, replace the filter.	✓	✓
Rotate the tires. Inspect the tire inflation pressures and the tire wear.	✓	✓
Visually inspect the brake lines and hoses for proper hook-up, binding, leaks, cracks, chafing, etc. Inspect the disc brake pads for wear and the rotors for surface condition. Inspect the drum brake linings for wear or cracks. Inspect other brake parts, including drums, wheel cylinders, calipers, parking brake, etc. Inspect the parking brake adjustment.	✓	✓
Inspect the engine coolant and the windshield washer fluid levels. Add fluid as needed.	✓	✓
Inspect the suspension and steering components. Inspect the front and rear suspension and the steering system for damaged, loose or missing parts, or signs of wear. Inspect the power steering lines and the hoses for proper hook-up, binding, leaks, cracks, chafing, etc.	--	✓
Visually inspect the coolant hoses and replace the hoses if they are cracked, swollen or deteriorated. Inspect all pipes, fittings and clamps; replace with GM parts as needed. To help ensure proper operation, a pressure test of the cooling system and pressure cap and cleaning the outside of the radiator and air conditioning condenser is recommended at least once a year.		✓
Inspect the wiper blades.	--	✓
Inspect the restraint system components. Ensure the safety belt reminder light and all the belts, buckles, latch plates, retractors and anchorages are working properly. Look for any other loose or damaged safety belt system parts. If you see anything that might keep a safety belt system from working correctly, repair or replaced the damaged part. Replace torn or frayed safety belts, refer to Operational and Functional Checks in Seat Belts. Inspect for any opened or broken air bag coverings, and repair or replace as needed. The air bag system does require regular maintenance.	--	✓

MAINTENANCE I AND II SERVICE SCHEDULES
2006-08 Ascender

Lubricate the body components.Lubricate all key lock cylinders, hood latch assemblies, secondary latches, pivots, spring anchor and release pawl, hood and door hinges, rear folding seats and liftgate hinges. Frequent lubrication may be required when exposed to a corrosive environment, refer to Fluid and Lubricant Recommendations . Applying dielectric silicone grease GM P/N 12345579 (Canadian P/N 1974984) or equivalent on the weatherstrips with a clean cloth.	--	✓
Inspect the transaxle fluid level and add fluid as needed.	--	✓
Inspect the suspension and steering components.Inspect the front and rear suspension and the steering system for damaged, loose or missing parts, or signs of wear. Inspect power steering lines and hoses for proper hook-up, binding, leaks, cracks, chafing, etc.	--	✓
Inspect the throttle system for interference or binding and for damaged or missing parts. Replace the parts as needed. Replace any components that have high effort or excessive wear. Do not lubricate the accelerator or the cruise control cables.	--	✓
Replace the passenger compartment air filter.	--	✓

22140_ASCE_C0013

PRECAUTIONS

Before servicing any vehicle, please be sure to read all of the following precautions, which deal with personal safety, prevention of component damage, and important points to take into consideration when servicing a motor vehicle:

• Never open, service or drain the radiator or cooling system when the engine is hot; serious burns can occur from the steam and hot coolant.

• Observe all applicable safety precautions when working around fuel. Whenever servicing the fuel system, always work in a well-ventilated area. Do not allow fuel spray or vapors to come in contact with a spark, open flame, or excessive heat (a hot drop light, for example). Keep a dry chemical fire extinguisher near the work area. Always keep fuel in a container specifically designed for fuel storage; also, always properly seal fuel containers to avoid the possibility of fire or explosion. Refer to the additional fuel system precautions later in this section.

• Fuel injection systems often remain pressurized, even after the engine has been turned **OFF**. The fuel system pressure must be relieved before disconnecting any fuel lines. Failure to do so may result in fire and/or personal injury.

• Brake fluid often contains polyglycol ethers and polyglycols. Avoid contact with the eyes and wash your hands thoroughly after handling brake fluid. If you do get brake fluid in your eyes, flush your eyes with clean, running water for 15 minutes. If eye irritation persists, or if you have taken brake fluid internally, IMMEDIATELY seek medical assistance.

• The EPA warns that prolonged contact with used engine oil may cause a number of skin disorders, including cancer. You should make every effort to minimize your exposure to used engine oil. Protective gloves should be worn when changing oil. Wash your hands and any other exposed skin areas as soon as possible after exposure to used engine oil. Soap and water, or waterless hand cleaner should be used.

• All new vehicles are now equipped with an air bag system, often referred to as a Supplemental Restraint System (SRS) or Supplemental Inflatable Restraint (SIR) system. The system must be disabled before performing service on or around system components, steering column, instrument panel components, wiring and sensors. Failure to follow safety and disabling procedures could result in accidental air bag deployment, possible personal injury and unnecessary system repairs.

• Always wear safety goggles when working with, or around, the air bag system. When carrying a non-deployed air bag, be sure the bag and trim cover are pointed away from your body. When placing a non-deployed air bag on a work surface, always face the bag and trim cover upward, away from the surface. This will reduce the motion of the module if it is accidentally deployed. Refer to the additional air bag system precautions later in this section.

• Clean, high quality brake fluid from a sealed container is essential to the safe and proper operation of the brake system. You should always buy the correct type of brake fluid for your vehicle. If the brake fluid becomes contaminated, completely flush the system with new fluid. Never reuse any brake fluid. Any brake fluid that is removed from the system should be discarded. Also, do not allow any brake fluid to come in contact with a painted surface; it will damage the paint.

• Never operate the engine without the proper amount and type of engine oil; doing so WILL result in severe engine damage.

• Timing belt maintenance is extremely important. Many models utilize an interference-type, non-freewheeling engine. If the timing belt breaks, the valves in the cylinder head may strike the pistons, causing potentially serious (also time-consuming and expensive) engine damage. Refer to the maintenance interval charts for the recommended replacement interval for the timing belt, and to the timing belt section for belt replacement and inspection.

• Disconnecting the negative battery cable on some vehicles may interfere with the functions of the on-board computer system(s) and may require the computer to undergo a relearning process once the negative battery cable is reconnected.

• When servicing drum brakes, only disassemble and assemble one side at a time, leaving the remaining side intact for reference.

• Only an MVAC-trained, EPA-certified automotive technician should service the air conditioning system or its components.

BRAKES

ANTI-LOCK BRAKE SYSTEM (ABS)

GENERAL INFORMATION

PRECAUTIONS

• Certain components within the ABS system are not intended to be serviced or repaired individually.

• Do not use rubber hoses or other parts not specifically specified for and ABS system. When using repair kits, replace all parts included in the kit. Partial or incorrect repair may lead to functional problems and require the replacement of components.

• Lubricate rubber parts with clean, fresh brake fluid to ease assembly. Do not use shop air to clean parts; damage to rubber components may result.

• Use only DOT 3 brake fluid from an unopened container.

• If any hydraulic component or line is removed or replaced, it may be necessary to bleed the entire system.

• A clean repair area is essential. Always clean the reservoir and cap thoroughly before removing the cap. The slightest amount of dirt in the fluid may plug an orifice and impair the system function. Perform repairs after components have been thoroughly cleaned; use only denatured alcohol to clean components. Do not allow ABS components to come into contact with any substance containing mineral oil; this includes used shop rags.

• The Anti-Lock control unit is a microprocessor similar to other computer units in the vehicle. Ensure that the ignition switch is **OFF** before removing or installing controller harnesses. Avoid static electricity discharge at or near the controller.

• If any arc welding is to be done on the vehicle, the control unit should be unplugged before welding operations begin.

BRAKES — BLEEDING THE BRAKE SYSTEM

BLEEDING PROCEDURE

BLEEDING PROCEDURE

1. Raise the vehicle in order to access the system bleed screws.
2. Bleed the system at the right rear wheel first.
3. Install a clear hose on the bleed screw.
4. Immerse the opposite end of the hose into a container partially filled with clean DOT 3 brake fluid.
5. Open the bleed screw ½ to 1 full turn.
6. Slowly depress the brake pedal. While the pedal is depressed to its full extent, tighten the bleed screw.
7. Release the brake pedal and wait 10–15 seconds for the master cylinder pistons to return to the home position.
8. Repeat the previous steps for the remaining wheels. The brake fluid which is present at each bleed screw should be clean and free of air.
9. This procedure may use more than a pint of fluid per wheel. Check the master cylinder fluid level every four to six strokes of the brake pedal in order to avoid running the system dry.
10. Press the brake pedal firmly and run the Scan Tool Automated Bleed Procedure. Release the brake pedal between each test.
11. Bleed all four wheels again using Steps 3–9. This will remove the remaining air from the brake system.
12. Evaluate the feel of the brake pedal before attempting to drive the vehicle.
13. Bleed the system as many times as necessary in order to obtain the appropriate feel of the pedal.

BRAKES — FRONT DISC BRAKES

✳✳ CAUTION

Dust and dirt accumulating on brake parts during normal use may contain asbestos fibers from production or aftermarket brake linings. Breathing excessive concentrations of asbestos fibers can cause serious bodily harm. Exercise care when servicing brake parts. Do not sand or grind brake lining unless equipment used is designed to contain the dust residue. Do not clean brake parts with compressed air or by dry brushing. Cleaning should be done by dampening the brake components with a fine mist of water, then wiping the brake components clean with a dampened cloth. Dispose of cloth and all residue containing asbestos fibers in an impermeable container with the appropriate label. Follow practices prescribed by the Occupational Safety and Health Administration (OSHA) and the Environmental Protection Agency (EPA) for the handling, processing, and disposing of dust or debris that may contain asbestos fibers.

BRAKE CALIPER

REMOVAL & INSTALLATION

1. Before servicing the vehicle, refer to the precautions in the beginning of this section.
2. Remove or disconnect the following:
   - ⅔ of the brake fluid from the master cylinder reservoir
   - Tire and wheel assembly
   - Caliper fluid line, then plug
   - Bolts retaining the caliper to the rotor
   - Caliper from the rotor
   - Disc brake pads from the caliper
   - Disc brake pad retaining clips from inside the caliper

To install:

3. Clean and lubricate the sleeves and bushings with silicon grease.
4. Install or connect the following:
   - Pads in the caliper
   - Caliper in position over the rotor
   - Mounting bolts and tighten to 31 ft. lbs. (42 Nm)
   - Fluid lines to the caliper and tighten to 33 ft. lbs. (45 Nm)
   - Wheel and tire assembly
5. Refill the master cylinder to the correct level. Bleed the brake system if the fluid lines were disconnected from the caliper.

DISC BRAKE PADS

REMOVAL & INSTALLATION

See Figures 1 and 2.

1. Before servicing the vehicle, refer to the precautions in the beginning of this section.
2. Remove or disconnect the following:
   - ⅔ of the brake fluid from the master cylinder
3. Place a C-clamp around the outer pad and caliper; tighten the C-clamp until the piston is fully compressed in the caliper.
   - Brake pads
   - Inboard pad and retaining spring from the caliper
   - Outboard pad from the caliper
   - Sleeves and bushings

To install:

4. Clean and lubricate the sleeves and bushing with silicone lubricant and install them in the caliper.
5. Clip the retaining spring onto the inboard pad and install the pad in the caliper.
6. Install or connect the following:
   - Outboard pad into the caliper
   - Caliper in position over the rotor and install the mounting bolts. Bend the tabs, on the outboard brake pad, over the caliper.
   - Wheel and tire assemblies
7. Refill the master cylinder and pump pedal to attain full brake pedal before Road-testing the vehicle.

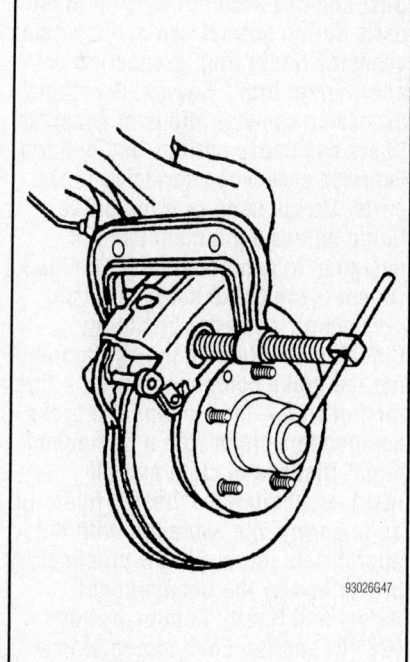

93026G47

Fig. 1 Compressing the caliper piston with a C-clamp

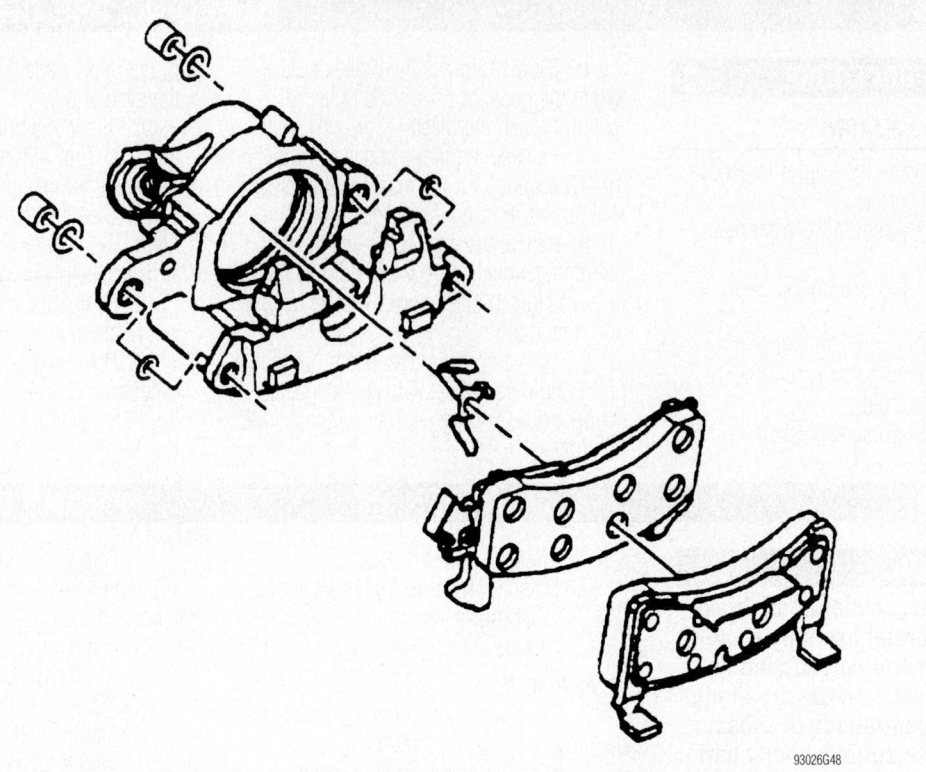

Fig. 2 Exploded view of the disc brake assembly

BRAKES

REAR DISC BRAKES

✴✴ CAUTION

Dust and dirt accumulating on brake parts during normal use may contain asbestos fibers from production or aftermarket brake linings. Breathing excessive concentrations of asbestos fibers can cause serious bodily harm. Exercise care when servicing brake parts. Do not sand or grind brake lining unless equipment used is designed to contain the dust residue. Do not clean brake parts with compressed air or by dry brushing. Cleaning should be done by dampening the brake components with a fine mist of water, then wiping the brake components clean with a dampened cloth. Dispose of cloth and all residue containing asbestos fibers in an impermeable container with the appropriate label. Follow practices prescribed by the Occupational Safety and Health Administration (OSHA) and the Environmental Protection Agency (EPA) for the handling, processing, and disposing of dust or debris that may contain asbestos fibers.

BRAKE CALIPER

REMOVAL & INSTALLATION

See Figure 3.

1. Before servicing the vehicle, refer to the precautions in the beginning of this section.
2. Raise and safely support the vehicle.
3. Remove or disconnect the following:

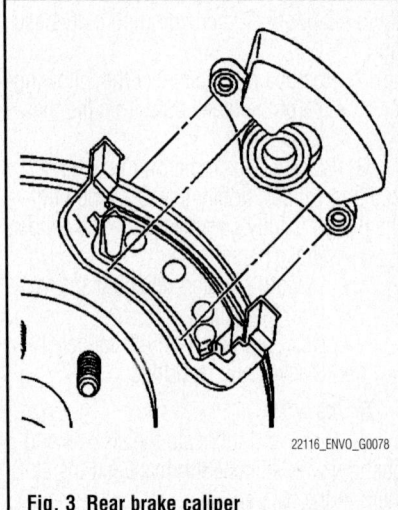

Fig. 3 Rear brake caliper

- Rear wheels
- Brake hose and cap line
- Retainers from caliper and remove caliper

To install:

4. Install or connect the following:
   - Brake pads if removed
   - Caliper over rotor, and onto mounts
   - Retainers, and tighten to 23 ft. lbs. or (31 Nm)
   - Brake hose, and tighten to 20 ft. lbs. (27 Nm)
5. Bleed brake system.
6. Install tires.
7. Refill the master cylinder and pump pedal to attain full brake pedal before Road-testing the vehicle.

DISC BRAKE PADS

REMOVAL & INSTALLATION

1. Before servicing the vehicle, refer to the precautions in the beginning of this section.
2. Remove or disconnect the following:
   - ⅔ of the brake fluid from the master cylinder
   - Wheels
3. Place a C-clamp around the outer

pad and caliper; tighten the C-clamp until the piston is fully compressed in the caliper.

- Top caliper retainer, and rotate caliper away from rotor
- Inboard pad and retaining spring from the caliper
- Outboard pad from the caliper

To install:

4. Clean and lubricate the sleeves and bushing with silicone lubricant
5. Install or connect the following:
   - Sleeves and bushings into the caliper
   - Clip the retaining spring onto the inboard pad and install the pad in the caliper
   - Outboard pad into the caliper
   - Caliper in position over the rotor and install the mounting bolts
   - Wheel and tire assemblies
6. Refill the master cylinder and pump pedal to attain full brake pedal before Road-testing the vehicle.

BRAKES — REAR DRUM BRAKES

✳✳ CAUTION

Dust and dirt accumulating on brake parts during normal use may contain asbestos fibers from production or aftermarket brake linings. Breathing excessive concentrations of asbestos fibers can cause serious bodily harm. Exercise care when servicing brake parts. Do not sand or grind brake lin-ing unless equipment used is designed to contain the dust residue. Do not clean brake parts with com-pressed air or by dry brushing. Cleaning should be done by dampen-ing the brake components with a fine mist of water, then wiping the brake components clean with a dampened cloth. Dispose of cloth and all residue containing asbestos fibers in an impermeable container with the appropriate label. Follow practices prescribed by the Occupational Safety and Health Administration (OSHA) and the Environmental Pro-tection Agency (EPA) for the han-dling, processing, and disposing of dust or debris that may contain asbestos fibers.

BRAKES — PARKING BRAKE

PARKING BRAKE SHOES

REMOVAL & INSTALLATION

See Figure 4.

1. Raise and support the rear of the vehicle safely using jackstands.
2. Remove the rear tire and wheel assembly.
3. Remove the caliper and rotor.
4. Disconnect the parking brake cable from the parking brake lever.
5. Remove the parking brake shoes assembly by sliding the shoe towards the hold-down spring until the shoe is discon-nected from the spring.
6. Remove the shoe from the actuation mechanism.
7. Clean all dirt, debris and dust from the parking brake assembly components using a clean rag.
8. Turn the adjustment screw to the fully home position in the notched adjust-ment nut, then back it off ¼ of a turn.
9. Align the slots in both the adjusting screw and the tappet to be parallel with the backing plate face.

To install:

10. Install a new parking brake shoe.
11. Position the shoe on the inboard side of the actuation mechanism.
12. Clip the shoe onto the hold-down spring. Make sure the shoe is central on the backing plate and has both tips located in the slots.
13. Manually check the parking brake for propor operation.

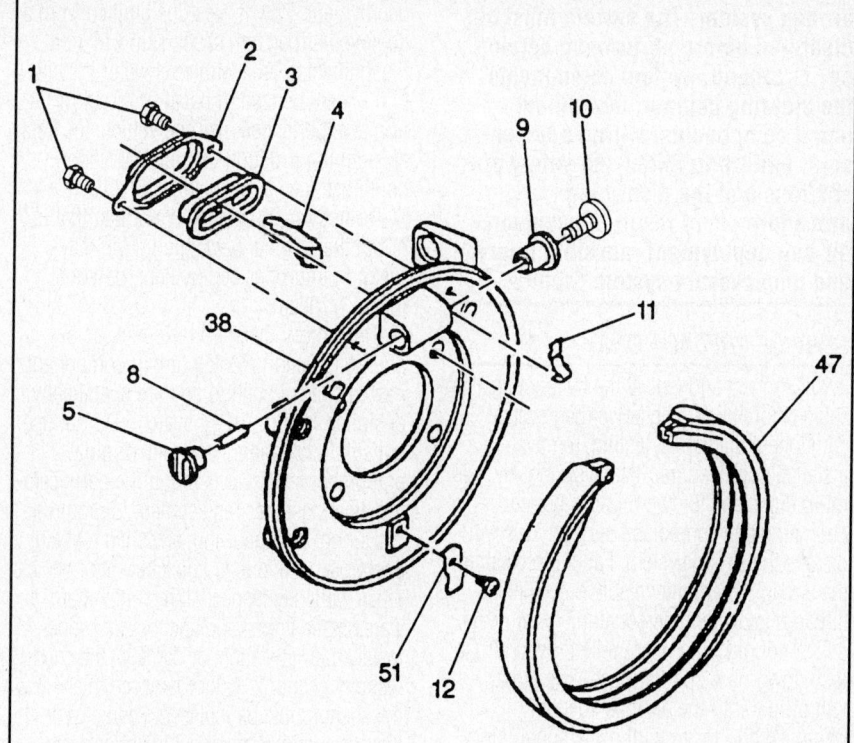

(1) Bolt/Screw, Retainer	(11) Pawl, Adjuster
(2) Retainer, Boot	(12) Bolt/Screw
(3) Boot	(38) Mounting Plate
(4) Lever	(47) Shoe and Lining
(8) Tappet	(49) Actuator
(9) Pushrod	(51) Clip
(10) Nut/Adjuster	

91119G03

Fig. 4 Exploded view of the rear parking brake assembly components

14. Attach the parking brake cable to the lever.

15. Adjust the parking brake shoe as outlined later in this section.

16. Install the caliper and the rotor.

17. Install the wheel and tire assembly.

18. Lower the vehicle and check for proper operation.

ADJUSTMENT

See Figure 5.

1. Raise and support the rear of the vehicle safely using jackstands.

2. Remove the rear tire and wheel assembly.

3. Remove the caliper and rotor.

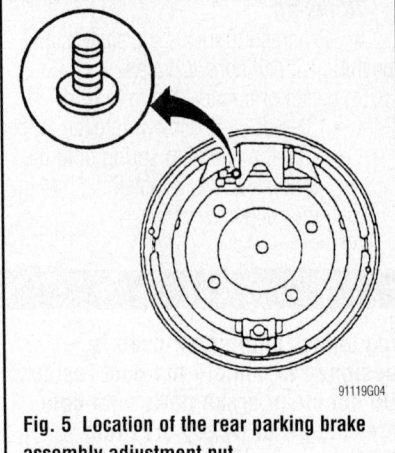

91119G04

Fig. 5 Location of the rear parking brake assembly adjustment nut

4. Disconnect the parking brake cable from the parking brake lever.

5. Adjust the shoe diameter using the adjuster nut. Turn the nut clockwise to increase the diameter until the rear wheel will not rotate forward without using excessive force. For location of the nut as refer to the accompanying illustration.

6. Attach the parking brake cable to the lever.

7. Install the caliper and the rotor.

8. Install the wheel and tire assembly.

9. Adjust the rear parking brake cables as outlined earlier in this section.

10. Lower the vehicle and check for proper operation.

CHASSIS ELECTRICAL

AIR BAG (SUPPLEMENTAL RESTRAINT SYSTEM)

GENERAL INFORMATION

✳✳ CAUTION

These vehicles are equipped with an air bag system. The system must be disarmed before performing service on, or around, system components, the steering column, instrument panel components, wiring and sensors. Failure to follow the safety precautions and the disarming procedure could result in accidental air bag deployment, possible injury and unnecessary system repairs.

SERVICE PRECAUTIONS

Disconnect and isolate the battery negative cable before beginning any airbag system component diagnosis, testing, removal, or installation procedures. Allow system capacitor to discharge for two minutes before beginning any component service. This will disable the airbag system. Failure to disable the airbag system may result in accidental airbag deployment, personal injury, or death.

Do not place an intact undeployed airbag face down on a solid surface. The airbag will propel into the air if accidentally deployed and may result in personal injury or death.

When carrying or handling an undeployed airbag, the trim side (face) of the airbag should be pointing towards the body to minimize possibility of injury if accidental deployment occurs. Failure to do this may result in personal injury or death.

Replace airbag system components with OEM replacement parts. Substitute parts may appear interchangeable, but internal differences may result in inferior occupant

protection. Failure to do so may result in occupant personal injury or death.

Wear safety glasses, rubber gloves, and long sleeved clothing when cleaning powder residue from vehicle after an airbag deployment. Powder residue emitted from a deployed airbag can cause skin irritation. Flush affected area with cool water if irritation is experienced. If nasal or throat irritation is experienced, exit the vehicle for fresh air until the irritation ceases. If irritation continues, see a physician.

Do not use a replacement airbag that is not in the original packaging. This may result in improper deployment, personal injury, or death.

The factory installed fasteners, screws and bolts used to fasten airbag components have a special coating and are specifically designed for the airbag system. Do not use substitute fasteners. Use only original equipment fasteners listed in the parts catalog when fastener replacement is required.

During, and following, any child restraint anchor service, due to impact event or vehicle repair, carefully inspect all mounting hardware, tether straps, and anchors for proper installation, operation, or damage. If a child restraint anchor is found damaged in any way, the anchor must be replaced. Failure to do this may result in personal injury or death.

Deployed and non-deployed airbags may or may not have live pyrotechnic material within the airbag inflator.

Do not dispose of driver/passenger/curtain airbags or seat belt tensioners unless you are sure of complete deployment. Refer to the Hazardous Substance Control System for proper disposal.

Dispose of deployed airbags and tensioners consistent with state, provincial, local, and federal regulations.

After any airbag component testing or service, do not connect the battery negative cable. Personal injury or death may result if the system test is not performed first.

If the vehicle is equipped with the Occupant Classification System (OCS), do not connect the battery negative cable before performing the OCS Verification Test using the scan tool and the appropriate diagnostic information. Personal injury or death may result if the system test is not performed properly.

Never replace both the Occupant Restraint Controller (ORC) and the Occupant Classification Module (OCM) at the same time. If both require replacement, replace one, then perform the Airbag System test before replacing the other.

Both the ORC and the OCM store Occupant Classification System (OCS) calibration data, which they transfer to one another when one of them is replaced. If both are replaced at the same time, an irreversible fault will be set in both modules and the OCS may malfunction and cause personal injury or death.

If equipped with OCS, the Seat Weight Sensor is a sensitive, calibrated unit and must be handled carefully. Do not drop or handle roughly. If dropped or damaged, replace with another sensor. Failure to do so may result in occupant injury or death.

If equipped with OCS, the front passenger seat must be handled carefully as well. When removing the seat, be careful when setting on floor not to drop. If dropped, the sensor may be inoperative, could result in occupant injury, or possibly death.

If equipped with OCS, when the passenger front seat is on the floor, no one should sit in the front passenger seat. This uneven force may damage the sensing ability of the seat weight sensors. If sat on and damaged, the sensor may be inoperative, could result in occupant injury, or possibly death.

DISARMING THE SYSTEM

Air Bag Fuse

1. Turn the steering wheel so that the vehicles wheels are pointing straight ahead.
2. Place the ignition in the OFF position.

➡The sensing and diagnostic module (SDM) may have more than one fused power input. To ensure there is no unwanted SIR deployment, personal injury, or unnecessary SIR system repairs, remove all fuses supplying power to the SDM. With all SDM fuses removed and the ignition switch in the ON position, the AIR BAG warning indicator illuminates. This is normal operation, and does not indicate a SIR system malfunction.

3. Locate and remove the fuse(s) supplying power to the SDM.
4. Wait 1 minute before working on the system.

Negative Battery Cable

1. Turn the steering wheel so that the vehicles wheels are pointing straight ahead.
2. Place the ignition in the OFF position.
3. Disconnect the negative battery cable.
4. Wait 1 minute before working on system.

ARMING THE SYSTEM

Air Bag Fuse

1. Place the ignition in the OFF position.
2. Install the fuse(s) supplying power to the SDM.
3. Turn the ignition switch to the ON position. The AIR BAG indicator will flash then turn OFF.

Negative Battery Cable

1. Place the ignition in the OFF position.
2. Connect the negative battery cable.
3. Turn the ignition switch to the ON position. The AIR BAG indicator will flash then turn OFF.

INFLATABLE RESTRAINT MODULE COIL CENTERING

See Figures 6 through 10.

✳✳ WARNING

A new inflatable restraint steering wheel module coil is pre-centered. Do not remove the centering tab from

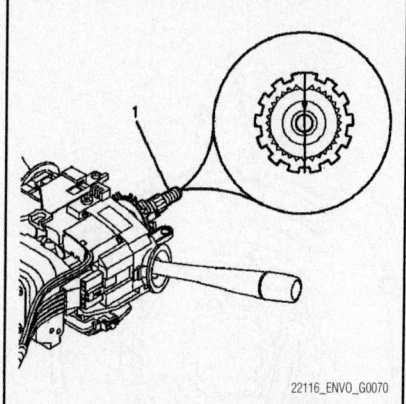

22116_ENVO_G0070

Fig. 6 Verify that the block tooth (1) of the steering shaft assembly is in the 12 o'clock position

the new inflatable restraint steering wheel module coil until installation is complete.

➡The new SIR coil assembly will be centered. Improper alignment of the SIR coil assembly may damage the unit, causing an inflatable restraint malfunction.

1. Verify the following conditions before centering the SIR coil:
- The wheels on the vehicle are straight ahead
- The block tooth (1) of the steering shaft assembly is in the 12 o'clock position
- The ignition switch assembly is in the LOCK position

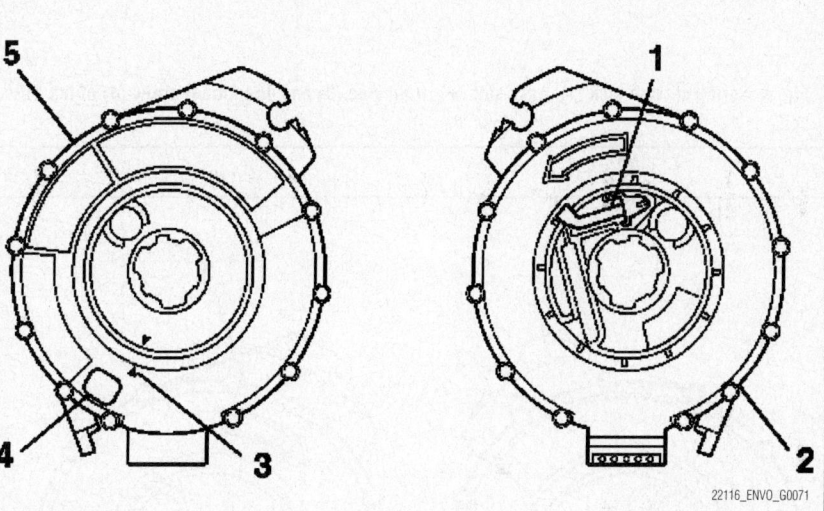

22116_ENVO_G0071

Fig. 7 Spring service lock (1), back side (2), alignment arrows (3), centering window (4) and front (5) of the SIR coil

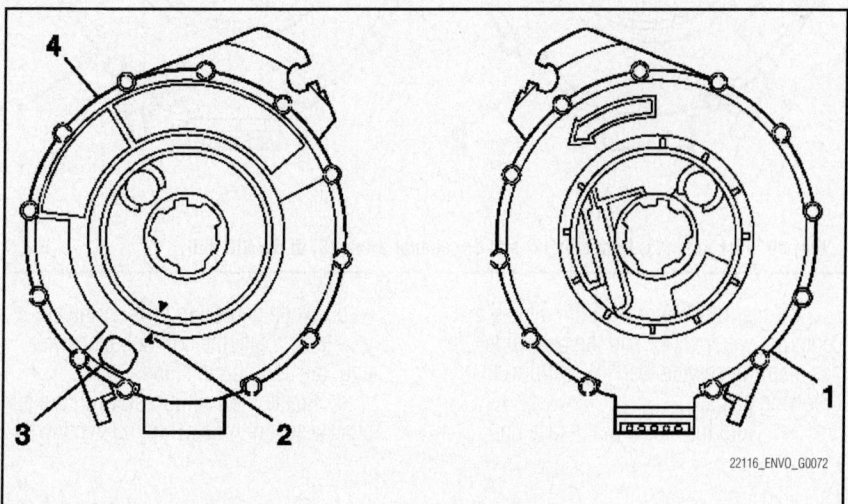

22116_ENVO_G0072

Fig. 8 Back side (1), alignment arrows (2), centering window (3) and front (4) of the SIR coil

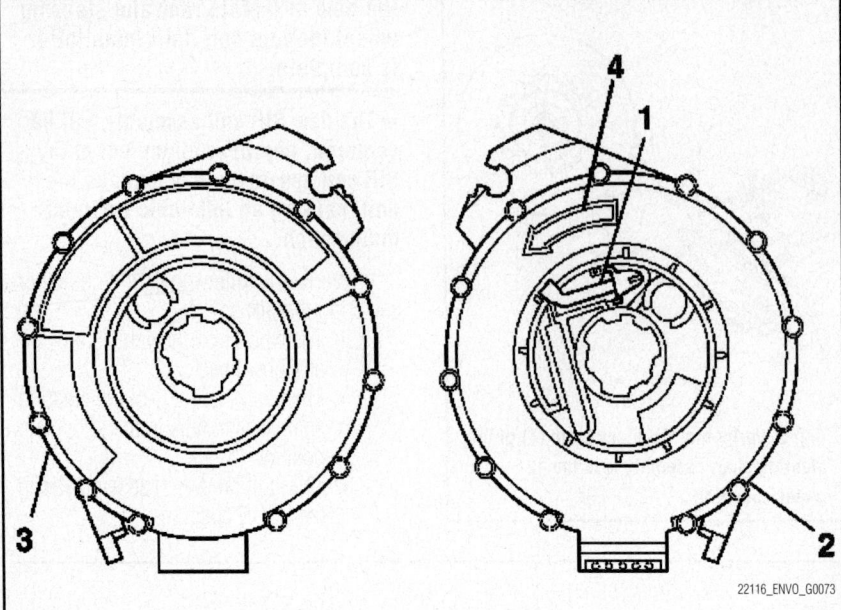

Fig. 9 Spring service lock (1), back side (2), front side (3) and directional arrow (4) of the SIR coil

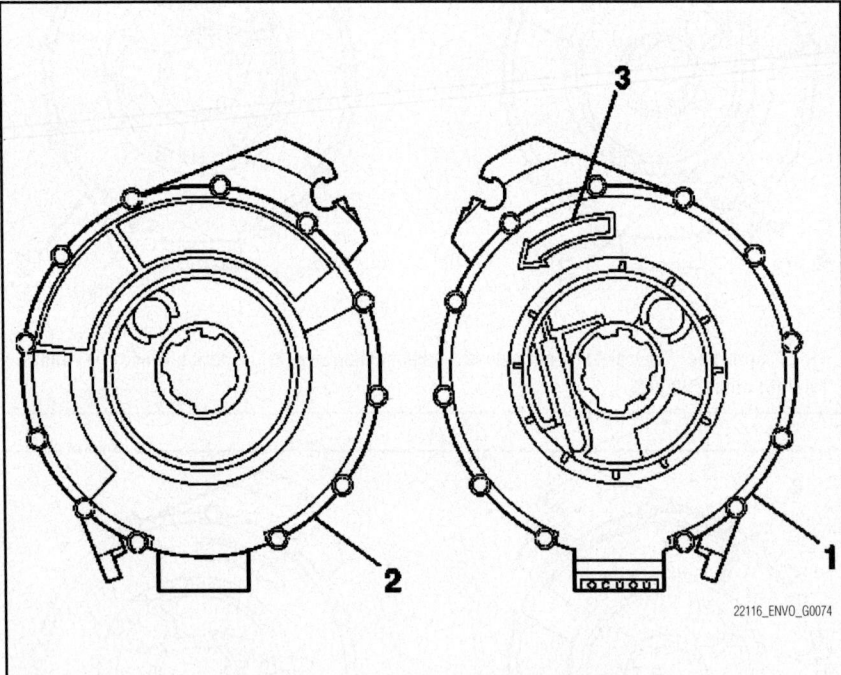

Fig. 10 Back side (1), front side (2) and directional arrow (3) of the SIR coil

2. If the front (5) of the SIR coil has a centering window (4), and the back side (2) has a spring service lock (1), perform the following steps:

a. Hold the coil with the face up.

b. While depressing the spring service lock, rotate the coil hub clockwise until the coil ribbon stops.

c. Rotate the coil hub slowly, counterclockwise, until the centering window appears yellow and both arrows (3) line up.

d. Release the spring service lock between the locking tab. The SIR coil is now centered.

e. Align the centered SIR coil with the horn tower and slide onto the steering shaft assembly.

3. If the front (4) of the SIR coil has a centering window (3) and the back side (1) has NO spring service lock, perform the following steps:

a. Hold the coil with the face up.

b. Rotate the coil hub clockwise until the coil ribbon stops.

c. Rotate the coil hub slowly, counterclockwise until the centering window appears yellow and both arrows (2) line up. This is the CENTER position.

d. While holding the coil hub in the CENTER position, align the coil with the horn tower and slide the coil onto the steering shaft assembly.

4. If no centering window is present on the front side (3) of the SIR coil, but a spring service lock (1) is on the back side (2), perform the following steps:

a. Hold the coil with the back side up.

b. While depressing the spring service lock, rotate the coil hub in the direction of the arrow (4) until the coil ribbon stops.

c. Still pressing the spring service lock, rotate the coil hub in the opposite direction 2½ revolutions.

d. Release the spring service lock between the locking tabs. The SIR coil is now centered.

e. Align the centered coil with the horn tower and slide the coil onto the steering shaft assembly.

5. If no centering window appears on the front side (2) of the SIR coil and no spring service lock exists on the back side (1), perform the following steps:

a. Hold the coil with the face up.

b. Rotate the coil hub in the direction of the arrow until the coil ribbon stops.

c. Rotate the coil hub, slowly, counterclockwise, for 2½ revolutions. This is the CENTER position.

d. While maintaining the coil hub in the CENTER position, align the centered coil with the horn tower and slide the coil onto the steering shaft assembly.

DRIVETRAIN

AUTOMATIC TRANSMISSION ASSEMBLY

REMOVAL & INSTALLATION

4.2L Engine

➡**This procedure requires the use of a Converter Holding Strap tool No. J 21366 to secure the torque converter to the transmission during removal and installation.**

1. Disconnect the negative battery.
2. Drain the transmission fluid.
3. Remove the filler tube nut and stud located on the right side of the engine.
4. Raise the vehicle.
5. If equipped with 2 wheel drive (2WD), remove the rear propeller shaft.
6. If equipped with 4 wheel drive (4WD), remove the transfer case.
7. Support the transmission with a transmission jack.
8. Remove the fuel tank shield if equipped.
9. Remove the transmission support.
10. Remove the transmission mount bolts and mount.
11. Remove the front exhaust pipe assembly.
12. Lower the transmission for access to the top and sides of the transmission.
13. Remove the range selector cable end from the transmission range selector lever ball stud and bracket.
14. Remove the transmission heat shield, transmission vent hose park/neutral position switch connector, and main connector from the transmission.
15. Remove the bolt that secures the fuel line bracket to the left side of the transmission.
16. Remove the flywheel-to-torque converter bolts. Be careful not to drop the bolts into the bell housing.
17. Disconnect the transmission oil cooler lines from the transmission. Plug the transmission oil cooler lines connectors in the transmission case.
18. Install a safety chain around the transmission.
19. Remove the bolt that secures the fuel line bracket to the bell housing.
20. Remove the bolts that secure the coolant pipe to the bell housing.
21. Remove the remaining nuts, studs and/or bolts that secure the transmission to the engine.
22. Install Converter Holding Strap tool No. J 21366 onto the transmission bell housing to hold the torque converter.

23. Pull the transmission straight back and remove it from the vehicle.

To install:

Installation is the reverse of removal, but please note the following important steps.

24. Make sure the torque converter is fully seated in the pump drive. If not, the transmission will not fit tightly to the rear of the engine block.
25. Raise the transmission into position and remove the torque converter holding strap. Carefully slide the transmission forward until the dowel pins are engaged while lining up the marks on the flywheel made during removal.
26. The torque converter should be flush with the flywheel and turn freely by hand.
27. Tighten the torque converter-to-flywheel bolts to 44 ft. lbs. (66 Nm).
28. Install the transmission-to-engine nuts, studs and or bolts. Tighten the studs and/or bolts to 37 ft. lbs. (50 Nm).
29. Tighten the bolts securing the heat shield to the transmission to 13 ft. lbs. (17 Nm).
30. Tighten the bolts and washers securing the transmission mount to 18 ft. lbs. (25 Nm).
31. Tighten the nut and washer securing the transmission mount to the transmission support to 35 ft. lbs. (46 Nm).
32. Refill the transmission with the proper amount and type of fluid.
33. Connect the negative battery cable. Start the vehicle and allow to warm while checking for leaks. Road test the vehicle to check for shift quality.

5.3L Engine

➡**This procedure requires the use of a Converter Holding Strap tool No. J 21366 to secure the torque converter to the transmission during removal and installation.**

1. Disconnect the negative battery.
2. Drain the transmission fluid.
3. Raise and support the vehicle.
4. Remove the rear propeller shaft.
5. Support the transmission with a jack.
6. Remove the nuts securing the transmission mount to the transmission support.
7. Remove the transmission support from the vehicle.
8. Remove the transmission mount.
9. Remove the front exhaust pipe assembly.
10. Lower the transmission to gain access to the top and sides of the transmission.

11. Remove the transfer case, if equipped.
12. Remove the range selector cable end from the transmission range selector lever ball stud and the bracket.
13. Remove the transmission heat shield.
14. Disconnect the transmission vent hose, the park/neutral position switch connectors, and the main electrical connector from the transmission.
15. Remove the transmission harness from the retainers.
16. Remove the bolt that secures the fuel line bracket to the left side of the transmission.
17. Remove the torque converter access plug.
18. Mark the flywheel and torque converter orientation for reassembly.
19. Remove the flywheel to torque converter bolts. Use care not to drop the bolts into the bell housing.
20. Disconnect the transmission oil cooler lines from the transmission.
21. Plug the transmission oil cooler line connectors in the transmission case.
22. Install a safety chain around the transmission.
23. Remove the nut that secures the filler tube to the bell housing.
24. Remove the transmission filler tube.
25. Remove the remaining nuts, studs and/or bolts that secure the transmission to the engine.
26. Install the J-21366 onto the transmission bell housing to retain the torque converter.
27. Pull the transmission straight back.
28. Remove the transmission from the vehicle.

To install:

29. Raise the transmission into place and remove the torque converter holding tool.
30. Slide the transmission straight onto the locating pins while lining up the marks on the flywheel and the torque converter made during removal. The torque converter must be flush onto the flywheel and rotate freely by hand.
31. Install nuts, studs and/or bolts securing the transmission to the engine and tighten to 37 ft. lbs. (50 Nm).
32. Install the fuel line retaining bracket to the transmission.
33. Install the flywheel-to-torque converter bolts and tighten to 44 ft. lbs. (66 Nm).

34. Install the torque converter access plug.

35. Remove the safety chain from the transmission.

36. Install the transmission filler tube.

37. Install the filler tube nut.

38. Install the transmission vent hose, fuel lines, and the wiring harness to the transmission.

39. Install the transmission harness to the retainers.

40. Install the heat shield to the transmission.

41. Install the bolts securing the heat shield to the transmission and tighten to 13 ft. lbs. (17 Nm).

42. Install the shift cable end to the transmission shift lever ball stud and bracket.

43. Install the transfer case, if equipped.

44. Install the front exhaust pipe assembly.

45. Install the transmission mount to the vehicle.

46. Install the bolts securing the transmission mount to the transmission and tighten to 18 ft. lbs. (25 Nm).

47. Install the transmission support to the vehicle.

48. Lower the transmission and remove the transmission jack.

49. Install the nuts securing the transmission mount to the transmission support and tighten to 35 ft. lbs. (46 Nm).

50. Install the rear propeller shaft.

51. Flush the transmission oil cooler and cooling lines at this time, if necessary.

52. Connect the transmission oil cooler lines to the transmission.

53. Lower the vehicle.

54. Connect the battery cable.

55. Fill the transmission to the proper level with DEXRON® III transmission fluid and check for leaks.

56. Road test the vehicle and check for proper operation.

TRANSFER CASE ASSEMBLY

REMOVAL & INSTALLATION

1. Before servicing the vehicle, refer to the precautions in the beginning of this section.

2. Disconnect the negative battery cable.

3. Raise and support the vehicle. Drain the transfer case.

4. Remove or disconnect the following:
- Fuel tank shield mounting bolts and shield
- Front and rear propeller shaft. Matchmark the shafts prior to removal.
- Fuel lines from the retainer
- Electrical harness from the retainers on the right and left sides
- Speed sensor electrical connectors
- Motor/encoder electrical connector
- Transfer case wiring harness
- Vent hose

5. Install a transmission jack to support the transfer case.
- Transfer case mounting bolts
- Transfer case from the vehicle
- Transfer case gasket and discard if damaged

To install:

6. Install or connect the following:

➡You must replace the transfer case gasket if it is damaged. Never use silicone sealant in place of, or with the transfer case gasket.

- Transfer case, using a new gasket if necessary
- Transfer case mounting bolts and tighten to 35 ft. lbs. (47 Nm)

7. Remove the transmission jack.

8. The remainder of installation is the reverse of removal.

9. Refill the transfer case.

FRONT HALFSHAFT

REMOVAL & INSTALLATION

See Figures 11 through 13.

1. Before servicing the vehicle, refer to the precautions in the beginning of this section.

2. Remove or disconnect the following:
- Front wheel

➡Place a drift through the caliper into the edge of the rotor to keep the rotor from turning when the nut is removed

- Wheel center cap, if equipped
- Halfshaft nut and discard. A new nut must be used for installation.
- Drift from the rotor
- Brake caliper and support it with a piece of wire to avoid damaging the brake hose
- Brake rotor

3. To remove the steering knuckle, remove or disconnect the following:
- Wheel hub and bearing

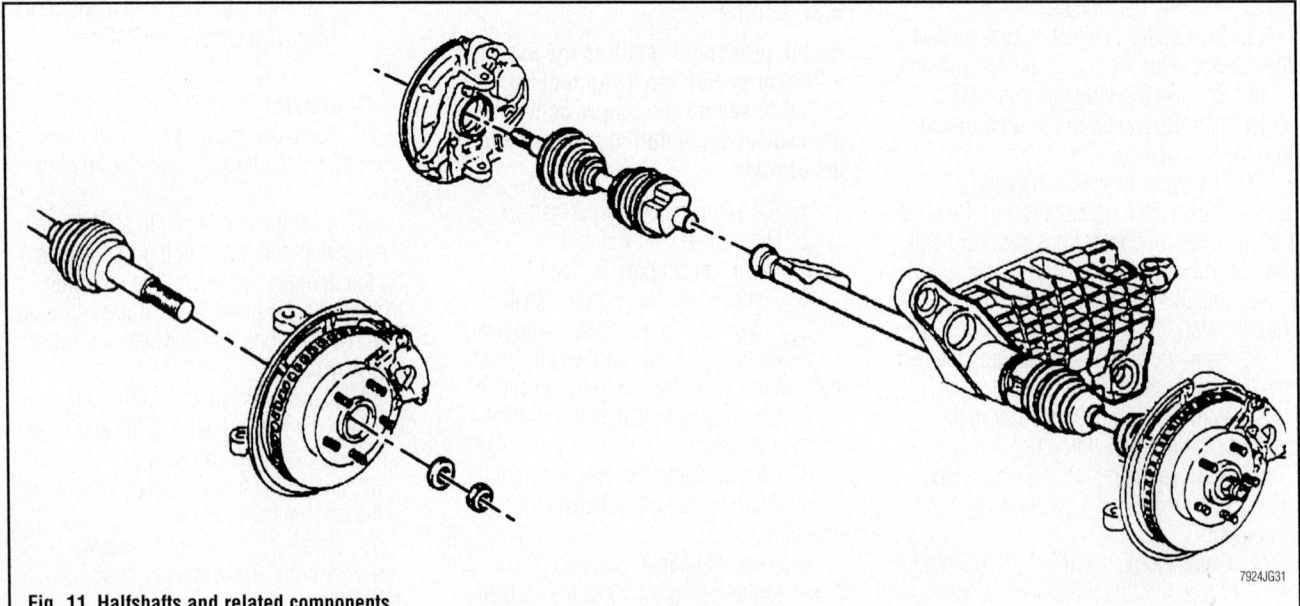

Fig. 11 Halfshafts and related components

7924JG31

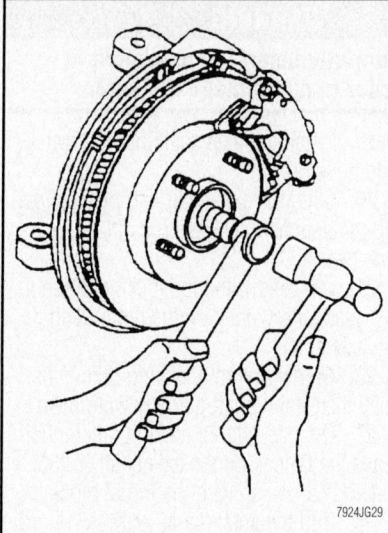

Fig. 12 Tap the halfshaft out of the hub without damaging the threads

- Outer tie rod retaining nut
- Outer tie rod end from the steering knuckle using a puller
- Brake hose bracket retaining bolts
- Brake hose bracket
- Anti-lock Brake System (ABS) wheel speed sensor wiring harness bracket, if necessary

- Upper control arm-to-steering knuckle pinch bolt and nut
- Upper control arm from the steering knuckle
- Lower ball joint retaining nut
- Steering knuckle from the control arm using a puller
- Steering knuckle

4. Remove the left side halfshaft from differential carrier, or right halfshaft from the clutch fork housing as follows:

   a. Place a brass drift against the tripot housing.

   b. Use a hammer to strike the drift outward from the case, striking hard enough to overcome the snapring tension holding the halfshaft.

5. Pull the halfshaft straight out of the differential carrier or clutch fork housing.

To install:

6. Install the halfshaft as follows:

   a. With both hands on the tripot housing, align the splines on the shaft with the differential carrier assembly (left) or clutch fork housing (right).

   b. Center the halfshaft into the differential carrier or clutch fork housing assembly seal.

   c. Firmly push the shaft straight into the differential carrier or clutch fork

housing assembly until the snapring is properly seated.

7. To install the steering knuckle, install or connect the following:

- Steering knuckle to the lower control arm
- Lower ball joint retaining nut and tighten to 81 ft. lbs. (110 Nm)
- Upper control arm to the steering knuckle
- Upper control arm pinch bolt and nut and tighten to 30 ft. lbs. (40 Nm)
- ABS wheel speed sensor harness bracket
- Brake hose bracket. Tighten the bolts to 7 ft. lbs. (10 Nm).
- Outer tie rod to the steering knuckle and tighten the nut to 33 ft. lbs. (45 Nm)
- Hub and bearing

8. Install or connect the following:

- New halfshaft nut and tighten to 103 ft. lbs. (140 Nm)
- Wheel

9. Lower the vehicle. Adjust the front toe.

REAR AXLE SHAFT, BEARING & SEAL

REMOVAL & INSTALLATION

See Figures 14 through 19.

1. Raise and safely support the rear of the vehicle securely on jackstands.

2. Remove the tire and wheel assembly.

3. Remove the brake caliper.

4. Remove the rear wheel speed sensor.

5. Remove the rear axle housing cover and the gasket.

6. Remove the pinion shaft locking bolt.

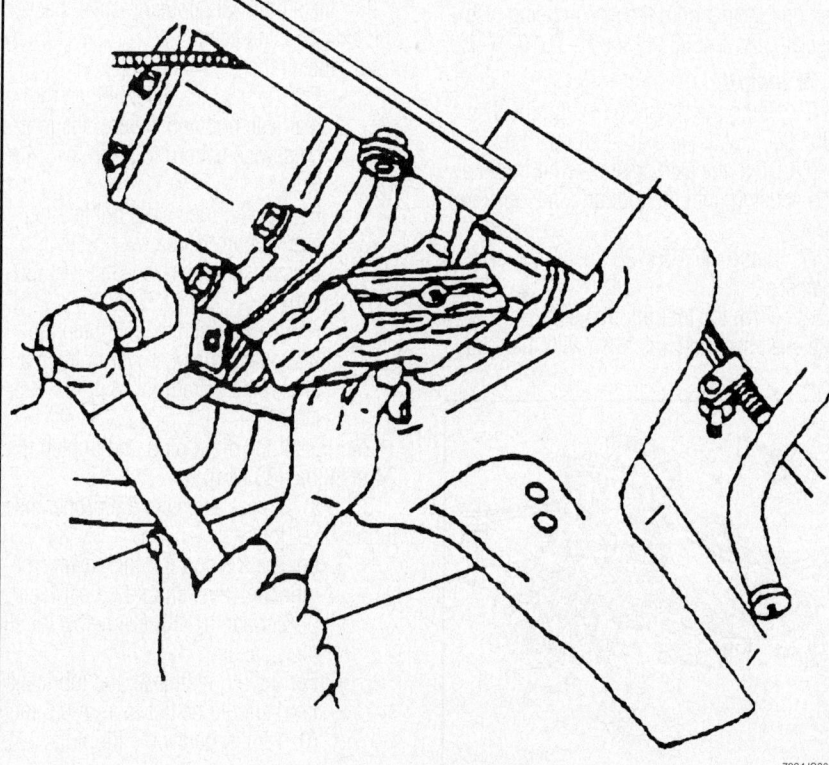

Fig. 13 Using a block of wood and a mallet, disengage the halfshaft from the differential assembly

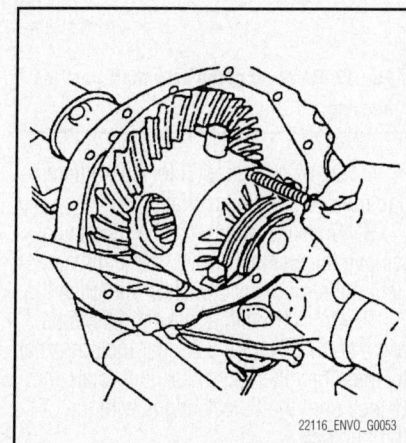

Fig. 14 Removal of the pinion shaft locking bolt

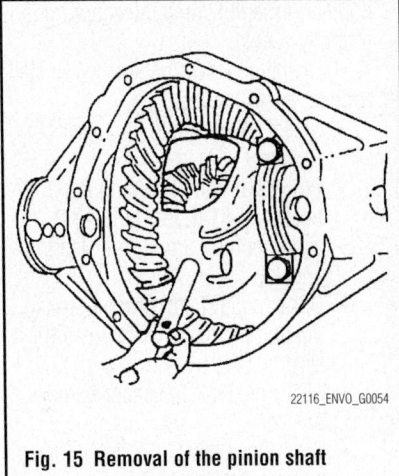

Fig. 15 Removal of the pinion shaft

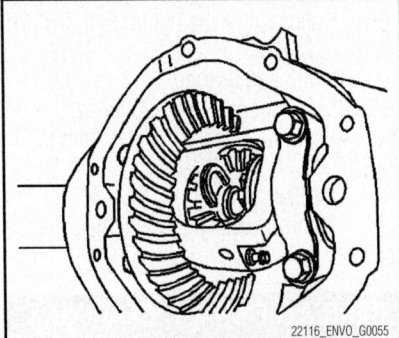

Fig. 16 Removal of the C-lock from the button end of the axle shaft

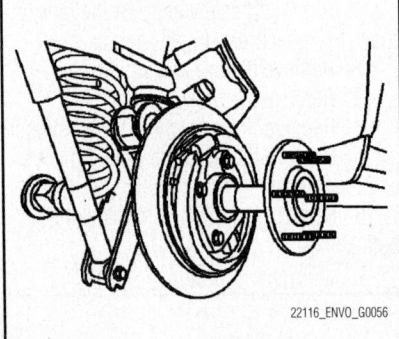

Fig. 17 Removal of the axle shaft from the housing

7. On axles without a locking differential, remove the pinion shaft.

8. On axles with a locking differential, remove the shaft part way. Rotate the case until the pinion shaft touches the housing.

9. On axles with a locking differential, use a screwdriver, or a similar tool, in order to enter the differential case and rotate the C-lock until the C-lock aligns with the thrust block.

10. Push the flange of the axle shaft toward the differential.

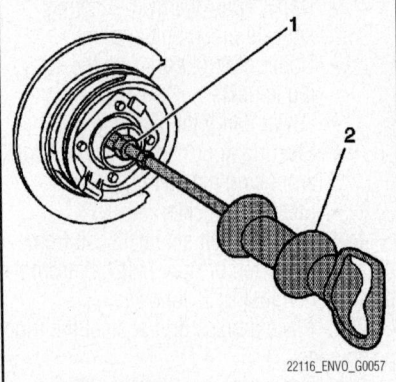

Fig. 18 Removal of the axle shaft seal and bearing together using special tools J-45857 (1) and J-2619-01 (2)

11. Remove the C-lock from the button end of the axle shaft.

➡**When removing the axle shaft, do not rotate the shaft. Rotating the shaft will misalign the gears. Misaligning the gears will make the installing of the axle shaft difficult.**

12. Remove the axle shaft from the housing. If the axle is difficult to remove, use the J-45859 (1) and the J-2619-01 (2) to remove the axle shaft from the housing.

13. To remove the seal only, use a suitable seal remover.

14. To remove the axle shaft seal and the bearing together from the axle housing, use special tools J-45857 (1) and J-2619-01 (2).

To install:

15. Using the J-23690 (1) and the J-8092 (2), install the axle shaft bearing.

16. Drive the axle shaft bearing into the axle housing until the tool bottoms against the tube.

17. Using the J-21128, install the axle shaft seal.

18. Drive the tool into the bore until the axle shaft seal bottoms flush with the tube.

Fig. 19 Installing the axle shaft bearing using J-23690 (1) and J-8092 (2)

✳✳ WARNING

Carefully insert the axle shaft in order to not damage the seal.

19. Install the axle shaft into the rear axle housing.

20. Slide the axle shaft into place allowing the splines to engage the differential side gear.

21. On axles without a locking differential, place the C-lock on the button end of the axle shaft.

22. On axles with a locking differential, keep the pinion shaft partially withdrawn.

23. On axles with a locking differential, place the C-lock on the axle shaft so that the ends are flush with the thrust block.

24. Pull the shaft flange outward in order to seat the C-lock in the differential gear.

25. Align the hole in the pinion shaft with the bolt hole in the differential case.

26. Install the new pinion shaft locking bolt:
- For the 8.0/8.6 inch axle, tighten the pinion shaft locking bolt to 27 ft. lbs. (36 Nm).
- For the 9.5 LD inch axle, tighten the pinion shaft locking bolt to 37 ft. lbs. (50 Nm).

➡**The axle housing gasket is reusable. Replace only if damaged.**

27. Install the axle housing cover gasket and axle housing cover.

28. Install the mounting bolts:
- For the 9.5 inch axle, tighten the rear axle housing cover bolts in a crosswise pattern to 30 ft. lbs. (40 Nm).
- For the 8.0 inch axle, tighten the rear axle housing cover bolts in a crosswise pattern to 20 ft. lbs. (30 Nm).
- For the 8.6 inch axle, tighten the rear axle housing cover bolts in a crosswise pattern to 18 ft. lbs. (25 Nm).

29. Install the drain plug and tighten to 24 inch lbs. (33 Nm).

30. Fill the rear axle with the proper axle lubricant as follows:
- For the 8.0 and 8.6 inch axles, the lubricant level should be between 0–0.4 inch (0–10mm) below the fill plug opening.
- For the 9.5 inch axle, the lubricant level should be between 0–0.5 inch (0–13mm) below the fill plug opening.

31. Install the brake caliper.

32. Install the rear wheel speed sensor.

33. Install the tire and wheel assembly.
34. Fill the rear axle with axle lubricant. Use the proper fluid.
35. Lower the vehicle.

REAR PINION SEAL

REMOVAL & INSTALLATION

See Figures 20 through 22.

1. Before servicing the vehicle, refer to the precautions in the beginning of this section.

➡**The following procedure requires the use of the Pinion Holding tool J-8614-10, the Pinion Flange Removal tool J-8614-1, J-8614-2, J-8614-3 and the Pinion Seal Installation tool J-23911 or J-33782.**

2. Remove or disconnect the following:
   - Driveshaft from the pinion flange. Matchmark the driveshaft prior to removal.
   - Driveshaft from the rear axle pinion flange and support the shaft up in body tunnel by wiring it to the exhaust pipe.

➡**If the U-joint bearings are not retained by a retainer strap, use a piece of tape to hold bearings on their journals.**

3. Mark the position of the pinion stem, flange and nut for reference.
4. Use an inch lbs. torque wrench to measure the amount of torque necessary to turn the pinion, then note this measurement as it is the combined pinion bearing, seal, carrier bearing, axle bearing and seal preload.
5. Remove or disconnect the following:
   - Pinion flange nut and washer, using a Pinion Holding tool J-8614-10

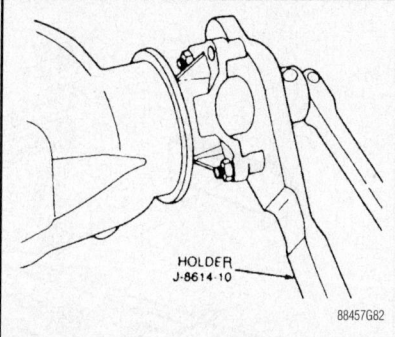

Fig. 20 Removing the pinion nut using a pinion holding fixture tool

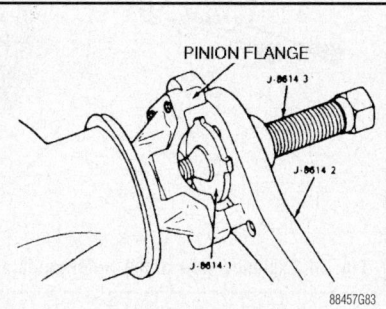

Fig. 21 A puller and adapter should be used to withdraw the pinion from the housing

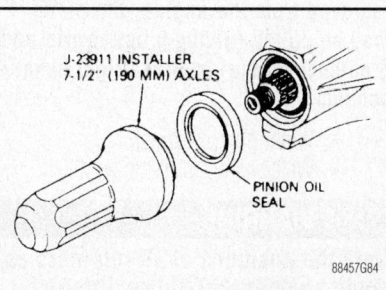

Fig. 22 Use the appropriately sized installation tool to drive the new seal into position.

and a Pinion Flange Removal tool J-8614-1, J-8614-2, J-8614-3, as applicable
   - Pinion flange
   - Pinion oil seal by driving it out of the differential with a blunt chisel; DO NOT damage the carrier

To install:

6. Examine the seal surface of pinion flange for tool marks, nicks or damage, such as a groove worn by the seal. If damaged, replace flange.
7. Examine the carrier bore and remove any burrs that might cause leaks around the O.D. of the seal.
8. Apply GM seal lubricant 1050169 to the outside diameter of the pinion flange and sealing lip of new seal.
9. Install or connect the following:
   - New pinion oil seal using a seal installer tool
   - Pinion flange and tighten nut to the same position as marked earlier. Tighten the nut a little at a time and turn the pinion flange several times after each tightening in order to set the rollers.
10. Measure the torque necessary to turn the pinion and compare this to the reading taken during removal. Tighten the nut additionally, as necessary to achieve the same preload as measured earlier.

➡**If fluid was lost from the differential housing during this procedure, be sure to check and add additional fluid, as necessary.**

11. Remove the support then align and secure the driveshaft assembly to the pinion flange.

➡**The original matchmarks MUST be aligned to assure proper shaft balance and prevent vibration.**

ENGINE COOLING

THERMOSTAT

REMOVAL & INSTALLATION

4.2L Engine

See Figure 23.

1. Remove the necessary coolant from the radiator.
2. Remove the alternator, as outlined in the Engine Electrical Section.
3. Loosen the outlet hose clamp at the thermostat housing. Remove the outlet hose from the thermostat housing.

4. Remove the thermostat housing bolts.
5. Remove the thermostat housing from the engine block.
6. Clean all of the surfaces of the thermostat housing.
7. Clean the sealing surface of the engine block.

To install:

8. Install the thermostat housing to the engine block.
9. Install the thermostat housing bolts and tighten to 89 inch lbs. (10 Nm).
10. Lubricate the inner diameter of the radiator hose with engine coolant.

11. Install the outlet hose to the thermostat housing. Secure the hose with the clamp.
12. Install the alternator.
13. Fill the cooling system with specified coolant and concentration.
14. Inspect all sealing surfaces for leaks after starting the engine.

5.3L Engine

➡**The thermostat is not serviceable separately. The water pump inlet and thermostat must be replaced as an assembly.**

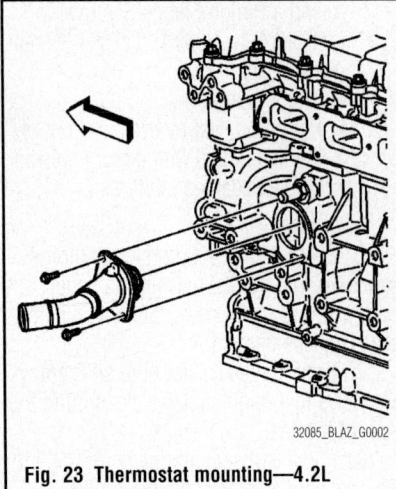

Fig. 23 Thermostat mounting—4.2L engine

1. Drain the cooling system to a level below the thermostat.
2. Remove the radiator outlet hose.
3. Remove the water pump inlet bolts.
4. Remove the water pump inlet and thermostat from the water pump.

To install:

5. Install the thermostat and thermostat housing to the water pump.
6. Install the thermostat housing bolts. Tighten the bolts to 11 ft. lbs. (15 Nm).
7. Install the radiator outlet hose.
8. Properly fill the engine cooling system and check for leaks.

WATER PUMP

REMOVAL & INSTALLATION

See Figures 24 and 25.

1. Before servicing the vehicle, refer to the precautions in the beginning of this section.
2. Disconnect the negative battery cable.
3. Drain the engine cooling system.
4. For 5.3L engines, loosen the air cleaner outlet duct clamps at the throttle body and Mass Airflow/Intake Air Temperature (MAF/IAT) sensor. Remove the bolt and air cleaner outlet duct.
5. Relieve the belt tension and remove the accessory drive belts or the serpentine drive belt, as applicable.
6. Remove or disconnect the following:
   • Upper fan shroud
   • Fan or fan and clutch assembly, as applicable
   • Water pump pulley; use a suitable tool to hold the pulley while removing the bolts
   • Coolant hose(s) from the water pump

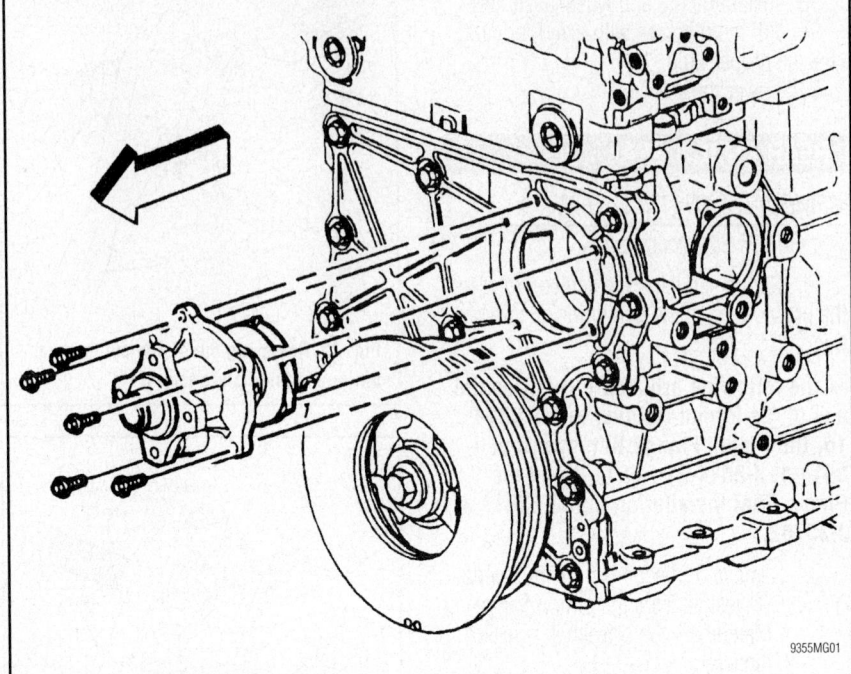

Fig. 24 Exploded view of the water pump assembly mounting—4.2L engine

➡For the hoses on some engines, removal may be easier if the hose is left attached until the pump is free from the block. Once the pump is removed from the engine, the pump may be pulled (giving a better grip and greater leverage) from the tight hose connection.

   • Water pump retainers
   • Water pump from the engine

❋❋ WARNING

Note the positions of all retainers as some engines will utilize different length fasteners in different locations and/or bolts and studs in different locations.

To install:

7. Clean the gasket mounting surfaces.

➡The water pumps on some of the engines covered may have been installed using sealer only, no gasket, at the factory. If a gasket is supplied with the replacement part, it should be used. Otherwise, a ⅛ in. (3mm) bead of RTV sealer should be used around the sealing surface of the pump.

8. Apply sealant to the water pump retainer threads.
9. Install or connect the following:
   • Water pump using a new gasket. Tighten the water pump retainers to

89 inch lbs. (10 Nm) for 4.2L engines. For 5.3L engines, tighten the bolts to 11 ft. lbs. (15 Nm), then to 22 ft. lbs. (30 Nm).
   • Coolant hose(s)
   • Water pump pulley. Tighten the pulley bolts to 18 ft. lbs. (25 Nm).
   • Fan or fan and clutch assembly
   • Serpentine drive belt (if equipped) by positioning the belt over the pulleys and carefully allow the tensioner back into contact with the belt.
   • V-belts (if equipped) and adjust the tension
   • Upper fan shroud
   • Negative battery cable
10. Refill the engine cooling system.
11. Run the engine and check for leaks.

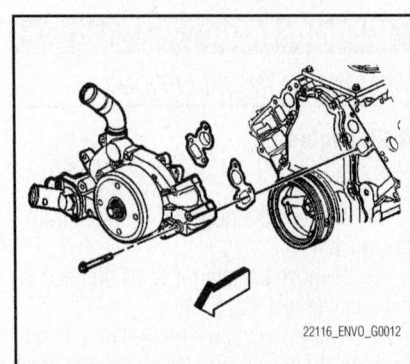

Fig. 25 Exploded view of the water pump assembly mounting—5.3L engine

ALTERNATOR

REMOVAL & INSTALLATION

4.2L Engine

See Figure 26.

1. Before servicing the vehicle, refer to the precautions in the beginning of this section.

2. Remove or disconnect the following:
   - Negative battery cable
   - Accessory belt
   - Positive battery cable nut from the generator
   - A/C line mounting bracket bolt at the engine lift hook
   - Right engine lift hook bolts
   - Engine lift hook
   - Mounting bolts
   - Alternator

To install:

3. Install or connect the following:
   - Alternator and loosely install the mounting blots
   - Tighten the alternator mounting bolts to 37 ft. lbs. (50 Nm)
   - Positive battery cable and secure with the nut; tighten the nut to 80 inch lbs. (9 Nm)
   - Engine lift hook and bolts; tighten the bolts to 37 ft. lbs. (50 Nm)
   - A/C line bracket to the lift hook, then tighten the retaining bolt to 89 inch lbs. (10 Nm)
   - Accessory belt
   - Negative battery cable

5.3L Engine

See Figure 27.

1. Before servicing the vehicle, refer to the precautions in the beginning of this section.

2. Remove or disconnect the following:
   - Negative battery cable
   - Accessory belt
   - Electrical connector
   - Terminal stud nut, after sliding boot down
   - Alternator cable
   - Mounting bolts
   - Alternator

To install:

3. Install or connect the following:
   - Alternator and loosely install the mounting bolts
   - Tighten the bolts to 37 ft. lbs. (50 Nm)
   - Alternator cable

- Terminal stud nut and tighten to 80 inch lbs. (9 Nm)
- Boot back over terminal stud

- Electrical connector
- Accessory belt
- Negative battery cable

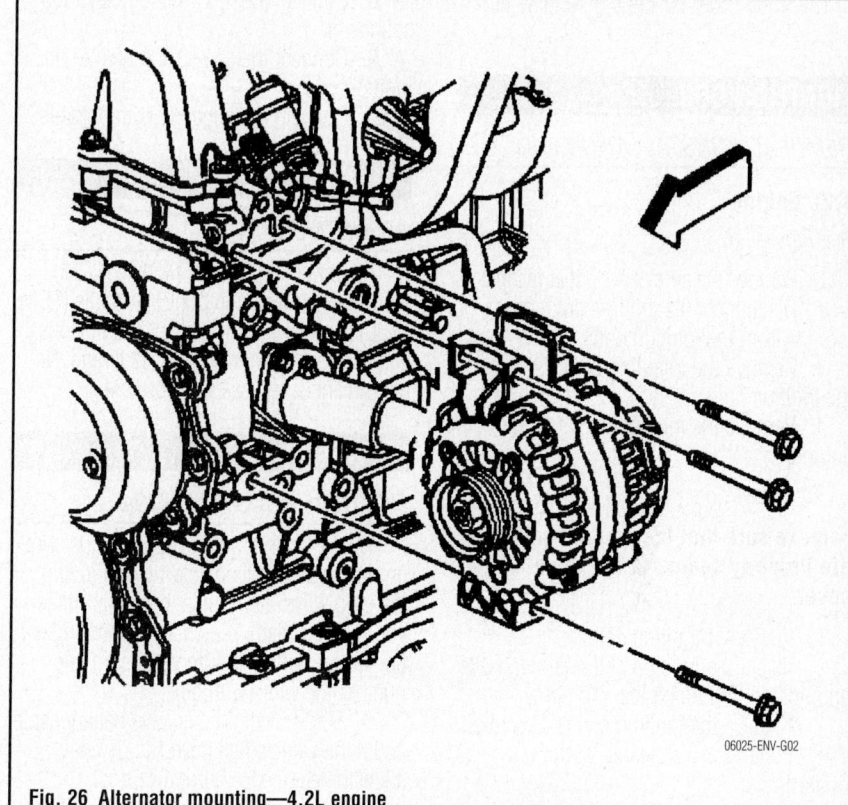

06025-ENV-G02

Fig. 26 Alternator mounting—4.2L engine

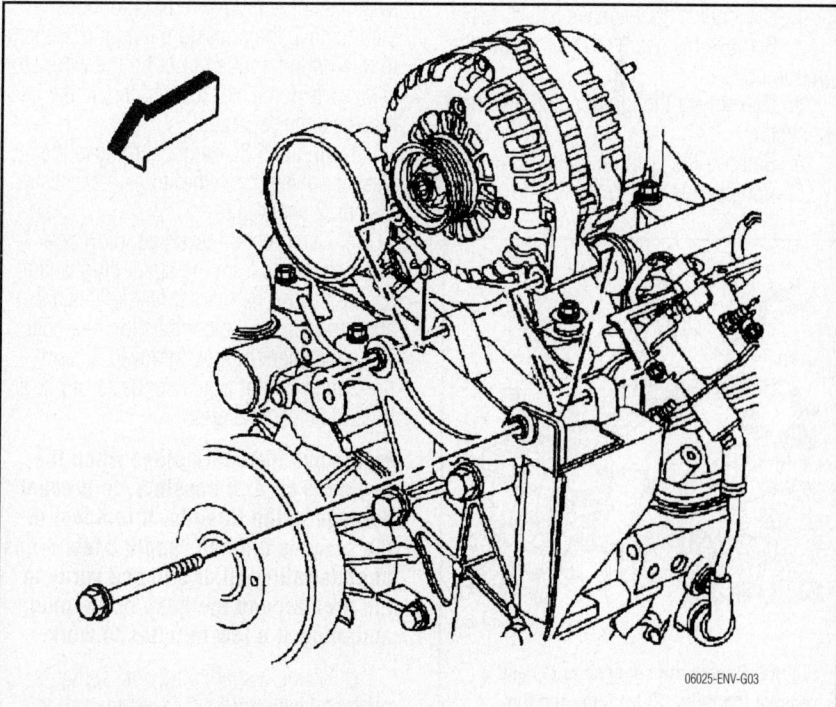

06025-ENV-G03

Fig. 27 Alternator mounting—5.3L engine

FIRING ORDER

The firing order for the 4.2L engine is 1-5-3-6-2-4.

The firing order for the 5.3L engine is 1-8-7-2-6-5-4-3.

IGNITION COIL

REMOVAL & INSTALLATION

4.2L Engine

See Figure 28.

1. Remove the air cleaner outlet resonator.
2. Disconnect the ignition coil connectors (1) from the ignition coils.
3. Remove the retaining bolts (2) from the ignition coils.
4. Remove the ignition coils (1) from the engine.

To install:

➡**Make sure that the ignition coil seals are properly seated to the valve cover.**

5. Install the ignition coil.
6. Install the ignition coil retaining bolts and tighten to 89 inch lbs. (10 Nm).
7. Replace the ignition coil connectors.
8. Install the air cleaner outlet resonator.

5.3L Engine

1. Disconnect the negative battery cable.
2. Remove the spark plug wire from the ignition coil.
3. Disconnect the ignition coil electrical connector.
4. Remove the ignition coil bolts.
5. Remove the ignition coil.

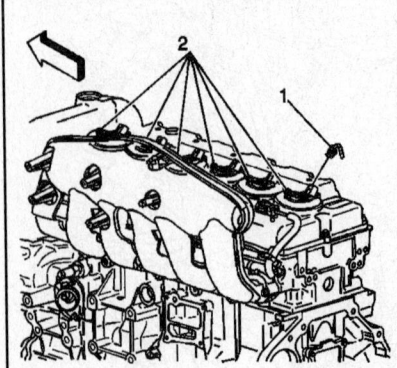

Fig. 28 Detach the connectors (1) and remove the bolts (2) from the ignition coils

To install:

6. Install the ignition coil.
7. Install the ignition coil bolts and tighten to 71 inch lbs. (8 Nm).
8. Connect the ignition coil electrical connector.
9. Connect the spark plug wire to the ignition coil.
10. Connect the negative battery cable.

IGNITION TIMING

ADJUSTMENT

The Powertrain Control Module (PCM) (4.2L) or the Electronic Control Module (ECM) (5.3L) controls all ignition system functions, and constantly corrects the spark timing. No adjustment is necessary or possible.

SPARK PLUGS

REMOVAL & INSTALLATION

When you're removing spark plugs, work on one at a time. Don't start by removing the plug wires all at once, because, unless you number them, they may become mixed up. Take a minute before you begin and number the wires with tape.

1. Disconnect the negative battery cable, and if the vehicle has been run recently, allow the engine to thoroughly cool.
2. On the 4.2L engine, remove the ignition coils. On the 5.3L engines, carefully twist the spark plug wire boot to loosen it, then remove the boot from the plug. Be sure to pull on the boot and not on the wire, otherwise the connector located inside the boot may become separated.
3. On the 5.3L engines, remove the washer solvent container to gain access to the No. 2 spark plug.
4. Using compressed air, blow any water or debris from the spark plug well to assure that no harmful contaminants are allowed to enter the combustion chamber when the spark plug is removed. If compressed air is not available, use a rag or a brush to clean the area.

➡**Remove the spark plugs when the engine is cold, if possible, to prevent damage to the threads. If removal of the plugs is difficult, apply a few drops of penetrating oil or silicone spray to the area around the base of the plug, and allow it a few minutes to work.**

5. Using a spark plug socket that is equipped with a rubber insert to properly hold the plug, turn the spark plug counter-clockwise to loosen and remove the spark plug from the bore.

✴✴ WARNING

Be sure not to use a flexible extension on the socket. Use of a flexible extension may allow a shear force to be applied to the plug. A shear force could break the plug off in the cylinder head, leading to costly and frustrating repairs.

To install:

6. Inspect the spark plug boot for tears or damage. If a damaged boot is found, the spark plug wire must be replaced.
7. Using a wire feeler gauge, check and adjust the spark plug gap. When using a gauge, the proper size should pass between the electrodes with a slight drag. The next larger size should not be able to pass while the next smaller size should pass freely.
8. Carefully thread the plug into the bore by hand. If resistance is felt before the plug is almost completely threaded, back the plug out and begin threading again. In small, hard to reach areas, an old spark plug wire and boot could be used as a threading tool. The boot will hold the plug while you twist the end of the wire and the wire is supple enough to twist before it would allow the plug to crossthread.

✴✴ WARNING

Do not use the spark plug socket to thread the plugs. Always carefully thread the plug by hand or using an old plug wire to prevent the possibility of crossthreading and damaging the cylinder head bore.

9. Carefully tighten the spark plug. If the plug you are installing is equipped with a crush washer, seat the plug, then tighten about 1/4 turn to crush the washer. If you are installing a tapered seat plug, tighten the plug to specifications provided by the vehicle or plug manufacturer.
10. On the 5.3L, install the washer solvent container to gain access to the No. 2 spark plug.
11. On the 4.2L engine, install the ignition coils. On the 5.3L, apply a small amount of silicone dielectric compound to the end of the spark plug lead or inside the spark plug boot to prevent sticking, then install the boot to the spark plug and push until it clicks into place. The click may be felt or heard, then gently pull back on the boot to assure proper contact.

ENGINE ELECTRICAL   **STARTING SYSTEM**

STARTER

REMOVAL & INSTALLATION

4.2L Engine

See Figure 29.

1. Before servicing the vehicle, refer to the precautions in the beginning of this section.
2. Disconnect the negative battery cable
3. Raise and safely support the vehicle.
4. Remove the left front tire and wheel assembly.
5. Working in the left fender area, disconnect the positive battery lead from the solenoid.
6. Remove or disconnect the following:
   • Starter mount bolt and nut
   • Starter motor

To install:

7. Install or connect the following:
   • Starter motor
   • Starter mounting bolt and nut. Tighten to 37 ft. lbs. (50 Nm).
   • Positive battery cable to the starter. Tighten the nut to 80 inch lbs. (9 Nm).
   • Left front tire and wheel assembly
8. Carefully lower the vehicle, then connect the negative battery cable.

5.3L Engine

See Figure 30.

1. Before servicing the vehicle, refer to the precautions in the beginning of this section.
2. Remove or disconnect the following:
   • Negative battery cable

   • Catalytic converter
   • Engine shield bolts and shield
   • Right transmission cover bolt
   • Starter bolts
   • Transmission cover and shield, after repositioning the starter
3. Position the starter down, with the terminals facing toward the front of the vehicle.

   • Starter solenoid nut
   • Starter lead from the solenoid stud
   • Starter lead nut
   • Positive cable from the starter stud
   • Starter

To install:

4. Install or connect the following:
   • Starter in the vehicle. Position the starter down , with the terminals facing toward the front of the vehicle.
   • Positive cable to the starter stud.
   • Starter lead nut and tighten to 80 inch lbs. (9 Nm)
   • Starter solenoid lead to the stud
   • Starter solenoid nut and tighten to 30 inch lbs. (3.4 Nm)
   • Install the shield and transmission cover, after repositioning the starter
5. Slide the starter rearward.
   • Starter bolts and tighten to 37 ft. lbs. (50 Nm)
   • Right transmission cover bolt and tighten to 80 inch lbs. (9 Nm)
   • Catalytic converter
   • Negative battery cable
6. Start the vehicle to check for proper operation.

06025-ENV-G06

Fig. 29 Starter mounting—4.2L engine

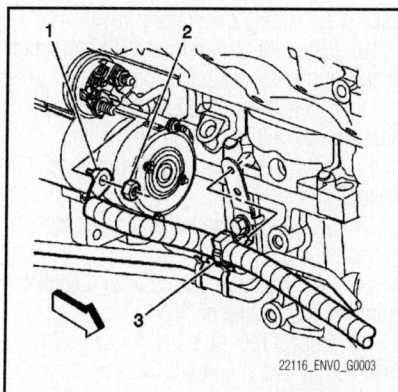

22116_ENVO_G0003

Fig. 30 View of the starter, positive cable (1) and starter lead nut (2)—5.3L engine

ENGINE MECHANICAL

➡Disconnecting the negative battery cable may interfere with the functions of the on board computer systems and may require the computer to undergo a relearning process, once the negative battery cable is reconnected.

ACCESSORY DRIVE BELTS

ACCESSORY BELT ROUTING

See Figures 31 and 32.

INSPECTION

Inspect the drive belt for signs of glazing or cracking. A glazed belt will be perfectly smooth from slippage, while a good belt will have a slight texture of fabric

42372-BLAZ-G01

Fig. 31 Accessory serpentine belt routing—4.2L engines

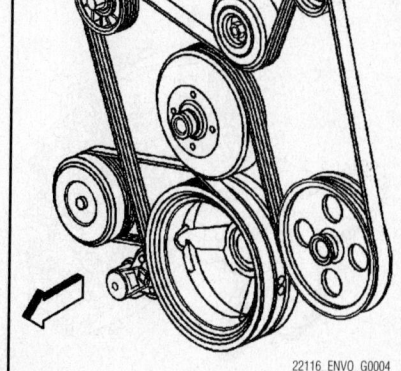

22116_ENVO_G0004

Fig. 32 Accessory drive belt and A/C belt routing—5.3L engines

visible. Cracks will usually start at the inner edge of the belt and run outward. All worn or damaged drive belts should be replaced immediately.

ADJUSTMENT

Serpentine belts are automatically tensioned by a system of idler and tensioner pulleys, thus require no adjustment. The serpentine belt tension can be checked by simply observing the belt acceptable belt wear range indicator located on the tensioner spindle. If the belt does not meet the specified range, it must be replaced.

➡ **A belt is considered "used" after 15 minutes of operation.**

REMOVAL & INSTALLATION

4.2L Engine

See Figure 33.

1. Install ⅜ inch breaker bar on the drive belt tensioner arm and turn the breaker bar clockwise enough to relieve the tension on the drive belt.
2. Remove the drive belt.
3. Release the tension on the tensioner arm.

To install:

4. Route the drive belt over all the pulleys except the drive belt tensioner pulley.
5. Install the ⅜ inch breaker bar on the drive belt tensioner arm and turn the breaker bar clockwise.
6. Install the drive belt over the drive belt tensioner pulley.
7. Slowly release the tension to the drive belt tensioner arm.
8. Inspect for proper installation of the drive belt on the pulleys.

5.3L Engine

ACCESSORY DRIVE BELT

See Figure 34.

Top of Form

1. Remove the air cleaner outlet duct.
2. Install a breaker bar with hex-head socket to the drive belt tensioner bolt.
3. Rotate the drive belt tensioner clockwise in order to relieve tension on the belt.
4. Remove the belt from the generator pulley.
5. Slowly release the tension on the drive belt tensioner.
6. Remove the breaker bar and socket and from the drive belt tensioner bolt.
7. Remove the belt from the remaining pulleys.
8. Clean and inspect the belt surfaces of all the pulleys.

To install:

9. Route the drive belt around all the pulleys except the generator pulley.
10. Install the breaker bar with hex-head socket to the belt tensioner bolt.
11. Rotate the belt tensioner clockwise in order to relieve the tension on the belt.
12. Install the drive belt on the generator pulley.
13. Slowly release the tension on the belt tensioner.
14. Remove the breaker bar and socket from the belt tensioner bolt.
15. Inspect the drive belt for proper installation and alignment.
16. Install the air cleaner outlet duct.

A/C COMPRESSOR BELT

See Figure 34.

1. Remove the accessory drive belt.
2. Raise the vehicle.
3. Install a ratchet into the square opening of the air conditioning (A/C) belt tensioner.
4. Rotate the A/C belt tensioner clockwise in order to relieve tension on the belt.
5. Remove the A/C belt from the pulleys.
6. Slowly release the tension on the A/C belt tensioner.
7. Remove the ratchet from the A/C belt tensioner.
8. Clean and inspect the belt surfaces of all the pulleys.

To install:

9. Install the A/C belt around the crankshaft balancer.
10. Install a ratchet into the square opening of the A/C drive belt tensioner.
11. Rotate the A/C belt tensioner clockwise in order to relieve tension on the belt.
12. Install the A/C belt over the idler pulley.
13. Install the A/C belt around the A/C compressor pulley.
14. Slowly release the tension on the A/C belt tensioner.
15. Remove the ratchet from the A/C belt tensioner.
16. Inspect the A/C belt for proper installation and alignment.
17. Lower the vehicle.
18. Install the accessory drive belt.

CAMSHAFT AND VALVE LIFTERS

REMOVAL & INSTALLATION

4.2L Engine

1. Before servicing the vehicle, refer to the precautions in the beginning of this section.
2. Disconnect the negative battery cable.
3. Discharge and recover the refrigerant from the air conditioning system, using the proper equipment.
4. Remove or disconnect the following:
   - Intake manifold
   - A/C line from the oil level indicator tube
   - A/C line from the accumulator
   - A/C bracket bolt from the engine lift hook
   - Engine lift bracket
   - Ignition control module electrical connectors
   - Ignition control module bolts and module

✳✳ WARNING

Be careful not to damage the clips that hold the harness housing in place.

42372-BLAZ-G01

Fig. 33 Accessory serpentine belt routing—4.2L engines

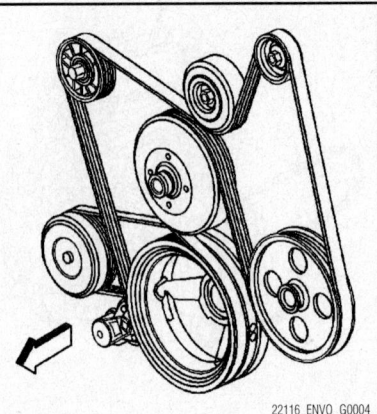

22116_ENVO_G0004

Fig. 34 Accessory drive belt and A/C belt routing—5.3L engine

- Engine electrical harness housing from the camshaft cover
- Fuel injection harness electrical connector
- Camshaft cover bolts and cover
- Exhaust and intake sprocket bolts

5. Install a suitable sprocket holding tool onto the cylinder head and adjust the horizontal bolts into the camshaft sprockets to maintain timing chain tension and avoid disturbing the timing chain components.

6. Carefully move the sprockets with the timing chain off of the camshafts.

➡ **Make sure to place the camshaft caps in a rack to keep them in order, so they may be installed in their original locations.**

7. Remove or disconnect the following:
- Camshaft cap bolts and caps
- Camshafts

To install:

8. Coat the camshaft journals with engine oil.
- Camshafts, in their original position
- Camshaft caps, in their original locations. Tighten the bolts to 106 inch lbs. (12 Nm).

9. Carefully place the camshaft sprockets back onto the camshafts and remove the holding tool.

10. Install or connect the following:
- Intake camshaft sprocket washer and bolt and the exhaust camshaft actuator bolt. Tighten the intake camshaft sprocket bolt to 22 ft. lbs. (30 Nm), plus an additional 135 degrees and the exhaust camshaft actuator bolt to 18 ft. lbs. (25 Nm), plus an additional 135 degrees.
- New camshaft cover seal
- New rubber ignition control module seals
- Camshaft cover and bolts. Tighten the bolts to 89 inch lbs. (10 Nm).
- Ignition control module. Tighten the bolts to 89 inch lbs. (10 Nm).
- Ignition control module electrical connectors
- Fuel injector electrical connectors
- Engine electrical harness housing
- A/C line bracket to the oil level indicator tube stud and secure with the nut. Tighten the nut to 62 inch lbs. (7 Nm).
- Engine lift bracket and secure the lift hook with the bolts. Tighten the bolts to 37 ft. lbs. (50 Nm).

- A/C line bracket to the engine lift bracket. Tighten the bolt to 89 inch lbs. (10 Nm).
- Intake manifold

11. Using the proper equipment, recharge the A/C system.

5.3L Engine

See Figures 35 through 37.

1. Before servicing the vehicle, refer to the precautions in the beginning of this section.

2. Disconnect the negative battery cable.

3. Discharge and recover the refrigerant from the air conditioning system, using the proper equipment.

4. Remove or disconnect the following:
- Condenser
- Cylinder head and gasket
- Valve lifter guide bolts
- Valve lifters and guide

➡ **If the lifters are stuck in the bores due to built up deposits, use Valve Lifter Remover tool No. J 3049-A or equivalent to remove the lifters**

- Valve lifters from the guide

➡ **Make sure to keep the lifters in order as you are removing them. They must be installed in their original locations.**

5. Clean and inspect the lifters for damage.
- Camshaft sensor bolt and sensor

6. Rotate the crankshaft until the timing marks on the crankshaft and camshaft sprockets are aligned.
- Camshaft sprocket bolts

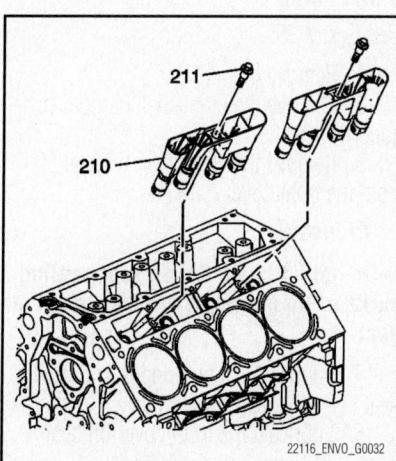

Fig. 35 Valve lifters and guides—5.3L engine

Do NOT turn the crankshaft after the timing chain has been removed to avoid damaging the pistons or valves!

- Camshaft sprocket and reposition the timing chain
- Camshaft retaining bolts and retainer
- Camshaft by installing three M8-1.25 x 4.0 in. (M8-1.25 x 1.00mm) bolts in the front of the camshaft to act as a handle; then, remove the camshaft while turning slightly from side to side, as necessary. Remove the bolts from the camshaft.

➡ **Take care not to damage the camshaft bearings when removing the camshaft.**

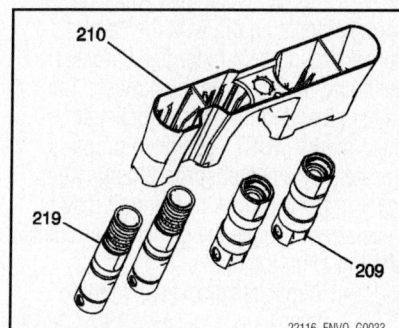

Fig. 36 Remove the lifters from the guides, making sure to keep them in order—5.3L engine

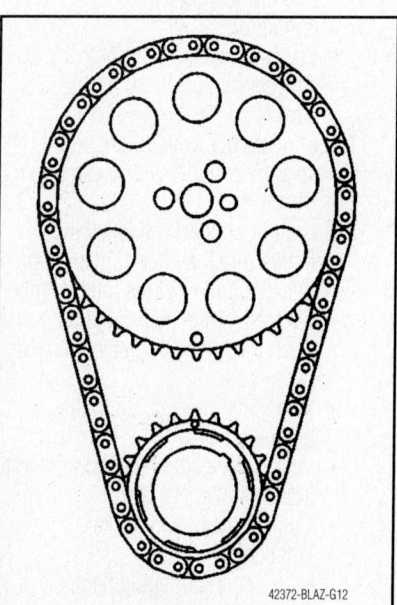

Fig. 37 Make sure the crankshaft and camshaft timing marks are aligned

7. Clean and inspect the camshaft and bearings.

To install:

➡**If the camshaft must be replaced, you must also replace the lifters.**

8. Lubricate the camshaft journals with clean engine oil.

9. Install or connect the following:
- Three bolts used during removal into the bolt hold in the front of the camshaft
- Camshaft carefully into the engine block, using the bolts as a handle. Remove the bolts.
- Camshaft retainer and bolts. Make sure the retaining plate is installed with the sealing gasket facing the engine block. Tighten the bolts to 18 ft. lbs. (25 Nm).

10. Properly locate the camshaft sprocket locating pin with the cam sprocket alignment hole. The sprocket teeth and timing chain must mesh. The camshaft and crankshaft sprocket alignment marks MUST be aligned properly. Locate the camshaft sprocket alignment mark in the 6 o'clock position. It may be necessary to rotate the camshaft or crankshaft to align the marks.
- Camshaft sprocket and timing chain
- Camshaft sprocket bolts and tighten to 26 ft. lbs. (35 Nm)
- Camshaft sensor O-ring, after making sure it is not damaged and lubricating it with clean engine oil
- Camshaft sensor and bolt. Torque the bolt to 18 ft. lbs. (25 Nm).

11. Lubricate the valve lifters and engine block lifter bores with clean engine oil.

12. Install or connect the following:
- Lifters into the lifter guides. Align the area on top of the lifter with the flat area in the lifter guide bore. Push the lifter completely into the guide bore.
- Valve lifters and guide to the engine block
- Valve lifter guide bolt and tighten to 106 inch lbs. (12 Nm)
- Cylinder head and gasket
- Condenser

13. Using the proper equipment, recharge the A/C system.

CRANKSHAFT FRONT SEAL

REMOVAL & INSTALLATION

4.2L Engine

See Figure 38.

➡**Do not damage the engine front cover or the crankshaft.**

1. Remove the crankshaft damper (balancer).

2. Pry out the crankshaft front oil seal using a suitable tool. Use the provided slots for prying out the seal.

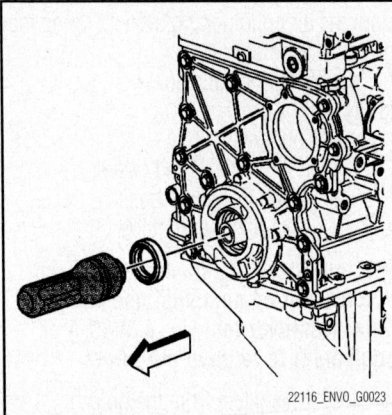

Fig. 38 Using the proper tool to install the crankshaft front oil seal—4.2L engine

To install:

3. Apply the engine oil to the outside diameter of the crankshaft front oil seal.

4. Use the special tool J44218 to install the front oil seal. Remove the J44218.

5. Install the crankshaft damper.

5.3L Engine

See Figure 39.

1. Remove the radiator.

2. Remove the crankshaft damper (balancer).

3. Remove the crankshaft oil seal (1) from the front cover.

To install:

➡**Do not lubricate the oil seal sealing surface. Do not reuse the crankshaft oil seal.**

4. Lubricate the outer edge of the oil seal (1) with clean engine oil.

5. Lubricate the front cover oil seal bore with clean engine oil.

6. Install the crankshaft front oil seal onto the J41478 guide.

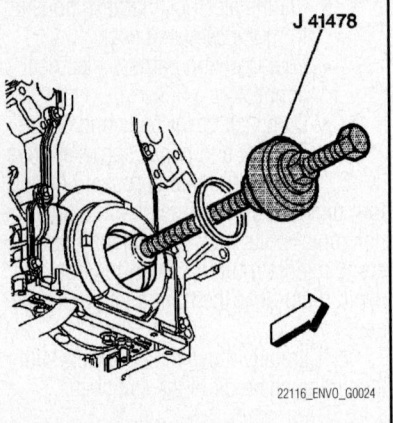

Fig. 39 Using the proper tool to install the crankshaft front oil seal—5.3L engine

7. Install J41478 threaded rod with nut, washer, guide, and oil seal into the end of the crankshaft.

8. Use J41478 in order to install the oil seal into the cover bore.

a. Use a wrench and hold the hex on the installer bolt.

b. Use a second wrench and rotate the installer nut clockwise until the seal bottoms in the cover bore.

c. Remove J41478.

d. Inspect the oil seal for proper installation.

9. The oil seal should be installed evenly and completely into the front cover bore.

10. Install the crankshaft damper.

11. Install the radiator.

CYLINDER HEAD

REMOVAL & INSTALLATION

4.2L Engine

See Figure 40.

1. Before servicing the vehicle, refer to the precautions in the beginning of this section.

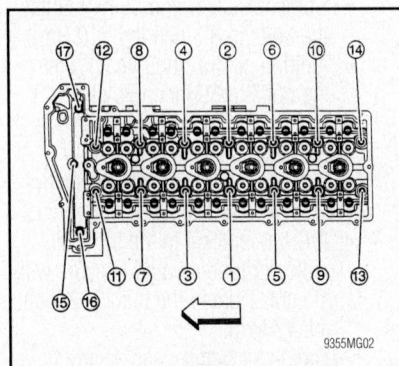

Fig. 40 Cylinder head bolt tightening sequence—4.2L engine

2. Disconnect the negative battery cable.

3. Drain the engine cooling system.

4. Remove or disconnect the following:
- Camshaft cover
- Exhaust manifold
- Front cover
- Cylinder head access hole plugs
- Timing chain tensioner shoe bolt and shoe
- Timing chain tensioner guide bolts and guide
- Timing chain and sprockets

5. Unfasten the cylinder head bolts by loosening them in the reverse of the torque sequence, then carefully remove the cylinder head.

6. Remove the cylinder head gasket.

To install:

7. Carefully clean and inspect the cylinder head and the gasket mounting surfaces.

➡The gasket surfaces on both the head and block must be clean of any foreign matter and free of nicks or heavy scratches. The cylinder bolt threads in the block and thread on the bolts must be cleaned (dirt will affect the bolt torque).

➡DO NOT apply sealer to composition steel-asbestos gaskets.

❋❋ WARNING

Make sure the number 1 cylinder is at Top Dead Center (TDC).

8. If using a steel only gasket, apply a thin and even coat of sealer to both sides of the gaskets.

9. Place a new gasket over the dowel pins with the bead or the words "This Side Up" facing upwards (as applicable), then carefully lower the cylinder head into position over the gasket and dowels.

10. Apply a coating of 12345493 or equivalent sealer to the threads of the cylinder head bolts, then thread the bolts into position until finger-tight.

11. Tighten the cylinder head bolts in sequence as follows:
   a. Tighten the long bolts (1-14), in sequence, to 30 ft. lbs. (40 Nm).
   b. Tighten the long bolts, in sequence, an additional 90 degrees.
   c. Tighten the long bolts, in sequence, an additional 60 degrees.
   d. Tighten the 2 long end bolts to 15 ft. lbs. (20 Nm).
   e. Tighten the 1 short end bolt to 13 ft. lbs. (18 Nm).

12. Install or connect the following:

- Cylinder head access hole plugs and tighten to 44 inch lbs. (5 Nm)
- Timing chain and sprockets
- Front cover
- Camshaft cover
- Exhaust manifold
- Negative battery cable

13. Properly refill the engine cooling system.

14. Run the engine to check for leaks.

5.3L Engine

LEFT SIDE

See Figures 41 through 43.

1. Before servicing the vehicle, refer to the precautions in the beginning of this section.

2. Drain the engine cooling system.

3. Remove or disconnect the following:
- Negative battery cable
- Alternator bracket
- Coolant air bleed pipe
- Left exhaust manifold
- Pushrods
- Auxiliary A/C bracket bolt, if equipped
- Cylinder head bolts. Discard the bolts
- Cylinder head
- Cylinder head gasket and discard

To install:

4. Carefully clean and inspect the cylinder head and the gasket mounting surfaces.

➡The gasket surfaces on both the head and block must be clean of any foreign matter and free of nicks or heavy scratches. The cylinder bolt threads in the block and thread on the bolts must be cleaned (dirt will affect the bolt torque).

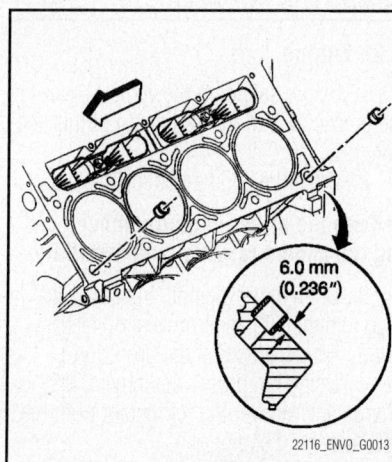

Fig. 41 Make sure the cylinder head locating pins are properly installed—5.3L engine

➡DO NOT apply any type sealer to the cylinder head gasket, unless otherwise specified.

5. Check the cylinder head locating pins for proper installation, location 0.236 in. (6.0mm), as shown.

6. Place a new gasket over the dowel pins. Inspect the displacement markings on the gasket for proper usage. When installed properly, the word "FRONT" on the left side, the tab on the gasket should be left of center or closer to the front of the engine.

7. Install or connect the following:
- Cylinder head

➡You must use new cylinder head bolts during reassembly. Do NOT reuse the old head bolts.

- NEW cylinder head bolts.

8. Tighten the cylinder head bolts in sequence as follows:
   a. Tighten the M11 bolts to 22 ft. lbs. (30 Nm).
   b. Tighten the M11 an additional 90 degrees.

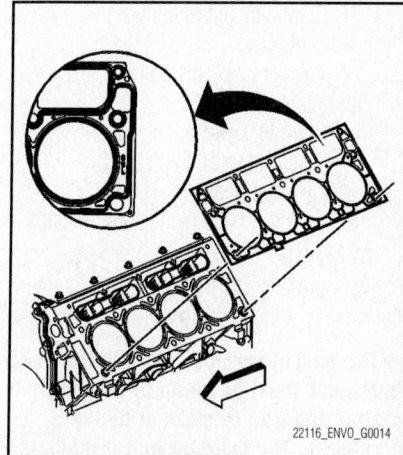

Fig. 42 Proper cylinder head gasket installation—5.3L engine

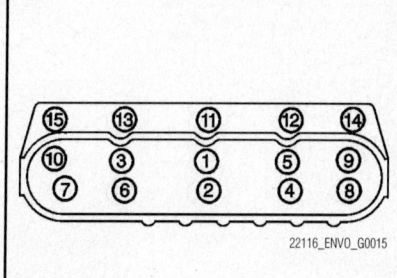

Fig. 43 Cylinder head bolt torque sequence—5.3L engine

c. Tighten M11 bolts, an additional 70 degrees.

d. Tighten the M8 bolts to 22 ft. lbs. (30 Nm). Tighten all the bolts beginning with the center bolt and working outward, alternating sides

9. Install or connect the following:
- Auxiliary A/C bracket, if equipped. Torque the bolt to 15 ft. lbs. (20 Nm).
- Pushrods
- Left exhaust manifold
- Coolant air bleed pipe
- Alternator bracket

10. Properly refill the engine cooling system.

11. Run the engine to check for leaks.

RIGHT SIDE

See Figure 44.

1. Before servicing the vehicle, refer to the precautions in the beginning of this section.

2. Drain the engine cooling system.

3. Remove or disconnect the following:
- Negative battery cable
- Oil level dipstick
- Coolant air bleed pipe
- Right exhaust manifold
- Pushrods
- Auxiliary A/C bracket nut, if equipped
- Cylinder head bolts 1, 2 and 3. Discard the bolts
- Cylinder head
- Cylinder head gasket and discard

To install:

4. Carefully clean and inspect the cylinder head and the gasket mounting surfaces.

➡**The gasket surfaces on both the head and block must be clean of any foreign matter and free of nicks or heavy scratches. The cylinder bolt threads in the block and thread on the bolts must be cleaned (dirt will affect the bolt torque).**

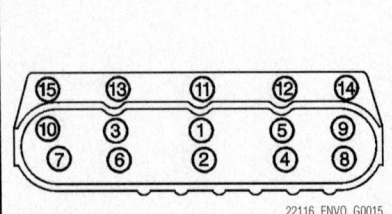

Fig. 44 Cylinder head bolt torque sequence—5.3L engine

➡**DO NOT apply any type sealer to the cylinder head gasket, unless otherwise specified.**

5. Check the cylinder head locating pins for proper installation, location (a) 0.327 in. (8.3mm), as shown.

6. Place a new gasket over the dowel pins. When installed properly, the word "FRONT" on the right side, the tab on the gasket should be right of center or closer.

7. Install or connect the following:
- Cylinder head

➡**You must use new cylinder head bolts during reassembly. Do NOT reuse the old head bolts.**

- NEW cylinder head bolts 1, 2 and 3.

8. Tighten the cylinder head bolts in sequence as follows:

a. Tighten the M11 bolts to 22 ft. lbs. (30 Nm).

b. Tighten the M11 an additional 90 degrees.

c. Tighten M11 bolts, an additional 70 degrees.

d. Tighten the M8 bolts to 22 ft. lbs. (30 Nm). Tighten all the bolts beginning with the center bolt and working outward, alternating sides

9. Install or connect the following:
- Auxiliary A/C bracket, if equipped. Torque the nut to 15 ft. lbs. (20 Nm).
- Pushrods
- Right exhaust manifold
- Coolant air bleed pipe
- Oil level dipstick

10. Properly refill the engine cooling system.

11. Run the engine to check for leaks.

ENGINE ASSEMBLY

REMOVAL & INSTALLATION

4.2L Engine

1. Before servicing the vehicle, refer to the precautions in the beginning of this section.

2. Drain the engine cooling system

➡**Keep the oil drain plug removed during the engine removal and installation.**

3. Drain the engine oil. Install a suitable plug into the oil pan to prevent oil leakage during the remainder of the procedure.

4. Using the proper equipment, discharge and recover the refrigerant from the A/C system, if equipped.

5. Remove or disconnect the following:
- Hood
- Negative battery cable

- Fuel system pressure
- Air cleaner assembly
- Throttle body
- Manifold Absolute Pressure (MAP) sensor
- Windshield washer solvent container
- Air intake baffle
- Grille
- Headlight housing
- Radiator support brace
- Hood
- A/C lines from the condenser
- Transmission cooler lines from the engine, not the radiator

6. Remove the cooling fan and shroud, tilting the radiator forward, and the cooling fan and shroud rearward for clearance.
- Accessory belt
- Power steering pump bolts; position the pump aside
- Heater hoses from the heater core
- Transmission filler tube bracket nut from the Air Injector Reactor (AIR) adapter
- AIR adapter

7. Install a suitable lift hook to the AIR adapter

8. Remove or disconnect the following:
- Oxygen (O$_2$) sensor connector
- A/C line from the accumulator
- Front axle actuator electrical connector
- Camshaft phaser actuator valve electrical connector
- Transmission cooler lines from the clips on the right side of the engine block
- Ignition coil harness connectors
- Harness retainer from the clips
- Power brake hose from the booster
- Powertrain Control Module (PCM)
- Fuel lines from the fuel pressure regulator. Cap the lines to avoid excessive fuel leakage.
- All harnesses from the engine harness bracket
- Engine harness bracket bolt and bracket
- Starter electrical connections
- A/C pressure sensor and clutch electrical connector
- Alternator electrical connector and battery lead
- Knock Sensor (KS), Crankshaft Position (CKP) and Camshaft Position (CMP) sensor electrical connectors
- 4 ground on the left side of the block

9. Raise and safely support the vehicle.
- Left and right side driveshafts
- Propeller shaft from the front axle pinion yoke

- Engine protection shield
- Exhaust pipe from the exhaust manifold. Slide the exhaust pipe backward slightly.
- Fuel tank shield, if equipped
- Torque converter access cover and bolts

10. Place a jack on the transmission fluid pan for support.

11. Remove the transmission support.

12. Lower the transmission enough to reach the top bell housing bolts.

13. Remove the top 4 bell housing bolts, there may be 2 harness clips that will need to be removed in order to have access to 2 of the top bolts.

14. Raise the transmission.

15. Reinstall the transmission support using only 2 through bolts.

16. Remove or disconnect the following:
- Remaining bell housing bolts (11 total)
- Left and right engine lower mount nuts
- Oil level sensor electrical connector
- Oil pressure switch electrical connector

17. Carefully lower the vehicle.

18. Remove the left, then the right upper engine mount nut.

19. Install a suitable engine hoist.

20. Raise the engine out of the compartment slowly, keeping the transmission supported.

21. Remove both engine mounts for clearance.

22. Continue raising the engine out of the vehicle.

23. Place the engine on a suitable engine stand.

To install:

24. Remove the engine from the engine stand.

25. Slowly install the engine into the engine compartment, aligning the engine mounts with the brackets.

26. When the engine mounts are aligned, install the engine mounts, putting the mount up through the engine mount brackets before inserting into the chassis mount brackets.

27. Lower the engine onto the mounts and install the upper engine mounting nuts. Tighten the nuts to 51 ft. lbs. (71 Nm).

28. Remove the engine hoist.

29. Lay the radiator into the radiator support, but do not install the radiator completely.

30. Raise and safely support the vehicle.

31. Install the lower bell housing bolts, except the top four.

32. Remove the 2 through bolts secure the transmission support, then lower the transmission.

33. Install the top 4 bell housing bolts and tighten all 11 bolts to 37 ft. lbs. (50 Nm).

34. Raise the transmission.

35. Install or connect the following:
- Transmission support
- 3 torque converter bolts and tighten to 44 ft. lbs. (60 Nm)
- Torque converter bolt cover
- Fuel tank shield, if equipped
- Engine protection shield
- Propeller shaft to the front axle pinion yoke
- Exhaust pipe to the manifold and tighten the bolts to 37 ft. lbs. (50 Nm)
- Oil level switch and oil pressure sender electrical connectors
- Oil pan drain plug and tighten to 19 ft. lbs. (26 Nm)
- Lower radiator hose
- Left and right wheel driveshafts

36. Lower the vehicle.
- 4 grounds on the left side of the block
- CMP, CKP and knock sensor electrical connectors
- Alternator and starter electrical connectors and battery leads. Torque the nuts to 80 inch lbs. (9 Nm).
- Fuel lines at the fuel pressure regulator
- Engine harness bracket and bolt. Torque the bolt to 37 ft. lbs. (50 Nm).
- Front differential vent hose, to the engine harness bracket
- PCM
- Power brake hose to the booster
- Harness retainer to its original location
- Ignition coil harness connectors
- Transmission cooler lines to clips on right side of engine block
- Camshaft phaser actuator valve electrical connector
- Front axle actuator electrical connector
- A/C line at the accumulator

37. Remove the lift hook.

38. Install or connect the following:
- AIR adapter and secure with the studs. Tighten to 18 ft. lbs. (25 Nm).
- Transmission filler tube bracket to AIR adapter stud and secure the bracket with the nut. Torque the nut to 89 inch lbs. (10 Nm).
- Heater hoses to the heater core
- Power steering pump and tighten the bolts to 18 ft. lbs. (25 Nm).

39. The remainder of installation is the reverse of removal, but please note the following important steps:

40. Connect the negative battery cable

41. Check all powertrain fluid levels and add, as necessary.

42. Refill the engine crankcase.

43. Refill the engine cooling system.

44. Perform the CKP System Variation Learn Procedure, as follows:

   a. Install a suitable scan tool and check for Diagnostic Trouble Codes (DTCs). If any DTCs, other than P1336 are set, resolve those codes first, before proceeding with this procedure.

   b. With the scan tool, select the crankshaft position variation learn procedure.

   c. Observe the fuel cut-off for the 4.2L engine.

   d. The scan tool will instruct you to perform certain steps, make sure you follow all directions given by the scan tool exactly.

   e. Enable the crankshaft position system variation learn procedure.

➡️**While the learn procedure is in progress, release the throttle immediately when the engine started to decelerate. The engine control is returned to the operator and the engine responds to throttle position after the learn procedure is complete.**

   f. Slowly increase the engine speed to the RPM that you observed.

   g. Immediately release the throttle when fuel cut-out is reached.

   h. The scan tool displays: Learn Status: Learned this ignition. If the scan tool does NOT display this message and not other DTCs set, you must perform further troubleshooting.

   i. Turn the ignition **OFF** for 30 seconds after the learn procedure has been completed successfully.

45. Start and run the engine, then check for leaks.

5.3L Engine

See Figure 45.

1. Before servicing the vehicle, refer to the precautions in the beginning of this section.

2. Drain the engine cooling system

3. Drain the engine oil.

4. Remove and recover the refrigerant, if equipped with A/C.

5. Remove or disconnect the following:
- Negative battery cable
- Hood
- Radiator

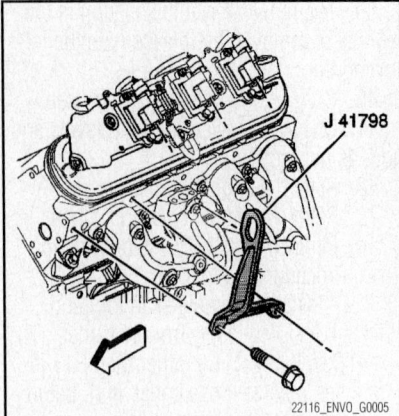

Fig. 45 If necessary, remove ignition coil(s) to install the engine lifting brackets—5.3L engine

- Radiator support brace
- Front axle, if 4WD
- Drive shafts
- Intake manifold
- Oil pressure sensor connector
- Oxygen (O_2) sensor connector
- Camshaft Position (CMP) sensor connector
- A/C compressor hose
- Rear auxiliary A/C compressor pipe fitting
- Rear auxiliary A/C compressor pipe nut and bolt. Tie the pipe out of the way.
- Engine Coolant Temperature (ECT) sensor
- Ground terminal bolt
- Retaining clips from the brackets
- A/C pressure switch electrical connector
- Retaining clip from the cylinder head
- Ground terminal bolts
- Starter
- Battery cable channel bolt
- Battery cable channel from the oil pan
- A/C compressor electrical connector

6. Collect all branches of the engine wiring harness, then position the harness out of the way.
- Alternator cable from the alternator
- Alternator bracket bolts, then position the bracket and alternator assembly aside
- Inlet and outlet hoses from the water outlet, using J 38185 to move the hose clamps
- Auxiliary heater inlet and outlet hose/pipe assembly from the heater water shutoff valve pipes
- Auxiliary heater inlet and outlet hoses/pipes from the water pump, using Hose Clamp Pliers J 38185

- Remove ignition coils, if necessary, to install Engine Lifting Brackets J 41798 to the cylinder heads

7. Install Engine Lifting Brackets J 41798 to the cylinder heads. Tighten the M8 bolts to 18 ft. lbs. (25 Nm) and the M10 bolts to 37 ft. lbs. (50 Nm).
- Catalytic converter
- 3 frame engine mount bracket bolts from the right and left sides
- Torque converter bolts
- Transmission oil level dipstick tube nut and tube
- Transmission bolt and stud on the right side
- Lower transmission bolt/studs
- 3 upper transmission bolts/studs

8. Install a suitable engine hoist to the engine lifting brackets.

9. Place a floor jack under the transmission for support.

10. Separate the engine from the transmission.

11. Remove the engine from the vehicle and place on a suitable engine stand.

12. Install Converter Holding Strap J 21366 to the transmission to hold the torque converter.

To install:

13. Remove Converter Holding Strap J 21366 from the transmission.

14. Attach the engine to a hoist and remove it from the engine stand

15. Install or connect the following:
- Engine into the vehicle. Match the transmission up to the engine, then remove the floor jack.
- 3 upper transmission bolts/studs and tighten to 37 ft. lbs. (50 Nm)
- Lower transmission bolts/studs and tighten to 37 ft. lbs. (50 Nm)
- Transmission bolt and stud on the right side and tighten to 37 ft. lbs. (50 Nm)
- Transmission oil level dipstick tube and nut. Torque to 89 inch lbs. (10 Nm).
- Torque converter bolts and tighten to 44 ft. lbs. (60 Nm)
- 3 frame engine mount bracket bolts to both the right and left sides. Torque the bolts to 37 ft. lbs. (50 Nm).
- Catalytic converter

16. Remove the engine lifting brackets from the cylinder heads
- Ignition coils, if removed, and tighten the bolts to 71 inch lbs. (8 Nm)
- Auxiliary heater inlet and outlet hoses
- Auxiliary heater inlet and outlet hose/pipe assembly to the heater water shutoff valve pipes

- Outlet and inlet hoses to the water outlet
- Bracket and alternator assembly. Tighten the bolts to 37 ft. lbs. (50 Nm).
- Cable to the alternator
- Position the engine wiring harness back over the engine
- A/C compressor electrical connector
- Battery cable channel to the oil pan and secure with the bolt. Torque to 106 inch lbs. (12 Nm).
- Starter
- Ground terminal bolts and tighten to 18 ft. lbs. (25 Nm)
- Retaining clip to the cylinder head
- A/C pressure switch electrical connector
- Retaining clips to the brackets
- Ground terminal bolt and tighten to 18 ft. lbs. (25 Nm)
- ECT sensor connector
- Rear auxiliary A/C compressor pipe nut and bolt. Torque to 15 ft. lbs. (20 Nm).
- A/C compressor hose
- Oil pressure sensor connector
- O_2 sensor connector
- CMP sensor connector
- Intake manifold
- Drive shafts
- Front axle, if removed
- Radiator support brace
- Radiator

17. Recharge the A/C system
- Negative battery cable
- Hood

18. Check all powertrain fluid levels and add, as necessary.

19. Refill the engine crankcase.

20. Refill the engine cooling system.

21. Start and run the engine, then check for leaks.

EXHAUST MANIFOLD

REMOVAL & INSTALLATION

4.2L Engine

See Figure 46.

1. Before servicing the vehicle, refer to the precautions in the beginning of this section.

2. Remove or disconnect the following:
- Negative battery cable

➡**It will be easier if the vehicle is only supported to a height where underhood access is still possible, the vehicle may be left in position for the entire procedure. If the vehicle is raised too high for underhood access, it will have**

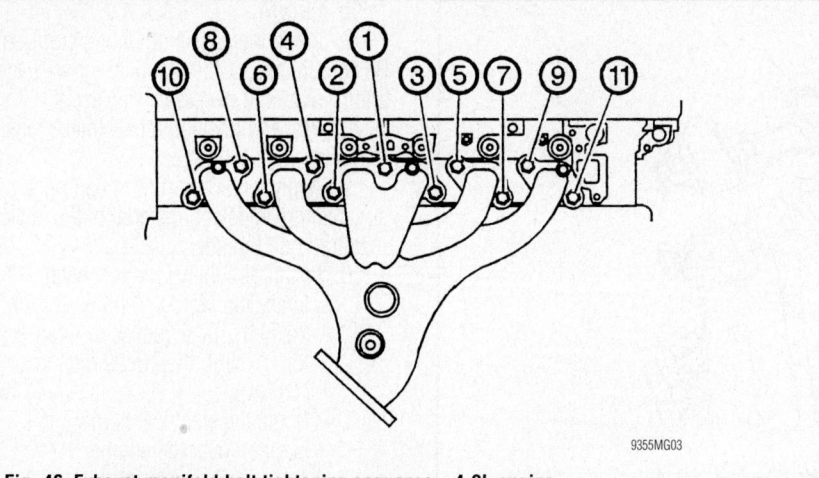

Fig. 46 Exhaust manifold bolt tightening sequence—4.2L engine

9355MG03

to lowered, raised and lowered again during the procedure.

- Air cleaner resonator outlet duct
- Transmission filler tube stud nut from the Air Injector Reactor (AIR) adapter and move the tube aside
- Oil level indicator tube
- Oxygen (O2) sensor from the exhaust manifold
- 4 manifold heat shield nuts and shield
- Exhaust pipe bolts from the exhaust manifold
- Exhaust manifold bolts, and manifold
- Old gaskets and discard

To install:

3. Using a putty knife, clean the gasket mounting surfaces. Inspect the exhaust manifold for distortion, cracks or damage; replace if necessary.

4. Apply a threadlock such as GM 12345493 to the threads of the manifold retainers prior to installation.

5. Install or connect the following:
- Exhaust manifold to the cylinder using a new gasket, then tighten the bolts, in 3 passes, in sequence, to 18 ft. lbs. (25 Nm)
- Heat shield studs, if necessary, and tighten to 89 inch lbs. (10 Nm)
- O2 sensor
- Exhaust manifold heat shield

➡**Apply a suitable anti-seize compound to the exhaust manifold heat shield nuts prior to installation.**

- Heat shield nuts and tighten to 44 inch lbs. (5 Nm)
- Exhaust pipe to the manifold with seal and retaining nuts. Tighten the nuts to 37 ft. lbs. (50 Nm).
- Oil level indicator tube

- Transmission filler tube back onto the AIR adapter block stud and secure with the nut. Tighten the bracket nut to 89 inch lbs. (10 Nm).
- Air cleaner resonator outlet duct
- Negative battery cable.

5.3L Engine

1. Before servicing the vehicle, refer to the precautions in the beginning of this section.

2. Remove or disconnect the following:
- Negative battery cable
- Spark plug wires from the spark plugs. Don't disconnect the wires from the ignition coil unless necessary for clearance.
- Exhaust manifold bolts, manifold and gasket. Discard the gasket.
- Heat shield bolt and shield, if necessary

To install:

3. Apply a 0.2 inch (5mm) bead of threadlock GM P/N 12345493, or equivalent to the threads of the exhaust manifold bolts. Do NOT apply sealer to the first 3 threads of the bolts.

4. Install or connect the following:
- New exhaust manifold gasket
- Exhaust manifold
- Exhaust manifold bolts. Tighten in two passes. First to 11 ft. lbs. (15 Nm), then to 18 ft. lbs. (25 Nm), starting with the center bolts and working outward.

5. Bend over the exposed edge of the gasket at the rear of the cylinder head using a flat punch or equivalent tool.
- Heat shield and bolts, if removed. Torque the bolts to 80 inch lbs. (9 Nm).
- Spark plug wires to the spark plugs
- Negative battery cable

INTAKE MANIFOLD

REMOVAL & INSTALLATION

4.2L Engine

1. Before servicing the vehicle, refer to the precautions in the beginning of this section.

2. Properly relieve the fuel system pressure.

3. Disconnect the negative battery cable.

4. Drain the engine cooling system.

5. Remove or disconnect the following:
- Throttle body
- Powertrain Control Module (PCM)
- All electrical harnesses from the engine harness bracket
- Front differential vent hose from the bracket clip
- Engine harness bracket bolt and bracket
- Manifold Absolute Pressure (MAP) sensor connector
- Crankcase ventilation hose
- Brake hose from the booster
- Alternator
- Intake manifold bolts and manifold.
- Manifold gasket

To install:

6. Clean the gasket mounting surfaces. Be sure to inspect the manifold for warpage and/or cracks. If necessary, replace it.

7. Properly position a new intake manifold gasket.

8. Install or connect the following:
- Intake manifold and bolts. Torque the bolts to 16 ft. lbs. (22 Nm).
- Alternator
- Brake hose to the booster
- Crankcase ventilation hose, lubricating the inner diameter first with 12345884, or equivalent lubricant
- MAP sensor electrical connector
- Engine harness bracket. Tighten the retaining bolt to 37 ft. lbs. (50 Nm).
- Front differential vent hose to the engine harness bracket clip
- All harnesses to their original locations onto the engine harness bracket
- PCM
- Throttle body
- Negative battery cable

9. Refill the engine cooling system.

5.3L Engine

See Figures 47 through 49.

➡**The intake manifold, throttle body, fuel rail and injectors can be removed as an assembly. If you are not servicing these components individually, remove the intake manifold as a complete assembly.**

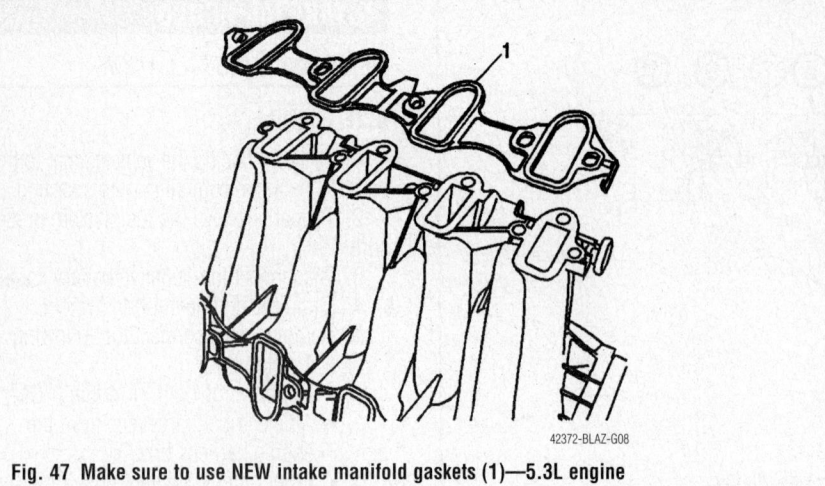

Fig. 47 Make sure to use NEW intake manifold gaskets (1)—5.3L engine

1. Before servicing the vehicle, refer to the precautions in the beginning of this section.

2. Properly relieve the fuel system pressure.

3. Disconnect the negative battery cable.

4. Drain the engine cooling system.

5. Remove or disconnect the following:
- Air cleaner outlet duct
- A/C compressor pressure switch electrical connector
- Harness clip from the cylinder head and fuel rail
- Mass Airflow/Intake Air Temperature sensor connector

6. Disconnect the electrical connectors from the following:
   a. Main coil
   b. Electronic Throttle Control (ETC)
   c. Fuel injectors. Matchmark the connectors, pull the Connector Position Assurance (CPA) retainer up 1 click. Push the tab on the connector in, then detach the injector connector.

- Alternator connector
- Evaporative emission (EVAP) purge solenoid electrical connector
- Knock Sensor (KS) electrical connector
- Main coil
- Fuel injector electrical connector
- Electrical harness clips from the fuel rail
- KS harness electrical connector from the intake manifold
- Positive Crankcase Ventilation (PCV) valve hose and valve
- Heater water shutoff valve actuator inlet hose from the intake manifold
- EVAP purge solenoid vent tube
- Vacuum brake booster hose from the rear of the intake manifold
- Upper engine wire harness retainer nut. Position the wire harness aside.
- Intake manifold bolts
- Intake manifold and gaskets. Discard the gaskets.

To install:

7. Clean the gasket mounting surfaces. Be sure to inspect the manifold for warpage and/or cracks. If necessary, replace it.

8. Properly position a new intake manifold gasket.

9. Apply a 0.20 in. (5mm) band of a suitable threadlocking material to the intake manifold bolt threads.

10. Install or connect the following:
- Intake manifold and bolts. Torque the bolts, in sequence to 44 inch lbs. (5 Nm), then to 89 inch lbs. (10 Nm).
- Route the electrical harness into position over the engine.
- Engine harness bracket nut and tighten to 89 inch lbs. (10 Nm)
- Vacuum brake booster hose to the rear of the intake manifold
- EVAP purge solenoid valve
- Heater water shutoff valve actuator inlet hose to the intake manifold
- PCV valve and hose
- EVAP purge solenoid, KS, MAP sensor, main coil & fuel injector electrical connectors
- Harness clips to the fuel rail
- Alternator electrical connector
- Main coil, ETC, fuel injector electrical connectors
- Electrical harness clips to the fuel rail
- A/C compressor pressure switch electrical connector
- Harness clip to the cylinder head
- Mass Airflow/Intake Air Temperature sensor connector
- Air cleaner outlet duct
- Fuel fill cap
- Negative battery cable

11. Refill the engine cooling system.

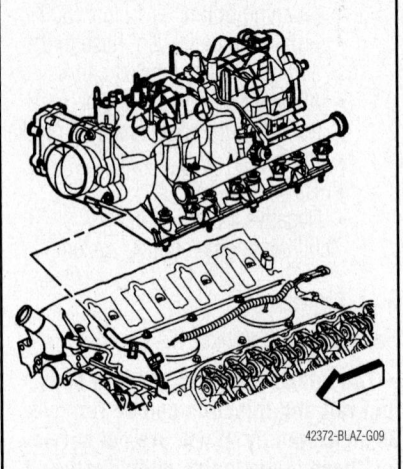

Fig. 48 Exploded view of the intake manifold—5.3L engine

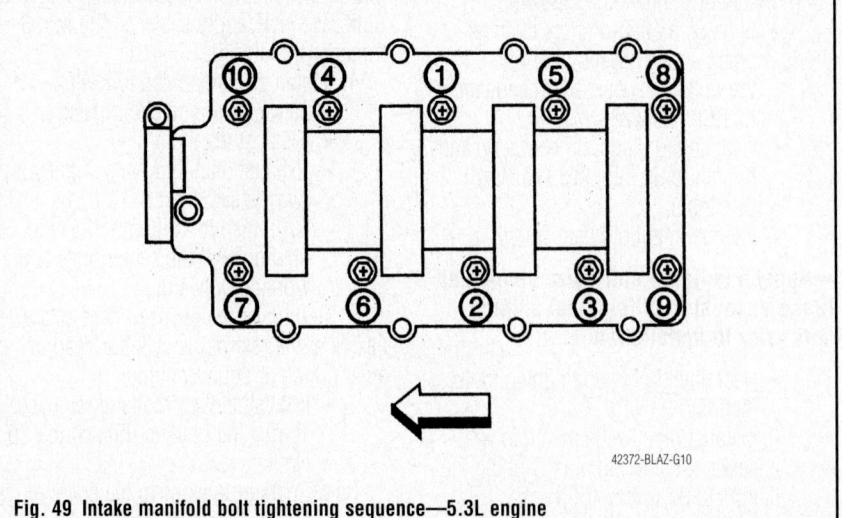

Fig. 49 Intake manifold bolt tightening sequence—5.3L engine

OIL PAN

REMOVAL & INSTALLATION

4.2L Engine

1. Before servicing the vehicle, refer to the precautions in the beginning of this section.

2. Disconnect the negative battery cable.

3. Remove or disconnect the following:
   - A/C compressor bottom bolts and loosen the top bolts
   - Oil dipstick and tube

4. Raise and safely support the vehicle.

5. Drain the engine crankcase oil.

6. Remove or disconnect the following:
   - Left and right front tire and wheel assemblies
   - Engine protection shield mounting bolts and shield
   - Front steering gear crossmember
   - Left and right driveshafts
   - Front drive axle clutch fork assembly
   - Prop shaft from the front axle pinion yoke
   - Unclip the transmission cooler lines from the engine block
   - Front differential bolts and position the differential aside
   - 4 transmission bell housing-to-oil pan bolts
   - Remaining oil pan bolts
   - Oil pan, by placing 2 oil pan bolts in the jack screws on the oil pan and tighten evenly to release the oil pan from the engine

To install:

7. Clean the gasket mounting surfaces.

➡The alignment between the rear of the oil pan and the rear of the block is critical. When the oil pan is installed it could be inadvertently shifted front or back a small amount which could cause a transmission alignment problem. The back to the oil pan needs to be flush with the engine block.

8. Apply a 0.12 in. (3mm) bead of sealant to engine block, rather than the oil pan.

➡The oil pan MUST be installed within 10 minutes of applying the sealant to the engine block.

9. Install or connect the following:
   - Oil pan, maneuvering it to clear the oil pump and screen assembly

➡After the bolts are installed, before tightening them to specifications, check the oil pan alignment. Use a straight edge on the back to the block and the oil pan transmission mounting surface.

   - Oil pan bolts; tighten the side bolts to 18 ft. lbs. (25 Nm) and the end bolts to 89 inch lbs. (10 Nm)
   - Transmission bell housing-to-oil pan bolts and tighten to 35 ft. lbs. (47 Nm)
   - A/C compressor bottom bolts. Tighten to 37 ft. lbs. (50 Nm)
   - Front differential bolts and tighten to 63 ft. lbs. (85 Nm)
   - Front drive axle and clutch fork assembly
   - Transmission cooler lines to block
   - Prop shaft to front differential
   - Steering gear crossmember
   - Left and right driveshaft
   - Oil pan drain plug. Tighten to 19 ft. lbs. (26 Nm)
   - Engine protection shield. Tighten the bolts to 18 ft. lbs. (25 Nm)
   - Left and right front wheel and tire assemblies

10. Carefully lower the vehicle.

11. Refill the crankcase with fresh oil. Start the engine, establish normal operating temperatures and check for leaks.

5.3L Engine

See Figures 50 through 52.

1. Before servicing the vehicle, refer to the precautions in the beginning of this section.

2. Disconnect the negative battery cable.

3. Drain the engine crankcase oil and differential oil.

4. Remove or disconnect the following:
   - Oil level dipstick
   - Front shock upper retaining nuts
   - Tires and wheels
   - Engine shield bolts and shield
   - Power steering gear
   - Left and right Antilock Brake System (ABS) wiring harnesses from the retainers
   - Wheel Speed Sensor (WSS) electrical connectors
   - Brake hose retaining bolts from the frame
   - Sway bar link pins from the lower control arm on both sides

5. Place an adjustable jackstand under the lower control arm.
   - Upper ball joint pinch bolt and nut
   - Upper control arm from the upper ball joint

6. Lower and remove the jackstand, letting the suspension hang.

   - Left driveshaft
   - Right driveshaft from the intermediate shaft bearing only. Do not remove the driveshaft from the steering knuckle. Position the driveshaft aside.

7. Using wire or hooks, secure the front shock modules to the frame. Do NOT let the shocks and steering knuckle hang without being supported.

8. Matchmark the position of the propeller shaft to the front axle pinion yoke.

9. Remove or disconnect the following:
   - Yoke retainer bolt and yoke retainers from the front axle pinion yoke. Wrap the bearing caps with tap to avoid losing the bearing rollers. Secure the propeller shaft to the frame.
   - Transmission oil cooler lines from the retainer
   - Transmission oil cooler line retaining bracket bolt and bracket

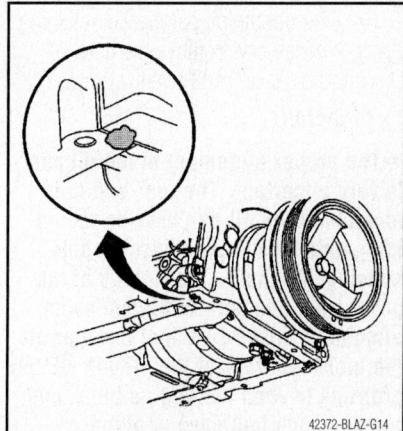

Fig. 50 Proper sealant application to the front cover gasket

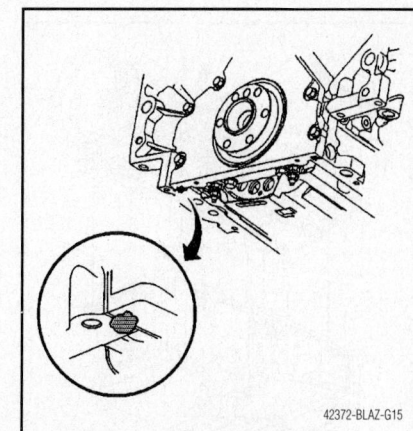

Fig. 51 Proper sealant application to the rear cover gasket

- Inner axle shaft
- Starter
- Flywheel inspection cover from the left side of the transmission
- Battery cable channel bolt from the front of the oil pan
- Battery cable channel from the oil pan
- Loosen the 2 upper A/C compressor bracket bolts
- 2 lower A/C compressor bracket bolts
- Front differential attachment bolts. Secure the front differential to the frame.
- 2 lower bellhousing bolts
- Oil pan bolts
- Oil pan by tilting the rear of the oil pan down to clear the transmission, pull the oil pan rearward past the front wire harness, then lower the oil pan clear of the vehicle

➡ **The oil pan gasket is reusable if it is not damaged.**

10. Drill out the oil pan gasket retaining rivets, if necessary. Remove the gaskets. Discard the gaskets and rivets.

To install:

➡ **The proper alignment of the oil pan is very important. The rear bolt hold location of the oil pan provide mounting points for the transmission bellhousing. To ensure the rigidity of the powertrain and correct transmission alignment, make sure that the rear of the block and rear of the oil pan NEVER protrude beyond the engine block and transmission bellhousing plane.**

➡ **If replacing the oil pan gasket, it is not necessary to rivet the NEW gasket to the pan.**

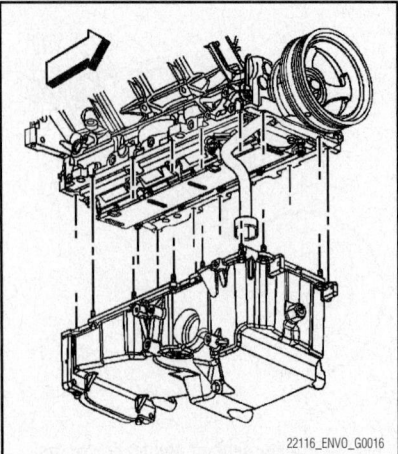

22116_ENVO_G0016

Fig. 52 Oil pan mounting—5.3L engine

11. Apply a 0.20 in. (5mm) bead of sealant 0.80 in. (20mm) long to the engine block. Apply the sealant directly onto the tabs of the front cover gasket that protrudes into the oil pan surface.

12. Apply a 0.20 in. (5mm) bead of sealant 0.80 in. (20mm) long to the engine block. Apply the sealant directly onto the tabs of the rear cover gasket that protrudes into the oil pan surface.

13. Pre-assemble the oil pan gasket and bolts to the pan. Install the gasket onto the pan. Install the oil pan bolts to the pan and through the gasket.

14. Install the oil pan, oil pan gasket and bolts to the engine block as an assembly.

15. Hand-start the bolts into the engine block snug-tight. Do not fully tighten yet.

16. Install the 2 lower bellhousing bolts and tighten to 37 ft. lbs. (50 Nm).

17. Tighten the 2 rear oil pan-to-rear cover bolts to 106 inch lbs. (12 Nm) and the remaining oil pan bolts to 18 ft. lbs. (25 Nm).

18. Release the differential from the frame and install to the oil pan. Install and tighten the bolts to 63 ft. lbs. (85 Nm).

19. Install or connect the following:
- 2 lower A/C compressor bracket bolts. Tighten the lower and upper compressor bolts to 37 ft. lbs. (50 Nm).
- Battery cable channel to the oil pan
- Battery cable channel bolt and tighten to 106 inch lbs. (12 Nm)
- Flywheel inspection cover to the left side of the transmission
- Starter
- Inner axle shaft
- Transmission oil cooler line retaining bracket and bolt. Torque the bolt to 80 inch lbs. (9 Nm).
- Transmission oil cooler lines to the retainer

20. Unhook the right driveshaft from the frame.
- Left and right driveshafts

21. Unsecure the shocks from the frame. Put adjustable jackstand under the lower control arm. Using the jackstand, raise the lower control arm and knuckle assembly in order to connect the upper ball joint to the upper control arm.
- Upper ball joint pinch nut and bolt and tighten to 30 ft. lbs. (40 Nm). Remove the jackstand.
- Sway bar link pins to the lower control arm on both sides
- Steering gear

22. Unsecure the prop shaft from the frame. Align the matchmarks on the prop shaft to the marks on the front axle pinion yoke.

- Propeller shaft to the front axle pinion yoke
- Yoke retainers and yoke retainer bolts to the front axle pinion yoke. Torque the bolts to 15 ft. lbs. (20 Nm).
- Brake hose retaining bolts to the frame and tighten to 18 ft. lbs. (25 Nm).
- WSS electrical connectors
- Left and right ABS wiring harnesses to the retainers
- Differential with oil
- Engine shield and bolts. Tighten the bolts to 18 ft. lbs. (25 Nm).
- Tires and wheels

23. Fill the engine with oil. Fill the power steering system with fluid.
- Upper shock nuts and tighten to 74 ft. lbs. (100 Nm).
- Oil dipstick
- Negative battery cable

OIL PUMP

REMOVAL & INSTALLATION

4.2L Engine

1. Before servicing the vehicle, refer to the precautions in the beginning of this section.

2. Remove or disconnect the following:
- Engine front cover
- Oil pump cover bolts
- Oil pump cover. Mark the inner and outer gears in relation to the pump housing.
- Inner and outer pump gears
- Oil pump pressure relief valve plug
- Oil pump pressure relief valve and spring

To install:

3. Install or connect the following:
- Oil pump pressure relief valve and spring
- Oil pump pressure relief valve plug. Tighten to 10 ft. lbs. (14 Nm).
- Oil pump outer and inner gears, as marked during removal
- Oil pump cover and bolts. Tighten the bolts to 89 inch lbs. (10 Nm).
- Front cover

5.3L Engine

See Figure 53.

1. Before servicing the vehicle, refer to the precautions in the beginning of this section.

2. Remove or disconnect the following:
- Oil pan
- Engine front cover

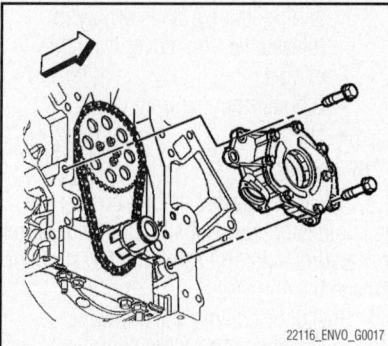

Fig. 53 Exploded view of the oil pump mounting—5.3L engine

- Oil pump screen bolt and nuts
- Oil pump screen with O-ring seal
- O-ring seal from the pump screen. Discard the O-ring seal.
- Remaining crankshaft oil deflector nuts
- Crankshaft oil deflector
- Oil pump bolts
- Oil pump

➡ **Do not let any dirt or debris into the oil pump or cap end.**

- Clean and inspect the oil pump.

To install:

3. Align the splined surfaces of the crankshaft sprocket and the oil pump drive gear and install the oil pump.

4. Install or connect the following:
- Oil pump onto the crankshaft sprocket until the pump housing contacts the face of the engine block
- Oil pump bolts and tighten to 18 ft. lbs. (25 Nm)
- Crankshaft oil deflector and nuts until snug
- New oil pump screen O-ring seal into the oil pump screen, after lubricating with clean engine oil

➡ **Push the oil pump screen tube completely into the oil pump prior to tightening the bolt. Do not let the bolt pull the tube into the pump.**

5. Align the oil pump screen mounting brackets with the correct crankshaft bearing cap studs.
- Oil pump screen
- Oil pump screen bolts and nuts. Tighten the bolts to 106 inch lbs. (12 Nm) and the nuts to 18 ft. lbs. (25 Nm).
- Engine front cover
- Oil pan

PISTON AND RING

POSITIONING

See Figures 59 and 60.

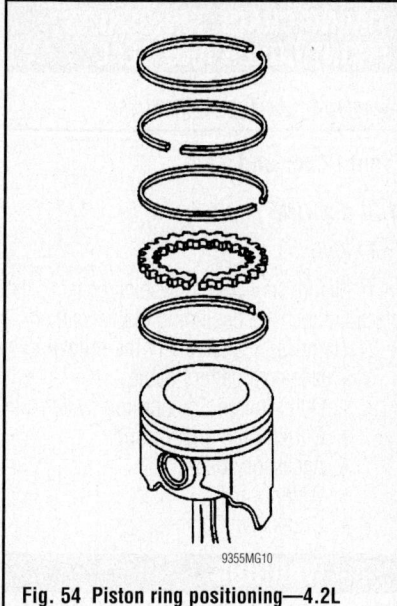

Fig. 54 Piston ring positioning—4.2L engine

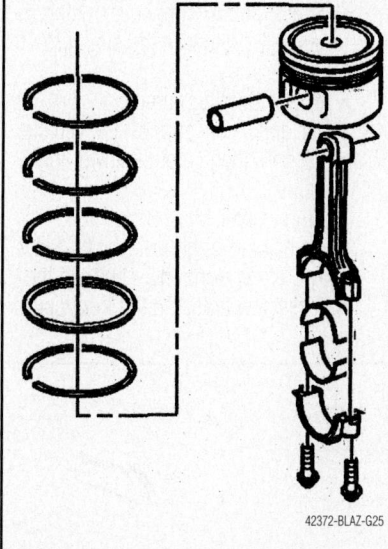

Fig. 55 Piston ring positioning—5.3L engine

REAR MAIN SEAL

REMOVAL & INSTALLATION

4.2L Engine

See Figure 56.

Please note that the transmission assembly must be removed to perform this procedure.

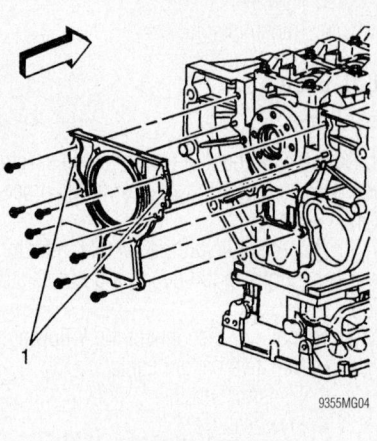

Fig. 56 Install 2 bolts into the jackscrew holes (1) to push the cover off of the block

1. Before servicing the vehicle, refer to the precautions in the beginning of this section.
2. Remove or disconnect the following:
- Negative battery cable
- Transmission
- Flywheel
- Crankshaft rear main seal housing bolts. Install 2 bolts into the jackscrew holes to release the cover from the block
- Crankshaft and rear main seal housing
- Rear main seal from the crankshaft snout

To install:

3. Install or connect the following:
- Rear main seal, using a suitable seal installation tool, then remove the tool
- Apply a 0.12 in. (3mm) bead of 12378521, or equivalent sealant to the rear mail seal housing
- Suitable cover alignment pins into the block

➡ **When you install a new seal, make sure to use the plastic installation sleeve supplies with the new seal. The sleeve should come off and be discarded after the seal is installed.**

4. Slide the crankshaft rear main seal housing over the alignment pins and crankshaft.

5. Install the crankshaft rear main seal housing bolts, except the 2 in place of the guide pins.

6. Remove the guide pins.

7. Install or connect the following:
- Remaining 2 crankshaft rear main seal housing bolts and tighten to 89 inch lbs. (10 Nm). Wipe off any excess sealant.

- Flywheel
- Transmission

5.3L Engine

See Figure 57.

Please note that the transmission assembly must be removed to perform this procedure.

1. Before servicing the vehicle, refer to the precautions in the beginning of this section.
2. Remove or disconnect the following:
   - Negative battery cable
   - Transmission
   - Flywheel
   - Crankshaft rear main oil seal from the rear cover

To install:

➡**The flywheel spacer (if applicable) must be removed prior to oil seal installation. Do not lubricate the oil seal Inside Diameter (ID) or crankshaft surface. Never reuse the rear main seal. Once it is removed, it must be replaced with a new seal.**

3. Lubricate the Outside Diameter (OD) of the rear main seal and the rear cover oil seal bore with clean engine oil. Do NOT let oil contact the seal surface or the crankshaft surface.
4. Install or connect the following:
   - Crankshaft Rear Oil Seal Installer Tool No. J 41479 tapered cone and bolts onto the rear of the crankshaft. Tighten the bolts until just snug, being careful not to overtighten.
   - Rear oil seal onto the tapered cone until the tool contacts the oil seal

5. Align the oil seal into the tool, Rotate the handle of the tool clockwise until the seal enters the rear cover and bottoms into the cover bore. Remove the tool.

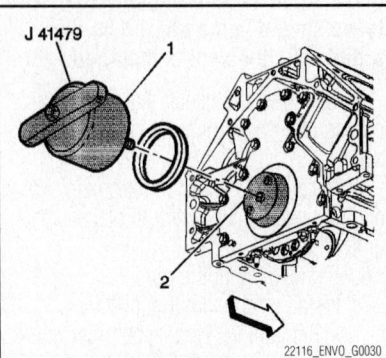

J 41479

Fig. 57 View of the rear main seal installation—5.3L engine

22116_ENV0_G0030

- Flywheel
- Transmission
- Negative battery cable

6. Start the engine and verify no oil leaks.

TIMING CHAIN, SPROCKETS, FRONT COVER AND SEAL

REMOVAL & INSTALLATION

Front Cover and Seal

4.2L ENGINE

See Figure 58.

1. Before servicing the vehicle, refer to the precautions in the beginning of this section.
2. Remove or disconnect the following:
   - Negative battery cable
   - Drain the engine cooling system.
   - Cooling fan and shroud
   - Accessory belt
   - Water pump
   - Crankshaft balancer

✼✼ WARNING

When removing the seal, be careful not to damage the front cover or crankshaft.

- Seal from the front cover, using a suitable prytool in the slots provided
- Power steering pump
3. Raise and safely support the vehicle.
   - Oil pan, then carefully lower the vehicle
   - 7mm center bolt
   - Remaining front cover bolts. Place two of the front cover bolts in the jackscrew holes on the front cover

and tighten the bolts evenly to release the front cover from the engine.
   - 2 bolts from the front cover
   - Oil pump

To install:

4. Clean the gasket mating surfaces of the engine and cover of all remaining gasket or sealer material. Be careful not to score or damage the surfaces.
5. Install or connect the following:
   - Suitable cover alignment pins, onto the engine

➡**The front cover MUST be installed within 10 minutes of applying the sealant.**

- Apply a 0.12 in. (3mm) beat of 12378521 or equivalent sealant to the trace grooves on the back side of the engine front cover. Apply sealant on the inside 3 bolt hole bosses on the cover also.
- Oil pump to the crankshaft splines
- Front cover and bolts, tighten the center bolt last. Tighten to 89 inch lbs. (10 Nm).

6. Remove the alignment pins and raise and safely support the vehicle. Install the oil pan, then lower the vehicle.
   - Power steering pump
   - Crankshaft balancer
   - Water pump
   - Accessory belt
   - Cooling fan and shroud
   - Negative battery cable

7. Properly refill the engine cooling system.
8. Run the engine until normal operating temperature has been reached, then check for leaks.

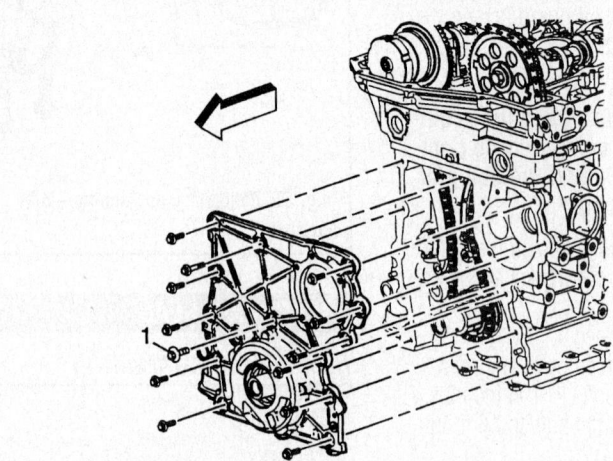

9355MG06

Fig. 58 Place 2 front cover bolts in the jackscrew holes on the cover and tighten to push the cover off of the engine

5.3L ENGINE

See Figures 59 and 60.

1. Before servicing the vehicle, refer to the precautions in the beginning of this section.
2. Properly discharge the A/C system.
3. Drain the engine cooling system.
4. Remove or disconnect the following:
   - Negative battery cable
   - A/C compressor and bracket
   - Water pump
   - Crankshaft balancer
   - Oil pan-to-front cover bolts
   - Front cover bolts
   - Front cover and gasket. Discard the gasket.
5. Clean and inspect the front cover.

To install:

6. Apply a 0.20 in. (5mm) bead of

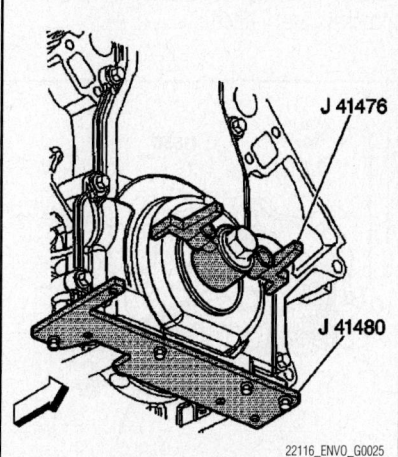

Fig. 59 Align the tapered legs of the tool with the machined alignment surfaces on the front cover—5.3L engine

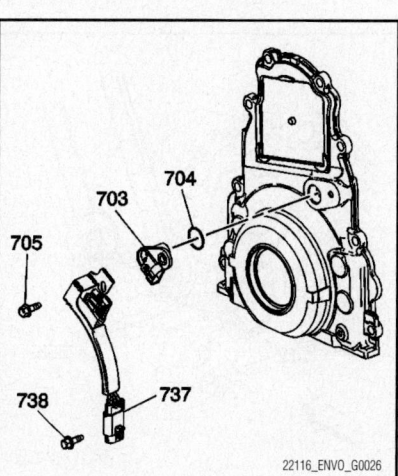

Fig. 60 Front cover seal installation using the proper tool—5.3L engine

sealant 0.80 in. (20mm) long to the oil pan-to-engine block junction.

7. Install or connect the following:
   - New front cover gasket and cover
   - Front cover bolts, finger-tight
   - Oil pan-to-front cover bolts, finger-tight
   - Front and Rear Cover Alignment Tool No. J 41476 to the front cover. Align the tapered legs of the tool with the machined alignment surfaces on the front cover
   - Crankshaft balancer bolt, finger-tight
   - Oil pan-to-front cover bolts to 18 ft. lbs. (25 Nm)
   - Front cover bolts to 18 ft. lbs. (25 Nm)
8. Remove the tool.
9. Install a NEW crankshaft front oil seal as follows:

   a. Remove the radiator for access.
   b. Remove the crankshaft balancer.
   c. Remove the crankshaft oil seal.
   d. Lubricate the outer edge ONLY of the NEW crankshaft oil seal with clean engine oil.
   e. Install the crankshaft front oil seal into the Crankshaft Front Seal Installation Tool No. J 41478 guide.
   f. Install the J 41478 threaded rod (with nut, washer, guide and oil seal) into the end of the crankshaft.
   g. Use J 41478 to install the oil seal into the cover bore. Use a wrench and hold the hex on the installer bolt. Use a second wrench to rotate the installer nut clockwise until the seal bottoms in the cover bore. Remove the tool.
   h. Check the seal for proper installation. It should be installed evenly

and completely into the front cover bore.

   i. Install the crankshaft balancer. Tighten the bolt to 37 ft. lbs. (50 Nm), plus an additional 140 degrees using a torque angle meter.
   j. Install the radiator.
10. Install or connect the following:
    - Water pump
    - A/C compressor and bracket
    - Cooling system with coolant
    - Negative battery cable
11. Properly recharge the A/C system

Timing Chain and Sprockets

4.2L ENGINE

See Figures 61 through 63.

➡ **The following procedure requires the use of the Crankshaft Holding tool No. J-44221 and a suitable torque angle meter.**

1. Before servicing the vehicle, refer to the precautions in the beginning of this section.
2. Remove or disconnect the following:
   - Camshaft cover
   - Timing chain (front) cover
   - Tension on the timing chain by moving the tensioner shoe in. Place a tee into the tension to hold the shoe in place.
   - Top chain guide bolts and guide
   - Exhaust camshaft position actuator bolt and actuator
   - Intake camshaft sprocket bolt and sprocket
   - Timing chain
   - Crankshaft sprocket
   - Cylinder head access hole plugs

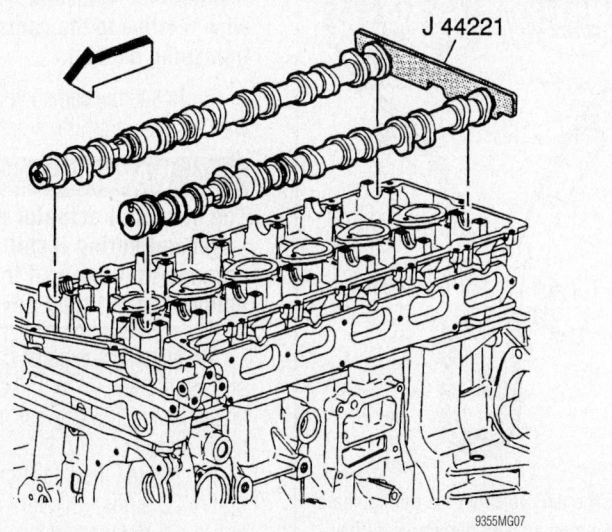

Fig. 61 Proper installation of the crankshaft holding tool with the No. 1 cylinder at TDC

- Timing chain tensioner shoe bolt and shoe
- Timing chain tensioner guide bolts and guide
- Timing chain tensioner bolts and tensioner

To install:

➡**Every seventh link of the timing chain is darkened to help in aligning the timing marks.**

3. Install or connect the following:
   - Timing chain tensioner and bolts. Tighten to 18 ft. lbs. (25 Nm).
   - Timing chain guide and bolts. Tighten to 89 inch lbs. (10 Nm).

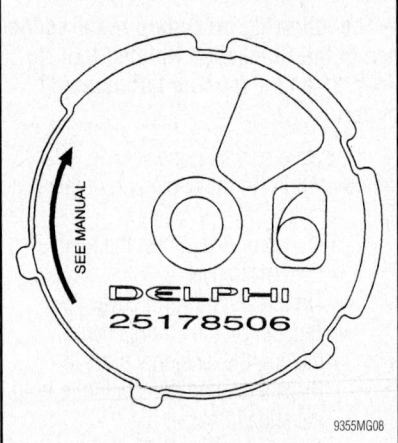

Fig. 62 Rotate the camshaft actuator clockwise

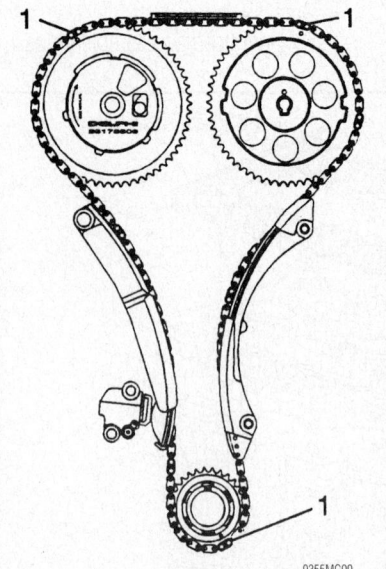

Fig. 63 The dark lines on the timing chain should be aligned with the marks on the sprockets

- Timing chain tensioner shoe and bolt. Tighten to 19 ft. lbs. (26 Nm).
- Cylinder head access hole plugs and tighten to 44 inch lbs. (5 Nm)
- Crankshaft Holding tool No. J-44221, or equivalent with the camshaft flats up and the No. 1 cylinder at Top Dead Center (TDC)
- Crankshaft sprocket
- Intake camshaft sprocket into the timing chain

4. Align the dark link of the timing chain with the timing mark on the intake camshaft sprocket. Feed the timing chain down through the opening in the head.
   - Timing chain onto the crankshaft sprocket. Align the dark link of the timing chain with the timing mark on the crankshaft sprocket.

➡**It may be necessary to remove the crankshaft holding tool to rotate and hold the camshaft hex to align the pin to the camshaft sprocket**

   - Intake camshaft sprocket onto the intake camshaft
   - Intake camshaft washer and bolt
   - Exhaust camshaft actuator into the timing chain. Align the dark link of the timing chain with the timing mark on the exhaust camshaft actuator.

➡**It may be necessary to remove the crankshaft holding tool to rotate and hold the camshaft hex to align the pin to the camshaft sprocket**

   - Exhaust camshaft actuator onto the exhaust camshaft

➡**Rotate the camshaft actuator clockwise relative to the camshaft prior to tightening the bolt.**

5. Rotate the camshaft actuator clockwise (as seen from the front of the vehicle).

✳✳ WARNING

The camshaft actuator must be fully advanced during installation. Engine damage may occur if the camshaft actuator is not fully advanced.

6. Install the exhaust camshaft actuator bolt and tighten to 18 ft. lbs. (25 Nm), plus an additional 135 degrees, using a torque angle meter.
7. Tighten the intake camshaft sprocket bolt to 22 ft. lbs. (30 Nm), plus an additional 135 degrees, using a torque angle meter.

8. Remove the tee from the timing chain tensioner to regain tension on the timing chain.
9. Remove the crankshaft holding tool. The dark lines on the timing chain should be aligned with the marks on the sprockets.
10. Install or connect the following:
    - Top chain guide
    - Suitable threadlock to the top chain guide bolt threads, then install and tighten to 89 inch lbs. (10 Nm)
    - Engine front cover
    - Camshaft cover

5.3L ENGINE

See Figures 64 through 66.

1. Before servicing the vehicle, refer to the precautions in the beginning of this section.
2. Remove the oil pump.
3. Rotate the crankshaft until the timing marks on the crankshaft and the camshaft sprockets are aligned.

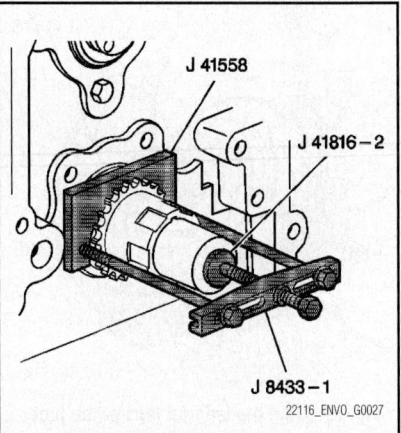

Fig. 64 Use the proper tools to remove the crankshaft sprocket—5.3L engine

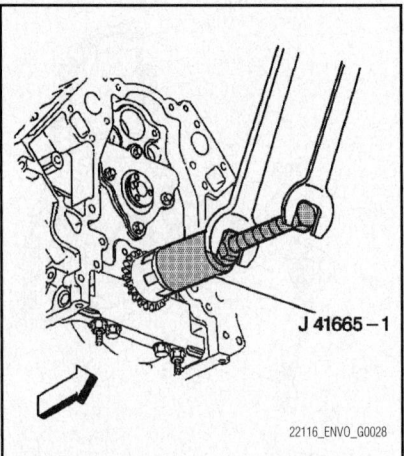

Fig. 65 Crankshaft sprocket installation—5.3L engine

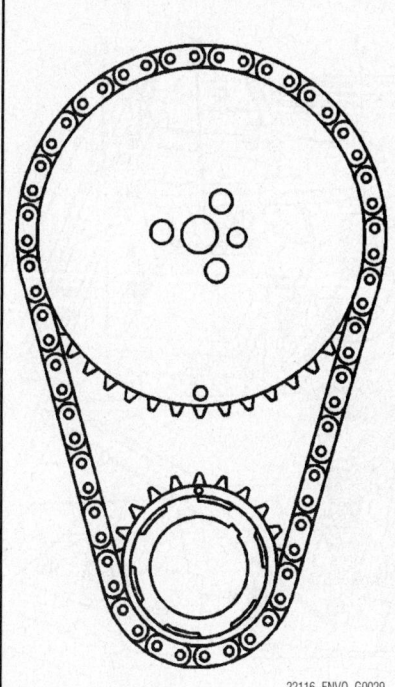

Fig. 66 Proper alignment of the timing marks for timing chain installation—5.3L engine

✳✳ **WARNING**

Do NOT turn the crankshaft after the timing chain has been removed to prevent damage to the pistons and valves.

4. Remove or disconnect the following:
   - Camshaft sprocket bolts
   - Camshaft sprocket and timing chain
   - Crankshaft sprocket using Pulley Puller No. J 8433, Crankshaft End Protector Tool No. J 41816-2 and Crankshaft Sprocket Removal Tool No. J 41558
   - Crankshaft sprocket key, if necessary
5. Clean and inspect the timing chain and sprockets.

To install:
6. Install or connect the following:
   - Key into the crankshaft keyway, if removed. Tap the key into the keyway until both ends of the key bottom into the crankshaft.
   - Crankshaft sprocket onto the front of the crankshaft. Align the crankshaft key with the sprocket keyway.
   - Crankshaft sprocket using Sprocket Installation Tool No. J 41665. Install the sprocket onto the crankshaft until fully seated against the crankshaft flange. Rotate the crankshaft sprocket until the alignment mark is in the 12 o'clock position.

➡**Properly locate the camshaft sprocket locating pin with the cam sprocket alignment hole. The sprocket teeth and timing chain must mesh. The camshaft and crankshaft sprocket alignment marks MUST be aligned properly. Locate the camshaft sprocket alignment mark in the 6 o'clock position. It may be necessary to rotate the camshaft or crankshaft to align the marks.**

   - Camshaft sprocket and timing chain
   - Camshaft sprocket bolts and tighten to 26 ft. lbs. (35 Nm)
   - Oil pump

VALVE LASH

ADJUSTMENT

The 4.2L and 5.3L engines do not require a periodic valve lash adjustment.

ENGINE PERFORMANCE & EMISSION CONTROL

COMPONENT LOCATIONS

See Figures 67 and 68.

ACCELERATOR PEDAL POSITION (APP) SENSOR

LOCATION

The accelerator pedal position (APP) sensor is mounted on the accelerator pedal assembly.

OPERATION

4.2L Engine

The APP sensor is mounted on the accelerator pedal assembly. The APP is actually 2 individual APP sensors within one housing. There are 2 separate signal, low reference, and 5-volt reference circuits. APP sensor 1 voltage increases as the accelerator pedal is depressed. APP sensor 2 voltage decreases as the accelerator pedal is depressed.

5.3L Engine

The accelerator pedal contains 2 individual accelerator pedal position (APP) sen-

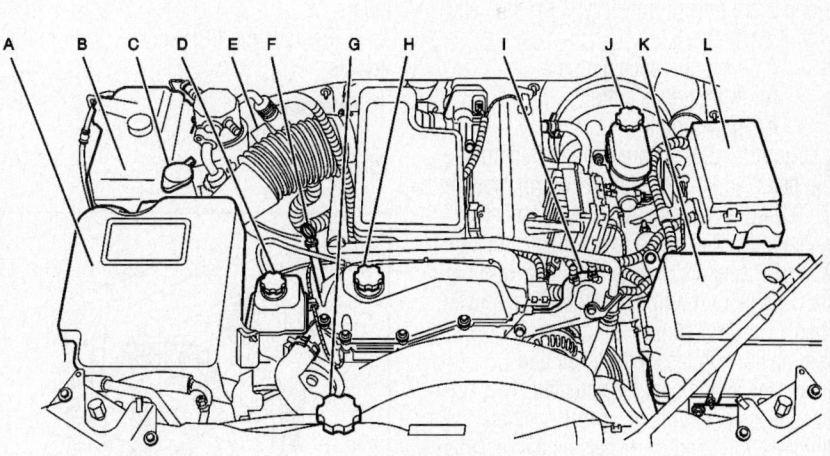

A. Engine air filter
B. Coolant recovery tank
C. Washer fluid reservoir
D. Power steering fluid reservoir
E. Transmission fluid dipstick (out of view)
F. Engine oil dipstick
G. Radiator cap
H. Engine oil fill cap
I. Remote negative (-) battery terminal (marked GND)
J. Brake master cylinder reservoir
K. Battery
L. Engine compartment fuse block

Fig. 67 Engine compartment component locations—4.2L engine

A. Engine coolant recovery tank
B. Engine air filter
C. Washer fluid reservoir
D. Engine oil dipstick
E. Transmission fluid dipstick
F. Engine oil fill cap
G. Radiator cap
H. Remote negative (-) terminal (marked GND)
I. Power steering fluid reservoir
J. Brake master cylinder reservoir
K. Engine compartment fuse block
L. Battery

22116_ENVO_G0142

Fig. 68 Engine compartment component locations—5.3L engine

sors within the assembly. The APP sensors 1 and 2 are potentiometer type sensors each with 3 circuits:

- A 5-volt reference circuit
- A low reference circuit
- A signal circuit

The APP sensors are used to determine the pedal angle. The engine control module (ECM) provides each APP sensor a 5-volt reference circuit and a low reference circuit. The APP sensors provide the ECM with signal voltage proportional to the pedal movement. The APP sensor 1 signal voltage at rest position is less than 1 volt and increases as the pedal is actuated. The APP sensor 2 signal voltage at rest position above 4 volts and decreases as the pedal is actuated.

REMOVAL & INSTALLATION

4.2L Engine

See Figure 69.

1. Disconnect the negative battery cable.
2. Disconnect the accelerator pedal position (APP) sensor electrical connector.

3. Remove the APP sensor retaining fasteners.
4. Remove the APP sensor (2) from the vehicle.

To install:

5. Install the APP sensor (2) to vehicle.

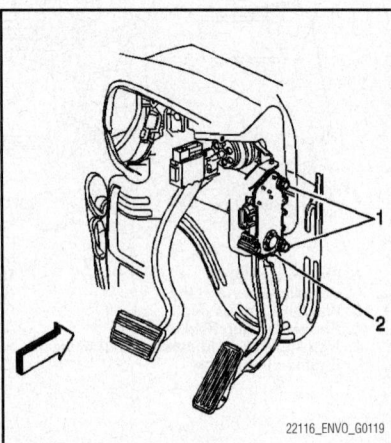

22116_ENVO_G0119

Fig. 69 Location of the accelerator pedal position (APP) sensor (2)

6. Install the APP sensor retaining fasteners (1) and tighten to 89 inch lbs. (10 Nm).
7. Connect the APP sensor electrical connector.
8. Connect the negative battery cable.

5.3L Engine

1. Disconnect the negative battery cable.
2. Disconnect the accelerator pedal position (APP) sensor electrical connector.
3. Remove the 3 APP sensor bolts (1).
4. Remove the APP sensor (2).

To install:

5. Install the APP sensor.
6. Install the 3 APP sensor bolts (1) and tighten to 15 ft. lbs. (20 Nm).
7. Connect the APP sensor electrical connector.
8. Verify that the vehicle meets the following conditions:

- The vehicle is not in a reduced engine power mode
- The ignition is ON

• The engine is OFF

9. Connect a scan tool in order to test for a proper throttle-opening and throttle-closing range.

10. Operate the accelerator pedal and monitor the throttle angles. The accelerator pedal should operate freely, without binding, between a closed throttle, and a wide open throttle.

11. Connect the negative battery cable.

CAMSHAFT POSITION (CMP) SENSOR

LOCATION

4.2L Engine

The Camshaft Position (CMP) sensor is located on the front right corner of the engine cylinder head.

5.3L Engine

The Camshaft Position (CMP) sensor is located on the front of the engine block, just above the crankshaft.

REMOVAL & INSTALLATION

4.2L Engine

See Figure 75.

1. Disconnect the negative battery cable.

2. Remove the Camshaft Position (CMP) sensor electrical connector (1).

3. Remove the CMP sensor retaining bolt.

To install:

4. Install the CMP sensor and tighten the CMP sensor bolt to 89 inch lbs. (10 Nm).

5. Install the CMP sensor electrical connector (1).

6. Connect the negative battery cable.

5.3L Engine

See Figure 71.

1. Disconnect the negative battery cable.

2. Remove the alternator bracket assembly.

3. Remove the Camshaft Position (CMP) sensor mounting bolts (1).

4. Remove the CMP sensor assembly (4, 5, 6) from the front cover (7).

5. Disconnect the CMP sensor jumper harness (2) and the engine harness (3) electrical connectors.

6. Remove the CMP sensor assembly (4, 5, 6).

7. Disconnect CMP sensor (5) from the jumper harness (4).

To install:

8. Reconnect the CMP sensor (5) and the jumper harness (4).

9. Install the O-ring (6) on the CMP sensor assembly (4, 5).

10. Reconnect the CMP sensor assembly (4, 5, 6) and the engine harness connector (3).

11. Install the CMP sensor assembly (4, 5, 6) in the front cover (7). Apply a small amount of clean motor oil to the O-ring (6).

12. Install the CMP sensor mounting bolts and tighten them to 18 ft. lbs. (25 Nm).

13. Install the alternator assembly.

14. Connect the negative battery cable.

CRANKSHAFT POSITION (CKP) SENSOR

LOCATION

4.2L Engine

The crankshaft position (CKP) sensor is located on the rear left bottom side of the engine block.

5.3L Engine

The crankshaft position (CKP) sensor is located on the rear right bottom side of the engine block.

REMOVAL & INSTALLATION

4.2L Engine

See Figure 72.

1. Disconnect the negative battery cable.

2. Raise and safely support the front of the vehicle securely on jackstands.

3. Disconnect the CKP sensor harness connector.

4. Remove the CKP sensor retaining bolt.

5. Remove the CKP sensor from the engine block.

To install:

6. Inspect the sensor O-ring for wear cracks or leakage and replace if necessary. Lubricate the new O-ring with engine oil before installation.

7. Install the CKP sensor into the engine block and tighten the bolt to 89 inch lbs. (10 Nm).

8. Install the CKP sensor retaining bolt.

9. Connect the CKP sensor harness connector.

10. Lower the vehicle.

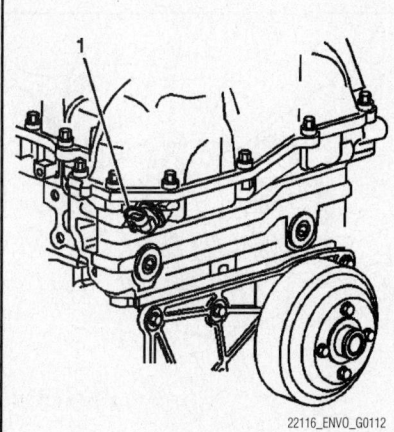

22116_ENVO_G0112

Fig. 70 Location of the Camshaft Position (CMP) sensor (1)—4.2L engine

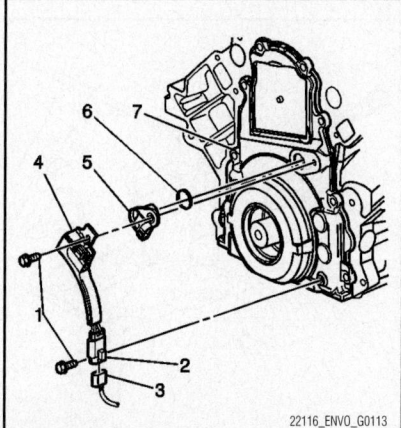

22116_ENVO_G0113

Fig. 71 Location of the Camshaft Position (CMP) sensor (5) and related components—5.3L engine

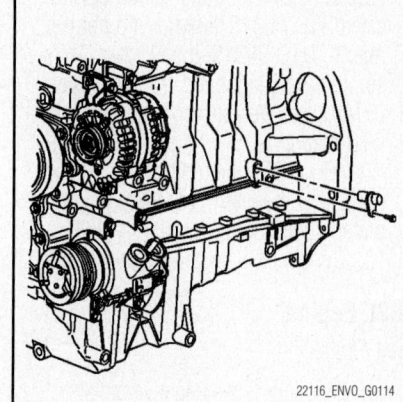

22116_ENVO_G0114

Fig. 72 Location of the Crankshaft Position (CKP) sensor—4.2L engine

11. Perform the crankshaft position system variation learn procedure as follows:

a. Install a scan tool.

b. Monitor the ECM for DTCs with a scan tool. If other DTCs are set, except DTC P0315, refer to Diagnostic Trouble Code (DTC) List - Vehicle for the applicable DTC that set.

c. With a scan tool, select the CKP system variation learn procedure and perform the following:

- Observe the fuel cut-off for the applicable engine
- Block the drive wheels
- Set the parking brake
- Place the vehicle's transmission in Park or Neutral
- Turn the A/C OFF
- Cycle the ignition from OFF to ON
- Apply and hold the brake pedal for the duration of the procedure
- Start and idle the engine
- Accelerate to wide open throttle (WOT). The engine should not accelerate beyond the calibrated fuel cut-off RPM value noted earlier. Release the throttle immediately if the value is exceeded.
- While the learn procedure is in progress, release the throttle immediately when the engine starts to decelerate. The engine control is returned to the operator and the engine responds to throttle position after the learn procedure is complete.
- Release the throttle when fuel cut-off occurs

d. The scan tool displays Learn Status: Learned this Ignition. If the scan tool indicates that DTC P0315 ran and passed, the CKP variation learn procedure is complete. If the scan tool indicates DTC P0315 failed or did not run, refer to DTC P0315. If any other DTCs set, refer to Diagnostic Trouble Code (DTC) List - Vehicle for the applicable DTC that set.

e. Turn OFF the ignition for 30 seconds after the learn procedure is completed successfully.

12. Connect the negative battery cable.

5.3L Engine

See Figure 73.

1. Disconnect the negative battery cable.

2. Remove the starter.

3. Disconnect the electrical connector (2) from the CKP sensor (1).

4. Clean the area around the CKP sensor before removal in order to avoid debris from entering the engine.

5. Remove the CKP sensor bolt.

6. Remove the CKP sensor.

To install:

7. Install the CKP sensor.

8. Install the CKP sensor bolt and tighten to 18 ft. lbs. (25 Nm).

9. Connect the electrical connector (2) to the CKP sensor (1).

10. Install the starter.

11. Perform the crankshaft position system variation learn procedure as follows:

a. Install a scan tool.

b. Monitor the ECM for DTCs with a scan tool. If other DTCs are set, except DTC P0315, refer to Diagnostic Trouble Code (DTC) List - Vehicle for the applicable DTC that set.

c. With a scan tool, select the CKP system variation learn procedure and perform the following:

- Accelerate to wide open throttle (WOT)
- Release throttle when fuel cut-off occurs
- Observe fuel cut-off for applicable engine
- Engine should not accelerate beyond calibrated RPM value
- Release throttle immediately if value is exceeded
- Block drive wheels
- Set parking brake
- DO NOT apply brake pedal
- Cycle ignition from OFF to ON
- Apply and hold brake pedal
- Start and idle engine
- Turn A/C OFF
- Vehicle must remain in Park or Neutral

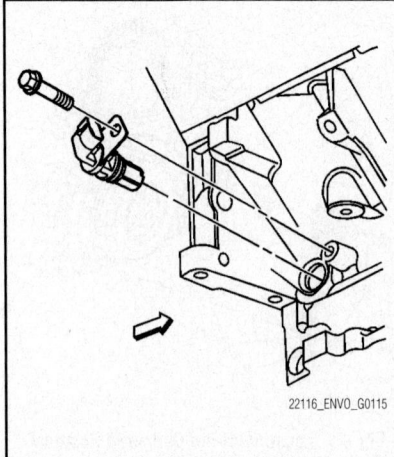

22116_ENVO_G0115

Fig. 73 Location CKP sensor—5.3L engine

d. The scan tool displays Learn Status: Learned this Ignition. If the scan tool indicates that DTC P0315 ran and passed, the CKP variation learn procedure is complete. If the scan tool indicates DTC P0315 failed or did not run, refer to DTC P0315. If any other DTCs set, refer to Diagnostic Trouble Code (DTC) List - Vehicle for the applicable DTC that set.

e. Turn OFF the ignition for 30 seconds after the learn procedure is completed successfully.

12. Connect the negative battery cable.

ENGINE CONTROL MODULE (ECM)

LOCATION

The Engine Control Module (ECM) is located in the front portion of the engine compartment

REMOVAL & INSTALLATION

4.2L Engine

Refer to Powertrain Control Module (PCM).

5.3L Engine

See Figure 74.

➡ **It is necessary to record the remaining engine oil life. If the replacement module is not programmed with the remaining engine oil life, the engine oil life will default to 100%. If the replacement module is not programmed with the remaining engine oil life, the engine oil will need to be changed at 5000 km (3,000 mi) from the last engine oil change.**

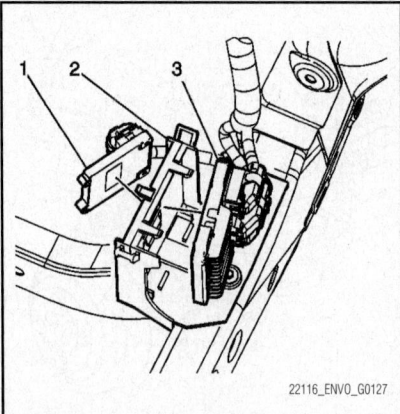

22116_ENVO_G0127

Fig. 74 Transmission Control Module (TCM) (1), ECM/TCM bracket (2) and Engine Control Module (3)—5.3L engine

1. Using a scan tool, retrieve the percentage of remaining engine oil. Record the remaining engine oil life.

2. Disconnect the negative battery cable.

3. Disconnect the cooling fan electrical connector for additional clearance while removing the ECM.

4. Depress the ECM/Transmission Control Module (TCM) cover retainers.

5. Remove the ECM/TCM cover from the ECM/TCM bracket.

✴✴ WARNING

Do not touch the connector pins or soldered components on the circuit board in order to prevent possible ElectroStatic Discharge (ESD) damage to the ECM.

➡ **It is not necessary to disconnect the ECM electrical connectors in order to remove the ECM from the ECM/TCM bracket. Only disconnect the electrical connectors if servicing of component requires disconnecting of the electrical connectors.**

✴✴ WARNING

Remove any debris from around the ECM connector surfaces before servicing the ECM. Inspect the ECM module connector gaskets when diagnosing or replacing the ECM. Ensure that the gaskets are installed correctly. The gaskets prevent contaminant intrusion into the ECM.

6. Disconnect the ECM electrical connectors from the ECM.

7. Release the bracket ECM retainers.

8. Tilt the ECM away from the ECM/TCM bracket.

9. Remove the ECM from the ECM bracket.

10. Only when replacement of the ECM/TCM bracket is necessary, remove the TCM.

11. Remove the ECM/TCM bracket retaining bolts.

12. Remove the ECM/TCM bracket from the vehicle frame.

To install:

13. If the ECM/TCM bracket was previously removed, install the ECM/TCM bracket to the vehicle frame.

14. Install the ECM/TCM bracket retaining bolts.

15. Tighten the ECM/TCM bracket bolts to 89 inch lbs. (10 Nm).

16. If the TCM was previously removed

from the ECM/TCM bracket, install the TCM.

17. Insert the ECM into the retaining slots of the ECM/TCM bracket.

18. Secure the ECM to the ECM/TCM mounting bracket ensuring the ECM retaining tabs are fully engaged.

19. Connect the ECM electrical connectors to the ECM if previously removed.

20. Install the ECM/TCM cover to the ECM/TCM bracket.

21. Ensure the ECM/TCM cover retainers are fully engaged with the ECM/TCM bracket.

22. Connect the cooling fan electrical connector.

23. Connect the negative battery cable.

24. If the ECM was replaced the replacement ECM must be programmed.

ENGINE COOLANT TEMPERATURE (ECT) SENSOR

LOCATION

4.2L Engine

The Engine Coolant Temperature (ECT) sensor is mounted on the top front of the engine cylinder head.

5.3L Engine

The Engine Coolant Temperature (ECT) sensor is mounted on the top front of the engine's left cylinder head.

REMOVAL & INSTALLATION

4.2L Engine

See Figure 75.

✴✴ WARNING

Use care when handling the coolant sensor. Damage to the coolant sensor will affect the operation of the fuel control system.

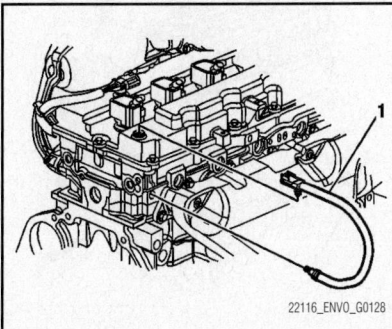

Fig. 75 Engine Coolant Temperature (ECT) sensor—4.2L engine

22116_ENVO_G0128

1. Turn the engine OFF.

2. Disconnect the negative battery terminal.

3. Drain coolant below the level of the Engine Coolant Temperature (ECT) sensor.

4. Disconnect the ECT sensor electrical connector (1).

5. Carefully remove the ECT sensor (1).

To install:

6. If installing the original sensor or a new sensor without sealant, apply thread sealer P/N 12346004 or equivalent.

7. Install the ECT sensor and tighten to 12 ft. lbs. (16 Nm).

8. Connect the ECT electrical connector (1).

9. Connect the negative battery terminal.

10. Refill the engine coolant.

5.3L Engine

See Figure 76.

✴✴ WARNING

Use care when handling the coolant sensor. Damage to the coolant sensor will affect the operation of the fuel control system.

1. Turn OFF the ignition.

2. Raise and suitably support the vehicle.

3. Drain the cooling system below the level of the Engine Coolant Temperature (ECT) sensor.

4. Lower the vehicle.

5. Disconnect the ECT sensor electrical connector (1).

6. Remove the ECT sensor.

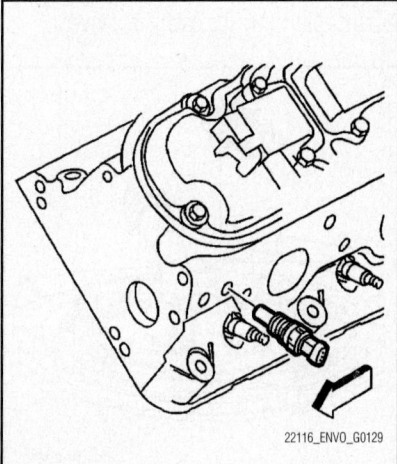

Fig. 76 Engine Coolant Temperature (ECT) sensor—5.3L engine

22116_ENVO_G0129

To install:

7. Coat the ECT sensor threads with sealer GM P/N 12346004 (Canadian P/N 10953480), or equivalent.

8. Install the ECT sensor and tighten to 15 ft. lbs. (20 Nm).

9. Connect the ECT sensor electrical connector (1).

10. Refill the engine coolant.

FUEL TANK PRESSURE (FTP) SENSOR

LOCATION

The Fuel Tank Pressure (FTP) sensor is located on top of the fuel tank, mounted on the fuel tank module.

REMOVAL & INSTALLATION

See Figure 77.

1. Disconnect the negative battery cable.

2. Remove the fuel tank.

3. Disconnect the fuel tank pressure harness connector.

4. Remove the Fuel Tank Pressure (FTP) sensor.

To install:

5. Install the new FTP seal.

6. Install the fuel tank pressure sensor.

7. Connect the fuel tank sensor harness connector.

8. Install the fuel tank.

9. Connect the negative battery cable.

HEATED OXYGEN (HO2S) SENSOR

LOCATION

The Heated Oxygen Sensors are located on each exhaust manifold and on each exhaust pipe after the catalytic converter.

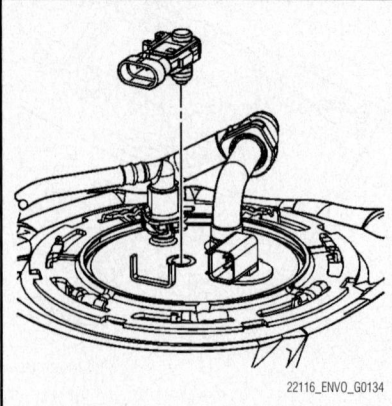

Fig. 77 Removal of the Fuel Tank Pressure (FTP) sensor

REMOVAL & INSTALLATION

See Figure 78.

1. Disconnect the negative battery cable.

2. Raise and safely support the vehicle, if necessary.

3. Disconnect the Heated Oxygen Sensor (HO2S) electrical connector.

4. Remove the HO2S using a J-39194-B.

To install:

5. Coat the threads of the heated oxygen sensor with the anti-seize compound P/N 5613695, or the equivalent if necessary.

6. Install the heated oxygen sensor using a J-39194-B and tighten to 30 ft. lbs. (41 Nm).

7. Connect the HO2S electrical connector.

8. Lower the vehicle, if raised.

9. Connect the negative battery cable.

INTAKE AIR TEMPERATURE (IAT) SENSOR

Refer to the Mass Air Flow/Intake Air Temperature (MAF/IAT) sensor procedure.

KNOCK SENSOR (KS)

LOCATION

4.2L Engine

The knock sensors (KS) are located on the left side of the engine block.

5.3L Engine

There are two (2) knock sensors (KS), one each located on the left and right middle sides of the engine block.

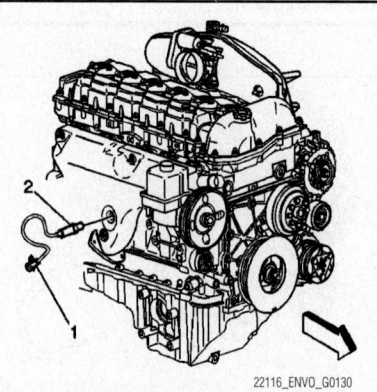

Fig. 78 Heated Oxygen Sensor (HO2S) 1 and electrical connector—4.2L engine shown

REMOVAL & INSTALLATION

4.2L Engine

See Figure 79.

1. Disconnect the negative battery cable.

2. Raise and safely support the front of the vehicle securely on jackstands.

3. Remove the knock sensor harness connector (4).

4. Remove the knock sensor retaining bolt (3).

5. Remove the appropriate knock sensor (1 or 2).

To install:

6. Install the knock sensor (1 or 2) and the bolt (3), then tighten the sensor to 18 ft. lbs. (25 Nm).

7. Connect the knock sensor harness connector (4).

8. Lower the vehicle.

9. Connect the negative battery cable.

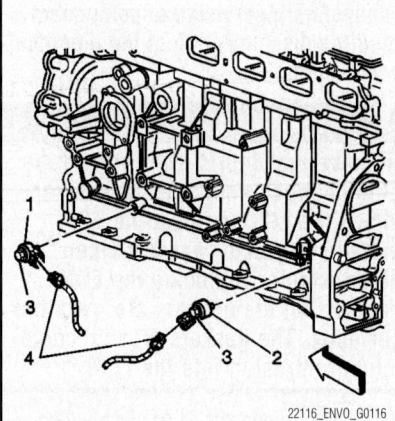

Fig. 79 Location of the knock sensors (1 and 2)—4.2L engine

5.3L Engine

LEFT SIDE

See Figure 80.

1. Disconnect the negative battery cable.

2. Remove the mounting bolt for the knock sensor 1.

3. Disconnect the electrical connector of the knock sensor from the engine harness (3).

4. Remove the knock sensor (2) from the engine block (4).

To install:

5. Reconnect the engine harness (3) and the knock sensor (2) electrical connectors.

6. Position the knock sensor 2 on the engine block (4).

7. Install the mounting bolt (1) for the knock sensor 2.

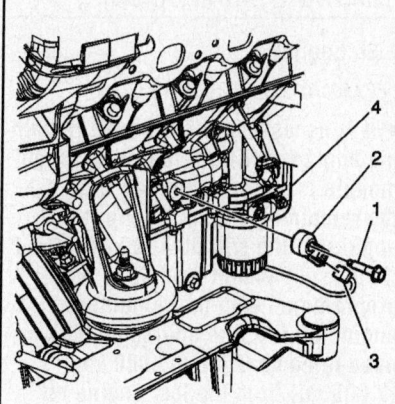

Fig. 80 Location of the left side knock sensor (2)—5.3L engine

8. Tighten the knock sensor mounting bolt (1) to 15 ft. lbs. (20 Nm).
9. Connect the negative battery cable.

RIGHT SIDE

See Figure 81.

1. Disconnect the negative battery cable.
2. Disconnect the electrical connector (3) from the knock sensor (2).
3. Remove the knock sensor bolt (1).
4. Remove the knock sensor (2) from the engine block.

To install:

5. Position the knock sensor (2) on the engine block.
6. Install the knock sensor bolt (1) and tighten to 15 ft. lbs. (20 Nm).
7. Connect the electrical connector (3) to the knock sensor (2).
8. Connect the negative battery cable.

MASS AIR FLOW/INTAKE AIR TEMPERATURE (MAF/IAT) SENSOR

LOCATION

The Mass Air Flow/Intake Air Temperature (MAF/IAT) sensor is mounted ahead of the throttle body on the air cleaner assembly.

4.2L Engine

See Figure 82.

➡Use care when handling the Mass Air Flow/Intake Air Temperature (MAF/IAT) sensor. Do not dent, puncture, or otherwise damage the honey cell located at the air inlet end of the MAF/IAT. Do not touch the sensing elements or allow anything including cleaning solvents and lubricants to come in contact with them. Use a small amount of a non-silicone based lubricant, on the air duct only, to aid in installation.

1. Disconnect the negative battery cable.
2. Disconnect the engine harness electrical connector (5) from the MAF/IAT sensor.
3. Remove the MAF/IAT sensor screws.
4. Remove the MAF/IAT sensor .

To install:

5. Install the MAF/IAT sensor.
6. Install the MAF/IAT sensor screws and tighten to 5 inch lbs. (0.6 Nm).
7. Connect the engine harness electrical connector (5) to the MAF/IAT sensor.
8. Connect the negative battery cable.

5.3L Engine

See Figure 83.

➡Use care when handling the Mass Air Flow/Intake Air Temperature (MAF/IAT) sensor. Do not dent, puncture, or otherwise damage the honey cell located at the air inlet end of the MAF/IAT. Do not touch the sensing elements or allow anything including cleaning solvents and lubricants to come in contact with them. Use a small amount of a non-silicone based lubricant, on the air duct only, to aid in installation.

1. Disconnect the negative battery cable.
2. Disconnect the MAF/IAT sensor electrical connector.
3. Loosen the clamps at the MAF/IAT sensor and the throttle body.
4. Remove the air cleaner outlet duct bolt.
5. Remove the air cleaner outlet duct.
6. Loosen the clamp attaching the MAF/IAT sensor to the air cleaner housing.
7. Remove the MAF/IAT sensor from the air cleaner housing.

To install:

➡The embossed arrow on the MAF/IAT sensor indicates the proper air flow direction. The arrow must point toward the engine.

8. Locate the air flow direction arrow (2) on the MAF/IAT sensor.
9. Install the MAF/IAT sensor on to the air cleaner housing.
10. Tighten the clamp securing the MAF/IAT sensor to the air cleaner housing and tighten the clamp to 62 inch lbs. (7 Nm).

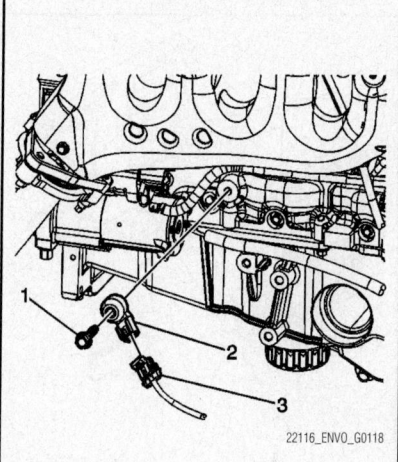

Fig. 81 Location of the right side knock sensor (2)—5.3L engine

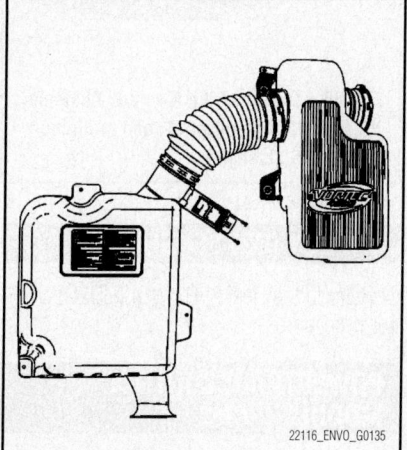

Fig. 82 Location of the Mass Air Flow/Intake Air Temperature (MAF/IAT) sensor—4.2L engine

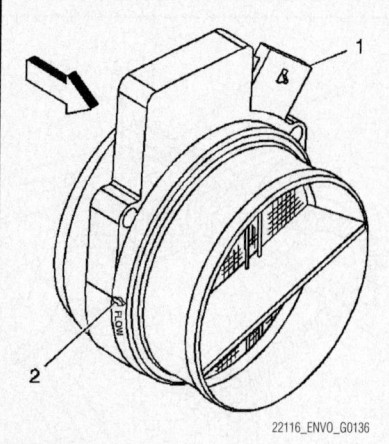

Fig. 83 The embossed arrow on the MAF/IAT sensor indicates the proper air flow direction. The arrow must point toward the engine—5.3L engine

11. Install the air cleaner outlet duct.

12. Install the air cleaner outlet duct bolt and tighten to 89 inch lbs. (10 Nm).

13. Tighten the clamps at the MAF/IAT sensor and the throttle body, then tighten the clamps to 62 inch lbs. (7 Nm).

14. Connect the MAF/IAT electrical connector.

15. Connect the negative battery cable.

MANIFOLD ABSOLUTE PRESSURE (MAP) SENSOR

LOCATION

4.2L Engine

The Manifold Absolute Pressure (MAP) sensor is located on top of the engine, near the firewall.

5.3L Engine

The Manifold Absolute Pressure (MAP) sensor is located on top of the engine, on top of the intake manifold plenum.

REMOVAL & INSTALLATION

4.2L Engine

See Figure 84.

1. Disconnect the negative battery cable.

2. Turn OFF the ignition.

3. Disconnect the Manifold Absolute Pressure (MAP) sensor electrical connector.

4. Press the retainer locking tabs inward, then pull the retainer (1) up to remove it.

5. Remove the MAP sensor (2).

6. Inspect the MAP sensor seal for damage, and replace as necessary.

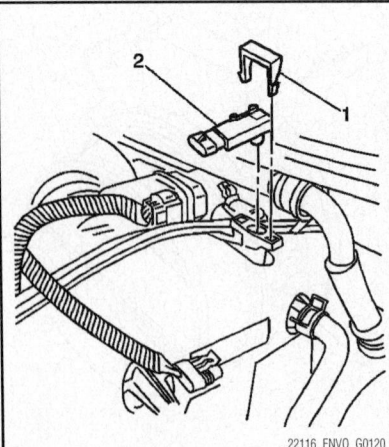

Fig. 84 Location of the retainer (1) and the Manifold Absolute Pressure (MAP) sensor (2)—4.2L engine

To install:

7. Install the MAP sensor (2).

8. Install the MAP sensor retainer (1).

9. Connect the electrical connector.

10. Connect the negative battery cable.

5.3L Engine

See Figure 85.

1. Disconnect the negative battery cable.

2. Disconnect the manifold absolute pressure (MAP) sensor electrical connector (3).

3. Remove the MAP sensor retaining clip (2) from the intake manifold.

4. Remove the MAP sensor (1) from the intake manifold.

To install:

5. Lightly coat the MAP sensor seal with clean engine oil before installing the sensor.

6. Install the MAP sensor (1). Push the MAP sensor into the intake manifold.

7. Install the MAP sensor retainer (2) to the intake manifold.

8. Connect the MAP sensor electrical connector (3).

9. Connect the negative battery cable.

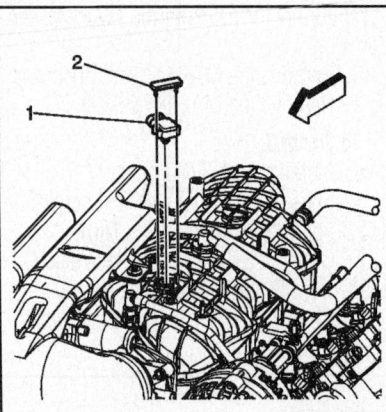

Fig. 85 Location of the Manifold Absolute Pressure (MAP) sensor (1) and retaining clip (2)—5.3L engine

OXYGEN (O2) SENSOR

Refer to the Heated Oxygen (HO2S) Sensor procedures.

POWERTRAIN CONTROL MODULE (PCM)

LOCATION

The Powertrain Control Module (PCM) is located in the engine compartment, mounted on the top left side of the engine.

REMOVAL & INSTALLATION

4.2L Engine

See Figures 86 through 88.

➡ It is necessary to record the remaining engine oil life. If the replacement module is not programmed with the remaining engine oil life, the engine oil life will default to 100%. If the replacement module is not programmed with the remaining engine oil life, the engine oil will need to be changed at 5000 km (3,000 mi) from the last engine oil change.

1. Using a scan tool, retrieve the percentage of remaining engine oil. Record the remaining engine oil life.

2. Disconnect the negative battery cable.

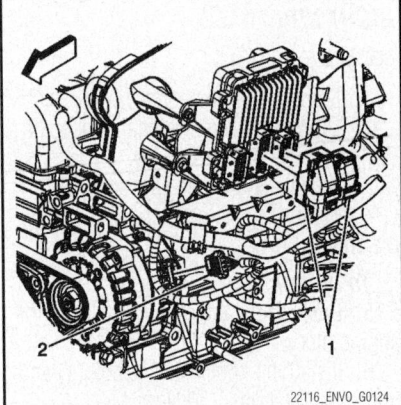

Fig. 86 Disconnect the instrument panel wiring harness electrical connector (1) from the PCM—4.2L engine

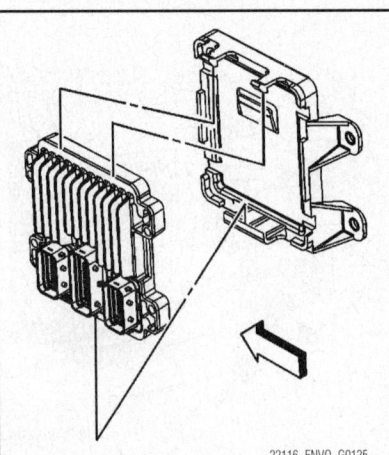

Fig. 87 Disengage the top retainers and remove the PCM from the bracket—4.2L engine

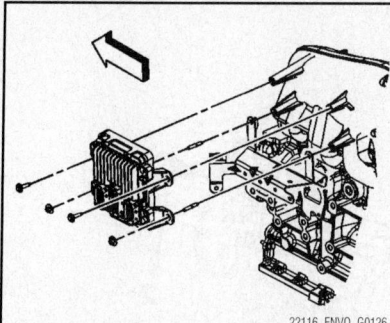

Fig. 88 Removal of the PCM and bracket, bolts and nuts—4.2L engine

✳✳ WARNING

In order to prevent internal damage to the PCM, the ignition must be OFF when disconnecting or reconnecting the PCM connector.

3. Disconnect the instrument panel wiring harness electrical connector (1) from the PCM.

4. Disconnect the engine wiring harness electrical connectors (1) from the PCM.

✳✳ WARNING

Do not touch the connector pins or soldered components on the circuit board in order to prevent possible ElectroStatic Discharge (ESD) damage to the PCM.

5. Disengage the top 2 retainers and remove the PCM from the bracket.

6. If the PCM and bracket require removal, perform the following steps:

   a. Remove the PCM bracket bolts and nuts.

   b. Remove the PCM bracket w/PCM from the studs.

To install:

7. If the PCM and bracket were removed, perform the following steps:

   a. Install the PCM bracket w/PCM to the studs.

   b. Install the PCM bracket bolts and nuts and tighten to 80 inch lbs. (9 Nm).

8. Set the PCM into the bottom retainer on the bracket and push the PCM rearward, engaging the 2 top retainers.

9. Connect the engine wiring harness electrical connectors (1) to the PCM.

10. Connect the I/P wiring harness electrical connector (1) to the PCM.

11. Connect the negative battery cable.

12. If a new PCM was installed, program the PCM.

13. Using a scan tool, set the remaining engine oil life.

5.3L Engine

Refer to Engine Control Module (ECM).

SECONDARY AIR INJECTION REACTION (AIR) SOLENOID VALVE

LOCATION

4.2L Engine

The secondary Air Injection Reaction (AIR) solenoid valve is located on the upper right side of the 4.2L engine.

REMOVAL & INSTALLATION

4.2L Engine

See Figure 89.

1. Remove the air cleaner outlet resonator.

2. Disconnect the electrical connector from the secondary Air Injection Reaction (AIR) solenoid valve.

3. Disconnect the AIR pump air outlet pipe from the AIR solenoid valve.

4. Remove the nut (1) securing the transmission fluid level indicator tube (2) to the AIR solenoid valve.

5. Remove the transmission fluid level indicator tube (2) from the AIR solenoid valve stud (3).

6. Remove the 2 AIR solenoid valve studs (3).

7. Remove the AIR solenoid valve (4) and the gasket (5) from the engine.

To install:

8. Install the AIR solenoid valve (4) and the gasket (5) to the engine.

9. Install the 2 AIR solenoid valve studs (3) and tighten the studs to 18 ft. lbs. (25 Nm).

10. Install the transmission fluid level indicator tube (2) to the AIR solenoid valve stud (3).

11. Install the nut (1) securing the transmission fluid level indicator tube (2) to the AIR solenoid valve and tighten the nut to 89 inch lbs. (10 Nm).

12. Connect the AIR pump air outlet pipe to the AIR solenoid valve.

13. Connect the electrical connector to the AIR solenoid valve.

14. Install the air cleaner outlet resonator.

THROTTLE POSITION SENSOR (TPS)

LOCATION

The Throttle Position Sensor (TPS) is an integral component of the throttle body assembly and cannot be serviced separately.

REMOVAL & INSTALLATION

Refer to the Throttle Body R & I procedure.

VEHICLE SPEED SENSOR (VSS)

LOCATION

The vehicle Speed Sensor (VSS) is located on the right rear side of the transmission case.

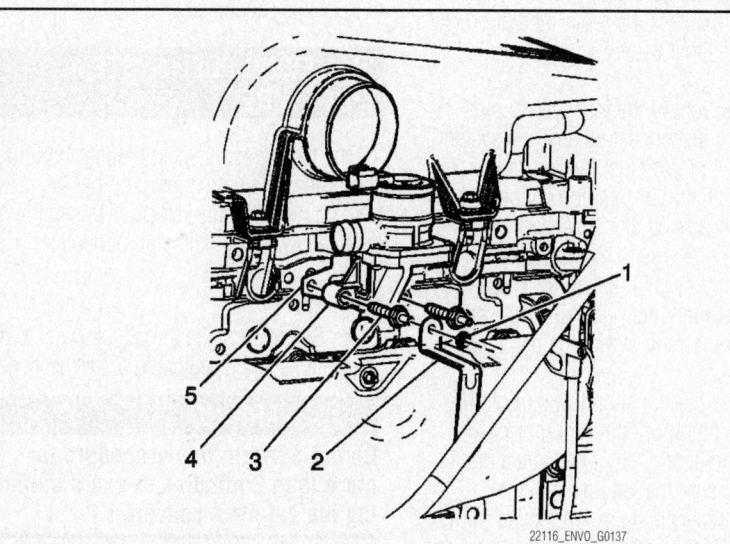

Fig. 89 Secondary Air Injection Reaction (AIR) solenoid valve and related components—4.2L engine

REMOVAL & INSTALLATION

See Figure 90.

1. Disconnect the negative battery cable.
2. Remove the harness connector.
3. Remove the bolt (2).
4. Remove the vehicle speed sensor (1).
5. Remove the O-ring seal (3).

To install:

6. Install the O-ring seal (3) on the vehicle speed sensor (1).
7. Coat the O-ring seal (3) with a thin film of transmission fluid.
8. Install the vehicle speed sensor (1) into the transmission case.
9. Install the bolt (2) and tighten to 97 inch lbs. (11 Nm).
10. Connect the wiring harness electrical connector to the vehicle speed sensor.
11. Refill the fluid as required.
12. Connect the negative battery cable.

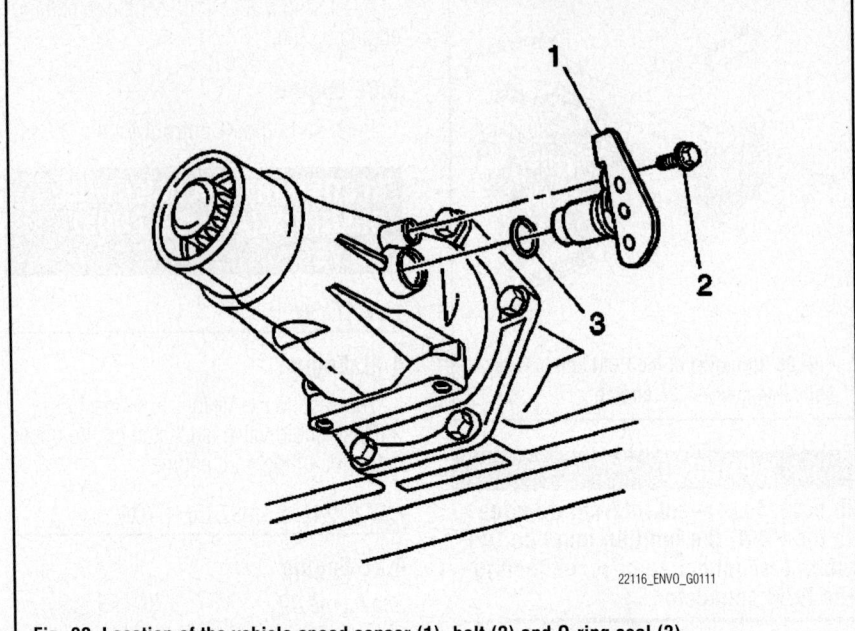

22116_ENVO_G0111

Fig. 90 Location of the vehicle speed sensor (1), bolt (2) and O-ring seal (3)

FUEL

FUEL SYSTEM SERVICE PRECAUTIONS

Safety is the most important factor when performing not only fuel system maintenance but any type of maintenance. Failure to conduct maintenance and repairs in a safe manner may result in serious personal injury or death. Maintenance and testing of the vehicle's fuel system components can be accomplished safely and effectively by adhering to the following rules and guidelines.

• To avoid the possibility of fire and personal injury, always disconnect the negative battery cable unless the repair or test procedure requires that battery voltage be applied.

• Always relieve the fuel system pressure prior to disconnecting any fuel system component (injector, fuel rail, pressure regulator, etc.), fitting or fuel line connection. Exercise extreme caution whenever relieving fuel system pressure to avoid exposing skin, face and eyes to fuel spray. Please be advised that fuel under pressure may penetrate the skin or any part of the body that it contacts.

• Always place a shop towel or cloth around the fitting or connection prior to loosening to absorb any excess fuel due to spillage. Ensure that all fuel spillage (should it occur) is quickly removed from engine surfaces. Ensure that all fuel soaked cloths or towels are deposited into a suitable waste container.

• Always keep a dry chemical (Class B) fire extinguisher near the work area.

• Do not allow fuel spray or fuel vapors to come into contact with a spark or open flame.

• Always use a back-up wrench when loosening and tightening fuel line connection fittings. This will prevent unnecessary stress and torsion to fuel line piping.

• Always replace worn fuel fitting O-rings with new Do not substitute fuel hose or equivalent where fuel pipe is installed.

Before servicing the vehicle, make sure to also refer to the precautions in the beginning of this section as well.

RELIEVING FUEL SYSTEM PRESSURE

The fuel systems operate under high fuel pressures. It is very important that the pressure be properly relieved prior to servicing the system or any of its components.

4.2L Engine

1. Before servicing the vehicle, refer to the precautions in the beginning of this section.

✳✳ WARNING

Do not perform this procedure for more than 2 minutes to avoid damaging the catalytic converter.

2. Loosen the fuel filler cap to release the fuel tank pressure.

GASOLINE FUEL INJECTION SYSTEM

3. Remove the fuel pump relay from the junction block.
4. Crank the engine, allowing it to start and stall.
5. Crank the engine for an additional 3 seconds to relieve any remaining fuel pressure.
6. Disconnect the negative battery cable to avoid repressurizing the fuel system.
7. Install the fuel pump relay in the junction block.
8. Tighten the fuel filler cap.
9. After you are finished working on the fuel system, connect the negative battery cable.

5.3L Engine

1. Disconnect the negative battery cable.
2. Install Fuel Pressure Gauge J 34730-1A or equivalent to the fuel pressure connection.
3. Loosen the fuel fill cap to relieve the fuel tank vapor pressure.
4. Open the valve on the fuel pressure gauge to bleed the system pressure. The fuel connections are now safe for servicing. Drain any fuel remaining in the gauge into an approved container. Once the system pressure is completely relieved, remove the fuel pressure gauge.

FUEL FILTER

REMOVAL & INSTALLATION

The fuel filter is contained in the fuel sender assembly inside the fuel tank. The paper filter element traps particles in the fuel that may damage the fuel injection system. The filter housing is made to withstand maximum fuel system pressure, exposure to fuel additives, and changes in temperature.

1. Before servicing the vehicle, refer to the precautions in the beginning of this section.

2. Properly relieve the fuel system pressure.

3. Disconnect the negative battery cable.

4. Remove the fuel filler cap, if not already done.

5. Remove the fuel pump assembly.

6. Remove the fuel filter from the fuel pump assembly. Replace the seals or O-rings.

To install:

7. Install the new fuel filter to the fuel pump assembly along with new seals or O-rings.

8. Install the fuel pump assembly.

9. Install the fuel filler cap.

10. Connect the negative battery cable.

11. Start the engine and check for leaks.

FUEL PUMP

REMOVAL & INSTALLATION

See Figure 91.

1. Before servicing the vehicle, refer to the precautions in the beginning of this section.

2. Properly relieve the fuel system pressure.

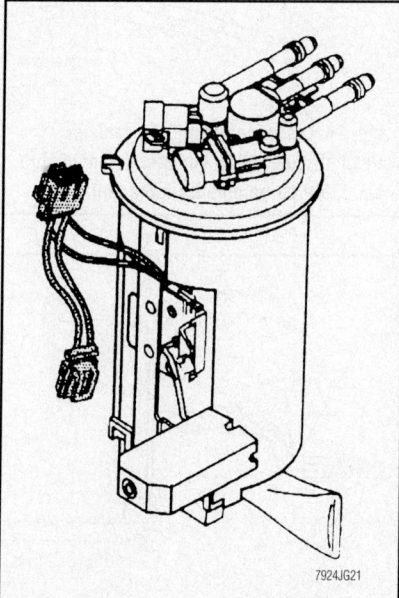

7924JG21

Fig. 91 View of the in-tank fuel pump assembly

3. Drain the fuel tank.
4. Support the fuel tank.
5. Remove or disconnect the following:
   - Negative battery cable
   - Filler neck from the tank
   - Shield from tank and tank straps
   - Fuel lines and vapor hose from pump
   - Electrical connection from fuel pump
   - Fuel tank
   - Fuel pump/sending unit assembly by turning the locking ring (located on top of the fuel tank) counterclockwise using a spanner wrench
   - Fuel pump from the fuel lever sending device

To install:

6. Install or connect the following:
   - Fuel pump in tank with new seal around opening

➡**The fuel pump strainer must be in a horizontal position when the fuel sender is installed in the tank. When installing the sender assembly, make sure that the fuel pump strainer does not block full travel of the float arm.**

   - Tank and connect fuel lines and vapor hose
   - Tank to the frame. Torque the fasteners to 33 ft. lbs. (45 Nm).
   - Shield
   - Fuel filler neck and clamp
   - Negative battery cable
7. Refill the tank.
8. Run the engine and check for leaks.

FUEL RAIL & INJECTORS

REMOVAL & INSTALLATION

4.2L Engine

1. Before servicing the vehicle, refer to the precautions in the beginning of this section.

2. Relieve the fuel system pressure. Refer to the fuel system relief procedure in this section.

3. Remove or disconnect the following:
   - Negative battery cable, if not done already
   - Intake manifold

➡**Clean the fuel rail assembly with a suitable spray cleaner before proceeding. Never soak the fuel rail in a cleaning solvent.**

   - Fuel pressure regulator vacuum line
   - Fuel feed and return pipes

   - Fuel injector in-line electrical connector
   - Fuel rail attaching bolts and fuel rail
   - Fuel injector harness connector from the fuel injectors
   - Injector retaining clip
   - Injector from the fuel rail
   - Retainer clip and O-ring seals from each end of the injector and discard

To install:

➡**Each injector is calibrated. When replacing the fuel injectors, be sure to replace it with the correct injector.**

4. Lubricate the new injector O-ring seats with engine oil.

5. Install or connect the following:
   - O-rings on the injector
   - New retainer clip on the injector

6. Push the fuel injector into the fuel rail socket, making sure the connector faces outward. The retainer clip locks to a flange on the fuel rail injector socket.
   - Fuel rail assembly. Tighten the bolts to 89 inch lbs. (10 Nm).
   - Fuel feed and return lines to the rail
   - Fuel injector electrical connectors
   - Fuel pressure regulator vacuum line
   - Intake manifold
   - Negative battery cable

7. Turn the ignition **ON** for 2 seconds and then turn it **OFF** for 10 seconds. Again turn the ignition **ON** and check for leaks.

5.3L Engine

See Figures 92 and 93.

1. Before servicing the vehicle, refer to the precautions in the beginning of this section.

2. Relieve the fuel system pressure. Refer to the fuel system relief procedure in this section.

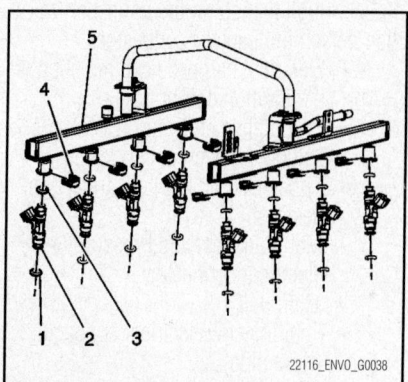

22116_ENVO_G0038

Fig. 92 Exploded view of the fuel rail mounting—5.3L engine

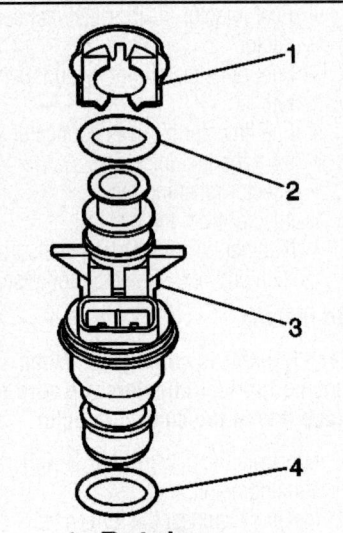

1. Retainer
2. O-ring seal
3. Fuel injector
4. O-ring seal

22116_ENVO_G0039

Fig. 93 Exploded view of the fuel injector (3), retainer (1) and O-ring seals (2, 4)—5.3L engine

3. Remove or disconnect the following:
- Negative battery cable, if not done already
- A/C compressor pressure switch electrical connector
- Wire harness from the clip on the cylinder head
- Mass Airflow/Intake Air Temperature (MAF/IAT) sensor connector
- Alternator electrical connector
- Right side electrical connectors from the coil main electrical harness, Electronic Throttle Control (ETC) and fuel injectors.

4. To detach the injector connector: Matchmark the connectors, pull the Connector Position Assurance (CPA) retainer up 1 click. Push the tab on the connector in, then detach the injector connector.
- Electrical harness from the clips on the ignition coil bracket
- Evaporative emission (EVAP) purge solenoid electrical connector
- Knock Sensor (KS) electrical connector
- Manifold Absolute Pressure (MAP) electrical connector
- Main coil
- Fuel injector electrical connector (right side)
- Electrical harness from the clips on the ignition coil bracket
- Upper engine wire harness retainer

nut. Position the wire harness aside.
- Fuel feed and return pipes from the rail
- Fuel pressure regulator vacuum line
- Fuel rail bolts
- Fuel rail, after cleaning with a spray-type cleaner

※※ WARNING

Be very careful when removing the fuel rail and injectors not to damage the connector terminals or injector spray tips

- Fuel injector from the fuel rail
- Fuel injector retainer clip and discard
- Fuel injector lower O-ring seals and discard

To install:

5. Install or connect the following:
- New O-ring seals on the injectors, after lubricating with clean engine oil
- New retainer clip on the injector
- Fuel injector by pushing it into the fuel rail socket
- Fuel rail
- Apply 0.20 (5mm) band of threadlock to the threads of the fuel rail bolts
- Fuel rail bolts and tighten to 89 inch lbs. (10 Nm)
- Fuel pressure regulator vacuum line
- Fuel feel and return pipes
- Route the upper electrical harness into position over the engine.
- Engine harness bracket nut and tighten to 89 inch lbs. (10 Nm)
- PCV valve and hose
- EVAP purge solenoid, KS, MAP sensor, main coil & fuel injector electrical connectors
- Harness to the clips on the ignition coil bracket
- Main coil, ETC, fuel injector electrical connectors
- Harness to the clips on the ignition coil bracket
- Alternator electrical connector
- MAF/IAT sensor connector
- Wire harness to the clip on the cylinder head
- A/C compressor switch electrical connector
- Air cleaner outlet duct
- Fuel fill cap
- Negative battery cable

6. Refill the engine cooling system.

FUEL TANK

REMOVAL & INSTALLATION

Standard Wheelbase Models

See Figures 94 through 96.

1. Relieve the fuel system pressure.
2. Disconnect the negative battery cable.
3. Raise and safely support the vehicle securely on jackstands.
4. Remove the 2 mounting bolts from the frame brace and remove the frame brace from the frame.
5. Remove the fuel tank shield to the frame retaining bolts and nut, then remove the fuel tank shield from the frame.
6. Drain the fuel tank as follows:
   a. Loosen the fuel fill hose clamp.
   b. Disconnect the fuel fill hose from the fuel tank.
   c. Use a hand or air operated pump device in order to drain as much fuel from the fuel tank as possible.
7. Disconnect the evaporative emission (EVAP) canister fresh air pipe.
8. Disconnect the EVAP canister solenoid pipe.

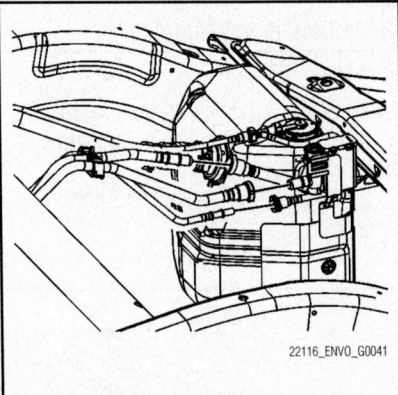

22116_ENVO_G0041

Fig. 94 Disconnect the EVAP canister fresh air pipe, EVAP canister solenoid pipe and EVAP purge pipe from the fuel tank

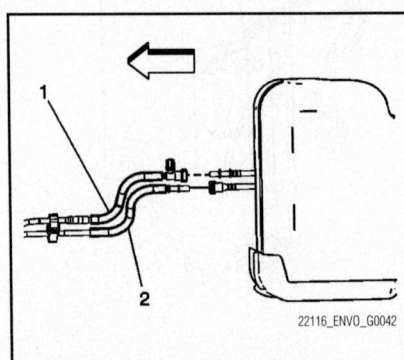

22116_ENVO_G0042

Fig. 95 Disconnect the EVAP pipe (1) and the fuel feed pipe (2) from the fuel tank

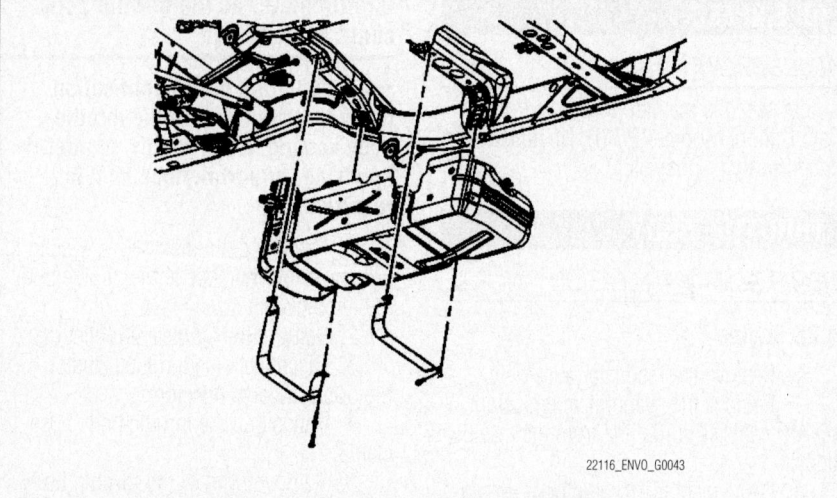

22116_ENVO_G0043

Fig. 96 Remove the fuel tank straps and carefully lower the fuel tank

9. Disconnect the EVAP purge pipe.

10. Disconnect the fuel filler pipe recirculation hose from the fuel tank.

11. Loosen the clamp securing the fuel fill pipe to the fuel tank.

12. Disconnect the fuel fill pipe from the fuel tank.

13. Disconnect the fuel feed pipe (2) and EVAP pipe (1) from the fuel tank.

14. Cap the fuel and EVAP pipes in order to prevent possible fuel system contamination.

15. Support the fuel tank.

16. Remove the fuel tank strap attaching bolts.

17. Remove the fuel tank straps.

18. Carefully lower the fuel tank.

19. Disconnect the EVAP vent valve electrical connector.

20. Disconnect the fuel tank pressure sensor electrical connector.

21. Disconnect the fuel sender electrical connector.

22. Remove the fuel tank.

23. Place the fuel tank in a suitable work area.

To install:

24. Support the fuel tank.

25. Connect the fuel sender electrical connector.

26. Connect the EVAP vent valve electrical connector.

27. Connect the fuel pressure sensor electrical connector.

28. Install the fuel tank straps.

29. Install the fuel tank strap attaching bolts and tighten to 24 ft. lbs. (32 Nm).

30. Remove the caps from the fuel and EVAP pipes.

31. Connect the fuel feed pipe (1) and the EVAP pipe (2) as follows:

a. Apply a few drops of clean engine oil to the male connection end.

b. Push both sides of the quick-connect fitting together in order to cause the retaining feature to snap into place.

c. Once installed, pull on both sides of the quick-connect fitting in order to make sure the connection is secure.

32. Connect the fuel fill pipe to the fuel tank and tighten the fuel fill hose clamp to 22 inch lbs. (2.5 Nm).

33. Connect the fuel filler pipe recirculation hose to the fuel tank as follows:

a. Apply a few drops of clean engine oil to the male connection end.

b. Push both sides of the quick-connect fitting together in order to cause the retaining feature to snap into place.

c. Once installed, pull on both sides of the quick-connect fitting in order to make sure the connection is secure.

34. Connect the EVAP purge pipe as follows:

a. Apply a few drops of clean engine oil to the male connection end.

b. Push both sides of the quick-connect fitting together in order to cause the retaining feature to snap into place.

c. Once installed, pull on both sides of the quick-connect fitting in order to make sure the connection is secure.

35. Connect the EVAP canister solenoid pipe as follows:

a. Apply a few drops of clean engine oil to the male connection end.

b. Push both sides of the quick-connect fitting together in order to cause the retaining feature to snap into place.

c. Once installed, pull on both sides of the quick-connect fitting in order to make sure the connection is secure.

36. Connect the EVAP canister fresh air pipe as follows:

a. Apply a few drops of clean engine oil to the male connection end.

b. Push both sides of the quick-connect fitting together in order to cause the retaining feature to snap into place.

c. Once installed, pull on both sides of the quick-connect fitting in order to make sure the connection is secure.

37. Lower the vehicle.

38. Refill the fuel tank.

39. Install the fuel filler cap.

40. Connect the negative battery cable.

41. Raise the vehicle.

42. Inspect for leaks as follows:

a. Turn ON the ignition, with the engine OFF for 10 seconds.

b. Turn OFF the ignition for 10 seconds.

c. Turn ON the ignition, with the engine OFF.

d. Inspect for fuel leaks.

43. Install the fuel tank shield, if equipped, to the frame and tighten the retaining bolts and nut to 24 ft. lbs. (32 Nm).

44. Install the frame brace to the frame using the 2 mounting bolts and tighten to 37 ft. lbs. (50 Nm).

45. Lower the vehicle.

Extended Wheelbase Models

1. Relieve the fuel system pressure.

2. Disconnect the negative battery cable.

3. Raise and safely support the vehicle securely on jackstands.

4. Remove the 2 mounting bolts from the frame brace and remove the frame brace from the frame.

5. Remove the fuel tank shield to the frame retaining bolts and nut, then remove the fuel tank shield from the frame.

6. Drain the fuel tank as follows:

a. Loosen the fuel fill hose clamp.

b. Disconnect the fuel fill hose from the fuel tank.

c. Use a hand or air operated pump device in order to drain as much fuel from the fuel tank as possible.

7. Loosen the fuel hose clamp at the fuel tank.

8. Separate the fuel hose from the fuel tank.

9. Disconnect the fuel feed pipe (1) and evaporative emission (EVAP) pipe (2).

10. Cap the fuel and EVAP pipes in order to prevent possible fuel system contamination.

11. With the aid of an assistant, support the fuel tank.

12. Remove the fuel tank strap bolts.

13. Remove the fuel tank straps.

14. Carefully lower the fuel tank.

15. Disconnect the fuel tank pressure sensor electrical connector.

16. Disconnect the EVAP vent valve electrical connector.

17. Disconnect the fuel tank module electrical connector.

18. Remove the fuel tank.

19. Place the fuel tank in a suitable work area.

To install:

20. With the aid of an assistant, position and support the fuel tank.

21. Connect the fuel sender electrical connector.

22. Connect the fuel tank pressure sensor electrical connector.

23. Connect the EVAP vent valve electrical connector.

24. Install the fuel tank straps.

25. Install the fuel tank strap bolts and tighten to 24 ft. lbs. (32 Nm).

26. Remove the caps from the fuel and EVAP pipes.

27. Connect the fuel feed pipe and EVAP pipe as follows:

   a. Apply a few drops of clean engine oil to the male connection end.

   b. Push both sides of the quick-connect fitting together in order to cause the retaining feature to snap into place.

   c. Once installed, pull on both sides of the quick-connect fitting in order to make sure the connection is secure.

28. Connect the fuel hose to the fuel tank and tighten the clamp to 22 inch lbs. (2.5 Nm).

29. Lower the vehicle.

30. Refill the fuel tank.

31. Install the fuel fill cap.

32. Connect the negative battery cable.

33. Inspect for leaks as follows:

   a. Turn ON the ignition, with the engine OFF for 10 seconds.

   b. Turn OFF the ignition for 10 seconds.

   c. Turn ON the ignition, with the engine OFF.

   d. Inspect for fuel leaks.

34. Install the fuel tank shield to the frame, if equipped, then install the retaining bolts and nut and tighten to 24 ft. lbs. (32 Nm).

35. Install the frame brace to the frame using the mounting bolts and tighten them to 37 ft. lbs. (50 Nm).

36. Lower the vehicle.

IDLE SPEED

ADJUSTMENT

Idle speed is maintained by the Powertrain Control Module (PCM). No adjustment is necessary or possible.

THROTTLE BODY

REMOVAL & INSTALLATION

4.2L Engine

1. Remove the resonator assembly.

2. Remove the evaporative emission (EVAP) canister purge line from the throttle body.

3. Disconnect the throttle body electrical connector.

4. Remove the throttle body assembly retaining bolts.

5. Remove the throttle body assembly and the gasket from the intake manifold.

6. Clean the gasket surface.

To install:

7. Install the throttle body assembly to the intake manifold with the gasket.

8. Add sealer GM P/N 12346004 (Canadian P/N 10953480) to the throttle control module bolt threads.

9. Install the throttle body assembly retaining bolts. Tighten the bolts to 89 inch lbs. (10 Nm).

10. Connect the throttle body electrical connector.

11. Install the EVAP canister purge line to the throttle body.

12. Install the resonator assembly.

5.3L Engine

✳✳ WARNING

Handle the electronic throttle control components carefully. Use cleanliness in order to prevent damage. Do not drop the electronic throttle control components. Do not roughly handle the electronic throttle control components. Do not immerse the electronic throttle control components in cleaning solvents of any type.

✳✳ WARNING

DO NOT for any reason, insert a screwdriver or other small hand tools into the throttle body to hold open the throttle plate, as the throttle body could be damaged.

➡**An eight digit part identification number is stamped on the throttle body casting. Refer to this number if servicing, or part replacement is required.**

1. Partially drain the cooling system in order to allow the hose at the throttle body to be removed.

2. Remove the air cleaner outlet duct.

3. Disconnect the throttle actuator motor electrical connector.

4. Reposition the throttle body hose clamp.

5. Remove both of the throttle body engine coolant hoses from the throttle body.

6. Remove the throttle body bolts and nuts.

➡**Do not reuse the throttle body gasket. Install a new gasket during assembly.**

7. Remove the throttle body and gasket. Discard the gasket.

To install:

8. Install a NEW throttle body gasket.

9. Install the throttle body.

10. Install the throttle body bolts and nuts. Tighten the bolts and nuts to 53 inch lbs. (6 Nm).

11. Connect the 2 throttle body engine coolant hoses to the throttle body.

12. Position the throttle body hose clamps.

13. Connect the throttle actuator motor electrical connector.

14. Install the air cleaner outlet duct.

15. Refill the cooling system.

16. Verify that the vehicle meets the following conditions:

   a. The vehicle is not in a reduced engine power mode.

   b. The ignition is **ON**.

   c. The engine is OFF.

17. Connect a scan tool in order to test for a proper throttle-opening and throttle-closing range.

18. Operate the accelerator pedal and monitor the throttle angles. The accelerator pedal should operate freely, without binding, between a closed throttle, and a wide open throttle (WOT).

19. Start the engine.

20. Inspect for coolant leaks.

HEATING & AIR CONDITIONING SYSTEM

BLOWER MOTOR

REMOVAL & INSTALLATION

Body VIN Type 3

See Figure 97.

1. Remove the HVAC module-auxiliary assembly.
2. Disconnect the electrical connectors (4) from the blower motor-auxiliary (3).
3. Remove the air outlet duct from the blower motor-auxiliary.
4. Remove the screws (2,5) from the blower motor-auxiliary.
5. Remove the blower motor-auxiliary (3).

To install:

6. Install the blower motor-auxiliary (3).
7. Install the retaining screws (2,5) to the blower motor-auxiliary. Tighten the screws to 88 inch lbs. (10 Nm).

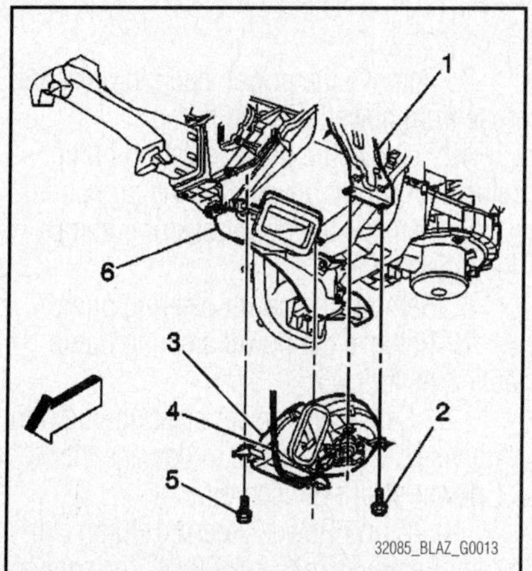

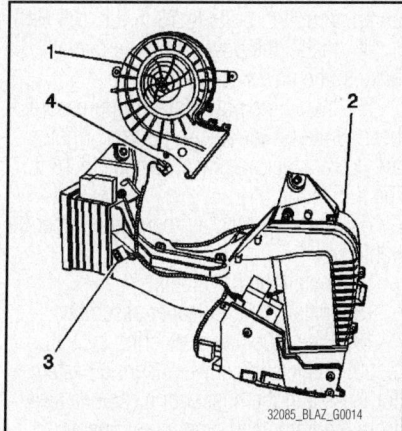

Fig. 97 Exploded view of the auxiliary blower motor (3) and related components—body VIN type 3 shown

32085_BLAZ_G0013

8. Install the air outlet duct to the blower motor-auxiliary.
9. Connect the electrical connectors (4) to the blower motor-auxiliary (3).
10. Install the HVAC module assembly.

Body VIN Type 6

See Figure 98.

1. Remove the right rear quarter trim panel.
2. Remove the retaining bolts (5) from the HVAC module-auxiliary (1).
3. Remove the retaining nuts (4) from the HVAC module-auxiliary (1) under the vehicle.
4. Disconnect the electrical connectors (3,4).
5. Remove the blower motor screws.
6. Remove the blower motor (1) from the HVAC module-auxiliary (2).

Fig. 98 View of the auxiliary blower motor (1), HVAC module (2) and related components—body VIN type 6 shown

32085_BLAZ_G0014

To install:

7. Install the blower motor-auxiliary (1) to the HVAC module-auxiliary (2).
8. Install the blower motor-auxiliary screws and tighten to 18 inch lbs. (2 Nm).
9. Connect the electrical connectors (3,4).
10. Install the retaining bolts (5) to the HVAC module-auxiliary.
11. Tighten the HVAC module-auxiliary retaining bolts (5). Tighten the bolts to 88 inch lbs. (10 Nm).
12. Install the retaining nuts (4) to the HVAC module-auxiliary (1) under the vehicle. Tighten the nuts to 88 inch lbs. (10 Nm).
13. Install the right rear quarter trim.

HEATER CORE

REMOVAL & INSTALLATION

1. Remove the HVAC module-auxiliary.
2. Remove the screws from heater core cover-auxiliary from the HVAC module-auxiliary.
3. Remove the heater core cover-auxiliary.
4. Remove the HVAC module pass thru seal-auxiliary.
5. Remove the heater core-auxiliary from the HVAC module-auxiliary.

To install:

6. Install the heater core-auxiliary to the HVAC module-auxiliary.
7. Install the HVAC module pass thru seal-auxiliary.
8. Install the heater core access cover-auxiliary to the HVAC module-auxiliary.
9. Install the screws to the heater core access cover-auxiliary and tighten to 18 inch lbs. (2 Nm).
10. Install the HVAC module-auxiliary.

STEERING

POWER RACK & PINION STEERING GEAR

REMOVAL & INSTALLATION

1. Before servicing the vehicle, refer to the precautions in the beginning of this section.
2. Raise and support the vehicle.
3. Position a fluid catch pan under the power steering gear.
4. Remove or disconnect the following:
   - Front tire and wheel assemblies
   - Outer tie rod retaining nuts

✷✷ WARNING

Do not try to separate a steering linkage joint by driving a wedge between the joint and the attached part. Doing this can cause seal damage and premature failure of the part.

- Outer tie rods from the steering knuckles using a suitable steering linkage and tie rod puller
- Lower intermediate shaft retaining bolt and shaft from the power steering gear
- Steering gear crossmember

- Feed and return fluid hoses from the steering gear. Immediately cap or plug all openings to prevent system contamination or excessive fluid loss.
5. Support the power steering gear.
- Power steering gear mounting bolts, then remove the gear from the vehicle
6. Loosen the outer tie rod jam nuts, then remove the outer tie rods from the inner tie rods and discard the jam nut.

To install:

7. Lubricate the inner tie rod threads with a suitable lubricant before installing the outer tie rod.

8. Install or connect the following:
- New jam nuts to the outer tie rods
- Outer tie rods to the inner tie rods
- Power steering gear to the vehicle. Tighten the retaining bolts to 81 ft. lbs. (110 Nm).

9. Remove the support from the power steering gear.
- Power steering hose(s) to the gear. Tighten the retaining bolt to 9 ft. lbs. (12 Nm).
- Steering gear crossmember
- Lower intermediate shaft to the power steering gear. Tighten the retaining bolt to 30 ft. lbs. (40 Nm).
- Outer tie rod ends to the steering knuckles. Tighten the retaining nuts to 33 ft. lbs. (45 Nm).
- Front tire and wheel assemblies

10. Remove the drain pan, then lower the vehicle.

11. Bleed the power steering system and adjust the front toe as necessary.

POWER STEERING PUMP

REMOVAL & INSTALLATION

4.2L Engine

See Figure 99.

1. Remove the air cleaner assembly.
2. Remove the drive belt.
3. Install a drain pan under the vehicle.
4. Disconnect the power steering pressure hose from the power steering pump.
5. Disconnect the power steering cooler hose from the power steering pump.
6. Disconnect the wiring harness from the wiring loom on the power steering pump.

7. Remove the power steering pump mounting bolts.
8. Remove the power steering pump.
9. Remove the power steering pump pulley.
10. Remove the power steering pump pulley using Power Steering Pump Pulley Remover tool no. J 25034-C, or equivalent.

To install:

11. Install the power steering pump pulley, as follows:
   a. Install the power steering pump pulley to the end of the power steering pump shaft.
   b. Install the power steering pump pulley to the power steering pump using Power Steering Pump Pulley Installer tool no. J 25033-C, or equivalent.
   c. Install the power steering pump pulley (1) flush against the end of the power steering pump shaft (2), with an allowable variance of 0.010 in. (0.25mm).
12. Install the power steering pump.
13. Install the power steering pump mounting bolts. Tighten the power steering pump mounting bolts to 18 ft. lbs. (25 Nm).
14. Install the power steering cooler hose to the power steering pump.
15. Install the power steering pressure hose to the power steering pump. Tighten the power steering pressure hose to 18 ft. lbs. (25 Nm).
16. Remove the drain pan from under the vehicle.
17. Install the drive belt.
18. Install the air cleaner assembly.
19. Bleed the power steering system.
20. Inspect the power steering system for leaks and the hoses for clearance away from the frame and other components.

5.3L Engine

See Figures 99 and 100.

1. Remove the drive belt.
2. Remove the PCM from PCM mounting bracket and move to the side.

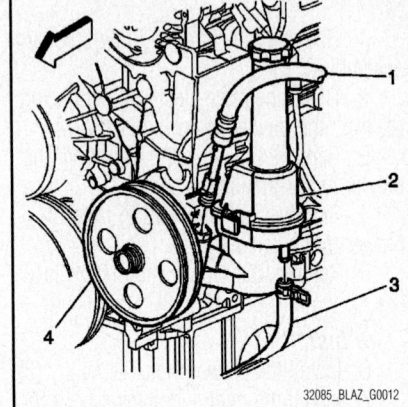
32085_BLAZ_G0012

Fig. 100 Remove the power steering pressure hose (1) and return hose (3) from power steering pump (2)

3. Remove the power steering pressure hose from power steering pump.
4. Remove the power steering pump return hose from power steering pump.
5. Remove the power steering pump mounting bolts.
6. Remove the power steering pump.
7. Remove the power steering pump pulley, as follows:
   a. Secure the power steering pump in a vise, taking care not to damage the power steering reservoir.
   b. Using Power Steering Pump Pulley Removal Tool J 25034-C, or equivalent, remove the power steering pump pulley.
8. If applicable, remove the power steering pump reservoir.

To install:

9. If applicable, install the power steering pump reservoir.
10. Install the power steering pump pulley, as follows:
   a. Install the power steering pump pulley to the end of the power steering pump shaft.
   b. Install the power steering pump pulley to the power steering pump using Power Steering Pump Pulley Installer tool no. J 25033-C, or equivalent.
   c. Install the power steering pump pulley (1) flush against the end of the power steering pump shaft (2), with an allowable variance of 0.010 in. (0.25mm).

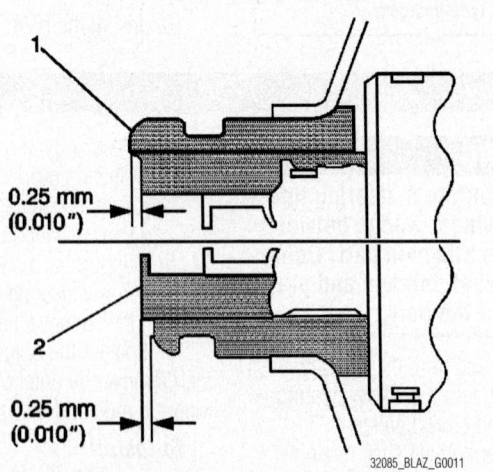

0.25 mm (0.010")

0.25 mm (0.010")

32085_BLAZ_G0011

Fig. 99 Install the power steering pump pulley (1) flush against the end of the power steering pump shaft (2), with an allowable variance of 0.010 in. (0.25mm)

11. Align the power steering pump with mounting bolt holes on engine block.

12. Install the power steering pump mounting bolts. Tighten the bolts to 18 ft. lbs. (25 Nm).

13. Attach the power steering pump return hose to power steering pump.

14. Attach the power steering pump pressure hose to power steering pump.

15. Tighten the fittings to 18 ft. lbs. (25 Nm).

16. Install the PCM to PCM mounting bracket.

17. Install the drive belt.

18. Bleed the power steering system.

SUSPENSION

COIL SPRING

REMOVAL & INSTALLATION

See Figure 101.

➡**This procedure requires the use of a suitable spring compressor.**

1. Before servicing the vehicle, refer to the precautions in the beginning of this section.

2. Remove or disconnect the following:
- Wheel
- Shock module
- Shock module yoke-to-shock absorber pinch bolt and nut

3. Spread the shock module yoke at the pinch bolt using a suitable flat-bladed tool.
- Shock module yoke from the shock absorber

4. Install pieces of heater hose or equivalent material to the shock module spring where the spring compressor contacts the lower part of the spring.

5. Install the shock module into the spring compressor.

➡**The spring is compressed when the shock absorber moves freely.**

6. Turn the spring compressor forcing screw until the coil spring is compressed.

7. Remove or disconnect the following:
- Shock absorber upper retaining nut
- Shock absorber from the shock module

8. Loosen the compressor forcing screw until the upper mounting plate and coil spring can be removed.
- Upper mounting plate and coil spring from the spring compressor

To install:

9. Install or connect the following:
- Coil spring and upper mounting plate to the spring compressor

10. Turn the compressor forcing screw until the coil spring is compressed.
- Shock absorber to the shock module. Tighten the retaining nut to 33 ft. lbs. (45 Nm)

11. Remove the shock module from the spring compressor. Remove the pieces of heater hose from the spring.
- Shock module yoke to the shock absorber
- Shock module yoke-to-shock pinch bolt and nut and tighten to 52 ft. lbs. (70 Nm)
- Shock module to the vehicle
- Tire and wheel

12. Lower the vehicle

LOWER BALL JOINT

REMOVAL & INSTALLATION

See Figures 102 through 104.

FRONT SUSPENSION

➡**This procedure requires the use of the following special tools: J 9519-E Lower Ball Joint Remover and Installer, J 34874 Booster Seal Remover/Installer, J 41435 Ball Joint Installer, J 45105-1 Ball Joint Flaring Adapter and J 45105-2 Receiver.**

1. On 4WD vehicles, remove the wheel center cap and drive axle nut.

2. Raise and support the vehicle.

3. Remove or disconnect the following:
- Tire and wheel
- Wheel hub and bearing, if necessary
- Outer tie rod retaining nut
- Out tie rod from the steering knuckle using a suitable puller
- Brake hose bracket retaining bolts and bracket
- Upper control arm-to-steering knuckle pinch bolt and nut
- Upper control arm from the steering knuckle
- Lower ball joint retaining nut
- Steering knuckle from the lower control arm using a suitable ball joint removal tool
- Steering knuckle from the vehicle
- Lower ball joint flange with a chisel

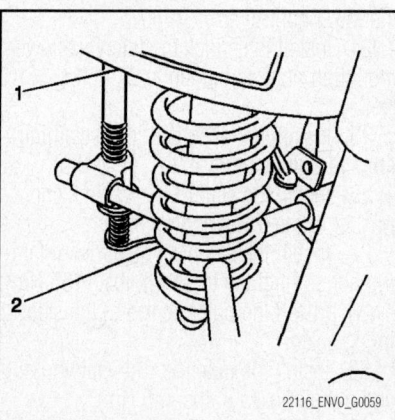

22116_ENVO_G0059

Fig. 101 Place pieces of heater hose to the spring where the compressor contacts the lower part of the spring

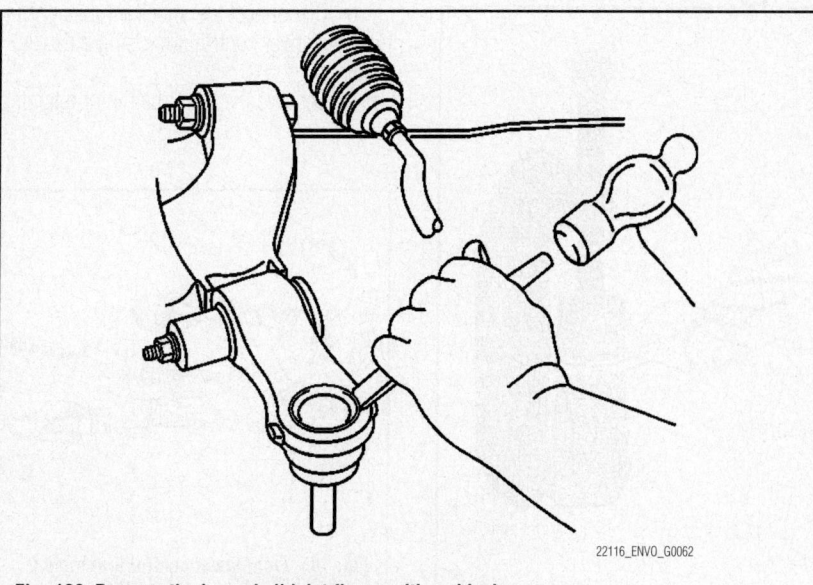

22116_ENVO_G0062

Fig. 102 Remove the lower ball joint flange with a chisel

4. Install tools J 9519-E and J 34874 to the lower ball joint, then use those tools to remove the lower ball joint from the lower control arm.

To install:

5. Install or connect the following:
- Lower ball joint to the lower control arm, using tools J 9519-E, J 41435 and J 45105-2

6. Remove the tools from the lower control arm.
- Tools J 9519-E and J 45105-1 to the lower ball joint

7. Flare the lower ball joint flange with J 9519-E and J 45105-1, then remove the tools from the lower ball joint.

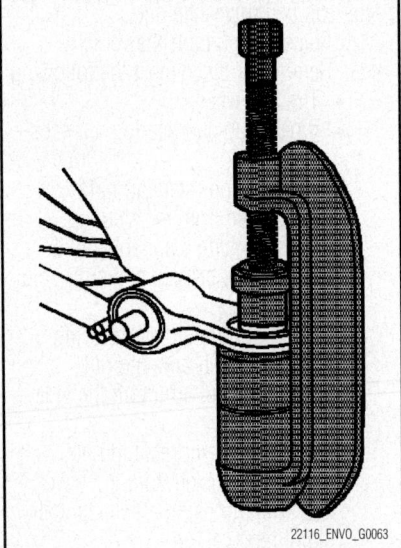

22116_ENVO_G0063

Fig. 103 Driving the lower joint from the control arm

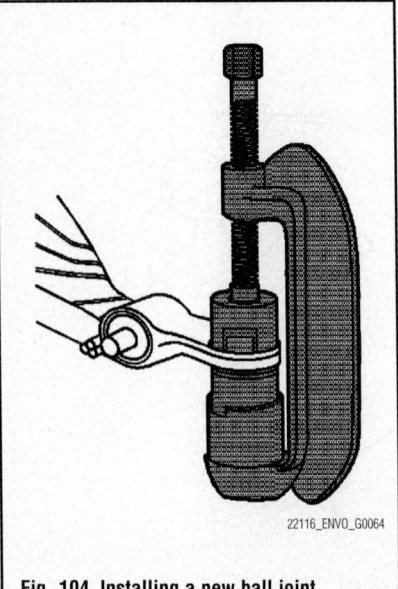

22116_ENVO_G0064

Fig. 104 Installing a new ball joint

- Steering knuckle to the lower control arm
- Lower ball joint retaining nut and tighten to 81 ft. lbs. (110 Nm)
- Upper control arm to the steering knuckle
- Upper control arm pinch bolt and nut and tighten to 30 ft. lbs. (41 Nm)
- Brake hose bracket to the steering knuckle
- Brake hose bracket retaining nuts and tighten to 7 ft. lbs. (10 Nm)
- Outer tie rod to the steering knuckle
- Outer tie rod retaining nut and tighten to 33 ft. lbs. (45 Nm)
- Wheel hub and bearing, if removed
- Tire and wheel

8. Lower the vehicle
- Drive axle nut, if 4WD, and tighten to 103 ft. lbs. (140 Nm)
- Wheel center cap, if removed

9. Check the front wheel alignment.

LOWER CONTROL ARM

REMOVAL & INSTALLATION

See Figure 105.

1. Raise and support the vehicle.
2. Remove the wheel and tire.
3. Remove the outer tie rod retaining nut.
4. Disconnect the outer tie rod from the steering knuckle using a tie rod puller.
5. Remove the stabilizer shaft link lower retaining nut.
6. Disconnect the stabilizer shaft link and washer from the lower control arm.
7. Remove the shock module yoke lower mounting nut.
8. Disconnect the shock module yoke from the lower control arm using a tie rod puller.
9. Remove the lower ball joint retaining nut.

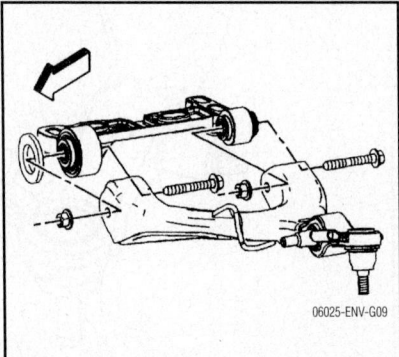

06025-ENV-G09

Fig. 105 Front lower control arm mounting

10. Disconnect the lower ball joint from steering knuckle using ball joint remover.
11. Remove the lower control arm-to-lower control arm bracket mounting nuts.
12. Note the direction the bolts are removed for installation.
13. Remove the lower control arm to lower control arm bracket mounting bolts.
14. Take care not to disengage the axle shaft from the transmission (4WD only).
15. Pivot the lower control arm outward and downward in order to disconnect the lower control arm from the lower control arm bracket.
16. Ensure that the spacer stays in position on the front control arm bracket front bushing.
17. Remove the lower control arm from the vehicle.

To install:

18. Position the lower control arm ball joint stud to the steering knuckle.
19. Ensure that the spacer stays in position on the front control arm bracket front bushing.
20. Pivot the lower control arm outward and upward in order to connect the lower control arm to the lower control arm bracket.
21. Install the lower control arm to lower control arm bracket mounting bolts.

➡**Ensure that the lower control arm is parallel to the lower control arm bracket during the installation and tightening of the lower control arm mounting bolts and nuts. This will ensure correct alignment of the lower control arm bushings.**

22. Install the lower control arm to lower control arm bracket mounting nuts and tighten to 96 ft. lbs. (130 Nm).
23. Connect the shock module yoke to the lower control arm.
24. Install the shock module yoke lower mounting nut and tighten to 81 ft. lbs. (110 Nm).
25. Install the lower ball joint retaining nut and tighten to 81 ft. lbs. (110 Nm).
26. Install the stabilizer shaft link and washer to the lower control arm.
27. Install the stabilizer shaft link retaining nut and tighten to 114 ft. lbs. (155 Nm).
28. Install the outer tie rod to the steering knuckle.
29. Install the outer tie rod retaining nut and tighten to 33 ft. lbs. (45 Nm).
30. Install the tire and wheel.
31. Lower the vehicle.
32. Check the front wheel alignment.

CONTROL ARM BUSHING REPLACEMENT

The control arm bushings are serviced with the control arm as an assembly.

LOWER CONTROL ARM BRACKET

REMOVAL & INSTALLATION

See Figure 106.

➡**This procedure requires the use of Steering Linkage and Tie Rod Puller tool No. J 24319-B and Ball Joint Remover tool No. J 43631.**

1. Before servicing the vehicle, refer to the precautions in the beginning of this section.
2. Raise the vehicle.
3. Remove or disconnect the following:
   - Tire and wheel
   - Outer tie rod retaining nut
   - Outer tie rod from the steering knuckle using Ball Joint Removal tool No. J 43631
   - Stabilizer shaft link lower nut, link and washer
   - Shock module yoke lower nut and shock module using Steering Linkage and Tie Rod Puller tool No. J 24319-B
   - Lower control arm-to-lower control arm bracket mounting bolts

➡**Make sure to note the direction that the bolts are removed for installation.**

   - Lower control arm-to-lower control arm bracket mounting bolts
   - Lower ball joint retaining nut
   - Lower ball joint from the steering knuckle using Ball Joint Removal tool No. J 43631

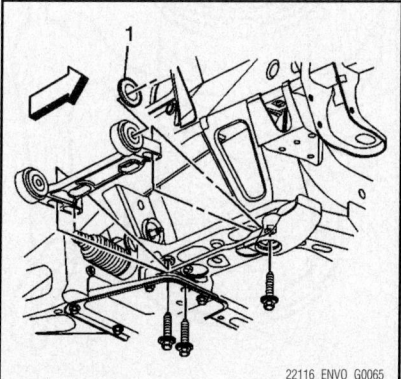

Fig. 106 Remove the lower control arm bracket from the vehicle

22116_ENV0_G0065

➡**On 4WD vehicles, make sure not to disengage the axle shaft from the transmission.**

4. Pivot the lower control arm out and down to disengage the lower control arm from the bracket, then remove the lower control arm from the knuckle.

➡**Note the position of the spacer (1) on the front bushing.**

5. Remove the lower control arm bracket mounting bolts from the frame.
6. Remove the lower control arm bracket from the vehicle.

To install:

➡**Ensure the spacer (1) is in the proper position on the front bushing.**

7. Install the lower control arm bracket to the vehicle.
8. Install the lower control arm bracket mounting bolts to the frame.
   - Tighten the front lower control arm bracket mounting bolt to 192 ft. lbs. (260 Nm).
   - Tighten the rear lower control arm bracket mounting bolts to 170 ft. lbs. (230 Nm).
9. Install or connect the following:
   - Lower control arm to the steering knuckle
   - Lower control to the bracket by pivoting it out and up

➡**During installation and tightening of the bolts and nuts, make sure that the lower control arm is parallel to the bracket. This is to maintain proper alignment of the lower control arm bushings.**

   - Lower control arm-to-bracket mounting bolts and tighten to 81 ft. lbs. (111 Nm)
   - Shock module yoke to the lower control arm
   - Shock module yoke lower mounting nut

➡**If it becomes necessary to replace the washer, use only an identical hardened steel, felt lined washer. Standard washers must not be used.**

   - Stabilizer shaft link and washer to the lower control arm
   - Stabilizer shaft link retaining bolt and tighten to 74 ft. lbs. (100 Nm)
   - Outer tie rod to the steering knuckle. Tighten the nuts to 33 ft. lbs. (45 Nm).
   - Tire and wheel
10. Lower the vehicle and check the front end alignment.

STABILIZER BAR

REMOVAL & INSTALLATION

1. Raise and support the vehicle.
2. Remove the stabilizer shaft links to the stabilizer shaft retaining nuts.
3. Remove the stabilizer shaft insulator clamp mounting bolts.
4. Remove the stabilizer shaft insulator clamp from the stabilizer shaft insulator.
5. Remove the stabilizer shaft insulators from the stabilizer shaft.
6. Remove the stabilizer shaft from the vehicle.

To install:

7. Install the stabilizer shaft to the vehicle.
8. Install the stabilizer shaft insulators to the stabilizer shaft.
9. Install the stabilizer shaft insulator clamp to the stabilizer shaft insulator.
10. Install the stabilizer shaft insulator clamp mounting bolts and tighten to 41 ft. lbs. (55 Nm).
11. Install the stabilizer shaft links to the stabilizer shaft and tighten to 74 ft. lbs. (100 Nm).
12. Lower the vehicle.

STEERING KNUCKLE

REMOVAL & INSTALLATION

1. Raise and support the vehicle.
2. Remove the tire and wheel.
3. On 4WD vehicles, remove wheel center cap, if equipped, and the drive axle nut and washer
4. Remove the wheel hub and bearing.
5. Remove the outer tie rod retaining nut.
6. Disconnect the outer tie rod from the steering knuckle using a tie rod puller.
7. Remove the brake hose bracket retaining bolts.
8. Remove the brake hose bracket from the steering knuckle.
9. Disconnect the ABS wheel speed sensor wiring harness bracket from the steering knuckle.
10. Remove the upper control arm to the steering knuckle pinch bolt and nut.
11. Disconnect the upper control arm from the steering knuckle.
12. Remove the lower ball joint retaining nut.
13. Remove the steering knuckle from the lower control arm.
14. Remove the steering knuckle from the vehicle.

To install:

15. Install the steering knuckle to the lower control arm.

16. Install the lower ball joint retaining nut and tighten to 81 ft. lbs. (110 Nm).

17. Connect the upper control arm to the steering knuckle.

18. Install upper control arm pinch bolt and nut and tighten to 30 ft. lbs. (40 Nm).

19. Connect the ABS wheel speed sensor wiring harness bracket to the steering knuckle.

20. Install the brake hose bracket to the steering knuckle.

21. Install the brake hose bracket retaining bolts and tighten to 89 inch lbs. (10 Nm).

22. Install the outer tie rod to the steering knuckle.

23. Install the new outer tie rod retaining nut and tighten to 33 ft. lbs. (45 Nm) on 2WD models, or 44 ft. lbs. (60 Nm) on 4WD models.

24. Install the wheel hub and bearing.

25. On 4WD vehicles, install the drive axle nut and tighten to 103 ft. lbs. (140 Nm), then install the center cap.

26. Install the tire and wheel.

27. Lower the vehicle.

28. Adjust the front toe.

STRUT/SHOCK MODULE

REMOVAL & INSTALLATION

See Figure 107.

➡**A "shock module", similar to a strut was used on these vehicles. This procedure requires the use of a suitable steering linkage and tie rod puller.**

1. Before servicing the vehicle, refer to the precautions in the beginning of this section.

2. Remove or disconnect the following:

- Shock module upper retaining nuts
- Tire and wheel
- Shock module-to-lower control arm retaining nut
- Shock module yoke from the lower control arm using a suitable puller
- Shock module from the shock tower and lower control arm

To install:

3. Install or connect the following:

- Shock module to the shock tower and lower control arm
- Shock module yoke to the lower control arm
- Shock module upper retaining nuts and tighten to 33 ft. lbs. (45 Nm)
- Shock module-to-lower control arm retaining nut and tighten to 81 ft. lbs. (110 Nm)
- Tire and wheel

UPPER CONTROL ARM

REMOVAL & INSTALLATION

1. Before servicing the vehicle, refer to the precautions in the beginning of this section.

2. Remove or disconnect the following:

- Tire and wheel assembly
- Upper ball joint-to-upper control arm pinch bolt and nut
- Upper control arm from the knuckle
- Anti-lock Brake System (ABS) wheel speed sensor wiring harness
- Upper control arm mounting bolts
- Upper control arm

To install:

3. Install or connect the following:

- Upper control arm and tighten the bolts to 111 ft. lbs. (150 Nm)
- ABS wheel speed sensor wiring harness
- Upper control arm to the steering knuckle
- Upper ball joint-to-upper control arm pinch bolt and nut and tighten to 30 ft. lbs. (40 Nm)
- Tire and wheel

4. Check the front wheel alignment.

CONTROL ARM BUSHING REPLACEMENT

The control arm bushings are serviced with the control arm as an assembly.

UPPER BALL JOINT

REMOVAL & INSTALLATION

See Figures 108 and 109.

➡**This procedure requires the use of the following special tools: J 9519-E Lower Ball Joint Remover and Installer, J 21474-01 Control Arm Bushing Set and J 45117 Ball Joint Installation Spacer.**

1. Raise and safely support the front of the vehicle securely on jackstands.

2. Remove the tire and wheel.

3. Remove the steering knuckle with wheel hub attached.

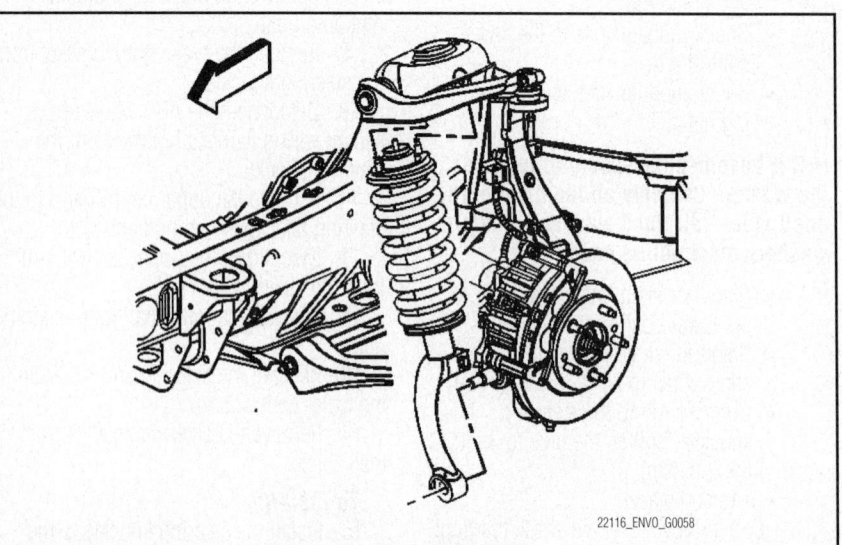

22116_ENVO_G0058

Fig. 107 View of the shock module used on the front suspension

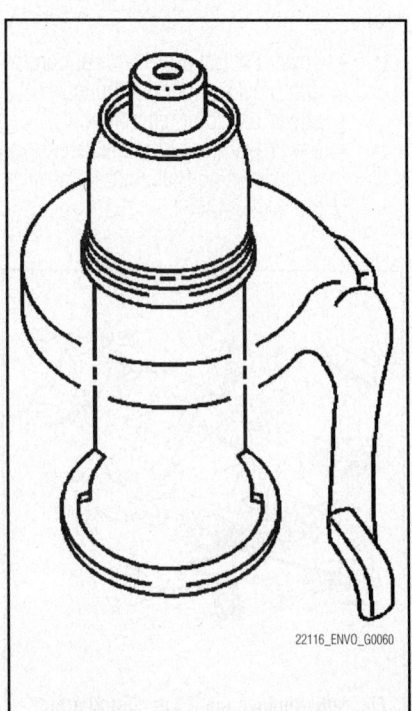

22116_ENVO_G0060

Fig. 108 Remove the upper ball joint boot

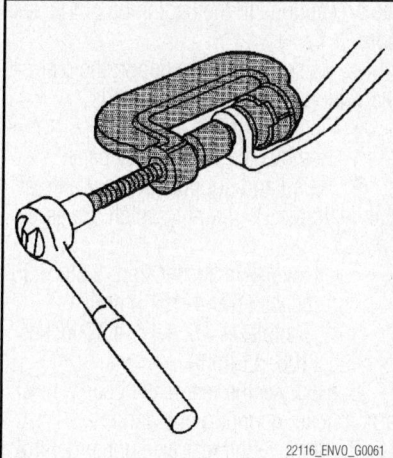

Fig. 109 Remove the upper ball joint from the steering knuckle using J-9519-E

22116_ENVO_G0061

4. Remove the upper ball joint retaining clip.

5. Remove the upper ball joint boot.

6. Remove the upper ball joint from the steering knuckle using J-9519-E.

To install:

7. Install the upper ball joint to steering knuckle using J-9519-E, J-21474-01, and J-45117 .

8. Install the upper ball joint retaining clip.

9. Install the steering knuckle with wheel hub attached.

10. Install the tire and wheel.

11. Lower the vehicle.

12. Check the front wheel alignment.

WHEEL HUB AND BEARINGS

REMOVAL & INSTALLATION

1. Before servicing the vehicle, refer to the precautions in the beginning of this section.

2. On 4WD vehicles, remove wheel center cap, if equipped, and the drive axle nut and washer

3. Raise and support the vehicle.

4. Remove or disconnect the following:
- Tire and wheel
- Caliper, leaving the fluid lines connected
- Brake rotor
- Halfshaft from the hub and bearing on 4WD vehicles. Place a brass drift against the outer edge of the halfshaft to protect the shaft threads. Use a hammer to sharply strike the brass drift, but to do not remove the halfshaft at this time.
- Wheel speed sensor
- Wheel hub and bearing-to-steering knuckle bolts and hub and bearing

➡**Lay the hub and bearing on the wheel studs on the outboard side. This will avoid damaging the bearing seal.**

- Splash shield from the steering knuckle
- Seal from the hub and bearing

To install:

5. Install or connect the following:
- Wheel hub and bearing seal
- Splash shield to the steering knuckle, making sure it's properly aligned
- Hub and bearing to the steering knuckle, aligning the threaded holes
- Hub and bearing bolts and tighten to 77 ft. lbs. (105 Nm)
- Wheel speed sensor. Tighten the bolt to 13 ft. lbs. (18 Nm).
- Rotor and brake caliper
- Tire and wheel

6. Lower the vehicle

7. On 4WD vehicles, install the drive axle nut and tighten to 103 ft. lbs. (140 Nm), then install the center cap.

ADJUSTMENT

The wheel bearings on these vehicles are not adjustable. If the bearings become loose or make noise, they must be replaced.

SUSPENSION

COIL SPRING

REMOVAL & INSTALLATION

1. Before servicing the vehicle, refer to the precautions in the beginning of this section.

2. Raise and support the vehicle.

3. Support the rear axle.

4. Remove the shock absorber lower mounting bolts.

➡**Do not lower the rear axle so the upper control arms contact the frame. This will damage the upper control arms.**

5. Lower the rear axle, then remove the coil springs.

➡**Be careful not to chip or scratch the coating of the coil springs when removing and installing the springs. Damaging the coating will cause premature failure of the coil springs.**

To install:

6. Install the coil springs, then raise the rear axle.

7. Install the shock absorber lower mounting bolts and tighten to 59 ft. lbs. (80 Nm).

8. Remove the rear axle support.

9. Lower the vehicle.

LOWER CONTROL ARM

REMOVAL & INSTALLATION

1. Raise and support the vehicle.

2. Remove the wheel and tire.

3. Raise and support the rear axle at the designed height of 5.33 in. (135.4mm).

4. Remove the rear axle lower control arm to the axle mounting nut and bolt.

5. Remove the rear axle lower control arm to the frame mounting nut and bolt.

6. Remove the lower control arm.

REAR SUSPENSION

To install:

7. Install the lower control arm.

8. Install the rear axle lower control arm to the frame mounting nut and bolt.

9. Install the rear axle lower control arm to the axle mounting bolt and nut and tighten to 74 ft. lbs. (100 Nm).

10. Remove the rear axle support.

11. Lower the vehicle.

SHOCK ABSORBER

REMOVAL & INSTALLATION

See Figure 110.

1. Before servicing the vehicle, refer to the precautions in the beginning of this section.

2. Properly support the rear axle assembly.

3. Remove or disconnect the following:
- Automatic level control air lines from the shock absorber, if equipped

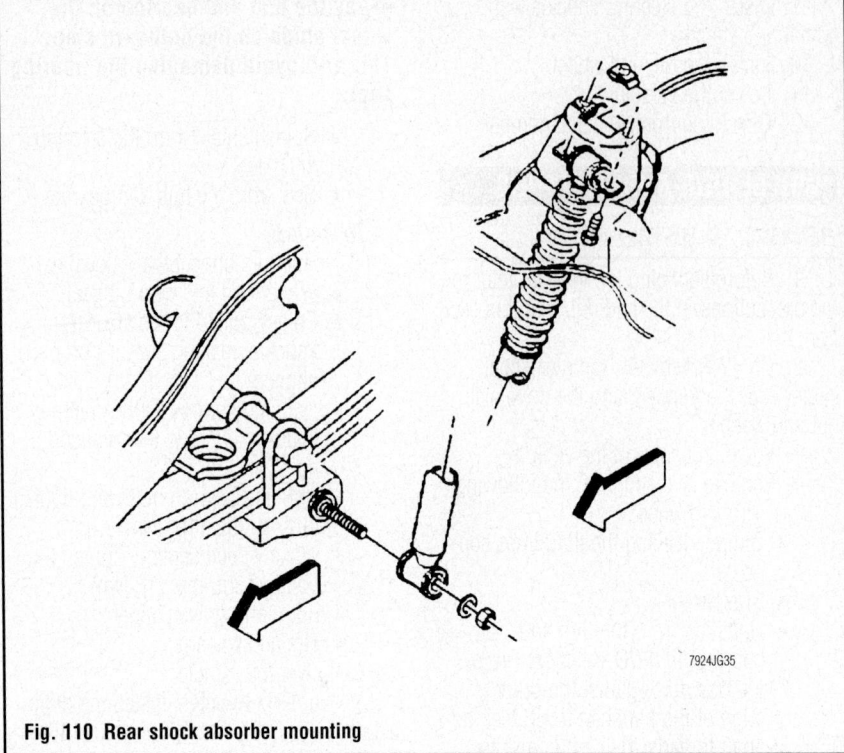

Fig. 110 Rear shock absorber mounting

- Shock absorber-to-frame retainer(s) at the top of the shock
- Shock-to-axle retainer(s) at the bottom of the shock
- Shock absorber

To install:

4. Install the shock in the vehicle and loosely install the upper mounting fasteners to retain it

5. Align the lower-end of the shock absorber with the axle mounting, then loosely install the retainers.

6. Tighten the upper and lower shock retainers to 63 ft. lbs. (85 Nm).

7. If equipped, attach the automatic level control air lines to the shock absorber.

STABILIZER BAR

REMOVAL & INSTALLATION

1. Raise and support the vehicle.
2. Remove the stabilizer shaft links to the stabilizer shaft retaining nuts.
3. Remove the stabilizer shaft insulator clamp mounting bolts.
4. Remove the stabilizer shaft insulator clamp from the stabilizer shaft insulator.
5. Remove the stabilizer shaft insulators from the stabilizer shaft.
6. Remove the stabilizer shaft from the vehicle.

To install:

7. Install the stabilizer shaft to the vehicle.

8. Install the stabilizer shaft insulators to the stabilizer shaft.

9. Install the stabilizer shaft insulator clamp to the stabilizer shaft insulator.

10. Install the stabilizer shaft insulator clamp mounting bolts and tighten to 52 ft. lbs. (70 Nm).

11. Install the stabilizer shaft links to the stabilizer shaft and tighten to 74 ft. lbs. (100 Nm).

12. Lower the vehicle.

UPPER CONTROL ARM

REMOVAL & INSTALLATION

See Figure 111.

1. Before servicing the vehicle, refer to the precautions in the beginning of this section.

2. Raise and safely support the rear of the vehicle securely on jackstands.

3. Remove the tire and wheel.

4. Remove the wheelhouse panel.

5. Raise and support the rear axle at the designed D - height, which are as follows:

- Except Air Suspension; 5.88–6.35 inches (149.4–161.4mm)
- Trailblazer SS; 4.17–4.49 inches (106–114mm)

6. Remove the rear axle upper control arm to axle mounting bolt and nut.

7. Remove the rear axle upper control arm to frame mounting bolt.

8. Remove the rear axle upper control arm.

To install:

9. Install the rear axle upper control arm.

10. Install the rear axle upper control arm to frame mounting bolt.

11. Install the rear axle upper control arm to axle mounting nut and bolt and tighten the bolts to 97 ft. lbs. (131 Nm).

12. Remove the rear axle support.

13. Install the wheelhouse panel.

14. Install the tire and wheel.

15. Lower the vehicle.

WHEEL BEARINGS

REMOVAL & INSTALLATION

For wheel bearing removal, refer to the procedure under Rear Axle Shaft, Bearing & Seal in the DRIVETRAIN section.

ADJUSTMENT

The wheel bearings on these vehicles are not adjustable. If the bearings become loose or make noise, they must be replaced.

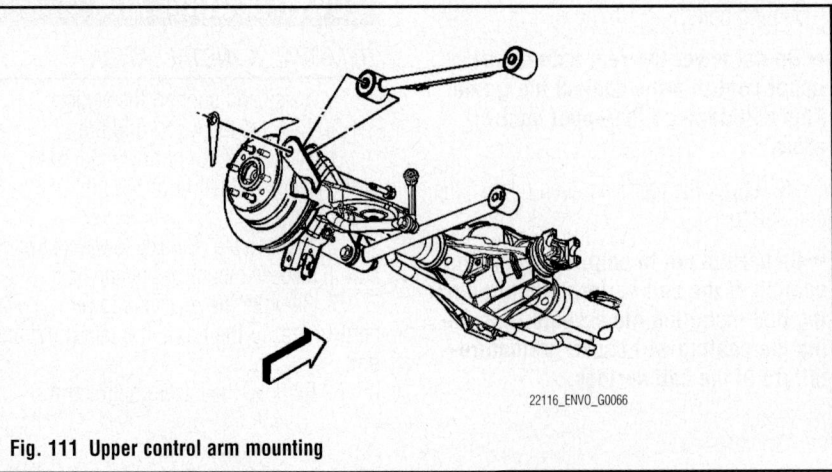

Fig. 111 Upper control arm mounting

ISUZU

Diagnostic Trouble Codes

15

DIAGNOSTIC TROUBLE CODES

OBD II VEHICLE APPLICATIONS

ISUZU

Ascender
2007-2008
• 4.2L I6 .... VIN S

i290
2007-08
• 2.9L 4-Cyl. .... VIN 9

i370
2007-08
• 3.7L 5-Cyl. .... VIN E

Gas Engine OBD II Trouble Code List (P0xxx Codes)

DTC	Trouble Code Title, Conditions & Possible Causes
DTC: P0101 **2T CCM, MIL:** YES **Years:** 2007, 2008 **Models:** Ascender, i290, i370 **Engines:** 2.9L VIN 9, 3.7L VIN E, 4.2L VIN S **Transmissions:** A/T, M/T	**Mass Airflow Sensor Circuit Range/Performance** DTC P0106, P0107, P0108, P0121, P0122 and P0123 not set, system voltage at 11-16v, engine running at a stable idle speed, throttle angle stable (±1%), Calculated airflow from 25-40 g/sec, conditions met for 1 second, and the PCM detected a MAF sensor frequency that was significantly higher or lower than a "predicted" MAF airflow based on throttle position and engine speed for 12.5 seconds over a 25 second period during the CCM Rationality test. **Possible Causes:** • Air leaks after the MAF sensor, or in the EGR or PCV system • Engine oil cap missing, engine oil dipstick not fully seated • MAF sensor is contaminated, dirty or out-of-calibration • MAF sensor ground circuit has high resistance • MAF minimum airflow rate to low at idle or during deceleration • MAP or TP sensor signal skewed, stuck or out of calibration • High signal interference (i.e., electrical noise from the ignition) • PCM has failed
DTC: P0102 **1T CCM, MIL:** YES **Years:** 2007, 2008 **Models:** Ascender, i290, i370 **Engines:** 2.9L VIN 9, 3.7L VIN E, 4.2L VIN S **Transmissions:** A/T, M/T	**MAF Sensor Circuit Low Frequency** Engine started, engine speed over 500 rpm for 10 seconds, system voltage over 11.5v, the PCM detected the MAF sensor frequency was less than 1000 Hz for a total of 50% of the last 100 samples in the CCM Rationality test (a sample is taken every cylinder event). **Possible Causes:** • MAF sensor signal is shorted to ground • MAF sensor power circuit is open • MAF sensor is contaminated, dirty or is damaged • PCM has failed
DTC: P0103 **1T CCM, MIL:** YES **Years:** 2007, 2008 **Models:** Ascender, i290, i370 **Engines:** 2.9L VIN 9, 3.7L VIN E, 4.2L VIN S **Transmissions:** A/T, M/T	**MAF Sensor Circuit High Frequency** Engine started, engine speed over 500 rpm for 10 seconds, system voltage over 11.5v, the PCM detected the MAF sensor frequency was more than 10,000 Hz for a total of 50% of the last 200 samples in the CCM Rationality test (a sample is taken every cylinder event). **Possible Causes:** • RFI or EMI interference from the Generator or Ignition system • RFI or EMI interference from an Ignition system component • MAF sensor is contaminated, dirty or is damaged • PCM has failed
DTC: P0106 **2T CCM, MIL:** YES **Years:** 2007, 2008 **Models:** Ascender, i290, i370 **Engines:** 2.9L VIN 9, 3.7L VIN E, 4.2L VIN S **Transmissions:** A/T, M/T	**MAP Sensor Signal Range/Performance** DTC P0121, P0122 and P0123 not set, engine speed stable (±100 rpm), throttle angle stable (±1%), IAC counts steady (±10 counts), EGR flow stable (±4%), no change in the A/C clutch, PSPS, Brake switch or TCC status, conditions met for 1 second, and the PCM detected the Actual MAP sensor value varied more than 10 kPa from the Expected MAP value for 10 seconds over a 20 second period. **Possible Causes:** • MAP sensor circuit open or shorted to ground (intermittent) • MAP sensor source vacuum line is leaking or restricted • MAP sensor source vacuum line is plugged at intake manifold • MAP sensor is damaged, out-of-calibration or has failed • PCM has failed
DTC: P0107 **1T CCM, MIL:** YES **Years:** 2007, 2008 **Models:** Ascender, i290, i370 **Engines:** 2.9L VIN 9, 3.7L VIN E, 4.2L VIN S **Transmissions:** A/T, M/T	**MAP Sensor Circuit Low Input** DTC P0121, P0122 and P0123 not set, engine started, system voltage from 11-16v, then with the engine speed below 1000 rpm and throttle angle over 1%, or with the engine speed over 1000 rpm and throttle angle over 2%, the PCM detected the MAP sensor was less than 0.04v (11 kPa) for 10 seconds over a 20 second period. **Possible Causes:** • MAP sensor circuit shorted to ground between sensor and PCM • MAP sensor power circuit is open or shorted to ground • MAP sensor is damaged or has failed • PCM has failed

DTC	Trouble Code Title, Conditions & Possible Causes
DTC: P0108 **1T CCM, MIL: YES** **Years:** 2007, 2008 **Models:** Ascender, i290, i370 Engines: 2.9L VIN 9, 3.7L VIN E, 4.2L VIN S **Transmissions:** A/T, M/T	**MAP Sensor Circuit High Input** DTC P0121, P0122 and P0123 not set, engine started, engine speed less than 1000 rpm and the throttle angle less than 3%, or with engine speed more than 1000 rpm and the throttle angle less than 10%, the PCM detected the MAP sensor was more than 4.40v (90 kPa) for 10 seconds over a 16 second period during the test. **Possible Causes:** • MAP sensor circuit is open between the sensor and the PCM • MAP sensor signal circuit is shorted to VREF or system power • MAP sensor ground circuit is open between sensor and PCM • MAP sensor is damaged or has failed • PCM has failed
DTC: P0112 **1T CCM, MIL: YES** **Years:** 2007, 2008 **Models:** Ascender, i290, i370 Engines: 2.9L VIN 9, 3.7L VIN E, 4.2L VIN S **Transmissions:** A/T, M/T	**Intake Air Temperature Sensor Circuit Low Input** DTC P0502 not set, engine started, engine runtime over 2 minutes, vehicle speed over 30 mph, and the PCM detected the IAT sensor indicated less than 0.10v (Scan Tool reads 298°F) for 12.5 seconds over a 20 second period during the CCM test. **Possible Causes:** • IAT sensor circuit shorted to ground between sensor and PCM • IAT sensor is damaged, out-of-calibration or has failed • PCM has failed
DTC: P0113 **1T CCM, MIL: YES** **Years:** 2007, 2008 **Models:** Ascender, i290, i370 **Engines:** 2.9L VIN 9, 3.7L VIN E, 4.2L VIN S **Transmissions:** A/T, M/T	**Intake Air Temperature Sensor Circuit High Input** DTC P0502 not set, engine runtime 4 minutes, vehicle speed over 20 mph, ECT sensor more than 140°F, MAF sensor less than 20 g/sec, and the PCM detected the IAT sensor indicated more than 4.90v (Scan Tool reads -38°F) for 12.5 second over a 25 second period during the CCM test. **Possible Causes:** • IAT sensor signal circuit open between the sensor and PCM • IAT sensor signal is shorted to VREF or system power (B+) • IAT sensor is damaged, out-of-calibration or has failed • PCM has failed
DTC: P0117 **1T CCM, MIL: YES** **Years:** 2007, 2008 **Models:** Ascender, i290, i370 **Engines:** 2.9L VIN 9, 3.7L VIN E, 4.2L VIN S **Transmissions:** A/T, M/T	**Engine Coolant Temperature Sensor Circuit Low Input** Engine started, engine runtime over 1 minute, and the PCM detected the ECT sensor indicated less than 0.10v (Scan Tool reads 302°F) for 6.25 seconds for 50 seconds over a 100 second period during the CCM test. **Possible Causes:** • ECT sensor circuit shorted to ground between sensor and PCM • ECT sensor is damaged, out-of-calibration or has failed • PCM has failed
DTC: P0118 **1T CCM, MIL: YES** **Years:** 2007, 2008 **Models:** Ascender, i290, i370 **Engines:** 2.9L VIN 9, 3.7L VIN E, 4.2L VIN S **Transmissions:** A/T, M/T	**Engine Coolant Temperature Sensor Circuit High Input** Engine started, engine runtime over 90 seconds, and the PCM detected the ECT sensor was more than 4.90v (Scan Tool reads -38°F) for 50 seconds over a 100 second period during the CCM test. **Possible Causes:** • ECT sensor signal circuit is open • ECT sensor signal circuit is shorted to VREF or system power • ECT sensor is damaged or has failed • PCM has failed
DTC: P0125 **2T CCM, MIL: YES** **Years:** 2007, 2008 **Models:** Ascender, i290, i370 **Engines:** 2.9L VIN 9, 3.7L VIN E, 4.2L VIN S **Transmissions:** A/T, M/T	**Insufficient Coolant Temperature For Closed Loop** DTC P0112, P0113, P0117, P0118, P1111, P1112, P1114 and P1115 not set, engine started, ECT and IAT sensors from 14-82°F at startup, then with the IAT sensor from 17-50°F at startup, the PCM detected the ECT signal did not reach 84°F after 20 minutes; or with the IAT sensor more than 50°F at startup, the PCM detected the ECT sensor signal was less than 84°F after 2 minutes, condition met at least 20 times during the CCM Rationality test. **Possible Causes:** • Inspect for low coolant level or an incorrect coolant mixture • Check the operation of the thermostat (it may be stuck open) • ECT sensor is damaged or out-of-calibration (it is "skewed") • ECT sensor signal circuit has high resistance • ECT sensor has failed • PCM has failed

DTC	Trouble Code Title, Conditions & Possible Causes

DTC: P0128
2T CCM, MIL: YES
Years: 2000, 2001, 2002, 2003, 2004, 2005
Models: Ascender, i290, i370
Engines: All
Transmissions: A/T, M/T
	Thermostat Malfunction
	DTC P0101, P0102, P0103, P0112, P0113, P0117, P0118 and P0502 not set, and the PCM detected under one of these cases:
	Cold Case Startup Conditions
	  • IAT sensor from 20-50°F, and the PCM detected it took over 263 seconds to reach a stabilized thermostat regulated temperature.
	  • Warm Case Startup Conditions
	  • IAT sensor from 50-128°F, and the PCM detected it took over 239 seconds to reach a stabilized thermostat regulated temperature.
	Possible Causes:
	  • Check the operation of the thermostat (it may be stuck open)
	  • ECT sensor is damaged or out-of-calibration (it is "skewed")
	  • PCM has failed

DTC: P0131
1T CCM, MIL: YES
Years: 2007, 2008
Models: Ascender, i290, i370
Engines: 2.9L VIN 9, 3.7L VIN E, 4.2L VIN S
Transmissions: A/T, M/T
	HO2S-11 (Bank 1 Sensor 1) Circuit Low Input
	DTC P0106, P0107, P0108, P0112, P0113, P0117, P0118, P0121, P0122, P0123, P0171, P0172 and P0300, P0301-P0306 not set, engine running in closed loop with the A/F ratio from 14.5-14.8:1, ECT sensor more than 140°F, throttle angle from 3-19% for 5 seconds, the PCM detected the HO2S-11 signal was less than 26 mv for 77 seconds over a 90 second period during the test.
	Possible Cause:
	  • HO2S signal circuit is open or shorted to ground
	  • HO2S is water or fuel contaminated, or it has failed
	  • PCM has failed

DTC: P0132
1T CCM, MIL: YES
Years: 2007, 2008
Models: Ascender, i290, i370
Engines: 2.9L VIN 9, 3.7L VIN E, 4.2L VIN S
Transmissions: A/T, M/T
	HO2S-11 (Bank 1 Sensor 1) Circuit High Input
	DTC P0106, P0107, P0108, P0112, P0113, P0117, P0118, P0121, P0122, P0123, P0171, P0172 and P0300, P0301-P0306 not set, engine running in closed loop with the A/F ratio from 14.5-14.8:1, throttle angle from 3-19% for 5 seconds, the PCM detected the HO2S-11 signal was less than 952 mv for 77 seconds over a 90 second period; or the HO2S-11 signal was more than 500 mv during Decel Fuel Cutoff mode for 3 seconds during the CCM test.
	Possible Causes:
	  • HO2S signal circuit shorted to system power (oil in connector)
	  • HO2S is water or fuel contaminated
	  • HO2S is damaged or has failed
	  • PCM has failed

DTC: P0133
2T OBD/O2S, MIL: YES
Years: 2007, 2008
Models: Ascender, i290, i370
Engines: 2.9L VIN 9, 3.7L VIN E, 4.2L VIN S
Transmissions: A/T, M/T
	HO2S-11 (Bank 1 Sensor 1) Slow Response
	DTC P0106, P0107, P0108, P0112, P0113, P0117, P0118, P0121, P0122, P0123, P0171, P0172 and P0300, P0301-P0306 not set, engine runtime 1 minute in closed loop, ECT sensor more than 122°F, engine speed from 1500-3000 rpm, MAF sensor from 9-42 g/sec, Purge duty cycle over 1%, conditions met for 3 seconds, then 90 seconds after entering closed loop, the PCM detected the HO2S-11 lean-to-rich average transition time was over 94 ms, or the rich-to-lean average transition time was over 105 ms during the test.
	Possible Causes:
	  • Exhaust leak present in the exhaust manifold or exhaust pipes
	  • HO2S element fuel contamination
	  • HO2S element has deteriorated
	  • PCM has failed

DTC: P0134
1T OBD/O2S, MIL: YES
Years: 2007, 2008
Models: Ascender, i290, i370
Engines: 2.9L VIN 9, 3.7L VIN E, 4.2L VIN S
Transmissions: A/T, M/T
	HO2S-11 (Bank 1 Sensor 1) Insufficient Activity Detected
	DTC P0106, P0107, P0108, P0112, P0113, P0117, P0118, P0121, P0122, P0123, P0171, P0172 and P0300, P0301-306 not set, system voltage from 11-16v, engine runtime over 40 seconds, then after the PCM determined the Oxygen Sensor Heater test passed, it detected the HO2S-11 signal remained from 400-500 mv for 77 seconds over a 90 second period in the Oxygen Sensor Monitor test.
	Possible Causes:
	  • Exhaust leak present in exhaust manifold or exhaust pipes
	  • HO2S element fuel contamination or has deteriorated
	  • HO2S signal circuit or the ground circuit has high resistance
	  • HO2S heater element has failed, or the heater circuit is open
	  • PCM has failed

DTC: P0135
2T OBD/O2S HTR1, MIL: YES
Years: 2007, 2008
Models: Ascender, i290, i370
Engines: 2.9L VIN 9, 3.7L VIN E, 4.2L VIN S
Transmissions: A/T, M/T
	HO2S-11 (Bank 1 Sensor 1) Heater Circuit Malfunction
	No HO2S-11 codes set, ECT and IAT sensors less than 90°F and within 14°F at startup, engine running, system voltage from 11-16v, then with the average Calculated airflow less than 15 g/sec during the test period, the PCM detected the HO2S-11 signal did not vary more than 150 mv from the bias voltage of 400 to 500 mv for up to 150 seconds during the Oxygen Sensor Heater Monitor test.
	Possible Causes:
	  • HO2S heater power circuit is open (check the 20A heater fuse)
	  • HO2S heater ground circuit is open
	  • HO2S heater element has high resistance or has failed
	  • PCM has failed

DTC	Trouble Code Title, Conditions & Possible Causes
DTC: P0137 **1T CCM, MIL: YES** **Years:** 2007, 2008 **Models:** Ascender, i290, i370 **Engines:** 2.9L VIN 9, 3.7L VIN E, 4.2L VIN S **Transmissions:** A/T, M/T	**HO2S-12 (Bank 1 Sensor 2) Circuit Low Input** DTC P0106, P0107, P0108, P0112, P0113, P0117, P0118, P0121, P0122, P0123, P0171, P0172 and P0300, P0301-P0306 not set, engine running in closed loop with the A/F ratio from 14.5-14.8:1, ECT sensor more than 140°F, throttle angle from 3-19% for 5 seconds, the PCM detected the HO2S-12 signal was less than 26 mv for 106 seconds over a 125 second period during the CCM test. **Possible Causes:** • HO2S signal circuit is open or shorted to ground • HO2S is water or fuel contaminated, or it has failed • PCM has failed
DTC: P0138 **1T CCM, MIL: YES** **Years:** 2007, 2008 **Models:** Ascender, i290, i370 **Engines:** 2.9L VIN 9, 3.7L VIN E, 4.2L VIN S Transmissions: A/T, M/T	**HO2S-12 (Bank 1 Sensor 2) Circuit High Input** DTC P0106, P0107, P0108, P0112, P0113, P0117, P0118, P0121, P0122, P0123, P0171, P0172 and P0300, P0301-P0306 not set, engine running in closed loop with the A/F ratio command at 14.5-14.8:1, throttle angle from 3-19% for 5 seconds, the PCM detected the HO2S-12 signal was less than 952 mv for 106 seconds over a 125 second period; or the HO2S-12 signal was more than 500 mv during Decel Fuel Cutoff mode for 3 seconds during the CCM test. **Possible Causes:** • HO2S signal circuit shorted to system power (oil in connector) • HO2S is water or fuel contaminated • HO2S is damaged or has failed • PCM has failed
DTC: P0140 **1T OBD/O2S, MIL: YES** **Years:** 2007, 2008 **Models:** Ascender, i290, i370 **Engines:** 2.9L VIN 9, 3.7L VIN E, 4.2L VIN S **Transmissions:** A/T, M/T	**HO2S-12 (Bank 1 Sensor 2) Insufficient Activity Detected** DTC P0106, P0107, P0108, P0112, P0113, P0117, P0118, P0121, P0122, P0123, P0171, P0172 and P0300, P0301-306 not set, system voltage from 11-16v, engine runtime over 40 seconds, then after the PCM determined the Oxygen Sensor Heater test passed, it detected the HO2S-12 signal remained from 426-474 mv for 105 seconds of a 125 second period in the Oxygen Sensor Monitor test. **Possible Causes:** • Exhaust leak present in exhaust manifold or exhaust pipes • HO2S element fuel contamination or has deteriorated • HO2S signal circuit or the ground circuit has high resistance • HO2S heater element has failed, or the heater circuit is open • PCM has failed
DTC: P0141 **2T OBD/O2S HTR1, MIL: YES** **Years:** 2007, 2008 **Models:** Ascender, i290, i370 **Engines:** 2.9L VIN 9, 3.7L VIN E, 4.2L VIN S **Transmissions:** A/T, M/T	**HO2S-12 (Bank 1 Sensor 2) Heater Circuit Malfunction** No HO2S-12 codes set, ECT and IAT sensors less than 90°F and within 11°F at startup, engine running, system voltage from 10-16v, then with the average Calculated airflow less than 23 g/sec during the test period, the PCM detected the HO2S-12 signal did not vary more than 150 mv from the bias voltage of 400 to 500 mv for up to 300 seconds during the Oxygen Sensor Heater Monitor test. **Possible Causes:** • HO2S heater power circuit is open (check the 20A heater fuse) • HO2S heater ground circuit is open • HO2S heater element has high resistance or has failed • PCM has failed
DTC: P0151 **1T CCM, MIL: YES** **Years:** 2007, 2008 **Models:** Ascender, i290, i370 **Engines:** 2.9L VIN 9, 3.7L VIN E, 4.2L VIN S **Transmissions:** A/T, M/T	**HO2S-21 (Bank 2 Sensor 1) Circuit Low Input** DTC P0106, P0107, P0108, P0112, P0113, P0117, P0118, P0121, P0122, P0123, P0171, P0172 and P0300, P0301-P0306 not set, engine running in closed loop with the A/F ratio from 14.5-14.8:1, ECT sensor more than 140°F, throttle angle from 3-19% for 5 seconds, the PCM detected the HO2S-21 signal was less than 22 mv for 77 seconds over a 90 second period during the CCM test. **Possible Causes:** • HO2S signal circuit is open or shorted to ground • HO2S is water or fuel contaminated • HO2S is damaged or has failed • PCM has failed
DTC: P0152 **1T CCM, MIL: YES** **Years:** 2007, 2008 **Models:** Ascender, i290, i370 **Engines:** 2.9L VIN 9, 3.7L VIN E, 4.2L VIN S **Transmissions:** A/T, M/T	**HO2S-21 (Bank 2 Sensor 1) Circuit High Input** DTC P0106, P0107, P0108, P0112, P0113, P0117, P0118, P0121, P0122, P0123, P0171, P0172 and P0300, P0301-P0306 not set, engine running in closed loop with the A/F ratio command at 14.5-14.8:1, throttle angle from 3-19% for 5 seconds, the PCM detected the HO2S-21 signal was less than 952 mv for 77 seconds over a 90 second period; or the HO2S-21 signal was more than 500 mv during Decel Fuel Cutoff mode for 3 seconds during the CCM test. **Possible Causes:** • HO2S signal circuit shorted to system power (oil in connector) • HO2S is water or fuel contaminated • HO2S is damaged or has failed • PCM has failed

DTC	Trouble Code Title, Conditions & Possible Causes
DTC: P0153 **2T OBD/O2S, MIL: YES** **Years:** 2007, 2008 **Models:** Ascender, i290, i370 **Engines:** 2.9L VIN 9, 3.7L VIN E, 4.2L VIN S **Transmissions:** A/T, M/T	**HO2S-21 (Bank 2 Sensor 1) Slow Response** DTC P0106, P0107, P0108, P0112, P0113, P0117, P0118, P0121, P0122, P0123, P0171, P0172 and P0300, P0301-P0306 not set, engine runtime 1 minute in closed loop, ECT sensor more than 122°F, engine speed from 1500-3000 rpm, MAF sensor from 9-42 g/sec, Purge duty cycle over 1%, conditions met for 3 seconds, then 90 seconds after entering closed loop, the PCM detected the HO2S-21 lean-to-rich average transition time was over 94 ms, or the rich-to-lean average transition time was over 105 ms during the Oxygen Sensor Monitor test. **Possible Causes:** • Exhaust leak present in the exhaust manifold or exhaust pipes • HO2S element fuel contamination • HO2S element has deteriorated • PCM has failed
DTC: P0154 **1T OBD/O2S, MIL: YES** **Years:** 2007, 2008 **Models:** Ascender, i290, i370 **Engines:** 2.9L VIN 9, 3.7L VIN E, 4.2L VIN S Transmissions: A/T, M/T	**HO2S-21 (Bank 2 Sensor 1) Insufficient Activity Detected** DTC P0106, P0107, P0108, P0112, P0113, P0117, P0118, P0121, P0122, P0123, P0171, P0172 and P0300, P0301-306 not set, system voltage from 11-16v, engine runtime over 40 seconds, then after the PCM determined the Oxygen Sensor Heater test passed, it detected the HO2S-21 signal remained from 400-500 mv for 77 seconds over a 90 second period in the Oxygen Sensor Monitor test. **Possible Causes:** • Exhaust leak present in exhaust manifold or exhaust pipes • HO2S element fuel contamination or has deteriorated • HO2S signal circuit or the ground circuit has high resistance • HO2S heater element has failed, or the heater circuit is open • PCM has failed
DTC: P0155 **2T OBD/O2S, MIL: YES** **Years:** 2007, 2008 **Models:** Ascender, i290, i370 **Engines:** 2.9L VIN 9, 3.7L VIN E, 4.2L VIN S **Transmissions:** A/T, M/T	**HO2S-21 (Bank 2 Sensor 1) Heater Circuit Malfunction** DTC P0151, P0152, P0153 and P0154 not set, ECT and IAT sensor signals less than 90°F, and within 11°F at startup, system voltage from 11-16v, throttle angle under 40%, average Calculated airflow less than 18 g/sec in the sample period, and the PCM detected the HO2S-21 signal did not vary more than 150 mv from the bias voltage (400-500 mv) for too long a period (maximum time is 120 seconds). **Possible Causes:** • HO2S power circuit is open (from the O2S heater fuse) • HO2S heater ground circuit is open • HO2S heater element has high resistance • HO2S heater element has failed (open or shorted) • PCM has failed
DTC: P0157 **1T CCM, MIL: YES** **Years:** 2007, 2008 **Models:** Ascender, i290, i370 **Engines:** 2.9L VIN 9, 3.7L VIN E, 4.2L VIN S **Transmissions:** A/T, M/T	**HO2S-22 (Bank 2 Sensor 2) Circuit Low Input** No related codes set, ECT sensor signal more than 140°F, engine running in closed loop with the A/F ratio at 14.5-14.8:1, throttle angle from 3-19%, and the PCM detected the HO2S-22 signal was less than 26 mv for 106 seconds out of a 125 second period; or that it was more than 400 mv in Power Enrichment Mode during the test. **Possible Causes:** • HO2S signal circuit is open or shorted to ground • HO2S is water or fuel contaminated • HO2S is damaged or has failed • PCM has failed
DTC: P0158 **1T CCM, MIL: YES** **Years:** 2007, 2008 **Models:** Ascender, i290, i370 **Engines:** 2.9L VIN 9, 3.7L VIN E, 4.2L VIN S **Transmissions:** A/T, M/T	**HO2S-22 (Bank 2 Sensor 2) Circuit High Input** DTC P0106, P0107, P0108, P0112, P0113, P0117, P0118, P0121, P0122, P0123, P0171, P0172 and P0300, P0301-P0306 not set, engine running in closed loop with the A/F ratio command at 14.5-14.8:1, throttle angle from 3-19% for 5 seconds, the PCM detected the HO2S-22 signal was less than 952 mv for 106 seconds over a 125 second period; or the HO2S-12 signal was more than 500 mv during Decel Fuel Cutoff mode for 3 seconds during the CCM test. **Possible Causes:** • HO2S signal circuit shorted to system power (oil in connector) • HO2S is water or fuel contaminated • HO2S is damaged or has failed • PCM has failed

DTC	Trouble Code Title, Conditions & Possible Causes
DTC: P0160 **1T OBD/O2S, MIL: YES** **Years:** 2007, 2008 **Models:** Ascender, i290, i370 **Engines:** 2.9L VIN 9, 3.7L VIN E, 4.2L VIN S **Transmissions:** A/T, M/T	**HO2S-22 (Bank 2 Sensor 2) Insufficient Activity Detected** DTC P0106, P0107, P0108, P0112, P0113, P0117, P0118, P0121, P0122, P0123, P0171, P0172 and P0300, P0301-P0306 not set, system voltage from 11-16v, engine runtime over 40 seconds, then after the PCM determined the Oxygen Sensor Heater test passed, it detected the HO2S-22 signal remained from 426-474 mv for 105 seconds of a 125 second period in the Oxygen Sensor Monitor test. **Possible Causes:** • Exhaust leak present in exhaust manifold or exhaust pipes • HO2S element has fuel contamination or has deteriorated • HO2S signal circuit or the ground circuit has high resistance • PCM has failed
DTC: P0161 **2T OBD/O2S HTR1, MIL: YES** **Years:** 2007, 2008 **Models:** Ascender, i290, i370 **Engines:** 2.9L VIN 9, 3.7L VIN E, 4.2L VIN S **Transmissions:** A/T, M/T	**HO2S-22 (Bank 2 Sensor 2) Heater Circuit Malfunction** No HO2S-22 codes set, ECT and IAT sensor less than 90°F, and within 11°F at startup, system voltage at 11-16v, average Calculated airflow during the test period less than 23 g/sec, and the PCM detected the HO2S-22 signal changed less than 150 mv from the bias voltage of 400-500 mv for up to 300 seconds during the test. **Possible Causes:** • HO2S power circuit is open (from the O2S heater fuse) • HO2S heater element has failed (it may be open or shorted) • PCM has failed
DTC: P0171 **2T OBD/FUEL, MIL: YES** **Years:** 2007, 2008 **Models:** Ascender, i290, i370 **Engines:** 2.9L VIN 9, 3.7L VIN E, 4.2L VIN S **Transmissions:** A/T, M/T	**Fuel System Too Lean (Bank 1)** DTC P0106, P0107, P0108, P0112, P0113, P0117, P0118, P0121, P0122, P0123, P0131, P0132, P0133, P0134, P0135, P0137, P0138, P0201-206, P0300, P0301=P0306, P0401, P0502, P0503, P0506, P0507, P1406 and P1441 not set, engine running in closed loop, system voltage from 11-16v, BARO sensor over 72.5 kPa, ECT sensor from 77-212°F, IAT sensor from -40 to 248°F, MAP sensor from 24-99 kPa, throttle angle less than 95%, VSS under 85 mph, engine speed from 400-6000 rpm, MAF sensor from 2-20 g/sec, Purge duty cycle over 0%, and the PCM detected the average of the Long Term fuel trim values was more than +20%. **Possible Causes:** • Air leaks after the MAF sensor, or in the EGR or PCV system • Base engine "mechanical" fault affecting one or more cylinders • Exhaust leaks before or near where the front HO2S is mounted • Fuel control sensor is out of calibration (i.e., ECT, IAT or MAP) • Fuel delivery system supplying too little fuel during cruise or idle periods (e.g., faulty fuel pump or dirty, restricted fuel filter) • Fuel injector (one or more) dirty or pressure regulator has failed • HO2S is contaminated, deteriorated or it has failed • Vehicle driven low on fuel or until it ran out of fuel
DTC: P0172 **2T OBD/FUEL, MIL: YES** **Years:** 2007, 2008 **Models:** Ascender, i290, i370 **Engines:** 2.9L VIN 9, 3.7L VIN E, 4.2L VIN S **Transmissions:** A/T, M/T	**Fuel System Too Rich (Bank 1)** DTC P0106, P0107, P0108, P0112, P0113, P0117, P0118, P0121, P0122, P0123, P0131, P0132, P0133, P0134, P0135, P0137, P0138, P0201-206, P0300, P0301=P0306, P0401, P0502, P0503, P0506, P0507, P1406 and P1441 not set, engine running in closed loop, BARO sensor over 72.5 kPa, ECT sensor from 77-212°F, IAT sensor from -40 to 248°F, MAP sensor from 24-99 kPa, throttle angle less than 95%, VSS under 85 mph, engine speed from 400-6000 rpm, MAF sensor from 2-20 g/sec, Purge duty cycle over 0%, and the PCM detected the average of the Long Term fuel trim values was more than -14% during the Fuel System Monitor test. **Possible Causes:** • Air leak at the exhaust pipe or manifold, or at air injection pipes • Base engine "mechanical" fault affecting one or more cylinders • EVAP system component has failed or canister fuel saturated • Fuel control sensor is out of calibration (i.e., ECT, IAT or MAP) • Fuel delivery system supplying too much fuel during cruise or idle periods (e.g., faulty fuel pump, or faulty pressure regulator) • Fuel injector(s) is leaking or stuck partially open (one or more) • HO2S is contaminated, deteriorated or it has failed

DTC	Trouble Code Title, Conditions & Possible Causes
DTC: P0174 **2T OBD/FUEL, MIL: YES** **Years:** 2007, 2008 **Models:** Ascender, i290, i370 **Engines:** 2.9L VIN 9, 3.7L VIN E, 4.2L VIN S **Transmissions:** A/T, M/T	**Fuel System Too Lean (Bank 2)** DTC P0106, P0107, P0108, P0112, P0113, P0117, P0118, P0121, P0122, P0123, P0131, P0132, P0133, P0134, P0135, P0137, P0138, P0201-206, P0300, P0301=P0306, P0401, P0502, P0503, P0506, P0507, P1406 and P1441 not set, engine running in closed loop, system voltage from 11-16v, BARO sensor over 72.5 kPa, ECT sensor from 77-212°F, IAT sensor from -40 to 248°F, MAP sensor from 24-99 kPa, throttle angle steady at less than 95%, VSS under 85 mph, engine speed from 400-6000 rpm, MAF sensor from 2-20 g/sec, Purge duty cycle over 0%, and the PCM detected the average of the Long Term fuel trim values was more than +20%. **Possible Causes:** • Air leaks after the MAF sensor, or in the EGR or PCV system • Base engine "mechanical" fault affecting one or more cylinders • Exhaust leaks before or near where the front HO2S is mounted • Fuel control sensor is out of calibration (i.e., ECT, IAT or MAP) • Fuel delivery system supplying too little fuel during cruise or idle periods (e.g., faulty fuel pump or dirty, restricted fuel filter) • Fuel injector (one or more) dirty or pressure regulator has failed • HO2S is contaminated, deteriorated or it has failed • Vehicle driven low on fuel or until it ran out of fuel
DTC: P0175 **2T OBD/FUEL, MIL: YES** **Years:** 2007, 2008 **Models:** Ascender, i290, i370 **Engines:** 2.9L VIN 9, 3.7L VIN E, 4.2L VIN S **Transmissions:** A/T, M/T	**Fuel System Too Rich (Bank 2)** DTC P0106, P0107, P0108, P0112, P0113, P0117, P0118, P0121, P0122, P0123, P0131, P0132, P0133, P0134, P0135, P0137, P0138, P0201-206, P0300, P0301=P0306, P0401, P0502, P0503, P0506, P0507, P1406 and P1441 not set, engine running in closed loop, BARO sensor over 72.5 kPa, ECT sensor from 77-212°F, IAT sensor from -40 to 248°F, MAP sensor from 24-99 kPa, throttle angle less than 95%, VSS under 85 mph, engine speed from 400-6000 rpm, MAF sensor from 2-20 g/sec, Purge duty cycle over 0%, and the PCM detected the average of the Long Term fuel trim values was more than -14% during the Fuel System Monitor test. **Possible Causes:** • Air leak at the exhaust pipe or manifold, or at air injection pipes • Base engine "mechanical" fault affecting one or more cylinders • EVAP system component has failed or canister fuel saturated • Fuel control sensor is out of calibration (i.e., ECT, IAT or MAP) • Fuel delivery system supplying too much fuel during cruise or idle periods (e.g., faulty fuel pump, or faulty pressure regulator) • Fuel injector(s) is leaking or stuck partially open (one or more) • HO2S is contaminated, deteriorated or it has failed
DTC: P0201 **1T CCM, MIL: YES** **Years:** 2007, 2008 **Models:** Ascender, i290, i370 **Engines:** 2.9L VIN 9, 3.7L VIN E, 4.2L VIN S **Transmissions:** A/T, M/T	**Injector Circuit Malfunction – Cylinder 1** Engine started, engine running, system voltage over 9v, and PCM detected the injector voltage for Cylinder 1 did not equal the system voltage with the injector commanded "off", or that the injector voltage did not equal zero (0) volts with the injector commanded "on". **Possible Causes:** • Fuel injector control circuit is open or shorted to ground • Fuel injector power circuit is open between injector and relay • Fuel Injector has failed • PCM has failed (injector driver circuit may be open or shorted)
DTC: P0202 **1T CCM, MIL: YES** **Years:** 2007, 2008 **Models:** Ascender, i290, i370 **Engines:** 2.9L VIN 9, 3.7L VIN E, 4.2L VIN S **Transmissions:** A/T, M/T	**Injector Circuit Malfunction – Cylinder 2** Engine started, engine running, system voltage over 9v, and PCM detected the injector voltage for Cylinder 2 did not equal the system voltage with the injector commanded "off", or that the injector voltage did not equal zero (0) volts with the injector commanded "on". **Possible Causes:** • Fuel injector control circuit is open or shorted to ground • Fuel injector power circuit open between injector and ECM fuse • Fuel Injector has failed • PCM has failed (injector driver circuit may be open or shorted)
DTC: P0203 **1T CCM, MIL: YES** **Years:** 2007, 2008 **Models:** Ascender, i290, i370 **Engines:** 2.9L VIN 9, 3.7L VIN E, 4.2L VIN S **Transmissions:** A/T, M/T	**Injector Circuit Malfunction – Cylinder 3** Engine started, engine running, system voltage over 9v, and PCM detected the injector voltage for Cylinder 3 did not equal the system voltage with the injector commanded "off", or that the injector voltage did not equal zero (0) volts with the injector commanded "on". **Possible Causes:** • Fuel injector control circuit is open or shorted to ground • Fuel injector power circuit open between injector and ECM fuse • Fuel Injector has failed • PCM has failed (injector driver circuit may be open or shorted)

DTC	Trouble Code Title, Conditions & Possible Causes
DTC: P0204 **1T CCM, MIL: YES** **Years:** 2007, 2008 **Models:** Ascender, i290, i370 **Engines:** 2.9L VIN 9, 3.7L VIN E, 4.2L VIN S **Transmissions:** A/T, M/T	**Injector Circuit Malfunction – Cylinder 4** Engine started, engine running, system voltage over 9v, and PCM detected the injector voltage for Cylinder 4 did not equal the system voltage with the injector commanded "off", or that the injector voltage did not equal zero (0) volts with the injector commanded "on". **Possible Causes:** • Fuel injector control circuit is open or shorted to ground • Fuel injector power circuit open between injector and ECM fuse • Fuel Injector has failed • PCM has failed (injector driver circuit may be open or shorted)
DTC: P0205 **1T CCM, MIL: YES** **Years:** 2007, 2008 **Models:** Ascender, i290, i370 **Engines:** 2.9L VIN 9, 3.7L VIN E, 4.2L VIN S **Transmissions:** A/T, M/T	**Injector Circuit Malfunction – Cylinder 5** Engine started, engine running, system voltage over 9v, and PCM detected the injector voltage for Cylinder 5 did not equal the system voltage with the injector commanded "off", or that the injector voltage did not equal zero (0) volts with the injector commanded "on". **Possible Causes:** • Fuel injector control circuit is open or shorted to ground • Fuel injector power circuit open between injector and ECM fuse • Fuel Injector has failed • PCM has failed (injector driver circuit may be open or shorted)
DTC: P0206 **1T CCM, MIL: YES** **Years:** 2007, 2008 **Models:** Ascender, i290, i370 **Engines:** 2.9L VIN 9, 3.7L VIN E, 4.2L VIN S **Transmissions:** A/T, M/T	**Injector Circuit Malfunction – Cylinder 6** Engine started, engine running, system voltage over 9v, and PCM detected the injector voltage for Cylinder 6 did not equal the system voltage with the injector commanded "off", or that the injector voltage did not equal zero (0) volts with the injector commanded "on". **Possible Causes:** • Fuel injector control circuit is open or shorted to ground • Fuel injector power circuit open between injector and ECM fuse • Fuel Injector has failed • PCM has failed (injector driver circuit may be open or shorted)
DTC: P0300 **2T CCM, MIL: YES** **Years:** 2007, 2008 **Models:** Ascender, i290, i370 **Engines:** 2.9L VIN 9, 3.7L VIN E, 4.2L VIN S **Transmissions:** A/T, M/T	**Multiple Cylinder Misfire Detected** DTC P0101, P0102, P0103, P0106, P0107, P0108, P0117, P0118, P0121, P0122, P0123, P0336, P0341, P0342, P0502 and P0503 not set, ECT sensor from 20-248°F, system voltage from 11-16v, engine speed from 800-5500 rpm, throttle angle stable (± 3%), and the PCM detected a crankshaft speed variation in one or more cylinders characteristic of a misfire condition during the Misfire Monitor test. **Note: If the misfire is severe, the MIL will flash on/off on the 1st trip!** **Possible Causes:** • Base engine mechanical fault that affects one or more cylinders • Fuel metering fault that affects more than one cylinder • Fuel pressure too low or too high, fuel supply contaminated • EVAP system problem or the EVAP canister is fuel saturated • EGR valve is stuck open or the PCV system has a vacuum leak • IC control circuit is shorted to ground (an intermittent fault) • Ignition system fault (a coil) that affects more than one cylinder • MAF sensor contamination (it can cause a very lean condition)
DTC: P0301 **2T CCM, MIL: YES** **Years:** 2007, 2008 **Models:** Ascender, i290, i370 **Engines:** 2.9L VIN 9, 3.7L VIN E, 4.2L VIN S **Transmissions:** A/T, M/T	**Misfire Detected - Cylinder 1** DTC P0101, P0102, P0103, P0106, P0107, P0108, P0117, P0118, P0121, P0122, P0123, P0336, P0341, P0342, P0502 and P0503 not set, ECT sensor from 20-248°F, system voltage from 11-16v, engine speed from 800-5500 rpm, throttle angle stable (± 3%), and the PCM detected a crankshaft speed variation in one cylinder characteristic of a misfire condition during the Misfire Diagnostic Monitor test. **Note: If the misfire is severe, the MIL will flash on/off on the 1st trip!** **Possible Causes:** • Base engine mechanical fault that affects only one cylinder • Fuel metering fault that affects only one cylinder • EGR valve is stuck open or the PCV system has a vacuum leak • Ignition system fault (i.e., a coil) that affects only one cylinder
DTC: P0302 **2T CCM, MIL: YES** **Years:** 2007, 2008 **Models:** Ascender, i290, i370 **Engines:** 2.9L VIN 9, 3.7L VIN E, 4.2L VIN S **Transmissions:** A/T, M/T	**Misfire Detected - Cylinder 2** DTC P0101, P0102, P0103, P0106, P0107, P0108, P0117, P0118, P0121, P0122, P0123, P0336, P0341, P0342, P0502 and P0503 not set, ECT sensor from 20-248°F, system voltage from 11-16v, engine speed from 800-5500 rpm, throttle angle stable (± 3%), and the PCM detected a crankshaft speed variation in one cylinder characteristic of a misfire condition during the Misfire Diagnostic Monitor test. **Note: If the misfire is severe, the MIL will flash on/off on the 1st trip!** **Possible Causes:** • Base engine mechanical fault that affects only one cylinder • Fuel metering fault that affects only one cylinder • EGR valve is stuck open or the PCV system has a vacuum leak • Ignition system fault (i.e., a coil) that affects only one cylinder

DTC	Trouble Code Title, Conditions & Possible Causes
DTC: P0303 **2T CCM, MIL: YES** **Years:** 2007, 2008 **Models:** Ascender, i290, i370 **Engines:** 2.9L VIN 9, 3.7L VIN E, 4.2L VIN S **Transmissions:** A/T, M/T	**Misfire Detected - Cylinder 3** DTC P0101, P0102, P0103, P0106, P0107, P0108, P0117, P0118, P0121, P0122, P0123, P0336, P0341, P0342, P0502 and P0503 not set, ECT sensor from 20-248°F, system voltage from 11-16v, engine speed from 800-5500 rpm, throttle angle stable ($\pm$ 3%), and the PCM detected a crankshaft speed variation in one cylinder characteristic of a misfire condition during the Misfire Diagnostic Monitor test. **Note: If the misfire is severe, the MIL will flash on/off on the 1st trip!** **Possible Causes:** • Base engine mechanical fault that affects only one cylinder • Fuel metering fault that affects only one cylinder • EGR valve is stuck open or the PCV system has a vacuum leak • Ignition system fault (i.e., a coil) that affects only one cylinder
DTC: P0304 **2T CCM, MIL: YES** **Years:** 2007, 2008 **Models:** Ascender, i290, i370 **Engines:** 2.9L VIN 9, 3.7L VIN E, 4.2L VIN S **Transmissions:** A/T, M/T	**Misfire Detected - Cylinder 4** DTC P0101, P0102, P0103, P0106, P0107, P0108, P0117, P0118, P0121, P0122, P0123, P0336, P0341, P0342, P0502 and P0503 not set, ECT sensor from 20-248°F, system voltage from 11-16v, engine speed from 800-5500 rpm, throttle angle stable ($\pm$ 3%), and the PCM detected a crankshaft speed variation in one cylinder characteristic of a misfire condition during the Misfire Diagnostic Monitor test. **Note: If the misfire is severe, the MIL will flash on/off on the 1st trip!** **Possible Causes:** • Base engine mechanical fault that affects only one cylinder • Fuel metering fault that affects only one cylinder • EGR valve is stuck open or the PCV system has a vacuum leak • Ignition system fault (i.e., a coil) that affects only one cylinder
DTC: P0305 **2T CCM, MIL: YES** **Years:** 2007, 2008 **Models:** Ascender, i370 **Engines:** 3.7L VIN E, 4.2L VIN S **Transmissions:** A/T, M/T	**Misfire Detected - Cylinder 5** DTC P0101, P0102, P0103, P0106, P0107, P0108, P0117, P0118, P0121, P0122, P0123, P0336, P0341, P0342, P0502 and P0503 not set, ECT sensor from 20-248°F, system voltage from 11-16v, engine speed from 800-5500 rpm, throttle angle stable ($\pm$ 3%), and the PCM detected a crankshaft speed variation in one cylinder characteristic of a misfire condition during the Misfire Diagnostic Monitor test. **Note: If the misfire is severe, the MIL will flash on/off on the 1st trip!** **Possible Causes:** • Base engine mechanical fault that affects only one cylinder • Fuel metering fault that affects only one cylinder • EGR valve is stuck open or the PCV system has a vacuum leak • Ignition system fault (i.e., a coil) that affects only one cylinder
DTC: P0306 **2T CCM, MIL: YES** **Years:** 2007, 2008 **Models:** Ascender, i370 **Engines:** 3.7L VIN E, 4.2L VIN S **Transmissions:** A/T, M/T	**Misfire Detected - Cylinder 6** DTC P0101, P0102, P0103, P0106, P0107, P0108, P0117, P0118, P0121, P0122, P0123, P0336, P0341, P0342, P0502 and P0503 not set, ECT sensor from 20-248°F, system voltage from 11-16v, engine speed from 800-5500 rpm, throttle angle stable ($\pm$ 3%), and the PCM detected a crankshaft speed variation in one cylinder characteristic of a misfire condition during the Misfire Diagnostic Monitor test. **Note: If the misfire is severe, the MIL will flash on/off on the 1st trip!** **Possible Causes:** • Base engine mechanical fault that affects only one cylinder • Fuel metering fault that affects only one cylinder • EGR valve is stuck open or the PCV system has a vacuum leak • Ignition system fault (i.e., a coil) that affects only one cylinder
DTC: P0336 **2T CCM, MIL: YES** **Years:** 2007, 2008 **Models:** Ascender, i290, i370 **Engines:** 2.9L VIN 9, 3.7L VIN E, 4.2L VIN S **Transmissions:** A/T, M/T	**Crankshaft Position 58X Sensor Circuit Malfunction** Engine started, and the PCM detected extra or missing pulses between consecutive Crankshaft Position (CKP) 58X sensor signals during 10 out of 100 revolutions during the CCM test. **Possible Causes:** • CKP sensor 58X signal circuit is open or shorted to ground • CKP sensor ground circuit is open or has high resistance • CKP sensor power circuit is open between the sensor and PCM • CKP sensor is damaged, or the reluctor wheel is damaged • PCM has failed (the Ignition module function is inside the PCM)

DTC	Trouble Code Title, Conditions & Possible Causes
DTC: P0337 **2T CCM, MIL: YES** **Years:** 2007, 2008 **Models:** Ascender, i290, i370 **Engines:** 2.9L VIN 9, 3.7L VIN E, 4.2L VIN S **Transmissions:** A/T, M/T	**Crankshaft Position 58X Sensor Circuit Low Input** DTC P0341 and P0342 not set, engine started, and the PCM did not detect any Crankshaft Position (CKP) 58X sensor pulses present between two (2) CMP sensor pulses, or it did not detect any CKP sensor pulses within 8 CMP sensor pulses during the CCM test. **Possible Causes:** • CKP sensor 58X signal circuit is open or shorted to ground • CKP sensor ground circuit is open or has high resistance • CKP sensor power circuit is open between the sensor and PCM • CKP sensor is damaged, or the reluctor wheel is damaged • PCM has failed (the Ignition module function is inside the PCM)
DTC: P0351 **1T CCM, MIL: YES** **Years:** 1996, 1997, 2007, 2008 **Models:** Ascender, i290, i370 **Engines:** 2.9L VIN 9, 3.7L VIN E, 4.2L VIN S **Transmissions:** A/T, M/T	**Ignition Control Module Circuit 1 Malfunction** Engine running with CKP 58X signals received, and the PCM detected the IC output signal did not equal 5v with the output commanded "on", or it did not equal 0v with the output commanded "off" in 20 tests over a 40 sample period during the CCM test. **Possible Causes:** • EST signal circuit is open between module and coil or the PCM • EST signal circuit shorted between module and coil or the PCM • Ignition Control module is damaged or has failed • PCM has failed
DTC: P0351 **2T CCM, MIL: YES** **Years:** 2000, 2001, 2002, 2003, 2004, 2005 **Models:** Ascender, i290, i370 **Engines:** 2.9L VIN 9, 3.7L VIN E, 4.2L VIN S **Transmissions:** A/T, M/T	**Ignition Control Module Circuit 1 Malfunction** Engine running with CKP 58X signals received, and the PCM detected the IC output signal did not equal 5v with the output commanded "on", or it did not equal 0v with the output commanded "off" in 20 tests over a 40 sample period during the CCM test. **Possible Causes:** • EST signal circuit is open between module and coil or the PCM • EST signal circuit shorted between module and coil or the PCM • ION module is damaged or has failed • PCM has failed
DTC: P0352 **1T CCM, MIL: YES** **Years:** 1996, 1997, 2007, 2008 **Models:** Ascender, i290, i370 **Engines:** 2.9L VIN 9, 3.7L VIN E, 4.2L VIN S **Transmissions:** A/T, M/T	**Ignition Control Module Circuit 2 Malfunction** Engine running with CKP 58X signals received, and the PCM detected the IC output signal did not equal 5v with the output commanded "on", or it did not equal 0v with the output commanded "off" in 20 tests over a 40 sample period during the CCM test. **Possible Causes:** • EST signal circuit is open between module and coil or the PCM • EST signal circuit shorted between module and coil or the PCM • Ignition Control module is damaged or has failed • PCM has failed
DTC: P0352 **1T CCM, MIL: YES** **Years:** 2000, 2001, 2002, 2003, 2004, 2005 **Models:** Ascender, i290, i370 **Engines:** 2.9L VIN 9, 3.7L VIN E, 4.2L VIN S **Transmissions:** A/T, M/T	**Ignition Control Module Circuit 2 Malfunction** Engine running with CKP 58X signals received, and the PCM detected the IC output signal did not equal 5v with the output commanded "on", or it did not equal 0v with the output commanded "off" in 20 tests over a 40 sample period during the CCM test. **Possible Causes:** • EST signal circuit is open between module and coil or the PCM • EST signal circuit shorted between module and coil or the PCM • ION module is damaged or has failed • PCM has failed
DTC: P0353 **1T CCM, MIL: YES** **Years:** 1996, 1997, 2007, 2008 **Models:** Ascender, i290, i370 **Engines:** 2.9L VIN 9, 3.7L VIN E, 4.2L VIN S **Transmissions:** A/T, M/T	**Ignition Control Module Circuit 3 Malfunction** Engine running with CKP 58X signals received, and the PCM detected the IC output signal did not equal 5v with the output commanded "on", or it did not equal 0v with the output commanded "off" in 20 tests over a 40 sample period during the CCM test. **Possible Causes:** • EST signal circuit is open between module and coil or the PCM • EST signal circuit shorted between module and coil or the PCM • Ignition Control module is damaged or has failed • PCM has failed
DTC: P0353 **1T CCM, MIL: YES** **Years:** 2000, 2001, 2002, 2003, 2004, 2005 **Models:** Ascender, i290, i370 **Engines:** 2.9L VIN 9, 3.7L VIN E, 4.2L VIN S **Transmissions:** A/T, M/T	**Ignition Control Module Circuit 3 Malfunction** Engine running with CKP 58X signals received, and the PCM detected the IC output signal did not equal 5v with the output commanded "on", or it did not equal 0v with the output commanded "off" in 20 tests over a 40 sample period during the CCM test. **Possible Causes:** • EST signal circuit is open between module and coil or the PCM • EST signal circuit shorted between module and coil or the PCM • ION module is damaged or has failed • PCM has failed

DTC	Trouble Code Title, Conditions & Possible Causes
DTC: P0355 **1T CCM, MIL: YES** **Years:** 1996, 1997, 2007, 2008 **Models:** Ascender, i290, i370 **Engines:** 2.9L VIN 9, 3.7L VIN E, 4.2L VIN S **Transmissions:** A/T, M/T	**Ignition Control Module Circuit 5 Malfunction** Engine running with CKP 58X signals received, and the PCM detected the IC output signal did not equal 5v with the output commanded "on", or it did not equal 0v with the output commanded "off" in 20 tests over a 40 sample period during the CCM test. **Possible Causes:** • EST signal circuit is open between module and coil or the PCM • EST signal circuit shorted between module and coil or the PCM • Ignition Control module is damaged or has failed • PCM has failed
DTC: P0355 **1T CCM, MIL: YES** **Years:** 2000, 2001, 2002, 2003, 2004, 2005 **Models:** Ascender, i290, i370 **Engines:** 2.9L VIN 9, 3.7L VIN E, 4.2L VIN S **Transmissions:** A/T, M/T	**Ignition Control Module Circuit 5 Malfunction** Engine running with CKP 58X signals received, and the PCM detected the IC output signal did not equal 5v with the output commanded "on", or it did not equal 0v with the output commanded "off" in 20 tests over a 40 sample period during the CCM test. **Possible Causes:** • EST signal circuit is open between module and coil or the PCM • EST signal circuit shorted between module and coil or the PCM • ION module is damaged or has failed • PCM has failed
DTC: P0356 **1T CCM, MIL: YES** **Years:** 1996, 1997, 1998, 1999 **Models:** Ascender, i290, i370 **Engines:** 2.9L VIN 9, 3.7L VIN E, 4.2L VIN S **Transmissions:** A/T, M/T	**Ignition Control Module Circuit 6 Malfunction** Engine running with CKP 58X signals received, and the PCM detected the IC output signal did not equal 5v with the output commanded "on", or it did not equal 0v with the output commanded "off" in 20 tests over a 40 sample period during the CCM test. **Possible Causes:** • EST signal circuit is open between module and coil or the PCM • EST signal circuit shorted between module and coil or the PCM • Ignition Control module is damaged or has failed • PCM has failed
DTC: P0356 **1T CCM, MIL: YES** **Years:** 2000, 2001, 2002, 2003, 2004, 2005 **Models:** Ascender, i290, i370 **Engines:** 2.9L VIN 9, 3.7L VIN E, 4.2L VIN S **Transmissions:** A/T, M/T	**Ignition Control Module Circuit 6 Malfunction** Engine running with CKP 58X signals received, and the PCM detected the IC output signal did not equal 5v with the output commanded "on", or it did not equal 0v with the output commanded "off" in 20 tests over a 40 sample period during the CCM test. **Possible Causes:** • EST signal circuit is open between module and coil or the PCM • EST signal circuit shorted between module and coil or the PCM • ION module is damaged or has failed • PCM has failed
DTC: P0401 **1T CCM, MIL: YES** **Years:** 2007, 2008 **Models:** Ascender, i290, i370 **Engines:** 2.9L VIN 9, 3.7L VIN E, 4.2L VIN S **Transmissions:** A/T, M/T	**Insufficient EGR System Flow Detected** No ECT, EGR Pintle Position, EVAP, IAC, IAT, MAP, Misfire, TP or VSS codes set, system voltage from 11-16v, ECT sensor more than 140°F, BARO sensor over 75 kPa, IAC position stable (±10 counts), A/C Clutch and TCC status unchanged, and VSS over 15 mph, then with the throttle closed (TP angle under 1%), EGR duty cycle under 1%, MAP sensor from 10-40 kPa (±2 kPa), engine speed from 1100-2000 rpm, the PCM detected the compensated MAP sensor signal indicated a value from 10.3-49.8 kPa during the EGR System test. **Possible Causes:** • Linear EGR valve "low" circuit is open or shorted to ground • Linear EGR valve "low" circuit is shorted to system power (B+) • EGR valve VREF (5-volt) is open between sensor and the PCM • EGR valve feedback circuit is open or shorted to ground • EGR valve is stuck closed, or partially open during the test • EGR exhaust flow path may be restricted • EGR valve is damaged, or has failed • PCM has failed
DTC: P0402 **2T CCM, MIL: YES** **Years:** 2007, 2008 **Models:** Ascender, i290, i370 **Engines:** 3.2L VIN W, 3.5L VIN X, 3.5L VIN Y **Transmissions:** A/T, M/T	**Excessive EGR System Excessive Flow Detected** Engine started, IAT sensor more than 38°F, engine running, system voltage at 11-16v, and the PCM detected the EGR position sensor signal indicated more than 21% over a 625 ms period during the EGR System flow test right after engine startup. **Possible Causes:** • Linear EGR valve control circuit is shorted to ground • EGR valve is stuck partially open during the initial startup test • EGR valve is damaged, or has excessive carbon buildup • PCM has failed

DTC	Trouble Code Title, Conditions & Possible Causes
DTC: P0404 **2T CCM, MIL: YES** **Years:** 2007, 2008 **Models:** Ascender, i290, i370 **Engines:** 3.2L VIN W, 3.5L VIN X, 3.5L VIN Y **Transmissions:** A/T, M/T	**EGR Pintle Position Sensor Circuit Range/Performance** Engine started, IAT sensor more than 38°F, engine speed less than 600 rpm, system voltage at 11-16v, then with the Desired EGR position at over 0%, the PCM detected the difference between the Actual and Desired EGR position was more than 15% for over 15 seconds. This fault must occur 3 times in a single trip to set a code. **Possible Causes:** • EGR valve is stuck partially open during the initial startup test • EGR valve is damaged, or has excessive carbon buildup • PCM has failed
DTC: P0405 **2T CCM, MIL: YES** **Years:** 2007, 2008 **Models:** Ascender, i290, i370 **Engines:** 3.2L VIN W, 3.5L VIN X, 3.5L VIN Y **Transmissions:** A/T, M/T	**EGR Pintle Position Sensor Circuit Low Input** Key on or engine running, system voltage from 11-16v, IAT sensor more than 140°F, and the PCM detected the EGR position sensor indicated less than 0.10v for 10 seconds during the CCM test. **Possible Causes:** • Linear EGR valve control circuit is shorted to ground • EGR valve is stuck partially open during the initial startup test • EGR valve is damaged, or has excessive carbon buildup • PCM has failed
DTC: P0406 **1T CCM, MIL: YES** **Years:** 2007, 2008 **Models:** Ascender, i290, i370 **Engines:** 3.2L VIN W, 3.5L VIN X, 3.5L VIN Y **Transmissions:** A/T, M/T	**EGR Pintle Position Sensor Circuit High Input** Engine started, engine running, system voltage from 11-16v, IAT sensor more than 41°F, and the PCM detected the EGR position sensor indicated more than 4.80v for 10 seconds during the test. **Possible Causes:** • EGR sensor ground circuit is open between sensor and PCM • EGR sensor is damaged or has failed • PCM has failed
DTC: P0420 **1T OBD/CAT1, MIL: YES** **Years:** 2007, 2008 **Models:** Ascender, i290, i370 **Engines:** 2.9L VIN 9, 3.7L VIN E, 4.2L VIN S **Transmissions:** A/T, M/T	**Catalyst Efficiency Below Normal (Bank 1)** DTC P0106, P0107, P0108, P0112, P0113, P0117, P0118, P0121, P0122, P0123, P0131, P0132, P0133, P0134, P0137, P0138, P0140, P0141, P0171, P0172, P0300, P0301-P0306, P0336, P0341, P0342, P0401, P0502, P0506 and P0507 not set, engine speed less than 3500 rpm in closed loop, ECT sensor more than 140°F, MAF sensor from 8-50 g/sec, engine load less than 99% (±8%), predicted Catalyst temperature over 750°F, vehicle speed from 16-75 mph, and the PCM detected the catalyst oxygen storage capacity was below an acceptable threshold during the Catalyst test. **Possible Causes:** • Air leaks at the exhaust manifold or in the exhaust pipes • Front HO2S or rear HO2S is contaminated with fuel or moisture • Front HO2S older (aged) than the rear HO2S (HO2S is lazy) • Front HO2S and/or the rear HO2S is loose in the mounting hole • Catalytic converter is damaged or has failed
DTC: P0430 **1T OBD/CAT2, MIL: YES** **Years:** 2007, 2008 **Models:** Ascender, i290, i370 **Engines:** 2.9L VIN 9, 3.7L VIN E, 4.2L VIN S **Transmissions:** A/T, M/T	**Catalyst Efficiency Below Normal (Bank 2)** DTC P0106, P0107, P0108, P0112, P0113, P0117, P0118, P0121, P0122, P0123, P0131, P0132, P0133, P0134, P0137, P0138, P0140, P0141, P0171, P0172, P0300, P0301-P0306, P0336, P0341, P0342, P0401, P0502, P0506 and P0507 not set, engine speed less than 3500 rpm in closed loop, ECT sensor more than 140°F, MAF sensor from 8-50 g/sec, engine load less than 99% (±8%), predicted Catalyst temperature over 750°F, vehicle speed from 16-75 mph, and the PCM detected the catalyst oxygen storage capacity was below an acceptable threshold during the Catalyst test. **Possible Causes:** • Air leaks at the exhaust manifold or in the exhaust pipes • Front HO2S or rear HO2S is contaminated with fuel or moisture • Front HO2S older (aged) than the rear HO2S (HO2S is lazy) • Front HO2S and/or the rear HO2S is loose in the mounting hole • Catalytic converter is damaged or has failed

DTC	Trouble Code Title, Conditions & Possible Causes
DTC: P0440 **2T CCM**, **MIL: YES** **Years:** 2007, 2008 **Models:** Ascender, i290, i370 **Engines:** 3.2L VIN W, 3.5L VIN X, 3.5L VIN Y **Transmissions:** A/T, M/T	**EVAP System Performance** DTC P0106, P0107, P0108, P0112, P0113, P0117, P0118, P0121, P0122, P0123, P1640 ad P1650 not set, ECT and IAT sensors less than 90°F and within 13°F at startup, ECT sensor over 39°F at startup, IAT sensor over 4°F at startup, system voltage from 11-16v, BARO sensor over 75 kPa, throttle angle from 7-30%, fuel level from 15-85% with minimum fuel slosh, vehicle speed under 75 mph, and the PCM determined it was unable to achieve or maintain vacuum in the system for 60-180 seconds during the EVAP Monitor leak test. **Possible Causes:** • Charcoal canister is loaded with fuel or moisture • ECT, IAT, MAP, VSS or TP sensor signals out-of-calibration • Fuel filler cap loose, cross-threaded, incorrect part or damaged • Fuel tank pressure sensor is damaged or has failed • Fuel tank or fuel tank sender assembly 'O' ring is leaking • Fuel tank vapor line(s) block, damaged or disconnected • Purge or Vent solenoid control circuit open or shorted to ground • Purge or Vent solenoid power circuit is open (check the fuse)
DTC: P0506 **1T CCM**, **MIL: YES** **Years:** 2007, 2008 **Models:** Ascender, i290, i370 **Engines:** 2.9L VIN 9, 3.7L VIN E, 4.2L VIN S **Transmissions:** A/T, M/T	**Idle Air Control System Low RPM** DTC P0106, P0107, P0108, P0112, P0113, P0117, P0118, P0121, P0122, P0123, P0125, P0131, P0132, P0133, P0134, P0200-206, P0300, P0301-P306, P0335, P0341, P0342, P0404, P0405, P0440, P0442, P0446, P0452, P0453, P0502, P0507, P0601, P0602, P0705, P1133, P1404 and P1441 not set, engine runtime over 125 seconds, system voltage from 11-16v, ECT sensor more than 122°F, IAT sensor more than -40°F, MAP sensor less than 40 kPa, Purge duty cycle over 10%, BARO sensor over 75 kPa, vehicle speed less than 2 mph with the throttle closed, and the PCM detected the Actual speed was 100-200 rpm below the Desired idle speed for 10 seconds based on the current engine coolant temperature. **Possible Causes:** • High resistance between the IAC 'A' high or low control circuits • Short to ground between the IAC 'B' high or low control circuits • IAC valve is damaged, dirty, sticking or has failed • The throttle plate is carbon fouled (it may need to be cleaned)
DTC: P0507 **1T CCM**, **MIL: YES** **Years:** 2007, 2008 **Models:** Ascender, i290, i370 **Engines:** 2.9L VIN 9, 3.7L VIN E, 4.2L VIN S **Transmissions:** A/T, M/T	**Idle Air Control System High RPM** DTC P0106, P0107, P0108, P0112, P0113, P0117, P0118, P0121, P0122, P0123, P0125, P0131, P0132, P0133, P0134, P0201-206, P0300, P0301-P0306, P0335, P0341, P0342, P0404, P0405, P0440, P0442, P0446, P0452, P0453, P0502, P0507, P0601, P0602, P0705, P1133, P1404 and P1441 not set, engine runtime over 125 seconds, system voltage from 11-16v, ECT sensor more than 122°F, IAT sensor more than -40°F, MAP sensor less than 40 kPa, Purge duty cycle over 10%, BARO sensor over 75 kPa, vehicle speed under 2 mph with the throttle closed, and the PCM detected the Actual speed was 100-200 rpm above the Desired idle speed for 10 seconds based on the current engine coolant temperature. **Possible Causes:** • High resistance between the IAC 'A' high or low control circuits • Short to ground between the IAC 'B' high or low control circuits • IAC valve is damaged, dirty, sticking or has failed • The throttle plate is carbon fouled (it may need to be cleaned)
DTC: P0711 **1T CCM**, **MIL: NO** **Years:** 2007, 2008 **Models:** Ascender, i290, i370 **Engines:** All Transmissions: A/T	**Transmission Fluid Temperature Sensor Performance** DTC P0722, P0723 and P1870 not set, engine started, system voltage from 11-16v, TFT sensor from -40°F to 69.8°F at startup, ECT sensor more than 150°F and has changed more than 90°F since startup, vehicle speed over 5 mph with the TCC slip speed over 120 rpm for 410 seconds, and the PCM detected the TFT sensor changed less than 2 counts since startup, or that its delta change was over 36°F at least 14 times during a 7 second period. **Possible Causes:** • TFT signal or ground circuit has a high resistance condition • TFT sensor is out-of-calibration (it may be skewed) • TFT sensor is damaged or has failed • PCM has failed

Gas Engine OBD II Trouble Code List (P1xxx Codes)

DTC	Trouble Code Title, Conditions & Possible Causes
DTC: P1106 **1T CCM, MIL: NO** **Years:** 2007, 2008 **Models:** Ascender, i290, i370 **Engines:** 2.9L VIN 9, 3.7L VIN E, 4.2L VIN S **Transmissions:** A/T, M/T	**MAP Sensor Circuit High Input (Intermittent)** DTC P0121, P0122 and P0123 not set, engine runtime 10 seconds, engine speed less than 1000 rpm with throttle angle less than 3%, or the engine speed is more than 1000 rpm with the throttle angle less than 10%, and the PCM detected an unexpected high value (over 80 kPa) on the MAP sensor circuit for 5 seconds of a 16 second period. **Note: For additional help with this code, view the Failure Records.** **Possible Causes:** • MAP sensor signal circuit is open (an intermittent fault) • MAP sensor ground circuit is open (an intermittent fault) • MAP sensor signal circuit is shorted to VREF or system power • MAP sensor is damaged or has failed
DTC: P1107 **1T CCM, MIL: NO** **Years:** 2007, 2008 **Models:** Ascender, i290, i370 **Engines:** 2.9L VIN 9, 3.7L VIN E, 4.2L VIN S **Transmissions:** A/T, M/T	**MAP Sensor Circuit Low Input (Intermittent)** DTC P0121, P0122, P0123 not set, engine running, engine speed less than 1000 rpm and throttle angle over 1%, or the engine speed more than 1000 rpm and throttle angle more than 2%, and the PCM detected an unexpected low value (below 11 kPa) on the MAP sensor circuit for 5 seconds of a 16 second period. **Note: For additional help with this code, view the Failure Records.** **Possible Causes:** • MAP sensor signal circuit shorted to ground (intermittent fault) • MAP VREF circuit open or shorted to ground (intermittent fault) • MAP sensor is damaged or has failed
DTC: P1111 **1T CCM, MIL: NO** **Years:** 2007, 2008 **Models:** Ascender, i290, i370 **Engines:** 2.9L VIN 9, 3.7L VIN E, 4.2L VIN S **Transmissions:** A/T, M/T	**IAT Sensor Circuit High Input (Intermittent)** Engine started, engine runtime over 4 minutes, ECT sensor more than 140°F, vehicle speed under 20 mph, MAF sensor less than 20 g/sec, and the PCM detected an unexpected high signal of over 4.90v (Scan Tool reads -38°F) on the IAT sensor circuit for 2.5 seconds over a 25 second period during the CCM test. **Note: For additional help with this code, view the Failure Records.** **Possible Causes:** • IAT sensor signal circuit is open (an intermittent fault) • IAT sensor is damaged or has failed (an intermittent fault) • PCM has failed
DTC: P1112 **1T CCM, MIL: NO** **Years:** 2007, 2008 **Models:** Ascender, i290, i370 **Engines:** 2.9L VIN 9, 3.7L VIN E, 4.2L VIN S **Transmissions:** A/T, M/T	**IAT Sensor Circuit Low Input (Intermittent)** Engine started, engine runtime over 4 minutes, ECT sensor more than 140°F, vehicle driven to a speed of over 20 mph, Calculated airflow less than 20 g/sec, and the PCM detected an unexpected low voltage condition of less than 0.10v (Scan Tool reads 298°F) on the IAT sensor circuit for 2.5 seconds over a 25 second period. **Note: For additional help with this code, view the Failure Records.** **Possible Causes:** • IAT sensor signal circuit is shorted to ground (intermittent fault) • IAT sensor is damaged or has failed (an intermittent fault) • PCM has failed
DTC: P1114 **1T CCM, MIL: NO** **Years:** 2007, 2008 **Models:** Ascender, i290, i370 **Engines:** 2.9L VIN 9, 3.7L VIN E, 4.2L VIN S **Transmissions:** A/T, M/T	**ECT Sensor Circuit Low Input (Intermittent)** Engine started, engine runtime more than 1 minute, and the PCM detected an unexpected low voltage condition of less than 0.10v (Scan Tool read 302°F) on the ECT sensor circuit for 10 seconds of a 100 second period. **Note: For additional help with this code, view the Failure Records.** **Possible Causes:** • ECT sensor signal circuit shorted to ground (intermittent fault) • ECT sensor is damaged or has failed • PCM has failed
DTC: P1115 **1T CCM, MIL: NO** **Years:** 2007, 2008 **Models:** Ascender, i290, i370 **Engines:** 2.9L VIN 9, 3.7L VIN E, 4.2L VIN S **Transmissions:** A/T, M/T	**ECT Sensor Circuit High Input (Intermittent)** Engine started, engine running for 1 minute, and the PCM detected an unexpected high voltage of over 4.90v (Scan Tool reads -38°F)] on the ECT sensor circuit for 10 seconds of a 100 second period. **Note: For additional help with this code, view the Failure Records.** **Possible Causes:** • ECT sensor signal circuit is open (an intermittent fault) • ECT sensor is damaged or has failed (an intermittent fault) • PCM has failed

DTC	Trouble Code Title, Conditions & Possible Causes
DTC: P1133 **2T OBD/O2S1, MIL: YES** **Years:** 2007, 2008 **Models:** Ascender, i290, i370 **Engines:** 2.9L VIN 9, 3.7L VIN E, 4.2L VIN S **Transmissions:** A/T, M/T	**HO2S-11 (Bank 1 Sensor 1) Insufficient Switching** DTC P0101, P0102, P0103, P0106, P0107, P0108, P0117, P0118, P0121, P0122, P0123, P0131, P0132, P0133, P0134, P0135, P0300, P0301-P0306, P0441 and P1441 not set, engine speed from 1500-3000 rpm in closed loop for 1 minute, system voltage from 11-16v, ECT sensor more than 122°F, MAF sensor from 9-42 g/sec, Purge duty cycle over 2%, conditions met for 3 seconds, and the PCM detected less than 23 rich-to-lean or lean-to-rich switches on the HO2S-11 signal circuit. **Possible Causes:** • Air leaks after the MAF sensor, or in the EGR or PCV system • Exhaust leaks before or near where the front HO2S is mounted • Fuel control sensor is out of calibration (i.e., ECT, IAT or MAP) • Fuel delivery system supplying too much or too little fuel during cruise or idle periods (e.g., faulty fuel pump, or dirty fuel filter) • Fuel injector (one or more) dirty, leaking or sticking • Fuel pressure regulator leaking, damaged or has failed • HO2S is contaminated, deteriorated or it has failed
DTC: P1134 **2T OBD/O2S1, MIL: YES** **Years:** 2007, 2008 **Models:** Ascender, i290, i370 **Engines:** 2.9L VIN 9, 3.7L VIN E, 4.2L VIN S **Transmissions:** A/T, M/T	**HO2S-11 (Bank 1 Sensor 1) Transition Time Ratio Error** DTC P0101, P0102, P0103, P0106, P0107, P0108, P0117, P0118, P0121, P0122, P0123, P0131, P0132, P0133, P0134, P0135, P0300, P0301-P0306, P0441 and P1441 not set, engine speed from 1500-3000 rpm in closed loop for 1 minute, system voltage from 11-16v, ECT sensor more than 122-167°F, MAF sensor at 18-42 g/sec, Purge duty cycle over 2%, conditions met for 3 seconds, and the PCM detected the HO2S-11 transition ratio from lean-to-rich and rich-to-lean was less than 0.44 or more than 3.8 during the test. **Possible Causes:** • Air leaks after the MAF sensor, or in the EGR or PCV system • Exhaust leaks before or near where the front HO2S is mounted • Fuel control sensor is out of calibration (i.e., ECT, IAT or MAP) • Fuel delivery system supplying too much or too little fuel during cruise or idle periods (e.g., faulty fuel pump, or dirty fuel filter) • Fuel injector (one or more) dirty, leaking or sticking • Fuel pressure regulator leaking, damaged or has failed • HO2S is contaminated, deteriorated or it has failed
DTC: P1153 **2T CCM, MIL: YES** **Years:** 2007, 2008 **Models:** Ascender, i290, i370 **Engines:** 2.9L VIN 9, 3.7L VIN E, 4.2L VIN S **Transmissions:** A/T, M/T	**HO2S-21 (Bank 2 Sensor 1) Insufficient Switching** DTC P0101, P0102, P0103, P0106, P0107, P0108, P0117, P0118, P0121, P0122, P0123, P0131, P0132, P0133, P0134, P0135, P0300, P0301-P0306, P0441 and P1441 not set, system voltage from 11-16v, engine speed from 1500-3000 rpm in closed loop for 1 minute, ECT sensor more than 122-167°F, MAF sensor from 9-42 g/sec, Purge duty cycle over 2%, conditions met for 3 seconds, and the PCM detected less than 27 rich-to-lean or lean-to-rich switches on the HO2S-21 signal circuit in the Oxygen Sensor Monitor test. **Possible Causes:** • Air leaks after the MAF sensor, or in the EGR or PCV system • Exhaust leaks before or near where the front HO2S is mounted • Fuel control sensor is out of calibration (i.e., ECT, IAT or MAP) • Fuel delivery system supplying too much or too little fuel during cruise or idle periods (e.g., faulty fuel pump, or dirty fuel filter) • Fuel injector (one or more) dirty, leaking or sticking • Fuel pressure regulator leaking, damaged or has failed • HO2S is contaminated, deteriorated or it has failed
DTC: P1154 **2T OBD/O2S1, MIL: YES** **Years:** 2007, 2008 **Models:** Ascender, i290, i370 **Engines:** 2.9L VIN 9, 3.7L VIN E, 4.2L VIN S **Transmissions:** A/T, M/T	**HO2S-21 (Bank 2 Sensor 1) Transition Time Ratio** DTC P0101, P0102, P0103, P0106, P0107, P0108, P0117, P0118, P0121, P0122, P0123, P0131, P0132, P0133, P0134, P0135, P0300, P0301-P0306, P0441 and P1441 not set, system voltage from 10-16v, engine speed from 1500-3000 rpm in closed loop for 1 minute, ECT sensor more than 122-167°F, MAF sensor from 9-42 g/sec, Purge duty cycle over 2%, conditions met for 3 seconds, then 90 seconds after entering closed loop control, the PCM detected the transition time ratio to switch from lean-to-rich or rich to lean from the HO2S-21 was less than 0.44 or more than 3.8 during the test. **Possible Causes:** • Air leaks at the exhaust manifold or exhaust pipes • HO2S signal circuit is open or shorted to ground (intermittent) • HO2S heater power circuit is open, or the heater has failed • HO2S contaminated with wrong fuel, has deteriorated or failed

DTC	Trouble Code Title, Conditions & Possible Causes
DTC: P1171 **1T CCM, MIL: YES** **Years:** 2007, 2008 **Models:** Ascender, i290, i370 **Engines:** 2.9L VIN 9, 3.7L VIN E, 4.2L VIN S **Transmissions:** A/T	**Fuel System Lean During Acceleration Detected** DTC P0131, P0132, P0133, P0134 and P1133 not set, ECT sensor more than 140°F, engine running in Power Enrichment mode in closed loop, and the PCM detected the HO2S-11 signal indicated less than 400 mv for 5 seconds during the Fuel System Monitor test. **Possible Causes:** • Air intake leaks in the engine, or in the PCV system (valve) • Air leaks at the EGR gasket, or at the EGR valve diaphragm • Base engine "mechanical" fault affecting one or more cylinders • Exhaust leaks before or near where the front HO2S is mounted • Fuel injectors (one or more) restricted (allowing too little fuel) • Fuel delivery system supplying too little fuel during acceleration periods (e.g., faulty fuel pump, dirty or restricted fuel filter) • Fuel control sensor out of calibration (i.e., IAT, MAF or MAP) • HO2S is contaminated, deteriorated or it has failed • Vehicle driven low on fuel or until it ran out of fuel
DTC: P1441 **2T CCM, MIL: YES** **Years:** 2007, 2008 **Models:** Ascender, i290, i370 **Engines:** 2.9L VIN 9, 3.7L VIN E, 4.2L VIN S **Transmissions:** A/T, M/T	**EVAP Vacuum Switch Circuit High Input** DTC P0106, P0107, P0108, P0112, P0113, P0121, P0122, P0123, P1640 and P1650 not set, system voltage at 11-16v, IAT sensor more than 32°F, fuel level from 15-85%, and the PCM detected a continuous "open" purge condition during the EVAP Monitor test. **Possible Causes:** • Vacuum switch signal circuit is open between switch and PCM • Vacuum switch ground circuit open between switch and ground • Vacuum switch is damaged or has failed • PCM has failed
DTC: P1625 **1T CCM, MIL: NO** **Years:** 2007, 2008 **Models:** Ascender, i290, i370 **Engines:** All **Transmissions:** A/T, M/T	**PCM Unexpected Reset Occurred** Key on, and the PCM detected a Clock or Computer Operating Properly (COP) reset or illegal software code interrupt occurred. **Note: For additional help with this code, view the Failure Records.** **Possible Causes:** • Perform a PCM Reset and retrieve the trouble codes
DTC: P1640 **1T CCM, MIL: NO** **Years:** 2007, 2008 **Models:** Ascender, i290, i370 **Engines:** 3.2L VIN W, 3.5L VIN X, 3.5L VIN Y **Transmissions:** A/T, M/T	**Output Driver Module 'A' Circuit Malfunction** DTC P1618 not set, engine started, engine running, system voltage over 13.2v for 4 seconds, and the PCM detected an open circuit condition and an unexpected high voltage condition on the Output Driver Module circuit (A/C Clutch Relay or Purge Solenoid) with the device "on" for 2.5 seconds during the CCM Rationality test. **Note: For additional help with this code, view the Failure Records.** **Possible Causes:** • One or more output device driver circuits connected to ODM 'A' has an open circuit condition • One or more output device driver circuits connected to ODM 'A' has a short-to-voltage condition • Check for an open power circuit to the related output devices • Disconnect the A/C clutch relay and Purge solenoid to find fault
DTC: P1650 **1T CCM, MIL: NO** **Years:** 2007, 2008 **Models:** Ascender, i290, i370 **Engines:** 3.2L VIN W, 3.5L VIN X, 3.5L VIN Y **Transmissions:** A/T, M/T	**Output Driver Module Circuit Malfunction** DTC P1618 not set, engine started, engine running, system voltage over 13.2v for 4 seconds, and the PCM detected the voltage on the Output Driver Module (ODM) circuit did not indicate less than 1.0 volt with the device commanded "on" for 0.5 seconds in the CCM test. **Note: For additional help with this code, view the Failure Records.** **Possible Causes:** • One or more output device driver circuits connected to ODM has an open circuit condition or has a short-to-voltage condition • Check for an open power circuit to the related output devices • Disconnect the A/C clutch relay and Purge solenoid to find fault
DTC: P1850 **1T CCM, MIL: NO** **Years:** 2007, 2008 **Models:** Ascender, i290, i370 **Engines:** 3.2L VIN W, 3.5L VIN X, 3.5L VIN Y **Transmissions:** A/T	**Kick Down Switch Circuit Low Input** DTC P0122 and P0123 not set, engine started, engine running, throttle position less than 70%, and the PCM detected the Kick Down Switch remained in "on" position during the CCM test. **Note: For additional help with this code, view the Failure Records.** **Possible Causes:** • Kick down switch signal circuit is shorted to ground • Kick down switch ground circuit open between switch and ground • Kick down switch is damaged, out-of-adjustment or has failed • PCM has failed

DTC	Trouble Code Title, Conditions & Possible Causes
DTC: P1850 **1T CCM**, **MIL: NO** **Years:** 2007, 2008 **Models:** Ascender, i290, i370 **Engines:** 3.2L VIN W, 3.5L VIN X, 3.5L VIN Y **Transmissions:** A/T	**A/T Brake Band Apply Solenoid Circuit Malfunction** Engine started, engine running, then with A/T Brake Band Apply solenoid commanded "on", and PCM detected the solenoid control signal was 12v, or with A/T Brake Band Apply solenoid commanded "off", the solenoid control signal was 1.34 to 1.56 seconds. **Note: The PCM controls this solenoid with a pulsewidth modulated (PWM) control signal.** **Possible Causes:** • A/T Brake band solenoid circuit is open or shorted to ground • A/T Brake band solenoid High circuit open or shorted to ground • A/T brake band apply solenoid is damaged or has failed • PCM has failed
DTC: P1860 **2T CCM**, **MIL: YES** **Years:** 2007, 2008 **Models:** Ascender, i290, i370 **Engines:** 3.2L VIN W, 3.5L VIN X, 3.5L VIN Y **Transmissions:** A/T	**Torque Converter Clutch PWM Solenoid Circuit Malfunction** DTC P0751, P0752, P0753, P0756, P0757 and P0758 not set, engine started, engine running, then with the TCC solenoid commanded "on", the PCM detected the solenoid control circuit signal was 12v, or with the TCC solenoid commanded "off", it detected the TCC control circuit signal was 0v for 1.25 seconds. **Possible Causes:** • TCC PWM solenoid control circuit open or shorted to ground • TCC PWM solenoid control circuit is shorted to power • TCC PWM solenoid is damaged or has failed • PCM has failed
DTC: P1870 **2T CCM**, **MIL: YES** **Years:** 2007, 2008 **Models:** Ascender, i290, i370 **Engines:** 3.2L VIN W, 3.5L VIN X, 3.5L VIN Y **Transmissions:** A/T	**Transmission Component Slipping Malfunction** DTC P0722, P0723, P0742, P0751, P0752, P0753, P0756, P0757, P0758, P1860 and P1870 not set, engine started, then driven to a speed of 15-58 mph, engine speed from 1000-3500 rpm, TP sensor signal from 15-99%, MAP sensor signal from 0-70 kPa, 50 < Engine Torque <300 Nm, gear selector in D4, TFT sensor signal from 68-302°F, speed ratio at 0.6-0.95, and the PCM detected the TCC slip speed was 250-800 rpm (event occurred 3 times within 7 seconds). **Possible Causes:** • Engine speed signal circuit open or shorted (intermittent fault) • Internal transmission component problem • TCC PWM or Shift Solenoids have failed (mechanical fault) • TR switch is damaged, out-of-adjustment or has failed

Commonly Used Abbreviations

2
2WD	Two Wheel Drive

4
4WD	Four Wheel Drive

A
A/C	Air Conditioning
ABDC	After Bottom Dead Center
ABS	Anti-lock Brakes
AC	Alternating Current
ACL	Air cleaner
ACT	Air Charge Temperature
AIR	Secondary Air Injection
ALCL	Assembly Line Communications Link
ALDL	Assembly Line Diagnostic Link
AT	Automatic Transaxle/Transmission
ATDC	After Top Dead Center
ATF	Automatic Transmission Fluid
ATS	Air Temperature Sensor
AWD	All Wheel Drive

B
BAP	Barometric Absolute Pressure
BARO	Barometric Pressure
BBDC	Before Bottom Dead Center
BCM	Body Control Module
BDC	Bottom Dead Center
BPT	Backpressure Transducer
BTDC	Before Top Dead Center
BVSV	Bimetallic Vacuum Switching Valve

C
CAC	Charge Air Cooler
CARB	California Air Resources Board
CAT	Catalytic Converter
CCC	Computer Command Control
CCCC	Computer Controlled Catalytic Converter
CCCI	Computer Controlled Coil Ignition
CCD	Computer Controlled Dwell
CDI	Capacitor Discharge Ignition
CEC	Computerized Engine Control
CFI	Continuous Fuel Injection
CIS	Continuous Injection System
CIS-E	Continuous Injection System - Electronic
CKP	Crankshaft Position
CL	Closed Loop
CMP	Camshaft Position
CPP	Clutch Pedal Position
CTOX	Continuous Trap Oxidizer System
CTP	Closed Throttle Position
CVC	Constant Vacuum Control
CYL	Cylinder

D
DBC	Dual Bed Catalyst
DC	Direct Current
DFI	Direct Fuel Injection
DIS	Distributorless Ignition System
DLC	Data Link Connector
DMM	Digital Multimeter
DOHC	Double Overhead Camshaft
DRB	Diagnostic Readout Box
DTC	Diagnostic Trouble Code
DTM	Diagnostic Test Mode
DVOM	Digital Volt/Ohmmeter

E
EBCM	Electronic Brake Control Module
ECM	Engine Control Module
ECT	Engine Coolant Temperature
ECU	Engine Control Unit or Electronic Control Unit
EDIS	Electronic Distributorless Ignition System
EEC	Electronic Engine Control
EEPROM	Electrically Erasable Programmable Read Only Memory
EFE	Early Fuel Evaporation
EGR	Exhaust Gas Recirculation
EGRT	Exhaust Gas Recirculation Temperature
EGRVC	EGR Valve Control
EPROM	Erasable Programmable Read Only Memory
EVAP	Evaporative Emissions
EVP	EGR Valve Position

F
FBC	Feedback Carburetor
FEEPROM	Flash Electrically Erasable Programmable Read Only Memory
FF	Flexible Fuel
FI	Fuel Injection
FT	Fuel Trim
FWD	Front Wheel Drive

G
GND	Ground

H
HAC	High Altitude Compensation
HEGO	Heated Exhaust Gas Oxygen sensor
HEI	High Energy Ignition
HO2 Sensor	Heated Oxygen Sensor

I
IAC	Idle Air Control
IAT	Intake Air Temperature
ICM	Ignition Control Module
IFI	Indirect Fuel Injection
IFS	Inertia Fuel Shutoff
ISC	Idle Speed Control
IVSV	Idle Vacuum Switching Valve

Commonly Used Abbreviations

K

KOEO	Key On, Engine Off
KOER	Key ON, Engine Running
KS	Knock Sensor

M

MAF	Mass Air Flow
MAP	Manifold Absolute Pressure
MAT	Manifold Air Temperature
MC	Mixture Control
MDP	Manifold Differential Pressure
MFI	Multiport Fuel Injection
MIL	Malfunction Indicator Lamp or Maintenance
MST	Manifold Surface Temperature
MVZ	Manifold Vacuum Zone

N

NVRAM	Nonvolatile Random Access Memory

O

O2 Sensor	Oxygen Sensor
OBD	On-Board Diagnostic
OC	Oxidation Catalyst
OHC	Overhead Camshaft
OL	Open Loop

P

P/S	Power Steering
PAIR	Pulsed Secondary Air Injection
PCM	Powertrain Control Module
PCS	Purge Control Solenoid
PCV	Positive Crankcase Ventilation
PIP	Profile Ignition Pick-up
PNP	Park/Neutral Position
PROM	Programmable Read Only Memory
PSP	Power Steering Pressure
PTO	Power Take-Off
PTOX	Periodic Trap Oxidizer System

R

RABS	Rear Anti-lock Brake System
RAM	Random Access Memory
ROM	Read Only Memory
RPM	Revolutions Per Minute
RWAL	Rear Wheel Anti-lock Brakes
RWD	Rear Wheel Drive

S

SBC	Single Bed Converter
SBEC	Single Board Engine Controller
SC	Supercharger
SCB	Supercharger Bypass
SFI	Sequential Multiport Fuel Injection
SIR	Supplemental Inflatable Restraint
SOHC	Single Overhead Camshaft
SPL	Smoke Puff Limiter
SPOUT	Spark Output
SRI	Service Reminder Indicator
SRS	Supplemental Restraint System
SRT	System Readiness Test
SSI	Solid State Ignition
ST	Scan Tool
STO	Self-Test Output

T

TAC	Thermostatic Air Clearner
TBI	Throttle Body Fuel Injection
TC	Turbocharger
TCC	Torque Converter Clutch
TCM	Transmission Control Module
TDC	Top Dead Center
TFI	Thick Film Ignition
TP	Throttle Position
TR Sensor	Transaxle/Transmission Range Sensor
TVV	Thermal Vacuum Valve
TWC	Three-way Catalytic Converter

V

VAF	Volume Air Flow, or Vane Air Flow
VAPS	Variable Assist Power Steering
VRV	Vacuum Regulator Valve
VSS	Vehicle Speed Sensor
VSV	Vacuum Switching Valve

W

WOT	Wide Open Throttle
WU-TWC	Warm Up Three-way Catalytic Converter

ENGLISH TO METRIC CONVERSION: TORQUE

To convert foot-pounds (ft. lbs.) to Newton-meters (Nm), multiply the number of ft. lbs. by 1.36

To convert Newton-meters (Nm) to foot-pounds (ft. lbs.), multiply the number of Nm by 0.7376

ft. lbs.	Nm	ft. lbs.	Nm	ft. lbs.	Nm	ft. lbs.	Nm
0.1	0.1	34	46.2	76	103.4	118	160.5
0.2	0.3	35	47.6	77	104.7	119	161.8
0.3	0.4	36	49.0	78	106.1	120	163.2
0.4	0.5	37	50.3	79	107.4	121	164.6
0.5	0.7	38	51.7	80	108.8	122	165.9
0.6	0.8	39	53.0	81	110.2	123	167.3
0.7	1.0	40	54.4	82	111.5	124	168.6
0.8	1.1	41	55.8	83	112.9	125	170.0
0.9	1.2	42	57.1	84	114.2	126	171.4
1	1.4	43	58.5	85	115.6	127	172.7
2	2.7	44	59.8	86	117.0	128	174.1
3	4.1	45	61.2	87	118.3	129	175.4
4	5.4	46	62.6	88	119.7	130	176.8
5	6.8	47	63.9	89	121.0	131	178.2
6	8.2	48	65.3	90	122.4	132	179.5
7	9.5	49	66.6	91	123.8	133	180.9
8	10.9	50	68.0	92	125.1	134	182.2
9	12.2	51	69.4	93	126.5	135	183.6
10	13.6	52	70.7	94	127.8	136	185.0
11	15.0	53	72.1	95	129.2	137	186.3
12	16.3	54	73.4	96	130.6	138	187.7
13	17.7	55	74.8	97	131.9	139	189.0
14	19.0	56	76.2	98	133.3	140	190.4
15	20.4	57	77.5	99	134.6	141	191.8
16	21.8	58	78.9	100	136.0	142	193.1
17	23.1	59	80.2	101	137.4	143	194.5
18	24.5	60	81.6	102	138.7	144	195.8
19	25.8	61	83.0	103	140.1	145	197.2
20	27.2	62	84.3	104	141.4	146	198.6
21	28.6	63	85.7	105	142.8	147	199.9
22	29.9	64	87.0	106	144.2	148	201.3
23	31.3	65	88.4	107	145.5	149	202.6
24	32.6	66	89.8	108	146.9	150	204.0
25	34.0	67	91.1	109	148.2	151	205.4
26	35.4	68	92.5	110	149.6	152	206.7
27	36.7	69	93.8	111	151.0	153	208.1
28	38.1	70	95.2	112	152.3	154	209.4
29	39.4	71	96.6	113	153.7	155	210.8
30	40.8	72	97.9	114	155.0	156	212.2
31	42.2	73	99.3	115	156.4	157	213.5
32	43.5	74	100.6	116	157.8	158	214.9
33	44.9	75	102.0	117	159.1	159	216.2

METRIC TO ENGLISH CONVERSION: TORQUE

To convert foot-pounds (ft. lbs.) to Newton-meters (Nm), multiply the number of ft. lbs. by 1.36

To convert Newton-meters (Nm) to foot-pounds (ft. lbs.), multiply the number of Nm by 0.7376

Nm	ft. lbs.	Nm	ft. lbs.	Nm	ft. lbs.	Nm	ft. lbs.	Nm	ft. lbs.
0.1	0.1	34	25.0	76	55.9	118	86.8	160	117.6
0.2	0.1	35	25.7	77	56.6	119	87.5	161	118.4
0.3	0.2	36	26.5	78	57.4	120	88.2	162	119.1
0.4	0.3	37	27.2	79	58.1	121	89.0	163	119.9
0.5	0.4	38	27.9	80	58.8	122	89.7	164	120.6
0.6	0.4	39	28.7	81	59.6	123	90.4	165	121.3
0.7	0.5	40	29.4	82	60.3	124	91.2	166	122.1
0.8	0.6	41	30.1	83	61.0	125	91.9	167	122.8
0.9	0.7	42	30.9	84	61.8	126	92.6	168	123.5
1	0.7	43	31.6	85	62.5	127	93.4	169	124.3
2	1.5	44	32.4	86	63.2	128	94.1	170	125.0
3	2.2	45	33.1	87	64.0	129	94.9	171	125.7
4	2.9	46	33.8	88	64.7	130	95.6	172	126.5
5	3.7	47	34.6	89	65.4	131	96.3	173	127.2
6	4.4	48	35.3	90	66.2	132	97.1	174	127.9
7	5.1	49	36.0	91	66.9	133	97.8	175	128.7
8	5.9	50	36.8	92	67.6	134	98.5	176	129.4
9	6.6	51	37.5	93	68.4	135	99.3	177	130.1
10	7.4	52	38.2	94	69.1	136	100.0	178	130.9
11	8.1	53	39.0	95	69.9	137	100.7	179	131.6
12	8.8	54	39.7	96	70.6	138	101.5	180	132.4
13	9.6	55	40.4	97	71.3	139	102.2	181	133.1
14	10.3	56	41.2	98	72.1	140	102.9	182	133.8
15	11.0	57	41.9	99	72.8	141	103.7	183	134.6
16	11.8	58	42.6	100	73.5	142	104.4	184	135.3
17	12.5	59	43.4	101	74.3	143	105.1	185	136.0
18	13.2	60	44.1	102	75.0	144	105.9	186	136.8
19	14.0	61	44.9	103	75.7	145	106.6	187	137.5
20	14.7	62	45.6	104	76.5	146	107.4	188	138.2
21	15.4	63	46.3	105	77.2	147	108.1	189	139.0
22	16.2	64	47.1	106	77.9	148	108.8	190	139.7
23	16.9	65	47.8	107	78.7	149	109.6	191	140.4
24	17.6	66	48.5	108	79.4	150	110.3	192	141.2
25	18.4	67	49.3	109	80.1	151	111.0	193	141.9
26	19.1	68	50.0	110	80.9	152	111.8	194	142.6
27	19.9	69	50.7	111	81.6	153	112.5	195	143.4
28	20.6	70	51.5	112	82.4	154	113.2	196	144.1
29	21.3	71	52.2	113	83.1	155	114.0	197	144.9
30	22.1	72	52.9	114	83.8	156	114.7	198	145.6
31	22.8	73	53.7	115	84.6	157	115.4	199	146.3
32	23.5	74	54.4	116	85.3	158	116.2	200	147.1
33	24.3	75	55.1	117	86.0	159	116.9	201	147.8

ENGLISH/METRIC CONVERSION: TEMPERATURE

To convert Fahrenheit (F°) to Celsius (C°), take F° temperature and subtract 32, multiply the result by 5 and divide the result by 9
To convert Celsius (C°) to Fahrenheit (F°), take C° temperature and multiply it by 9, divide the result by 5 and add 32

F°	C°	F°	C°	C°	F°	C°	F°
-40	-40.0	150	65.6	-38	-36.4	46	114.8
-35	-37.2	155	68.3	-36	-32.8	48	118.4
-30	-34.4	160	71.1	-34	-29.2	50	122
-25	-31.7	165	73.9	-32	-25.6	52	125.6
-20	-28.9	170	76.7	-30	-22	54	129.2
-15	-26.1	175	79.4	-28	-18.4	56	132.8
-10	-23.3	180	82.2	-26	-14.8	58	136.4
-5	-20.6	185	85.0	-24	-11.2	60	140
0	-17.8	190	87.8	-22	-7.6	62	143.6
1	-17.2	195	90.6	-20	-4	64	147.2
2	-16.7	200	93.3	-18	-0.4	66	150.8
3	-16.1	205	96.1	-16	3.2	68	154.4
4	-15.6	210	98.9	-14	6.8	70	158
5	-15.0	212	100.0	-12	10.4	72	161.6
10	-12.2	215	101.7	-10	14	74	165.2
15	-9.4	220	104.4	-8	17.6	76	168.8
20	-6.7	225	107.2	-6	21.2	78	172.4
25	-3.9	230	110.0	-4	24.8	80	176
30	-1.1	235	112.8	-2	28.4	82	179.6
35	1.7	240	115.6	0	32	84	183.2
40	4.4	245	118.3	2	35.6	86	186.8
45	7.2	250	121.1	4	39.2	88	190.4
50	10.0	255	123.9	6	42.8	90	194
55	12.8	260	126.7	8	46.4	92	197.6
60	15.6	265	129.4	10	50	94	201.2
65	18.3	270	132.2	12	53.6	96	204.8
70	21.1	275	135.0	14	57.2	98	208.4
75	23.9	280	137.8	16	60.8	100	212
80	26.7	285	140.6	18	64.4	102	215.6
85	29.4	290	143.3	20	68	104	219.2
90	32.2	295	146.1	22	71.6	106	222.8
95	35.0	300	148.9	24	75.2	108	226.4
100	37.8	305	151.7	26	78.8	110	230
105	40.6	310	154.4	28	82.4	112	233.6
110	43.3	315	157.2	30	86	114	237.2
115	46.1	320	160.0	32	89.6	116	240.8
120	48.9	325	162.8	34	93.2	118	244.4
125	51.7	330	165.6	36	96.8	120	248
130	54.4	335	168.3	38	100.4	122	251.6
135	57.2	340	171.1	40	104	124	255.2
140	60.0	345	173.9	42	107.6	126	258.8
145	62.8	350	176.7	44	111.2	128	262.4

LENGTH CONVERSION

To convert inches (in.) to millimeters (mm), multiply the number of inches by 25.4
To convert millimeters (mm) to inches (in.), multiply the number of millimeters by 0.04

Inches	Millimeters	Inches	Millimeters	Inches	Millimeters	Inches	Millimeters
0.0001	0.00254	0.005	0.1270	0.09	2.286	4	101.6
0.0002	0.00508	0.006	0.1524	0.1	2.54	5	127.0
0.0003	0.00762	0.007	0.1778	0.2	5.08	6	152.4
0.0004	0.01016	0.008	0.2032	0.3	7.62	7	177.8
0.0005	0.01270	0.009	0.2286	0.4	10.16	8	203.2
0.0006	0.01524	0.01	0.254	0.5	12.70	9	228.6
0.0007	0.01778	0.02	0.508	0.6	15.24	10	254.0
0.0008	0.02032	0.03	0.762	0.7	17.78	11	279.4
0.0009	0.02286	0.04	1.016	0.8	20.32	12	304.8
0.001	0.0254	0.05	1.270	0.9	22.86	13	330.2
0.002	0.0508	0.06	1.524	1	25.4	14	355.6
0.003	0.0762	0.07	1.778	2	50.8	15	381.0
0.004	0.1016	0.08	2.032	3	76.2	16	406.4

ENGLISH/METRIC CONVERSION: LENGTH

To convert inches (in.) to millimeters (mm), multiply the number of inches by 25.4

To convert millimeters (mm) to inches (in.), multiply the number of millimeters by 0.04

Inches		Millimeters	Inches		Millimeters	Inches		Millimeters
Fraction	Decimal	Decimal	Fraction	Decimal	Decimal	Fraction	Decimal	Decimal
1/64	0.016	0.397	11/32	0.344	8.731	11/16	0.688	17.463
1/32	0.031	0.794	23/64	0.359	9.128	45/64	0.703	17.859
3/64	0.047	1.191	3/8	0.375	9.525	23/32	0.719	18.256
1/16	0.063	1.588	25/64	0.391	9.922	47/64	0.734	18.653
5/64	0.078	1.984	13/32	0.406	10.319	3/4	0.750	19.050
3/32	0.094	2.381	27/64	0.422	10.716	49/64	0.766	19.447
7/64	0.109	2.778	7/16	0.438	11.113	25/32	0.781	19.844
1/8	0.125	3.175	29/64	0.453	11.509	51/64	0.797	20.241
9/64	0.141	3.572	15/32	0.469	11.906	13/16	0.813	20.638
5/32	0.156	3.969	31/64	0.484	12.303	53/64	0.828	21.034
11/64	0.172	4.366	1/2	0.500	12.700	27/32	0.844	21.431
3/16	0.188	4.763	33/64	0.516	13.097	55/64	0.859	21.828
13/64	0.203	5.159	17/32	0.531	13.494	7/8	0.875	22.225
7/32	0.219	5.556	35/64	0.547	13.891	57/64	0.891	22.622
15/64	0.234	5.953	9/16	0.563	14.288	29/32	0.906	23.019
1/4	0.250	6.350	37/64	0.578	14.684	59/64	0.922	23.416
17/64	0.266	6.747	19/32	0.594	15.081	15/16	0.938	23.813
9/32	0.281	7.144	39/64	0.609	15.478	61/64	0.953	24.209
19/64	0.297	7.541	5/8	0.625	15.875	31/32	0.969	24.606
5/16	0.313	7.938	41/64	0.641	16.272	63/64	0.984	25.003
21/64	0.328	8.334	21/32	0.656	16.669	1/1	1.000	25.400
			43/64	0.672	17.066			

Chilton labor times have been a trusted standard in the industry for over 60 years. That legacy continues in the Chilton 2008 Labor Guide Manuals and CD-ROM. Chilton labor times take into account the real world environment in which technicians work. Three labor times are available for many models: Chilton's standard and severe service times, plus OEM warranty time. Chilton labor times are accepted by most insurance and extended warranty companies. Vehicle makes and models conform to current Automotive Aftermarket Industry Association (AAIA) standards. Get your hands on the newest version of a classic.

Hardcover Manuals are 8 1/2" x 11", ©2007

Chilton 2008 Labor Guide Manuals
978-1-4283-2035-2 (1-4283-2035-0) Part No.142035

Chilton 2008 Labor Guide CD-ROM
978-1-4283-2041-3 (1-4283-2041-5) Part No.142041

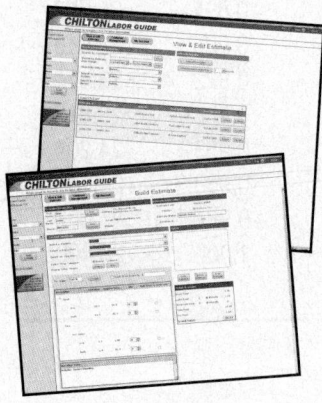

Labor Guide CD-ROM Benefits:

- save time with automatically calculated labor charges, taxes, & parts as total job estimated
- create professional estimates for your customers and worksheets for your technicians, printing them whenever needed
- keep track of customers, prior estimates, and your own parts or package jobs with less paper
- estimate all your vehicles with one tool- labor times from 1981 through current domestic and imported vehicles

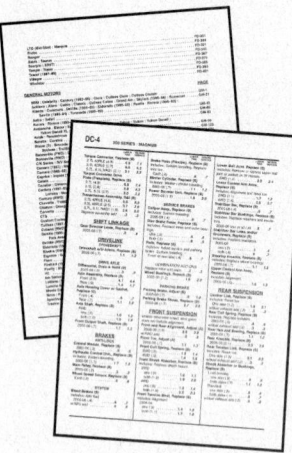

Labor Guide CD-ROM Benefits:

- enjoy quicker referencing than ever by using separate, easier-to-handle, domestic and imported vehicle manuals filling more than 2500 pages
- find the times fast by using:
  - tabs that display contents by manufacturer and model
  - two indexes--labor operations and systems--in each model group
  - manufacturers arranged alphabetically, and page numbering that includes manufacturer code so you know where you are in the book
- manufacturers are arranged alphabetically within each volume
- estimate all your vehicles with one tool- labor times from 1981 through current domestic and imported vehicles

Chilton 2008® Service Manuals include 11 manuals covering DaimlerChrysler, Ford, General Motors, Asian, and European vehicles. Users will be expertly provided with the most currently available information to assist in daily activities. These new, reliable, and comprehensive manuals provide essential information, allowing users to accurately and efficiently diagnose and repair. Step-by-step procedures and helpful illustrations provide easy references for jobs. These new service manuals cover 2006 and 2007 models, plus any available 2008 models.

Service Manual Benefits:

- multi-volume manual set, organized by vehicle manufacturer, provides more than 2000 pages of expertly written content
- access new year, make, and model information without repeating previous edition's content
- comprehensive, technically detailed content—including exploded view illustrations, diagnostics and specification charts—arranged alphabetically by model group for quick, easy access

Chilton 2008 DaimlerChrysler Service Manuals (2 volume set)—ISBN 978-1-4283-2204-2 (1-4283-2204-3) Part No. 142204
Chilton 2008 Ford Service Manuals (2 volume set)—ISBN 978-1-4283-2208-0 (1-4283-2208-6) Part No. 142208
Chilton 2008 General Motors Service Manuals (2 volume set)—ISBN 978-1-4283-2211-0 (1-4283-2211-6) Part No. 142211
Chilton 2008 Asian Service Manuals (4 volume set)—ISBN 978-1-4283-2214-1 (1-4283-2214-0) Part No. 142214
Chilton 2008 European Service Manual—ISBN 978-1-4283-2220-2 (1-4283-2220-5) Part No. 142220

Manuals are 8 1/2" x 11"

The *Chilton® Perennial Editions* contain repair and maintenance information for popular mechanical systems that may not be available elsewhere. They offer a wide range of repair information on cars, trucks, vans, and SUVs dating back to the early 1960s, and as current as 2002. Information for 1993 and later model years includes scheduled maintenance interval charts.

Benefits:

- covers the most common vehicle models found in the repair aftermarket today
- allows quick understanding of systems using exploded-view illustrations, diagrams, and charts
- simplify tough jobs with easy-to-follow removal and installation instructions for heater core and other components
- gives complete coverage of repair procedures from drive train to chassis and associated components

Auto Repair Manual, 1998-2002, 1,426 pages
ISBN 978-0-8019-9362-6 (0-8019-9362-8) Part No. 9362
Auto Repair Manual, 1993-1997, 2,064 pages
ISBN 978-0-8019-7919-4 (0-8019-7919-6) Part No. 7919
Auto Repair Manual, 1980-1987, 1,344 pages
ISBN 978-0-8019-7670-4 (0-8019-7670-7) Part No. 7670
Import Car Repair Manual, 1998-2002, 1,792 pps
ISBN 978-0-8019-9363-3 ISBN 0-8019-9363-6/Part No. 9363
Import Car Repair Manual, 1993-1997, 2,080 pps
ISBN 978-0-8019-7920-0 (0-8019-7920-X) Part No. 7920
Import Car Repair Manual, 1988-1992, 1,632 pages
ISBN 978-0-8019-7907-1 (0-8019-7907-2) Part No. 7907
Import Car Repair Manual, 1980-1987, 1,488 pages
ISBN 978-0-8019-7672-8 (0-8019-7672-3) Part No. 7672

Truck & Van Repair Manual, 1998-2002, 1,408 pages
ISBN 978-0-8019-9364-0 (0-8019-9364-4) Part No. 9364
Truck & Van Repair Manual, 1993-1997, 2,096 pages
ISBN 978-0-8019-7921-7 (0-8019-7921-8) Part No. 7921
Truck & Van Repair Manual, 1991-1995, 1,664 pages
ISBN 978-0-8019-7911-8 (0-8019-7911-0) Part No. 7911
Truck & Van Repair Manual, 1986-1990, 1,536 pages
ISBN 978-0-8019-7902-6 (0-8019-7902-1) Part No. 7902
Truck & Van Repair Manual, 1979-1986, 1,440 pages
ISBN 978-0-8019-7655-1 (0-8019-7655-3) Part No. 7655

SUV Repair Manual, 1998-2002, 1,292 pages
ISBN 978-0-8019-9365-7 (0-8019-9365-2)Part No. 9365

Hardcover manuals are 8 1/2" x 11".

Chilton Collector's Editions—*Reference Manuals for Vintage Vehicles*
Auto Repair Manual, 1964-1971, ISBN 978-0-8019-5974-5 (0-8019-5974-8) Part No. 5974,
Truck & Van Repair Manual, 1971-1978, ISBN 978-0-8019-7012-2 (0-8019-7012-1) Part No. 7012

Chilton Timing Belts, 1985-2005

Chilton
ISBN 978-1-4018-9880-9 (1-4018-9880-7) Part No. 129880

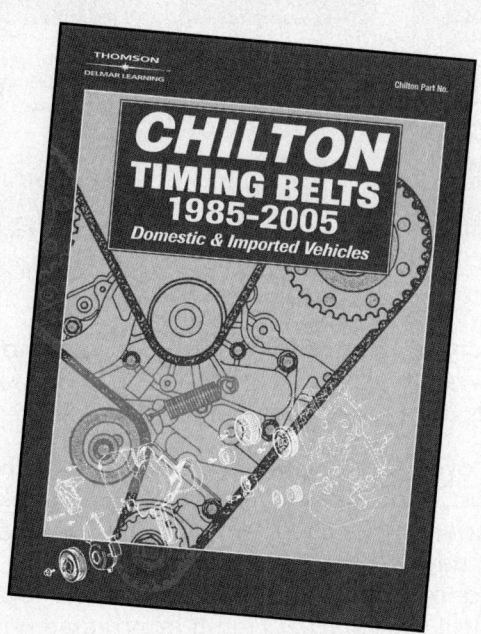

Timing belt procedures can represent increased profits for automotive repair shops and service stations, and this manual contains all the information automotive technicians need to properly service timing belts on domestic and imported cars, vans, and light trucks through 2005 models. Clear, straightforward procedures, illustrations, and specifications help to communicate 20 years of vehicle applications for fast, accurate inspection, replacement, and tensioning of timing belts. Readers will learn step-by-step how to perform key procedures both quickly and safely, while learning the correct labor time to charge for the service. OEM-recommended replacement intervals for proper maintenance of customer's vehicles are also featured.

Benefits:

- detailed illustrations clearly demonstrate important concepts, such as how to correctly align camshaft and crankshaft timing marks, and how to simplify serpentine belt installation
- readers are made aware of potential hazards and time-wasting practices that can impede safe and profitable service procedures
- special tools are identified so that completing the service is as easy and quick as possible

544 pp, 8 1/2" x 11", softcover, ©2006

ALSO AVAILABLE:
Quick-Reference Manuals
Chilton
The Chilton Professional Series offers Quick-Reference Manuals for the automotive professional, providing complete coverage on repair and maintenance, adjustments, and diagnostic procedures for specific systems and components.

Benefits:
- step-by-step procedures
- easy-to-use manufacturer and model indexing
- detailed illustrations and exploded views
- handy specifications or data charts

Heater Core Service 1990-2000,
13-Digit ISBN 978-0-8019-9311-4 Part No. 9311
(10-Digit ISBN 0-8019-9311-3) 560 pp
Brake Specifications and Service 1990-2000
13-Digit ISBN 978-0-8019-9312-1 Part No. 9312
(10-Digit ISBN 0-8019-9312-1) 520 pp

Electric Cooling Fans, Accessory Drive Belts & Water Pumps, 1995-1999,
13-Digit ISBN 978-0-8019-9126-4/Part No. 9126
(10-Digit ISBN 0-8019-9126-9) 312 pp
Powertrain Codes & Oxygen Sensors, 1990-1999,
13-Digit ISBN 978-0-8019-9127-1/Part No. 9127
(10-Digit ISBN 0-8019-9127-7) 400 pp

This pioneering eight-book series offers automotive service shop owners and those wanting to be shop owners the necessary business and customer service skills to run a successful automotive service facility.

The series covers three main topical areas: personnel management, business management, and sales and marketing. Each book provides a framework to help technicians make consistent, high-quality, and productive service a part of every day shop operations. According to the author, "Great performance coupled with increased customer loyalty, trust, and operational excellence will almost always result in increased profits."

Automotive Service Management Series Benefits:

- real-world approach reflects author's experience as a fourth generation technician, a repair & service company owner, and an automotive industry trainer
- all-inclusive coverage spans from designing an automotive repair facility floor plan through financial management techniques, customer/staff relations, and more
- length of each book makes it easy to incorporate this series into workshops, seminars, and training/education courses
- information is available "as is" or for customization

Total Customer Relationship Management
        ISBN 978-1-4018-2657-4 (1-4018-2657-1) Part No. 22657
From Intent to Implementation
        ISBN 978-1-4018-2658-1 (1-4018-2658-X) Part No. 22658
Operational Excellence
        ISBN 978-1-4018-2659-8 (1-4018-2659-8) Part No. 22659
Building a Team
        ISBN 978-4018-2660-4 (1-4018-2660-1) Part No. 22660
The High Performance Shop
        ISBN 978-1-4018-2661-1 (1-4018-2661-X) Part No. 22661
Safety Communications
        ISBN 978-1-4018-2662-8 (1-4018-2662-8) Part No. 22662
Managing Dollars with Sense
        ISBN 978-1-4018-2663-5 (1-4018-2663-6) Part No. 22663
Operations Management
        ISBN 978-1-4018-2665-9 (1-4018-2665-2) Part No. 22665
Entire Set of 8 Books
        ISBN 978-1-4018-2499-0 (1-4018-2499-4) Part No. 2499

Softcover manuals are 8 1/2" x 11", ©2003

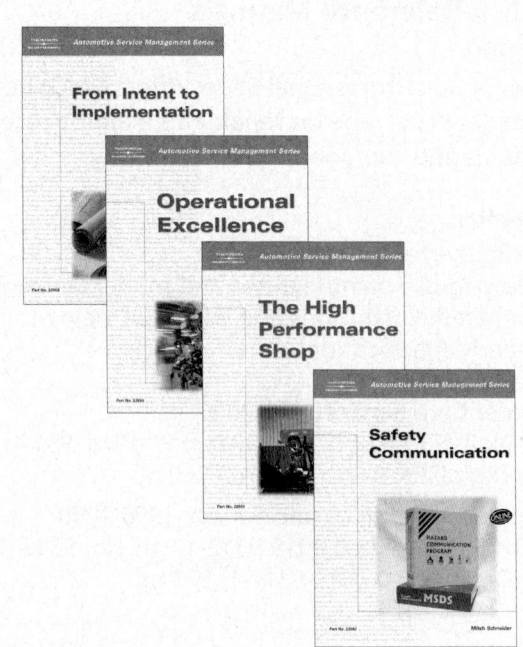

ASE CERTIFICATION TEST PREPARATION

WE SUPPORT
PROFESSIONAL CERTIFICATION
THROUGH THE
National Institute for
AUTOMOTIVE
SERVICE
EXCELLENCE

**You Deserve The Best
When You Are Putting
Your Skills To The Test!**

ASE Test Preparation

(A1) Engine Repair, 4E	1-4180-3878-4
(A2) Transmissions and Transaxles, 4E	1-4180-3879-2
(A3) Manual Drive Train and Axles, 4E	1-4180-3880-6
(A4) Suspension and Steering, 4E	1-4180-3881-4
(A5) Brakes, 4E	1-4180-3882-2
(A6) Electrical-Electronic Systems, 4E	1-4180-3883-0
(A7) Heating and Air Conditioning, 4E	1-4180-3884-9
(A8) Engine Performance, 4E	1-4180-3885-7
(L1) Advanced Engine Performance, 4E	1-4180-3888-1
(X1) Exhaust Systems, 4E	1-4180-3886-5
(P2) Parts Specialist, 4E	1-4180-3887-3
(C1) Service Consultant, 2E	1-4180-3889-X
(B2) Painting and Refinishing, 3E	1-4018-3664-X
(B3) Non Structural Analysis and Damage Repair, 3E	1-4018-3665-8
(B4) Structural Analysis and Damage Repair, 3E	1-4018-3666-6
(B5) Mechanical and Electrical Components, 3E	1-4018-3667-4
(B6) Damage Analysis and Estimation, 3E	1-4018-3668-2
(M1) Cylinder Head Specialist	0-7668-6280-1
(M2) Cylinder Block Specialist	0-7668-6281-X
(M3) Assembly Specialist	0-7668-6282-8
(T1) Gasoline Engines, 4E	1-4180-4828-3
(T2) Diesel Engines, 4E	1-4180-4829-1
(T3) Drive Train, 4E	1-4180-4830-5
(T4) Brakes, 4E	1-4180-4831-3
(T5) Suspension and Steering, 4E	1-4180-4832-1
(T6) Electrical and Electronic Systems, 4E	1-4180-4834-8
(T7) Heating, Ventilation, and Air Conditioning, 4E	1-4180-4835-6
(T8) Preventative Maintenance, 4E	1-4180-4836-4
(S2) Diesel Engines	1-4018-1822-6
(S4) Brakes	1-4018-1824-2
(S5) Suspension and Steering	1-4018-1825-0
(H2) Diesel Engines	1-4180-4998-0
(H3) Drive Train	1-4354-5376-X
(H4) Brakes	1-4180-4998-0
(H5) Suspension & Steering	1-4283-4011-4
(H6) Electrical-Electronic Systems	1-4180-4999-9
(H7) Electrical/Electronic	1-4180-4999-7

ASE Test Preparation in Spanish

Spanish (A1) Engine Repair	1-4018-1014-4
Spanish (A2) Transmissions and Transaxle	1-4018-1015-2
Spanish (A3) Manual Drive Train and Axles	1-4018-1016-0
Spanish (A4) Suspension and Steering	1-4018-1017-9
Spanish (A5) Brakes	1-4018-1018-7
Spanish (A6) Electrical-Electronic Systems	1-4018-1019-5
Spanish (A7) Heating and Air Conditioning	1-4018-1020-9
Spanish (A8) Engine Performance	1-4018-1021-7
Spanish (L1) Advanced Engine Performance	1-4018-1022-5
Spanish (X1) Exhaust Systems	1-4018-1024-1
Spanish (P2) Parts Specialist	1-4018-1023-3
Spanish (B2) Painting and Refinishing	1-4018-9255-8
Spanish (B3) Non-Structural Analysis and Damage Repair	1-4018-2544-3
Spanish (B4) Structural Analysis and Damage Repair	1-4018-9131-4
Spanish (B5) Mechanical and Electrical Components	1-4018-7759-1
Spanish (B6) Damage Analysis and Estimation	1-4018-6573-9

*Switch between English & Spanish at the click of a button!

Online ASE Test Preparation
Place your order online at www.techniciantestprep.com

*Online (A1) Automotive Engine Repair	1-4180-1305-6
*Online (A2) Automatic Transmissions & Transaxles	1-4180-1306-4
*Online (A3) Automotive Manual Drive Trains & Axles	1-4180-1307-2
*Online (A4) Automotive Suspension & Steering	1-4180-1308-0
*Online (A5) Automotive Brakes	1-4180-1309-9
*Online (A6) Automotive Electrical/Electronic Systems	1-4180-1310-2
*Online (A7) Automotive Heating & Air Conditioning	1-4180-1311-0
*Online (A8) Automotive Engine Performance	1-4180-1312-9
*Online (X1) Exhaust Systems	1-4180-1313-7
*Online (P2) Automobile Parts Specialist	1-4180-1314-5
*Online (L1) Automotive Advance Engine Performance	1-4180-1315-3
*Online (C1) Service Consultant	1-4180-1316-1
Online (T1) Gasoline Engines	1-4018-7897-0
Online (T2) Diesel Engines	1-4018-7898-9
Online (T3) Drive Train	1-4018-7900-4
Online (T4) Brakes	1-4018-7901-2
Online (T5) Suspension & Steering	1-4018-7903-9
Online (T6) Electrical & Electronics	1-4180-1879-1
Online (T7) Heating, Ventilation, & Air Conditioning	1-4180-1880-5
Online (T8) Preventative Maintenance	1-4018-7906-3

Complete Series

	IBSN
ASE Test Preparation Manuals for Automotive (A1-A8, X1, P2, L1, C1)	1-4180-3954-3
ASE Test Preparation Manuals for Automotive (A1-A8 & L1)	1-4180-6139-5
ASE Test Preparation Manuals for Automotive (A1-A8, L1, & P2)	1-4180-4197-1
ASE Test Preparation Manuals for Automotive (A1-A8)	1-4180-6237-5
ASE Test Preparation Manuals for Automotive (A1-A8, L1, P2 & X1)	1-4180-6335-5
Online ASE Test Preparation for Automotive (A1-A8, X1, P2, L1, C1)	1-4180-1344-7
Place your order online at www.techniciantestprep.com	
ASE Test Preparation Manuals for Medium/Heavy Duty Truck (T1-T8)	1-4180-4934-4
Online ASE Test Preparation for Medium/Heavy Duty Truck Package (T1-T8)	1-4180-0611-4
Place your order online at www.techniciantestprep.com	
ASE Test Preparation Manuals for Collision (B2-B6)	1-4018-5120-7
ASE Test Preparation Manuals for Collision in Spanish (B2-B6))	1-4018-4155-4
ASE Test Preparation Manuals for Engine Machinist (M1-M3)	0-7668-6283-6